8e

Organic Chemistry

8e

Organic Chemistry

John McMurry
Cornell University

BROOKS/COLE
CENGAGE Learning™

Australia • Brazil • Japan • Korea • Mexico • Singapore • Spain • United Kingdom • United States

***Organic Chemistry*, Eighth Edition**
John McMurry

Executive Editor: Lisa Lockwood

Senior Developmental Editor: Sandra Kiselica

Assistant Editor: Elizabeth Woods

Senior Media Editor: Lisa Weber

Media Editor: Stephanie Van Camp

Senior Marketing Manager: Barb Bartoszek

Marketing Assistant: Kevin Carroll

Marketing Communications Manager: Linda Yip

Content Project Manager: Teresa L. Trego

Art Director: John Walker

Print Buyer: Karen Hunt

Rights Acquisitions Specialist: Dean Dauphinais

Production Service: Graphic World, Inc.

Text Designer: RHDG Riezebos Holzbaur

Photo Researcher: The Bill Smith Group/ Chis Althof

Copy Editor: Graphic World, Inc.

OWL producers: Stephen Battisti, Cindy Stein, David Hart (Center for Educational Software Development, University of Massachusetts, Amherst)

Illustrator: Graphic World, Inc.

Cover Designer: John Walker

Cover Image: © Raul Touzon/Getty Images

Compositor: Graphic World, Inc.

Library of Congress Control Number: 2010936830

Student Edition:

ISBN-13: 978-0-8400-5444-9

ISBN-10: 0-8400-5444-0

Brooks/Cole
20 Davis Drive
Belmont, CA 94002-3098
USA

Cengage Learning is a leading provider of customized learning solutions with office locations around the globe, including Singapore, the United Kingdom, Australia, Mexico, Brazil, and Japan. Locate your local office at **www.cengage.com/global.**

Cengage Learning products are represented in Canada by Nelson Education, Ltd.

To learn more about Brooks/Cole, visit **www.cengage.com/brookscole**

Purchase any of our products at your local college store or at our preferred online store **www.cengagebrain.com.**

Printed in Canada
1 2 3 4 5 6 7 14 13 12 11 10

Brief Contents

Contents

3 Organic Compounds: Alkanes and Their Stereochemistry 74

4 Organic Compounds: Cycloalkanes and Their Stereochemistry 108

5 Stereochemistry at Tetrahedral Centers 142

11 Reactions of Alkyl Halides: Nucleophilic Substitutions and Eliminations 372

12 Structure Determination: Mass Spectrometry and Infrared Spectroscopy 424

13 Structure Determination: Nuclear Magnetic Resonance Spectroscopy 456

Preview of Carbonyl Chemistry 712

19 Aldehydes and Ketones: Nucleophilic Addition Reactions 722

20 Carboxylic Acids and Nitriles 778

31 Synthetic Polymers 1242

Preface

I love writing, and I love explaining organic chemistry. This book is now in its eighth edition, but I'm still going over every word and every explanation, updating a thousand small details and trying to improve everything. My aim is always to refine the features that made earlier editions so successful, while adding new ones.

CHANGES AND ADDITIONS FOR THIS EIGHTH EDITION

- End-of-chapter problems are now grouped by topic so that students can focus on specific subjects.
- Figure references and cross-references are identified by color to better tie the text to nearby illustrations and to previous material.
- Many new problems at the ends of chapters have been added, with a particular emphasis on biologically related topics.
- Coverage of stereochemistry at tetrahedral centers has been moved forward to Chapter 5.

Specific changes within individual chapters include:

- *Chapter 2—Polar Covalent Bonds; Acids and Bases.* A new end-of-chapter *A Deeper Look* dealing with dental anesthetics derived from cocaine has been added.
- *Chapter 5—Stereochemistry at Tetrahedral Centers.* This crucial topic, so important for understanding biological chemistry, has been moved forward from its previous position in Chapter 9 to give it more prominence.
- *Chapter 6—An Overview of Organic Reactions.* All bond-dissociation data have been updated, and the discussion of energy changes in multistep reactions has been enhanced.
- *Chapter 7—Alkenes: Structure and Reactivity.* A new end-of-chapter *A Deeper Look* dealing with bioprospecting and the hunt for biologically active natural products has been added.
- *Chapter 8—Alkenes: Reactions and Synthesis.* New Sections 8.12 and 8.13 dealing with the stereochemistry of addition reactions to chiral and achiral alkenes have been added.
- *Chapter 10—Organohalides.* A discussion of synthesizing alkyl fluorides from alcohols has been added to Section 10.5 to reflect the increasing importance of fluorinated compounds in pharmaceutical chemistry. Coverage of the palladium-catalyzed Suzuki–Miyaura coupling reaction has been added to Section 10.7.
- *Chapter 12—Structure Determination: Mass Spectrometry and Infrared Spectroscopy.* A new end-of-chapter *A Deeper Look* dealing with X-ray crystallography has been added.
- *Chapter 13—Structure Determination: Nuclear Magnetic Resonance Spectroscopy.* Integration of NMR spectra is presented in a digital format rather than in the outdated stair-step format.

- *Chapter 16—Chemistry of Benzene: Electrophilic Aromatic Substitution.* Coverage of aromatic fluorination has been added to Section 16.2 to reflect the increasing importance of fluorinated compounds in pharmaceutical chemistry.
- *Chapter 17—Alcohols and Phenols.* The standard method for oxidizing alcohols has been changed to using the Dess–Martin periodinane to reflect current laboratory practice.
- *Chapter 18—Ethers and Epoxides; Thiols and Sulfides.* Coverage of the Claisen rearrangement has been expanded, and a biological example of the reaction has been added in Section 18.4. Epoxide opening by amine nucleophiles has been added to Section 18.6.
- *Chapter 19—Aldehydes and Ketones: Nucleophilic Addition Reactions.* The Tollens' oxidation has been deleted, and the mechanism of the Wittig reaction has been updated.
- *Chapter 22—Carbonyl Alpha-Substitution Reactions.* A new end-of-chapter *A Deeper Look* dealing with barbiturates has been added.
- *Chapter 25—Biomolecules: Carbohydrates.* Section 25.11 on cell-surface carbohydrates has been changed to focus on influenza viruses such as avian flu and swine flu.
- *Chapter 28—Biomolecules: Nucleic Acids.* The chapter material has been extensively updated.
- *Chapter 29—The Organic Chemistry of Metabolic Pathways.* A new end-of-chapter *A Deeper Look* dealing with statin drugs has been added.
- *Chapter 31—Synthetic Polymers.* A new Section 31.5 on olefin metathesis polymerization has been added.

FEATURES

- The "Why This Chapter?" section is a short paragraph that appears at the end of the introduction to every chapter and tells students why the material about to be covered is important.
- Key Ideas are highlighted. These include topics pivotal to students' development in organic chemistry, such as Proposing a Mechanism for a Reaction and the Rules of Resonance. These Key Ideas are further reinforced in end-of-chapter problems marked with a ▲ icon.

- The Exercises are assignable in OWL for Organic Chemistry, an online homework assessment tool in which students can practice and test their knowledge.
- Each Worked Example includes a Strategy and a detailed Solution and is followed by problems for students to try on their own. This book has more than 1800 in-text and end-of-chapter problems.
- An overview chapter, *A Preview of Carbonyl Chemistry,* follows Chapter 18 and highlights the idea that studying organic chemistry requires both summarizing and looking ahead.
- The Visualizing Chemistry Problems that begin the exercises at the end of each chapter offer students an opportunity to see chemistry in a different way by visualizing molecules rather than by simply interpreting structural formulas.

- Applied essays called *A Deeper Look* complement the text and highlight applications to the chemistry. They include, "Where Do Drugs Come From?" in Chapter 6 and "Molecular Mechanics" in Chapter 4.
- Summaries and Key Word lists help students by outlining the key concepts of the chapter.
- Summaries of Reactions at the ends of appropriate chapters bring together the key reactions from the chapter in one complete list.

Ancillaries for Students and Instructors

A COMPLETE ANCILLARY PACKAGE

Study Guide and Solutions Manual by Susan McMurry, provides answers and clear, step-by-step explanations to all in-text and end-of-chapter exercises. ISBN: 0-8400-5445-9

Companion Website has study tools, including a glossary and flashcards. Accessible from **www.cengagebrain.com**.

OWL for Organic Chemistry
OWL with eBook 24-Month Instant Access ISBN-13: 978-0-495-91561-4
OWL with eBook 6-Month Instant Access ISBN-13: 978-1-111-67700-8

Online Web Learning

By Steve Hixson and Peter Lillya of the University of Massachusetts, Amherst, and William Vining of the State University of New York at Oneonta. End-of chapter questions by David W. Brown, Florida Gulf Coast University. **OWL** Online Web Learning offers more assignable, gradable content (including end-of-chapter questions specific to this textbook) and more reliability and flexibility than any other system. OWL's powerful course management tools allow instructors to control due dates, number of attempts, and whether students see answers or receive feedback on how to solve problems. OWL includes the **Cengage YouBook**, a Flash-based eBook that is interactive and customizable. It features a text edit tool that allows instructors to modify the textbook narrative as needed. With the Cengage YouBook, instructors can quickly re-order entire sections and chapters or hide any content they don't teach to create an eBook that perfectly matches their syllabus. Instructors can further customize the Cengage YouBook by publishing web links. It also includes animated figures, video clips, highlighting, notes, and more.

Developed by chemistry instructors for teaching chemistry, OWL is the only system specifically designed to support **mastery learning**, where students work as long as they need to master each chemical concept and skill. OWL has already helped hundreds of thousands of students master chemistry through a wide range of assignment types, including tutorials and algorithmically generated homework questions that provide instant, answer-specific feedback.

OWL is continually enhanced with online learning tools to address the various learning styles of today's students such as:

- **Quick Prep** review courses that help students learn essential skills to succeed in General and Organic Chemistry
- **Jmol** molecular visualization program for rotating molecules and measuring bond distances and angles

In addition, when you become an OWL user, you can expect service that goes far beyond the ordinary. To learn more or to see a demo, please contact your Cengage Learning representative or visit us at **www.cengage.com/owl**.

CengageBrain.com App. Students can prepare for class anytime and anywhere using the CengageBrain.com application developed specifically for the Apple iPhone® and iPod touch®, which allows students to access free study materials—book-specific quizzes, flashcards, related Cengage Learning materials and more—so they can study the way they want, when they want to . . . even on the go. To learn more about this complimentary application, please visit **www.cengagebrain.com**.

Visit CengageBrain.com. To access these and additional course materials, please visit **www.cengagebrain.com**. At the CengageBrain.com home page, search for the ISBN (from the back cover of your book) using the search box at the top of the page. This will take you to the product page where these resources can be found. (Instructors can log in at **login.cengage.com**.)

***Pushing Electrons: A Guide for Students of Organic Chemistry*, third edition,** by Daniel P. Weeks. A workbook designed to help students learn techniques of electron pushing, its programmed approach emphasizes repetition and active participation. ISBN: 0-03-020693-6

Spartan Model Electronic Modeling Kit. This set of easy-to-use builders allows for the construction and 3-D manipulation of molecules of any size or complexity—from a hydrogen atom to DNA and everything in between. This kit includes the SpartanModel software on CD-ROM, an extensive molecular database, 3-D glasses, and a Tutorial and Users Guide that includes a wealth of activities to help you get the most out of your course. ISBN: 1-111-67490-6

PowerLecture Instructor CD/DVD Package, ExamView®, and Test Bank. This cross-platform lecture support CD-ROM for instructors contains PowerPoint Lecture Slides, Image Libraries, JoinIn (clicker) questions, Test Bank, and ExamView Computerized Testing. The PowerPoint slides feature updated content and art, as well as worked-out problems to help illustrate concept applications. ExamView and Test Bank now feature more question variety and range of difficulty. ISBN: 0-8400-5451-3

Understanding the Principles of Organic Chemistry: A Laboratory Course, by Pedersen and Myers. Class-tested by thousands of students and using simple equipment and green chemistry ideas, this laboratory manual includes 36 experiments that introduce traditional as well as recently developed synthetic methods. Offering up-to-date and novel experiments not found in other lab manuals, this innovative book focuses on safety, gives students practice in the basic techniques used in the organic lab, and includes microscale experiments, many drawn from the recent literature. ISBN: 1-111-42816-6

ACKNOWLEDGMENTS

I thank all the people who helped to shape this book and its message. At Cengage they include Lisa Lockwood, executive editor; Sandra Kiselica, senior development editor; Elizabeth Woods, associate editor; Teresa Trego, content project manager; Lisa Weber and Stephanie Van Camp, media editors; and Barb Bartoszek, marketing manager.

Above all, I thank my wife, Susan, who has once again read and improved the entire manuscript and has supported this project through all eight editions. I couldn't do it without her.

I am grateful to my colleagues who reviewed the manuscript for this book. They include:

Reviewers of the Eighth Edition

Andrew Bolig, San Francisco State University
Indraneel Ghosh, University of Arizona
Stephen Godleski, State University of New York, Brockport
Gordon Gribble, Dartmouth College
Matthew E. Hart, Grand Valley State University
Darren Johnson, University of Oregon
Ernest G. Nolen, Colgate University
Douglas C. Smith, California State University, San Bernadino
Gary Sulikowski, Vanderbilt University
Richard Weiss, Georgetown University
Yan Zhao, Iowa State University

Reviewers of the Seventh Edition

Arthur W. Bull, Oakland University
Robert Coleman, Ohio State University
Nicholas Drapela, Oregon State University
Christopher Hadad, Ohio State University
Eric J. Kantorowski, California Polytechnic State University
James J. Kiddle, Western Michigan University
Joseph B. Lambert, Northwestern University
Dominic McGrath, University of Arizona
Thomas A. Newton, University of Southern Maine
Michael Rathke, Michigan State University
Laren M. Tolbert, Georgia Institute of Technology

1

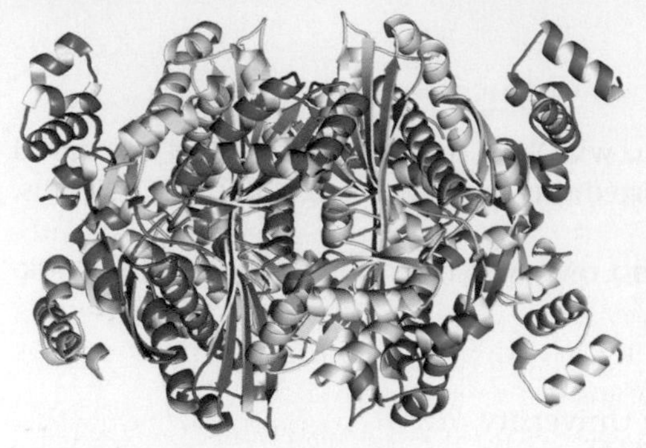

The enzyme HMG-CoA reductase, shown here as a so-called ribbon model, catalyzes a crucial step in the body's synthesis of cholesterol. Understanding how this enzyme functions has led to the development of drugs credited with saving millions of lives.

Structure and Bonding

OWL Sign in to OWL for Organic Chemistry at **www.cengage.com/owl** to view tutorials and simulations, develop problem-solving skills, and complete online homework assigned by your professor.

What is organic chemistry, and why should you study it? The answers to these questions are all around you. Every living organism is made of organic chemicals. The proteins that make up your hair, skin, and muscles; the DNA that controls your genetic heritage; the foods that nourish you; and the medicines that heal you are all organic chemicals. Anyone with a curiosity about life and living things, and anyone who wants to be a part of the remarkable advances now occurring in medicine and the biological sciences, must first understand organic chemistry. Look at the following drawings for instance, which show the chemical structures of some molecules whose names might be familiar to you. Although the drawings may appear unintelligible at this point, don't worry. Before long, they'll make perfectly good sense, and you'll soon be drawing similar structures for any substance you're interested in.

Rofecoxib (Vioxx)

Atorvastatin (Lipitor)

Oxycodone (OxyContin)

Cholesterol

Benzylpenicillin

The foundations of organic chemistry date from the mid-1700s, when chemistry was evolving from an alchemist's art into a modern science. Little was known about chemistry at that time, and the behavior of the "organic" substances isolated from plants and animals seemed different from that of the "inorganic" substances found in minerals. Organic compounds were generally low-melting solids and were usually more difficult to isolate, purify, and work with than high-melting inorganic compounds.

To many chemists, the simplest explanation for the difference in behavior between organic and inorganic compounds was that organic compounds contained a peculiar "vital force" as a result of their origin in living sources. Because of this vital force, chemists believed, organic compounds could not be prepared and manipulated in the laboratory as could inorganic compounds. As early as 1816, however, this vitalistic theory received a heavy blow when Michel Chevreul found that soap, prepared by the reaction of alkali with animal fat, could be separated into several pure organic compounds, which he termed *fatty acids*. For the first time, one organic substance (fat) was converted into others (fatty acids plus glycerin) without the intervention of an outside vital force.

$$\text{Animal fat} \xrightarrow[H_2O]{NaOH} \text{Soap} + \text{Glycerin}$$

$$\text{Soap} \xrightarrow{H_3O^+} \text{"Fatty acids"}$$

Little more than a decade later, the vitalistic theory suffered still further when Friedrich Wöhler discovered in 1828 that it was possible to convert the "inorganic" salt ammonium cyanate into the "organic" substance urea, which had previously been found in human urine.

$$NH_4^+ \ {}^-OCN \xrightarrow{\text{Heat}} H_2N{-}C({=}O){-}NH_2$$

Ammonium cyanate **Urea**

By the mid-1800s, the weight of evidence was clearly against the vitalistic theory and it was clear that there was no fundamental difference between organic and inorganic compounds. The same fundamental principles explain the behaviors of all substances, regardless of origin or complexity. The only distinguishing characteristic of organic chemicals is that all contain the element carbon.

Organic chemistry, then, is the study of carbon compounds. But why is carbon special? Why, of the more than 50 million presently known chemical compounds, do most of them contain carbon? The answers to these questions come from carbon's electronic structure and its consequent position in the periodic table **(Figure 1.1)**. As a group 4A element, carbon can share four valence electrons and form four strong covalent bonds. Furthermore, carbon atoms can bond to one another, forming long chains and rings. Carbon, alone of all elements, is able to form an immense diversity of compounds, from the simple methane, with one carbon atom, to the staggeringly complex DNA, which can have more than *100 million* carbons.

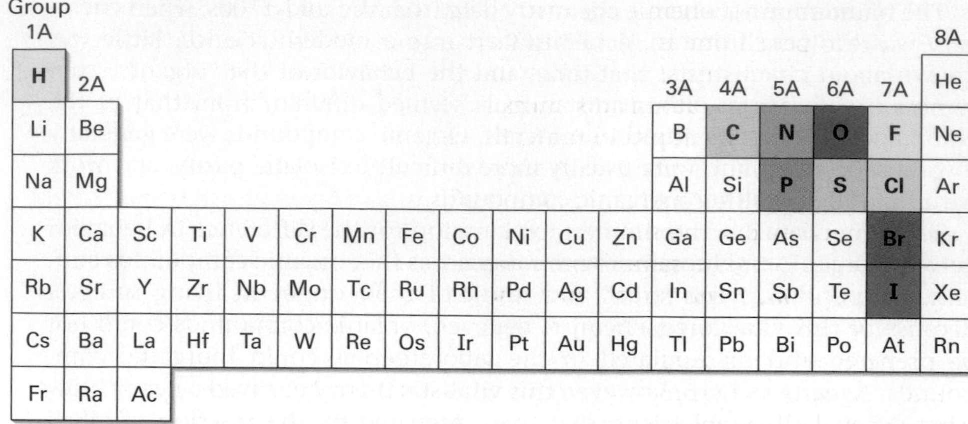

Figure 1.1 The position of **carbon** in the periodic table. Other elements commonly found in organic compounds are shown in the colors typically used to represent them.

Not all carbon compounds are derived from living organisms of course. Modern chemists have developed a remarkably sophisticated ability to design and synthesize new organic compounds in the laboratory—medicines, dyes, polymers, and a host of other substances. Organic chemistry touches the lives of everyone; its study can be a fascinating undertaking.

Why This Chapter? We'll ease into the study of organic chemistry by first reviewing some ideas about atoms, bonds, and molecular geometry that you may recall from your general chemistry course. Much of the material in this chapter and the next is likely to be familiar to you, but it's nevertheless a good idea to make sure you understand it before going on.

1.1 Atomic Structure: The Nucleus

As you probably know from your general chemistry course, an atom consists of a dense, positively charged nucleus surrounded at a relatively large distance by negatively charged electrons **(Figure 1.2)**. The nucleus consists of subatomic particles called protons, which are positively charged, and neutrons, which are electrically neutral. Because an atom is neutral overall, the number of positive protons in the nucleus and the number of negative electrons surrounding the nucleus are the same.

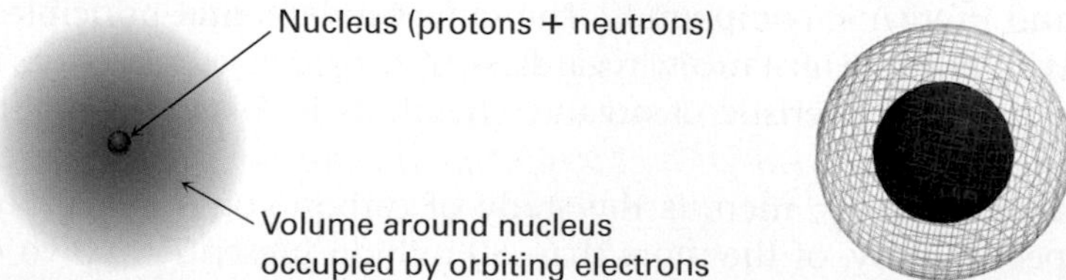

Figure 1.2 A schematic view of an atom. The dense, positively charged nucleus contains most of the atom's mass and is surrounded by negatively charged electrons. The three-dimensional view on the right shows calculated electron-density surfaces. Electron density increases steadily toward the nucleus and is 40 times greater at the blue solid surface than at the gray mesh surface.

Although extremely small—about 10^{-14} to 10^{-15} meter (m) in diameter—the nucleus nevertheless contains essentially all the mass of the atom. Electrons have negligible mass and circulate around the nucleus at a distance of approximately 10^{-10} m. Thus, the diameter of a typical atom is about 2×10^{-10} m, or

200 picometers (pm), where 1 pm = 10^{-12} m. To give you an idea of how small this is, a thin pencil line is about 3 million carbon atoms wide. Many organic chemists and biochemists, particularly in the United States, still use the unit *angstrom* (Å) to express atomic distances, where 1 Å – 100 pm = 10^{-10} m, but we'll stay with the SI unit picometer in this book.

A specific atom is described by its atomic number (*Z*), which gives the number of protons (or electrons) it contains, and its mass number (*A*), which gives the total number of protons plus neutrons in its nucleus. All the atoms of a given element have the same atomic number—1 for hydrogen, 6 for carbon, 15 for phosphorus, and so on—but they can have different mass numbers depending on how many neutrons they contain. Atoms with the same atomic number but different mass numbers are called **isotopes**.

The weighted average mass in atomic mass units (amu) of an element's naturally occurring isotopes is called the element's atomic mass (or atomic weight)—1.008 amu for hydrogen, 12.011 amu for carbon, 30.974 amu for phosphorus, and so on. Atomic masses of the elements are given in the periodic table in the front of this book.

1.2 Atomic Structure: Orbitals

How are the electrons distributed in an atom? You might recall from your general chemistry course that, according to the quantum mechanical model, the behavior of a specific electron in an atom can be described by a mathematical expression called a *wave equation*—the same type of expression used to describe the motion of waves in a fluid. The solution to a wave equation is called a *wave function*, or **orbital**, and is denoted by the Greek letter psi (ψ).

By plotting the square of the wave function, ψ^2, in three-dimensional space, an orbital describes the volume of space around a nucleus that an electron is most likely to occupy. You might therefore think of an orbital as looking like a photograph of the electron taken at a slow shutter speed. In such a photo, the orbital would appear as a blurry cloud, indicating the region of space where the electron has been. This electron cloud doesn't have a sharp boundary, but for practical purposes we can set the limits by saying that an orbital represents the space where an electron spends 90% to 95% of its time.

What do orbitals look like? There are four different kinds of orbitals, denoted *s*, *p*, *d*, and *f*, each with a different shape. Of the four, we'll be concerned primarily with *s* and *p* orbitals because these are the most common in organic and biological chemistry. An *s* orbital is spherical, with the nucleus at its center; a *p* orbital is dumbbell-shaped; and four of the five *d* orbitals are cloverleaf-shaped, as shown in **Figure 1.3**. The fifth *d* orbital is shaped like an elongated dumbbell with a doughnut around its middle.

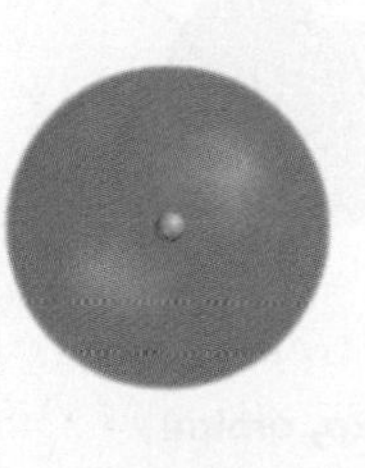

An *s* orbital

A *p* orbital

A *d* orbital

Figure 1.3 Representations of *s*, *p*, and *d* orbitals. An *s* orbital is spherical, a *p* orbital is dumbbell-shaped, and four of the five *d* orbitals are cloverleaf-shaped. **Different lobes** of *p* and *d* orbitals are often drawn for convenience as teardrops, but their actual shape is more like that of a doorknob, as indicated.

The orbitals in an atom are organized into different **electron shells**, centered around the nucleus and having successively larger size and energy. Different shells contain different numbers and kinds of orbitals, and each orbital within a shell can be occupied by two electrons. The first shell contains only a single *s* orbital, denoted 1*s*, and thus holds only 2 electrons. The second shell contains one 2*s* orbital and three 2*p* orbitals and thus holds a total of 8 electrons. The third shell contains a 3*s* orbital, three 3*p* orbitals, and five 3*d* orbitals, for a total capacity of 18 electrons. These orbital groupings and their energy levels are shown in **Figure 1.4**.

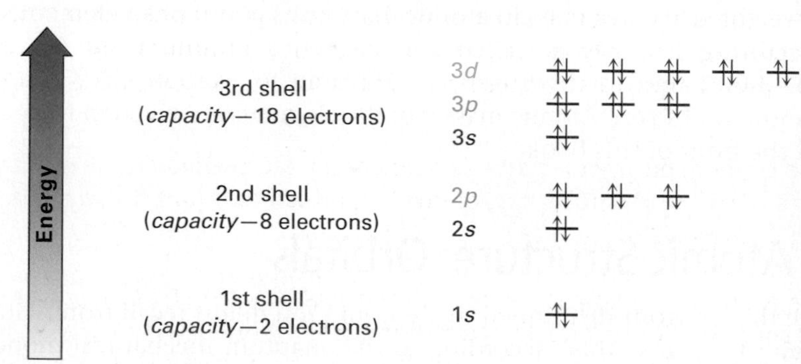

Figure 1.4 The energy levels of electrons in an atom. The first shell holds a maximum of 2 electrons in one **1*s*** orbital; the second shell holds a maximum of 8 electrons in one **2*s*** and three **2*p*** orbitals; the third shell holds a maximum of 18 electrons in one **3*s***, three **3*p***, and five **3*d*** orbitals; and so on. The two electrons in each orbital are represented by up and down arrows, ↑↓. Although not shown, the energy level of the 4*s* orbital falls between 3*p* and 3*d*.

The three different *p* orbitals within a given shell are oriented in space along mutually perpendicular directions, denoted p_x, p_y, and p_z. As shown in **Figure 1.5**, the two lobes of each *p* orbital are separated by a region of zero electron density called a **node**. Furthermore, the two orbital regions separated by the node have different algebraic signs, + and −, in the wave function, as represented by the different colors in Figure 1.5. We'll see in **Section 1.11** that these algebraic signs of different orbital lobes have important consequences with respect to chemical bonding and chemical reactivity.

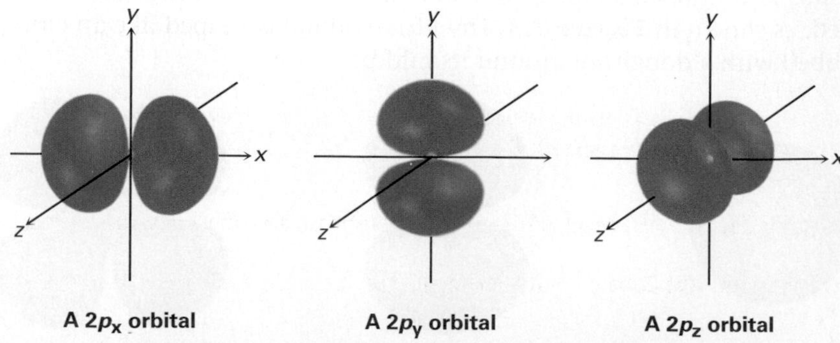

Figure 1.5 Shapes of the 2*p* orbitals. Each of the three mutually perpendicular, dumbbell-shaped orbitals has **two lobes** separated by a node. The two lobes have different algebraic signs in the corresponding wave function, as indicated by the different colors.

1.3 Atomic Structure: Electron Configurations

The lowest-energy arrangement, or **ground-state electron configuration**, of an atom is a listing of the orbitals occupied by its electrons. We can predict this arrangement by following three rules.

RULE 1
The lowest-energy orbitals fill up first, according to the order $1s \rightarrow 2s \rightarrow 2p \rightarrow 3s \rightarrow 3p \rightarrow 4s \rightarrow 3d$, a statement called the aufbau principle. Note that the $4s$ orbital lies between the $3p$ and $3d$ orbitals in energy.

RULE 2
Electrons act in some ways as if they were spinning around an axis, somewhat as the earth spins. This spin can have two orientations, denoted as up (↑) and down (↓). Only two electrons can occupy an orbital, and they must be of opposite spin, a statement called the Pauli exclusion principle.

RULE 3
If two or more empty orbitals of equal energy are available, one electron occupies each with spins parallel until all orbitals are half-full, a statement called Hund's rule.

Some examples of how these rules apply are shown in Table 1.1. Hydrogen, for instance, has only one electron, which must occupy the lowest-energy orbital. Thus, hydrogen has a $1s$ ground-state configuration. Carbon has six electrons and the ground-state configuration $1s^2\ 2s^2\ 2p_x^{\ 1}\ 2p_y^{\ 1}$, and so forth. Note that a superscript is used to represent the number of electrons in a particular orbital.

Table 1.1 Ground-State Electron Configurations of Some Elements

Element	Atomic number	Configuration	Element	Atomic number	Configuration
Hydrogen	1	$1s$ ↑	Phosphorus	15	$3p$ ↑ ↑ ↑
					$3s$ ↑↓
Carbon	6	$2p$ ↑ ↑ —			$2p$ ↑↓ ↑↓ ↑↓
		$2s$ ↑↓			$2s$ ↑↓
		$1s$ ↑↓			$1s$ ↑↓

Problem 1.1
Give the ground-state electron configuration for each of the following elements:
(a) Oxygen **(b)** Nitrogen **(c)** Sulfur

Problem 1.2
How many electrons does each of the following elements have in its outermost electron shell?
(a) Magnesium **(b)** Cobalt **(c)** Selenium

1.4 Development of Chemical Bonding Theory

By the mid-1800s, the new science of chemistry was developing rapidly and chemists had begun to probe the forces holding compounds together. In 1858, August Kekulé and Archibald Couper independently proposed that, in all organic compounds, carbon is *tetravalent*—it always forms four bonds when it joins other elements to form stable compounds. Furthermore, said Kekulé, carbon atoms can bond to one another to form extended chains of linked atoms. In 1865, Kekulé provided another major advance when he suggested that carbon chains can double back on themselves to form *rings* of atoms.

Although Kekulé and Couper were correct in describing the tetravalent nature of carbon, chemistry was still viewed in a two-dimensional way until 1874. In that year, Jacobus van't Hoff and Joseph Le Bel added a third dimension to our ideas about organic compounds when they proposed that the four bonds of carbon are not oriented randomly but have specific spatial directions. Van't Hoff went even further and suggested that the four atoms to which carbon is bonded sit at the corners of a regular tetrahedron, with carbon in the center.

A representation of a tetrahedral carbon atom is shown in **Figure 1.6**. Note the conventions used to show three-dimensionality: solid lines represent bonds in the plane of the page, the heavy wedged line represents a bond coming out of the page toward the viewer, and the dashed line represents a bond receding back behind the page, away from the viewer. These representations will be used throughout the text.

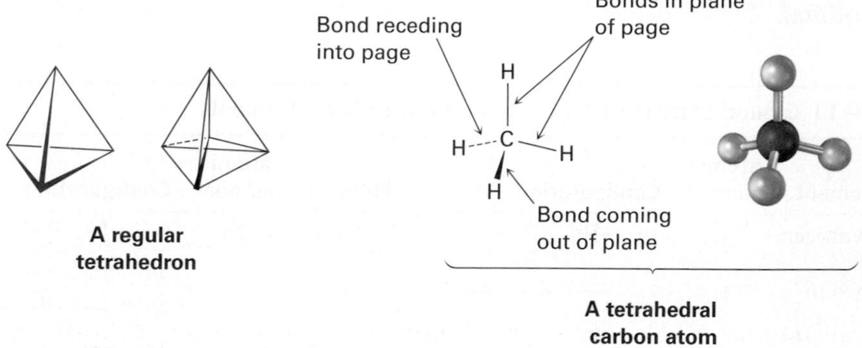

Figure 1.6 A representation of a tetrahedral carbon atom. The solid lines represent bonds in the plane of the paper, the heavy wedged line represents a bond coming out of the plane of the page, and the dashed line represents a bond going back behind the plane of the page.

Why, though, do atoms bond together, and how can bonds be described electronically? The *why* question is relatively easy to answer: atoms bond together because the compound that results is more stable and lower in energy than the separate atoms. Energy—usually as heat—always flows out of the chemical system when a bond forms. Conversely, energy must be put into the chemical system to break a bond. Making bonds always releases energy, and breaking bonds always absorbs energy. The *how* question is more difficult. To answer it, we need to know more about the electronic properties of atoms.

We know through observation that eight electrons (an electron *octet*) in an atom's outermost shell, or **valence shell**, impart special stability to the noble-gas elements in group 8A of the periodic table: Ne (2 + *8*); Ar (2 + 8 + *8*); Kr (2 + 8 + 18 +*8*). We also know that the chemistry of main-group elements is governed by their tendency to take on the electron configuration of the nearest noble gas. The alkali metals in group 1A, for example, achieve a noble-gas configuration by losing the single *s* electron from their valence shell to form a cation, while the halogens in group 7A achieve a noble-gas configuration by gaining a *p* electron to fill their valence shell and form an anion. The resultant ions are held together in compounds like $Na^+ Cl^-$ by an electrostatic attraction that we call an *ionic bond.*

But how do elements closer to the middle of the periodic table form bonds? Look at methane, CH_4, the main constituent of natural gas, for example. The bonding in methane is not ionic because it would take too much energy for carbon ($1s^2\ 2s^2\ 2p^2$) either to gain or lose four electrons to achieve a noble-gas configuration. As a result, carbon bonds to other atoms, not by gaining or losing electrons, but by sharing them. Such a shared-electron bond, first proposed in 1916 by G. N. Lewis, is called a **covalent bond**. The neutral collection of atoms held together by covalent bonds is called a **molecule**.

A simple way of indicating the covalent bonds in molecules is to use what are called *Lewis structures,* or **electron-dot structures**, in which the valence-shell electrons of an atom are represented as dots. Thus, hydrogen has one dot representing its 1*s* electron, carbon has four dots ($2s^2\ 2p^2$), oxygen has six dots ($2s^2\ 2p^4$), and so on. A stable molecule results whenever a noble-gas configuration is achieved for all the atoms—eight dots (an octet) for main-group atoms or two dots for hydrogen. Simpler still is the use of *Kekulé structures,* or **line-bond structures**, in which a two-electron covalent bond is indicated as a line drawn between atoms.

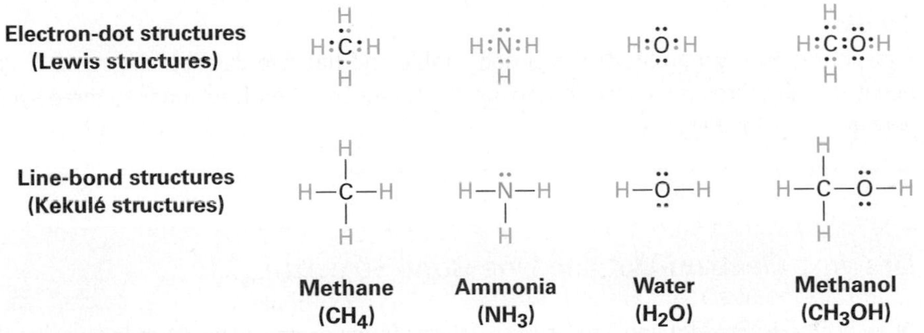

Electron-dot structures (Lewis structures)

Line-bond structures (Kekulé structures)

Methane (CH_4) **Ammonia (NH_3)** **Water (H_2O)** **Methanol (CH_3OH)**

The number of covalent bonds an atom forms depends on how many additional valence electrons it needs to reach a noble-gas configuration. Hydrogen has one valence electron (1*s*) and needs one more to reach the helium configuration ($1s^2$), so it forms one bond. Carbon has four valence electrons ($2s^2\ 2p^2$) and needs four more to reach the neon configuration ($2s^2\ 2p^6$), so it forms four bonds. Nitrogen has five valence electrons ($2s^2\ 2p^3$), needs three more, and forms three bonds; oxygen has six valence electrons ($2s^2\ 2p^4$), needs two more, and forms two bonds; and the halogens have seven valence electrons, need one more, and form one bond.

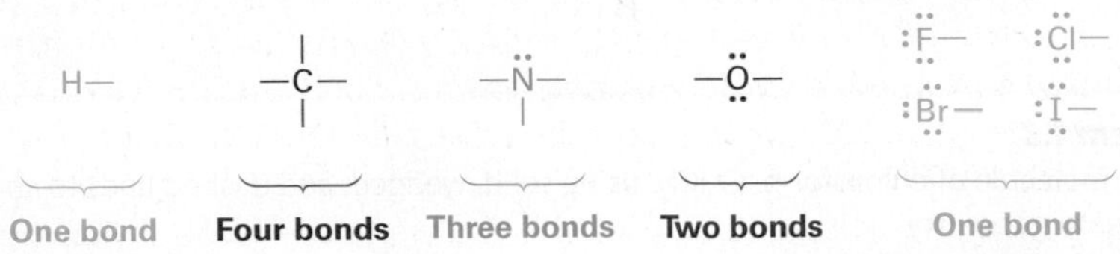

H— (One bond) —C— (**Four bonds**) —N— (Three bonds) —O— (**Two bonds**) F—, Cl—, Br—, I— (One bond)

Valence electrons that are not used for bonding are called **lone-pair electrons**, or *nonbonding electrons*. The nitrogen atom in ammonia, NH_3, for instance, shares six valence electrons in three covalent bonds and has its remaining two valence electrons in a nonbonding lone pair. As a time-saving shorthand, nonbonding electrons are often omitted when drawing line-bond structures, but you still have to keep them in mind since they're often crucial in chemical reactions.

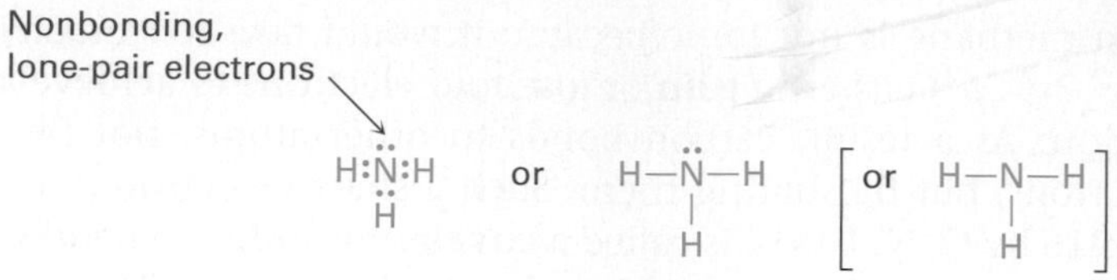

Nonbonding, lone-pair electrons

H:N̈:H or H—N̈—H [or H—N—H]

Ammonia

Worked Example 1.1 Predicting the Number of Bonds Formed by an Atom

How many hydrogen atoms does phosphorus bond to in forming phosphine, $PH_?$?

Strategy

Identify the periodic group of phosphorus, and tell from that how many electrons (bonds) are needed to make an octet.

Solution

Phosphorus is in group 5A of the periodic table and has five valence electrons. It thus needs to share three more electrons to make an octet and therefore bonds to three hydrogen atoms, giving PH_3.

Worked Example 1.2 Drawing Electron-Dot and Line-Bond Structures

Draw both electron-dot and line-bond structures for chloromethane, CH_3Cl.

Strategy

Remember that a bond—that is, a pair of shared electrons—is represented as a line between atoms.

Solution

Hydrogen has one valence electron, carbon has four valence electrons, and chlorine has seven valence electrons. Thus, chloromethane is represented as

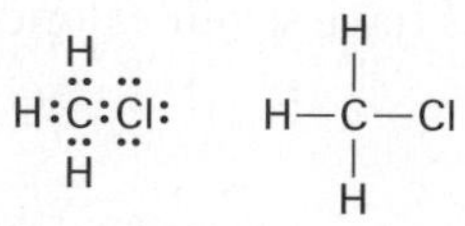

H:C:Cl: H—C—Cl **Chloromethane**

Problem 1.3

Draw a molecule of chloroform, $CHCl_3$, using solid, wedged, and dashed lines to show its tetrahedral geometry.

Problem 1.4
Convert the following representation of ethane, C_2H_6, into a conventional drawing that uses solid, wedged, and dashed lines to indicate tetrahedral geometry around each carbon (gray = C, ivory = H).

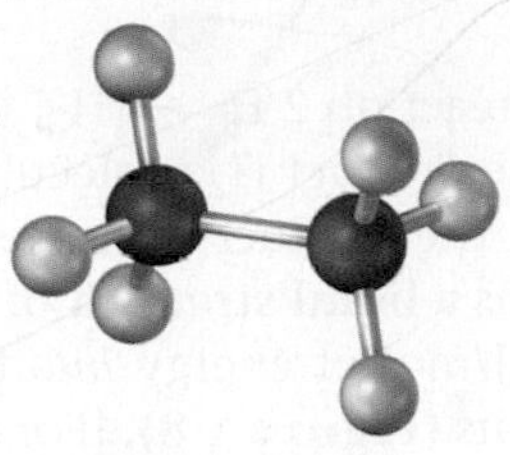
Ethane

Problem 1.5
What are likely formulas for the following substances?
(a) $CCl_?$ **(b)** $AlH_?$ **(c)** $CH_?Cl_2$ **(d)** $SiF_?$ **(e)** $CH_3NH_?$

Problem 1.6
Write line-bond structures for the following substances, showing all nonbonding electrons:
(a) $CHCl_3$, chloroform **(b)** H_2S, hydrogen sulfide
(c) CH_3NH_2, methylamine **(d)** CH_3Li, methyllithium

Problem 1.7
Why can't an organic molecule have the formula C_2H_7?

1.5 Describing Chemical Bonds: Valence Bond Theory

How does electron sharing lead to bonding between atoms? Two models have been developed to describe covalent bonding: *valence bond theory* and *molecular orbital theory*. Each model has its strengths and weaknesses, and chemists tend to use them interchangeably depending on the circumstances. Valence bond theory is the more easily visualized of the two, so most of the descriptions we'll use in this book derive from that approach.

According to **valence bond theory**, a covalent bond forms when two atoms approach each other closely and a singly occupied orbital on one atom overlaps a singly occupied orbital on the other atom. The electrons are now paired in the overlapping orbitals and are attracted to the nuclei of both atoms, thus bonding the atoms together. In the H_2 molecule, for instance, the H–H bond results from the overlap of two singly occupied hydrogen 1*s* orbitals.

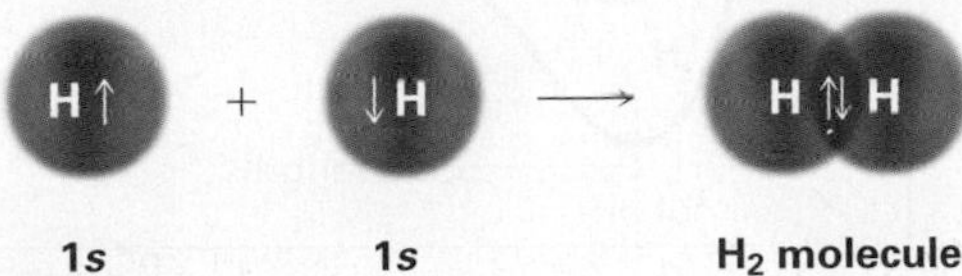

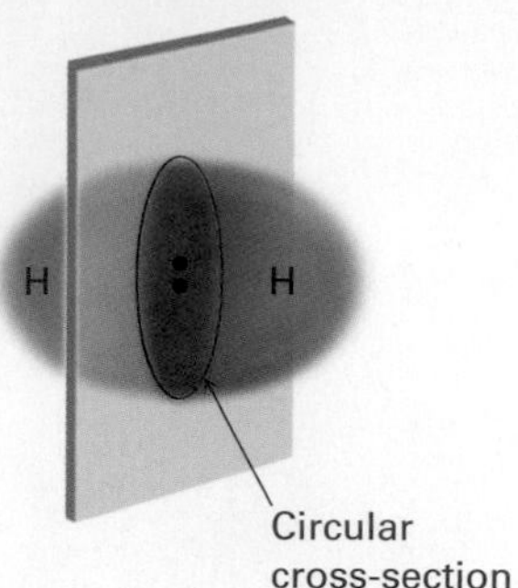

Figure 1.7 The cylindrical symmetry of the H–H σ bond in an H_2 molecule. The intersection of a plane cutting through the σ bond is a circle.

The overlapping orbitals in the H_2 molecule have the elongated egg shape we might get by pressing two spheres together. If a plane were to pass through the middle of the bond, the intersection of the plane and the overlapping orbitals would be a circle. In other words, the H–H bond is cylindrically symmetrical, as shown in **Figure 1.7**. Such bonds, which are formed by the head-on overlap of two atomic orbitals along a line drawn between the nuclei, are called **sigma (σ) bonds.**

During the bond-forming reaction 2 H· → H_2, 436 kJ/mol (104 kcal/mol) of energy is released. Because the product H_2 molecule has 436 kJ/mol less energy than the starting 2 H· atoms, the product is more stable than the reactant and we say that the H–H bond has a **bond strength** of 436 kJ/mol. In other words, we would have to put 436 kJ/mol of energy *into* the H–H bond to break the H_2 molecule apart into H atoms **(Figure 1.8)**. [For convenience, we'll generally give energies in both kilocalories (kcal) and the SI unit kilojoules (kJ): 1 kJ = 0.2390 kcal; 1 kcal = 4.184 kJ.]

Figure 1.8 Relative energy levels of two H atoms and the H_2 molecule. The H_2 molecule has 436 kJ/mol (104 kcal/mol) less energy than the two H atoms, so 436 kJ/mol of energy is **released** when the H–H bond **forms**. Conversely, 436 kJ/mol is **absorbed** when the H–H bond **breaks**.

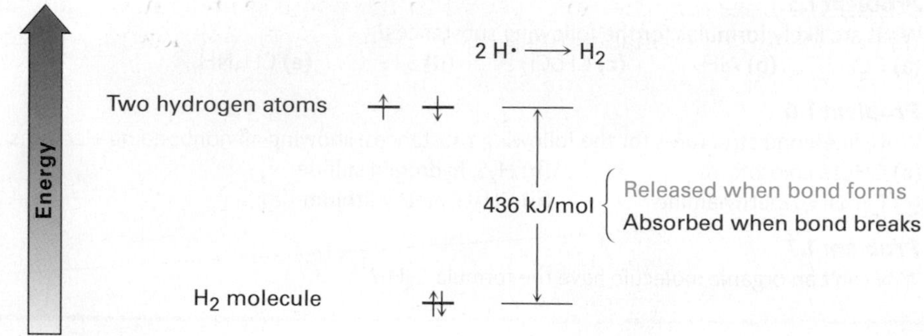

How close are the two nuclei in the H_2 molecule? If they are too close, they will repel each other because both are positively charged, yet if they're too far apart, they won't be able to share the bonding electrons. Thus, there is an optimum distance between nuclei that leads to maximum stability **(Figure 1.9)**. Called the **bond length**, this distance is 74 pm in the H_2 molecule. Every covalent bond has both a characteristic bond strength and bond length.

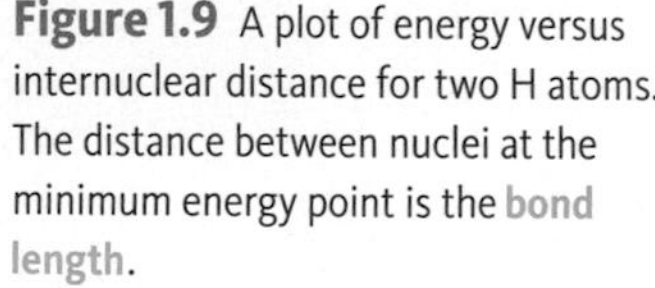

Figure 1.9 A plot of energy versus internuclear distance for two H atoms. The distance between nuclei at the minimum energy point is the **bond length**.

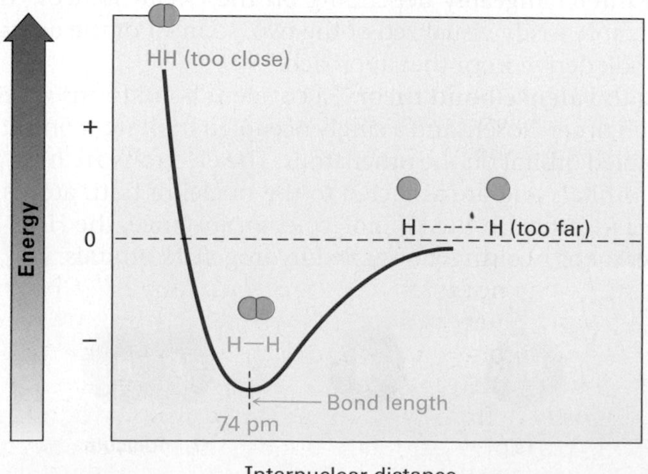

1.6 sp^3 Hybrid Orbitals and the Structure of Methane

The bonding in the hydrogen molecule is fairly straightforward, but the situation is more complicated in organic molecules with tetravalent carbon atoms. Take methane, CH_4, for instance. As we've seen, carbon has four valence electrons ($2s^2\ 2p^2$) and forms four bonds. Because carbon uses two kinds of orbitals for bonding, 2*s* and 2*p*, we might expect methane to have two kinds of C–H bonds. In fact, though, all four C–H bonds in methane are identical and are spatially oriented toward the corners of a regular tetrahedron (Figure 1.6). How can we explain this?

An answer was provided in 1931 by Linus Pauling, who showed mathematically how an *s* orbital and three *p* orbitals on an atom can combine, or *hybridize,* to form four equivalent atomic orbitals with tetrahedral orientation. Shown in **Figure 1.10**, these tetrahedrally oriented orbitals are called ***sp*^{3} hybrids**. Note that the superscript 3 in the name sp^3 tells how many of each type of atomic orbital combine to form the hybrid, not how many electrons occupy it.

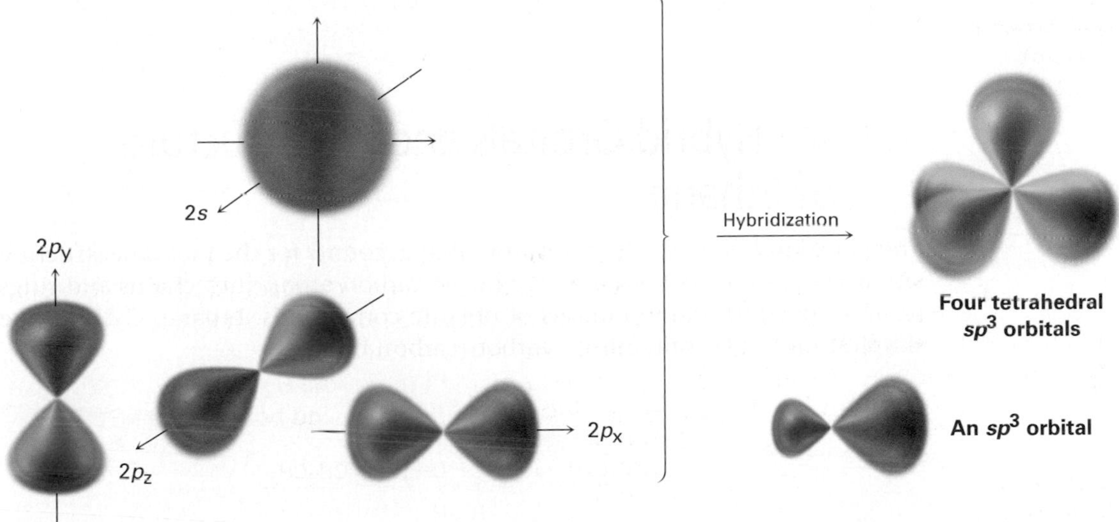

Figure 1.10 Four sp^3 hybrid orbitals, oriented to the corners of a regular tetrahedron, are formed by combination of an ***s* orbital** and three *p* orbitals **(red/blue)**. The sp^3 hybrids have two lobes and are unsymmetrical about the nucleus, giving them a directionality and allowing them to form strong bonds when they overlap an orbital from another atom.

The concept of hybridization explains how carbon forms four equivalent tetrahedral bonds but not why it does so. The shape of the hybrid orbital suggests the answer. When an *s* orbital hybridizes with three *p* orbitals, the resultant sp^3 hybrid orbitals are unsymmetrical about the nucleus. One of the two lobes is larger than the other and can therefore overlap more effectively with an orbital from another atom to form a bond. As a result, sp^3 hybrid orbitals form stronger bonds than do unhybridized *s* or *p* orbitals.

The asymmetry of sp^3 orbitals arises because, as noted previously, the two lobes of a p orbital have different algebraic signs, + and −, in the wave function. Thus, when a p orbital hybridizes with an s orbital, the positive p lobe adds to the s orbital but the negative p lobe subtracts from the s orbital. The resultant hybrid orbital is therefore unsymmetrical about the nucleus and is strongly oriented in one direction.

When each of the four identical sp^3 hybrid orbitals of a carbon atom overlaps with the 1s orbital of a hydrogen atom, four identical C–H bonds are formed and methane results. Each C–H bond in methane has a strength of 439 kJ/mol (105 kcal/mol) and a length of 109 pm. Because the four bonds have a specific geometry, we also can define a property called the **bond angle**. The angle formed by each H–C–H is 109.5°, the so-called tetrahedral angle. Methane thus has the structure shown in **Figure 1.11**.

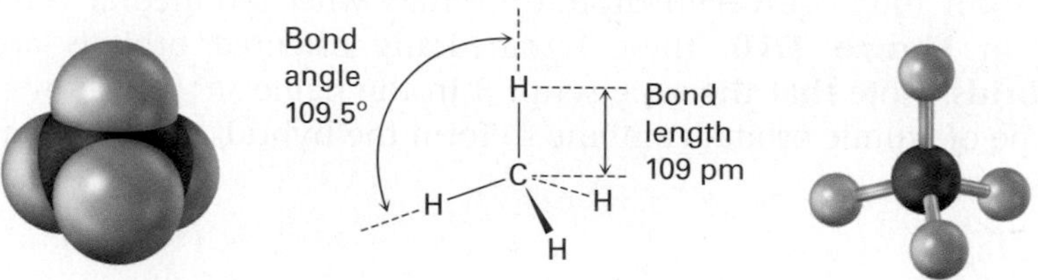

Figure 1.11 The structure of methane, showing its 109.5° bond angles.

1.7 sp^3 Hybrid Orbitals and the Structure of Ethane

The same kind of orbital hybridization that accounts for the methane structure also accounts for the bonding together of carbon atoms into chains and rings to make possible many millions of organic compounds. Ethane, C_2H_6, is the simplest molecule containing a carbon–carbon bond.

```
  H H          H   H
H:C:C:H      H—C—C—H      CH3CH3
  H H          H   H
```

Some representations of ethane

We can picture the ethane molecule by imagining that the two carbon atoms bond to each other by σ overlap of an sp^3 hybrid orbital from each **(Figure 1.12)**. The remaining three sp^3 hybrid orbitals on each carbon overlap with the 1s orbitals of three hydrogens to form the six C–H bonds. The C–H bonds in ethane are similar to those in methane, although a bit weaker—421 kJ/mol (101 kcal/mol) for ethane versus 439 kJ/mol for methane. The C–C bond is 154 pm long and has a strength of 377 kJ/mol (90 kcal/mol). All the bond angles of ethane are near, although not exactly at, the tetrahedral value of 109.5°.

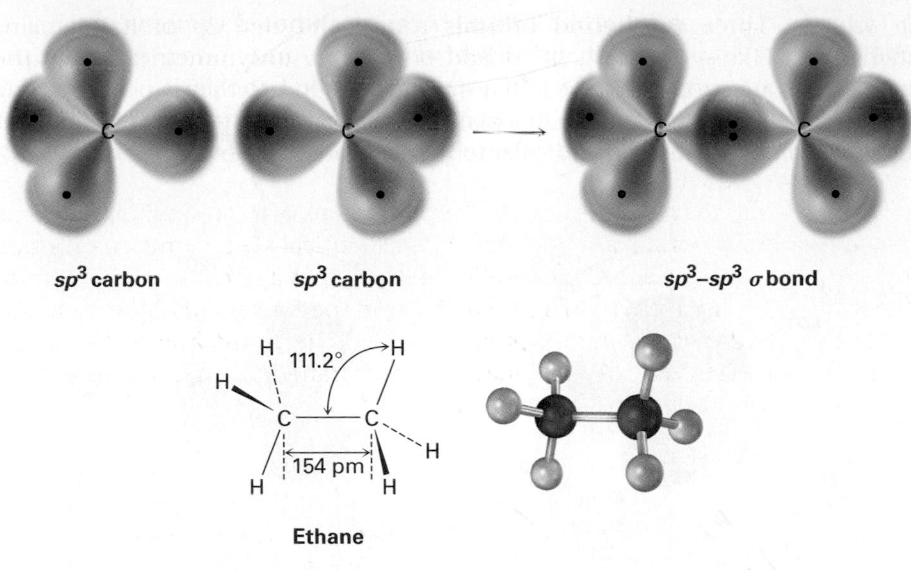

Figure 1.12 The structure of ethane. The carbon–carbon bond is formed by σ overlap of sp^3 hybrid orbitals. For clarity, the smaller lobes of the sp^3 hybrid orbitals are not shown.

Problem 1.8
Draw a line-bond structure for propane, $CH_3CH_2CH_3$. Predict the value of each bond angle, and indicate the overall shape of the molecule.

Problem 1.9
Convert the following molecular model of hexane, a component of gasoline, into a line-bond structure (gray = C, ivory = H).

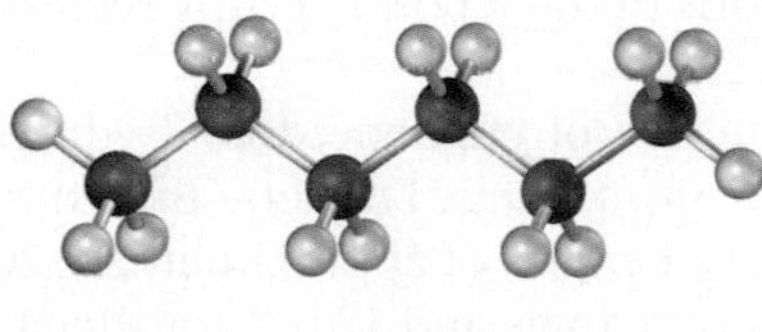

Hexane

1.8 sp^2 Hybrid Orbitals and the Structure of Ethylene

The bonds we've seen in methane and ethane are called *single bonds* because they result from the sharing of one electron pair between bonded atoms. It was recognized nearly 150 years ago, however, that carbon atoms can also form *double bonds* by sharing *two* electron pairs between atoms or *triple bonds* by sharing *three* electron pairs. Ethylene, for instance, has the structure $H_2C{=}CH_2$ and contains a carbon–carbon double bond, while acetylene has the structure $HC{\equiv}CH$ and contains a carbon–carbon triple bond.

How are multiple bonds described by valence bond theory? When we discussed sp^3 hybrid orbitals in **Section 1.6**, we said that the four valence-shell atomic orbitals of carbon combine to form four equivalent sp^3 hybrids. Imagine instead that the 2*s* orbital combines with only *two* of the three available

$2p$ orbitals. Three **sp^2 hybrid orbitals** result, and one $2p$ orbital remains unchanged. Like sp^3 hybrids, sp^2 hybrid orbitals are unsymmetrical about the nucleus and are strongly oriented in a specific direction so they can form strong bonds. The three sp^2 orbitals lie in a plane at angles of 120° to one another, with the remaining *p* orbital perpendicular to the sp^2 plane, as shown in **Figure 1.13**.

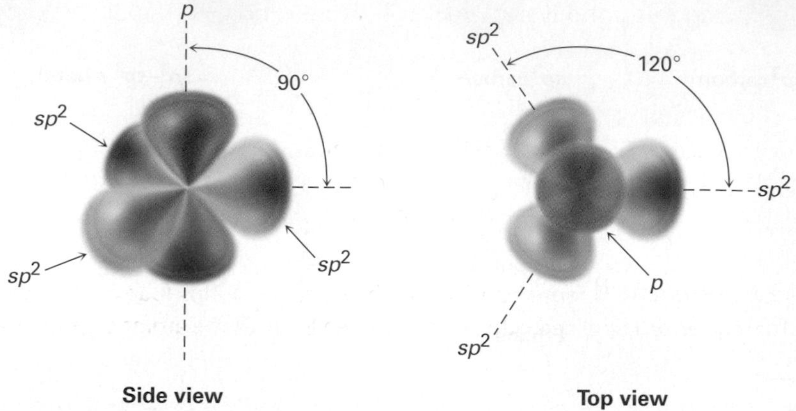

Figure 1.13 sp^2 Hybridization. The three equivalent **sp^2 hybrid orbitals** lie in a plane at angles of 120° to one another, and a single unhybridized *p* orbital **(red/blue)** is perpendicular to the sp^2 plane.

When two carbons with sp^2 hybridization approach each other, they form a strong σ bond by sp^2–sp^2 head-on overlap. At the same time, the unhybridized *p* orbitals interact by sideways overlap to form what is called a **pi (π) bond**. The combination of an sp^2–sp^2 σ bond and a $2p$–$2p$ π bond results in the sharing of four electrons and the formation of a carbon–carbon double bond **(Figure 1.14)**. Note that the electrons in the σ bond occupy the region centered between nuclei, while the electrons in the π bond occupy regions above and below a line drawn between nuclei.

To complete the structure of ethylene, four hydrogen atoms form σ bonds with the remaining four sp^2 orbitals. Ethylene thus has a planar structure, with H–C–H and H–C–C bond angles of approximately 120°. (The actual values are 117.4° for the H–C–H bond angle and 121.3° for the H–C–C bond angle.) Each C–H bond has a length of 108.7 pm and a strength of 464 kJ/mol (111 kcal/mol).

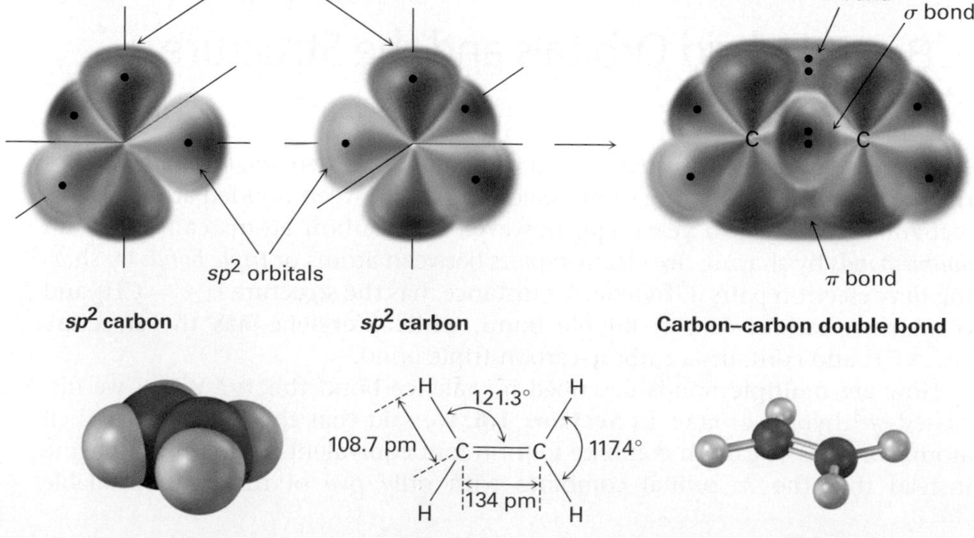

Figure 1.14 The structure of ethylene. One part of the double bond in ethylene results from σ (head-on) overlap of **sp^2 orbitals**, and the other part results from π (sideways) overlap of unhybridized *p* orbitals **(red/blue)**. The π bond has regions of electron density above and below a line drawn between nuclei.

As you might expect, the carbon–carbon double bond in ethylene is both shorter and stronger than the single bond in ethane because it has four electrons bonding the nuclei together rather than two. Ethylene has a C=C bond length of 134 pm and a strength of 728 kJ/mol (174 kcal/mol) versus a C–C length of 154 pm and a strength of 377 kJ/mol for ethane. The carbon–carbon double bond is less than twice as strong as a single bond because the sideways overlap in the π part of the double bond is not as great as the head-on overlap in the σ part.

Worked Example 1.3 Drawing Electron-Dot and Line-Bond Structures

Commonly used in biology as a tissue preservative, formaldehyde, CH_2O, contains a carbon–*oxygen* double bond. Draw electron-dot and line-bond structures of formaldehyde, and indicate the hybridization of the carbon orbitals.

Strategy

We know that hydrogen forms one covalent bond, carbon forms four, and oxygen forms two. Trial and error, combined with intuition, is needed to fit the atoms together.

Solution

There is only one way that two hydrogens, one carbon, and one oxygen can combine:

Electron-dot structure **Line-bond structure**

Like the carbon atoms in ethylene, the carbon atom in formaldehyde is in a double bond and its orbitals are therefore sp^2-hybridized.

Problem 1.10

Draw a line-bond structure for propene, $CH_3CH{=}CH_2$. Indicate the hybridization of the orbitals on each carbon, and predict the value of each bond angle.

Problem 1.11

Draw a line-bond structure for 1,3-butadiene, $H_2C{=}CH{-}CH{=}CH_2$. Indicate the hybridization of the orbitals on each carbon, and predict the value of each bond angle.

Problem 1.12

Following is a molecular model of aspirin (acetylsalicylic acid). Identify the hybridization of the orbitals on each carbon atom in aspirin, and tell which atoms have lone pairs of electrons (gray = C, red = O, ivory = H).

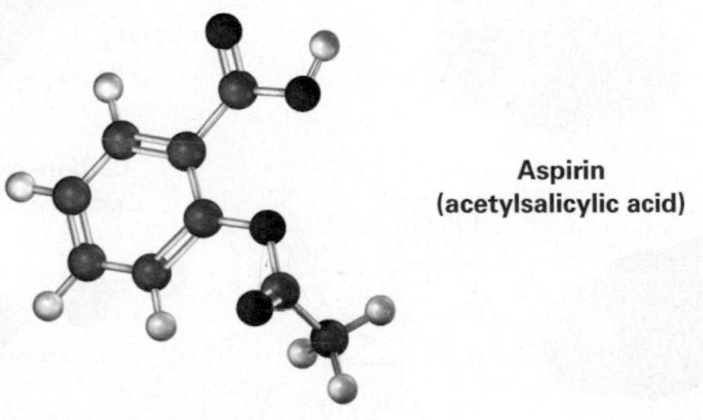

Aspirin (acetylsalicylic acid)

1.9 *sp* Hybrid Orbitals and the Structure of Acetylene

In addition to forming single and double bonds by sharing two and four electrons, respectively, carbon also can form a *triple* bond by sharing six electrons. To account for the triple bond in a molecule such as acetylene, H—C≡C—H, we need a third kind of hybrid orbital, an ***sp* hybrid**. Imagine that, instead of combining with two or three *p* orbitals, a carbon 2*s* orbital hybridizes with only a single *p* orbital. Two *sp* hybrid orbitals result, and two *p* orbitals remain unchanged. The two *sp* orbitals are oriented 180° apart on the *x*-axis, while the remaining two *p* orbitals are perpendicular on the *y*-axis and the *z*-axis, as shown in **Figure 1.15**.

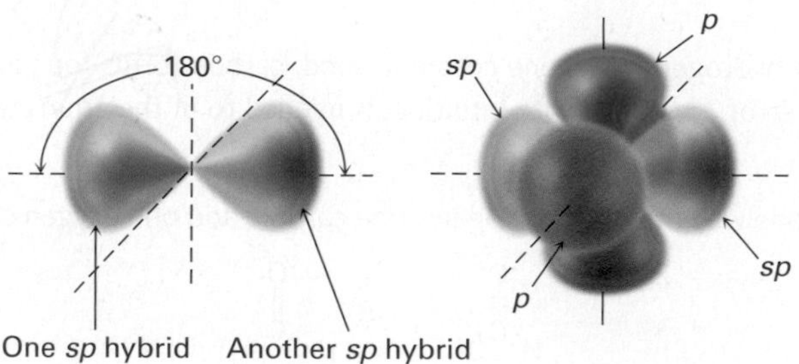

Figure 1.15 *sp* Hybridization. The two *sp* hybrid orbitals are oriented 180° away from each other, perpendicular to the two remaining *p* orbitals **(red/blue)**.

When two *sp* carbon atoms approach each other, *sp* hybrid orbitals on each carbon overlap head-on to form a strong *sp–sp* σ bond. At the same time, the p_Z orbitals from each carbon form a p_Z–p_Z π bond by sideways overlap, and the p_Y orbitals overlap similarly to form a p_Y–p_Y π bond. The net effect is the sharing of six electrons and formation of a carbon–carbon triple bond. The two remaining *sp* hybrid orbitals each form a σ bond with hydrogen to complete the acetylene molecule **(Figure 1.16)**.

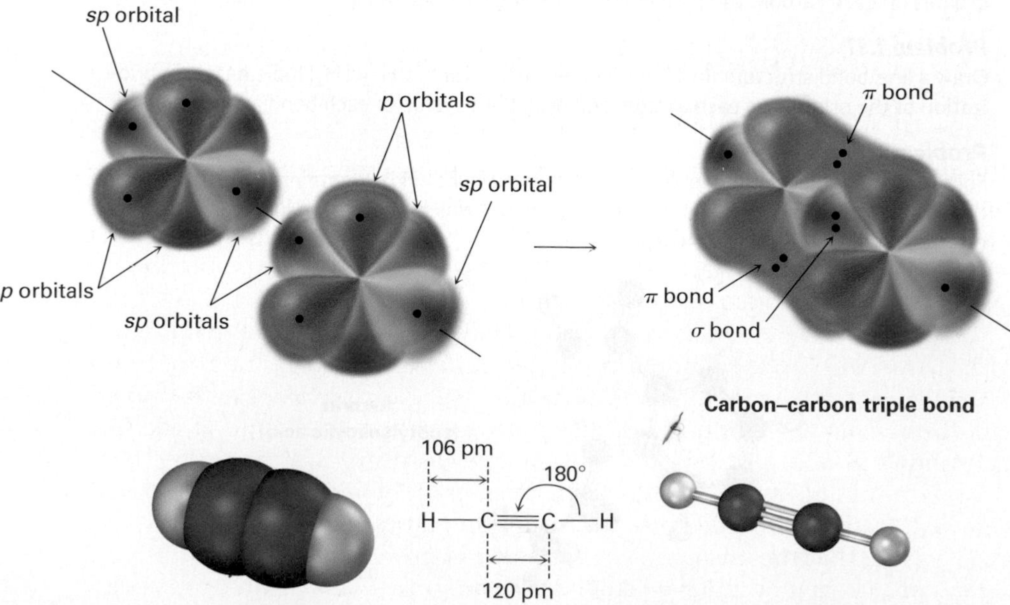

Figure 1.16 The structure of acetylene. The two carbon atoms are joined by one *sp–sp* σ bond and two *p–p* π bonds.

As suggested by *sp* hybridization, acetylene is a linear molecule with H–C–C bond angles of 180°. The C–H bonds have a length of 106 pm and a strength of 558 kJ/mol (133 kcal/mol). The C–C bond length in acetylene is 120 pm, and its strength is about 965 kJ/mol (231 kcal/mol), making it the shortest and strongest of any carbon–carbon bond. A comparison of *sp*, sp^2, and sp^3 hybridization is given in Table 1.2.

Table 1.2 Comparison of C–C and C–H Bonds in Methane, Ethane, Ethylene, and Acetylene

Molecule	Bond	Bond strength (kJ/mol)	Bond strength (kcal/mol)	Bond length (pm)
Methane, CH_4	(sp^3) C—H	439	105	109
Ethane, CH_3CH_3	(sp^3) C—C (sp^3)	377	90	154
	(sp^3) C—H	421	101	109
Ethylene, $H_2C{=}CH_2$	(sp^2) C=C (sp^2)	728	174	134
	(sp^2) C—H	464	111	109
Acetylene, $HC{\equiv}CH$	(sp) C≡C (sp)	965	231	120
	(sp) C—H	558	133	106

Problem 1.13

Draw a line-bond structure for propyne, $CH_3C{\equiv}CH$. Indicate the hybridization of the orbitals on each carbon, and predict a value for each bond angle.

1.10 Hybridization of Nitrogen, Oxygen, Phosphorus, and Sulfur

The valence-bond concept of orbital hybridization described in the previous four sections is not limited to carbon. Covalent bonds formed by other elements can also be described using hybrid orbitals. Look, for instance, at the nitrogen atom in methylamine (CH_3NH_2), an organic derivative of ammonia (NH_3) and the substance responsible for the odor of rotting fish.

The experimentally measured H–N–H bond angle in methylamine is 107.1°, and the C–N–H bond angle is 110.3°, both of which are close to the 109.5° tetrahedral angle found in methane. We therefore assume that nitrogen forms four sp^3-hybridized orbitals, just as carbon does. One of the four sp^3 orbitals is occupied by two nonbonding electrons, and the other three hybrid orbitals have one electron each. Overlap of these three half-filled nitrogen orbitals with half-filled orbitals from other atoms (C or H) gives methylamine. Note that the unshared lone pair of electrons in the fourth sp^3 hybrid orbital of nitrogen occupies as much space as an N–H bond does and is very important to the chemistry of methylamine and other nitrogen-containing organic molecules.

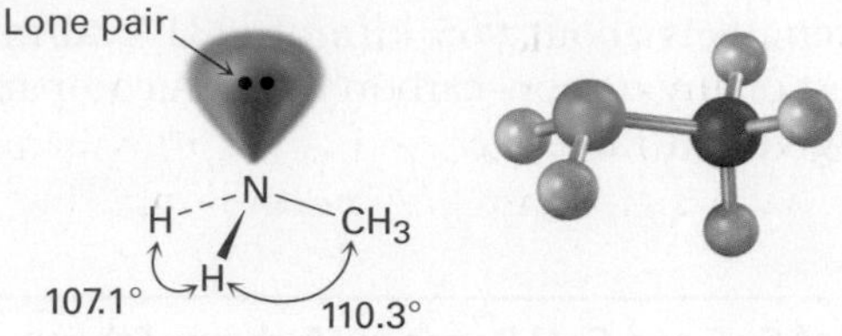

Lone pair

H, N, CH_3, H

107.1° 110.3°

Methylamine

Like the carbon atom in methane and the nitrogen atom in methylamine, the oxygen atom in methanol (methyl alcohol) and many other organic molecules can be described as sp^3-hybridized. The C–O–H bond angle in methanol is 108.5°, very close to the 109.5° tetrahedral angle. Two of the four sp^3 hybrid orbitals on oxygen are occupied by nonbonding electron lone pairs, and two are used to form bonds.

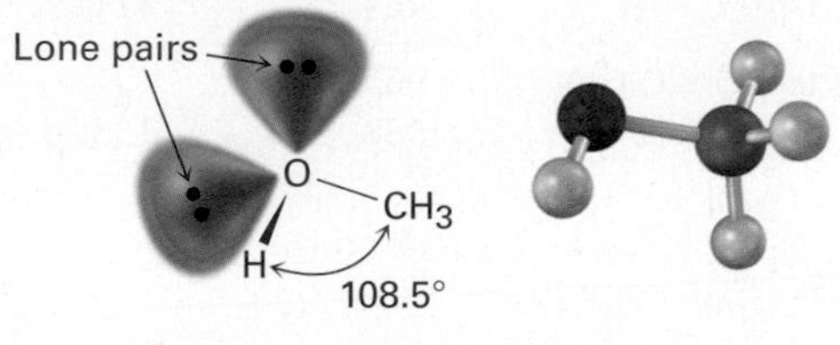

Methanol
(methyl alcohol)

Phosphorus and sulfur are the third-row analogs of nitrogen and oxygen, and the bonding in both can be described using hybrid orbitals. Because of their positions in the third row, however, both phosphorus and sulfur can expand their outer-shell octets and form more than the typical number of covalent bonds. Phosphorus, for instance, often forms five covalent bonds, and sulfur often forms four.

Phosphorus is most commonly encountered in biological molecules in *organophosphates,* compounds that contain a phosphorus atom bonded to four oxygens, with one of the oxygens also bonded to carbon. Methyl phosphate, $CH_3OPO_3{}^{2-}$, is the simplest example. The O–P–O bond angle in such compounds is typically in the range 110 to 112°, implying sp^3 hybridization for the phosphorus orbitals.

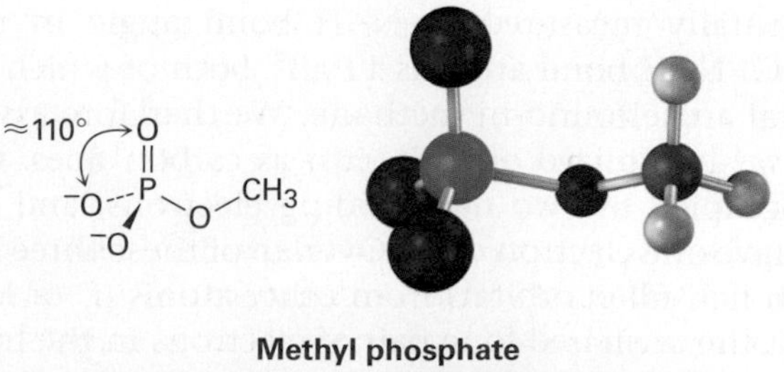

≈110°

Methyl phosphate
(an organophosphate)

Sulfur is most commonly encountered in biological molecules either in compounds called *thiols,* which have a sulfur atom bonded to one hydrogen and one carbon, or in *sulfides,* which have a sulfur atom bonded to two carbons. Produced by some bacteria, methanethiol (CH_3SH) is the simplest example of a thiol, and dimethyl sulfide [$(CH_3)_2S$] is the simplest example of a sulfide. Both can be described by approximate sp^3 hybridization around sulfur, although both have significant deviation from the 109.5° tetrahedral angle.

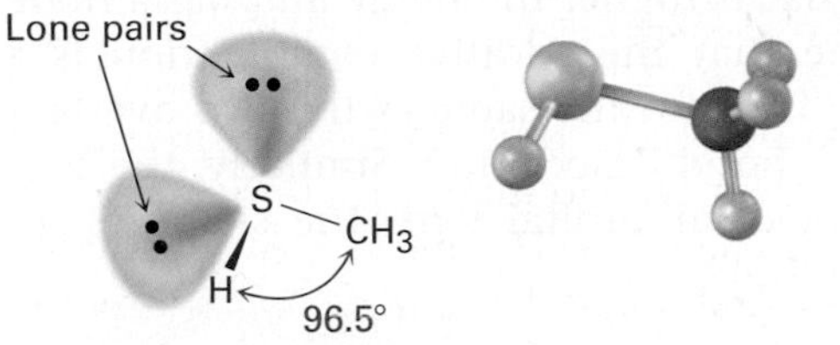

Lone pairs
96.5°
Methanethiol

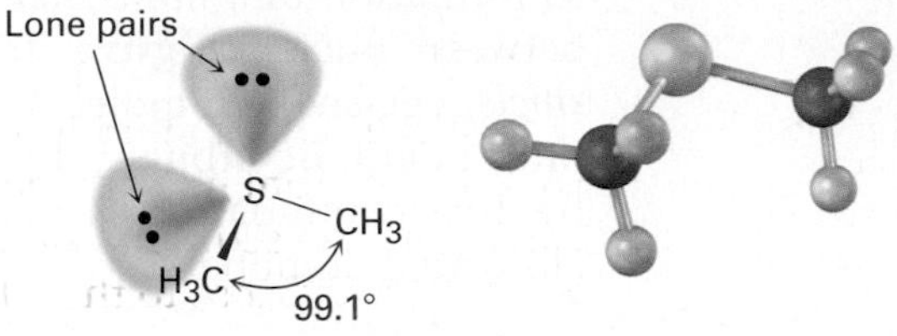

Lone pairs
99.1°
Dimethyl sulfide

Problem 1.14

Identify all nonbonding lone pairs of electrons in the following molecules, and tell what geometry you expect for each of the indicated atoms.

(a) The oxygen atom in dimethyl ether, $CH_3{-}O{-}CH_3$

(b) The nitrogen atom in trimethylamine, $H_3C{-}\underset{\displaystyle CH_3}{\underset{|}{N}}{-}CH_3$

(c) The phosphorus atom in phosphine, PH_3

(d) The sulfur atom in the amino acid methionine, $CH_3{-}S{-}CH_2CH_2\underset{\displaystyle NH_2}{\underset{|}{C}}H\overset{\displaystyle O}{\overset{\|}{C}}OH$

1.11 Describing Chemical Bonds: Molecular Orbital Theory

We said in **Section 1.5** that chemists use two models for describing covalent bonds: valence bond theory and molecular orbital theory. Having now seen the valence bond approach, which uses hybrid atomic orbitals to account for geometry and assumes the overlap of atomic orbitals to account for electron sharing, let's look briefly at the molecular orbital approach to bonding. We'll return to the topic in Chapters 14, 15, and 30 for a more in-depth discussion.

Molecular orbital (MO) theory describes covalent bond formation as arising from a mathematical combination of atomic orbitals (wave functions) on different atoms to form *molecular orbitals,* so called because they belong to the entire *molecule* rather than to an individual atom. Just as an *atomic* orbital, whether unhybridized or hybridized, describes a region of space around an atom where an electron is likely to be found, so a molecular orbital describes a region of space in a *molecule* where electrons are most likely to be found.

Like an atomic orbital, a molecular orbital has a specific size, shape, and energy. In the H_2 molecule, for example, two singly occupied 1*s* atomic orbitals combine to form two molecular orbitals. There are two ways for the orbital combination to occur—an additive way and a subtractive way. The additive combination leads to formation of a molecular orbital that is lower in energy and roughly egg-shaped, while the subtractive combination leads to formation of a molecular orbital that is higher in energy and has a node between nuclei **(Figure 1.17)**. Note that the additive combination is a *single,* egg-shaped, molecular orbital; it is not the same as the two overlapping 1*s* atomic orbitals of the valence bond description. Similarly, the subtractive combination is a single molecular orbital with the shape of an elongated dumbbell.

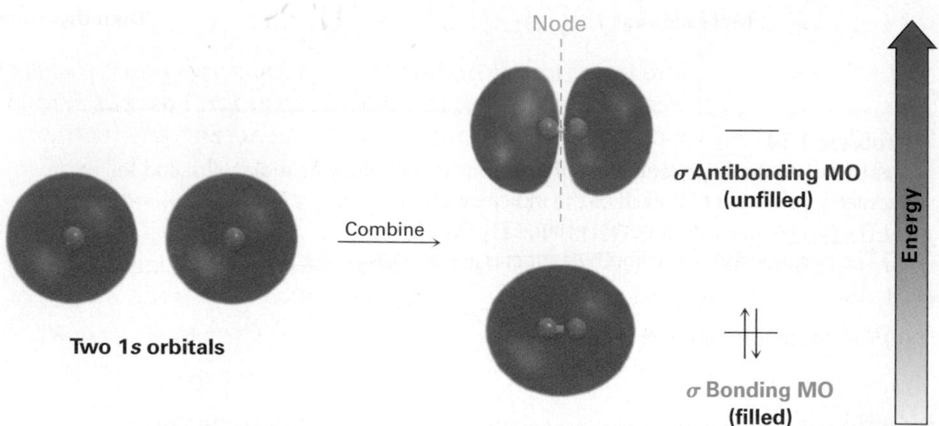

Figure 1.17 Molecular orbitals of H_2. Combination of two hydrogen 1*s* atomic orbitals leads to two H_2 molecular orbitals. The lower-energy, **bonding MO** is filled, and the higher-energy, **antibonding MO** is unfilled.

The additive combination is lower in energy than the two hydrogen 1*s* atomic orbitals and is called a **bonding MO** because electrons in this MO spend most of their time in the region between the two nuclei, thereby bonding the atoms together. The subtractive combination is higher in energy than the two hydrogen 1*s* orbitals and is called an **antibonding MO** because any electrons it contains *can't* occupy the central region between the nuclei, where there is a node, and can't contribute to bonding. The two nuclei therefore repel each other.

Just as bonding and antibonding σ molecular orbitals result from the head-on combination of two *s* atomic orbitals in H_2, so bonding and antibonding π molecular orbitals result from the sideways combination of two *p* atomic orbitals in ethylene. As shown in **Figure 1.18**, the lower-energy, π bonding MO has no node between nuclei and results from combination of *p* orbital lobes with the same algebraic sign. The higher-energy, π antibonding MO has a node between nuclei and results from combination of lobes with opposite algebraic signs. Only the bonding MO is occupied; the higher-energy, antibonding MO is vacant. We'll see in Chapters 14, 15, and 30 that molecular orbital theory is particularly useful for describing π bonds in compounds that have more than one double bond.

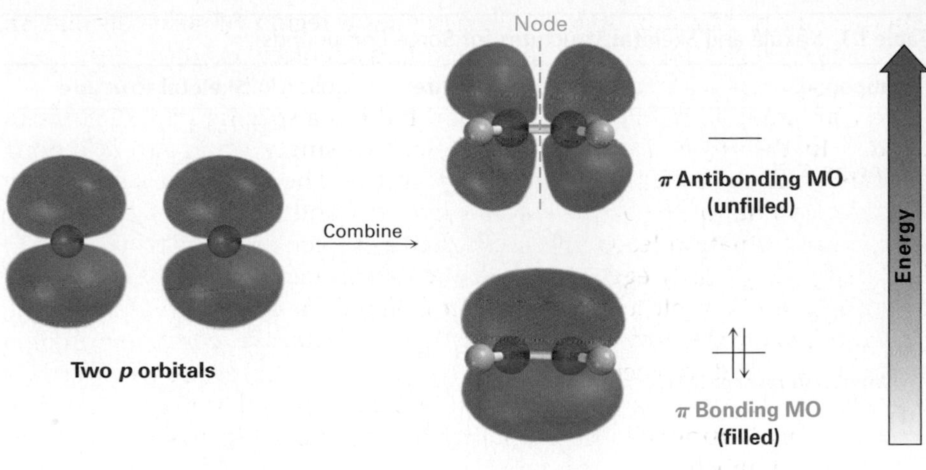

Figure 1.18 A molecular orbital description of the C–C π bond in ethylene. The lower-energy, **π bonding MO** results from an additive combination of *p* orbital lobes with the same algebraic sign and is filled. The higher-energy, **π antibonding MO** results from a subtractive combination of *p* orbital lobes with the opposite algebraic signs and is unfilled.

1.12 Drawing Chemical Structures

Let's cover just one more point before ending this introductory chapter. In the structures we've been drawing until now, a line between atoms has represented the two electrons in a covalent bond. Drawing every bond and every atom is tedious, however, so chemists have devised several shorthand ways for writing structures. In **condensed structures**, carbon–hydrogen and carbon–carbon single bonds aren't shown; instead, they're understood. If a carbon has three hydrogens bonded to it, we write CH_3; if a carbon has two hydrogens bonded to it, we write CH_2; and so on. The compound called 2-methylbutane, for example, is written as follows:

```
           H
           |
         H-C-H
           |                       Condensed structures
   H   H   |   H             CH3     /          \
   |   |   |   |              |     ↙            ↘
 H-C---C---C---C-H  =  CH3CH2CHCH3   or   CH3CH2CH(CH3)2
   |   |   |   |
   H   H   H   H

     2-Methylbutane
```

Notice that the horizontal bonds between carbons aren't shown in condensed structures—the CH_3, CH_2, and CH units are simply placed next to each other—but the vertical carbon–carbon bond in the first of the condensed structures drawn above is shown for clarity. Notice also in the second of the condensed structures that the two CH_3 units attached to the CH carbon are grouped together as $(CH_3)_2$.

Even simpler than condensed structures are **skeletal structures** such as those shown in Table 1.3. The rules for drawing skeletal structures are straightforward.

RULE 1

Carbon atoms aren't usually shown. Instead, a carbon atom is assumed to be at each intersection of two lines (bonds) and at the end of each line. Occasionally, a carbon atom might be indicated for emphasis or clarity.

Table 1.3 Kekulé and Skeletal Structures for Some Compounds

Compound	Kekulé structure	Skeletal structure
Isoprene, C_5H_8	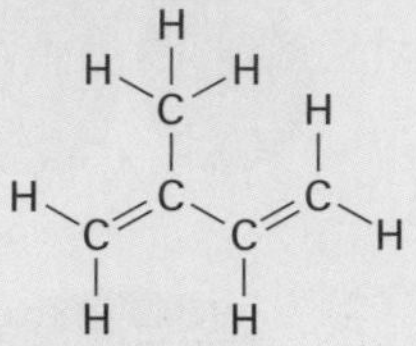	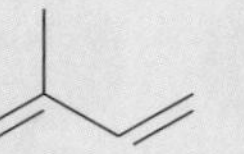
Methylcyclohexane, C_7H_{14}	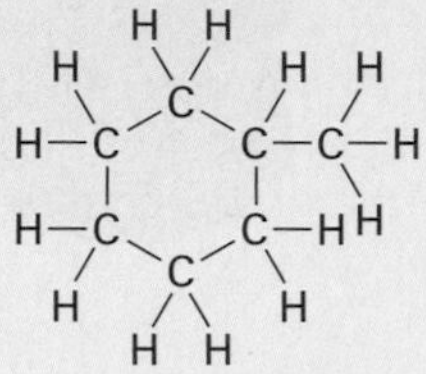	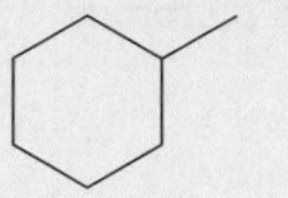
Phenol, C_6H_6O	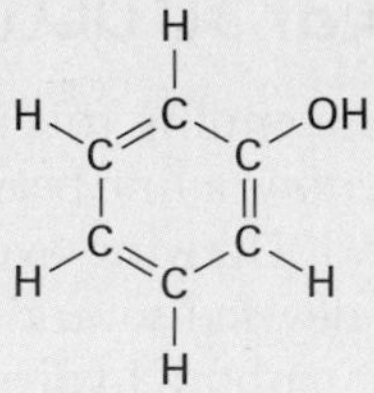	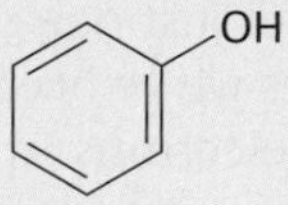 OH

RULE 2

Hydrogen atoms bonded to carbon aren't shown. Because carbon always has a valence of 4, we mentally supply the correct number of hydrogen atoms for each carbon.

RULE 3

Atoms other than carbon and hydrogen *are* shown.

One further comment: although such groupings as $-CH_3$, $-OH$, and $-NH_2$ are usually written with the C, O, or N atom first and the H atom second, the order of writing is sometimes inverted to H_3C-, $HO-$, and H_2N- if needed to make the bonding connections in a molecule clearer. Larger units such as $-CH_2CH_3$ are not inverted, though; we don't write H_3CH_2C- because it would be confusing. There are, however, no well-defined rules that cover all cases; it's largely a matter of preference.

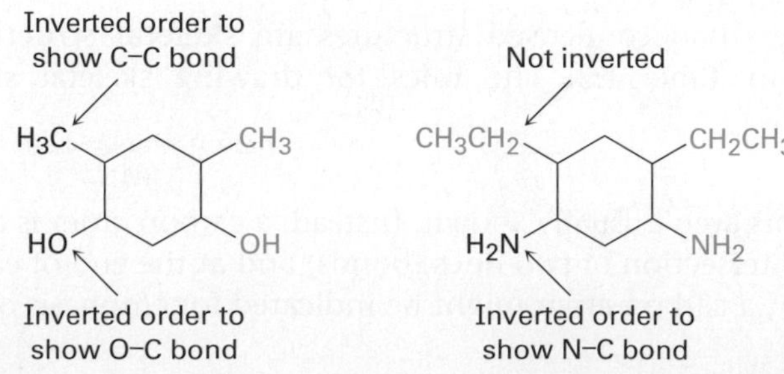

Worked Example 1.4 — Interpreting a Line-Bond Structure

Carvone, a substance responsible for the odor of spearmint, has the following structure. Tell how many hydrogens are bonded to each carbon, and give the molecular formula of carvone.

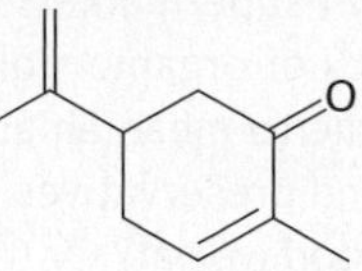

Carvone

Strategy

The end of a line represents a carbon atom with 3 hydrogens, CH_3; a two-way intersection is a carbon atom with 2 hydrogens, CH_2; a three-way intersection is a carbon atom with 1 hydrogen, CH; and a four-way intersection is a carbon atom with no attached hydrogens.

Solution

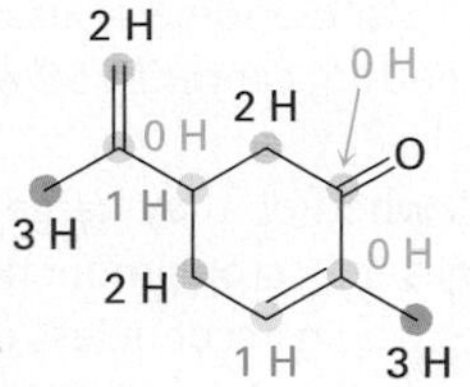

Carvone ($C_{10}H_{14}O$)

Problem 1.15

Tell how many hydrogens are bonded to each carbon in the following compounds, and give the molecular formula of each substance:

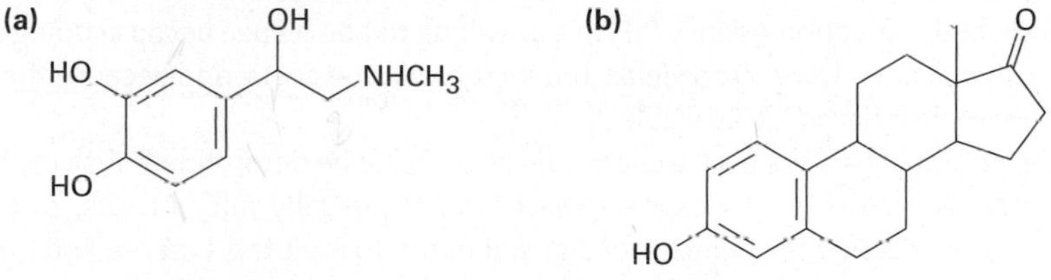

(a) **Adrenaline**

(b) **Estrone (a hormone)**

Problem 1.16

Propose skeletal structures for compounds that satisfy the following molecular formulas. There is more than one possibility in each case.

(a) C_5H_{12} **(b)** C_2H_7N **(c)** C_3H_6O **(d)** C_4H_9Cl

Problem 1.17

The following molecular model is a representation of *para*-aminobenzoic acid (PABA), the active ingredient in many sunscreens. Indicate the positions of the multiple bonds, and draw a skeletal structure (gray = C, red = O, blue = N, ivory = H).

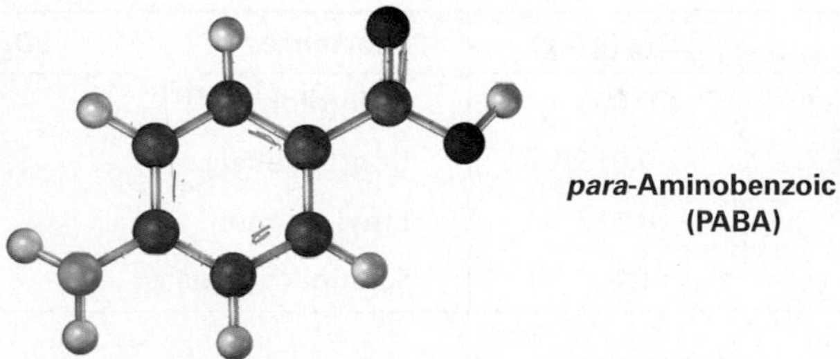

***para*-Aminobenzoic acid (PABA)**

© David R. Frazier Photolibrary, Inc./Alamy

How dangerous is the pesticide being sprayed on this crop?

Organic Foods: Risk versus Benefit

A DEEPER LOOK

Contrary to what you may hear in supermarkets or on television, *all* foods are organic—that is, complex mixtures of organic molecules. Even so, when applied to food, the word *organic* has come to mean an absence of synthetic chemicals, typically pesticides, antibiotics, and preservatives. How concerned should we be about traces of pesticides in the food we eat? Or toxins in the water we drink? Or pollutants in the air we breathe?

Life is not risk-free—we all take many risks each day without even thinking about it. We decide to ride a bike rather than drive, even though there is a ten times greater likelihood per mile of dying in a bicycling accident than in a car. We decide to walk down stairs rather than take an elevator, even though 7000 people die from falls each year in the United States. Some of us decide to smoke cigarettes, even though it increases our chance of getting cancer by 50%. But what about risks from chemicals like pesticides?

One thing is certain: without pesticides, whether they target weeds (herbicides), insects (insecticides), or molds and fungi (fungicides), crop production would drop significantly, food prices would increase, and famines would occur in less developed parts of the world. Take the herbicide atrazine, for instance. In the United States alone, approximately 100 million pounds of atrazine are used each year to kill weeds in corn, sorghum, and sugarcane fields, greatly improving the yields of these crops. Nevertheless, the use of atrazine continues to be a concern because traces persist in the environment. Indeed, heavy atrazine exposure *can* pose health risks to humans and some animals, but the United States Environmental Protection Agency (EPA) is unwilling to ban its use because doing so would result in significantly lower crop yields and increased food costs, and because there is no suitable alternative herbicide available.

How can the potential hazards from a chemical like atrazine be determined? Risk evaluation of chemicals is carried out by exposing test animals, usually mice or rats, to the chemical and then monitoring the animals for signs of harm. To limit the expense and time needed, the amounts administered are typically hundreds or thousands of times greater than those a person might normally encounter. The results obtained in animal tests are then distilled into a single number called an LD_{50}, the amount of substance per kilogram body weight that is a lethal dose for 50% of the test animals. For atrazine, the LD_{50} value is between 1 and 4 g/kg depending on the animal species. Aspirin, for comparison, has an LD_{50} of 1.1 g/kg, and ethanol (ethyl alcohol) has an LD_{50} of 10.6 g/kg.

Table 1.4 lists values for some other familiar substances. The lower the value, the more toxic the substance. Note, though, that LD_{50} values tell only about the effects of heavy

Table 1.4 Some LD_{50} Values

Substance	LD_{50} (g/kg)	Substance	LD_{50} (g/kg)
Strychnine	0.005	Chloroform	1.2
Arsenic trioxide	0.015	Iron(II) sulfate	1.5
DDT	0.115	Ethyl alcohol	10.6
Aspirin	1.1	Sodium cyclamate	17

(continued)

(continued)

exposure for a relatively short time. They say nothing about the risks of long-term exposure, such as whether the substance can cause cancer or interfere with development in the unborn.

So, should we still use atrazine? All decisions involve tradeoffs, and the answer is rarely obvious. Does the benefit of increased food production outweigh possible health risks of a pesticide? Do the beneficial effects of a new drug outweigh a potentially dangerous side effect in a small number of users? Different people will have different opinions, but an honest evaluation of facts is surely the best way to start. At present, atrazine is approved for continued use in the United States because the EPA believes that the benefits of increased food production outweigh possible health risks. At the same time, though, the use of atrazine is being phased out in Europe.

Summary

The purpose of this chapter has been to get you up to speed—to review some ideas about atoms, bonds, and molecular geometry. As we've seen, **organic chemistry** is the study of carbon compounds. Although a division into organic and inorganic chemistry occurred historically, there is no scientific reason for the division.

An atom consists of a positively charged nucleus surrounded by one or more negatively charged electrons. The electronic structure of an atom can be described by a quantum mechanical wave equation, in which electrons are considered to occupy **orbitals** around the nucleus. Different orbitals have different energy levels and different shapes. For example, *s* orbitals are spherical and *p* orbitals are dumbbell-shaped. The **ground-state electron configuration** of an atom can be found by assigning electrons to the proper orbitals, beginning with the lowest-energy ones.

A **covalent bond** is formed when an electron pair is shared between atoms. According to **valence bond theory**, electron sharing occurs by overlap of two atomic orbitals. According to **molecular orbital (MO) theory**, bonds result from the mathematical combination of atomic orbitals to give molecular orbitals, which belong to the entire molecule. Bonds that have a circular cross-section and are formed by head-on interaction are called **sigma (σ) bonds**; bonds formed by sideways interaction of *p* orbitals are called **pi (π) bonds**.

In the valence bond description, carbon uses hybrid orbitals to form bonds in organic molecules. When forming only single bonds with tetrahedral geometry, carbon uses four equivalent **sp^3 hybrid orbitals**. When forming a double bond with planar geometry, carbon uses three equivalent **sp^2 hybrid orbitals** and one unhybridized *p* orbital. When forming a triple bond with linear geometry, carbon uses two equivalent ***sp* hybrid orbitals** and two unhybridized *p* orbitals. Other atoms such as nitrogen, phosphorus, oxygen, and sulfur also use hybrid orbitals to form strong, oriented bonds.

Organic molecules are usually drawn using either condensed structures or skeletal structures. In **condensed structures**, carbon–carbon and carbon–hydrogen bonds aren't shown. In **skeletal structures**, only the bonds and not the atoms are shown. A carbon atom is assumed to be at the ends and at the junctions of lines (bonds), and the correct number of hydrogens is mentally supplied.

Key words

antibonding MO, 20
bond angle, 12
bond length, 10
bond strength, 10
bonding MO, 20
condensed structure, 21
covalent bond, 7
electron-dot structure, 7
electron shell, 4
ground-state electron configuration, 5
isotope, 3
line-bond structure, 7
lone-pair electrons, 8
molecular orbital (MO) theory, 19
molecule, 7
node, 4
orbital, 3
organic chemistry, 1
pi (π) bond, 14
sigma (σ) bond, 10
skeletal structure, 21
***sp* hybrid orbital, 16**
sp^2 hybrid orbital, 14
sp^3 hybrid orbital, 11
valence bond theory, 9
valence shell, 6

Working Problems

There's no surer way to learn organic chemistry than by working problems. Although careful reading and rereading of this text are important, reading alone isn't enough. You must also be able to use the information you've read and be able to apply your knowledge in new situations. Working problems gives you practice at doing this.

Each chapter in this book provides many problems of different sorts. The in-chapter problems are placed for immediate reinforcement of ideas just learned, while end-of-chapter problems provide additional practice and are of several types. They begin with a short section called "Visualizing Chemistry," which helps you "see" the microscopic world of molecules and provides practice for working in three dimensions. After the visualizations are many "Additional Problems," which are organized by topic. Early problems are primarily of the drill type, providing an opportunity for you to practice your command of the fundamentals. Later problems tend to be more thought-provoking, and some are real challenges.

As you study organic chemistry, take the time to work the problems. Do the ones you can, and ask for help on the ones you can't. If you're stumped by a particular problem, check the accompanying *Study Guide and Solutions Manual* for an explanation that will help clarify the difficulty. Working problems takes effort, but the payoff in knowledge and understanding is immense.

Exercises

OWL Interactive versions of these problems are assignable in OWL for Organic Chemistry.

Visualizing Chemistry

(Problems 1.1–1.17 appear within the chapter.)

1.18 Convert each of the following molecular models into a skeletal structure, and give the formula of each. Only the connections between atoms are shown; multiple bonds are not indicated (gray = C, red = O, blue = N, ivory = H).

(a)

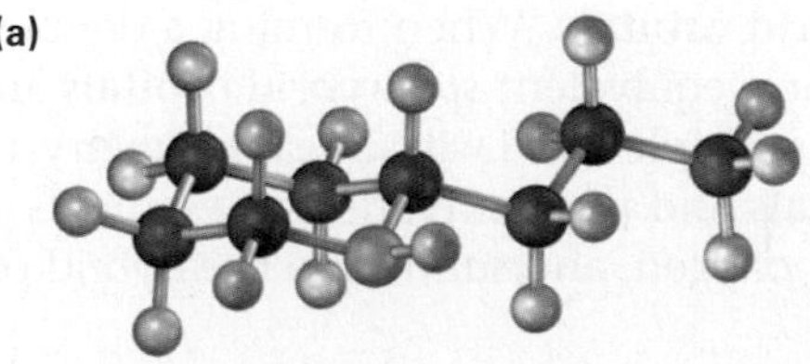

Coniine (the toxic substance in poison hemlock)

(b)

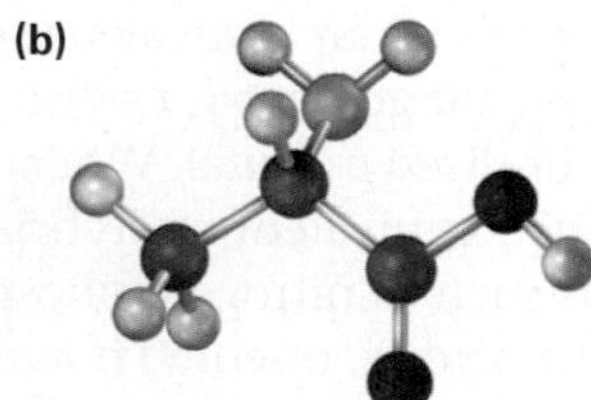

Alanine (an amino acid)

1.19 The following model is a representation of citric acid, the key substance in the so-called citric acid cycle by which food molecules are metabolized in the body. Only the connections between atoms are shown; multiple bonds are not indicated. Complete the structure by indicating the positions of multiple bonds and lone-pair electrons (gray = C, red = O, ivory = H).

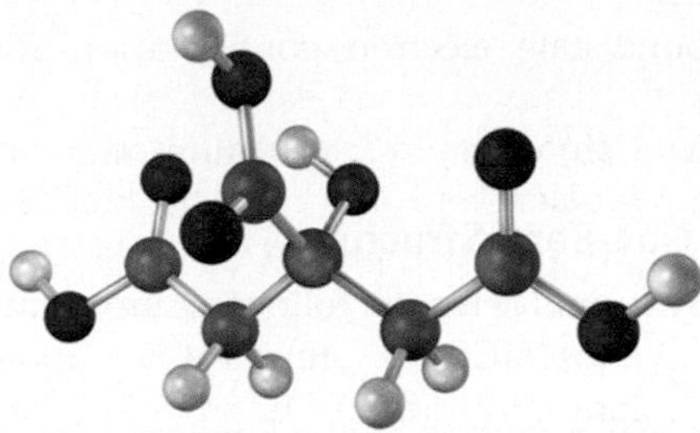

1.20 The following model is a representation of acetaminophen, a pain reliever sold in drugstores under a variety of names, including Tylenol. Identify the hybridization of each carbon atom in acetaminophen, and tell which atoms have lone pairs of electrons (gray = C, red = O, blue = N, ivory = H).

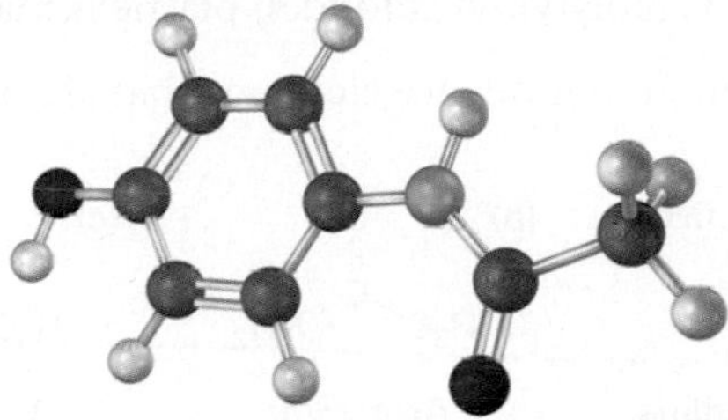

1.21 The following model is a representation of aspartame, $C_{14}H_{18}N_2O_5$, known commercially under many names, including NutraSweet. Only the connections between atoms are shown; multiple bonds are not indicated. Complete the structure for aspartame, and indicate the positions of multiple bonds (gray = C, red = O, blue = N, ivory = H).

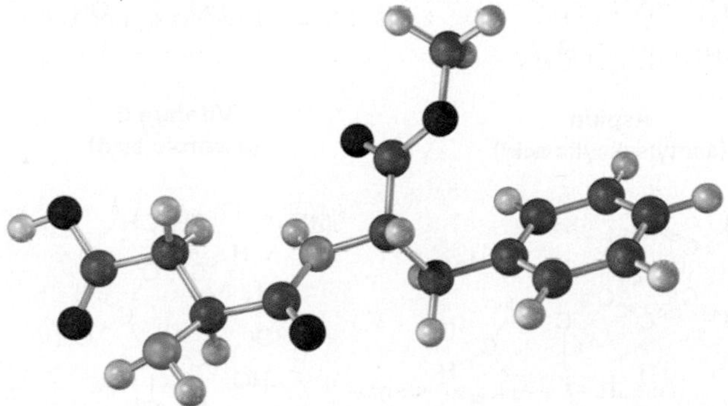

Additional Problems

Electron Configurations

1.22 How many valence electrons does each of the following dietary trace elements have?
(a) Zinc (b) Iodine (c) Silicon (d) Iron

1.23 Give the ground-state electron configuration for each of the following elements:
(a) Potassium (b) Arsenic (c) Aluminum (d) Germanium

Electron-Dot and Line-Bond Structures

1.24 What are likely formulas for the following molecules?
(a) $NH_?OH$ (b) $AlCl_?$ (c) $CF_2Cl_?$ (d) $CH_?O$

1.25 Why can't molecules with the following formulas exist?
(a) CH_5 (b) C_2H_6N (c) $C_3H_5Br_2$

1.26 Draw an electron-dot structure for acetonitrile, C_2H_3N, which contains a carbon–nitrogen triple bond. How many electrons does the nitrogen atom have in its outer shell? How many are bonding, and how many are nonbonding?

1.27 Draw a line-bond structure for vinyl chloride, C_2H_3Cl, the starting material from which PVC [poly(vinyl chloride)] plastic is made.

1.28 Fill in any nonbonding valence electrons that are missing from the following structures:

(a) $H_3C-S-S-CH_3$
Dimethyl disulfide

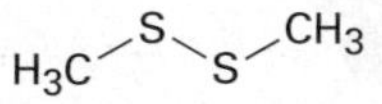

(b) $H_3C-C(=O)-NH_2$
Acetamide

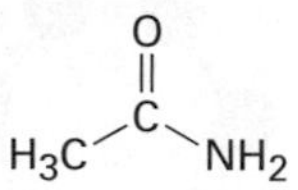

(c) $H_3C-C(=O)-O^-$
Acetate ion

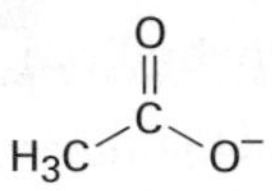

1.29 Convert the following line-bond structures into molecular formulas:

(a)
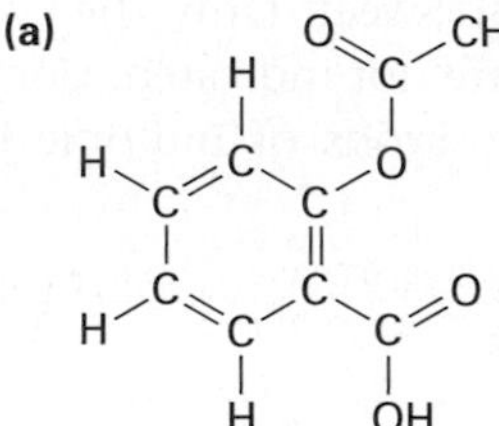
Aspirin (acetylsalicylic acid)

(b) 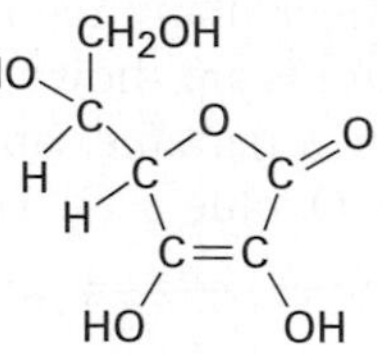
Vitamin C (ascorbic acid)

(c)
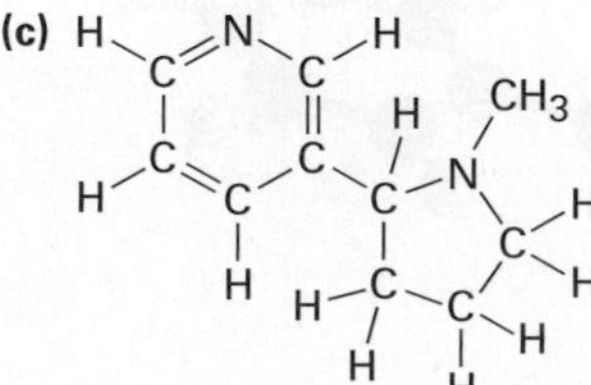
Nicotine

(d) 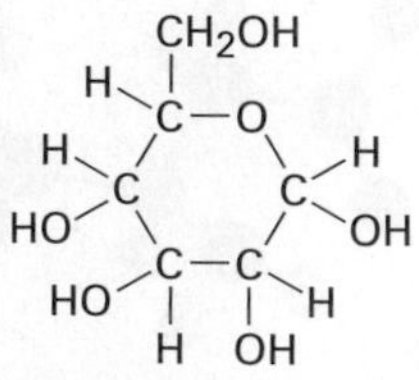
Glucose

1.30 Convert the following molecular formulas into line-bond structures that are consistent with valence rules:

(a) C_3H_8 (b) CH_5N
(c) C_2H_6O (2 possibilities) (d) C_3H_7Br (2 possibilities)
(e) C_2H_4O (3 possibilities) (f) C_3H_9N (4 possibilities)

1.31 Draw a three-dimensional representation of the oxygen-bearing carbon atom in ethanol, CH_3CH_2OH, using the standard convention of solid, wedged, and dashed lines.

1.32 Oxaloacetic acid, an important intermediate in food metabolism, has the formula $C_4H_4O_5$ and contains three C=O bonds and two O–H bonds. Propose two possible structures.

1.33 Draw structures for the following molecules, showing lone pairs:

(a) Acrylonitrile, C_3H_3N, which contains a carbon–carbon double bond and a carbon–nitrogen triple bond
(b) Ethyl methyl ether, C_3H_8O, which contains an oxygen atom bonded to two carbons
(c) Butane, C_4H_{10}, which contains a chain of four carbon atoms
(d) Cyclohexene, C_6H_{10}, which contains a ring of six carbon atoms and one carbon–carbon double bond

1.34 Potassium methoxide, $KOCH_3$, contains both covalent and ionic bonds. Which do you think is which?

Hybridization

1.35 What is the hybridization of each carbon atom in acetonitrile (Problem 1.26)?

1.36 What kind of hybridization do you expect for each carbon atom in the following molecules?

(a) Propane, $CH_3CH_2CH_3$

(b) 2-Methylpropene, $CH_3C(CH_3){=}CH_2$

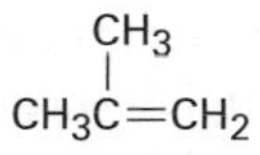

(c) 1-Butene-3-yne, $H_2C{=}CH{-}C{\equiv}CH$

(d) Acetic acid, $CH_3C(=O)OH$

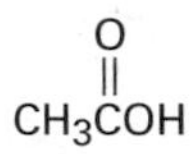

1.37 What is the shape of benzene, and what hybridization do you expect for each carbon?

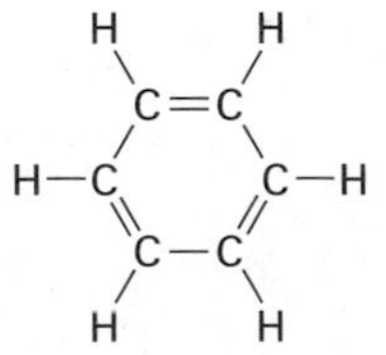

Benzene (C_6H_6: six-membered ring of C atoms with alternating double bonds, each C bearing one H)

1.38 What bond angles do you expect for each of the following, and what kind of hybridization do you expect for the central atom in each?

(a) $H_2N{-}CH_2{-}C({=}O){-}OH$

Glycine (an amino acid)

(b) H, N, C, H ring structure

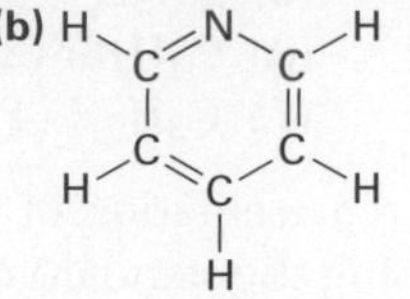

Pyridine

(c) $CH_3{-}CH(OH){-}C({=}O){-}OH$

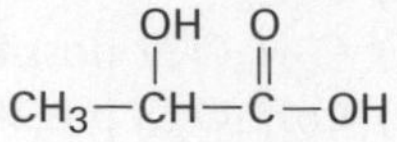

Lactic acid (in sour milk)

1.39 Propose structures for molecules that meet the following descriptions:

(a) Contains two sp^2-hybridized carbons and two sp^3-hybridized carbons
(b) Contains only four carbons, all of which are sp^2-hybridized
(c) Contains two sp-hybridized carbons and two sp^2-hybridized carbons

1.40 What kind of hybridization do you expect for each carbon atom in the following molecules?

(a) H, O, H, $CH_2{-}CH_3$, H_2N, CH_2, $\overset{+}{N}$, $CH_2{-}CH_3$, Cl^-

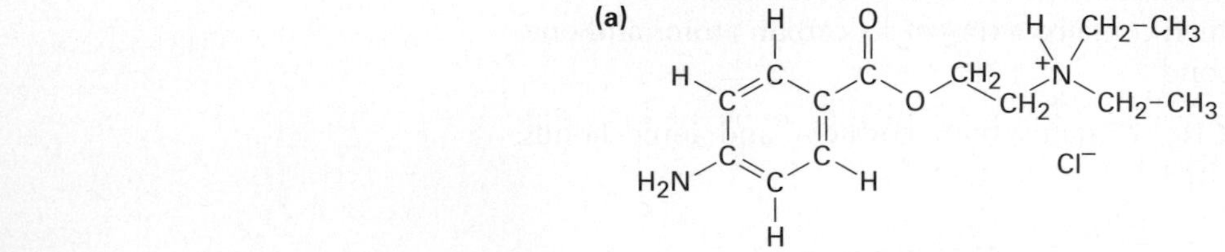

Procaine

(b) CH_2OH, HO, O, O, H, HO, OH

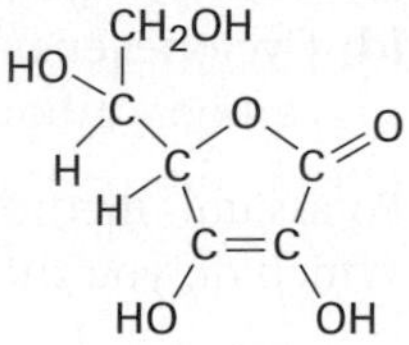

Vitamin C (ascorbic acid)

1.41 Pyridoxal phosphate, a close relative of vitamin B_6, is involved in a large number of metabolic reactions. Tell the hybridization, and predict the bond angles for each nonterminal atom.

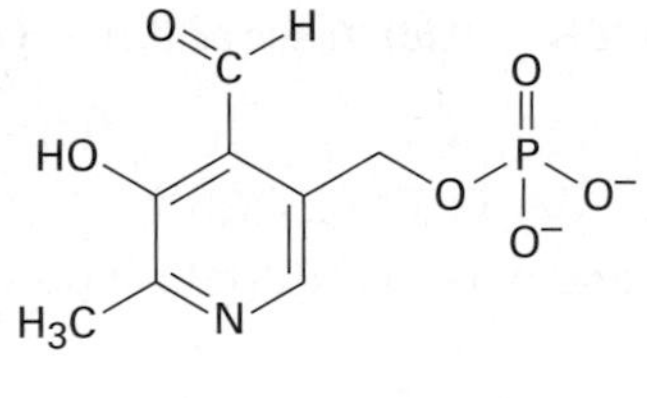

O, H, HO, H_3C, N, O, P, O^-, O^-

Pyridoxal phosphate

Skeletal Structures

1.42 Convert the following structures into skeletal drawings:

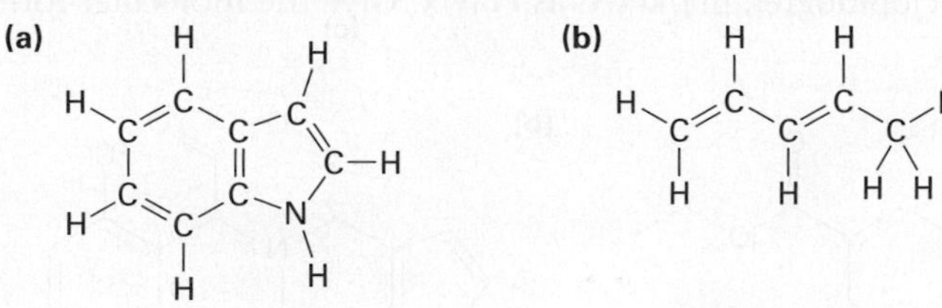

(a) **Indole** (b) **1,3-Pentadiene**

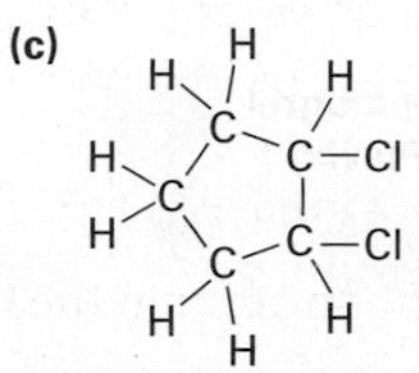

(c) **1,2-Dichlorocyclopentane**

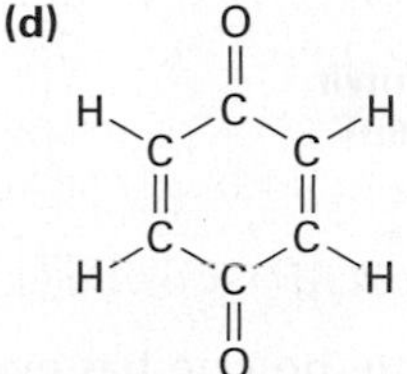

(d) **Benzoquinone**

1.43 Tell the number of hydrogens bonded to each carbon atom in the following substances, and give the molecular formula of each:

(a) (b) 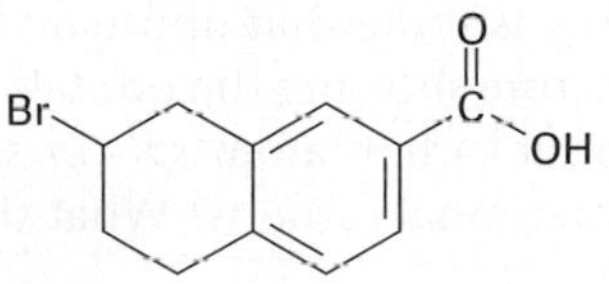 (c)

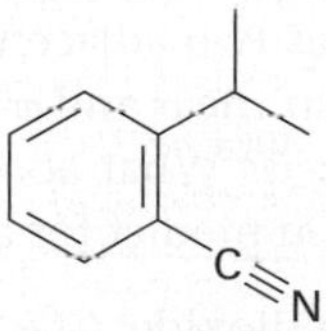

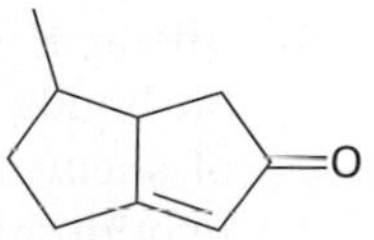

1.44 Quetiapine, marketed as Seroquel, is a heavily prescribed antipsychotic drug used in the treatment of schizophrenia and bipolar disorder. Convert the following representation into a skeletal structure, and give the molecular formula of quetiapine.

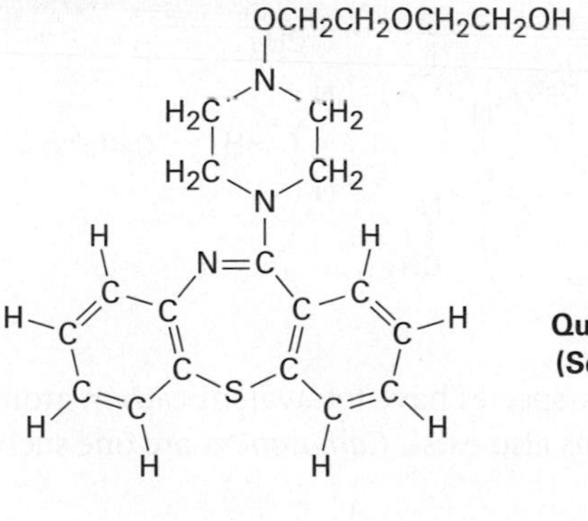

Quetiapine (Seroquel)

1.45 Tell the number of hydrogens bonded to each carbon atom in **(a)** the anti-influenza agent oseltamivir, marketed as Tamiflu, and **(b)** the platelet aggregation inhibitor clopidogrel, marketed as Plavix. Give the molecular formula of each.

(a) NH_2 **Oseltamivir (Tamiflu)**

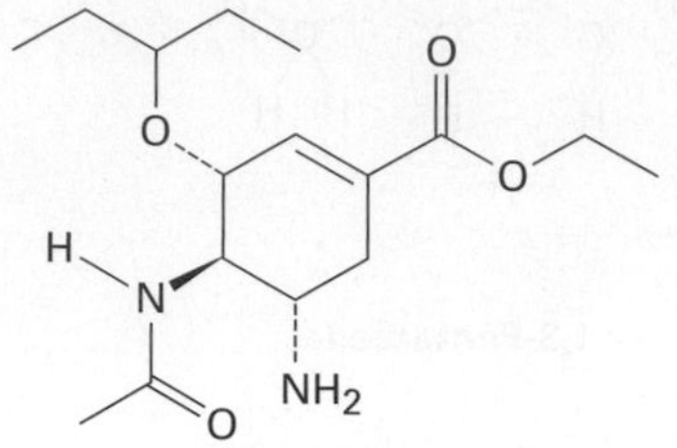

(b) **Clopidogrel (Plavix)**

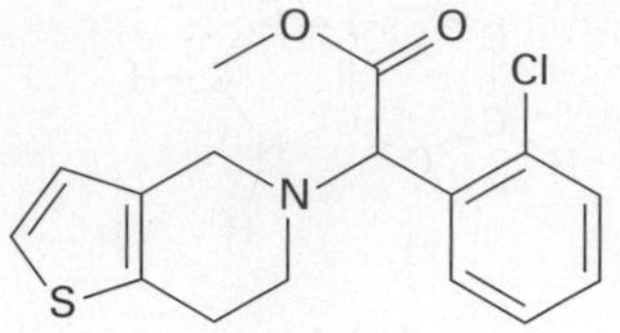

General Problems

1.46 Why do you suppose no one has ever been able to make cyclopentyne as a stable molecule?

Cyclopentyne

1.47 Allene, $H_2C{=}C{=}CH_2$, is somewhat unusual in that it has two adjacent double bonds. Draw a picture showing the orbitals involved in the σ and π bonds of allene. Is the central carbon atom sp^2- or sp-hybridized? What about the hybridization of the terminal carbons? What shape do you predict for allene?

1.48 Allene (see Problem 1.47) is related structurally to carbon dioxide, CO_2. Draw a picture showing the orbitals involved in the σ and π bonds of CO_2, and identify the likely hybridization of carbon.

1.49 Complete the electron-dot structure of caffeine, showing all lone-pair electrons, and identify the hybridization of the indicated atoms.

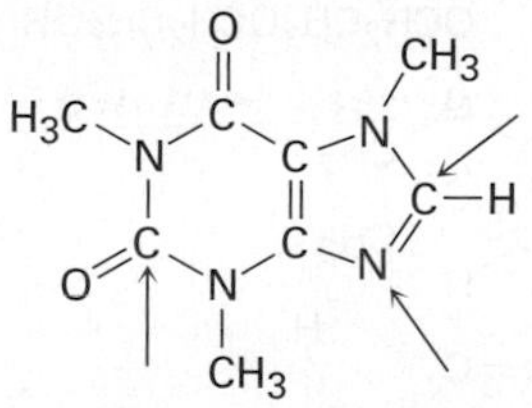

Caffeine

1.50 Most stable organic species have tetravalent carbon atoms, but species with trivalent carbon atoms also exist. *Carbocations* are one such class of compounds.

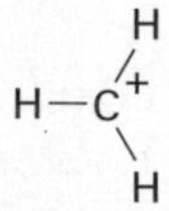

A carbocation

(a) How many valence electrons does the positively charged carbon atom have?
(b) What hybridization do you expect this carbon atom to have?
(c) What geometry is the carbocation likely to have?

1.51 A *carbanion* is a species that contains a negatively charged, trivalent carbon.

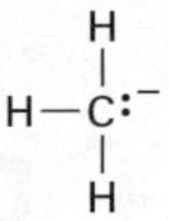

A carbanion

(a) What is the electronic relationship between a carbanion and a trivalent nitrogen compound such as NH_3?
(b) How many valence electrons does the negatively charged carbon atom have?
(c) What hybridization do you expect this carbon atom to have?
(d) What geometry is the carbanion likely to have?

1.52 Divalent carbon species called *carbenes* are capable of fleeting existence. For example, methylene, $:CH_2$, is the simplest carbene. The two unshared electrons in methylene can be either paired in a single orbital or unpaired in different orbitals. Predict the type of hybridization you expect carbon to adopt in singlet (spin-paired) methylene and triplet (spin-unpaired) methylene. Draw a picture of each, and identify the valence orbitals on carbon.

1.53 There are two different substances with the formula C_4H_{10}. Draw both, and tell how they differ.

1.54 There are two different substances with the formula C_3H_6. Draw both, and tell how they differ.

1.55 There are two different substances with the formula C_2H_6O. Draw both, and tell how they differ.

1.56 There are three different substances that contain a carbon–carbon double bond and have the formula C_4H_8. Draw them, and tell how they differ.

1.57 Among the most common over-the-counter drugs you might find in a medicine cabinet are mild pain relievers such ibuprofen (Advil, Motrin), naproxen (Aleve), and acetaminophen (Tylenol).

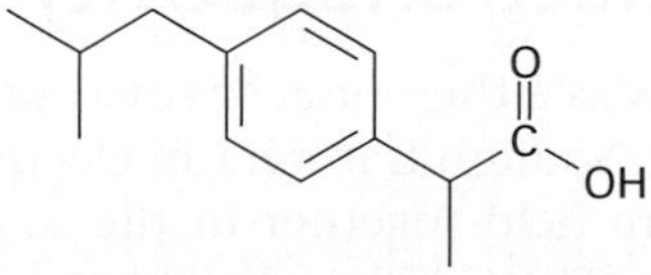

Ibuprofen

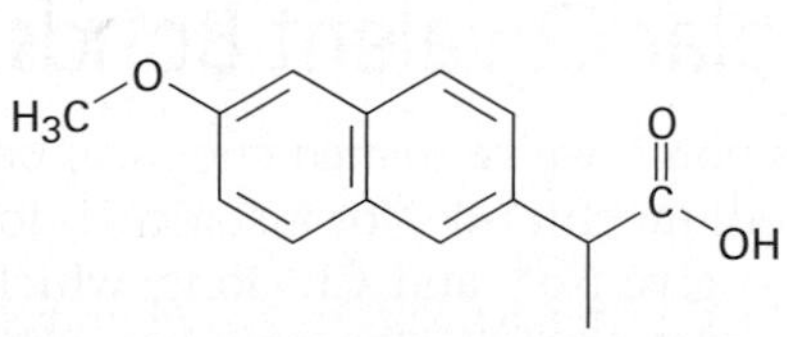

Naproxen

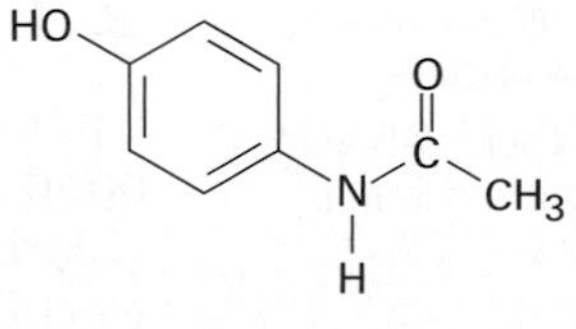

Acetaminophen

(a) How many sp^3-hybridized carbons does each molecule have?
(b) How many sp^2-hybridized carbons does each molecule have?
(c) Can you spot any similarities in their structures?

2

The opium poppy is the source of morphine, one of the first "vegetable alkali," or *alkaloids*, to be isolated.

Polar Covalent Bonds; Acids and Bases

We saw in the last chapter how covalent bonds between atoms are described, and we looked at the valence bond model, which uses hybrid orbitals to account for the observed shapes of organic molecules. Before going on to a systematic study of organic chemistry, however, we still need to review a few fundamental topics. In particular, we need to look more closely at how electrons are distributed in covalent bonds and at some of the consequences that arise when the electrons in a bond are not shared equally between atoms.

Why This Chapter? Understanding organic and biological chemistry means knowing not just what happens but also why and how it happens at the molecular level. In this chapter, we'll look at some of the ways that chemists describe and account for chemical reactivity, thereby providing a foundation to understand the specific reactions discussed in subsequent chapters. Topics such as bond polarity, the acid–base behavior of molecules, and hydrogen-bonding are a particularly important part of that foundation.

2.1 Polar Covalent Bonds: Electronegativity

Up to this point, we've treated chemical bonds as either ionic or covalent. The bond in sodium chloride, for instance, is ionic. Sodium transfers an electron to chlorine to give Na^+ and Cl^- ions, which are held together in the solid by electrostatic attractions between unlike charges. The C–C bond in ethane, however, is covalent. The two bonding electrons are shared equally by the two equivalent carbon atoms, resulting in a symmetrical electron distribution in the bond. Most bonds, however, are neither fully ionic nor fully covalent but are somewhere between the two extremes. Such bonds are called **polar covalent bonds**, meaning that the bonding electrons are attracted more strongly by one atom than the other so that the electron distribution between atoms is not symmetrical **(Figure 2.1)**.

OWL Sign in to OWL for Organic Chemistry at **www.cengage.com/owl** to view tutorials and simulations, develop problem-solving skills, and complete online homework assigned by your professor.

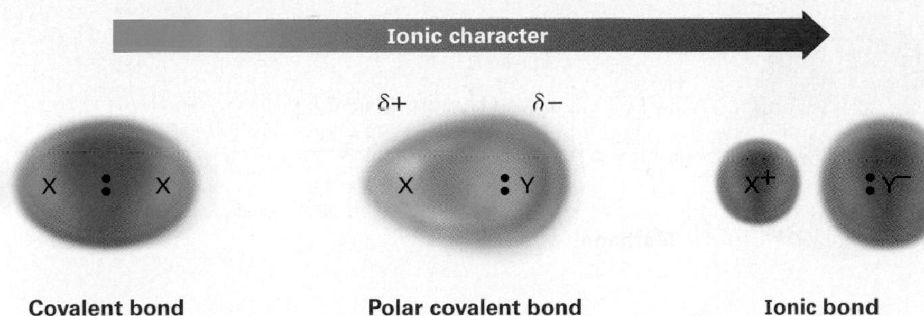

Figure 2.1 The continuum in bonding from covalent to ionic is a result of an unequal distribution of bonding electrons between atoms. The symbol δ (lowercase Greek delta) means *partial* charge, either partial positive (δ+) for the electron-poor atom or partial negative (δ−) for the electron-rich atom.

Bond polarity is due to differences in **electronegativity (EN)**, the intrinsic ability of an atom to attract the shared electrons in a covalent bond. As shown in **Figure 2.2**, electronegativities are based on an arbitrary scale, with fluorine the most electronegative (EN = 4.0) and cesium the least (EN = 0.7). Metals on the left side of the periodic table attract electrons weakly and have lower electronegativities, while oxygen, nitrogen, and halogens on the right side of the periodic table attract electrons strongly and have higher electronegativities. Carbon, the most important element in organic compounds, has an electronegativity value of 2.5.

H 2.1																	He
Li 1.0	Be 1.6											B 2.0	C 2.5	N 3.0	O 3.5	F 4.0	Ne
Na 0.9	Mg 1.2											Al 1.5	Si 1.8	P 2.1	S 2.5	Cl 3.0	Ar
K 0.8	Ca 1.0	Sc 1.3	Ti 1.5	V 1.6	Cr 1.6	Mn 1.5	Fe 1.8	Co 1.9	Ni 1.9	Cu 1.9	Zn 1.6	Ga 1.6	Ge 1.8	As 2.0	Se 2.4	Br 2.8	Kr
Rb 0.8	Sr 1.0	Y 1.2	Zr 1.4	Nb 1.6	Mo 1.8	Tc 1.9	Ru 2.2	Rh 2.2	Pd 2.2	Ag 1.9	Cd 1.7	In 1.7	Sn 1.8	Sb 1.9	Te 2.1	I 2.5	Xe
Cs 0.7	Ba 0.9	La 1.0	Hf 1.3	Ta 1.5	W 1.7	Re 1.9	Os 2.2	Ir 2.2	Pt 2.2	Au 2.4	Hg 1.9	Tl 1.8	Pb 1.9	Bi 1.9	Po 2.0	At 2.1	Rn

Figure 2.2 Electronegativity values and trends. Electronegativity generally increases from left to right across the periodic table and decreases from top to bottom. The values are on an arbitrary scale, with F = 4.0 and Cs = 0.7. Elements in **red** are the most electronegative, those in yellow are medium, and those in green are the least electronegative.

As a rough guide, bonds between atoms whose electronegativities differ by less than 0.5 are nonpolar covalent, bonds between atoms whose electronegativities differ by 0.5–2 are polar covalent, and bonds between atoms whose electronegativities differ by more than 2 are largely ionic. Carbon–hydrogen bonds, for example, are relatively nonpolar because carbon (EN = 2.5) and hydrogen (EN = 2.1) have similar electronegativities. Bonds between carbon and *more* electronegative elements such as oxygen (EN = 3.5) and nitrogen (EN = 3.0), by contrast, are polarized so that the bonding electrons are drawn away from carbon toward the electronegative atom. This leaves carbon with a partial positive charge, denoted by δ+, and the electronegative atom with a partial negative charge, δ− (δ is the lowercase Greek letter delta). An example is the C O bond in methanol, CH_3OH **(Figure 2.3a)**. Bonds between carbon and *less* electronegative elements are polarized so that carbon bears a partial negative charge and the other atom bears a partial positive charge. An example is the C–Li bond in methyllithium, CH_3Li **(Figure 2.3b)**.

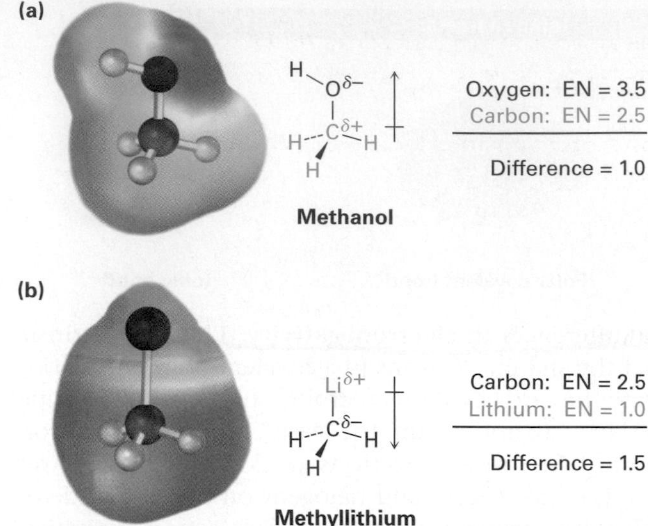

Figure 2.3 **(a)** Methanol, CH_3OH, has a polar covalent C–O bond, and **(b)** methyllithium, CH_3Li, has a polar covalent C–Li bond. The computer-generated representations, called electrostatic potential maps, use color to show calculated charge distributions, ranging from **red (electron-rich; δ−)** to **blue (electron-poor; δ+)**.

Note in the representations of methanol and methyllithium in Figure 2.3 that a crossed arrow $\leftarrow\!\!\!+$ is used to indicate the direction of bond polarity. By convention, *electrons are displaced in the direction of the arrow.* The tail of the arrow (which looks like a plus sign) is electron-poor (δ+), and the head of the arrow is electron-rich (δ−).

Note also in Figure 2.3 that calculated charge distributions in molecules can be displayed visually with what are called *electrostatic potential maps,* which use color to indicate electron-rich (red; δ−) and electron-poor (blue; δ+) regions. In methanol, oxygen carries a partial negative charge and is colored red, while the carbon and hydrogen atoms carry partial positive charges and are colored blue-green. In methyllithium, lithium carries a partial positive charge (blue), while carbon and the hydrogen atoms carry partial negative charges (red). Electrostatic potential maps are useful because they show at a glance the electron-rich and electron-poor atoms in molecules. We'll make frequent use of these maps throughout the text and will see many examples of how electronic structure correlates with chemical reactivity.

When speaking of an atom's ability to polarize a bond, we often use the term *inductive effect*. An **inductive effect** is simply the shifting of electrons in a σ bond in response to the electronegativity of nearby atoms. Metals, such as lithium and magnesium, inductively donate electrons, whereas reactive nonmetals, such as oxygen and nitrogen, inductively withdraw electrons. Inductive effects play a major role in understanding chemical reactivity, and we'll use them many times throughout this text to explain a variety of chemical observations.

Problem 2.1

Which element in each of the following pairs is more electronegative?

(a) Li or H **(b)** B or Br **(c)** Cl or I **(d)** C or H

Problem 2.2

Use the δ+/δ− convention to indicate the direction of expected polarity for each of the bonds indicated.

(a) $H_3C–Cl$ **(b)** $H_3C–NH_2$ **(c)** $H_2N–H$
(d) $H_3C–SH$ **(e)** $H_3C–MgBr$ **(f)** $H_3C–F$

Problem 2.3

Use the electronegativity values shown in Figure 2.2 to rank the following bonds from least polar to most polar: $H_3C{-}Li$, $H_3C{-}K$, $H_3C{-}F$, $H_3C{-}MgBr$, $H_3C{-}OH$

Problem 2.4

Look at the following electrostatic potential map of chloromethane, and tell the direction of polarization of the C–Cl bond:

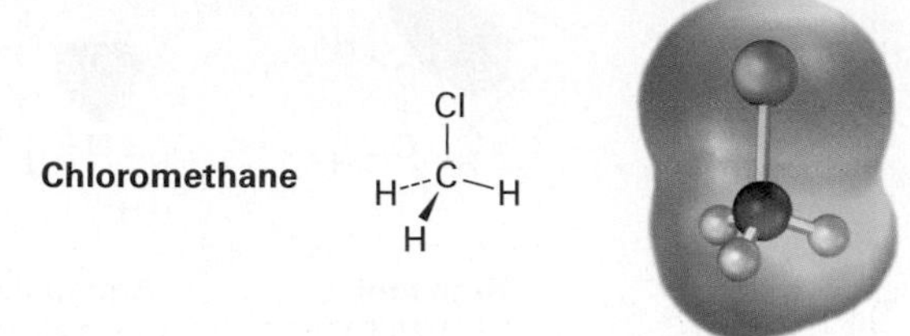

2.2 Polar Covalent Bonds: Dipole Moments

Just as individual bonds are often polar, molecules as a whole are often polar also. Molecular polarity results from the vector summation of all individual bond polarities and lone-pair contributions in the molecule. As a practical matter, strongly polar substances are often soluble in polar solvents like water, whereas less polar substances are insoluble in water.

Net molecular polarity is measured by a quantity called the *dipole moment* and can be thought of in the following way: assume that there is a center of mass of all positive charges (nuclei) in a molecule and a center of mass of all negative charges (electrons). If these two centers don't coincide, then the molecule has a net polarity.

The **dipole moment**, μ (Greek mu), is defined as the magnitude of the charge Q at either end of the molecular dipole times the distance r between the charges, $\mu = Q \times r$. Dipole moments are expressed in *debyes* (D), where 1 D = 3.336×10^{-30} coulomb meter (C · m) in SI units. For example, the unit charge on an electron is 1.60×10^{-19} C. Thus, if one positive charge and one negative charge are separated by 100 pm (a bit less than the length of a typical covalent bond), the dipole moment is 1.60×10^{-29} C · m, or 4.80 D.

$$\mu = Q \times r$$

$$\mu = (1.60 \times 10^{-19}\,\text{C})(100 \times 10^{-12}\,\text{m})\left(\frac{1\ \text{D}}{3.336 \times 10^{-30}\ \text{C} \cdot \text{m}}\right) = 4.80\ \text{D}$$

Dipole moments for some common substances are given in Table 2.1. Of the compounds shown in the table, sodium chloride has the largest dipole moment (9.00 D) because it is ionic. Even small molecules like water (μ = 1.85 D), methanol (CH_3OH; μ = 1.70 D), and ammonia (μ = 1.47 D), have substantial dipole moments, however, both because they contain strongly electronegative atoms (oxygen and nitrogen) and because all three molecules have lone-pair electrons. The lone-pair electrons on oxygen and nitrogen atom stick out into space away from the positively charged nuclei, giving rise to a considerable charge separation and making a large contribution to the dipole moment.

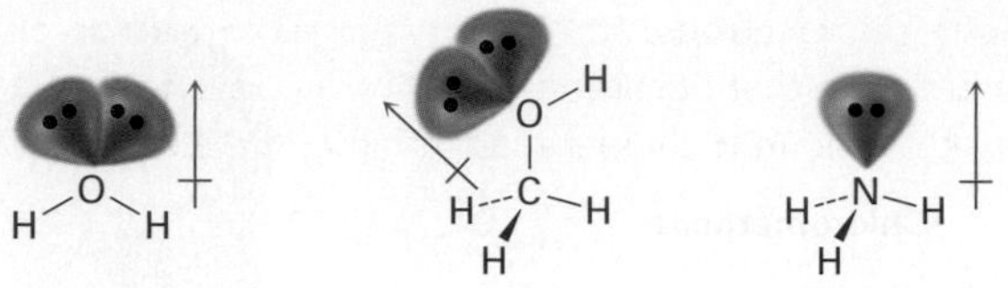

Water (μ = 1.85 D) **Methanol** (μ = 1.70 D) **Ammonia** (μ = 1.47 D)

Table 2.1 Dipole Moments of Some Compounds

Compound	Dipole moment (D)	Compound	Dipole moment (D)
NaCl	9.00	NH_3	1.47
CH_2O	2.33	CH_3NH_2	1.31
CH_3Cl	1.87	CO_2	0
H_2O	1.85	CH_4	0
CH_3OH	1.70	CH_3CH_3	0
CH_3CO_2H	1.70	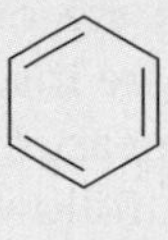 Benzene	0
CH_3SH	1.52		

In contrast with water, methanol, and ammonia, molecules such as carbon dioxide, methane, ethane, and benzene have zero dipole moments. Because of the symmetrical structures of these molecules, the individual bond polarities and lone-pair contributions exactly cancel.

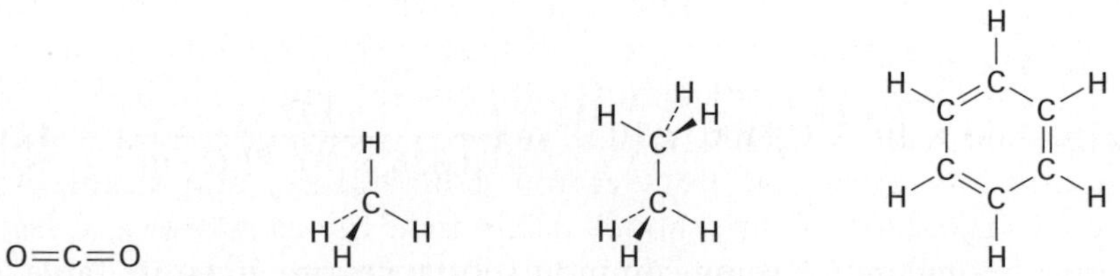

O=C=O

Carbon dioxide (μ = 0) **Methane** (μ = 0) **Ethane** (μ = 0) **Benzene** (μ = 0)

Worked Example 2.1

Predicting the Direction of a Dipole Moment

Make a three-dimensional drawing of methylamine, CH_3NH_2, a substance responsible for the odor of rotting fish, and show the direction of its dipole moment ($\mu = 1.31$).

Strategy

Look for any lone-pair electrons, and identify any atom with an electronegativity substantially different from that of carbon. (Usually, this means O, N, F, Cl, or Br.) Electron density will be displaced in the general direction of the electronegative atoms and the lone pairs.

Solution

Methylamine contains an electronegative nitrogen atom with a lone-pair electrons. The dipole moment thus points generally from $-CH_3$ toward the lone pair.

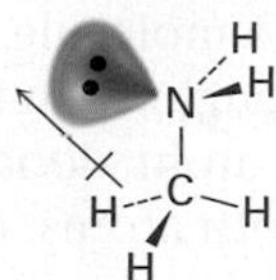

Methylamine
(μ = 1.31)

Problem 2.5

Ethylene glycol, $HOCH_2CH_2OH$, has zero dipole moment even though carbon–oxygen bonds are strongly polar and oxygen has two lone-pairs of electrons. Explain.

Problem 2.6

Make three-dimensional drawings of the following molecules, and predict whether each has a dipole moment. If you expect a dipole moment, show its direction.

(a) $H_2C{=}CH_2$ **(b)** $CHCl_3$ **(c)** CH_2Cl_2 **(d)** $H_2C{=}CCl_2$

2.3 Formal Charges

Closely related to the ideas of bond polarity and dipole moment is the concept of assigning *formal charges* to specific atoms within a molecule, particularly atoms that have an apparently "abnormal" number of bonds. Look at dimethyl sulfoxide (CH_3SOCH_3), for instance, a solvent commonly used for preserving biological cell lines at low temperature. The sulfur atom in dimethyl sulfoxide has three bonds rather than the usual two and has a formal positive charge. The oxygen atom, by contrast, has one bond rather than the usual two and has a formal negative charge. Note that an electrostatic potential map of dimethyl sulfoxide shows the oxygen as negative (red) and the sulfur as relatively positive (blue), in accordance with the formal charges.

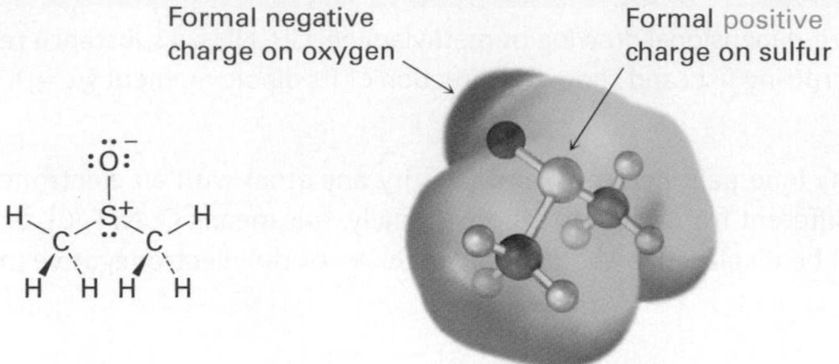

Dimethyl sulfoxide

Formal charges, as the name suggests, are a formalism and don't imply the presence of actual ionic charges in a molecule. Instead, they're a device for electron "bookkeeping" and can be thought of in the following way: a typical covalent bond is formed when each atom donates one electron. Although the bonding electrons are shared by both atoms, each atom can still be considered to "own" one electron for bookkeeping purposes. In methane, for instance, the carbon atom owns one electron in each of the four C–H bonds, for a total of four. Because a neutral, isolated carbon atom has four valence electrons, and because the carbon atom in methane still owns four, the methane carbon atom is neutral and has no formal charge.

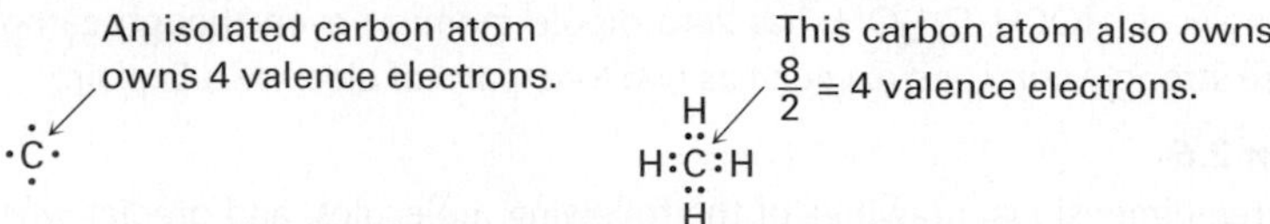

The same is true for the nitrogen atom in ammonia, which has three covalent N–H bonds and two nonbonding electrons (a lone pair). Atomic nitrogen has five valence electrons, and the ammonia nitrogen also has five—one in each of three shared N–H bonds plus two in the lone pair. Thus, the nitrogen atom in ammonia has no formal charge.

An isolated nitrogen atom owns 5 valence electrons.

This nitrogen atom also owns $\frac{6}{2} + 2 = 5$ valence electrons.

The situation is different in dimethyl sulfoxide. Atomic sulfur has six valence electrons, but the dimethyl sulfoxide sulfur owns only *five*—one in each of the two S–C single bonds, one in the S–O single bond, and two in a lone pair. Thus, the sulfur atom has formally lost an electron and therefore has a positive charge. A similar calculation for the oxygen atom shows that it has formally gained an electron and has a negative charge. Atomic oxygen has six valence electrons, but the oxygen in dimethyl sulfoxide has seven—one in the O–S bond and two in each of three lone pairs.

For sulfur:

Sulfur valence electrons = 6
Sulfur bonding electrons = 6
Sulfur nonbonding electrons = 2

Formal charge = 6 − 6/2 − 2 = +1

For oxygen:

Oxygen valence electrons = 6
Oxygen bonding electrons = 2
Oxygen nonbonding electrons = 6

Formal charge = 6 − 2/2 − 6 = −1

To express the calculations in a general way, the **formal charge** on an atom is equal to the number of valence electrons in a neutral, isolated atom minus the number of electrons owned by that bonded atom in a molecule. The number of electrons in the bonded atom, in turn, is equal to half the number of bonding electrons plus the nonbonding, lone-pair electrons.

$$\textbf{Formal charge} = \begin{pmatrix}\text{Number of}\\ \text{valence electrons}\\ \text{in free atom}\end{pmatrix} - \begin{pmatrix}\text{Number of}\\ \text{valence electrons}\\ \text{in bonded atom}\end{pmatrix}$$

$$= \begin{pmatrix}\text{Number of}\\ \text{valence electrons}\\ \text{in free atom}\end{pmatrix} - \left(\frac{\text{Number of bonding electrons}}{2}\right) - \begin{pmatrix}\text{Number of}\\ \text{nonbonding}\\ \text{electrons}\end{pmatrix}$$

A summary of commonly encountered formal charges and the bonding situations in which they occur is given in Table 2.2. Although only a bookkeeping device, formal charges often give clues about chemical reactivity, so it's helpful to be able to identify and calculate them correctly.

Table 2.2 A Summary of Common Formal Charges

Atom	C	C	C	N	N	O	O	S	S	P
Structure	—Ċ— (three bonds, one nonbonding electron)	—C⁺— (three bonds)	—C⁻— (three bonds, lone pair)	—N⁺— (four bonds)	—N⁻— (two bonds, two lone pairs)	—O⁺— (three bonds, lone pair)	—O:⁻ (one bond, three lone pairs)	—S⁺— (three bonds, lone pair)	—S:⁻ (one bond, three lone pairs)	—P⁺— (four bonds)
Valence electrons	4	4	4	5	5	6	6	6	6	5
Number of bonds	3	3	3	4	2	3	1	3	1	4
Number of nonbonding electrons	1	0	2	0	4	2	6	2	6	0
Formal charge	0	+1	−1	+1	−1	+1	−1	+1	−1	+1

Problem 2.7

Calculate formal charges for the nonhydrogen atoms in the following molecules:

(a) Diazomethane, $H_2C{=}N{=}\ddot{N}{:}$ **(b)** Acetonitrile oxide, $H_3C{-}C{\equiv}N{-}\ddot{\underset{\cdot\cdot}{O}}{:}$

(c) Methyl isocyanide, $H_3C{-}N{\equiv}C{:}$

Problem 2.8

Organic phosphate groups occur commonly in biological molecules. Calculate formal charges on the four O atoms in the methyl phosphate dianion.

```
  ┌      H      :O:    ┐2–
  │      |   ··  ||  ·· │
  │   H—C—O—P—O:  │
  │      |   ··  |   ·· │
  │      H      :O:    │
  └              ··     ┘
```

Methyl phosphate ion

2.4 Resonance

Most substances can be represented unambiguously by the Kekulé line-bond structures we've been using up to this point, but an interesting problem sometimes arises. Look at the acetate ion, for instance. When we draw a line-bond structure for acetate, we need to show a double bond to one oxygen and a single bond to the other. But which oxygen is which? Should we draw a double bond to the "top" oxygen and a single bond to the "bottom" oxygen, or vice versa?

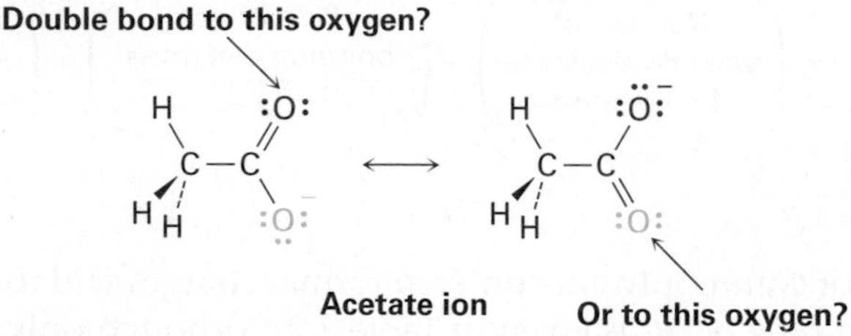

Although the two oxygen atoms in the acetate ion appear different in line-bond structures, experiments show that they are equivalent. Both carbon–oxygen bonds, for example, are 127 pm in length, midway between the length of a typical C–O single bond (135 pm) and a typical C=O double bond (120 pm). In other words, *neither* of the two structures for acetate is correct by itself. The true structure is intermediate between the two, and an electrostatic potential map shows that both oxygen atoms share the negative charge and have equal electron densities (red).

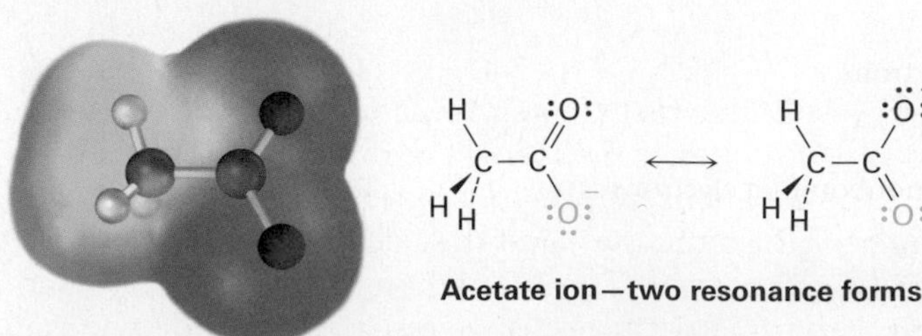

Acetate ion—two resonance forms

The two individual line-bond structures for acetate ion are called **resonance forms**, and their special resonance relationship is indicated by the double-headed arrow between them. *The only difference between resonance forms is the placement of the π and nonbonding valence electrons*. The atoms themselves occupy exactly the same place in both resonance forms, the connections between atoms are the same, and the three-dimensional shapes of the resonance forms are the same.

A good way to think about resonance forms is to realize that a substance like the acetate ion is the same as any other. Acetate doesn't jump back and forth between two resonance forms, spending part of the time looking like one and part of the time looking like the other. Rather, acetate has a single unchanging structure that we say is a **resonance hybrid** of the two individual forms and has characteristics of both. The only "problem" with acetate is that we can't draw it accurately using a familiar line-bond structure—line-bond structures just don't work well for resonance hybrids. The difficulty, however, is with the *representation* of acetate on paper, not with acetate itself.

Resonance is a very useful concept that we'll return to on numerous occasions throughout the rest of this book. We'll see in Chapter 15, for instance, that the six carbon–carbon bonds in aromatic compounds, such as benzene, are equivalent and that benzene is best represented as a hybrid of two resonance forms. Although each individual resonance form seems to imply that benzene has alternating single and double bonds, neither form is correct by itself. The true benzene structure is a hybrid of the two individual forms, and all six carbon–carbon bonds are equivalent. This symmetrical distribution of electrons around the molecule is evident in an electrostatic potential map.

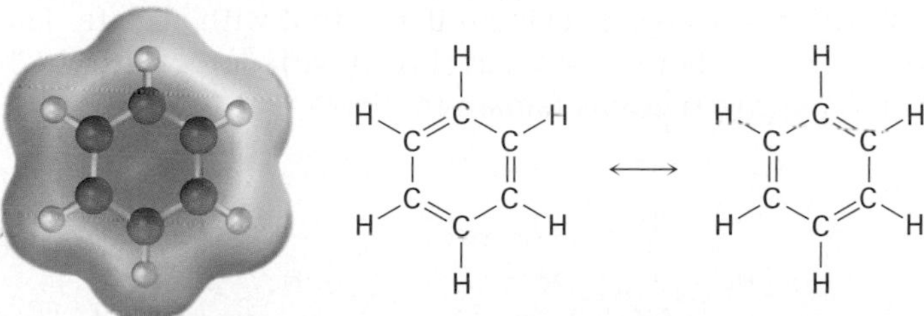

Benzene (two resonance forms)

2.5 Rules for Resonance Forms

When first dealing with resonance forms, it's useful to have a set of guidelines that describe how to draw and interpret them. The following rules should be helpful:

RULE 1

Individual resonance forms are imaginary, not real. The real structure is a composite, or resonance hybrid, of the different forms. Species such as the acetate ion and benzene are no different from any other. They have single, unchanging structures, and they do not switch back and forth between resonance forms. The only difference between these and other substances is in the way they must be represented in drawings on paper.

Key IDEAS

Test your knowledge of Key Ideas by answering end-of-chapter exercises marked with ▲.

RULE 2

Resonance forms differ only in the placement of their π or nonbonding electrons. Neither the position nor the hybridization of any atom changes from one resonance form to another. In the acetate ion, for instance, the carbon atom is sp^2-hybridized and the oxygen atoms remain in exactly the same place in both resonance forms. Only the positions of the π electrons in the C=O bond and the lone-pair electrons on oxygen differ from one form to another. This movement of electrons from one resonance structure to another can be indicated by using curved arrows. *A curved arrow always indicates the movement of electrons, not the movement of atoms.* An arrow shows that a pair of electrons moves *from* the atom or bond at the tail of the arrow *to* the atom or bond at the head of the arrow.

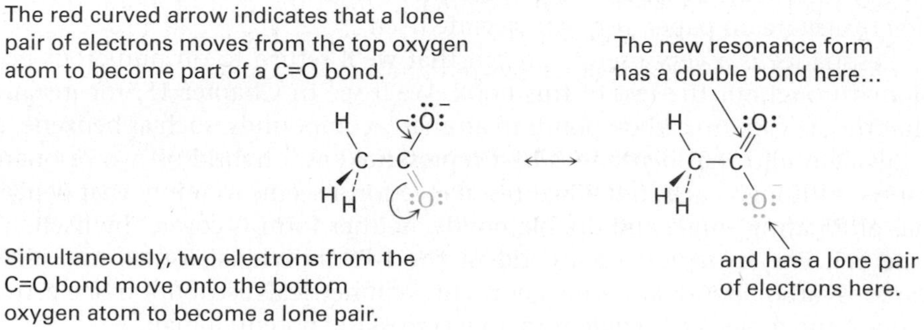

The red curved arrow indicates that a lone pair of electrons moves from the top oxygen atom to become part of a C=O bond.

The new resonance form has a double bond here...

Simultaneously, two electrons from the C=O bond move onto the bottom oxygen atom to become a lone pair.

and has a lone pair of electrons here.

The situation with benzene is similar to that with acetate. The π electrons in the double bonds move, as shown with curved arrows, but the carbon and hydrogen atoms remain in place.

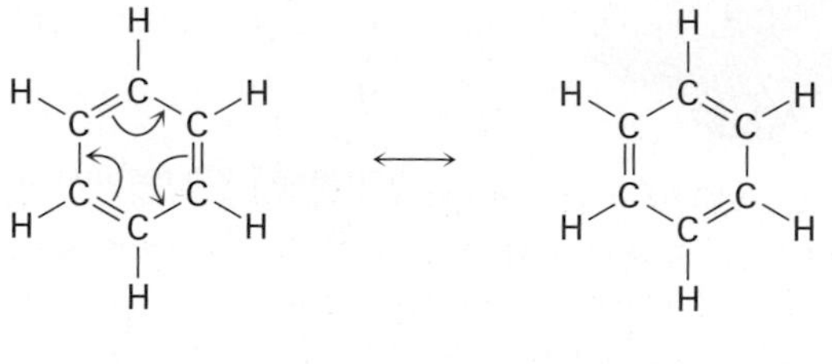

RULE 3

Different resonance forms of a substance don't have to be equivalent. As an example, we'll see in Chapter 22 that a compound such as acetone, which contains a C=O bond, can be converted into its anion by reaction with a strong base. The resultant anion has two resonance forms. One form contains a carbon–*oxygen* double bond and has a negative charge on *carbon;* the other contains a carbon–*carbon* double bond and has a negative charge on *oxygen.* Even though the two resonance forms aren't equivalent, both contribute to the overall resonance hybrid.

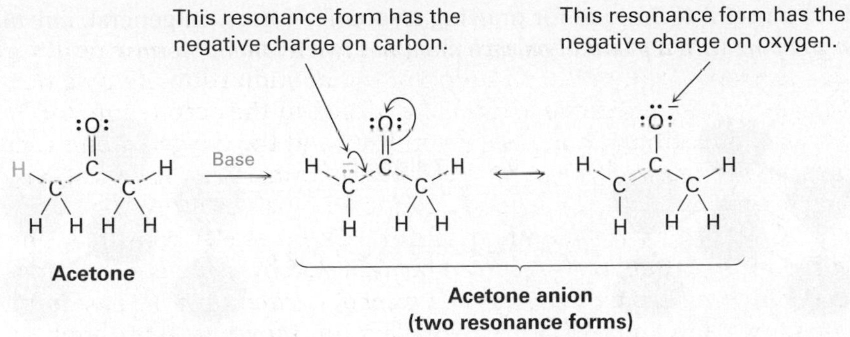

This resonance form has the negative charge on carbon.

This resonance form has the negative charge on oxygen.

Acetone

Base

Acetone anion
(two resonance forms)

When two resonance forms are nonequivalent, the actual structure of the resonance hybrid resembles the more stable form more than it resembles the less stable form. Thus, we might expect the true structure of the acetone anion to be more like that of the form that places the negative charge on the electronegative oxygen atom rather than on carbon.

RULE 4

Resonance forms obey normal rules of valency. A resonance form is like any other structure: the octet rule still applies to second-row, main-group atoms. For example, one of the following structures for the acetate ion is not a valid resonance form because the carbon atom has five bonds and ten valence electrons:

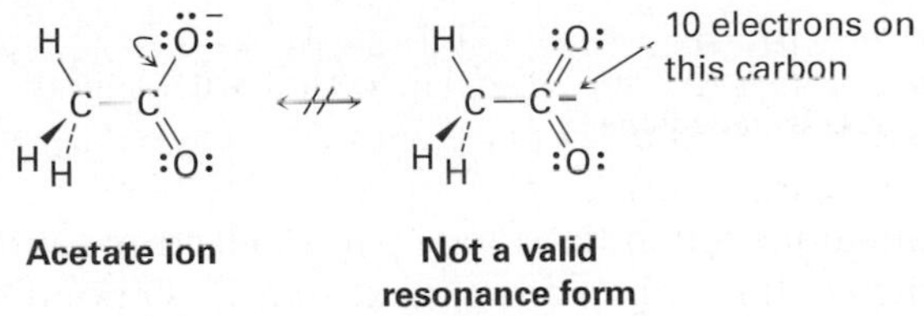

10 electrons on this carbon

Acetate ion

Not a valid resonance form

RULE 5

The resonance hybrid is more stable than any individual resonance form. In other words, resonance leads to stability. Generally speaking, the larger the number of resonance forms, the more stable a substance is because its electrons are spread out over a larger part of the molecule and are closer to more nuclei. We'll see in Chapter 15, for instance, that a benzene ring is more stable because of resonance than might otherwise be expected.

2.6 Drawing Resonance Forms

Look back at the resonance forms of the acetate ion and the acetone anion shown in the previous section. The pattern seen there is a common one that leads to a useful technique for drawing resonance forms. In general, *any three-atom grouping with a p orbital on each atom has two resonance forms:*

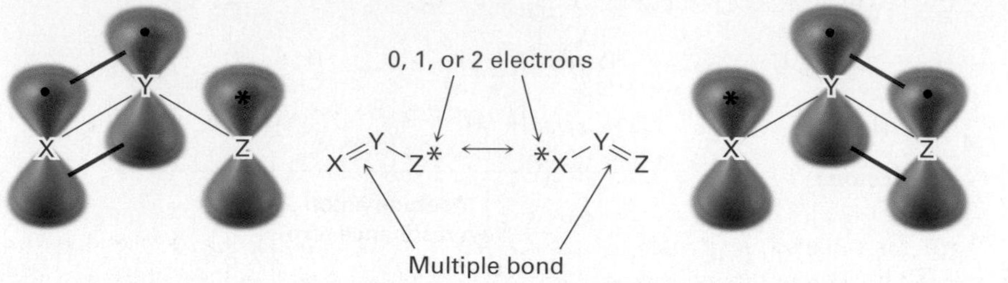

0, 1, or 2 electrons

X=Y–Z* ⟷ *X–Y=Z

Multiple bond

The atoms X, Y, and Z in the general structure might be C, N, O, P, S, or others, and the asterisk (*) might mean that the *p* orbital on atom Z is vacant, that it contains a single electron, or that it contains a lone pair of electrons. The two resonance forms differ simply by an exchange in position of the multiple bond and the asterisk from one end of the three-atom grouping to the other.

By learning to recognize such three-atom groupings within larger structures, resonance forms can be systematically generated. Look, for instance, at the anion produced when H^+ is removed from 2,4-pentanedione by reaction with a base. How many resonance structures does the resultant anion have?

Base

2,4-Pentanedione

The 2,4-pentanedione anion has a lone pair of electrons and a formal negative charge on the central carbon atom, next to a C=O bond on the left. The O=C–C:$^-$ grouping is a typical one for which two resonance structures can be drawn.

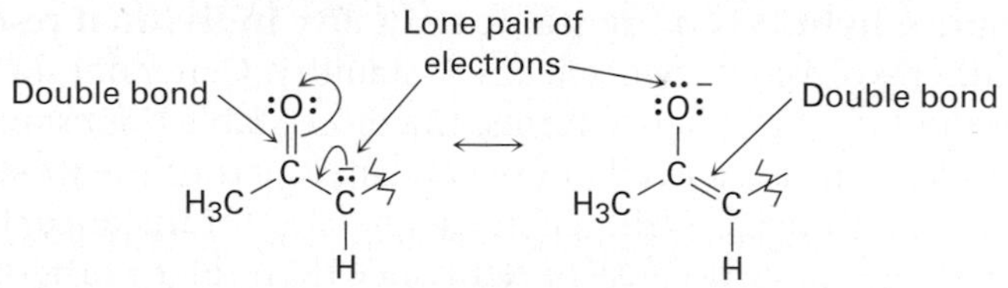

Lone pair of electrons

Double bond

Double bond

Just as there is a C=O bond to the left of the lone pair, there is a second C=O bond to the right. Thus, we can draw a total of three resonance structures for the 2,4-pentanedione anion.

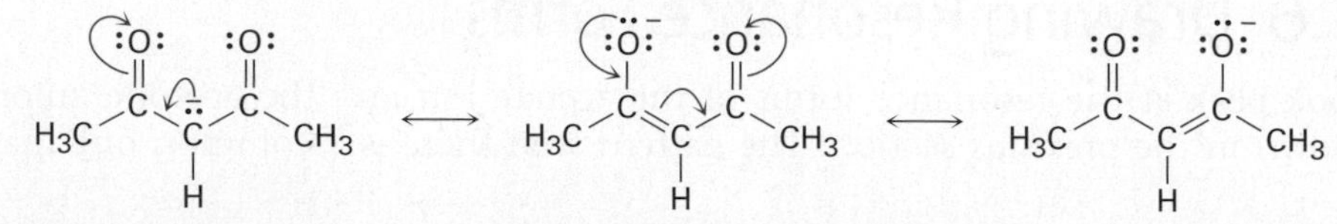

Drawing Resonance Forms for an Anion

Worked Example 2.2

Draw three resonance structures for the carbonate ion, CO_3^{2-}.

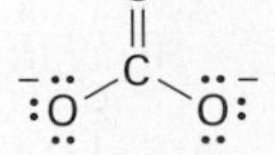

Carbonate ion

Strategy

Look for three-atom groupings that contain a multiple bond next to an atom with a *p* orbital. Then exchange the positions of the multiple bond and the electrons in the *p* orbital. In the carbonate ion, each of the singly bonded oxygen atoms with its lone pairs and negative charge is next to the C=O double bond, giving the grouping O=C–O:⁻.

Solution

Exchanging the position of the double bond and an electron lone pair in each grouping generates three resonance structures.

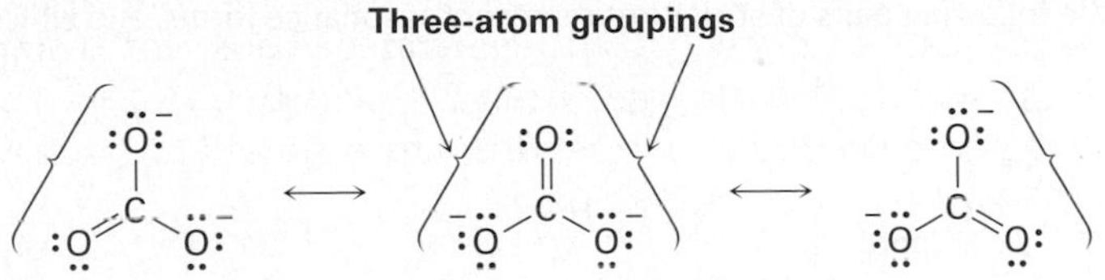

Three-atom groupings

Drawing Resonance Forms for a Radical

Worked Example 2.3

Draw three resonance forms for the pentadienyl radical, where a *radical* is a substance that contains a single, unpaired electron in one of its orbitals, denoted by a dot (·).

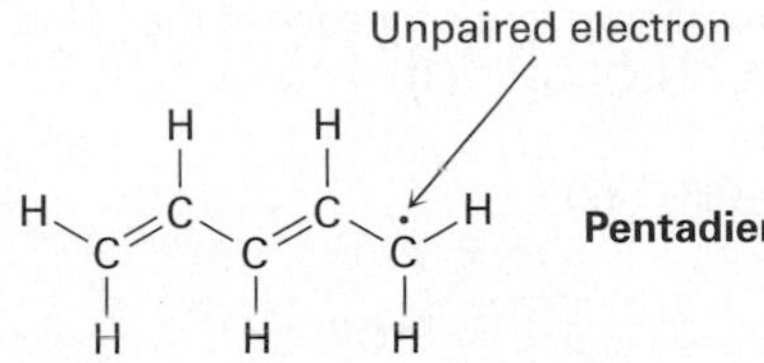

Unpaired electron

Pentadienyl radical

Strategy

Find the three-atom groupings that contain a multiple bond next to a *p* orbital.

Solution

The unpaired electron is on a carbon atom next to a C=C bond, giving a typical three-atom grouping that has two resonance forms.

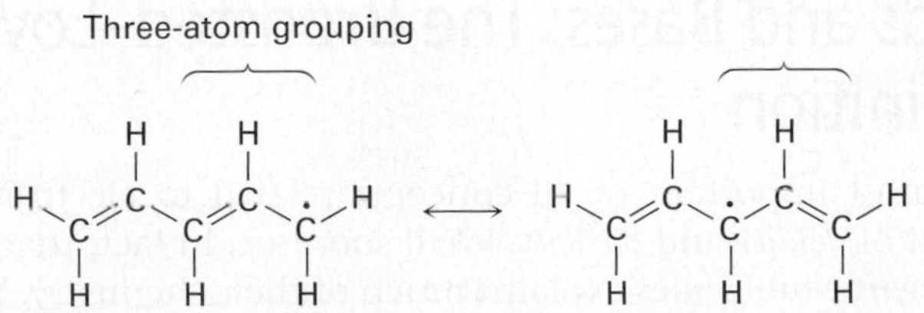

Three-atom grouping

In the second resonance form, the unpaired electron is next to another double bond, giving another three-atom grouping and leading to another resonance form.

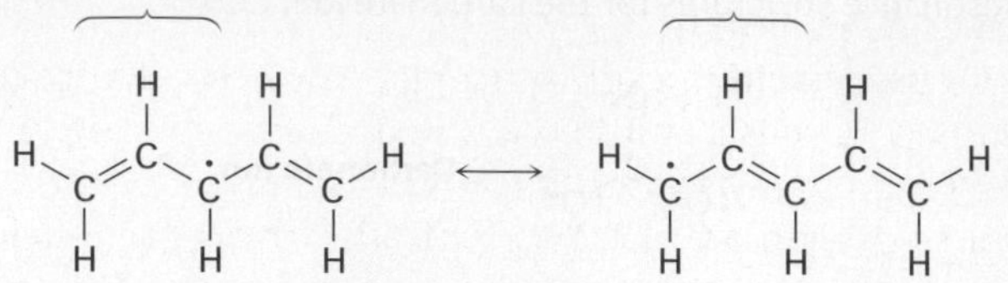

Three-atom grouping

Thus, the three resonance forms for the pentadienyl radical are:

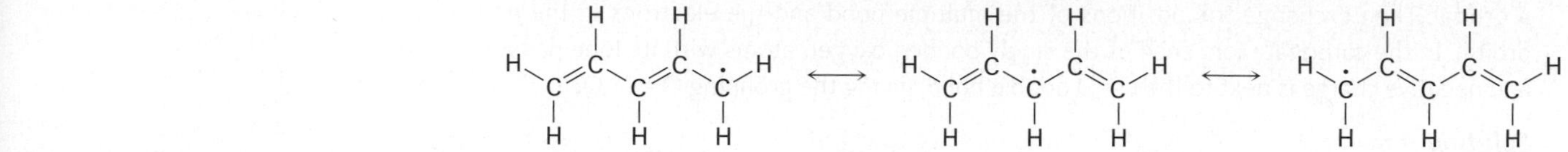

Problem 2.9

Which of the following pairs of structures represent resonance forms, and which do not? Explain.

(a) and

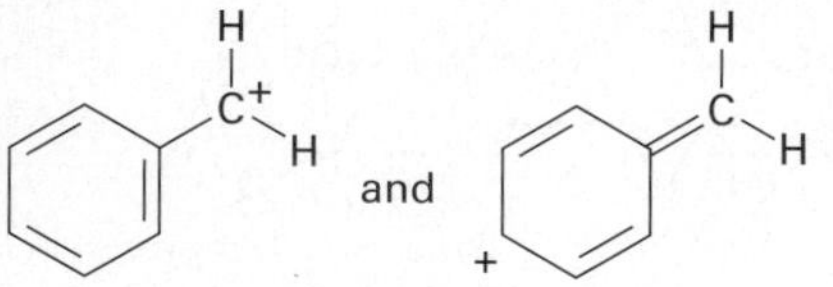

(b) and

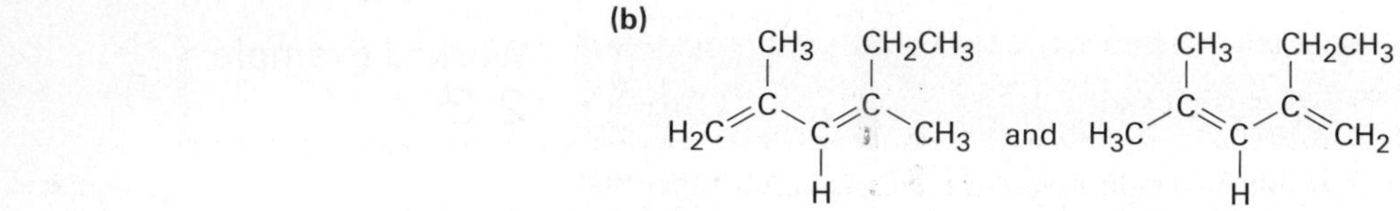

Problem 2.10

Draw the indicated number of resonance forms for each of the following species:

(a) The methyl phosphate anion, $CH_3OPO_3^{2-}$ (3)

(b) The nitrate anion, NO_3^- (3)

(c) The allyl cation, $H_2C{=}CH{-}CH_2^+$ (2)

(d) The benzoate anion (4)

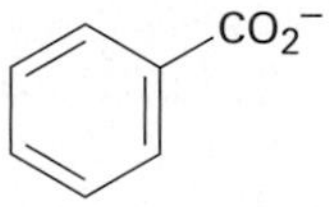

CO_2^-

2.7 Acids and Bases: The Brønsted–Lowry Definition

Perhaps the most important of all concepts related to electronegativity and polarity is that of *acidity* and *basicity*. We'll soon see, in fact, that the acid–base behavior of organic molecules explains much of their chemistry. You may recall from a course in general chemistry that two definitions of acidity are frequently used: the *Brønsted–Lowry definition* and the *Lewis definition*. We'll look at the Brønsted–Lowry definition in this and the following three sections and then discuss the Lewis definition in **Section 2.11**.

A **Brønsted–Lowry acid** is a substance that donates a hydrogen ion, H^+, and a **Brønsted–Lowry base** is a substance that accepts a hydrogen ion. (The name *proton* is often used as a synonym for H^+ because loss of the valence electron from a neutral hydrogen atom leaves only the hydrogen nucleus—a proton.) When gaseous hydrogen chloride dissolves in water, for example, a polar HCl molecule acts as an acid and donates a proton, while a water molecule acts as a base and accepts the proton, yielding chloride ion (Cl^-) and hydronium ion (H_3O^+). This and other acid–base reactions are reversible, so we'll write them with double, forward-and-backward arrows.

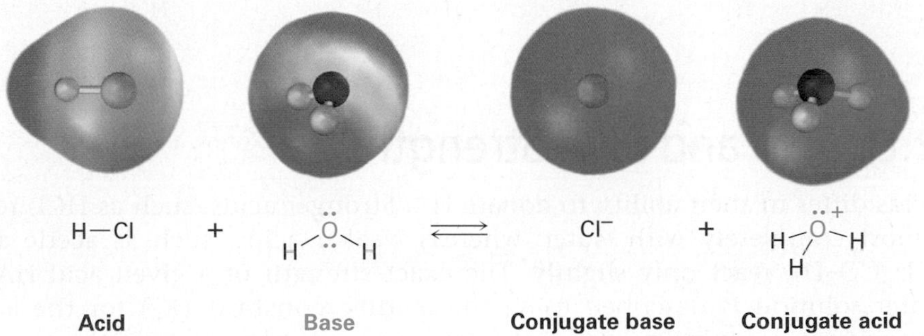

Chloride ion, the product that results when the acid HCl loses a proton, is called the **conjugate base** of the acid, and hydronium ion, the product that results when the base H_2O gains a proton, is called the **conjugate acid** of the base. Other common mineral acids such as H_2SO_4 and HNO_3 behave similarly, as do organic acids such as acetic acid, CH_3CO_2H.

In a general sense,

$$\underset{\text{Acid}}{H{-}A} + \underset{\text{Base}}{:B} \rightleftharpoons \underset{\text{Conjugate base}}{:A^-} + \underset{\text{Conjugate acid}}{H{-}B^+}$$

For example:

$$\underset{\text{Acid}}{H_3C{-}C({=}O){-}\ddot{O}{-}H} + \underset{\text{Base}}{:\ddot{O}{-}H^-} \rightleftharpoons \underset{\text{Conjugate base}}{H_3C{-}C({=}O){-}\ddot{O}:^-} + \underset{\text{Conjugate acid}}{H{-}\ddot{O}{-}H}$$

$$\underset{\text{Acid}}{H{-}\ddot{O}{-}H} + \underset{\text{Base}}{H{-}\ddot{N}H{-}H} \rightleftharpoons \underset{\text{Conjugate base}}{H{-}\ddot{O}:^-} + \underset{\text{Conjugate acid}}{NH_4^+}$$

Notice that water can act either as an acid or as a base, depending on the circumstances. In its reaction with HCl, water is a base that accepts a proton to give the hydronium ion, H_3O^+. In its reaction with ammonia (NH_3), however, water is an acid that donates a proton to give ammonium ion (NH_4^+) and hydroxide ion, HO^-.

Problem 2.11

Nitric acid (HNO_3) reacts with ammonia (NH_3) to yield ammonium nitrate. Write the reaction, and identify the acid, the base, the conjugate acid product, and the conjugate base product.

2.8 Acid and Base Strength

Acids differ in their ability to donate H^+. Stronger acids, such as HCl, react almost completely with water, whereas weaker acids, such as acetic acid (CH_3CO_2H), react only slightly. The exact strength of a given acid HA in water solution is described using the **acidity constant (K_a)** for the acid-dissociation equilibrium. Remember from general chemistry that the concentration of solvent is ignored in the equilibrium expression and that brackets [ ] around a substance refer to the concentration of the enclosed species in moles per liter.

$$HA + H_2O \rightleftharpoons A^- + H_3O^+$$

$$K_a = \frac{[H_3O^+][A^-]}{[HA]}$$

Stronger acids have their equilibria toward the right and thus have larger acidity constants, whereas weaker acids have their equilibria toward the left and have smaller acidity constants. The range of K_a values for different acids is enormous, running from about 10^{15} for the strongest acids to about 10^{-60} for the weakest. The common inorganic acids such as H_2SO_4, HNO_3, and HCl have K_a's in the range of 10^2 to 10^9, while organic acids generally have K_a's in the range of 10^{-5} to 10^{-15}. As you gain more experience, you'll develop a rough feeling for which acids are "strong" and which are "weak" (always remembering that the terms are relative).

Acid strengths are normally expressed using pK_a values rather than K_a values, where the **pK_a** is the negative common logarithm of the K_a:

$$pK_a = -\log K_a$$

A *stronger* acid (larger K_a) has a *smaller* pK_a, and a *weaker* acid (smaller K_a) has a *larger* pK_a. Table 2.3 lists the pK_a's of some common acids in order of their strength, and a more comprehensive table is given in Appendix B.

Table 2.3 Relative Strengths of Some Common Acids and Their Conjugate Bases

	Acid	Name	pK_a	Conjugate base	Name	
Weaker acid	CH_3CH_2OH	Ethanol	16.00	$CH_3CH_2O^-$	Ethoxide ion	Stronger base
	H_2O	Water	15.74	HO^-	Hydroxide ion	
	HCN	Hydrocyanic acid	9.31	CN^-	Cyanide ion	
	$H_2PO_4^-$	Dihydrogen phosphate ion	7.21	HPO_4^{2-}	Hydrogen phosphate ion	
	CH_3CO_2H	Acetic acid	4.76	$CH_3CO_2^-$	Acetate ion	
	H_3PO_4	Phosphoric acid	2.16	$H_2PO_4^-$	Dihydrogen phosphate ion	
	HNO_3	Nitric acid	−1.3	NO_3^-	Nitrate ion	
Stronger acid	HCl	Hydrochloric acid	−7.0	Cl^-	Chloride ion	Weaker base

Notice that the pK_a value shown in Table 2.3 for water is 15.74, which results from the following calculation. Because water is both the acid and the solvent, the equilibrium expression is

$$\underset{\text{(acid)}}{H_2O} + \underset{\text{(solvent)}}{H_2O} \rightleftharpoons OH^- + H_3O^+$$

$$K_a = \frac{[H_3O^+][A^-]}{[HA]} = \frac{[H_3O^+][OH^-]}{[H_2O]} = \frac{[1.0 \times 10^{-7}][1.0 \times 10^{-7}]}{[55.4]} = 1.8 \times 10^{-16}$$

$$pK_a = 15.74$$

The numerator in this expression is the so-called ion-product constant for water, $K_w = [H_3O^+][OH^-] = 1.00 \times 10^{-14}$, and the denominator is the molar concentration of pure water, $[H_2O] = 55.4$ M at 25 °C. The calculation is artificial in that the concentration of "solvent" water is ignored while the concentration of "acid" water is not, but it is nevertheless useful for making a comparison of water with other weak acids on a similar footing.

Notice also in Table 2.3 that there is an inverse relationship between the acid strength of an acid and the base strength of its conjugate base. A *strong* acid has a *weak* conjugate base, and a *weak* acid has a *strong* conjugate base. To understand this inverse relationship, think about what is happening to the acidic hydrogen in an acid–base reaction. A strong acid is one that loses H^+ easily, meaning that its conjugate base holds the H^+ weakly and is therefore a weak base. A weak acid is one that loses H^+ with difficulty, meaning that its conjugate base holds the proton tightly and is therefore a strong base. The fact that HCl is a strong acid, for example, means that Cl^- does not hold H^+ tightly and is thus a weak base. Water, on the other hand, is a weak acid, meaning that OH^- holds H^+ tightly and is a strong base.

Problem 2.12

The amino acid phenylalanine has pK_a = 1.83, and tryptophan has pK_a = 2.83. Which is the stronger acid?

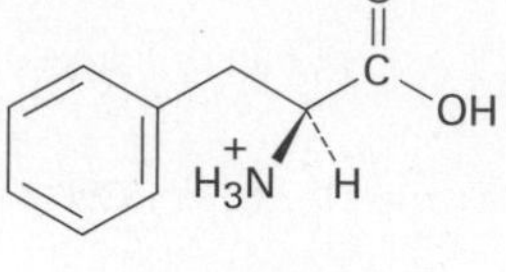

Phenylalanine
(pK_a = 1.83)

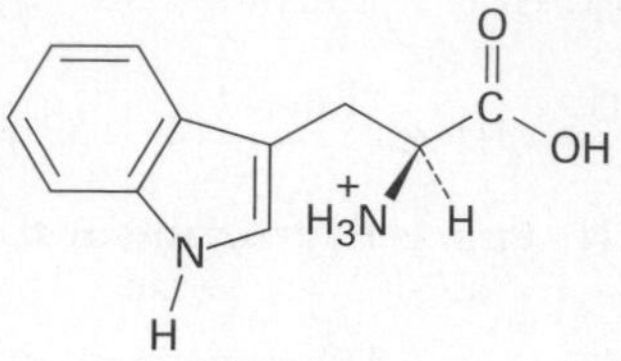

Tryptophan
(pK_a = 2.83)

Problem 2.13

Amide ion, H_2N^-, is a much stronger base than hydroxide ion, HO^-. Which is the stronger acid, NH_3 or H_2O? Explain.

2.9 Predicting Acid–Base Reactions from pK_a Values

Compilations of pK_a values like those in Table 2.3 and Appendix B are useful for predicting whether a given acid–base reaction will take place because H^+ will always go *from* the stronger acid *to* the stronger base. That is, an acid will donate a proton to the conjugate base of a weaker acid, and the conjugate base of a weaker acid will remove the proton from a stronger acid. Since water (pK_a = 15.74) is a weaker acid than acetic acid (pK_a = 4.76), for example, hydroxide ion holds a proton more tightly than acetate ion does. Hydroxide ion will therefore react to a large extent with acetic acid, CH_3CO_2H, to yield acetate ion and H_2O.

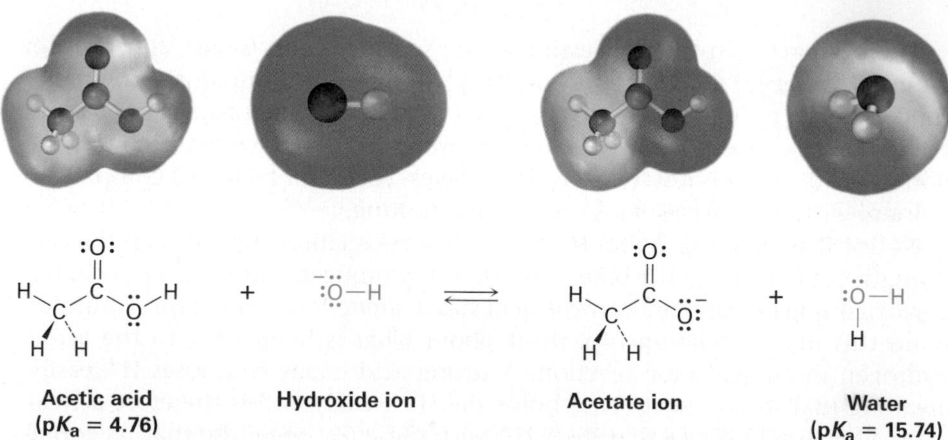

Another way to predict acid–base reactivity is to remember that the product conjugate acid in an acid–base reaction must be weaker and less reactive than the starting acid and the product conjugate base must be weaker and less reactive than the starting base. In the reaction of acetic acid with hydroxide ion, for example, the product conjugate acid (H_2O) is weaker than the starting acid (CH_3CO_2H), and the product conjugate base ($CH_3CO_2^-$) is weaker than the starting base (OH^-).

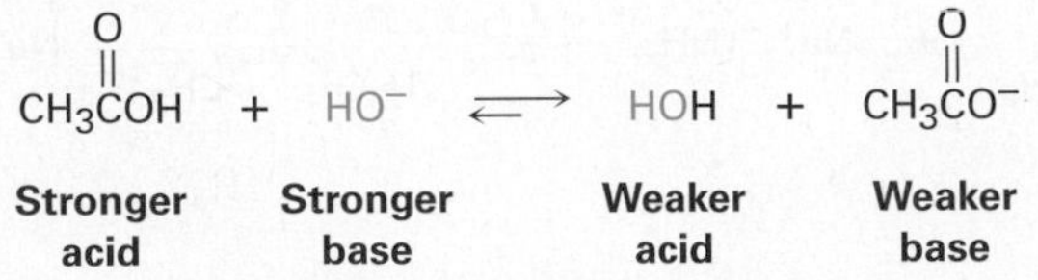

$$\underset{\textbf{Stronger acid}}{CH_3\overset{O}{\overset{\|}{C}}OH} + \underset{\textbf{Stronger base}}{HO^-} \rightleftharpoons \underset{\textbf{Weaker acid}}{HOH} + \underset{\textbf{Weaker base}}{CH_3\overset{O}{\overset{\|}{C}}O^-}$$

Worked Example 2.4

Predicting Acid Strengths from pK_a Values

Water has pK_a = 15.74, and acetylene has pK_a = 25. Which is the stronger acid? Does hydroxide ion react to a significant extent with acetylene?

$$\underset{\textbf{Acetylene}}{H{-}C{\equiv}C{-}H} + OH^- \xrightarrow{?} H{-}C{\equiv}C{:}^- + H_2O$$

Strategy

In comparing two acids, the one with the lower pK_a is stronger. Thus, water is a stronger acid than acetylene and gives up H^+ more easily.

Solution

Because water is a stronger acid and gives up H^+ more easily than acetylene does, the HO^- ion must have less affinity for H^+ than the $HC{\equiv}C{:}^-$ ion has. In other words, the anion of acetylene is a stronger base than hydroxide ion, and the reaction will not proceed significantly as written.

Worked Example 2.5

Calculating K_a from pK_a

According to the data in Table 2.3, acetic acid has pK_a = 4.76. What is its K_a?

Strategy

Since pK_a is the negative logarithm of K_a, it's necessary to use a calculator with an ANTILOG or INV LOG function. Enter the value of the pK_a (4.76), change the sign (−4.76), and then find the antilog (1.74×10^{-5}).

Solution

$K_a = 1.74 \times 10^{-5}$.

Problem 2.14

Will either of the following reactions take place to a significant extent as written, according to the data in Table 2.3?

(a) $HCN + CH_3CO_2^-\,Na^+ \xrightarrow{?} Na^+\,^-CN + CH_3CO_2H$

(b) $CH_3CH_2OH + Na^+\,^-CN \xrightarrow{?} CH_3CH_2O^-\,Na^+ + HCN$

Problem 2.15

Ammonia, NH_3, has p$K_a \approx 36$, and acetone has p$K_a \approx 19$. Will the following reaction take place to a significant extent?

$$H_3C{-}C({=}O){-}CH_3 \;+\; Na^{+}\,{}^{-}{:}\ddot{N}H_2 \;\xrightarrow{?}\; H_3C{-}C({=}O){-}CH_2{:}^{-}\;Na^{+} \;+\; \ddot{N}H_3$$

Acetone

Problem 2.16

What is the K_a of HCN if its pK_a = 9.31?

2.10 Organic Acids and Organic Bases

Many of the reactions we'll be seeing in future chapters, including practically all biological reactions, involve organic acids and organic bases. Although it's too early to go into the details of these processes now, you might keep the following generalities in mind:

Organic Acids

Organic acids are characterized by the presence of a positively polarized hydrogen atom (blue in electrostatic potential maps) and are of two main kinds: those acids such as methanol and acetic acid that contain a hydrogen atom bonded to an electronegative oxygen atom (O–H) and those such as acetone **(Section 2.5)** that contain a hydrogen atom bonded to a carbon atom next to a C=O bond (O=C–C–H).

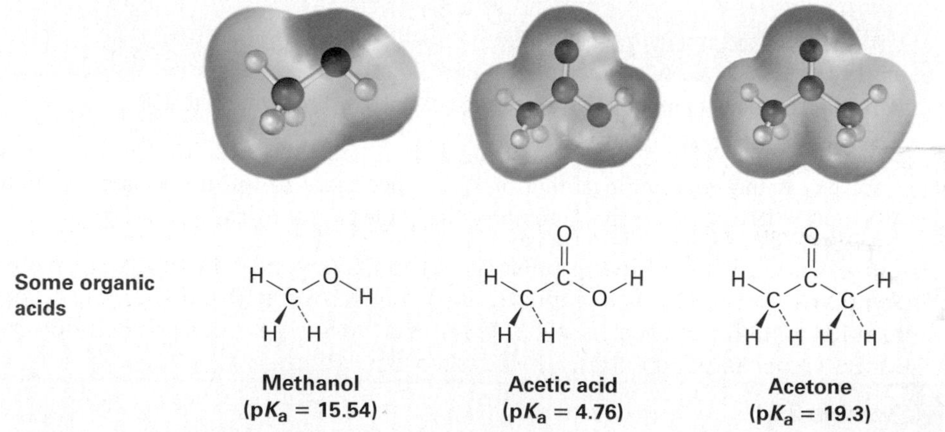

Methanol contains an O–H bond and is a weak acid, while acetic acid also contains an O–H bond and is a somewhat stronger acid. In both cases, acidity is due to the fact that the conjugate base resulting from loss of H^+ is stabilized by having its negative charge on a strongly electronegative oxygen atom. In addition, the conjugate base of acetic acid is stabilized by resonance **(Sections 2.4 and 2.5)**.

Anion is stabilized by having negative charge on a highly electronegative atom.

Anion is stabilized both by having negative charge on a highly electronegative atom and by resonance.

The acidity of acetone and other compounds with C=O bonds is due to the fact that the conjugate base resulting from loss of H^+ is stabilized by resonance. In addition, one of the resonance forms stabilizes the negative charge by placing it on an electronegative oxygen atom.

Anion is stabilized both by resonance and by having negative charge on a highly electronegative atom.

Electrostatic potential maps of the conjugate bases from methanol, acetic acid, and acetone are shown in **Figure 2.4**. As you might expect, all three show a substantial amount of negative charge (red) on oxygen.

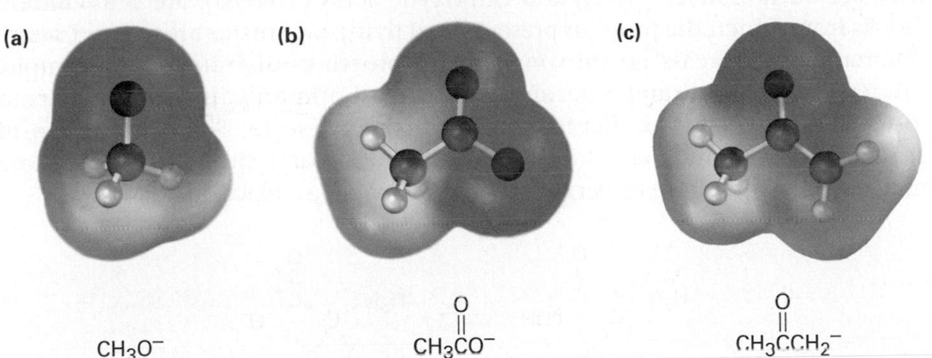

Figure 2.4 Electrostatic potential maps of the conjugate bases of **(a)** methanol, **(b)** acetic acid, and **(c)** acetone. The electronegative oxygen atoms stabilize the negative charge in all three.

Compounds called *carboxylic acids*, which contain the $-CO_2H$ grouping, occur abundantly in all living organisms and are involved in almost all metabolic pathways. Acetic acid, pyruvic acid, and citric acid are examples. You might note that at the typical pH of 7.3 found within cells, carboxylic acids are usually dissociated and exist as their carboxylate anions, $-CO_2^-$.

Acetic acid **Pyruvic acid** **Citric acid**

Organic Bases

Organic bases are characterized by the presence of an atom (reddish in electrostatic potential maps) with a lone pair of electrons that can bond to H^+. Nitrogen-containing compounds such as methylamine are the most common organic bases and are involved in almost all metabolic pathways, but oxygen-containing compounds can also act as bases when reacting with a sufficiently strong acid. Note that some oxygen-containing compounds can act both as acids and as bases depending on the circumstances, just as water can. Methanol and acetone, for instance, act as *acids* when they donate a proton but as *bases* when their oxygen atom accepts a proton.

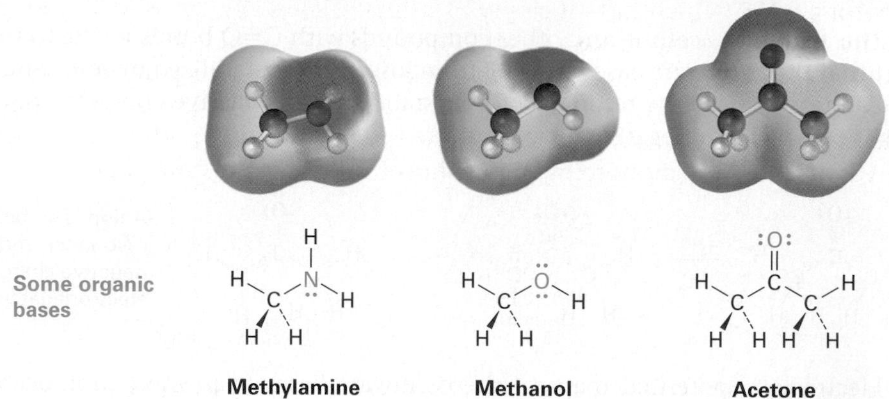

Some organic bases

We'll see in Chapter 26 that substances called *amino acids,* so-named because they are both amines ($-NH_2$) and carboxylic acids ($-CO_2H$), are the building blocks from which the proteins present in all living organisms are made. Twenty different amino acids go into making up proteins—alanine is an example. Interestingly, alanine and other amino acids exist primarily in a doubly charged form called a *zwitterion* rather than in the uncharged form. The zwitterion form arises because amino acids have both acidic and basic sites within the same molecule and therefore undergo an internal acid–base reaction.

$$H_2N-CH(CH_3)-C(=O)-OH \rightleftharpoons H_3\overset{+}{N}-CH(CH_3)-C(=O)-O^-$$

Alanine (uncharged form) **Alanine (zwitterion form)**

2.11 Acids and Bases: The Lewis Definition

The Lewis definition of acids and bases is broader and more encompassing than the Brønsted–Lowry definition because it's not limited to substances that donate or accept just protons. A **Lewis acid** is a substance that *accepts an electron pair,* and a **Lewis base** is a substance that *donates an electron pair.* The donated electron pair is shared between the acid and the base in a covalent bond.

Filled orbital — Vacant orbital

B: + A ⟶ B—A

Lewis base **Lewis acid**

Lewis Acids and the Curved Arrow Formalism

The fact that a Lewis acid is able to accept an electron pair means that it must have either a vacant, low-energy orbital or a polar bond to hydrogen so that it can donate H^+ (which has an empty 1*s* orbital). Thus, the Lewis definition of acidity includes many species in addition to H^+. For example, various metal cations, such as Mg^{2+}, are Lewis acids because they accept a pair of electrons when they form a bond to a base. We'll also see in later chapters that certain metabolic reactions begin with an acid–base reaction between Mg^{2+} as a Lewis acid and an organic diphosphate or triphosphate ion as the Lewis base.

Mg^{2+} + (organodiphosphate ion) ⟶ (acid–base complex with Mg^{2+})

Lewis acid — **Lewis base (an organodiphosphate ion)** — **Acid–base complex**

In the same way, compounds of group 3A elements, such as BF_3 and $AlCl_3$, are Lewis acids because they have unfilled valence orbitals and can accept electron pairs from Lewis bases, as shown in **Figure 2.5**. Similarly, many transition-metal compounds, such as $TiCl_4$, $FeCl_3$, $ZnCl_2$, and $SnCl_4$, are Lewis acids.

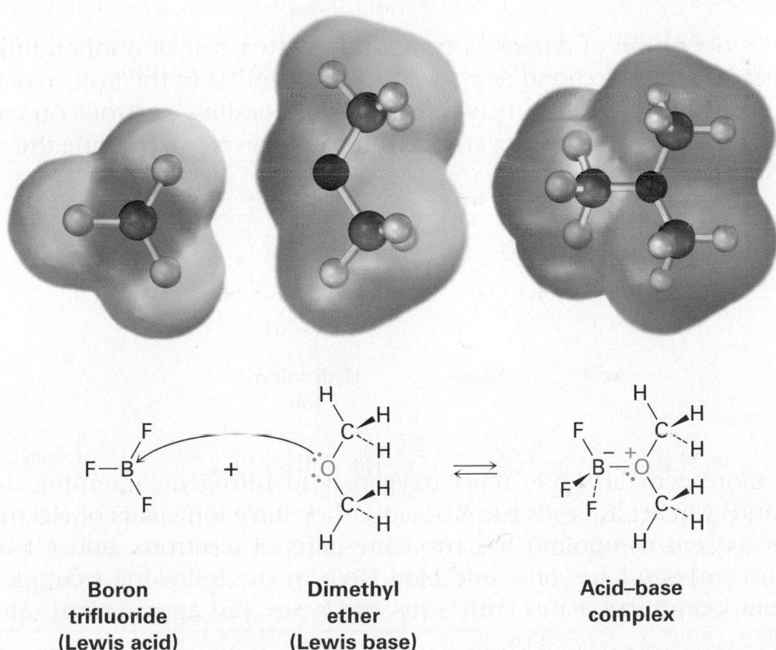

Figure 2.5 The reaction of boron trifluoride, a Lewis acid, with dimethyl ether, a Lewis base. The Lewis acid accepts a pair of electrons, and the Lewis base donates a pair of nonbonding electrons. Note how the movement of electrons *from* the Lewis base *to* the Lewis acid is indicated by a curved arrow. Note also how, in electrostatic potential maps, the boron becomes **more negative** after reaction because it has gained electrons and the oxygen atom becomes **more positive** because it has donated electrons.

Look closely at the acid–base reaction in Figure 2.5, and note how it is shown. Dimethyl ether, the Lewis base, donates an electron pair to a vacant valence orbital of the boron atom in BF_3, a Lewis acid. The direction of electron-pair flow from the base to acid is shown using curved arrows, just as the direction of electron flow in going from one resonance structure to another was shown using curved arrows in **Section 2.5**. *A curved arrow always means that a pair of electrons moves* from *the atom at the tail of the arrow* to *the atom at the head of the arrow.* We'll use this curved-arrow notation throughout the remainder of this text to indicate electron flow during reactions.

Some further examples of Lewis acids follow:

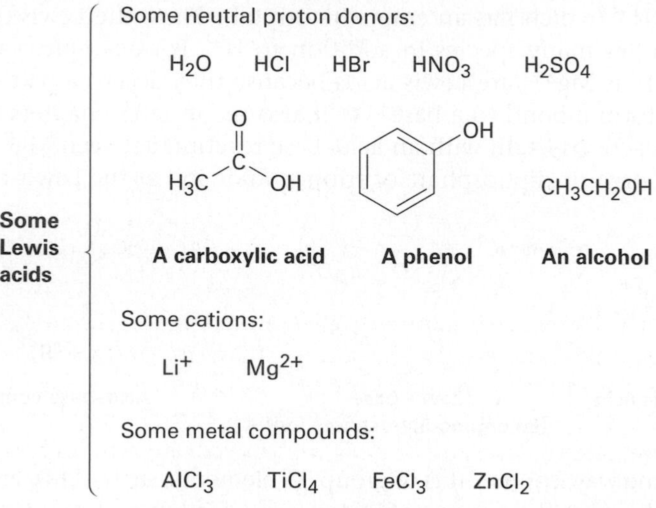

Lewis Bases

The Lewis definition of a base—a compound with a pair of nonbonding electrons that it can use to bond to a Lewis acid—is similar to the Brønsted–Lowry definition. Thus, H_2O, with its two pairs of nonbonding electrons on oxygen, acts as a Lewis base by donating an electron pair to an H^+ in forming the hydronium ion, H_3O^+.

$$\underset{\text{Acid}}{Cl{-}H} + \underset{\text{Base}}{:\ddot{O}H_2} \rightleftharpoons \underset{\text{Hydronium ion}}{H{-}\overset{+}{\ddot{O}}H_2} + Cl^-$$

In a more general sense, most oxygen- and nitrogen-containing organic compounds can act as Lewis bases because they have lone pairs of electrons. A divalent oxygen compound has two lone pairs of electrons, and a trivalent nitrogen compound has one lone pair. Note in the following examples that some compounds can act as both acids and bases, just as water can. Alcohols and carboxylic acids, for instance, act as acids when they donate an H^+ but as bases when their oxygen atom accepts an H^+.

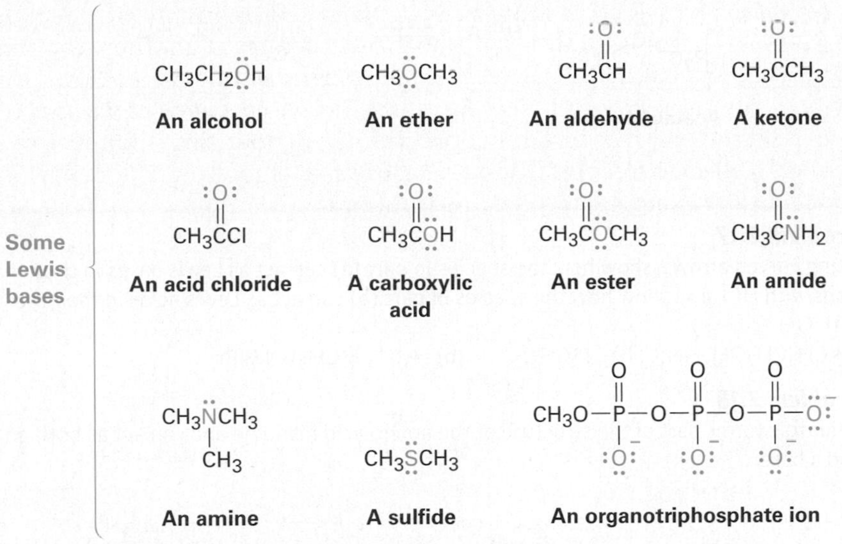

Some Lewis bases

$CH_3CH_2\ddot{O}H$ **An alcohol**; $CH_3\ddot{O}CH_3$ **An ether**; CH_3CH (with :O: double bond) **An aldehyde**; CH_3CCH_3 (with :O: double bond) **A ketone**

CH_3CCl **An acid chloride**; CH_3COH **A carboxylic acid**; CH_3COCH_3 **An ester**; CH_3CNH_2 **An amide**

$CH_3\ddot{N}CH_3$ (with CH_3) **An amine**; $CH_3\ddot{S}CH_3$ **A sulfide**; $CH_3O{-}P{-}O{-}P{-}O{-}P{-}\ddot{O}:^-$ **An organotriphosphate ion**

Notice in the list of Lewis bases just given that some compounds, such as carboxylic acids, esters, and amides, have more than one atom with a lone pair of electrons and can therefore react at more than one site. Acetic acid, for example, can be protonated either on the doubly bonded oxygen atom or on the singly bonded oxygen atom. Reaction normally occurs only once in such instances, and the more stable of the two possible protonation products is formed. For acetic acid, protonation by reaction with sulfuric acid occurs on the doubly bonded oxygen because that product is stabilized by two resonance forms.

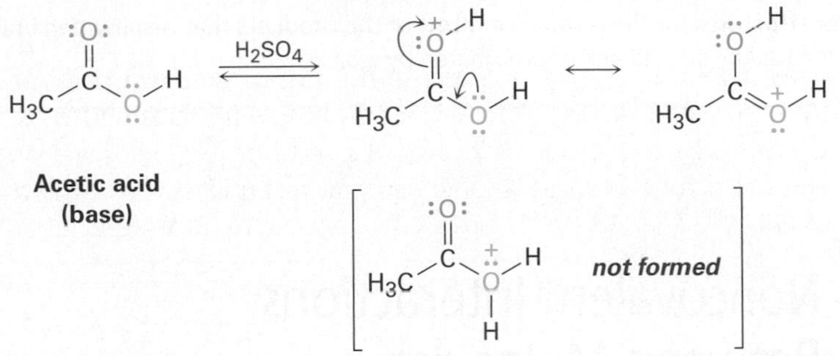

H_2SO_4

Acetic acid (base)

not formed

Using Curved Arrows to Show Electron Flow

Worked Example 2.6

Using curved arrows, show how acetaldehyde, CH_3CHO, can act as a Lewis base.

Strategy

A Lewis base donates an electron pair to a Lewis acid. We therefore need to locate the electron lone pairs on acetaldehyde and use a curved arrow to show the movement of a pair toward the H atom of the acid.

Solution

H_3C–C(=:O:)–H + H–A ⇌ H_3C–C(=$\overset{+}{O}$–H)–H + :A^-

Acetaldehyde

Problem 2.17

Using curved arrows, show how the species in part **(a)** can act as Lewis bases in their reactions with HCl, and show how the species in part **(b)** can act as Lewis acids in their reaction with OH^-.

(a) CH_3CH_2OH, $HN(CH_3)_2$, $P(CH_3)_3$ **(b)** H_3C^+, $B(CH_3)_3$, $MgBr_2$

Problem 2.18

Imidazole forms part of the structure of the amino acid histidine and can act as both an acid and a base.

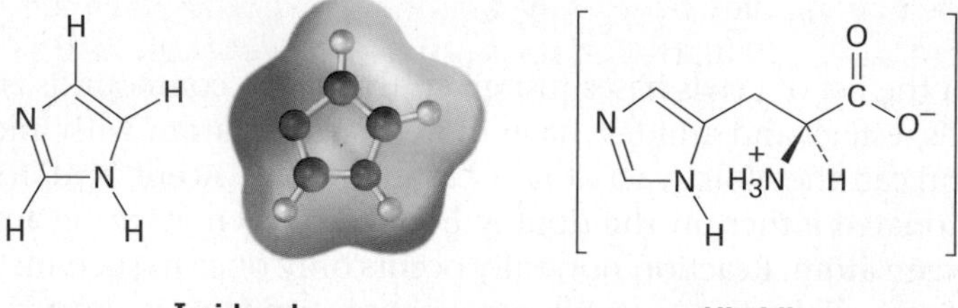

Imidazole **Histidine**

(a) Look at the electrostatic potential map of imidazole, and identify the most acidic hydrogen atom and the most basic nitrogen atom.

(b) Draw structures for the resonance forms of the products that result when imidazole is protonated by an acid and deprotonated by a base.

2.12 Noncovalent Interactions Between Molecules

When thinking about chemical reactivity, chemists usually focus their attention on bonds, the covalent interactions between atoms *within* molecules. Also important, however, particularly in large biomolecules like proteins and nucleic acids, are a variety of interactions *between* molecules that strongly affect molecular properties. Collectively called either *intermolecular forces, van der Waals forces,* or **noncovalent interactions**, they are of several different types: dipole–dipole forces, dispersion forces, and hydrogen bonds.

Dipole–dipole forces occur between polar molecules as a result of electrostatic interactions among dipoles. The forces can be either attractive or repulsive depending on the orientation of the molecules—attractive when unlike charges are together and repulsive when like charges are together. The attractive geometry is lower in energy and therefore predominates **(Figure 2.6)**.

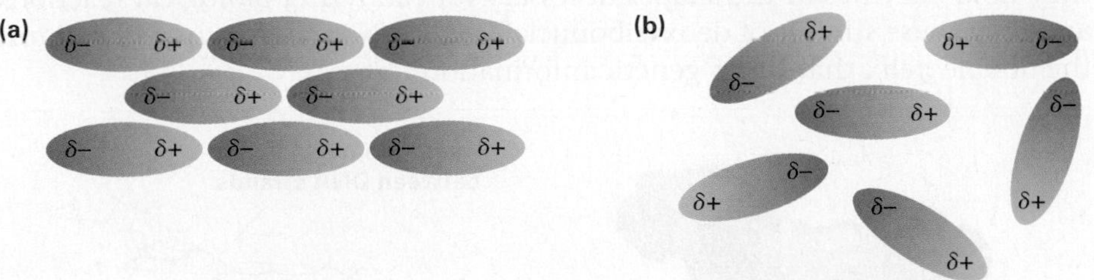

Figure 2.6 Dipole–dipole forces cause polar molecules **(a)** to attract one another when they orient with unlike charges together, but **(b)** to repel one another when they orient with like charges together.

Dispersion forces occur between all neighboring molecules and arise because the electron distribution within molecules is constantly changing. Although uniform on a time-averaged basis, the electron distribution even in nonpolar molecules is likely to be nonuniform at any given instant. One side of a molecule may, by chance, have a slight excess of electrons relative to the opposite side, giving the molecule a temporary dipole. This temporary dipole in one molecule causes a nearby molecule to adopt a temporarily opposite dipole, with the result that a tiny attraction is induced between the two **(Figure 2.7)**. Temporary molecular dipoles have only a fleeting existence and are constantly changing, but their cumulative effect is often strong enough to hold molecules close together so that a substance is a liquid or solid rather than a gas.

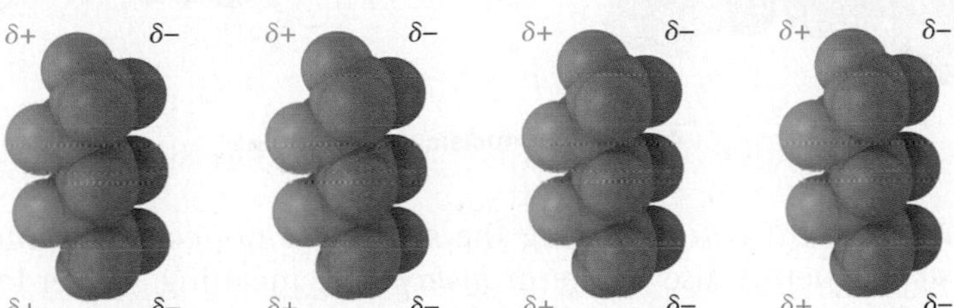

Figure 2.7 Attractive dispersion forces in nonpolar molecules are caused by temporary dipoles, as shown in these models of pentane, C_5H_{12}.

Perhaps the most important noncovalent interaction in biological molecules is the **hydrogen bond**, an attractive interaction between a hydrogen bonded to an electronegative O or N atom and an unshared electron pair on another O or N atom. In essence, a hydrogen bond is a very strong dipole–dipole interaction involving polarized O–H or N–H bonds. Electrostatic potential maps of water and ammonia clearly show the positively polarized hydrogens (blue) and the negatively polarized oxygens and nitrogens (red).

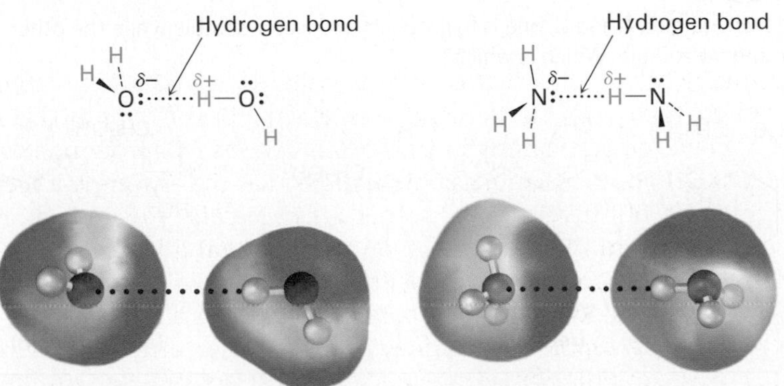

Hydrogen bonding has enormous consequences for living organisms. Hydrogen bonds cause water to be a liquid rather than a gas at ordinary temperatures, they hold enzymes in the shapes necessary for catalyzing biological reactions, and they cause strands of deoxyribonucleic acid (DNA) to pair up and coil into the double helix that stores genetic information.

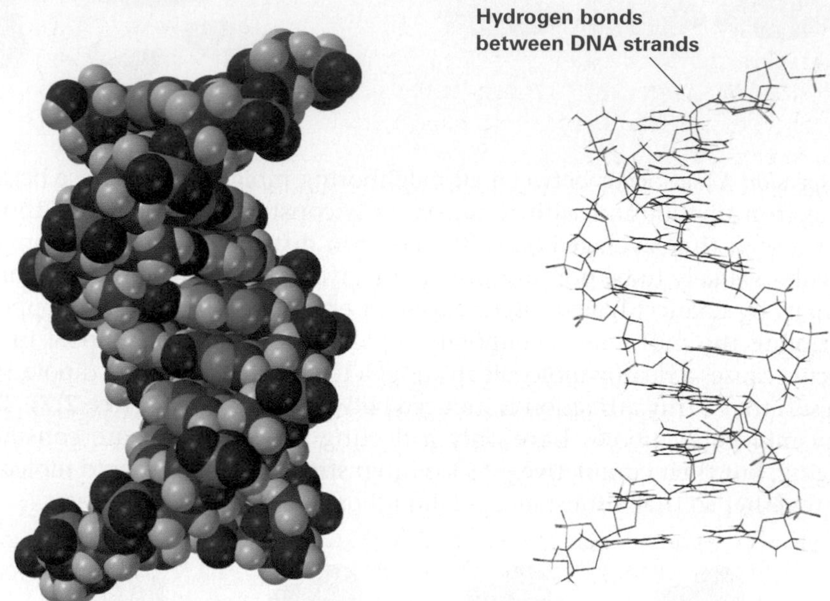

A deoxyribonucleic acid segment

One further point before leaving the subject of noncovalent interactions: biochemists frequently use the term *hydrophilic,* meaning "water-loving," to describe a substance that is strongly attracted to water and the term *hydrophobic,* meaning "water-fearing," to describe a substance that is not strongly attracted to water. Hydrophilic substances, such as table sugar, usually have a number of ionic charges or polar –OH groups in their structure so they can form hydrogen bonds, whereas hydrophobic substances, such as vegetable oil, do not have groups that form hydrogen bonds, so their attraction to water is limited to weak dispersion forces.

Problem 2.19

Of the two vitamins A and C, one is hydrophilic and water-soluble while the other is hydrophobic and fat-soluble. Which is which?

H_3C CH_3 CH_3 CH_3 CH_2OH CH_3

Vitamin A
(retinol)

CH_2OH H HO O O HO OH

Vitamin C
(ascorbic acid)

A DEEPER LOOK Alkaloids: From Cocaine to Dental Anesthetics

The coca bush *Erythroxylon coca*, native to upland rain forest areas of Colombia, Ecuador, Peru, Bolivia, and western Brazil, is the source of the alkaloid cocaine.

Just as ammonia (NH_3) is a weak base, there are a large number of nitrogen-containing organic compounds called *amines* that are also weak bases. In the early days of organic chemistry, basic amines derived from natural sources were known as vegetable alkali, but they are now called *alkaloids.* More than 20,000 alkaloids are known. Their study provided much of the impetus for the growth of organic chemistry in the nineteenth century and remains today an active and fascinating area of research.

Alkaloids vary widely in structure, from the simple to the enormously complex. The odor of rotting fish, for example, is caused largely by methylamine, CH_3NH_2, a simple relative of ammonia in which one of the NH_3 hydrogens has been replaced by an organic CH_3 group. In fact, the use of lemon juice to mask fish odors is simply an acid–base reaction of the citric acid in lemons with methylamine base in the fish.

Many alkaloids have pronounced biological properties, and approximately 50% of the pharmaceutical agents used today are derived from naturally occurring amines. As just three examples, morphine, an analgesic agent, is obtained from the opium poppy *Papaver somniferum*. Ephedrine, a bronchodilator, decongestant, and appetite suppressant, is obtained from the Chinese plant *Ephedra sinica.* Cocaine, both an anesthetic and a stimulant, is obtained from the coca bush *Erythroxylon coca,* endemic to the upland rain forest areas of central South America. (And yes, there really was a small amount of cocaine in the original Coca-Cola recipe, although it was removed in 1906.)

Morphine **Ephedrine** **Cocaine**

Cocaine itself is no longer used as a medicine because it is too addictive, but its anesthetic properties provoked a search for related but nonaddictive compounds. This search ultimately resulted in the synthesis of the "caine" anesthetics that are commonly used today in dental and surgical anesthesia. Procaine, the first such compound, was synthesized in 1898 and marketed under the name Novocain. It was rapidly adopted and remains in use today as a topical anesthetic. Other related compounds with different activity profiles followed: Lidocaine, marketed as Xylocaine, was introduced in 1943, and mepivacaine (Carbocaine) in the early 1960s. More recently, bupivacaine (Marcaine) and prilocaine (Citanest) have gained popularity. Both are quick-acting, but the effects of

(continued)

(continued)

bupivacaine last for 3 to 6 hours while those of prilocaine fade after 45 minutes. Note some structural similarity of all the caines to cocaine itself.

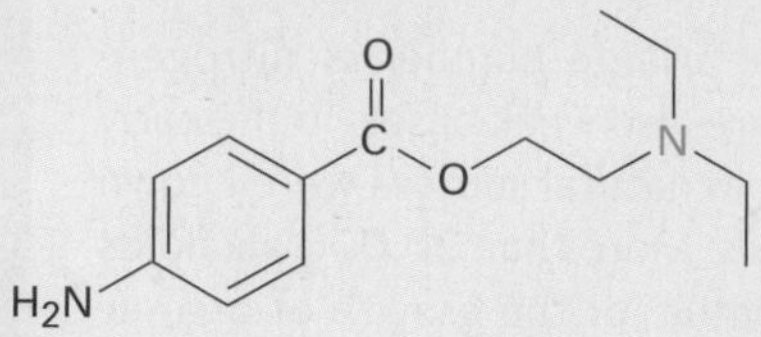

Procaine (Novocain)

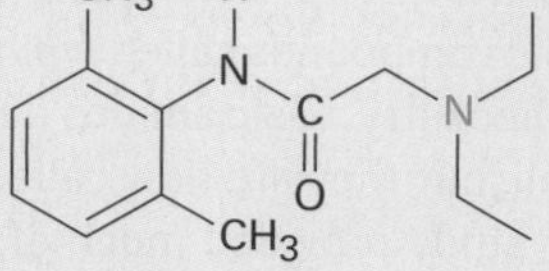

Lidocaine (Xylocaine)

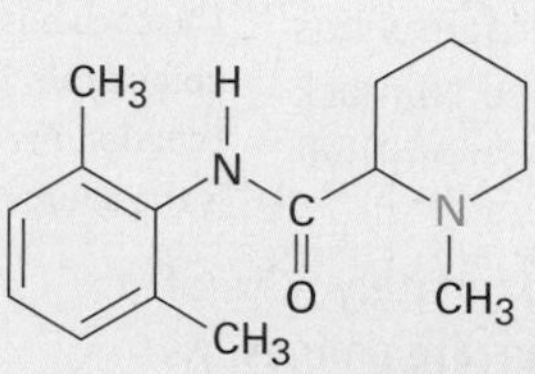

Mepivacaine (Carbocaine)

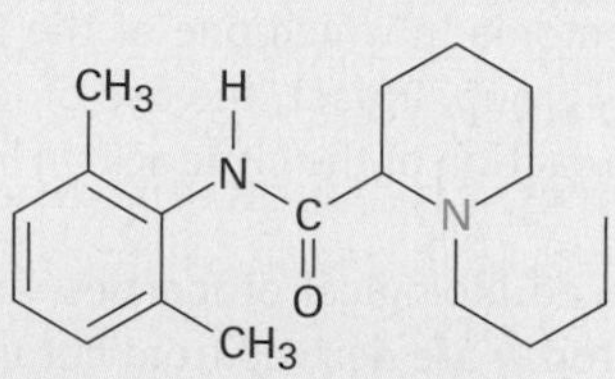

Bupivacaine (Marcaine)

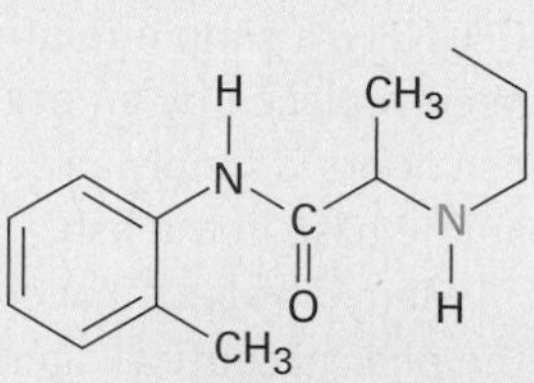

Prilocaine (Citanest)

A recent report from the U.S. National Academy of Sciences estimates than less than 1% of all living species have been characterized. Thus, alkaloid chemistry remains today an active area of research, and innumerable substances with potentially useful properties remain to be discovered. Undoubtedly even the caine anesthetics will become obsolete at some point, perhaps supplanted by newly discovered alkaloids.

Key words

acidity constant (K_a), 50
Brønsted–Lowry acid, 49
Brønsted–Lowry base, 49
conjugate acid, 49
conjugate base, 49
dipole moment (μ), 37
electronegativity (EN), 35
formal charge, 41
hydrogen bond, 61
inductive effect, 36
Lewis acid, 56
Lewis base, 56
noncovalent interaction, 60
pK_a, 50
polar covalent bond, 34
resonance form, 43
resonance hybrid, 43

Summary

Understanding both organic and biological chemistry means knowing not just what happens but also why and how it happens at the molecular level. In this chapter, we've reviewed some of the ways that chemists describe and account for chemical reactivity, thereby providing a foundation for understanding the specific reactions that will be discussed in subsequent chapters.

Organic molecules often have **polar covalent bonds** as a result of unsymmetrical electron sharing caused by differences in the **electronegativity** of atoms. A carbon–oxygen bond is polar, for example, because oxygen attracts the shared electrons more strongly than carbon does. Carbon–hydrogen bonds are relatively nonpolar. Many molecules as a whole are also polar owing to the presence of individual polar bonds and electron lone pairs. The polarity of a molecule is measured by its **dipole moment**, μ.

Plus (+) and minus (−) signs are often used to indicate the presence of **formal charges** on atoms in molecules. Assigning formal charges to specific atoms is a bookkeeping technique that makes it possible to keep track of the valence electrons around an atom and offers some clues about chemical reactivity.

Some substances, such as acetate ion and benzene, can't be represented by a single line-bond structure and must be considered as a **resonance hybrid** of two or more structures, neither of which is correct by itself. The only difference between two **resonance forms** is in the location of their π and nonbonding electrons. The nuclei remain in the same places in both structures, and the hybridization of the atoms remains the same.

Acidity and basicity are closely related to the ideas of polarity and electronegativity. A **Brønsted–Lowry acid** is a compound that can donate a proton (hydrogen ion, H^+), and a **Brønsted–Lowry base** is a compound that can accept a proton. The strength of a Brønsted–Lowry acid or base is expressed by its **acidity constant, K_a,** or by the negative logarithm of the acidity constant, **pK_a**. The larger the pK_a, the weaker the acid. More useful is the Lewis definition of acids and bases. A **Lewis acid** is a compound that has a low-energy empty orbital that can accept an electron pair; Mg^{2+}, BF_3, $AlCl_3$, and H^+ are examples. A **Lewis base** is a compound that can donate an unshared electron pair; NH_3 and H_2O are examples. Most organic molecules that contain oxygen and nitrogen can act as Lewis bases toward sufficiently strong acids.

A variety of **noncovalent interactions** have a significant effect on the properties of large biomolecules. **Hydrogen bonding**—the attractive interaction between a positively polarized hydrogen atom bonded to an oxygen or nitrogen atom with an unshared electron pair on another O or N atom, is particularly important in giving proteins and nucleic acids their shapes.

Exercises

Visualizing Chemistry

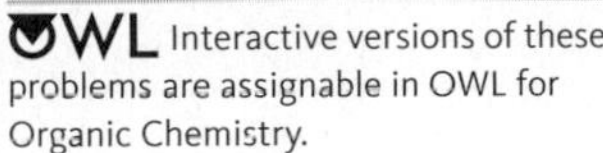

OWL Interactive versions of these problems are assignable in OWL for Organic Chemistry.

▲ denotes problems linked to the Key Ideas in this chapter.

(Problems 2.1–2.19 appear within the chapter.)

2.20 Fill in the multiple bonds in the following model of naphthalene, $C_{10}H_8$ (gray = C, ivory = H). How many resonance structures does naphthalene have? Draw them.

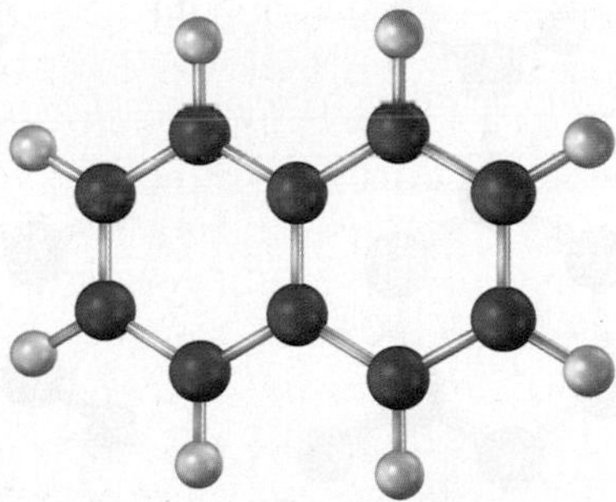

▲ Problems linked to Key Ideas in this chapter

2.21 The following model is a representation of ibuprofen, a common over-the-counter pain reliever. Indicate the positions of the multiple bonds, and draw a skeletal structure (gray = C, red = O, ivory = H).

2.22 *cis*-1,2-Dichloroethylene and *trans*-dichloroethylene are *isomers,* compounds with the same formula but different chemical structures. Look at the following electrostatic potential maps, and tell whether either compound has a dipole moment.

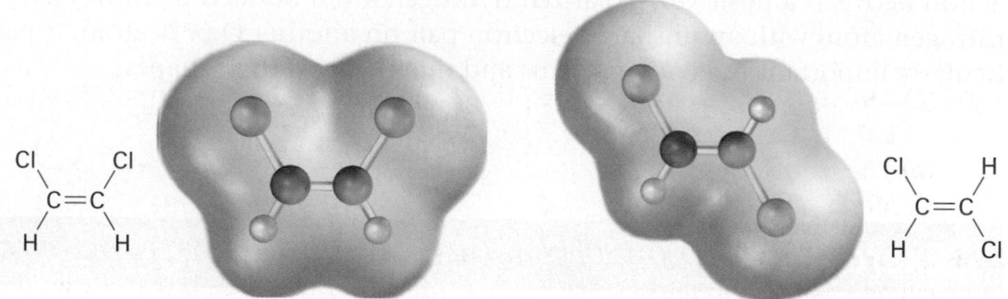

***cis*-1,2-Dichloroethylene** ***trans*-1,2-Dichloroethylene**

2.23 The following molecular models are representations of **(a)** adenine and **(b)** cytosine, constituents of DNA (deoxyribonucleic acid). Indicate the positions of multiple bonds and lone pairs for both, and draw skeletal structures (gray = C, red = O, blue = N, ivory = H).

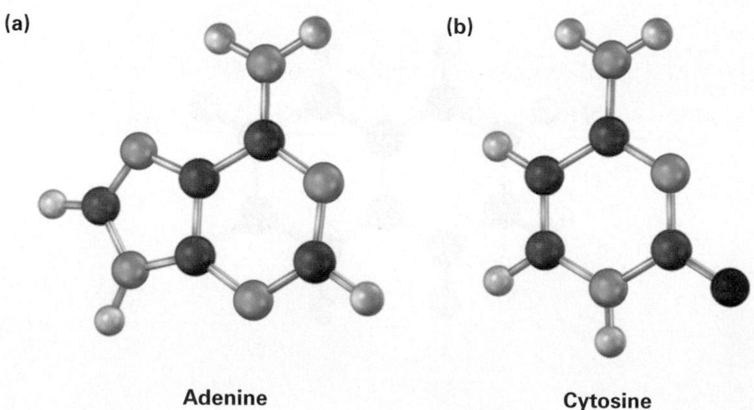

Adenine **Cytosine**

▲ Problems linked to Key Ideas in this chapter

Additional Problems

Electronegativity and Dipole Moments

2.24 Identify the most electronegative element in each of the following molecules:
(a) CH_2FCl (b) $FCH_2CH_2CH_2Br$
(c) $HOCH_2CH_2NH_2$ (d) CH_3OCH_2Li

2.25 Use the electronegativity table given in Figure 2.2 on page 35 to predict which bond in each of the following pairs is more polar, and indicate the direction of bond polarity for each compound.
(a) H_3C–Cl or Cl–Cl (b) H_3C–H or H–Cl
(c) HO–CH_3 or $(CH_3)_3Si$–CH_3 (d) H_3C–Li or Li–OH

2.26 Which of the following molecules has a dipole moment? Indicate the expected direction of each.

(a) 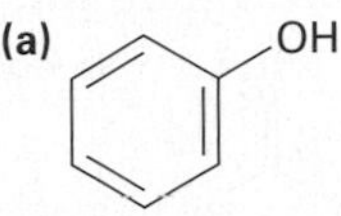

(b) 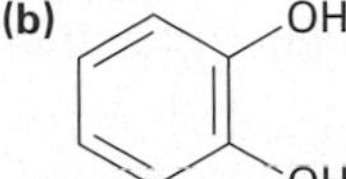

(c) 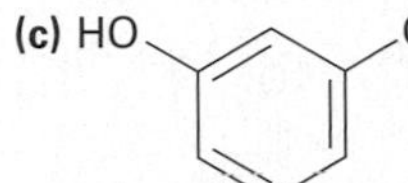

(d) 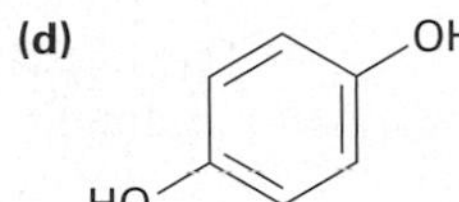

2.27 (a) The H–Cl bond length is 136 pm. What would the dipole moment of HCl be if the molecule were 100% ionic, H^+ Cl^-?
(b) The actual dipole moment of HCl is 1.08 D. What is the percent ionic character of the H–Cl bond?

2.28 Phosgene, $Cl_2C{=}O$, has a smaller dipole moment than formaldehyde, $H_2C{=}O$, even though it contains electronegative chlorine atoms in place of hydrogen. Explain.

2.29 Fluoromethane (CH_3F, $\mu = 1.81$ D) has a smaller dipole moment than chloromethane (CH_3Cl, $\mu = 1.87$ D) even though fluorine is more electronegative than chlorine. Explain.

2.30 Methanethiol, CH_3SH, has a substantial dipole moment ($\mu = 1.52$) even though carbon and sulfur have identical electronegativities. Explain.

Formal Charges

2.31 Calculate the formal charges on the atoms shown in red.

(a) $(CH_3)_2\ddot{O}BF_3$ (b) $H_2\ddot{C}{-}N{\equiv}N{:}$ (c) $H_2C{=}N{=}\ddot{N}{:}$

(d) $:\ddot{O}{=}\ddot{O}{-}\ddot{O}:$ (e) 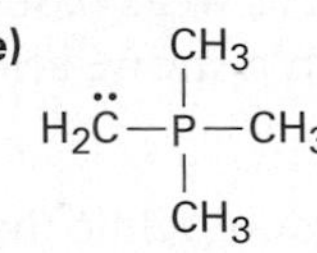

(f) 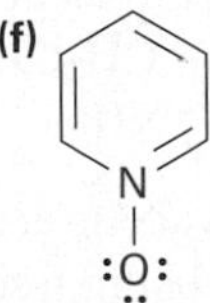

2.32 Assign formal charges to the atoms in each of the following molecules:

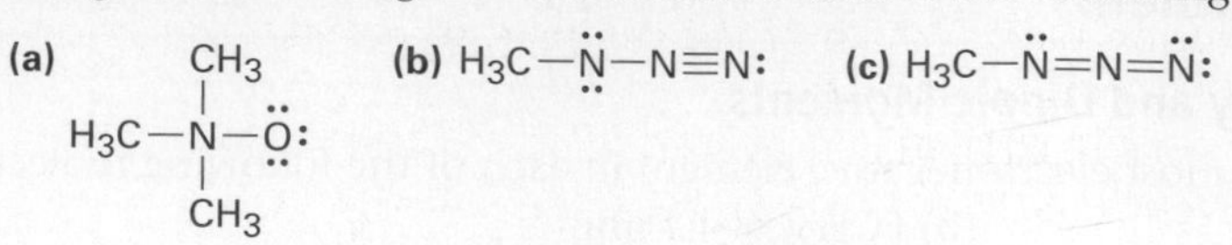

(a) $H_3C-N(CH_3)_2-\ddot{O}:$ (b) $H_3C-\ddot{N}-N\equiv N:$ (c) $H_3C-\ddot{N}=N=\ddot{N}:$

Resonance

2.33 Which of the following pairs of structures represent resonance forms?

(a) and (b) and (c) and (d) and

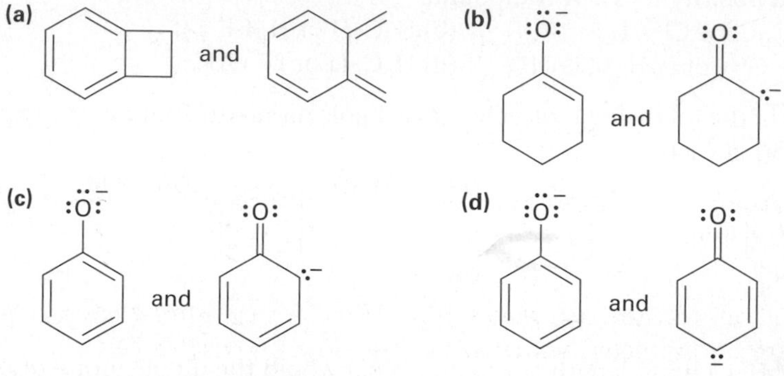

2.34 ▲ Draw as many resonance structures as you can for the following species:

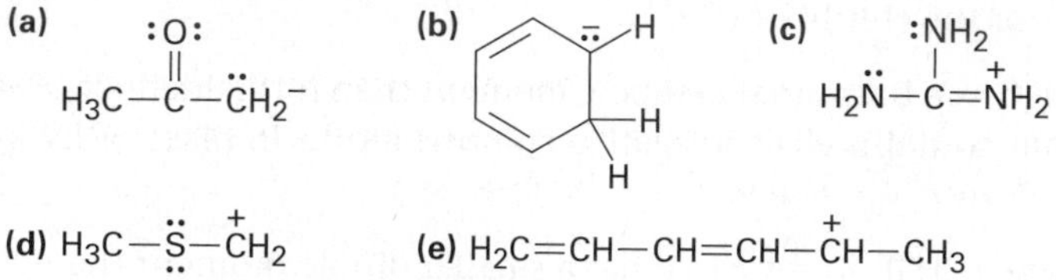

(a) $H_3C-C(=O)-\ddot{C}H_2^-$ (b) (c) $H_2\ddot{N}-C(\ddot{N}H_2)=\overset{+}{N}H_2$

(d) $H_3C-\ddot{S}-\overset{+}{C}H_2$ (e) $H_2C=CH-CH=CH-\overset{+}{C}H-CH_3$

2.35 1,3-Cyclobutadiene is a rectangular molecule with two shorter double bonds and two longer single bonds. Why do the following structures *not* represent resonance forms?

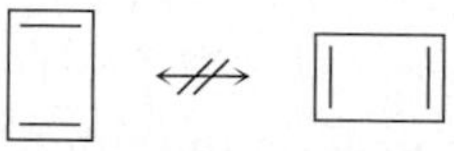

Acids and Bases

2.36 Alcohols can act either as weak acids or as weak bases, just as water can. Show the reaction of methanol, CH_3OH, with a strong acid such as HCl and with a strong base such as Na^+ $^-NH_2$.

2.37 ▲ The O–H hydrogen in acetic acid is more acidic than any of the C]H hydrogens. Explain this result using resonance structures.

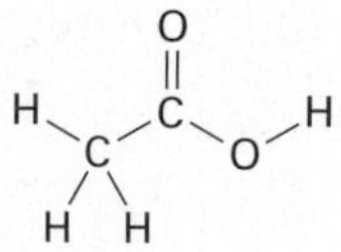

Acetic acid

▲ Problems linked to Key Ideas in this chapter

2.38 Draw electron-dot structures for the following molecules, indicating any unshared electron pairs. Which of the compounds are likely to act as Lewis acids and which as Lewis bases?
(a) $AlBr_3$ (b) $CH_3CH_2NH_2$ (c) BH_3
(d) HF (e) CH_3SCH_3 (f) $TiCl_4$

2.39 Write the products of the following acid–base reactions:
(a) $CH_3OH + H_2SO_4 \rightleftarrows$?
(b) $CH_3OH + NaNH_2 \rightleftarrows$?
(c) $CH_3NH_3^+ \; Cl^- + NaOH \rightleftarrows$?

2.40 Rank the following substances in order of increasing acidity:

$CH_3\overset{O}{\overset{\|}{C}}CH_3$	$CH_3\overset{O}{\overset{\|}{C}}CH_2\overset{O}{\overset{\|}{C}}CH_3$	C_6H_5–OH	$CH_3\overset{O}{\overset{\|}{C}}OH$
Acetone ($\mathbf{p}K_\mathbf{a}$ **= 19.3)**	**2,4-Pentanedione** ($\mathbf{p}K_\mathbf{a}$ **= 9)**	**Phenol** ($\mathbf{p}K_\mathbf{a}$ **= 9.9)**	**Acetic acid** ($\mathbf{p}K_\mathbf{a}$ **= 4.76)**

2.41 Which, if any, of the substances in Problem 2.40 is a strong enough acid to react almost completely with NaOH? (The pK_a of H_2O is 15.74.)

2.42 The ammonium ion (NH_4^+, pK_a = 9.25) has a lower pK_a than the methylammonium ion ($CH_3NH_3^+$, pK_a = 10.66). Which is the stronger base, ammonia (NH_3) or methylamine (CH_3NH_2)? Explain.

2.43 Is *tert*-butoxide anion a strong enough base to react significantly with water? In other words, can a solution of potassium *tert*-butoxide be prepared in water? The pK_a of *tert*-butyl alcohol is approximately 18.

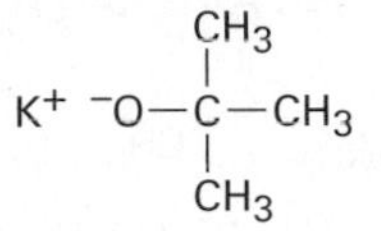

$K^+ \; ^-O-C(CH_3)_3$ **Potassium *tert*-butoxide**

2.44 Predict the structure of the product formed in the reaction of the organic base pyridine with the organic acid acetic acid, and use curved arrows to indicate the direction of electron flow.

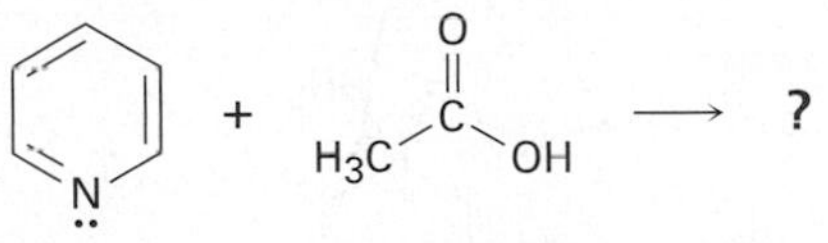

C_5H_5N: + $H_3C-\overset{O}{\overset{\|}{C}}-OH \longrightarrow$?

Pyridine **Acetic acid**

2.45 Calculate K_a values from the following pK_a's:
(a) Acetone, pK_a = 19.3 (b) Formic acid, pK_a = 3.75

2.46 Calculate pK_a values from the following K_a's:
(a) Nitromethane, $K_a = 5.0 \times 10^{-11}$ (b) Acrylic acid, $K_a = 5.6 \times 10^{-5}$

▲ Problems linked to Key Ideas in this chapter

2.47 What is the pH of a 0.050 M solution of formic acid, $pK_a = 3.75$?

2.48 Sodium bicarbonate, $NaHCO_3$, is the sodium salt of carbonic acid (H_2CO_3), $pK_a = 6.37$. Which of the substances shown in Problem 2.40 will react significantly with sodium bicarbonate?

General Problems

2.49 Maleic acid has a dipole moment, but the closely related fumaric acid, a substance involved in the citric acid cycle by which food molecules are metabolized, does not. Explain.

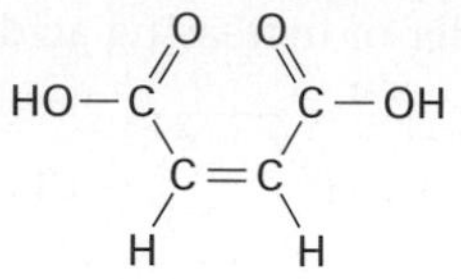

Maleic acid 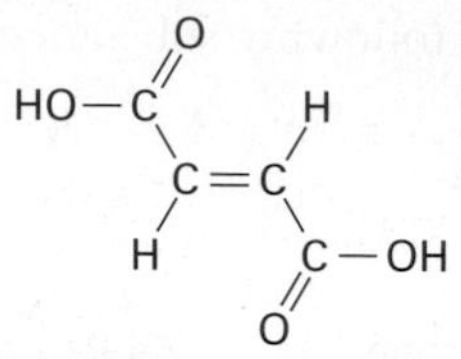 **Fumaric acid**

2.50 Assume that you have two unlabeled bottles, one of which contains phenol ($pK_a = 9.9$) and one of which contains acetic acid ($pK_a = 4.76$). In light of your answer to Problem 2.48, suggest a simple way to determine what is in each bottle.

2.51 Identify the acids and bases in the following reactions:

(a) $CH_3OH + H^+ \longrightarrow CH_3\overset{+}{O}H_2$

(b) $CH_3COCH_3 + TiCl_4 \longrightarrow$ $H_3C-\overset{+}{O}(\overset{-}{TiCl_4})=C-CH_3$

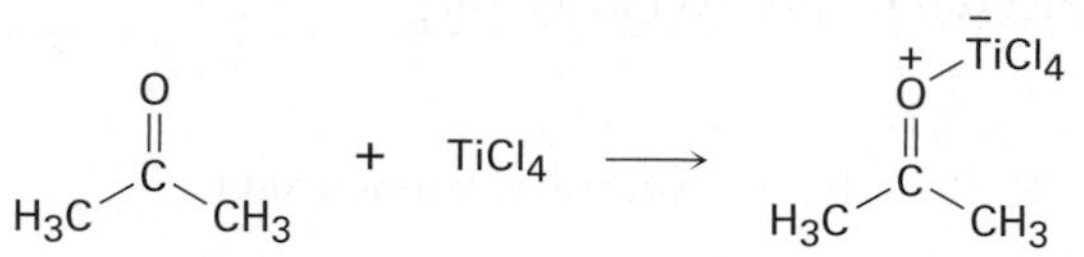

(c) cyclohexanone + NaH $\longrightarrow$ enolate carbanion Na^+ + H_2

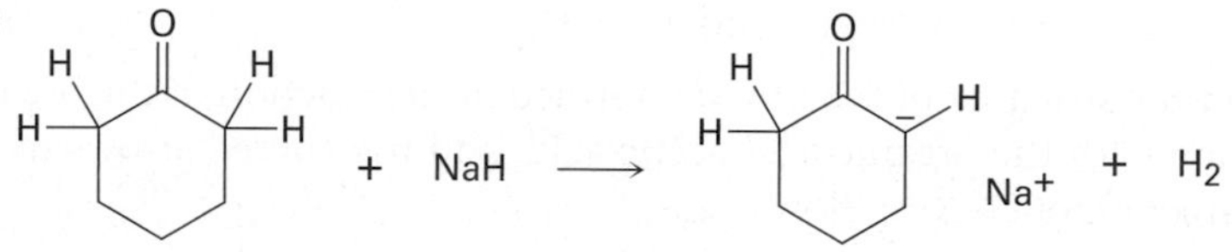

(d) morpholine + BH_3 $\longrightarrow$ morpholine–BH_3 adduct ($\overset{+}{N}$–$\overset{-}{B}H_3$)

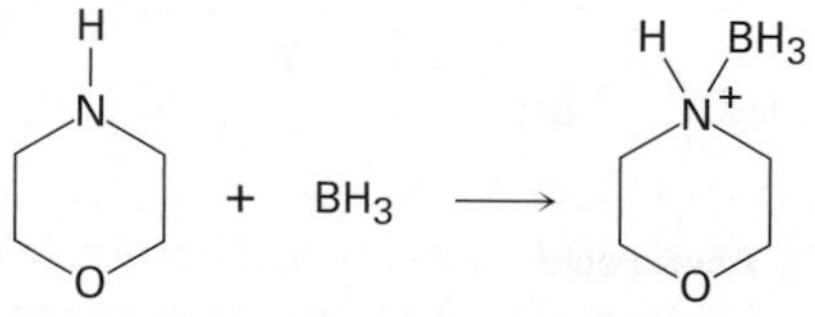

▲ Problems linked to Key Ideas in this chapter

2.52 ▲ Which of the following pairs represent resonance structures?

(a) $CH_3C{\equiv}\overset{+}{N}{-}\ddot{\underset{..}{O}}{:}^{-}$ and $CH_3\overset{+}{C}{=}\ddot{N}{-}\ddot{\underset{..}{O}}{:}^{-}$

(b) $CH_3C(=\!O){-}\ddot{\underset{..}{O}}{:}^{-}$ and $:\!\overset{-}{C}H_2C(=\!O){-}\ddot{\underset{..}{O}}{-}H$

(c) $C_6H_5C(=\!O)\overset{+}{N}H_3$ and $C_6H_5C(=\overset{+}{O}H)\ddot{N}H_2$

(d) $CH_2{=}\overset{+}{N}(O^{-})_2$ and $:\!\overset{-}{C}H_2{-}\overset{+}{N}(=\!O)(O^{-})$

2.53 ▲ Draw as many resonance structures as you can for the following species, adding appropriate formal charges to each:

(a) Nitromethane, $H_3C{-}\overset{+}{N}(=\!O)({-}\ddot{\underset{..}{O}}{:}^{-})$

(b) Ozone, $:\ddot{O}{=}\overset{+}{\ddot{O}}{-}\ddot{\underset{..}{O}}{:}^{-}$

(c) Diazomethane, $H_2C{=}\overset{+}{N}{=}\overset{-}{\ddot{N}}{:}$

2.54 Carbocations, which contain a trivalent, positively charged carbon atom, react with water to give alcohols:

$$H_3C{-}\overset{+}{C}H{-}CH_3 \xrightarrow{H_2O} H_3C{-}CH(OH){-}CH_3 + H^+$$

A carbocation → An alcohol

How can you account for the fact that the following carbocation gives a mixture of *two* alcohols on reaction with water?

$$H_3C{-}\overset{+}{C}H{-}CH{=}CH_2 \xrightarrow{H_2O} H_3C{-}CH(OH){-}CH{=}CH_2 + H_3C{-}CH{=}CH{-}CH_2OH$$

2.55 We'll see in the next chapter that organic molecules can be classified according to the *functional groups* they contain, where a functional group is a collection of atoms with a characteristic chemical reactivity. Use the electronegativity values given in Figure 2.2 on page 35 to predict the direction of polarization of the following functional groups.

(a) Ketone: $C{=}O$

(b) Alcohol: $C{-}OH$

(c) Amide: $C(=\!O){-}NH_2$

(d) Nitrile: $—C{\equiv}N$

▲ Problems linked to Key Ideas in this chapter

2.56 The *azide* functional group (Problem 2.55), such as occurs in azidobenzene, contains three adjacent nitrogen atoms. One resonance structures for azidobenzene is shown. Draw three additional resonance structures, and assign appropriate formal charges to the atoms in all four.

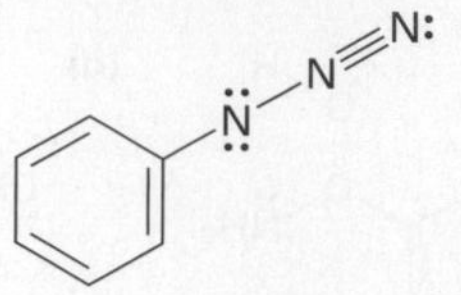

Azidobenzene

2.57 Phenol, C_6H_5OH, is a stronger acid than methanol, CH_3OH, even though both contain an O–H bond. Draw the structures of the anions resulting from loss of H^+ from phenol and methanol, and use resonance structures to explain the difference in acidity.

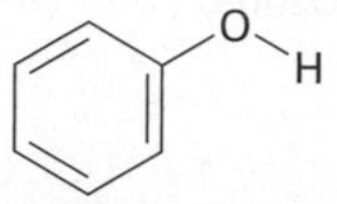

Phenol (pK_a = 9.89) 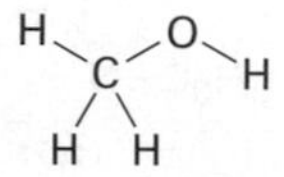 **Methanol (pK_a = 15.54)**

2.58 Thiamin diphosphate (TPP), a derivative of vitamin B_1 required for glucose metabolism, is a weak acid that can be deprotonated by base. Assign formal charges to the appropriate atoms in both TPP and its deprotonation product.

Thiamin diphosphate (TPP)

▲ Problems linked to Key Ideas in this chapter

3

The bristlecone pine is the oldest living organism on Earth. The waxy coating on its needles contains a mixture of organic compounds called alkanes, the subject of this chapter.

Organic Compounds: Alkanes and Their Stereochemistry

According to *Chemical Abstracts,* the publication that abstracts and indexes the chemical literature, there are more than 50 million known organic compounds. Each of these compounds has its own physical properties, such as melting point and boiling point, and each has its own chemical reactivity.

Chemists have learned through years of experience that organic compounds can be classified into families according to their structural features and that the members of a given family often have similar chemical behavior. Instead of 40 million compounds with random reactivity, there are a few dozen families of organic compounds whose chemistry is reasonably predictable. We'll study the chemistry of specific families throughout much of this book, beginning in this chapter with a look at the simplest family, the *alkanes*.

Why This Chapter? Alkanes are relatively unreactive and not often involved in chemical reactions, but they nevertheless provide a useful vehicle for introducing some important general ideas. In this chapter, we'll use alkanes to introduce the basic approach to naming organic compounds and to take an initial look at some of the three-dimensional aspects of molecules, a topic of particular importance in understanding biological organic chemistry.

3.1 Functional Groups

The structural features that make it possible to classify compounds into families are called *functional groups*. A **functional group** is a group of atoms within a molecule that has a characteristic chemical behavior. Chemically, a given functional group behaves in nearly the same way in every molecule it's a part of. For example, compare ethylene, a plant hormone that causes fruit to ripen, with menthene, a much more complicated molecule found in peppermint oil. Both substances contain a carbon–carbon double-bond functional group, and both therefore react with Br_2 in the same way to give a product in which a Br atom

OWL Sign in to OWL for Organic Chemistry at **www.cengage.com/owl** to view tutorials and simulations, develop problem-solving skills, and complete online homework assigned by your professor.

has added to each of the double-bond carbons **(Figure 3.1)**. This example is typical: *the chemistry of every organic molecule, regardless of size and complexity, is determined by the functional groups it contains.*

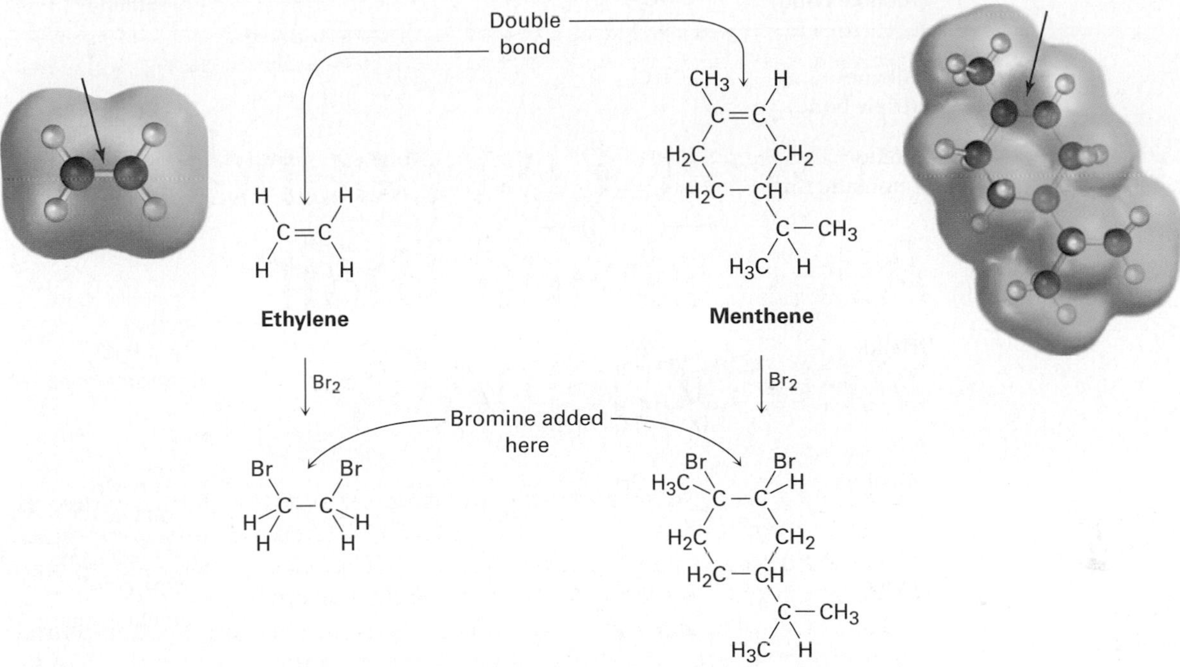

Figure 3.1 The reactions of ethylene and menthene with **bromine**. In both molecules, the carbon–carbon double-bond functional group has a similar polarity pattern, so both molecules react with Br_2 in the same way. The size and complexity of the molecules are not important.

Look at Table 3.1 on pages 76 and 77, which lists many of the common functional groups and gives simple examples of their occurrence. Some functional groups have only carbon–carbon double or triple bonds; others have halogen atoms; and still others contain oxygen, nitrogen, or sulfur. Much of the chemistry you'll be studying is the chemistry of these functional groups.

Functional Groups with Carbon–Carbon Multiple Bonds

Alkenes, alkynes, and arenes (aromatic compounds) all contain carbon–carbon multiple bonds. *Alkenes* have a double bond, *alkynes* have a triple bond, and *arenes* have alternating double and single bonds in a six-membered ring of carbon atoms. Because of their structural similarities, these compounds also have chemical similarities.

Table 3.1 Structures of Some Common Functional Groups

Name	Structure*	Name ending	Example
Alkene (double bond)	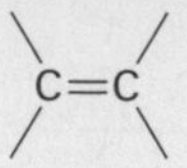 $C{=}C$	*-ene*	$H_2C{=}CH_2$ Ethene
Alkyne (triple bond)	$-C{\equiv}C-$	*-yne*	$HC{\equiv}CH$ Ethyne
Arene (aromatic ring)	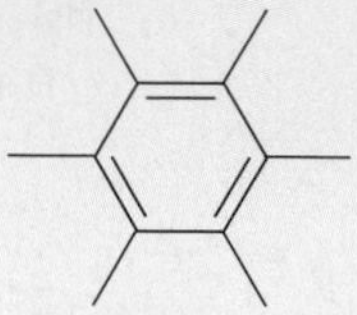	None	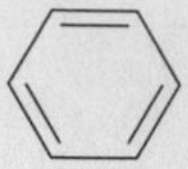 Benzene
Halide	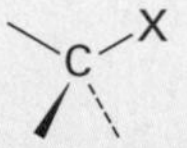 $C-X$ (X = F, Cl, Br, I)	None	CH_3Cl Chloromethane
Alcohol	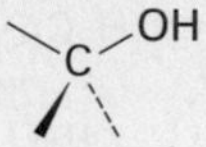 $C-OH$	*-ol*	CH_3OH Methanol
Ether	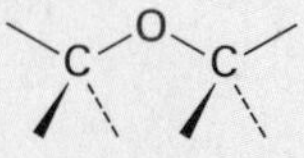 $C-O-C$	*ether*	CH_3OCH_3 Dimethyl ether
Monophosphate	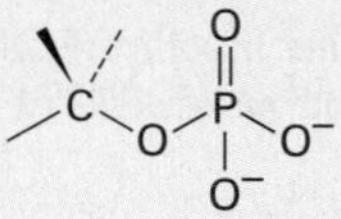 $C-O-P(=O)(O^-)-O^-$	*phosphate*	$CH_3OPO_3{}^{2-}$ Methyl phosphate
Diphosphate	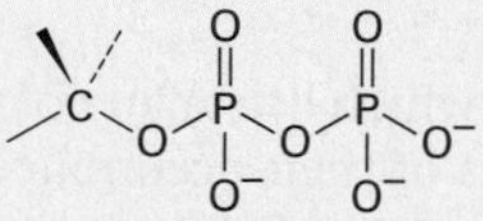 $C-O-P(=O)(O^-)-O-P(=O)(O^-)-O^-$	*diphosphate*	$CH_3OP_2O_6{}^{3-}$ Methyl diphosphate
Amine	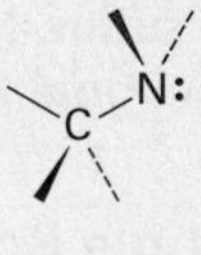 $C-N:$	*-amine*	CH_3NH_2 Methylamine
Imine (Schiff base)	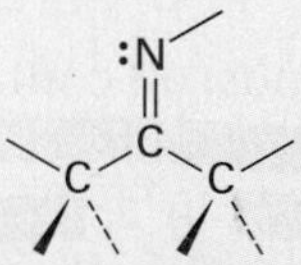 $:N{=}C(C)C$	None	$CH_3C(=NH)CH_3$ Acetone imine
Nitrile	$-C{\equiv}N$	*-nitrile*	$CH_3C{\equiv}N$ Ethanenitrile
Thiol	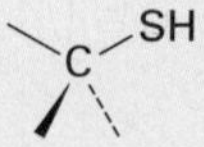 $C-SH$	*-thiol*	CH_3SH Methanethiol

*The bonds whose connections aren't specified are assumed to be attached to carbon or hydrogen atoms in the rest of the molecule.

Continued

Table 3.1 Structures of Some Common Functional Groups *(continued)*

Name	Structure*	Name ending	Example
Sulfide	C–S–C	*sulfide*	CH_3SCH_3 Dimethyl sulfide
Disulfide	C–S–S–C	*disulfide*	CH_3SSCH_3 Dimethyl disulfide
Sulfoxide	$C\overset{O^-}{\overset{\vert}{S^+}}C$	*sulfoxide*	$CH_3\overset{O^-}{\overset{\vert +}{S}}CH_3$ Dimethyl sulfoxide
Aldehyde	$\overset{O}{\overset{\Vert}{C}}H$	*-al*	$CH_3\overset{O}{\overset{\Vert}{C}}H$ Ethanal
Ketone	$C\overset{O}{\overset{\Vert}{C}}C$	*-one*	$CH_3\overset{O}{\overset{\Vert}{C}}CH_3$ Propanone
Carboxylic acid	$C\overset{O}{\overset{\Vert}{C}}OH$	-oic acid	$CH_3\overset{O}{\overset{\Vert}{C}}OH$ Ethanoic acid
Ester	$C\overset{O}{\overset{\Vert}{C}}OC$	*-oate*	$CH_3\overset{O}{\overset{\Vert}{C}}OCH_3$ Methyl ethanoate
Thioester	$C\overset{O}{\overset{\Vert}{C}}SC$	*-thioate*	$CH_3\overset{O}{\overset{\Vert}{C}}SCH_3$ Methyl ethanethioate
Amide	$C\overset{O}{\overset{\Vert}{C}}\ddot{N}$	*-amide*	$CH_3\overset{O}{\overset{\Vert}{C}}NH_2$ Ethanamide
Acid chloride	$C\overset{O}{\overset{\Vert}{C}}Cl$	*-oyl chloride*	$CH_3\overset{O}{\overset{\Vert}{C}}Cl$ Ethanoyl chloride
Carboxylic acid anhydride	$C\overset{O}{\overset{\Vert}{C}}O\overset{O}{\overset{\Vert}{C}}C$	*-oic anhydride*	$CH_3\overset{O}{\overset{\Vert}{C}}O\overset{O}{\overset{\Vert}{C}}CH_3$ Ethanoic anhydride

*The bonds whose connections aren't specified are assumed to be attached to carbon or hydrogen atoms in the rest of the molecule.

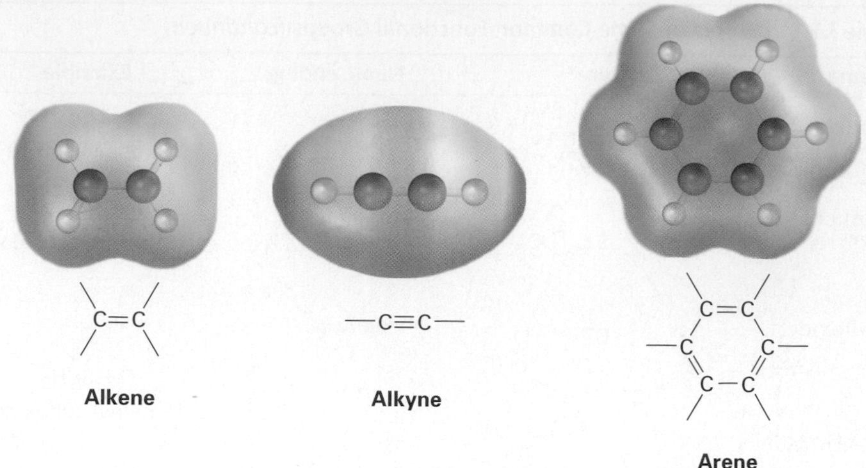

Functional Groups with Carbon Singly Bonded to an Electronegative Atom Alkyl halides (haloalkanes), alcohols, ethers, alkyl phosphates, amines, thiols, sulfides, and disulfides all have a carbon atom singly bonded to an electronegative atom—halogen, oxygen, nitrogen, or sulfur. Alkyl halides have a carbon atom bonded to halogen (−X), alcohols have a carbon atom bonded to the oxygen of a hydroxyl group (−OH), ethers have two carbon atoms bonded to the same oxygen, organophosphates have a carbon atom bonded to the oxygen of a phosphate group ($-OPO_3^{2-}$), amines have a carbon atom bonded to a nitrogen, thiols have a carbon atom bonded to the sulfur of an −SH group, sulfides have two carbon atoms bonded to the same sulfur, and disulfides have carbon atoms bonded to two sulfurs that are joined together. In all cases, the bonds are polar, with the carbon atom bearing a partial positive charge (δ+) and the electronegative atom bearing a partial negative charge (δ−).

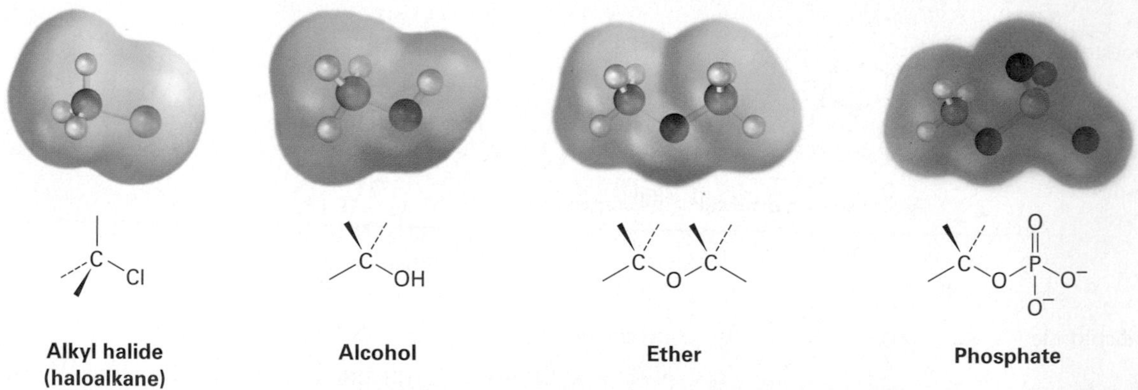

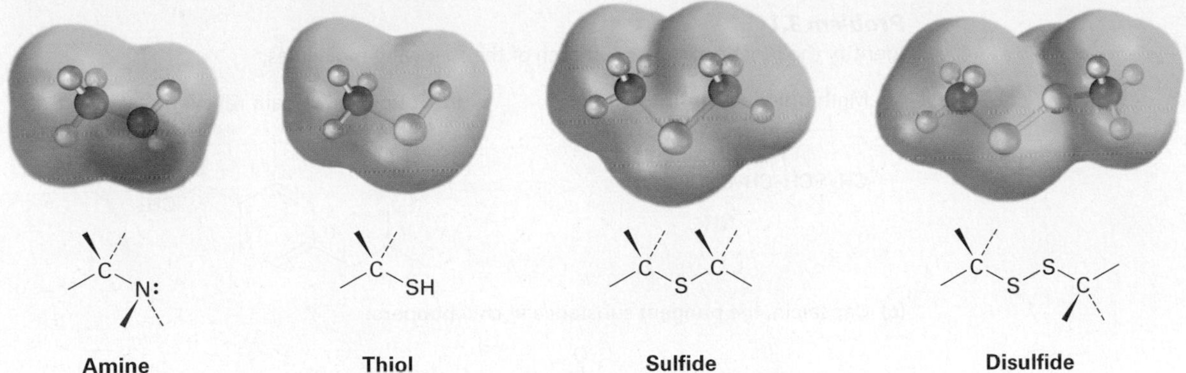

Amine Thiol Sulfide Disulfide

Functional Groups with a Carbon–Oxygen Double Bond (Carbonyl Groups) The *carbonyl group,* C=O (pronounced car-bo-**neel**) is common to many of the families listed in Table 3.1. Carbonyl groups are present in a large majority of organic compounds and in practically all biological molecules. These compounds behave similarly in many respects but differ depending on the identity of the atoms bonded to the carbonyl-group carbon. Aldehydes have at least one hydrogen bonded to the C=O, ketones have two carbons bonded to the C=O, carboxylic acids have an –OH group bonded to the C=O, esters have an ether-like oxygen bonded to the C=O, thioesters have a sulfide-like sulfur bonded to the C=O, amides have an amine-like nitrogen bonded to the C=O, acid chlorides have a chlorine bonded to the C=O, and so on. The carbonyl carbon atom bears a partial positive charge ($\delta+$), and the oxygen bears a partial negative charge ($\delta-$).

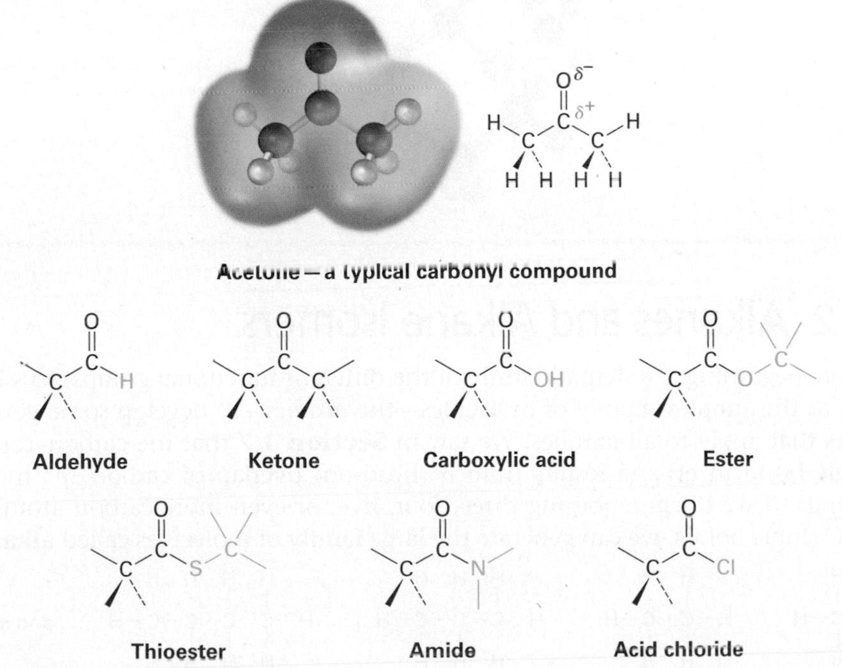

Acetone—a typical carbonyl compound

Problem 3.1

Identify the functional groups in each of the following molecules:

(a) Methionine, an amino acid:

$$CH_3SCH_2CH_2\underset{NH_2}{\underset{|}{C}}H\overset{O}{\overset{\|}{C}}OH$$

(b) Ibuprofen, a pain reliever:

CO_2H, CH_3

(c) Capsaicin, the pungent substance in chili peppers:

H_3C, O, HO, N, H, O, CH_3, CH_3

Problem 3.2

Propose structures for simple molecules that contain the following functional groups:

(a) Alcohol **(b)** Aromatic ring **(c)** Carboxylic acid
(d) Amine **(e)** Both ketone and amine **(f)** Two double bonds

Problem 3.3

Identify the functional groups in the following model of arecoline, a veterinary drug used to control worms in animals. Convert the drawing into a line-bond structure and a molecular formula (red = O, blue = N).

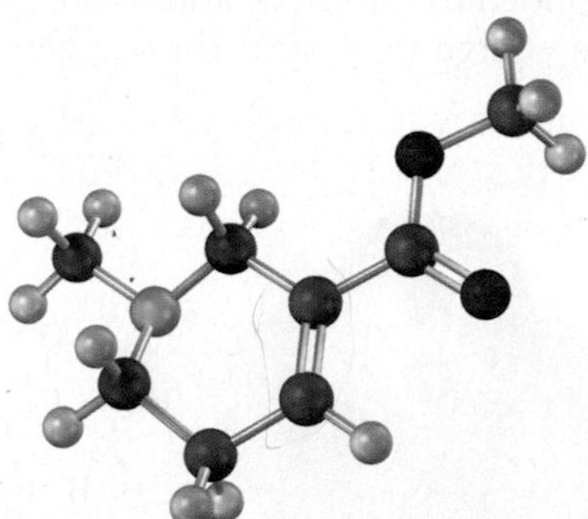

3.2 Alkanes and Alkane Isomers

Before beginning a systematic study of the different functional groups, let's look first at the simplest family of molecules—the *alkanes*—to develop some general ideas that apply to all families. We saw in **Section 1.7** that the carbon–carbon single bond in ethane results from σ (head-on) overlap of carbon sp^3 hybrid orbitals. If we imagine joining three, four, five, or even more carbon atoms by C–C single bonds, we can generate the large family of molecules called **alkanes**.

$$\underset{\textbf{Methane}}{CH_4} \qquad \underset{\textbf{Ethane}}{H_3C\text{–}CH_3} \qquad \underset{\textbf{Propane}}{H_3C\text{–}CH_2\text{–}CH_3} \qquad \underset{\textbf{Butane}}{H_3C\text{–}CH_2\text{–}CH_2\text{–}CH_3} \quad \ldots \text{and so on}$$

Alkanes are often described as *saturated hydrocarbons:* **hydrocarbons** because they contain only carbon and hydrogen; **saturated** because they have only C–C and C–H single bonds and thus contain the maximum possible number of hydrogens per carbon. They have the general formula C_nH_{2n+2}, where n is an integer. Alkanes are also occasionally called **aliphatic** compounds, a name derived from the Greek *aleiphas,* meaning "fat." We'll see in **Section 27.1** that many animal fats contain long carbon chains similar to alkanes.

$$
\begin{array}{l}
CH_2O\overset{O}{\overset{\|}{C}}CH_2CH_2CH_2CH_2CH_2CH_2CH_2CH_2CH_2CH_2CH_2CH_2CH_2CH_2CH_2CH_2CH_3 \\
|\\
CHO\overset{O}{\overset{\|}{C}}CH_2CH_2CH_2CH_2CH_2CH_2CH_2CH_2CH_2CH_2CH_2CH_2CH_2CH_2CH_2CH_2CH_3 \\
|\\
CH_2O\overset{O}{\overset{\|}{C}}CH_2CH_2CH_2CH_2CH_2CH_2CH_2CH_2CH_2CH_2CH_2CH_2CH_2CH_2CH_2CH_2CH_3
\end{array}
$$

A typical animal fat

Think about the ways that carbon and hydrogen might combine to make alkanes. With one carbon and four hydrogens, only one structure is possible: methane, CH_4. Similarly, there is only one combination of two carbons with six hydrogens (ethane, CH_3CH_3) and only one combination of three carbons with eight hydrogens (propane, $CH_3CH_2CH_3$). When larger numbers of carbons and hydrogens combine, however, more than one structure is possible. For example, there are *two* substances with the formula C_4H_{10}: the four carbons can all be in a row (butane), or they can branch (isobutane). Similarly, there are three C_5H_{12} molecules, and so on for larger alkanes.

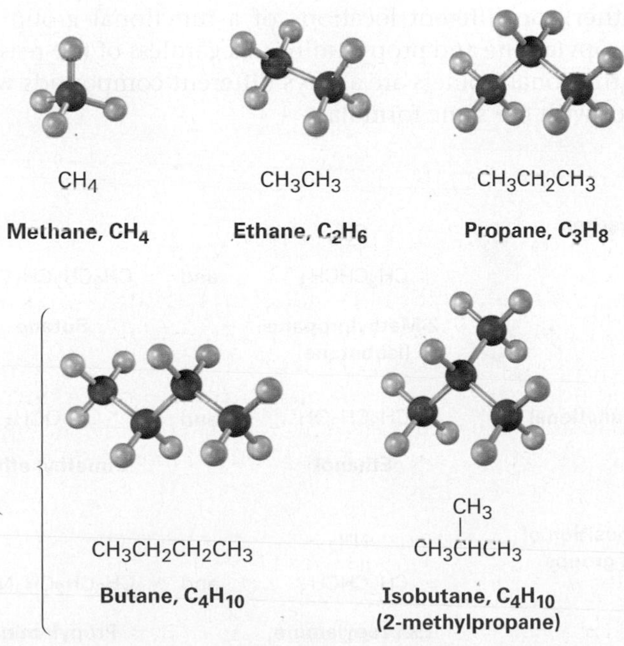

Methane, CH_4 **Ethane, C_2H_6** **Propane, C_3H_8**

Butane, C_4H_{10} **Isobutane, C_4H_{10} (2-methylpropane)**

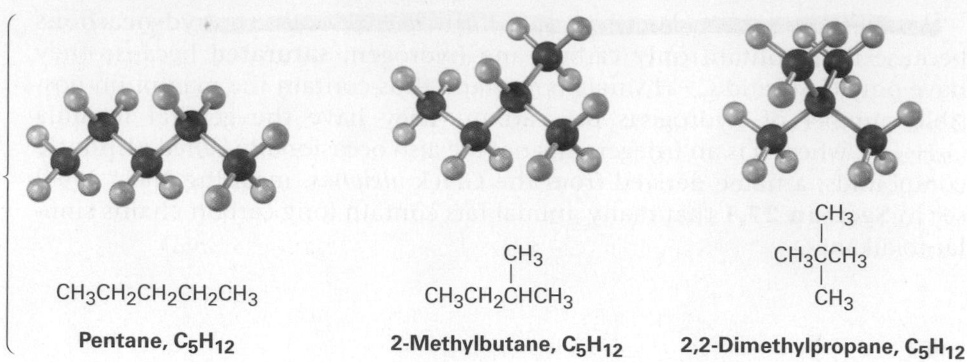

Pentane, C_5H_{12} **2-Methylbutane, C_5H_{12}** **2,2-Dimethylpropane, C_5H_{12}**

Compounds like butane and pentane, whose carbons are all connected in a row, are called **straight-chain alkanes**, or *normal alkanes*. Compounds like 2-methylpropane (isobutane), 2-methylbutane, and 2,2-dimethylpropane, whose carbon chains branch, are called **branched-chain alkanes**.

Compounds like the two C_4H_{10} molecules and the three C_5H_{12} molecules, which have the same formula but different structures, are called *isomers,* from the Greek *isos* + *meros,* meaning "made of the same parts." **Isomers** are compounds that have the same numbers and kinds of atoms but differ in the way the atoms are arranged. Compounds like butane and isobutane, whose atoms are connected differently, are called **constitutional isomers**. We'll see shortly that other kinds of isomers are also possible, even among compounds whose atoms are connected in the same order. As Table 3.2 shows, the number of possible alkane isomers increases dramatically as the number of carbon atoms increases.

Table 3.2 Number of Alkane Isomers

Formula	Number of isomers
C_6H_{14}	5
C_7H_{16}	9
C_8H_{18}	18
C_9H_{20}	35
$C_{10}H_{22}$	75
$C_{15}H_{32}$	4,347
$C_{20}H_{42}$	366,319
$C_{30}H_{62}$	4,111,846,763

Constitutional isomerism is not limited to alkanes—it occurs widely throughout organic chemistry. Constitutional isomers may have different carbon skeletons (as in isobutane and butane), different functional groups (as in ethanol and dimethyl ether), or different locations of a functional group along the chain (as in isopropylamine and propylamine). Regardless of the reason for the isomerism, constitutional isomers are always different compounds with different properties but with the same formula.

Different carbon skeletons **C_4H_{10}**	$CH_3CH(CH_3)CH_3$ **2-Methylpropane (isobutane)**	and	$CH_3CH_2CH_2CH_3$ **Butane**
Different functional groups **C_2H_6O**	CH_3CH_2OH **Ethanol**	and	CH_3OCH_3 **Dimethyl ether**
Different position of functional groups **C_3H_9N**	$CH_3CH(NH_2)CH_3$ **Isopropylamine**	and	$CH_3CH_2CH_2NH_2$ **Propylamine**

A given alkane can be drawn in many ways. For example, the straight-chain, four-carbon alkane called butane can be represented by any of the structures shown in **Figure 3.2**. These structures don't imply any particular three-dimensional geometry for butane; they indicate only the connections among atoms. In practice, as noted in **Section 1.12**, chemists rarely draw all the bonds in a molecule and usually refer to butane by the condensed structure, $CH_3CH_2CH_2CH_3$ or $CH_3(CH_2)_2CH_3$. Still more simply, butane can be represented as n-C_4H_{10}, where n denotes *normal* (straight-chain) butane.

$CH_3—CH_2—CH_2—CH_3$ $CH_3CH_2CH_2CH_3$ $CH_3(CH_2)_2CH_3$

Figure 3.2 Some representations of butane, C_4H_{10}. The molecule is the same regardless of how it's drawn. These structures imply only that butane has a continuous chain of four carbon atoms; they do not imply any specific geometry.

Straight-chain alkanes are named according to the number of carbon atoms they contain, as shown in Table 3.3. With the exception of the first four compounds—methane, ethane, propane, and butane—whose names have historical roots, the alkanes are named based on Greek numbers. The suffix *-ane* is added to the end of each name to indicate that the molecule identified is an alkane. Thus, pent*ane* is the five-carbon alkane, hex*ane* is the six-carbon alkane, and so on. We'll soon see that these alkane names form the basis for naming all other organic compounds, so at least the first ten should be memorized.

Table 3.3 Names of Straight-Chain Alkanes

Number of carbons (n)	Name	Formula (C_nH_{2n+2})	Number of carbons (n)	Name	Formula (C_nH_{2n+2})
1	Methane	CH_4	9	Nonane	C_9H_{20}
2	Ethane	C_2H_6	10	Decane	$C_{10}H_{22}$
3	Propane	C_3H_8	11	Undecane	$C_{11}H_{24}$
4	Butane	C_4H_{10}	12	Dodecane	$C_{12}H_{26}$
5	Pentane	C_5H_{12}	13	Tridecane	$C_{13}H_{28}$
6	Hexane	C_6H_{14}	20	Icosane	$C_{20}H_{42}$
7	Heptane	C_7H_{16}	30	Triacontane	$C_{30}H_{62}$
8	Octane	C_8H_{18}			

Drawing the Structures of Isomers

Worked Example 3.1

Propose structures for two isomers with the formula C_2H_7N.

Strategy

We know that carbon forms four bonds, nitrogen forms three, and hydrogen forms one. Write down the carbon atoms first, and then use a combination of trial and error plus intuition to put the pieces together.

Solution

There are two isomeric structures. One has the connection C–C–N, and the other has the connection C–N–C.

These pieces . . . 2 –C– 1 –N– 7 H–

give . . .

these structures. $CH_3CH_2NH_2$ (H–C–C–N–H with H's) and CH_3NHCH_3 (H–C–N–C–H with H's)

Problem 3.4

Draw structures of the five isomers of C_6H_{14}.

Problem 3.5

Propose structures that meet the following descriptions:

(a) Two isomeric esters with the formula $C_5H_{10}O_2$

(b) Two isomeric nitriles with the formula C_4H_7N

(c) Two isomeric disulfides with the formula $C_4H_{10}S_2$

Problem 3.6

How many isomers are there with the following descriptions?

(a) Alcohols with the formula C_3H_8O

(b) Bromoalkanes with the formula C_4H_9Br

(c) Thioesters with the formula C_4H_8OS

3.3 Alkyl Groups

If you imagine removing a hydrogen atom from an alkane, the partial structure that remains is called an **alkyl group**. Alkyl groups are not stable compounds themselves, they are simply parts of larger compounds. Alkyl groups are named by replacing the *-ane* ending of the parent alkane with an *-yl* ending. For example, removal of a hydrogen from methane, CH_4, generates a *methyl* group, $-CH_3$, and removal of a hydrogen from ethane, CH_3CH_3, generates an *ethyl* group, $-CH_2CH_3$. Similarly, removal of a hydrogen atom from the end carbon of any straight-chain alkane gives the series of straight-chain alkyl groups shown in Table 3.4. Combining an alkyl group with any of the functional groups listed earlier makes it possible to generate and name many thousands of compounds. For example:

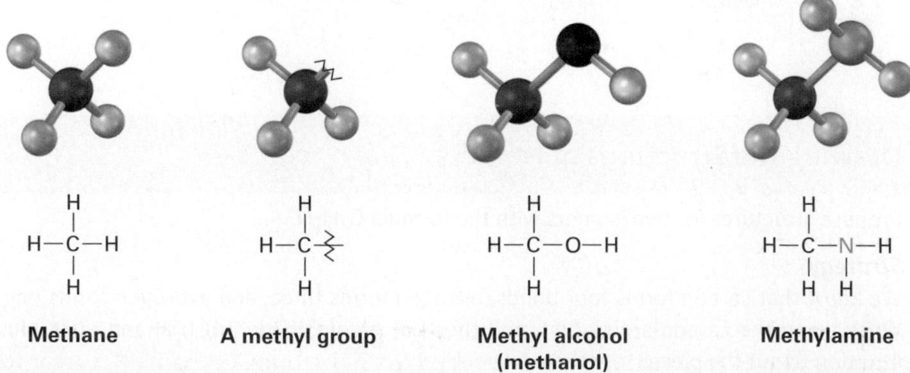

Methane **A methyl group** **Methyl alcohol (methanol)** **Methylamine**

Table 3.4 Some Straight-Chain Alkyl Groups

Alkane	Name	Alkyl group	Name (abbreviation)
CH_4	Methane	$-CH_3$	Methyl (Me)
CH_3CH_3	Ethane	$-CH_2CH_3$	Ethyl (Et)
$CH_3CH_2CH_3$	Propane	$-CH_2CH_2CH_3$	Propyl (Pr)
$CH_3CH_2CH_2CH_3$	Butane	$-CH_2CH_2CH_2CH_3$	Butyl (Bu)
$CH_3CH_2CH_2CH_2CH_3$	Pentane	$-CH_2CH_2CH_2CH_2CH_3$	Pentyl, or amyl

Just as straight-chain alkyl groups are generated by removing a hydrogen from an end carbon, branched alkyl groups are generated by removing a hydrogen atom from an internal carbon. Two 3-carbon alkyl groups and four 4-carbon alkyl groups are possible **(Figure 3.3)**.

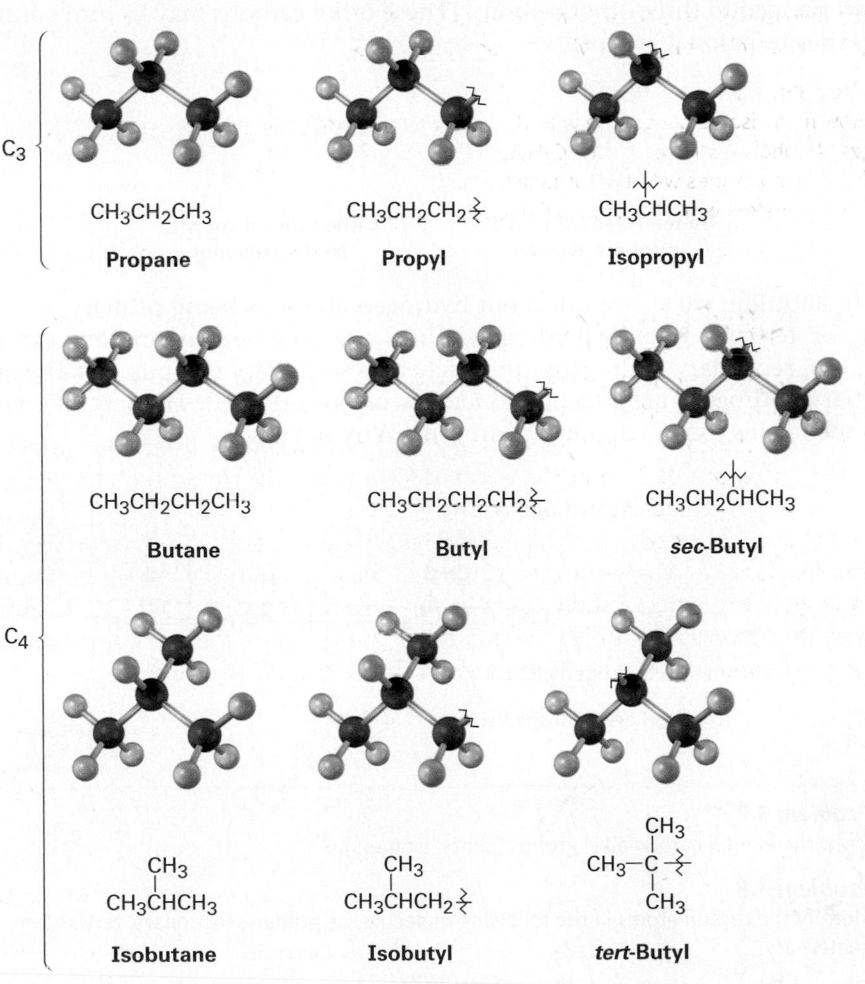

Figure 3.3 Alkyl groups generated from straight-chain alkanes.

One further comment about naming alkyl groups: the prefixes *sec-* (for secondary) and *tert-* (for tertiary) used for the C_4 alkyl groups in Figure 3.3 refer to *the number of other carbon atoms attached to the branching carbon atom.* There are four possibilities: primary (1°), secondary (2°), tertiary (3°), and quaternary (4°).

RCH_3

Primary **carbon (1°) is bonded to one other carbon.**

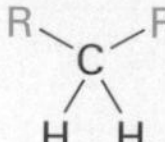

Secondary **carbon (2°) is bonded to two other carbons.**

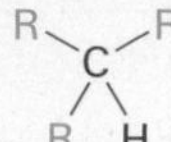

Tertiary **carbon (3°) is bonded to three other carbons.**

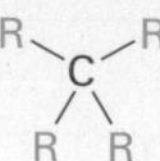

Quaternary **carbon (4°) is bonded to four other carbons.**

The symbol **R** is used here and throughout organic chemistry to represent a *generalized* organic group. The R group can be methyl, ethyl, propyl, or any of a multitude of others. You might think of **R** as representing the **R**est of the molecule, which isn't specified.

The terms *primary, secondary, tertiary,* and *quaternary* are routinely used in organic chemistry, and their meanings need to become second nature. For example, if we were to say, "Citric acid is a tertiary alcohol," we would mean that it has an alcohol functional group (–OH) bonded to a carbon atom that is itself bonded to three other carbons. (These other carbons may in turn connect to other functional groups.)

R_3C–OH (R–C(OH)(R)–R)

General class of tertiary alcohols, R_3COH

HO_2CCH_2–C(OH)(CO_2H)–CH_2CO_2H

Citric acid—a specific tertiary alcohol

In addition, we also speak about hydrogen atoms as being primary, secondary, or tertiary. Primary hydrogen atoms are attached to primary carbons (RCH_3), secondary hydrogens are attached to secondary carbons (R_2CH_2), and tertiary hydrogens are attached to tertiary carbons (R_3CH). There is, of course, no such thing as a quaternary hydrogen. (Why not?)

Primary hydrogens (CH_3)

$CH_3CH_2CH(CH_3)CH_3$ = [H–C–C–C(CH$_3$)–C–H, fully expanded]

Secondary hydrogens (CH_2)

A tertiary hydrogen (CH)

Problem 3.7

Draw the eight 5-carbon alkyl groups (pentyl isomers).

Problem 3.8

Identify the carbon atoms in the following molecules as primary, secondary, tertiary, or quaternary:

(a) $CH_3CH(CH_3)CH_2CH_2CH_3$

(b) $CH_3CH_2CH(CH(CH_3)CH_3)CH_2CH_3$

(c) $CH_3CH(CH_3)CH_2C(CH_3)_2CH_3$

Problem 3.9

Identify the hydrogen atoms on the compounds shown in Problem 3.8 as primary, secondary, or tertiary.

Problem 3.10

Draw structures of alkanes that meet the following descriptions:

(a) An alkane with two tertiary carbons
(b) An alkane that contains an isopropyl group
(c) An alkane that has one quaternary and one secondary carbon

3.4 Naming Alkanes

In earlier times, when relatively few pure organic chemicals were known, new compounds were named at the whim of their discoverer. Thus, urea (CH_4N_2O) is a crystalline substance isolated from urine; morphine ($C_{17}H_{19}NO_3$) is an analgesic (painkiller) named after Morpheus, the Greek god of dreams; and acetic acid, the primary organic constituent of vinegar, is named from the Latin word for vinegar, *acetum*.

As the science of organic chemistry slowly grew in the 19th century, so too did the number of known compounds and the need for a systematic method of naming them. The system of nomenclature we'll use in this book is that devised by the International Union of Pure and Applied Chemistry (IUPAC, usually spoken as **eye**-you-pac).

A chemical name typically has four parts in the IUPAC system of nomenclature: prefix, parent, locant, and suffix. The prefix identifies the various **substituent** groups in the molecule, the parent selects a main part of the molecule and tells how many carbon atoms are in that part, the locants give the positions of the functional groups and substituents, and the suffix identifies the primary functional group.

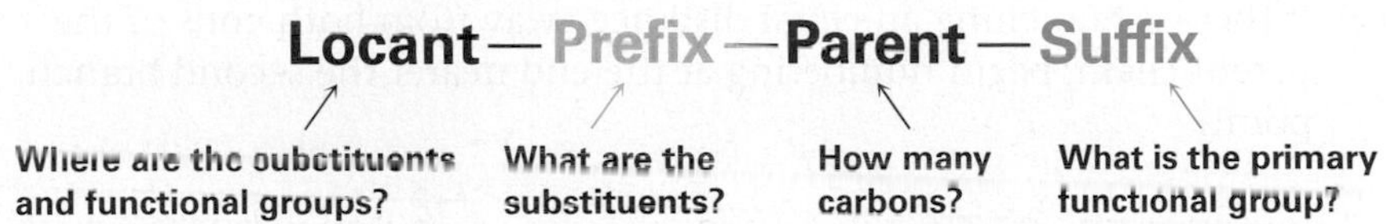

As we cover new functional groups in later chapters, the applicable IUPAC rules of nomenclature will be given. In addition, Appendix A at the back of this book gives an overall view of organic nomenclature and shows how compounds that contain more than one functional group are named. (If preferred, you can study that appendix now.) For the present, let's see how to name branched-chain alkanes and learn some general rules that are applicable to all compounds.

All but the most complex branched-chain alkanes can be named by following four steps. For a very few compounds, a fifth step is needed.

STEP 1

Find the parent hydrocarbon.

(a) Find the longest continuous chain of carbon atoms in the molecule, and use the name of that chain as the parent name. The longest chain may not always be apparent from the manner of writing; you may have to "turn corners."

```
          CH2CH3
          |
CH3CH2CH2CH—CH3        Named as a substituted hexane

   CH3
   |
   CH2
   |
CH3—CHCH—CH2CH3        Named as a substituted heptane
        |
        CH2CH2CH3
```

(b) If two different chains of equal length are present, choose the one with the larger number of branch points as the parent.

```
   CH3                               CH3
   |                                 |
CH3CHCHCH2CH2CH3                  CH3CH—CHCH2CH2CH3
     |                                  |
     CH2CH3                             CH2CH3
```

Named as a hexane with *two* substituents — *NOT* — as a hexane with *one* substituent

STEP 2

Number the atoms in the longest chain.

(a) Beginning at the end nearer the first branch point, number each carbon atom in the parent chain.

```
      2   1                                  6   7
      CH2CH3                                 CH2CH3
      |                                      |
CH3—CHCH—CH2CH3          NOT           CH3—CHCH—CH2CH3
    3  |4                                  5  |4
       CH2CH2CH3                              CH2CH2CH3
       5   6   7                              3   2   1
```

The first branch occurs at C3 in the proper system of numbering, not at C4.

(b) If there is branching an equal distance away from both ends of the parent chain, begin numbering at the end nearer the second branch point.

```
     8   9                                         2   1
     CH2CH3    CH3 CH2CH3                          CH2CH3    CH3 CH2CH3
     |         |   |                               |         |   |
CH3—CHCH2CH2CH—CHCH2CH3          NOT          CH3—CHCH2CH2CH—CHCH2CH3
    7 6  5  4   3 2  1                            3 4  5  6   7 8  9
```

STEP 3

Identify and number the substituents.

(a) Assign a number, or *locant,* to each substituent to locate its point of attachment to the parent chain.

```
     9   8
     CH3CH2     H3C CH2CH3
     |            | |
CH3—CHCH2CH2CHCHCH2CH3          Named as a nonane
    7  6  5  4 3 2  1
```

Substituents: On C3, CH_2CH_3 (3-ethyl)
On C4, CH_3 (4-methyl)
On C7, CH_3 (7-methyl)

(b) If there are two substituents on the same carbon, give both the same number. There must be as many numbers in the name as there are substituents.

```
        CH3  CH3
       4|    |
CH3CH2CCH2CHCH3        Named as a hexane
 6   5  |3   2  1
        CH2CH3
```

Substituents:	On C2, CH_3	(2-methyl)
	On C4, CH_3	(4-methyl)
	On C4, CH_2CH_3	(4-ethyl)

STEP 4

Write the name as a single word.

Use hyphens to separate the different prefixes, and use commas to separate numbers. If two or more different substituents are present, cite them in alphabetical order. If two or more identical substituents are present on the parent chain, use one of the multiplier prefixes *di-*, *tri-*, *tetra-*, and so forth, but don't use these prefixes for alphabetizing. Full names for some of the examples we have been using follow.

```
        2   1
        CH2CH3
        |
CH3CH2CH2CH—CH3
 6   5  4   3
```

3-Methylhexane

```
        8   9
        CH2CH3    CH3  CH2CH3
        |         |    |
CH3—CHCH2CH2CH—CHCH2CH3
     7  6   5   4   3 2   1
```

3-Ethyl-4,7-dimethylnonane

```
       CH3
       |
CH3CHCHCH2CH2CH3
 1   2 |3 4  5  6
       CH2CH3
```

3-Ethyl-2-methylhexane

```
     2   1
     CH2CH3
     |
CH3CHCHCH2CH3
    3 4|
       CH2CH2CH3
       5   6   7
```

4-Ethyl-3-methylheptane

```
        CH3  CH3
       4|    |
CH3CH2CCH2CHCH3
 6   5  |3   2  1
        CH2CH3
```

4-Ethyl-2,4-dimethylhexane

STEP 5

Name a complex substituent as though it were itself a compound.

In some particularly complex cases, a fifth step is necessary. It occasionally happens that a substituent on the main chain has sub-branching. In the following case, for instance, the substituent at C6 is a three-carbon chain with a methyl sub-branch. To name the compound fully, the complex substituent must first be named.

```
     CH3                  CH3
  1 2|  3  4    5   6     |
CH3CHCHCH2CH2CH—CH2CHCH3
        |        |
       CH3      CH2CH2CH2CH3
                 7   8   9  10
```

Named as a 2,3,6-trisubstituted decane

```
[         CH3     ]
[         |       ]
[  ⌇—CH2CHCH3     ]
[     1   2  3    ]
```

A 2-methylpropyl group

Number the branched substituent beginning at its point of its attachment to the main chain, and identify it—in this case, a 2-methylpropyl group. The substituent is treated as a whole and is alphabetized according to the first letter of its complete name, including any numerical prefix. It is set off in parentheses when naming the entire molecule.

```
       CH3                   CH3
   1  2|  3  4    5   6       |
   CH3CHCHCH2CH2CH—CH2CHCH3
           |             |
          CH3         CH2CH2CH2CH3
                       7   8   9  10
```

2,3-Dimethyl-6-(2-methylpropyl)decane

As a further example:

```
                       CH3
            4    3     |2  1
            CH2CH2CHCH3
   9    8     7    6   5|
   CH3CH2CH2CH2CH—CHCHCH3
                     |     |
                   H3C  CH3
```

5-(1,2-Dimethylpropyl)-2-methylnonane

```
     1   2   3
   —CHCHCH3
      |   |
    H3C  CH3
```

A 1,2-dimethylpropyl group

For historical reasons, some of the simpler branched-chain alkyl groups also have nonsystematic, common names, as noted earlier.

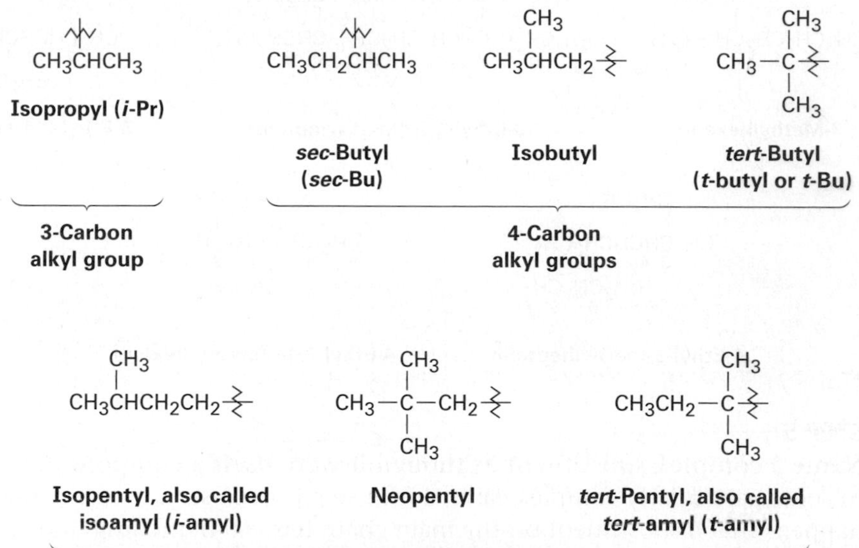

The common names of these simple alkyl groups are so well entrenched in the chemical literature that IUPAC rules make allowance for them. Thus, the following compound is properly named either 4-(1-methylethyl)heptane or 4-isopropylheptane. There's no choice but to memorize these common names; fortunately, there are only a few of them.

```
            CH3CHCH3
                |
   CH3CH2CH2CHCH2CH2CH3
```

4-(1-Methylethyl)heptane or 4-Isopropylheptane

When writing an alkane name, the nonhyphenated prefix iso- is considered part of the alkyl-group name for alphabetizing purposes, but the hyphenated and italicized prefixes *sec-* and *tert-* are not. Thus, isopropyl and isobutyl are listed alphabetically under *i*, but *sec*-butyl and *tert*-butyl are listed under *b*.

Naming Alkanes

Worked Example 3.2

What is the IUPAC name of the following alkane?

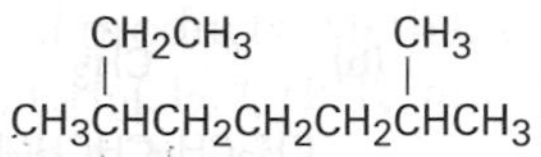

```
CH2CH3        CH3
|             |
CH3CHCH2CH2CH2CHCH3
```

Strategy

Find the longest continuous carbon chain in the molecule, and use that as the parent name. This molecule has a chain of eight carbons—octane—with two methyl substituents. (You have to turn corners to see it.) Numbering from the end nearer the first methyl substituent indicates that the methyls are at C2 and C6

Solution

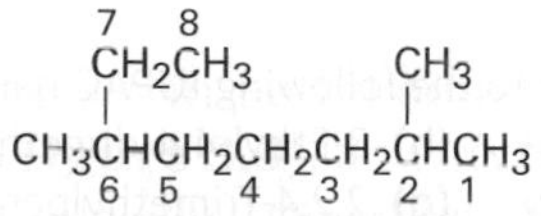

```
7  8
CH2CH3        CH3
|             |
CH3CHCH2CH2CH2CHCH3
   6  5  4  3  2 1
```

2,6-Dimethyloctane

Converting a Chemical Name into a Structure

Worked Example 3.3

Draw the structure of 3-isopropyl-2-methylhexane.

Strategy

This is the reverse of Worked Example 3.2 and uses a reverse strategy. Look at the parent name (hexane), and draw its carbon structure.

C—C—C—C—C—C **Hexane**

Next, find the substituents (3-isopropyl and 2-methyl), and place them on the proper carbons.

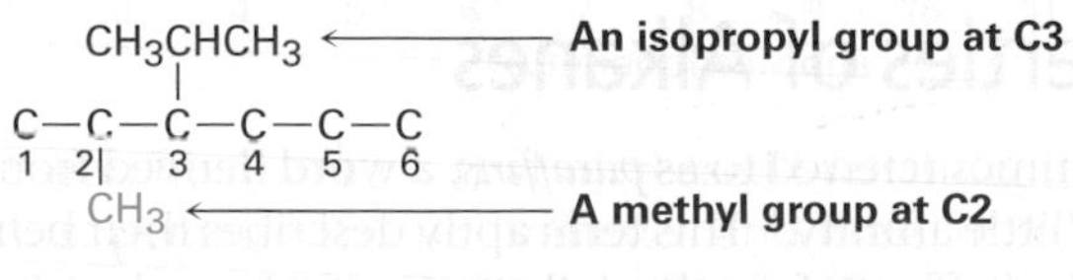

```
    CH3CHCH3 <———— An isopropyl group at C3
        |
C—C—C—C—C—C
1 2| 3  4 5 6
  CH3 <———— A methyl group at C2
```

Finally, add hydrogens to complete the structure.

boils at 125.7 °C, whereas isooctane (2,2,4-trimethylpentane) boils at 99.3 °C. Branched-chain alkanes are lower-boiling because they are more nearly spherical than straight-chain alkanes, have smaller surface areas, and consequently have smaller dispersion forces.

3.6 Conformations of Ethane

Up to now, we've viewed molecules primarily in a two-dimensional way and have given little thought to any consequences that might arise from the spatial arrangement of atoms in molecules. Now it's time to add a third dimension to our study. **Stereochemistry** is the branch of chemistry concerned with the three-dimensional aspects of molecules. We'll see on many occasions in future chapters that the exact three-dimensional structure of a molecule is often crucial to determining its properties and biological behavior.

We know from **Section 1.5** that σ bonds are cylindrically symmetrical. In other words, the intersection of a plane cutting through a carbon–carbon single-bond orbital looks like a circle. Because of this cylindrical symmetry, rotation is possible around carbon–carbon bonds in open-chain molecules. In ethane, for instance, rotation around the C−C bond occurs freely, constantly changing the spatial relationships between the hydrogens on one carbon and those on the other **(Figure 3.5)**.

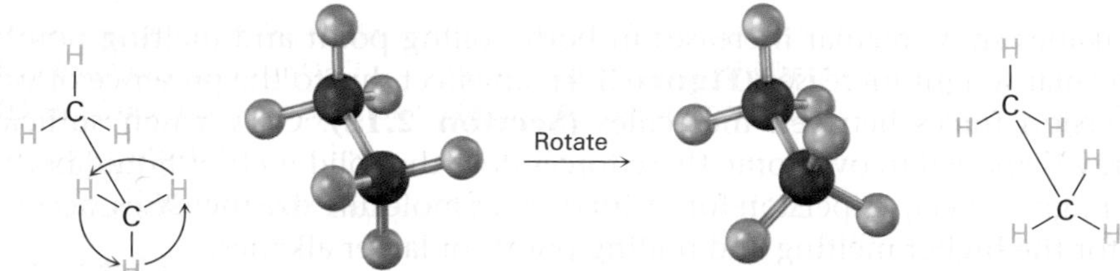

Figure 3.5 Rotation occurs around the carbon–carbon single bond in ethane because of σ bond cylindrical symmetry.

The different arrangements of atoms that result from bond rotation are called **conformations**, and molecules that have different arrangements are called **conformational isomers, or conformers.** Unlike constitutional isomers, however, different conformers often can't be isolated because they interconvert too rapidly.

Conformational isomers are represented in two ways, as shown in **Figure 3.6**. A *sawhorse representation* views the carbon–carbon bond from an oblique angle and indicates spatial orientation by showing all C−H bonds. A **Newman projection** views the carbon–carbon bond directly end-on and represents the two carbon atoms by a circle. Bonds attached to the front carbon are represented by lines to the center of the circle, and bonds attached to the rear carbon are represented by lines to the edge of the circle.

Back carbon

Front carbon

Sawhorse representation

Newman projection

Figure 3.6 A sawhorse representation and a Newman projection of ethane. The sawhorse representation views the molecule from an oblique angle, while the Newman projection views the molecule end-on. Note that the molecular model of the Newman projection appears at first to have six atoms attached to a single carbon. Actually, the front carbon, with three attached green atoms, is directly in front of the rear carbon, with three attached **red atoms**.

Despite what we've just said, we actually don't observe *perfectly* free rotation in ethane. Experiments show that there is a small (12 kJ/mol; 2.9 kcal/mol) barrier to rotation and that some conformations are more stable than others. The lowest-energy, most stable conformation is the one in which all six C–H bonds are as far away from one another as possible—**staggered** when viewed end-on in a Newman projection. The highest-energy, least stable conformation is the one in which the six C–H bonds are as close as possible—**eclipsed** in a Newman projection. At any given instant, about 99% of ethane molecules have an approximately staggered conformation and only about 1% are near the eclipsed conformation.

Rotate rear carbon 60°

4.0 kJ/mol

4.0 kJ/mol

4.0 kJ/mol

Ethane—staggered conformation

Ethane—eclipsed conformation

The extra 12 kJ/mol of energy present in the eclipsed conformation of ethane is called **torsional strain**. Its cause has been the subject of controversy, but the major factor is an interaction between C–H bonding orbitals on one carbon with antibonding orbitals on the adjacent carbon, which stabilizes the staggered conformation relative to the eclipsed one. Because the total strain of 12 kJ/mol arises from three equal hydrogen–hydrogen eclipsing interactions, we can assign a value of approximately 4.0 kJ/mol (1.0 kcal/mol) to each single interaction. The barrier to rotation that results can be represented on a graph of potential energy versus degree of rotation in which the angle between C–H bonds on front and back carbons as viewed end-on (the *dihedral angle*) goes full circle from 0 to 360°. Energy minima occur at staggered conformations, and energy maxima occur at eclipsed conformations, as shown in **Figure 3.7**.

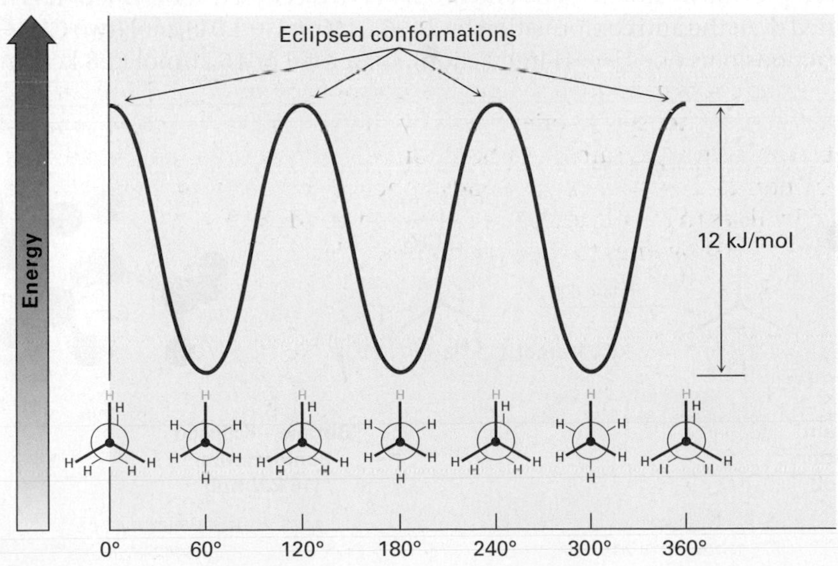

Figure 3.7 A graph of potential energy versus bond rotation in ethane. The staggered conformations are 12 kJ/mol lower in energy than the eclipsed conformations.

3.7 Conformations of Other Alkanes

Propane, the next higher member in the alkane series, also has a torsional barrier that results in hindered rotation around the carbon–carbon bonds. The barrier is slightly higher in propane than in ethane—a total of 14 kJ/mol (3.4 kcal/mol) versus 12 kJ/mol.

The eclipsed conformation of propane has three interactions—two ethane-type hydrogen–hydrogen interactions and one additional hydrogen–methyl interaction. Since each eclipsing H⟷H interaction is the same as that in ethane and thus has an energy "cost" of 4.0 kJ/mol, we can assign a value of $14 - (2 \times 4.0) = 6.0$ kJ/mol (1.4 kcal/mol) to the eclipsing H⟷CH_3 interaction **(Figure 3.8)**.

Staggered propane

Eclipsed propane

Figure 3.8 Newman projections of propane showing staggered and eclipsed conformations. The staggered conformer is lower in energy by 14 kJ/mol.

The conformational situation becomes more complex for larger alkanes because not all staggered conformations have the same energy and not all eclipsed conformations have the same energy. In butane, for instance, the lowest-energy arrangement, called the **anti conformation**, is the one in which the two methyl groups are as far apart as possible—180° away from each other. As rotation around the C2–C3 bond occurs, an eclipsed conformation is reached in which there are two CH_3⟷H interactions and one H⟷H interaction. Using the energy values derived previously from ethane and propane, this eclipsed conformation is more strained than the anti conformation by 2×6.0 kJ/mol + 4.0 kJ/mol (two CH_3⟷H interactions plus one H⟷H interaction), for a total of 16 kJ/mol (3.8 kcal/mol).

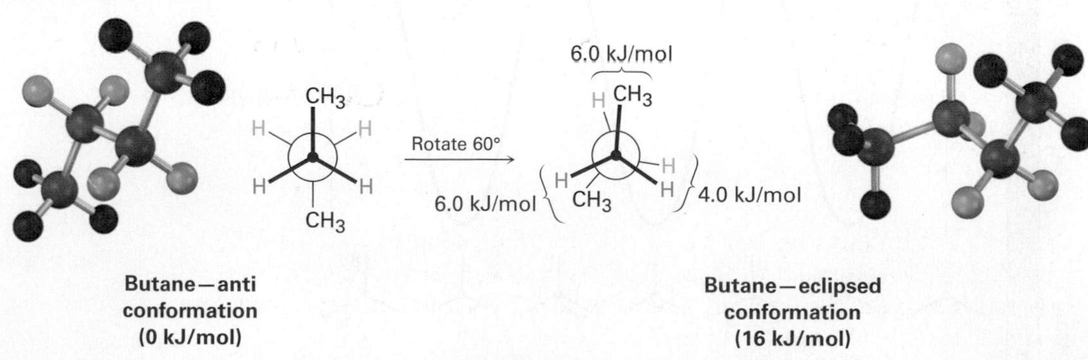

As bond rotation continues, an energy minimum is reached at the staggered conformation where the methyl groups are 60° apart. Called the **gauche conformation**, it lies 3.8 kJ/mol (0.9 kcal/mol) higher in energy than the anti conformation even though it has no eclipsing interactions. This energy difference occurs because the hydrogen atoms of the methyl groups are near one another in the gauche conformation, resulting in what is called *steric strain.* **Steric strain** is the repulsive interaction that occurs when atoms are forced closer together than their atomic radii allow. It's the result of trying to force two atoms to occupy the same space.

Steric strain
3.8 kJ/mol
Rotate 60°

Butane—eclipsed conformation (16 kJ/mol)

Butane—gauche conformation (3.8 kJ/mol)

As the dihedral angle between the methyl groups approaches 0°, an energy maximum is reached at a second eclipsed conformation. Because the methyl groups are forced even closer together than in the gauche conformation, both torsional strain and steric strain are present. A total strain energy of 19 kJ/mol (4.5 kcal/mol) has been estimated for this conformation, making it possible to calculate a value of 11 kJ/mol (2.6 kcal/mol) for the $CH_3 \longleftrightarrow CH_3$ eclipsing interaction: total strain of 19 kJ/mol less the strain of two $H \longleftrightarrow H$ eclipsing interactions (2 × 4.0 kcal/mol) equals 11 kJ/mol.

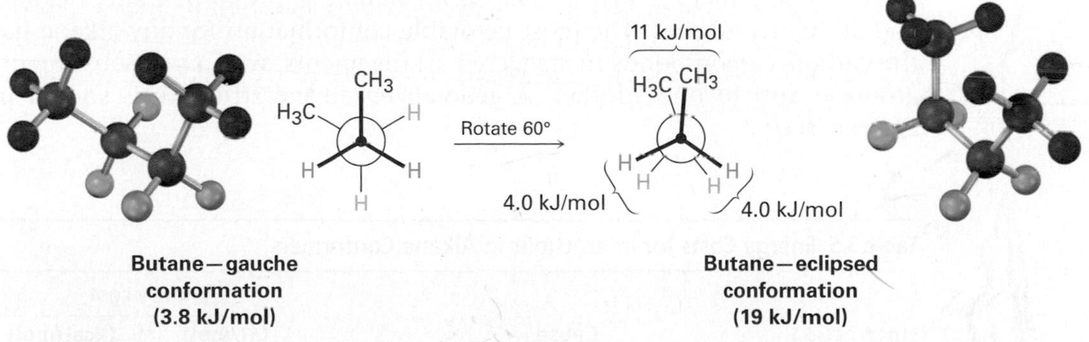

Butane—gauche conformation (3.8 kJ/mol)

Butane—eclipsed conformation (19 kJ/mol)

After 0°, the rotation becomes a mirror image of what we've already seen: another gauche conformation is reached, another eclipsed conformation, and finally a return to the anti conformation. A plot of potential energy versus rotation about the C2–C3 bond is shown in **Figure 3.9**.

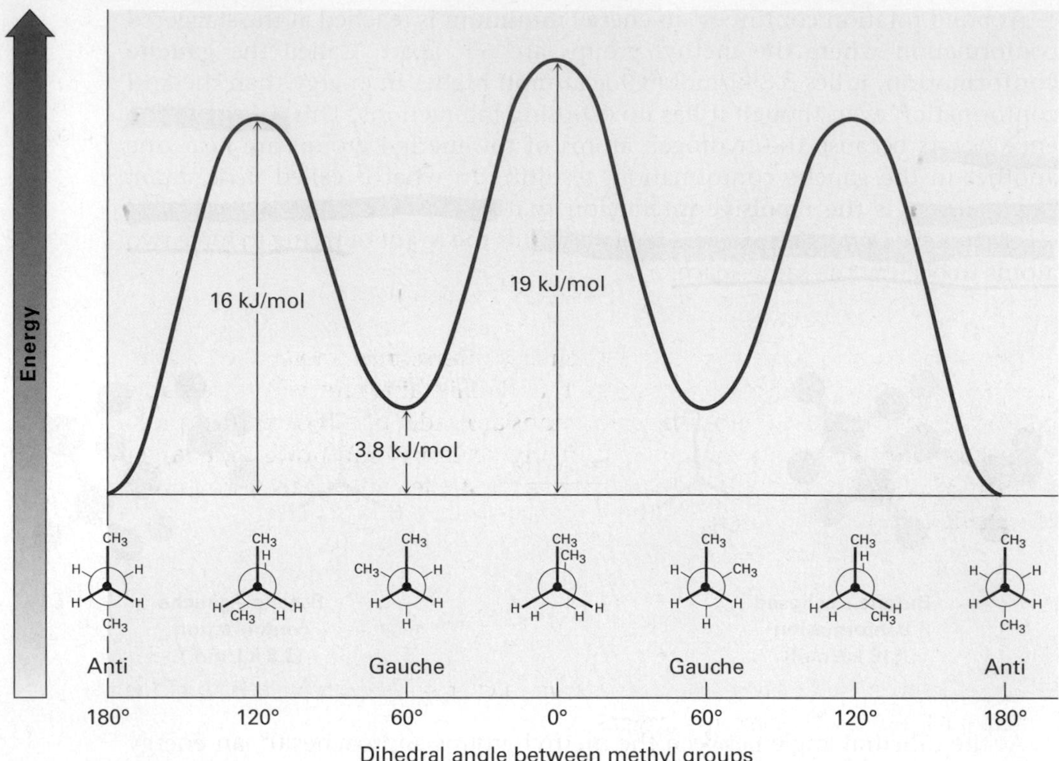

Figure 3.9 A plot of potential energy versus rotation for the C2–C3 bond in butane. The energy maximum occurs when the two methyl groups eclipse each other, and the energy minimum occurs when the two methyl groups are 180° apart (anti).

The notion of assigning definite energy values to specific interactions within a molecule is a very useful one that we'll return to in the next chapter. A summary of what we've seen thus far is given in Table 3.5.

The same principles just developed for butane apply to pentane, hexane, and all higher alkanes. The most favorable conformation for any alkane has the carbon–carbon bonds in staggered arrangements, with large substituents arranged anti to one another. A generalized alkane structure is shown in **Figure 3.10**.

Table 3.5 Energy Costs for Interactions in Alkane Conformers

Interaction	Cause	Energy cost (kJ/mol)	Energy cost (kcal/mol)
H⟷H eclipsed	Torsional strain	4.0	1.0
H⟷CH_3 eclipsed	Mostly torsional strain	6.0	1.4
CH_3⟷CH_3 eclipsed	Torsional and steric strain	11	2.6
CH_3⟷CH_3 gauche	Steric strain	3.8	0.9

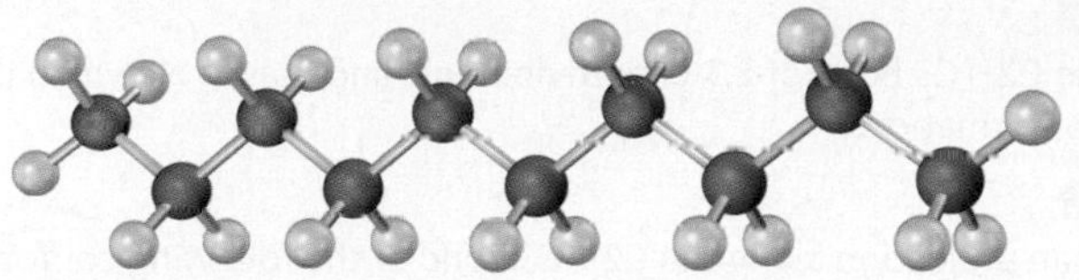

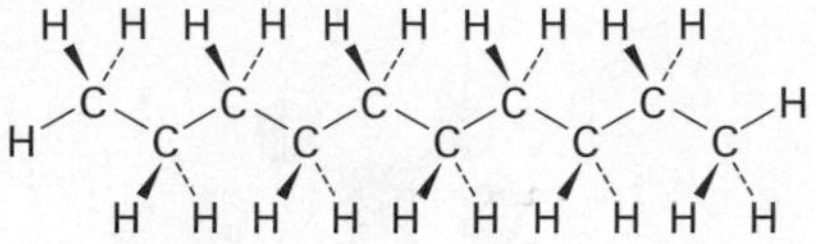

Figure 3.10 The most stable alkane conformation is the one in which all substituents are staggered and the carbon–carbon bonds are arranged anti, as shown in this model of decane.

One final point: saying that one particular conformer is "more stable" than another doesn't mean the molecule adopts and maintains only the more stable conformation. At room temperature, rotations around σ bonds occur so rapidly that all conformers are in equilibrium. At any given instant, however, a larger percentage of molecules will be found in a more stable conformation than in a less stable one.

Drawing Newman Projections

Worked Example 3.4

Sight along the C1–C2 bond of 1-chloropropane, and draw Newman projections of the most stable and least stable conformations.

Strategy

The most stable conformation of a substituted alkane is generally a staggered one in which large groups have an anti relationship. The least stable conformation is generally an eclipsed one in which large groups are as close as possible.

Solution

Most stable (staggered) **Least stable (eclipsed)**

Problem 3.15

Make a graph of potential energy versus angle of bond rotation for propane, and assign values to the energy maxima.

Problem 3.16

Sight along the C2–C1 bond, 2-methylpropane (isobutane) and

(a) draw a Newman projection of the most stable conformation.

(b) draw a Newman projection of the least stable conformation.

(c) make a graph of energy versus angle of rotation around the C2–C1 bond.

(d) Since an H⟷H eclipsing interaction costs 4.0 kJ/mol and an H⟷CH_3 eclipsing interaction costs 6.0 kJ/mol, assign relative values to the maxima and minima in your graph.

Problem 3.17

Sight along the C2—C3 bond of 2,3-dimethylbutane, and draw a Newman projection of the most stable conformation.

Problem 3.18

Draw a Newman projection along the C2—C3 bond of the following conformation of 2,3-dimethylbutane, and calculate a total strain energy:

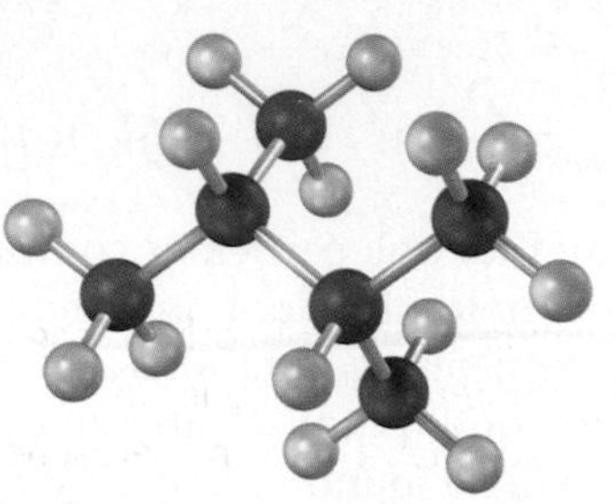

Gasoline is a finite resource. It won't be around forever.

Gasoline | A DEEPER LOOK

British Foreign Minister Ernest Bevin once said that "The Kingdom of Heaven runs on righteousness, but the Kingdom of Earth runs on alkanes." (Actually, he said "runs on oil" not "runs on alkanes," but they're essentially the same.) By far, the major sources of alkanes are the world's natural gas and petroleum deposits. Laid down eons ago, these deposits are thought to be derived primarily from the decomposition of tiny single-celled marine organisms called foraminifera. *Natural gas* consists chiefly of methane but also contains ethane, propane, and butane. *Petroleum* is a complex mixture of hydrocarbons that must be separated into fractions and then further refined before it can be used.

The petroleum era began in August 1859, when the world's first oil well was drilled by Edwin Drake near Titusville, Pennsylvania. The petroleum was distilled into fractions according to boiling point, but it was high-boiling kerosene, or lamp oil, rather than gasoline that was primarily sought. Literacy was becoming widespread at the time, and people wanted better light for reading than was available from candles. Gasoline was too volatile for use in lamps and was initially considered a waste by-product. The world has changed greatly since those early days, however, and it is now gasoline rather than lamp oil that is prized.

Petroleum refining begins by fractional distillation of crude oil into three principal cuts according to boiling point (bp): straight-run gasoline (bp 30–200 °C), kerosene (bp 175–300 °C), and heating oil, or diesel fuel (bp 275–400 °C). Further distillation under reduced pressure then yields lubricating oils and waxes and leaves a tarry residue of asphalt. The distillation of crude oil is only the first step in gasoline production, however. Straight-run gasoline turns out to be a poor fuel in automobiles because of engine knock, an uncontrolled combustion that can occur in a hot engine.

The *octane number* of a fuel is the measure by which its antiknock properties are judged. It was recognized long ago that straight-chain hydrocarbons are far more prone to induce

(continued)

(continued)

engine knock than are highly branched compounds. Heptane, a particularly bad fuel, is assigned a base value of 0 octane number, and 2,2,4-trimethylpentane, commonly known as isooctane, has a rating of 100.

$$CH_3CH_2CH_2CH_2CH_2CH_2CH_3$$

Heptane
(octane number = 0)

$$\begin{array}{c} CH_3\ CH_3 \\ |\quad\ \ | \\ CH_3CCH_2CHCH_3 \\ | \\ CH_3 \end{array}$$

2,2,4-Trimethylpentane
(octane number = 100)

Because straight-run gasoline burns so poorly in engines, petroleum chemists have devised numerous methods for producing higher-quality fuels. One of these methods, *catalytic cracking*, involves taking the high-boiling kerosene cut (C_{11}–C_{14}) and "cracking" it into smaller branched molecules suitable for use in gasoline. Another process, called *reforming*, is used to convert C_6–C_8 alkanes to aromatic compounds such as benzene and toluene, which have substantially higher octane numbers than alkanes. The final product that goes in your tank has an approximate composition of 15% C_4–C_8 straight-chain alkanes, 25% to 40% C_4–C_{10} branched-chain alkanes, 10% cyclic alkanes, 10% straight-chain and cyclic alkenes, and 25% arenes (aromatics).

Summary

Even though alkanes are relatively unreactive and rarely involved in chemical reactions, they nevertheless provide a useful vehicle for introducing some important general ideas. In this chapter, we've used alkanes to introduce the basic approach to naming organic compounds and to take an initial look at some of the three-dimensional aspects of molecules.

A **functional group** is a group of atoms within a larger molecule that has a characteristic chemical reactivity. Because functional groups behave in approximately the same way in all molecules where they occur, the chemical reactions of an organic molecule are largely determined by its functional groups.

Alkanes are a class of **saturated hydrocarbons** with the general formula C_nH_{2n+2}. They contain no functional groups, are relatively inert, and can be either **straight-chain** (*normal*) or **branched**. Alkanes are named by a series of IUPAC rules of nomenclature. Compounds that have the same chemical formula but different structures are called **isomers**. More specifically, compounds such as butane and isobutane, which differ in their connections between atoms, are called **constitutional isomers**.

Carbon–carbon single bonds in alkanes are formed by σ overlap of carbon sp^3 hybrid orbitals. Rotation is possible around σ bonds because of their cylindrical symmetry, and alkanes therefore exist in a large number of rapidly interconverting **conformations**. **Newman projections** make it possible to visualize the spatial consequences of bond rotation by sighting directly along a carbon–carbon bond axis. Not all alkane conformations are equally stable. The **staggered** conformation of ethane is 12 kJ/mol (2.9 kcal/mol) more stable than the **eclipsed** conformation because of **torsional strain**. In general, any alkane is most stable when all its bonds are staggered.

Key words

aliphatic, 81
alkane, 80
alkyl group, 84
anti conformation, 96
branched-chain alkane, 82
conformation, 94
conformers, 94
constitutional isomers, 82
eclipsed conformation, 95
functional group, 74
gauche conformation, 97
hydrocarbon, 81
isomers, 82
Newman projection, 94
R group, 86
saturated, 81
staggered conformation, 95
stereochemistry, 94
steric strain, 97
straight-chain alkane, 82
substituent, 87
torsional strain, 95

Exercises

OWL Interactive versions of these problems are assignable in OWL for Organic Chemistry.

Visualizing Chemistry

(Problems 3.1–3.18 appear within the chapter.)

3.19 Identify the functional groups in the following substances, and convert each drawing into a molecular formula (red = O, blue = N).

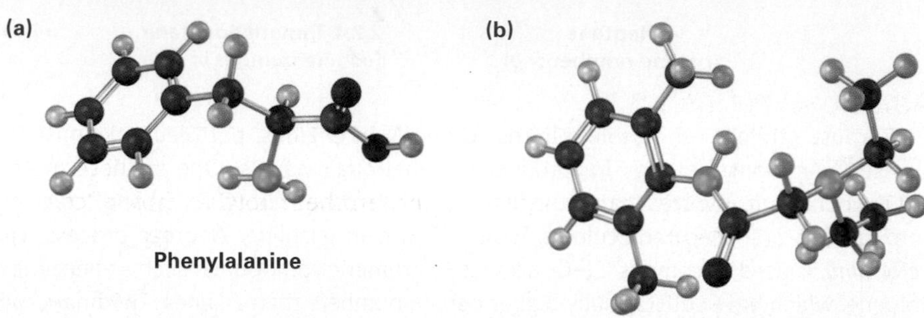

3.20 Give IUPAC names for the following alkanes, and convert each drawing into a skeletal structure:

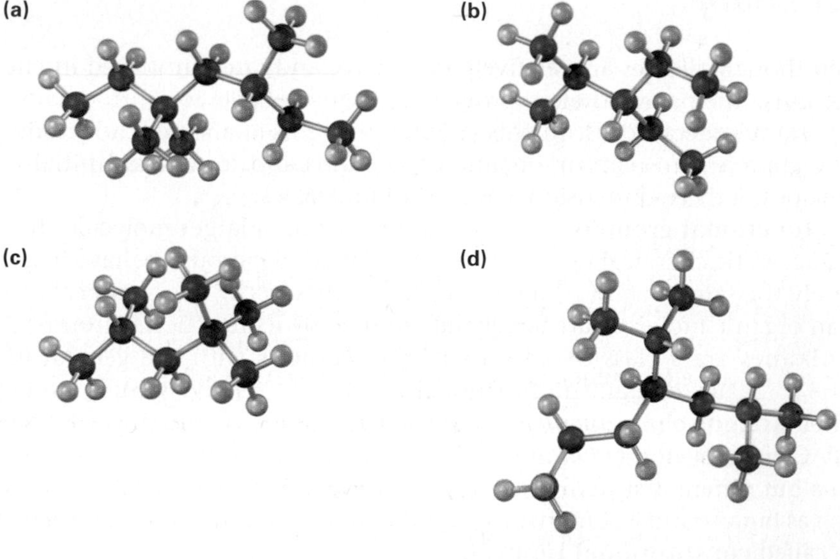

3.21 Draw a Newman projection along the C2–C3 bond of the following conformation of 2-butanol.

Additional Problems

Functional Groups

3.22 Locate and identify the functional groups in the following molecules.

(a) CH_2OH $NHCH_3$

(b) 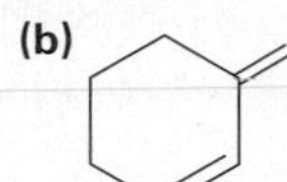

(c) H N C=O CH_3

(d) O ‖ CH_3CHCOH | NH_2

(e) O

(f) Cl O

3.23 Propose structures that meet the following descriptions:

(a) A ketone with five carbons (b) A four-carbon amide
(c) A five-carbon ester (d) An aromatic aldehyde
(e) A keto ester (f) An amino alcohol

3.24 Propose structures for the following:

(a) A ketone, C_4H_8O (b) A nitrile, C_5H_9N
(c) A dialdehyde, $C_4H_6O_2$ (d) A bromoalkene, $C_6H_{11}Br$
(e) An alkane, C_6H_{14} (f) A *cyclic* saturated hydrocarbon, C_6H_{12}
(g) A diene (dialkene), C_5H_8 (h) A keto alkene, C_5H_8O

3.25 Predict the hybridization of the carbon atom in each of the following functional groups:

(a) Ketone (b) Nitrile (c) Carboxylic acid

3.26 Draw the structures of the following molecules:

(a) *Biacetyl,* $C_4H_6O_2$, a substance with the aroma of butter; it contains no rings or carbon–carbon multiple bonds.

(b) *Ethylenimine,* C_2H_5N, a substance used in the synthesis of melamine polymers; it contains no multiple bonds.

(c) *Glycerol,* $C_3H_8O_3$, a substance isolated from fat and used in cosmetics; it has an –OH group on each carbon.

Isomers

3.27 Draw structures that meet the following descriptions (there are many possibilities):

(a) Three isomers with the formula C_8H_{18}

(b) Two isomers with the formula $C_4H_8O_2$

3.28 Draw structures of the nine isomers of C_7H_{16}.

3.29 In each of the following sets, which structures represent the same compound and which represent different compounds?

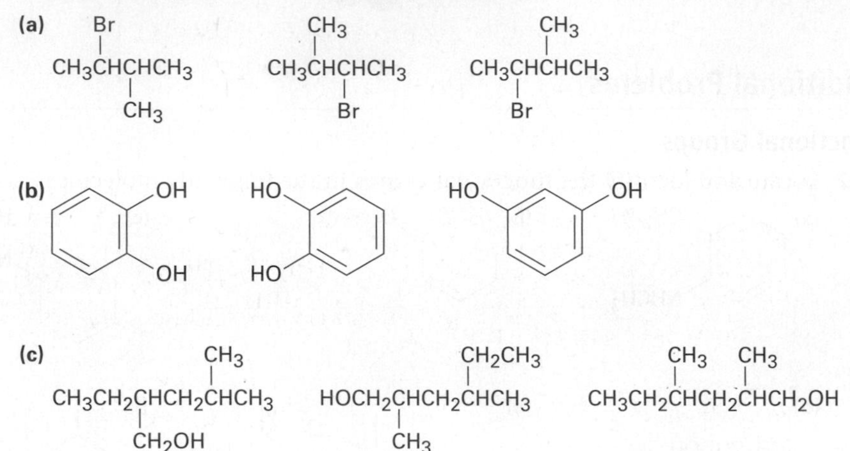

3.30 There are seven constitutional isomers with the formula $C_4H_{10}O$. Draw as many as you can.

3.31 Draw as many compounds as you can that fit the following descriptions:

(a) Alcohols with formula $C_4H_{10}O$ (b) Amines with formula $C_5H_{13}N$

(c) Ketones with formula $C_5H_{10}O$ (d) Aldehydes with formula $C_5H_{10}O$

(e) Esters with formula $C_4H_8O_2$ (f) Ethers with formula $C_4H_{10}O$

3.32 Draw compounds that contain the following:

(a) A primary alcohol (b) A tertiary nitrile

(c) A secondary thiol (d) Both primary and secondary alcohols

(e) An isopropyl group (f) A quaternary carbon

Naming Compounds

3.33 Draw and name all monobromo derivatives of pentane, $C_5H_{11}Br$.

3.34 Draw and name all monochloro derivatives of 2,5-dimethylhexane, $C_8H_{17}Cl$.

3.35 Draw structures for the following:

(a) 2-Methylheptane (b) 4-Ethyl-2,2-dimethylhexane

(c) 4-Ethyl-3,4-dimethyloctane (d) 2,4,4-Trimethylheptane

(e) 3,3-Diethyl-2,5-dimethylnonane (f) 4-Isopropyl-3-methylheptane

3.36 Draw a compound that:

(a) Has only primary and tertiary carbons

(b) Has no secondary or tertiary carbons

(c) Has four secondary carbons

3.37 Draw a compound that:
(a) Has nine primary hydrogens
(b) Has only primary hydrogens

3.38 Give IUPAC names for the following compounds:

(a)
```
   CH3
   |
CH3CHCH2CH2CH3
```

(b)
```
       CH3
       |
CH3CH2CCH3
       |
       CH3
```

(c)
```
   H3C CH3
    |  |
CH3CHCCH2CH2CH3
      |
      CH3
```

(d)
```
      CH2CH3    CH3
      |         |
CH3CH2CHCH2CH2CHCH3
```

(e)
```
         CH3  CH2CH3
         |    |
CH3CH2CH2CHCH2CCH3
              |
              CH3
```

(f)
```
   H3C CH3
    |  |
CH3C—CCH2CH2CH3
    |  |
   H3C CH3
```

3.39 Name the five isomers of C_6H_{14}.

3.40 Explain why each of the following names is incorrect:
(a) 2,2-Dimethyl-6-ethylheptane (b) 4-Ethyl-5,5-dimethylpentane
(c) 3-Ethyl-4,4-dimethylhexane (d) 5,5,6-Trimethyloctane
(e) 2-Isopropyl-4-methylheptane

3.41 Propose structures and give IUPAC names for the following:
(a) A diethyldimethylhexane (b) A (3-methylbutyl)-substituted alkane

Conformations

3.42 Consider 2-methylbutane (isopentane). Sighting along the C2–C3 bond:
(a) Draw a Newman projection of the most stable conformation.
(b) Draw a Newman projection of the least stable conformation.
(c) If a $CH_3 \longleftrightarrow CH_3$ eclipsing interaction costs 11 kJ/mol (2.5 kcal/mol) and a $CH_3 \longleftrightarrow CH_3$ gauche interaction costs 3.8 kJ/mol (0.9 kcal/mol), make a quantitative plot of energy versus rotation about the C2–C3 bond.

3.43 What are the relative energies of the three possible staggered conformations around the C2–C3 bond in 2,3-dimethylbutane? (See Problem 3.42.)

3.44 Construct a qualitative potential-energy diagram for rotation about the C–C bond of 1,2-dibromoethane. Which conformation would you expect to be most stable? Label the anti and gauche conformations of 1,2-dibromoethane.

3.45 Which conformation of 1,2-dibromoethane (Problem 3.44) would you expect to have the largest dipole moment? The observed dipole moment of 1,2-dibromoethane is $\mu = 1.0$ D. What does this tell you about the actual conformation of the molecule?

3.46 Draw the most stable conformation of pentane, using wedges and dashes to represent bonds coming out of the paper and going behind the paper, respectively.

3.47 Draw the most stable conformation of 1,4-dichlorobutane, using wedges and dashes to represent bonds coming out of the paper and going behind the paper, respectively.

General Problems

3.48 For each of the following compounds, draw an isomer that has the same functional groups.

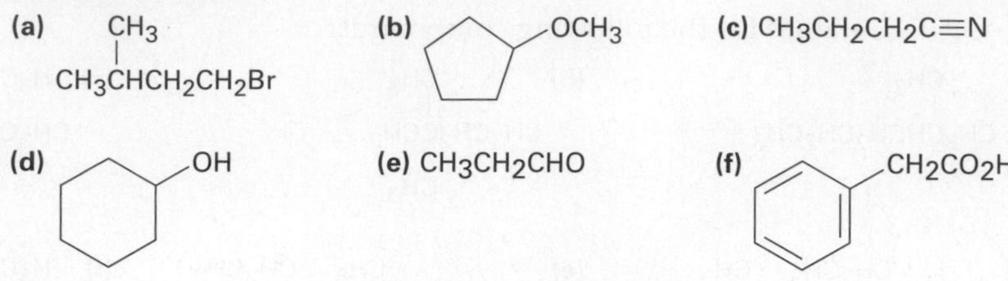

3.49 Malic acid, $C_4H_6O_5$, has been isolated from apples. Because this compound reacts with 2 molar equivalents of base, it is a dicarboxylic acid.
(a) Draw at least five possible structures.
(b) If malic acid is a secondary alcohol, what is its structure?

3.50 Formaldehyde, $H_2C{=}O$, is known to all biologists because of its usefulness as a tissue preservative. When pure, formaldehyde *trimerizes* to give trioxane, $C_3H_6O_3$, which, surprisingly enough, has no carbonyl groups. Only one monobromo derivative ($C_3H_5BrO_3$) of trioxane is possible. Propose a structure for trioxane.

3.51 The barrier to rotation about the C–C bond in bromoethane is 15 kJ/mol (3.6 kcal/mol).
(a) What energy value can you assign to an H$\longleftrightarrow$Br eclipsing interaction?
(b) Construct a quantitative diagram of potential energy versus bond rotation for bromoethane.

3.52 Increased substitution around a bond leads to increased strain. Take the four substituted butanes listed below, for example. For each compound, sight along the C2–C3 bond and draw Newman projections of the most stable and least stable conformations. Use the data in Table 3.5 to assign strain energy values to each conformation. Which of the eight conformations is most strained? Which is least strained?
(a) 2-Methylbutane (b) 2,2-Dimethylbutane
(c) 2,3-Dimethylbutane (d) 2,2,3-Trimethylbutane

3.53 The cholesterol-lowering agents called *statins,* such as simvastatin (Zocor) and pravastatin (Pravachol), are among the most widely prescribed drugs in the world, with annual sales estimated at approximately $15 billion. Identify the functional groups in both, and tell how the two substances differ.

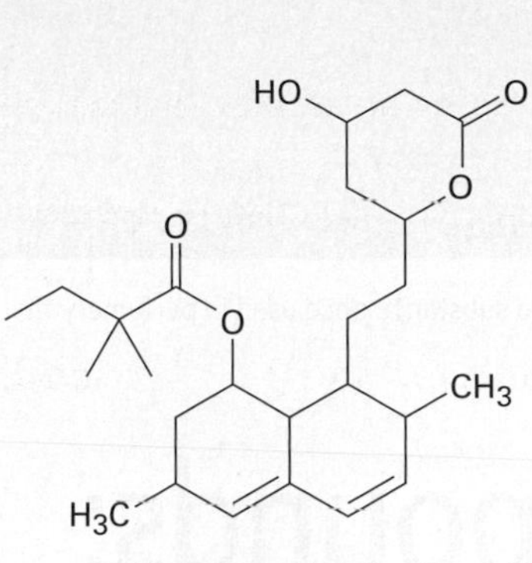

Simvastatin (Zocor)

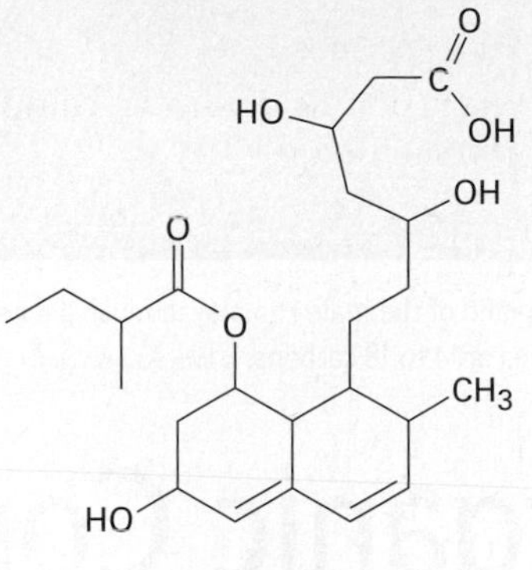

Pravastatin (Pravachol)

3.54 We'll look in the next chapter at *cycloalkanes*—saturated cyclic hydrocarbons—and we'll see that the molecules generally adopt puckered, nonplanar conformations. Cyclohexane, for instance, has a puckered shape like a lounge chair rather than a flat shape. Why?

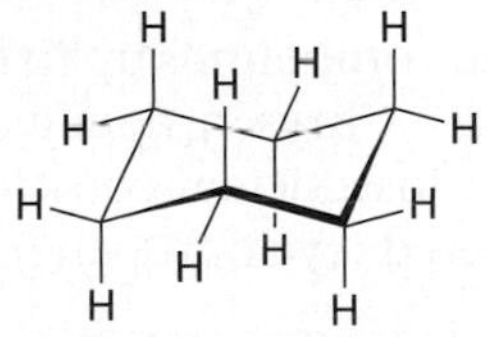

Nonplanar cyclohexane 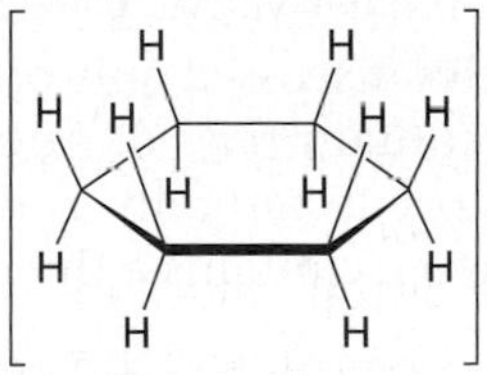 **Planar cyclohexane**

3.55 We'll see in the next chapter that there are two isomeric substances both named 1,2-dimethylcyclohexane. Explain.

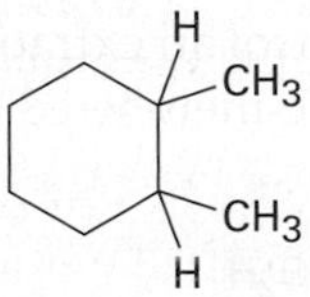

1,2-Dimethylcyclohexane

4

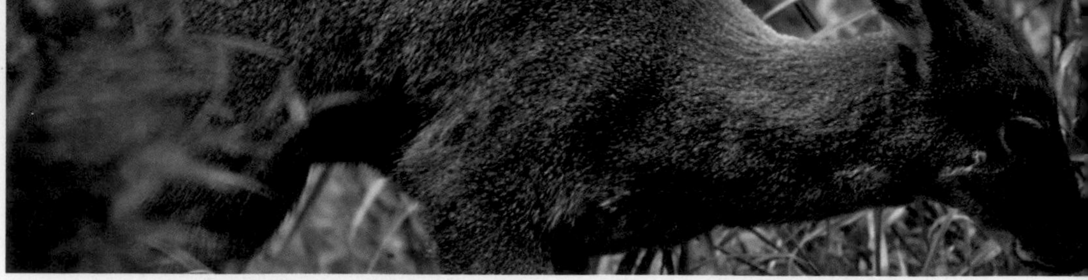

The musk gland of the male Himalayan musk deer secretes a substance once used in perfumery that contains cycloalkanes of 14 to 18 carbons. © Indiapicture/Alamy

Organic Compounds: Cycloalkanes and Their Stereochemistry

Although we've discussed only open-chain compounds up to now, most organic compounds contain *rings* of carbon atoms. Chrysanthemic acid, for instance, whose esters occur naturally as the active insecticidal constituents of chrysanthemum flowers, contains a three-membered (cyclopropane) ring.

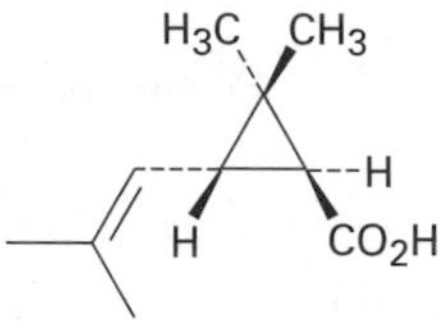

H_3C CH_3 H H CO_2H

Chrysanthemic acid

Prostaglandins, potent hormones that control an extraordinary variety of physiological functions in humans, contain a five-membered (cyclopentane) ring.

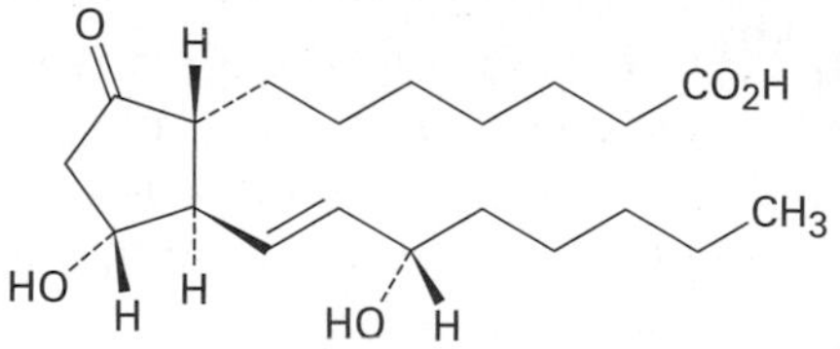

O H CO_2H CH_3 HO H H HO H

Prostaglandin E_1

Steroids, such as cortisone, contain four rings joined together—3 six-membered (cyclohexane) and 1 five-membered. We'll discuss steroids and their properties in more detail in **Sections 27.6 and 27.7**.

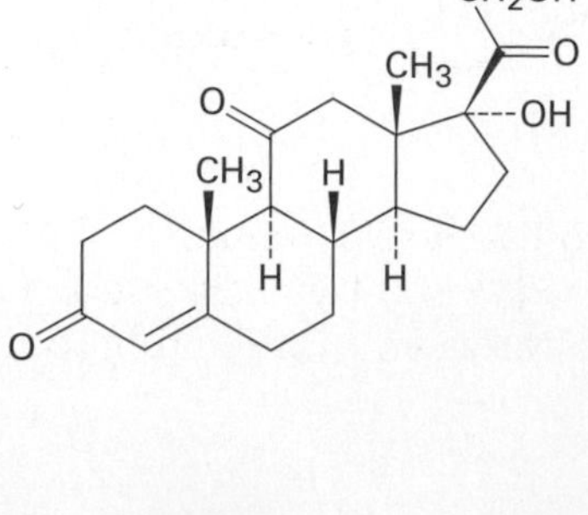

CH_2OH O CH_3 OH O CH_3 H H H O

Cortisone

OWL Sign in to OWL for Organic Chemistry at **www.cengage.com/owl** to view tutorials and simulations, develop problem-solving skills, and complete online homework assigned by your professor.

Why This Chapter? We'll see numerous instances in future chapters where the chemistry of a given functional group is affected by being in a ring rather than an open chain. Because cyclic molecules are so commonly encountered in most pharmaceuticals and in all classes of biomolecules, including proteins, lipids, carbohydrates, and nucleic acids, it's important to understand the consequences of cyclic structures.

4.1 Naming Cycloalkanes

Saturated cyclic hydrocarbons are called **cycloalkanes**, or **alicyclic** compounds (**ali**phatic **cyclic**). Because cycloalkanes consist of rings of $-CH_2-$units, they have the general formula $(CH_2)_n$, or C_nH_{2n}, and can be represented by polygons in skeletal drawings.

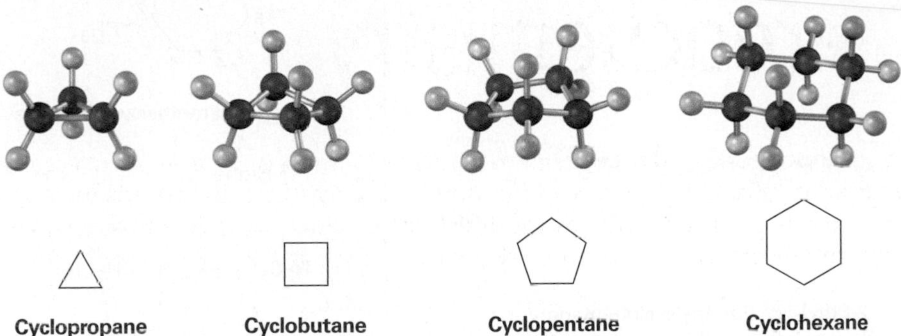

Substituted cycloalkanes are named by rules similar to those we saw in the previous chapter for open-chain alkanes **(Section 3.4)**. For most compounds, there are only two steps.

STEP 1

Find the parent.

Count the number of carbon atoms in the ring and the number in the largest substituent. If the number of carbon atoms in the ring is equal to or greater than the number in the substituent, the compound is named as an alkyl-substituted cycloalkane. If the number of carbon atoms in the largest substituent is greater than the number in the ring, the compound is named as a cycloalkyl-substituted alkane. For example:

$-CH_3$ $\quad$ $-CH_2CH_2CH_2CH_3$

3 carbons $\quad$ 4 carbons

Methylcyclopentane $\quad$ **1-Cyclopropylbutane**

STEP 2

Number the substituents, and write the name.

For an alkyl- or halo-substituted cycloalkane, choose a point of attachment as carbon 1 and number the substituents on the ring so that the *second* substituent has as low a number as possible. If ambiguity still exists, number so that the third or fourth substituent has as low a number as possible, until a point of difference is found.

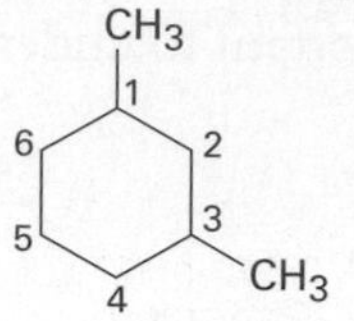

1,3-Dimethylcyclohexane NOT **1,5-Dimethylcyclohexane**

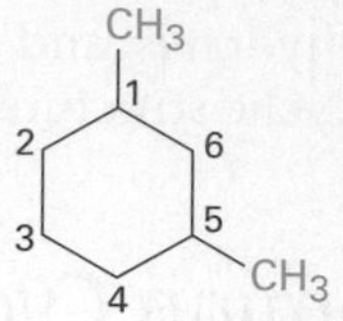

Lower Higher

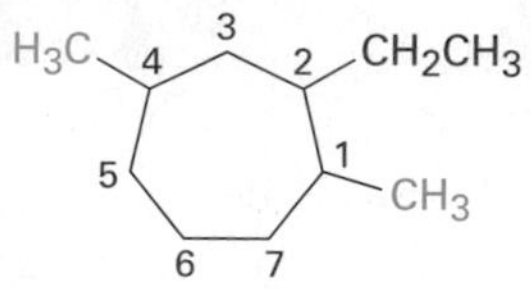

2-Ethyl-1,4-dimethylcycloheptane NOT **1-Ethyl-2,6-dimethylcycloheptane** / **3-Ethyl-1,4-dimethylcycloheptane**

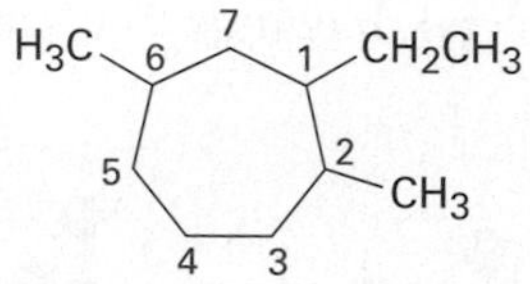

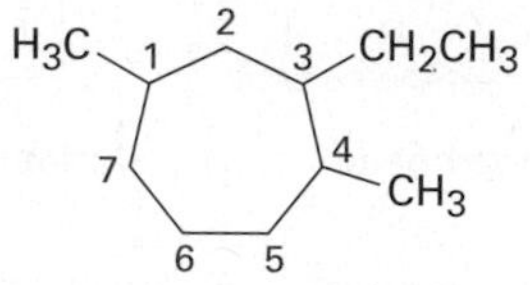

Lower Lower Higher Higher

(a) When two or more different alkyl groups that could potentially receive the same numbers are present, number them by alphabetical priority, ignoring numerical prefixes such as di- and tri-.

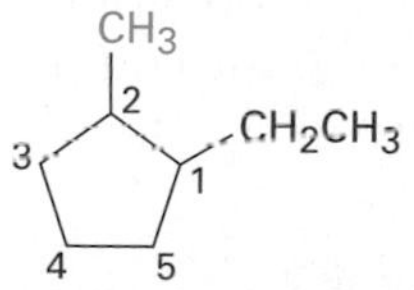

1-Ethyl-2-methylcyclopentane NOT **2-Ethyl-1-methylcyclopentane**

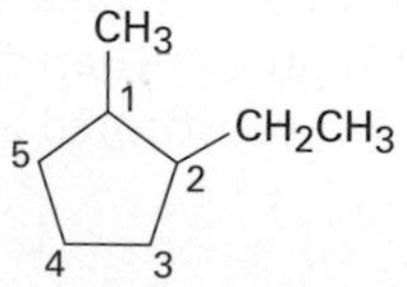

(b) If halogens are present, treat them just like alkyl groups.

1-Bromo-2-methylcyclobutane NOT **2-Bromo-1-methylcyclobutane**

Some additional examples follow:

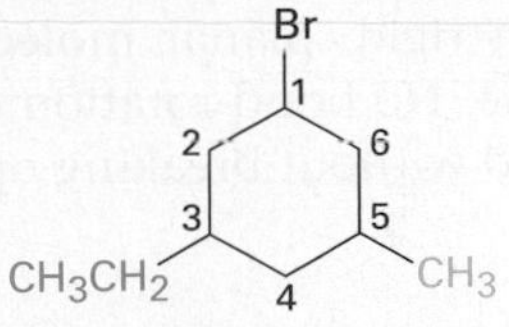

Br, 1, 2, 6, 3, 5, 4, CH_3CH_2, CH_3

1-Bromo-3-ethyl-5-methyl-cyclohexane

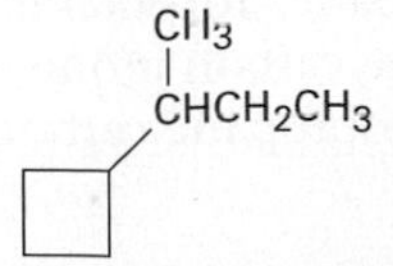

CH_3, $CHCH_2CH_3$

(1-Methylpropyl)cyclobutane or *sec*-butylcyclobutane

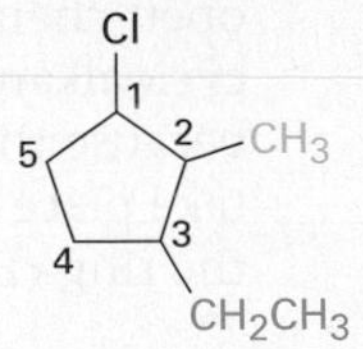

Cl, 1, 2, 3, 4, 5, CH_3, CH_2CH_3

1-Chloro-3-ethyl-2-methyl-cyclopentane

Problem 4.1

Give IUPAC names for the following cycloalkanes:

(a) CH_3, CH_3

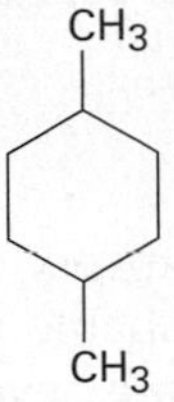

(b) $CH_2CH_2CH_3$, CH_3

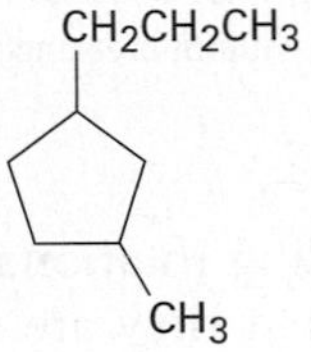

(c)

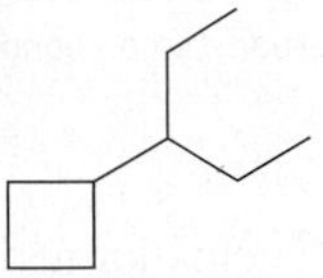

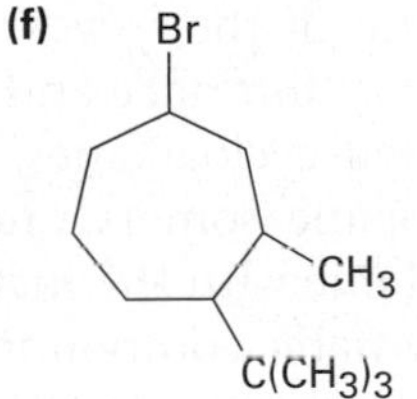

(d) CH_2CH_3, Br

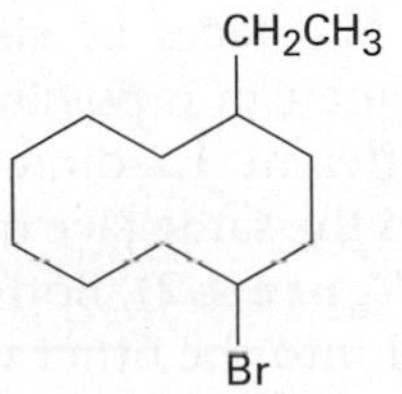

(e) CH_3, $CH(CH_3)_2$

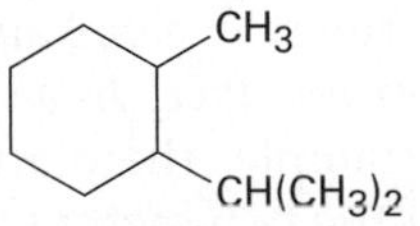

(f) Br, CH_3, $C(CH_3)_3$

Problem 4.2

Draw structures corresponding to the following IUPAC names:

(a) 1,1-Dimethylcyclooctane **(b)** 3-Cyclobutylhexane
(c) 1,2-Dichlorocyclopentane **(d)** 1,3-Dibromo-5-methylcyclohexane

Problem 4.3

Name the following cycloalkane:

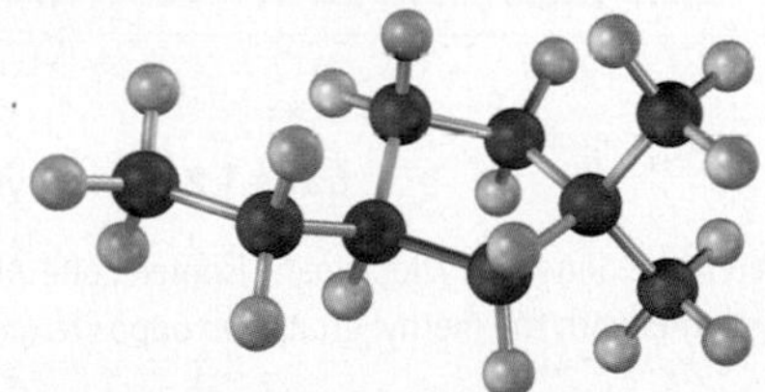

4.2 Cis–Trans Isomerism in Cycloalkanes

In many respects, the chemistry of cycloalkanes is like that of open-chain alkanes: both are nonpolar and fairly inert. There are, however, some important differences. One difference is that cycloalkanes are less flexible than open-chain alkanes. In contrast with the relatively free rotation around single bonds in open-chain alkanes **(Sections 3.6 and 3.7)**, there is much less freedom in cycloalkanes. Cyclopropane, for example, must be a rigid, planar molecule because three points (the carbon atoms) define a plane. No bond rotation can take place around a cyclopropane carbon–carbon bond without breaking open the ring **(Figure 4.1)**.

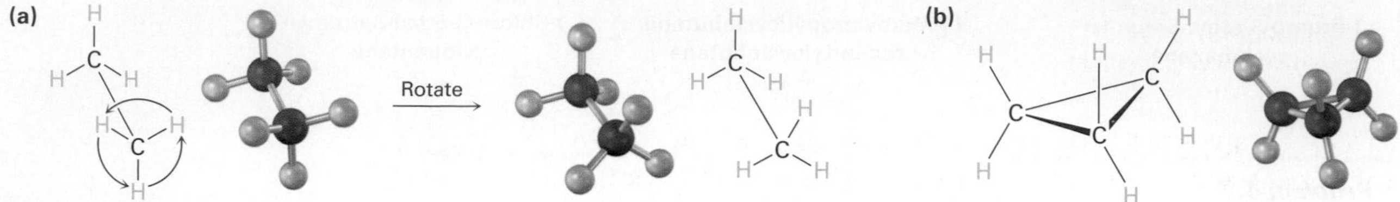

Figure 4.1 **(a)** Rotation occurs around the carbon–carbon bond in ethane, but **(b)** no rotation is possible around the carbon–carbon bonds in cyclopropane without breaking open the ring.

Larger cycloalkanes have increasing rotational freedom, and the very large rings (C_{25} and up) are so floppy that they are nearly indistinguishable from open-chain alkanes. The common ring sizes (C_3–C_7), however, are severely restricted in their molecular motions.

Because of their cyclic structures, cycloalkanes have two faces as viewed edge-on, a "top" face and a "bottom" face. As a result, isomerism is possible in substituted cycloalkanes. For example, there are two different 1,2-dimethylcyclopropane isomers, one with the two methyl groups on the same face of the ring and one with the methyl groups on opposite faces **(Figure 4.2)**. Both isomers are stable compounds, and neither can be converted into the other without breaking and reforming chemical bonds.

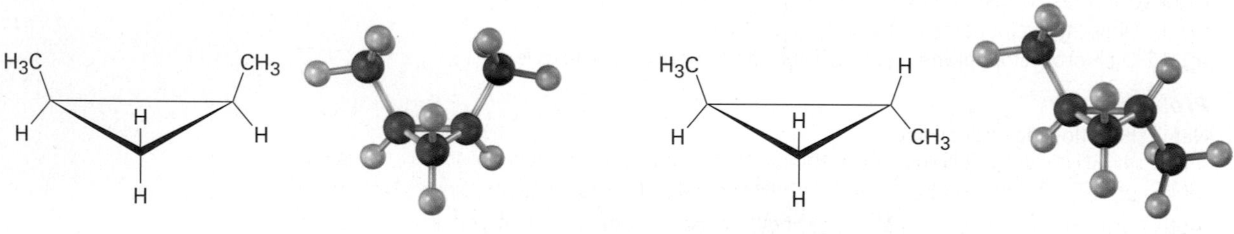

Figure 4.2 There are two different 1,2-dimethylcyclopropane isomers, one with the methyl groups on the same face of the ring (cis) and the other with the methyl groups on opposite faces of the ring (trans). The two isomers do not interconvert.

Unlike the constitutional isomers butane and isobutane, which have their atoms connected in a different order **(Section 3.2)**, the two 1,2-dimethylcyclopropanes have the same order of connections but differ in the spatial orientation of the atoms. Such compounds, which have their atoms connected in the same order but differ in three-dimensional orientation, are called stereochemical isomers, or **stereoisomers**. More generally, the term **stereochemistry** is used to refer to the three-dimensional aspects of chemical structure and reactivity.

Constitutional isomers (different connections between atoms)

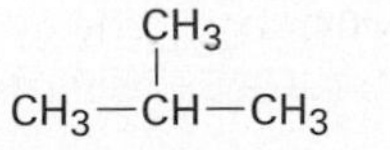

$CH_3-CH(CH_3)-CH_3$ and $CH_3-CH_2-CH_2-CH_3$

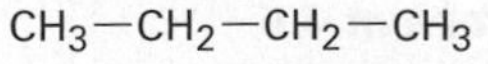

Stereoisomers (same connections but different three-dimensional geometry)

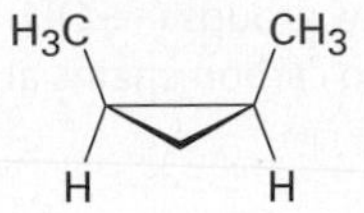

H3C, CH3, H, H and H3C, H, H, CH3

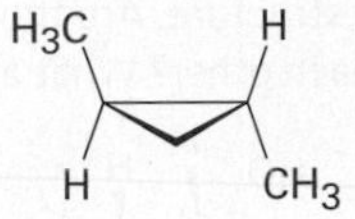

The 1,2-dimethylcyclopropanes are members of a subclass of stereoisomers called **cis–trans isomers.** The prefixes *cis-* (Latin "on the same side") and *trans-* (Latin "across") are used to distinguish between them. Cis–trans isomerism is a common occurrence in substituted cycloalkanes and in many cyclic biological molecules.

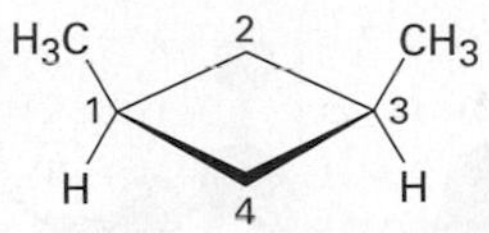

***cis*-1,3-Dimethylcyclobutane**

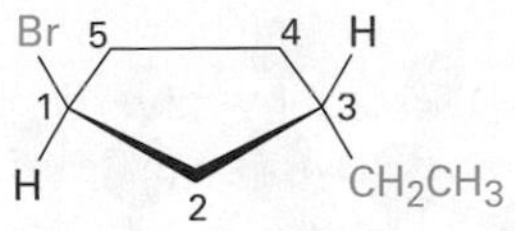

***trans*-1-Bromo-3-ethylcyclopentane**

Naming Cycloalkanes

Worked Example 4.1

Name the following substances, including the *cis-* or *trans-* prefix:

(a) H, CH3, H3C, H (b) H, Cl, Cl, H

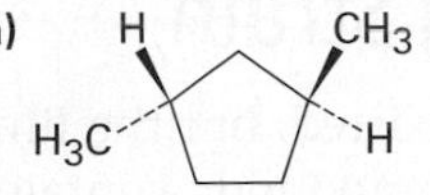

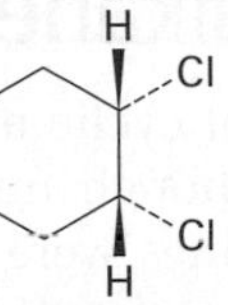

Strategy

In these views, the ring is roughly in the plane of the page, a wedged bond protrudes out of the page, and a dashed bond recedes into the page. Two substituents are cis if they are both out of or both into the page, and they are trans if one is out of and one is into the page.

Solution

(a) *trans*-1,3-Dimethylcyclopentane **(b)** *cis*-1,2-Dichlorocyclohexane

Problem 4.4

Name the following substances, including the *cis-* or *trans-* prefix:

(a) H, CH3, Cl, H (b) H3C, CH2CH3, H, H

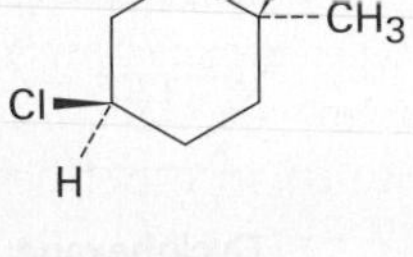

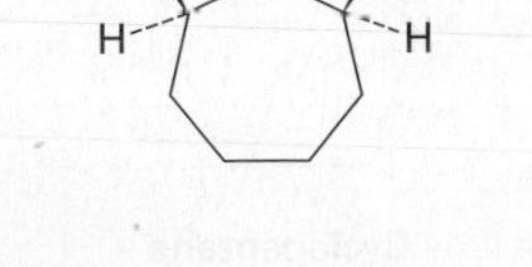

Problem 4.5

Draw the structures of the following molecules:

(a) *trans*-1-Bromo-3-methylcyclohexane **(b)** *cis*-1,2-Dimethylcyclobutane

(c) *trans*-1-*tert*-Butyl-2-ethylcyclohexane

Problem 4.6

Prostaglandin $F_{2\alpha}$, a hormone that causes uterine contraction during childbirth, has the following structure. Are the two hydroxyl groups (—OH) on the cyclopentane ring cis or trans to each other? What about the two carbon chains attached to the ring?

Prostaglandin $F_{2\alpha}$

Problem 4.7

Name the following substances, including the *cis*- or *trans*- prefix (red-brown = Br):

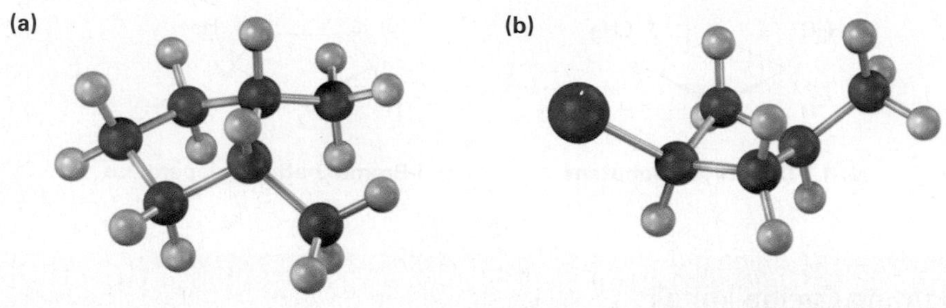

4.3 Stability of Cycloalkanes: Ring Strain

Chemists in the late 1800s knew that cyclic molecules existed, but the limitations on ring size were unclear. Although numerous compounds containing five-membered and six-membered rings were known, smaller and larger ring sizes had not been prepared, despite many efforts.

A theoretical interpretation of this observation was proposed in 1885 by Adolf von Baeyer, who suggested that small and large rings might be unstable due to **angle strain**—the strain induced in a molecule when bond angles are forced to deviate from the ideal 109° tetrahedral value. Baeyer based his suggestion on the simple geometric notion that a three-membered ring (cyclopropane) should be an equilateral triangle with bond angles of 60° rather than 109°, a four-membered ring (cyclobutane) should be a square with bond angles of 90°, a five-membered ring should be a regular pentagon with bond angles of 108°, and so on. Continuing this argument, large rings should be strained by having bond angles that are much greater than 109°.

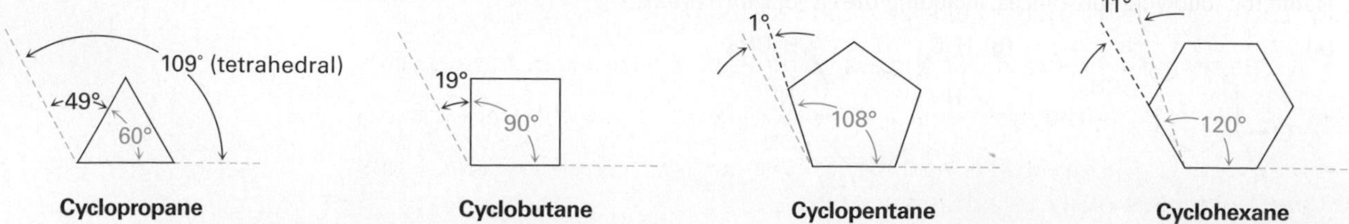

What are the facts? To measure the amount of strain in a compound, we have to measure the total energy of the compound and then subtract the energy of a strain-free reference compound. The difference between the two values should represent the amount of extra energy in the molecule due to strain. The simplest experimental way to do this for a cycloalkane is to measure its *heat of combustion,* the amount of heat released when the compound burns completely with oxygen. The more energy (strain) the compound contains, the more energy (heat) is released on combustion.

$$(CH_2)_n + 3n/2\ O_2 \longrightarrow n\ CO_2 + n\ H_2O + \textbf{Heat}$$

Because the heat of combustion of a cycloalkane depends on size, we need to look at heats of combustion per CH_2 unit. Subtracting a reference value derived from a strain-free acyclic alkane and then multiplying by the number of CH_2 units in the ring gives the overall strain energy. **Figure 4.3** shows the results.

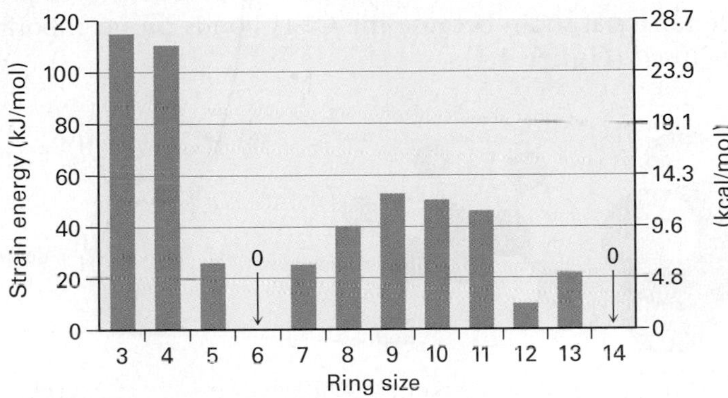

Figure 4.3 Cycloalkane strain energies, calculated by taking the difference between cycloalkane heat of combustion per CH_2 and acyclic alkane heat of combustion per CH_2, and multiplying by the number of CH_2 units in a ring. Small and medium rings are strained, but cyclohexane rings and very large rings are strain-free.

The data in Figure 4.3 show that Baeyer's theory is only partially correct. Cyclopropane and cyclobutane are indeed strained, just as predicted, but cyclopentane is more strained than predicted, and cyclohexane is strain-free. Cycloalkanes of intermediate size have only modest strain, and rings of 14 carbons or more are strain-free. Why is Baeyer's theory wrong?

Baeyer's theory is wrong for the simple reason that he assumed all cycloalkanes to be flat. In fact, as we'll see in the next section, most cycloalkanes are *not* flat; they adopt puckered three-dimensional conformations that allow bond angles to be nearly tetrahedral. As a result, angle strain occurs only in three- and four-membered rings, which have little flexibility. For most ring sizes, particularly the medium-ring (C_7–C_{11}) cycloalkanes, torsional strain caused by H⟷H eclipsing interactions on adjacent carbons **(Section 3.6)** and steric strain caused by the repulsion between nonbonded atoms that approach too closely **(Section 3.7)** are the most important factors. Thus, three kinds of strain contribute to the overall energy of a cycloalkane.

- **Angle strain**—the strain due to expansion or compression of bond angles
- **Torsional strain**—the strain due to eclipsing of bonds on neighboring atoms
- **Steric strain**—the strain due to repulsive interactions when atoms approach each other too closely

Problem 4.8

Each H ⟷ H eclipsing interaction in ethane costs about 4.0 kJ/mol. How many such interactions are present in cyclopropane? What fraction of the overall 115 kJ/mol (27.5 kcal/mol) strain energy of cyclopropane is due to torsional strain?

Problem 4.9

cis-1,2-Dimethylcyclopropane has more strain than *trans*-1,2-dimethylcyclopropane. How can you account for this difference? Which of the two compounds is more stable?

4.4 Conformations of Cycloalkanes

Cyclopropane

Cyclopropane is the most strained of all rings, primarily because of the angle strain caused by its 60° C–C–C bond angles. In addition, cyclopropane has considerable torsional strain because the C–H bonds on neighboring carbon atoms are eclipsed **(Figure 4.4)**.

Figure 4.4 The structure of cyclopropane, showing the eclipsing of neighboring C–H bonds that gives rise to torsional strain. Part **(b)** is a Newman projection along a C–C bond.

How can the hybrid-orbital model of bonding account for the large distortion of bond angles from the normal 109° tetrahedral value to 60° in cyclopropane? The answer is that cyclopropane has *bent bonds*. In an unstrained alkane, maximum bonding is achieved when two atoms have their overlapping orbitals pointing directly toward each other. In cyclopropane, though, the orbitals can't point directly toward each other; rather, they overlap at a slight angle. The result is that cyclopropane bonds are weaker and more reactive than typical alkane bonds—255 kJ/mol (61 kcal/mol) for a C–C bond in cyclopropane versus 370 kJ/mol (88 kcal/mol) for a C–C bond in open-chain propane.

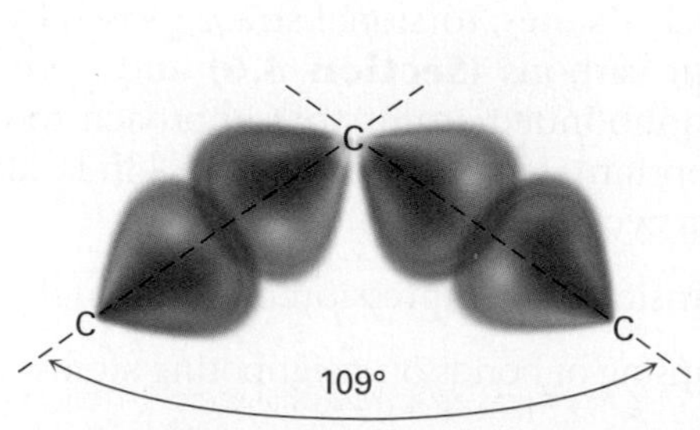

Typical alkane C–C bonds

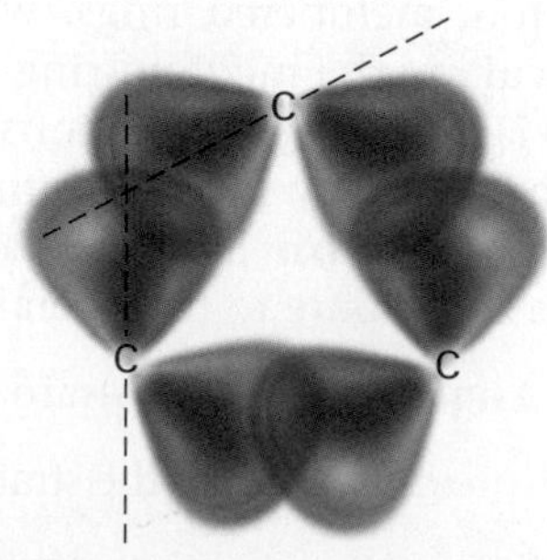

Typical bent cyclopropane C–C bonds

Cyclobutane

Cyclobutane has less angle strain than cyclopropane but has more torsional strain because of its larger number of ring hydrogens. As a result, the total strain for the two compounds is nearly the same—110 kJ/mol (26.4 kcal/mol) for cyclobutane versus 115 kJ/mol (27.5 kcal/mol) for cyclopropane. Cyclobutane is not quite flat but is slightly bent so that one carbon atom lies about 25° above the plane of the other three **(Figure 4.5)**. The effect of this slight bend is to *increase* angle strain but to *decrease* torsional strain, until a minimum-energy balance between the two opposing effects is achieved.

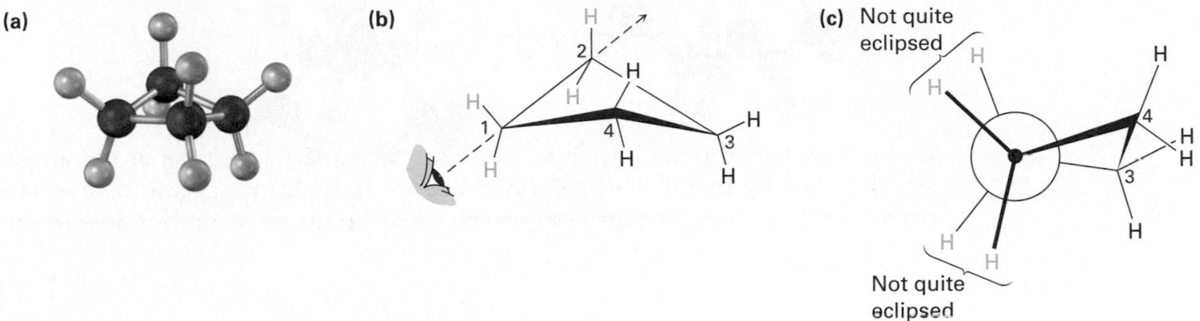

Figure 4.5 The conformation of cyclobutane. Part **(c)** is a Newman projection along a C–C bond, showing that neighboring C–H bonds are not quite eclipsed.

Cyclopentane

Cyclopentane was predicted by Baeyer to be nearly strain-free, but it actually has a total strain energy of 26 kJ/mol (6.2 kcal/mol). Although planar cyclopentane has practically no angle strain, it has a large amount of torsional strain. Cyclopentane therefore twists to adopt a puckered, nonplanar conformation that strikes a balance between increased angle strain and decreased torsional strain. Four of the cyclopentane carbon atoms are in approximately the same plane, with the fifth carbon atom bent out of the plane. Most of the hydrogens are nearly staggered with respect to their neighbors **(Figure 4.6)**.

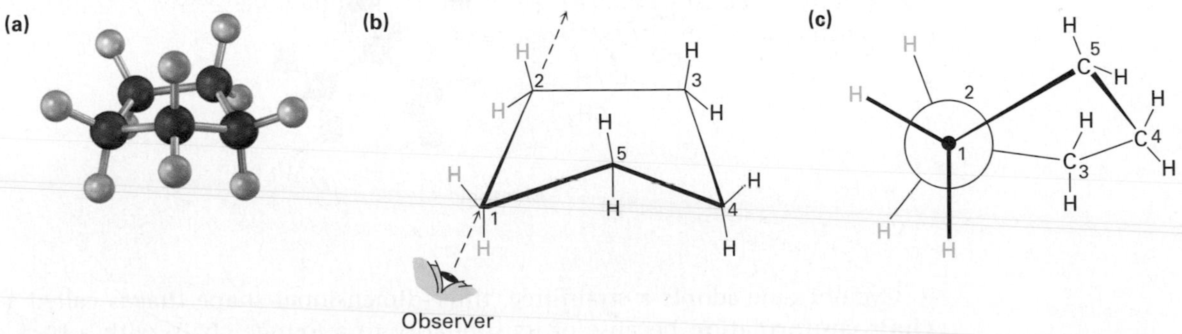

Figure 4.6 The conformation of cyclopentane. Carbons 1, 2, 3, and 4 are nearly planar, but carbon 5 is out of the plane. Part **(c)** is a Newman projection along the C1–C2 bond, showing that neighboring C–H bonds are nearly staggered.

Problem 4.10

How many H⟷H eclipsing interactions would be present if cyclopentane were planar? Assuming an energy cost of 4.0 kJ/mol for each eclipsing interaction, how much torsional strain would planar cyclopentane have? Since the measured total strain of cyclopentane is 26 kJ/mol, how much of the torsional strain is relieved by puckering?

Problem 4.11

Two conformations of *cis*-1,3-dimethylcyclobutane are shown. What is the difference between them, and which do you think is likely to be more stable?

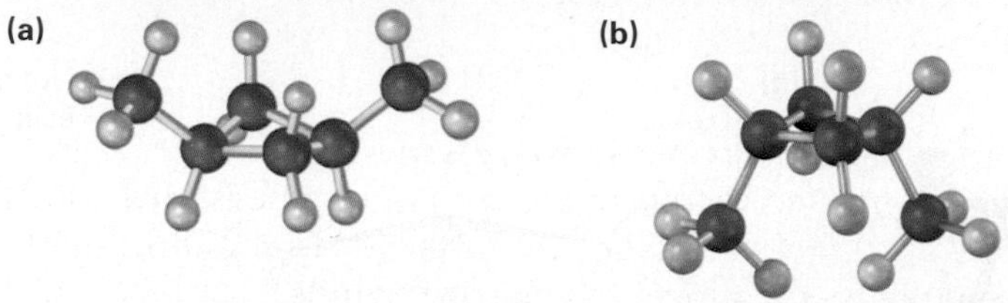

4.5 Conformations of Cyclohexane

Substituted cyclohexanes are the most common cycloalkanes and occur widely in nature. A large number of compounds, including steroids and many pharmaceutical agents, have cyclohexane rings. The flavoring agent menthol, for instance, has three substituents on a six-membered ring.

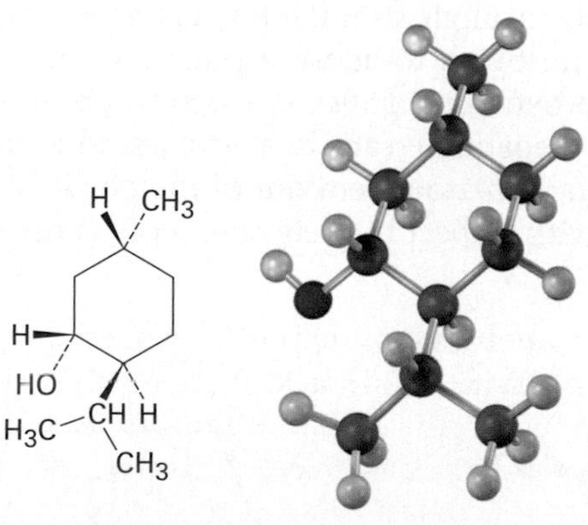

Menthol

Cyclohexane adopts a strain-free, three-dimensional shape that is called a **chair conformation** because of its similarity to a lounge chair, with a back, seat, and footrest **(Figure 4.7)**. Chair cyclohexane has neither angle strain nor torsional strain—all C–C–C bond angles are near the 109.5° tetrahedral value, and all neighboring C–H bonds are staggered.

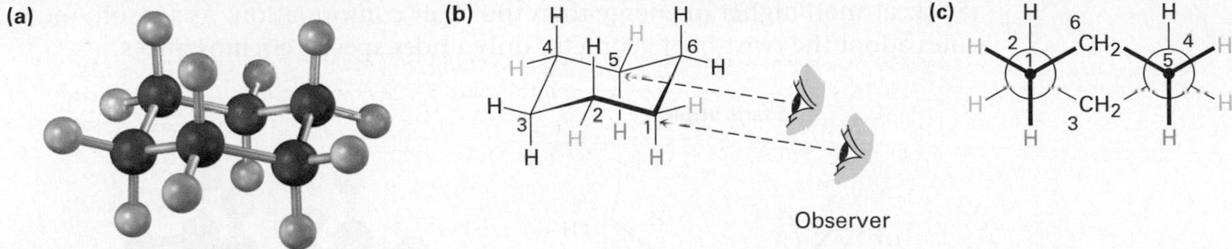

Figure 4.7 The strain-free chair conformation of cyclohexane. All C—C—C bond angles are 111.5°, close to the ideal 109.5° tetrahedral angle, and all neighboring C—H bonds are staggered.

The easiest way to visualize chair cyclohexane is to build a molecular model. (In fact, do it now if you have access to a model kit.) Two-dimensional drawings like that in Figure 4.7 are useful, but there's no substitute for holding, twisting, and turning a three-dimensional model in your own hands.

The chair conformation of cyclohexane can be drawn in three steps.

STEP 1
Draw two parallel lines, slanted downward and slightly offset from each other. This means that four of the cyclohexane carbons lie in a plane.

STEP 2
Place the topmost carbon atom above and to the right of the plane of the other four, and connect the bonds.

STEP 3
Place the bottommost carbon atom below and to the left of the plane of the middle four, and connect the bonds. Note that the bonds to the bottommost carbon atom are parallel to the bonds to the topmost carbon.

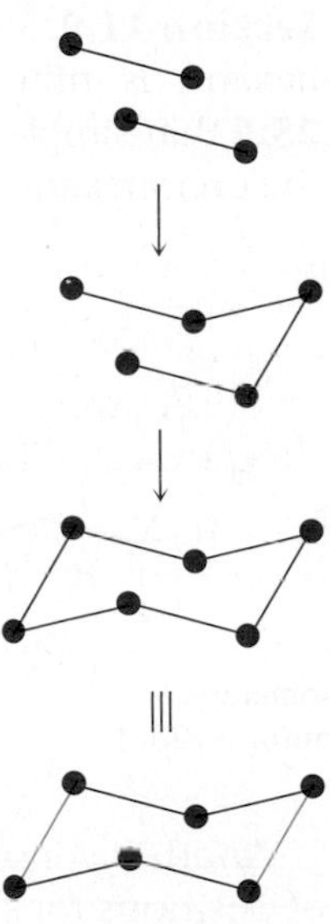

When viewing cyclohexane, it's helpful to remember that the lower bond is in front and the upper bond is in back. If this convention is not defined, an optical illusion can make it appear that the reverse is true. For clarity, all cyclohexane rings drawn in this book will have the front (lower) bond heavily shaded to indicate nearness to the viewer.

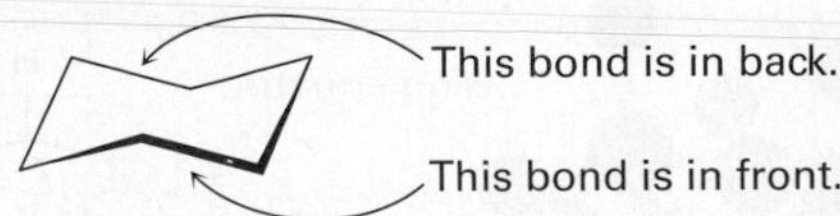

In addition to the chair conformation of cyclohexane, an alternative called the **twist-boat conformation** is also nearly free of angle strain. It does, however, have both steric strain and torsional strain and is about 23 kJ/mol

(5.5 kcal/mol) higher in energy than the chair conformation. As a result, molecules adopt the twist-boat geometry only under special circumstances.

Steric strain

H ↔ H
H ↔ H
Torsional strain

Twist-boat cyclohexane (23 kJ/mol strain)

4.6 Axial and Equatorial Bonds in Cyclohexane

The chair conformation of cyclohexane leads to many consequences. We'll see in **Section 11.9**, for instance, that the chemical behavior of many substituted cyclohexanes is influenced by their conformation. In addition, we'll see in **Section 25.5** that simple carbohydrates, such as glucose, adopt a conformation based on the cyclohexane chair and that their chemistry is directly affected as a result.

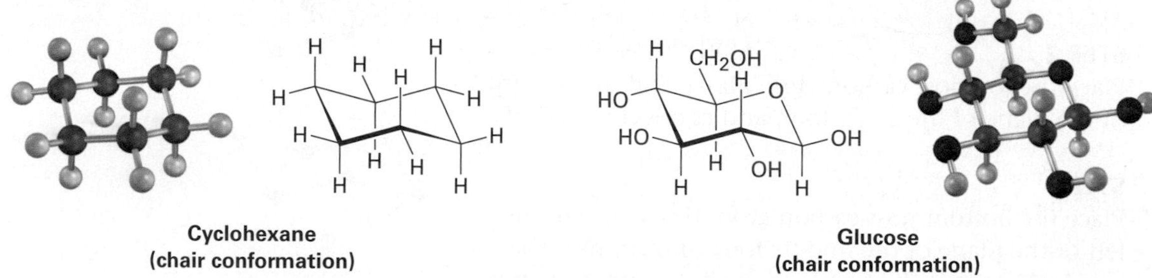

Cyclohexane (chair conformation)

Glucose (chair conformation)

Another consequence of the chair conformation is that there are two kinds of positions for substituents on the cyclohexane ring: *axial* positions and *equatorial* positions **(Figure 4.8)**. The six **axial** positions are perpendicular to the ring, parallel to the ring axis, and the six **equatorial** positions are in the rough plane of the ring, around the ring equator.

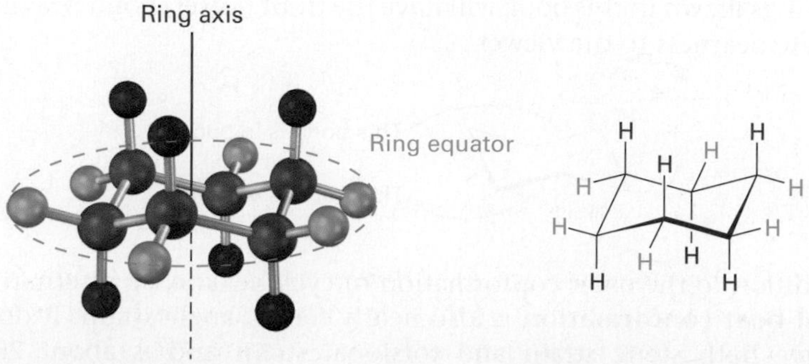

Figure 4.8 **Axial** and equatorial positions in chair cyclohexane. The six axial hydrogens are parallel to the ring axis, and the six equatorial hydrogens are in a band around the ring equator.

As shown in Figure 4.8, each carbon atom in chair cyclohexane has one axial and one equatorial hydrogen. Furthermore, each face of the ring has three axial and three equatorial hydrogens in an alternating arrangement. For example, if the top face of the ring has axial hydrogens on carbons 1, 3, and 5, then it has equatorial hydrogens on carbons 2, 4, and 6. Exactly the reverse is true for the bottom face: carbons 1, 3, and 5 have equatorial hydrogens, but carbons 2, 4, and 6 have axial hydrogens **(Figure 4.9)**.

Figure 4.9 Alternating **axial** and **equatorial** positions in chair cyclohexane, as shown in a view looking directly down the ring axis. Each carbon atom has one axial and one equatorial position, and each face has alternating axial and equatorial positions.

Note that we haven't used the words *cis* and *trans* in this discussion of cyclohexane conformation. Two hydrogens on the same face of the ring are always cis, regardless of whether they're axial or equatorial and regardless of whether they're adjacent. Similarly, two hydrogens on opposite faces of the ring are always trans.

Axial and equatorial bonds can be drawn following the procedure in **Figure 4.10**. Look at a molecular model as you practice.

Axial bonds: The six axial bonds, one on each carbon, are parallel and alternate up–down.

Equatorial bonds: The six equatorial bonds, one on each carbon, come in three sets of two parallel lines. Each set is also parallel to two ring bonds. Equatorial bonds alternate between sides around the ring.

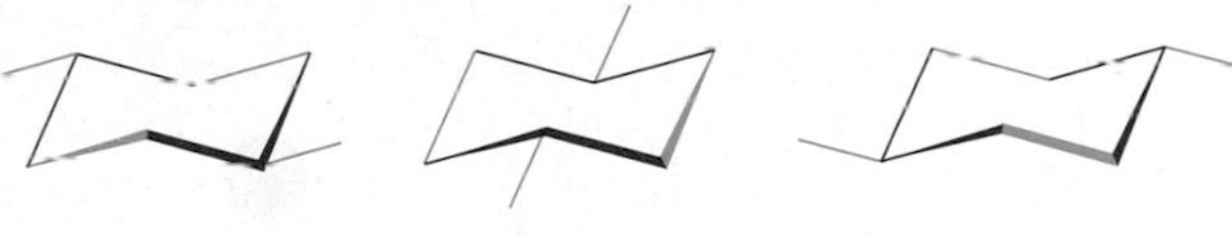

Completed cyclohexane

Figure 4.10 A procedure for drawing axial and equatorial bonds in chair cyclohexane.

Because chair cyclohexane has two kinds of positions—axial and equatorial—we might expect to find two isomeric forms of a monosubstituted cyclohexane. In fact, we don't. There is only *one* methylcyclohexane, *one* bromocyclohexane, *one* cyclohexanol (hydroxycyclohexane), and so on, because cyclohexane rings are *conformationally mobile* at room temperature. Different chair conformations readily interconvert, exchanging axial and equatorial positions. This interconversion, usually called a **ring-flip**, is shown in **Figure 4.11**.

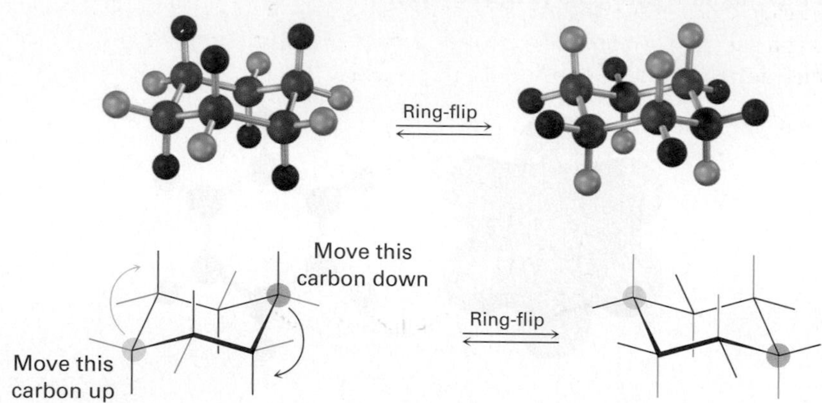

Figure 4.11 A ring-flip in chair cyclohexane interconverts axial and equatorial positions. What is **axial** in the starting structure becomes equatorial in the ring-flipped structure, and what is **equatorial** in the starting structure is axial after ring-flip.

As shown in Figure 4.11, a chair cyclohexane can be ring-flipped by keeping the middle four carbon atoms in place while folding the two end carbons in opposite directions. In so doing, an axial substituent in one chair form becomes an equatorial substituent in the ring-flipped chair form and vice versa. For example, axial bromocyclohexane becomes equatorial bromocyclohexane after ring-flip. Since the energy barrier to chair–chair interconversion is only about 45 kJ/mol (10.8 kcal/mol), the process is rapid at room temperature and we see what appears to be a single structure rather than distinct axial and equatorial isomers.

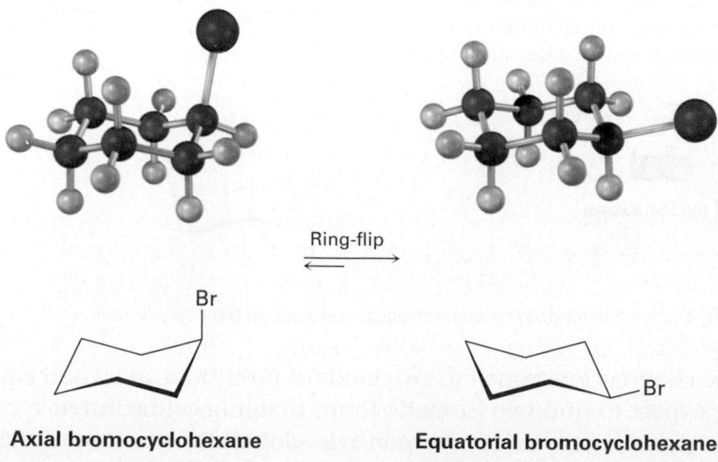

Drawing the Chair Conformation of a Substituted Cyclohexane

Worked Example 4.2

Draw 1,1-dimethylcyclohexane in a chair conformation, indicating which methyl group in your drawing is axial and which is equatorial.

Strategy

Draw a chair cyclohexane ring using the procedure in Figure 4.10, and then put two methyl groups on the same carbon. The methyl group in the rough plane of the ring is equatorial, and the one directly above or below the ring is axial.

Solution

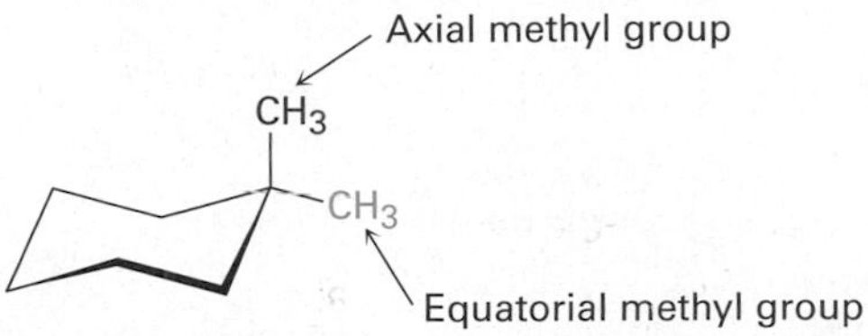

Problem 4.12

Draw two different chair conformations of cyclohexanol (hydroxycyclohexane), showing all hydrogen atoms. Identify each position as axial or equatorial.

Problem 4.13

Draw two different chair conformations of *trans*-1,4-dimethylcyclohexane, and label all positions as axial or equatorial.

Problem 4.14

Identify each of the colored positions—red, blue, and green—as axial or equatorial. Then carry out a ring-flip, and show the new positions occupied by each color.

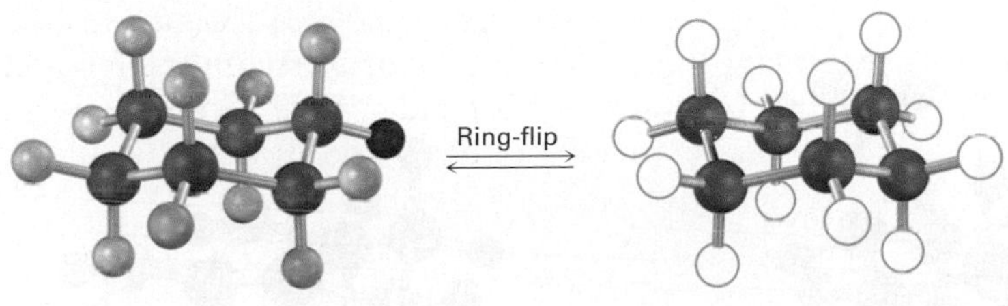

4.7 Conformations of Monosubstituted Cyclohexanes

Even though cyclohexane rings flip rapidly between chair conformations at room temperature, the two conformations of a monosubstituted cyclohexane aren't equally stable. In methylcyclohexane, for instance, the equatorial conformation is more stable than the axial conformation by 7.6 kJ/mol (1.8 kcal/mol). The same is true of other monosubstituted cyclohexanes: a substituent is almost always more stable in an equatorial position than in an axial position.

Key IDEAS

Test your knowledge of Key Ideas by answering end-of-chapter exercises marked with ▲.

You might recall from your general chemistry course that it's possible to calculate the percentages of two isomers at equilibrium using the equation

$\Delta E = -RT \ln K$, where ΔE is the energy difference between isomers, R is the gas constant [8.315 J/(K·mol)], T is the Kelvin temperature, and K is the equilibrium constant between isomers. For example, an energy difference of 7.6 kJ/mol means that about 95% of methylcyclohexane molecules have the methyl group equatorial at any given instant and only 5% have the methyl group axial. **Figure 4.12** plots the relationship between energy and isomer percentages.

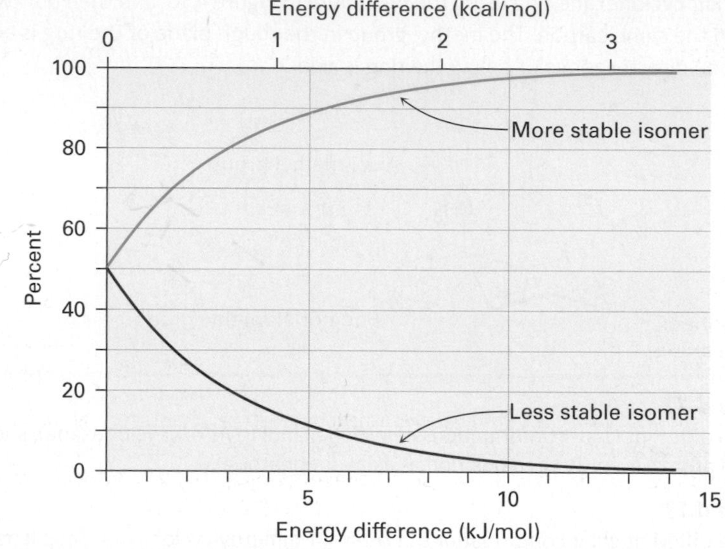

Figure 4.12 A plot of the percentages of two isomers at equilibrium versus the energy difference between them. The curves are calculated using the equation $\Delta E = -RT \ln K$.

The energy difference between axial and equatorial conformations is due to steric strain caused by **1,3-diaxial interactions**. The axial methyl group on C1 is too close to the axial hydrogens three carbons away on C3 and C5, resulting in 7.6 kJ/mol of steric strain **(Figure 4.13)**.

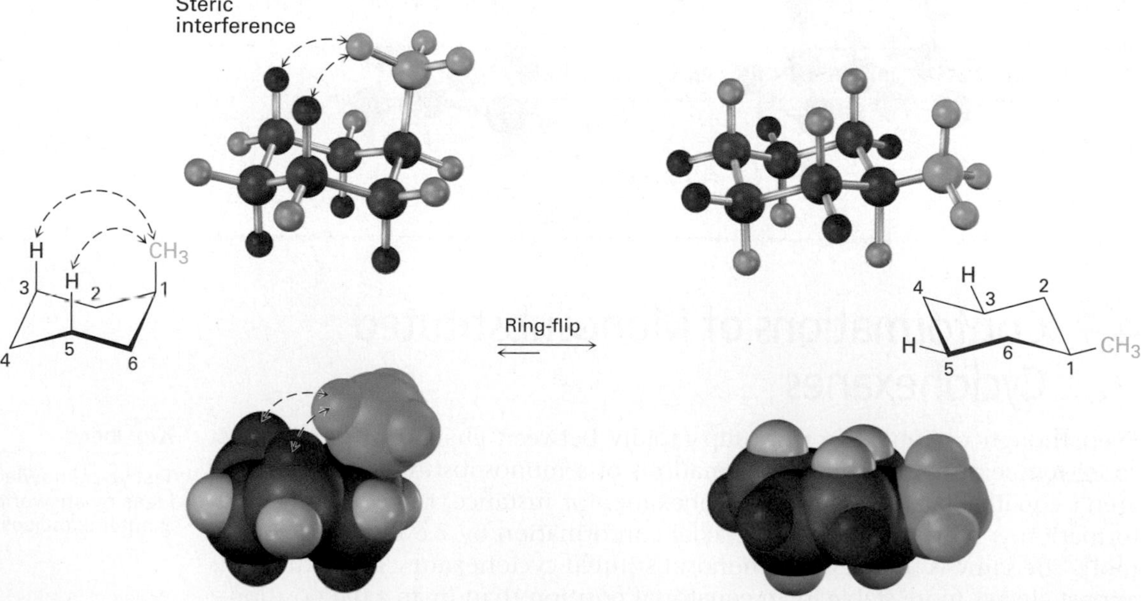

Figure 4.13 Interconversion of axial and equatorial methylcyclohexane, as represented in several formats. The equatorial conformation is more stable than the axial conformation by 7.6 kJ/mol.

The 1,3-diaxial steric strain in substituted methylcyclohexane is already familiar—we saw it previously as the steric strain between methyl groups in gauche butane. Recall from **Section 3.7** that gauche butane is less stable than anti butane by 3.8 kJ/mol (0.9 kcal/mol) because of steric interference between hydrogen atoms on the two methyl groups. Comparing a four-carbon fragment of axial methylcyclohexane with gauche butane shows that the steric interaction is the same in both cases **(Figure 4.14)**. Because axial methylcyclohexane has two such interactions, it has 2 × 3.8 = 7.6 kJ/mol of steric strain. Equatorial methylcyclohexane has no such interactions and is therefore more stable.

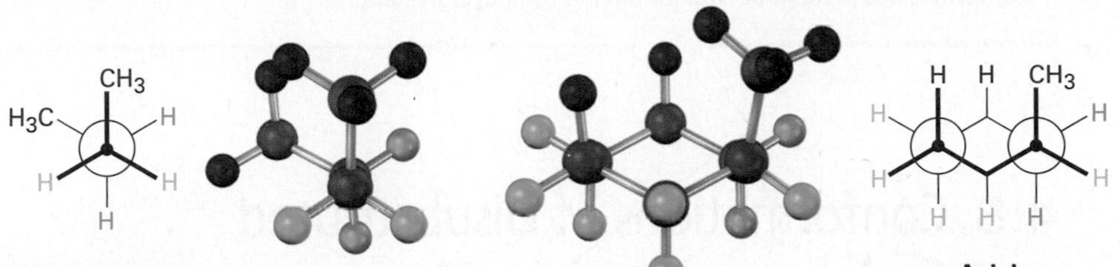

Figure 4.14 The origin of 1,3-diaxial interactions in methylcyclohexane. The steric strain between an **axial methyl group** and **an axial hydrogen atom** three carbons away is identical to the steric strain in gauche butane. Note that the $-CH_3$ group in methylcyclohexane moves slightly away from a true axial position to minimize the strain.

The exact amount of 1,3-diaxial steric strain in a given substituted cyclohexane depends on the nature and size of the substituent, as indicated in Table 4.1. Not surprisingly, the amount of steric strain increases through the series $H_3C- < CH_3CH_2- < (CH_3)_2CH- << (CH_3)_3C-$, paralleling the increasing size of the alkyl groups. Note that the values in Table 4.1 refer to 1,3-diaxial interactions of the substituent with a single hydrogen atom. These values must be doubled to arrive at the amount of strain in a monosubstituted cyclohexane.

Table 4.1 Steric Strain in Monosubstituted Cyclohexanes

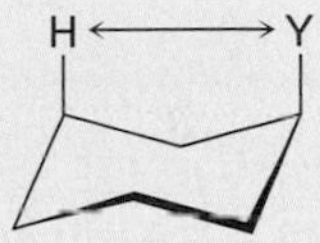

	1,3-Diaxial strain	
Y	(kJ/mol)	(kcal/mol)
F	0.5	0.12
Cl, Br	1.0	0.25
OH	2.1	0.5
CH_3	3.8	0.9
CH_2CH_3	4.0	0.95
$CH(CH_3)_2$	4.6	1.1
$C(CH_3)_3$	11.4	2.7
C_6H_5	6.3	1.5
CO_2H	2.9	0.7
CN	0.4	0.1

Problem 4.15

What is the energy difference between the axial and equatorial conformations of cyclohexanol (hydroxycyclohexane)?

Problem 4.16

Why do you suppose an axial cyano (–CN) substituent causes practically no 1,3-diaxial steric strain (0.4 kJ/mol)? Use molecular models to help with your answer.

Problem 4.17

Look at Figure 4.12 on page 124, and estimate the percentages of axial and equatorial conformations present at equilibrium in bromocyclohexane.

4.8 Conformations of Disubstituted Cyclohexanes

Monosubstituted cyclohexanes are always more stable with their substituent in an equatorial position, but the situation in disubstituted cyclohexanes is more complex because the steric effects of both substituents must be taken into account. All steric interactions in both possible chair conformations must be analyzed before deciding which conformation is favored.

Let's look at 1,2-dimethylcyclohexane as an example. There are two isomers, *cis*-1,2-dimethylcyclohexane and *trans*-1,2-dimethylcyclohexane, which must be considered separately. In the cis isomer, both methyl groups are on the same face of the ring and the compound can exist in either of the two chair conformations shown in **Figure 4.15**. (It may be easier for you to see whether a compound is cis- or trans-disubstituted by first drawing the ring as a flat representation and then converting to a chair conformation.)

***cis*-1,2-Dimethylcyclohexane**

One gauche interaction (3.8 kJ/mol)
Two $CH_3 \leftrightarrow H$ diaxial interactions (7.6 kJ/mol)
Total strain: 3.8 + 7.6 = 11.4 kJ/mol

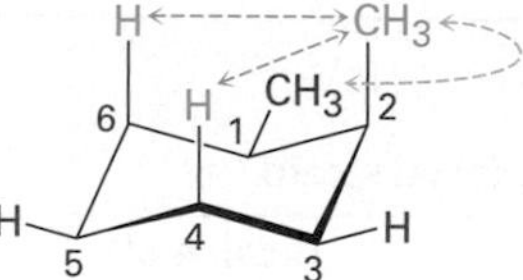

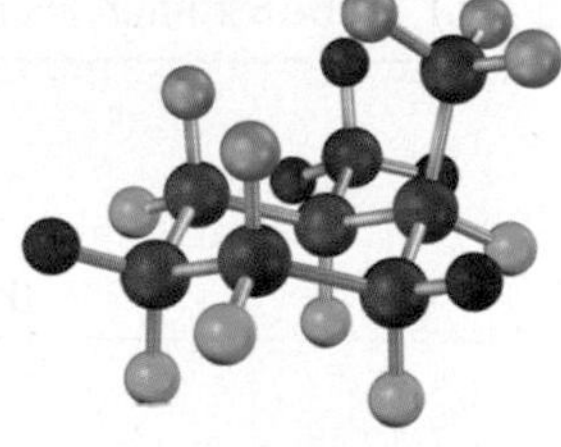

Ring-flip

One gauche interaction (3.8 kJ/mol)
Two $CH_3 \leftrightarrow H$ diaxial interactions (7.6 kJ/mol)
Total strain: 3.8 + 7.6 = 11.4 kJ/mol

H
CH3
H
H
5
6
1
H
3
CH3
4
2

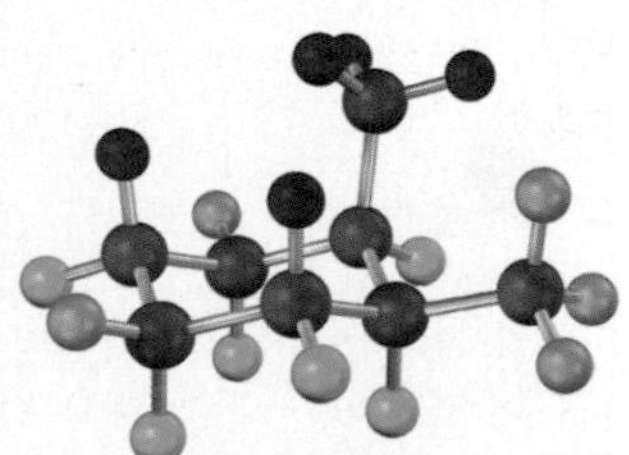

Figure 4.15 Conformations of *cis*-1,2-dimethylcyclohexane. The two chair conformations are equal in energy because each has one axial methyl group and one equatorial methyl group.

Both chair conformations of *cis*-1,2-dimethylcyclohexane have one axial methyl group and one equatorial methyl group. The top conformation in Figure 4.15 has an axial methyl group at C2, which has 1,3 diaxial interactions with hydrogens on C4 and C6. The ring-flipped conformation has an axial methyl group at C1, which has 1,3-diaxial interactions with hydrogens on C3 and C5. In addition, both conformations have gauche butane interactions between the two methyl groups. The two conformations are equal in energy, with a total steric strain of 3 × 3.8 kJ/mol = 11.4 kJ/mol (2.7 kcal/mol).

In *trans*-1,2-dimethylcyclohexane, the two methyl groups are on opposite faces of the ring and the compound can exist in either of the two chair conformations shown in **Figure 4.16**. The situation here is quite different from that of the cis isomer. The top conformation in Figure 4.16 has both methyl groups equatorial and therefore has only a gauche butane interaction between them (3.8 kJ/mol) but no 1,3-diaxial interactions. The ring-flipped conformation, however, has both methyl groups axial. The axial methyl group at C1 interacts with axial hydrogens at C3 and C5, and the axial methyl group at C2 interacts with axial hydrogens at C4 and C6. These four 1,3-diaxial interactions produce a steric strain of 4 × 3.8 kJ/mol = 15.2 kJ/mol and make the diaxial conformation 15.2 − 3.8 = 11.4 kJ/mol less favorable than the diequatorial conformation. We therefore predict that *trans*-1,2-dimethylcyclohexane will exist almost exclusively in the diequatorial conformation.

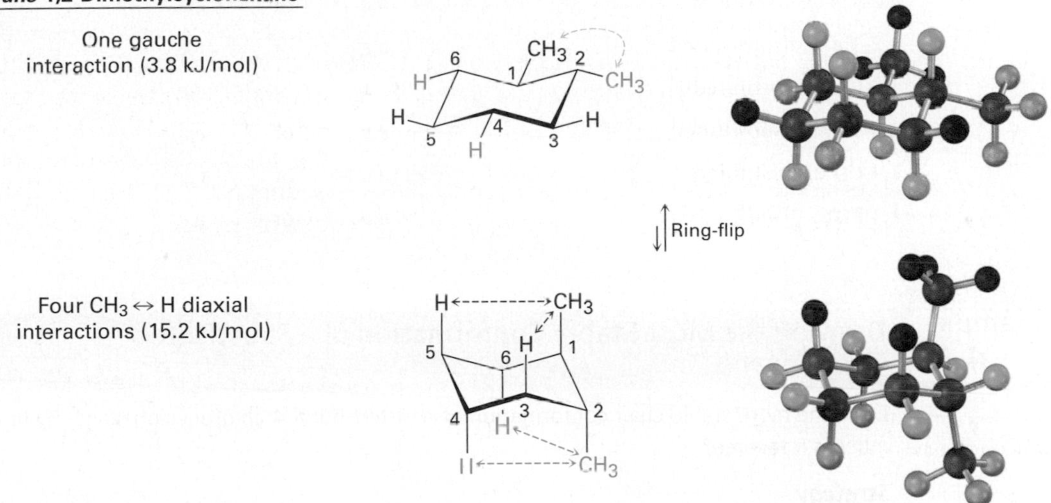

Figure 4.16 Conformations of *trans*-1,2-dimethylcyclohexane. The conformation with both methyl groups equatorial (top) is favored by 11.4 kJ/mol (2.7 kcal/mol) over the conformation with both methyl groups axial (bottom).

The same kind of **conformational analysis** just carried out for *cis*- and *trans*-1,2-dimethylcyclohexane can be done for any substituted cyclohexane, such as *cis*-1-*tert*-butyl-4-chlorocyclohexane (see Worked Example 4.3). As you might imagine, though, the situation becomes more complex as the number of substituents increases. For instance, compare glucose with mannose, a carbohydrate present in seaweed. Which do you think is more strained? In glucose, all substituents on the six-membered ring are equatorial, while in mannose, one of the –OH groups is axial, making mannose more strained.

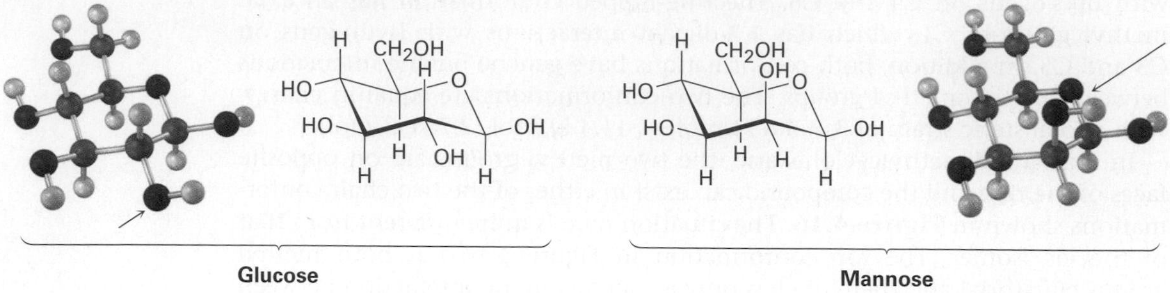

Glucose

Mannose

A summary of the various axial and equatorial relationships among substituent groups in the different possible cis and trans substitution patterns for disubstituted cyclohexanes is given in Table 4.2.

Table 4.2 Axial and Equatorial Relationships in Cis- and Trans-Disubstituted Cyclohexanes

Cis/trans substitution pattern	Axial/equatorial relationships		
1,2-Cis disubstituted	a,e	or	e,a
1,2-Trans disubstituted	a,a	or	e,e
1,3-Cis disubstituted	a,a	or	e,e
1,3-Trans disubstituted	a,e	or	e,a
1,4-Cis disubstituted	a,e	or	e,a
1,4-Trans disubstituted	a,a	or	e,e

Worked Example 4.3 Drawing the Most Stable Conformation of a Substituted Cyclohexane

Draw the more stable chair conformation of *cis*-1-*tert*-butyl-4-chlorocyclohexane. By how much is it favored?

Strategy

Draw the two possible chair conformations, and calculate the strain energy in each. Remember that equatorial substituents cause less strain than axial substituents.

Solution

First draw the two chair conformations of the molecule:

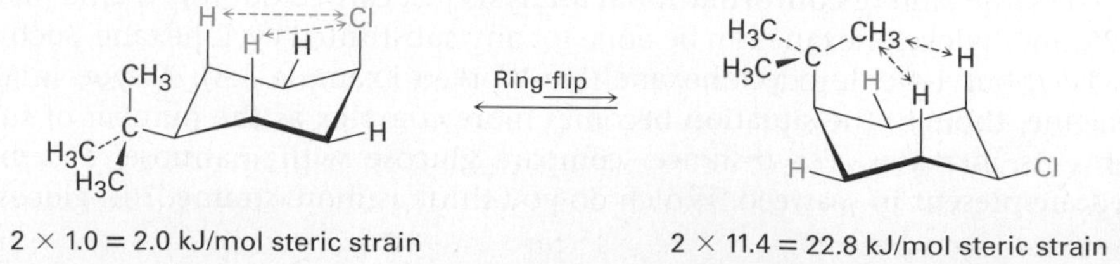

Ring-flip

$2 \times 1.0 = 2.0$ kJ/mol steric strain

$2 \times 11.4 = 22.8$ kJ/mol steric strain

In the conformation on the left, the *tert*-butyl group is equatorial and the chlorine is axial. In the conformation on the right, the *tert*-butyl group is axial and the chlorine is equatorial. These conformations aren't of equal energy because an axial *tert*-butyl substituent and an axial chloro substituent produce different amounts of steric strain. Table 4.1 shows that the 1,3-diaxial interaction between a hydrogen and a *tert*-butyl group costs 11.4 kJ/mol (2.7 kcal/mol), whereas the interaction between a hydrogen and a chlorine costs only 1.0 kJ/mol (0.25 kcal/mol). An axial *tert*-butyl group therefore produces (2 × 11.4 kJ/mol) − (2 × 1.0 kJ/mol) = 20.8 kJ/mol (4.9 kcal/mol) more steric strain than does an axial chlorine, and the compound preferentially adopts the conformation with the chlorine axial and the *tert*-butyl equatorial.

Problem 4.18

Draw the more stable chair conformation of the following molecules, and estimate the amount of strain in each:

(a) *trans*-1-Chloro-3-methylcyclohexane **(b)** *cis*-1-Ethyl-2-methylcyclohexane
(c) *cis*-1-Bromo-4-ethylcyclohexane **(d)** *cis*-1-*tert*-Butyl-4-ethylcyclohexane

Problem 4.19

Identify each substituent in the following compound as axial or equatorial, and tell whether the conformation shown is the more stable or less stable chair form (green = Cl):

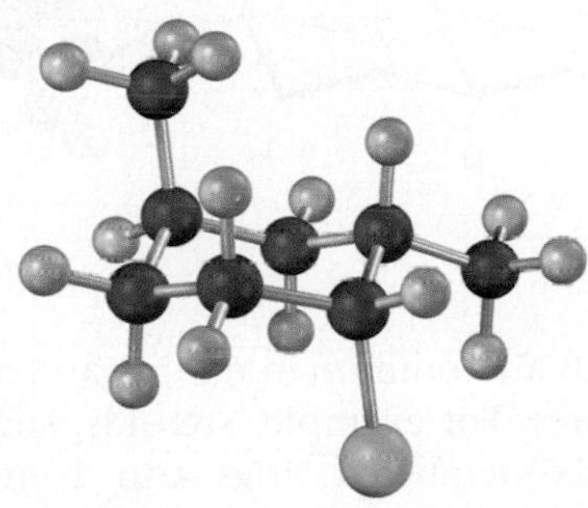

4.9 Conformations of Polycyclic Molecules

The final point we'll consider about cycloalkane stereochemistry is to see what happens when two or more cycloalkane rings are fused together along a common bond to construct a **polycyclic** molecule—for example, decalin.

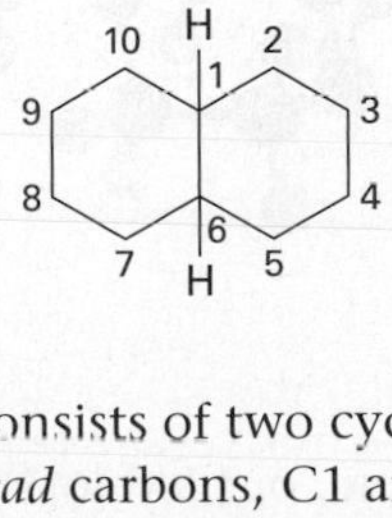

Decalin—two fused cyclohexane rings

Decalin consists of two cyclohexane rings joined to share two carbon atoms (the *bridgehead* carbons, C1 and C6) and a common bond. Decalin can exist in either of two isomeric forms, depending on whether the rings are trans fused or cis fused. In *cis*-decalin, the hydrogen atoms at the bridgehead carbons are on the same face of the rings; in *trans*-decalin, the bridgehead hydrogens are on opposite faces. **Figure 4.17** shows how both compounds can be represented using chair cyclohexane conformations. Note that *cis*- and *trans*-decalin are not interconvertible by ring-flips or other rotations. They are cis–trans stereoisomers and have the same relationship to each other that *cis*- and *trans*-1,2-dimethylcyclohexane have.

Figure 4.17 Representations of *cis*- and *trans*-decalin. The **hydrogen atoms** at the bridgehead carbons are on the same face of the rings in the cis isomer but on opposite faces in the trans isomer.

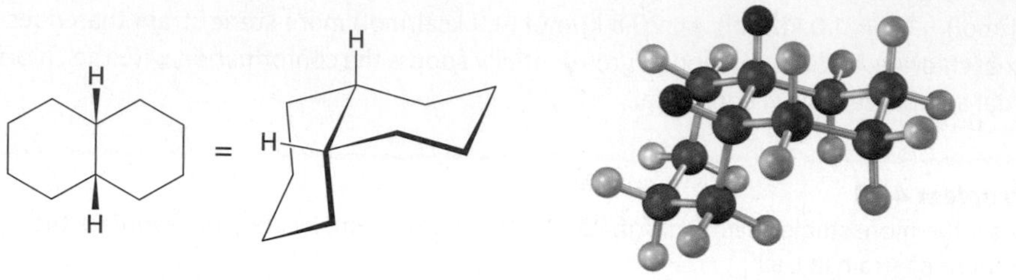

cis-Decalin

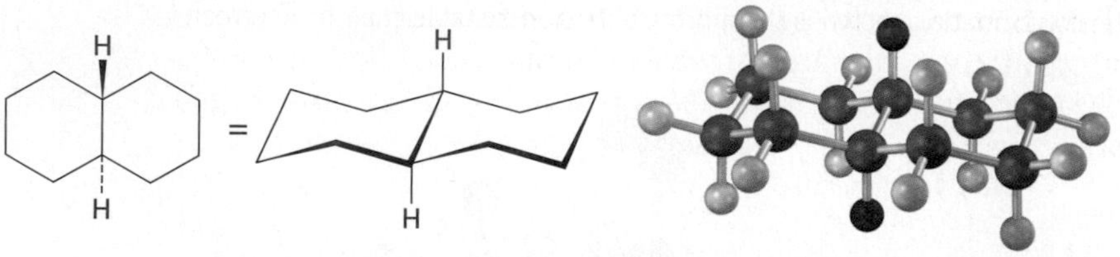

trans-Decalin

Polycyclic compounds are common in nature, and many valuable substances have fused-ring structures. For example, steroids, such as the male hormone testosterone, have 3 six-membered rings and 1 five-membered ring fused together. Although steroids look complicated compared with cyclohexane or decalin, the same principles that apply to the conformational analysis of simple cyclohexane rings apply equally well (and often better) to steroids.

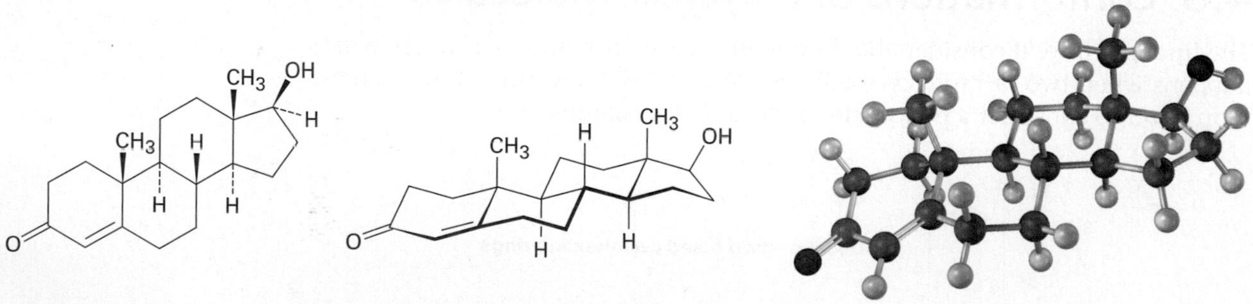

Testosterone (a steroid)

Another common ring system is the norbornane, or bicyclo[2.2.1]heptane, structure. Like decalin, norbornane is a *bicycloalkane,* so called because *two* rings would have to be broken open to generate an acyclic structure. Its systematic name, bicyclo[2.2.1]heptane, reflects the fact that the molecule has seven carbons, is bicyclic, and has three "bridges" of 2, 2, and 1 carbon atoms connecting the two bridgehead carbons.

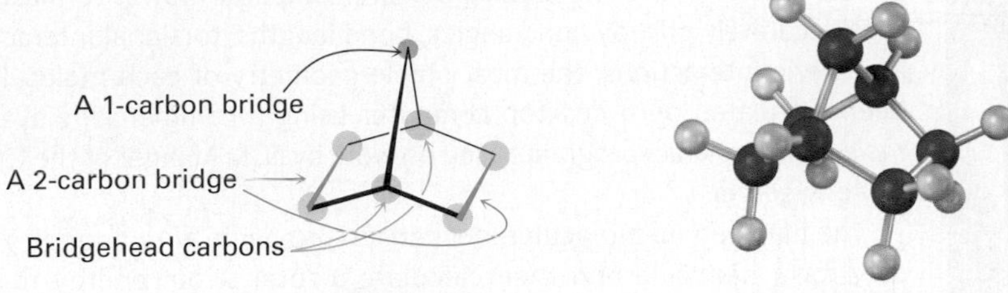

Norbornane (bicyclo[2.2.1]heptane)

Norbornane has a conformationally locked boat cyclohexane ring **(Section 4.5)** in which carbons 1 and 4 are joined by an additional CH_2 group. Note how, in drawing this structure, a break in the rear bond indicates that the vertical bond crosses in front of it. Making a molecular model is particularly helpful when trying to see the three-dimensionality of norbornane.

Substituted norbornanes, such as camphor, are found widely in nature, and many have been important historically in developing organic structural theories.

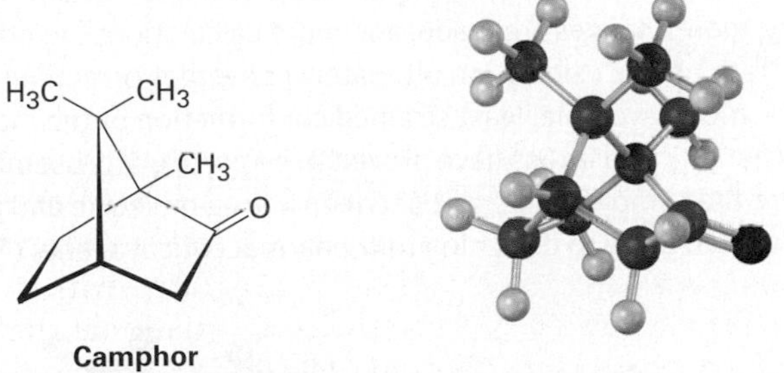

Camphor

Problem 4.20

Which isomer is more stable, *cis*-decalin or *trans*-decalin? Explain.

Problem 4.21

Look at the following structure of the female hormone estrone, and tell whether each of the two indicated ring-fusions is cis or trans.

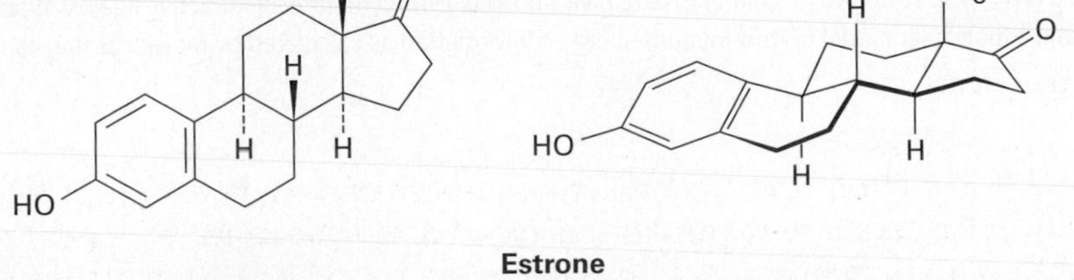

CH3 O H H H HO

CH3 H O HO H H

Estrone

Computer programs make it possible to portray accurate representations of molecular geometry.

Molecular Mechanics | A DEEPER LOOK

All the structural models in this book are computer-drawn. To make sure they accurately portray bond angles, bond lengths, torsional interactions, and steric interactions, the most stable geometry of each molecule has been calculated on a desktop computer using a commercially available *molecular mechanics* program based on work by N. L. Allinger of the University of Georgia.

The idea behind molecular mechanics is to begin with a rough geometry for a molecule and then calculate a total strain energy for that starting geometry, using mathematical equations that assign values to specific kinds of molecular interactions. Bond angles that are too large or too small cause angle strain; bond lengths that are too short or too long cause stretching or compressing strain; unfavorable eclipsing interactions around single bonds cause torsional strain; and nonbonded atoms that approach each other too closely cause steric, or *van der Waals,* strain.

$$E_{\text{total}} = E_{\text{bond stretching}} + E_{\text{angle strain}} + E_{\text{torsional strain}} + E_{\text{van der Waals}}$$

After calculating a total strain energy for the starting geometry, the program automatically changes the geometry slightly in an attempt to lower strain—perhaps by lengthening a bond that is too short or decreasing an angle that is too large. Strain is recalculated for the new geometry, more changes are made, and more calculations are done. After dozens or hundreds of iterations, the calculation ultimately converges on a minimum energy that corresponds to the most favorable, least strained conformation of the molecule.

Molecular mechanics calculations have proven to be particularly useful in pharmaceutical research, where the complementary fit between a drug molecule and a receptor molecule in the body is often a key to designing new pharmaceutical agents **(Figure 4.18)**.

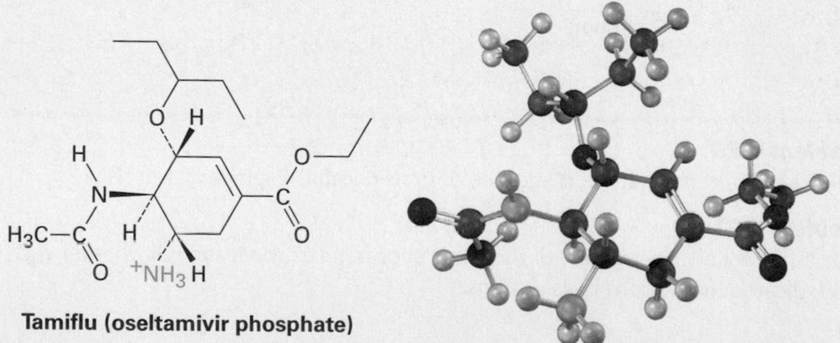

Figure 4.18 The structure of Tamiflu (oseltamivir phosphate), an antiviral agent active against type A influenza, and a molecular model of its minimum-energy conformation as calculated by molecular mechanics.

Summary

Cyclic molecules are so commonly encountered throughout organic and biological chemistry that it's important to understand the consequences of their cyclic structures. Thus, we've taken a close look at cyclic structures in this chapter.

A **cycloalkane** is a saturated cyclic hydrocarbon with the general formula C_nH_{2n}. In contrast to open-chain alkanes, where nearly free rotation occurs around C–C bonds, rotation is greatly reduced in cycloalkanes. Disubstituted cycloalkanes can therefore exist as **cis–trans isomers**. The cis isomer has both substituents on the same face of the ring; the trans isomer has substituents on opposite faces. Cis–trans isomers are just one kind of **stereoisomers**—compounds that have the same connections between atoms but different three-dimensional arrangements.

Not all cycloalkanes are equally stable. Three kinds of strain contribute to the overall energy of a cycloalkane: (1) **angle strain** is the resistance of a bond angle to compression or expansion from the normal 109° tetrahedral value, (2) torsional strain is the energy cost of having neighboring C–H bonds eclipsed rather than staggered, and (3) steric strain is the repulsive interaction that arises when two groups attempt to occupy the same space.

Cyclopropane (115 kJ/mol strain) and cyclobutane (110.4 kJ/mol strain) have both angle strain and torsional strain. Cyclopentane is free of angle strain but has a substantial torsional strain due to its large number of eclipsing interactions. Both cyclobutane and cyclopentane pucker slightly away from planarity to relieve torsional strain.

Cyclohexane is strain-free because it adopts a puckered **chair conformation**, in which all bond angles are near 109° and all neighboring C–H bonds are staggered. Chair cyclohexane has two kinds of positions: **axial** and **equatorial**. Axial positions are oriented up and down, parallel to the ring axis, while equatorial positions lie in a belt around the equator of the ring. Each carbon atom has one axial and one equatorial position.

Chair cyclohexanes are conformationally mobile and can undergo a **ring-flip**, which interconverts axial and equatorial positions. Substituents on the ring are more stable in the equatorial position because axial substituents cause **1,3-diaxial interactions**. The amount of 1,3-diaxial steric strain caused by an axial substituent depends on its size.

Key words

Exercises

OWL Interactive versions of these problems are assignable in OWL for Organic Chemistry.

▲ denotes problems linked to the Key Ideas in this chapter.

Visualizing Chemistry

(Problems 4.1–4.21 appear within the chapter.)

4.22 Name the following cycloalkanes:

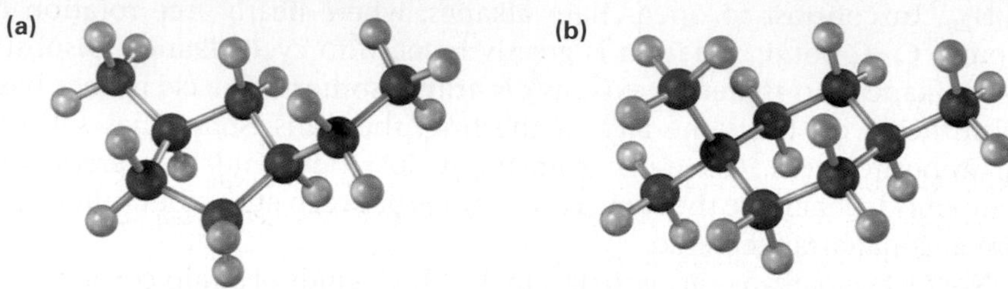

4.23 Name the following compound, identify each substituent as axial or equatorial, and tell whether the conformation shown is the more stable or less stable chair form (green = Cl):

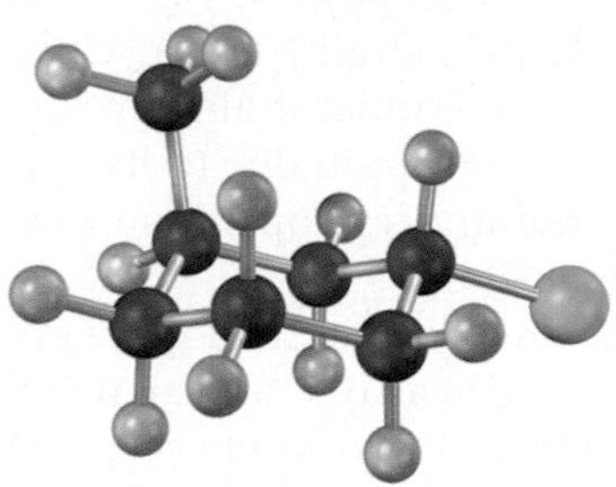

4.24 ▲ A trisubstituted cyclohexane with three substituents—red, green, and blue—undergoes a ring-flip to its alternative chair conformation. Identify each substituent as axial or equatorial, and show the positions occupied by the three substituents in the ring-flipped form.

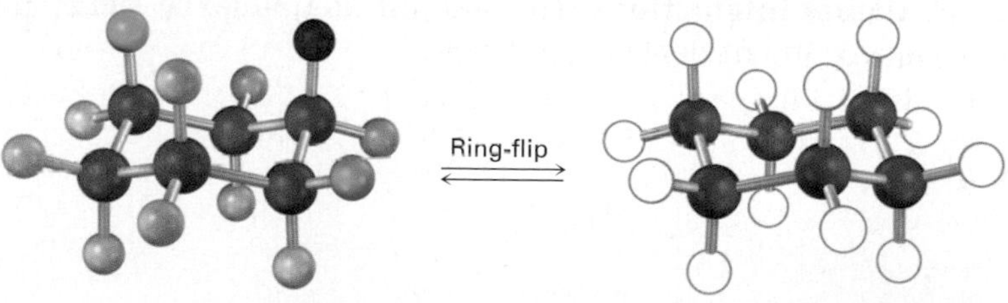

▲ Problems linked to Key Ideas in this chapter

4.25 The following cyclohexane derivative has three substituents—red, green, and blue. Identify each substituent as axial or equatorial, and identify each pair of relationships (red–blue, red–green, and blue–green) as cis or trans.

4.26 Glucose exists in two forms having a 36:64 ratio at equilibrium. Draw a skeletal structure of each, describe the difference between them, and tell which of the two you think is more stable (red = O).

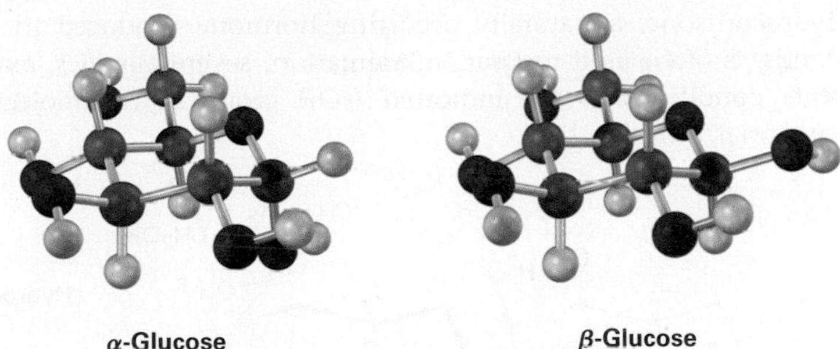

Additional Problems

Cycloalkane Isomers

4.27 Draw the five cycloalkanes with the formula C_5H_{10}.

4.28 Draw two constitutional isomers of *cis*-1,2-dibromocyclopentane.

4.29 Draw a stereoisomer of *trans*-1,3-dimethylcyclobutane.

4.30 Tell whether the following pairs of compounds are identical, constitutional isomers, stereoisomers, or unrelated.

(a) *cis*-1,3-Dibromocyclohexane and *trans*-1,4-dibromocyclohexane

(b) 2,3-Dimethylhexane and 2,3,3-trimethylpentane

(c)

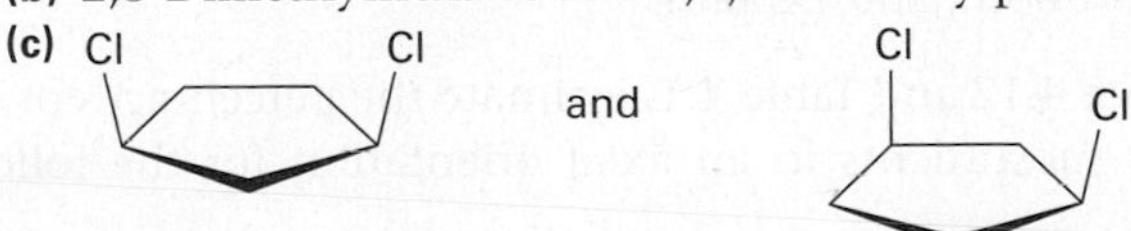

▲ Problems linked to Key Ideas in this chapter

4.31 Draw three isomers of *trans*-1,2-dichlorocyclobutane, and label them as either constitutional isomers or stereoisomers.

4.32 Identify each pair of relationships among the –OH groups in glucose (red–blue, red–green, red–black, blue–green, blue–black, green–black) as cis or trans.

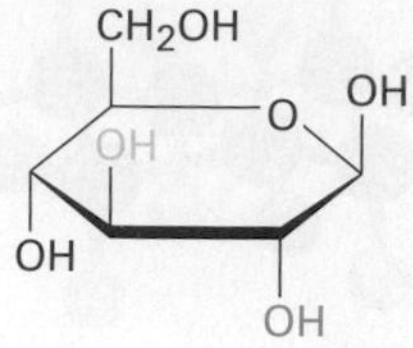

CH_2OH O OH OH OH OH **Glucose**

4.33 ▲ Draw 1,3,5-trimethylcyclohexane using a hexagon to represent the ring. How many cis–trans stereoisomers are possible?

Cycloalkane Conformation and Stability

4.34 Hydrocortisone, a naturally occurring hormone produced in the adrenal glands, is often used to treat inflammation, severe allergies, and numerous other conditions. Is the indicated –OH group in the molecule axial or equatorial?

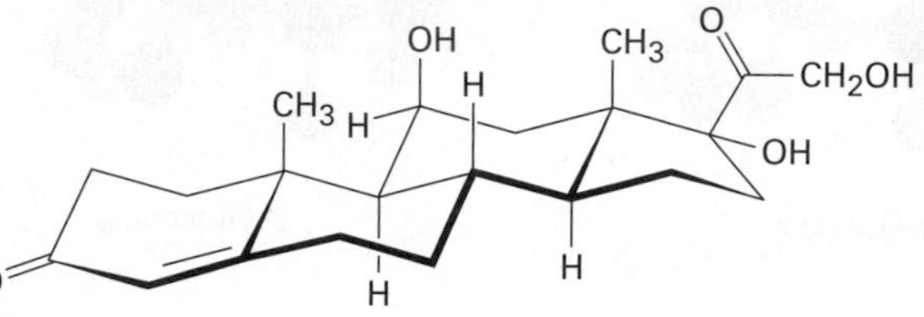

OH CH_3 O H CH_2OH CH_3 H OH O H H **Hydrocortisone**

4.35 A 1,2-cis disubstituted cyclohexane, such as *cis*-1,2-dichlorocyclohexane, must have one group axial and one group equatorial. Explain.

4.36 A 1,2-trans disubstituted cyclohexane must have either both groups axial or both groups equatorial. Explain.

4.37 Why is a 1,3-cis disubstituted cyclohexane more stable than its trans isomer?

4.38 Which is more stable, a 1,4-trans disubstituted cyclohexane or its cis isomer?

4.39 *cis*-1,2-Dimethylcyclobutane is less stable than its trans isomer, but *cis*-1,3-dimethylcyclobutane is more stable than its trans isomer. Draw the most stable conformations of both, and explain.

4.40 From the data in Figure 4.12 and Table 4.1, estimate the percentages of molecules that have their substituents in an axial orientation for the following compounds:

(a) Isopropylcyclohexane
(b) Fluorocyclohexane
(c) Cyclohexanecarbonitrile, $C_6H_{11}CN$

▲ Problems linked to Key Ideas in this chapter

4.41 ▲ Assume that you have a variety of cyclohexanes substituted in the positions indicated. Identify the substituents as either axial or equatorial. For example, a 1,2-cis relationship means that one substituent must be axial and one equatorial, whereas a 1,2-trans relationship means that both substituents are axial or both are equatorial.

(a) 1,3-Trans disubstituted (b) 1,4-Cis disubstituted
(c) 1,3-Cis disubstituted (d) 1,5-Trans disubstituted
(e) 1,5-Cis disubstituted (f) 1,6-Trans disubstituted

Cyclohexane Conformational Analysis

4.42 Draw the two chair conformations of *cis*-1-chloro-2-methylcyclohexane. Which is more stable, and by how much?

4.43 Draw the two chair conformations of *trans*-1-chloro-2-methylcyclohexane. Which is more stable?

4.44 Galactose, a sugar related to glucose, contains a six-membered ring in which all the substituents except the −OH group indicated below in red are equatorial. Draw galactose in its more stable chair conformation.

$HOCH_2$ O OH HO OH OH

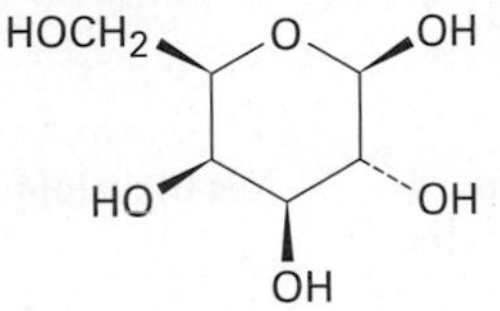

Galactose

4.45 Draw the two chair conformations of menthol, and tell which is more stable.

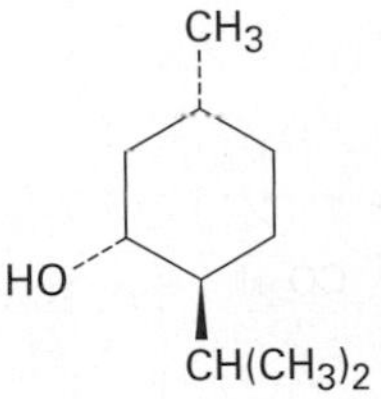

CH_3 HO $CH(CH_3)_2$

Menthol

4.46 There are four cis–trans isomers of menthol (Problem 4.45), including the one shown. Draw the other three.

4.47 ▲ The diaxial conformation of *cis*-1,3-dimethylcyclohexane is approximately 23 kJ/mol (5.4 kcal/mol) less stable than the diequatorial conformation. Draw the two possible chair conformations, and suggest a reason for the large energy difference.

4.48 Approximately how much steric strain does the 1,3-diaxial interaction between the two methyl groups introduce into the diaxial conformation of *cis*-1,3-dimethylcyclohexane? (See Problem 4.47.)

4.49 In light of your answer to Problem 4.48, draw the two chair conformations of 1,1,3-trimethylcyclohexane and estimate the amount of strain energy in each. Which conformation is favored?

▲ Problems linked to Key Ideas in this chapter

4.50 One of the two chair structures of *cis*-1-chloro-3-methylcyclohexane is more stable than the other by 15.5 kJ/mol (3.7 kcal/mol). Which is it? What is the energy cost of a 1,3-diaxial interaction between a chlorine and a methyl group?

General Problems

4.51 We saw in Problem 4.20 that *cis*-decalin is less stable than *trans*-decalin. Assume that the 1,3-diaxial interactions in *cis*-decalin are similar to those in axial methylcyclohexane [that is, one $CH_2 \leftrightarrow H$ interaction costs 3.8 kJ/mol (0.9 kcal/mol)], and calculate the magnitude of the energy difference between *cis*- and *trans*-decalin.

4.52 Using molecular models as well as structural drawings, explain why *trans*-decalin is rigid and cannot ring-flip whereas *cis*-decalin can easily ring-flip.

4.53 *trans*-Decalin is more stable than its cis isomer, but *cis*-bicyclo[4.1.0]heptane is more stable than its trans isomer. Explain.

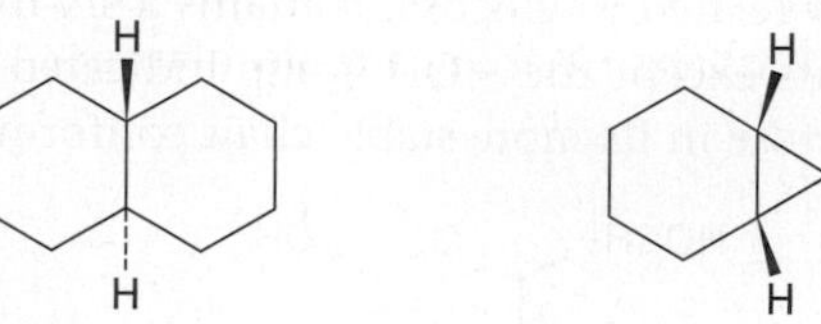

***trans*-Decalin** ***cis*-Bicyclo[4.1.0]heptane**

4.54 As mentioned in Problem 3.53, the statin drugs, such as simvastatin (Zocor), pravastatin (Pravachol), and atorvastatin (Lipitor) are the most widely prescribed drugs in the world.

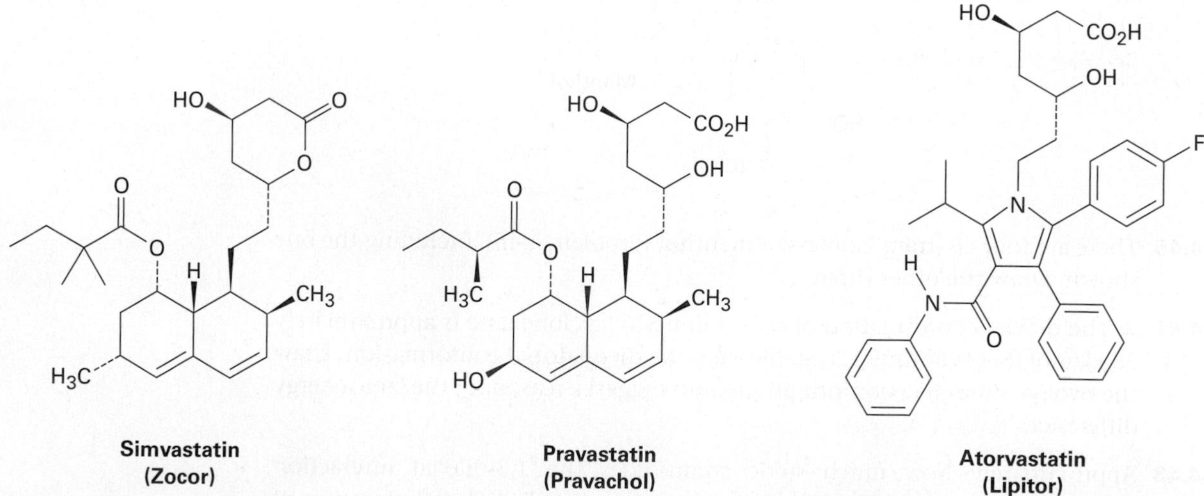

Simvastatin (Zocor) **Pravastatin (Pravachol)** **Atorvastatin (Lipitor)**

(a) Are the two indicated bonds on simvastatin cis or trans?
(b) What are the cis/trans relationships among the three indicated bonds on pravastatin?
(c) Why can't the three indicated bonds on atorvastatin be identified as cis or trans?

4.55 ▲ *myo*-Inositol, one of the isomers of 1,2,3,4,5,6-hexahydroxycyclohexane, acts as a growth factor in both animals and microorganisms. Draw the most stable chair conformation of *myo*-inositol.

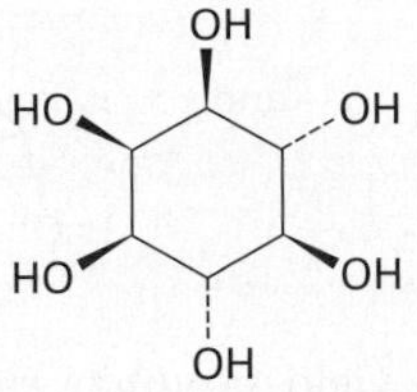

myo-Inositol

4.56 How many cis–trans stereoisomers of *myo*-inositol (Problem 4.55) are there? Draw the structure of the most stable isomer.

4.57 The German chemist J. Bredt proposed in 1935 that bicycloalkenes such as 1-norbornene, which have a double bond to the bridgehead carbon, are too strained to exist. Explain. (Making a molecular model will be helpful.)

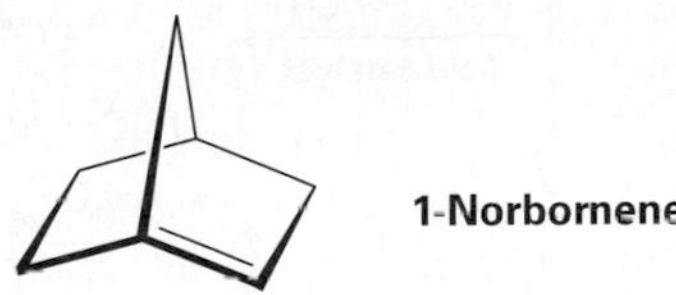

1-Norbornene

4.58 Tell whether each of the following substituents on a steroid is axial or equatorial. (A substituent that is "up" is on the top face of the molecule as drawn, and a substituent that is "down" is on the bottom face.)

(a) Substituent up at C3

(b) Substituent down at C7

(c) Substituent down at C11

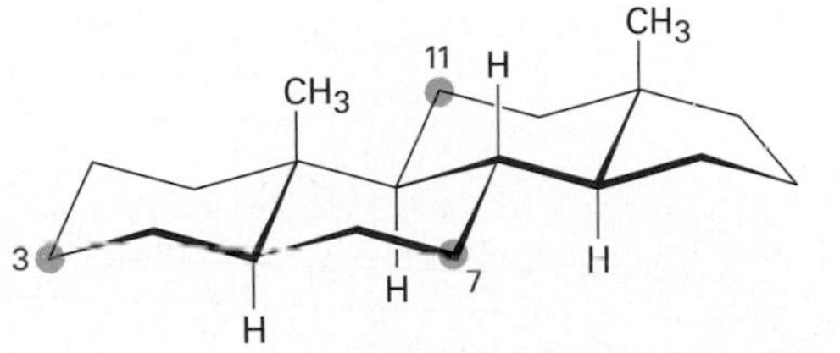

4.59 Amantadine is an antiviral agent that is active against influenza type A infection. Draw a three-dimensional representation of amantadine, showing the chair cyclohexane rings.

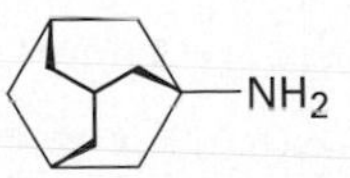

Amantadine

▲ Problems linked to Key Ideas in this chapter

4.60 Here's a difficult one. There are two different substances named *trans*-1,2-dimethylcyclopentane. What is the relationship between them? (We'll explore this kind of isomerism in the next chapter.)

4.61 Ketones react with alcohols to yield products called *acetals*. Why does the all-cis isomer of 4-*tert*-butyl-1,3-cyclohexanediol react readily with acetone and an acid catalyst to form an acetal, but other stereoisomers do not react? In formulating your answer, draw the more stable chair conformations of all four stereoisomers and the product acetal from each.

H C(CH3)3 H HO HO H — H_3C–CO–CH_3, Acid catalyst → H C(CH3)3 H O H3C H3C O H + H_2O

An acetal

4.62 Alcohols undergo an *oxidation* reaction to yield carbonyl compounds on treatment with CrO_3. For example, 2-*tert*-butylcyclohexanol gives 2-*tert*-butylcyclohexanone. If axial −OH groups are generally more reactive than their equatorial isomers, which do you think reacts faster, the cis isomer of 2-*tert*-butylcyclohexanol or the trans isomer? Explain.

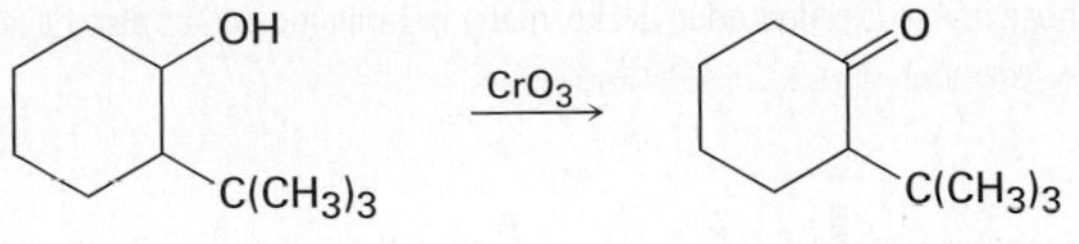

OH
CrO_3
O
$C(CH_3)_3$
$C(CH_3)_3$

2-*tert*-Butylcyclohexanol **2-*tert*-Butylcyclohexanone**

▲ Problems linked to Key Ideas in this chapter

5

Like the mountain whose image is reflected in a lake, many organic molecules also have mirror-image counterparts.

Stereochemistry at Tetrahedral Centers

Are you right-handed or left-handed? You may not spend much time thinking about it, but handedness plays a surprisingly large role in your daily activities. Many musical instruments, such as oboes and clarinets, have a handedness to them; the last available softball glove always fits the wrong hand; left-handed people write in a "funny" way. The reason for these difficulties is that our hands aren't identical; rather, they're *mirror images*. When you hold a *left* hand up to a mirror, the image you see looks like a *right* hand. Try it.

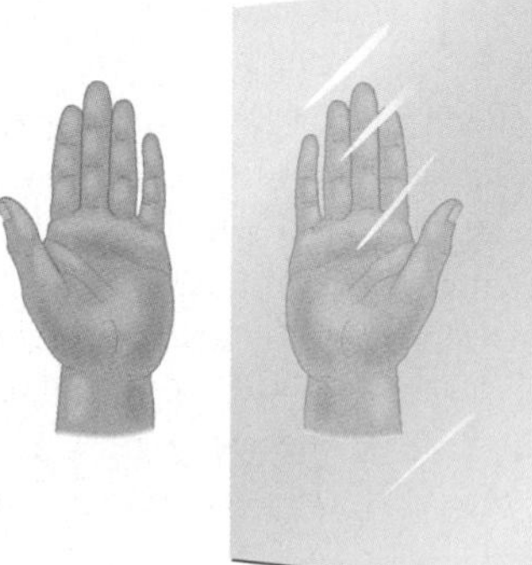

Left hand Right hand

Handedness is also important in organic and biological chemistry, where it arises primarily as a consequence of the tetrahedral stereochemistry of sp^3-hybridized carbon atoms. Many drugs and almost all the molecules in our bodies—amino acids, carbohydrates, nucleic acids, and many more—are handed. Furthermore, molecular handedness makes possible the precise interactions between enzymes and their substrates that are involved in the hundreds of thousands of chemical reactions on which life is based.

Why This Chapter? Understanding the causes and consequences of molecular handedness is crucial to understanding organic and biological chemistry. The subject can be a bit complex at first, but the material covered in this chapter nevertheless forms the basis for much of the remainder of the book.

OWL Sign in to OWL for Organic Chemistry at **www.cengage.com/owl** to view tutorials and simulations, develop problem-solving skills, and complete online homework assigned by your professor.

5.1 Enantiomers and the Tetrahedral Carbon

What causes molecular handedness? Look at generalized molecules of the type CH_3X, CH_2XY, and CHXYZ shown in **Figure 5.1**. On the left are three molecules, and on the right are their images reflected in a mirror. The CH_3X and CH_2XY molecules are identical to their mirror images and thus are not handed. If you make a molecular model of each molecule and its mirror image, you find that you can superimpose one on the other so that all atoms coincide. The CHXYZ molecule, by contrast, is *not* identical to its mirror image. You can't superimpose a model of the molecule on a model of its mirror image for the same reason that you can't superimpose a left hand on a right hand: they simply aren't the same.

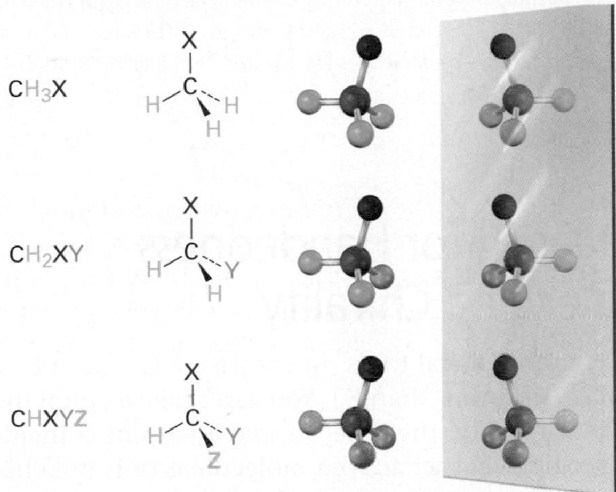

Figure 5.1 Tetrahedral carbon atoms and their mirror images. Molecules of the type CH_3X and CH_2XY are identical to their mirror images, but a molecule of the type CHXYZ is not. A CHXYZ molecule is related to its mirror image in the same way that a right hand is related to a left hand.

Molecules that are not identical to their mirror images are kinds of stereoisomers called **enantiomers** (Greek *enantio,* meaning "opposite"). Enantiomers are related to each other as a right hand is related to a left hand and result whenever a tetrahedral carbon is bonded to four different substituents (one need not be H). For example, lactic acid (2-hydroxypropanoic acid) exists as a pair of enantiomers because there are four different groups (–H, –OH, $-CH_3$, $-CO_2H$) bonded to the central carbon atom. The enantiomers are called (+)-lactic acid and (−)-lactic acid. Both are found in sour milk, but only the (+) enantiomer occurs in muscle tissue.

$CH_3-C(H)(OH)-CO_2H$ X–C(H)(Y)–Z

Lactic acid: a molecule of general formula CHXYZ

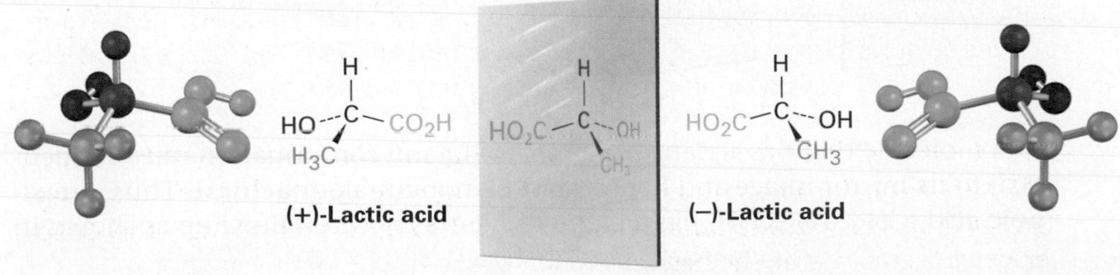

No matter how hard you try, you can't superimpose a molecule of (+)-lactic acid on a molecule of (−)-lactic acid. If any two groups match up, say $-H$ and $-CO_2H$, the remaining two groups don't match **(Figure 5.2)**.

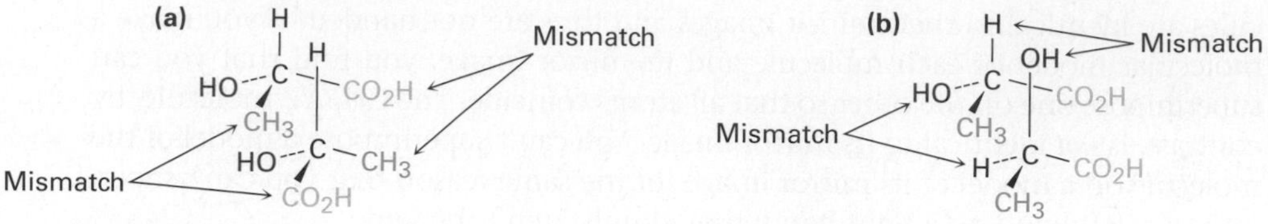

Figure 5.2 Attempts at superimposing the mirror-image forms of lactic acid. **(a)** When the $-H$ and $-OH$ substituents match up, the $-CO_2H$ and $-CH_3$ substituents don't; **(b)** when $-CO_2H$ and $-CH_3$ match up, $-H$ and $-OH$ don't. Regardless of how the molecules are oriented, they aren't identical.

5.2 The Reason for Handedness in Molecules: Chirality

Key **IDEAS**

Test your knowledge of Key Ideas by answering end-of-chapter exercises marked with ▲.

A molecule that is not identical to its mirror image is said to be **chiral** (**ky**-ral, from the Greek *cheir,* meaning "hand"). You can't take a chiral molecule and its enantiomer and place one on the other so that all atoms coincide.

How can you predict whether a given molecule is or is not chiral? *A molecule is not chiral if it has a plane of symmetry.* A plane of symmetry is a plane that cuts through the middle of a molecule (or any object) in such a way that one half of the molecule or object is a mirror image of the other half. A laboratory flask, for example, has a plane of symmetry. If you were to cut the flask in half, one half would be a mirror image of the other half. A hand, however, does not have a plane of symmetry. One "half" of a hand is not a mirror image of the other half **(Figure 5.3)**.

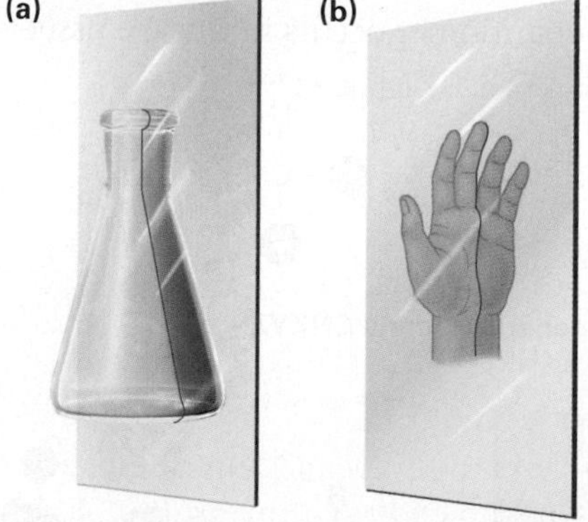

Figure 5.3 The meaning of *symmetry plane.* **(a)** An object like the flask has a symmetry plane cutting through it so that right and left halves are mirror images. **(b)** An object like a hand has no symmetry plane; the right "half" of a hand is not a mirror image of the left half.

A molecule that has a plane of symmetry in any conformation must be identical to its mirror image and hence must be nonchiral, or **achiral**. Thus, propanoic acid, $CH_3CH_2CO_2H$, has a plane of symmetry when lined up as shown in

Figure 5.4 and is achiral, while lactic acid, $CH_3CH(OH)CO_2H$, has no plane of symmetry in any conformation and is chiral.

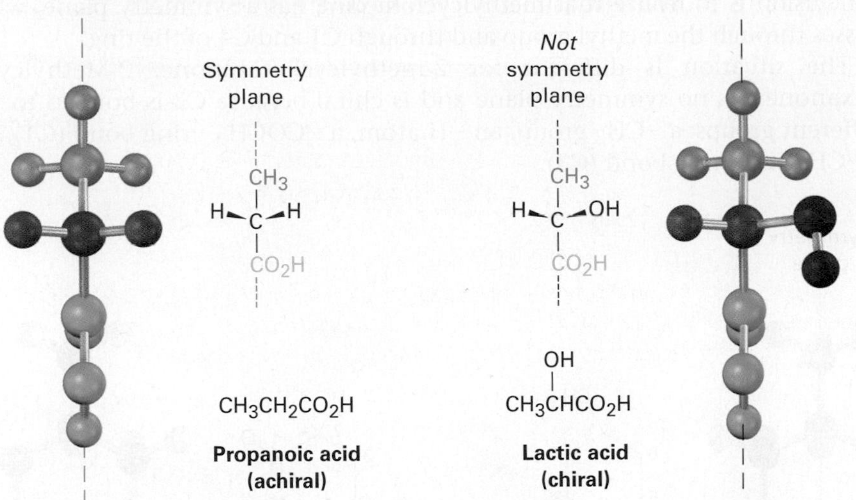

Figure 5.4 The achiral propanoic acid molecule versus the chiral lactic acid molecule. Propanoic acid has a plane of symmetry that makes one side of the molecule a mirror image of the other side. Lactic acid has no such symmetry plane.

The most common, although not the only, cause of chirality in an organic molecule is the presence of a tetrahedral carbon atom bonded to four different groups—for example, the central carbon atom in lactic acid. Such carbons are referred to as **chirality centers**, although other terms such as *stereocenter, asymmetric center,* and *stereogenic center* have also been used. Note that *chirality* is a property of the entire molecule, whereas a chirality *center* is the *cause* of chirality.

Detecting a chirality center in a complex molecule takes practice because it's not always immediately apparent that four different groups are bonded to a given carbon. The differences don't necessarily appear right next to the chirality center. For example, 5-bromodecane is a chiral molecule because four different groups are bonded to C5, the chirality center (marked with an asterisk). A butyl substituent is similar to a pentyl substituent, but it isn't identical. The difference isn't apparent until four carbon atoms away from the chirality center, but there's still a difference.

$$CH_3CH_2CH_2CH_2CH_2\overset{Br}{\underset{H}{C^*}}CH_2CH_2CH_2CH_3$$

5-Bromodecane (chiral)

Substituents on carbon 5

—H

—Br

$-CH_2CH_2CH_2CH_3$ (butyl)

$-CH_2CH_2CH_2CH_2CH_3$ (pentyl)

As other possible examples, look at methylcyclohexane and 2-methylcyclohexanone. Methylcyclohexane is achiral because no carbon atom in the molecule is bonded to four different groups. You can immediately eliminate all $-CH_2-$ carbons and the $-CH_3$ carbon from consideration, but what about C1 on the ring? The C1 carbon atom is bonded to a $-CH_3$ group, to an –H atom, and to C2 and C6 of the ring. Carbons 2 and 6 are equivalent, however, as are carbons

3 and 5. Thus, the C6–C5–C4 "substituent" is equivalent to the C2–C3–C4 substituent, and methylcyclohexane is achiral. Another way of reaching the same conclusion is to realize that methylcyclohexane has a symmetry plane, which passes through the methyl group and through C1 and C4 of the ring.

The situation is different for 2-methylcyclohexanone. 2-Methylcyclohexanone has no symmetry plane and is chiral because C2 is bonded to four different groups: a $-CH_3$ group, an $-H$ atom, a $-COCH_2-$ ring bond (C1), and a $-CH_2CH_2-$ ring bond (C3).

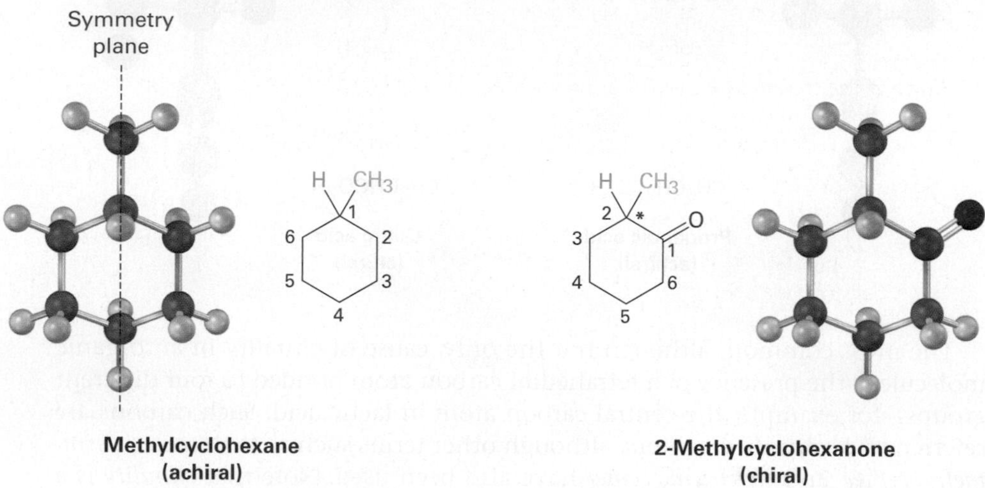

Methylcyclohexane (achiral)

2-Methylcyclohexanone (chiral)

Several more examples of chiral molecules are shown below. Check for yourself that the labeled carbons are chirality centers. You might note that carbons in $-CH_2-$, $-CH_3$, C=O, C=C, and C≡C groups can't be chirality centers. (Why not?)

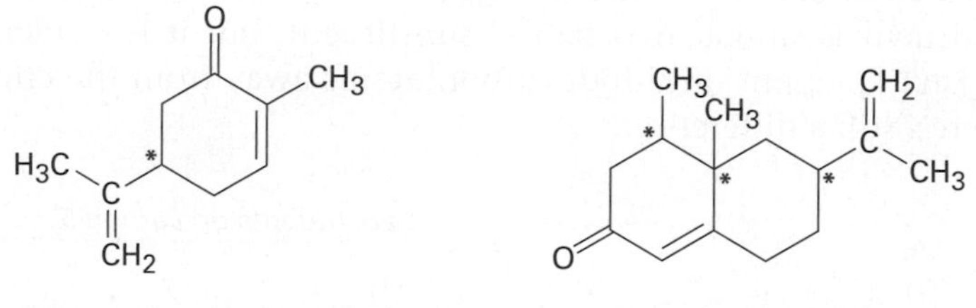

Carvone (spearmint oil)

Nootkatone (grapefruit oil)

Worked Example 5.1 Drawing the Three-Dimensional Structure of a Chiral Molecule

Draw the structure of a chiral alcohol.

Strategy

An alcohol is a compound that contains the $-OH$ functional group. To make an alcohol chiral, we need to have four different groups bonded to a single carbon atom, say $-H$, $-OH$, $-CH_3$, and $-CH_2CH_3$.

Solution

$$\begin{array}{c} OH \\ | \\ CH_3CH_2-\overset{}{C^*}-CH_3 \\ | \\ H \end{array}$$

2 Butanol (chiral)

Problem 5.1

Which of the following objects are chiral?

(a) Soda can **(b)** Screwdriver **(c)** Screw **(d)** Shoe

Problem 5.2

Which of the following molecules are chiral? Identify the chirality center(s) in each.

(a) $CH_2CH_2CH_3$, N, H

Coniine (poison hemlock)

(b) H CH_3, HO, H, H

Menthol (flavoring agent)

(c) CH_3O, N—CH_3, H, H

Dextromethorphan (cough suppressant)

Problem 5.3

Alanine, an amino acid found in proteins, is chiral. Draw the two enantiomers of alanine using the standard convention of solid, wedged, and dashed lines.

$$\begin{array}{c} NH_2 \\ | \\ CH_3CHCO_2H \end{array}$$

Alanine

Problem 5.4

Identify the chirality centers in the following molecules (green = Cl, yellow-green = F):

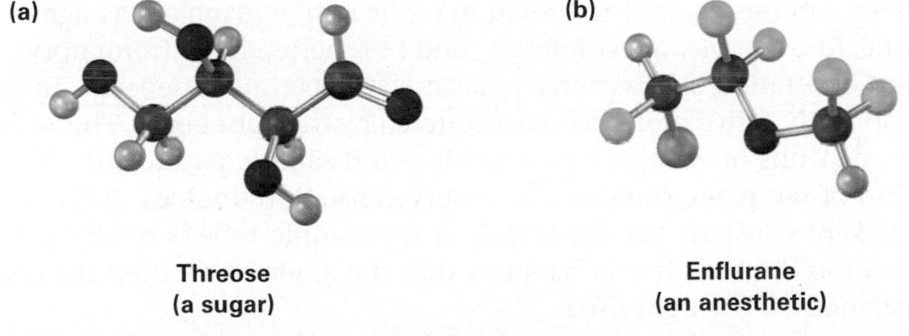

Threose (a sugar)

Enflurane (an anesthetic)

5.3 Optical Activity

The study of chirality originated in the early 19th century during investigations by the French physicist Jean-Baptiste Biot into the nature of *plane-polarized light*. A beam of ordinary light consists of electromagnetic waves that oscillate in an infinite number of planes at right angles to the direction of light travel. When a beam of ordinary light passes through a device called a *polarizer,* however, only the light waves oscillating in a single plane pass through and the light is said to be plane-polarized. Light waves in all other planes are blocked out.

Biot made the remarkable observation that when a beam of plane-polarized light passes through a solution of certain organic molecules, such as sugar or camphor, the plane of polarization is *rotated* through an angle, α. Not all organic substances exhibit this property, but those that do are said to be **optically active**.

The angle of rotation can be measured with an instrument called a *polarimeter,* represented in **Figure 5.5**. A solution of optically active organic molecules is placed in a sample tube, plane-polarized light is passed through the tube, and rotation of the polarization plane occurs. The light then goes through a second polarizer called the *analyzer*. By rotating the analyzer until the light passes through *it,* we can find the new plane of polarization and can tell to what extent rotation has occurred.

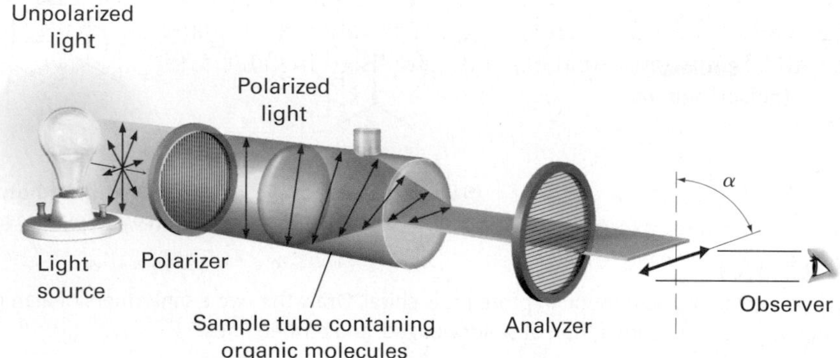

Figure 5.5 Schematic representation of a polarimeter. Plane-polarized light passes through a solution of optically active molecules, which rotate the plane of polarization.

In addition to determining the extent of rotation, we can also find the direction. From the vantage point of the observer looking directly at the analyzer, some optically active molecules rotate polarized light to the left (counterclockwise) and are said to be **levorotatory**, whereas others rotate polarized light to the right (clockwise) and are said to be **dextrorotatory**. By convention, rotation to the left is given a minus sign (−) and rotation to the right is given a plus sign (+). (−)-Morphine, for example, is levorotatory, and (+)-sucrose is dextrorotatory.

The extent of rotation observed in a polarimetry experiment depends on the number of optically active molecules encountered by the light beam. This number, in turn, depends on sample concentration and sample pathlength. If the concentration of sample is doubled, the observed rotation doubles. If the concentration is kept constant but the length of the sample tube is doubled, the observed rotation doubles. It also happens that the angle of rotation depends on the wavelength of the light used.

To express optical rotations in a meaningful way so that comparisons can be made, we have to choose standard conditions. The **specific rotation, $[\alpha]_D$**, of a compound is defined as the observed rotation when light of 589.6 nanometer (nm; 1 nm = 10^{-9} m) wavelength is used with a sample pathlength l of 1 decimeter (dm; 1 dm = 10 cm) and a sample concentration c of 1 g/cm^3. (Light of 589.6 nm, the so-called sodium D line, is the yellow light emitted from common sodium street lamps.)

$$[\alpha]_D \times \frac{\text{Observed rotation (degrees)}}{\text{Pathlength, } l \text{ (dm)} \times \text{Concentration, } c \text{ (g/cm}^3\text{)}} = \frac{\alpha}{l \times c}$$

Table 5.1 Specific Rotation of Some Organic Molecules

Compound	$[\alpha]_D$	Compound	$[\alpha]_D$
Penicillin V	+233	Cholesterol	−31.5
Sucrose	+66.47	Morphine	−132
Camphor	+44.26	Cocaine	−16
Chloroform	0	Acetic acid	0

When optical rotation data are expressed in this standard way, the specific rotation, $[\alpha]_D$, is a physical constant characteristic of a given optically active compound. For example, (+)-lactic acid has $[\alpha]_D = +3.82$, and (−)-lactic acid has $[\alpha]_D = -3.82$. That is, the two enantiomers rotate plane-polarized light to exactly the same extent but in opposite directions. Note that the units of specific rotation are $[(\text{deg} \cdot \text{cm}^2)/\text{g}]$ but that values are usually expressed without the units. Some additional examples are listed in Table 5.1.

Worked Example 5.2 Calculating an Optical Rotation

A 1.20 g sample of cocaine, $[\alpha]_D = -16$, was dissolved in 7.50 mL of chloroform and placed in a sample tube having a pathlength of 5.00 cm. What was the observed rotation?

Cocaine

Strategy

Since $[\alpha]_D = \dfrac{\alpha}{l \times c}$

Then $\alpha = l \times c \times [\alpha]_D$

where $[\alpha]_D = -16$; $l = 5.00\text{ cm} = 0.500\text{ dm}$; $c = 1.20\text{ g}/7.50\text{ cm}^3 = 0.160\text{ g/cm}^3$

Solution

$\alpha = (-16)\,(0.500)\,(0.160) = -1.3°$.

Problem 5.5

Is cocaine (Worked Example 5.2) dextrorotatory or levorotatory?

Problem 5.6

A 1.50 g sample of coniine, the toxic extract of poison hemlock, was dissolved in 10.0 mL of ethanol and placed in a sample cell with a 5.00 cm pathlength. The observed rotation at the sodium D line was +1.21°. Calculate $[\alpha]_D$ for coniine.

5.4 Pasteur's Discovery of Enantiomers

Little was done after Biot's discovery of optical activity until 1848, when Louis Pasteur began work on a study of crystalline tartaric acid salts derived from wine. On crystallizing a concentrated solution of sodium ammonium tartrate below 28 °C, Pasteur made the surprising observation that two distinct kinds of crystals precipitated. Furthermore, the two kinds of crystals were nonsuperimposable mirror images and were related in the same way that a right hand is related to a left hand.

Working carefully with tweezers, Pasteur was able to separate the crystals into two piles, one of "right-handed" crystals and one of "left-handed" crystals, like those shown in **Figure 5.6**. Although the original sample, a 50:50 mixture of right and left, was optically inactive, solutions of the crystals from each of the sorted piles were optically active and their specific rotations were equal in amount but opposite in sign.

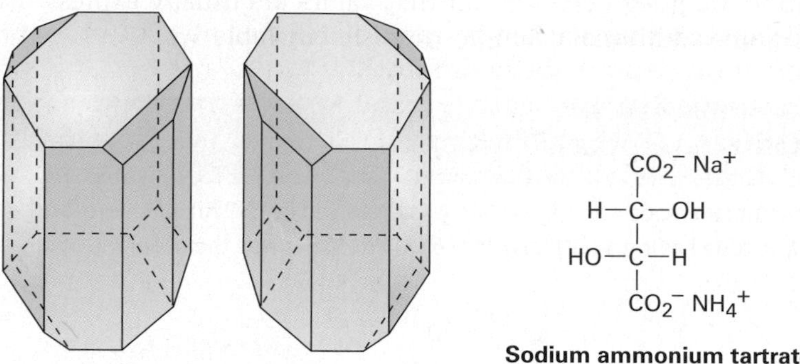

Figure 5.6 Drawings of sodium ammonium tartrate crystals taken from Pasteur's original sketches. One of the crystals is dextrorotatory in solution, and the other is levorotatory.

Pasteur was far ahead of his time. Although the structural theory of Kekulé had not yet been proposed, Pasteur explained his results by speaking of the molecules themselves, saying, "There is no doubt that [in the *dextro* tartaric acid] there exists an asymmetric arrangement having a nonsuperimposable image. It is no less certain that the atoms of the *levo* acid have precisely the inverse asymmetric arrangement." Pasteur's vision was extraordinary, for it was not until 25 years later that his ideas regarding the asymmetric carbon atom were confirmed.

Today, we would describe Pasteur's work by saying that he had discovered enantiomers. Enantiomers, also called *optical isomers,* have identical physical properties, such as melting point and boiling point, but differ in the direction in which their solutions rotate plane-polarized light.

5.5 Sequence Rules for Specifying Configuration

Key IDEAS

Test your knowledge of Key Ideas by answering end-of-chapter exercises marked with ▲.

Structural drawings provide a visual representation of stereochemistry, but a written method for indicating the three-dimensional arrangement, or **configuration**, of substituents at a chirality center is also needed. The method used employs a set of *sequence rules* to rank the four groups attached to the chirality center and then looks at the handedness with which those groups are attached.

Called the **Cahn–Ingold–Prelog rules** after the chemists who proposed them, the sequence rules are as follows:

RULE 1

Look at the four atoms directly attached to the chirality center, and rank them according to atomic number. The atom with the highest atomic number has the highest ranking (first), and the atom with the lowest atomic number (usually hydrogen) has the lowest ranking (fourth). When different isotopes of the same element are compared, such as deuterium (^{2}H) and protium (^{1}H), the heavier isotope ranks higher than the lighter isotope. Thus, atoms commonly found in organic compounds have the following order.

Atomic number	35		17		16		15		8		7		6		(2)		(1)	
Higher ranking	Br	>	Cl	>	S	>	P	>	O	>	N	>	C	>	^{2}H	>	^{1}H	Lower ranking

RULE 2

If a decision can't be reached by ranking the first atoms in the substituent, look at the second, third, or fourth atoms away from the chirality center until the first difference is found. A $-CH_2CH_3$ substituent and a $-CH_3$ substituent are equivalent by rule 1 because both have carbon as the first atom. By rule 2, however, ethyl ranks higher than methyl because ethyl has a *carbon* as its highest second atom, while methyl has only *hydrogen* as its second atom. Look at the following pairs of examples to see how the rule works:

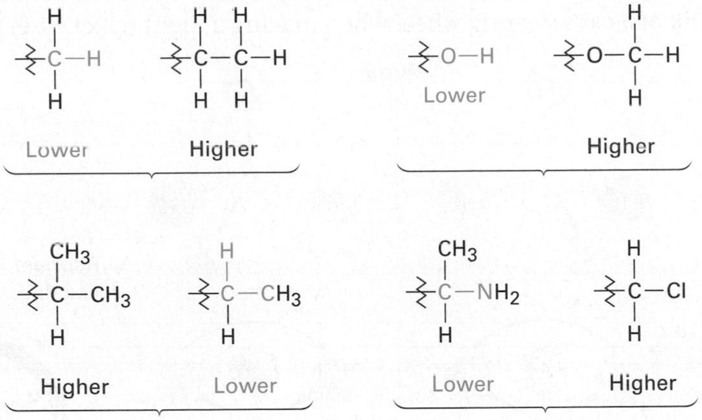

RULE 3

Multiple-bonded atoms are equivalent to the same number of single-bonded atoms. For example, an aldehyde substituent (—CH=O), which has a carbon atom *doubly* bonded to *one* oxygen, is equivalent to a substituent having a carbon atom *singly* bonded to *two* oxygens:

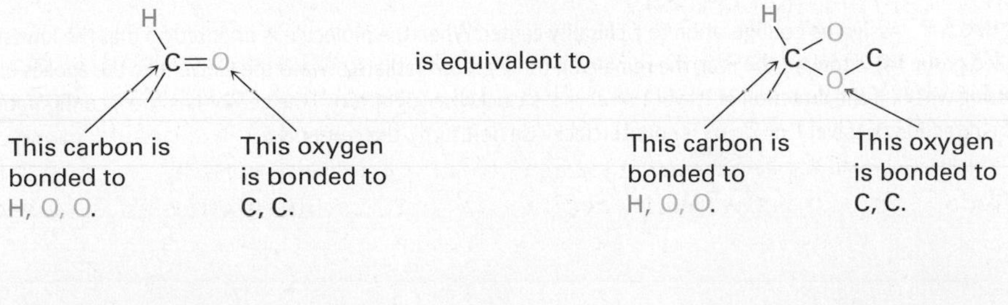

As further examples, the following pairs are equivalent:

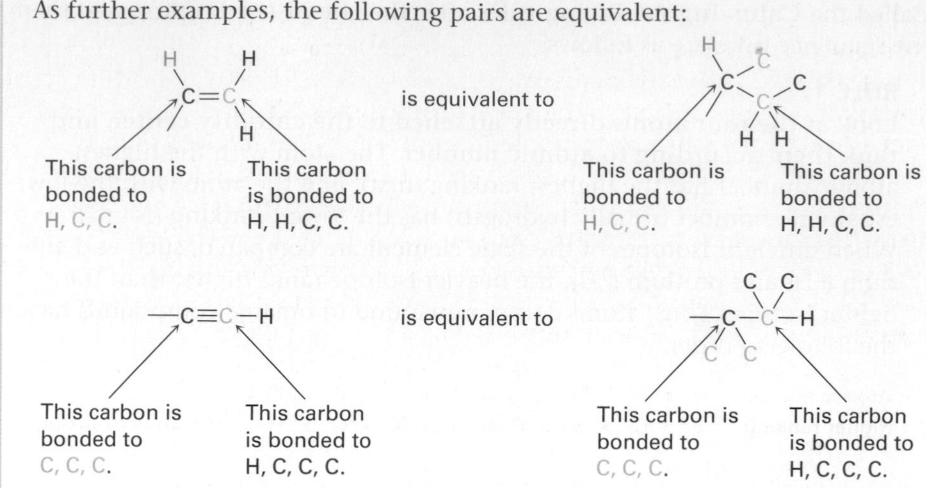

Having ranked the four groups attached to a chiral carbon, we describe the stereochemical configuration around the carbon by orienting the molecule so that the group with the lowest ranking (4) points directly back, away from us. We then look at the three remaining substituents, which now appear to radiate toward us like the spokes on a steering wheel **(Figure 5.7)**. If a curved arrow drawn from the highest to second-highest to third-highest ranked substituent (1 → 2 → 3) is clockwise, we say that the chirality center has the ***R* configuration** (Latin *rectus,* meaning "right"). If an arrow from 1 → 2 → 3 is counterclockwise, the chirality center has the ***S* configuration** (Latin *sinister,* meaning "left"). To remember these assignments, think of a car's steering wheel when making a *R*ight (clockwise) turn.

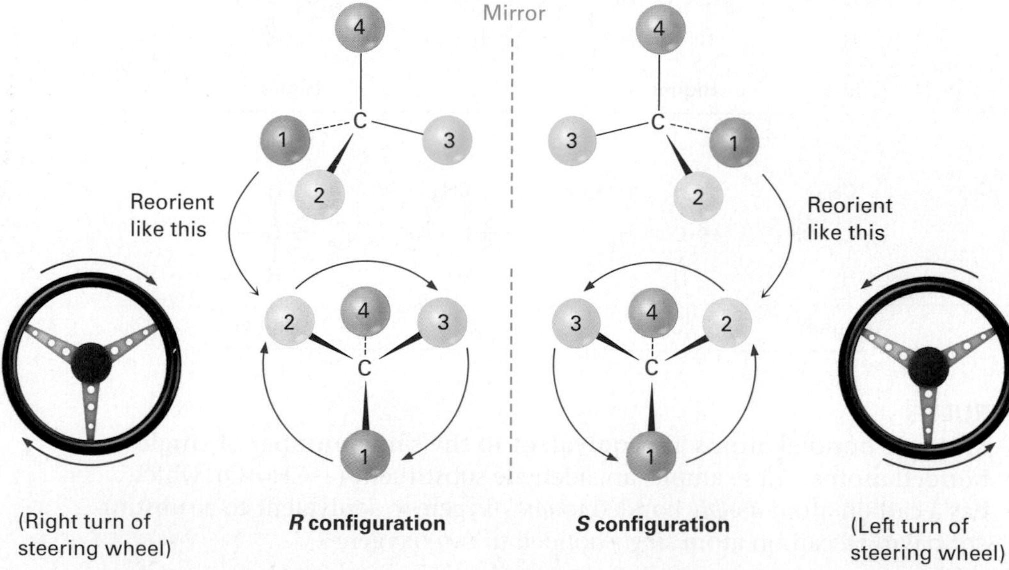

Figure 5.7 Assigning configuration to a chirality center. When the molecule is oriented so that the lowest-ranked group (**4**) is toward the rear, the remaining three groups radiate toward the viewer like the spokes of a steering wheel. If the direction of travel **1** → **2** → **3** is clockwise (right turn), the center has the *R* configuration. If the direction of travel **1** → **2** → **3** is counterclockwise (left turn), the center is *S*.

Look at (−)-lactic acid in **Figure 5.8** for an example of how to assign configuration. Sequence rule 1 says that $-OH$ is ranked 1 and $-H$ is ranked 4, but it doesn't allow us to distinguish between $-CH_3$ and $-CO_2H$ because both groups have carbon as their first atom. Sequence rule 2, however, says that $-CO_2H$ ranks higher than $-CH_3$ because O (the highest second atom in $-CO_2H$) outranks H (the highest second atom in $-CH_3$). Now, turn the molecule so that the fourth-ranked group ($-H$) is oriented toward the rear, away from the observer. Since a curved arrow from 1 ($-OH$) to 2 ($-CO_2H$) to 3 ($-CH_3$) is clockwise (right turn of the steering wheel), (−)-lactic acid has the *R* configuration. Applying the same procedure to (+)-lactic acid leads to the opposite assignment.

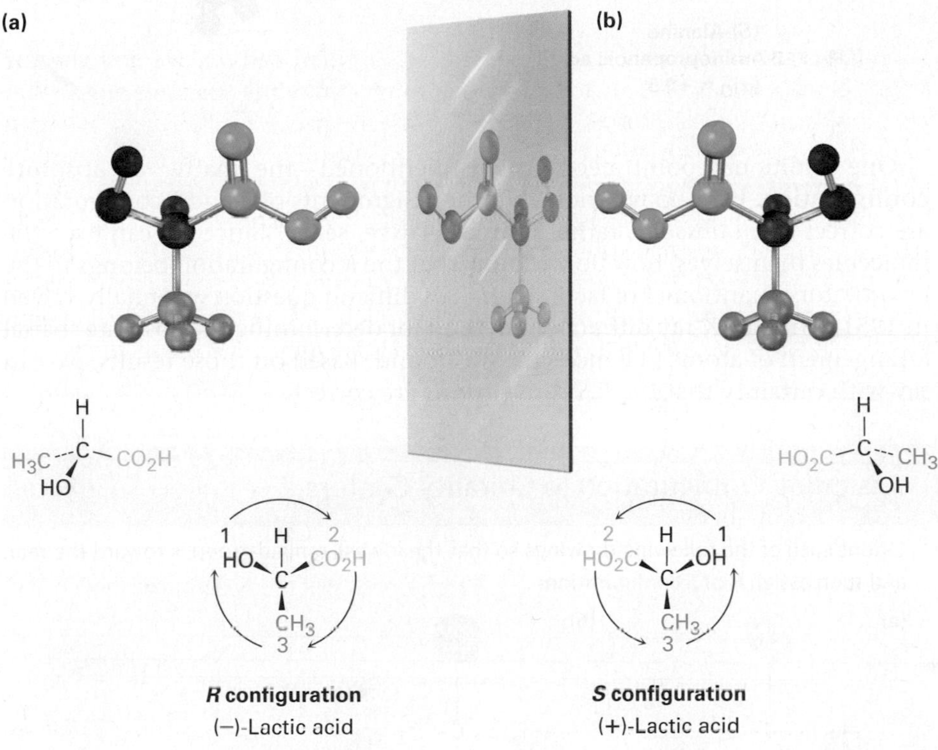

Figure 5.8 Assigning configuration to **(a)** (*R*)-(−)-lactic acid and **(b)** (*S*)-(+)-lactic acid.

Further examples are provided by naturally occurring (−)-glyceraldehyde and (+)-alanine, which both have the S configuration as shown in **Figure 5.9**. Note that the sign of optical rotation, (+) or (−), is not related to the *R,S* designation. (*S*)-Glyceraldehyde happens to be levorotatory (−), and (*S*)-alanine happens to be dextrorotatory (+). There is no simple correlation between *R,S* configuration and direction or magnitude of optical rotation.

Figure 5.9 Assigning configuration to **(a)** (−)-glyceraldehyde. **(b)** (+)-alanine. Both happen to have the *S* configuration, although one is levorotatory and the other is dextrorotatory.

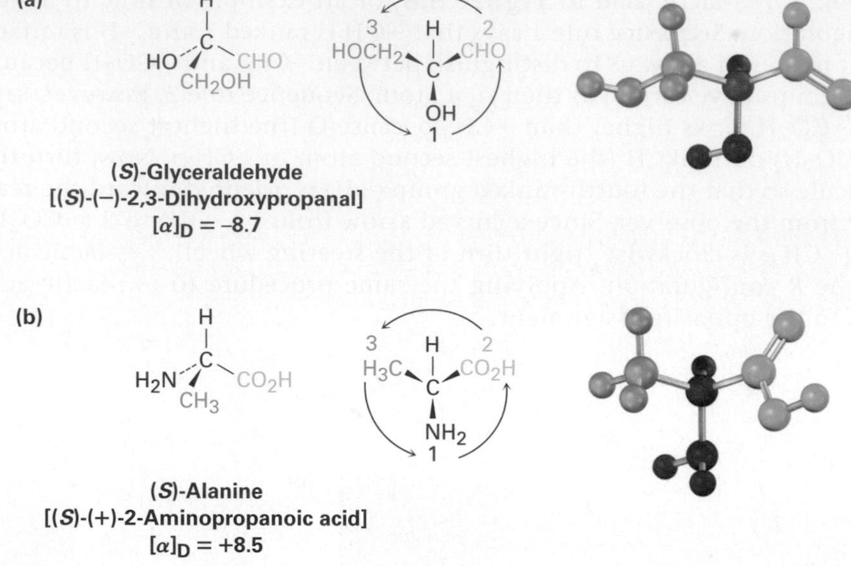

One additional point needs to be mentioned—the matter of **absolute configuration**. How do we know that the assignments of *R* and *S* configuration are correct in an *absolute,* rather than a relative, sense? Since we can't see the molecules themselves, how do we know that the *R* configuration belongs to the levorotatory enantiomer of lactic acid? This difficult question was finally solved in 1951, when an X-ray diffraction method for determining the absolute spatial arrangement of atoms in a molecule was found. Based on those results, we can say with certainty that the *R,S* conventions are correct.

Worked Example 5.3 Assigning Configuration to Chirality Centers

Orient each of the following drawings so that the lowest-ranked group is toward the rear, and then assign *R* or *S* configuration:

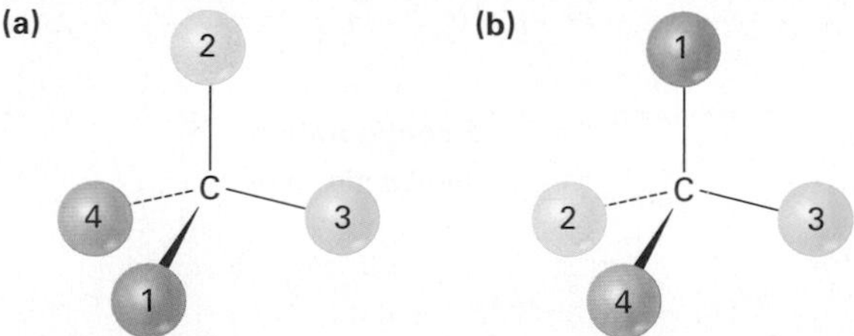

Strategy

It takes practice to be able to visualize and orient a chirality center in three dimensions. You might start by indicating where the observer must be located—180° opposite the lowest-ranked group. Then imagine yourself in the position of the observer, and redraw what you would see.

Solution

In **(a)**, you would be located in front of the page toward the top right of the molecule, and you would see group 2 to your left, group 3 to your right, and group 1 below you. This corresponds to an *R* configuration.

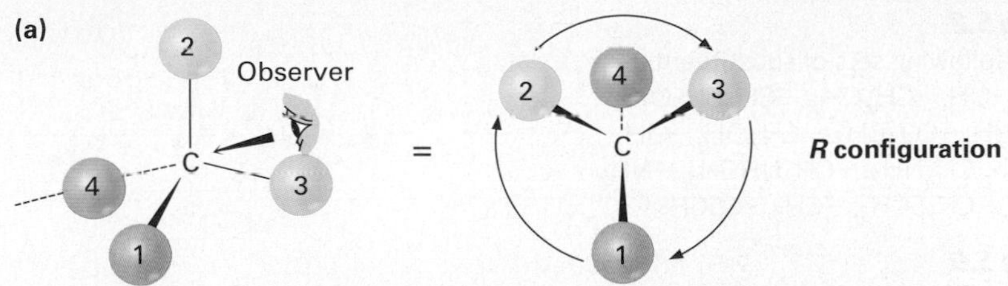

In **(b)**, you would be located behind the page toward the top left of the molecule from your point of view, and you would see group 3 to your left, group 1 to your right, and group 2 below you. This also corresponds to an *R* configuration.

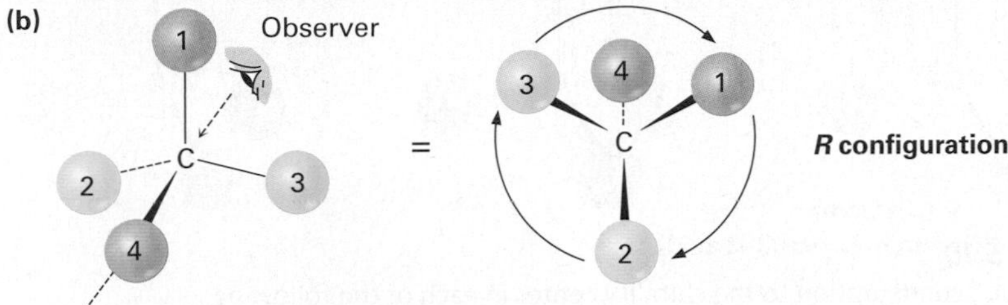

Drawing the Three-Dimensional Structure of a Specific Enantiomer

Worked Example 5.4

Draw a tetrahedral representation of (*R*)-2-chlorobutane.

Strategy

Begin by ranking the four substituents bonded to the chirality center: (1) $-Cl$, (2) $-CH_2CH_3$, (3) $-CH_3$, (4) $-H$. To draw a tetrahedral representation of the molecule, orient the lowest-ranked group ($-H$) away from you and imagine that the other three groups are coming out of the page toward you. Then place the remaining three substituents such that the direction of travel 1 → 2 → 3 is clockwise (right turn), and tilt the molecule toward you to bring the rear hydrogen into view. Using molecular models is a great help in working problems of this sort.

Solution

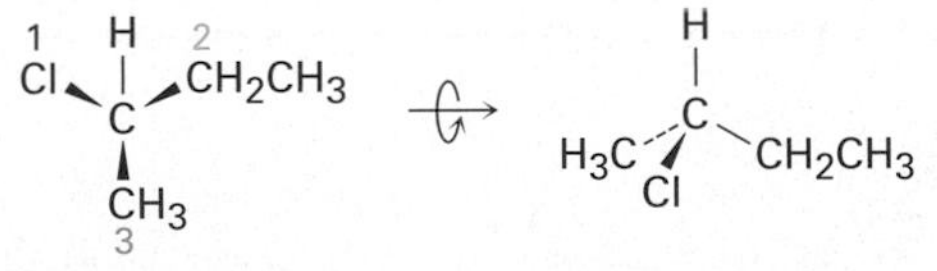

(*R*)-2-Chlorobutane

Problem 5.7

Which member in each of the following sets ranks higher?

(a) $-H$ or $-Br$
(b) $-Cl$ or $-Br$
(c) $-CH_3$ or $-CH_2CH_3$
(d) $-NH_2$ or $-OH$
(e) $-CH_2OH$ or $-CH_3$
(f) $-CH_2OH$ or $-CH{=}O$

Problem 5.8

Rank the following sets of substituents:

(a) $-H, -OH, -CH_2CH_3, -CH_2CH_2OH$
(b) $-CO_2H, -CO_2CH_3, -CH_2OH, -OH$
(c) $-CN, -CH_2NH_2, -CH_2NHCH_3, -NH_2$
(d) $-SH, -CH_2SCH_3, -CH_3, -SSCH_3$

Problem 5.9

Orient each of the following drawings so that the lowest-ranked group is toward the rear, and then assign *R* or *S* configuration:

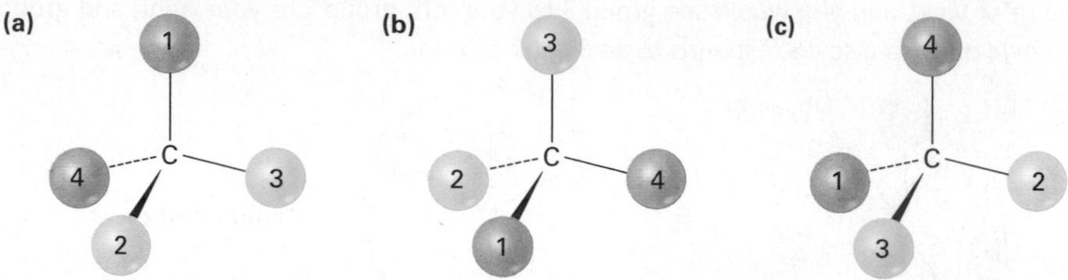

Problem 5.10

Assign *R* or *S* configuration to the chirality center in each of the following molecules:

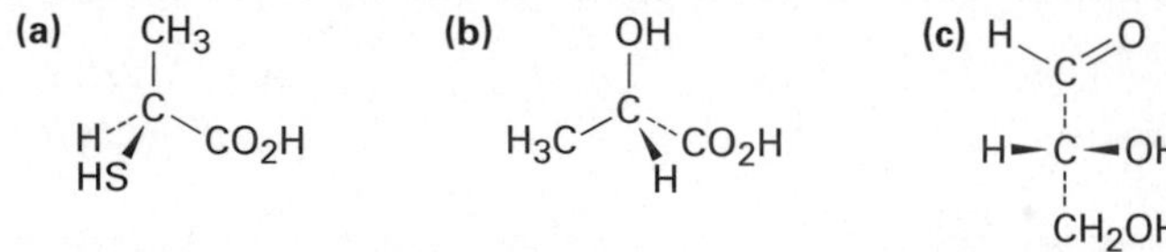

Problem 5.11

Draw a tetrahedral representation of (*S*)-2-pentanol (2-hydroxypentane).

Problem 5.12

Assign *R* or *S* configuration to the chirality center in the following molecular model of the amino acid methionine (blue = N, yellow = S):

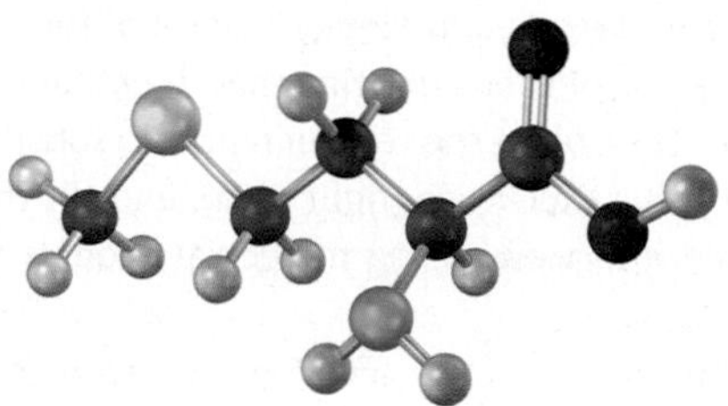

5.6 Diastereomers

Molecules like lactic acid, alanine, and glyceraldehyde are relatively simple because each has only one chirality center and only two stereoisomers. The situation becomes more complex, however, with molecules that have more than one chirality center. As a general rule, a molecule with *n* chirality centers can have up to 2^n stereoisomers (although it may have fewer, as we'll see below). Take the amino acid threonine (2-amino-3-hydroxybutanoic acid), for example. Since threonine has two chirality centers (C2 and C3), there are four possible stereoisomers, as shown in **Figure 5.10**. Check for yourself that the *R,S* configurations are correct.

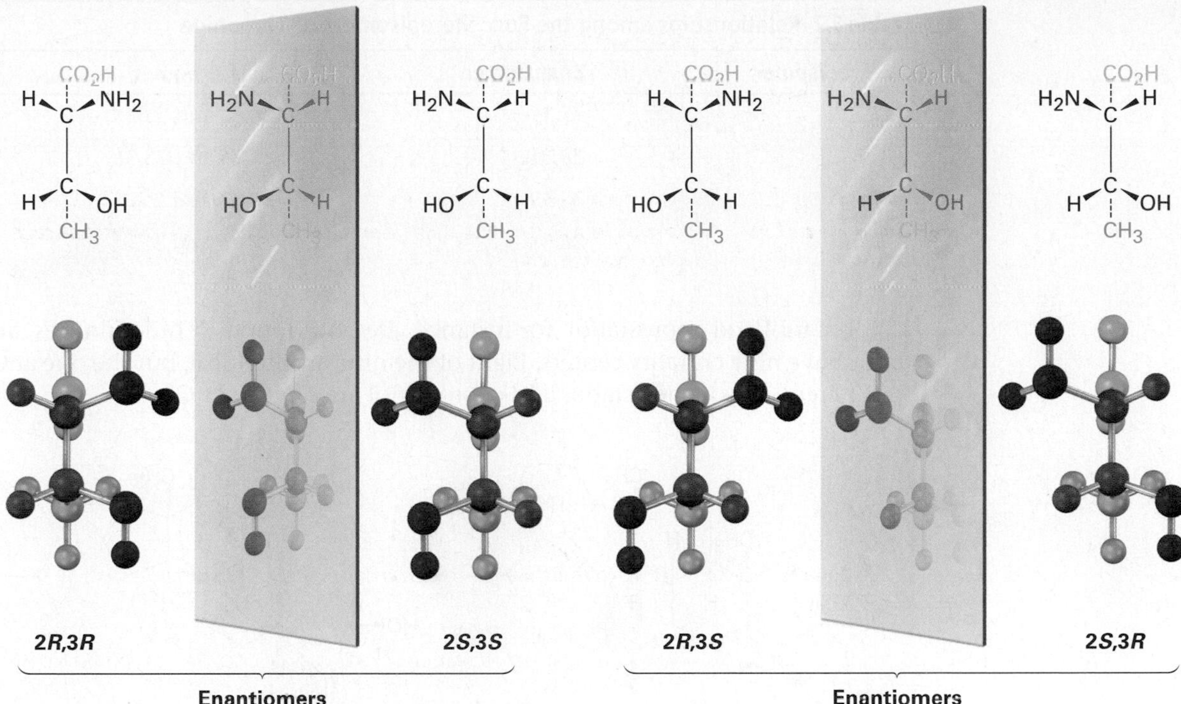

Figure 5.10 The four stereoisomers of 2-amino-3-hydroxybutanoic acid.

The four stereoisomers of 2-amino-3-hydroxybutanoic acid can be grouped into two pairs of enantiomers. The 2*R*,3*R* stereoisomer is the mirror image of 2*S*,3*S*, and the 2*R*,3*S* stereoisomer is the mirror image of 2*S*,3*R*. But what is the relationship between any two molecules that are not mirror images? What, for instance, is the relationship between the 2*R*,3*R* isomer and the 2*R*,3*S* isomer? They are stereoisomers, yet they aren't enantiomers. To describe such a relationship, we need a new term—*diastereomer.*

Diastereomers are stereoisomers that are not mirror images. Since we used the right-hand/left-hand analogy to describe the relationship between two enantiomers, we might extend the analogy by saying that the relationship between diastereomers is like that of hands from different people. Your hand and your friend's hand look *similar,* but they aren't identical and they aren't mirror images. The same is true of diastereomers: they're similar, but they aren't identical and they aren't mirror images.

Note carefully the difference between enantiomers and diastereomers: enantiomers have opposite configurations at *all* chirality centers, whereas diastereomers have opposite configurations at *some* (one or more) chirality centers but the same configuration at others. A full description of the four stereoisomers of threonine is given in Table 5.2. Of the four, only the 2*S*,3*R* isomer, $[\alpha]_D = -28.3$, occurs naturally in plants and animals and is an essential human nutrient. This result is typical: most biological molecules are chiral, and usually only one stereoisomer is found in nature.

In the special case where two diastereomers differ at only one chirality center but are the same at all others, we say that the compounds are **epimers.**

Table 5.2 Relationships among the Four Stereoisomers of Threonine

Stereoisomer	Enantiomer	Diastereomer
2*R*,3*R*	2*S*,3*S*	2*R*,3*S* and 2*S*,3*R*
2*S*,3*S*	2*R*,3*R*	2*R*,3*S* and 2*S*,3*R*
2*R*,3*S*	2*S*,3*R*	2*R*,3*R* and 2*S*,3*S*
2*S*,3*R*	2*R*,3*S*	2*R*,3*R* and 2*S*,3*S*

Cholestanol and coprostanol, for instance, are both found in human feces, and both have nine chirality centers. Eight of the nine are identical, but the one at C5 is different. Thus, cholestanol and coprostanol are *epimeric* at C5.

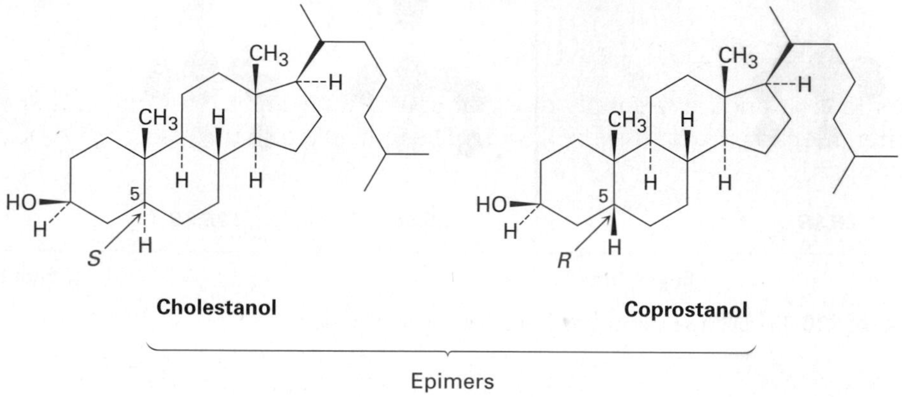

Problem 5.13

One of the following molecules **(a)**–**(d)** is D-erythrose 4-phosphate, an intermediate in the Calvin photosynthetic cycle by which plants incorporate CO_2 into carbohydrates. If D-erythrose 4-phosphate has *R* stereochemistry at both chirality centers, which of the structures is it? Which of the remaining three structures is the enantiomer of D-erythrose 4-phosphate, and which are diastereomers?

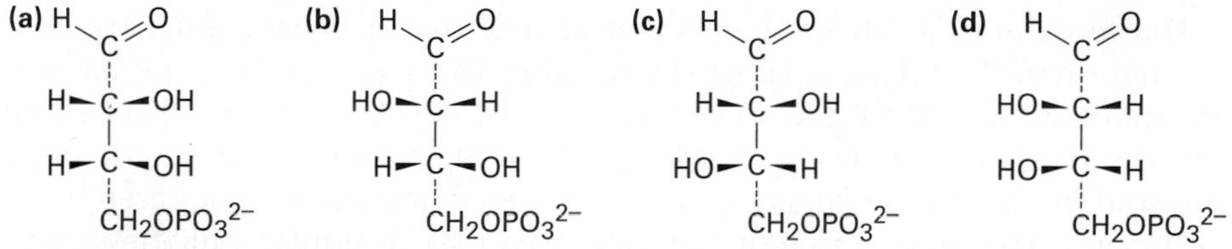

Problem 5.14

How many chirality centers does morphine have? How many stereoisomers of morphine are possible in principle?

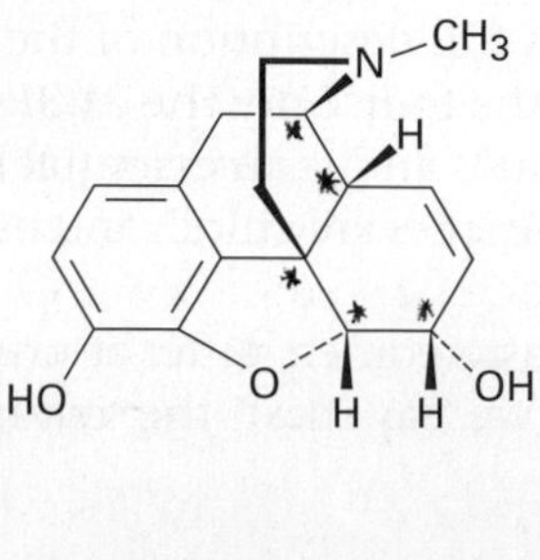

Problem 5.15

Assign *R,S* configuration to each chirality center in the following molecular model of the amino acid isoleucine (blue = N):

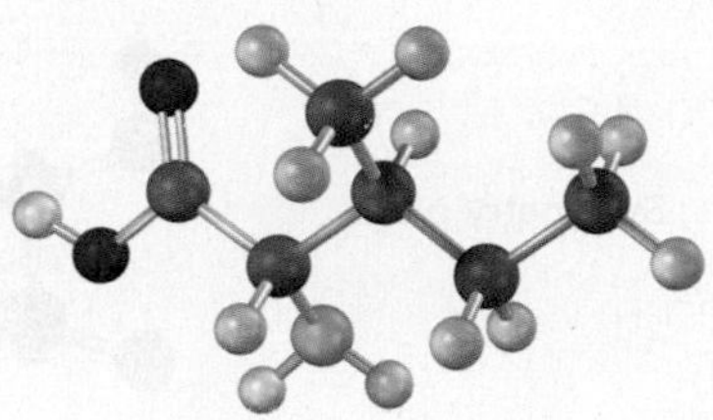

5.7 Meso Compounds

Let's look at another example of a compound with more than one chirality center: the tartaric acid used by Pasteur. The four stereoisomers can be drawn as follows:

Mirror | Mirror

2R,3R	**2S,3S**	**2R,3S**	**2S,3R**
1 CO_2H	1 CO_2H	1 CO_2H	1 CO_2H
H, OH on 2C	HO, H on 2C	H, OH on 2C	HO, H on 2C
HO, H on 3C	H, OH on 3C	H, OH on 3C	HO, H on 3C
4 CO_2H	4 CO_2H	4 CO_2H	4 CO_2H

The 2*R*,3*R* and 2*S*,3*S* structures are nonsuperimposable mirror images and therefore represent a pair of enantiomers. A close look at the 2*R*,3*S* and 2*S*,3*R* structures, however, shows that they *are* superimposable, and thus identical, as can be seen by rotating one structure 180°.

2R,3S	Rotate 180°	**2S,3R**
1 CO_2H		1 CO_2H
H, OH on 2C	→	HO, H on 2C
H, OH on 3C		HO, H on 3C
4 CO_2H		4 CO_2H

Identical

The 2*R*,3*S* and 2*S*,3*R* structures are identical because the molecule has a plane of symmetry and is therefore achiral. The symmetry plane cuts through the C2–C3 bond, making one half of the molecule a mirror image of the other half (**Figure 5.11**). Because of the plane of symmetry, the molecule is achiral, despite the fact that it has two chirality centers. Compounds that are achiral, yet contain chirality centers, are called **meso compounds (me-zo)**. Thus, tartaric acid exists in three stereoisomeric forms: two enantiomers and one meso form.

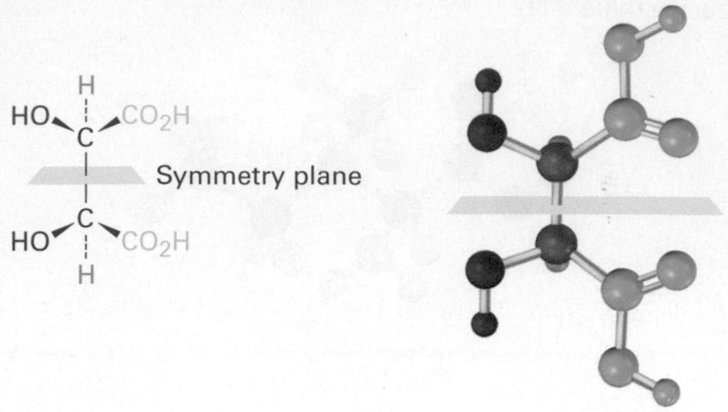

Figure 5.11 A symmetry plane through the C2–C3 bond of *meso*-tartaric acid makes the molecule achiral.

Symmetry plane

Some physical properties of the three stereoisomers are listed in Table 5.3. The (+)- and (−)-tartaric acids have identical melting points, solubilities, and densities, but they differ in the sign of their rotation of plane-polarized light. The meso isomer, by contrast, is diastereomeric with the (+) and (−) forms. It has no mirror-image relationship to (+)- and (−)-tartaric acids, is a different compound altogether, and has different physical properties.

Table 5.3 Some Properties of the Stereoisomers of Tartaric Acid

Stereoisomer	Melting point (°C)	$[\alpha]_D$	Density (g/cm^3)	Solubility at 20 °C (g/100 mL H_2O)
(+)	168–170	+12	1.7598	139.0
(−)	168–170	−12	1.7598	139.0
Meso	146–148	0	1.6660	125.0

Worked Example 5.5 Distinguishing Chiral Compounds from Meso Compounds

Does *cis*-1,2-dimethylcyclobutane have any chirality centers? Is it chiral?

Strategy

To see whether a chirality center is present, look for a carbon atom bonded to four different groups. To see whether the molecule is chiral, look for the presence or absence of a symmetry plane. Not all molecules with chirality centers are chiral overall—meso compounds are an exception.

Solution

A look at the structure of *cis*-1,2-dimethylcyclobutane shows that both methyl-bearing ring carbons (C1 and C2) are chirality centers. Overall, though, the compound is achiral because there is a symmetry plane bisecting the ring between C1 and C2. Thus, the molecule is a meso compound.

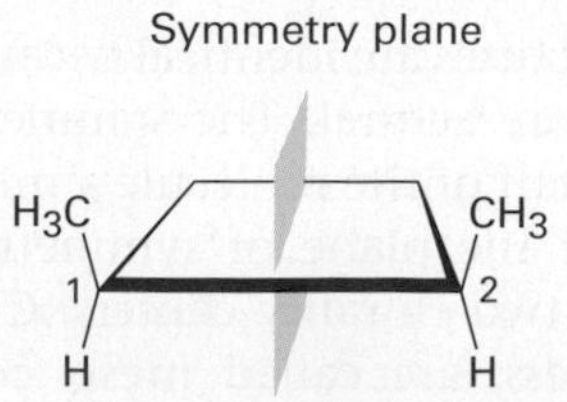

Symmetry plane

Problem 5.16

Which of the following structures represent meso compounds?

(a) 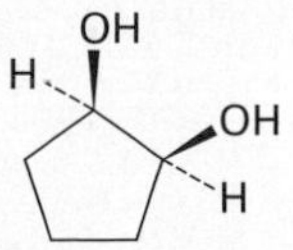 (b) 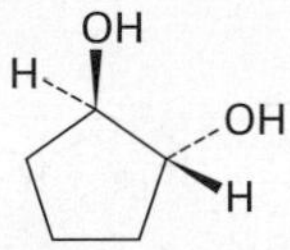 (c) 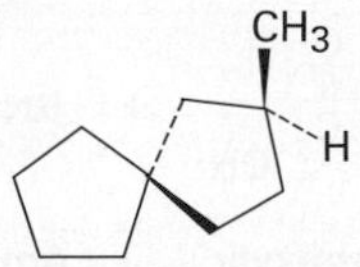 (d)

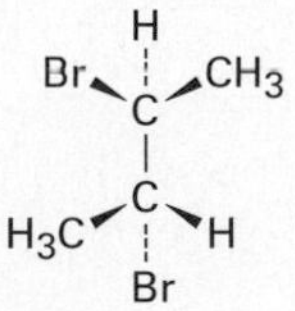

Problem 5.17

Which of the following have a meso form? (Recall that the *-ol* suffix refers to an alcohol, ROH.)

(a) 2,3-Butanediol **(b)** 2,3-Pentanediol **(c)** 2,4-Pentanediol

Problem 5.18

Does the following structure represent a meso compound? If so, indicate the symmetry plane.

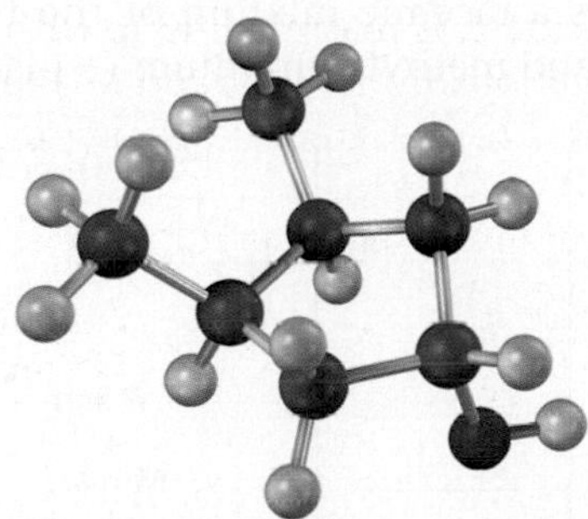

5.8 Racemic Mixtures and the Resolution of Enantiomers

To end this discussion of stereoisomerism, let's return for a last look at Pasteur's pioneering work, described in **Section 5.4**. Pasteur took an optically inactive tartaric acid salt and found that he could crystallize from it two optically active forms having what we would now call the 2*R*,3*R* and 2*S*,3*S* configurations. But what was the optically inactive form he started with? It couldn't have been *meso*-tartaric acid, because *meso*-tartaric acid is a different chemical compound and can't interconvert with the two chiral enantiomers without breaking and re-forming chemical bonds.

The answer is that Pasteur started with a 50:50 *mixture* of the two chiral tartaric acid enantiomers. Such a mixture is called a **racemate (raa-suh-mate)**, or *racemic mixture*, and is denoted by either the symbol (±) or the prefix *d,l* to indicate an equal mixture of dextrorotatory and levorotatory forms. Racemates show no optical rotation because the (+) rotation from one enantiomer exactly cancels the (−) rotation from the other. Through luck, Pasteur was able to separate, or **resolve**, racemic tartaric acid into its (+) and (−) enantiomers. Unfortunately, the fractional crystallization technique he used doesn't work for most racemates, so other methods are needed.

The most common method of resolution uses an acid–base reaction between the racemate of a chiral carboxylic acid (RCO_2H) and an amine base (RNH_2) to yield an ammonium salt:

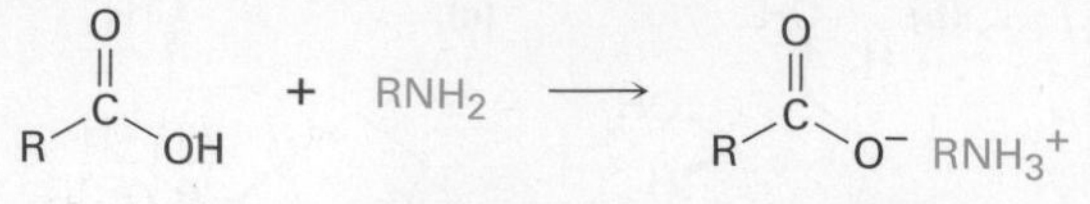

$$RC(=O)OH + RNH_2 \longrightarrow RC(=O)O^- \ RNH_3^+$$

Carboxylic acid **Amine base** **Ammonium salt**

To understand how this method of resolution works, let's see what happens when a racemic mixture of chiral acids, such as (+)- and (−)-lactic acids, reacts with an achiral amine base, such as methylamine, CH_3NH_2. Stereochemically, the situation is analogous to what happens when left and right hands (chiral) pick up a ball (achiral). Both left and right hands pick up the ball equally well, and the products—ball in right hand versus ball in left hand—are mirror images. In the same way, both (+)- and (−)-lactic acid react with methylamine equally well, and the product is a racemic mixture of the two enantiomers methylammonium (+)-lactate and methylammonium (−)-lactate **(Figure 5.12)**.

Figure 5.12 Reaction of racemic lactic acid with achiral methylamine leads to a racemic mixture of ammonium salts.

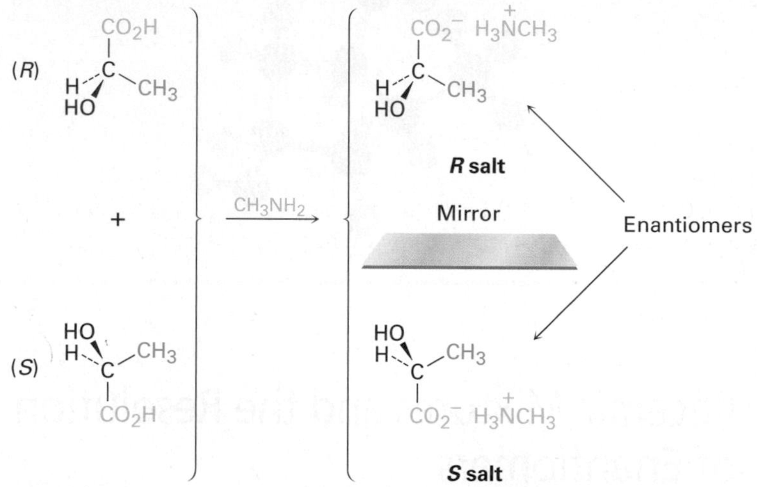

Now let's see what happens when the racemic mixture of (+)- and (−)-lactic acids reacts with a single enantiomer of a chiral amine base, such as (*R*)-1-phenylethylamine. Stereochemically, the situation is analogous to what happens when left and right hands (chiral) put on a right-handed glove (*also chiral*). Left and right hands don't put on the right-handed glove in the same way, so the products—right hand in right glove versus left hand in right glove—are not mirror images; they're similar but different.

In the same way, (+)- and (−)-lactic acids react with (*R*)-1-phenylethylamine to give two different products **(Figure 5.13)**. (*R*)-Lactic acid reacts with (*R*)-1-phenylethylamine to give the *R,R* salt, and (*S*)-lactic acid reacts with the *R* amine to give the *S,R* salt. *The two salts are diastereomers*. They have different chemical and physical properties, and it may therefore be possible to separate them by crystallization or some other means. Once separated, acidification of the two diastereomeric salts with a strong acid then allows us to isolate the two pure enantiomers of lactic acid and to recover the chiral amine for reuse.

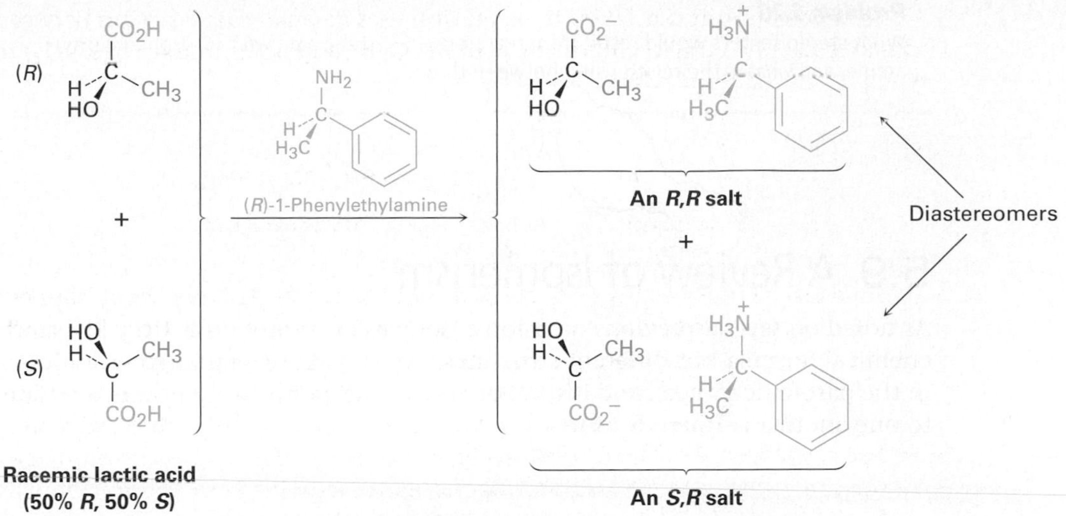

(R)-lactic acid + (S)-lactic acid (Racemic lactic acid) reacting with (R)-1-Phenylethylamine to give An R,R salt + An S,R salt (Diastereomers)

Racemic lactic acid
(50% *R*, 50% *S*)

(R)-1-Phenylethylamine

An *R,R* salt

An *S,R* salt

Diastereomers

Figure 5.13 Reaction of racemic lactic acid with (*R*)-1-phenylethylamine yields a mixture of diastereomeric ammonium salts, which have different properties and can be separated.

Worked Example 5.6

Predicting the Chirality of a Reaction Product

We'll see in **Section 21.3** that carboxylic acids (RCO_2H) react with alcohols (R′OH) to form esters (RCO_2R'). Suppose that (±)-lactic acid reacts with CH_3OH to form the ester, methyl lactate. What stereochemistry would you expect the product(s) to have? What is the relationship of the products?

$$CH_3CH(OH)COOH + CH_3OH \xrightarrow{\text{Acid catalyst}} CH_3CH(OH)COOCH_3 + H_2O$$

Lactic acid **Methanol** **Methyl lactate**

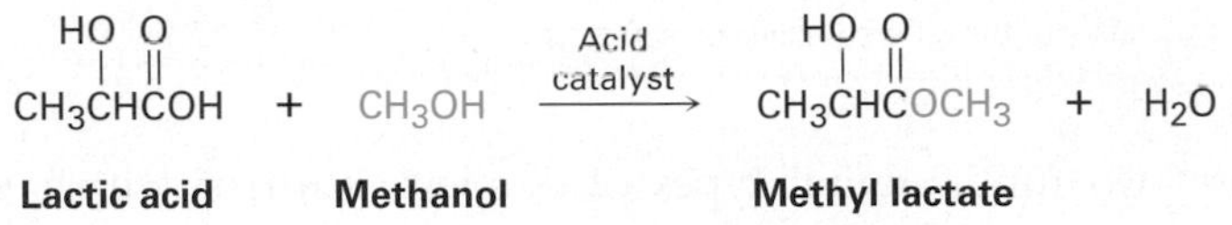

Solution

Reaction of a racemic acid with an achiral alcohol such as methanol yields a racemic mixture of mirror-image (enantiomeric) products.

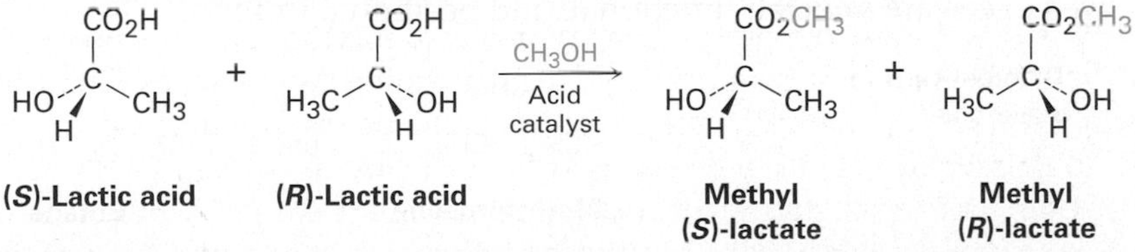

(S)-Lactic acid + (R)-Lactic acid $\xrightarrow{CH_3OH,\ \text{Acid catalyst}}$ Methyl (S)-lactate + Methyl (R)-lactate

(*S*)-Lactic acid **(*R*)-Lactic acid** **Methyl (*S*)-lactate** **Methyl (*R*)-lactate**

Problem 5.19

Suppose that acetic acid (CH_3CO_2H) reacts with (*S*)-2-butanol to form an ester (see Worked Example 5.6). What stereochemistry would you expect the product(s) to have? What is the relationship of the products?

$$CH_3COOH + CH_3CH(OH)CH_2CH_3 \xrightarrow{\text{Acid catalyst}} CH_3COOCH(CH_3)CH_2CH_3 + H_2O$$

Acetic acid **2-Butanol** ***sec*-Butyl acetate**

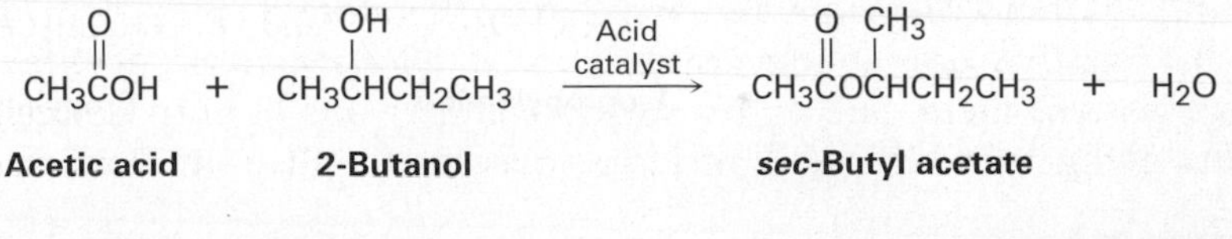

Problem 5.20

What stereoisomers would result from reaction of (±)-lactic acid with (*S*)-1-phenylethylamine, and what is the relationship between them?

5.9 A Review of Isomerism

As noted on several previous occasions, isomers are compounds with the same chemical formula but different structures. We've seen several kinds of isomers in the past few chapters, and it's a good idea at this point to see how they relate to one another **(Figure 5.14)**.

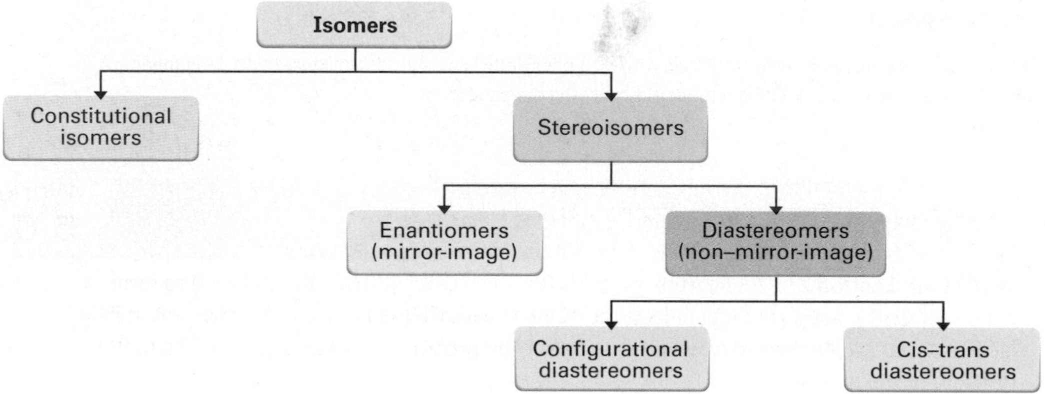

Figure 5.14 A summary of the different kinds of isomers.

There are two fundamental types of isomers, both of which we've now encountered: constitutional isomers and stereoisomers.

Constitutional isomers **(Section 3.2)** are compounds whose atoms are connected differently. Among the kinds of constitutional isomers we've seen are skeletal, functional, and positional isomers.

Different carbon skeletons	$CH_3CH(CH_3)CH_3$ **2-Methylpropane**	and	$CH_3CH_2CH_2CH_3$ **Butane**
Different functional groups	CH_3CH_2OH **Ethyl alcohol**	and	CH_3OCH_3 **Dimethyl ether**
Different position of functional groups	$CH_3CH(NH_2)CH_3$ **Isopropylamine**	and	$CH_3CH_2CH_2NH_2$ **Propylamine**

Stereoisomers (Section 4.2) are compounds whose atoms are connected in the same order but with a different spatial arrangement. Among the kinds of stereoisomers we've seen are enantiomers, diastereomers, and cis–trans isomers of cycloalkanes. Actually, cis–trans isomers are just a subclass of diastereomers because they are non–mirror-image stereoisomers:

Enantiomers (nonsuperimposable mirror-image stereoisomers)

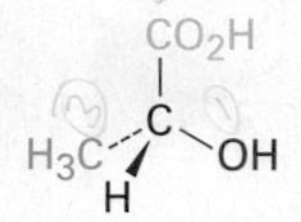

(*R*)-Lactic acid 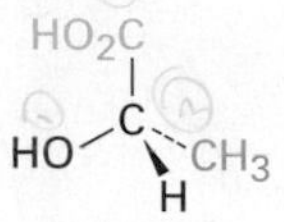 **(*S*)-Lactic acid**

Diastereomers (nonsuperimposable non–mirror-image stereoisomers)

Configurational diastereomers

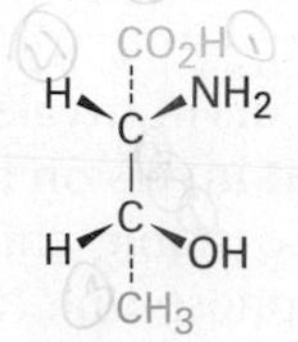

2*R*,3*R*-2-Amino-3-hydroxybutanoic acid 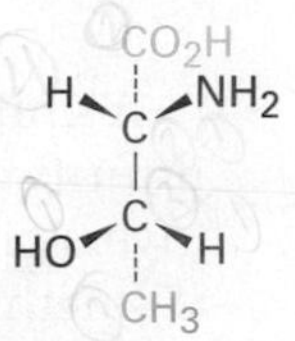 **2*R*,3*S*-2-Amino-3-hydroxybutanoic acid**

Cis–trans diastereomers (substituents on same side or opposite side of double bond or ring)

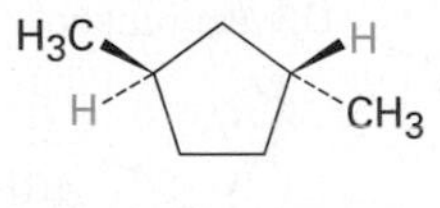

***trans*-1,3-Dimethylcyclopentane** and ***cis*-1,3-Dimethylcyclopentane**

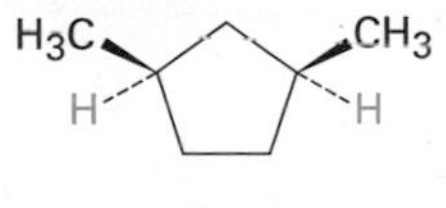

Problem 5.21

What kinds of isomers are the following pairs?

(a) (*S*)-5-Chloro-2-hexene and chlorocyclohexane

(b) (2*R*,3*R*)-Dibromopentane and (2*S*,3*R*)-dibromopentane

5.10 Chirality at Nitrogen, Phosphorus, and Sulfur

Although the most common cause of chirality is the presence of four different substituents bonded to a tetrahedral atom, that atom doesn't necessarily have to be carbon. Nitrogen, phosphorus, and sulfur are all commonly encountered in organic molecules, and all can be chirality centers. We know, for instance, that trivalent nitrogen is tetrahedral, with its lone pair of electrons acting as the fourth "substituent" **(Section 1.10)**. Is trivalent nitrogen chiral? Does a compound such as ethylmethylamine exist as a pair of enantiomers?

The answer is both yes and no. Yes in principle, but no in practice. Most trivalent nitrogen compounds undergo a rapid umbrella-like inversion that interconverts enantiomers, so we can't isolate individual enantiomers except in special cases.

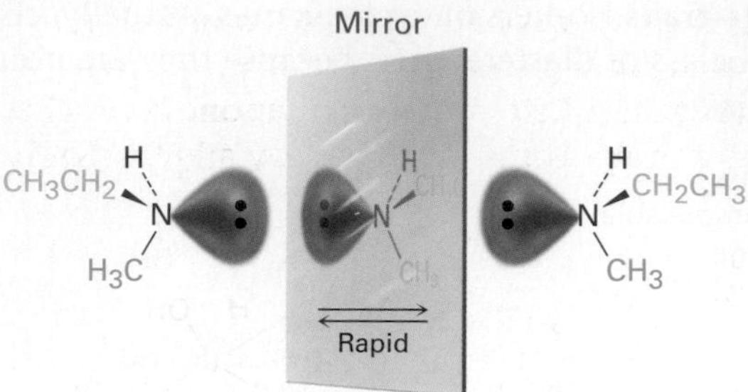

A similar situation occurs in trivalent phosphorus compounds, or *phosphines*. It turns out, though, that inversion at phosphorus is substantially slower than inversion at nitrogen, so stable chiral phosphines *can* be isolated. (*R*)- and (*S*)-methylpropylphenylphosphine, for example, are configurationally stable for several hours at 100 °C. We'll see the importance of phosphine chirality in **Section 26.7** in connection with the synthesis of chiral amino acids.

(*R*)-Methylpropylphenylphosphine (configurationally stable)

Divalent sulfur compounds are achiral, but trivalent sulfur compounds called *sulfonium salts* (R_3S^+) can be chiral. Like phosphines, sulfonium salts undergo relatively slow inversion, so chiral sulfonium salts are configurationally stable and can be isolated. Perhaps the best known example is the coenzyme *S*-adenosylmethionine, the so-called biological methyl donor, which is involved in many metabolic pathways as a source of CH_3 groups. (The "*S*" in the name *S*-adenosylmethionine stands for *sulfur* and means that the adenosyl group is attached to the sulfur atom of the amino acid methionine.) The molecule has *S* stereochemistry at sulfur and is configurationally stable for several days at room temperature. Its *R* enantiomer is also known but is not biologically active.

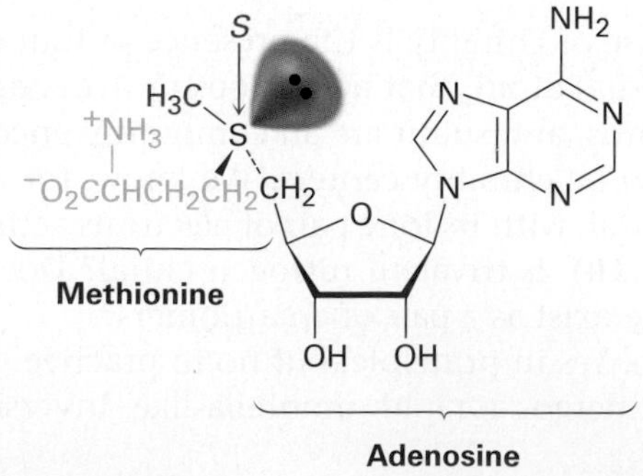

(*S*)-*S*-Adenosylmethionine

Methionine

Adenosine

5.11 Prochirality

Closely related to the concept of chirality, and particularly important in biological chemistry, is the notion of *prochirality*. A molecule is said to be **prochiral** if it can be converted from achiral to chiral in a single chemical step. For instance, an unsymmetrical ketone like 2-butanone is prochiral because it can be converted to the chiral alcohol 2-butanol by addition of hydrogen, as we'll see in **Section 17.4**.

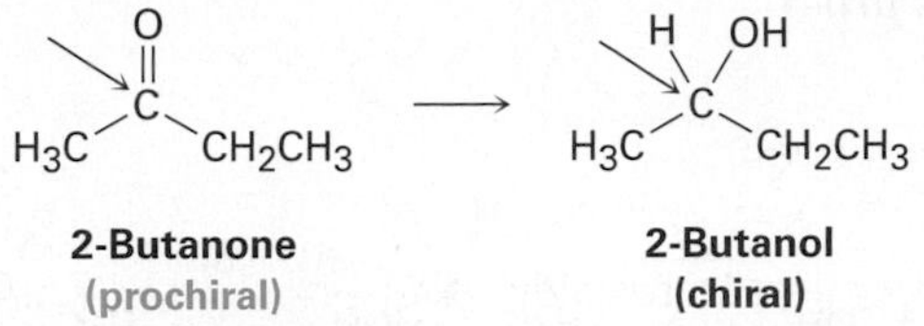

2-Butanone (prochiral) → **2-Butanol** (chiral)

Which enantiomer of 2-butanol is produced depends on which face of the planar carbonyl group undergoes reaction. To distinguish between the possibilities, we use the stereochemical descriptors *Re* and *Si*. Rank the three groups attached to the trigonal, sp^2-hybridized carbon, and imagine curved arrows from the highest to second-highest to third-highest ranked substituents. The face on which the arrows curve clockwise is designated ***Re*** (similar to *R*), and the face on which the arrows curve counterclockwise is designated ***Si*** (similar to *S*). In this particular example, addition of hydrogen from the *Re* faces gives (*S*)-butan-2-ol, and addition from the *Si* face gives (*R*)-butan-2-ol.

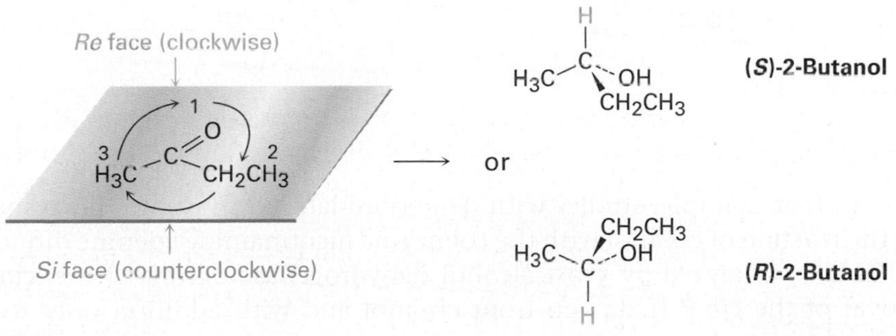

Re face (clockwise)

Si face (counterclockwise)

(*S*)-2-Butanol

or

(*R*)-2-Butanol

In addition to compounds with planar, sp^2-hybridized atoms, compounds with tetrahedral, sp^3-hybridized atoms can also be prochiral. An sp^3-hybridized atom is said to be a **prochirality center** if, by changing one of its attached groups, it becomes a chirality center. The $-CH_2OH$ carbon atom of ethanol, for instance, is a prochirality center because changing one of its attached $-H$ atoms converts it into a chirality center.

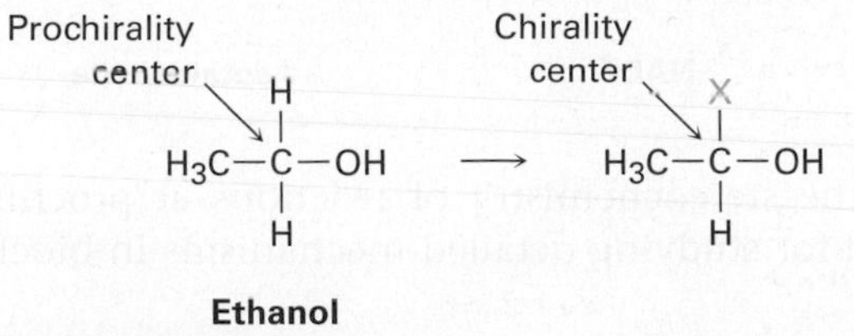

Prochirality center → Chirality center

Ethanol

To distinguish between the two identical atoms (or groups of atoms) on a prochirality center, we imagine a change that will raise the ranking of one atom over the other without affecting its rank with respect to other attached groups. On the $-CH_2OH$ carbon of ethanol, for instance, we might imagine replacing one of the ^1H atoms (protium) by ^2H (deuterium). The newly introduced ^2H atom ranks higher than the remaining ^1H atom, but it remains lower than other groups attached to the carbon. Of the two identical atoms in the original compound, that atom whose replacement leads to an *R* chirality center is said to be ***pro-R*** and that atom whose replacement leads to an *S* chirality center is ***pro-S***.

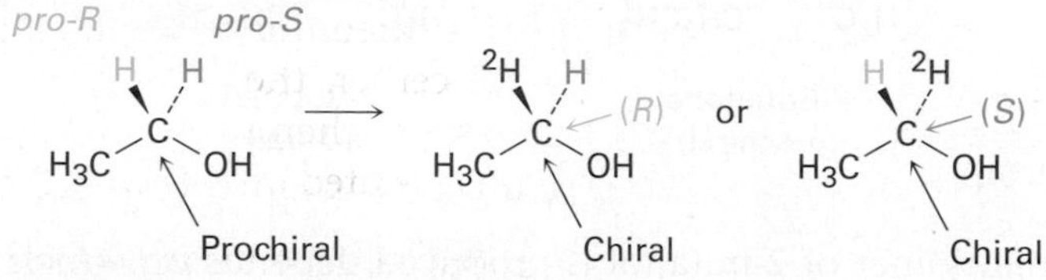

A large number of biological reactions involve prochiral compounds. One of the steps in the citric acid cycle by which food is metabolized, for instance, is the addition of H_2O to fumarate to give malate. Addition of $-OH$ occurs on the *Si* face of a fumarate carbon and gives (*S*)-malate as product.

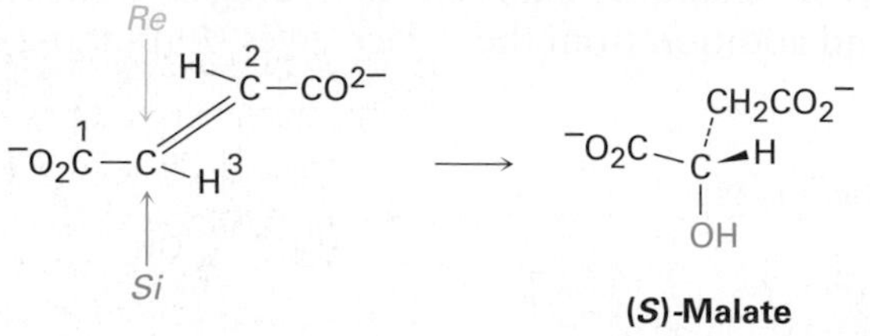

As another example, studies with deuterium-labeled substrates have shown that the reaction of ethanol with the coenzyme nicotinamide adenine dinucleotide (NAD^+) catalyzed by yeast alcohol dehydrogenase occurs with exclusive removal of the *pro-R* hydrogen from ethanol and with addition only to the *Re* face of NAD^+.

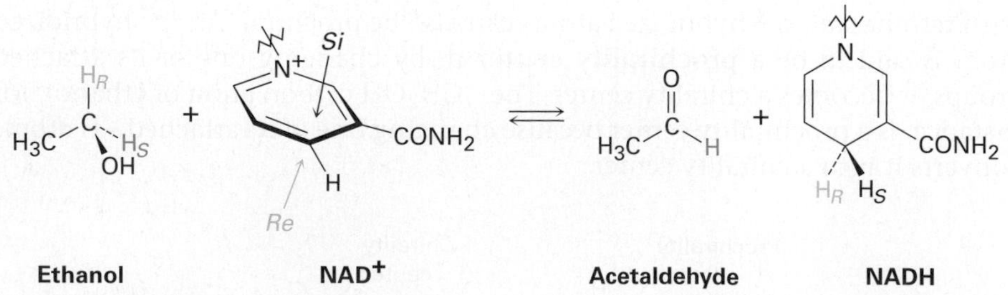

Determining the stereochemistry of reactions at prochirality centers is a powerful method for studying detailed mechanisms in biochemical reactions.

As just one example, the conversion of citrate to (*cis*)-aconitate in the citric acid cycle has been shown to occur with loss of a *pro-R* hydrogen, implying that the OH and H groups leave from opposite sides of the molecule.

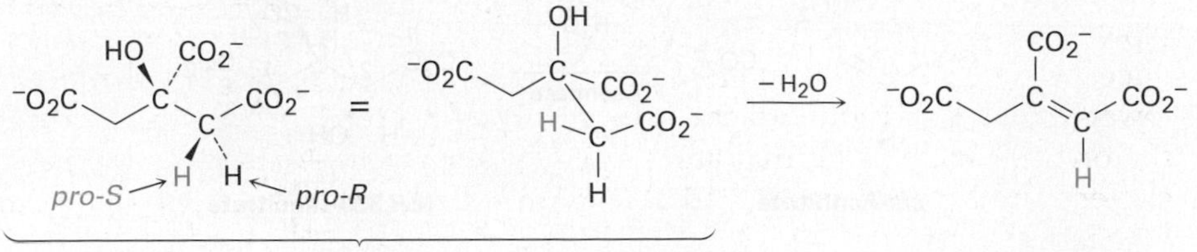

Citrate $\xrightarrow{-H_2O}$ *cis*-Aconitate

Note that when drawing compounds like threonine, cholestanol, and coprostanol, which have more than one chiral center, the wedges and dashes in a structure are used only to imply *relative* stereochemistry within the molecule rather than absolute stereochemistry, unless stated otherwise.

Problem 5.22

Identify the indicated hydrogens in the following molecules as *pro-R* or *pro-S*:

(a) (*S*)-Glyceraldehyde

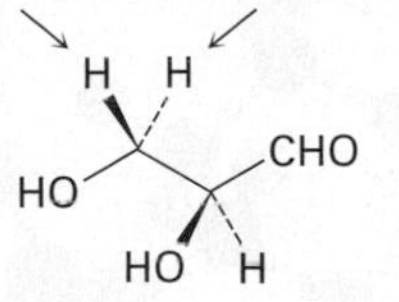

(b) Phenylalanine

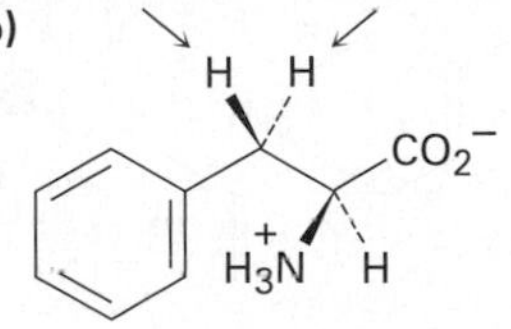

Problem 5.23

Identify the indicated faces of carbon atoms in the following molecules as *Re* or *Si*:

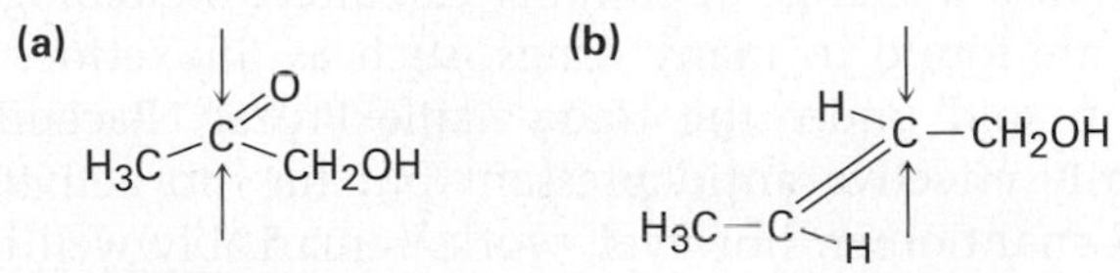

(a) Hydroxyacetone

(b) Crotyl alcohol

Problem 5.24

The lactic acid that builds up in tired muscles is formed from pyruvate. If the reaction occurs with addition of hydrogen to the *Re* face of pyruvate, what is the stereochemistry of the product?

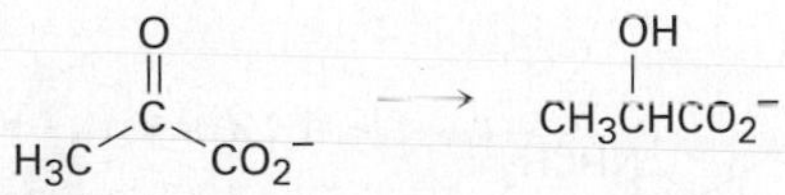

Pyruvate → Lactate

Problem 5.25

The aconitase-catalyzed addition of water to *cis*-aconitate in the citric acid cycle occurs with the following stereochemistry. Does the addition of the OH group occur on the *Re* or the *Si* face of the substrate? What about the addition of the H? Do the H and OH groups adds from the same side of the double bond or from opposite sides?

^-O_2C–C(H)=C(CO_2^-)–CH_2–CO_2^- $\xrightarrow[\text{Aconitase}]{H_2O}$ ^-O_2C–CH(OH)–CH(CO_2^-)–CH_2–CO_2^-

cis-Aconitate **(2*R*,3*S*)-Isocitrate**

5.12 Chirality in Nature and Chiral Environments

Although the different enantiomers of a chiral molecule have the same physical properties, they usually have different biological properties. For example, the (+) enantiomer of limonene has the odor of oranges and lemons, but the (−) enantiomer has the odor of pine trees.

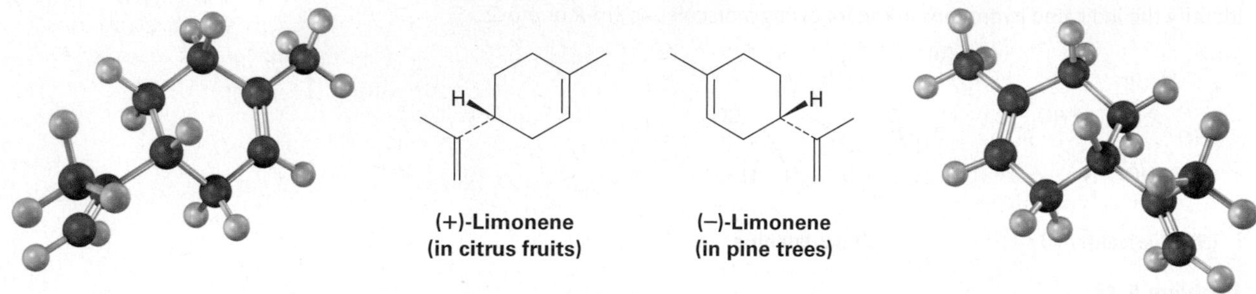

(+)-Limonene (in citrus fruits) **(−)-Limonene (in pine trees)**

More dramatic examples of how a change in chirality can affect the biological properties of a molecule are found in many drugs, such as fluoxetine, a heavily prescribed medication sold under the trade name Prozac. Racemic fluoxetine is an extraordinarily effective antidepressant but has no activity against migraine. The pure *S* enantiomer, however, works remarkably well in preventing migraine. Other examples of how chirality affects biological properties are given in *A Deeper Look* at the end of this chapter.

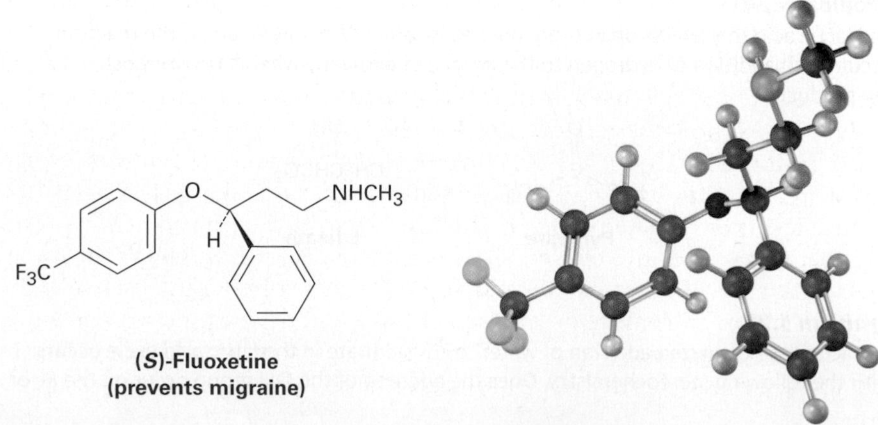

(*S*)-Fluoxetine (prevents migraine)

Why do different enantiomers have different biological properties? To have a biological effect, a substance typically must fit into an appropriate receptor that has an exactly complementary shape. But because biological receptors are chiral, only one enantiomer of a chiral substrate can fit in, just as only a right hand can fit into right-handed glove. The mirror-image enantiomer will be a misfit, like a left hand in a right-handed glove. A representation of the interaction between a chiral molecule and a chiral biological receptor is shown in **Figure 5.15**: one enantiomer fits the receptor perfectly, but the other does not.

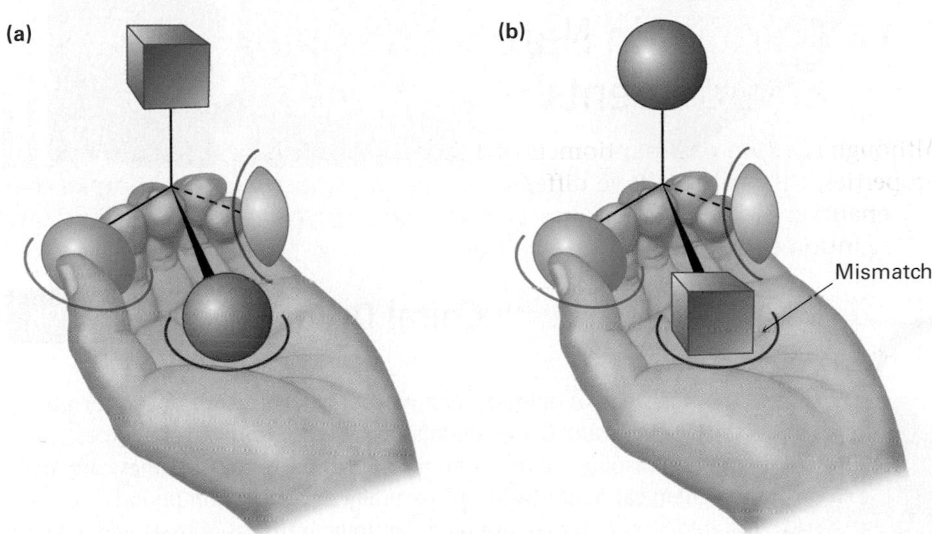

Figure 5.15 Imagine that a left hand interacts with a chiral object, much as a biological receptor interacts with a chiral molecule. **(a)** One enantiomer fits into the hand perfectly: **green thumb**, **red palm**, and **gray pinkie finger**, with the **blue substituent exposed.** **(b)** The other enantiomer, however, can't fit into the hand. When the green thumb and gray pinkie finger interact appropriately, the palm holds a blue substituent rather than a red one, with the **red substituent exposed.**

The hand-in-glove fit of a chiral substrate into a chiral receptor is relatively straightforward, but it's less obvious how a prochiral substrate can undergo a selective reaction. Take the reaction of ethanol with NAD^+ catalyzed by yeast alcohol dehydrogenase. As we saw at the end of **Section 5.11**, the reaction occurs with exclusive removal of the *pro-R* hydrogen from ethanol and with addition only to the *Re* face of the NAD^+ carbon.

We can understand this result by imagining that the chiral enzyme receptor again has three binding sites, as was previously the case in Figure 5.15. When green and gray substituents of a prochiral substrate are held appropriately, however, only one of the two red substituents—say, the *pro-S* one—is also held while the other, *pro-R,* substituent is exposed for reaction.

We describe the situation by saying that the receptor provides a **chiral environment** for the substrate. In the absence of a chiral environment, the two red substituents are chemically identical, but in the presence of the chiral environment, they are chemically distinctive **(Figure 5.16a)**. The situation is similar to what happens when you pick up a coffee mug. By itself, the mug has a plane of symmetry and is achiral. When you pick up the mug, however, your hand provides a chiral environment so that one side becomes much more accessible and easier to drink from than the other **(Figure 5.16b)**.

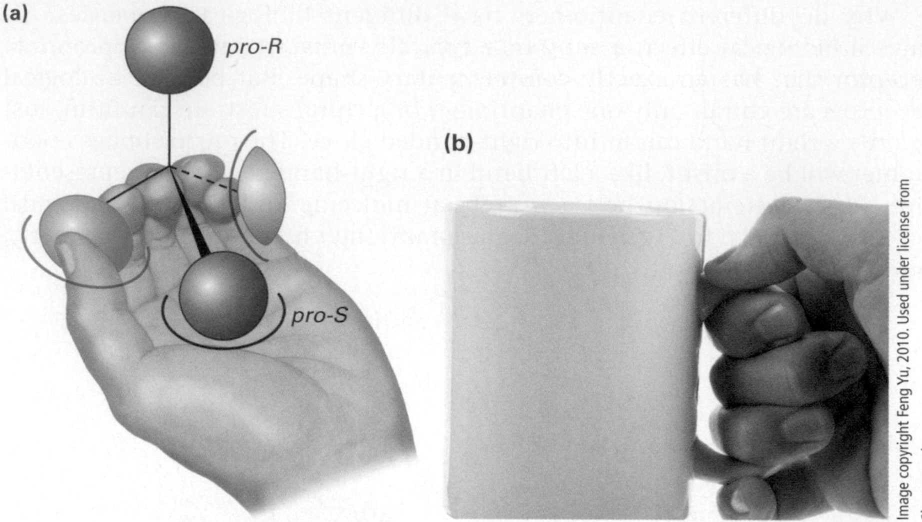

Figure 5.16 **(a)** When a prochiral molecule is held in a chiral environment, the **two seemingly identical substituents** are distinguishable. **(b)** Similarly, when an achiral coffee mug is held in the chiral environment of your hand, it's much easier to drink from one side than the other because the two sides of the mug are now distinguishable.

Chiral Drugs | A DEEPER LOOK

The *S* enantiomer of ibuprofen soothes the aches and pains of athletic injuries much more effectively than the *R* enantiomer.

The hundreds of different pharmaceutical agents approved for use by the U.S. Food and Drug Administration come from many sources. Many drugs are isolated directly from plants or bacteria, and others are made by chemical modification of naturally occurring compounds. An estimated 33%, however, are made entirely in the laboratory and have no relatives in nature.

Those drugs that come from natural sources, either directly or after chemical modification, are usually chiral and are generally found only as a single enantiomer rather than as a racemate. Penicillin V, for example, an antibiotic isolated from the *Penicillium* mold, has the 2*S*,5*R*,6*R* configuration. Its enantiomer, which does not occur naturally but can be made in the laboratory, has no antibiotic activity.

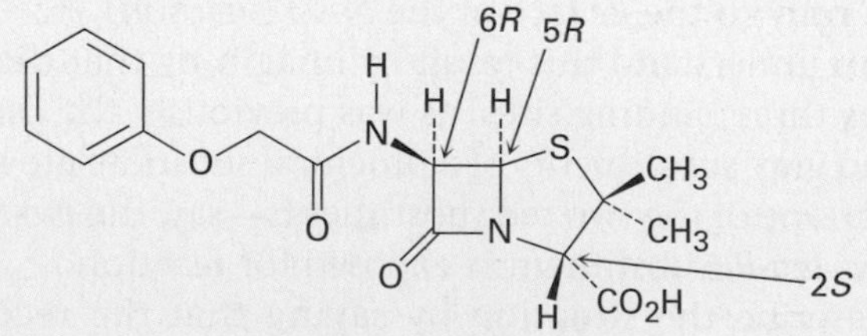

Penicillin V (2*S*,5*R*,6*R* configuration)

In contrast to drugs from natural sources, those drugs that are made entirely in the laboratory either are achiral or, if chiral, are often produced and sold as racemates. Ibuprofen, for example, has one chirality center and is sold commercially under such trade names as Advil, Nuprin, and Motrin as a 50 : 50 mixture of *R* and *S*. It turns out, however, that only

(continued)

(continued)

the *S* enantiomer is active as an analgesic and anti-inflammatory agent. The *R* enantiomer of ibuprofen is inactive, although it is slowly converted in the body to the active *S* form.

H CO_2H C CH_3

(*S*)-Ibuprofen
(an active analgesic agent)

Not only is it chemically wasteful to synthesize and administer an enantiomer that does not serve the intended purpose, many instances are now known where the presence of the "wrong" enantiomer in a racemic mixture either affects the body's ability to utilize the "right" enantiomer or has unintended pharmacological effects of its own. The presence of (*R*)-ibuprofen in the racemic mixture, for instance, slows the rate at which the *S* enantiomer takes effect in the body, from 12 minutes to 38 minutes.

To get around this problem, pharmaceutical companies attempt to devise methods of *enantioselective synthesis*, which allow them to prepare only a single enantiomer rather than a racemic mixture. Viable methods have been developed for the preparation of (*S*)-ibuprofen, which is now being marketed in Europe. We'll look further into enantioselective synthesis in the Chapter 19 *A Deeper Look.*

Summary

In this chapter, we've looked at some of the causes and consequences of molecular handedness—a topic of particular importance in understanding biological chemistry. The subject can be a bit complex but is so important that it's worthwhile spending the time needed to become familiar with it.

An object or molecule that is not superimposable on its mirror image is said to be **chiral**, meaning "handed." A chiral molecule is one that does not have a plane of symmetry cutting through it so that one half is a mirror image of the other half. The most common cause of chirality in organic molecules is the presence of a tetrahedral, sp^3-hybridized carbon atom bonded to four different groups—a so-called **chirality center**. Chiral compounds can exist as a pair of nonsuperimposable mirror-image stereoisomers called **enantiomers**. Enantiomers are identical in all physical properties except for their **optical activity**, or direction in which they rotate plane-polarized light.

The stereochemical **configuration** of a chirality center can be specified as either ***R*** (*rectus*) or ***S*** (*sinister*) by using the **Cahn–Ingold–Prelog rules**. First

Key words

absolute configuration, 154
achiral, 144
Cahn–Ingold–Prelog rules, 151
chiral, 144
chiral environment, 171
chirality center, 145
configuration, 150
dextrorotatory, 148
diastereomers, 157
enantiomers, 143
epimers, 157
levorotatory, 148

Key words—cont'd

meso compound, 159
optically active, 148
***pro-R* configuration, 168**
***pro-S* configuration, 168**
prochiral, 167
prochirality center, 167
***R* configuration, 152**
racemate, 161
***Re* face, 167**
resolution, 161
***S* configuration, 152**
***Si* face, 167**
specific rotation, $[\alpha]_D$, 148

rank the four substituents on the chiral carbon atom, and then orient the molecule so that the lowest-ranked group points directly back. If a curved arrow drawn in the direction of decreasing rank ($1 \rightarrow 2 \rightarrow 3$) for the remaining three groups is clockwise, the chirality center has the *R* configuration. If the direction is counterclockwise, the chirality center has the *S* configuration.

Some molecules have more than one chirality center. Enantiomers have opposite configuration at all chirality centers, whereas **diastereomers** have the same configuration in at least one center but opposite configurations at the others. **Epimers** are diastereomers that differ in configuration at only one chirality center. A compound with *n* chirality centers can have a maximum of 2n stereoisomers.

Meso compounds contain chirality centers but are achiral overall because they have a plane of symmetry. Racemic mixtures, or **racemates**, are 50:50 mixtures of (+) and (−) enantiomers. Racemates and individual diastereomers differ in their physical properties, such as solubility, melting point, and boiling point.

A molecule is **prochiral** if it can be converted from achiral to chiral in a single chemical step. A prochiral sp^2-hybridized atom has two faces, described as either ***Re*** or ***Si***. An sp^3-hybridized atom is a **prochirality center** if, by changing one of its attached atoms, a chirality center results. The atom whose replacement leads to an *R* chirality center is ***pro-R***, and the atom whose replacement leads to an *S* chirality center is ***pro-S***.

Exercises

OWL Interactive versions of these problems are assignable in OWL for Organic Chemistry.

▲ denotes problems linked to the Key Ideas in this chapter.

Visualizing Chemistry

(Problems 5.1–5.25 appear within the chapter.)

5.26 Which of the following structures are identical? (Green = Cl.)

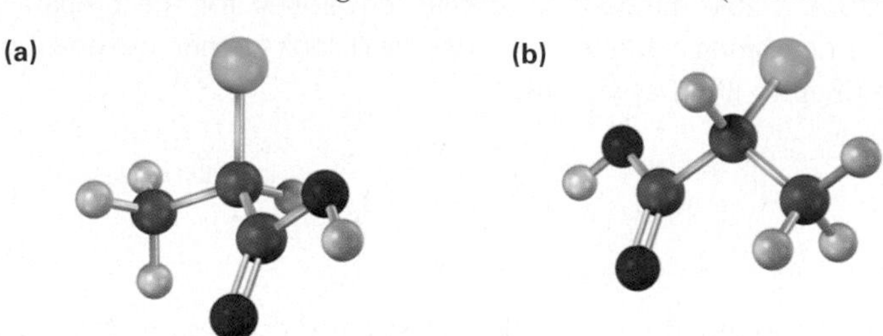

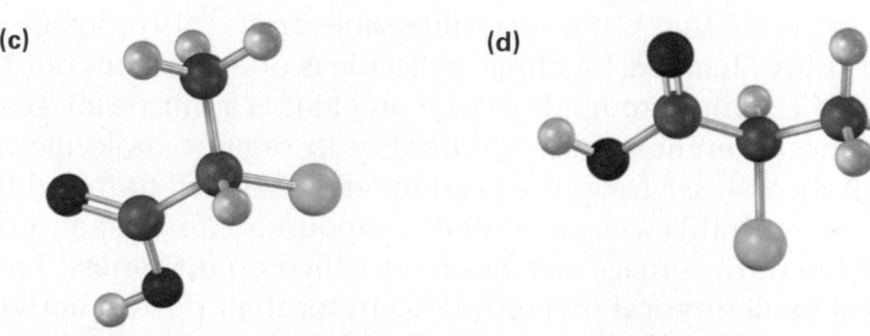

▲ Problems linked to Key Ideas in this chapter

5.27 ▲ Assign *R* or *S* configuration to the chirality centers in the following molecules (blue = N):

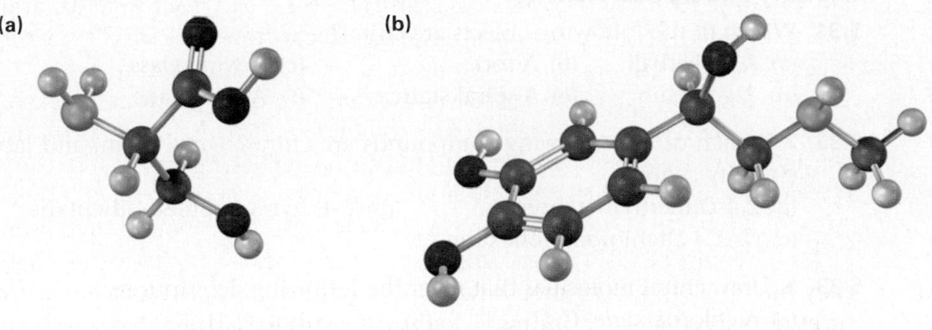

5.28 Which, if any, of the following structures represent meso compounds? (Blue = N, green = Cl.)

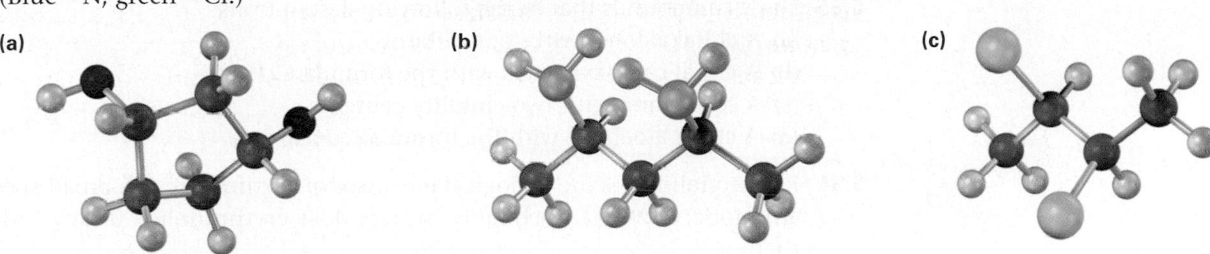

5.29 ▲ Assign *R* or *S* configuration to each chirality center in pseudoephedrine, an over-the-counter decongestant found in cold remedies (blue = N).

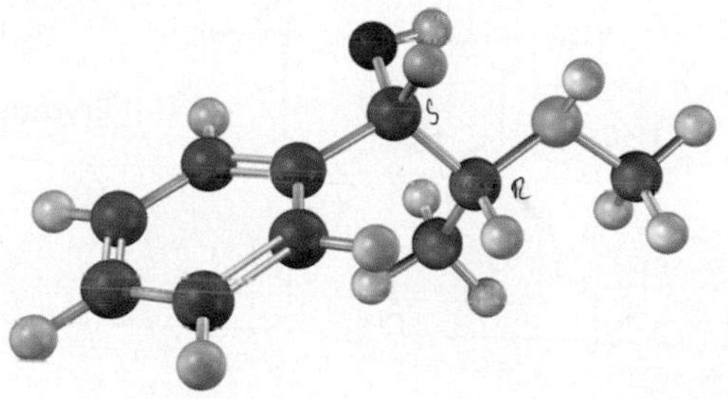

5.30 Orient each of the following drawings so that the lowest-ranked group is toward the rear, and then assign *R* or *S* configuration:

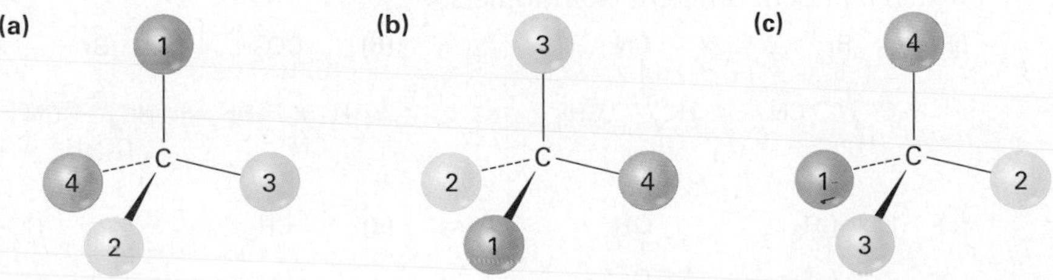

▲ Problems linked to Key Ideas in this chapter

Additional Problems

Chirality and Optical Activity

5.31 Which of the following objects are chiral?
(a) A basketball (b) A fork (c) A wine glass
(d) A golf club (e) A spiral staircase (f) A snowflake

5.32 ▲ Which of the following compounds are chiral? Draw them, and label the chirality centers.
(a) 2,4-Dimethylheptane (b) 5-Ethyl-3,3-dimethylheptane
(c) *cis*-1,4-Dichlorocyclohexane

5.33 ▲ Draw chiral molecules that meet the following descriptions:
(a) A chloroalkane, $C_5H_{11}Cl$ (b) An alcohol, $C_6H_{14}O$
(c) An alkene, C_6H_{12} (d) An alkane, C_8H_{18}

5.34 ▲ Eight alcohols have the formula $C_5H_{12}O$. Draw them. Which are chiral?

5.35 Draw compounds that fit the following descriptions:
(a) A chiral alcohol with four carbons
(b) A chiral carboxylic acid with the formula $C_5H_{10}O_2$
(c) A compound with two chirality centers
(d) A chiral aldehyde with the formula C_3H_5BrO

5.36 Erythronolide B is the biological precursor of erythromycin, a broad-spectrum antibiotic. How many chirality centers does erythronolide B have? Identify them.

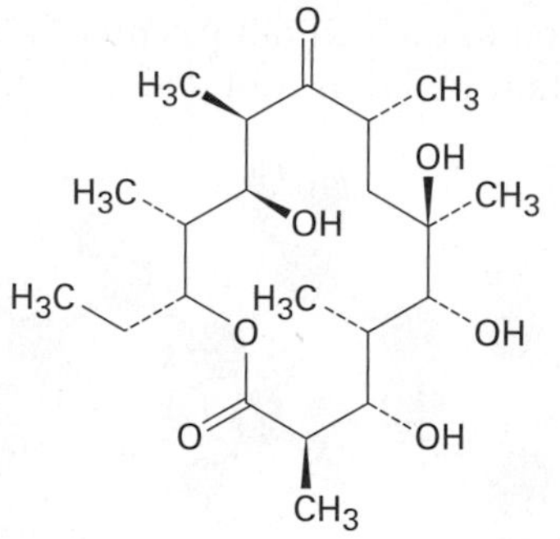

O
H3C CH3
OH
H3C CH3
OH
H3C H3C OH
O
O OH
CH3

Erythronolide B

Assigning Configuration to Chirality Centers

5.37 Which of the following pairs of structures represent the same enantiomer, and which represent different enantiomers?

(a) Br; H_3C–C–CN; H — CN; H–C–CH_3; Br

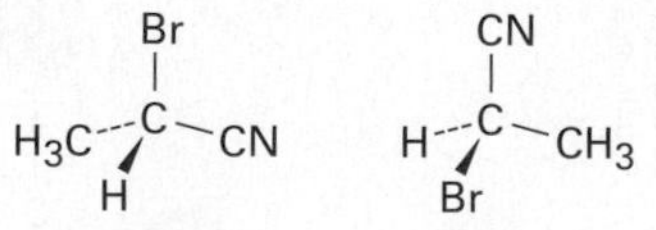

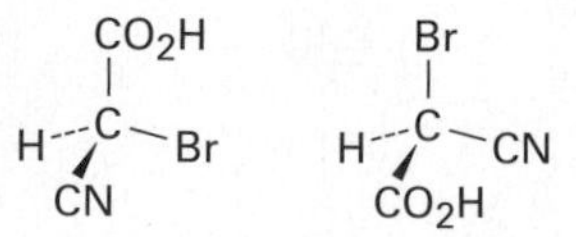

(b) CO_2H; H–C–Br; CN — Br; H–C–CN; CO_2H

(c) CH_3; H–C–OH; CH_3CH_2 — OH; H–C–CH_2CH_3; H_3C

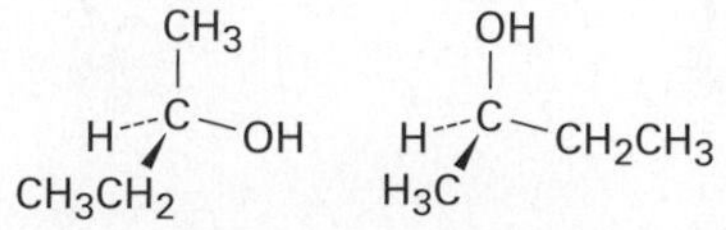

(d) CH_3; H–C–CO_2H; H_2N — CO_2H; H_3C–C–H; H_2N

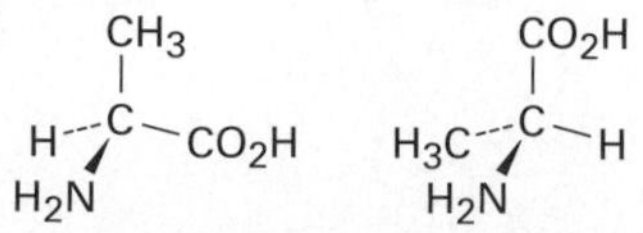

▲ Problems linked to Key Ideas in this chapter

5.38 What is the relationship between the specific rotations of (2*R*,3*R*)-dichloropentane and (2*S*,3*S*)-dichloropentane? Between (2*R*,3*S*)-dichloropentane and (2*R*,3*R*)-dichloropentane?

5.39 What is the stereochemical configuration of the enantiomer of (2*S*,4*R*)-2,4-octanediol? (A diol is a compound with two –OH groups.)

5.40 What are the stereochemical configurations of the two diastereomers of (2*S*,4*R*)-2,4-octanediol? (A diol is a compound with two –OH groups.)

5.41 Orient each of the following drawings so that the lowest-ranked group is toward the rear, and then assign *R* or *S* configuration:

(a)

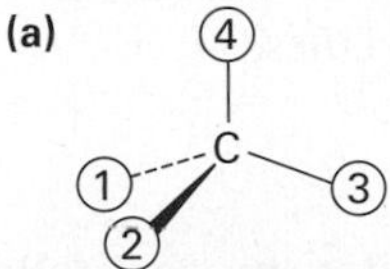

(b)

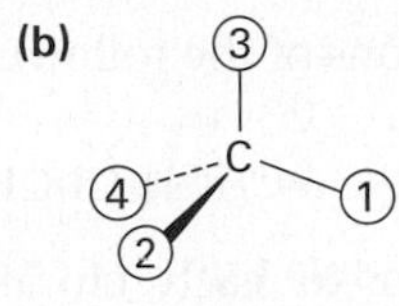

(c)

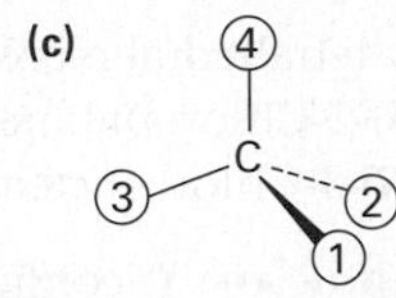

5.42 Assign Cahn–Ingold–Prelog rankings to the following sets of substituents:

(a) $-CH{=}CH_2$, $-CH(CH_3)_2$, $-C(CH_3)_3$, $-CH_2CH_3$

(b) $-C{\equiv}CH$, $-CH{=}CH_2$, $-C(CH_3)_3$,

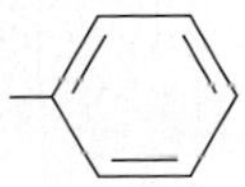

(c) $-CO_2CH_3$, $-COCH_3$, $-CH_2OCH_3$, $-CH_2CH_3$

(d) $-C{\equiv}N$, $-CH_2Br$, $-CH_2CH_2Br$, $-Br$

5.43 Assign *R* or *S* configurations to the chirality centers in the following molecules:

(a) H OH

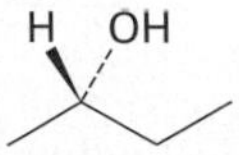

(b)

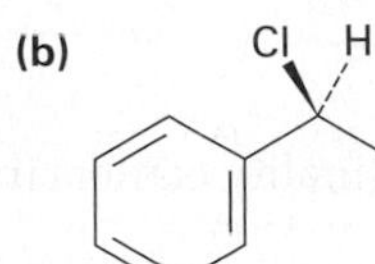

(c) H OCH_3 $HOCH_2$ CO_2H

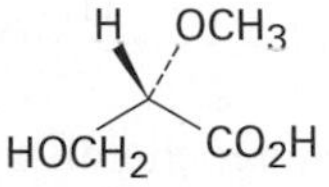

5.44 Assign *R* or *S* configuration to each chirality center in the following molecules:

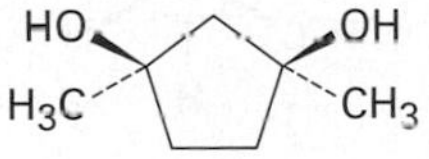

(a) OH H Cl H

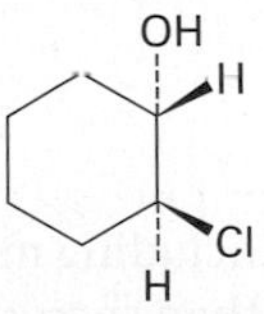

(b) H CH_3 CH_3CH_2 H

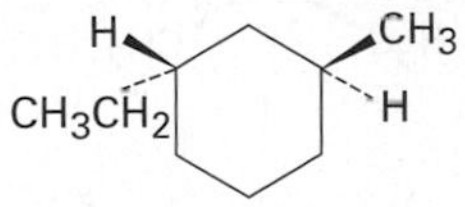

(c) HO OH H_3C CH_3

5.45 Assign *R* or *S* configuration to each chirality center in the following biological molecules:

(a) Biotin $CH_2CH_2CH_2CH_2CO_2^-$

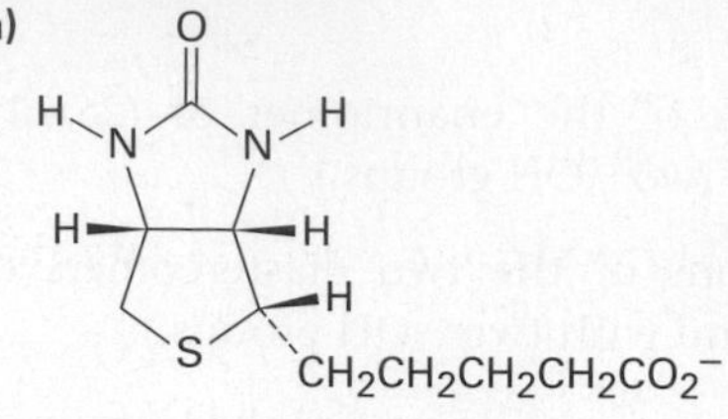

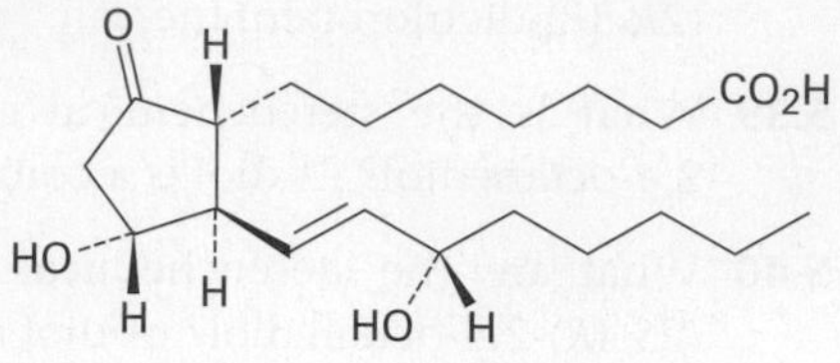

(b) Prostaglandin E_1 CO_2H

5.46 Draw tetrahedral representations of the following molecules:

(a) (*S*)-2-Chlorobutane

(b) (*R*)-3-Chloro-1-pentene [$H_2C{=}CHCH(Cl)CH_2CH_3$]

5.47 Assign *R* or *S* configuration to each chirality center in the following molecules:

(a) H Br H_3C CH_3 H Br

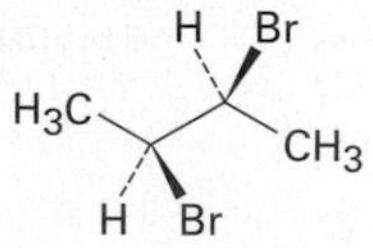

(b) H H H NH_2 CO_2H H OH

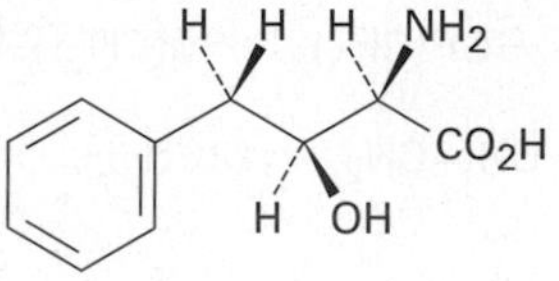

5.48 Assign *R* or *S* configurations to the chirality centers in ascorbic acid (vitamin C).

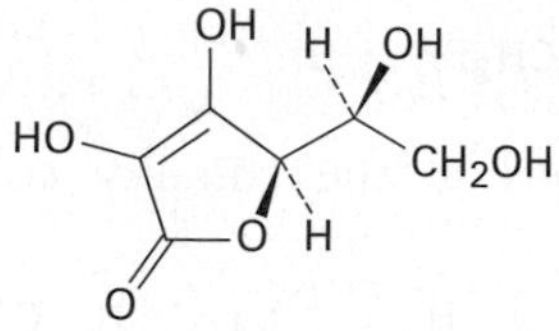

OH H OH HO CH_2OH O H O **Ascorbic acid**

5.49 Assign *R* or *S* stereochemistry to the chirality centers in the following Newman projections:

(a) Cl H CH_3 H_3C H H

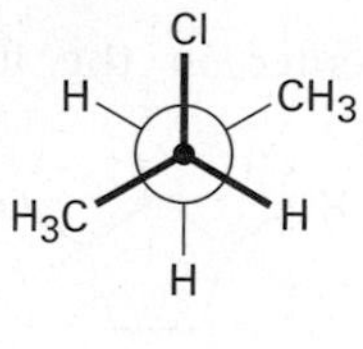

(b) H H_3C OH H_3C CH_3 H

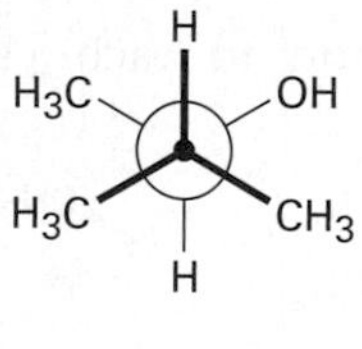

5.50 Xylose is a common sugar found in many types of wood, including maple and cherry. Because it is much less prone to cause tooth decay than sucrose, xylose has been used in candy and chewing gum. Assign *R* or *S* configurations to the chirality centers in xylose.

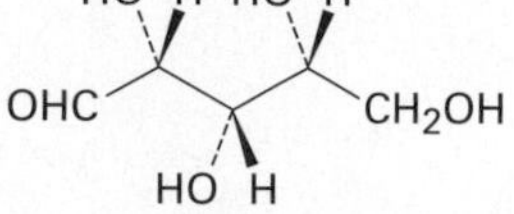

HO H HO H OHC CH_2OH HO H **(+)-Xylose**

▲ Problems linked to Key Ideas in this chapter

Meso Compounds

5.51 Draw examples of the following:

(a) A meso compound with the formula C_8H_{18}

(b) A meso compound with the formula C_9H_{20}

(c) A compound with two chirality centers, one *R* and the other *S*

5.52 Draw the meso form of each of the following molecules, and indicate the plane of symmetry in each:

(a) 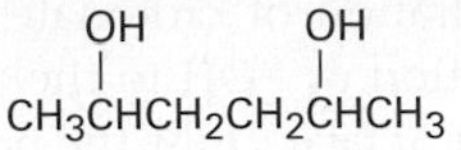 $CH_3CH(OH)CH_2CH_2CH(OH)CH_3$

(b) 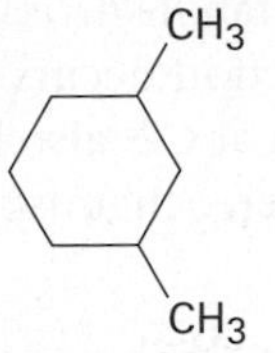 CH_3, CH_3 (dimethylcyclohexane)

(c) 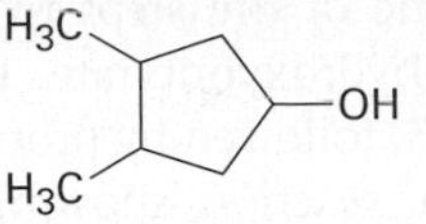 H_3C, OH, H_3C (cyclopentane)

5.53 Draw the structure of a meso compound that has five carbons and three chirality centers.

5.54 Ribose, an essential part of ribonucleic acid (RNA), has the following structure:

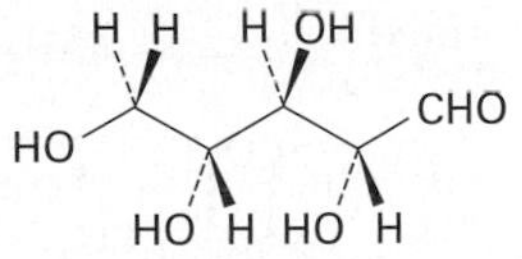

HO, H H, H OH, CHO, HO H, HO H — **Ribose**

(a) How many chirality centers does ribose have? Identify them.

(b) How many stereoisomers of ribose are there?

(c) Draw the structure of the enantiomer of ribose.

(d) Draw the structure of a diastereomer of ribose.

5.55 On reaction with hydrogen gas with a platinum catalyst, ribose (Problem 5.54) is converted into ribitol. Is ribitol optically active or inactive? Explain.

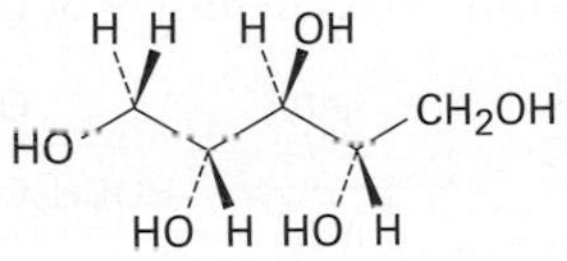

HO, H H, H OH, CH_2OH, HO H, HO H — **Ribitol**

Prochirality

5.56 Identify the indicated hydrogens in the following molecules as *pro-R* or *pro-S*:

(a) 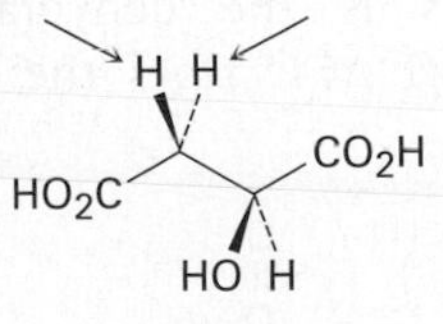 H H, HO_2C, CO_2H, HO H — **Malic acid**

(b) 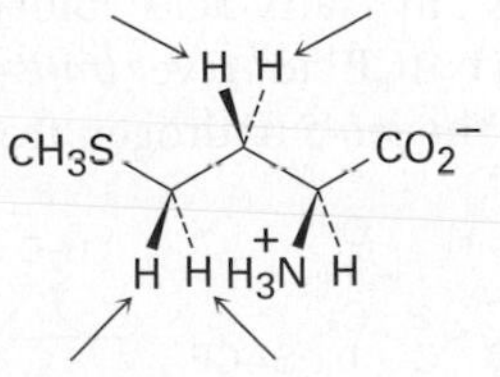 H H, CH_3S, CO_2^-, H H $H_3\overset{+}{N}$ H — **Methionine**

(c) 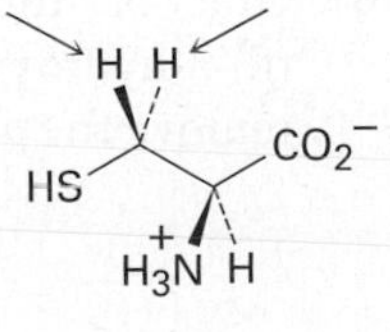 H H, HS, CO_2^-, $H_3\overset{+}{N}$ H — **Cysteine**

▲ Problems linked to Key Ideas in this chapter

5.57 Identify the indicated faces in the following molecules as *Re* or *Si:*

(a) Pyruvate

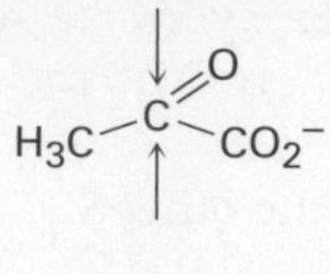

$H_3C-C(=O)-CO_2^-$

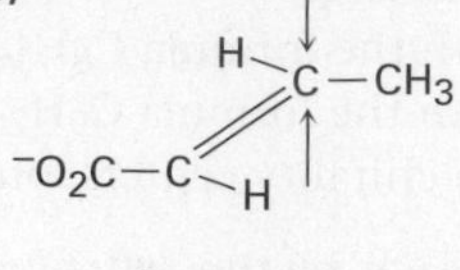

(b) Crotonate

$H-C(CH_3)=C(H)-CO_2^-$

5.58 One of the steps in fat metabolism is the hydration of crotonate to yield 3-hydroxybutyrate. The reaction occurs by addition of –OH to the *Si* face at C3, followed by protonation at C2, also from the *Si* face. Draw the product of the reaction, showing the stereochemistry of each step.

Crotonate ⟶ 3-Hydroxybutyrate ($CH_3CH(OH)CH_2CO_2^-$)

5.59 The dehydration of citrate to yield *cis*-aconitate, a step in the citric acid cycle, involves the *pro-R* "arm" of citrate rather than the *pro-S* arm. Which of the following two products is formed?

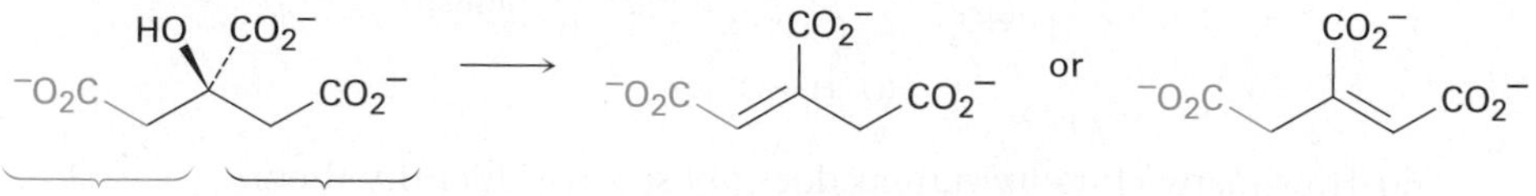

Citrate ⟶ *cis*-Aconitate or *cis*-Aconitate

5.60 The first step in the metabolism of glycerol, formed by digestion of fats, is phosphorylation of the *pro-R* $-CH_2OH$ group by reaction with adenosine triphosphate (ATP) to give the corresponding glycerol phosphate plus adenosine diphosphate (ADP). Show the stereochemistry of the product.

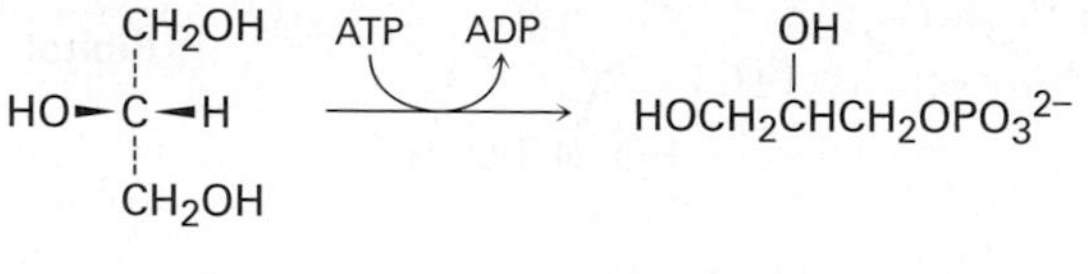

Glycerol ($HO-C(H)(CH_2OH)_2$) —ATP → ADP⟶ Glycerol phosphate ($HOCH_2CH(OH)CH_2OPO_3^{2-}$)

5.61 One of the steps in fatty-acid biosynthesis is the dehydration of (*R*)-3-hydroxybutyryl ACP to give *trans*-crotonyl ACP. Does the reaction remove the *pro-R* or the *pro-S* hydrogen from C2?

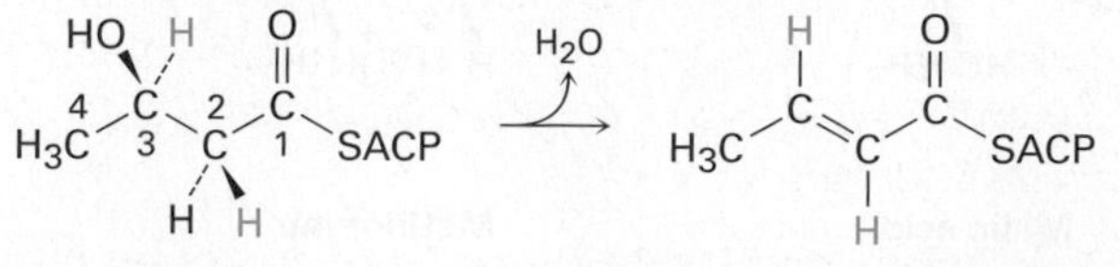

(*R*)-3-Hydroxybutyryl ACP —H_2O⟶ *trans*-Crotonyl ACP

▲ Problems linked to Key Ideas in this chapter

General Problems

5.62 Draw all possible stereoisomers of 1,2-cyclobutanedicarboxylic acid, and indicate the interrelationships. Which, if any, are optically active? Do the same for 1,3-cyclobutanedicarboxylic acid.

5.63 Draw tetrahedral representations of the two enantiomers of the amino acid cysteine, $HSCH_2CH(NH_2)CO_2H$, and identify each as *R* or *S*.

5.64 The naturally occurring form of the amino acid cysteine (Problem 5.63) has the *S* configuration at its chirality center. On treatment with a mild oxidizing agent, two cysteines join to give cystine, a disulfide. Assuming that the chirality center is not affected by the reaction, is cystine optically active? Explain.

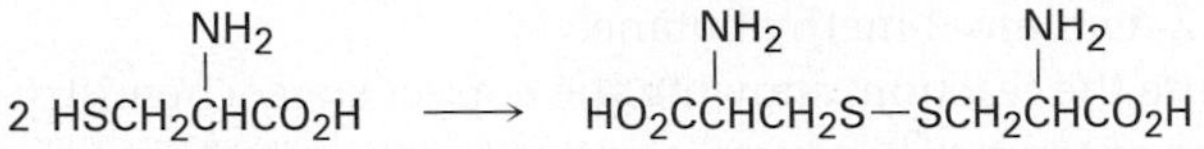

$$2\ HSCH_2CH(NH_2)CO_2H \longrightarrow HO_2CCH(NH_2)CH_2S{-}SCH_2CH(NH_2)CO_2H$$

Cysteine **Cystine**

5.65 Draw tetrahedral representations of the following molecules:

(a) The 2*S*,3*R* enantiomer of 2,3-dibromopentane

(b) The meso form of 3,5-heptanediol

5.66 Assign *R,S* configurations to the chiral centers in cephalexin, trade-named Keflex, the most widely prescribed antibiotic in the United States.

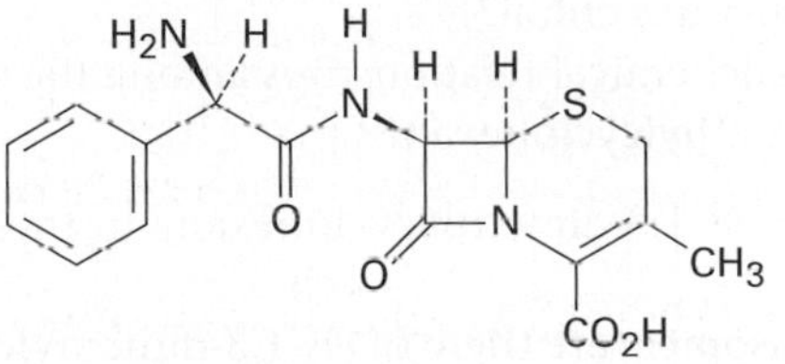

Cephalexin

5.67 Chloramphenicol, a powerful antibiotic isolated in 1949 from the *Streptomyces venezuelae* bacterium, is active against a broad spectrum of bacterial infections and is particularly valuable against typhoid fever. Assign *R,S* configurations to the chirality centers in chloramphenicol.

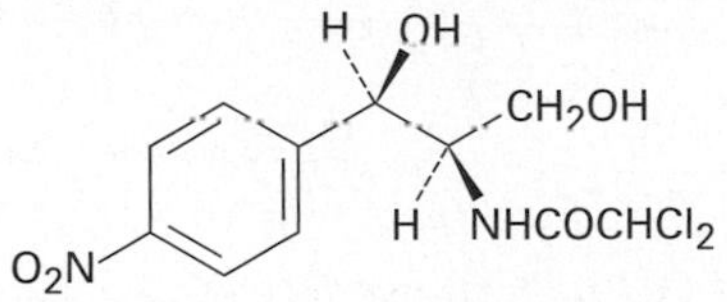

Chloramphenicol

5.68 *Allenes* are compounds with adjacent carbon–carbon double bonds. Many allenes are chiral, even though they don't contain chirality centers. Mycomycin, for example, a naturally occurring antibiotic isolated from the bacterium *Nocardia acidophilus,* is chiral and has $[\alpha]_D = -130$. Explain why mycomycin is chiral.

$$HC{\equiv}C{-}C{\equiv}C{-}CH{=}C{=}CH{-}CH{=}CH{-}CH{=}CH{-}CH_2CO_2H$$

Mycomycin

▲ Problems linked to Key Ideas in this chapter

5.69 Long before chiral allenes were known (Problem 5.68), the resolution of 4-methylcyclohexylideneacetic acid into two enantiomers had been carried out. Why is it chiral? What geometric similarity does it have to allenes?

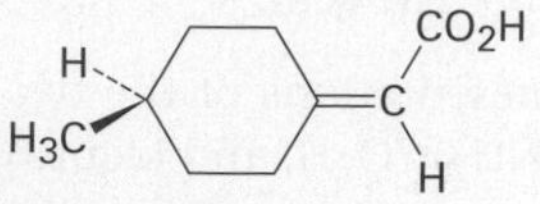

4-Methylcyclohexylideneacetic acid

5.70 (*S*)-1-Chloro-2-methylbutane undergoes light-induced reaction with Cl_2 to yield a mixture of products, among which are 1,4-dichloro-2-methylbutane and 1,2-dichloro-2-methylbutane.

(a) Write the reaction, showing the correct stereochemistry of the reactant.

(b) One of the two products is optically active, but the other is optically inactive. Which is which?

5.71 How many stereoisomers of 2,4-dibromo-3-chloropentane are there? Draw them, and indicate which are optically active.

5.72 Draw both *cis*- and *trans*-1,4-dimethylcyclohexane in their more stable chair conformations.

(a) How many stereoisomers are there of *cis*-1,4-dimethylcyclohexane, and how many of *trans*-1,4-dimethylcyclohexane?

(b) Are any of the structures chiral?

(c) What are the stereochemical relationships among the various stereoisomers of 1,4-dimethylcyclohexane?

5.73 Draw both *cis*- and *trans*-1,3-dimethylcyclohexane in their more stable chair conformations.

(a) How many stereoisomers are there of *cis*-1,3-dimethylcyclohexane, and how many of *trans*-1,3-dimethylcyclohexane?

(b) Are any of the structures chiral?

(c) What are the stereochemical relationships among the various stereoisomers of 1,3-dimethylcyclohexane?

5.74 *cis*-1,2-Dimethylcyclohexane is optically inactive even though it has two chirality centers. Explain.

5.75 We'll see in Chapter 11 that alkyl halides react with hydrosulfide ion (HS^-) to give a product whose stereochemistry is *inverted* from that of the reactant.

$$\text{C—Br} \xrightarrow{HS^-} \text{HS—C} + Br^-$$

An alkyl bromide

Draw the reaction of (*S*)-2-bromobutane with HS^- ion to yield 2-butanethiol, $CH_3CH_2CH(SH)CH_3$. Is the stereochemistry of the product *R* or *S*?

5.76 Ketones react with sodium acetylide (the sodium salt of acetylene, $Na^{+-}:C\equiv CH$) to give alcohols. For example, the reaction of sodium acetylide with 2-butanone yields 3-methyl-1-pentyn-3-ol:

$$H_3C-C(=O)-CH_2CH_3 \xrightarrow[\text{2. } H_3O^+]{\text{1. } Na^{+-}:C\equiv CH} HC\equiv C-C(CH_3)(OH)-CH_2CH_3$$

2-Butanone **3-Methyl-1-pentyn-3-ol**

(a) Is the product chiral?

(b) Assuming that the reaction takes place with equal likelihood from both *Re* and *Si* faces of the carbonyl group, is the product optically active? Explain.

5.77 Imagine that a reaction similar to that in Problem 5.76 is carried out between sodium acetylide and (*R*)-2-phenylpropanal to yield 4-phenyl-1-pentyn-3-ol:

$$\xrightarrow[\text{2. } H_3O^+]{\text{1. } Na^{+-}:C\equiv CH}$$

(*R*)-2-Phenylpropanal **4-Phenyl-1-pentyn-3-ol**

(a) Is the product chiral?

(b) Draw both major and minor reaction products, assuming that the reaction takes place preferentially from the *Re* face of the carbonyl group. Is the product mixture optically active? Explain.

▲ Problems linked to Key Ideas in this chapter

6

Many chemical reactions are like these balanced rocks. They need a shove of energy to get them started moving.

An Overview of Organic Reactions

OWL Sign in to OWL for Organic Chemistry at **www.cengage.com/owl** to view tutorials and simulations, develop problem-solving skills, and complete online homework assigned by your professor.

When first approached, organic chemistry might seem overwhelming. It's not so much that any one part is difficult to understand, it's that there are so many parts: tens of millions of compounds, dozens of functional groups, and an apparently endless number of reactions. With study, though, it becomes evident that there are only a few fundamental ideas that underlie all organic reactions. Far from being a collection of isolated facts, organic chemistry is a beautifully logical subject that is unified by a few broad themes. When these themes are understood, learning organic chemistry becomes much easier and memorization is minimized. The aim of this book is to describe the themes and clarify the patterns that unify organic chemistry.

Why This Chapter? All chemical reactions, whether they take place in the laboratory or in living organisms, follow the same "rules." Reactions in living organisms often look more complex than laboratory reactions because of the size of the biomolecules and the involvement of biological catalysts called *enzymes,* but the principles governing all reactions are the same.

To understand both organic and biological chemistry, it's necessary to know not just *what* occurs but also *why* and *how* chemical reactions take place. In this chapter, we'll start with an overview of the fundamental kinds of organic reactions, we'll see why reactions occur, and we'll see how reactions can be described. Once this background is out of the way, we'll then be ready to begin studying the details of organic chemistry.

6.1 Kinds of Organic Reactions

Organic chemical reactions can be organized broadly in two ways—by *what kinds* of reactions occur and by *how* those reactions occur. Let's look first at the kinds of reactions that take place. There are four general types of organic reactions: *additions, eliminations, substitutions,* and *rearrangements.*

* **Addition reactions** occur when two reactants add together to form a single product with no atoms "left over." An example that we'll be

studying soon is the reaction of an alkene, such as ethylene, with HBr to yield an alkyl bromide.

These two reactants . . . $H_2C{=}CH_2$ + H—Br ⟶ $H{-}CH_2{-}CH_2Br$ . . . add to give this product.

Ethylene (an alkene) **Bromoethane (an alkyl halide)**

* **Elimination reactions** are, in a sense, the opposite of addition reactions. They occur when a single reactant splits into two products, often with formation of a small molecule such as water or HBr. An example is the acid-catalyzed reaction of an alcohol to yield water and an alkene.

This one reactant . . . $H{-}CH_2{-}CH_2OH$ ⇌ (Acid catalyst) $H_2C{=}CH_2$ + H_2O . . . gives these two products.

Ethanol (an alcohol) **Ethylene (an alkene)**

* **Substitution reactions** occur when two reactants exchange parts to give two new products. An example is the reaction of an ester such as methyl acetate with water to yield a carboxylic acid plus an alcohol. Similar reactions occur in many biological pathways, including the metabolism of dietary fats.

These two reactants... $H_3C{-}C(=O){-}O{-}CH_3$ + H—O—H ⟶ (Acid catalyst) $H_3C{-}C(=O){-}O{-}H$ + H—O—CH_3 ...give these two products.

Methyl acetate (an ester) **Acetic acid (a carboxylic acid)** **Methanol (an alcohol)**

* **Rearrangement reactions** occur when a single reactant undergoes a reorganization of bonds and atoms to yield an isomeric product. An example is the conversion of dihydroxyacetone phosphate into its constitutional isomer glyceraldehyde 3-phosphate, a step in the glycolysis pathway by which carbohydrates are metabolized.

This reactant... $^{2-}O_3PO{-}CH_2{-}C(=O){-}CH_2{-}OH$ ⟶ $^{2-}O_3PO{-}CH_2{-}CH(OH){-}CH{=}O$ ...gives this isomeric product.

Dihydroxyacetone phosphate **Glyceraldehyde 3-phosphate**

Problem 6.1

Classify each of the following reactions as an addition, elimination, substitution, or rearrangement:

(a) $CH_3Br + KOH \rightarrow CH_3OH + KBr$

(b) $CH_3CH_2Br \rightarrow H_2C{=}CH_2 + HBr$

(c) $H_2C{=}CH_2 + H_2 \rightarrow CH_3CH_3$

6.2 How Organic Reactions Occur: Mechanisms

Having looked at the kinds of reactions that take place, let's now see how reactions occur. An overall description of how a reaction occurs is called a **reaction mechanism**. A mechanism describes in detail exactly what takes place at each stage of a chemical transformation—which bonds are broken and in what order, which bonds are formed and in what order, and what the relative rates of the steps are. A complete mechanism must also account for all reactants used and all products formed.

All chemical reactions involve bond-breaking and bond-making. When two molecules come together, react, and yield products, specific bonds in the reactant molecules are broken and specific bonds in the product molecules are formed. Fundamentally, there are two ways in which a covalent two-electron bond can break. A bond can break in an electronically *symmetrical* way so that one electron remains with each product fragment, or a bond can break in an electronically *unsymmetrical* way so that both bonding electrons remain with one product fragment, leaving the other with a vacant orbital. The symmetrical cleavage is said to be *homolytic,* and the unsymmetrical cleavage is said to be *heterolytic*.

We'll develop the point in more detail later, but you might note for now that the movement of *one* electron in the symmetrical process is indicated using a half-headed, or "fishhook," arrow (⌒), whereas the movement of *two* electrons in the unsymmetrical process is indicated using a full-headed curved arrow (↷).

A : B ⟶ A· + ·B — Symmetrical bond-breaking (radical): one bonding electron stays with each product.

A : B ⟶ A^+ + $:B^-$ — Unsymmetrical bond-breaking (polar): two bonding electrons stay with one product.

Just as there are two ways in which a bond can break, there are two ways in which a covalent two-electron bond can form. A bond can form in an electronically symmetrical way if one electron is donated to the new bond by each reactant or in an unsymmetrical way if both bonding electrons are donated by one reactant.

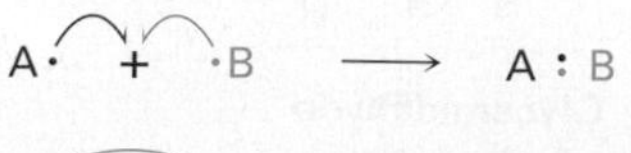

Symmetrical bond-making (radical): one bonding electron is donated by each reactant.

A^+ + $:B^-$ ⟶ A : B — Unsymmetrical bond-making (polar): two bonding electrons are donated by one reactant.

Processes that involve symmetrical bond-breaking and bond-making are called **radical reactions.** A **radical**, often called a *"free radical,"* is a neutral chemical species that contains an odd number of electrons and thus has a single, unpaired electron in one of its orbitals. Processes that involve unsymmetrical bond-breaking and bond-making are called **polar reactions**. Polar reactions involve species that have an even number of electrons and thus have only electron pairs in their orbitals. Polar processes are by far the more common reaction type in both organic and biological chemistry, and a large part of this book is devoted to their description.

In addition to polar and radical reactions, there is a third, less commonly encountered process called a *pericyclic reaction.* Rather than explain pericyclic reactions now, though, we'll look at them more carefully in Chapter 30.

6.3 Radical Reactions

Radical reactions are not as common as polar reactions but are nevertheless important in some industrial processes and biological pathways. Let's see briefly how they occur.

A radical is highly reactive because it contains an atom with an odd number of electrons (usually seven) in its valence shell, rather than a stable, noble-gas octet. A radical can achieve a valence-shell octet in several ways. For example, the radical might abstract an atom and one bonding electron from another reactant, leaving behind a new radical. The net result is a radical substitution reaction.

Unpaired electron → Rad· + A:B ⟶ Rad:A + ·B ← Unpaired electron

Reactant radical | **Substitution product** | **Product radical**

Alternatively, a reactant radical might add to a double bond, taking one electron from the double bond and yielding a new radical. The net result is a radical addition reaction.

Unpaired electron → Rad· + C=C ⟶ Rad–C–C· ← Unpaired electron

Reactant radical | **Alkene** | **Addition product radical**

An example of an industrially useful radical reaction is the chlorination of methane to yield chloromethane. This substitution reaction is the first step in the preparation of the solvents dichloromethane (CH_2Cl_2) and chloroform ($CHCl_3$).

$$CH_4 + Cl{-}Cl \xrightarrow{\text{Light}} CH_3Cl + H{-}Cl$$

Methane | **Chlorine** | **Chloromethane**

Like many radical reactions in the laboratory, methane chlorination requires three kinds of steps: *initiation, propagation,* and *termination.*

Initiation Irradiation with ultraviolet light begins the reaction by breaking the relatively weak Cl–Cl bond of a small number of Cl_2 molecules to give a few reactive chlorine radicals.

$$:\ddot{Cl}:\ddot{Cl}: \xrightarrow{\text{Light}} 2\ :\ddot{Cl}\cdot$$

Propagation Once produced, a reactive chlorine radical collides with a methane molecule in a propagation step, abstracting a hydrogen atom to give HCl and a methyl radical ($\cdot CH_3$). This methyl radical reacts further with Cl_2 in a second propagation step to give the product chloromethane plus a new chlorine radical, which cycles back and repeats the first propagation step. Thus, once the sequence has been initiated, it becomes a self-sustaining cycle of repeating steps (a) and (b), making the overall process a *chain reaction.*

$$\text{(a)}\ :\ddot{Cl}\cdot + H:CH_3 \longrightarrow H:\ddot{Cl}: + \cdot CH_3$$

$$\text{(b)}\ :\ddot{Cl}:\ddot{Cl}: + \cdot CH_3 \longrightarrow :\ddot{Cl}\cdot + :\ddot{Cl}:CH_3$$

Termination Occasionally, two radicals might collide and combine to form a stable product. When that happens, the reaction cycle is broken and the chain is ended. Such termination steps occur infrequently, however, because the concentration of radicals in the reaction at any given moment is very small. Thus, the likelihood that two radicals will collide is also small.

$$:\ddot{Cl}\cdot + \cdot\ddot{Cl}: \longrightarrow :\ddot{Cl}:\ddot{Cl}:$$

$$:\ddot{Cl}\cdot + \cdot CH_3 \longrightarrow :\ddot{Cl}:CH_3$$

$$H_3C\cdot + \cdot CH_3 \longrightarrow H_3C:CH_3$$

Possible termination steps

As a biological example of a radical reaction, look at the synthesis of *prostaglandins,* a large class of molecules found in virtually all body tissues and fluids. A number of pharmaceuticals are based on or derived from prostaglandins, including medicines that induce labor during childbirth, reduce intraocular pressure in glaucoma, control bronchial asthma, and help treat congenital heart defects.

Prostaglandin biosynthesis is initiated by abstraction of a hydrogen atom from arachidonic acid by an iron–oxygen radical, thereby generating a new, carbon radical in a substitution reaction. Don't be intimidated by the size of the molecules; focus on the changes occurring in each step. (To help you do that, the unchanged part of the molecule is "ghosted," with only the reactive part clearly visible.)

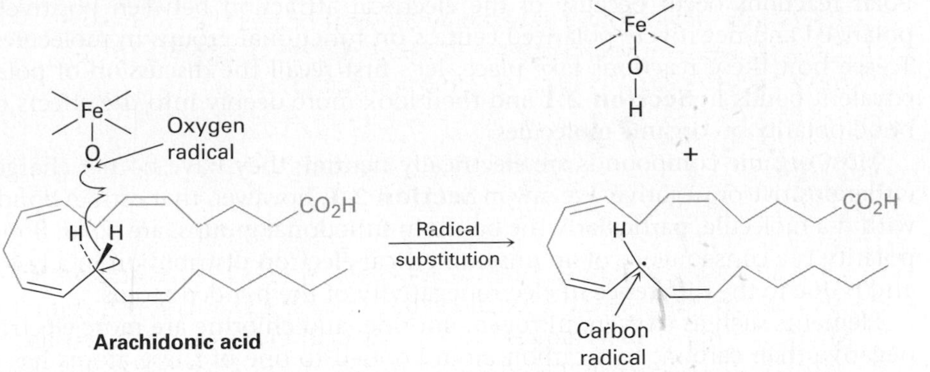

Fe Oxygen radical O H H CO_2H Radical substitution Fe O H + H CO_2H Carbon radical

Arachidonic acid

Following the initial abstraction of a hydrogen atom, the carbon radical then reacts with O_2 to give an oxygen radical, which reacts with a C=C bond within the same molecule in an addition reaction. Several further transformations ultimately yield prostaglandin H_2.

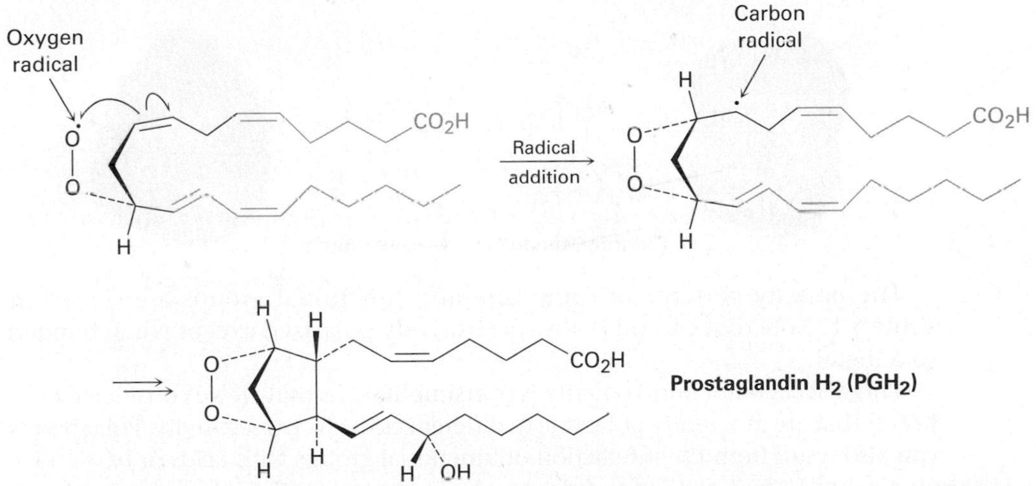

Oxygen radical O O H CO_2H Radical addition Carbon radical H O O H CO_2H

H H O O H H H OH CO_2H **Prostaglandin H_2 (PGH_2)**

Problem 6.2

Radical chlorination of alkanes is not generally useful because mixtures of products often result when more than one kind of C–H bond is present in the substrate. Draw and name all monochloro substitution products $C_6H_{13}Cl$ you might obtain by reaction of 2-methylpentane with Cl_2.

Problem 6.3

Using a curved fishhook arrow, propose a mechanism for formation of the cyclopentane ring of prostaglandin H_2.

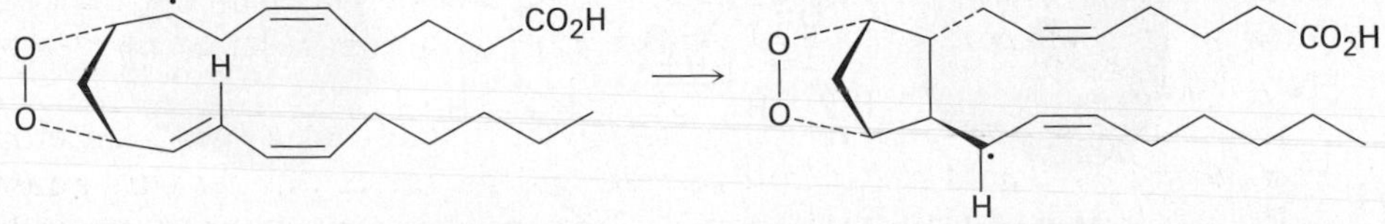

O O H CO_2H → O O H CO_2H

6.4 Polar Reactions

Polar reactions occur because of the electrical attraction between positively polarized and negatively polarized centers on functional groups in molecules. To see how these reactions take place, let's first recall the discussion of polar covalent bonds in **Section 2.1** and then look more deeply into the effects of bond polarity on organic molecules.

Most organic compounds are electrically neutral; they have no net charge, either positive or negative. We saw in **Section 2.1**, however, that certain bonds within a molecule, particularly the bonds in functional groups, are polar. Bond polarity is a consequence of an unsymmetrical electron distribution in a bond and is due to the difference in electronegativity of the bonded atoms.

Elements such as oxygen, nitrogen, fluorine, and chlorine are more electronegative than carbon, so a carbon atom bonded to one of these atoms has a partial positive charge ($\delta+$). Conversely, metals are less electronegative than carbon, so a carbon atom bonded to a metal has a partial negative charge ($\delta-$). Electrostatic potential maps of chloromethane and methyllithium illustrate these charge distributions, showing that the carbon atom in chloromethane is electron-poor (blue) while the carbon in methyllithium is electron-rich (red).

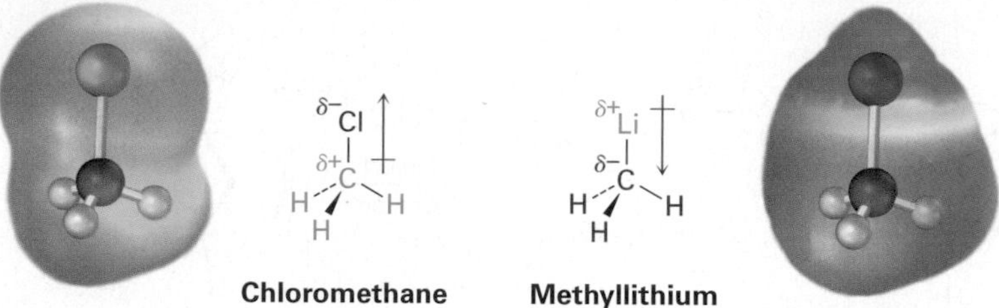

Chloromethane **Methyllithium**

The polarity patterns of some common functional groups are shown in Table 6.1. Note that carbon is always positively polarized except when bonded to a metal.

This discussion of bond polarity is oversimplified in that we've considered only bonds that are inherently polar due to differences in electronegativity. Polar bonds can also result from the interaction of functional groups with acids or bases. Take an alcohol such as methanol, for example. In neutral methanol, the carbon atom is somewhat electron-poor because the electronegative oxygen attracts the electrons in the C–O bond. On protonation of the methanol oxygen by an acid, however, a full positive charge on oxygen attracts the electrons in the C–O bond much more strongly and makes the carbon much more electron-poor. We'll see numerous examples throughout this book of reactions that are catalyzed by acids because of the resultant increase in bond polarity on protonation.

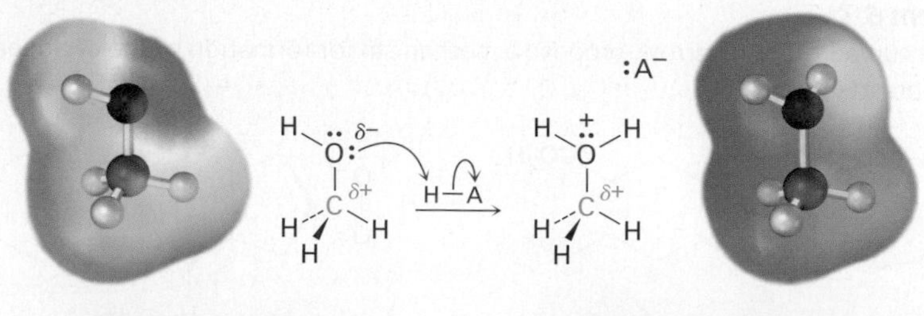

Methanol—weakly electron-poor carbon

Protonated methanol—strongly electron-poor carbon

Table 6.1 Polarity Patterns in Some Common Functional Groups

Compound type	Functional group structure	Compound type	Functional group structure
Alcohol	$\overset{\delta+}{C}-\overset{\delta-}{OH}$	Carbonyl	$\overset{\delta+}{C}=\overset{\delta-}{O}$
Alkene	C=C Symmetrical, nonpolar	Carboxylic acid	$-\overset{\delta+}{C}(=\overset{\delta-}{O})-\overset{\delta-}{OH}$
Alkyl halide	$\overset{\delta+}{C}-\overset{\delta-}{X}$	Carboxylic acid chloride	$-\overset{\delta+}{C}(=\overset{\delta-}{O})-\overset{\delta-}{Cl}$
Amine	$\overset{\delta+}{C}-\overset{\delta-}{NH_2}$	Thioester	$-\overset{\delta+}{C}(=\overset{\delta-}{O})-\overset{\delta-}{S}-C$
Ether	$\overset{\delta+}{C}-\overset{\delta-}{O}-\overset{\delta+}{C}$	Aldehyde	$-\overset{\delta+}{C}(=\overset{\delta-}{O})-H$
Thiol	$\overset{\delta+}{C}-\overset{\delta-}{SH}$	Ester	$-\overset{\delta+}{C}(=\overset{\delta-}{O})-\overset{\delta-}{O}-C$
Nitrile	$-\overset{\delta+}{C}\equiv\overset{\delta-}{N}$	Ketone	$-\overset{\delta+}{C}(=\overset{\delta-}{O})-C$
Grignard reagent	$\overset{\delta-}{C}-\overset{\delta+}{MgBr}$		
Alkyllithium	$\overset{\delta-}{C}-\overset{\delta+}{Li}$		

Yet a further consideration is the *polarizability* (as opposed to polarity) of atoms in a molecule. As the electric field around a given atom changes because of changing interactions with solvent or other polar molecules nearby, the electron distribution around that atom also changes. The measure of this response to an external electrical influence is called the polarizability of the atom. Larger atoms with more loosely held electrons are more polarizable, and smaller atoms with fewer, tightly held electrons are less polarizable. Thus, sulfur is more polarizable than oxygen, and iodine is more polarizable than chlorine. The effect of this higher polarizability for sulfur and iodine is that carbon–sulfur and carbon–iodine bonds, although nonpolar according to electronegativity values (Figure 2.2 on page 35), nevertheless usually react as if they were polar.

$$\overset{\delta-}{S}H-\overset{\delta+}{C} \qquad \overset{\delta-}{I}-\overset{\delta+}{C}$$

What does functional-group polarity mean with respect to chemical reactivity? Because unlike charges attract, the fundamental characteristic of all polar organic reactions is that electron-rich sites react with electron-poor sites. Bonds are made when an electron-rich atom donates a pair of electrons to an electron-poor atom, and bonds are broken when one atom leaves with both electrons from the former bond.

As we saw in **Section 2.11**, chemists indicate the movement of an electron pair during a polar reaction by using a curved, full-headed arrow. A curved arrow shows where electrons move when reactant bonds are broken and product bonds are formed. It means that an electron pair moves *from* the atom (or bond) at the tail of the arrow *to* the atom at the head of the arrow during the reaction.

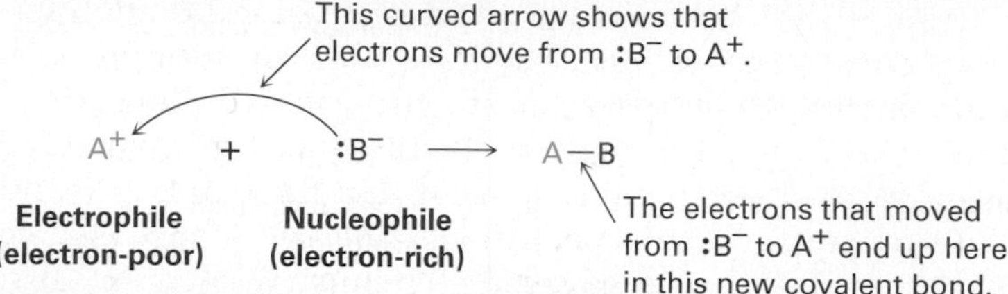

In referring to the electron-rich and electron-poor species involved in polar reactions, chemists use the words *nucleophile* and *electrophile*. A **nucleophile** is a substance that is "nucleus-loving." (Remember that a nucleus is positively charged.) A nucleophile has a negatively polarized, electron-rich atom and can form a bond by donating a pair of electrons to a positively polarized, electron-poor atom. Nucleophiles can be either neutral or negatively charged; ammonia, water, hydroxide ion, and chloride ion are examples. An **electrophile**, by contrast, is "electron-loving." An electrophile has a positively polarized, electron-poor atom and can form a bond by accepting a pair of electrons from a nucleophile. Electrophiles can be either neutral or positively charged. Acids (H^+ donors), alkyl halides, and carbonyl compounds are examples **(Figure 6.1)**.

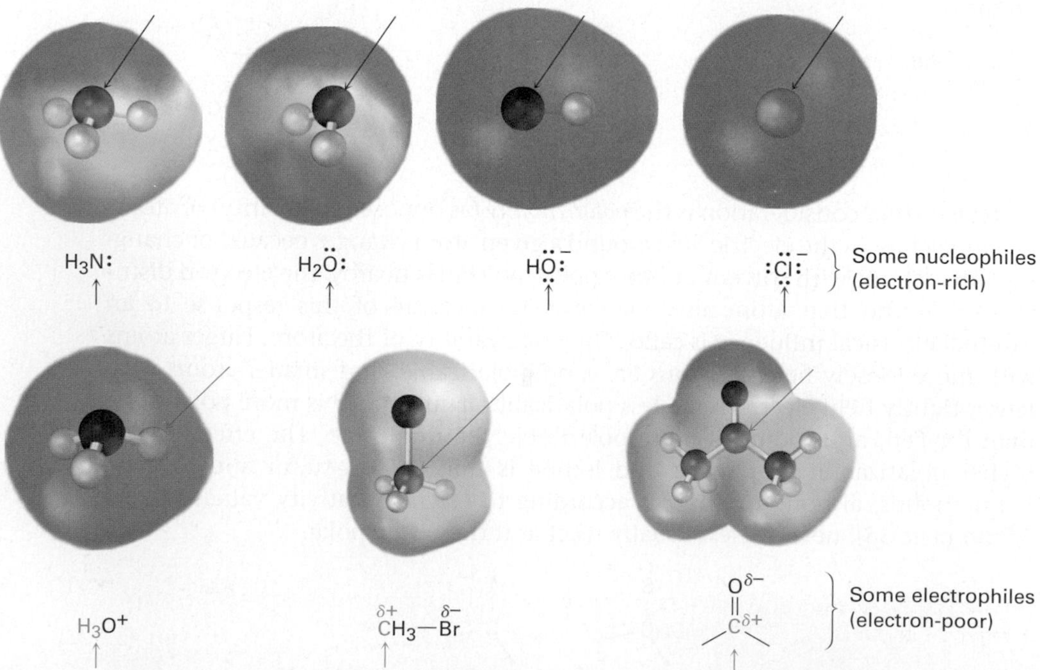

Figure 6.1 Some nucleophiles and electrophiles. Electrostatic potential maps identify the nucleophilic **(negative)** and electrophilic **(positive)** atoms.

Note that neutral compounds can often react either as nucleophiles or as electrophiles, depending on the circumstances. After all, if a compound is neutral yet has an electron-*rich* nucleophilic site, it must also have a corresponding electron-*poor* electrophilic site. Water, for instance, acts as an electrophile when it donates H^+ but acts as a nucleophile when it donates a nonbonding pair of electrons. Similarly, a carbonyl compound acts as an electrophile when it reacts at its positively polarized carbon atom, yet acts as a nucleophile when it reacts at its negatively polarized oxygen atom.

If the definitions of nucleophiles and electrophiles sound similar to those given in **Section 2.11** for Lewis acids and Lewis bases, that's because there is indeed a correlation. Lewis bases are electron donors and behave as nucleophiles, whereas Lewis acids are electron acceptors and behave as electrophiles. Thus, much of organic chemistry is explainable in terms of acid–base reactions. The main difference is that the words *acid* and *base* are used broadly in all fields of chemistry, while the words *nucleophile* and *electrophile* are used primarily in organic chemistry when bonds to carbon are involved.

Worked Example 6.1

Identifying Electrophiles and Nucleophiles

Which of the following species is likely to behave as a nucleophile and which as an electrophile?

(a) NO_2^+ **(b)** CN^- **(c)** CH_3NH_2 **(d)** $(CH_3)_3S^+$

Strategy

A nucleophile has an electron-rich site, either because it is negatively charged or because it has a functional group containing an atom that has a lone pair of electrons. An electrophile has an electron-poor site, either because it is positively charged or because it has a functional group containing an atom that is positively polarized.

Solution

(a) NO_2^+ (nitronium ion) is likely to be an electrophile because it is positively charged.

(b) $:C{\equiv}N^-$ (cyanide ion) is likely to be a nucleophile because it is negatively charged.

(c) CH_3NH_2 (methylamine) might be either a nucleophile or an electrophile depending on the circumstances. The lone pair of electrons on the nitrogen atom makes methylamine a potential nucleophile, while positively polarized N–H hydrogens make methylamine a potential acid (electrophile).

(d) $(CH_3)_3S^+$ (trimethylsulfonium ion) is likely to be an electrophile because it is positively charged.

Problem 6.4

Which of the following species are likely to be nucleophiles and which electrophiles? Which might be both?

(a) CH_3Cl (b) CH_3S^- (c) N-methylimidazole (ring: N, N–CH_3) (d) $CH_3CH{=}O$ ($CH_3\overset{O}{\overset{\|}{C}}H$)

Problem 6.5

An electrostatic potential map of boron trifluoride is shown. Is BF_3 likely to be a nucleophile or an electrophile? Draw a Lewis structure for BF_3, and explain your answer.

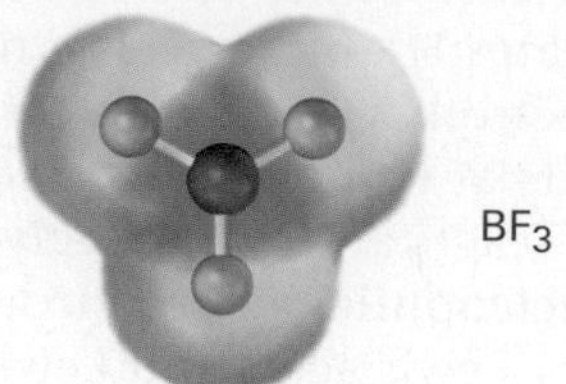

6.5 An Example of a Polar Reaction: Addition of HBr to Ethylene

Let's look at a typical polar process—the addition reaction of an alkene, such as ethylene, with hydrogen bromide. When ethylene is treated with HBr at room temperature, bromoethane is produced. Overall, the reaction can be formulated as

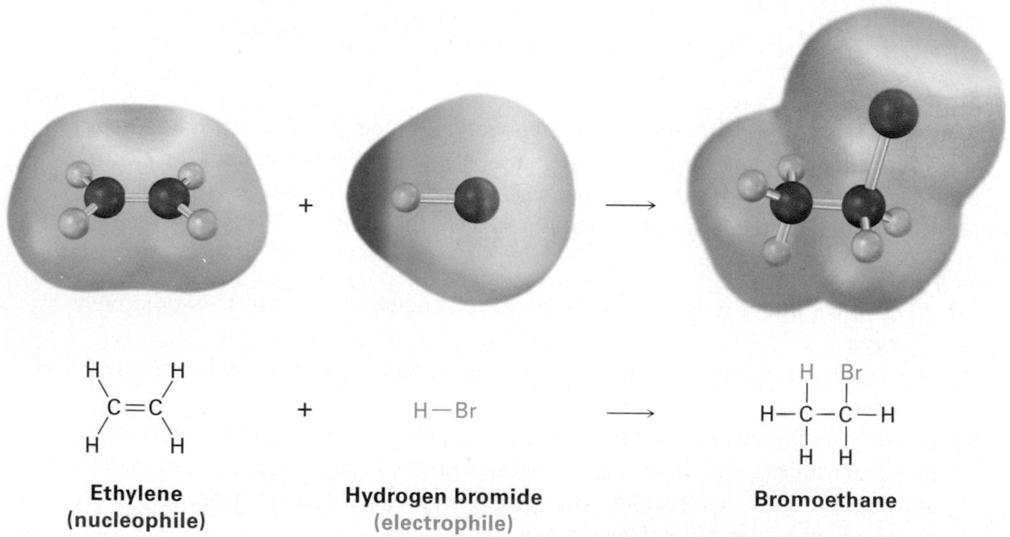

The reaction is an example of a polar reaction type known as an *electrophilic addition reaction* and can be understood using the general ideas discussed in the previous section. Let's begin by looking at the two reactants.

What do we know about ethylene? We know from **Section 1.8** that a carbon–carbon double bond results from orbital overlap of two sp^2-hybridized carbon atoms. The σ part of the double bond results from sp^2–sp^2 overlap, and the π part results from p–p overlap.

What kind of chemical reactivity might we expect of a C=C bond? We know that alkanes, such as ethane, are relatively inert because all valence electrons are tied up in strong, nonpolar, C–C and C–H bonds. Furthermore, the bonding electrons in alkanes are relatively inaccessible to approaching reactants because they are sheltered in σ bonds between nuclei. The electronic situation in alkenes is quite different, however. For one thing, double bonds have a greater electron density than single bonds—four electrons in a double bond versus only two in a single bond. In addition, the electrons in the π bond are accessible to approaching reactants because they are located above and below the plane of the double bond rather than being sheltered between the nuclei **(Figure 6.2)**. As a result, the double bond is nucleophilic and the chemistry of alkenes is dominated by reactions with electrophiles.

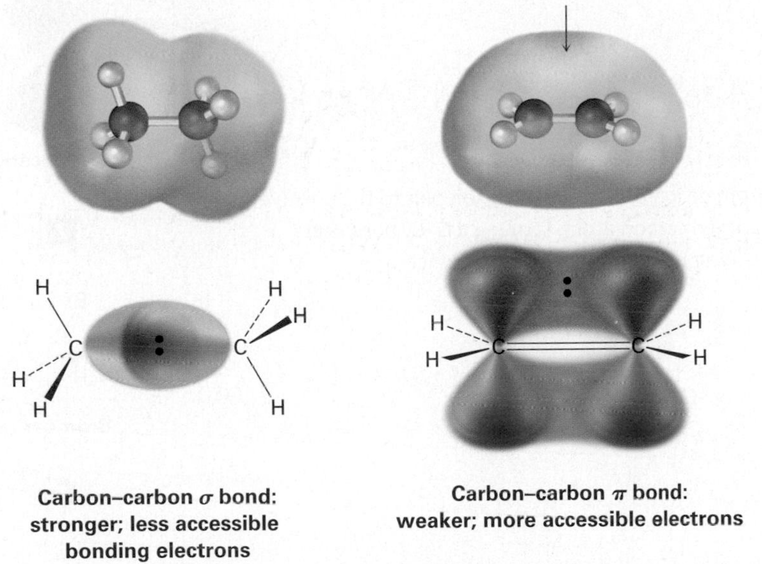

Figure 6.2 A comparison of carbon–carbon single and double bonds. A double bond is both more accessible to approaching reactants than a single bond and more electron-rich (more nucleophilic). An electrostatic potential map of ethylene indicates that the double bond is the region of **highest negative charge**.

What about the second reactant, HBr? As a strong acid, HBr is a powerful proton (H^+) donor and electrophile. Thus, the reaction between HBr and ethylene is a typical electrophile–nucleophile combination, characteristic of all polar reactions.

We'll see more details about alkene electrophilic addition reactions shortly, but for the present we can imagine the reaction as taking place by the pathway shown in **Figure 6.3**. The reaction begins when the alkene nucleophile donates a pair of electrons from its C=C bond to HBr to form a new C–H bond plus Br^-, as indicated by the path of the curved arrows in the first step of Figure 6.3. One curved arrow begins at the middle of the double bond (the source of the electron pair) and points to the hydrogen atom in HBr (the atom to which a bond will form). This arrow indicates that a new C–H bond forms using electrons from the former C=C bond. Simultaneously, a second curved arrow begins in the middle of the H–Br bond and points to the Br, indicating that the H–Br bond breaks and the electrons remain with the Br atom, giving Br^-.

Figure 6.3 | MECHANISM

The electrophilic addition reaction of ethylene and HBr. The reaction takes place in two steps, both of which involve electrophile–nucleophile interactions.

1 A hydrogen atom on the electrophile HBr is attacked by π electrons from the nucleophilic double bond, forming a new C–H bond. This leaves the other carbon atom with a + charge and a vacant *p* orbital. Simultaneously, two electrons from the H–Br bond move onto bromine, giving bromide anion.

2 Bromide ion donates an electron pair to the positively charged carbon atom, forming a C–Br bond and yielding the neutral addition product.

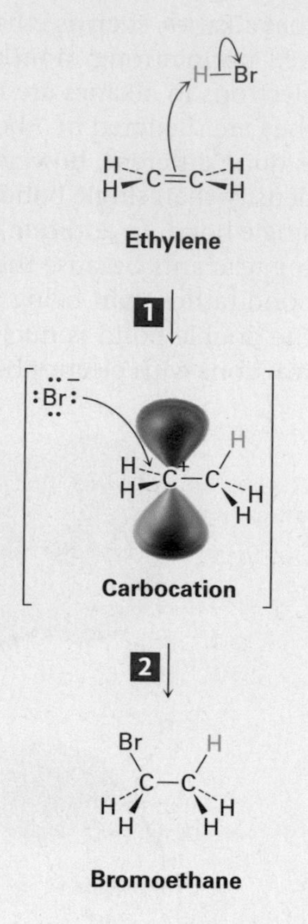

When one of the alkene carbon atoms bonds to the incoming hydrogen, the other carbon atom, having lost its share of the double-bond electrons, now has only six valence electrons and is left with a positive charge. This positively charged species—a carbon-cation, or **carbocation**—is itself an electrophile that can accept an electron pair from nucleophilic Br^- anion in a second step, forming a C–Br bond and yielding the observed addition product. Once again, a curved arrow in Figure 6.3 shows the electron-pair movement from Br^- to the positively charged carbon.

The electrophilic addition of HBr to ethylene is only one example of a polar process; there are many others that we'll study in detail in later chapters. But regardless of the details of individual reactions, all polar reactions take place between an electron-poor site and an electron-rich site and involve the donation of an electron pair from a nucleophile to an electrophile.

Problem 6.6

What product would you expect from reaction of cyclohexene with HBr? With HCl?

[cyclohexene] + HBr ⟶ ?

Problem 6.7

Reaction of HBr with 2-methylpropene yields 2-bromo-2-methylpropane. What is the structure of the carbocation formed during the reaction? Show the mechanism of the reaction.

$$(H_3C)_2C{=}CH_2 + HBr \longrightarrow CH_3{-}C(CH_3)_2{-}Br$$

2-Methylpropene **2-Bromo-2-methylpropane**

6.6 Using Curved Arrows in Polar Reaction Mechanisms

It takes practice to use curved arrows properly in reaction mechanisms, but there are a few rules and a few common patterns you should look for that will help you become more proficient:

Key **IDEAS**

Test your knowledge of Key Ideas by answering end-of-chapter exercises marked with ▲.

RULE 1

Electrons move *from* a nucleophilic source (Nu: or Nu:⁻) *to* an electrophilic sink (E or E⁺). The nucleophilic source must have an electron pair available, usually either as a lone pair or in a multiple bond. For example:

Electrons usually flow *from* one of these nucleophiles.

:O: → E   N: → E   C:⁻ → E   C=C → E

The electrophilic sink must be able to accept an electron pair, usually because it has either a positively charged atom or a positively polarized atom in a functional group. For example:

Electrons usually flow *to* one of these electrophiles.

Nu: → C⁺   Nu: → C(δ+)—Halogen(δ−)   Nu: → H(δ+)—O(δ−)   Nu: → C(δ+)=O(δ−)

RULE 2

The nucleophile can be either negatively charged or neutral. If the nucleophile is negatively charged, the atom that donates an electron pair becomes neutral. For example:

Negatively charged   Neutral

$$CH_3{-}\ddot{O}{:}^- + H{-}Br{:} \longrightarrow CH_3{-}\ddot{O}(H){:} + {:}Br{:}^-$$

If the nucleophile is neutral, the atom that donates the electron pair acquires a positive charge. For example:

Neutral   Positively charged

$$H_2C{=}CH_2 + H{-}Br \longrightarrow H_2\overset{+}{C}{-}CH_3 + {:}Br{:}^-$$

RULE 3

The electrophile can be either positively charged or neutral. If the electrophile is positively charged, the atom bearing that charge becomes neutral after accepting an electron pair. For example:

Positively charged — Neutral

$$H_2C{=}CH_2 + H{-}\ddot{O}^{+}H_2 \longrightarrow H_2C^{+}{-}CH_3 + \ddot{O}H_2$$

If the electrophile is neutral, the atom that ultimately accepts the electron pair acquires a negative charge. For this to happen, however, the negative charge must be stabilized by being on an electronegative atom such as oxygen, nitrogen, or a halogen. Carbon and hydrogen do not typically stabilize a negative charge. For example:

Neutral — Negatively charged

$$H_2C{=}CH_2 + H{-}Br \longrightarrow H_2C^{+}{-}CH_3 + Br^{-}$$

The result of Rules 2 and 3 together is that charge is conserved during the reaction. A negative charge in one of the reactants gives a negative charge in one of the products, and a positive charge in one of the reactants gives a positive charge in one of the products.

RULE 4

The octet rule must be followed. That is, no second-row atom can be left with ten electrons (or four for hydrogen). If an electron pair moves *to* an atom that already has an octet (or two for hydrogen), another electron pair must simultaneously move *from* that atom to maintain the octet. When two electrons move from the C=C bond of ethylene to the hydrogen atom of H_3O^+, for instance, two electrons must leave that hydrogen. This means that the H–O bond must break and the electrons must stay with the oxygen, giving neutral water.

This hydrogen already has two electrons. When another electron pair moves to the hydrogen from the double bond, the electron pair in the H–O bond must leave.

$$H_2C{=}CH_2 + H{-}\ddot{O}^{+}H_2 \longrightarrow H_2C^{+}{-}CH_3 + \ddot{O}H_2$$

Worked Example 6.2 gives another example of drawing curved arrows.

Using Curved Arrows in Reaction Mechanisms

Worked Example 6.2

Add curved arrows to the following polar reaction to show the flow of electrons:

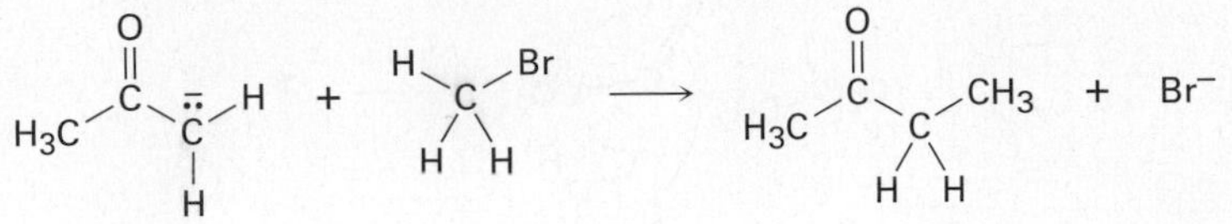

$$H_3C-C(=O)-\ddot{C}^-H_2 + CH_3Br \longrightarrow H_3C-C(=O)-CH_2-CH_3 + Br^-$$

Strategy

Look at the reaction, and identify the bonding changes that have occurred. In this case, a C–Br bond has broken and a C–C bond has formed. The formation of the C–C bond involves donation of an electron pair from the nucleophilic carbon atom of the reactant on the left to the electrophilic carbon atom of CH_3Br, so we draw a curved arrow originating from the lone pair on the negatively charged C atom and pointing to the C atom of CH_3Br. At the same time that the C–C bond forms, the C–Br bond must break so that the octet rule is not violated. We therefore draw a second curved arrow from the C–Br bond to Br. The bromine is now a stable Br^- ion.

Solution

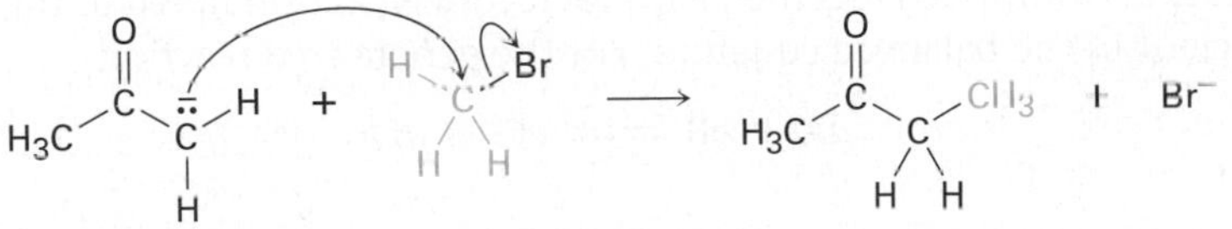

$$H_3C-C(=O)-\ddot{C}^-H_2 + CH_3Br \longrightarrow H_3C-C(=O)-CH_2-CH_3 + Br^-$$

Problem 6.8

Add curved arrows to the following polar reactions to indicate the flow of electrons in each:

(a)

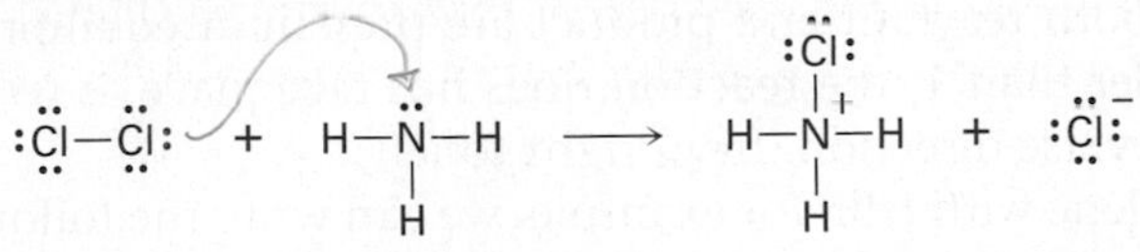

$$:\ddot{Cl}-\ddot{Cl}: + H-\ddot{N}H-H \longrightarrow H-\overset{+}{N}(Cl)H-H + :\ddot{Cl}:^-$$

(b)

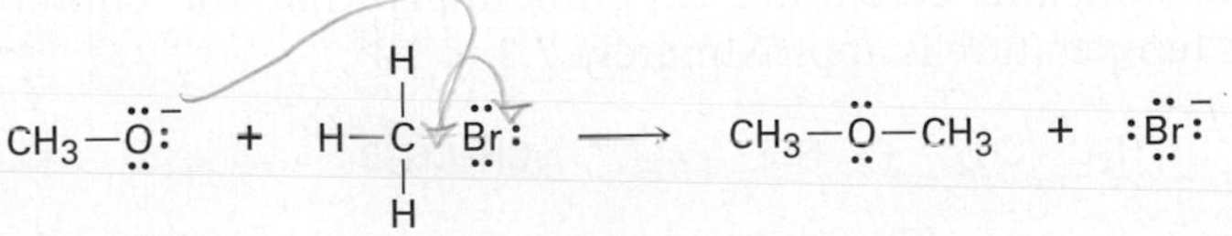

$$CH_3-\ddot{O}:^- + H-C(H)(H)-\ddot{Br}: \longrightarrow CH_3-\ddot{O}-CH_3 + :\ddot{Br}:^-$$

(c)

$$H_3C-C(-\ddot{O}:^-)(Cl)-OCH_3 \longrightarrow H_3C-C(=O:)-OCH_3 + :\ddot{Cl}:^-$$

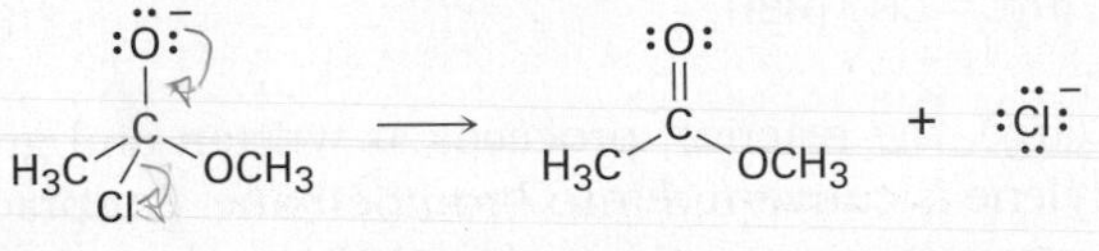

Problem 6.9

Predict the products of the following polar reaction, a step in the citric acid cycle for food metabolism, by interpreting the flow of electrons indicated by the curved arrows:

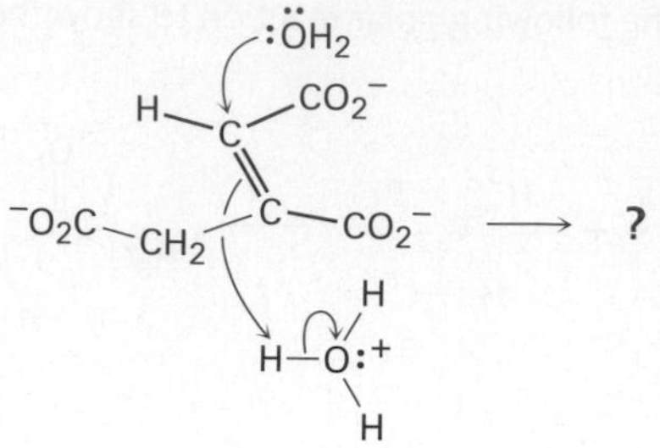

$$^{-}O_2C-CH_2-C(CO_2^-)=C(H)(CO_2^-) + :\ddot{O}H_2 + H_3O^+ \longrightarrow ?$$

6.7 Describing a Reaction: Equilibria, Rates, and Energy Changes

Every chemical reaction can go in either forward or reverse direction. Reactants can go forward to products, and products can revert to reactants. As you may remember from your general chemistry course, the position of the resulting chemical equilibrium is expressed by an equation in which K_{eq}, the equilibrium constant, is equal to the product concentrations multiplied together, divided by the reactant concentrations multiplied together, with each concentration raised to the power of its coefficient in the balanced equation. For the generalized reaction

$$aA + bB \rightleftarrows cC + dD$$

we have

$$K_{eq} = \frac{[C]^c\ [D]^d}{[A]^a\ [B]^b}$$

The value of the equilibrium constant tells which side of the reaction arrow is energetically favored. If K_{eq} is much larger than 1, then the product concentration term $[C]^c\ [D]^d$ is much larger than the reactant concentration term $[A]^a\ [B]^b$, and the reaction proceeds as written from left to right. If K_{eq} is near 1, appreciable amounts of both reactant and product are present at equilibrium. And if K_{eq} is much smaller than 1, the reaction does not take place as written but instead goes in the reverse direction, from right to left.

In the reaction of ethylene with HBr, for example, we can write the following equilibrium expression and determine experimentally that the equilibrium constant at room temperature is approximately 7.1×10^7:

$$H_2C{=}CH_2 + HBr \rightleftarrows CH_3CH_2Br$$

$$K_{eq} = \frac{[CH_3CH_2Br]}{[H_2C{=}CH_2]\ [HBr]} = 7.1 \times 10^7$$

Because K_{eq} is relatively large, the reaction proceeds as written and greater than 99.999 99% of the ethylene is converted into bromoethane. For practical purposes, an equilibrium constant greater than about 10^3 means that the amount of reactant left over will be barely detectable (less than 0.1%).

What determines the magnitude of the equilibrium constant? For a reaction to have a favorable equilibrium constant and proceed as written, the energy of the products must be lower than the energy of the reactants. In other words, energy must be released. The situation is analogous to that of a rock poised precariously in a high-energy position near the top of a hill. When it rolls downhill, the rock releases energy until it reaches a more stable, low-energy position at the bottom.

The energy change that occurs during a chemical reaction is called the **Gibbs free-energy change (ΔG)**, which is equal to the free energy of the products minus the free energy of the reactants: $\Delta G = G_{products} - G_{reactants}$. For a favorable reaction, ΔG has a negative value, meaning that energy is lost by the chemical system and released to the surroundings, usually as heat. Such reactions are said to be **exergonic**. For an unfavorable reaction, ΔG has a positive value, meaning that energy is absorbed by the chemical system *from* the surroundings. Such reactions are said to be **endergonic**.

You might also recall from general chemistry that the *standard* free-energy change for a reaction is denoted $\Delta G°$, where the superscript ° means that the reaction is carried out under standard conditions, with pure substances in their most stable form at 1 atm pressure and a specified temperature, usually 298 K. For biological reactions, the standard free-energy change is symbolized $\Delta G°'$ and refers to a reaction carried out at pH = 7.0 with solute concentrations of 1.0 M.

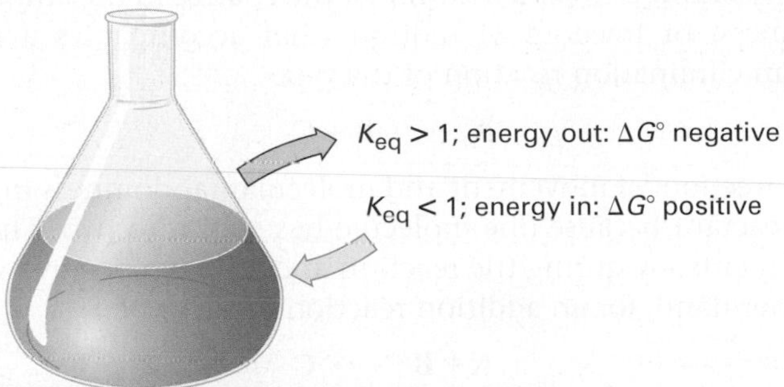

Because the equilibrium constant, K_{eq}, and the standard free-energy change, $\Delta G°$, both measure whether a reaction is favorable, they are mathematically related by the equation

$$\Delta G° = -RT \ln K_{eq} \quad \text{or} \quad K_{eq} = e^{-\Delta G°/RT}$$

where $R = 8.314\ \text{J/(K} \cdot \text{mol)} = 1.987\ \text{cal/(K} \cdot \text{mol)}$
T = Kelvin temperature
$e = 2.718$
$\ln K_{eq}$ = natural logarithm of K_{eq}

For example, the reaction of ethylene with HBr has $K_{eq} = 7.1 \times 10^7$, so $\Delta G° = -44.8$ kJ/mol (-10.7 kcal/mol) at 298 K:

$$K_{eq} = 7.1 \times 10^7 \quad \text{and} \quad \ln K_{eq} = 18.08$$

$$\Delta G° = -RT \ln K_{eq} = -[8.314\ \text{J/(K} \cdot \text{mol)}]\ (298\ \text{K})\ (18.08)$$

$$= -44{,}800\ \text{J/mol} = -44.8\ \text{kJ/mol}$$

The free-energy change ΔG is made up of two terms, an *enthalpy* term, ΔH, and a temperature-dependent *entropy* term, $T\Delta S$. Of the two terms, the enthalpy term is often larger and more dominant.

$$\Delta G° = \Delta H° - T\Delta S°$$

For the reaction of ethylene with HBr at room temperature (298 K), the approximate values are

$$H_2C{=}CH_2 \;+\; HBr \;\rightleftharpoons\; CH_3CH_2Br \qquad \begin{cases} \Delta G° = -44.8 \text{ kJ/mol} \\ \Delta H° = -84.1 \text{ kJ/mol} \\ \Delta S° = -0.132 \text{ kJ/(K} \cdot \text{mol)} \\ K_{eq} = 7.1 \times 10^7 \end{cases}$$

The **enthalpy change, ΔH,** also called the **heat of reaction**, is a measure of the change in total bonding energy during a reaction. If ΔH is negative, as in the reaction of HBr with ethylene, the products have less energy than the reactants. Thus, the products are more stable and have stronger bonds than the reactants, heat is released, and the reaction is said to be **exothermic**. If ΔH is positive, the products are less stable and have weaker bonds than the reactants, heat is absorbed, and the reaction is said to be **endothermic**. For example, if a reaction breaks reactant bonds with a total strength of 380 kJ/mol and forms product bonds with a total strength of 400 kJ/mol, then ΔH for the reaction is −20 kJ/mol and the reaction is exothermic.

The **entropy change, ΔS,** is a measure of the change in the amount of molecular randomness, or freedom of motion, that accompanies a reaction. For example, in an elimination reaction of the type

$$A \longrightarrow B + C$$

there is more freedom of movement and molecular randomness in the products than in the reactant because one molecule has split into two. Thus, there is a net increase in entropy during the reaction and ΔS has a positive value.

On the other hand, for an addition reaction of the type

$$A + B \longrightarrow C$$

the opposite is true. Because such reactions restrict the freedom of movement of two molecules by joining them together, the product has less randomness than the reactants and ΔS has a negative value. The reaction of ethylene and HBr to yield bromoethane, which has $\Delta S° = -0.132$ kJ/(K · mol), is an example. Table 6.2 describes the thermodynamic terms more fully.

Knowing the value of K_{eq} for a reaction is useful, but it's important to realize the limitations. An equilibrium constant tells only the *position* of the equilibrium, or how much product is theoretically possible. It doesn't tell the *rate* of reaction, or how fast the equilibrium is established. Some reactions are extremely slow even though they have favorable equilibrium constants. Gasoline is stable at room temperature, for instance, because the rate of its reaction with oxygen is slow at 298 K. Only at higher temperatures, such as contact with a lighted match, does gasoline react rapidly with oxygen and undergoes complete conversion to the equilibrium products water and carbon dioxide. Rates (*how fast* a reaction occurs) and equilibria (*how much* a reaction occurs) are entirely different.

Rate ⟶ **Is the reaction fast or slow?**

Equilibrium ⟶ **In what direction does the reaction proceed?**

Table 6.2 Explanation of Thermodynamic Quantities: $\Delta G° = \Delta H° - T\Delta S°$

Term	Name	Explanation
$\Delta G°$	Gibbs free-energy change	The energy difference between reactants and products. When $\Delta G°$ is negative, the reaction is **exergonic**, has a favorable equilibrium constant, and can occur spontaneously. When $\Delta G°$ is positive, the reaction is **endergonic**, has a unfavorable equilibrium constant, and cannot occur spontaneously.
$\Delta H°$	Enthalpy change	The heat of reaction, or difference in strength between the bonds broken in a reaction and the bonds formed. When $\Delta H°$ is negative, the reaction releases heat and is **exothermic**. When $\Delta H°$ is positive, the reaction absorbs heat and is **endothermic**.
$\Delta S°$	Entropy change	The change in molecular randomness during a reaction. When $\Delta S°$ is negative, randomness decreases. When $\Delta S°$ is positive, randomness increases.

Problem 6.10

Which reaction is more energetically favored, one with $\Delta G° = -44$ kJ/mol or one with $\Delta G° = +44$ kJ/mol?

Problem 6.11

Which reaction is likely to be more exergonic, one with $K_{eq} = 1000$ or one with $K_{eq} = 0.001$?

6.8 Describing a Reaction: Bond Dissociation Energies

We've just seen that heat is released (negative ΔH) when a bond is formed because the products are more stable and have stronger bonds than the reactants. Conversely, heat is absorbed (positive ΔH) when a bond is broken because the products are less stable and have weaker bonds than the reactants. The amount of energy needed to break a given bond to produce two radical fragments when the molecule is in the gas phase at 25 °C is a quantity called *bond strength*, or **bond dissociation energy (*D*)**.

$$\text{A : B} \xrightarrow[\text{energy}]{\text{Bond dissociation}} \text{A}\cdot \ + \ \cdot\text{B}$$

Each specific bond has its own characteristic strength, and extensive tables of data are available. For example, a C–H bond in methane has a bond dissociation energy $D = 439.3$ kJ/mol (105.0 kcal/mol), meaning that 439.3 kJ/mol must be added to break a C–H bond of methane to give the two radical fragments $\cdot CH_3$ and $\cdot$H. Conversely, 439.3 kJ/mol of energy is released when a methyl radical and a hydrogen atom combine to form methane. Table 6.3 lists some other bond strengths.

Think again about the connection between bond strengths and chemical reactivity. In an exothermic reaction, more heat is released than is absorbed. But because making bonds in the products releases heat and breaking bonds in the reactants absorbs heat, the bonds in the products must be stronger than the

Table 6.3 Some Bond Dissociation Energies, *D*

Bond	*D* (kJ/mol)	Bond	*D* (kJ/mol)	Bond	*D* (kJ/mol)
H—H	436	$(CH_3)_3C$—I	227	$(CH_3)_2CH$—CH_3	369
H—F	570	$H_2C{=}CH$—H	464	$(CH_3)_3C$—CH_3	363
H—Cl	431	$H_2C{=}CH$—Cl	396	$H_2C{=}CH$—CH_3	426
H—Br	366	$H_2C{=}CHCH_2$—H	369	$H_2C{=}CHCH_2$—CH_3	318
H—I	298	$H_2C{=}CHCH_2$—Cl	298	$H_2C{=}CH_2$	728
Cl—Cl	242	C_6H_5—H	472	C_6H_5—CH_3	427
Br—Br	194	C_6H_5—Cl	400	$C_6H_5CH_2$—CH_3	325
I—I	152	$C_6H_5CH_2$—H	375	$CH_3C(=O)$—H	374
CH_3—H	439	$C_6H_5CH_2$—Cl	300	HO—H	497
CH_3—Cl	350	C_6H_5—Br	336	HO—OH	211
CH_3—Br	294	C_6H_5—OH	464	CH_3O—H	440
CH_3—I	239	HC≡C—H	558	CH_3S—H	366
CH_3—OH	385	CH_3—CH_3	377	C_2H_5O—H	441
CH_3—NH_2	386	C_2H_5—CH_3	370	$CH_3C(=O)$—CH_3	352
C_2H_5—H	421			CH_3CH_2O—CH_3	355
C_2H_5—Cl	352			NH_2—H	450
C_2H_5—Br	293			H—CN	528
C_2H_5—I	233				
C_2H_5—OH	391				
$(CH_3)_2CH$—H	410				
$(CH_3)_2CH$—Cl	354				
$(CH_3)_2CH$—Br	299				
$(CH_3)_3C$—H	400				
$(CH_3)_3C$—Cl	352				
$(CH_3)_3C$—Br	293				

bonds in the reactants. In other words, exothermic reactions are favored by products with strong bonds and by reactants with weak, easily broken bonds.

Sometimes, particularly in biochemistry, reactive substances that undergo highly exothermic reactions, such as ATP (adenosine triphosphate), are referred to as "energy-rich" or "high-energy" compounds. Such a label doesn't mean that ATP is special or different from other compounds, it only means that ATP has relatively weak bonds that require a relatively small amount of heat to break, thus leading to a larger release of heat when a strong new bond forms in a reaction. When a typical organic phosphate such as glycerol 3-phosphate reacts with water, for instance, only 9 kJ/mol of heat is released ($\Delta H = -9$ kJ/mol), but when ATP reacts with water, 30 kJ/mol of heat is released ($\Delta H = -30$ kJ/mol). The difference between the two reactions is due to the fact that the bond broken in ATP is substantially weaker than the bond broken in glycerol 3-phosphate. We'll see the metabolic importance of this reaction in later chapters.

$\Delta H°' = -9$ kJ/mol

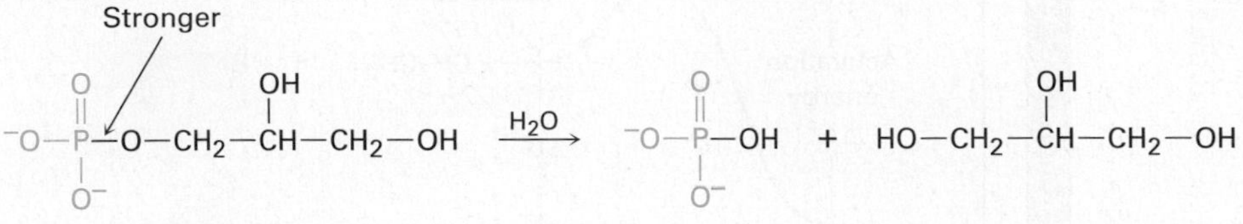

Stronger

$$^{-}O-P(=O)(O^{-})-O-CH_2-CH(OH)-CH_2-OH \xrightarrow{H_2O} {}^{-}O-P(=O)(O^{-})-OH + HO-CH_2-CH(OH)-CH_2-OH$$

Glycerol 3-phosphate — Glycerol

$\Delta H°' = -30$ kJ/mol

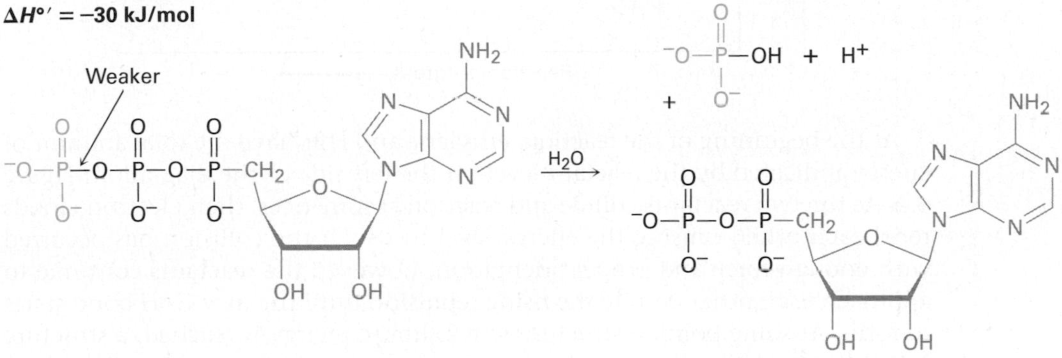

Weaker

Adenosine triphosphate (ATP) $\xrightarrow{H_2O}$ ${}^{-}O-P(=O)(O^{-})-OH + H^+$ + Adenosine diphosphate (ADP)

6.9 Describing a Reaction: Energy Diagrams and Transition States

For a reaction to take place, reactant molecules must collide and reorganization of atoms and bonds must occur. Let's again look at the addition reaction of HBr and ethylene.

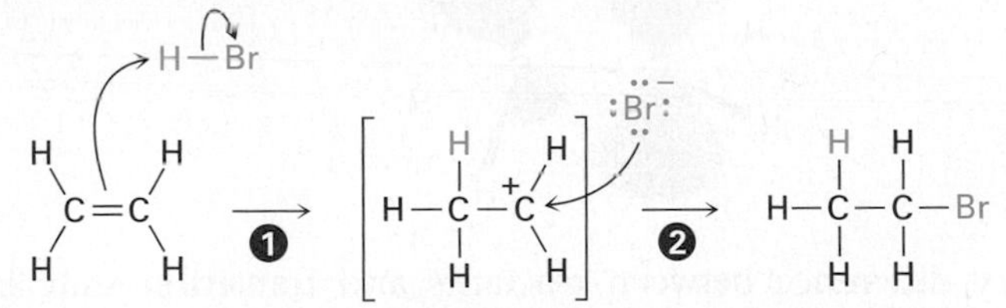

$$H_2C{=}CH_2 + H{-}Br \xrightarrow{❶} [H_3C{-}CH_2^{+}]\ \text{(Carbocation)} + :Br:^{-} \xrightarrow{❷} H_3C{-}CH_2{-}Br$$

Carbocation

As the reaction proceeds, ethylene and HBr must approach each other, the ethylene π bond and the H–Br bond must break, a new C–H bond must form in step ❶, and a new C–Br bond must form in step ❷.

To depict graphically the energy changes that occur during a reaction, chemists use energy diagrams, such as that shown in **Figure 6.4**. The vertical axis of the diagram represents the total energy of all reactants, and the horizontal axis, called the *reaction coordinate*, represents the progress of the reaction from beginning to end. Let's see how the addition of HBr to ethylene can be described in an energy diagram.

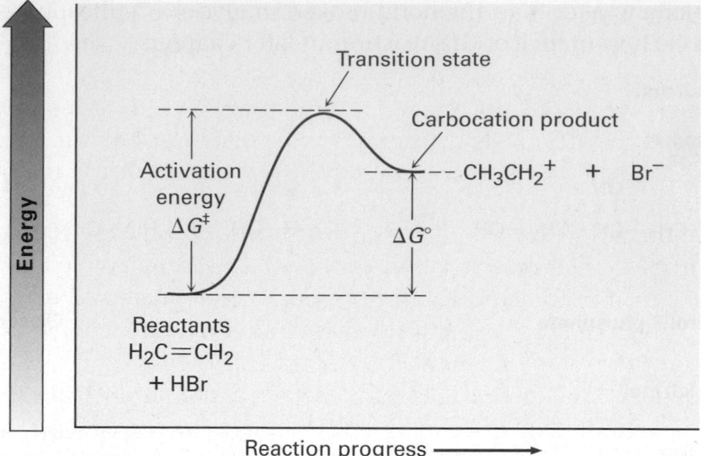

Figure 6.4 An energy diagram for the first step in the reaction of ethylene with HBr. The energy difference between reactants and transition state, $\Delta G^{\ddagger}$, defines the reaction rate. The energy difference between reactants and carbocation product, $\Delta G°$, defines the position of the equilibrium.

At the beginning of the reaction, ethylene and HBr have the total amount of energy indicated by the reactant level on the left side of the diagram in Figure 6.4. As the two reactants collide and reaction commences, their electron clouds repel each other, causing the energy level to rise. If the collision has occurred with enough force and proper orientation, however, the reactants continue to approach each other despite the rising repulsion until the new C–H bond starts to form. At some point, a structure of maximum energy is reached, a structure called the *transition state.*

The **transition state** represents the highest-energy structure involved in this step of the reaction. It is unstable and can't be isolated, but we can nevertheless imagine it to be an activated complex of the two reactants in which both the C=C π bond and H–Br bond are partially broken and the new C–H bond is partially formed **(Figure 6.5)**.

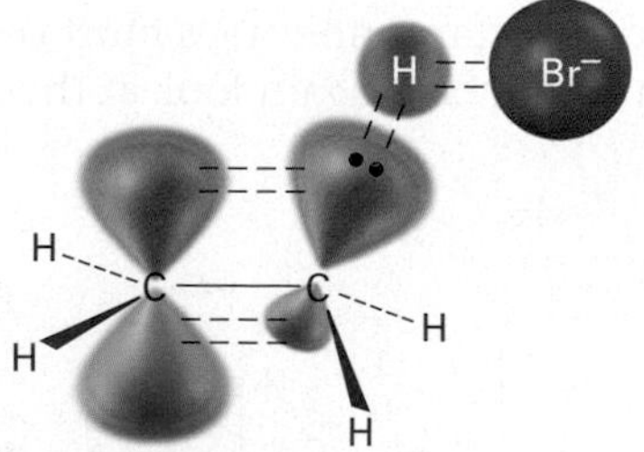

Figure 6.5 A hypothetical transition-state structure for the first step of the reaction of ethylene with HBr. The C=C π bond and H–Br bond are just beginning to break, and the C–H bond is just beginning to form.

The energy difference between reactants and transition state is called the **activation energy, $\Delta G^{\ddagger}$**, and determines how rapidly the reaction occurs at a given temperature. (The double-dagger superscript, ‡, always refers to the transition state.) A large activation energy results in a slow reaction because few collisions occur with enough energy for the reactants to reach the transition state. A small activation energy results in a rapid reaction because almost all collisions occur with enough energy for the reactants to reach the transition state.

As an analogy, you might think of reactants that need enough energy to climb the activation barrier to the transition state as similar to hikers who need enough energy to climb to the top of a mountain pass. If the pass is a high one, the hikers need a lot of energy and surmount the barrier with difficulty. If the pass is low, however, the hikers need less energy and reach the top easily.

As a rough generalization, many organic reactions have activation energies in the range 40 to 150 kJ/mol (10–35 kcal/mol). The reaction of ethylene with HBr, for example, has an activation energy of approximately 140 kJ/mol (34 kcal/mol). Reactions with activation energies less than 80 kJ/mol take place at or below room temperature, while reactions with higher activation energies normally require a higher temperature to give the reactants enough energy to climb the activation barrier.

Once the transition state is reached, the reaction can either continue on to give the carbocation product or revert back to reactants. When reversion to reactants occurs, the transition-state structure comes apart and an amount of free energy corresponding to $-\Delta G^\ddagger$ is released. When the reaction continues on to give the carbocation, the new C–H bond forms fully and an amount of energy corresponding to the difference between transition state and carbocation product is released. The net energy change for the step, $\Delta G°$, is represented in the diagram as the difference in level between reactant and product. Since the carbocation is higher in energy than the starting alkene, the step is endergonic, has a positive value of $\Delta G°$, and absorbs energy.

Not all energy diagrams are like that shown for the reaction of ethylene and HBr. Each reaction has its own energy profile. Some reactions are fast (small $\Delta G^\ddagger$) and some are slow (large $\Delta G^\ddagger$); some have a negative $\Delta G°$, and some have a positive $\Delta G°$. **Figure 6.6** illustrates some different possibilities.

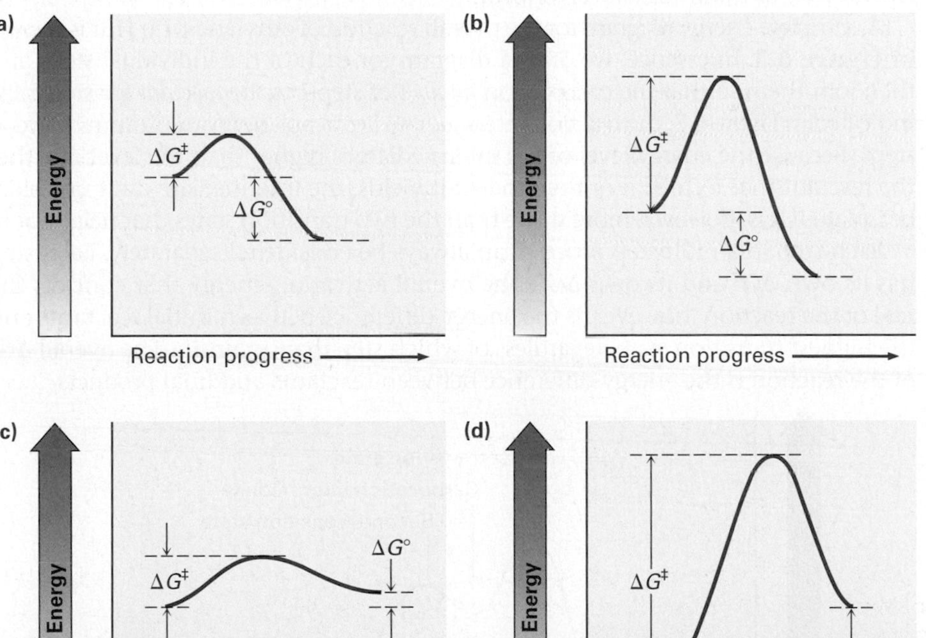

Figure 6.6 Some hypothetical energy diagrams: **(a)** a fast exergonic reaction (small $\Delta G^\ddagger$, negative $\Delta G°$); **(b)** a slow exergonic reaction (large $\Delta G^\ddagger$, negative $\Delta G°$); **(c)** a fast endergonic reaction (small $\Delta G^\ddagger$, small positive $\Delta G°$); **(d)** a slow endergonic reaction (large $\Delta G^\ddagger$, positive $\Delta G°$).

Problem 6.12

Which reaction is faster, one with $\Delta G^\ddagger = +45$ kJ/mol or one with $\Delta G^\ddagger = +70$ kJ/mol?

6.10 Describing a Reaction: Intermediates

How can we describe the carbocation formed in the first step of the reaction of ethylene with HBr? The carbocation is clearly different from the reactants, yet it isn't a transition state and it isn't a final product.

Reaction intermediate

We call the carbocation, which exists only transiently during the course of the multistep reaction, a **reaction intermediate**. As soon as the intermediate is formed in the first step by reaction of ethylene with H^+, it reacts further with Br^- in a second step to give the final product, bromoethane. This second step has its own activation energy ($\Delta G^‡$), its own transition state, and its own energy change ($\Delta G°$). We can picture the second transition state as an activated complex between the electrophilic carbocation intermediate and the nucleophilic bromide anion, in which Br^- donates a pair of electrons to the positively charged carbon atom as the new C–Br bond just starts to form.

A complete energy diagram for the overall reaction of ethylene with HBr is shown in **Figure 6.7**. In essence, we draw a diagram for each of the individual steps and then join them so that the carbocation *product* of step 1 is the *reactant* for step 2. As indicated in Figure 6.7, the reaction intermediate lies at an energy minimum between steps. Because the energy level of the intermediate is higher than the level of either the reactant that formed it or the product it yields, the intermediate can't normally be isolated. It is, however, more stable than the two transition states that neighbor it.

Each step in a multistep process can always be considered separately. Each step has its own $\Delta G^‡$ and its own $\Delta G°$. The overall activation energy that controls the rate of the reaction, however, is the energy difference between initial reactants and the highest transition state, regardless of which step that occurs in. The overall $\Delta G°$ of the reaction is the energy difference between reactants and final products.

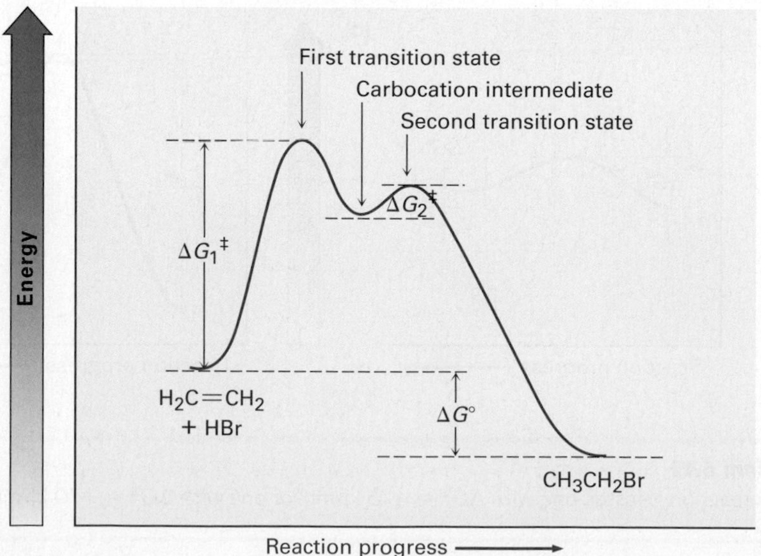

Figure 6.7 An energy diagram for the reaction of ethylene with HBr. Two separate steps are involved, each with its own activation energy ($\Delta G^‡$) and free-energy change ($\Delta G°$). The overall $\Delta G^‡$ for the complete reaction is the energy difference between reactants and the highest transition state (which corresponds to $\Delta G_1^‡$ in this case), and the overall $\Delta G°$ for the reaction is the energy difference between reactants and final products.

The biological reactions that take place in living organisms have the same energy requirements as reactions that take place in the laboratory and can be described in similar ways. They are, however, constrained by the fact that they must have low enough activation energies to occur at moderate temperatures, and they must release energy in relatively small amounts to avoid overheating the organism. These constraints are generally met through the use of large, structurally complex, enzyme catalysts that change the mechanism of a reaction to an alternative pathway that proceeds through a series of small steps rather than one or two large steps. Thus, a typical energy diagram for a biological reaction might look like that in **Figure 6.8**.

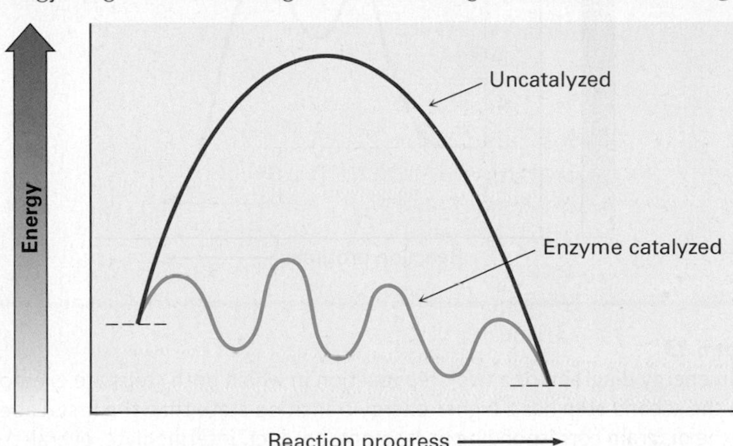

Figure 6.8 An energy diagram for a typical, **enzyme-catalyzed biological reaction** versus an **uncatalyzed laboratory reaction**. The biological reaction involves many steps, each of which has a relatively small activation energy and small energy change. The end result is the same, however.

Worked Example 6.3 Drawing an Energy Diagram for a Reaction

Sketch an energy diagram for a one-step reaction that is fast and highly exergonic.

Strategy

A fast reaction has a small $\Delta G^\ddagger$, and a highly exergonic reaction has a large negative $\Delta G°$.

Solution

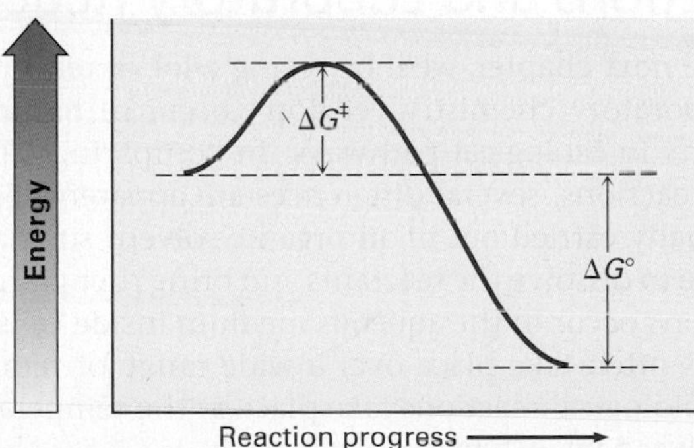

Worked Example 6.4 Drawing an Energy Diagram for a Reaction

Sketch an energy diagram for a two-step exergonic reaction whose second step has a higher-energy transition state than its first step. Show $\Delta G^\ddagger$ and $\Delta G°$ for the overall reaction.

Strategy

A two-step reaction has two transition states and an intermediate between them. The $\Delta G^{\ddagger}$ for the overall reaction is the energy change between reactants and the highest-energy transition state—the second one in this case. An exergonic reaction has a negative overall $\Delta G°$.

Solution

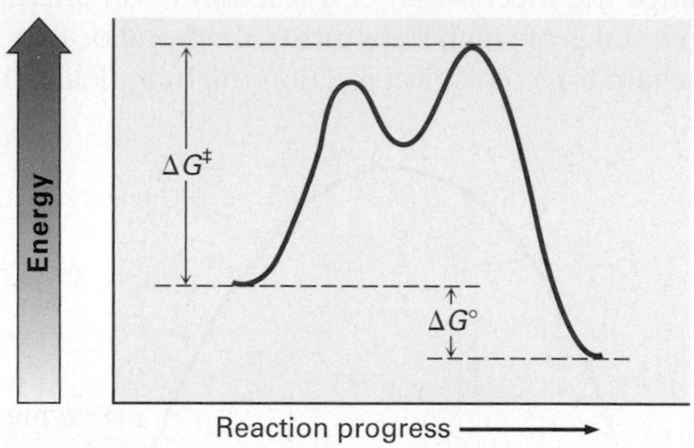

Problem 6.13

Sketch an energy diagram for a two-step reaction in which both steps are exergonic and in which the second step has a higher-energy transition state than the first. Label the parts of the diagram corresponding to reactant, product, intermediate, overall $\Delta G^{\ddagger}$, and overall $\Delta G°$.

6.11 A Comparison Between Biological Reactions and Laboratory Reactions

Beginning in the next chapter, we'll be seeing a lot of reactions, some that are important in laboratory chemistry yet don't occur in nature and others that have counterparts in biological pathways. In comparing laboratory reactions with biological reactions, several differences are apparent. For one, laboratory reactions are usually carried out in an organic solvent such as diethyl ether or dichloromethane to dissolve the reactants and bring them into contact, whereas biological reactions occur in the aqueous medium inside cells. For another, laboratory reactions often take place over a wide range of temperatures without catalysts, while biological reactions take place at the temperature of the organism and are catalyzed by enzymes.

We'll look at enzymes in more detail in **Section 26.10**, but you may already be aware that an enzyme is a large, globular, protein molecule that contains in its structure a protected pocket called its *active site*. The active site is lined by acidic or basic groups as needed for catalysis and has precisely the right shape to bind and hold a substrate molecule in the orientation necessary for reaction. **Figure 6.9** shows a molecular model of hexokinase, along with an X-ray crystal structure of the glucose substrate and adenosine diphosphate (ADP) bound in the active site. Hexokinase is an enzyme that catalyzes the initial step of glucose metabolism—the transfer of a phosphate group from ATP to glucose, giving glucose 6-phosphate and ADP. The structures of ATP and ADP were shown at the end of **Section 6.8**.

$$\text{Glucose} \xrightarrow[\text{Hexokinase}]{\text{ATP} \;\; \text{ADP}} \text{Glucose 6-phosphate}$$

Note how the hexokinase-catalyzed phosphorylation reaction of glucose is written. It's common when writing biological equations to show only the structures of the primary reactant and product, while abbreviating the structures of various biological "reagents" and by-products such as ATP and ADP. A curved arrow intersecting the straight reaction arrow indicates that ATP is also a reactant and ADP also a product.

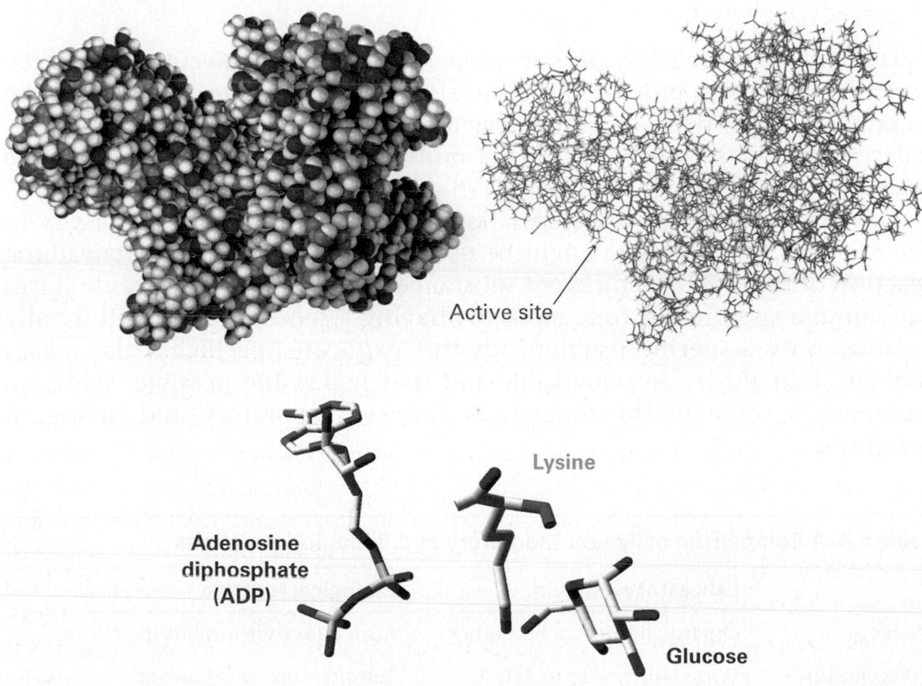

Figure 6.9 Models of hexokinase in space-filling and wire-frame formats, showing the cleft that contains the active site where substrate binding and reaction catalysis occur. At the bottom is an X-ray crystal structure of the enzyme active site, showing the positions of both glucose and ADP as well as a lysine amino acid that acts as a base to deprotonate glucose.

Yet another difference between laboratory and biological reactions is that laboratory reactions are often done using relatively small, simple reagents such as Br_2, HCl, $NaBH_4$, CrO_3, and so forth, while biological reactions usually involve relatively complex "reagents" called *coenzymes*. In the hexokinase-catalyzed phosphorylation of glucose just shown, ATP is the coenzyme. As another example, compare the H_2 molecule, a laboratory reagent that adds to a carbon–carbon double bond to yield an alkane, with the reduced nicotinamide adenine dinucleotide (NADH) molecule, a coenzyme that effects an analogous addition of hydrogen to a double bond in many biological pathways. Of all the atoms in the entire coenzyme, only the one hydrogen atom shown in red is transferred to the double-bond substrate.

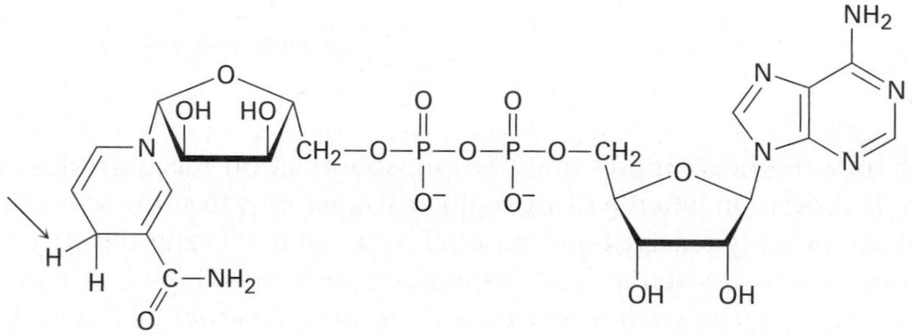

Reduced nicotinamide adenine dinucleotide, NADH (a coenzyme)

Don't be intimidated by the size of the ATP or NADH molecule; most of the structure is there to provide an overall shape for binding to the enzyme and to provide appropriate solubility behavior. When looking at biological molecules, focus on the small part of the molecule where the chemical change takes place.

One final difference between laboratory and biological reactions is in their specificity. A catalyst might be used in the laboratory to catalyze the reaction of thousands of different substances, but an enzyme, because it can only bind a specific substrate molecule having a specific shape, will usually catalyze only a specific reaction. It's this exquisite specificity that makes biological chemistry so remarkable and that makes life possible. Table 6.4 summarizes some of the differences between laboratory and biological reactions.

Table 6.4 A Comparison of Typical Laboratory and Biological Reactions

	Laboratory reaction	Biological reaction
Solvent	Organic liquid, such as ether	Aqueous environment in cells
Temperature	Wide range; −80 to 150 °C	Temperature of organism
Catalyst	Either none, or very simple	Large, complex enzymes needed
Reagent size	Usually small and simple	Relatively complex coenzymes
Specificity	Little specificity for substrate	Very high specificity for substrate

A DEEPER LOOK Where Do Drugs Come From?

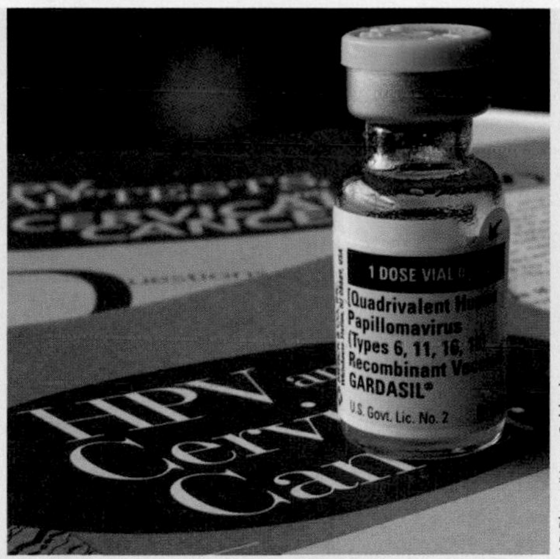

AP Photo/Harry Cabluck

Introduced in June, 2006, Gardasil is the first vaccine ever approved for the prevention of cancer. Where do new drugs like this come from?

It has been estimated that major pharmaceutical companies in the United States spend some \$33 billion per year on drug research and development, while government agencies and private foundations spend another \$28 billion. What does this money buy? For the period 1981 to 2008, the money resulted in a total of 989 new molecular entities (NMEs)—new biologically active chemical substances approved for sale as drugs by the U.S. Food and Drug Administration (FDA). That's an average of only 35 new drugs each year, spread over all diseases and conditions, and the number is steadily falling. In 2008, only 20 NMEs were approved.

Where do the new drugs come from? According to a study carried out at the U.S. National Cancer Institute, only about 33% of new drugs are entirely synthetic and completely unrelated to any naturally occurring substance. The remaining 67% take their lead, to a greater or lesser extent, from nature. Vaccines and genetically engineered proteins of biological origin account for 15% of NMEs, but most new drugs come from *natural products,* a catchall term generally taken to mean small molecules found in bacteria, plants, and other living organisms. Unmodified natural products isolated directly from the producing organism account for 24% of NMEs, while natural products that have been chemically modified in the laboratory account for the remaining 28%.

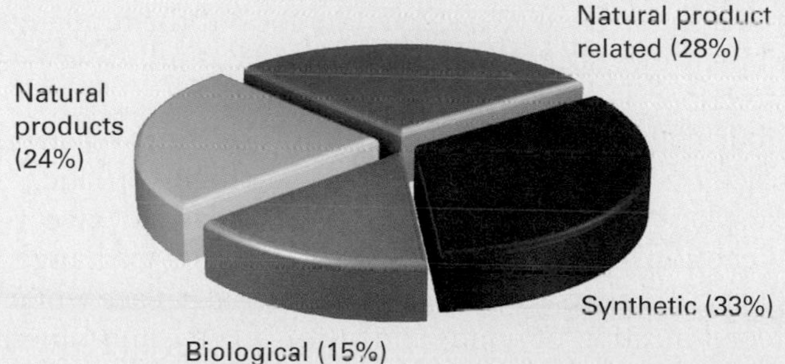

Many years of work go into screening many thousands of substances to identify a single compound that might ultimately gain approval as an NME. But after that single compound has been identified, the work has just begun because it takes an average of 9 to 10 years for a drug to make it through the approval process. First, the safety of the drug in animals must be demonstrated and an economical method of manufacture must be devised. With these preliminaries out of the way, an Investigational New Drug (IND) application is submitted to the FDA for permission to begin testing in humans.

Human testing takes 5 to 7 years and is divided into three phases. Phase I clinical trials are carried out on a small group of healthy volunteers to establish safety and look for side effects. Several months to a year are needed, and only about 70% of drugs pass at this point. Phase II clinical trials next test the drug for 1 to 2 years in several hundred patients with the target disease or condition, looking both for safety and for efficacy, and only about 33% of the original group pass. Finally, phase III trials are undertaken on a large sample of patients to document definitively the drug's safety, dosage, and efficacy. If the

(continued)

(continued)

drug is one of the 25% of the original group that make it to the end of phase III, all the data are then gathered into a New Drug Application (NDA) and sent to the FDA for review and approval, which can take another 2 years. Ten years have elapsed and at least $500 million has been spent, with only a 20% success rate for the drugs that began testing. Finally, though, the drug will begin to appear in medicine cabinets. The following timeline shows the process.

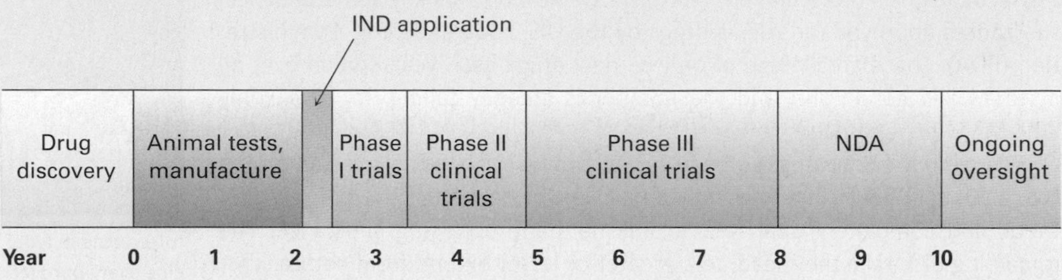

Summary

Key words

activation energy ($\Delta G^{\ddagger}$), 206
addition reaction, 184
bond dissociation energy (D), 203
carbocation, 196
electrophile, 192
elimination reaction, 185
endergonic, 201
endothermic, 202
enthalpy change (ΔH), 202
entropy change (ΔS), 202
exergonic, 201
exothermic, 202
Gibbs free-energy change (ΔG), 201
heat of reaction, 202
nucleophile, 192
polar reaction, 187
radical, 187
radical reaction, 187
reaction intermediate, 208
reaction mechanism, 186
rearrangement reaction, 185
substitution reaction, 185
transition state, 206

All chemical reactions, whether in the laboratory or in living organisms, follow the same "rules." To understand both organic and biological chemistry, it's necessary to know not just *what* occurs but also *why* and *how* chemical reactions take place. In this chapter, we've taken a brief look at the fundamental kinds of organic reactions, we've seen why reactions occur, and we've seen how reactions can be described.

There are four common kinds of reactions: **addition reactions** take place when two reactants add together to give a single product; **elimination reactions** take place when one reactant splits apart to give two products; **substitution reactions** take place when two reactants exchange parts to give two new products; and **rearrangement reactions** take place when one reactant undergoes a reorganization of bonds and atoms to give an isomeric product.

A full description of how a reaction occurs is called its **mechanism**. There are two general kinds of mechanisms by which most reactions take place: **radical** mechanisms and **polar** mechanisms. Polar reactions, the more common type, occur because of an attractive interaction between a **nucleophilic** (electron-rich) site in one molecule and an **electrophilic** (electron-poor) site in another molecule. A bond is formed in a polar reaction when the nucleophile donates an electron pair to the electrophile. This movement of electrons is indicated by a curved arrow showing the direction of electron travel from the nucleophile to the electrophile. Radical reactions involve species that have an odd number of electrons. A bond is formed when each reactant donates one electron.

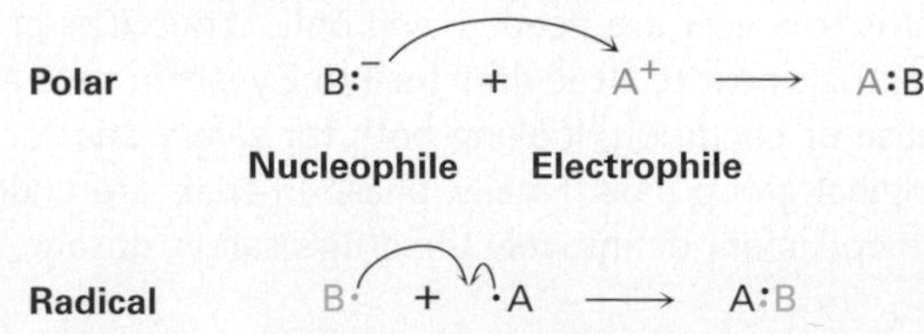

The energy changes that take place during reactions can be described by considering both rates (how fast the reactions occur) and equilibria (how much the reactions occur). The position of a chemical equilibrium is determined by the value of the **free-energy change (ΔG)** for the reaction, where $\Delta G = \Delta H - T\Delta S$. The **enthalpy** term (ΔH) corresponds to the net change in strength of chemical bonds broken and formed during reaction; the **entropy** term (ΔS) corresponds to the change in the amount of molecular randomness during the reaction. Reactions that have negative values of ΔG release energy, are said to be **exergonic**, and have favorable equilibria. Reactions that have positive values of ΔG absorb energy, are said to be **endergonic**, and have unfavorable equilibria.

A reaction can be described pictorially using an energy diagram that follows the reaction course from reactant through transition state to product. The **transition state** is an activated complex occurring at the highest-energy point of a reaction. The amount of energy needed by reactants to reach this high point is the **activation energy, $\Delta G^\ddagger$**. The higher the activation energy, the slower the reaction.

Many reactions take place in more than one step and involve the formation of a **reaction intermediate**. An intermediate is a species that lies at an energy minimum between steps on the reaction curve and is formed briefly during the course of a reaction.

Exercises

Visualizing Chemistry

OWL Interactive versions of these problems are assignable in OWL for Organic Chemistry.

▲ denotes problems linked to the Key Ideas in this chapter.

(Problems 6.1–6.13 appear within the chapter.)

6.14 The following alkyl halide can be prepared by addition of HBr to two different alkenes. Draw the structures of both (reddish-brown = Br).

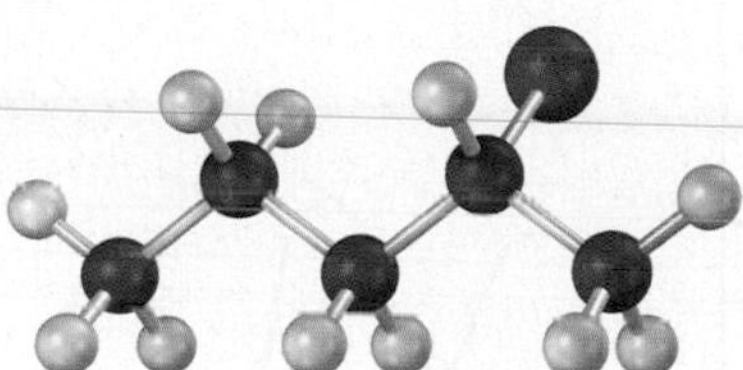

6.15 The following structure represents the carbocation intermediate formed in the addition reaction of HBr to two different alkenes. Draw the structures of both.

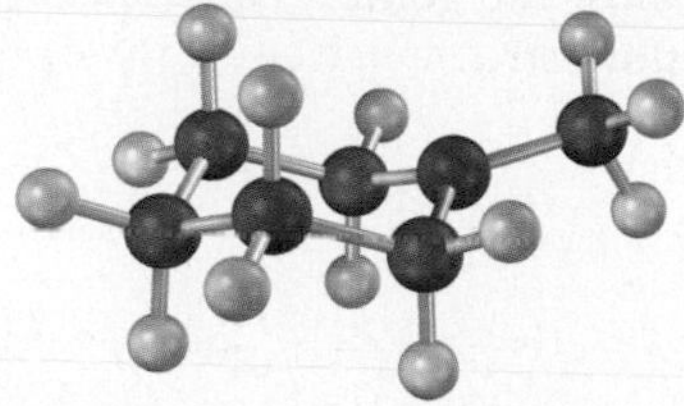

▲ Problems linked to Key Ideas in this chapter

6.16 Electrostatic potential maps of **(a)** formaldehyde (CH_2O) and **(b)** methanethiol (CH_3SH) are shown. Is the formaldehyde carbon atom likely to be electrophilic or nucleophilic? What about the methanethiol sulfur atom? Explain.

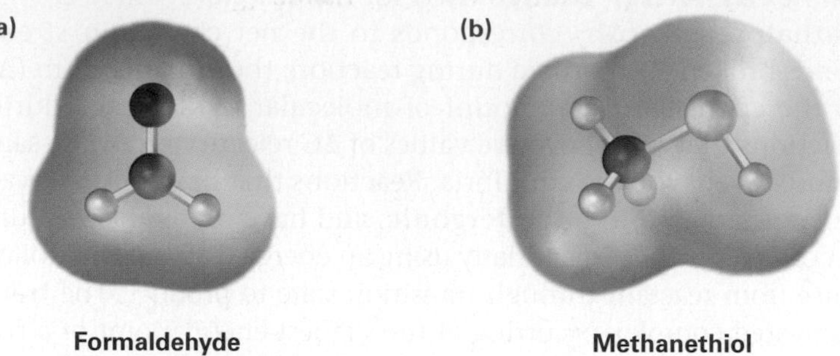

6.17 Look at the following energy diagram:

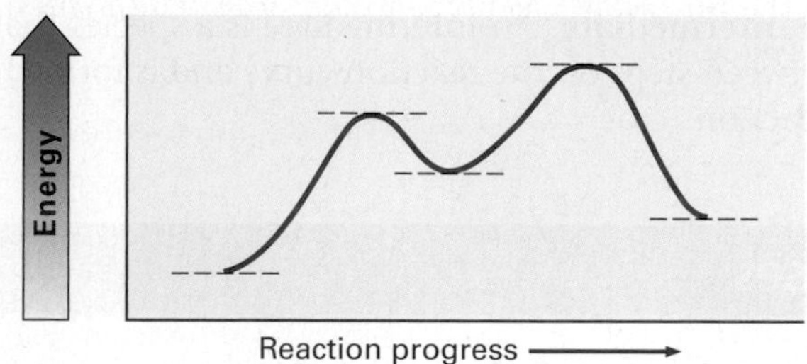

(a) Is $\Delta G°$ for the reaction positive or negative? Label it on the diagram.
(b) How many steps are involved in the reaction?
(c) How many transition states are there? Label them on the diagram.

6.18 Look at the following energy diagram for an enzyme-catalyzed reaction:

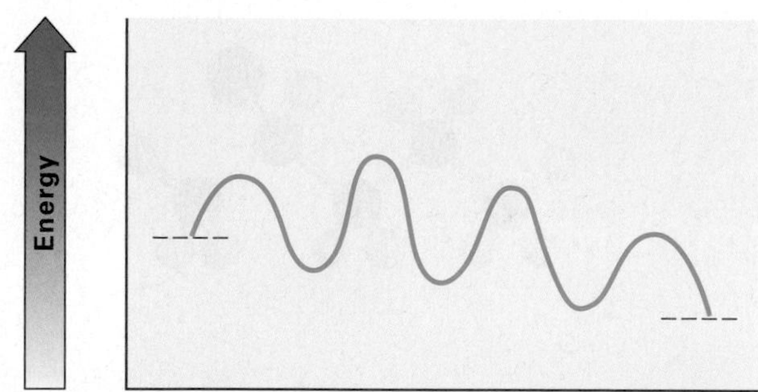

(a) How many steps are involved?
(b) Which step is most exergonic?
(c) Which step is slowest?

▲ Problems linked to Key Ideas in this chapter

Additional Problems

Polar Reactions

6.19 Identify the functional groups in the following molecules, and show the polarity of each:

(a) $CH_3CH_2C{\equiv}N$

(b) (cyclopentyl)–OCH_3

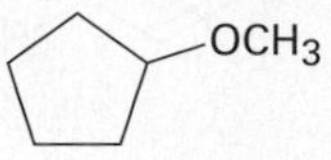

(c) $CH_3\overset{O}{\overset{\|}{C}}CH_2\overset{O}{\overset{\|}{C}}OCH_3$

(d) (cyclohexa-2,5-diene-1,4-dione)

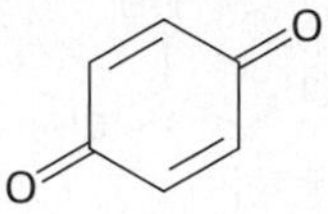

(e) (but-2-enamide, NH_2)

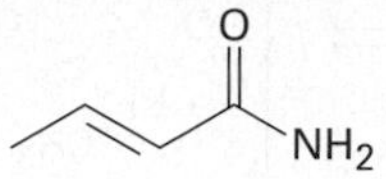

(f) (benzaldehyde, C(=O)H)

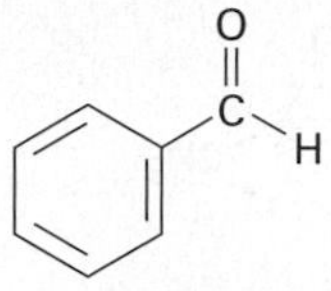

6.20 Identify the following reactions as additions, eliminations, substitutions, or rearrangements:

(a) CH_3CH_2Br + NaCN ⟶ CH_3CH_2CN (+ NaBr)

(b) (cyclohexanol)–OH $\xrightarrow[\text{catalyst}]{\text{Acid}}$ (cyclohexene) (+ H_2O)

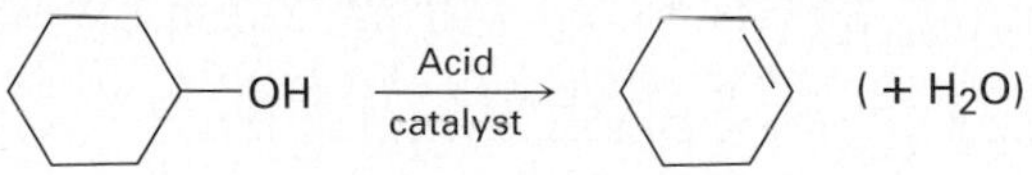

(c) (cyclopentadiene) + (methyl vinyl ketone) $\xrightarrow{\text{Heat}}$ (bicyclic ketone product)

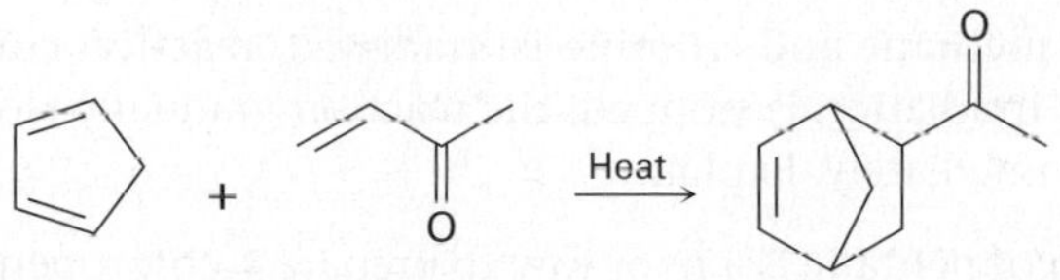

(d) (cyclohexane) + $O_2N{-}NO_2$ $\xrightarrow{\text{Light}}$ (nitrocyclohexane, NO_2) (+ HNO_2)

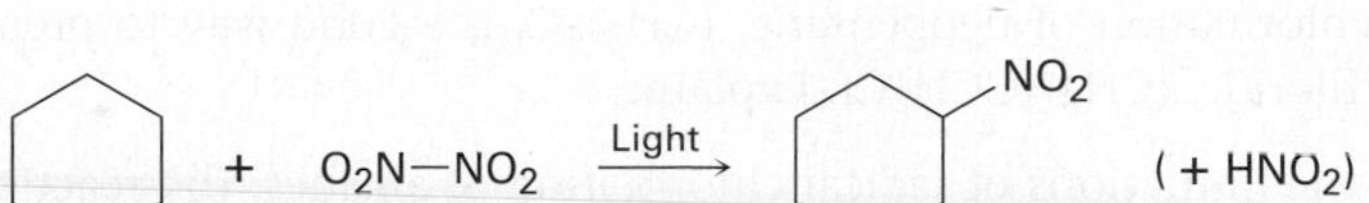

6.21 Identify the likely electrophilic and nucleophilic sites in each of the following molecules:

(a) Testosterone (structure with CH_3, OH, H labels)

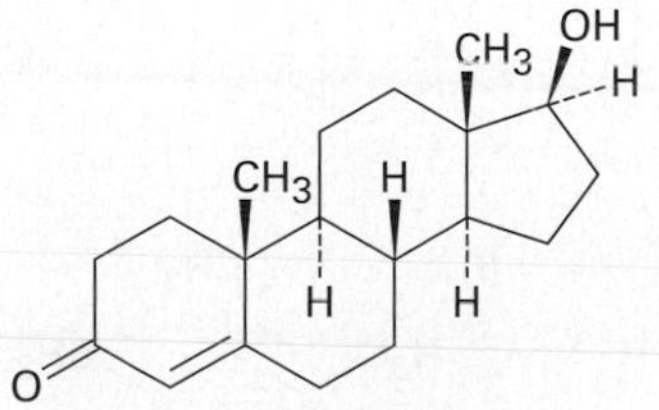

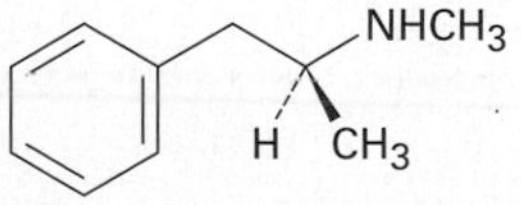

(b) Amphetamine (structure with $NHCH_3$, H, CH_3 labels)

▲ Problems linked to Key Ideas in this chapter

6.22 ▲ Add curved arrows to the following polar reactions to indicate the flow of electrons in each:

(a)

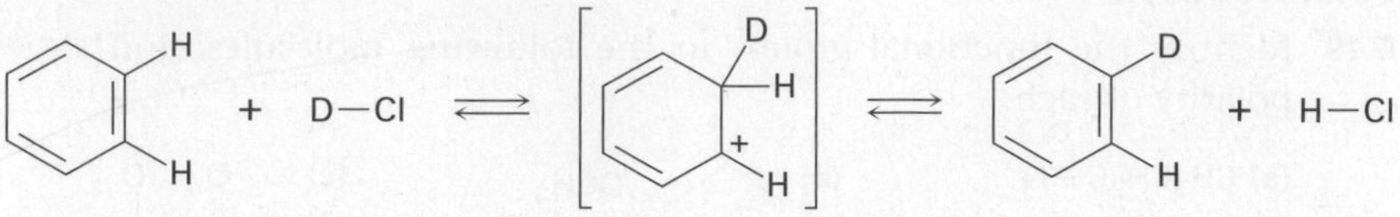

+ D—Cl ⇌ [ ] ⇌ + H—Cl

(b)

+ H—Cl ⇌ [ ] ⇌

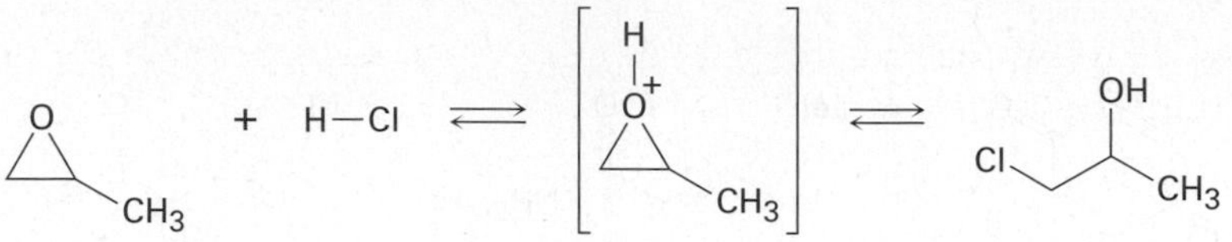

6.23 ▲ Follow the flow of electrons indicated by the curved arrows in each of the following polar reactions, and predict the products that result:

(a) ⇌ ?

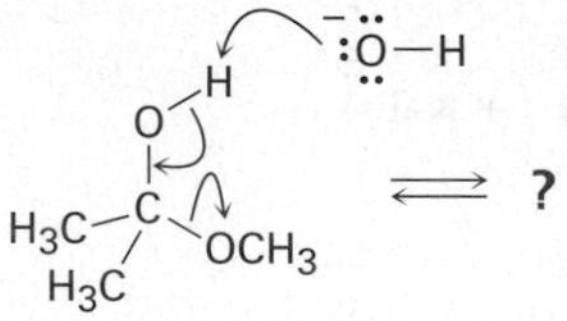

(b) ⇌ ?

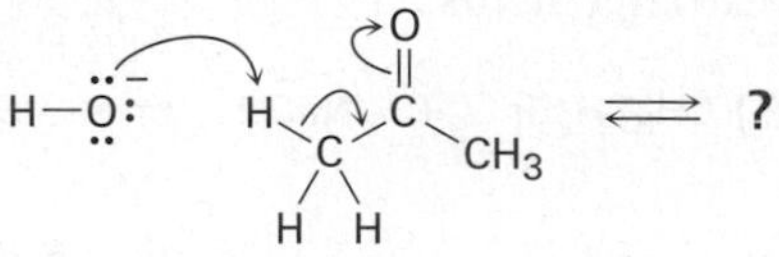

Radical Reactions

6.24 When a mixture of methane and chlorine is irradiated, reaction commences immediately. When irradiation is stopped, the reaction gradually slows down but does not stop immediately. Explain.

6.25 Radical chlorination of pentane is a poor way to prepare 1-chloropentane, but radical chlorination of neopentane, $(CH_3)_4C$, is a good way to prepare neopentyl chloride, $(CH_3)_3CCH_2Cl$. Explain.

6.26 Despite the limitations of radical chlorination of alkanes, the reaction is still useful for synthesizing certain halogenated compounds. For which of the following compounds does radical chlorination give a single monochloro product?

(a) CH_3CH_3

(b) $CH_3CH_2CH_3$

(c)

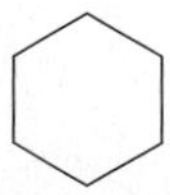

(d) $CH_3C(CH_3)_2CH_2CH_3$

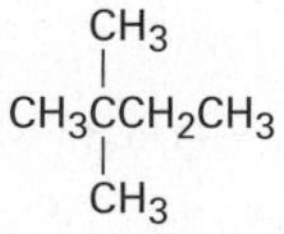

(e) $CH_3C{\equiv}CCH_3$

(f)

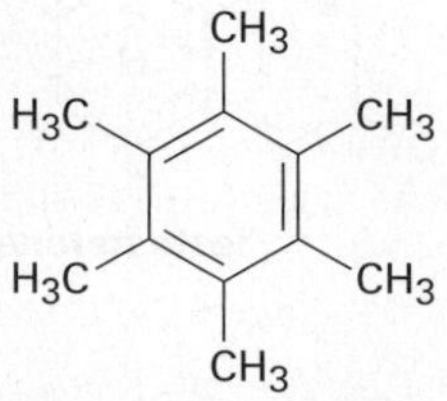

▲ Problems linked to Key Ideas in this chapter

Energy Diagrams and Reaction Mechanisms

6.27 What is the difference between a transition state and an intermediate?

6.28 Draw an energy diagram for a one-step reaction with $K_{eq} < 1$. Label the parts of the diagram corresponding to reactants, products, transition state, $\Delta G°$, and $\Delta G^‡$. Is $\Delta G°$ positive or negative?

6.29 Draw an energy diagram for a two-step reaction with $K_{eq} > 1$. Label the overall $\Delta G°$, transition states, and intermediate. Is $\Delta G°$ positive or negative?

6.30 Draw an energy diagram for a two-step exergonic reaction whose second step is faster than its first step.

6.31 Draw an energy diagram for a reaction with $K_{eq} = 1$. What is the value of $\Delta G°$ in this reaction?

6.32 The addition of water to ethylene to yield ethanol has the following thermodynamic parameters:

$$H_2C{=}CH_2 + H_2O \rightleftharpoons CH_3CH_2OH \quad \begin{cases} \Delta H° = -44 \text{ kJ/mol} \\ \Delta S° = -0.12 \text{ kJ/(K} \cdot \text{mol)} \\ K_{eq} = 24 \end{cases}$$

(a) Is the reaction exothermic or endothermic?
(b) Is the reaction favorable (spontaneous) or unfavorable (nonspontaneous) at room temperature (298 K)?

6.33 When isopropylidenecyclohexane is treated with strong acid at room temperature, isomerization occurs by the mechanism shown below to yield 1-isopropylcyclohexene:

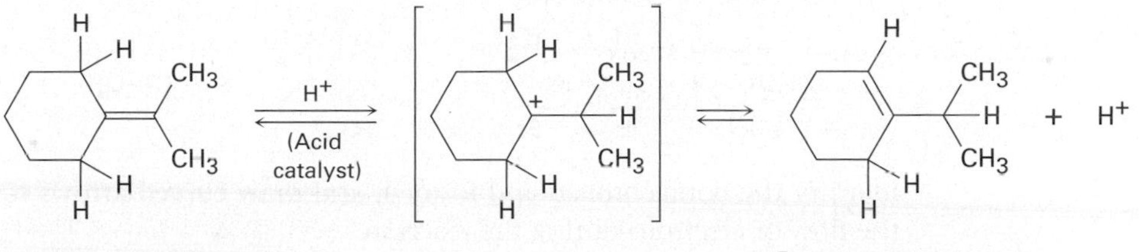

H^+ (Acid catalyst)

Isopropylidenecyclohexane **1-Isopropylcyclohexene**

At equilibrium, the product mixture contains about 30% isopropylidenecyclohexane and about 70% 1-isopropylcyclohexene.
(a) What is an approximate value of K_{eq} for the reaction?
(b) Since the reaction occurs slowly at room temperature, what is its approximate $\Delta G^‡$?
(c) Draw an energy diagram for the reaction.

6.34 ▲ Add curved arrows to the mechanism shown in Problem 6.33 to indicate the electron movement in each step.

General Problems

6.35 2-Chloro-2-methylpropane reacts with water in three steps to yield 2-methyl-2-propanol. The first step is slower than the second, which in turn is much

▲ Problems linked to Key Ideas in this chapter

slower than the third. The reaction takes place slowly at room temperature, and the equilibrium constant is near 1.

$$(CH_3)_3C{-}Cl \rightleftharpoons \left[(CH_3)_3C^+ \xrightleftharpoons{H_2O} (CH_3)_3C{-}O^+H_2 \right] \xrightleftharpoons{H_2O} (CH_3)_3C{-}O{-}H + H_3O^+ + Cl^-$$

2-Chloro-2-methylpropane → **2-Methyl-2-propanol**

(a) Give approximate values for $\Delta G^\ddagger$ and ΔG° that are consistent with the above information.

(b) Draw an energy diagram for the reaction, labeling all points of interest and making sure that the relative energy levels on the diagram are consistent with the information given.

6.36 ▲ Add curved arrows to the mechanism shown in Problem 6.35 to indicate the electron movement in each step.

6.37 The reaction of hydroxide ion with chloromethane to yield methanol and chloride ion is an example of a general reaction type called a *nucleophilic substitution reaction:*

$$HO^- + CH_3Cl \rightleftharpoons CH_3OH + Cl^-$$

The value of ΔH° for the reaction is -75 kJ/mol, and the value of ΔS° is $+54$ J/(K·mol). What is the value of ΔG° (in kJ/mol) at 298 K? Is the reaction exothermic or endothermic? Is it exergonic or endergonic?

6.38 Methoxide ion (CH_3O^-) reacts with bromoethane in a single step according to the following equation:

$$CH_3\ddot{\underset{..}{O}}{:}^- + H_3C{-}CH_2Br \longrightarrow H_2C{=}CH_2 + CH_3OH + {:}\ddot{\underset{..}{Br}}{:}^-$$

Identify the bonds broken and formed, and draw curved arrows to represent the flow of electrons during the reaction.

6.39 ▲ Ammonia reacts with acetyl chloride (CH_3COCl) to give acetamide (CH_3CONH_2). Identify the bonds broken and formed in each step of the reaction, and draw curved arrows to represent the flow of electrons in each step.

$$H_3C{-}C({=}\ddot{O}{:}){-}Cl \xrightarrow{:NH_3} H_3C{-}C({-}\ddot{O}{:}^-)(Cl){-}NH_3^+ \longrightarrow H_3C{-}C({=}\ddot{O}{:}){-}NH_3^+ \longrightarrow$$

Acetyl chloride

$$\xrightarrow{:NH_3} H_3C{-}C({=}\ddot{O}{:}){-}\ddot{N}H_2 + NH_4^+\ Cl^-$$

Acetamide

▲ Problems linked to Key Ideas in this chapter

6.40 The naturally occurring molecule α-terpineol is biosynthesized by a route that includes the following step:

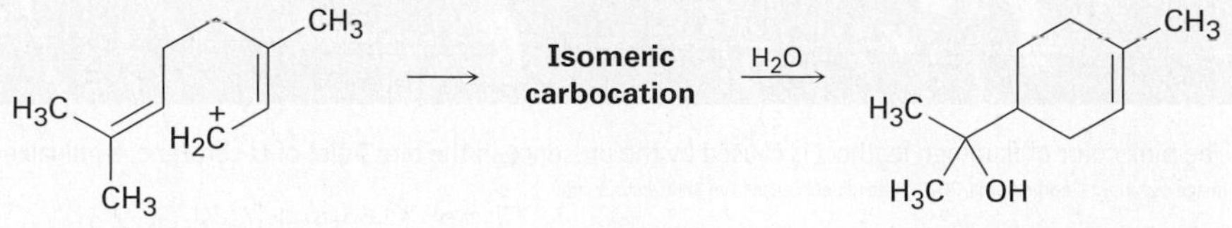

Carbocation → Isomeric carbocation → (H_2O) → α-Terpineol

(a) Propose a likely structure for the isomeric carbocation intermediate.
(b) Show the mechanism of each step in the biosynthetic pathway, using curved arrows to indicate electron flow.

6.41 Predict the product(s) of each of the following biological reactions by interpreting the flow of electrons as indicated by the curved arrows:

(a) → ?

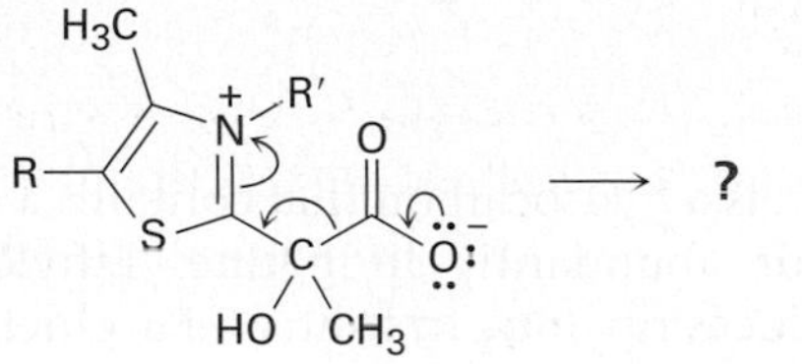

(b) → ?

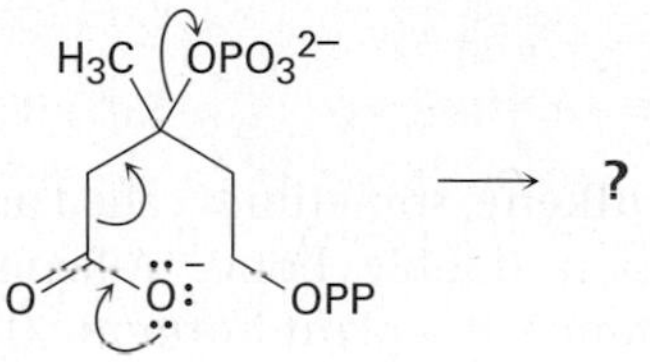

(c) → ?

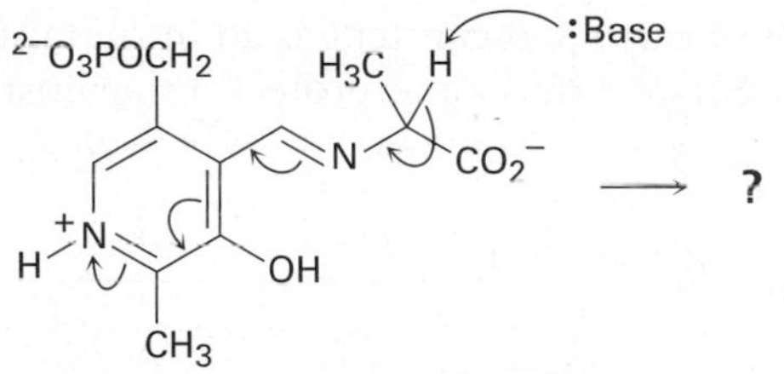

6.42 Reaction of 2-methylpropene with HBr might, in principle, lead to a mixture of two alkyl bromide addition products. Name them, and draw their structures.

6.43 Draw the structures of the two carbocation intermediates that might form during the reaction of 2-methylpropene with HBr (Problem 6.42). We'll see in the next chapter that the stability of carbocations depends on the number of alkyl substituents attached to the positively charged carbon—the more alkyl substituents there are, the more stable the cation. Which of the two carbocation intermediates you drew is more stable?

▲ Problems linked to Key Ideas in this chapter

7

The pink color of flamingo feathers is caused by the presence in the bird's diet of β-carotene, a polyalkene.

Alkenes: Structure and Reactivity

An **alkene**, sometimes called an *olefin,* is a hydrocarbon that contains a carbon–carbon double bond. Alkenes occur abundantly in nature. Ethylene, for instance, is a plant hormone that induces ripening in fruit, and α-pinene is the major component of turpentine. Life itself would be impossible without such alkenes as β-carotene, a compound that contains 11 double bonds. An orange pigment responsible for the color of carrots, β-carotene is an important dietary source of vitamin A and is thought to offer some protection against certain types of cancer.

$H_2C=CH_2$

Ethylene

α-Pinene

β-Carotene
(orange pigment and vitamin A precursor)

Why This Chapter? Carbon–carbon double bonds are present in most organic and biological molecules, so a good understanding of their behavior is needed. In this chapter, we'll look at some consequences of alkene stereoisomerism and then focus on the broadest and most general class of alkene reactions, the electrophilic addition reaction.

OWL Sign in to OWL for Organic Chemistry at **www.cengage.com/owl** to view tutorials and simulations, develop problem-solving skills, and complete online homework assigned by your professor.

7.1 Industrial Preparation and Use of Alkenes

Ethylene and propylene, the simplest alkenes, are the two most important organic chemicals produced industrially. Approximately 127 million metric tons of ethylene and 54 million metric tons of propylene are produced worldwide each year for use in the synthesis of polyethylene, polypropylene, ethylene glycol, acetic acid, acetaldehyde, and a host of other substances **(Figure 7.1)**.

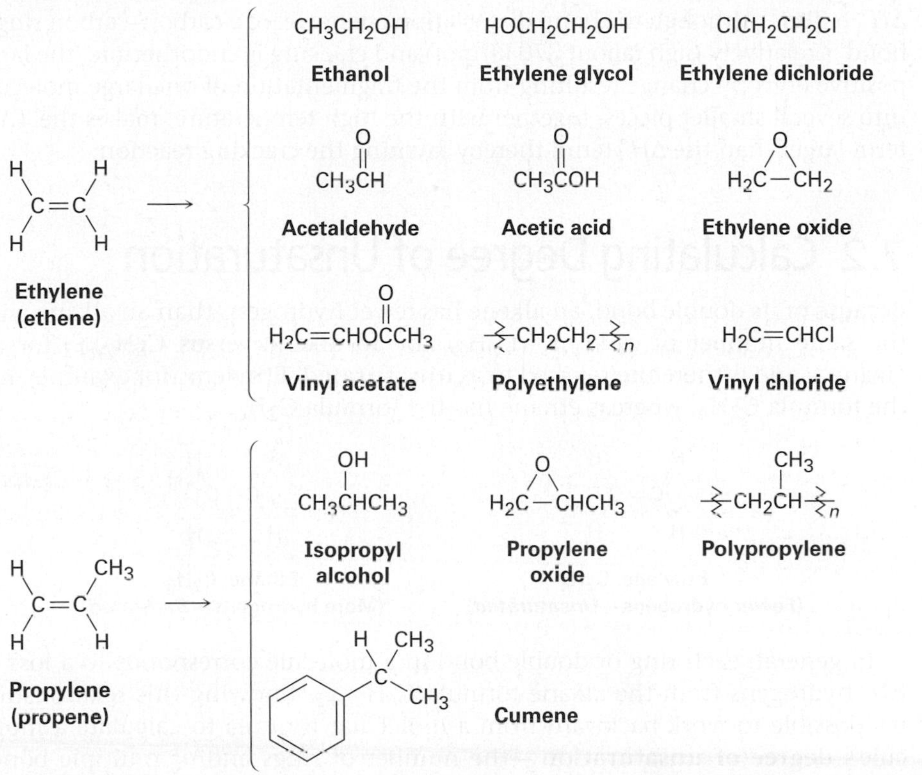

Ethylene (ethene) $H_2C=CH_2$ →
- CH_3CH_2OH Ethanol
- $HOCH_2CH_2OH$ Ethylene glycol
- $ClCH_2CH_2Cl$ Ethylene dichloride
- CH_3CHO Acetaldehyde
- CH_3COOH Acetic acid
- H_2C-CH_2 (O bridged) Ethylene oxide
- $H_2C=CHOCOCH_3$ Vinyl acetate
- $-(CH_2CH_2)_n-$ Polyethylene
- $H_2C=CHCl$ Vinyl chloride

Propylene (propene) $H_2C=CHCH_3$ →
- $CH_3CH(OH)CH_3$ Isopropyl alcohol
- $H_2C-CHCH_3$ (O bridged) Propylene oxide
- $-(CH_2CH(CH_3))_n-$ Polypropylene
- $C_6H_5CH(CH_3)_2$ Cumene

Figure 7.1 Compounds derived industrially from ethylene and propylene.

Ethylene, propylene, and butene are synthesized industrially by steam cracking of light (C_2–C_8) alkanes.

$$CH_3(CH_2)_nCH_3 \quad [n = 0\text{–}6] \xrightarrow{850\text{–}900\ °C,\ \text{steam}} H_2 + H_2C{=}CH_2 + CH_3CH{=}CH_2 + CH_3CH_2CH{=}CH_2$$

Steam cracking takes place without a catalyst at temperatures up to 900 °C. The process is complex, although it undoubtedly involves radical reactions. The high-temperature reaction conditions cause spontaneous homolytic breaking of C–C and C–H bonds, with resultant formation of smaller fragments. We might imagine, for instance, that a molecule of butane splits into two ethyl radicals, each of which then loses a hydrogen atom to generate two molecules of ethylene.

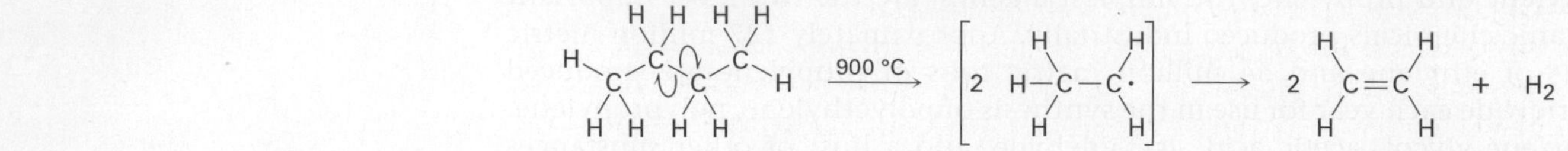

Steam cracking is an example of a reaction whose energetics are dominated by entropy ($\Delta S°$) rather than by enthalpy ($\Delta H°$) in the free-energy equation $\Delta G° = \Delta H° - T\Delta S°$. Although the bond dissociation energy D for a carbon–carbon single bond is relatively high (about 370 kJ/mol) and cracking is endothermic, the large positive entropy change resulting from the fragmentation of one large molecule into several smaller pieces, together with the high temperature, makes the $T\Delta S°$ term larger than the $\Delta H°$ term, thereby favoring the cracking reaction.

7.2 Calculating Degree of Unsaturation

Because of its double bond, an alkene has fewer hydrogens than an alkane with the same number of carbons—C_nH_{2n} for an alkene versus C_nH_{2n+2} for an alkane—and is therefore referred to as **unsaturated**. Ethylene, for example, has the formula C_2H_4, whereas ethane has the formula C_2H_6.

Ethylene: C_2H_4
(Fewer hydrogens—*Unsaturated*)

Ethane: C_2H_6
(More hydrogens—*Saturated*)

In general, each ring or double bond in a molecule corresponds to a loss of two hydrogens from the alkane formula C_nH_{2n+2}. Knowing this relationship, it's possible to work backward from a molecular formula to calculate a molecule's **degree of unsaturation**—the number of rings and/or multiple bonds present in the molecule.

Let's assume that we want to find the structure of an unknown hydrocarbon. A molecular weight determination on the unknown yields a value of 82 amu, which corresponds to a molecular formula of C_6H_{10}. Since the saturated C_6 alkane (hexane) has the formula C_6H_{14}, the unknown compound has two fewer pairs of hydrogens ($H_{14} - H_{10} = H_4 = 2\ H_2$) so its degree of unsaturation is 2. The unknown therefore contains two double bonds, one ring and one double bond, two rings, or one triple bond. There's still a long way to go to establish structure, but the simple calculation has told us a lot about the molecule.

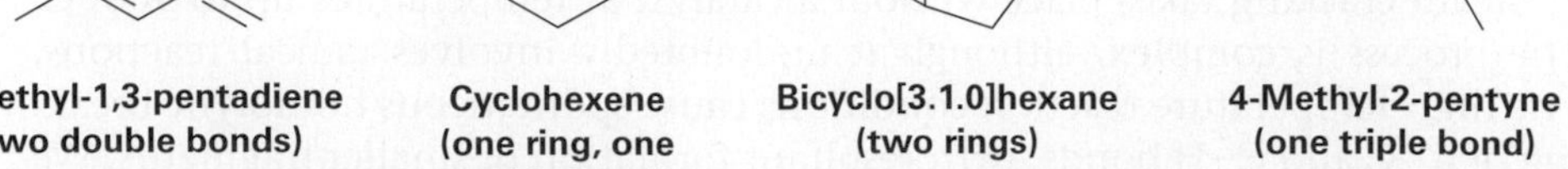

4-Methyl-1,3-pentadiene
(two double bonds)

Cyclohexene
(one ring, one double bond)

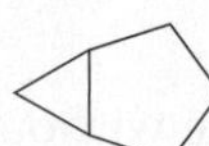

Bicyclo[3.1.0]hexane
(two rings)

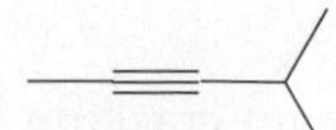

4-Methyl-2-pentyne
(one triple bond)

C_6H_{10}

Similar calculations can be carried out for compounds containing elements other than just carbon and hydrogen.

- **Organohalogen compounds (C, H, X, where X = F, Cl, Br, or I)** A halogen substituent acts as a replacement for hydrogen in an organic molecule, so we can add the number of halogens and hydrogens to arrive at an equivalent hydrocarbon formula from which the degree of unsaturation can be found. For example, the formula $C_4H_6Br_2$ is equivalent to the hydrocarbon formula C_4H_8 and thus corresponds to one degree of unsaturation.

Replace 2 Br by 2 H

$BrCH_2CH{=}CHCH_2Br$ = $HCH_2CH{=}CHCH_2H$

$C_4H_6Br_2$ = "C_4H_8" One unsaturation: one double bond

Add

- **Organooxygen compounds (C, H, O)** Oxygen forms two bonds, so it doesn't affect the formula of an equivalent hydrocarbon and can be ignored when calculating the degree of unsaturation. You can convince yourself of this by seeing what happens when an oxygen atom is inserted into an alkane bond: C–C becomes C–O–C or C–H becomes C–O–H, and there is no change in the number of hydrogen atoms. For example, the formula C_5H_8O is equivalent to the hydrocarbon formula C_5H_8 and thus corresponds to two degrees of unsaturation.

O removed from here

$H_2C{=}CHCH{=}CHCH_2OH$ = $H_2C{=}CHCH{=}CHCH_2{-}H$

C_5H_8O = "C_5H_8" Two unsaturations: two double bonds

- **Organonitrogen compounds (C, H, N)** Nitrogen forms three bonds, so an organonitrogen compound has one more hydrogen than a related hydrocarbon. We therefore subtract the number of nitrogens from the number of hydrogens to arrive at the equivalent hydrocarbon formula. Again, you can convince yourself of this by seeing what happens when a nitrogen atom is inserted into an alkane bond: C–C becomes C–NH–C or C–H becomes C–NH_2, meaning that one additional hydrogen atom has been added. We must therefore subtract this extra hydrogen atom to arrive at the equivalent hydrocarbon formula. For example, the formula C_5H_9N is equivalent to C_5H_8 and thus has two degrees of unsaturation.

Removed

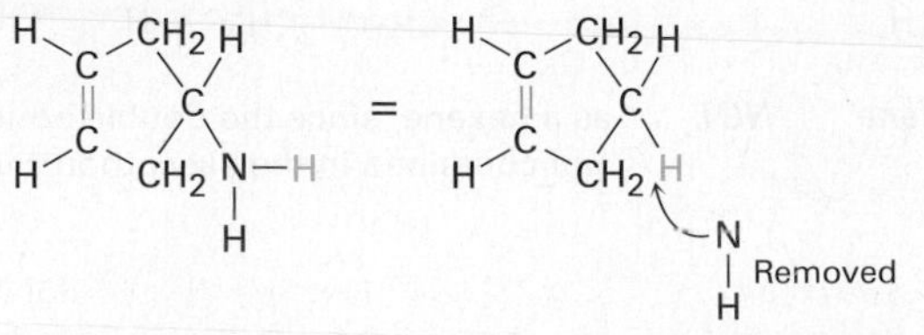

C_5H_9N = "C_5H_8" Two unsaturations: one ring and one double bond

To summarize:

- **Add** the number of **halogens** to the number of hydrogens.
- **Ignore** the number of **oxygens**.
- **Subtract** the number of **nitrogens** from the number of hydrogens.

Problem 7.1

Calculate the degree of unsaturation in each of the following formulas, and then draw as many structures as you can for each:

(a) C_4H_8 **(b)** C_4H_6 **(c)** C_3H_4

Problem 7.2

Calculate the degree of unsaturation in each of the following formulas:

(a) C_6H_5N **(b)** $C_6H_5NO_2$ **(c)** $C_8H_9Cl_3$
(d) $C_9H_{16}Br_2$ **(e)** $C_{10}H_{12}N_2O_3$ **(f)** $C_{20}H_{32}ClN$

Problem 7.3

Diazepam, marketed as an antianxiety medication under the name Valium, has three rings, eight double bonds, and the formula $C_{16}H_?ClN_2O$. How many hydrogens does diazepam have? (Calculate the answer; don't count hydrogens in the structure.)

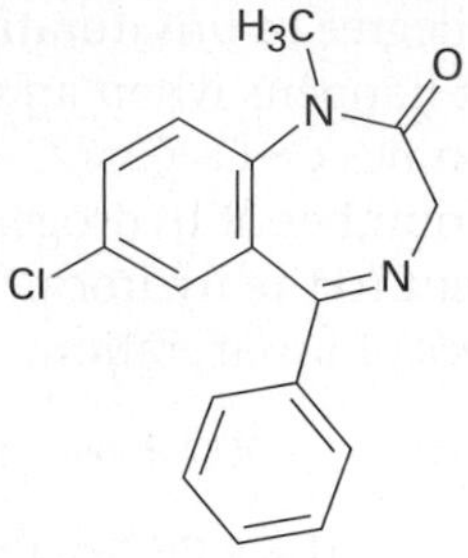

H_3C, O, N, Cl, N

Diazepam

7.3 Naming Alkenes

Alkenes are named using a series of rules similar to those for alkanes **(Section 3.4)**, with the suffix *-ene* used instead of *-ane* to identify the functional group. There are three steps.

STEP 1

Name the parent hydrocarbon. Find the longest carbon chain containing the double bond, and name the compound accordingly, using the suffix *-ene:*

$CH_3CH_2(CH_3CH_2CH_2)C{=}CH_2$ $CH_3CH_2(CH_3CH_2CH_2)C{=}CH_2$

Named as a *pentene* *NOT* as a hexene, since the double bond is not contained in the six-carbon chain

STEP 2

Number the carbon atoms in the chain. Begin at the end nearer the double bond or, if the double bond is equidistant from the two ends, begin at the end nearer the first branch point. This rule ensures that the double-bond carbons receive the lowest possible numbers.

$$\underset{6\ \ 5\ \ 4\ \ 3\ \ \ 2\ \ 1}{CH_3CH_2CH_2CH{=}CHCH_3} \qquad \underset{1\ \ 2\ \ 3\ \ \ 4\ \ 5\ \ 6}{CH_3\overset{CH_3}{\overset{|}{C}}HCH{=}CHCH_2CH_3}$$

STEP 3

Write the full name. Number the substituents according to their positions in the chain, and list them alphabetically. Indicate the position of the double bond by giving the number of the first alkene carbon and placing that number directly before the parent name. If more than one double bond is present, indicate the position of each and use one of the suffixes *-diene*, *-triene*, and so on.

$$\underset{6\ \ 5\ \ 4\ \ 3\ \ \ 2\ \ 1}{CH_3CH_2CH_2CH{=}CHCH_3} \qquad \underset{1\ \ 2\ \ 3\ \ \ 4\ \ 5\ \ 6}{CH_3\overset{CH_3}{\overset{|}{C}}HCH{=}CHCH_2CH_3}$$

2-Hexene **2-Methyl-3-hexene**

$$(CH_3CH_2)(\underset{5\ \ 4\ \ 3}{CH_3CH_2CH_2})\overset{2}{C}{=}\overset{1}{C}H_2 \qquad \underset{1\ \ \ 2\ \ \ 3\ \ \ \ 4}{H_2C{=}\overset{CH_3}{\overset{|}{C}}{-}CH{=}CH_2}$$

2-Ethyl-1-pentene **2-Methyl-1,3-butadiene**

We should also note that IUPAC changed their naming recommendations in 1993 to place the locant indicating the position of the double bond immediately before the *-ene* suffix rather than before the parent name: but-2-ene rather than 2-butene, for instance. This change has not been widely accepted by the chemical community in the United States, however, so we'll stay with the older but more commonly used names. Be aware, though, that you may occasionally encounter the newer system.

$$\underset{7\ \ 6\ \ \ 5\ \ 4\ \ \ \ 3\ \ 2\ \ 1}{CH_3CH_2\overset{CH_3}{\overset{|}{C}}HCH{=}CH\overset{CH_3}{\overset{|}{C}}HCH_3} \qquad \underset{1\ \ \ 2\ \ 3\ \ 4\ \ \ \ 5\ \ 6}{H_2C{=}CH\overset{CH_2CH_2CH_3}{\overset{|}{C}}HCH{=}CHCH_3}$$

Older naming system:	**2,5-Dimethyl-3-heptene**	**3-Propyl-1,4-hexadiene**
(Newer naming system:	**2,5-Dimethylhept-3-ene**	**3-Propylhexa-1,4-diene)**

Cycloalkenes are named similarly, but because there is no chain end to begin from, we number the cycloalkene so that the double bond is between C1 and C2 and the first substituent has as low a number as possible. It's not necessary to indicate the position of the double bond in the name because it's always between C1 and C2. As with open-chain alkenes, newer but not yet widely accepted naming rules place the locant immediately before the suffix in a diene.

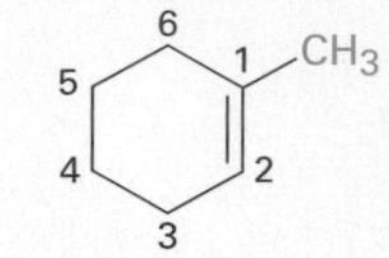

1-Methylcyclohexene

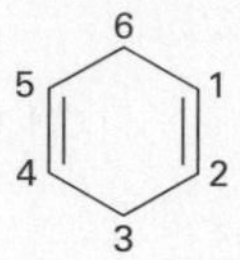

1,4-Cyclohexadiene (New: Cyclohexa-1,4-diene)

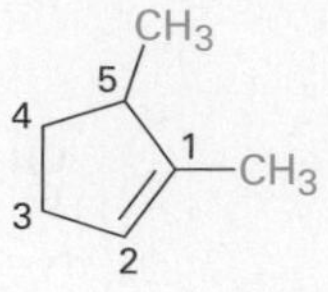

1,5-Dimethylcyclopentene

For historical reasons, there are a few alkenes whose names are firmly entrenched in common usage but don't conform to the rules. For example, the alkene derived from ethane should be called *ethene,* but the name *ethylene* has been used so long that it is accepted by IUPAC. Table 7.1 lists several other common names that are often used and are recognized by IUPAC. Note also that a $=CH_2$ substituent is called a **methylene group**, a $H_2C{=}CH-$ substituent is called a **vinyl group**, and a $H_2C{=}CHCH_2-$ substituent is called an **allyl group.**

$H_2C{=}$ — **A methylene group**

$H_2C{=}CH-$ — **A vinyl group**

$H_2C{=}CH{-}CH_2-$ — **An allyl group**

Table 7.1 Common Names of Some Alkenes

Compound	Systematic name	Common name
$H_2C{=}CH_2$	Ethene	Ethylene
$CH_3CH{=}CH_2$	Propene	Propylene
$CH_3C(CH_3){=}CH_2$	2-Methylpropene	Isobutylene
$H_2C{=}C(CH_3){-}CH{=}CH_2$	2-Methyl-1,3-butadiene	Isoprene

Problem 7.4

Give IUPAC names for the following compounds:

(a) $H_2C{=}CHCH(CH_3)C(CH_3)_2CH_3$ (printed with H_3C and CH_3 above, CH_3 below)

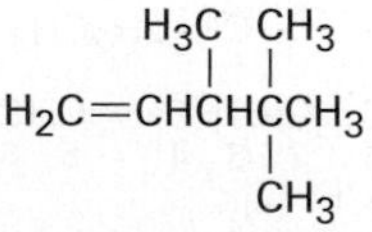

(b) $CH_3CH_2CH{=}C(CH_3)CH_2CH_3$

(c) $CH_3CH{=}CHCH(CH_3)CH{=}CHCH(CH_3)CH_3$

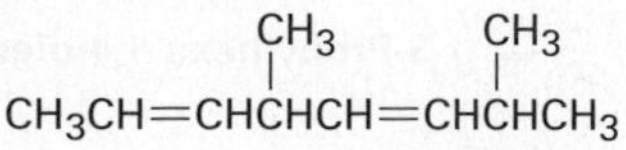

(d) $CH_3CH_2CH_2CH{=}CHCH(CH(CH_3)CH_2CH_3)CH_2CH_3$ (printed with $CH_3CHCH_2CH_3$ above)

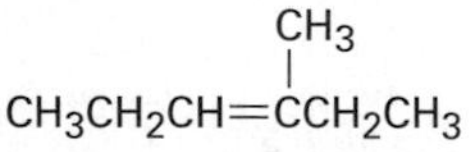

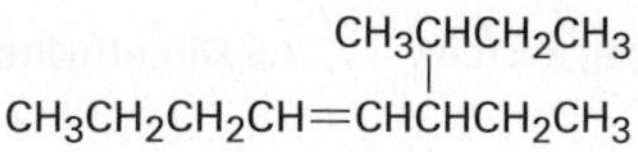

Problem 7.5

Draw structures corresponding to the following IUPAC names:

(a) 2-Methyl-1,5-hexadiene

(b) 3-Ethyl-2,2-dimethyl-3-heptene

(c) 2,3,3-Trimethyl-1,4,6-octatriene

(d) 3,4-Diisopropyl-2,5-dimethyl-3-hexene

Problem 7.6

Name the following cycloalkenes:

(a) CH_3, CH_3 **(b)** CH_3, CH_3 **(c)** $CH(CH_3)_2$

Problem 7.7

Change the following old names to new, post-1993 names, and draw the structure of each compound:

(a) 2,5,5-Trimethyl-2-hexene **(b)** 2,3-Dimethyl-1,3-cyclohexadiene

7.4 Cis–Trans Isomerism in Alkenes

We saw in Chapter 1 that the carbon–carbon double bond can be described in two ways. In valence bond language **(Section 1.8)**, the carbons are sp^2-hybridized and have three equivalent hybrid orbitals that lie in a plane at angles of 120° to one another. The carbons form a σ bond by head-on overlap of sp^2 orbitals and a π bond by sideways overlap of unhybridized p orbitals oriented perpendicular to the sp^2 plane, as shown in Figure 1.14 on page 14.

In molecular orbital language **(Section 1.11)**, interaction between the p orbitals leads to one bonding and one antibonding π molecular orbital. The π bonding MO has no node between nuclei and results from a combination of p orbital lobes with the same algebraic sign. The π antibonding MO has a node between nuclei and results from a combination of lobes with different algebraic signs, as shown in Figure 1.18, page 21.

Although essentially free rotation is possible around single bonds **(Section 3.6)**, the same is not true of double bonds. For rotation to occur around a double bond, the π bond must break and re-form **(Figure 7.2)**. Thus, the barrier to double-bond rotation must be at least as great as the strength of the π bond itself, an estimated 350 kJ/mol (84 kcal/mol). Recall that the barrier to bond rotation in ethane is only 12 kJ/mol.

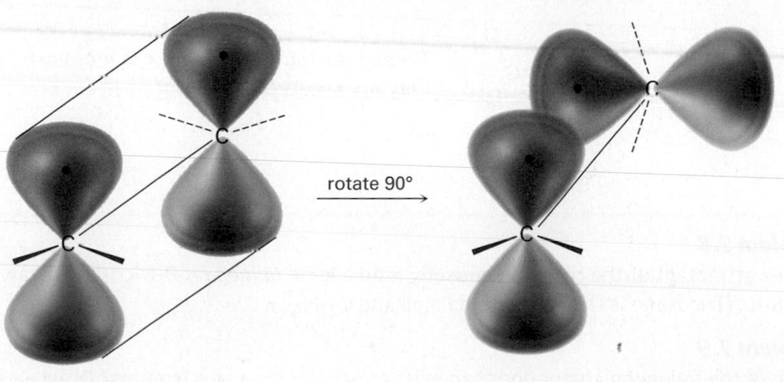

Figure 7.2 The π bond must break for rotation to take place around a carbon–carbon double bond.

The lack of rotation around carbon–carbon double bonds is of more than just theoretical interest; it also has chemical consequences. Imagine the situation for a disubstituted alkene such as 2-butene. (*Disubstituted* means that two substituents other than hydrogen are bonded to the double-bond carbons.) The two methyl groups in 2-butene can be either on the same side of the double bond or on opposite sides, a situation similar to that in disubstituted cycloalkanes **(Section 4.2)**.

Since bond rotation can't occur, the two 2-butenes can't spontaneously interconvert; they are different, isolable compounds. As with disubstituted cycloalkanes, we call such compounds *cis–trans stereoisomers*. The compound with substituents on the same side of the double bond is called *cis*-2-butene, and the isomer with substituents on opposite sides is *trans*-2-butene **(Figure 7.3)**.

H_3C, CH_3 / C=C / H, H — ***cis*-2-Butene**

H, CH_3 / C=C / H_3C, H — ***trans*-2-Butene**

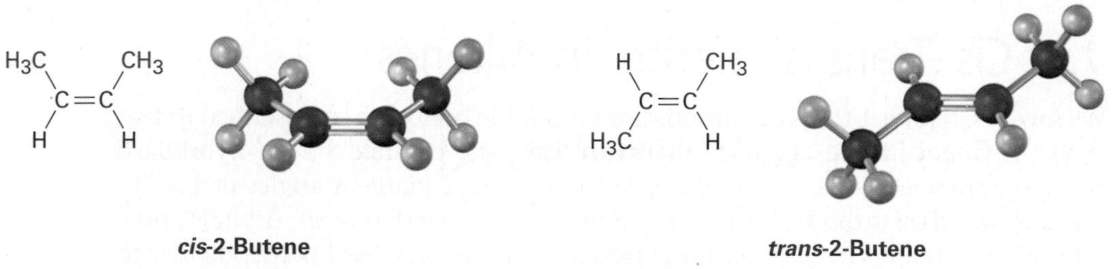

Figure 7.3 Cis and trans isomers of 2-butene. The cis isomer has the two methyl groups on the same side of the double bond, and the trans isomer has the methyl groups on opposite sides.

Cis–trans isomerism is not limited to disubstituted alkenes. It can occur whenever both double-bond carbons are attached to two different groups. If one of the double-bond carbons is attached to two identical groups, however, then cis–trans isomerism is not possible **(Figure 7.4)**.

Figure 7.4 The requirement for cis–trans isomerism in alkenes. Compounds that have one of their carbons bonded to two identical groups can't exist as cis–trans isomers. Only when both carbons are bonded to two different groups is cis–trans isomerism possible.

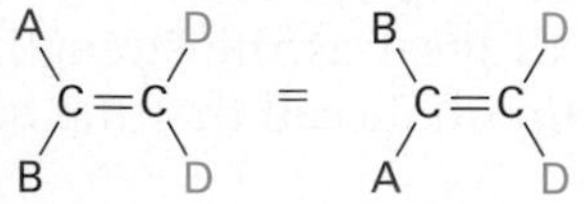

(A, B)C=C(D, D) = (B, A)C=C(D, D) — These two compounds are identical; they are not cis–trans isomers.

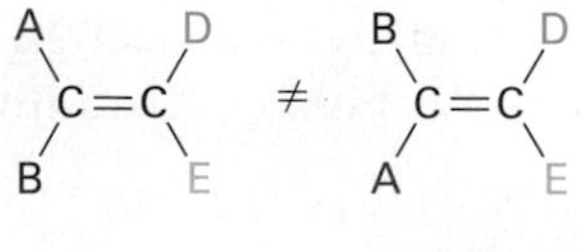

(A, B)C=C(D, E) ≠ (B, A)C=C(D, E) — These two compounds are not identical; they are cis–trans isomers.

Problem 7.8

The sex attractant of the common housefly is an alkene named *cis*-9-tricosene. Draw its structure. (Tricosane is the straight-chain alkane $C_{23}H_{48}$.)

Problem 7.9

Which of the following compounds can exist as pairs of cis–trans isomers? Draw each cis–trans pair, and indicate the geometry of each isomer.

(a) $CH_3CH{=}CH_2$ **(b)** $(CH_3)_2C{=}CHCH_3$
(c) $CH_3CH_2CH{=}CHCH_3$ **(d)** $(CH_3)_2C{=}C(CH_3)CH_2CH_3$
(e) $ClCH{=}CHCl$ **(f)** $BrCH{=}CHCl$

Problem 7.10

Name the following alkenes, including the cis or trans designation:

(a) **(b)**

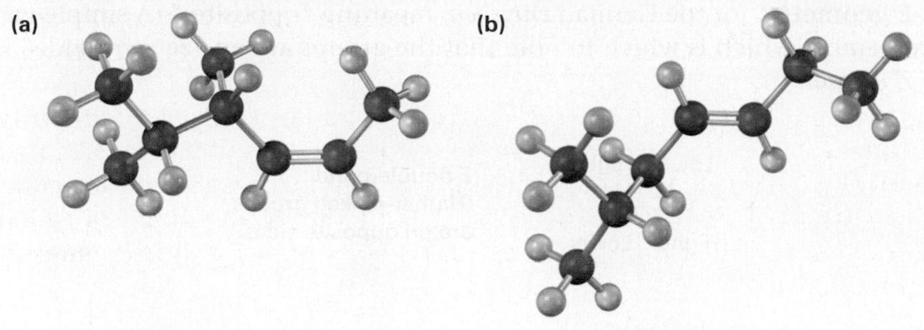

7.5 Alkene Stereochemistry and the *E,Z* Designation

Key IDEAS

Test your knowledge of Key Ideas by answering end-of-chapter exercises marked with ▲.

The cis–trans naming system used in the previous section works only with disubstituted alkenes—compounds that have two substituents other than hydrogen on the double bond. With trisubstituted and tetrasubstituted double bonds, a more general method is needed for describing double-bond geometry. (*Trisubstituted* means three substituents other than hydrogen on the double bond; *tetrasubstituted* means four substituents other than hydrogen.)

The method used for describing alkene stereochemistry is called the ***E,Z* system** and employs the same Cahn–Ingold–Prelog sequence rules given in **Section 5.5** for specifying the configuration of a chirality center. Let's briefly review the sequence rules and then see how they're used to specify double-bond geometry. For a more thorough review, you should reread **Section 5.5**.

RULE 1

Considering each of the double-bond carbons separately, look at the two substituents attached and rank them according to the atomic number of the first atom in each. An atom with higher atomic number ranks higher than an atom with lower atomic number.

RULE 2

If a decision can't be reached by ranking the first atoms in the two substituents, look at the second, third, or fourth atoms away from the double-bond until the first difference is found.

RULE 3

Multiple-bonded atoms are equivalent to the same number of single-bonded atoms.

Once the two groups attached to each doubly bonded carbon atom have been ranked as either higher or lower, look at the entire molecule. If the higher-ranked groups on each carbon are on the same side of the double bond, the alkene is said to have **Z geometry**, for the German *zusammen,* meaning "together." If the higher-ranked groups are on opposite sides, the alkene has ***E* geometry**, for the German *entgegen,* meaning "opposite." (A simple way to remember which is which to note that the groups are on "ze zame zide" in the *Z* isomer.)

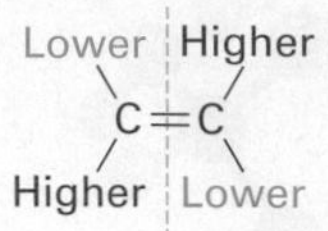

Lower | Higher; C=C; Higher | Lower

E double bond (Higher-ranked groups are on opposite sides.)

Higher | Higher; C=C; Lower | Lower

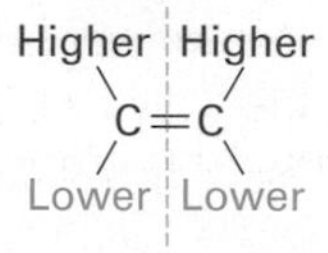

Z double bond (Higher-ranked groups are on the same side.)

For further practice, work through each of the following examples to convince yourself that the assignments are correct:

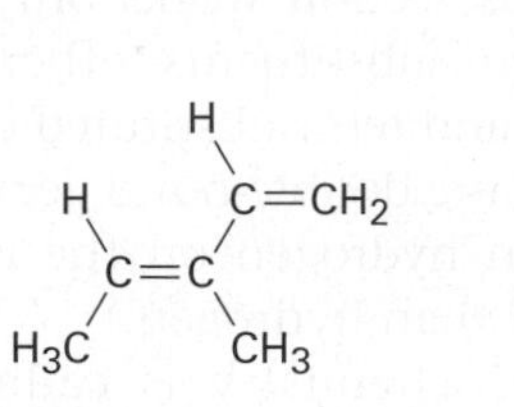

(*E*)-3-Methyl-1,3-pentadiene

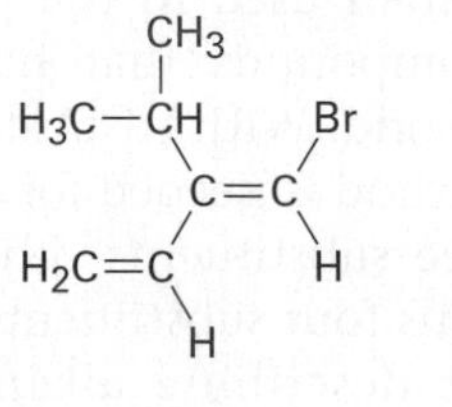

(*E*)-1-Bromo-2-isopropyl-1,3-butadiene

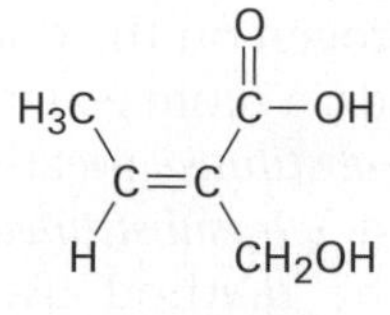

(*Z*)-2-Hydroxymethyl-2-butenoic acid

Worked Example 7.1 Assigning *E* and *Z* Configurations to Substituted Alkenes

Assign *E* or *Z* configuration to the double bond in the following compound:

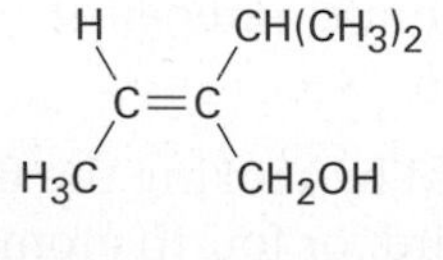

H, $CH(CH_3)_2$; C=C; H_3C, CH_2OH

Strategy

Look at the two substituents connected to each double-bond carbon, and determine their ranking using the Cahn–Ingold–Prelog rules. Then see whether the two higher-ranked groups are on the same or opposite sides of the double bond.

Solution

The left-hand carbon has $-H$ and $-CH_3$ substituents, of which $-CH_3$ ranks higher by sequence rule 1. The right-hand carbon has $-CH(CH_3)_2$ and $-CH_2OH$ substituents, which are equivalent by rule 1. By rule 2, however, $-CH_2OH$ ranks higher than $-CH(CH_3)_2$ because the substituent $-CH_2OH$ has an *oxygen* as its highest second atom, but $-CH(CH_3)_2$ has a *carbon* as its highest second atom. The two higher-ranked groups are on the same side of the double bond, so we assign *Z* configuration.

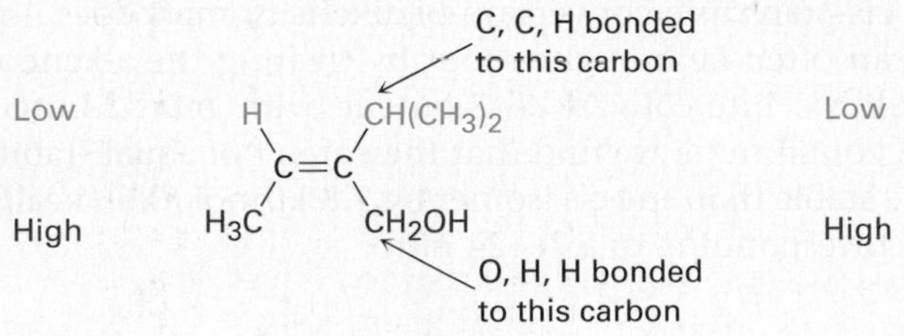

Low H $CH(CH_3)_2$ Low
C=C
High H_3C CH_2OH High

C, C, H bonded to this carbon

O, H, H bonded to this carbon

***Z* configuration**

Problem 7.11

Which member in each of the following sets ranks higher?

(a) $-H$ or $-CH_3$ **(b)** $-Cl$ or $-CH_2Cl$
(c) $-CH_2CH_2Br$ or $-CH{=}CH_2$ **(d)** $-NHCH_3$ or $-OCH_3$
(e) $-CH_2OH$ or $-CH{=}O$ **(f)** $-CH_2OCH_3$ or $-CH{=}O$

Problem 7.12

Rank the substituents in each of the following sets according to the sequence rules:

(a) $-CH_3$, $-OH$, $-H$, $-Cl$
(b) $-CH_3$, $-CH_2CH_3$, $-CH{=}CH_2$, $-CH_2OH$
(c) $-CO_2H$, $-CH_2OH$, $-C{\equiv}N$, $-CH_2NH_2$
(d) $-CH_2CH_3$, $-C{\equiv}CH$, $-C{\equiv}N$, $-CH_2OCH_3$

Problem 7.13

Assign *E* or *Z* configuration to the following alkenes:

(a) 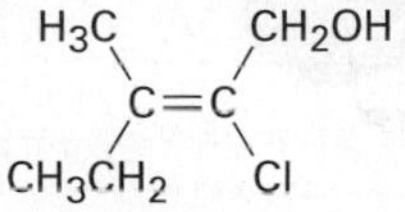 H_3C, CH_2OH / C=C / CH_3CH_2, Cl

(b) 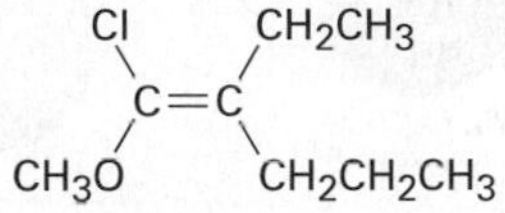 Cl, CH_2CH_3 / C=C / CH_3O, $CH_2CH_2CH_3$

(c) CH_3 (ring), CO_2H / C=C / CH_2OH

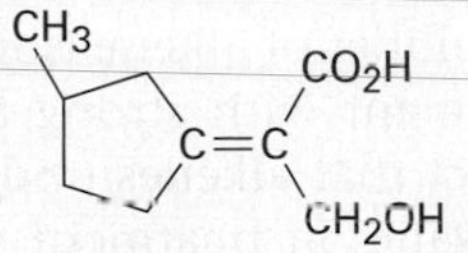

(d) 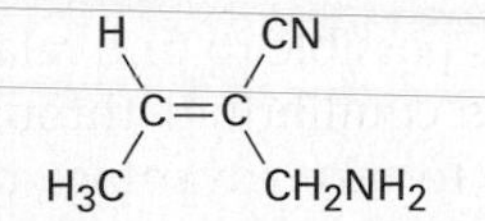 H, CN / C=C / H_3C, CH_2NH_2

Problem 7.14

Assign stereochemistry (*E* or *Z*) to the double bond in the following compound, and convert the drawing into a skeletal structure (red = O):

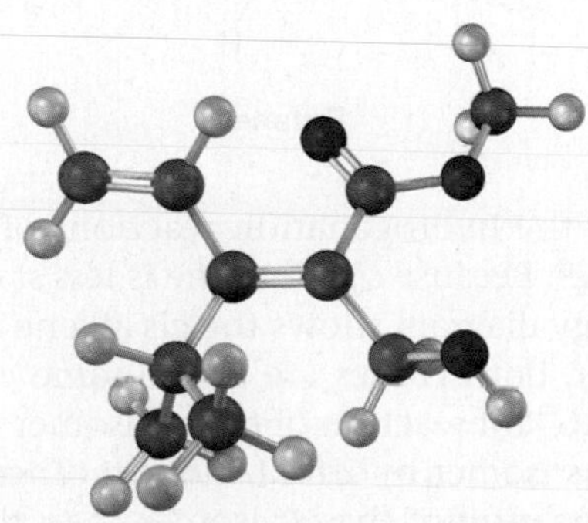

7.6 Stability of Alkenes

Although the cis–trans interconversion of alkene isomers does not occur spontaneously, it can often be brought about by treating the alkene with a strong acid catalyst. If we interconvert *cis*-2-butene with *trans*-2-butene and allow them to reach equilibrium, we find that they aren't of equal stability. The trans isomer is more stable than the cis isomer by 2.8 kJ/mol (0.66 kcal/mol) at room temperature, corresponding to a 76:24 ratio.

$$\underset{\text{Trans (76\%)}}{(H)(H_3C)C{=}C(CH_3)(H)} \xrightleftharpoons[\text{catalyst}]{\text{Acid}} \underset{\text{Cis (24\%)}}{(H_3C)(H)C{=}C(CH_3)(H)}$$

Cis alkenes are less stable than their trans isomers because of steric strain between the two larger substituents on the same side of the double bond. This is the same kind of steric interference that we saw previously in the axial conformation of methylcyclohexane **(Section 4.7)**.

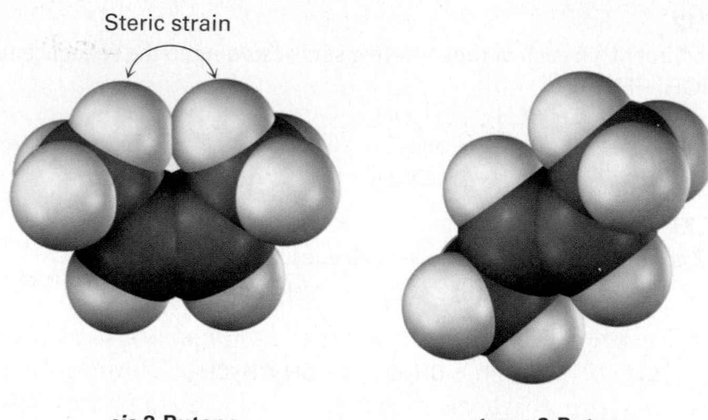

***cis*-2-Butene** **_trans_-2-Butene**

Although it's sometimes possible to find relative stabilities of alkene isomers by establishing a cis–trans equilibrium through treatment with strong acid, a more general method is to take advantage of the fact that alkenes undergo a *hydrogenation* reaction to give the corresponding alkane on treatment with H_2 gas in the presence of a catalyst such as palladium or platinum.

$$\underset{\textit{trans}\text{-2-Butene}}{(H)(H_3C)C{=}C(CH_3)(H)} \xrightarrow[\text{Pd}]{H_2} \underset{\text{Butane}}{H_3C{-}CH_2{-}CH_2{-}CH_3} \xleftarrow[\text{Pd}]{H_2} \underset{\textit{cis}\text{-2-Butene}}{(H_3C)(H)C{=}C(CH_3)(H)}$$

Energy diagrams for the hydrogenation reactions of *cis*- and *trans*-2-butene are shown in **Figure 7.5**. Because *cis*-2-butene is less stable than *trans*-2-butene by 2.8 kJ/mol, the energy diagram shows the cis alkene at a higher energy level. After reaction, however, both curves are at the same energy level (butane). It therefore follows that $\Delta G°$ for reaction of the cis isomer must be larger than $\Delta G°$ for reaction of the trans isomer by 2.8 kJ/mol. In other words, more energy is released in the hydrogenation of the cis isomer than the trans isomer because the cis isomer has more energy to begin with.

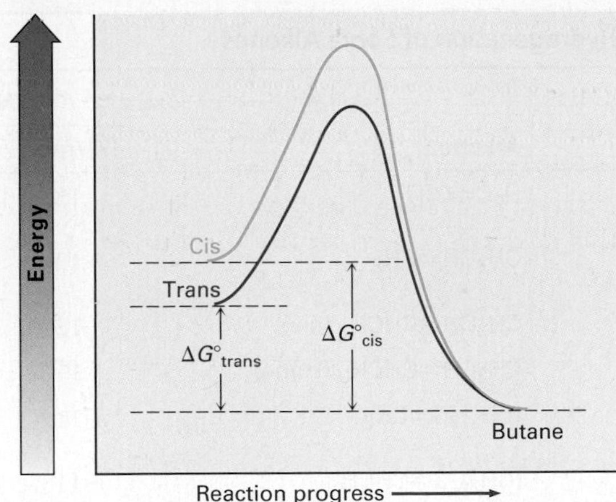

Figure 7.5 Energy diagrams for hydrogenation of *cis*- and *trans*-2-butene. The cis isomer is higher in energy than the trans isomer by about 2.8 kJ/mol and therefore releases more energy in the reaction.

If we were to measure the so-called heats of hydrogenation ($\Delta H°_{hydrog}$) for two double-bond isomers and find their difference, we could determine the relative stabilities of cis and trans isomers without having to measure an equilibrium position. *cis*-2-Butene, for instance, has $\Delta H°_{hydrog}$ = −120 kJ/mol (−28.6 kcal/mol), while *trans*-2-butene has $\Delta H°_{hydrog}$ = −116 kJ/mol (−27.6 kcal/mol)—a difference of 4 kJ/mol.

H_3C CH_3 C=C H H

Cis isomer
$\Delta H°_{hydrog}$ **= −120 kJ/mol**

H CH_3 C=C H_3C H

Trans isomer
$\Delta H°_{hydrog}$ **= 116 kJ/mol**

The 4 kJ/mol energy difference between the 2-butene isomers calculated from heats of hydrogenation agrees reasonably well with the 2.8 kcal/mol energy difference calculated from equilibrium data, but the numbers aren't exactly the same for two reasons. First, there is probably some experimental error, since heats of hydrogenation are difficult to measure accurately. Second, heats of reaction and equilibrium constants don't measure exactly the same thing. Heats of reaction measure enthalpy changes, $\Delta H°$, whereas equilibrium constants measure free-energy changes, $\Delta G°$, so we might expect a slight difference between the two.

Table 7.2 lists some representative data for the hydrogenation of different alkenes and shows that alkenes become more stable with increasing substitution. That is, alkenes follow the stability order:

Tetrasubstituted > Trisubstituted > Disubstituted > Monosubstituted

$R_2C{=}CR_2$ > $R_2C{=}CHR$ > RHC=CHR (trans) ≈ $R_2C{=}CH_2$ > RHC=CH_2

The stability order of substituted alkenes is due to a combination of two factors. One is a stabilizing interaction between the C=C π bond and adjacent C–H σ bonds on substituents. In valence-bond language, the interaction is called **hyperconjugation**. In a molecular orbital description, there is a bonding MO that extends over the four-atom C=C–C–H grouping, as shown in **Figure 7.6**. The more substituents that are present on the double bond, the more hyperconjugation there is and the more stable the alkene.

Table 7.2 Heats of Hydrogenation of Some Alkenes

Substitution	Alkene	$\Delta H°_{hydrog}$ (kJ/mol)	$\Delta H°_{hydrog}$ (kcal/mol)
Ethylene	$H_2C{=}CH_2$	−137	−32.8
Monosubstituted	$CH_3CH{=}CH_2$	−126	−30.1
Disubstituted	$CH_3CH{=}CHCH_3$ (cis)	−120	−28.6
	$CH_3CH{=}CHCH_3$ (trans)	−116	−27.6
	$(CH_3)_2C{=}CH_2$	−119	−28.4
Trisubstituted	$(CH_3)_2C{=}CHCH_3$	−113	−26.9
Tetrasubstituted	$(CH_3)_2C{=}C(CH_3)_2$	−111	−26.6

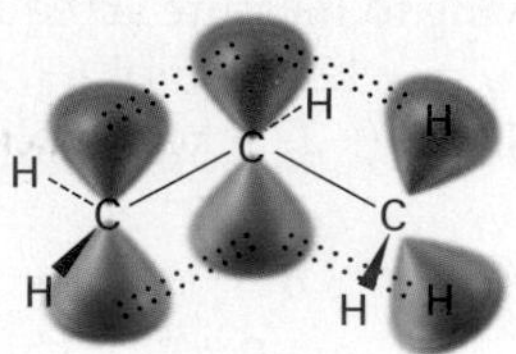

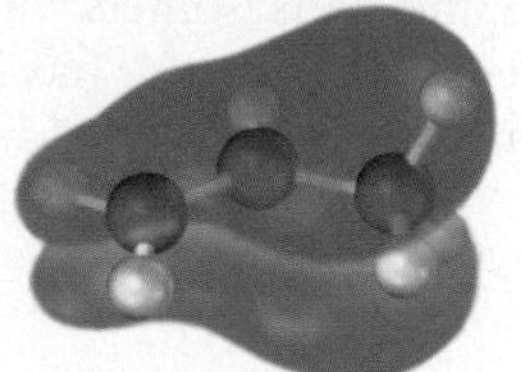

Figure 7.6 Hyperconjugation is a stabilizing interaction between the C=C π bond and adjacent C–H σ bonds on substituents. The more substituents there are, the greater the stabilization of the alkene.

A second factor that contributes to alkene stability involves bond strengths. A bond between an sp^2 carbon and an sp^3 carbon is somewhat stronger than a bond between two sp^3 carbons. Thus, in comparing 1-butene and 2-butene, the monosubstituted isomer has one sp^3–sp^3 bond and one sp^3–sp^2 bond, while the disubstituted isomer has two sp^3–sp^2 bonds. More highly substituted alkenes always have a higher ratio of sp^3–sp^2 bonds to sp^3–sp^3 bonds than less highly substituted alkenes and are therefore more stable.

sp^3–sp^2 sp^2–sp^3

$CH_3{-}CH{=}CH{-}CH_3$

2-Butene (more stable)

sp^3–sp^3 sp^3–sp^2

$CH_3{-}CH_2{-}CH{=}CH_2$

1-Butene (less stable)

Problem 7.15

Name the following alkenes, and tell which compound in each pair is more stable:

(a) $H_2C{=}CHCH_2CH_3$ or $H_2C{=}C(CH_3)CH_3$

(b) cis: H, H / H_3C, $CH_2CH_2CH_3$ C=C or trans: H, $CH_2CH_2CH_3$ / H_3C, H C=C

(c) 1-methylcyclohexene (CH_3) or 3-methylcyclohexene (CH_3)

7.7 Electrophilic Addition Reactions of Alkenes

Before beginning a detailed discussion of alkene reactions, let's review briefly some conclusions from the previous chapter. We said in **Section 6.5** that alkenes behave as nucleophiles (Lewis bases) in polar reactions, donating a pair of electrons from their electron-rich C=C bond to an electrophile (Lewis acid). For example, reaction of 2-methylpropene with HBr yields 2-bromo-2-methylpropane. A careful study of this and similar reactions by Christopher Ingold and others in the 1930s led to the generally accepted mechanism shown in **Figure 7.7** for **electrophilic addition reactions**.

Figure 7.7 | MECHANISM

Mechanism of the electrophilic addition of HBr to 2-methylpropene. The reaction occurs in two steps, protonation and bromide addition, and involves a carbocation intermediate.

1 A hydrogen atom on the electrophile HBr is attacked by π electrons from the nucleophilic double bond, forming a new C–H bond. This leaves the other carbon atom with a + charge and a vacant p orbital. Simultaneously, two electrons from the H–Br bond move onto bromine, giving bromide anion.

2 Bromide ion donates an electron pair to the positively charged carbon atom, forming a C–Br bond and yielding the neutral addition product.

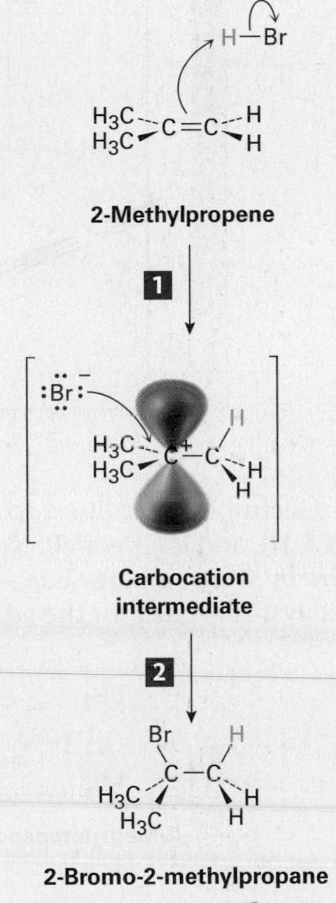

The reaction begins with an attack on the hydrogen of the electrophile HBr by the electrons of the nucleophilic π bond. Two electrons from the π bond form a new σ bond between the entering hydrogen and an alkene carbon, as shown by the curved arrow at the top of Figure 7.7. The carbocation intermediate that results is itself an electrophile, which can accept an electron pair from nucleophilic Br^- ion to form a C–Br bond and yield a neutral addition product.

An energy diagram for the overall electrophilic addition reaction **(Figure 7.8)** has two peaks (transition states) separated by a valley (carbocation intermediate). The energy level of the intermediate is higher than that of the starting alkene, but the reaction as a whole is exergonic (negative $\Delta G°$). The first step, protonation of the alkene to yield the intermediate cation, is relatively slow but, once formed, the cation intermediate rapidly reacts further to yield the final alkyl bromide product. The relative rates of the two steps are indicated in Figure 7.8 by the fact that $\Delta G^{\ddagger}_1$ is larger than $\Delta G^{\ddagger}_2$.

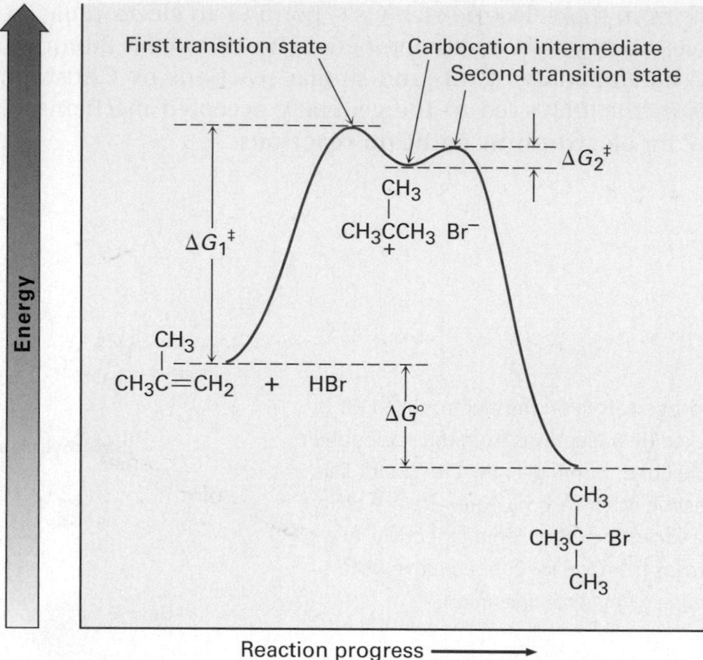

Figure 7.8 Energy diagram for the two-step electrophilic addition of HBr to 2-methylpropene. The first step is slower than the second step.

Electrophilic addition to alkenes is successful not only with HBr but with HCl, HI, and H_2O as well. Note that HI is usually generated in the reaction mixture by treating potassium iodide with phosphoric acid and that a strong acid catalyst is needed for the addition of water.

$$(CH_3)_2C{=}CH_2 + HCl \xrightarrow{\text{Ether}} CH_3{-}CCl(CH_3){-}CH_3$$

2-Methylpropene → **2-Chloro-2-methylpropane (94%)**

$$CH_3CH_2CH_2CH{=}CH_2 \xrightarrow[H_3PO_4]{KI} CH_3CH_2CH_2CHICH_3$$

1-Pentene → **(HI)** → **2-Iodopentane**

$$\text{1-Methylcyclohexene} \xrightarrow[H_2SO_4\ \text{catalyst}]{H_2O} \text{1-Methylcyclohexanol}$$

1-Methylcyclohexene → **1-Methylcyclohexanol**

Writing Organic Reactions

This is a good time to mention that organic reaction equations are sometimes written in different ways to emphasize different points. In describing a laboratory process, for instance, the reaction of 2-methylpropene with HCl might be written in the format A + B → C to emphasize that both reactants are equally important for the purposes of the discussion. The solvent and notes about other reaction conditions such as temperature are written either above or below the reaction arrow.

$$(CH_3)_2C{=}CH_2 + HCl \xrightarrow[25\ ^\circ C]{Ether} CH_3{-}C(CH_3)_2{-}Cl$$

Solvent

2-Methylpropene

2-Chloro-2-methylpropane

Alternatively, we might write the same reaction in a format to emphasize that 2-methylpropene is the reactant whose chemistry is of greater interest. The second reactant, HCl, is placed above the reaction arrow together with notes about solvent and reaction conditions.

$$(CH_3)_2C{=}CH_2 \xrightarrow[Ether,\ 25\ ^\circ C]{HCl} CH_3{-}C(CH_3)_2{-}Cl$$

Reactant

Solvent

2-Methylpropene

2-Chloro-2-methylpropane

In describing a biological process, the reaction is usually written to show only the structures of the primary reactant and product, while abbreviating the structures of various biological "reagents" and by-products by using a curved arrow that intersects the straight reaction arrow. As discussed in **Section 6.11**, the reaction of glucose with ATP to give glucose 6-phosphate plus ADP would be written as

Glucose $\xrightarrow[Hexokinase]{ATP \quad ADP}$ Glucose 6-phosphate

Glucose

Glucose 6-phosphate

Worked Example 7.2 Predicting the Product of an Electrophilic Addition Reaction

What product would you expect from reaction of HCl with 1-ethylcyclopentene?

$$\text{1-ethylcyclopentene } (CH_2CH_3) + HCl \longrightarrow ?$$

Strategy

When solving a problem that asks you to predict a reaction product, begin by looking at the functional group(s) in the reactants and deciding what kind of reaction is likely to occur. In the present instance, the reactant is an alkene that will probably undergo an electrophilic addition reaction with HCl. Next, recall what you know about electrophilic addition reactions and use your knowledge to predict the product. You know that electrophilic addition reactions follow Markovnikov's rule, so H^+ will add to the double-bond carbon that has one alkyl group (C2 on the ring) and the Cl will add to the double-bond carbon that has two alkyl groups (C1 on the ring).

Solution

The expected product is 1-chloro-1-ethylcyclopentane.

2 alkyl groups on this carbon (C1)

1 alkyl group on this carbon (C2)

$$\text{1-ethylcyclopentene} + HCl \longrightarrow \text{1-chloro-1-ethylcyclopentane}$$

1-Chloro-1-ethylcyclopentane

Worked Example 7.3 Synthesizing a Specific Compound

What alkene would you start with to prepare the following alkyl halide? There may be more than one possibility.

$$? \longrightarrow CH_3CH_2C(Cl)(CH_3)CH_2CH_2CH_3$$

Strategy

When solving a problem that asks how to prepare a given product, *always work backward.* Look at the product, identify the functional group(s) it contains, and ask yourself, "How can I prepare that functional group?" In the present instance, the product is a tertiary alkyl chloride, which can be prepared by reaction of an alkene with HCl. The carbon atom bearing the –Cl atom in the product must be one of the double-bond carbons in the reactant. Draw and evaluate all possibilities.

Solution

There are three possibilities, any one of which could give the desired product according to Markovnikov's rule.

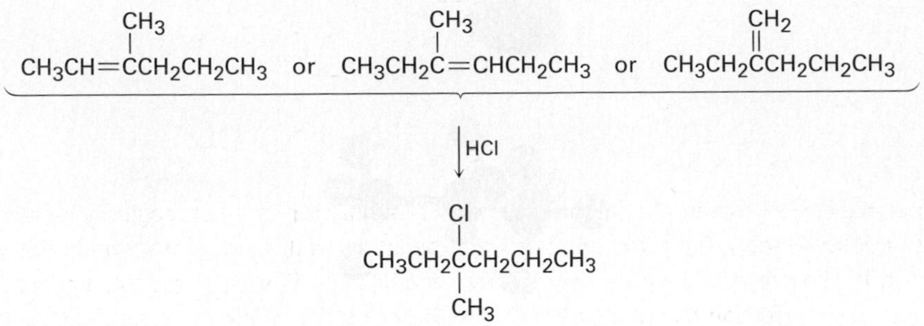

$$\underbrace{\overset{\displaystyle CH_3}{CH_3CH{=}\overset{|}{C}CH_2CH_2CH_3} \quad \text{or} \quad \overset{\displaystyle CH_3}{CH_3CH_2\overset{|}{C}{=}CHCH_2CH_3} \quad \text{or} \quad \overset{\displaystyle CH_2}{CH_3CH_2\overset{\|}{C}CH_2CH_2CH_3}}$$

$$\downarrow HCl$$

$$CH_3CH_2\overset{\displaystyle Cl}{\underset{\displaystyle CH_3}{C}}CH_2CH_2CH_3$$

Problem 7.16

Predict the products of the following reactions:

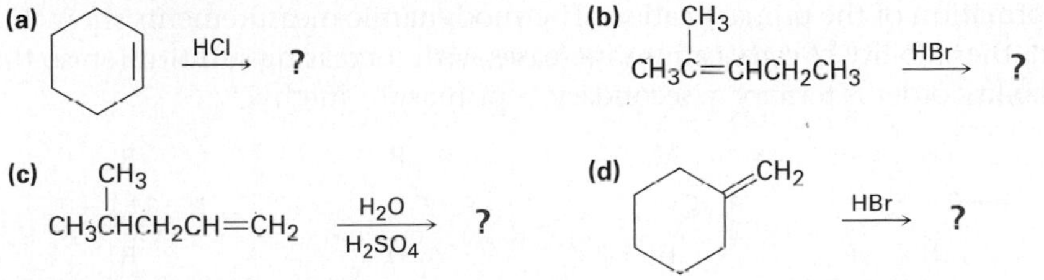

(a) cyclohexene $\xrightarrow{HCl}$?

(b) $CH_3\overset{\displaystyle CH_3}{\overset{|}{C}}{=}CHCH_2CH_3 \xrightarrow{HBr}$?

(c) $CH_3\overset{\displaystyle CH_3}{\overset{|}{C}H}CH_2CH{=}CH_2 \xrightarrow[H_2SO_4]{H_2O}$?

(Addition of H_2O occurs.)

(d) methylenecyclohexane ($=CH_2$) $\xrightarrow{HBr}$?

Problem 7.17

What alkenes would you start with to prepare the following products?

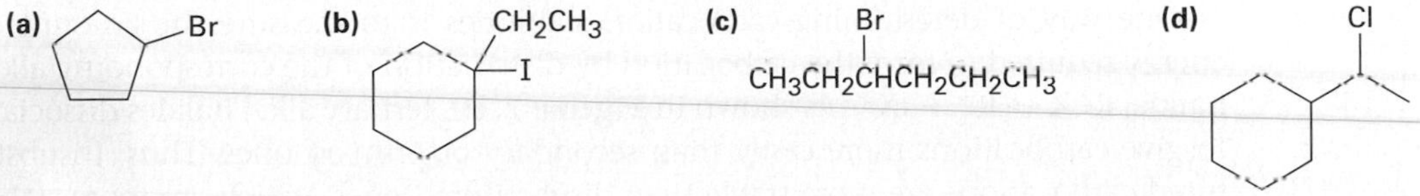

(a) bromocyclopentane (cyclopentyl–Br)

(b) 1-ethyl-1-iodocyclohexane (CH_2CH_3, I)

(c) $CH_3CH_2\overset{\displaystyle Br}{\overset{|}{C}H}CH_2CH_2CH_3$

(d) (1-chloroethyl)cyclohexane (Cl)

7.9 Carbocation Structure and Stability

To understand why Markovnikov's rule works, we need to learn more about the structure and stability of carbocations and about the general nature of reactions and transition states. The first point to explore involves structure.

A great deal of experimental evidence has shown that carbocations are planar. The trivalent carbon is sp^2-hybridized, and the three substituents are oriented toward the corners of an equilateral triangle, as indicated in **Figure 7.9**. Because there are only six valence electrons on carbon and all six are used in the three σ bonds, the p orbital extending above and below the plane is unoccupied.

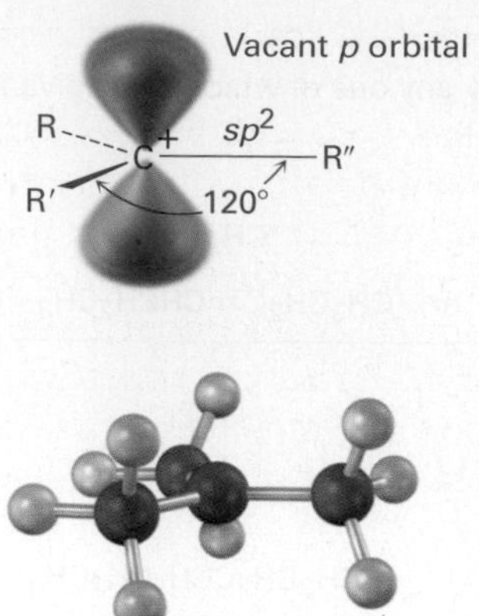

Figure 7.9 The structure of a carbocation. The trivalent carbon is sp^2-hybridized and has a vacant *p* orbital perpendicular to the plane of the carbon and three attached groups.

The second point to explore involves carbocation stability. 2-Methylpropene might react with H^+ to form a carbocation having three alkyl substituents (a tertiary ion, 3°), or it might react to form a carbocation having one alkyl substituent (a primary ion, 1°). Since the tertiary alkyl chloride, 2-chloro-2-methylpropane, is the only product observed, formation of the tertiary cation is evidently favored over formation of the primary cation. Thermodynamic measurements show that, indeed, the stability of carbocations increases with increasing substitution so that the stability order is tertiary > secondary > primary > methyl.

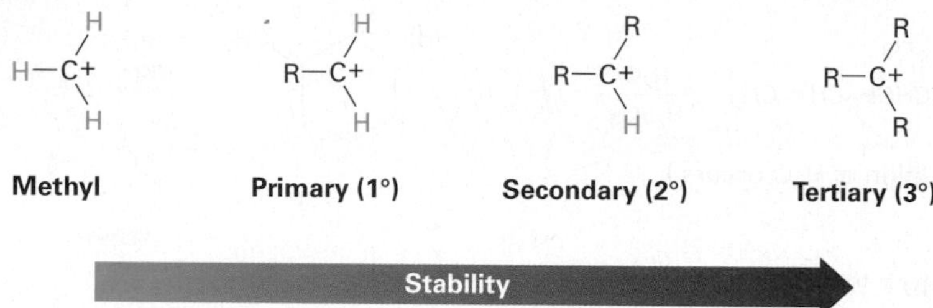

One way of determining carbocation stabilities is to measure the amount of energy required to form the carbocation by dissociation of the corresponding alkyl halide, $R{-}X \rightarrow R^+ + :X^-$. As shown in **Figure 7.10**, tertiary alkyl halides dissociate to give carbocations more easily than secondary or primary ones. Thus, trisubstituted carbocations are more stable than disubstituted ones, which are more stable than monosubstituted ones. The data in Figure 7.10 are taken from measurements made in the gas phase, but a similar stability order is found for carbocations in solution. The dissociation enthalpies are much lower in solution because polar solvents can stabilize the ions, but the order of carbocation stability remains the same.

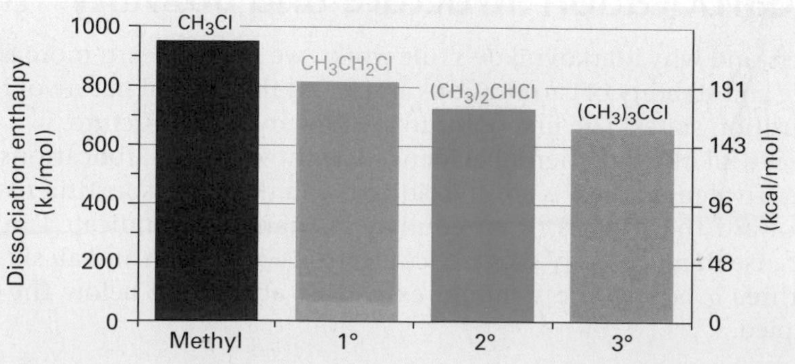

Figure 7.10 A plot of dissociation enthalpy versus substitution pattern for the gas-phase dissociation of alkyl chlorides to yield carbocations. More highly substituted alkyl halides dissociate more easily than less highly substituted ones.

Why are more highly substituted carbocations more stable than less highly substituted ones? There are at least two reasons. Part of the answer has to do with inductive effects, and part has to do with hyperconjugation. Inductive effects, discussed in **Section 2.1** in connection with polar covalent bonds, result from the shifting of electrons in a σ bond in response to the electronegativity of nearby atoms. In the present instance, electrons from a relatively larger and more polarizable alkyl group can shift toward a neighboring positive charge more easily than the electron from a hydrogen. Thus, the more alkyl groups there are attached to the positively charged carbon, the more electron density shifts toward the charge and the more inductive stabilization of the cation occurs **(Figure 7.11)**.

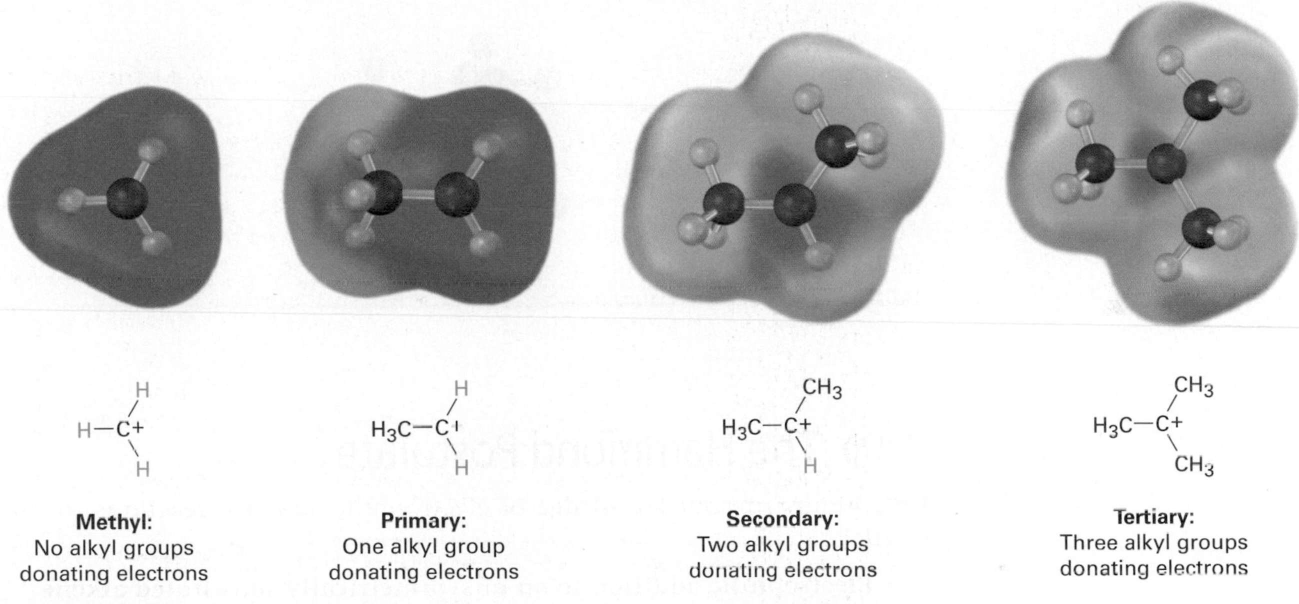

Figure 7.11 A comparison of inductive stabilization for methyl, primary, secondary, and tertiary carbocations. The more alkyl groups that are bonded to the positively charged carbon, the more electron density shifts toward the charge, making the charged carbon less electron poor (blue in electrostatic potential maps).

Hyperconjugation, discussed in **Section 7.6** in connection with the stabilities of substituted alkenes, is the stabilizing interaction between a p orbital and properly oriented C–H σ bonds on neighboring carbons that are roughly parallel to the p orbital. The more alkyl groups there are on the carbocation, the more possibilities there are for hyperconjugation and the more stable the carbocation. **Figure 7.12** shows the molecular orbital for the ethyl carbocation, $CH_3CH_2^+$, and indicates the difference between the C–H bond perpendicular to the cation p orbital and the two C–H bonds more nearly parallel to the cation p orbital. Only the roughly parallel C–H bonds are oriented properly to take part in hyperconjugation.

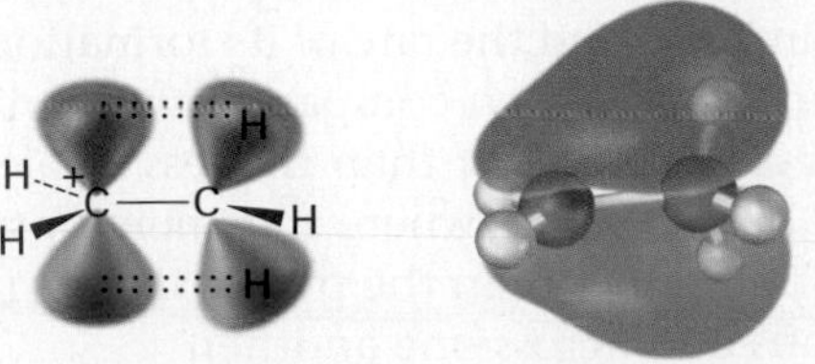

Figure 7.12 Stabilization of the ethyl carbocation, $CH_3CH_2^+$, through hyperconjugation. Interaction of neighboring C–H σ bonds with the vacant p orbital stabilizes the cation and lowers its energy. The molecular orbital shows that only the two C–H bonds more nearly parallel to the cation p orbital are oriented properly. The C–H bond perpendicular to the cation p orbital cannot take part.

Problem 7.18

Show the structures of the carbocation intermediates you would expect in the following reactions:

(a) $CH_3CH_2C(CH_3){=}CHCH(CH_3)CH_3 \xrightarrow{HBr} ?$

(b) (cyclopentylidene)$=CHCH_3 \xrightarrow{HI} ?$

Problem 7.19

Draw a skeletal structure of the following carbocation. Identify it as primary, secondary, or tertiary, and identify the hydrogen atoms that have the proper orientation for hyperconjugation in the conformation shown.

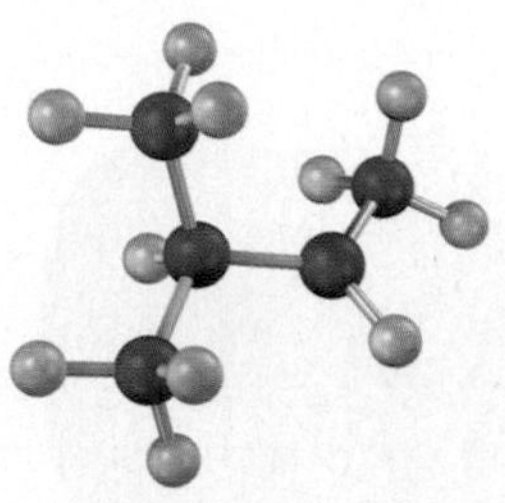

7.10 The Hammond Postulate

Let's summarize our knowledge of electrophilic addition reactions to this point:

- **Electrophilic addition to an unsymmetrically substituted alkene gives the more highly substituted carbocation intermediate.** A more highly substituted carbocation forms faster than a less highly substituted one and, once formed, rapidly goes on to give the final product.
- **A more highly substituted carbocation is more stable than a less highly substituted one.** That is, the stability order of carbocations is tertiary > secondary > primary > methyl.

What we have not yet seen is how these two points are related. Why does the *stability* of the carbocation intermediate affect the *rate* at which it's formed and thereby determine the structure of the final product? After all, carbocation stability is determined by the free-energy change $\Delta G°$, but reaction rate is determined by the activation energy $\Delta G^{\ddagger}$. The two quantities aren't directly related.

Although there is no simple quantitative relationship between the stability of a carbocation intermediate and the rate of its formation, there *is* an intuitive relationship. It's generally true when comparing two similar reactions that the more stable intermediate forms faster than the less stable one. The situation is shown graphically in **Figure 7.13**, where the energy profile in part (a) represents the typical situation rather than the profile in part (b). That is, the curves for two similar reactions don't cross one another.

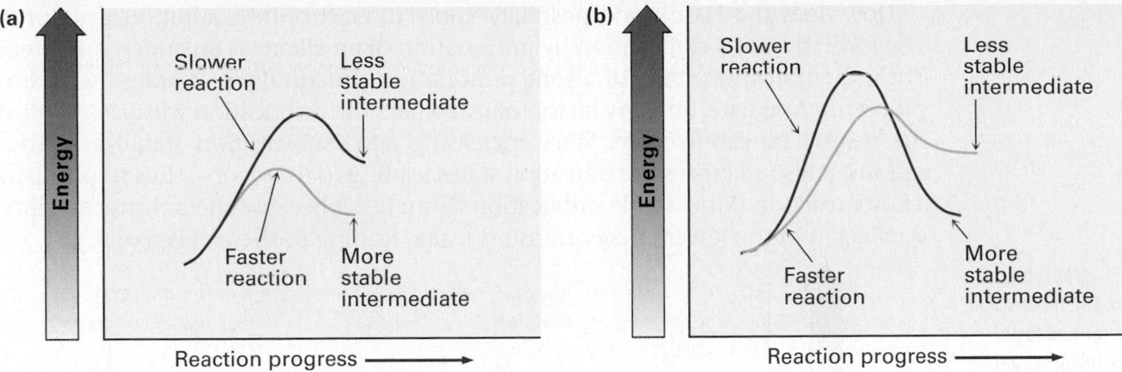

Figure 7.13 Energy diagrams for two similar competing reactions. In **(a)**, the faster reaction yields the more stable intermediate. In **(b)**, the slower reaction yields the more stable intermediate. The curves shown in **(a)** represent the typical situation.

Called the **Hammond postulate**, the explanation of the relationship between reaction rate and intermediate stability goes like this: Transition states represent energy maxima. They are high-energy activated complexes that occur transiently during the course of a reaction and immediately go on to a more stable species. Although we can't actually observe transition states because they have no finite lifetime, the Hammond postulate says that we can get an idea of a particular transition state's structure by looking at the structure of the nearest stable species. Imagine the two cases shown in **Figure 7.14**, for example. The reaction profile in part (a) shows the energy curve for an endergonic reaction step, and the profile in part (b) shows the curve for an exergonic step.

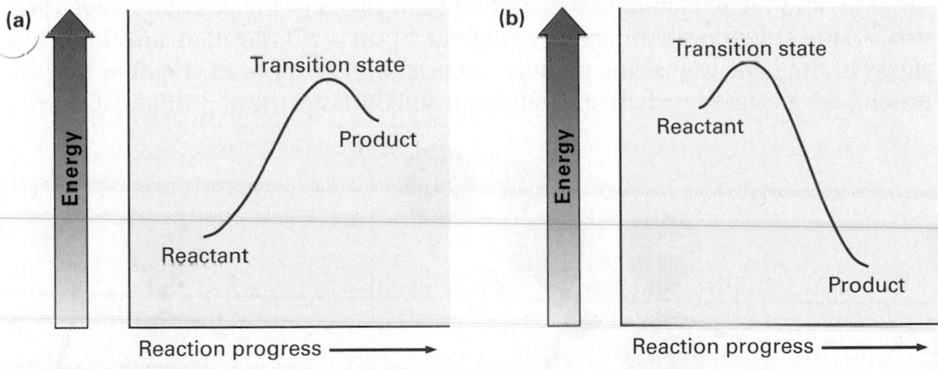

Figure 7.14 Energy diagrams for endergonic and exergonic steps. **(a)** In an endergonic step, the energy levels of transition state and *product* are closer. **(b)** In an exergonic step, the energy levels of transition state and *reactant* are closer.

In an endergonic reaction (Figure 7.14a), the energy level of the transition state is closer to that of the product than to that of the reactant. Since the transition state is closer energetically to the product, we make the natural assumption that it's also closer structurally. In other words, *the transition state for an endergonic reaction step structurally resembles the product of that step*. Conversely, the transition state for an exergonic reaction (Figure 7.14b) is closer energetically, and thus structurally, to the reactant than to the product. We therefore say that *the transition state for an exergonic reaction step structurally resembles the reactant for that step.*

Hammond postulate

The structure of a transition state resembles the structure of the nearest stable species. Transition states for endergonic steps structurally resemble products, and transition states for exergonic steps structurally resemble reactants.

How does the Hammond postulate apply to electrophilic addition reactions? The formation of a carbocation by protonation of an alkene is an endergonic step. Thus, the transition state for alkene protonation structurally resembles the carbocation intermediate, and any factor that stabilizes the carbocation will also stabilize the nearby transition state. Since increasing alkyl substitution stabilizes carbocations, it also stabilizes the transition states leading to those ions, thus resulting in a faster reaction. More stable carbocations form faster because their greater stability is reflected in the lower-energy transition state leading to them **(Figure 7.15)**.

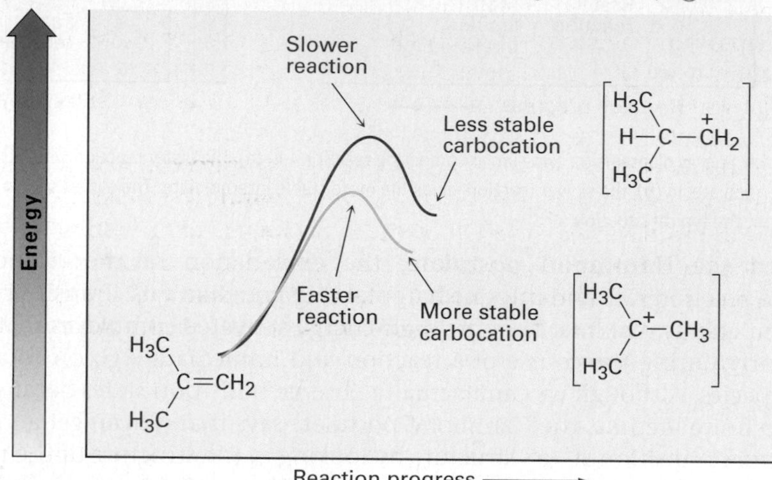

Figure 7.15 Energy diagrams for carbocation formation. The more stable tertiary carbocation is formed faster (green curve) because its increased stability lowers the energy of the transition state leading to it.

We can imagine the transition state for alkene protonation to be a structure in which one of the alkene carbon atoms has almost completely rehybridized from sp^2 to sp^3 and the remaining alkene carbon bears much of the positive charge **(Figure 7.16)**. This transition state is stabilized by hyperconjugation and inductive effects in the same way as the product carbocation. The more alkyl groups that are present, the greater the extent of stabilization and the faster the transition state forms.

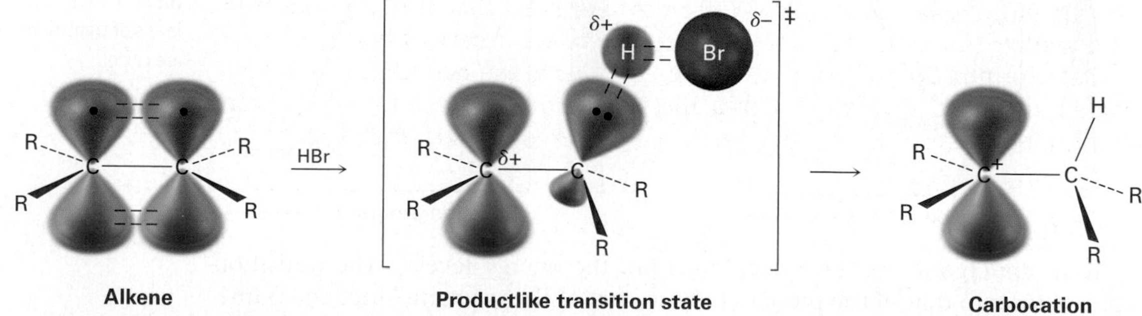

Figure 7.16 The hypothetical structure of a transition state for alkene protonation. The transition state is closer in both energy and structure to the carbocation than to the alkene. Thus, an increase in carbocation stability (lower $\Delta G°$) also causes an increase in transition-state stability (lower $\Delta G^{\ddagger}$), thereby increasing the rate of its formation.

Problem 7.20

What about the second step in the electrophilic addition of HCl to an alkene—the reaction of chloride ion with the carbocation intermediate? Is this step exergonic or endergonic? Does the transition state for this second step resemble the reactant (carbocation) or product (alkyl chloride)? Make a rough drawing of what the transition-state structure might look like.

7.11 Evidence for the Mechanism of Electrophilic Additions: Carbocation Rearrangements

How do we know that the carbocation mechanism for electrophilic addition reactions of alkenes is correct? The answer is that we *don't* know it's correct; at least we don't know with complete certainty. Although an incorrect reaction mechanism can be disproved by demonstrating that it doesn't account for observed data, a correct reaction mechanism can never be entirely proven. The best we can do is to show that a proposed mechanism is consistent with all known facts. If enough facts are accounted for, the mechanism is probably correct.

One of the best pieces of evidence supporting the carbocation mechanism for the electrophilic addition reaction was discovered during the 1930s by F. C. Whitmore of the Pennsylvania State University, who found that structural rearrangements often occur during the reaction of HX with an alkene. For example, reaction of HCl with 3-methyl-1-butene yields a substantial amount of 2-chloro-2-methylbutane in addition to the "expected" product, 2-chloro-3-methylbutane.

3-Methyl-1- butene + HCl ⟶ 2-Chloro-3-methylbutane (approx. 50%) + 2-Chloro-2-methylbutane (approx. 50%)

If the reaction takes place in a single step, it would be difficult to account for rearrangement, but if the reaction takes place in several steps, rearrangement is more easily explained. Whitmore suggested that it is a carbocation intermediate that undergoes rearrangement. The secondary carbocation intermediate formed by protonation of 3-methyl-1-butene rearranges to a more stable tertiary carbocation by a **hydride shift**—the shift of a hydrogen atom and its electron pair (a hydride ion, :H^-) between neighboring carbons.

3-Methyl-1- butene + H—Cl ⟶ [A 2° carbocation ⇌ (Hydride shift) A 3° carbocation]

A 2° carbocation —Cl^-→ 2-Chloro-3-methylbutane

A 3° carbocation —Cl^-→ 2-Chloro-2-methylbutane

Carbocation rearrangements can also occur by the shift of an alkyl group with its electron pair. For example, reaction of 3,3-dimethyl-1-butene with HCl leads to an equal mixture of unrearranged 2-chloro-3,3-dimethylbutane and rearranged 2-chloro-2,3-dimethylbutane. In this instance, a secondary carbocation rearranges to a more stable tertiary carbocation by the shift of a methyl group.

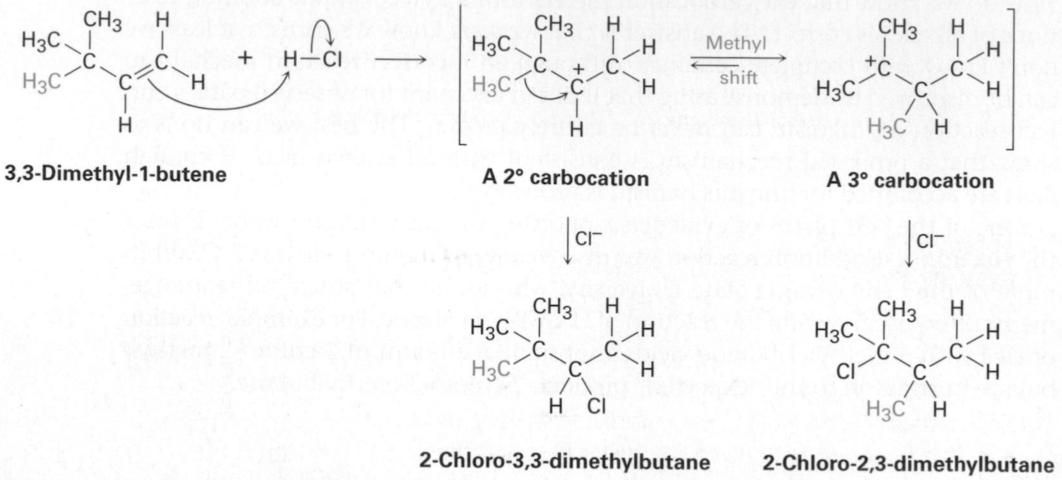

3,3-Dimethyl-1-butene + H–Cl → A 2° carbocation ⇌ (Methyl shift) A 3° carbocation

A 2° carbocation + Cl^- → 2-Chloro-3,3-dimethylbutane

A 3° carbocation + Cl^- → 2-Chloro-2,3-dimethylbutane

Note the similarities between the two carbocation rearrangements: in both cases, a group (:H^- or :CH_3^-) moves to an adjacent positively charged carbon, taking its bonding electron pair with it. Also in both cases, a less stable carbocation rearranges to a more stable ion. Rearrangements of this kind are a common feature of carbocation chemistry and are particularly important in the biological pathways by which steroids and related substances are synthesized. An example is the following hydride shift that occurs during the biosynthesis of cholesterol.

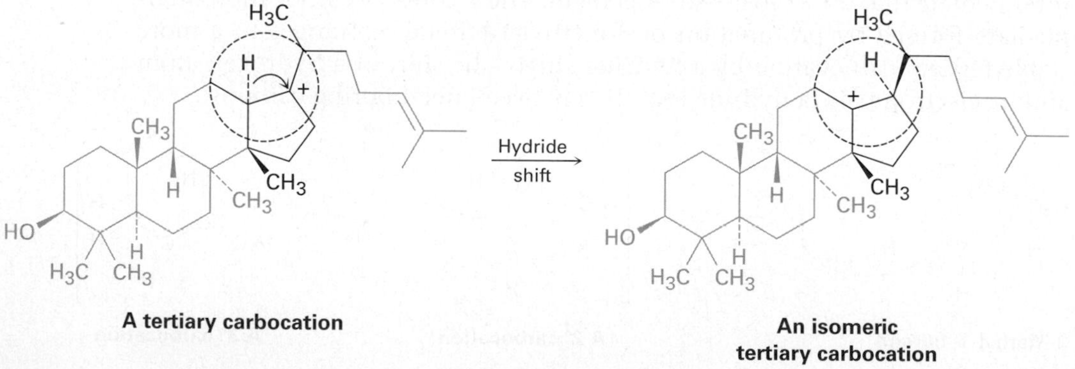

A tertiary carbocation → (Hydride shift) An isomeric tertiary carbocation

A word of advice that we've noted before and will repeat on occasion: biological molecules are often larger and more complex in appearance than the molecules chemists work with in the laboratory, but don't be intimidated. When looking at *any* chemical transformation, whether biochemical or not, focus on the part of the molecule where the change is occurring and don't worry about the rest. The tertiary carbocation just pictured looks complicated, but all the chemistry is taking place in the small part of the molecule inside the red circle.

Problem 7.21

On treatment with HBr, vinylcyclohexane undergoes addition and rearrangement to yield 1-bromo-1-ethylcyclohexane. Using curved arrows, propose a mechanism to account for this result.

Vinylcyclohexane $\xrightarrow{HBr}$ 1-Bromo-1-ethylcyclohexane (CH_2CH_3, Br)

Vinylcyclohexane **1-Bromo-1-ethylcyclohexane**

A DEEPER LOOK Bioprospecting: Hunting for Natural Products

Rapamycin, an immunosuppressant natural product used during organ transplants, was originally isolated from a soil sample found on Easter Island, or Rapa Nui, an island 2200 miles off the coast of Chile known for its giant Moai statues.

Most people know the names of the common classes of biomolecules—proteins, carbohydrates, lipids, and nucleic acids—but there are far more kinds of compounds in living organisms than just those four. All living organisms also contain a vast diversity of substances usually grouped under the heading *natural products.* The term natural product really refers to *any* naturally occurring substance but is generally taken to mean a so-called secondary metabolite—a small molecule that is not essential to the growth and development of the producing organism and is not classified by structure.

It has been estimated that well over 300,000 secondary metabolites exist, and it's thought that their primary function is to increase the likelihood of an organism's survival by repelling or attracting other organisms. Alkaloids, such as morphine; antibiotics, such as erythromycin and the penicillins; and immunosuppressive agents, such as rapamycin (sirolimus) prescribed for liver transplant recipients, are examples.

Rapamycin (Sirolimus)

(continued)

(continued)

Where do these natural products come from, and how are they found? Although most chemists and biologists spend most of their time in the laboratory, a few spend their days scuba diving on South Pacific islands or trekking through the rainforests of South America and Southeast Asia at work as bioprospectors. Their job is to hunt for new and unusual natural products that might be useful as drugs.

As noted in the Chapter 6 *A Deeper Look,* more than half of all new drug candidates come either directly or indirectly from natural products. Morphine from the opium poppy, prostaglandin E_1 from sheep prostate glands, erythromycin A from a *Streptomyces erythreus* bacterium cultured from a Philippine soil sample, and benzylpenicillin from the mold *Penicillium notatum* are examples. The immunosuppressive agent rapamycin, whose structure is shown on the previous page, was first isolated from a *Streptomyces hygroscopicus* bacterium found in a soil sample from Easter Island (Rapa Nui), located 2200 miles off the coast of Chile.

With less than 1% of living organisms yet investigated, bioprospectors have a lot of work to do. But there is a race going on. Rainforests throughout the world are being destroyed at an alarming rate, causing many species of both plants and animals to become extinct before they can even be examined. Fortunately, the governments in many countries seem aware of the problem, but there is as yet no international treaty on biodiversity that could help preserve vanishing species.

Summary

Key words

alkene ($R_2C{=}CR_2$), 222
allyl group, 228
degree of unsaturation, 224
***E* geometry, 232**
***E,Z* system, 231**
electrophilic addition reaction, 237
Hammond postulate, 247
hydride shift, 249
hyperconjugation, 235
Markovnikov's rule, 240
methylene group, 228
regiospecific, 240
unsaturated, 224
vinyl group, 228
***Z* geometry, 232**

Carbon–carbon double bonds are present in most organic and biological molecules, so a good understanding of their behavior is needed. In this chapter, we've looked at some consequences of alkene stereoisomerism and at the details of the broadest and most general class of alkene reactions—the electrophilic addition reaction.

An **alkene** is a hydrocarbon that contains a carbon–carbon double bond. Because they contain fewer hydrogens than alkanes with the same number of carbons, alkenes are said to be **unsaturated**.

Because rotation around the double bond can't occur, substituted alkenes can exist as cis–trans stereoisomers. The geometry of a double bond can be specified by application of the Cahn–Ingold–Prelog sequence rules, which rank the substituents on each double-bond carbon. If the higher-ranking groups on each carbon are on the same side of the double bond, the geometry is ***Z*** (*zusammen,* "together"); if the higher-ranking groups on each carbon are on opposite sides of the double bond, the geometry is ***E*** (*entgegen,* "apart").

Alkene chemistry is dominated by **electrophilic addition reactions**. When HX reacts with an unsymmetrically substituted alkene, **Markovnikov's rule** predicts that the H will add to the carbon having fewer alkyl substituents and the X group will add to the carbon having more alkyl substituents. Electrophilic additions to alkenes take place through carbocation intermediates formed by reaction of the nucleophilic alkene π bond with electrophilic H^+. Carbocation stability follows the order

$$\begin{array}{ccccccc} \text{Tertiary (3°)} & > & \text{Secondary (2°)} & > & \text{Primary (1°)} & > & \text{Methyl} \\ R_3C^+ & > & R_2CH^+ & > & RCH_2^+ & > & CH_3^+ \end{array}$$

Markovnikov's rule can be restated by saying that, in the addition of HX to an alkene, the more stable carbocation intermediate is formed. This result is explained by the **Hammond postulate**, which says that the transition state of an exergonic reaction step structurally resembles the reactant, whereas the transition state of an endergonic reaction step structurally resembles the product. Since an alkene protonation step is endergonic, the stability of the more highly substituted carbocation is reflected in the stability of the transition state leading to its formation.

Evidence in support of a carbocation mechanism for electrophilic additions comes from the observation that structural rearrangements often take place during reaction. Rearrangements occur by shift of either a hydride ion, **:H$^-$** (a **hydride shift**), or an alkyl anion, **:R$^-$**, from a carbon atom to the neighboring positively charged carbon. The result is isomerization of a less stable carbocation to a more stable one.

Exercises

Visualizing Chemistry

OWL Interactive versions of these problems are assignable in OWL for Organic Chemistry.

▲ denotes problems linked to the Key Ideas in this chapter.

(Problems 7.1–7.21 appear within the chapter.)

7.22 Name the following alkenes, and convert each drawing into a skeletal structure:

(a) **(b)**

7.23 Assign *E* or *Z* stereochemistry to the double bonds in each of the following alkenes, and convert each drawing into a skeletal structure (red = O, green = Cl):

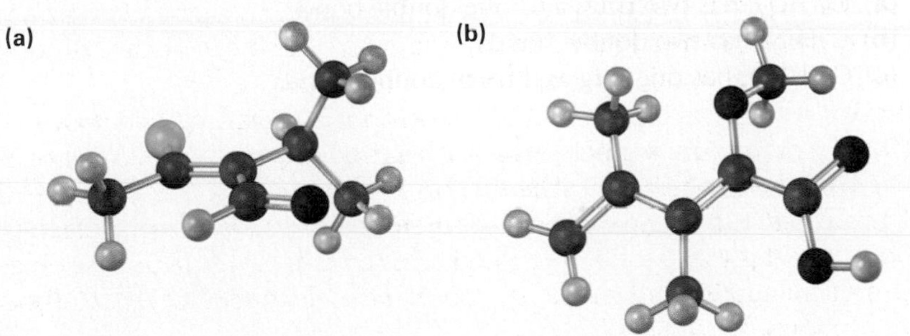

7.24 The following carbocation is an intermediate in the electrophilic addition reaction of HCl with two different alkenes. Identify both, and tell which C–H bonds in the carbocation are aligned for hyperconjugation with the vacant *p* orbital on the positively charged carbon.

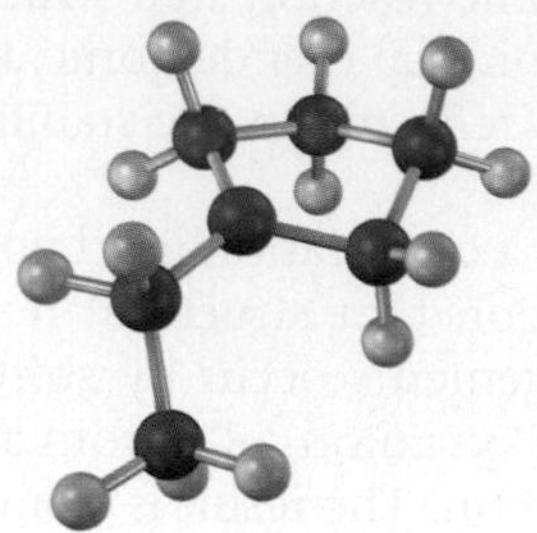

7.25 The following alkyl bromide can be made by HBr addition to three different alkenes. Show their structures.

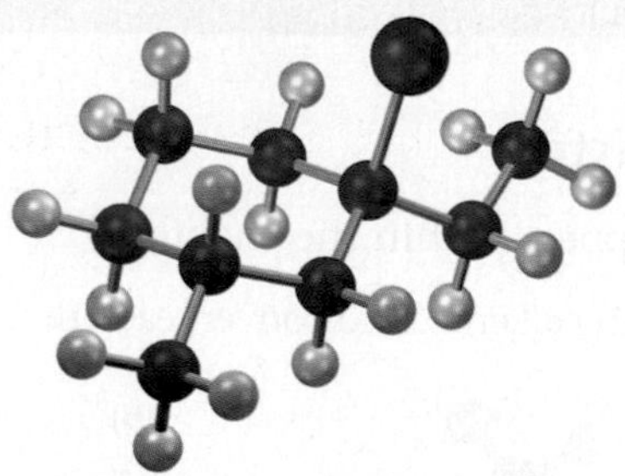

Additional Problems

Calculating a Degree of Unsaturation

7.26 Calculate the degree of unsaturation in the following formulas, and draw five possible structures for each:

(a) $C_{10}H_{16}$ (b) C_8H_8O (c) $C_7H_{10}Cl_2$
(d) $C_{10}H_{16}O_2$ (e) $C_5H_9NO_2$ (f) $C_8H_{10}ClNO$

7.27 How many hydrogens does each of the following compounds have?

(a) $C_8H_?O_2$, has two rings and one double bond
(b) $C_7H_?N$, has two double bonds
(c) $C_9H_?NO$, has one ring and three double bonds

▲ Problems linked to Key Ideas in this chapter

7.28 Loratadine, marketed as an antiallergy medication under the name Claritin, has four rings, eight double bonds, and the formula $C_{22}H_?ClN_2O_2$. How many hydrogens does loratadine have? (Calculate your answer; don't count hydrogens in the structure.)

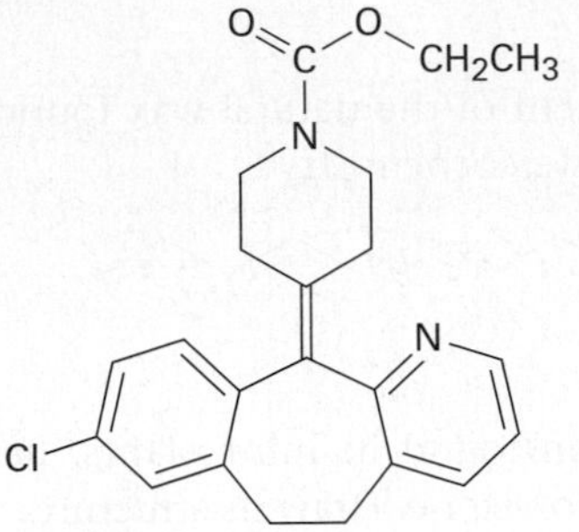

O=C(–O–CH_2CH_3)–N ... N ... Cl

Loratadine

Naming Alkenes

7.29 Name the following alkenes:

(a) CH_3 / H, $CHCH_2CH_3$ / C=C / H_3C, H

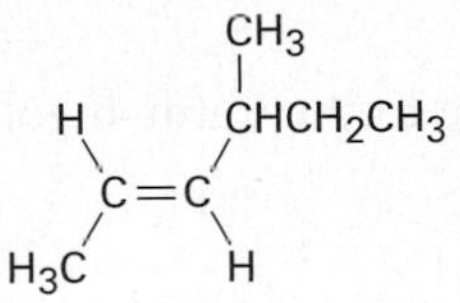

(b) CH_3, CH_2CH_3 / $CH_3CHCH_2CH_2CH$, CH_3 / C=C / H_3C, H

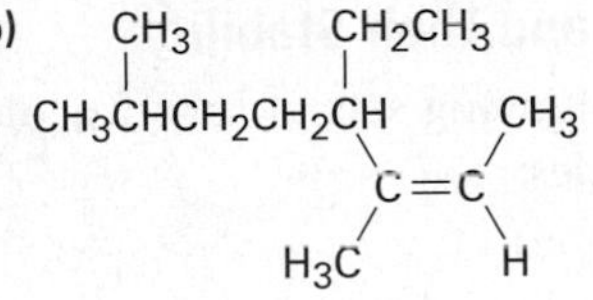

(c) CH_2CH_3 / $H_2C{=}CCH_2CH_3$

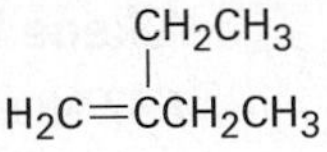

(d) H, CH_3 / H_3C, C=C / $H_2C{=}CHCHCH$, H / CH_3

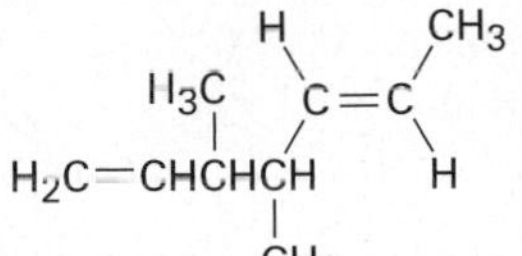

(e) H, H / H_3C, C=C / C=C, CH_3 / $CH_3CH_2CH_2$, CH_3

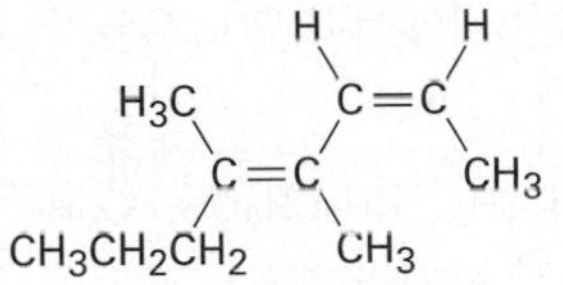

(f) $H_2C{=}C{=}CHCH_3$

7.30 Draw structures corresponding to the following systematic names:

(a) (4*E*)-2,4-Dimethyl-1,4-hexadiene
(b) *cis*-3,3-Dimethyl-4-propyl-1,5-octadiene
(c) 4-Methyl-1,2-pentadiene
(d) (3*E*,5*Z*)-2,6-Dimethyl-1,3,5,7-octatetraene
(e) 3-Butyl-2-heptene
(f) *trans*-2,2,5,5-Tetramethyl-3-hexene

7.31 Name the following cycloalkenes:

(a) 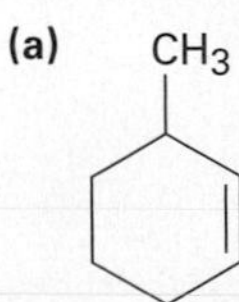

(b) 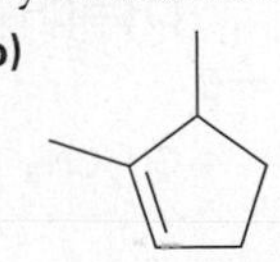

(c) 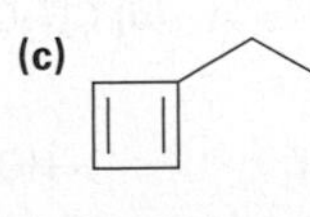

(d) 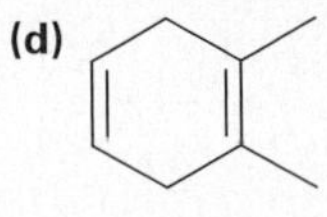

(e) 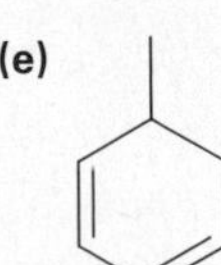

(f)

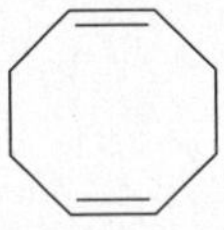

▲ Problems linked to Key Ideas in this chapter

7.32 Ocimene is a triene found in the essential oils of many plants. What is its IUPAC name, including stereochemistry?

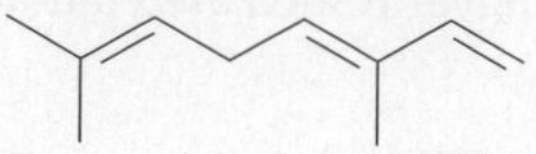

Ocimene

7.33 α-Farnesene is a constituent of the natural wax found on apples. What is its IUPAC name, including stereochemistry?

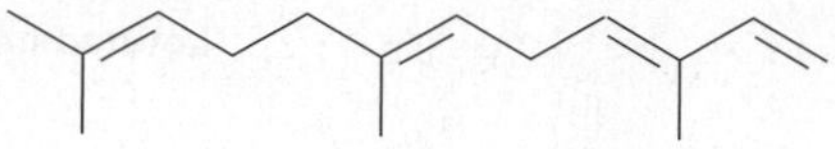

α-Farnesene

7.34 Menthene, a hydrocarbon found in mint plants, has the systematic name 1-isopropyl-4-methylcyclohexene. Draw its structure.

7.35 Draw and name the six alkene isomers, C_5H_{10}, including *E,Z* isomers.

7.36 Draw and name the 17 alkene isomers, C_6H_{12}, including *E,Z* isomers.

Alkene Isomers and Their Stability

7.37 Rank the following sets of substituents according to the Cahn–Ingold–Prelog sequence rules:

(a) $-CH_3$, $-Br$, $-H$, $-I$

(b) $-OH$, $-OCH_3$, $-H$, $-CO_2H$

(c) $-CO_2H$, $-CO_2CH_3$, $-CH_2OH$, $-CH_3$

(d) $-CH_3$, $-CH_2CH_3$, $-CH_2CH_2OH$, $-C(=O)CH_3$

(e) $-CH{=}CH_2$, $-CN$, $-CH_2NH_2$, $-CH_2Br$

(f) $-CH{=}CH_2$, $-CH_2CH_3$, $-CH_2OCH_3$, $-CH_2OH$

7.38 ▲ Assign *E* or *Z* configuration to each of the following compounds:

(a) $HOCH_2$, H_3C / C=C / CH_3, H

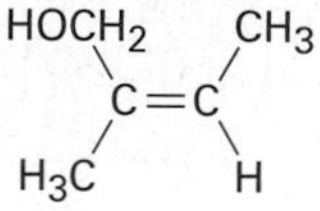

(b) HO_2C, Cl / C=C / H, OCH_3

(c) NC, CH_3CH_2 / C=C / CH_3, CH_2OH

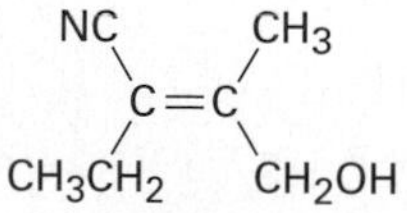

(d) CH_3O_2C, HO_2C / C=C / $CH{=}CH_2$, CH_2CH_3

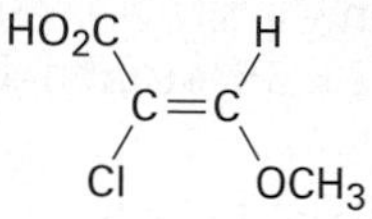

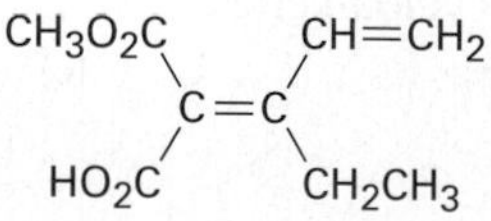

▲ Problems linked to Key Ideas in this chapter

7.39 ▲ Which of the following *E,Z* designations are correct, and which are incorrect?

(a) CH_3-substituted cyclohexylidene: C=C with CO_2H and H — *Z*

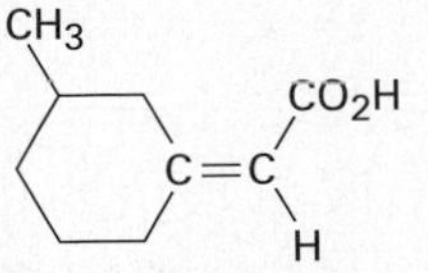

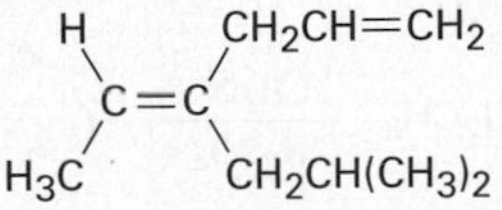

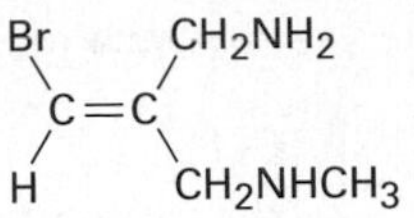

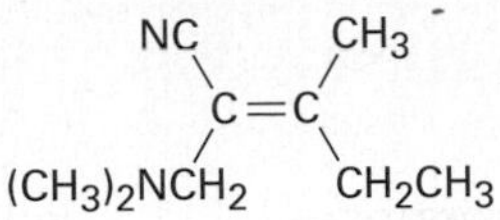

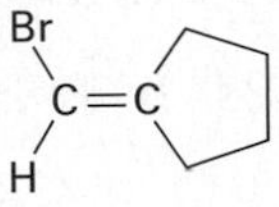

(b) H and H_3C / $CH_2CH{=}CH_2$ and $CH_2CH(CH_3)_2$ — *E*

(c) Br and H / CH_2NH_2 and CH_2NHCH_3 — *Z*

(d) NC and $(CH_3)_2NCH_2$ / CH_3 and CH_2CH_3 — *E*

(e) Br and H / cyclopentylidene — *Z*

(f) $HOCH_2$ and CH_3OCH_2 / CO_2H and $COCH_3$ — *E*

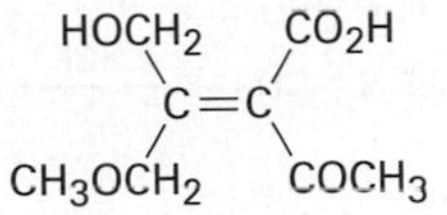

7.40 *trans*-2-Butene is more stable than *cis*-2-butene by only 4 kJ/mol, but *trans*-2,2,5,5-tetramethyl-3-hexene is more stable than its cis isomer by 39 kJ/mol. Explain.

7.41 Cyclodecene can exist in both cis and trans forms, but cyclohexene cannot. Explain. (Making molecular models is helpful.)

7.42 Normally, a trans alkene is *more* stable than its cis isomer. *trans*-Cyclooctene, however, is *less* stable than *cis*-cyclooctene by 38.5 kJ/mol. Explain.

7.43 *trans*-Cyclooctene is less stable than *cis*-cyclooctene by 38.5 kJ/mol, but *trans*-cyclononene is less stable than *cis*-cyclononene by only 12.2 kJ/mol. Explain.

7.44 Tamoxifen, a drug used in the treatment of breast cancer, and clomiphene, a drug used as a fertility treatment, have similar structures but very different effects. Assign *E* or *Z* configuration to the double bonds in both compounds.

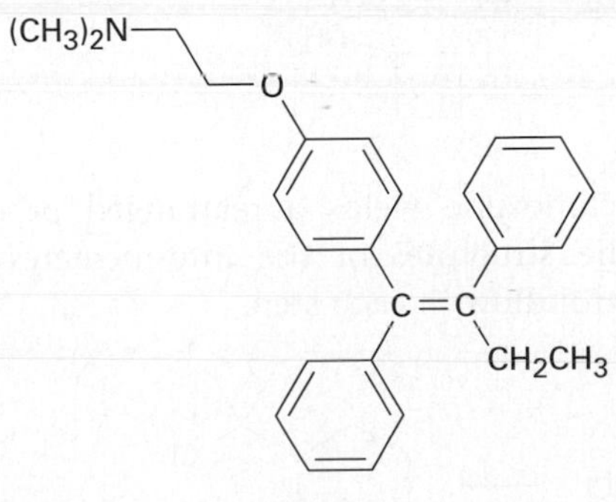

$(CH_3)_2N$— ... O ... C=C ... CH_2CH_3

Tamoxifen
(anticancer)

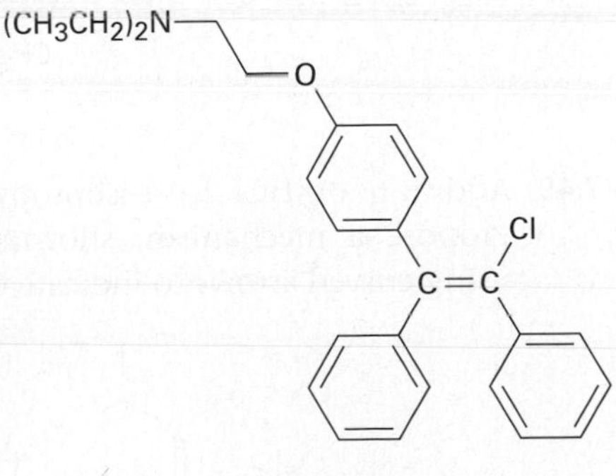

$(CH_3CH_2)_2N$— ... O ... C=C ... Cl

Clomiphene
(fertility treatment)

▲ Problems linked to Key Ideas in this chapter

Carbocations and Electrophilic Addition Reactions

7.45 Predict the major product in each of the following reactions:

(a) $CH_3CH_2CH{=}C(CH_3)CH_2CH_3 \xrightarrow[H_2SO_4]{H_2O}$?

(Addition of H_2O occurs.)

(b) 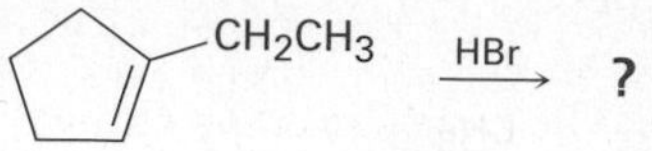 1-ethylcyclopentene (CH_2CH_3 on ring) $\xrightarrow{HBr}$?

(c) 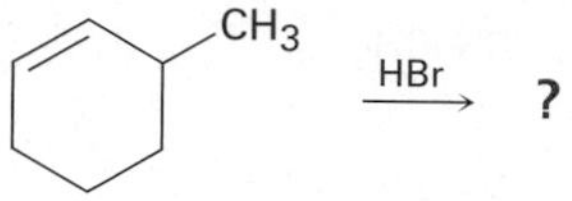 3-methylcyclohexene (CH_3 on ring) $\xrightarrow{HBr}$?

(d) $H_2C{=}CHCH_2CH_2CH_2CH{=}CH_2 \xrightarrow{2\ HCl}$?

7.46 ▲ Predict the major product from addition of HBr to each of the following alkenes:

(a) 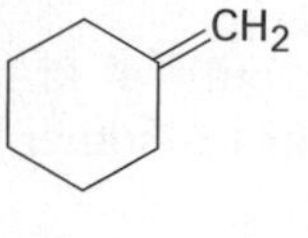 methylenecyclohexane ($=CH_2$) **(b)** 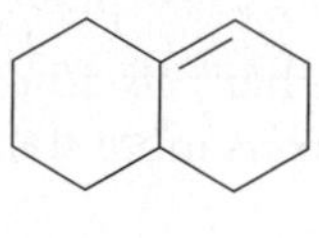 bicyclic alkene (octahydronaphthalene) **(c)** $CH_3CH{=}CHCH(CH_3)CH_3$

7.47 ▲ Alkenes can be converted into alcohols by acid-catalyzed addition of water. Assuming that Markovnikov's rule is valid, predict the major alcohol product from each of the following alkenes.

(a) $CH_3CH_2C(CH_3){=}CHCH_3$ **(b)** 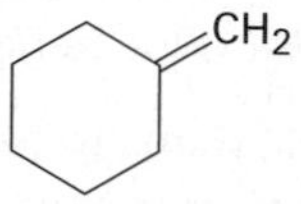 methylenecyclohexane ($=CH_2$) **(c)** $CH_3CH(CH_3)CH_2CH{=}CH_2$

7.48 Each of the following carbocations can rearrange to a more stable ion. Propose structures for the likely rearrangement products.

(a) $CH_3CH_2CH_2CH_2^+$ **(b)** $CH_3\overset{+}{C}HCH(CH_3)CH_3$ **(c)** 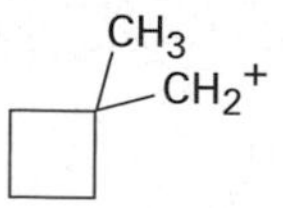 cyclobutane bearing CH_3 and CH_2^+ on the same carbon

7.49 Addition of HCl to 1-isopropylcyclohexene yields a rearranged product. Propose a mechanism, showing the structures of the intermediates and using curved arrows to indicate electron flow in each step.

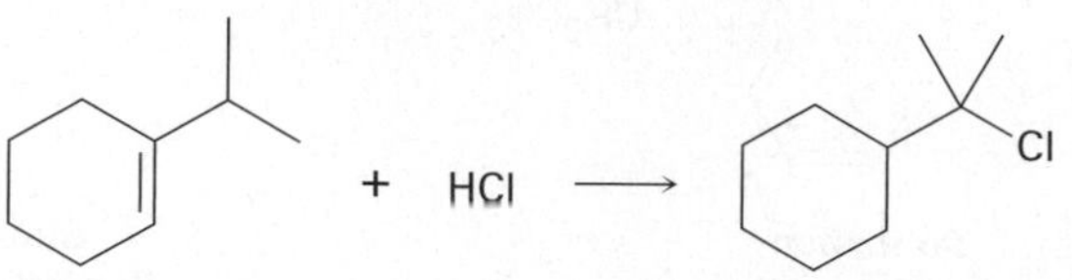

1-isopropylcyclohexene + HCl ⟶ cyclohexyl–C(CH_3)_2Cl

▲ Problems linked to Key Ideas in this chapter

General Problems

7.50 Allene (1,2-propadiene), $H_2C{=}C{=}CH_2$, has two adjacent double bonds. What kind of hybridization must the central carbon have? Sketch the bonding π orbitals in allene. What shape do you predict for allene?

7.51 The heat of hydrogenation for allene (Problem 7.50) to yield propane is -295 kJ/mol, and the heat of hydrogenation for a typical monosubstituted alkene such as propene is -126 kJ/mol. Is allene more stable or less stable than you might expect for a diene? Explain.

7.52 Retin A, or retinoic acid, is a medication commonly used to reduce wrinkles and treat severe acne. How many different isomers arising from double-bond isomerizations are possible?

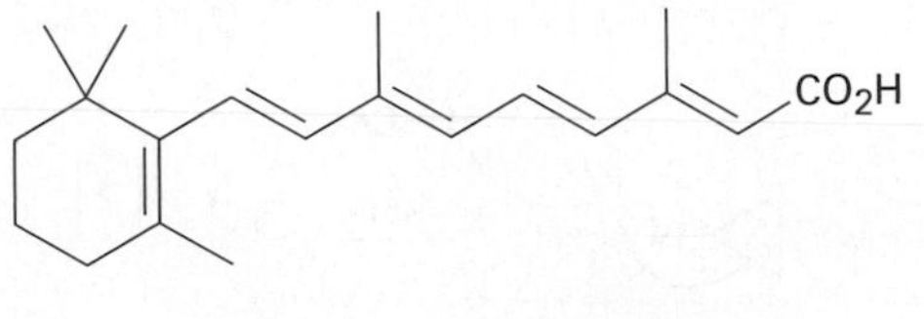

CO_2H

Retin A (retinoic acid)

7.53 Fucoserratene and ectocarpene are sex pheromones produced by marine brown algae. What are their systematic names? (Ectocarpene is a bit difficult; make your best guess, and then check your answer in the *Study Guide and Solutions Manual*.)

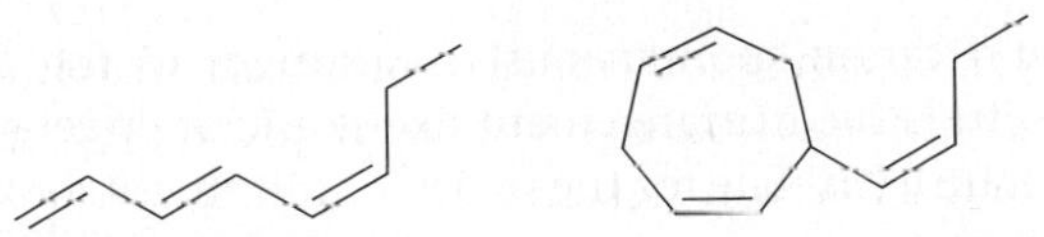

Fucoserratene **Ectocarpene**

7.54 ▲ *tert*-Butyl esters $[RCO_2C(CH_3)_3]$ are converted into carboxylic acids (RCO_2H) by reaction with trifluoroacetic acid, a reaction useful in protein synthesis (Section 26.7). Assign *E,Z* designation to the double bonds of both reactant and product in the following scheme, and explain why there is an apparent change of double-bond stereochemistry:

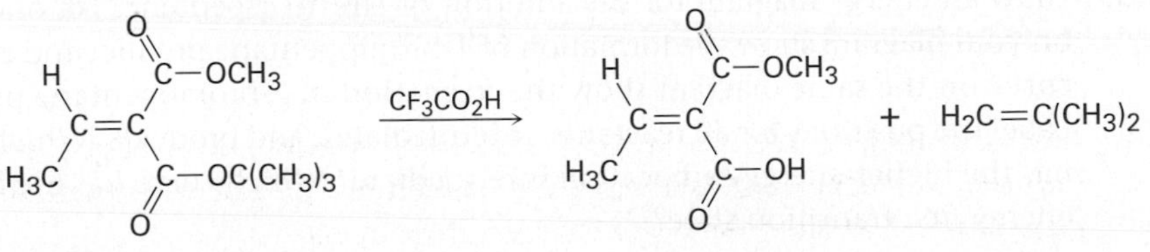

$$(H)(H_3C)C{=}C(COOCH_3)(COOC(CH_3)_3) \xrightarrow{CF_3CO_2H} (H)(H_3C)C{=}C(COOCH_3)(COOH) + H_2C{=}C(CH_3)_2$$

▲ Problems linked to Key Ideas in this chapter

7.55 Addition of HCl to 1-isopropenyl-1-methylcyclopentane yields 1-chloro-1,2,2-trimethylcyclohexane. Propose a mechanism, showing the structures of the intermediates and using curved arrows to indicate electron flow in each step.

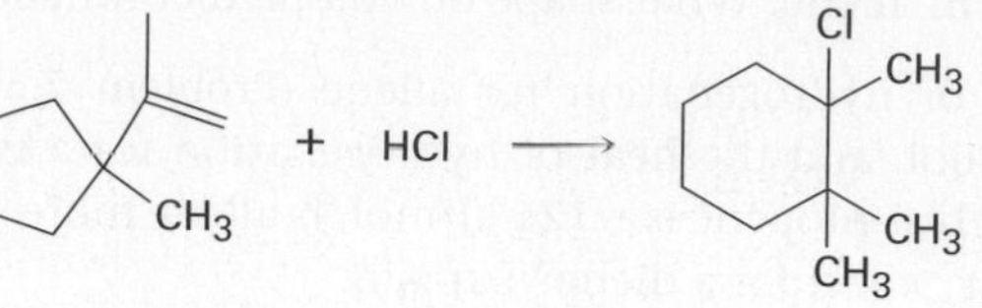

CH_3 + HCl ⟶ Cl, CH_3, CH_3, CH_3

7.56 Vinylcyclopropane reacts with HBr to yield a rearranged alkyl bromide. Follow the flow of electrons as represented by the curved arrows, show the structure of the carbocation intermediate in brackets, and show the structure of the final product.

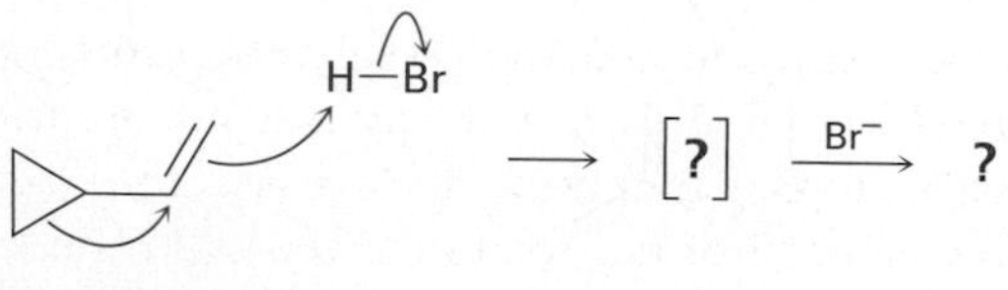

H—Br ⟶ [?] $\xrightarrow{Br^-}$?

Vinylcyclopropane

7.57 Calculate the degree of unsaturation in each of the following formulas:

(a) Cholesterol, $C_{27}H_{46}O$
(b) DDT, $C_{14}H_9Cl_5$
(c) Prostaglandin E_1, $C_{20}H_{34}O_5$
(d) Caffeine, $C_8H_{10}N_4O_2$
(e) Cortisone, $C_{21}H_{28}O_5$
(f) Atropine, $C_{17}H_{23}NO_3$

7.58 The isobutyl cation spontaneously rearranges to the *tert*-butyl cation by a hydride shift. Is the rearrangement exergonic or endergonic? Draw what you think the transition state for the hydride shift might look like according to the Hammond postulate.

H_3C CH_3 C H C+ H H ⟶ CH_3 +C H_3C CH_3

Isobutyl cation **_tert_-Butyl cation**

7.59 Draw an energy diagram for the addition of HBr to 1-pentene. Let one curve on your diagram show the formation of 1-bromopentane product and another curve on the same diagram show the formation of 2-bromopentane product. Label the positions for all reactants, intermediates, and products. Which curve has the higher-energy carbocation intermediate? Which curve has the higher-energy first transition state?

7.60 Make sketches of the transition-state structures involved in the reaction of HBr with 1-pentene (Problem 7.59). Tell whether each structure resembles reactant or product.

7.61 Limonene, a fragrant hydrocarbon found in lemons and oranges, is biosynthesized from geranyl diphosphate by the following pathway. Add curved arrows to show the mechanism of each step. Which step involves an alkene

▲ Problems linked to Key Ideas in this chapter

electrophilic addition? (The ion $OP_2O_6{}^{4-}$ is the diphosphate ion, and "Base" is an unspecified base in the enzyme that catalyzes the reaction.)

+ $OP_2O_6{}^{4-}$

$OP_2O_6{}^{3-}$ → + → + (H) —Base→

Geranyl diphosphate

Limonene

7.62 *epi*-Aristolochene, a hydrocarbon found in both pepper and tobacco, is biosynthesized by the following pathway. Add curved arrows to show the mechanism of each step. Which steps involve alkene electrophilic addition(s), and which involve carbocation rearrangement(s)? (The abbreviation H–A stands for an unspecified acid, and "Base" is an unspecified base in the enzyme.)

H_3C, CH_3, H —H–A (acid)→ CH_3, +, H → CH_3, +, H, CH_3, H →

CH_3, H, +, H, CH_3 → H, +, H, CH_3, H, CH_3 —Base→ H, CH_3, H, CH_3

***epi*-Aristolochene**

7.63 Aromatic compounds such as benzene react with alkyl chlorides in the presence of $AlCl_3$ catalyst to yield alkylbenzenes. The reaction occurs through a carbocation intermediate, formed by reaction of the alkyl chloride with $AlCl_3$ ($R{-}Cl + AlCl_3 \rightarrow R^+ + AlCl_4{}^-$). How can you explain the observation that reaction of benzene with 1-chloropropane yields isopropylbenzene as the major product?

$$\text{C}_6\text{H}_6 + CH_3CH_2CH_2Cl \xrightarrow{AlCl_3} \text{C}_6\text{H}_5\text{CH}(CH_3)CH_3$$

7.64 Reaction of 2,3-dimethyl-1-butene with HBr leads to an alkyl bromide, $C_6H_{13}Br$. On treatment of this alkyl bromide with KOH in methanol, elimination of HBr occurs and a hydrocarbon that is isomeric with the starting alkene is formed. What is the structure of this hydrocarbon, and how do you think it is formed from the alkyl bromide?

▲ Problems linked to Key Ideas in this chapter

In biological pathways, dehydrations rarely occur with isolated alcohols. Instead, they normally take place on substrates in which the –OH is positioned two carbons away from a carbonyl group. In the biosynthesis of fats, for instance, β-hydroxybutyryl ACP is converted by dehydration to *trans*-crotonyl ACP, where ACP is an abbreviation for *acyl carrier protein*. We'll see the reason for this requirement in **Section 11.10**.

$$H_3C\text{–}CH(OH)\text{–}CH_2\text{–}C(=O)\text{–}ACP \longrightarrow H_3C\text{–}CH{=}CH\text{–}C(=O)\text{–}ACP + H_2O$$

β-Hydroxybutyryl ACP **trans-Crotonyl ACP**

Problem 8.1

One problem with elimination reactions is that mixtures of products are often formed. For example, treatment of 2-bromo-2-methylbutane with KOH in ethanol yields a mixture of two alkene products. What are their likely structures?

Problem 8.2

How many alkene products, including *E,Z* isomers, might be obtained by dehydration of 3-methyl-3-hexanol with aqueous sulfuric acid?

$$CH_3CH_2CH_2C(OH)(CH_3)CH_2CH_3 \xrightarrow{H_2SO_4} ?$$

3-Methyl-3-hexanol

8.2 Halogenation of Alkenes: Addition of X_2

Bromine and chlorine add rapidly to alkenes to yield 1,2-dihalides, a process called *halogenation*. For example, more than 18 million tons 1,2-dichloroethane (ethylene dichloride) is synthesized worldwide each year, much of it by addition of Cl_2 to ethylene. The product is used both as a solvent and as starting material for the manufacture of poly(vinyl chloride), PVC. Fluorine is too reactive and difficult to control for most laboratory applications, and iodine does not react with most alkenes.

$$H_2C{=}CH_2 + Cl_2 \longrightarrow H\text{–}CHCl\text{–}CHCl\text{–}H$$

Ethylene **1,2-Dichloroethane (ethylene dichloride)**

Based on what we've seen thus far, a possible mechanism for the reaction of bromine with alkenes might involve electrophilic addition of Br^+ to the alkene, giving a carbocation intermediate that could undergo further reaction with Br^- to yield the dibromo addition product.

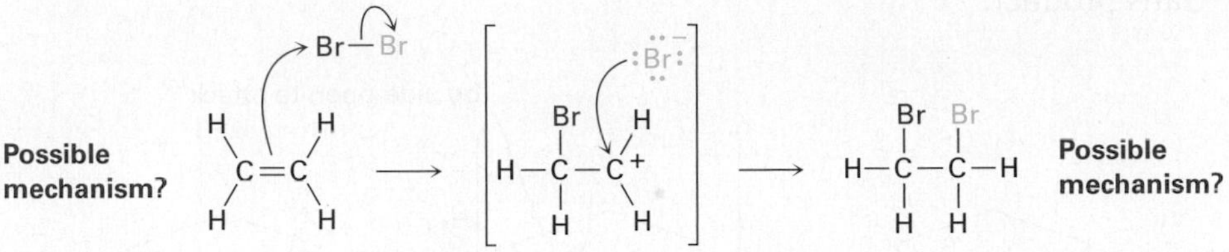

Possible mechanism?

Although this mechanism seems plausible, it's not fully consistent with known facts. In particular, it doesn't explain the *stereochemistry* of the addition reaction. That is, the mechanism doesn't tell which product stereoisomer is formed.

When the halogenation reaction is carried out on a cycloalkene, such as cyclopentene, only the *trans* stereoisomer of the dihalide addition product is formed rather than the mixture of cis and trans isomers that might have been expected if a planar carbocation intermediate were involved. We say that the reaction occurs with **anti stereochemistry**, meaning that the two bromine atoms come from opposite faces of the double bond—one from the top face and one from the bottom face.

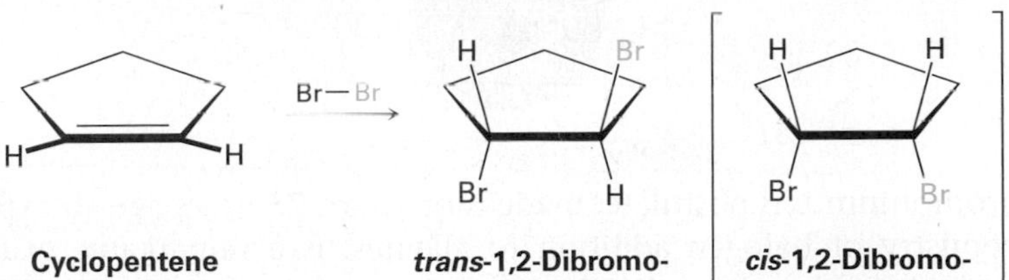

Cyclopentene — Br—Br → ***trans*-1,2-Dibromo-cyclopentane** (sole product) — [***cis*-1,2-Dibromo-cyclopentane** ***(Not formed)***]

An explanation for the observed stereochemistry of addition was suggested in 1937 by George Kimball and Irving Roberts, who proposed that the reaction intermediate is not a carbocation but is instead a **bromonium ion, R_2Br^+**, formed by electrophilic addition of Br^+ to the alkene. (Similarly, a *chloronium ion* contains a positively charged, divalent chlorine, R_2Cl^+.) The bromonium ion is formed in a single step by interaction of the alkene with Br_2 and simultaneous loss of Br^-.

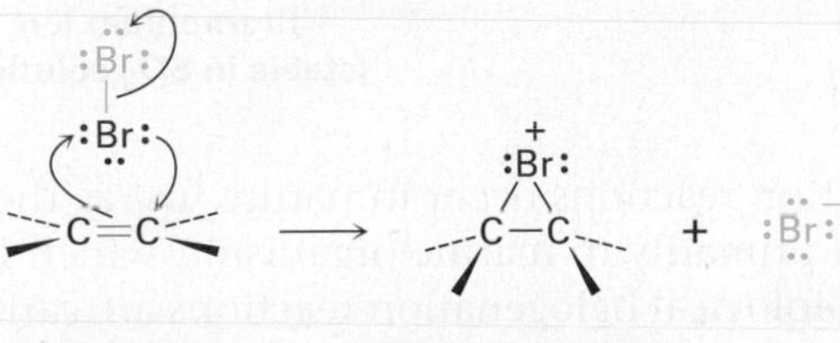

An alkene → **A bromonium ion** + $:\ddot{Br}:^-$

How does the formation of a bromonium ion account for the observed anti stereochemistry of addition to cyclopentene? If a bromonium ion is formed as an intermediate, we can imagine that the large bromine atom might "shield" one side of the molecule. Reaction with Br^- ion in the second step could then occur only from the opposite, unshielded side to give trans product.

Top side open to attack

Bottom side shielded from attack

Cyclopentene **Bromonium ion intermediate** ***trans*-1,2-Dibromocyclopentane**

The bromonium ion postulate, made more than 75 years ago to explain the stereochemistry of halogen addition to alkenes, is a remarkable example of deductive logic in chemistry. Arguing from experimental results, chemists were able to make a hypothesis about the intimate mechanistic details of alkene electrophilic reactions. Subsequently, strong evidence supporting the mechanism came from the work of George Olah, who prepared and studied *stable* solutions of cyclic bromonium ions in liquid SO_2. There's no question that bromonium ions exist.

SbF_5 / Liquid SO_2 → SbF_6^-

Bromonium ion (stable in SO_2 solution)

Alkene halogenation reactions occur in nature just as they do in the laboratory but are limited primarily to marine organisms, which live in a halide-rich environment. The biological halogenation reactions are carried out by enzymes called *haloperoxidases,* which use H_2O_2 to oxidize Br^- or Cl^- ions to a biological equivalent of Br^+ or Cl^+. Electrophilic addition to the double bond of a substrate molecule then yields a bromonium or chloronium ion intermediate just as in the laboratory, and reaction with another halide ion completes the process. Halomon, for example, an anticancer pentahalide isolated from red alga, is thought to arise by a route that involves twofold addition of BrCl through the corresponding bromonium ions.

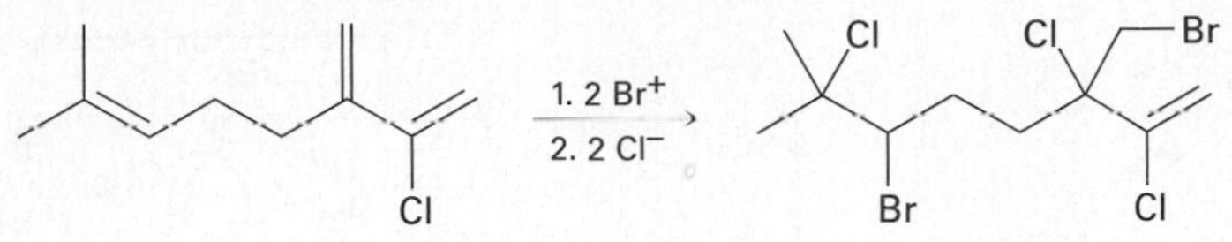

1. 2 Br^+
2. 2 Cl^-

Halomon

Problem 8.3
What product would you expect to obtain from addition of Cl_2 to 1,2-dimethylcyclohexene? Show the stereochemistry of the product.

Problem 8.4
Addition of HCl to 1,2-dimethylcyclohexene yields a mixture of two products. Show the stereochemistry of each, and explain why a mixture is formed.

8.3 Halohydrins from Alkenes: Addition of HOX

Another example of an electrophilic addition is the reaction of alkenes with the hypohalous acids HO–Cl or HO–Br to yield 1,2-halo alcohols, called **halohydrins**. Halohydrin formation doesn't take place by direct reaction of an alkene with HOBr or HOCl, however. Rather, the addition is done indirectly by reaction of the alkene with either Br_2 or Cl_2 in the presence of water.

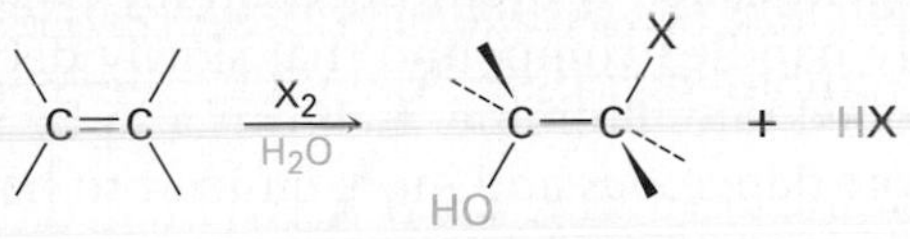

X_2, H_2O; + HX

An alkene   A halohydrin

We saw in the previous section that when Br_2 reacts with an alkene, the cyclic bromonium ion intermediate reacts with the only nucleophile present, Br^- ion. If the reaction is carried out in the presence of an additional nucleophile, however, the intermediate bromonium ion can be intercepted by the added nucleophile and diverted to a different product. In the presence of a high concentration of water, for instance, water competes with Br^- ion as nucleophile and reacts with the bromonium ion intermediate to yield a bromohydrin. The net effect is addition of HO–Br to the alkene by the pathway shown in **Figure 8.1**.

Figure 8.1 | MECHANISM

Bromohydrin formation by reaction of an alkene with Br_2 in the presence of water. Water acts as a nucleophile in step 2 to react with the intermediate bromonium ion.

1 Reaction of the alkene with Br_2 yields a bromonium ion intermediate, as previously discussed.

2 Water acts as a nucleophile, using a lone pair of electrons to open the bromonium ion ring and form a bond to carbon. Since oxygen donates its electrons in this step, it now has the positive charge.

3 Loss of a proton (H^+) from oxygen then gives H_3O^+ and the neutral bromohydrin addition product.

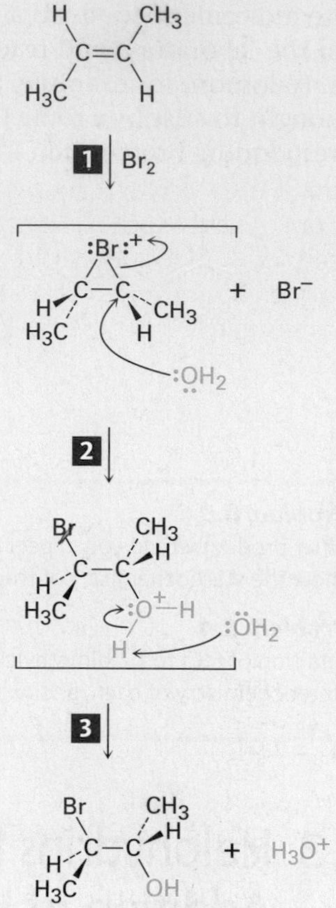

3-Bromo-2-butanol (a bromohydrin)

In practice, few alkenes are soluble in water, and bromohydrin formation is often carried out in a solvent such as aqueous dimethyl sulfoxide, CH_3SOCH_3 (DMSO), using a reagent called *N*-bromosuccinimide (NBS) as a source of Br_2. NBS is a stable, easily handled compound that slowly decomposes in water to yield Br_2 at a controlled rate. Bromine itself can also be used in the addition reaction, but it is more dangerous and more difficult to handle than NBS.

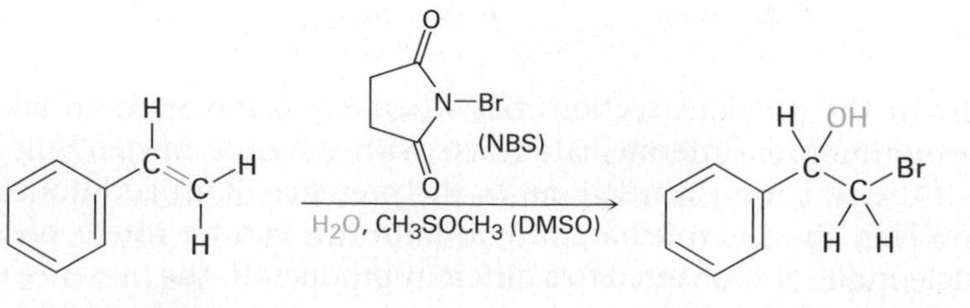

(NBS), H_2O, CH_3SOCH_3 (DMSO)

Styrene → **2-Bromo-1-phenylethanol (70%)**

Note that the aromatic ring in the above example does not react with Br_2 under the conditions used, even though it appears to contain three carbon–carbon double bonds. As we'll see in Chapter 15, aromatic rings are a good deal more stable and less reactive than might be expected.

There are a number of biological examples of halohydrin formation, particularly in marine organisms. As with halogenation (**Section 8.2**), halohydrin formation is carried out by haloperoxidases, which function by oxidizing Br^- or Cl^- ions to the corresponding HOBr or HOCl bonded to a metal atom in the enzyme. Electrophilic addition to the double bond of a substrate molecule then yields a bromonium or chloronium ion intermediate, and reaction with water gives the halohydrin. For example:

$$C_6H_5CH{=}CHCH_2OH \xrightarrow[\text{Bromoperoxidase}]{H_2O_2,\ Br^-,\ pH = 3} C_6H_5CH(OH)CH(Br)CH_2OH$$

Problem 8.5

What product would you expect from the reaction of cyclopentene with NBS and water? Show the stereochemistry.

Problem 8.6

When an unsymmetrical alkene such as propene is treated with *N*-bromosuccinimide in aqueous dimethyl sulfoxide, the major product has the bromine atom bonded to the less highly substituted carbon atom. Is this Markovnikov or non-Markovnikov orientation? Explain.

$$CH_3CH{=}CH_2 \xrightarrow{Br_2,\ H_2O} CH_3CH(OH)CH_2Br$$

8.4 Hydration of Alkenes: Addition of H_2O by Oxymercuration

Water adds to alkenes to yield alcohols, a process called *hydration*. The reaction takes place on treatment of the alkene with water and a strong acid catalyst, such as H_2SO_4, by a mechanism similar to that of HX addition. Thus, as shown in **Figure 8.2**, protonation of an alkene double bond yields a carbocation intermediate, which reacts with water to yield a protonated alcohol product, ROH_2^+. Loss of H^+ from this protonated alcohol gives the neutral alcohol and regenerates the acid catalyst.

Figure 8.2 | MECHANISM

Mechanism of the acid-catalyzed hydration of an alkene to yield an alcohol. Protonation of the alkene gives a carbocation intermediate, which reacts with water. The initial product is then deprotonated.

1 A hydrogen atom on the electrophile H_3O^+ is attacked by π electrons from the nucleophilic double bond, forming a new C–H bond. This leaves the other carbon atom with a + charge and a vacant *p* orbital. Simultaneously, two electrons from the H–O bond move onto oxygen, giving neutral water.

2 The nucleophile H_2O donates an electron pair to the positively charged carbon atom, forming a C–O bond and leaving a positive charge on oxygen in the protonated alcohol addition product.

3 Water acts as a base to remove H^+, regenerating H_3O^+ and yielding the neutral alcohol addition product.

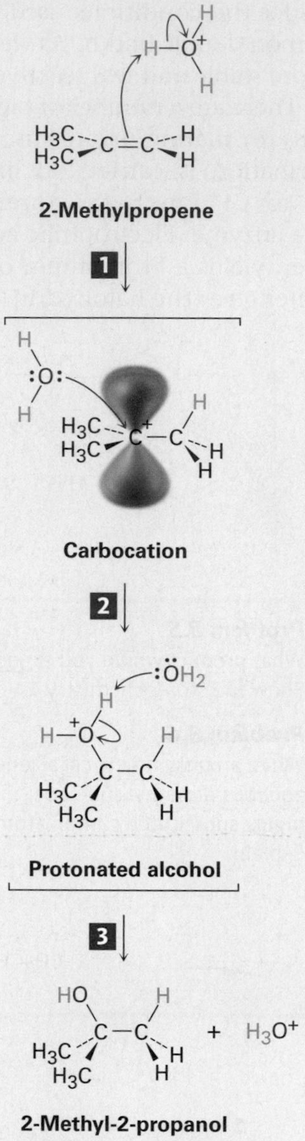

Acid-catalyzed alkene hydration is particularly suited to large-scale industrial procedures, and approximately 300,000 tons of ethanol is manufactured each year in the United States by hydration of ethylene. The reaction is of little value in the typical laboratory, however, because it requires high temperatures—250 °C in the case of ethylene—and strongly acidic conditions.

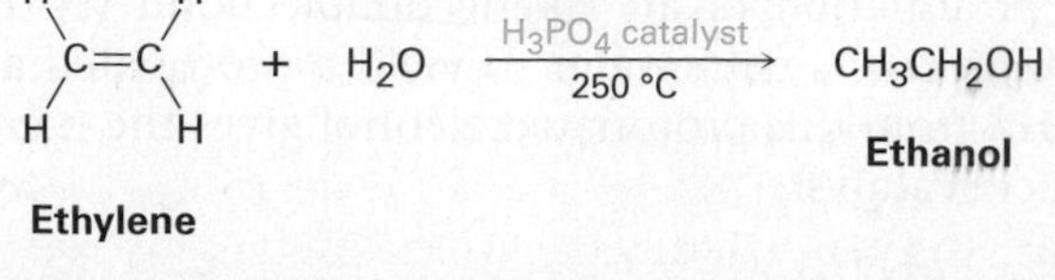

$$H_2C{=}CH_2 + H_2O \xrightarrow[250\ °C]{H_3PO_4\ \text{catalyst}} CH_3CH_2OH$$

Ethylene → **Ethanol**

Acid-catalyzed hydration of isolated double bonds, although known, is also uncommon in biological pathways. More frequently, biological hydrations require that the double bond be adjacent to a carbonyl group for reaction to proceed. Fumarate, for instance, is hydrated to give malate as one step in the citric acid cycle of food metabolism. Note that the requirement for an adjacent carbonyl group in the addition of water is the same as that we saw in **Section 8.1** for the elimination of water. We'll see the reason for the requirement in **Section 19.13**, but might note for now that the reaction is not an electrophilic addition but instead occurs through a mechanism that involves formation of an anion intermediate followed by protonation by an acid HA.

H_2O, pH = 7.4, Fumarase; HA

Fumarate **Anion intermediate** **Malate**

When it comes to circumventing problems like those with acid-catalyzed alkene hydrations, laboratory chemists have a great advantage over the cellular "chemists" in living organisms. Laboratory chemists are not constrained to carry out their reactions in water solution; they can choose from any of a large number of solvents. Laboratory reactions don't need to be carried out at a fixed temperature; they can take place over a wide range of temperatures. And laboratory reagents aren't limited to containing carbon, oxygen, nitrogen, and a few other elements; they can contain any element in the periodic table.

In the laboratory, alkenes are often hydrated by the **oxymercuration–demercuration** procedure. Oxymercuration involves electrophilic addition of Hg^{2+} to the alkene on reaction with mercury(II) acetate [$(CH_3CO_2)_2Hg$, often abbreviated $Hg(OAc)_2$] in aqueous tetrahydrofuran (THF) solvent. When the intermediate organomercury compound is then treated with sodium borohydride, $NaBH_4$, demercuration occurs to produce an alcohol. For example:

1. $Hg(OAc)_2$, H_2O/THF
2. $NaBH_4$

1-Methylcyclopentene **1-Methylcyclopentanol (92%)**

Alkene oxymercuration is closely analogous to halohydrin formation. The reaction is initiated by electrophilic addition of Hg^{2+} (mercuric) ion to the alkene to give an intermediate *mercurinium ion,* whose structure resembles that of a bromonium ion **(Figure 8.3)**. Nucleophilic addition of water as in halohydrin formation, followed by loss of a proton, then yields a stable organomercury product. The final step, demercuration of the organomercury compound by reaction with sodium borohydride, is complex and involves radicals. Note that the regiochemistry of the reaction corresponds to Markovnikov addition of water; that is, the –OH group attaches to the more highly substituted carbon atom, and the –H attaches to the less highly substituted carbon. The hydrogen that replaces mercury in the demercuration step can attach from either side of the molecule depending on the exact circumstances.

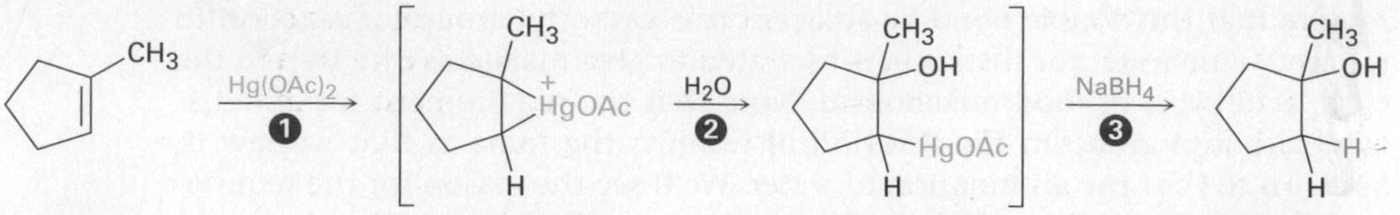

Figure 8.3 Mechanism of the oxymercuration of an alkene to yield an alcohol. (1) Electrophilic addition of Hg^{2+} gives a mercurinium ion, which (2) reacts with water as in halohydrin formation. Loss of a proton gives an organomercury product, and (3) reaction with $NaBH_4$ removes the mercury. The product of the reaction is the more highly substituted alcohol, corresponding to Markovnikov regiochemistry.

Problem 8.7

What products would you expect from oxymercuration–demercuration of the following alkenes?

(a) $CH_3CH_2CH_2CH{=}CH_2$ **(b)** $CH_3C(CH_3){=}CHCH_2CH_3$

Problem 8.8

From what alkenes might the following alcohols have been prepared?

(a) $CH_3C(OH)(CH_3)CH_2CH_2CH_2CH_3$ **(b)** (structure: cyclohexyl–CH(OH)–CH3)

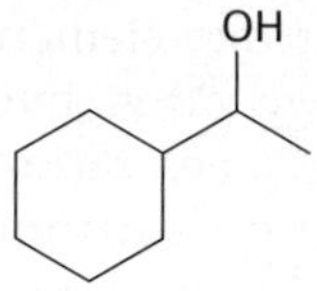

8.5 Hydration of Alkenes: Addition of H_2O by Hydroboration

In addition to the oxymercuration–demercuration method, which yields the Markovnikov product, a complementary method that yields the non-Markovnikov product is also useful. Discovered in 1959 by H.C. Brown and called **hydroboration**, the reaction involves addition of a B–H bond of borane, BH_3, to an alkene to yield an organoborane intermediate, RBH_2. Oxidation of the organoborane by reaction with basic hydrogen peroxide, H_2O_2, then gives an alcohol. For example:

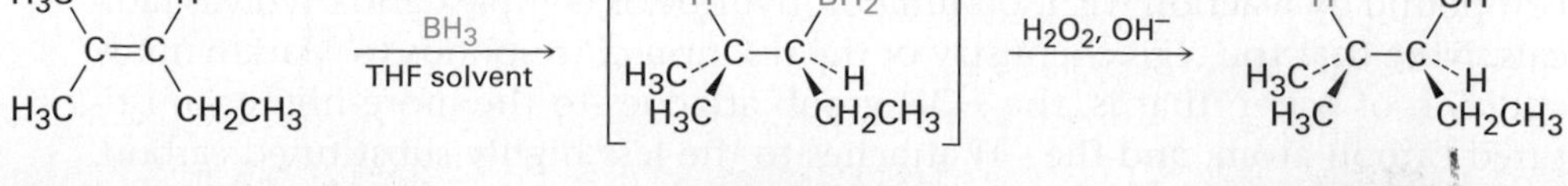

Borane is very reactive as a Lewis acid because the boron atom has only six electrons in its valence shell. In tetrahydrofuran solution, BH_3 accepts an electron pair from a solvent molecule in a Lewis acid–base reaction to complete its octet and form a stable BH_3–THF complex.

Electrophilic

Borane THF BH_3–THF complex

When an alkene reacts with BH_3 in THF solution, rapid addition to the double bond occurs three times and a trialkylborane, R_3B, is formed. For example, 1 molar equivalent of BH_3 adds to 3 molar equivalents of cyclohexene to yield tricyclohexylborane. When tricyclohexylborane is then treated with aqueous hydrogen H_2O_2 in basic solution, an oxidation takes place. The three C–B bonds are broken, –OH groups bond to the three carbons, and 3 equivalents of cyclohexanol are produced. The net effect of the two-step hydroboration–oxidation sequence is hydration of the alkene double bond.

3 Cyclohexene $\xrightarrow[\text{THF solvent}]{BH_3}$ Tricyclohexylborane $\xrightarrow[H_2O,\ NaOH]{H_2O_2}$ 3 Cyclohexanol (87%) + $B(OH)_3$

One of the features that makes the hydroboration reaction so useful is the regiochemistry that results when an unsymmetrical alkene is hydroborated. For example, hydroboration–oxidation of 1-methylcyclopentene yields *trans*-2-methylcyclopentanol. Boron and hydrogen add to the alkene from the same face of the double bond—that is, with **syn stereochemistry**, the opposite of anti—with boron attaching to the less highly substituted carbon. During the oxidation step, the boron is replaced by an –OH with the same stereochemistry, resulting in an overall syn non-Markovnikov addition of water. This stereochemical result is particularly useful because it is complementary to the Markovnikov regiochemistry observed for oxymercuration–demercuration.

1-Methylcyclopentene $\xrightarrow[\text{THF solvent}]{BH_3}$ [Organoborane Intermediate] $\xrightarrow{H_2O_2,\ OH^-}$ *trans*-2-Methylcyclopentanol (85% yield)

Why does alkene hydroboration take place with syn, non-Markovnikov regiochemistry to yield the less highly substituted alcohol? Hydroboration differs from many other alkene addition reactions in that it occurs in a single step without a carbocation intermediate **(Figure 8.4)**. Because both C–H and C–B bonds form at the same time and from the same face of the alkene, syn stereochemistry results. Non-Markovnikov regiochemistry occurs because attachment of boron is favored at the less sterically crowded carbon atom of the alkene rather than at the more crowded carbon.

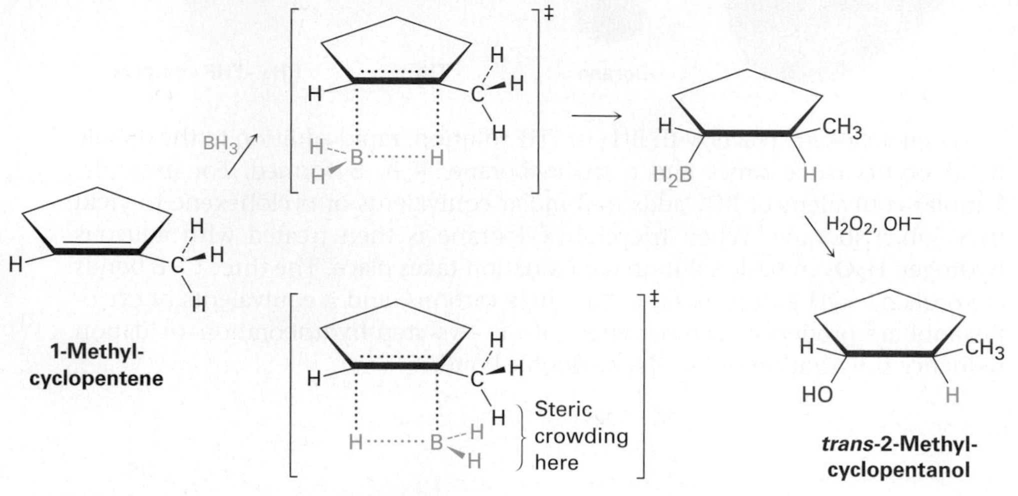

Figure 8.4 Mechanism of alkene hydroboration. The reaction occurs in a single step in which both C–H and C–B bonds form at the same time and on the same face of the double bond. The lower energy, more rapidly formed transition state is the one with less steric crowding, leading to non-Markovnikov regiochemistry.

Worked Example 8.1

Predicting the Products Formed in a Reaction

What products would you obtain from reaction of 2,4-dimethyl-2-pentene with:
(a) BH_3, followed by H_2O_2, OH^- **(b)** $Hg(OAc)_2$, followed by $NaBH_4$

Strategy

When predicting the product of a reaction, you have to recall what you know about the kind of reaction being carried out and then apply that knowledge to the specific case you're dealing with. In the present instance, recall that the two methods of hydration—hydroboration–oxidation and oxymercuration–demercuration—give complementary products. Hydroboration–oxidation occurs with syn stereochemistry and gives the non-Markovnikov addition product; oxymercuration–demercuration gives the Markovnikov product.

Solution

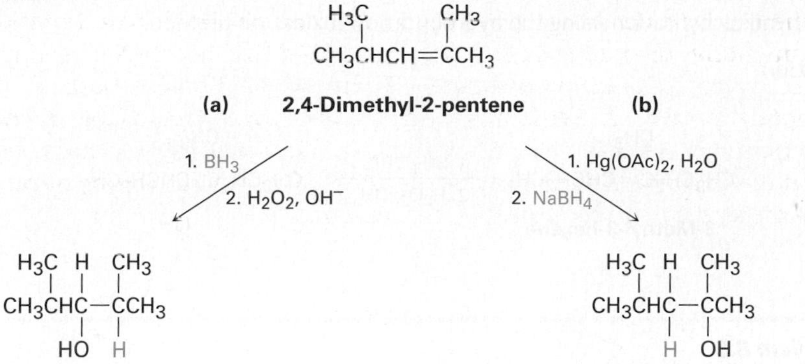

$$\underset{\textbf{2,4-Dimethyl-2-pentene}}{\overset{\displaystyle H_3C\quad\ CH_3}{CH_3CHCH{=}CCH_3}}$$

(a) 1. BH_3, 2. H_2O_2, OH^- → 2,4-Dimethyl-3-pentanol: $CH_3CH(CH_3)CH(OH)CH(CH_3)CH_3$ [drawn as H₃C H CH₃ / CH₃CHC—CCH₃ / HO H]

(b) 1. $Hg(OAc)_2$, H_2O, 2. $NaBH_4$ → 2,4-Dimethyl-2-pentanol: $CH_3CH(CH_3)CH_2C(OH)(CH_3)CH_3$ [drawn as H₃C H CH₃ / CH₃CHC—CCH₃ / H OH]

2,4-Dimethyl-3-pentanol **2,4-Dimethyl-2-pentanol**

Synthesizing an Alcohol

Worked Example 8.2

How might you prepare the following alcohol?

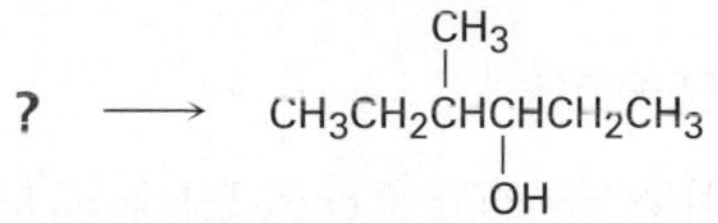

$$? \longrightarrow CH_3CH_2CH(CH_3)CH(OH)CH_2CH_3$$

Strategy

Problems that require the synthesis of a specific target molecule should always be worked backward. Look at the target, identify its functional group(s), and ask yourself "What are the methods for preparing that functional group?" In the present instance, the target molecule is a secondary alcohol (R_2CHOH), and we've seen that alcohols can be prepared from alkenes by either hydroboration–oxidation or oxymercuration. The –OH bearing carbon in the product must have been a double-bond carbon in the alkene reactant, so there are two possibilities: 4-methyl-2-hexene and 3-methyl-3-hexene.

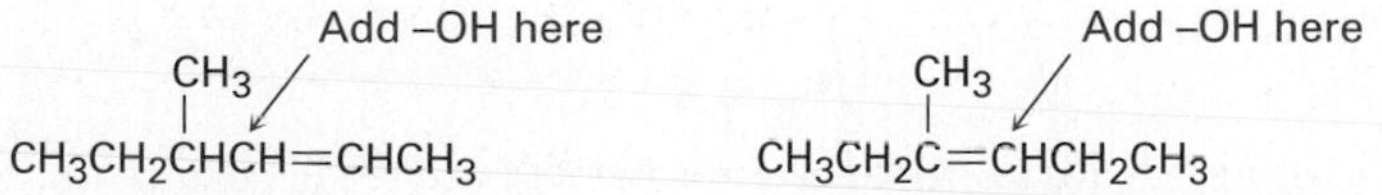

$$CH_3CH_2CH(CH_3)CH{=}CHCH_3 \qquad CH_3CH_2C(CH_3){=}CHCH_2CH_3$$

Add –OH here (arrow to C3 in each)

4-Methyl-2-hexene **3-Methyl-3-hexene**

4-Methyl-2-hexene has a disubstituted double bond, RCH=CHR′, and will probably give a mixture of two alcohols with either hydration method since Markovnikov's rule does not apply to symmetrically substituted alkenes. 3-Methyl-3-hexene, however, has a trisubstituted double bond, and should give only the desired product on non-Markovnikov hydration using the hydroboration–oxidation method.

Solution

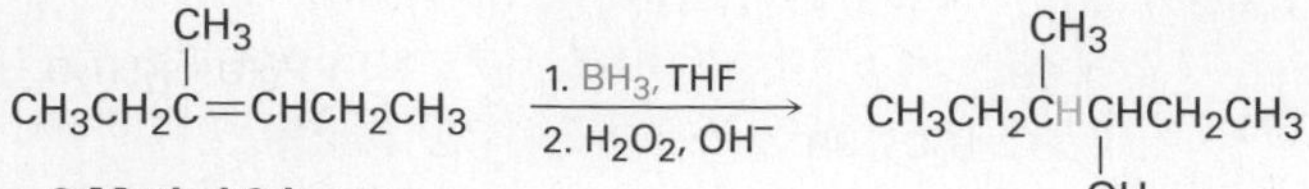

$$\underset{\textbf{3-Methyl-3-hexene}}{CH_3CH_2\overset{\overset{CH_3}{|}}{C}{=}CHCH_2CH_3} \xrightarrow[\text{2. } H_2O_2,\ OH^-]{\text{1. } BH_3,\ THF} CH_3CH_2\overset{\overset{CH_3}{|}}{C}H\underset{\underset{OH}{|}}{C}HCH_2CH_3$$

Problem 8.9

Show the structures of the products you would obtain by hydroboration–oxidation of the following alkenes:

(a) 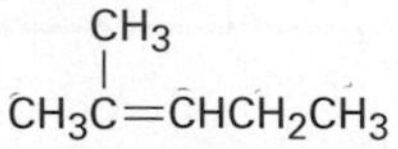 $CH_3\overset{\overset{CH_3}{|}}{C}{=}CHCH_2CH_3$

(b) 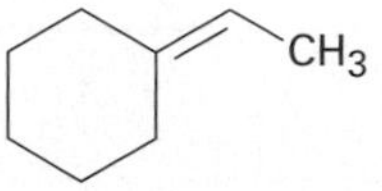 (cyclohexylidene)–CH_3

Problem 8.10

What alkenes might be used to prepare the following alcohols by hydroboration–oxidation?

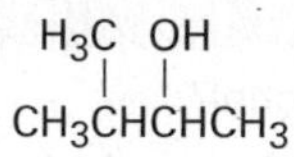

(a) 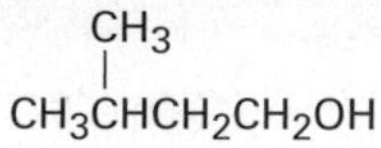 $CH_3\overset{\overset{CH_3}{|}}{C}HCH_2CH_2OH$

(b) $CH_3\overset{\overset{H_3C}{|}}{C}H\overset{\overset{OH}{|}}{C}HCH_3$

(c) (cyclohexyl)–CH_2OH

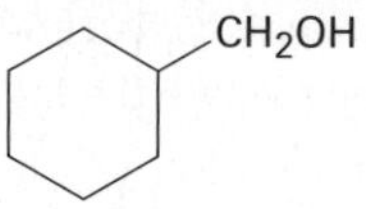

Problem 8.11

The following cycloalkene gives a mixture of two alcohols on hydroboration followed by oxidation. Draw the structures of both, and explain the result.

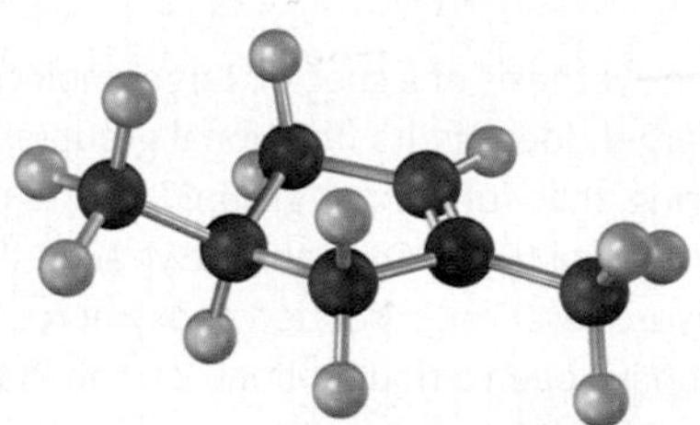

8.6 Reduction of Alkenes: Hydrogenation

Alkenes react with H_2 in the presence of a metal catalyst such as palladium or platinum to yield the corresponding saturated alkane addition products. We describe the result by saying that the double bond has been **hydrogenated**, or *reduced*. Note that the word *reduction* is used somewhat differently in organic chemistry from what you might have learned previously. In general chemistry, a reduction is defined as the gain of one or more electrons by an atom. In organic chemistry, however, a **reduction** is a reaction that results in a gain of electron density by carbon, caused either by bond formation between carbon and a less electronegative atom—usually hydrogen—or by bond-breaking between carbon and a more electronegative atom—usually oxygen, nitrogen, or a halogen. We'll explore the topic in more detail in **Section 10.8**.

Reduction Increases electron density on carbon by:
- forming this: C–H
- or breaking one of these: C–O C–N C–X

A reduction:

$$\text{C=C} + H_2 \xrightarrow{\text{Catalyst}} \text{H}_3\text{C–CH}_3\text{ (H–C–C–H)}$$

An alkene **An alkane**

Platinum and palladium are the most common laboratory catalysts for alkene hydrogenations. Palladium is normally used as a very fine powder "supported" on an inert material such as charcoal (Pd/C) to maximize surface area. Platinum is normally used as PtO_2, a reagent known as *Adams' catalyst* after its discoverer, Roger Adams.

Catalytic hydrogenation, unlike most other organic reactions, is a *heterogeneous* process rather than a homogeneous one. That is, the hydrogenation reaction does not occur in a homogeneous solution but instead takes place on the surface of solid catalyst particles. Hydrogenation usually occurs with syn stereochemistry: both hydrogens add to the double bond from the same face.

$$\xrightarrow[\text{CH}_3\text{CO}_2\text{H solvent}]{\text{H}_2\text{, PtO}_2}$$

1,2-Dimethylcyclohexene **cis-1,2-Dimethylcyclohexane (82%)**

As shown in **Figure 8.5**, hydrogenation begins with adsorption of H_2 onto the catalyst surface. Complexation between catalyst and alkene then occurs as a vacant orbital on the metal interacts with the filled alkene π orbital. In the final steps, hydrogen is inserted into the double bond and the saturated product diffuses away from the catalyst. The stereochemistry of hydrogenation is syn because both hydrogens add to the double bond from the same catalyst surface.

Figure 8.5 | MECHANISM

Mechanism of alkene hydrogenation. The reaction takes place with syn stereochemistry on the surface of insoluble catalyst particles.

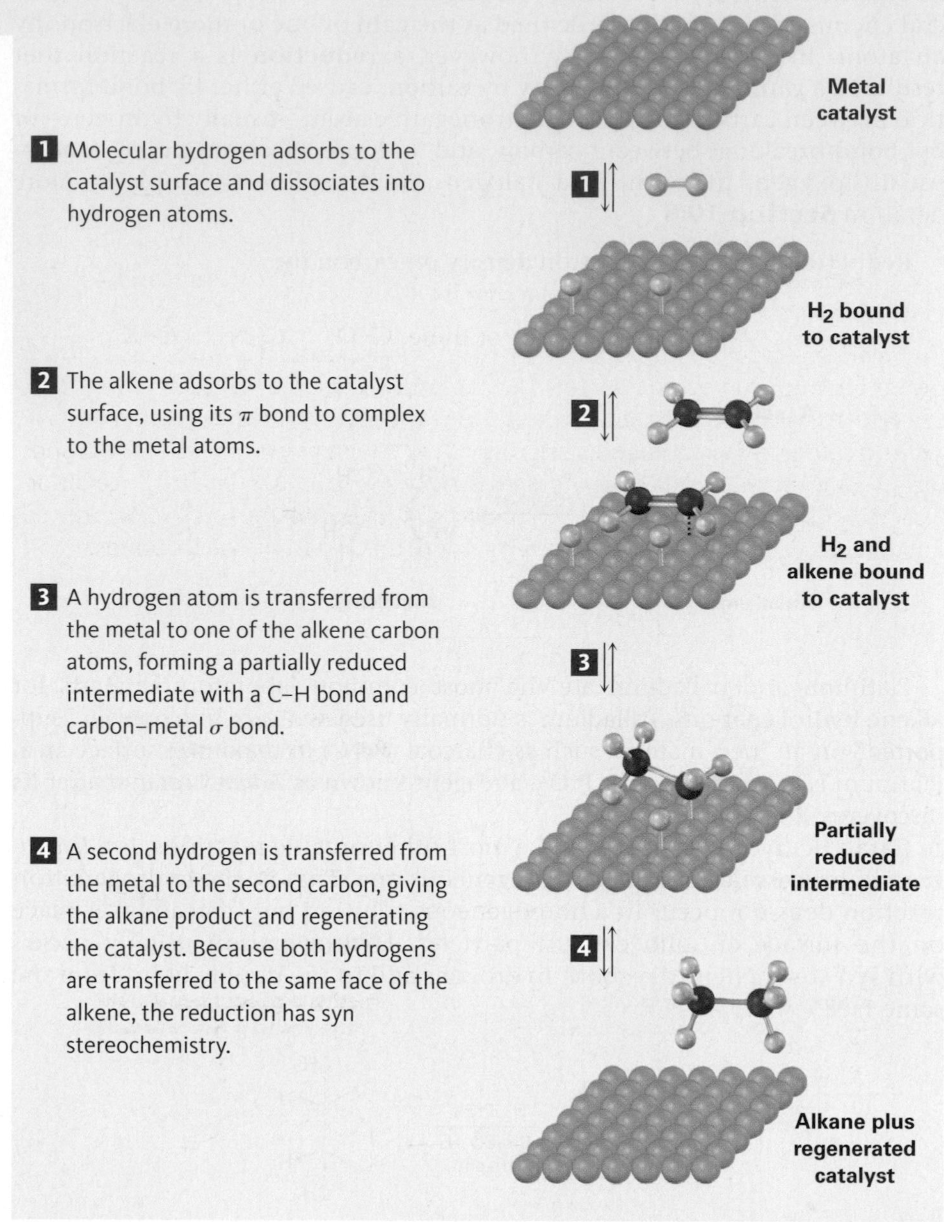

An interesting feature of catalytic hydrogenation is that the reaction is extremely sensitive to the steric environment around the double bond. As a result, the catalyst usually approaches only the more accessible face of an alkene, giving rise to a single product. In α-pinene, for example, one of the methyl groups attached to the four-membered ring hangs over the top face of the double bond and blocks approach of the hydrogenation catalyst from that side. Reduction therefore occurs exclusively from the bottom face to yield the product shown.

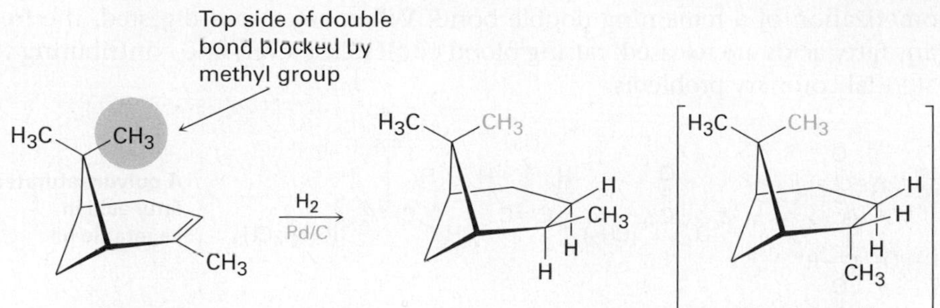

Top side of double bond blocked by methyl group

H_3C CH_3 CH_3 $\xrightarrow[\text{Pd/C}]{H_2}$ H_3C CH_3 H CH_3 H H [H_3C CH_3 H H H CH_3]

α-Pinene

(Not formed)

Alkenes are much more reactive than most other unsaturated functional groups toward catalytic hydrogenation, and the reaction is therefore quite selective. Other functional groups, such as aldehydes, ketones, esters, and nitriles, often survive alkene hydrogenation conditions unchanged, although reaction with these groups does occur under more vigorous conditions. Note particularly in the hydrogenation of methyl 3-phenylpropenoate shown below that the aromatic ring is not reduced by hydrogen and palladium even though it contains apparent double bonds.

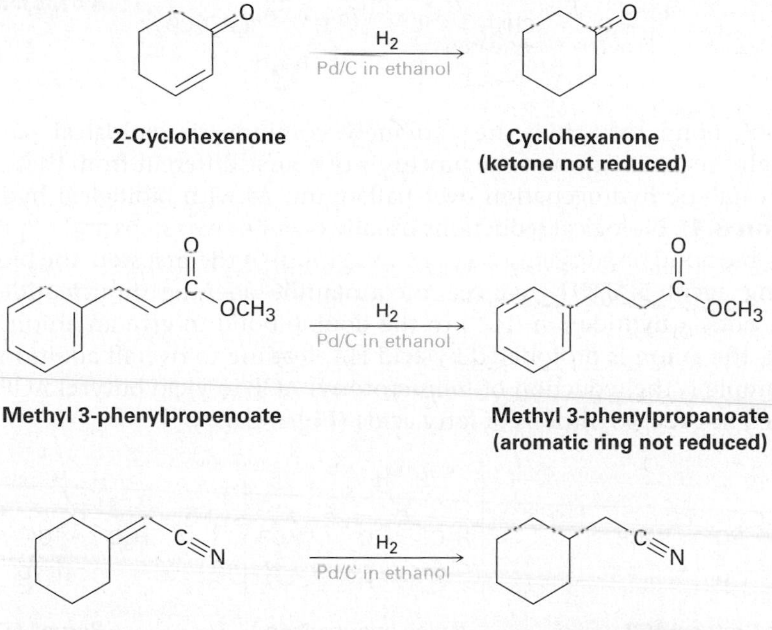

$\xrightarrow[\text{Pd/C in ethanol}]{H_2}$

2-Cyclohexenone → **Cyclohexanone (ketone not reduced)**

OCH_3 $\xrightarrow[\text{Pd/C in ethanol}]{H_2}$ OCH_3

Methyl 3-phenylpropenoate → **Methyl 3-phenylpropanoate (aromatic ring not reduced)**

C≡N $\xrightarrow[\text{Pd/C in ethanol}]{H_2}$ C≡N

Cyclohexylideneacetonitrile → **Cyclohexylacetonitrile (nitrile not reduced)**

In addition to its usefulness in the laboratory, catalytic hydrogenation is also important in the food industry, where unsaturated vegetable oils are reduced on a large scale to produce the saturated fats used in margarine and cooking products **(Figure 8.6)**. As we'll see in **Section 27.1**, vegetable oils are triesters of glycerol, $HOCH_2CH(OH)CH_2OH$, with three long-chain carboxylic acids called *fatty acids*. The fatty acids are generally polyunsaturated, and their double bonds have cis stereochemistry. Complete hydrogenation yields the corresponding saturated fatty acids, but incomplete hydrogenation often results in partial cis–trans isomerization of a remaining double bond. When eaten and digested, the free trans fatty acids are released, raising blood cholesterol levels and contributing to potential coronary problems.

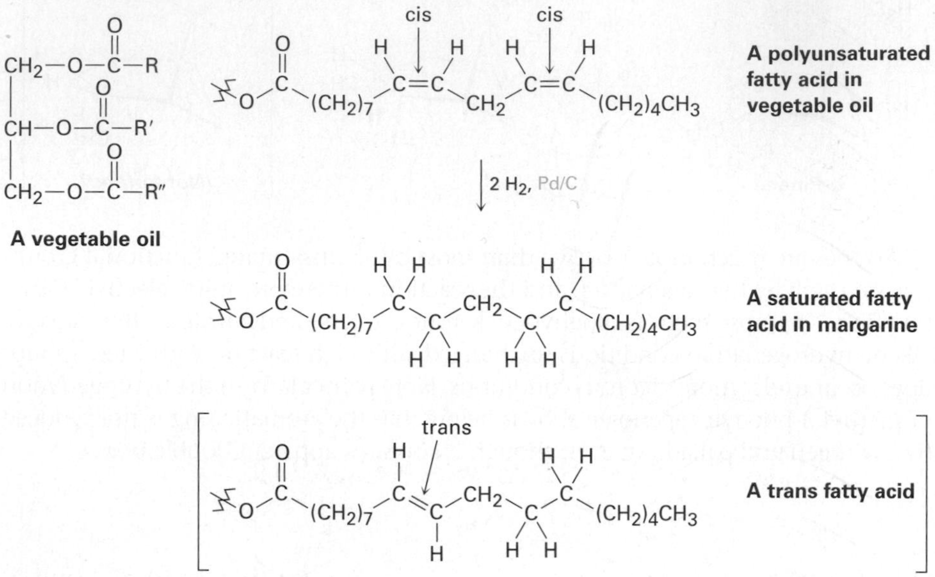

Figure 8.6 Catalytic hydrogenation of polyunsaturated fats leads to saturated products, along with a small amount of isomerized trans fats.

A vegetable oil

A polyunsaturated fatty acid in vegetable oil

$2\ H_2$, Pd/C

A saturated fatty acid in margarine

A trans fatty acid

Double-bond reductions are extremely common in biological pathways, although the mechanism of the process is of course different from that of laboratory catalytic hydrogenation over palladium. As with biological hydrations **(Section 8.4)**, biological reductions usually occur in two steps and require that the double bond be adjacent to a carbonyl group. In the first step, the biological reducing agent NADPH (reduced nicotinamide adenine dinucleotide phosphate), adds a hydride ion ($H:^-$) to the double bond to give an anion. In the second, the anion is protonated by acid HA, leading to overall addition of H_2. An example is the reduction of *trans*-crotonyl ACP to yield butyryl ACP, a step involved in the biosynthesis of fatty acids **(Figure 8.7)**.

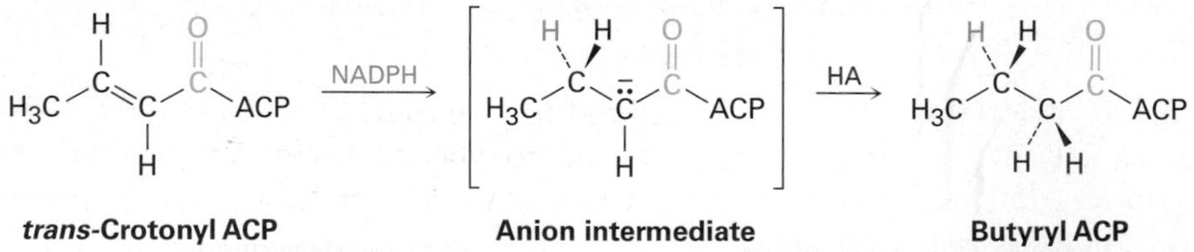

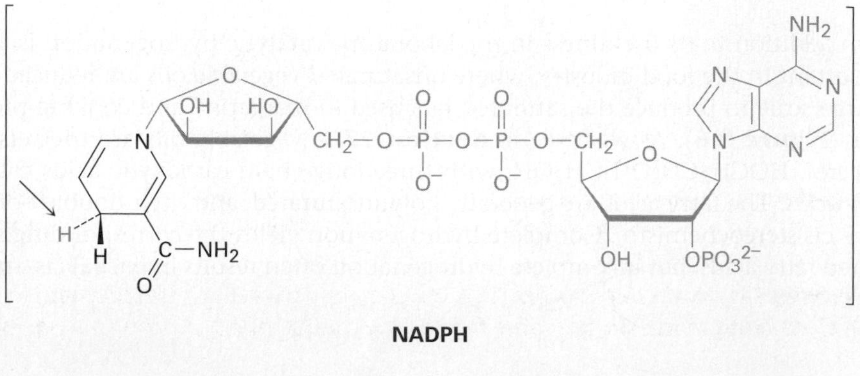

Figure 8.7 Reduction of the carbon–carbon double bond in *trans*-crotonyl ACP, a step in the biosynthesis of fatty acids. **One hydrogen** is delivered from NADPH as a hydride ion, $H:^-$; the **other hydrogen** is delivered by protonation of the anion intermediate with an acid, HA.

trans-Crotonyl ACP → (NADPH) → Anion intermediate → (HA) → Butyryl ACP

NADPH

Problem 8.12

What product would you obtain from catalytic hydrogenation of the following alkenes?

(a) $CH_3C(CH_3){=}CHCH_2CH_3$

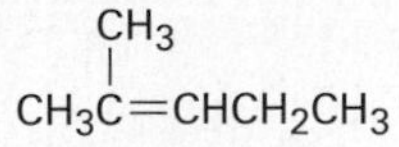

(b) [3,3-dimethylcyclopentene structure: CH_3, CH_3]

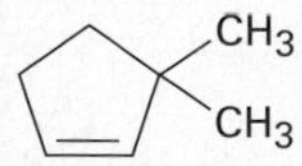

(c) [cyclohexene bearing $C(CH_3)_3$: H_3C, CH_3, C, CH_3 = H, CH_3, C, CH_3, CH_3]

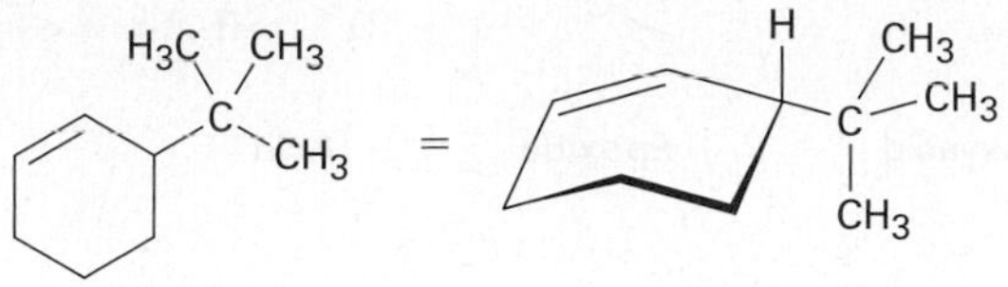

8.7 Oxidation of Alkenes: Epoxidation and Hydroxylation

Like the word *reduction* used in the previous section for the addition of hydrogen to a double bond, the word *oxidation* has a slightly different meaning in organic chemistry from what you might have previously learned. In general chemistry, an oxidation is defined as the loss of one or more electrons by an atom. In organic chemistry, however, an **oxidation** is a reaction that results in a loss of electron density by carbon, caused either by bond formation between carbon and a more electronegative atom—usually oxygen, nitrogen, or a halogen—or by bond-breaking between carbon and a less electronegative atom—usually hydrogen. Note that an *oxidation* often adds oxygen, while a *reduction* often adds hydrogen.

Oxidation Decreases electron density on carbon by:
- forming one of these: C–O C–N C–X
- or breaking this: C–H

In the laboratory, alkenes are oxidized to give *epoxides* on treatment with a peroxyacid, RCO_3H, such as *meta*-chloroperoxybenzoic acid. An **epoxide**, also called an *oxirane,* is a cyclic ether with an oxygen atom in a three-membered ring. For example:

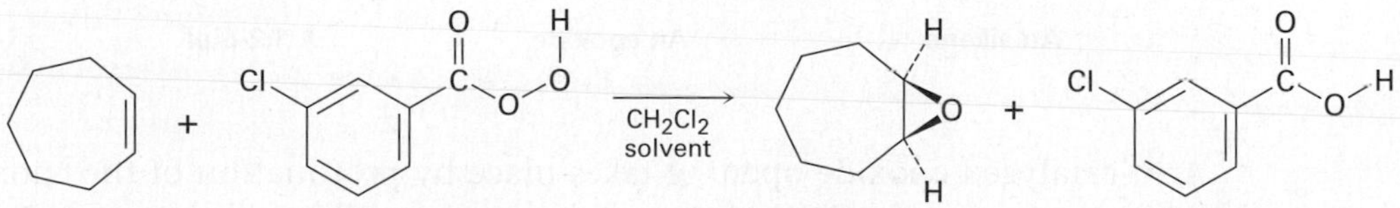

Cycloheptene + *meta*-Chloroperoxybenzoic acid $\xrightarrow[\text{solvent}]{CH_2Cl_2}$ 1,2-Epoxycycloheptane + *meta*-Chlorobenzoic acid

Peroxyacids transfer an oxygen atom to the alkene with syn stereochemistry—both C–O bonds form on the same face of the double bond—through a one-step mechanism without intermediates. The oxygen atom farthest from the carbonyl group is the one transferred.

Alkene Peroxyacid Epoxide Acid

Another method for the synthesis of epoxides is through the use of halohydrins, prepared by electrophilic addition of HO–X to alkenes **(Section 8.3)**. When a halohydrin is treated with base, HX is eliminated and an epoxide is produced.

Cyclohexene → (Cl_2, H_2O) → *trans*-2-Chlorocyclohexanol → (NaOH, H_2O) → 1,2-Epoxycyclohexane (73%) + H_2O + NaCl

Epoxides undergo an acid-catalyzed ring-opening reaction with water (a *hydrolysis*) to give the corresponding 1,2-dialcohol, or *diol,* also called a **glycol**. Thus, the net result of the two-step alkene epoxidation/hydrolysis is **hydroxylation**—the addition of an –OH group to each of the two double-bond carbons. In fact, approximately 18 million metric tons of ethylene glycol, $HOCH_2CH_2OH$, most of it used for automobile antifreeze, is produced worldwide each year by epoxidation of ethylene followed by hydrolysis.

An alkene → (Epoxidation) → An epoxide → (H_3O^+) → A 1,2-diol

Acid-catalyzed epoxide opening takes place by protonation of the epoxide to increase its reactivity, followed by nucleophilic addition of water. This nucleophilic addition is analogous to the final step of alkene bromination, in which a cyclic bromonium ion is opened by a nucleophile **(Section 8.2)**. That is, a *trans*-1,2-diol results when an epoxycycloalkane is opened by aqueous acid, just as a *trans*-1,2-dibromide results when a cycloalkene is brominated. We'll look at epoxide chemistry in more detail in **Section 18.6**.

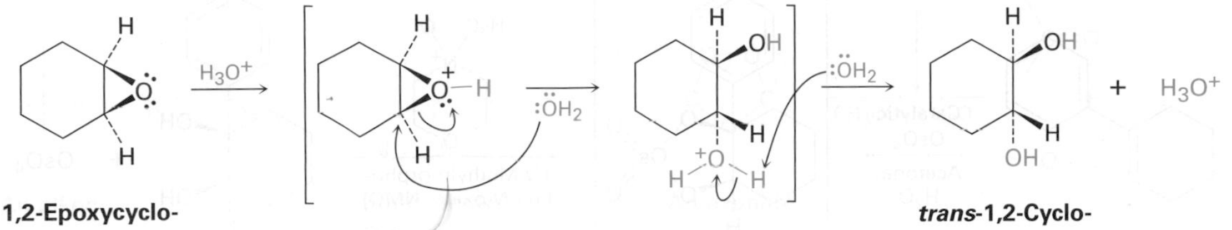

H H_3O^+ H :$\ddot{O}H_2$ H OH :$\ddot{O}H_2$ H OH + H_3O^+

1,2-Epoxycyclohexane

trans-1,2-Cyclohexanediol (86%)

Recall the following:

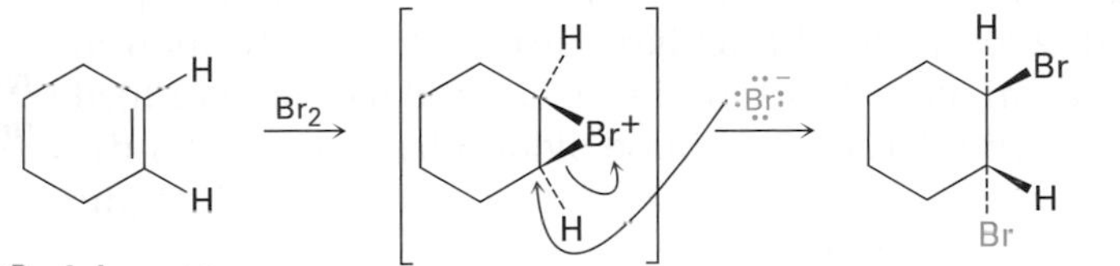

Br_2 :$\ddot{Br}$:$^-$

Cyclohexene

trans-1,2-Dibromocyclohexane

Hydroxylation can be carried out directly without going through an intermediate epoxide by treating an alkene with osmium tetroxide, OsO_4. The reaction occurs with syn stereochemistry and does not involve a carbocation intermediate. Instead, it takes place through an intermediate cyclic *osmate,* which is formed in a single step by addition of OsO_4 to the alkene. This cyclic osmate is then cleaved using aqueous sodium bisulfite, $NaHSO_3$.

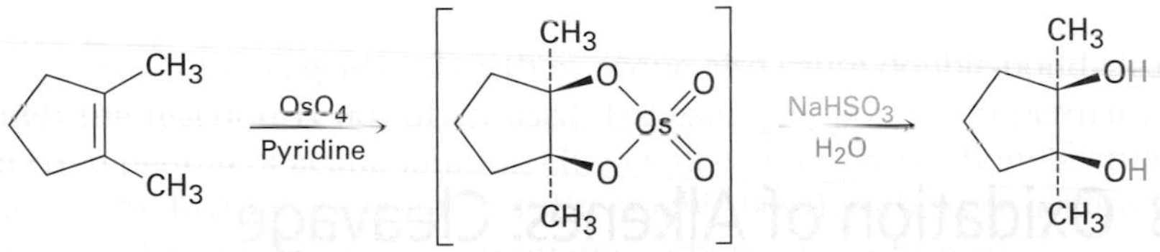

CH_3 OsO_4 Pyridine CH_3 $NaHSO_3$ H_2O CH_3 OH

1,2-Dimethylcyclopentene

A cyclic osmate intermediate

cis-1,2-Dimethyl-1,2-cyclopentanediol (87%)

Because OsO_4 is both very expensive and *very* toxic, the reaction is usually carried out using only a small, catalytic amount of OsO_4 in the presence of a stoichiometric amount of a safe and inexpensive co-oxidant such as *N*-methylmorpholine *N*-oxide, abbreviated NMO. The initially formed osmate intermediate reacts rapidly with NMO to yield the product diol plus compound is formed. As indicated in the following examples, the cleavage reaction takes place through a cyclic periodate intermediate.

A 1,2-diol —(HIO_4, H_2O, THF)→ [Cyclic periodate intermediate] → 6-Oxoheptanal (86%)

A 1,2-diol —(HIO_4, H_2O, THF)→ [Cyclic periodate intermediate] → 2 Cyclopentanone (81%)

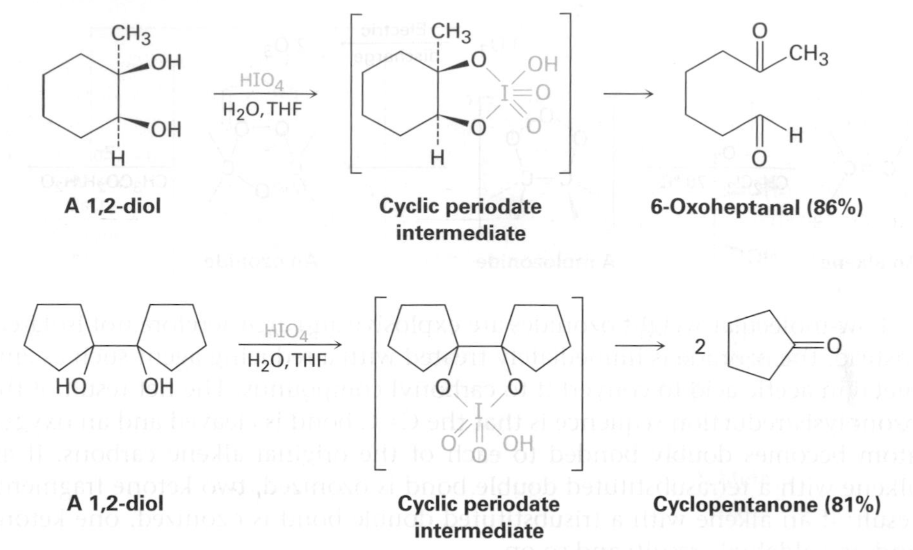

Worked Example 8.3 Predicting the Reactant in an Ozonolysis Reaction

What alkene would yield a mixture of cyclopentanone and propanal on treatment with ozone followed by reduction with zinc?

$$? \xrightarrow[\text{2. Zn, acetic acid}]{\text{1. } O_3} \text{cyclopentanone} + CH_3CH_2\overset{O}{\overset{\|}{C}}H$$

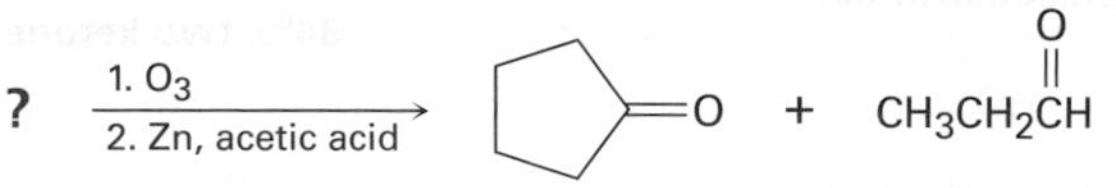

Strategy

Reaction of an alkene with ozone, followed by reduction with zinc, cleaves the C=C bond and gives two carbonyl-containing fragments. That is, the C=C bond becomes two C=O bonds. Working backward from the carbonyl-containing products, the alkene precursor can be found by removing the oxygen from each product and joining the two carbon atoms to form a double bond.

Solution

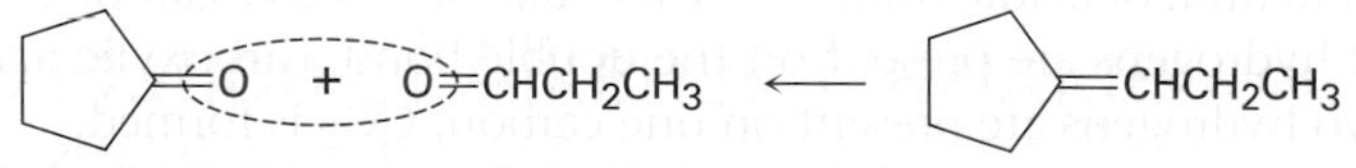

Cyclopentanone (C=O) + O=CHCH$_2$CH$_3$ ← cyclopentylidene=CHCH$_2$CH$_3$

Problem 8.15

What products would you expect from reaction of 1-methylcyclohexene with the following reagents?

(a) Aqueous acidic $KMnO_4$ **(b)** O_3, followed by Zn, CH_3CO_2H

Problem 8.16

Propose structures for alkenes that yield the following products on reaction with ozone followed by treatment with Zn:

(a) $(CH_3)_2C{=}O + H_2C{=}O$ **(b)** 2 equiv $CH_3CH_2CH{=}O$

8.9 Addition of Carbenes to Alkenes: Cyclopropane Synthesis

Yet another kind of alkene addition is the reaction with a *carbene* to yield a cyclopropane. A **carbene, R_2C:**, is a neutral molecule containing a divalent carbon with only six electrons in its valence shell. It is therefore highly reactive and is generated only as a reaction intermediate, rather than as an isolable molecule. Because they're electron-deficient, carbenes behave as electrophiles and react with nucleophilic C=C bonds. The reaction occurs in a single step without intermediates.

$$>C{=}C< \; + \; R_2C{:} \longrightarrow \text{cyclopropane}$$

An alkene **A carbene** **A cyclopropane**

One of the simplest methods for generating a substituted carbene is by treatment of chloroform, $CHCl_3$, with a strong base such as KOH. As shown in **Figure 8.8**, loss of a proton from $CHCl_3$ gives the trichloromethanide anion, $^{-}{:}CCl_3$, which spontaneously expels a Cl^- ion to yield dichlorocarbene, $:CCl_2$.

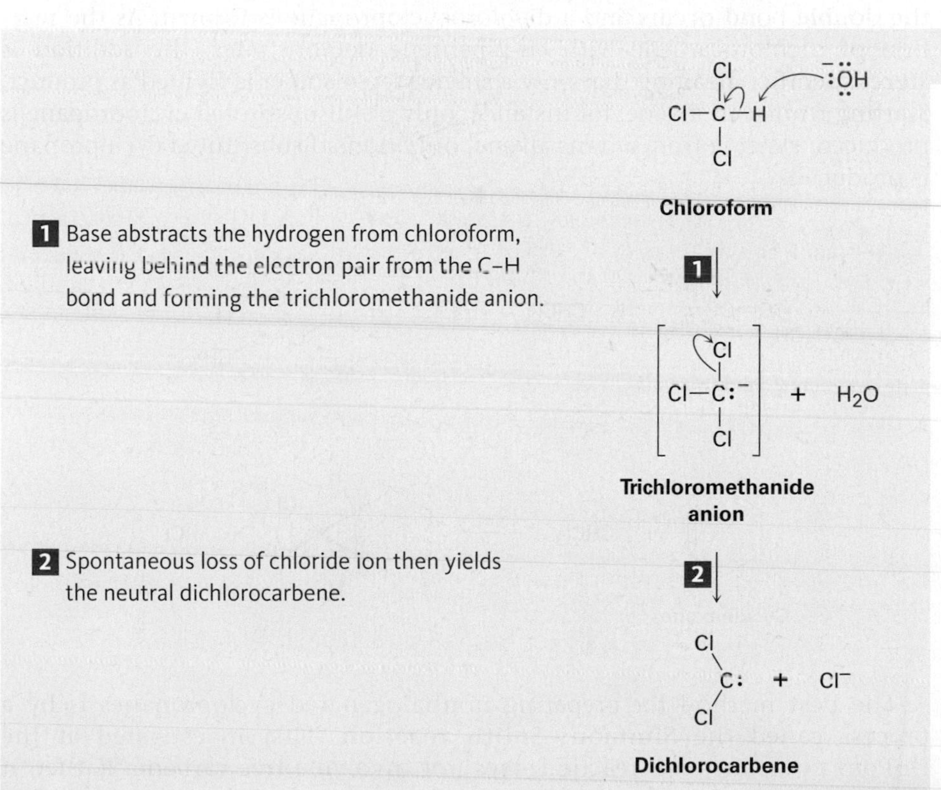

Figure 8.8 | MECHANISM

Mechanism of the formation of dichlorocarbene by reaction of chloroform with strong base. Deprotonation of $CHCl_3$ gives the trichloromethanide anion, $^{-}{:}CCl_3$, which spontaneously expels a Cl^- ion.

The dichlorocarbene carbon atom is sp^2-hybridized, with a vacant p orbital extending above and below the plane of the three atoms and with an unshared pair of electrons occupying the third sp^2 lobe. Note that this electronic description of dichlorocarbene is similar to that of a carbocation **(Section 7.9)** with respect to both the sp^2 hybridization of carbon and the vacant p orbital. Electrostatic potential maps further show this similarity **(Figure 8.9)**.

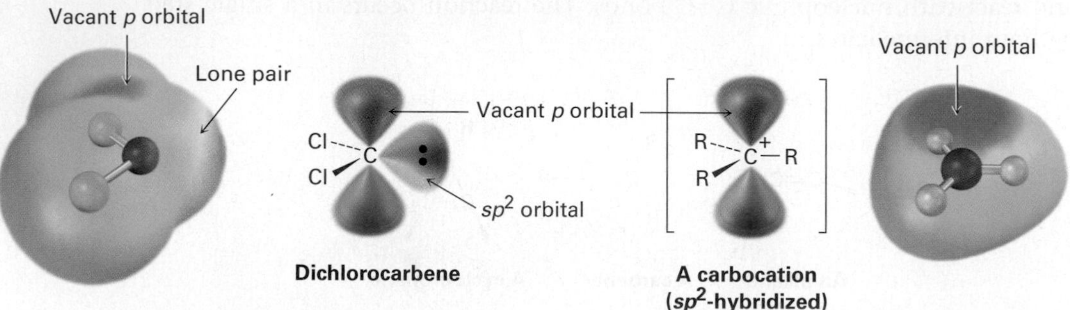

Figure 8.9 The structure of dichlorocarbene. Electrostatic potential maps show how the **positive region** coincides with the empty p orbital in both dichlorocarbene and a carbocation (CH_3^+). The **negative region** in the dichlorocarbene map coincides with the lone-pair electrons.

If dichlorocarbene is generated in the presence of an alkene, addition to the double bond occurs and a dichlorocyclopropane is formed. As the reaction of dichlorocarbene with *cis*-2-pentene demonstrates, the addition is **stereospecific**, meaning that only a single stereoisomer is formed as product. Starting from a cis alkene, for instance, only cis-disubstituted cyclopropane is produced; starting from a trans alkene, only trans-disubstituted cyclopropane is produced.

$$\text{cis-2-Pentene} + CHCl_3 \xrightarrow{KOH} \text{cis-1,1-dichloro-2-ethyl-3-methylcyclopropane} + KCl$$

cis-2-Pentene

$$\text{Cyclohexene} + CHCl_3 \xrightarrow{KOH} \text{7,7-dichlorobicyclo[4.1.0]heptane} + KCl$$

Cyclohexene

The best method for preparing nonhalogenated cyclopropanes is by a process called the **Simmons–Smith reaction**. First investigated at the DuPont company, this reaction does not involve a free carbene. Rather, it utilizes a *carbenoid*—a metal-complexed reagent with carbene-like reactivity. When diiodomethane is treated with a specially prepared zinc–copper mix, (iodomethyl)zinc iodide, ICH_2ZnI, is formed. In the presence of an alkene, (iodomethyl)zinc iodide transfers a CH_2 group to the double bond and yields the cyclopropane. For example, cyclohexene reacts cleanly and in good yield to give the corresponding cyclopropane. Although we won't discuss the mechanistic details, carbene addition to an alkene is one of a general class of reactions called *cycloadditions,* which we'll study more carefully in Chapter 30.

$$CH_2I_2 + Zn(Cu) \longrightarrow ICH_2{-}ZnI \quad [\text{":}CH_2\text{"}]$$

Diiodomethane **(Iodomethyl)zinc iodide (a carbenoid)**

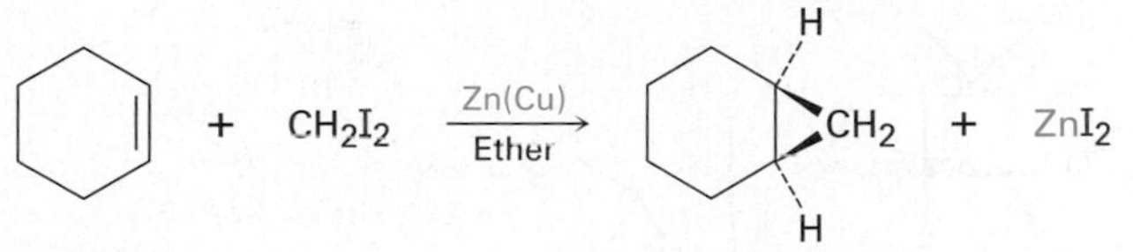

Cyclohexene + CH_2I_2 $\xrightarrow[\text{Ether}]{\text{Zn(Cu)}}$ Bicyclo[4.1.0]heptane (H, CH_2, H) + ZnI_2

Cyclohexene **Bicyclo[4.1.0]heptane (92%)**

Problem 8.17

What products would you expect from the following reactions?

(a) Methylenecyclohexane ($=CH_2$) + $CHCl_3$ $\xrightarrow{\text{KOH}}$?

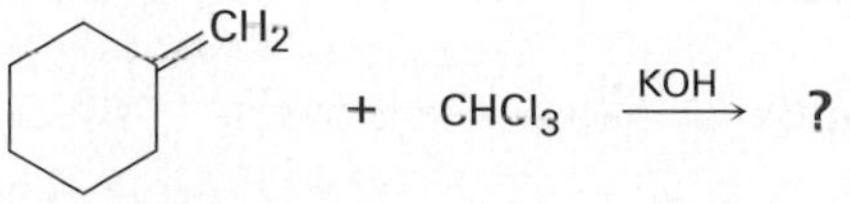

(b) $CH_3CH(CH_3)CH_2CH{=}CHCH_3$ + CH_2I_2 $\xrightarrow{\text{Zn(Cu)}}$?

8.10 Radical Additions to Alkenes: Chain-Growth Polymers

In our brief introduction to radical reactions in **Section 6.3**, we said that radicals can add to C=C bonds, taking one electron from the double bond and leaving one behind to yield a new radical. Let's now look at the process in more detail, focusing on the industrial synthesis of alkene polymers. A **polymer** is simply a large—sometimes *very* large—molecule built up by repetitive bonding together of many smaller molecules, called **monomers**.

Nature makes wide use of biological polymers. Cellulose, for instance, is a polymer built of repeating glucose monomer units; proteins are polymers built of repeating amino acid monomers; and nucleic acids are polymers built of repeating nucleotide monomers.

Cellulose—a glucose polymer

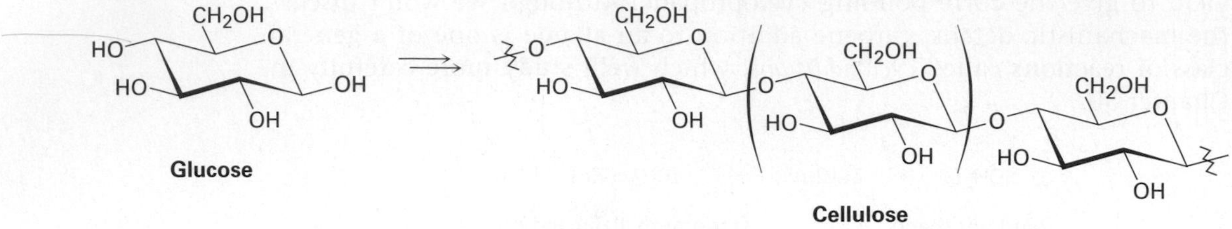

Glucose

Cellulose

Protein—an amino acid polymer

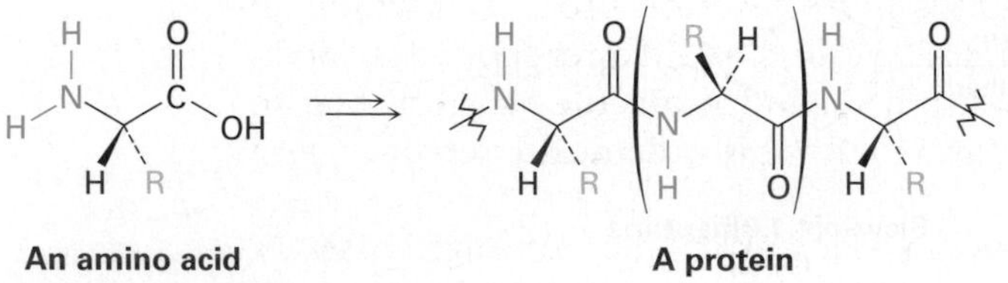

An amino acid

A protein

Nucleic acid—a nucleotide polymer

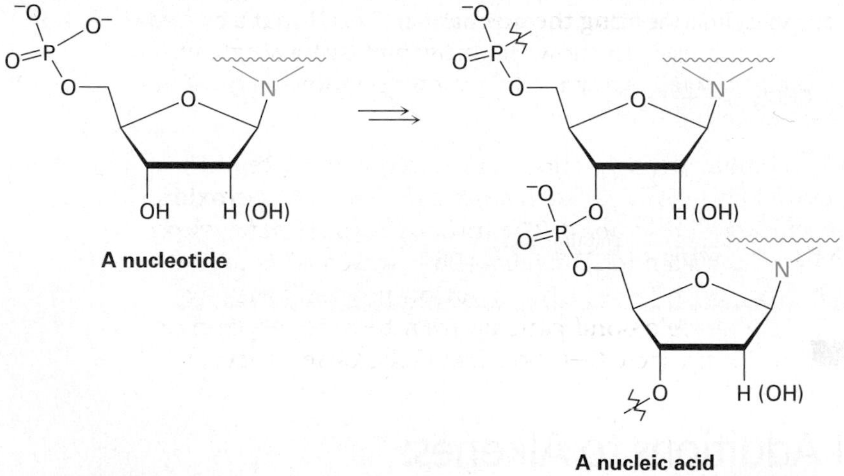

A nucleotide

A nucleic acid

Synthetic polymers, such as polyethylene, are chemically much simpler than biopolymers, but there is still a great diversity to their structures and properties, depending on the identity of the monomers and on the reaction conditions used for polymerization. The simplest synthetic polymers are those that result when an alkene is treated with a small amount of a suitable catalyst. Ethylene, for example, yields polyethylene, an enormous alkane that may have a molecular weight up to *6 million* amu and may contain as many as 200,000 monomer units incorporated into a gigantic hydrocarbon chain. Worldwide production of polyethylene is approximately 80 million metric tons per year.

Polyethylene—a synthetic alkene polymer

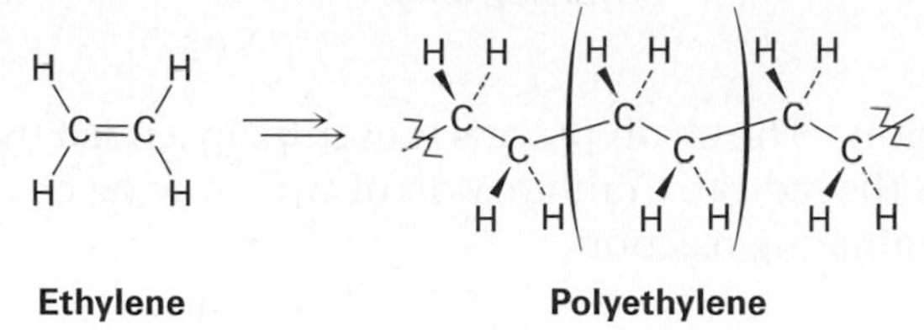

Ethylene

Polyethylene

Polyethylene and other simple alkene polymers are called **chain-growth polymers** because they are formed in a chain reaction process in which an initiator adds to a carbon–carbon double bond to yield a reactive intermediate. The intermediate then reacts with a second molecule of monomer to yield a new intermediate, which reacts with a third monomer unit, and so on.

Historically, ethylene polymerization was carried out at high pressure (1000–3000 atm) and high temperature (100–250 °C) in the presence of a radical initiator such as benzoyl peroxide, although other catalysts and reaction conditions are now used. The key step is the addition of a radical to the ethylene double bond, a reaction similar in many respects to what takes place in the addition of an electrophile. In writing the mechanism, recall that a curved half-arrow, or "fishhook" ⌒, is used to show the movement of a single electron, as opposed to the full curved arrow used to show the movement of an electron pair in a polar reaction.

- **Initiation** The polymerization reaction is initiated when a few radicals are generated on heating a small amount of benzoyl peroxide catalyst to break the weak O–O bond. The initially formed benzoyloxy radical loses CO_2 and gives a phenyl radical (Ph·), which adds to the C=C bond of ethylene to start the polymerization process. One electron from the ethylene double bond pairs up with the odd electron on the phenyl radical to form a new C–C bond, and the other electron remains on carbon.

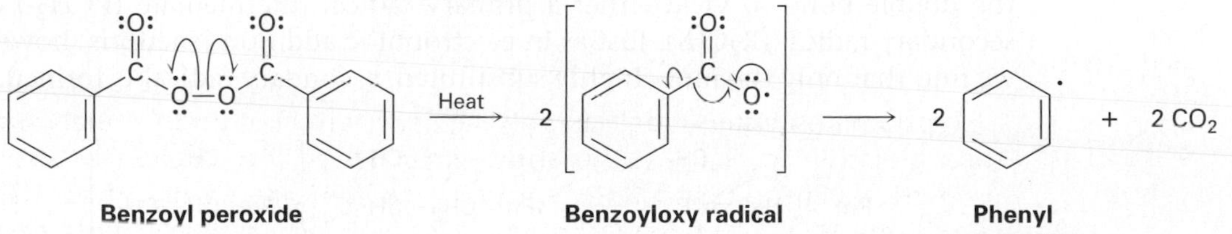

Heat → 2 [ ] → 2 + 2 CO_2

Benzoyl peroxide

Benzoyloxy radical

Phenyl radical (Ph·)

$$\text{Ph}\cdot + \text{H}_2\text{C}{=}\text{CH}_2 \longrightarrow \text{Ph}{-}\text{CH}_2\text{CH}_2\cdot$$

- **Propagation** Polymerization occurs when the carbon radical formed in the initiation step adds to another ethylene molecule to yield another radical. Repetition of the process for hundreds or thousands of times builds the polymer chain.

$$Ph{-}CH_2CH_2\cdot \;\; H_2C{=}CH_2 \longrightarrow Ph{-}CH_2CH_2CH_2CH_2\cdot \xrightarrow{\text{Repeat many times}} Ph{-}(CH_2CH_2)_nCH_2CH_2\cdot$$

- **Termination** The chain process is eventually ended by a reaction that consumes the radical. Combination of two growing chains is one possible chain-terminating reaction.

$$2\ R{-}CH_2CH_2\cdot \longrightarrow R{-}CH_2CH_2CH_2CH_2{-}R$$

Ethylene is not unique in its ability to form a polymer. Many substituted ethylenes, called *vinyl monomers,* also undergo polymerization to yield polymers with substituent groups regularly spaced on alternating carbon atoms along the chain. Propylene, for example, yields polypropylene, and styrene yields polystyrene.

$$H_2C{=}CHCH_3 \longrightarrow \left(CH_2CH(CH_3)CH_2CH(CH_3)CH_2CH(CH_3)CH_2CH(CH_3)\right)$$

Propylene **Polypropylene**

$$H_2C{=}CH{-}C_6H_5 \longrightarrow \left(CH_2CH(C_6H_5)CH_2CH(C_6H_5)CH_2CH(C_6H_5)CH_2CH(C_6H_5)\right)$$

Styrene **Polystyrene**

When an unsymmetrically substituted vinyl monomer such as propylene or styrene is polymerized, the radical addition steps can take place at either end of the double bond to yield either a primary radical intermediate ($RCH_2\cdot$) or a secondary radical ($R_2CH\cdot$). Just as in electrophilic addition reactions, however, we find that only the more highly substituted, secondary radical is formed.

$$Ph\cdot \;\; H_2C{=}CH(CH_3) \longrightarrow Ph{-}CH_2{-}CH(CH_3)\cdot \quad \left[\,Ph{-}CH(CH_3){-}CH_2\cdot\,\right]$$

Secondary radical **Primary radical (Not formed)**

Table 8.1 shows some commercially important alkene polymers, their uses, and the vinyl monomers from which they are made.

Table 8.1 Some Alkene Polymers and Their Uses

Monomer	Formula	Trade or common name of polymer	Uses
Ethylene	$H_2C{=}CH_2$	Polyethylene	Packaging, bottles
Propene (propylene)	$H_2C{=}CHCH_3$	Polypropylene	Moldings, rope, carpets
Chloroethylene (vinyl chloride)	$H_2C{=}CHCl$	Poly(vinyl chloride) Tedlar	Insulation, films, pipes
Styrene	$H_2C{=}CHC_6H_5$	Polystyrene	Foam, moldings
Tetrafluoroethylene	$F_2C{=}CF_2$	Teflon	Gaskets, nonstick coatings
Acrylonitrile	$H_2C{=}CHCN$	Orlon, Acrilan	Fibers
Methyl methacrylate	$H_2C{=}C(CH_3)CO_2CH_3$	Plexiglas, Lucite	Paint, sheets, moldings
Vinyl acetate	$H_2C{=}CHOCOCH_3$	Poly(vinyl acetate)	Paint, adhesives, foams

Predicting the Structure of a Polymer

Worked Example 8.4

Show the structure of poly(vinyl chloride), a polymer made from $H_2C{=}CHCl$, by drawing several repeating units.

Strategy

Mentally break the carbon–carbon double bond in the monomer unit, and form single bonds by connecting numerous units together.

Solution

The general structure of poly(vinyl chloride) is

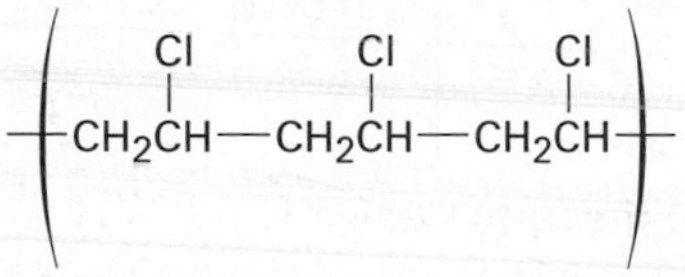

$$\left(-CH_2\underset{Cl}{CH}-CH_2\underset{Cl}{CH}-CH_2\underset{Cl}{CH}- \right)$$

Problem 8.18

Show the monomer units you would use to prepare the following polymers:

(a)

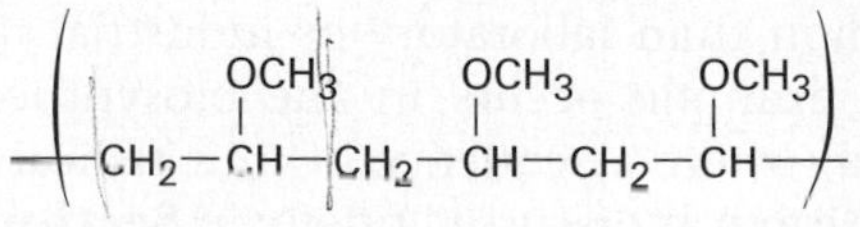

$$\left(-CH_2-\underset{OCH_3}{CH}-CH_2-\underset{OCH_3}{CH}-CH_2-\underset{OCH_3}{CH}- \right)$$

(b)

$$\left(-\underset{Cl}{CH}-\underset{Cl}{CH}-\underset{Cl}{CH}-\underset{Cl}{CH}-\underset{Cl}{CH}-\underset{Cl}{CH}- \right)$$

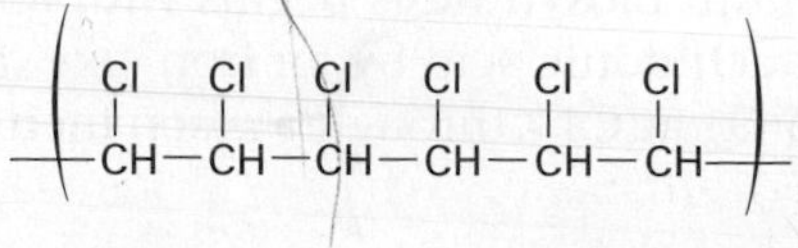

Problem 8.19

One of the chain-termination steps that sometimes occurs to interrupt polymerization is the following reaction between two radicals. Propose a mechanism for the reaction, using fishhook arrows to indicate electron flow.

$$2\ \text{-}CH_2\dot{C}H_2 \longrightarrow \text{-}CH_2CH_3 + \text{-}CH{=}CH_2$$

8.11 Biological Additions of Radicals to Alkenes

The same high reactivity of radicals that makes possible the alkene polymerization we saw in the previous section also makes it difficult to carry out controlled radical reactions on complex molecules. As a result, there are severe limitations on the usefulness of radical addition reactions in the laboratory. In contrast to an electrophilic addition, where reaction occurs once and the reactive cation intermediate is rapidly quenched by a nucleophile, the reactive intermediate in a radical reaction is not usually quenched. Instead, it reacts again and again in a largely uncontrollable way.

Electrophilic addition (Intermediate is quenched, so reaction stops.)

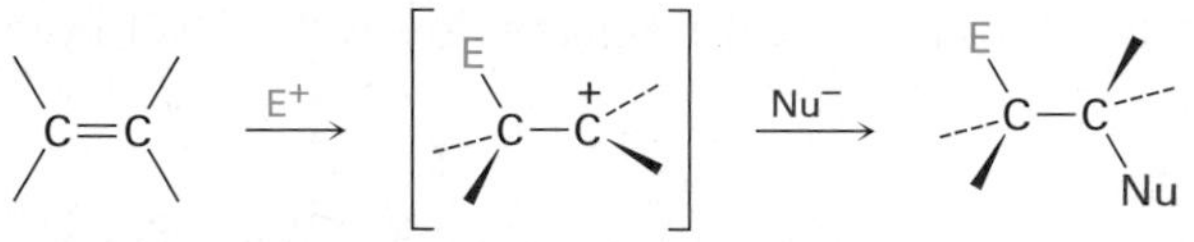

C=C —E⁺→ [E–C–C⁺] —Nu⁻→ E–C–C–Nu

Radical addition (Intermediate is not quenched, so reaction does not stop.)

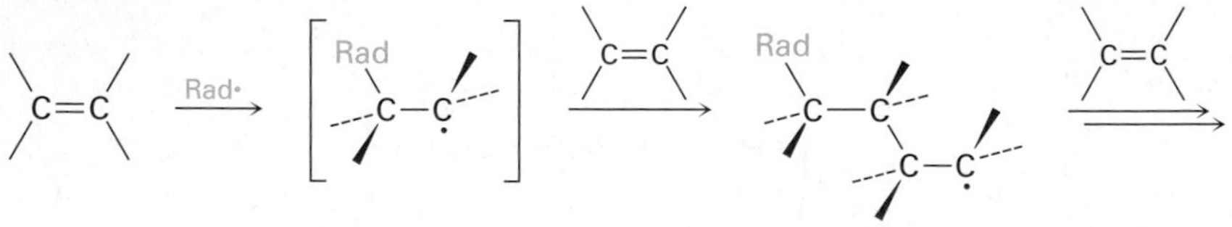

C=C —Rad•→ [Rad–C–C•] —C=C→ Rad–C–C–C–C• —C=C→

In biological reactions, the situation is different from that in the laboratory. Only one substrate molecule at a time is present in the active site of the enzyme where reaction takes place, and that molecule is held in a precise position, with other necessary reacting groups nearby. As a result, biological radical reactions are more controlled and more common than laboratory or industrial radical reactions. A particularly impressive example occurs in the biosynthesis of prostaglandins from arachidonic acid, where a sequence of four radical additions take place. The reaction mechanism was discussed briefly in **Section 6.3**.

As shown in **Figure 8.10**, prostaglandin biosynthesis begins with abstraction of a hydrogen atom from C13 of arachidonic acid by an iron–oxy radical to give a carbon radical that reacts with O_2 at C11 through a resonance form.

The oxygen radical that results adds to the C8–C9 double bond to give a carbon radical at C8, which adds to the C12–C13 double bond and gives a carbon radical at C13. A resonance form of this carbon radical adds at C15 to a second O_2 molecule, completing the prostaglandin skeleton. Reduction of the O–O bond then gives prostaglandin H_2, called PGH_2. The pathway looks complicated, but the entire process is catalyzed with exquisite control by a single enzyme.

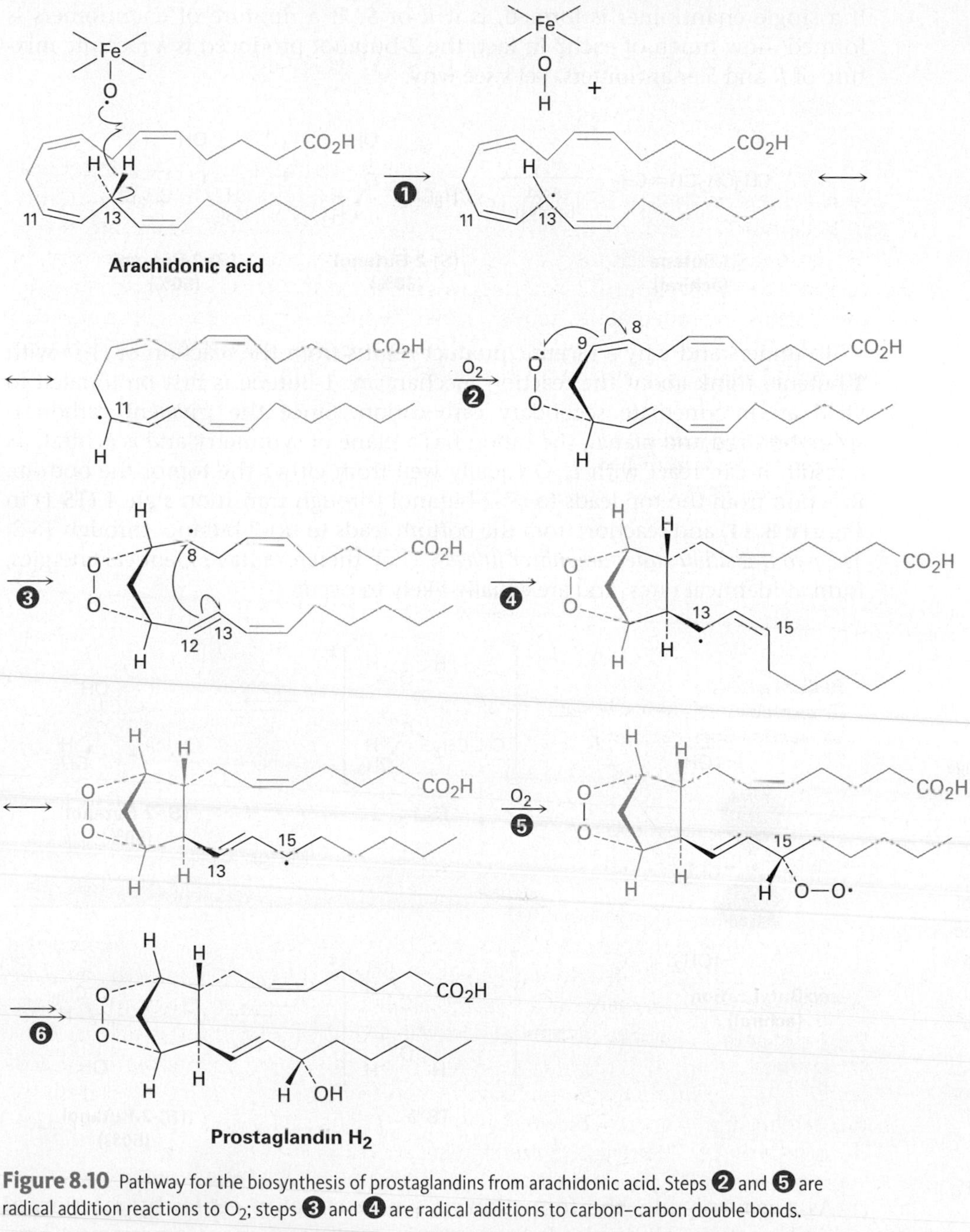

Figure 8.10 Pathway for the biosynthesis of prostaglandins from arachidonic acid. Steps 2 and 5 are radical addition reactions to O_2; steps 3 and 4 are radical additions to carbon–carbon double bonds.

8.12 Reaction Stereochemistry: Addition of H_2O to an Achiral Alkene

Most of the biochemical reactions that take place in the body, as well as many organic reactions in the laboratory, yield products with chirality centers. For example, acid-catalyzed addition of H_2O to 1-butene in the laboratory yields 2-butanol, a chiral alcohol. What is the stereochemistry of this chiral product? If a single enantiomer is formed, is it *R* or *S*? If a mixture of enantiomers is formed, how much of each? In fact, the 2-butanol produced is a racemic mixture of *R* and *S* enantiomers. Let's see why.

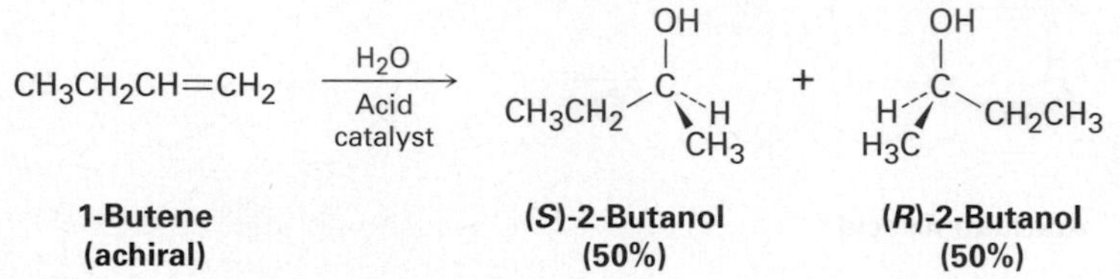

$CH_3CH_2CH{=}CH_2$ $\xrightarrow[\text{Acid catalyst}]{H_2O}$ (S)-2-Butanol (50%) + (R)-2-Butanol (50%)

1-Butene (achiral) **(*S*)-2-Butanol (50%)** **(*R*)-2-Butanol (50%)**

To understand why a racemic product results from the reaction of H_2O with 1-butene, think about the reaction mechanism. 1-Butene is first protonated to yield an intermediate secondary carbocation. Since the trivalent carbon is sp^2-hybridized and planar, the cation has a plane of symmetry and is achiral. As a result, it can react with H_2O equally well from either the top or the bottom. Reaction from the top leads to (*S*)-2-butanol through transition state 1 (TS 1) in **Figure 8.11**, and reaction from the bottom leads to (*R*)-2-butanol through TS 2. *The two transition states are mirror images.* They therefore have identical energies, form at identical rates, and are equally likely to occur.

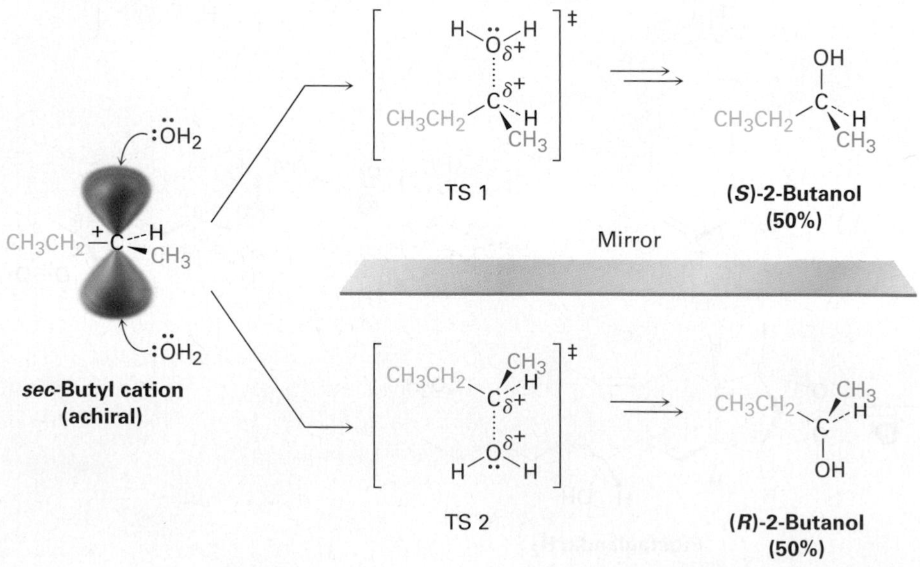

Figure 8.11 Reaction of H_2O with the carbocation resulting from protonation of 1-butene. Reaction from the top leads to *S* product and is the mirror image of reaction from the bottom, which leads to *R* product. Because they are energetically identical, they are equally likely and lead to a racemic mixture of products. The dotted C···O bond in the transition state indicates partial bond formation.

As a general rule, the formation of a new chirality center by reaction of achiral reactants always leads to a racemic mixture of enantiomeric products. Put another way, optical activity can't appear from nowhere; an optically active product can only result by starting with an optically active reactant or chiral environment **(Section 5.12)**.

In contrast to laboratory reactions, enzyme-catalyzed biological reactions often give a single enantiomer of a chiral product, even when the substrate is achiral. One step in the citric acid cycle of food metabolism, for instance, is the aconitase-catalyzed addition of water to (*Z*)-aconitate (usually called *cis*-aconitate) to give isocitrate.

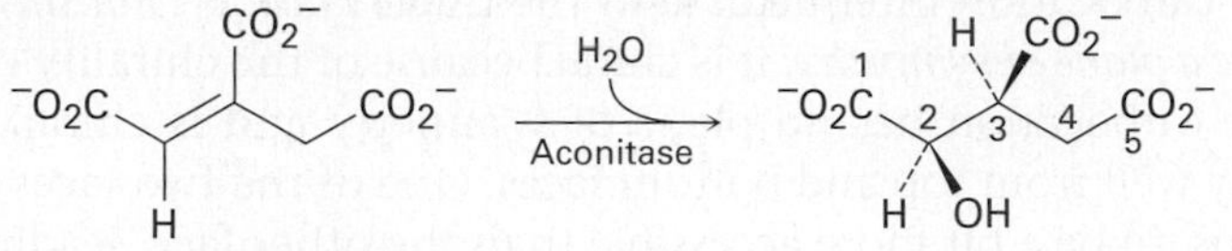

cis-Aconitate (achiral) → (H_2O, Aconitase) → **(2R,3S)-Isocitrate**

Even though *cis*-aconitate is achiral, only the (2*R*,3*S*) enantiomer of the product is formed. As discussed in **Sections 5.11 and 5.12**, *cis*-aconitate is a prochiral molecule, which is held in a chiral environment by the aconitase enzyme during the reaction. In that chiral environment, the two faces of the double bond are chemically distinct, and addition occurs on only the *Re* face at C2.

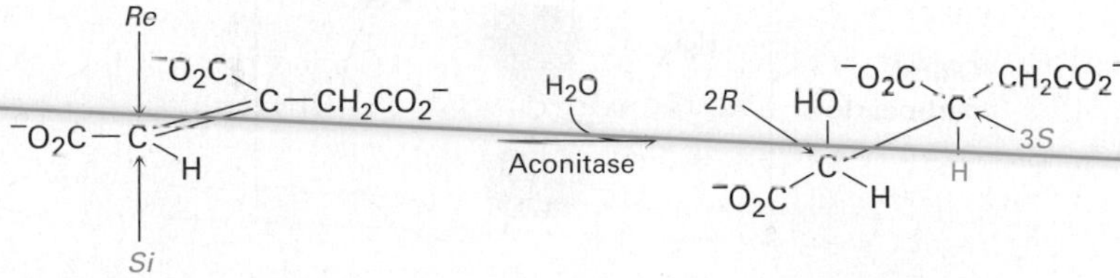

cis-Aconitate → (H_2O, Aconitase) → **(2R,3S)-Isocitrate**

8.13 Reaction Stereochemistry: Addition of H_2O to a Chiral Alkene

The reaction discussed in the previous section involves an addition to an achiral reactant and forms an optically inactive, racemic mixture of two enantiomeric products. What would happen, though, if we were to carry out the reaction on a *single* enantiomer of a *chiral* reactant? For example, what stereochemical result would be obtained from addition of H_2O to a chiral alkene, such as (*R*)-4-methyl-1-hexene? The product of the reaction, 4-methyl-2-hexanol, has two chirality centers and so has four possible stereoisomers.

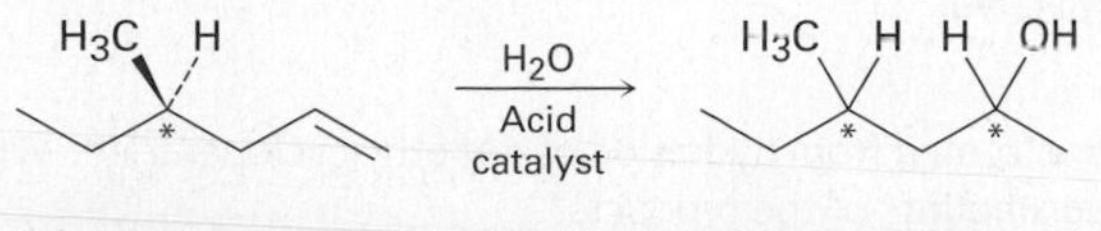

(R)-4-Methyl-1-hexene (chiral) → (H_2O, Acid catalyst) → **4-Methyl-2-hexanol (chiral)**

Let's think about the two chirality centers separately. What about the configuration at C4, the methyl-bearing carbon atom? Since C4 has the *R* configuration in the starting material and this chirality center is unaffected by the reaction, its configuration is unchanged. Thus, the configuration at C4 in the product remains *R* (assuming that the relative rankings of the four attached groups are not changed by the reaction).

What about the configuration at C2, the newly formed chirality center? As shown in **Figure 8.12**, the stereochemistry at C2 is established by reaction of H_2O with a carbocation intermediate in the usual manner. *But this carbocation does not have a plane of symmetry;* it is chiral because of the chirality center at C4. Because the carbocation has no plane of symmetry and is chiral, it does not react equally well from top and bottom faces. One of the two faces is likely, for steric reasons, to be a bit more accessible than the other face, leading to a mixture of *R* and *S* products in some ratio other than 50:50. Thus, two diastereomeric products, (2*R*,4*R*)-4-methyl-2-hexanol and (2*S*,4*R*)-4-methyl-2-hexanol, are formed in unequal amounts, and the mixture is optically active.

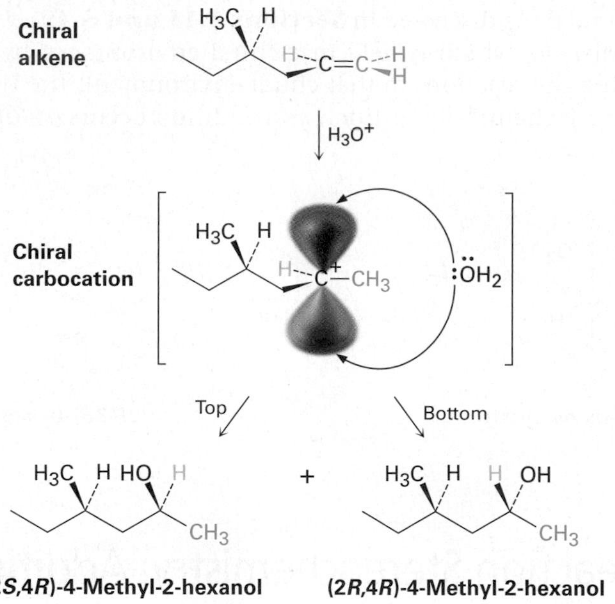

Figure 8.12 Stereochemistry of the acid-catalyzed addition of H_2O to the chiral alkene, (*R*)-4-methyl-1-hexene. A mixture of diastereomeric 2*R*,4*R* and 2*S*,4*R* products is formed in unequal amounts because reaction of the chiral carbocation intermediate is not equally likely from top and bottom. The product mixture is optically active.

Chiral alkene

H_3O^+

Chiral carbocation

Top

Bottom

(2*S*,4*R*)-4-Methyl-2-hexanol + **(2*R*,4*R*)-4-Methyl-2-hexanol**

As a general rule, the formation of a new chirality center by the reaction of a chiral reactant leads to unequal amounts of diastereomeric products. If the chiral reactant is optically active because only one enantiomer is used rather than a racemic mixture, then the products are also optically active.

Problem 8.20

What products are formed from acid-catalyzed hydration of racemic (±)-4-methyl-1-hexene? What can you say about the relative amounts of the products? Is the product mixture optically active?

Problem 8.21

What products are formed from hydration of 4-methylcyclopentene? What can you say about the relative amounts of the products?

A DEEPER LOOK Terpenes: Naturally Occurring Alkenes

The wonderful fragrance of leaves from the California bay laurel is due primarily to myrcene, a simple terpene.

Ever since its discovery in Persia around 1000 A.D., it has been known that *steam distillation,* the codistillation of plant materials with water, produces a fragrant mixture of liquids called *essential oils.* The resulting oils have long been used as medicines, spices, and perfumes, and their investigation played a major role in the emergence of organic chemistry as a science during the 19th century.

Chemically, plant essential oils consist largely of mixtures of compounds called *terpenoids*—small organic molecules with an immense diversity of structure. More than 35,000 different terpenoids are known. Some are open-chain molecules, and others contain rings; some are hydrocarbons, and others contain oxygen. Hydrocarbon terpenoids, in particular, are known as *terpenes,* and all contain double bonds. For example:

H_3C CH_3 CH_3

Myrcene (oil of bay) **α-Pinene (turpentine)**

CH_3 CH_3 H_3C CH_3 CH_2 CH_3

Humulene (oil of hops) **β-Santalene (sandalwood oil)**

Regardless of their apparent structural differences, all terpenoids are related. According to a formalism called the *isoprene rule,* they can be thought of as arising from head-to-tail joining of 5-carbon isoprene units (2-methyl-1,3-butadiene). Carbon 1 is the head of the isoprene unit, and carbon 4 is the tail. For example, myrcene contains two isoprene units joined head to tail, forming an 8-carbon chain with two 1-carbon branches. α-Pinene similarly contains two isoprene units assembled into a more complex cyclic structure, and humulene contains three isoprene units. See if you can identify the isoprene units in α-pinene, humulene, and β-santalene.

Head Tail 1 2 3 4

Isoprene → **Myrcene**

(continued)

(continued)

Terpenes (and terpenoids) are further classified according to the number of 5-carbon units they contain. Thus, *monoterpenes* are 10-carbon substances derived from two isoprene units, *sesquiterpenes* are 15-carbon molecules derived from three isoprene units, *diterpenes* are 20-carbon substances derived from four isoprene units, and so on. Monoterpenes and sesquiterpenes are found primarily in plants, but the higher terpenoids occur in both plants and animals, and many have important biological roles. The triterpenoid lanosterol, for instance, is the biological precursor from which all steroid hormones are made.

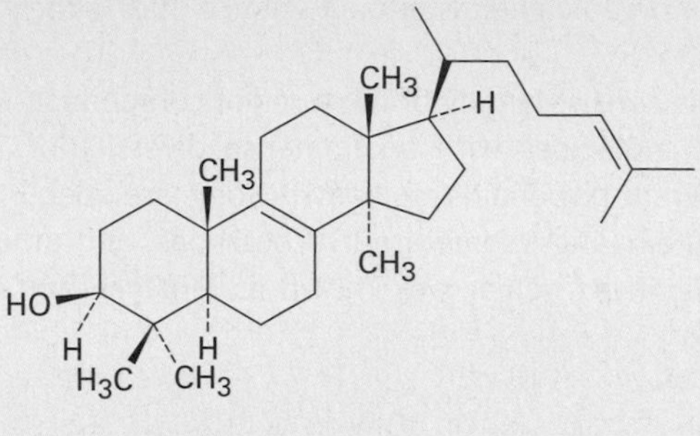

CH_3 H CH_3 CH_3 HO H H H_3C CH_3

Lanosterol (a triterpene, C_{30})

Isoprene itself is not the true biological precursor of terpenoids. Nature instead uses two "isoprene equivalents"—isopentenyl diphosphate and dimethylallyl diphosphate—which are themselves made by two different routes depending on the organism. Lanosterol, in particular, is biosynthesized from acetic acid by a complex pathway that has been worked out in great detail. We'll look at the subject more closely in **Sections 27.5 and 27.7**.

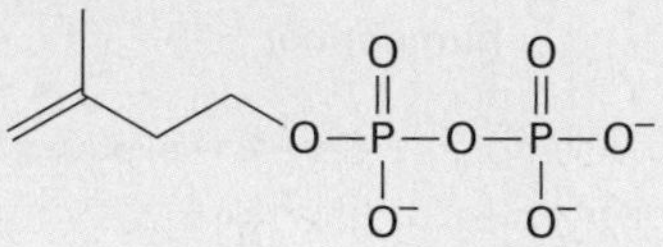

Isopentenyl diphosphate 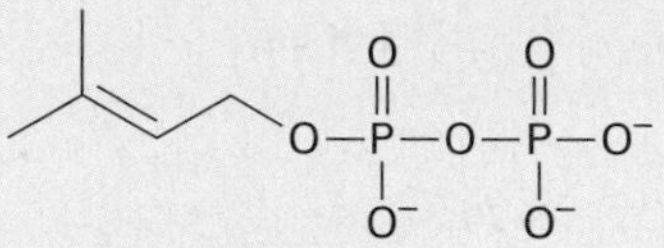 **Dimethylallyl diphosphate**

Summary

Key words

anti stereochemistry, 265
bromonium ion, 265
carbene, 287
chain-growth polymer, 291
epoxide, 281
glycol, 282
halohydrin, 267
hydroboration, 272
hydrogenation, 276
hydroxylation, 282
monomer, 289
oxidation, 281
oxymercuration–demercuration, 271
ozonide, 284
polymer, 289

With the background needed to understand organic reactions now covered, this chapter has begun the systematic description of major functional groups.

Alkenes are generally prepared by an *elimination reaction,* such as *dehydrohalogenation,* the elimination of HX from an alkyl halide, or *dehydration,* the elimination of water from an alcohol. The flip side of that elimination reaction to prepare alkenes is the addition of various substances to the alkene double bond to give saturated products.

HCl, HBr, and HI add to alkenes by a two-step electrophilic addition mechanism. Initial reaction of the nucleophilic double bond with H^+ gives a carbocation intermediate, which then reacts with halide ion. Bromine and chlorine add to alkenes via three-membered-ring **bromonium ion** or chloronium ion intermediates to give addition products having **anti stereochemistry**. If water is present during the halogen addition reaction, a **halohydrin** is formed.

Hydration of an alkene—the addition of water—is carried out by either of two procedures, depending on the product desired. **Oxymercuration–demercuration** involves electrophilic addition of Hg^{2+} to an alkene, followed by trapping of the cation intermediate with water and subsequent treatment with $NaBH_4$. **Hydroboration** involves addition of borane (BH_3) followed by oxidation of the intermediate organoborane with alkaline H_2O_2. The two hydration methods are complementary: oxymercuration–demercuration gives the product of Markovnikov addition, whereas hydroboration–oxidation gives the product with non-Markovnikov **syn stereochemistry**.

Alkenes are **reduced** by addition of H_2 in the presence of a catalyst such as platinum or palladium to yield alkanes, a process called catalytic **hydrogenation**. Alkenes are also **oxidized** by reaction with a peroxyacid to give **epoxides**, which can be converted into trans-1,2-diols by acid-catalyzed hydrolysis. The corresponding cis-1,2-diols can be made directly from alkenes by **hydroxylation** with OsO_4. Alkenes can also be cleaved to produce carbonyl compounds by reaction with ozone, followed by reduction with zinc metal. In addition, alkenes react with divalent substances called **carbenes, R_2C:**, to give cyclopropanes. Nonhalogenated cyclopropanes are best prepared by treatment of the alkene with CH_2I_2 and zinc–copper, a process called the **Simmons–Smith reaction**.

Alkene **polymers**—large molecules resulting from repetitive bonding together of many hundreds or thousands of small **monomer** units—are formed by chain-reaction polymerization of simple alkenes. Polyethylene, polypropylene, and polystyrene are examples. As a general rule, radical addition reactions are not common in the laboratory but occur much more frequently in biological pathways.

Many reactions give chiral products. If the reactants are optically inactive, the products are also optically inactive. If one or both of the reactants is optically active, the products can also be optically active.

Key words—cont'd

reduction, 277
Simmons–Smith reaction, 288
stereospecific, 288
syn stereochemistry, 273

Learning Reactions

What's seven times nine? Sixty-three, of course. You didn't have to stop and figure it out; you knew the answer immediately because you long ago learned the multiplication tables. Learning the reactions of organic chemistry requires the same approach: reactions have to be learned for immediate recall if they are to be useful.

Different people take different approaches to learning reactions. Some people make flashcards; others find studying with friends to be helpful. To help guide your study, most chapters in this book end with a summary of the reactions just presented. In addition, the accompanying *Study Guide and Solutions Manual* has several appendixes that organize organic reactions from other viewpoints. Fundamentally, though, there are no shortcuts. Learning organic chemistry does take effort.

Summary of Reactions

Note: **No stereochemistry is implied unless specifically indicated with wedged, solid, and dashed lines.**

1. Addition reactions of alkenes
   (a) Addition of HCl, HBr, and HI (Sections 7.7 and 7.8)
   Markovnikov regiochemistry occurs, with H adding to the less highly substituted alkene carbon and halogen adding to the more highly substituted carbon.

$$\mathrm{>C{=}C<} \xrightarrow[\text{Ether}]{\text{HX}} \mathrm{-\overset{H}{C}-\overset{X}{C}-}$$

(continued)

(b) Addition of halogens Cl_2 and Br_2 (Section 8.2)
Anti addition is observed through a halonium ion intermediate.

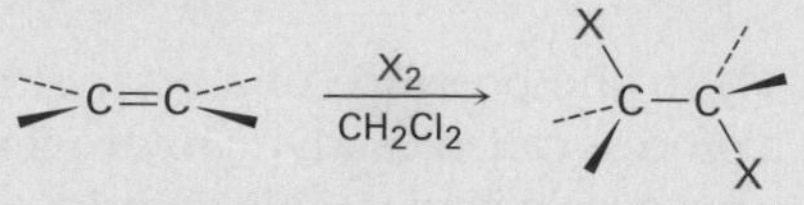

$$\text{C=C} \xrightarrow[\text{CH}_2\text{Cl}_2]{\text{X}_2} \text{X–C–C–X (anti)}$$

(c) Halohydrin formation (Section 8.3)
Markovnikov regiochemistry and anti stereochemistry occur.

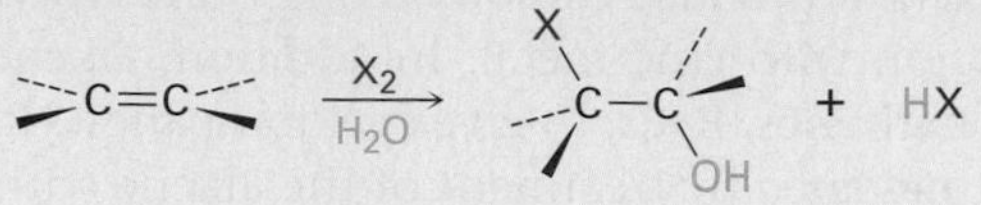

$$\text{C=C} \xrightarrow[\text{H}_2\text{O}]{\text{X}_2} \text{X–C–C–OH} + \text{HX}$$

(d) Addition of water by oxymercuration–demercuration (Section 8.4)
Markovnikov regiochemistry occurs.

$$\text{C=C} \xrightarrow[\text{2. NaBH}_4]{\text{1. Hg(OAc)}_2\text{, H}_2\text{O/THF}} \text{HO–C–C–H}$$

(e) Addition of water by hydroboration–oxidation (Section 8.5)
Non-Markovnikov syn addition occurs.

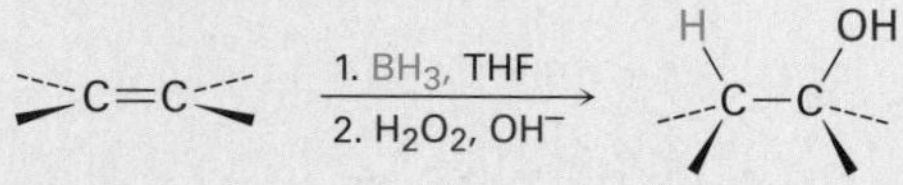

$$\text{C=C} \xrightarrow[\text{2. H}_2\text{O}_2\text{, OH}^-]{\text{1. BH}_3\text{, THF}} \text{H–C–C–OH}$$

(f) Catalytic hydrogenation (Section 8.6)
Syn addition occurs.

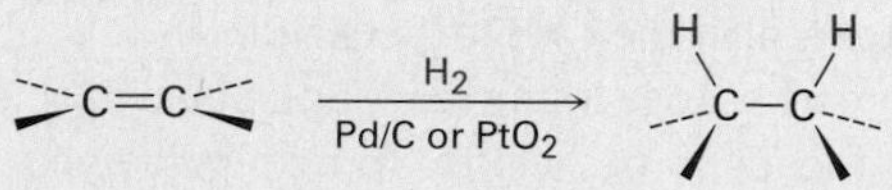

$$\text{C=C} \xrightarrow[\text{Pd/C or PtO}_2]{\text{H}_2} \text{H–C–C–H}$$

(g) Epoxidation with a peroxyacid (Section 8.7)
Syn addition occurs.

$$\text{C=C} \xrightarrow{\text{RCOOH (O)}} \text{C–C (epoxide, O bridging)}$$

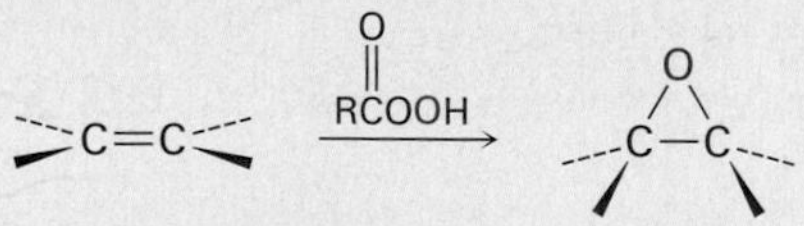

(h) Hydroxylation with OsO_4 (Section 8.7)
Syn addition occurs.

$$\text{C=C} \xrightarrow[\text{2. NaHSO}_3\text{, H}_2\text{O or OsO}_4\text{, NMO}]{\text{1. OsO}_4} \text{HO–C–C–OH}$$

(i) Addition of carbenes to yield cyclopropanes (Section 8.9)

(1) Dichlorocarbene addition

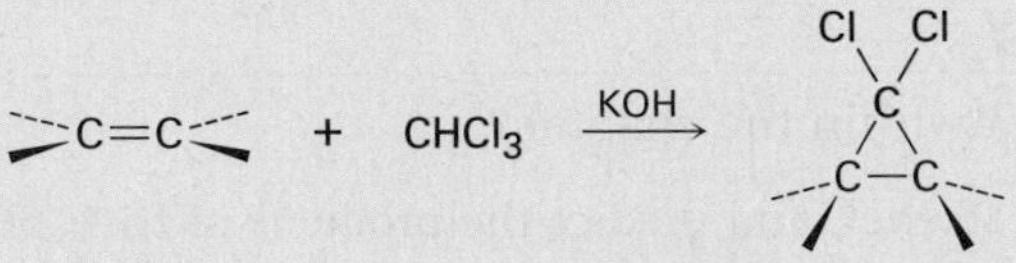

$>C{=}C< \; + \; CHCl_3 \xrightarrow{KOH}$ (cyclopropane with CCl_2)

(2) Simmons–Smith reaction

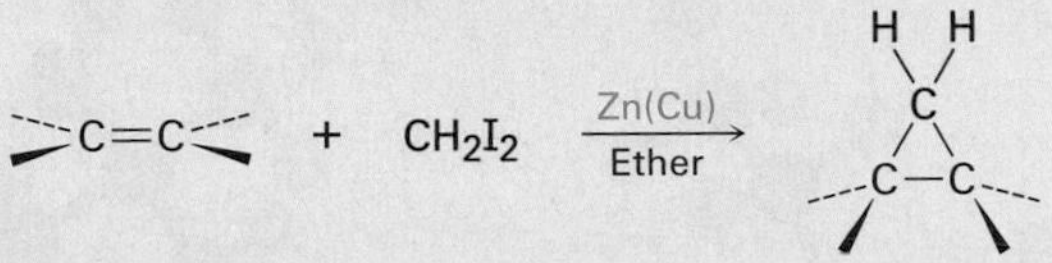

$>C{=}C< \; + \; CH_2I_2 \xrightarrow[\text{Ether}]{\text{Zn(Cu)}}$ (cyclopropane with CH_2)

2. Hydroxylation by acid-catalyzed epoxide hydrolysis (Section 8.7)

Anti stereochemistry occurs.

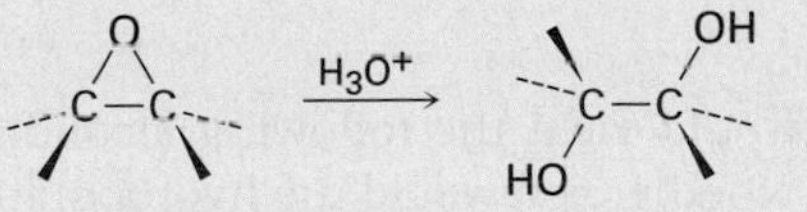

$\text{epoxide} \xrightarrow{H_3O^+}$ anti 1,2-diol (HO, OH)

3. Oxidative cleavage of alkenes (Section 8.8)

(a) Reaction with ozone followed by zinc in acetic acid

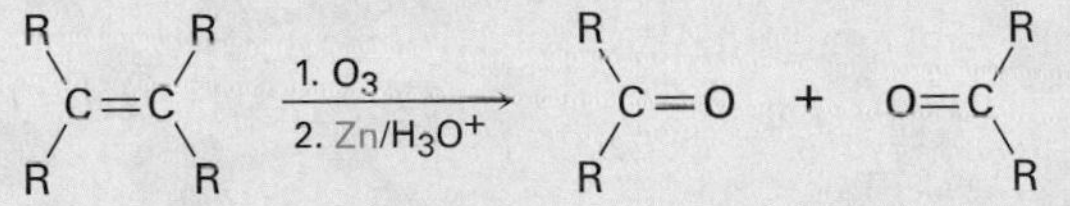

$R_2C{=}CR_2 \xrightarrow[\text{2. Zn/}H_3O^+]{\text{1. }O_3} R_2C{=}O \; + \; O{=}CR_2$

(b) Reaction with $KMnO_4$ in acidic solution

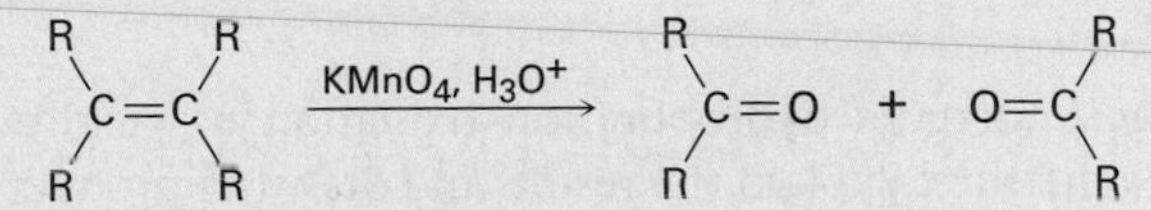

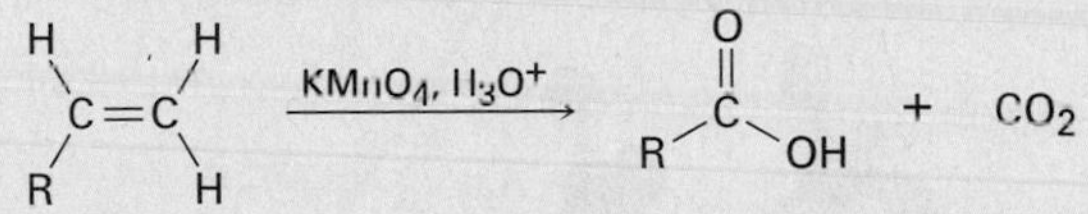

$R_2C{=}CR_2 \xrightarrow{KMnO_4,\, H_3O^+} R_2C{=}O \; + \; O{=}CR_2$

$RHC{=}CH_2 \xrightarrow{KMnO_4,\, H_3O^+} RCOOH \; + \; CO_2$

4. Cleavage of 1,2-diols (Section 8.8)

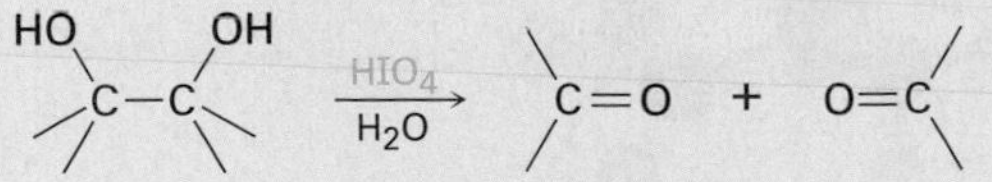

$>C(OH){-}C(OH)< \xrightarrow[H_2O]{HIO_4} >C{=}O \; + \; O{=}C<$

Exercises

OWL Interactive versions of these problems are assignable in OWL for Organic Chemistry.

Visualizing Chemistry

(Problems 8.1–8.21 appear within the chapter.)

8.22 Name the following alkenes, and predict the products of their reaction with (1) *meta*-chloroperoxybenzoic acid, (2) $KMnO_4$ in aqueous acid, and (3) O_3, followed by Zn in acetic acid:

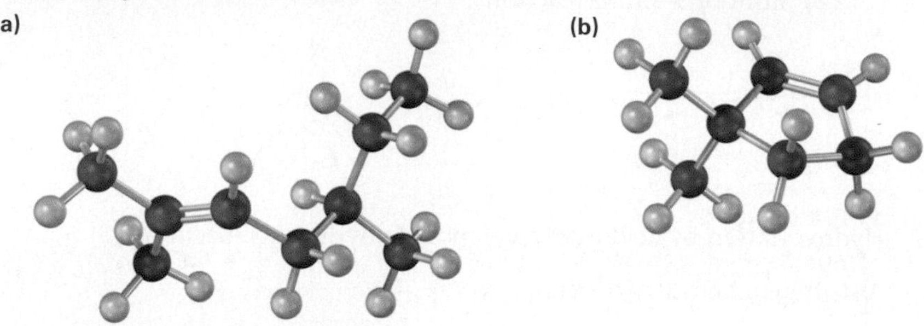

8.23 Draw the structures of alkenes that would yield the following alcohols on hydration (red = O). Tell in each case whether you would use hydroboration–oxidation or oxymercuration–demercuration.

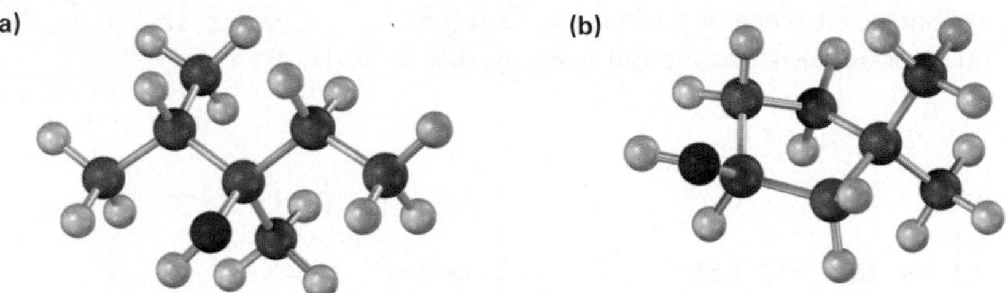

8.24 The following alkene undergoes hydroboration–oxidation to yield a single product rather than a mixture. Explain the result, and draw the product showing its stereochemistry.

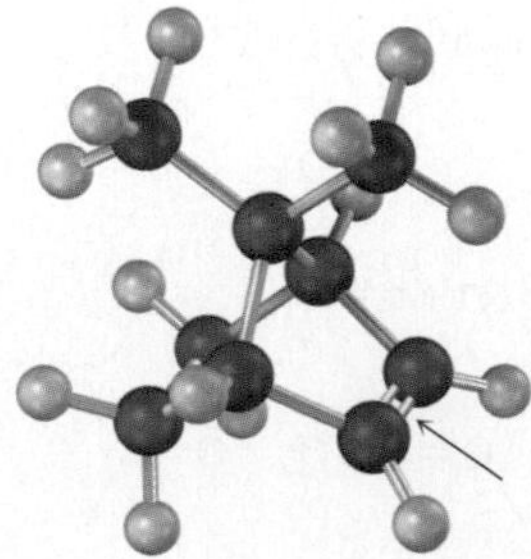

8.25 From what alkene was the following 1,2-diol made, and what method was used, epoxide hydrolysis or OsO_4?

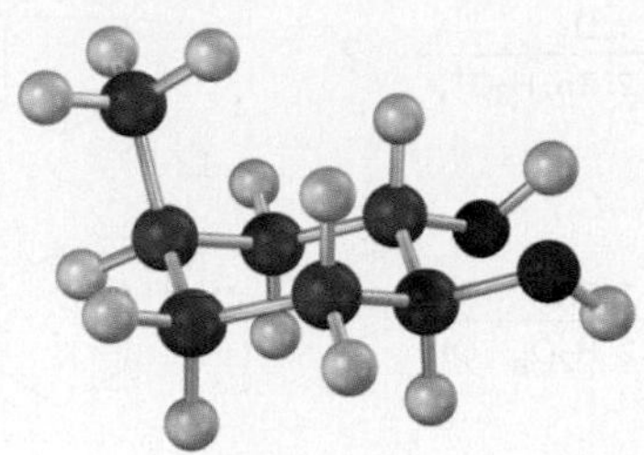

Additional Problems

Reactions of Alkenes

8.26 Predict the products of the following reactions (the aromatic ring is unreactive in all cases). Indicate regiochemistry when relevant.

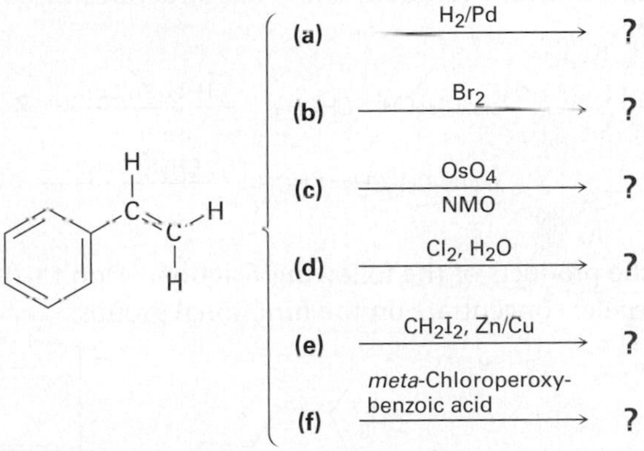

Starting material: styrene (C_6H_5–CH=CH_2)

(a) $\xrightarrow{H_2/Pd}$?

(b) $\xrightarrow{Br_2}$?

(c) $\xrightarrow[NMO]{OsO_4}$?

(d) $\xrightarrow{Cl_2,\ H_2O}$?

(e) $\xrightarrow{CH_2I_2,\ Zn/Cu}$?

(f) $\xrightarrow{\textit{meta}\text{-Chloroperoxybenzoic acid}}$?

8.27 Suggest structures for alkenes that give the following reaction products. There may be more than one answer for some cases.

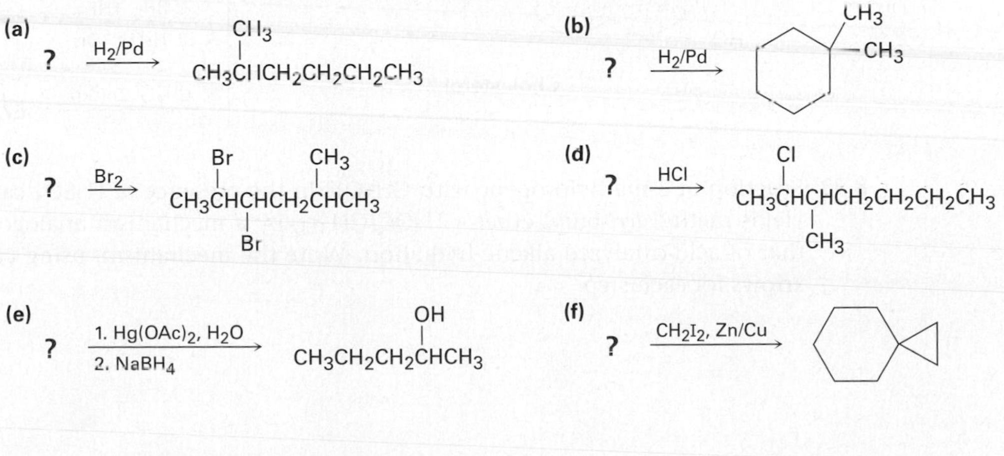

(a) ? $\xrightarrow{H_2/Pd}$ $CH_3CHCH_2CH_2CH_2CH_3$ (with CH_3 on C2)

(b) ? $\xrightarrow{H_2/Pd}$ 1,1-dimethylcyclohexane (ring carbon bearing two CH_3 groups)

(c) ? $\xrightarrow{Br_2}$ $CH_3CHCHCH_2CHCH_3$ (Br on C2, Br on C3, CH_3 on C5)

(d) ? $\xrightarrow{HCl}$ $CH_3CHCHCH_2CH_2CH_2CH_3$ (Cl on C2, CH_3 on C3)

(e) ? $\xrightarrow[2.\ NaBH_4]{1.\ Hg(OAc)_2,\ H_2O}$ $CH_3CH_2CH_2CHCH_3$ (OH on C4)

(f) ? $\xrightarrow{CH_2I_2,\ Zn/Cu}$ spiro cyclohexane–cyclopropane

8.28 Predict the products of the following reactions, showing both regiochemistry and stereochemistry where appropriate:

(a) [1-methylcyclohexene, CH$_3$, H] $\xrightarrow[\text{2. Zn, } H_3O^+]{\text{1. } O_3}$?

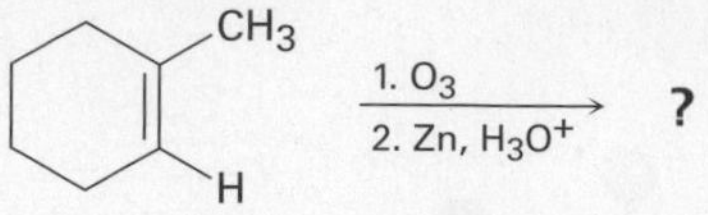

(b) [cyclohexene] $\xrightarrow[H_3O^+]{KMnO_4}$?

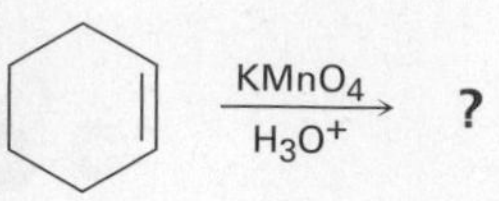

(c) [1-methylcyclohexene, CH$_3$] $\xrightarrow[\text{2. } H_2O_2,\ ^-OH]{\text{1. } BH_3}$?

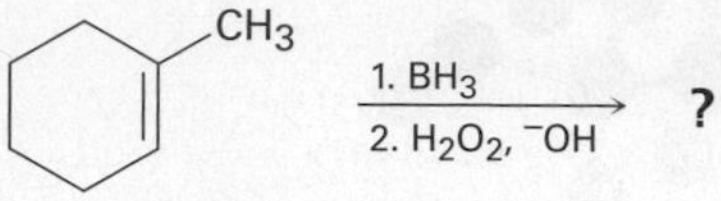

(d) [1-methylcyclohexene, CH$_3$] $\xrightarrow[\text{2. } NaBH_4]{\text{1. } Hg(OAc)_2,\ H_2O}$?

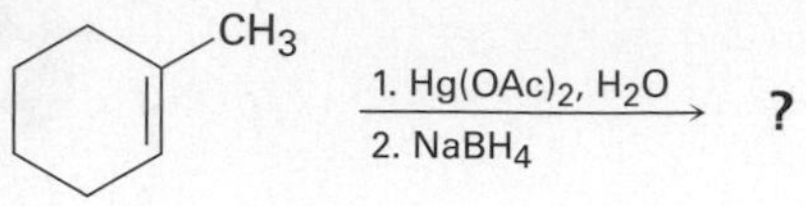

8.29 Which reaction would you expect to be faster, addition of HBr to cyclohexene or to 1-methylcyclohexene? Explain.

8.30 What product will result from hydroboration–oxidation of 1-methylcyclopentene with deuterated borane, BD_3? Show both the stereochemistry (spatial arrangement) and the regiochemistry (orientation) of the product.

8.31 The cis and trans isomers of 2-butene give different cyclopropane products in the Simmons–Smith reaction. Show the structures of both, and explain the difference.

$$\textit{cis}\text{-}CH_3CH{=}CHCH_3 \xrightarrow{CH_2I_2,\ Zn(Cu)} ?$$

$$\textit{trans}\text{-}CH_3CH{=}CHCH_3 \xrightarrow{CH_2I_2,\ Zn(Cu)} ?$$

8.32 Predict the products of the following reactions. Don't worry about the size of the molecule; concentrate on the functional groups.

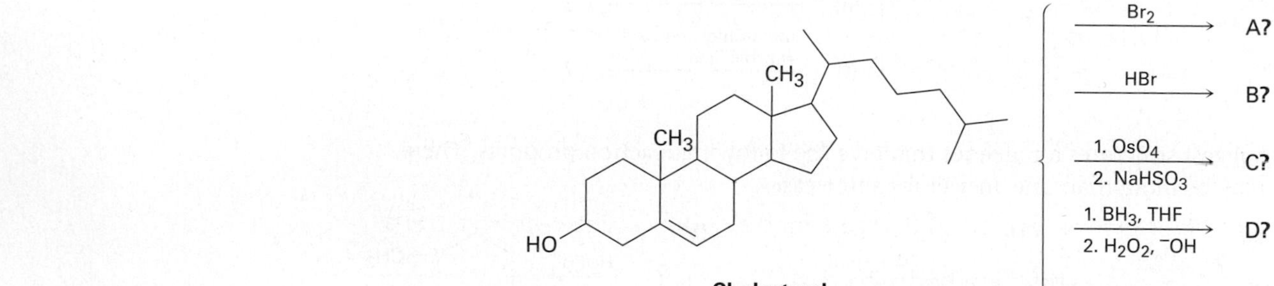

[Structure: cholesterol, with HO, CH$_3$, CH$_3$ labels]

Cholesterol

$\xrightarrow{Br_2}$ A?

$\xrightarrow{HBr}$ B?

$\xrightarrow[\text{2. } NaHSO_3]{\text{1. } OsO_4}$ C?

$\xrightarrow[\text{2. } H_2O_2,\ ^-OH]{\text{1. } BH_3\text{, THF}}$ D?

$\xrightarrow{CH_2I_2,\ Zn(Cu)}$ E?

8.33 Reaction of 2-methylpropene with CH_3OH in the presence of H_2SO_4 catalyst yields methyl *tert*-butyl ether, $CH_3OC(CH_3)_3$, by a mechanism analogous to that of acid-catalyzed alkene hydration. Write the mechanism, using curved arrows for each step.

8.34 Addition of HCl to 1-methoxycyclohexene yields 1-chloro-1-methoxycyclohexane as the sole product. Use resonance structures of the carbocation intermediate to explain why none of the other regioisomer is formed.

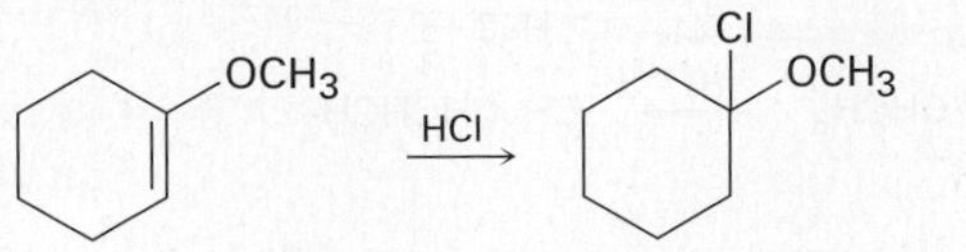

1-Methoxycyclohexene $\xrightarrow{HCl}$ 1-Chloro-1-methoxycyclohexane

Synthesis Using Alkenes

8.35 How would you carry out the following transformations? Tell the reagents you would use in each case.

(a) Cyclopentene $\xrightarrow{?}$ cis-1,2-cyclopentanediol (H, OH, OH, H)

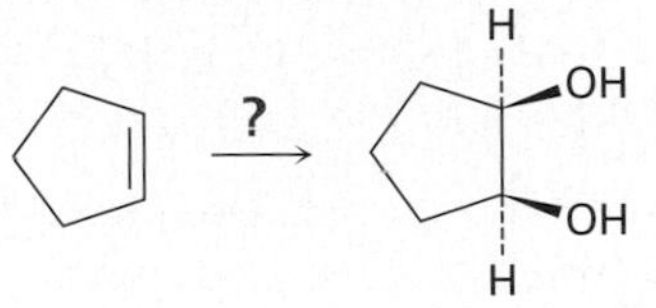

(b) Cyclopentene $\xrightarrow{?}$ cyclopentanol (OH)

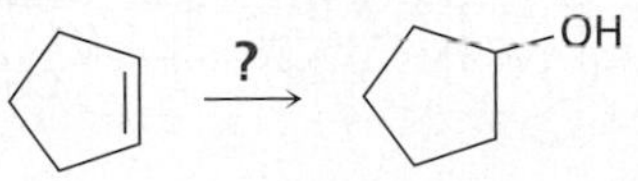

(c) Cyclopentene $\xrightarrow{?}$ dichlorocyclopropane-fused cyclopentane (H, Cl, Cl, H)

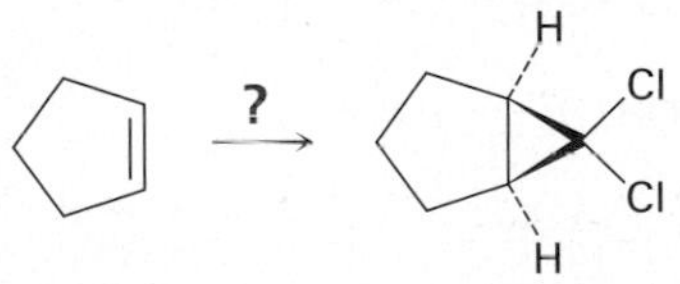

(d) 1-Methylcyclohexanol (CH_3, OH) $\xrightarrow{?}$ 1-methylcyclohexene (CH_3)

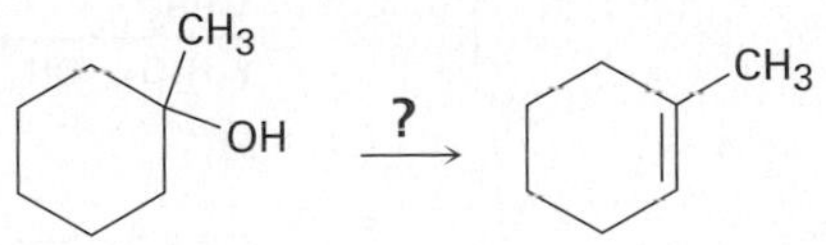

(e) $CH_3CH{=}CHCH(CH_3)CH_3 \xrightarrow{?} CH_3CHO + CH_3CH(CH_3)CHO$

(f) $CH_3C(CH_3){=}CH_2 \xrightarrow{?} CH_3CH(CH_3)CH_2OH$

8.36 Draw the structure of an alkene that yields only acetone, $(CH_3)_2C{=}O$, on ozonolysis followed by treatment with Zn.

8.37 Show the structures of alkenes that give the following products on oxidative cleavage with $KMnO_4$ in acidic solution:

(a) $CH_3CH_2CO_2H + CO_2$

(b) $(CH_3)_2C{=}O + CH_3CH_2CH_2CO_2H$

(c) 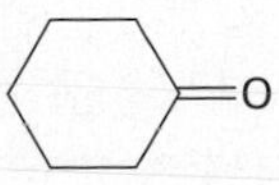 Cyclohexanone ($=$O) $+ (CH_3)_2C{=}O$

(d) $CH_3CH_2C(=O)CH_2CH_2CH_2CH_2CO_2H$

8.38 In planning the synthesis of one compound from another, it's just as important to know what *not* to do as to know what to do. The following reactions all have serious drawbacks to them. Explain the potential problems of each.

(a) $CH_3C(CH_3){=}CHCH_3 \xrightarrow{HI} CH_3CH(CH_3)CH(I)CH_3$

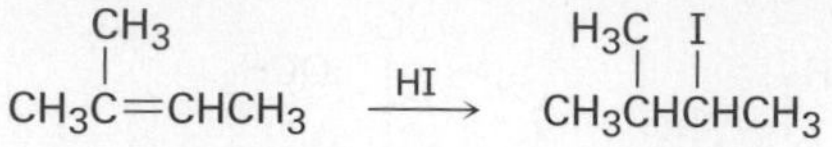

(b) Cyclopentene $\xrightarrow[\text{2. NaHSO}_3]{\text{1. OsO}_4}$ trans-1,2-cyclopentanediol (H, OH / H, OH)

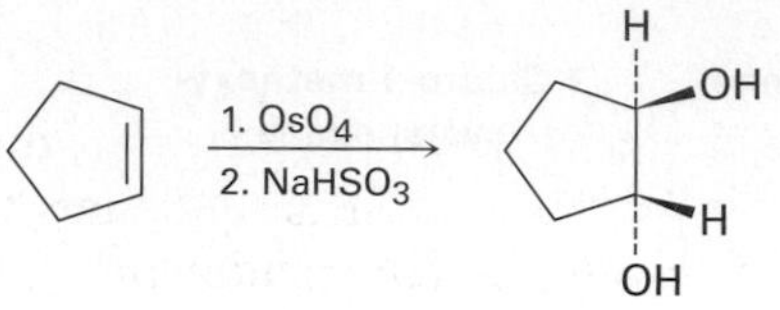

(c) Cyclohexene $\xrightarrow[\text{2. Zn}]{\text{1. O}_3}$ CHO, CHO

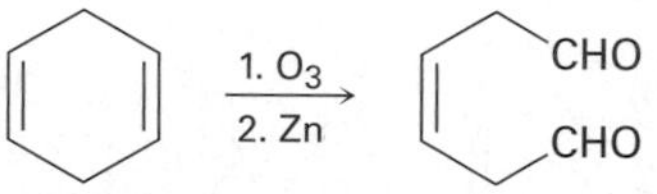

(d) 1-Methylcyclohexene (CH_3) $\xrightarrow[\text{2. H}_2\text{O}_2,\ ^-\text{OH}]{\text{1. BH}_3}$ product (H, CH_3 / H, OH)

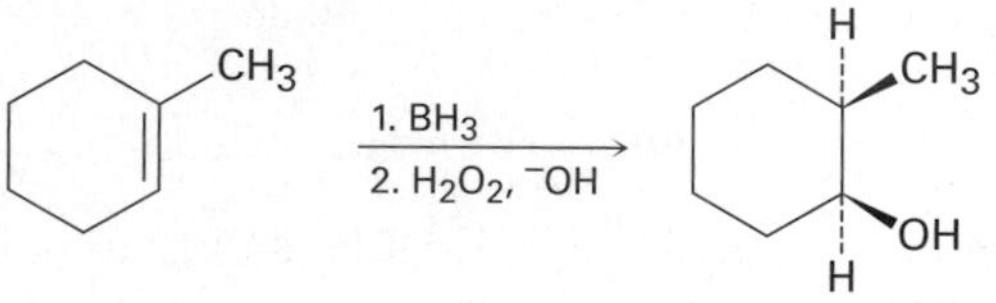

8.39 Which of the following alcohols could *not* be made selectively by hydroboration–oxidation of an alkene? Explain.

(a) $CH_3CH_2CH_2CH(OH)CH_3$

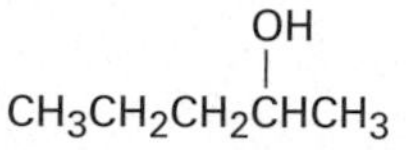

(b) $(CH_3)_2CHC(OH)(CH_3)_2$

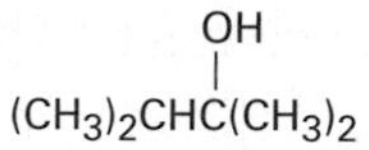

(c) Cyclohexane ring with H, CH_3 / OH, H

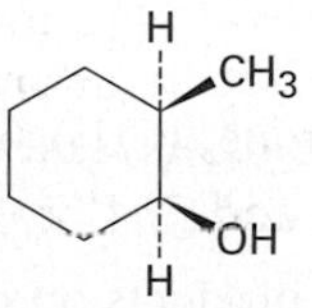

(d) Cyclohexane ring with OH, CH_3 / H, H

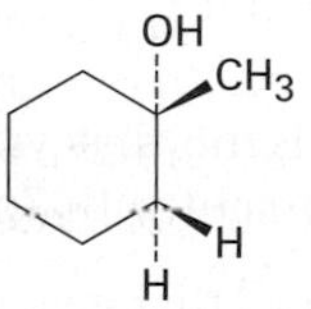

Polymers

8.40 Plexiglas, a clear plastic used to make many molded articles, is made by polymerization of methyl methacrylate. Draw a representative segment of Plexiglas.

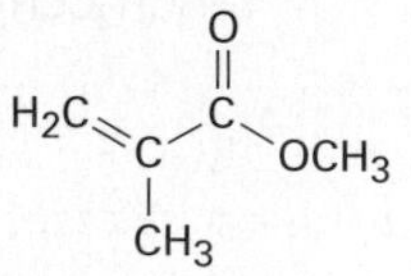

$H_2C{=}C(CH_3)C(=O)OCH_3$ **Methyl methacrylate**

8.41 Poly(vinyl pyrrolidone), prepared from *N*-vinylpyrrolidone, is used both in cosmetics and as a synthetic blood substitute. Draw a representative segment of the polymer.

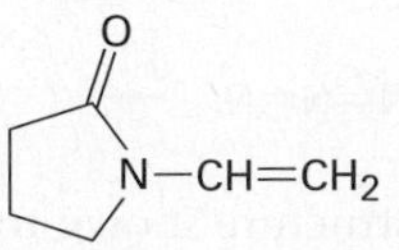

N—CH=CH$_2$ **N-Vinylpyrrolidone**

8.42 When a single alkene monomer, such as ethylene, is polymerized, the product is a *homopolymer.* If a mixture of two alkene monomers is polymerized, however, a *copolymer* often results. The following structure represents a segment of a copolymer called *Saran.* What two monomers were copolymerized to make Saran?

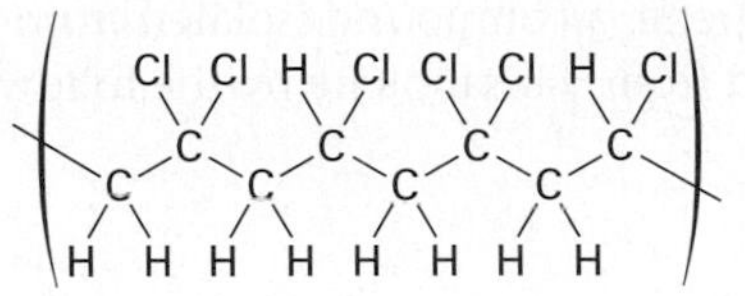

Saran

General Problems

8.43 Compound A has the formula $C_{10}H_{16}$. On catalytic hydrogenation over palladium, it reacts with only 1 molar equivalent of H_2. Compound A also undergoes reaction with ozone, followed by zinc treatment, to yield a symmetrical diketone, **B** ($C_{10}H_{16}O_2$).

(a) How many rings does **A** have?

(b) What are the structures of **A** and **B**?

(c) Write the reactions.

8.44 An unknown hydrocarbon **A** with the formula C_6H_{12} reacts with 1 molar equivalent of H_2 over a palladium catalyst. Hydrocarbon **A** also reacts with OsO_4 to give diol **B**. When oxidized with $KMnO_4$ in acidic solution, **A** gives two fragments. One fragment is propanoic acid, $CH_3CH_2CO_2H$, and the other fragment is ketone **C**. What are the structures of **A**, **B**, and **C**? Write all reactions, and show your reasoning.

8.45 Using an oxidative cleavage reaction, explain how you would distinguish between the following two isomeric dienes:

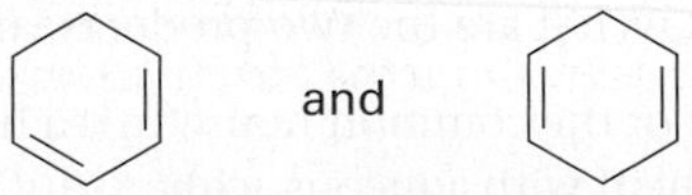

and

8.46 Compound A, $C_{10}H_{18}O$, undergoes reaction with dilute H_2SO_4 at 50 °C to yield a mixture of two alkenes, $C_{10}H_{16}$. The major alkene product, **B**, gives only cyclopentanone after ozone treatment followed by reduction with zinc in acetic acid. Identify **A** and **B**, and write the reactions.

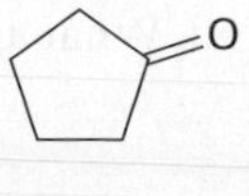

Cyclopentanone

8.47 Iodine azide, IN_3, adds to alkenes by an electrophilic mechanism similar to that of bromine. If a monosubstituted alkene such as 1-butene is used, only one product results:

$$CH_3CH_2CH{=}CH_2 + I{-}N{=}N{=}N \longrightarrow CH_3CH_2\underset{}{C}H(N{=}N{=}N)CH_2I$$

(a) Add lone-pair electrons to the structure shown for IN_3, and draw a second resonance form for the molecule.

(b) Calculate formal charges for the atoms in both resonance structures you drew for IN_3 in part (a).

(c) In light of the result observed when IN_3 adds to 1-butene, what is the polarity of the $I{-}N_3$ bond? Propose a mechanism for the reaction using curved arrows to show the electron flow in each step.

8.48 10-Bromo-α-chamigrene, a compound isolated from marine algae, is thought to be biosynthesized from γ-bisabolene by the following route:

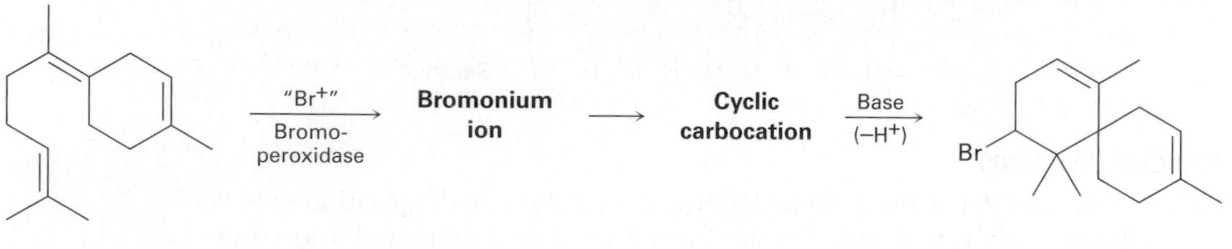

γ-Bisabolene —"Br+" / Bromo-peroxidase→ **Bromonium ion** → **Cyclic carbocation** —Base (–H+)→ **10-Bromo-α-chamigrene**

Draw the structures of the intermediate bromonium and cyclic carbocation, and propose mechanisms for all three steps.

8.49 Draw the structure of a hydrocarbon that absorbs 2 molar equivalents of H_2 on catalytic hydrogenation and gives only butanedial on ozonolysis.

$$HCCH_2CH_2CH \text{ (with C=O at each end)}$$ **Butanedial**

8.50 Simmons–Smith reaction of cyclohexene with diiodomethane gives a single cyclopropane product, but the analogous reaction of cyclohexene with 1,1-diiodoethane gives (in low yield) a mixture of two isomeric methylcyclopropane products. What are the two products, and how do they differ?

8.51 The sex attractant of the common housefly is a hydrocarbon with the formula $C_{23}H_{46}$. On treatment with aqueous acidic $KMnO_4$, two products are obtained, $CH_3(CH_2)_{12}CO_2H$ and $CH_3(CH_2)_7CO_2H$. Propose a structure.

8.52 Compound **A** has the formula C_8H_8. It reacts rapidly with $KMnO_4$ to give CO_2 and a carboxylic acid, **B** ($C_7H_6O_2$), but reacts with only 1 molar equivalent of H_2 on catalytic hydrogenation over a palladium catalyst. On hydrogenation under conditions that reduce aromatic rings, 4 equivalents of H_2 are taken up and hydrocarbon **C** (C_8H_{16}) is produced. What are the structures of **A**, **B**, and **C**? Write the reactions.

8.53 Isolated from marine algae, prelaureatin is thought to be biosynthesized from laurediol by the following route. Propose a mechanism.

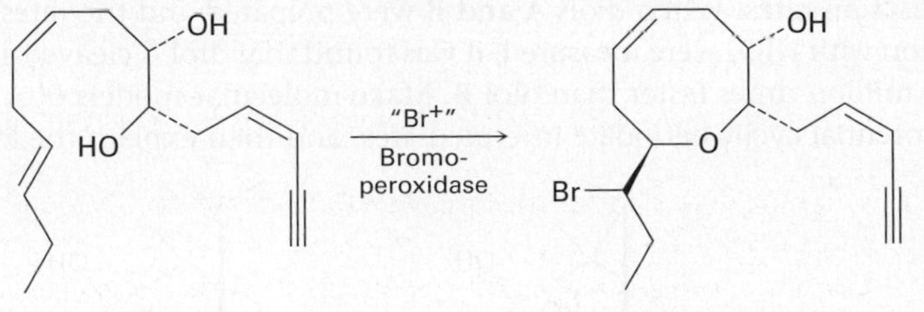

"Br^+"
Bromo-
peroxidase

Laurediol **Prelaureatin**

8.54 How would you distinguish between the following pairs of compounds using simple chemical tests? Tell what you would do and what you would see.
(a) Cyclopentene and cyclopentane **(b)** 2-Hexene and benzene

8.55 Dichlorocarbene can be generated by heating sodium trichloroacetate. Propose a mechanism for the reaction, and use curved arrows to indicate the movement of electrons in each step. What relationship does your mechanism bear to the base-induced elimination of HCl from chloroform?

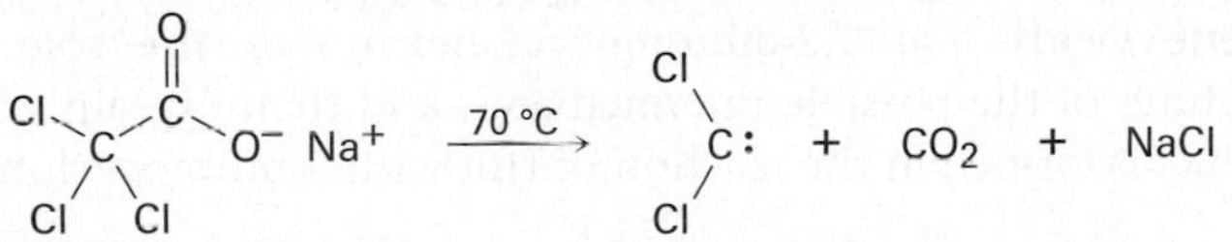

$$Cl_3C{-}CO_2^-\ Na^+ \xrightarrow{70\ °C} :CCl_2 + CO_2 + NaCl$$

8.56 α-Terpinene, $C_{10}H_{16}$, is a pleasant-smelling hydrocarbon that has been isolated from oil of marjoram. On hydrogenation over a palladium catalyst, α-terpinene reacts with 2 molar equivalents of H_2 to yield a hydrocarbon, $C_{10}H_{20}$. On ozonolysis, followed by reduction with zinc and acetic acid, α-terpinene yields two products, glyoxal and 6-methyl-2,5-heptanedione.

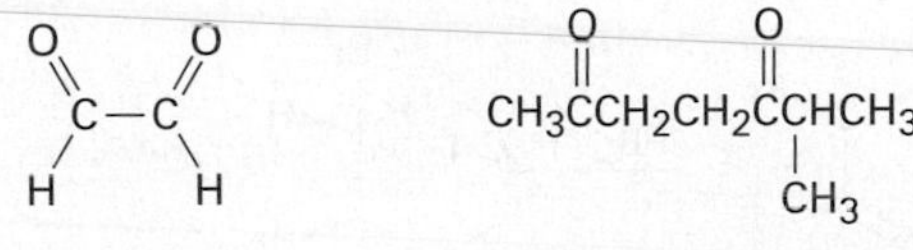

Glyoxal **6-Methyl-2,5-heptanedione**

(a) How many degrees of unsaturation does α-terpinene have?
(b) How many double bonds and how many rings does it have?
(c) Propose a structure for α-terpinene.

8.57 Evidence that cleavage of 1,2-diols by HIO_4 occurs through a five-membered cyclic periodate intermediate is based on *kinetic data*—the measurement of reaction rates. When diols **A** and **B** were prepared and the rates of their reaction with HIO_4 were measured, it was found that diol **A** cleaved approximately 1 million times faster than diol **B**. Make molecular models of **A** and **B** and of potential cyclic periodate intermediates, and then explain the kinetic results.

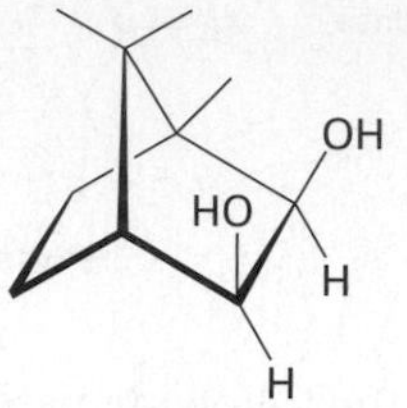

HO OH H H — **A** (*cis* diol)

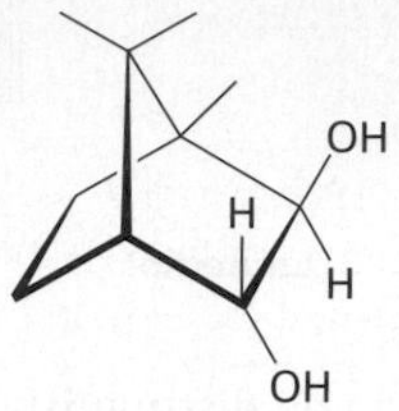

OH H H OH — **B** (*trans* diol)

8.58 Reaction of HBr with 3-methylcyclohexene yields a mixture of four products: *cis*- and *trans*-1-bromo-3-methylcyclohexane and *cis*- and *trans*-1-bromo-2-methylcyclohexane. The analogous reaction of HBr with 3-bromocyclohexene yields *trans*-1,2-dibromocyclohexane as the sole product. Draw structures of the possible intermediates, and then explain why only a single product is formed in the reaction of HBr with 3-bromocyclohexene.

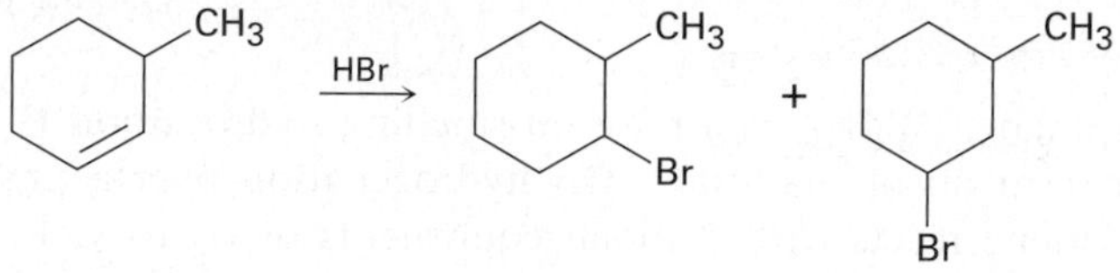

CH_3 —HBr→ CH_3, Br (**cis, trans**) + CH_3, Br (**cis, trans**)

Br —HBr→ Br, H, Br, H

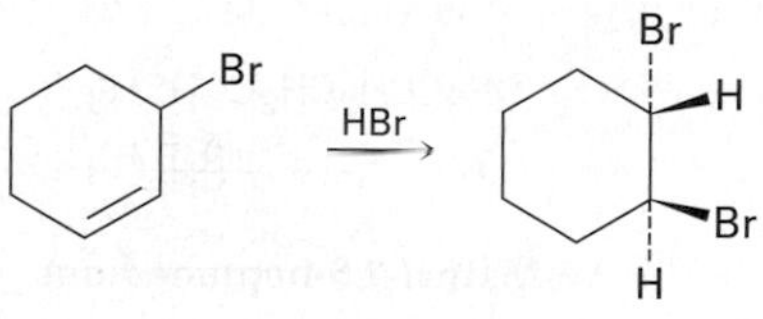

8.59 Reaction of cyclohexene with mercury(II) acetate in CH_3OH rather than H_2O, followed by treatment with $NaBH_4$, yields cyclohexyl methyl ether rather than cyclohexanol. Suggest a mechanism.

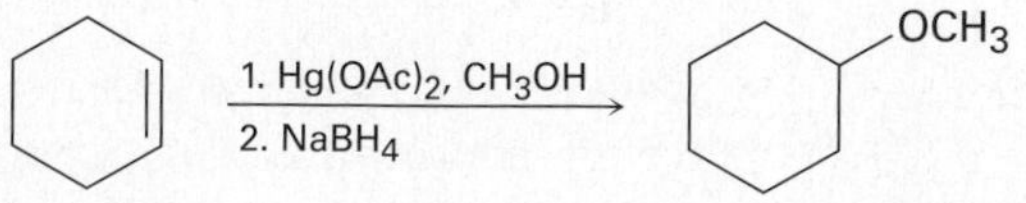

Cyclohexene —1. $Hg(OAc)_2$, CH_3OH; 2. $NaBH_4$→ OCH_3 **Cyclohexyl methyl ether**

8.60 Use your general knowledge of alkene chemistry to suggest a mechanism for the following reaction.

CO_2CH_3 $\xrightarrow{Hg(OAc)_2}$ CO_2CH_3

AcO—Hg

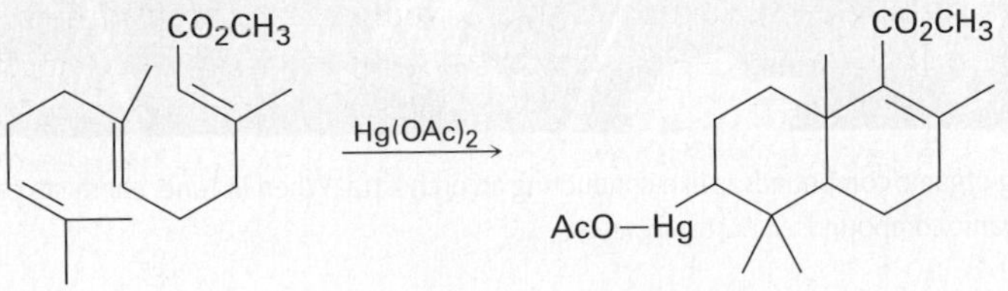

8.61 Treatment of 4-penten-1-ol with aqueous Br_2 yields a cyclic bromo ether rather than the expected bromohydrin, Suggest a mechanism, using curved arrows to show electron movement.

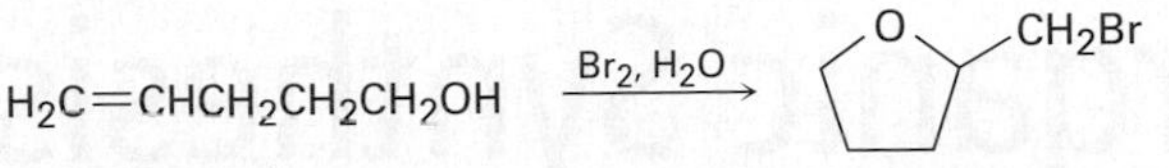

$$H_2C{=}CHCH_2CH_2CH_2OH \xrightarrow{Br_2,\ H_2O} \text{(O-ring)}{-}CH_2Br$$

4-Penten-1-ol **2-(Bromomethyl)tetrahydrofuran**

8.62 Hydroboration of 2-methyl-2-pentene at 25 °C followed by oxidation with alkaline H_2O_2 yields 2-methyl-3-pentanol, but hydroboration at 160 °C followed by oxidation yields 4-methyl-1-pentanol. Suggest a mechanism.

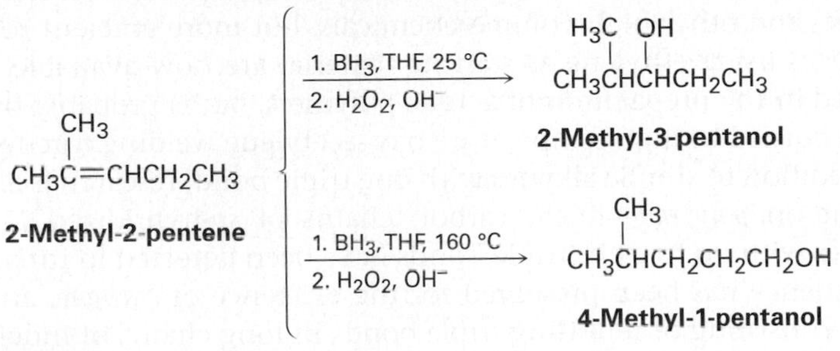

2-Methyl-2-pentene, $CH_3C(CH_3){=}CHCH_2CH_3$:

- 1. BH_3, THF, 25 °C; 2. H_2O_2, OH^- → $CH_3CH(CH_3)CH(OH)CH_2CH_3$ **2-Methyl-3-pentanol**
- 1. BH_3, THF, 160 °C; 2. H_2O_2, OH^- → $CH_3CH(CH_3)CH_2CH_2CH_2OH$ **4-Methyl-1-pentanol**

8.63 We'll see in the next chapter that alkynes undergo many of the same reactions that alkenes do. What product might you expect from each of the following reactions?

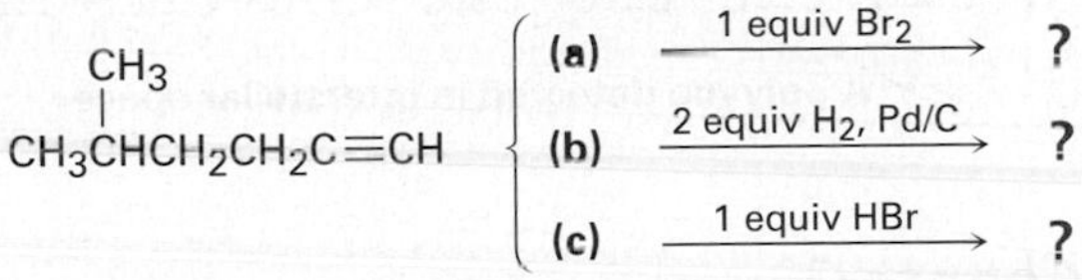

$CH_3CH(CH_3)CH_2CH_2C{\equiv}CH$

- **(a)** $\xrightarrow{\text{1 equiv } Br_2}$?
- **(b)** $\xrightarrow{\text{2 equiv } H_2,\ Pd/C}$?
- **(c)** $\xrightarrow{\text{1 equiv HBr}}$?

8.64 Hydroxylation of *cis*-2-butene with OsO_4 yields a different product than hydroxylation of *trans*-2-butene. Draw the structure, show the stereochemistry of each product, and explain the difference between them.

8.65 Compound **A**, $C_{11}H_{16}O$, was found to be an optically active alcohol. Despite its apparent unsaturation, no hydrogen was absorbed on catalytic reduction over a palladium catalyst. On treatment of **A** with dilute sulfuric acid, dehydration occurred and an optically inactive alkene **B**, $C_{11}H_{14}$, was produced as the major product. Alkene **B**, on ozonolysis, gave two products. One product was identified as propanal, CH_3CH_2CHO. Compound **C**, the other product, was shown to be a ketone, C_8H_8O. How many degrees of unsaturation does **A** have? Write the reactions, and identify **A**, **B**, and **C**.

9

Synthesizing organic compounds is like conducting an orchestra. When in tune, chemists can create highly complex organic compounds. © Olaf Doering/Alamy

Alkynes: An Introduction to Organic Synthesis

An **alkyne** is a hydrocarbon that contains a carbon–carbon triple bond. Acetylene, H—C≡C—H, the simplest alkyne, was once widely used in industry as the starting material for the preparation of acetaldehyde, acetic acid, vinyl chloride, and other high-volume chemicals, but more efficient routes to these substances using ethylene as starting material are now available. Acetylene is still used in the preparation of acrylic polymers, but is probably best known as the gas burned in high-temperature oxy–acetylene welding torches.

In addition to simple alkynes with one triple bond, research is also being carried out on *polyynes*—linear carbon chains of *sp*-hybridized carbon atoms. Polyynes with up to eight triple bonds have been detected in interstellar space, and evidence has been presented for the existence of *carbyne,* an allotrope of carbon consisting of repeating triple bonds in long chains of indefinite length. The electronic properties of polyynes are being explored for potential use in nanotechnology applications.

H—C≡C—C≡C—C≡C—C≡C—C≡C—C≡C—C≡C—H

A polyyne detected in interstellar space

Why This Chapter? Alkynes are less common than alkenes, both in the laboratory and in living organisms, so we won't cover them in great detail. The real importance of this chapter is that we'll use alkyne chemistry as a vehicle to begin looking at some of the general strategies used in organic synthesis—the construction of complex molecules in the laboratory. Without the ability to design and synthesize new molecules in the laboratory, many of the medicines we take for granted would not exist and few new ones would be made.

9.1 Naming Alkynes

Alkyne nomenclature follows the general rules for hydrocarbons discussed in **Sections 3.4 and 7.3**. The suffix *-yne* is used, and the position of the triple bond is indicated by giving the number of the first alkyne carbon in the chain.

OWL Sign in to OWL for Organic Chemistry at **www.cengage.com/owl** to view tutorials and simulations, develop problem-solving skills, and complete online homework assigned by your professor.

Numbering the main chain begins at the end nearer the triple bond so that the triple bond receives as low a number as possible.

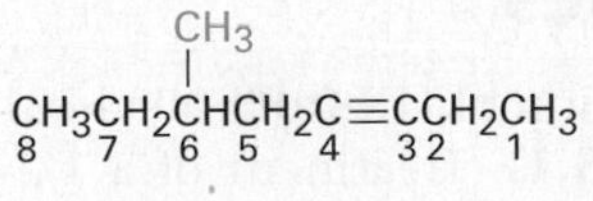

Begin numbering at the end nearer the triple bond.

6-Methyl-3-octyne

(New: 6-Methyloct-3-yne)

Compounds with more than one triple bond are called diynes, triynes, and so forth; compounds containing both double and triple bonds are called enynes (not ynenes). Numbering of an enyne chain starts from the end nearer the first multiple bond, whether double or triple. When there is a choice in numbering, double bonds receive lower numbers than triple bonds. For example:

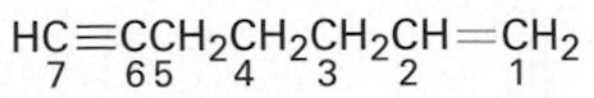

1-Hepten-6-yne

(New: Hept-1-en-6-yne)

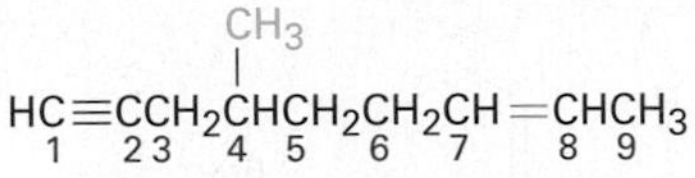

4-Methyl-7-nonen-1-yne

(New: 4-Methylnon-7-en-1-yne)

As with alkyl and alkenyl substituents derived from alkanes and alkenes, respectively, alkynyl groups are also possible.

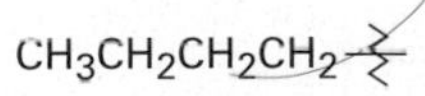

Butyl
(an alkyl group)

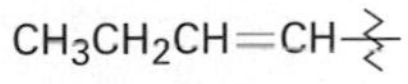

1-Butenyl
(a vinylic group)

(New: But-1-enyl)

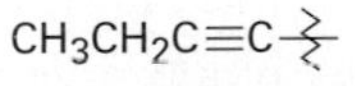

1-Butynyl
(an alkynyl group)

(New: But-1-ynyl)

Problem 9.1

Name the following compounds:

(a)

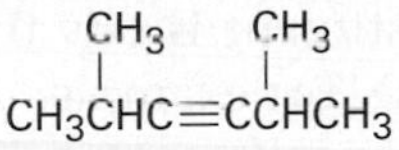

(b)

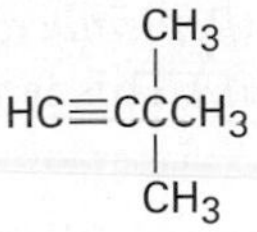

(c)

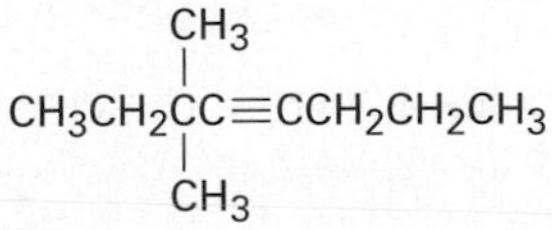

(d)

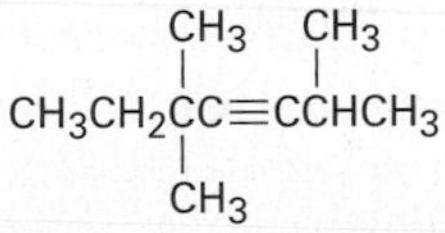

(e)

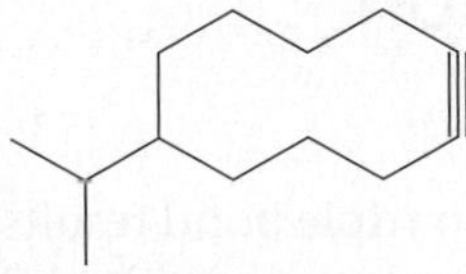

(f) $CH_3CH{=}CHCH{=}CHC{\equiv}CCH_3$

Problem 9.2

There are seven isomeric alkynes with the formula C_6H_{10}. Draw and name them.

9.2 Preparation of Alkynes: Elimination Reactions of Dihalides

Alkynes can be prepared by the elimination of HX from alkyl halides in much the same manner as alkenes **(Section 8.1)**. Treatment of a 1,2-dihaloalkane (a *vicinal* dihalide) with an excess amount of a strong base such as KOH or $NaNH_2$ results in a twofold elimination of HX and formation of an alkyne. As with the elimination of HX to form an alkene, we'll defer a full discussion of this topic and the relevant reaction mechanisms until Chapter 11.

The starting vicinal dihalides are themselves readily available by addition of Br_2 or Cl_2 to alkenes. Thus, the overall halogenation/dehydrohalogenation sequence makes it possible to go from an alkene to an alkyne. For example, diphenylethylene is converted into diphenylacetylene by reaction with Br_2 and subsequent base treatment.

Br_2, CH_2Cl_2

1,2-Diphenylethylene (stilbene) → **1,2-Dibromo-1,2-diphenylethane (a vicinal dibromide)**

2 KOH, ethanol

→ **Diphenylacetylene (85%)** + $2\ H_2O$ + 2 KBr

The twofold dehydrohalogenation takes place through a vinylic halide intermediate, which suggests that vinylic halides themselves should give alkynes when treated with strong base. (*Remember:* A *vinylic* substituent is one that is attached to a double-bond carbon.) This is indeed the case. For example:

$H_3C(Cl)C{=}CH(CH_2OH)$ → 1. $2\ NaNH_2$; 2. H_3O^+ → $CH_3C{\equiv}CCH_2OH$

(*Z*)-3-Chloro-2-buten-1-ol → **2-Butyn-1-ol**

9.3 Reactions of Alkynes: Addition of HX and X_2

You might recall from **Section 1.9** that a carbon–carbon triple bond results from the interaction of two *sp*-hybridized carbon atoms. The two *sp* hybrid orbitals of carbon lie at an angle of 180° to each other along an axis perpendicular to the axes of the two unhybridized $2p_y$ and $2p_z$ orbitals. When two *sp*-hybridized carbons approach each other, one *sp*–*sp* σ bond and two *p*–*p* π bonds are formed.

The two remaining *sp* orbitals form bonds to other atoms at an angle of 180° from the carbon–carbon bond. Thus, acetylene is a linear molecule with H–C≡C bond angles of 180° (**Figure 9.1**). The length of the C≡C bond is 120 pm, and its strength is approximately 965 kJ/mol (231 kcal/mol), making it the shortest and strongest known carbon–carbon bond.

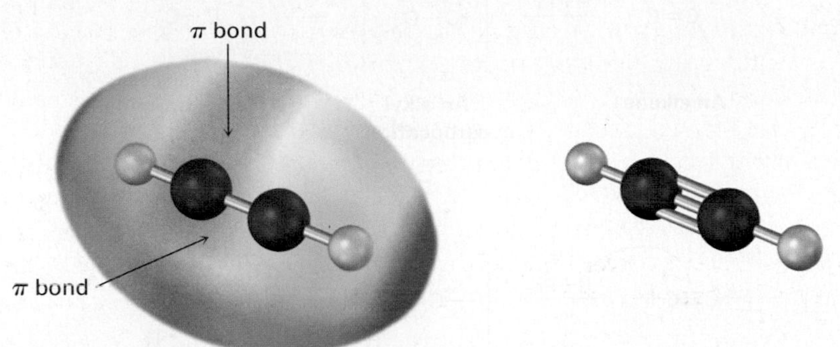

Figure 9.1 The structure of acetylene, H—C≡C—H. The H—C≡C bond angles are 180°, and the C≡C bond length is 120 pm. The electrostatic potential map shows that the π bonds create a **negative belt** around the molecule.

As a general rule, electrophiles undergo addition reactions with alkynes much as they do with alkenes. Take the reaction of alkynes with HX, for instance. The reaction often can be stopped after addition of 1 equivalent of HX, but reaction with an excess of HX leads to a dihalide product. For example, reaction of 1-hexyne with 2 equivalents of HBr yields 2,2-dibromohexane. As the following examples indicate, the regiochemistry of addition follows Markovnikov's rule, with halogen adding to the more highly substituted side of the alkyne bond and hydrogen adding to the less highly substituted side. Trans stereochemistry of H and X normally, although not always, results in the product.

$$CH_3CH_2CH_2CH_2C{\equiv}CH \xrightarrow{HBr} CH_3CH_2CH_2CH_2C(Br){=}CH_2 \xrightarrow{HBr} CH_3CH_2CH_2CH_2CBr_2CH_3$$

1-Hexyne → **2-Bromo-1-hexene** → **2,2-Dibromohexane**

$$CH_3CH_2C{\equiv}CCH_2CH_3 \xrightarrow[CH_3CO_2H]{HCl} (Cl)(CH_3CH_2)C{=}C(CH_2CH_3)(H) \xrightarrow[CH_3CO_2H]{HCl} CH_3CH_2CCl_2CH_2CH_2CH_3$$

3-Hexyne → **(*Z*)-3-Chloro-3-hexene** → **3,3-Dichlorohexane**

Bromine and chlorine also add to alkynes to give addition products, and trans stereochemistry again results.

$$CH_3CH_2C{\equiv}CH \xrightarrow[CH_2Cl_2]{Br_2} (Br)(CH_3CH_2)C{=}C(H)(Br) \xrightarrow[CH_2Cl_2]{Br_2} CH_3CH_2CBr_2{-}CHBr_2$$

1-Butyne → **(*E*)-1,2-Dibromo-1-butene** → **1,1,2,2-Tetrabromobutane**

The mechanism of alkyne additions is similar but not identical to that of alkene additions. When an electrophile such as HBr adds to an alkene, the reaction takes place in two steps and involves an alkyl carbocation intermediate

(Sections 7.7 and 7.8). If HBr were to add by the same mechanism to an *alkyne*, an analogous *vinylic* carbocation would be formed as the intermediate.

An alkene → (H–Br) → [An alkyl carbocation] + Br^- → An alkyl bromide

An alkyne → (H–Br) → [A vinylic carbocation] + Br^- → A vinylic bromide

A vinylic carbocation has an *sp*-hybridized carbon and generally forms less readily than an alkyl carbocation **(Figure 9.2)**. As a rule, a secondary vinylic carbocation forms about as readily as a primary alkyl carbocation, but a primary vinylic carbocation is so difficult to form that there is no clear evidence it even exists. Thus, many alkyne additions occur through more complex mechanistic pathways.

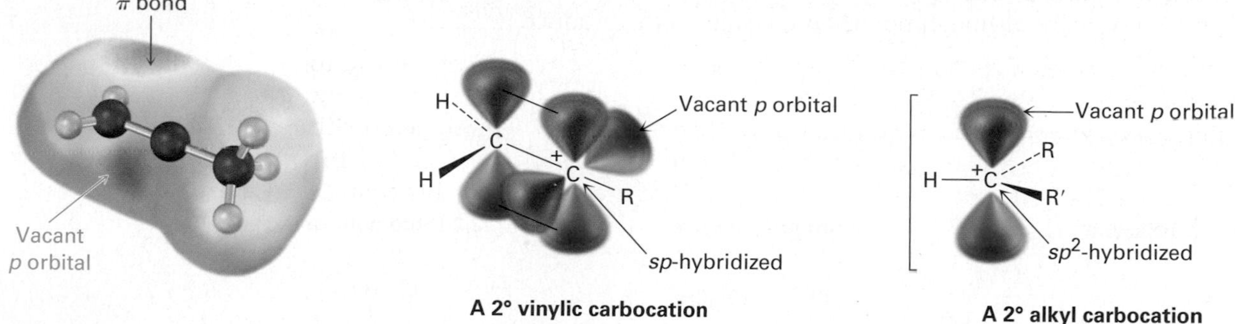

Figure 9.2 The structure of a secondary vinylic carbocation. The cationic carbon atom is *sp*-hybridized and has a vacant *p* orbital perpendicular to the plane of the π bond orbitals. Only one R group is attached to the positively charged carbon rather than two, as in a secondary alkyl carbocation. The electrostatic potential map shows that the **most positive regions** coincide with lobes of the vacant *p* orbital and are perpendicular to the **most negative regions** associated with the π bond.

Problem 9.3

What products would you expect from the following reactions?

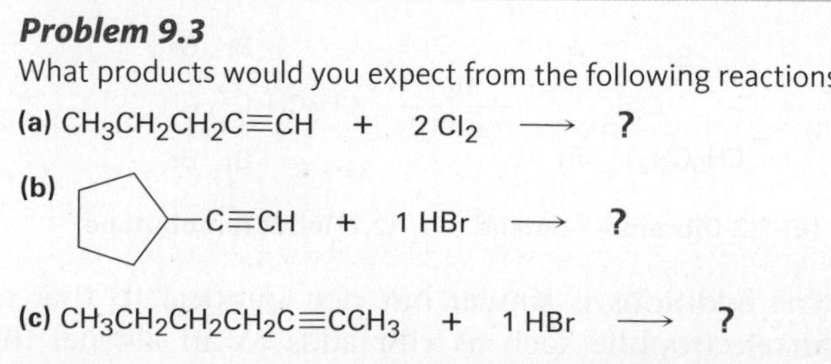

(a) $CH_3CH_2CH_2C{\equiv}CH + 2\ Cl_2 \longrightarrow$?

(b) (cyclopentyl)–C≡CH + 1 HBr ⟶ ?

(c) $CH_3CH_2CH_2CH_2C{\equiv}CCH_3 + 1\ HBr \longrightarrow$?

9.4 Hydration of Alkynes

Like alkenes **(Sections 8.4 and 8.5)**, alkynes can be hydrated by either of two methods. Direct addition of water catalyzed by mercury(II) ion yields the Markovnikov product, and indirect addition of water by a hydroboration–oxidation sequence yields the non-Markovnikov product.

Mercury(II)-Catalyzed Hydration of Alkynes

Alkynes don't react directly with aqueous acid but will undergo hydration readily in the presence of mercury(II) sulfate as a Lewis acid catalyst. The reaction occurs with Markovnikov regiochemistry, so the –OH group adds to the more highly substituted carbon and the –H attaches to the less highly substituted one.

$$CH_3CH_2CH_2CH_2C{\equiv}CH \xrightarrow[HgSO_4]{H_2O,\ H_2SO_4} \left[CH_3CH_2CH_2CH_2C(OH){=}CH_2 \right] \longrightarrow CH_3CH_2CH_2CH_2C({=}O)CH_3$$

1-Hexyne — An enol — 2-Hexanone (78%)

Interestingly, the product actually isolated from alkyne hydration is not the vinylic alcohol, or **enol** (*ene* + *ol*), but is instead a ketone. Although the enol is an intermediate in the reaction, it immediately rearranges to a ketone by a process called *keto–enol tautomerism.* The individual keto and enol forms are said to be **tautomers**, a word used to describe two isomers that under spontaneous interconversion accompanied by the change in position of a hydrogen. With few exceptions, the keto–enol tautomeric equilibrium lies on the side of the ketone; enols are almost never isolated. We'll look more closely at this equilibrium in **Section 22.1**.

Enol tautomer (less favored) ⇌ Keto tautomer (more favored)

As shown in **Figure 9.3**, the mechanism of the mercury(II)-catalyzed alkyne hydration reaction is analogous to the oxymercuration reaction of alkenes **(Section 8.4)**. Electrophilic addition of mercury(II) ion to the alkyne gives a vinylic cation, which reacts with water and loses a proton to yield a mercury-containing enol intermediate. In contrast with alkene oxymercuration, however, no treatment with $NaBH_4$ is necessary to remove the mercury. The acidic reaction conditions alone are sufficient to effect replacement of mercury by hydrogen. Tautomerization then gives the ketone.

Figure 9.3 | MECHANISM

Mechanism of the mercury(II)-catalyzed hydration of an alkyne to yield a ketone. The reaction occurs through initial formation of an intermediate enol, which tautomerizes to the ketone.

1 The alkyne uses a pair of electrons to attack the electrophilic mercury(II) ion, yielding a mercury-containing vinylic carbocation intermediate.

2 Nucleophilic attack of water on the carbocation forms a C–O bond and yields a protonated mercury-containing enol.

3 Abstraction of H^+ from the protonated enol by water gives an organomercury compound.

4 Replacement of Hg^{2+} by H^+ occurs to give a neutral enol.

5 The enol undergoes tautomerization to give the final ketone product.

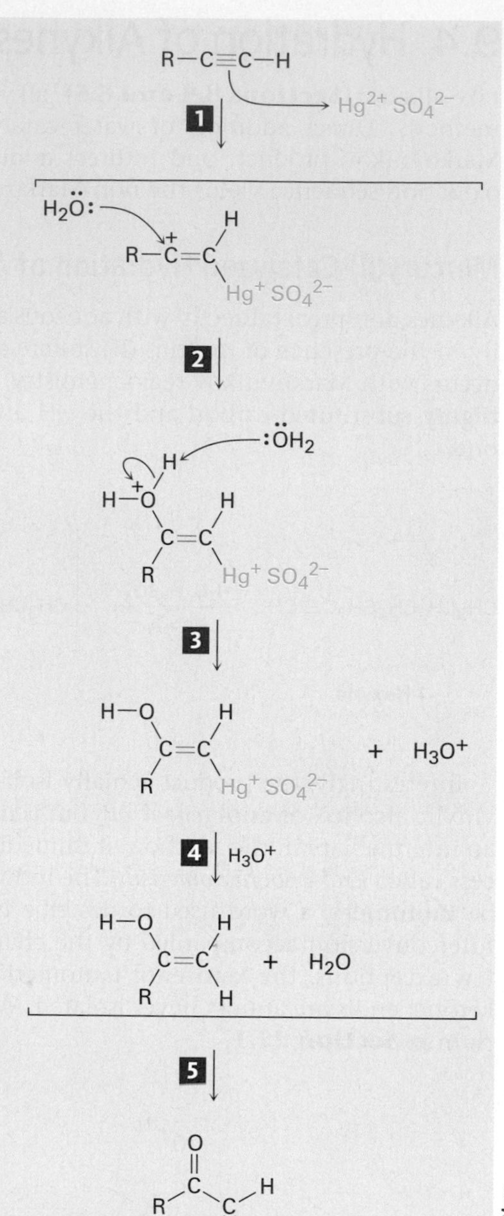

A mixture of both possible ketones results when an unsymmetrically substituted internal alkyne (RC≡CR′) is hydrated. The reaction is therefore most useful when applied to a terminal alkyne (RC≡CH) because only a methyl ketone is formed.

An internal alkyne

$$\text{R—C≡C—R}' \xrightarrow[\text{HgSO}_4]{\text{H}_3\text{O}^+} \text{RC(=O)CH}_2\text{R}' + \text{RCH}_2\text{C(=O)R}'$$

Mixture

A terminal alkyne

$$R{-}C{\equiv}C{-}H \xrightarrow[HgSO_4]{H_3O^+} R{-}\overset{\overset{O}{\|}}{C}{-}CH_3$$

A methyl ketone

Problem 9.4

What product would you obtain by hydration of the following alkynes?

(a) $CH_3CH_2CH_2C{\equiv}CCH_2CH_2CH_3$ **(b)** $CH_3\overset{\overset{CH_3}{|}}{C}HCH_2C{\equiv}CCH_2CH_2CH_3$

Problem 9.5

What alkynes would you start with to prepare the following ketones?

(a) $CH_3CH_2CH_2\overset{\overset{O}{\|}}{C}CH_3$ **(b)** $CH_3CH_2\overset{\overset{O}{\|}}{C}CH_2CH_3$

Hydroboration–Oxidation of Alkynes

Borane adds rapidly to an alkyne just as it does to an alkene, and the resulting vinylic borane can be oxidized by H_2O_2 to yield an enol. Tautomerization then gives either a ketone or an aldehyde, depending on the structure of the alkyne reactant. Hydroboration–oxidation of an internal alkyne such as 3-hexyne gives a ketone, and hydroboration–oxidation of a terminal alkyne gives an aldehyde. Note that the relatively unhindered terminal alkyne undergoes two additions, giving a doubly hydroborated intermediate. Oxidation with H_2O_2 at pH 8 then replaces both boron atoms by oxygen and generates the aldehyde.

An internal alkyne

$$3\ CH_3CH_2C{\equiv}CCH_2CH_3 \xrightarrow[THF]{BH_3} [\text{A vinylic borane}] \xrightarrow[H_2O,\ NaOH]{H_2O_2} 3\ [(H)(CH_3CH_2)C{=}C(OH)(CH_2CH_3)] \longrightarrow 3\ CH_3CH_2CH_2\overset{\overset{O}{\|}}{C}CH_2CH_3$$

A vinylic borane (B bonded to three $(CH_3CH_2)C{=}C(H)(CH_2CH_3)$ groups)

An enol

3-Hexanone

A terminal alkyne

$$CH_3CH_2CH_2CH_2C{\equiv}CH \xrightarrow[THF]{BH_3} CH_3CH_2CH_2CH_2CH_2{-}CH(BR_2)_2 \xrightarrow[H_2O,\ pH\ 8]{H_2O_2} CH_3CH_2CH_2CH_2CH_2\overset{\overset{O}{\|}}{C}H$$

1-Hexyne **Hexanal (70%)**

The hydroboration–oxidation sequence is complementary to the direct, mercury(II)-catalyzed hydration reaction of a terminal alkyne because different products result. Direct hydration with aqueous acid and mercury(II) sulfate leads to a methyl ketone, whereas hydroboration–oxidation of the same terminal alkyne leads to an aldehyde.

$$R{-}C{\equiv}C{-}H \xrightarrow[HgSO_4]{H_2O,\ H_2SO_4} R{-}\overset{O}{\overset{\|}{C}}{-}CH_3$$

A terminal alkyne → **A methyl ketone**

$$R{-}C{\equiv}C{-}H \xrightarrow[2.\ H_2O_2]{1.\ BH_3,\ THF} R{-}CH_2{-}\overset{O}{\overset{\|}{C}}{-}H$$

A terminal alkyne → **An aldehyde**

Problem 9.6

What alkyne would you start with to prepare each of the following compounds by a hydroboration–oxidation reaction?

(a) $C_6H_5{-}CH_2\overset{O}{\overset{\|}{C}}H$

(b) $CH_3\overset{CH_3}{\overset{|}{C}}HCH_2\overset{O}{\overset{\|}{C}}\underset{CH_3}{\underset{|}{C}}HCH_3$

Problem 9.7

How would you prepare the following carbonyl compounds starting from an alkyne (reddish brown = Br)?

(a) (b)

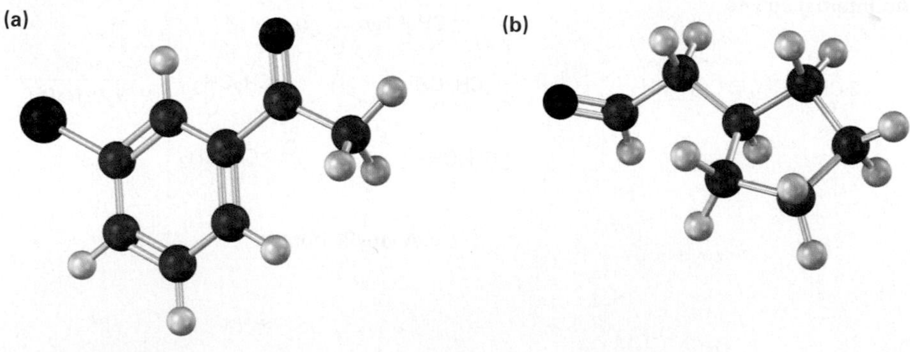

9.5 Reduction of Alkynes

Alkynes are reduced to alkanes by addition of H_2 over a metal catalyst. The reaction occurs in two steps through an alkene intermediate, and measurements show that the first step in the reaction is more exothermic than the second step.

$$HC{\equiv}CH \xrightarrow[\text{Catalyst}]{H_2} H_2C{=}CH_2 \qquad \Delta H^\circ_{hydrog} = -176 \text{ kJ/mol } (-42 \text{ kcal/mol})$$

$$H_2C{=}CH_2 \xrightarrow[\text{Catalyst}]{H_2} CH_3{-}CH_3 \qquad \Delta H^\circ_{hydrog} = -137 \text{ kJ/mol } (-33 \text{ kcal/mol})$$

Complete reduction to the alkane occurs when palladium on carbon (Pd/C) is used as catalyst, but hydrogenation can be stopped at the alkene stage if the less active *Lindlar catalyst* is used. The Lindlar catalyst is a finely divided palladium metal that has been precipitated onto a calcium carbonate support and then deactivated by treatment with lead acetate and quinoline, an aromatic amine. The hydrogenation occurs with syn stereochemistry **(Section 8.5)**, giving a cis alkene product.

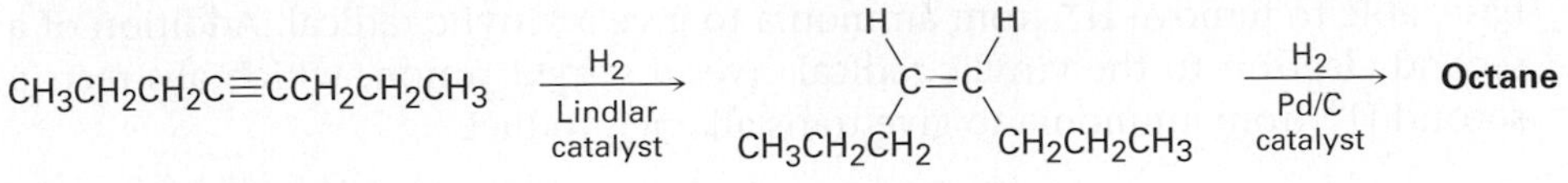

$$CH_3CH_2CH_2C{\equiv}CCH_2CH_2CH_3 \xrightarrow[\text{Lindlar catalyst}]{H_2} \text{cis-}CH_3CH_2CH_2CH{=}CHCH_2CH_2CH_3 \xrightarrow[\text{Pd/C catalyst}]{H_2} \text{Octane}$$

4-Octyne **cis-4-Octene**

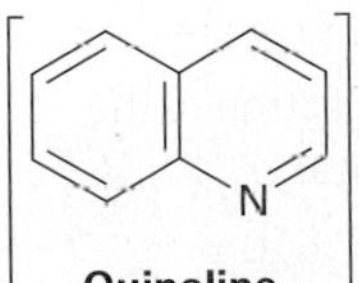

Quinoline

The alkyne hydrogenation reaction has been explored extensively by the Hoffmann–LaRoche pharmaceutical company, where it is used in the commercial synthesis of vitamin A. The cis isomer of vitamin A produced initially on hydrogenation is converted to the trans isomer by heating.

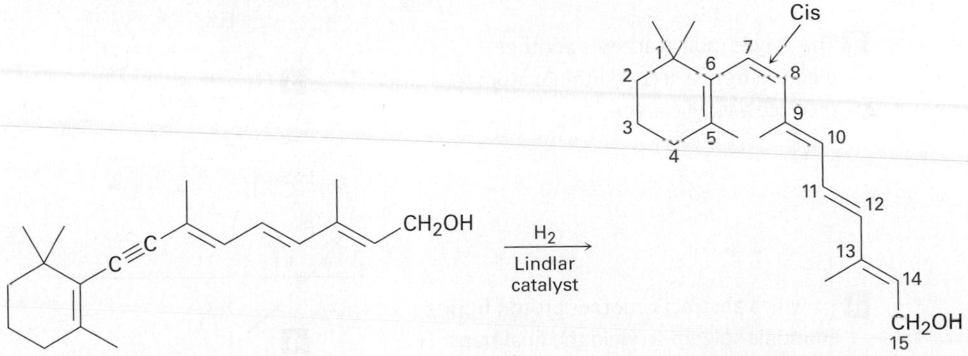

CH_2OH $\xrightarrow[\text{Lindlar catalyst}]{H_2}$ Cis, 1, 2, 3, 4, 5, 6, 7, 8, 9, 10, 11, 12, 13, 14, CH_2OH 15

7-*cis*-Retinol (7-*cis*-vitamin A; vitamin A has a trans double bond at C7)

An alternative method for the conversion of an alkyne to an alkene uses sodium or lithium metal as the reducing agent in liquid ammonia as solvent. This method is complementary to the Lindlar reduction because it produces trans rather than cis alkenes. For example, 5-decyne gives *trans*-5-decene on treatment with lithium in liquid ammonia.

$$CH_3CH_2CH_2CH_2C{\equiv}CCH_2CH_2CH_2CH_3 \xrightarrow[NH_3]{Li} (CH_3CH_2CH_2CH_2)HC{=}CH(CH_2CH_2CH_2CH_3)$$

5-Decyne → ***trans*-5-Decene (78%)**

Alkali metals dissolve in liquid ammonia at −33 °C to produce a deep blue solution containing the metal cation and ammonia-solvated electrons. When an alkyne is then added to the solution, reduction occurs by the mechanism shown in **Figure 9.4**. An electron first adds to the triple bond to yield an intermediate anion radical—a species that is both an anion (has a negative charge) and a radical (has an odd number of electrons). This anion radical is a strong base, able to remove H^+ from ammonia to give a vinylic radical. Addition of a second electron to the vinylic radical gives a vinylic anion, which abstracts a second H^+ from ammonia to give trans alkene product.

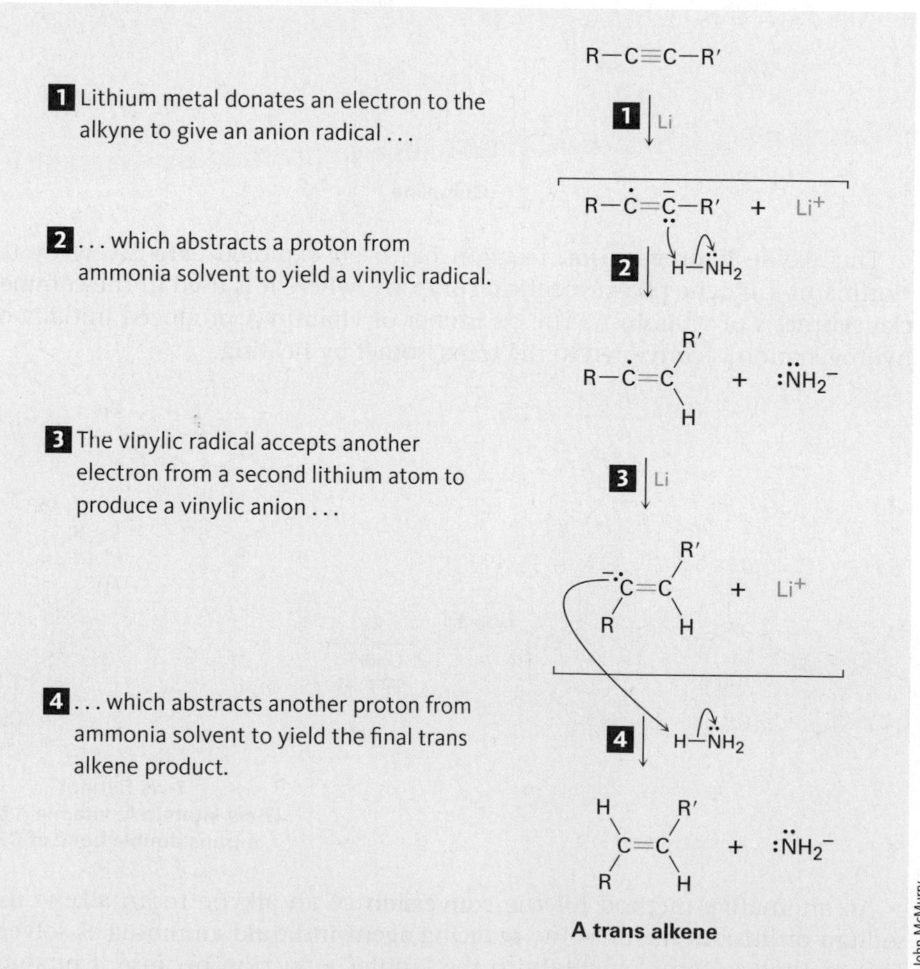

Figure 9.4 | MECHANISM

Mechanism of the lithium/ammonia reduction of an alkyne to produce a trans alkene.

Trans stereochemistry of the alkene product is established during the second reduction step (**3**) when the less hindered trans vinylic anion is formed from the vinylic radical. Vinylic radicals undergo rapid cis–trans equilibration, but vinylic anions equilibrate much less rapidly. Thus, the more stable trans vinylic anion is formed rather than the less stable cis anion and is then protonated without equilibration.

Problem 9.8
Using any alkyne needed, how would you prepare the following alkenes?
(a) *trans*-2-Octene **(b)** *cis*-3-Heptene **(c)** 3-Methyl-1-pentene

9.6 Oxidative Cleavage of Alkynes

Alkynes, like alkenes, can be cleaved by reaction with powerful oxidizing agents such as ozone or $KMnO_4$, although the reaction is of little value and we mention it only for completeness. A triple bond is generally less reactive than a double bond, and yields of cleavage products are sometimes low. The products obtained from cleavage of an internal alkyne are carboxylic acids; from a terminal alkyne, CO_2 is formed as one product.

An internal alkyne

$$\text{R}-\text{C}\equiv\text{C}-\text{R}' \xrightarrow{\text{KMnO}_4 \text{ or O}_3} \text{RCOOH} + \text{HOOC}\text{R}'$$

A terminal alkyne

$$\text{R}-\text{C}\equiv\text{C}-\text{H} \xrightarrow{\text{KMnO}_4 \text{ or O}_3} \text{RCOOH} + \text{O}=\text{C}=\text{O}$$

9.7 Alkyne Acidity: Formation of Acetylide Anions

The most striking difference between alkenes and alkynes is that terminal alkynes are relatively acidic. When a terminal alkyne is treated with a strong base, such as sodium amide, $Na^+ \, {}^-NH_2$, the terminal hydrogen is removed and an **acetylide anion** is formed.

$$\underset{\textbf{A terminal alkyne}}{\text{R}-\text{C}\equiv\text{C}-\text{H}} \xrightarrow{:\ddot{\text{N}}\text{H}_2^- \text{ Na}^+} \underset{\textbf{An acetylide anion}}{\text{R}-\text{C}\equiv\text{C}:^- \text{ Na}^+} + :\text{NH}_3$$

According to the Brønsted–Lowry definition **(Section 2.7)**, an acid is a substance that donates H^+. Although we usually think of oxyacids (H_2SO_4, HNO_3) or halogen acids (HCl, HBr) in this context, any compound containing a hydrogen atom can be an acid under the right circumstances. By measuring dissociation constants of different acids and expressing the results as pK_a values, an acidity order can be established. Recall from **Section 2.8** that a lower pK_a corresponds to a stronger acid and a higher pK_a corresponds to a weaker acid.

Where do hydrocarbons lie on the acidity scale? As the data in Table 9.1 show, both methane ($pK_a \approx 60$) and ethylene ($pK_a = 44$) are very weak acids and thus do not react with any of the common bases. Acetylene, however, has $pK_a = 25$ and can be deprotonated by the conjugate base of any acid whose pK_a is greater than 25. Amide ion (NH_2^-), for example, the conjugate base of ammonia ($pK_a = 35$), is often used to deprotonate terminal alkynes.

Table 9.1 Acidity of Simple Hydrocarbons

Family	Example	K_a	pK_a	
Alkyne	$HC{\equiv}CH$	10^{-25}	25	Stronger acid ↑
Alkene	$H_2C{=}CH_2$	10^{-44}	44	
Alkane	CH_4	10^{-60}	60	Weaker acid

Why are terminal alkynes more acidic than alkenes or alkanes? In other words, why are acetylide anions more stable than vinylic or alkyl anions? The simplest explanation involves the hybridization of the negatively charged carbon atom. An acetylide anion has an *sp*-hybridized carbon, so the negative charge resides in an orbital that has 50% "*s* character." A vinylic anion has an sp^2-hybridized carbon with 33% *s* character, and an alkyl anion (sp^3) has only 25% *s* character. Because *s* orbitals are nearer the positive nucleus and lower in energy than *p* orbitals, the negative charge is stabilized to a greater extent in an orbital with higher *s* character **(Figure 9.5)**.

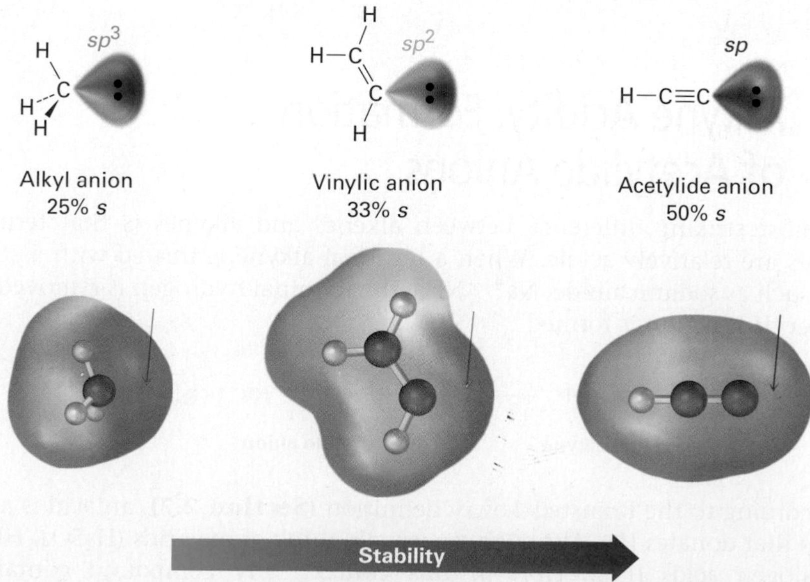

Figure 9.5 A comparison of alkyl, vinylic, and acetylide anions. The acetylide anion, with *sp* hybridization, has more *s* character and is more stable. Electrostatic potential maps show that placing the negative charge closer to the carbon nucleus makes carbon appear less negative (red).

Problem 9.9

The pK_a of acetone, CH_3COCH_3, is 19.3. Which of the following bases is strong enough to deprotonate acetone?

(a) KOH (pK_a of H_2O = 15.7) **(b)** Na^+ $^-C{\equiv}CH$ (pK_a of C_2H_2 = 25)

(c) $NaHCO_3$ (pK_a of H_2CO_3 = 6.4) **(d)** $NaOCH_3$ (pK_a of CH_3OH = 15.6)

9.8 Alkylation of Acetylide Anions

The negative charge and unshared electron pair on carbon make an acetylide anion strongly nucleophilic. As a result, an acetylide anion can react with electrophiles, such as alkyl halides, in a process that replaces the halide and yields a new alkyne product.

$$H{-}C{\equiv}C{:}^-\ Na^+ \ + \ CH_3Br \longrightarrow H{-}C{\equiv}C{-}CH_3 \ + \ NaBr$$

Acetylide anion **Propyne**

We won't study the details of this substitution reaction until Chapter 11 but for now can picture it as happening by the pathway shown in **Figure 9.6**. The nucleophilic acetylide ion uses an electron pair to form a bond to the positively polarized, electrophilic carbon atom of bromomethane. As the new C–C bond forms, Br^- departs, taking with it the electron pair from the former C–Br bond and yielding propyne as product. We call such a reaction an **alkylation** because a new alkyl group has become attached to the starting alkyne.

Figure 9.6 | MECHANISM

A mechanism for the alkylation reaction of acetylide anion with bromomethane to give propyne.

1 The nucleophilic acetylide anion uses its electron lone pair to form a bond to the positively polarized, electrophilic carbon atom of bromomethane. As the new C–C bond begins to form, the C–Br bond begins to break in the transition state.

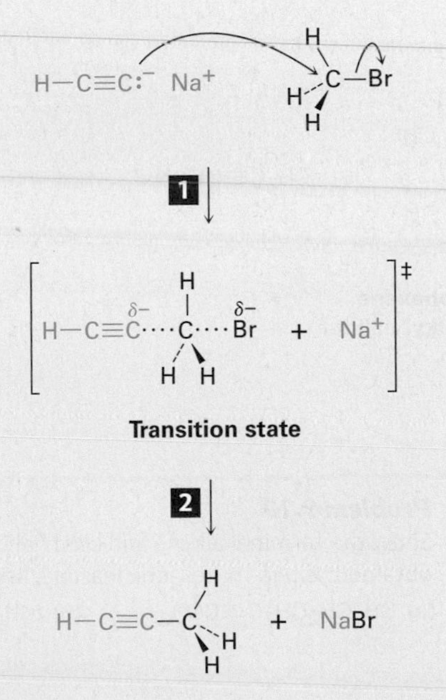

2 The new C–C bond is fully formed and the old C–Br bond is fully broken at the end of the reaction.

Alkyne alkylation is not limited to acetylene itself. Any terminal alkyne can be converted into its corresponding anion and then alkylated by treatment with an alkyl halide, yielding an internal alkyne. For example, conversion of 1-hexyne into its anion, followed by reaction with 1-bromobutane, yields 5-decyne.

$$CH_3CH_2CH_2CH_2C{\equiv}CH \xrightarrow[\text{2. } CH_3CH_2CH_2CH_2Br]{\text{1. } NaNH_2,\ NH_3} CH_3CH_2CH_2CH_2C{\equiv}CCH_2CH_2CH_2CH_3$$

1-Hexyne → **5-Decyne (76%)**

Because of its generality, acetylide alkylation is a good method for preparing substituted alkynes from simpler precursors. A terminal alkyne can be prepared by alkylation of acetylene itself, and an internal alkyne can be prepared by further alkylation of a terminal alkyne.

$$H{-}C{\equiv}C{-}H \xrightarrow{NaNH_2} [H{-}C{\equiv}C{:}^-\ Na^+] \xrightarrow{RCH_2Br} H{-}C{\equiv}C{-}CH_2R$$

Acetylene → **A terminal alkyne**

$$R{-}C{\equiv}C{-}H \xrightarrow{NaNH_2} [R{-}C{\equiv}C{:}^-\ Na^+] \xrightarrow{R'CH_2Br} R{-}C{\equiv}C{-}CH_2R'$$

A terminal alkyne → **An internal alkyne**

The alkylation reaction is limited to the use of primary alkyl bromides and alkyl iodides because acetylide ions are sufficiently strong bases to cause elimination instead of substitution when they react with secondary and tertiary alkyl halides. For example, reaction of bromocyclohexane with propyne anion yields the elimination product cyclohexene rather than the substitution product 1-propynylcyclohexane.

Bromocyclohexane (a secondary alkyl halide) + $CH_3C{\equiv}C{:}^-\ Na^+$ → Cyclohexene + $CH_3C{\equiv}CH$ + NaBr

↛ 1-propynylcyclohexane ($C{\equiv}C{-}CH_3$ on cyclohexane) — *Not formed*

Bromocyclohexane
(a secondary alkyl halide)

Cyclohexene

Not formed

Problem 9.10

Show the terminal alkyne and alkyl halide from which the following products can be obtained. If two routes look feasible, list both.

(a) $CH_3CH_2CH_2C{\equiv}CCH_3$ **(b)** $(CH_3)_2CHC{\equiv}CCH_2CH_3$ **(c)** cyclohexyl–$C{\equiv}CCH_3$

Problem 9.11

How would you prepare *cis*-2-butene starting from propyne, an alkyl halide, and any other reagents needed? This problem can't be worked in a single step. You'll have to carry out more than one reaction.

9.9 An Introduction to Organic Synthesis

There are many reasons for carrying out the laboratory synthesis of an organic compound. In the pharmaceutical industry, new molecules are designed and synthesized in the hope that some might be useful new drugs. In the chemical industry, syntheses are done to devise more economical routes to known compounds. In academic laboratories, the synthesis of extremely complex molecules is sometimes done just for the intellectual challenge involved in mastering so difficult a subject. The successful synthesis route is a highly creative work that is sometimes described by such subjective terms as *elegant* or *beautiful.*

In this book, too, we will often devise syntheses of molecules from simpler precursors, but our purpose is to learn. The ability to plan a successful multistep synthetic sequence requires a working knowledge of the uses and limitations of many different organic reactions. Furthermore, it requires the practical ability to fit together the steps in a sequence such that each reaction does only what is desired without causing changes elsewhere in the molecule. Planning a synthesis makes you approach a chemical problem in a logical way, draw on your knowledge of chemical reactions, and organize that knowledge into a workable plan—it helps you learn organic chemistry.

There's no secret to planning an organic synthesis: all it takes is a knowledge of the different reactions and some practice. The only real trick is to *work backward* in what is often called a **retrosynthetic** direction. Don't look at a potential starting material and ask yourself what reactions it might undergo. Instead, look at the final product and ask, "What was the immediate precursor of that product?" For example, if the final product is an alkyl halide, the immediate precursor might be an alkene, to which you could add HX. If the final product is a cis alkene, the immediate precursor might be an alkyne, which you could hydrogenate using the Lindlar catalyst. Having found an immediate precursor, work backward again, one step at a time, until you get back to the starting material. You have to keep the starting material in mind, of course, so that you can work back to it, but you don't want that starting material to be your main focus.

Let's work several examples of increasing complexity.

Devising a Synthesis Route

Worked Example 9.1

Synthesize *cis*-2-hexene from 1-pentyne and an alkyl halide. More than one step is needed.

$$CH_3CH_2CH_2C{\equiv}CH \;+\; RX \;\longrightarrow\; CH_3CH_2CH_2(H)C{=}C(H)CH_3$$

1-Pentyne; Alkyl halide; *cis*-2-Hexene

Strategy

When undertaking any synthesis problem, you should look at the product, identify the functional groups it contains, and then ask yourself how those functional groups can be prepared. Always work retrosynthetically, one step at a time.

The product in this case is a cis-disubstituted alkene, so the first question is, "What is an immediate precursor of a cis-disubstituted alkene?" We know that an alkene can be prepared from an alkyne by reduction and that the right choice of experimental conditions will allow us to prepare either a trans-disubstituted alkene (using lithium in liquid ammonia) or a cis-disubstituted alkene (using catalytic hydrogenation over the Lindlar catalyst). Thus, reduction of 2-hexyne by catalytic hydrogenation using the Lindlar catalyst should yield *cis*-2-hexene.

$$CH_3CH_2CH_2C{\equiv}CCH_3 \xrightarrow[\text{Lindlar catalyst}]{H_2} (CH_3CH_2CH_2)HC{=}CH(CH_3)$$

2-Hexyne → ***cis*-2-Hexene**

Next ask, "What is an immediate precursor of 2-hexyne?" We've seen that an internal alkyne can be prepared by alkylation of a terminal alkyne anion. In the present instance, we're told to start with 1-pentyne and an alkyl halide. Thus, alkylation of the anion of 1-pentyne with iodomethane should yield 2-hexyne.

$$CH_3CH_2CH_2C{\equiv}CH + NaNH_2 \xrightarrow{\text{In } NH_3} CH_3CH_2CH_2C{\equiv}C{:}^- \ Na^+$$

1-Pentyne

$$CH_3CH_2CH_2C{\equiv}C{:}^- \ Na^+ + CH_3I \xrightarrow{\text{In THF}} CH_3CH_2CH_2C{\equiv}CCH_3$$

2-Hexyne

Solution

cis-2-Hexene can be synthesized from the given starting materials in three steps.

$$CH_3CH_2CH_2C{\equiv}CH \xrightarrow[\text{2. } CH_3I\text{, THF}]{\text{1. } NaNH_2\text{, } NH_3} CH_3CH_2CH_2C{\equiv}CCH_3 \xrightarrow[\text{Lindlar catalyst}]{H_2} (CH_3CH_2CH_2)HC{=}CH(CH_3)$$

1-Pentyne → **2-Hexyne** → ***cis*-2-Hexene**

Worked Example 9.2 Devising a Synthesis Route

Synthesize 2-bromopentane from acetylene and an alkyl halide. More than one step is needed.

$$HC{\equiv}CH + RX \xrightarrow{?} CH_3CH_2CH_2CH(Br)CH_3$$

Acetylene + **Alkyl halide** → **2-Bromopentane**

Strategy

Identify the functional group in the product (an alkyl bromide) and work the problem retrosynthetically. What is an immediate precursor of an alkyl bromide? Perhaps an alkene plus HBr. Of the two possibilities, Markovnikov addition of HBr to 1-pentene looks like a better choice than addition to 2-pentene because the latter reaction would give a mixture of isomers.

$$\begin{matrix} CH_3CH_2CH_2CH{=}CH_2 \\ \text{or} \\ CH_3CH_2CH{=}CHCH_3 \end{matrix} \xrightarrow[\text{Ether}]{\text{HBr}} CH_3CH_2CH_2\underset{}{\overset{\overset{\displaystyle Br}{|}}{C}}HCH_3$$

What is an immediate precursor of an alkene? Perhaps an alkyne, which could be reduced.

$$CH_3CH_2CH_2C{\equiv}CH \xrightarrow[\text{Lindlar catalyst}]{H_2} CH_3CH_2CH_2CH{=}CH_2$$

What is an immediate precursor of a terminal alkyne? Perhaps sodium acetylide and an alkyl halide.

$$Na^+ \; :\bar{C}{\equiv}CH \; + \; BrCH_2CH_2CH_3 \longrightarrow CH_3CH_2CH_2C{\equiv}CH$$

Solution

The desired product can be synthesized in four steps from acetylene and 1-bromopropane.

$$\underset{\textbf{Acetylene}}{HC{\equiv}CH} \xrightarrow[\text{2. } CH_3CH_2CH_2Br\text{, THF}]{\text{1. } NaNH_2\text{, } NH_3} \underset{\textbf{1-Pentyne}}{CH_3CH_2CH_2C{\equiv}CH} \xrightarrow[\text{Lindlar catalyst}]{H_2} \underset{\textbf{1-Pentene}}{CH_3CH_2CH_2CH{=}CH_2}$$

$$\xrightarrow{\text{HBr, ether}} \underset{\textbf{2-Bromopentane}}{CH_3CH_2CH_2\underset{\underset{\displaystyle Br}{|}}{C}HCH_3}$$

Devising a Synthesis Route

Worked Example 9.3

Synthesize 5-methyl-1-hexanol (5-methyl-1-hydroxyhexane) from acetylene and an alkyl halide.

$$\underset{\textbf{Acetylene}}{HC{\equiv}CH} + \underset{\textbf{Alkyl halide}}{RX} \Longrightarrow \underset{\textbf{5-Methyl-1-hexanol}}{CH_3\overset{\overset{\displaystyle CH_3}{|}}{C}HCH_2CH_2CH_2CH_2OH}$$

Strategy

What is an immediate precursor of a primary alcohol? Perhaps a terminal alkene, which could be hydrated with non-Markovnikov regiochemistry by reaction with borane followed by oxidation with H_2O_2.

$$\underset{\text{CH}_3\text{CHCH}_2\text{CH}_2\text{CH}{=}\text{CH}_2}{\overset{\text{CH}_3}{|}} \xrightarrow[\text{2. H}_2\text{O}_2\text{, NaOH}]{\text{1. BH}_3} \underset{\text{CH}_3\text{CHCH}_2\text{CH}_2\text{CH}_2\text{CH}_2\text{OH}}{\overset{\text{CH}_3}{|}}$$

What is an immediate precursor of a terminal alkene? Perhaps a terminal alkyne, which could be reduced.

$$\underset{\text{CH}_3\text{CHCH}_2\text{CH}_2\text{C}{\equiv}\text{CH}}{\overset{\text{CH}_3}{|}} \xrightarrow[\text{Lindlar catalyst}]{\text{H}_2} \underset{\text{CH}_3\text{CHCH}_2\text{CH}_2\text{CH}{=}\text{CH}_2}{\overset{\text{CH}_3}{|}}$$

What is an immediate precursor of 5-methyl-1-hexyne? Perhaps acetylene and 1-bromo-3-methylbutane.

$$\text{HC}{\equiv}\text{CH} \xrightarrow{\text{NaNH}_2} \text{Na}^+\ {}^-\text{C}{\equiv}\text{CH} \xrightarrow{\underset{\text{CH}_3\text{CHCH}_2\text{CH}_2\text{Br}}{\overset{\text{CH}_3}{|}}} \underset{\text{CH}_3\text{CHCH}_2\text{CH}_2\text{C}{\equiv}\text{CH}}{\overset{\text{CH}_3}{|}}$$

Solution

The synthesis can be completed in four steps from acetylene and 1-bromo-3-methylbutane:

$$\underset{\textbf{Acetylene}}{\text{HC}{\equiv}\text{CH}} \xrightarrow[\text{2. CH}_3\text{CH(CH}_3\text{)CH}_2\text{CH}_2\text{Br}]{\text{1. NaNH}_2} \underset{\textbf{5-Methyl-1-hexyne}}{\underset{\text{CH}_3\text{CHCH}_2\text{CH}_2\text{C}{\equiv}\text{CH}}{\overset{\text{CH}_3}{|}}} \xrightarrow[\text{Lindlar catalyst}]{\text{H}_2} \underset{\textbf{5-Methyl-1-hexene}}{\underset{\text{CH}_3\text{CHCH}_2\text{CH}_2\text{CH}{=}\text{CH}_2}{\overset{\text{CH}_3}{|}}}$$

$$\xrightarrow[\text{2. H}_2\text{O}_2\text{, NaOH}]{\text{1. BH}_3} \underset{\textbf{5-Methyl-1-hexanol}}{\underset{\text{CH}_3\text{CHCH}_2\text{CH}_2\text{CH}_2\text{CH}_2\text{OH}}{\overset{\text{CH}_3}{|}}}$$

Problem 9.12

Beginning with 4-octyne as your only source of carbon, and using any inorganic reagents necessary, how would you synthesize the following compounds?

(a) *cis*-4-Octene **(b)** Butanal **(c)** 4-Bromooctane
(d) 4-Octanol **(e)** 4,5-Dichlorooctane **(f)** Butanoic acid

Problem 9.13

Beginning with acetylene and any alkyl halide needed, how would you synthesize the following compounds?

(a) Decane **(b)** 2,2-Dimethylhexane **(c)** Hexanal **(d)** 2-Heptanone

A DEEPER LOOK The Art of Organic Synthesis

Vitamin B_{12} has been synthesized from scratch in the laboratory, but the bacteria growing on sludge from municipal sewage plants do a much better job.

If you think some of the synthesis problems at the end of this chapter are hard, try devising a synthesis of vitamin B_{12} starting only from simple substances you can buy in a chemical catalog. This extraordinary achievement was reported in 1973 as the culmination of a collaborative effort headed by Robert B. Woodward of Harvard University and Albert Eschenmoser of the Swiss Federal Institute of Technology in Zürich. More than 100 graduate students and postdoctoral associates contributed to the work, which took more than a decade to complete.

Vitamin B_{12}

Why put such extraordinary effort into the laboratory synthesis of a molecule so easily obtained from natural sources? There are many reasons. On a basic human level, a chemist might be motivated primarily by the challenge, much as a climber might be challenged by the ascent of a difficult peak. Beyond the pure challenge, the completion of a difficult synthesis is also valuable for the way in which it establishes new standards and raises the field to a new level. If vitamin B_{12} can be made, then why can't any molecule found in nature be made? Indeed, the decades that have passed since the work of Woodward and Eschenmoser have seen the laboratory synthesis of many enormously complex and valuable substances. Sometimes these substances—for instance, the anticancer compound paclitaxel, trade named Taxol—are not easily available in nature, so laboratory synthesis is the only method for obtaining larger quantities.

Paclitaxel (Taxol)

(continued)

(continued)

But perhaps the most important reason for undertaking a complex synthesis is that, in so doing, new reactions and new chemistry are discovered. It invariably happens in synthesis that a point is reached at which the planned route fails. At such a time, the only alternatives are to quit or to devise a way around the difficulty. New reactions and new principles come from such situations, and it is in this way that the science of organic chemistry grows richer. In the synthesis of vitamin B_{12}, for example, unexpected findings emerged that led to the understanding of an entire new class of reactions—the *pericyclic* reactions that are the subject of Chapter 30 in this book. From synthesizing vitamin B_{12} to understanding pericyclic reactions—no one could have possibly predicted such a link at the beginning of the synthesis, but that is the way of science.

Summary

Key words

acetylide anion, 325
alkylation, 327
alkyne (RC≡CR), 314
enol, 319
retrosynthetic, 329
tautomer, 319

Alkynes are less common than alkenes, both in the laboratory and in living organisms, so we haven't covered them in great detail. The real importance of this chapter is that alkyne chemistry is a useful vehicle to look at the general strategies used in organic synthesis—the construction of complex molecules in the laboratory.

An **alkyne** is a hydrocarbon that contains a carbon–carbon triple bond. Alkyne carbon atoms are *sp*-hybridized, and the triple bond consists of one *sp–sp* σ bond and two *p–p* π bonds. There are relatively few general methods of alkyne synthesis. Two good ones are the alkylation of an acetylide anion with a primary alkyl halide and the twofold elimination of HX from a vicinal dihalide.

The chemistry of alkynes is dominated by electrophilic addition reactions, similar to those of alkenes. Alkynes react with HBr and HCl to yield vinylic halides and with Br_2 and Cl_2 to yield 1,2-dihalides (vicinal dihalides). Alkynes can be hydrated by reaction with aqueous sulfuric acid in the presence of mercury(II) catalyst. The reaction leads to an intermediate **enol** that immediately **tautomerizes** to yield a ketone. Because the addition reaction occurs with Markovnikov regiochemistry, a methyl ketone is produced from a terminal alkyne. Alternatively, hydroboration–oxidation of a terminal alkyne yields an aldehyde.

Alkynes can be reduced to yield alkenes and alkanes. Complete reduction of the triple bond over a palladium hydrogenation catalyst yields an alkane; partial reduction by catalytic hydrogenation over a Lindlar catalyst yields a cis alkene. Reduction of the alkyne with lithium in ammonia yields a trans alkene.

Terminal alkynes are weakly acidic. The alkyne hydrogen can be removed by a strong base such as $Na^+ \ {}^-NH_2$ to yield an **acetylide anion**. An acetylide anion acts as a nucleophile and can displace a halide ion from a primary alkyl halide in an **alkylation** reaction. Acetylide anions are more stable than either alkyl anions or vinylic anions because their negative charge is in a hybrid orbital with 50% *s* character, allowing the charge to be closer to the nucleus.

Summary of Reactions

1. Preparation of alkynes
   (a) Dehydrohalogenation of vicinal dihalides (Section 9.2)

$$\text{R—CH(Br)—CH(Br)—R}' \xrightarrow[\text{or 2 NaNH}_2\text{, NH}_3]{\text{2 KOH, ethanol}} \text{R—C}\equiv\text{C—R}' + 2\,H_2O + 2\,KBr$$

$$\text{R—CH}=\text{CBr—R}' \xrightarrow[\text{or NaNH}_2\text{, NH}_3]{\text{KOH, ethanol}} \text{R—C}\equiv\text{C—R}' + H_2O + KBr$$

   (b) Alkylation of acetylide anions (Section 9.8)

$$\underset{\textbf{Acetylene}}{HC\equiv CH} \xrightarrow{NaNH_2} HC\equiv C^-\,Na^+ \xrightarrow{RCH_2Br} \underset{\textbf{A terminal alkyne}}{HC\equiv CCH_2R}$$

$$\underset{\textbf{A terminal alkyne}}{RC\equiv CH} \xrightarrow{NaNH_2} RC\equiv C^-\,Na^+ \xrightarrow{R'CH_2Br} \underset{\textbf{An internal alkyne}}{RC\equiv CCH_2R'}$$

2. Reactions of alkynes
   (a) Addition of HCl and HBr (Section 9.3)

$$\text{R—C}\equiv\text{C—R} \xrightarrow[\text{Ether}]{HX} \text{(X)(R)C}=\text{C(R)(H)} \xrightarrow[\text{Ether}]{HX} \text{R—CX}_2\text{—CH}_2\text{—R}$$

   (b) Addition of Cl_2 and Br_2 (Section 9.3)

$$\text{R—C}\equiv\text{C—R}' \xrightarrow[CH_2Cl_2]{X_2} \text{(X)(R)C}=\text{C(R}')\text{(X)} \xrightarrow[CH_2Cl_2]{X_2} \text{R—CX}_2\text{—CX}_2\text{—R}'$$

   (c) Hydration (Section 9.4)
      (1) Mercuric sulfate catalyzed

$$\text{R—C}\equiv\text{CH} \xrightarrow[HgSO_4]{H_2SO_4,\ H_2O} \underset{\textbf{An enol}}{\left[\text{R—C(OH)}=\text{CH}_2\right]} \longrightarrow \underset{\textbf{A methyl ketone}}{\text{R—C(}=\text{O)—CH}_3}$$

(continued)

(2) Hydroboration–oxidation

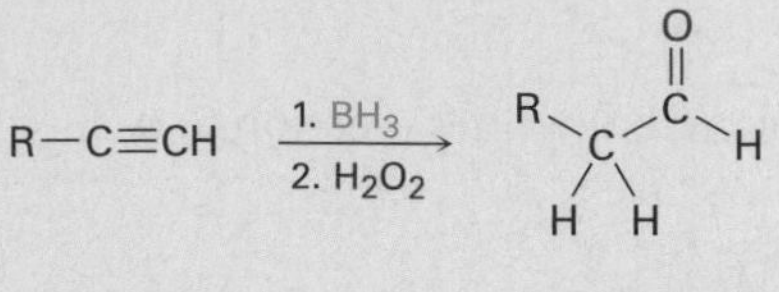

$$R-C\equiv CH \xrightarrow[2.\ H_2O_2]{1.\ BH_3} R-CH_2-CHO$$

An aldehyde

(d) Reduction (Section 9.5)

(1) Catalytic hydrogenation

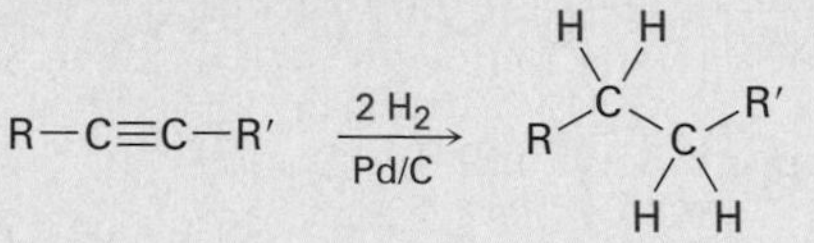

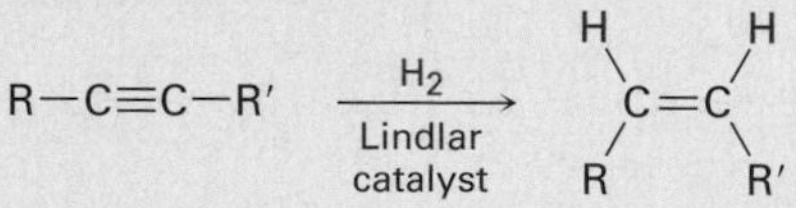

$$R-C\equiv C-R' \xrightarrow[Pd/C]{2\ H_2} R-CH_2-CH_2-R'$$

$$R-C\equiv C-R' \xrightarrow[\text{Lindlar catalyst}]{H_2} \text{cis-}RCH=CHR'$$

A cis alkene

(2) Lithium in liquid ammonia

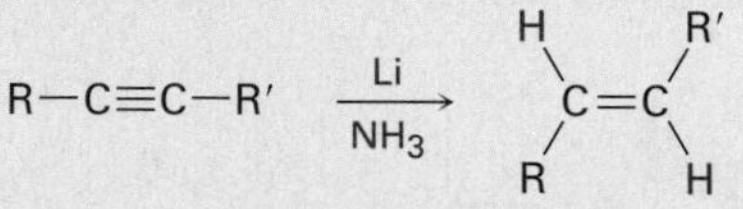

$$R-C\equiv C-R' \xrightarrow[NH_3]{Li} \text{trans-}RCH=CHR'$$

A trans alkene

(e) Conversion into acetylide anions (Section 9.7)

$$R-C\equiv C-H \xrightarrow[NH_3]{NaNH_2} R-C\equiv C:^-\ Na^+ \ + \ NH_3$$

Exercises

Visualizing Chemistry

OWL Interactive versions of these problems are assignable in OWL for Organic Chemistry.

(Problems 9.1–9.13 appear within the chapter.)

9.14 Name the following alkynes, and predict the products of their reaction with (1) H_2 in the presence of a Lindlar catalyst and (2) H_3O^+ in the presence of $HgSO_4$:

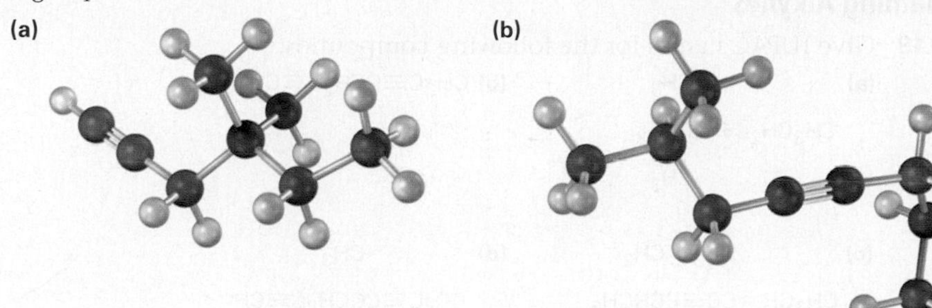

9.15 From what alkyne might each of the following substances have been made? (Green = Cl.)

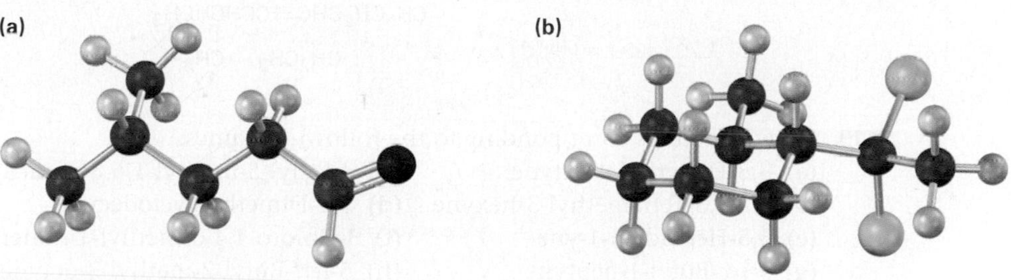

9.16 How would you prepare the following substances, starting from any compounds having four carbons or fewer?

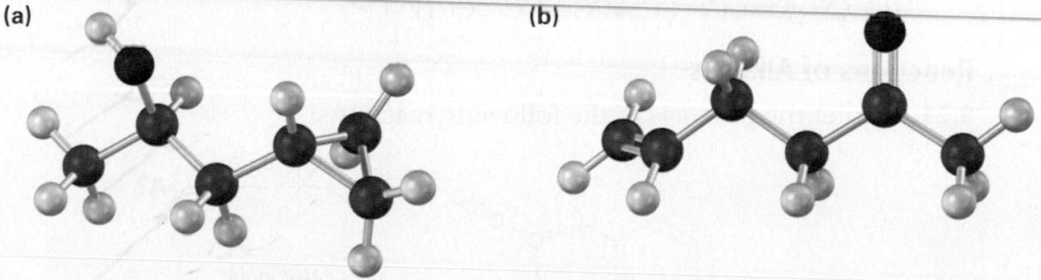

9.17 The following cycloalkyne is too unstable to exist. Explain.

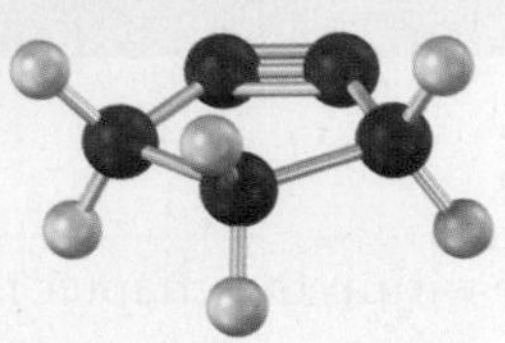

Additional Problems

Naming Alkynes

9.18 Give IUPAC names for the following compounds:

(a)

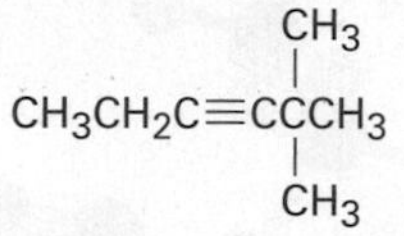

$$\begin{array}{c} CH_3 \\ | \\ CH_3CH_2C{\equiv}CCCH_3 \\ | \\ CH_3 \end{array}$$

(methyl groups on the carbon after C≡C)

(b) $CH_3C{\equiv}CCH_2C{\equiv}CCH_2CH_3$

(c)

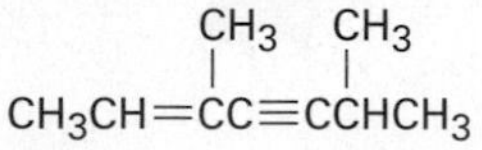

$$\begin{array}{c} \quad CH_3 \quad CH_3 \\ CH_3CH{=}CC{\equiv}CCHCH_3 \end{array}$$

(d)

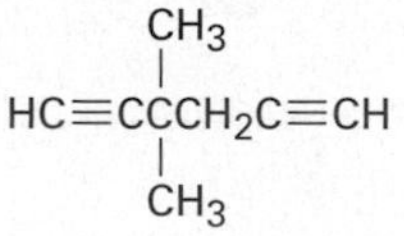

$$\begin{array}{c} CH_3 \\ | \\ HC{\equiv}CCCH_2C{\equiv}CH \\ | \\ CH_3 \end{array}$$

(e) $H_2C{=}CHCH{=}CHC{\equiv}CH$

(f)

$$\begin{array}{c} CH_2CH_3 \\ | \\ CH_3CH_2CHC{\equiv}CCHCHCH_3 \\ | \qquad\quad | \\ CH_2CH_3 \quad CH_3 \end{array}$$

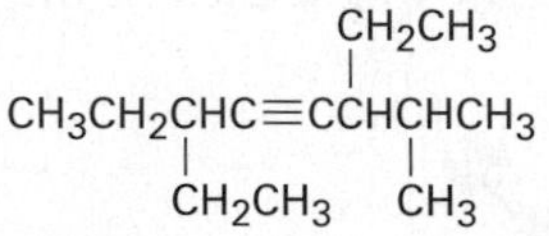

9.19 Draw structures corresponding to the following names:

(a) 3,3-Dimethyl-4-octyne
(b) 3-Ethyl-5-methyl-1,6,8-decatriyne
(c) 2,2,5,5-Tetramethyl-3-hexyne
(d) 3,4-Dimethylcyclodecyne
(e) 3,5-Heptadien-1-yne
(f) 3-Chloro-4,4-dimethyl-1-nonen-6-yne
(g) 3-*sec*-Butyl-1-heptyne
(h) 5-*tert*-Butyl-2-methyl-3-octyne

9.20 The following two hydrocarbons have been isolated from various plants in the sunflower family. Name them according to IUPAC rules.

(a) $CH_3CH{=}CHC{\equiv}CC{\equiv}CCH{=}CHCH{=}CHCH{=}CH_2$ (all trans)
(b) $CH_3C{\equiv}CC{\equiv}CC{\equiv}CC{\equiv}CC{\equiv}CCH{=}CH_2$

Reactions of Alkynes

9.21 Predict the products of the following reactions:

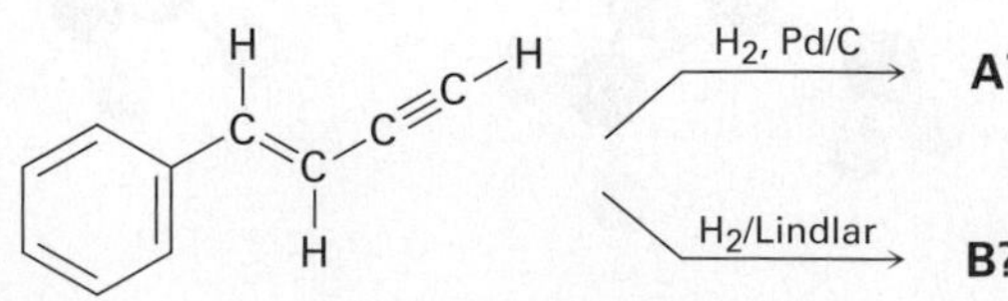

9.22 Predict the products from reaction of 1-hexyne with the following reagents:

(a) 1 equiv HBr
(b) 1 equiv Cl_2
(c) H_2, Lindlar catalyst
(d) $NaNH_2$ in NH_3, then CH_3Br
(e) H_2O, H_2SO_4, $HgSO_4$
(f) 2 equiv HCl

9.23 Predict the products from reaction of 5-decyne with the following reagents:
(a) H_2, Lindlar catalyst (b) Li in NH_3
(c) 1 equiv Br_2 (d) BH_3 in THF, then H_2O_2, OH^-
(e) H_2O, H_2SO_4, $HgSO_4$ (f) Excess H_2, Pd/C catalyst

9.24 Predict the products from reaction of 2-hexyne with the following reagents:
(a) 2 equiv Br_2 (b) 1 equiv HBr (c) Excess HBr
(d) Li in NH_3 (e) H_2O, H_2SO_4, $HgSO_4$

9.25 Propose structures for hydrocarbons that give the following products on oxidative cleavage by $KMnO_4$ or O_3:

(a) CO_2 + $CH_3(CH_2)_5CO_2H$

(b) CH_3CO_2H + $C_6H_5CO_2H$

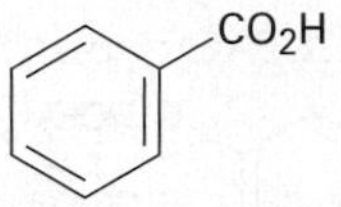

(c) $HO_2C(CH_2)_8CO_2H$

(d) CH_3CHO + $CH_3C(=O)CH_2CH_2CO_2H$ + CO_2

(e) $HC(=O)CH_2CH_2CH_2CH_2C(=O)CO_2H$ + CO_2

9.26 Identify the reagents a–c in the following scheme:

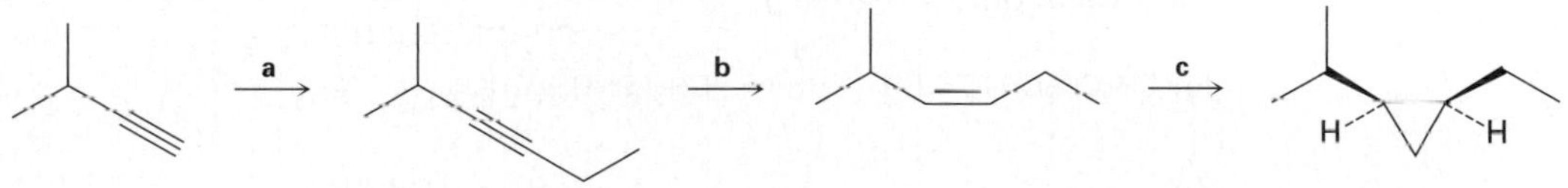

[Scheme: 3-methyl-1-butyne →(a) 2-methyl-3-hexyne →(b) cis-2-methyl-3-hexene →(c) cis-disubstituted cyclopropane]

Organic Synthesis

9.27 How would you carry out the following conversions? More than one step may be needed in some instances.

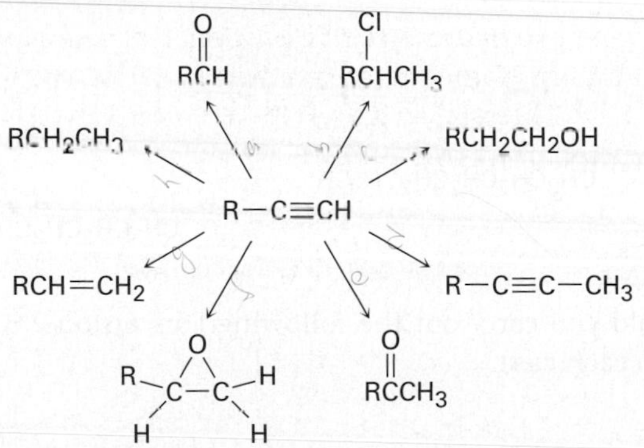

[Scheme: R–C≡CH converted to each of: RCH_2CH_3; $RC(=O)H$ (RCH); $RCHClCH_3$; RCH_2CH_2OH; $RCH{=}CH_2$; cis epoxide R(H)C–C(H)H with O bridge; $RC(=O)CH_3$; $R{-}C{\equiv}C{-}CH_3$]

9.28 How would you carry out the following reactions?

(a) $CH_3CH_2C{\equiv}CH \xrightarrow{?} CH_3CH_2C(=O)CH_3$

(b) $CH_3CH_2C{\equiv}CH \xrightarrow{?} CH_3CH_2CH_2CHO$

(c) $C_6H_5C{\equiv}CH \xrightarrow{?} C_6H_5C{\equiv}C{-}CH_3$

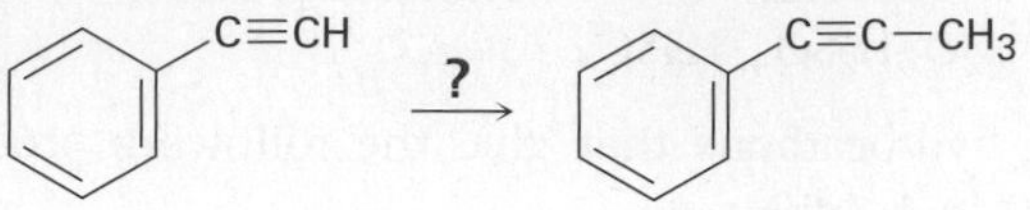

(d) $C_6H_5C{\equiv}CCH_3 \xrightarrow{?}$ $C_6H_5CH{=}CHCH_3$ (phenyl and CH_3 cis; the two H's cis)

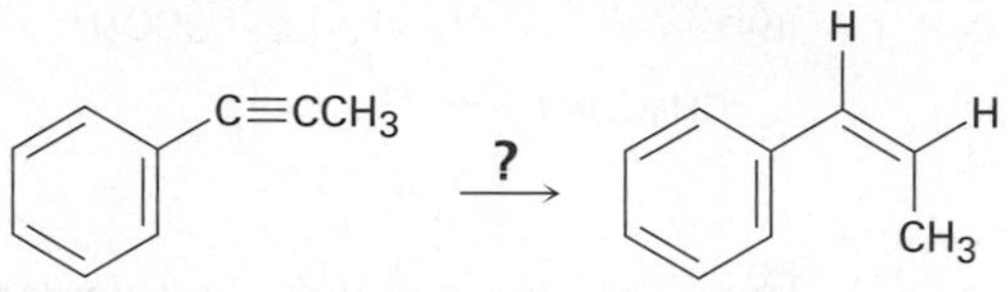

(e) $CH_3CH_2C{\equiv}CH \xrightarrow{?} CH_3CH_2CO_2H$

(f) $CH_3CH_2CH_2CH_2CH{=}CH_2 \xrightarrow[\text{(2 steps)}]{?} CH_3CH_2CH_2CH_2C{\equiv}CH$

9.29 Each of the following syntheses requires more than one step. How would you carry them out?

(a) $CH_3CH_2CH_2C{\equiv}CH \xrightarrow{?} CH_3CH_2CH_2CHO$

(b) $(CH_3)_2CHCH_2C{\equiv}CH \xrightarrow{?}$ $(CH_3)_2CHCH_2CH{=}CHCH_2CH_3$ (trans: H and CH_2CH_3 on one carbon, $(CH_3)_2CHCH_2$ and H on the other)

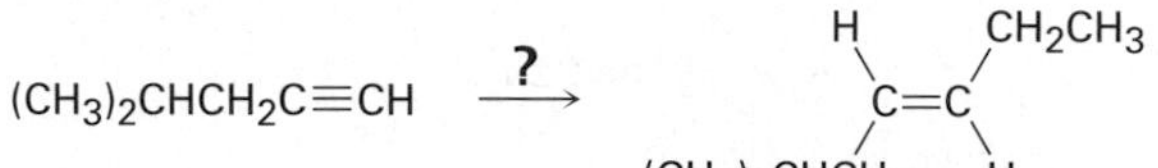

9.30 How would you carry out the following transformation? More than one step is needed.

$CH_3CH_2CH_2CH_2C{\equiv}CH \xrightarrow{?}$ cis-1-butyl-2-methylcyclopropane (CH_2 ring carbon bearing H, H; ring carbons bearing H and $CH_3CH_2CH_2CH_2$, and H and CH_3)

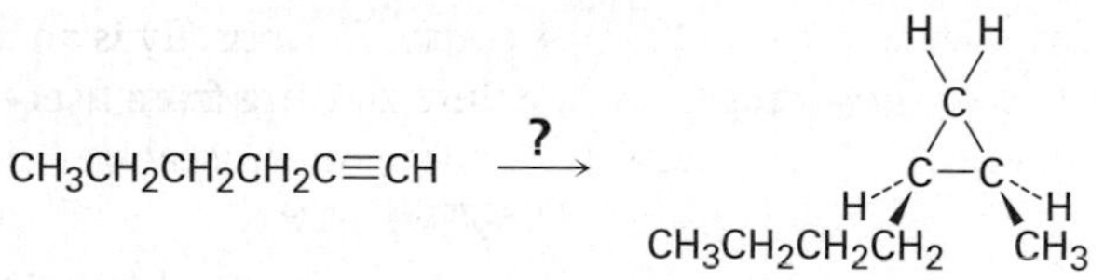

9.31 How would you carry out the following conversions? More than one step is needed in each case.

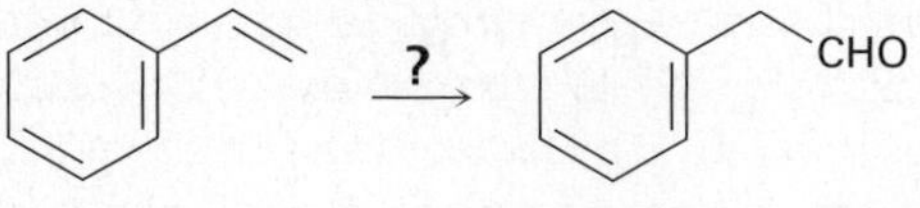

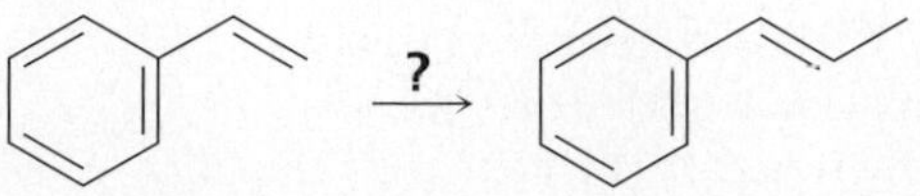

$C_6H_5CH{=}CH_2 \xrightarrow{?} C_6H_5CH_2CHO$

$C_6H_5CH{=}CH_2 \xrightarrow{?} C_6H_5CH{=}CHCH_3$

9.32 Synthesize the following compounds using 1-butyne as the only source of carbon, along with any inorganic reagents you need. More than one step may be needed.

(a) 1,1,2,2-Tetrachlorobutane (b) 1,1-Dichloro-2-ethylcyclopropane

9.33 How would you synthesize the following compounds from acetylene and any alkyl halides with four or fewer carbons? More than one step may be needed.

(a) $CH_3CH_2CH_2C{\equiv}CH$ (b) $CH_3CH_2C{\equiv}CCH_2CH_3$

(c) $CH_3CH(CH_3)CH_2CH{=}CH_2$ (d) $CH_3CH_2CH_2C({=}O)CH_2CH_2CH_2CH_3$

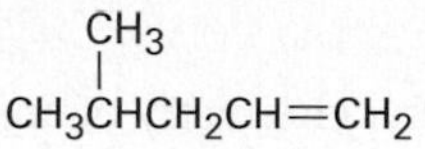

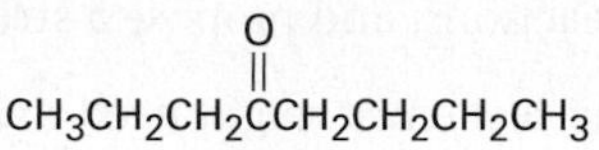

(e) $CH_3CH_2CH_2CH_2CH_2CHO$

9.34 How would you carry out the following reactions to introduce deuterium into organic molecules?

(a) $CH_3CH_2C{\equiv}CCH_2CH_3 \xrightarrow{?}$ (D)(C₂H₅)C=C(D)(C₂H₅), with both D on the same side

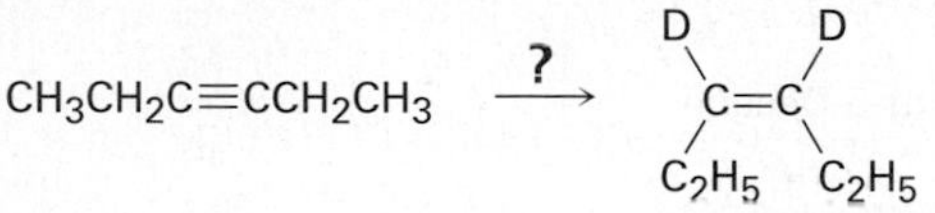

(b) $CH_3CH_2C{\equiv}CCH_2CH_3 \xrightarrow{?}$ (D)(C₂H₅)C=C(C₂H₅)(D), with D and C₂H₅ on opposite sides

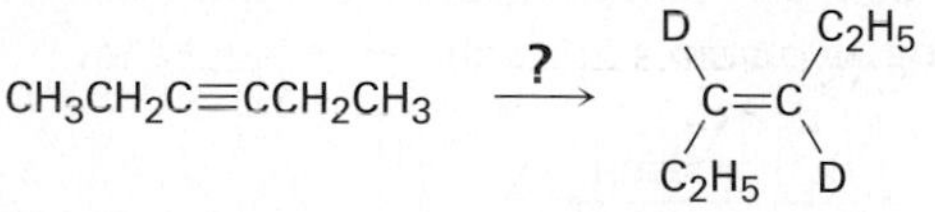

(c) $CH_3CH_2CH_2C{\equiv}CH \xrightarrow{?} CH_3CH_2CH_2C{\equiv}CD$

(d) $C_6H_5C{\equiv}CH \xrightarrow{?} C_6H_5CD{=}CD_2$

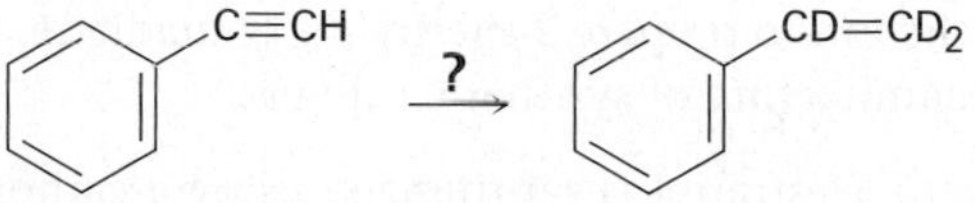

9.35 How would you prepare cyclodecyne starting from acetylene and any alkyl halide needed?

9.36 The sex attractant given off by the common housefly is an alkene named *muscalure.* Propose a synthesis of muscalure starting from acetylene and any alkyl halides needed. What is the IUPAC name for muscalure?

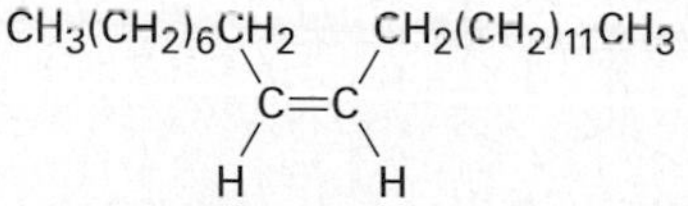

$CH_3(CH_2)_6CH_2$ and $CH_2(CH_2)_{11}CH_3$ on one carbon each of C=C, with H on each carbon (cis) — **Muscalure**

General Problems

9.37 A hydrocarbon of unknown structure has the formula C_8H_{10}. On catalytic hydrogenation over the Lindlar catalyst, 1 equivalent of H_2 is absorbed. On hydrogenation over a palladium catalyst, 3 equivalents of H_2 are absorbed.

(a) How many degrees of unsaturation are present in the unknown?
(b) How many triple bonds are present?
(c) How many double bonds are present?
(d) How many rings are present?
(e) Draw a structure that fits the data.

9.38 Compound **A** (C_9H_{12}) absorbed 3 equivalents of H_2 on catalytic reduction over a palladium catalyst to give **B** (C_9H_{18}). On ozonolysis, compound **A** gave, among other things, a ketone that was identified as cyclohexanone. On treatment with $NaNH_2$ in NH_3, followed by addition of iodomethane, compound **A** gave a new hydrocarbon, **C** ($C_{10}H_{14}$). What are the structures of **A**, **B**, and **C**?

9.39 Hydrocarbon **A** has the formula $C_{12}H_8$. It absorbs 8 equivalents of H_2 on catalytic reduction over a palladium catalyst. On ozonolysis, only two products are formed: oxalic acid (HO_2CCO_2H) and succinic acid ($HO_2CCH_2CH_2CO_2H$). Write the reactions, and propose a structure for **A**.

9.40 Occasionally, a chemist might need to *invert* the stereochemistry of an alkene—that is, to convert a cis alkene to a trans alkene, or vice versa. There is no one-step method for doing an alkene inversion, but the transformation can be carried out by combining several reactions in the proper sequence. How would you carry out the following reactions?

(a) *trans*-5-Decene $\xrightarrow{?}$ *cis*-5-Decene

(b) *cis*-5-Decene $\xrightarrow{?}$ *trans*-5-Decene

9.41 Organometallic reagents such as sodium acetylide undergo an addition reaction with ketones, giving alcohols:

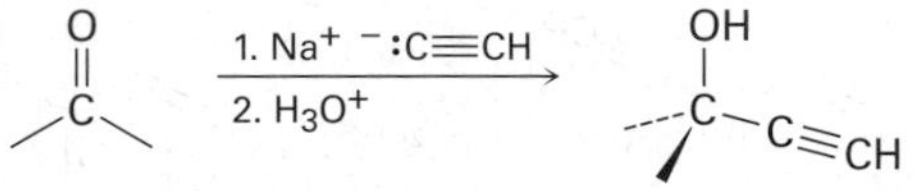

O=C< $\xrightarrow[\text{2. } H_3O^+]{\text{1. } Na^+ \ ^-:C{\equiv}CH}$ C(OH)–C≡CH

How might you use this reaction to prepare 2-methyl-1,3-butadiene, the starting material used in the manufacture of synthetic rubber?

9.42 The oral contraceptive agent Mestranol is synthesized using a carbonyl addition reaction like that shown in Problem 9.41. Draw the structure of the ketone needed.

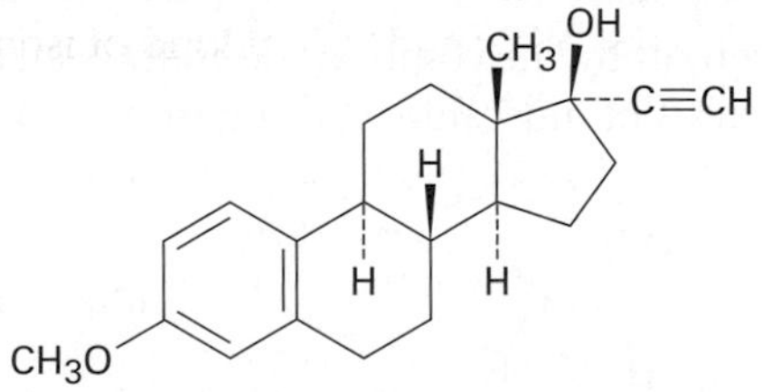

CH3 OH C≡CH H H H CH3O

Mestranol

9.43 1-Octen-3-ol, a potent mosquito attractant commonly used in mosquito traps, can be prepared in two steps from hexanal, $CH_3CH_2CH_2CH_2CH_2CHO$. The first step is an acetylide-addition reaction like that described in Problem 9.41. What is the structure of the product from the first step, and how can it be converted into 1-octen-3-ol?

$CH_3CH_2CH_2CH_2CH_2CH(OH)CH{=}CH_2$ **1-Octen-3-ol**

9.44 Erythrogenic acid, $C_{18}H_{26}O_2$, is an acetylenic fatty acid that turns a vivid red on exposure to light. On catalytic hydrogenation over a palladium catalyst, 5 equivalents of H_2 are absorbed, and stearic acid, $CH_3(CH_2)_{16}CO_2H$, is produced. Ozonolysis of erythrogenic acid gives four products: formaldehyde, CH_2O; oxalic acid, HO_2CCO_2H; azelaic acid, $HO_2C(CH_2)_7CO_2H$; and the aldehyde acid $OHC(CH_2)_4CO_2H$. Draw two possible structures for erythrogenic acid, and suggest a way to tell them apart by carrying out some simple reactions.

9.45 Hydrocarbon **A** has the formula C_9H_{12} and absorbs 3 equivalents of H_2 to yield **B**, C_9H_{18}, when hydrogenated over a Pd/C catalyst. On treatment of **A** with aqueous H_2SO_4 in the presence of mercury(II), two isomeric ketones, **C** and **D**, are produced. Oxidation of **A** with $KMnO_4$ gives a mixture of acetic acid (CH_3CO_2H) and the tricarboxylic acid **E**. Propose structures for compounds **A–D**, and write the reactions.

CH_2CO_2H

$HO_2CCH_2CHCH_2CO_2H$

E

9.46 Terminal alkynes react with Br_2 and water to yield bromo ketones. For example:

C≡CH (on benzene ring) $\xrightarrow{Br_2,\ H_2O}$ (benzene ring)–C(=O)–CH_2Br

Propose a mechanism for the reaction. To what reaction of alkenes is the process analogous?

9.47 A *cumulene* is a compound with three adjacent double bonds. Draw an orbital picture of a cumulene. What kind of hybridization do the two central carbon atoms have? What is the geometric relationship of the substituents on one end to the substituents on the other end? What kind of isomerism is possible? Make a model to help see the answer.

$$R_2C{=}C{=}C{=}CR_2$$

A cumulene

9.48 Reaction of acetone with D_3O^+ yields hexadeuterioacetone. That is, all the hydrogens in acetone are exchanged for deuterium. Review the mechanism of mercuric ion–catalyzed alkyne hydration, and then propose a mechanism for this deuterium incorporation.

$$H_3C{-}C({=}O){-}CH_3 \xrightarrow{D_3O^+} D_3C{-}C({=}O){-}CD_3$$

Acetone **Hexadeuterioacetone**

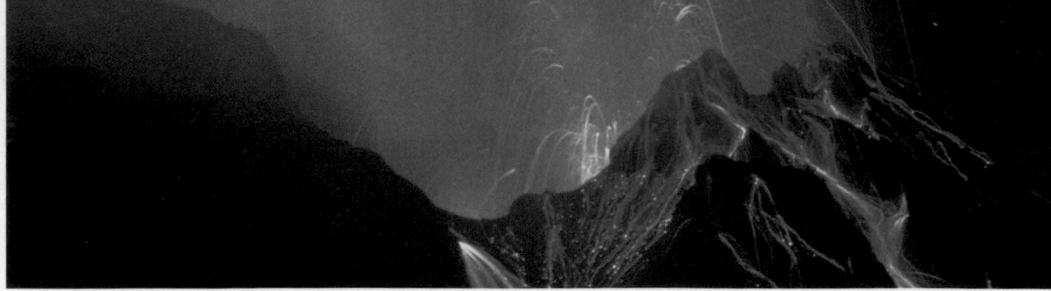

The gases released during volcanic eruptions contain large amounts of organohalides, including chloromethane, chloroform, dichlorodifluoromethane, and many others.

10 Organohalides

Now that we've covered the chemistry of hydrocarbons, it's time to start looking at more complex substances that contain elements in addition to C and H. We'll begin by discussing the chemistry of **organohalides**, compounds that contain one or more halogen atoms.

Halogen-substituted organic compounds are widespread in nature, and more than 5000 organohalides have been found in algae and various other marine organisms. Chloromethane, for example, is released in large amounts by ocean kelp, as well as by forest fires and volcanoes. Halogen-containing compounds also have a vast array of industrial applications, including their use as solvents, inhaled anesthetics in medicine, refrigerants, and pesticides.

Trichloroethylene (a solvent) | **Halothane** (an inhaled anesthetic) | **Dichlorodifluoromethane** (a refrigerant)

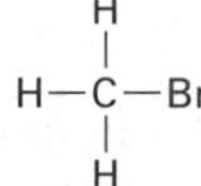

Bromomethane (a fumigant)

Still other halo-substituted compounds are used as medicines and food additives. The nonnutritive sweetener sucralose, marketed as Splenda, contains four chlorine atoms, for instance. Sucralose is about 600 times as sweet as sucrose, so only 1 mg is equivalent to an entire teaspoon of table sugar.

Sucralose

A large variety of organohalides are known. The halogen might be bonded to an alkynyl group (C≡C–X), a vinylic group (C=C–X), an aromatic ring (Ar–X), or an alkyl group. We'll be concerned in this chapter, however, primarily with **alkyl halides**, compounds with a halogen atom bonded to a saturated, sp^3-hybridized carbon atom.

OWL Sign in to OWL for Organic Chemistry at **www.cengage.com/owl** to view tutorials and simulations, develop problem-solving skills, and complete online homework assigned by your professor.

Why This Chapter? Alkyl halides are encountered less frequently than their oxygen-containing relatives and are not often involved in the biochemical pathways of terrestrial organisms, but some of the *kinds* of reactions they undergo—nucleophilic substitutions and eliminations—*are* encountered frequently. Thus, alkyl halide chemistry acts as a relatively simple model for many mechanistically similar but structurally more complex reactions found in biomolecules. We'll begin in this chapter with a look at how to name and prepare alkyl halides, and we'll see several of their reactions. Then in the next chapter, we'll make a detailed study of the substitution and elimination reactions of alkyl halides—two of the most important and well-studied reaction types in organic chemistry.

10.1 Names and Properties of Alkyl Halides

Although commonly called *alkyl halides,* halogen-substituted alkanes are named systematically as *haloalkanes* **(Section 3.4)**, treating the halogen as a substituent on a parent alkane chain. There are three steps:

STEP 1

Find the longest chain, and name it as the parent. If a double or triple bond is present, the parent chain must contain it.

STEP 2

Number the carbons of the parent chain beginning at the end nearer the first substituent, whether alkyl or halo. Assign each substituent a number according to its position on the chain.

$CH_3\overset{CH_3}{C}HCH_2\underset{CH_3}{C}H\overset{Br}{C}HCH_2CH_3$ (carbons numbered 1–7 from left)

5-Bromo-2,4-dimethylheptane

$CH_3\overset{Br}{C}HCH_2\underset{CH_3}{C}H\overset{CH_3}{C}HCH_2CH_3$ (carbons numbered 1–7 from left)

2-Bromo-4,5-dimethylheptane

If different halogens are present, number all and list them in alphabetical order when writing the name.

$BrCH_2CH_2\overset{Cl}{C}H\underset{CH_3}{C}HCH_3$ (carbons numbered 1–5 from left)

1-Bromo-3-chloro-4-methylpentane

STEP 3

If the parent chain can be properly numbered from either end by step 2, begin at the end nearer the substituent that has alphabetical precedence.

$CH_3\overset{CH_3}{C}HCH_2CH_2\overset{Br}{C}HCH_3$ (carbons numbered 6–1 from left)

2-Bromo-5-methylhexane
(*Not* 5-bromo-2-methylhexane)

In addition to their systematic names, many simple alkyl halides are also named by identifying first the alkyl group and then the halogen. For example, CH_3I can be called either iodomethane or methyl iodide. Such names are well entrenched in the chemical literature and in daily usage, but they won't be used in this book.

CH_3I
Iodomethane (or methyl iodide)

Cl
|
CH_3CHCH_3
2-Chloropropane (or isopropyl chloride)

Br-cyclohexane
Bromocyclohexane (or cyclohexyl bromide)

Halogens increase in size going down the periodic table, so the lengths of the corresponding carbon–halogen bonds increase accordingly (Table 10.1). In addition, C–X bond strengths decrease going down the periodic table. As we've been doing thus far, we'll continue to use the abbreviation X to represent any of the halogens F, Cl, Br, or I.

Table 10.1 A Comparison of the Halomethanes

		Bond strength		
Halomethane	**Bond length (pm)**	**(kJ/mol)**	**(kcal/mol)**	**Dipole moment (*D*)**
CH_3F	139	460	110	1.85
CH_3Cl	178	350	84	1.87
CH_3Br	193	294	70	1.81
CH_3I	214	239	57	1.62

In our discussion of bond polarity in functional groups in **Section 6.4**, we noted that halogens are more electronegative than carbon. The C–X bond is therefore polar, with the carbon atom bearing a slight positive charge ($\delta+$) and the halogen a slight negative charge ($\delta-$). This polarity results in a substantial dipole moment for all the halomethanes (Table 10.1) and implies that the alkyl halide C–X carbon atom should behave as an electrophile in polar reactions. We'll soon see that this is indeed the case.

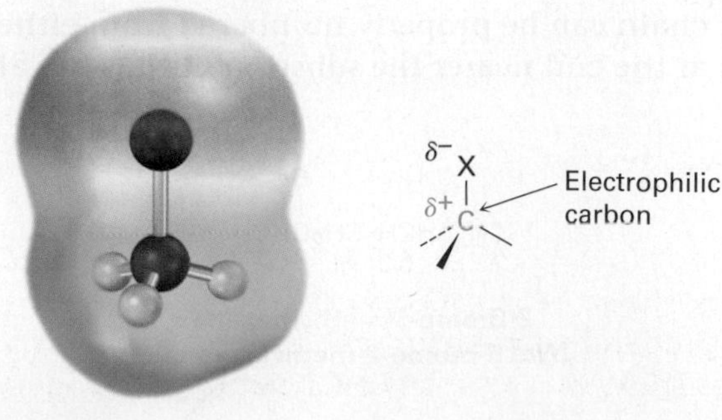

Problem 10.1

Give IUPAC names for the following alkyl halides:

(a) $CH_3CH_2CH_2CH_2I$

(b) $CH_3\overset{CH_3}{\overset{|}{C}}HCH_2CH_2Cl$

(c) $BrCH_2CH_2CH_2\overset{CH_3}{\underset{CH_3}{C}}CH_2Br$

(d) $CH_3\overset{CH_3}{\underset{Cl}{C}}CH_2CH_2Cl$

(e) $CH_3\overset{I}{C}H\overset{CH_2CH_2Cl}{C}HCH_2CH_3$

(f) $CH_3\overset{Br}{C}HCH_2CH_2\overset{Cl}{C}HCH_3$

Problem 10.2

Draw structures corresponding to the following IUPAC names:

(a) 2-Chloro-3,3-dimethylhexane
(b) 3,3-Dichloro-2-methylhexane
(c) 3-Bromo-3-ethylpentane
(d) 1,1-Dibromo-4-isopropylcyclohexane
(e) 4-*sec*-Butyl-2-chlorononane
(f) 1,1-Dibromo-4-*tert*-butylcyclohexane

10.2 Preparing Alkyl Halides from Alkanes: Radical Halogenation

Simple alkyl halides can sometimes be prepared by reaction of an alkane with Cl_2 or Br_2 in the presence of light through a radical chain-reaction pathway **(Section 6.3)**. The mechanism is shown in **Figure 10.1** for chlorination.

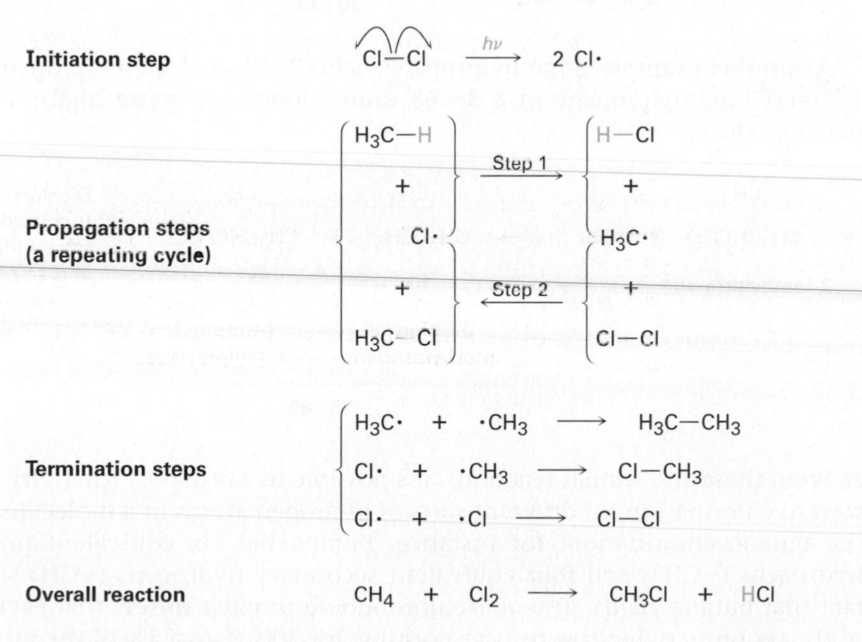

Figure 10.1 Mechanism of the radical chlorination of methane. Three kinds of steps are required: initiation, propagation, and termination. The propagation steps are a repeating cycle, with Cl· a reactant in step 1 and a product in step 2, and with $\cdot CH_3$ a product in step 1 and a reactant in step 2. (The symbol *hν* shown in the initiation step is the standard way of indicating irradiation with light.)

Recall from **Section 6.3** that radical substitution reactions require three kinds of steps: *initiation, propagation,* and *termination.* Once an initiation step has started the process by producing radicals, the reaction continues in a self-sustaining cycle. The cycle requires two repeating propagation steps in which a radical, the halogen, and the alkane yield alkyl halide product plus more radical to carry on the chain. The chain is occasionally terminated by the combination of two radicals.

Although interesting from a mechanistic point of view, alkane halogenation is a poor synthetic method for preparing alkyl halides because mixtures of products invariably result. For example, chlorination of methane does not stop cleanly at the monochlorinated stage but continues to give a mixture of dichloro, trichloro, and even tetrachloro products.

$$CH_4 + Cl_2 \xrightarrow{h\nu} CH_3Cl + HCl$$

$$CH_3Cl \xrightarrow{Cl_2} CH_2Cl_2 + HCl$$

$$CH_2Cl_2 \xrightarrow{Cl_2} CHCl_3 + HCl$$

$$CHCl_3 \xrightarrow{Cl_2} CCl_4 + HCl$$

The situation is even worse for chlorination of alkanes that have more than one sort of hydrogen. For example, chlorination of butane gives two monochlorinated products in a 30:70 ratio in addition to dichlorobutane, trichlorobutane, and so on.

$$\underset{\textbf{Butane}}{CH_3CH_2CH_2CH_3} + Cl_2 \xrightarrow{h\nu} \underset{\textbf{1-Chlorobutane}}{CH_3CH_2CH_2CH_2Cl} + \underset{\textbf{2-Chlorobutane}}{CH_3CH_2CH(Cl)CH_3} + \text{Dichloro-, trichloro-, tetrachloro-, and so on}$$

30 : 70

As another example, 2-methylpropane yields 2-chloro-2-methylpropane and 1-chloro-2-methylpropane in a 35:65 ratio, along with more highly chlorinated products.

$$\underset{\textbf{2-Methylpropane}}{CH_3CH(CH_3)CH_3} + Cl_2 \xrightarrow{h\nu} \underset{\textbf{2-Chloro-2-methylpropane}}{CH_3C(CH_3)(Cl)CH_3} + \underset{\textbf{1-Chloro-2-methylpropane}}{CH_3CH(CH_3)CH_2Cl} + \text{Dichloro-, trichloro-, tetrachloro-, and so on}$$

35 : 65

From these and similar reactions, it's possible to calculate a reactivity order toward chlorination for different sorts of hydrogen atoms in a molecule. Take the butane chlorination, for instance. Butane has six equivalent primary hydrogens ($-CH_3$) and four equivalent secondary hydrogens ($-CH_2-$). The fact that butane yields 30% of 1-chlorobutane product means that each one of the six primary hydrogens is responsible for $30\% \div 6 = 5\%$ of the product. Similarly, the fact that 70% of 2-chlorobutane is formed means that each of the four secondary hydrogens is responsible for $70\% \div 4 = 17.5\%$ of the product. Thus, a secondary hydrogen reacts $17.5\% \div 5\% = 3.5$ times as often as a primary hydrogen.

A similar calculation for the chlorination of 2-methylpropane indicates that each of the nine primary hydrogens accounts for 65% ÷ 9 = 7.2% of the product, while the single tertiary hydrogen (R_3CH) accounts for 35% of the product. Thus, a tertiary hydrogen is 35 ÷ 7.2 = 5 times as reactive as a primary hydrogen toward chlorination.

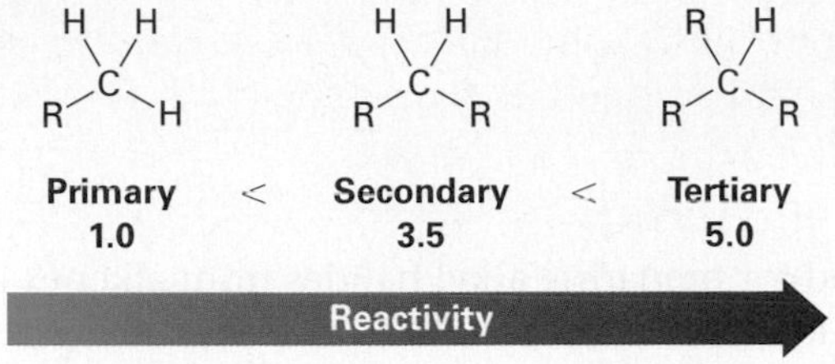

Primary < Secondary < Tertiary
1.0 3.5 5.0
Reactivity

The observed reactivity order of alkane hydrogens toward radical chlorination can be explained by looking at the bond dissociation energies given previously in Table 6.3 on page 204. The data show that a tertiary C–H bond (400 kJ/mol; 96 kcal/mol) is weaker than a secondary C–H bond (410 kJ/mol; 98 kcal/mol), which is in turn weaker than a primary C–H bond (421 kJ/mol; 101 kcal/mol). Since less energy is needed to break a tertiary C–H bond than to break a primary or secondary C–H bond, the resultant tertiary radical is more stable than a primary or secondary radical.

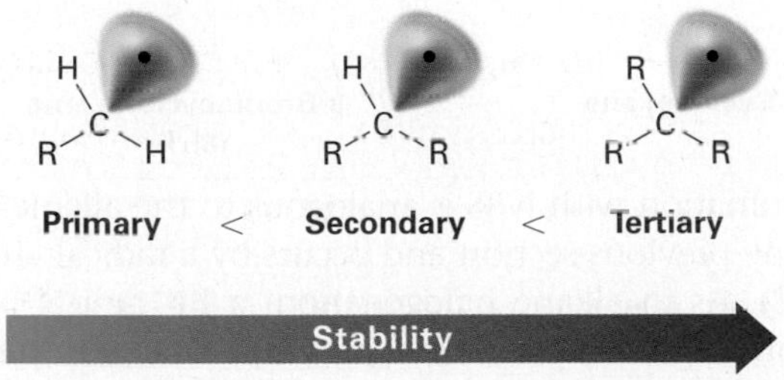

Primary < Secondary < Tertiary
Stability

Problem 10.3
Draw and name all monochloro products you would expect to obtain from radical chlorination of 2-methylpentane. Which, if any, are chiral?

Problem 10.4
Taking the relative reactivities of 1°, 2°, and 3° hydrogen atoms into account, what product(s) would you expect to obtain from monochlorination of 2-methylbutane? What would the approximate percentage of each product be? (Don't forget to take into account the number of each sort of hydrogen.)

10.3 Preparing Alkyl Halides from Alkenes: Allylic Bromination

We've already seen several methods for preparing alkyl halides from alkenes, including the reactions of HX and X_2 with alkenes in electrophilic addition reactions **(Sections 7.7 and 8.2)**. The hydrogen halides HCl, HBr, and HI react with alkenes by a polar mechanism to give the product of Markovnikov addition. Bromine and chlorine undergo anti addition through halonium ion intermediates to give 1,2-dihalogenated products.

X_2 HX

X = Cl or Br X = Cl, Br, or I

Another method for preparing alkyl halides from alkenes is by reaction with *N*-bromosuccinimide (abbreviated NBS) in the presence of light to give products resulting from substitution of hydrogen by bromine at the position next to the double bond—the **allylic** position. Cyclohexene, for example, gives 3-bromocyclohexene.

N—Br (NBS)

Allylic positions

$h\nu$, CCl_4

Cyclohexene **3-Bromocyclohexene (85%)**

This allylic bromination with NBS is analogous to the alkane chlorination reaction discussed in the previous section and occurs by a radical chain reaction pathway **(Figure 10.2)**. As in alkane halogenation, a Br· radical abstracts an allylic hydrogen atom, forming an allylic radical plus HBr. The HBr then reacts with NBS to form Br_2, which in turn reacts with the allylic radical to yield the brominated product and a Br· radical that cycles back into the first step and carries on the chain.

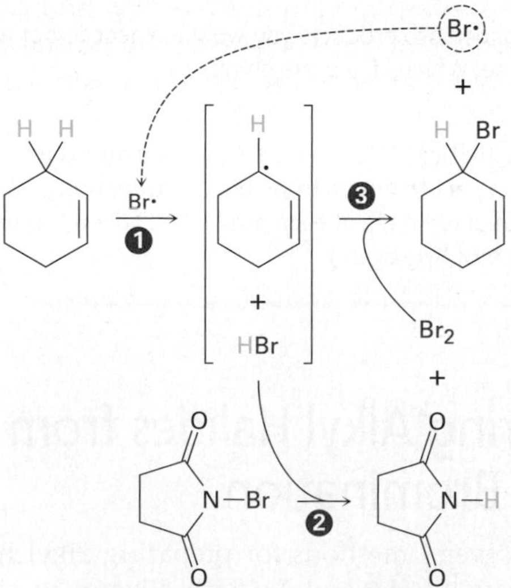

Figure 10.2 Mechanism of allylic bromination of an alkene with NBS. The process is a radical chain reaction in which (1) a Br· radical abstracts an allylic hydrogen atom of the alkene and gives an allylic radical plus HBr. (2) The HBr then reacts with NBS to form Br_2, which (3) reacts with the allylic radical to yield the bromoalkene product and a Br· radical that carries on the chain.

Why does bromination with NBS occur exclusively at an allylic position rather than elsewhere in the molecule? The answer, once again, is found by looking at bond dissociation atom energies to see the relative stabilities of various kinds of radicals. Although a typical secondary alkyl C–H bond has a strength of about 410 kJ/mol (98 kcal/mol) and a typical vinylic C–H bond has a strength of 465 kJ/mol (111 kcal/mol), an *allylic* C–H bond has a strength of only about 370 kJ/mol (88 kcal/mol). An allylic radical is therefore more stable than a typical alkyl radical with the same substitution by about 40 kJ/mol (9 kcal/mol).

Allylic
370 kJ/mol (88 kcal/mol)
Alkyl
410 kJ/mol (98 kcal/mol)
Vinylic
465 kJ/mol (111 kcal/mol)

We can thus expand the stability ordering to include vinylic and allylic radicals.

Vinylic < Methyl < Primary < Secondary < Tertiary < Allylic

Stability

10.4 Stability of the Allyl Radical: Resonance Revisited

To see why an allylic radical is so stable, look at the orbital picture in **Figure 10.3**. The radical carbon atom with an unpaired electron can adopt sp^2 hybridization, placing the unpaired electron in a *p* orbital and giving a structure that is electronically symmetrical. The *p* orbital on the central carbon can therefore overlap equally well with a *p* orbital on either of the two neighboring carbons.

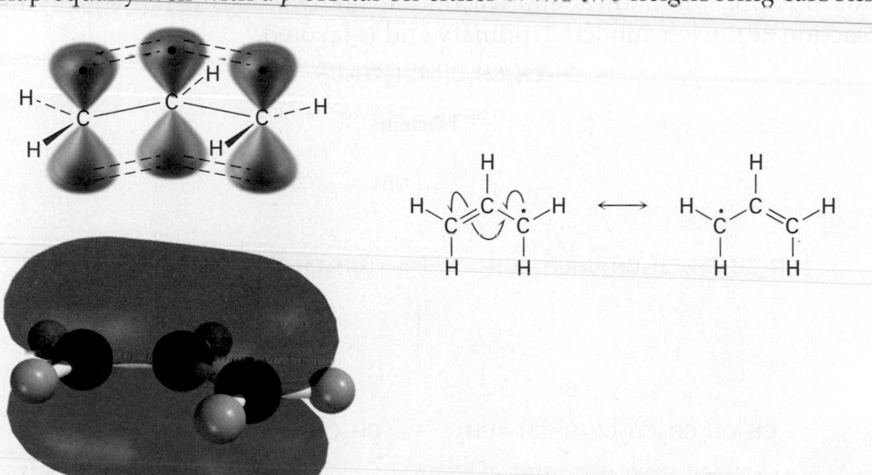

Figure 10.3 An orbital view of the allyl radical. The *p* orbital on the central carbon can overlap equally well with a *p* orbital on either neighboring carbon, giving rise to two equivalent resonance structures.

Because the allyl radical is electronically symmetrical, it has two resonance forms—one with the unpaired electron on the left and the double bond on the right and another with the unpaired electron on the right and the double bond on the left. Neither structure is correct by itself; the true structure of the allyl radical is a resonance hybrid of the two. (You might want to review **Sections 2.4–2.6** to brush up on resonance.) As noted in **Section 2.5**, the greater the number of resonance forms, the greater the stability of a compound because bonding electrons are attracted to more nuclei. An allyl radical, with two resonance forms, is therefore more stable than a typical alkyl radical, which has only a single structure.

In molecular orbital terms, the stability of the allyl radical is due to the fact that the unpaired electron is **delocalized**, or spread out, over an extended π orbital network rather than localized at only one site, as shown by the computer-generated MO in Figure 10.3. This delocalization is particularly apparent in the so-called spin density surface in **Figure 10.4**, which shows the calculated location of the unpaired electron. The two terminal carbons share the unpaired electron equally.

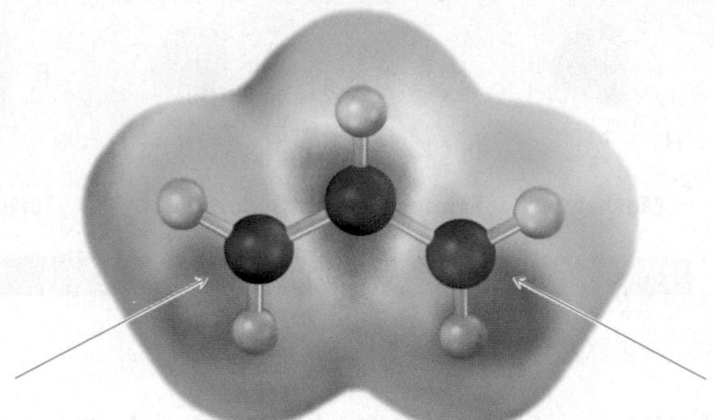

Figure 10.4 The spin density surface of the allyl radical locates the position of the **unpaired electron** and shows that it is equally shared between the two terminal carbons.

In addition to its effect on stability, delocalization of the unpaired electron in the allyl radical has other chemical consequences. Because the unpaired electron is delocalized over both ends of the π orbital system, reaction with Br_2 can occur at either end. As a result, allylic bromination of an unsymmetrical alkene often leads to a mixture of products. For example, bromination of 1-octene gives a mixture of 3-bromo-1-octene and 1-bromo-2-octene. The two products are not formed in equal amounts, however, because the intermediate allylic radical is not symmetrical and reaction at the two ends is not equally likely. Reaction at the less hindered, primary end is favored.

$$CH_3CH_2CH_2CH_2CH_2CH_2CH{=}CH_2$$

1-Octene

$\downarrow$ NBS, $h\nu$, CCl_4

$$\left[CH_3CH_2CH_2CH_2CH_2\dot{C}HCH{=}CH_2 \longleftrightarrow CH_3CH_2CH_2CH_2CH_2CH{=}CH\dot{C}H_2 \right]$$

$\downarrow$

$$CH_3CH_2CH_2CH_2CH_2CH(Br)CH{=}CH_2 \quad + \quad CH_3CH_2CH_2CH_2CH_2CH{=}CHCH_2Br$$

3-Bromo-1-octene (17%)

1-Bromo-2-octene (83%)
(53 : 47 trans : cis)

The products of allylic bromination reactions are useful for conversion into dienes by dehydrohalogenation with base. Cyclohexene can be converted into 1,3-cyclohexadiene, for example.

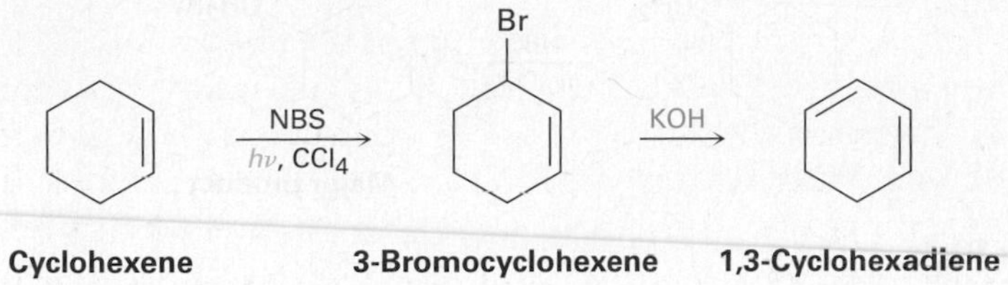

NBS, $h\nu$, CCl_4 → KOH →

Cyclohexene 3-Bromocyclohexene 1,3-Cyclohexadiene

Worked Example 10.1 Predicting the Product of an Allylic Bromination Reaction

What products would you expect from reaction of 4,4-dimethylcyclohexene with NBS?

Strategy

Draw the alkene reactant, and identify the allylic positions. In this case, there are two different allylic positions; we'll label them **A** and **B**. Now abstract an allylic hydrogen from each position to generate the two corresponding allylic radicals. Each of the two allylic radicals can add a Br atom at either end (**A** or **A′**; **B** or **B′**), to give a mixture of up to four products. Draw and name the products. In the present instance, the "two" products from reaction at position **B** are identical, so only three products are formed in this reaction.

Solution

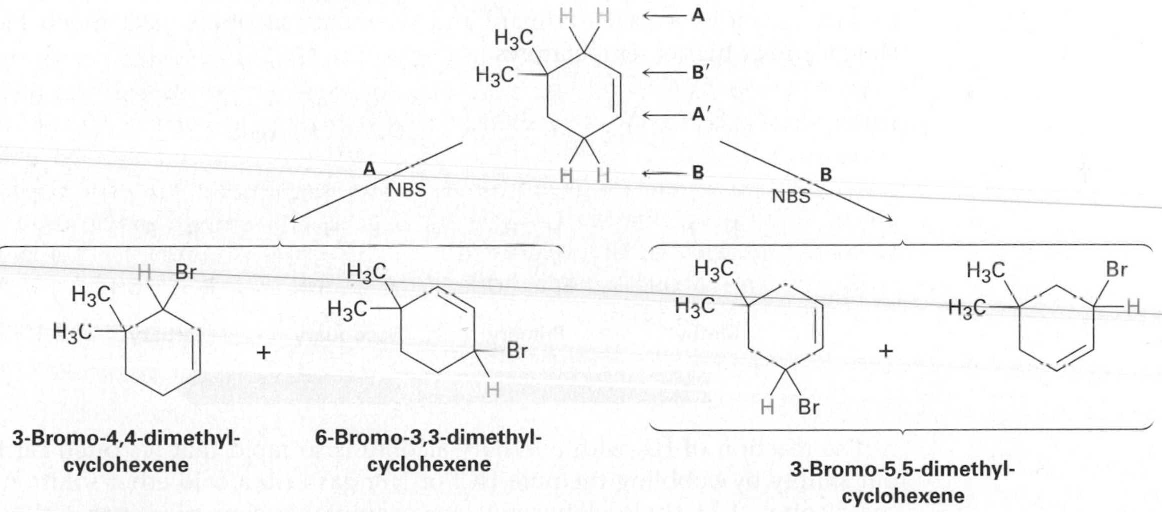

A NBS B NBS

3-Bromo-4,4-dimethylcyclohexene + 6-Bromo-3,3-dimethylcyclohexene

3-Bromo-5,5-dimethylcyclohexene

Problem 10.5

Draw three resonance forms for the cyclohexadienyl radical.

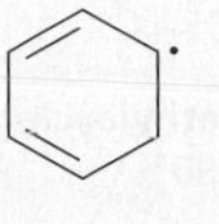

Cyclohexadienyl radical

Problem 10.6

The major product of the reaction of methylenecyclohexane with *N*-bromosuccinimide is 1-(bromomethyl)cyclohexene. Explain.

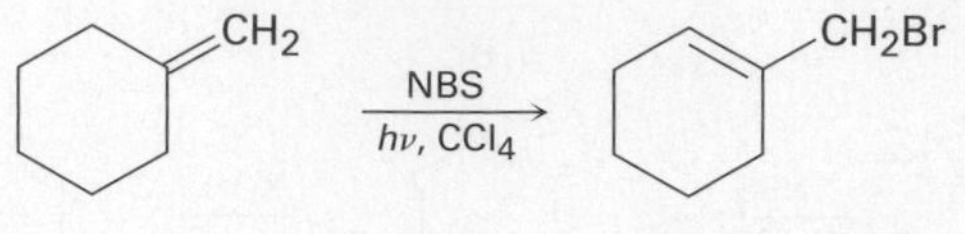

CH_2 —NBS, $h\nu$, CCl_4→ CH_2Br

Major product

Problem 10.7

What products would you expect from reaction of the following alkenes with NBS? If more than one product is formed, show the structures of all.

(a) CH_3 **(b)** $CH_3CH(CH_3)CH{=}CHCH_2CH_3$

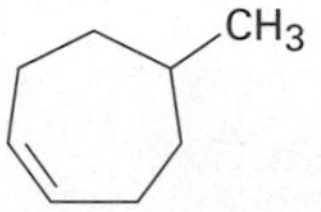

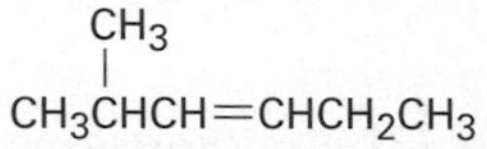

10.5 Preparing Alkyl Halides from Alcohols

The most generally useful method for preparing alkyl halides is to make them from alcohols, which themselves can be obtained from carbonyl compounds as we'll see in **Sections 17.4 and 17.5**. Because of the importance of the process, many different methods have been developed to transform alcohols into alkyl halides. The simplest method is to treat the alcohol with HCl, HBr, or HI. For reasons that will be discussed in **Section 11.5**, the reaction works best with tertiary alcohols, R_3COH. Primary and secondary alcohols react much more slowly and at higher temperatures.

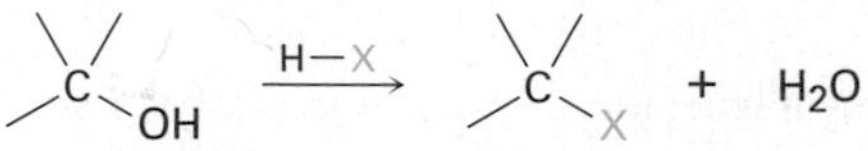

$$\text{C–OH} \xrightarrow{\text{H–X}} \text{C–X} + H_2O$$

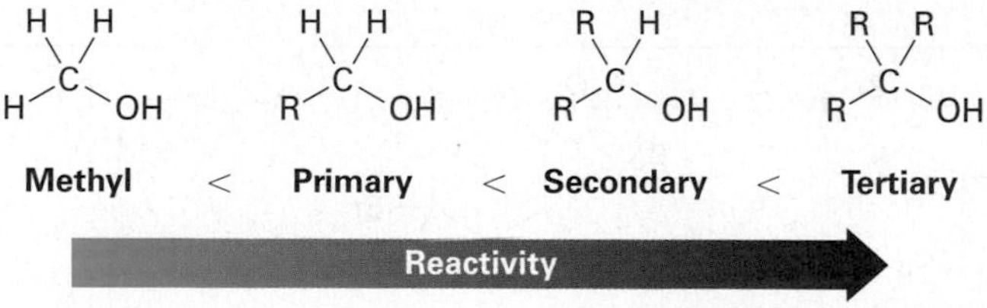

H H / H–C–OH   H H / R–C–OH   R H / R–C–OH   R R / R–C–OH

Methyl < **Primary** < **Secondary** < **Tertiary**

Reactivity

The reaction of HX with a tertiary alcohol is so rapid that it's often carried out simply by bubbling the pure HCl or HBr gas into a cold ether solution of the alcohol. 1-Methylcyclohexanol, for example, is converted into 1-chloro-1-methylcyclohexane by treating with HCl.

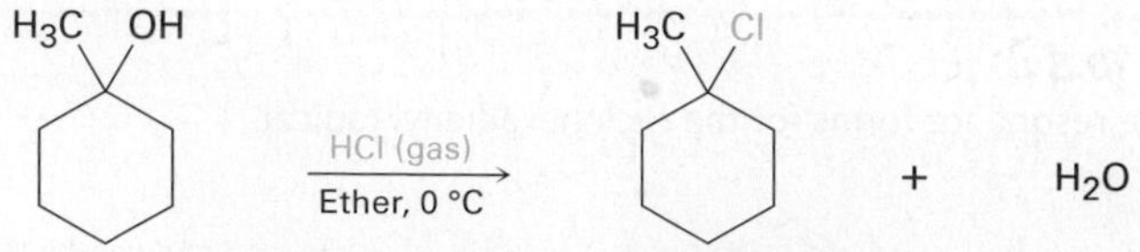

H_3C OH —HCl (gas), Ether, 0 °C→ H_3C Cl + H_2O

1-Methylcyclohexanol   **1-Chloro-1-methylcyclohexane (90%)**

Primary and secondary alcohols are best converted into alkyl halides by treatment with either thionyl chloride ($SOCl_2$) or phosphorus tribromide (PBr_3). These reactions, which normally take place readily under mild conditions, are less acidic and less likely to cause acid-catalyzed rearrangements than the HX method.

Benzoin $\xrightarrow[\text{Pyridine}]{SOCl_2}$ (86%) + SO_2 + HCl

$$3\ CH_3CH_2CH(OH)CH_3 \xrightarrow[\text{Ether, 35 °C}]{PBr_3} 3\ CH_3CH_2CH(Br)CH_3 + H_3PO_3$$

2-Butanol → 2-Bromobutane (86%)

As the preceding examples indicate, the yields of these $SOCl_2$ and PBr_3 reactions are generally high and other functional groups such as ethers, carbonyls, and aromatic rings don't usually interfere. We'll look at the mechanisms of these and other related substitution reactions in **Section 11.3**.

Alkyl fluorides can also be prepared from alcohols. Numerous alternative reagents are used for the reaction, including diethylaminosulfur trifluoride [$(CH_3CH_2)_2NSF_3$] and HF in pyridine solvent.

Cyclohexanol $\xrightarrow[\text{Pyridine}]{HF}$ Fluorocyclohexane (99%)

Problem 10.8

How would you prepare the following alkyl halides from the corresponding alcohols?

(a) $CH_3C(Cl)(CH_3)CH_3$ **(b)** $CH_3CH(Br)CH_2CH(CH_3)CH_3$ **(c)** $BrCH_2CH_2CH_2CH_2CH(CH_3)CH_3$ **(d)** F (3,3-dimethylcyclopentyl fluoride structure)

10.6 Reactions of Alkyl Halides: Grignard Reagents

Alkyl halides, RX, react with magnesium metal in ether or tetrahydrofuran (THF) solvent to yield alkylmagnesium halides, RMgX. The products, called **Grignard reagents** after their discoverer, Victor Grignard, are examples of *organometallic* compounds because they contain a carbon–metal bond. In addition to alkyl halides, Grignard reagents can also be made from alkenyl (vinylic) and aryl (aromatic) halides. The halogen can be Cl, Br, or I, although chlorides are less reactive than bromides and iodides. Organofluorides rarely react with magnesium.

$$\left.\begin{array}{l}1^\circ\text{ alkyl}\\2^\circ\text{ alkyl}\\3^\circ\text{ alkyl}\\\text{alkenyl}\\\text{aryl}\end{array}\right\} \longrightarrow \text{R—X} \longleftarrow \left\{\begin{array}{l}\text{Cl}\\\text{Br}\\\text{I}\end{array}\right.$$

$$\text{R—X} \xrightarrow[\text{Ether or THF}]{\text{Mg}} \text{R—Mg—X}$$

As you might expect from the discussion of electronegativity and bond polarity in **Section 6.4**, the carbon–magnesium bond is polarized, making the carbon atom of Grignard reagents both nucleophilic and basic. An electrostatic potential map of methylmagnesium iodide, for instance, indicates the electron-rich (red) character of the carbon bonded to magnesium.

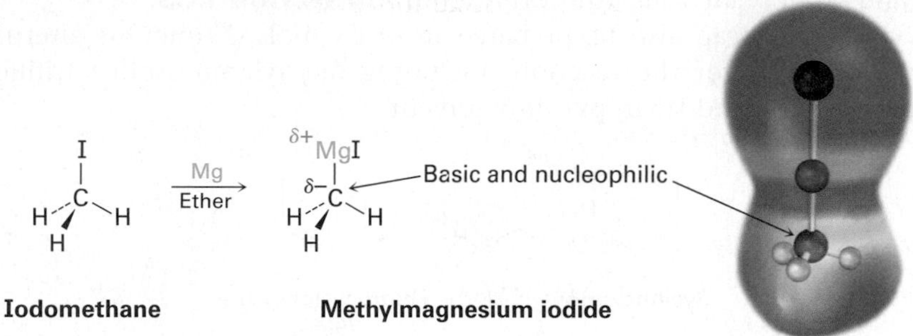

Iodomethane **Methylmagnesium iodide**

A Grignard reagent is formally the magnesium salt, $R_3C^{-\,+}MgX$, of a carbon acid, $R_3C—H$, and is thus a carbon anion, or **carbanion**. But because hydrocarbons are such weak acids, with pK_a's in the range 44 to 60 **(Section 9.7)**, carbon anions are very strong bases. Grignard reagents must therefore be protected from atmospheric moisture to prevent their being protonated and destroyed in an acid–base reaction: R—Mg—X + H_2O → R—H + HO—Mg—X.

$$CH_3CH_2CH_2CH_2CH_2CH_2Br \xrightarrow[\text{Ether}]{\text{Mg}} CH_3CH_2CH_2CH_2CH_2CH_2MgBr \xrightarrow{H_2O} CH_3CH_2CH_2CH_2CH_2CH_3$$

1-Bromohexane **1-Hexylmagnesium bromide** **Hexane**

Grignard reagents themselves don't occur in living organisms, but they are useful carbon-based nucleophiles in several important laboratory reactions, which we'll look at in detail in Chapter 17. In addition, they act as a simple model for other, more complex carbon-based nucleophiles that *are* important in biological chemistry. We'll see many examples in Chapter 29.

Problem 10.9

How strong a base would you expect a Grignard reagent to be? Look at Table 9.1 on page 326, and predict whether the following reactions will occur as written. (The pK_a of NH_3 is 35.)

(a) $CH_3MgBr + H{-}C{\equiv}C{-}H \rightarrow CH_4 + H{-}C{\equiv}C{-}MgBr$

(b) $CH_3MgBr + NH_3 \rightarrow CH_4 + H_2N{-}MgBr$

Problem 10.10

How might you replace a halogen substituent by a deuterium atom if you wanted to prepare a deuterated compound?

$$CH_3CH(Br)CH_2CH_3 \xrightarrow{?} CH_3CH(D)CH_2CH_3$$

10.7 Organometallic Coupling Reactions

Many other kinds of organometallic compounds can be prepared in a manner similar to that of Grignard reagents. For instance, alkyllithium reagents, RLi, can be prepared by the reaction of an alkyl halide with lithium metal. Alkyllithiums are both nucleophiles and strong bases, and their chemistry is similar in many respects to that of alkylmagnesium halides.

$$CH_3CH_2CH_2CH_2Br \xrightarrow[\text{Pentane}]{2\ Li} CH_3CH_2CH_2CH_2Li + LiBr$$

1-Bromobutane → **Butyllithium**

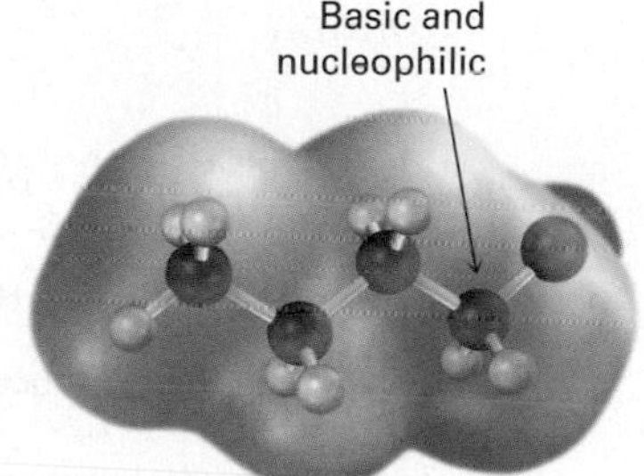

One particularly valuable reaction of alkyllithiums is in making lithium diorganocopper compounds, R_2CuLi, by reaction with copper(I) iodide in diethyl ether as solvent. Called **Gilman reagents**, lithium diorganocopper compounds are useful because they undergo a *coupling* reaction with organochlorides, bromides, and iodides (but not fluorides). One of the alkyl groups from the Gilman reagent replaces the halogen of the organohalide, forming a new carbon–carbon bond and yielding a hydrocarbon product. Lithium dimethylcopper, for instance, reacts with 1-iododecane to give undecane in 90% yield.

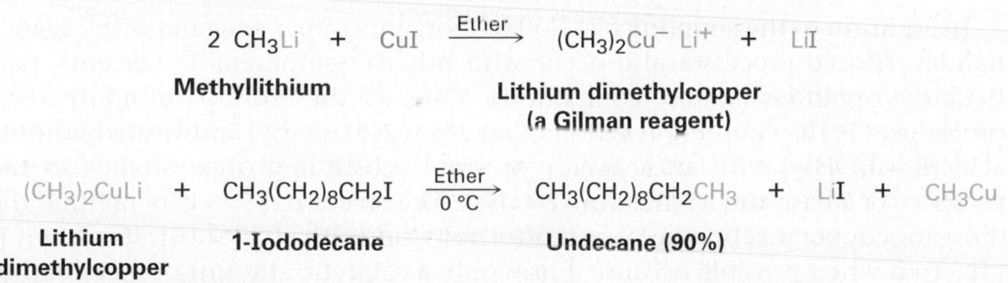

This organometallic coupling reaction is useful in organic synthesis because it forms carbon–carbon bonds, thereby making possible the preparation of larger molecules from smaller ones. As the following examples indicate, the coupling reaction can be carried out on aryl and vinylic halides as well as on alkyl halides.

$$(n\text{-}C_7H_{15})HC{=}CHI + (n\text{-}C_4H_9)_2CuLi \longrightarrow (n\text{-}C_7H_{15})HC{=}CH(C_4H_9\text{-}n) + n\text{-}C_4H_9Cu + LiI$$

***trans*-1-Iodo-1-nonene** → ***trans*-5-Tridecene (71%)**

$$C_6H_5I + (CH_3)_2CuLi \longrightarrow C_6H_5CH_3 + CH_3Cu + LiI$$

Iodobenzene → **Toluene (91%)**

An organocopper coupling reaction is carried out commercially to synthesize muscalure, (9*Z*)-tricosene, the sex attractant secreted by the common housefly. Minute amounts of muscalure greatly increase the lure of insecticide-treated fly bait and provide an effective and species-specific means of insect control.

$$CH_3(CH_2)_7CH{=}CH(CH_2)_7CH_2Br \xrightarrow{[CH_3(CH_2)_4]_2CuLi} CH_3(CH_2)_7CH{=}CH(CH_2)_{12}CH_3$$

***cis*-1-Bromo-9-octadecene** → **Muscalure (9*Z*-tricosene)**

The mechanism of the coupling reaction involves initial formation of a triorganocopper intermediate, followed by coupling and loss of RCu. The coupling is not a typical polar nucleophilic substitution reaction of the sort considered in the next chapter.

$$R{-}X + [R'{-}Cu{-}R']^- \, Li^+ \longrightarrow \left[\begin{array}{c} R \\ | \\ R'{-}Cu{-}R' \end{array}\right] \longrightarrow R{-}R' + R'{-}Cu$$

In addition to the coupling reaction of diorganocopper reagents with organohalides, related processes also occur with other organometallic reagents, particularly organopalladium compounds. One of the most commonly used procedures is the coupling reaction of an aromatic or vinyl substituted boronic acid [R—B(OH)$_2$] with an aromatic or vinyl substituted organohalide in the presence of a base and a palladium catalyst. The reaction is less general than the diorganocopper reaction because it does not work with alkyl substrates, but it is preferred when possible because it uses only a catalytic amount of metal rather than a full equivalent and because palladium compounds are less toxic than copper compounds. For example:

CH_3, OH, H_3C–B, OH, CH_3 + I–Ph $\xrightarrow[\text{CaCO}_3,\ \text{THF}]{\text{Pd(PPh}_3)_4}$ H_3C, CH_3, CH_3

[Ph = Phenyl,]

A *biaryl* compound (92%)

Called the *Suzuki–Miyaura reaction,* the process is particularly useful for preparing so-called biaryl compounds, which have two aromatic rings joined together. A large number of commonly used drugs fit this description, so the Suzuki–Miyaura reaction is much-used in the pharmaceutical industry. As an example, valsartan, marketed as Diovan, is a widely prescribed antihypertensive agent whose synthesis begins with a Suzuki–Miyaura coupling of *ortho*-chlorobenzonitrile with *para*-methylbenzeneboronic acid.

H_3C, $B(OH)_2$ — ***para*-Methylbenzeneboronic acid** + Cl, N≡C — ***ortho*-Chlorobenzonitrile** $\xrightarrow[K_2CO_3]{\text{Pd catalyst}}$ H_3C, N≡C $\Longrightarrow$ O, H, $CH(CH_3)_2$, $CH_3CH_2CH_2CH_2$–C–N, CO_2H, N, N, N, N–H — **Valsartan (Diovan)**

Shown in a simplified form in **Figure 10.5**, the mechanism of the Suzuki–Miyaura reaction involves initial reaction of the aromatic halide with the palladium catalyst to form an organopalladium intermediate, followed by reaction of that intermediate with the aromatic boronic acid. The resultant diorganopalladium complex then decomposes to the coupled biaryl product plus regenerated catalyst.

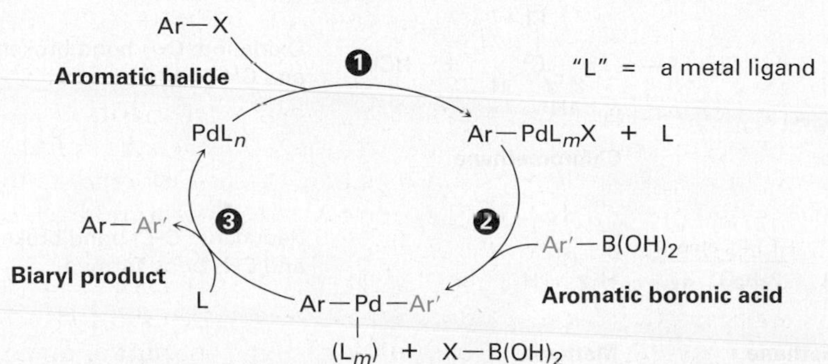

Figure 10.5 Mechanism of the Suzuki–Miyaura coupling reaction of an aromatic boronic acid with an aromatic halide to give a biaryl. The reaction takes place by (1) reaction of the aromatic halide, ArX, with the catalyst to form an organopalladium intermediate, followed by (2) reaction with the aromatic boronic acid. (3) Subsequent decomposition of the diarylpalladium intermediate gives the biaryl product.

Problem 10.11

How would you carry out the following transformations using an organocopper coupling reaction? More than one step is required in each case.

(a) Cyclohexene $\xrightarrow{?}$ 3-methylcyclohexene (CH_3)

(b) $CH_3CH_2CH_2CH_2Br \xrightarrow{?} CH_3CH_2CH_2CH_2CH_2CH_2CH_2CH_3$

(c) $CH_3CH_2CH_2CH{=}CH_2 \xrightarrow{?} CH_3CH_2CH_2CH_2CH_2CH_2CH_2CH_2CH_2CH_3$

10.8 Oxidation and Reduction in Organic Chemistry

We've pointed out on several occasions that some of the reactions discussed in this and earlier chapters are either oxidations or reductions. As noted in **Section 8.7**, an organic oxidation results in a loss of electron density by carbon, caused either by bond formation between carbon and a more electronegative atom (usually O, N, or a halogen) or by bond-breaking between carbon and a less electronegative atom (usually H). Conversely, an organic reduction results in a gain of electron density by carbon, caused either by bond formation between carbon and a less electronegative atom or by bond-breaking between carbon and a more electronegative atom **(Section 8.6)**.

Oxidation Decreases electron density on carbon by:
- forming one of these: C–O C–N C–X
- or breaking this: C–H

Reduction Increases electron density on carbon by:
- forming this: C–H
- or breaking one of these: C–O C–N C–X

Based on these definitions, the chlorination reaction of methane to yield chloromethane is an oxidation because a C–H bond is broken and a C–Cl bond is formed. The conversion of an alkyl chloride to an alkane via a Grignard reagent followed by protonation is a reduction, however, because a C–Cl bond is broken and a C–H bond is formed.

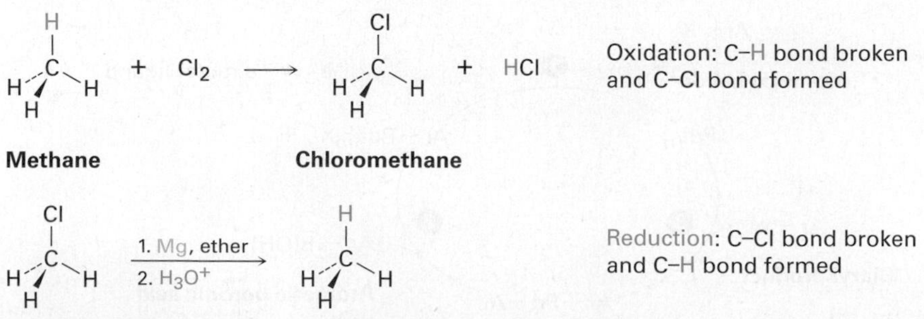

As other examples, the reaction of an alkene with Br_2 to yield a 1,2-dibromide is an oxidation because two C–Br bonds are formed, but the reaction of an alkene with HBr to yield an alkyl bromide is neither an oxidation nor a reduction because both a C–H and a C–Br bond are formed.

$$H_2C{=}CH_2 + Br_2 \longrightarrow BrCH_2{-}CH_2Br$$

Ethylene → **1,2-Dibromoethane**

Oxidation: Two new bonds formed between carbon and a more electronegative element

$$H_2C{=}CH_2 + HBr \longrightarrow HCH_2{-}CH_2Br$$

Ethylene → **Bromoethane**

Neither oxidation nor reduction: One new C–H bond and one new C–Br bond formed

A list of compounds of increasing oxidation level is shown in **Figure 10.6**. Alkanes are at the lowest oxidation level because they have the maximum possible number of C–H bonds per carbon, and CO_2 is at the highest level because it has the maximum possible number of C–O bonds per carbon. Any reaction that converts a compound from a lower level to a higher level is an oxidation, any reaction that converts a compound from a higher level to a lower level is a reduction, and any reaction that doesn't change the level is neither an oxidation nor a reduction.

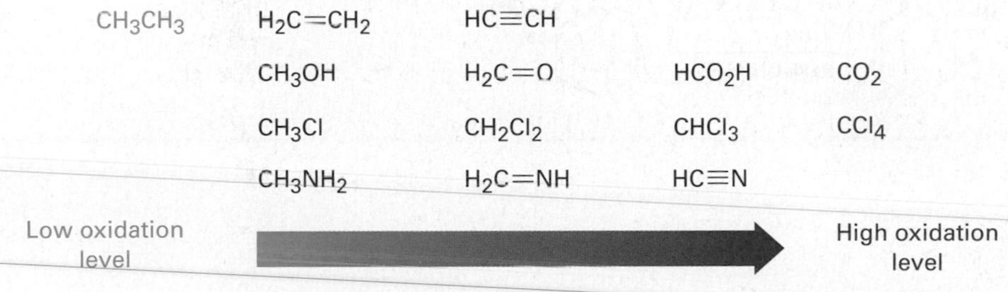

Figure 10.6 Oxidation levels of some common types of compounds.

Worked Example 10.2 shows how to compare the oxidation levels of different compounds with the same number of carbon atoms.

Worked Example 10.2

Comparing Oxidation Levels

Rank the following compounds in order of increasing oxidation level:

$$CH_3CH{=}CH_2 \qquad CH_3CH(OH)CH_3 \qquad CH_3C(=O)CH_3 \qquad CH_3CH_2CH_3$$

Strategy

Compounds that have the same number of carbon atoms can be compared by adding the number of C–O, C–N, and C–X bonds in each and then subtracting the number of C–H bonds. The larger the resultant value, the higher the oxidation level.

Solution

The first compound (propene) has six C—H bonds, giving an oxidation level of −6; the second (2-propanol) has one C—O bond and seven C—H bonds, giving an oxidation level of −6; the third (acetone) has two C—O bonds and six C—H bonds, giving an oxidation level of −4; and the fourth (propane) has eight C—H bonds, giving an oxidation level of −8. Thus, the order of increasing oxidation level is

$$CH_3CH_2CH_3 \;<\; CH_3CH{=}CH_2 \;=\; CH_3\overset{\displaystyle OH}{\overset{|}{C}}HCH_3 \;<\; CH_3\overset{\displaystyle O}{\overset{\|}{C}}CH_3$$

Problem 10.12

Rank each of the following series of compounds in order of increasing oxidation level:

(a) O Cl

(b) CH_3CN $CH_3CH_2NH_2$ $H_2NCH_2CH_2NH_2$

Problem 10.13

Tell whether each of the following reactions is an oxidation, a reduction, or neither.

(a)

$$CH_3CH_2\overset{\displaystyle O}{\overset{\|}{C}}H \xrightarrow[H_2O]{NaBH_4} CH_3CH_2CH_2OH$$

(b) 1. BH_3 2. NaOH, H_2O_2 OH

Marine corals secrete organohalogen compounds that act as a feeding deterrent to fish.

Naturally Occurring Organohalides

A DEEPER LOOK

As recently as 1970, only about 30 naturally occurring organohalides were known. It was simply assumed that chloroform, halogenated phenols, chlorinated aromatic compounds called PCBs, and other such substances found in the environment were industrial pollutants. Now, a bit more than a third of a century later, the situation is quite different. More than 5000 organohalides have been found to occur naturally, and tens of thousands more surely exist. From a simple compound like chloromethane to an extremely complex one like the antibiotic vancomycin, a remarkably diverse range of organohalides exists in plants, bacteria, and animals. Many even have valuable physiological activity. The pentahalogenated alkene halomon, for

(continued)

(continued)

instance, has been isolated from the red alga *Portieria hornemannii* and found to have anti-cancer activity against several human tumor cell lines.

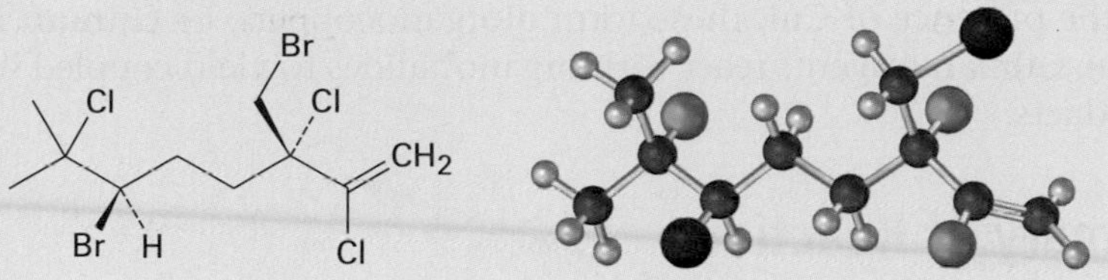

Br
Cl
Cl
CH_2
Br H
Cl

Halomon

Some naturally occurring organohalides are produced in massive quantities. Forest fires, volcanoes, and marine kelp release up to *5 million tons* of CH_3Cl per year, for example, while annual industrial emissions total about 26,000 tons. Termites are thought to release as much as 10^8 kg of chloroform per year. A detailed examination of the Okinawan acorn worm *Ptychodera flava* found that the 64 million worms living in a 1 km^2 study area excreted nearly 8000 pounds per year of bromophenols and bromoindoles, compounds previously thought to be nonnatural pollutants.

Why do organisms produce organohalides, many of which are undoubtedly toxic? The answer seems to be that many organisms use organohalogen compounds for self-defense, either as feeding deterrents, as irritants to predators, or as natural pesticides. Marine sponges, coral, and sea hares, for example, release foul-tasting organohalides that deter fish, starfish, and other predators from eating them. Even humans appear to produce halogenated compounds as part of their defense against infection. The human immune system contains a peroxidase enzyme capable of carrying out halogenation reactions on fungi and bacteria, thereby killing the pathogen. And most remarkable of all, even free chlorine—Cl_2—has been found to be present in humans.

Much remains to be learned—only a few hundred of the more than 500,000 known species of marine organisms have been examined—but it is clear that organohalides are an integral part of the world around us.

Summary

Alkyl halides are not often found in terrestrial organisms, but the kinds of reactions they undergo are among the most important and well-studied reaction types in organic chemistry. In this chapter, we saw how to name and prepare alkyl halides, and we'll soon make a detailed study of their substitution and elimination reactions.

Simple alkyl halides can be prepared by radical halogenation of alkanes, but mixtures of products usually result. The reactivity order of alkanes toward halogenation is identical to the stability order of radicals: $R_3C\cdot > R_2CH\cdot > RCH_2\cdot$. Alkyl halides can also be prepared from alkenes by reaction with *N*-bromosuccinimide (NBS) to give the product of **allylic** bromination. The NBS bromination of alkenes takes place through an intermediate allylic radical, which is stabilized by resonance.

Alcohols react with HX to form alkyl halides, but the reaction works well only for tertiary alcohols, R_3COH. Primary and secondary alkyl halides are normally

Key words

alkyl halide, 344
allylic, 350
carbanion, 356
delocalized, 352
Gilman reagent (LiR_2Cu), 357
Grignard reagent (RMgX), 355
organohalide, 344

prepared from alcohols using either $SOCl_2$, PBr_3, or HF in pyridine. Alkyl halides react with magnesium in ether solution to form organomagnesium halides, called **Grignard reagents (RMgX)**, which are both nucleophilic and strongly basic.

Alkyl halides also react with lithium metal to form organolithium reagents, RLi. In the presence of CuI, these form diorganocoppers, or **Gilman reagents (LiR_2Cu)**. Gilman reagents react with organohalides to yield coupled hydrocarbon products.

Summary of Reactions

1. Preparation of alkyl halides
   (a) From alkenes by allylic bromination (Section 10.3)

   NBS; $h\nu$, CCl_4

   (b) From alcohols (Section 10.5)
   (1) Reaction with HX

   HX; Ether

   Reactivity order: 3° > 2° > 1°

   (2) Reaction of 1° and 2° alcohols with $SOCl_2$

   $SOCl_2$; Pyridine

   (3) Reaction of 1° and 2° alcohols with PBr_3

   PBr_3; Ether

   (4) Reaction of 1° and 2° alcohols with HF–pyridine

   HF; Pyridine

   Pyridine

2. Reactions of alkyl halides
   (a) Formation of Grignard (organomagnesium) reagents (Section 10.6)

   $$\text{R–X} \xrightarrow[\text{Ether}]{\text{Mg}} \text{R–Mg–X}$$

(continued)

(b) Formation of Gilman (diorganocopper) reagents (Section 10.7)

$$\text{R—X} \xrightarrow[\text{Pentane}]{\text{2 Li}} \text{R—Li} + \text{LiX}$$

$$2\ \text{R—Li} + \text{CuI} \xrightarrow{\text{In ether}} [\text{R—Cu—R}]^- \text{Li}^+ + \text{LiI}$$

(c) Organometallic coupling (Section 10.7)

(1) Diorganocopper reaction

$$\text{R}_2\text{CuLi} + \text{R}'\text{—X} \xrightarrow{\text{In ether}} \text{R—R}' + \text{RCu} + \text{LiX}$$

(2) Palladium-catalyzed Suzuki–Miyaura reaction

$$\text{C}_6\text{H}_5\text{—B(OH)}_2 + \text{I—C}_6\text{H}_5 \xrightarrow[\text{CaCO}_3,\ \text{THF}]{\text{Pd(PPh}_3)_4} \text{C}_6\text{H}_5\text{—C}_6\text{H}_5$$

Exercises

Visualizing Chemistry

(Problems 10.1–10.13 appear within the chapter.)

OWL Interactive versions of these problems are assignable in OWL for Organic Chemistry.

10.14 Give IUPAC names for the following alkyl halides (green = Cl):

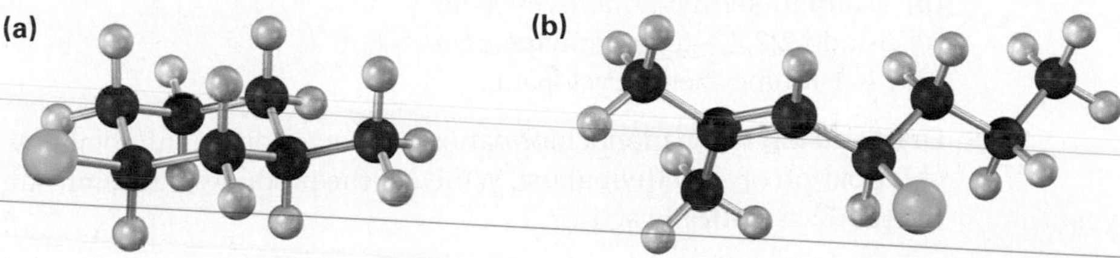

10.15 Show the product(s) of reaction of the following alkenes with NBS:

(a)

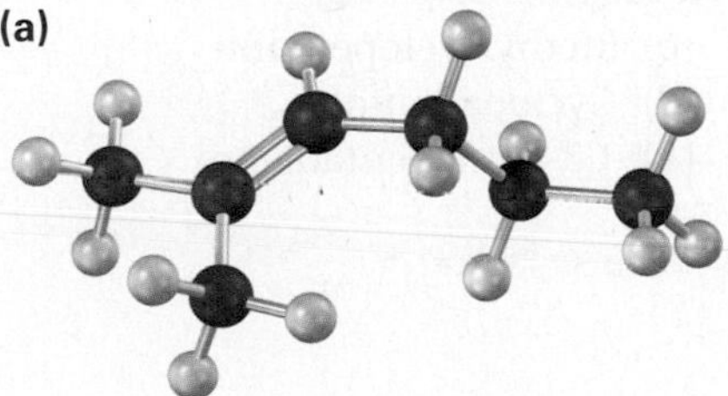

(b)

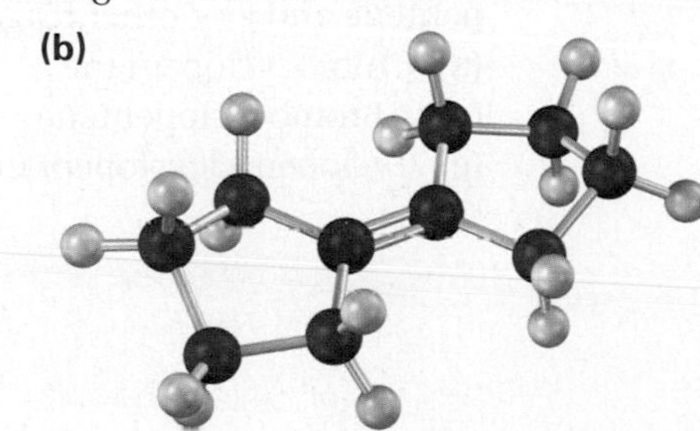

10.16 The following alkyl bromide can be prepared by reaction of the alcohol (*S*)-2-pentanol with PBr_3. Name the compound, assign (*R*) or (*S*) stereochemistry, and tell whether the reaction of the alcohol occurs with retention of the same stereochemistry or with a change in stereochemistry (reddish brown = Br).

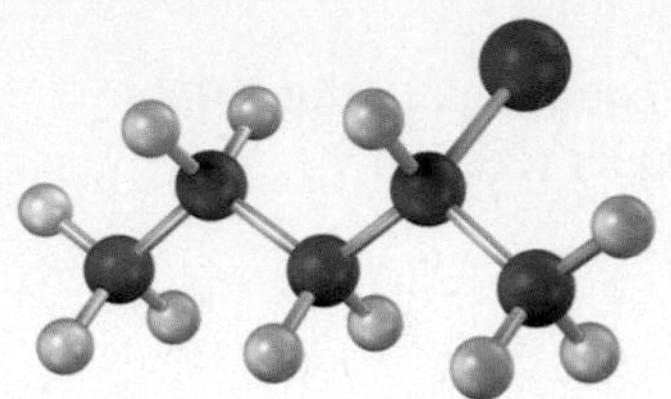

Additional Problems

Naming Alkyl Halides

10.17 Name the following alkyl halides:

(a) H_3C Br Br CH_3 (substituents on C2, C3, C4, C6)
$CH_3CHCHCHCH_2CHCH_3$

(b) I (on C5)
$CH_3CH{=}CHCH_2CHCH_3$

(c) Br (on C2), Cl CH_3 (on C4, C5), CH_3 (on C2)
$CH_3CCH_2CHCHCH_3$

(d) CH_2Br (on C3)
$CH_3CH_2CHCH_2CH_2CH_3$

(e) $ClCH_2CH_2CH_2C{\equiv}CCH_2Br$

10.18 Draw structures corresponding to the following IUPAC names:
(a) 2,3-Dichloro-4-methylhexane
(b) 4-Bromo-4-ethyl-2-methylhexane
(c) 3-Iodo-2,2,4,4-tetramethylpentane
(d) *cis*-1-Bromo-2-ethylcyclopentane

10.19 Draw and name the monochlorination products you might obtain by radical chlorination of 2-methylbutane. Which of the products are chiral? Are any of the products optically active?

Synthesizing Alkyl Halides

10.20 How would you prepare the following compounds, starting with cyclopentene and any other reagents needed?
(a) Chlorocyclopentane
(b) Methylcyclopentane
(c) 3-Bromocyclopentene
(d) Cyclopentanol
(e) Cyclopentylcyclopentane
(f) 1,3-Cyclopentadiene

10.21 Predict the product(s) of the following reactions:

(a) 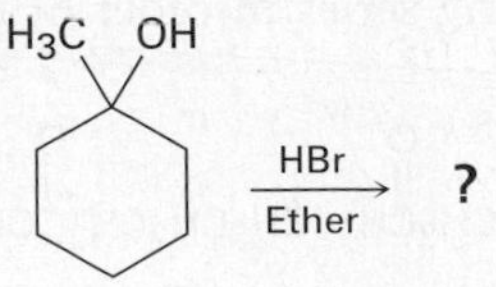 1-Methylcyclohexanol (H_3C, OH) $\xrightarrow[\text{Ether}]{\text{HBr}}$?

(b) $CH_3CH_2CH_2CH_2OH \xrightarrow{SOCl_2}$?

(c) 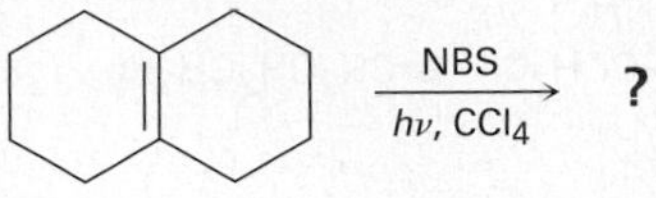 (bicyclic alkene) $\xrightarrow[h\nu,\ CCl_4]{\text{NBS}}$?

(d) 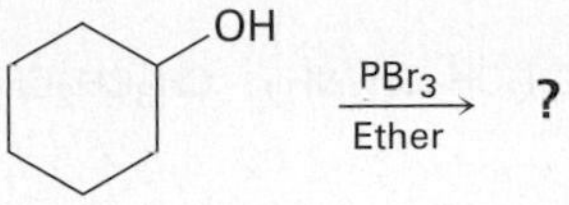 Cyclohexylmethanol (OH) $\xrightarrow[\text{Ether}]{PBr_3}$?

(e) $CH_3CH_2CHBrCH_3 \xrightarrow[\text{Ether}]{\text{Mg}}$ A? $\xrightarrow{H_2O}$ B?

(f) $CH_3CH_2CH_2CH_2Br \xrightarrow[\text{Pentane}]{\text{Li}}$ A? $\xrightarrow{\text{CuI}}$ B?

(g) $CH_3CH_2CH_2CH_2Br$ + $(CH_3)_2CuLi \xrightarrow{\text{Ether}}$?

10.22 A chemist requires a large amount of 1-bromo-2-pentene as starting material for a synthesis and decides to carry out an NBS allylic bromination reaction. What is wrong with the following synthesis plan? What side products would form in addition to the desired product?

$$CH_3CH_2CH{=}CHCH_3 \xrightarrow[h\nu,\ CCl_4]{\text{NBS}} CH_3CH_2CH{=}CHCH_2Br$$

10.23 What product(s) would you expect from the reaction of 1-methylcyclohexene with NBS? Would you use this reaction as part of a synthesis?

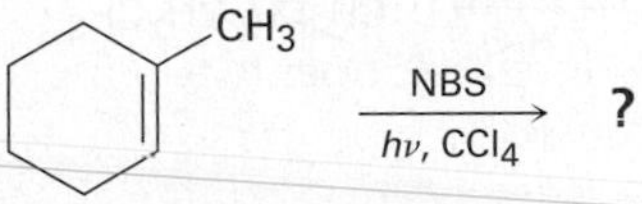

1-Methylcyclohexene (CH_3) $\xrightarrow[h\nu,\ CCl_4]{\text{NBS}}$?

10.24 What product(s) would you expect from the reaction of 1,4-hexadiene with NBS? What is the structure of the most stable radical intermediate?

10.25 What product would you expect from the reaction of 1-phenyl-2-butene with NBS? Explain.

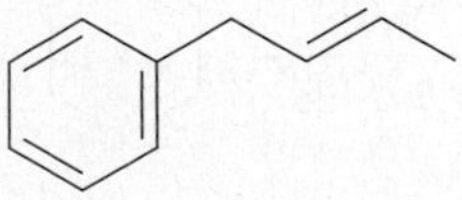

1-Phenyl-2-butene

Oxidation and Reduction

10.26 Rank the compounds in each of the following series in order of increasing oxidation level:

(a)

$CH_3CH{=}CHCH_3$ $\quad CH_3CH_2CH{=}CH_2$ $\quad CH_3CH_2CH_2\overset{\overset{O}{\|}}{C}H$ $\quad CH_3CH_2CH_2\overset{\overset{O}{\|}}{C}OH$

(b)

$CH_3CH_2CH_2NH_2$ $\quad CH_3CH_2CH_2Br$ $\quad CH_3\overset{\overset{O}{\|}}{C}CH_2Cl$ $\quad BrCH_2CH_2CH_2Cl$

10.27 Which of the following compounds have the same oxidation level, and which have different levels?

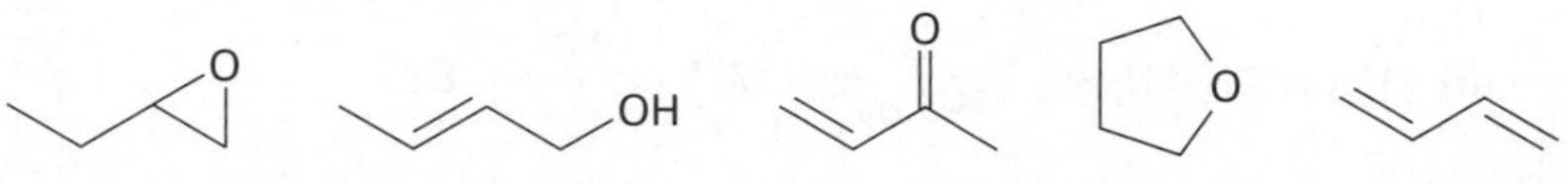

1 2 (OH) 3 (O) 4 (O) 5

10.28 Tell whether each of the following reactions is an oxidation, a reduction, or neither:

(a)

$$CH_3CH_2OH \xrightarrow{CrO_3} CH_3\overset{\overset{O}{\|}}{C}H$$

(b)

$$H_2C{=}CH\overset{\overset{O}{\|}}{C}CH_3 + NH_3 \longrightarrow H_2NCH_2CH_2\overset{\overset{O}{\|}}{C}CH_3$$

(c)

$$CH_3CH_2\overset{\overset{Br}{|}}{C}HCH_3 \xrightarrow[2.\ H_2O]{1.\ Mg} CH_3CH_2CH_2CH_3$$

General Problems

10.29 Alkylbenzenes such as toluene (methylbenzene) react with NBS to give products in which bromine substitution has occurred at the position next to the aromatic ring (the *benzylic* position). Explain, based on the bond dissociation energies in Table 6.3 on page 204.

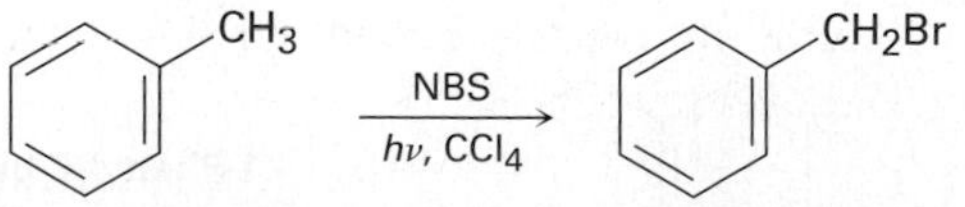

$$C_6H_5CH_3 \xrightarrow[h\nu,\ CCl_4]{NBS} C_6H_5CH_2Br$$

10.30 Draw resonance structures for the benzyl radical, $C_6H_5CH_2\cdot$, the intermediate produced in the NBS bromination reaction of toluene (Problem 10.29).

10.31 Draw resonance structures for the following species:

(a) $CH_3CH{=}CHCH{=}CHCH{=}CH\overset{+}{C}H_2$ **(b)** 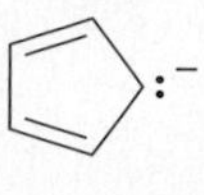 cyclopentadienyl anion $:^-$ **(c)** $CH_3C{\equiv}\overset{+}{N}{-}\ddot{\underset{..}{O}}:^-$

10.32 (*S*)-3-Methylhexane undergoes radical bromination to yield optically inactive 3-bromo-3-methylhexane as the major product. Is the product chiral? What conclusions can you draw about the radical intermediate?

10.33 Assume that you have carried out a radical chlorination reaction on (*R*)-2-chloropentane and have isolated (in low yield) 2,4-dichloropentane. How many stereoisomers of the product are formed and in what ratio? Are any of the isomers optically active? (See Problem 10.32.)

10.34 How would you carry out the following syntheses?

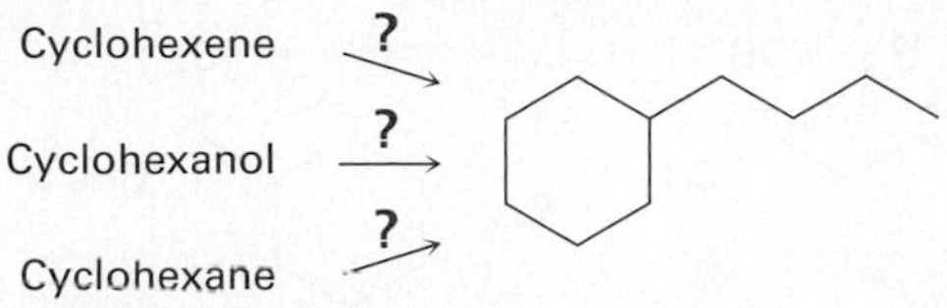

Cyclohexene $\xrightarrow{?}$
Cyclohexanol $\xrightarrow{?}$
Cyclohexane $\xrightarrow{?}$

10.35 The syntheses shown here are unlikely to occur as written. What is wrong with each?

(a) $CH_3CH_2CH_2F \xrightarrow[\text{2. } H_3O^+]{\text{1. Mg}} CH_3CH_2CH_3$

(b) CH_2, CH_3 $\xrightarrow[h\nu,\ CCl_4]{\text{NBS}}$ Br, CH_2, CH_3

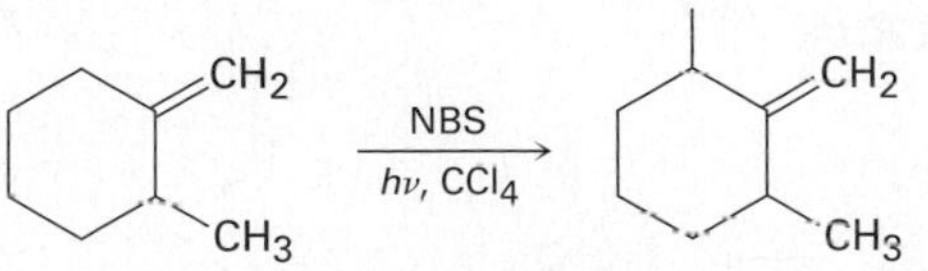

(c) F $\xrightarrow[\text{Ether}]{(CH_3)_2CuLi}$ CH_3

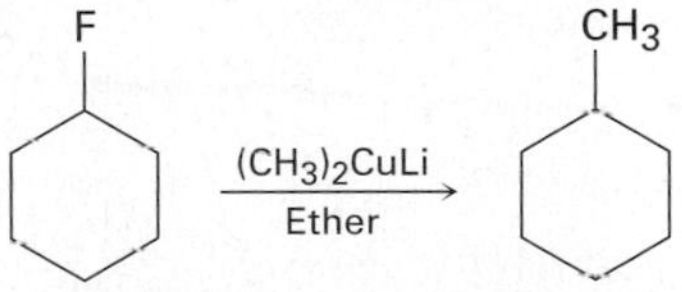

10.36 Why do you suppose it's not possible to prepare a Grignard reagent from a bromo alcohol such as 4-bromo-1-pentanol? Give another example of a molecule that is unlikely to form a Grignard reagent.

$CH_3CH(Br)CH_2CH_2CH_2OH \xrightarrow{Mg}\!\!\!\!\!/\ \ CH_3CH(MgBr)CH_2CH_2CH_2OH$

10.37 Addition of HBr to a double bond with an ether (–OR) substituent occurs regiospecifically to give a product in which the –Br and –OR are bonded to the same carbon. Draw the two possible carbocation intermediates in this electrophilic addition reaction, and explain using resonance why the observed product is formed.

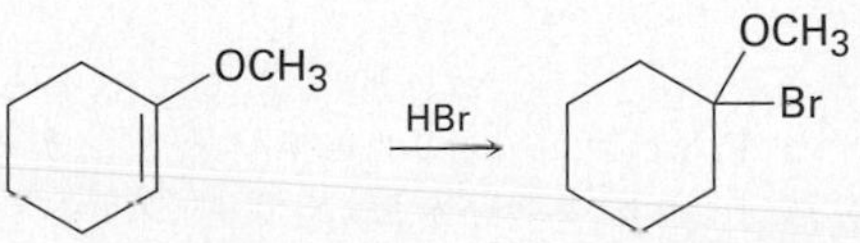

OCH_3 $\xrightarrow{HBr}$ OCH_3, Br

10.38 Alkyl halides can be reduced to alkanes by a radical reaction with tributyltin hydride, $(C_4H_9)_3SnH$, in the presence of light ($h\nu$). Propose a radical chain mechanism by which the reaction might occur. The initiation step is the light-induced homolytic cleavage of the Sn–H bond to yield a tributyltin radical.

$$R{-}X + (C_4H_9)_3SnH \xrightarrow{h\nu} R{-}H + (C_4H_9)_3SnX$$

10.39 Identify the reagents **a–c** in the following scheme:

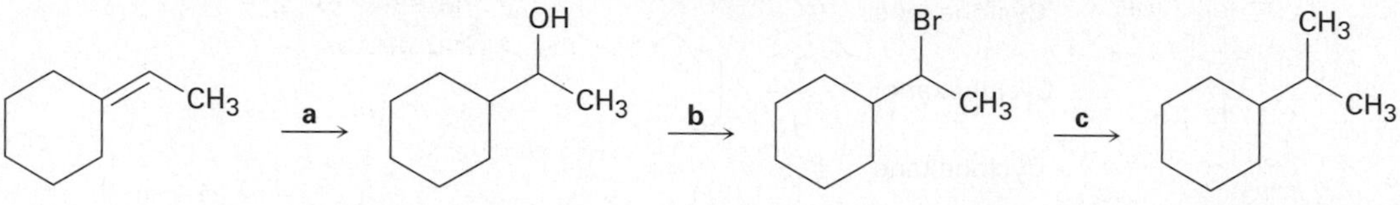

CH3 —**a**→ OH, CH3 —**b**→ Br, CH3 —**c**→ CH3, CH3

10.40 Tertiary alkyl halides, R_3CX, undergo spontaneous dissociation to yield a carbocation, R_3C^+, plus halide ion. Which do you think reacts faster, $(CH_3)_3CBr$ or $H_2C{=}CHC(CH_3)_2Br$? Explain.

10.41 In light of the fact that tertiary alkyl halides undergo spontaneous dissociation to yield a carbocation plus halide ion (Problem 10.40), propose a mechanism for the following reaction.

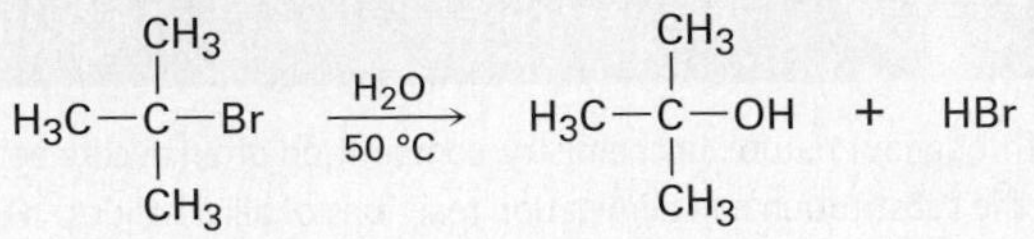

$$H_3C-C(CH_3)_2-Br \xrightarrow[50\ ^\circ C]{H_2O} H_3C-C(CH_3)_2-OH \; + \; HBr$$

10.42 Carboxylic acids (RCO_2H; $pK_a \approx 5$) are approximately 10^{11} times more acidic than alcohols (ROH; $pK_a \approx 16$). In other words, a carboxylate ion (RCO_2^-) is more stable than an alkoxide ion (RO^-). Explain, using resonance.

10.43 How might you use a Suzuki–Miyaura coupling to prepare the following biaryl compound? Show the two potential reaction partners.

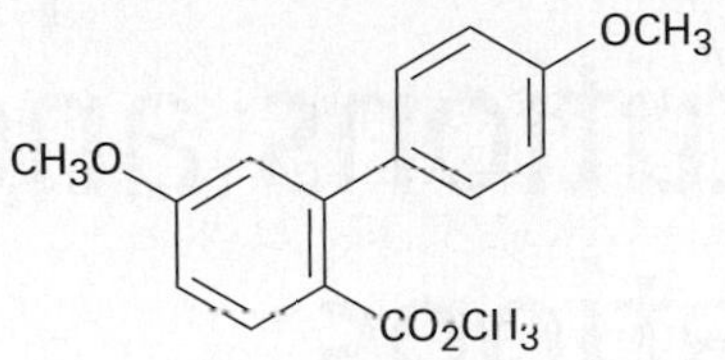

OCH_3 CH_3O CO_2CH_3

Competition occurs throughout nature. In chemistry, competition often occurs between alternative reaction pathways, such as in the substitution and elimination reactions of alkyl halides. Cheryl Ann Quigley/Shutterstock

11 Reactions of Alkyl Halides: Nucleophilic Substitutions and Eliminations

We saw in the preceding chapter that the carbon–halogen bond in an alkyl halide is polar and that the carbon atom is electron-poor. Thus, alkyl halides are electrophiles, and much of their chemistry involves polar reactions with nucleophiles and bases. Alkyl halides do one of two things when they react with a nucleophile/base, such as hydroxide ion: either they undergo *substitution* of the X group by the nucleophile, or they undergo *elimination* of HX to yield an alkene.

Substitution

$$\text{H–C–C–Br} + OH^- \longrightarrow \text{H–C–C–OH} + Br^-$$

Elimination

$$\text{H–C–C–Br} + OH^- \longrightarrow \text{C=C} + H_2O + Br^-$$

Why This Chapter? Nucleophilic substitution and base-induced elimination are two of the most widely occurring and versatile reaction types in organic chemistry, both in the laboratory and in biological pathways. We'll look at them closely in this chapter to see how they occur, what their characteristics are, and how they can be used. We'll begin with substitution reactions.

11.1 The Discovery of Nucleophilic Substitution Reactions

The discovery of the nucleophilic substitution reaction of alkyl halides dates back to work carried out in 1896 by the German chemist Paul Walden. Walden found that the pure enantiomeric (+)- and (−)-malic acids could be interconverted through a series of simple substitution reactions. When Walden treated (−)-malic acid with PCl_5, he isolated (+)-chlorosuccinic acid. This, on treatment with wet Ag_2O, gave (+)-malic acid. Similarly, reaction of (+)-malic acid with PCl_5 gave (−)-chlorosuccinic acid, which was converted into (−)-malic acid when treated with wet Ag_2O. The full cycle of reactions is shown in **Figure 11.1**.

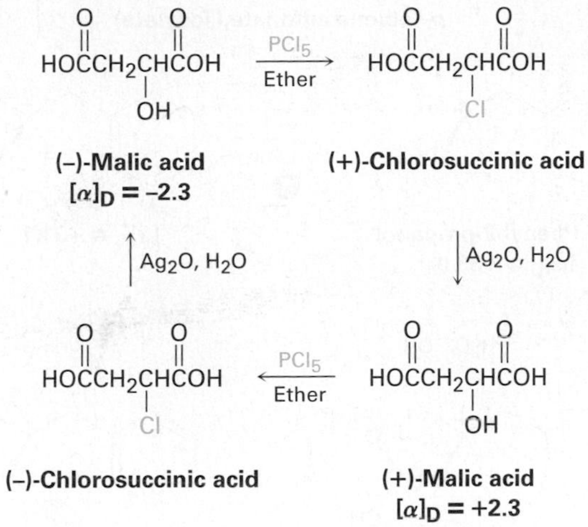

Figure 11.1 Walden's cycle of reactions interconverting (+)- and (−)-malic acids.

At the time, the results were astonishing. The eminent chemist Emil Fischer called Walden's discovery "the most remarkable observation made in the field of optical activity since the fundamental observations of Pasteur." Because (−)-malic acid was converted into (+)-malic acid, *some reactions in the cycle must have occurred with a change, or inversion, in configuration at the chirality center*. But which ones, and how? (Remember from **Section 5.5** that the direction of light rotation and the configuration of a chirality center aren't directly related. You can't tell by looking at the sign of rotation whether a change in configuration has occurred during a reaction.)

Today, we refer to the transformations taking place in Walden's cycle as **nucleophilic substitution reactions** because each step involves the substitution of one nucleophile (chloride ion, Cl^-, or hydroxide ion, HO^-) by another. Nucleophilic substitution reactions are one of the most common and versatile reaction types in organic chemistry.

$$\text{R—X} + \text{Nu:}^- \longrightarrow \text{R—Nu} + \text{X:}^-$$

Following the work of Walden, further investigations were undertaken during the 1920s and 1930s to clarify the mechanism of nucleophilic substitution reactions and to find out how inversions of configuration occur. Among the first series studied was one that interconverted the two enantiomers of 1-phenyl-2-propanol **(Figure 11.2)**. Although this particular series of reactions involves nucleophilic substitution of an alkyl *p*-toluenesulfonate (called a *tosylate*) rather than an alkyl halide, exactly the same type of reaction is involved as that studied by Walden. For all practical purposes, the entire tosylate group acts as if it were simply a halogen substituent. (In fact, when you see a tosylate substituent in a molecule, do a mental substitution and tell yourself that you're dealing with an alkyl halide.)

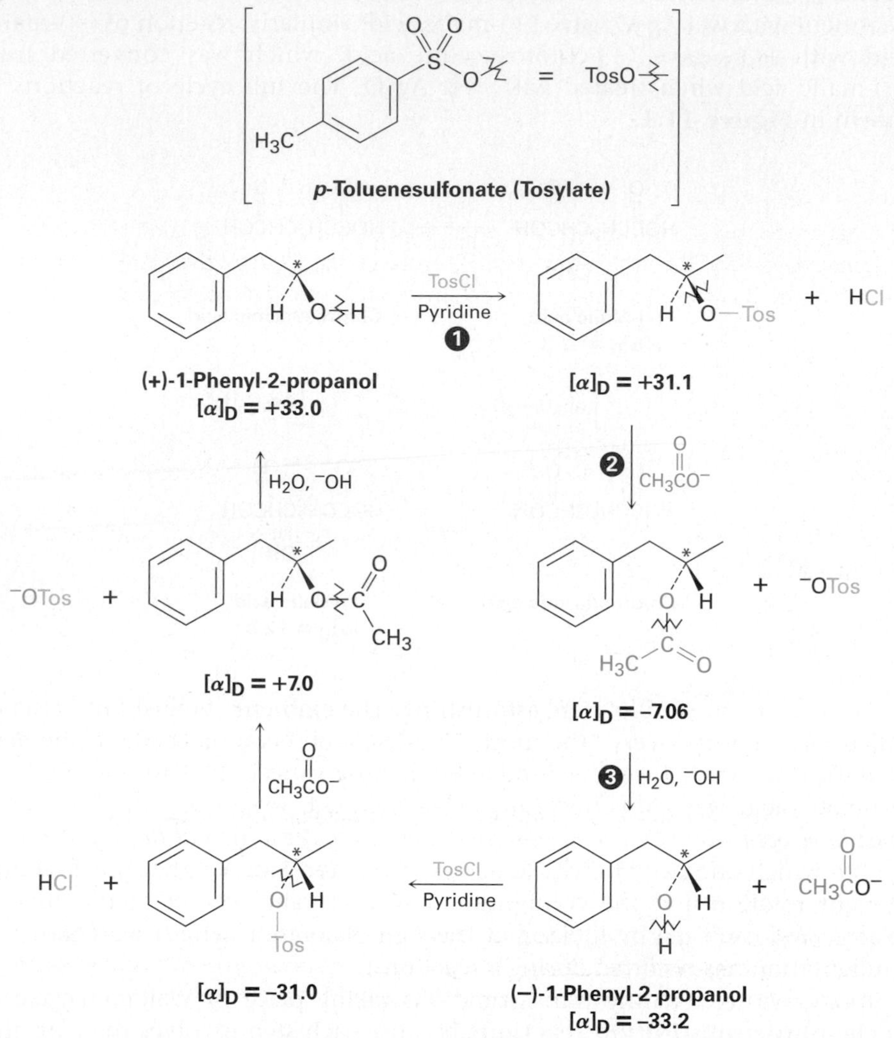

Figure 11.2 A Walden cycle interconverting (+) and (−) enantiomers of 1-phenyl-2-propanol. Chirality centers are marked by asterisks, and the bonds broken in each reaction are indicated by red wavy lines. The inversion of chirality occurs in step 2, where acetate ion substitutes for tosylate ion.

H_3C TosO

p-Toluenesulfonate (Tosylate)

TosCl, Pyridine (1)

(+)-1-Phenyl-2-propanol $[\alpha]_D = +33.0$

$[\alpha]_D = +31.1$ + HCl

(2) CH_3CO^-

H_2O, ^-OH

^-OTos + ... $[\alpha]_D = +7.0$

$[\alpha]_D = -7.06$ + ^-OTos

CH_3CO^-

(3) H_2O, ^-OH

HCl + ... $[\alpha]_D = -31.0$

TosCl, Pyridine

(−)-1-Phenyl-2-propanol $[\alpha]_D = -33.2$ + CH_3CO^-

In the three-step reaction sequence shown in Figure 11.2, (+)-1-phenyl-2-propanol is interconverted with its (−) enantiomer, so at least one of the three steps must involve an inversion of configuration at the chirality center. Step 1, formation of a tosylate, occurs by breaking the O–H bond of the alcohol rather than the C–O bond to the chiral carbon, so the configuration around carbon is unchanged. Similarly, step 3, hydroxide-ion cleavage of the acetate, takes place without breaking the C–O bond at the chirality center. *The inversion of stereochemical configuration must therefore take place in step 2, the nucleophilic substitution of tosylate ion by acetate ion.*

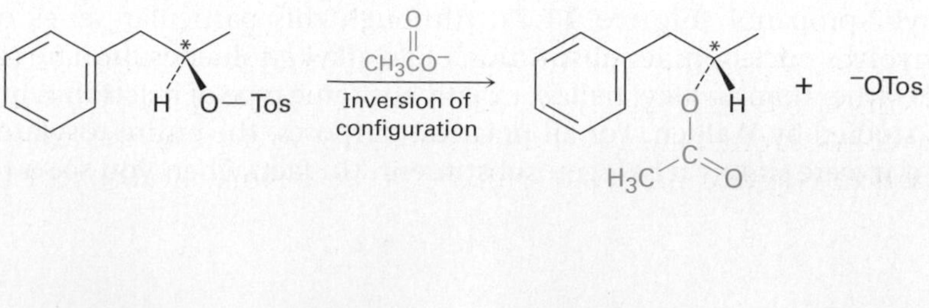

CH_3CO^-, Inversion of configuration → + ^-OTos

From this and nearly a dozen other series of similar reactions, workers concluded that the nucleophilic substitution reaction of a primary or secondary alkyl halide or tosylate always proceeds with inversion of configuration. (Tertiary alkyl halides and tosylates, as we'll see shortly, give different stereochemical results and react by a different mechanism.)

Worked Example 11.1

Predicting the Stereochemistry of a Nucleophilic Substitution Reaction

What product would you expect from a nucleophilic substitution reaction of (*R*)-1-bromo-1-phenylethane with cyanide ion, $^-C{\equiv}N$, as nucleophile? Show the stereochemistry of both reactant and product, assuming that inversion of configuration occurs.

Br

Na^+ $^-C{\equiv}N$ → ?

Strategy

Draw the *R* enantiomer of the reactant, and then change the configuration of the chirality center while replacing the ^-Br with a ^-CN.

Solution

H Br → $^-C{\equiv}N$ → N≡C H

(*R*)-1-Bromo-1-phenylethane **(*S*)-2-Phenylpropanenitrile**

Problem 11.1

What product would you expect to obtain from a nucleophilic substitution reaction of (*S*)-2-bromohexane with acetate ion, $CH_3CO_2^-$? Assume that inversion of configuration occurs, and show the stereochemistry of both reactant and product.

11.2 The S_N2 Reaction

In every chemical reaction, there is a direct relationship between the rate at which the reaction occurs and the concentrations of the reactants. When we measure this relationship, we measure the **kinetics** of the reaction. For example, let's look at the kinetics of a simple nucleophilic substitution—the reaction of CH_3Br with OH^- to yield CH_3OH plus Br^-.

$$HO:^- + CH_3{-}Br: \longrightarrow HO{-}CH_3 + :Br:^-$$

At a given temperature, solvent, and concentration of reactants, the substitution occurs at a certain rate. If we double the concentration of OH^-, the frequency of encounter between the reaction partners doubles and we find that the reaction rate also doubles. Similarly, if we double the concentration of

CH_3Br, the reaction rate again doubles. We call such a reaction, in which the rate is linearly dependent on the concentrations of two species, a **second-order reaction.** Mathematically, we can express this second-order dependence of the nucleophilic substitution reaction by setting up a *rate equation.* As either [RX] or [$^-$OH] changes, the rate of the reaction changes proportionately.

$$\text{Reaction rate} = \text{Rate of disappearance of reactant}$$

$$= k \times [\text{RX}] \times [^-\text{OH}]$$

where [RX] = CH_3Br concentration in molarity

[$^-$OH] = $^-$OH concentration in molarity

k = A constant value (the rate constant)

A mechanism that accounts for both the inversion of configuration and the second-order kinetics that are observed with nucleophilic substitution reactions was suggested in 1937 by the British chemists E. D. Hughes and Christopher Ingold, who formulated what they called the **S_N2 reaction**—short for *substitution, nucleophilic, bimolecular.* (*Bimolecular* means that two molecules, nucleophile and alkyl halide, take part in the step whose kinetics are measured.)

The essential feature of the S_N2 mechanism is that it takes place in a single step without intermediates when the incoming nucleophile reacts with the alkyl halide or tosylate (the *substrate*) from a direction opposite the group that is displaced (the *leaving group*). As the nucleophile comes in on one side of the substrate and bonds to the carbon, the halide or tosylate departs from the other side, thereby inverting the stereochemical configuration. The process is shown in **Figure 11.3** for the reaction of (*S*)-2-bromobutane with HO^- to give (*R*)-2-butanol.

Figure 11.3 | MECHANISM

The mechanism of the S_N2 reaction. The reaction takes place in a single step when the incoming nucleophile approaches from a direction 180° away from the leaving halide ion, thereby inverting the stereochemistry at carbon.

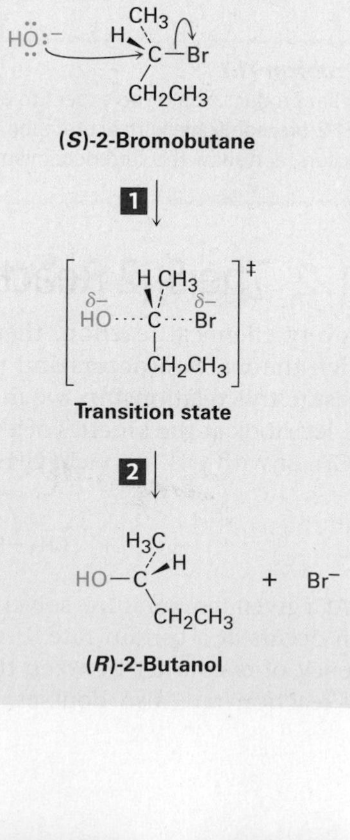

As shown in Figure 11.3, the S_N2 reaction occurs when an electron pair on the nucleophile Nu:$^-$ forces out the group X:$^-$, which takes with it the electron pair from the former C–X bond. This occurs through a transition state in which the new Nu–C bond is partially forming at the same time that the old C–X bond is partially breaking and in which the negative charge is shared by both the incoming nucleophile and the outgoing halide ion. The transition state for this inversion has the remaining three bonds to carbon in a planar arrangement **(Figure 11.4)**.

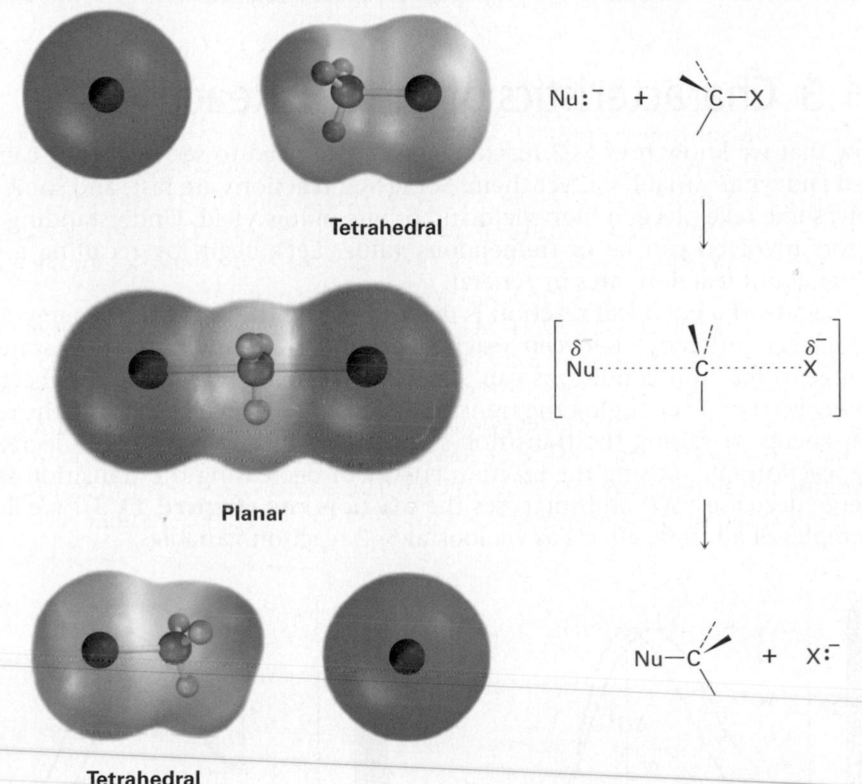

Figure 11.4 The transition state of an S_N2 reaction has a planar arrangement of the carbon atom and the remaining three groups. Electrostatic potential maps show that **negative charge** is delocalized in the transition state.

The mechanism proposed by Hughes and Ingold is fully consistent with experimental results, explaining both stereochemical and kinetic data. Thus, the requirement for backside approach of the entering nucleophile from a direction 180° away from the departing X group causes the stereochemistry of the substrate to invert, much like an umbrella turning inside out in the wind. The Hughes–Ingold mechanism also explains why second-order kinetics are found: the S_N2 reaction occurs in a single step that involves both alkyl halide and nucleophile. Two molecules are involved in the step whose rate is measured.

Problem 11.2

What product would you expect to obtain from S_N2 reaction of OH^- with (*R*)-2-bromobutane? Show the stereochemistry of both reactant and product.

Problem 11.3

Assign configuration to the following substance, and draw the structure of the product that would result on nucleophilic substitution reaction with HS^- (reddish brown = Br):

11.3 Characteristics of the S_N2 Reaction

***Key* IDEAS**

Test your knowledge of Key Ideas by answering end-of-chapter exercises marked with ▲.

Now that we know how S_N2 reactions occur, we need to see how they can be used and what variables affect them. Some S_N2 reactions are fast, and some are slow; some take place in high yield and others in low yield. Understanding the factors involved can be of tremendous value. Let's begin by recalling a few things about reaction rates in general.

The rate of a chemical reaction is determined by the activation energy $\Delta G^\ddagger$, the energy difference between reactant ground state and transition state. A change in reaction conditions can affect $\Delta G^\ddagger$ either by changing the reactant energy level or by changing the transition-state energy level. Lowering the reactant energy or raising the transition-state energy increases $\Delta G^\ddagger$ and decreases the reaction rate; raising the reactant energy or decreasing the transition-state energy decreases $\Delta G^\ddagger$ and increases the reaction rate **(Figure 11.5)**. We'll see examples of all these effects as we look at S_N2 reaction variables.

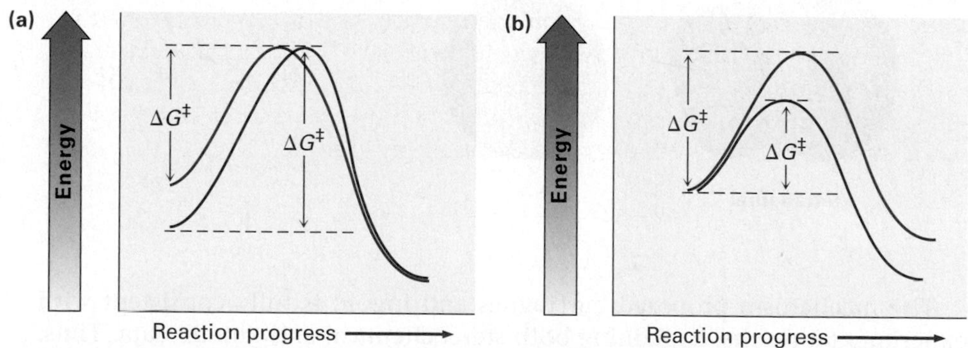

Figure 11.5 The effects of changes in reactant and transition-state energy levels on reaction rate. **(a)** A higher reactant energy level (red curve) corresponds to a faster reaction (smaller $\Delta G^\ddagger$). **(b)** A higher transition-state energy level (red curve) corresponds to a slower reaction (larger $\Delta G^\ddagger$).

The Substrate: Steric Effects in the S_N2 Reaction

The first S_N2 reaction variable to look at is the structure of the substrate. Because the S_N2 transition state involves partial bond formation between the incoming nucleophile and the alkyl halide carbon atom, it seems reasonable that a hindered, bulky substrate should prevent easy approach of the nucleophile, making bond formation difficult. In other words, the transition state for reaction of a sterically hindered substrate, whose carbon atom is "shielded" from approach of the incoming nucleophile, is higher in energy and forms more slowly than the corresponding transition state for a less hindered substrate **(Figure 11.6)**.

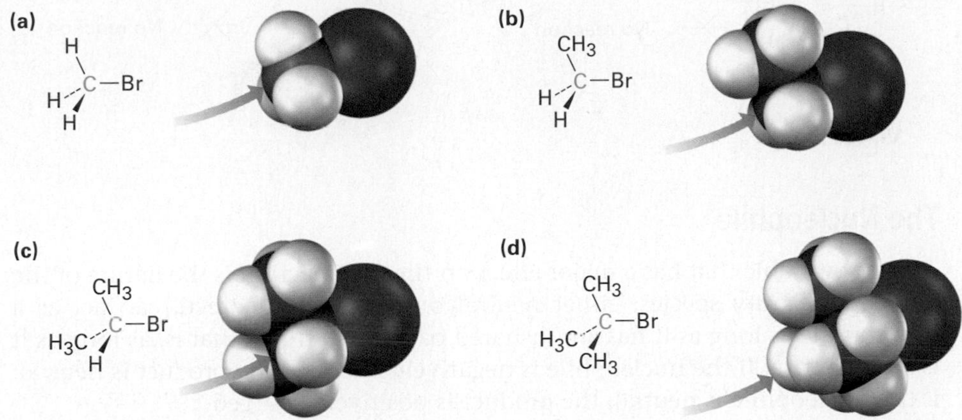

Figure 11.6 Steric hindrance to the S_N2 reaction. As the models indicate, the carbon atom in **(a)** bromomethane is readily accessible, resulting in a fast S_N2 reaction. The carbon atoms in **(b)** bromoethane (primary), **(c)** 2-bromopropane (secondary), and **(d)** 2-bromo-2-methylpropane (tertiary) are successively more hindered, resulting in successively slower S_N2 reactions.

As Figure 11.6 shows, the difficulty of nucleophile approach increases as the three substituents bonded to the halo-substituted carbon atom increase in size. Methyl halides are by far the most reactive substrates in S_N2 reactions, followed by primary alkyl halides such as ethyl and propyl. Alkyl branching at the reacting center, as in isopropyl halides (2°), slows the reaction greatly, and further branching, as in *tert*-butyl halides (3°), effectively halts the reaction. Even branching one carbon removed from the reacting center, as in 2,2-dimethylpropyl *(neopentyl)* halides, greatly slows nucleophilic displacement. As a result, S_N2 reactions occur only at relatively unhindered sites and are normally useful only with methyl halides, primary halides, and a few simple secondary halides. Relative reactivities for some different substrates are as follows:

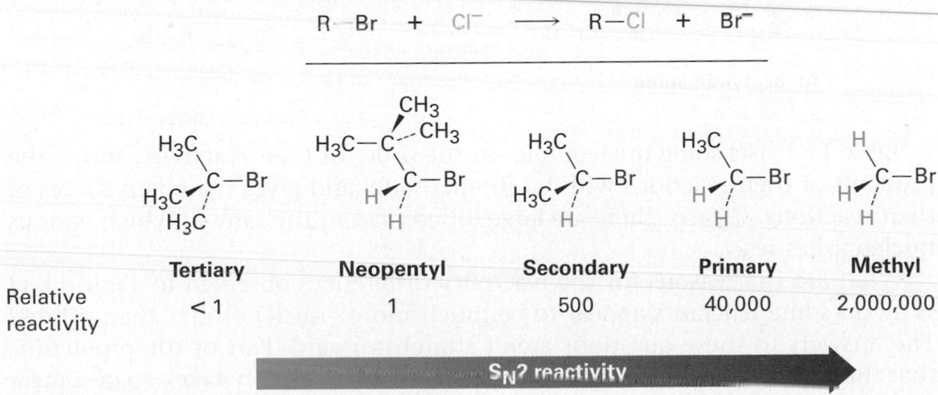

Vinylic halides ($R_2C{=}CRX$) and aryl halides are not shown on this reactivity list because they are unreactive toward S_N2 displacement. This lack of reactivity is due to steric factors: the incoming nucleophile would have to approach in the plane of the carbon–carbon double bond and burrow through part of the molecule to carry out a backside displacement.

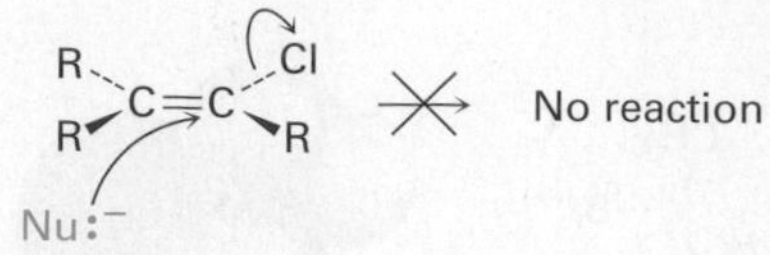

R₂C=C(Cl)R + Nu:⁻ ⟶ No reaction

Vinylic halide

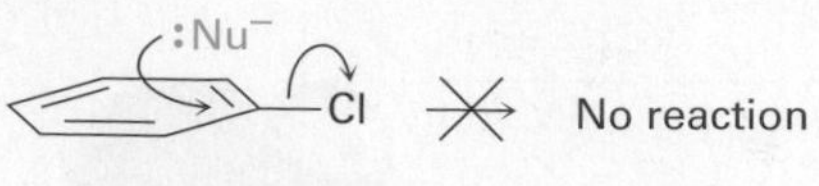

Aryl–Cl + :Nu⁻ ⟶ No reaction

Aryl halide

The Nucleophile

Another variable that has a major effect on the S_N2 reaction is the nature of the nucleophile. Any species, either neutral or negatively charged, can act as a nucleophile as long as it has an unshared pair of electrons; that is, as long as it is a Lewis base. If the nucleophile is negatively charged, the product is neutral; if the nucleophile is neutral, the product is positively charged.

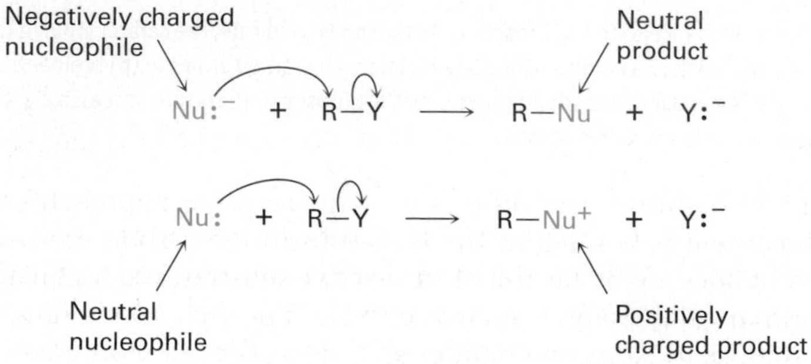

Negatively charged nucleophile → Neutral product

$$Nu{:}^- + R{-}Y \longrightarrow R{-}Nu + Y{:}^-$$

Neutral nucleophile → Positively charged product

$$Nu{:} + R{-}Y \longrightarrow R{-}Nu^+ + Y{:}^-$$

A wide array of substances can be prepared using nucleophilic substitution reactions. In fact, we've already seen examples in previous chapters. The reaction of an acetylide anion with an alkyl halide discussed in **Section 9.8**, for instance, is an S_N2 reaction in which the acetylide nucleophile displaces a halide leaving group.

$$R{-}C{\equiv}C{:}^- + CH_3Br \xrightarrow[\text{reaction}]{S_N2} R{-}C{\equiv}C{-}CH_3 + Br^-$$

An acetylide anion

Table 11.1 lists some nucleophiles in the order of their reactivity, shows the products of their reactions with bromomethane, and gives the relative rates of their reactions. Clearly, there are large differences in the rates at which various nucleophiles react.

What are the reasons for the reactivity differences observed in Table 11.1? Why do some reactants appear to be much more "nucleophilic" than others? The answers to these questions aren't straightforward. Part of the problem is that the term *nucleophilicity* is imprecise. The term is usually taken to be a measure of the affinity of a nucleophile for a carbon atom in the S_N2 reaction, but the reactivity of a given nucleophile can change from one reaction to the next. The exact nucleophilicity of a species in a given reaction depends on the substrate, the solvent, and even the reactant concentrations. Detailed explanations

Table 11.1 Some S_N2 Reactions with Bromomethane

$Nu:^- + CH_3Br \rightarrow CH_3Nu + Br^-$

Nucleophile		Product		Relative rate of reaction
Formula	**Name**	**Formula**	**Name**	
H_2O	Water	$CH_3OH_2^+$	Methylhydronium ion	1
$CH_3CO_2^-$	Acetate	$CH_3CO_2CH_3$	Methyl acetate	500
NH_3	Ammonia	$CH_3NH_3^+$	Methylammonium ion	700
Cl^-	Chloride	CH_3Cl	Chloromethane	1,000
HO^-	Hydroxide	CH_3OH	Methanol	10,000
CH_3O^-	Methoxide	CH_3OCH_3	Dimethyl ether	25,000
I^-	Iodide	CH_3I	Iodomethane	100,000
^-CN	Cyanide	CH_3CN	Acetonitrile	125,000
HS^-	Hydrosulfide	CH_3SH	Methanethiol	125,000

for the observed nucleophilicities aren't always simple, but some trends can be detected in the data of Table 11.1.

- **Nucleophilicity roughly parallels basicity** when comparing nucleophiles that have the same reacting atom. Thus, OH^- is both more basic and more nucleophilic than acetate ion, $CH_3CO_2^-$, which in turn is more basic and more nucleophilic than H_2O. Since "nucleophilicity" is usually taken as the affinity of a Lewis base for a carbon atom in the S_N2 reaction and "basicity" is the affinity of a base for a proton, it's easy to see why there might be a correlation between the two kinds of behavior.
- **Nucleophilicity usually increases going down a column of the periodic table.** Thus, HS^- is more nucleophilic than HO^-, and the halide reactivity order is $I^- > Br^- > Cl^-$. Going down the periodic table, elements have their valence electrons in successively larger shells where they are successively farther from the nucleus, less tightly held, and consequently more reactive. The matter is complex, though, and the nucleophilicity order can change depending on the solvent.
- **Negatively charged nucleophiles are usually more reactive than neutral ones.** As a result, S_N2 reactions are often carried out under basic conditions rather than neutral or acidic conditions.

Problem 11.4

What product would you expect from S_N2 reaction of 1-bromobutane with each of the following?

(a) NaI **(b)** KOH **(c)** $H-C\equiv C-Li$ **(d)** NH_3

Problem 11.5

Which substance in each of the following pairs is more reactive as a nucleophile? Explain.

(a) $(CH_3)_2N^-$ or $(CH_3)_2NH$ **(b)** $(CH_3)_3B$ or $(CH_3)_3N$ **(c)** H_2O or H_2S

The Leaving Group

Still another variable that can affect the S_N2 reaction is the nature of the group displaced by the incoming nucleophile. Because the leaving group is expelled with a negative charge in most S_N2 reactions, the best leaving groups are those that best stabilize the negative charge in the transition state. The greater the extent of charge stabilization by the leaving group, the lower the energy of the transition state and the more rapid the reaction. But as we saw in **Section 2.8**, those groups that best stabilize a negative charge are also the weakest bases. Thus, weak bases such as Cl^-, Br^-, and tosylate ion make good leaving groups, while strong bases such as OH^- and NH_2^- make poor leaving groups.

	OH^-, NH_2^-, OR^-	F^-	Cl^-	Br^-	I^-	$TosO^-$
Relative reactivity	<<1	1	200	10,000	30,000	60,000

Leaving group reactivity →

It's just as important to know which are poor leaving groups as to know which are good, and the preceding data clearly indicate that F^-, HO^-, RO^-, and H_2N^- are not displaced by nucleophiles. In other words, alkyl fluorides, alcohols, ethers, and amines do not typically undergo S_N2 reactions. To carry out an S_N2 reaction with an alcohol, it's necessary to convert the ^-OH into a better leaving group. This, in fact, is just what happens when a primary or secondary alcohol is converted into either an alkyl chloride by reaction with $SOCl_2$ or an alkyl bromide by reaction with PBr_3 **(Section 10.5)**.

A 1° or 2° alcohol: C—OH (with H) → ($SOCl_2$, ether) → [C—O—S(=O)—Cl] A chlorosulfite → (Cl^-, S_N2) → Cl—C (with H) An alkyl chloride

A 1° or 2° alcohol → (PBr_3, ether) → [C—O—PBr_2] A dibromophosphite → (Br^-, S_N2) → Br—C (with H) An alkyl bromide

Alternatively, an alcohol can be made more reactive toward nucleophilic substitution by treating it with *para*-toluenesulfonyl chloride to form a tosylate. As noted previously, tosylates are even more reactive than halides in nucleophilic substitutions. Note that tosylate formation does not change the configuration of the oxygen-bearing carbon because the C–O bond is not broken.

A 1° or 2° alcohol: C—OH (with H) → (p-$CH_3C_6H_4SO_2Cl$; Ether, pyridine) → C—O—SO_2—C_6H_4—CH_3 A tosylate

The one general exception to the rule that ethers don't typically undergo S_N2 reactions occurs with epoxides, the three-membered cyclic ethers that we saw in **Section 8.7**. Epoxides, because of the angle strain in the three-membered ring, are much more reactive than other ethers. They react with aqueous acid to give 1,2-diols, as we saw in **Section 8.7**, and they react readily with many other nucleophiles as well. Propene oxide, for instance, reacts with HCl to give 1-chloro-2-propanol by S_N2 backside attack on the less hindered primary carbon atom. We'll look at the process in more detail in **Section 18.6**.

Propene oxide → (H—Cl) → [protonated epoxide + Cl^-] → CH_3CHCH_2Cl (OH on C2) **1-Chloro-2-propanol**

Problem 11.6
Rank the following compounds in order of their expected reactivity toward S_N2 reaction:

CH_3Br, CH_3OTos, $(CH_3)_3CCl$, $(CH_3)_2CHCl$

The Solvent

The rates of S_N2 reactions are strongly affected by the solvent. Protic solvents—those that contain an –OH or –NH group—are generally the worst for S_N2 reactions, while polar aprotic solvents, which are polar but don't have an –OH or –NH group, are the best.

Protic solvents, such as methanol and ethanol, slow down S_N2 reactions by **solvation** of the reactant nucleophile. The solvent molecules hydrogen bond to the nucleophile and form a cage around it, thereby lowering its energy and reactivity.

RO—H···X^-···H—OR (with H—OR above and below)

A solvated anion (reduced nucleophilicity due to enhanced ground-state stability)

In contrast with protic solvents, which decrease the rates of S_N2 reactions by lowering the ground-state energy of the nucleophile, polar aprotic solvents increase the rates of S_N2 reactions by raising the ground-state energy of the nucleophile. Acetonitrile (CH_3CN), dimethylformamide [$(CH_3)_2NCHO$, abbreviated DMF], dimethyl sulfoxide [$(CH_3)_2SO$, abbreviated DMSO], and hexamethylphosphoramide {[$(CH_3)_2N]_3PO$, abbreviated HMPA} are particularly useful. These solvents can dissolve many salts because of their high polarity, but they tend to solvate metal cations rather than nucleophilic anions. As a result, the bare unsolvated anions have a greater nucleophilicity and S_N2 reactions take place at correspondingly faster rates. For instance, a rate increase of 200,000 has been observed on changing from methanol to HMPA for the reaction of azide ion with 1-bromobutane.

$$CH_3CH_2CH_2CH_2{-}Br \;+\; N_3^- \longrightarrow CH_3CH_2CH_2CH_2{-}N_3 \;+\; Br^-$$

Solvent	CH_3OH	H_2O	DMSO	DMF	CH_3CN	HMPA
Relative reactivity	1	7	1300	2800	5000	200,000

Solvent reactivity →

Problem 11.7

Organic solvents like benzene, ether, and chloroform are neither protic nor strongly polar. What effect would you expect these solvents to have on the reactivity of a nucleophile in S_N2 reactions?

A Summary of S_N2 Reaction Characteristics

The effects on S_N2 reactions of the four variables—substrate structure, nucleophile, leaving group, and solvent—are summarized in the following statements and in the energy diagrams of **Figure 11.7**:

Substrate Steric hindrance raises the energy of the S_N2 transition state, increasing $\Delta G^‡$ and decreasing the reaction rate **(Figure 11.7a)**. As a result, S_N2 reactions are best for methyl and primary substrates. Secondary substrates react slowly, and tertiary substrates do not react by an S_N2 mechanism.

Nucleophile Basic, negatively charged nucleophiles are less stable and have a higher ground-state energy than neutral ones, decreasing $\Delta G^‡$ and increasing the S_N2 reaction rate **(Figure 11.7b)**.

Leaving group Good leaving groups (more stable anions) lower the energy of the transition state, decreasing $\Delta G^‡$ and increasing the S_N2 reaction rate **(Figure 11.7c)**.

Solvent Protic solvents solvate the nucleophile, thereby lowering its ground-state energy, increasing $\Delta G^‡$, and decreasing the S_N2 reaction rate. Polar aprotic solvents surround the accompanying cation but not the nucleophilic anion, thereby raising the groundstate energy of the nucleophile, decreasing $\Delta G^‡$, and increasing the reaction rate **(Figure 11.7d)**.

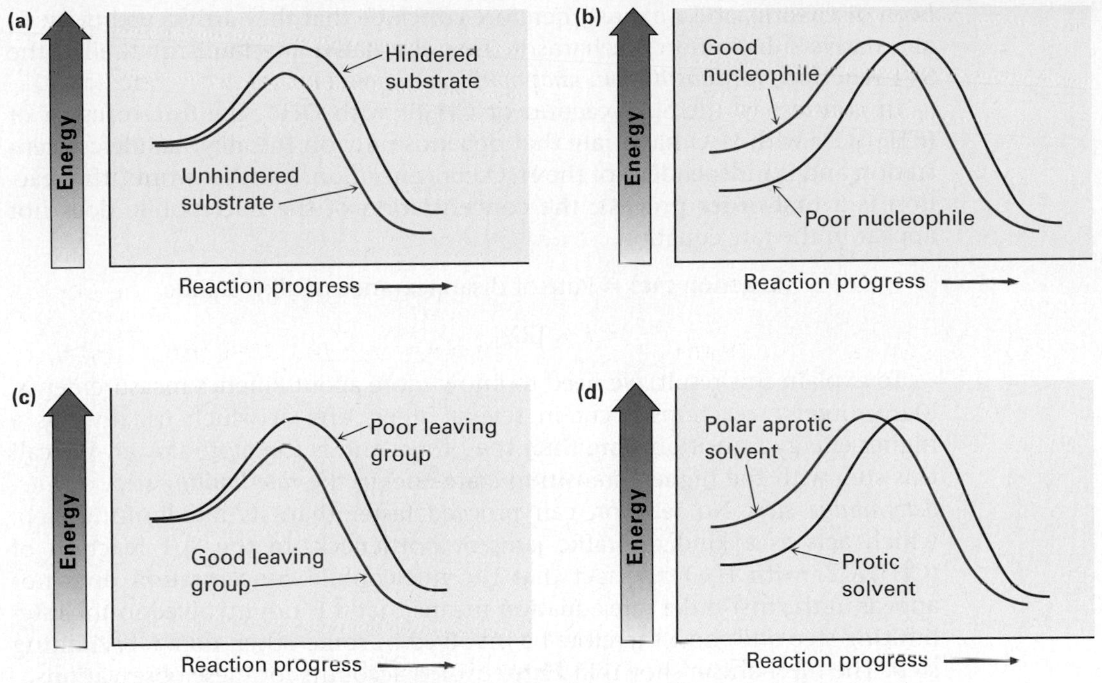

Figure 11.7 Energy diagrams showing the effects of **(a)** substrate, **(b)** nucleophile, **(c)** leaving group, and **(d)** solvent on S_N2 reaction rates. Substrate and leaving group effects are felt primarily in the transition state. Nucleophile and solvent effects are felt primarily in the reactant ground state.

11.4 The S_N1 Reaction

Most nucleophilic substitutions take place by the S_N2 pathway just discussed. The reaction is favored when carried out with an unhindered substrate and a negatively charged nucleophile in a polar aprotic solvent, but is disfavored when carried out with a hindered substrate and a neutral nucleophile in a protic solvent. You might therefore expect the reaction of a tertiary substrate (hindered) with water (neutral, protic) to be among the slowest of substitution reactions. Remarkably, however, the opposite is true. The reaction of the tertiary halide $(CH_3)_3CBr$ with H_2O to give the alcohol 2-methyl-2-propanol is more than *1 million times* as fast as the corresponding reaction of CH_3Br to give methanol.

$$R{-}Br \ + \ H_2O \ \longrightarrow \ R{-}OH \ + \ HBr$$

	CH_3Br	CH_3CH_2Br	$(CH_3)_2CHBr$	$(CH_3)_3CBr$
	Methyl	**Primary**	**Secondary**	**Tertiary**
Relative reactivity	< 1	1	12	1,200,000

Reactivity →

What's going on here? Clearly, a nucleophilic substitution reaction is occurring—a halogen is replacing a hydroxyl group—yet the reactivity order seems backward. These reactions can't be taking place by the S_N2 mechanism we've been discussing, and we must therefore conclude that they are occurring by an alternative substitution mechanism. This alternative mechanism is called the **S_N1 reaction**, for *substitution, nucleophilic, unimolecular.*

In contrast to the S_N2 reaction of CH_3Br with OH^-, the S_N1 reaction of $(CH_3)_3CBr$ with H_2O has a rate that depends only on the alkyl halide concentration and is independent of the H_2O concentration. In other words, the reaction is a **first-order process**; the concentration of the nucleophile does not appear in the rate equation.

$$\text{Reaction rate} = \text{Rate of disappearance of alkyl halide}$$
$$= k \times [\text{RX}]$$

To explain this result, we need to know more about kinetics measurements. Many organic reactions occur in several steps, one of which usually has a higher-energy transition state than the others and is therefore slower. We call this step with the highest transition-state energy the *rate-limiting step,* or *rate-determining step.* No reaction can proceed faster than its rate-limiting step, which acts as a kind of traffic jam, or bottleneck. In the S_N1 reaction of $(CH_3)_3CBr$ with H_2O, the fact that the nucleophile concentration does not appear in the first-order rate equation means that it is not involved in the rate-limiting step and must therefore be involved in some other, non–rate-limiting step. The mechanism shown in **Figure 11.8** accounts for these observations.

Figure 11.8 | MECHANISM

The mechanism of the S_N1 reaction of 2-bromo-2-methylpropane with H_2O involves three steps. Step **1**—the spontaneous, unimolecular dissociation of the alkyl bromide to yield a carbocation—is rate-limiting.

1 Spontaneous dissociation of the alkyl bromide occurs in a slow, rate-limiting step to generate a carbocation intermediate plus bromide ion.

2 The carbocation intermediate reacts with water as nucleophile in a fast step to yield protonated alcohol as product.

3 Loss of a proton from the protonated alcohol intermediate then gives the neutral alcohol product.

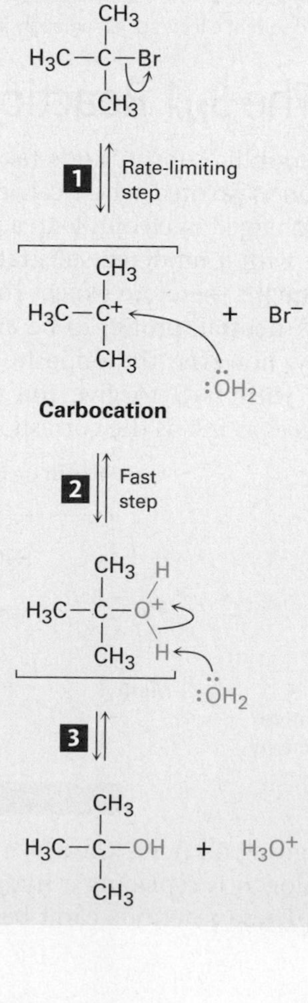

$(CH_3)_3C{-}Br$

1 Rate-limiting step

$H_3C{-}C^+(CH_3)_2$ + Br^- $:\ddot{O}H_2$

Carbocation

2 Fast step

$(CH_3)_3C{-}O^+H_2$ $:\ddot{O}H_2$

3

$(CH_3)_3C{-}OH$ + H_3O^+

Unlike what happens in an S_N2 reaction, where the leaving group is displaced at the same time the incoming nucleophile approaches, an S_N1 reaction takes place by loss of the leaving group *before* the nucleophile approaches. 2-Bromo-2-methylpropane spontaneously dissociates to the *tert*-butyl carbocation plus Br^- in a slow rate-limiting step, and the intermediate carbocation is then immediately trapped by the nucleophile water in a faster second step. Water is not a reactant in the step whose rate is measured. The energy diagram is shown in **Figure 11.9**.

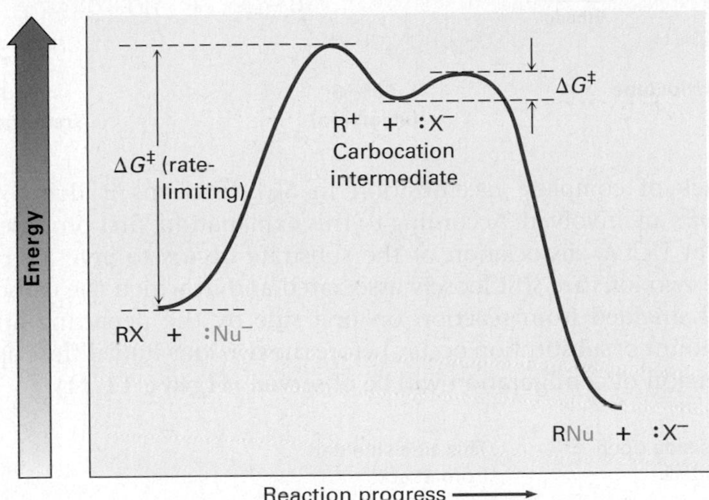

Figure 11.9 An energy diagram for an S_N1 reaction. The rate-limiting step is the spontaneous dissociation of the alkyl halide to give a carbocation intermediate. Reaction of the carbocation with a nucleophile then occurs in a second, faster step.

Because an S_N1 reaction occurs through a carbocation intermediate, its stereochemical outcome is different from that of an S_N2 reaction. Carbocations, as we've seen, are planar, sp^2-hybridized, and achiral. Thus, if we carry out an S_N1 reaction on one enantiomer of a chiral reactant and go through an achiral carbocation intermediate, the product must lose its optically activity **(Section 8.12)**. That is, the symmetrical intermediate carbocation can react with a nucleophile equally well from either side, leading to a racemic, 50:50 mixture of enantiomers **(Figure 11.10)**.

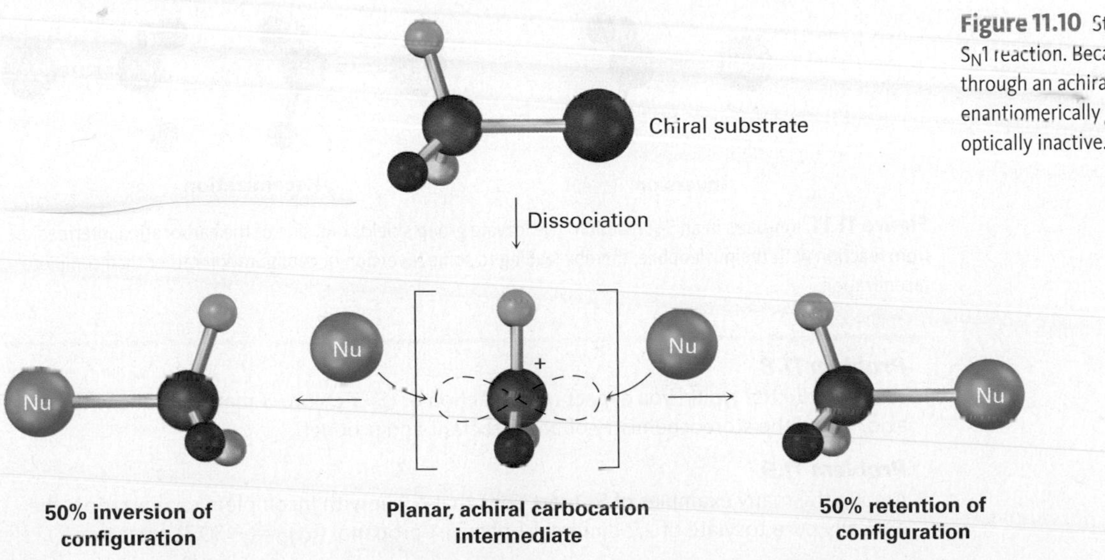

Figure 11.10 Stereochemistry of the S_N1 reaction. Because the reaction goes through an achiral intermediate, an enantiomerically pure reactant gives an optically inactive. racemic product.

The conclusion that S_N1 reactions on enantiomerically pure substrates should give racemic products is nearly, but not exactly, what is found. In fact, few S_N1 displacements occur with complete racemization. Most give a minor (0–20%) excess of inversion. The reaction of (*R*)-6-chloro-2,6-dimethyloctane with H_2O, for example, leads to an alcohol product that is approximately 80% racemized and 20% inverted (80% *R,S* + 20% *S* is equivalent to 40% *R* + 60% S).

H_3C, CH_2CH_3, Cl—C, $CH_2CH_2CH_2CH(CH_3)CH_3$ $\xrightarrow[\text{Ethanol}]{H_2\ddot{O}:}$ CH_3CH_2, CH_3, C—OH, $CH_3CH(CH_3)CH_2CH_2CH_2$ + H_3C, CH_2CH_3, HO—C, $CH_2CH_2CH_2CH(CH_3)CH_3$

(*R*)-6-Chloro-2,6-dimethyloctane | **60% *S* (inversion)** | **40% *R* (retention)**

This lack of complete racemization in S_N1 reactions is due to the fact that *ion pairs* are involved. According to this explanation, first proposed by Saul Winstein at UCLA, dissociation of the substrate occurs to give a structure in which the two ions are still loosely associated and in which the carbocation is effectively shielded from reaction on one side by the departing anion. If a certain amount of substitution occurs before the two ions fully diffuse apart, then a net inversion of configuration will be observed **(Figure 11.11)**.

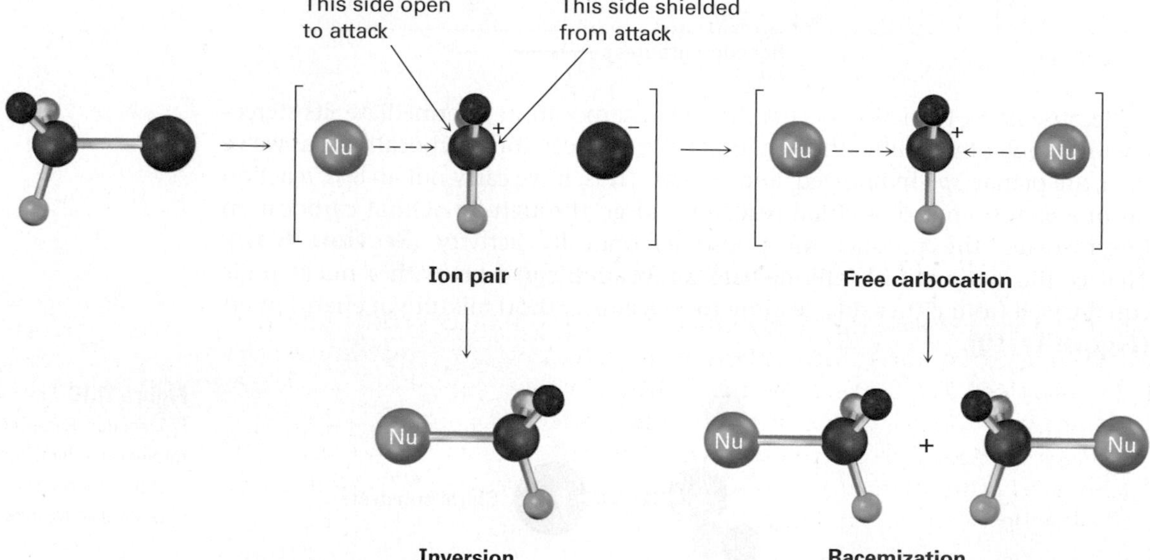

Figure 11.11 Ion-pairs in an S_N1 reaction. The leaving group shields one side of the carbocation intermediate from reaction with the nucleophile, thereby leading to some inversion of configuration rather than complete racemization.

Problem 11.8

What product(s) would you expect from reaction of (*S*)-3-chloro-3-methyloctane with acetic acid? Show the stereochemistry of both reactant and product.

Problem 11.9

Among the many examples of S_N1 reactions that occur with incomplete racemization, the optically pure tosylate of 2,2-dimethyl-1-phenyl-1-propanol ($[\alpha]_D = -30.3$) gives the corresponding acetate ($[\alpha]_D = +5.3$) when heated in acetic acid. If complete inversion had occurred, the optically pure acetate would have had $[\alpha]_D = +53.6$. What percentage racemization and what percentage inversion occurred in this reaction?

H C(CH_3)$_3$ — C — OTos $\xrightarrow{CH_3COOH}$ H C(CH_3)$_3$ — C — O — C(=O) — CH_3 + TosOH

$[\alpha]_D = -30.3$

Observed $[\alpha]_D = +5.3$
(optically pure $[\alpha]_D = +53.6$)

Problem 11.10
Assign configuration to the following substrate, and show the stereochemistry and identity of the product you would obtain by S_N1 reaction with water (reddish brown = Br):

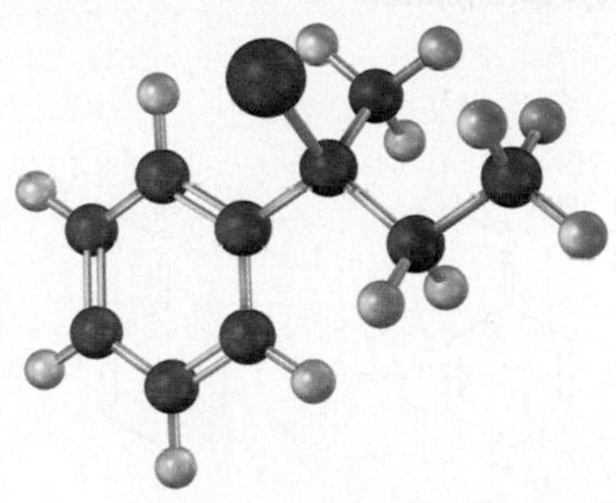

11.5 Characteristics of the S_N1 Reaction

Just as the S_N2 reaction is strongly influenced by the structure of the substrate, the leaving group, the nucleophile, and the solvent, the S_N1 reaction is similarly influenced. Factors that lower $\Delta G^\ddagger$, either by lowering the energy level of the transition state or by raising the energy level of the ground state, favor faster S_N1 reactions. Conversely, factors that raise $\Delta G^\ddagger$, either by raising the energy level of the transition state or by lowering the energy level of the reactant, slow down the S_N1 reaction.

Key **IDEAS**

Test your knowledge of Key Ideas by answering end-of-chapter exercises marked with ▲.

The Substrate

According to the Hammond postulate **(Section 7.10)**, any factor that stabilizes a high-energy intermediate also stabilizes the transition state leading to that intermediate. Since the rate-limiting step in an S_N1 reaction is the spontaneous, unimolecular dissociation of the substrate to yield a carbocation, the reaction is favored whenever a stabilized carbocation intermediate is formed. The more stable the carbocation intermediate, the faster the S_N1 reaction.

We saw in **Section 7.9** that the stability order of alkyl carbocations is $3° > 2° > 1° > CH_3$. To this list we must also add the resonance-stabilized allyl and benzyl cations. Just as allylic radicals are unusually stable because the unpaired electron can be delocalized over an extended π orbital system **(Section 10.4)**, so allylic and benzylic carbocations are unusually stable. (The word **benzylic** means "next to an aromatic ring.") As **Figure 11.12** indicates, an allylic cation has two resonance forms. In one form the double bond is on the "left"; in the other form it's on the "right." A benzylic cation has five resonance forms, all of which contribute to the overall resonance hybrid.

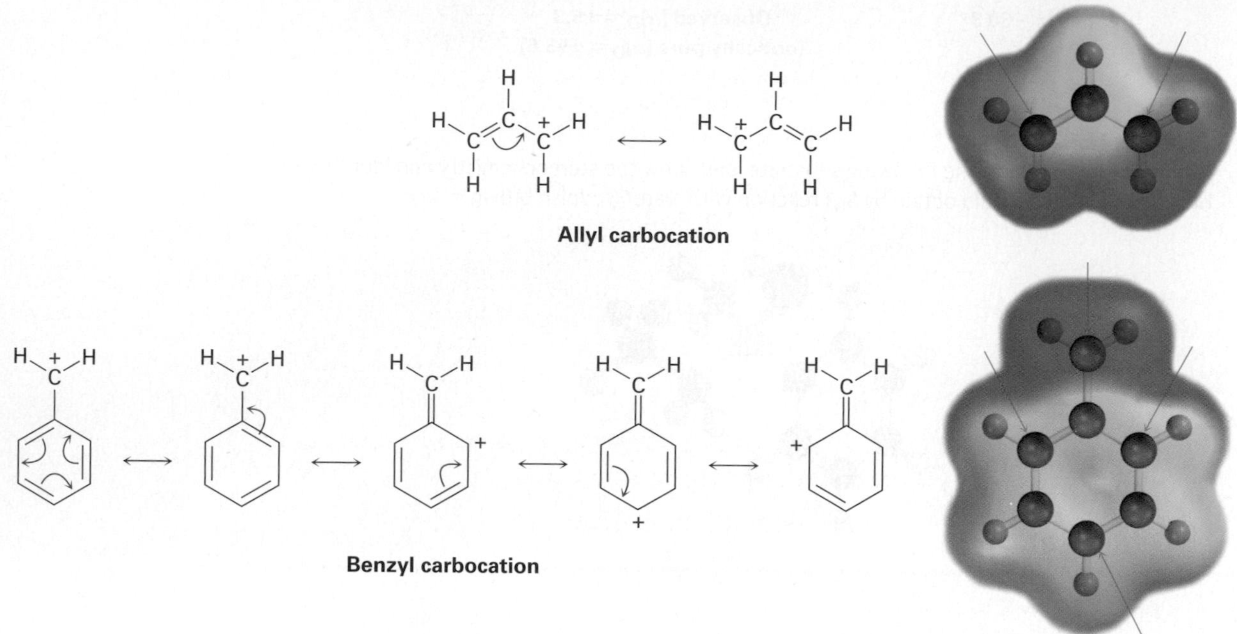

Figure 11.12 Resonance forms of allylic and benzylic carbocations. The positive charge is delocalized over the π system in both. **Electron-poor atoms** are indicated by blue arrows.

Because of resonance stabilization, a primary allylic or benzylic carbocation is about as stable as a secondary alkyl carbocation and a secondary allylic or benzylic carbocation is about as stable as a tertiary alkyl carbocation. This stability order of carbocations is the same as the order of S_N1 reactivity for alkyl halides and tosylates.

Methyl < **Primary** < **Allylic** ≈ **Benzylic** ≈ **Secondary** < **Tertiary**

Carbocation stability

We should also note parenthetically that primary allylic and benzylic substrates are particularly reactive in S_N2 reactions as well as in S_N1 reactions. Allylic and benzylic C–X bonds are about 50 kJ/mol (12 kcal/mol) weaker than the corresponding saturated bonds and are therefore more easily broken.

CH_3CH_2-Cl	$H_2C{=}CHCH_2-Cl$	$C_6H_5CH_2-Cl$
338 kJ/mol (81 kcal/mol)	289 kJ/mol (69 kcal/mol)	293 kJ/mol (70 kcal/mol)

Problem 11.11

Rank the following substances in order of their expected S_N1 reactivity:

CH_3CH_2Br $\qquad H_2C{=}CHCH(Br)CH_3 \qquad H_2C{=}CHBr \qquad CH_3CH(Br)CH_3$

Problem 11.12

3-Bromo-1-butene and 1-bromo-2-butene undergo S_N1 reaction at nearly the same rate even though one is a secondary halide and the other is primary. Explain.

The Leaving Group

We said during the discussion of S_N2 reactivity that the best leaving groups are those that are most stable; that is, those that are the conjugate bases of strong acids. An identical reactivity order is found for the S_N1 reaction because the leaving group is directly involved in the rate-limiting step. Thus, the S_N1 reactivity order is

$$HO^- < Cl^- < Br^- < I^- \approx TosO^- \quad H_2O$$

Leaving group reactivity →

Note that in the S_N1 reaction, which is often carried out under acidic conditions, neutral water is sometimes the leaving group. This occurs, for example, when an alkyl halide is prepared from a tertiary alcohol by reaction with HBr or HCl **(Section 10.5)**. As shown in **Figure 11.13**, the alcohol is first protonated and then spontaneously loses H_2O to generate a carbocation, which reacts with halide ion to give the alkyl halide. Knowing that an S_N1 reaction is involved in the conversion of alcohols to alkyl halides explains why the reaction works well only for tertiary alcohols. Tertiary alcohols react fastest because they give the most stable carbocation intermediates.

Figure 11.13 | MECHANISM

The mechanism of the S_N1 reaction of a tertiary alcohol with HBr to yield an alkyl halide. Neutral water is the leaving group (step **2**).

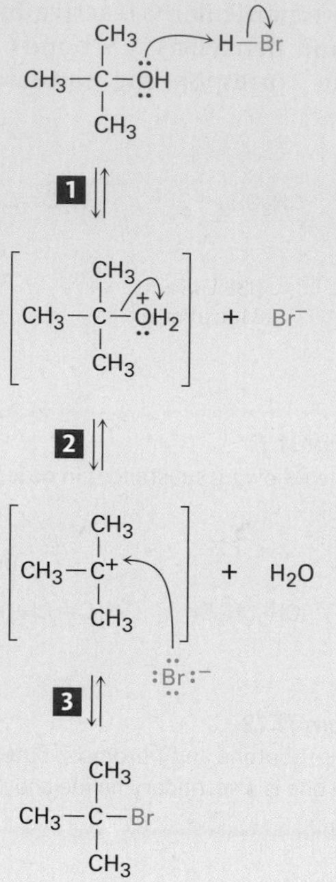

$$(CH_3)_3C\text{–}\ddot{O}H + H\text{–}Br$$

1 The –OH group is first protonated by HBr.

$$\left[(CH_3)_3C\text{–}\overset{+}{O}H_2\right] + Br^-$$

2 Spontaneous dissociation of the protonated alcohol occurs in a slow, rate-limiting step to yield a carbocation intermediate plus water.

Carbocation

$$\left[(CH_3)_3C^+\right] + H_2O$$

3 The carbocation intermediate reacts with bromide ion in a fast step to yield the neutral substitution product.

$$:\ddot{Br}:^- \longrightarrow (CH_3)_3C\text{–}Br$$

The Nucleophile

The nature of the nucleophile plays a major role in the S_N2 reaction but does not affect an S_N1 reaction. Because the S_N1 reaction occurs through a rate-limiting step in which the added nucleophile has no part, the nucleophile can't affect the reaction rate. The reaction of 2-methyl-2-propanol with HX, for instance, occurs at the same rate regardless of whether X is Cl, Br, or I. Furthermore, neutral nucleophiles are just as effective as negatively charged ones, so S_N1 reactions frequently occur under neutral or acidic conditions.

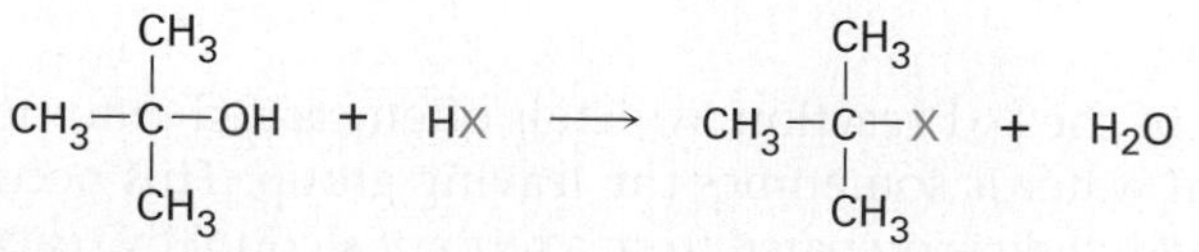

$$(CH_3)_3C\text{–}OH + HX \longrightarrow (CH_3)_3C\text{–}X + H_2O$$

2-Methyl-2-propanol (Same rate for X = Cl, Br, I)

The Solvent

What about the solvent? Do solvents have the same effect in S_N1 reactions that they have in S_N2 reactions? The answer is both yes and no. Yes, solvents have a large effect on S_N1 reactions, but no, the reasons for the effects on S_N1 and S_N2 reactions are not the same. Solvent effects in the S_N2 reaction are due largely to stabilization or destabilization of the nucleophile *reactant,* while solvent effects in the S_N1 reaction are due largely to stabilization or destabilization of the *transition state.*

The Hammond postulate says that any factor stabilizing the intermediate carbocation should increase the rate of an S_N1 reaction. Solvation of the carbocation—the interaction of the ion with solvent molecules—has just such an effect. Solvent molecules orient around the carbocation so that the electron-rich ends of the solvent dipoles face the positive charge **(Figure 11.14)**, thereby lowering the energy of the ion and favoring its formation.

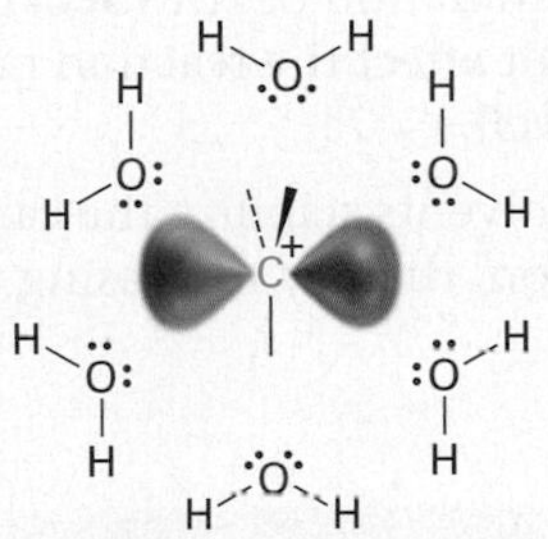

Figure 11.14 Solvation of a carbocation by water. The electron-rich oxygen atoms of solvent molecules orient around the positively charged carbocation and thereby stabilize it.

The properties of a solvent that contribute to its ability to stabilize ions by solvation are related to the solvent's polarity. S_N1 reactions take place much more rapidly in strongly polar solvents, such as water and methanol, than in less polar solvents, such as ether and chloroform. In the reaction of 2-chloro-2-methylpropane, for example, a rate increase of 100,000 is observed on going from ethanol (less polar) to water (more polar). The rate increases on going from a hydrocarbon solvent to water are so large they can't be measured accurately.

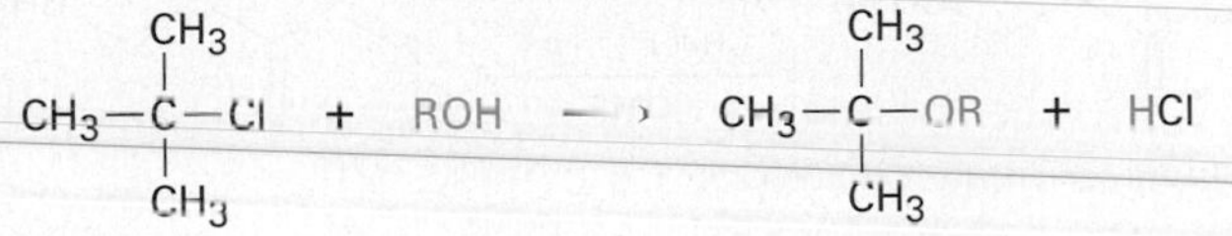

$$(CH_3)_3C{-}Cl + ROH \longrightarrow (CH_3)_3C{-}OR + HCl$$

	Ethanol	40% Water/ 60% Ethanol	80% Water/ 20% Ethanol	Water
Relative reactivity	1	100	14,000	100,000

Solvent reactivity →

It should be emphasized again that both the S_N1 and the S_N2 reaction show solvent effects, but that they do so for different reasons. S_N2 reactions are *disfavored* in protic solvents because the *ground-state energy* of the nucleophile is lowered by solvation. S_N1 reactions are *favored* in protic solvents because the *transition-state energy* leading to carbocation intermediate is lowered by solvation.

A Summary of S_N1 Reaction Characteristics

The effects on S_N1 reactions of the four variables—substrate, leaving group, nucleophile, and solvent—are summarized in the following statements:

Substrate The best substrates yield the most stable carbocations. As a result, S_N1 reactions are best for tertiary, allylic, and benzylic halides.

Leaving group Good leaving groups increase the reaction rate by lowering the energy level of the transition state for carbocation formation.

Nucleophile The nucleophile must be nonbasic to prevent a competitive elimination of HX **(Section 11.7)**, but otherwise does not affect the reaction rate. Neutral nucleophiles work well.

Solvent Polar solvents stabilize the carbocation intermediate by solvation, thereby increasing the reaction rate.

Worked Example 11.2

Predicting the Mechanism of a Nucleophilic Substitution Reaction

Predict whether each of the following substitution reactions is likely to be S_N1 or S_N2:

(a) Cl ... $\xrightarrow[CH_3CO_2H,\ H_2O]{CH_3CO_2^-\ Na^+}$ OAc

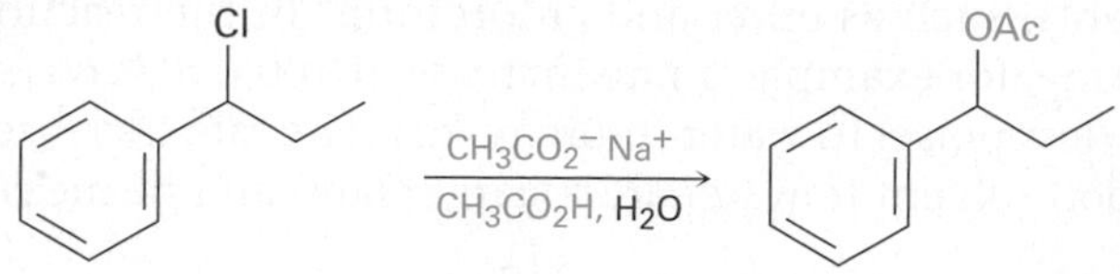

(b) CH_2Br ... $\xrightarrow[DMF]{CH_3CO_2^-\ Na^+}$ CH_2OAc

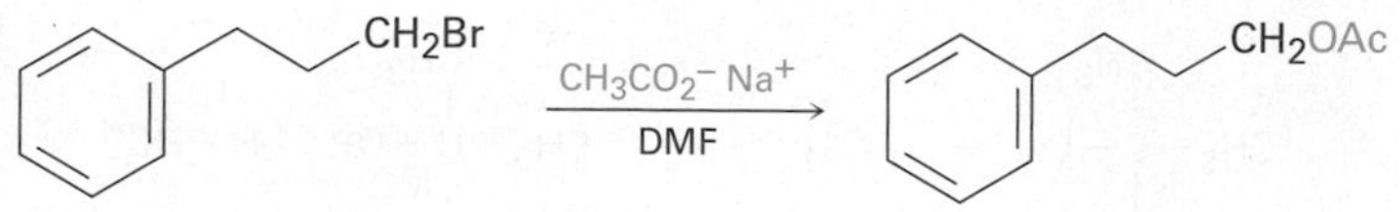

Strategy

Look at the substrate, leaving group, nucleophile, and solvent. Then decide from the summaries at the ends of **Sections 11.3 and 11.5** whether an S_N1 or an S_N2 reaction is favored. S_N1 reactions are favored by tertiary, allylic, or benzylic substrates, by good leaving groups, by nonbasic nucleophiles, and by protic solvents. S_N2 reactions are favored by primary substrates, by good leaving groups, by good nucleophiles, and by polar aprotic solvents.

Solution

(a) This is likely to be an S_N1 reaction because the substrate is secondary and benzylic, the nucleophile is weakly basic, and the solvent is protic.

(b) This is likely to be an S_N2 reaction because the substrate is primary, the nucleophile is a reasonably good one, and the solvent is polar aprotic.

Problem 11.13

Predict whether each of the following substitution reactions is likely to be S_N1 or S_N2:

(a) [cyclohexenyl alcohol] $\xrightarrow[CH_3OH]{HCl}$ [cyclohexenyl chloride]

(b) $H_2C{=}C(CH_3)CH_2Br \xrightarrow[CH_3CN]{Na^+\ ^-SCH_3} H_2C{=}C(CH_3)CH_2SCH_3$

11.6 Biological Substitution Reactions

Both S_N1 and S_N2 reactions are well-known in biological chemistry, particularly in the pathways for biosynthesis of the many thousands of plant-derived substances called *terpenoids,* which we'll discuss in **Section 27.5**. Unlike what typically happens in the laboratory, however, the substrate in a biological substitution reaction is usually an organodiphosphate rather than an alkyl halide. Thus, the leaving group is the diphosphate ion, abbreviated PP_i, rather than a halide ion. In fact, it's useful to think of the diphosphate group as the "biological equivalent" of a halogen. The dissociation of an organodiphosphate in a biological reaction is typically assisted by complexation to a divalent metal cation such as Mg^{2+} to help neutralize charge and make the diphosphate a better leaving group.

$R{-}Cl \xrightarrow{\text{Dissociation}} R^+ + Cl^-$

An alkyl chloride

$R{-}OPOPO^- \ (\cdot Mg^{2+}) \xrightarrow{\text{Dissociation}} R^+ + {}^-OPOPO^- \ (\cdot Mg^{2+}) \ (PP_i)$

An organodiphosphate **Diphosphate ion**

Two S_N1 reactions occur during the biosynthesis of geraniol, a fragrant alcohol found in roses and used in perfumery. Geraniol biosynthesis begins with dissociation of dimethylallyl diphosphate to give an allylic carbocation, which reacts with isopentenyl diphosphate **(Figure 11.15)**. From the viewpoint of isopentenyl diphosphate, the reaction is an electrophilic alkene addition, but from the viewpoint of dimethylallyl diphosphate, the process is an S_N1 reaction in which the carbocation intermediate reacts with a double bond as the nucleophile.

Following this initial S_N1 reaction, loss of the *pro-R* hydrogen gives geranyl diphosphate, itself an allylic diphosphate that dissociates a second time. Reaction of the geranyl carbocation with water in a second S_N1 reaction, followed by loss of a proton, then yields geraniol.

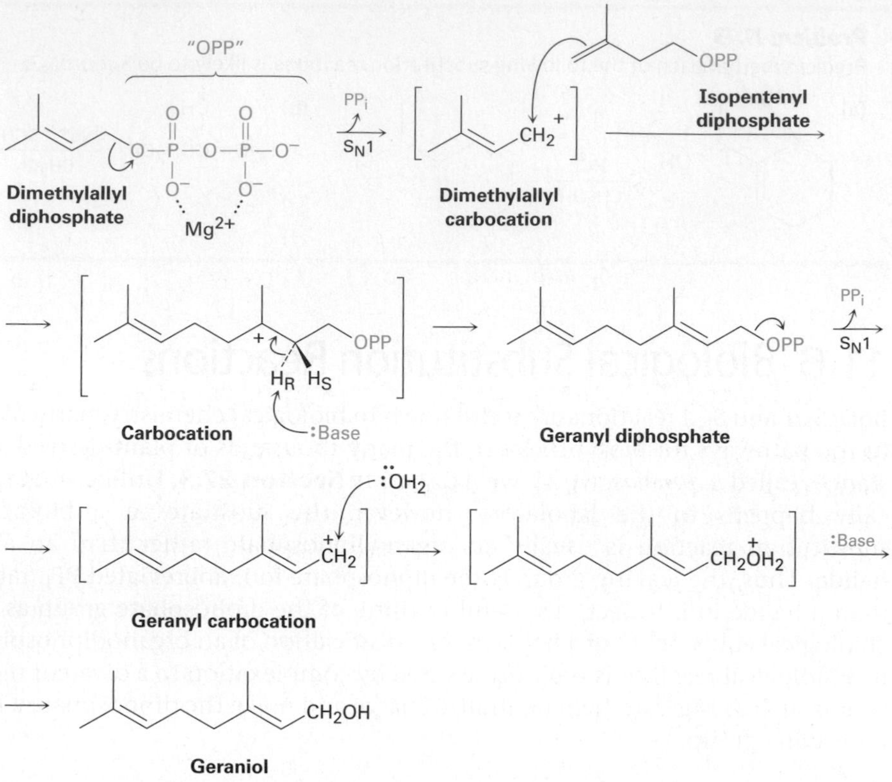

Figure 11.15 Biosynthesis of geraniol from dimethylallyl diphosphate. Two S_N1 reactions occur, both with diphosphate ion as the leaving group.

S_N2 reactions are involved in almost all biological methylations, which transfer a $-CH_3$ group from an electrophilic donor to a nucleophile. The donor is *S*-adenosylmethionine (abbreviated SAM), which contains a positively charged sulfur (a sulfonium ion, **Section 5.12**), and the leaving group is the neutral *S*-adenosylhomocysteine molecule. In the biosynthesis of epinephrine (adrenaline) from norepinephrine, for instance, the nucleophilic nitrogen atom of norepinephrine attacks the electrophilic methyl carbon atom of S-adenosylmethionine in an S_N2 reaction, displacing *S*-adenosylhomocysteine **(Figure 11.16)**. In effect, *S*-adenosylmethionine is simply a biological equivalent of CH_3Cl.

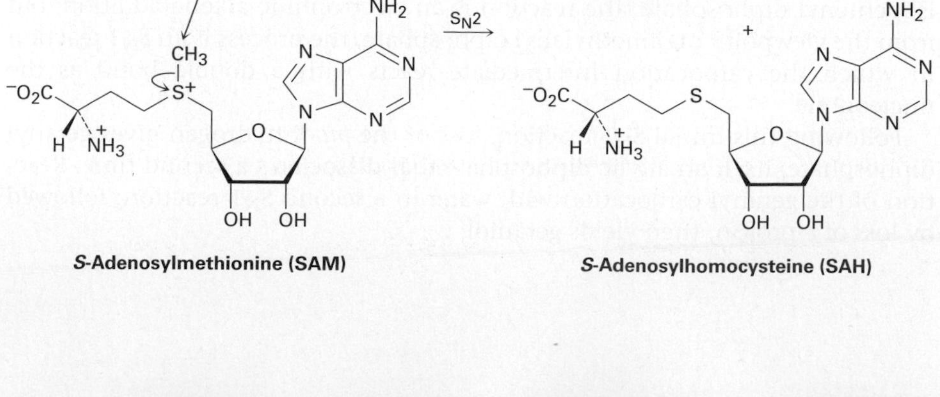

Figure 11.16 The biosynthesis of epinephrine from norepinephrine occurs by an S_N2 reaction with *S*-adenosylmethionine.

Problem 11.14
Review the mechanism of geraniol biosynthesis shown in Figure 11.15, and propose a mechanism for the biosynthesis of limonene from linalyl diphosphate.

OPP ⟹

Linalyl diphosphate **Limonene**

11.7 Elimination Reactions: Zaitsev's Rule

We said at the beginning of this chapter that two kinds of reactions can take place when a nucleophile/Lewis base reacts with an alkyl halide. The nucleophile can either substitute for the halide by reaction at carbon or can cause elimination of HX by reaction at a neighboring hydrogen:

Key **IDEAS**

Test your knowledge of Key Ideas by answering end-of-chapter exercises marked with ▲.

Substitution

H–C–C–Br + OH^- ⟶ H–C–C–OH + Br^-

Elimination

H–C–C–Br + OH^- ⟶ C=C + H_2O + Br^-

Elimination reactions are more complex than substitution reactions for several reasons. One is the problem of regiochemistry. What products result by loss of HX from an unsymmetrical halide? In fact, elimination reactions almost always give mixtures of alkene products, and the best we can usually do is to predict which will be the major product.

According to **Zaitsev's rule**, formulated in 1875 by the Russian chemist Alexander Zaitsev, base-induced elimination reactions generally (although not always) give the more stable alkene product—that is, the alkene with more alkyl substituents on the double-bond carbons. In the following two cases, for example, the more highly substituted alkene product predominates.

ZAITSEV'S RULE
In the elimination of HX from an alkyl halide, the more highly substituted alkene product predominates.

$CH_3CH_2CH(Br)CH_3$ $\xrightarrow[CH_3CH_2OH]{CH_3CH_2O^-\ Na^+}$ $CH_3CH{=}CHCH_3$ + $CH_3CH_2CH{=}CH_2$

2-Bromobutane — **2-Butene (81%)** — **1-Butene (19%)**

$CH_3CH_2C(Br)(CH_3)CH_3$ $\xrightarrow[CH_3CH_2OH]{CH_3CH_2O^-\ Na^+}$ $CH_3CH{=}C(CH_3)CH_3$ + $CH_3CH_2C(CH_3){=}CH_2$

2-Bromo-2-methylbutane — **2-Methyl-2-butene (70%)** — **2-Methyl-1-butene (30%)**

Another factor that complicates a study of elimination reactions is that they can take place by different mechanisms, just as substitutions can. We'll consider three of the most common mechanisms—the E1, E2, and E1cB reactions—which differ in the timing of C–H and C–X bond-breaking.

In the E1 reaction, the C–X bond breaks first to give a carbocation intermediate that undergoes subsequent base abstraction of H^+ to yield the alkene. In the E2 reaction, base-induced C–H bond cleavage is simultaneous with C–X bond cleavage, giving the alkene in a single step. In the E1cB reaction (cB for "conjugate base"), base abstraction of the proton occurs first, giving a carbanion ($R:^-$) intermediate. This anion, the conjugate base of the reactant "acid," then undergoes loss of X^- in a subsequent step to give the alkene. All three mechanisms occur frequently in the laboratory, but the E1cB mechanism predominates in biological pathways.

E1 Reaction: C–X bond breaks first to give a carbocation intermediate, followed by base removal of a proton to yield the alkene.

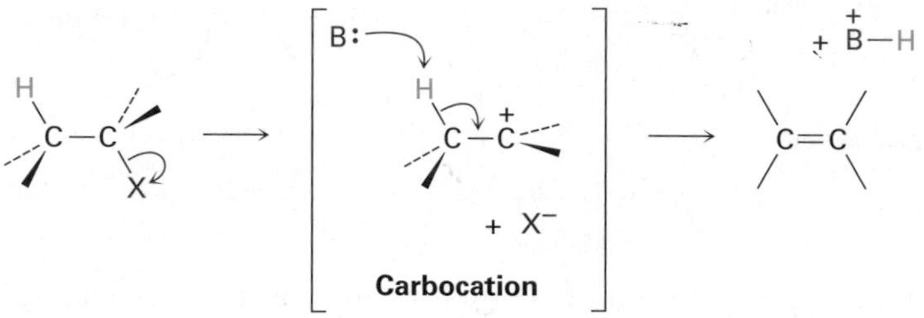

E2 Reaction: C–H and C–X bonds break simultaneously, giving the alkene in a single step without intermediates.

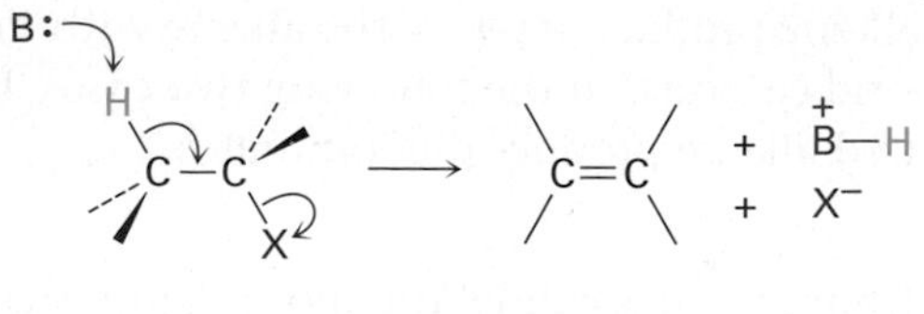

E1cB Reaction: C–H bond breaks first, giving a carbanion intermediate that loses X^- to form the alkene.

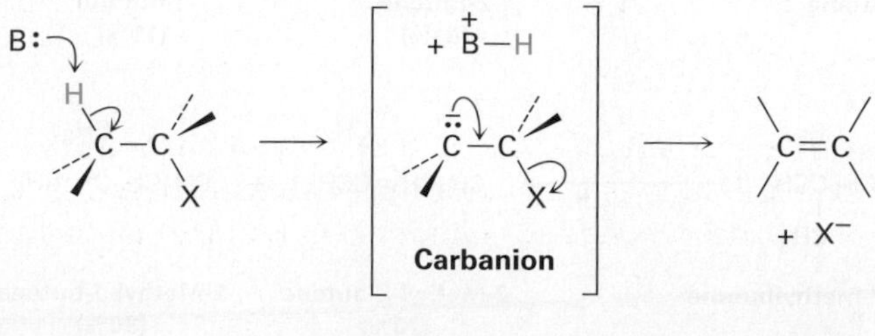

Predicting the Product of an Elimination Reaction

Worked Example 11.3

What product would you expect from reaction of 1-chloro-1 methylcyclohexane with KOH in ethanol?

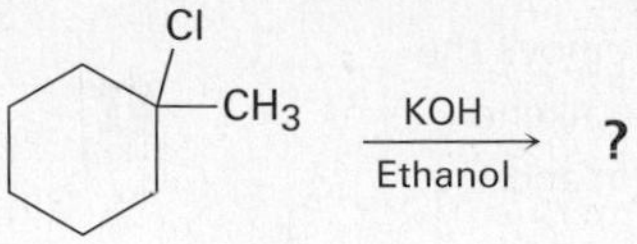

[Reaction scheme: 1-chloro-1-methylcyclohexane (Cl, CH_3) $\xrightarrow[\text{Ethanol}]{\text{KOH}}$?]

Strategy

Treatment of an alkyl halide with a strong base such as KOH yields an alkene. To find the products in a specific case, locate the hydrogen atoms on each carbon next to the leaving group, and then generate the potential alkene products by removing HX in as many ways as possible. The major product will be the one that has the most highly substituted double bond—in this case, 1-methylcyclohexene.

Solution

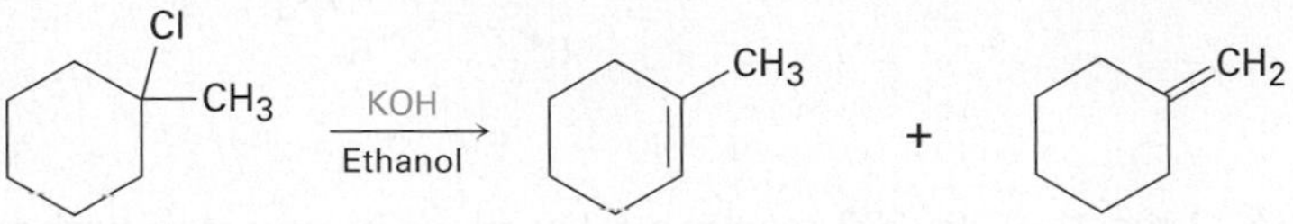

[Reaction scheme: 1-Chloro-1-methylcyclohexane $\xrightarrow[\text{Ethanol}]{\text{KOH}}$ 1-Methylcyclohexene (major) + Methylenecyclohexane (minor)]

1-Chloro-1-methylcyclohexane — **1-Methylcyclohexene (major)** — **Methylenecyclohexane (minor)**

Problem 11.15

Ignoring double-bond stereochemistry, what products would you expect from elimination reactions of the following alkyl halides? Which product will be the major product in each case?

(a) 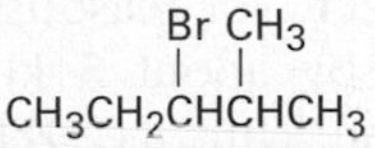 $CH_3CH_2CH(Br)CH(CH_3)CH_3$ (Br CH3 / CH3CH2CHCHCH3)

(b) 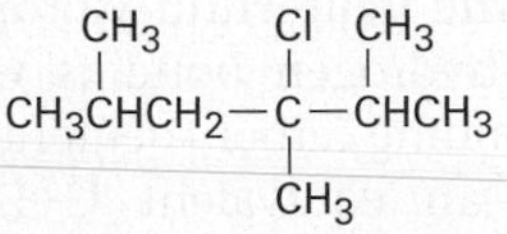 $CH_3CH(CH_3)CH_2-C(Cl)(CH_3)-CH(CH_3)CH_3$ (CH3 Cl CH3 / CH3CHCH2—C—CHCH3 / CH3)

(c) Cyclohexyl—CH(Br)CH3 (Br / —CHCH3)

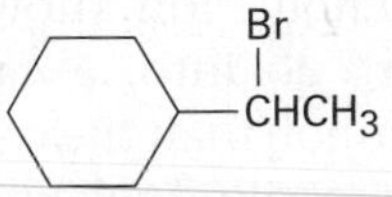

Problem 11.16

What alkyl halides might the following alkenes have been made from?

(a) 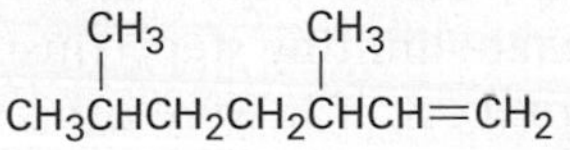 $CH_3CH(CH_3)CH_2CH_2CH(CH_3)CH{=}CH_2$ (CH3 CH3 / CH3CHCH2CH2CHCH=CH2)

(b) 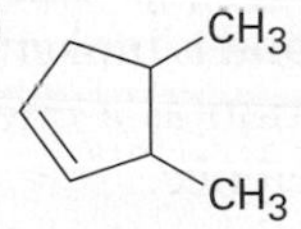 3,4-Dimethylcyclopentene (ring with two CH_3 groups)

11.8 The E2 Reaction and the Deuterium Isotope Effect

The **E2 reaction** (for *elimination, bimolecular*) occurs when an alkyl halide is treated with a strong base, such as hydroxide ion or alkoxide ion (RO^-). It is the most commonly occurring pathway for elimination and can be formulated as shown in **Figure 11.17**.

Figure 11.17 | MECHANISM

Mechanism of the E2 reaction of an alkyl halide. The reaction takes place in a single step through a transition state in which the double bond begins to form at the same time the H and X groups are leaving.

1 Base (B:) attacks a neighboring hydrogen and begins to remove the H at the same time as the alkene double bond starts to form and the X group starts to leave.

2 Neutral alkene is produced when the C–H bond is fully broken and the X group has departed with the C–X bond electron pair.

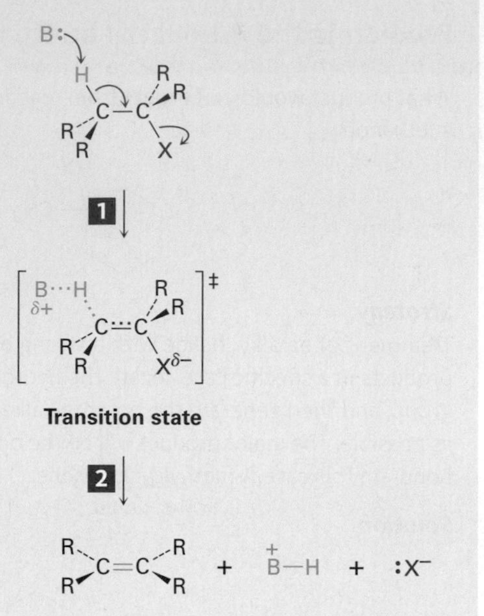

Like the S_N2 reaction, the E2 reaction takes place in one step without intermediates. As the base begins to abstract H^+ from a carbon next to the leaving group, the C–H bond begins to break, a C=C bond begins to form, and the leaving group begins to depart, taking with it the electron pair from the C–X bond. Among the pieces of evidence supporting this mechanism is that E2 reactions show second-order kinetics and follow the rate law: rate = k × [RX] × [Base]. That is, both base and alkyl halide take part in the rate-limiting step.

A second piece of evidence in support of the E2 mechanism is provided by a phenomenon known as the **deuterium isotope effect**. For reasons that we won't go into, a carbon–hydrogen bond is weaker by about 5 kJ/mol (1.2 kcal/mol) than the corresponding carbon–deuterium bond. Thus, a C–H bond is more easily broken than an equivalent C–D bond, and the rate of C–H bond cleavage is faster. For instance, the base-induced elimination of HBr from 1-bromo-2-phenylethane proceeds 7.11 times as fast as the corresponding elimination of DBr from 1-bromo-2,2-dideuterio-2-phenylethane. This result tells us that the C–H (or C–D) bond is broken in the rate-limiting step, consistent with our picture of the E2 reaction as a one-step process. If it were otherwise, we couldn't measure a rate difference.

(D) (D)
(H) (H)
C–CH_2Br —Base→ (D)(H) C=CH_2

(H)—**Faster reaction**
(D)—**Slower reaction**

Yet a third piece of mechanistic evidence involves the stereochemistry of E2 eliminations. As shown by a large number of experiments, E2 reactions occur with *periplanar* geometry, meaning that all four reacting atoms—the hydrogen, the two carbons, and the leaving group—lie in the same plane. Two such geometries are possible: **syn periplanar** geometry, in which the H and the X are on the same side of the molecule, and **anti periplanar** geometry, in which the H and the X are on opposite sides of the molecule. Of the two, anti periplanar geometry is energetically preferred because it allows the substituents on the two carbons to adopt a staggered relationship, whereas syn geometry requires that the substituents be eclipsed.

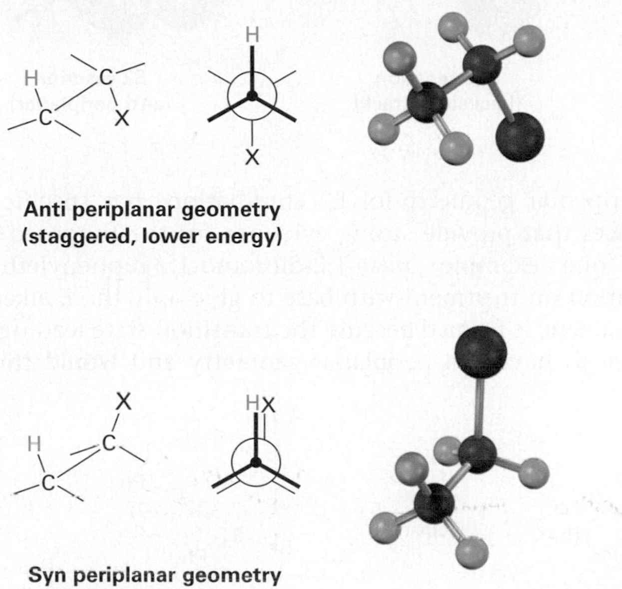

What's so special about periplanar geometry? Because the sp^3 σ orbitals in the reactant C–H and C–X bonds must overlap and become p π orbitals in the alkene product, there must also be some overlap in the transition state. This can occur most easily if all the orbitals are in the same plane to begin with—that is, if they're periplanar **(Figure 11.18)**.

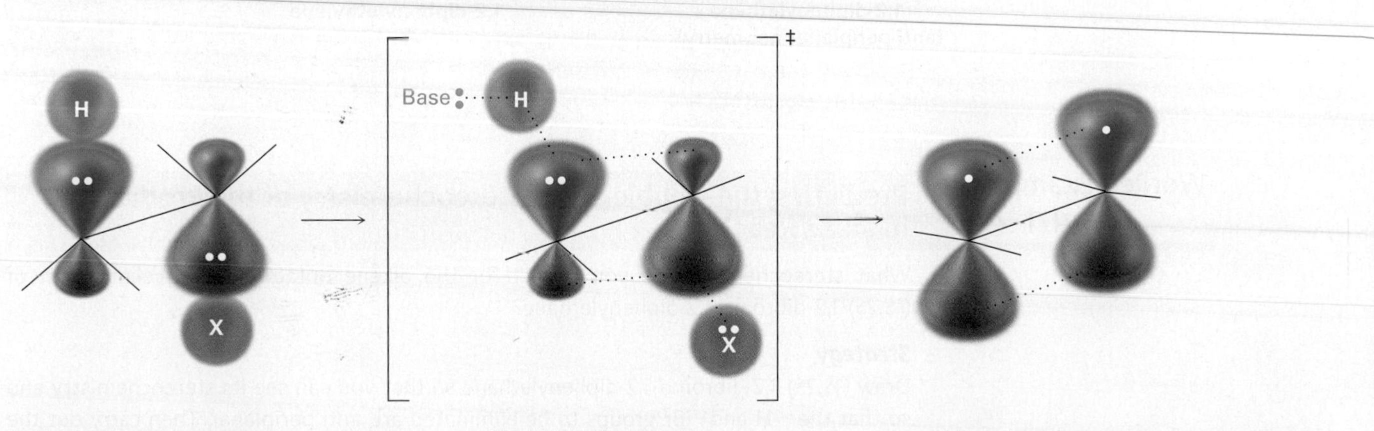

Figure 11.18 The transition state for the E2 reaction of an alkyl halide with base. Overlap of the developing p orbitals in the transition state requires periplanar geometry of the reactant.

You can think of E2 elimination reactions with periplanar geometry as being similar to S_N2 reactions with 180° geometry. In an S_N2 reaction, an electron pair from the incoming nucleophile pushes out the leaving group on the opposite side of the molecule. In an E2 reaction, an electron pair from a neighboring C–H bond pushes out the leaving group on the opposite side of the molecule.

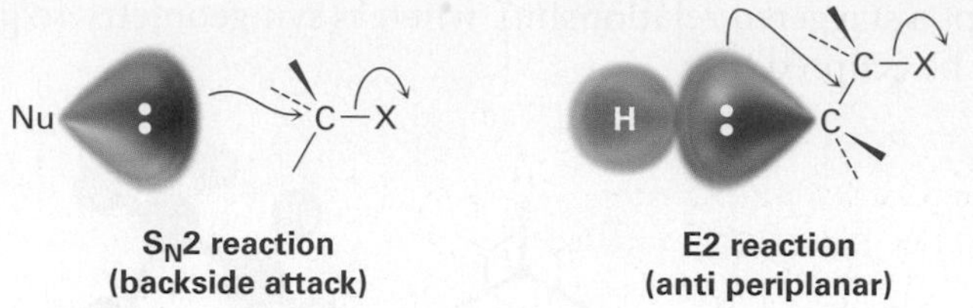

S_N2 reaction (backside attack)

E2 reaction (anti periplanar)

Anti periplanar geometry for E2 eliminations has specific stereochemical consequences that provide strong evidence for the proposed mechanism. To take just one example, *meso*-1,2-dibromo-1,2-diphenylethane undergoes E2 elimination on treatment with base to give only the *E* alkene. None of the isomeric *Z* alkene is formed because the transition state leading to the *Z* alkene would have to have syn periplanar geometry and would thus be higher in energy.

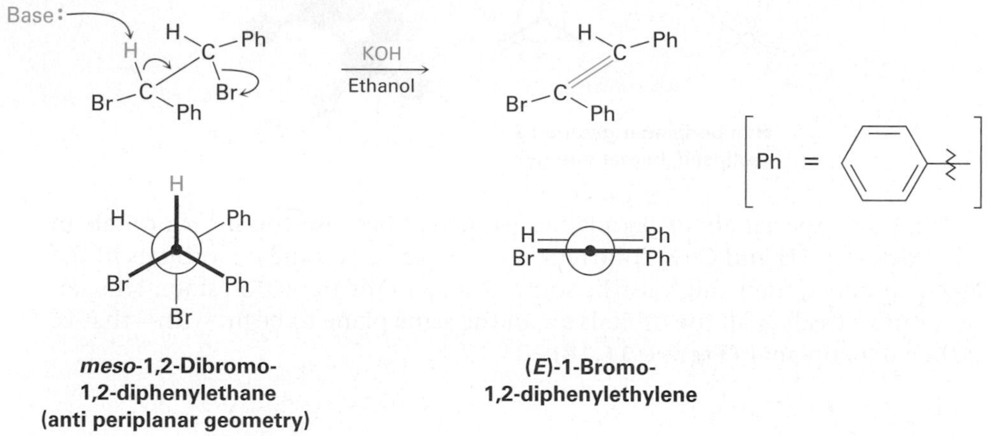

meso-1,2-Dibromo-1,2-diphenylethane (anti periplanar geometry)

(*E*)-1-Bromo-1,2-diphenylethylene

Worked Example 11.4 Predicting the Double-Bond Stereochemistry of the Product in an E2 Reaction

What stereochemistry do you expect for the alkene obtained by E2 elimination of (1*S*,2*S*)-1,2-dibromo-1,2-diphenylethane?

Strategy

Draw (1*S*,2*S*)-1,2-dibromo-1,2-diphenylethane so that you can see its stereochemistry and so that the –H and –Br groups to be eliminated are anti periplanar. Then carry out the elimination while keeping all substituents in approximately their same positions, and see what alkene results.

Solution

Anti periplanar elimination of HBr gives (*Z*)-1-bromo-1,2-diphenylethylene.

B: H Ph H C Br C Br Ph → Ph C H Br C Ph

Problem 11.17

What stereochemistry do you expect for the alkene obtained by E2 elimination of (1*R*,2*R*)-1,2-dibromo-1,2-diphenylethane? Draw a Newman projection of the reacting conformation.

Problem 11.18

What stereochemistry do you expect for the trisubstituted alkene obtained by E2 elimination of the following alkyl halide on treatment with KOH? (Reddish brown = Br.)

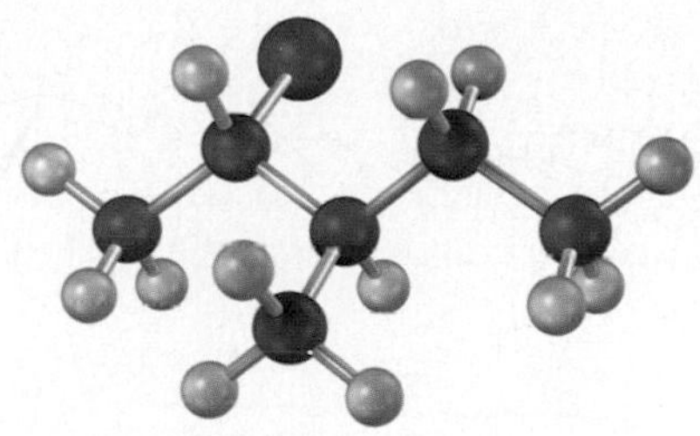

11.9 The E2 Reaction and Cyclohexane Conformation

Anti periplanar geometry for E2 reactions is particularly important in cyclohexane rings, where chair geometry forces a rigid relationship between the substituents on neighboring carbon atoms **(Section 4.8)**. The anti periplanar requirement for E2 reactions overrides Zaitsev's rule and can be met in cyclohexanes only if the hydrogen and the leaving group are trans diaxial **(Figure 11.19)**. If either the leaving group or the hydrogen is equatorial, E2 elimination can't occur.

Axial chlorine: H and Cl are anti periplanar

H H Cl = H H Cl —Base, E_2 reaction→ + HCl

Equatorial chlorine: H and Cl are not anti periplanar

H H Cl H H = H H H Cl H —Base→ No reaction from this conformation

Figure 11.19 The geometric requirement for an E2 reaction in a substituted cyclohexane. The leaving group and the hydrogen must both be axial for anti periplanar elimination to occur.

The elimination of HCl from the isomeric menthyl and neomenthyl chlorides shown in **Figure 11.20** gives a good illustration of this trans-diaxial requirement. Neomenthyl chloride undergoes elimination of HCl on reaction with ethoxide ion 200 times as fast as menthyl chloride. Furthermore, neomenthyl chloride yields 3-menthene as the major alkene product, whereas menthyl chloride yields 2-menthene.

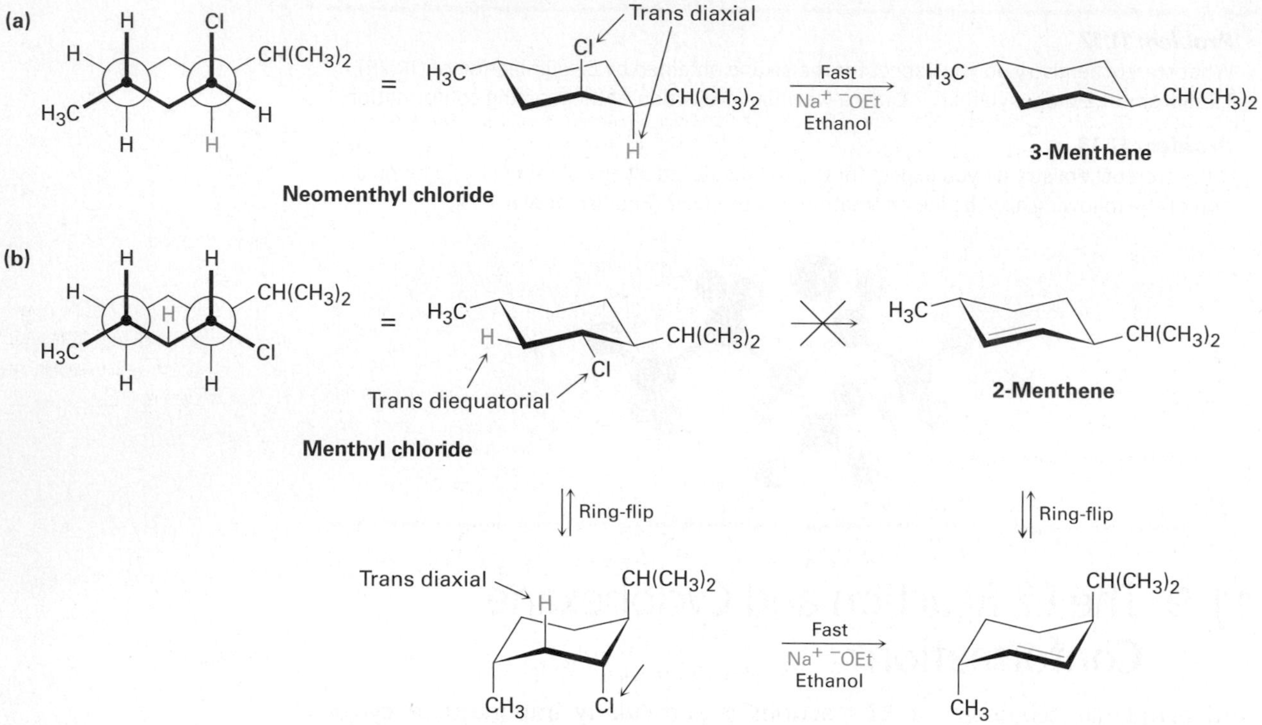

Figure 11.20 Dehydrochlorination of menthyl and neomenthyl chlorides. **(a)** Neomenthyl chloride loses HCl directly from its more stable conformation, but **(b)** menthyl chloride must first ring-flip to a higher energy conformation before HCl loss can occur. The abbreviation "Et" represents an ethyl group.

The difference in reactivity between the isomeric menthyl chlorides is due to the difference in their conformations. Neomenthyl chloride has the conformation shown in Figure 11.20a, with the methyl and isopropyl groups equatorial and the chlorine axial—a perfect geometry for E2 elimination. Loss of the hydrogen atom at C4 occurs easily to yield the more substituted alkene product, 3-menthene, as predicted by Zaitsev's rule.

Menthyl chloride, by contrast, has a conformation in which all three substituents are equatorial (Figure 11.20b). To achieve the necessary geometry for elimination, menthyl chloride must first ring-flip to a higher-energy chair conformation, in which all three substituents are axial. E2 elimination then occurs with loss of the only trans-diaxial hydrogen available, leading to the non-Zaitsev product 2-menthene. The net effect of the simple change in chlorine stereochemistry is a 200-fold change in reaction rate and a complete change of product. The chemistry of the molecule is controlled by its conformation.

Problem 11.19

Which isomer would you expect to undergo E2 elimination faster, *trans*-1-bromo-4-*tert*-butylcyclohexane or *cis*-1-bromo-4-*tert*-butylcyclohexane? Draw each molecule in its more stable chair conformation, and explain your answer.

11.10 The E1 and E1cB Reactions

The E1 Reaction

Just as the E2 reaction is analogous to the S_N2 reaction, the S_N1 reaction has a close analog called the **E1 reaction** (for *elimination, unimolecular*). The E1 reaction can be formulated as shown in **Figure 11.21** for the elimination of HCl from 2-chloro-2-methylpropane.

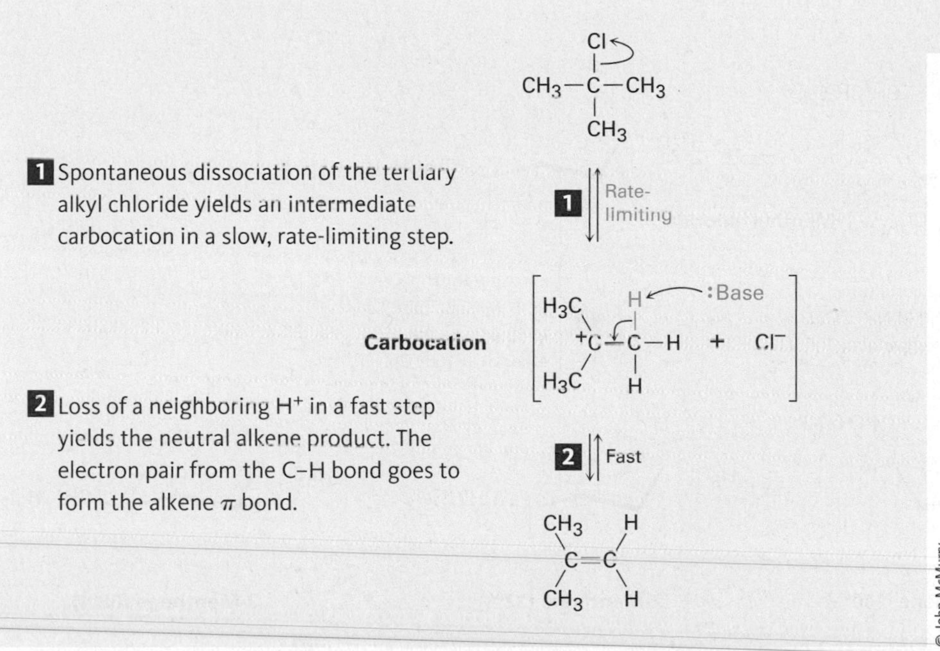

Figure 11.21 | MECHANISM

Mechanism of the E1 reaction. Two steps are involved, the first of which is rate-limiting, and a carbocation intermediate is present.

E1 eliminations begin with the same unimolecular dissociation to give a carbocation that we saw in the S_N1 reaction, but the dissociation is followed by loss of H^+ from the adjacent carbon rather than by substitution. In fact, the E1 and S_N1 reactions normally occur together whenever an alkyl halide is treated in a protic solvent with a nonbasic nucleophile. Thus, the best E1 substrates are also the best S_N1 substrates, and mixtures of substitution and elimination products are usually obtained. For example, when 2-chloro-2-methylpropane is warmed to 65 °C in 80% aqueous ethanol, a 64:36 mixture of 2-methyl-2-propanol (S_N1) and 2-methylpropene (E1) results.

$$(CH_3)_3C{-}Cl \xrightarrow[65\ °C]{H_2O,\ ethanol} (CH_3)_3C{-}OH + (CH_3)_2C{=}CH_2$$

2-Chloro-2-methylpropane → **2-Methyl-2-propanol (64%)** + **2-Methylpropene (36%)**

Much evidence has been obtained in support of the E1 mechanism. For example, E1 reactions show first-order kinetics, consistent with a rate-limiting, unimolecular dissociation process. Furthermore, E1 reactions show no deuterium isotope effect because rupture of the C–H (or C–D) bond occurs after the rate-limiting step rather than during it. Thus, we can't measure a rate difference between a deuterated and nondeuterated substrate.

A final piece of evidence involves the stereochemistry of elimination. Unlike the E2 reaction, where anti periplanar geometry is required, there is no geometric requirement on the E1 reaction because the halide and the hydrogen are lost in separate steps. We might therefore expect to obtain the more stable (Zaitsev's rule) product from E1 reaction, which is just what we find. To return to a familiar example, menthyl chloride loses HCl under E1 conditions in a polar solvent to give a mixture of alkenes in which the Zaitsev product, 3-menthene, predominates **(Figure 11.22)**.

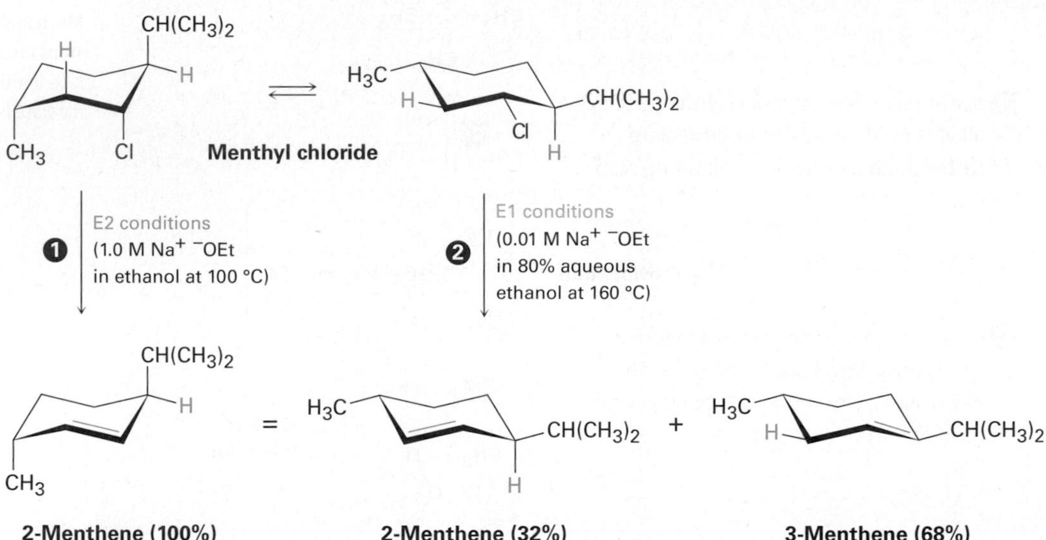

Figure 11.22 Elimination reactions of menthyl chloride. E2 conditions (1, strong base in 100% ethanol) lead to 2-menthene through an anti periplanar elimination, whereas E1 conditions (2, dilute base in 80% aqueous ethanol) lead to a mixture of 2-menthene and 3-menthene.

The E1cB Reaction

In contrast to the E1 reaction, which involves a carbocation intermediate, the **E1cB reaction** takes place through a carbanion intermediate. Base-induced abstraction of a proton in a slow, rate-limiting step gives an anion, which expels a leaving group on the adjacent carbon. The reaction is particularly common in substrates that have a poor leaving group, such as −OH, two carbons removed from a carbonyl group, HO—C—CH—C=O. The poor leaving group disfavors the alternative E1 and E2 possibilities, and the carbonyl group makes the adjacent hydrogen unusually acidic by resonance stabilization of the anion intermediate. We'll look at this acidifying effect of a carbonyl group in **Section 22.5**.

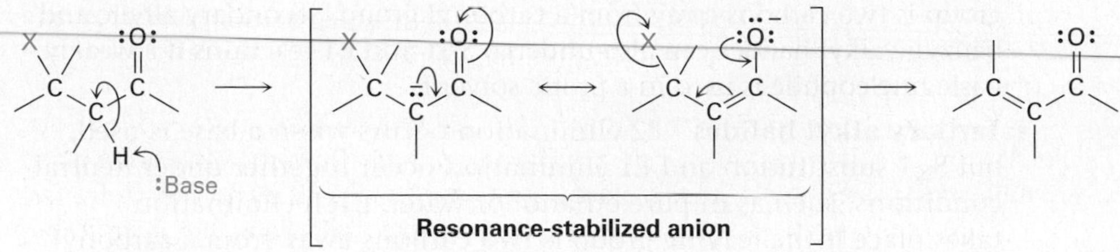

Resonance-stabilized anion

11.11 Biological Elimination Reactions

All three elimination reactions—E2, E1, and E1cB—occur in biological pathways, but the E1cB mechanism is particularly common. The substrate is usually an alcohol rather than an alkyl halide, and the H atom removed is usually adjacent to a carbonyl group, just as in laboratory reactions. Thus, 3-hydroxy carbonyl compounds are frequently converted to unsaturated carbonyl compounds by elimination reactions. A typical example occurs during the biosynthesis of fats when a 3-hydroxybutyryl thioester is dehydrated to the corresponding unsaturated (crotonyl) thioester. The base in this reaction is a histidine amino acid in the enzyme, and loss of the –OH group is assisted by simultaneous protonation.

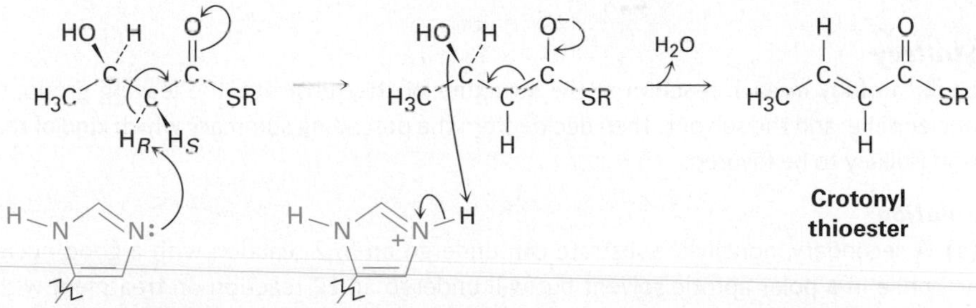

3-Hydroxybutyryl thioester

Crotonyl thioester

11.12 A Summary of Reactivity: S_N1, S_N2, E1, E1cB, and E2

S_N1, S_N2, E1, E1cB, E2—how can you keep it all straight and predict what will happen in any given case? Will substitution or elimination occur? Will the reaction be bimolecular or unimolecular? There are no rigid answers to these questions, but it's possible to recognize some trends and make some generalizations.

- **Primary alkyl halides** S_N2 substitution occurs if a good nucleophile is used, E2 elimination occurs if a strong, sterically hindered base is used, and E1cB elimination occurs if the leaving group is two carbons away from a carbonyl group.

- **Secondary alkyl halides** S_N2 substitution occurs if a weakly basic nucleophile is used in a polar aprotic solvent, E2 elimination predominates if a strong base is used, and E1cB elimination takes place if the leaving group is two carbons away from a carbonyl group. Secondary allylic and benzylic alkyl halides can also undergo S_N1 and E1 reactions if a weakly basic nucleophile is used in a protic solvent.
- **Tertiary alkyl halides** E2 elimination occurs when a base is used, but S_N1 substitution and E1 elimination occur together under neutral conditions, such as in pure ethanol or water. E1cB elimination takes place if the leaving group is two carbons away from a carbonyl group.

Worked Example 11.5 Predicting the Product and Mechanism of Reactions

Tell whether each of the following reactions is likely to be S_N1, S_N2, E1, E1cB, or E2, and predict the product of each:

(a) Cl (cyclopentyl chloride) $\xrightarrow[CH_3OH]{Na^+ \ ^-OCH_3}$?

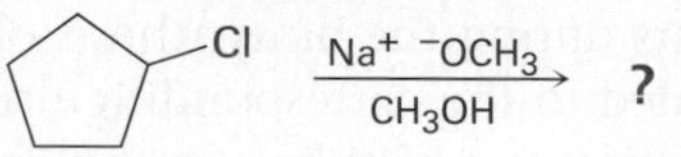

(b) Br (1-bromo-1-phenylethane) $\xrightarrow[H_2O]{HCO_2H}$?

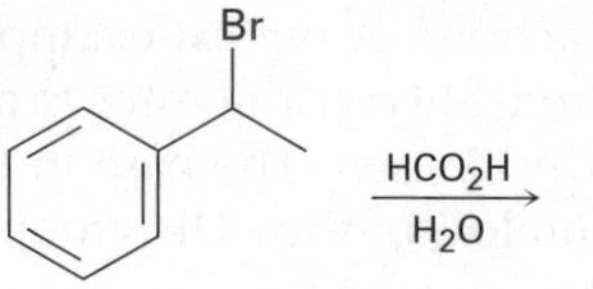

Strategy

Look carefully in each reaction at the structure of the substrate, the leaving group, the nucleophile, and the solvent. Then decide from the preceding summary which kind of reaction is likely to be favored.

Solution

(a) A secondary, nonallylic substrate can undergo an S_N2 reaction with a good nucleophile in a polar aprotic solvent but will undergo an E2 reaction on treatment with a strong base in a protic solvent. In this case, E2 reaction is likely to predominate.

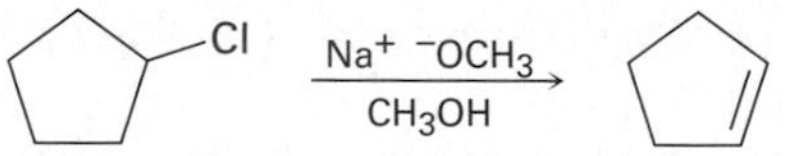

Cl $\xrightarrow[CH_3OH]{Na^+ \ ^-OCH_3}$ (cyclopentene) **E2 reaction**

(b) A secondary benzylic substrate can undergo an S_N2 reaction on treatment with a nonbasic nucleophile in a polar aprotic solvent and will undergo an E2 reaction on treatment with a base. Under protic conditions, such as aqueous formic acid (HCO_2H), an S_N1 reaction is likely, along with some E1 reaction.

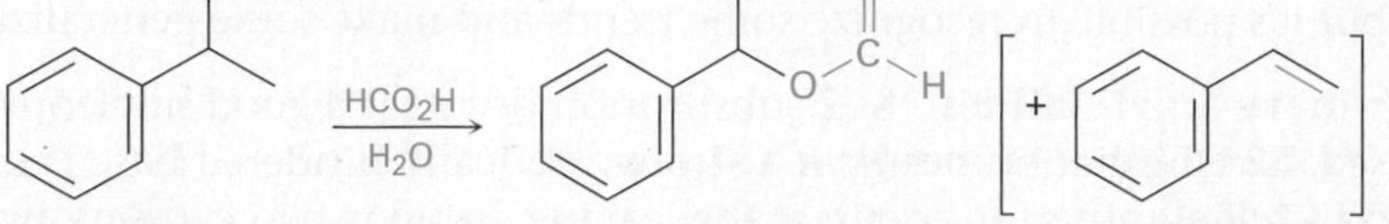

Br $\xrightarrow[H_2O]{HCO_2H}$ O–C(=O)–H (formate ester) + [styrene]

S_N1 **E1**

Problem 11.20

Tell whether each of the following reactions is likely to be S_N1, S_N2, E1, E1cB, or E2:

(a) $CH_3CH_2CH_2CH_2Br \xrightarrow[\text{THF}]{NaN_3} CH_3CH_2CH_2CH_2N{=}N{=}N$

(b) $CH_3CH_2CH(Cl)CH_2CH_3 \xrightarrow[\text{Ethanol}]{KOH} CH_3CH_2CH{=}CHCH_3$

(c) 1-Cl, 1-CH_3 cyclohexane $\xrightarrow{CH_3CO_2H}$ 1-$OCCH_3$ (C=O), 1-CH_3 cyclohexane

(d) 2-(1-hydroxycyclohexyl)cyclohexanone (O, OH) $\xrightarrow[\text{Ethanol}]{NaOH}$ 2-cyclohexylidenecyclohexanone (O)

A DEEPER LOOK Green Chemistry

Let's hope disasters like this are never repeated.

Organic chemistry in the 20th century changed the world, giving us new medicines, insecticides, adhesives, textiles, dyes, building materials, composites, and all manner of polymers. But these advances did not come without a cost: every chemical process produces wastes that must be dealt with, including reaction solvents and toxic by-products that might evaporate into the air or be leached into groundwater if not disposed of properly. Even apparently harmless by-products must be safely buried or otherwise sequestered. As always, there's no such thing as a free lunch; with the good also comes the bad.

It may never be possible to make organic chemistry completely benign, but awareness of the environmental problems caused by many chemical processes has grown dramatically in recent years, giving rise to a movement called *green chemistry*. Green chemistry is the design and implementation of chemical products and processes that reduce waste and attempt to eliminate the generation of hazardous substances. There are 12 principles of green chemistry:

Prevent waste – Waste should be prevented rather than treated or cleaned up after it has been created.

Maximize atom economy – Synthetic methods should maximize the incorporation of all materials used in a process into the final product so that waste is minimized.

Use less hazardous processes – Synthetic methods should use reactants and generate wastes with minimal toxicity to health and the environment.

(continued)

(continued)

Design safer chemicals – Chemical products should be designed to have minimal toxicity.

Use safer solvents – Minimal use should be made of solvents, separation agents, and other auxiliary substances in a reaction.

Design for energy efficiency – Energy requirements for chemical processes should be minimized, with reactions carried out at room temperature if possible.

Use renewable feedstocks – Raw materials should come from renewable sources when feasible.

Minimize derivatives – Syntheses should be designed with minimal use of protecting groups to avoid extra steps and reduce waste.

Use catalysis – Reactions should be catalytic rather than stoichiometric.

Design for degradation – Products should be designed to be biodegradable at the end of their useful lifetimes.

Monitor pollution in real time – Processes should be monitored in real time for the formation of hazardous substances.

Prevent accidents – Chemical substances and processes should minimize the potential for fires, explosions, or other accidents.

The foregoing 12 principles won't all be met in most real-world applications, but they provide a worthy goal to aim for and they can make chemists think more carefully about the environmental implications of their work. Real success stories are already occurring, and more are in progress. Approximately 7 million pounds per year of ibuprofen (6 billion tablets!) is now made by a "green" process that produces approximately 99% less waste than the process it replaces. Only three steps are needed, the anhydrous HF solvent used in the first step is recovered and reused, and the second and third steps are catalytic.

Summary

The reaction of an alkyl halide or tosylate with a nucleophile/base results either in *substitution* or in *elimination*. The resultant nucleophilic substitution and base-induced elimination reactions are two of the most widely occurring and versatile reaction types in organic chemistry, both in the laboratory and in biological pathways.

Nucleophilic substitutions are of two types: **S_N2 reactions** and **S_N1 reactions**. In the S_N2 reaction, the entering nucleophile approaches the halide from a direction 180° away from the leaving group, resulting in an umbrella-like inversion of configuration at the carbon atom. The reaction is kinetically **second-order** and is strongly inhibited by increasing steric bulk of the reactants. Thus, S_N2 reactions are favored for primary and secondary substrates.

In the S_N1 reaction, the substrate spontaneously dissociates to a carbocation in a slow **rate-limiting step**, followed by a rapid reaction with the nucleophile. As a result, S_N1 reactions are kinetically **first-order** and take place with substantial racemization of configuration at the carbon atom. They are most favored for tertiary substrates. Both S_N1 and S_N2 reactions occur in biological pathways, although the leaving group is typically a diphosphate ion rather than a halide.

Eliminations of alkyl halides to yield alkenes occur by three mechanisms: **E2 reactions**, **E1 reactions**, and **E1cB reactions**, which differ in the timing of C–H and C–X bond-breaking. In the E2 reaction, C–H and C–X bond-breaking occur simultaneously when a base abstracts H^+ from one carbon at the same time the leaving group departs from the neighboring carbon. The reaction takes place preferentially through an **anti periplanar** transition state in which the four reacting atoms—hydrogen, two carbons, and leaving group—are in the same plane. The reaction shows second-order kinetics and a **deuterium isotope effect**, and occurs when a secondary or tertiary substrate is treated with a strong base. These elimination reactions usually give a mixture of alkene products in which the more highly substituted alkene predominates (**Zaitsev's rule**).

In the E1 reaction, C–X bond-breaking occurs first. The substrate dissociates to yield a carbocation in the slow rate-limiting step before losing H^+ from an adjacent carbon in a second step. The reaction shows first-order kinetics and no deuterium isotope effect and occurs when a tertiary substrate reacts in polar, nonbasic solution.

In the E1cB reaction, C–H bond-breaking occurs first. A base abstracts a proton to give a carbanion, followed by loss of the leaving group from the adjacent carbon in a second step. The reaction is favored when the leaving group is two carbons removed from a carbonyl, which stabilizes the intermediate anion by resonance. Biological elimination reactions typically occur by this E1cB mechanism.

In general, substrates react in the following way:

RCH_2X (primary)	⟶	Mostly S_N2 substitution
R_2CHX (secondary)	⟶	S_N2 substitution with nonbasic nucleophiles E2 elimination with strong bases
R_3CX (tertiary)	⟶	Mostly E2 elimination (S_N1 substitution and E1 elimination in nonbasic solvents)

Key words

Summary of Reactions

1. Nucleophilic substitutions
   (a) S_N1 reaction of 3°, allylic, and benzylic halides (Sections 11.4 and 11.5)

$$R_3C{-}X \longrightarrow \left[R_3C^+\right] \xrightarrow{:Nu^-} R_3C{-}Nu + :X^-$$

   (b) S_N2 reaction of 1° and simple 2° halides (Sections 11.2 and 11.3)

$$Nu: + {-}C{-}X \longrightarrow Nu{-}C{-} + X:^-$$

2. Eliminations
   (a) E1 reaction (Section 11.10)

$$H{-}C{-}C(X)(R){-}R \longrightarrow \left[H{-}C{-}C^+(R){-}R\right] \longrightarrow C{=}C(R)(R) + HX$$

   (b) E1cB reaction (Section 11.10)

$$X{-}C{-}C(H){-}C(=O){-} \xrightarrow{:Base} \left[X{-}C{-}\ddot{C}^-{-}C(=O){-}\right] \longrightarrow C{=}C{-}C{=}O + HX$$

   (c) E2 reaction (Section 11.8)

$$Base: + H{-}C{-}C{-}Br \xrightarrow[\text{Ethanol}]{\text{KOH}} C{=}C$$

Exercises

Visualizing Chemistry

(Problems 11.1–11.20 appear within the chapter.)

11.21 Write the product you would expect from reaction of each of the following alkyl halides with (1) $Na^+ \ {}^-SCH_3$ and (2) $Na^+ \ {}^-OH$ (green = Cl):

(a)

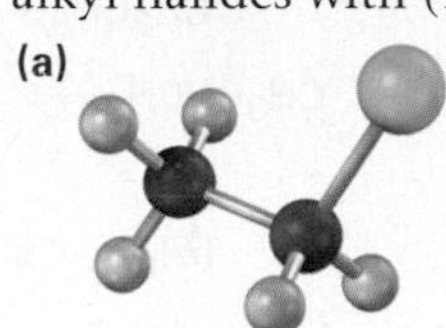

(b)

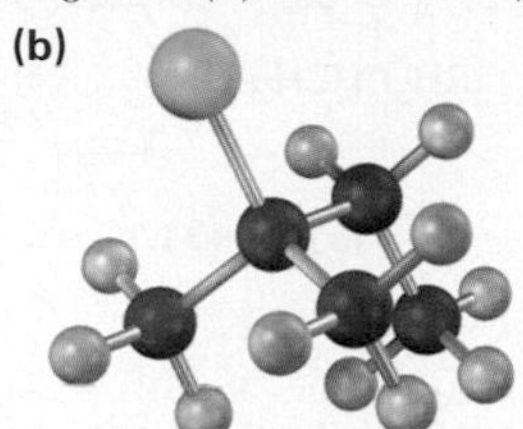

(c)

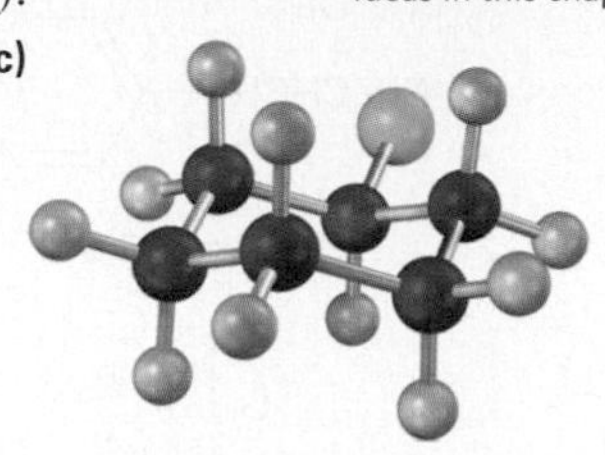

OWL Interactive versions of these problems are assignable in OWL for Organic Chemistry.

▲ denotes problems linked to the Key Ideas in this chapter.

11.22 From what alkyl bromide was the following alkyl acetate made by S_N2 reaction? Write the reaction, showing all stereochemistry.

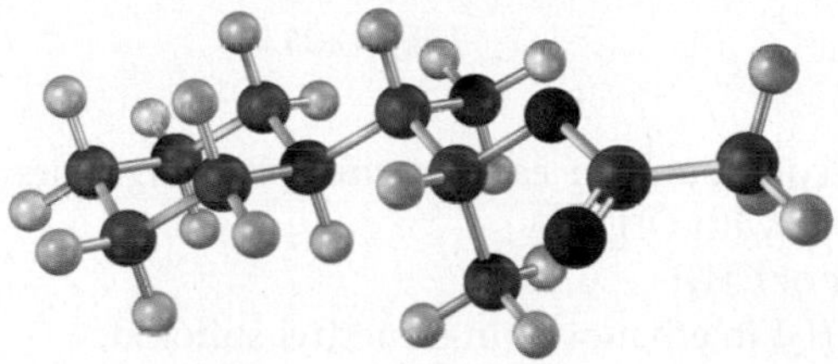

11.23 Assign *R* or *S* configuration to the following molecule, write the product you would expect from S_N2 reaction with NaCN, and assign *R* or *S* configuration to the product (green = Cl):

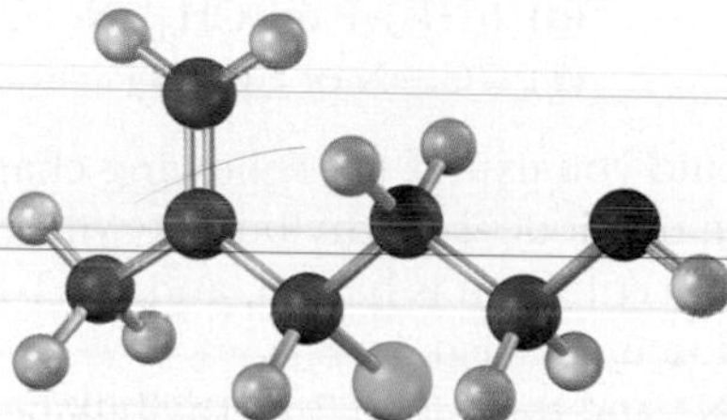

11.24 Draw the structure and assign *Z* or *E* stereochemistry to the product you expect from E2 reaction of the following molecule with NaOH (green = Cl):

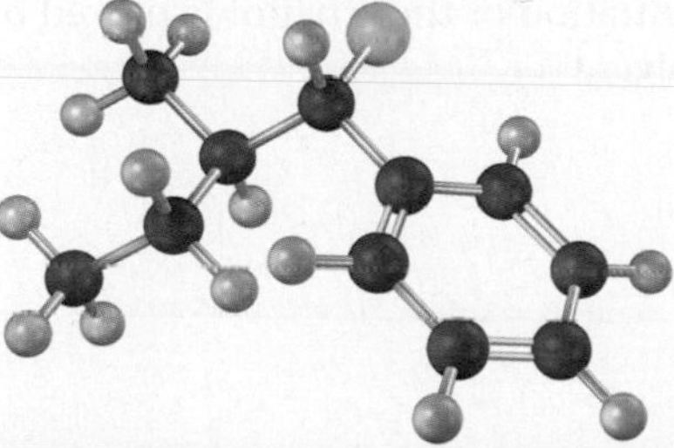

Additional Problems

Nucleophilic Substitution Reactions

11.25 Draw all isomers of C_4H_9Br, name them, and arrange them in order of decreasing reactivity in the S_N2 reaction.

11.26 The following Walden cycle has been carried out. Explain the results, and indicate where Walden inversion is occurring.

$CH_3CH(OH)CH_2-C_6H_5$ $[\alpha]_D = +33.0$ $\xrightarrow{TosCl}$ $CH_3CH(OTos)CH_2-C_6H_5$ $[\alpha]_D = +31.1$ $\xrightarrow[Heat]{CH_3CH_2OH}$ $CH_3CH(OCH_2CH_3)CH_2-C_6H_5$ $[\alpha]_D = -19.9$

$CH_3CH(OH)CH_2-C_6H_5$ $\xrightarrow{K}$ $[CH_3CH(O^-\,K^+)CH_2-C_6H_5]$ $\xrightarrow{CH_3CH_2Br}$ $CH_3CH(OCH_2CH_3)CH_2-C_6H_5$ $[\alpha]_D = +23.5$

11.27 ▲ Which compound in each of the following pairs will react faster in an S_N2 reaction with OH^-?
(a) CH_3Br or CH_3I
(b) CH_3CH_2I in ethanol or in dimethyl sulfoxide
(c) $(CH_3)_3CCl$ or CH_3Cl
(d) $H_2C{=}CHBr$ or $H_2C{=}CHCH_2Br$

11.28 Which reactant in each of the following pairs is more nucleophilic? Explain.
(a) $^-NH_2$ or NH_3 **(b)** H_2O or $CH_3CO_2^-$
(c) BF_3 or F^- **(d)** $(CH_3)_3P$ or $(CH_3)_3N$
(e) I^- or Cl^- **(f)** $^-C{\equiv}N$ or $^-OCH_3$

11.29 What effect would you expect the following changes to have on the rate of the S_N2 reaction of 1-iodo-2-methylbutane with cyanide ion?
(a) The CN^- concentration is halved, and the 1-iodo-2-methylbutane concentration is doubled.
(b) Both the CN^- and the 1-iodo-2-methylbutane concentrations are tripled.

11.30 What effect would you expect the following changes to have on the rate of the reaction of ethanol with 2-iodo-2-methylbutane?
(a) The concentration of the halide is tripled.
(b) The concentration of the ethanol is halved by adding diethyl ether as an inert solvent.

▲ Problems linked to Key Ideas in this chapter

11.31 How might you prepare each of the following molecules using a nucleophilic substitution reaction at some step?

(a) $CH_3C{\equiv}CCH(CH_3)CH_3$ (b) $CH_3{-}O{-}C(CH_3)_2CH_3$ (c) $CH_3CH_2CH_2CH_2CN$ (d) $CH_3CH_2CH_2NH_2$

11.32 ▲ Which reaction in each of the following pairs would you expect to be faster?

(a) The S_N2 displacement by I^- on CH_3Cl or on CH_3OTos
(b) The S_N2 displacement by $CH_3CO_2^-$ on bromoethane or on bromocyclohexane
(c) The S_N2 displacement on 2-bromopropane by $CH_3CH_2O^-$ or by CN^-
(d) The S_N2 displacement by $HC{\equiv}C^-$ on bromomethane in benzene or in acetonitrile

11.33 Predict the product and give the stereochemistry resulting from reaction of each of the following nucleophiles with (*R*)-2-bromooctane:

(a) ^-CN (b) $CH_3CO_2^-$ (c) CH_3S^-

11.34 (*R*)-2-Bromooctane undergoes racemization to give (±)-2-bromooctane when treated with NaBr in dimethyl sulfoxide. Explain.

Elimination Reactions

11.35 Propose structures for compounds that fit the following descriptions:

(a) An alkyl halide that gives a mixture of three alkenes on E2 reaction
(b) An organohalide that will not undergo nucleophilic substitution
(c) An alkyl halide that gives the non-Zaitsev product on E2 reaction
(d) An alcohol that reacts rapidly with HCl at 0 °C

11.36 What products would you expect from the reaction of 1-bromopropane with each of the following?

(a) $NaNH_2$ (b) $KOC(CH_3)_3$ (c) NaI
(d) NaCN (e) $NaC{\equiv}CH$ (f) Mg, then H_2O

11.37 1-Chloro-1,2-diphenylethane can undergo E2 elimination to give either *cis*- or *trans*-1,2-diphenylethylene (stilbene). Draw Newman projections of the reactive conformations leading to both possible products, and suggest a reason why the trans alkene is the major product.

$$C_6H_5CHClCH_2C_6H_5 \xrightarrow{^-OCH_3} C_6H_5CH{=}CHC_6H_5$$

1-Chloro-1,2-diphenylethane — *trans*-1,2-Diphenylethylene

11.38 Predict the major alkene product of the following E1 reaction:

$$CH_3CH(CH_3)CBr(CH_3)CH_2CH_3 \xrightarrow[\text{Heat}]{\text{HOAc}} ?$$

▲ Problems linked to Key Ideas in this chapter

11.39 There are eight diastereomers of 1,2,3,4,5,6-hexachlorocyclohexane. Draw each in its more stable chair conformation. One isomer loses HCl in an E2 reaction nearly 1000 times more slowly than the others. Which isomer reacts so slowly, and why?

General Problems

11.40 The reactions shown below are unlikely to occur as written. Tell what is wrong with each, and predict the actual product.

(a) $CH_3CH(Br)CH_2CH_3 \xrightarrow[(CH_3)_3COH]{K^+\ ^-OC(CH_3)_3} CH_3CH(OC(CH_3)_3)CH_2CH_3$

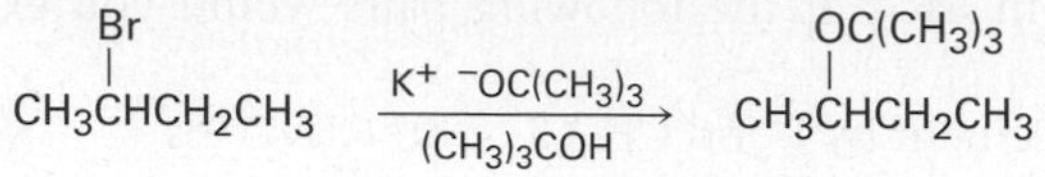

(b) Fluorocyclohexane $\xrightarrow{Na^+\ ^-OH}$ Cyclohexanol (OH)

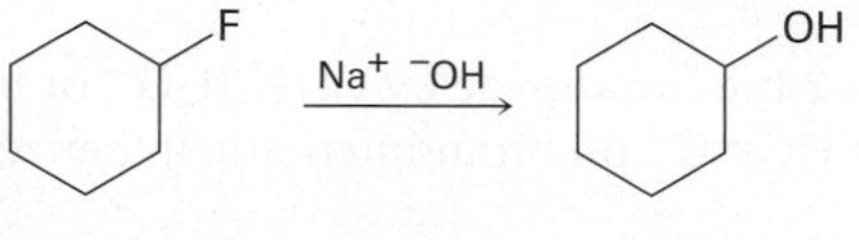

(c) 1-Methylcyclohexanol (OH, CH_3) $\xrightarrow[\text{Pyridine (a base)}]{SOCl_2}$ 1-Chloro-1-methylcyclohexane (Cl, CH_3)

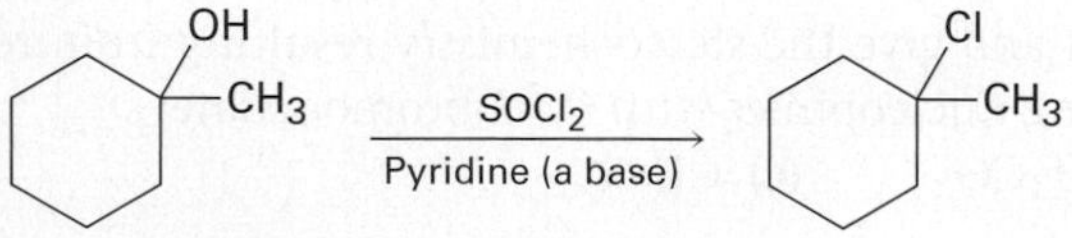

11.41 ▲ Order each of the following sets of compounds with respect to S_N1 reactivity:

(a) $H_3C{-}C(CH_3)_2{-}Cl$ $C_6H_5C(CH_3)_2Cl$ (H₃C, CH₃, C, Cl on benzene ring) $CH_3CH_2CH(NH_2)CH_3$

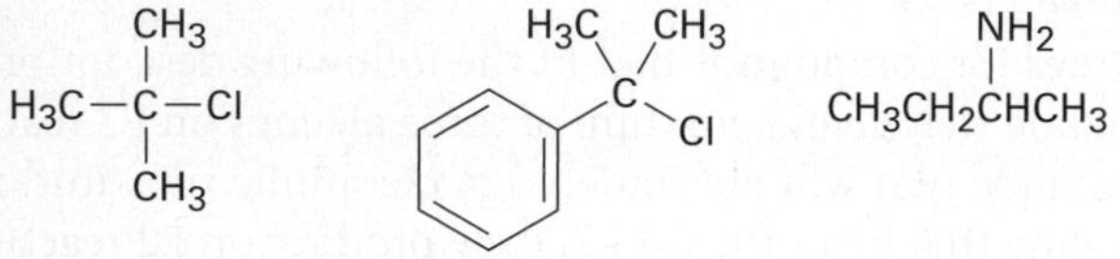

(b) $(CH_3)_3CCl$ $(CH_3)_3CBr$ $(CH_3)_3COH$

(c) $C_6H_5CH_2Br$ $C_6H_5CH(Br)CH_3$ $(C_6H_5)_3CBr$

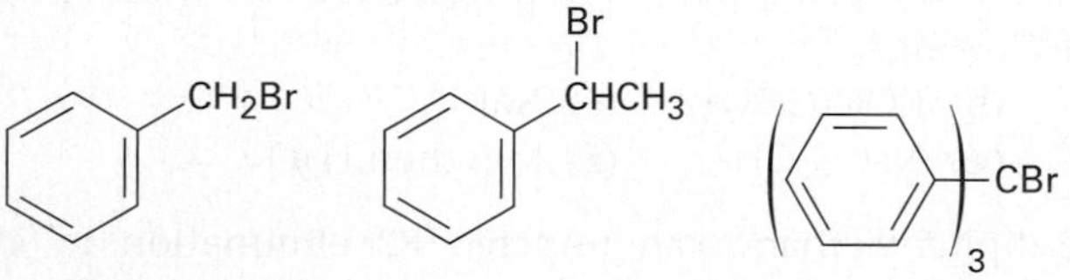

11.42 ▲ Order each of the following sets of compounds with respect to S_N2 reactivity:

(a) $H_3C{-}C(CH_3)_2{-}Cl$ $CH_3CH_2CH_2Cl$ $CH_3CH_2CH(Cl)CH_3$

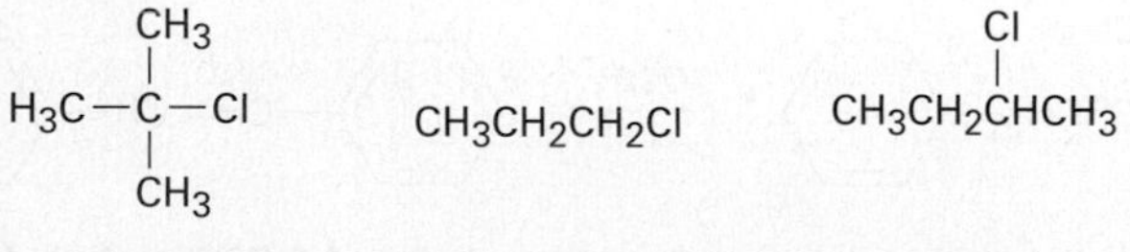

(b) $CH_3CH(CH_3)CH(Br)CH_3$ $CH_3CH(CH_3)CH_2Br$ $CH_3C(CH_3)_2CH_2Br$

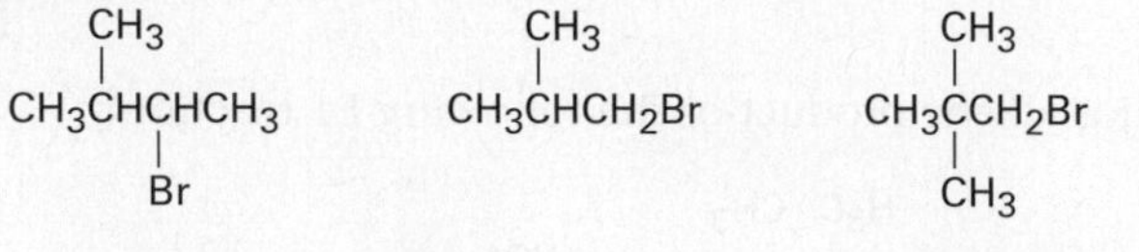

(c) $CH_3CH_2CH_2OCH_3$ $CH_3CH_2CH_2OTos$ $CH_3CH_2CH_2Br$

▲ Problems linked to Key Ideas in this chapter

11.43 Reaction of the following *S* tosylate with cyanide ion yields a nitrile product that also has *S* stereochemistry. Explain.

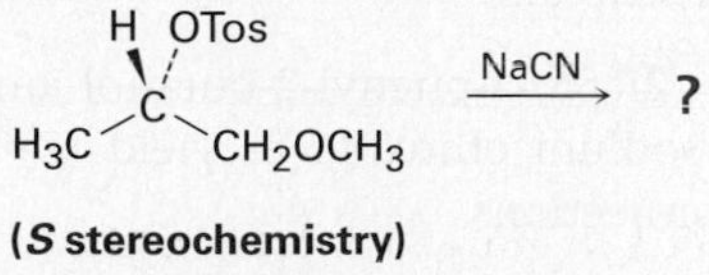

H OTos
H_3C–C–CH_2OCH_3 $\xrightarrow{\text{NaCN}}$?

(*S* stereochemistry)

11.44 Ethers can often be prepared by S_N2 reaction of alkoxide ions, RO^-, with alkyl halides. Suppose you wanted to prepare cyclohexyl methyl ether. Which of the two possible routes shown below would you choose? Explain.

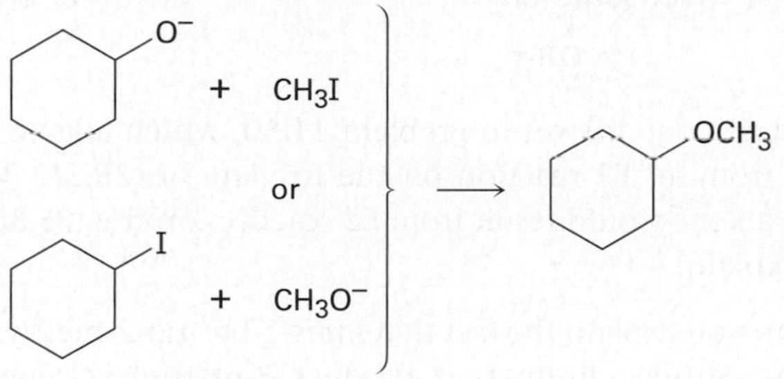

Cyclohexyl–O^- + CH_3I

or

Cyclohexyl–I + CH_3O^-

$\longrightarrow$ Cyclohexyl–OCH_3

11.45 We saw in Section 8.7 that bromohydrins are converted into epoxides when treated with base. Propose a mechanism, using curved arrows to show the electron flow.

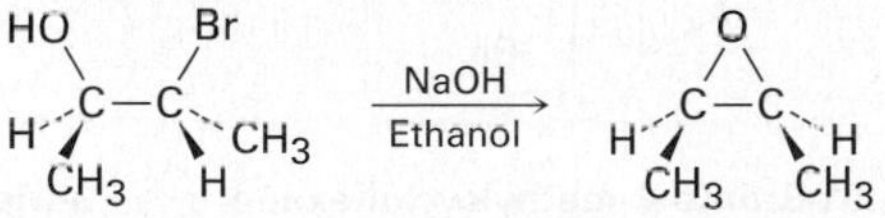

HO, Br; H, CH_3; CH_3, H $\xrightarrow[\text{Ethanol}]{\text{NaOH}}$ O; H, CH_3; H, CH_3

11.46 Show the stereochemistry of the epoxide (see Problem 11.45) you would obtain by formation of a bromohydrin from *trans*-2-butene, followed by treatment with base.

11.47 In light of your answer to Problem 11.45, what product might you expect from treatment of 4-bromo-1-butanol with base?

$BrCH_2CH_2CH_2CH_2OH \xrightarrow{\text{Base}}$?

11.48 ▲ The following tertiary alkyl bromide does not undergo a nucleophilic substitution reaction by either S_N1 or S_N2 mechanisms. Explain.

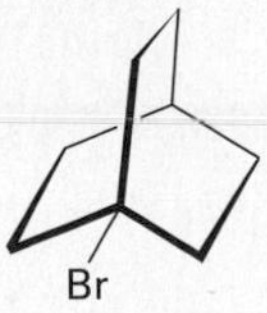

Br

▲ Problems linked to Key Ideas in this chapter

11.49 In addition to not undergoing substitution reactions, the alkyl bromide shown in Problem 11.48 also fails to undergo an elimination reaction when treated with base. Explain.

11.50 The tosylate of (2*R*,3*S*)-3-phenyl-2-butanol undergoes E2 elimination on treatment with sodium ethoxide to yield (*Z*)-2-phenyl-2-butene. Explain, using Newman projections.

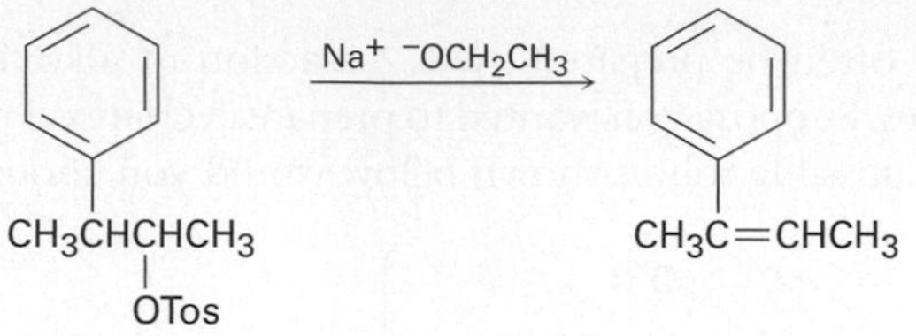

$CH_3CHCHCH_3$ (OTos) $\xrightarrow{Na^+\ ^-OCH_2CH_3}$ $CH_3C{=}CHCH_3$

11.51 In light of your answer to Problem 11.50, which alkene, *E* or *Z*, would you expect from an E2 reaction on the tosylate of (2*R*,3*R*)-3-phenyl-2-butanol? Which alkene would result from E2 reaction on the (2*S*,3*R*) and (2*S*,3*S*) tosylates? Explain.

11.52 How can you explain the fact that *trans*-1-bromo-2-methylcyclohexane yields the non-Zaitsev elimination product 3-methylcyclohexene on treatment with base?

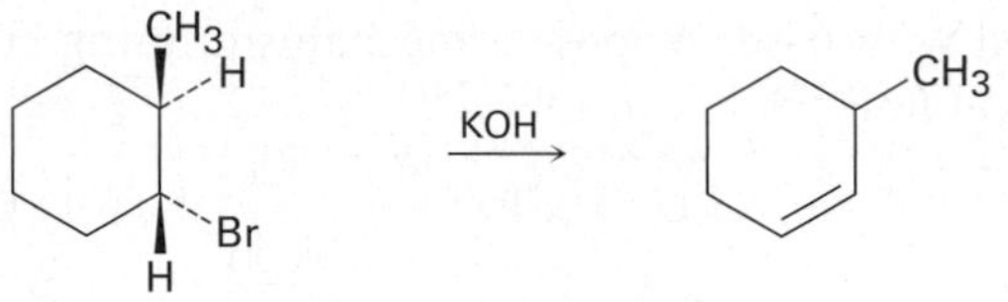

$\xrightarrow{KOH}$

***trans*-1-Bromo-2-methylcyclohexane** **3-Methylcyclohexene**

11.53 Predict the product(s) of the following reaction, indicating stereochemistry where necessary:

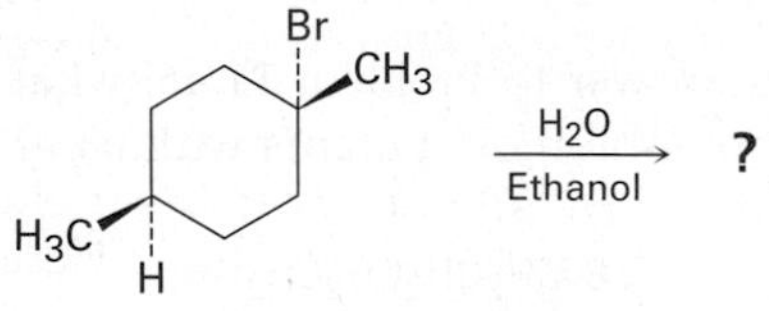

$\xrightarrow[\text{Ethanol}]{H_2O}$?

▲ Problems linked to Key Ideas in this chapter

11.54 Metabolism of *S*-adenosylhomocysteine (Section 11.6) involves the following sequence. Propose a mechanism for the second step.

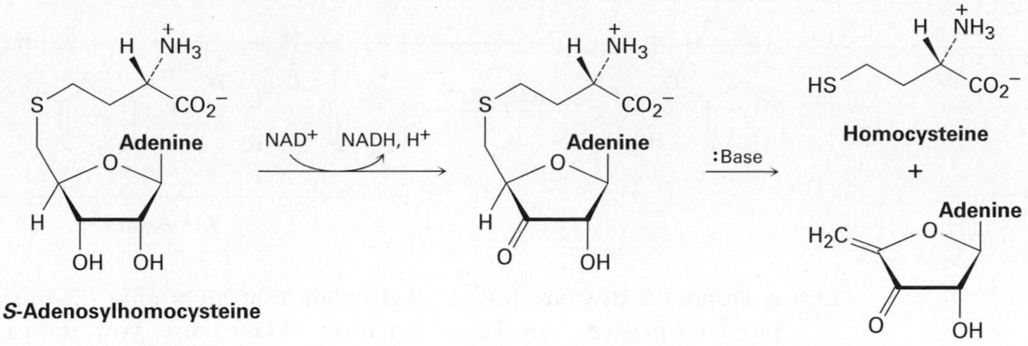

S-Adenosylhomocysteine —(NAD⁺ → NADH, H⁺)→ [3-keto intermediate] —(:Base)→ Homocysteine + [enone with Adenine]

11.55 Reaction of iodoethane with CN^- yields a small amount of *isonitrile*, $CH_3CH_2N{\equiv}C$, along with the nitrile $CH_3CH_2C{\equiv}N$ as the major product. Write electron-dot structures for both products, assign formal charges as necessary, and propose mechanisms to account for their formation.

11.56 ▲ Alkynes can be made by dehydrohalogenation of vinylic halides in a reaction that is essentially an E2 process. In studying the stereochemistry of this elimination, it was found that (*Z*)-2-chloro-2-butenedioic acid reacts 50 times as fast as the corresponding *E* isomer. What conclusion can you draw about the stereochemistry of eliminations in vinylic halides? How does this result compare with eliminations of alkyl halides?

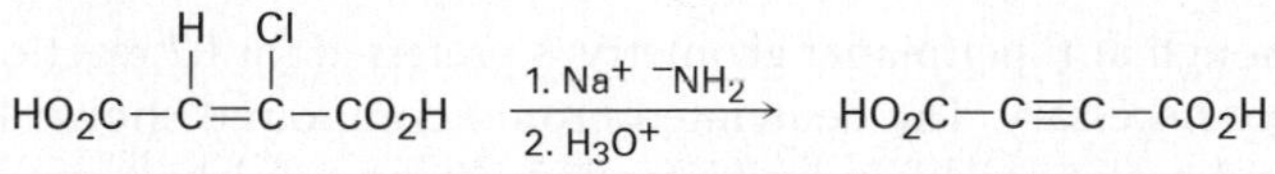

$$HO_2C{-}C(H){=}C(Cl){-}CO_2H \xrightarrow[2.\ H_3O^+]{1.\ Na^+\ {}^-NH_2} HO_2C{-}C{\equiv}C{-}CO_2H$$

11.57 (*S*)-2-Butanol slowly racemizes on standing in dilute sulfuric acid. Explain.

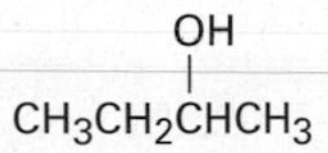

$CH_3CH_2CH(OH)CH_3$ **2-Butanol**

11.58 Reaction of HBr with (*R*)-3-methyl-3-hexanol leads to racemic 3-bromo-3-methylhexane. Explain.

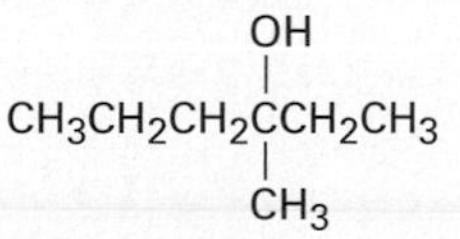

$CH_3CH_2CH_2C(OH)(CH_3)CH_2CH_3$ **3-Methyl-3-hexanol**

▲ Problems linked to Key Ideas in this chapter

11.59 Treatment of 1-bromo-2-deuterio-2-phenylethane with strong base leads to a mixture of deuterated and nondeuterated phenylethylenes in an approximately 7:1 ratio. Explain.

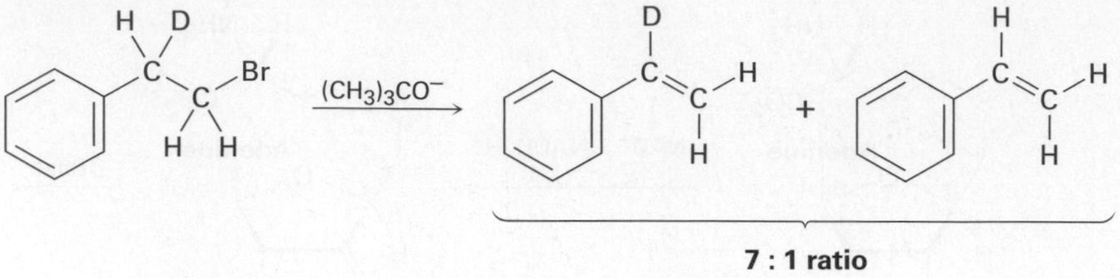

$(CH_3)_3CO^-$

7 : 1 ratio

11.60 ▲ Propose a structure for an alkyl halide that gives only (*E*)-3-methyl-2-phenyl-2-pentene on E2 elimination. Make sure you indicate the stereochemistry.

11.61 One step in the urea cycle for ridding the body of ammonia is the conversion of argininosuccinate to the amino acid arginine plus fumarate. Propose a mechanism for the reaction, and show the structure of arginine.

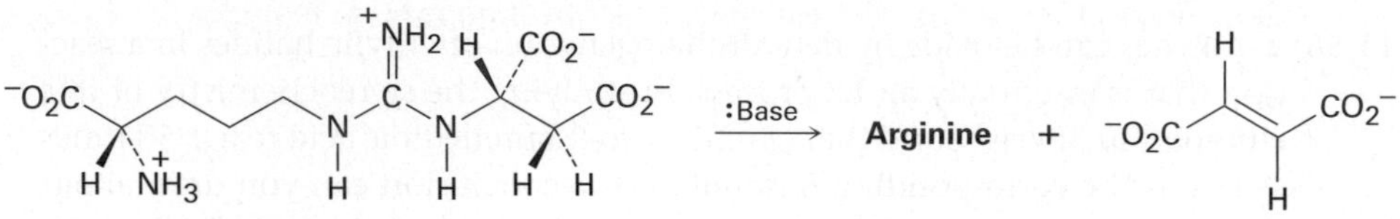

:Base → **Arginine** + Fumarate

Argininosuccinate **Fumarate**

11.62 Although anti periplanar geometry is preferred for E2 reactions, it isn't absolutely necessary. The deuterated bromo compound shown here reacts with strong base to yield an undeuterated alkene. Clearly, a syn elimination has occurred. Make a molecular model of the reactant, and explain the result.

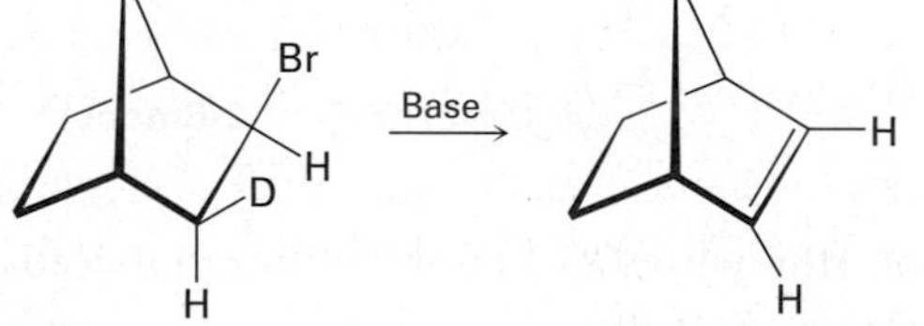

Base →

▲ Problems linked to Key Ideas in this chapter

11.63 In light of your answer to Problem 11.62, explain why one of the following isomers undergoes E2 reaction approximately 100 times as fast as the other. Which isomer is more reactive, and why?

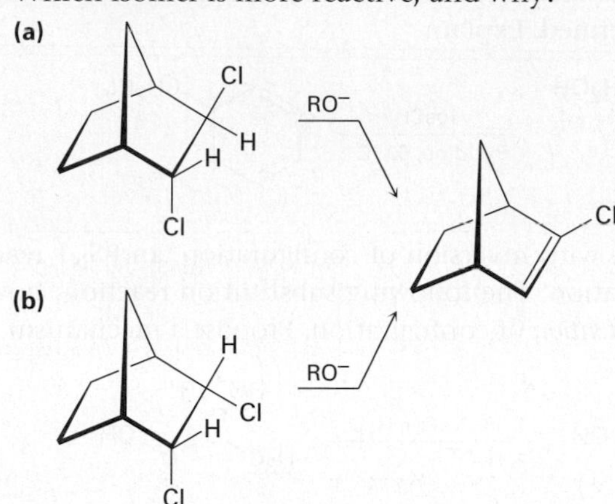

11.64 Methyl esters (RCO_2CH_3) undergo a cleavage reaction to yield carboxylate ions plus iodomethane on heating with LiI in dimethylformamide:

$$C_6H_{11}\text{–}C(=O)OCH_3 \xrightarrow[\text{DMF}]{\text{LiI}} C_6H_{11}\text{–}C(=O)O^-\ Li^+ \ + \ CH_3I$$

The following evidence has been obtained: (1) The reaction occurs much faster in DMF than in ethanol. (2) The corresponding ethyl ester ($RCO_2CH_2CH_3$) cleaves approximately 10 times more slowly than the methyl ester. Propose a mechanism for the reaction. What other kinds of experimental evidence could you gather to support your hypothesis?

11.65 The reaction of 1-chlorooctane with $CH_3CO_2^-$ to give octyl acetate is greatly accelerated by adding a small quantity of iodide ion. Explain.

11.66 Compound **X** is optically inactive and has the formula $C_{16}H_{16}Br_2$. On treatment with strong base, **X** gives hydrocarbon **Y**, $C_{16}H_{14}$. Compound **Y** absorbs 2 equivalents of hydrogen when reduced over a palladium catalyst and reacts with ozone to give two fragments. One fragment, **Z**, is an aldehyde with formula C_7H_6O. The other fragment is glyoxal, $(CHO)_2$. Write the reactions involved, and suggest structures for **X**, **Y**, and **Z**. What is the stereochemistry of **X**?

▲ Problems linked to Key Ideas in this chapter

11.67 When a primary alcohol is treated with *p*-toluenesulfonyl chloride at room temperature in the presence of an organic base such as pyridine, a tosylate is formed. When the same reaction is carried out at higher temperature, an alkyl chloride is often formed. Explain.

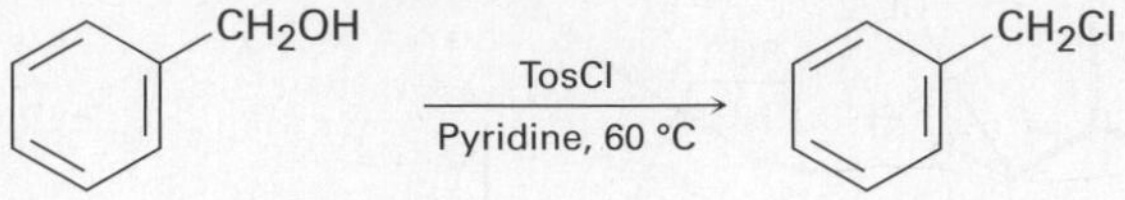

CH_2OH TosCl Pyridine, 60 °C CH_2Cl

11.68 S_N2 reactions take place with inversion of configuration, and S_N1 reactions take place with racemization. The following substitution reaction, however, occurs with complete *retention* of configuration. Propose a mechanism.

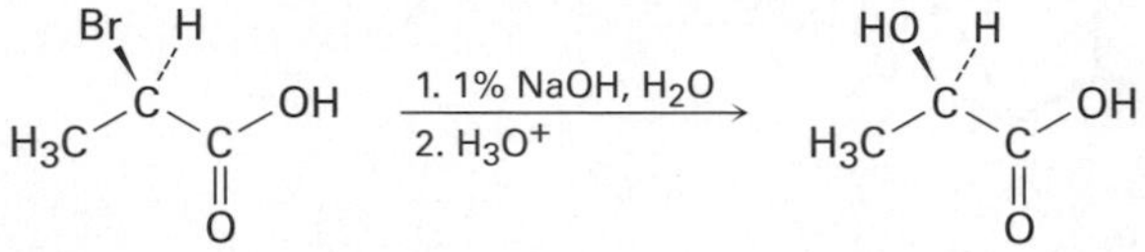

Br H H_3C C C OH O 1. 1% NaOH, H_2O 2. H_3O^+ HO H H_3C C C OH O

11.69 Propose a mechanism for the following reaction, an important step in the laboratory synthesis of proteins:

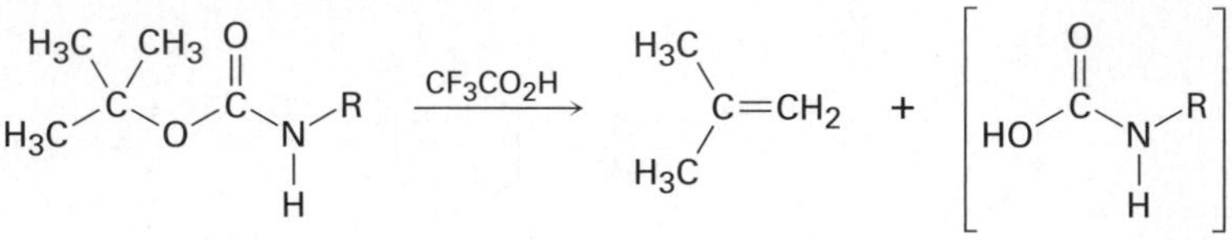

H_3C CH_3 O H_3C C O C N R H CF_3CO_2H H_3C C=CH_2 H_3C + [HO C N R H O]

11.70 The amino acid methionine is formed by a methylation reaction of homocysteine with *N*-methyltetrahydrofolate. The stereochemistry of the reaction has been probed by carrying out the transformation using a donor with a "chiral methyl group" that contains protium (H), deuterium (D), and tritium (T) isotopes of hydrogen. Does the methylation reaction occur with inversion or retention of configuration?

HS CO_2^- $H_3\overset{+}{N}$ H **Homocysteine** Methionine synthase H C S D T CO_2^- $H_3\overset{+}{N}$ H **Methionine**

T H C D O H N N NHAr H_2N N H + H N

***N*-Methyltetrahydrofolate** **Tetrahydrofolate**

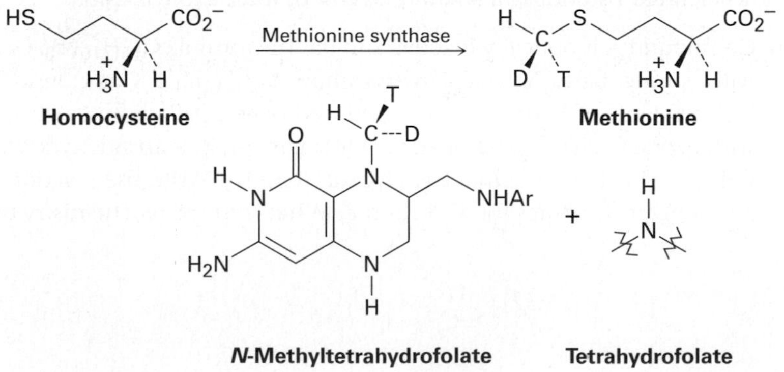

▲ Problems linked to Key Ideas in this chapter

11.71 Amines are converted into alkenes by a two-step process called the *Hofmann elimination*. S_N2 reaction of the amine with an excess of CH_3I in the first step yields an intermediate that undergoes E2 reaction when treated with silver oxide as base. Pentylamine, for example, yields 1-pentene. Propose a structure for the intermediate, and explain why it undergoes ready elimination.

$$CH_3CH_2CH_2CH_2CH_2NH_2 \xrightarrow[\text{2. } Ag_2O,\ H_2O]{\text{1. Excess } CH_3I} CH_3CH_2CH_2CH{=}CH_2$$

11.72 The antipsychotic drug flupentixol is prepared by the following scheme:

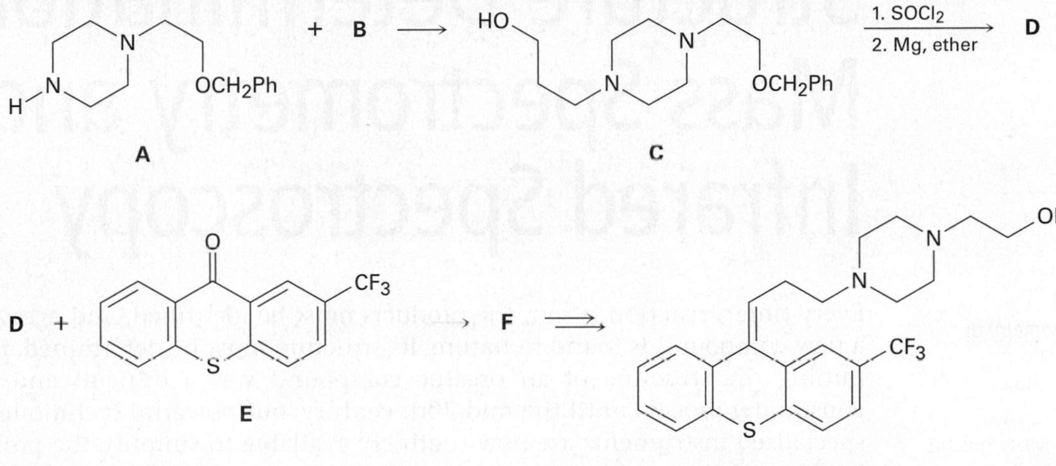

A + B ⟶ C $\xrightarrow[\text{2. Mg, ether}]{\text{1. } SOCl_2}$ D

D + E ⟶ F ⟹ Flupentixol

(A: H–N piperazine with OCH₂Ph side chain; C: HO– chain on piperazine with OCH₂Ph; E: thioxanthone with O, S, CF₃; Flupentixol: piperazine with N–CH₂CH₂OH, CF₃, S)

(a) What alkyl chloride **B** reacts with amine **A** to form **C**?

(b) Compound **C** is treated with $SOCl_2$, and the product is allowed to react with magnesium metal to give a Grignard reagent **D**. What is the structure of **D**?

(c) We'll see in Section 19.7 that Grignard reagents add to ketones, such as E, to give tertiary alcohols, such as F. Because of the newly formed chirality center, compound F exists as a pair of enantiomers. Draw both, and assign *R,S* configuration.

(d) Two stereoisomers of flupentixol are subsequently formed from F, but only one is shown. Draw the other isomer, and identify the type of stereoisomerism.

12

More than a thousand different chemical compounds have been isolated from coffee. Their structures were determined using various spectroscopic techniques. © webphotographeer/iStockphoto

Structure Determination: Mass Spectrometry and Infrared Spectroscopy

Every time a reaction is run, the products must be identified, and every time a new compound is found in nature, its structure must be determined. Determining the structure of an organic compound was a difficult and time-consuming process until the mid-20th century, but powerful techniques and specialized instruments are now routinely available to simplify the problem. In this and the next two chapters, we'll look at four such techniques—mass spectrometry (MS), infrared (IR) spectroscopy, ultraviolet spectroscopy (UV), and nuclear magnetic resonance spectroscopy (NMR)—and we'll see the kind of information that can be obtained from each.

Mass spectrometry	What is the size and formula?
Infrared spectroscopy	What functional groups are present?
Ultraviolet spectroscopy	Is a conjugated π electron system present?
Nuclear magnetic resonance spectroscopy	What is the carbon–hydrogen framework?

Why This Chapter? Finding the structures of new molecules, whether small ones synthesized in the laboratory or large proteins and nucleic acids found in living organisms, is central to progress in chemistry and biochemistry. We can only scratch the surface of structure determination in this book, but after reading this and the following two chapters, you should have a good idea of the range of structural techniques available and of how and when each is used.

12.1 Mass Spectrometry of Small Molecules: Magnetic-Sector Instruments

At its simplest, **mass spectrometry (MS)** is a technique for measuring the mass, and therefore the molecular weight (MW), of a molecule. In addition, it's often possible to gain structural information about a molecule by measuring the masses of the fragments produced when molecules are broken apart.

More than 20 different kinds of commercial mass spectrometers are available depending on the intended application, but all have three basic parts: an *ionization source* in which sample molecules are given an electrical charge, a *mass analyzer* in which ions are separated by their mass-to-charge ratio, and a *detector* in which the separated ions are observed and counted.

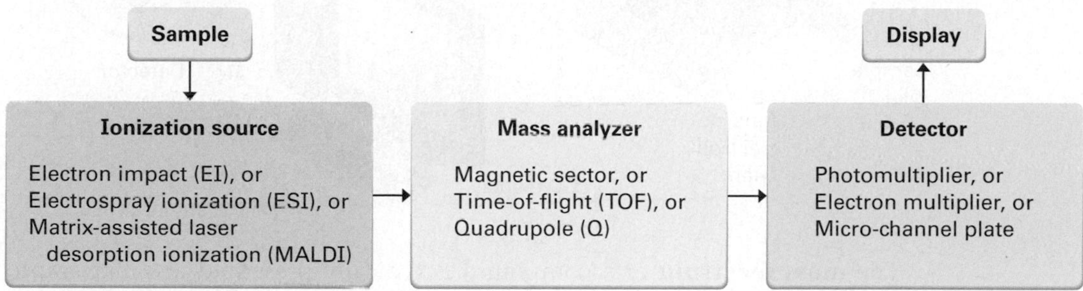

Among the most common mass spectrometers used for routine purposes in the laboratory is the electron-impact, magnetic-sector instrument shown schematically in **Figure 12.1**. A small amount of sample is vaporized into the ionization source, where it is bombarded by a stream of high-energy electrons. The energy of the electron beam can be varied but is commonly around 70 electron volts (eV), or 6700 kJ/mol. When a high-energy electron strikes an organic molecule, it dislodges a valence electron from the molecule, producing a *cation radical*—*cation* because the molecule has lost an electron and now has a positive charge; *radical* because the molecule now has an odd number of electrons.

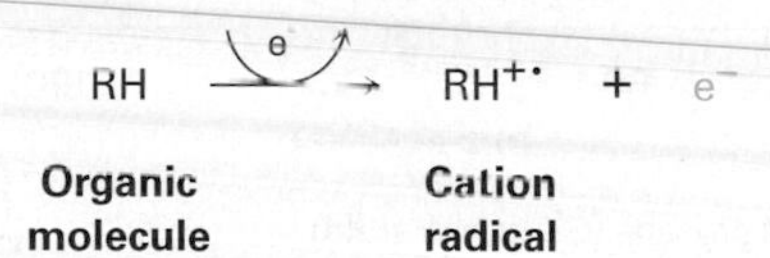

Electron bombardment transfers so much energy that most of the cation radicals fragment after formation. They fly apart into smaller pieces, some of which retain the positive charge and some of which are neutral. The fragments then flow through a curved pipe in a strong magnetic field, which deflects them into different paths according to their mass-to-charge ratio (m/z). Neutral fragments are not deflected by the magnetic field and are lost on the walls of the pipe, but positively charged fragments are sorted by the mass spectrometer onto a detector, which records them as peaks at the various m/z ratios. Since the number of charges z on each ion is usually 1, the value of m/z for each ion is simply its mass m. Masses up to approximately 2500 atomic mass units (amu) can be analyzed.

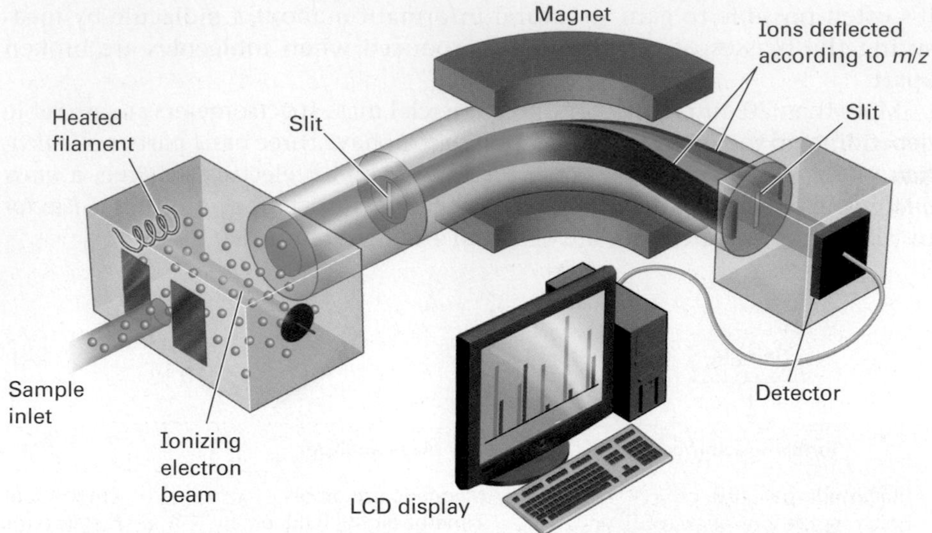

Figure 12.1 A representation of an electron-ionization, magnetic-sector mass spectrometer. Molecules are ionized by collision with high-energy electrons, causing some of the molecules to fragment. Passage of the charged fragments through a magnetic field then sorts them according to their mass.

The **mass spectrum** of a compound is typically presented as a bar graph, with masses (*m*/*z* values) on the *x* axis and intensity, or relative abundance of ions of a given *m*/*z* striking the detector, on the *y* axis. The tallest peak, assigned an intensity of 100%, is called the **base peak**, and the peak that corresponds to the unfragmented cation radical is called the **parent peak**, or the *molecular ion* (M^+). **Figure 12.2** shows the mass spectrum of propane.

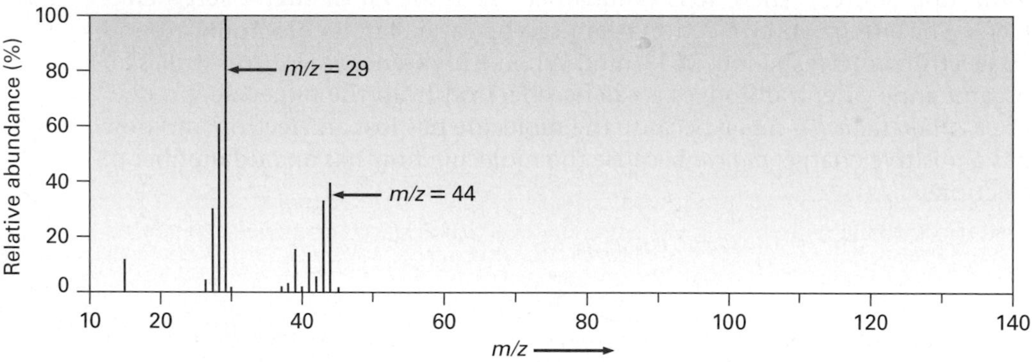

Figure 12.2 Mass spectrum of propane (C_3H_8; MW = 44).

Mass spectral fragmentation patterns are usually complex, and the molecular ion is often not the base peak. The mass spectrum of propane in Figure 12.2, for instance, shows a molecular ion at $m/z = 44$ that is only about 30% as high as the base peak at $m/z = 29$. In addition, many other fragment ions are present.

12.2 Interpreting Mass Spectra

What kinds of information can we get from a mass spectrum? The most obvious information is the molecular weight of the sample, which in itself can be invaluable. If we were given samples of hexane (MW = 86), 1-hexene (MW = 84), and 1-hexyne (MW = 82), for example, mass spectrometry would easily distinguish them.

Some instruments, called *double-focusing mass spectrometers,* have such high resolution that they provide exact mass measurements accurate to 5 ppm, or about 0.0005 amu, making it possible to distinguish between two formulas with the same nominal mass. For example, both C_5H_{12} and C_4H_8O have MW=72, but they differ slightly beyond the decimal point: C_5H_{12} has an exact mass of 72.0939 amu, whereas C_4H_8O has an exact mass of 72.0575 amu. A high-resolution instrument can easily distinguish between them. Note, however, that exact mass measurements refer to molecules with specific isotopic compositions. Thus, the sum of the exact atomic masses of the specific isotopes in a molecule is measured—1.007 83 amu for ^1H, 12.000 00 amu for ^{12}C, 14.003 07 amu for ^{14}N, 15.994 91 amu for ^{16}O, and so on—rather than the sum of the average atomic masses of elements as found on a periodic table.

Unfortunately, not every compound shows a molecular ion in its electron-impact mass spectrum. Although M^+ is usually easy to identify if it's abundant, some compounds, such as 2,2-dimethylpropane, fragment so easily that no molecular ion is observed **(Figure 12.3)**. In such cases, alternative "soft" ionization methods that do not use electron bombardment can prevent or minimize fragmentation.

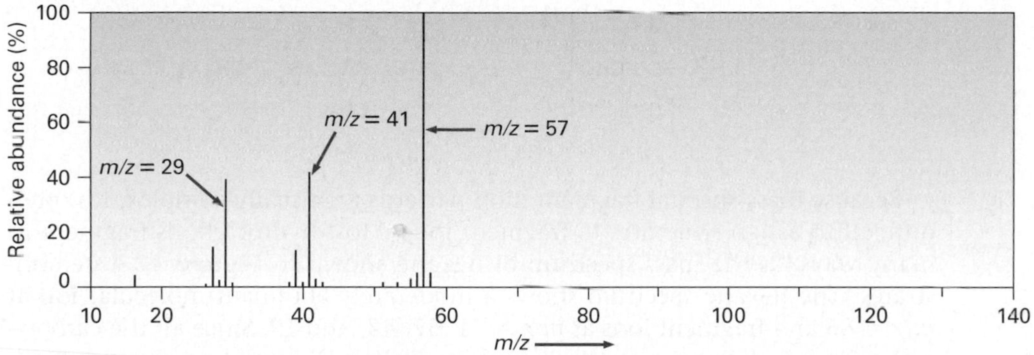

Figure 12.3 Mass spectrum of 2,2-dimethylpropane (C_5H_{12}; MW = 72). No molecular ion is observed when electron-impact ionization is used. What do you think is the formula and structure of the M^+ peak at $m/z = 57$?

Knowing the molecular weight makes it possible to narrow greatly the choices of molecular formula. For example, if the mass spectrum of an unknown compound shows a molecular ion at m/z = 110, the molecular formula is likely to be C_8H_{14}, $C_7H_{10}O$, $C_6H_6O_2$, or $C_6H_{10}N_2$. There are always a number of molecular formulas possible for all but the lowest molecular weights, and a computer can easily generate a list of choices.

A further point about mass spectrometry, noticeable in the spectra of both propane (Figure 12.2) and 2,2-dimethylpropane (Figure 12.3), is that the peak for the molecular ion is not at the highest m/z value. There is also a small peak at M+1 because of the presence of different isotopes in the molecules. Although ^{12}C is the most abundant carbon isotope, a small amount (1.10% natural abundance) of ^{13}C is also present. Thus, a certain percentage of the molecules analyzed in the mass spectrometer are likely to contain a ^{13}C atom, giving rise to the observed M+1 peak. In addition, a small amount of ^2H (deuterium; 0.015% natural abundance) is present, making a further contribution to the M+1 peak.

Mass spectrometry would be useful even if molecular weight and formula were the only information that could be obtained, but in fact we can get much more. For one thing, the mass spectrum of a compound serves as a kind of "molecular fingerprint." Each organic compound fragments in a unique way depending on its structure, and the likelihood of two compounds having identical mass spectra is small. Thus, it's sometimes possible to identify an unknown by computer-based matching of its mass spectrum to one of the more than 592,000 spectra recorded in a data base called the *Registry of Mass Spectral Data*.

It's also possible to derive structural information about a molecule by interpreting its fragmentation pattern. Fragmentation occurs when the high-energy cation radical flies apart by spontaneous cleavage of a chemical bond. One of the two fragments retains the positive charge and is a carbocation, while the other fragment is a neutral radical.

Not surprisingly, the positive charge often remains with the fragment that is best able to stabilize it. In other words, a relatively stable carbocation is often formed during fragmentation. For example, 2,2-dimethylpropane tends to fragment in such a way that the positive charge remains with the *tert*-butyl group. 2,2-Dimethylpropane therefore has a base peak at m/z = 57, corresponding to $C_4H_9^+$ (Figure 12.3).

$$\left[(CH_3)_3C{-}CH_3 \right]^{+\cdot} \longrightarrow H_3C{-}\overset{CH_3}{\underset{CH_3}{C^+}} + \cdot CH_3$$

m/z = 57

Because mass-spectral fragmentation patterns are usually complex, it's often difficult to assign structures to fragment ions. Most hydrocarbons fragment in many ways, as the mass spectrum of hexane shown in **Figure 12.4** demonstrates. The hexane spectrum shows a moderately abundant molecular ion at m/z = 86 and fragment ions at m/z = 71, 57, 43, and 29. Since all the carbon–carbon bonds of hexane are electronically similar, all break to a similar extent, giving rise to the observed mixture of ions.

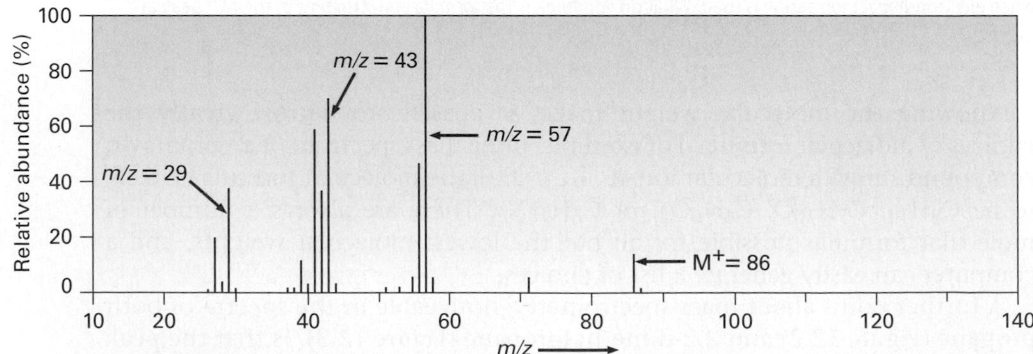

Figure 12.4 Mass spectrum of hexane (C_6H_{14}; MW = 86). The base peak is at m/z = 57, and numerous other ions are present.

Figure 12.5 shows how the hexane fragments might arise. The loss of a methyl radical from the hexane cation radical (M^+ = 86) gives rise to a fragment of mass 71; the loss of an ethyl radical accounts for a fragment of mass 57; the loss of a propyl radical accounts for a fragment of mass 43; and the loss of a butyl radical accounts for a fragment of mass 29. With practice, it's sometimes possible to analyze the fragmentation pattern of an unknown compound and work backward to a structure that is compatible with the data.

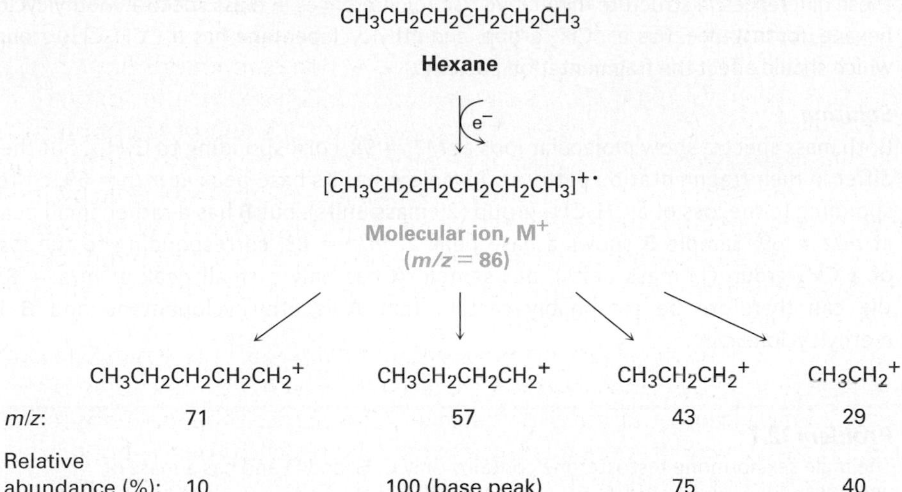

Figure 12.5 Fragmentation of hexane in a mass spectrometer.

We'll see in the next section and in later chapters that specific functional groups, such as alcohols, ketones, aldehydes, and amines, show specific kinds of mass spectral fragmentations that can be interpreted to provide structural information.

Using Mass Spectra to Identify Compounds

Worked Example 12.1

Assume that you have two unlabeled samples, one of methylcyclohexane and the other of ethylcyclopentane. How could you use mass spectrometry to tell them apart? The mass spectra of both are shown in **Figure 12.6**.

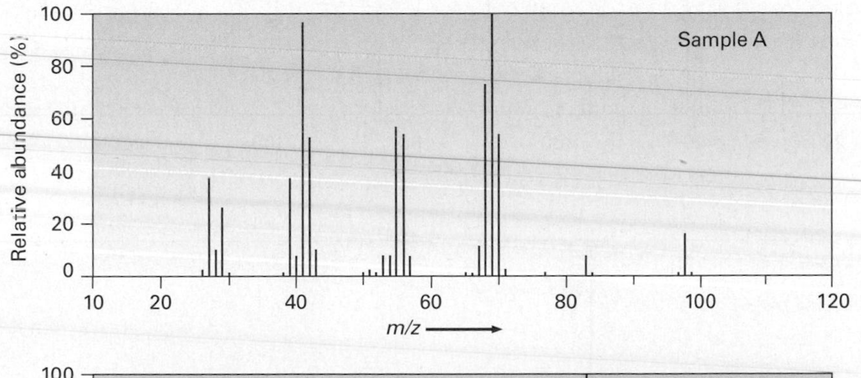

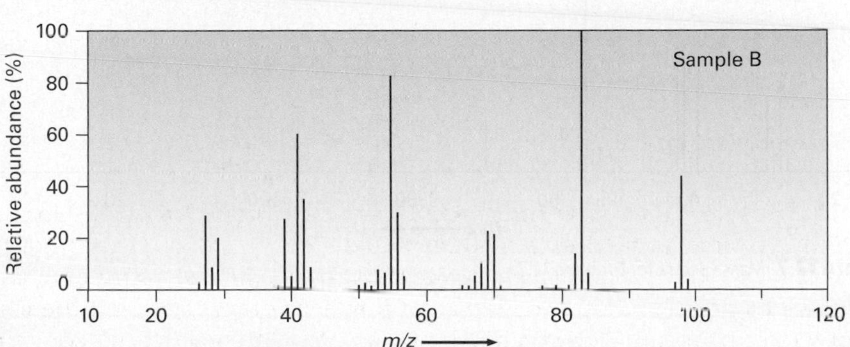

Figure 12.6 Mass spectra of unlabeled samples **A** and **B** for Worked Example 12.1.

Strategy

Look at the possible structures and decide on how they differ. Then think about how any of these differences in structure might give rise to differences in mass spectra. Methylcyclohexane, for instance, has a $-CH_3$ group, and ethylcyclopentane has a $-CH_2CH_3$ group, which should affect the fragmentation patterns.

Solution

Both mass spectra show molecular ions at $M^+ = 98$, corresponding to C_7H_{14}, but they differ in their fragmentation patterns. Sample **A** has its base peak at $m/z = 69$, corresponding to the loss of a CH_2CH_3 group (29 mass units), but **B** has a rather small peak at $m/z = 69$. Sample **B** shows a base peak at $m/z = 83$, corresponding to the loss of a CH_3 group (15 mass units), but sample **A** has only a small peak at $m/z = 83$. We can therefore be reasonably certain that **A** is ethylcyclopentane and **B** is methylcyclohexane.

Problem 12.1

The male sex hormone testosterone contains only C, H, and O and has a mass of 288.2089 amu as determined by high-resolution mass spectrometry. What is the likely molecular formula of testosterone?

Problem 12.2

Two mass spectra are shown in **Figure 12.7**. One spectrum is that of 2-methyl-2-pentene; the other is of 2-hexene. Which is which? Explain.

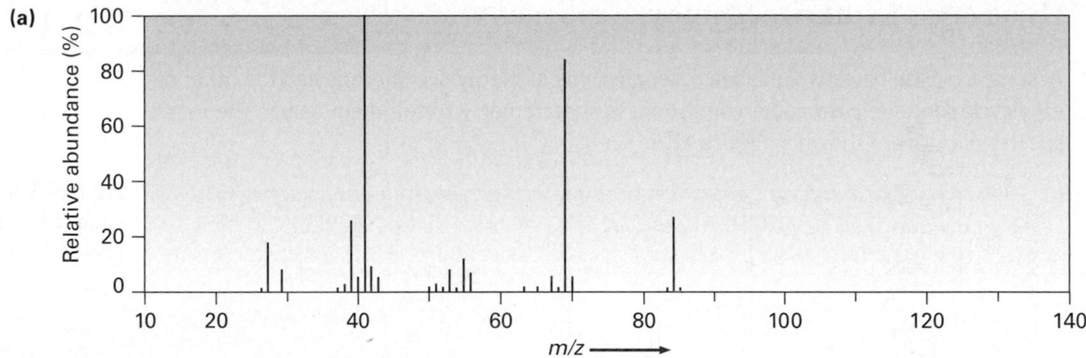

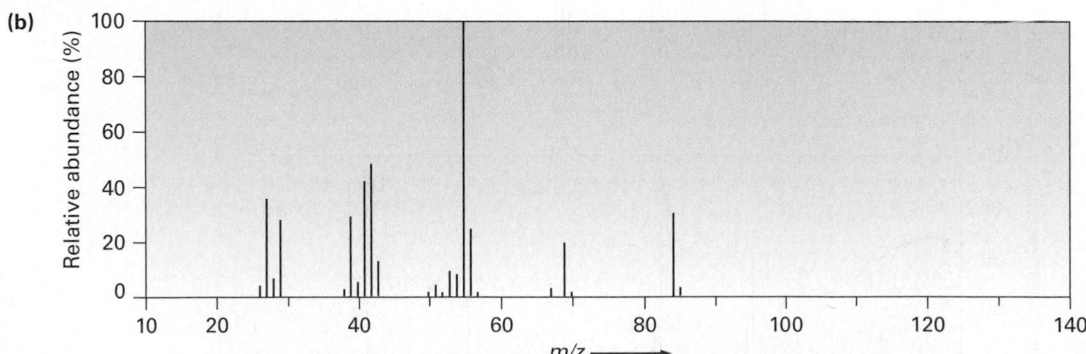

Figure 12.7 Mass spectra for Problem 12.2.

12.3 Mass Spectrometry of Some Common Functional Groups

As each functional group is discussed in future chapters, mass-spectral fragmentations characteristic of that group will be described. As a preview, though, we'll point out some distinguishing features of several common functional groups.

Alcohols

Alcohols undergo fragmentation in the mass spectrometer by two pathways: *alpha cleavage* and *dehydration.* In the α-cleavage pathway, a C–C bond nearest the hydroxyl group is broken, yielding a neutral radical plus a resonance-stabilized, oxygen-containing cation.

$$\left[RCH_2 \text{—} C \text{—} OH \right]^{+\cdot} \xrightarrow{\text{Alpha cleavage}} RCH_2\cdot + \left[:\ddot{O}H\text{—}C^+ \longleftrightarrow {}^+\ddot{O}H\text{=}C \right]$$

In the dehydration pathway, water is eliminated, yielding an alkene radical cation with a mass 18 units less than M^+.

$$\left[H\text{—}C\text{—}C\text{—}OH \right]^{+\cdot} \xrightarrow{\text{Dehydration}} H_2O + \left[C\text{=}C \right]^{+\cdot}$$

Amines

Aliphatic amines undergo a characteristic α cleavage in the mass spectrometer, similar to that observed for alcohols. A C–C bond nearest the nitrogen atom is broken, yielding an alkyl radical and a resonance-stabilized, nitrogen-containing cation.

$$\left[RCH_2 \text{—} C \text{—} NR_2 \right]^{+\cdot} \xrightarrow{\text{Alpha cleavage}} RCH_2\cdot + \left[:NR_2\text{—}C^+ \longleftrightarrow {}^+NR_2\text{=}C \right]$$

Carbonyl Compounds

Ketones and aldehydes that have a hydrogen on a carbon three atoms away from the carbonyl group undergo a characteristic mass-spectral cleavage called the *McLafferty rearrangement.* The hydrogen atom is transferred to the carbonyl oxygen, a C–C bond is broken, and a neutral alkene fragment is produced. The charge remains with the oxygen-containing fragment.

$$\left[H\text{—}C\text{—}C\text{—}C\text{—}C(\text{=}O)\text{—}R \right]^{+\cdot} \xrightarrow{\text{McLafferty rearrangement}} C\text{=}C + \left[H\text{—}O\text{—}C\text{=}C \right]^{+\cdot}$$

In addition, ketones and aldehydes frequently undergo α cleavage of the bond between the carbonyl group and the neighboring carbon. Alpha cleavage yields a neutral radical and a resonance-stabilized acyl cation.

$$\left[R-\overset{\overset{\large O}{\|}}{C}-R' \right]^{+\cdot} \xrightarrow[\text{cleavage}]{\text{Alpha}} R\cdot \;+\; \left[:\overset{..}{O}=C^{+}-R' \longleftrightarrow :O^{+}\equiv C-R' \right]$$

Worked Example 12.2 Identifying Fragmentation Patterns in a Mass Spectrum

The mass spectrum of 2-methyl-3-pentanol is shown in **Figure 12.8**. What fragments can you identify?

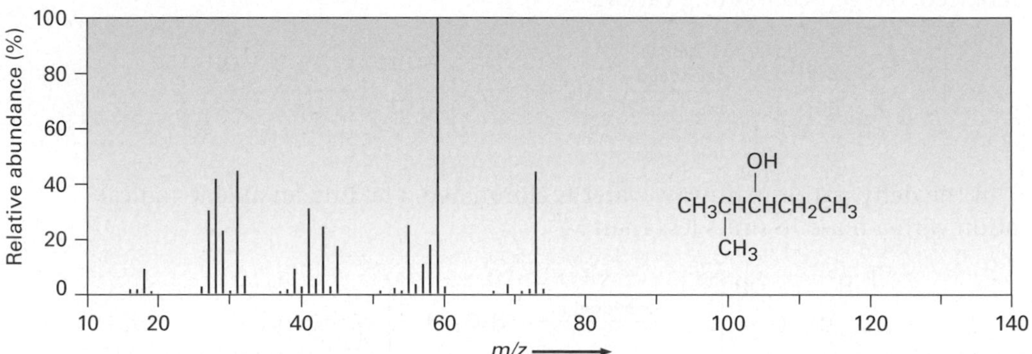

Figure 12.8 Mass spectrum of 2-methyl-3-pentanol, Worked Example 12.2.

Strategy

Calculate the mass of the molecular ion, and identify the functional groups in the molecule. Then write the fragmentation processes you might expect, and compare the masses of the resultant fragments with those peaks present in the spectrum.

Solution

2-Methyl-3-pentanol, an open-chain alcohol, has $M^+ = 102$ and might be expected to fragment by α cleavage and by dehydration. These processes would lead to fragment ions of $m/z = 84$, 73, and 59. Of the three expected fragments, dehydration is not observed (no $m/z = 84$ peak), but both α cleavages take place ($m/z = 73, 59$).

Loss of C_3H_7 ($M^+ - 43$) by alpha cleavage gives a peak of mass 59.

Loss of C_2H_5 ($M^+ - 29$) by alpha cleavage gives a peak of mass 73.

$M^+ = 102$ OH

Problem 12.3

What are the masses of the charged fragments produced in the following cleavage pathways?

(a) Alpha cleavage of 2-pentanone ($CH_3COCH_2CH_2CH_3$)

(b) Dehydration of cyclohexanol (hydroxycyclohexane)

(c) McLafferty rearrangement of 4-methyl-2-pentanone [$CH_3COCH_2CH(CH_3)_2$]

(d) Alpha cleavage of triethylamine [$(CH_3CH_2)_3N$]

Problem 12.4

List the masses of the parent ion and of several fragments you might expect to find in the mass spectrum of the following molecule:

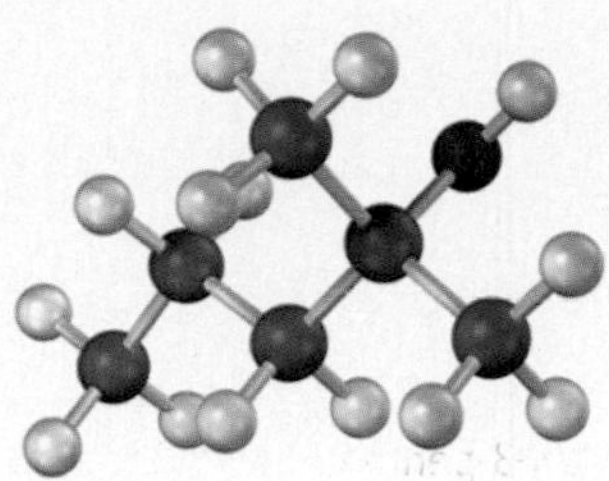

12.4 Mass Spectrometry in Biological Chemistry: Time-of-Flight (TOF) Instruments

Most biochemical analyses by MS use either electrospray ionization (ESI) or matrix-assisted laser desorption ionization (MALDI), typically linked to a time-of-flight (TOF) mass analyzer. Both ESI and MALDI are soft ionization methods that produce charged molecules with little fragmentation, even with biological samples of very high molecular weight.

In an ESI source, the sample is dissolved in a polar solvent and sprayed through a steel capillary tube. As it exits the tube, it is subjected to a high voltage that causes it to become protonated by removing one or more H^+ ions from the solvent. The volatile solvent is then evaporated, giving variably protonated sample molecules ($M+H_n{}^{n+}$). In a MALDI source, the sample is adsorbed onto a suitable matrix compound, such as 2,5-dihydroxybenzoic acid, which is ionized by a short burst of laser light. The matrix compound then transfers the energy to the sample and protonates it, forming $M+H_n{}^{n+}$ ions.

Following ion formation, the variably protonated sample molecules are electrically focused into a small packet with a narrow spatial distribution, and the packet is given a sudden kick of energy by an accelerator electrode. Since each molecule in the packet is given the same energy, $E = mv^2/2$, it begins moving with a velocity that depends on the square root of its mass, $v = \sqrt{2E/m}$. Lighter molecules move faster, and heavier molecules move slower. The analyzer itself—the *drift tube*—is simply an electrically grounded metal tube inside which the different charged molecules become separated as they move along at different velocities and take different amounts of time to complete their flight.

The TOF technique is considerably more sensitive than the magnetic sector alternative, and protein samples of up to 100 kilodaltons (100,000 amu) can be separated with a mass accuracy of 3 ppm. **Figure 12.9** shows a MALDI–TOF spectrum of chicken egg-white lysozyme, MW = 14,306.7578 daltons. (Biochemists generally use the unit *dalton,* abbreviated Da, instead of amu.)

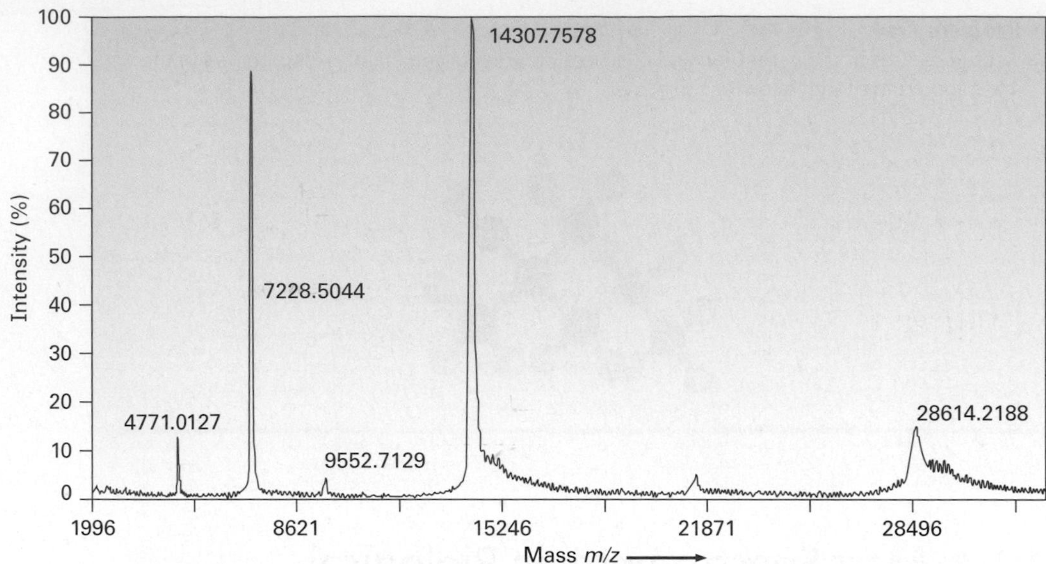

Figure 12.9 MALDI–TOF mass spectrum of chicken egg-white lysozyme. The peak at 14,307.7578 daltons (amu) is due to the monoprotonated protein, $M+H^+$, and the peak at 28,614.2188 daltons is due to an impurity formed by dimerization of the protein. Other peaks are various protonated species, $M+H_n{}^{n+}$.

12.5 Spectroscopy and the Electromagnetic Spectrum

Infrared, ultraviolet, and nuclear magnetic resonance spectroscopies differ from mass spectrometry in that they are nondestructive and involve the interaction of molecules with electromagnetic energy rather than with an ionizing source. Before beginning a study of these techniques, however, let's briefly review the nature of radiant energy and the electromagnetic spectrum.

Visible light, X rays, microwaves, radio waves, and so forth are all different kinds of *electromagnetic radiation.* Collectively, they make up the **electromagnetic spectrum**, shown in **Figure 12.10**. The electromagnetic spectrum is arbitrarily divided into regions, with the familiar visible region accounting for only a small portion, from 3.8×10^{-7} m to 7.8×10^{-7} m in wavelength. The visible region is flanked by the infrared and ultraviolet regions.

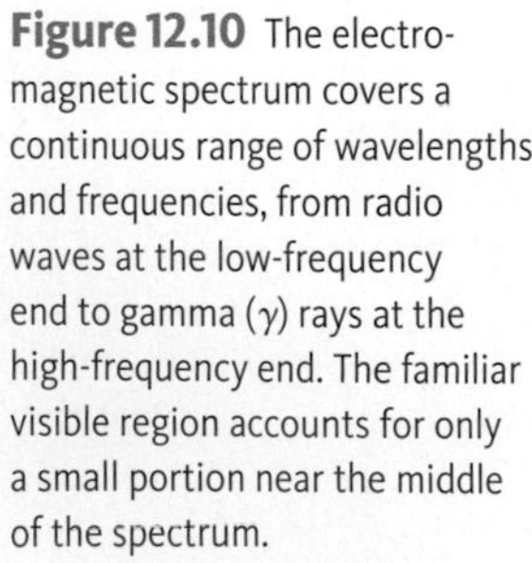

Figure 12.10 The electromagnetic spectrum covers a continuous range of wavelengths and frequencies, from radio waves at the low-frequency end to gamma (γ) rays at the high-frequency end. The familiar visible region accounts for only a small portion near the middle of the spectrum.

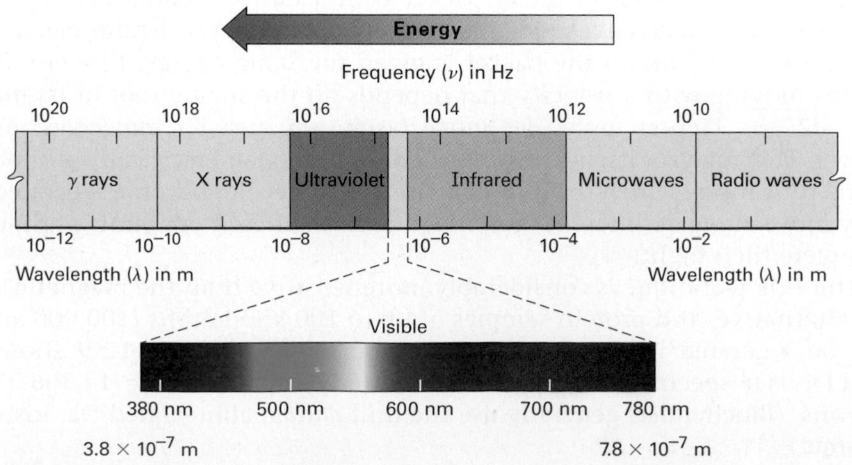

Electromagnetic radiation is often said to have dual behavior. In some respects, it has the properties of a particle, called a *photon,* yet in other respects it behaves as an energy wave. Like all waves, electromagnetic radiation is characterized by a *wavelength,* a *frequency,* and an *amplitude* **(Figure 12.11)**. The **wavelength, λ** (Greek lambda), is the distance from one wave maximum to the next. The **frequency, ν** (Greek nu), is the number of waves that pass by a fixed point per unit time, usually given in reciprocal seconds (s^{-1}), or **hertz, Hz** (1 Hz = 1 s^{-1}). The **amplitude** is the height of a wave, measured from midpoint to peak. The intensity of radiant energy, whether a feeble glow or a blinding glare, is proportional to the square of the wave's amplitude.

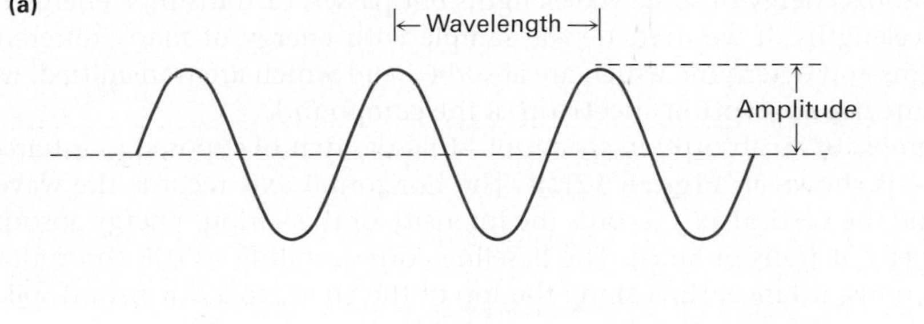

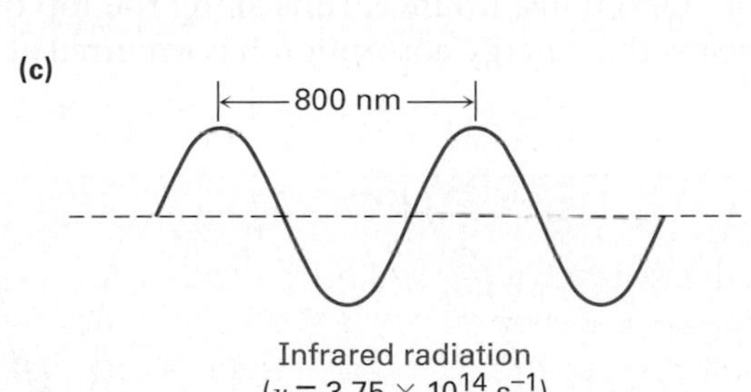

Figure 12.11 Electromagnetic waves are characterized by a wavelength, a frequency, and an amplitude. **(a)** Wavelength (λ) is the distance between two successive wave maxima. Amplitude is the height of the wave measured from the center. **(b)–(c)** What we perceive as different kinds of electromagnetic radiation are simply waves with different wavelengths and frequencies.

Multiplying the wavelength of a wave in meters (m) by its frequency in reciprocal seconds (s^{-1}) gives the speed of the wave in meters per second (m/s). The rate of travel of all electromagnetic radiation in a vacuum is a constant value, commonly called the "speed of light" and abbreviated *c.* Its numerical value is defined as exactly 2.997 924 58 × 10^8 m/s, usually rounded off to 3.00 × 10^8 m/s.

$$\text{Wavelength} \times \text{Frequency} = \text{Speed}$$

$$\lambda\ (\text{m}) \times \nu\ (\text{s}^{-1}) = c\ (\text{m/s})$$

$$\lambda = \frac{c}{\nu} \quad \text{or} \quad \nu = \frac{c}{\lambda}$$

Just as matter comes only in discrete units called atoms, electromagnetic energy is transmitted only in discrete amounts called *quanta.* The amount of energy ϵ corresponding to 1 quantum of energy (1 photon) of a given frequency ν is expressed by the Planck equation

$$\varepsilon = h\nu = \frac{hc}{\lambda}$$

where h = Planck's constant (6.62 × 10^{-34} J · s = 1.58 × 10^{-34} cal · s).

The Planck equation says that the energy of a given photon varies directly with its frequency ν but inversely with its wavelength λ. High frequencies and short wavelengths correspond to high-energy radiation such as gamma rays; low frequencies and long wavelengths correspond to low-energy radiation such as radio waves. Multiplying ϵ by Avogadro's number N_A gives the same equation in more familiar units, where E represents the energy of Avogadro's number (one "mole") of photons of wavelength λ:

$$E = \frac{N_A hc}{\lambda} = \frac{1.20 \times 10^{-4}\ \text{kJ/mol}}{\lambda\ (\text{m})} \quad \text{or} \quad \frac{2.86 \times 10^{-5}\ \text{kcal/mol}}{\lambda\ (\text{m})}$$

When an organic compound is exposed to a beam of electromagnetic radiation, it absorbs energy of some wavelengths but passes, or transmits, energy of other wavelengths. If we irradiate the sample with energy of many different wavelengths and determine which are absorbed and which are transmitted, we can measure the **absorption spectrum** of the compound.

An example of an absorption spectrum—that of ethanol exposed to infrared radiation—is shown in **Figure 12.12**. The horizontal axis records the wavelength, and the vertical axis records the intensity of the various energy absorptions in percent transmittance. The baseline corresponding to 0% absorption (or 100% transmittance) runs along the top of the chart, so a downward spike means that energy absorption has occurred at that wavelength.

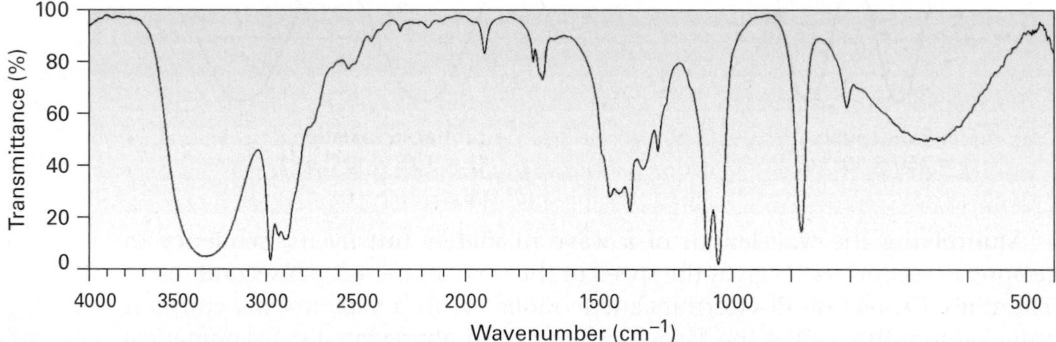

Figure 12.12 An infrared absorption spectrum of ethanol, CH_3CH_2OH. A transmittance of 100% means that all the energy is passing through the sample, whereas a lower transmittance means that some energy is being absorbed. Thus, each downward spike corresponds to an energy absorption.

The energy a molecule gains when it absorbs radiation must be distributed over the molecule in some way. With infrared radiation, the absorbed energy causes bonds to stretch and bend more vigorously. With ultraviolet radiation, the energy causes an electron to jump from a lower-energy orbital to a higher-energy one. Different radiation frequencies affect molecules in different ways, but each provides structural information when the results are interpreted.

There are many kinds of spectroscopies, which differ according to the region of the electromagnetic spectrum used. We'll look at three: infrared spectroscopy, ultraviolet spectroscopy, and nuclear magnetic resonance spectroscopy. Let's begin by seeing what happens when an organic sample absorbs infrared energy.

Correlating Energy and Frequency of Radiation

Worked Example 12.3

Which is higher in energy, FM radio waves with a frequency of 1.015×10^8 Hz (101.5 MHz) or visible green light with a frequency of 5×10^{14} Hz?

Strategy

Remember the equations $\epsilon = h\nu$ and $\epsilon = hc/\lambda$, which say that energy increases as frequency increases and as wavelength decreases.

Solution

Since visible light has a higher frequency than radio waves, it is higher in energy.

Problem 12.5

Which has higher energy, infrared radiation with $\lambda = 1.0 \times 10^{-6}$ m or an X ray with $\lambda = 3.0 \times 10^{-9}$ m? Radiation with $\nu = 4.0 \times 10^9$ Hz or with $\lambda = 9.0 \times 10^{-6}$ m?

Problem 12.6

It's useful to develop a feeling for the amounts of energy that correspond to different parts of the electromagnetic spectrum. Calculate the energies in kJ/mol of each of the following kinds of radiation:

(a) A gamma ray with $\lambda = 5.0 \times 10^{-11}$ m
(b) An X ray with $\lambda = 3.0 \times 10^{-9}$ m
(c) Ultraviolet light with $\nu = 6.0 \times 10^{15}$ Hz
(d) Visible light with $\nu = 7.0 \times 10^{14}$ Hz
(e) Infrared radiation with $\lambda = 2.0 \times 10^{-5}$ m
(f) Microwave radiation with $\nu = 1.0 \times 10^{11}$ Hz

12.6 Infrared Spectroscopy

The **infrared (IR)** region of the electromagnetic spectrum covers the range from just above the visible (7.8×10^{-7} m) to approximately 10^{-4} m, but only the midportion from 2.5×10^{-6} m to 2.5×10^{-5} m is used by organic chemists (**Figure 12.13**). Wavelengths within the IR region are usually given in micrometers ($1\ \mu m = 10^{-6}$ m), and frequencies are given in wavenumbers rather than in hertz. The **wavenumber** ($\tilde{\nu}$) is the reciprocal of the wavelength in centimeters and is therefore expressed in units of cm^{-1}.

$$\text{Wavenumber:} \quad \tilde{\nu}\ (cm^{-1}) = \frac{1}{\lambda\ (cm)}$$

Thus, the useful IR region is from 4000 to 400 cm^{-1}, corresponding to energies of 48.0 kJ/mol to 4.80 kJ/mol (11.5–1.15 kcal/mol).

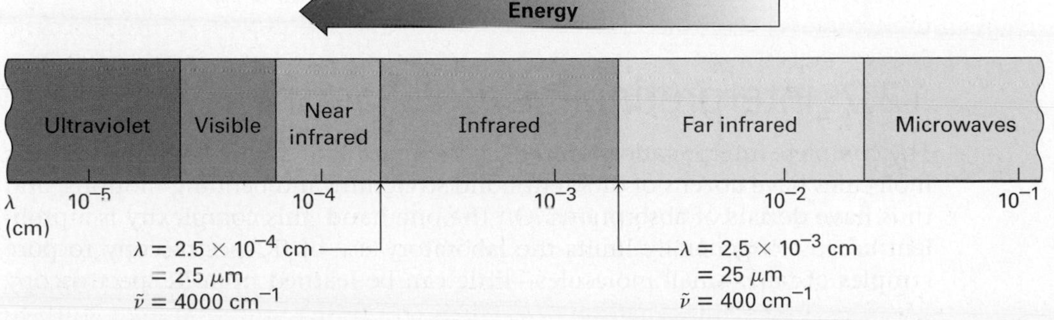

Figure 12.13 The infrared and adjacent regions of the electromagnetic spectrum.

Why does an organic molecule absorb some wavelengths of IR radiation but not others? All molecules have a certain amount of energy and are in constant motion. Their bonds stretch and contract, atoms wag back and forth, and other molecular vibrations occur. Some of the kinds of allowed vibrations are shown below:

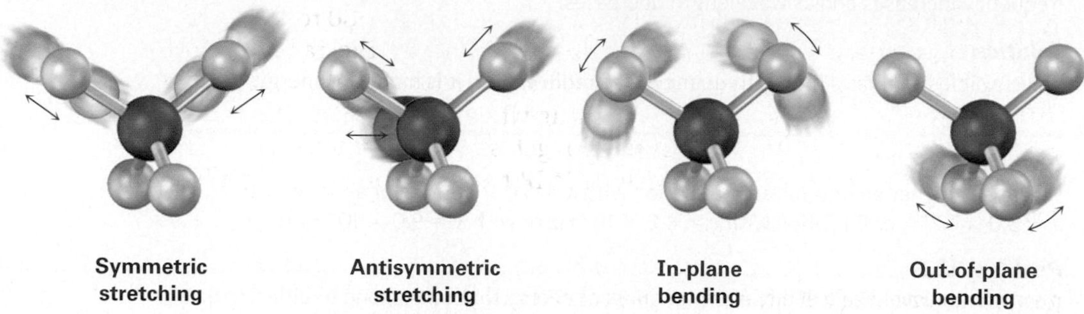

The amount of energy a molecule contains is not continuously variable but is *quantized*. That is, a molecule can stretch or bend only at specific frequencies corresponding to specific energy levels. Take bond stretching, for example. Although we usually speak of bond lengths as if they were fixed, the numbers given are really averages. In fact, a typical C–H bond with an average bond length of 110 pm is actually vibrating at a specific frequency, alternately stretching and contracting as if there were a spring connecting the two atoms.

When a molecule is irradiated with electromagnetic radiation, energy is absorbed if the frequency of the radiation matches the frequency of the vibration. The result of this energy absorption is an increased amplitude for the vibration; in other words, the "spring" connecting the two atoms stretches and compresses a bit further. Since each frequency absorbed by a molecule corresponds to a specific molecular motion, we can find what kinds of motions a molecule has by measuring its IR spectrum. By then interpreting those motions, we can find out what kinds of bonds (functional groups) are present in the molecule.

IR spectrum → What molecular motions? → What functional groups?

12.7 Interpreting Infrared Spectra

The complete interpretation of an IR spectrum is difficult because most organic molecules have dozens of different bond stretching and bending motions, and thus have dozens of absorptions. On the one hand, this complexity is a problem because it generally limits the laboratory use of IR spectroscopy to pure samples of fairly small molecules—little can be learned from IR spectroscopy about large, complex biomolecules. On the other hand, the complexity is useful because an IR spectrum acts as a unique fingerprint of a compound. In fact, the complex region of the IR spectrum from 1500 cm^{-1} to around 400 cm^{-1} is called the *fingerprint region.* If two samples have identical IR spectra, they are almost certainly identical compounds.

Fortunately, we don't need to interpret an IR spectrum fully to get useful structural information. Most functional groups have characteristic IR absorption bands that don't change from one compound to another. The C=O absorption of a ketone is almost always in the range 1680 to 1750 cm^{-1}; the O–H absorption of an alcohol is almost always in the range 3400 to 3650 cm^{-1}; the C=C absorption of an alkene is almost always in the range 1640 to 1680 cm^{-1}; and so forth. By learning where characteristic functional-group absorptions occur, it's possible to get structural information from IR spectra. Table 12.1 lists the characteristic IR bands of some common functional groups.

Table 12.1 Characteristic IR Absorptions of Some Functional Groups

Functional Group	Absorption (cm^{-1})	Intensity
Alkane		
C–H	2850–2960	Medium
Alkene		
=C–H	3020–3100	Medium
C=C	1640–1680	Medium
Alkyne		
≡C–H	3300	Strong
C≡C	2100–2260	Medium
Alkyl halide		
C–Cl	600–800	Strong
C–Br	500–600	Strong
Alcohol		
O–H	3400–3650	Strong, broad
C–O	1050–1150	Strong
Arene		
C–H	3030	Weak
Aromatic ring	1660–2000	Weak
	1450–1600	Medium
Amine		
N–H	3300–3500	Medium
C–N	1030–1230	Medium
Carbonyl compound		
C=O	1670–1780	Strong
Aldehyde	1730	Strong
Ketone	1715	Strong
Ester	1735	Strong
Amide	1690	Strong
Carboxylic acid	1710	Strong
Carboxylic acid		
O–H	2500–3100	Strong, broad
Nitrile		
C≡N	2210–2260	Medium
Nitro		
NO_2	1540	Strong

Look at the IR spectra of hexane, 1-hexene, and 1-hexyne in **Figure 12.14** to see an example of how IR spectroscopy can be used. Although all three IR spectra contain many peaks, there are characteristic absorptions of the C=C and C≡C functional groups that allow the three compounds to be distinguished. Thus, 1-hexene shows a characteristic C=C absorption at 1660 cm^{-1} and a vinylic =C–H absorption at 3100 cm^{-1}, whereas 1-hexyne has a C≡C absorption at 2100 cm^{-1} and a terminal alkyne ≡C–H absorption at 3300 cm^{-1}.

(a)

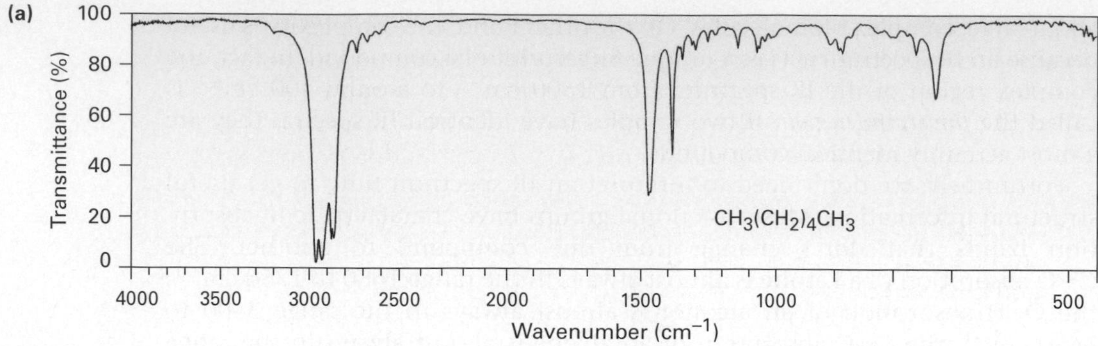

(b)

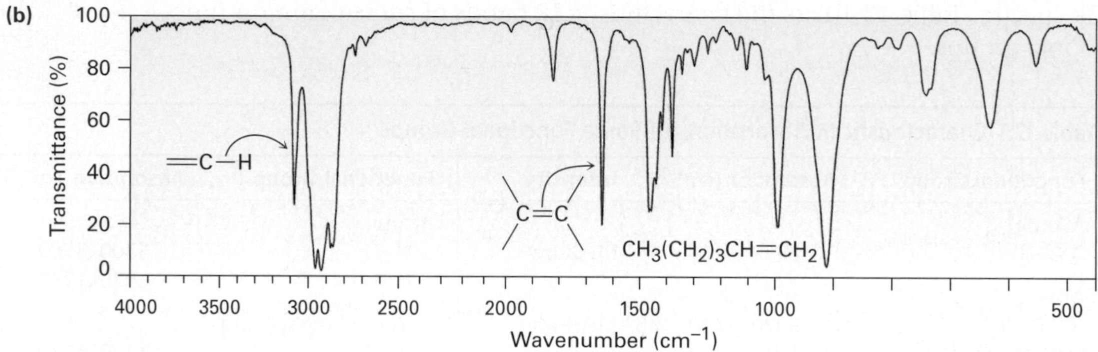

(c)

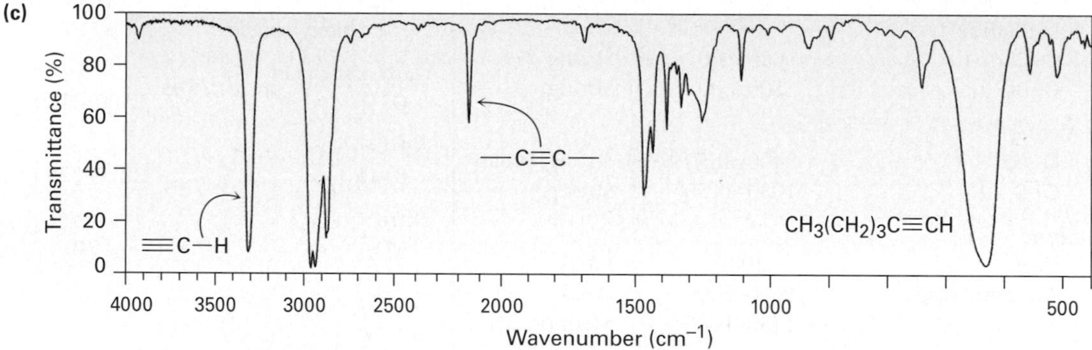

Figure 12.14 IR spectra of **(a)** hexane, **(b)** 1-hexene, and **(c)** 1-hexyne. Spectra like these are easily obtained on submilligram amounts of material in a few minutes using commercially available instruments.

It helps in remembering the position of specific IR absorptions to divide the IR region from 4000 cm^{-1} to 400 cm^{-1} into four parts, as shown in **Figure 12.15**.

- The region from 4000 to 2500 cm^{-1} corresponds to absorptions caused by N–H, C–H, and O–H single-bond stretching motions. N–H and O–H bonds absorb in the 3300 to 3600 cm^{-1} range; C–H bond stretching occurs near 3000 cm^{-1}.

- The region from 2500 to 2000 cm^{-1} is where triple-bond stretching occurs. Both C≡N and C≡C bonds absorb here.
- The region from 2000 to 1500 cm^{-1} is where double bonds (C=O, C=N, and C=C) absorb. Carbonyl groups generally absorb in the range 1680 to 1750 cm^{-1}, and alkene stretching normally occurs in the narrow range 1640 to 1680 cm^{-1}.
- The region below 1500 cm^{-1} is the fingerprint portion of the IR spectrum. A large number of absorptions due to a variety of C–C, C–O, C–N, and C–X single-bond vibrations occur here.

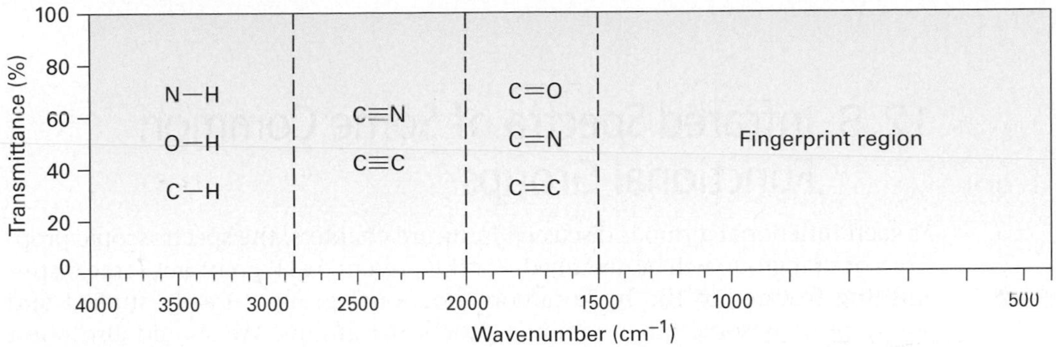

Figure 12.15 The four regions of the infrared spectrum: single bonds to hydrogen, triple bonds, double bonds, and fingerprint.

Why do different functional groups absorb where they do? As noted previously, a good analogy is that of two weights (atoms) connected by a spring (a bond). Short, strong bonds vibrate at a higher energy and higher frequency than do long, weak bonds, just as a short, strong spring vibrates faster than a long, weak spring. Thus, triple bonds absorb at a higher frequency than double bonds, which in turn absorb at a higher frequency than single bonds. In addition, springs connecting small weights vibrate faster than springs connecting large weights. Thus, C–H, O–H, and N–H bonds vibrate at a higher frequency than bonds between heavier C, O, and N atoms.

Distinguishing Isomeric Compounds by IR Spectroscopy

Worked Example 12.4

Acetone (CH_3COCH_3) and 2-propen-1-ol ($H_2C{=}CHCH_2OH$) are isomers. How could you distinguish them by IR spectroscopy?

Strategy

Identify the functional groups in each molecule, and refer to Table 12.1.

Solution

Acetone has a strong C=O absorption at 1715 cm^{-1}, while 2-propen-1-ol has an –OH absorption at 3500 cm^{-1} and a C=C absorption at 1660 cm^{-1}.

Problem 12.7

What functional groups might the following molecules contain?

(a) A compound with a strong absorption at 1710 cm^{-1}

(b) A compound with a strong absorption at 1540 cm^{-1}

(c) A compound with strong absorptions at 1720 cm^{-1} and at 2500–3100 cm^{-1}

Problem 12.8

How might you use IR spectroscopy to distinguish between the following pairs of isomers?

(a) CH_3CH_2OH and CH_3OCH_3 **(b)** Cyclohexane and 1-hexene

(c) $CH_3CH_2CO_2H$ and $HOCH_2CH_2CHO$

12.8 Infrared Spectra of Some Common Functional Groups

As each functional group is discussed in future chapters, the spectroscopic properties of that group will be described. For the present, we'll point out some distinguishing features of the hydrocarbon functional groups already studied and briefly preview some other common functional groups. We should also point out, however, that in addition to interpreting absorptions that *are* present in an IR spectrum, it's also possible to get structural information by noticing which absorptions are *not* present. If the spectrum of a compound has no absorptions at 3300 and 2150 cm^{-1}, the compound is not a terminal alkyne; if the spectrum has no absorption near 3400 cm^{-1}, the compound is not an alcohol; and so on.

Alkanes

The IR spectrum of an alkane is fairly uninformative because no functional groups are present and all absorptions are due to C–H and C–C bonds. Alkane C–H bonds show a strong absorption from 2850 to 2960 cm^{-1}, and saturated C–C bonds show a number of bands in the 800 to 1300 cm^{-1} range. Since most organic compounds contain saturated alkane-like portions, most organic compounds have these characteristic IR absorptions. The C–H and C–C bands are clearly visible in the three spectra shown in Figure 12.14.

Alkanes —C—H 2850–2960 cm^{-1}

—C—C— 800–1300 cm^{-1}

Alkenes

Alkenes show several characteristic stretching absorptions. Vinylic =C–H bonds absorb from 3020 to 3100 cm^{-1}, and alkene C=C bonds usually absorb near 1650 cm^{-1}, although in some cases the peaks can be rather small and difficult to see clearly. Both absorptions are visible in the 1-hexene spectrum in Figure 12.14b.

Monosubstituted and disubstituted alkenes have characteristic =C–H out-of-plane bending absorptions in the 700 to 1000 cm^{-1} range, thereby allowing the substitution pattern on a double bond to be determined. Monosubstituted alkenes such as 1-hexene show strong characteristic bands at 910 and 990 cm^{-1}, and 2,2-disubstituted alkenes ($R_2C{=}CH_2$) have an intense band at 890 cm^{-1}.

Alkenes	=C–H	3020–3100 cm^{-1}
	C=C	1640–1680 cm^{-1}
	$RCH{=}CH_2$	910 and 990 cm^{-1}
	$R_2C{=}CH_2$	890 cm^{-1}

Alkynes

Alkynes show a C≡C stretching absorption at 2100 to 2260 cm^{-1}, an absorption that is much more intense for terminal alkynes than for internal alkynes. In fact, symmetrically substituted triple bonds like that in 3-hexyne show no absorption at all, for reasons we won't go into. Terminal alkynes such as 1-hexyne also have a characteristic ≡C–H stretch at 3300 cm^{-1} (Figure 12.14c). This band is diagnostic for terminal alkynes because it is fairly intense and quite sharp.

Alkynes	–C≡C–	2100–2260 cm^{-1}
	≡C–H	3300 cm^{-1}

Aromatic Compounds

Aromatic compounds, such as benzene, have a weak C–H stretching absorption at 3030 cm^{-1}, just to the left of a typical saturated C–H band. In addition, they have a series of weak absorptions in the 1660 to 2000 cm^{-1} range and a series of medium-intensity absorptions in the 1450 to 1600 cm^{-1} region. These latter absorptions are due to complex molecular motions of the entire ring. The IR spectrum of phenylacetylene, shown in Figure 12.17 at the end of this section, gives an example.

Aromatic compounds	C–H	3030 cm^{-1} (weak)
	(benzene ring)	1660–2000 cm^{-1} (weak) 1450–1600 cm^{-1} (medium)

Alcohols

The O–H functional group of alcohols is easy to spot. Alcohols have a characteristic band in the range 3400 to 3650 cm^{-1} that is usually broad and intense. If present, it's hard to miss this band or to confuse it with anything else.

Alcohols	–O–H	3400–3650 cm^{-1} (broad, intense)

Amines

The N–H functional group of amines is also easy to spot in the IR, with a characteristic absorption in the 3300 to 3500 cm^{-1} range. Although alcohols absorb in the same range, an N–H absorption is much sharper and less intense than an O–H band.

Amines —N—H 3300–3500 cm^{-1} (sharp, medium intensity)

Carbonyl Compounds

Carbonyl functional groups are the easiest to identify of all IR absorptions because of their sharp, intense peak in the range 1670 to 1780 cm^{-1}. Most important, the exact position of absorption within the range can often be used to identify the exact kind of carbonyl functional group—aldehyde, ketone, ester, and so forth.

Aldehydes Saturated aldehydes absorb at 1730 cm^{-1}; aldehydes next to either a double bond or an aromatic ring absorb at 1705 cm^{-1}.

Aldehydes CH_3CH_2CHO 1730 cm^{-1}; $CH_3CH{=}CHCHO$ 1705 cm^{-1}; benzaldehyde (C_6H_5CHO) 1705 cm^{-1}

Ketones Saturated open-chain ketones and six-membered cyclic ketones absorb at 1715 cm^{-1}, five-membered cyclic ketones absorb at 1750 cm^{-1}, and ketones next to a double bond or an aromatic ring absorb at 1690 cm^{-1}.

Ketones CH_3COCH_3 1715 cm^{-1}; cyclopentanone 1750 cm^{-1}; $CH_3CH{=}CHCOCH_3$ 1690 cm^{-1}; $C_6H_5COCH_3$ 1690 cm^{-1}

Esters Saturated esters absorb at 1735 cm^{-1}; esters next to either an aromatic ring or a double bond absorb at 1715 cm^{-1}.

Esters CH_3COOCH_3 1735 cm^{-1}; $CH_3CH{=}CHCOOCH_3$ 1715 cm^{-1}; $C_6H_5COOCH_3$ 1715 cm^{-1}

Predicting IR Absorptions of Compounds

Worked Example 12.5

Where might the following compounds have IR absorptions?

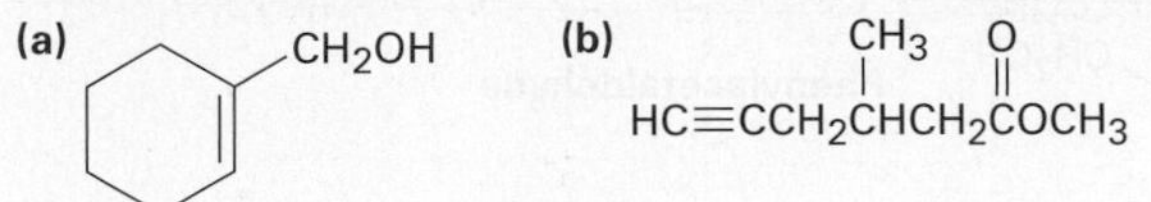

Strategy

Identify the functional groups in each molecule, and then check Table 12.1 to see where those groups absorb.

Solution

(a) *Absorptions:* 3400–3650 cm^{-1} (O–H), 3020–3100 cm^{-1} (=C–H), 1640–1680 cm^{-1} (C=C). This molecule has an alcohol O–H group and an alkene double bond.

(b) *Absorptions:* 3300 cm^{-1} (≡C–H), 2100–2260 cm^{-1} (C≡C), 1735 cm^{-1} (C=O). This molecule has a terminal alkyne triple bond and a saturated ester carbonyl group.

Identifying Functional Groups from an IR Spectrum

Worked Example 12.6

The IR spectrum of an unknown compound is shown in **Figure 12.16**. What functional groups does the compound contain?

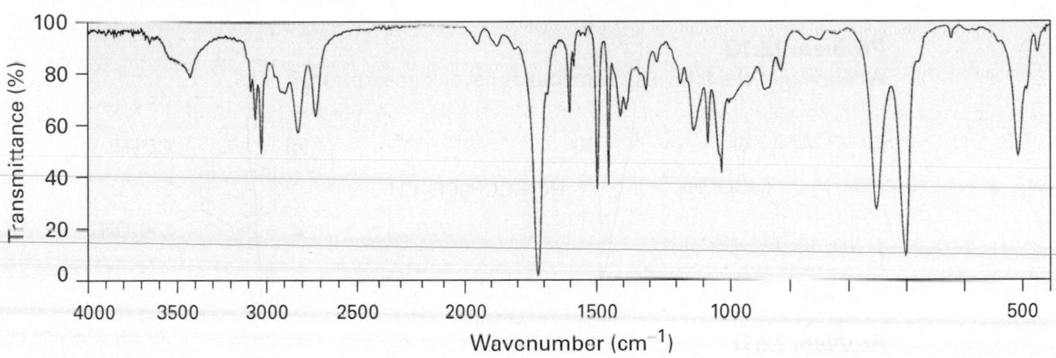

Figure 12.16 IR spectrum for Worked Example 12.6.

Strategy

All IR spectra have many absorptions, but those useful for identifying specific functional groups are usually found in the region from 1500 cm^{-1} to 3300 cm^{-1}. Pay particular attention to the carbonyl region (1670 to 1780 cm^{-1}), the aromatic region (1660 to 2000 cm^{-1}), the triple-bond region (2000 to 2500 cm^{-1}), and the C–H region (2500 to 3500 cm^{-1}).

Solution

The spectrum shows an intense absorption at 1725 cm^{-1} due to a carbonyl group (perhaps an aldehyde, –CHO), a series of weak absorptions from 1800 to 2000 cm^{-1} characteristic of aromatic compounds, and a C–H absorption near 3030 cm^{-1}, also characteristic of aromatic compounds. In fact, the compound is phenylacetaldehyde.

O
‖
CH_2CH **Phenylacetaldehyde**

Problem 12.9

The IR spectrum of phenylacetylene is shown in **Figure 12.17**. What absorption bands can you identify?

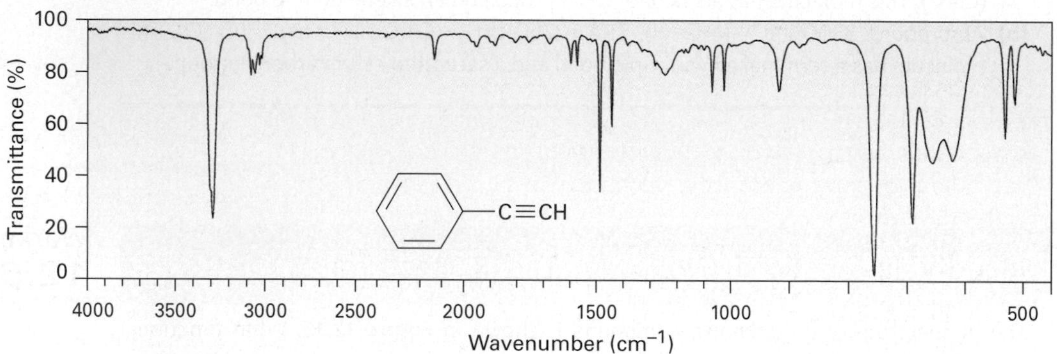

Figure 12.17 The IR spectrum of phenylacetylene, Problem 12.9.

Problem 12.10

Where might the following compounds have IR absorptions?

(a) O ‖ $COCH_3$

(b) O ‖ $HC{\equiv}CCH_2CH_2CH$

(c) CO_2H, CH_2OH

Problem 12.11

Where might the following compound have IR absorptions?

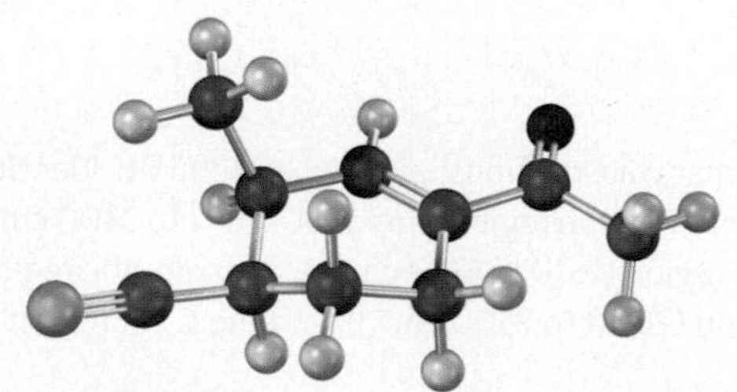

A DEEPER LOOK X-Ray Crystallography

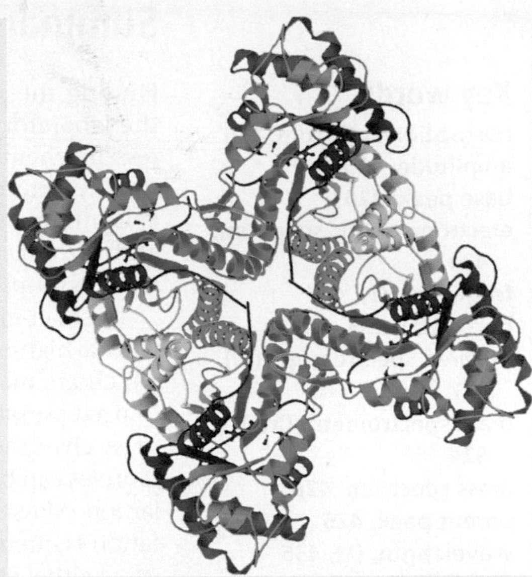

The structure of human muscle fructose-1,6-bisphosphate aldolase, as determined by X-ray crystallography and downloaded from the Protein Data Bank, 1ALD.

The various spectroscopic techniques described in this and the next two chapters are enormously important in chemistry and have been fine-tuned to such a degree that the structure of almost any molecule can be found. Nevertheless, wouldn't it be nice if you could simply look at a molecule and "see" its structure with your eyes?

Determining the three-dimensional shape of an object around you is easy—you just look at it, let your eyes focus the light rays reflected from the object, and let your brain assemble the data into a recognizable image. If the object is small, you use a microscope and let the microscope lens focus the visible light. Unfortunately, there is a limit to what you can see, even with the best optical microscope. Called the diffraction limit, you can't see anything smaller than the wavelength of light you are using for the observation. Visible light has wavelengths of several hundred nanometers, but atoms in molecules have dimension on the order of 0.1 nm. Thus, to "see" a molecule—whether a small one in the laboratory or a large, complex enzyme with a molecular weight in the tens of thousands—you need wavelengths in the 0.1 nm range, which corresponds to X rays.

Let's say that we want to determine the structure and shape of an enzyme or other biological molecule. The technique used is called *X-ray crystallography.* First, the molecule is crystallized (which often turns out to be the most difficult and time-consuming part of the entire process) and a small crystal with a dimension of 0.4 to 0.5 mm on its longest axis is glued to the end of a glass fiber. The fiber and attached crystal are then mounted in an instrument called an X-ray diffractometer, which consists of a radiation source, a sample positioning and orienting device that can rotate the crystal in any direction, a detector, and a controlling computer.

Once mounted in the diffractometer, the crystal is irradiated with X rays, usually so-called CuK_α radiation with a wavelength of 0.154 nm. When the X rays strike the enzyme crystal, they interact with electrons in the molecule and are scattered into a diffraction pattern which, when detected and visualized, appears as a series of intense spots against a null background.

Manipulation of the diffraction pattern to extract three-dimensional molecular data is a complex process, but the final result is that an electron-density map of the molecule is produced. Because electrons are largely localized around atoms, any two centers of electron density located within bonding distance of each other are assumed to represent bonded atoms, leading to a recognizable chemical structure. So important is this structural information for biochemistry that an online database of more than 66,000 biological substances has been created. Operated by Rutgers University and funded by the U.S. National Science Foundation, the Protein Data Bank (PDB) is a worldwide repository for processing and distributing three-dimensional structural data for biological macromolecules. We'll see how to access the PDB in the Chapter 26 *A Deeper Look.*

Summary

Key words

absorption spectrum, 436
amplitude, 435
base peak, 426
electromagnetic spectrum, 434
frequency, (ν), 435
hertz (Hz), 435
infrared spectroscopy (IR), 437
mass spectrometry (MS), 424
mass spectrum, 426
parent peak, 426
wavelength, (λ), 435
wavenumber ($\tilde{\nu}$), 437

Finding the structure of a new molecule, whether a small one synthesized in the laboratory or a large protein found in living organisms, is central to progress in chemistry and biochemistry. The structure of an organic molecule is usually determined using spectroscopic methods, including mass spectrometry and infrared spectroscopy. **Mass spectrometry (MS)** tells the molecular weight and formula of a molecule; **infrared (IR) spectroscopy** identifies the functional groups present in the molecule.

In small-molecule mass spectrometry, molecules are first ionized by collision with a high-energy electron beam. The ions then fragment into smaller pieces, which are magnetically sorted according to their mass-to-charge ratio (m/z). The ionized sample molecule is called the *molecular ion,* M^+, and measurement of its mass gives the molecular weight of the sample. Structural clues about unknown samples can be obtained by interpreting the fragmentation pattern of the molecular ion. Mass-spectral fragmentations are usually complex, however, and interpretation is often difficult. In biological mass spectrometry, molecules are protonated using either electrospray ionization (ESI) or matrix-assisted laser desorption ionization (MALDI), and the protonated molecules are separated by time-of-flight (TOF).

Infrared spectroscopy involves the interaction of a molecule with **electromagnetic radiation**. When an organic molecule is irradiated with infrared energy, certain **frequencies** are absorbed by the molecule. The frequencies absorbed correspond to the amounts of energy needed to increase the amplitude of specific molecular vibrations such as bond stretchings and bendings. Since every functional group has a characteristic combination of bonds, every functional group has a characteristic set of infrared absorptions. For example, the terminal alkyne ≡C–H bond absorbs IR radiation of 3300 cm^{-1} frequency, and the alkene C=C bond absorbs in the range 1640 to 1680 cm^{-1}. By observing which frequencies of infrared radiation are absorbed by a molecule and which are not, it's possible to determine the functional groups a molecule contains.

Exercises

OWL Interactive versions of these problems are assignable in OWL for Organic Chemistry.

Visualizing Chemistry

(Problems 12.1–12.11 appear within the chapter.)

12.12 Where in the IR spectrum would you expect each of the following molecules to absorb?

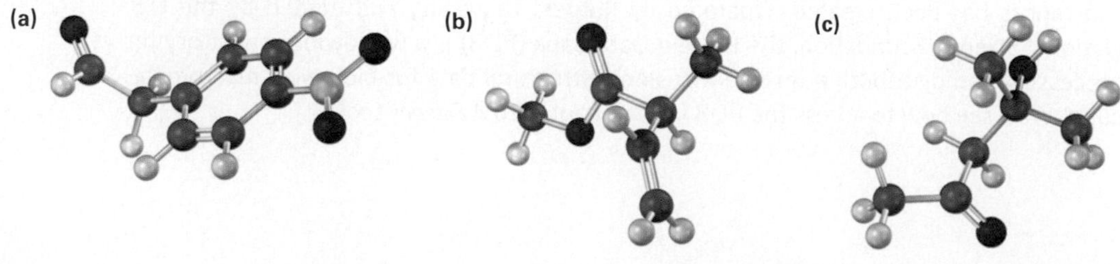

12.13 Show the structures of the likely fragments you would expect in the mass spectra of the following molecules:

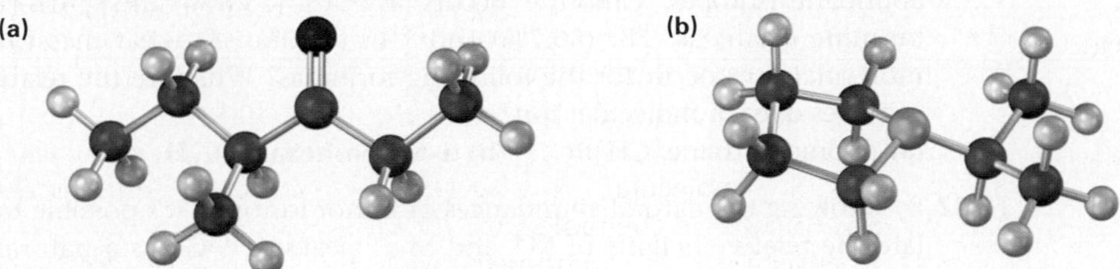

Additional Problems

Mass Spectrometry

12.14 Propose structures for compounds that fit the following mass-spectral data:
(a) A hydrocarbon with $M^+ = 132$ (b) A hydrocarbon with $M^+ = 166$
(c) A hydrocarbon with $M^+ = 84$

12.15 Write molecular formulas for compounds that show the following molecular ions in their high-resolution mass spectra, assuming that C, H, N, and O might be present. The exact atomic masses are: 1.007 83 (^1H) 12.000 00 (^{12}C), 14.003 07 (^{14}N), 15.994 91 (^{16}O).
(a) $M^+ = 98.0844$ (b) $M^+ = 123.0320$

12.16 Camphor, a saturated monoketone from the Asian camphor tree, is used among other things as a moth repellent and as a constituent of embalming fluid. If camphor has $M^+ = 152.1201$ by high-resolution mass spectrometry, what is its molecular formula? How many rings does camphor have?

12.17 The *nitrogen rule* of mass spectrometry says that a compound containing an odd number of nitrogens has an odd-numbered molecular ion. Conversely, a compound containing an even number of nitrogens has an even-numbered M^+ peak. Explain.

12.18 In light of the nitrogen rule mentioned in Problem 12.17, what is the molecular formula of pyridine, $M^+ = 79$?

12.19 Nicotine is a diamino compound isolated from dried tobacco leaves. Nicotine has two rings and $M^+ = 162.1157$ by high-resolution mass spectrometry. Give a molecular formula for nicotine, and calculate the number of double bonds.

12.20 The hormone cortisone contains C, H, and O, and shows a molecular ion at $M^+ = 360.1937$ by high-resolution mass spectrometry. What is the molecular formula of cortisone? (The degree of unsaturation of cortisone is 8.)

12.21 Halogenated compounds are particularly easy to identify by their mass spectra because both chlorine and bromine occur naturally as mixtures of two abundant isotopes. Chlorine occurs as ^{35}Cl (75.8%) and ^{37}Cl (24.2%); bromine occurs as ^{79}Br (50.7%) and ^{81}Br (49.3%). At what masses do the molecular ions occur for the following formulas? What are the relative percentages of each molecular ion?

(a) Bromomethane, CH_3Br **(b)** 1-Chlorohexane, $C_6H_{13}Cl$

12.22 By knowing the natural abundances of minor isotopes, it's possible to calculate the relative heights of M^+ and M+1 peaks. If ^{13}C has a natural abundance of 1.10%, what are the relative heights of the M^+ and M+1 peaks in the mass spectrum of benzene, C_6H_6?

12.23 Propose structures for compounds that fit the following data:

(a) A ketone with $M^+ = 86$ and fragments at $m/z = 71$ and $m/z = 43$

(b) An alcohol with $M^+ = 88$ and fragments at $m/z = 73$, $m/z = 70$, and $m/z = 59$

12.24 2-Methylpentane (C_6H_{14}) has the mass spectrum shown. Which peak represents M^+? Which is the base peak? Propose structures for fragment ions of $m/z = 71$, 57, 43, and 29. Why does the base peak have the mass it does?

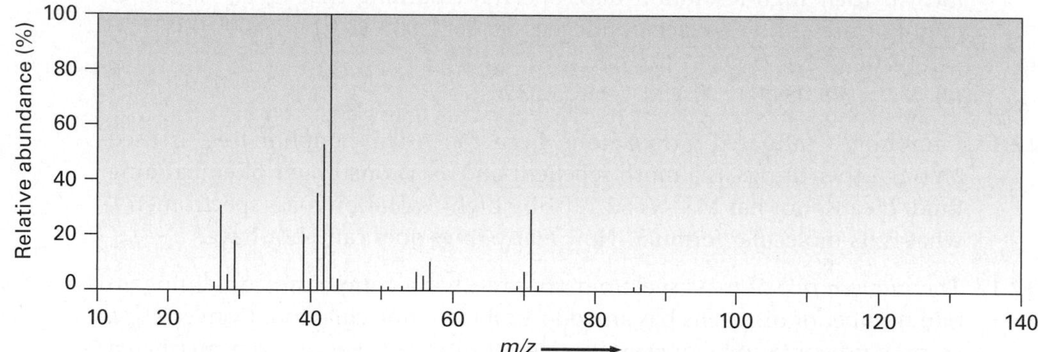

12.25 Assume that you are in a laboratory carrying out the catalytic hydrogenation of cyclohexene to cyclohexane. How could you use a mass spectrometer to determine when the reaction is finished?

12.26 What fragments might you expect in the mass spectra of the following compounds?

(a) **(b)** **(c)**

Infrared Spectroscopy

12.27 How might you use IR spectroscopy to distinguish among the three isomers 1-butyne, 1,3-butadiene, and 2-butyne?

12.28 Would you expect two enantiomers such as (*R*)-2-bromobutane and (*S*)-2-bromobutane to have identical or different IR spectra? Explain.

12.29 Would you expect two diastereomers such as *meso*-2,3-dibromobutane and (2*R*,3*R*)-dibromobutane to have identical or different IR spectra? Explain.

12.30 Propose structures for compounds that meet the following descriptions:

(a) C_5H_8, with IR absorptions at 3300 and 2150 cm^{-1}
(b) C_4H_8O, with a strong IR absorption at 3400 cm^{-1}
(c) C_4H_8O, with a strong IR absorption at 1715 cm^{-1}
(d) C_8H_{10}, with IR absorptions at 1600 and 1500 cm^{-1}

12.31 How could you use infrared spectroscopy to distinguish between the following pairs of isomers?

(a) $HC{\equiv}CCH_2NH_2$ and $CH_3CH_2C{\equiv}N$
(b) CH_3COCH_3 and CH_3CH_2CHO

12.32 Two infrared spectra are shown. One is the spectrum of cyclohexane, and the other is the spectrum of cyclohexene. Identify them, and explain your answer.

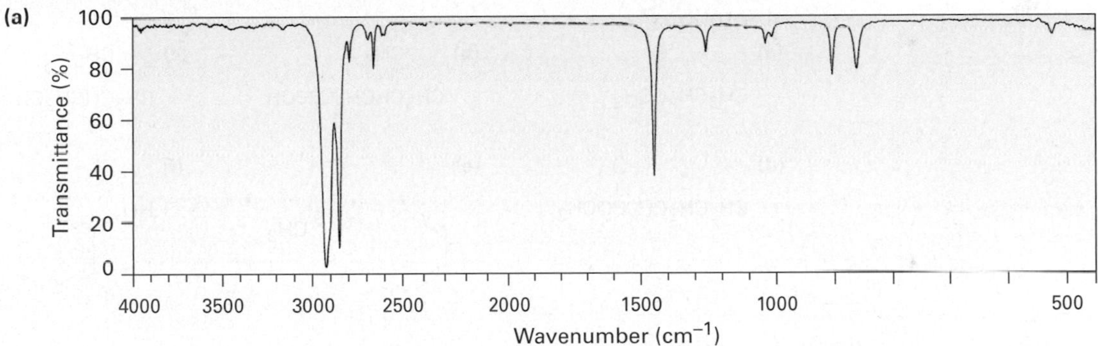

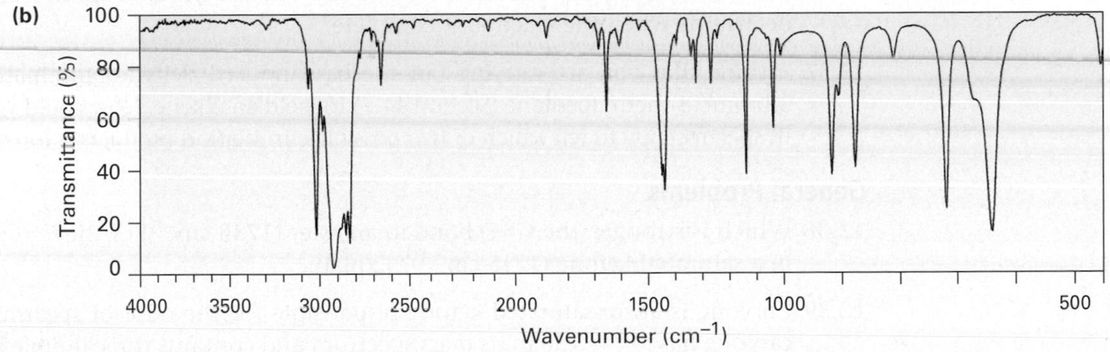

12.33 At what approximate positions might the following compounds show IR absorptions?

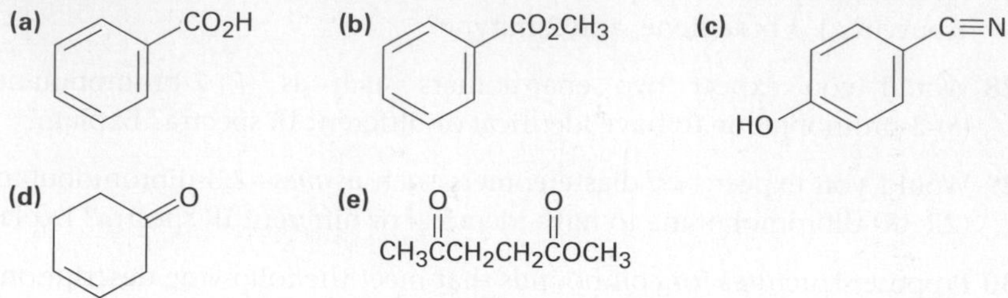

12.34 How would you use infrared spectroscopy to distinguish between the following pairs of constitutional isomers?

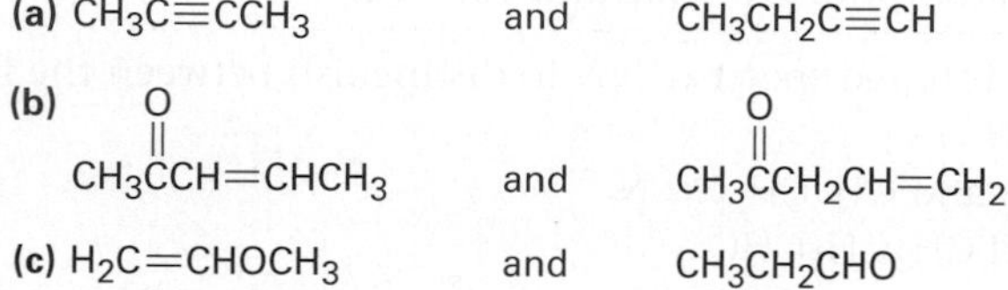

12.35 At what approximate positions might the following compounds show IR absorptions?

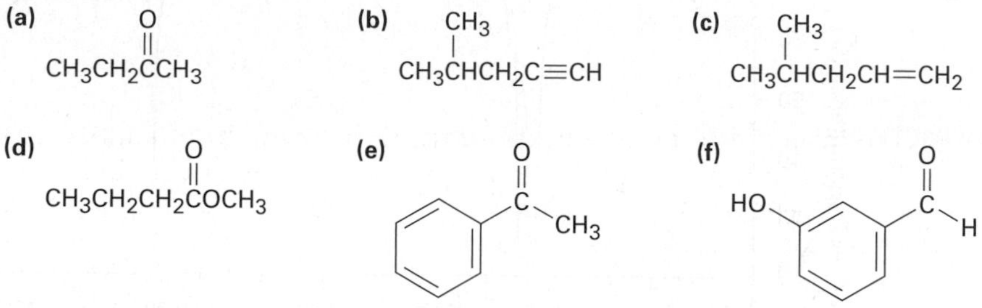

12.36 Assume that you are carrying out the dehydration of 1-methylcyclohexanol to yield 1-methylcyclohexene. How could you use infrared spectroscopy to determine when the reaction is complete?

12.37 Assume that you are carrying out the base-induced dehydrobromination of 3-bromo-3-methylpentane (Section 11.7) to yield an alkene. How could you use IR spectroscopy to tell which of two possible elimination products is formed?

General Problems

12.38 Which is stronger, the C=O bond in an ester (1735 cm^{-1}) or the C=O bond in a saturated ketone (1715 cm^{-1})? Explain.

12.39 Carvone is an unsaturated ketone responsible for the odor of spearmint. If carvone has $M^+ = 150$ in its mass spectrum and contains three double bonds and one ring, what is its molecular formula?

12.40 Carvone (Problem 12.39) has an intense infrared absorption at 1690 cm^{-1}. What kind of ketone does carvone contain?

12.41 The mass spectrum **(a)** and the infrared spectrum **(b)** of an unknown hydrocarbon are shown. Propose as many structures as you can.

(a)

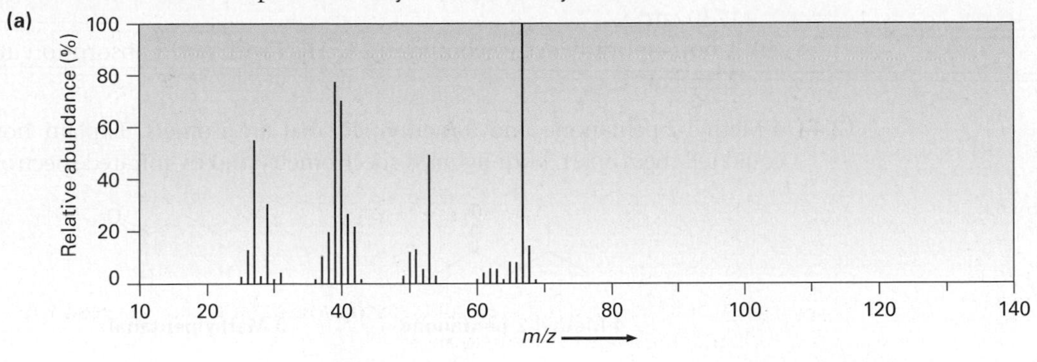

(b)

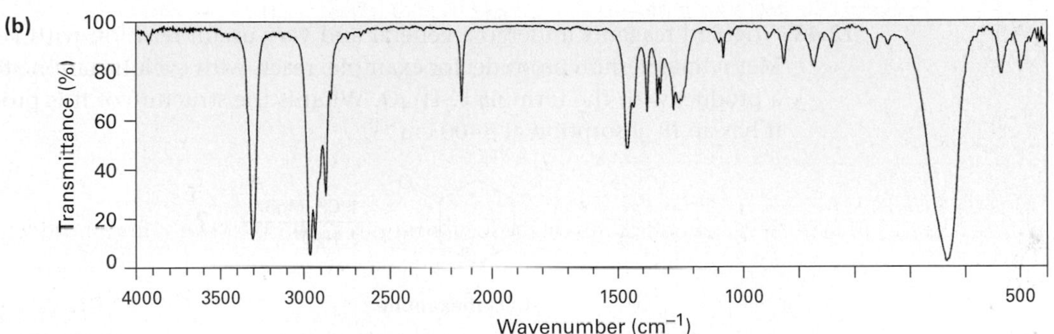

12.42 The mass spectrum **(a)** and the infrared spectrum **(b)** of another unknown hydrocarbon are shown. Propose as many structures as you can.

(a)

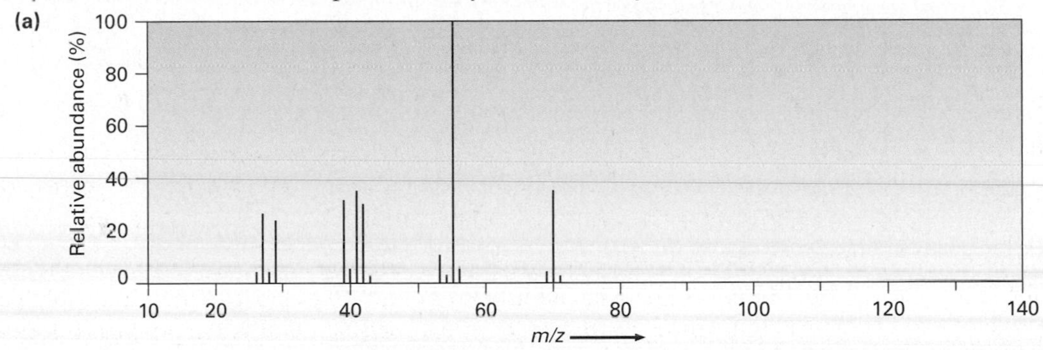

(b)

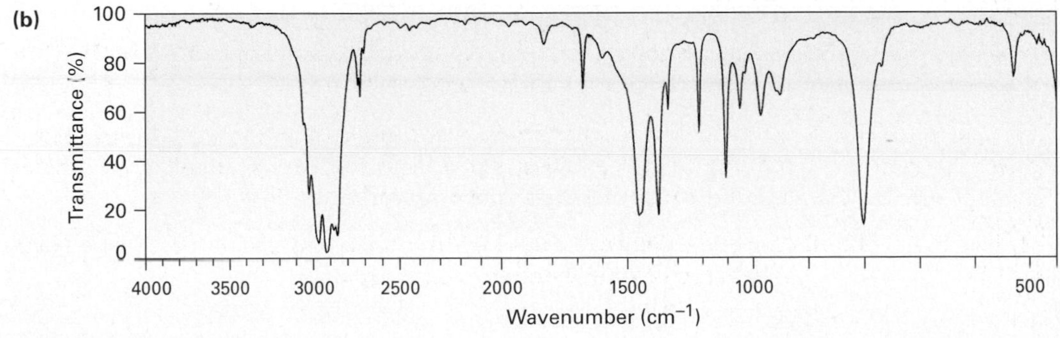

12.43 Propose structures for compounds that meet the following descriptions:

(a) An optically active compound $C_5H_{10}O$ with an IR absorption at 1730 cm^{-1}

(b) A non–optically active compound C_5H_9N with an IR absorption at 2215 cm^{-1}

12.44 4-Methyl-2-pentanone and 3-methylpentanal are isomers. Explain how you could tell them apart, both by mass spectrometry and by infrared spectroscopy.

O

H

4-Methyl-2-pentanone **3-Methylpentanal**

12.45 Grignard reagents undergo a general and very useful reaction with ketones. Methylmagnesium bromide, for example, reacts with cyclohexanone to yield a product with the formula $C_7H_{14}O$. What is the structure of this product if it has an IR absorption at 3400 cm^{-1}?

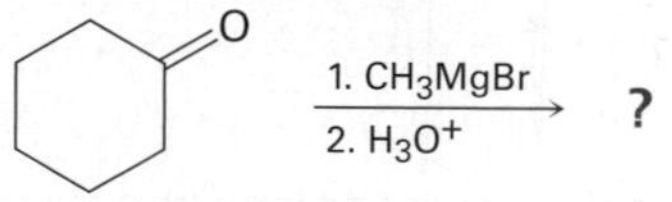

O

1. CH_3MgBr
2. H_3O^+
?

Cyclohexanone

12.46 Ketones undergo a reduction when treated with sodium borohydride, $NaBH_4$. What is the structure of the compound produced by reaction of 2-butanone with $NaBH_4$ if it has an IR absorption at 3400 cm^{-1} and $M^+ = 74$ in the mass spectrum?

$$CH_3CH_2\overset{O}{\overset{\|}{C}}CH_3 \xrightarrow[2.\ H_3O^+]{1.\ NaBH_4} ?$$

2-Butanone

12.47 Nitriles, $R{-}C{\equiv}N$, undergo a hydrolysis reaction when heated with aqueous acid. What is the structure of the compound produced by hydrolysis of propanenitrile, $CH_3CH_2C{\equiv}N$, if it has IR absorptions at 2500–3100 cm^{-1} and 1710 cm^{-1} and has $M^+ = 74$?

13

NMR spectroscopy is an invaluable aid in carrying out the design and synthesis of new drugs.

Structure Determination: Nuclear Magnetic Resonance Spectroscopy

OWL Sign in to OWL for Organic Chemistry at **www.cengage.com/owl** to view tutorials and simulations, develop problem-solving skills, and complete online homework assigned by your professor.

Nuclear magnetic resonance (NMR) spectroscopy is the most valuable spectroscopic technique available to organic chemists. It's the method of structure determination that organic chemists turn to first.

We saw in Chapter 12 that mass spectrometry gives a molecule's formula and infrared spectroscopy identifies a molecule's functional groups. Nuclear magnetic resonance spectroscopy complements these other techniques by mapping a molecule's carbon–hydrogen framework. Taken together, mass spectrometry, IR, and NMR make it possible to determine the structures of even very complex molecules.

Mass spectrometry	Molecular size and formula
Infrared spectroscopy	Functional groups
NMR spectroscopy	Map of carbon–hydrogen framework

Why This Chapter? The opening sentence in this chapter says it all: NMR is by far the most valuable spectroscopic technique for structure determination. Although we'll just give an overview of the subject in this chapter, focusing on NMR applications to small molecules, more advanced NMR techniques are also used in biological chemistry to study protein structure and folding.

13.1 Nuclear Magnetic Resonance Spectroscopy

Many kinds of nuclei behave as if they were spinning about an axis, much as the earth spins daily. Because they're positively charged, these spinning nuclei act like tiny magnets and interact with an external magnetic field, denoted B_0. Not all nuclei act this way, but fortunately for organic chemists, both the proton (^{1}H) and the ^{13}C nucleus do have spins. (In speaking about NMR, the words *proton* and *hydrogen* are often used interchangeably, since a hydrogen nucleus is just a proton.) Let's see what the consequences of nuclear spin are and how we can use the results.

In the absence of an external magnetic field, the spins of magnetic nuclei are oriented randomly. When a sample containing these nuclei is placed between the poles of a strong magnet, however, the nuclei adopt specific orientations, much as a compass needle orients in the earth's magnetic field. A spinning ^{1}H or ^{13}C nucleus can orient so that its own tiny magnetic field is aligned either with (parallel to) or against (antiparallel to) the external field. The two orientations don't have the same energy, however, and aren't equally likely. The parallel orientation is slightly lower in energy by an amount that depends on the strength of the external field, making this spin state very slightly favored over the antiparallel orientation **(Figure 13.1)**.

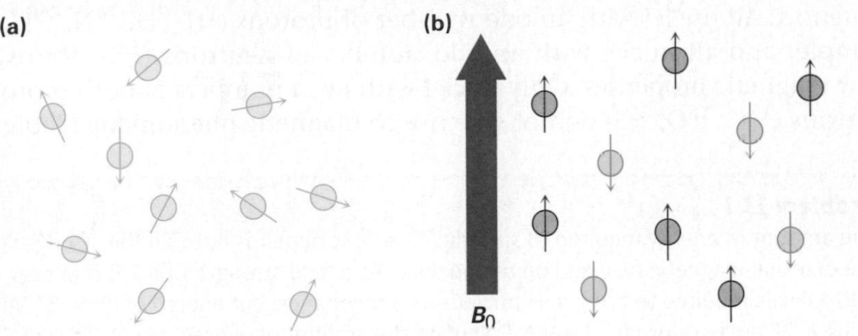

Figure 13.1 **(a)** Nuclear spins are oriented randomly in the absence of an external magnetic field but **(b)** have a specific orientation in the presence of an external field, B_0. Some of the spins (red) are aligned parallel to the external field while others (blue) are antiparallel. The parallel spin state is slightly lower in energy and therefore favored.

If the oriented nuclei are irradiated with electromagnetic radiation of the proper frequency, energy absorption occurs and the lower-energy state "spin-flips" to the higher-energy state. When this spin-flip occurs, the magnetic nuclei are said to be in resonance with the applied radiation—hence the name *nuclear magnetic resonance.*

The exact frequency necessary for resonance depends both on the strength of the external magnetic field and on the identity of the nuclei. If a very strong magnetic field is applied, the energy difference between the two spin states is larger and higher-frequency (higher-energy) radiation is required for a spin-flip. If a weaker magnetic field is applied, less energy is required to effect the transition between nuclear spin states **(Figure 13.2)**.

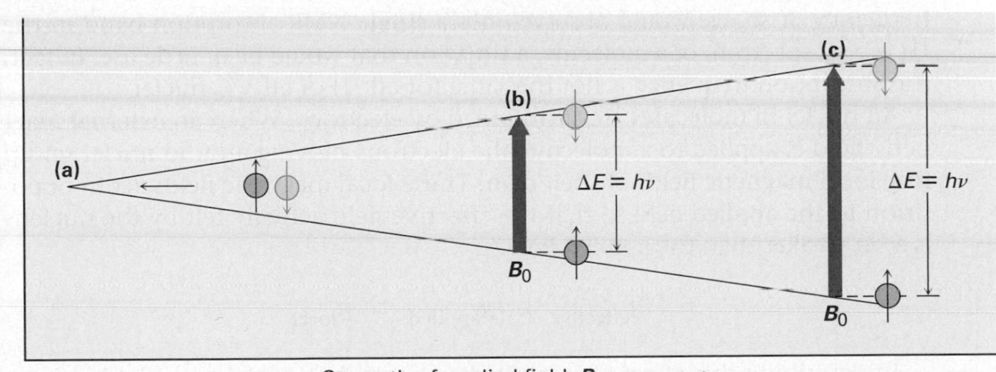

Figure 13.2 The energy difference ΔE between nuclear spin states depends on the strength of the applied magnetic field. Absorption of energy with frequency ν converts a nucleus from a lower spin state to a higher spin state. **(a)** Spin states have equal energies in the absence of an applied magnetic field but **(b)** have unequal energies in the presence of a magnetic field. At $\nu = 200$ MHz, $\Delta E = 8.0 \times 10^{-5}$ kJ/mol (1.9×10^{-5} kcal/mol). **(c)** The energy difference between spin states is greater at larger applied fields. At $\nu = 500$ MHz, $\Delta E = 2.0 \times 10^{-4}$ kJ/mol.

Table 13.1 The NMR Behavior of Some Common Nuclei

Magnetic nuclei	Nonmagnetic nuclei
^{1}H	^{12}C
^{13}C	^{16}O
^{2}H	^{32}S
^{14}N	
^{19}F	
^{31}P	

In practice, superconducting magnets that produce enormously powerful fields up to 21.2 tesla (T) are sometimes used, but field strengths in the range of 4.7 to 7.0 T are more common. At a magnetic field strength of 4.7 T, so-called radiofrequency (rf) energy in the 200 MHz range (1 MHz = 10^6 Hz) brings a ^{1}H nucleus into resonance, and rf energy of 50 MHz brings a ^{13}C nucleus into resonance. At the highest field strength currently available in commercial instruments (21.2 T), 900 MHz energy is required for ^{1}H spectroscopy. These energies needed for NMR are much smaller than those required for IR spectroscopy; 200 MHz rf energy corresponds to only 8.0×10^{-5} kJ/mol versus the 4.8 to 48 kJ/mol needed for IR spectroscopy.

^{1}H and ^{13}C nuclei are not unique in their ability to exhibit the NMR phenomenon. All nuclei with an odd number of protons (^{1}H, ^{2}H, ^{14}N, ^{19}F, ^{31}P, for example) and all nuclei with an odd number of neutrons (^{13}C, for example) show magnetic properties. Only nuclei with even numbers of both protons and neutrons (^{12}C, ^{16}O, ^{32}S) do not give rise to magnetic phenomena (Table 13.1).

Problem 13.1

The amount of energy required to spin-flip a nucleus depends both on the strength of the external magnetic field and on the nucleus. At a field strength of 4.7 T, rf energy of 200 MHz is required to bring a ^{1}H nucleus into resonance, but energy of only 187 MHz will bring a ^{19}F nucleus into resonance. Calculate the amount of energy required to spin-flip a ^{19}F nucleus. Is this amount greater or less than that required to spin-flip a ^{1}H nucleus?

Problem 13.2

Calculate the amount of energy required to spin-flip a proton in a spectrometer operating at 300 MHz. Does increasing the spectrometer frequency from 200 to 300 MHz increase or decrease the amount of energy necessary for resonance?

13.2 The Nature of NMR Absorptions

From the description thus far, you might expect all ^{1}H nuclei in a molecule to absorb energy at the same frequency and all ^{13}C nuclei to absorb at the same frequency. If so, we would observe only a single NMR absorption band in the ^{1}H or ^{13}C spectrum of a molecule, a situation that would be of little use. In fact, the absorption frequency is not the same for all ^{1}H or all ^{13}C nuclei.

All nuclei in molecules are surrounded by electrons. When an external magnetic field is applied to a molecule, the electrons moving around nuclei set up tiny local magnetic fields of their own. These local magnetic fields act in opposition to the applied field so that the effective field actually felt by the nucleus is a bit weaker than the applied field.

$$\boldsymbol{B}_{\text{effective}} = \boldsymbol{B}_{\text{applied}} - \boldsymbol{B}_{\text{local}}$$

In describing this effect of local fields, we say that nuclei are **shielded** from the full effect of the applied field by the surrounding electrons. Because each chemically distinct nucleus in a molecule is in a slightly different electronic environment, each nucleus is shielded to a slightly different extent and the effective magnetic field felt by each is slightly different. These tiny differences in the effective magnetic fields experienced by different nuclei can be detected, and we thus see a distinct NMR signal for each chemically distinct ^{13}C or ^{1}H nucleus in a molecule. As a result, an NMR spectrum effectively maps the carbon–hydrogen framework of an organic molecule. With practice, it's possible to read the map and derive structural information.

Figure 13.3 shows both the ^{1}H and the ^{13}C NMR spectra of methyl acetate, $CH_3CO_2CH_3$. The horizontal axis shows the effective field strength felt by the nuclei, and the vertical axis indicates the intensity of absorption of rf energy. Each peak in the NMR spectrum corresponds to a chemically distinct ^{1}H or ^{13}C nucleus in the molecule. Note that NMR spectra are formatted with the zero absorption line at the bottom, whereas IR spectra are formatted with the zero absorption line at the top; **Section 12.5**. Note also that ^{1}H and ^{13}C spectra can't be observed simultaneously on the same spectrometer because different amounts of energy are required to spin-flip the different kinds of nuclei. The two spectra must be recorded separately.

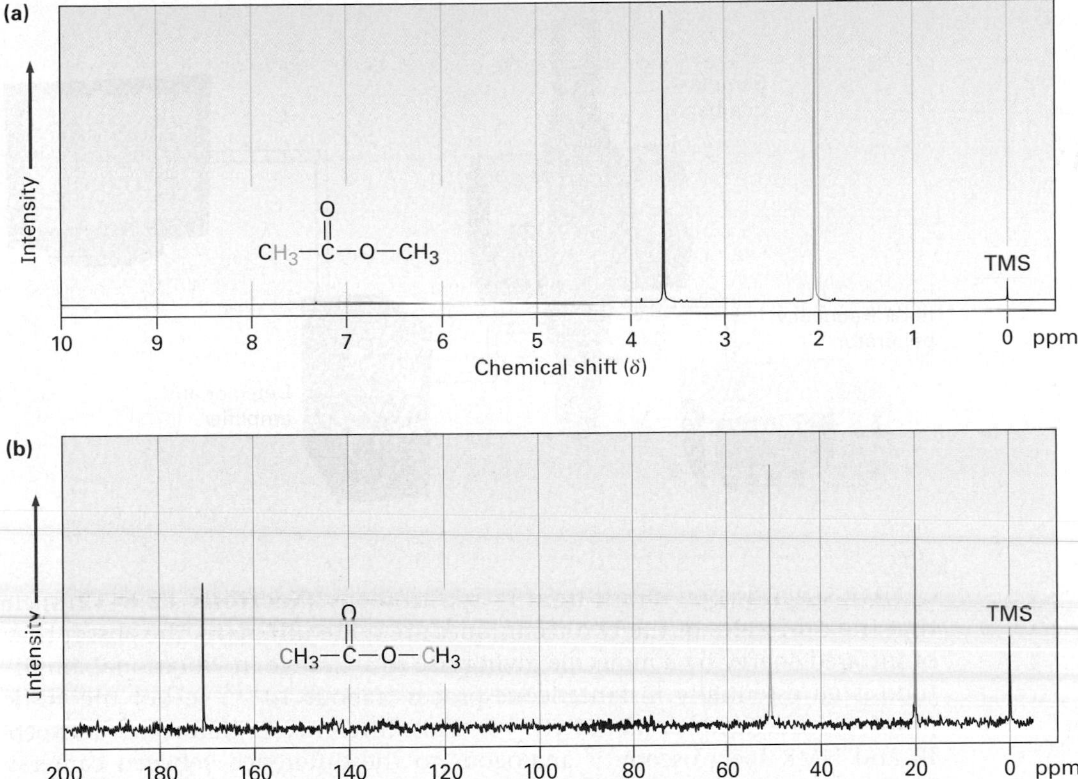

Figure 13.3 **(a)** The ^{1}H NMR spectrum and **(b)** the ^{13}C NMR spectrum of methyl acetate, $CH_3CO_2CH_3$. The small peak labeled "TMS" at the far right of each spectrum is a calibration peak, as explained in the next section.

The ^{13}C spectrum of methyl acetate in Figure 13.3b shows three peaks, one for each of the three chemically distinct carbon atoms in the molecule. The ^{1}H NMR spectrum in Figure 13.3a shows only two peaks, however, even though methyl acetate has six hydrogens. One peak is due to the $CH_3C{=}O$ hydrogens, and the other to the $-OCH_3$ hydrogens. Because the three hydrogens in each methyl group have the same electronic environment, they are shielded to the same extent and are said to be equivalent. Chemically equivalent nuclei always show a single absorption. The two methyl groups themselves, however, are nonequivalent, so the two sets of hydrogens absorb at different positions.

The operation of a basic NMR spectrometer is illustrated in **Figure 13.4**. An organic sample is dissolved in a suitable solvent (usually deuteriochloroform, $CDCl_3$, which has no hydrogens) and placed in a thin glass tube between the poles of a magnet. The strong magnetic field causes the ^1H and ^{13}C nuclei in the molecule to align in one of the two possible orientations, and the sample is irradiated with rf energy. If the frequency of the rf irradiation is held constant and the strength of the applied magnetic field is varied, each nucleus comes into resonance at a slightly different field strength. A sensitive detector monitors the absorption of rf energy, and the electronic signal is then amplified and displayed as a peak.

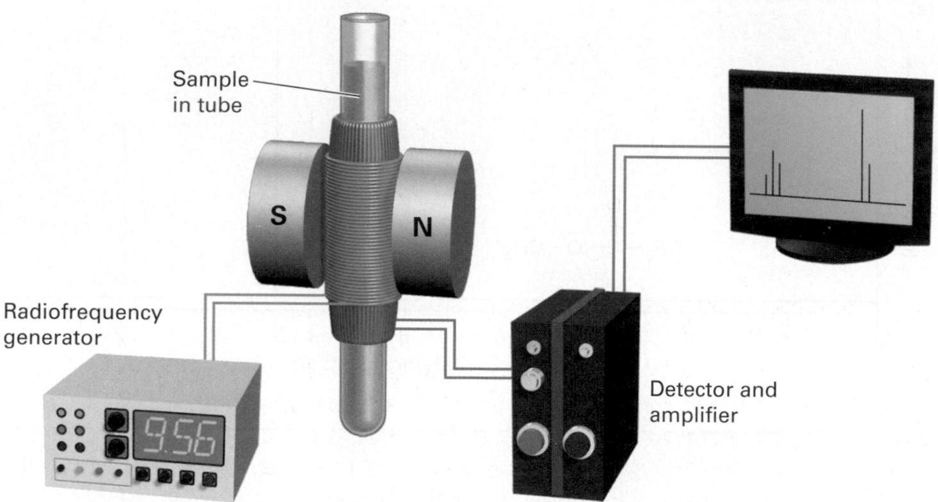

Figure 13.4 Schematic operation of a basic NMR spectrometer. A thin glass tube containing the sample solution is placed between the poles of a strong magnet and irradiated with rf energy.

NMR spectroscopy differs from IR spectroscopy **(Sections 12.6–12.8)** in that the timescales of the two techniques are quite different. The absorption of infrared energy by a molecule giving rise to a change in vibrational amplitude is an essentially instantaneous process (about 10^{-13} s), but the NMR process is much slower (about 10^{-3} s). This difference in timescales between IR and NMR spectroscopy is analogous to the difference between cameras operating at very fast and very slow shutter speeds. The fast camera (IR) takes an instantaneous picture and freezes the action. If two rapidly interconverting species are present, IR spectroscopy records the spectrum of both. The slow camera (NMR), however, takes a blurred, time-averaged picture. If two species interconverting faster than 10^3 times per second are present in a sample, NMR records only a single, averaged spectrum, rather than separate spectra of the two discrete species.

Because of this blurring effect, NMR spectroscopy can be used to measure the rates and activation energies of very fast processes. In cyclohexane, for example, a ring-flip **(Section 4.6)** occurs so rapidly at room temperature that axial and equatorial hydrogens can't be distinguished by NMR; only a single, averaged ^{1}H NMR absorption is seen for cyclohexane at 25 °C. At −90 °C, however, the ring-flip is slowed down enough that two absorption peaks are seen, one for the six axial hydrogens and one for the six equatorial hydrogens. Knowing the temperature and the rate at which signal blurring begins to occur, it's possible to calculate that the activation energy for the cyclohexane ring-flip is 45 kJ/mol (10.8 kcal/mol).

E_{act} = 45 kJ/mol

^{1}H NMR: 1 peak at 25 °C
2 peaks at −90 °C

Problem 13.3

2-Chloropropene shows signals for three kinds of protons in its ^{1}H NMR spectrum. Explain.

13.3 Chemical Shifts

NMR spectra are displayed on charts that show the applied field strength increasing from left to right **(Figure 13.5)**. Thus, the left part of the chart is the low-field, or **downfield**, side, and the right part is the high-field, or **upfield**, side. Nuclei that absorb on the downfield side of the chart require a lower field strength for resonance, implying that they have less shielding. Nuclei that absorb on the upfield side require a higher field strength for resonance, implying that they have more shielding.

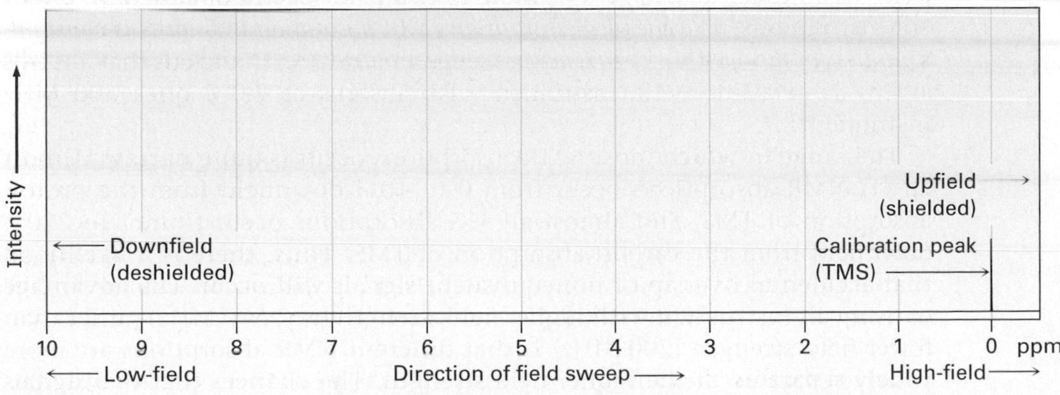

Figure 13.5 The NMR chart. The downfield, deshielded side is on the left, and the upfield, shielded side is on the right. The tetramethylsilane (TMS) absorption is used as reference point.

To define the position of an absorption, the NMR chart is calibrated and a reference point is used. In practice, a small amount of tetramethylsilane [TMS; $(CH_3)_4Si$] is added to the sample so that a reference absorption peak is produced when the spectrum is run. TMS is used as reference for both ^{1}H and ^{13}C measurements because it produces in both a single peak that occurs upfield of other absorptions normally found in organic compounds. The ^{1}H and ^{13}C spectra of methyl acetate in Figure 13.3 have the TMS reference peak indicated.

The position on the chart at which a nucleus absorbs is called its **chemical shift.** The chemical shift of TMS is set as the zero point, and other absorptions normally occur downfield, to the left on the chart. NMR charts are calibrated using an arbitrary scale called the **delta (δ) scale**, where 1 δ equals 1 part-per-million (1 ppm) of the spectrometer operating frequency. For example, if we were measuring the ^{1}H NMR spectrum of a sample using an instrument operating at 200 MHz, 1 δ would be 1 millionth of 200,000,000 Hz, or 200 Hz. If we were measuring the spectrum using a 500 MHz instrument, 1 δ = 500 Hz. The following equation can be used for any absorption:

$$\delta = \frac{\text{Observed chemical shift (number of Hz away from TMS)}}{\text{Spectrometer frequency in MHz}}$$

Although this method of calibrating NMR charts may seem complex, there's a good reason for it. As we saw earlier, the rf frequency required to bring a given nucleus into resonance depends on the spectrometer's magnetic field strength. But because there are many different kinds of spectrometers with many different magnetic field strengths available, chemical shifts given in frequency units (Hz) vary from one instrument to another. Thus, a resonance that occurs at 120 Hz downfield from TMS on one spectrometer might occur at 600 Hz downfield from TMS on another spectrometer with a more powerful magnet.

By using a system of measurement in which NMR absorptions are expressed in relative terms (parts per million relative to spectrometer frequency) rather than absolute terms (Hz), it's possible to compare spectra obtained on different instruments. *The chemical shift of an NMR absorption in δ units is constant, regardless of the operating frequency of the spectrometer.* A ^{1}H nucleus that absorbs at 2.0 δ on a 200 MHz instrument also absorbs at 2.0 δ on a 500 MHz instrument.

The range in which most NMR absorptions occur is quite narrow. Almost all ^{1}H NMR absorptions occur from 0 to 10 δ downfield from the proton absorption of TMS, and almost all ^{13}C absorptions occur from 1 to 220 δ downfield from the carbon absorption of TMS. Thus, there is a likelihood that accidental overlap of nonequivalent signals will occur. The advantage of using an instrument with higher field strength (say, 500 MHz) rather than lower field strength (200 MHz) is that different NMR absorptions are more widely separated at the higher field strength. The chances that two signals will accidentally overlap are therefore lessened, and interpretation of spectra becomes easier. For example, two signals that are only 20 Hz apart at 200 MHz (0.1 ppm) are 50 Hz apart at 500 MHz (still 0.1 ppm).

Problem 13.4

The following ^{1}H NMR peaks were recorded on a spectrometer operating at 200 MHz. Convert each into δ units.

(a) $CHCl_3$; 1454 Hz **(b)** CH_3Cl; 610 Hz
(c) CH_3OH; 693 Hz **(d)** CH_2Cl_2; 1060 Hz

Problem 13.5

When the ^{1}H NMR spectrum of acetone, CH_3COCH_3, is recorded on an instrument operating at 200 MHz, a single sharp resonance at 2.1 δ is seen.

(a) How many hertz downfield from TMS does the acetone resonance correspond to?
(b) If the ^{1}H NMR spectrum of acetone were recorded at 500 MHz, what would the position of the absorption be in δ units?
(c) How many hertz downfield from TMS does this 500 MHz resonance correspond to?

13.4 ^{13}C NMR Spectroscopy: Signal Averaging and FT–NMR

Everything we've said thus far about NMR spectroscopy applies to both ^{1}H and ^{13}C spectra. Now, though, let's focus only on ^{13}C spectroscopy because it's much easier to interpret. What we learn now about interpreting ^{13}C spectra will simplify the subsequent discussion of ^{1}H spectra.

In some ways, it's surprising that carbon NMR is even possible. After all, ^{12}C, the most abundant carbon isotope, has no nuclear spin and can't be seen by NMR. Carbon-13 is the only naturally occurring carbon isotope with a nuclear spin, but its natural abundance is only 1.1%. Thus, only about 1 of every 100 carbons in an organic sample is observable by NMR. The problem of low abundance has been overcome, however, by the use of *signal averaging* and *Fourier-transform NMR* (**FT–NMR**). Signal averaging increases instrument sensitivity, and FT–NMR increases instrument speed.

The low natural abundance of ^{13}C means that any individual NMR spectrum is extremely "noisy." That is, the signals are so weak that they are cluttered with random background electronic noise, as shown in **Figure 13.6a** on the next page. If, however, hundreds or thousands of individual runs are added together by a computer and then averaged, a greatly improved spectrum results (Figure 13.6b). Background noise, because of its random nature, averages to zero, while the nonzero signals stand out clearly. Unfortunately, the value of signal averaging is limited when using the method of NMR spectrometer operation described in **Section 13.2**, because it takes about 5 to 10 minutes to obtain a single spectrum. Thus, a faster way to obtain spectra is needed if signal averaging is to be used.

In the method of NMR spectrometer operation described in **Section 13.2**, the rf frequency is held constant while the strength of the magnetic field is varied so that all signals in the spectrum are recorded sequentially. In the FT–NMR technique used by modern spectrometers, however, all the signals are recorded simultaneously. A sample is placed in a magnetic field of constant strength and is irradiated with a short pulse of rf energy that covers the entire range of useful frequencies. All ^{1}H or ^{13}C nuclei in the sample resonate at once, giving a complex, composite signal that is mathematically manipulated using so-called Fourier transforms and then displayed in the usual way. Because all resonance signals are collected at once, it takes only a few seconds rather than a few minutes to record an entire spectrum.

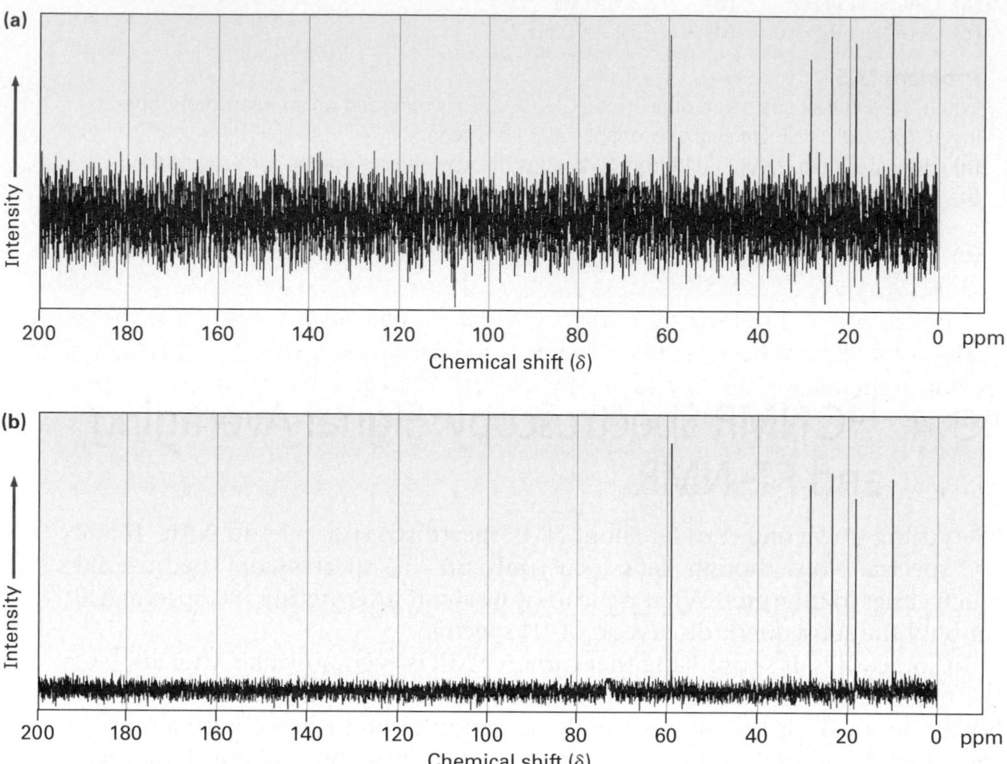

Figure 13.6 Carbon-13 NMR spectra of 1-pentanol, $CH_3CH_2CH_2CH_2CH_2OH$. Spectrum **(a)** is a single run, showing the large amount of background noise. Spectrum **(b)** is an average of 200 runs.

Combining the speed of FT–NMR with the sensitivity enhancement of signal averaging is what gives modern NMR spectrometers their power. Literally thousands of spectra can be taken and averaged in a few hours, resulting in sensitivity so high that a ^{13}C NMR spectrum can be obtained on less than 0.1 mg of sample and a ^{1}H spectrum can be recorded on only a few *micro*grams.

13.5 Characteristics of ^{13}C NMR Spectroscopy

At its simplest, ^{13}C NMR makes it possible to count the number of different carbon atoms in a molecule. Look at the ^{13}C NMR spectra of methyl acetate and 1-pentanol shown previously in Figures 13.3b and 13.6b. In each case, a single sharp resonance line is observed for each different carbon atom.

Most ^{13}C resonances are between 0 and 220 ppm downfield from the TMS reference line, with the exact chemical shift of each ^{13}C resonance dependent on that carbon's electronic environment within the molecule. **Figure 13.7** shows the correlation of chemical shift with environment.

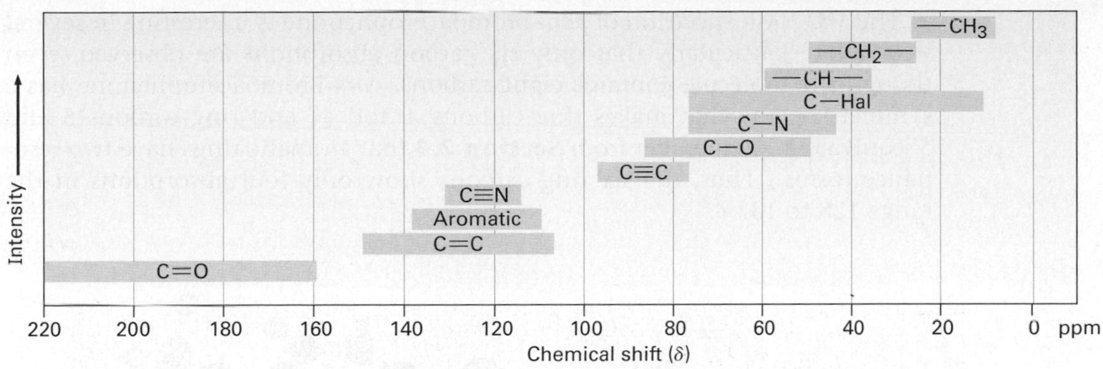

Figure 13.7 Chemical shift correlations for ^{13}C NMR.

The factors that determine chemical shifts are complex, but it's possible to make some generalizations from the data in Figure 13.7. One trend is that a carbon's chemical shift is affected by the electronegativity of nearby atoms. Carbons bonded to oxygen, nitrogen, or halogen absorb downfield (to the left) of typical alkane carbons. Because electronegative atoms attract electrons, they pull electrons away from neighboring carbon atoms, causing those carbons to be deshielded and to come into resonance at a lower field.

Another trend is that *sp*^{3}-hybridized carbons generally absorb from 0 to 90 δ, while *sp*^{2} carbons absorb from 110 to 220 δ. Carbonyl carbons (C=O) are particularly distinct in ^{13}C NMR and are always found at the low-field end of the spectrum, from 160 to 220 δ. **Figure 13.8** shows the ^{13}C NMR spectra of 2-butanone and *para*-bromoacetophenone and indicates the peak assignments. Note that the C=O carbons are at the left edge of the spectrum in each case.

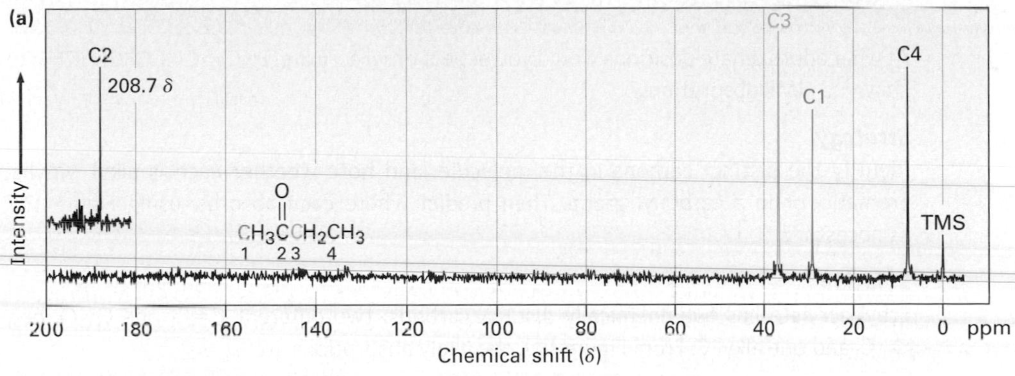

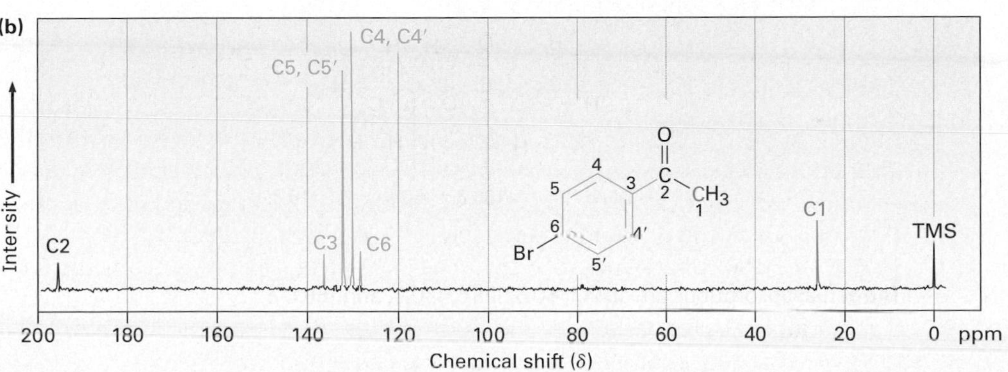

Figure 13.8 Carbon-13 NMR spectra of **(a)** 2-butanone and **(b)** *para*-bromoacetophenone.

The ^{13}C NMR spectrum of *para*-bromoacetophenone is interesting in several ways. Note particularly that only six carbon absorptions are observed, even though the molecule contains eight carbons. *para*-Bromoacetophenone has a symmetry plane that makes ring carbons 4 and 4′, and ring carbons 5 and 5′ equivalent. (Remember from **Section 2.4** that aromatic rings have two resonance forms.) Thus, the six ring carbons show only four absorptions in the range 128 to 137 δ.

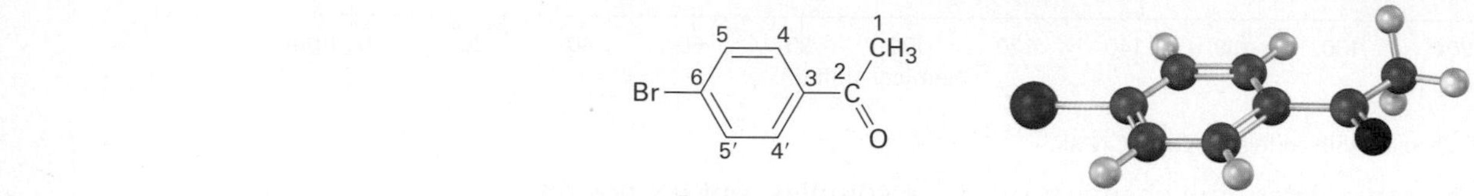

***para*-Bromoacetophenone**

A second interesting point about both spectra in Figure 13.8 is that the peaks aren't uniform in size. Some peaks are larger than others even though they are one-carbon resonances (except for the two 2-carbon peaks of *para*-bromoacetophenone). This difference in peak size is a general feature of ^{13}C NMR spectra.

Worked Example 13.1 Predicting Chemical Shifts in ^{13}C NMR Spectra

At what approximate positions would you expect ethyl acrylate, $H_2C{=}CHCO_2CH_2CH_3$, to show ^{13}C NMR absorptions?

Strategy

Identify the distinct carbons in the molecule, and note whether each is alkyl, vinylic, aromatic, or in a carbonyl group. Then predict where each absorbs, using Figure 13.7 as necessary.

Solution

Ethyl acrylate has five chemically distinct carbons: two different C=C, one C=O, one O–C, and one alkyl C. From Figure 13.7, the likely absorptions are

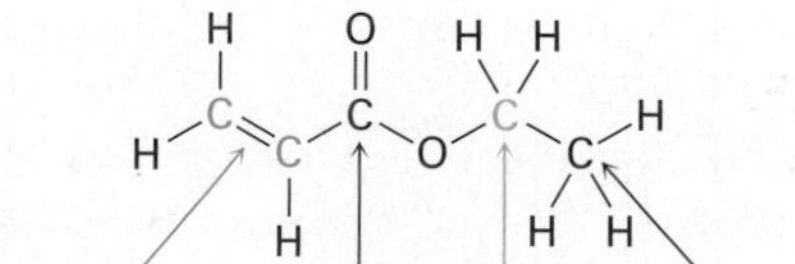

The actual absorptions are at 14.1, 60.5, 128.5, 130.3, and 166.0 δ.

Problem 13.6

Predict the number of carbon resonance lines you would expect in the ^{13}C NMR spectra of the following compounds:

(a) Methylcyclopentane
(b) 1-Methylcyclohexene
(c) 1,2-Dimethylbenzene
(d) 2-Methyl-2-butene
(e)

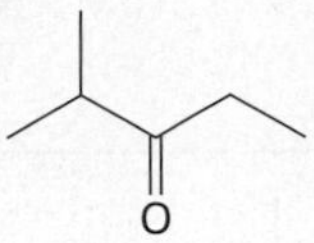

(f)

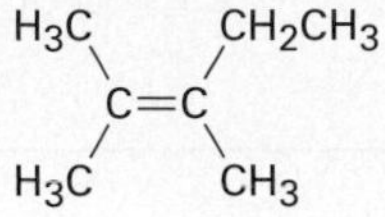

Problem 13.7

Propose structures for compounds that fit the following descriptions:

(a) A hydrocarbon with seven lines in its ^{13}C NMR spectrum
(b) A six carbon compound with only five lines in its ^{13}C NMR spectrum
(c) A four-carbon compound with three lines in its ^{13}C NMR spectrum

Problem 13.8

Assign the resonances in the ^{13}C NMR spectrum of methyl propanoate, $CH_3CH_2CO_2CH_3$ **(Figure 13.9)**.

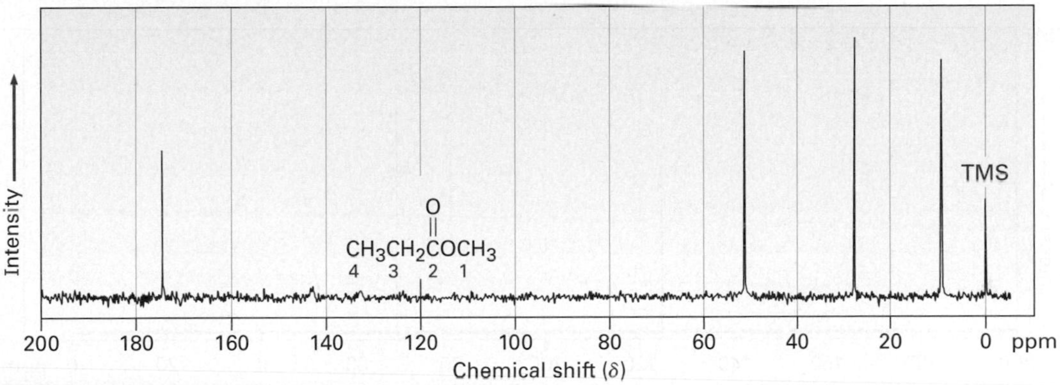

Figure 13.9 ^{13}C NMR spectrum of methyl propanoate, Problem 13.8.

13.6 DEPT ^{13}C NMR Spectroscopy

Numerous techniques developed in recent years have made it possible to obtain enormous amounts of information from ^{13}C NMR spectra. Among these techniques is one called DEPT–NMR, for *distortionless enhancement by polarization transfer,* which makes it possible to distinguish among signals due to CH_3, CH_2, CH, and quaternary carbons. That is, the number of hydrogens attached to each carbon in a molecule can be determined.

A DEPT experiment is usually done in three stages, as shown in **Figure 13.10** for 6-methyl-5-hepten-2-ol. The first stage is to run an ordinary spectrum (called a broadband-decoupled spectrum) to locate the chemical shifts of all carbons. Next, a second spectrum called a DEPT-90 is run, using special conditions under which only signals due to CH carbons appear. Signals due to CH_3, CH_2, and quaternary carbons are absent. Finally, a third spectrum called a DEPT-135 is run, using conditions under which CH_3 and CH resonances appear as positive signals, CH_2 resonances appear as *negative* signals—that is, as peaks below the baseline—and quaternary carbons are again absent.

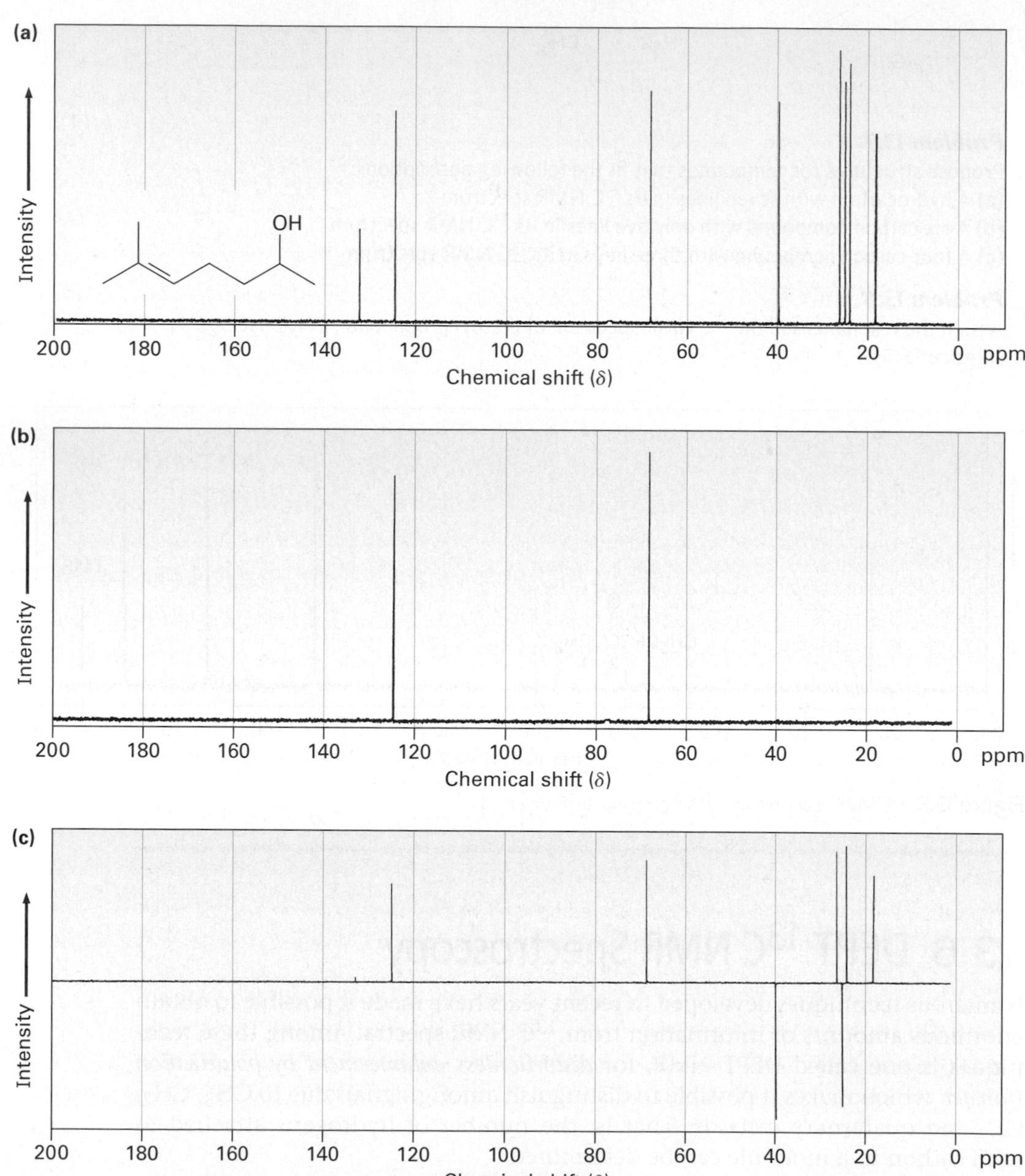

Figure 13.10 DEPT–NMR spectra for 6-methyl-5-hepten-2-ol. Part **(a)** is an ordinary broadband-decoupled spectrum, which shows signals for all eight carbons. Part **(b)** is a DEPT-90 spectrum, which shows only signals for the two CH carbons. Part **(c)** is a DEPT-135 spectrum, which shows positive signals for the two CH and three CH_3 carbons and negative signals for the two CH_2 carbons.

Putting together the information from all three spectra makes it possible to tell the number of hydrogens attached to each carbon. The CH carbons are identified in the DEPT-90 spectrum, the CH_2 carbons are identified as the negative peaks in the DEPT-135 spectrum, the CH_3 carbons are identified by subtracting the CH peaks from the positive peaks in the DEPT-135 spectrum, and quaternary carbons are identified by subtracting all peaks in the DEPT-135 spectrum from the peaks in the broadband-decoupled spectrum.

Broadband-decoupled	DEPT-90	DEPT-135
C, CH, CH_2, CH_3	CH	CH_3, CH are positive CH_2 is negative

C	Subtract DEPT-135 from broadband-decoupled spectrum
CH	DEPT-90
CH_2	Negative DEPT-135
CH_3	Subtract DEPT-90 from positive DEPT-135

Assigning a Chemical Structure from a ^{13}C NMR Spectrum

Worked Example 13.2

Propose a structure for an alcohol, $C_4H_{10}O$, that has the following ^{13}C NMR spectral data:

Broadband decoupled ^{13}C NMR: 19.0, 31.7, 69.5 δ;
DEPT-90: 31.7 δ;
DEPT-135: positive peak at 19.0 δ, negative peak at 69.5 δ.

Strategy

As noted in **Section 7.2**, it usually helps with compounds of known formula but unknown structure to calculate the compound's degree of unsaturation. In the present instance, a formula of $C_4H_{10}O$ corresponds to a saturated, open-chain molecule.

To gain information from the ^{13}C data, let's begin by noting that the unknown alcohol has four carbon atoms, yet has only three NMR absorptions, which implies that two of the carbons must be equivalent. Looking at chemical shifts, two of the absorptions are in the typical alkane region (19.0 and 31.7 δ), while one is in the region of a carbon bonded to an electronegative atom (69.5 δ)—oxygen in this instance. The DEPT-90 spectrum tells us that the alkyl carbon at 31.7 δ is tertiary (CH); the DEPT-135 spectrum tells us that the alkyl carbon at 19.0 δ is a methyl (CH_3) and that the carbon bonded to oxygen (69.5 δ) is secondary (CH_2). The two equivalent carbons are probably both methyls bonded to the same tertiary carbon, $(CH_3)_2CH$–. We can now put the pieces together to propose a structure: 2-methyl-1-propanol.

Solution

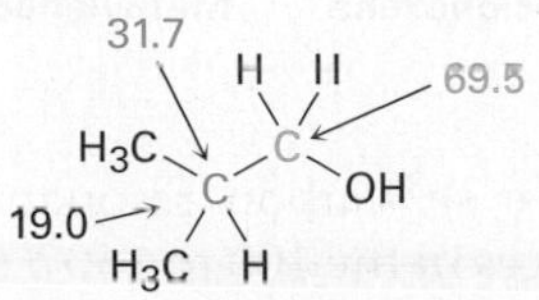

31.7
H H 69.5
H_3C C C OH
19.0
H_3C H

2-Methyl-1-propanol

Problem 13.9

Assign a chemical shift to each carbon in 6-methyl-5-hepten-2-ol (Figure 13.10).

Problem 13.10

Estimate the chemical shift of each carbon in the following molecule. Predict which carbons will appear in the DEPT-90 spectrum, which will give positive peaks in the DEPT-135 spectrum, and which will give negative peaks in the DEPT-135 spectrum.

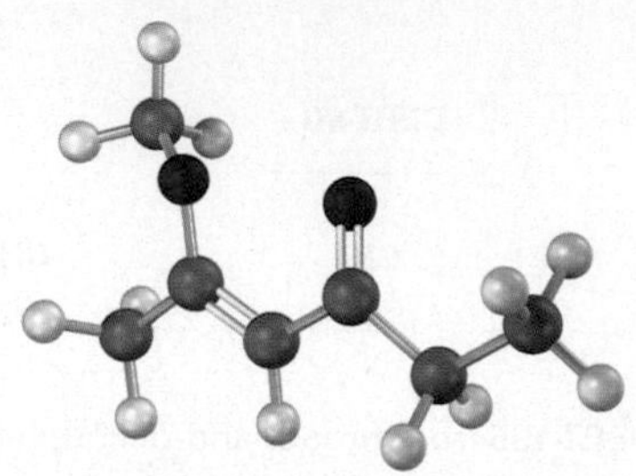

Problem 13.11

Propose a structure for an aromatic hydrocarbon, $C_{11}H_{16}$, that has the following ^{13}C NMR spectral data:

Broadband decoupled: 29.5, 31.8, 50.2, 125.5, 127.5, 130.3, 139.8 δ

DEPT-90: 125.5, 127.5, 130.3 δ

DEPT-135: positive peaks at 29.5, 125.5, 127.5, 130.3 δ; negative peak at 50.2 δ

13.7 Uses of ^{13}C NMR Spectroscopy

The information derived from ^{13}C NMR spectroscopy is extraordinarily useful for structure determination. Not only can we count the number of nonequivalent carbon atoms in a molecule, we can also get information about the electronic environment of each carbon and find how many protons each is attached to. As a result, we can answer many structural questions that go unanswered by IR spectroscopy or mass spectrometry.

Here's an example: how do we know that the E2 reaction of an alkyl halide follows Zaitsev's rule **(Section 11.7)**? Does treatment of 1-chloro-1-methylcyclohexane with a strong base give predominantly the trisubstituted alkene 1-methylcyclohexene or the disubstituted alkene methylenecyclohexane?

H_3C Cl — KOH / Ethanol → CH_3 or CH_2 ?

1-Chloro-1-methylcyclohexane **1-Methylcyclohexene** **Methylenecyclohexane**

1-Methylcyclohexene will have five sp^3-carbon resonances in the 20 to 50 δ range and two sp^2-carbon resonances in the 100 to 150 δ range. Methylenecyclohexane, however, because of its symmetry, will have only three sp^3-carbon resonance peaks and two sp^2-carbon peaks. The spectrum of the actual reaction product, shown in **Figure 13.11**, clearly identifies 1-methylcyclohexene as the product of this E2 reaction.

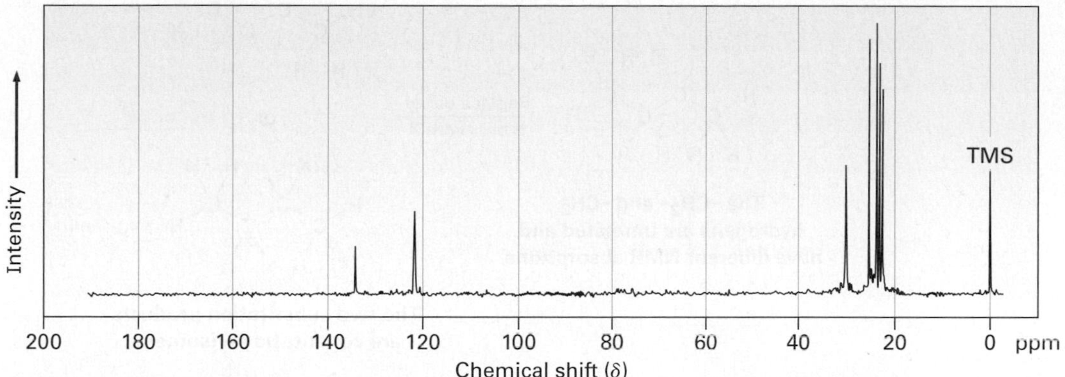

Figure 13.11 The ^{13}C NMR spectrum of 1-methylcyclohexene, the E2 reaction product from treatment of 1-chloro-1-methylcyclohexane with base.

Problem 13.12

We saw in **Section 9.3** that addition of HBr to a terminal alkyne leads to the Markovnikov addition product, with the Br bonding to the more highly substituted carbon. How could you use ^{13}C NMR to identify the product of the addition of 1 equivalent of HBr to 1-hexyne?

13.8 ^{1}H NMR Spectroscopy and Proton Equivalence

Having looked at ^{13}C spectra, let's now focus on ^{1}H NMR spectroscopy. Because each electronically distinct hydrogen in a molecule has its own unique absorption, one use of ^{1}H NMR is to find out how many kinds of electronically nonequivalent hydrogens are present. In the ^{1}H NMR spectrum of methyl acetate shown previously in Figure 13.3a on page 459, for instance, there are two signals, corresponding to the two kinds of nonequivalent protons present, $CH_3C{=}O$ protons and $-OCH_3$ protons.

For relatively small molecules, a quick look at the structure is often enough to decide how many kinds of protons are present and thus how many NMR absorptions might appear. If in doubt, though, the equivalence or nonequivalence of two protons can be determined by comparing the structures that would be formed if each hydrogen were replaced by an X group. There are four possibilities.

- One possibility is that the protons are chemically unrelated and thus nonequivalent. If so, the products formed on substitution of H by X would be different constitutional isomers. In butane, for instance, the $-CH_3$ protons are different from the $-CH_2-$ protons. They therefore give different products on substitution by X than the $-CH_2$ protons and would likely show different NMR absorptions.

Replace either H or H with X

or

The $-CH_2-$ and $-CH_3$ hydrogens are unrelated and have different NMR absorptions.

The two substitution products are constitutional isomers.

- A second possibility is that the protons are chemically identical and thus electronically equivalent. If so, the same product would be formed regardless of which H is substituted by X. In butane, for instance, the six $-CH_3$ hydrogens on C1 and C4 are identical, would give the identical structure on substitution by X, and would show the identical NMR absorption. Such protons are said to be **homotopic**.

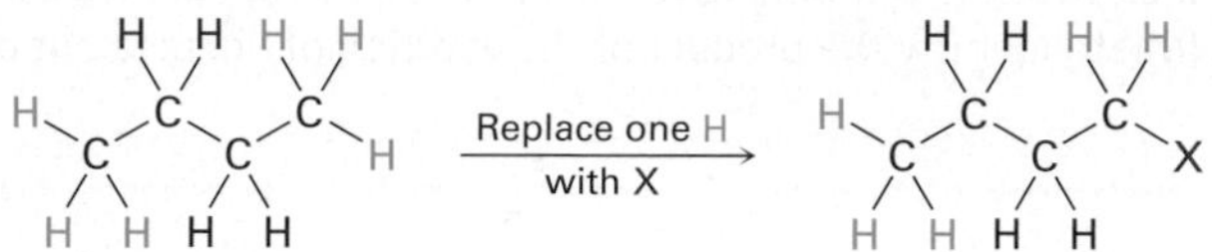

The six $-CH_3$ hydrogens are homotopic and have the same NMR absorption.

Only one substitution product is possible.

- The third possibility is a bit more subtle. Although they might at first seem homotopic, the two $-CH_2-$ hydrogens on C2 in butane (and the two $-CH_2-$ hydrogens on C3) are in fact not identical. Substitution by X of a hydrogen at C2 (or C3) would form a new chirality center, so different enantiomers **(Section 5.1)** would result depending on whether the *pro-R* or *pro-S* hydrogen were substituted for **(Section 5.11)**. Such hydrogens, whose substitution by X would lead to different enantiomers, are said to be **enantiotopic**. Enantiotopic hydrogens, even though not identical, are nevertheless electronically equivalent and thus have the same NMR absorption.

pro-S *pro-R*

Replace either H or H with X

or

The two hydrogens on C2 (and on C3) are enantiotopic and have the same NMR absorption.

The two possible substitution products are enantiomers.

- The fourth possibility arises in chiral molecules, such as (*R*)-2-butanol. The two $-CH_2-$ hydrogens at C3 are neither homotopic nor enantiotopic. Since substitution of a hydrogen at C3 would form a second chirality center, different diastereomers **(Section 5.6)** would result depending on whether the *pro-R* or *pro-S* hydrogen were substituted for. Such hydrogens, whose substitution by X leads to different diastereomers, are said to be **diastereotopic**. Diastereotopic hydrogens are neither chemically nor electronically equivalent. They are completely different and would likely show different NMR absorptions.

Replace either H or H with X

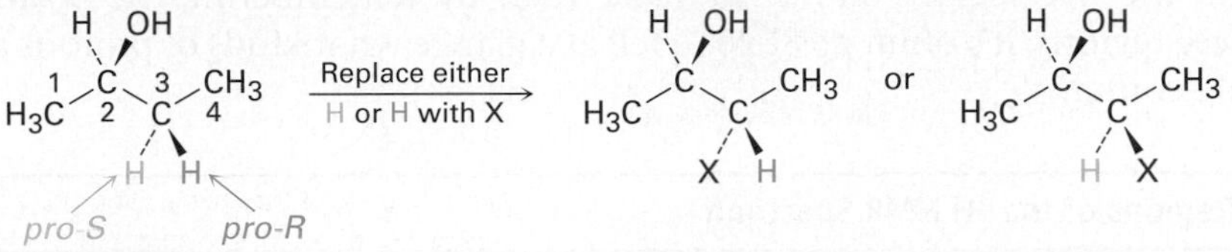

The two hydrogens on C3 are diastereotopic and have different NMR absorptions.

The two possible substitution products are diastereomers.

Problem 13.13

Identify the indicated sets of protons as unrelated, homotopic, enantiotopic, or diastereotopic:

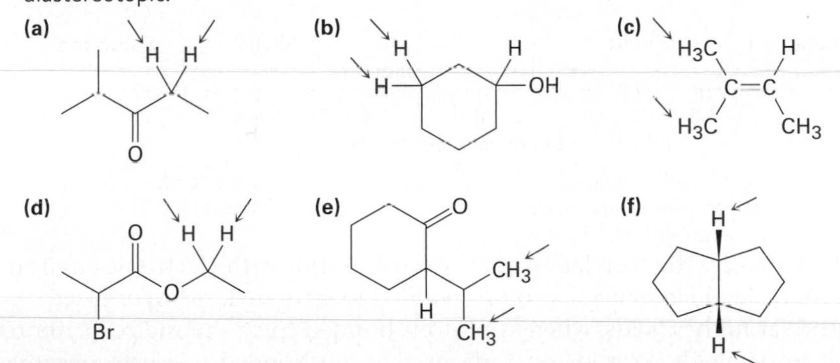

Problem 13.14

How many kinds of electronically nonequivalent protons are present in each of the following compounds, and thus how many NMR absorptions might you expect in each?

(a) CH_3CH_2Br **(b)** $CH_3OCH_2CH(CH_3)_2$ **(c)** $CH_3CH_2CH_2NO_2$
(d) Methylbenzene **(e)** 2-Methyl-1-butene **(f)** *cis*-3-Hexene

Problem 13.15

How many absorptions would you expect (*S*)-malate, an intermediate in carbohydrate metabolism, to have in its ^{1}H NMR spectrum? Explain.

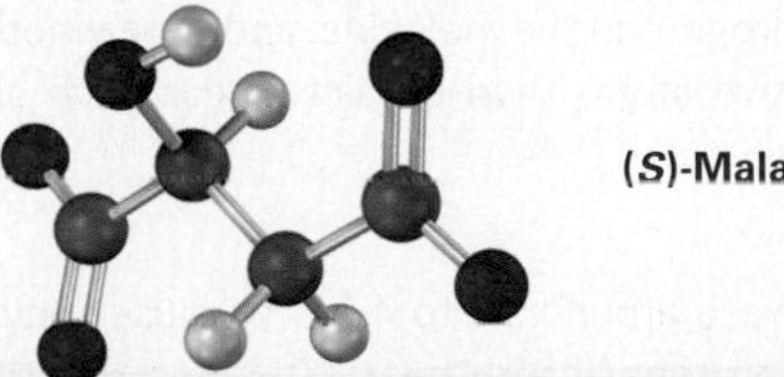

(*S*)-Malate

13.9 Chemical Shifts in ^{1}H NMR Spectroscopy

As we said previously, differences in chemical shifts are caused by the small local magnetic fields of electrons surrounding the different nuclei. Nuclei that are more strongly shielded by electrons require a higher applied field to bring them into resonance and therefore absorb on the right side of the NMR chart. Nuclei that are less strongly shielded need a lower applied field for resonance and therefore absorb on the left of the NMR chart.

Most ^{1}H chemical shifts fall within the range 0 to 10 δ, which can be divided into the five regions shown in Table 13.2. By remembering the positions of these regions, it's often possible to tell at a glance what kinds of protons a molecule contains.

Table 13.2 Regions of the ^{1}H NMR Spectrum

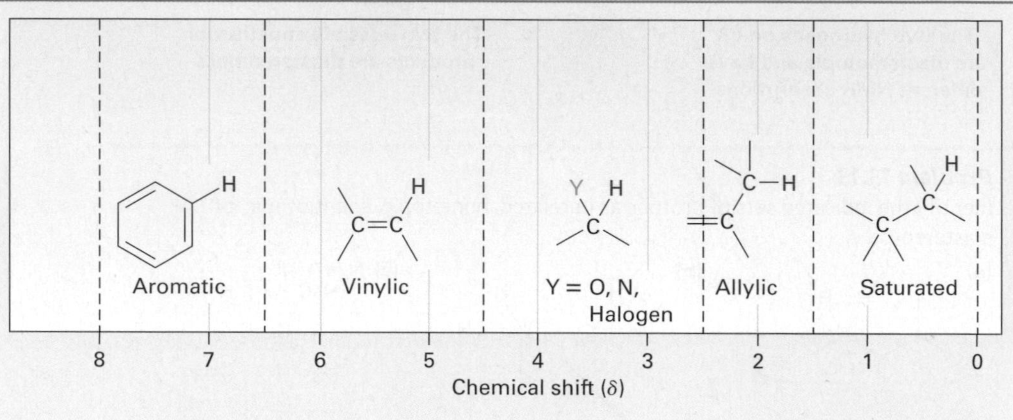

Table 13.3 shows the correlation of ^{1}H chemical shift with electronic environment in more detail. In general, protons bonded to saturated, *sp*^{3}-hybridized carbons absorb at higher fields, whereas protons bonded to *sp*^{2}-hybridized carbons absorb at lower fields. Protons on carbons that are bonded to electronegative atoms, such as N, O, or halogen, also absorb at lower fields.

Worked Example 13.3 Predicting Chemical Shifts in ^{1}H NMR Spectra

Methyl 2,2-dimethylpropanoate $(CH_3)_3CCO_2CH_3$ has two peaks in its ^{1}H NMR spectrum. What are their approximate chemical shifts?

Strategy

Identify the types of hydrogens in the molecule, and note whether each is alkyl, vinylic, or next to an electronegative atom. Then predict where each absorbs, using Table 13.3 if necessary.

Solution

The $-OCH_3$ protons absorb around 3.5 to 4.0 δ because they are on carbon bonded to oxygen. The $(CH_3)_3C-$ protons absorb near 1.0 δ because they are typical alkane-like protons.

Table 13.3 Correlation of ^1H Chemical Shift with Environment

Type of hydrogen		Chemical shift (δ)	Type of hydrogen		Chemical shift (δ)
Reference	$Si(CH_3)_4$	0	Alcohol	—C—O—H	2.5–5.0
Alkyl (primary)	$—CH_3$	0.7–1.3	Alcohol, ether	H—C—O—	3.3–4.5
Alkyl (secondary)	$—CH_2—$	1.2–1.6	Vinylic	C=C—H	4.5–6.5
Alkyl (tertiary)	—CH—	1.4–1.8	Aryl	Ar—H	6.5–8.0
Allylic	C=C—C—H	1.6–2.2	Aldehyde	O=C—H	9.7–10.0
Methyl ketone	$O=C—CH_3$	2.0–2.4	Carboxylic acid	O=C—O—H	11.0–12.0
Aromatic methyl	$Ar—CH_3$	2.4–2.7			
Alkynyl	—C≡C—H	2.5–3.0			
Alkyl halide	H—C—Hal	2.5–4.0			

Problem 13.16

Each of the following compounds has a single ^1H NMR peak. Approximately where would you expect each compound to absorb?

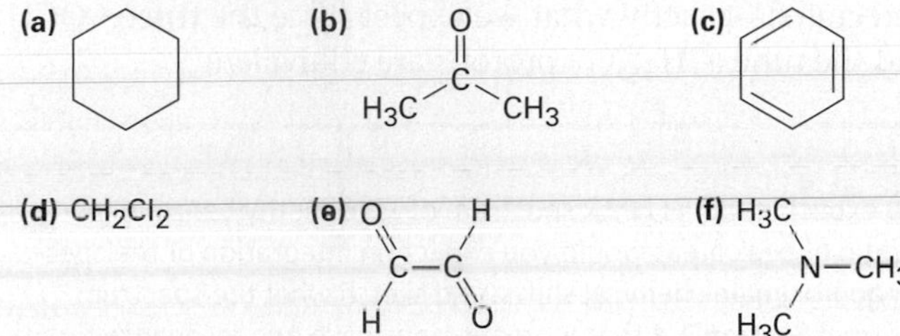

Problem 13.17

Identify the different kinds of nonequivalent protons in the following molecule, and tell where you would expect each to absorb:

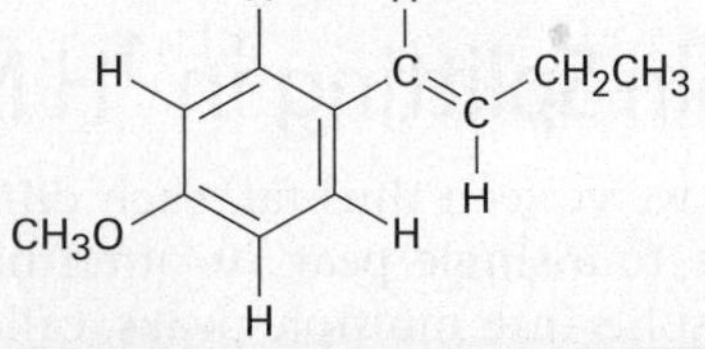

H H
H C CH2CH3
C
H
CH3O H
H

13.10 Integration of ^{1}H NMR Absorptions: Proton Counting

Look at the ^{1}H NMR spectrum of methyl 2,2-dimethylpropanoate in **Figure 13.12**. There are two peaks, corresponding to the two kinds of protons, but the peaks aren't the same size. The peak at 1.2 δ, due to the $(CH_3)_3C$– protons, is larger than the peak at 3.7 δ, due to the $–OCH_3$ protons.

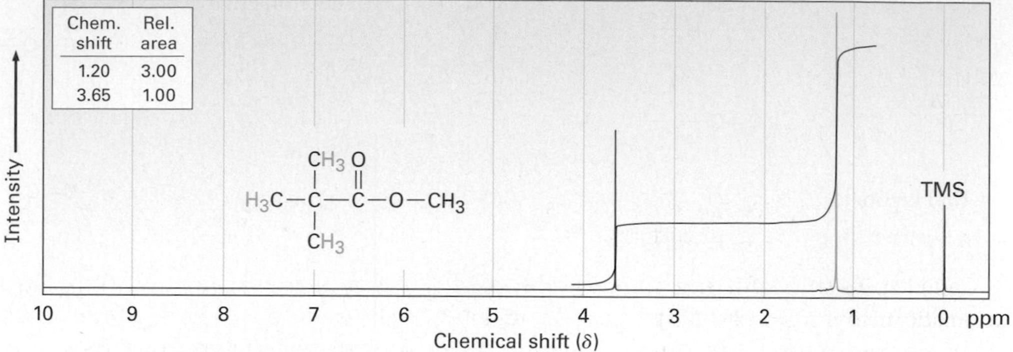

Figure 13.12 The ^{1}H NMR spectrum of methyl 2,2-dimethylpropanoate. Integrating the peaks in a stair-step manner shows that they have a 1:3 ratio, corresponding to the ratio of the numbers of protons (3:9) responsible for each peak. Modern instruments give a direct digital readout of relative peak areas.

The area under each peak is proportional to the number of protons causing that peak. By electronically measuring, or **integrating**, the area under each peak, it's possible to measure the relative numbers of the different kinds of protons in a molecule.

Modern NMR instruments provide a digital readout of relative peak areas, but an older, more visual method displays the integrated peak areas as a stair-step line, with the height of each step proportional to the area under the peak, and therefore proportional to the relative number of protons causing the peak. For example, the two steps for the peaks in methyl 2,2-dimethylpropanoate are found to have a 1:3 (or 3:9) height ratio when integrated—exactly what we expect since the three $–OCH_3$ protons are equivalent and the nine $(CH_3)_3C$– protons are equivalent.

Problem 13.18

How many peaks would you expect in the ^{1}H NMR spectrum of 1,4-dimethylbenzene (*para*-xylene, or *p*-xylene)? What ratio of peak areas would you expect on integration of the spectrum? Refer to Table 13.3 for approximate chemical shifts, and sketch what the spectrum would look like. (Remember from **Section 2.4** that aromatic rings have two resonance forms.)

CH_3

H_3C

***p*-Xylene**

13.11 Spin–Spin Splitting in ^{1}H NMR Spectra

In the ^{1}H NMR spectra we've seen thus far, each different kind of proton in a molecule has given rise to a single peak. It often happens, though, that the absorption of a proton splits into multiple peaks, called a **multiplet**. For example, in the ^{1}H NMR spectrum of bromoethane shown in **Figure 13.13**, the $-CH_2Br$ protons appear as four peaks (a *quartet*) centered at 3.42 δ and the $-CH_3$ protons appear as three peaks (a *triplet*) centered at 1.68 δ.

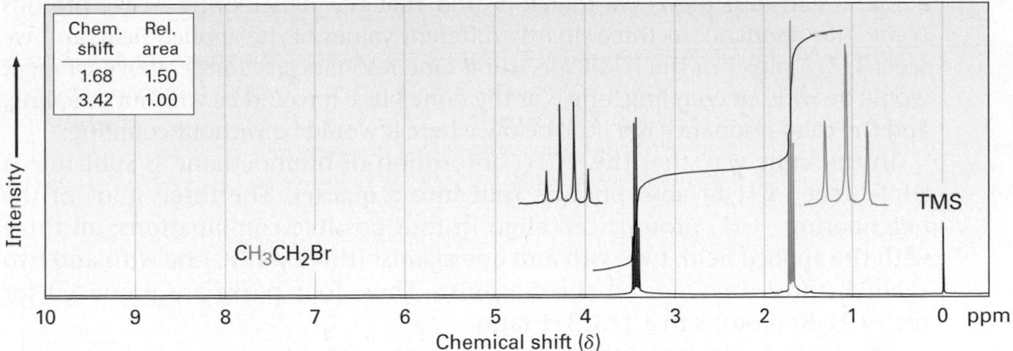

Figure 13.13 The ^{1}H NMR spectrum of bromoethane, CH_3CH_2Br. The $-CH_2Br$ protons appear as a quartet at 3.42 δ, and the $-CH_3$ protons appear as a triplet at 1.68 δ.

Called **spin–spin splitting**, multiple absorptions of a nucleus are caused by the interaction, or **coupling**, of the spins of nearby nuclei. In other words, the tiny magnetic field produced by one nucleus affects the magnetic field felt by a neighboring nucleus. Look at the $-CH_3$ protons in bromoethane, for example. The three equivalent $-CH_3$ protons are neighbored by two other magnetic nuclei—the two protons on the adjacent $-CH_2Br$ group. Each of the neighboring $-CH_2Br$ protons has its own nuclear spin, which can align either with or against the applied field, producing a tiny effect that is felt by the $-CH_3$ protons.

There are three ways in which the spins of the two $-CH_2Br$ protons can align, as shown in **Figure 13.14**. If both proton spins align with the applied field, the total effective field felt by the neighboring $-CH_3$ protons is slightly larger than it would otherwise be. Consequently, the applied field necessary to cause resonance is slightly reduced. Alternatively, if one of the $-CH_2Br$ proton spins aligns with the field and one aligns against the field, there is no effect on the neighboring $-CH_3$ protons. (This arrangement can occur in two ways, depending on which of the two proton spins aligns which way.) Finally, if both $-CH_2Br$ proton spins align against the applied field, the effective field felt by the $-CH_3$ protons is slightly smaller than it would otherwise be and the applied field needed for resonance is slightly increased.

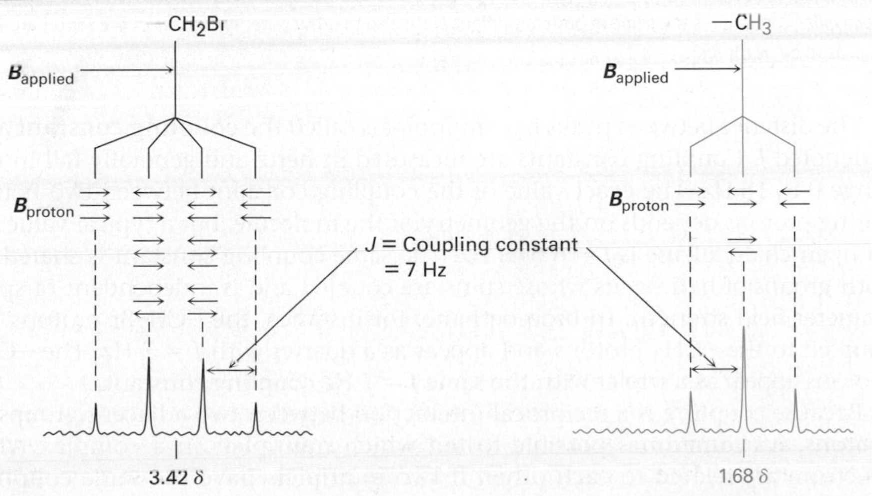

Figure 13.14 The origin of spin–spin splitting in bromoethane. The nuclear spins of neighboring protons, indicated by horizontal arrows, align either with or against the applied field, causing the splitting of absorptions into multiplets.

Any given molecule has only one of the three possible alignments of $-CH_2Br$ spins, but in a large collection of molecules, all three spin states are represented in a 1:2:1 statistical ratio. We therefore find that the neighboring $-CH_3$ protons come into resonance at three slightly different values of the applied field, and we see a 1:2:1 triplet in the NMR spectrum. One resonance is a little above where it would be without coupling, one is at the same place it would be without coupling, and the third resonance is a little below where it would be without coupling.

In the same way that the $-CH_3$ absorption of bromoethane is split into a triplet, the $-CH_2Br$ absorption is split into a quartet. The three spins of the neighboring $-CH_3$ protons can align in four possible combinations: all three with the applied field, two with and one against (three ways), one with and two against (three ways), or all three against. Thus, four peaks are produced for the $-CH_2Br$ protons in a 1:3:3:1 ratio.

As a general rule, called the ***n* + 1 rule**, protons that have *n* equivalent neighboring protons show $n + 1$ peaks in their NMR spectrum. For example, the spectrum of 2-bromopropane in **Figure 13.15** shows a doublet at 1.71 δ and a seven-line multiplet, or septet, at 4.28 δ. The septet is caused by splitting of the –CHBr– proton signal by six equivalent neighboring protons on the two methyl groups ($n = 6$ leads to $6 + 1 = 7$ peaks). The doublet is due to signal splitting of the six equivalent methyl protons by the single –CHBr– proton ($n = 1$ leads to 2 peaks). Integration confirms the expected 6:1 ratio.

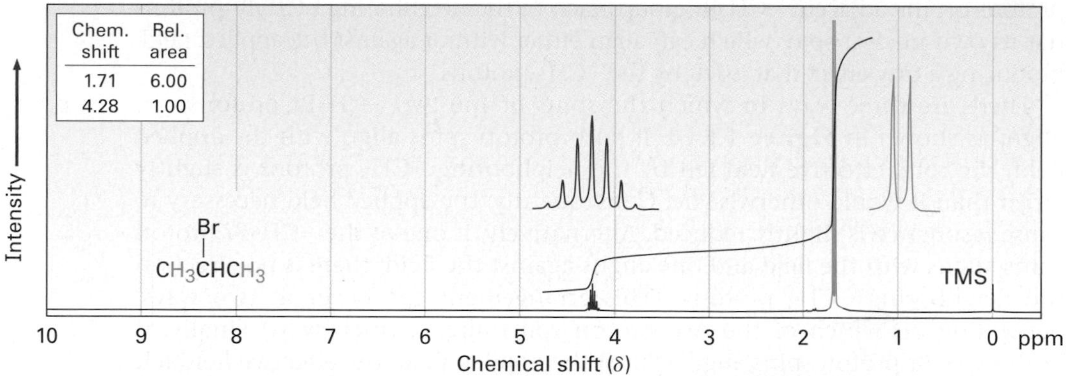

Figure 13.15 The ^{1}H NMR spectrum of 2-bromopropane. The $-CH_3$ proton signal at 1.71 δ is split into a doublet, and the –CHBr– proton signal at 4.28 δ is split into a septet. Note that the distance between peaks—the *coupling constant*—is the same in both multiplets. Note also that the outer two peaks of the septet are so small as to be nearly lost.

The distance between peaks in a multiplet is called the **coupling constant** and is denoted *J*. Coupling constants are measured in hertz and generally fall in the range 0 to 18 Hz. The exact value of the coupling constant between two neighboring protons depends on the geometry of the molecule, but a typical value for an open-chain alkane is $J = 6$ to 8 Hz. The same coupling constant is shared by both groups of hydrogens whose spins are coupled and is independent of spectrometer field strength. In bromoethane, for instance, the $-CH_2Br$ protons are coupled to the $-CH_3$ protons and appear as a quartet with $J = 7$ Hz. The $-CH_3$ protons appear as a triplet with the same $J = 7$ Hz coupling constant.

Because coupling is a reciprocal interaction between two adjacent groups of protons, it's sometimes possible to tell which multiplets in a complex NMR spectrum are related to each other. If two multiplets have the same coupling constant, they are probably related, and the protons causing those multiplets are therefore adjacent in the molecule.

The most commonly observed coupling patterns and the relative intensities of lines in their multiplets are listed in Table 13.4. Note that it's not possible for a given proton to have five equivalent neighboring protons. (Why not?) A six-line multiplet, or sextet, is therefore found only when a proton has five non-equivalent neighboring protons that coincidentally happen to be coupled with an identical coupling constant *J*.

Table 13.4 Some Common Spin Multiplicities

Number of equivalent adjacent protons	Multiplet	Ratio of intensities
0	Singlet	1
1	Doublet	1:1
2	Triplet	1:2:1
3	Quartet	1:3:3:1
4	Quintet	1:4:6:4:1
6	Septet	1:6:15:20:15:6:1

Spin–spin splitting in ^{1}H NMR can be summarized in three rules.

RULE 1

Chemically equivalent protons don't show spin–spin splitting. The equivalent protons may be on the same carbon or on different carbons, but their signals don't split.

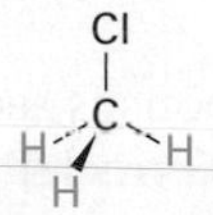

Three C–H protons are chemically equivalent; no splitting occurs.

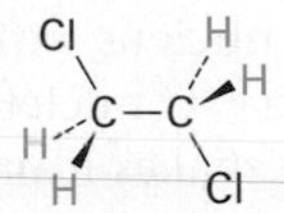

Four C–H protons are chemically equivalent; no splitting occurs.

RULE 2

The signal of a proton with *n* equivalent neighboring protons is split into a multiplet of *n* + 1 peaks with coupling constant *J*. Protons that are farther than two carbon atoms apart don't usually couple, although they sometimes show small coupling when they are separated by a π bond.

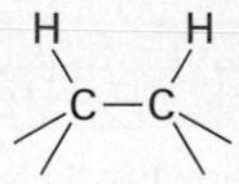

Splitting observed

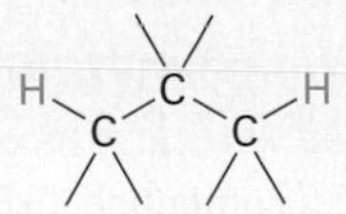

Splitting not usually observed

RULE 3

Two groups of protons coupled to each other have the same coupling constant, *J*.

The spectrum of *para*-methoxypropiophenone in **Figure 13.16** further illustrates the three rules. The downfield absorptions at 6.91 and 7.93 δ are due to the four aromatic ring protons. There are two kinds of aromatic protons, each of which gives a signal that is split into a doublet by its neighbor. The $-OCH_3$ signal is unsplit and appears as a sharp singlet at 3.84 δ. The $-CH_2-$ protons next to the carbonyl group appear at 2.93 δ in the region expected for protons on carbon next to an unsaturated center, and their signal is split into a quartet by coupling with the protons of the neighboring methyl group. The methyl protons appear as a triplet at 1.20 δ in the usual upfield region.

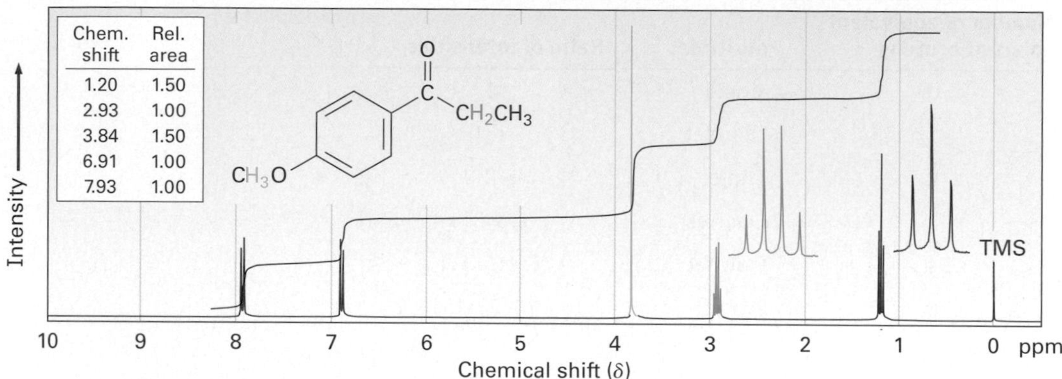

Figure 13.16 The ¹H NMR spectrum of *para*-methoxypropiophenone.

One further question needs to be answered before leaving the topic of spin–spin splitting. Why is spin–spin splitting seen only for ^1H NMR? Why is there no splitting of *carbon* signals into multiplets in ^{13}C NMR? After all, you might expect that the spin of a given ^{13}C nucleus would couple with the spin of an adjacent magnetic nucleus, either ^{13}C or ^1H.

No coupling of a ^{13}C nucleus with nearby carbons is seen because their low natural abundance makes it unlikely that two ^{13}C nuclei will be adjacent. No coupling of a ^{13}C nucleus with nearby hydrogens is seen because ^{13}C spectra, as previously noted **(Section 13.6)**, are normally recorded using broadband decoupling. At the same time that the sample is irradiated with a pulse of rf energy to cover the carbon resonance frequencies, it is also irradiated by a second band of rf energy covering all the hydrogen resonance frequencies. This second irradiation makes the hydrogens spin-flip so rapidly that their local magnetic fields average to zero and no coupling with carbon spins occurs.

Worked Example 13.4 Assigning a Chemical Structure from a ¹H NMR Spectrum

Propose a structure for a compound, $C_5H_{12}O$, that fits the following ^1H NMR data: 0.92 δ (3 H, triplet, $J = 7$ Hz), 1.20 δ (6 H, singlet), 1.50 δ (2 H, quartet, $J = 7$ Hz), 1.64 δ (1 H, broad singlet).

Strategy

As noted in Worked Example 13.2, it's best to begin solving structural problems by calculating a molecule's degree of unsaturation. In the present instance, a formula of $C_5H_{12}O$ corresponds to a saturated, open-chain molecule, either an alcohol or an ether.

To interpret the NMR information, let's look at each absorption individually. The three-proton absorption at 0.92 δ is due to a methyl group in an alkane-like environment, and the triplet splitting pattern implies that the CH_3 is next to a CH_2. Thus, our molecule contains an ethyl group, CH_3CH_2–. The six-proton singlet at 1.20 δ is due to two equivalent alkane-like methyl groups attached to a carbon with no hydrogens, $(CH_3)_2C$, and the two-proton quartet at 1.50 δ is due to the CH_2 of the ethyl group. All 5 carbons and 11 of the 12 hydrogens in the molecule are now accounted for. The remaining hydrogen, which appears as a broad one-proton singlet at 1.64 δ, is probably due to an OH group, since there is no other way to account for it. Putting the pieces together gives the structure: 2-methyl-2-butanol.

Solution

1.20 δ — CH_3; 1.50 δ

CH_3—C—CH_2CH_3 ⟵ 0.92 δ **2-Methyl-2-butanol**

OH ⟵ 1.64 δ

Problem 13.19

Predict the splitting patterns you would expect for each proton in the following molecules:

(a) $CHBr_2CH_3$ **(b)** $CH_3OCH_2CH_2Br$ **(c)** $ClCH_2CH_2CH_2Cl$

(d)

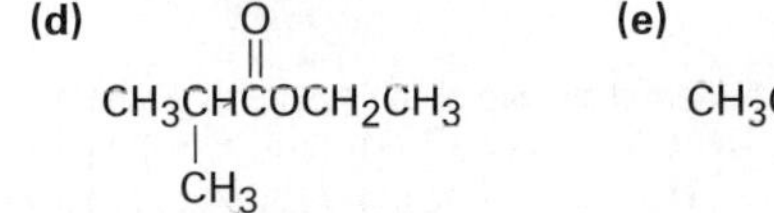

(e)

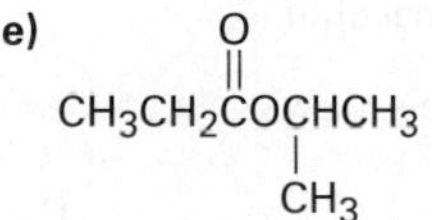

(f)

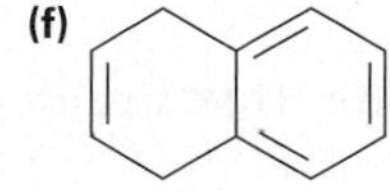

Problem 13.20

Draw structures for compounds that meet the following descriptions:

(a) C_2H_6O; one singlet **(b)** C_3H_7Cl; one doublet and one septet

(c) $C_4H_8Cl_2O$; two triplets **(d)** $C_4H_8O_2$; one singlet, one triplet, and one quartet

Problem 13.21

The integrated ^1H NMR spectrum of a compound of formula $C_4H_{10}O$ is shown in **Figure 13.17**. Propose a structure.

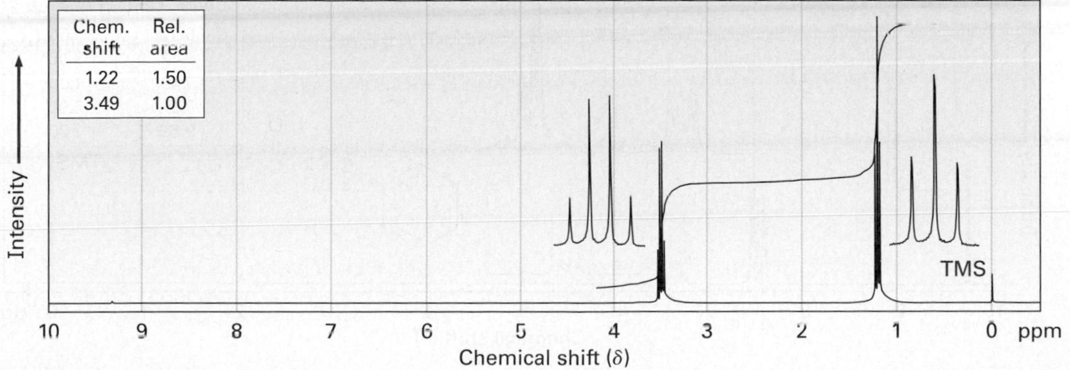

Figure 13.17 An integrated ^1H NMR spectrum for Problem 13.21.

13.12 More Complex Spin–Spin Splitting Patterns

In the ^{1}H NMR spectra we've seen so far, the chemical shifts of different protons have been distinct and the spin–spin splitting patterns have been straightforward. It often happens, however, that different kinds of hydrogens in a molecule have accidentally overlapping signals. The spectrum of toluene (methylbenzene) in **Figure 13.18**, for example, shows that the five aromatic ring protons give a complex, overlapping pattern, even though they aren't all equivalent.

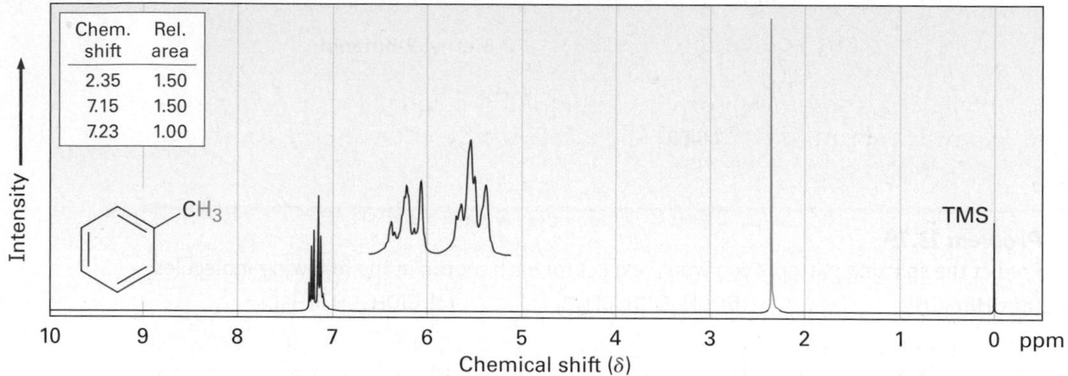

Figure 13.18 The ^{1}H NMR spectrum of toluene, showing the accidental overlap of the five nonequivalent aromatic ring protons.

Yet another complication in ^{1}H NMR spectroscopy arises when a signal is split by two or more *nonequivalent* kinds of protons, as is the case with *trans*-cinnamaldehyde, isolated from oil of cinnamon **(Figure 13.19)**. Although the $n + 1$ rule predicts splitting caused by equivalent protons, splittings caused by nonequivalent protons are more complex.

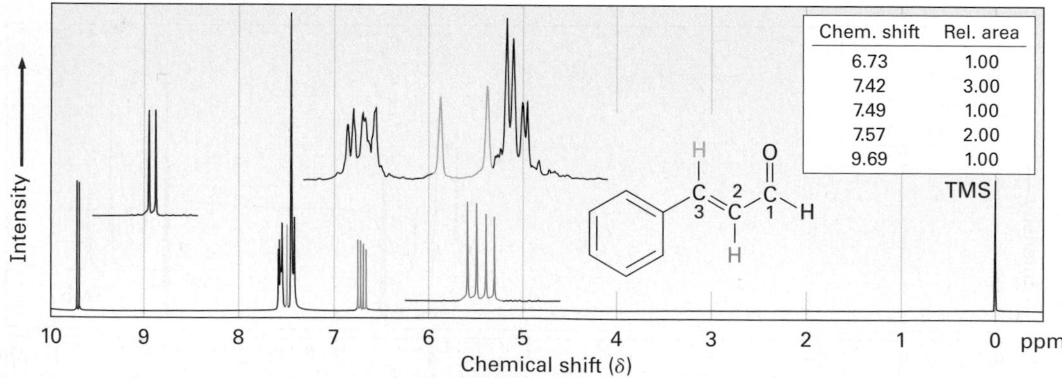

Figure 13.19 The ^{1}H NMR spectrum of *trans*-cinnamaldehyde. The signal of the proton at C2 (blue) is split into four peaks—a doublet of doublets—by the two nonequivalent neighboring protons.

To understand the ^{1}H NMR spectrum of *trans*-cinnamaldehyde, we have to isolate the different parts and look at the signal of each proton individually.

- The five aromatic proton signals (black in Figure 13.19) overlap into a complex pattern with a large peak at 7.42 δ and a broad absorption at 7.57 δ.
- The aldehyde proton signal at C1 (red) appears in the normal downfield position at 9.69 δ and is split into a doublet with $J = 6$ Hz by the adjacent proton at C2.
- The vinylic proton at C3 (green) is next to the aromatic ring and is therefore shifted downfield from the normal vinylic region. This C3 proton signal appears as a doublet centered at 7.49 δ. Because it has one neighbor proton at C2, its signal is split into a doublet, with $J = 12$ Hz.
- The C2 vinylic proton signal (blue) appears at 6.73 δ and shows an interesting, four-line absorption pattern. It is coupled to the two non-equivalent protons at C1 and C3 with two different coupling constants: $J_{1\text{-}2} = 6$ Hz and $J_{2\text{-}3} = 12$ Hz.

A good way to understand the effect of multiple coupling such as occurs for the C2 proton of *trans*-cinnamaldehyde is to draw a *tree diagram*, like that in **Figure 13.20**. The diagram shows the individual effect of each coupling constant on the overall pattern. Coupling with the C3 proton splits the signal of the C2 proton in *trans*-cinnamaldehyde into a doublet with $J = 12$ Hz. Further coupling with the aldehyde proton then splits each peak of the doublet into new doublets with $J = 6$ Hz, and we therefore observe a four-line spectrum for the C2 proton.

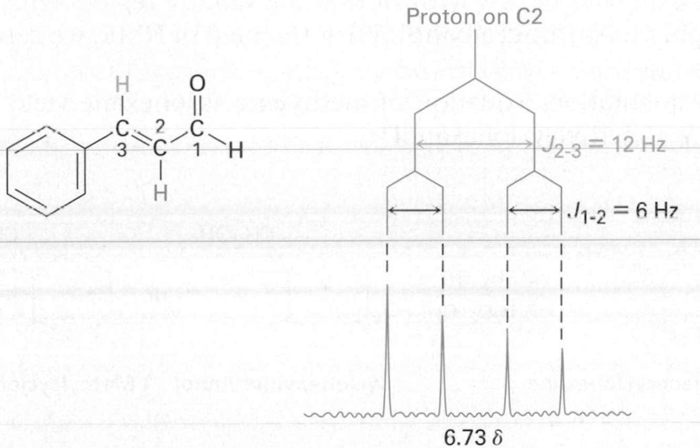

Figure 13.20 A tree diagram for the C2 proton of *trans*-cinnamaldehyde shows how it is coupled to the C1 and C3 protons with different coupling constants.

One further point evident in the cinnamaldehyde spectrum is that the four peaks of the C2 proton signal are not all the same size. The two left-hand peaks are somewhat larger than the two right-hand peaks. Such a size difference occurs whenever coupled nuclei have similar chemical shifts in this case, 7.49 δ for the C3 proton and 6.73 δ for the C2 proton. The peaks nearer the signal of the coupled partner are always larger, and the peaks farther from the signal of the coupled partner are always smaller. Thus, the left-hand peaks of the C2 proton multiplet at 6.73 δ are closer to the C3 proton absorption at 7.49 δ and are larger than the right-hand peaks. At the same time, the right-hand peak of the C3 proton doublet at 7.49 δ is larger than the left-hand peak because it is closer to the C2 proton multiplet at 6.73 δ. This skewing effect on multiplets can often be useful because it tells where to look in the spectrum to find the coupled partner: look in the direction of the larger peaks.

Problem 13.22

3-Bromo-1-phenyl-1-propene shows a complex NMR spectrum in which the vinylic proton at C2 is coupled with both the C1 vinylic proton (J = 16 Hz) and the C3 methylene protons (J = 8 Hz). Draw a tree diagram for the C2 proton signal, and account for the fact that a five-line multiplet is observed.

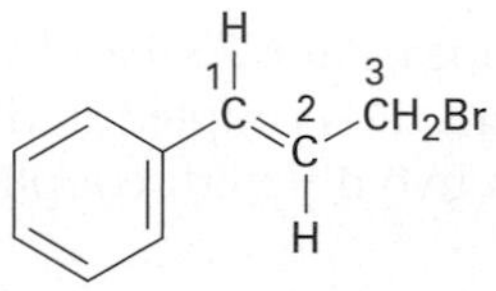

H, 1, 3, C, 2, C, CH_2Br, H

3-Bromo-1-phenyl-1-propene

13.13 Uses of ^{1}H NMR Spectroscopy

NMR is used to help identify the product of nearly every reaction run in the laboratory. For example, we said in **Section 8.5** that hydroboration–oxidation of alkenes occurs with non-Markovnikov regiochemistry to yield the less highly substituted alcohol. With the help of NMR, we can now prove this statement.

Does hydroboration–oxidation of methylenecyclohexane yield cyclohexylmethanol or 1-methylcyclohexanol?

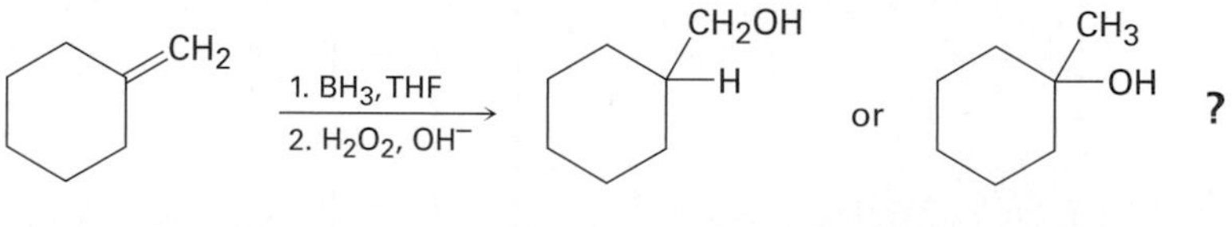

CH_2 — 1. BH_3, THF; 2. H_2O_2, OH^- → CH_2OH, H or CH_3, OH ?

Methylenecyclohexane **Cyclohexylmethanol** **1-Methylcyclohexanol**

The ^{1}H NMR spectrum of the reaction product is shown in **Figure 13.21a**. The spectrum shows a 2-proton peak at 3.40 δ, indicating that the product has a $-CH_2-$ group bonded to an electronegative oxygen atom ($-CH_2OH$). Furthermore, the spectrum shows no large three-proton singlet absorption near 1 δ, where we would expect the signal of a quaternary $-CH_3$ group to appear. (Figure 13.21b gives the spectrum of 1-methylcyclohexanol, the alternative product.) Thus, it's clear that cyclohexylmethanol is the reaction product.

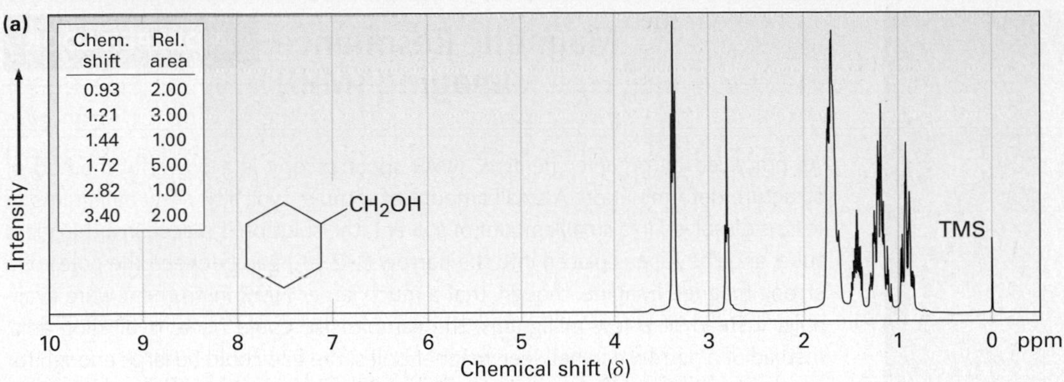

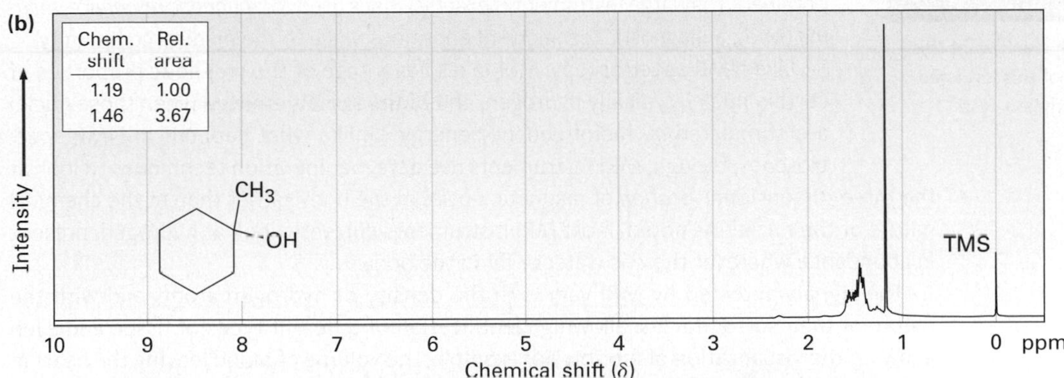

Figure 13.21 **(a)** The ¹H NMR spectrum of cyclohexylmethanol, the product from hydroboration–oxidation of methylenecyclohexane, and **(b)** the ¹H NMR spectrum of 1-methylcyclohexanol, the possible alternative reaction product.

Problem 13.23

How could you use ¹H NMR to determine the regiochemistry of electrophilic addition to alkenes? For example, does addition of HCl to 1-methylcyclohexene yield 1-chloro-1-methylcyclohexane or 1-chloro-2-methylcyclohexane?

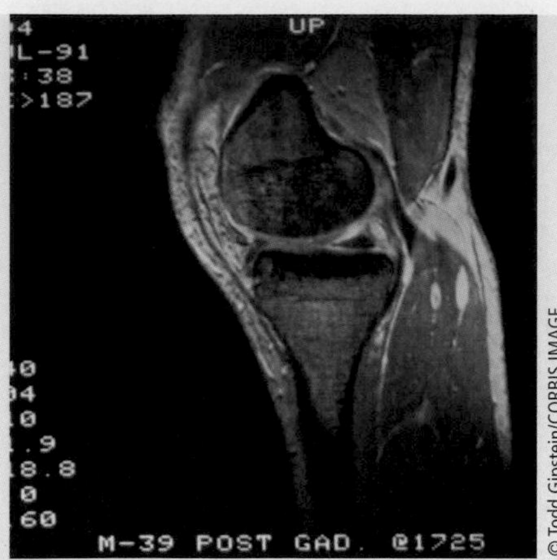

If you're a runner, you really don't want this to happen to you. The MRI of this left knee shows the presence of a ganglion cyst.

Magnetic Resonance Imaging (MRI)

A DEEPER LOOK

As practiced by organic chemists, NMR spectroscopy is a powerful method of structure determination. A small amount of sample, typically a few milligrams or less, is dissolved in a small amount of solvent, the solution is placed in a thin glass tube, and the tube is placed into the narrow (1–2 cm) gap between the poles of a strong magnet. Imagine, though, that a much larger NMR instrument were available. Instead of a few milligrams, the sample size could be tens of kilograms; instead of a narrow gap between magnet poles, the gap could be large enough for a whole person to climb into so that an NMR spectrum of body parts could be obtained. That large instrument is exactly what's used for *magnetic resonance imaging (MRI)*, a diagnostic technique of enormous value to the medical community.

Like NMR spectroscopy, MRI takes advantage of the magnetic properties of certain nuclei, typically hydrogen, and of the signals emitted when those nuclei are stimulated by radiofrequency energy. Unlike what happens in NMR spectroscopy, though, MRI instruments use data manipulation techniques to look at the three-dimensional *location* of magnetic nuclei in the body rather than at the chemical nature of the nuclei. As noted, most MRI instruments currently look at hydrogen, present in abundance wherever there is water or fat in the body.

The signals detected by MRI vary with the density of hydrogen atoms and with the nature of their surroundings, allowing identification of different types of tissue and even allowing the visualization of motion. For example, the volume of blood leaving the heart in a single stroke can be measured, and heart motion can be observed. Soft tissues that don't show up well on X rays can be seen clearly, allowing diagnosis of brain tumors, strokes, and other conditions. The technique is also valuable in diagnosing damage to knees or other joints and is a noninvasive alternative to surgical explorations.

Several types of atoms in addition to hydrogen can be detected by MRI, and the applications of images based on ^{31}P atoms are being explored. The technique holds great promise for studies of metabolism.

Summary

Key words

chemical shift, 462
coupling, 477
coupling constant (*J*), 478
delta (δ) scale, 462
diastereotopic, 473
downfield, 461
enantiotopic, 472
FT–NMR, 463
homotopic, 472
integration, 476
multiplet, 476
***n* + 1 rule, 478**

Nuclear magnetic resonance spectroscopy, or **NMR**, is the most valuable of the numerous spectroscopic techniques used for structure determination. Although we focused in this chapter on NMR applications to small molecules, more advanced NMR techniques are also used in biological chemistry to study protein structure and folding.

When magnetic nuclei, such as ^{1}H and ^{13}C, are placed in a strong magnetic field, their spins orient either with or against the field. On irradiation with radiofrequency (rf) waves, energy is absorbed and the nuclei "spin-flip" from the lower energy state to the higher energy state. This absorption of rf energy is detected, amplified, and displayed as an NMR spectrum.

Each electronically distinct ^{1}H or ^{13}C nucleus in a molecule comes into resonance at a slightly different value of the applied field, thereby producing a unique absorption signal. The exact position of each peak is called the **chemical shift**. Chemical shifts are caused by electrons setting up tiny local magnetic fields that **shield** a nearby nucleus from the applied field.

The NMR chart is calibrated in **delta units (δ)**, where 1 δ = 1 ppm of spectrometer frequency. Tetramethylsilane (TMS) is used as a reference point because it shows both ^{1}H and ^{13}C absorptions at unusually high values of the applied magnetic field. The TMS absorption occurs at the right-hand (**upfield**) side of the chart and is arbitrarily assigned a value of 0 δ.

^{13}C spectra are run on Fourier-transform NMR (**FT–NMR**) spectrometers using broadband decoupling of proton spins so that each chemically distinct carbon shows a single unsplit resonance line. As with ^{1}H NMR, the chemical shift of each ^{13}C signal provides information about a carbon's chemical environment in the sample. In addition, the number of protons attached to each carbon can be determined using the DEPT–NMR technique.

In ^{1}H NMR spectra, the area under each absorption peak can be electronically **integrated** to determine the relative number of hydrogens responsible for each peak. In addition, neighboring nuclear spins can **couple**, causing the **spin–spin splitting** of NMR peaks into **multiplets**. The NMR signal of a hydrogen neighbored by n equivalent adjacent hydrogens splits into $n + 1$ peaks (the **n + 1 rule**) with **coupling constant J**.

Key words—cont'd

nuclear magnetic resonance (NMR) spectroscopy, 456
shielding, 458
spin–spin splitting, 477
upfield, 461

Exercises

Visualizing Chemistry

(Problems 13.1–13.23 appear within the chapter.)

OWL Interactive versions of these problems are assignable in OWL for Organic Chemistry.

13.24 Into how many peaks would you expect the ^{1}H NMR signals of the indicated protons to be split? (Green = Cl.)

(a)

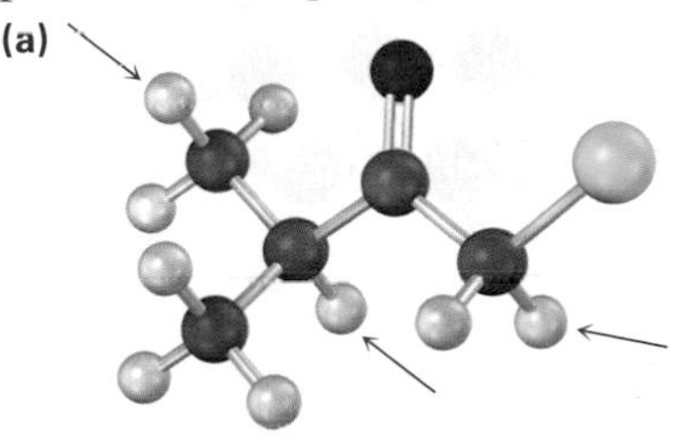

(b)

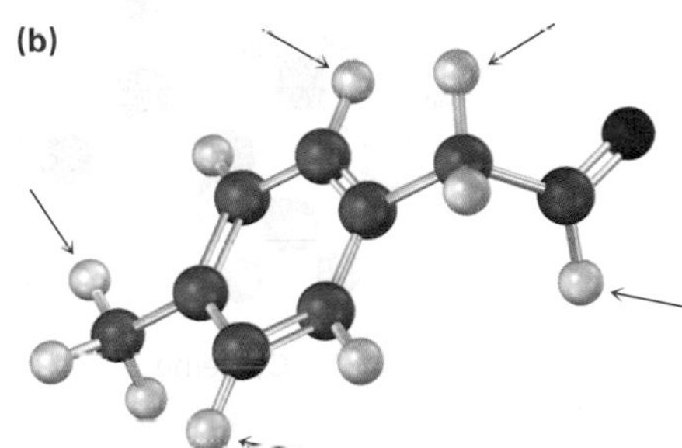

13.25 How many absorptions would you expect the following compound to have in its ^{1}H and ^{13}C NMR spectra?

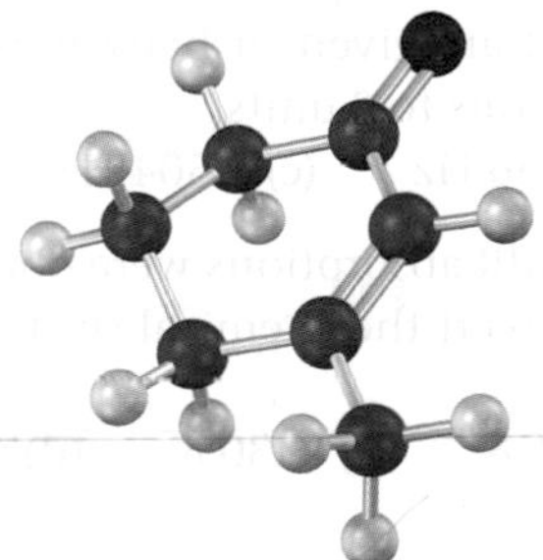

C2–C3 single bond in 1,3-butadiene, for instance, has a length of 147 pm, some 6 pm shorter than the C2–C3 bond in butane (153 pm).

147 pm	153 pm
$H_2C{=}CH{-}CH{=}CH_2$	$CH_3{-}CH_2{-}CH_2{-}CH_3$
1,3-Butadiene	**Butane**

Another distinctive property of conjugated dienes is their unusual stability, as evidenced by their heats of hydrogenation (Table 14.1). We saw in **Section 7.6** that monosubstituted alkenes, such as 1-butene, have $\Delta H°_{hydrog}$ near −126 kJ/mol (−30.1 kcal/mol), whereas disubstituted alkenes, such as 2-methylpropene, have $\Delta H°_{hydrog}$ near −119 kJ/mol (−28.4 kcal/mol), approximately 7 kJ/mol less negative. We concluded from these data that more highly substituted alkenes are more stable than less substituted ones. That is, more highly substituted alkenes release less heat on hydrogenation because they contain less energy to start with. A similar conclusion can be drawn for conjugated dienes.

Table 14.1 Heats of Hydrogenation for Some Alkenes and Dienes

		$\Delta H°_{hydrog}$	
Alkene or diene	**Product**	**(kJ/mol)**	**(kcal/mol)**
$CH_3CH_2CH{=}CH_2$	$CH_3CH_2CH_2CH_3$	−126	−30.1
$CH_3C(CH_3){=}CH_2$	$CH_3CH(CH_3)CH_3$	−119	−28.4
$H_2C{=}CHCH_2CH{=}CH_2$	$CH_3CH_2CH_2CH_2CH_3$	−253	−60.5
$H_2C{=}CH{-}CH{=}CH_2$	$CH_3CH_2CH_2CH_3$	−236	−56.4
$H_2C{=}CH{-}C(CH_3){=}CH_2$	$CH_3CH_2CH(CH_3)CH_3$	−229	−54.7

Because a monosubstituted alkene has a $\Delta H°_{hydrog}$ of approximately −126 kJ/mol, we might expect that a compound with two monosubstituted double bonds would have a $\Delta H°_{hydrog}$ approximately twice that value, or −252 kJ/mol. Nonconjugated dienes, such as 1,4-pentadiene ($\Delta H°_{hydrog}$ = −253 kJ/mol), meet this expectation, but the conjugated diene 1,3-butadiene ($\Delta H°_{hydrog}$ = −236 kJ/mol) does not. 1,3-Butadiene is approximately 16 kJ/mol (3.8 kcal/mol) more stable than expected.

	$\Delta H°_{hydrog}$ (kJ/mol)	
$H_2C{=}CHCH_2CH{=}CH_2$	−126 + (−126) = −252	Expected
1,4-Pentadiene	−253	Observed
	1	Difference
$H_2C{=}CHCH{=}CH_2$	−126 + (−126) = −252	Expected
1,3-Butadiene	−236	Observed
	−16	Difference

What accounts for the stability of conjugated dienes? According to valence bond theory **(Sections 1.5 and 1.8)**, the stability is due to orbital hybridization. Typical C–C single bonds like those in alkanes result from σ overlap of sp^3 orbitals on both carbons, but in a conjugated diene, the central C–C single bond results from σ overlap of sp^2 orbitals on both carbons. Since sp^2 orbitals have more s character (33% s) than sp^3 orbitals (25% s), the electrons in sp^2 orbitals are closer to the nucleus and the bonds they form are somewhat shorter and stronger. Thus, the "extra" stability of a conjugated diene results in part from the greater amount of s character in the orbitals forming the C–C single bond.

CH_3–CH_2–CH_2–CH_3 H_2C=CH–CH=CH_2

Bonds formed by overlap of sp^3 orbitals   **Bond formed by overlap of sp^2 orbitals**

According to molecular orbital theory **(Section 1.11)**, the stability of a conjugated diene arises because of an interaction between the π orbitals of the two double bonds. To review briefly, when two p atomic orbitals combine to form a π bond, two π molecular orbitals (MOs) result. One is lower in energy than the starting p orbitals and is therefore bonding; the other is higher in energy, has a node between nuclei, and is antibonding. The two π electrons occupy the low-energy, bonding orbital, resulting in formation of a stable bond between atoms **(Figure 14.1)**.

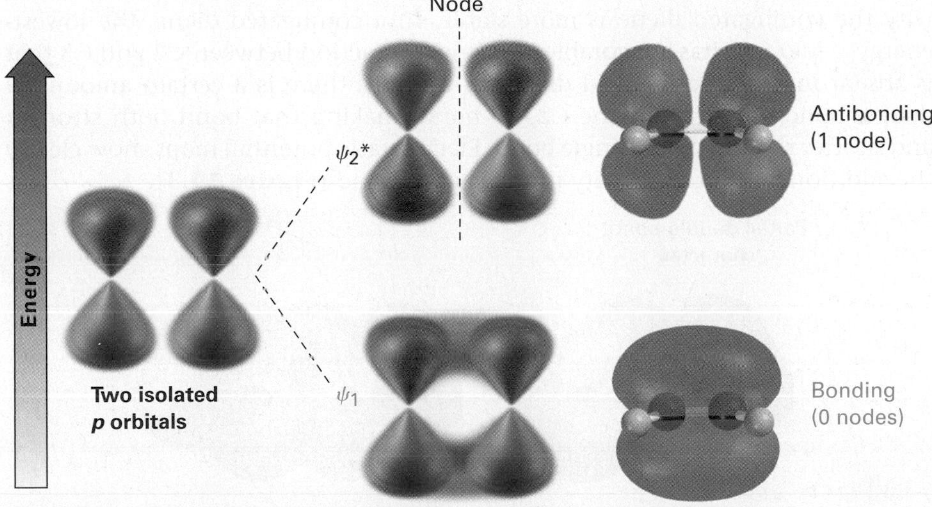

Figure 14.1 Two p orbitals combine to form two π molecular orbitals. Both electrons occupy the low-energy, bonding orbital, leading to a net lowering of energy and formation of a stable bond. The asterisk on $\psi_2{}^*$ indicates an antibonding orbital.

Now let's combine four adjacent p atomic orbitals, as occurs in a conjugated diene. In so doing, we generate a set of four π molecular orbitals, two of which are bonding and two of which are antibonding **(Figure 14.2)**. The four π electrons occupy the two bonding orbitals, leaving the antibonding orbitals vacant.

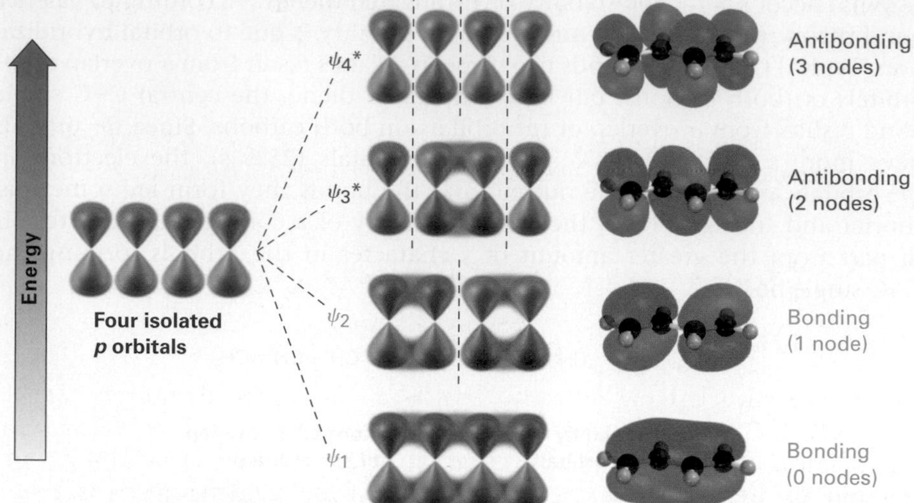

Figure 14.2 Four π molecular orbitals in 1,3-butadiene. Note that the number of nodes between nuclei increases as the energy level of the orbital increases.

The lowest-energy π molecular orbital (denoted ψ_1, Greek psi) has no nodes between the nuclei and is therefore bonding. The π MO of next lowest energy, ψ_2, has one node between nuclei and is also bonding. Above ψ_1 and ψ_2 in energy are the two antibonding π MOs, ψ_3^* and ψ_4^*. (The asterisks indicate antibonding orbitals.) Note that the number of nodes between nuclei increases as the energy level of the orbital increases. The ψ_3^* orbital has two nodes between nuclei, and ψ_4^*, the highest-energy MO, has three nodes between nuclei.

Comparing the π molecular orbitals of 1,3-butadiene (two conjugated double bonds) with those of 1,4-pentadiene (two isolated double bonds) shows why the conjugated diene is more stable. In a conjugated diene, the lowest-energy π MO (ψ_1) has a favorable bonding interaction between C2 and C3 that is absent in a nonconjugated diene. As a result, there is a certain amount of double-bond character to the C2–C3 bond, making that bond both stronger and shorter than a typical single bond. Electrostatic potential maps show clearly the additional electron density in the central bond **(Figure 14.3)**.

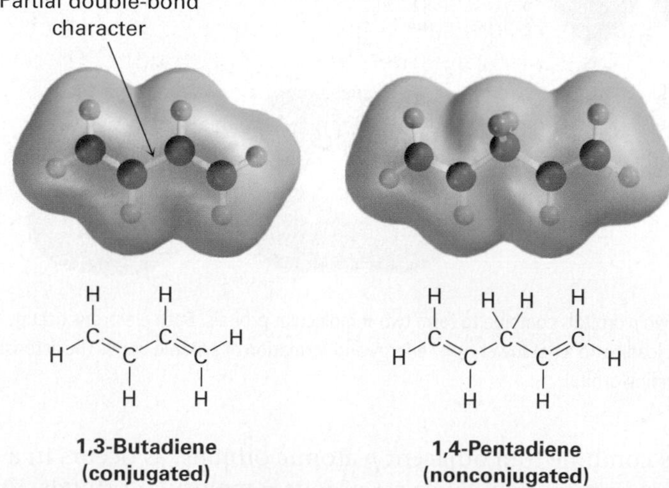

Figure 14.3 Electrostatic potential maps of 1,3-butadiene (conjugated) and 1,4-pentadiene (nonconjugated) show additional **electron density** in the central C–C bond of 1,3-butadiene, corresponding to partial double-bond character.

In describing 1,3-butadiene, we say that the π electrons are spread out, or *delocalized,* over the entire π framework rather than localized between two specific nuclei. Delocalization allows the bonding electrons to be closer to more nuclei, thus leading to lower energy and greater stability.

Problem 14.1

Allene, $H_2C{=}C{=}CH_2$, has a heat of hydrogenation of −298 kJ/mol (−71.3 kcal/mol). Rank a conjugated diene, a nonconjugated diene, and an allene in order of stability.

14.2 Electrophilic Additions to Conjugated Dienes: Allylic Carbocations

One of the most striking differences between conjugated dienes and typical alkenes is their behavior in electrophilic addition reactions. To review briefly, the addition of an electrophile to a carbon–carbon double bond is a general reaction of alkenes **(Section 7.7)**. Markovnikov regiochemistry is found because the more stable carbocation is formed as an intermediate. Thus, addition of HCl to 2-methylpropene yields 2-chloro-2-methylpropane rather than 1-chloro-2-methylpropane, and addition of 2 mol equiv of HCl to the nonconjugated diene 1,4-pentadiene yields 2,4-dichloropentane.

$$CH_3C(CH_3){=}CH_2 \xrightarrow[\text{Ether}]{\text{HCl}} \left[CH_3\overset{+}{C}(CH_3)CH_3\right] \longrightarrow CH_3C(Cl)(CH_3)CH_3$$

2-Methylpropene **Tertiary carbocation** **2-Chloro-2-methylpropane**

$$H_2C{=}CHCH_2CH{=}CH_2 \xrightarrow[\text{Ether}]{\text{HCl}} CH_3CH(Cl)CH_2CH(Cl)CH_3$$

1,4-Pentadiene (nonconjugated) **2,4-Dichloropentane**

Conjugated dienes also undergo electrophilic addition reactions readily, but mixtures of products are invariably obtained. Addition of HBr to 1,3-butadiene, for instance, yields a mixture of two products (not counting cis–trans isomers). 3-Bromo-1-butene is the typical Markovnikov product of **1,2 addition** to a double bond, but 1-bromo-2-butene appears unusual. The double bond in this product has moved to a position between carbons 2 and 3, and HBr has added to carbons 1 and 4, a result described as **1,4 addition**.

$$H_2C{=}CH{-}CH{=}CH_2 \xrightarrow{\text{HBr}} H_2C{=}CH{-}CHBr{-}CH_3 \;+\; BrCH_2{-}CH{=}CH{-}CH_3$$

1,3-Butadiene

3-Bromo-1-butene (71%; 1,2-addition)

1-Bromo-2-butene (29%; 1,4-addition)

Many other electrophiles besides HBr add to conjugated dienes, and mixtures of products are usually formed. For example, Br_2 adds to 1,3-butadiene to give a mixture of 3,4-dibromo-1-butene and 1,4-dibromo-2-butene.

Br_2, 20 °C

1,3-Butadiene | **3,4-Dibromo-1-butene (55%; 1,2-addition)** + **1,4-Dibromo-2-butene (45%; 1,4-addition)**

How can we account for the formation of 1,4-addition products? The answer is that allylic carbocations are involved as intermediates (recall that the word *allylic* means "next to a double bond"). When 1,3-butadiene reacts with an electrophile such as H^+, two carbocation intermediates are possible—a primary nonallylic carbocation and a secondary allylic cation. Because an allylic cation is stabilized by resonance between two forms **(Section 11.5)**, it is more stable and forms faster than a nonallylic carbocation.

1,3-Butadiene — HBr →

Secondary, allylic (Br^-)

Primary, nonallylic (Br^-)
(not formed)

When the allylic cation reacts with Br^- to complete the electrophilic addition, reaction can occur either at C1 or at C3 because both carbons share the positive charge **(Figure 14.4)**. Thus, a mixture of 1,2- and 1,4-addition products results. You might recall that a similar product mixture was seen for NBS bromination of alkenes in **Section 10.3**, a reaction that proceeds through an allylic *radical*.

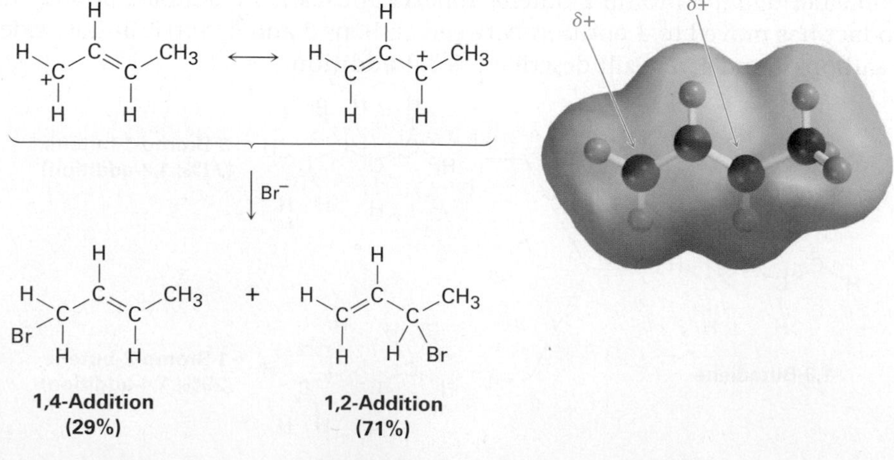

Figure 14.4 An electrostatic potential map of the allylic carbocation produced by protonation of 1,3-butadiene shows that the positive charge is shared by carbons 1 and 3. Reaction of Br^- with the more positive carbon (C3) gives predominantly the 1,2-addition product.

Predicting the Product of an Electrophilic Addition Reaction of a Conjugated Diene

Worked Example 14.1

Give the structures of the likely products from reaction of 1 equivalent of HCl with 2-methyl-1,3-cyclohexadiene. Show both 1,2 and 1,4 adducts.

Strategy

Electrophilic addition of HCl to a conjugated diene involves the formation of allylic carbocation intermediates. Thus, the first step is to protonate the two ends of the diene and draw the resonance forms of the two allylic carbocations that result. Then allow each resonance form to react with Cl^-, generating a maximum of four possible products.

In the present instance, protonation of the C1–C2 double bond gives a carbocation that can react further to give the 1,2 adduct 3-chloro-3-methylcyclohexene and the 1,4 adduct 3-chloro-1-methylcyclohexene. Protonation of the C3–C4 double bond gives a symmetrical carbocation, whose two resonance forms are equivalent. Thus, the 1,2 adduct and the 1,4 adduct have the same structure: 6-chloro-1-methylcyclohexene. Of the two possible modes of protonation, the first is more likely because it yields a more stable, tertiary allylic cation rather than a less-stable, secondary allylic cation.

Solution

HCl

2-Methyl-1,3-cyclohexadiene

1,2 1,4 1,2 and 1,4

3-Chloro-3-methylcyclohexene 3-Chloro-1-methylcyclohexene 6-Chloro-1-methylcyclohexene

Problem 14.2

Give the structures of both 1,2 and 1,4 adducts resulting from reaction of 1 equivalent of HCl with 1,3-pentadiene.

Problem 14.3

Look at the possible carbocation intermediates produced during addition of HCl to 1,3-pentadiene (Problem 14.2), and predict which 1,2 adduct predominates. Which 1,4 adduct predominates?

Problem 14.4

Give the structures of both 1,2 and 1,4 adducts resulting from reaction of 1 equivalent of HBr with the following compound:

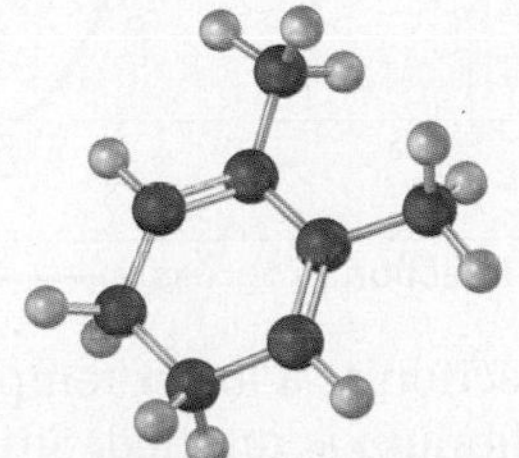

14.3 Kinetic versus Thermodynamic Control of Reactions

***Key* IDEAS**

Test your knowledge of Key Ideas by answering end-of-chapter exercises marked with ▲.

Electrophilic addition to a conjugated diene at or below room temperature normally leads to a mixture of products in which the 1,2 adduct predominates over the 1,4 adduct. When the same reaction is carried out at higher temperatures, though, the product ratio often changes and the 1,4 adduct predominates. For example, addition of HBr to 1,3-butadiene at 0 °C yields a 71:29 mixture of 1,2 and 1,4 adducts, but the same reaction carried out at 40 °C yields a 15:85 mixture. Furthermore, when the product mixture formed at 0 °C is heated to 40 °C in the presence of HBr, the ratio of adducts slowly changes from 71:29 to 15:85. Why?

HBr

1,3-Butadiene **1,2-Adduct** **1,4-Adduct**

	1,2-Adduct	1,4-Adduct
At 0 °C:	71%	29%
At 40 °C:	15%	85%

To understand the effect of temperature on product distribution, let's briefly review what we said in **Section 6.7** about rates and equilibria. Imagine a reaction that can give either or both of two products, **B** and **C**.

A → B, A → C

Let's assume that **B** forms faster than **C** (in other words, $\Delta G^{\ddagger}_{B} < \Delta G^{\ddagger}_{C}$) but that **C** is more stable than **B** (in other words, $\Delta G^{\circ}_{C} > \Delta G^{\circ}_{B}$). An energy diagram for the two processes might look like that shown in **Figure 14.5**.

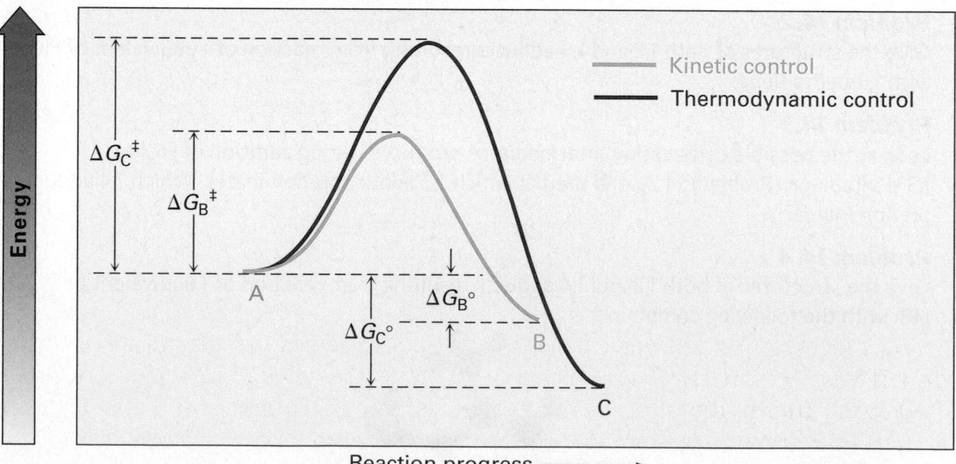

Figure 14.5 An energy diagram for two competing reactions in which the less stable product **B** forms faster than the more stable product **C**.

Let's first carry out the reaction at a lower temperature so that both processes are irreversible and no equilibrium is reached. Since **B** forms faster than **C**, **B** is the major product. It doesn't matter that **C** is more stable than **B**, because the two are not in equilibrium. *The product of an irreversible reaction depends only on relative rates, not on stability.* Such reactions are said to be under **kinetic control**.

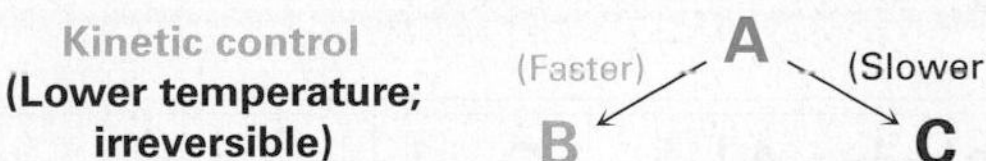

Now let's carry out the same reaction at some higher temperature so that both processes are readily reversible and an equilibrium is reached. Since **C** is more stable than **B**, **C** is the major product obtained. It doesn't matter that **C** forms more slowly than **B**, because the two are in equilibrium. *The product of a readily reversible reaction depends only on stability, not on relative rates.* Such reactions are said to be under equilibrium control, or **thermodynamic control**.

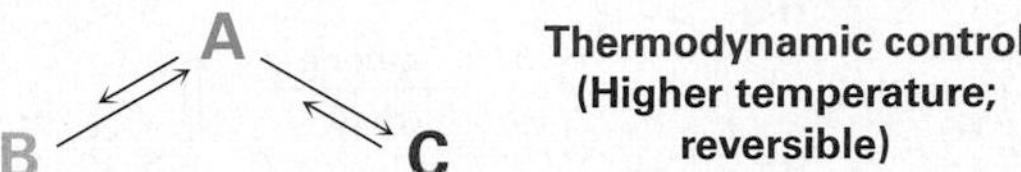

We can now explain the effect of temperature on electrophilic addition reactions of conjugated dienes. At low temperature (0 °C), HBr adds to 1,3-butadiene under kinetic control to give a 71:29 mixture of products, with the more rapidly formed 1,2 adduct predominating. Since these conditions don't allow the reaction to reach equilibrium, the product that forms faster predominates. At higher temperature (40 °C), however, the reaction occurs under thermodynamic control to give a 15:85 mixture of products, with the more stable 1,4 adduct predominating. The higher temperature allows the addition process to become reversible, so an equilibrium mixture of products results. **Figure 14.6** shows the situation in an energy diagram.

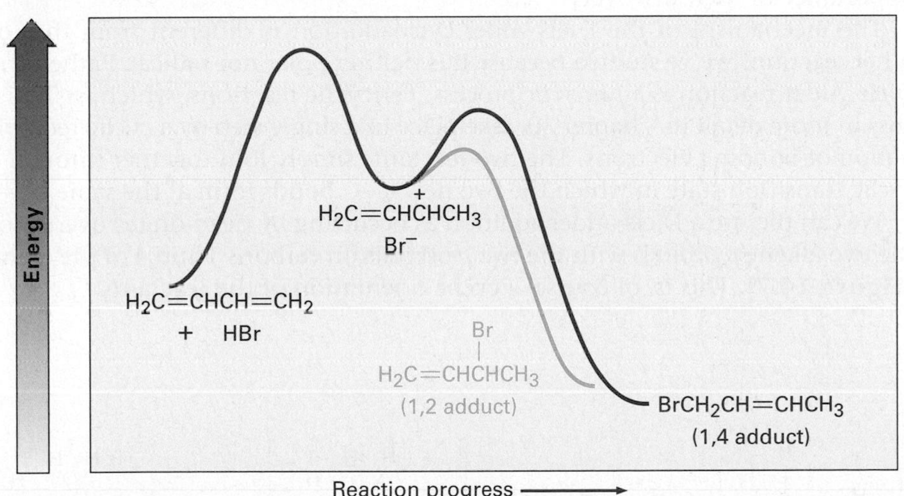

Figure 14.6 Energy diagram for the electrophilic addition of HBr to 1,3-butadiene. The 1,2 adduct is the kinetic product because it forms faster, but the 1,4 adduct is the thermodynamic product because it is more stable.

The electrophilic addition of HBr to 1,3-butadiene is a good example of how a change in experimental conditions can change the product of a reaction. The concept of thermodynamic control versus kinetic control is a useful one that we can sometimes take advantage of in the laboratory.

Problem 14.5

The 1,2 adduct and the 1,4 adduct formed by reaction of HBr with 1,3-butadiene are in equilibrium at 40 °C. Propose a mechanism by which the interconversion of products takes place.

Problem 14.6
Why do you suppose 1,4 adducts of 1,3-butadiene are generally more stable than 1,2 adducts?

14.4 The Diels–Alder Cycloaddition Reaction

Perhaps the most striking difference between conjugated and nonconjugated dienes is that conjugated dienes undergo an addition reaction with alkenes to yield substituted cyclohexene products. For example, 1,3-butadiene and 3-buten-2-one give 3-cyclohexenyl methyl ketone.

1,3-Butadiene + 3-Buten-2-one $\xrightarrow[\text{Heat}]{\text{Toluene}}$ 3-Cyclohexenyl methyl ketone (96%)

This process, named the **Diels–Alder cycloaddition reaction** after its discoverers, is extremely useful in the laboratory because it forms two carbon–carbon bonds in a single step and is one of the few general methods available for making cyclic molecules. (As the name implies, a *cycloaddition* reaction is one in which two reactants add together to give a cyclic product.) The 1950 Nobel Prize in Chemistry was awarded to Diels and Alder in recognition of the importance of their discovery.

The mechanism of the Diels–Alder cycloaddition is different from that of other reactions we've studied because it is neither polar nor radical. Rather, the Diels–Alder reaction is a *pericyclic* process. Pericyclic reactions, which we'll discuss in more detail in Chapter 30, take place in a single step by a cyclic redistribution of bonding electrons. The two reactants simply join together through a cyclic transition state in which the two new C–C bonds form at the same time.

We can picture a Diels–Alder addition as occurring by head-on (σ) overlap of the two alkene p orbitals with the two p orbitals on carbons 1 and 4 of the diene **(Figure 14.7)**. This is, of course, a cyclic orientation of the reactants.

Figure 14.7 Mechanism of the Diels–Alder cycloaddition reaction. The reaction occurs in a single step through a cyclic transition state in which the two new C–C bonds form simultaneously.

In the Diels–Alder transition state, the two alkene carbons and carbons 1 and 4 of the diene rehybridize from sp^2 to sp^3 to form two new single bonds, while carbons 2 and 3 of the diene remain sp^2-hybridized to form the new double bond in the cyclohexene product. We'll study this mechanism in more detail in **Section 30.5** but will concentrate for the present on learning about the characteristics and uses of the Diels–Alder reaction.

14.5 Characteristics of the Diels–Alder Reaction

The Dienophile

The Diels–Alder cycloaddition reaction occurs most rapidly if the alkene component, or **dienophile** ("diene lover"), has an electron-withdrawing substituent group. Thus, ethylene itself reacts sluggishly, but propenal, ethyl propenoate, maleic anhydride, benzoquinone, propenenitrile, and similar compounds are highly reactive. Note also that alkynes, such as methyl propynoate, can act as Diels–Alder dienophiles.

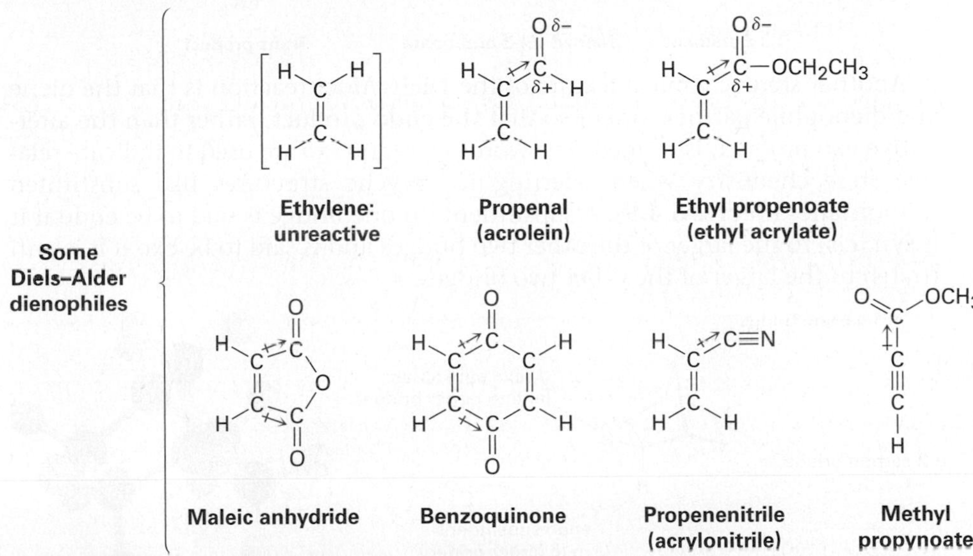

In all cases, the double or triple bond of the dienophile is adjacent to the positively polarized carbon of an electron-withdrawing substituent. As a result, the double-bond carbons in these substances are substantially less electron-rich than the carbons in ethylene, as indicated by the electrostatic potential maps in **Figure 14.8**.

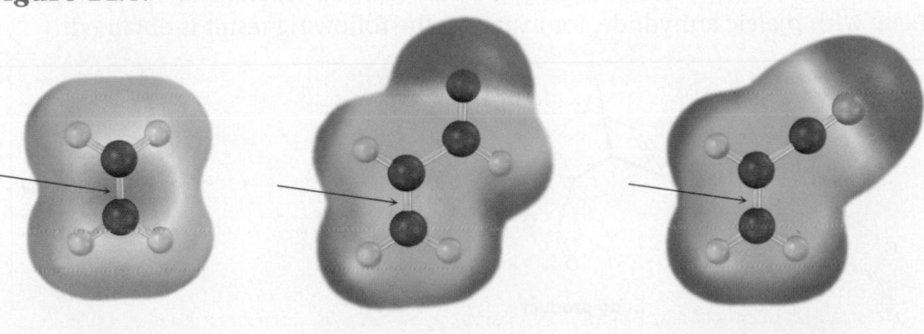

Figure 14.8 Electrostatic potential maps of ethylene, propenal, and propenenitrile show that electron-withdrawing groups make the double-bond carbons **less electron-rich**.

One of the most useful features of the Diels–Alder reaction is that it is *stereospecific,* meaning that a single product stereoisomer is formed. Furthermore, the stereochemistry of the dienophile is retained. If we carry out the cycloaddition with methyl *cis*-2-butenoate, only the cis-substituted cyclohexene product is formed. With methyl *trans*-2-butenoate, only the trans-substituted cyclohexene product is formed.

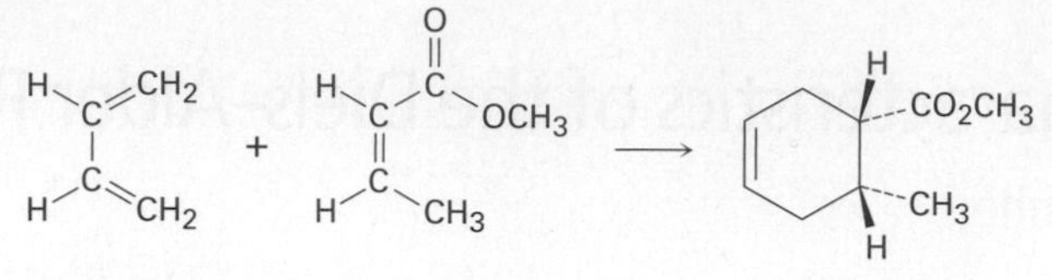

1,3-Butadiene **Methyl (*Z*)-2-butenoate** **Cis product**

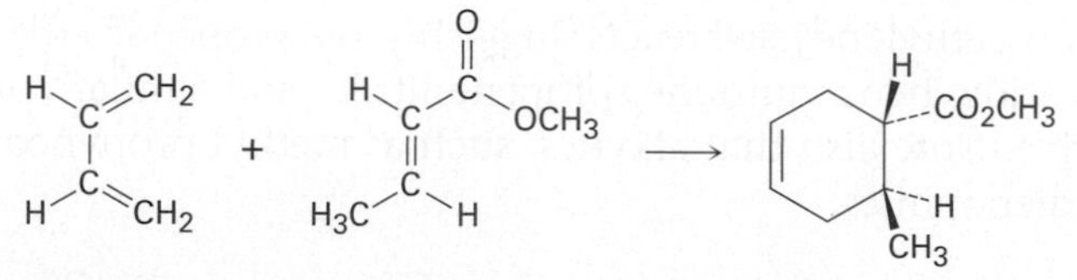

1,3-Butadiene **Methyl (*E*)-2-butenoate** **Trans product**

Another stereochemical feature of the Diels–Alder reaction is that the diene and dienophile partners orient so that the endo product, rather than the alternative exo product, is formed. The words *endo* and *exo* are used to indicate relative stereochemistry when referring to bicyclic structures like substituted norbornanes **(Section 4.9)**. A substituent on one bridge is said to be endo if it is syn (cis) to the larger of the other two bridges and is said to be exo if it is anti (trans) to the larger of the other two bridges.

A 1-carbon bridge

A 2-carbon bridge

exo substituent (anti to larger bridge)

endo substituent (syn to larger bridge)

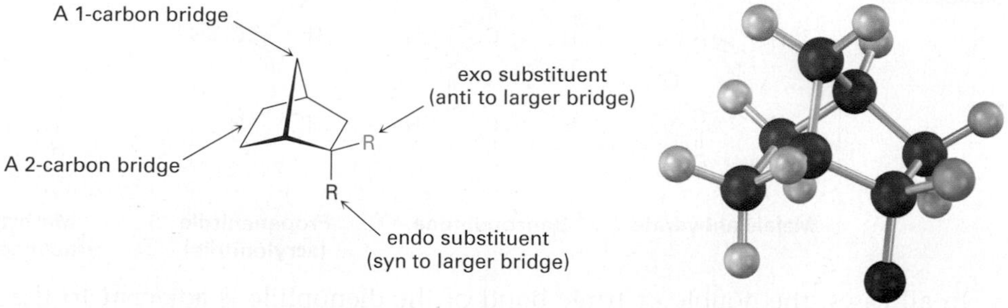

Endo products result from Diels–Alder reactions because the amount of orbital overlap between diene and dienophile is greater when the reactants lie directly on top of one another so that the electron-withdrawing substituent on the dienophile is underneath the diene double bonds. In the reaction of 1,3-cyclopentadiene with maleic anhydride, for instance, the following result is obtained:

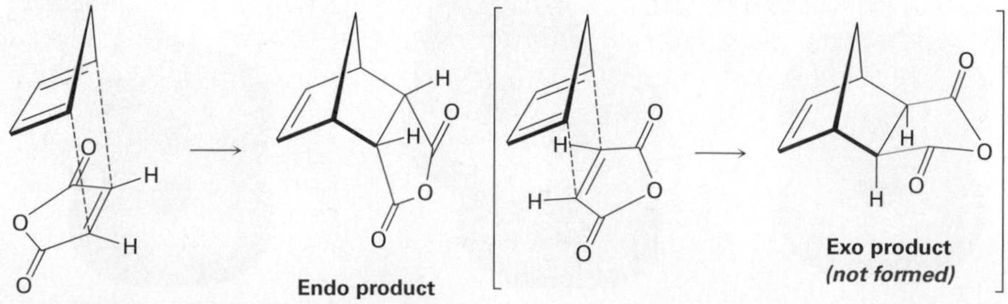

Maleic anhydride **Endo product** **Exo product** ***(not formed)***

Predicting the Product of a Diels–Alder Reaction

Worked Example 14.2

Predict the product of the following Diels–Alder reaction:

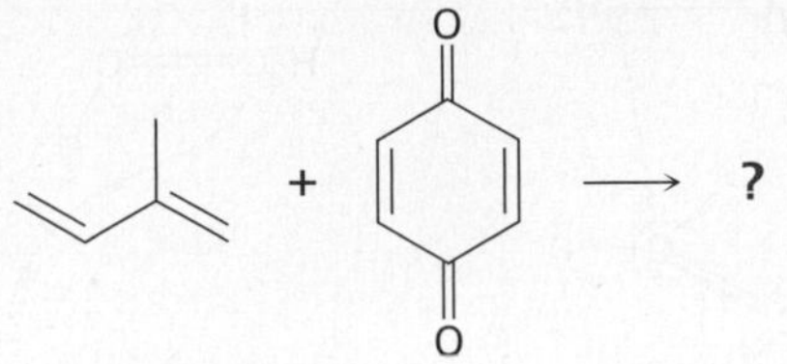

Strategy

Draw the diene so that the ends of its two double bonds are near the dienophile double bond. Then form two single bonds between the partners, convert the three double bonds into single bonds, and convert the former single bond of the diene into a double bond. Because the dienophile double bond is cis to begin with, the two attached hydrogens must remain cis in the product.

Solution

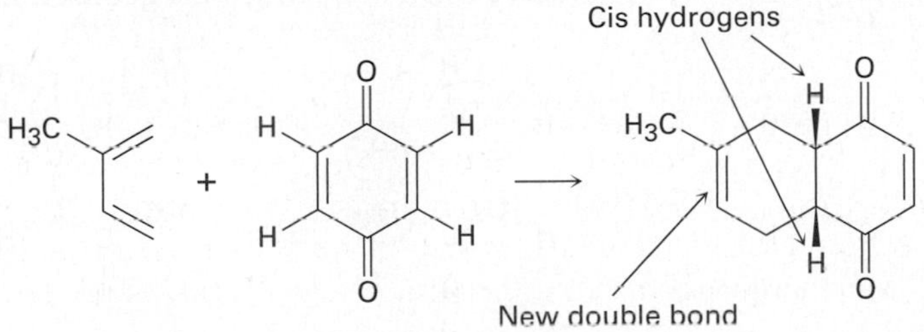

Cis hydrogens

H_3C H H H H H_3C H H O O O O

New double bond

Problem 14.7

Predict the product of the following Diels–Alder reaction:

H, $C{-}OCH_3$, $C{=}C$, H_3C, H

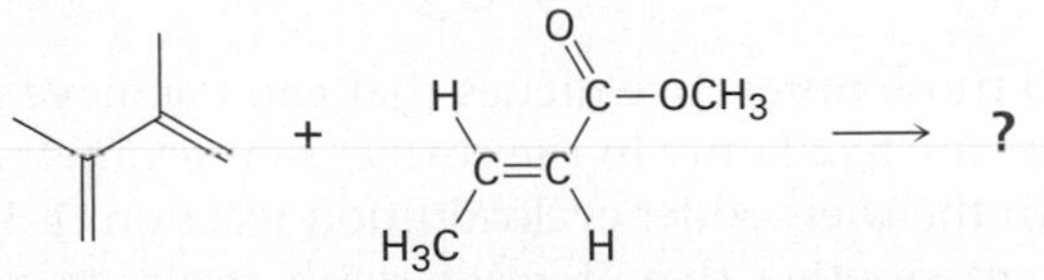

The Diene

Just as the dienophile component has certain constraints that affect its reactivity, so too does the conjugated diene component. The diene must adopt what is called an *s-cis conformation,* meaning "cis-like" about the single bond, to undergo a Diels–Alder reaction. Only in the *s*-cis conformation are carbons 1 and 4 of the diene close enough to react through a cyclic transition state.

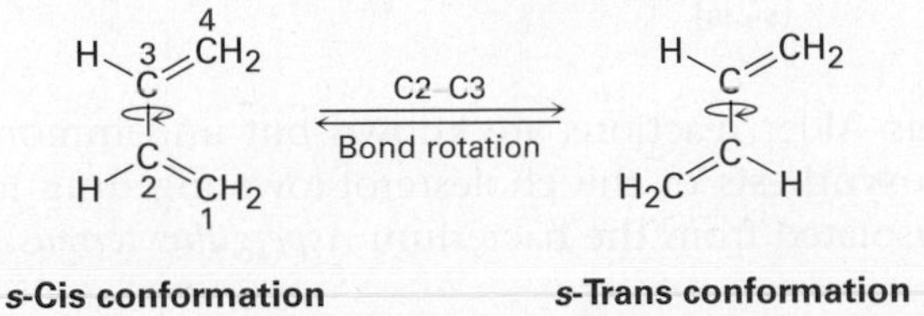

C2–C3 Bond rotation

***s*-Cis conformation** ***s*-Trans conformation**

In the alternative *s*-trans conformation, the ends of the diene partner are too far apart to overlap with the dienophile *p* orbitals.

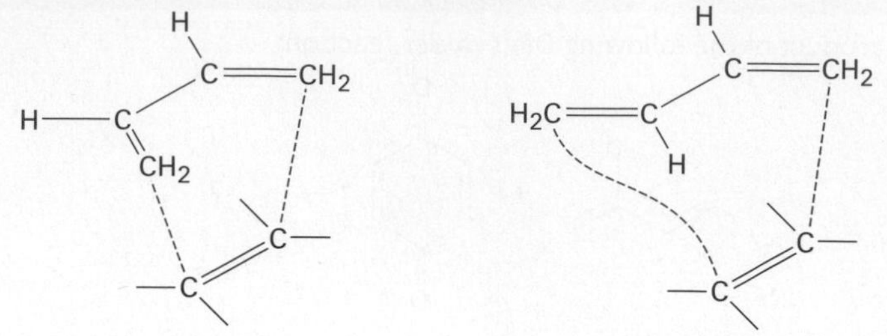

Successful reaction **No reaction (ends too far apart)**

Two examples of dienes that can't adopt an *s*-cis conformation, and thus don't undergo Diels–Alder reactions, are shown in **Figure 14.9**. In the bicyclic diene, the double bonds are rigidly fixed in an *s*-trans arrangement by geometric constraints of the rings. In (2*Z*,4*Z*)-hexadiene, steric strain between the two methyl groups prevents the molecule from adopting *s*-cis geometry.

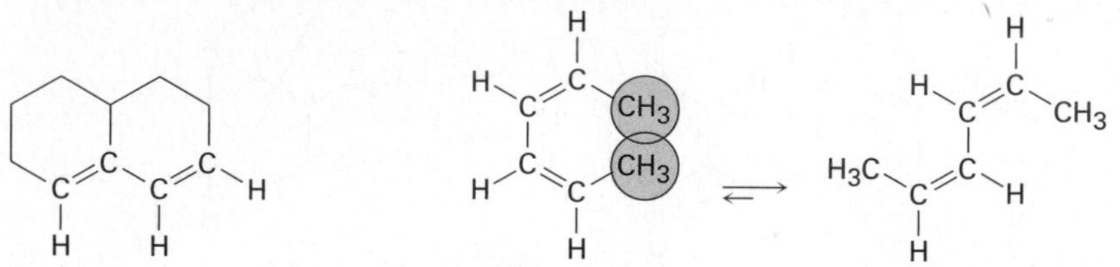

A bicyclic diene (rigid *s*-trans diene) **Severe steric strain in *s*-cis form** **(2*Z*,4*Z*)-Hexadiene (*s*-trans, more stable)**

Figure 14.9 Two dienes that can't achieve an *s*-cis conformation and thus can't undergo Diels–Alder reactions.

In contrast to those unreactive dienes that can't achieve an *s*-cis conformation, other dienes are fixed only in the correct *s*-cis geometry and are therefore highly reactive in the Diels–Alder cycloaddition reaction. 1,3-Cyclopentadiene, for example, is so reactive that it reacts with itself. At room temperature, 1,3-cyclopentadiene *dimerizes*. One molecule acts as diene and a second molecule acts as dienophile in a self-Diels–Alder reaction.

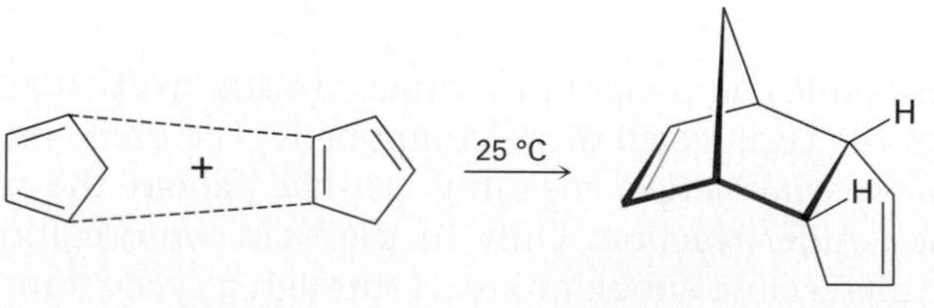

1,3-Cyclopentadiene (*s*-cis) 25 °C **Bicyclopentadiene**

Biological Diels–Alder reactions are known but uncommon. One example occurs in the biosynthesis of the cholesterol-lowering drug lovastatin (trade name Mevacor) isolated from the bacterium *Aspergillus terreus*. The key step is the *intramolecular* Diels–Alder reaction of a triene, in which the diene and dienophile components are within the same molecule.

Lovastatin

Problem 14.8

Which of the following alkenes would you expect to be good Diels–Alder dienophiles?

(a) $H_2C{=}CHCCl$ (with C=O)

(b) $H_2C{=}CHCH_2CH_2COCH_3$ (with C=O)

(c)

(d)

(e)

Problem 14.9

Which of the following dienes have an s-cis conformation, and which have an s-trans conformation? Of the s-trans dienes, which can readily rotate to s-cis?

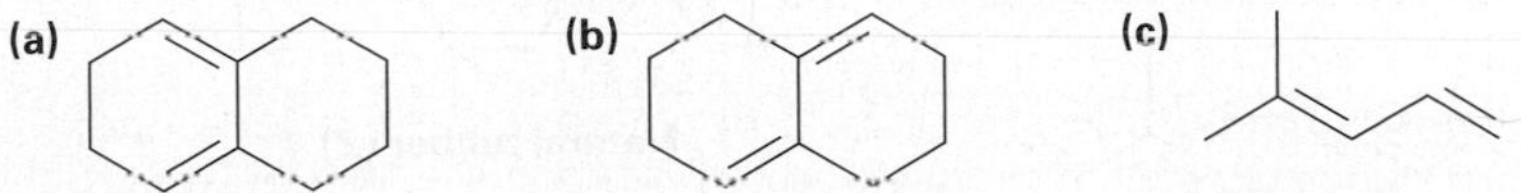

Problem 14.10

Predict the product of the following Diels–Alder reaction:

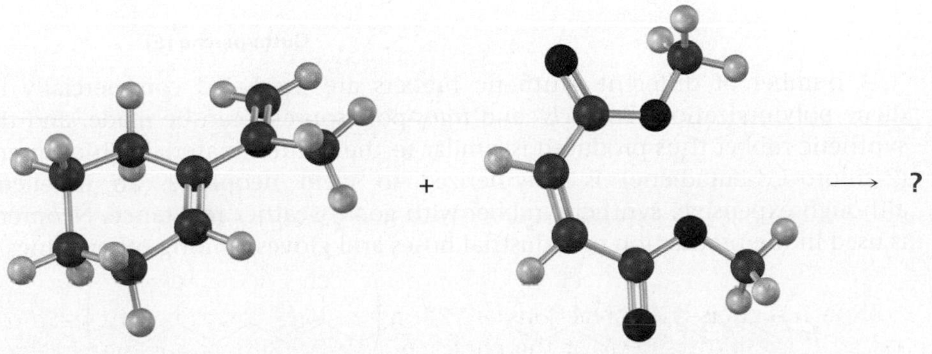

14.6 Diene Polymers: Natural and Synthetic Rubbers

Conjugated dienes can be polymerized just as simple alkenes can **(Section 8.10)**. Diene polymers are structurally more complex than simple alkene polymers, though, because double bonds remain every four carbon atoms along the chain, leading to the possibility of cis–trans isomers. The initiator (In) for the reaction can be either a radical, as occurs in ethylene polymerization, or an acid. Note that the polymerization is a 1,4 addition of the growing chain to a conjugated diene monomer.

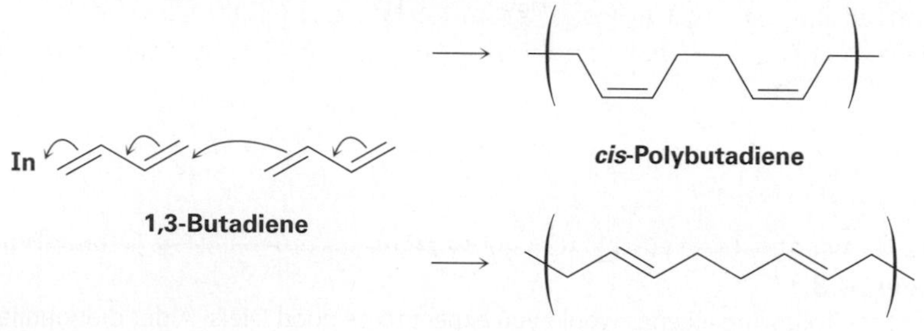

In

1,3-Butadiene

***cis*-Polybutadiene**

***trans*-Polybutadiene**

Rubber is a naturally occurring diene polymer of isoprene (2-methyl-1,3-butadiene) and is produced by more than 400 different plants. The major source is the so-called rubber tree, *Hevea brasiliensis,* from which the crude material, called *latex,* is harvested as it drips from a slice made through the bark. The double bonds of rubber have *Z* stereochemistry, but gutta-percha, the *E* isomer of rubber, also occurs naturally. Harder and more brittle than rubber, gutta-percha has a variety of minor applications, including occasional use in dentistry and as the covering on golf balls.

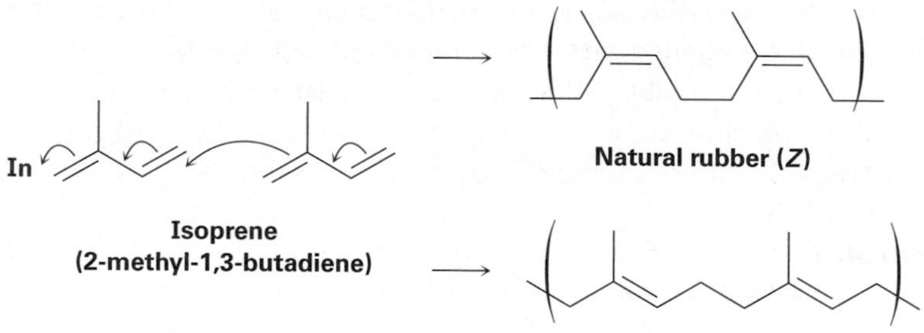

In

Isoprene
(2-methyl-1,3-butadiene)

Natural rubber (*Z*)

Gutta-percha (*E*)

A number of different synthetic rubbers are produced commercially by diene polymerization. Both *cis*- and *trans*-polyisoprene can be made, and the synthetic rubber thus produced is similar to the natural material. Chloroprene (2-chloro-1,3-butadiene) is polymerized to yield neoprene, an excellent, although expensive, synthetic rubber with good weather resistance. Neoprene is used in the production of industrial hoses and gloves, among other things.

Cl Cl

In

Cl Cl

Chloroprene
(2-chloro-1,3-butadiene)

Neoprene (*Z*)

Both natural and synthetic rubbers are too soft and tacky to be useful until they are hardened by heating with elemental sulfur, a process called *vulcanization.* Vulcanization cross-links the rubber chains together by forming carbon–sulfur bonds between them, thereby hardening and stiffening the polymer. The exact degree of hardening can be varied, yielding material soft enough for automobile tires or hard enough for bowling balls *(ebonite).*

The unusual ability of rubber to stretch and then contract to its original shape is due to the irregular shapes of the polymer chains caused by the double bonds. These double bonds introduce bends and kinks into the polymer chains, thereby preventing neighboring chains from nestling together. When stretched, the randomly coiled chains straighten out and orient along the direction of the pull but are kept from sliding over one another by the cross-links. When the stretch is released, the polymer reverts to its original random state.

Problem 14.11

Draw a segment of the polymer that might be prepared from 2-phenyl-1,3-butadiene.

Problem 14.12

Show the mechanism of the acid-catalyzed polymerization of 1,3-butadiene.

14.7 Structure Determination in Conjugated Systems: Ultraviolet Spectroscopy

Mass spectrometry, infrared spectroscopy, and nuclear magnetic resonance spectroscopy are techniques of structure determination applicable to all organic molecules. In addition to these three generally useful methods, there is a fourth—**ultraviolet (UV) spectroscopy**—that is applicable only to conjugated compounds. UV is less commonly used than the other three spectroscopic techniques because of the specialized information it gives, so we'll mention it only briefly.

Mass spectrometry	Molecular size and formula
IR spectroscopy	Functional groups present
NMR spectroscopy	Carbon–hydrogen framework
UV spectroscopy	**Nature of conjugated π electron system**

The ultraviolet region of the electromagnetic spectrum extends from the short-wavelength end of the visible region (4×10^{-7} m) to the long-wavelength end of the X-ray region (10^{-8} m), but the narrow range from 2×10^{-7} m to 4×10^{-7} m is the part of greatest interest to organic chemists. Absorptions in this region are usually measured in nanometers (nm), where 1 nm $= 10^{-9}$ m. Thus, the ultraviolet range of interest is from 200 to 400 nm **(Figure 14.10)**.

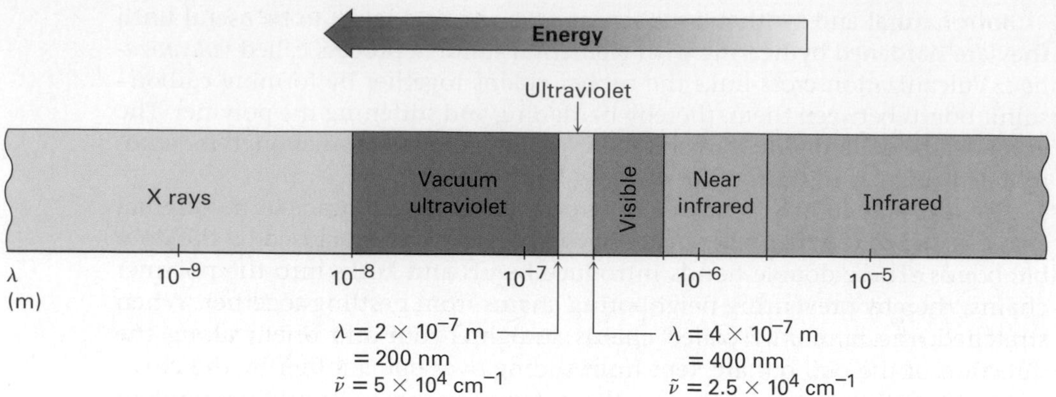

Figure 14.10 The ultraviolet (UV) and neighboring regions of the electromagnetic spectrum.

We saw in **Section 12.5** that when an organic molecule is irradiated with electromagnetic energy, the radiation either passes through the sample or is absorbed, depending on its energy. With IR irradiation, the energy absorbed corresponds to the amount necessary to increase molecular vibrations. With UV radiation, the energy absorbed corresponds to the amount necessary to promote an electron from a lower-energy orbital to a higher-energy one in a conjugated molecule. The conjugated diene 1,3-butadiene, for instance, has four π molecular orbitals, as shown previously in Figure 14.2 on page 504. The two lower-energy, bonding MOs are occupied in the ground state, and the two higher-energy, antibonding MOs are unoccupied.

On irradiation with ultraviolet light (*hν*), 1,3-butadiene absorbs energy and a π electron is promoted from the **highest occupied molecular orbital**, or **HOMO**, to the **lowest unoccupied molecular orbital**, or **LUMO**. Since the electron is promoted from a bonding π molecular orbital to an antibonding π^* molecular orbital, we call this a $\pi \rightarrow \pi^*$ excitation (read as "pi to pi star"). The energy gap between the HOMO and the LUMO of 1,3-butadiene is such that UV light of 217 nm wavelength is required to accomplish the $\pi \rightarrow \pi^*$ electronic transition **(Figure 14.11)**.

Figure 14.11 Ultraviolet excitation of 1,3-butadiene results in the promotion of an electron from ψ_2, the highest occupied molecular orbital (HOMO), to $\psi_3{}^*$, the lowest unoccupied molecular orbital (LUMO).

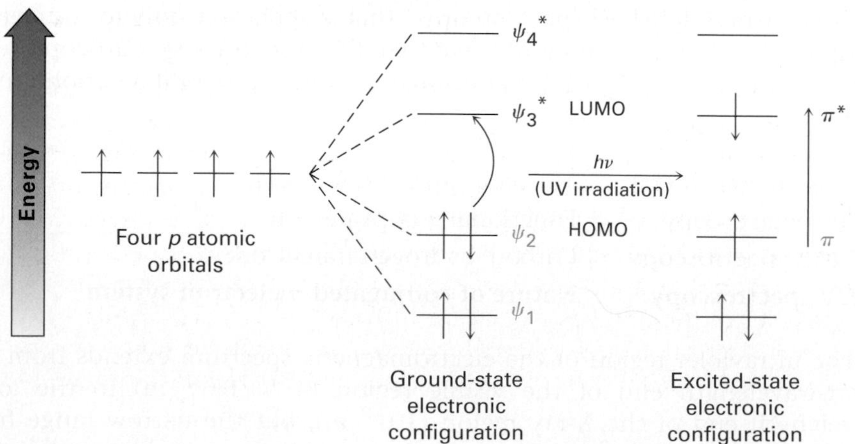

An ultraviolet spectrum is recorded by irradiating the sample with UV light of continuously changing wavelength. When the wavelength corresponds to the energy level required to excite an electron to a higher level, energy is absorbed. This absorption is detected and displayed on a chart that plots wavelength versus *absorbance (A)*, defined as

$$A = \log \frac{I_0}{I}$$

where I_0 is the intensity of the incident light and I is the intensity of the light transmitted through the sample.

Note that UV spectra differ from IR spectra in the way they are presented. For historical reasons, IR spectra are usually displayed so that the baseline corresponding to zero absorption runs across the top of the chart and a valley indicates an absorption, whereas UV spectra are displayed with the baseline at the bottom of the chart so that a peak indicates an absorption **(Figure 14.12)**.

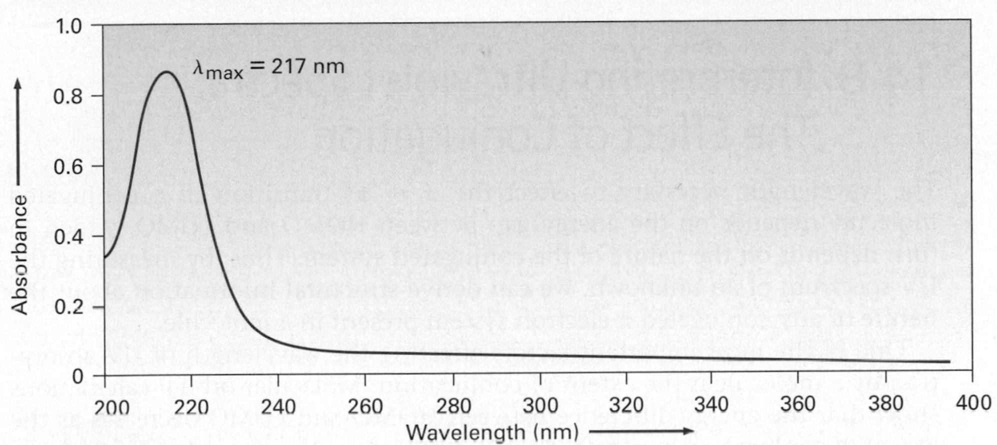

Figure 14.12 The ultraviolet spectrum of 1,3-butadiene, λ_{max} = 217 nm.

The amount of UV light absorbed is expressed as the sample's **molar absorptivity (ϵ)**, defined by the equation

$$\varepsilon = \frac{A}{c \times l}$$

where A = Absorbance
c = Concentration in mol/L
l = Sample pathlength in cm

Molar absorptivity is a physical constant, characteristic of the particular substance being observed and thus characteristic of the particular π electron system in the molecule. Typical values for conjugated dienes are in the range ϵ = 10,000 to 25,000. The units for molar absorptivity, L/(mol · cm), are usually dropped.

A particularly important use of this equation comes from rearranging it to the form $c = A/(\epsilon \cdot l)$, which lets us measure the concentration of a sample in solution when A, ϵ, and l are known. As an example, β-carotene, the pigment responsible for the orange color of carrots, has ϵ = 138,000 L/(mol · cm). If a sample of β-carotene is placed in a cell with a pathlength of 1.0 cm and the UV absorbance reads 0.37, then the concentration of β-carotene in the sample is

$$c = \frac{A}{\varepsilon l} = \frac{0.37}{\left(1.38 \times 10^5 \frac{\text{L}}{\text{mol} \cdot \text{cm}}\right)(1.00 \text{ cm})}$$

$$= 2.7 \times 10^{-6} \text{ mol/L}$$

Unlike IR and NMR spectra, which show many absorptions for a given molecule, UV spectra are usually quite simple—often only a single peak. The peak is usually broad, and we identify its position by noting the wavelength at the very top of the peak—λ_{max}, read as "lambda max."

Problem 14.13

Calculate the energy range of electromagnetic radiation in the UV region of the spectrum from 200 to 400 nm (see **Section 12.5**). How does this value compare with the values calculated previously for IR and NMR spectroscopy?

Problem 14.14

If pure vitamin A has $\lambda_{max} = 325$ ($\epsilon = 50{,}100$), what is the vitamin A concentration in a sample whose absorbance at 325 nm is $A = 0.735$ in a cell with a pathlength of 1.00 cm?

14.8 Interpreting Ultraviolet Spectra: The Effect of Conjugation

The wavelength necessary to effect the $\pi \rightarrow \pi^*$ transition in a conjugated molecule depends on the energy gap between HOMO and LUMO, which in turn depends on the nature of the conjugated system. Thus, by measuring the UV spectrum of an unknown, we can derive structural information about the nature of any conjugated π electron system present in a molecule.

One of the most important factors affecting the wavelength of UV absorption by a molecule is the extent of conjugation. Molecular orbital calculations show that the energy difference between HOMO and LUMO decreases as the extent of conjugation increases. Thus, 1,3-butadiene absorbs at $\lambda_{max} = 217$ nm, 1,3,5-hexatriene absorbs at $\lambda_{max} = 258$ nm, and 1,3,5,7-octatetraene absorbs at $\lambda_{max} = 290$ nm. (Remember: longer wavelength means lower energy.)

Other kinds of conjugated systems, such as conjugated enones and aromatic rings, also have characteristic UV absorptions that are useful in structure determination. The UV absorption maxima of some representative conjugated molecules are given in Table 14.2.

Table 14.2 Ultraviolet Absorptions of Some Conjugated Molecules

Name	Structure	λ_{max} (nm)
2-Methyl-1,3-butadiene	$H_2C=C(CH_3)-CH=CH_2$	220
1,3-Cyclohexadiene	[ring structure]	256
1,3,5-Hexatriene	$H_2C=CH-CH=CH-CH=CH_2$	258
1,3,5,7-Octatetraene	$H_2C=CH-CH=CH-CH=CH-CH=CH_2$	290
3-Buten-2-one	$H_2C=CH-C(=O)-CH_3$	219
Benzene	[ring structure]	203

Problem 14.15

Which of the following compounds would you expect to show ultraviolet absorptions in the 200 to 400 nm range?

(a) (b) (c) CN

(d) O OH O O **Aspirin**

(e) CH_3 O

(f) N H **Indole**

14.9 Conjugation, Color, and the Chemistry of Vision

Why are some organic compounds colored while others aren't? β-Carotene, the pigment in carrots, is purple-orange, for instance, while cholesterol is colorless. The answer involves both the chemical structures of colored molecules and the way we perceive light.

The visible region of the electromagnetic spectrum is adjacent to the ultraviolet region, extending from approximately 400 to 800 nm. Colored compounds have such extended systems of conjugation that their "UV" absorptions extend into the visible region. β-Carotene, for example, has 11 double bonds in conjugation, and its absorption occurs at λ_{max} = 455 nm **(Figure 14.13)**.

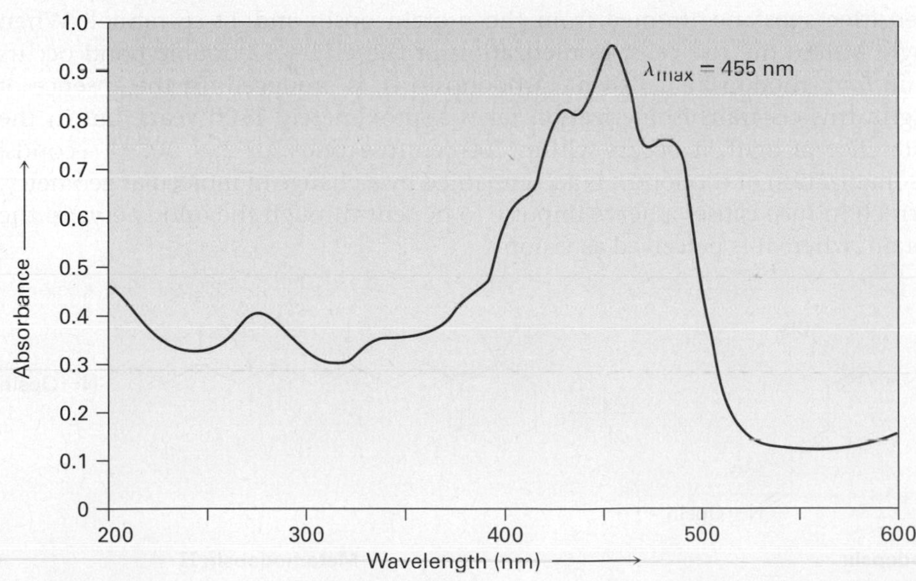

Figure 14.13 Ultraviolet spectrum of β-carotene, a conjugated molecule with 11 double bonds. The absorption occurs in the visible region.

"White" light from the sun or from a lamp consists of all wavelengths in the visible region. When white light strikes β-carotene, the wavelengths from 400 to 500 nm (blue) are absorbed while all other wavelengths are transmitted and can reach our eyes. We therefore see the white light with the blue removed, and we perceive a yellow-orange color for β-carotene.

Conjugation is crucial not only for the colors we see in organic molecules but also for the light-sensitive molecules on which our visual system is based. The key substance for vision is dietary β-carotene, which is converted to vitamin A by enzymes in the liver, oxidized to an aldehyde called 11-*trans*-retinal, and then isomerized by a change in geometry of the C11–C12 double bond to produce 11-*cis*-retinal.

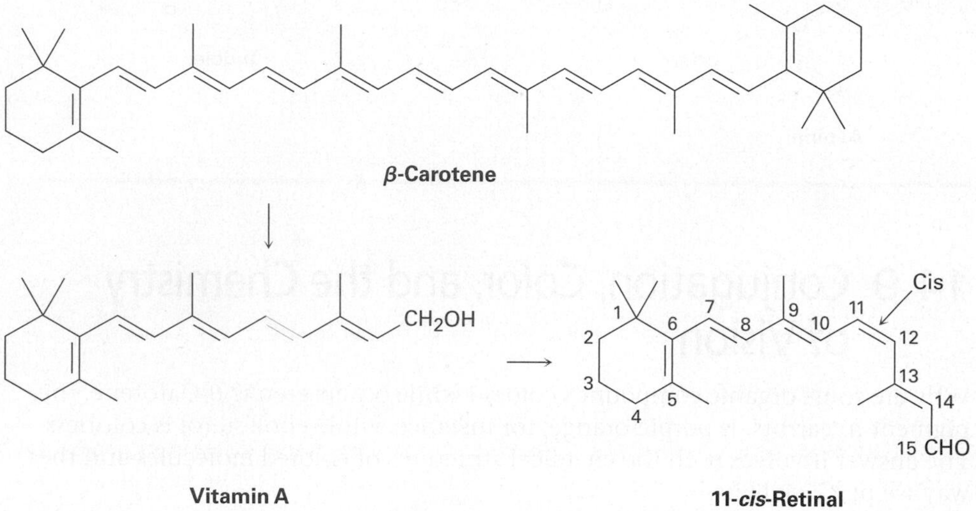

β-Carotene

Vitamin A

11-*cis*-Retinal

There are two main types of light-sensitive receptor cells in the retina of the human eye, *rod* cells and *cone* cells. The 3 million or so rod cells are primarily responsible for seeing in dim light, whereas the 100 million cone cells are responsible for seeing in bright light and for the perception of bright colors. In the rod cells of the eye, 11-*cis*-retinal is converted into rhodopsin, a light-sensitive substance formed from the protein opsin and 11-*cis*-retinal. When light strikes the rod cells, isomerization of the C11–C12 double bond occurs and *trans*-rhodopsin, called metarhodopsin II, is produced. In the absence of light, this cis–trans isomerization takes approximately 1100 years, but in the presence of light, it occurs within 200 femtoseconds, or 2×10^{-13} seconds! Isomerization of rhodopsin is accompanied by a change in molecular geometry, which in turn causes a nerve impulse to be sent through the optic nerve to the brain, where it is perceived as vision.

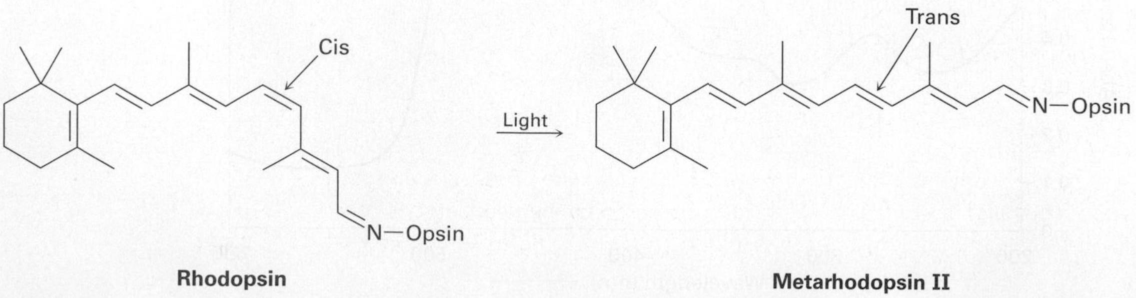

Rhodopsin

Metarhodopsin II

Metarhodopsin II is then recycled back into rhodopsin by a multistep sequence involving cleavage to all-*trans*-retinal and cis–trans isomerization back to 11-*cis*-retinal.

A DEEPER LOOK Photolithography

Manufacturing the ultrathin circuitry on this computer chip depends on the organic chemical reactions of special polymers.

Fifty years ago, someone interested in owning a computer would have paid approximately $150,000 for 16 megabytes of random-access memory that would have occupied a volume the size of a small desk. Today, anyone can buy 60 times as much computer memory for $20 and fit the small chip into their shirt pocket. The difference between then and now is due to improvements in *photolithography,* the process by which integrated-circuit chips are made.

Photolithography begins by coating a layer of SiO_2 onto a silicon wafer and further coating with a thin (0.5–1.0 μm) film of a light-sensitive organic polymer called a *resist.* A *mask* is then used to cover those parts of the chip that will become a circuit, and the wafer is irradiated with UV light. The nonmasked sections of the polymer undergo a chemical change when irradiated that makes them more soluble than the masked, unirradiated sections. On washing the irradiated chip with solvent, solubilized polymer is selectively removed from the irradiated areas, exposing the SiO_2 underneath. This SiO_2 is then chemically etched away by reaction with hydrofluoric acid, leaving behind a pattern of polymer-coated SiO_2. Further washing removes the remaining polymer, leaving a positive image of the mask in the form of exposed ridges of SiO_2 **(Figure 14.14)**. Additional cycles of coating, masking, and etching then produce the completed chips.

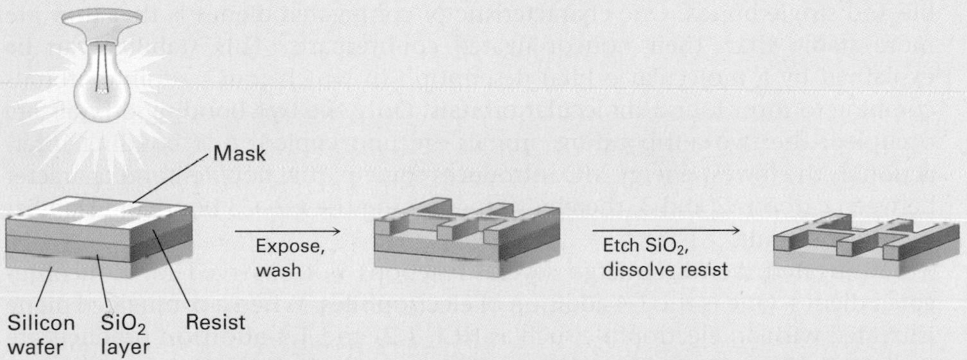

Figure 14.14 Outline of the photolithography process for producing integrated circuit chips.

The polymer resist currently used in chip manufacturing is based on the two-component *diazoquinone–novolac system.* Novolac resin is a soft, relatively low-molecular-weight polymer made from methylphenol and formaldehyde, while the diazoquinone is a bicyclic (two-ring) molecule containing a diazo group (=N=N) adjacent to a ketone carbonyl (C=O). The diazoquinone–novolac mix is relatively insoluble when fresh, but on exposure to ultraviolet light and water vapor, the diazoquinone component undergoes reaction to yield N_2 and a carboxylic acid, which can be washed away with dilute base. Novolac–diazoquinone

(continued)

(continued)

technology is capable of producing features as small as 0.5 μm (5×10^{-7} m), but still further improvements in miniaturization are being developed.

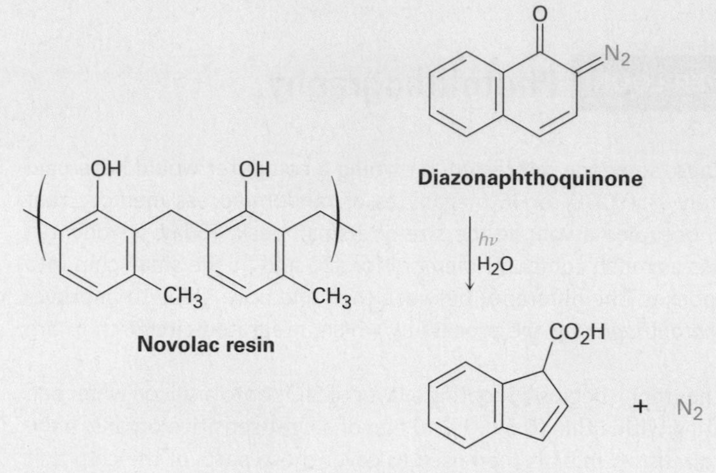

Diazonaphthoquinone

$h\nu$, H_2O → CO_2H + N_2

Novolac resin

Summary

Key words

- 1,2 addition, 505
- 1,4 addition, 505
- conjugated, 500
- Diels–Alder cycloaddition reaction, 510
- dienophile, 511
- highest occupied molecular orbital (HOMO), 518
- kinetic control, 509
- lowest unoccupied molecular orbital (LUMO), 518
- molar absorptivity (ϵ), 519
- thermodynamic control, 509
- ultraviolet (UV) spectroscopy, 517

The unsaturated compounds we've looked at previously have had only one double bond, but many compounds have numerous sites of unsaturation, which gives them some distinctive properties. Many such compounds are common in nature, including pigments and hormones.

A **conjugated** diene or other compound is one that contains alternating double and single bonds. One characteristic of conjugated dienes is that they are more stable than their nonconjugated counterparts. This stability can be explained by a molecular orbital description in which four *p* atomic orbitals combine to form four π molecular orbitals. Only the two bonding orbitals are occupied; the two antibonding orbitals are unoccupied. A π bonding interaction in the lowest-energy MO introduces some partial double-bond character between carbons 2 and 3, thereby strengthening the C2–C3 bond and stabilizing the molecule.

Conjugated dienes undergo several reactions not observed for nonconjugated dienes. One is the 1,4 addition of electrophiles. When a conjugated diene is treated with an electrophile such as HCl, **1,2-** and **1,4-addition** products are formed. Both are formed from the same resonance-stabilized allylic carbocation intermediate and are produced in varying amounts depending on the reaction conditions. The 1,2 adduct is usually formed faster and is said to be the product of **kinetic control**. The 1,4 adduct is usually more stable and is said to be the product of **thermodynamic control**.

Another reaction unique to conjugated dienes is the **Diels–Alder cycloaddition**. Conjugated dienes react with electron-poor alkenes (**dienophiles**) in a single step through a cyclic transition state to yield a cyclohexene product. The reaction is stereospecific, meaning that only a single product stereoisomer is formed, and can occur only if the diene is able to adopt an *s*-cis conformation.

Ultraviolet (UV) spectroscopy is a method of structure determination applicable specifically to conjugated π-electron systems. When a conjugated molecule is irradiated with ultraviolet light, energy absorption occurs and a π electron is promoted from the **highest occupied molecular orbital (HOMO)** to the **lowest unoccupied molecular orbital (LUMO)**. For 1,3-butadiene, radiation of λ_{max} = 217 nm is required. The greater the extent of conjugation, the less the energy needed and the longer the wavelength of radiation required.

Summary of Reactions

1. Electrophilic addition reactions (Sections 14.2 and 14.3)

HBr

2. Diels–Alder cycloaddition reaction (Sections 14.4 and 14.5)

Toluene, Heat

A diene | A dienophile | A cyclohexene

Exercises

Visualizing Chemistry

(Problems 14.1–14.15 appear within the chapter.)

OWL Interactive versions of these problems are assignable in OWL for Organic Chemistry.

▲ denotes problems linked to the Key Ideas in this chapter.

14.16 Show the structures of all possible adducts of the following diene with 1 equivalent of HCl:

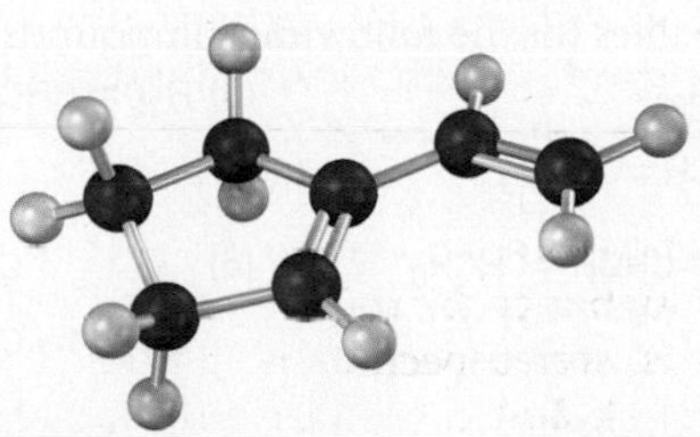

14.17 Show the product of the Diels–Alder reaction of the following diene with 3-buten-2-one, $H_2C{=}CHCOCH_3$. Make sure you show the full stereochemistry of the reaction product.

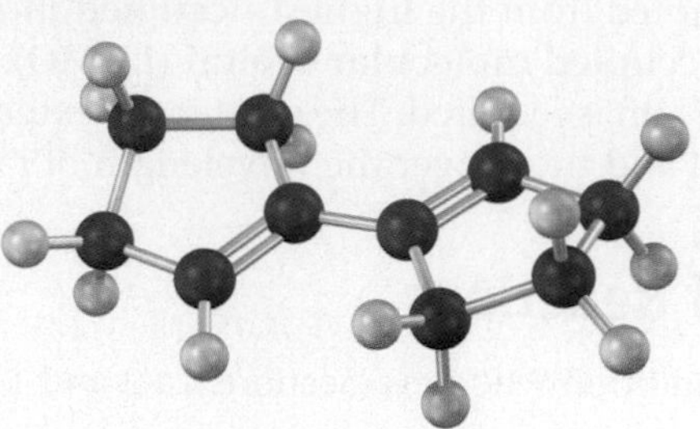

14.18 The following diene does not undergo Diels–Alder reactions. Explain.

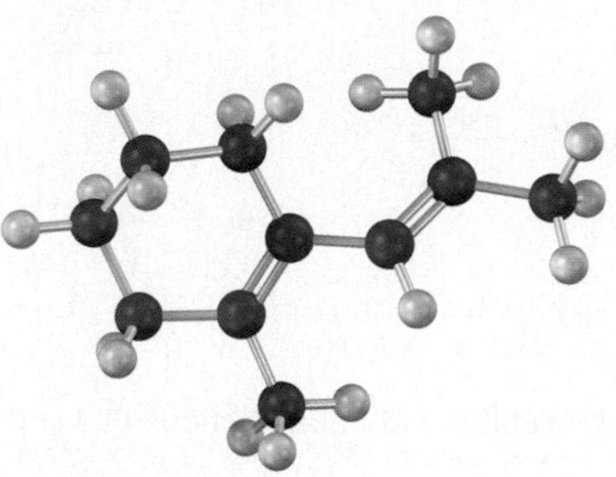

14.19 The following model is that of an allylic carbocation intermediate formed by protonation of a conjugated diene with HBr. Show the structure of the diene and the structures of the final reaction products.

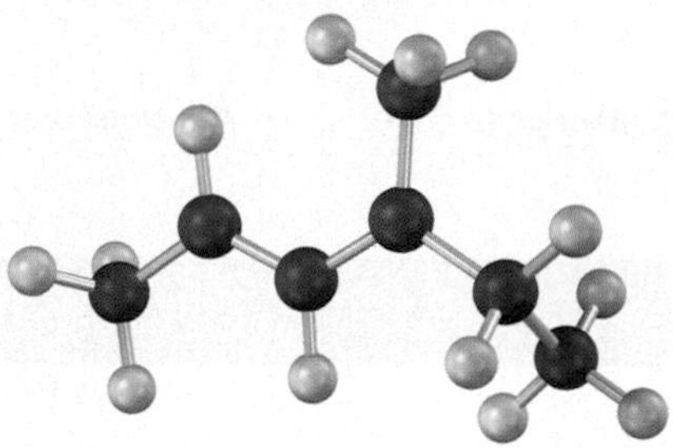

Additional Problems

Conjugated Dienes

14.20 Give IUPAC names for the following compounds:

(a)
```
        CH3
        |
CH3CH=CCH=CHCH3
```

(b) $H_2C{=}CHCH{=}CHCH{=}CHCH_3$

(c) $CH_3CH{=}C{=}CHCH{=}CHCH_3$

(d)
```
        CH2CH2CH3
        |
CH3CH=CCH=CH2
```

▲ Problems linked to Key Ideas in this chapter

14.21 Draw and name the six possible diene isomers of formula C_5H_8. Which of the six are conjugated dienes?

14.22 What product(s) would you expect to obtain from reaction of 1,3-cyclohexadiene with each of the following?

(a) 1 mol Br_2 in CH_2Cl_2
(b) O_3 followed by Zn
(c) 1 mol HCl in ether
(d) 1 mol DCl in ether
(e) 3-Buten-2-one ($H_2C{=}CHCOCH_3$)
(f) Excess OsO_4, followed by $NaHSO_3$

14.23 Electrophilic addition of Br_2 to isoprene (2-methyl-1,3-butadiene) yields the following product mixture:

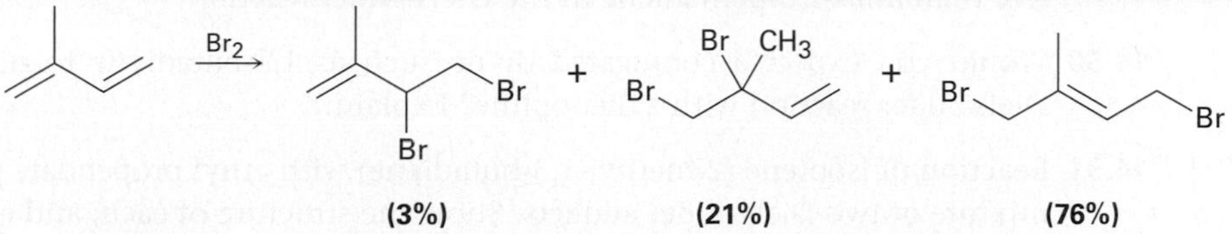

Br_2 → Br, Br (3%) + Br CH3, Br (21%) + Br, Br (76%)

Of the 1,2-addition products, explain why 3,4-dibromo-3-methyl-1-butene (21%) predominates over 3,4-dibromo-2-methyl-1-butene (3%).

14.24 Propose a structure for a conjugated diene that gives the same product from both 1,2 and 1,4 addition of HBr.

14.25 Draw the possible products resulting from addition of 1 equivalent of HCl to 1-phenyl-1,3-butadiene. Which would you expect to predominate, and why?

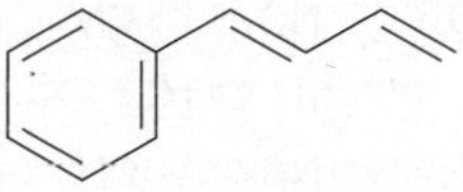

1-Phenyl-1,3-butadiene

Diels–Alder Reactions

14.26 Predict the products of the following Diels–Alder reactions:

(a) + O, O, O ⟶ ?

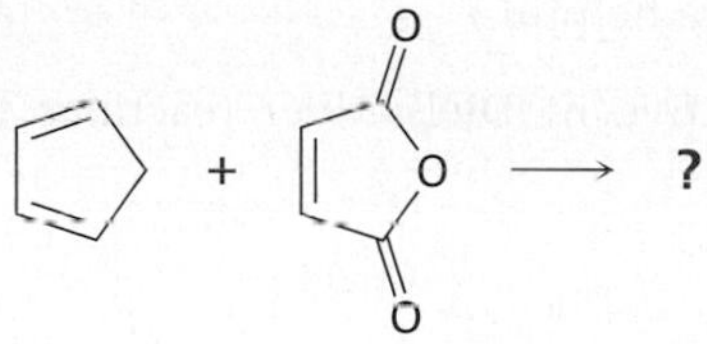

(b) + O, O ⟶ ?

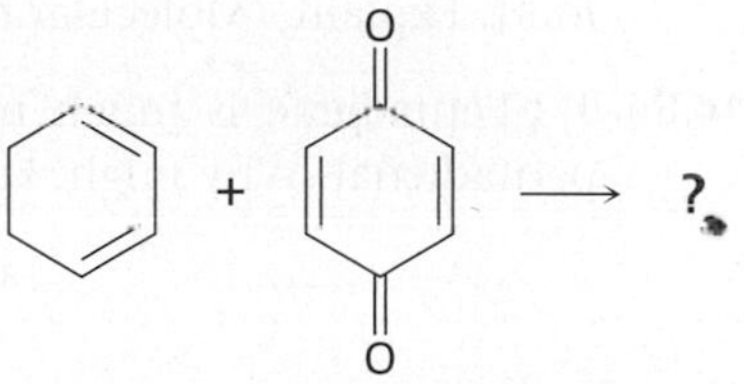

14.27 2,3-Di-*tert*-butyl-1,3-butadiene does not undergo Diels–Alder reactions. Explain.

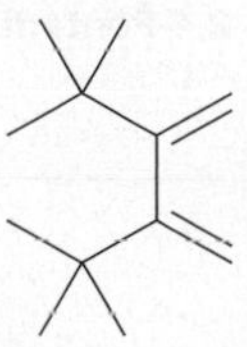

2,3-Di-*tert*-butyl-1,3-butadiene

▲ Problems linked to Key Ideas in this chapter

14.28 Show the structure, including stereochemistry, of the product from the following Diels–Alder reaction:

$COCH_3$ (C=O) + CH_3OC (C=O) ⟶ ?

14.29 How can you account for the fact that *cis*-1,3-pentadiene is much less reactive than *trans*-1,3-pentadiene in the Diels–Alder reaction?

14.30 Would you expect a conjugated diyne such as 1,3-butadiyne to undergo Diels–Alder reaction with a dienophile? Explain.

14.31 Reaction of isoprene (2-methyl-1,3-butadiene) with ethyl propenoate gives a mixture of two Diels–Alder adducts. Show the structure of each, and explain why a mixture is formed.

+ $CO_2CH_2CH_3$ ⟶ ?

14.32 Rank the following dienophiles in order of their expected reactivity in the Diels–Alder reaction.

$(H)_2C{=}C(H)(CH_3)$ $(H)_2C{=}C(H)(CHO)$ $(NC)_2C{=}C(CN)_2$ $(H_3C)_2C{=}C(CH_3)_2$

14.33 1,3-Cyclopentadiene is very reactive in Diels–Alder cycloaddition reactions, but 1,3-cyclohexadiene is less reactive and 1,3-cycloheptadiene is nearly inert. Explain. (Molecular models are helpful.)

14.34 1,3-Pentadiene is much more reactive in Diels–Alder reactions than 2,4-pentadienal. Why might this be?

1,3-Pentadiene   **2,4-Pentadienal**

▲ Problems linked to Key Ideas in this chapter

14.35 How could you use Diels–Alder reactions to prepare the following products? Show the starting diene and dienophile in each case.

(a)

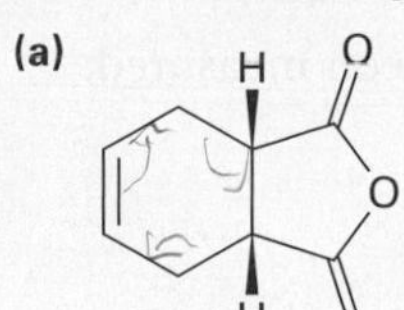

(b)

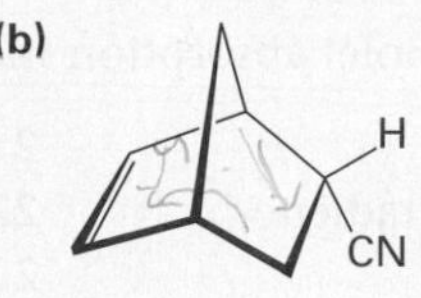

(c)

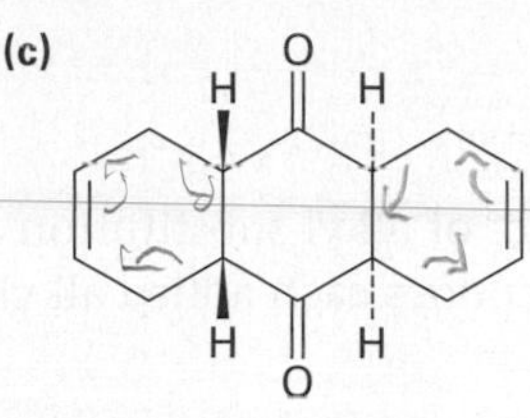

(d) CO_2CH_3

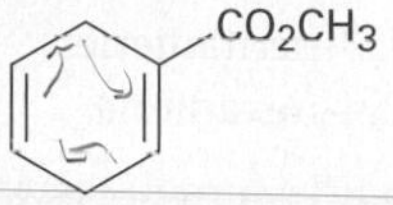

Diene Polymers

14.36 Diene polymers contain occasional vinyl branches along the chain. How do you think these branches might arise?

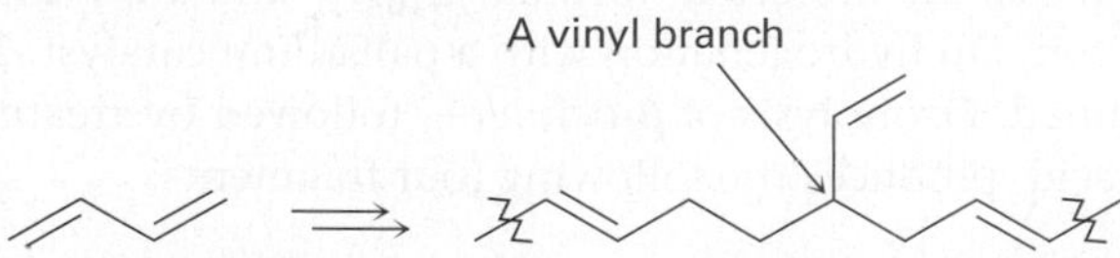

14.37 Tires whose sidewalls are made of natural rubber tend to crack and weather rapidly in areas around cities where high levels of ozone and other industrial pollutants are found. Explain.

14.38 ▲ 1,3-Cyclopentadiene polymerizes slowly at room temperature to yield a polymer that has no double bonds except on the ends. On heating, the polymer breaks down to regenerate 1,3-cyclopentadiene. Propose a structure for the product.

UV Spectroscopy

14.39 Which of the following compounds would you expect to have a $\pi \rightarrow \pi^*$ UV absorption in the 200 to 400 nm range?

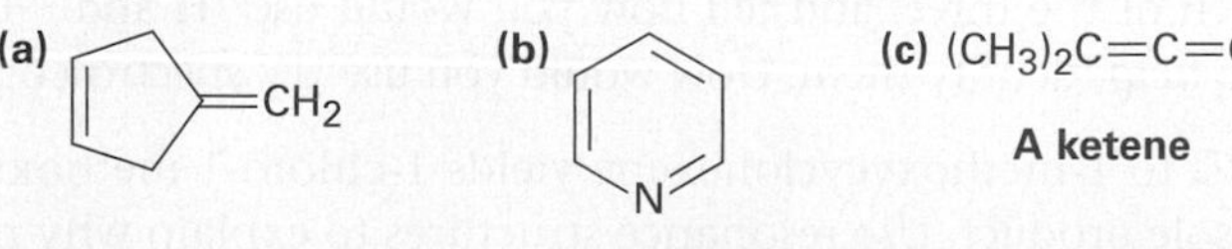

(c) $(CH_3)_2C{=}C{=}O$

A ketene

▲ Problems linked to Key Ideas in this chapter

14.40 Would you expect allene, $H_2C{=}C{=}CH_2$, to show a UV absorption in the 200 to 400 nm range? Explain.

14.41 The following ultraviolet absorption maxima have been measured:

1,3-Butadiene	217 nm
2-Methyl-1,3-butadiene	220 nm
1,3-Pentadiene	223 nm
2,3-Dimethyl-1,3-butadiene	226 nm
2,4-Hexadiene	227 nm
2,4-Dimethyl-1,3-pentadiene	232 nm
2,5-Dimethyl-2,4-hexadiene	240 nm

What conclusion can you draw about the effect of alkyl substitution on UV absorption maxima? Approximately what effect does each added alkyl group have?

14.42 1,3,5-Hexatriene has λ_{max} = 258 nm. In light of your answer to Problem 14.41, approximately where would you expect 2,3-dimethyl-1,3,5-hexatriene to absorb?

14.43 β-Ocimene is a pleasant-smelling hydrocarbon found in the leaves of certain herbs. It has the molecular formula $C_{10}H_{16}$ and a UV absorption maximum at 232 nm. On hydrogenation with a palladium catalyst, 2,6-dimethyloctane is obtained. Ozonolysis of β-ocimene, followed by treatment with zinc and acetic acid, produces the following four fragments:

$CH_3C(=O)CH_3$	$HC(=O)H$	$CH_3C(=O){-}C(=O)H$	$HC(=O)CH_2C(=O)H$
Acetone	**Formaldehyde**	**Pyruvaldehyde**	**Malonaldehyde**

(a) How many double bonds does β-ocimene have?
(b) Is β-ocimene conjugated or nonconjugated?
(c) Propose a structure for β-ocimene.
(d) Write the reactions, showing starting material and products.

General Problems

14.44 Treatment of 3,4-dibromohexane with strong base leads to loss of 2 equivalents of HBr and formation of a product with formula C_6H_{10}. Three products are possible. Name each of the three, and tell how you would use ^1H and ^{13}C NMR spectroscopy to help identify them. How would you use UV spectroscopy?

14.45 Addition of HCl to 1-methoxycyclohexene yields 1-chloro-1-methoxycyclohexane as the sole product. Use resonance structures to explain why none of the other regioisomer is formed.

1-methoxycyclohexene (OCH_3) $\xrightarrow{HCl}$ 1-chloro-1-methoxycyclohexane (OCH_3, Cl)

▲ Problems linked to Key Ideas in this chapter

14.46 Aldrin, a chlorinated insecticide now banned for use in the United States, can be made by Diels–Alder reaction of hexachloro-1,3-cyclopentadiene with norbornadiene. What is the structure of aldrin?

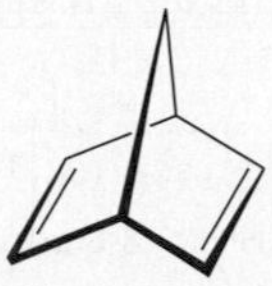

Norbornadiene

14.47 Norbornadiene (Problem 14.46) can be prepared by reaction of chloroethylene with 1,3-cyclopentadiene, followed by treatment of the product with sodium ethoxide. Write the overall scheme, and identify the two kinds of reactions.

14.48 ▲ We've seen that the Diels–Alder cycloaddition reaction is a one-step, pericyclic process that occurs through a cyclic transition state. Propose a mechanism for the following reaction:

Heat → + $H_2C{=}CH_2$

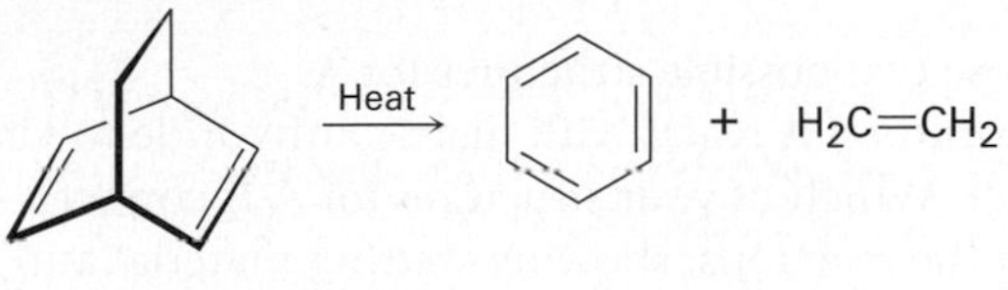

14.49 In light of your answer to Problem 14.48, propose a mechanism for the following reaction:

CO_2CH_3 — C≡C — CO_2CH_3; O; O; Heat → CO_2CH_3, CO_2CH_3 + CO_2

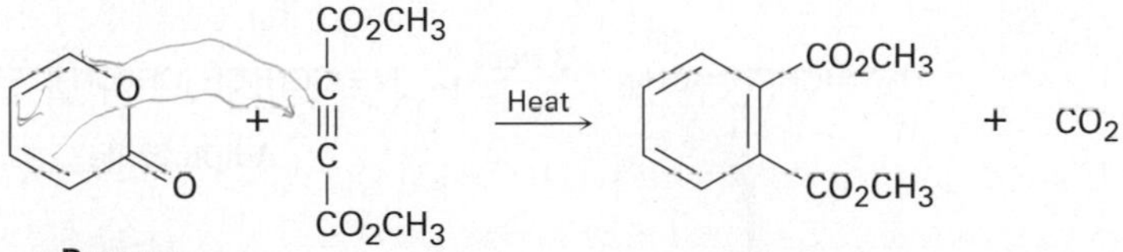

α-Pyrone

14.50 The triene shown here reacts with 2 equivalents of maleic anhydride to yield a product with the formula $C_{17}H_{16}O_6$. Predict a structure for the product.

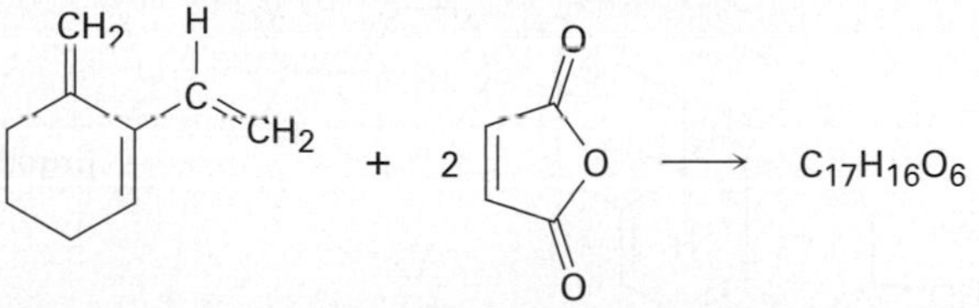

CH_2; H; C; CH_2 + 2 O, O, O → $C_{17}H_{16}O_6$

14.51 Myrcene, $C_{10}H_{16}$, is found in oil of bay leaves and is isomeric with β-ocimene (Problem 14.43). It has an ultraviolet absorption at 226 nm and can be catalytically hydrogenated to yield 2,6-dimethyloctane. On ozonolysis followed by zinc/acetic acid treatment, myrcene yields formaldehyde, acetone, and 2-oxopentanedial:

$$\overset{O}{\overset{\|}{H C}}CH_2CH_2\overset{O}{\overset{\|}{C}}-\overset{O}{\overset{\|}{C}}H \quad \textbf{2-Oxopentanedial}$$

Propose a structure for myrcene, and write the reactions, showing starting material and products.

14.52 Hydrocarbon **A**, $C_{10}H_{14}$, has a UV absorption at $\lambda_{max} = 236$ nm and gives hydrocarbon **B**, $C_{10}H_{18}$, on catalytic hydrogenation. Ozonolysis of **A** followed by zinc/acetic acid treatment yields the following diketo dialdehyde:

$$\overset{O}{\overset{\|}{H C}}CH_2CH_2CH_2\overset{O}{\overset{\|}{C}}-\overset{O}{\overset{\|}{C}}CH_2CH_2CH_2\overset{O}{\overset{\|}{C}}H$$

(a) Propose two possible structures for **A**.
(b) Hydrocarbon **A** reacts with maleic anhydride to yield a Diels–Alder adduct. Which of your structures for **A** is correct?
(c) Write the reactions, showing starting material and products.

14.53 Adiponitrile, a starting material used in the manufacture of nylon, can be prepared in three steps from 1,3-butadiene. How would you carry out this synthesis?

$$H_2C{=}CHCH{=}CH_2 \xrightarrow{\text{3 steps}} N{\equiv}CCH_2CH_2CH_2CH_2C{\equiv}N$$

Adiponitrile

14.54 Ergosterol, a precursor of vitamin D, has $\lambda_{max} = 282$ nm and molar absorptivity $\epsilon = 11{,}900$. What is the concentration of ergosterol in a solution whose absorbance $A = 0.065$ with a sample pathlength $l = 1.00$ cm?

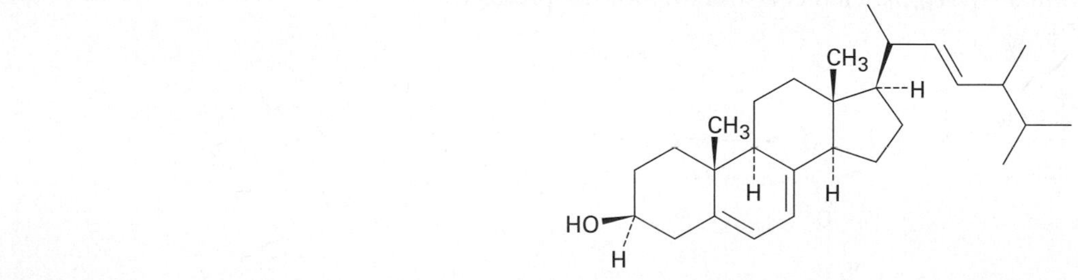

Ergosterol ($C_{28}H_{44}O$)

▲ Problems linked to Key Ideas in this chapter

14.55 ▲ Dimethyl butynedioate undergoes a Diels–Alder reaction with (2*E*,4*E*)-hexadiene. Show the structure and stereochemistry of the product.

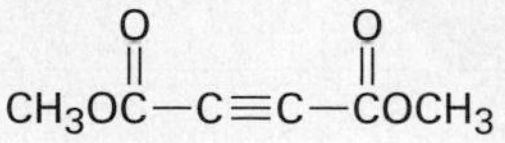

$CH_3OC(=O)-C\equiv C-C(=O)OCH_3$ **Dimethyl butynedioate**

14.56 Dimethyl butynedioate also undergoes a Diels–Alder reaction with (2*E*,4*Z*)-hexadiene, but the stereochemistry of the product is different from that of the (2*E*,4*E*) isomer (Problem 14.55). Explain.

14.57 How would you carry out the following synthesis (more than one step is required)? What stereochemical relationship between the $-CO_2CH_3$ group attached to the cyclohexane ring and the –CHO groups would your synthesis produce?

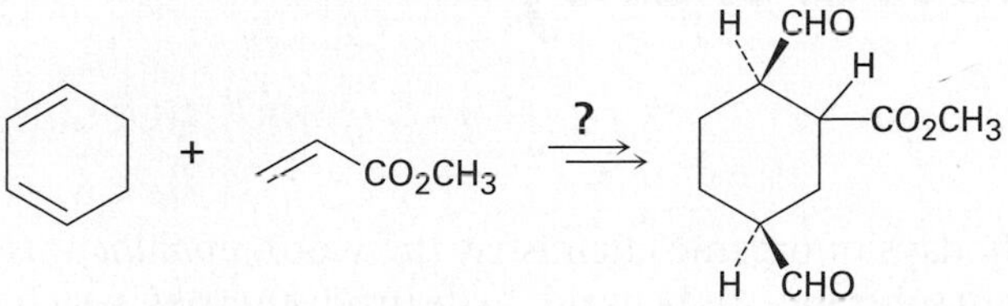

Cyclohexadiene + $CH_2=CHCO_2CH_3$ $\xrightarrow{?}$ cyclohexane bearing H, CHO; H, CO_2CH_3; H, CHO

14.58 The double bond of an *enamine* (alk*ene* + *amine*) is much more nucleophilic than a typical alkene double bond. Assuming that the nitrogen atom in an enamine is sp^2-hybridized, draw an orbital picture of an enamine, and explain why the double bond is electron-rich.

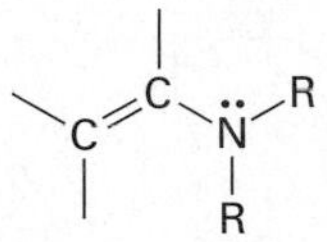

An enamine ($C=C-\ddot{N}R_2$)

14.59 Benzene has an ultraviolet absorption at λ_{max} = 204 nm, and para-toluidine has λ_{max} = 235 nm. How do you account for this difference?

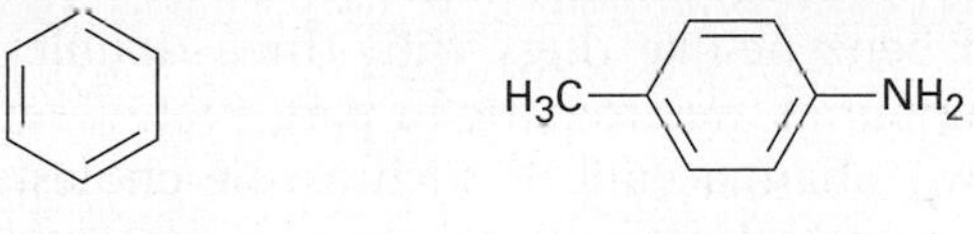

Benzene (λ_{max} = 204 nm) — $H_3C-C_6H_4-NH_2$ ***p*-Toluidine** (λ_{max} = 235 nm)

▲ Problems linked to Key Ideas in this chapter

15

A fennel plant is an aromatic herb used in cooking. A phenyl group—pronounced exactly the same way—is the characteristic structural unit of "aromatic" organic compounds.

Benzene and Aromaticity

In the early days of organic chemistry, the word *aromatic* was used to describe such fragrant substances as benzaldehyde (from cherries, peaches, and almonds), toluene (from Tolu balsam), and benzene (from coal distillate). It was soon realized, however, that substances grouped as aromatic differed from most other organic compounds in their chemical behavior.

Benzene **Benzaldehyde** **Toluene**

Today, the association of aromaticity with fragrance has long been lost, and we now use the word **aromatic** to refer to the class of compounds that contain six-membered benzene-like rings with three double bonds. Many naturally occurring compounds are aromatic in part, including steroids such as estrone and well-known pharmaceuticals such as the cholesterol-lowering drug atorvastatin, marketed as Lipitor. Benzene itself causes a depressed white blood cell count (leukopenia) on prolonged exposure and should not be used as a laboratory solvent.

Estrone

Atorvastatin (Lipitor)

OWL Sign in to OWL for Organic Chemistry at **www.cengage.com/owl** to view tutorials and simulations, develop problem-solving skills, and complete online homework assigned by your professor.

Why This Chapter? The reactivity of substituted aromatic compounds, more than that of any other class of substances, is intimately tied to their exact structure. As a result, aromatic compounds provide an extraordinarily sensitive probe for studying the relationship between structure and reactivity. We'll examine that relationship in this and the next chapter, and we'll find that the lessons learned are applicable to all other organic compounds, including such particularly important substances as the nucleic acids that control our genetic makeup.

15.1 Sources and Names of Aromatic Compounds

Simple aromatic hydrocarbons come from two main sources: coal and petroleum. Coal is an enormously complex mixture made up primarily of large arrays of benzene-like rings joined together. Thermal breakdown of coal occurs when it is heated to 1000 °C in the absence of air, and a mixture of volatile products called *coal tar* boils off. Fractional distillation of coal tar yields benzene, toluene, xylene (dimethylbenzene), naphthalene, and a host of other aromatic compounds **(Figure 15.1)**.

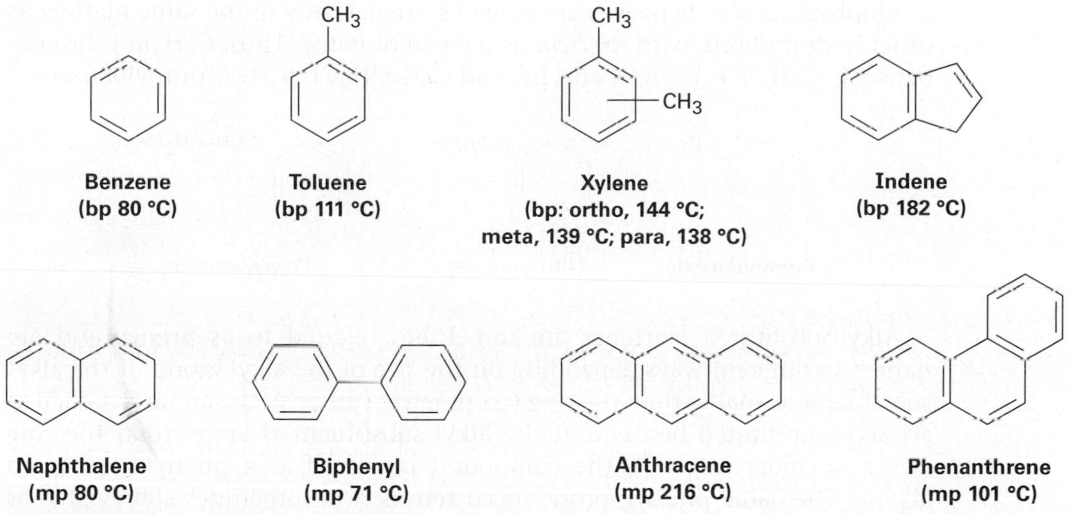

Figure 15.1 Some aromatic hydrocarbons found in coal tar.

Unlike coal, petroleum contains few aromatic compounds and consists largely of alkanes (Chapter 3 *A Deeper Look*). During petroleum refining, however, aromatic molecules are formed when alkanes are passed over a catalyst at about 500 °C under high pressure.

Aromatic substances, more than any other class of organic compounds, have acquired a large number of nonsystematic names. IUPAC rules discourage the use of most such names but do allow for some of the more widely used ones to be retained (Table 15.1). Thus, methylbenzene is known commonly as *toluene;* hydroxybenzene, as *phenol;* aminobenzene, as *aniline;* and so on.

Table 15.1 Common Names of Some Aromatic Compounds

Structure	Name	Structure	Name
CH_3	Toluene (bp 111 °C)	CHO	Benzaldehyde (bp 178 °C)
OH	Phenol (mp 43 °C)	CO_2H	Benzoic acid (mp 122 °C)
NH_2	Aniline (bp 184 °C)	CH_3, CH_3	ortho-Xylene (bp 144 °C)
O, C, CH_3	Acetophenone (mp 21 °C)	H, C, C, H, H	Styrene (bp 145 °C)

Monosubstituted benzenes are named systematically in the same manner as other hydrocarbons, with *-benzene* as the parent name. Thus, C_6H_5Br is bromobenzene, $C_6H_5NO_2$ is nitrobenzene, and $C_6H_5CH_2CH_2CH_3$ is propylbenzene.

Br NO$_2$ $CH_2CH_2CH_3$

Bromobenzene **Nitrobenzene** **Propylbenzene**

Alkyl-substituted benzenes are sometimes referred to as **arenes** and are named in different ways depending on the size of the alkyl group. If the alkyl substituent is smaller than the ring (six or fewer carbons), the arene is named as an alkyl-substituted benzene. If the alkyl substituent is larger than the ring (seven or more carbons), the compound is named as a phenyl-substituted alkane. The name **phenyl**, pronounced **fen**-nil and sometimes abbreviated as Ph or *Φ* (Greek *phi*), is used for the $-C_6H_5$ unit when the benzene ring is considered as a substituent. The word is derived from the Greek *pheno* ("I bear light"), commemorating the discovery of benzene by Michael Faraday in 1825 from the oily residue left by the illuminating gas used in London street lamps. In addition, the name **benzyl** is used for the $C_6H_5CH_2-$ group.

1CH_3 $CHCH_2CH_2CH_2CH_2CH_3$ 2 3 4 5 6 7 CH_2

A phenyl group **2-Phenylheptane** **A benzyl group**

Disubstituted benzenes are named using one of the prefixes **ortho** ***(o)***, **meta** ***(m)***, or **para** ***(p)***. An ortho-disubstituted benzene has its two substituents in a 1,2 relationship on the ring, a meta-disubstituted benzene has its two substituents in a 1,3 relationship, and a para-disubstituted benzene has its substituents in a 1,4 relationship.

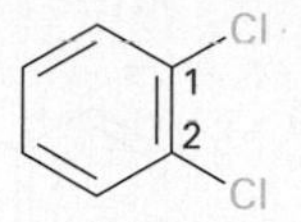

***ortho*-Dichlorobenzene**
1,2 disubstituted

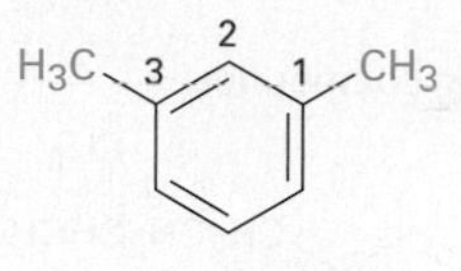

***meta*-Dimethylbenzene**
(*meta*-xylene)
1,3 disubstituted

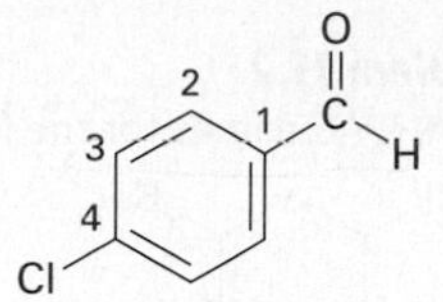

***para*-Chlorobenzaldehyde**
1,4 disubstituted

The ortho, meta, para system of nomenclature is also useful when discussing reactions. For example, we might describe the reaction of bromine with toluene by saying, "Reaction occurs at the para position"—in other words, at the position para to the methyl group already present on the ring.

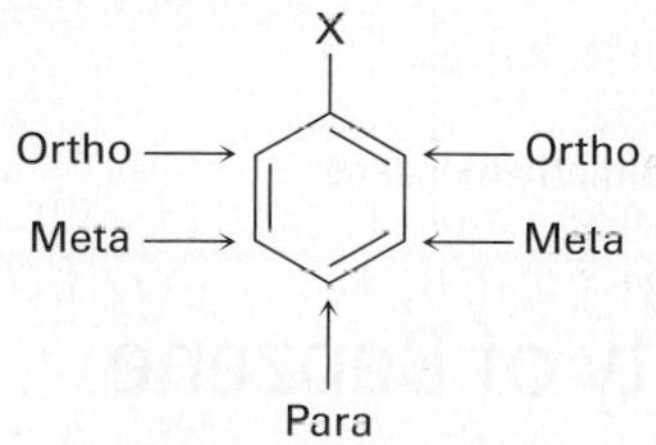

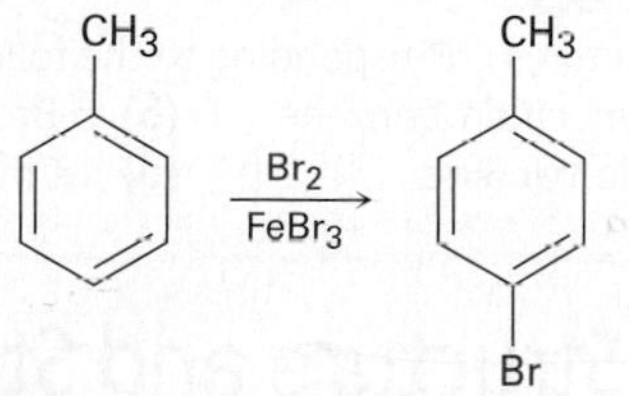

Ortho, Meta, Para

$\xrightarrow[FeBr_3]{Br_2}$

Toluene ***p*-Bromotoluene**

As with cycloalkanes **(Section 4.1)**, benzenes with more than two substituents are named by choosing a point of attachment as carbon 1 and numbering the substituents on the ring so that the second substituent has as low a number as possible. If ambiguity still exists, number so that the third or fourth substituent has as low a number as possible, until a point of difference is found. The substituents are listed alphabetically when writing the name.

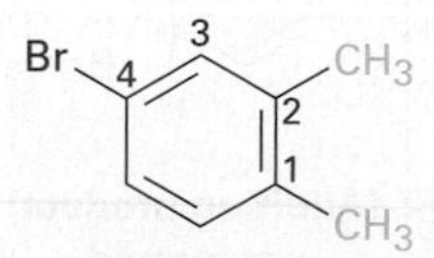

4-Bromo-1,2-dimethylbenzene 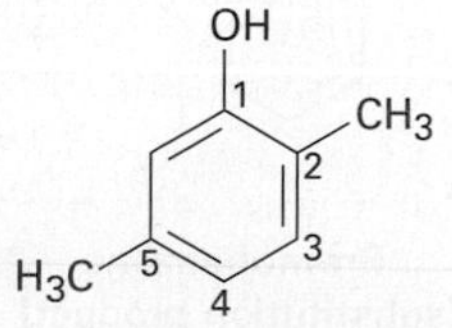 **2,5-Dimethylphenol** 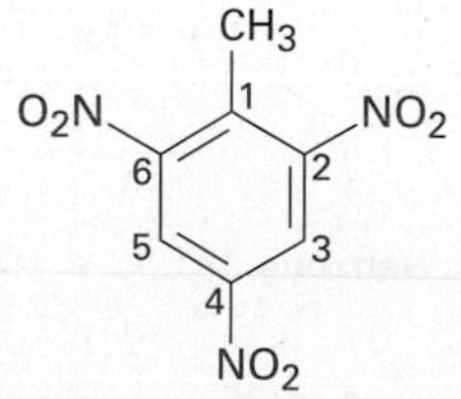 **2,4,6-Trinitrotoluene (TNT)**

Note in the second and third examples shown that *-phenol* and *-toluene* are used as the parent names rather than *-benzene*. Any of the monosubstituted aromatic compounds shown in Table 15.1 can serve as a parent name, with the principal substituent (–OH in phenol or $-CH_3$ in toluene) attached to C1 on the ring.

Problem 15.1

Tell whether the following compounds are ortho-, meta-, or para-disubstituted:

(a) 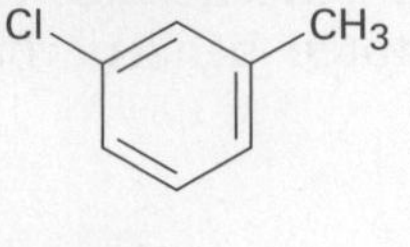 Cl, CH_3 **(b)** 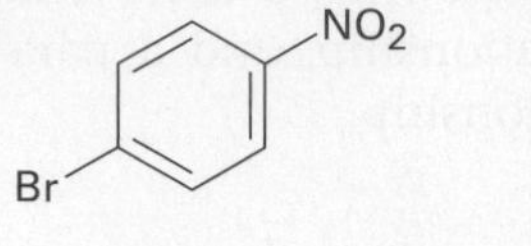 NO_2, Br **(c)** 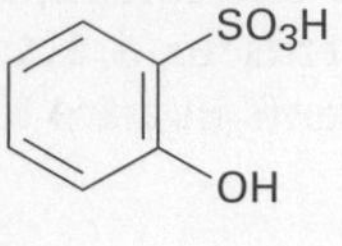 SO_3H, OH

Problem 15.2

Give IUPAC names for the following compounds:

(a) 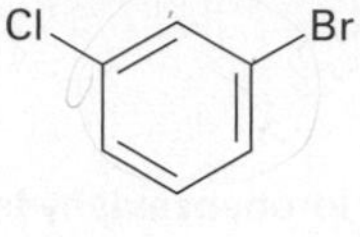 Cl, Br **(b)** 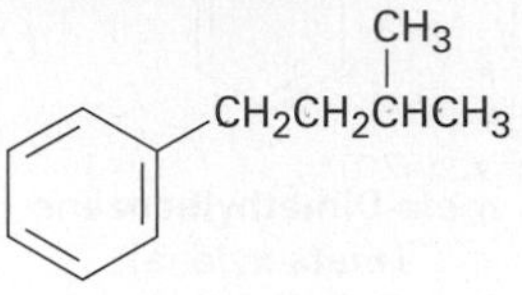 CH_3 / $CH_2CH_2CHCH_3$ **(c)** 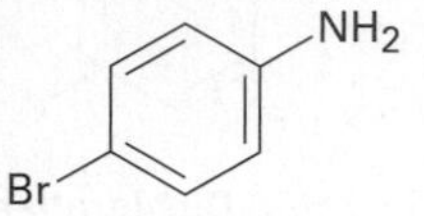 NH_2, Br

(d) 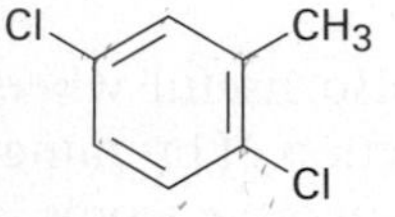 Cl, CH_3, Cl **(e)** 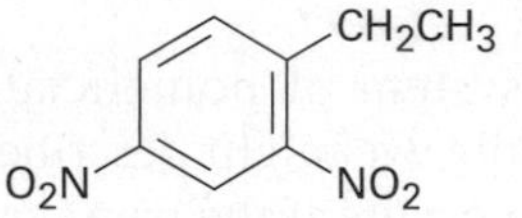 CH_2CH_3, O_2N, NO_2 **(f)** 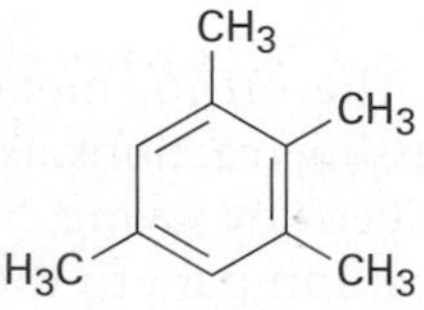 CH_3, CH_3, H_3C, CH_3

Problem 15.3

Draw structures corresponding to the following IUPAC names:

(a) *p*-Bromochlorobenzene **(b)** *p*-Bromotoluene

(c) *m*-Chloroaniline **(d)** 1-Chloro-3,5-dimethylbenzene

15.2 Structure and Stability of Benzene

Benzene (C_6H_6) has six fewer hydrogens than the corresponding six-carbon cycloalkane (C_6H_{12}) and is clearly unsaturated, usually being represented as a six-membered ring with alternating double and single bonds. Yet it has been known since the mid-1800s that benzene is much less reactive than typical alkenes and fails to undergo typical alkene addition reactions. Cyclohexene, for instance, reacts rapidly with Br_2 and gives the addition product 1,2-dibromocyclohexane, but benzene reacts only slowly with Br_2 and gives the *substitution* product C_6H_5Br.

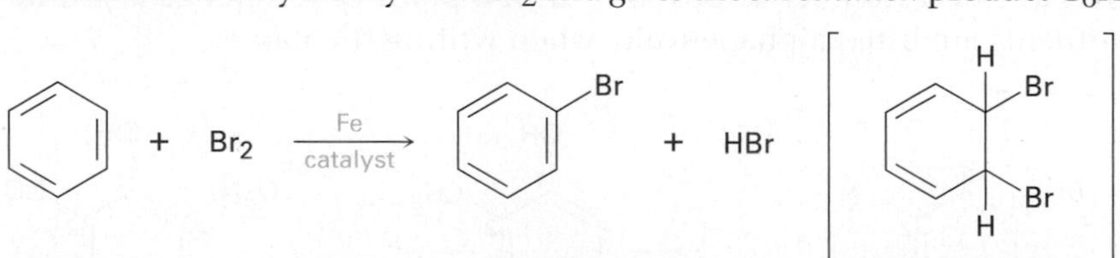

Benzene + Br_2 $\xrightarrow[\text{catalyst}]{\text{Fe}}$ Bromobenzene (substitution product) + HBr

[H, Br, Br, H] **(Addition product) *not* formed**

Benzene **Bromobenzene (substitution product)**

We can get a quantitative idea of benzene's stability by measuring heats of hydrogenation **(Section 7.6)**. Cyclohexene, an isolated alkene, has $\Delta H^\circ_{\text{hydrog}} = -118$ kJ/mol (-28.2 kcal/mol), and 1,3-cyclohexadiene, a conjugated diene, has $\Delta H^\circ_{\text{hydrog}} = -230$ kJ/mol (-55.0 kcal/mol). As noted in **Section 14.1**, this value for 1,3-cyclohexadiene is a bit less than twice that for cyclohexene because conjugated dienes are more stable than isolated dienes.

Carrying the process one step further, we might expect $\Delta H^\circ_{\text{hydrog}}$ for "cyclohexatriene" (benzene) to be a bit less than -356 kJ/mol, or three times the cyclohexene value. The actual value, however, is −206 kJ/mol, some 150 kJ/mol (36 kcal/mol) less than expected. Since 150 kJ/mol less heat than expected is released during hydrogenation of benzene, benzene must have 150 kJ/mol less energy to begin with. In other words, benzene is more stable than expected by 150 kJ/mol **(Figure 15.2)**.

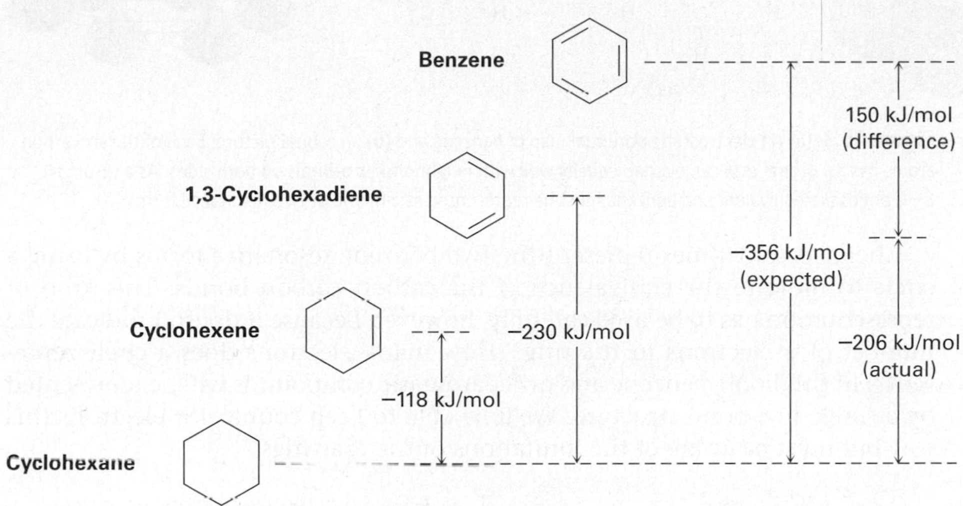

Figure 15.2 A comparison of the heats of hydrogenation for cyclohexene, 1,3-cyclohexadiene, and benzene. Benzene is 150 kJ/mol (36 kcal/mol) more stable than might be expected for "cyclohexatriene."

Further evidence for the unusual nature of benzene is that all its carbon–carbon bonds have the same length—139 pm—intermediate between typical single (154 pm) and double (134 pm) bonds. In addition, an electrostatic potential map shows that the electron density in all six C–C bonds is identical. Thus, benzene is a planar molecule with the shape of a regular hexagon. All C–C–C bond angles are 120°, all six carbon atoms are sp^2-hybridized, and each carbon has a *p* orbital perpendicular to the plane of the six-membered ring.

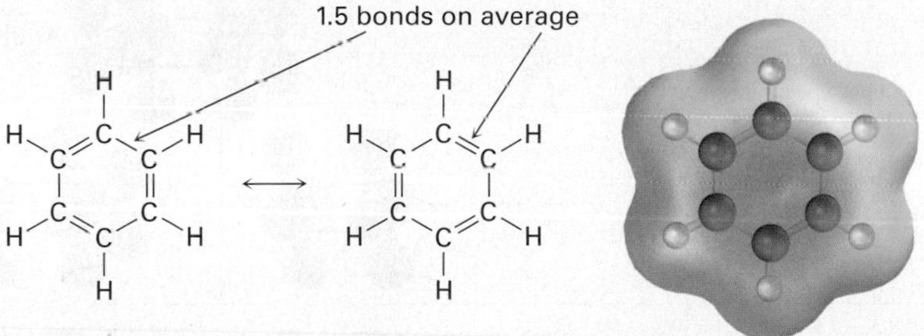

Because all six carbon atoms and all six *p* orbitals in benzene are equivalent, it's impossible to define three localized π bonds in which a given *p* orbital overlaps only one neighboring *p* orbital. Rather, each *p* orbital overlaps equally well with both neighboring *p* orbitals, leading to a picture of benzene in which all six π electrons are free to move about the entire ring **(Figure 15.3b)**. In resonance terms **(Sections 2.4 and 2.5)**, benzene is a hybrid of two equivalent forms. Neither form is correct by itself; the true structure of benzene is somewhere in between the two resonance forms but is impossible to draw with our usual conventions. Because of this resonance, benzene is more stable and less reactive than a typical alkene.

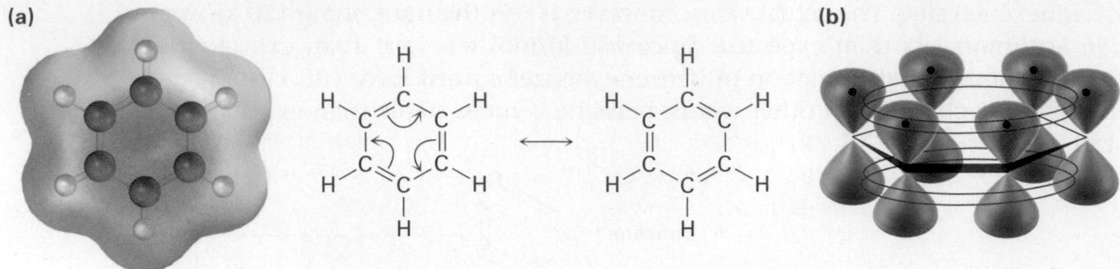

Figure 15.3 **(a)** An electrostatic potential map of benzene and **(b)** an orbital picture. Each of the six carbon atoms has a *p* orbital that can overlap equally well with neighboring *p* orbitals on both sides. As a result, all C—C bonds are equivalent and benzene must be represented as a hybrid of two resonance forms.

Chemists sometimes represent the two benzene resonance forms by using a circle to indicate the equivalence of the carbon–carbon bonds. This kind of representation has to be used carefully, however, because it doesn't indicate the number of π electrons in the ring. (How many electrons does a circle represent?) In this book, benzene and other aromatic compounds will be represented by a single line-bond structure. We'll be able to keep count of π electrons this way but must be aware of the limitations of the drawings.

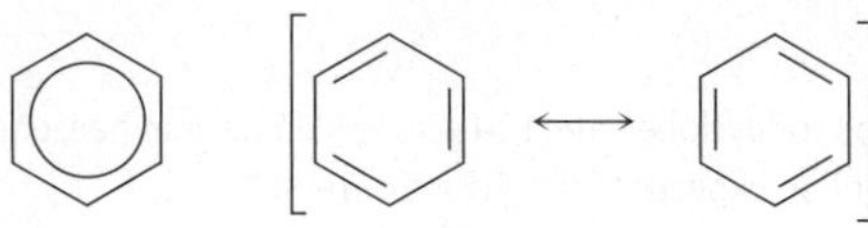

Alternative representations of benzene. The "circle" representation must be used carefully since it doesn't indicate the number of π electrons in the ring.

Having just seen a resonance description of benzene, let's now look at the alternative molecular orbital description. We can construct π molecular orbitals for benzene just as we did for 1,3-butadiene in **Section 14.1**. If six *p* atomic orbitals combine in a cyclic manner, six benzene molecular orbitals result, as shown in **Figure 15.4**. The three low-energy molecular orbitals, denoted ψ_1, ψ_2, and ψ_3, are bonding combinations, and the three high-energy orbitals are antibonding.

Figure 15.4 The six benzene π molecular orbitals. The bonding orbitals ψ_2 and ψ_3 have the same energy and are said to be degenerate, as are the antibonding orbitals ψ_4* and ψ_5*. The orbitals ψ_3 and ψ_4* have no π electron density on two carbons because of a node passing through these atoms.

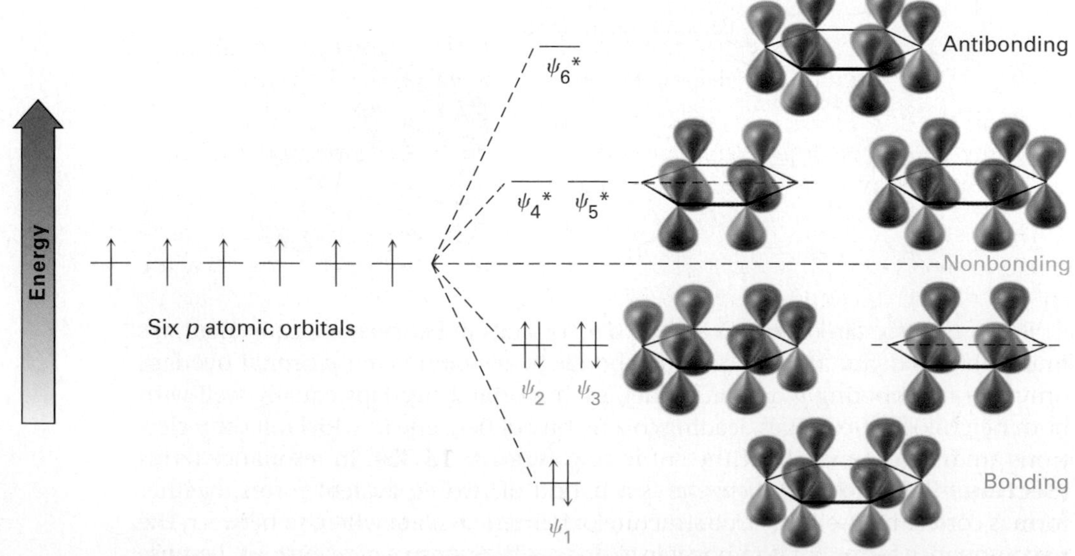

Note that the two bonding orbitals ψ_2 and ψ_3 have the same energy, as do the two antibonding orbitals ψ_4* and ψ_5*. Such orbitals with the same energy are said to be *degenerate*. Note also that the two orbitals ψ_3 and ψ_4* have nodes passing through ring carbon atoms, thereby leaving no π electron density on these carbons. The six *p* electrons of benzene occupy the three bonding molecular orbitals and are delocalized over the entire conjugated system, leading to the observed 150 kJ/mol stabilization of benzene.

Problem 15.4

Pyridine is a flat, hexagonal molecule with bond angles of 120°. It undergoes substitution rather than addition and generally behaves like benzene. Draw a picture of the π orbitals of pyridine to explain its properties. Check your answer by looking ahead to **Section 15.5.**

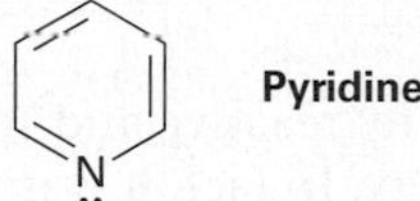

Pyridine

15.3 Aromaticity and the Hückel 4*n* + 2 Rule

Let's list what we've said thus far about benzene and, by extension, about other benzene-like aromatic molecules.

- Benzene is cyclic and conjugated.
- Benzene is unusually stable, having a heat of hydrogenation 150 kJ/mol less negative than we might expect for a conjugated cyclic triene.
- Benzene is planar and has the shape of a regular hexagon. All bond angles are 120°, all carbon atoms are sp^2-hybridized, and all carbon–carbon bond lengths are 139 pm.
- Benzene undergoes substitution reactions that retain the cyclic conjugation rather than electrophilic addition reactions that would destroy the conjugation.
- Benzene can be described as a resonance hybrid whose structure is intermediate between two line-bond structures.

This list would seem to be a good description of benzene and other aromatic molecules, but it isn't enough. Something else, called the **Hückel 4*n* + 2 rule**, is needed to complete a description of aromaticity. According to a theory devised in 1931 by the German physicist Erich Hückel, a molecule is aromatic only if it has a planar, monocyclic system of conjugation and contains a total of $4n + 2$ *π electrons*, where *n* is an integer ($n = 0, 1, 2, 3, \ldots$). In other words, only molecules with 2, 6, 10, 14, 18, . . . π electrons can be aromatic. Molecules with $4n$ π electrons (4, 8, 12, 16, . . .) can't be aromatic, even though they may be cyclic, planar, and apparently conjugated. In fact, planar, conjugated molecules with $4n$ π electrons are said to be **antiaromatic** because delocalization of their π electrons would lead to their destabilization.

Let's look at several examples to see how the Hückel $4n + 2$ rule works.

- **Cyclobutadiene** has four π electrons and is antiaromatic. The π electrons are localized in two double bonds rather than delocalized around the ring, as indicated by an electrostatic potential map.

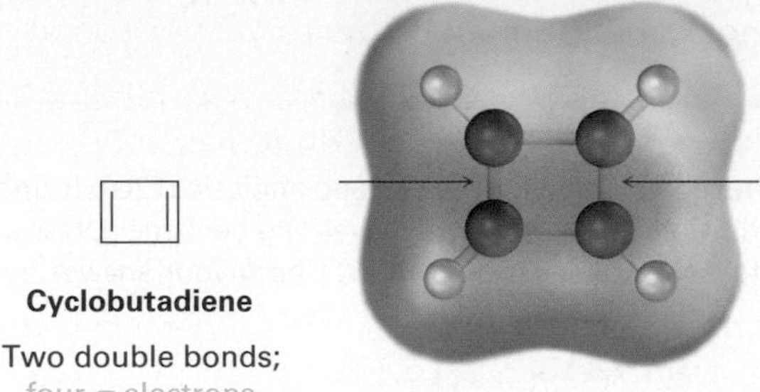

Cyclobutadiene is highly reactive and shows none of the properties associated with aromaticity. In fact, it was not even prepared until 1965, when Rowland Pettit of the University of Texas was able to make it at low temperature. Even at −78 °C, however, cyclobutadiene is so reactive that it dimerizes by a Diels–Alder reaction. One molecule behaves as a diene and the other as a dienophile.

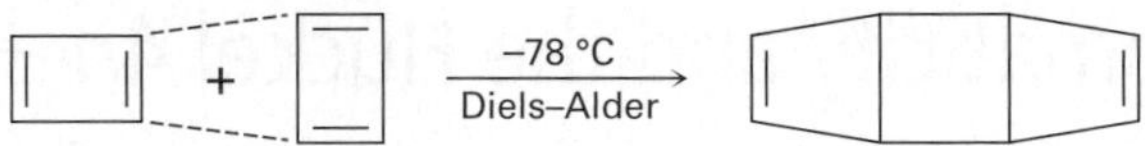

- **Benzene** has six π electrons ($4n + 2 = 6$ when $n = 1$) and is aromatic.

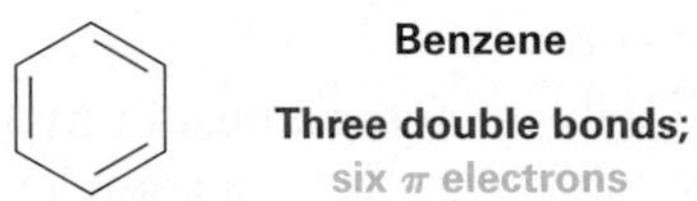

- **Cyclooctatetraene** has eight π electrons and is not aromatic. The π electrons are localized into four double bonds rather than delocalized around the ring, and the molecule is tub-shaped rather than planar.

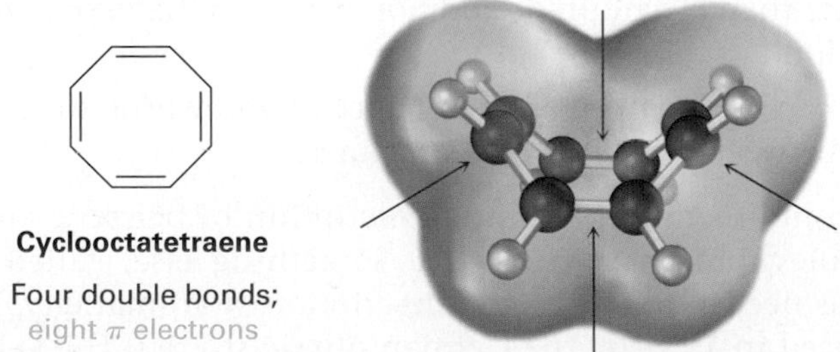

Chemists in the early 1900s believed that the only requirement for aromaticity was the presence of a cyclic conjugated system. It was therefore expected that cyclooctatetraene, as a close analog of benzene, would also prove to be unusually stable. The facts, however, proved otherwise. When cyclooctatetraene was first prepared in 1911 by the German chemist Richard Willstätter, it was found not to be particularly stable but to resemble an open-chain polyene in its reactivity.

In fact, cyclooctatetraene is not even conjugated because neighboring p orbitals don't have the necessary parallel alignment for overlap. The π electrons are localized in four discrete C=C bonds rather than delocalized around the ring. X-ray studies show that the C–C single bonds are 147 pm long and the double bonds are 134 pm long. In addition, the ^{1}H NMR spectrum shows a single sharp resonance line at 5.7 δ, a value characteristic of an alkene rather than an aromatic molecule.

What's so special about $4n + 2$ π electrons? Why do 2, 6, 10, 14 . . . π electrons lead to aromatic stability, while other numbers of electrons do not? The answer comes from molecular orbital theory. When the energy levels of molecular orbitals for cyclic conjugated molecules are calculated, it turns out that there is always a single lowest-lying MO, above which the MOs come in degenerate pairs. Thus, when electrons fill the various molecular orbitals, it takes two electrons, or one pair, to fill the lowest-lying orbital and four electrons, or two pairs, to fill each of n succeeding energy levels—a total of $4n + 2$. Any other number would leave an energy level partially filled.

Key IDEAS

Test your knowledge of Key Ideas by answering end-of-chapter exercises marked with ▲.

The six π molecular orbitals of benzene were shown previously in Figure 15.4, and their relative energies are shown again in **Figure 15.5**. The lowest-energy MO, ψ_1, occurs singly and contains two electrons. The next two lowest-energy orbitals, ψ_2 and ψ_3, are degenerate, and it therefore takes four electrons to fill both. The result is a stable six-π-electron aromatic molecule with filled bonding orbitals.

Energy
Six p atomic orbitals
ψ_6^*
ψ_4^* ψ_5^*
ψ_2 ψ_3
ψ_1

Figure 15.5 Energy levels of the six benzene π molecular orbitals. There is a single, lowest-energy orbital, above which the orbitals come in degenerate pairs.

Problem 15.5

To be aromatic, a molecule must have $4n + 2$ π electrons and must have cyclic conjugation. 1,3,5,7,9-Cyclodecapentaene fulfills one of these criteria but not the other and has resisted all attempts at synthesis. Explain.

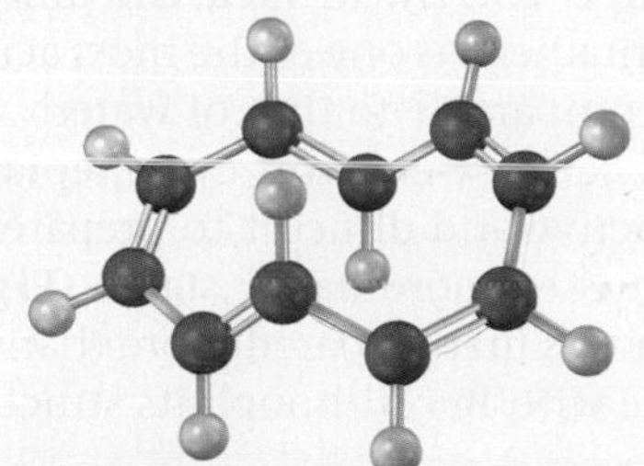

15.4 Aromatic Ions

According to the Hückel criteria for aromaticity, a molecule must be cyclic, conjugated (nearly planar with a *p* orbital on each atom), and have $4n + 2$ π electrons. Nothing in this definition says that the number of π electrons must be the same as the number of atoms in the ring or that the substance must be neutral. In fact, the numbers can be different and the substance can be an ion. Thus, both the cyclopentadienyl anion and the cycloheptatrienyl cation are aromatic even though both are ions and neither contains a six-membered ring.

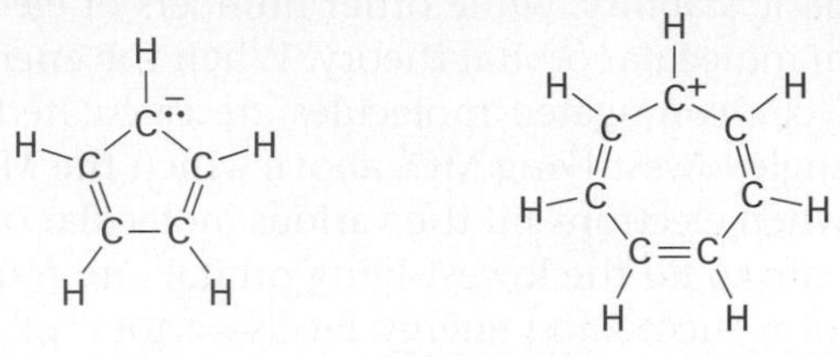

Cyclopentadienyl anion **Cycloheptatrienyl cation**

Six π electrons; aromatic ions

To see why the cyclopentadienyl anion and the cycloheptatrienyl cation are aromatic, imagine starting from the related neutral hydrocarbons, 1,3-cyclopentadiene and 1,3,5-cycloheptatriene, and removing one hydrogen from the saturated CH_2 carbon in each. If that carbon then rehybridizes from sp^3 to sp^2, the resultant products would be fully conjugated, with a *p* orbital on every carbon. There are three ways in which the hydrogen might be removed.

- The hydrogen can be removed with *both* electrons ($H:^-$) from the C–H bond, leaving a carbocation as product.
- The hydrogen can be removed with *one* electron (H·) from the C–H bond, leaving a carbon radical as product.
- The hydrogen can be removed with *no* electrons (H^+) from the C–H bond, leaving a carbanion as product.

All the potential products formed by removing a hydrogen from 1,3-cyclopentadiene and from 1,3,5-cycloheptatriene can be drawn with numerous resonance structures, but Hückel's rule predicts that only the six-π-electron cyclopentadienyl anion and cycloheptatrienyl cation should be aromatic. The other products are predicted by the $4n + 2$ rule to be unstable and antiaromatic **(Figure 15.6)**.

In practice, both the four-π-electron cyclopentadienyl cation and the five-π-electron cyclopentadienyl radical are highly reactive and difficult to prepare. Neither shows any sign of the stability expected for an aromatic system. The six-π-electron cyclopentadienyl anion, by contrast, is easily prepared and remarkably stable **(Figure 15.7a)**. In fact, the anion is so stable and easily formed that 1,3-cyclopentadiene is one of the most acidic hydrocarbons known, with $pK_a = 16$, a value comparable to that of water!

In the same way, the seven-π-electron cycloheptatrienyl radical and eight-π-electron anion are reactive and difficult to prepare, while the six-π-electron cycloheptatrienyl cation is extraordinarily stable **(Figure 15.7b)**. In fact, the cycloheptatrienyl cation was first prepared more than a century ago by reaction of 1,3,5-cycloheptatriene with Br_2, although its structure was not recognized at the time.

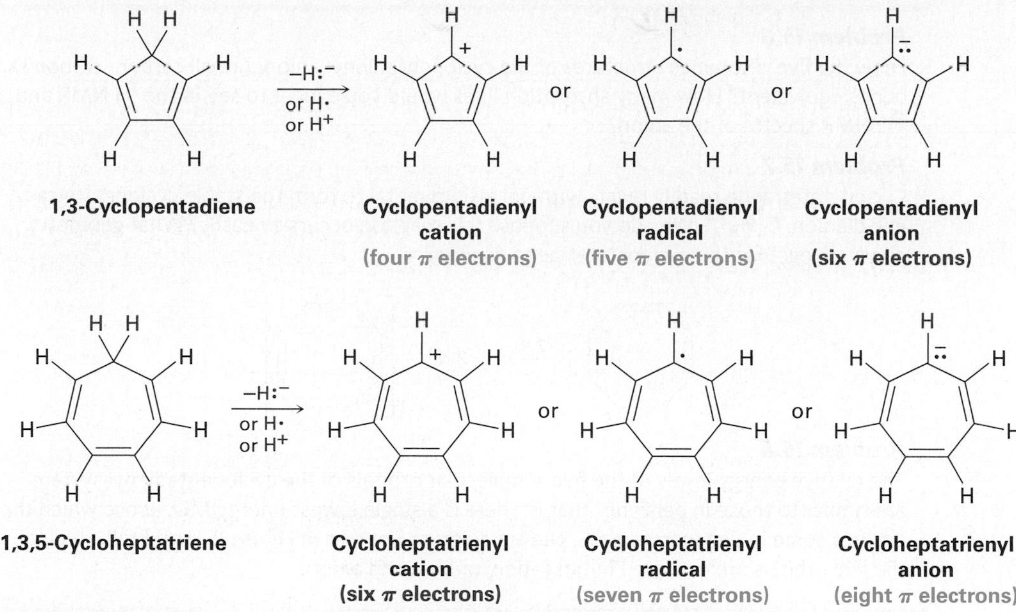

Figure 15.6 The aromatic six-π-electron cyclopentadienyl anion can be formed by removing a hydrogen ion (H^+) from the CH_2 group of 1,3-cyclopentadiene. Similarly, the aromatic six-π-electron cycloheptatrienyl cation can be generated by removing a hydride ion ($H:^-$) from the CH_2 group of 1,3,5-cycloheptatriene.

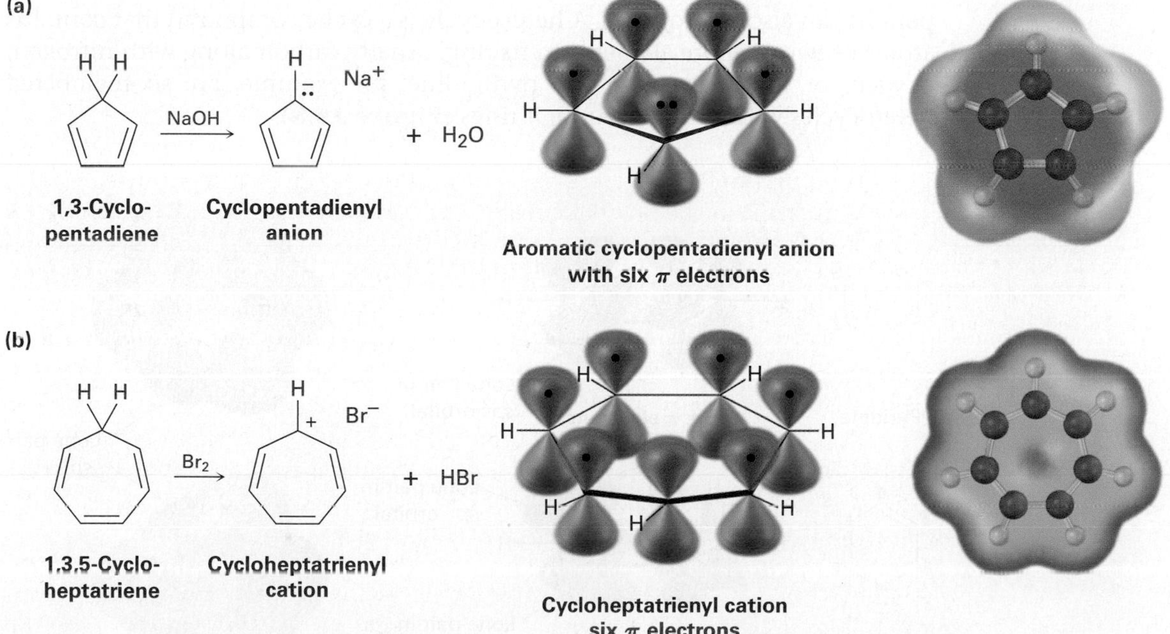

Figure 15.7 **(a)** The aromatic cyclopentadienyl anion, showing the cyclic conjugation and six π electrons in five *p* orbitals, and **(b)** the aromatic cycloheptatrienyl cation, showing the cyclic conjugation and six π electrons in seven *p* orbitals. Electrostatic potential maps indicate that both ions are symmetrical, with the charge equally shared among all atoms in each ring.

Problem 15.6

Draw the five resonance structures of the cyclopentadienyl anion. Are all carbon–carbon bonds equivalent? How many absorption lines would you expect to see in the ^{1}H NMR and ^{13}C NMR spectra of the anion?

Problem 15.7

Cyclooctatetraene readily reacts with potassium metal to form the stable cyclooctatetraene dianion, $C_8H_8^{2-}$. Why do you suppose this reaction occurs so easily? What geometry do you expect for the cyclooctatetraene dianion?

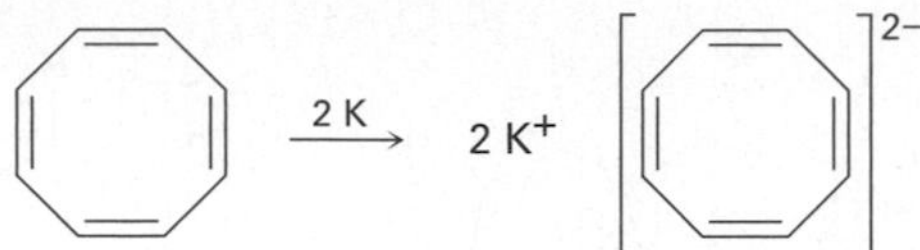

Problem 15.8

The relative energy levels of the five π molecular orbitals of the cyclopentadienyl system are similar to those in benzene. That is, there is a single lowest-energy MO, above which the orbitals come in degenerate pairs. Draw a diagram like that in Figure 15.5, and tell which of the five orbitals are occupied in the cation, radical, and anion.

15.5 Aromatic Heterocycles: Pyridine and Pyrrole

Look back once again at the definition of aromaticity in **Section 15.3**: . . . a cyclic, conjugated molecule containing $4n + 2$ π electrons. Nothing in this definition says that the atoms in the ring must be *carbon*. In fact, *heterocyclic* compounds can also be aromatic. A **heterocycle** is a cyclic compound that contains atoms of two or more elements in its ring, usually carbon along with nitrogen, oxygen, or sulfur. Pyridine and pyrimidine, for example, are six-membered heterocycles with nitrogen in their rings **(Figure 15.8)**.

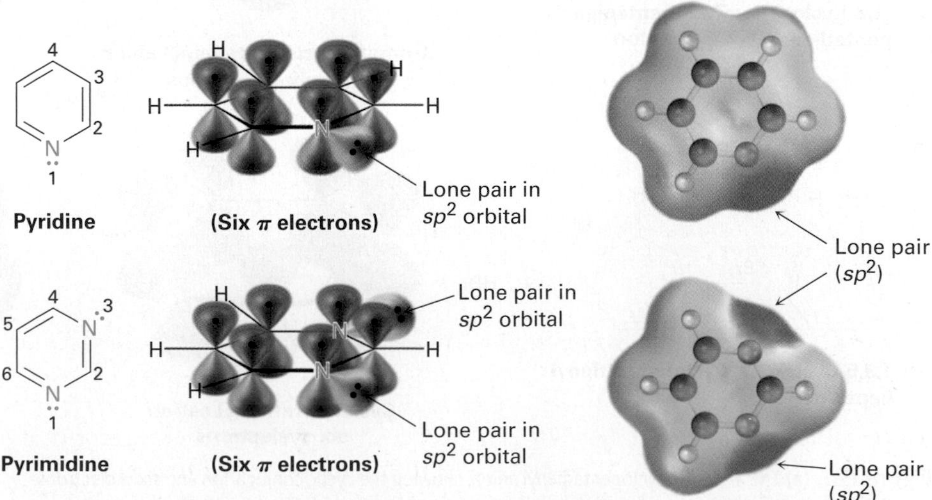

Figure 15.8 Pyridine and pyrimidine are nitrogen-containing aromatic heterocycles with π electron arrangements like that of benzene. Both have a lone pair of electrons on nitrogen in an sp^2 orbital in the plane of the ring.

Pyridine is much like benzene in its π electron structure. Each of the five sp^2-hybridized carbons has a p orbital perpendicular to the plane of the ring, and each p orbital contains one π electron. The nitrogen atom is also sp^2-hybridized and has one electron in a p orbital, bringing the total to six π electrons. The nitrogen lone-pair electrons (red in an electrostatic potential map) are in an sp^2 orbital in the plane of the ring and are not part of the aromatic π system. Pyrimidine, also shown in Figure 15.8, is a benzene analog that has two nitrogen atoms in a six-membered, unsaturated ring. Both nitrogens are sp^2-hybridized, and each contributes one electron to the aromatic π system.

Pyrrole (spelled with two *r*'s and one *l*) and imidazole are *five*-membered heterocycles, yet both have *six* π electrons and are aromatic. In pyrrole, each of the four sp^2-hybridized carbons contributes one π electron and the sp^2-hybridized nitrogen atom contributes the two from its lone pair, which occupies a p orbital **(Figure 15.9)**. Imidazole, also shown in Figure 15.9, is an analog of pyrrole that has two nitrogen atoms in a five-membered, unsaturated ring. Both nitrogens are sp^2-hybridized, but one is in a double bond and contributes only one electron to the aromatic π system while the other is not in a double bond and contributes two from its lone pair.

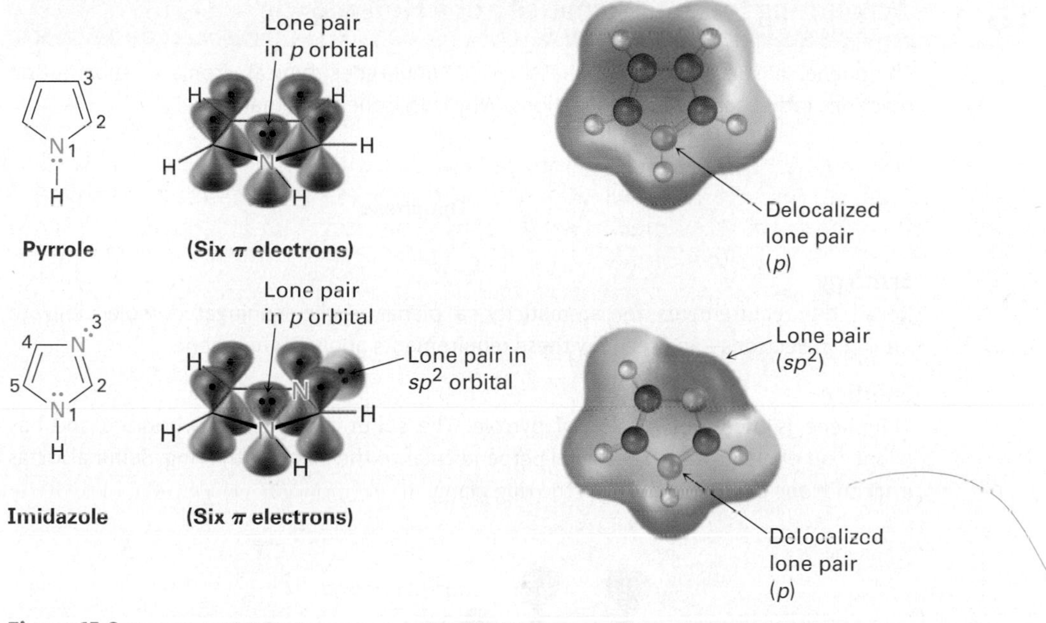

Figure 15.9 Pyrrole and imidazole are five-membered, nitrogen-containing heterocycles but have six π electron arrangements like that of the cyclopentadienyl anion. Both have a lone pair of electrons on nitrogen in a p orbital perpendicular to the ring.

Note that nitrogen atoms have different roles depending on the structure of the molecule. The nitrogen atoms in pyridine and pyrimidine are both in double bonds and contribute only one π electron to the aromatic sextet, just as a carbon atom in benzene does. The nitrogen atom in pyrrole, however, is not in a double bond and contributes two π electrons (its lone pair) to the aromatic sextet. In imidazole, both kinds of nitrogen are present in the same molecule—a double-bonded "pyridine-like" nitrogen that contributes one π electron and a "pyrrole-like" nitrogen that contributes two.

Pyrimidine and imidazole rings are particularly important in biological chemistry. Pyrimidine, for instance, is the parent ring system in cytosine, thymine, and uracil, three of the five heterocyclic amine bases found in nucleic acids. An aromatic imidazole ring is present in histidine, one of the twenty amino acids found in proteins.

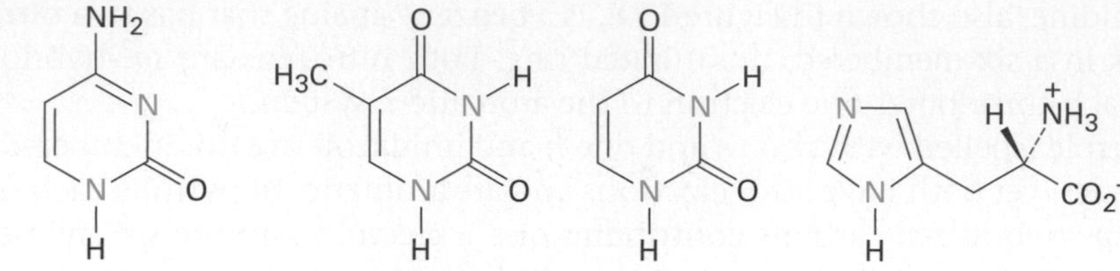

Cytosine (in DNA and RNA) — **Thymine** (in DNA) — **Uracil** (in RNA) — **Histidine** (an amino acid)

Worked Example 15.1 Accounting for the Aromaticity of a Heterocycle

Thiophene, a sulfur-containing heterocycle, undergoes typical aromatic substitution reactions rather than addition reactions. Why is thiophene aromatic?

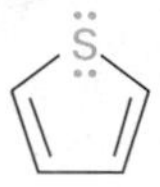

Thiophene

Strategy

Recall the requirements for aromaticity—a planar, cyclic, conjugated molecule with $4n + 2$ π electrons—and see how these requirements apply to thiophene.

Solution

Thiophene is the sulfur analog of pyrrole. The sulfur atom is sp^2-hybridized and has a lone pair of electrons in a *p* orbital perpendicular to the plane of the ring. Sulfur also has a second lone pair of electrons in the ring plane.

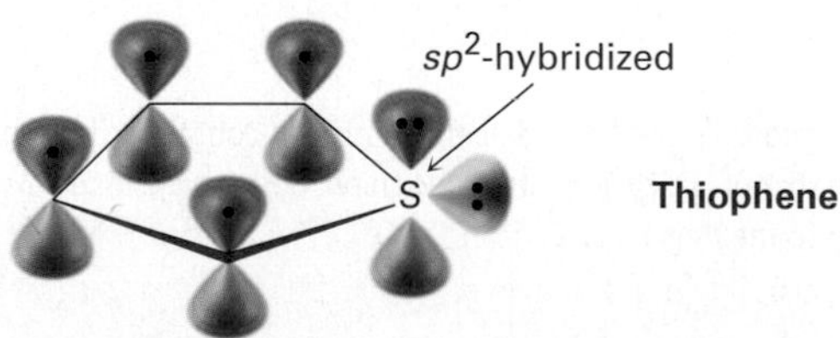

Problem 15.9

Draw an orbital picture of furan to show how the molecule is aromatic.

Furan

Problem 15.10

Thiamin, or vitamin B_1, contains a positively charged five-membered nitrogen–sulfur heterocycle called a *thiazolium* ring. Explain why the thiazolium ring is aromatic.

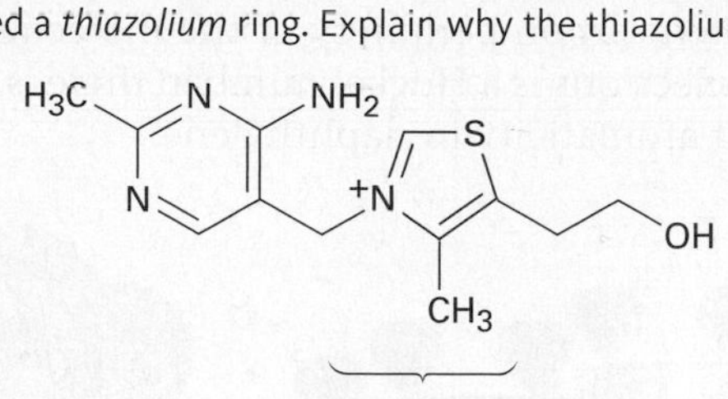

H_3C N NH_2 S +N OH CH_3

Thiamin

Thiazolium ring

15.6 Polycyclic Aromatic Compounds

The Hückel rule is strictly applicable only to monocyclic compounds, but the general concept of aromaticity can be extended to include *polycyclic* aromatic compounds. Naphthalene, with two benzene-like rings fused together; anthracene, with three rings; benzo[a]pyrene, with five rings; and coronene, with six rings, are all well-known aromatic hydrocarbons. Benzo[a]pyrene is particularly interesting because it is one of the cancer-causing substances found in tobacco smoke.

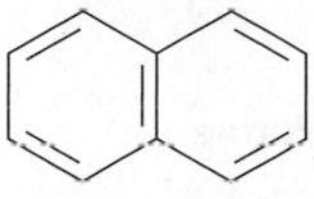

Naphthalene 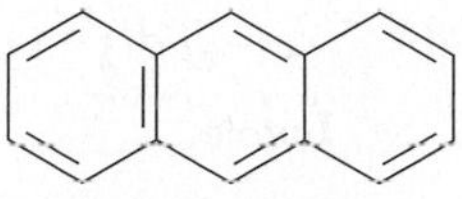 **Anthracene** 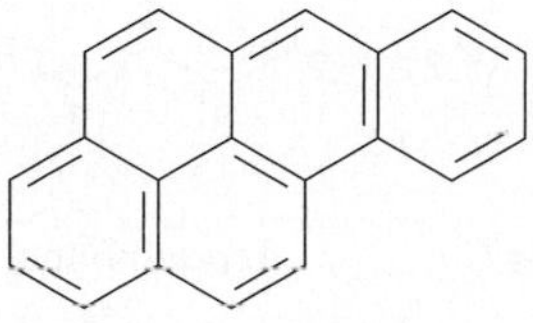 **Benzo[a]pyrene** 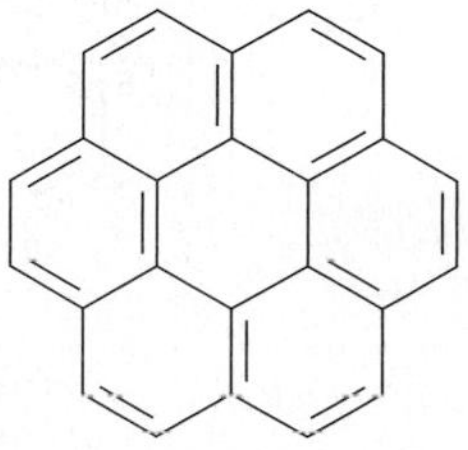 **Coronene**

All polycyclic aromatic hydrocarbons can be represented by a number of different resonance forms. Naphthalene, for instance, has three.

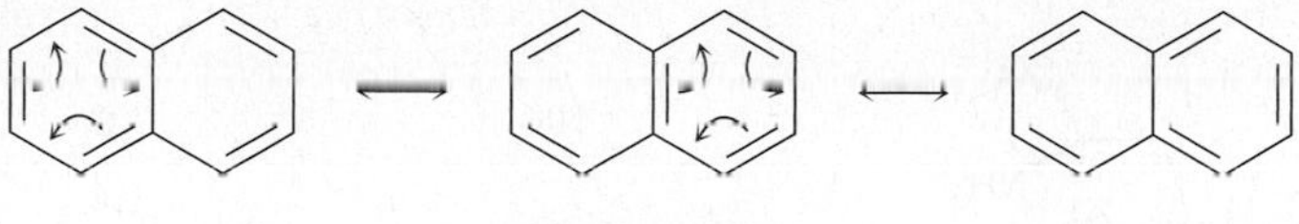

Naphthalene

Naphthalene and other polycyclic aromatic hydrocarbons show many of the chemical properties associated with aromaticity. Thus, measurement of its heat of hydrogenation shows an aromatic stabilization energy of approximately 250 kJ/mol (60 kcal/mol). Furthermore, naphthalene reacts slowly with electrophiles such as Br_2 to give substitution products rather than double-bond addition products.

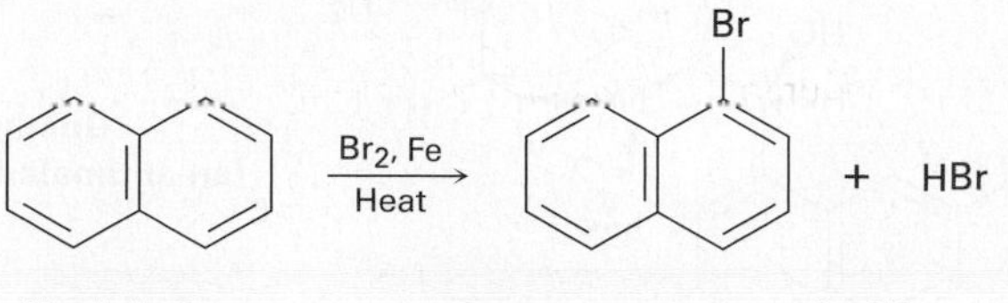

$\xrightarrow[\text{Heat}]{Br_2,\ Fe}$ Br + HBr

Naphthalene **1-Bromonaphthalene (75%)**

The aromaticity of naphthalene is explained by the orbital picture in **Figure 15.10**. Naphthalene has a cyclic, conjugated π electron system, with p orbital overlap both around the ten-carbon periphery of the molecule and across the central bond. Since ten π electrons is a Hückel number, there is π electron delocalization and consequent aromaticity in naphthalene.

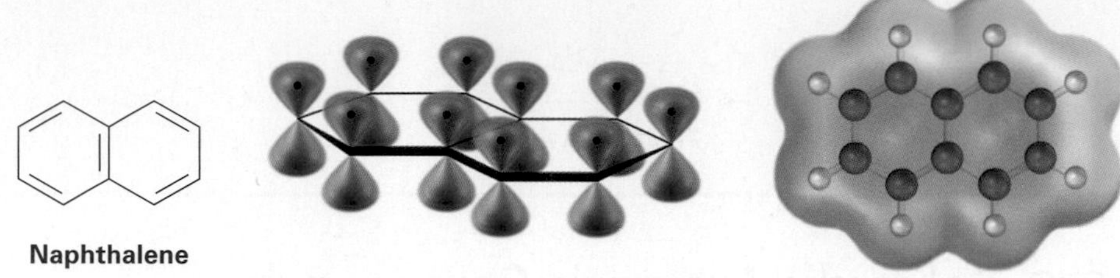

Figure 15.10 An orbital picture and electrostatic potential map of naphthalene, showing that the ten π electrons are fully delocalized throughout both rings.

Just as there are heterocyclic analogs of benzene, there are also many heterocyclic analogs of naphthalene. Among the most common are quinoline, isoquinoline, indole, and purine. Quinoline, isoquinoline, and purine all contain pyridine-like nitrogens that are part of a double bond and contribute one electron to the aromatic π system. Indole and purine both contain pyrrole-like nitrogens that contribute two π electrons.

Quinoline **Isoquinoline** **Indole** **Purine**

Among the many biological molecules that contain polycyclic aromatic rings, the amino acid tryptophan contains an indole ring and the antimalarial drug quinine contains a quinoline ring. Adenine and guanine, two of the five heterocyclic amine bases found in nucleic acids, have rings based on purine.

Tryptophan (an amino acid)

Adenine (in DNA and RNA)

Guanine (in DNA and RNA)

Quinine (an antimalarial agent)

Problem 15.11

Azulene, a beautiful blue hydrocarbon, is an isomer of naphthalene. Is azulene aromatic? Draw a second resonance form of azulene in addition to that shown.

Azulene

Problem 15.12

How many electrons does each of the four nitrogen atoms in purine contribute to the aromatic π system?

Purine

15.7 Spectroscopy of Aromatic Compounds

Infrared Spectroscopy

Aromatic rings show a characteristic C–H stretching absorption at 3030 cm^{-1} and a series of peaks in the 1450 to 1600 cm^{-1} range of the infrared spectrum. The aromatic C–H band at 3030 cm^{-1} generally has low intensity and occurs just to the left of a typical saturated C–H band.

As many as four absorptions are sometimes observed in the 1450 to 1600 cm^{-1} region because of complex molecular motions of the ring itself. Two bands, one at 1500 cm^{-1} and one at 1600 cm^{-1}, are usually the most intense. In addition, aromatic compounds show weak absorptions in the 1660 to 2000 cm^{-1} region and strong absorptions in the 690 to 900 cm^{-1} range due to C–H out-of-plane bending. The exact position of both sets of absorptions is diagnostic of the substitution pattern of the aromatic ring.

Monosubstituted:	690–710 cm^{-1}	***m*-Disubstituted:**	690–710 cm^{-1}
	730–770 cm^{-1}		810–850 cm^{-1}
***o*-Disubstituted:**	735–770 cm^{-1}	***p*-Disubstituted:**	810–840 cm^{-1}

The IR spectrum of toluene in **Figure 15.11** shows these characteristic absorptions.

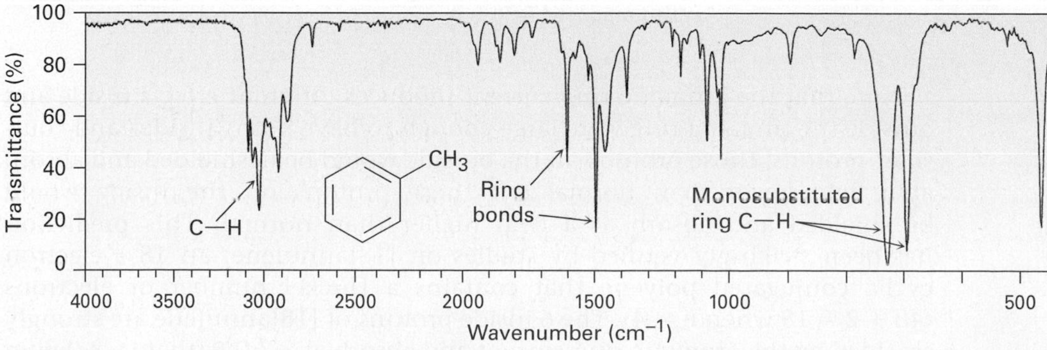

Figure 15.11 The infrared spectrum of toluene.

Ultraviolet Spectroscopy

Aromatic rings are detectable by ultraviolet spectroscopy because they contain a conjugated π electron system. In general, aromatic compounds show a series of bands, with a fairly intense absorption near 205 nm and a less intense absorption in the 255 to 275 nm range. The presence of these bands in the ultraviolet spectrum of a molecule is a sure indication of an aromatic ring.

Nuclear Magnetic Resonance Spectroscopy

Hydrogens directly bonded to an aromatic ring are easily identifiable in the ^{1}H NMR spectrum. Aromatic hydrogens are strongly deshielded by the ring and absorb between 6.5 and 8.0 δ. The spins of nonequivalent aromatic protons on substituted rings often couple with each other, giving rise to spin–spin splitting patterns that can identify the substitution of the ring.

Much of the difference in chemical shift between aromatic protons (6.5–8.0 δ) and vinylic protons (4.5–6.5 δ) is due to a property of aromatic rings called *ring-current.* When an aromatic ring is oriented perpendicular to a strong magnetic field, the delocalized π electrons circulate around the ring, producing a small local magnetic field. This induced field opposes the applied field in the middle of the ring but reinforces the applied field outside the ring **(Figure 15.12)**. Aromatic protons therefore experience an effective magnetic field greater than the applied field and come into resonance at a lower applied field.

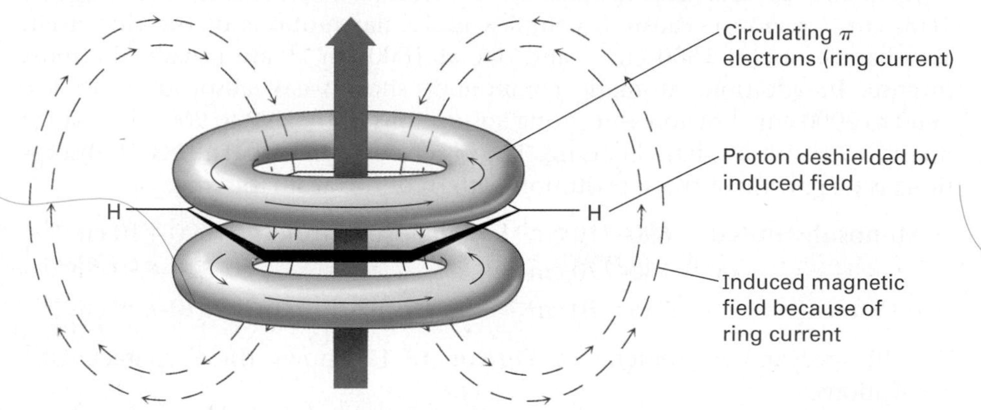

Figure 15.12 The origin of aromatic ring-current. Aromatic protons are deshielded by the induced magnetic field caused by delocalized π electrons circulating around the aromatic ring.

Note that the aromatic ring-current produces different effects inside and outside the ring. If a ring were large enough to have both "inside" and "outside" protons, those protons on the outside would be deshielded and absorb at a field lower than normal but those protons on the inside would be shielded and absorb at a field higher than normal. This prediction has been strikingly verified by studies on [18]annulene, an 18-π-electron cyclic conjugated polyene that contains a Hückel number of electrons ($4n + 2 = 18$ when n = 4). The 6 inside protons of [18]annulene are strongly shielded by the aromatic ring-current and absorb at -3.0 δ (that is, 3.0 ppm *upfield* from TMS, off the normal chart), while the 12 outside protons are strongly deshielded and absorb in the typical aromatic region at 9.3 ppm downfield from TMS.

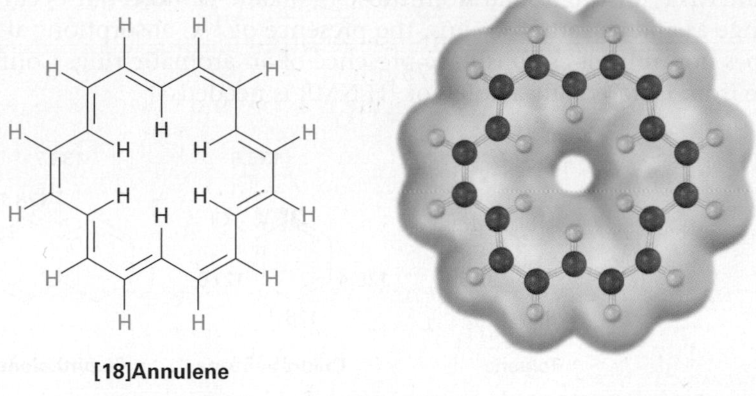

[18]Annulene

Inside H: −3.0 δ
Outside H: 9.3 δ

The presence of a ring-current is characteristic of all Hückel aromatic molecules and is a good test of aromaticity. For example, benzene, a six-π-electron aromatic molecule, absorbs at 7.37 δ because of its ring-current, but cyclooctatetraene, an eight-π-electron nonaromatic molecule, absorbs at 5.78 δ.

Hydrogens on carbon next to aromatic rings—*benzylic* hydrogens—also show distinctive absorptions in the NMR spectrum. Benzylic protons normally absorb downfield from other alkane protons in the region from 2.3 to 3.0 δ.

Benzylic protons, 2.3–3.0 δ

Aryl protons, 6.5–8.0 δ

The ^{1}H NMR spectrum of *p*-bromotoluene, shown in **Figure 15.13**, displays many of the features just discussed. The aromatic protons appear as two doublets at 7.02 and 7.45 δ, and the benzylic methyl protons absorb as a sharp singlet at 2.29 δ. Integration of the spectrum shows the expected 2:2:3 ratio of peak areas.

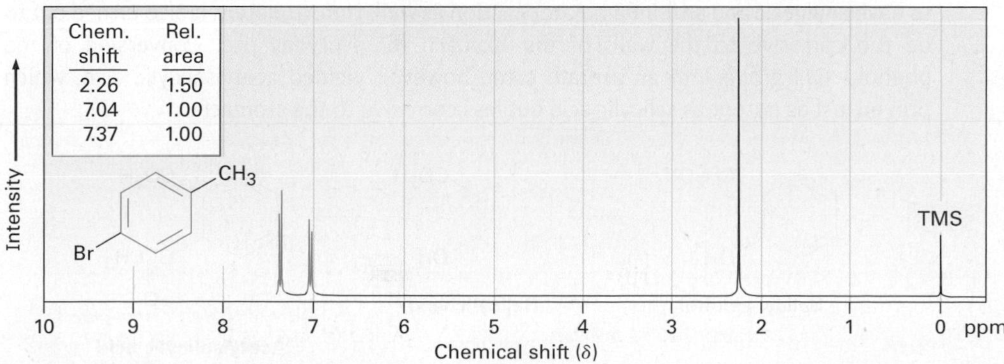

Figure 15.13 The ^{1}H NMR spectrum of *p*-bromotoluene.

The carbon atoms in an aromatic ring absorb in the range 110 to 140 δ in the ^{13}C NMR spectrum, as indicated by the examples in **Figure 15.14**. These resonances are easily distinguished from those of alkane carbons but occur in the same range as alkene carbons. Thus, the presence of ^{13}C absorptions at 110 to 140 δ does not in itself establish the presence of an aromatic ring. Confirming evidence from infrared, ultraviolet, or ^{1}H NMR is needed.

Figure 15.14 Some ^{13}C NMR absorptions of aromatic compounds (δ units).

Benzene: 128.4

Toluene: 137.7, 21.3 (CH_3), 129.3, 128.5, 125.6

Chlorobenzene: 133.8 (Cl), 127.6, 128.4, 125.4

Naphthalene: 133.7, 128.1, 126.0

Many athletes rely on NSAIDs to help with pain and soreness.

Aspirin, NSAIDs, and COX-2 Inhibitors

A DEEPER LOOK

Whatever the cause—whether tennis elbow, a sprained ankle, or a wrenched knee—pain and inflammation seem to go together. They are, however, different in their origin, and powerful drugs are available for treating each separately. Codeine, for example, is a powerful *analgesic*, or pain reliever, used in the management of debilitating pain, while cortisone and related steroids are potent *anti-inflammatory* agents, used for treating arthritis and other crippling inflammations. For minor pains and inflammation, both problems are often treated at the same time by using a common over-the-counter medication called an *NSAID*, or *nonsteroidal anti-inflammatory drug.*

The most common NSAID is aspirin, or acetylsalicylic acid, whose use goes back to the late 1800s. It had been known from before the time of Hippocrates in 400 BC that fevers could be lowered by chewing the bark of willow trees. The active agent in willow bark was found in 1827 to be an aromatic compound called *salicin*, which could be converted by reaction with water into salicyl alcohol and then oxidized to give salicylic acid. Salicylic acid turned out to be even more effective than salicin for reducing fevers and to have analgesic and anti-inflammatory action as well. Unfortunately, it also turned out to be too corrosive to the walls of the stomach for everyday use. Conversion of the phenol –OH group into an acetate ester, however, yielded acetylsalicylic acid, which proved just as potent as salicylic acid but less corrosive to the stomach.

Salicyl alcohol (CH_2OH, OH) ⟶ **Salicylic acid** (CO_2H, OH) ⟶ **Acetylsalicylic acid (aspirin)** (CO_2H, $OCCH_3$ with C=O)

(continued)

(continued)

Although extraordinary in its effect, aspirin is also more dangerous than commonly believed. A dose of only about 15 g can be fatal to a small child, and aspirin can cause stomach bleeding and allergic reactions in long-term users. Even more serious is a condition called *Reye's syndrome,* a potentially fatal reaction to aspirin sometimes seen in children recovering from the flu. As a result of these problems, numerous other NSAIDs have been developed in the last several decades, most notably ibuprofen and naproxen.

Like aspirin, both ibuprofen and naproxen are relatively simple aromatic compounds containing a side-chain carboxylic acid group. Ibuprofen, sold under the names Advil, Nuprin, Motrin, and others, has roughly the same potency as aspirin but is less prone to cause stomach upset. Naproxen, sold under the names Aleve and Naprosyn, also has about the same potency as aspirin but remains active in the body six times longer.

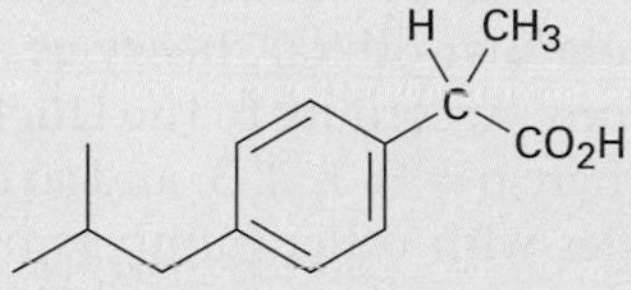

H CH_3 C CO_2H

Ibuprofen
(Advil, Nuprin, Motrin)

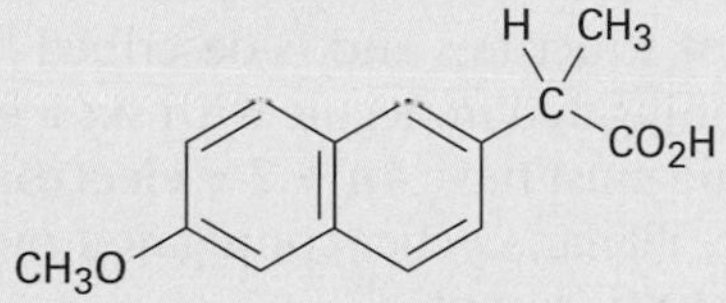

H CH_3 C CO_2H CH_3O

Naproxen
(Aleve, Naprosyn)

Aspirin and other NSAIDs function by blocking the cyclooxygenase (COX) enzymes that carry out the body's synthesis of prostaglandins **(Sections 8.11 and 27.4)**. There are two forms of the enzyme: COX-1, which carries out the normal physiological production of prostaglandins, and COX-2, which mediates the body's response to arthritis and other inflammatory conditions. Unfortunately, both COX-1 and COX-2 enzymes are blocked by aspirin, ibuprofen, and other NSAIDs, thereby shutting down not only the response to inflammation but also various protective functions, including the control mechanism for production of acid in the stomach.

Medicinal chemists have devised a number of drugs that act as selective inhibitors of the COX-2 enzyme. Inflammation is thereby controlled without blocking protective functions. Originally heralded as a breakthrough in arthritis treatment, the first generation of COX-2 inhibitors, including Vioxx, Celebrex, and Bextra, turned out to cause potentially serious heart problems, particularly in elderly or compromised patients. The second generation of COX-2 inhibitors promises to be safer but will be closely scrutinized for side effects before gaining approval.

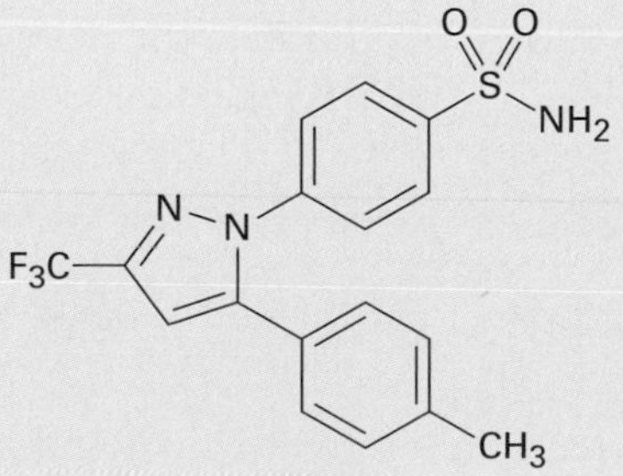

O O S NH_2 N N F_3C CH_3

Celecoxib
(Celebrex)

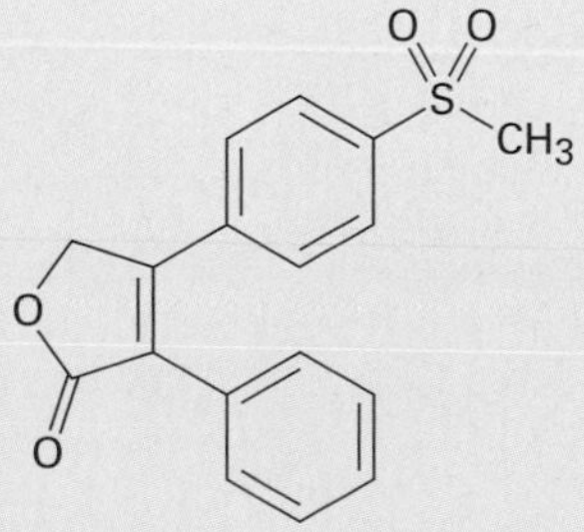

O O S CH_3 O O

Rofecoxib
(Vioxx)

Summary

Key words

antiaromatic, 541
arene, 536
aromatic, 534
benzyl, 536
heterocycle, 546
Hückel $4n + 2$ rule, 541
meta (*m*), 537
ortho (*o*), 537
para (*p*), 537
phenyl, 536

Aromatic rings are a common part of many biological structures and are particularly important in nucleic acid chemistry and in the chemistry of several amino acids. In this chapter, we've seen how and why aromatic compounds are different from such apparently related compounds as cycloalkenes.

The word **aromatic** is used for historical reasons to refer to the class of compounds related structurally to benzene. Aromatic compounds are systematically named according to IUPAC rules, but many common names are also used. Disubstituted benzenes are named as **ortho** (1,2 disubstituted), **meta** (1,3 disubstituted), or **para** (1,4 disubstituted) derivatives. The C_6H_5- unit itself is referred to as a **phenyl** group, and the $C_6H_5CH_2-$ unit is a **benzyl** group.

Benzene is described by valence-bond theory as a resonance hybrid of two equivalent structures and is described by molecular orbital theory as a planar, cyclic, conjugated molecule with six π electrons. According to the **Hückel rule**, a molecule must have **$4n + 2$ π electrons**, where n = 0, 1, 2, 3, and so on, to be aromatic. Planar, cyclic, conjugated molecules with other numbers of π electrons are **antiaromatic**.

Other kinds of substances besides benzene-like compounds can also be aromatic. The cyclopentadienyl anion and the cycloheptatrienyl cation, for instance, are aromatic ions. Pyridine and pyrimidine are six-membered, nitrogen-containing, aromatic **heterocycles**. Pyrrole and imidazole are five-membered, nitrogen-containing heterocycles. Naphthalene, quinoline, indole, and many others are polycyclic aromatic compounds.

Aromatic compounds have the following characteristics:

- Aromatic compounds are cyclic, planar, and conjugated.
- Aromatic compounds are unusually stable. Benzene, for instance, has a heat of hydrogenation 150 kJ/mol less than we might expect for a cyclic triene.
- Aromatic compounds react with electrophiles to give substitution products, in which cyclic conjugation is retained, rather than addition products, in which conjugation is destroyed.
- Aromatic compounds have $4n + 2$ π electrons, which are delocalized over the ring.

Exercises

Visualizing Chemistry

(Problems 15.1–15.12 appear within the chapter.)

OWL Interactive versions of these problems are assignable in OWL for Organic Chemistry.

▲ denotes problems linked to the Key Ideas in this chapter.

15.13 Give IUPAC names for the following substances (red = O, blue = N):

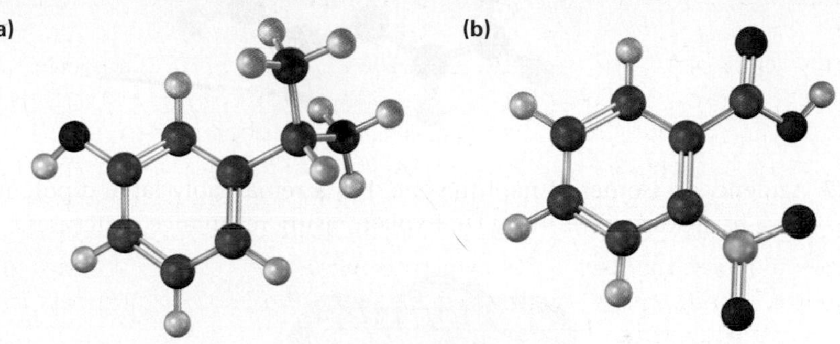

15.14 ▲ All-cis cyclodecapentaene is a stable molecule that shows a single absorption in its ^{1}H NMR spectrum at 5.67 δ. Tell whether it is aromatic, and explain its NMR spectrum.

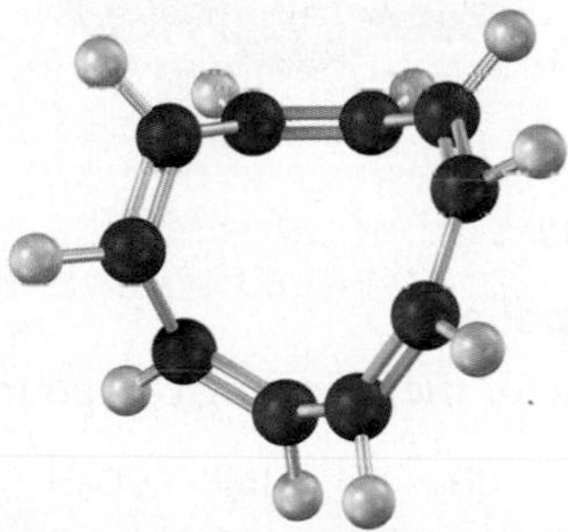

15.15 ▲ 1,6-Methanonaphthalene has an interesting ^{1}H NMR spectrum in which the eight hydrogens around the perimeter absorb at 6.9 to 7.3 δ, while the two CH_2 protons absorb at -0.5 δ. Tell whether it is aromatic, and explain its NMR spectrum.

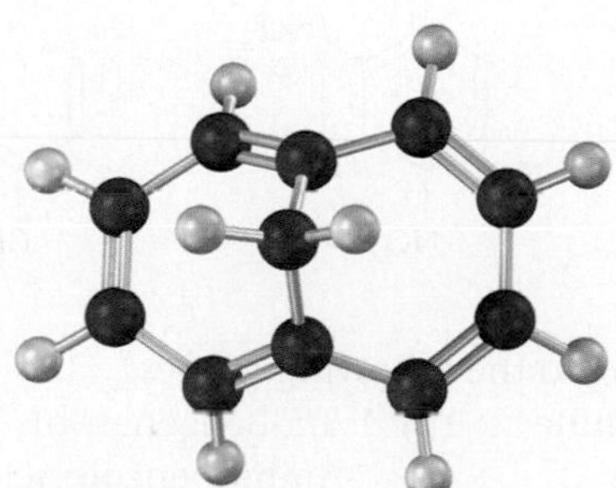

1,6-Methanonaphthalene

15.16 The following molecular model is that of a carbocation. Draw two resonance structures for the carbocation, indicating the positions of the double bonds.

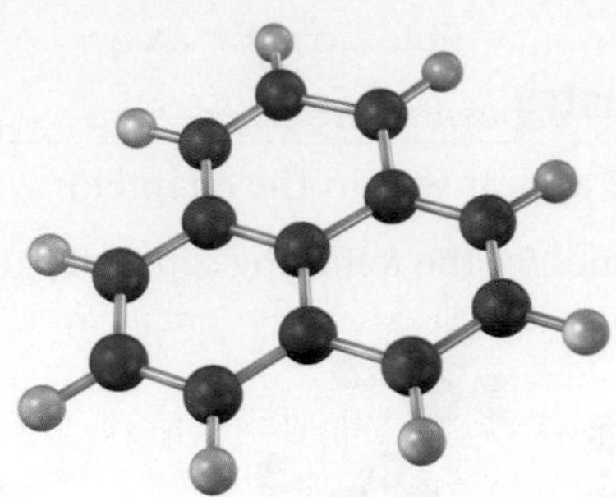

15.17 Azulene, an isomer of naphthalene, has a remarkably large dipole moment for a hydrocarbon ($\mu = 1.0$ D). Explain, using resonance structures.

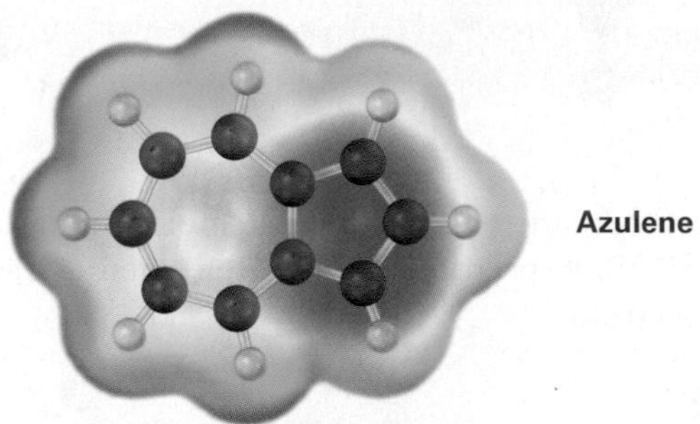

Additional Problems

Naming Aromatic Compounds

15.18 Give IUPAC names for the following compounds:

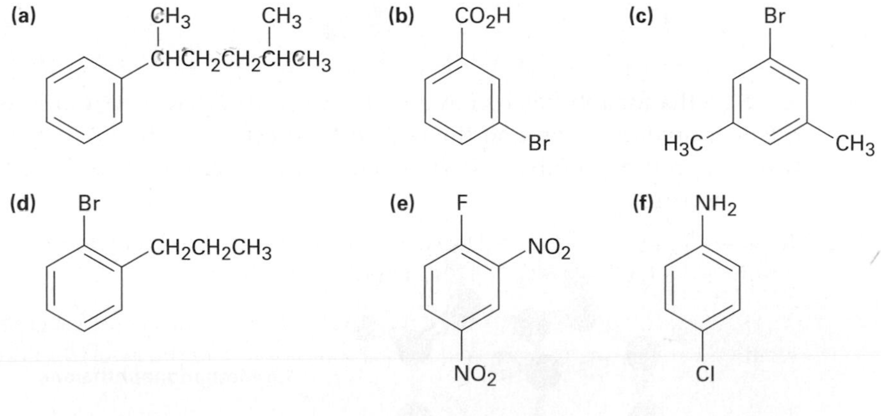

(a) $C_6H_5CH(CH_3)CH_2CH_2CH(CH_3)CH_3$ — labels: CH_3, CH_3, $CHCH_2CH_2CHCH_3$

(b) CO_2H, Br

(c) Br, H_3C, CH_3

(d) Br, $CH_2CH_2CH_3$

(e) F, NO_2, NO_2

(f) NH_2, Cl

15.19 Draw structures corresponding to the following names:

(a) 3-Methyl-1,2-benzenediamine
(b) 1,3,5-Benzenetriol
(c) 3-Methyl-2-phenylhexane
(d) *o*-Aminobenzoic acid
(c) *m*-Bromophenol
(f) 2,4,6-Trinitrophenol (picric acid)

▲ Problems linked to Key Ideas in this chapter

15.20 Draw and name all possible isomers of the following:
(a) Dinitrobenzene (b) Bromodimethylbenzene (c) Trinitrophenol

15.21 Draw and name all possible aromatic compounds with the formula C_7H_7Cl.

15.22 Draw and name all possible aromatic compounds with the formula C_8H_9Br. (There are 14.)

Structure of Aromatic Compounds

15.23 ▲ Propose structures for aromatic hydrocarbons that meet the following descriptions:
(a) C_9H_{12}; gives only one $C_9H_{11}Br$ product on substitution of a hydrogen on the aromatic ring with bromine
(b) $C_{10}H_{14}$; gives only one $C_{10}H_{13}Cl$ product on substitution of a hydrogen on the aromatic ring with chlorine
(c) C_8H_{10}; gives three C_8H_9Br products on substitution of a hydrogen on the aromatic ring with bromine
(d) $C_{10}H_{14}$; gives two $C_{10}H_{13}Cl$ products on substitution of a hydrogen on the aromatic ring with chlorine

15.24 Look at the three resonance structures of naphthalene shown in Section 15.6, and account for the fact that not all carbon–carbon bonds have the same length. The C1–C2 bond is 136 pm long, whereas the C2–C3 bond is 139 pm long.

15.25 Anthracene has four resonance structures, one of which is shown. Draw the other three.

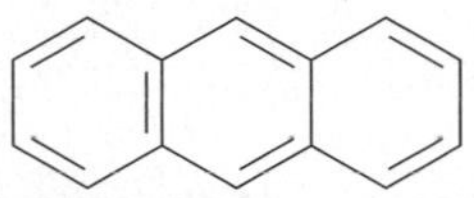

Anthracene

15.26 Phenanthrene has five resonance structures, one of which is shown. Draw the other four.

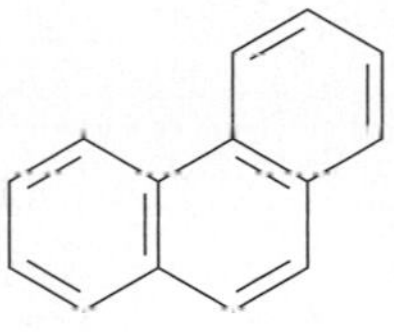

Phenanthrene

15.27 Look at the five resonance structures for phenanthrene (Problem 15.26), and predict which of its carbon–carbon bonds is shortest.

15.28 In 1932, A. A. Levine and A. G. Cole studied the ozonolysis of *o*-xylene and isolated three products: glyoxal, 2,3-butanedione, and pyruvaldehyde:

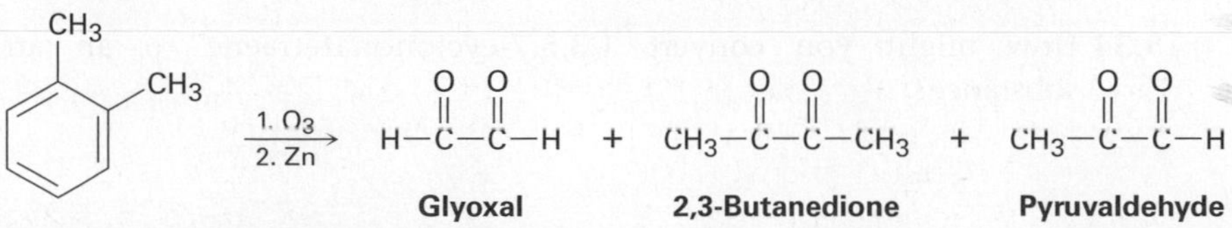

$$o\text{-xylene} \xrightarrow[\text{2. Zn}]{\text{1. } O_3} \underset{\textbf{Glyoxal}}{H-\overset{O}{\overset{\|}{C}}-\overset{O}{\overset{\|}{C}}-H} + \underset{\textbf{2,3-Butanedione}}{CH_3-\overset{O}{\overset{\|}{C}}-\overset{O}{\overset{\|}{C}}-CH_3} + \underset{\textbf{Pyruvaldehyde}}{CH_3-\overset{O}{\overset{\|}{C}}-\overset{O}{\overset{\|}{C}}-H}$$

▲ Problems linked to Key Ideas in this chapter

In what ratio would you expect the three products to be formed if *o*-xylene is a resonance hybrid of two structures? The actual ratio found was 3 parts glyoxal, 1 part 2,3-butanedione, and 2 parts pyruvaldehyde. What conclusions can you draw about the structure of *o*-xylene?

Aromaticity and Hückel's Rule

15.29 3-Chlorocyclopropene, on treatment with $AgBF_4$, gives a precipitate of AgCl and a stable solution of a product that shows a single ^{1}H NMR absorption at 11.04 δ. What is a likely structure for the product, and what is its relation to Hückel's rule?

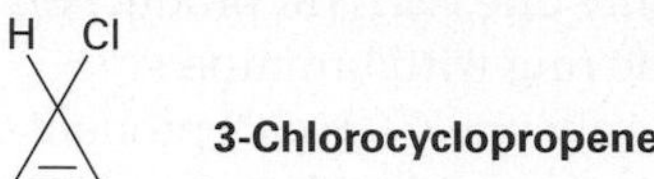

15.30 Draw an energy diagram for the three molecular orbitals of the cyclopropenyl system (C_3H_3). How are these three molecular orbitals occupied in the cyclopropenyl anion, cation, and radical? Which of the three substances is aromatic according to Hückel's rule?

15.31 Cyclopropanone is highly reactive because of its large amount of angle strain. but methylcyclopropenone, although even more strained than cyclopropanone, is nevertheless quite stable and can even be distilled. Explain, taking the polarity of the carbonyl group into account.

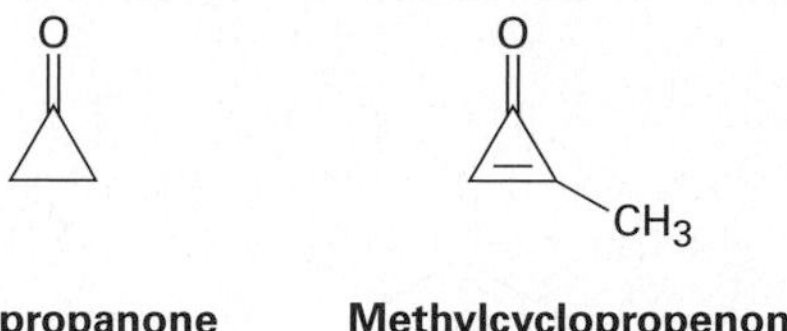

15.32 Cycloheptatrienone is stable, but cyclopentadienone is so reactive that it can't be isolated. Explain, taking the polarity of the carbonyl group into account.

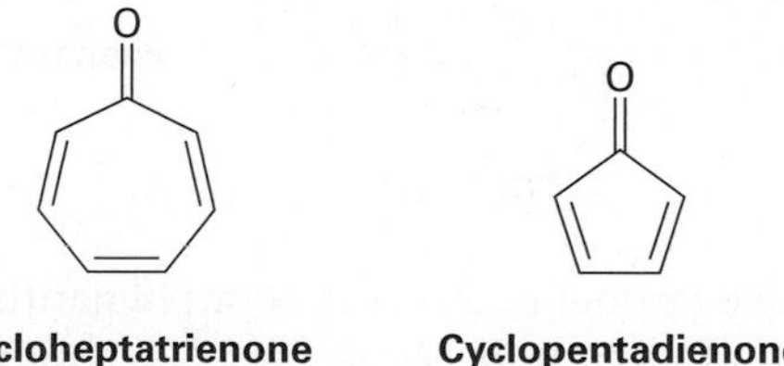

15.33 Which would you expect to be most stable, cyclononatetraenyl radical, cation, or anion?

15.34 How might you convert 1,3,5,7-cyclononatetraene to an aromatic substance?

▲ Problems linked to Key Ideas in this chapter

15.35 Calicene, like azulene (Problem 15.17), has an unusually large dipole moment for a hydrocarbon. Explain, using resonance structures.

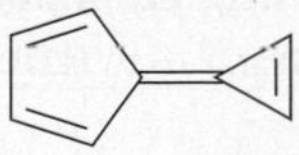

Calicene

15.36 Pentalene is a most elusive molecule that has been isolated only at liquid-nitrogen temperature. The pentalene dianion, however, is well known and quite stable. Explain.

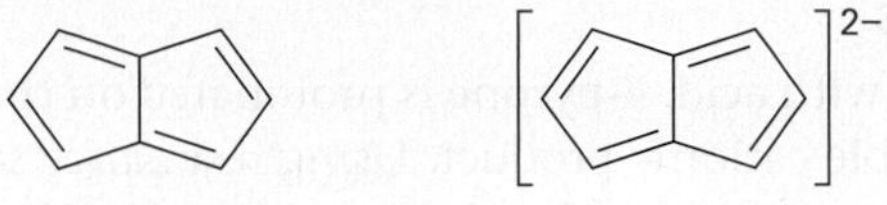

2–

Pentalene **Pentalene dianion**

15.37 Indole is an aromatic heterocycle that has a benzene ring fused to a pyrrole ring. Draw an orbital picture of indole.

(a) How many π electrons does indole have?

(b) What is the electronic relationship of indole to naphthalene?

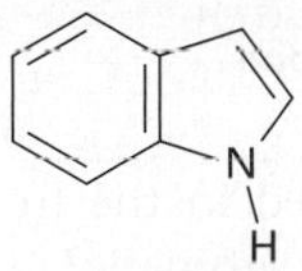

N
H
Indole

15.38 Ribavirin, an antiviral agent used against hepatitis C and viral pneumonia, contains a 1,2,4-triazole ring. Why is the ring aromatic?

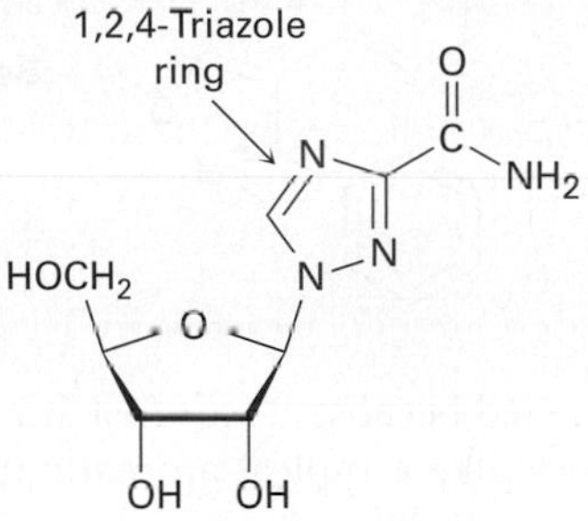

1,2,4-Triazole ring
O
C
NH$_2$
N
N
N
HOCH$_2$
O
OH OH
Ribavirin

Spectroscopy

15.39 Compound A, C_8H_{10}, yields three substitution products, C_8H_9Br, on reaction with Br_2. Propose two possible structures for A. The ^{1}H NMR spectrum of A shows a complex four-proton multiplet at 7.0 δ and a six-proton singlet at 2.30 δ. What is the structure of A?

15.40 What is the structure of a hydrocarbon that has $M^+ = 120$ in its mass spectrum and has the following ^{1}H NMR spectrum?

7.25 δ (5 H, broad singlet); 2.90 δ (1 H, septet, J = 7 Hz); 1.22 δ (6 H, doublet, J = 7 Hz)

▲ Problems linked to Key Ideas in this chapter

15.41 Propose structures for compounds that fit the following descriptions:

(a) $C_{10}H_{14}$
^1H NMR: 7.18 δ (4 H, broad singlet); 2.70 δ (4 H, quartet, $J = 7$ Hz); 1.20 δ (6 H, triplet, $J = 7$ Hz)
IR: 745 cm^{-1}

(b) $C_{10}H_{14}$
^1H NMR: 7.0 δ (4 H, broad singlet); 2.85 δ (1 H, septet, $J = 8$ Hz); 2.28 δ (3 H, singlet); 1.20 δ (6 H, doublet, $J = 8$ Hz)
IR: 825 cm^{-1}

General Problems

15.42 On reaction with acid, 4-pyrone is protonated on the carbonyl-group oxygen to give a stable cationic product. Using resonance structures and the Hückel $4n + 2$ rule, explain why the protonated product is so stable.

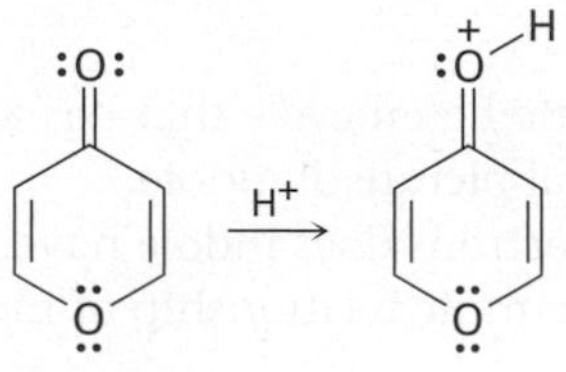

H^+

4-Pyrone

15.43 Bextra, a COX-2 inhibitor used in the treatment of arthritis, contains an isoxazole ring. Why is the ring aromatic?

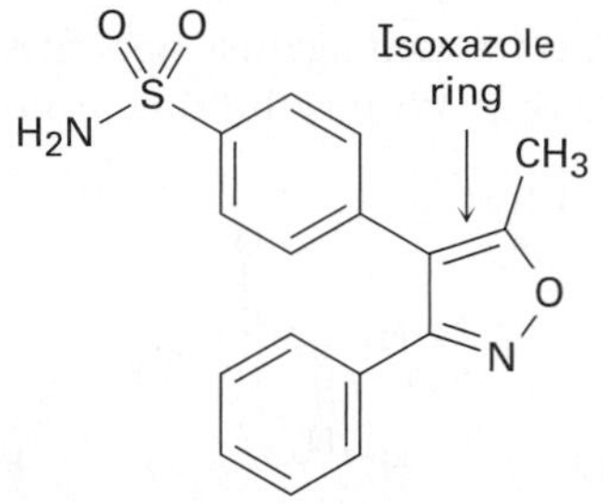

Isoxazole ring

Bextra

15.44 *N*-Phenylsydnone, so-named because it was first studied at the University of Sydney, Australia, behaves like a typical aromatic molecule. Explain, using the Hückel $4n + 2$ rule.

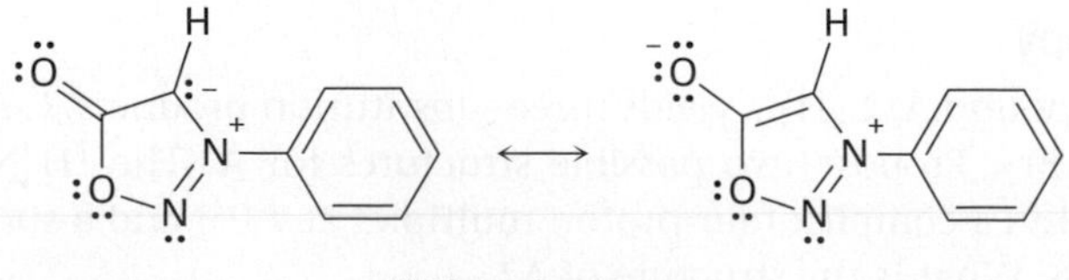

***N*-Phenylsydnone**

▲ Problems linked to Key Ideas in this chapter

15.45 Show the relative energy levels of the seven π molecular orbitals of the cycloheptatrienyl system. Tell which of the seven orbitals are filled in the cation, radical, and anion, and account for the aromaticity of the cycloheptatrienyl cation.

15.46 1-Phenyl-2-butene has an ultraviolet absorption at $\lambda_{max} = 208$ nm ($\epsilon = 8000$). On treatment with a small amount of strong acid, isomerization occurs and a new substance with $\lambda_{max} = 250$ nm ($\epsilon = 15{,}800$) is formed. Propose a structure for this isomer, and suggest a mechanism for its formation.

15.47 Propose structures for aromatic compounds that have the following ^1H NMR spectra:

(a) C_8H_9Br
IR: 820 cm^{-1}

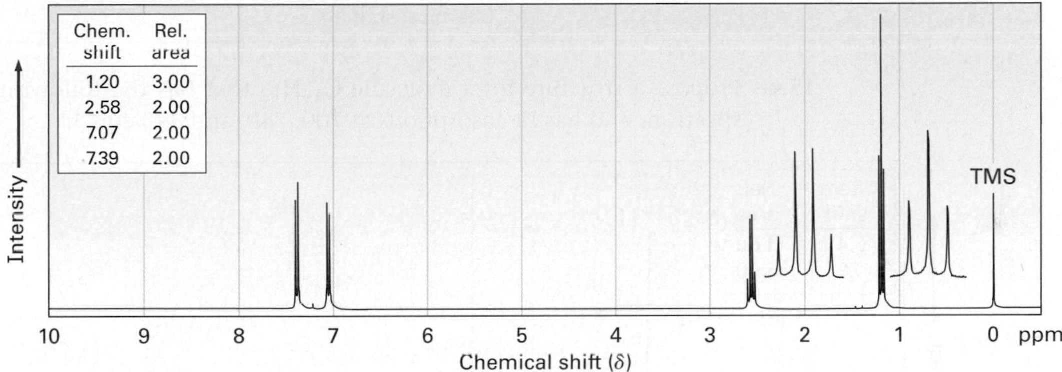

(b) C_9H_{12}
IR: 750 cm^{-1}

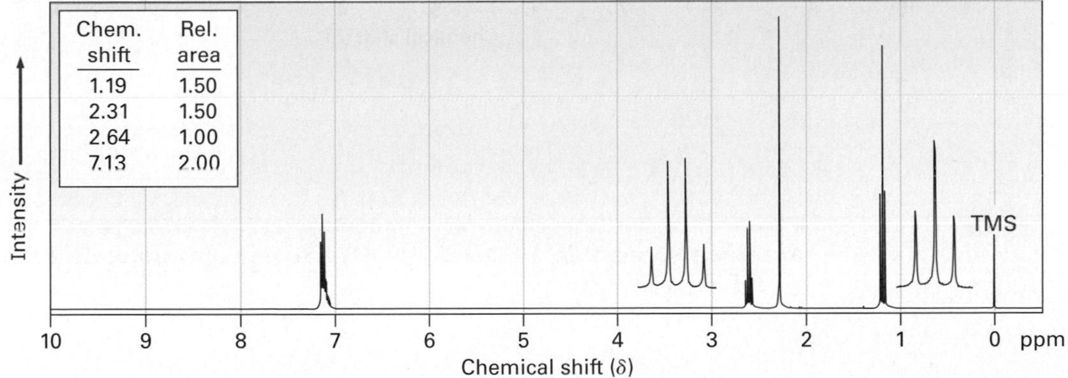

▲ Problems linked to Key Ideas in this chapter

(c) $C_{11}H_{16}$
IR: 820 cm^{-1}

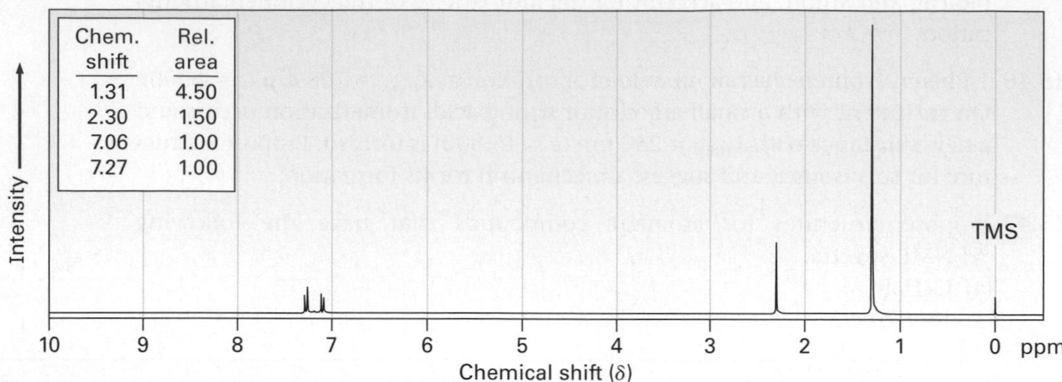

15.48 Propose a structure for a molecule $C_{14}H_{12}$ that has the following ^1H NMR spectrum and has IR absorptions at 700, 740, and 890 cm^{-1}:

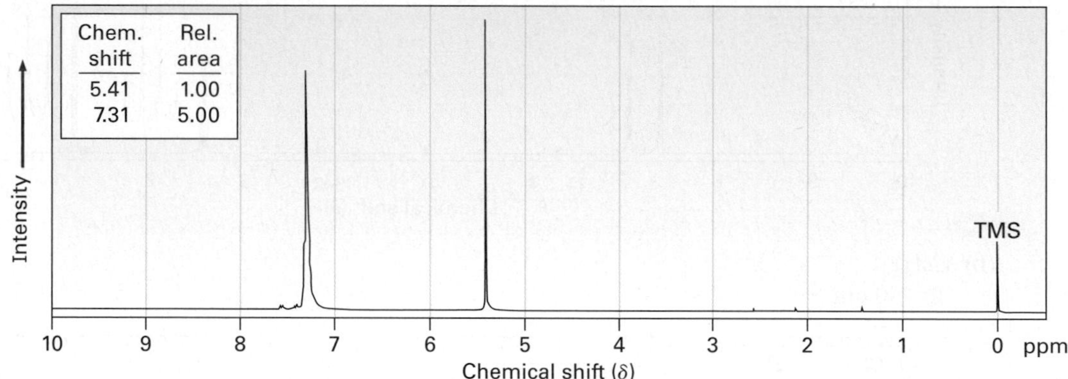

15.49 Aromatic substitution reactions occur by addition of an electrophile such as Br^+ to the aromatic ring to yield an allylic carbocation intermediate, followed by loss of H^+. Show the structure of the intermediate formed by reaction of benzene with Br^+.

15.50 The substitution reaction of toluene with Br_2 can, in principle, lead to the formation of three isomeric bromotoluene products. In practice, however, only *o*- and *p*-bromotoluene are formed in substantial amounts. The meta isomer is not formed. Draw the structures of the three possible carbocation intermediates (Problem 15.49), and explain why ortho and para products predominate over meta.

16

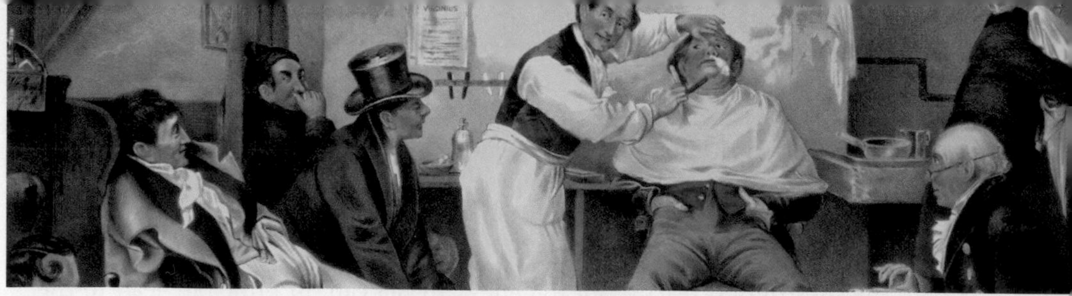

In the 19th and early-20th centuries, benzene was used as an aftershave lotion because of its pleasant smell and as a solvent to decaffeinate coffee beans. Neither is a good idea. © Niday Picture Library/Alamy

Chemistry of Benzene: Electrophilic Aromatic Substitution

OWL Sign in to OWL for Organic Chemistry at **www.cengage.com/owl** to view tutorials and simulations, develop problem-solving skills, and complete online homework assigned by your professor.

In the preceding chapter, we looked at *aromaticity*—the stability associated with benzene and related compounds that contain a cyclic conjugated system of $4n + 2$ π electrons. In this chapter, we'll look at some of the unique reactions that aromatic molecules undergo.

The most common reaction of aromatic compounds is **electrophilic aromatic substitution**, in which an electrophile (E^+) reacts with an aromatic ring and substitutes for one of the hydrogens. The reaction is characteristic of all aromatic rings, not just benzene and substituted benzenes. In fact, the ability of a compound to undergo electrophilic substitution is a good test of aromaticity.

$$C_6H_6 + E^+ \longrightarrow C_6H_5E + H^+$$

Many different substituents can be introduced onto an aromatic ring through electrophilic substitution reactions. To list some possibilities, an aromatic ring can be substituted by a halogen (–Cl, –Br, I), a nitro group ($-NO_2$), a sulfonic acid group ($-SO_3H$), a hydroxyl group (–OH), an alkyl group (–R), or an acyl group (–COR). Starting from only a few simple materials, it's possible to prepare many thousands of substituted aromatic compounds.

Why This Chapter? This chapter continues the coverage of aromatic molecules begun in the preceding chapter, but we'll shift focus to concentrate on reactions, looking at the relationship between aromatic structure and reactivity. This relationship is critical to an understanding of how many biological molecules and pharmaceutical agents are synthesized and why they behave as they do.

16.1 Electrophilic Aromatic Substitution Reactions: Bromination

Before seeing how electrophilic aromatic substitutions occur, let's briefly recall what we said in Chapter 7 about electrophilic alkene additions. When a reagent such as HCl adds to an alkene, the electrophilic hydrogen approaches the π electrons of the double bond and forms a bond to one carbon, leaving a positive charge at the other carbon. This carbocation intermediate then reacts with the nucleophilic Cl^- ion to yield the addition product.

An electrophilic aromatic substitution reaction begins in a similar way, but there are a number of differences. One difference is that aromatic rings are less reactive toward electrophiles than alkenes are. For example, Br_2 in CH_2Cl_2 solution reacts instantly with most alkenes but does not react with benzene at room temperature. For bromination of benzene to take place, a catalyst such as $FeBr_3$ is needed. The catalyst makes the Br_2 molecule more electrophilic by polarizing it to give an $FeBr_4^-Br^+$ species that reacts as if it were Br^+. The polarized Br_2 molecule then reacts with the nucleophilic benzene ring to yield a nonaromatic carbocation intermediate that is doubly allylic **(Section 11.5)** and has three resonance forms.

$$Br{-}Br \; + \; FeBr_3 \longrightarrow Br^+ \; ^-FeBr_4$$

Although more stable than a typical alkyl carbocation because of resonance, the intermediate in electrophilic aromatic substitution is nevertheless much less stable than the starting benzene ring itself, with its 150 kJ/mol (36 kcal/mol) of aromatic stability. Thus, the reaction of an electrophile with a benzene ring is endergonic, has a substantial activation energy, and is rather slow. **Figure 16.1** shows an energy diagram comparing the reaction of an electrophile with an alkene and with benzene. The benzene reaction is slower (higher $\Delta G^{\ddagger}$) because the starting material is more stable.

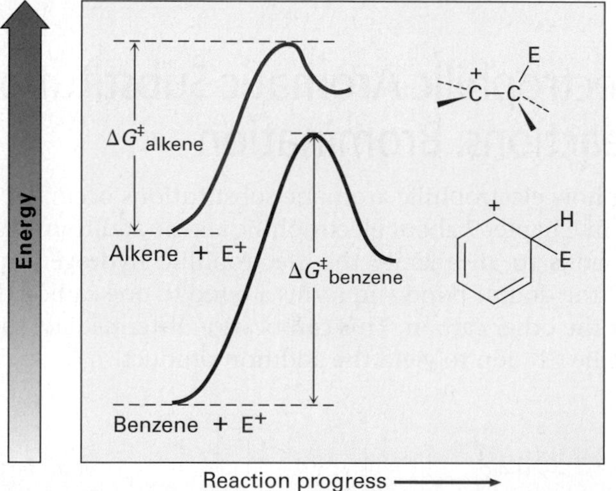

Figure 16.1 A comparison of the reactions of an electrophile (E^+) with an alkene and with benzene: $\Delta G^{\ddagger}_{alkene} < \Delta G^{\ddagger}_{benzene}$. The benzene reaction is slower than the alkene reaction because of the stability of the aromatic ring.

Another difference between alkene addition and aromatic substitution occurs after the carbocation intermediate has formed. Instead of adding Br^- to give an addition product, the carbocation intermediate loses H^+ from the bromine-bearing carbon to give a substitution product. Note that this loss of H^+ is similar to what occurs in the second step of an E1 reaction **(Section 11.10)**. The net effect of reaction of Br_2 with benzene is the substitution of H^+ by Br^+ by the overall mechanism shown in **Figure 16.2**.

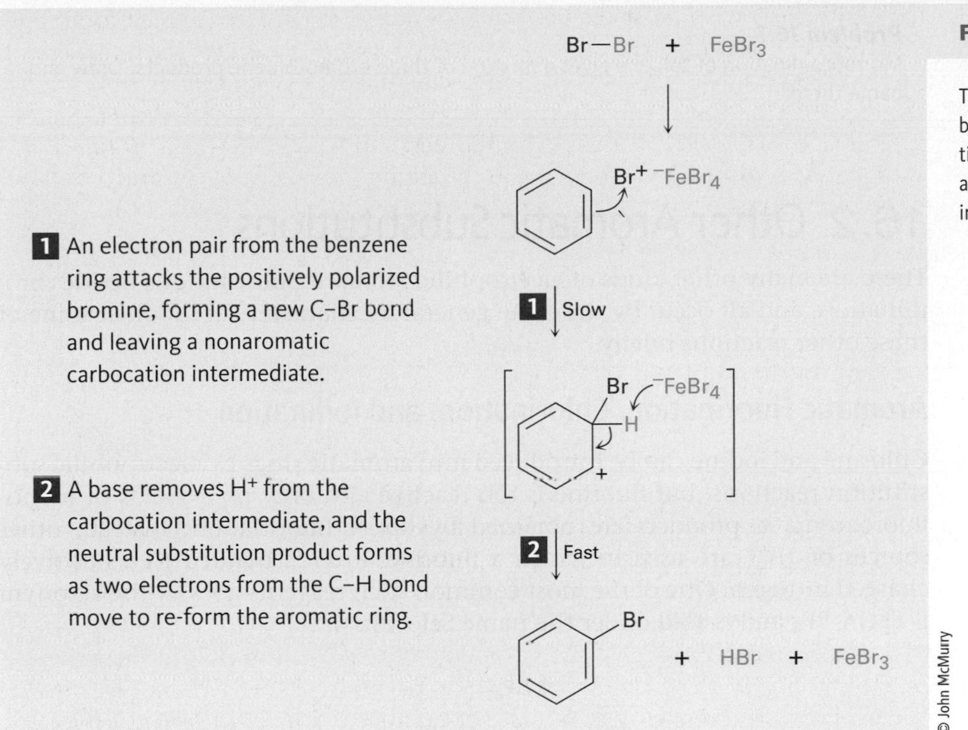

Figure 16.2 | MECHANISM

The mechanism of the electrophilic bromination of benzene. The reaction occurs in two steps and involves a resonance-stabilized carbocation intermediate.

Why does the reaction of Br_2 with benzene take a different course than its reaction with an alkene? The answer is straightforward. If *addition* occurred, the 150 kJ/mol stabilization energy of the aromatic ring would be lost and the overall reaction would be endergonic. When *substitution* occurs, though, the stability of the aromatic ring is retained and the reaction is exergonic. An energy diagram for the overall process is shown in **Figure 16.3**.

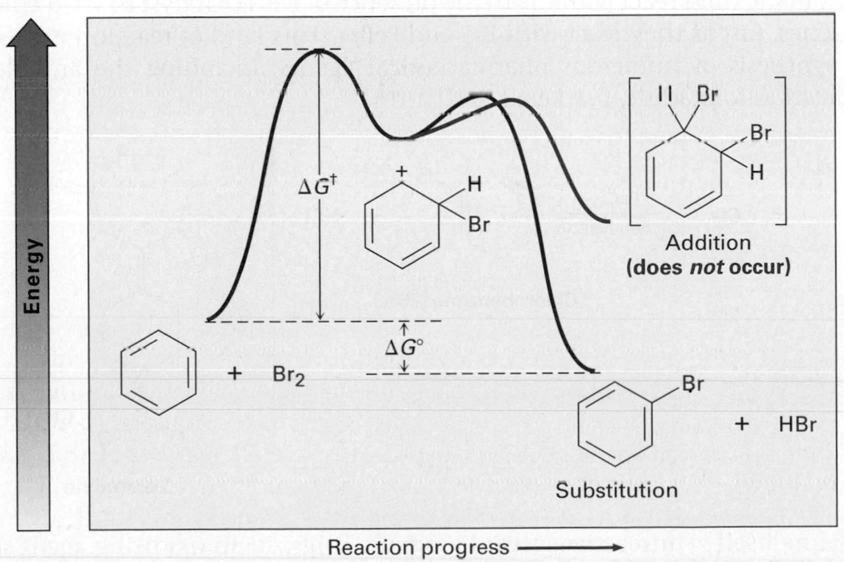

Figure 16.3 An energy diagram for the electrophilic bromination of benzene. Because the stability of the aromatic ring is retained, the overall process is exergonic.

Problem 16.1

Monobromination of toluene gives a mixture of three bromotoluene products. Draw and name them.

16.2 Other Aromatic Substitutions

There are many other kinds of electrophilic aromatic substitutions besides bromination, and all occur by the same general mechanism. Let's look at some of these other reactions briefly.

Aromatic Fluorination, Chlorination, and Iodination

Chlorine and iodine can be introduced into aromatic rings by electrophilic substitution reactions, but fluorine is too reactive and only poor yields of monofluoroaromatic products are obtained by direct fluorination. Instead, other sources of "F^+" are used in which a fluorine atom is bonded to a positively charged nitrogen. One of the most common such reagents goes by the acronym F-TEDA-BF_4 and is sold under the name Selectfluor.

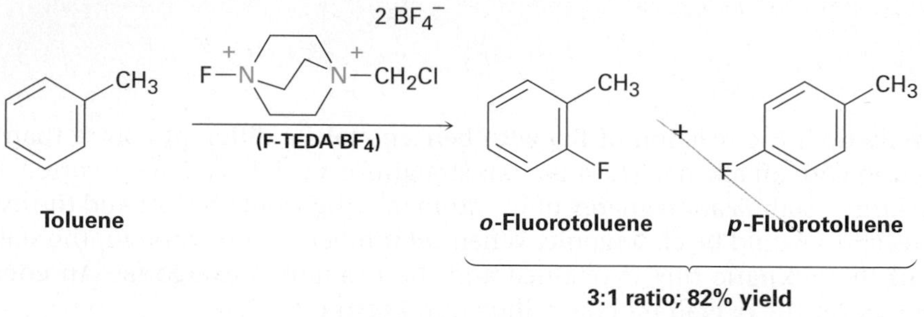

$2\ BF_4^-$

CH_3 F—N^+ N^+—CH_2Cl (F-TEDA-BF_4) CH_3 F + F CH_3

Toluene

o-Fluorotoluene *p*-Fluorotoluene

3:1 ratio; 82% yield

Aromatic rings react with Cl_2 in the presence of $FeCl_3$ catalyst to yield chlorobenzenes, just as they react with Br_2 and $FeBr_3$. This kind of reaction is used in the synthesis of numerous pharmaceutical agents, including the antiallergy medication loratadine, marketed as Claritin.

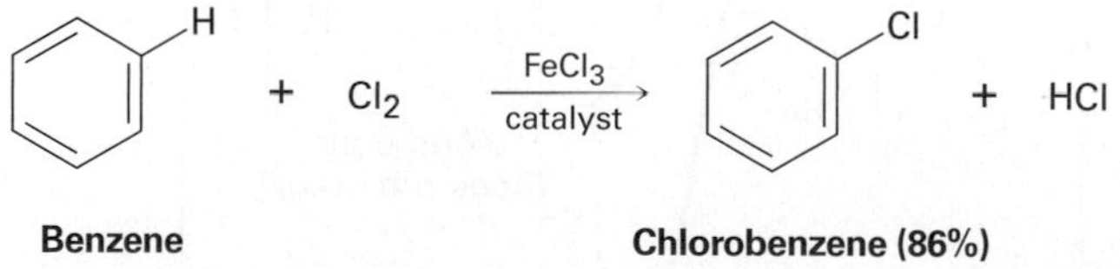

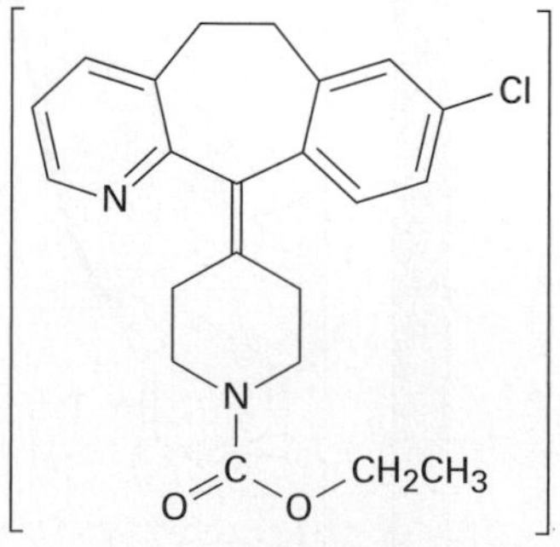

H + Cl_2 $\xrightarrow[\text{catalyst}]{FeCl_3}$ Cl + HCl

Benzene

Chlorobenzene (86%)

Cl N N O C O CH_2CH_3

Loratadine

Iodine itself is unreactive toward aromatic rings, so an oxidizing agent such as hydrogen peroxide or a copper salt such as $CuCl_2$ must be added to the reaction. These substances accelerate the iodination reaction by oxidizing I_2 to a more powerful electrophilic species that reacts as if it were I^+. The aromatic ring then reacts with I^+ in the typical way, yielding a substitution product.

$$I_2 + 2\,Cu^{2+} \longrightarrow 2\,I^+ + 2\,Cu^+$$

Electrophilic aromatic halogenations also occur in the biosynthesis of many naturally occurring molecules, particularly those produced by marine organisms. In humans, the best-known example occurs in the thyroid gland during the biosynthesis of thyroxine, a thyroid hormone involved in regulating growth and metabolism. The amino acid tyrosine is first iodinated by thyroid peroxidase, and two of the iodinated tyrosine molecules then couple. The electrophilic iodinating agent is an I^+ species, perhaps hypoiodous acid (HIO), that is formed from iodide ion by oxidation with H_2O_2.

Aromatic Nitration

Aromatic rings are nitrated by reaction with a mixture of concentrated nitric and sulfuric acids. The electrophile is the nitronium ion, NO_2^+, which is formed from HNO_3 by protonation and loss of water. The nitronium ion reacts with benzene to yield a carbocation intermediate, and loss of H^+ from this intermediate gives the neutral substitution product, nitrobenzene **(Figure 16.4)**.

Figure 16.4 The mechanism of electrophilic nitration of an aromatic ring. An electrostatic potential map of the reactive electrophile NO_2^+ shows that the **nitrogen atom is most positive.**

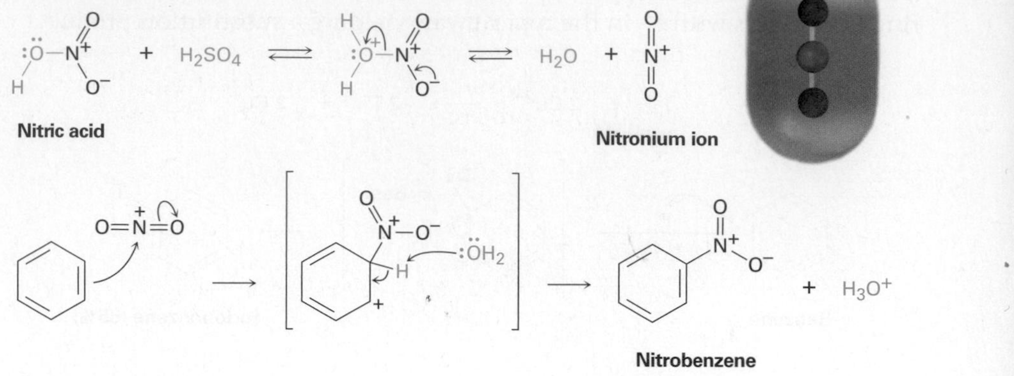

Nitric acid

Nitronium ion

Nitrobenzene

Electrophilic nitration of an aromatic ring does not occur in nature but is particularly important in the laboratory because the nitro-substituted product can be reduced by reagents such as iron, tin, or $SnCl_2$ to yield an *arylamine,* $ArNH_2$. Attachment of an amino group to an aromatic ring by the two-step nitration/reduction sequence is a key part of the industrial synthesis of many dyes and pharmaceutical agents. We'll discuss this reduction and other reactions of aromatic nitrogen compounds in Chapter 24.

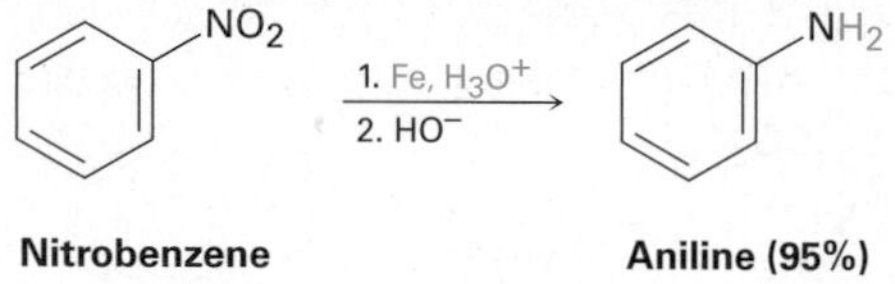

1. Fe, H_3O^+
2. HO^-

Nitrobenzene

Aniline (95%)

Aromatic Sulfonation

Aromatic rings can be sulfonated by reaction with fuming sulfuric acid, a mixture of H_2SO_4 and SO_3. The reactive electrophile is either HSO_3^+ or neutral SO_3, depending on reaction conditions, and substitution occurs by the same two-step mechanism seen previously for bromination and nitration **(Figure 16.5)**. Note, however, that the sulfonation reaction is readily reversible; it can occur either forward or backward, depending on the reaction conditions. Sulfonation is favored in strong acid, but desulfonation is favored in hot, dilute aqueous acid.

Figure 16.5 The mechanism of electrophilic sulfonation of an aromatic ring. An electrostatic potential map of the reactive electrophile $HOSO_2^+$ shows that **sulfur and hydrogen are the most positive atoms.**

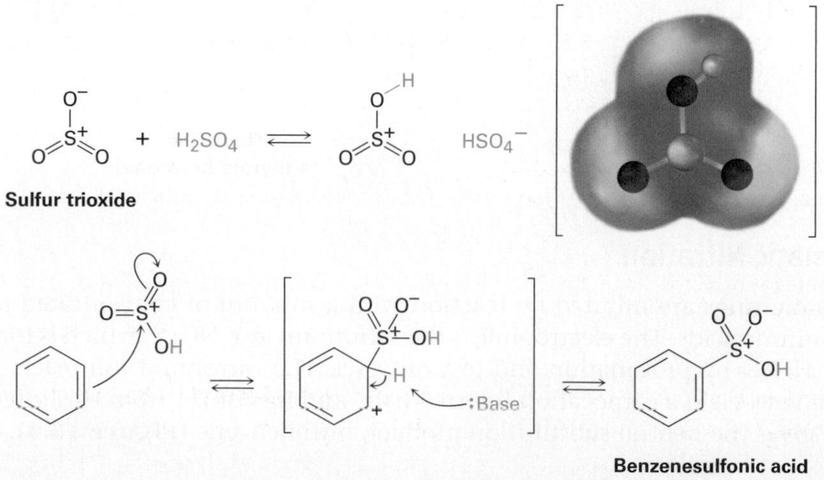

Sulfur trioxide

Benzenesulfonic acid

Aromatic sulfonation does not occur naturally but is widely used in the preparation of dyes and pharmaceutical agents. For example, the sulfa drugs, such as sulfanilamide, were among the first clinically useful antibiotics. Although largely replaced today by more effective agents, sulfa drugs are still used in the treatment of meningitis and urinary tract infections. These drugs are prepared commercially by a process that involves aromatic sulfonation as the key step.

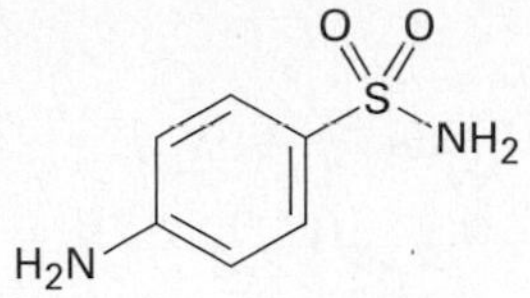

Sulfanilamide (an antibiotic)

Aromatic Hydroxylation

Direct hydroxylation of an aromatic ring to yield a hydroxybenzene (a *phenol*) is difficult and rarely done in the laboratory but occurs much more frequently in biological pathways. An example is the hydroxylation of *p*-hydroxyphenylacetate to give 3,4-dihydroxyphenylacetate. The reaction is catalyzed by *p*-hydroxyphenylacetate-3-hydroxylase and requires molecular oxygen plus the coenzyme reduced flavin adenine dinucleotide, abbreviated $FADH_2$.

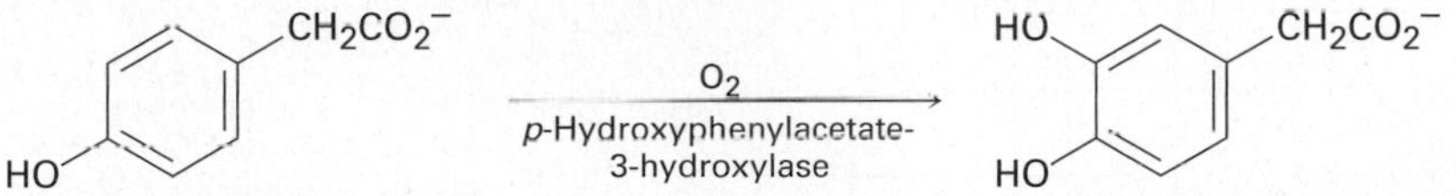

O_2, *p*-Hydroxyphenylacetate-3-hydroxylase

***p*-Hydroxyphenylacetate** → **3,4-Dihydroxyphenylacetate**

By analogy with other electrophilic aromatic substitutions, you might expect that an electrophilic oxygen species acting as an "OH^+ equivalent" is needed for the hydroxylation reaction. That is just what happens, as shown in **Figure 16.6** on the next page. Protonation of FAD hydroperoxide, RO–OH, occurs to give an intermediate that reacts as if it were HO^+ ($RO–OH + H^+ \rightarrow ROH + OH^+$ and carries out an electrophilic aromatic substitution reaction. The FAD hydroperoxide is itself formed by reaction of $FADH_2$ with O_2.

Problem 16.2

Propose a mechanism for the electrophilic fluorination of benzene with F-TEDA-BF_4.

Problem 16.3

How many products might be formed on chlorination of *o*-xylene (*o*-dimethylbenzene), *m*-xylene, and *p*-xylene?

Problem 16.4

When benzene is treated with D_2SO_4, deuterium slowly replaces all six hydrogens in the aromatic ring. Explain.

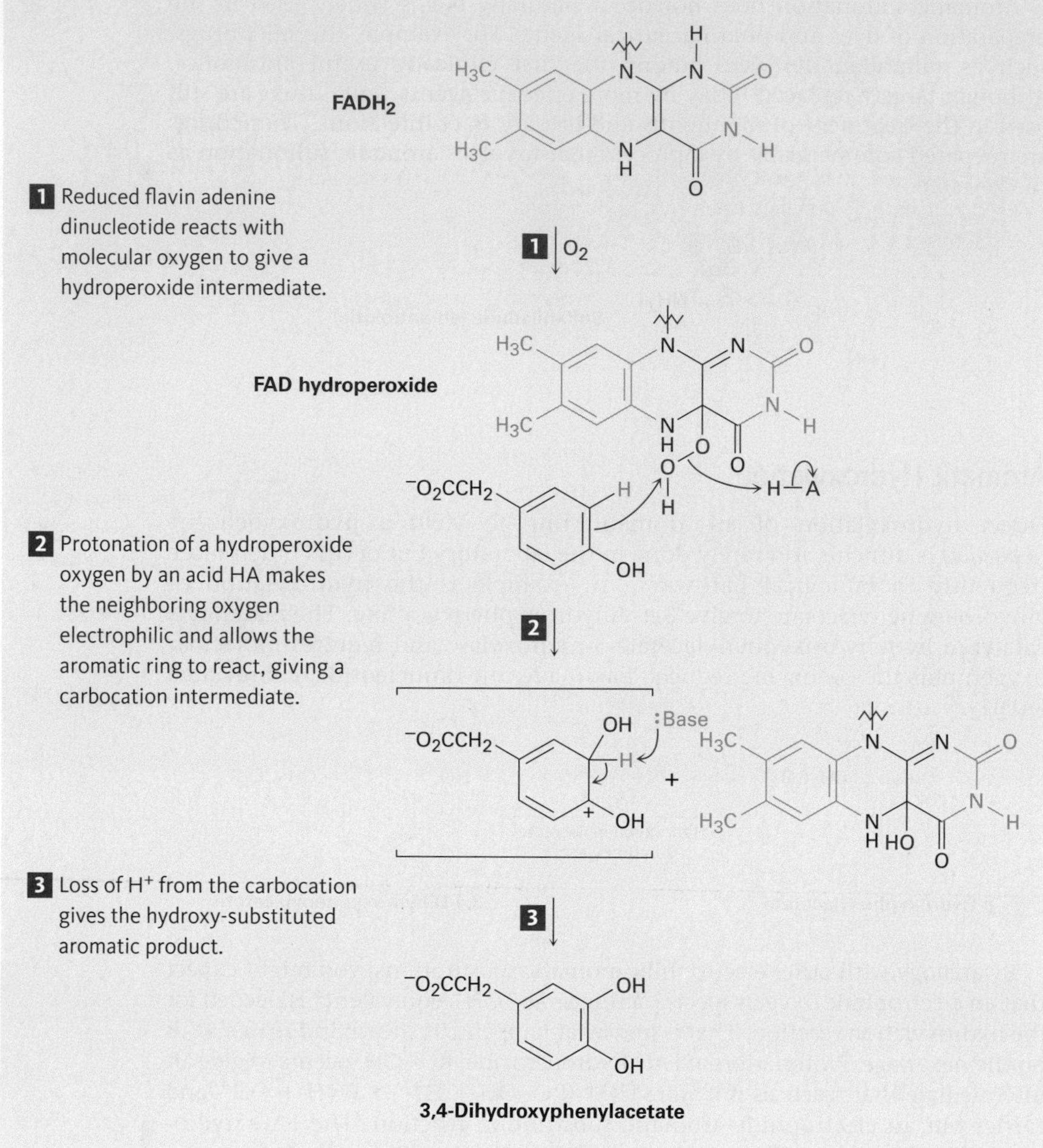

Figure 16.6 | **MECHANISM**

Mechanism of the electrophilic hydroxylation of *p*-hydroxyphenylacetate, by reaction with FAD hydroperoxide. The hydroxylating species is an "OH^+ equivalent" that arises by protonation of FAD hydroperoxide.

16.3 Alkylation and Acylation of Aromatic Rings: The Friedel–Crafts Reaction

Among the most useful electrophilic aromatic substitution reactions in the laboratory is **alkylation**—the introduction of an alkyl group onto the benzene ring. Called the **Friedel–Crafts reaction** after its discoverers, the reaction is carried out by treating the aromatic compound with an alkyl chloride, RCl, in the presence of $AlCl_3$ to generate a carbocation electrophile, R^+. Aluminum chloride catalyzes the reaction by helping the alkyl halide to dissociate in much the same way that $FeBr_3$ catalyzes aromatic brominations by polarizing Br_2 **(Section 16.1)**. Loss of H^+ then completes the reaction **(Figure 16.7)**.

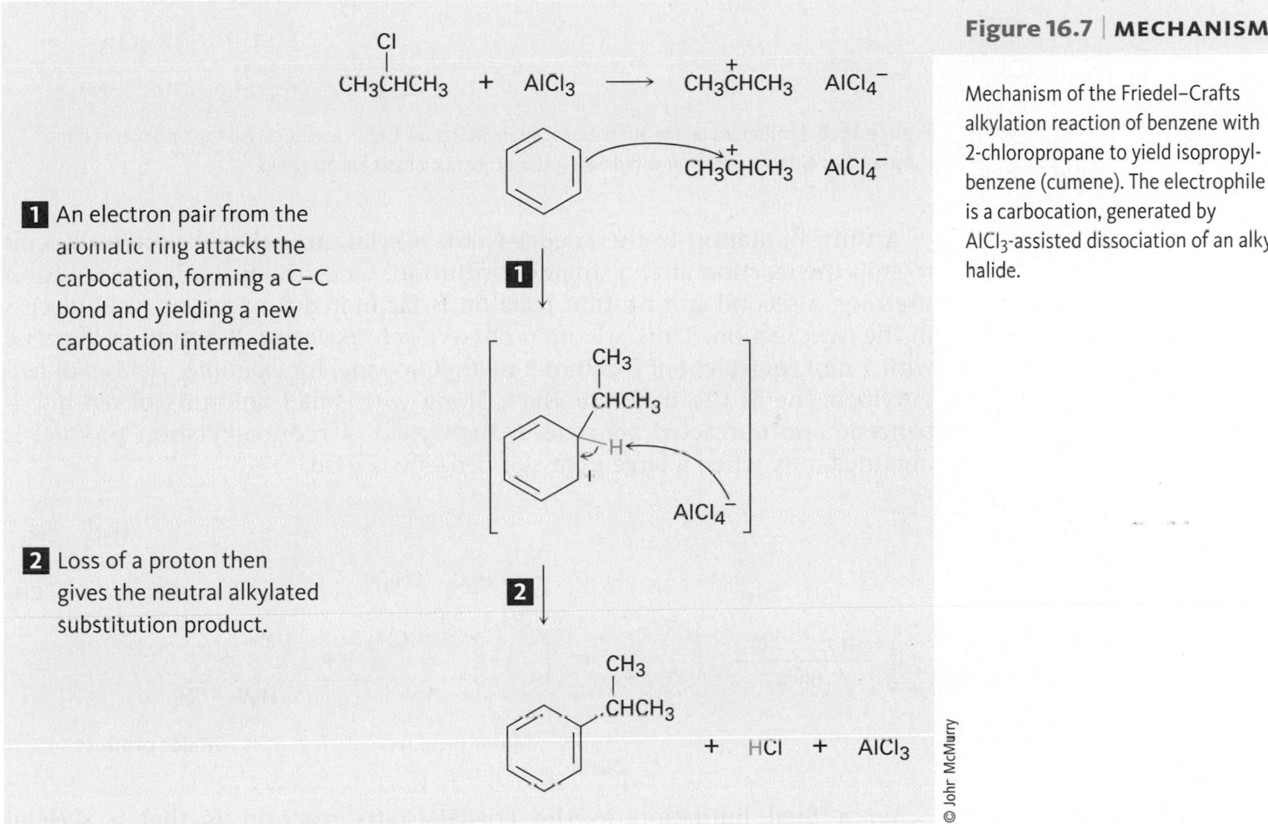

Figure 16.7 | **MECHANISM**

Mechanism of the Friedel–Crafts alkylation reaction of benzene with 2-chloropropane to yield isopropylbenzene (cumene). The electrophile is a carbocation, generated by $AlCl_3$-assisted dissociation of an alkyl halide.

Despite its utility, the Friedel–Crafts alkylation has several limitations. For one thing, only *alkyl* halides can be used. Aromatic (aryl) halides and vinylic halides don't react because aryl and vinylic carbocations are too high in energy to form under Friedel–Crafts conditions.

An aryl halide (Cl)   (Cl) **A vinylic halide**

Not reactive

Another limitation is that Friedel–Crafts reactions don't succeed on aromatic rings that are substituted either by a strongly electron-withdrawing group such as carbonyl (C=O) or by a basic amino group that can be protonated. We'll see in the next section that the presence of a substituent group already on a ring can have a dramatic effect on that ring's subsequent reactivity toward further electrophilic substitution. Rings that contain any of the substituents listed in **Figure 16.8** do not undergo Friedel–Crafts alkylation.

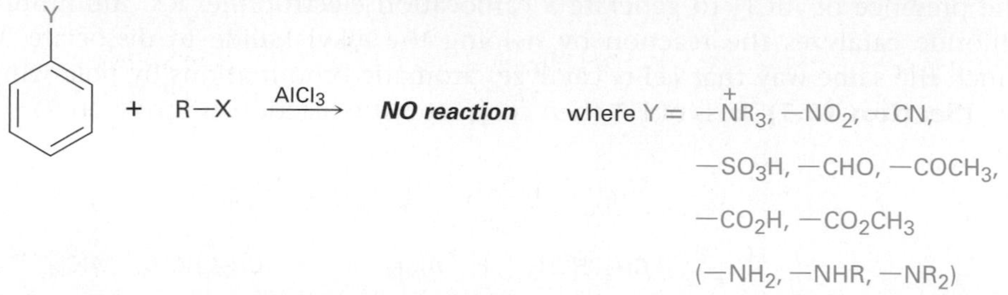

Figure 16.8 Limitations on the aromatic substrate in Friedel–Crafts reactions. No reaction occurs if the substrate has either an electron-withdrawing substituent or a basic amino group.

A third limitation to the Friedel–Crafts alkylation is that it's often difficult to stop the reaction after a single substitution. Once the first alkyl group is on the ring, a second substitution reaction is facilitated for reasons we'll discuss in the next section. Thus, we often observe *polyalkylation.* Reaction of benzene with 1 mol equivalent of 2-chloro-2-methylpropane, for example, yields *p*-di-*tert*-butylbenzene as the major product, along with small amounts of *tert*-butylbenzene and unreacted benzene. A high yield of monoalkylation product is obtained only when a large excess of benzene is used.

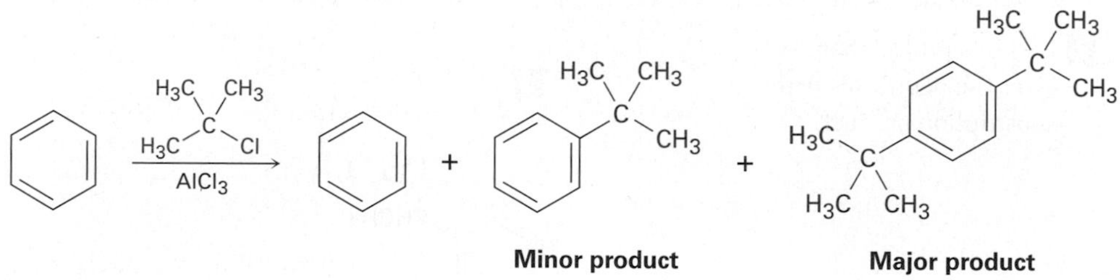

Yet a final limitation to the Friedel–Crafts reaction is that a skeletal rearrangement of the alkyl carbocation electrophile sometimes occurs during reaction, particularly when a primary alkyl halide is used. Treatment of benzene with 1-chlorobutane at 0 °C, for instance, gives an approximately 2:1 ratio of rearranged (*sec*-butyl) to unrearranged (butyl) products.

The carbocation rearrangements that accompany Friedel–Crafts reactions are like those that accompany electrophilic additions to alkenes **(Section 7.11)** and occur either by hydride shift or alkyl shift. For example, the relatively unstable primary butyl carbocation produced by reaction of 1-chlorobutane with $AlCl_3$ rearranges to the more stable secondary butyl carbocation by shift of a hydrogen atom and its electron pair (a hydride ion, $H:^-$) from C2 to C1. Similarly, alkylation of benzene with 1-chloro-2,2-dimethylpropane yields

(1,1-dimethylpropyl)benzene. The initially formed primary carbocation rearranges to a tertiary carbocation by shift of a methyl group and its electron pair from C2 to C1.

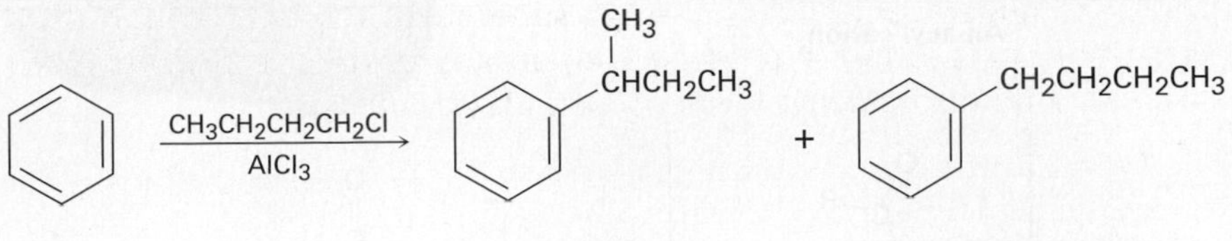

Benzene $\xrightarrow[AlCl_3]{CH_3CH_2CH_2CH_2Cl}$ sec-Butylbenzene (65%) ($C_6H_5CH(CH_3)CH_2CH_3$) + Butylbenzene (35%) ($C_6H_5CH_2CH_2CH_2CH_3$)

Benzene — ***sec*-Butylbenzene (65%)** — **Butylbenzene (35%)**

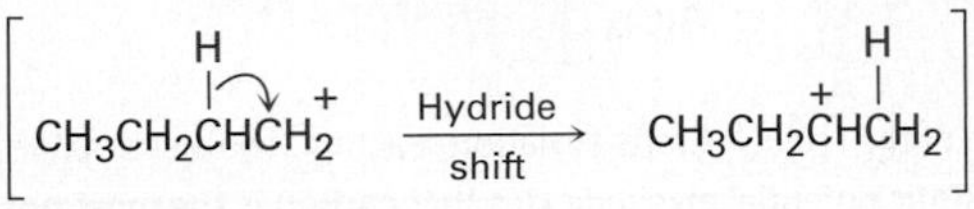

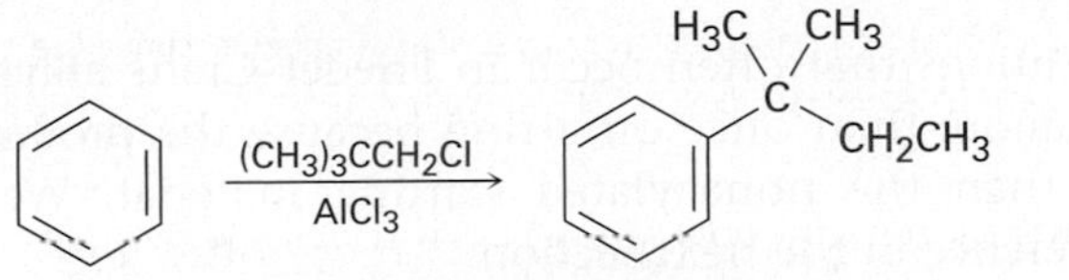

$$\left[CH_3CH_2\overset{H}{C}H\overset{+}{C}H_2 \xrightarrow[\text{shift}]{\text{Hydride}} CH_3CH_2\overset{+}{C}HCH_2 \right]$$

(H shown on the cation carbon after the shift: $CH_3CH_2\overset{+}{C}H\overset{H}{C}H_2$)

Benzene $\xrightarrow[AlCl_3]{(CH_3)_3CCH_2Cl}$ (1,1-Dimethylpropyl)benzene ($C_6H_5C(CH_3)_2CH_2CH_3$)

Benzene — **(1,1-Dimethylpropyl)benzene**

$$\left[CH_3-\underset{CH_3}{\overset{CH_3}{C}}-\overset{+}{C}H_2 \xrightarrow[\text{shift}]{\text{Alkyl}} CH_3-\underset{CH_3}{\overset{+}{C}}-CH_2CH_3 \right]$$

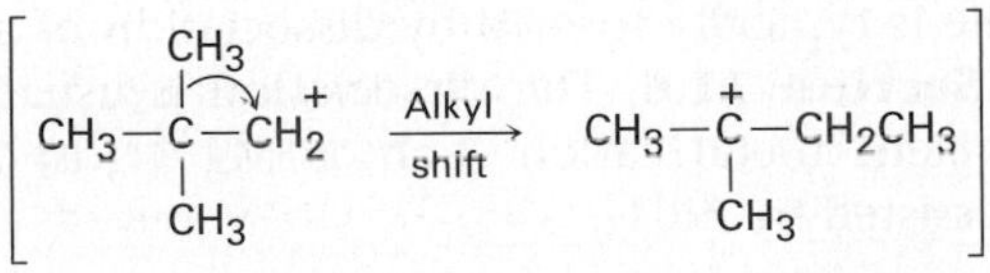

Just as an aromatic ring is alkylated by reaction with an alkyl chloride, it is **acylated** by reaction with a carboxylic acid chloride, RCOCl, in the presence of $AlCl_3$. That is, an **acyl group** (–COR; pronounced **a**-sil) is substituted onto the aromatic ring. For example, reaction of benzene with acetyl chloride yields the ketone, acetophenone.

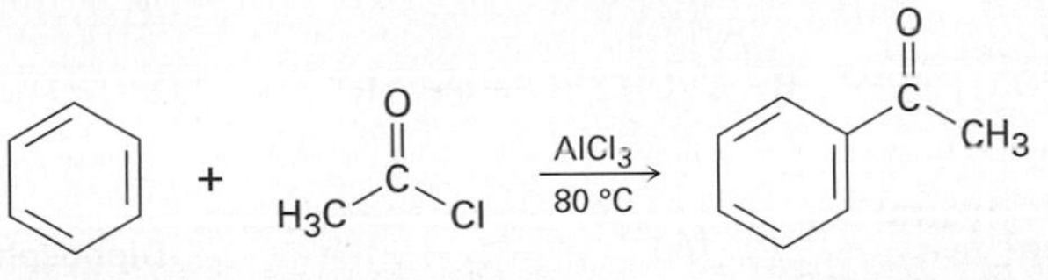

Benzene + $H_3C-\overset{O}{\overset{\|}{C}}-Cl$ $\xrightarrow[80\ °C]{AlCl_3}$ $C_6H_5-\overset{O}{\overset{\|}{C}}-CH_3$

Benzene — **Acetyl chloride** — **Acetophenone (95%)**

The mechanism of the Friedel–Crafts acylation reaction is similar to that of Friedel–Crafts alkylation, and the same limitations on the aromatic substrate noted previously in Figure 16.8 for alkylation also apply to acylation. The reactive electrophile is a resonance-stabilized acyl cation, generated by reaction between the acyl chloride and $AlCl_3$ (**Figure 16.9**). As the resonance structures in the figure indicate, an acyl cation is stabilized by interaction of the vacant orbital on carbon with lone-pair electrons on the neighboring oxygen. Because of this stabilization, no carbocation rearrangement occurs during acylation.

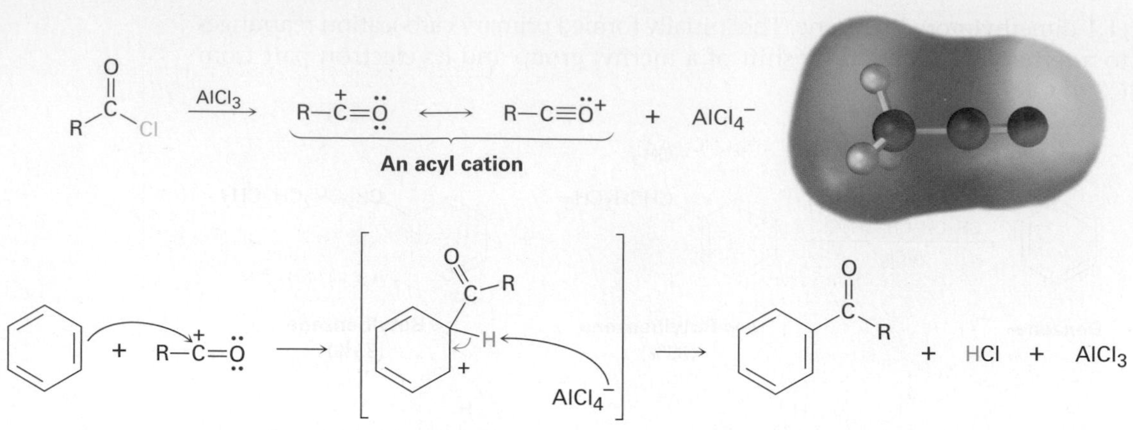

$$\text{RCOCl} \xrightarrow{AlCl_3} \underbrace{R{-}\overset{+}{C}{=}\ddot{O}: \longleftrightarrow R{-}C{\equiv}\ddot{O}^{+}}_{\text{An acyl cation}} + AlCl_4^-$$

Figure 16.9 Mechanism of the Friedel–Crafts acylation reaction. The electrophile is a resonance-stabilized acyl cation, whose electrostatic potential map indicates that **carbon is the most positive atom.**

Unlike the multiple substitutions that often occur in Friedel–Crafts alkylations, acylations never occur more than once on a ring because the product acylbenzene is less reactive than the nonacylated starting material. We'll account for this reactivity difference in the next section.

Aromatic alkylations occur in numerous biological pathways, although there is of course no $AlCl_3$ present in living systems to catalyze the reaction. Instead, the carbocation electrophile is typically formed by dissociation of an organodiphosphate, as we saw in **Section 11.6**. The dissociation is usually assisted by complexation to a divalent metal cation such as Mg^{2+} just as dissociation of an alkyl chloride is assisted by $AlCl_3$.

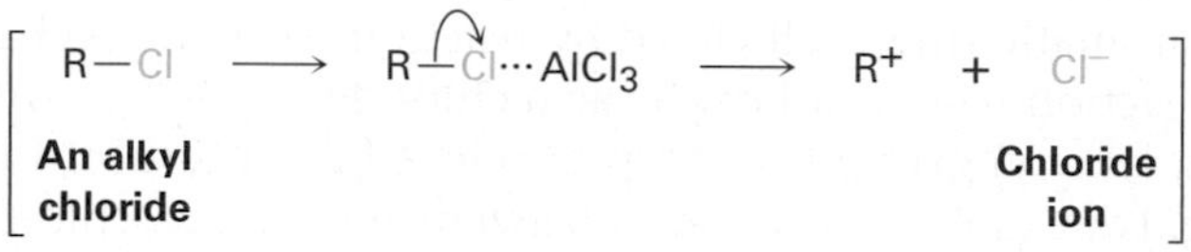

$$R{-}Cl \longrightarrow R{-}Cl\cdots AlCl_3 \longrightarrow R^+ + Cl^-$$

An alkyl chloride — Chloride ion

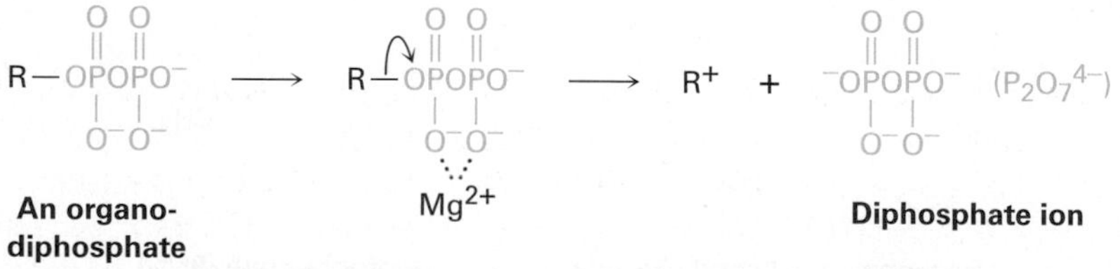

$$R{-}OPO_2OPO_3^{3-} \longrightarrow R{-}OPO_2OPO_3^{3-}\cdots Mg^{2+} \longrightarrow R^+ + {}^-OPO_2OPO_3^{3-}\ (P_2O_7^{4-})$$

An organodiphosphate — Mg^{2+} — Diphosphate ion

An example of a biological Friedel–Crafts reaction occurs during the biosynthesis of phylloquinone, or vitamin K_1, the human blood-clotting factor. Phylloquinone is formed by reaction of 1,4-dihydroxynaphthoic acid with phytyl diphosphate. Phytyl diphosphate first dissociates to a resonance-stabilized allylic carbocation, which then substitutes onto the aromatic ring in the typical way. Several further transformations lead to phylloquinone **(Figure 16.10)**.

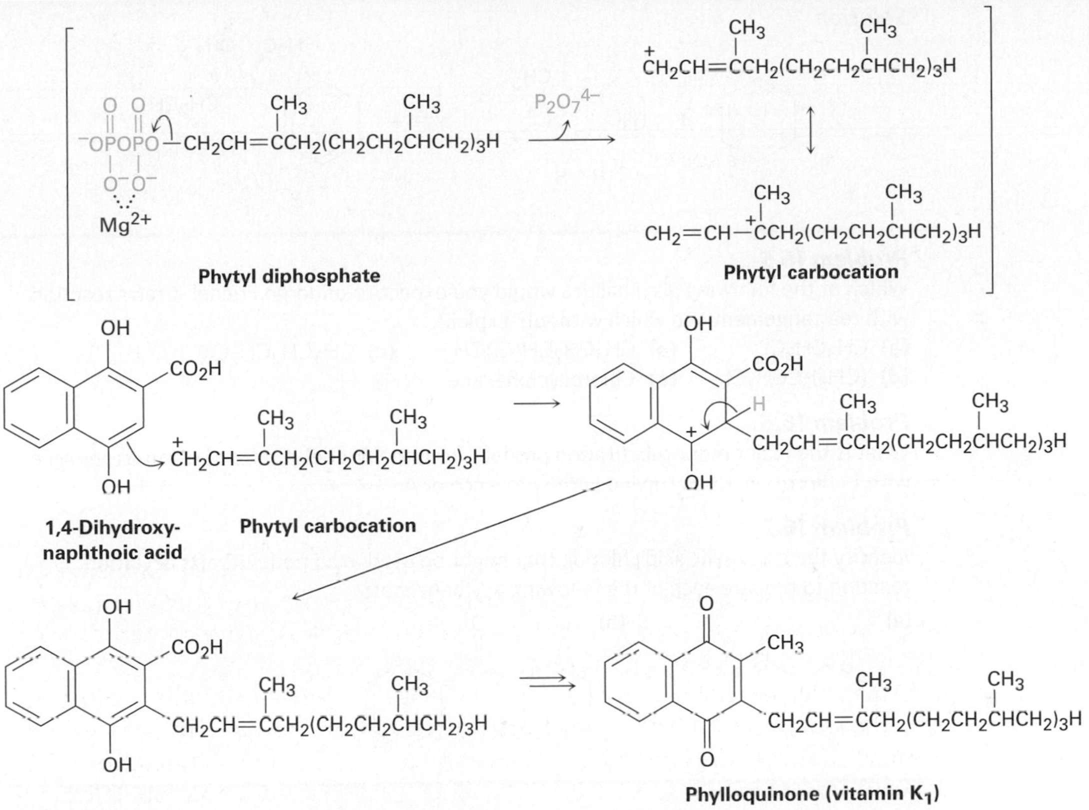

Phytyl diphosphate

Phytyl carbocation

1,4-Dihydroxy-naphthoic acid

Phytyl carbocation

Phylloquinone (vitamin K_1)

Figure 16.10 Biosynthesis of phylloquinone (vitamin K_1) from 1,4-dihydroxynaphthoic acid. The key step that joins the 20-carbon phytyl side chain to the aromatic ring is a Friedel–Crafts-like electrophilic substitution reaction with a diphosphate ion as the leaving group.

Predicting the Product of a Carbocation Rearrangement

Worked Example 16.1

The Friedel–Crafts reaction of benzene with 2-chloro-3-methylbutane in the presence of $AlCl_3$ occurs with a carbocation rearrangement. What is the structure of the product?

Strategy

A Friedel–Crafts reaction involves initial formation of a carbocation, which can rearrange by either a hydride shift or an alkyl shift to give a more stable carbocation. Draw the initial carbocation, assess its stability, and see if the shift of a hydride ion or an alkyl group from a neighboring carbon will result in increased stability. In the present instance, the initial carbocation is a secondary one that can rearrange to a more stable tertiary one by a hydride shift.

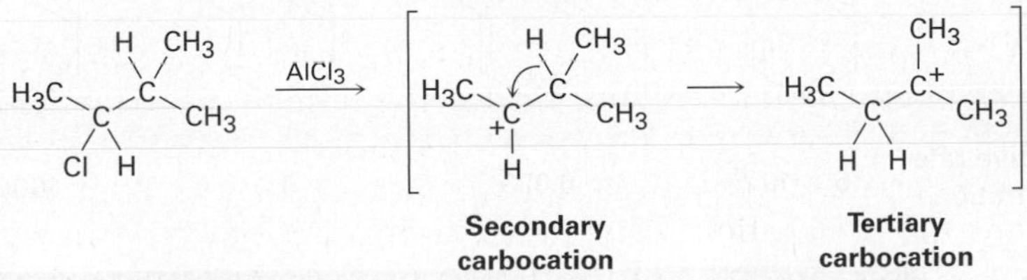

$AlCl_3$

Secondary carbocation

Tertiary carbocation

Use this more stable tertiary carbocation to complete the Friedel–Crafts reaction.

Solution

Benzene + $H_3C{-}CH_2{-}C^+(CH_3){-}CH_3$ ⟶ $C_6H_5{-}C(CH_3)_2{-}CH_2CH_3$

Problem 16.5

Which of the following alkyl halides would you expect to undergo Friedel–Crafts reaction with rearrangement and which without? Explain.

(a) CH_3CH_2Cl **(b)** $CH_3CH_2CH(Cl)CH_3$ **(c)** $CH_3CH_2CH_2Cl$
(d) $(CH_3)_3CCH_2Cl$ **(e)** Chlorocyclohexane

Problem 16.6

What is the major monosubstitution product from the Friedel–Crafts reaction of benzene with 1-chloro-2-methylpropane in the presence of $AlCl_3$?

Problem 16.7

Identify the carboxylic acid chloride that might be used in a Friedel–Crafts acylation reaction to prepare each of the following acylbenzenes:

(a) O **(b)** O

16.4 Substituent Effects in Substituted Aromatic Rings

Only one product can form when an electrophilic substitution occurs on benzene, but what would happen if we were to carry out a reaction on an aromatic ring that already has a substituent? A substituent already present on the ring has two effects.

- **Substituents affect the *reactivity* of the aromatic ring.** Some substituents activate the ring, making it more reactive than benzene, and some deactivate the ring, making it less reactive than benzene. In aromatic nitration, for instance, an $-OH$ substituent makes the ring 1000 times more reactive than benzene, while an $-NO_2$ substituent makes the ring more than 10 million times less reactive.

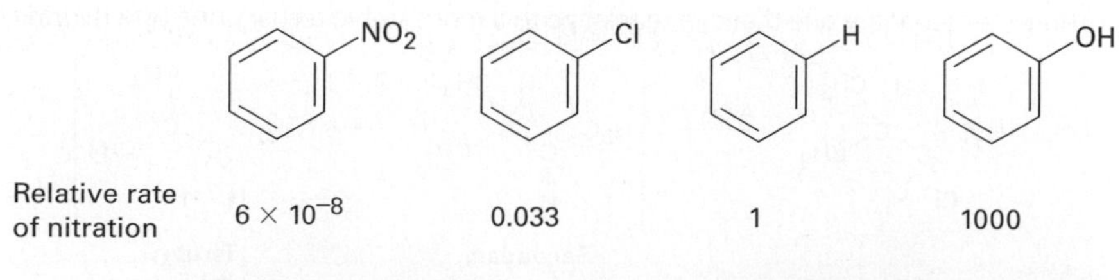

- **Substituents affect the *orientation* of the reaction.** The three possible disubstituted products—ortho, meta, and para—are usually not formed in equal amounts. Instead, the nature of the substituent already present on the benzene ring determines the position of the second substitution. An –OH group directs substitution toward the ortho and para positions, for instance, while a carbonyl group such as –CHO directs substitution primarily toward the meta position. Table 16.1 lists experimental results for the nitration of some substituted benzenes.

Table 16.1 Orientation of Nitration in Substituted Benzenes

Y–benzene $\xrightarrow[H_2SO_4,\ 25\ °C]{HNO_3}$ Y–benzene–NO_2

	Product (%)				Product (%)		
	Ortho	Meta	Para		Ortho	Meta	Para
Meta-directing deactivators				**Ortho- and para-directing deactivators**			
$-\overset{+}{N}(CH_3)_3$	2	87	11	–F	13	1	86
$-NO_2$	7	91	2	–Cl	35	1	64
$-CO_2H$	22	76	2	–Br	43	1	56
–CN	17	81	2	–I	45	1	54
$-CO_2CH_3$	28	66	6	**Ortho- and para-directing activators**			
$-COCH_3$	26	72	2	$-CH_3$	63	3	34
–CHO	19	72	9	–OH	50	0	50
				$-NHCOCH_3$	19	2	79

Substituents can be classified into three groups, as shown in **Figure 16.11**: *ortho- and para-directing activators, ortho- and para-directing deactivators,* and *meta-directing deactivators*. There are no meta-directing activators. Notice how the directing effect of a group correlates with its reactivity. All meta-directing groups are strongly deactivating, and most ortho- and para-directing groups are activating. The halogens are unique in being ortho- and para-directing but weakly deactivating.

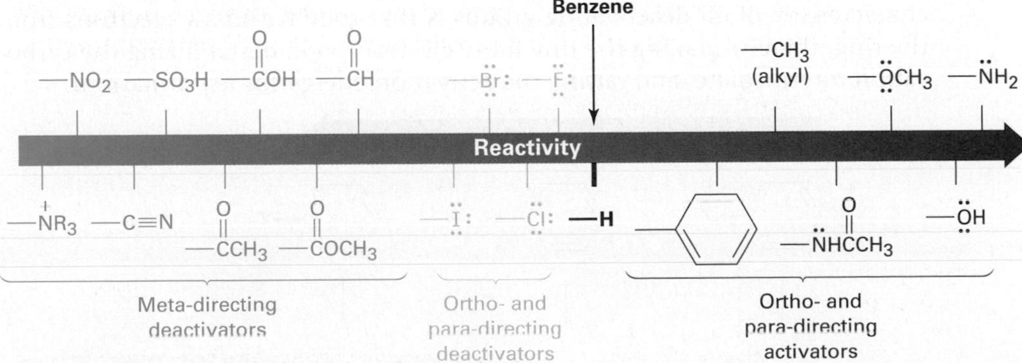

Figure 16.11 Classification of substituent effects in electrophilic aromatic substitution. All activating groups are ortho- and para-directing, and all deactivating groups other than halogen are meta-directing. The halogens are unique in being deactivating but ortho- and para-directing.

Worked Example 16.2 Predicting the Product of an Electrophilic Aromatic Substitution Reaction

Predict the major product of the sulfonation of toluene.

Strategy

Identify the substituent present on the ring, and decide whether it is ortho- and para-directing or meta-directing. According to Figure 16.11, an alkyl substituent is ortho- and para-directing, so sulfonation of toluene will give primarily a mixture of *o*-toluenesulfonic acid and *p*-toluenesulfonic acid.

Solution

$$\text{Toluene} \xrightarrow[H_2SO_4]{SO_3} o\text{-Toluenesulfonic acid} + p\text{-Toluenesulfonic acid}$$

Toluene (CH_3) — **o-Toluenesulfonic acid** (CH_3, SO_3H) — **p-Toluenesulfonic acid** (HO_3S, CH_3)

Problem 16.8

Rank the compounds in each of the following groups in order of their reactivity to electrophilic substitution:

(a) Nitrobenzene, phenol, toluene, benzene
(b) Phenol, benzene, chlorobenzene, benzoic acid
(c) Benzene, bromobenzene, benzaldehyde, aniline

Problem 16.9

Predict the major products of the following reactions:

(a) Nitration of bromobenzene **(b)** Bromination of nitrobenzene
(c) Chlorination of phenol **(d)** Bromination of aniline

16.5 An Explanation of Substituent Effects

Activation and Deactivation of Aromatic Rings

What makes a group either activating or deactivating? The common characteristic of all activating groups is that they donate electrons to the ring, thereby making the ring more electron-rich, stabilizing the carbocation intermediate, and lowering the activation energy for its formation. Conversely, the common characteristic of all deactivating groups is that they withdraw electrons from the ring, thereby making the ring more electron-poor, destabilizing the carbocation intermediate, and raising the activation energy for its formation.

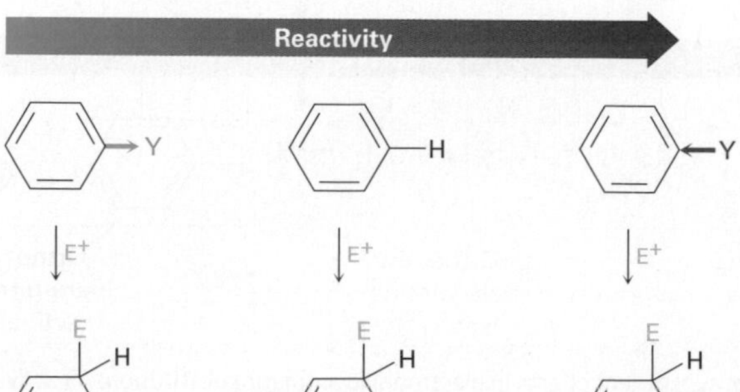

Y withdraws electrons; carbocation intermediate is less stable, and ring is less reactive.

Y donates electrons; carbocation intermediate is more stable, and ring is more reactive.

Compare the electrostatic potential maps of benzaldehyde (deactivated), chlorobenzene (weakly deactivated), and phenol (activated) with that of benzene. As shown in **Figure 16.12**, the ring is more positive (yellow-green) when an electron-withdrawing group such as –CHO or –Cl is present and more negative (red) when an electron-donating group such as –OH is present.

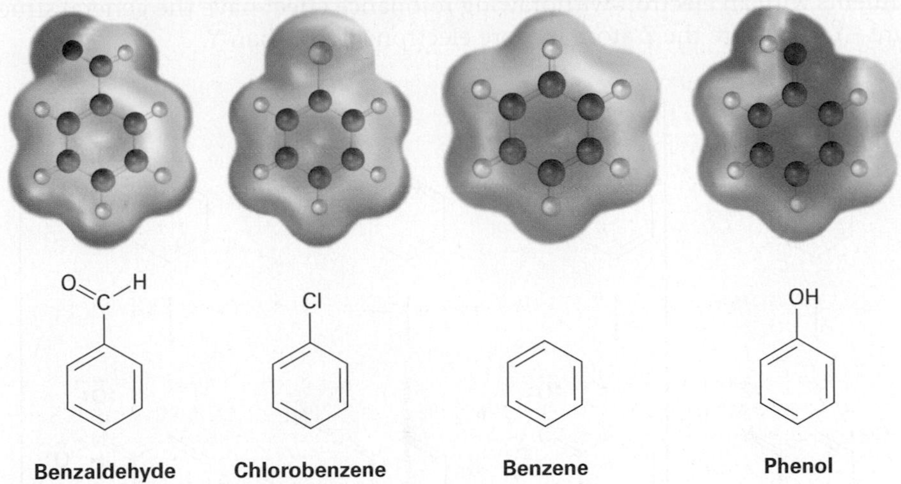

Figure 16.12 Electrostatic potential maps of benzene and several substituted benzenes show that an electron-withdrawing group (–CHO or –Cl) makes the ring more **electron-poor**, while an electron-donating group (–OH) makes the ring more **electron-rich**.

The withdrawal or donation of electrons by a substituent group is controlled by an interplay of *inductive effects* and *resonance effects*. As we saw in **Section 2.1**, an **inductive effect** is the withdrawal or donation of electrons through a σ bond due to electronegativity. Halogens, hydroxyl groups, carbonyl groups, cyano groups, and nitro groups inductively withdraw electrons through the σ bond linking the substituent to a benzene ring. The effect is most pronounced in halobenzenes and phenols, in which the electronegative atom is directly attached to the ring, but is also significant in carbonyl compounds, nitriles, and nitro compounds, in which the electronegative atom is farther removed. Alkyl groups, on the other hand, inductively donate electrons. This is the same hyperconjugative donating effect that causes alkyl substituents to stabilize alkenes **(Section 7.6)** and carbocations **(Section 7.9)**.

Inductive electron withdrawal

Inductive electron donation

A **resonance effect** is the withdrawal or donation of electrons through a π bond due to the overlap of a *p* orbital on the substituent with a *p* orbital on the aromatic ring. Carbonyl, cyano, and nitro substituents, for example, withdraw electrons from the aromatic ring by resonance. The π electrons flow from the ring to the substituent, leaving a positive charge in the ring. Note that substituents with an electron-withdrawing resonance effect have the general structure –Y=Z, where the Z atom is more electronegative than Y.

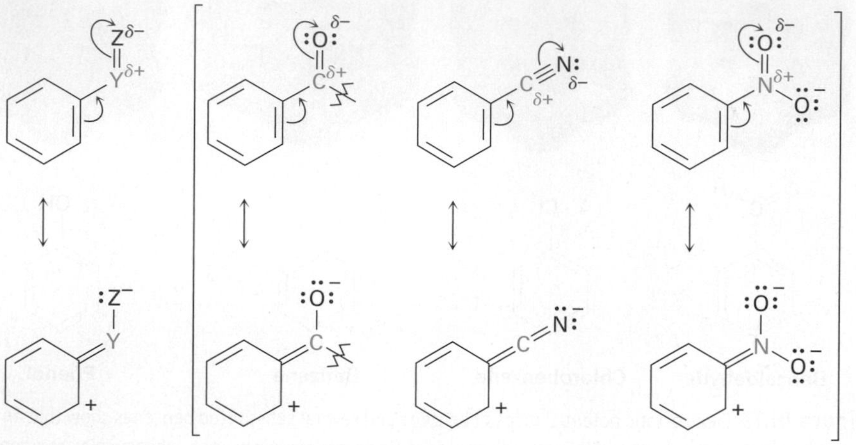

Resonance electron-withdrawing group

Conversely, halogen, hydroxyl, alkoxyl (–OR), and amino substituents donate electrons to the aromatic ring by resonance. Lone-pair electrons flow from the substituents to the ring, placing a negative charge in the ring. Substituents with an electron-donating resonance effect have the general structure $-\ddot{Y}$, where the Y atom has a lone pair of electrons available for donation to the ring.

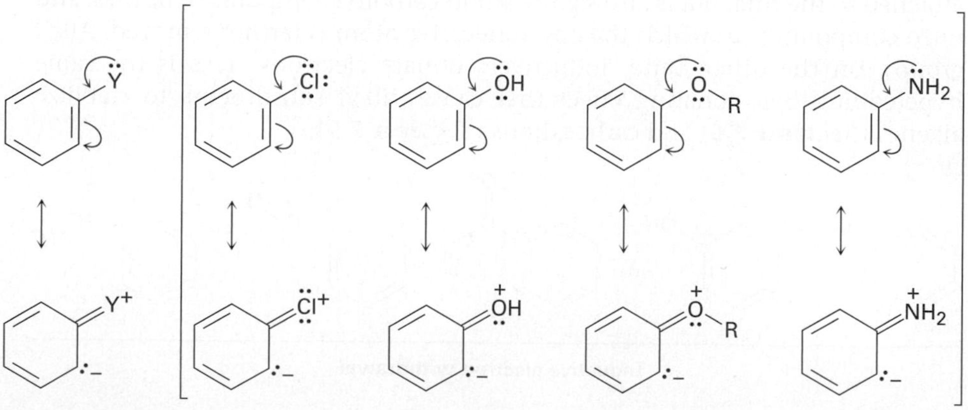

Resonance electron-donating group

One further point: inductive effects and resonance effects don't necessarily act in the same direction. Halogen, hydroxyl, alkoxyl, and amino substituents, for instance, have electron-withdrawing inductive effects because of the electronegativity of the –X, –O, or –N atom bonded to the aromatic ring but have electron-donating resonance effects because of the lone-pair electrons on those –X, –O, or –N atoms. When the two effects act in opposite directions, the stronger effect dominates. Thus, hydroxyl, alkoxyl, and amino substituents are activators because their stronger electron-donating resonance effect outweighs their weaker electron-withdrawing inductive effect. Halogens, however, are deactivators because their stronger electron-withdrawing inductive effect outweighs their weaker electron-donating resonance effect.

Problem 16.10

Use Figure 16.11 to explain why Friedel–Crafts alkylations often give polysubstitution but Friedel–Crafts acylations do not.

$$\text{Benzene} \xrightarrow[\text{AlCl}_3]{\text{CH}_3\text{Cl}} \text{C}_6\text{H}_5\text{CH}_3 + \text{H}_3\text{C–C}_6\text{H}_4\text{–CH}_3$$

(Product mixture)

$$\text{Benzene} \xrightarrow[\text{AlCl}_3]{\text{CH}_3\text{CCl (C=O)}} \text{C}_6\text{H}_5\text{CCH}_3 \text{ (C=O)}$$

(Sole product)

Problem 16.11

An electrostatic potential map of (trifluoromethyl)benzene, $C_6H_5CF_3$, is shown. Would you expect (trifluoromethyl)benzene to be more reactive or less reactive than toluene toward electrophilic substitution? Explain.

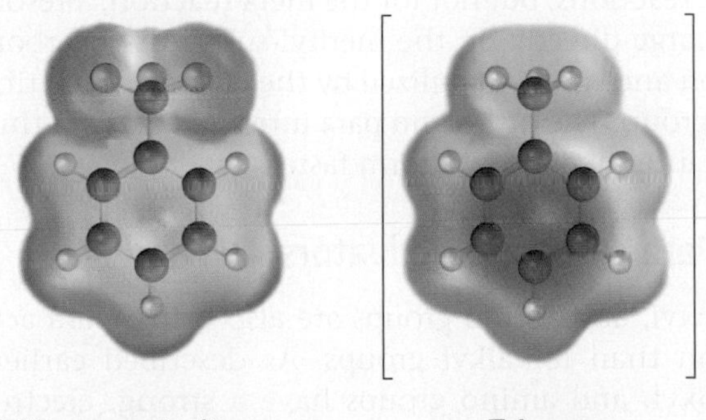

(Trifluoromethyl)benzene **Toluene**

Ortho- and Para-Directing Activators: Alkyl Groups

Inductive and resonance effects account not only for reactivity but also for the orientation of electrophilic aromatic substitutions. Take alkyl groups, for instance, which have an electron-donating inductive effect and are ortho and para directors. The results of toluene nitration are shown in **Figure 16.13**.

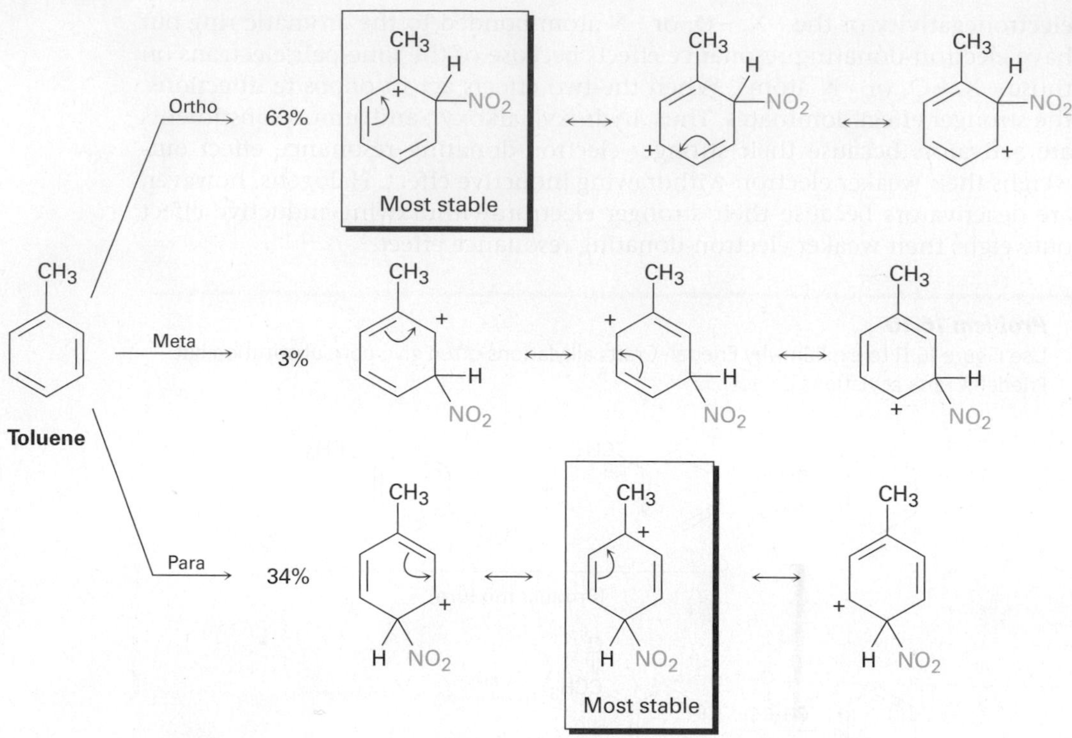

Figure 16.13 Carbocation intermediates in the nitration of toluene. Ortho and para intermediates are more stable than the meta intermediate because the positive charge is on a tertiary carbon rather than a secondary carbon.

Nitration of toluene might occur either ortho, meta, or para to the methyl group, giving the three carbocation intermediates shown in Figure 16.13. Although all three intermediates are resonance-stabilized, the ortho and para intermediates are more stabilized than the meta intermediate. For both the ortho and para reactions, but not for the meta reaction, a resonance form places the positive charge directly on the methyl-substituted carbon, where it is in a tertiary position and can be stabilized by the electron-donating inductive effect of the methyl group. The ortho and para intermediates are thus lower in energy than the meta intermediate and form faster.

Ortho- and Para-Directing Activators: OH and NH_2

Hydroxyl, alkoxyl, and amino groups are also ortho–para activators, but for a different reason than for alkyl groups. As described earlier in this section, hydroxyl, alkoxyl, and amino groups have a strong, electron-donating resonance effect that outweighs a weaker electron-withdrawing inductive effect. When phenol is nitrated, for instance, reaction can occur either ortho, meta, or para to the –OH group, giving the carbocation intermediates shown in **Figure 16.14**. The ortho and para intermediates are more stable than the meta intermediate because they have more resonance forms, including one particularly favorable form that allows the positive charge to be stabilized by electron donation from the substituent oxygen atom. The intermediate from meta reaction has no such stabilization.

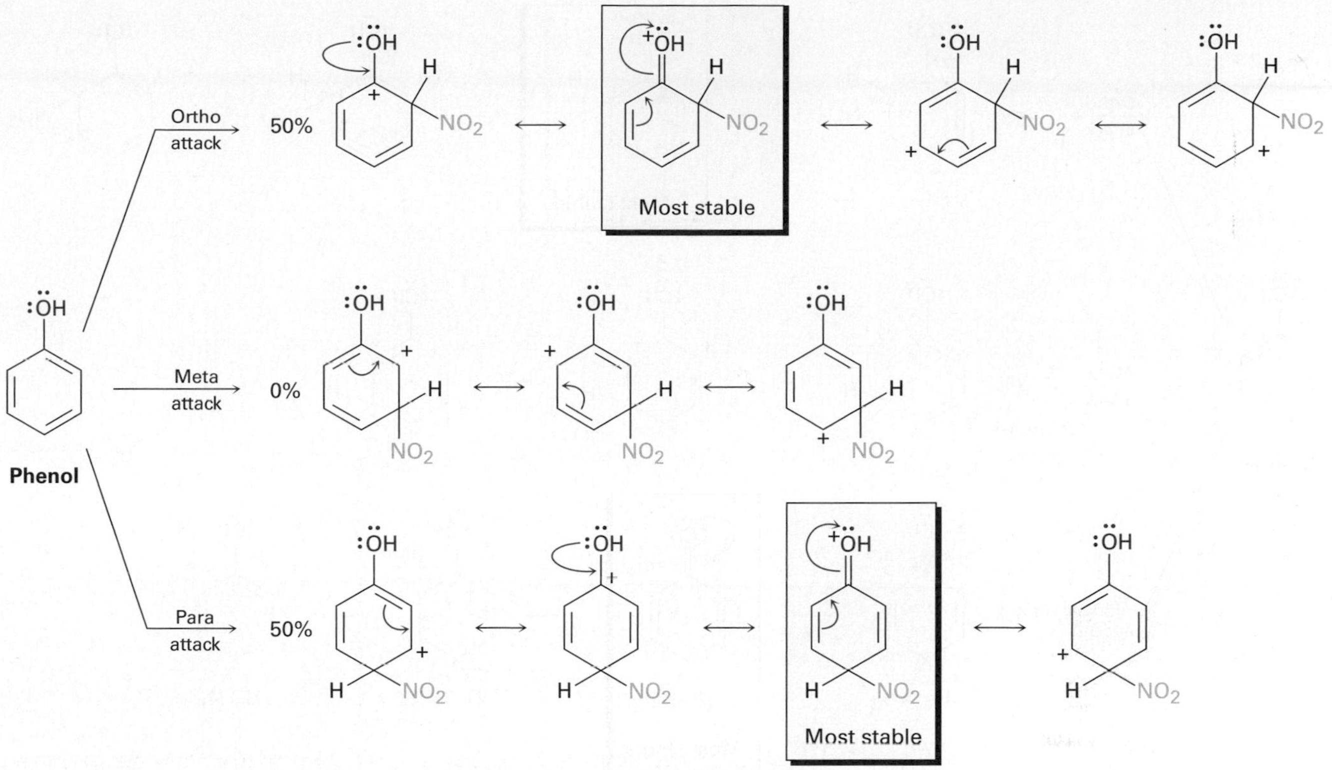

Figure 16.14 Carbocation intermediates in the nitration of phenol. The ortho and para intermediates are more stable than the meta intermediate because they have more resonance forms, including one particularly favorable form that involves electron donation from the oxygen atom.

Problem 16.12
Acetanilide is less reactive than aniline toward electrophilic substitution. Explain.

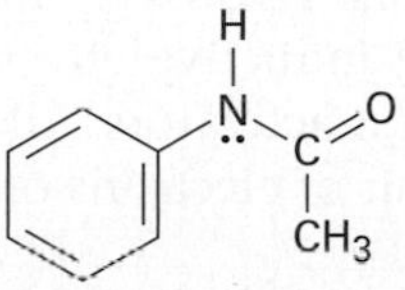

Ortho- and Para-Directing Deactivators: Halogens

Halogens are deactivating because their stronger electron-withdrawing inductive effect outweighs their weaker electron-donating resonance effect. Although weak, that electron-donating resonance effect is nevertheless felt only at the ortho and para positions and not at the meta position **(Figure 16.15)**. Thus, a halogen substituent can stabilize the positive charge of the carbocation intermediates from ortho and para reaction in the same way that hydroxyl and amino substituents can. The meta intermediate, however, has no such stabilization and is therefore formed more slowly.

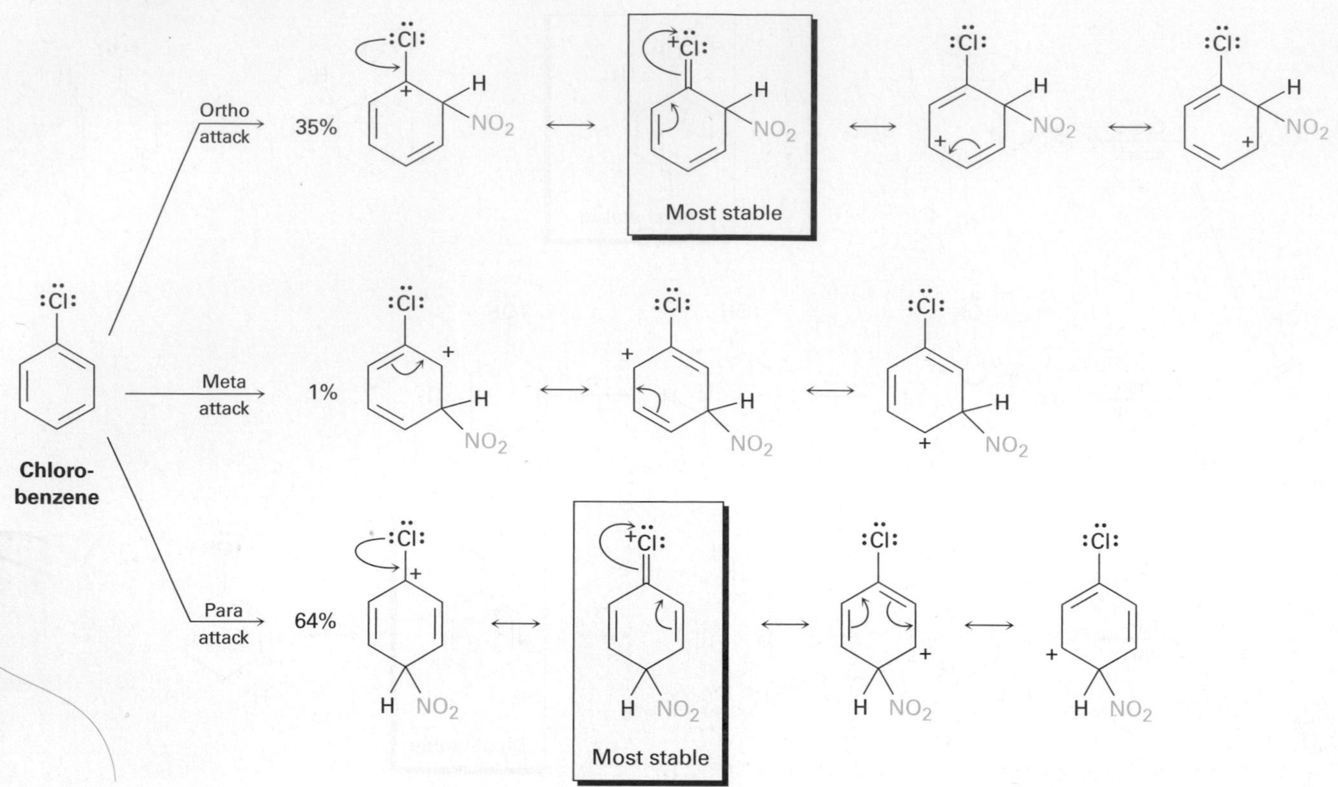

Figure 16.15 Carbocation intermediates in the nitration of chlorobenzene. The ortho and para intermediates are more stable than the meta intermediate because of electron donation of the halogen lone-pair electrons.

Note again that halogens, hydroxyl, alkoxyl, and amino groups all withdraw electrons inductively but donate electrons by resonance. Halogens have a stronger electron-withdrawing inductive effect but a weaker electron-donating resonance effect and are thus deactivators. Hydroxyl, alkoxyl, and amino groups have a weaker electron-withdrawing inductive effect but a stronger electron-donating resonance effect and are thus activators. All are ortho and para directors, however, because of the lone pair of electrons on the atom bonded to the aromatic ring.

Meta-Directing Deactivators

The influence of meta-directing substituents can be explained using the same kinds of arguments used for ortho and para directors. Look at the nitration of benzaldehyde, for instance **(Figure 16.16)**. Of the three possible carbocation intermediates, the meta intermediate has three favorable resonance forms, but the ortho and para intermediates have only two. In both ortho and para intermediates, the third resonance form is unfavorable because it places the positive charge directly on the carbon that bears the aldehyde group, where it is disfavored by a repulsive interaction with the positively polarized carbon atom of the C=O group. Hence, the meta intermediate is more favored and is formed faster than the ortho and para intermediates.

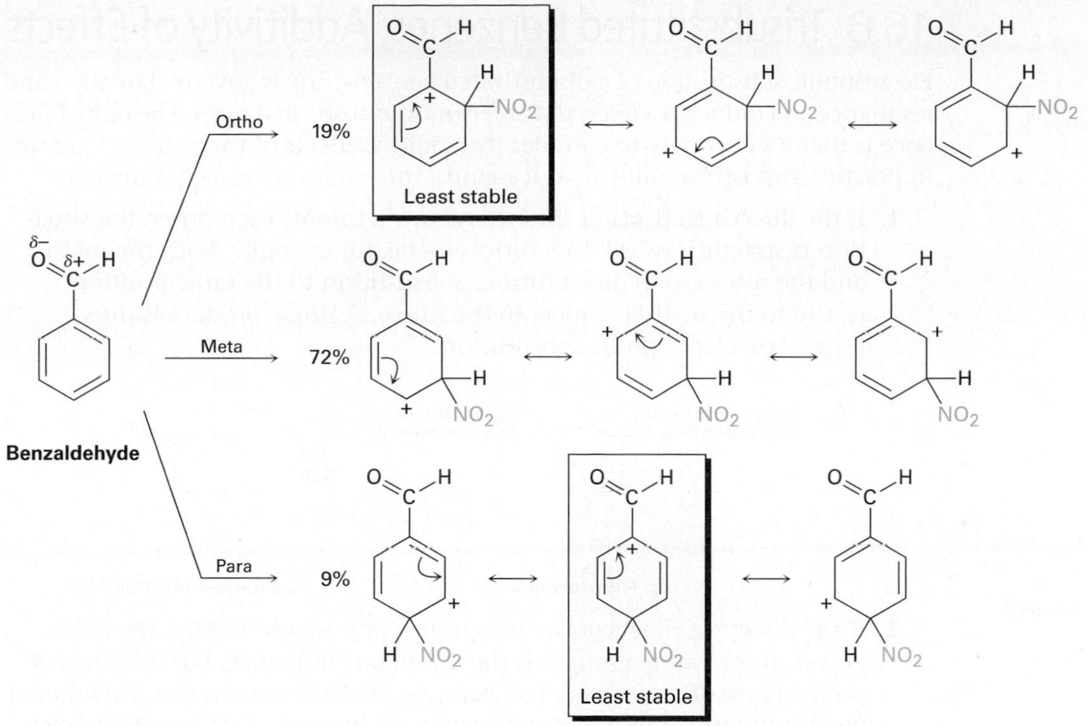

Ortho 19%
Least stable
Meta 72%
Benzaldehyde
Para 9%
Least stable

Figure 16.16 Carbocation intermediates in the nitration of benzaldehyde. The ortho and para intermediates are less stable than the meta intermediate. The meta intermediate is more favorable than ortho and para intermediates because it has three favorable resonance forms rather than two.

In general, any substituent that has a positively polarized atom ($\delta+$) directly attached to the ring will make one of the resonance forms of the ortho and para intermediates unfavorable and will thus act as a meta director.

Problem 16.13

Draw resonance structures for the intermediates from reaction of an electrophile at the ortho, meta, and para positions of nitrobenzene. Which intermediates are most stable?

A Summary of Substituent Effects in Aromatic Substitution

Key IDEAS

Test your knowledge of Key Ideas by answering end-of-chapter exercises marked with ▲.

A summary of the activating and directing effects of substituents in electrophilic aromatic substitution is shown in Table 16.2.

Table 16.2 Substituent Effects in Electrophilic Aromatic Substitution

Substituent	Reactivity	Orienting effect	Inductive effect	Resonance effect
$-CH_3$	Activating	Ortho, para	Weak donating	—
$-OH$, $-NH_2$	Activating	Ortho, para	Weak withdrawing	Strong donating
$-F$, $-Cl$, $-Br$, $-I$	Deactivating	Ortho, para	Strong withdrawing	Weak donating
$-NO_2$, $-CN$, $-CHO$, $-CO_2R$, $-COR$, $-CO_2H$	Deactivating	Meta	Strong withdrawing	Strong withdrawing

16.6 Trisubstituted Benzenes: Additivity of Effects

Electrophilic substitution of a disubstituted benzene ring is governed by the same resonance and inductive effects that affect monosubstituted rings. The only difference is that it's necessary to consider the additive effects of two different groups. In practice, this isn't as difficult as it sounds; three rules are usually sufficient.

1. If the directing effects of the two groups reinforce each other, the situation is straightforward. In *p*-nitrotoluene, for example, both the methyl and the nitro group direct further substitution to the same position (ortho to the methyl = meta to the nitro). A single product is thus formed on electrophilic substitution.

CH_3 directs here. NO_2 directs here. — CH_3 directs here. NO_2 directs here. — $\xrightarrow[FeBr_3]{Br_2}$

p-Nitrotoluene — **2-Bromo-4-nitrotoluene**

2. If the directing effects of the two groups oppose each other, the more powerful activating group has the dominant influence, but mixtures of products are often formed. For example, bromination of *p*-methylphenol yields primarily 2-bromo-4-methylphenol because $-OH$ is a more powerful activator than $-CH_3$.

OH directs here. — OH directs here. — CH_3 directs here. — CH_3 directs here. — $\xrightarrow[H_2SO_4]{HNO_3}$

p-Methylphenol — **4-Methyl-2-nitrophenol**

3. Further substitution rarely occurs between the two groups in a meta-disubstituted compound because this site is too hindered. Aromatic rings with three adjacent substituents must therefore be prepared by some other route, such as by substitution of an ortho-disubstituted compound.

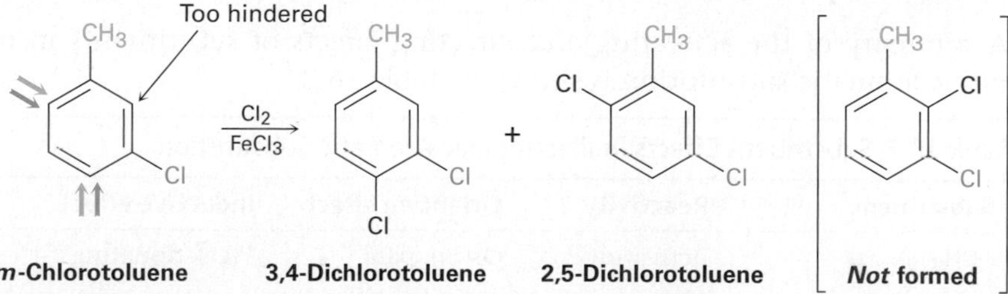

But:

$\xrightarrow[FeCl_3]{Cl_2}$ +

o-Nitrotoluene — **2-Chloro-6-nitrotoluene** — **4-Chloro-2-nitrotoluene**

Worked Example 16.3

Predicting the Product of Substitution on a Disubstituted Benzene

What product would you expect from bromination of *p*-methylbenzoic acid?

Strategy

Identify the two substituents present on the ring, decide the directing effect of each and, if necessary, decide which substituent is the stronger activator. In the present case, the carboxyl group ($-CO_2H$) is a meta director and the methyl group is an ortho and para director. Both groups direct bromination to the position next to the methyl group, yielding 3-bromo-4-methylbenzoic acid.

Solution

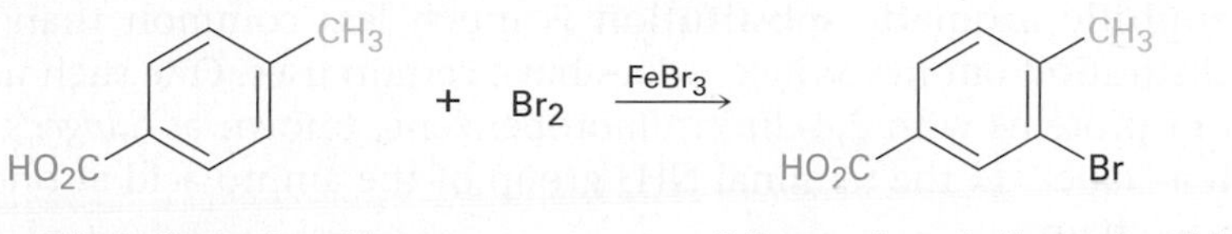

CH_3, HO_2C + Br_2 $\xrightarrow{FeBr_3}$ CH_3, HO_2C, Br

p-Methylbenzoic acid **3-Bromo-4-methylbenzoic acid**

Problem 16.14

At what position would you expect electrophilic substitution to occur in each of the following substances?

(a) 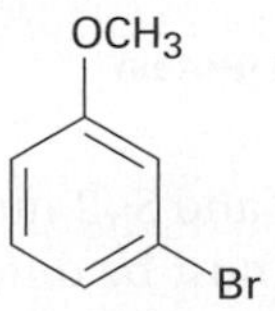 OCH_3, Br

(b) 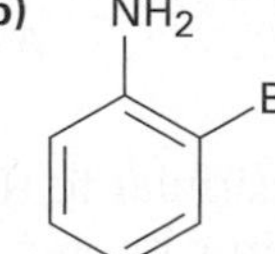

(c) 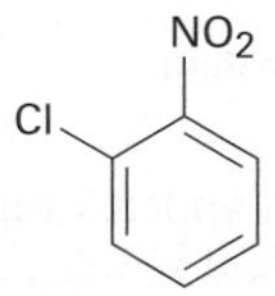 NO_2, Cl

Problem 16.15

Show the major product(s) from reaction of the following substances with (1) CH_3CH_2Cl, $AlCl_3$ and (2) HNO_3, H_2SO_4:

(a)

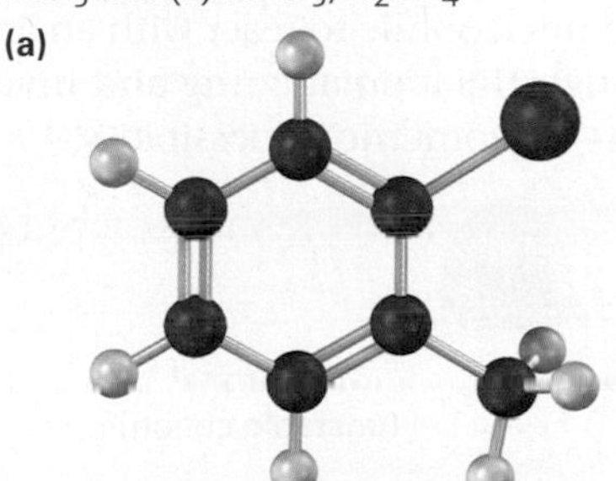

(b)

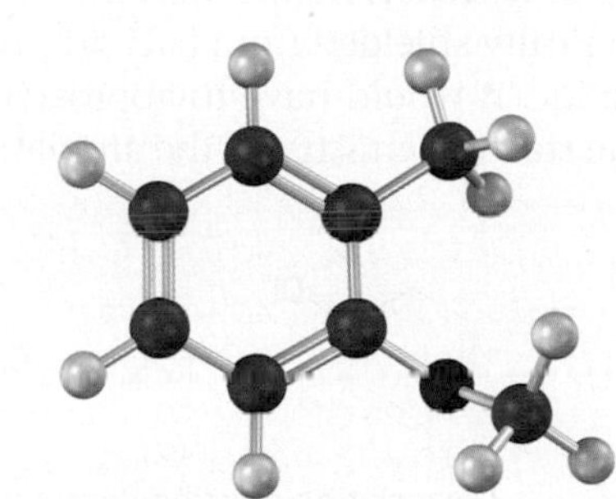

16.7 Nucleophilic Aromatic Substitution

Although aromatic substitution reactions usually occur by an *electrophilic* mechanism, aryl halides that have electron-withdrawing substituents can also undergo a *nucleophilic* substitution reaction. For example, 2,4,6-trinitrochlorobenzene reacts with aqueous NaOH at room temperature to give 2,4,6-trinitrophenol. The nucleophile OH^- has substituted for Cl^-.

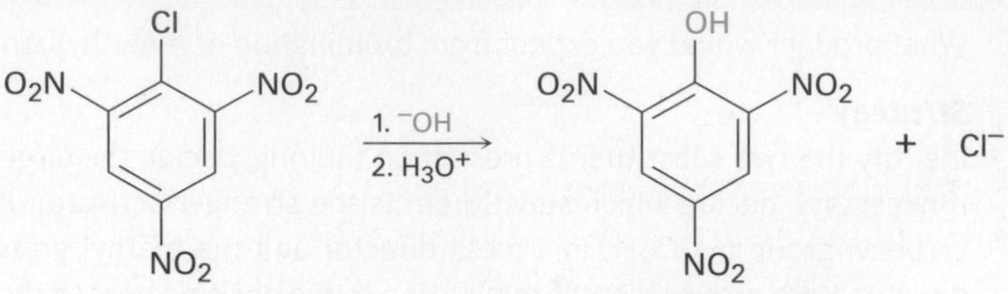

Cl, O_2N, NO_2, NO_2 → (1. ^-OH; 2. H_3O^+) → OH, O_2N, NO_2, NO_2 + Cl^-

2,4,6-Trinitrochlorobenzene **2,4,6-Trinitrophenol (100%)**

Nucleophilic aromatic substitution is much less common than electrophilic substitution but nevertheless does have certain uses. One such use is the reaction of proteins with 2,4-dinitrofluorobenzene, known as *Sanger's reagent,* to attach a "label" to the terminal NH_2 group of the amino acid at one end of the protein chain.

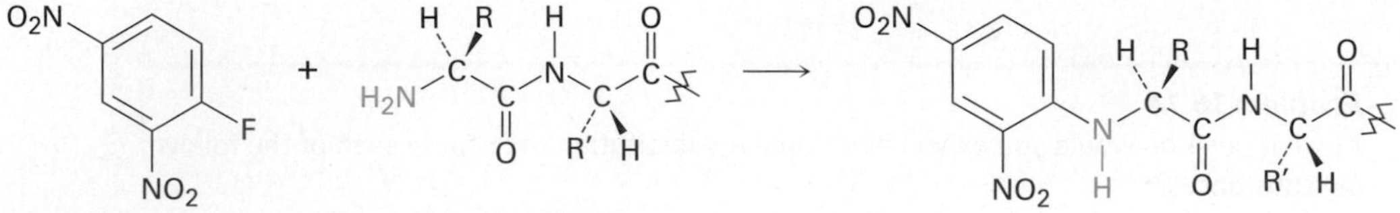

2,4-Dinitrofluorobenzene + **A protein** → **A labeled protein**

Although the reaction appears superficially similar to the S_N1 and S_N2 nucleophilic substitutions of alkyl halides discussed in Chapter 11, it must be different because aryl halides are inert to both S_N1 and S_N2 conditions. S_N1 reactions don't occur with aryl halides because dissociation of the halide is energetically unfavorable due to the instability of the potential aryl cation product. S_N2 reactions don't occur with aryl halides because the halo-substituted carbon of the aromatic ring is sterically shielded from backside approach. For a nucleophile to react with an aryl halide, it would have to approach directly through the aromatic ring and invert the stereochemistry of the aromatic ring carbon—a geometric impossibility.

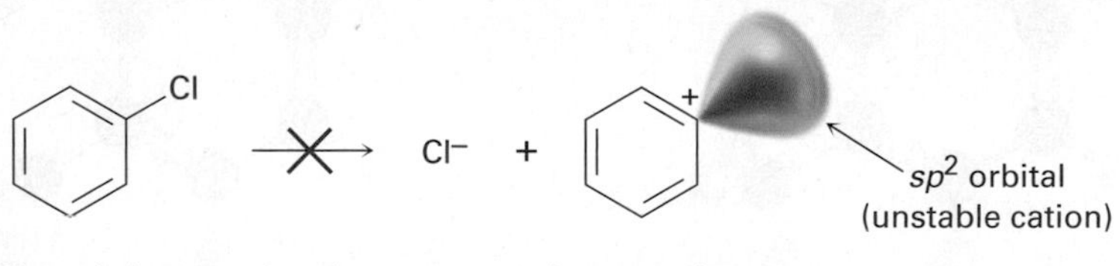

Cl ⟶✕ Cl^- + (aryl cation) — *sp*[2] orbital (unstable cation)

Dissociation reaction does not occur because the aryl cation is unstable; therefore, no S_N1 reaction.

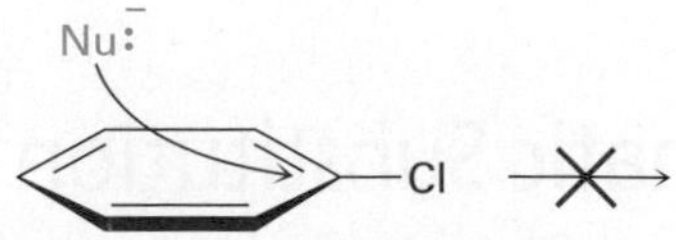

Nu:⁻ — Cl ⟶✕

Backside displacement is sterically blocked; therefore, no S_N2 reaction.

Nucleophilic substitutions on an aromatic ring proceed by the mechanism shown in **Figure 16.17**. The nucleophile first adds to the electron-deficient aryl halide, forming a resonance-stabilized, negatively charged intermediate called a Meisenheimer complex after its discoverer. Halide ion is then eliminated.

Figure 16.17 | MECHANISM

Mechanism of nucleophilic aromatic substitution. The reaction occurs in two steps and involves a resonance-stabilized carbanion intermediate.

1 Nucleophilic addition of hydroxide ion to the electron-poor aromatic ring takes place, yielding a stabilized carbanion intermediate.

2 The carbanion intermediate undergoes elimination of chloride ion in a second step to give the substitution product.

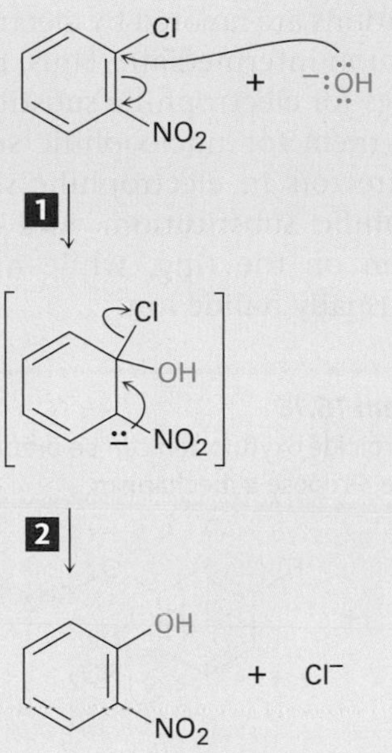

Nucleophilic aromatic substitution occurs only if the aromatic ring has an electron-withdrawing substituent in a position ortho or para to the leaving group to stabilize the anion intermediate through resonance **(Figure 16.18)**. A meta substituent offers no such resonance stabilization. Thus, *p*-chloronitrobenzene and *o*-chloronitrobenzene react with hydroxide ion at 130 °C to yield substitution products, but *m*-chloronitrobenzene is inert to OH^-.

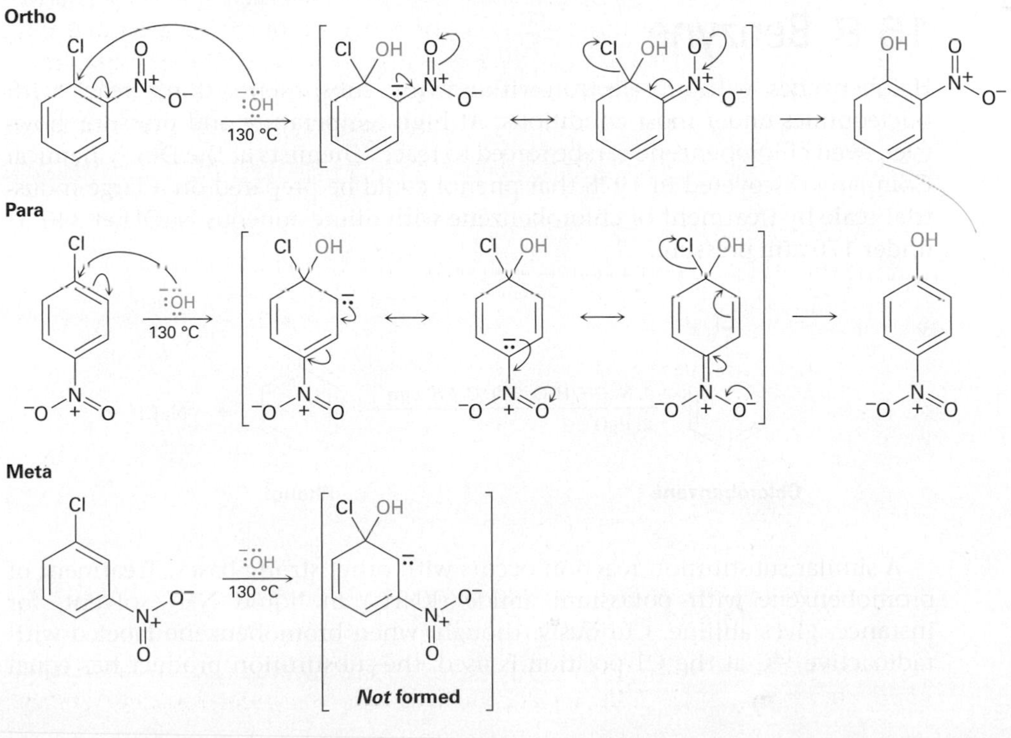

Ortho

Para

Meta

***Not* formed**

Figure 16.18 Nucleophilic aromatic substitution on nitrochlorobenzenes. Only in the ortho and para intermediates is the negative charge stabilized by a resonance interaction with the nitro group, so only the ortho and para isomers undergo reaction.

Note the differences between electrophilic and nucleophilic aromatic substitutions. Electrophilic substitutions are favored by electron-*donating* substituents, which stabilize a carbocation intermediate, while nucleophilic substitutions are favored by electron-*withdrawing* substituents, which stabilize a carbanion intermediate. Thus, the electron-withdrawing groups that *deactivate* rings for electrophilic substitution (nitro, carbonyl, cyano, and so forth) *activate* them for nucleophilic substitution. What's more, these groups are meta directors in electrophilic substitution but are ortho–para directors in nucleophilic substitution. And finally, electrophilic substitutions replace hydrogen on the ring, while nucleophilic substitutions replace a leaving group, usually halide ion.

Problem 16.16

The herbicide oxyfluorfen can be prepared by reaction between a phenol and an aryl fluoride. Propose a mechanism.

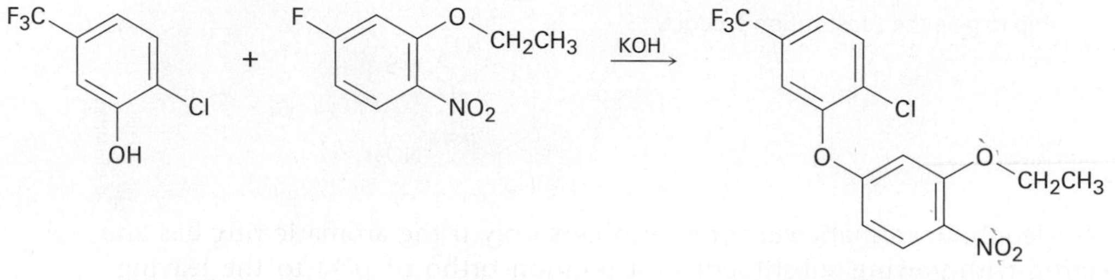

F_3C Cl OH + F O CH_2CH_3 NO_2 $\xrightarrow{KOH}$ F_3C Cl O O CH_2CH_3 NO_2

Oxyfluorfen

16.8 Benzyne

Halobenzenes without electron-withdrawing substituents don't react with nucleophiles under most conditions. At high temperature and pressure, however, even chlorobenzene can be forced to react. Chemists at the Dow Chemical Company discovered in 1928 that phenol could be prepared on a large industrial scale by treatment of chlorobenzene with dilute aqueous NaOH at 340 °C under 170 atm pressure.

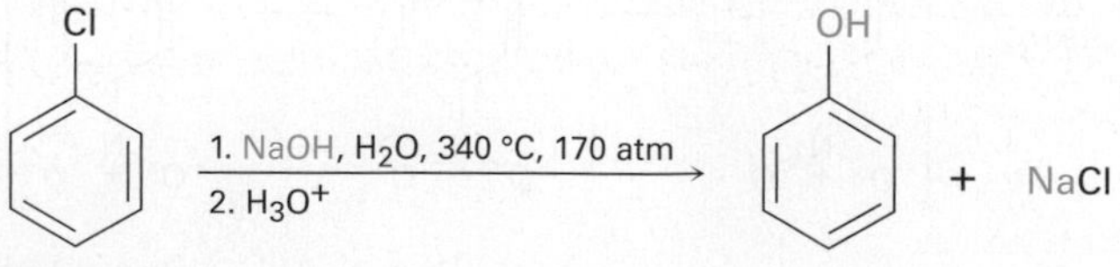

Cl $\xrightarrow[2.\ H_3O^+]{1.\ NaOH,\ H_2O,\ 340\ °C,\ 170\ atm}$ OH + NaCl

Chlorobenzene **Phenol**

A similar substitution reaction occurs with other strong bases. Treatment of bromobenzene with potassium amide (KNH_2) in liquid NH_3 solvent, for instance, gives aniline. Curiously, though, when bromobenzene labeled with radioactive ^{14}C at the C1 position is used, the substitution product has equal amounts of the label at both C1 and C2, implying the presence of a symmetrical reaction intermediate in which C1 and C2 are equivalent.

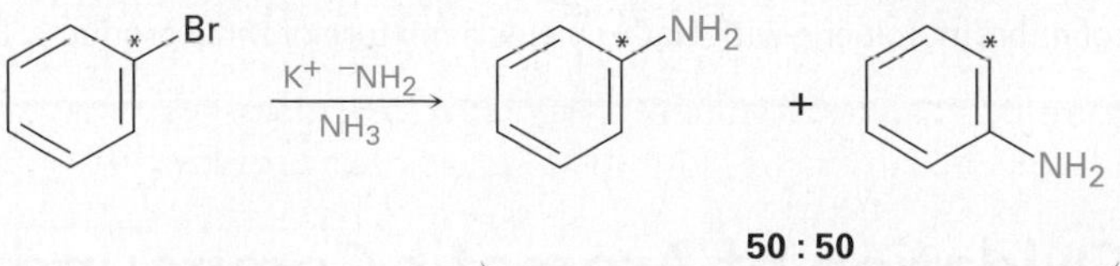

Bromobenzene $\xrightarrow[NH_3]{K^+ \ ^-NH_2}$ Aniline (50 : 50)

Further mechanistic evidence comes from trapping experiments. When bromobenzene is treated with KNH_2 in the presence of a conjugated diene, such as furan, a Diels–Alder reaction **(Section 14.4)** occurs, implying that the symmetrical intermediate is a **benzyne**, formed by elimination of HBr from bromobenzene. Benzyne is too reactive to be isolated as a pure compound but, in the presence of water, addition occurs to give the phenol. In the presence of a diene, Diels–Alder cycloaddition takes place.

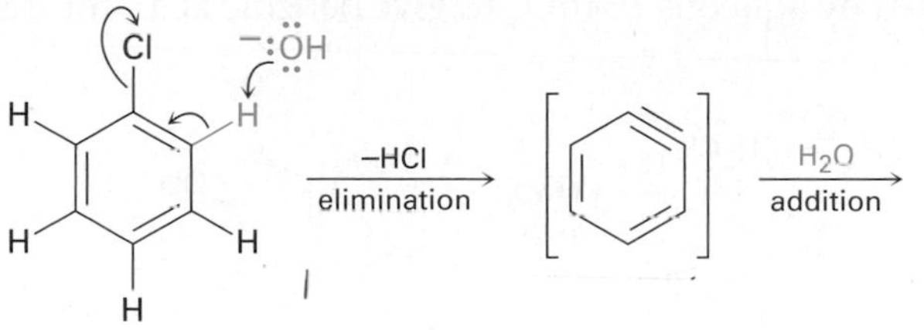

Chlorobenzene $\xrightarrow[\text{elimination}]{-HCl}$ Benzyne $\xrightarrow[\text{addition}]{H_2O}$ Phenol

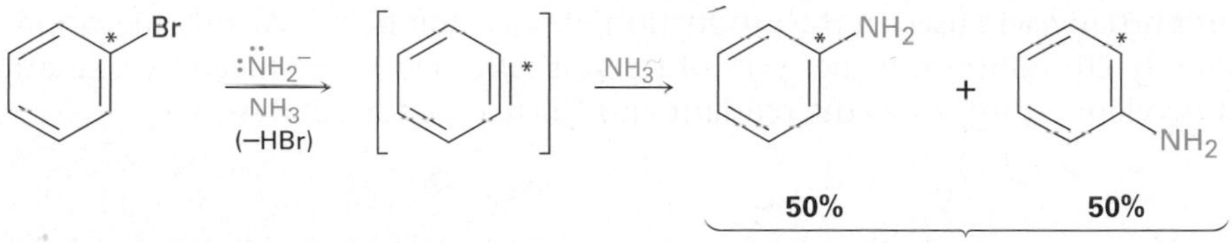

Bromobenzene $\xrightarrow[NH_3 \ (-HBr)]{:NH_2^-}$ Benzyne (symmetrical) $\xrightarrow{NH_3}$ Aniline (50% + 50%)

The electronic structure of benzyne, shown in **Figure 16.19**, is that of a highly distorted alkyne. Although a typical alkyne triple bond uses *sp*-hybridized carbon atoms, the benzyne triple bond uses sp^2-hybridized carbons. Furthermore, a typical alkyne triple bond has two mutually perpendicular π bonds formed by *p–p* overlap, but the benzyne triple bond has one π bond formed by *p–p* overlap and one π bond formed by sp^2–sp^2 overlap. The latter π bond is in the plane of the ring and is very weak.

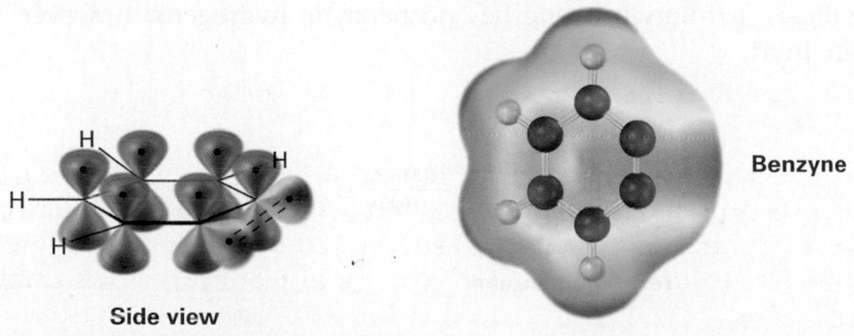

Figure 16.19 An orbital picture and electrostatic potential map of benzyne. The benzyne carbons are sp^2-hybridized, and the "third" bond results from weak overlap of two adjacent sp^2 orbitals.

Problem 16.17

Treatment of *p*-bromotoluene with NaOH at 300 °C yields a mixture of *two* products, but treatment of *m*-bromotoluene with NaOH yields a mixture of *three* products. Explain.

16.9 Oxidation of Aromatic Compounds

Oxidation of Alkyl Side Chains

Despite its unsaturation, the benzene ring is inert to strong oxidizing agents such as $KMnO_4$ and $Na_2Cr_2O_7$, reagents that will cleave alkene carbon–carbon bonds **(Section 8.8)**. It turns out, however, that the presence of the aromatic ring has a dramatic effect on alkyl side chains. Alkyl side chains react rapidly with oxidizing agents and are converted into carboxyl groups, $-CO_2H$. The net effect is conversion of an alkylbenzene into a benzoic acid, Ar–R → Ar–CO_2H. Butylbenzene is oxidized by aqueous $KMnO_4$ to give benzoic acid, for instance.

$CH_2CH_2CH_2CH_3$ — $\xrightarrow[H_2O]{KMnO_4}$ — O, C, OH

Butylbenzene **Benzoic acid (85%)**

A similar oxidation is employed industrially for the preparation of the terephthalic acid used in the production of polyester fibers. Worldwide, approximately 20 million tons per year of terephthalic acid is produced by oxidation of *p*-xylene, using air as the oxidant and Co(III) salts as catalyst.

CH_3, CH_3 — $\xrightarrow[Co(III)]{O_2}$ — O, C, OH; O, C, OH

***p*-Xylene** **Terephthalic acid**

The mechanism of side-chain oxidation is complex and involves reaction of C–H bonds at the position next to the aromatic ring to form intermediate benzylic radicals. *tert*-Butylbenzene has no benzylic hydrogens, however, and is therefore inert.

H_3C, CH_3, C, CH_3 — $\xrightarrow[H_2O]{KMnO_4}$ — ***No reaction***

***tert*-Butylbenzene**

Analogous side-chain oxidations occur in various biosynthetic pathways. The neurotransmitter norepinephrine, for instance, is biosynthesized from dopamine by a benzylic hydroxylation reaction. The process is catalyzed by the copper-containing enzyme dopamine β-monooxygenase and occurs by a radical mechanism. A copper–oxygen species in the enzyme first abstracts the *pro*-R benzylic hydrogen to give a radical, and a hydroxyl is then transferred from copper to carbon.

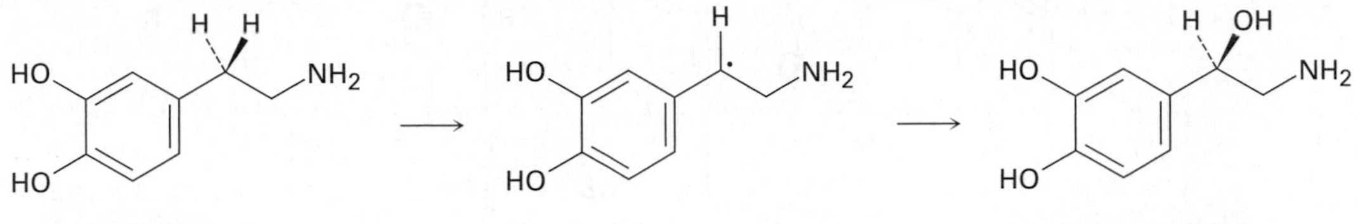

Dopamine → → Norepinephrine

Problem 16.18

What aromatic products would you obtain from the $KMnO_4$ oxidation of the following substances?

(a) 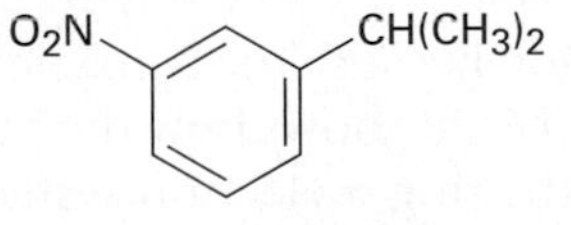 O_2N–C_6H_4–$CH(CH_3)_2$ **(b)** 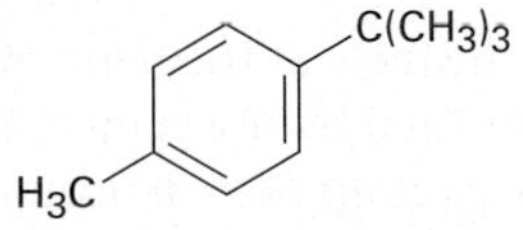 H_3C–C_6H_4–$C(CH_3)_3$

Bromination of Alkylbenzene Side Chains

Side-chain bromination at the benzylic position occurs when an alkylbenzene is treated with *N*-bromosuccinimide (NBS). For example, propylbenzene gives (1-bromopropyl)benzene in 97% yield on reaction with NBS in the presence of benzoyl peroxide, $(PhCO_2)_2$, as a radical initiator. Bromination occurs exclusively in the benzylic position next to the aromatic ring and does not give a mixture of products.

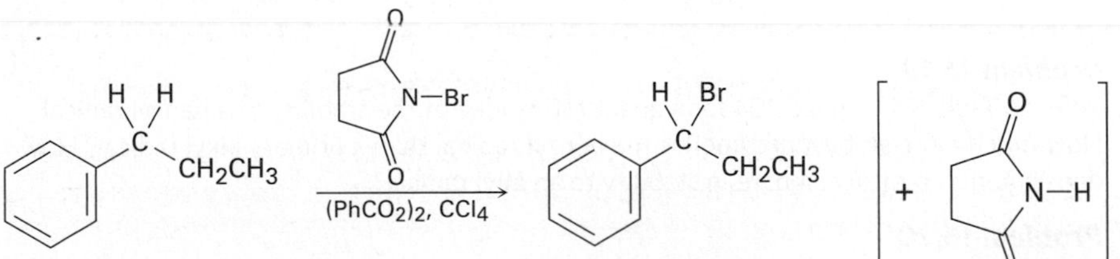

$$\text{Propylbenzene} \xrightarrow[(PhCO_2)_2,\ CCl_4]{\text{NBS}} \text{(1-Bromopropyl)benzene (97\%)} + \text{succinimide}$$

Propylbenzene **(1-Bromopropyl)benzene (97%)**

The mechanism of benzylic bromination is similar to that discussed in **Section 10.3** for allylic bromination of alkenes. Abstraction of a benzylic hydrogen atom first generates an intermediate benzylic radical, which then reacts with Br_2 in step 2 to yield product and a Br· radical that cycles back into the reaction to carry on the chain. The Br_2 necessary for reaction with the benzylic radical is produced in step 3 by a concurrent reaction of HBr with NBS.

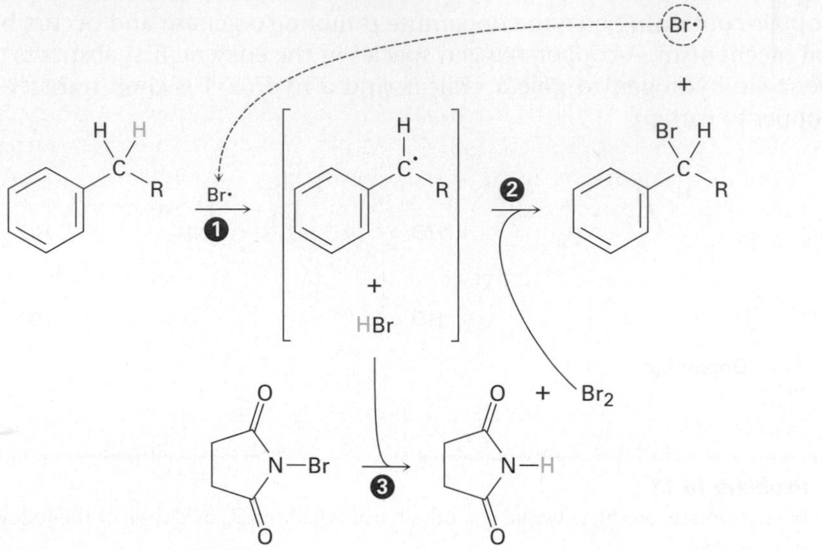

Reaction occurs exclusively at the benzylic position because the benzylic radical intermediate is stabilized by resonance. **Figure 16.20** shows how the benzyl radical is stabilized by overlap of its *p* orbital with the ring π electron system.

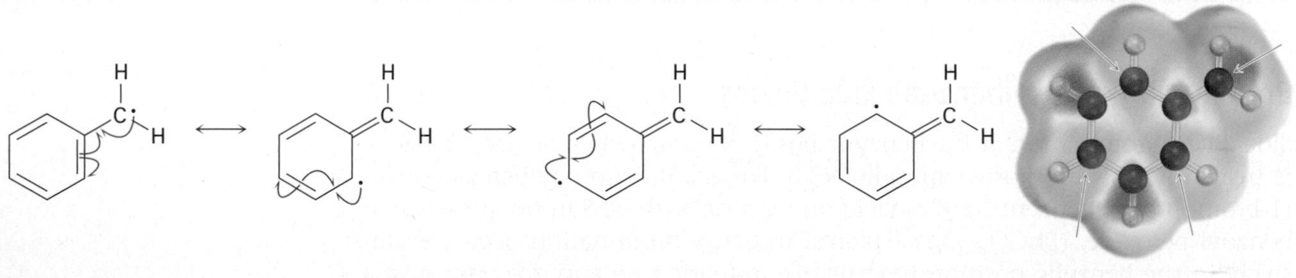

Figure 16.20 A resonance-stabilized benzylic radical. The spin-density surface shows that the unpaired electron is shared by the ortho and para carbons of the ring.

Problem 16.19

Refer to Table 6.3 on page 204 for a quantitative idea of the stability of a benzyl radical. How much more stable (in kJ/mol) is the benzyl radical than a primary alkyl radical? How does a benzyl radical compare in stability to an allyl radical?

Problem 16.20

Styrene, the simplest alkenylbenzene, is prepared commercially for use in plastics manufacture by catalytic dehydrogenation of ethylbenzene. How might you prepare styrene from benzene using reactions you've studied?

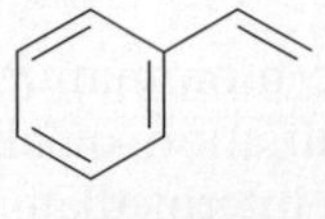

Styrene

16.10 Reduction of Aromatic Compounds

Catalytic Hydrogenation of Aromatic Rings

Just as aromatic rings are generally inert to oxidation, they're also inert to catalytic hydrogenation under conditions that reduce typical alkene double bonds. As a result, it's possible to reduce an alkene double bond selectively in the presence of an aromatic ring. For example, 4-phenyl-3-buten-2-one is reduced to 4-phenyl-2-butanone using a palladium catalyst at room temperature and atmospheric pressure. Neither the benzene ring nor the ketone carbonyl group is affected.

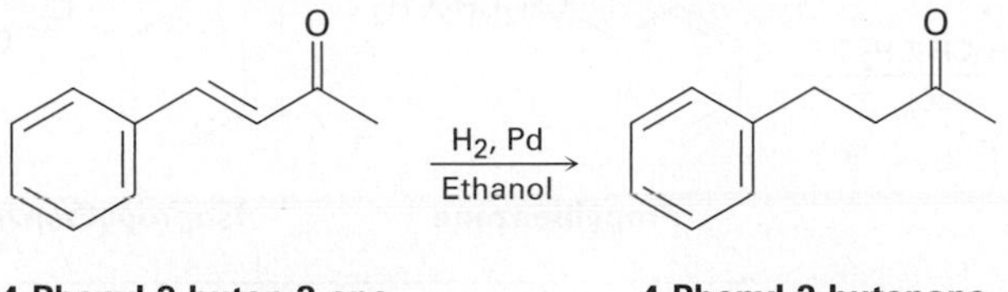

O O

H_2, Pd

Ethanol

4-Phenyl-3-buten-2-one **4-Phenyl-2-butanone (100%)**

To hydrogenate an aromatic ring, it's necessary either to use a platinum catalyst with hydrogen gas at several hundred atmospheres pressure or to use a more effective catalyst such as rhodium on carbon. Under these conditions, aromatic rings are converted into cyclohexanes. For example, *o*-xylene yields 1,2-dimethylcyclohexane, and 4-*tert*-butylphenol gives 4-*tert*-butylcyclohexanol.

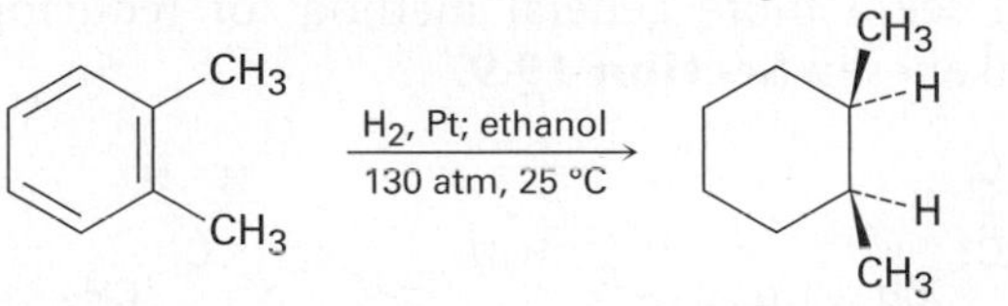

CH_3 CH_3 CH_3 CH_3 H H

H_2, Pt; ethanol

130 atm, 25 °C

***o*-Xylene** ***cis*-1,2-Dimethylcyclohexane**

H_3C CH_3 C CH_3 HO

H_2, Rh/C; ethanol

1 atm, 25 °C

H_3C CH_3 C CH_3 H HO H

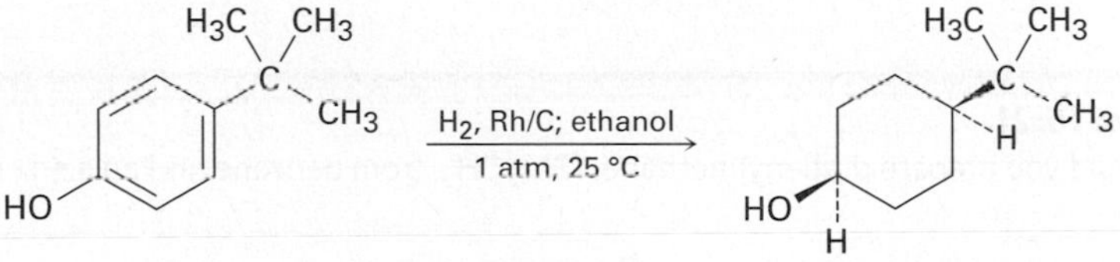

4-*tert*-Butylphenol ***cis*-4-*tert*-Butylcyclohexane**

Reduction of Aryl Alkyl Ketones

In the same way that an aromatic ring activates a neighboring (benzylic) C–H toward oxidation, it also activates a benzylic carbonyl group toward reduction. Thus, an aryl alkyl ketone prepared by Friedel–Crafts acylation of an aromatic ring can be converted into an alkylbenzene by catalytic hydrogenation over a palladium catalyst. Propiophenone, for instance, is reduced to propylbenzene by catalytic hydrogenation. Since the net effect of Friedel–Crafts acylation followed by reduction is the preparation of a primary alkylbenzene, this two-step sequence of reactions makes it possible to circumvent the carbocation rearrangement problems associated with direct Friedel–Crafts alkylation using a primary alkyl halide (**Section 16.3**).

CH_3CH_2CCl (O), $AlCl_3$ → Propiophenone (95%) → H_2/Pd → Propylbenzene (100%)

$CH_3CH_2CH_2Cl$, $AlCl_3$ → Propylbenzene + Isopropylbenzene

Propiophenone (95%) **Propylbenzene (100%)**

Propylbenzene **Isopropylbenzene**

Mixture of two products

The conversion of a carbonyl group into a methylene group ($C{=}O \rightarrow CH_2$) by catalytic hydrogenation is limited to aryl alkyl ketones; dialkyl ketones are not reduced under these conditions. Furthermore, the catalytic reduction of aryl alkyl ketones is not compatible with the presence of a nitro substituent on the aromatic ring because a nitro group is reduced to an amino group under the reaction conditions. We'll see a more general method for reducing ketone carbonyl groups to yield alkanes in **Section 19.9**.

m-Nitroacetophenone → H_2, Pd/C, Ethanol → *m*-Ethylaniline

Problem 16.21

How would you prepare diphenylmethane, $(Ph)_2CH_2$, from benzene and an acid chloride?

16.11 Synthesis of Polysubstituted Benzenes

One of the surest ways to learn organic chemistry is to work synthesis problems. The ability to plan a successful multistep synthesis of a complex molecule requires a working knowledge of the uses and limitations of a great many organic reactions. Not only must you know *which* reactions to use, you must also know *when* to use them because the order in which reactions are carried out is often critical to the success of the overall scheme.

The ability to plan a sequence of reactions in the right order is particularly important in the synthesis of substituted aromatic rings, where the introduction of a new substituent is strongly affected by the directing effects of other substituents. Planning syntheses of substituted aromatic compounds is therefore a good way to gain confidence in using the many reactions learned in the past few chapters.

During our previous discussion of strategies for working synthesis problems in **Section 9.9**, we said that it's usually best to work a problem backward, or *retrosynthetically*. Look at the target molecule and ask yourself, "What is an immediate precursor of this compound?" Choose a likely answer and continue working backward, one step at a time, until you arrive at a simple starting material. Let's try some examples.

Synthesizing a Polysubstituted Benzene

Worked Example 16.4

Synthesize 4-bromo-2-nitrotoluene from benzene.

Strategy

Draw the target molecule, identify the substituents, and recall how each group can be introduced separately. Then plan retrosynthetically.

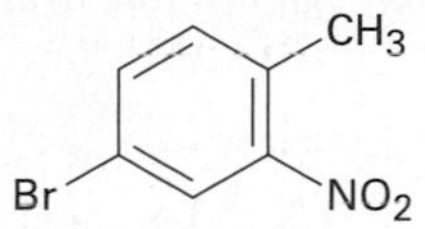

CH_3 Br NO_2

4-Bromo-2-nitrotoluene

The three substituents on the ring are a bromine, a methyl group, and a nitro group. A bromine can be introduced by bromination with $Br_2/FeBr_3$, a methyl group can be introduced by Friedel–Crafts alkylation with $CH_3Cl/AlCl_3$, and a nitro group can be introduced by nitration with HNO_3/H_2SO_4.

Solution

Ask yourself, "What is an immediate precursor of the target?" The final step will involve introduction of one of three groups—bromine, methyl, or nitro—so we have to consider three possibilities. Of the three, the bromination of *o*-nitrotoluene could be used because the activating methyl group would dominate the deactivating nitro group and direct bromination to the right position. Unfortunately, a mixture of product isomers would be formed. A Friedel–Crafts reaction can't be used as the final step because this reaction doesn't work on a nitro-substituted (strongly deactivated) benzene. The best precursor of the desired product is probably *p*-bromotoluene, which can be nitrated ortho to the activating methyl group to give a single product.

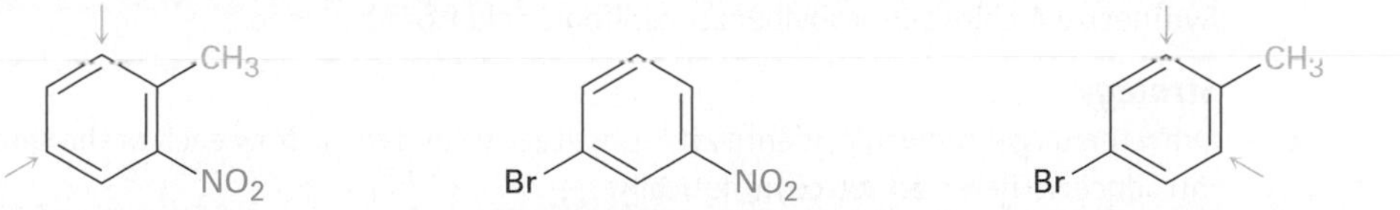

CH_3 NO_2

***o*-Nitrotoluene**

This ring will give a mixture of isomers on bromination.

Br NO_2

***m*-Bromonitrobenzene**

This deactivated ring will not undergo a Friedel–Crafts reaction.

CH_3 Br

***p*-Bromotoluene**

This ring will give only the desired isomer on nitration.

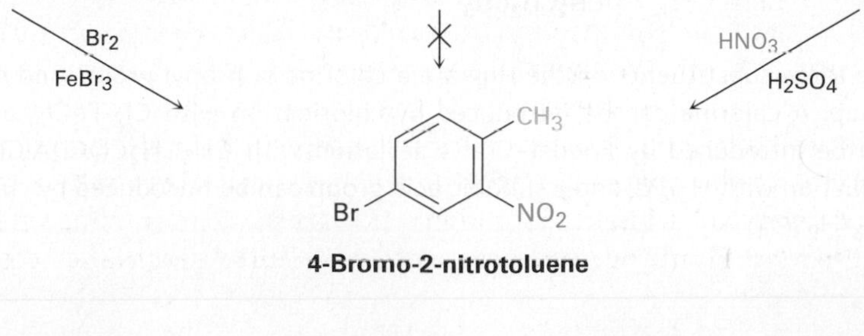

Br_2 / $FeBr_3$

HNO_3 / H_2SO_4

CH_3 Br NO_2

4-Bromo-2-nitrotoluene

Next ask, "What is an immediate precursor of *p*-bromotoluene?" Perhaps toluene is an immediate precursor because the methyl group would direct bromination to the ortho and para positions. Alternatively, bromobenzene might be an immediate precursor because we could carry out a Friedel–Crafts methylation and obtain a mixture of ortho and para products. Both answers are satisfactory, although both would also lead unavoidably to a product mixture that would have to be separated.

CH_3 Br_2 $FeBr_3$ CH_3 CH_3Cl $AlCl_3$ Br Br

Toluene ***p*-Bromotoluene (+ ortho isomer)** **Bromobenzene**

"What is an immediate precursor of toluene?" Benzene, which could be methylated in a Friedel–Crafts reaction. Alternatively, "What is an immediate precursor of bromobenzene?" Benzene, which could be brominated.

The retrosynthetic analysis has provided two valid routes from benzene to 4-bromo-2-nitrotoluene.

CH_3Cl $AlCl_3$ CH_3 Br_2 $FeBr_3$ **Toluene** CH_3 HNO_3 H_2SO_4 CH_3 Br Br NO_2

Benzene Br_2 $FeBr_3$ CH_3Cl $AlCl_3$ ***p*-Bromotoluene** **4-Bromo-2-nitrotoluene** Br

Bromobenzene

Worked Example 16.5 Synthesizing a Polysubstituted Benzene

Synthesize 4-chloro-2-propylbenzenesulfonic acid from benzene.

Strategy

Draw the target molecule, identify its substituents, and recall how each of the three can be introduced. Then plan retrosynthetically.

SO_3H **4-Chloro-2-propylbenzenesulfonic acid** Cl $CH_2CH_2CH_3$

The three substituents on the ring are a chlorine, a propyl group, and a sulfonic acid group. A chlorine can be introduced by chlorination with $Cl_2/FeCl_3$, a propyl group can be introduced by Friedel–Crafts acylation with $CH_3CH_2COCl/AlCl_3$ followed by reduction with H_2/Pd, and a sulfonic acid group can be introduced by sulfonation with SO_3/H_2SO_4.

Solution

"What is an immediate precursor of the target?" The final step will involve introduction of one of three groups—chlorine, propyl, or sulfonic acid—so we have to consider three possibilities. Of the three, the chlorination of *o*-propylbenzenesulfonic acid can't be used because the reaction would occur at the wrong position. Similarly, a Friedel–Crafts reaction can't be used as the final step because this reaction doesn't work on sulfonic acid-substituted (strongly deactivated) benzenes. Thus, the immediate precursor of the desired product is probably *m*-chloropropylbenzene, which can be sulfonated to give a mixture of product isomers that must then be separated.

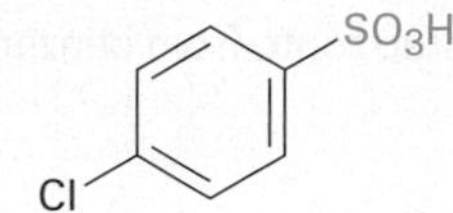

o-Propylbenzene-sulfonic acid
This ring will give the wrong isomer on chlorination.

p-Chlorobenzene-sulfonic acid
This deactivated ring will not undergo a Friedel–Crafts reaction.

***m*-Chloropropylbenzene**
This ring will give the desired product on sulfonation.

SO_3, H_2SO_4

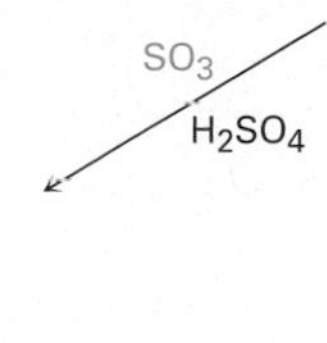

4-Chloro-2-propylbenzenesulfonic acid

"What is an immediate precursor of *m*-chloropropylbenzene?" Because the two substituents have a meta relationship, the first substituent placed on the ring must be a meta director so that the second substitution will take place at the proper position. Furthermore, because primary alkyl groups such as propyl can't be introduced directly by Friedel–Crafts alkylation, the precursor of *m*-chloropropylbenzene is probably *m*-chloropropiophenone, which could be catalytically reduced.

***m*-Chloropropiophenone** $\xrightarrow[\text{Pd, C}]{H_2}$ ***m*-Chloropropylbenzene**

"What is an immediate precursor of *m*-chloropropiophenone?" Propiophenone, which could be chlorinated in the meta position.

Propiophenone $\xrightarrow[FeCl_3]{Cl_2}$ ***m*-Chloropropiophenone**

"What is an immediate precursor of propiophenone?" Benzene, which could undergo Friedel–Crafts acylation with propanoyl chloride and $AlCl_3$.

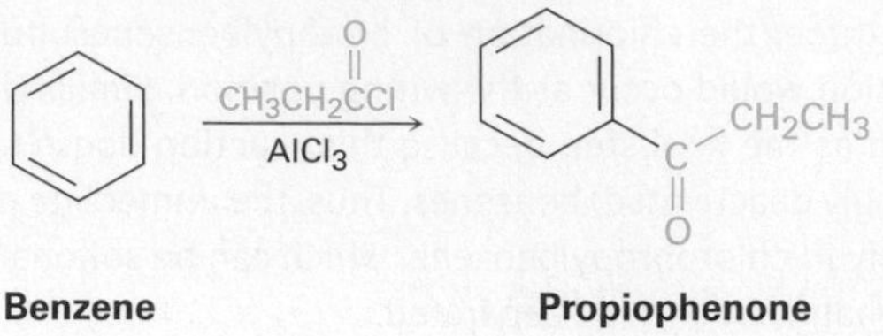

CH_3CH_2CCl (O), $AlCl_3$

Benzene → **Propiophenone** (CH_2CH_3, C=O)

The final synthesis is a four-step route from benzene:

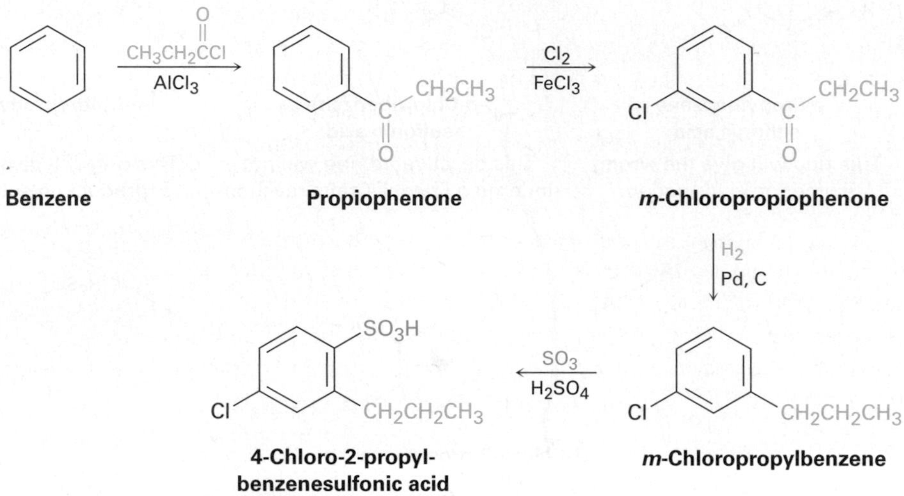

Benzene —(CH_3CH_2CCl (O), $AlCl_3$)→ **Propiophenone** —(Cl_2, $FeCl_3$)→ **m-Chloropropiophenone** —(H_2, Pd, C)→ **m-Chloropropylbenzene** —(SO_3, H_2SO_4)→ **4-Chloro-2-propylbenzenesulfonic acid**

Cl, CH_2CH_3, C, O, SO_3H, $CH_2CH_2CH_3$

Planning an organic synthesis has been compared with playing chess. There are no tricks; all that's required is a knowledge of the allowable moves (the organic reactions) and the discipline to plan ahead, carefully evaluating the consequences of each move. Practicing may not be easy, but it's a great way to learn organic chemistry.

Problem 16.22

How might you synthesize the following substances from benzene?

(a) *m*-Chloronitrobenzene
(b) *m*-Chloroethylbenzene
(c) 4-Chloro-1-nitro-2-propylbenzene
(d) 3-Bromo-2-methylbenzenesulfonic acid

Problem 16.23

In planning a synthesis, it's as important to know what not to do as to know what to do. As written, the following reaction schemes have flaws in them. What is wrong with each?

(a) CN (benzonitrile) —(1. CH_3CH_2COCl, $AlCl_3$; 2. HNO_3, H_2SO_4)→ CN, O_2N, CH_2CH_3, C=O

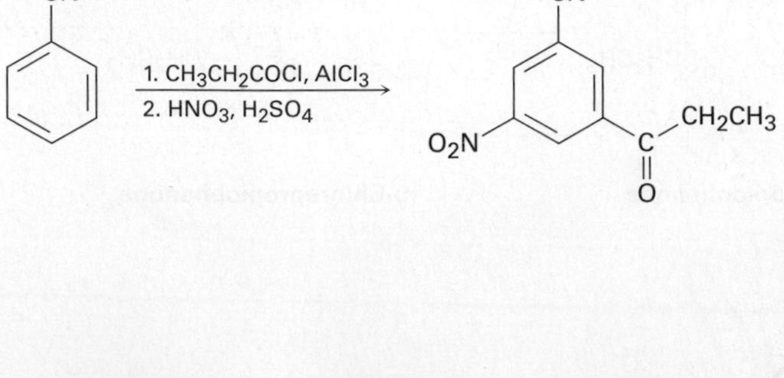

(b) Chlorobenzene $\xrightarrow[\text{2. } Cl_2,\ FeCl_3]{\text{1. } CH_3CH_2CH_2Cl,\ AlCl_3}$ 1,2-dichloro-4-propylbenzene ($CH_3CH_2CH_2$, Cl, Cl)

A DEEPER LOOK Combinatorial Chemistry

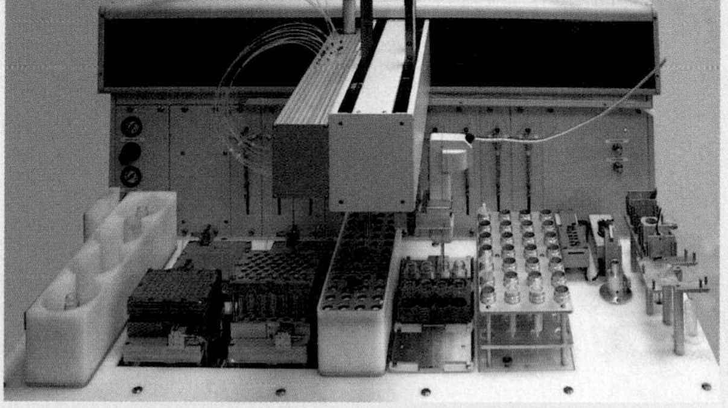

Organic chemistry by robot means no spilled flasks!

Traditionally, organic compounds have been synthesized one at a time. This works well for preparing large amounts of a few substances, but it doesn't work so well for preparing small amounts of a great many substances. This latter goal is particularly important in the pharmaceutical industry, where vast numbers of structurally similar compounds must be screened to find the optimum drug candidate.

To speed the process of drug discovery, *combinatorial chemistry* has been developed to prepare what are called *combinatorial libraries,* in which anywhere from a few dozen to several hundred thousand substances are prepared simultaneously. Among the early successes of combinatorial chemistry is the development of a benzodiazepine library, a class of aromatic compounds much used as antianxiety agents.

(Structure: benzodiazepine core with substituents R_1, R_2, R_3, R_4, N, N, O)

Benzodiazepine library (R_1–R_4 are various organic substituents)

Two main approaches to combinatorial chemistry are used—*parallel synthesis* and *split synthesis.* In parallel synthesis, each compound is prepared independently. Typically, a reactant is first linked to the surface of polymer beads, which are then placed into small wells on a 96-well glass plate. Programmable robotic instruments add different sequences of building blocks to the different wells, thereby making 96 different products. When the reaction sequences are complete, the polymer beads are washed and their products are released.

In split synthesis, the initial reactant is again linked to the surface of polymer beads, which are then divided into several groups. A different building block is added to each group of beads, the different groups are combined, and the reassembled mix is again split to form new groups. Another building block is added to each group, the groups are again combined and redivided, and the process continues. If, for example, the beads are divided into four groups at each step, the number of compounds increases in the progression $4 \rightarrow 16 \rightarrow 64 \rightarrow 256$. After 10 steps, more than 1 million compounds have been prepared **(Figure 16.21)**.

(continued)

(continued)

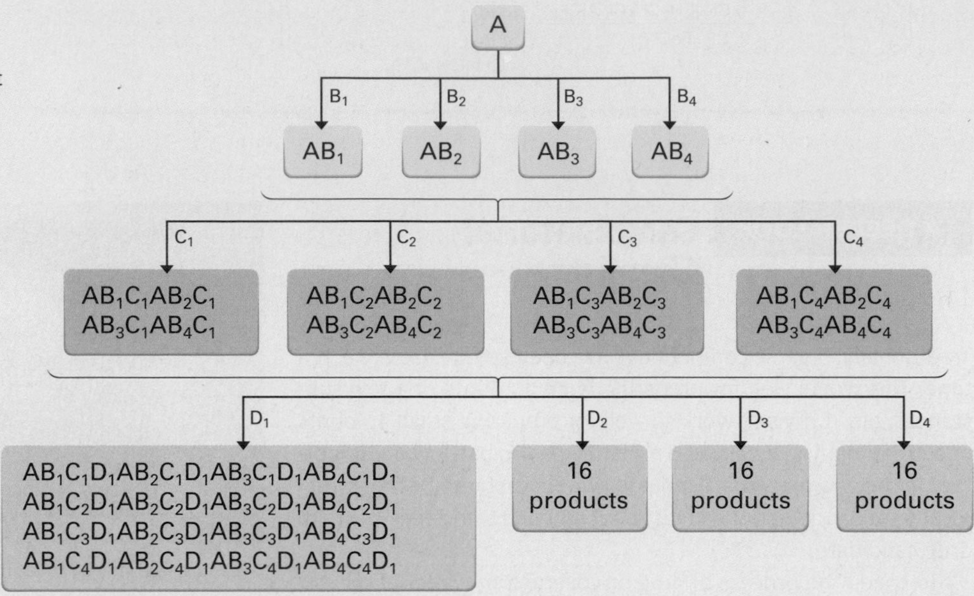

Figure 16.21 The results of split combinatorial synthesis. Assuming that 4 different building blocks are used at each step, 64 compounds result after 3 steps, and more than one million compounds result after 10 steps.

Of course, with so many different final products mixed together, the problem is to identify them. What structure is linked to what bead? Several approaches to this problem have been developed, all of which involve the attachment of encoding labels to each polymer bead to keep track of the chemistry each has undergone. Encoding labels used thus far have included proteins, nucleic acids, halogenated aromatic compounds, and even computer chips.

Summary

Key words

acyl group, 577
acylation, 577
alkylation, 575
benzyne, 595
electrophilic aromatic substitution, 566
Friedel–Crafts reaction, 575
inductive effect, 583
nucleophilic aromatic substitution, 592
resonance effect, 584

We've continued the coverage of aromatic molecules in this chapter, shifting focus to concentrate on reactions. In particular, we've looked at the relationship between aromatic structure and reactivity, a relationship critical to understanding how numerous biological molecules and pharmaceutical agents are synthesized and why they behave as they do.

An **electrophilic aromatic substitution reaction** takes place in two steps—initial reaction of an electrophile, E^+, with the aromatic ring, followed by loss of H^+ from the resonance-stabilized carbocation intermediate to regenerate the aromatic ring.

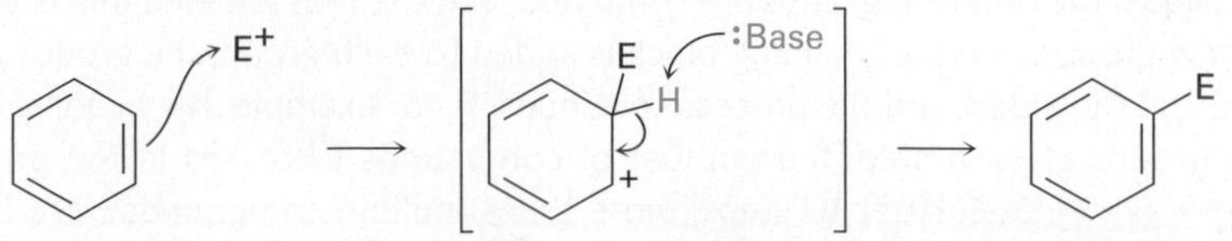

Many variations of the reaction can be carried out, including halogenation, nitration, and sulfonation. **Friedel–Crafts alkylation** and **acylation** reactions, which involve reaction of an aromatic ring with carbocation electrophiles, are particularly useful. They are limited, however, by the fact that the aromatic ring must be at least as reactive as a halobenzene. In addition, polyalkylation and carbocation rearrangements often occur in Friedel–Crafts alkylation.

Substituents on the benzene ring affect both the reactivity of the ring toward further substitution and the orientation of that substitution. Groups can be classified as ortho- and para-directing activators, ortho- and para-directing deactivators, or meta-directing deactivators. Substituents influence aromatic rings by a combination of resonance and inductive effects. **Resonance effects** are transmitted through π bonds; **inductive effects** are transmitted through σ bonds.

Halobenzenes undergo **nucleophilic aromatic substitution** through either of two mechanisms. If the halobenzene has a strongly electron-withdrawing substituent in the ortho or para position, substitution occurs by addition of a nucleophile to the ring, followed by elimination of halide from the intermediate anion. If the halobenzene is not activated by an electron-withdrawing substituent, substitution can occur by elimination of HX to give a **benzyne**, followed by addition of a nucleophile.

The benzylic position of an alkylbenzene can be brominated by reaction with *N*-bromosuccinimide, and the entire side chain can be degraded to a carboxyl group by oxidation with aqueous $KMnO_4$. Aromatic rings can also be reduced to cyclohexanes by hydrogenation over a platinum or rhodium catalyst, and aryl alkyl ketones are reduced to alkylbenzenes by hydrogenation over a platinum catalyst.

Summary of Reactions

1. Electrophilic aromatic substitution
   (a) Fluorination (Section 16.2)

   $2\ BF_4^-$; $F-\overset{+}{N}$ ... $\overset{+}{N}-CH_2Cl$ (F-TEDA-BF_4): benzene → fluorobenzene

   (b) Bromination (Section 16.1)

   benzene $+ Br_2 \xrightarrow{FeBr_3}$ bromobenzene $+ HBr$

   (c) Chlorination (Section 16.2)

   benzene $\xrightarrow{Cl_2,\ FeCl_3}$ chlorobenzene $+ HCl$

(continued)

(d) Iodination (Section 16.2)

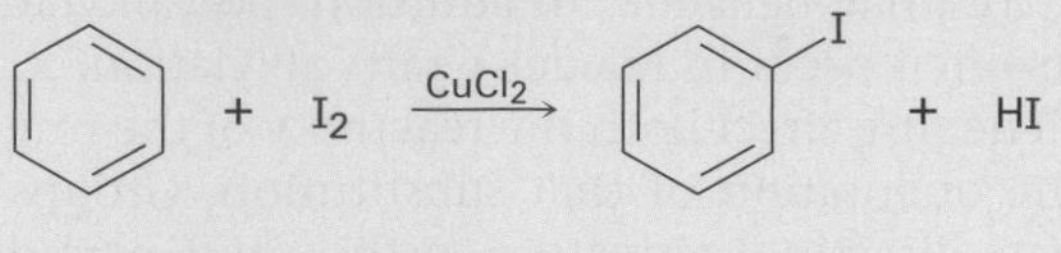

$$C_6H_6 + I_2 \xrightarrow{CuCl_2} C_6H_5I + HI$$

(e) Nitration (Section 16.2)

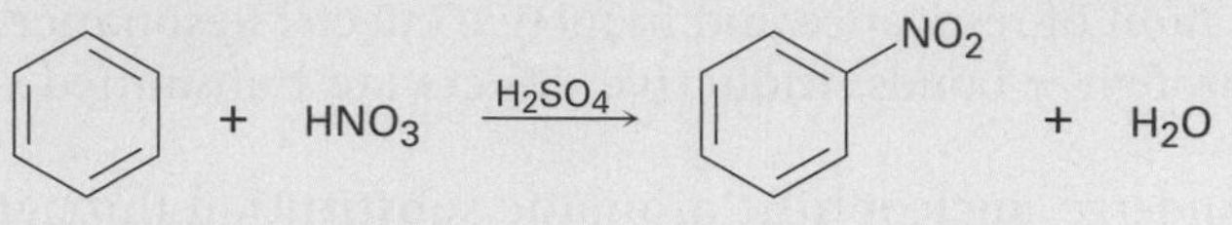

$$C_6H_6 + HNO_3 \xrightarrow{H_2SO_4} C_6H_5NO_2 + H_2O$$

(f) Sulfonation (Section 16.2)

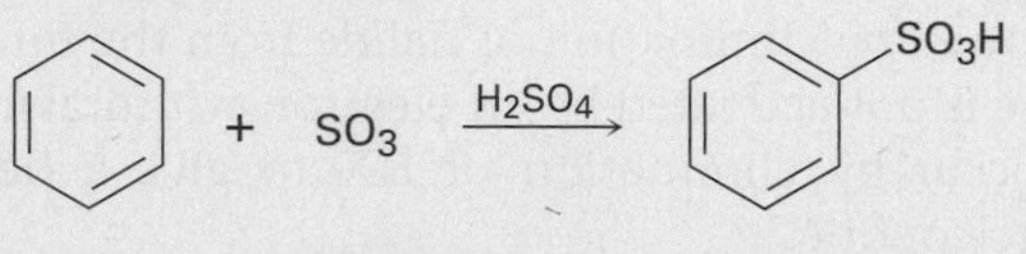

$$C_6H_6 + SO_3 \xrightarrow{H_2SO_4} C_6H_5SO_3H$$

(g) Friedel–Crafts alkylation (Section 16.3)

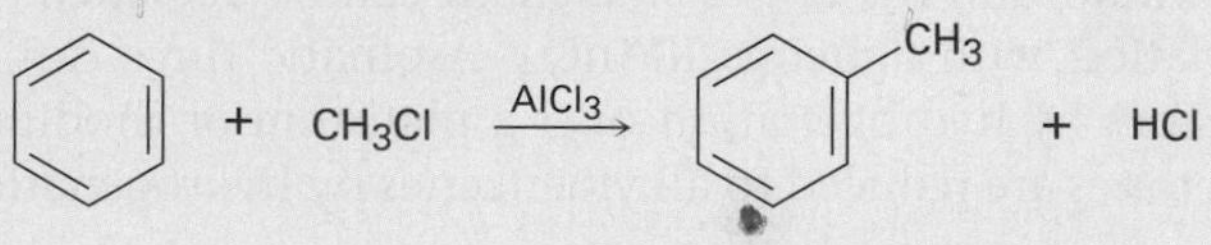

$$C_6H_6 + CH_3Cl \xrightarrow{AlCl_3} C_6H_5CH_3 + HCl$$

Aromatic ring. Must be at least as reactive as a halobenzene.
Alkyl halide. Primary alkyl halides undergo carbocation rearrangement.

(h) Friedel–Crafts acylation (Section 16.3)

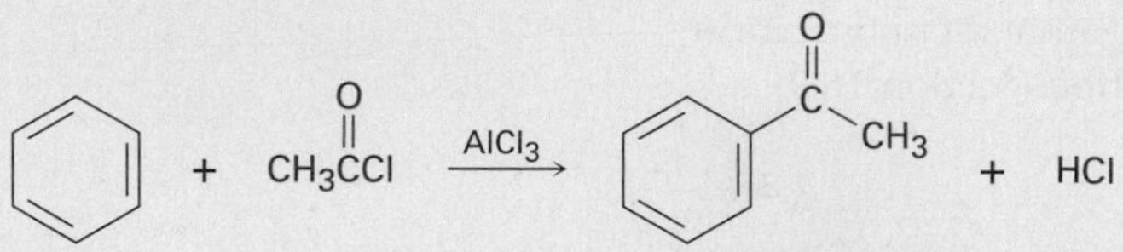

$$C_6H_6 + CH_3COCl \xrightarrow{AlCl_3} C_6H_5COCH_3 + HCl$$

2. Reduction of aromatic nitro groups (Section 16.2)

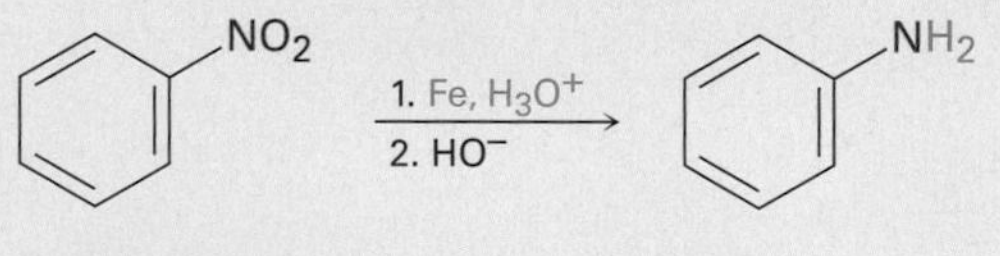

$$C_6H_5NO_2 \xrightarrow[2.\ HO^-]{1.\ Fe,\ H_3O^+} C_6H_5NH_2$$

3. Nucleophilic aromatic substitution
(a) By addition to activated aryl halides (Section 16.7)

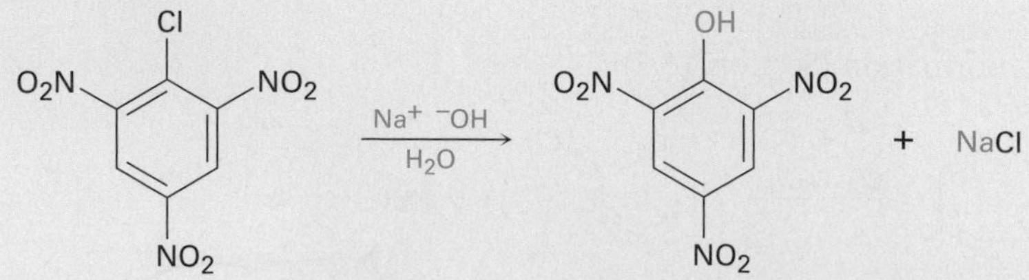

$$\text{2,4,6-}(O_2N)_3C_6H_2Cl \xrightarrow[H_2O]{Na^+\ ^-OH} \text{2,4,6-}(O_2N)_3C_6H_2OH + NaCl$$

(b) By formation of benzyne intermediate from unactivated aryl halide (Section 16.8)

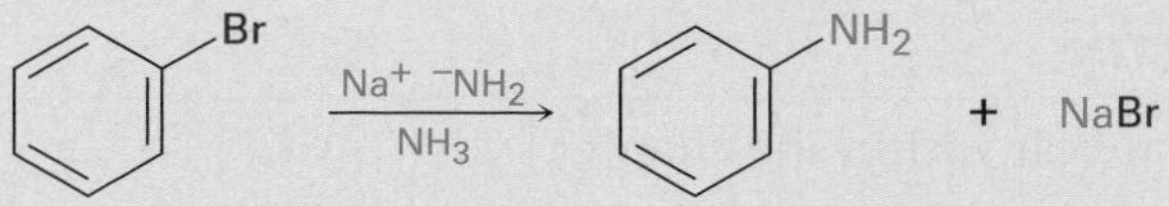

Br, $Na^+ \; ^-NH_2$ / NH_3 → NH_2 + NaBr

4. Oxidation of alkylbenzene side chain (Section 16.9)

CH_3, $KMnO_4$ / H_2O → CO_2H

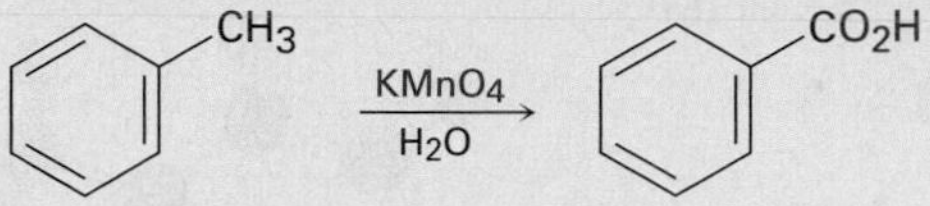

5. Benzylic bromination of alkylbenzene side chain (Section 16.9)

CH_3, O / N—Br / O, $(BzO)_2$, CCl_4 → CH_2Br

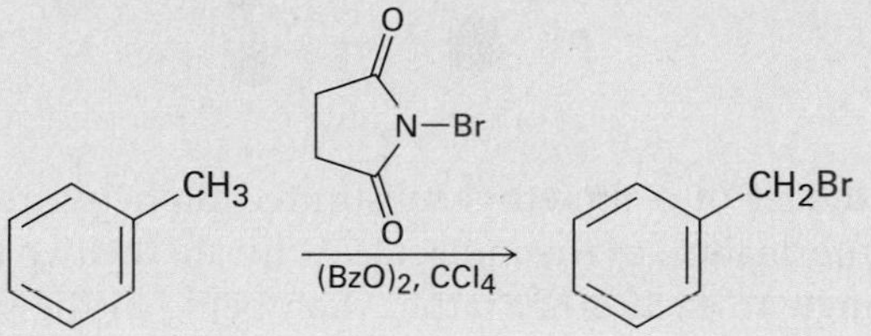

6. Catalytic hydrogenation of aromatic ring (Section 16.10)

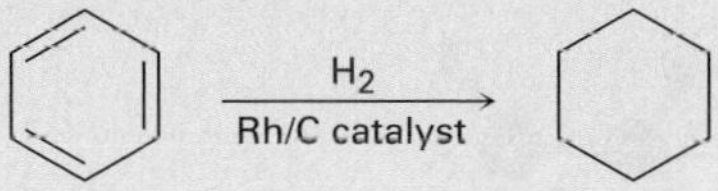

H_2 / Rh/C catalyst →

7. Reduction of aryl alkyl ketones (Section 16.10)

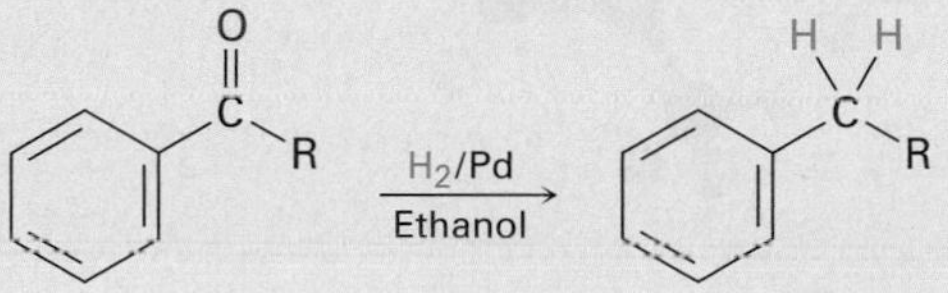

O, C, R, H_2/Pd / Ethanol → H H, C, R

Exercises

OWL Interactive versions of these problems are assignable in OWL for Organic Chemistry.

▲ denotes problems linked to the Key Ideas in this chapter.

Visualizing Chemistry

(Problems 16.1–16.23 appear within the chapter.)

16.24 Draw the product from reaction of each of the following substances with (1) Br_2, $FeBr_3$ and (2) CH_3COCl, $AlCl_3$.

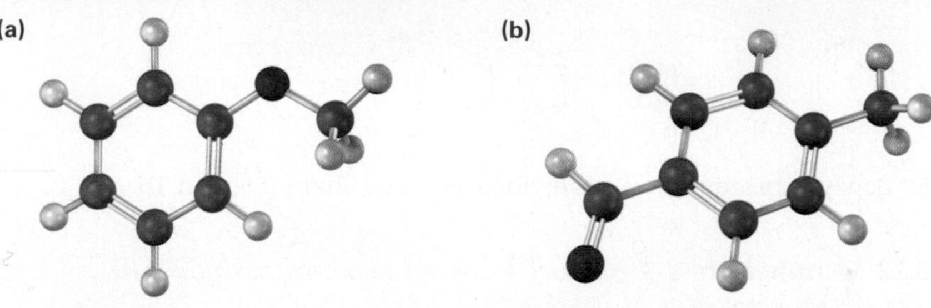

16.25 The following molecular model of a dimethyl-substituted biphenyl represents the lowest-energy conformation of the molecule. Why are the two benzene rings tilted at a 63° angle to each other rather than being in the same plane so that their *p* orbitals can overlap? Why doesn't complete rotation around the single bond joining the two rings occur?

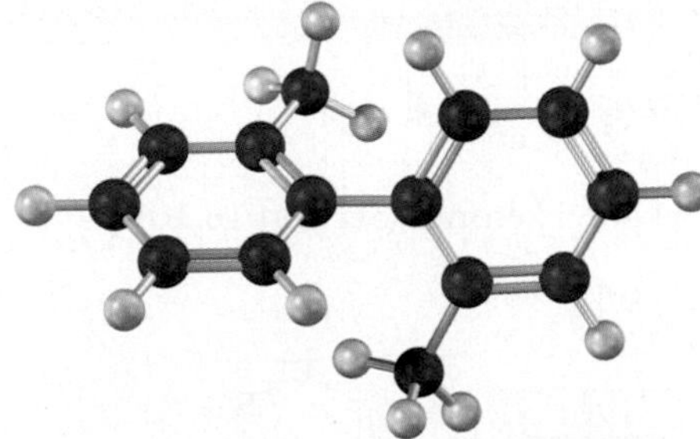

16.26 How would you synthesize the following compound starting from benzene? More than one step is needed.

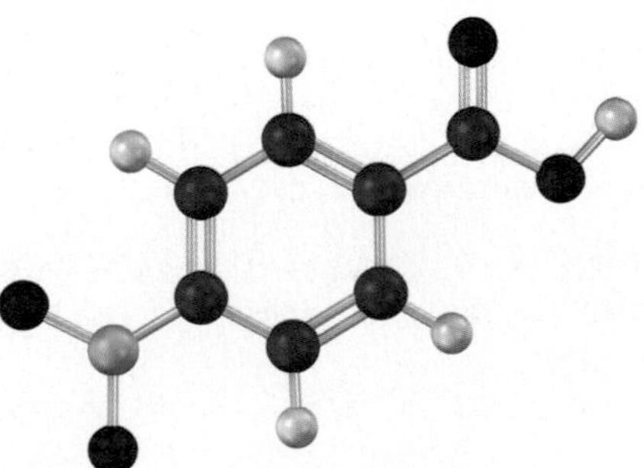

16.27 The following compound can't be synthesized using the methods discussed in this chapter. Why not?

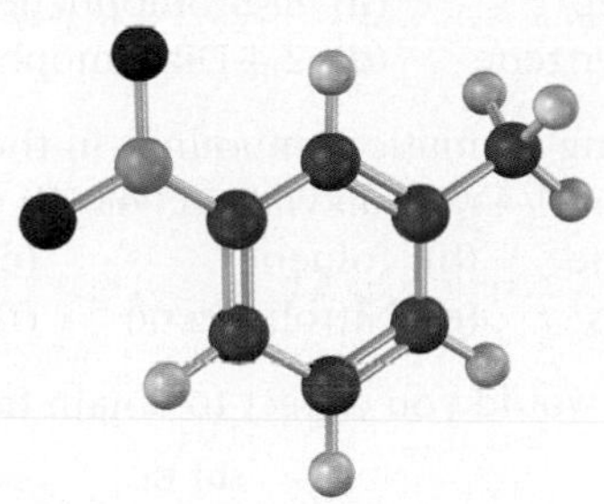

Additional Problems

Reactivity and Orientation of Electrophilic Substitutions

16.28 Identify each of the following groups as an activator or deactivator and as an *o,p*-director or *m*-director:

(a) $-N(CH_3)_2$ **(b)** cyclopentyl **(c)** $-OCH_2CH_3$ **(d)** cyclohexylcarbonyl ($-C(=O)-C_6H_{11}$)

16.29 Predict the major product(s) of nitration of the following substances. Which react faster than benzene, and which slower?

(a) Bromobenzene (b) Benzonitrile (c) Benzoic acid
(d) Nitrobenzene (e) Benzenesulfonic acid (f) Methoxybenzene

16.30 ▲ Rank the compounds in each group according to their reactivity toward electrophilic substitution.

(a) Chlorobenzene, *o*-dichlorobenzene, benzene
(b) *p*-Bromonitrobenzene, nitrobenzene, phenol
(c) Fluorobenzene, benzaldehyde, *o*-xylene
(d) Benzonitrile, *p*-methylbenzonitrile, *p*-methoxybenzonitrile

16.31 ▲ Predict the major monoalkylation products you would expect to obtain from reaction of the following substances with chloromethane and $AlCl_3$:

(a) Bromobenzene (b) *m*-Bromophenol
(c) *p*-Chloroaniline (d) 2,4-Dichloronitrobenzene
(e) 2,4-Dichlorophenol (f) Benzoic acid
(g) *p*-Methylbenzenesulfonic acid (h) 2,5-Dibromotoluene

16.32 Name and draw the major product(s) of electrophilic chlorination of the following compounds:

(a) *m*-Nitrophenol (b) *o*-Xylene
(c) *p*-Nitrobenzoic acid (d) *p*-Bromobenzenesulfonic acid

▲ Problems linked to Key Ideas in this chapter

16.33 Predict the major product(s) you would obtain from sulfonation of the following compounds:
(a) Fluorobenzene (b) *m*-Bromophenol
(c) *m*-Dichlorobenzene (d) 2,4-Dibromophenol

16.34 Rank the following aromatic compounds in the expected order of their reactivity toward Friedel–Crafts alkylation. Which compounds are unreactive?
(a) Bromobenzene (b) Toluene (c) Phenol
(d) Aniline (e) Nitrobenzene (f) *p*-Bromotoluene

16.35 What product(s) would you expect to obtain from the following reactions?

(a) (o-nitroacetophenone: ring with C(=O)CH3 and NO2) $\xrightarrow{H_2/Pd}$?

(b) (m-dibromobenzene: Br, Br) $\xrightarrow[\text{2. Fe, } H_3O^+]{\text{1. } HNO_3,\ H_2SO_4}$?

(c) (tetralin) $\xrightarrow[H_2O]{KMnO_4}$?

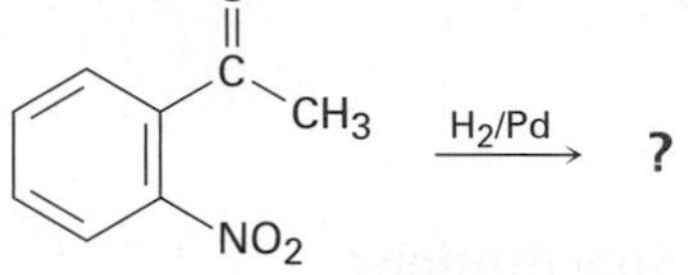

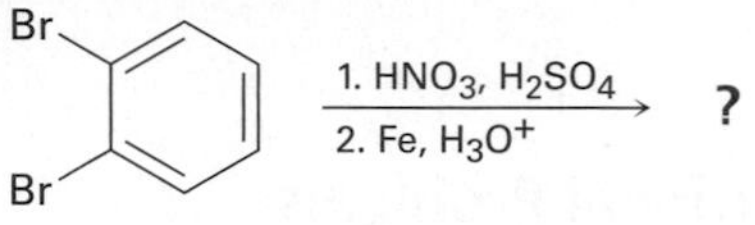

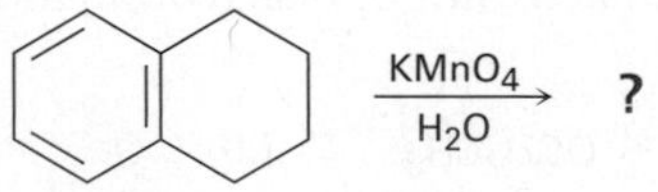

(d) (p-chloroanisole: Cl, OCH_3) $\xrightarrow[AlCl_3]{CH_3CH_2CH_2Cl}$?

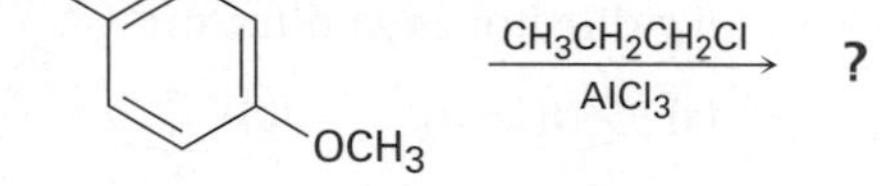

16.36 Predict the major product(s) of the following reactions:

(a) (chlorobenzene: Cl) $\xrightarrow[AlCl_3]{CH_3CH_2Cl}$?

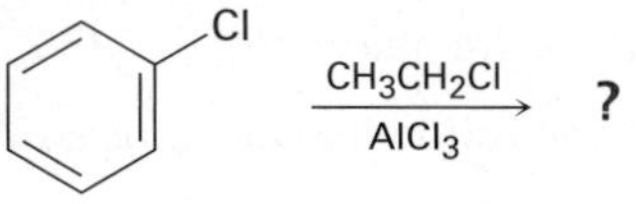

(b) (diphenyl ether: O) $\xrightarrow[AlCl_3]{CH_3CH_2COCl}$?

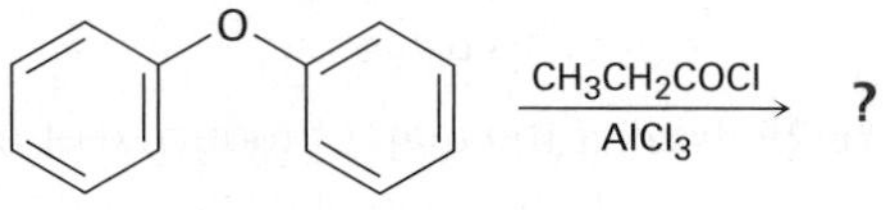

(c) (benzoic acid: CO_2H) $\xrightarrow[H_2SO_4]{HNO_3}$?

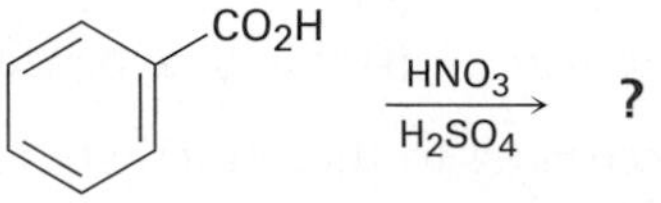

(d) ($N(CH_2CH_3)_2$ on benzene) $\xrightarrow[H_2SO_4]{SO_3}$?

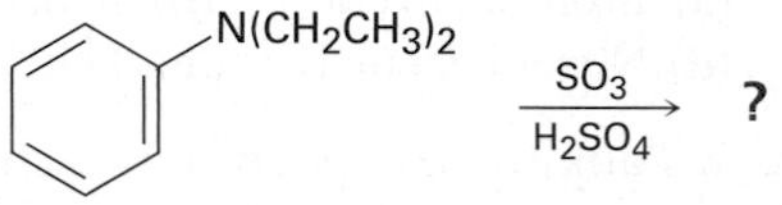

Mechanisms of Electrophilic Substitutions

16.37 Aromatic iodination can be carried out with a number of reagents, including iodine monochloride, ICl. What is the direction of polarization of ICl? Propose a mechanism for the iodination of an aromatic ring with ICl.

16.38 The sulfonation of an aromatic ring with SO_3 and H_2SO_4 is reversible. That is, heating benzenesulfonic acid with H_2SO_4 yields benzene. Show the mechanism of the desulfonation reaction. What is the electrophile?

16.39 The carbocation electrophile in a Friedel–Crafts reaction can be generated in ways other than by reaction of an alkyl chloride with $AlCl_3$. For example, reaction of benzene with 2-methylpropene in the presence of H_3PO_4 yields *tert*-butylbenzene. Propose a mechanism for this reaction.

16.40 The *N,N,N*-trimethylammonium group, $-\overset{+}{N}(CH_3)_3$, is one of the few groups that is a meta-directing deactivator yet has no electron-withdrawing resonance effect. Explain.

▲ Problems linked to Key Ideas in this chapter

16.41 The nitroso group, –N=O, is one of the few nonhalogens that is an ortho- and para-directing deactivator. Explain by drawing resonance structures of the carbocation intermediates in ortho, meta, and para electrophilic reaction on nitrosobenzene, $C_6H_5N{=}O$.

16.42 Triphenylmethane can be prepared by reaction of benzene and chloroform in the presence of $AlCl_3$. Propose a mechanism for the reaction.

+ $CHCl_3$ $\xrightarrow{AlCl_3}$

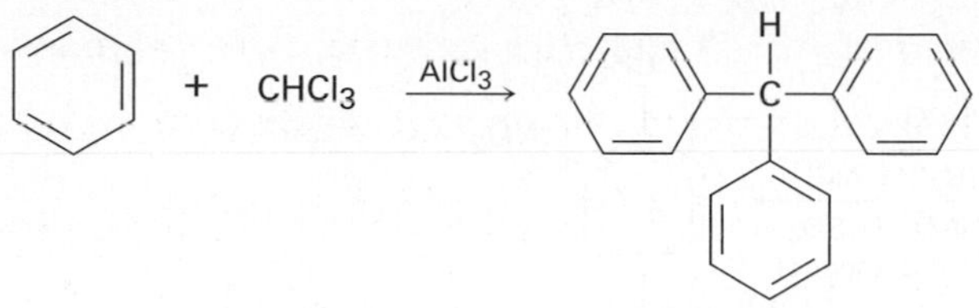

16.43 Using resonance structures of the intermediates, explain why bromination of biphenyl occurs at ortho and para positions rather than at meta.

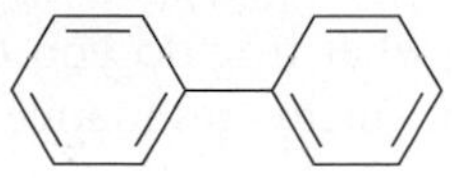

Biphenyl

16.44 Benzene and alkyl-substituted benzenes can be hydroxylated by reaction with H_2O_2 in the presence of an acidic catalyst. What is the structure of the reactive electrophile? Propose a mechanism for the reaction.

$\xrightarrow[CF_3SO_3H \text{ catalyst}]{H_2O_2}$ OH

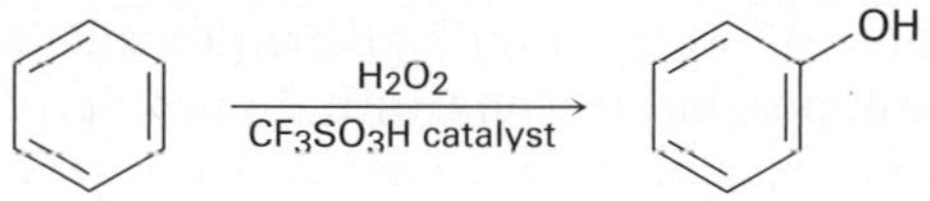

Organic Synthesis

16.45 How would you synthesize the following substances starting from benzene or phenol? Assume that ortho- and para-substitution products can be separated.
(a) *o*-Bromobenzoic acid (b) *p*-Methoxytoluene
(c) 2,4,6-Trinitrobenzoic acid (d) *m*-Bromoaniline

16.46 Starting with benzene as your only source of aromatic compounds, how would you synthesize the following substances? Assume that you can separate ortho and para isomers if necessary.
(a) *p*-Chloroacetophenone (b) *m*-Bromonitrobenzene
(c) *o*-Bromobenzenesulfonic acid (d) *m*-Chlorobenzenesulfonic acid

16.47 Starting with either benzene or toluene, how would you synthesize the following substances? Assume that ortho and para isomers can be separated.
(a) 2-Bromo-4-nitrotoluene (b) 1,3,5-Trinitrobenzene
(c) 2,4,6-Tribromoaniline (d) *m*-Fluorobenzoic acid

▲ Problems linked to Key Ideas in this chapter

16.48 As written, the following syntheses have flaws. What is wrong with each?

(a) CH_3 (benzene) → 1. Cl_2, $FeCl_3$; 2. $KMnO_4$ → CO_2H, Cl

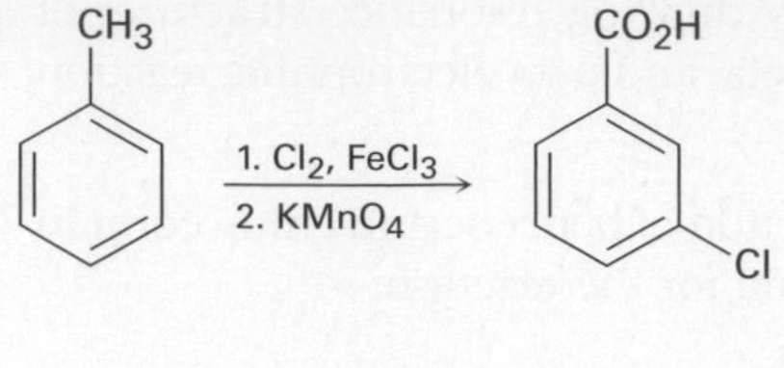

(b) Cl (benzene) → 1. HNO_3, H_2SO_4; 2. CH_3Cl, $AlCl_3$; 3. Fe, H_3O^+; 4. NaOH, H_2O → Cl, CH_3, NH_2

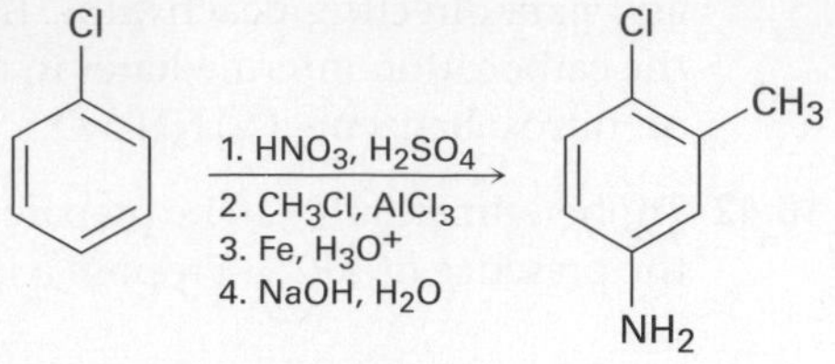

(c) CH_3 (benzene) → 1. $CH_3\overset{O}{\overset{\|}{C}}Cl$, $AlCl_3$; 2. HNO_3, H_2SO_4; 3. H_2/Pd; ethanol → CH_3, NO_2, CH_2CH_3

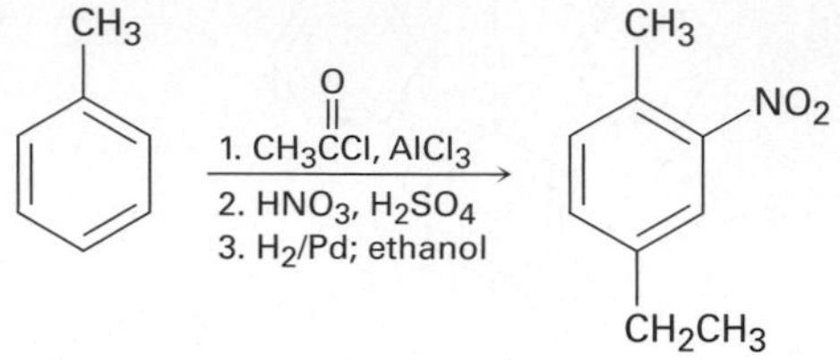

General Problems

16.49 ▲ At what position and on what ring do you expect nitration of 4-bromobiphenyl to occur? Explain, using resonance structures of the potential intermediates.

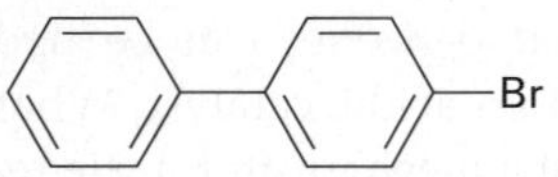

Br **4-Bromobiphenyl**

16.50 ▲ Electrophilic substitution on 3-phenylpropanenitrile occurs at the ortho and para positions, but reaction with 3-phenylpropenenitrile occurs at the meta position. Explain, using resonance structures of the intermediates.

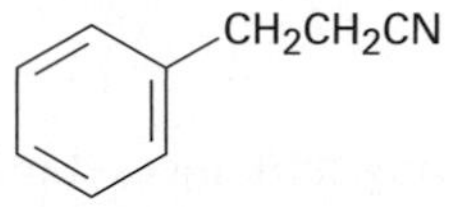

CH_2CH_2CN CN

3-Phenylpropanenitrile 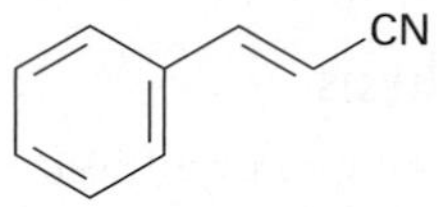 **3-Phenylpropenenitrile**

16.51 Addition of HBr to 1-phenylpropene yields only (1-bromopropyl)benzene. Propose a mechanism for the reaction, and explain why none of the other regioisomer is produced.

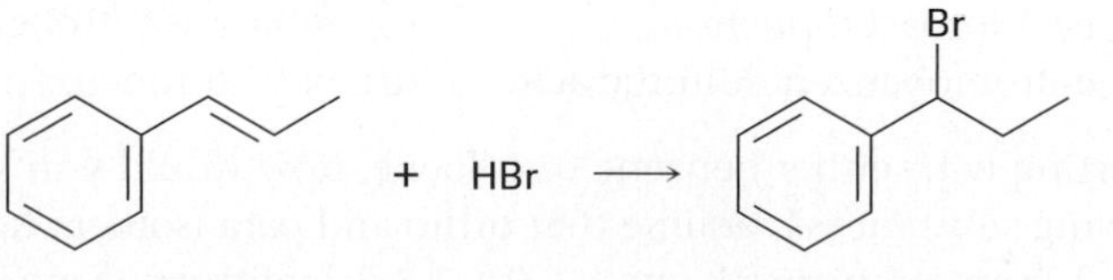

+ HBr ⟶ Br

▲ Problems linked to Key Ideas in this chapter

16.52 At what position, and on what ring, would you expect the following substances to undergo electrophilic substitution?

(a)

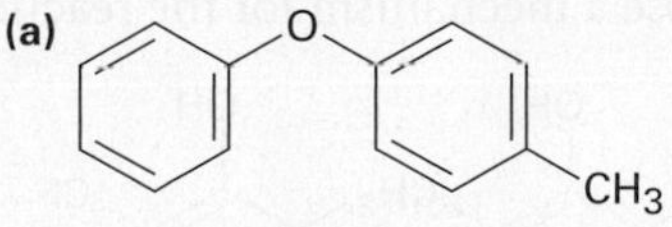

(b)

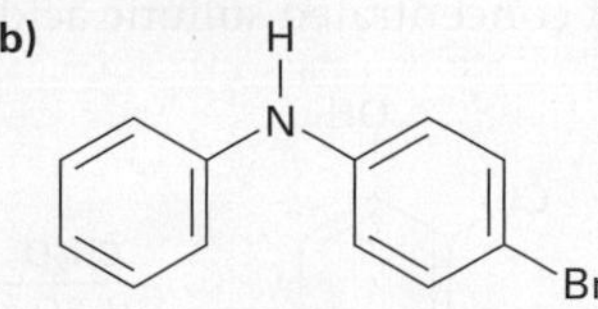

(c)

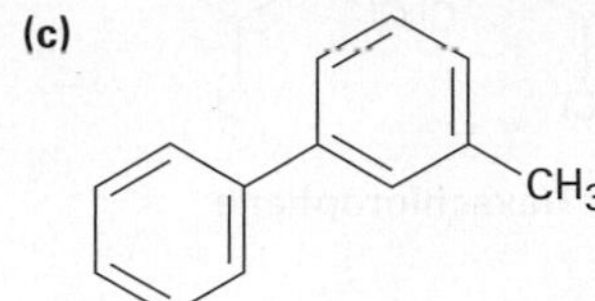

(d)

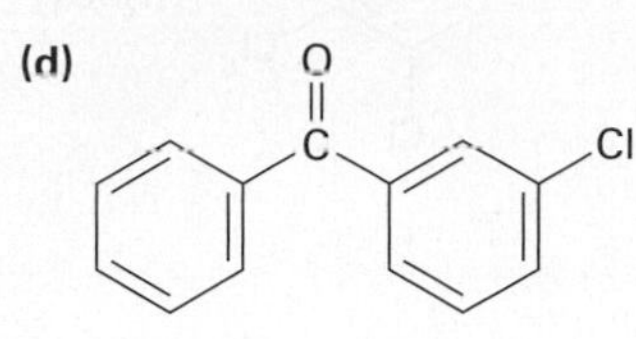

16.53 At what position, and on what ring, would you expect bromination of benzanilide to occur? Explain by drawing resonance structures of theintermediates.

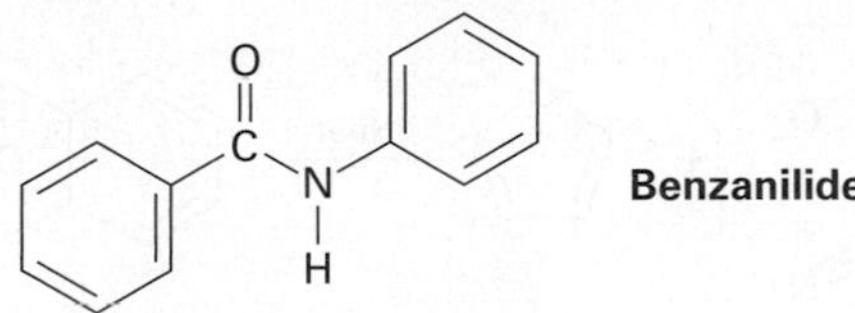

Benzanilide

16.54 Would you expect the Friedel–Crafts reaction of benzene with (*R*)-2-chlorobutane to yield optically active or racemic product? Explain.

16.55 How would you synthesize the following substances starting from benzene?

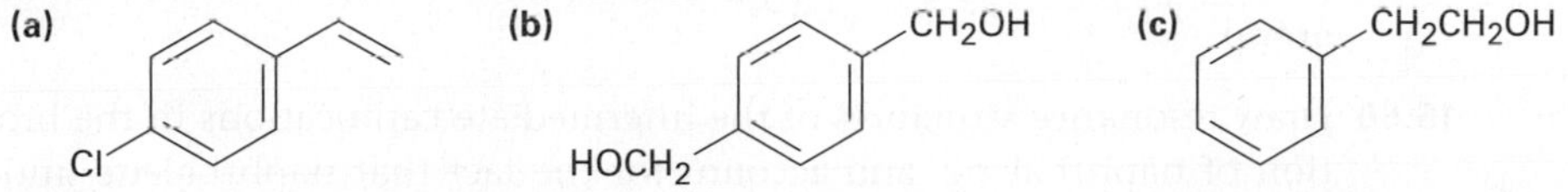

(a)

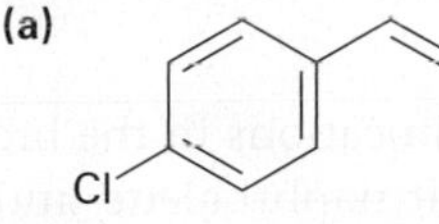

(b)

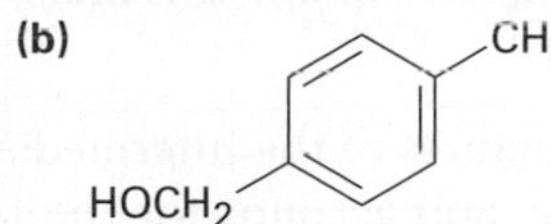

(c) CH_2CH_2OH

16.56 The compound MON-0585 is a nontoxic, biodegradable larvicide that is highly selective against mosquito larvae. Synthesize MON-0585 using either benzene or phenol as a source of the aromatic rings.

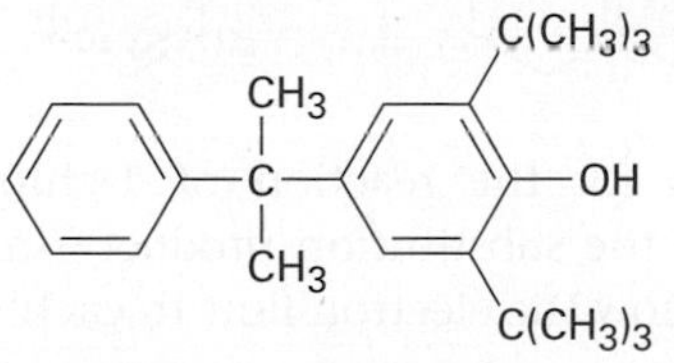

$C(CH_3)_3$ CH_3 C OH CH_3 $C(CH_3)_3$

MON-0585

▲ Problems linked to Key Ideas in this chapter

16.57 Hexachlorophene, a substance used in the manufacture of germicidal soaps, is prepared by reaction of 2,4,5-trichlorophenol with formaldehyde in the presence of concentrated sulfuric acid. Propose a mechanism for the reaction.

OH, Cl, Cl, Cl $\xrightarrow[H_2SO_4]{CH_2O}$ OH, OH, Cl, CH_2, Cl, Cl Cl, Cl, Cl

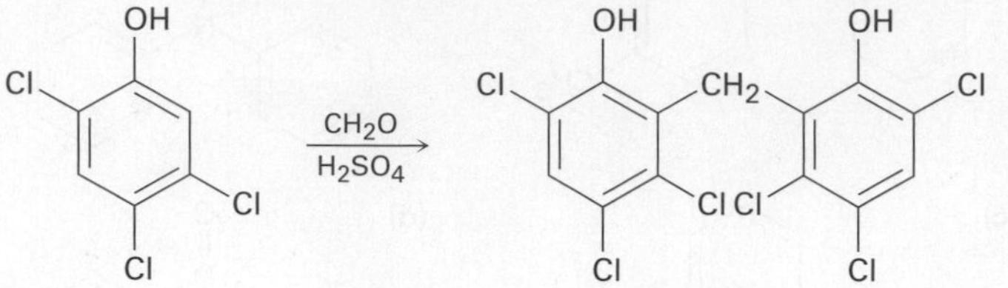

Hexachlorophene

16.58 Benzenediazonium carboxylate decomposes when heated to yield N_2, CO_2, and a reactive substance that can't be isolated. When benzenediazonium carboxylate is heated in the presence of furan, the following reaction is observed:

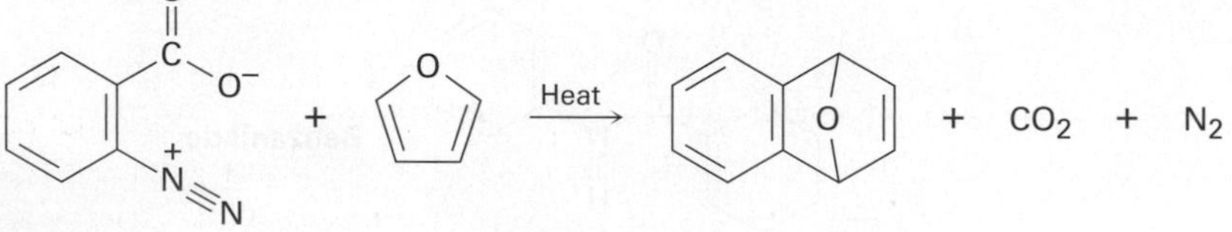

$$\xrightarrow{\text{Heat}} + CO_2 + N_2$$

What intermediate is involved in this reaction? Propose a mechanism for its formation.

16.59 Phenylboronic acid, $C_6H_5B(OH)_2$, is nitrated to give 15% ortho-substitution product and 85% meta. Explain the meta-directing effect of the $-B(OH)_2$ group.

16.60 Draw resonance structures of the intermediate carbocations in the bromination of naphthalene, and account for the fact that naphthalene undergoes electrophilic substitution at C1 rather than C2.

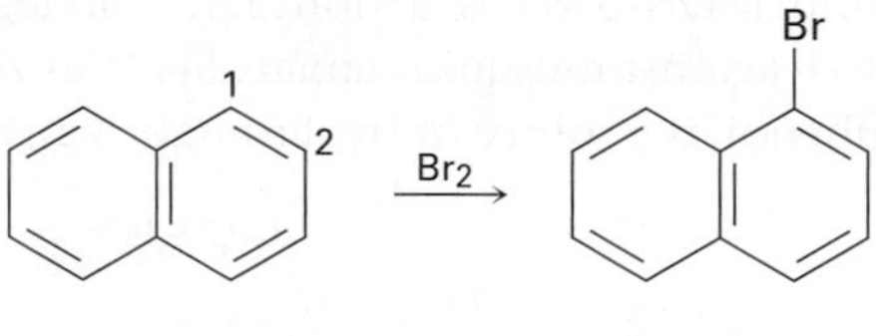

$$\xrightarrow{Br_2}$$

16.61 Propose a mechanism for the reaction of 1-chloroanthraquinone with methoxide ion to give the substitution product 1-methoxyanthraquinone. Use curved arrows to show the electron flow in each step.

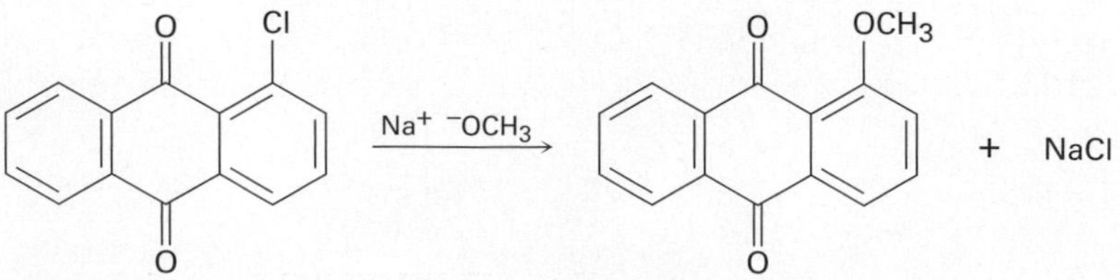

$$\xrightarrow{Na^+ \ ^-OCH_3} + NaCl$$

1-Chloroanthraquinone **1-Methoxyanthraquinone**

▲ Problems linked to Key Ideas in this chapter

16.62 4-Chloropyridine undergoes reaction with dimethylamine to yield 4-dimethylaminopyridine. Propose a mechanism for the reaction.

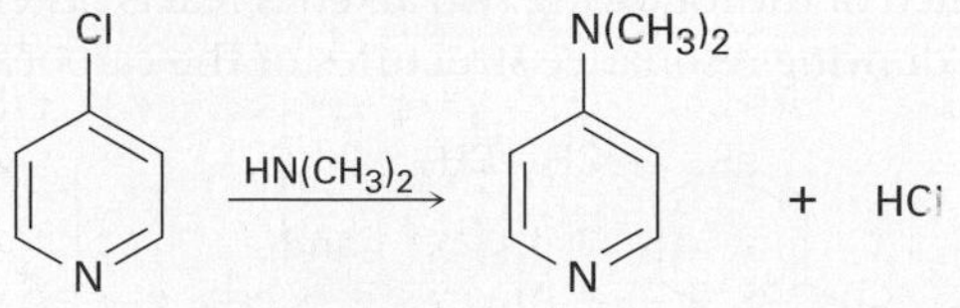

16.63 *p*-Bromotoluene reacts with potassium amide to give a mixture of *m*- and *p*-methylaniline. Explain.

16.64 Propose a mechanism to account for the reaction of benzene with 2,2,5,5-tetramethyltetrahydrofuran.

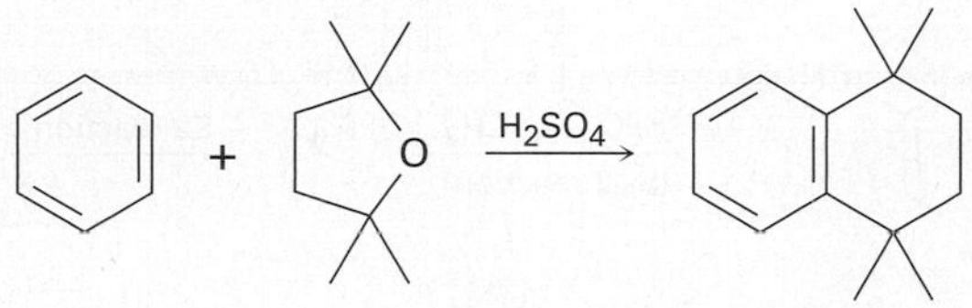

16.65 Propose a mechanism to account for the following reaction:

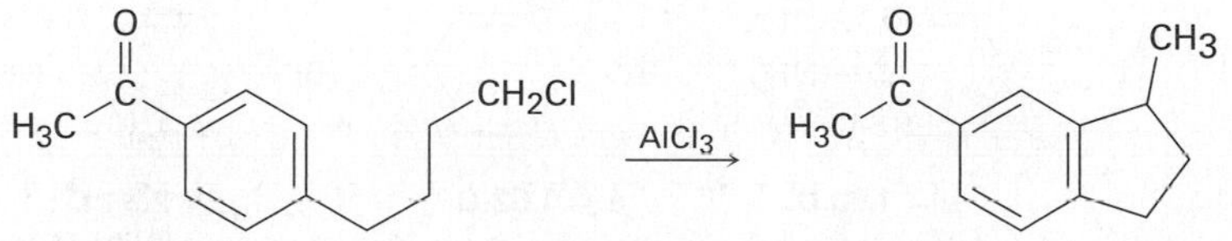

16.66 In the *Gatterman–Koch reaction,* a formyl group (–CHO) is introduced directly onto a benzene ring. For example, reaction of toluene with CO and HCl in the presence of mixed $CuCl/AlCl_3$ gives p-methylbenzaldehyde. Propose a mechanism.

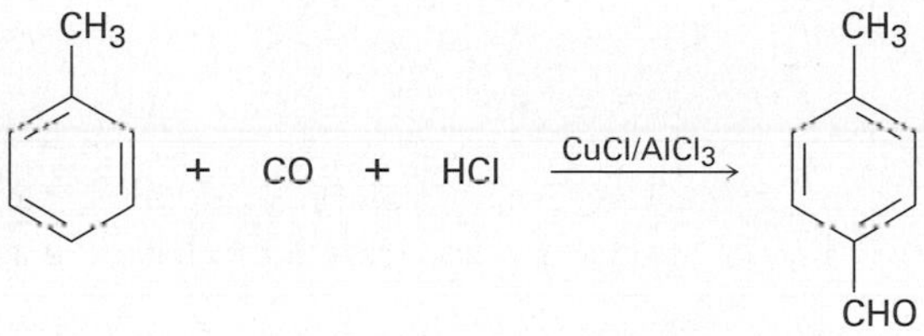

16.67 Treatment of *p-tert*-butylphenol with a strong acid such as H_2SO_4 yields phenol and 2-methylpropene. Propose a mechanism.

16.68 How would you synthesize the following compounds from benzene? Assume that ortho and para isomers can be separated.

(a)

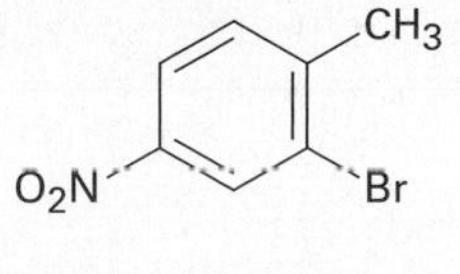

(b)

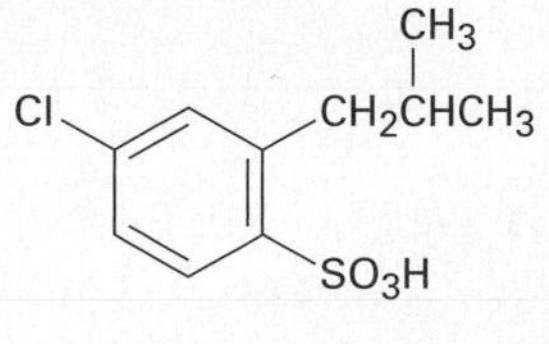

▲ Problems linked to Key Ideas in this chapter

16.69 You know the mechanism of HBr addition to alkenes, and you know the effects of various substituent groups on aromatic substitution. Use this knowledge to predict which of the following two alkenes reacts faster with HBr. Explain your answer by drawing resonance structures of the carbocation intermediates.

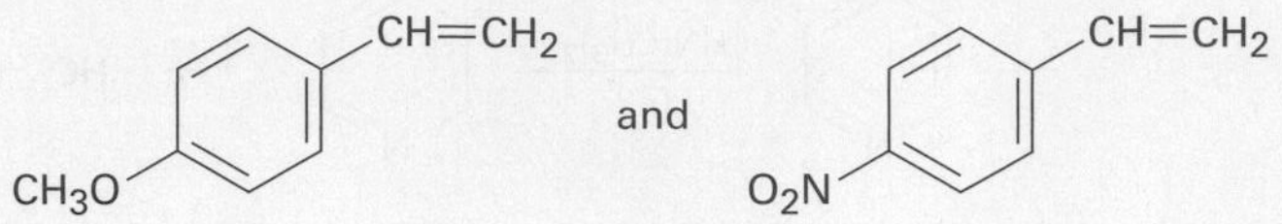

CH_3O–C_6H_4–CH=CH_2 and O_2N–C_6H_4–CH=CH_2

16.70 Benzyl bromide is converted into benzaldehyde by heating in dimethyl sulfoxide. Propose a structure for the intermediate, and show the mechanisms of the two steps in the reaction.

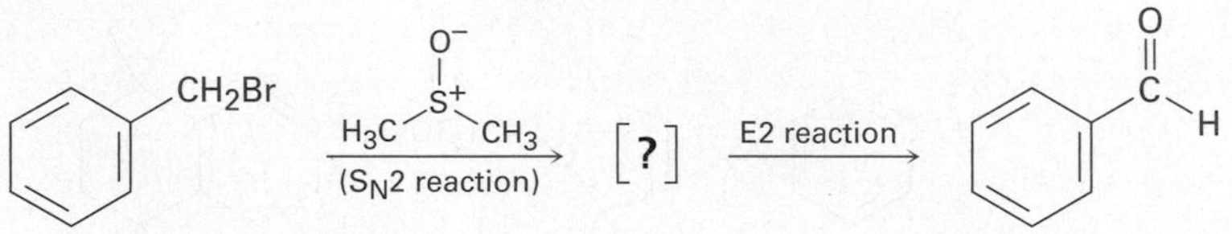

$C_6H_5CH_2Br$ —[H_3C–$S^+(O^-)$–CH_3 (S_N2 reaction)]→ [?] —[E2 reaction]→ C_6H_5–CHO

16.71 Use your knowledge of directing effects, along with the following data, to deduce the directions of the dipole moments in aniline and bromobenzene.

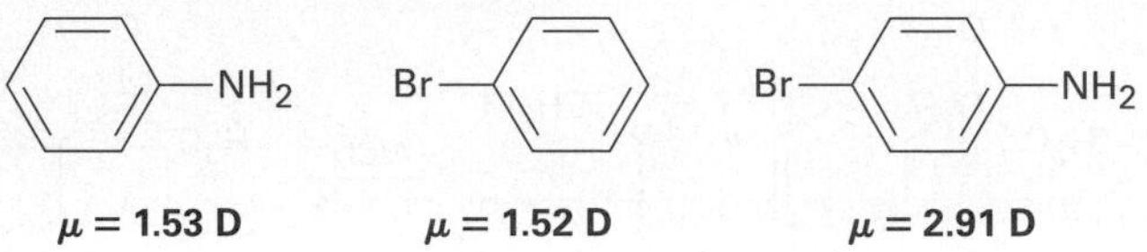

Aniline ($C_6H_5NH_2$)	Bromobenzene (C_6H_5Br)	Br–C_6H_4–NH_2
μ = 1.53 D	μ = 1.52 D	μ = 2.91 D

16.72 Identify the reagents represented by the letters **a–e** in the following scheme:

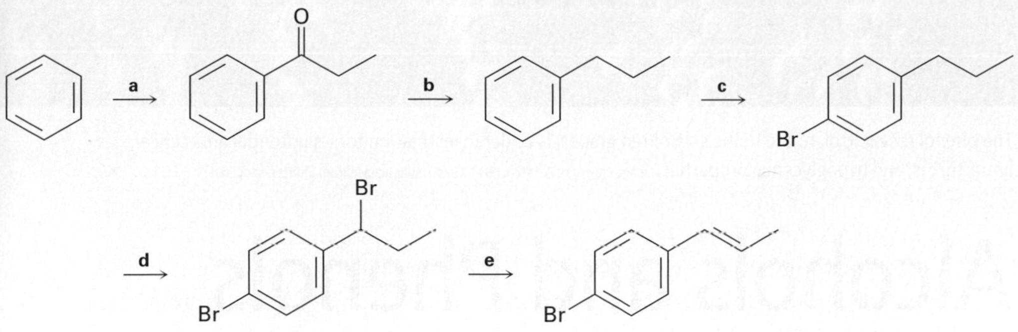

a, b, c, d, e; O; Br; Br; Br; Br

16.73 Phenols (ArOH) are relatively acidic, and the presence of a substituent group on the aromatic ring has a large effect. The pK_a of unsubstituted phenol, for example, is 9.89, while that of *p*-nitrophenol is 7.15. Draw resonance structures of the corresponding phenoxide anions and explain the data.

16.74 Would you expect *p*-methylphenol to be more acidic or less acidic than unsubstituted phenol? Explain. (See Problem 16.73.)

▲ Problems linked to Key Ideas in this chapter

17

The phenol resveratrol, found in the skin of red grapes, is under investigation for its potential anti-cancer, antiarthritis, and hypoglycemic properties.

Alcohols and Phenols

Alcohols and **phenols** can be thought of as organic derivatives of water in which one of the water hydrogens is replaced by an organic group: H–O–H versus R–O–H and Ar–O–H. In practice, the group name *alcohol* is restricted to compounds that have their –OH group bonded to a saturated, sp^3-hybridized carbon atom, while compounds with their –OH group bonded to a vinylic, sp^2-hybridized carbon are called *enols*. We'll look at enols in Chapter 22.

An alcohol **A phenol** [**An enol**]

Alcohols occur widely in nature and have many industrial and pharmaceutical applications. Methanol, for instance, is one of the most important of all industrial chemicals. Historically, methanol was prepared by heating wood in the absence of air and thus came to be called *wood alcohol*. Today, approximately 40 million metric tons (13 billion gallons) of methanol is manufactured worldwide each year, most of it by catalytic reduction of carbon monoxide with hydrogen gas. Methanol is toxic to humans, causing blindness in small doses (15 mL) and death in larger amounts (100–250 mL). Industrially, it is used both as a solvent and as a starting material for production of formaldehyde (CH_2O) and acetic acid (CH_3CO_2H).

$$CO + 2\,H_2 \xrightarrow[\text{Zinc oxide/chromia catalyst}]{400\ ^\circ C} CH_3OH$$

Ethanol was one of the first organic chemicals to be prepared and purified. Its production by fermentation of grains and sugars has been carried out for perhaps 9000 years, and its purification by distillation goes back at least as far as the 12th century. Today, approximately 55 million metric tons (18 billion gallons) of ethanol is produced worldwide each year, most of it by fermentation of corn, barley, sorghum, and other plant sources. Essentially the entire amount is used for automobile fuel.

OWL Sign in to OWL for Organic Chemistry at **www.cengage.com/owl** to view tutorials and simulations, develop problem-solving skills, and complete online homework assigned by your professor.

Ethanol for industrial use as a solvent or chemical intermediate is largely obtained by acid-catalyzed hydration of ethylene at high temperature.

$$H_2C{=}CH_2 \xrightarrow[\substack{H_3PO_4 \\ 250\ ^\circ C}]{H_2O} CH_3CH_2OH$$

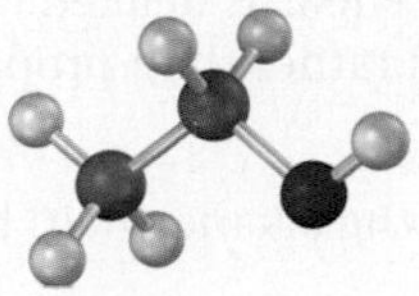

Phenols occur widely throughout nature and also serve as intermediates in the industrial synthesis of products as diverse as adhesives and antiseptics. Phenol itself is a general disinfectant found in coal tar; methyl salicylate is a flavoring agent found in oil of wintergreen; and the urushiols are the allergenic constituents of poison oak and poison ivy. Note that the word *phenol* is the name both of the specific compound (hydroxybenzene) and of a class of compounds.

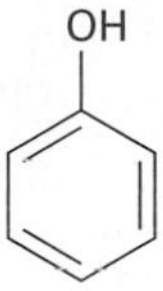

OH | OH, CO_2CH_3 | OH, OH, R

Phenol (also known as carbolic acid) 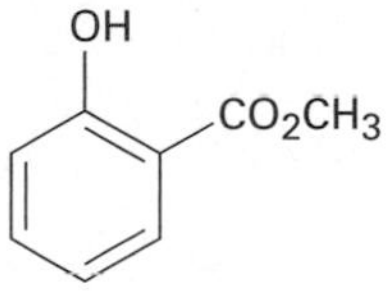 | **Methyl salicylate** 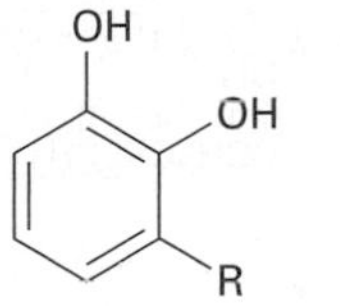 | **Urushiols (R = different C_{15} alkyl and alkenyl chains)**

Why This Chapter? Up to this point, we've focused on developing some general ideas of organic reactivity, on looking at the chemistry of hydrocarbons and alkyl halides, and on seeing some of the tools used in structural studies. With that background, it's now time to begin a study of the oxygen-containing functional groups that lie at the heart of organic and biological chemistry. We'll look at alcohols in this chapter and then move on to carbonyl compounds in Chapters 19 through 23.

17.1 Naming Alcohols and Phenols

Alcohols are classified as primary (1°), secondary (2°), or tertiary (3°), depending on the number of organic groups bonded to the hydroxyl-bearing carbon.

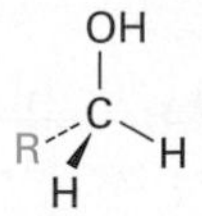

OH, C, R, H, H | OH, C, R, H, R | OH, C, R, R, R

A primary (1°) alcohol 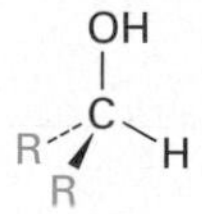 | **A secondary (2°) alcohol** 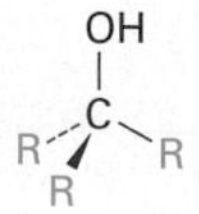 | **A tertiary (3°) alcohol**

Simple alcohols are named by the IUPAC system as derivatives of the parent alkane, using the suffix *-ol*.

RULE 1

Select the longest carbon chain containing the hydroxyl group, and derive the parent name by replacing the *-e* ending of the corresponding alkane with *-ol*. The *-e* is deleted to prevent the occurrence of two adjacent vowels: propanol rather than propaneol, for example.

RULE 2

Number the alkane chain beginning at the end nearer the hydroxyl group.

RULE 3

Number the substituents according to their position on the chain, and write the name, listing the substituents in alphabetical order and identifying the position to which the –OH is bonded. Note that in naming *cis*-1,4-cyclohexanediol, the final *-e* of cyclohexane is not deleted because the next letter, *d*, is not a vowel; that is, cyclohexanediol rather than cyclohexandiol. Also, as with alkenes **(Section 7.3)**, newer IUPAC naming recommendations place the locant immediately before the suffix rather than before the parent.

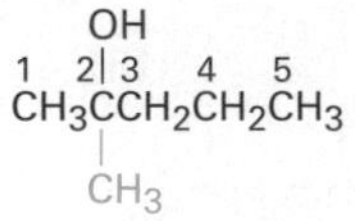

$CH_3C(OH)(CH_3)CH_2CH_2CH_3$

2-Methyl-2-pentanol
(New: 2-Methylpentan-2-ol)

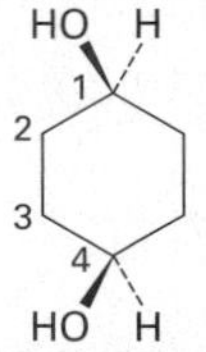

cis-1,4-Cyclohexanediol
(New: cis-Cyclohexane-1,4-diol)

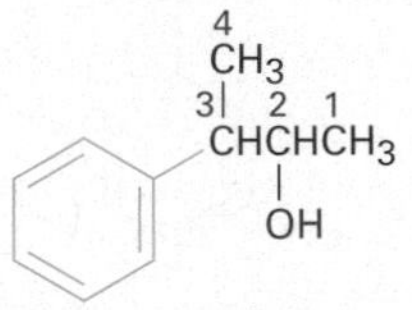

$C_6H_5CH(CH_3)CH(OH)CH_3$

3-Phenyl-2-butanol
(New: 3-Phenylbutan-2-ol)

Some simple and widely occurring alcohols have common names that are accepted by IUPAC. For example:

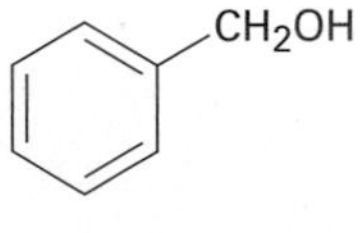

$C_6H_5CH_2OH$	$H_2C{=}CHCH_2OH$	$(CH_3)_3COH$	$HOCH_2CH_2OH$	$HOCH_2CH(OH)CH_2OH$
Benzyl alcohol (phenylmethanol)	**Allyl alcohol (2-propen-1-ol)**	***tert*-Butyl alcohol (2-methyl-2-propanol)**	**Ethylene glycol (1,2-ethanediol)**	**Glycerol (1,2,3-propanetriol)**

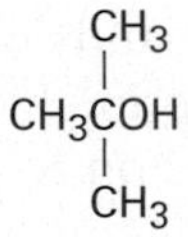

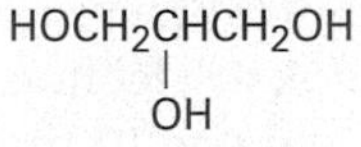

Phenols are named as described previously for aromatic compounds according to the rules discussed in **Section 15.1**. Note that *-phenol* is used as the parent name rather than *-benzene*.

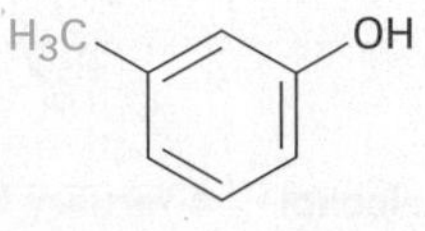

m-Methylphenol
(m-Cresol)

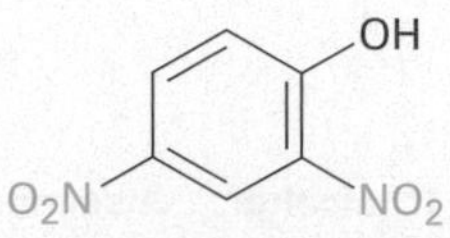

2,4-Dinitrophenol

Problem 17.1

Give IUPAC names for the following compounds:

(a) $CH_3CH(OH)CH_2CH(OH)CH(CH_3)CH_3$

(b) $C_6H_5CH_2CH_2C(OH)(CH_3)CH_3$

(c) HO (cyclohexane ring) CH_3, CH_3

(d) H, Br, H, OH (cyclopentane ring)

(e) H_3C, OH, Br (benzene ring)

(f) (cyclopentene ring) OH

Problem 17.2

Draw structures corresponding to the following IUPAC names:

(a) (*Z*)-2-Ethyl-2-buten-1-ol **(b)** 3-Cyclohexen-1-ol
(c) *trans*-3-Chlorocycloheptanol **(d)** 1,4-Pentanediol
(e) 2,6-Dimethylphenol **(f)** *o*-(2-Hydroxyethyl)phenol

17.2 Properties of Alcohols and Phenols

Alcohols and phenols have nearly the same geometry around the oxygen atom as water. The R–O–H bond angle has an approximately tetrahedral value (108.5° in methanol, for instance), and the oxygen atom is sp^3-hybridized.

Also like water, alcohols and phenols have higher boiling points than might be expected because of hydrogen-bonding **(Section 2.12)**. A positively polarized –OH hydrogen atom from one molecule is attracted to a lone pair of electrons on the electronegative oxygen atom of another molecule, resulting in a weak force that holds the molecules together **(Figure 17.1)**. These intermolecular attractions must be overcome for a molecule to break free from the liquid and enter the vapor state, so the boiling temperature is raised. For example, 1-propanol (MW = 60), butane (MW = 58), and chloroethane (MW = 65) have similar molecular weights, yet 1-propanol boils at 97 °C, compared with −0.5 °C for the alkane and 12.5 °C for the chloroalkane.

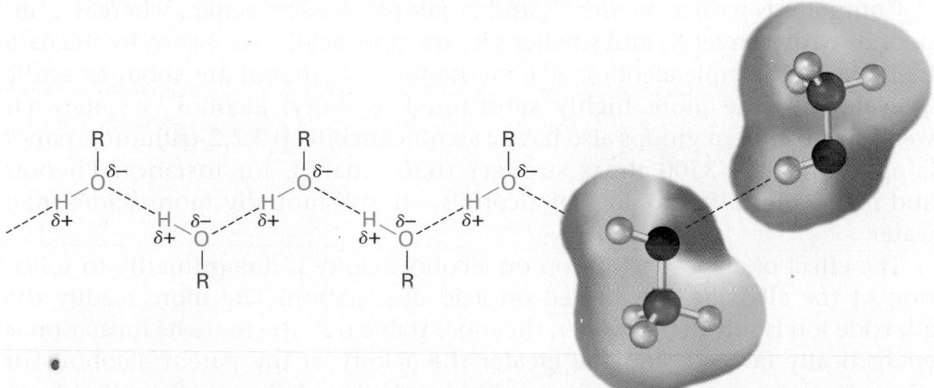

Figure 17.1 Hydrogen-bonding in alcohols and phenols. Attraction between a positively polarized OH hydrogen and a negatively polarized oxygen holds molecules together. The electrostatic potential map of methanol shows the **positively polarized** O–H hydrogen and the **negatively polarized** oxygen.

Another similarity with water is that alcohols and phenols are both weakly basic and weakly acidic. As weak bases, they are reversibly protonated by strong acids to yield oxonium ions, ROH_2^+.

An alcohol

An oxonium ion

$$\left[\text{or ArOH} + \text{HX} \rightleftharpoons \text{Ar}\overset{+}{\text{O}}\text{H}_2\ \text{X}^-\right]$$

As weak acids, they dissociate slightly in dilute aqueous solution by donating a proton to water, generating H_3O^+ and an **alkoxide ion, RO^-**, or a **phenoxide ion, ArO^-**.

An alcohol

An alkoxide ion

A phenol

A phenoxide ion

Recall from the earlier discussion of acidity in **Sections 2.7–2.11** that the strength of any acid HA in water can be expressed by an acidity constant, K_a.

$$K_a = \frac{[A^-][H_3O^+]}{[HA]} \qquad pK_a = -\log K_a$$

Compounds with a smaller K_a and larger pK_a are less acidic, whereas compounds with a larger K_a and smaller pK_a are more acidic. As shown by the data in Table 17.1, simple alcohols like methanol and ethanol are about as acidic as water but the more highly substituted *tert*-butyl alcohol is somewhat weaker. Substituent groups also have a significant effect: 2,2,2-trifluoroethanol is approximately 3700 times stronger than ethanol, for instance. Phenols and *thiols,* the sulfur analogs of alcohols, are substantially more acidic than water.

The effect of alkyl substitution on alcohol acidity is due primarily to solvation of the alkoxide ion formed on acid dissociation. The more readily the alkoxide ion is solvated by water, the more stable it is, the more its formation is energetically favored, and the greater the acidity of the parent alcohol. For example, the oxygen atom of an unhindered alkoxide ion, such as that from methanol, is sterically accessible and is easily solvated by water. The oxygen

Table 17.1 Acidity Constants of Some Alcohols and Phenols

Compound	pK_a	
$(CH_3)_3COH$	18.00	Weaker acid
CH_3CH_2OH	16.00	
H_2O	15.74	
CH_3OH	15.54	
CF_3CH_2OH	12.43	
p-Aminophenol	10.46	
CH_3SH	10.3	
p-Methylphenol	10.17	
Phenol	9.89	
p-Chlorophenol	9.38	
p-Nitrophenol	7.15	Stronger acid

atom of a hindered alkoxide ion, however, such as that from *tert*-butyl alcohol, is less easily solvated and is therefore less stabilized.

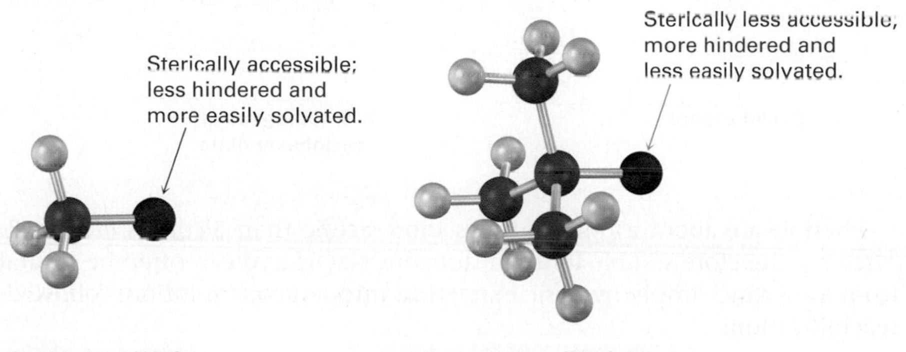

Methoxide ion, CH_3O^- (pK_a = 15.54)

***tert*-Butoxide ion, $(CH_3)_3CO^-$ (pK_a = 18.00)**

Inductive effects **(Section 16.5)** are also important in determining alcohol acidities. Electron-withdrawing halogen substituents, for instance, stabilize an alkoxide ion by spreading out the charge over a larger volume, thus making the alcohol more acidic. Compare, for example, the acidities of ethanol (pK_a = 16.00) and 2,2,2-trifluoroethanol (pK_a = 12.43), or of *tert*-butyl alcohol (pK_a = 18.0) and nonafluoro-*tert*-butyl alcohol (pK_a = 5.4).

Electron-withdrawing groups stabilize the alkoxide ion and lower the pK_a.

$(F_3C)_3C{-}O^-$ versus $(H_3C)_3C{-}O^-$

pK_a = 5.4 pK_a = 18.0

Because alcohols are weak acids, they don't react with weak bases, such as amines or bicarbonate ion, and they react to only a limited extent with metal hydroxides, such as NaOH. Alcohols do, however, react with alkali metals and with strong bases such as sodium hydride (NaH), sodium amide ($NaNH_2$), and Grignard reagents (RMgX). Alkoxides are themselves bases that are frequently used as reagents in organic chemistry. They are named systematically by adding the *-ate* suffix to the name of the alcohol. Methanol becomes methanolate, for instance.

$$2\ (CH_3)_3C{-}OH + 2\ K \longrightarrow 2\ (CH_3)_3C{-}O^-\ K^+ + H_2$$

***tert*-Butyl alcohol (2-methyl-2-propanol)** → **Potassium *tert*-butoxide (potassium 2-methyl-2-propanolate)**

$$CH_3OH + NaH \longrightarrow CH_3O^-\ Na^+ + H_2$$

Methanol → **Sodium methoxide (sodium methanolate)**

$$CH_3CH_2OH + NaNH_2 \longrightarrow CH_3CH_2O^-\ Na^+ + NH_3$$

Ethanol → **Sodium ethoxide (sodium ethanolate)**

$$C_6H_{11}OH + CH_3MgBr \longrightarrow C_6H_{11}O^-\ {}^+MgBr + CH_4$$

Cyclohexanol → **Bromomagnesium cyclohexanolate**

Phenols are about a million times more acidic than alcohols (Table 17.1). They are therefore soluble in dilute aqueous NaOH and can often be separated from a mixture simply by basic extraction into aqueous solution, followed by reacidification.

$$C_6H_5OH + NaOH \longrightarrow C_6H_5O^-\ Na^+ + H_2O$$

Phenol → **Sodium phenoxide (sodium phenolate)**

Phenols are more acidic than alcohols because the phenoxide anion is resonance-stabilized. Delocalization of the negative charge over the ortho and para positions of the aromatic ring results in increased stability of the phenoxide anion relative to undissociated phenol and in a consequently lower $\Delta G°$ for dissociation. **Figure 17.2** compares electrostatic potential maps of an alkoxide ion (CH_3O^-) with phenoxide ion and shows how the negative charge in phenoxide ion is delocalized from oxygen to the ring.

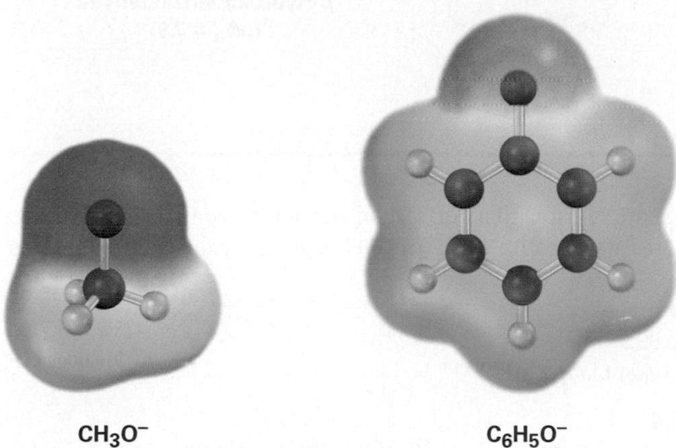

Figure 17.2 The resonance-stabilized phenoxide ion is more stable than an alkoxide ion. Electrostatic potential maps show how the **negative charge** is concentrated on oxygen in the methoxide ion but is spread over the aromatic ring in the phenoxide ion.

Substituted phenols can be either more acidic or less acidic than phenol itself, depending on whether the substituent is electron-withdrawing or electron-donating **(Section 16.5)**. Phenols with an electron-withdrawing substituent are more acidic because these substituents delocalize the negative charge; phenols with an electron-donating substituent are less acidic because these substituents concentrate the charge. The acidifying effect of an electron-withdrawing substituent is particularly noticeable in phenols with a nitro group at the ortho or para position.

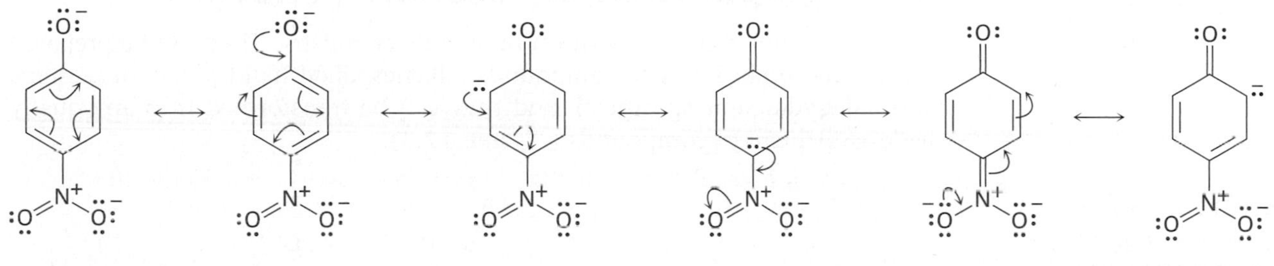

Predicting the Relative Acidity of a Substituted Phenol

Worked Example 17.1

Is *p*-hydroxybenzaldehyde more acidic or less acidic than phenol?

Strategy

Identify the substituent on the aromatic ring, and decide whether it is electron-donating or electron-withdrawing. Electron-withdrawing substituents make the phenol more acidic by stabilizing the phenoxide anion, and electron-donating substituents make the phenol less acidic by destabilizing the anion.

Solution

We saw in **Section 16.5** that a carbonyl group is electron-withdrawing. Thus, *p*-hydroxybenzaldehyde is more acidic ($pK_a = 7.9$) than phenol ($pK_a = 9.89$).

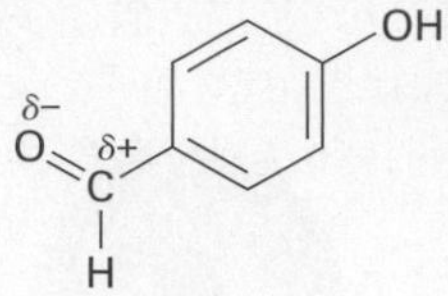

p-Hydroxybenzaldehyde ($pK_a = 7.9$)

Problem 17.3

The following data for isomeric four-carbon alcohols show that there is a decrease in boiling point with increasing substitution of the OH-bearing carbon. How might you account for this trend?

1-Butanol, bp 117.5 °C
2-Butanol, bp 99.5 °C
2-Methyl-2-propanol, bp 82.2 °C

Problem 17.4

Rank the following substances in order of increasing acidity:

(a) $(CH_3)_2CHOH$, $HC{\equiv}CH$, $(CF_3)_2CHOH$, CH_3OH
(b) Phenol, *p*-methylphenol, *p*-(trifluoromethyl)phenol
(c) Benzyl alcohol, phenol, *p*-hydroxybenzoic acid

Problem 17.5

p-Nitrobenzyl alcohol is more acidic than benzyl alcohol, but *p*-methoxybenzyl alcohol is less acidic. Explain.

17.3 Preparation of Alcohols: A Review

Alcohols occupy a central position in organic chemistry. They can be prepared from many other kinds of compounds (alkenes, alkyl halides, ketones, esters, and aldehydes, among others), and they can be transformed into an equally wide assortment of compounds **(Figure 17.3)**.

Figure 17.3 The central position of alcohols in organic chemistry. Alcohols can be prepared from, and converted into, many other kinds of compounds.

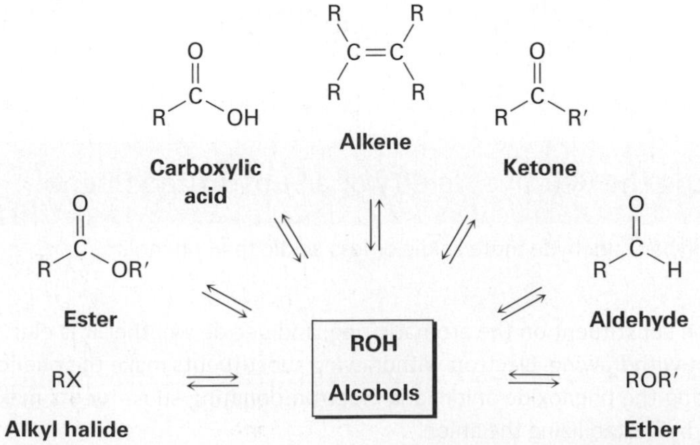

We've already seen several methods of alcohol synthesis:

- Alcohols can be prepared by hydration of alkenes. Because the direct hydration of alkenes with aqueous acid is generally a poor reaction in the laboratory, two indirect methods are commonly used. Hydroboration–oxidation yields the product of syn, non-Markovnikov hydration **(Section 8.5)**, whereas oxymercuration–demercuration yields the product of Markovnikov hydration **(Section 8.4)**.

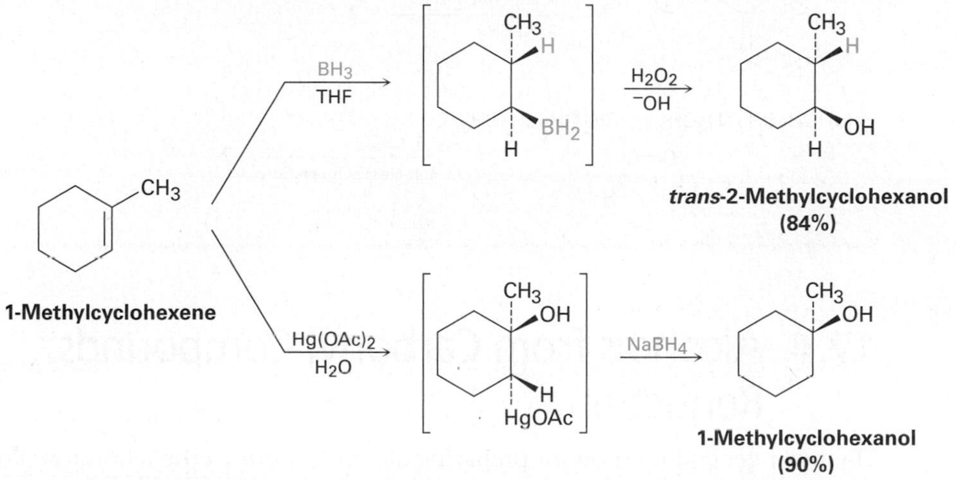

1-Methylcyclohexene

BH_3, THF → [intermediate: CH_3, H, BH_2, H] → H_2O_2, ^{-}OH → ***trans*-2-Methylcyclohexanol (84%)**

$Hg(OAc)_2$, H_2O → [intermediate: CH_3, OH, H, HgOAc] → $NaBH_4$ → **1-Methylcyclohexanol (90%)**

- 1,2-Diols can be prepared either by direct hydroxylation of an alkene with OsO_4 followed by reduction with $NaHSO_3$ or by acid-catalyzed hydrolysis of an epoxide **(Section 8.7)**. The OsO_4 reaction occurs with syn stereochemistry to give a cis diol, and epoxide opening occurs with anti stereochemistry to give a trans diol.

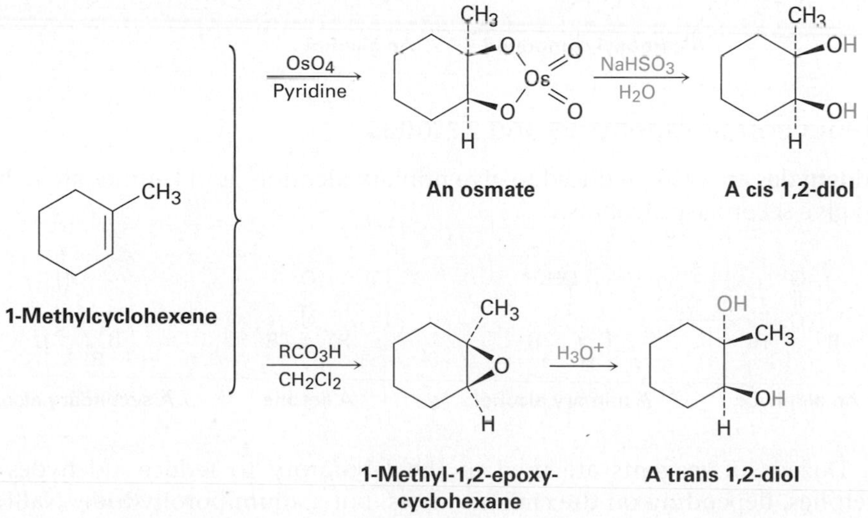

1-Methylcyclohexene

OsO_4, Pyridine → **An osmate** → $NaHSO_3$, H_2O → **A cis 1,2-diol**

RCO_3H, CH_2Cl_2 → **1-Methyl-1,2-epoxycyclohexane** → H_3O^+ → **A trans 1,2-diol**

Problem 17.6

Predict the products of the following reactions:

(a) $CH_3CH_2CH{=}C(CH_3)CH_3$ $\xrightarrow[\text{2. NaOH, } H_2O_2]{\text{1. } BH_3}$?

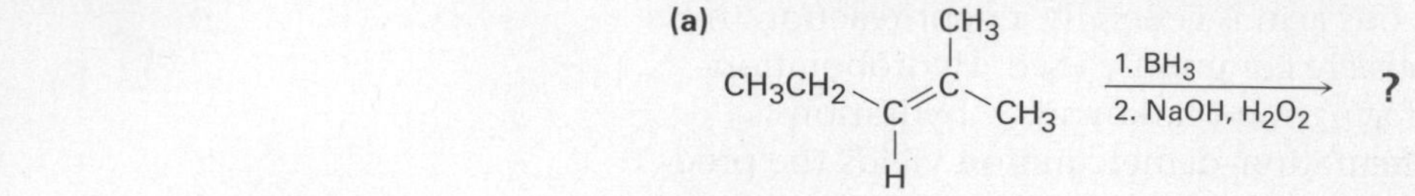

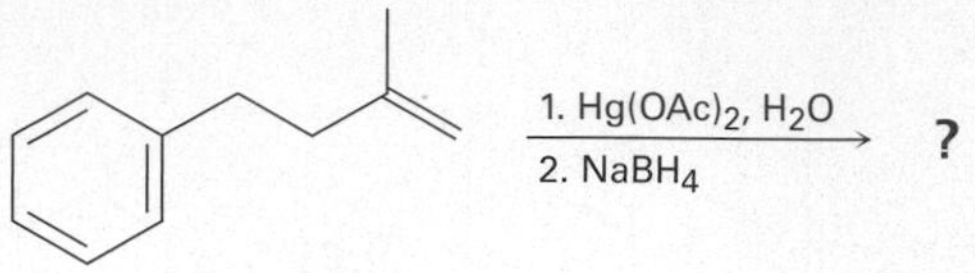

(b) (3-methyl-3-butenyl)benzene $\xrightarrow[\text{2. } NaBH_4]{\text{1. Hg(OAc)}_2\text{, } H_2O}$?

(c) $CH_3CH_2CH_2CH_2CH{=}CHCH_2CH_2CH_2CH_3$ (cis, H and H on same side) $\xrightarrow[\text{2. } NaHSO_3\text{, } H_2O]{\text{1. } OsO_4}$?

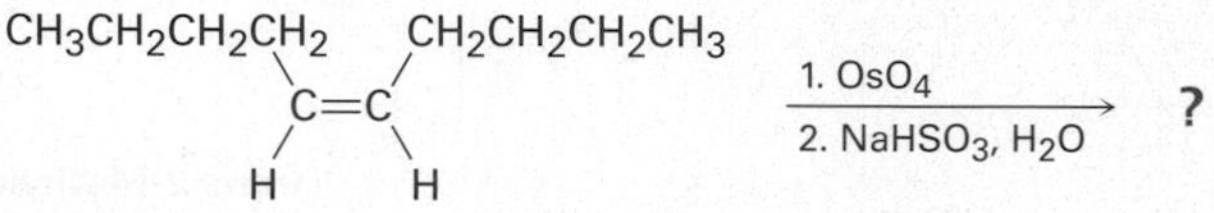

17.4 Alcohols from Carbonyl Compounds: Reduction

The most general method for preparing alcohols, both in the laboratory and in living organisms, is by the reduction of a carbonyl compound. Just as reduction of an alkene adds hydrogen to a C=C bond to give an alkane **(Section 8.6)**, reduction of a carbonyl compound adds hydrogen to a C=O bond to give an alcohol. All kinds of carbonyl compounds can be reduced, including aldehydes, ketones, carboxylic acids, and esters.

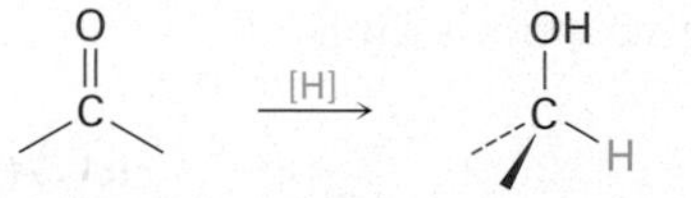

$$\text{C=O} \xrightarrow{[H]} \text{H–C–OH}$$

where [H] is a reducing agent

A carbonyl compound **An alcohol**

Reduction of Aldehydes and Ketones

Aldehydes are easily reduced to give primary alcohols, and ketones are reduced to give secondary alcohols.

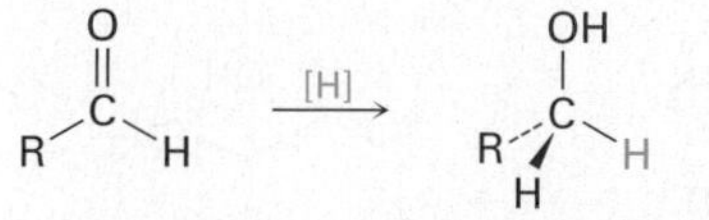

$$RCHO \xrightarrow{[H]} RCH_2OH \qquad RCOR' \xrightarrow{[H]} RR'CHOH$$

An aldehyde **A primary alcohol** 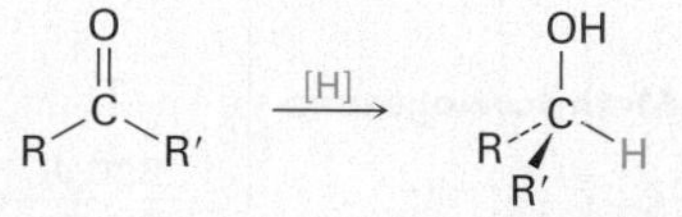 **A ketone** **A secondary alcohol**

Dozens of reagents are used in the laboratory to reduce aldehydes and ketones, depending on the circumstances, but sodium borohydride, $NaBH_4$, is usually chosen because of its safety and ease of handling. Sodium borohydride is a white, crystalline solid that can be weighed in the open atmosphere and used in either water or alcohol solution.

Aldehyde reduction

$$\mathrm{CH_3CH_2CH_2CHO} \xrightarrow[\text{2. } \mathrm{H_3O^+}]{\text{1. } \mathrm{NaBH_4}\text{, ethanol}} \mathrm{CH_3CH_2CH_2CH_2OH}$$

Butanal → **1-Butanol (85%)** **(a 1° alcohol)**

Ketone reduction

Dicyclohexyl ketone —(1. $NaBH_4$, ethanol; 2. H_3O^+)→ **Dicyclohexylmethanol (88%)** **(a 2° alcohol)**

Lithium aluminum hydride, $LiAlH_4$, is another reducing agent often used for reduction of aldehydes and ketones. A grayish powder that is soluble in ether and tetrahydrofuran, $LiAlH_4$ is much more reactive than $NaBH_4$ but also more dangerous. It reacts violently with water and decomposes explosively when heated above 120 °C.

2-Cyclohexenone —(1. $LiAlH_4$, ether; 2. H_3O^+)→ **2-Cyclohexenol (94%)**

We'll defer a detailed discussion of the mechanisms of these reductions until Chapter 19. For the moment, we'll simply note that they involve the addition of a nucleophilic hydride ion (:H^-) to the positively polarized, electrophilic carbon atom of the carbonyl group. The initial product is an alkoxide ion, which is protonated by addition of H_3O^+ in a second step to yield the alcohol product.

A carbonyl compound —(:H^-)→ **An alkoxide ion intermediate** —(H_3O^+)→ **An alcohol**

In living organisms, aldehyde and ketone reductions are carried out by either of the coenzymes NADH (reduced nicotinamide adenine dinucleotide) or NADPH (reduced nicotinamide adenine dinucleotide phosphate). Although these biological "reagents" are much more complex structurally than $NaBH_4$ or $LiAlH_4$, the mechanisms of laboratory and biological reactions are similar. The coenzyme acts as a hydride-ion donor to give an alkoxide anion, and the intermediate anion is then protonated by acid. An example is the reduction of acetoacetyl ACP to β-hydroxybutyryl ACP, a step in the biological synthesis of fats **(Figure 17.4)**. Note that the *pro-R* hydrogen of NADPH is the one transferred in this example. Enzyme-catalyzed reactions usually occur with high specificity, although it's not usually possible to predict the stereochemical result before the fact.

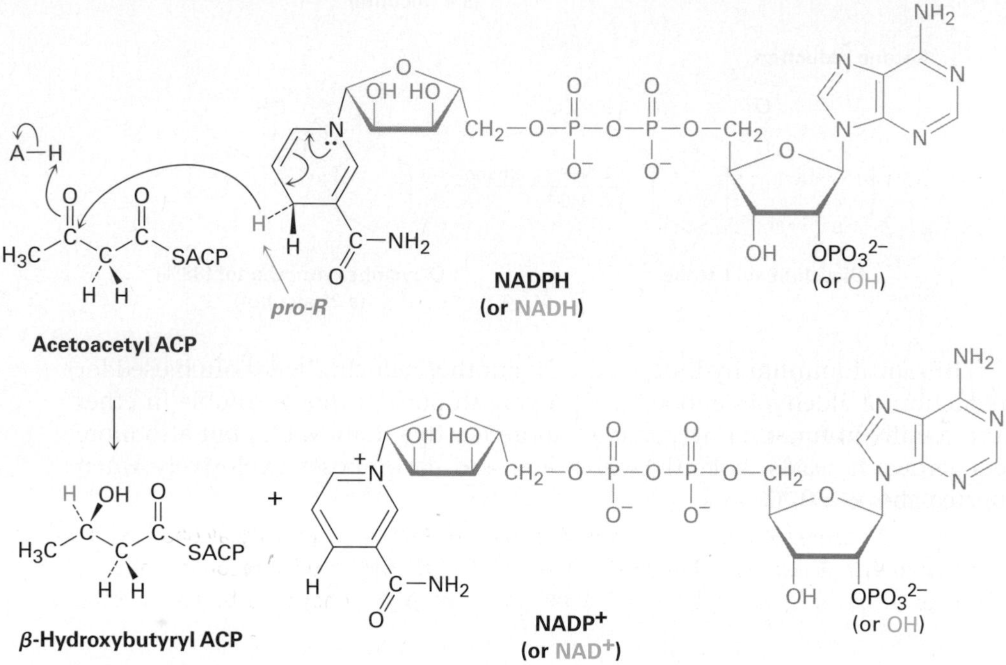

Figure 17.4 The biological reduction of a ketone (acetoacetyl ACP) to an alcohol (β-hydroxybutyryl ACP) by NADPH.

Reduction of Carboxylic Acids and Esters

Carboxylic acids and esters are reduced to give primary alcohols.

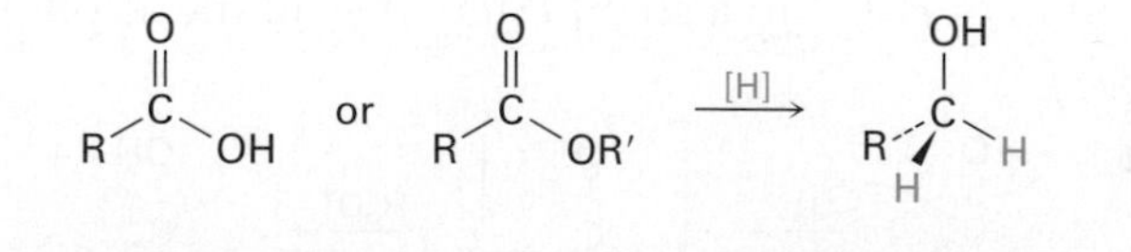

These reactions aren't as rapid as the reductions of aldehydes and ketones. $NaBH_4$ reduces esters very slowly and does not reduce carboxylic acids at all. Instead, carboxylic acid and ester reductions are usually carried out with the more reactive reducing agent $LiAlH_4$. All carbonyl groups, including acids, esters, ketones, and aldehydes, are reduced by $LiAlH_4$. Note that one hydrogen atom is delivered to the carbonyl carbon atom during aldehyde and ketone reductions but that two hydrogens become bonded to the former carbonyl carbon during carboxylic acid and ester reductions. We'll defer a discussion of the mechanisms of these reactions until Chapter 21.

Carboxylic acid reduction

$$CH_3(CH_2)_7CH{=}CH(CH_2)_7\overset{O}{\overset{\|}{C}}OH \xrightarrow[2.\ H_3O^+]{1.\ LiAlH_4,\ ether} CH_3(CH_2)_7CH{=}CH(CH_2)_7CH_2OH$$

9-Octadecenoic acid (oleic acid) → **9-Octadecen-1-ol (87%)**

Ester reduction

$$CH_3CH_2CH{=}CH\overset{O}{\overset{\|}{C}}OCH_3 \xrightarrow[2.\ H_3O^+]{1.\ LiAlH_4,\ ether} CH_3CH_2CH{=}CHCH_2OH + CH_3OH$$

Methyl 2-pentenoate → **2-Penten-1-ol (91%)**

Worked Example 17.2 Identifying a Reactant, Given the Product

What carbonyl compounds would you reduce to obtain the following alcohols?

(a) $CH_3CH_2\overset{CH_3}{\overset{|}{C}}HCH_2\overset{OH}{\overset{|}{C}}HCH_3$ **(b)** (4-methyl-2-penten-1-ol skeletal structure) CH_2OH

Strategy

Identify the target alcohol as primary, secondary, or tertiary. A primary alcohol can be prepared by reduction of an aldehyde, an ester, or a carboxylic acid; a secondary alcohol can be prepared by reduction of a ketone; and a tertiary alcohol can't be prepared by reduction.

Solution

(a) The target molecule is a secondary alcohol, which can be prepared only by reduction of a ketone. Either $NaBH_4$ or $LiAlH_4$ can be used.

$$CH_3CH_2\overset{CH_3}{\overset{|}{C}}HCH_2\overset{O}{\overset{\|}{C}}CH_3 \xrightarrow[2.\ H_3O^+]{1.\ NaBH_4\ or\ LiAlH_4} CH_3CH_2\overset{CH_3}{\overset{|}{C}}HCH_2\overset{OH}{\overset{|}{C}}HCH_3$$

(b) The target molecule is a primary alcohol, which can be prepared by reduction of an aldehyde, an ester, or a carboxylic acid. $LiAlH_4$ is needed for the ester and carboxylic acid reductions.

(skeletal) CHO — 1. $NaBH_4$ or $LiAlH_4$; 2. H_3O^+ →
(skeletal) CO_2CH_3 — 1. $LiAlH_4$; 2. H_3O^+ → (skeletal) CH_2OH
(skeletal) CO_2H — 1. $LiAlH_4$; 2. H_3O^+ →

Problem 17.7

What reagent would you use to accomplish each of the following reactions?

(a) $CH_3\overset{O}{\overset{\|}{C}}CH_2CH_2\overset{O}{\overset{\|}{C}}OCH_3 \xrightarrow{?} CH_3\overset{OH}{\overset{|}{C}}HCH_2CH_2\overset{O}{\overset{\|}{C}}OCH_3$

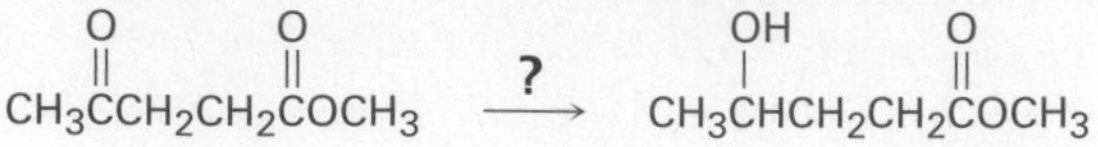

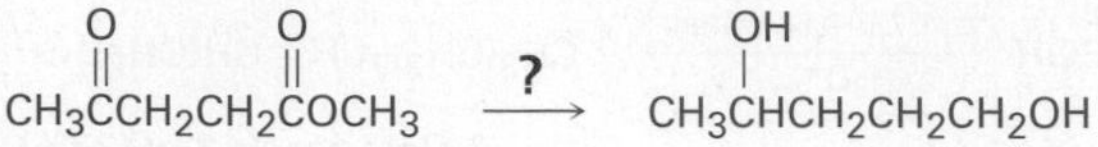

(b) $CH_3\overset{O}{\overset{\|}{C}}CH_2CH_2\overset{O}{\overset{\|}{C}}OCH_3 \xrightarrow{?} CH_3\overset{OH}{\overset{|}{C}}HCH_2CH_2CH_2OH$

(c) O ? OH

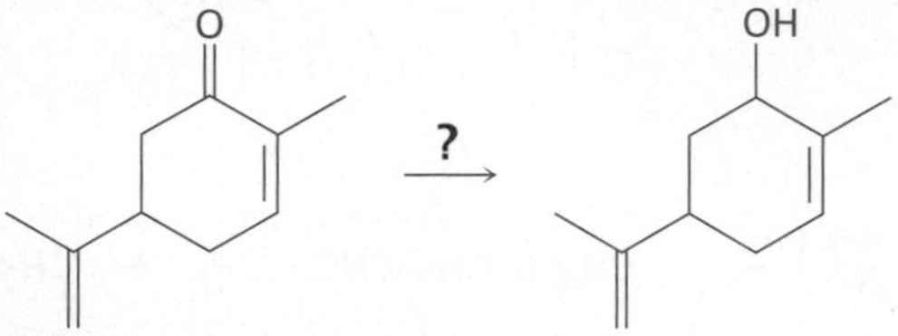

Problem 17.8

What carbonyl compounds give the following alcohols on reduction with $LiAlH_4$? Show all possibilities.

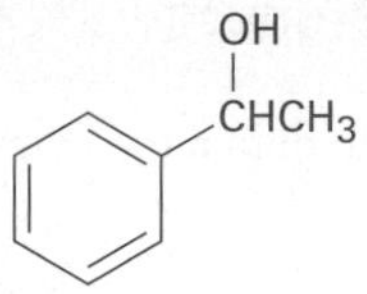

(a) CH_2OH (b) OH CHCH₃ (c) OH H (d) $(CH_3)_2CHCH_2OH$

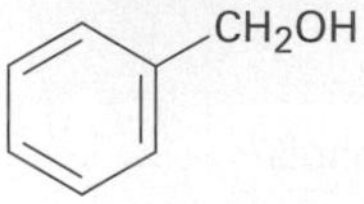

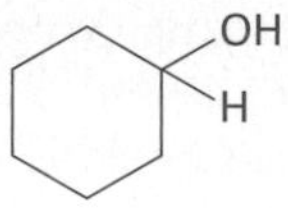

17.5 Alcohols from Carbonyl Compounds: Grignard Reaction

Grignard reagents (RMgX), prepared by reaction of organohalides with magnesium **(Section 10.6)**, react with carbonyl compounds to yield alcohols in much the same way that hydride reducing agents do. Just as carbonyl reduction involves addition of a hydride ion nucleophile to the C=O bond, Grignard reaction involves addition of a carbanion nucleophile ($R:^{-}\,^{+}MgX$).

$$R{-}X + Mg \longrightarrow \overset{\delta-}{R}{-}\overset{\delta+}{MgX}$$

A Grignard reagent

R = 1°, 2°, or 3° alkyl, aryl, or vinylic
X = Cl, Br, I

$$\overset{O}{\overset{\|}{C}} \xrightarrow[\text{2. } H_3O^+]{\text{1. RMgX, ether}} \overset{OH}{\overset{|}{C}}{-}R + HOMgX$$

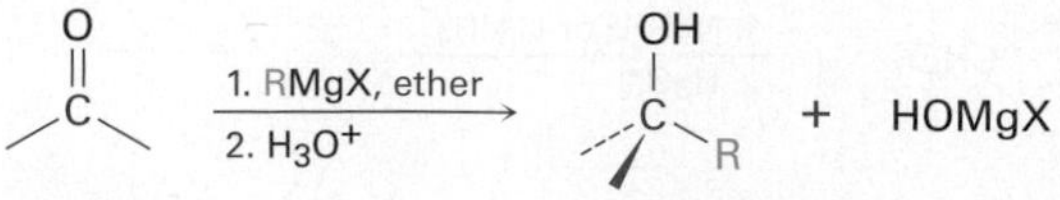

The reaction of Grignard reagents with carbonyl compounds has no direct counterpart in biological chemistry because organomagnesium compounds are too strongly basic to exist in an aqueous medium. Nevertheless, the reaction is worth knowing about for two reasons. First, the reaction is an unusually broad and useful method of alcohol synthesis and demonstrates again the relative freedom with which chemists can operate in the laboratory. Second, the reaction *does* have an indirect biological counterpart, for we'll see in Chapter 23 that the addition of stabilized carbon nucleophiles to carbonyl compounds is used in almost all metabolic pathways as the major process for forming carbon–carbon bonds.

As examples of their addition to carbonyl compounds, Grignard reagents react with formaldehyde, $H_2C{=}O$, to give primary alcohols, with aldehydes to give secondary alcohols, and with ketones to give tertiary alcohols.

Formaldehyde reaction

Cyclohexylmagnesium bromide (MgBr) + Formaldehyde ($H_2C{=}O$) $\xrightarrow[\text{2. } H_3O^+]{\text{1. Mix in ether}}$ Cyclohexylmethanol (65%) (CH_2OH)

Cyclohexylmagnesium bromide | **Formaldehyde** | **Cyclohexylmethanol (65%) (a 1° alcohol)**

Aldehyde reaction

Phenylmagnesium bromide (MgBr) + CH_3CHCH_2CH (with CH_3 and O) $\xrightarrow[\text{2. } H_3O^+]{\text{1. Mix in ether}}$ CH_3CHCH_2CH–phenyl (with CH_3 and OH)

Phenylmagnesium bromide | **3-Methylbutanal** | **3-Methyl-1-phenyl-1-butanol (73%) (a 2° alcohol)**

Ketone reaction

CH_3CH_2MgBr + cyclohexanone $\xrightarrow[\text{2. } H_3O^+]{\text{1. Mix in ether}}$ 1-ethylcyclohexanol (OH, CH_2CH_3)

Ethylmagnesium bromide | **Cyclohexanone** | **1-Ethylcyclohexanol (89%) (a 3° alcohol)**

Esters react with Grignard reagents to yield tertiary alcohols in which two of the substituents bonded to the hydroxyl-bearing carbon have come from the Grignard reagent, just as $LiAlH_4$ reduction of an ester adds two hydrogens.

$CH_3CH_2CH_2CH_2C(=O)OCH_2CH_3$ $\xrightarrow[\text{2. } H_3O^+]{\text{1. 2 } CH_3MgBr}$ $CH_3CH_2CH_2CH_2C(CH_3)_2OH$ + CH_3CH_2OH

Ethyl pentanoate | **2-Methyl-2-hexanol (85%) (a 3° alcohol)**

Carboxylic acids don't give addition products with Grignard reagents because the acidic carboxyl hydrogen reacts with the basic Grignard reagent to yield a hydrocarbon and the magnesium salt of the acid.

$$\left[\text{R–X} + \text{Mg} \longrightarrow \text{R–MgX}\right]$$

$$\text{R}'\text{–C(=O)–OH} \xrightarrow{\text{RMgX, ether}} \text{R}'\text{–C(=O)–O}^- \ {}^+\text{MgBr} + \text{R–H}$$

A carboxylic acid → **A carboxylic acid salt**

The Grignard reaction, although useful, also has limitations. One major problem is that a Grignard reagent can't be prepared from an organohalide if other reactive functional groups are present in the same molecule. For example, a compound that is both an alkyl halide and a ketone can't form a Grignard reagent because it would react with itself. Similarly, a compound that is both an alkyl halide and a carboxylic acid, an alcohol, or an amine can't form a Grignard reagent because the acidic $RCO_2\mathbf{H}$, $RO\mathbf{H}$, or $RN\mathbf{H}_2$ hydrogen present in the same molecule would react with the basic Grignard reagent as rapidly as it forms. In general, Grignard reagents can't be prepared from alkyl halides that contain the following functional groups (FG):

Br—(Molecule)—FG

where FG = —OH, —NH, —SH, —CO_2H — The Grignard reagent is protonated by these groups.

FG = —C(=O)H, —C(=O)R, —C(=O)NR_2, —C≡N, —NO_2, —SO_2R — The Grignard reagent adds to these groups.

As with the reduction of carbonyl compounds discussed in the previous section, we'll defer a detailed treatment of the mechanism of Grignard reactions until Chapter 19. For the moment, it's sufficient to note that Grignard reagents act as nucleophilic carbanions (**:**R^-) and that their addition to a carbonyl compound is analogous to the addition of hydride ion. The intermediate is an alkoxide ion, which is protonated by addition of H_3O^+ in a second step.

$$\text{C}^{\delta+}\text{=O}^{\delta-} \xrightarrow{:\text{R}^-} \left[\text{–C(O}^-\text{)–R}\right] \xrightarrow{\text{H}_3\text{O}^+} \text{–C(OH)–R}$$

A carbonyl compound → **An alkoxide ion intermediate** → **An alcohol**

Using a Grignard Reaction to Synthesize an Alcohol

Worked Example 17.3

How could you use the addition of a Grignard reagent to a ketone to synthesize 2-phenyl-2-butanol?

Strategy

Draw the product, and identify the three groups bonded to the alcohol carbon atom. One of the three will have come from the Grignard reagent, and the remaining two will have come from the ketone.

Solution

2-Phenyl-2-butanol has a methyl group, an ethyl group, and a phenyl group ($-C_6H_5$) attached to the alcohol carbon atom. Thus, the possibilities are addition of ethylmagnesium bromide to acetophenone, addition of methylmagnesium bromide to propiophenone, and addition of phenylmagnesium bromide to 2-butanone.

Acetophenone — 1. CH_3CH_2MgBr 2. H_3O^+ → **2-Phenyl-2-butanol**

Propiophenone — 1. CH_3MgBr 2. H_3O^+ → **2-Phenyl-2-butanol**

2-Butanone ($H_3C-CO-CH_2CH_3$) — 1. C_6H_5MgBr 2. H_3O^+ → **2-Phenyl-2-butanol**

Using a Grignard Reaction to Synthesize an Alcohol

Worked Example 17.4

How could you use the reaction of a Grignard reagent with a carbonyl compound to synthesize 2-methyl-2-pentanol?

Strategy

Draw the product, and identify the three groups bonded to the alcohol carbon atom. If the three groups are all different, the starting carbonyl compound must be a ketone. If two of the three groups are identical, the starting carbonyl compound might be either a ketone or an ester.

Solution

In the present instance, the product is a tertiary alcohol with two methyl groups and one propyl group. Starting from a ketone, the possibilities are addition of methylmagnesium bromide to 2-pentanone and addition of propylmagnesium bromide to acetone.

$$CH_3CH_2CH_2C(=O)CH_3 \xrightarrow[\text{2. } H_3O^+]{\text{1. } CH_3MgBr} CH_3CH_2CH_2C(CH_3)_2OH$$

2-Pentanone → **2-Methyl-2-pentanol**

$$H_3CC(=O)CH_3 \xrightarrow[\text{2. } H_3O^+]{\text{1. } CH_3CH_2CH_2MgBr} CH_3CH_2CH_2C(CH_3)_2OH$$

Acetone → **2-Methyl-2-pentanol**

Starting from an ester, the only possibility is addition of methylmagnesium bromide to an ester of butanoic acid, such as methyl butanoate.

$$CH_3CH_2CH_2C(=O)OCH_3 \xrightarrow[\text{2. } H_3O^+]{\text{1. 2 } CH_3MgBr} CH_3CH_2CH_2C(CH_3)_2OH + CH_3OH$$

Methyl butanoate → **2-Methyl-2-pentanol**

Problem 17.9

Show the products obtained from addition of methylmagnesium bromide to the following compounds:

(a) Cyclopentanone **(b)** Benzophenone (diphenyl ketone) **(c)** 3-Hexanone

Problem 17.10

Use a Grignard reaction to prepare the following alcohols:

(a) 2-Methyl-2-propanol **(b)** 1-Methylcyclohexano **(c)** 3-Methyl-3-pentanol
(d) 2-Phenyl-2-butanol **(e)** Benzyl alcohol **(f)** 4-Methyl-1-pentanol

Problem 17.11

Use the reaction of a Grignard reagent with a carbonyl compound to synthesize the following compound:

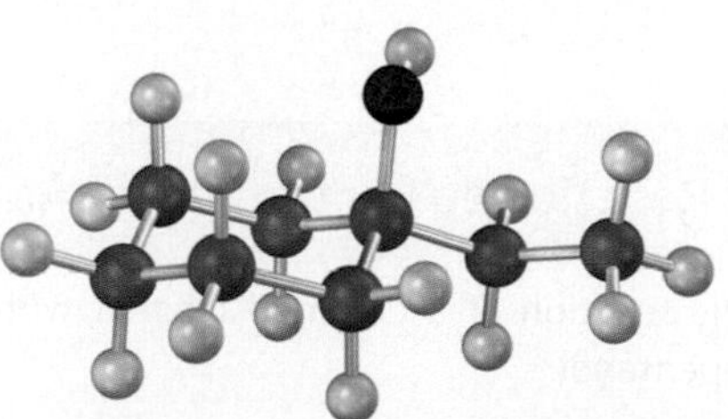

17.6 Reactions of Alcohols

We've already seen several reactions of alcohols—their conversion into alkyl halides and tosylates in **Section 10.5** and their dehydration to give alkenes in **Section 8.1**—although without mechanistic details. Let's now look at those details.

Conversion of Alcohols into Alkyl Halides

Tertiary alcohols react with either HCl or HBr at 0 °C by an S_N1 mechanism through a carbocation intermediate. Primary and secondary alcohols are much more resistant to acid, however, and are best converted into halides by treatment with either $SOCl_2$ or PBr_3 through an S_N2 mechanism.

The reaction of a tertiary alcohol with HX takes place by an S_N1 mechanism when acid protonates the hydroxyl oxygen atom, water is expelled to generate a carbocation, and the cation reacts with nucleophilic halide ion to give the alkyl halide product.

$$R_3C{-}OH \xrightarrow[\text{or HBr}]{\text{HCl}} \left[R_3C{-}\overset{+}{O}H_2\right] \xrightarrow[S_N1]{-H_2O} \left[R_3C^{+}\right] \xrightarrow[\text{or Br}^-]{\text{Cl}^-} R_3C{-}Cl\ (Br)$$

A 3° alcohol — A carbocation — An alkyl chloride or bromide

The reactions of primary and secondary alcohols with $SOCl_2$ and PBr_3 take place by S_N2 mechanisms. Hydroxide ion itself is too poor a leaving group to be displaced by nucleophiles in S_N2 reactions, but reaction of an alcohol with $SOCl_2$ or PBr_3 converts the –OH into a much better leaving group, either a chlorosulfite (–OSOCl) or a dibromophosphite (–$OPBr_2$), that is readily expelled by backside nucleophilic substitution.

$$\text{A 1° or 2° alcohol: } HC{-}OH \xrightarrow[\text{ether}]{SOCl_2} \left[Cl^- \; HC{-}O{-}\overset{O}{\overset{\|}{S}}{-}Cl\right] \xrightarrow{S_N2} Cl{-}CH + SO_2 + HCl$$

A chlorosulfite — An alkyl chloride

$$HC{-}OH \xrightarrow[\text{ether}]{PBr_3} \left[Br^- \; HC{-}O{-}PBr_2\right] \xrightarrow{S_N2} Br{-}CH + Br_2POH$$

A dibromophosphite — An alkyl bromide

Conversion of Alcohols into Tosylates

Alcohols react with *p*-toluenesulfonyl chloride (tosyl chloride, *p* TosCl) in pyridine solution to yield alkyl tosylates, ROTos **(Section 11.1)**. Only the O–H bond of the alcohol is broken in this reaction; the C–O bond remains intact, so no change of configuration occurs if the oxygen is attached to a chirality center.

The resultant alkyl tosylates behave much like alkyl halides, undergoing both S_N1 and S_N2 substitution reactions.

An alcohol + *p*-Toluenesulfonyl chloride $\xrightarrow{\text{Pyridine}}$ A tosylate (ROTos) + Pyridine·HCl

One of the most important reasons for using tosylates in S_N2 reactions is stereochemical. The S_N2 reaction of an alcohol via an alkyl halide proceeds with *two* inversions of configuration—one to make the halide from the alcohol and one to substitute the halide—and yields a product with the same stereochemistry as the starting alcohol. The S_N2 reaction of an alcohol via a tosylate, however, proceeds with only one inversion and yields a product of opposite stereochemistry to the starting alcohol. **Figure 17.5** shows a series of reactions on the *R* enantiomer of 2-octanol that illustrates these stereochemical relationships.

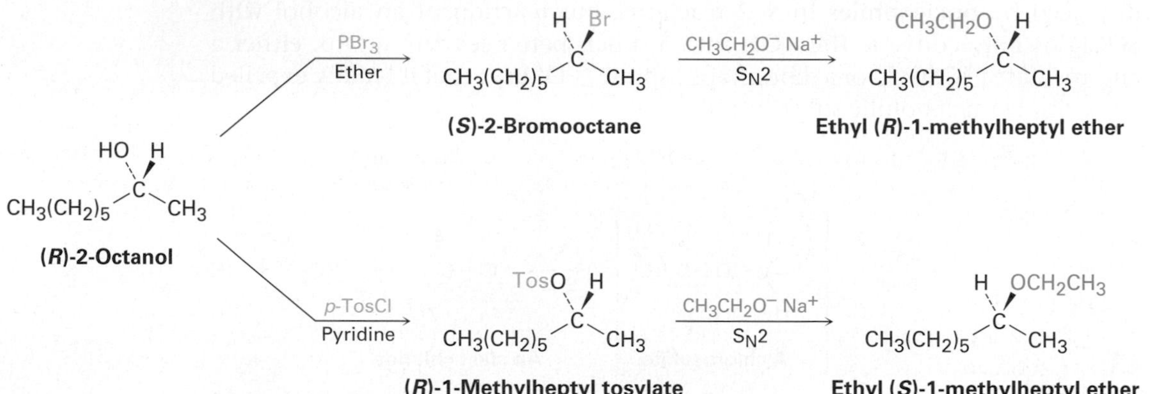

Figure 17.5 Stereochemical consequences of S_N2 reactions on derivatives of (*R*)-2-octanol. Substitution through the halide gives a product with the same stereochemistry as the starting alcohol; substitution through the tosylate gives a product with opposite stereochemistry to the starting alcohol.

Problem 17.12

How would you carry out the following transformation, a step used in the commercial synthesis of (*S*)-ibuprofen?

Dehydration of Alcohols to Yield Alkenes

A third important reaction of alcohols, both in the laboratory and in biological pathways, is their dehydration to give alkenes. Because of the usefulness of the reaction, a number of ways have been devised for carrying out dehydrations. One method that works particularly well for tertiary alcohols is the acid-catalyzed reaction discussed in **Section 8.1**. For example, treatment of 1-methylcyclohexanol with warm aqueous sulfuric acid in a solvent such as tetrahydrofuran results in loss of water and formation of 1-methylcyclohexene.

H_3C OH — $\xrightarrow[50\ °C]{H_3O^+,\ THF}$ — CH_3

1-Methylcyclohexanol **1-Methylcyclohexene (91%)**

Acid-catalyzed dehydrations usually follow Zaitsev's rule **(Section 11.7)** and yield the more stable alkene as the major product. Thus, 2-methyl-2-butanol gives primarily 2-methyl-2-butene (trisubstituted double bond) rather than 2-methyl-1-butene (disubstituted double bond).

$$H_3C-C(CH_3)(OH)-CH_2CH_3 \xrightarrow[25\ °C]{H_3O^+,\ THF} (CH_3)_2C{=}CHCH_3 + CH_2{=}C(CH_3)-CH_2CH_3$$

2-Methyl-2-butanol **2-Methyl-2-butene (trisubstituted)** **2-Methyl-1-butene (disubstituted)**

Major product **Minor product**

The reaction is an E1 process **(Section 11.10)** and occurs by the three-step mechanism shown in **Figure 17.6** on the next page. Protonation of the alcohol oxygen is followed by unimolecular loss of water to generate a carbocation intermediate and final loss of a proton from the neighboring carbon atom to complete the process. As usual for E1 reactions, tertiary alcohols react fastest because they lead to stabilized, tertiary carbocation intermediates. Secondary alcohols can be made to react, but the conditions are severe (75% H_2SO_4, 100 °C) and sensitive molecules don't survive.

To circumvent the need for strong acid and allow the dehydration of secondary alcohols in a gentler way, reagents have been developed that are effective under mild, basic conditions. One such reagent, phosphorus oxychloride ($POCl_3$) in the basic amine solvent pyridine, is often able to effect the dehydration of secondary and tertiary alcohols at 0 °C.

CH_3 OH — $\xrightarrow[Pyridine,\ 0\ °C]{POCl_3}$ — CH_3 H

1-Methylcyclohexanol **1-Methylcyclohexene (96%)**

Figure 17.6 | MECHANISM

Mechanism of the acid-catalyzed dehydration of a tertiary alcohol to yield an alkene. The process is an E1 reaction and involves a carbocation intermediate.

Alcohol dehydrations carried out with $POCl_3$ in pyridine take place by an E2 mechanism, as shown in **Figure 17.7**. Because hydroxide ion is a poor leaving group **(Section 11.3)**, direct E2 elimination of water from an alcohol does not occur. On reaction with $POCl_3$, however, the –OH group is converted into a dichlorophosphate ($-OPOCl_2$), which is a good leaving group and is readily eliminated. Pyridine is both the reaction solvent and the base that removes a neighboring proton in the E2 elimination step.

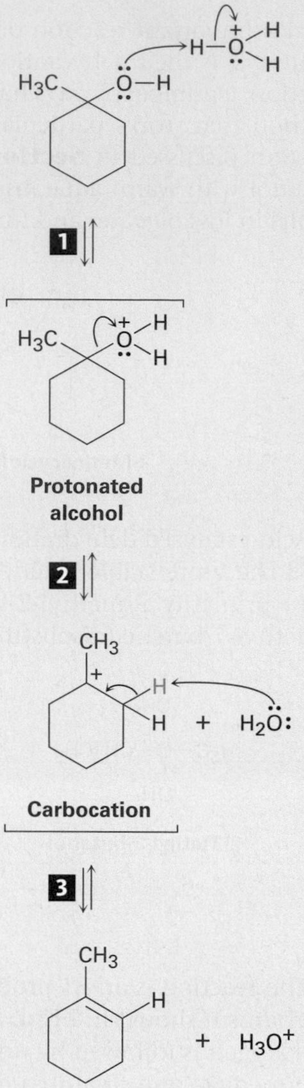

Figure 17.7 | MECHANISM

Mechanism of the dehydration of secondary and tertiary alcohols by reaction with $POCl_3$ in pyridine. The reaction is an E2 process.

1 The alcohol hydroxyl group reacts with $POCl_3$ to form a dichlorophosphate intermediate.

2 E2 elimination then occurs by the usual one-step mechanism as the amine base pyridine abstracts a proton from the neighboring carbon at the same time that the dichlorophosphate group is leaving.

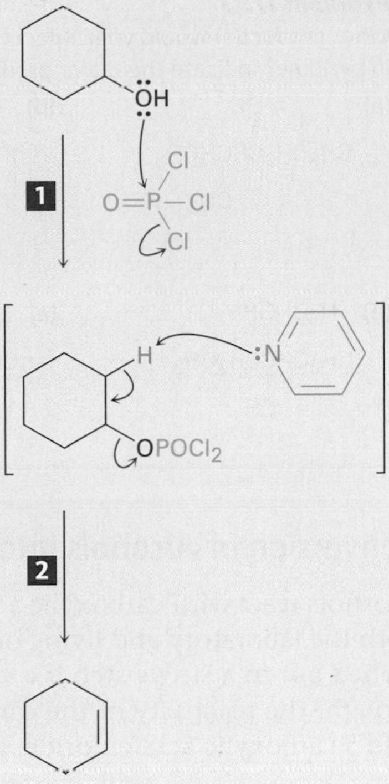

As noted previously in **Section 11.11**, biological dehydrations are also common and usually occur by an E1cB mechanism on a substrate in which the –OH group is two carbons away from a carbonyl group. An example occurs in the biosynthesis of the aromatic amino acid tyrosine. A base (:B) first abstracts a proton from the carbon adjacent to the carbonyl group, and the anion intermediate then expels the –OH group with simultaneous protonation by an acid (HA) to form water.

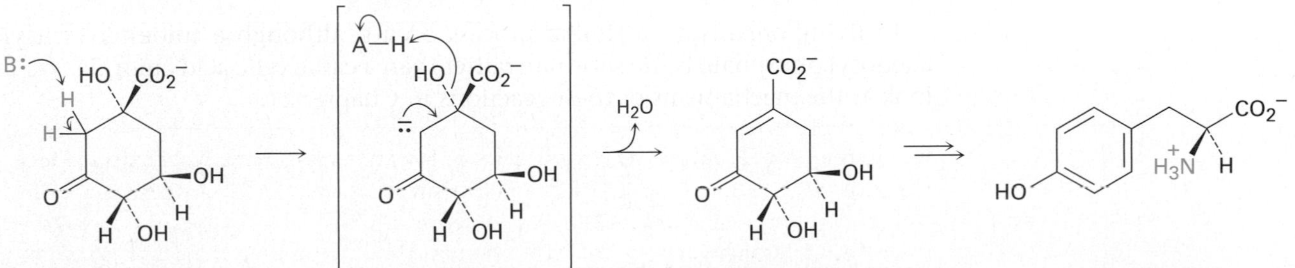

5-Dehydroquinate → [**Anion intermediate**] → (H_2O) **5-Dehydroshikimate** ⇉ **Tyrosine**

Problem 17.13

What product(s) would you expect from dehydration of the following alcohols with $POCl_3$ in pyridine? Indicate the major product in each case.

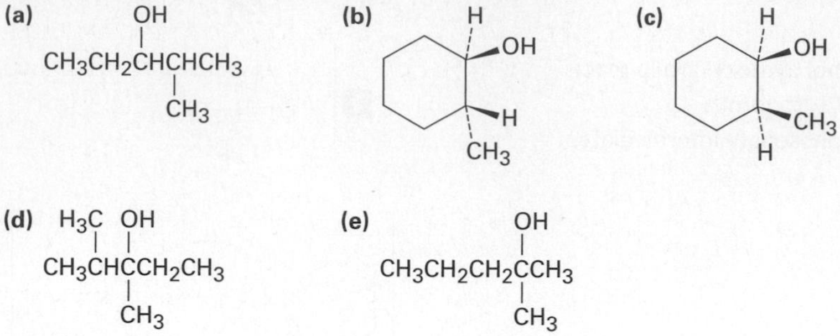

(a) $CH_3CH_2CH(OH)CH(CH_3)CH_3$
(b) cyclohexane with H, OH and H, CH_3 substituents
(c) cyclohexane with H, OH and CH_3, H substituents
(d) $CH_3CH(CH_3)C(OH)(CH_3)CH_2CH_3$
(e) $CH_3CH_2CH_2C(OH)(CH_3)CH_3$

Conversion of Alcohols into Esters

Alcohols react with carboxylic acids to give esters, a reaction that is common in both the laboratory and living organisms. In the laboratory, the reaction can be carried out in a single step if a strong acid is used as catalyst. More frequently, though, the reactivity of the carboxylic acid is enhanced by first converting it into a carboxylic acid chloride, which then reacts with the alcohol.

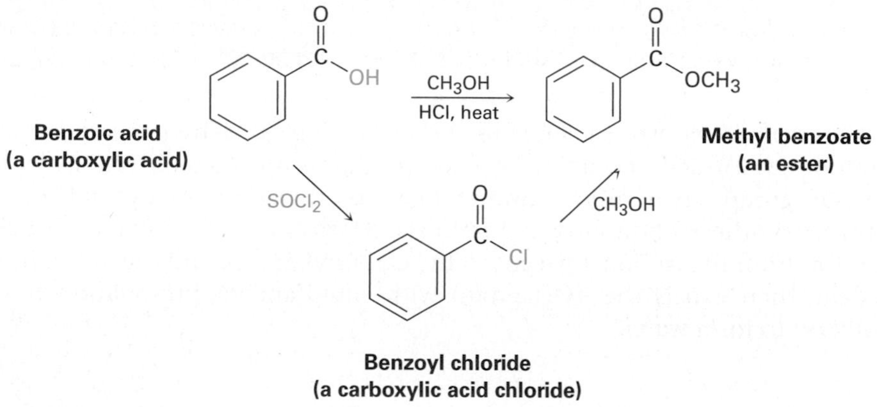

Benzoic acid (a carboxylic acid) —(CH_3OH, HCl, heat)→ **Methyl benzoate (an ester)**

Benzoic acid —($SOCl_2$)→ **Benzoyl chloride (a carboxylic acid chloride)** —(CH_3OH)→ Methyl benzoate

In living organisms, a similar process occurs, although a thioester or acyl adenosyl phosphate is the substrate rather than a carboxylic acid chloride. We'll look at the mechanisms of these reactions in Chapter 21.

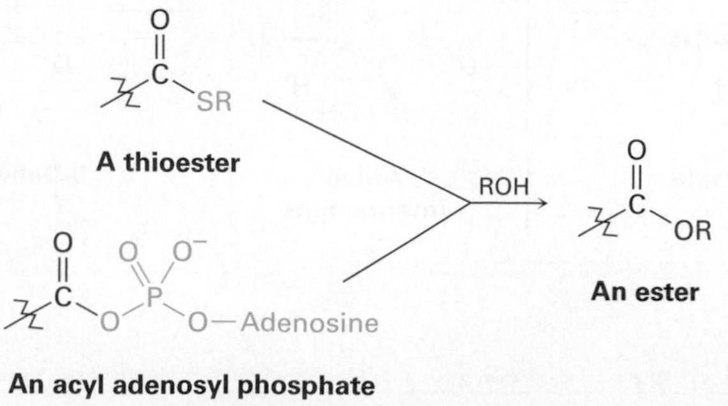

A thioester / **An acyl adenosyl phosphate** —(ROH)→ **An ester**

17.7 Oxidation of Alcohols

Perhaps the most valuable reaction of alcohols is their oxidation to give carbonyl compounds—the opposite of the reduction of carbonyl compounds to give alcohols. Primary alcohols yield aldehydes or carboxylic acids, secondary alcohols yield ketones, but tertiary alcohols don't normally react with most oxidizing agents.

Primary alcohol — $RCH_2OH \xrightarrow{[O]} RCHO$ **An aldehyde** $\xrightarrow{[O]} RCOOH$ **A carboxylic acid**

Secondary alcohol — $RR'CHOH \xrightarrow{[O]} RCOR'$ **A ketone**

Tertiary alcohol — $RR'R''COH \xrightarrow{[O]}$ *No reaction*

The oxidation of a primary or secondary alcohol can be accomplished by any of a large number of reagents, including $KMnO_4$, CrO_3, and $Na_2Cr_2O_7$. Which reagent is used in a specific case depends on such factors as cost, convenience, reaction yield, and alcohol sensitivity. For example, the large-scale oxidation of a simple, inexpensive alcohol such as cyclohexanol might best be done with a cheap oxidant such as $Na_2Cr_2O_7$. On the other hand, the small-scale oxidation of a delicate and expensive polyfunctional alcohol might best be done with one of several mild and high-yielding reagents, regardless of cost.

Primary alcohols are oxidized to either aldehydes or carboxylic acids, depending on the reagents chosen and the conditions used. Older methods were often based on Cr(VI) reagents such as CrO_3 or $Na_2Cr_2O_7$, but a more common current choice for preparing an aldehyde from a primary alcohol in the laboratory is to use the I(V)-containing *Dess–Martin periodinane* in dichloromethane solvent.

Geraniol (CH_2OH) $\xrightarrow[CH_2Cl_2]{\text{Dess–Martin periodinane}}$ **Geranial (84%)**

Dess–Martin periodinane (AcO, OAc, OAc on I; O; O)

–OAc = acetate ($-O-C(=O)-CH_3$)

Most other commonly used oxidizing agents, such as chromium trioxide (CrO_3) in aqueous acid, oxidize primary alcohols directly to carboxylic acids. An aldehyde is involved as an intermediate in this reaction but can't usually be isolated because it is further oxidized too rapidly.

$$CH_3(CH_2)_8CH_2OH \xrightarrow[H_3O^+,\ \text{acetone}]{CrO_3} CH_3(CH_2)_8COOH$$

1-Decanol → **Decanoic acid (93%)**

Secondary alcohols are easily oxidized to give ketones. For a sensitive or costly alcohol, the Dess–Martin procedure is often used because the reaction is nonacidic and occurs at lower temperatures. For a large-scale oxidation, however, an inexpensive reagent such as $Na_2Cr_2O_7$ in aqueous acetic acid might be used.

$$\text{4-}tert\text{-Butylcyclohexanol} \xrightarrow[H_2O,\ CH_3CO_2H,\ \text{heat}]{Na_2Cr_2O_7} \text{4-}tert\text{-Butylcyclohexanone (91\%)}$$

4-*tert*-Butylcyclohexanol → **4-*tert*-Butylcyclohexanone (91%)**

All these oxidations occur by a mechanism that is closely related to the E2 reaction **(Section 11.8)**. In the Dess–Martin oxidation, for instance, the first step involves a substitution reaction between the alcohol and the I(V) reagent to form a new periodinane intermediate, followed by expulsion of reduced I(III) as the leaving group. Similarly, when a Cr(VI) reagent, such as CrO_3, is the oxidant, reaction with the alcohol gives a chromate intermediate followed by expulsion of a reduced Cr(IV) species. Although we usually think of the E2 reaction as a means of generating a carbon–*carbon* double bond by elimination of a halide leaving group, the reaction is also useful for generating a carbon–*oxygen* double bond by elimination of a reduced iodine or metal as the leaving group.

Biological alcohol oxidations are the opposite of biological carbonyl reductions and are carried by the coenzymes NAD^+ and $NADP^+$. A base removes the –OH proton, and the alkoxide ion transfers a hydride ion to the coenzyme. An example is the oxidation of *sn*-glycerol 3-phosphate to dihydroxyacetone phosphate, a step in the biological metabolism of fats **(Figure 17.8)**. Note that addition occurs exclusively on the *Re* face of the NAD^+ ring, adding a hydrogen with *pro-R* stereochemistry.

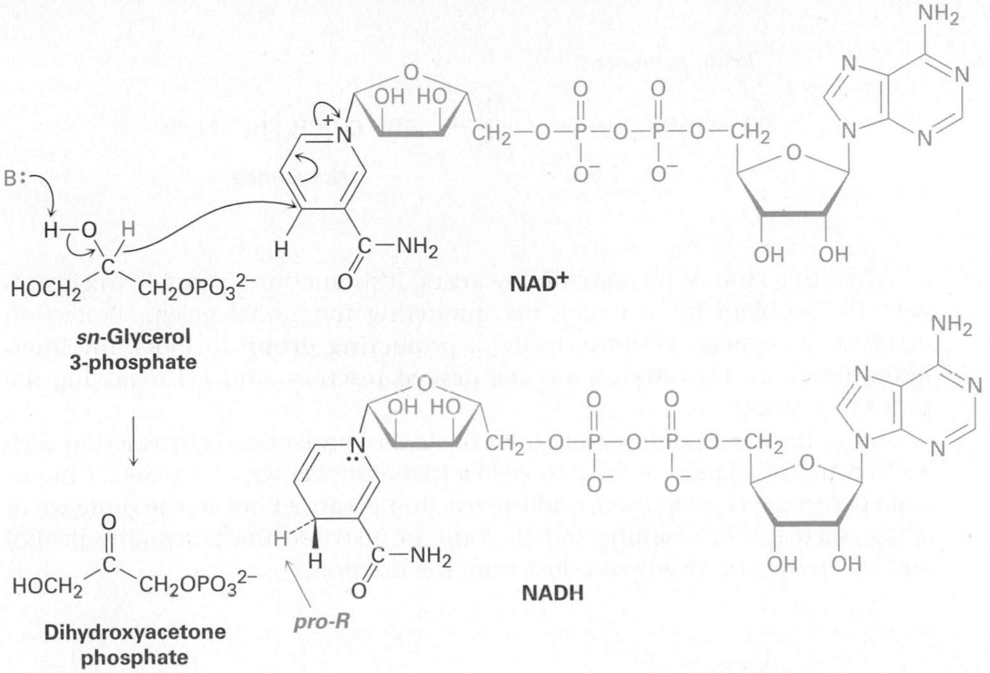

Figure 17.8 The biological oxidation of an alcohol (*sn*-glycerol 3-phosphate) to give a ketone (dihydroxyacetone phosphate). This mechanism is the exact opposite of the ketone reduction shown previously in Figure 17.4.

Problem 17.14

What alcohols would give the following products on oxidation?

(a) 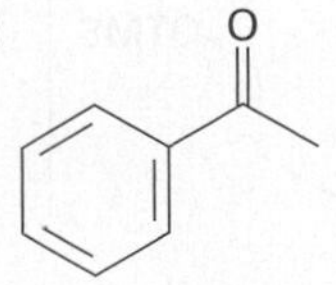 **(b)** $CH_3CH(CH_3)CHO$ **(c)**

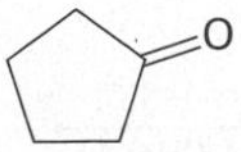

Problem 17.15

What products would you expect from oxidation of the following compounds with CrO_3 in aqueous acid? With the Dess–Martin periodinane?

(a) 1-Hexanol **(b)** 2-Hexanol **(c)** Hexanal

17.8 Protection of Alcohols

It often happens, particularly during the synthesis of complex molecules, that one functional group in a molecule interferes with an intended reaction on another functional group elsewhere in the same molecule. We saw earlier in this chapter, for instance, that a Grignard reagent can't be prepared from a halo alcohol because the C–Mg bond is not compatible with the presence of an acidic –OH group in the same molecule.

Acidic hydrogen

$$HO{-}CH_2CH_2CH_2{-}Br \xrightarrow[\text{Ether}]{\text{Mg}} \!\!\!\!\!\!\!\times\;\; HO{-}CH_2CH_2CH_2{-}MgBr$$

Not formed

When this kind of incompatibility arises, it's sometimes possible to circumvent the problem by *protecting* the interfering functional group. Protection involves three steps: (1) introducing a **protecting group** to block the interfering function, (2) carrying out the desired reaction, and (3) removing the protecting group.

One of the more common methods of alcohol protection is by reaction with a chlorotrialkylsilane, Cl–SiR_3, to yield a trialkylsilyl ether, R′–O–SiR_3. Chlorotrimethylsilane is often used, and the reaction is carried out in the presence of a base, such as triethylamine, to help form the alkoxide anion from the alcohol and to remove the HCl by-product from the reaction.

$:N(CH_2CH_3)_3$

$$R{-}O{-}H + (H_3C)_3Si{-}Cl \longrightarrow R{-}O{-}Si(CH_3)_3 + (CH_3CH_2)_3NH^+\ Cl^-$$

An alcohol | **Chlorotrimethylsilane** | **A trimethylsilyl (TMS) ether**

For example:

$$\text{Cyclohexanol} \xrightarrow[(CH_3CH_2)_3N]{(CH_3)_3SiCl} \text{Cyclohexyl–O–Si(CH}_3)_3\ \ [\text{Cyclohexyl–OTMS}]$$

Cyclohexanol | **Cyclohexyl trimethylsilyl ether (94%)**

The ether-forming step is an S_N2-like reaction of the alkoxide ion on the silicon atom, with concurrent loss of the leaving chloride anion. Unlike most

S_N2 reactions, though, this reaction takes place at a *tertiary* center—a trialkyl-substituted silicon atom. The reaction occurs because silicon, a third-row atom, is larger than carbon and forms longer bonds. The three methyl substituents attached to silicon thus offer less steric hindrance to reaction than they do in the analogous *tert*-butyl chloride.

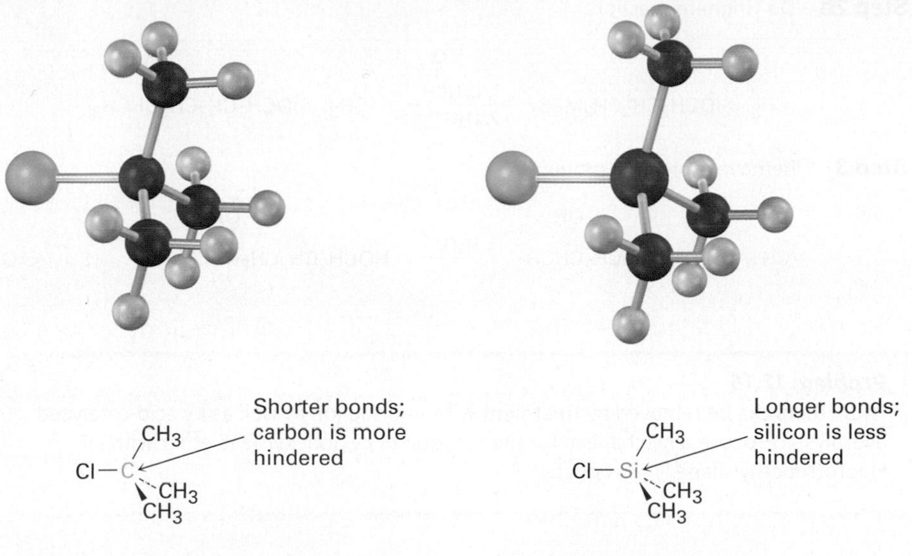

Shorter bonds; carbon is more hindered

Longer bonds; silicon is less hindered

C–C bond length: 154 pm

C–Si bond length: 195 pm

Like most other ethers, which we'll study in the next chapter, TMS ethers are relatively unreactive. They have no acidic hydrogens and don't react with oxidizing agents, reducing agents, or Grignard reagents. They do, however, react with aqueous acid or with fluoride ion to regenerate the alcohol.

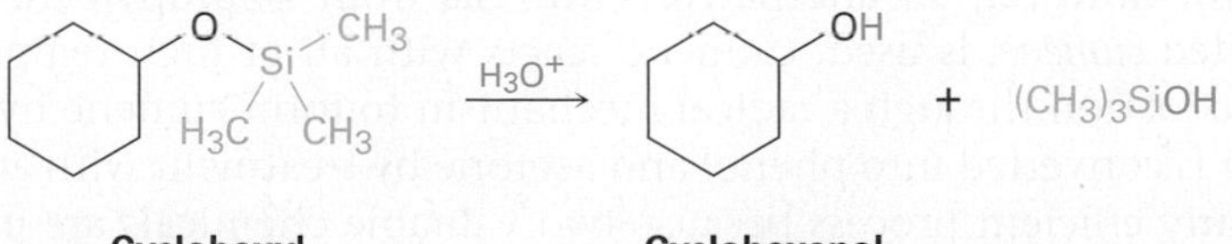

$$\text{Cyclohexyl trimethylsilyl ether} \xrightarrow{H_3O^+} \text{Cyclohexanol} + (CH_3)_3SiOH$$

Cyclohexyl trimethylsilyl ether **Cyclohexanol**

To now solve the problem posed at the beginning of this section, it's possible to use a halo alcohol in a Grignard reaction by employing a protection sequence. For example, we can add 3-bromo-1-propanol to acetaldehyde by the route shown in **Figure 17.9**.

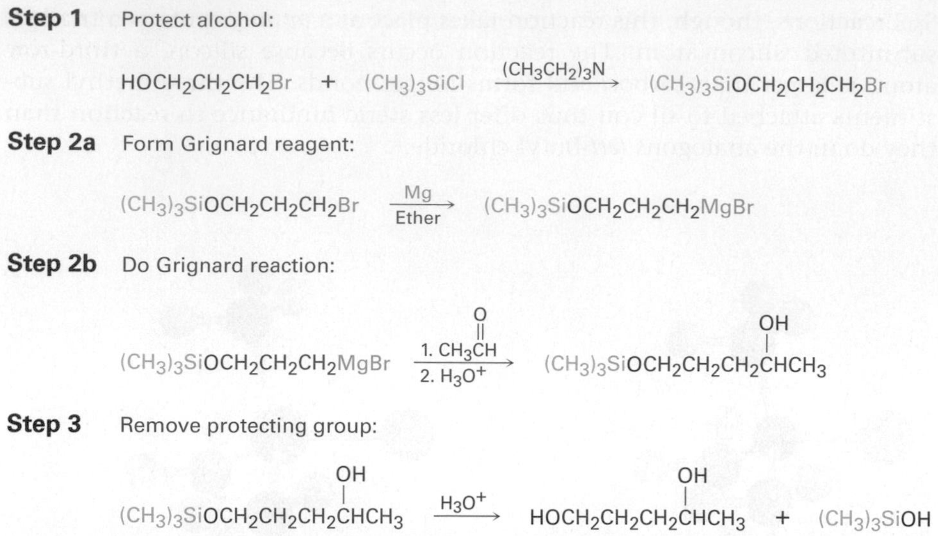

Figure 17.9 Use of a TMS-protected alcohol during a Grignard reaction.

Step 1 Protect alcohol:

$$HOCH_2CH_2CH_2Br + (CH_3)_3SiCl \xrightarrow{(CH_3CH_2)_3N} (CH_3)_3SiOCH_2CH_2CH_2Br$$

Step 2a Form Grignard reagent:

$$(CH_3)_3SiOCH_2CH_2CH_2Br \xrightarrow[\text{Ether}]{\text{Mg}} (CH_3)_3SiOCH_2CH_2CH_2MgBr$$

Step 2b Do Grignard reaction:

$$(CH_3)_3SiOCH_2CH_2CH_2MgBr \xrightarrow[\text{2. } H_3O^+]{\text{1. } CH_3CHO} (CH_3)_3SiOCH_2CH_2CH_2CH(OH)CH_3$$

Step 3 Remove protecting group:

$$(CH_3)_3SiOCH_2CH_2CH_2CH(OH)CH_3 \xrightarrow{H_3O^+} HOCH_2CH_2CH_2CH(OH)CH_3 + (CH_3)_3SiOH$$

Problem 17.16

TMS ethers can be removed by treatment with fluoride ion as well as by acid-catalyzed hydrolysis. Propose a mechanism for the reaction of cyclohexyl TMS ether with LiF. Fluorotrimethylsilane is a product.

17.9 Phenols and Their Uses

Historically, the outbreak World War I provided a stimulus for the industrial preparation of large amounts of synthetic phenol, which was needed as a raw material to manufacture the explosive, picric acid (2,4,6-trinitrophenol). Today, approximately 8 million metric tons of phenol is manufactured worldwide each year for use in such products as Bakelite resin and adhesives for binding plywood.

Phenol was manufactured for many years by the Dow process, in which chlorobenzene reacts with NaOH at high temperature and pressure **(Section 16.8)**. Now, however, an alternative synthesis from isopropylbenzene, commonly called *cumene,* is used. Cumene reacts with air at high temperature by benzylic oxidation through a radical mechanism to form cumene hydroperoxide, which is converted into phenol and acetone by treatment with acid. This is a particularly efficient process because two valuable chemicals are prepared at the same time.

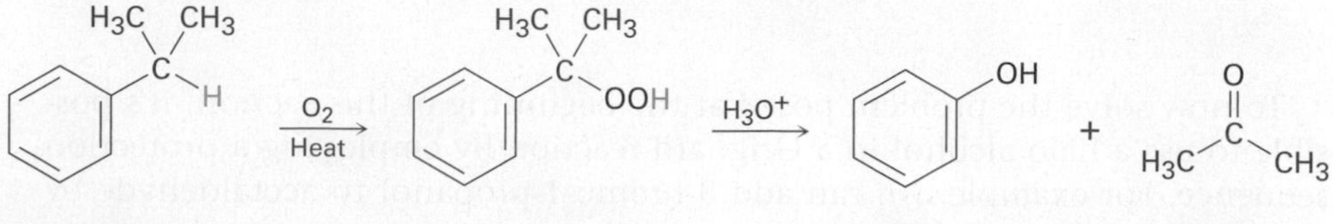

$$C_6H_5C(CH_3)_2H \xrightarrow[\text{Heat}]{O_2} C_6H_5C(CH_3)_2OOH \xrightarrow{H_3O^+} C_6H_5OH + CH_3COCH_3$$

Cumene (isopropylbenzene) → **Cumene hydroperoxide** → **Phenol** + **Acetone**

As shown in **Figure 17.10**, the reaction occurs by protonation of oxygen followed by shift of the phenyl group from carbon to oxygen with simultaneous loss of water. Readdition of water then yields an intermediate called a *hemiacetal*—a compound that contains one –OR group and one –OH group bonded to the same carbon atom—which breaks down to phenol and acetone.

Figure 17.10 | MECHANISM

Mechanism of the formation of phenol by acid-catalyzed rearrangement of cumene hydroperoxide.

1 Protonation of the hydroperoxy group on the terminal oxygen atom gives an oxonium ion . . .

2 . . . which undergoes rearrangement by migration of the phenyl ring from carbon to oxygen, expelling water as the leaving group and giving a carbocation.

3 Nucleophilic addition of water to the carbocation yields another oxonium ion . . .

4 . . . which rearranges by a proton shift from one oxygen to another.

5 Elimination of phenol gives acetone as co-product and regenerates the acid catalyst.

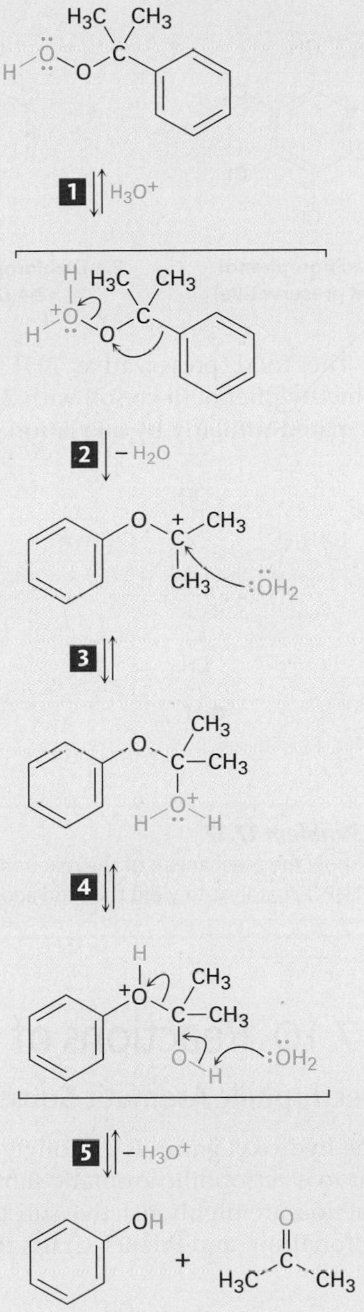

In addition to its use in making resins and adhesives, phenol is also the starting material for the synthesis of chlorinated phenols and the food preservatives BHT (butylated hydroxytoluene) and BHA (butylated hydroxyanisole). Pentachlorophenol, a widely used wood preservative, is prepared by reaction of phenol with excess Cl_2. The herbicide 2,4-D (2,4-dichlorophenoxyacetic acid) is prepared from 2,4-dichlorophenol, and the hospital antiseptic agent hexachlorophene is prepared from 2,4,5-trichlorophenol.

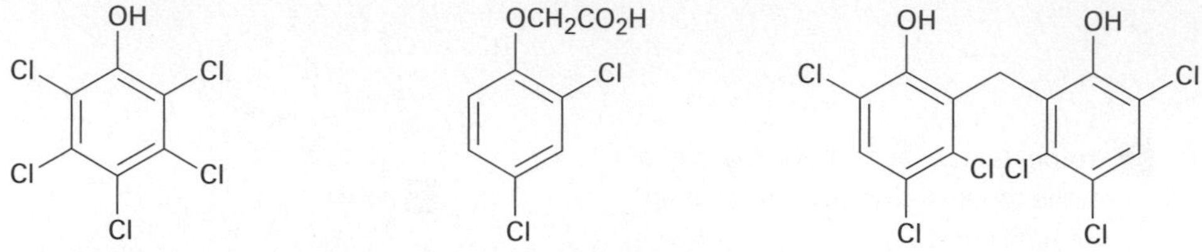

Pentachlorophenol (wood preservative) **2,4-Dichlorophenoxyacetic acid, 2,4-D (herbicide)** **Hexachlorophene (antiseptic)**

The food preservative BHT is prepared by Friedel–Crafts alkylation of *p*-methylphenol (*p*-cresol) with 2-methylpropene in the presence of acid; BHA is prepared similarly by alkylation of *p*-methoxyphenol.

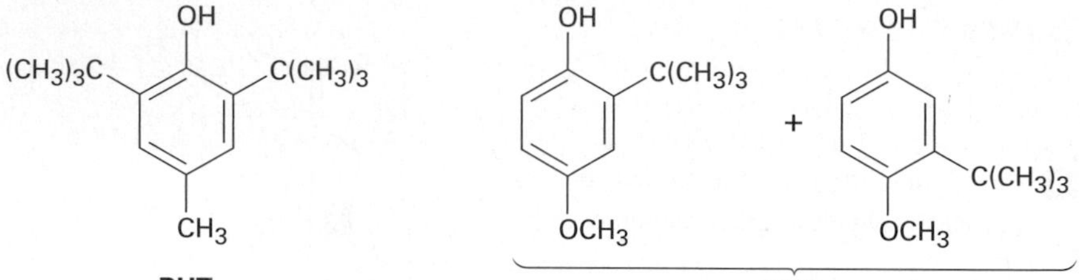

BHT **BHA**

Problem 17.17

Show the mechanism of the reaction of *p*-methylphenol with 2-methylpropene and H_3PO_4 catalyst to yield the food additive BHT.

17.10 Reactions of Phenols

Electrophilic Aromatic Substitution Reactions

The hydroxyl group is a strongly activating, ortho- and para-directing substituent in electrophilic aromatic substitution reactions **(Section 16.4)**. As a result, phenols are highly reactive substrates for electrophilic halogenation, nitration, sulfonation, and Friedel–Crafts reactions.

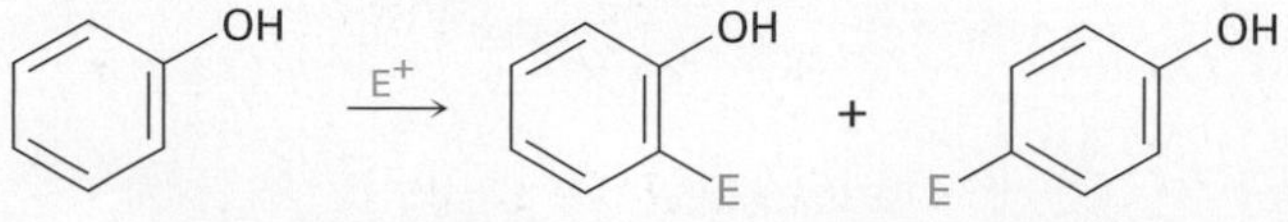

Oxidation of Phenols: Quinones

Phenols don't undergo oxidation in the same way that alcohols do because they don't have a hydrogen atom on the hydroxyl-bearing carbon. Instead, oxidation of a phenol yields a cyclohexa-2,5-diene-1,4-dione, or **quinone**. Many different oxidizing agents will accomplish the transformation, with $Na_2Cr_2O_7$ a common choice for simple phenols and potassium nitrosodisulfonate [$(KSO_3)_2NO$], called Fremy's salt, used in more complex cases.

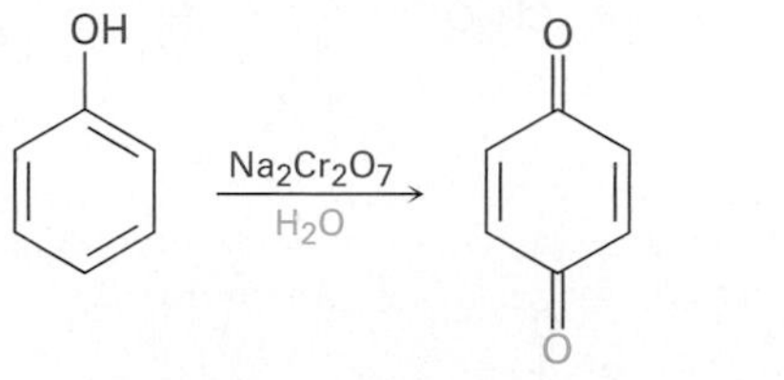

OH

$Na_2Cr_2O_7$
H_2O

O

O

Phenol **Benzoquinone (79%)**

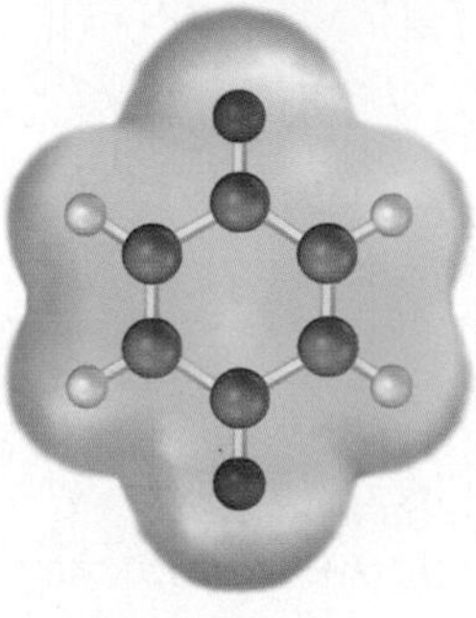

Quinones are a valuable class of compounds because of their oxidation–reduction, or *redox,* properties. They can be easily reduced to **hydroquinones** (*p*-dihydroxybenzenes) by reagents such as $NaBH_4$ and $SnCl_2$, and hydroquinones can be easily reoxidized back to quinones by $Na_2Cr_2O_7$.

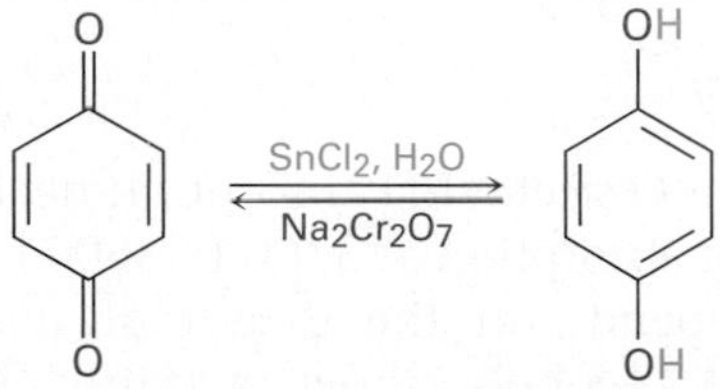

O OH

$SnCl_2$, H_2O
$Na_2Cr_2O_7$

O OH

Benzoquinone **Hydroquinone**

The redox properties of quinones are crucial to the functioning of living cells, where compounds called *ubiquinones* act as biochemical oxidizing agents to mediate the electron-transfer processes involved in energy production. Ubiquinones, also called *coenzymes Q,* are components of the cells of all aerobic organisms, from the simplest bacterium to humans. They are so named because of their ubiquitous occurrence in nature.

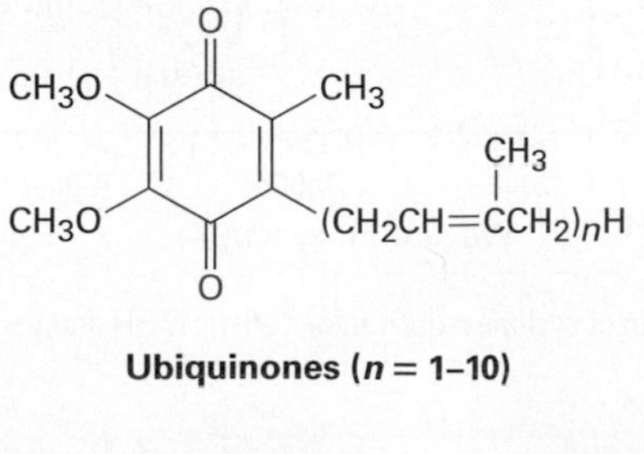

O

CH_3O CH_3

CH_3

CH_3O $(CH_2CH{=}CCH_2)_nH$

O

Ubiquinones (n = 1–10)

Ubiquinones function within the mitochondria of cells to mediate the respiration process in which electrons are transported from the biological reducing agent NADH to molecular oxygen. Through a complex series of steps, the ultimate result is a cycle whereby NADH is oxidized to NAD^+, O_2 is reduced to water, and energy is produced. Ubiquinone acts only as an intermediary and is itself unchanged.

Step 1

NADH + H^+ + [quinone: O, CH_3O, CH_3, CH_3O, R, O] ⟶ [hydroquinone: OH, CH_3O, CH_3, CH_3O, R, OH] + NAD^+

Reduced form **Oxidized form**

Step 2

[hydroquinone: OH, CH_3O, CH_3, CH_3O, R, OH] + $\frac{1}{2}O_2$ ⟶ [quinone: O, CH_3O, CH_3, CH_3O, R, O] + H_2O

Net change: NADH + $\frac{1}{2}O_2$ + H^+ ⟶ NAD^+ + H_2O

17.11 Spectroscopy of Alcohols and Phenols

Infrared Spectroscopy

Alcohols have a strong C–O stretching absorption near 1050 cm^{-1} and a characteristic O–H stretching absorption at 3300 to 3600 cm^{-1}. The exact position of the O–H stretch depends on the extent of hydrogen-bonding in the molecule. Unassociated alcohols show a fairly sharp absorption near 3600 cm^{-1}, whereas hydrogen-bonded alcohols show a broader absorption in the 3300 to 3400 cm^{-1} range. The hydrogen-bonded hydroxyl absorption appears at 3350 cm^{-1} in the IR spectrum of cyclohexanol **(Figure 17.11)**.

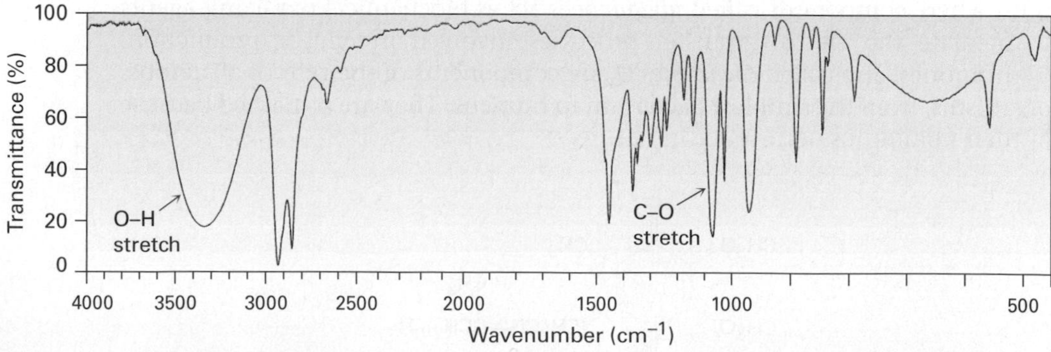

Figure 17.11 Infrared spectrum of cyclohexanol. Characteristic O—H and C—O stretching absorptions are indicated.

Phenols also show a characteristic broad IR absorption at 3500 cm^{-1} due to the –OH group, as well as the usual 1500 and 1600 cm^{-1} aromatic bands **(Figure 17.12)**. In phenol itself, the monosubstituted aromatic-ring peaks at 690 and 760 cm^{-1} are visible.

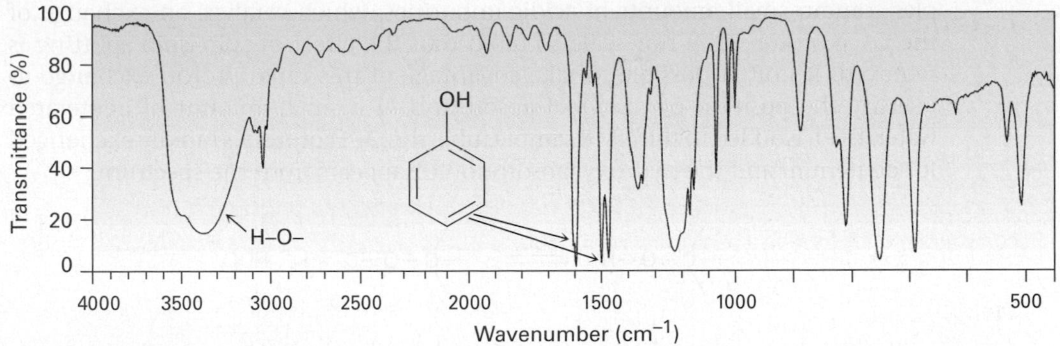

Figure 17.12 Infrared spectrum of phenol.

Problem 17.18

Assume that you need to prepare 5-cholesten-3-one from cholesterol. How could you use IR spectroscopy to tell whether the reaction was successful? What differences would you look for in the IR spectra of starting material and product?

CrO_3, H_3O^+

Cholesterol → **5-Cholestene-3-one**

Nuclear Magnetic Resonance Spectroscopy

Carbon atoms bonded to electron-withdrawing –OH groups are deshielded and absorb at a lower field in the ^{13}C NMR spectrum than do typical alkane carbons. Most alcohol carbon absorptions fall in the range 50 to 80 δ, as the following data illustrate for cyclohexanol:

OH: 69.5 δ, 35.5 δ, 24.4 δ, 25.9 δ

Alcohols also show characteristic absorptions in the ^{1}H NMR spectrum. Hydrogens on the oxygen-bearing carbon atom are deshielded by the electron-withdrawing effect of the nearby oxygen, and their absorptions occur in the range 3.4 to 4.5 δ. Spin–spin splitting, however, is not usually observed between the O–H proton of an alcohol and the neighboring protons on carbon. Most samples contain small amounts of acidic impurities, which catalyze an exchange of the O–H proton on a timescale so rapid that the effect of spin–spin splitting is removed. It's often possible to take advantage of this rapid proton exchange to identify the position of the O–H absorption. If a small amount of deuterated water, D_2O, is added to the NMR sample tube, the O–H proton is rapidly exchanged for deuterium and the hydroxyl absorption disappears from the spectrum.

$$\text{—C—O—H} \xrightleftharpoons{D_2O} \text{—C—O—D} + \text{HDO}$$

Typical spin–spin splitting *is* observed between protons on the oxygen-bearing carbon and other neighbors. For example, the signal of the two $-CH_2O-$ protons in 1-propanol is split into a triplet by coupling with the neighboring $-CH_2-$ protons **(Figure 17.13)**.

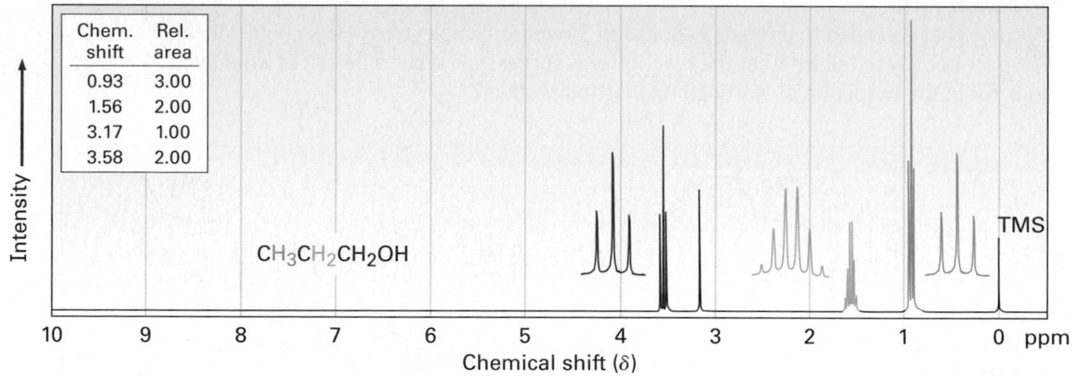

Figure 17.13 ^{1}H NMR spectrum of 1-propanol. The protons on the oxygen-bearing carbon are split into a triplet at 3.58 δ.

Phenols, like all aromatic compounds, show ^{1}H NMR absorptions near 7 to 8 δ, the expected position for aromatic-ring protons **(Section 15.7)**. In addition, phenol O–H protons absorb at 3 to 8 δ. In neither case are these absorptions uniquely diagnostic for phenols, since other kinds of protons absorb in the same range.

Problem 17.19

When the ^{1}H NMR spectrum of an alcohol is run in dimethyl sulfoxide (DMSO) solvent rather than in chloroform, exchange of the O–H proton is slow and spin–spin splitting is seen between the O–H proton and C–H protons on the adjacent carbon. What spin multiplicities would you expect for the hydroxyl protons in the following alcohols?

(a) 2-Methyl-2-propanol **(b)** Cyclohexanol **(c)** Ethanol
(d) 2-Propanol **(e)** Cholesterol **(f)** 1-Methylcyclohexanol

Mass Spectrometry

As noted previously in **Section 12.3**, alcohols undergo fragmentation in the mass spectrometer by two characteristic pathways, alpha cleavage and dehydration. In the alpha-cleavage pathway, a C–C bond nearest the hydroxyl group is broken, yielding a neutral radical plus a resonance-stabilized, oxygen-containing cation.

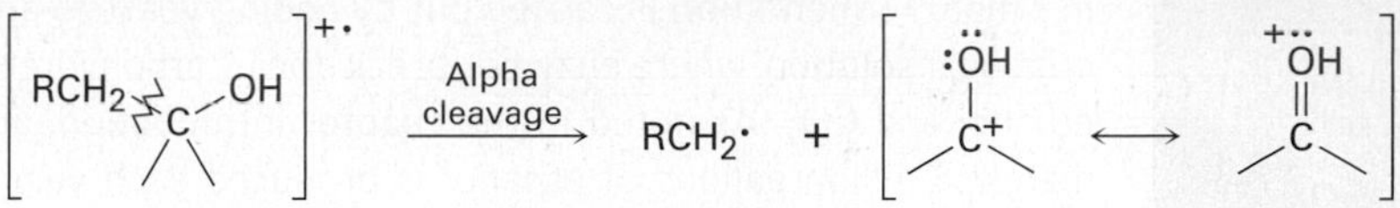

$$[RCH_2{-}C{-}OH]^{+\bullet} \xrightarrow{\text{Alpha cleavage}} RCH_2^{\bullet} + [\,{:}\ddot{O}H{-}C^{+} \longleftrightarrow {}^{+}\ddot{O}H{=}C\,]$$

In the dehydration pathway, water is eliminated, yielding an alkene radical cation.

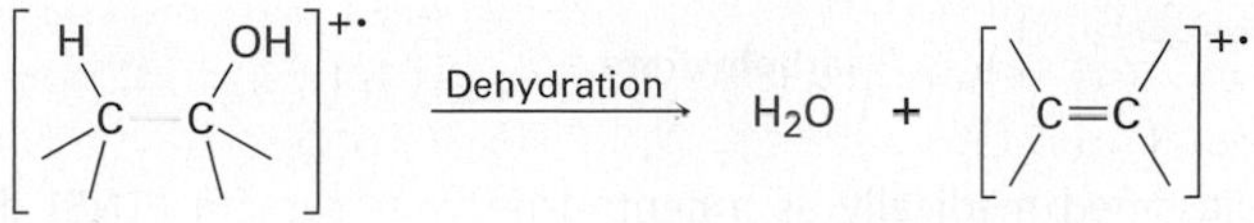

$$[H{-}C{-}C{-}OH]^{+\bullet} \xrightarrow{\text{Dehydration}} H_2O + [C{=}C]^{+\bullet}$$

Both fragmentation modes are apparent in the mass spectrum of 1-butanol (**Figure 17.14**). The peak at $m/z = 56$ is due to loss of water from the molecular ion, and the peak at $m/z = 31$ is due to an alpha cleavage.

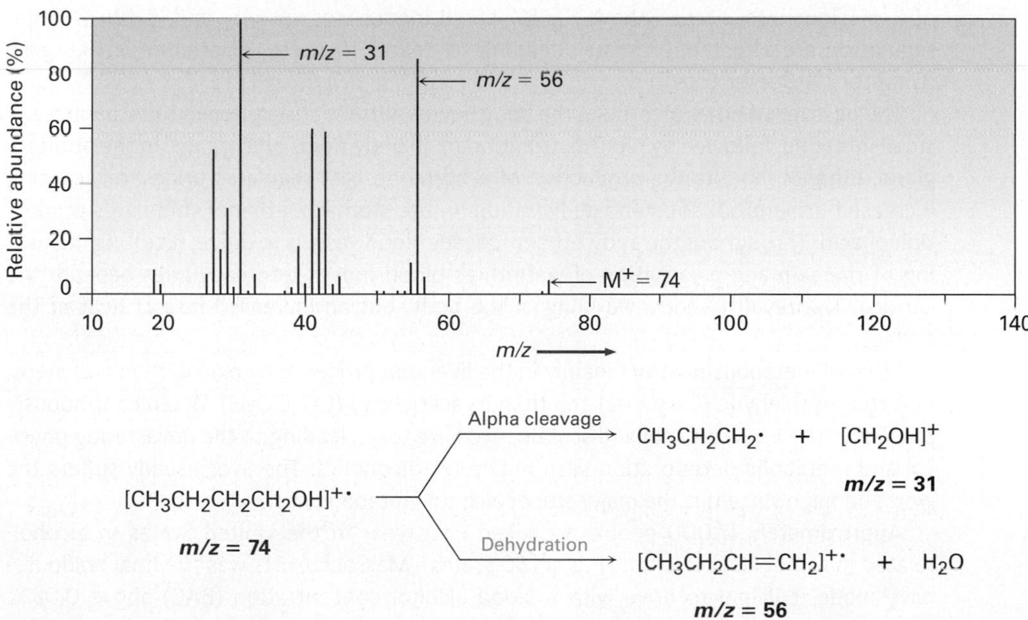

Figure 17.14 Mass spectrum of 1-butanol ($M^+ = 74$). Dehydration gives a peak at $m/z = 56$, and fragmentation by alpha cleavage gives a peak at $m/z = 31$.

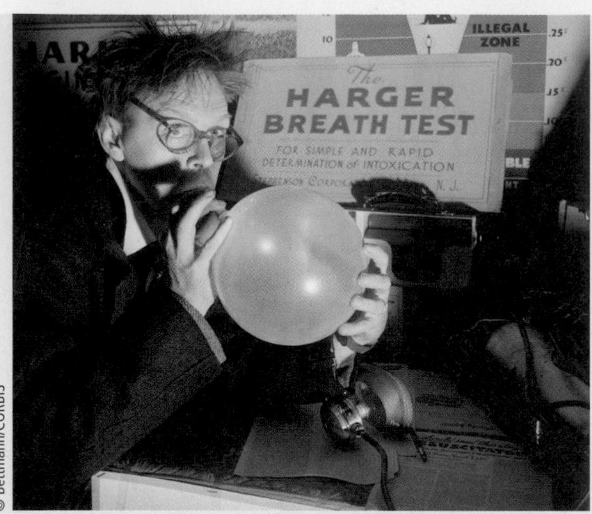

The Harger Drunkometer was the first breath analyzer, introduced in 1938 to help convict drunk drivers.

Ethanol: Chemical, Drug, Poison

A DEEPER LOOK

The production of ethanol by fermentation of grains and sugars is one of the oldest known organic reactions, going back at least 8000 years in the Middle East and perhaps as many as 9000 years in China. Fermentation is carried out by adding yeast to an aqueous sugar solution, where enzymes break down carbohydrates into ethanol and CO_2. As noted in the chapter introduction, approximately 4 billion gallons of ethanol is produced each year in the United States by fermentation, with essentially the entire amount used to make E90 automobile fuel.

$$\underset{\text{A carbohydrate}}{C_6H_{12}O_6} \xrightarrow{\text{Yeast}} 2\ CH_3CH_2OH + 2\ CO_2$$

Ethanol is classified medically as a central nervous system (CNS) depressant. Its effects—that is, being drunk—resemble the human response to anesthetics. There is an initial excitability and increase in sociable behavior, but this results from depression of inhibition rather than from stimulation. At a blood alcohol concentration of 0.1% to 0.3%, motor coordination is affected, accompanied by loss of balance, slurred speech, and amnesia. When blood alcohol concentration rises to 0.3% to 0.4%, nausea and loss of consciousness occur. Above 0.6%, spontaneous respiration and cardiovascular regulation are affected, ultimately leading to death. The LD_{50} of ethanol is 10.6 g/kg (Chapter 1 *A Deeper Look*).

The passage of ethanol through the body begins with its absorption in the stomach and small intestine, followed by rapid distribution to all body fluids and organs. In the pituitary gland, ethanol inhibits the production of a hormone that regulates urine flow, causing increased urine production and dehydration. In the stomach, ethanol stimulates production of acid. Throughout the body, ethanol causes blood vessels to dilate, resulting in flushing of the skin and a sensation of warmth as blood moves into capillaries beneath the surface. The result is not a warming of the body, but an increased loss of heat at the surface.

Ethanol metabolism occurs mainly in the liver and proceeds by oxidation in two steps, first to acetaldehyde (CH_3CHO) and then to acetic acid (CH_3CO_2H). When continuously present in the body, ethanol and acetaldehyde are toxic, leading to the devastating physical and metabolic deterioration seen in chronic alcoholics. The liver usually suffers the worst damage since it is the major site of alcohol metabolism.

Approximately 17,000 people are killed each year in the United States in alcohol-related automobile accidents. Thus, all 50 states—Massachusetts was the final holdout—have made it illegal to drive with a blood alcohol concentration (BAC) above 0.08%. Fortunately, simple tests have been devised for measuring blood alcohol concentration. The *Breathalyzer test* measures alcohol concentration in expired air by the color change that occurs when the bright orange oxidizing agent potassium dichromate ($K_2Cr_2O_7$) is reduced to blue-green chromium(III). The *Intoxilyzer* test uses IR spectroscopy to measure blood alcohol levels in expired air. Just breathe into the machine, and let the spectrum tell the tale.

Summary

In previous chapters, we focused on developing general ideas of organic reactivity, looking at the chemistry of hydrocarbons and alkyl halides and seeing some of the tools used in structural studies. With that accomplished, we have now begun in this chapter to study of the oxygen-containing functional groups that lie at the heart of organic and biological chemistry.

Alcohols are among the most versatile of all organic compounds. They occur widely in nature, are important industrially, and have an unusually rich chemistry. The most widely used methods of alcohol synthesis start with carbonyl compounds. Aldehydes, esters, and carboxylic acids are reduced by reaction with $LiAlH_4$ to give primary alcohols (RCH_2OH); ketones are reduced to yield secondary alcohols (R_2CHOH).

Alcohols are also prepared by reaction of carbonyl compounds with Grignard reagents, RMgX. Addition of a Grignard reagent to formaldehyde yields a primary alcohol, addition to an aldehyde yields a secondary alcohol, and addition to a ketone or an ester yields a tertiary alcohol. The Grignard reaction is limited by the fact that Grignard reagents can't be prepared from alkyl halides that contain reactive functional groups in the same molecule. This problem can sometimes be avoided by **protecting** the interfering functional group. Alcohols are often protected by formation of trimethylsilyl (TMS) ethers.

Alcohols undergo many reactions and can be converted into many other functional groups. They can be dehydrated to give alkenes by treatment with $POCl_3$ and can be transformed into alkyl halides by treatment with PBr_3 or $SOCl_2$. Furthermore, alcohols are weakly acidic ($pK_a \approx 16–18$) and react with strong bases and with alkali metals to form **alkoxide anions**, which are used frequently in organic synthesis. Perhaps the most important reactions of alcohols is their oxidation to carbonyl compounds. Primary alcohols yield either aldehydes or carboxylic acids, secondary alcohols yield ketones, but tertiary alcohols are not normally oxidized.

Phenols are aromatic counterparts of alcohols but are more acidic ($pK_a \approx 10$) because the corresponding **phenoxide anions** are resonance stabilized by delocalization of the negative charge into the aromatic ring. Substitution of the aromatic ring by an electron-withdrawing group increases phenol acidity, and substitution by an electron-donating group decreases acidity. Phenols can be oxidized to **quinones**, and quinones can be reduced back to **hydroquinones**.

Key words

alcohol (ROH), 620
alkoxide ion (RO^-), 624
hydroquinone, 653
phenol (ArOH), 620
phenoxide ion (ArO^-), 624
protecting group, 648
quinone, 653

Summary of Reactions

1. Synthesis of alcohols
   (a) Reduction of carbonyl compounds (Section 17.4)
      (1) Aldehydes

$$RCHO \xrightarrow[2.\ H_3O^+]{1.\ NaBH_4 \text{ or } LiAlH_4} RCH_2OH$$

Primary alcohol

(continued)

(2) Ketones

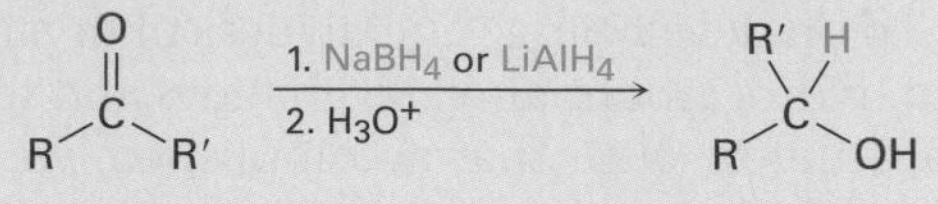

$$\mathrm{R{-}C({=}O){-}R'} \xrightarrow[\text{2. } H_3O^+]{\text{1. } NaBH_4 \text{ or } LiAlH_4} \mathrm{R{-}CH(R'){-}OH}$$

Secondary alcohol

(3) Esters

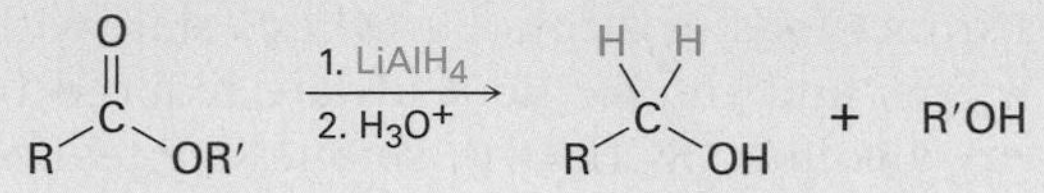

$$\mathrm{R{-}C({=}O){-}OR'} \xrightarrow[\text{2. } H_3O^+]{\text{1. } LiAlH_4} \mathrm{R{-}CH_2{-}OH} + \mathrm{R'OH}$$

Primary alcohol

(4) Carboxylic acids

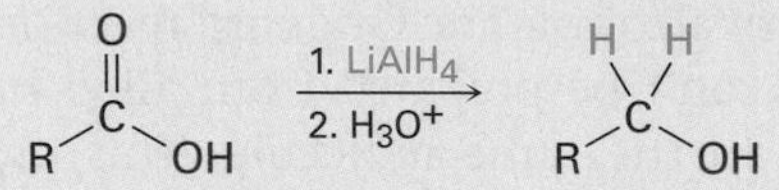

$$\mathrm{R{-}C({=}O){-}OH} \xrightarrow[\text{2. } H_3O^+]{\text{1. } LiAlH_4} \mathrm{R{-}CH_2{-}OH}$$

Primary alcohol

(b) Grignard addition to carbonyl compounds (Section 17.5)

(1) Formaldehyde

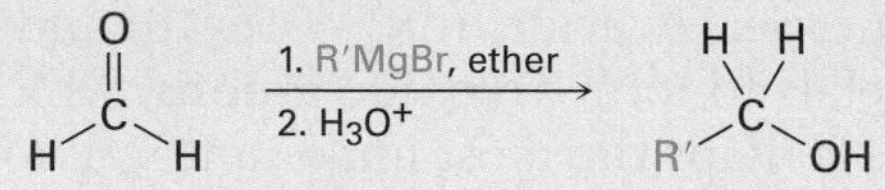

$$\mathrm{H{-}C({=}O){-}H} \xrightarrow[\text{2. } H_3O^+]{\text{1. } R'MgBr\text{, ether}} \mathrm{R'{-}CH_2{-}OH}$$

Primary alcohol

(2) Aldehydes

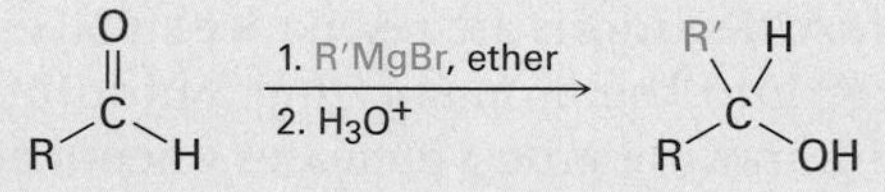

$$\mathrm{R{-}C({=}O){-}H} \xrightarrow[\text{2. } H_3O^+]{\text{1. } R'MgBr\text{, ether}} \mathrm{R{-}CH(R'){-}OH}$$

Secondary alcohol

(3) Ketones

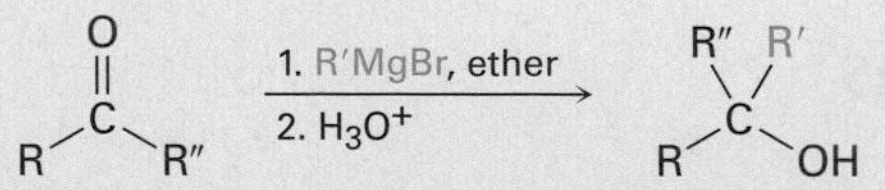

$$\mathrm{R{-}C({=}O){-}R''} \xrightarrow[\text{2. } H_3O^+]{\text{1. } R'MgBr\text{, ether}} \mathrm{R{-}C(R'')(R'){-}OH}$$

Tertiary alcohol

(4) Esters

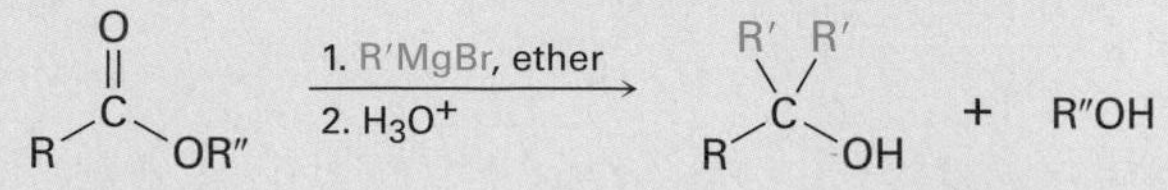

$$\mathrm{R{-}C({=}O){-}OR''} \xrightarrow[\text{2. } H_3O^+]{\text{1. } R'MgBr\text{, ether}} \mathrm{R{-}C(R')(R'){-}OH} + \mathrm{R''OH}$$

Tertiary alcohol

2. Reactions of alcohols
   (a) Dehydration (Section 17.6)
      (1) Tertiary alcohols

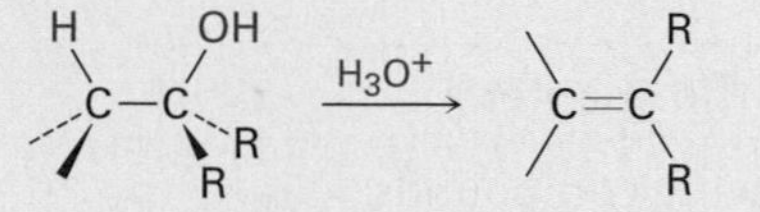

$$\text{H–C–C(OH)(R)(R)} \xrightarrow{H_3O^+} \text{C=C(R)(R)}$$

      (2) Secondary and tertiary alcohols

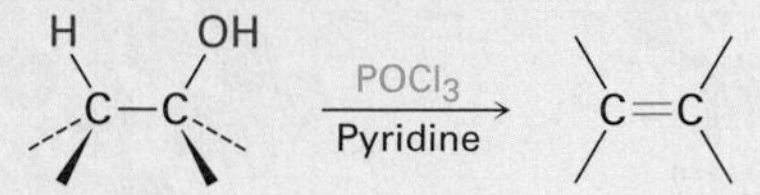

$$\text{H–C–C–OH} \xrightarrow[\text{Pyridine}]{POCl_3} \text{C=C}$$

   (b) Oxidation (Section 17.7)
      (1) Primary alcohols

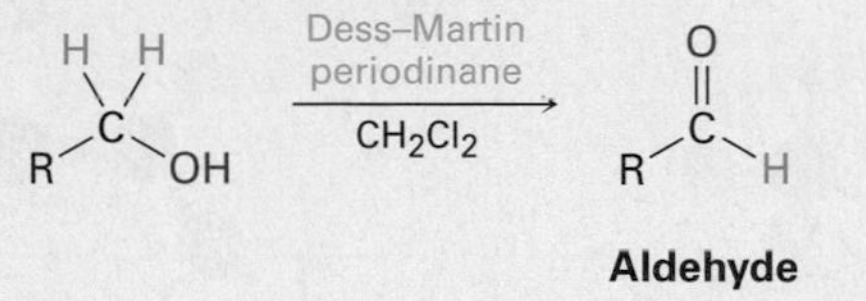

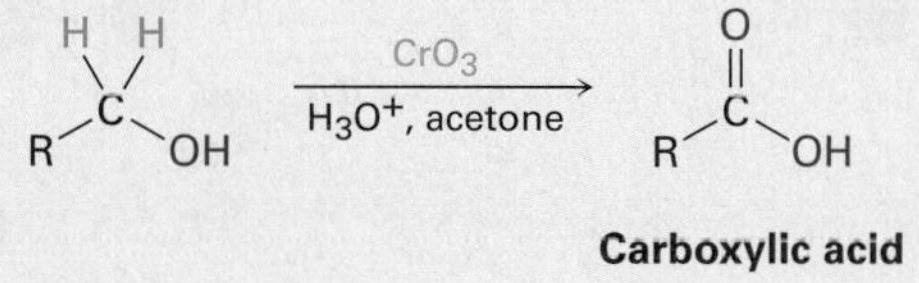

$$RCH_2OH \xrightarrow[CH_2Cl_2]{\text{Dess–Martin periodinane}} RCHO$$

Aldehyde

$$RCH_2OH \xrightarrow[H_3O^+,\ \text{acetone}]{CrO_3} RCOOH$$

Carboxylic acid

      (2) Secondary alcohols

$$RCH(OH)R' \xrightarrow[CH_2Cl_2]{\text{Dess–Martin periodinane}} RCOR'$$

Ketone

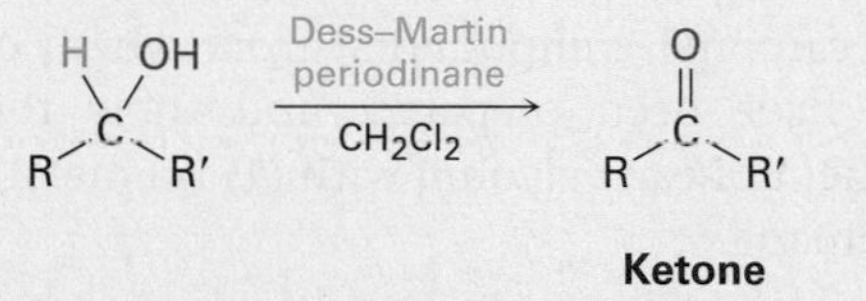

3. Oxidation of phenols to quinones (Section 17.10)

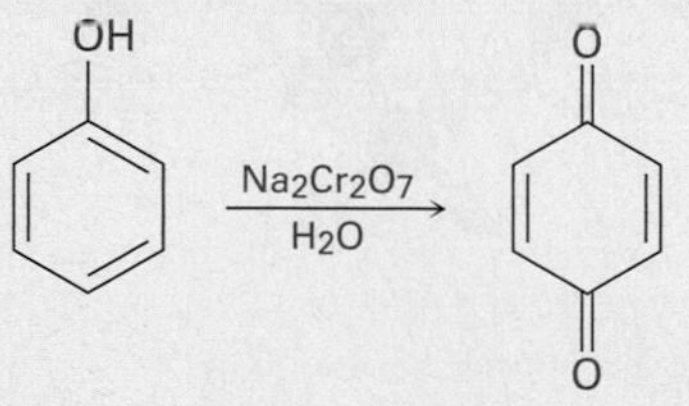

$$\text{Phenol (OH)} \xrightarrow[H_2O]{Na_2Cr_2O_7} \text{Quinone (O, O)}$$

Exercises

OWL Interactive versions of these problems are assignable in OWL for Organic Chemistry.

Visualizing Chemistry

(Problems 17.1–17.19 appear within the chapter.)

17.20 Give IUPAC names for the following compounds:

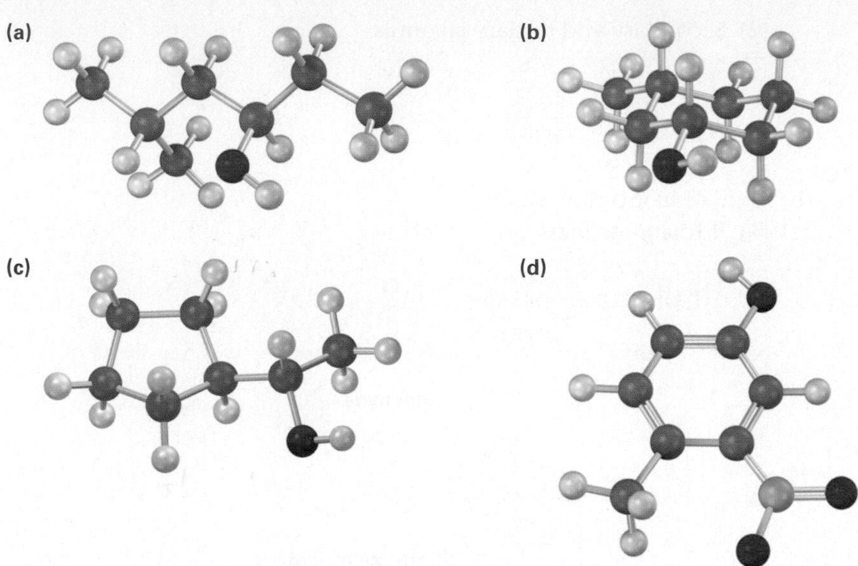

17.21 Draw the structure of the carbonyl compound(s) from which each of the following alcohols might have been prepared, and show the products you would obtain by treatment of each alcohol with (1) Na metal, (2) $SOCl_2$, and (3) Dess–Martin periodinane.

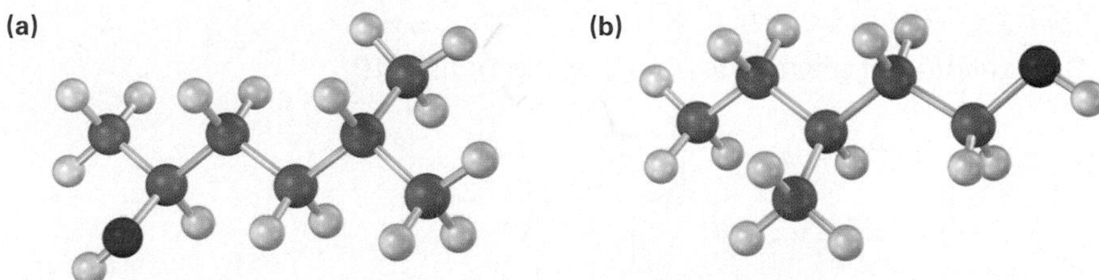

17.22 Predict the product from reaction of the following substance (reddish brown = Br) with:

(a) PBr_3 (b) Aqueous H_2SO_4 (c) $SOCl_2$
(d) Dess–Martin periodinane (e) Br_2, $FeBr_3$

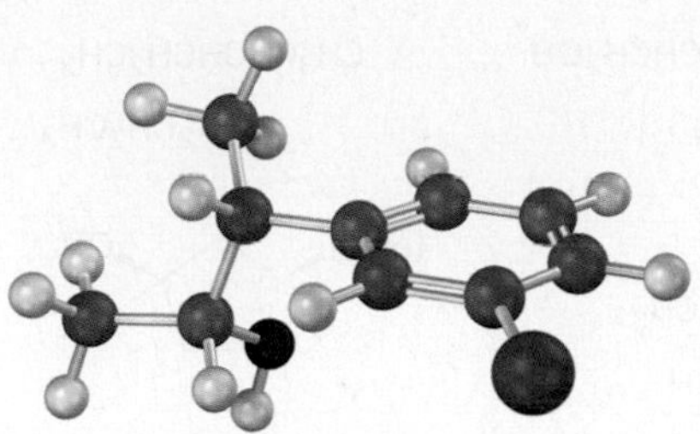

17.23 Predict the product from reaction of the following substance with:

(a) $NaBH_4$; then H_3O^+ (b) $LiAlH_4$; then H_3O^+
(c) 2 CH_3CH_2MgBr; then H_3O^+

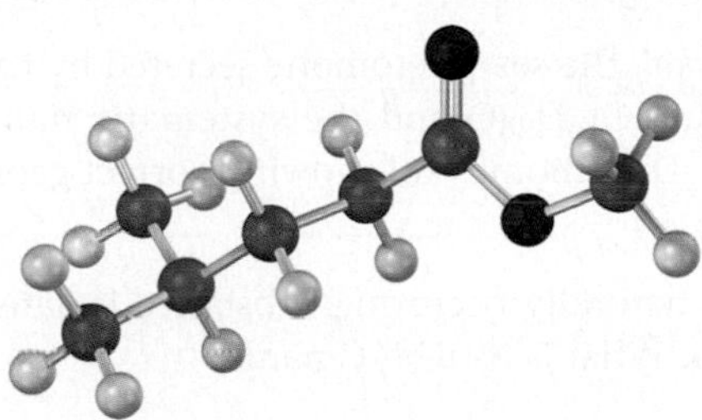

17.24 Name and assign *R* or *S* stereochemistry to the product(s) you would obtain by reaction of the following substance with ethylmagnesium bromide. Is the product chiral? Is it optically active? Explain.

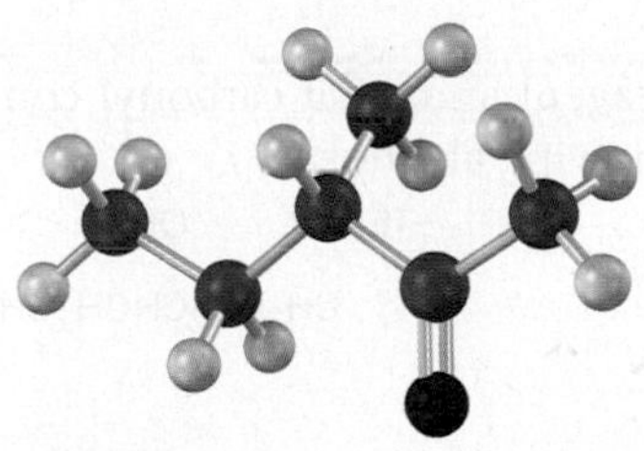

Additional Problems

Naming Alcohols

17.25 Give IUPAC names for the following compounds:

(a) 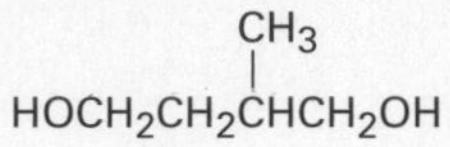 $HOCH_2CH_2CH(CH_3)CH_2OH$

(b) 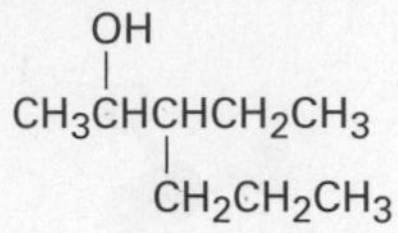 $CH_3CH(OH)CH(CH_2CH_2CH_3)CH_2CH_3$

(c) 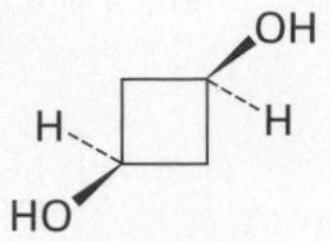 [cyclobutane with OH and H on two carbons, H and HO on the opposite carbon]

(d) 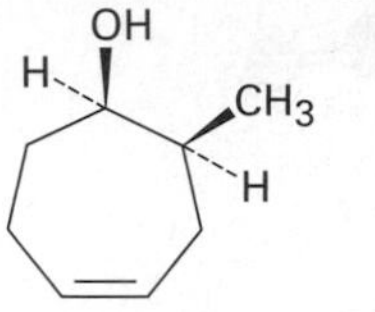 [cycloheptene with OH, H, CH3, H substituents]

(e) 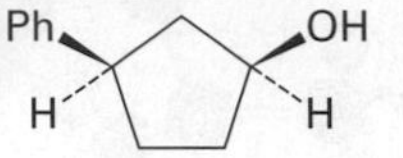 [cyclopentane with Ph, H, OH, H substituents]

(f) 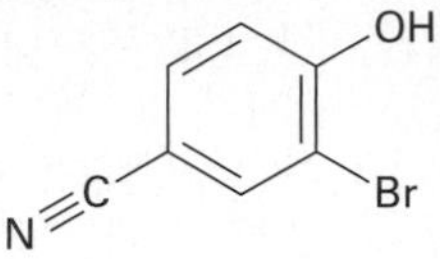 [benzene ring with OH, Br, and N≡C– substituents]

17.26 Draw and name the eight isomeric alcohols with formula $C_5H_{12}O$.

17.27 Which of the eight alcohols you identified in Problem 17.26 react with CrO_3 in aqueous acid? Show the products you would expect from each reaction.

17.28 Named *bombykol,* the sex pheromone secreted by the female silkworm moth has the formula $C_{16}H_{28}O$ and the systematic name (10*E*,12*Z*)-10,12-hexadecadien-1-ol. Draw bombykol showing correct geometry for the two double bonds.

17.29 *Carvacrol* is a naturally occurring substance isolated from oregano, thyme, and marjoram. What is its IUPAC name?

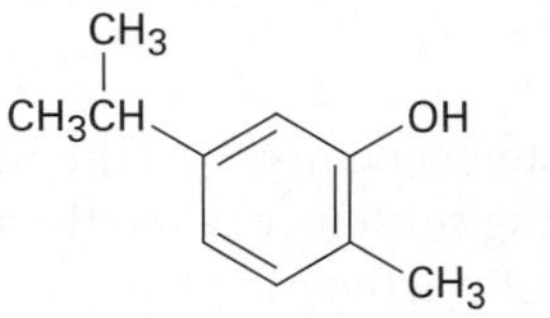

[Structure: benzene ring with $CH_3CH(CH_3)$–, OH, and CH_3 substituents] **Carvacrol**

Synthesizing Alcohols

17.30 What Grignard reagent and what carbonyl compound might you start with to prepare the following alcohols?

(a) 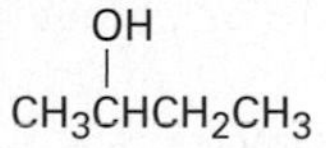 $CH_3CH(OH)CH_2CH_3$

(b) 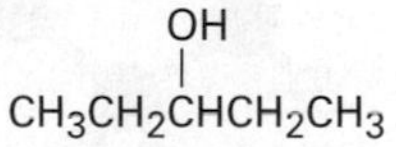 $CH_3CH_2CH(OH)CH_2CH_3$

(c) 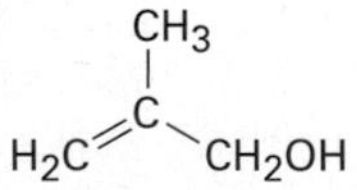 $H_2C{=}C(CH_3)CH_2OH$

(d) 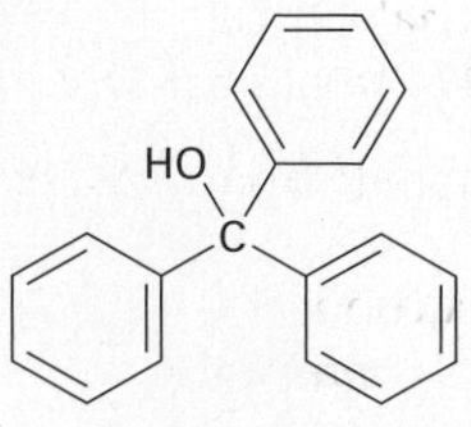 $(C_6H_5)_3C{-}OH$

(e) 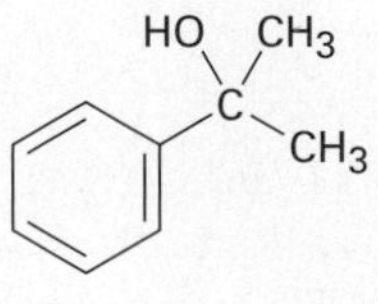 $C_6H_5C(OH)(CH_3)_2$

(f) 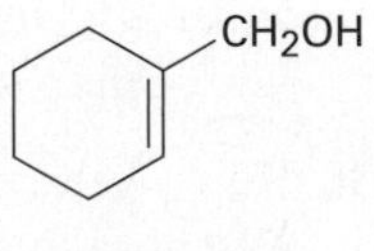 [cyclohexene with CH_2OH substituent]

17.31 What carbonyl compounds would you reduce to prepare the following alcohols? List all possibilities.

(a)

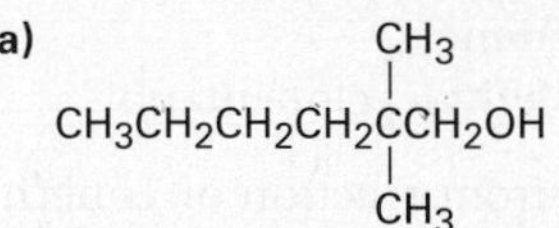

(b)

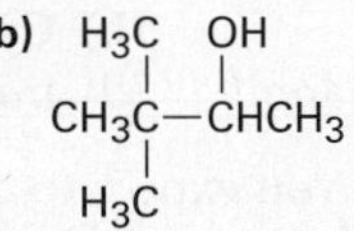

(c) (cyclohexyl)CH(OH)CH$_2$CH$_3$

17.32 What carbonyl compounds might you start with to prepare the following compounds by Grignard reaction? List all possibilities.

(a) 2-Methyl-2-propanol **(b)** 1-Ethylcyclohexanol
(c) 3-Phenyl-3-pentanol **(d)** 2-Phenyl-2-pentanol

(e)

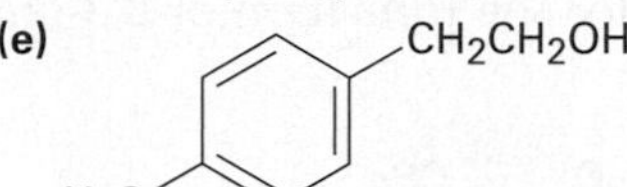

(f)

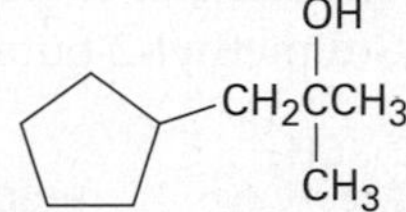

17.33 How would you synthesize the following alcohols, starting with benzene and other alcohols of six or fewer carbons as your only organic reagents?

(a)

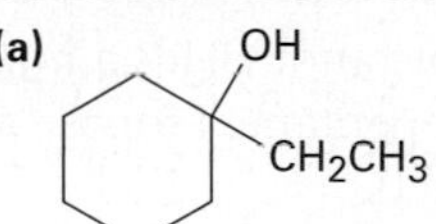

(b)

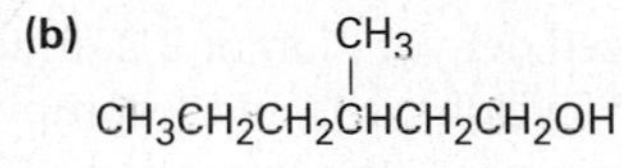

(c)

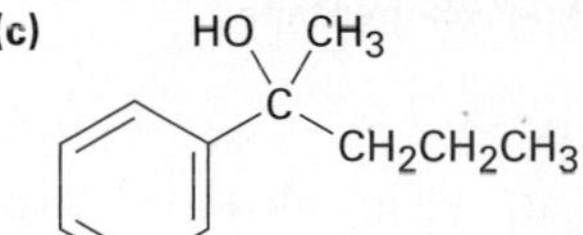

(d) CH$_3$CH(CH$_3$)CH$_2$CH(OH)CH$_2$CH$_3$

Reactions of Alcohols

17.34 What products would you obtain from reaction of 1-pentanol with the following reagents?

(a) PBr_3 (b) $SOCl_2$
(c) CrO_3, H_2O, H_2SO_4 (d) Dess–Martin periodinane

17.35 How would you prepare the following compounds from 2-phenylethanol? More than one step may be required.

(a) Styrene ($PhCH{=}CH_2$) (b) Phenylacetaldehyde ($PhCH_2CHO$)
(c) Phenylacetic acid ($PhCH_2CO_2H$) (d) Benzoic acid
(e) Ethylbenzene (f) Benzaldehyde
(g) 1-Phenylethanol (h) 1-Bromo-2-phenylethane

17.36 How would you prepare the following compounds from 1-phenylethanol? More than one step may be required.

(a) Acetophenone ($PhCOCH_3$) (b) Benzyl alcohol
(c) *m*-Bromobenzoic acid (d) 2-Phenyl-2-propanol

17.37 How would you prepare the following substances from cyclopentanol? More than one step may be required.

(a) Cyclopentanone (b) Cyclopentene

(c) 1-Methylcyclopentanol (d) *trans*-2-Methylcyclopentanol

17.38 What products would you expect to obtain from reaction of 1-methylcyclohexanol with the following reagents?

(a) HBr (b) NaH (c) H_2SO_4 (d) $Na_2Cr_2O_7$

Mechanisms

17.39 Evidence for the intermediate carbocations in the acid-catalyzed dehydration of alcohols comes from the observation that rearrangements sometimes occur. Propose a mechanism to account for the formation of 2,3-dimethyl-2-butene from 3,3-dimethyl-2-butanol.

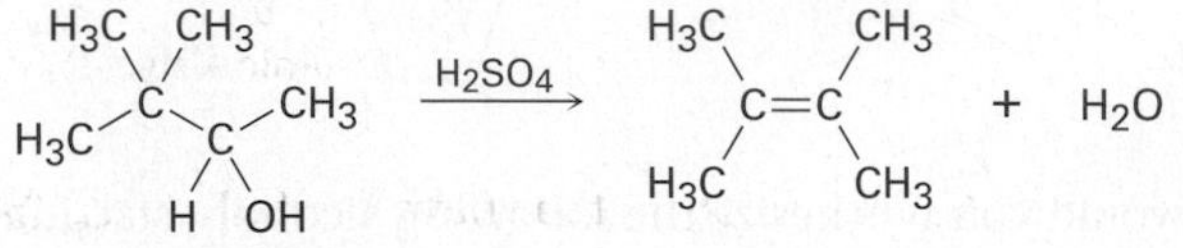

H_3C CH_3 / H_3C–C–C(CH_3)(H)(OH) $\xrightarrow{H_2SO_4}$ $(H_3C)_2C{=}C(CH_3)_2$ + H_2O

17.40 Acid-catalyzed dehydration of 2,2-dimethylcyclohexanol yields a mixture of 1,2-dimethylcyclohexene and isopropylidenecyclopentane. Propose a mechanism to account for the formation of both products.

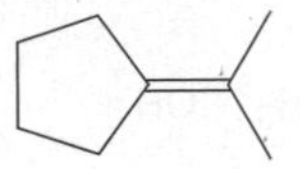

Isopropylidenecyclopentane

17.41 Epoxides react with Grignard reagents to yield alcohols. Propose a mechanism.

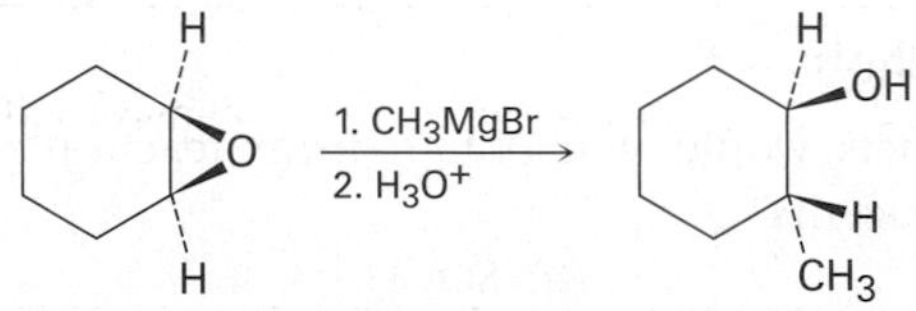

H, O, H $\xrightarrow[\text{2. } H_3O^+]{\text{1. } CH_3MgBr}$ H, OH, H, CH_3

17.42 Treatment of the following epoxide with aqueous acid produces a carbocation intermediate that reacts with water to give a diol product. Show the structure of the carbocation, and propose a mechanism for the second step.

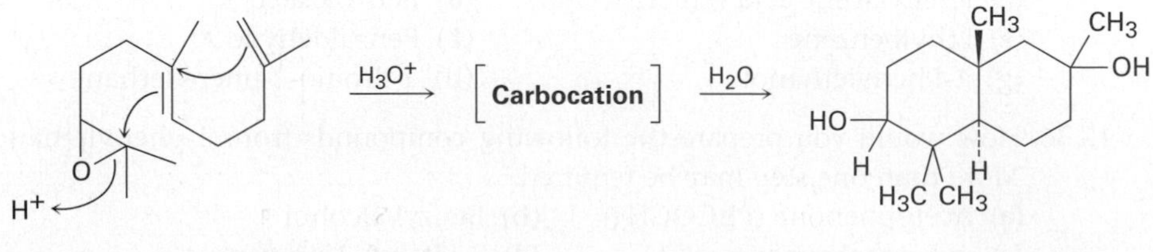

O, H^+ $\xrightarrow{H_3O^+}$ [**Carbocation**] $\xrightarrow{H_2O}$ CH_3, CH_3, OH, HO, H, H, H_3C CH_3

17.43 Reduction of 2-butanone with $NaBH_4$ yields 2-butanol. Is the product chiral? Is it optically active? Explain.

Spectroscopy

17.44 The following ^1H NMR spectrum is that of an alcohol, $C_8H_{10}O$. Propose a structure.

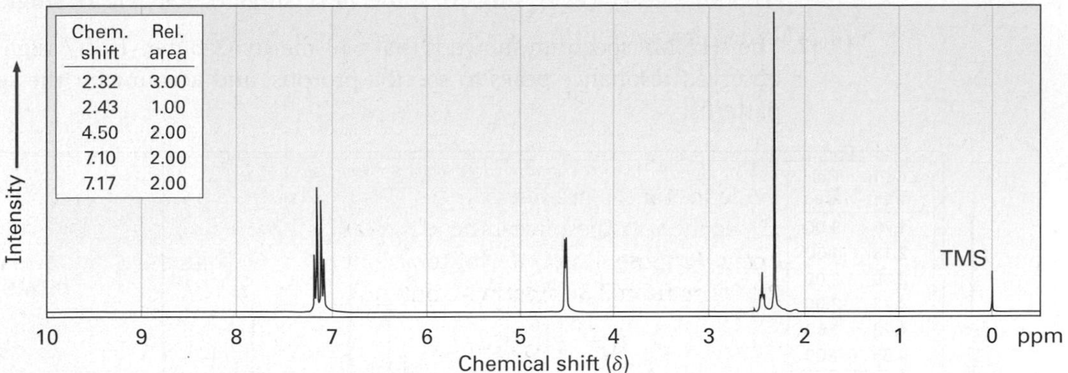

17.45 Propose structures for alcohols that have the following ^1H NMR spectra:

(a) $C_5H_{12}O$

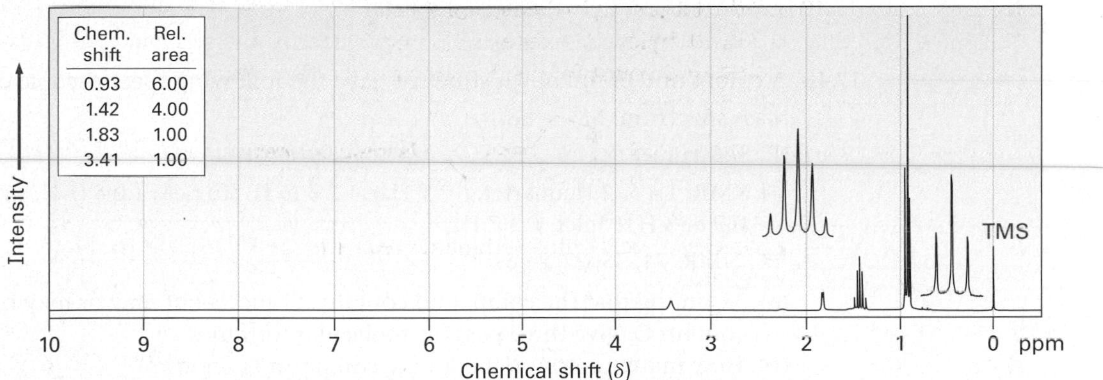

(b) $C_8H_{10}O$

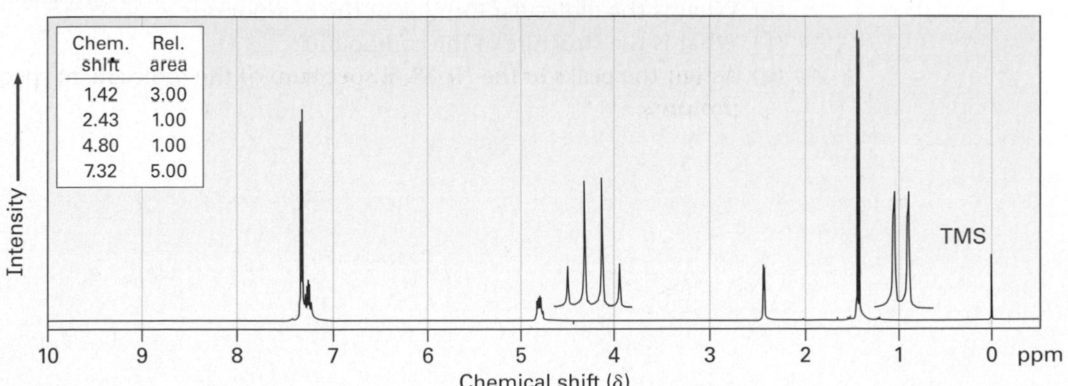

17.46 Propose a structure consistent with the following spectral data for a compound $C_8H_{18}O_2$:

IR: 3350 cm^{-1}

^{1}H NMR: 1.24 δ (12 H, singlet); 1.56 δ (4 H, singlet); 1.95 δ (2 H, singlet)

17.47 The ^{1}H NMR spectrum shown is that of 3-methyl-3-buten-1-ol. Assign all the observed resonance peaks to specific protons, and account for the splitting patterns.

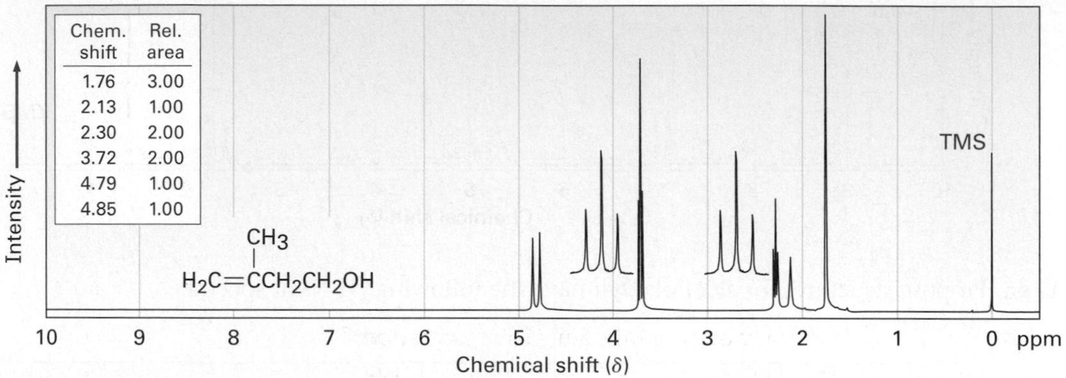

17.48 A compound of unknown structure gave the following spectroscopic data:

Mass spectrum: M$^+$ = 88.1

IR: 3600 cm^{-1}

^{1}H NMR: 1.4 δ (2 H, quartet, J = 7 Hz); 1.2 δ (6 H, singlet); 1.0 δ (1 H, singlet); 0.9 δ (3 H, triplet, J = 7 Hz)

^{13}C NMR: 74, 35, 27, 25 δ

(a) Assuming that the compound contains C and H but may or may not contain O, give three possible molecular formulas.
(b) How many protons (H) does the compound contain?
(c) What functional group(s) does the compound contain?
(d) How many carbons does the compound contain?
(e) What is the molecular formula of the compound?
(f) What is the structure of the compound?
(g) Assign the peaks in the ^{1}H NMR spectrum of the molecule to specific protons.

17.49 Propose a structure for a compound $C_{15}H_{24}O$ that has the following ^1H NMR spectrum. The peak marked by an asterisk disappears when D_2O is added to the sample.

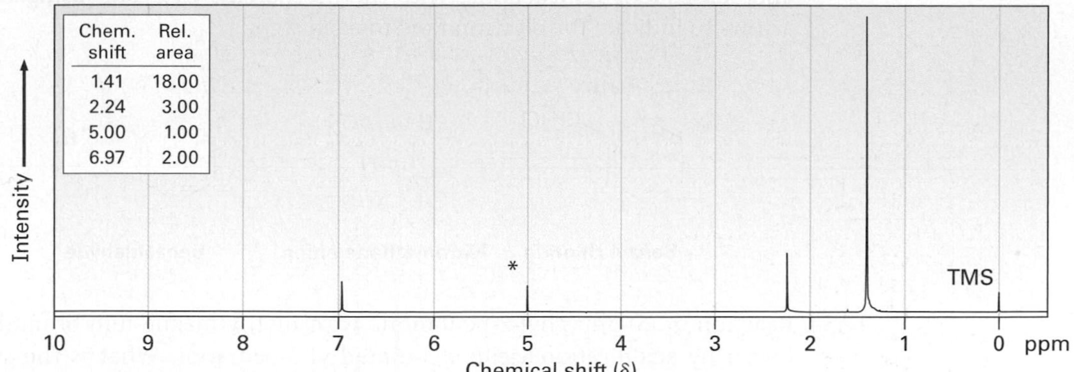

General Problems

17.50 How would you carry out the following transformations?

(a) CO_2H $\xrightarrow{?}$ CO_2H

(b) CO_2H $\xrightarrow{?}$ CH_2OH

(c) CO_2H $\xrightarrow{?}$ CH_2SH

17.51 Benzoquinone is an excellent dienophile in the Diels–Alder reaction. What product would you expect from reaction of benzoquinone with 1 equivalent of 1,3-butadiene? From reaction with 2 equivalents of 1,3-butadiene?

17.52 Rank the following substituted phenols in order of increasing acidity, and explain your answer:

OH   F, OH CH_3O, OH   N≡C, OH

17.53 Benzyl chloride can be converted into benzaldehyde by treatment with nitromethane and base. The reaction involves initial conversion of nitromethane into its anion, followed by S_N2 reaction of the anion with benzyl chloride and subsequent E2 reaction. Write the mechanism in detail, using curved arrows to indicate the electron flow in each step.

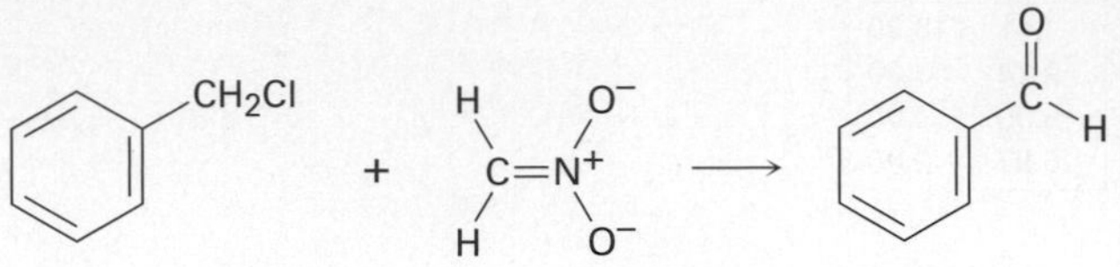

CH_2Cl H O^- C=N^+ H O^- O C H

Benzyl chloride **Nitromethane anion** **Benzaldehyde**

17.54 Reaction of (*S*)-3-methyl-2-pentanone with methylmagnesium bromide followed by acidification yields 2,3-dimethyl-2-pentanol. What is the stereochemistry of the product? Is the product optically active?

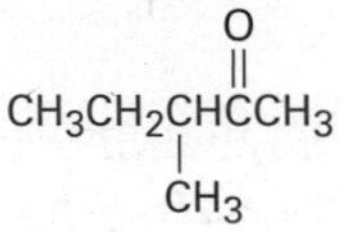

$CH_3CH_2CHCCH_3$ (with O on C, CH_3 on CH) **3-Methyl-2-pentanone**

17.55 Testosterone is one of the most important male steroid hormones. When testosterone is dehydrated by treatment with acid, rearrangement occurs to yield the product shown. Propose a mechanism to account for this reaction.

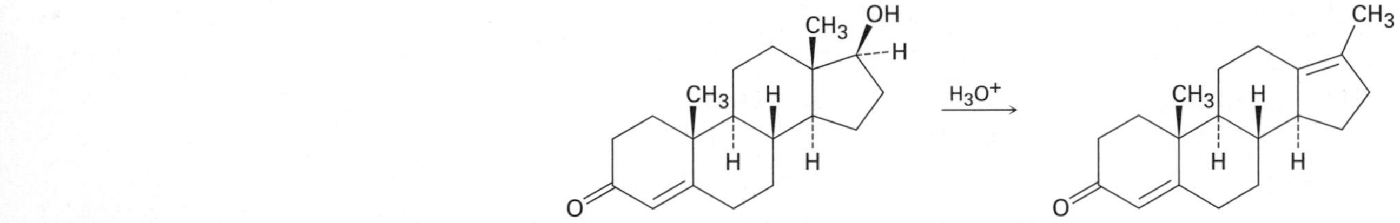

CH_3 OH H CH_3 H H H O $\xrightarrow{H_3O^+}$ CH_3 CH_3 H H H O

Testosterone

17.56 Starting from testosterone (Problem 17.55), how would you prepare the following substances?

(a)

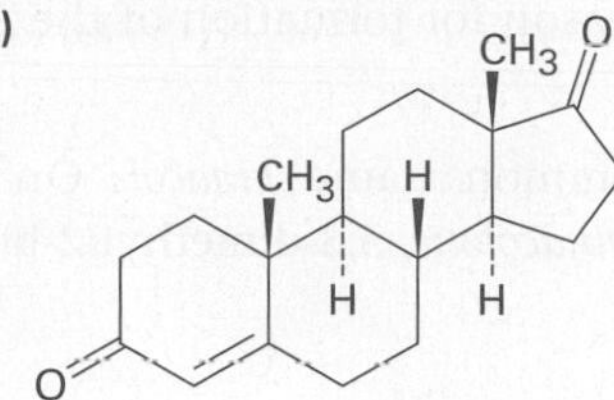

(b)

(c)

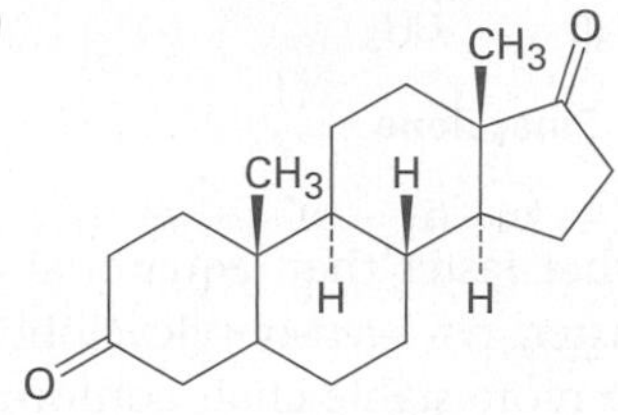

(d)

17.57 *p*-Nitrophenol and 2,6-dimethyl-4-nitrophenol both have pK_a = 7.15, but 3,5-dimethyl-4-nitrophenol has pK_a = 8.25. Why is 3,5-dimethyl-4-nitrophenol so much less acidic?

pK_a = 7.15 **pK_a = 7.15** **pK_a = 8.25**

17.58 Compound A, $C_{10}H_{18}O$, undergoes reaction with dilute H_2SO_4 at 25 °C to yield a mixture of two alkenes, $C_{10}H_{16}$. The major alkene product, **B**, gives only cyclopentanone after ozone treatment followed by reduction with zinc in acetic acid. Write the reactions involved, and identify A and B.

17.59 Compound A, $C_5H_{10}O$, is one of the basic building blocks of nature. All steroids and many other naturally occurring compounds are built from compound A. Spectroscopic analysis of A yields the following information:

IR: 3400 cm^{-1}; 1640 cm^{-1}

^{1}H NMR: 1.63 δ (3 H, singlet); 1.70 δ (3 H, singlet); 3.83 δ (1 H, broad singlet); 4.15 δ (2 H, doublet, *J* = 7 Hz); 5.70 δ (1 H, triplet, *J* = 7 Hz)

(a) How many double bonds and/or rings does A have?

(b) From the IR spectrum, what is the identity of the oxygen-containing functional group?

(c) What kinds of protons are responsible for the NMR absorptions listed?

(d) Propose a structure for A.

17.60 Dehydration of *trans*-2-methylcyclopentanol with $POCl_3$ in pyridine yields predominantly 3-methylcyclopentene. Is the stereochemistry of this dehydration syn or anti? Can you suggest a reason for formation of the observed product? (Make molecular models!)

17.61 2,3-Dimethyl-2,3-butanediol has the common name *pinacol.* On heating with aqueous acid, pinacol rearranges to *pinacolone,* 3,3-dimethyl-2-butanone. Suggest a mechanism for this reaction.

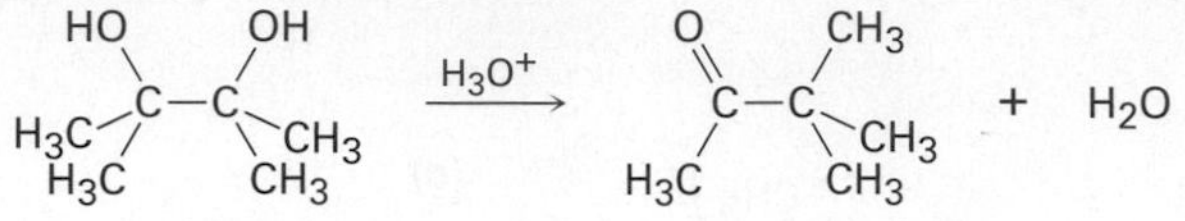

$$(CH_3)_2C(OH)\text{–}C(OH)(CH_3)_2 \xrightarrow{H_3O^+} CH_3C(=O)\text{–}C(CH_3)_3 + H_2O$$

Pinacol **Pinacolone**

17.62 As a rule, axial alcohols oxidize somewhat faster than equatorial alcohols. Which would you expect to oxidize faster, *cis*-4-*tert*-butylcyclohexanol or *trans*-4-*tert*-butylcyclohexanol? Draw the more stable chair conformation of each molecule.

17.63 Propose a synthesis of bicyclohexylidene, starting from cyclohexanone as the only source of carbon.

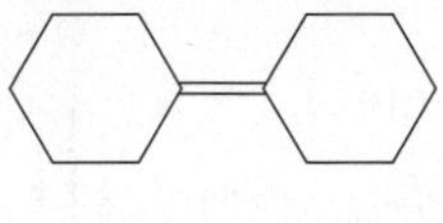

Bicyclohexylidene

17.64 A problem often encountered in the oxidation of primary alcohols to acids is that esters are sometimes produced as by-products. For example, oxidation of ethanol yields acetic acid and ethyl acetate:

$$CH_3CH_2OH \xrightarrow{CrO_3} CH_3\overset{O}{\overset{\|}{C}}OH + CH_3\overset{O}{\overset{\|}{C}}OCH_2CH_3$$

Propose a mechanism to account for the formation of ethyl acetate. Take into account the reversible reaction between aldehydes and alcohols:

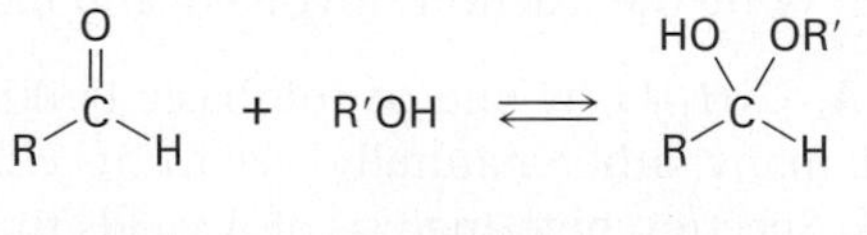

$$R\overset{O}{\overset{\|}{C}}H + R'OH \rightleftharpoons RCH(OH)(OR')$$

17.65 Identify the reagents **a–f** in the following scheme:

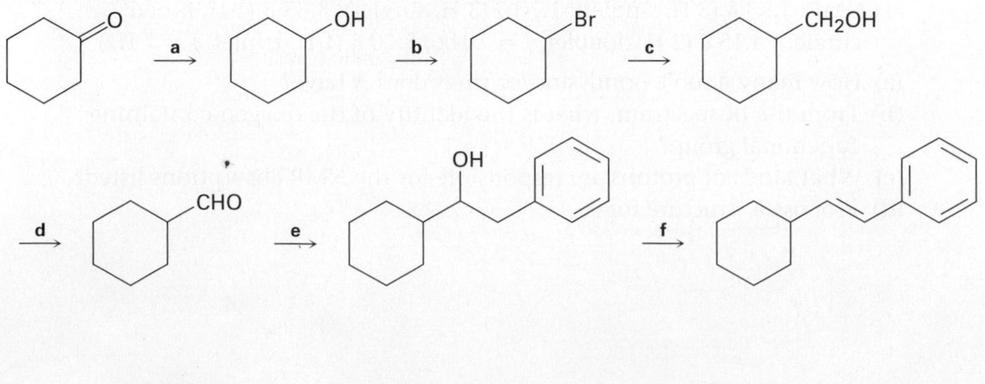

Cyclohexanone $\xrightarrow{a}$ cyclohexanol (OH) $\xrightarrow{b}$ bromocyclohexane (Br) $\xrightarrow{c}$ cyclohexylmethanol (CH_2OH) $\xrightarrow{d}$ cyclohexanecarbaldehyde (CHO) $\xrightarrow{e}$ 1-cyclohexyl-2-phenylethanol (OH) $\xrightarrow{f}$ 1-cyclohexyl-2-phenylethene

17.66 Galactose, a constituent of the disaccharide lactose found in dairy products, is metabolized by a pathway that includes the isomerization of UDP-galactose to UDP-glucose, where UDP = uridylyl diphosphate. The enzyme responsible for the transformation uses NAD^+ as cofactor. Propose a mechanism.

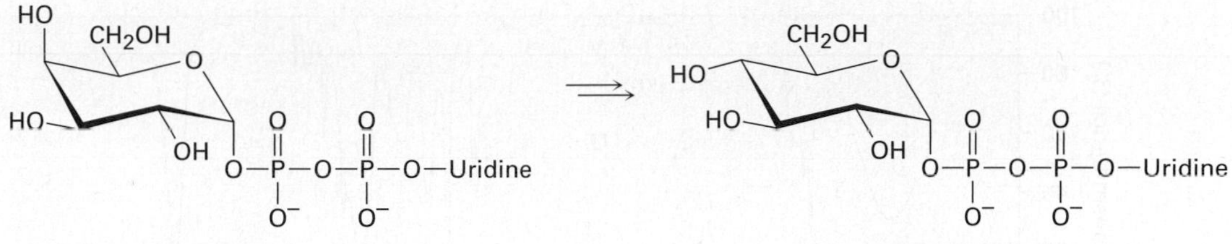

UDP-galactose ⇄ UDP-glucose

17.67 Propose structures for alcohols that have the following ^1H NMR spectra:

(a) $C_9H_{12}O$

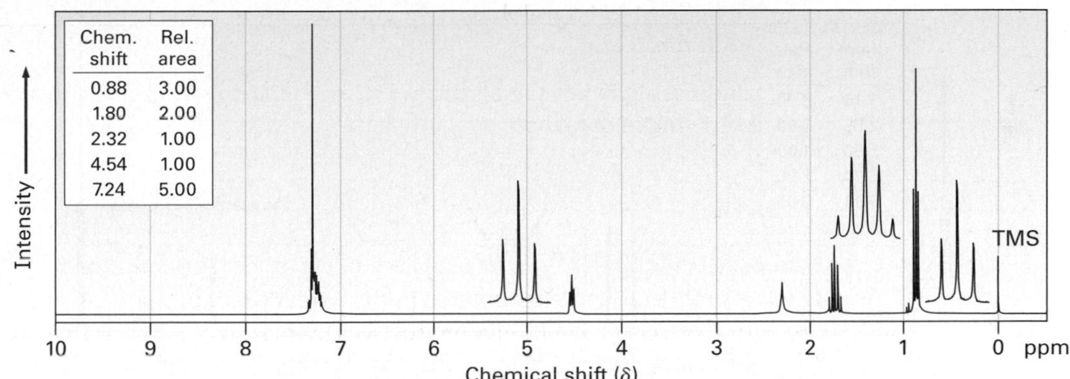

(b) $C_8H_{10}O_2$

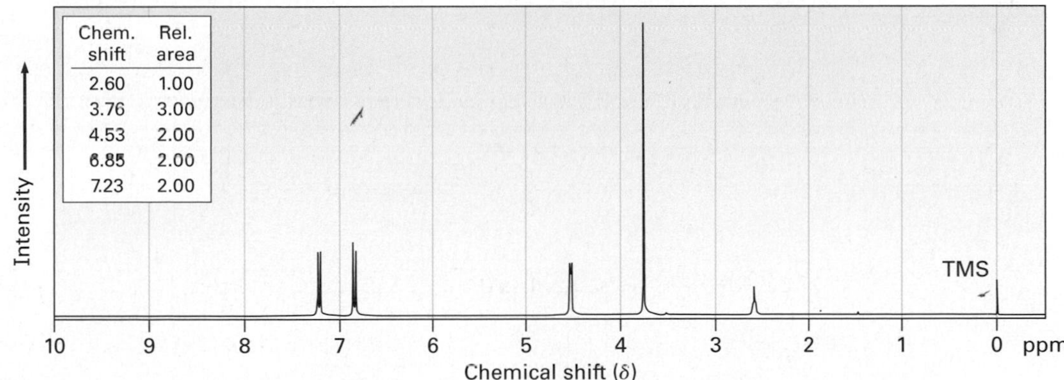

17.68 Compound **A**, $C_8H_{10}O$, has the IR and ^1H NMR spectra shown. Propose a structure consistent with the observed spectra, and assign each peak in the NMR spectrum. Note that the absorption at 5.5 δ disappears when D_2O is added.

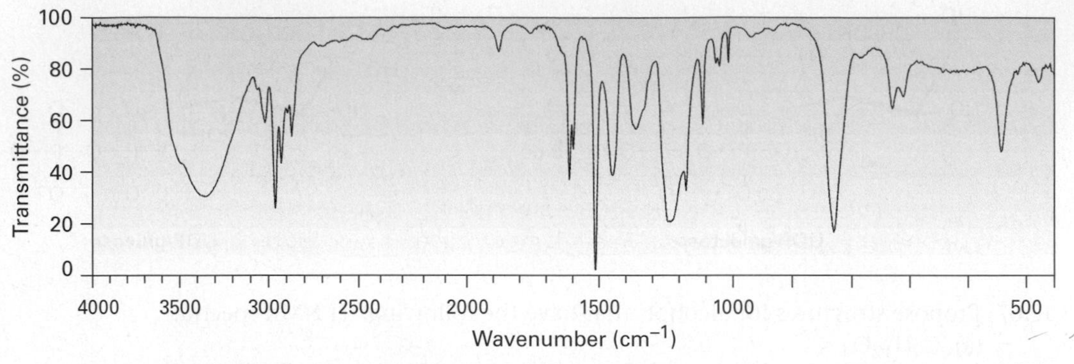

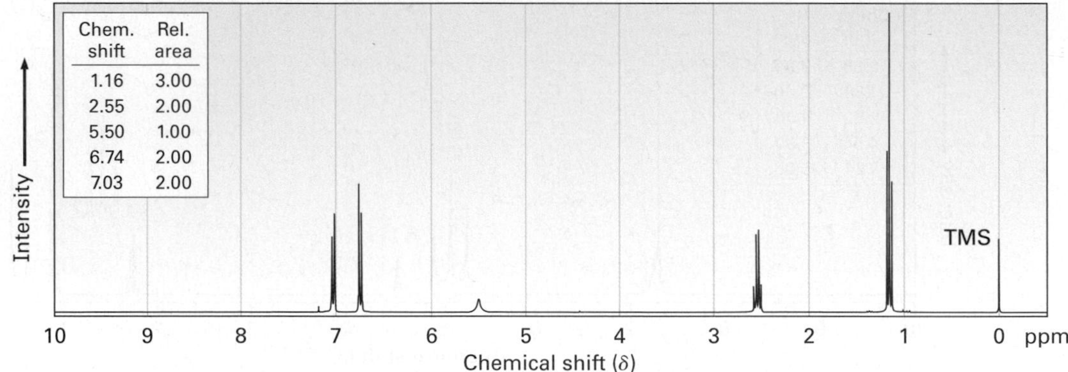

17.69 The reduction of carbonyl compounds by reaction with hydride reagents ($H:^-$) and the Grignard addition by reaction with organomagnesium halides ($R:^-$ ^+MgBr) are examples of *nucleophilic carbonyl addition reactions.* What analogous product do you think might result from reaction of cyanide ion with a ketone?

$$\text{(ketone) } \xrightarrow[H_3O^+]{CN^-} \ ?$$

17.70 Ethers can be prepared by reaction of an alkoxide or phenoxide ion with a primary alkyl halide. Anisole, for instance, results from reaction of sodium phenoxide with iodomethane. What kind of reaction is occurring? Show the mechanism.

$$C_6H_5O^-\ Na^+ + CH_3I \longrightarrow C_6H_5OCH_3$$

Sodium phenoxide → **Anisole**

18

The appalling and unforgettable odor of skunks is due to a mixture of several simple thiols. Comstock Images/ Jupiterimages

Ethers and Epoxides; Thiols and Sulfides

Ethers (R–O–R′), like the alcohols we saw in the preceding chapter, are also organic derivatives of water, but they have two organic groups bonded to the same oxygen atom rather than one. The organic groups might be alkyl, aryl, or vinylic, and the oxygen atom might be in an open chain or a ring.

Perhaps the most well-known ether is diethyl ether, which has a long history of medicinal use as an anesthetic and industrial use as a solvent. Other useful ethers include anisole, a pleasant-smelling aromatic ether used in perfumery, and tetrahydrofuran (THF), a cyclic ether often used as a solvent.

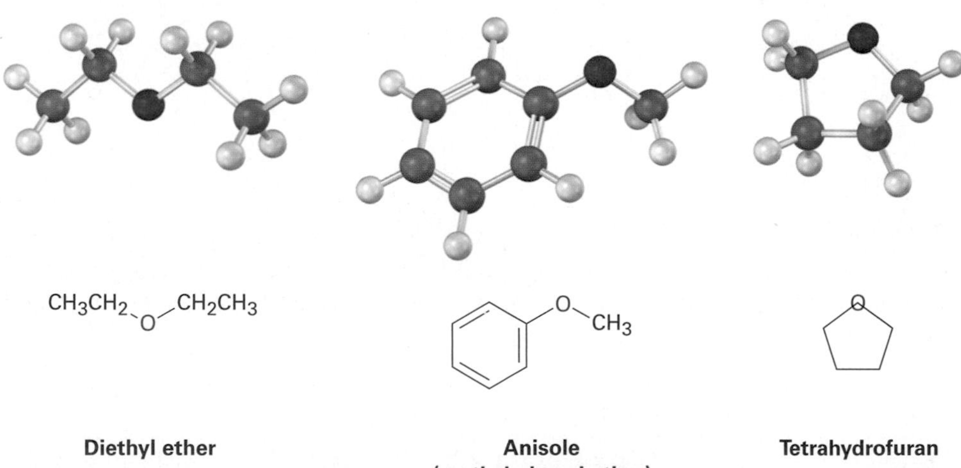

Thiols (R–S–H) and **sulfides (R–S–R′)** are sulfur analogs of alcohols and ethers, respectively. Both functional groups are found in various biomolecules, although not as commonly as their oxygen-containing relatives.

Why This Chapter? This chapter finishes the coverage of functional groups with C–O and C–S single bonds that was begun in Chapter 17. We'll focus primarily on ethers and take only a brief look at thiols and sulfides before going on to an extensive coverage of compounds with C=O double bonds in Chapters 19 through 23.

OWL Sign in to OWL for Organic Chemistry at **www.cengage.com/owl** to view tutorials and simulations, develop problem-solving skills, and complete online homework assigned by your professor.

18.1 Names and Properties of Ethers

Simple ethers with no other functional groups are named by identifying the two organic substituents and adding the word *ether.*

Isopropyl methyl ether **Ethyl phenyl ether**

If other functional groups are present, the ether part is considered an *alkoxy* substituent. For example:

***p*-Dimethoxybenzene** **4-*tert*-Butoxy-1-cyclohexene**

Like alcohols, ethers have nearly the same geometry as water. The R–O–R bonds have an approximately tetrahedral bond angle (112° in dimethyl ether), and the oxygen atom is sp^3-hybridized.

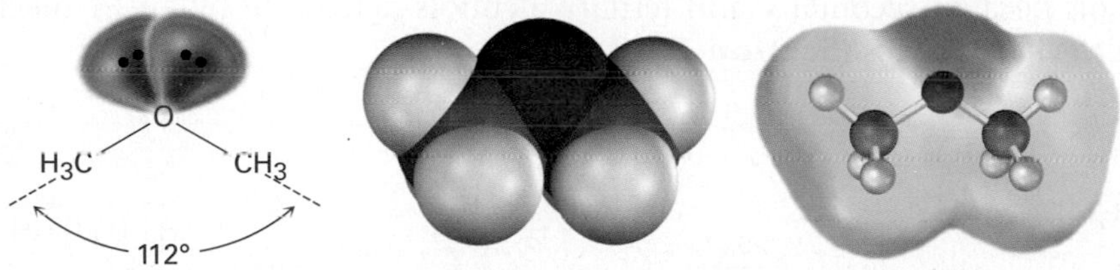

The electronegative oxygen atom gives ethers a slight dipole moment, and the boiling points of ethers are often slightly higher than the boiling points of comparable alkanes. Table 18.1 compares the boiling points of some common ethers and the corresponding hydrocarbons.

Table 18.1 Comparison of Boiling Points of Ethers and Hydrocarbons

Ether	Boiling point °C	Hydrocarbon	Boiling point °C
CH_3OCH_3	−25	$CH_3CH_2CH_3$	−45
$CH_3CH_2OCH_2CH_3$	34.6	$CH_3CH_2CH_2CH_2CH_3$	36
O	65		49
OCH_3	158	CH_2CH_3	136

Ethers are relatively stable and unreactive in many respects, but some ethers react slowly with the oxygen in air to give *peroxides,* compounds that contain an O–O bond. The peroxides from low-molecular-weight ethers such as diisopropyl ether and tetrahydrofuran are explosive and extremely dangerous, even in tiny amounts. Ethers are very useful as solvents in the laboratory, but they must always be used cautiously and should not be stored for long periods of time.

Problem 18.1

Name the following ethers:

(a) $CH_3CHOCHCH_3$ (with CH_3 on each CH)

(b) cyclopentyl–$OCH_2CH_2CH_3$

(c) Br–benzene–OCH_3 (para)

(d) cyclohexenyl–OCH_3

(e) $CH_3CHCH_2OCH_2CH_3$ (with CH_3 on CH)

(f) $H_2C{=}CHCH_2OCH{=}CH_2$

18.2 Synthesis of Ethers

Diethyl ether and other simple symmetrical ethers are prepared industrially by the sulfuric acid–catalyzed reaction of alcohols. The reaction occurs by S_N2 displacement of water from a protonated ethanol molecule by the oxygen atom of a second ethanol. Unfortunately, the method is limited to use with primary alcohols because secondary and tertiary alcohols dehydrate by an E1 mechanism to yield alkenes **(Section 17.6)**.

$CH_3CH_2OH \xrightarrow{H_3O^+} [CH_3CH_2OH_2^+] \xrightarrow[S_N2]{CH_3CH_2OH} [CH_3CH_2O^+(H)CH_2CH_3] + H_2O \longrightarrow CH_3CH_2OCH_2CH_3 + H_3O^+$

The Williamson Ether Synthesis

The most generally useful method of preparing ethers is by the *Williamson ether synthesis,* in which an alkoxide ion reacts with a primary alkyl halide or tosylate in an S_N2 reaction. As we saw earlier in **Section 17.2**, the alkoxide ion is normally prepared by reaction of an alcohol with a strong base such as sodium hydride, NaH.

Cyclopentanol $\xrightarrow[THF]{NaH}$ Alkoxide ion ($O^-\ Na^+$) $\xrightarrow{CH_3-I}$ Cyclopentyl methyl ether (74%)

Cyclopentanol **Alkoxide ion** **Cyclopentyl methyl ether (74%)**

A useful variation of the Williamson synthesis involves silver oxide, Ag_2O, as a mild base rather than NaH. Under these conditions, the free alcohol reacts directly with alkyl halide, so there is no need to preform the metal alkoxide intermediate. Sugars react particularly well; glucose, for example, reacts with excess iodomethane in the presence of Ag_2O to generate a pentaether in 85% yield.

$$\alpha\text{-D-Glucose} \xrightarrow[Ag_2O]{CH_3I} \alpha\text{-D-Glucose pentamethyl ether (85\%)} + AgI$$

α-D-Glucose **α-D-Glucose pentamethyl ether (85%)**

Because the Williamson synthesis is an S_N2 reaction, it is subject to all the usual constraints, as discussed in **Section 11.3**. Primary halides and tosylates work best because competitive E2 elimination can occur with more hindered substrates. Unsymmetrical ethers should therefore be synthesized by reaction between the more hindered alkoxide partner and less hindered halide partner rather than vice versa. For example, *tert*-butyl methyl ether, a substance used in the 1990s as an octane booster in gasoline, is best prepared by reaction of *tert*-butoxide ion with iodomethane rather than by reaction of methoxide ion with 2-chloro-2-methylpropane.

$$(CH_3)_3C\text{–}O^- + CH_3\text{–}I \longrightarrow (CH_3)_3C\text{–}O\text{–}CH_3 + I^-$$

***tert*-Butoxide** **Iodomethane** ***tert*-Butyl methyl ether**

$$CH_3O^- + H_2C\text{–}C(CH_3)_2Cl \longrightarrow H_2C{=}C(CH_3)_2 + CH_3OH + Cl^-$$

2-Chloro-2-methylpropane **2-Methylpropene**

Problem 18.2

Why do you suppose only symmetrical ethers are prepared by the sulfuric acid–catalyzed dehydration procedure? What product(s) would you expect if ethanol and 1-propanol were allowed to react together? In what ratio would the products be formed if the two alcohols were of equal reactivity?

Problem 18.3

How would you prepare the following ethers using a Williamson synthesis?

(a) Methyl propyl ether **(b)** Anisole (methyl phenyl ether)
(c) Benzyl isopropyl ether **(d)** Ethyl 2,2-dimethylpropyl ether

Alkoxymercuration of Alkenes

We saw in **Section 8.4** that alkenes react with water in the presence of mercuric acetate to yield a hydroxymercuration product. Subsequent treatment with $NaBH_4$ breaks the C–Hg bond and yields the alcohol. A similar **alkoxymercuration** reaction occurs when an alkene is treated with an *alcohol* in the presence of mercuric acetate or, even better, mercuric trifluoroacetate, $(CF_3CO_2)_2Hg$. Demercuration by reaction with $NaBH_4$ then yields an ether. The net result is Markovnikov addition of the alcohol to the alkene.

Styrene $\xrightarrow[CH_3OH]{(CF_3CO_2)_2Hg}$ [intermediate with OCH_3 and HgO_2CCF_3] $\xrightarrow{NaBH_4}$ 1-Methoxy-1-phenylethane (97%)

Cyclohexene $\xrightarrow[2.\ NaBH_4]{1.\ (CF_3CO_2)_2Hg,\ CH_3CH_2OH}$ Cyclohexyl ethyl ether (100%)

The mechanism of the alkoxymercuration reaction is similar to that described in **Section 8.4** for hydroxymercuration. The reaction is initiated by electrophilic addition of Hg^{2+} to the alkene, followed by reaction of the intermediate cation with alcohol and reduction of the C–Hg bond by $NaBH_4$. A variety of alcohols and alkenes can be used in the alkoxymercuration reaction. Primary, secondary, and even tertiary alcohols react well, but ditertiary ethers can't be prepared because of steric hindrance to reaction.

Worked Example 18.1 Synthesizing an Ether

How would you prepare ethyl phenyl ether? Use whichever method you think is more appropriate, the Williamson synthesis or the alkoxymercuration reaction.

Strategy

Draw the target ether, identify the two groups attached to oxygen, and recall the limitations of the two methods for preparing ethers. The Williamson synthesis uses an S_N2 reaction and requires that one of the two groups attached to oxygen be either secondary or (preferably) primary. The alkoxymercuration reaction requires that one of the two groups come from an alkene precursor. Ethyl phenyl ether could be made by either method.

Ethyl phenyl ether (C_6H_5–O–CH_2CH_3)

Primary carbon; compatible with Williamson method

Alkene derived; compatible with alkoxymercuration method

Solution

Phenol → (1. NaOH; 2. CH_3CH_2Br) or (1. $(CF_3CO_2)_2Hg$, $H_2C{=}CH_2$; 2. $NaBH_4$) → Ethyl phenyl ether

Phenol **Ethyl phenyl ether**

Problem 18.4

Review the mechanism of oxymercuration shown in Figure 8.3 on page 272, and then write the mechanism of the alkoxymercuration reaction of 1-methylcyclopentene with ethanol. Use curved arrows to show the electron flow in each step.

Problem 18.5

How would you prepare the following ethers? Use whichever method you think is more appropriate, the Williamson synthesis or the alkoxymercuration reaction.

(a) Butyl cyclohexyl ether **(b)** Benzyl ethyl ether ($C_6H_5CH_2OCH_2CH_3$)
(c) *sec*-Butyl *tert*-butyl ether **(d)** Tetrahydrofuran

Problem 18.6

Rank the following halides in order of their reactivity in the Williamson synthesis:

(a) Bromoethane, 2-bromopropane, bromobenzene
(b) Chloroethane, bromoethane, 1-iodopropene

18.3 Reactions of Ethers: Acidic Cleavage

Ethers are unreactive to many reagents used in organic chemistry, a property that accounts for their wide use as reaction solvents. Halogens, dilute acids, bases, and nucleophiles have no effect on most ethers. In fact, ethers undergo only one truly general reaction—they are cleaved by strong acids. Aqueous HBr and HI both work well, but HCl does not cleave ethers.

Ethyl phenyl ether → (HBr, H_2O; Reflux) → Phenol + CH_3CH_2Br

Ethyl phenyl ether **Phenol** **Bromoethane**

Acidic ether cleavages are typical nucleophilic substitution reactions and take place by either S_N1 or S_N2 mechanisms depending on the structure of the substrate. Ethers with only primary and secondary alkyl groups react by an S_N2 mechanism, in which I^- or Br^- attacks the protonated ether at the less hindered site. This usually results in a selective cleavage into a single alcohol and a single alkyl halide. For example, ethyl isopropyl ether yields exclusively isopropyl alcohol and iodoethane on cleavage by HI because nucleophilic attack by iodide ion occurs at the less hindered primary site rather than at the more hindered secondary site.

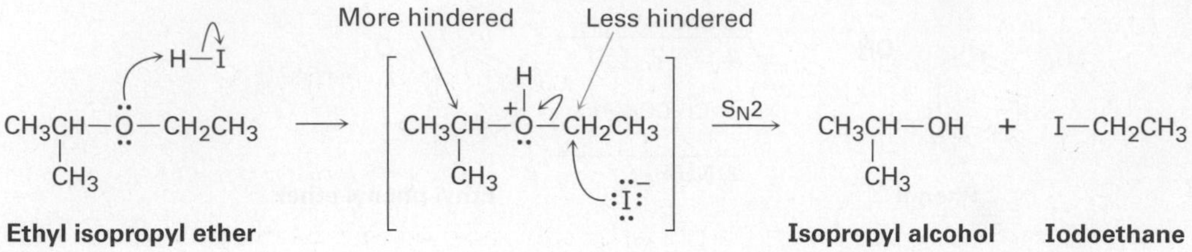

More hindered Less hindered

$$CH_3CH(CH_3){-}\ddot{O}{-}CH_2CH_3 \xrightarrow{H{-}I} \left[CH_3CH(CH_3){-}\overset{+}{O}H{-}CH_2CH_3 \quad :\ddot{I}:^-\right] \xrightarrow{S_N2} CH_3CH(CH_3){-}OH + I{-}CH_2CH_3$$

Ethyl isopropyl ether **Isopropyl alcohol** **Iodoethane**

Ethers with a tertiary, benzylic, or allylic group cleave by either an S_N1 or E1 mechanism because these substrates can produce stable intermediate carbocations. These reactions are often fast and take place at moderate temperatures. *tert*-Butyl ethers, for example, react by an E1 mechanism on treatment with trifluoroacetic acid at 0 °C. We'll see in **Section 26.7** that the reaction is often used in the laboratory synthesis of peptides.

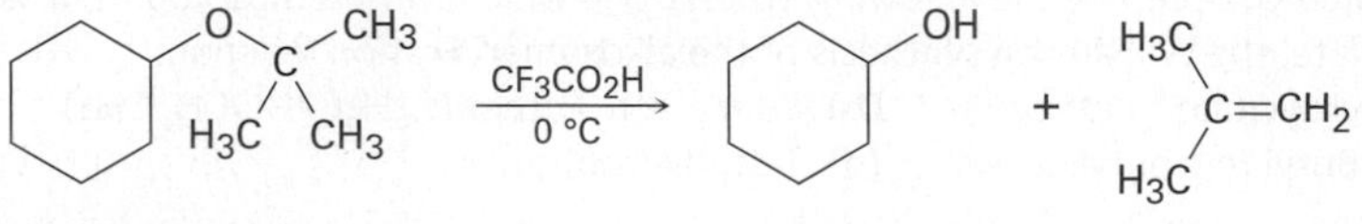

$$C_6H_{11}{-}O{-}C(CH_3)_3 \xrightarrow[0\ °C]{CF_3CO_2H} C_6H_{11}{-}OH + (H_3C)_2C{=}CH_2$$

***tert*-Butyl cyclohexyl ether** **Cyclohexanol (90%)** **2-Methylpropene**

Worked Example 18.2 Predicting the Product of an Ether Cleavage Reaction

Predict the products of the following reaction:

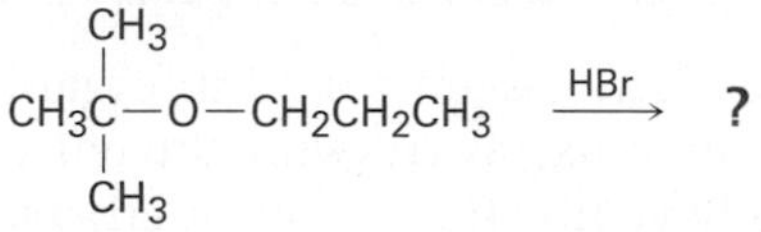

$$CH_3C(CH_3)_2{-}O{-}CH_2CH_2CH_3 \xrightarrow{HBr} ?$$

Strategy

Identify the substitution pattern of the two groups attached to oxygen—in this case a tertiary alkyl group and a primary alkyl group. Then recall the guidelines for ether cleavages. An ether with only primary and secondary alkyl groups usually undergoes cleavage by S_N2 attack of a nucleophile on the less hindered alkyl group, but an ether with a tertiary alkyl group usually undergoes cleavage by an S_N1 mechanism. In this case, an S_N1 cleavage of the tertiary C—O bond will occur, giving 1-propanol and a tertiary alkyl bromide. In addition, a competitive E1 reaction leading to alkene might occur.

Solution

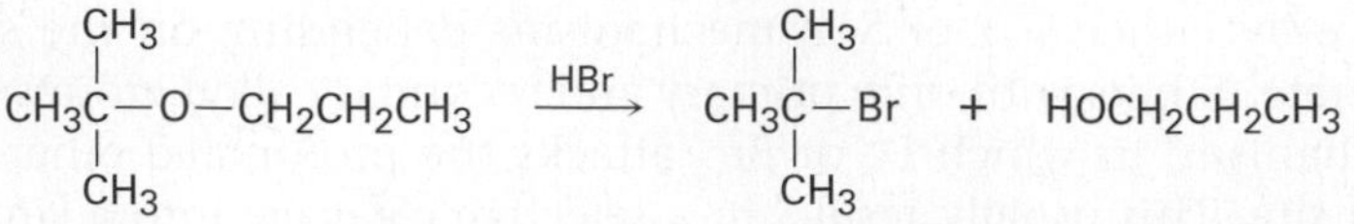

$$CH_3C(CH_3)_2{-}O{-}CH_2CH_2CH_3 \xrightarrow{HBr} CH_3C(CH_3)_2{-}Br + HOCH_2CH_2CH_3$$

***tert*-Butyl propyl ether** **2-Bromo-2-methylpropane** **1-Propanol**

Problem 18.7

Predict the products of the following reactions:

(a) [structure: phenyl tert-alkyl methyl ether] $\xrightarrow{HBr}$?

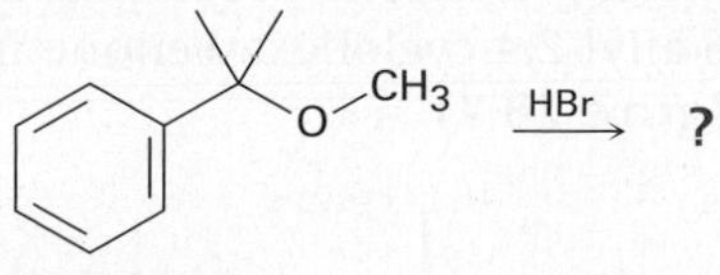

(b) $CH_3CH_2CH(CH_3)-O-CH_2CH_2CH_3 \xrightarrow{HBr}$?

Problem 18.8

Write the mechanism of the acid-induced cleavage of *tert*-butyl cyclohexyl ether to yield cyclohexanol and 2-methylpropene.

Problem 18.9

Why are HI and HBr more effective than HCl in cleaving ethers? (See **Section 11.3.**)

18.4 Reactions of Ethers: Claisen Rearrangement

Unlike the acid-induced ether cleavage reaction discussed in the previous section, which is general to all ethers, the **Claisen rearrangement** is specific to allyl aryl ethers ($H_2C{=}CHCH_2{-}O{-}Ar$) and allyl vinyl ethers ($H_2C{=}CHCH_2{-}O{-}CH{=}CH_2$). Treatment of a phenoxide ion with 3-bromopropene (allyl bromide) results in a Williamson ether synthesis and formation of an allyl aryl ether. Heating the allyl aryl ether to 200–250 °C then effects Claisen rearrangement, leading to an *o*-allylphenol. The net result is alkylation of the phenol in an ortho position.

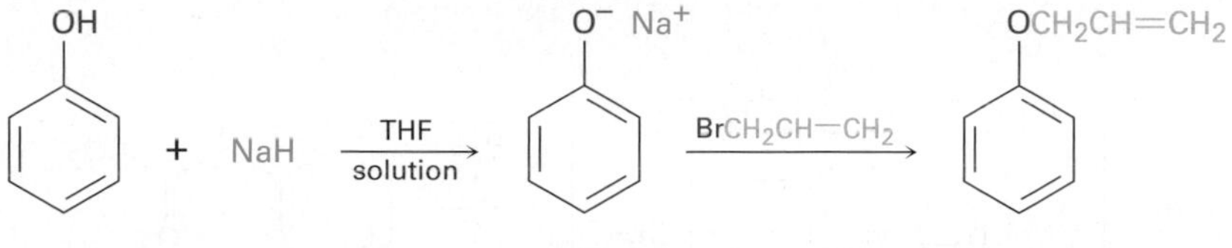

Phenol (OH) + NaH $\xrightarrow[\text{solution}]{\text{THF}}$ Sodium phenoxide ($O^-\ Na^+$) $\xrightarrow{BrCH_2CH{=}CH_2}$ Allyl phenyl ether ($OCH_2CH{=}CH_2$)

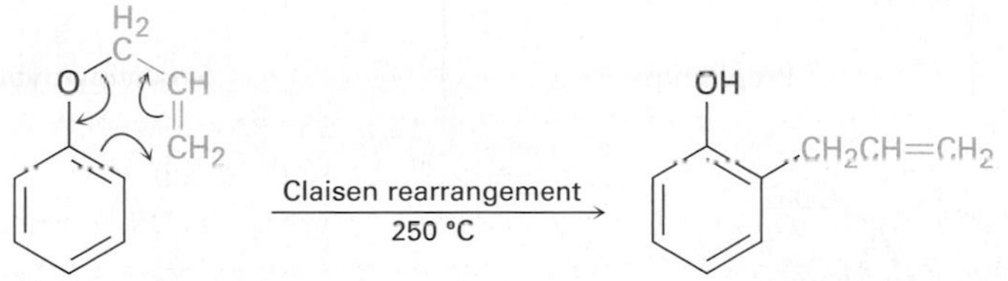

Allyl phenyl ether $\xrightarrow[\text{250 °C}]{\text{Claisen rearrangement}}$ *o*-Allylphenol (OH, $CH_2CH{=}CH_2$)

A similar rearrangement takes place with allyl vinyl ethers, leading to a so-called γ,δ-unsaturated ketone or aldehyde.

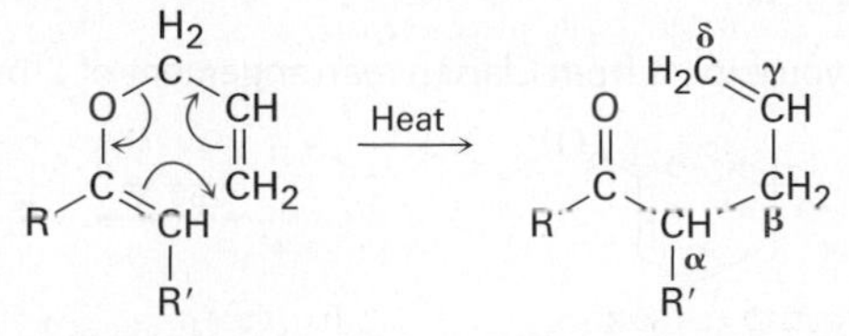

An allyl vinyl ether $\xrightarrow{\text{Heat}}$ A γ,δ unsaturated ketone

[structures: $R-C(=CH-R')-O-CH_2-CH{=}CH_2$ → $R-C(=O)-CH(R')_\alpha-CH_2{}_\beta-CH_\gamma{=}CH_2{}_\delta$]

Like the Diels–Alder reaction discussed in **Sections 14.4 and 14.5**, the Claisen rearrangement reaction takes place in a single step through a pericyclic mechanism in which a reorganization of bonding electrons occurs through a six-membered, cyclic transition state. The 6-allyl-2,4-cyclohexadienone intermediate then isomerizes to *o*-allylphenol **(Figure 18.1)**.

Figure 18.1 The mechanism of the Claisen rearrangement. The C–O bond-breaking and C–C bond-making occur simultaneously.

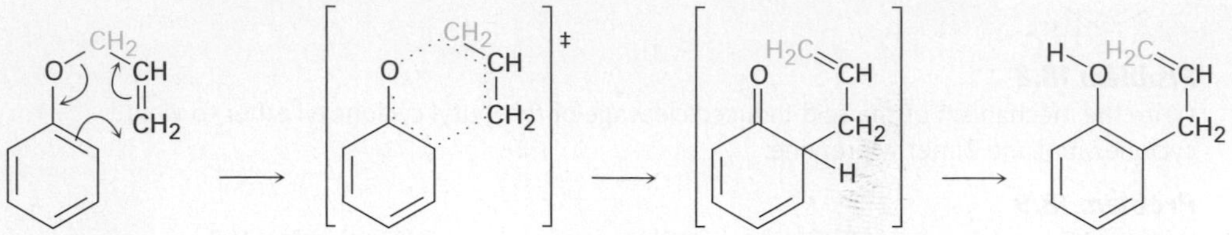

Allyl phenyl ether → **Transition state** → **Intermediate (6-allyl-2,4-cyclohexadienone)** → ***o*-Allylphenol**

Evidence for this mechanism comes from the observation that the rearrangement takes place with an inversion of the allyl group. That is, allyl phenyl ether containing a ^{14}C label on the allyl *ether* carbon atom yields *o*-allylphenol in which the label is on the *terminal* vinylic carbon (green in Figure 18.1). We'll look at the reaction in more detail in **Section 30.8**.

Claisen rearrangements are uncommon in biological pathways, but a well-studied example does occur during biosynthesis of the amino acids phenylalanine and tyrosine. Both phenylalanine and tyrosine arise from a precursor called prephenate, which is itself formed by a biological Claisen rearrangement of the allylic vinyl ether chorismate.

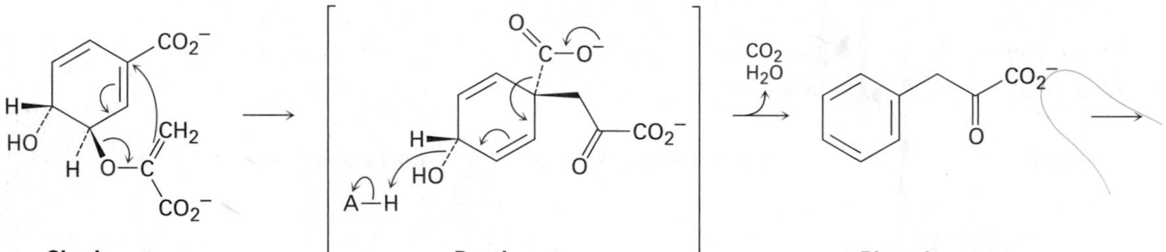

Chorismate → **Prephenate** → (CO_2, H_2O) **Phenylpyruvate** → (Glutamate, α-Ketoglutarate) **Phenylalanine**

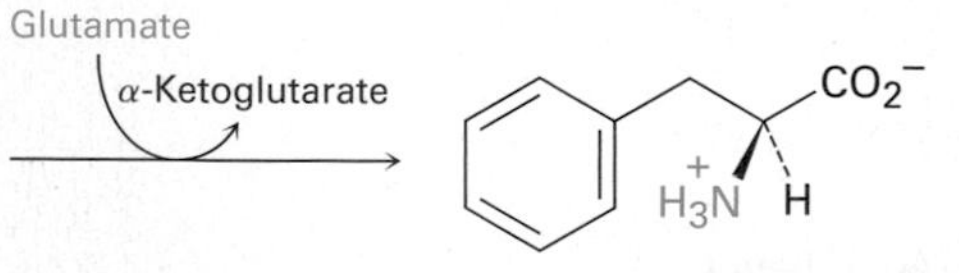

Problem 18.10

What product would you expect from Claisen rearrangement of 2-butenyl phenyl ether?

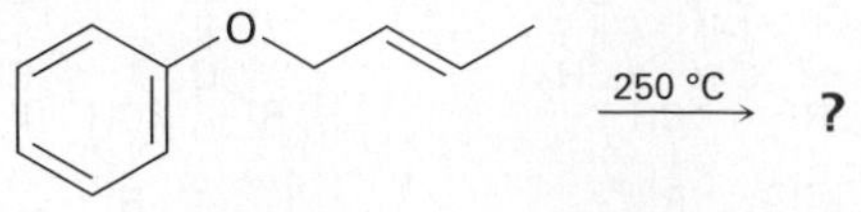

2-Butenyl phenyl ether → (250 °C) ?

18.5 Cyclic Ethers: Epoxides

For the most part, cyclic ethers behave like acyclic ethers. The chemistry of the ether functional group is the same, whether it's in an open chain or in a ring. Common cyclic ethers such as tetrahydrofuran and dioxane, for example, are often used as solvents because of their inertness, yet they can be cleaved by strong acids.

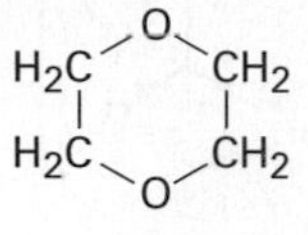

1,4-Dioxane 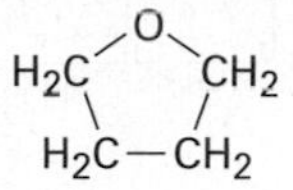 **Tetrahydrofuran**

The one group of cyclic ethers that behaves differently from open-chain ethers are the three-membered-ring compounds called *epoxides,* or *oxiranes,* which we saw in **Section 8.7**. The strain of the three-membered ring gives epoxides unique chemical reactivity.

Ethylene oxide, the simplest epoxide, is an intermediate in the manufacture of both ethylene glycol, used for automobile antifreeze, and polyester polymers. Approximately 18 million metric tons of ethylene oxide is produced worldwide each year, most of it by air oxidation of ethylene over a silver oxide catalyst at 300 °C. This process is not useful for other epoxides, however, and is of little value in the laboratory.

Note that the name *ethylene oxide* is not a systematic one because the *-ene* ending implies the presence of a double bond in the molecule. The name is frequently used, however, because ethylene oxide is derived *from* ethylene by addition of an oxygen atom. Other simple epoxides are named similarly. The systematic name for ethylene oxide is 1,2-epoxyethane.

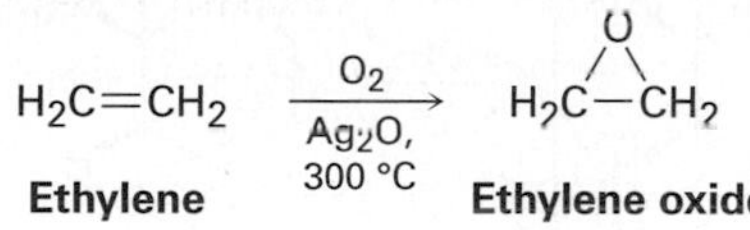

$$H_2C{=}CH_2 \xrightarrow[\text{Ag}_2\text{O, 300 °C}]{O_2} H_2C{-}CH_2 \text{ (with bridging O)}$$

Ethylene **Ethylene oxide**

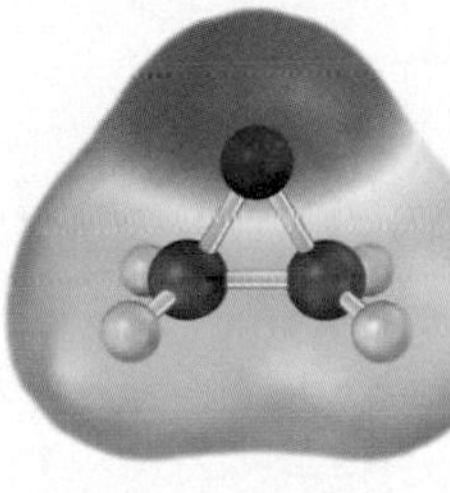

In the laboratory, as we saw in **Section 8.7**, epoxides are usually prepared by treatment of an alkene with a peroxyacid (RCO_3H), typically *m*-chloroperoxybenzoic acid.

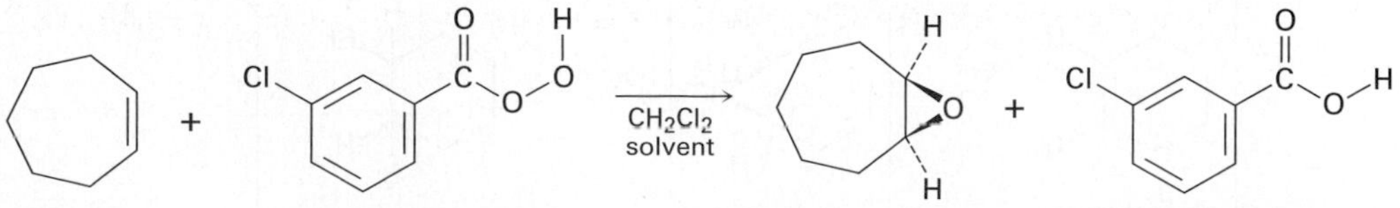

Cycloheptene + ***meta*-Chloroperoxybenzoic acid** $\xrightarrow[\text{solvent}]{CH_2Cl_2}$ **1,2-Epoxycycloheptane** + ***meta*-Chlorobenzoic acid**

Epoxides can also be prepared from halohydrins, themselves produced by electrophilic addition of HO–X to alkenes **(Section 8.3)**. When a halohydrin is treated with base, HX is eliminated and an epoxide is produced by an *intramolecular* Williamson ether synthesis. That is, the nucleophilic alkoxide ion and the electrophilic alkyl halide are in the same molecule.

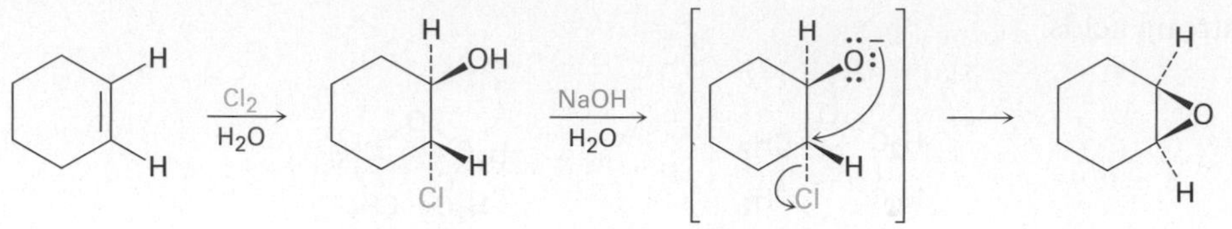

Cl_2, H_2O; NaOH, H_2O

Cyclohexene **trans-2-Chloro-cyclohexanol** **1,2-Epoxy-cyclohexane**

Problem 18.11

Reaction of *cis*-2-butene with *m*-chloroperoxybenzoic acid yields an epoxide different from that obtained by reaction of the trans isomer. Explain.

18.6 Reactions of Epoxides: Ring-Opening

Acid-Catalyzed Epoxide Opening

Epoxides are cleaved by treatment with acid just as other ethers are, but under much milder conditions because of ring strain. As we saw in **Section 8.7**, dilute aqueous acid at room temperature is sufficient to cause the hydrolysis of epoxides to give 1,2-diols, also called *vicinal glycols*. (The word *vicinal* means "adjacent," and a *glycol* is a diol.) The epoxide cleavage takes place by S_N2-like backside attack of a nucleophile on the protonated epoxide, giving a trans-1,2-diol as product.

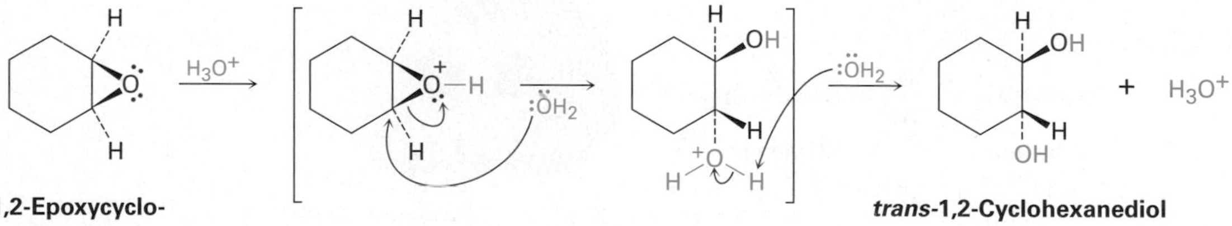

H_3O^+; $:\ddot{O}H_2$; $:\ddot{O}H_2$; $+ H_3O^+$

1,2-Epoxycyclo-hexane ***trans*-1,2-Cyclohexanediol (86%)**

Recall the following:

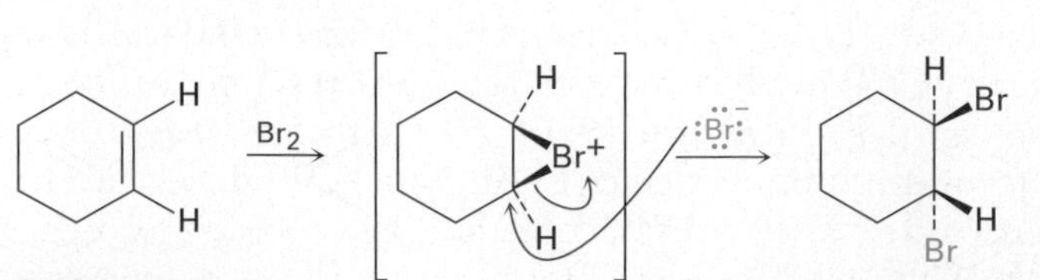

Br_2; $:\ddot{Br}:^-$

Cyclohexene ***trans*-1,2-Dibromo-cyclohexane**

Epoxides can also be opened by reaction with acids other than H_3O^+. If anhydrous HX is used, for instance, an epoxide is converted into a trans halohydrin.

HX
Ether

A trans 2-halocyclohexanol

where X = F, Br, Cl, or I

The regiochemistry of acid-catalyzed ring-opening depends on the epoxide's structure, and a mixture of products is often formed. When both epoxide carbon atoms are either primary or secondary, attack of the nucleophile occurs primarily at the less highly substituted site—an S_N2-like result. When one of the epoxide carbon atoms is tertiary, however, nucleophilic attack occurs primarily at the *more* highly substituted site—an S_N1-like result. Thus, 1,2-epoxypropane reacts with HCl to give primarily 1-chloro-2-propanol, but 2-methyl-1,2-epoxypropane gives 2-chloro-2-methyl-1-propanol as the major product.

Secondary Primary

HCl
Ether

1,2-Epoxypropane **1-Chloro-2-propanol (90%)** **2-Chloro-1-propanol (10%)**

Tertiary Primary

HCl
Ether

2-Methyl-1,2-epoxypropane **2-Chloro-2-methyl-1-propanol (60%)** **1-Chloro-2-methyl-2-propanol (40%)**

The mechanisms of these acid-catalyzed epoxide openings are more complex than they at first appear. They seem to be neither purely S_N1 nor S_N2 but instead to be midway between the two extremes and to have characteristics of both. Take the reaction of 1,2-epoxy-1-methylcyclohexane with HBr shown in **Figure 18.2**, for instance. The reaction yields only a single stereoisomer of 2-bromo-2-methylcyclohexanol in which the –Br and –OH groups are trans, an S_N2-like result caused by backside displacement of the epoxide oxygen. But the fact that Br^- attacks the more hindered tertiary side of the epoxide rather than the less hindered secondary side is an S_N1-like result in which the more stable, tertiary carbocation is involved.

Evidently, the transition state for acid-catalyzed epoxide opening has an S_N2-like geometry but also has a large amount of S_N1-like carbocationic character. Since the positive charge in the protonated epoxide is shared by the more highly substituted carbon atom, backside attack of Br^- occurs at the more highly substituted site.

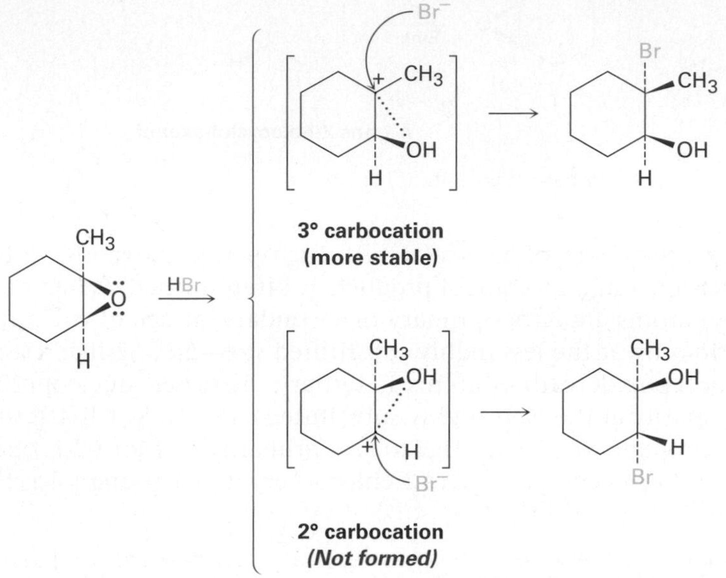

Figure 18.2 Ring-opening of 1,2-epoxy-1-methylcyclohexane with HBr. There is a high degree of S_N1-like carbocation character in the transition state, which leads to backside attack of the nucleophile at the tertiary center and to formation of a product isomer that has —Br and —OH groups trans.

3° carbocation (more stable)

2° carbocation *(Not formed)*

Worked Example 18.3 Predicting the Product of Epoxide Ring-Opening

Predict the major product of the following reaction:

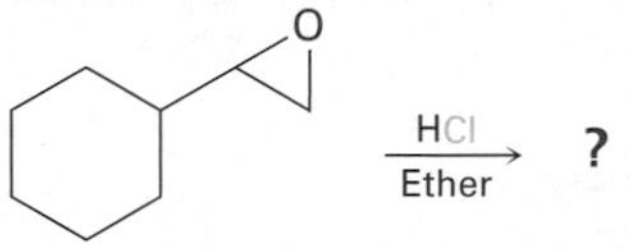

Strategy

Identify the substitution pattern of the two epoxide carbon atoms—in this case, one carbon is secondary and one is primary. Then recall the guidelines for epoxide cleavages. An epoxide with only primary and secondary carbons usually undergoes cleavage by S_N2-like attack of a nucleophile on the less hindered carbon, but an epoxide with a tertiary carbon atom usually undergoes cleavage by backside attack on the more hindered carbon. In this case, an S_N2 cleavage of the primary C—O epoxide bond will occur.

Solution

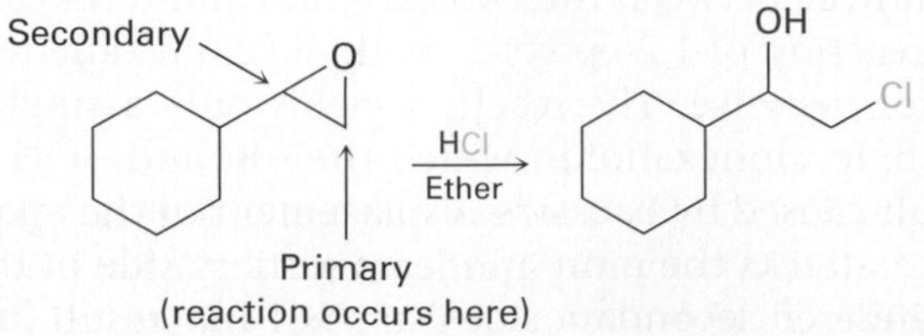

Secondary

Primary (reaction occurs here)

Problem 18.12

Predict the major product of each of the following reactions:

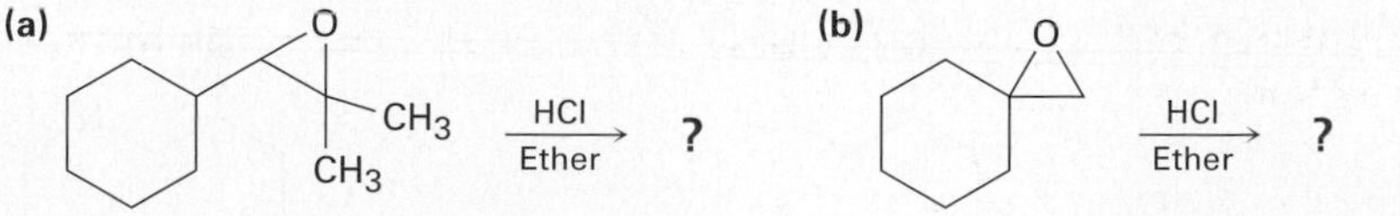

(a) O, CH_3, CH_3 $\xrightarrow[\text{Ether}]{\text{HCl}}$?

(b) O $\xrightarrow[\text{Ether}]{\text{HCl}}$?

Problem 18.13

How would you prepare the following diols?

(a) (b)

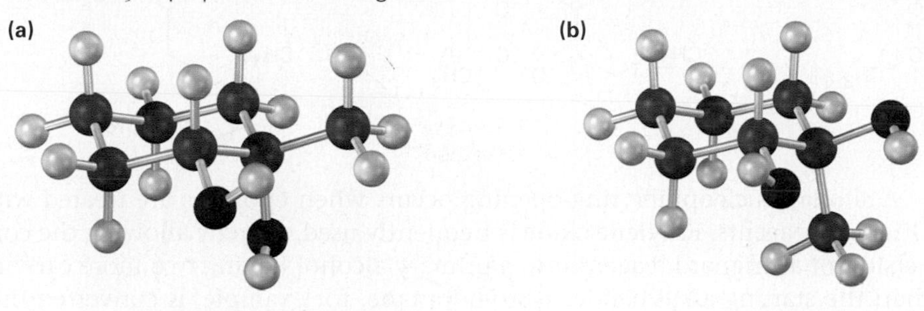

Base-Catalyzed Epoxide Opening

Unlike other ethers, epoxide rings can be cleaved by bases and nucleophiles as well as by acid. Although an ether oxygen is normally a poor leaving group in an S_N2 reaction **(Section 11.3)**, the strain of the three-membered ring causes epoxides to react with hydroxide ion at elevated temperatures.

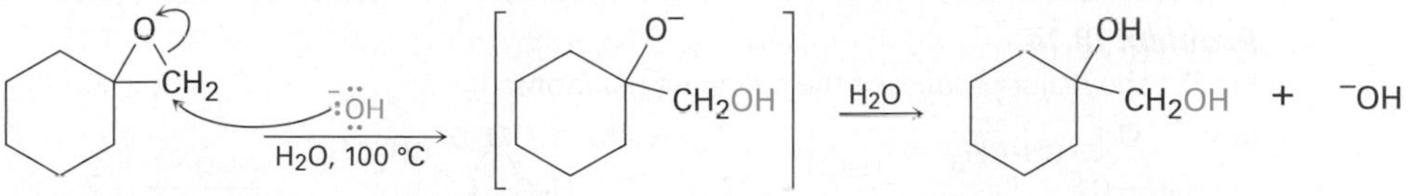

O, CH_2 $\xrightarrow[H_2O,\ 100\ °C]{:\ddot{O}H}$ [O^-, CH_2OH] $\xrightarrow{H_2O}$ OH, CH_2OH + ^-OH

Methylenecyclohexane oxide

1-Hydroxymethyl-cyclohexanol (70%)

Base-catalyzed epoxide opening is a typical S_N2 reaction in which attack of the nucleophile takes place at the less hindered epoxide carbon. For example, 1,2-epoxypropane reacts with ethoxide ion exclusively at the less highly substituted, primary, carbon to give 1-ethoxy-2-propanol.

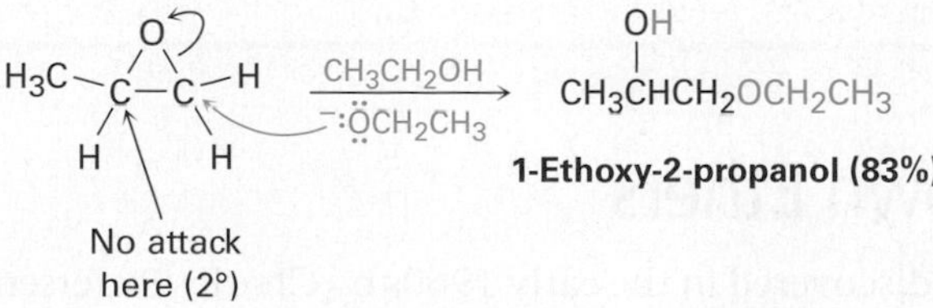

$H_3C-C-C-H$ (O; H, H) $\xrightarrow[^-:\ddot{O}CH_2CH_3]{CH_3CH_2OH}$ $CH_3\overset{OH}{\overset{|}{C}}HCH_2OCH_2CH_3$

No attack here (2°)

1-Ethoxy-2-propanol (83%)

Many different nucleophiles can be used for epoxide opening, including amines (RNH_2 or R_2NH) and Grignard reagents (RMgX). An example of an amine reacting with an epoxide occurs in the commercial synthesis of metoprolol, a so-called β-blocker that is used for treatment of cardiac arrhythmias, hypertension, and heart attacks. β-blockers are among the most widely prescribed drugs in the world.

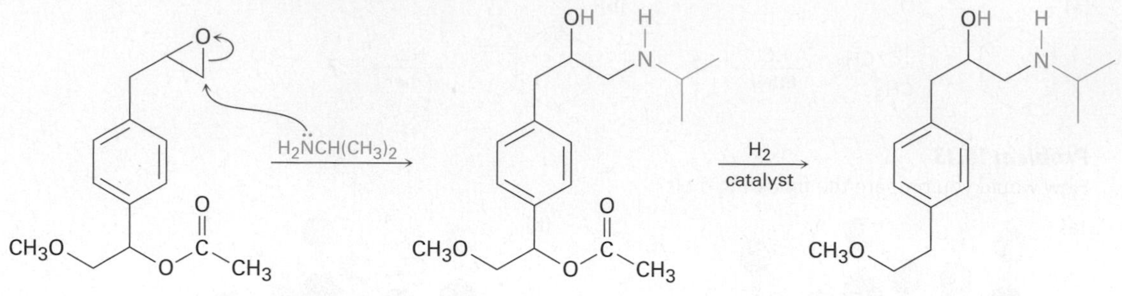

$H_2\ddot{N}CH(CH_3)_2$ → ; H_2, catalyst →

Metoprolol

A similar nucleophilic ring-opening occurs when epoxides are treated with Grignard reagents. Ethylene oxide is frequently used, thereby allowing the conversion of a Grignard reagent into a primary alcohol having two more carbons than the starting alkyl halide. 1-Bromobutane, for example, is converted into 1-hexanol by reaction of its Grignard reagent with ethylene oxide.

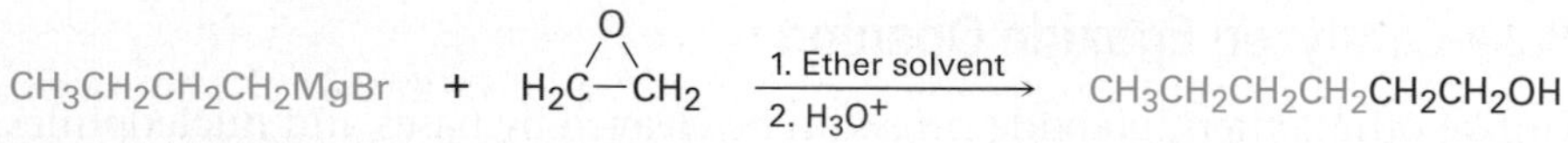

$$CH_3CH_2CH_2CH_2MgBr + H_2C\overset{O}{-}CH_2 \xrightarrow[2.\ H_3O^+]{1.\ \text{Ether solvent}} CH_3CH_2CH_2CH_2CH_2CH_2OH$$

Butylmagnesium bromide **Ethylene oxide** **1-Hexanol (62%)**

Problem 18.14

Predict the major product of the following reactions:

(a) $H_2C\overset{O}{-}C(CH_2CH_3)(CH_3) \xrightarrow[H_2O^{18}]{NaOH}$?

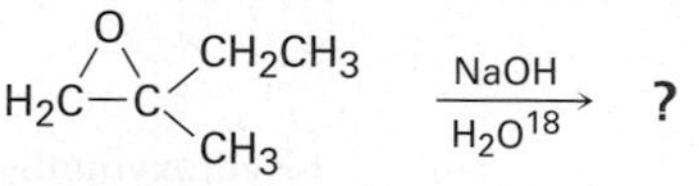

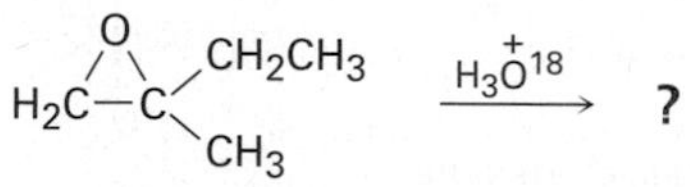

(b) $H_2C\overset{O}{-}C(CH_2CH_3)(CH_3) \xrightarrow{H_3\overset{+}{O}^{18}}$?

(c) $H_3C(H)C\overset{O}{-}C(CH_2CH_3)(CH_3) \xrightarrow[2.\ H_3O^+]{1.\ C_6H_5MgBr}$?

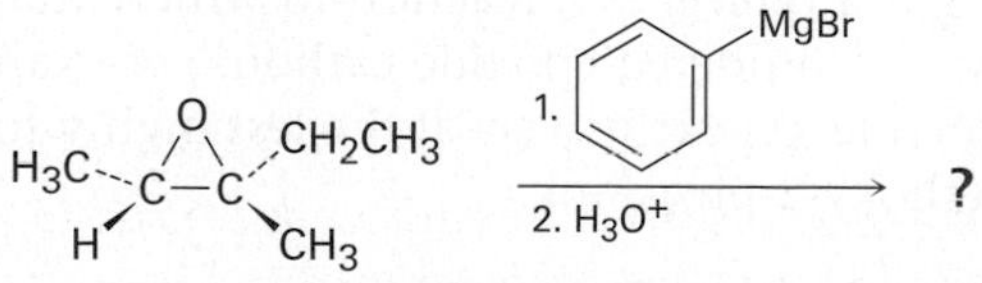

18.7 Crown Ethers

Crown ethers, discovered in the early 1960s by Charles Pedersen at the DuPont Company, are a relatively recent addition to the ether family. They are named according to the general format *x*-crown-*y*, where *x* is the total number of atoms in the ring and *y* is the number of oxygen atoms. Thus, 18-crown-6 ether is an 18-membered ring containing 6 ether oxygen atoms. Note the size and negative (red) character of the crown ether cavity in the following electrostatic potential map.

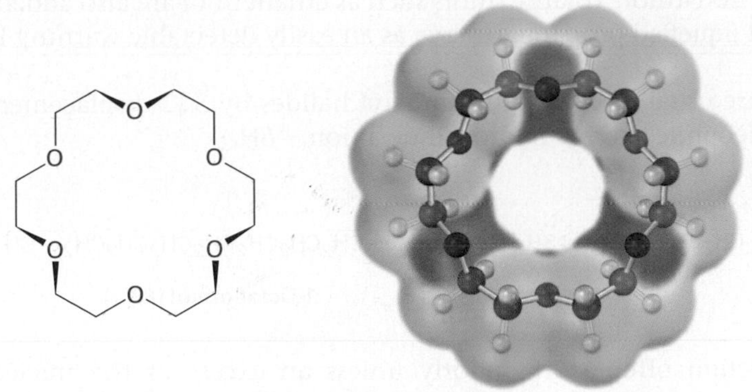

18-Crown-6 ether

The importance of crown ethers derives from their ability to sequester specific metal cations in the center of the polyether cavity. 18-Crown-6, for example, binds strongly with potassium ion. As a result, a solution of 18-crown-6 in a nonpolar organic solvent will dissolve many potassium salts. Potassium permanganate, $KMnO_4$, dissolves in toluene in the presence of 18-crown-6, for instance, and the resulting solution is a valuable reagent for oxidizing alkenes.

The effect of using a crown ether to dissolve an inorganic salt in a hydrocarbon or ether solvent is similar to the effect of dissolving the salt in a polar aprotic solvent such as DMSO, DMF, or HMPA **(Section 11.3)**. In both cases, the metal cation is strongly solvated, leaving the anion bare. Thus, the S_N2 reactivity of an anion is tremendously enhanced in the presence of a crown ether.

Problem 18.15
15-Crown-5 and 12-crown-4 ethers complex Na^+ and Li^+, respectively. Make models of these crown ethers, and compare the sizes of the cavities.

18.8 Thiols and Sulfides

Thiols

Thiols, sometimes called *mercaptans,* are sulfur analogs of alcohols. They are named by the same system used for alcohols, with the suffix *-thiol* used in place of *-ol*. The –SH group itself is referred to as a **mercapto group**. Like alcohols, thiols are weakly acidic; the pK_a of CH_3SH, for instance, is 10.3. Unlike alcohols, however, thiols don't typically form hydrogen bonds because the sulfur atom is not sufficiently electronegative.

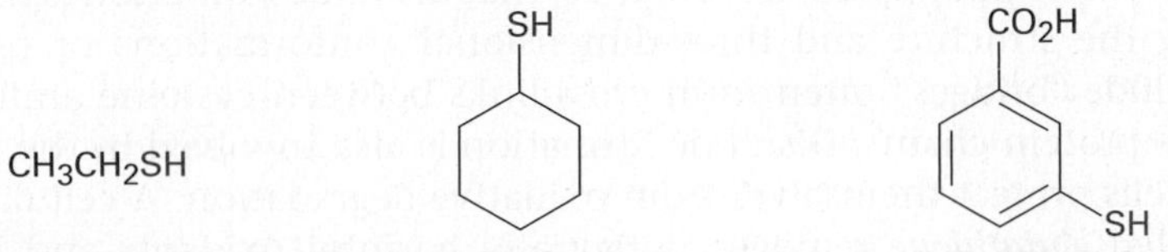

Ethanethiol   Cyclohexanethiol   *m*-Mercaptobenzoic acid

The most striking characteristic of thiols is their appalling odor. Skunk scent, for instance, is caused primarily by the simple thiols 3-methyl-1-butanethiol and 2-butene-1-thiol. Volatile thiols such as ethanethiol are also added to natural gas and liquefied propane to serve as an easily detectable warning in case of leaks.

Thiols are usually prepared from alkyl halides by S_N2 displacement with a sulfur nucleophile such as hydrosulfide anion, ^-SH.

$$CH_3CH_2CH_2CH_2CH_2CH_2CH_2CH_2{-}Br \;+\; {:}\ddot{S}H^- \longrightarrow CH_3CH_2CH_2CH_2CH_2CH_2CH_2CH_2{-}SH \;+\; Br^-$$

1-Bromooctane **1-Octanethiol (83%)**

The reaction often works poorly unless an excess of the nucleophile is used because the product thiol can undergo a second S_N2 reaction with alkyl halide to give a sulfide as a by-product. To circumvent this problem, thiourea, $(NH_2)_2C{=}S$, is often used as the nucleophile in the preparation of a thiol from an alkyl halide. The reaction occurs by displacement of the halide ion to yield an intermediate alkyl isothiourea salt, which is hydrolyzed by subsequent reaction with aqueous base.

$$CH_3CH_2CH_2CH_2CH_2CH_2CH_2CH_2{-}Br \;+\; (H_2N)_2C{=}\ddot{S}{:} \longrightarrow \left[CH_3CH_2CH_2CH_2CH_2CH_2CH_2CH_2{-}\overset{+}{S}{=}C(NH_2){-}NH_2\right] Br^-$$

1-Bromooctane **Thiourea**

$$\xrightarrow{H_2O,\ NaOH} CH_3CH_2CH_2CH_2CH_2CH_2CH_2CH_2{-}SH \;+\; H_2N{-}C({=}O){-}NH_2$$

1-Octanethiol (83%) **Urea**

Thiols can be oxidized by Br_2 or I_2 to yield **disulfides (RSSR′)**. The reaction is easily reversed, and a disulfide can be reduced back to a thiol by treatment with zinc and acid.

$$2\,R{-}SH \underset{Zn,\ H^+}{\overset{I_2}{\rightleftharpoons}} R{-}S{-}S{-}R \;+\; 2\,HI$$

A thiol **A disulfide**

This thiol–disulfide interconversion is a key part of numerous biological processes. We'll see in Chapter 26, for instance, that disulfide formation is involved in defining the structure and three-dimensional conformations of proteins, where disulfide "bridges" often form cross-links between cysteine amino acid units in the protein chains. Disulfide formation is also involved in the process by which cells protect themselves from oxidative degradation. A cellular component called *glutathione* removes potentially harmful oxidants and is itself oxidized to glutathione disulfide in the process. Reduction back to the thiol requires the coenzyme reduced flavin adenine dinucleotide, abbreviated $FADH_2$.

H_2O_2 / $FADH_2$

Glutathione (GSH) **Glutathione disulfide (GSSG)**

Sulfides

Sulfides are the sulfur analogs of ethers just as thiols are the sulfur analogs of alcohols. Sulfides are named by following the same rules used for ethers, with *sulfide* used in place of *ether* for simple compounds and *alkylthio* used in place of *alkoxy* for more complex substances.

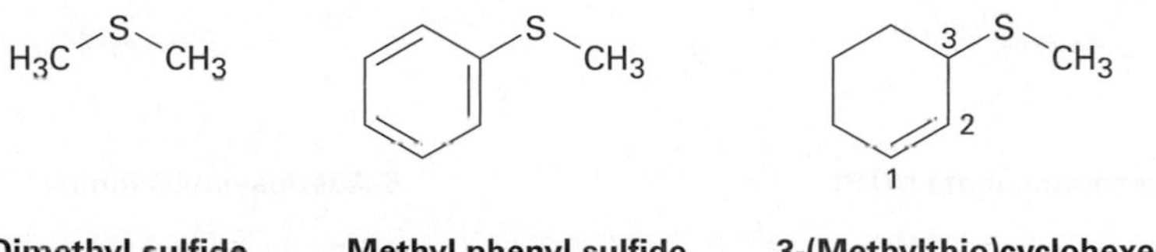

Dimethyl sulfide **Methyl phenyl sulfide** **3-(Methylthio)cyclohexene**

Treatment of a thiol with a base, such as NaH, gives the corresponding **thiolate ion (RS^-)**, which undergoes reaction with a primary or secondary alkyl halide to give a sulfide. The reaction occurs by an S_N2 mechanism, analogous to the Williamson synthesis of ethers **(Section 18.2)**.

$$\text{C}_6\text{H}_5\text{S}^- \text{Na}^+ + CH_3\text{–I} \longrightarrow \text{C}_6\text{H}_5\text{–S–}CH_3 + \text{NaI}$$

Sodium benzenethiolate **Methyl phenyl sulfide (96%)**

Despite their close structural similarity, sulfides and ethers differ substantially in their chemistry. Because the valence electrons on sulfur are farther from the nucleus and are less tightly held than those on oxygen ($3p$ electrons versus $2p$ electrons), sulfur compounds are more nucleophilic than their oxygen analogs. Unlike dialkyl ethers, dialkyl sulfides react rapidly with primary alkyl halides by an S_N2 mechanism to give **sulfonium ions (R_3S^+)**.

$$CH_3-\ddot{S}-CH_3 + CH_3-I \xrightarrow{THF} CH_3-\overset{CH_3}{\overset{|+}{S}}-CH_3\ I^-$$

Dimethyl sulfide **Iodomethane** **Trimethylsulfonium iodide**

The most common example of this process in living organisms is the reaction of the amino acid methionine with adenosine triphosphate (ATP; **Section 6.8**) to give *S*-adenosylmethionine. The reaction is somewhat unusual in that the biological leaving group in this S_N2 process is the *triphosphate* ion rather than the more frequently seen diphosphate ion **(Section 11.6)**.

Methionine + **Adenosine triphosphate (ATP)** $\xrightarrow{S_N2}$ **Triphosphate ion** + ***S*-Adenosylmethionine**

Sulfonium ions are themselves useful alkylating agents because a nucleophile can attack one of the groups bonded to the positively charged sulfur, displacing a neutral sulfide as leaving group. We saw an example in **Section 11.6** (Figure 11.16 on page 396) in which *S*-adenosylmethionine transferred a methyl group to norepinephrine to give adrenaline.

Another difference between sulfides and ethers is that sulfides are easily oxidized. Treatment of a sulfide with hydrogen peroxide, H_2O_2, at room temperature yields the corresponding **sulfoxide (R_2SO)**, and further oxidation of the sulfoxide with a peroxyacid yields a **sulfone (R_2SO_2)**.

$$\text{Methyl phenyl sulfide} \xrightarrow[H_2O,\ 25\ °C]{H_2O_2} \text{Methyl phenyl sulfoxide} \xrightarrow{CH_3CO_3H} \text{Methyl phenyl sulfone}$$

Methyl phenyl sulfide **Methyl phenyl sulfoxide** **Methyl phenyl sulfone**

Dimethyl sulfoxide (DMSO) is a particularly well-known sulfoxide that is often used as a polar aprotic solvent. It must be handled with care, however, because it has a remarkable ability to penetrate the skin, carrying along whatever is dissolved in it.

Dimethyl sulfoxide (a polar aprotic solvent)

Problem 18.16

Name the following compounds:

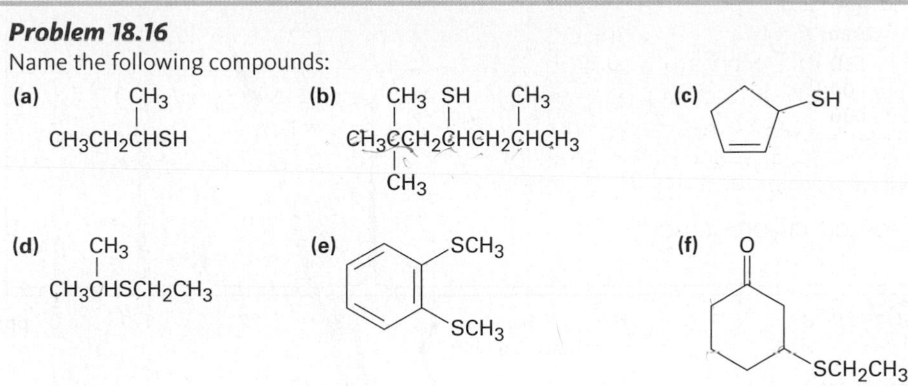

Problem 18.17

2-Butene-1-thiol is one component of skunk spray. How would you synthesize this substance from methyl 2-butenoate? From 1,3-butadiene?

$$CH_3CH{=}CHCOCH_3 \longrightarrow CH_3CH{=}CHCH_2SH$$

Methyl 2-butenoate **2-Butene-1-thiol**

18.9 Spectroscopy of Ethers

Infrared Spectroscopy

Ethers are difficult to identify by IR spectroscopy. Although they show an absorption due to C–O single-bond stretching in the range 1050 to 1150 cm^{-1}, many other kinds of absorptions occur in the same range. **Figure 18.3** shows the IR spectrum of diethyl ether and identifies the C–O stretch.

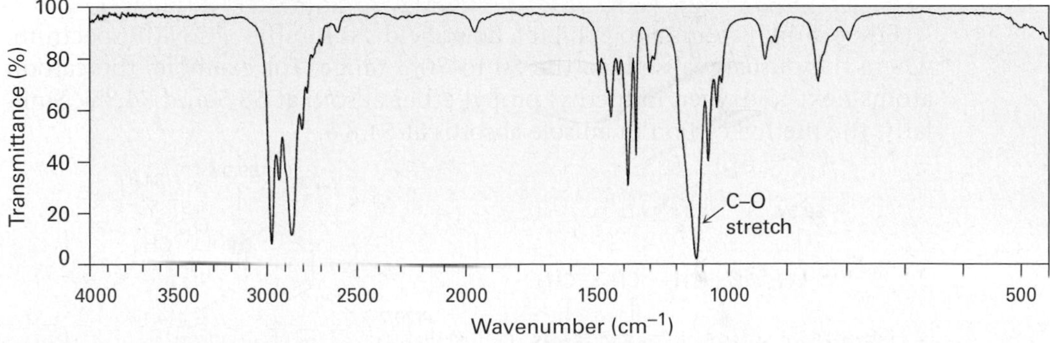

Figure 18.3 The infrared spectrum of diethyl ether, $CH_3CH_2OCH_2CH_3$.

Nuclear Magnetic Resonance Spectroscopy

Hydrogens on carbon next to an ether oxygen are shifted downfield from the normal alkane resonance and show ^1H NMR absorptions in the region 3.4 to 4.5 δ. This downfield shift is clearly seen in the spectrum of dipropyl ether shown in **Figure 18.4**.

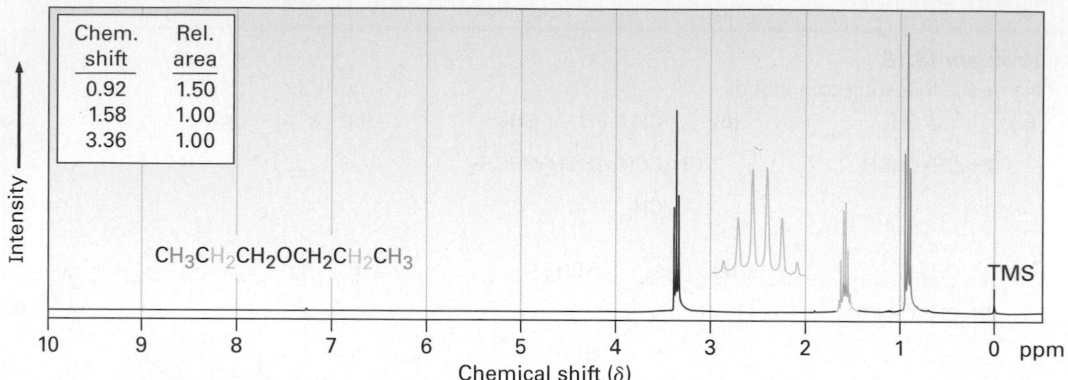

Figure 18.4 The ^1H NMR spectrum of dipropyl ether. Protons on carbon next to oxygen are shifted downfield to 3.4 δ.

Epoxides absorb at a slightly higher field than other ethers and show characteristic resonances at 2.5 to 3.5 δ in their ^1H NMR spectra, as indicated for 1,2-epoxypropane in **Figure 18.5**.

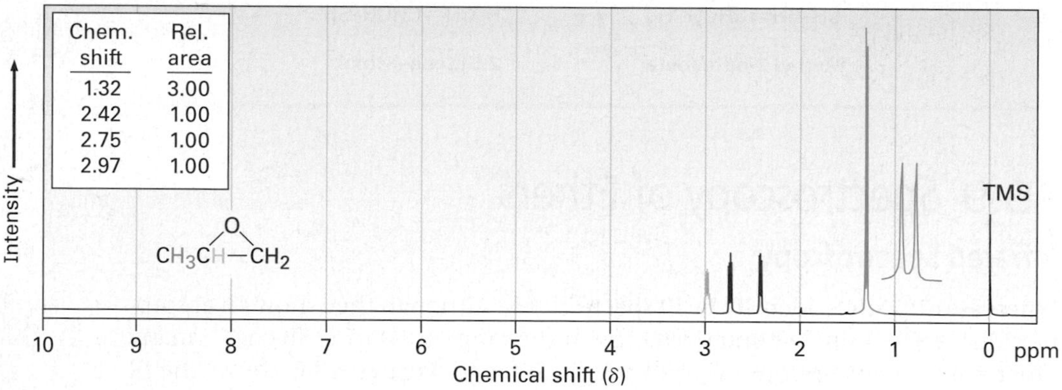

Figure 18.5 The ^1H NMR spectrum of 1,2-epoxypropane.

Ether carbon atoms also exhibit a downfield shift in the ^{13}C NMR spectrum, where they usually absorb in the 50 to 80 δ range. For example, the carbon atoms next to oxygen in methyl propyl ether absorb at 58.5 and 74.8 δ. Similarly, the methyl carbon in anisole absorbs at 54.8 δ.

58.5 δ 74.8 δ
$CH_3—O—CH_2—CH_2—CH_3$
23.3 δ 10.7 δ

159.9 δ 54.8 δ
O
CH_3
120.7 δ 114.1 δ
129.5 δ

Problem 18.18

The ^1H NMR spectrum shown is that of a cyclic ether with the formula C_4H_8O. Propose a structure.

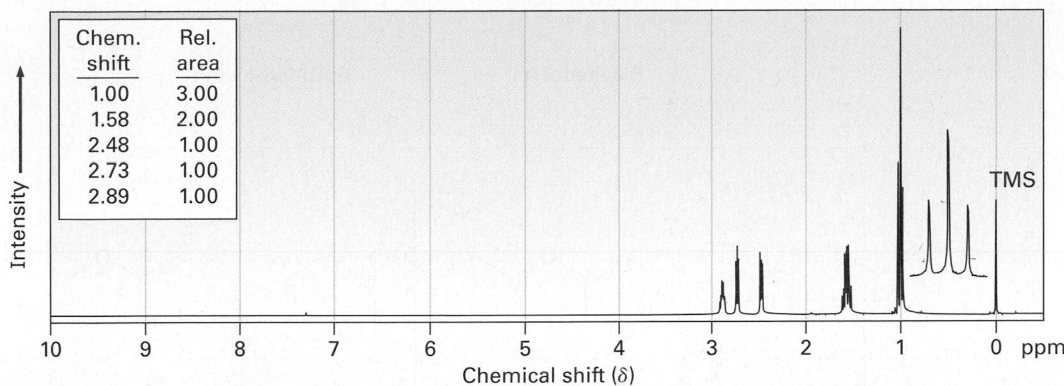

A DEEPER LOOK Epoxy Resins and Adhesives

Few nonchemists know exactly what an epoxide is, but practically everyone has used an "epoxy glue" for household repairs or an epoxy resin for a protective coating. Worldwide, approximately 15 billion dollars' worth of epoxies are used annually for a vast number of adhesive and coating applications, including many in the aerospace industry. Much of the new Boeing 787 Dreamliner, for instance, is held together with epoxy-based adhesives.

Epoxy resins and adhesives generally consist of two components that are mixed just prior to use. One component is a liquid "prepolymer," and the second is a "curing agent" that reacts with the prepolymer and causes it to solidify.

The most widely used epoxy resins and adhesives are based on a prepolymer made from bisphenol A and epichlorohydrin. On treatment with base, bisphenol A is converted into its anion, which acts as a nucleophile in an S_N2 reaction with epichlorohydrin. Each epichlorohydrin molecule can react with two molecules of bisphenol A, once by S_N2 displacement of chloride ion and once by nucleophilic opening of the epoxide ring. At the same time, each bisphenol A molecule can react with two epichlorohydrins, leading to a long polymer chain. Each end of a prepolymer chain has an unreacted epoxy group, and each chain has numerous secondary alcohol groups spaced regularly along its midsection.

Kayaks are often made of a high-strength polymer coated with epoxy resin.

(continued)

(continued)

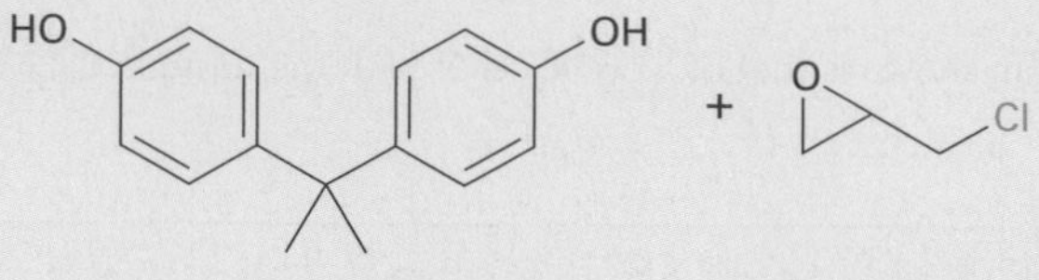

When the epoxide is to be used, a basic curing agent such as a tertiary amine, R_3N, is added to cause the individual prepolymer chains to link together. This cross-linking of chains is simply a base-catalyzed, S_N2 epoxide ring-opening of an –OH group in the middle of one chain with an epoxide group on the end of another chain. The result of such cross-linking is formation of a vast, three-dimensional tangle that has enormous strength and chemical resistance.

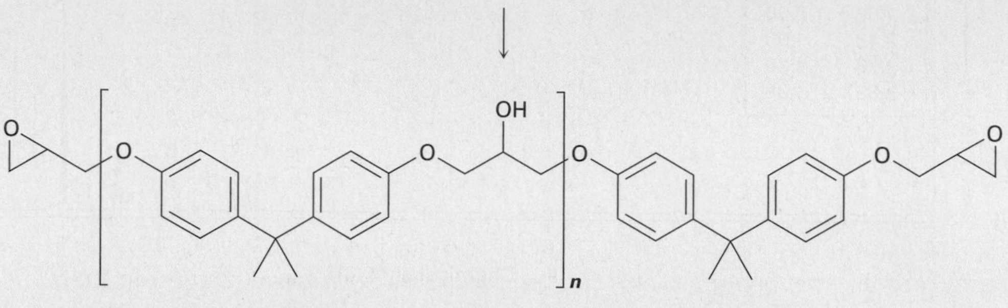

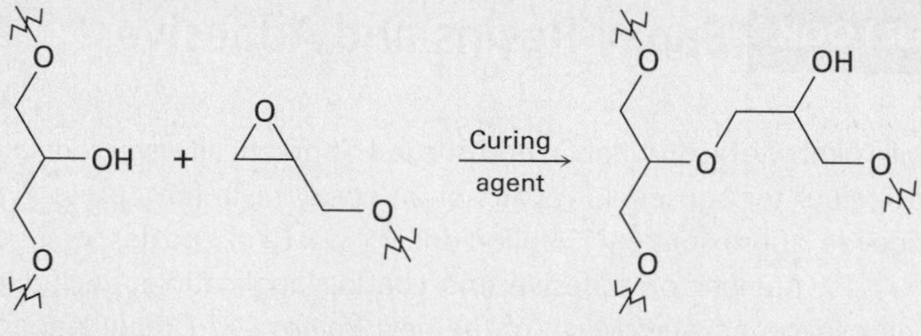

Key words

alkoxymercuration, 680
Claisen rearrangement, 683
crown ether, 690
disulfide (RSSR′), 692
ether (ROR′), 676
mercapto group (–SH), 691
sulfide (RSR′), 676
sulfone (R_2SO_2), 694
sulfonium ion (R_3S^+), 694
sulfoxide (R_2SO), 694
thiol (RSH), 676
thiolate ion (RS^-), 693

Summary

This chapter has finished the coverage of functional groups with C–O and C–S single bonds, focusing primarily on ethers, epoxides, thiols, and sulfides. **Ethers** are compounds that have two organic groups bonded to the same oxygen atom, ROR′. The organic groups can be alkyl, vinylic, or aryl, and the oxygen atom can be in a ring or in an open chain. Ethers are prepared by either the Williamson ether synthesis, which involves S_N2 reaction of an alkoxide ion with a primary alkyl halide, or the **alkoxymercuration** reaction, which involves Markovnikov addition of an alcohol to an alkene.

Ethers are inert to most reagents but react with strong acids to give cleavage products. Both HI and HBr are often used. The cleavage reaction takes place by an S_N2 mechanism at the less highly substituted site if only primary and secondary alkyl groups are bonded to the ether oxygen, but by an S_N1 or E1 mechanism if one of the alkyl groups bonded to oxygen is tertiary. Allyl aryl ethers and allyl vinyl ethers undergo **Claisen rearrangement** to give *o*-allylphenols and γ,δ-unsaturated ketones, respectively.

Epoxides are cyclic ethers with a three-membered, oxygen-containing ring. Because of the strain in the ring, epoxides undergo a cleavage reaction with both acids and bases. Acid-catalyzed ring-opening occurs with a regiochemistry that depends on the structure of the epoxide. Cleavage of the C–O bond at the less highly substituted site occurs if both epoxide carbons are primary or secondary, but cleavage of the C–O bond to the more highly substituted site occurs if one of the epoxide carbons is tertiary. Base-catalyzed epoxide ring-opening occurs by S_N2 reaction of a nucleophile at the less hindered epoxide carbon.

Thiols, the sulfur analogs of alcohols, are usually prepared by S_N2 reaction of an alkyl halide with thiourea. Mild oxidation of a thiol yields a **disulfide**, and mild reduction of a disulfide gives back the thiol. **Sulfides**, the sulfur analogs of ethers, are prepared by an S_N2 reaction between a thiolate anion and a primary or secondary alkyl halide. Sulfides are more nucleophilic than ethers and can be alkylated by reaction with a primary alkyl halide to yield a **sulfonium ion**. Sulfides can also be oxidized to **sulfoxides** and to **sulfones**.

Summary of Reactions

1. Synthesis of ethers (Section 18.2)
   (a) Williamson ether synthesis

$$RO^- + R'CH_2X \longrightarrow ROCH_2R' + X^-$$

   (b) Alkoxymercuration/demercuration

$$\text{C=C} \xrightarrow[\text{2. NaBH}_4]{\text{1. ROH, }(CF_3CO_2)_2Hg} \text{H–C–C–OR}$$

2. Reactions of ethers
   (a) Cleavage by HBr or HI (Section 18.3)

$$R{-}O{-}R' \xrightarrow[H_2O]{HX} RX + R'OH$$

   (b) Claisen rearrangement (Section 18.4)

$$C_6H_5OCH_2CH{=}CH_2 \xrightarrow{250\ °C} o\text{-}(CH_2CH{=}CH_2)C_6H_4OH$$

$$\text{R–C(=CH–R')–O–CH}_2\text{–CH=CH}_2 \xrightarrow{\text{Heat}} \text{R–C(=O)–CH(R')}_{\alpha}\text{–CH}_{2\,\beta}\text{–CH}_{\gamma}\text{=CH}_{2\,\delta}$$

(continued)

(c) Acid-catalyzed epoxide opening (Section 18.6)

$$\text{epoxide} \xrightarrow{H_3O^+} \text{trans-1,2-diol (HO–C–C–OH)}$$

$$\text{epoxide} \xrightarrow{HBr} \text{HO–C–C–Br}$$

(d) Base-catalyzed epoxide opening (Section 18.6)

$$\text{epoxide} \xrightarrow{RO^-,\ ROH} \text{HO–C–C–OR}$$

$$RMgX + H_2C\overset{O}{—}CH_2 \xrightarrow[2.\ H_3O^+]{1.\ \text{Ether solvent}} RCH_2CH_2OH$$

3. Synthesis of thiols (Section 18.8)

$$RCH_2Br \xrightarrow[2.\ H_2O,\ NaOH]{1.\ (H_2N)_2C{=}S} RCH_2SH$$

4. Oxidation of thiols to disulfides (Section 18.8)

$$2\ RSH \xrightarrow{I_2,\ H_2O} RS{-}SR$$

5. Synthesis of sulfides (Section 18.8)

$$RS^- + R'CH_2Br \longrightarrow RSCH_2R' + Br^-$$

6. Oxidation of sulfides to sulfoxides and sulfones (Section 18.8)

$$R{-}\ddot{S}{-}R' \xrightarrow{H_2O_2} R{-}\overset{O}{\overset{\|}{S}}{-}R'$$

$$R{-}\overset{O}{\overset{\|}{S}}{-}R' \xrightarrow{RCO_3H} R{-}S(=O)_2{-}R'$$

Exercises

Visualizing Chemistry

(Problems 18.1–18.18 appear within the chapter.)

OWL Interactive versions of these problems are assignable in OWL for Organic Chemistry.

18.19 Give IUPAC names for the following compounds (reddish brown = Br; yellow = S):

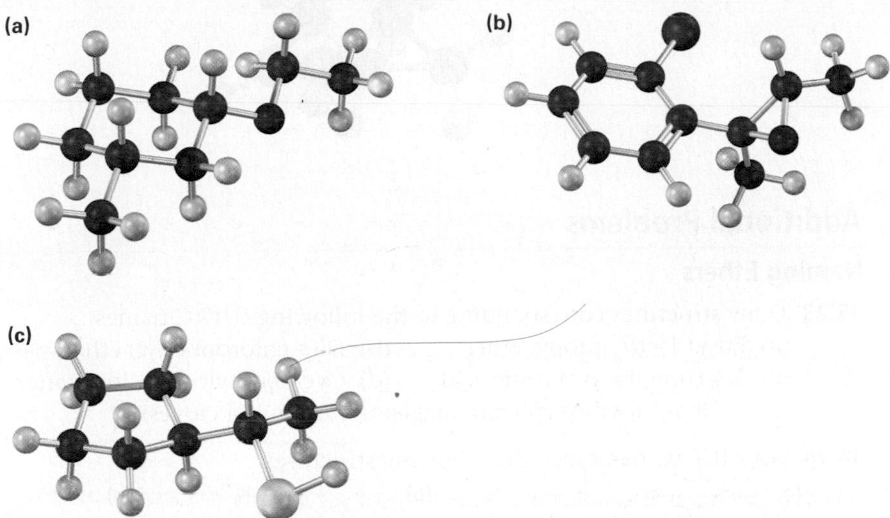

18.20 Show the product, including stereochemistry, that would result from reaction of the following epoxide with HBr:

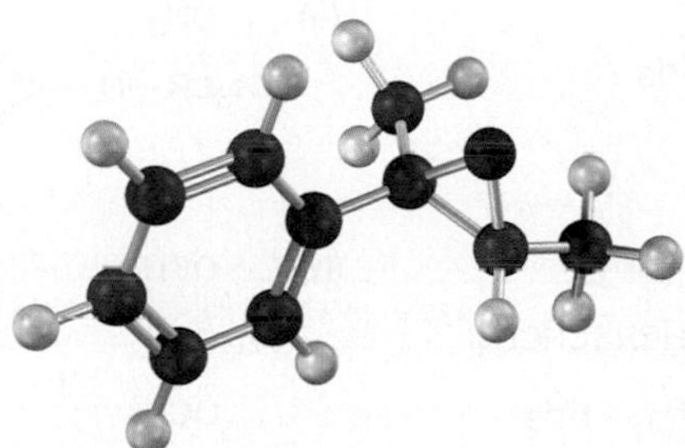

18.21 Show the product, including stereochemistry, of the following reaction:

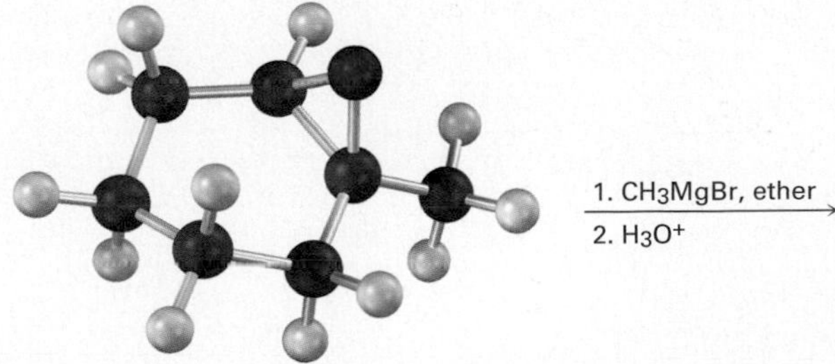

18.22 Treatment of the following alkene with a peroxyacid yields an epoxide different from that obtained by reaction with aqueous Br_2 followed by base treatment. Propose structures for the two epoxides, and explain the result.

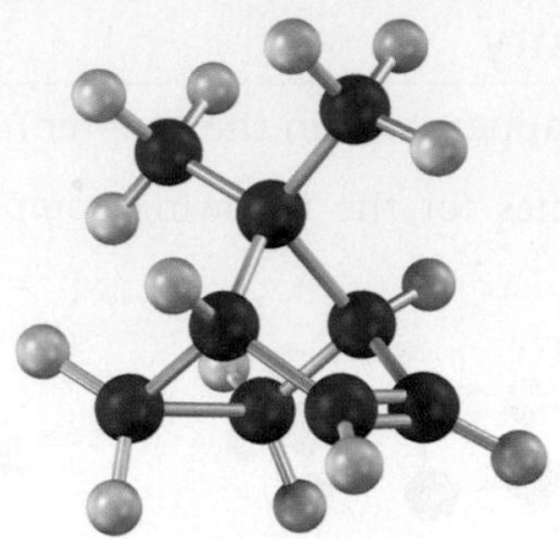

Additional Problems

Naming Ethers

18.23 Draw structures corresponding to the following IUPAC names:

(a) Ethyl 1-ethylpropyl ether
(b) Di(*p*-chlorophenyl) ether
(c) 3,4-Dimethoxybenzoic acid
(d) Cyclopentyloxycyclohexane
(e) 4-Allyl-2-methoxyphenol (eugenol; from oil of cloves)

18.24 Give IUPAC names for the following structures:

(a) S

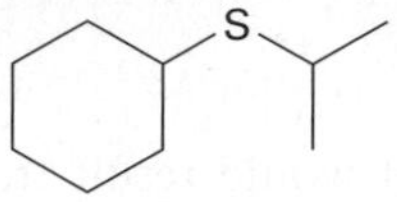

(b) OCH_3, OCH_3

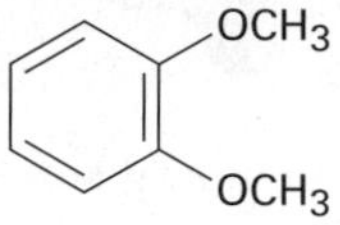

(c) O

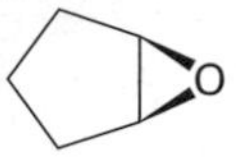

(d) 1 O, 2 CH_3, 3, 4, 5

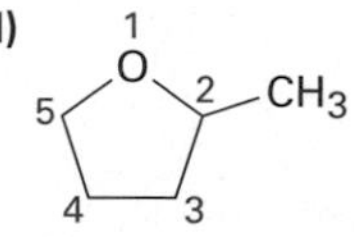

(e) CH_3; $CH_3CH—O—$

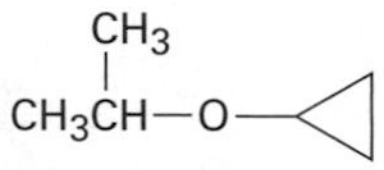

(f) SH, NO_2

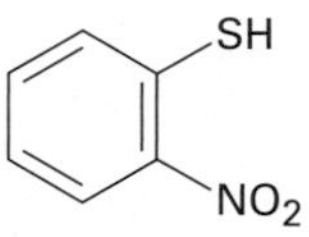

(g) CH_3 CH_3; $CH_3CH_2CHCHCHSCHCH_3$; CH_3 CH_3

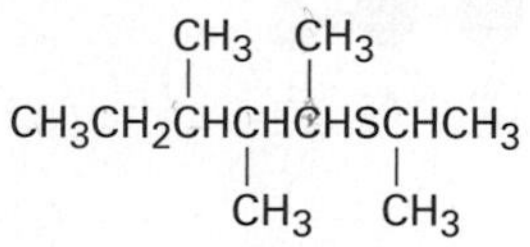

(h) OCH_3; CH_3CCH_3; OCH_3

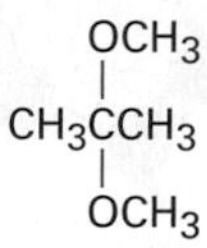

(i) SCH_3; SCH_3

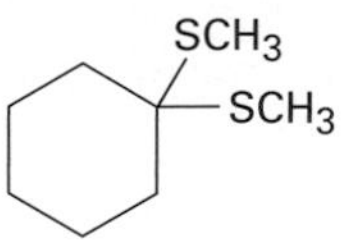

Synthesizing Ethers

18.25 How would you prepare the following ethers?

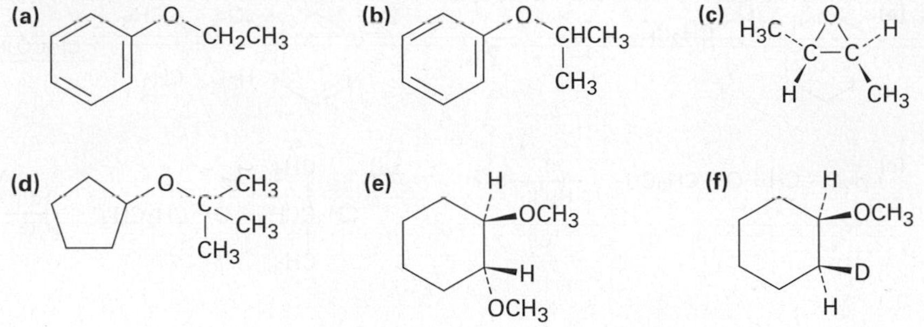

(a) (b) (c) (d) (e) (f)

18.26 How would you prepare the following compounds from 1-phenylethanol?

(a) Methyl 1-phenylethyl ether (b) Phenylepoxyethane
(c) *tert*-Butyl 1-phenylethyl ether (d) 1-Phenylethanethiol

18.27 *tert*-Butyl ethers can be prepared by the reaction of an alcohol with 2-methylpropene in the presence of an acid catalyst. Propose a mechanism for this reaction.

18.28 Treatment of *trans*-2-chlorocyclohexanol with NaOH yields 1,2-epoxycyclohexane, but reaction of the cis isomer under the same conditions yields cyclohexanone. Propose mechanisms for both reactions, and explain why the different results are obtained.

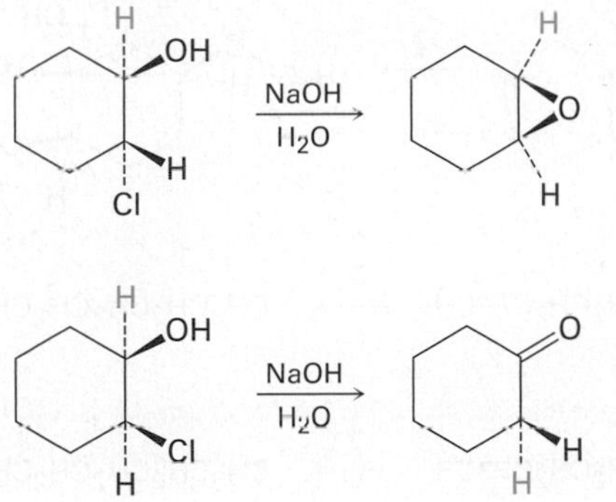

NaOH, H_2O

NaOH, H_2O

Reactions of Ethers and Epoxides

18.29 Predict the products of the following ether cleavage reactions:

(a) Cyclohexyl–O–CH_2CH_3 $\xrightarrow[H_2O]{HI}$?

(b) Phenyl–O–$C(CH_3)_3$ $\xrightarrow{CF_3CO_2H}$?

(c) $H_2C{=}CH{-}O{-}CH_2CH_3$ $\xrightarrow[H_2O]{HI}$?

(d) $CH_3C(CH_3)_2CH_2{-}O{-}CH_2CH_3$ $\xrightarrow[H_2O]{HI}$?

18.30 How would you carry out the following transformations? More than one step may be required.

(a) Cyclohexene $\xrightarrow{?}$ Cyclohexyl–OCH_2CH_3

(b) H_3C, H / H, OCH_3 substituted cyclohexane $\xrightarrow{?}$ H_3C, H / Br, H substituted cyclohexane

(c) $(H_3C)_3C$–cyclohexene $\xrightarrow{?}$ $(H_3C)_3C$, H / OH, H / OH, H substituted cyclohexane

(d) $CH_3CH_2CH_2CH_2C{\equiv}CH$ $\xrightarrow{?}$ $CH_3CH_2CH_2CH_2CH_2CH_2OCH_3$

(e) $CH_3CH_2CH_2CH_2C{\equiv}CH$ $\xrightarrow{?}$ $CH_3CH_2CH_2CH_2CH(OCH_3)CH_3$

18.31 What product would you expect from cleavage of tetrahydrofuran with HI?

18.32 Write the mechanism of the hydrolysis of *cis*-5,6-epoxydecane by reaction with aqueous acid. What is the stereochemistry of the product, assuming normal backside S_N2 attack?

18.33 What is the stereochemistry of the product from acid-catalyzed hydrolysis of *trans*-5,6-epoxydecane? How does the product differ from that formed in Problem 18.32?

18.34 Acid-catalyzed hydrolysis of a 1,2-epoxycyclohexane produces a trans-diaxial 1,2-diol. What product would you expect to obtain from acidic hydrolysis of *cis*-3-*tert*-butyl-1,2-epoxycyclohexane? (Recall that the bulky *tert*-butyl group locks the cyclohexane ring into a specific conformation.)

18.35 Imagine that you have treated (2*R*,3*R*)-2,3-epoxy-3-methylpentane with aqueous acid to carry out a ring-opening reaction.

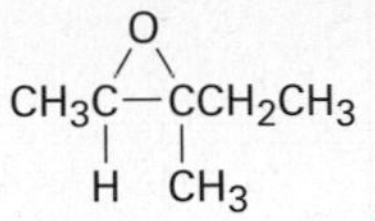

$$CH_3\overset{O}{C}H-\underset{CH_3}{C}CH_2CH_3$$

2,3-Epoxy-3-methylpentane (no stereochemistry implied)

(a) Draw the epoxide, showing stereochemistry.
(b) Draw and name the product, showing stereochemistry.
(c) Is the product chiral? Explain.
(d) Is the product optically active? Explain.

18.36 Epoxides are reduced by treatment with lithium aluminum hydride to yield alcohols. Propose a mechanism for this reaction.

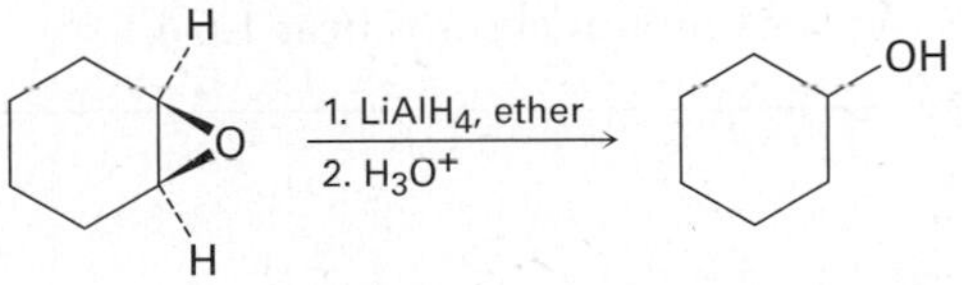

1,2-Epoxycyclohexane (H, O, H) → 1. $LiAlH_4$, ether; 2. H_3O^+ → cyclohexanol (OH)

18.37 Show the structure and stereochemistry of the alcohol that would result if 1,2-epoxycyclohexane were reduced with lithium aluminum deuteride, $LiAlD_4$ (Problem 18.36).

Spectroscopy

18.38 The red fox (*Vulpes vulpes*) uses a chemical communication system based on scent marks in urine. One component of fox urine is a sulfide whose mass spectrum has $M^+ = 116$. IR spectroscopy shows an intense band at 890 cm^{-1}, and ^1H NMR spectroscopy reveals the following peaks:

1.74 δ (3 H, singlet); 2.11 δ (3 H, singlet); 2.27 δ (2 H, triplet, J = 4.2 Hz); 2.57 δ (2 H, triplet, J = 4.2 Hz); 4.73 δ (2 H, broad)

Propose a structure consistent with these data. [*Note:* $(CH_3)_2S$ absorbs at 2.1 δ.]

18.39 Anethole, $C_{10}H_{12}O$, a major constituent of the oil of anise, has the ^1H NMR spectrum shown. On oxidation with $Na_2Cr_2O_7$, anethole yields *p*-methoxybenzoic acid. What is the structure of anethole? Assign all peaks in the NMR spectrum, and account for the observed splitting patterns.

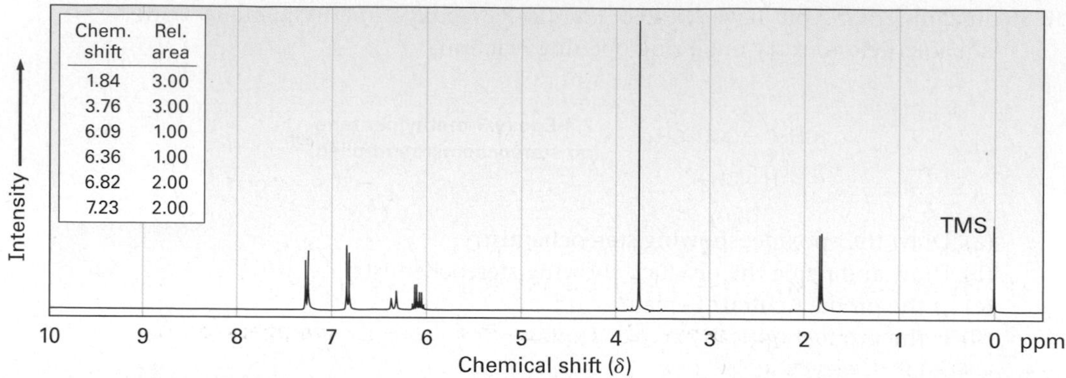

Chem. shift	Rel. area
1.84	3.00
3.76	3.00
6.09	1.00
6.36	1.00
6.82	2.00
7.23	2.00

18.40 Propose structures for compounds that have the following ^1H NMR spectra:

(a) $C_5H_{12}S$ (An –SH proton absorbs near 1.6 δ.)

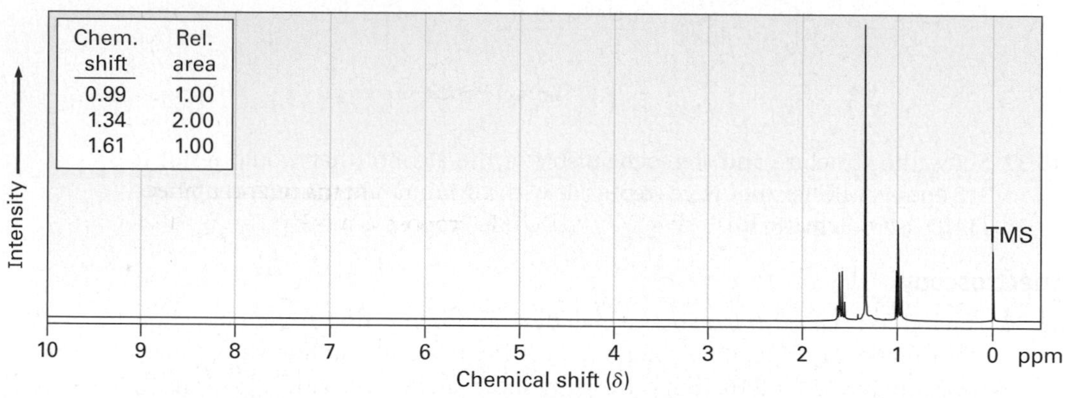

Chem. shift	Rel. area
0.99	1.00
1.34	2.00
1.61	1.00

(b) $C_9H_{11}BrO$

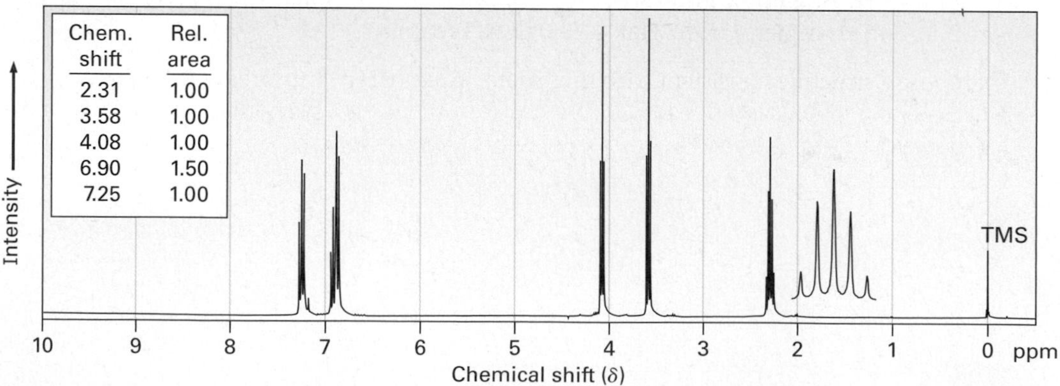

Chem. shift	Rel. area
2.31	1.00
3.58	1.00
4.08	1.00
6.90	1.50
7.25	1.00

General Problems

18.41 Predict the products of the following reactions:

(a) $C_6H_5OCH_2CH(CH_3)CH_3 \xrightarrow{HBr}$?

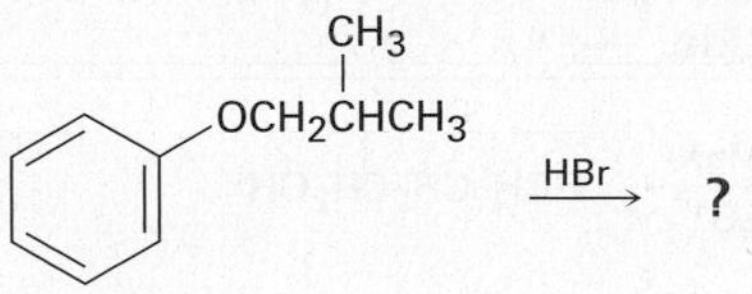

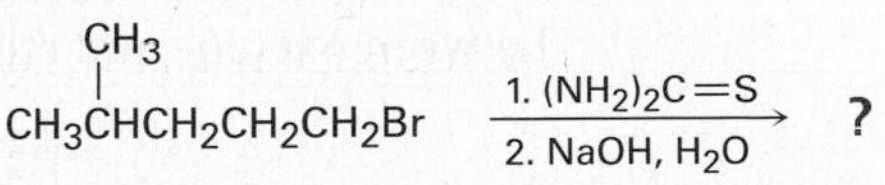

(b) $CH_3CH(CH_3)CH_2CH_2CH_2Br \xrightarrow[2.\ NaOH,\ H_2O]{1.\ (NH_2)_2C{=}S}$?

(c) Cyclopentyl–SH $\xrightarrow{Br_2}$?

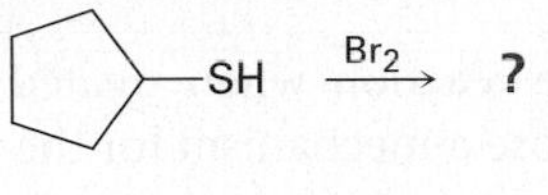

(d) 3-(Ethylthio)cyclohexene (SCH_2CH_3) $\xrightarrow{H_2O_2,\ H_2O}$?

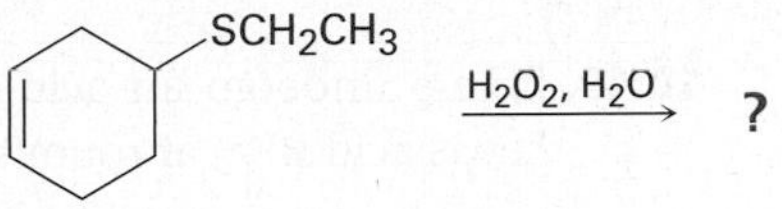

18.42 How would you synthesize anethole (Problem 18.39) from phenol?

18.43 How could you prepare benzyl phenyl ether from benzene and phenol? More than one step is required.

18.44 When 2-methyl-2,5-pentanediol is treated with sulfuric acid, dehydration occurs and 2,2-dimethyltetrahydrofuran is formed. Suggest a mechanism for this reaction. Which of the two oxygen atoms is most likely to be eliminated, and why?

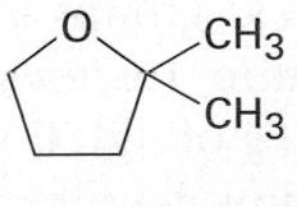

2,2-Dimethyltetrahydrofuran

18.45 Methyl aryl ethers, such as anisole, are cleaved to iodomethane and a phenoxide ion by treatment with LiI in hot DMF. Propose a mechanism for this reaction.

18.46 *Meerwein's reagent,* triethyloxonium tetrafluoroborate, is a powerful ethylating agent that converts alcohols into ethyl ethers at neutral pH. Show the reaction of Meerwein's reagent with cyclohexanol, and account for the fact that trialkyloxonium salts are much more reactive alkylating agents than alkyl iodides.

$(CH_3CH_2)_3O^+\ BF_4^-$ **Meerwein's reagent**

18.47 Safrole, a substance isolated from oil of sassafras, is used as a perfumery agent. Propose a synthesis of safrole from catechol (1,2-benzenediol).

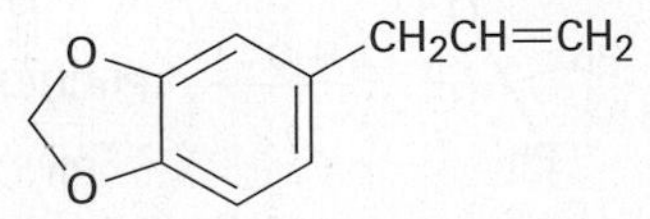

$CH_2CH{=}CH_2$ **Safrole**

18.48 Grignard reagents react with oxetane, a four-membered cyclic ether, to yield primary alcohols, but the reaction is much slower than the corresponding reaction with ethylene oxide. Suggest a reason for the difference in reactivity between oxetane and ethylene oxide.

Oxetane $\xrightarrow[\text{2. } H_3O^+]{\text{1. RMgX}}$ $RCH_2CH_2CH_2OH$

Oxetane

18.49 Ethers undergo an acid-catalyzed cleavage reaction when treated with the Lewis acid BBr_3 at room temperature. Propose a mechanism for the reaction.

$C_6H_5OCH_3 \xrightarrow[\text{2. } H_2O]{\text{1. } BBr_3} C_6H_5OH + CH_3Br$

18.50 The *Zeisel method* is an old analytical procedure for determining the number of methoxyl groups in a compound. A weighed amount of the compound is heated with concentrated HI, ether cleavage occurs, and the iodomethane product is distilled off and passed into an alcohol solution of $AgNO_3$, where it reacts to form a precipitate of silver iodide. The AgI is then collected and weighed, and the percentage of methoxyl groups in the sample is thereby determined. For example, 1.06 g of vanillin, the material responsible for the characteristic odor of vanilla, yields 1.60 g of AgI. If vanillin has a molecular weight of 152, how many methoxyl groups does it contain?

18.51 Disparlure, $C_{19}H_{38}O$, is a sex attractant released by the female gypsy moth, *Lymantria dispar.* The ^1H NMR spectrum of disparlure shows a large absorption in the alkane region, 1 to 2 δ, and a triplet at 2.8 δ. Treatment of disparlure, first with aqueous acid and then with $KMnO_4$, yields two carboxylic acids identified as undecanoic acid and 6-methylheptanoic acid. ($KMnO_4$ cleaves 1,2-diols to yield carboxylic acids.) Neglecting stereochemistry, propose a structure for disparlure. The actual compound is a chiral molecule with 7*R*,8*S* stereochemistry. Draw disparlure, showing the correct stereochemistry.

18.52 How would you synthesize racemic disparlure (Problem 18.51) from compounds having ten or fewer carbons?

18.53 Treatment of 1,1-diphenyl-1,2-epoxyethane with aqueous acid yields diphenylacetaldehyde as the major product. Propose a mechanism for the reaction.

$Ph_2C(O)CH_2$ (epoxide) $\xrightarrow{H_3O^+}$ Ph_2CHCHO

18.54 How would you prepare *o*-hydroxyphenylacetaldehyde from phenol? More than one step is required.

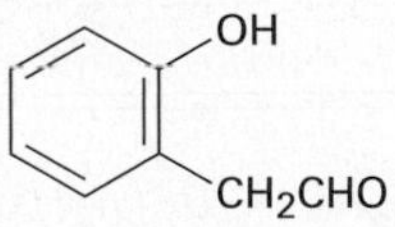

OH

CH_2CHO

***o*-Hydroxyphenylacetaldehyde**

18.55 Identify the reagents **a–e** in the following scheme:

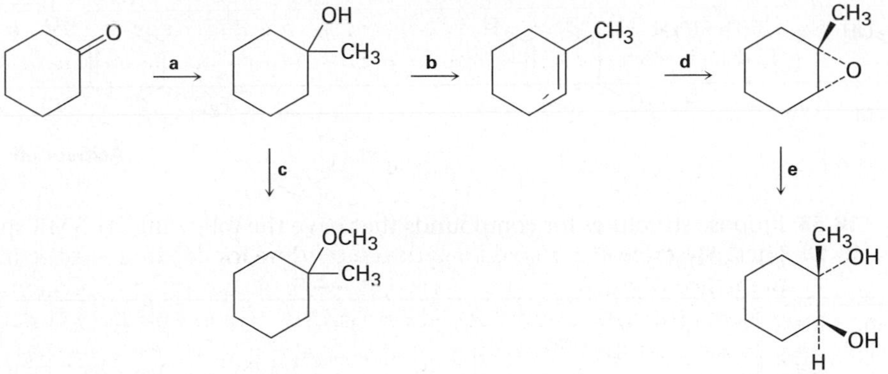

OH, CH_3, a, b, CH_3, d, CH_3, O, c, e, OCH_3, CH_3, CH_3, OH, OH, H

18.56 Fluoxetine, a heavily prescribed antidepressant marketed under the name Prozac, can be prepared by a route that begins with reaction between a phenol and an alkyl chloride.

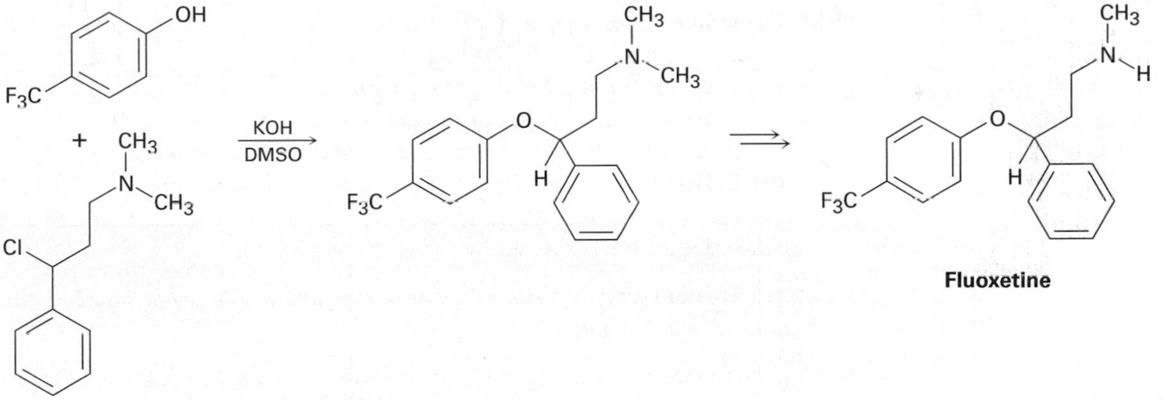

OH, F_3C, +, CH_3, N, CH_3, Cl, KOH, DMSO, CH_3, N, CH_3, O, H, F_3C, CH_3, N, H, O, H, F_3C

Fluoxetine

(a) The rate of the reaction depends on both phenol and alkyl halide. Is this an S_N1 or an S_N2 reaction? Show the mechanism.

(b) The physiologically active enantiomer of fluoxetine has (*S*) stereochemistry. Based on your answer in part **(a)**, draw the structure of the alkyl chloride you would need, showing the correct stereochemistry.

18.57 The herbicide acifluorfen can be prepared by a route that begins with reaction between a phenol and an aryl fluoride. Propose a mechanism.

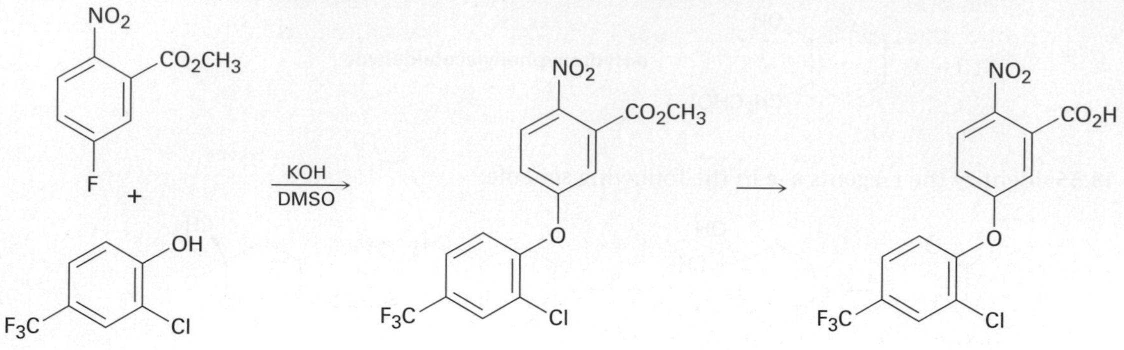

$$\text{(4-F-2-}CO_2CH_3\text{-nitrobenzene)} + \text{(2-Cl-4-}CF_3\text{-phenol)} \xrightarrow[\text{DMSO}]{\text{KOH}} \text{(diaryl ether, }CO_2CH_3\text{)} \longrightarrow \text{(diaryl ether, }CO_2H\text{)}$$

Acifluorfen

18.58 Propose structures for compounds that have the following ^1H NMR spectra:

(a) $C_4H_{10}O_2$

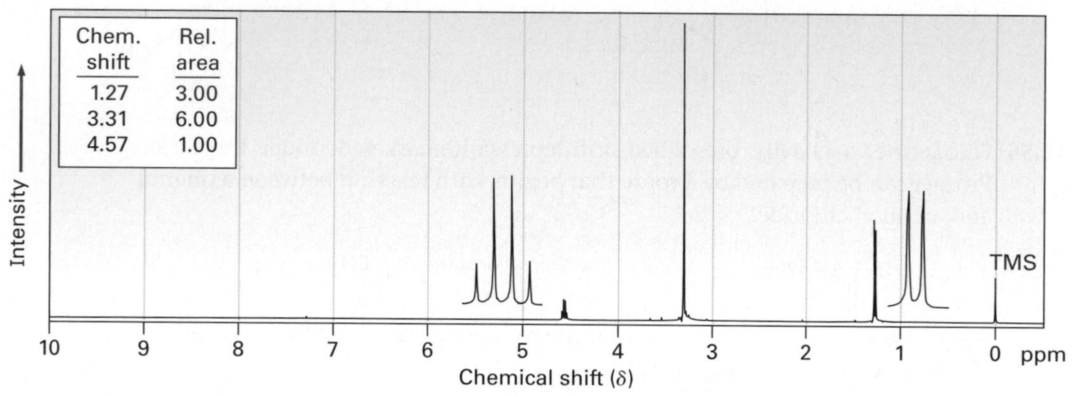

(b) $C_9H_{10}O$

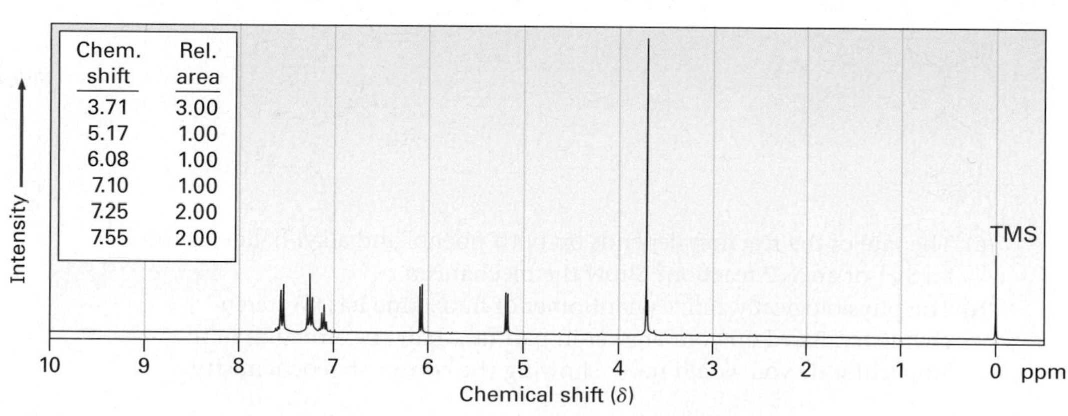

18.59 Aldehydes and ketones undergo acid-catalyzed reaction with alcohols to yield *hemiacetals,* compounds that have one alcohol-like oxygen and one ether-like oxygen bonded to the same carbon. Further reaction of a hemiacetal with alcohol then yields an *acetal,* a compound that has two ether-like oxygens bonded to the same carbon.

$$\mathrm{C{=}O} + \mathrm{ROH} \xrightarrow[\text{catalyst}]{\mathrm{H^+}} \mathrm{C(OR)(OH)} \xrightarrow[\mathrm{H^+}]{\mathrm{ROH}} \mathrm{C(OR)(OR)} + \mathrm{H_2O}$$

A hemiacetal **An acetal**

(a) Show the structures of the hemiacetal and acetal you would obtain by reaction of cyclohexanone with ethanol.

(b) Propose a mechanism for the conversion of a hemiacetal into an acetal.

18.60 We saw in Section 17.4 that ketones react with $NaBH_4$ to yield alcohols. We'll also see in Section 22.3 that ketones react with Br_2 to yield α-bromo ketones. Perhaps surprisingly, treatment with $NaBH_4$ of the α-bromo ketone from acetophenone yields an epoxide rather than a bromo alcohol. Show the structure of the epoxide, and explain its formation.

$$\mathrm{C_6H_5COCH_3} \xrightarrow{\mathrm{Br_2}} \mathrm{C_6H_5COCH_2Br} \xrightarrow{\mathrm{NaBH_4}} \text{Epoxide}$$

Acetophenone **An α-bromo ketone**

18.61 Propose a mechanism to account for the following transformation. What two kinds of reactions are occurring?

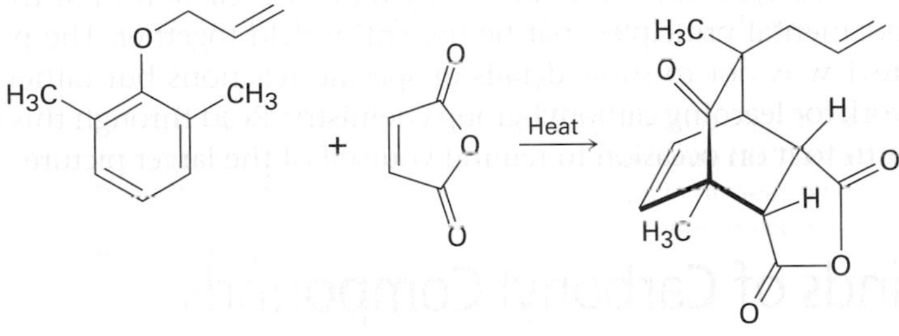

II. Nature of the Carbonyl Group

The carbon–oxygen double bond of a carbonyl group is similar in many respects to the carbon–carbon double bond of an alkene. The carbonyl carbon atom is sp^2-hybridized and forms three σ bonds. The fourth valence electron remains in a carbon p orbital and forms a π bond to oxygen by overlap with an oxygen p orbital. The oxygen atom also has two nonbonding pairs of electrons, which occupy its remaining two orbitals.

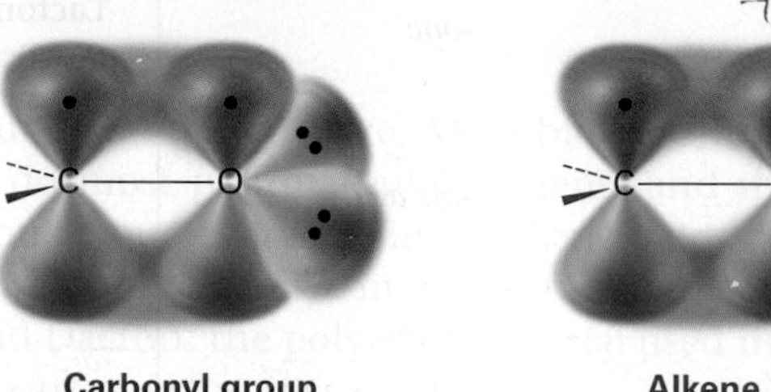

Carbonyl group | Alkene

Like alkenes, carbonyl compounds are planar about the double bond and have bond angles of approximately 120°. **Figure 1** shows the structure of acetaldehyde and indicates its bond lengths and angles. As you might expect, the carbon–oxygen double bond is both shorter (122 pm versus 143 pm) and stronger [732 kJ/mol (175 kcal/mol) versus 385 kJ/mol (92 kcal/mol)] than a C–O single bond.

Figure 1 Structure of acetaldehyde.

Bond angle	(°)	Bond length	(pm)
H—C—C	118	C=O	122
C—C=O	121	C—C	150
H—C=O	121	OC—H	109

Electron-rich

Electron-poor

As indicated by the electrostatic potential map in Figure 1, the carbon–oxygen double bond is strongly polarized because of the high electronegativity of oxygen relative to carbon. Thus, the carbonyl carbon atom carries a partial positive charge, is an electrophilic (Lewis acidic) site, and reacts with nucleophiles. Conversely, the carbonyl oxygen atom carries a partial negative charge, is a nucleophilic (Lewis basic) site, and reacts with electrophiles. We'll see in the next five chapters that the majority of carbonyl-group reactions can be rationalized by simple polarity arguments.

III. General Reactions of Carbonyl Compounds

Both in the laboratory and in living organisms, most reactions of carbonyl compounds take place by one of four general mechanisms: *nucleophilic addition, nucleophilic acyl substitution, alpha substitution,* and *carbonyl condensation.*

These mechanisms have many variations, just as alkene electrophilic addition reactions and S_N2 reactions do, but the variations are much easier to learn when the fundamental features of the mechanisms are made clear. Let's see what the four mechanisms are and what kinds of chemistry carbonyl compounds undergo.

Nucleophilic Addition Reactions of Aldehydes and Ketones (Chapter 19)

The most common reaction of aldehydes and ketones is the **nucleophilic addition reaction**, in which a nucleophile, $:Nu^-$, adds to the electrophilic carbon of the carbonyl group. Since the nucleophile uses an electron pair to form a new bond to carbon, two electrons from the carbon–oxygen double bond must move toward the electronegative oxygen atom to give an alkoxide anion. The carbonyl carbon rehybridizes from sp^2 to sp^3 during the reaction, and the alkoxide ion product therefore has tetrahedral geometry.

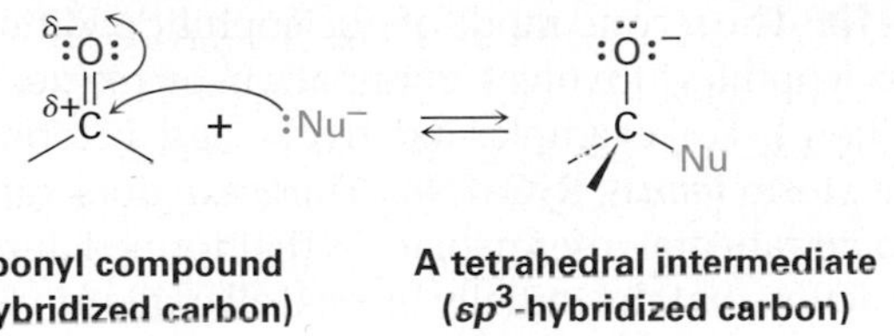

A carbonyl compound (sp^2-hybridized carbon)

A tetrahedral intermediate (sp^3-hybridized carbon)

Once formed, and depending on the nature of the nucleophile, the tetrahedral alkoxide intermediate can undergo either of two further reactions, as shown in **Figure 2**. Often, the tetrahedral alkoxide intermediate is simply protonated by water or acid to form an alcohol product. Alternatively, the tetrahedral intermediate can be protonated and expel the oxygen to form a new double bond between the carbonyl carbon and the nucleophile. We'll study both processes in detail in Chapter 19.

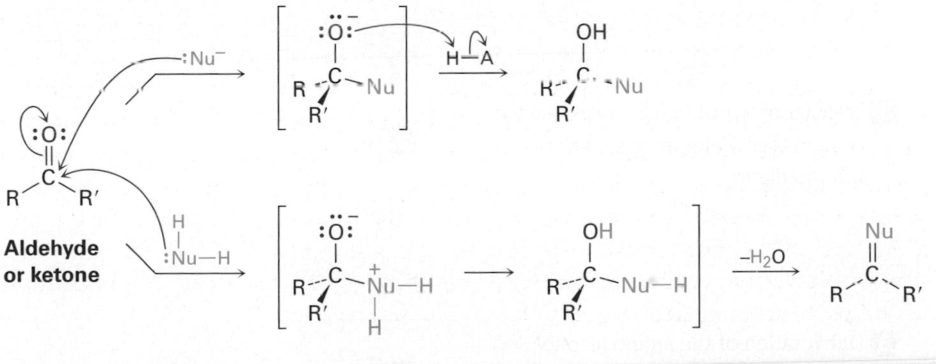

Aldehyde or ketone

Figure 2 The addition reaction of an aldehyde or a ketone with a nucleophile. Depending on the nucleophile, either an alcohol or a compound with a C=Nu double bond is formed.

Formation of an Alcohol The simplest reaction of a tetrahedral alkoxide intermediate is protonation to yield an alcohol. We've already seen two examples of this kind of process during reduction of aldehydes and ketones with hydride reagents such as $NaBH_4$ and $LiAlH_4$ **(Section 17.4)** and during Grignard reactions **(Section 17.5)**. During a reduction, the nucleophile that adds to the carbonyl group is a hydride ion, H:$^-$, while during a Grignard reaction, the nucleophile is a carbanion, $R_3C:^-$.

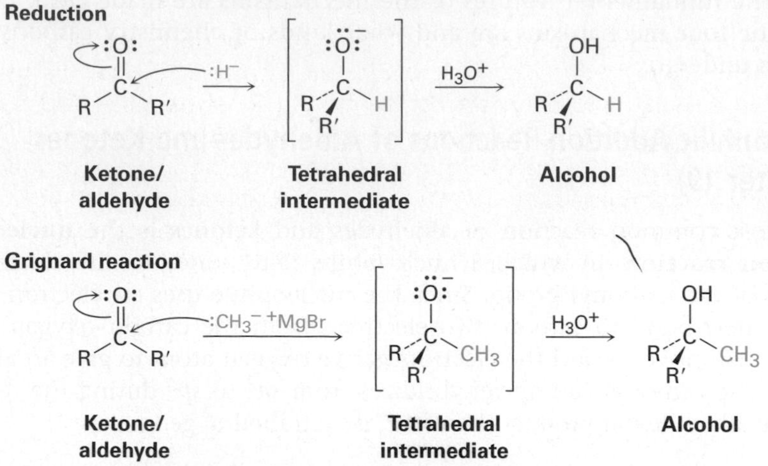

Formation of C=Nu The second mode of nucleophilic addition, which often occurs with amine nucleophiles, involves elimination of oxygen and formation of a C=Nu double bond. For example, aldehydes and ketones react with primary amines, RNH_2, to form *imines*, $R_2C{=}NR'$. These reactions proceed through exactly the same kind of tetrahedral intermediate as that formed during hydride reduction and Grignard reaction, but the initially formed alkoxide ion is not isolated. Instead, it is protonated and then loses water to form an imine, as shown in **Figure 3**.

Figure 3 | MECHANISM

Formation of an imine, $R_2C{=}NR'$, by reaction of an amine with an aldehyde or a ketone.

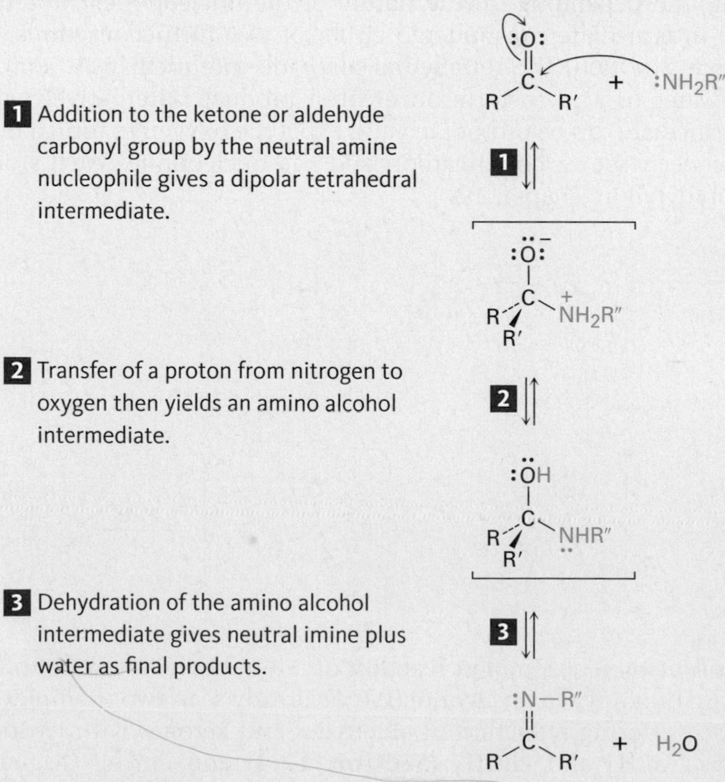

Nucleophilic Acyl Substitution Reactions of Carboxylic Acid Derivatives (Chapter 21)

The second fundamental reaction of carbonyl compounds, **nucleophilic acyl substitution**, is related to the nucleophilic addition reaction just discussed but occurs only with carboxylic acid derivatives rather than with aldehydes and ketones. When the carbonyl group of a carboxylic acid derivative reacts with a nucleophile, addition occurs in the usual way, but the initially formed tetrahedral alkoxide intermediate is not isolated. Because carboxylic acid derivatives have a leaving group bonded to the carbonyl-group carbon, the tetrahedral intermediate can react further by expelling the leaving group and forming a new carbonyl compound:

Carboxylic acid derivative → **Tetrahedral intermediate** → $RCONu$ + $:Y^-$

[Y = –OR (ester), –Cl (acid chloride), $-NH_2$ (amide), or –OCOR (acid anhydride)]

The net effect of nucleophilic acyl substitution is the replacement of the leaving group by the entering nucleophile. We'll see in Chapter 21, for instance, that acid chlorides are rapidly converted into esters by treatment with alkoxide ions **(Figure 4)**.

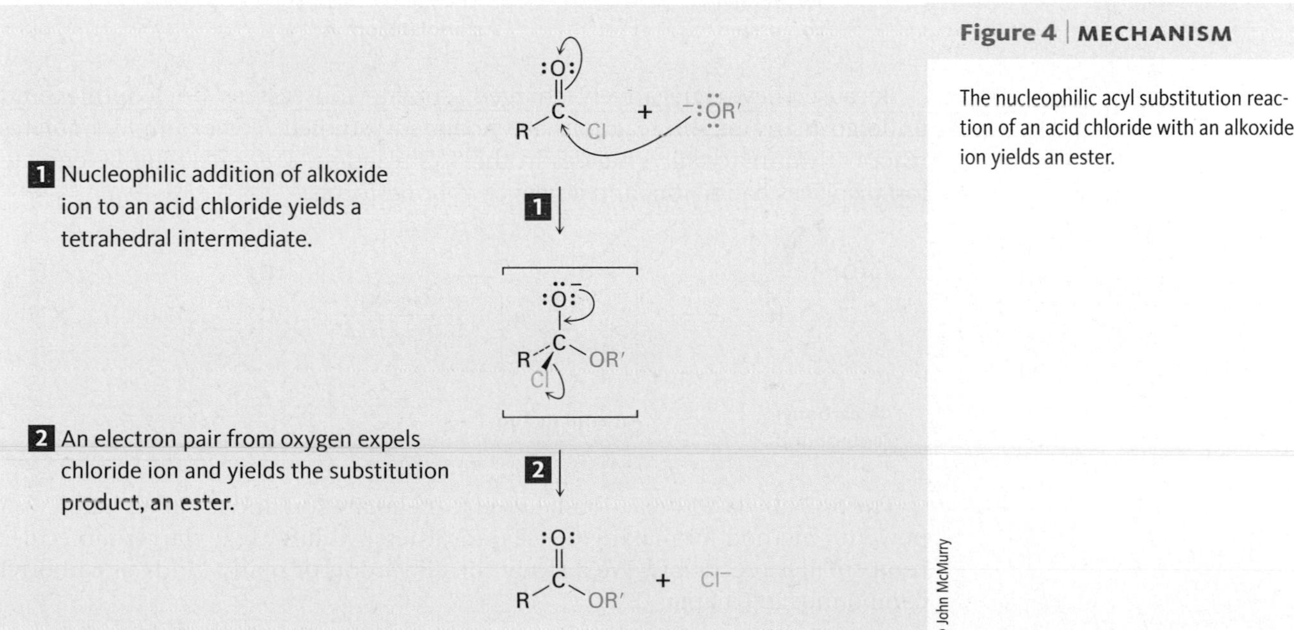

Figure 4 | MECHANISM

The nucleophilic acyl substitution reaction of an acid chloride with an alkoxide ion yields an ester.

Alpha-Substitution Reactions (Chapter 22)

The third major reaction of carbonyl compounds, **alpha substitution**, occurs at the position next to the carbonyl group—the alpha (α) position. This reaction, which takes place with all carbonyl compounds regardless of structure, results in the substitution of an α hydrogen by an electrophile through the formation of an intermediate *enol* or *enolate ion:*

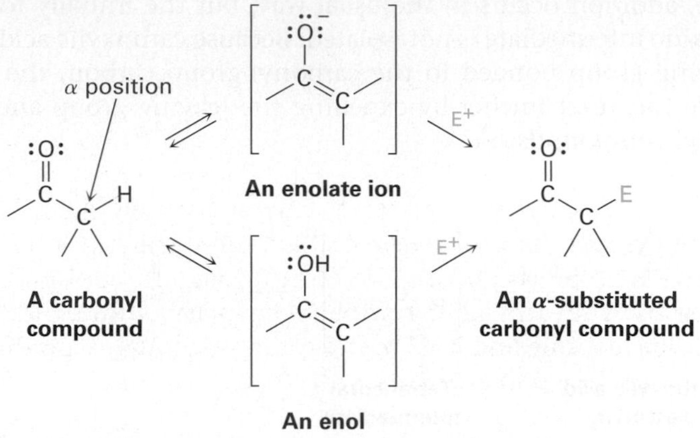

α position

A carbonyl compound

An enolate ion

An enol

E^+

An α-substituted carbonyl compound

For reasons that we'll explore in Chapter 22, the presence of a carbonyl group renders the hydrogens on the α carbon acidic. Carbonyl compounds therefore react with strong base to yield enolate ions.

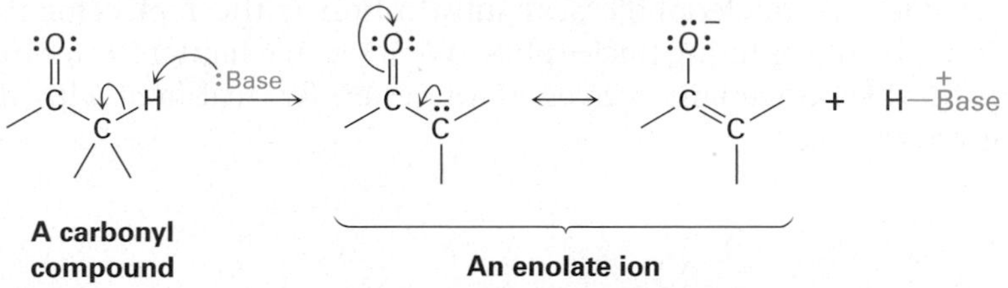

:Base

+ H—$\overset{+}{\text{Base}}$

A carbonyl compound

An enolate ion

Because they're negatively charged, enolate ions act as nucleophiles and undergo many of the reactions we've already studied. For example, enolates react with primary alkyl halides in the S_N2 reaction. The nucleophilic enolate ion displaces halide ion, and a new C–C bond forms:

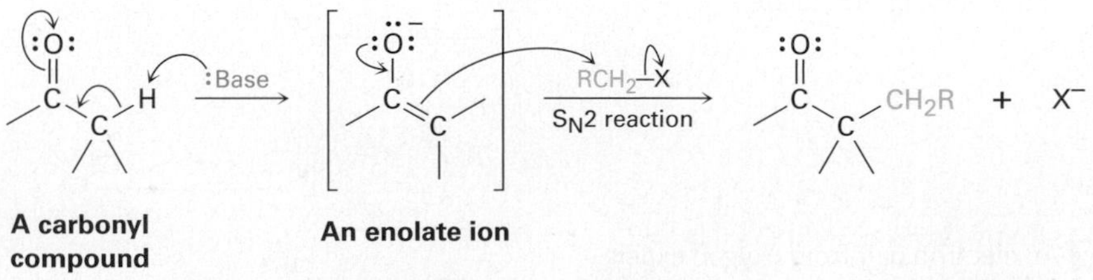

:Base

RCH_2–X

S_N2 reaction

CH_2R + X^-

A carbonyl compound

An enolate ion

The S_N2 alkylation reaction between an enolate ion and an alkyl halide is a powerful method for making C–C bonds, thereby building up larger molecules from smaller precursors. We'll study the alkylation of many kinds of carbonyl compounds in Chapter 22.

Carbonyl Condensation Reactions (Chapter 23)

The fourth and last fundamental reaction of carbonyl groups, **carbonyl condensation**, takes place when two carbonyl compounds react with each other. When acetaldehyde is treated with base, for instance, two molecules combine to yield the hydroxy aldehyde product known as *aldol* (*ald*ehyde + alcohol):

$$\mathrm{H_3C{-}CHO} + \mathrm{H_3C{-}CHO} \xrightleftharpoons{\text{NaOH}} \mathrm{H_3C{-}CH(OH){-}CH_2{-}CHO}$$

Two acetaldehydes **Aldol**

Although the carbonyl condensation reaction appears different from the three processes already discussed, it's actually quite similar. A carbonyl condensation reaction is simply a combination of a nucleophilic addition step and an α-substitution step. The initially formed enolate ion of one acetaldehyde molecule acts as a nucleophile and adds to the carbonyl group of another acetaldehyde molecule, as shown in **Figure 5**.

Figure 5 | MECHANISM

A carbonyl condensation reaction between two molecules of acetaldehyde yields a hydroxy aldehyde product.

1 Base abstracts an acidic alpha hydrogen from one acetaldehyde molecule, yielding a resonance-stabilized enolate ion.

2 The enolate ion adds as a nucleophile to the carbonyl group of a second acetaldehyde, producing a tetrahedral alkoxide ion.

3 The tetrahedral intermediate is protonated by solvent to yield the neutral aldol product and regenerate the base catalyst.

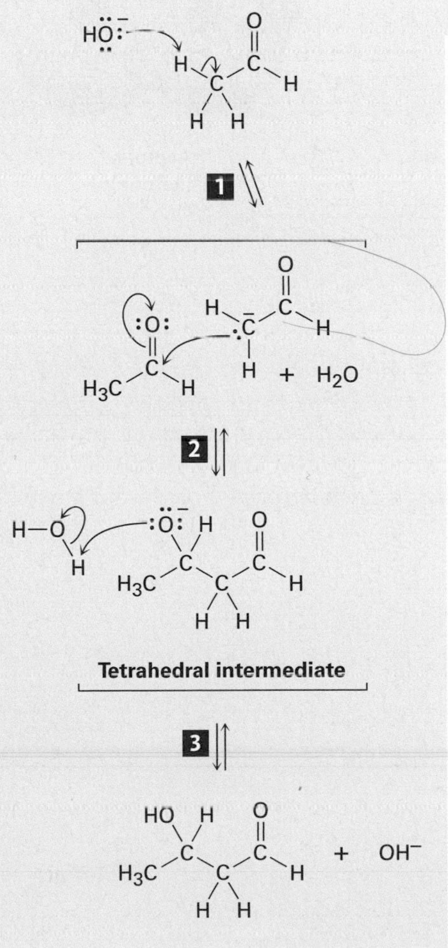

IV. Summary

To a great extent, the chemistry of living organisms is the chemistry of carbonyl compounds. We have not looked at the details of specific carbonyl reactions in this short preview but rather have laid the groundwork for the next five chapters. All the carbonyl-group reactions we'll be studying in Chapters 19 through 23 fall into one of the four fundamental categories discussed in this preview. Knowing where we'll be heading should help you keep matters straight in understanding this most important of all functional groups.

Problems

1. Judging from the following electrostatic potential maps, which kind of carbonyl compound has the more electrophilic carbonyl carbon atom, a ketone or an acid chloride? Which has the more nucleophilic carbonyl oxygen atom? Explain.

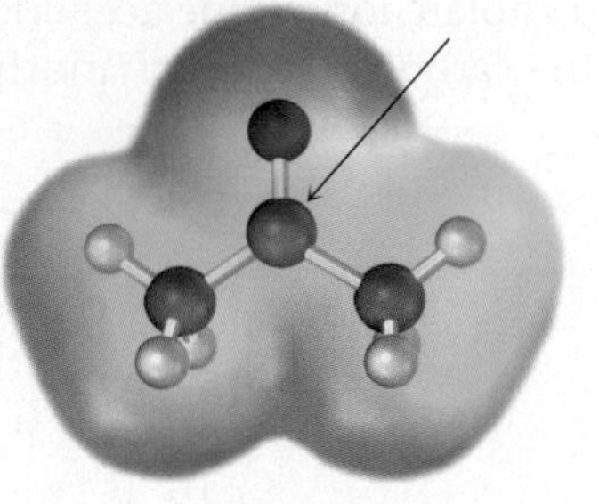

Acetone (ketone)

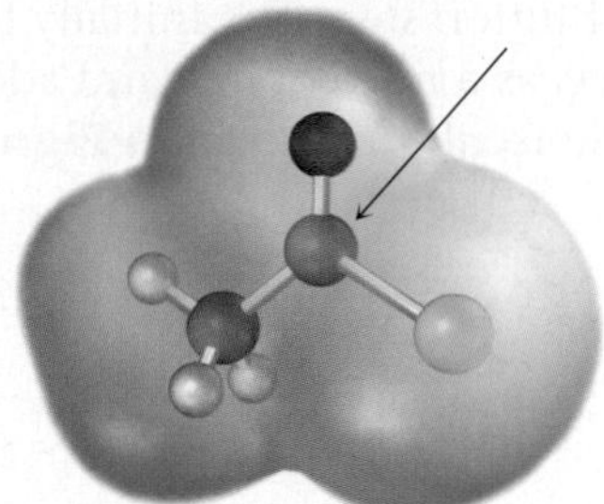

Acetyl chloride (acid chloride)

2. Predict the product formed by nucleophilic addition of cyanide ion (CN^-) to the carbonyl group of acetone, followed by protonation to give an alcohol:

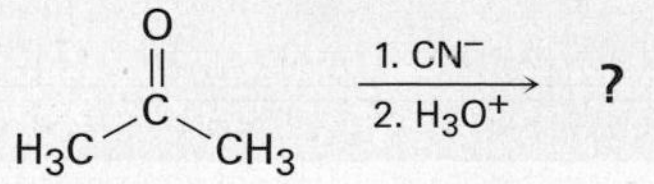

Acetone

3. Identify each of the following reactions as a nucleophilic addition, nucleophilic acyl substitution, an α substitution, or a carbonyl condensation:

(a)

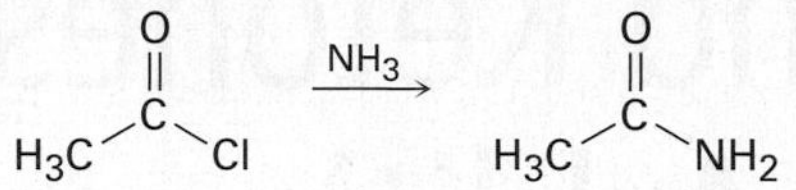

(b)

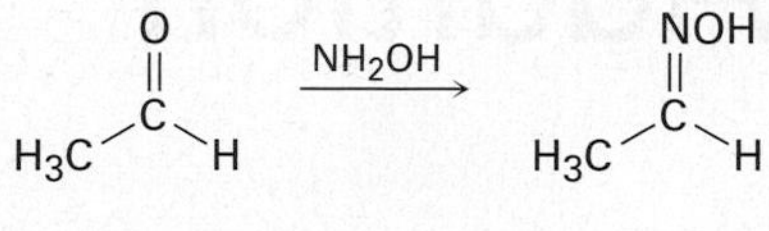

(c)

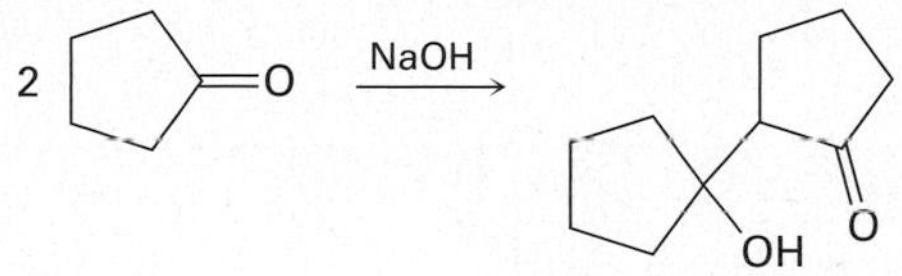

Few flowers are more beautiful or more fragrant than roses. Their perfumed odor is due to several simple organic compounds, including the ketone β-damascenone. Image copyright Veronika Trofer, 2010. Used under license from Shutterstock.com

19 Aldehydes and Ketones: Nucleophilic Addition Reactions

Aldehydes (RCHO) and **ketones (R_2CO)** are among the most widely occurring of all compounds. In nature, many substances required by living organisms are aldehydes or ketones. The aldehyde pyridoxal phosphate, for instance, is a coenzyme involved in a large number of metabolic reactions; the ketone hydrocortisone is a steroid hormone secreted by the adrenal glands to regulate fat, protein, and carbohydrate metabolism.

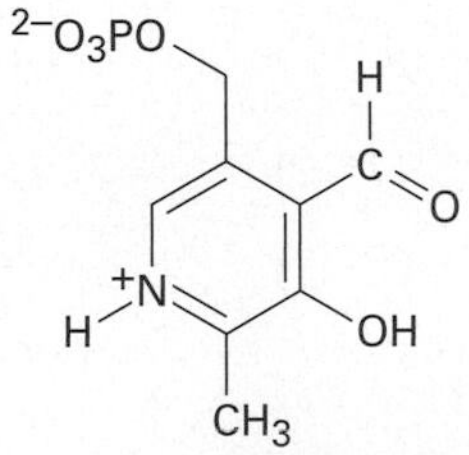

Pyridoxal phosphate (PLP)

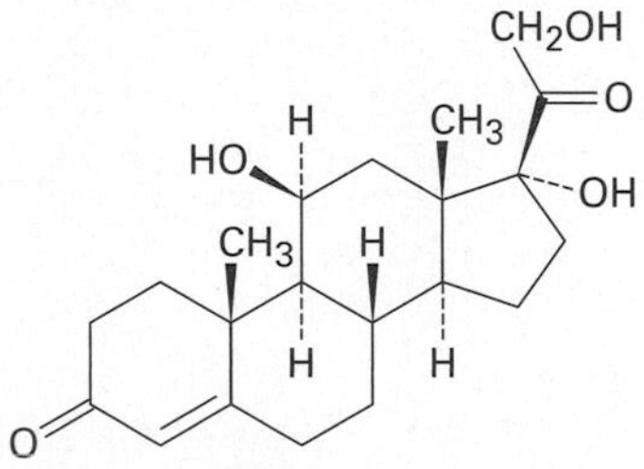

Hydrocortisone

In the chemical industry, simple aldehydes and ketones are produced in large quantities for use as solvents and as starting materials to prepare a host of other compounds. For example, more than 23 million tons per year of formaldehyde, $H_2C{=}O$, is produced worldwide for use in building insulation materials and in the adhesive resins that bind particle board and plywood. Acetone, $(CH_3)_2C{=}O$, is widely used as an industrial solvent; approximately 3.3 million tons per year is produced worldwide. Formaldehyde is synthesized industrially by catalytic oxidation of methanol, and one method of acetone preparation involves oxidation of 2-propanol.

OWL Sign in to OWL for Organic Chemistry at **www.cengage.com/owl** to view tutorials and simulations, develop problem-solving skills, and complete online homework assigned by your professor.

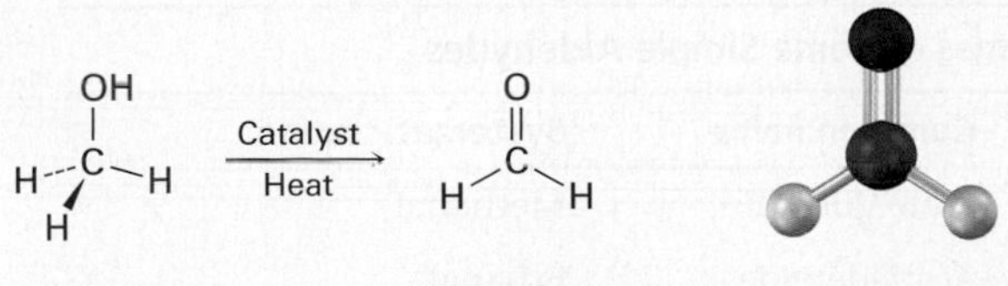

OH — C — H (Methanol) →(Catalyst, Heat) H — C(=O) — H (Formaldehyde)

Methanol **Formaldehyde**

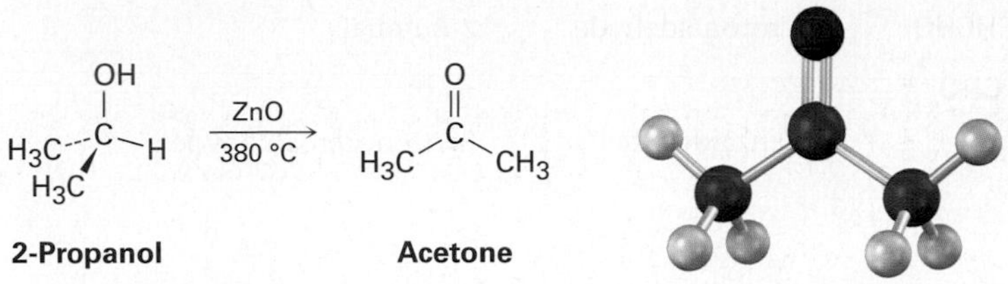

H_3C — C(OH)(H) — H_3C →(ZnO, 380 °C) H_3C — C(=O) — CH_3

2-Propanol **Acetone**

Why This Chapter? Much of organic chemistry is the chemistry of carbonyl compounds. Aldehydes and ketones, in particular, are intermediates in the synthesis of many pharmaceutical agents, in almost all biological pathways, and in numerous industrial processes, so an understanding of their properties and reactions is essential. We'll look in this chapter at some of their most important reactions.

19.1 Naming Aldehydes and Ketones

Aldehydes are named by replacing the terminal *-e* of the corresponding alkane name with *-al*. The parent chain must contain the –CHO group, and the –CHO carbon is numbered as carbon 1. Note in the following examples that the longest chain in 2-ethyl-4-methylpentanal is actually a hexane, but this chain does not include the –CHO group and thus is not the parent.

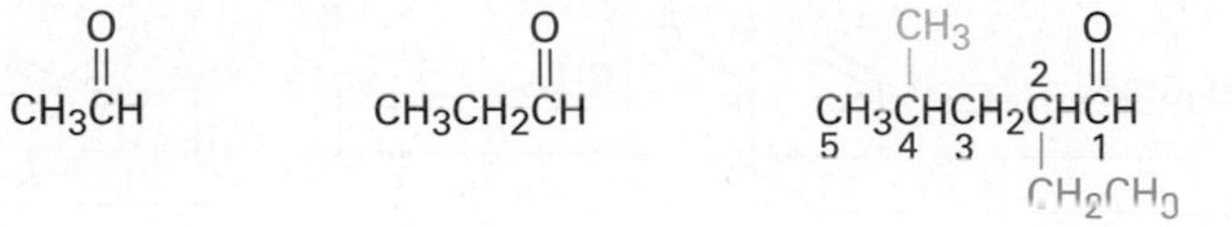

$CH_3CH(=O)$ — **Ethanal (acetaldehyde)**

$CH_3CH_2CH(=O)$ — **Propanal (propionaldehyde)**

$\overset{5}{C}H_3\overset{4}{C}H(CH_3)\overset{3}{C}H_2\overset{2}{C}H(CH_2CH_3)\overset{1}{C}H(=O)$ — **2-Ethyl-4-methylpentanal**

For cyclic aldehydes in which the –CHO group is directly attached to a ring, the suffix *-carbaldehyde* is used.

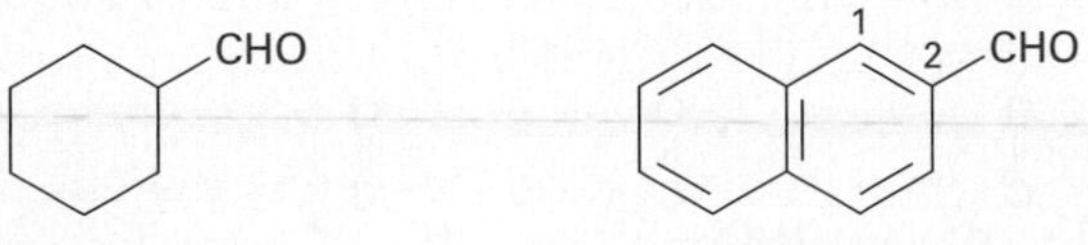

Cyclohexanecarbaldehyde **2-Naphthalenecarbaldehyde**

A few simple and well-known aldehydes have common names that are recognized by IUPAC. Several that you might encounter are listed in Table 19.1.

Table 19.1 Common Names of Some Simple Aldehydes

Formula	Common name	Systematic name
HCHO	Formaldehyde	Methanal
CH_3CHO	Acetaldehyde	Ethanal
$H_2C{=}CHCHO$	Acrolein	Propenal
$CH_3CH{=}CHCHO$	Crotonaldehyde	2-Butenal
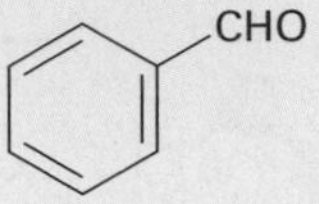 C_6H_5CHO	Benzaldehyde	Benzenecarbaldehyde

Ketones are named by replacing the terminal *-e* of the corresponding alkane name with *-one.* The parent chain is the longest one that contains the ketone group, and the numbering begins at the end nearer the carbonyl carbon. As with alkenes **(Section 7.3)** and alcohols **(Section 17.1)**, the locant is placed before the parent name in older rules but before the suffix in newer IUPAC recommendations. For example:

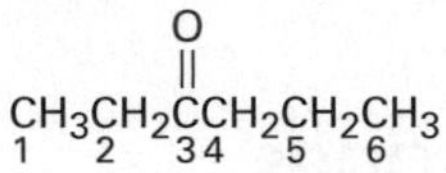

$CH_3CH_2C(O)CH_2CH_2CH_3$ (numbered 1 2 3 4 5 6)
3-Hexanone
(New: Hexan-3-one)

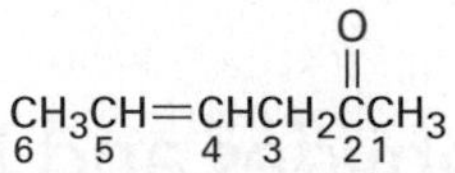

$CH_3CH{=}CHCH_2C(O)CH_3$ (numbered 6 5 4 3 2 1)
4-Hexen-2-one
(New: Hex-4-en-2-one)

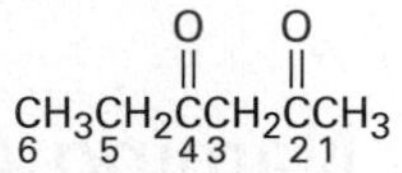

$CH_3CH_2C(O)CH_2C(O)CH_3$ (numbered 6 5 4 3 2 1)
2,4-Hexanedione
(New: Hexane-2,4-dione)

A few ketones are allowed by IUPAC to retain their common names.

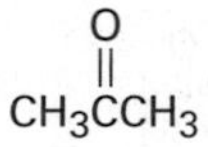

$CH_3C(O)CH_3$ **Acetone**

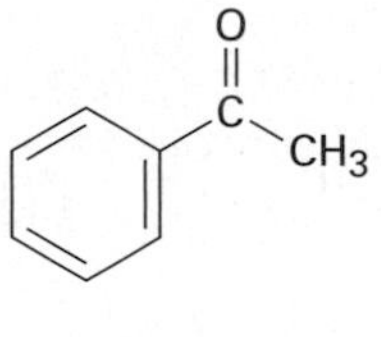

$C_6H_5C(O)CH_3$ **Acetophenone**

$C_6H_5C(O)C_6H_5$ **Benzophenone**

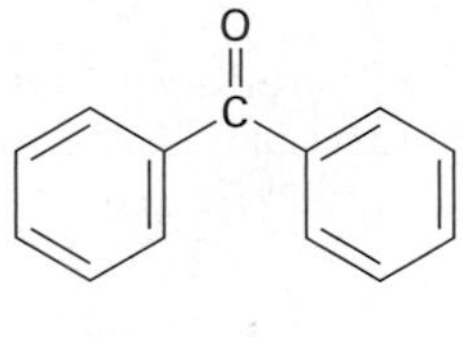

When it's necessary to refer to the R–C=O as a substituent, the name **acyl** (**a**-sil) **group** is used and the name ending *-yl* is attached. Thus, $-COCH_3$ is an *acetyl* group, –CHO is a *formyl* group, –COAr is an *aroyl* group, and $-COC_6H_5$ is a *benzoyl* group.

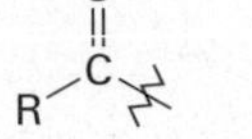

$R-C(O)-$ **An acyl group**

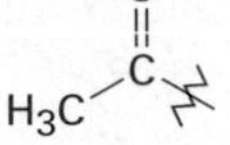

$H_3C-C(O)-$ **Acetyl**

$H-C(O)-$ **Formyl**

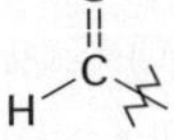

$C_6H_5-C(O)-$ **Benzoyl**

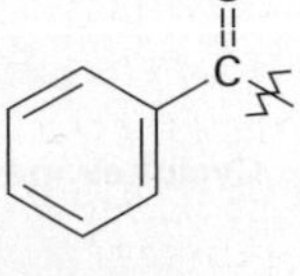

If other functional groups are present and the doubly bonded oxygen is considered a substituent on a parent chain, the prefix *oxo-* is used. For example:

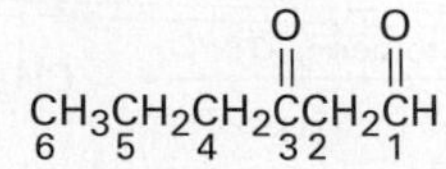

$$\underset{6\ \ 5\ \ 4\ \ 3\ 2\ \ 1}{CH_3CH_2CH_2\overset{O}{\overset{\|}{C}}CH_2\overset{O}{\overset{\|}{C}}H}\quad \textbf{3-Oxohexanal}$$

Problem 19.1

Name the following aldehydes and ketones:

(a) $CH_3CH_2\overset{O}{\overset{\|}{C}}\underset{CH_3}{\underset{|}{C}}HCH_3$ **(b)** $C_6H_5CH_2CH_2CHO$ **(c)** $CH_3\overset{O}{\overset{\|}{C}}CH_2CH_2CH_2\overset{O}{\overset{\|}{C}}CH_2CH_3$

(d) (cyclohexane with CH_3, H, H, CHO) **(e)** $CH_3CH{=}CHCH_2CH_2\overset{O}{\overset{\|}{C}}H$ **(f)** (cyclohexanone with H_3C, H, H, CH_3)

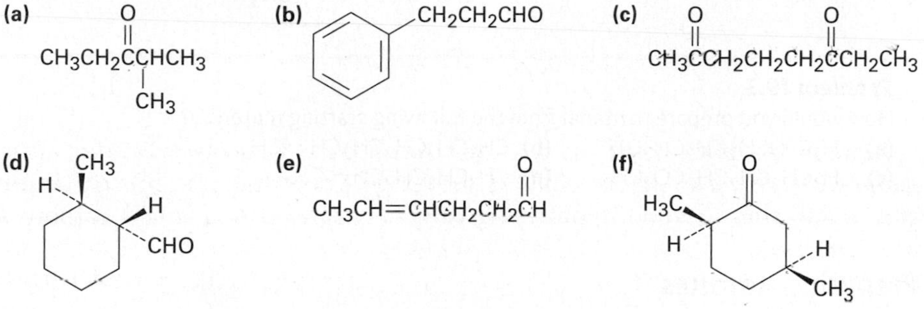

Problem 19.2

Draw structures corresponding to the following names:

(a) 3-Methylbutanal **(b)** 4-Chloro-2-pentanone
(c) Phenylacetaldehyde **(d)** *cis*-3-*tert*-Butylcyclohexanecarbaldehyde
(e) 3-Methyl-3-butenal **(f)** 2-(1-Chloroethyl)-5-methylheptanal

19.2 Preparing Aldehydes and Ketones

Preparing Aldehydes

One of the best methods of aldehyde synthesis is by oxidation of primary alcohols, as we saw in **Section 17.7**. The reaction is often carried out using the Dess–Martin periodinane reagent in dichloromethane solvent at room temperature:

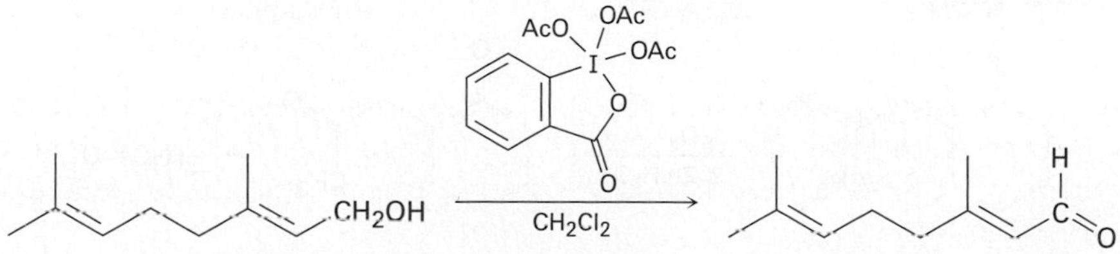

Geraniol —(Dess–Martin periodinane: AcO, OAc, OAc; CH_2Cl_2)→ Geranial (84%)

Geraniol **Geranial (84%)**

A second method of aldehyde synthesis is one that we'll mention here just briefly and then return to in **Section 21.6**. Certain carboxylic acid derivatives can be *partially* reduced to yield aldehydes. The partial reduction of an ester by diisobutylaluminum hydride (DIBAH, or DIBAL-H), for instance, is an important laboratory-scale method of aldehyde synthesis, and mechanistically related processes also occur in biological pathways. The reaction is normally carried out at −78 °C (dry-ice temperature) in toluene solution.

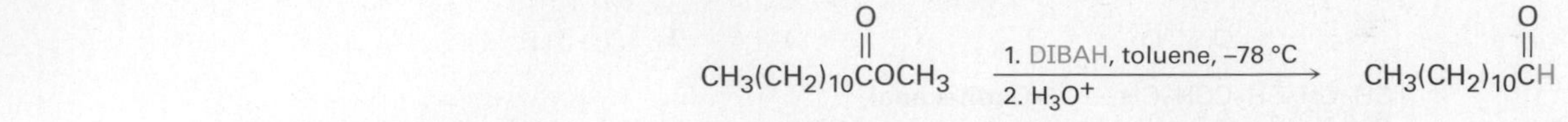

$$\underset{\textbf{Methyl dodecanoate}}{CH_3(CH_2)_{10}\overset{O}{\overset{\|}{C}}OCH_3} \xrightarrow[\text{2. } H_3O^+]{\text{1. DIBAH, toluene, −78 °C}} \underset{\textbf{Dodecanal (88\%)}}{CH_3(CH_2)_{10}\overset{O}{\overset{\|}{C}}H}$$

where DIBAH = $CH_3\underset{CH_3}{CH}CH_2-\overset{H}{Al}-CH_2\underset{CH_3}{CH}CH_3$

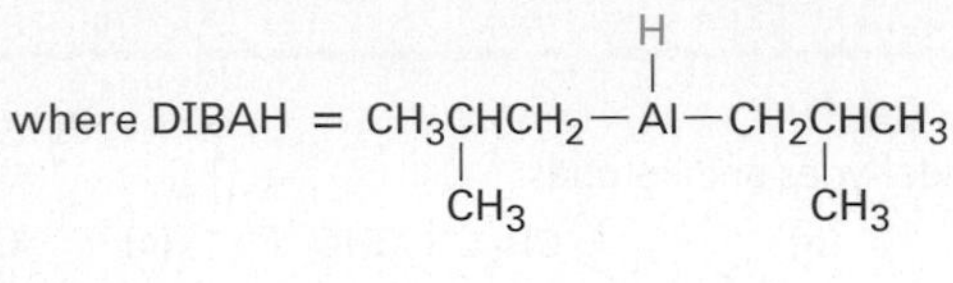

Problem 19.3

How would you prepare pentanal from the following starting materials?

(a) $CH_3CH_2CH_2CH_2CH_2OH$ **(b)** $CH_3CH_2CH_2CH_2CH{=}CH_2$

(c) $CH_3CH_2CH_2CH_2CO_2CH_3$ **(d)** $CH_3CH_2CH_2CH{=}CH_2$

Preparing Ketones

For the most part, methods of ketone synthesis are similar to those for aldehydes. Secondary alcohols are oxidized by a variety of reagents to give ketones **(Section 17.7)**. The choice of oxidant depends on such factors as reaction scale, cost, and acid or base sensitivity of the alcohol. Either the Dess–Martin periodinane or a Cr(VI) regent such as CrO_3 is a common choice.

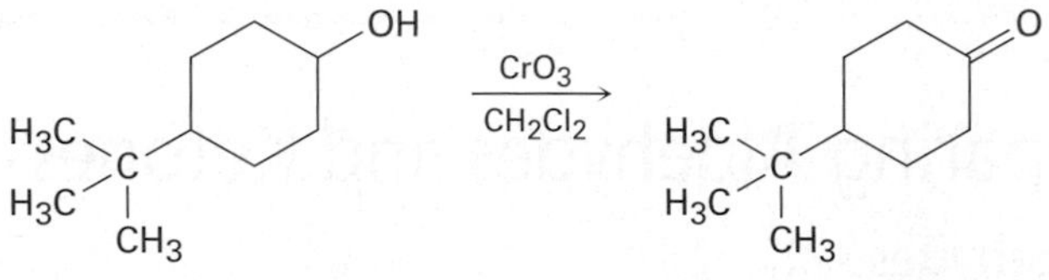

$\xrightarrow[CH_2Cl_2]{CrO_3}$

4-*tert*-Butylcyclohexanol **4-*tert*-Butylcyclohexanone (90%)**

Other methods include the ozonolysis of alkenes in which one of the unsaturated carbon atoms is disubstituted **(Section 8.8)** and Friedel–Crafts acylation of an aromatic ring with an acid chloride in the presence of $AlCl_3$ catalyst **(Section 16.3)**.

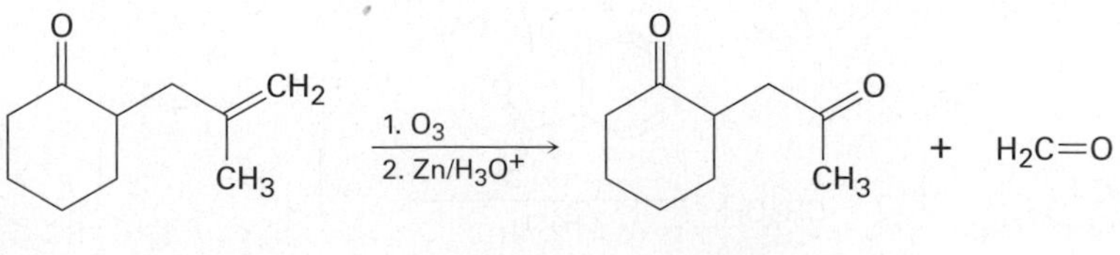

$\xrightarrow[\text{2. Zn/}H_3O^+]{\text{1. } O_3}$ + $H_2C{=}O$

70%

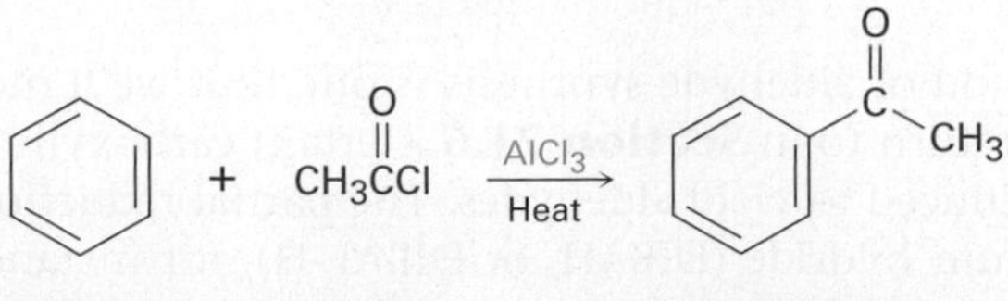

Benzene + **Acetyl chloride** $CH_3\overset{O}{\overset{\|}{C}}Cl$ $\xrightarrow[\text{Heat}]{AlCl_3}$ **Acetophenone (95%)**

In addition to those methods already discussed, ketones can also be prepared from certain carboxylic acid derivatives, just as aldehydes can. Among the most useful reactions of this sort is that between an acid chloride and a lithium diorganocopper reagent such as we saw in **Section 10.7**. We'll discuss this reaction in more detail in **Section 21.4**.

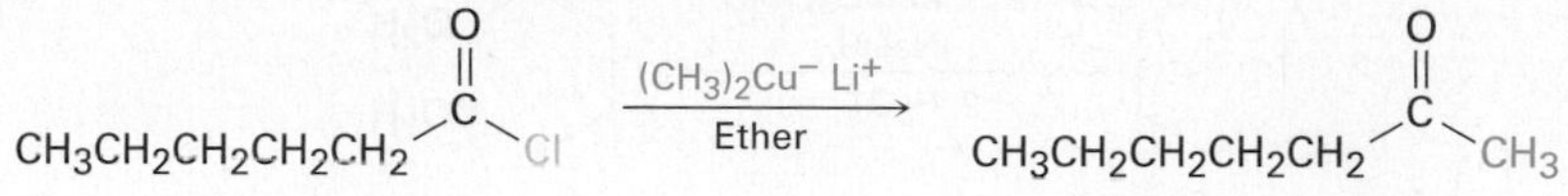

$$CH_3CH_2CH_2CH_2CH_2C(=O)Cl \xrightarrow[\text{Ether}]{(CH_3)_2Cu^- \, Li^+} CH_3CH_2CH_2CH_2CH_2C(=O)CH_3$$

Hexanoyl chloride → **2-Heptanone (81%)**

Problem 19.4

How would you carry out the following reactions? More than one step may be required.

(a) 3-Hexyne → 3-Hexanone
(b) Benzene → *m*-Bromoacetophenone
(c) Bromobenzene → Acetophenone
(d) 1-Methylcyclohexene → 2-Methylcyclohexanone

19.3 Oxidation of Aldehydes and Ketones

Aldehydes are easily oxidized to yield carboxylic acids, but ketones are generally inert toward oxidation. The difference is a consequence of structure: aldehydes have a –CHO proton that can be abstracted during oxidation, but ketones do not.

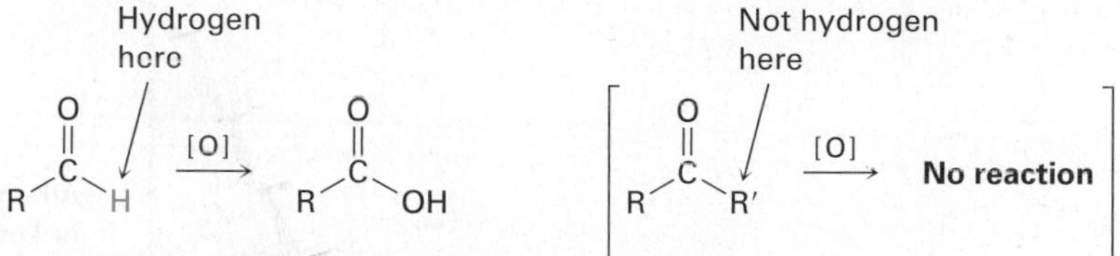

$$RC(=O)H \xrightarrow{[O]} RC(=O)OH$$

Hydrogen here (the H of the aldehyde)

An aldehyde → **A carboxylic acid**

$$\left[RC(=O)R' \xrightarrow{[O]} \text{No reaction} \right]$$

Not hydrogen here (the R′ of the ketone)

A ketone

Many oxidizing agents, including $KMnO_4$ and hot HNO_3, convert aldehydes into carboxylic acids, but CrO_3 in aqueous acid is a more common choice. The oxidation occurs rapidly at room temperature and generally results in good yields.

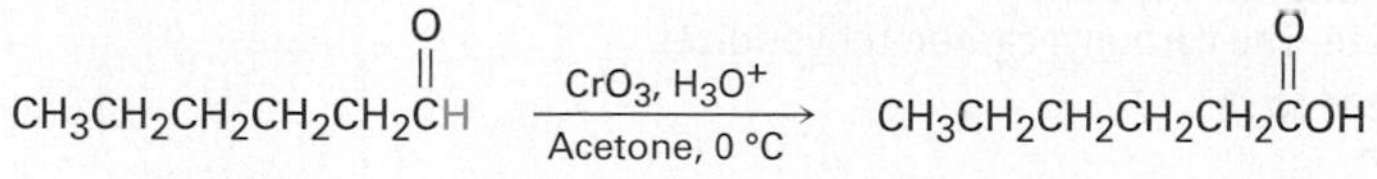

$$CH_3CH_2CH_2CH_2CH_2CHO \xrightarrow[\text{Acetone, 0 °C}]{CrO_3,\ H_3O^+} CH_3CH_2CH_2CH_2CH_2COOH$$

Hexanal → **Hexanoic acid (85%)**

Aldehyde oxidations occur through intermediate 1,1-diols, or *hydrates,* which are formed by a reversible nucleophilic addition of water to the carbonyl group. Even though formed to only a small extent at equilibrium, the hydrate reacts like any typical primary or secondary alcohol and is oxidized to a carbonyl compound **(Section 17.7)**.

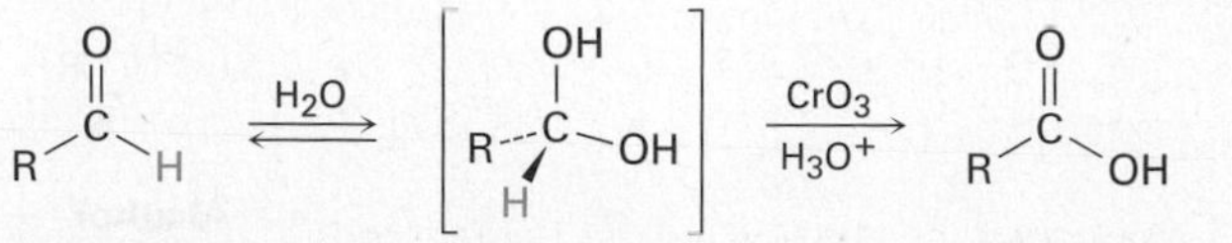

$$RC(=O)H \overset{H_2O}{\rightleftharpoons} \left[RCH(OH)OH \right] \xrightarrow[H_3O^+]{CrO_3} RC(=O)OH$$

An aldehyde → **A hydrate** → **A carboxylic acid**

Ketones are inert to most oxidizing agents but undergo a slow cleavage reaction of the C–C bond next to the carbonyl group when treated with hot alkaline $KMnO_4$. The reaction is not often used and is mentioned here only for completeness.

Cyclohexanone $\xrightarrow[\text{2. } H_3O^+]{\text{1. } KMnO_4,\ H_2O,\ NaOH}$ Hexanedioic acid (79%)

Cyclohexanone **Hexanedioic acid (79%)**

19.4 Nucleophilic Addition Reactions of Aldehydes and Ketones

As we saw in the *Preview of Carbonyl Chemistry,* the most general reaction of aldehydes and ketones is the **nucleophilic addition reaction**. As shown in **Figure 19.1**, a nucleophile, :Nu$^-$, approaches the carbonyl group from an angle of about 105° opposite the carbonyl oxygen and forms a bond to the electrophilic C=O carbon atom. At the same time, rehybridization of the carbonyl carbon from sp^2 to sp^3 occurs, an electron pair from the C=O bond moves toward the electronegative oxygen atom, and a tetrahedral alkoxide ion intermediate is produced. Protonation of the alkoxide by addition of acid then gives an alcohol.

Figure 19.1 | MECHANISM

A nucleophilic addition reaction to an aldehyde or ketone. The nucleophile approaches the carbonyl group from an angle of approximately 75° to the plane of the sp^2 orbitals, the carbonyl carbon rehybridizes from sp^2 to sp^3, and an alkoxide ion is formed. Protonation by addition of acid then gives an alcohol.

1 An electron pair from the nucleophile adds to the electrophilic carbon of the carbonyl group, pushing an electron pair from the C=O bond onto oxygen and giving an alkoxide ion intermediate. The carbonyl carbon rehybridizes from sp^2 to sp^3.

2 Protonation of the alkoxide anion intermediate gives the neutral alcohol addition product.

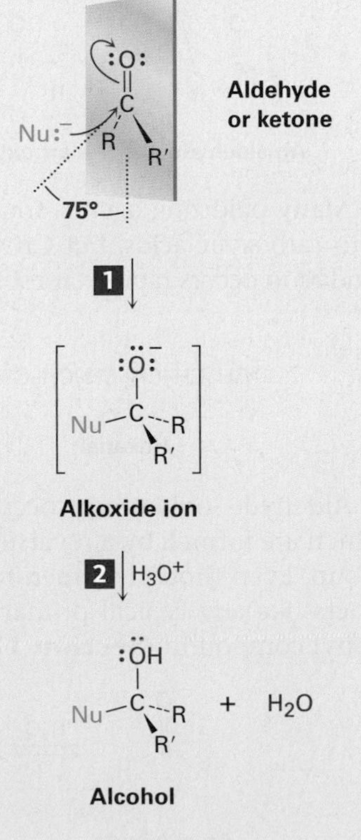

The nucleophile can be either negatively charged (:Nu$^-$) or neutral (:Nu). If it's neutral, however, it usually carries a hydrogen atom that can subsequently be eliminated, :Nu–H. For example:

Some negatively charged nucleophiles

- $H\ddot{O}:^-$ (hydroxide ion)
- $H:^-$ (hydride ion)
- $R_3C:^-$ (a carbanion)
- $R\ddot{O}:^-$ (an alkoxide ion)
- $N{\equiv}C:^-$ (cyanide ion)

Some neutral nucleophiles

- $H\ddot{O}H$ (water)
- $R\ddot{O}H$ (an alcohol)
- $H_3N:$ (ammonia)
- $R\ddot{N}H_2$ (an amine)

Nucleophilic additions to aldehydes and ketones have two general variations, as shown in **Figure 19.2**. In one variation, the tetrahedral intermediate is protonated by water or acid to give an alcohol as the final product. In the second variation, the carbonyl oxygen atom is protonated and then eliminated as HO$^-$ or H_2O to give a product with a C=Nu double bond.

Aldehyde or ketone

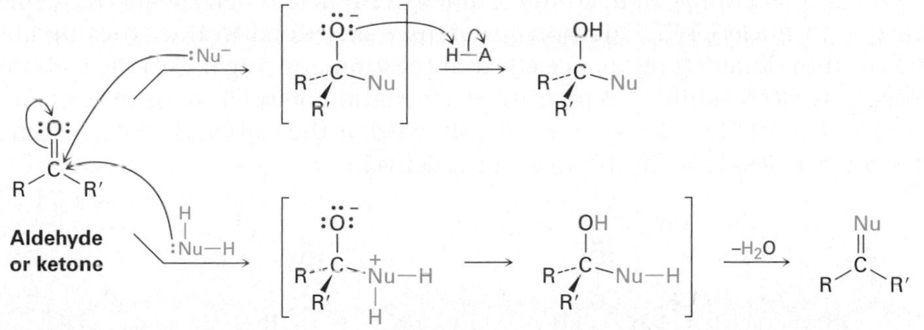

Figure 19.2 Two general reaction pathways following addition of a nucleophile to an aldehyde or ketone. The top pathway leads to an alcohol product; the bottom pathway leads to a product with a C=Nu double bond.

Aldehydes are generally more reactive than ketones in nucleophilic addition reactions for both steric and electronic reasons. Sterically, the presence of only one large substituent bonded to the C=O carbon in an aldehyde versus two large substituents in a ketone means that a nucleophile is able to approach an aldehyde more readily. Thus, the transition state leading to the tetrahedral intermediate is less crowded and lower in energy for an aldehyde than for a ketone **(Figure 19.3)**.

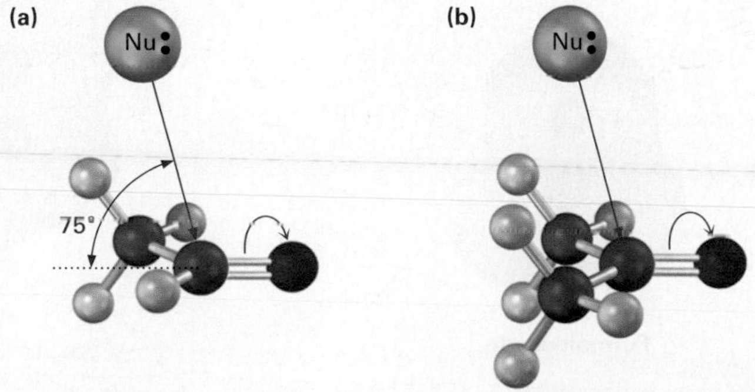

Figure 19.3 **(a)** Nucleophilic addition to an aldehyde is sterically less hindered because only one relatively large substituent is attached to the carbonyl-group carbon. **(b)** A ketone, however, has two large substituents and is more hindered. The approach of the nucleophile is along the C=O bond at an angle of about 75° to the plane of the carbon sp^2 orbitals.

Electronically, aldehydes are more reactive than ketones because of the greater polarization of aldehyde carbonyl groups. To see this polarity difference, recall the stability order of carbocations **(Section 7.9)**. A primary carbocation is higher in energy and thus more reactive than a secondary carbocation because it has only one alkyl group inductively stabilizing the positive charge rather than two. In the same way, an aldehyde has only one alkyl group inductively stabilizing the partial positive charge on the carbonyl carbon rather than two, is a bit more electrophilic, and is therefore more reactive than a ketone.

1° carbocation
(less stable, more reactive)

2° carbocation
(more stable, less reactive)

Aldehyde
(less stabilization of δ+, more reactive)

Ketone
(more stabilization of δ+, less reactive)

One further comparison: aromatic aldehydes, such as benzaldehyde, are less reactive in nucleophilic addition reactions than aliphatic aldehydes because the electron-donating resonance effect of the aromatic ring makes the carbonyl group less electrophilic. Comparing electrostatic potential maps of formaldehyde and benzaldehyde, for example, shows that the carbonyl carbon atom is less positive (less blue) in the aromatic aldehyde.

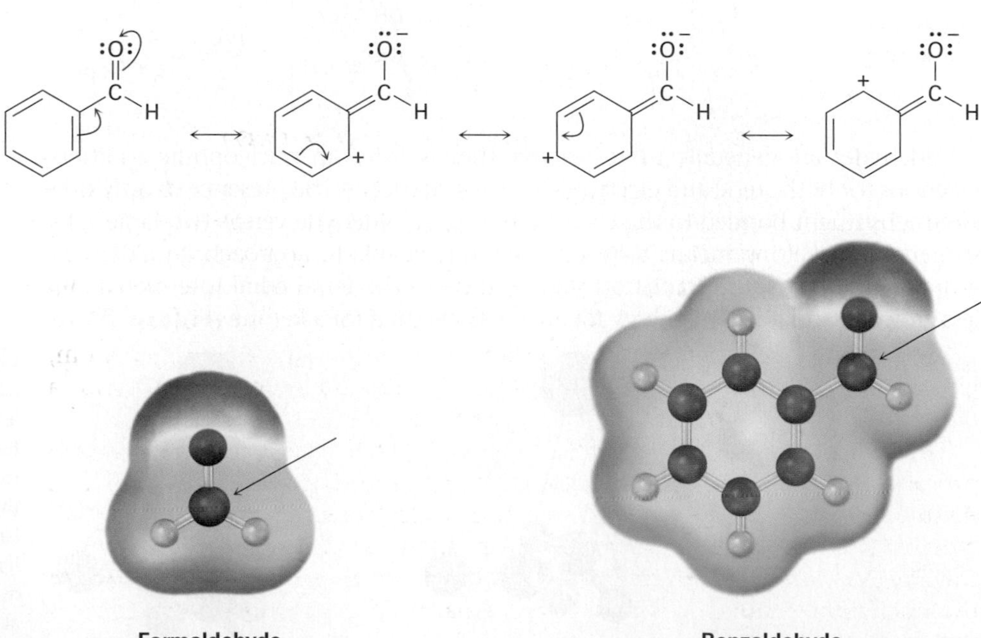

Formaldehyde

Benzaldehyde

Problem 19.5

Treatment of an aldehyde or ketone with cyanide ion ($^{-}$:C≡N), followed by protonation of the tetrahedral alkoxide ion intermediate, gives a *cyanohydrin*. Show the structure of the cyanohydrin obtained from cyclohexanone.

Problem 19.6

p-Nitrobenzaldehyde is more reactive toward nucleophilic additions than *p*-methoxybenzaldehyde. Explain.

19.5 Nucleophilic Addition of H_2O: Hydration

Aldehydes and ketones react with water to yield 1,1-diols, or *geminal (gem)* diols. The hydration reaction is reversible, and a gem diol can eliminate water to regenerate an aldehyde or ketone.

$$H_3C{-}C({=}O){-}CH_3 + H_2O \rightleftharpoons (H_3C)_2C(OH)_2$$

Acetone (99.9%) **Acetone hydrate (0.1%)**

The position of the equilibrium between a gem diol and an aldehyde or ketone depends on the structure of the carbonyl compound. The equilibrium generally favors the carbonyl compound for steric reasons, but the gem diol is favored for a few simple aldehydes. For example, an aqueous solution of formaldehyde consists of 99.9% gem diol and 0.1% aldehyde at equilibrium, whereas an aqueous solution of acetone consists of only about 0.1% gem diol and 99.9% ketone.

$$H{-}C({=}O){-}H + H_2O \rightleftharpoons H_2C(OH)_2$$

Formaldehyde (0.1%) **Formaldehyde hydrate (99.9%)**

The nucleophilic addition of water to an aldehyde or ketone is slow under neutral conditions but is catalyzed by both base and acid. Under basic conditions **(Figure 19.4a)**, the nucleophile is negatively charged (OH^-) and uses a pair of its electrons to form a bond to the electrophilic carbon atom of the C=O group. At the same time, the C=O carbon atom rehybridizes from sp^2 to sp^3 and two electrons from the C=O π bond are pushed onto the oxygen atom, giving an alkoxide ion. Protonation of the alkoxide ion by water then yields a neutral addition product plus regenerated OH^-.

Under acidic conditions **(Figure 19.4b)**, the carbonyl oxygen atom is first protonated by H_3O^+ to make the carbonyl group more strongly electrophilic. A neutral nucleophile, H_2O, then uses a pair of electrons to bond to the carbon atom of the C=O group, and two electrons from the C=O π bond move onto the oxygen atom. The positive charge on oxygen is thereby neutralized, while the nucleophile gains a positive charge. Finally, deprotonation by water gives the neutral addition product and regenerates the H_3O^+ catalyst.

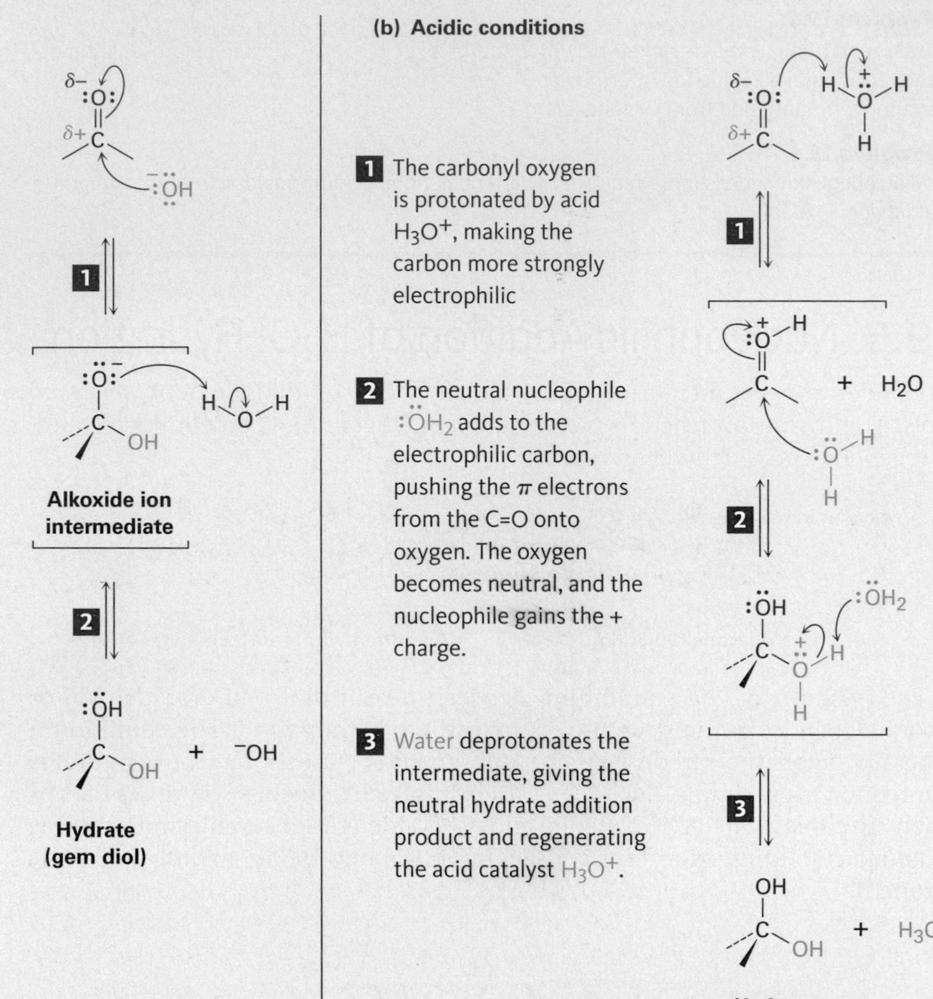

Figure 19.4 | MECHANISM

The mechanism of a nucleophilic addition reaction of aldehydes and ketones under both basic and acidic conditions. **(a)** Under basic conditions, a negatively charged nucleophile adds to the carbonyl group to give an alkoxide ion intermediate, which is subsequently protonated. **(b)** Under acidic conditions, protonation of the carbonyl group occurs first, followed by addition of a neutral nucleophile and subsequent deprotonation.

Note the key difference between the base-catalyzed and acid-catalyzed reactions. The base-catalyzed reaction takes place rapidly because water is converted into hydroxide ion, a much better *nucleophile*. The acid-catalyzed reaction takes place rapidly because the carbonyl compound is converted by protonation into a much better *electrophile*.

The hydration reaction just described is typical of what happens when an aldehyde or ketone is treated with a nucleophile of the type H–Y, where the Y atom is electronegative and can stabilize a negative charge (oxygen, halogen, or sulfur, for instance). In such reactions, the nucleophilic addition is reversible, with the equilibrium generally favoring the carbonyl reactant rather than the tetrahedral addition product. In other words, treatment of an aldehyde or ketone with CH_3OH, H_2O, HCl, HBr, or H_2SO_4 does not normally lead to a stable alcohol addition product.

$$RC(=O)R' + H{-}Y \rightleftharpoons RR'C(OH)Y$$

Favored when
$Y = -OCH_3, -OH, -Br, -Cl, HSO_4^-$

Problem 19.7
When dissolved in water, trichloroacetaldehyde exists primarily as its hydrate, called chloral hydrate. Show the structure of chloral hydrate.

Problem 19.8
The oxygen in water is primarily (99.8%) ^{16}O, but water enriched with the heavy isotope ^{18}O is also available. When an aldehyde or ketone is dissolved in ^{18}O-enriched water, the isotopic label becomes incorporated into the carbonyl group. Explain.

$$R_2C{=}O + H_2O \rightleftharpoons R_2C{=}O + H_2O \qquad \text{where } O = {}^{18}O$$

19.6 Nucleophilic Addition of HCN: Cyanohydrin Formation

Aldehydes and unhindered ketones undergo a nucleophilic addition reaction with HCN to yield **cyanohydrins, RCH(OH)C≡N.** Studies carried out in the early 1900s by Arthur Lapworth showed that cyanohydrin formation is reversible and base-catalyzed. Reaction occurs slowly when pure HCN is used but rapidly when a small amount of base is added to generate the nucleophilic cyanide ion, CN^-. Addition of CN^- takes place by a typical nucleophilic addition pathway, yielding a tetrahedral intermediate that is protonated by HCN to give cyanohydrin product plus regenerated CN^-.

:C≡N⁻ ⇌ ; HCN ⇌ ; + :C≡N⁻

Benzaldehyde **Tetrahedral intermediate** **Mandelonitrile (88%)**

Cyanohydrin formation is somewhat unusual because it is one of the few examples of the addition of a protic acid (H–Y) to a carbonyl group. As noted in the previous section, protic acids such as H_2O, HBr, HCl, and H_2SO_4 don't normally yield carbonyl addition products because the equilibrium constants are unfavorable. With HCN, however, the equilibrium favors the cyanohydrin adduct.

Cyanohydrin formation is useful because of the further chemistry that can be carried out on the product. For example, a nitrile (R–C≡N) can be reduced with

$LiAlH_4$ to yield a primary amine (RCH_2NH_2) and can be hydrolyzed by hot aqueous acid to yield a carboxylic acid. Thus, cyanohydrin formation provides a method for transforming an aldehyde or ketone into a different functional group.

Benzaldehyde —HCN→ Mandelonitrile —1. $LiAlH_4$, THF; 2. H_2O→ 2-Amino-1-phenylethanol

Mandelonitrile —H_3O^+, heat→ Mandelic acid (90%)

Problem 19.9

Cyclohexanone forms a cyanohydrin in good yield but 2,2,6-trimethylcyclohexanone does not. Explain.

19.7 Nucleophilic Addition of Hydride and Grignard Reagents: Alcohol Formation

Addition of Hydride Reagents: Reduction

We saw in **Section 17.4** that the most common method for preparing alcohols, both in the laboratory and in living organisms, is by the reduction of carbonyl compounds. Aldehydes are reduced with sodium borohydride ($NaBH_4$) to give primary alcohols, and ketones are reduced similarly to give secondary alcohols.

Aldehyde —$NaBH_4$, Ethanol→ 1° Alcohol

Ketone —$NaBH_4$, Ethanol→ 2° Alcohol

Carbonyl reduction occurs by a typical nucleophilic addition mechanism under basic conditions, as shown previously in Figure 19.4a. Although the details of carbonyl-group reductions are complex, $LiAlH_4$ and $NaBH_4$ act as if they were donors of hydride ion nucleophile, $:H^-$, and the initially formed alkoxide ion intermediate is then protonated by addition of aqueous acid. The reaction is effectively irreversible because the reverse process would require expulsion of a very poor leaving group.

$R_2C{=}O$ —"$:H^-$" from $NaBH_4$→ [alkoxide] —H_3O^+→ alcohol + H_2O

Addition of Grignard Reagents, RMgX

Just as aldehydes and ketones undergo nucleophilic addition with hydride ion to give alcohols, they undergo a similar addition with Grignard reagent nucleophiles, R:$^-$ $^+$MgX. Aldehydes give secondary alcohols on reaction with Grignard reagents in ether solution, and ketones give tertiary alcohols.

Aldehyde $\xrightarrow[\text{2. } H_3O^+]{\text{1. R''MgX}}$ 2° Alcohol Ketone $\xrightarrow[\text{2. } H_3O^+]{\text{1. R''MgX}}$ 3° Alcohol

As shown in **Figure 19.5**, a Grignard reaction begins with an acid–base complexation of Mg^{2+} to the carbonyl oxygen atom of the aldehyde or ketone, thereby making the carbonyl group a better electrophile. Nucleophilic addition of R:$^-$ then produces a tetrahedral magnesium alkoxide intermediate, and protonation by addition of water or dilute aqueous acid in a separate step yields the neutral alcohol. Like reduction, Grignard additions are effectively irreversible because a carbanion is too poor a leaving group to be expelled in a reversal step.

Figure 19.5 | MECHANISM

1 The Lewis acid Mg^{2+} first forms an acid–base complex with the basic oxygen atom of the aldehyde or ketone, thereby making the carbonyl group a better acceptor.

2 Nucleophilic addition of an alkyl group :R$^-$ to the aldehyde or ketone produces a tetrahedral magnesium alkoxide intermediate . . .

3 . . . which undergoes hydrolysis when water is added in a separate step. The final product is a neutral alcohol.

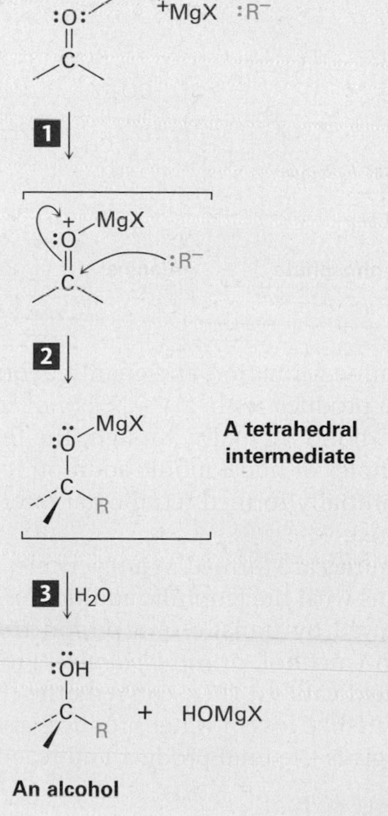

Mechanism of the Grignard reaction. Complexation of the carbonyl oxygen with the Lewis acid Mg^{2+} and subsequent nucleophilic addition of a carbanion to an aldehyde or ketone is followed by protonation of the alkoxide intermediate to yield an alcohol.

19.8 Nucleophilic Addition of Amines: Imine and Enamine Formation

Primary amines, RNH_2, add to aldehydes and ketones to yield **imines, $R_2C{=}NR$**. Secondary amines, R_2NH, add similarly to yield **enamines, $R_2N{-}CR{=}CR_2$** (*ene* + *amine* = unsaturated amine).

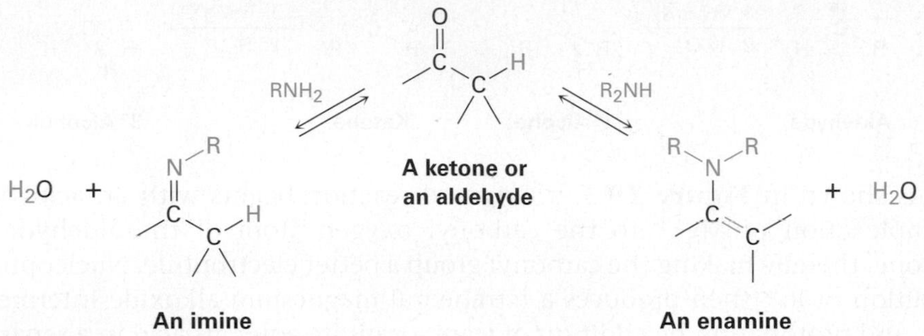

H_2O + An imine ⇌ (RNH_2) A ketone or an aldehyde ⇌ (R_2NH) An enamine + H_2O

Imines are particularly common as intermediates in many biological pathways, where they are often called **Schiff bases**. The amino acid alanine, for instance, is metabolized in the body by reaction with the aldehyde pyridoxal phosphate (PLP), a derivative of vitamin B_6, to yield a Schiff base that is further degraded.

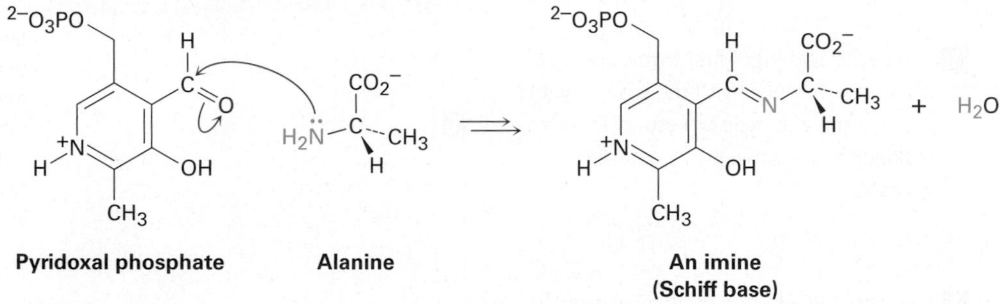

Pyridoxal phosphate + Alanine ⟶ An imine (Schiff base) + H_2O

Imine formation and enamine formation appear different because one leads to a product with a C=N bond and the other leads to a product with a C=C bond. Actually, though, the reactions are quite similar. Both are typical examples of nucleophilic addition reactions in which water is eliminated from the initially formed tetrahedral intermediate and a new C=Nu double bond is formed.

Imines are formed in a reversible, acid-catalyzed process **(Figure 19.6)** that begins with nucleophilic addition of the primary amine to the carbonyl group, followed by transfer of a proton from nitrogen to oxygen to yield a neutral amino alcohol, or *carbinolamine*. Protonation of the carbinolamine oxygen by an acid catalyst then converts the –OH into a better leaving group ($-OH_2^+$), and E1-like loss of water produces an iminium ion. Loss of a proton from nitrogen gives the final product and regenerates the acid catalyst.

Figure 19.6 | MECHANISM

Mechanism of imine formation by reaction of an aldehyde or ketone with a primary amine. The key step is the initial nucleophilic addition to yield a carbinolamine intermediate, which then loses water to give the imine.

Ketone/aldehyde

1 Nucleophilic attack on the ketone or aldehyde by the lone-pair electrons of an amine leads to a dipolar tetrahedral intermediate.

1 $:NH_2R$

2 A proton is then transferred from nitrogen to oxygen, yielding a neutral carbinolamine.

2 Proton transfer

Carbinolamine

3 Acid catalyst protonates the hydroxyl oxygen.

3 H_3O^+

4 The nitrogen lone-pair electrons expel water, giving an iminium ion.

4 $-H_2O$

Iminium ion

5 Loss of H^+ from nitrogen then gives the neutral imine product.

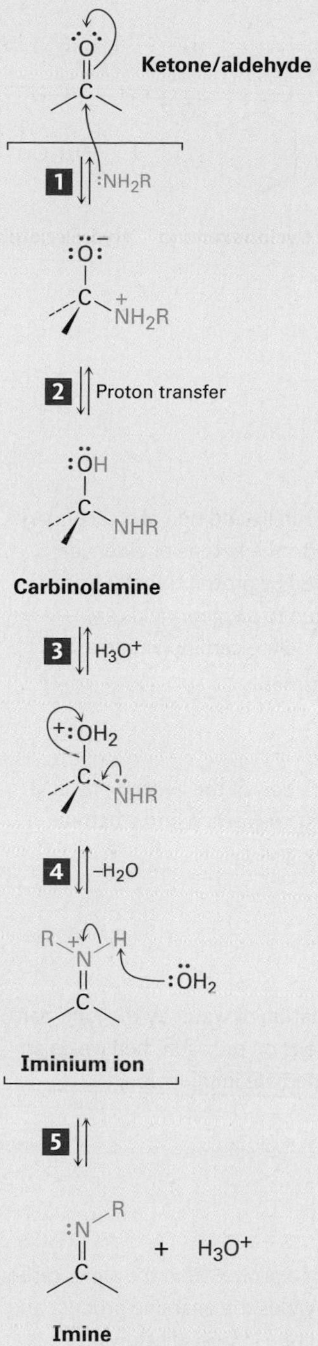

5

Imine + H_3O^+

Imine formation with such reagents as hydroxylamine and 2,4-dinitrophenylhydrazine is sometimes useful because the products of these reactions—*oximes* and *2,4-dinitrophenylhydrazones (2,4-DNPs),* respectively—are often crystalline and easy to handle. Such crystalline derivatives are occasionally prepared as a means of purifying and characterizing liquid ketones or aldehydes.

Oxime

Cyclohexanone + NH_2OH ⟶ Cyclohexanone oxime + H_2O

Cyclohexanone **Hydroxylamine** **Cyclohexanone oxime (mp 90 °C)**

Figure 19.7 | MECHANISM

Mechanism of enamine formation by reaction of an aldehyde or ketone with a secondary amine, R_2NH. The iminium ion intermediate formed in step 3 has no hydrogen attached to N and so must lose H^+ from the carbon two atoms away.

1 Nucleophilic addition of a secondary amine to the ketone or aldehyde, followed by proton transfer from nitrogen to oxygen, yields an intermediate carbinolamine in the normal way.

1 R_2NH

2 Protonation of the hydroxyl by acid catalyst converts it into a better leaving group.

2 H_3O^+

3 Elimination of water by the lone-pair electrons on nitrogen then yields an intermediate iminium ion.

3 $-H_2O$

4 Loss of a proton from the alpha carbon atom yields the enamine product and regenerates the acid catalyst.

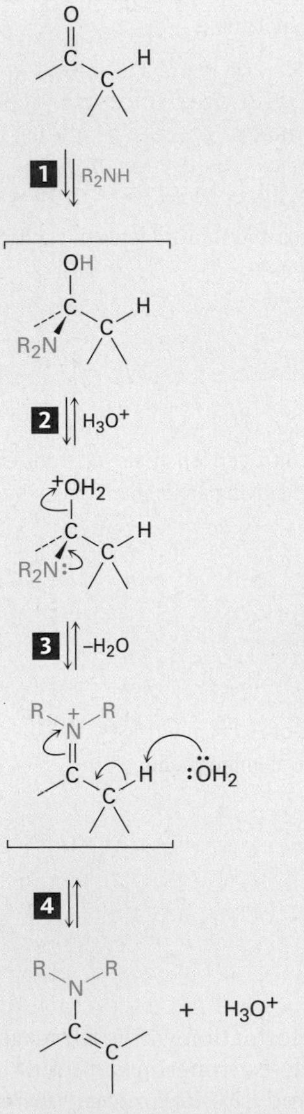

+ H_3O^+

Enamine

2,4-Dinitrophenylhydrazone

Acetone + 2,4-Dinitrophenylhydrazine → Acetone 2,4-dinitrophenylhydrazone (mp 126 °C) + H_2O

Reaction of an aldehyde or ketone with a secondary amine, R_2NH, rather than a primary amine yields an enamine. As shown in **Figure 19.7**, the process is identical to imine formation up to the iminium ion stage, but at this point there is no proton on nitrogen that can be lost to form a neutral imine product. Instead, a proton is lost from the *neighboring* carbon (the α carbon), yielding an enamine.

Imine and enamine formation are slow at both high pH and low pH but reach a maximum rate at a weakly acidic pH around 4 to 5. For example, the profile of pH versus rate shown in **Figure 19.8** for the reaction between acetone and hydroxylamine, NH_2OH, indicates that the maximum reaction rate is obtained at pH 4.5.

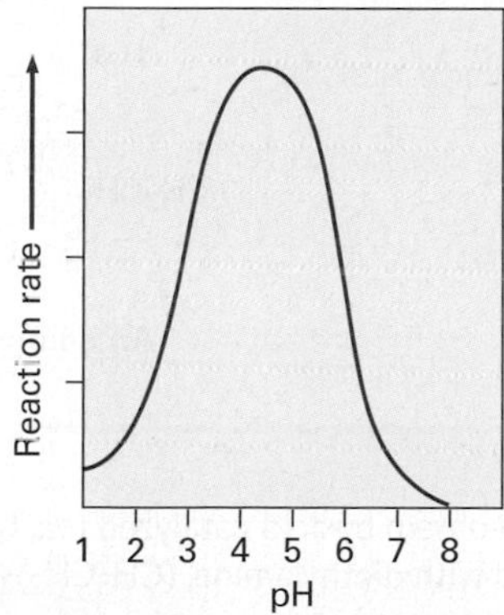

Figure 19.8 Dependence on pH of the rate of reaction between acetone and hydroxylamine: $(CH_3)_2C{=}O + NH_2OH \rightarrow (CH_3)_2C{=}NOH + H_2O$.

We can explain the observed pH dependence of imine formation by looking at the individual steps in the mechanism. As indicated in Figure 19.7, an acid catalyst is required in step 3 to protonate the intermediate carbinolamine, thereby converting the –OH into a better leaving group. Thus, reaction will be slow if not enough acid is present (that is, at high pH). On the other hand, if too much acid is present (low pH), the basic amine nucleophile is completely protonated, so the initial nucleophilic addition step can't occur.

Evidently, a pH of 4.5 represents a compromise between the need for *some* acid to catalyze the rate-limiting dehydration step but *not too much* acid so as to avoid complete protonation of the amine. Each individual nucleophilic addition reaction has its own requirements, and reaction conditions must be optimized to obtain maximum reaction rates.

Worked Example 19.1 Predicting the Product of Reaction between a Ketone and an Amine

Show the products you would obtain by acid-catalyzed reaction of 3-pentanone with methylamine, CH_3NH_2, and with dimethylamine, $(CH_3)_2NH$.

Strategy

An aldehyde or ketone reacts with a primary amine, RNH_2, to yield an imine, in which the carbonyl oxygen atom has been replaced by the =N–R group of the amine. Reaction of the same aldehyde or ketone with a secondary amine, R_2NH, yields an enamine, in which the oxygen atom has been replaced by the $-NR_2$ group of the amine and the double bond has moved to a position between the former carbonyl carbon and the neighboring carbon.

Solution

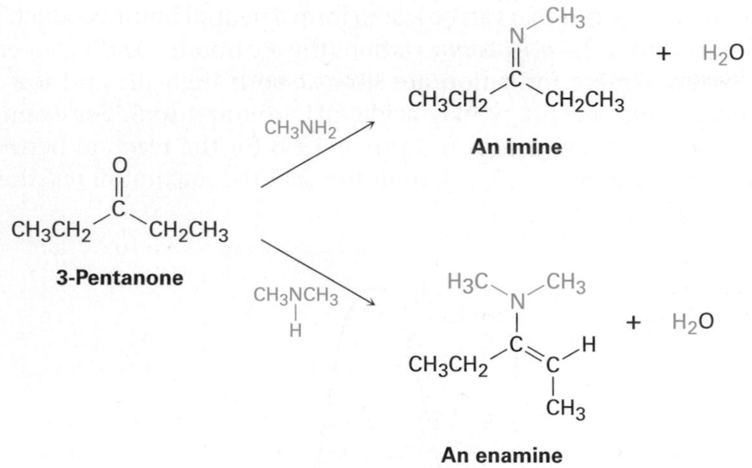

3-Pentanone ($CH_3CH_2C(=O)CH_2CH_3$) $\xrightarrow{CH_3NH_2}$ $CH_3CH_2C(=NCH_3)CH_2CH_3$ + H_2O

An imine

3-Pentanone $\xrightarrow{CH_3NHCH_3}$ $CH_3CH_2C(N(CH_3)_2)=CHCH_3$ + H_2O

3-Pentanone

An enamine

Problem 19.10

Show the products you would obtain by acid-catalyzed reaction of cyclohexanone with ethylamine, $CH_3CH_2NH_2$, and with diethylamine, $(CH_3CH_2)_2NH$.

Problem 19.11

Imine formation is reversible. Show all the steps involved in the acid-catalyzed reaction of an imine with water (hydrolysis) to yield an aldehyde or ketone plus primary amine.

Problem 19.12

Draw the following molecule as a skeletal structure, and show how it can be prepared from a ketone and an amine.

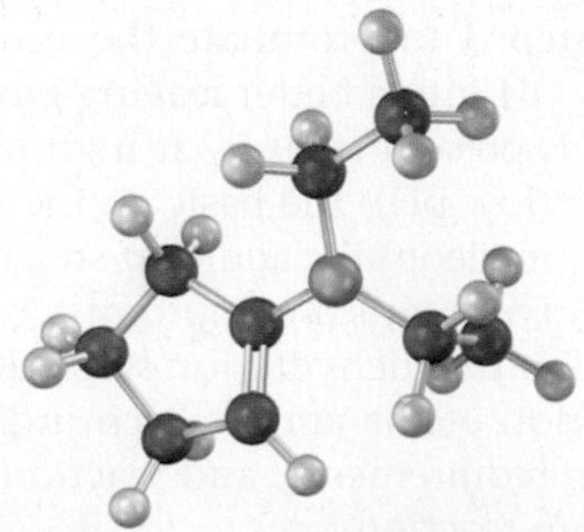

19.9 Nucleophilic Addition of Hydrazine: The Wolff–Kishner Reaction

A useful variant of the imine-forming reaction just discussed involves the treatment of an aldehyde or ketone with hydrazine, H_2NNH_2, in the presence of KOH. Called the **Wolff–Kishner reaction**, the process is a useful and general method for converting an aldehyde or ketone into an alkane, $R_2C{=}O \rightarrow R_2CH_2$.

Propiophenone $\xrightarrow[KOH]{H_2NNH_2}$ Propylbenzene (82%) + N_2 + H_2O

Propiophenone **Propylbenzene (82%)**

Cyclopropanecarbaldehyde $\xrightarrow[KOH]{H_2NNH_2}$ Methylcyclopropane (72%) + N_2 + H_2O

Cyclopropane-carbaldehyde **Methylcyclo-propane (72%)**

As shown in **Figure 19.9**, the Wolff–Kishner reaction involves formation of a *hydrazone* intermediate, $R_2C{=}NNH_2$, followed by base-catalyzed double-bond migration, loss of N_2 gas to give a carbanion, and protonation to give the alkane product. The double-bond migration takes place when base removes one of the weakly acidic NH protons in step 2 to generate a hydrazone anion, which has an allylic resonance structure that places the double bond between nitrogens and the negative charge on carbon. Reprotonation then occurs on carbon to generate the double-bond rearrangement product. The next step—loss of nitrogen and formation of an alkyl anion—is driven by the large thermodynamic stability of the N_2 molecule.

Note that the Wolff–Kishner reduction accomplishes the same overall transformation as the catalytic hydrogenation of an acylbenzene to yield an alkylbenzene **(Section 16.10)**. The Wolff–Kishner reduction is more general and more useful than catalytic hydrogenation, however, because it works well with both alkyl and aryl ketones.

Problem 19.13

Show how you could prepare the following compounds from 4-methyl-3-penten-2-one, $(CH_3)_2C{=}CHCOCH_3$.

(a) $CH_3CH(CH_3)CH_2C(=O)CH_3$ **(b)** $CH_3C(CH_3){=}CHCH_2CH_3$ **(c)** $CH_3CH(CH_3)CH_2CH_2CH_3$

Figure 19.9 | MECHANISM

Mechanism of the Wolff–Kishner reduction of an aldehyde or ketone to yield an alkane.

1 Reaction of the aldehyde or ketone with hydrazine yields a hydrazone in the normal way.

2 Base abstracts a weakly acidic N–H proton, yielding a hydrazone anion. This anion has a resonance form that places the negative charge on carbon and the double bond between nitrogens.

3 Protonation of the hydrazone anion takes place on carbon to yield a neutral intermediate.

4 Deprotonation of the remaining weakly acidic N–H occurs with simultaneous loss of nitrogen to give a carbanion . . .

5 . . . which is protonated to give the alkane product.

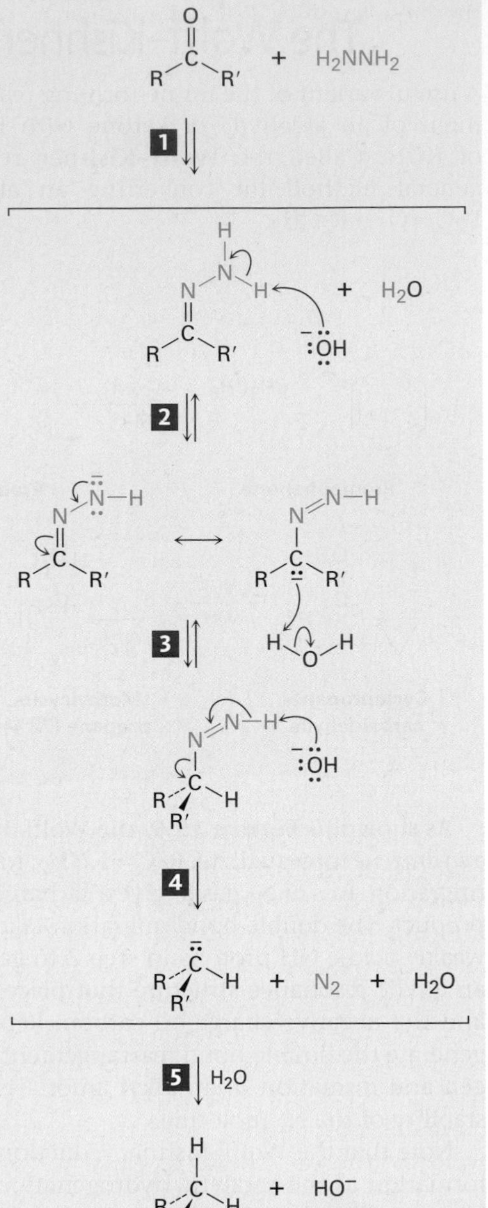

© John McMurry

19.10 Nucleophilic Addition of Alcohols: Acetal Formation

Aldehydes and ketones react reversibly with 2 equivalents of an alcohol in the presence of an acid catalyst to yield **acetals**, **$R_2C(OR')_2$**, frequently called *ketals* if derived from a ketone. Cyclohexanone, for instance, reacts with methanol in the presence of HCl to give the corresponding dimethyl acetal.

$$\text{Cyclohexanone} \xrightarrow[\text{HCl catalyst}]{2\ CH_3OH} \text{Cyclohexanone dimethyl acetal} + H_2O$$

Cyclohexanone — Cyclohexanone dimethyl acetal

Acetal formation is similar to the hydration reaction discussed in **Section 19.5**. Like water, alcohols are weak nucleophiles that add to aldehydes and ketones only slowly under neutral conditions. Under acidic conditions, however, the reactivity of the carbonyl group is increased by protonation, so addition of an alcohol occurs rapidly.

A neutral carbonyl group is moderately electrophilic because of the polarity of the C–O bond.

A protonated carbonyl group is strongly electrophilic because of the positive charge on carbon.

As shown in **Figure 19.10**, nucleophilic addition of an alcohol to the carbonyl group initially yields a hydroxy ether called a **hemiacetal**, analogous to the gem diol formed by addition of water. Hemiacetals are formed reversibly, with the equilibrium normally favoring the carbonyl compound. In the presence of acid, however, a further reaction occurs. Protonation of the –OH group, followed by an E1-like loss of water, leads to an oxonium ion, $R_2C{=}OR^+$, which undergoes a second nucleophilic addition of alcohol to yield the protonated acetal. Loss of a proton completes the reaction.

Because all the steps in acetal formation are reversible, the reaction can be driven either forward (from carbonyl compound to acetal) or backward (from acetal to carbonyl compound), depending on the conditions. The forward reaction is favored by conditions that remove water from the medium and thus drive the equilibrium to the right. In practice, this is often done by distilling off water as it forms. The reverse reaction is favored by treating the acetal with a large excess of aqueous acid to drive the equilibrium to the left.

Acetals are useful because they can act as protecting groups for aldehydes and ketones in the same way that trimethylsilyl ethers act as protecting groups for alcohols **(Section 17.8)**. As we saw previously, it sometimes happens that one functional group interferes with intended chemistry elsewhere in a complex molecule. For example, if we wanted to reduce only the ester group of ethyl 4-oxopentanoate, the ketone would interfere. Treatment of the starting keto ester with $LiAlH_4$ would reduce both the keto and the ester groups to give a diol product.

$$CH_3\overset{O}{\overset{\|}{C}}CH_2CH_2\overset{O}{\overset{\|}{C}}OCH_2CH_3 \xrightarrow{?} CH_3\overset{O}{\overset{\|}{C}}CH_2CH_2CH_2OH$$

Ethyl 4-oxopentanoate — 5-Hydroxy-2-pentanone

Figure 19.10 | MECHANISM

Mechanism of acid-catalyzed acetal formation by reaction of an aldehyde or ketone with an alcohol.

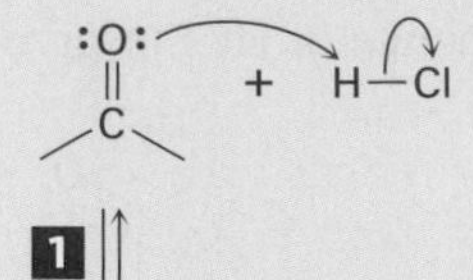

1 Protonation of the carbonyl oxygen strongly polarizes the carbonyl group and . . .

2 . . . activates the carbonyl group for nucleophilic attack by oxygen lone-pair electrons from the alcohol.

3 Loss of a proton yields a neutral hemiacetal tetrahedral intermediate.

Hemiacetal

4 Protonation of the hemiacetal hydroxyl converts it into a good leaving group.

5 Dehydration yields an intermediate oxonium ion.

6 Addition of a second equivalent of alcohol gives a protonated acetal.

7 Loss of a proton yields the neutral acetal product.

Acetal

By protecting the keto group as an acetal, however, the problem can be circumvented. Like other ethers, acetals are unreactive to bases, hydride reducing agents, Grignard reagents, and catalytic hydrogenation conditions, but they are cleaved by acid. Thus, we can accomplish the selective reduction of the ester group in ethyl 4-oxopentanoate by first converting the keto group to an acetal, then reducing the ester with $LiAlH_4$, and then removing the acetal by treatment with aqueous acid. (In practice, it's often convenient to use 1 equivalent of a diol such as ethylene glycol as the alcohol and to form a *cyclic* acetal. The mechanism of cyclic acetal formation using 1 equivalent of ethylene glycol is exactly the same as that using 2 equivalents of methanol or other monoalcohol.)

OCH_2CH_3 — $HOCH_2CH_2OH$, Acid catalyst → OCH_2CH_3 + H_2O

Ethyl 4-oxopentanoate

Can't be done directly

1. $LiAlH_4$
2. H_3O^+

$HOCH_2CH_2OH$ + CH_2OH ← H_3O^+ — CH_2OH + CH_3CH_2OH

5-Hydroxy-2-pentanone

Acetal and hemiacetal groups are particularly common in carbohydrate chemistry. Glucose, for instance, is a polyhydroxy aldehyde that undergoes an *internal* nucleophilic addition reaction and exists primarily as a cyclic hemiacetal.

HÖ: H H OH O HOCH₂ C H H OH H OH ⇌ CH₂OH HO HO OH OH

Glucose—open chain **Glucose—cyclic hemiacetal**

Predicting the Product of Reaction between a Ketone and an Alcohol

Worked Example 19.2

Show the structure of the acetal you would obtain by acid-catalyzed reaction of 2-pentanone with 1,3-propanediol.

Strategy

Acid-catalyzed reaction of an aldehyde or ketone with 2 equivalents of a monoalcohol or 1 equivalent of a diol yields an acetal, in which the carbonyl oxygen atom is replaced by two —OR groups from the alcohol.

Solution

$$CH_3CH_2CH_2C(=O)CH_3 \xrightarrow[\text{H}^+ \text{ catalyst}]{HOCH_2CH_2CH_2OH} \text{cyclic acetal} + H_2O$$

2-Pentanone

Problem 19.14

Show all the steps in the acid-catalyzed formation of a cyclic acetal from ethylene glycol and an aldehyde or ketone.

Problem 19.15

Identify the carbonyl compound and the alcohol that were used to prepare the following acetal:

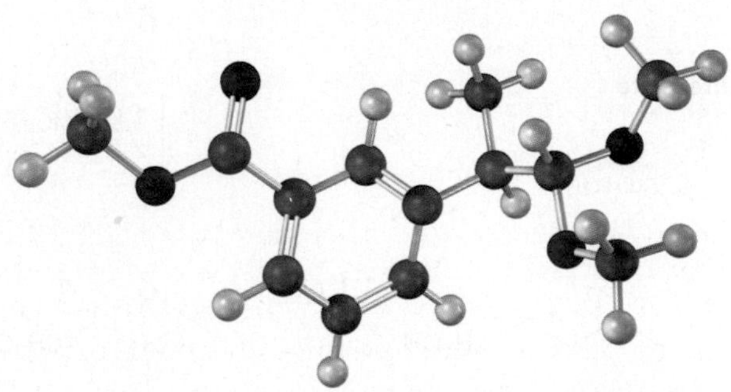

19.11 Nucleophilic Addition of Phosphorus Ylides: The Wittig Reaction

Aldehydes and ketones are converted into alkenes by means of a nucleophilic addition called the **Wittig reaction**. The reaction has no direct biological counterpart but is important both because of its wide use in the laboratory and drug manufacture and because of its mechanistic similarity to reactions of the coenzyme thiamin diphosphate, which we'll see in **Section 29.6**.

In the Wittig reaction, a triphenylphosphorus *ylide,* $R_2\overset{-}{C}-\overset{+}{P}Ph_3$, also called a *phosphorane* and sometimes written in the resonance form $R_2C{=}PPh_3$, adds to an aldehyde or ketone to yield a four-membered cyclic intermediate called an *oxaphosphetane.* The oxaphosphetane is not isolated, but instead spontaneously decomposes to give an alkene plus triphenylphosphine oxide, $O{=}PPh_3$. In effect, the oxygen atom of the aldehyde or ketone and the $R_2C{=}$ bonded to phosphorus exchange places. (An **ylide**—pronounced **ill**-id—is a neutral, dipolar compound with adjacent plus and minus charges.)

$$\underset{\text{Aldehyde/ketone}}{>C{=}O} + \underset{\text{Ylide}}{CHR{=}PPh_3} \rightleftharpoons \underset{\text{Oxaphosphetane}}{\left[\begin{array}{c} O-PPh_3 \\ | \quad | \\ -C-C(R)(H) \end{array}\right]} \longrightarrow \underset{\text{Alkene}}{>C{=}C(R)(H)} + \underset{\text{Triphenylphosphine oxide}}{O{=}PPh_3}$$

The initial addition step appears to take place by different pathways depending on the structure of the reactants and the exact experimental conditions. One pathway involves a one-step cycloaddition process analogous to the Diels–Alder cycloaddition reaction **(Section 14.4)**. The other pathway involves a nucleophilic addition reaction to give a dipolar intermediate called a *betaine* (**bay**-ta-een), which undergoes ring closure.

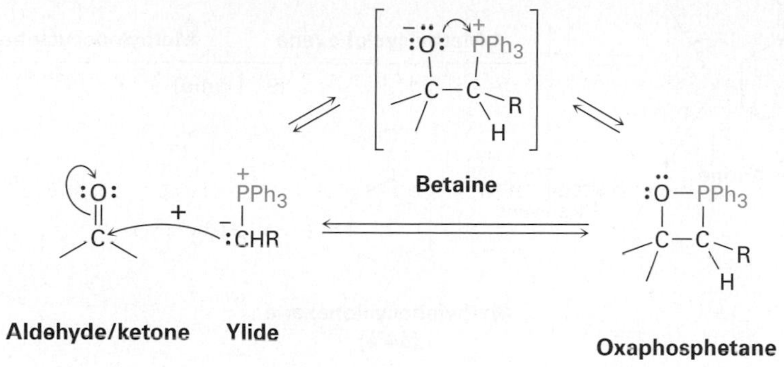

Betaine

Aldehyde/ketone Ylide

Oxaphosphetane

The phosphorus ylides necessary for Wittig reaction are easily prepared by S_N2 reaction of primary (and some secondary) alkyl halides with triphenylphosphine, $(Ph)_3P$, followed by treatment with base. Triphenylphosphine is a good nucleophile in S_N2 reactions, and yields of the resultant alkyltriphenylphosphonium salts are high. Because of the positive charge on phosphorus, the hydrogen on the neighboring carbon is weakly acidic and can be removed by a strong base such as butyllithium (BuLi) to generate the neutral ylide. For example:

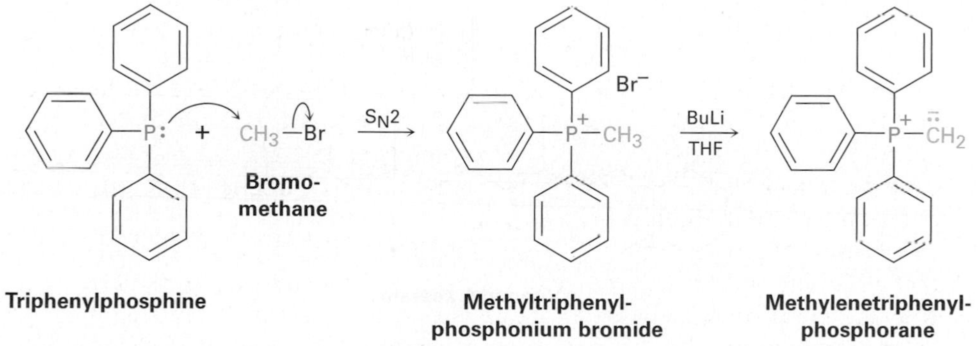

Triphenylphosphine + CH_3—Br (Bromomethane) $\xrightarrow{S_N2}$ Methyltriphenylphosphonium bromide $\xrightarrow[THF]{BuLi}$ Methylenetriphenylphosphorane

The Wittig reaction is extremely general, and a great many monosubstituted, disubstituted, and trisubstituted alkenes can be prepared from the appropriate combination of phosphorane and aldehyde or ketone. Tetrasubstituted alkenes can't be prepared, however, because of steric hindrance during the reaction.

The real value of the Wittig reaction is that it yields a pure alkene of predictable structure. The C=C bond in the product is always exactly where the C=O group was in the reactant, and no alkene isomers (except E,Z isomers) are formed. For example, Wittig reaction of cyclohexanone with methylenetriphenylphosphorane yields only the single alkene product methylenecyclohexane. By contrast, addition of methylmagnesium bromide to cyclohexanone, followed by dehydration with $POCl_3$, yields a roughly 9:1 mixture of two alkenes.

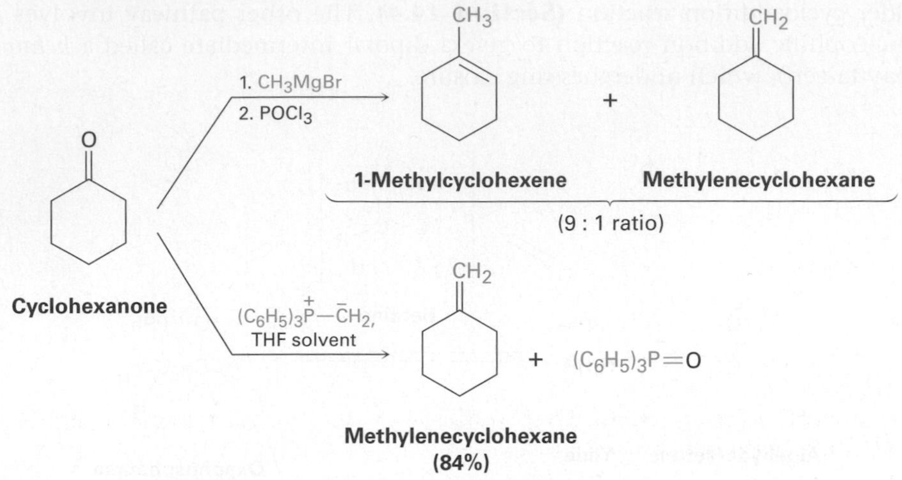

Cyclohexanone

1. CH_3MgBr
2. $POCl_3$

1-Methylcyclohexene + Methylenecyclohexane

(9 : 1 ratio)

$(C_6H_5)_3\overset{+}{P}—\overset{-}{C}H_2$, THF solvent

Methylenecyclohexane (84%) + $(C_6H_5)_3P=O$

Wittig reactions are used commercially in the synthesis of numerous pharmaceutical agents. For example, the German chemical company BASF prepares vitamin A by Wittig reaction between a 15-carbon ylide and a 5-carbon aldehyde.

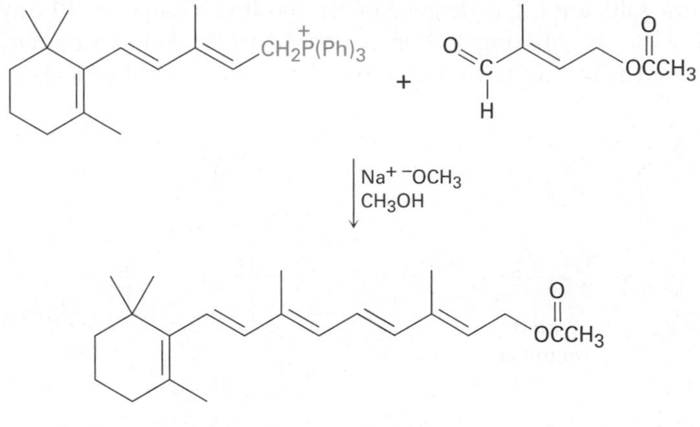

$CH_2\overset{+}{P}(Ph)_3$ + $OCH–C(H)=...–OCCH_3$

$Na^+ \ ^-OCH_3$, CH_3OH

Vitamin A acetate

Worked Example 19.3 Synthesizing an Alkene Using a Wittig Reaction

What carbonyl compound and what phosphorus ylide might you use to prepare 3-ethyl-2-pentene?

Strategy

An aldehyde or ketone reacts with a phosphorus ylide to yield an alkene in which the oxygen atom of the carbonyl reactant is replaced by the $=CR_2$ of the ylide. Preparation of the phosphorus ylide itself usually involves S_N2 reaction of a primary alkyl halide with triphenylphosphine, so the ylide is typically primary, $RCH{=}P(Ph)_3$. This means that the disubstituted alkene carbon in the product comes from the carbonyl reactant, while the monosubstituted alkene carbon comes from the ylide.

Solution

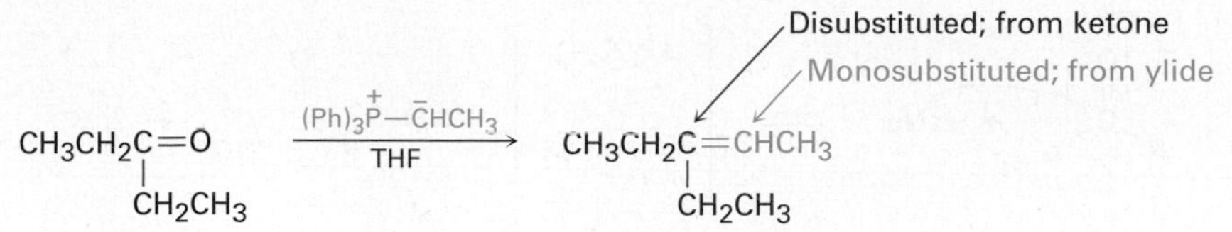

$$\underset{\textbf{3-Pentanone}}{CH_3CH_2\underset{\displaystyle CH_2CH_3}{\underset{|}{C}}{=}O} \xrightarrow[\text{THF}]{(Ph)_3\overset{+}{P}{-}\overset{-}{C}HCH_3} \underset{\textbf{3-Ethyl-2-pentene}}{CH_3CH_2\underset{\displaystyle CH_2CH_3}{\underset{|}{C}}{=}CHCH_3}$$

Disubstituted; from ketone
Monosubstituted; from ylide

Problem 19.16
What carbonyl compound and what phosphorus ylide might you use to prepare each of the following compounds?

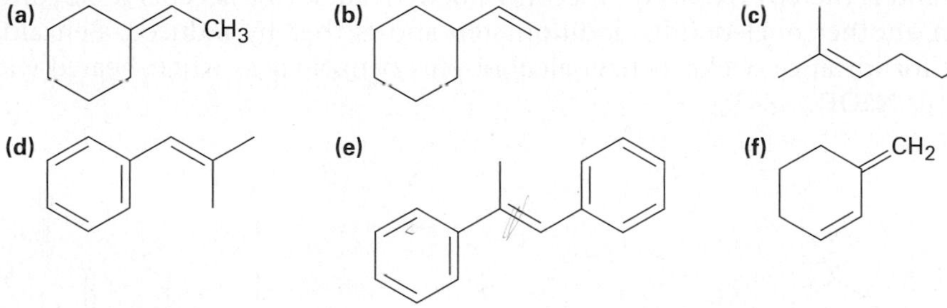

(a) CH_3 (b) (c)

(d) (e) (f) CH_2

Problem 19.17
β-Carotene, a yellow food-coloring agent and dietary source of vitamin A can be prepared by a *double* Wittig reaction between 2 equivalents of β-ionylideneacetaldehyde and a *diylide*. Show the structure of the β-carotene product.

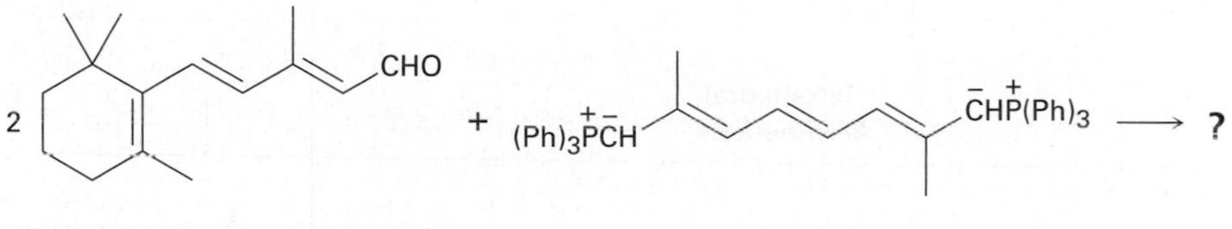

2 [CHO] + $(Ph)_3\overset{+}{P}\overset{-}{C}H$ ... $\overset{-}{C}H\overset{+}{P}(Ph)_3$ ⟶ ?

β-Ionylideneacetaldehyde **A diylide**

19.12 Biological Reductions

As a general rule, nucleophilic addition reactions are characteristic only of aldehydes and ketones, not of carboxylic acid derivatives. The reason for the difference is structural. As discussed previously in the *Preview of Carbonyl Compounds* and shown in **Figure 19.11**, the tetrahedral intermediate produced by addition of a nucleophile to a carboxylic acid derivative can eliminate a leaving group, leading to a net nucleophilic acyl substitution reaction.

The tetrahedral intermediate produced by addition of a nucleophile to an aldehyde or ketone, however, has only alkyl or hydrogen substituents and thus can't usually expel a leaving group. One exception to this rule, however, is the **Cannizzaro reaction**, discovered in 1853.

Figure 19.11 Carboxylic acid derivatives have an electronegative substituent Y = —Br, —Cl, —OR, —NR_2 that can be expelled as a leaving group from the tetrahedral intermediate formed by nucleophilic addition. Aldehydes and ketones have no such leaving group and thus do not usually undergo this reaction.

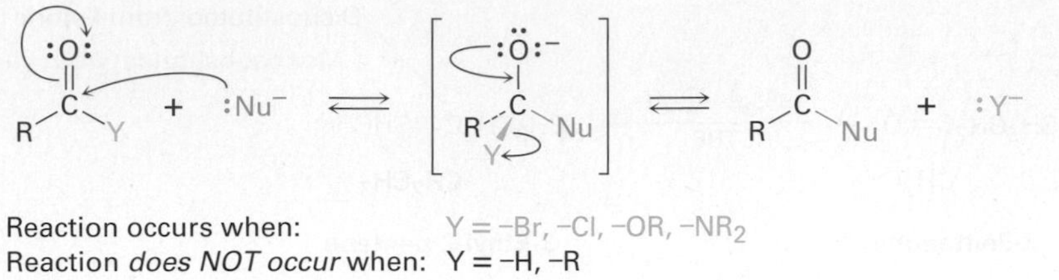

Reaction occurs when: Y = –Br, –Cl, –OR, –NR_2

Reaction *does NOT occur* when: Y = –H, –R

The Cannizzaro reaction takes place by nucleophilic addition of OH^- to an aldehyde to give a tetrahedral intermediate, *which expels hydride ion as a leaving group* and is thereby oxidized. A second aldehyde molecule accepts the hydride ion in another nucleophilic addition step and is thereby reduced. Benzaldehyde, for instance, yields benzyl alcohol plus benzoic acid when heated with aqueous NaOH.

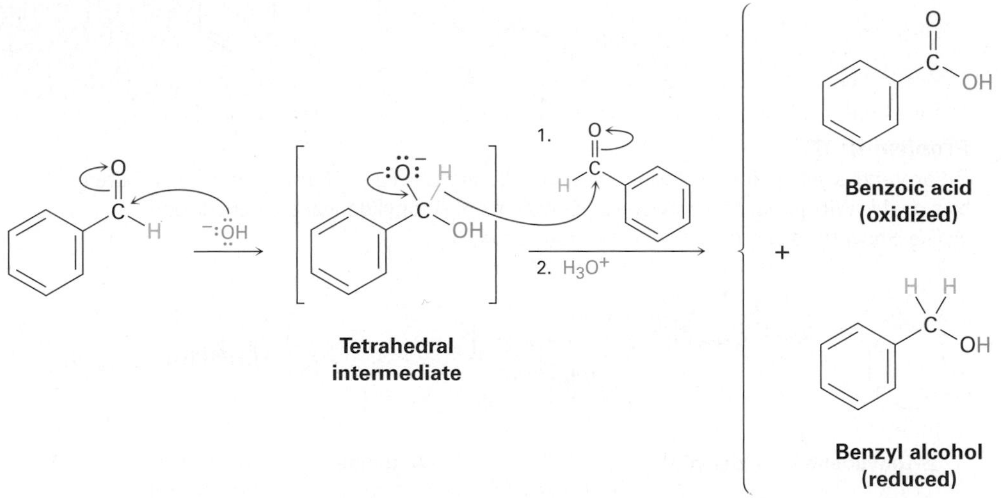

Tetrahedral intermediate

Benzoic acid (oxidized)

Benzyl alcohol (reduced)

The Cannizzaro reaction is little used today but is interesting mechanistically because it is a simple laboratory analogy for the primary biological pathway by which carbonyl reductions occur in living organisms. In nature, as we saw in **Section 17.4**, one of the most important reducing agents is NADH, reduced nicotinamide adenine dinucleotide. NADH donates H^- to aldehydes and ketones, thereby reducing them, in much the same way that the tetrahedral alkoxide intermediate in a Cannizzaro reaction does. The electron lone pair on a nitrogen atom of NADH expels H^- as leaving group, which adds to a carbonyl group in another molecule **(Figure 19.12)**. As an example, pyruvate is converted during intense muscle activity to (*S*)-lactate, a reaction catalyzed by lactate dehydrogenase.

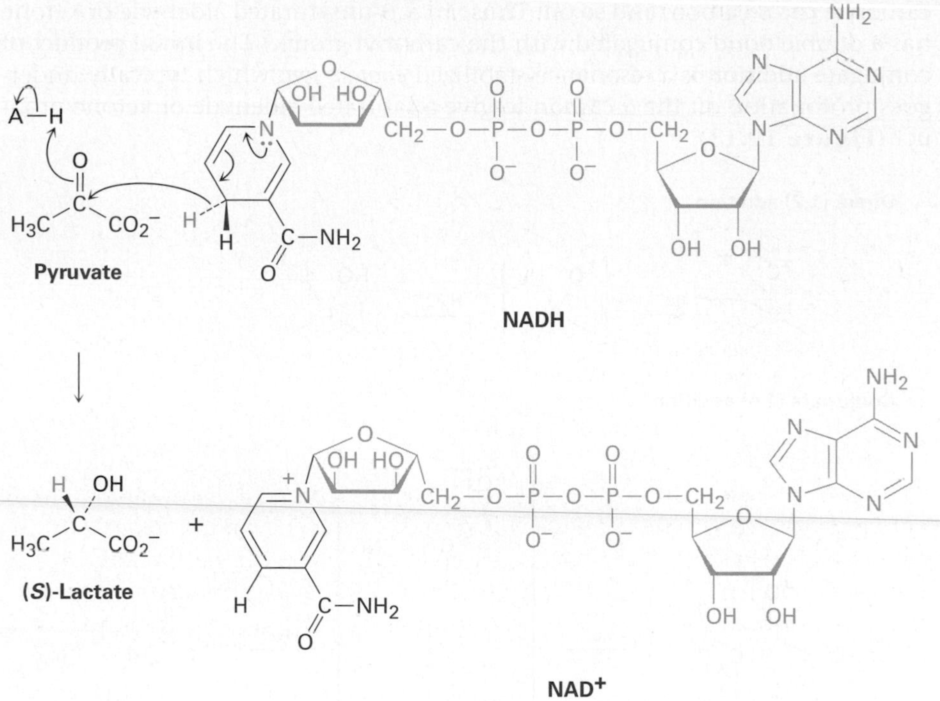

Pyruvate

NADH

(*S*)-Lactate

NAD+

Figure 19.12 Mechanism of biological aldehyde and ketone reductions by the coenzyme NADH. The key step is an expulsion of hydride ion from NADH and donation to the carbonyl group.

Problem 19.18

When *o*-phthalaldehyde is treated with base, *o*-(hydroxymethyl)benzoic acid is formed. Show the mechanism of this reaction.

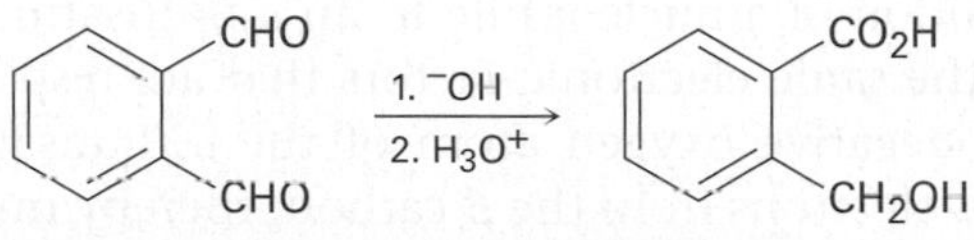

1. ^{-}OH
2. H_3O^+

o-Phthalaldehyde **o-(Hydroxymethyl)benzoic acid**

Problem 19.19

What is the stereochemistry of the pyruvate reduction shown in Figure 19.12? Does NADH lose its *pro-R* or *pro-S* hydrogen? Does addition occur to the *Si* face or *Re* face of pyruvate? (Review **Section 5.11**.)

19.13 Conjugate Nucleophilic Addition to α,β-Unsaturated Aldehydes and Ketones

All the reactions we've been discussing to this point have involved the addition of a nucleophile directly to the carbonyl group, a so-called **1,2 addition.** Closely related to this direct addition is the **conjugate addition**, or **1,4 addition**, of a nucleophile to the C=C bond of an α,β-unsaturated aldehyde or ketone. (The carbon atom next to a carbonyl group is often called the α carbon, the next carbon is the β carbon, and so on. Thus, an α,β-unsaturated aldehyde or ketone has a double bond conjugated with the carbonyl group.) The initial product of conjugate addition is a resonance-stabilized *enolate ion,* which typically undergoes protonation on the α carbon to give a saturated aldehyde or ketone product (**Figure 19.13**).

Figure 19.13 A comparison of direct (1,2) and conjugate (1,4) nucleophilic addition reactions. In the conjugate addition, a nucleophile adds to the β carbon of an α,β-unsaturated aldehyde or ketone and protonation occurs on the α carbon.

Direct (1,2) addition

Conjugate (1,4) addition

α,β-Unsaturated aldehyde/ketone

Enolate ion

Saturated aldehyde/ketone

The conjugate addition of a nucleophile to an α,β-unsaturated aldehyde or ketone is caused by the same electronic factors that are responsible for direct addition. The electronegative oxygen atom of the α,β-unsaturated carbonyl compound withdraws electrons from the β carbon, thereby making it electron-poor and more electrophilic than a typical alkene carbon.

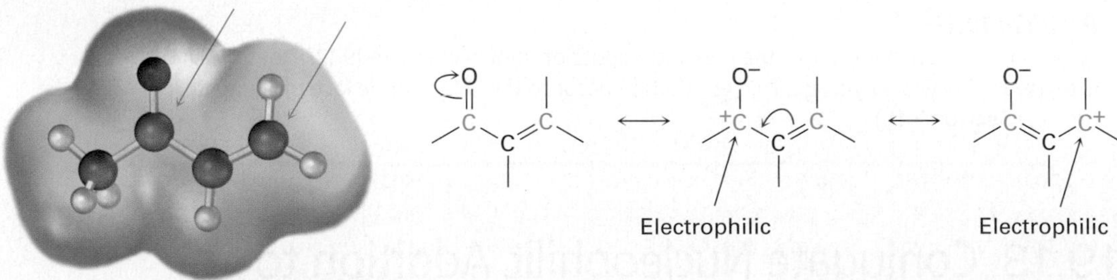

As noted, conjugate addition of a nucleophile to the β carbon of an α,β-unsaturated aldehyde or ketone leads to an enolate ion intermediate, which is protonated on the α carbon to give the saturated product (Figure 19.13). The net effect is addition of the nucleophile to the C=C bond, with the carbonyl group itself unchanged. In fact, of course, the carbonyl group is crucial to the success of the reaction. The C=C bond would not be activated for addition, and no reaction would occur, without the carbonyl group.

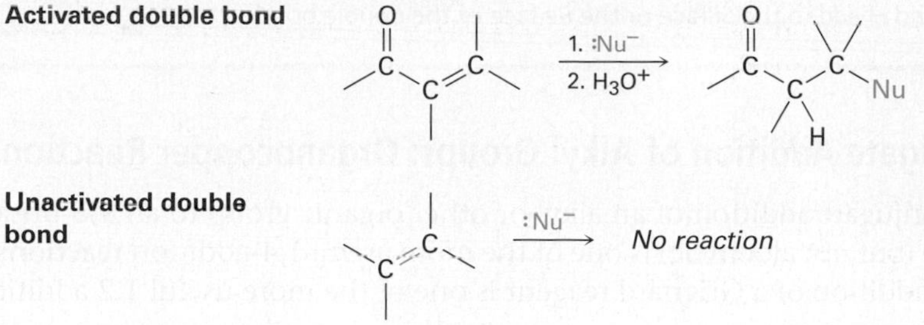

Activated double bond

1. :Nu⁻
2. H_3O^+

Unactivated double bond

:Nu⁻ → *No reaction*

Conjugate Addition of Amines

Both primary and secondary amines add to α,β-unsaturated aldehydes and ketones to yield β-amino aldehydes and ketones rather than the alternative imines. Under typical reaction conditions, both modes of addition occur rapidly. But because the reactions are reversible, they generally proceed with thermodynamic control rather than kinetic control **(Section 14.3)**, so the more stable conjugate addition product is often obtained to the complete exclusion of the less stable direct addition product.

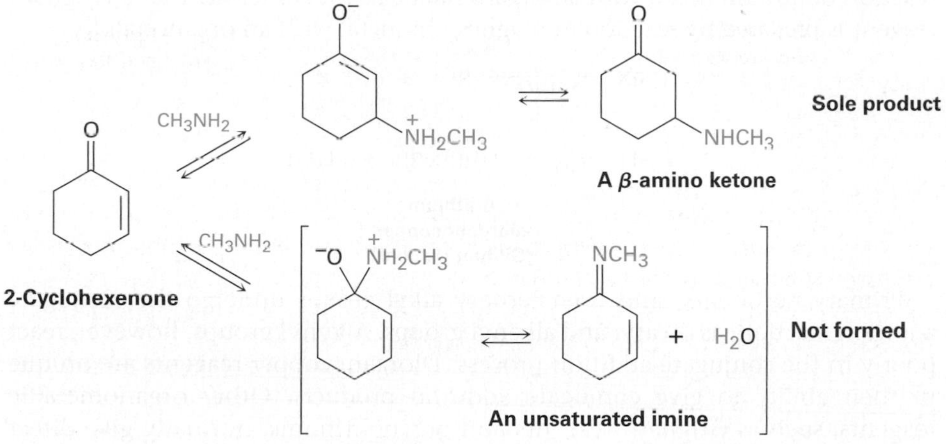

CH_3NH_2 — $\overset{+}{N}H_2CH_3$ — NHCH$_3$ — **Sole product**

A β-amino ketone

2-Cyclohexenone

CH_3NH_2 — $^-$O $\overset{+}{N}H_2CH_3$ — NCH_3 + H_2O — **Not formed**

An unsaturated imine

Conjugate Addition of Water

Water can add reversibly to α,β-unsaturated aldehydes and ketones to yield β-hydroxy aldehydes and ketones, although the position of the equilibrium generally favors unsaturated reactant rather than saturated adduct. Related additions to α,β-unsaturated carboxylic acids occur in numerous biological pathways, such as the citric acid cycle of food metabolism in which *cis*-aconitate is converted into isocitrate by conjugate addition of water to a double bond.

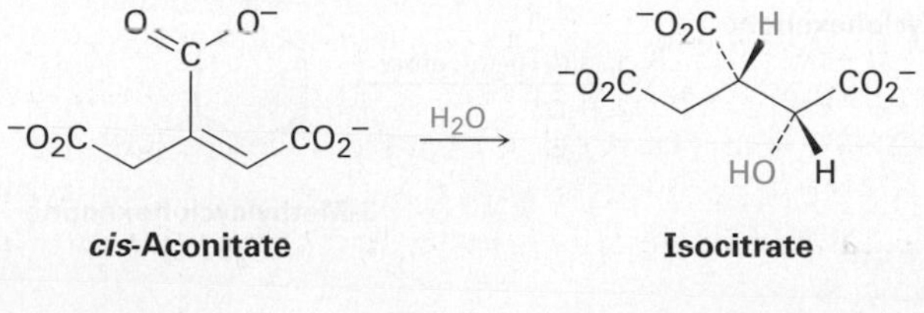

H_2O →

***cis*-Aconitate** **Isocitrate**

Problem 19.20

Assign *R* or *S* stereochemistry to the two chirality centers in isocitrate, and tell whether OH and H add to the *Si* face or the *Re* face of the double bond.

Conjugate Addition of Alkyl Groups: Organocopper Reactions

The conjugate addition of an alkyl or other organic group to an α,β-unsaturated ketone (but not aldehyde) is one of the more useful 1,4-addition reactions, just as direct addition of a Grignard reagent is one of the more useful 1,2 additions.

1. " :R⁻ "
2. H_3O^+

α,β-Unsaturated ketone

Conjugate addition of an organic group is carried out by treating the α,β-unsaturated ketone with a lithium diorganocopper reagent, R_2CuLi. As we saw in **Section 10.7**, lithium diorganocopper (Gilman) reagents are prepared by reaction between 1 equivalent of copper(I) iodide and 2 equivalents of an organolithium regent, RLi. The organolithium reagent, in turn, is formed by reaction of lithium metal with an organohalide in the same way that a Grignard reagent is prepared by reaction of magnesium metal with an organohalide.

$$RX \xrightarrow[\text{Pentane}]{\text{2 Li}} RLi + Li^+ X^-$$

$$2\ RLi \xrightarrow[\text{Ether}]{\text{CuI}} Li^+(R\bar{Cu}R) + Li^+ I^-$$

A lithium diorganocopper (Gilman reagent)

Primary, secondary, and even tertiary alkyl groups undergo the conjugate addition reaction, as do aryl and alkenyl groups. Alkynyl groups, however, react poorly in the conjugate addition process. Diorganocopper reagents are unique in their ability to give conjugate addition products. Other organometallic reagents, such as Grignard reagents and organolithiums, normally give direct carbonyl addition on reaction with α,β-unsaturated ketones.

1. CH_3MgBr, ether or CH_3Li
2. H_3O^+

1-Methyl-2-cyclohexen-1-ol (95%)

2-Cyclohexenone

1. $Li(CH_3)_2Cu$, ether
2. H_3O^+

3-Methylcyclohexanone (97%)

The mechanism of the reaction is thought to involve conjugate nucleophilic addition of the diorganocopper anion, R_2Cu^-, to the unsaturated ketone to give a copper-containing intermediate. Transfer of an R group from copper to carbon, followed by elimination of a neutral organocopper species, RCu, gives the final product.

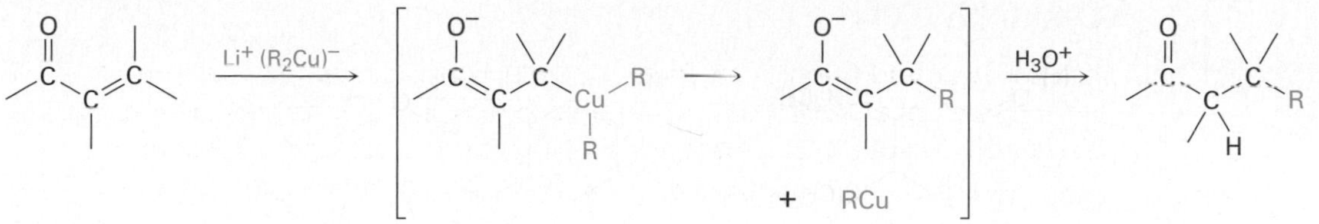

Worked Example 19.4 Synthesis Using a Conjugate Addition Reaction

How might you use a conjugate addition reaction to prepare 2-methyl-3-propylcyclopentanone?

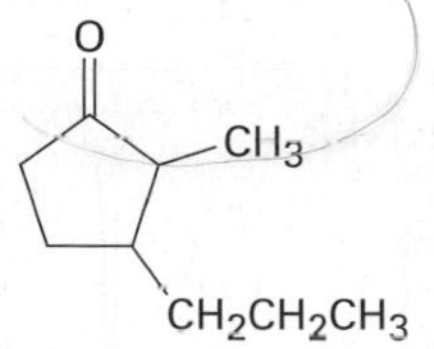

2-Methyl-3-propylcyclopentanone

Strategy

A ketone with a substituent group in its β position might be prepared by a conjugate addition of that group to an α,β-unsaturated ketone. In the present instance, the target molecule has a propyl substituent on the β carbon and might therefore be prepared from 2-methyl-2-cyclopentenone by reaction with lithium dipropylcopper.

Solution

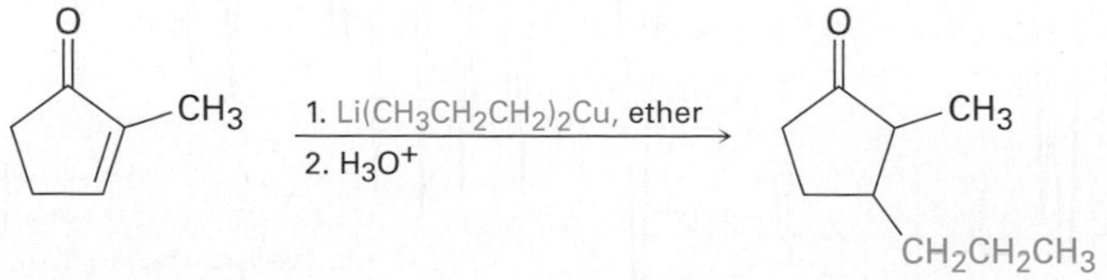

2-Methyl-2-cyclopentenone **2-Methyl-3-propylcyclopentanone**

Problem 19.21

Treatment of 2-cyclohexenone with HCN/KCN yields a saturated keto nitrile rather than an unsaturated cyanohydrin. Show the structure of the product, and propose a mechanism for the reaction.

Problem 19.22

How might conjugate addition reactions of lithium diorganocopper reagents be used to synthesize the following compounds?

(a) $CH_3CH_2CH_2CH_2CH_2\overset{O}{\overset{\|}{C}}CH_3$

(b) [3,3-dimethylcyclohexanone: cyclohexanone ring with CH_3 and CH_3 on the same carbon]

(c) [cyclohexanone ring bearing CH_2CH_3 and $C(CH_3)_3$ substituents]

(d) [bicyclic ketone (decalone) bearing $CH{=}CH_2$ at the ring-fusion carbon]

19.14 Spectroscopy of Aldehydes and Ketones

Infrared Spectroscopy

Aldehydes and ketones show a strong C=O bond absorption in the IR region from 1660 to 1770 cm^{-1}, as the spectra of benzaldehyde and cyclohexanone demonstrate **(Figure 19.14)**. In addition, aldehydes show two characteristic C–H absorptions in the range 2720 to 2820 cm^{-1}.

(a)

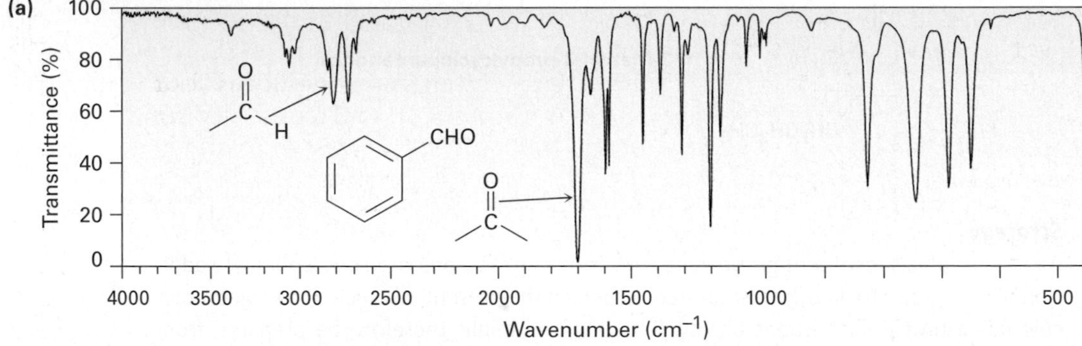

(b)

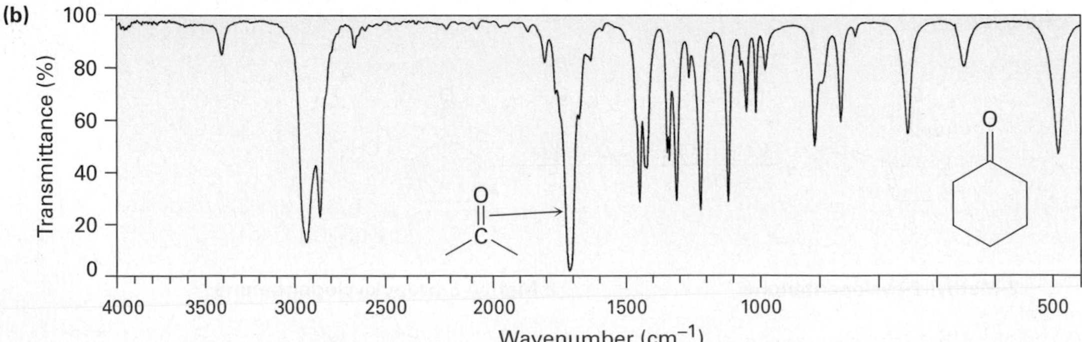

Figure 19.14 Infrared spectra of **(a)** benzaldehyde and **(b)** cyclohexanone.

The exact position of the C=O absorption is diagnostic of the nature of the carbonyl group. As the data in Table 19.2 indicate, saturated aldehydes usually

Table 19.2 Infrared Absorptions of Some Aldehydes and Ketones

Carbonyl type	Example	Absorption (cm^{-1})
Saturated aldehyde	CH_3CHO	1730
Aromatic aldehyde	PhCHO	1705
α,β-Unsaturated aldehyde	$H_2C{=}CHCHO$	1705
Saturated ketone	CH_3COCH_3	1715
Cyclohexanone		1715
Cyclopentanone		1750
Cyclobutanone		1785
Aromatic ketone	$PhCOCH_3$	1690
α,β-Unsaturated ketone	$H_2C{=}CHCOCH_3$	1685

show carbonyl absorptions near 1730 cm^{-1} in the IR spectrum, but conjugation of the aldehyde to an aromatic ring or a double bond lowers the absorption by 25 cm^{-1} to near 1705 cm^{-1}. Saturated aliphatic ketones and cyclohexanones both absorb near 1715 cm^{-1}, and conjugation with a double bond or an aromatic ring again lowers the absorption by 30 cm^{-1} to 1685 to 1690 cm^{-1}. Angle strain in the carbonyl group caused by reducing the ring size of cyclic ketones to four or five raises the absorption position.

The values given in Table 19.2 are remarkably constant from one aldehyde or ketone to another. As a result, IR spectroscopy is a powerful tool for identifying the kind of a carbonyl group in a molecule of unknown structure. An unknown that shows an IR absorption at 1730 cm^{-1} is almost certainly an aldehyde rather than a ketone; an unknown that shows an IR absorption at 1750 cm^{-1} is almost certainly a cyclopentanone, and so on.

Problem 19.23

How might you use IR spectroscopy to determine whether reaction between 2-cyclohexenone and lithium dimethylcopper gives the direct addition product or the conjugate addition product?

Problem 19.24

Where would you expect each of the following compounds to absorb in the IR spectrum?

(a) 4-Penten-2-one
(b) 3-Penten-2-one
(c) 2,2-Dimethylcyclopentanone
(d) *m*-Chlorobenzaldehyde
(e) 3-Cyclohexenone
(f) 2-Hexenal

Nuclear Magnetic Resonance Spectroscopy

Aldehyde protons (RCHO) absorb near 10 δ in the ^1H NMR spectrum and are very distinctive because no other absorptions occur in this region. The aldehyde proton shows spin–spin coupling with protons on the neighboring carbon, with coupling constant $J \approx 3$ Hz. Acetaldehyde, for example, shows a quartet at 9.8 δ for the aldehyde proton, indicating that there are three protons neighboring the –CHO group **(Figure 19.15)**.

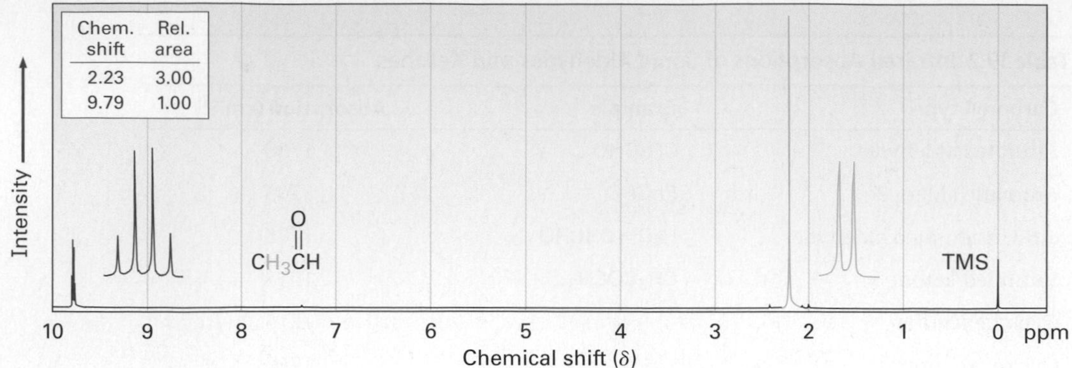

Figure 19.15 ^{1}H NMR spectrum of acetaldehyde. The absorption of the aldehyde proton appears at 9.8 δ and is split into a quartet.

Hydrogens on the carbon next to a carbonyl group are slightly deshielded and normally absorb near 2.0 to 2.3 δ. The acetaldehyde methyl group in Figure 19.15, for instance, absorbs at 2.20 δ. Methyl ketones are particularly distinctive because they always show a sharp three-proton singlet near 2.1 δ.

The carbonyl-group carbon atoms of aldehydes and ketones have characteristic ^{13}C NMR resonances in the range 190 to 215 δ. Since no other kinds of carbons absorb in this range, the presence of an NMR absorption near 200 δ is clear evidence for a carbonyl group. Saturated aldehyde or ketone carbons usually absorb in the region from 200 to 215 δ, while aromatic and α,β-unsaturated carbonyl carbons absorb in the 190 to 200 δ region.

CH_3CHO: 31, 200

Benzaldehyde: 136.5, 192, 134, 130,129

$CH_3CCH_2CH_3$ (O): 29.5, 209, 37, 8

Cyclohexanone: 211, 25, 27, 42

Acetophenone: 137, 198, 26.5, 133, 128.5,128

Mass Spectrometry

Aliphatic aldehydes and ketones that have hydrogens on their gamma (γ) carbon atoms undergo a characteristic mass spectral cleavage called the *McLafferty rearrangement.* A hydrogen atom is transferred from the γ carbon to the carbonyl oxygen, the bond between the α and β carbons is broken, and a neutral alkene fragment is produced. The charge remains with the oxygen-containing fragment.

McLafferty rearrangement

In addition to fragmentation by the McLafferty rearrangement, aldehydes and ketones also undergo cleavage of the bond between the carbonyl group and the α carbon, called an *α cleavage*. Alpha cleavage yields a neutral radical and a resonance-stabilized acyl cation.

$$\left[RC(=O)R' \right]^{+\cdot} \xrightarrow{\text{Alpha cleavage}} R\cdot \; + \; \left[:\!O\!:\!=\!C^{+}\!-\!R' \longleftrightarrow :\!O^{+}\!\equiv\!C\!-\!R' \right]$$

Fragment ions from both McLafferty rearrangement and α cleavage are visible in the mass spectrum of 5-methyl-2-hexanone shown in **Figure 19.16**. McLafferty rearrangement with loss of 2-methylpropene yields a fragment with m/z = 58. Alpha cleavage occurs primarily at the more substituted side of the carbonyl group, leading to a $[CH_3CO]^+$ fragment with m/z = 43.

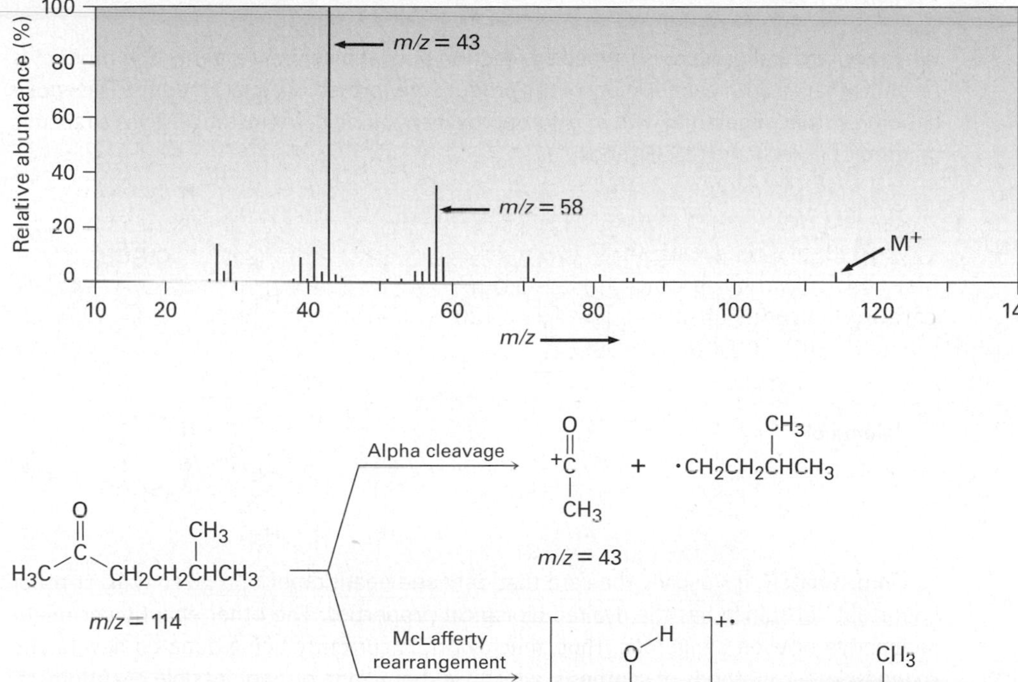

Figure 19.16 Mass spectrum of 5-methyl-2-hexanone. The peak at m/z = 58 is due to McLafferty rearrangement. The abundant peak at m/z = 43 is due to α cleavage at the more highly substituted side of the carbonyl group. Note that the peak due to the molecular ion is very small.

Problem 19.25

How might you use mass spectrometry to distinguish between the following pairs of isomers?

(a) 3-Methyl-2-hexanone and 4-methyl-2-hexanone
(b) 3-Heptanone and 4-heptanone
(c) 2-Methylpentanal and 3-methylpentanal

Problem 19.26

Tell the prominent IR absorptions and mass spectral peaks you would expect for the following compound:

Enantioselective Synthesis

A DEEPER LOOK

Whenever a chiral product is formed by reaction between achiral reagents, the product is racemic; that is, both enantiomers of the product are formed in equal amounts. The epoxidation reaction of geraniol with *m*-chloroperoxybenzoic acid, for instance, gives a racemic mixture of (2*R*,3*S*) and (2*S*,3*R*) epoxides.

Geraniol (CH_2OH) → (Cl–C_6H_4–CO_3H) → (*S*,*R*) epoxide (H_3C, CH_2OH) 50% + (*R*,*S*) epoxide (H_3C, CH_2OH) 50%

Unfortunately, it's usually the case that only one enantiomer of a given drug or other important substance has the desired biological properties. The other enantiomer might be inactive or even dangerous. Thus, much work is currently being done on developing *enantioselective* methods of synthesis, which yield only one of two possible enantiomers. So important has enantioselective synthesis become that the 2001 Nobel Prize in Chemistry was awarded to three pioneers in the field: William S. Knowles, K. Barry Sharpless, and Ryoji Noyori.

Several approaches to enantioselective synthesis have been taken, but the most efficient are those that use chiral catalysts to temporarily hold a substrate molecule in an unsymmetrical environment—the same strategy that nature uses when catalyzing reactions with chiral enzymes. While in that unsymmetrical environment, the substrate may be more open to reaction on one side than on another, leading to an excess of one enantiomeric product over another. As an analogy, think about picking up a coffee mug in your right hand to take a drink. The mug by itself is achiral, but as soon as you pick it up by the handle, it becomes chiral. One side of the mug now faces toward you so you can drink from it, but the other side faces away. The two sides are different, with one side much more accessible to you than the other.

(continued)

(continued)

Among the thousands of enantioselective reactions now known, one of the most useful is the so-called Sharpless epoxidation, in which an allylic alcohol, such as geraniol, is treated with *tert*-butyl hydroperoxide, $(CH_3)_3C$—OOH, in the presence of titanium tetraisopropoxide and diethyl tartrate (DET) as a chiral auxiliary reagent. When the (*R*,*R*) tartrate is used, geraniol is converted into its 2*R*,3*S* epoxide with 98% selectivity, whereas use of the (*S*,*S*) tartrate gives the 2*S*,3*R* epoxide enantiomer. We say that the major product in each case is formed with an *enantiomeric excess* of 96%, meaning that 4% of the product is racemic (2% 2*R*,3*S* plus 2% 2*S*,3*R*) and an extra 96% of a single enantiomer is formed. The mechanistic details by which the chiral catalyst works are a bit complex, although it appears that a chiral complex of two tartrate molecules with one titanium is involved.

A substance made from the tartaric acid found at the bottom of this wine vat catalyzes enantioselective reactions.

(*R*,*R*)-Diethyl tartrate **(*S*,*S*)-Diethyl tartrate**

Geraniol

$(CH_3)_3C$–OOH, $Ti[OCH(CH_3)_2]_4$, (*R*,*R*)-DET → **2*R*,3*S* isomer—98%** + 2% 2*S*,3*R*

$(CH_3)_3C$–OOH, $Ti[OCH(CH_3)_2]_4$, (*S*,*S*)-DET → **2*S*,3*R* isomer—98%** + 2% 2*R*,3*S*

Summary

Aldehydes and ketones are among the most important of all functional groups, both in the chemical industry and in biological pathways. In this chapter, we've looked at some of their typical reactions. Aldehydes are normally prepared in the laboratory by oxidation of primary alcohols or by partial reduction of esters. Ketones are similarly prepared by oxidation of secondary alcohols.

The **nucleophilic addition reaction** is the most common general reaction type for aldehydes and ketones. Many different kinds of products can be prepared by nucleophilic additions. Aldehydes and ketones are reduced by $NaBH_4$ or $LiAlH_4$ to yield primary and secondary alcohols, respectively. Addition of Grignard reagents to aldehydes and ketones also gives alcohols (secondary and tertiary, respectively), and addition of HCN yields **cyanohydrins**. Primary amines add to carbonyl compounds yielding **imines**, or **Schiff bases**, and secondary

Key words

acetal [$R_2C(OR')_2$], 742
acyl group, 724
1,2-addition, 751
1,4-addition, 751
aldehyde (RCHO), 722
Cannizzaro reaction, 750
conjugate addition, 751
cyanohydrin [RCH(OH)C≡N], 733

Key words—cont'd

amines yield **enamines**. Reaction of an aldehyde or ketone with hydrazine and base gives an alkane (the **Wolff–Kishner reaction**). Alcohols add to carbonyl groups to yield **acetals**, which are valuable as protecting groups. Phosphorus **ylides** add to aldehydes and ketones in the **Wittig reaction** to give alkenes.

α,β-Unsaturated aldehydes and ketones often react with nucleophiles to give the product of **conjugate addition**, or **1,4 addition**. Particularly useful are the conjugate addition of an amine and the conjugate addition of an organic group by reaction with a diorganocopper reagent.

IR spectroscopy is helpful for identifying aldehydes and ketones. Carbonyl groups absorb in the IR range 1660 to 1770 cm^{-1}, with the exact position highly diagnostic of the kind of carbonyl group present in the molecule. ^{13}C NMR spectroscopy is also useful for aldehydes and ketones because their carbonyl carbons show resonances in the 190 to 215 δ range. ^1H NMR is useful for aldehyde $-CHO$ protons, which absorb near 10 δ. Aldehydes and ketones undergo two characteristic kinds of fragmentation in the mass spectrometer: α cleavage and McLafferty rearrangement.

Summary of Reactions

1. Preparation of aldehydes (Section 19.2)
   (a) Oxidation of primary alcohols (Section 17.7)

$$RCH(OH)H \xrightarrow[CH_2Cl_2]{\text{Dess–Martin periodinane}} RC(=O)H$$

   (b) Partial reduction of esters (Section 19.2)

$$RC(=O)OR' \xrightarrow[2.\ H_3O^+]{1.\ \text{DIBAH, toluene}} RC(=O)H + R'OH$$

2. Preparation of ketones
   (a) Oxidation of secondary alcohols (Section 17.7)

$$RCH(OH)R' \xrightarrow[\text{or } CrO_3]{\text{Periodinane}} RC(=O)R'$$

   (b) Diorganocopper reaction with acid chlorides (Section 19.2)

$$RC(=O)Cl + R'_2CuLi \xrightarrow{\text{Ether}} RC(=O)R'$$

3. Oxidation of aldehydes (Section 19.3)

$$RC(=O)H \xrightarrow{CrO_3,\ H_3O^+} RC(=O)OH$$

(continued)

4. Nucleophilic addition reactions of aldehydes and ketones

(a) Addition of hydride to give alcohols: reduction (Section 19.7)

$$RC(=O)R' \xrightarrow[\text{2. } H_3O^+]{\text{1. } NaBH_4\text{, ethanol}} RC(H)(OH)R'$$

(b) Addition of Grignard reagents to give alcohols (Section 19.7)

$$RC(=O)R' \xrightarrow[\text{2. } H_3O^+]{\text{1. } R''MgX\text{, ether}} RC(R'')(OH)R'$$

(c) Addition of HCN to give cyanohydrins (Section 19.6)

$$RC(=O)R' \xrightleftharpoons{HCN} RC(CN)(OH)R'$$

(d) Addition of primary amines to give imines (Section 19.8)

$$RC(=O)R' \xrightleftharpoons{R''NH_2} RC(=NR'')R' + H_2O$$

(e) Addition of secondary amines to give enamines (Section 19.8)

$$RC(=O)C(H)< \xrightleftharpoons{HNR'_2} RC(NR'_2)=C< + H_2O$$

(f) Wolff–Kishner reaction to give alkanes (Section 19.9)

$$RC(=O)R' \xrightarrow[KOH]{H_2NNH_2} RC(H)(H)R' + N_2 + H_2O$$

(g) Addition of alcohols to give acetals (Section 19.10)

$$RC(=O)R' + 2\ R''OH \xrightarrow[\text{catalyst}]{\text{Acid}} RC(OR'')(OR'')R' + H_2O$$

(h) Addition of phosphorus ylides to give alkenes: Wittig reaction (Section 19.11)

$$RC(=O)R' + (C_6H_5)_3\overset{+}{P}-\overset{-}{C}HR'' \xrightarrow{THF} (R)(R')C=C(R'')(H) + (C_6H_5)_3P=O$$

5. Conjugate additions to α,β-unsaturated aldehydes and ketones (Section 19.13)

(a) Conjugate addition of amines

$$RC(=O)C=C< \xrightleftharpoons{R'NH_2} RC(=O)C(H)-C(NHR')<$$

(continued)

(b) Conjugate addition of water

$$\mathrm{-C(=O)-C{=}C<} \;\underset{}{\overset{H_2O}{\rightleftharpoons}}\; \mathrm{-C(=O)-CH-C-OH}$$

(c) Conjugate addition of alkyl groups by diorganocopper reaction

$$\mathrm{R-C(=O)-C{=}C<} \;\xrightarrow[\text{2. } H_3O^+]{\text{1. } R'_2CuLi\text{, ether}}\; \mathrm{R-C(=O)-CH-C-R'}$$

Exercises

OWL Interactive versions of these problems are assignable in OWL for Organic Chemistry.

Visualizing Chemistry

(Problems 19.1–19.26 appear within the chapter.)

19.27 Each of the following substances can be prepared by a nucleophilic addition reaction between an aldehyde or ketone and a nucleophile. Identify the reactants from which each was prepared. If the substance is an acetal, identify the carbonyl compound and the alcohol; if it is an imine, identify the carbonyl compound and the amine; and so forth.

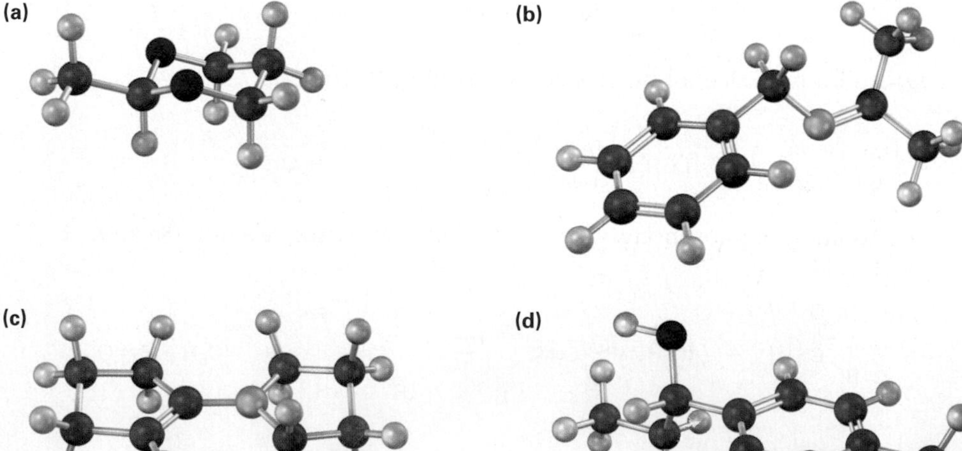

19.28 The following molecular model represents a tetrahedral intermediate resulting from addition of a nucleophile to an aldehyde or ketone. Identify the reactants, and write the structure of the final product when the nucleophilic addition reaction is complete.

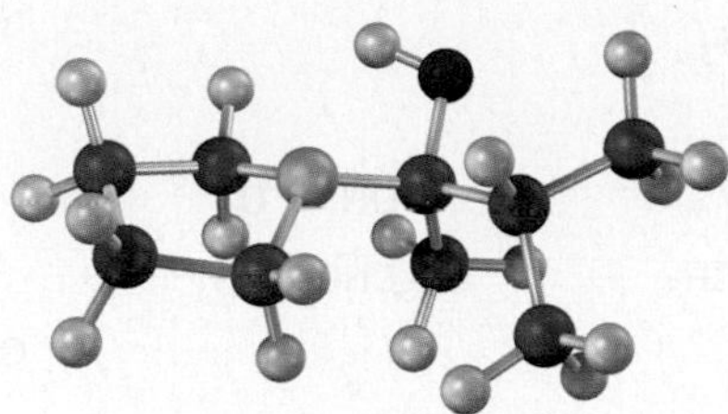

19.29 The enamine prepared from acetone and dimethylamine is shown in its lowest-energy form.

(a) What is the geometry and hybridization of the nitrogen atom?

(b) What orbital on nitrogen holds the lone pair of electrons?

(c) What is the geometric relationship between the *p* orbitals of the double bond and the nitrogen orbital that holds the lone pair? Why do you think this geometry represents the minimum energy?

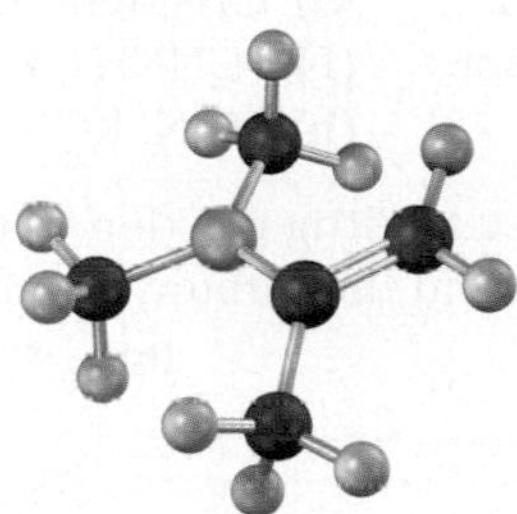

Additional Problems

Naming Aldehydes and Ketones

19.30 Draw structures corresponding to the following names:

(a) Bromoacetone

(b) (*S*)-2-Hydroxypropanal

(c) 2-Methyl-3-heptanone

(d) (2*S*,3*R*)-2,3,4-Trihydroxybutanal

(e) 2,2,4,4-Tetramethyl-3-pentanone

(f) 4-Methyl-3-penten-2-one

(g) Butanedial

(h) 3-Phenyl-2-propenal

(i) 6,6-Dimethyl-2,4-cyclohexadienone

(j) *p*-Nitroacetophenone

19.31 Draw and name the seven aldehydes and ketones with the formula $C_5H_{10}O$. Which are chiral?

19.32 Give IUPAC names for the following compounds:

(a) 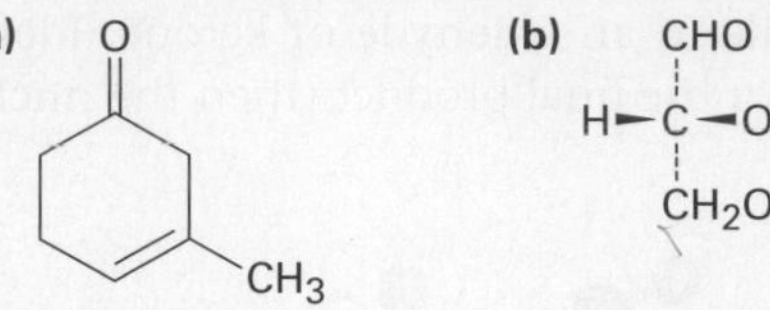

(b)

CHO
H—C—OH
CH₂OH

(c) 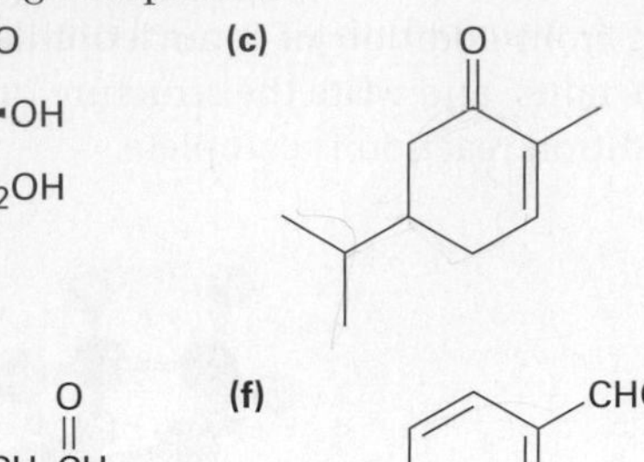

(d)

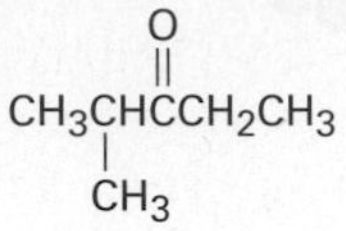

```
       O
       ‖
CH3CHCCH2CH3
    |
    CH3
```

(e)

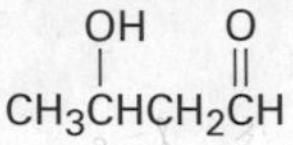

```
  OH   O
  |    ‖
CH3CHCH2CH
```

(f) OHC–C₆H₄–CHO

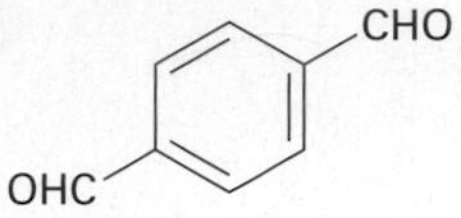

19.33 Draw structures of compounds that fit the following descriptions:

(a) An α,β-unsaturated ketone, C_6H_8O
(b) An α-diketone
(c) An aromatic ketone, $C_9H_{10}O$
(d) A diene aldehyde, C_7H_8O

Reactions of Aldehydes and Ketones

19.34 Predict the products of the reaction of (1) phenylacetaldehyde and (2) acetophenone with the following reagents:

(a) $NaBH_4$, then H_3O^+
(b) Dess–Martin reagent
(c) NH_2OH, HCl catalyst
(d) CH_3MgBr, then H_3O^+
(e) 2 CH_3OH, HCl catalyst
(f) H_2NNH_2, KOH
(g) $(C_6H_5)_3P{=}CH_2$
(h) HCN, KCN

19.35 Show how you might use a Wittig reaction to prepare the following alkenes. Identify the alkyl halide and the carbonyl components.

(a) 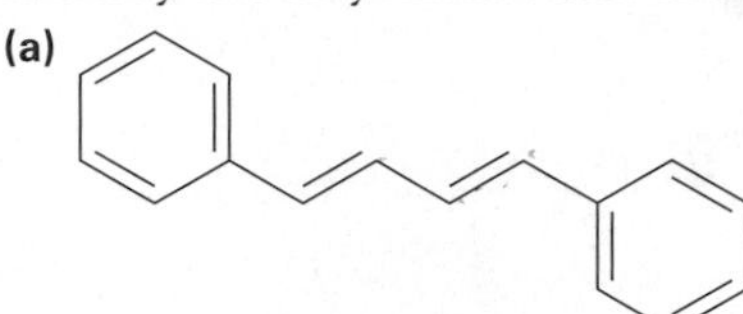

(b) 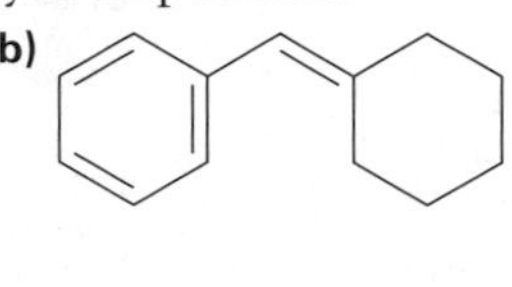

19.36 How would you use a Grignard reaction on an aldehyde or ketone to synthesize the following compounds?

(a) 2-Pentanol
(b) 1-Butanol
(c) 1-Phenylcyclohexanol
(d) Diphenylmethanol

19.37 How might you carry out the following selective transformations? One of the two schemes requires a protection step. (Recall from Section 19.4 that aldehydes are more reactive than ketones toward nucleophilic addition.)

(a)

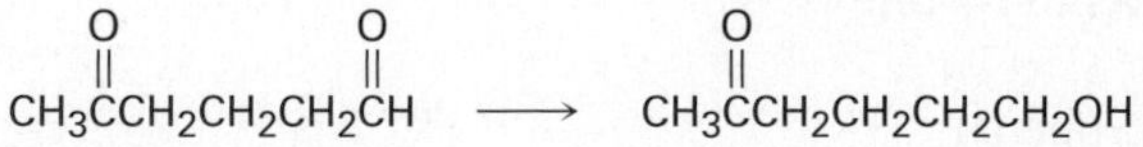

```
   O              O              O
   ‖              ‖              ‖
CH3CCH2CH2CH2CH  ——→  CH3CCH2CH2CH2CH2OH
```

(b)

```
   O              O          OH           O
   ‖              ‖          |            ‖
CH3CCH2CH2CH2CH  ——→  CH3CHCH2CH2CH2CH
```

19.38 How would you prepare the following substances from 2-cyclohexenone? More than one step may be needed.

(a)

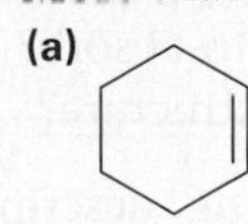

(b)

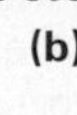

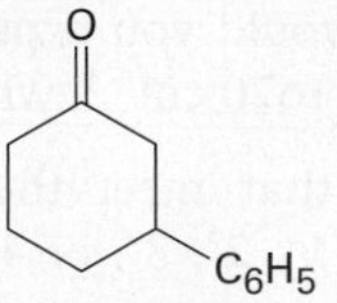

C_6H_5

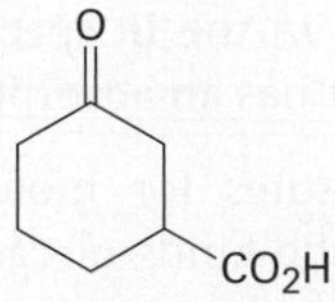

(c) CO_2H

(d) 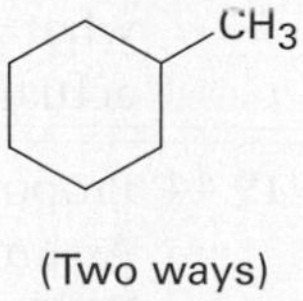 CH_3 (Two ways)

19.39 How would you synthesize the following substances from benzaldehyde and any other reagents needed?

(a)

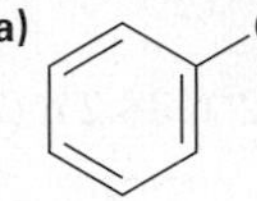

(b)

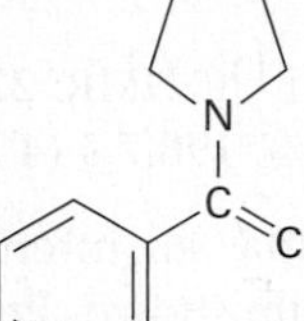

(c)

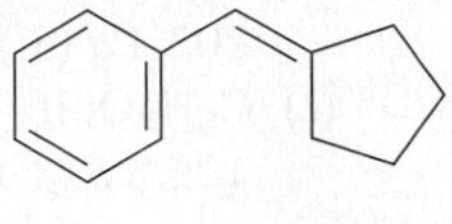

19.40 Carvone is the major constituent of spearmint oil. What products would you expect from reaction of carvone with the following reagents?

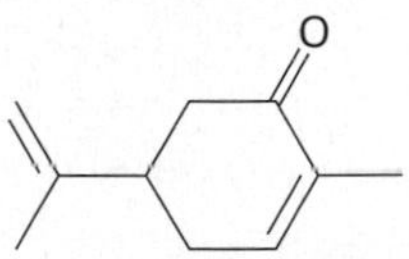

Carvone

(a) $(CH_3)_2Cu^- Li^+$, then H_3O^+
(b) $LiAlH_4$, then H_3O^+
(c) CH_3NH_2
(d) C_6H_5MgBr, then H_3O^+
(e) H_2/Pd
(f) CrO_3, H_3O^+
(g) $(C_6H_5)_3\overset{+}{P}\overset{-}{C}HCH_3$
(h) $HOCH_2CH_2OH$, HCl

19.41 How would you synthesize the following compounds from cyclohexanone?

(a) 1-Methylcyclohexene
(b) 2-Phenylcyclohexanone
(c) *cis*-1,2-Cyclohexanediol
(d) 1-Cyclohexylcyclohexanol

Spectroscopy

19.42 At what position would you expect to observe IR absorptions for the following molecules?

(a)

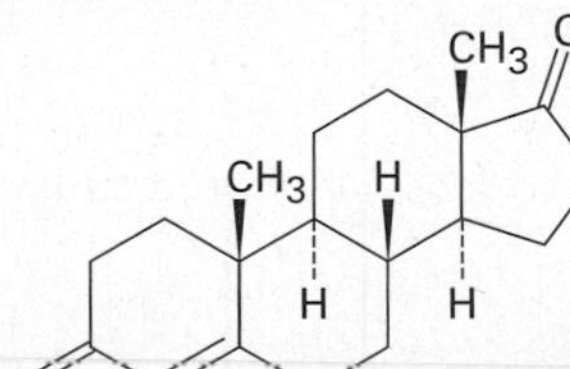

4-Androstene-3,17-dione

(b)

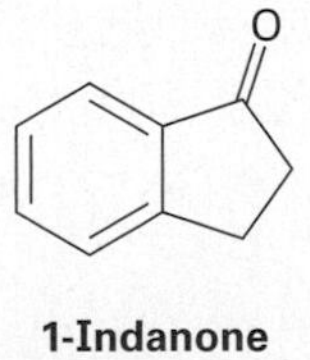

1-Indanone

(c)

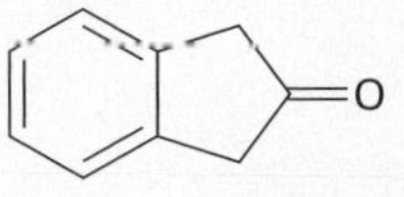

(d) 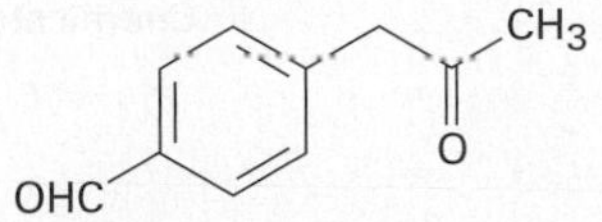 CH_3 OHC

19.43 Acid-catalyzed dehydration of 3-hydroxy-3-phenylcyclohexanone leads to an unsaturated ketone. What possible structures are there for the product? At what position in the IR spectrum would you expect each to absorb? If the actual product has an absorption at 1670 cm^{-1}, what is its structure?

19.44 Propose structures for molecules that meet the following descriptions. Assume that the kinds of carbons (1°, 2°, 3°, or 4°) have been assigned by DEPT-NMR.

(a) $C_6H_{12}O$; IR: 1715 cm^{-1}; ^{13}C NMR: 8.0 δ (1°), 18.5 δ (1°), 33.5 δ (2°), 40.6 δ (3°), 214.0 δ (4°)

(b) $C_5H_{10}O$; IR: 1730 cm^{-1}; ^{13}C NMR: 22.6 δ (1°), 23.6 δ (3°), 52.8 δ (2°), 202.4 δ (3°)

(c) C_6H_8O; IR: 1680 cm^{-1}; ^{13}C NMR: 22.9 δ (2°), 25.8 δ (2°), 38.2 δ (2°), 129.8 δ (3°), 150.6 δ (3°), 198.7 δ (4°)

19.45 Compound A, $C_8H_{10}O_2$, has an intense IR absorption at 1750 cm^{-1} and gives the ^{13}C NMR spectrum shown. Propose a structure for A.

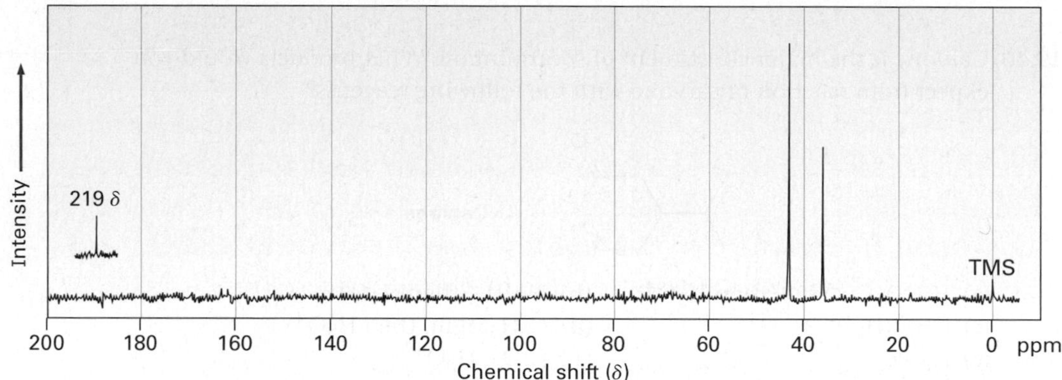

19.46 Propose structures for ketones or aldehydes that have the following ^1H NMR spectra:

(a) C_4H_7ClO
IR: 1715 cm^{-1}

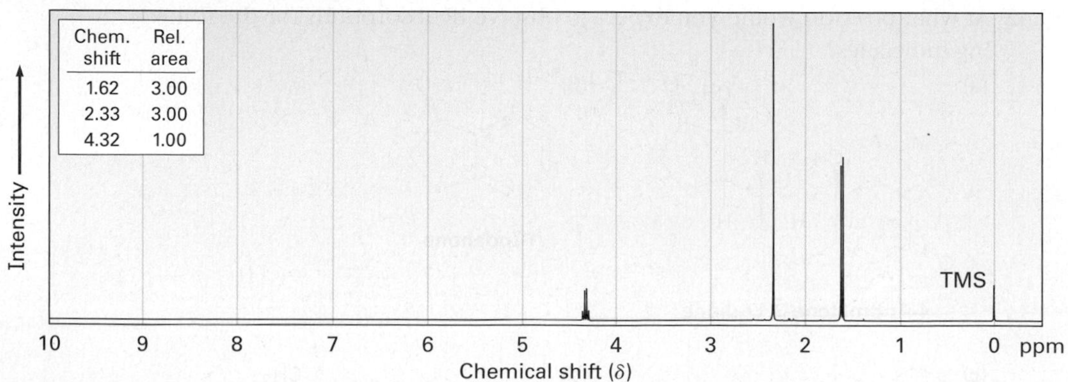

(b) $C_7H_{14}O$
IR: 1710 cm^{-1}

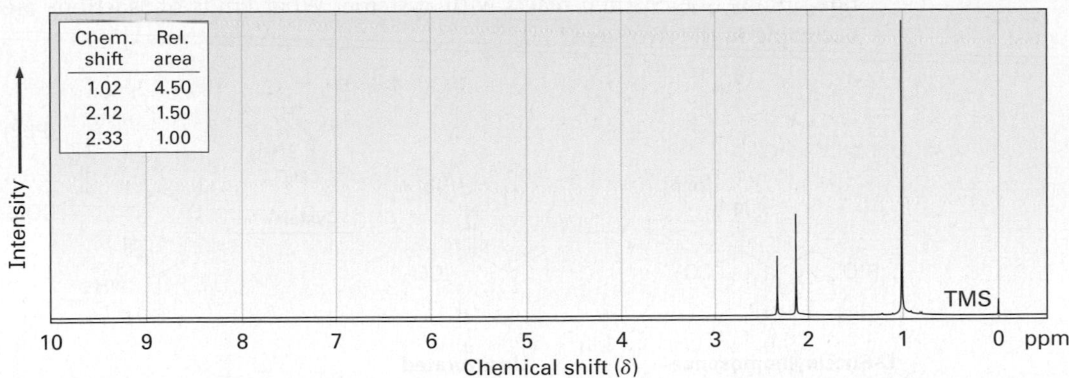

General Problems

19.47 Aldehydes can be prepared by the Wittig reaction using (methoxymethylene)-triphenylphosphorane as the Wittig reagent and then hydrolyzing the product with acid. For example,

Cyclohexanone + $(C_6H_5)_3\overset{+}{P}-\overset{-}{C}HOCH_3$ ⟶ (methoxymethylene)cyclohexane ($=C(H)OCH_3$) $\xrightarrow{H_3O^+}$ cyclohexanecarbaldehyde (CHO)

(Methoxymethylene)-triphenylphosphorane

(a) How would you prepare the necessary phosphorane?
(b) Propose a mechanism for the hydrolysis step.

19.48 When 4-hydroxybutanal is treated with methanol in the presence of an acid catalyst, 2-methoxytetrahydrofuran is formed. Explain.

$$HOCH_2CH_2CH_2CHO \xrightarrow[HCl]{CH_3OH} \text{2-methoxytetrahydrofuran (ring O, } OCH_3)$$

19.49 The S_N2 reaction of (dibromomethyl)benzene, $C_6H_5CHBr_2$, with NaOH yields benzaldehyde rather than (dihydroxymethyl)benzene, $C_6H_5CH(OH)_2$. Explain.

19.50 Reaction of 2-butanone with HCN yields a chiral product. What stereochemistry does the product have? Is it optically active?

19.51 One of the steps in the metabolism of fats is the reaction of an unsaturated acyl CoA with water to give a β-hydroxyacyl CoA. Propose a mechanism.

$$RCH_2CH_2CH{=}CH\overset{O}{\overset{\|}{C}}SCoA \xrightarrow{H_2O} RCH_2CH_2\overset{OH}{\overset{|}{C}H}-CH_2\overset{O}{\overset{\|}{C}}SCoA$$

Unsaturated acyl CoA   **β-Hydroxyacyl CoA**

19.52 The amino acid methionine is biosynthesized by a multistep route that includes reaction of an imine of pyridoxal phosphate (PLP) to give an unsaturated imine, which then reacts with cysteine. What kinds of reactions are occurring in the two steps?

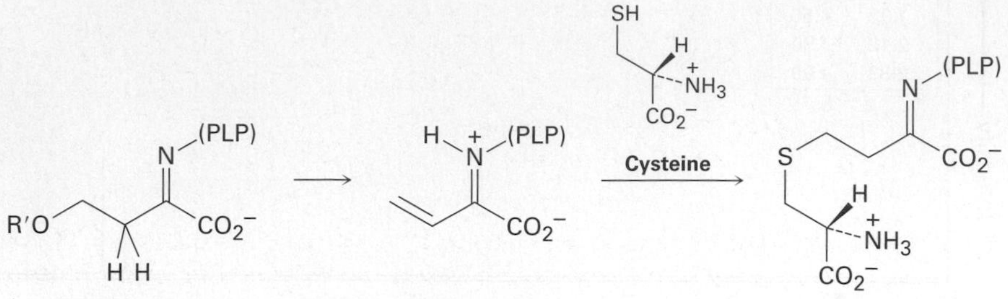

SH

H

$\overset{+}{N}H_3$

CO_2^-

(PLP)

N

(PLP)

H + (PLP)

N

Cysteine

S

CO_2^-

H

R'O

CO_2^-

CO_2^-

$\overset{+}{N}H_3$

H H

CO_2^-

***O*-Succinylhomoserine–PLP imine**

Unsaturated imine

19.53 Each of the following reaction schemes contains one or more flaws. What is wrong in each case? How would you correct each scheme?

(a) 1. CH_3MgBr 2. H_3O^+ → ; 1. $LiAlH_4$ 2. H_3O^+ → ; CH_3 ; CH_3

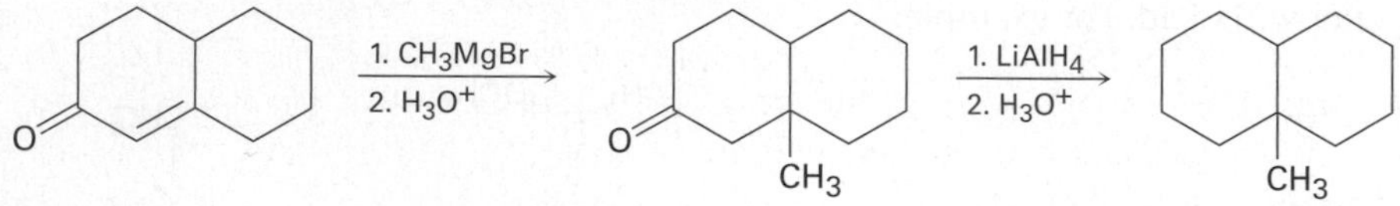

(b) $C_6H_5CH{=}CHCH_2OH \xrightarrow[H_3O^+]{CrO_3} C_6H_5CH{=}CHCHO \xrightarrow{H^+,\ CH_3OH} C_6H_5CH{=}CHCH(OCH_3)_2$

(c) $CH_3COCH_3 \xrightarrow[\text{Ethanol}]{HCN,\ KCN} CH_3C(OH)(CN)CH_3 \xrightarrow{H_3O^+} CH_3C(OH)(CH_2NH_2)CH_3$

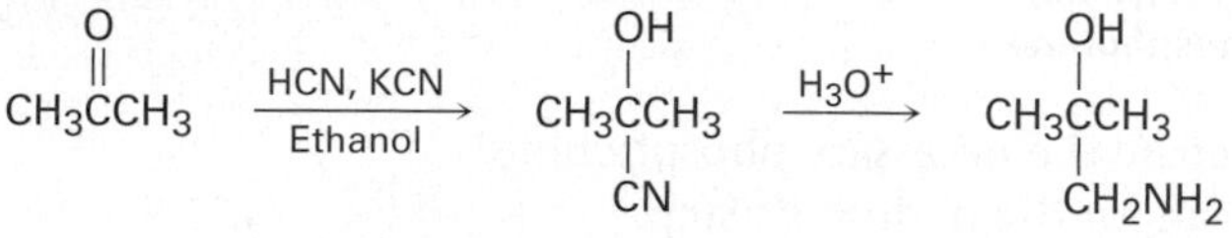

19.54 6-Methyl-5-hepten-2-one is a constituent of lemongrass oil. How could you synthesize this substance from methyl 4-oxopentanoate?

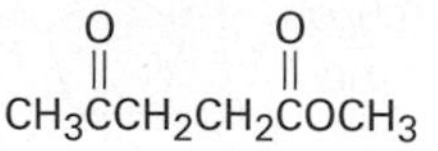

$CH_3COCH_2CH_2COOCH_3$ **Methyl 4-oxopentanoate**

19.55 Aldehydes and ketones react with thiols to yield *thioacetals* just as they react with alcohols to yield acetals. Predict the product of the following reaction, and propose a mechanism:

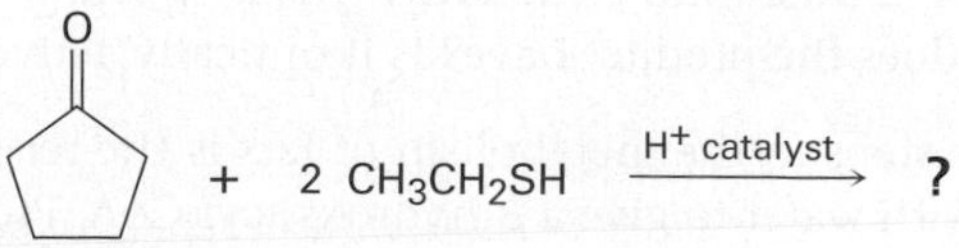

(cyclopentanone) + 2 CH_3CH_2SH $\xrightarrow{H^+ \text{ catalyst}}$?

19.56 Ketones react with dimethylsulfonium methylide to yield epoxides. Suggest a mechanism for the reaction.

Cyclohexanone + $\ddot{C}H_2\overset{+}{S}(CH_3)_2$ $\xrightarrow[\text{solvent}]{\text{DMSO}}$ (epoxide) + $(CH_3)_2S$

Dimethylsulfonium methylide

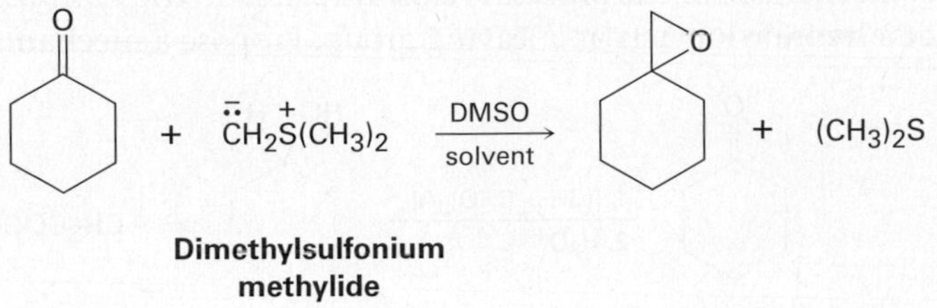

19.57 When cyclohexanone is heated in the presence of a large amount of acetone cyanohydrin and a small amount of base, cyclohexanone cyanohydrin and acetone are formed. Propose a mechanism.

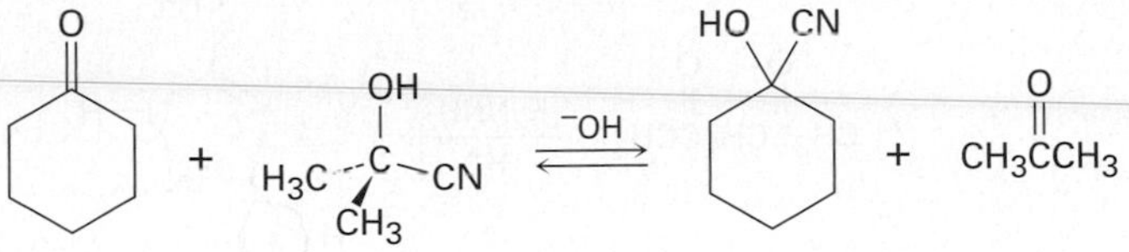

Cyclohexanone + $H_3C{-}C(OH)(CH_3){-}CN$ $\overset{^-OH}{\rightleftharpoons}$ cyclohexanone cyanohydrin (HO, CN) + $CH_3\overset{O}{\overset{\|}{C}}CH_3$

19.58 Tamoxifen is a drug used in the treatment of breast cancer. How would you prepare tamoxifen from benzene, the following ketone, and any other reagents needed?

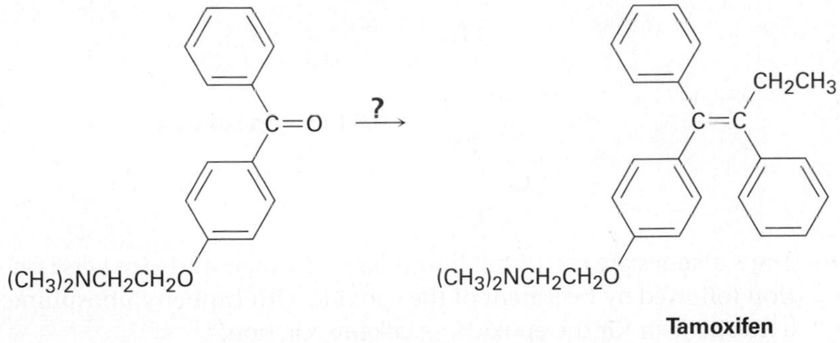

Ketone ($C{=}O$; $(CH_3)_2NCH_2CH_2O$) $\xrightarrow{?}$ Tamoxifen ($C{=}C$, CH_2CH_3; $(CH_3)_2NCH_2CH_2O$)

Tamoxifen

19.59 Paraldehyde, a sedative and hypnotic agent, is prepared by treatment of acetaldehyde with an acidic catalyst. Propose a mechanism for the reaction.

$3\ CH_3\overset{O}{\overset{\|}{C}}H$ $\xrightarrow[\text{catalyst}]{H^+}$ Paraldehyde (H_3C, O, CH_3, O, O, CH_3)

Paraldehyde

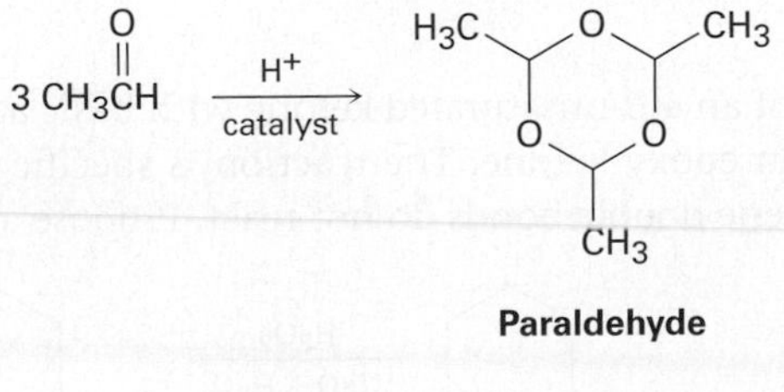

19.60 The Meerwein–Ponndorf–Verley reaction involves reduction of a ketone by treatment with an excess of aluminum triisopropoxide, $[(CH_3)_2CHO]_3Al$. The mechanism of the process is closely related to the Cannizzaro reaction in that a hydride ion acts as a leaving group. Propose a mechanism.

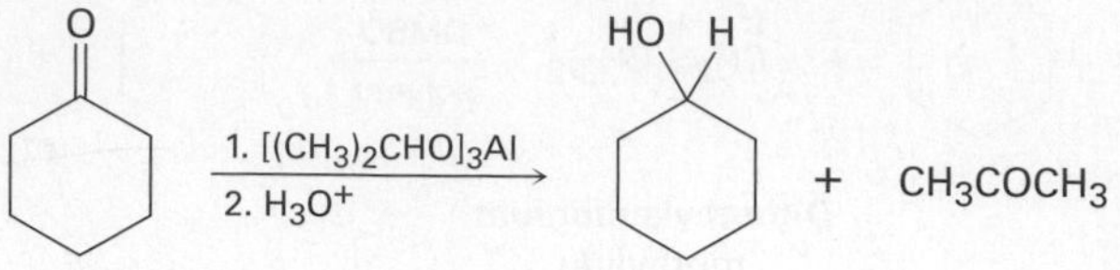

$$\text{Cyclohexanone} \xrightarrow[\text{2. } H_3O^+]{\text{1. } [(CH_3)_2CHO]_3Al} \text{Cyclohexanol (HO, H)} + CH_3COCH_3$$

19.61 Propose a mechanism to account for the formation of 3,5-dimethylpyrazole from hydrazine and 2,4-pentanedione. Look carefully to see what has happened to each carbonyl carbon in going from starting material to product.

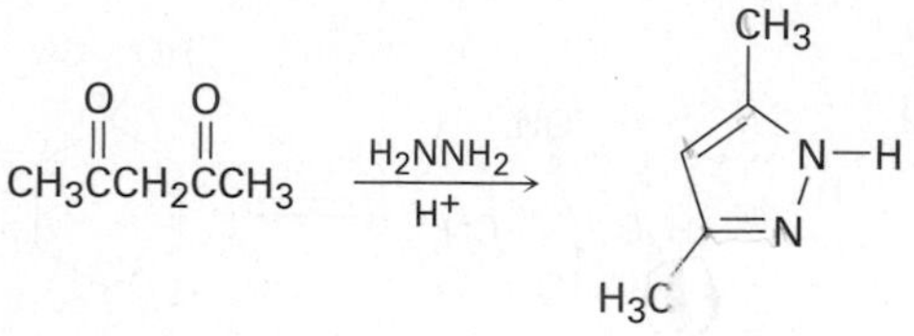

$$CH_3\overset{O}{\overset{\|}{C}}CH_2\overset{O}{\overset{\|}{C}}CH_3 \xrightarrow[H^+]{H_2NNH_2} \text{3,5-Dimethylpyrazole}$$

2,4-Pentanedione **3,5-Dimethylpyrazole**

19.62 In light of your answer to Problem 19.61, propose a mechanism for the formation of 3,5-dimethylisoxazole from hydroxylamine and 2,4-pentanedione.

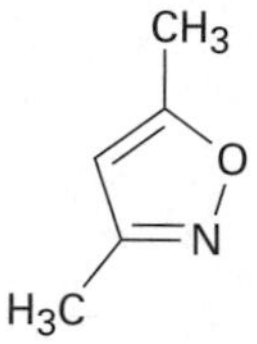

3,5-Dimethylisoxazole

19.63 Trans alkenes are converted into their cis isomers and vice versa on epoxidation followed by treatment of the epoxide with triphenylphosphine. Propose a mechanism for the epoxide → alkene reaction.

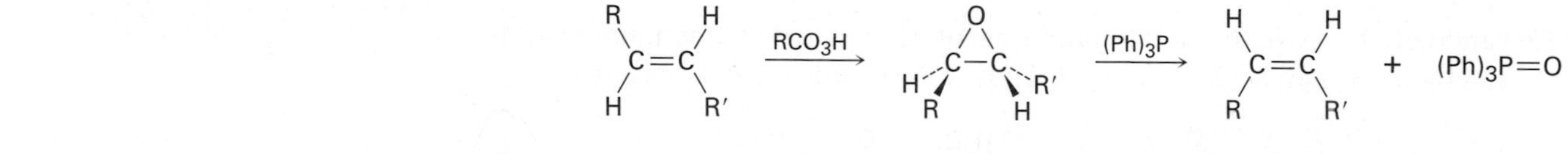

$$\text{trans-RCH=CHR'} \xrightarrow{RCO_3H} \text{epoxide} \xrightarrow{(Ph)_3P} \text{cis-RCH=CHR'} + (Ph)_3P{=}O$$

19.64 Treatment of an α,β-unsaturated ketone with basic aqueous hydrogen peroxide yields an epoxy ketone. The reaction is specific to unsaturated ketones; isolated alkene double bonds do not react. Propose a mechanism.

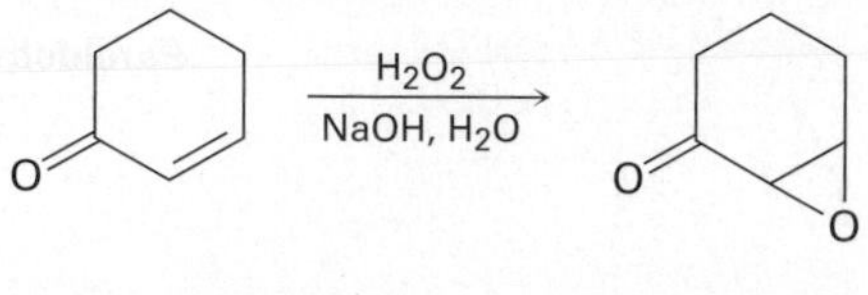

$$\text{2-Cyclohexenone} \xrightarrow[NaOH,\ H_2O]{H_2O_2} \text{2,3-Epoxycyclohexanone}$$

19.65 One of the biological pathways by which an amine is converted to a ketone involves two steps: (1) oxidation of the amine by NAD^+ to give an imine and (2) hydrolysis of the imine to give a ketone plus ammonia. Glutamate, for instance, is converted by this process into α-ketoglutarate. Show the structure of the imine intermediate, and propose mechanisms for both steps.

^-O_2C–CH$_2$CH$_2$–CH(NH_3^+)–CO_2^- (H, $\overset{+}{N}H_3$) $\xrightarrow{NAD^+}$ Imine $\xrightarrow{H_2O}$ ^-O_2C–CH$_2$CH$_2$–C(=O)–CO_2^- + NH_3

Glutamate **α-Ketoglutarate**

19.66 Compound **A**, MW = 86, shows an IR absorption at 1730 cm^{-1} and a very simple ^1H NMR spectrum with peaks at 9.7 δ (1 H, singlet) and 1.2 δ (9 H, singlet). Propose a structure for **A**.

19.67 Compound **B** is isomeric with **A** (Problem 19.66) and shows an IR peak at 1715 cm^{-1}. The ^1H NMR spectrum of **B** has peaks at 2.4 δ (1 H, septet, J = 7 Hz), 2.1 δ (3 H, singlet), and 1.2 δ (6 H, doublet, J = 7 Hz). What is the structure of **B**?

19.68 The ^1H NMR spectrum shown is that of a compound with formula $C_9H_{10}O$. How many double bonds and/or rings does this compound contain? If the unknown has an IR absorption at 1690 cm^{-1}, what is a likely structure?

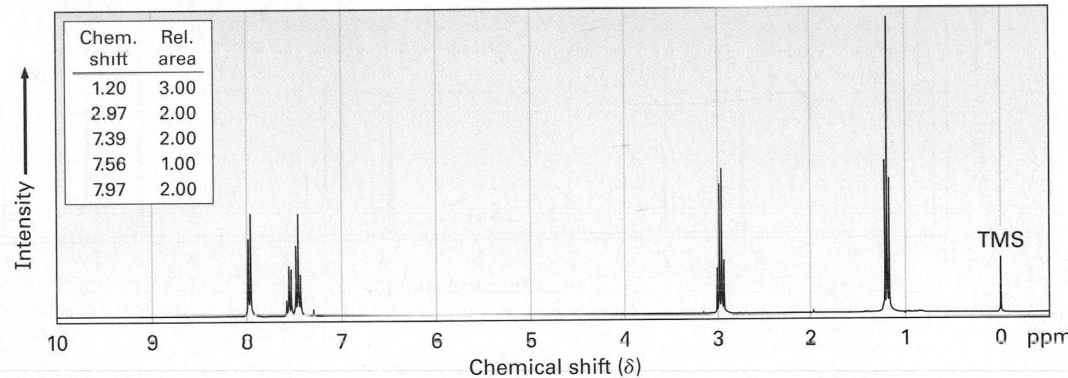

19.69 The ^1H NMR spectrum shown is that of a compound isomeric with the one in Problem 19.68. This isomer has an IR absorption at 1730 cm^{-1}. Propose a structure. [*Note:* Aldehyde protons (CHO) often show low coupling constants to adjacent hydrogens, so the splitting of aldehyde signals is not always apparent.]

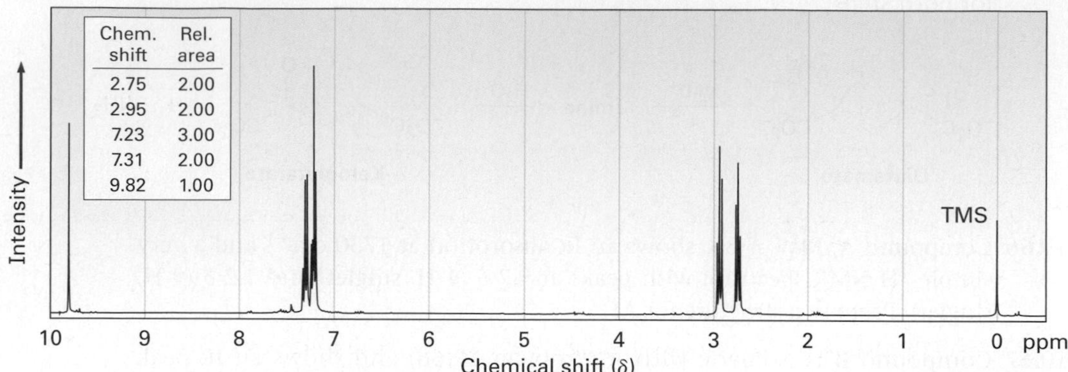

19.70 Propose structures for ketones or aldehydes that have the following ^1H NMR spectra:

(a) $C_9H_{10}O_2$
IR: 1695 cm^{-1}

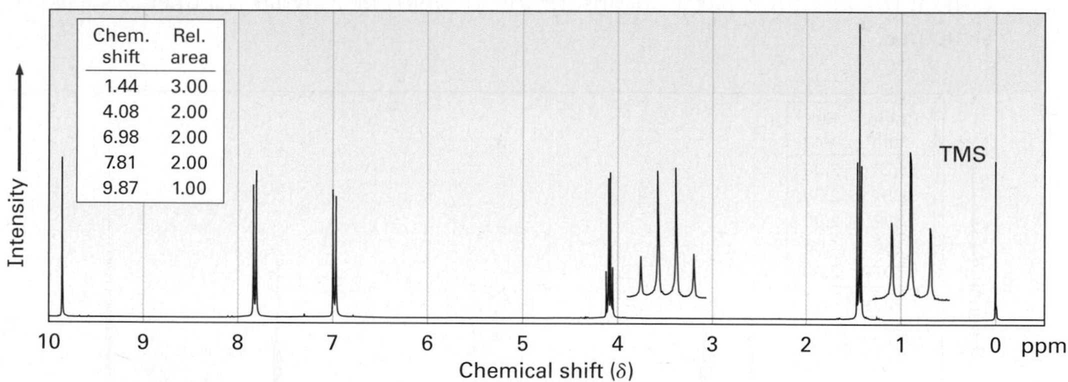

(b) C_4H_6O
IR: 1690 cm^{-1}

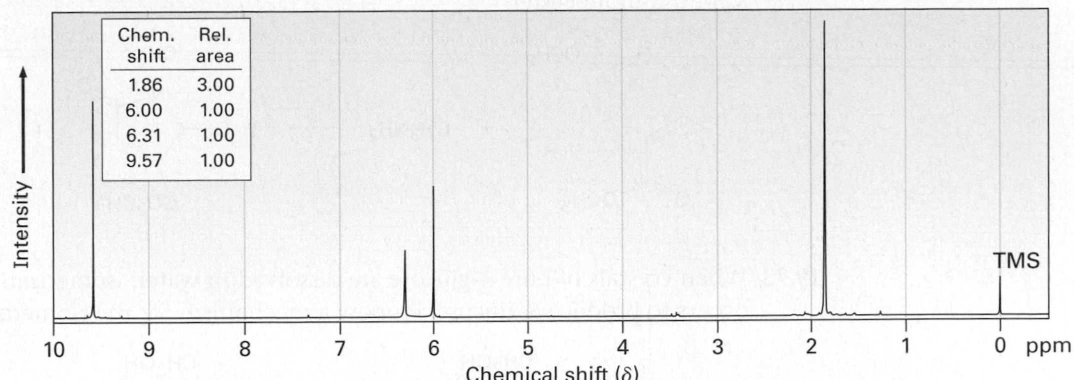

19.71 Propose structures for ketones or aldehydes that have the following ^{1}H NMR spectra.

(a) $C_{10}H_{12}O$
IR: 1710 cm^{-1}

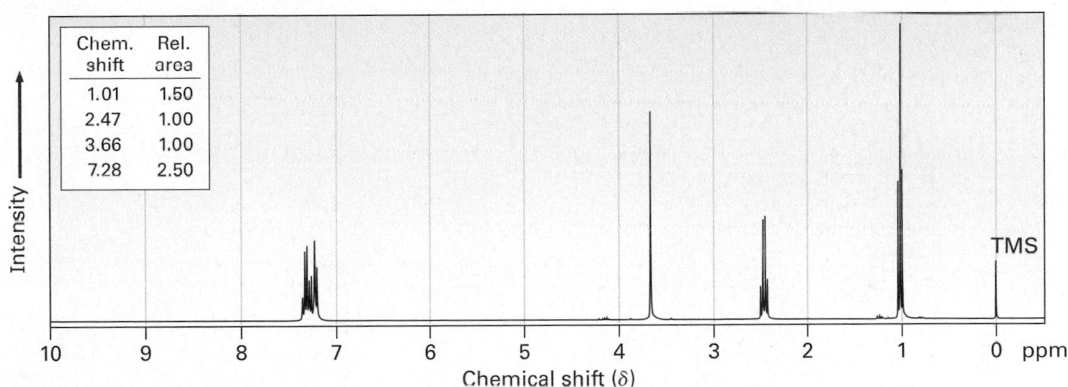

(b) $C_6H_{12}O_3$
IR: 1715 cm^{-1}

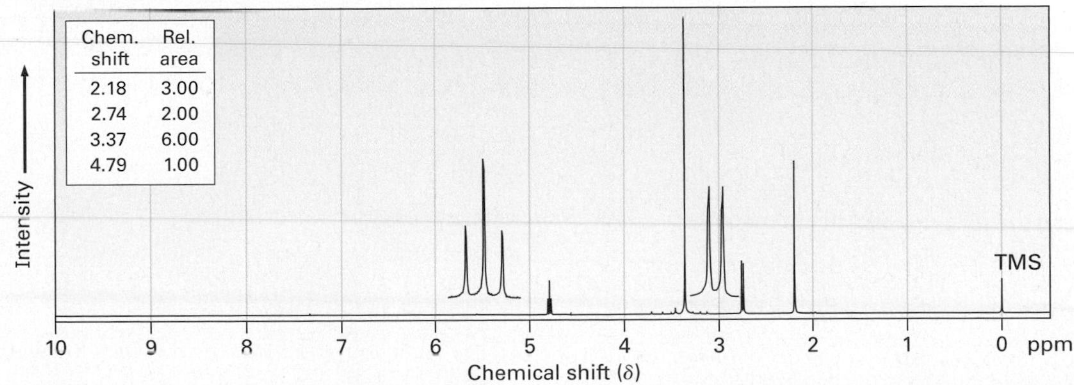

19.72 Primary amines react with esters to yield amides: $RCO_2R' + R''NH_2 \rightarrow RCONHR'' + R'OH$. Propose a mechanism for the following reaction of an α,β-unsaturated ester.

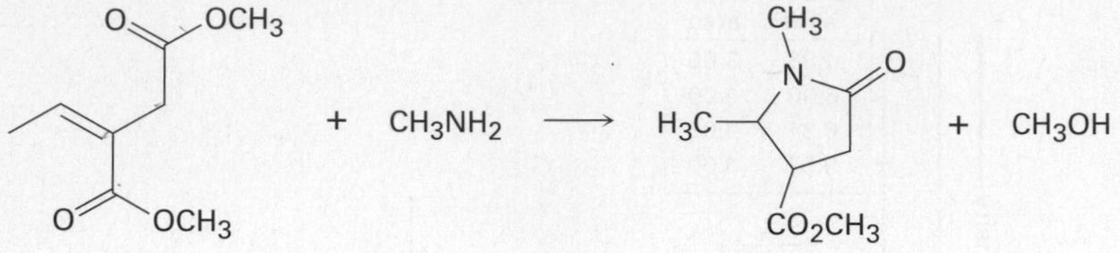

$$\text{(diester)} + CH_3NH_2 \longrightarrow \text{(N-methyl lactam with } H_3C \text{ and } CO_2CH_3\text{)} + CH_3OH$$

19.73 When crystals of pure α-glucose are dissolved in water, isomerization slowly occurs to produce β-glucose. Propose a mechanism for the isomerization.

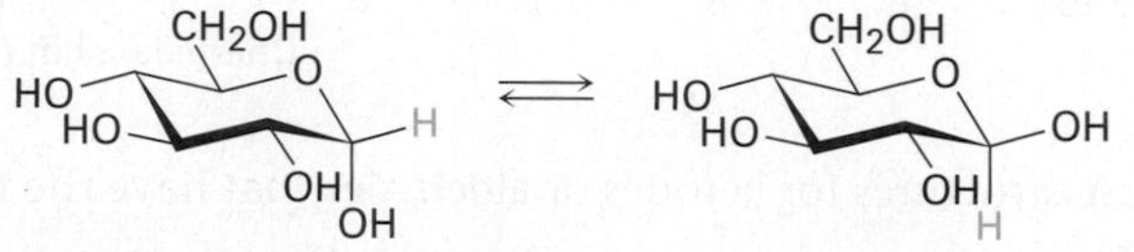

α-Glucose ⇌ β-Glucose

19.74 When glucose (Problem 19.73) is treated with $NaBH_4$, reaction occurs to yield *sorbitol,* a polyalcohol commonly used as a food additive. Show how this reduction occurs.

Glucose (ring structure with CH_2OH, O, HO, HO, OH, OH, H) $\xrightarrow[H_2O]{NaBH_4}$ $HOCH_2CH(OH)CH(OH)CH(OH)CH(OH)CH_2OH$

Glucose **Sorbitol**

19.75 The Wharton reaction converts an epoxy ketone to an allylic alcohol by reaction with hydrazine. Propose a mechanism. (*Hint:* Review the Wolff–Kishner reaction in Section 19.9.)

Epoxy ketone (bicyclic, with CH_3, O, O) + H_2NNH_2 $\xrightarrow[CH_3CO_2H]{CH_3CO_2^-\ ^+Na}$ allylic alcohol (bicyclic, with CH_3, OH) + N_2

20

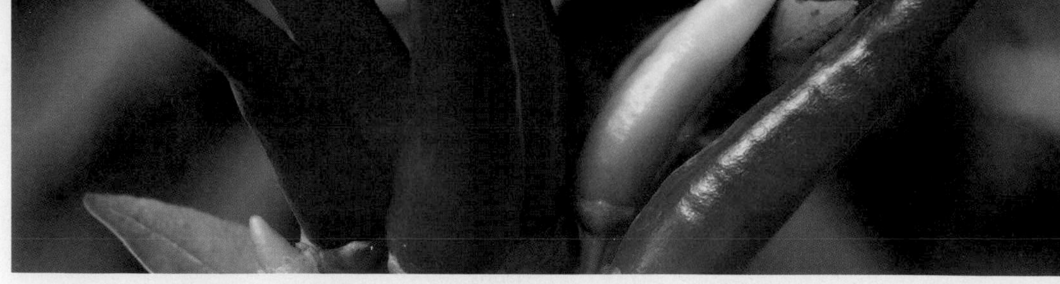

The burning sensation produced by touching or eating chili peppers is due to capsaicin, a carboxylic acid derivative called an amide. Image copyright Marie C. Fields, 2010. Used under license from Shutterstock.com

Carboxylic Acids and Nitriles

Carboxylic acids, RCO_2H, occupy a central place among carbonyl compounds. Not only are they valuable in themselves, they also serve as starting materials for preparing numerous carboxylic acid derivatives such as acid chlorides, esters, amides, and thioesters. In addition, carboxylic acids are present in the majority of biological pathways.

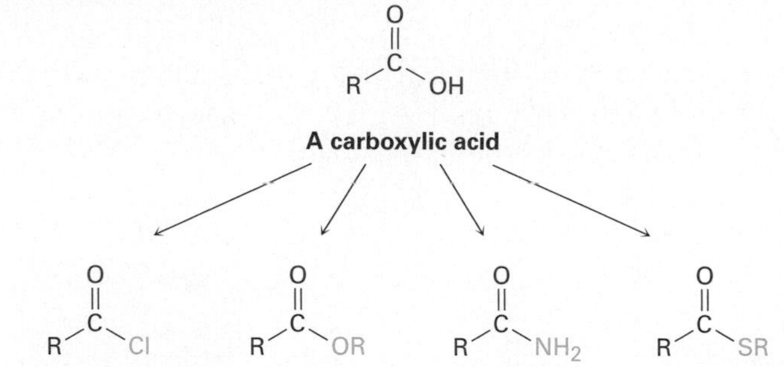

A carboxylic acid

An acid chloride | An ester | An amide | A thioester

A great many carboxylic acids are found in nature: acetic acid, CH_3CO_2H, is the chief organic component of vinegar; butanoic acid, $CH_3CH_2CH_2CO_2H$, is responsible for the rancid odor of sour butter; and hexanoic acid (caproic acid), $CH_3(CH_2)_4CO_2H$, is responsible for the unmistakable aroma of goats and dirty gym socks (the name comes from the Latin *caper,* meaning "goat"). Other examples are cholic acid, a major component of human bile, and long-chain aliphatic acids such as palmitic acid, $CH_3(CH_2)_{14}CO_2H$, a biological precursor of fats and vegetable oils.

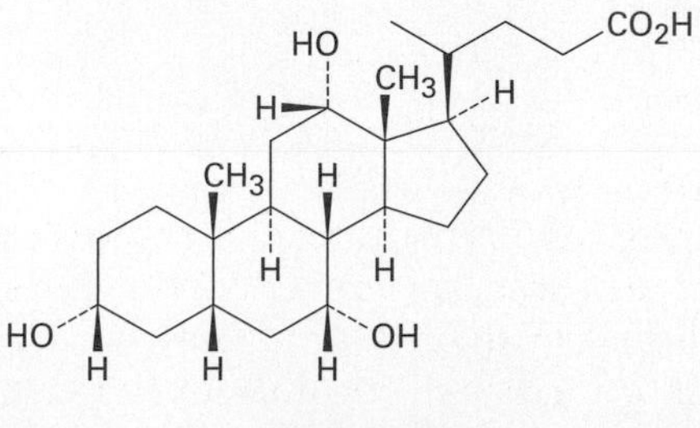

Cholic acid

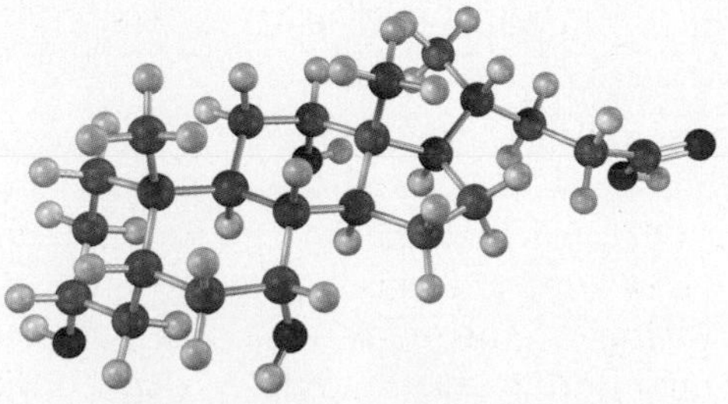

OWL Sign in to OWL for Organic Chemistry at **www.cengage.com/owl** to view tutorials and simulations, develop problem-solving skills, and complete online homework assigned by your professor.

Approximately 5 million metric tons of acetic acid is produced worldwide each year for a variety of purposes, including preparation of the vinyl acetate polymer used in paints and adhesives. About 20% of the acetic acid synthesized industrially is obtained by oxidation of acetaldehyde. Much of the remaining 80% is prepared by the rhodium-catalyzed reaction of methanol with carbon monoxide.

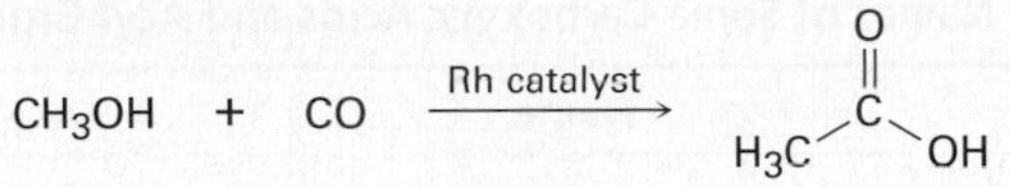

$$CH_3OH + CO \xrightarrow{\text{Rh catalyst}} H_3C\text{–}C(=O)\text{–}OH$$

Why This Chapter? Carboxylic acids are present in many industrial processes and most biological pathways and are the starting materials from which other acyl derivatives are made. Thus, an understanding of their properties and reactions is fundamental to understanding organic chemistry. We'll look both at acids and at their close relatives, nitriles ($RC{\equiv}N$), in this chapter and at carboxylic acid derivatives in the next chapter.

20.1 Naming Carboxylic Acids and Nitriles

Carboxylic Acids, RCO_2H

Simple carboxylic acids derived from open-chain alkanes are systematically named by replacing the terminal *-e* of the corresponding alkane name with *-oic acid*. The $-CO_2H$ carbon atom is numbered C1.

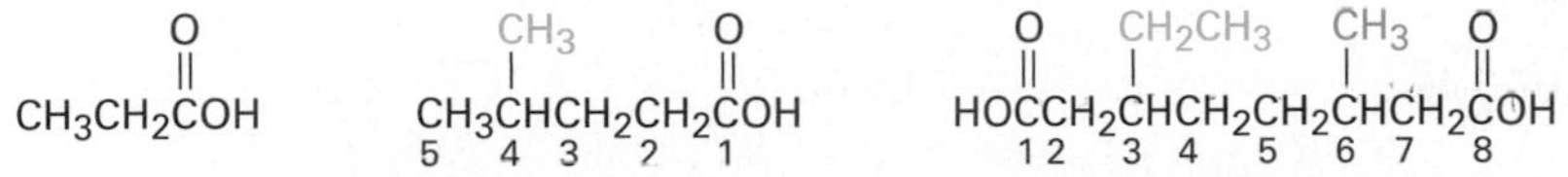

CH_3CH_2COOH — **Propanoic acid**

$CH_3CH(CH_3)CH_2CH_2COOH$ (C5 C4 C3 C2 C1) — **4-Methylpentanoic acid**

$HOOCCH_2CH(CH_2CH_3)CH_2CH_2CH(CH_3)CH_2COOH$ (C1–C8) — **3-Ethyl-6-methyloctanedioic acid**

Compounds that have a $-CO_2H$ group bonded to a ring are named using the suffix *-carboxylic acid*. The CO_2H carbon is attached to C1 in this system and is not itself numbered. As a substituent, the CO_2H group is called a **carboxyl group**.

***trans*-4-Hydroxycyclohexanecarboxylic acid**

1-Cyclopentenecarboxylic acid

Because many carboxylic acids were among the first organic compounds to be isolated and purified, a large number of common names exist (Table 20.1). Biological chemists, in particular, make frequent use of these names, so you may find yourself referring back to this list on occasion. We'll use systematic names in this book, with a few exceptions such as formic (methanoic) acid and acetic (ethanoic) acid, whose names are accepted by IUPAC and are so well known that it makes little sense to refer to them any other way.

Also listed in Table 20.1 are the names of acyl groups $\left(R{-}\overset{\overset{O}{\|}}{C}{-}\right)$ derived from the parent acids. Except for the eight entries at the top of Table 20.1, whose names have a *-yl* ending, all other acyl groups are named using an *-oyl* ending.

Table 20.1 Common Names of Some Carboxylic Acids and Acyl Groups

Structure	Name	Acyl group
HCO_2H	Formic	Formyl
CH_3CO_2H	Acetic	Acetyl
$CH_3CH_2CO_2H$	Propionic	Propionyl
$CH_3CH_2CH_2CO_2H$	Butyric	Butyryl
HO_2CCO_2H	Oxalic	Oxalyl
$HO_2CCH_2CO_2H$	Malonic	Malonyl
$HO_2CCH_2CH_2CO_2H$	Succinic	Succinyl
$HO_2CCH_2CH_2CH_2CO_2H$	Glutaric	Glutaryl
$HO_2CCH_2CH_2CH_2CH_2CO_2H$	Adipic	Adipoyl
$H_2C{=}CHCO_2H$	Acrylic	Acryloyl
$HO_2CCH{=}CHCO_2H$	Maleic (cis)	Maleoyl
	Fumaric (trans)	Fumaroyl
$HOCH_2CO_2H$	Glycolic	Glycoloyl
$CH_3\overset{\overset{OH}{\vert}}{C}HCO_2H$	Lactic	Lactoyl
$CH_3\overset{\overset{O}{\|}}{C}CO_2H$	Pyruvic	Pyruvoyl
$HOCH_2\overset{\overset{OH}{\vert}}{C}HCO_2H$	Glyceric	Gylceroyl
$HO_2C\overset{\overset{OH}{\vert}}{C}HCH_2CO_2H$	Malic	Maloyl
$HO_2C\overset{\overset{O}{\|}}{C}CH_2CO_2H$	Oxaloacetic	Oxaloacetyl
CO_2H	Benzoic	Benzoyl
CO_2H, CO_2H	Phthalic	Phthaloyl

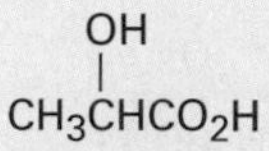

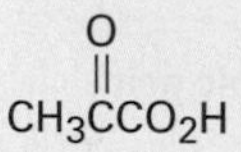

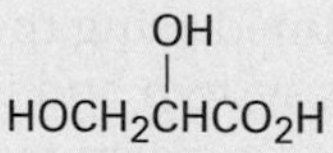

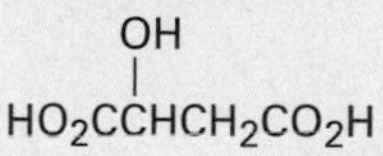

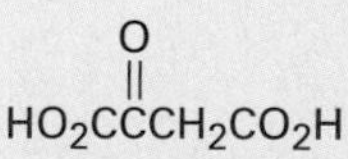

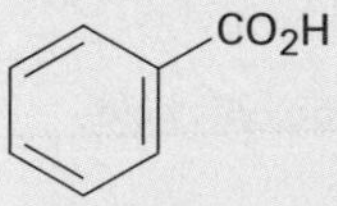

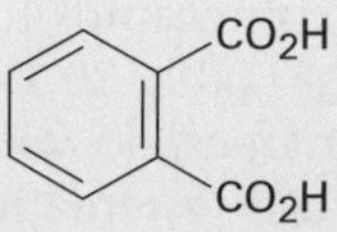

Nitriles, RC≡N

Compounds containing the –C≡N functional group are called **nitriles** and undergo some chemistry similar to that of carboxylic acids. Simple open-chain nitriles are named by adding *-nitrile* as a suffix to the alkane name, with the nitrile carbon numbered C1.

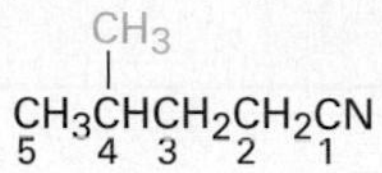

$$\underset{5\quad 4\quad 3\quad\ 2\quad\ 1}{CH_3\overset{\overset{\displaystyle CH_3}{|}}{C}HCH_2CH_2CN}$$ **4-Methylpentanenitrile**

Nitriles can also be named as derivatives of carboxylic acids by replacing the *-ic acid* or *-oic acid* ending with *-onitrile,* or by replacing the *-carboxylic acid* ending with *-carbonitrile.* The nitrile carbon atom is attached to C1 but is not itself numbered.

$CH_3C{\equiv}N$ — **Acetonitrile (from acetic acid)**

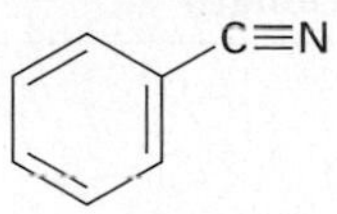

C≡N (on benzene ring) — **Benzonitrile (from benzoic acid)**

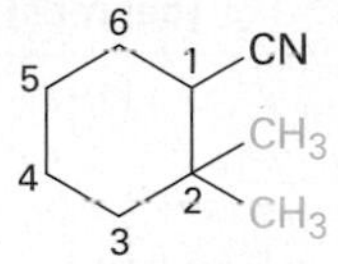

CN, CH3, CH3 (ring positions 1–6) — **2,2-Dimethylcyclohexanecarbonitrile (from 2,2-dimethylcyclohexanecarboxylic acid)**

If another carboxylic acid derivative is present in the same molecule, the prefix *cyano-* is used for the C≡N group.

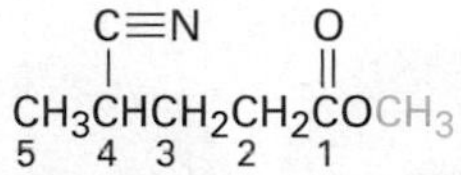

$$\underset{5\quad 4\quad 3\quad\ 2\quad\ 1}{CH_3\overset{\overset{\displaystyle C\equiv N}{|}}{C}HCH_2CH_2\overset{\overset{\displaystyle O}{\|}}{C}OCH_3}$$ **Methyl 4-cyanopentanoate**

Problem 20.1

Give IUPAC names for the following compounds:

(a) $CH_3\overset{\overset{\displaystyle CH_3}{|}}{C}HCH_2\overset{\overset{\displaystyle O}{\|}}{C}OH$

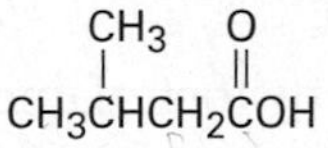

(b) $CH_3\overset{\overset{\displaystyle Br}{|}}{C}HCH_2CH_2\overset{\overset{\displaystyle O}{\|}}{C}OH$

(c) $CH_3CH_2\overset{\overset{\displaystyle CO_2H}{|}}{C}HCH_2CH_2CH_3$

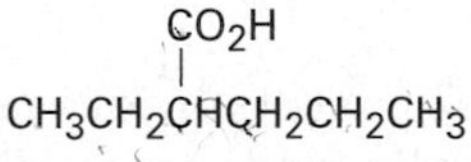

(d) H, H, C=C, H_3C, $CH_2CH_2\overset{\overset{\displaystyle O}{\|}}{C}OH$

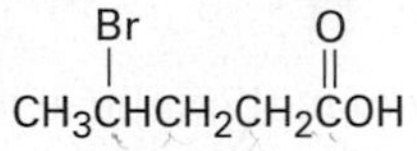

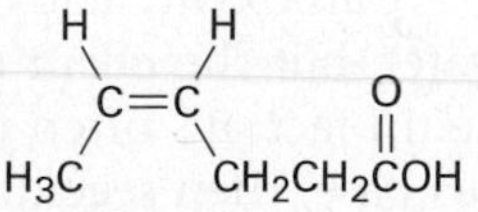

(e) $CH_3\overset{\overset{\displaystyle CH_3}{|}}{C}HCH_2\overset{\overset{\displaystyle CN}{|}}{C}HCH_3$

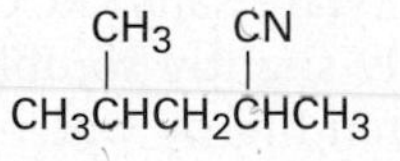

(f) H, H, HO_2C, CO_2H (cyclopentane)

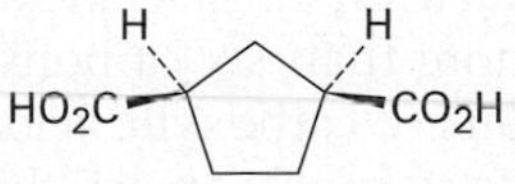

Problem 20.2

Draw structures corresponding to the following IUPAC names:

(a) 2,3-Dimethylhexanoic acid
(b) 4-Methylpentanoic acid
(c) *trans*-1,2-Cyclobutanedicarboxylic acid
(d) *o*-Hydroxybenzoic acid
(e) (9*Z*,12*Z*)-9,12-Octadecadienoic acid
(f) 2-Pentenenitrile

20.2 Structure and Properties of Carboxylic Acids

Carboxylic acids are similar in some respects to both ketones and alcohols. Like ketones, the carboxyl carbon is sp^2-hybridized, and carboxylic acid groups are therefore planar with C–C=O and O=C–O bond angles of approximately 120° (Table 20.2).

Table 20.2 Physical Parameters for Acetic Acid

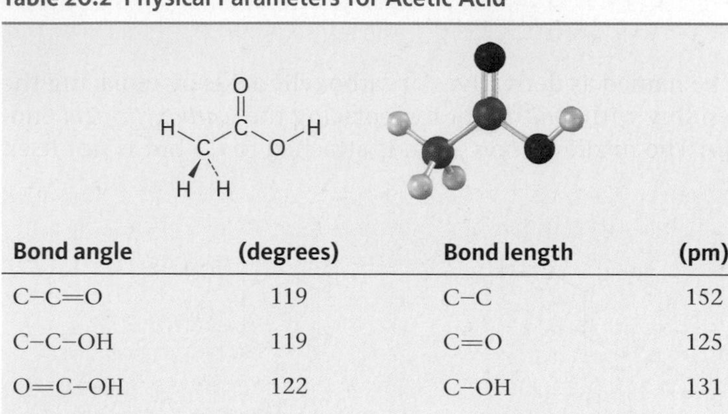

Bond angle	(degrees)	Bond length	(pm)
C–C=O	119	C–C	152
C–C–OH	119	C=O	125
O=C–OH	122	C–OH	131

Like alcohols, carboxylic acids are strongly associated because of hydrogen-bonding. Most carboxylic acids exist as cyclic dimers held together by two hydrogen bonds. This strong hydrogen-bonding has a noticeable effect on boiling points, making carboxylic acids much higher boiling than the corresponding alcohols. Acetic acid, for instance, has a boiling point of 117.9 °C, versus 78.3 °C for ethanol, even though both compounds have two carbons.

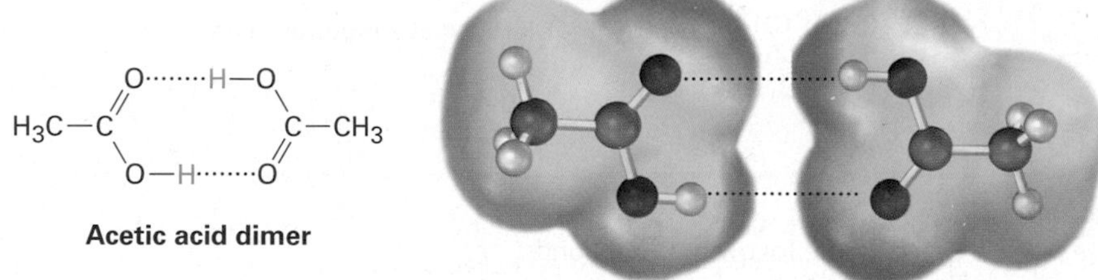

Acetic acid dimer

The most obvious property of carboxylic acids is implied by their name: carboxylic acids are *acidic*. They therefore react with bases such as NaOH and $NaHCO_3$ to give metal carboxylate salts, $RCO_2^-\ M^+$. Carboxylic acids with more than six carbons are only slightly soluble in water, but the alkali metal salts of carboxylic acids are often highly water-soluble. In fact, it's often possible to purify an acid by extracting its salt into aqueous base, then reacidifying and extracting the pure acid back into an organic solvent.

$$RCOOH + NaOH \xrightarrow{H_2O} RCOO^-\ Na^+ + H_2O$$

A carboxylic acid (water-insoluble) → **A carboxylic acid salt (water-soluble)**

Like other Brønsted–Lowry acids discussed in **Section 2.7**, carboxylic acids dissociate slightly in dilute aqueous solution to give H_3O^+ and the corresponding carboxylate anions, RCO_2^-. The extent of dissociation is given by an acidity constant, K_a.

$$RCOOH + H_2O \rightleftharpoons RCOO^- + H_3O^+$$

$$K_a = \frac{[RCO_2^-][H_3O^+]}{[RCO_2H]} \quad \text{and} \quad pK_a = -\log K_a$$

A list of K_a values for various carboxylic acids is given in Table 20.3. For most, K_a is approximately 10^{-4} to 10^{-5}. Acetic acid, for instance, has $K_a = 1.75 \times 10^{-5}$ at 25 °C, which corresponds to a pK_a of 4.76. In practical terms, a K_a value near 10^{-5} means that only about 0.1% of the molecules in a 0.1 M solution are dissociated, as opposed to the 100% dissociation found with strong mineral acids like HCl.

Table 20.3 Acidity of Some Carboxylic Acids

Structure	K_a	pK_a	
CF_3CO_2H	0.59	0.23	**Stronger acid**
HCO_2H	1.77×10^{-4}	3.75	
$HOCH_2CO_2H$	1.5×10^{-4}	3.84	
$C_6H_5CO_2H$	6.46×10^{-5}	4.19	
$H_2C{=}CHCO_2H$	5.6×10^{-5}	4.25	
CH_3CO_2H	1.75×10^{-5}	4.76	
$CH_3CH_2CO_2H$	1.34×10^{-5}	4.87	
CH_3CH_2OH (ethanol)	(1.00×10^{-16})	(16.00)	**Weaker acid**

Although much weaker than mineral acids, carboxylic acids are nevertheless much stronger acids than alcohols and phenols. The K_a of ethanol, for example, is approximately 10^{-16}, making ethanol a weaker acid than acetic acid by a factor of 10^{11}.

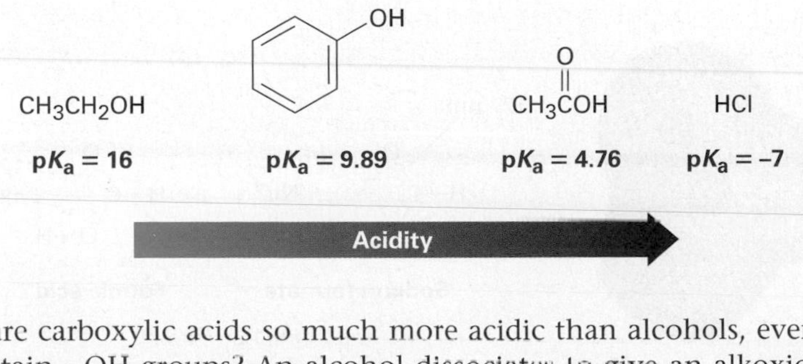

CH_3CH_2OH	Phenol (C_6H_5OH)	CH_3COH (acetic acid)	HCl
$pK_a = 16$	**$pK_a = 9.89$**	**$pK_a = 4.76$**	**$pK_a = -7$**

Acidity →

Why are carboxylic acids so much more acidic than alcohols, even though both contain –OH groups? An alcohol dissociates to give an alkoxide ion, in which the negative charge is localized on a single electronegative atom. A carboxylic acid, however, gives a carboxylate ion, in which the negative charge is delocalized over two equivalent oxygen atoms **(Figure 20.1)**. In resonance terms **(Section 2.4)**, a carboxylate ion is a stabilized resonance hybrid of two equivalent structures. Since a carboxylate ion is more stable than an alkoxide ion, it is lower in energy and more favored in the dissociation equilibrium.

Figure 20.1 An alkoxide ion has its charge localized on one oxygen atom and is less stable, while a carboxylate ion has the charge spread equally over both oxygens and is therefore more stable.

Ethanol $\overset{H_2O}{\rightleftharpoons}$ H_3O^+ + **Ethoxide ion (localized charge)**

Acetic acid $\overset{H_2O}{\rightleftharpoons}$ H_3O^+ + **Acetate ion (delocalized charge)**

Experimental evidence for the equivalence of the two carboxylate oxygens comes from X-ray crystallographic studies on sodium formate. Both carbon–oxygen bonds are 127 pm in length, midway between the C=O double bond (120 pm) and C–O single bond (134 pm) of formic acid. An electrostatic potential map of the formate ion also shows how the negative charge (red) is spread equally over both oxygens.

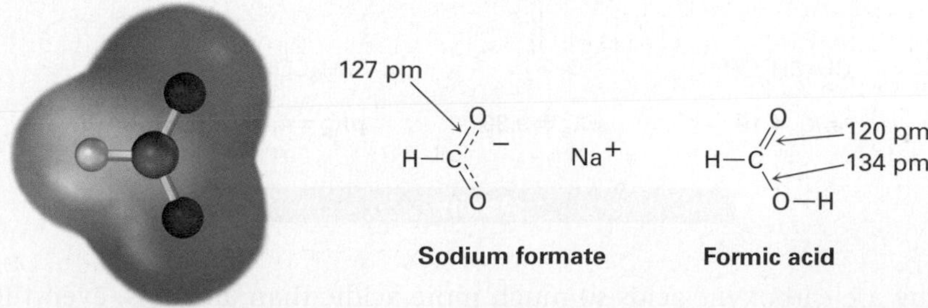

Problem 20.3

Assume you have a mixture of naphthalene and benzoic acid that you want to separate. How might you take advantage of the acidity of one component in the mixture to effect a separation?

Problem 20.4

The K_a for dichloroacetic acid is 3.32×10^{-2}. Approximately what percentage of the acid is dissociated in a 0.10 M aqueous solution?

20.3 Biological Acids and the Henderson–Hasselbalch Equation

In acidic solution, at low pH, a carboxylic acid is completely undissociated and exists entirely as RCO_2H. In basic solution, at high pH, a carboxylic acid is completely dissociated and exists entirely as RCO_2^-. Inside living cells, however, the pH is neither acidic nor basic but is instead buffered to nearly neutral pH—in humans, to pH = 7.3, a value often referred to as *physiological pH.* In what form, then, do carboxylic acids exist inside cells? The question is an important one for understanding the acid catalysts so often found in biological reactions.

If the pK_a value of a given acid and the pH of the medium are known, the percentages of dissociated and undissociated forms can be calculated using the **Henderson–Hasselbalch equation.**

For any acid HA, we have

$$pK_a = -\log \frac{[H_3O^+][A^-]}{[HA]} = -\log [H_3O^+] - \log \frac{[A^-]}{[HA]}$$

$$= pH - \log \frac{[A^-]}{[HA]}$$

which can be rearranged to give

$$pH = pK_a + \log \frac{[A^-]}{[HA]} \qquad \textbf{Henderson–Hasselbalch equation}$$

$$\text{so} \qquad \log \frac{[A^-]}{[HA]} = pH - pK_a$$

This equation says that the logarithm of the concentration of dissociated acid $[A^-]$ divided by the concentration of undissociated acid [HA] is equal to the pH of the solution minus the pK_a of the acid. Thus, if we know both the pH of the solution and the pK_a of the acid, we can calculate the ratio of $[A^-]$ to [HA]. Furthermore, when pH = pK_a, the two forms HA and A^- are present in equal amounts because log 1 = 0.

As an example of how to use the Henderson–Hasselbalch equation, let's find out what species are present in a 0.0010 M solution of acetic acid at pH = 7.3.

According to Table 20.3, the pK_a of acetic acid is 4.76. From the Henderson–Hasselbalch equation, we have

$$\log \frac{[A^-]}{[HA]} = pH - pK_a = 7.3 - 4.76 = 2.54$$

$$\frac{[A^-]}{[HA]} = \text{antilog}\,(2.54) = 3.5 \times 10^2 \quad \text{so} \quad [A^-] = (3.5 \times 10^2)\,[HA]$$

In addition, we know that

$$[A^-] + [HA] = 0.0010\ M$$

Solving the two simultaneous equations gives $[A^-] = 0.0010$ M and $[HA] = 3 \times 10^{-6}$ M. In other words, at a physiological pH of 7.3, essentially 100% of acetic acid molecules in a 0.0010 M solution are dissociated to the acetate ion.

What is true for acetic acid is also true for other carboxylic acids: at the physiological pH that exists inside cells, carboxylic acids are almost entirely dissociated. To reflect this fact, we always refer to cellular carboxylic acids by the name of their anion—acetate, lactate, citrate, and so forth, rather than acetic acid, lactic acid, and citric acid.

Problem 20.5

Calculate the percentages of dissociated and undissociated forms present in the following solutions:

(a) 0.0010 M glycolic acid ($HOCH_2CO_2H$; pK_a = 3.83) at pH = 4.50

(b) 0.0020 M propanoic acid (pK_a = 4.87) at pH = 5.30

20.4 Substituent Effects on Acidity

The listing of pK_a values shown previously in Table 20.3 indicates that there are substantial differences in acidity from one carboxylic acid to another. For example, trifluoroacetic acid ($K_a = 0.59$) is 33,000 times as strong as acetic acid ($K_a = 1.75 \times 10^{-5}$). How can we account for such differences?

Because the dissociation of a carboxylic acid is an equilibrium process, any factor that stabilizes the carboxylate anion relative to undissociated carboxylic acid will drive the equilibrium toward increased dissociation and result in increased acidity. For instance, three electron-withdrawing fluorine atoms delocalize the negative charge in the trifluoroacetate anion, thereby stabilizing the ion and increasing the acidity of CF_3CO_2H. In the same way, glycolic acid ($HOCH_2CO_2H$; pK_a = 3.83) is stronger than acetic acid because of the electron-withdrawing effect of the electronegative oxygen atom.

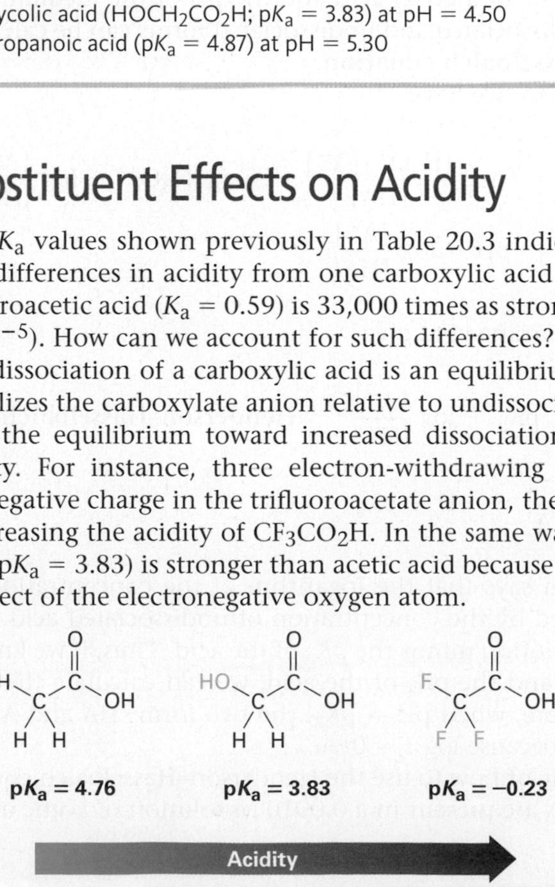

$H_3C{-}CO_2H$ (H–C(H)(H)–C(=O)–OH): **pK_a = 4.76**

$HOCH_2{-}CO_2H$ (HO–C(H)(H)–C(=O)–OH): **pK_a = 3.83**

$F_3C{-}CO_2H$ (F–C(F)(F)–C(=O)–OH): **pK_a = −0.23**

Acidity →

Because inductive effects operate through σ bonds and are dependent on distance, the effect of halogen substitution decreases as the substituent moves farther from the carboxyl. Thus, 2-chlorobutanoic acid has pK_a = 2.86, 3-chlorobutanoic acid has pK_a = 4.05, and 4-chlorobutanoic acid has pK_a = 4.52, similar to that of butanoic acid itself.

$ClCH_2CH_2CH_2COOH$ (pK_a = 4.52) — $CH_3CH(Cl)CH_2COOH$ (pK_a = 4.05) — $CH_3CH_2CH(Cl)COOH$ (pK_a = 2.86)

Acidity →

Substituent effects on acidity are also found in substituted benzoic acids. We said during the discussion of electrophilic aromatic substitution in **Section 16.5** that substituents on the aromatic ring strongly affect reactivity. Aromatic rings with electron-donating groups are activated toward further electrophilic substitution, and aromatic rings with electron-withdrawing groups are deactivated. Exactly the same effects are noticed on the acidity of substituted benzoic acids (Table 20.4).

Table 20.4 Substituent Effects on Acidity of *p*-Substituted Benzoic Acids

(Structure: Y–C_6H_4–COOH)

	Y	$K_a \times 10^{-5}$	pK_a	
Stronger acid ↑	$-NO_2$	39	3.41	Deactivating groups
	–CN	28	3.55	
	–CHO	18	3.75	
	–Br	11	3.96	
	–Cl	10	4.0	
	–H	6.46	4.19	
	$-CH_3$	4.3	4.34	Activating groups
	$-OCH_3$	3.5	4.46	
Weaker acid	–OH	3.3	4.48	

As Table 20.4 shows, an electron-donating (activating) group such as methoxy decreases acidity by destabilizing the carboxylate anion and an electron-withdrawing (deactivating) group such as nitro increases acidity by stabilizing the carboxylate anion.

p-Methoxybenzoic acid (pK_a = 4.46) — **Benzoic acid** (pK_a = 4.19) — **p-Nitrobenzoic acid** (pK_a = 3.41)

Acidity →

Because it's much easier to measure the acidity of a substituted benzoic acid than it is to determine the relative reactivity of an aromatic ring toward electrophilic substitution, the correlation between the two effects is useful for predicting reactivity. If we want to know the effect of a certain substituent on electrophilic reactivity, we can simply find the acidity of the corresponding benzoic acid. Worked Example 20.1 gives an illustration.

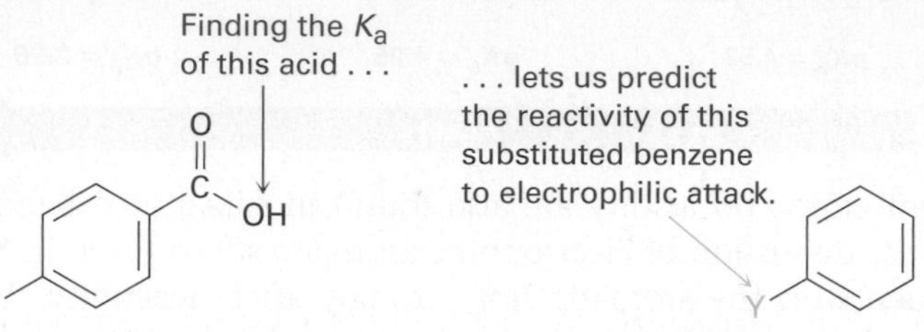

Finding the K_a of this acid . . .

. . . lets us predict the reactivity of this substituted benzene to electrophilic attack.

Worked Example 20.1 Predicting the Effect of a Substituent on the Reactivity of an Aromatic Ring toward Electrophilic Substitution

The pK_a of *p*-(trifluoromethyl)benzoic acid is 3.6. Is the trifluoromethyl substituent an activating or deactivating group in electrophilic aromatic substitution?

Strategy

Decide whether *p*-(trifluoromethyl)benzoic acid is stronger or weaker than benzoic acid. A substituent that strengthens the acid is a deactivating group because it withdraws electrons, and a substituent that weakens the acid is an activating group because it donates electrons.

Solution

A pK_a of 3.6 means that *p*-(trifluoromethyl)benzoic acid is stronger than benzoic acid, whose pK_a is 4.19. Thus, the trifluoromethyl substituent favors dissociation by helping stabilize the negative charge. Trifluoromethyl must therefore be an electron-withdrawing, deactivating group.

Problem 20.6

Which would you expect to be a stronger acid, the lactic acid found in tired muscles or acetic acid? Explain.

$CH_3CH(OH)COOH$ **Lactic acid**

Problem 20.7

Dicarboxylic acids have two dissociation constants, one for the initial dissociation into a monoanion and one for the second dissociation into a dianion. For oxalic acid, $HO_2C—CO_2H$, the first ionization constant is pK_{a1} = 1.2 and the second ionization constant is pK_{a2} = 4.2. Why is the second carboxyl group so much less acidic than the first?

Problem 20.8

The pK_a of *p*-cyclopropylbenzoic acid is 4.45. Is cyclopropylbenzene likely to be more reactive or less reactive than benzene toward electrophilic bromination? Explain.

Problem 20.9

Rank the following compounds in order of increasing acidity. Don't look at a table of pK_a data to help with your answer.

(a) Benzoic acid, *p*-methylbenzoic acid, *p*-chlorobenzoic acid

(b) *p*-Nitrobenzoic acid, acetic acid, benzoic acid

20.5 Preparing Carboxylic Acids

Let's review briefly some of the methods for preparing carboxylic acids that we've seen in previous chapters.

- Oxidation of a substituted alkylbenzene with $KMnO_4$ or $Na_2Cr_2O_7$ gives a substituted benzoic acid **(Section 16.9)**. Both primary and secondary alkyl groups can be oxidized, but tertiary groups are not affected.

$$O_2N-C_6H_4-CH_3 \xrightarrow[H_2O,\ 95\ °C]{KMnO_4} O_2N-C_6H_4-COOH$$

***p*-Nitrotoluene** → ***p*-Nitrobenzoic acid (88%)**

- Oxidation of a primary alcohol or an aldehyde yields a carboxylic acid **(Sections 17.7 and 19.3)**. Primary alcohols are often oxidized with CrO_3 in aqueous acid, and aldehydes are similarly oxidized.

$$CH_3CH(CH_3)CH_2CH_2CH_2OH \xrightarrow[H_3O^+]{CrO_3} CH_3CH(CH_3)CH_2CH_2COOH$$

4-Methyl-1-pentanol → **4-Methylpentanoic acid**

$$CH_3CH_2CH_2CH_2CH_2CHO \xrightarrow[H_3O^+]{CrO_3} CH_3CH_2CH_2CH_2CH_2COOH$$

Hexanal → **Hexanoic acid**

Hydrolysis of Nitriles

Carboxylic acids can be prepared from nitriles on heating with aqueous acid or base by a mechanism that we'll see in **Section 20.7**. Since nitriles themselves are usually made by S_N2 reaction of a primary or secondary alkyl halide with CN^-, the two-step sequence of cyanide displacement followed by nitrile hydrolysis is a good way to make a carboxylic acid from an alkyl halide ($RBr \rightarrow RC{\equiv}N \rightarrow RCO_2H$). Note that the product acid has one more carbon than the starting alkyl halide. An example occurs in one commercial route for the synthesis of the nonsteroidal anti-inflammatory drug ibuprofen. (See Chapter 15 *A Deeper Look.*)

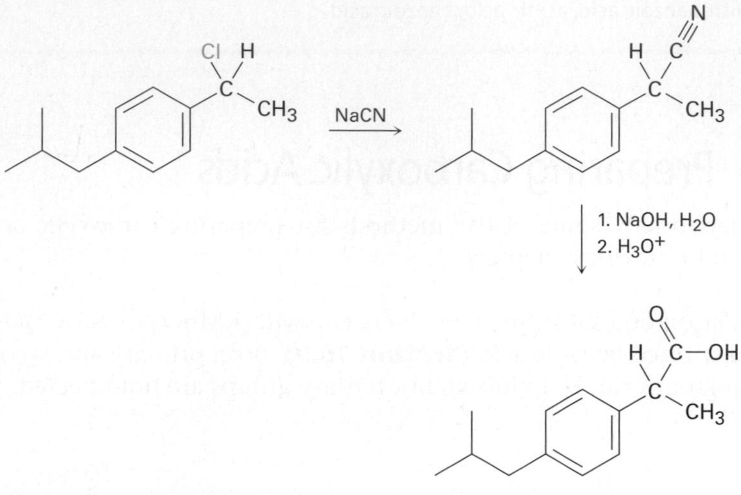

Ibuprofen

Carboxylation of Grignard Reagents

Another method for preparing carboxylic acids is by reaction of a Grignard reagent with CO_2 to yield a metal carboxylate, followed by protonation to give the carboxylic acid. This **carboxylation** reaction is usually carried out by bubbling a stream of dry CO_2 gas through a solution of the Grignard reagent. The organomagnesium halide adds to a C=O bond of carbon dioxide in a typical nucleophilic carbonyl addition reaction, and protonation of the carboxylate by addition of aqueous HCl in a separate step then gives the free carboxylic acid. For example

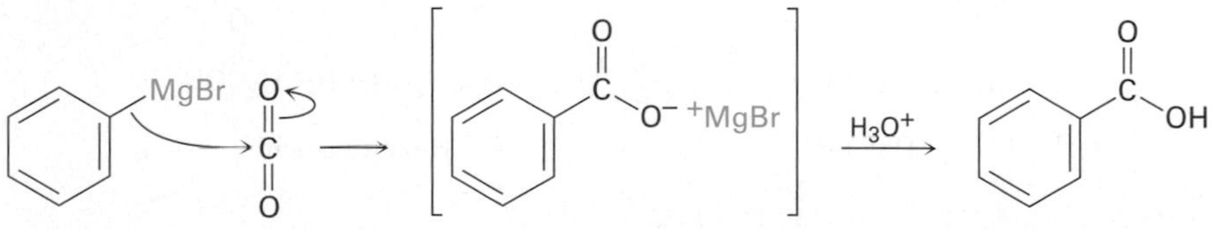

Phenylmagnesium bromide

Benzoic acid

As noted previously, there are no Grignard reagents inside living cells, but there are other types of stabilized carbanions that are often carboxylated. One of the initial steps in fatty-acid biosynthesis, for instance, involves formation of a carbanion from acetyl CoA, followed by carboxylation to yield malonyl CoA.

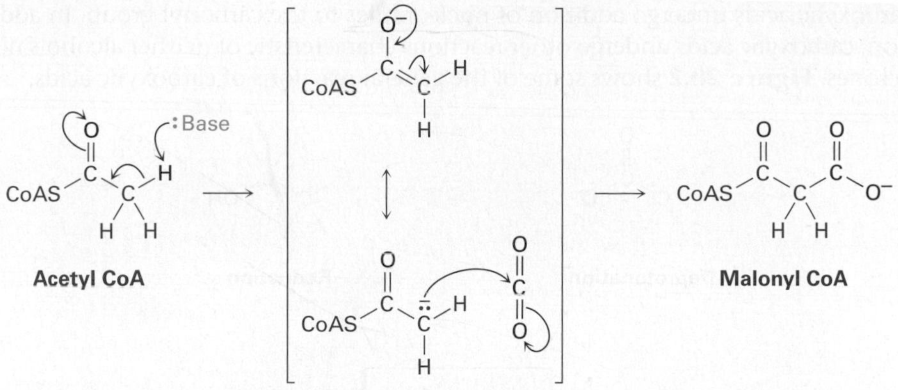

Devising a Synthesis Route for a Carboxylic Acid

Worked Example 20.2

How would you prepare phenylacetic acid ($PhCH_2CO_2H$) from benzyl bromide ($PhCH_2Br$)?

Strategy

We've seen two methods for preparing carboxylic acids from alkyl halides: (1) cyanide ion displacement followed by hydrolysis and (2) formation of a Grignard reagent followed by carboxylation. The first method involves an S_N2 reaction and is therefore limited to use with primary and some secondary alkyl halides. The second method involves formation of a Grignard reagent and is therefore limited to use with organic halides that have no acidic hydrogens or reactive functional groups elsewhere in the molecule. In the present instance, either method would work well.

Solution

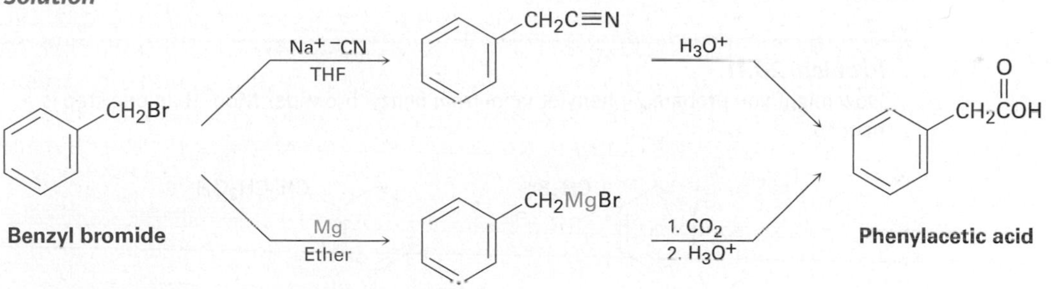

Problem 20.10

How would you prepare the following carboxylic acids?

(a) $(CH_3)_3CCO_2H$ from $(CH_3)_3CCl$

(b) $CH_3CH_2CH_2CO_2H$ from $CH_3CH_2CH_2Br$

20.6 Reactions of Carboxylic Acids: An Overview

We commented earlier in this chapter that carboxylic acids are similar in some respects to both alcohols and ketones. Like alcohols, carboxylic acids can be deprotonated to give anions, which are good nucleophiles in S_N2 reactions. Like ketones, carboxylic acids undergo addition of nucleophiles to the carbonyl group. In addition, carboxylic acids undergo other reactions characteristic of neither alcohols nor ketones. **Figure 20.2** shows some of the general reactions of carboxylic acids.

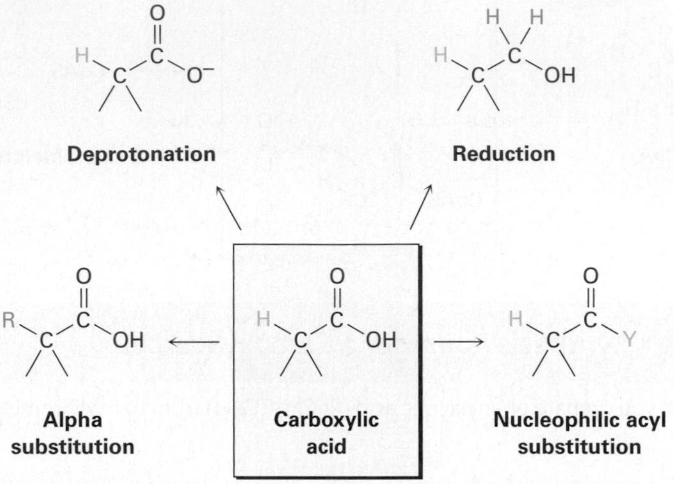

Figure 20.2 Some general reactions of carboxylic acids.

Deprotonation

Reduction

Alpha substitution

Carboxylic acid

Nucleophilic acyl substitution

Reactions of carboxylic acids can be grouped into the four categories indicated in Figure 20.2. Of the four, we've already discussed the acidic behavior of carboxylic acids in **Sections 20.2 through 20.4**, and we mentioned reduction by treatment of the acid with $LiAlH_4$ in **Section 17.4**. The remaining two categories are examples of fundamental carbonyl-group reaction mechanisms—nucleophilic acyl substitution and α substitution—that will be discussed in detail in Chapters 21 and 22.

Problem 20.11
How might you prepare 2-phenylethanol from benzyl bromide? More than one step is needed.

$C_6H_5CH_2Br \xrightarrow{?} C_6H_5CH_2CH_2OH$

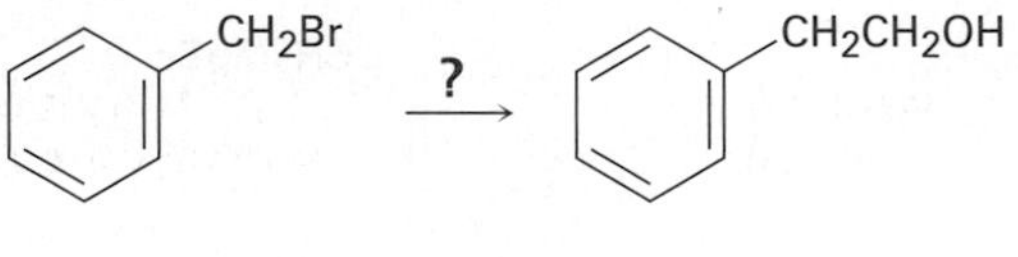

Problem 20.12
How might you carry out the following transformation? More than one step is needed.

$C_5H_9CH_2OH \xrightarrow{?} C_5H_9CH_2CH_2OH$

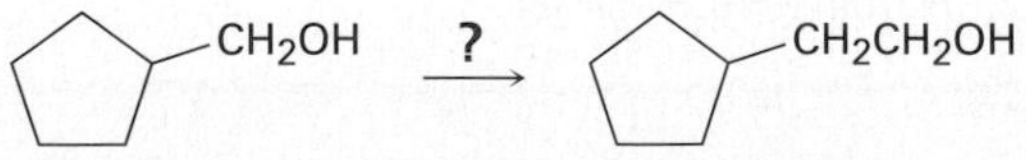

20.7 Chemistry of Nitriles

Nitriles are analogous to carboxylic acids in that both have a carbon atom with three bonds to an electronegative atom and both contain a π bond. Thus, some reactions of nitriles and carboxylic acids are similar. Both kinds of compounds are electrophiles, for instance, and both undergo nucleophilic addition reactions.

$$R{-}C{\equiv}N \qquad R{-}C({=}O){-}OH$$

A nitrile—three bonds to nitrogen **An acid—three bonds to two oxygens**

Nitriles occur infrequently in living organisms, although several hundred examples are known. Cyanocycline A, for instance, has been isolated from the bacterium *Streptomyces lavendulae* and found to have both antimicrobial and antitumor activity. In addition, more than 1000 compounds called *cyanogenic glycosides* are known. Derived primarily from plants, cyanogenic glycosides contain a sugar with an acetal carbon, one oxygen of which is bonded to a nitrile-bearing carbon (sugar–O–C–CN). On hydrolysis with aqueous acid, the acetal is cleaved **(Section 19.10)**, generating a cyanohydrin (HO–C–CN), which releases hydrogen cyanide. It's thought that the primary function of cyanogenic glycosides is to protect the plant by poisoning any animal foolish enough to eat it. Lotaustralin from the cassava plant is an example.

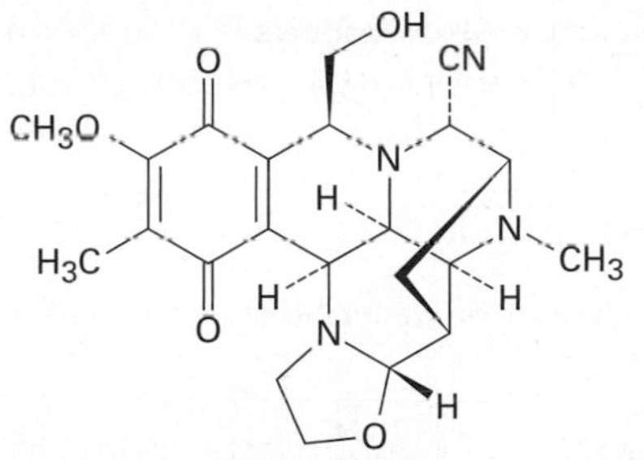

Cyanocycline A

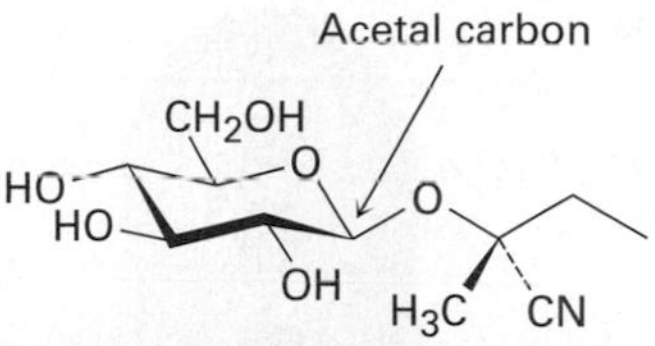

Acetal carbon

Lotaustralin (a cyanogenic glycoside)

Preparation of Nitriles

The simplest method of nitrile preparation is the S_N2 reaction of CN^- with a primary or secondary alkyl halide, as discussed in **Section 20.5**. Another method for preparing nitriles is by dehydration of a primary amide, $RCONH_2$. Thionyl chloride is often used for the reaction, although other dehydrating agents such as $POCl_3$ also work.

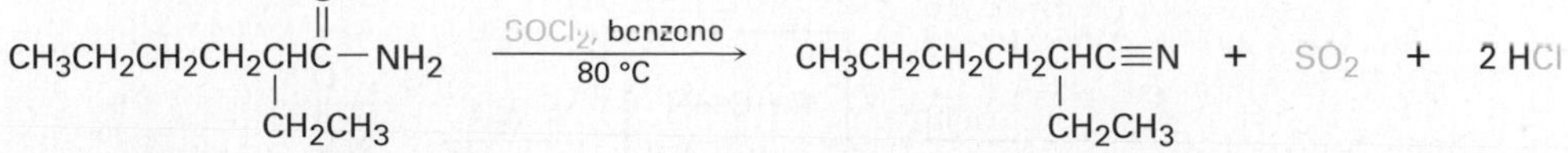

$$CH_3CH_2CH_2CH_2CH(CH_2CH_3)C(=O){-}NH_2 \xrightarrow[80\ °C]{SOCl_2,\ benzene} CH_3CH_2CH_2CH_2CH(CH_2CH_3)C{\equiv}N + SO_2 + 2\ HCl$$

2-Ethylhexanamide **2-Ethylhexanenitrile (94%)**

The dehydration occurs by initial reaction of $SOCl_2$ on the nucleophilic amide oxygen atom, followed by deprotonation and a subsequent E2-like elimination reaction.

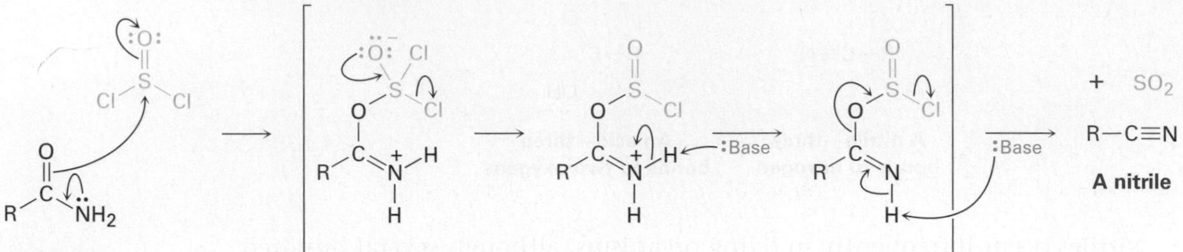

An amide

A nitrile

Both methods of nitrile synthesis—S_N2 displacement by CN^- on an alkyl halide and amide dehydration—are useful, but the synthesis from amides is more general because it is not limited by steric hindrance.

Reactions of Nitriles

Like a carbonyl group, a nitrile group is strongly polarized and has an electrophilic carbon atom. Nitriles therefore react with nucleophiles to yield sp^2-hybridized imine anions in a reaction analogous to the formation of an sp^3-hybridized alkoxide ion by nucleophilic addition to a carbonyl group.

Carbonyl compound

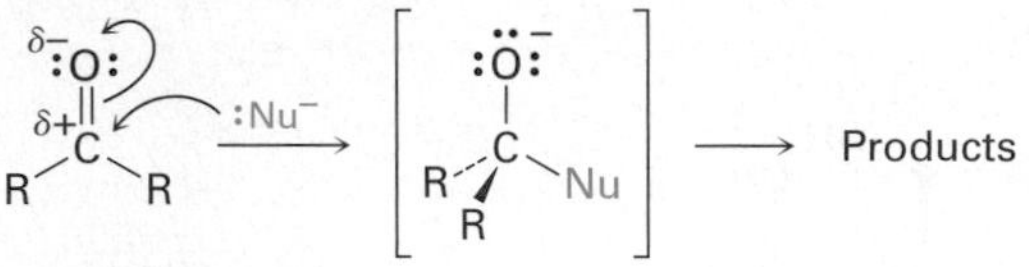

Products

Nitrile

Products

Imine anion

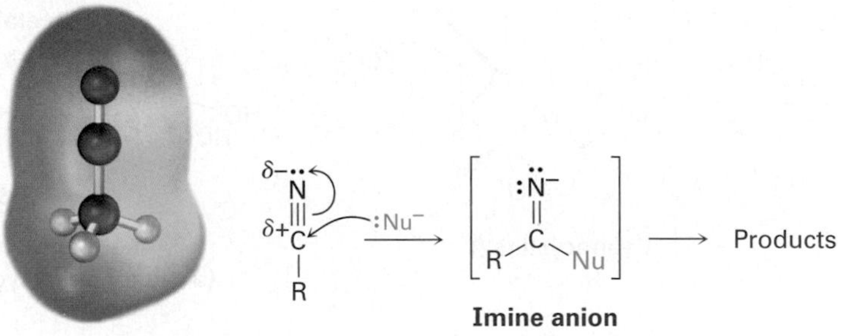

Some general reactions of nitriles are shown in **Figure 20.3**.

Figure 20.3 Some reactions of nitriles.

Amide

Carboxylic acid

Nitrile

Amine

Ketone

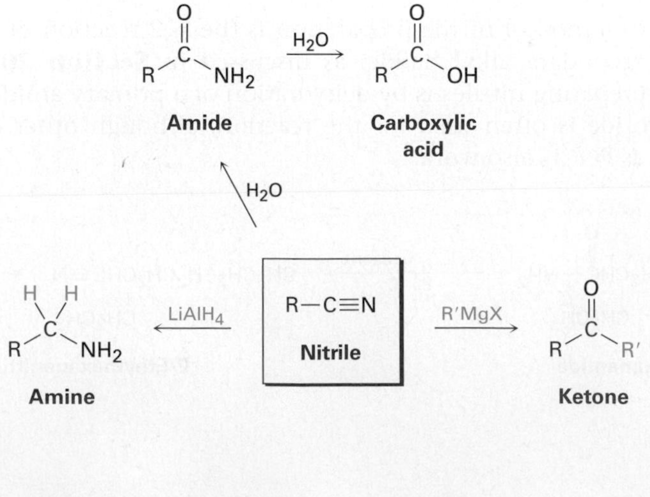

Hydrolysis: Conversion of Nitriles into Carboxylic Acids Among the most useful reactions of nitriles is their hydrolysis to yield first an amide and then a carboxylic acid plus ammonia or an amine. The reaction occurs in either basic or acidic aqueous solution:

$$\underset{\textbf{A nitrile}}{R{-}C{\equiv}N} \xrightarrow[\text{or NaOH, } H_2O]{H_3O^+} \underset{\textbf{An amide}}{RC(=O)NH_2} \xrightarrow[\text{or NaOH, } H_2O]{H_3O^+} \underset{\textbf{A carboxylic acid}}{RC(=O)OH} + NH_3$$

As shown in **Figure 20.4**, base-catalyzed nitrile hydrolysis involves nucleophilic addition of hydroxide ion to the polar C≡N bond to give an imine anion in a process similar to nucleophilic addition to a polar C=O bond to give an alkoxide anion. Protonation then gives a hydroxy imine, which tautomerizes **(Section 9.4)** to an amide in a step similar to the tautomerization of an enol to a ketone. Further hydrolysis gives a carboxylate ion.

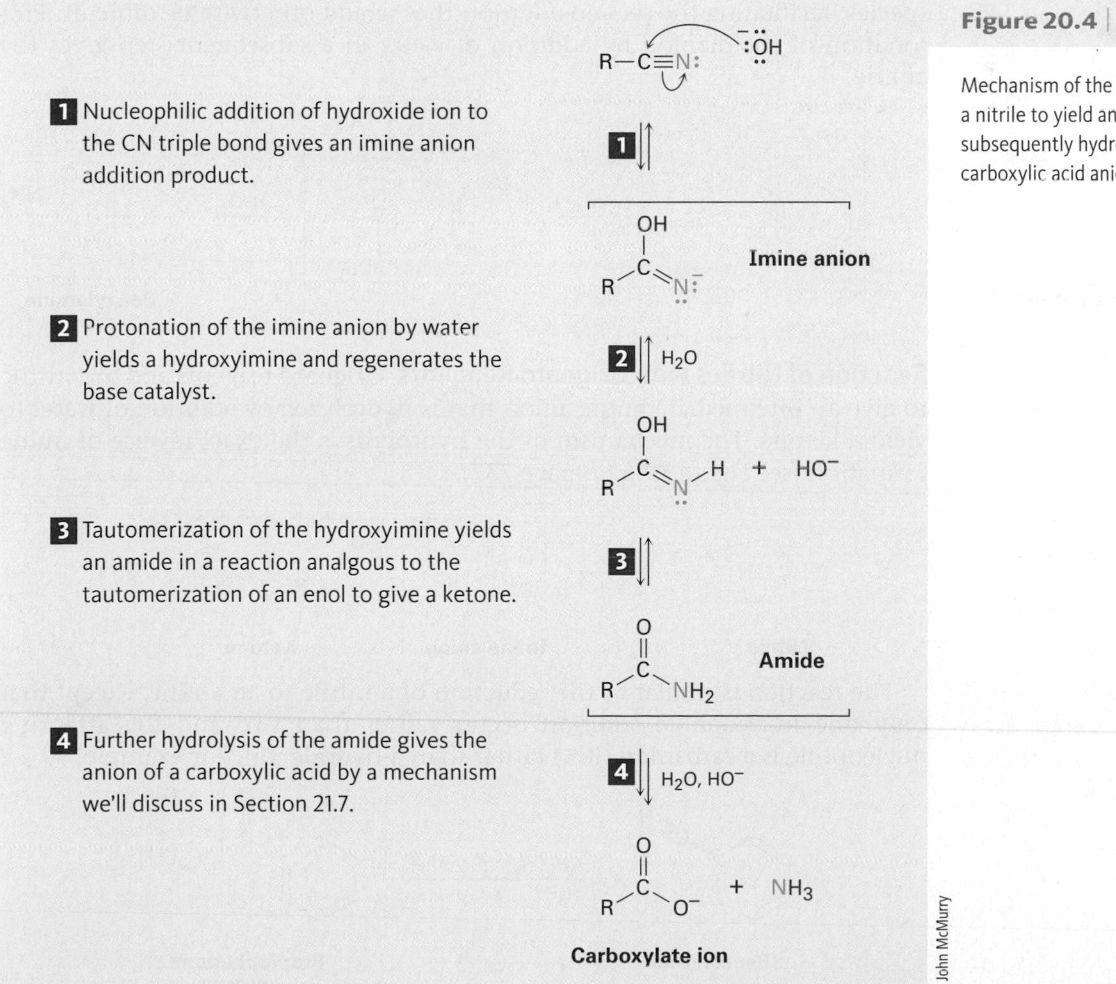

Figure 20.4 | MECHANISM

Mechanism of the basic hydrolysis of a nitrile to yield an amide, which is subsequently hydrolyzed further to a carboxylic acid anion.

The further hydrolysis of the amide intermediate takes place by a nucleophilic addition of hydroxide ion to the amide carbonyl group, which yields a tetrahedral alkoxide ion. Expulsion of amide ion, NH_2^-, as leaving group gives the carboxylate ion, thereby driving the reaction toward products. Subsequent acidification in a separate step yields the carboxylic acid. We'll look at this process in more detail in **Section 21.7**.

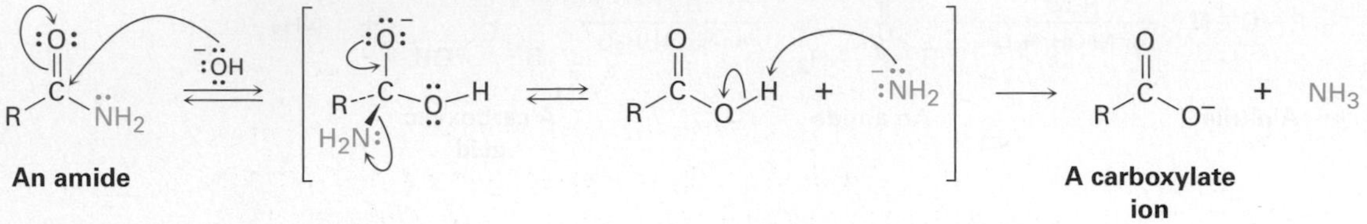

An amide

A carboxylate ion

Reduction: Conversion of Nitriles into Amines Reduction of a nitrile with $LiAlH_4$ gives a primary amine, RNH_2. The reaction occurs by nucleophilic addition of hydride ion to the polar C≡N bond, yielding an imine anion, which still contains a C=N bond and therefore undergoes a second nucleophilic addition of hydride to give a *dianion*. Both monoanion and dianion intermediates are undoubtedly stabilized by Lewis acid–base complexation to an aluminum species, facilitating the second addition that would otherwise be difficult. Protonation of the dianion by addition of water in a subsequent step gives the amine.

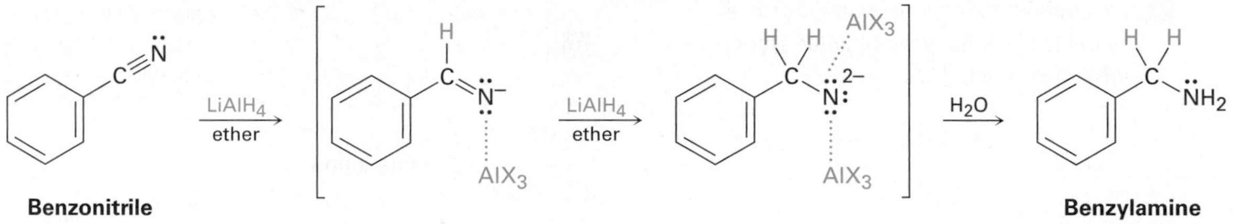

Benzonitrile

Benzylamine

Reaction of Nitriles with Grignard Reagents Grignard reagents add to a nitrile to give an intermediate imine anion that is hydrolyzed by addition of water to yield a ketone. The mechanism of the hydrolysis is the exact reverse of imine formation (see Figure 19.6 on page 737).

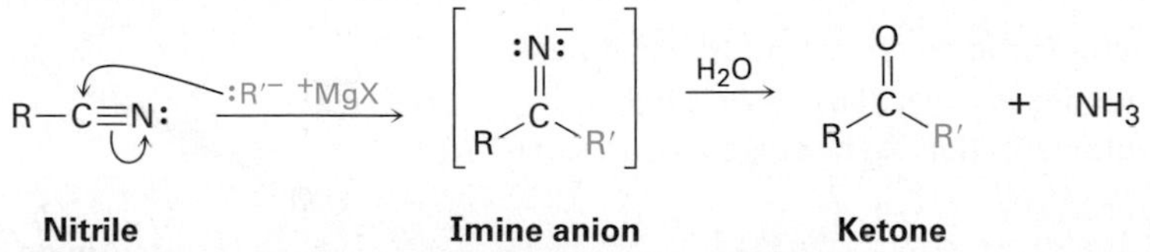

Nitrile

Imine anion

Ketone

The reaction is similar to the reduction of a nitrile to an amine, except that only one nucleophilic addition occurs rather than two and the attacking nucleophile is a carbanion (R:⁻) rather than a hydride ion. For example:

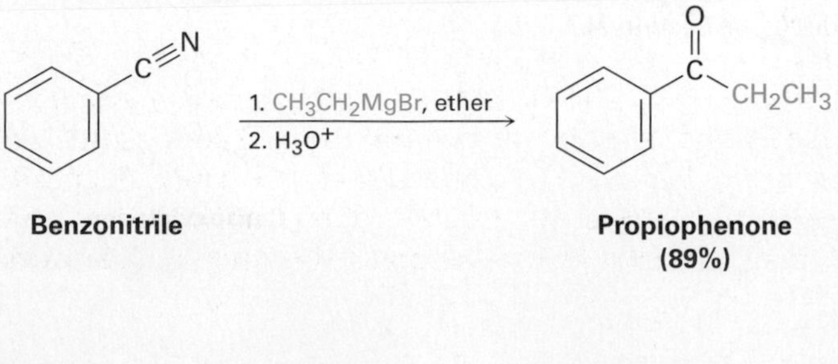

Benzonitrile

Propiophenone (89%)

Synthesizing a Ketone from a Nitrile

Worked Example 20.3

How would you prepare 2-methyl-3-pentanone from a nitrile?

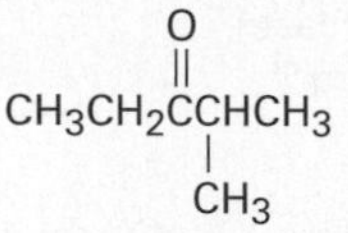

$$\mathrm{CH_3CH_2\overset{O}{\overset{\|}{C}}\underset{CH_3}{\underset{|}{C}}HCH_3} \quad \textbf{2-Methyl-3-pentanone}$$

Strategy

A ketone results from the reaction between a Grignard reagent and a nitrile, with the C$\equiv$N carbon of the nitrile becoming the carbonyl carbon. Identify the two groups attached to the carbonyl carbon atom in the product. One will come from the Grignard reagent, and the other will come from the nitrile.

Solution

There are two possibilities.

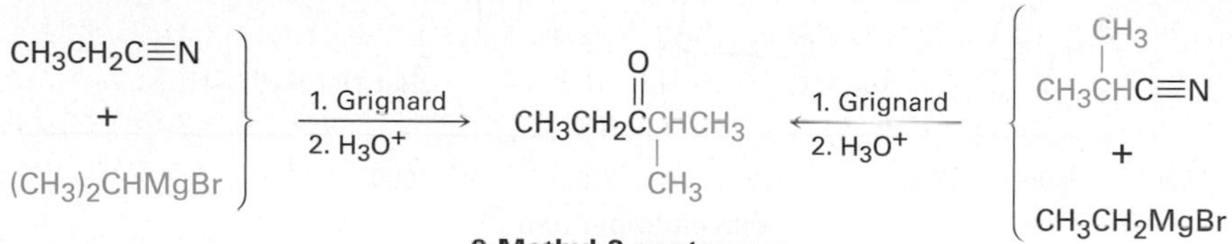

$$\left.\begin{matrix} \mathrm{CH_3CH_2C{\equiv}N} \\ + \\ \mathrm{(CH_3)_2CHMgBr} \end{matrix}\right\} \xrightarrow[\text{2. } \mathrm{H_3O^+}]{\text{1. Grignard}} \mathrm{CH_3CH_2\overset{O}{\overset{\|}{C}}\underset{CH_3}{\underset{|}{C}}HCH_3} \xleftarrow[\text{2. } \mathrm{H_3O^+}]{\text{1. Grignard}} \left\{\begin{matrix} \mathrm{CH_3\overset{CH_3}{\overset{|}{C}}HC{\equiv}N} \\ + \\ \mathrm{CH_3CH_2MgBr} \end{matrix}\right.$$

2-Methyl-3-pentanone

Problem 20.13

How would you prepare the following carbonyl compounds from a nitrile?

(a) $\mathrm{CH_3CH_2\overset{O}{\overset{\|}{C}}CH_2CH_3}$

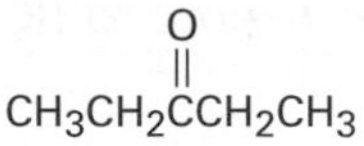

(b) $p\text{-}\mathrm{O_2N{-}C_6H_4{-}\overset{O}{\overset{\|}{C}}{-}CH_3}$

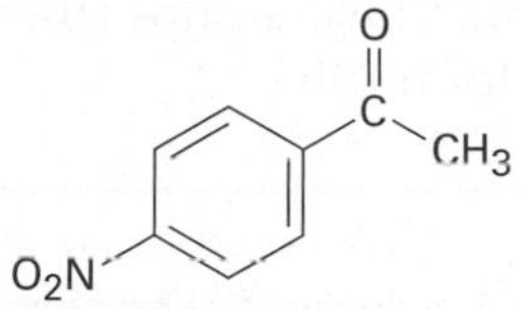

Problem 20.14

How would you prepare 1-phenyl-2-butanone, $C_6H_5CH_2COCH_2CH_3$, from benzyl bromide, $C_6H_5CH_2Br$? More than one step is required.

20.8 Spectroscopy of Carboxylic Acids and Nitriles

Infrared Spectroscopy

Carboxylic acids have two characteristic IR absorptions that make the $-CO_2H$ group easily identifiable. The O–H bond of the carboxyl group gives rise to a very broad absorption over the range 2500 to 3300 cm^{-1}, and the C=O bond shows an absorption between 1710 and 1760 cm^{-1}. The exact position of C=O absorption depends both on the structure of the molecule and on whether the acid is free (monomeric) or hydrogen-bonded (dimeric). Free carboxyl groups absorb at 1760 cm^{-1}, but the more commonly encountered dimeric carboxyl groups absorb in a broad band centered around 1710 cm^{-1}.

Free carboxyl (uncommon), 1760 cm^{-1}: R—C(=O)—O—H

Associated carboxyl (usual case), 1710 cm^{-1}: R—C(=O)—O—H······O=C(—R)—O—H······O (dimer)

Both the broad O–H absorption and the C=O absorption at 1710 cm^{-1} (dimeric) are identified in the IR spectrum of butanoic acid shown in **Figure 20.5**.

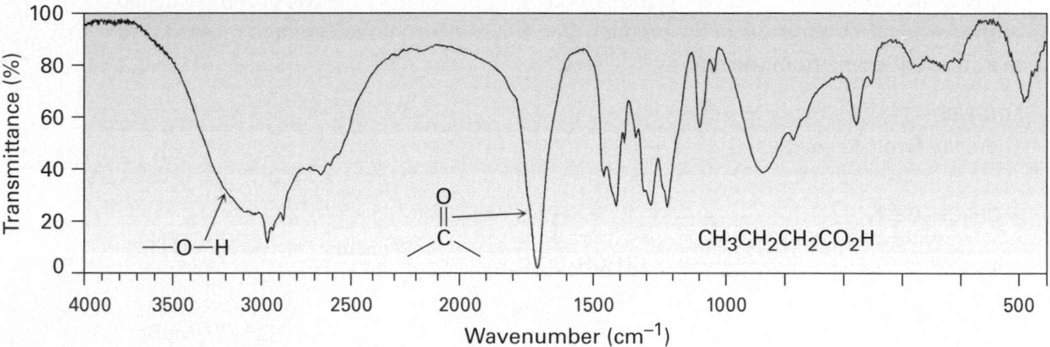

Figure 20.5 IR spectrum of butanoic acid, $CH_3CH_2CH_2CO_2H$.

Nitriles show an intense and easily recognizable C≡N bond absorption near 2250 cm^{-1} for saturated compounds and 2230 cm^{-1} for aromatic and conjugated molecules. Few other functional groups absorb in this region, so IR spectroscopy is highly diagnostic for nitriles.

Problem 20.15

Cyclopentanecarboxylic acid and 4-hydroxycyclohexanone have the same formula ($C_6H_{10}O_2$), and both contain an —OH and a C=O group. How could you distinguish between them by IR spectroscopy?

Nuclear Magnetic Resonance Spectroscopy

Carboxyl carbon atoms absorb in the range 165 to 185 δ in the ^{13}C NMR spectrum, with aromatic and α,β-unsaturated acids near the upfield end of the range (~165 δ) and saturated aliphatic acids near the downfield end (~185 δ). Nitrile carbons absorb in the range 115 to 130 δ.

CH_3CH_2COH (C=O): 9, 25, 181

Benzoic acid (C_6H_5COOH): 134, 128,130, 129, 173

$CH_3CH{=}CHCOH$ (C=O): 18, 122, 148, 172

$CH_3CH_2C{\equiv}N$: 10, 11, 121

In the ^1H NMR spectrum, the acidic $-CO_2H$ proton normally absorbs as a singlet near 12 δ. As with alcohols **(Section 17.11)**, the $-CO_2H$ proton can be replaced by deuterium when D_2O is added to the sample tube, causing the absorption to disappear from the NMR spectrum. **Figure 20.6** shows the ^1H NMR spectrum of phenylacetic acid. Note that the carboxyl proton absorption occurs at 12.0 δ.

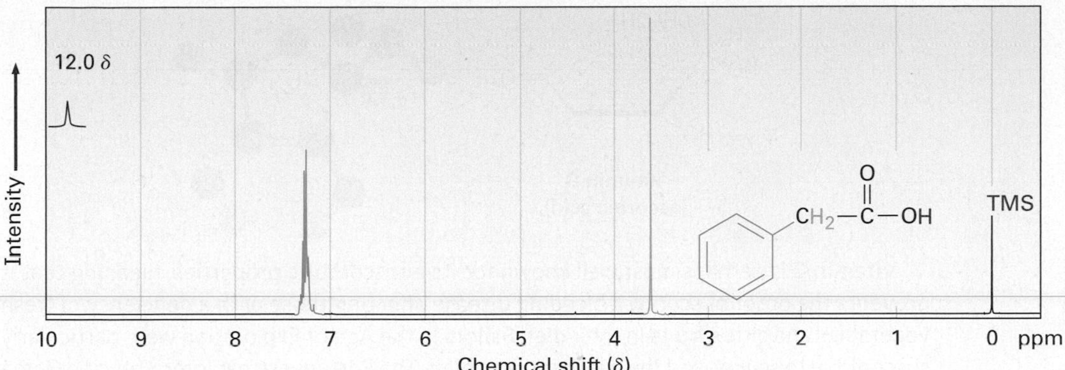

Figure 20.6 Proton NMR spectrum of phenylacetic acid, $C_6H_5CH_2CO_2H$.

Problem 20.16

How could you distinguish between the isomers cyclopentanecarboxylic acid and 4-hydroxycyclohexanone by ^1H and ^{13}C NMR spectroscopy? (See Problem 20.15.)

A DEEPER LOOK Vitamin C

The word *vitamin,* despite its common usage, is actually an imprecise term. Generally speaking, a vitamin is an organic substance that a given organism requires in small amounts to live and grow but is unable to synthesize and must obtain in its diet. Thus, to be considered a vitamin, only a *small* amount of the substance is needed—anywhere from a few micrograms to 100 mg or so per day. Dietary substances needed in larger amounts, such as some amino acids and unsaturated fats, are not considered vitamins.

Furthermore, different organisms need different vitamins. More than 4000 species of mammals can synthesize ascorbic acid in their bodies, for instance, but humans are not among them. Ascorbic acid is therefore a human vitamin—what we all know as vitamin C—and must be obtained in our diet. Small amounts of more than a dozen other substances are similarly required by humans: retinol (vitamin A), thiamine (vitamin B_1), and tocopherol (vitamin E), for instance.

Vitamin C is surely the best known of all human vitamins. It was the first to be discovered (1928), the first to be structurally characterized (1933), and the first to be synthesized in the laboratory (1933). Over 110,000 metric tons of vitamin C is synthesized worldwide each year, more than the total amount of all other vitamins

In addition to the hazards of weather, participants in early polar expeditions often suffered from scurvy, caused by a dietary vitamin C deficiency.

(continued)

(continued)

combined. In addition to its use as a vitamin supplement, vitamin C is used as a food preservative, a "flour improver" in bakeries, and an animal food additive.

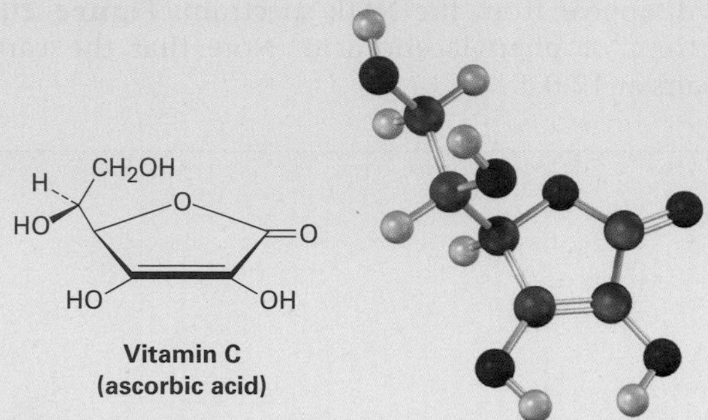

Vitamin C
(ascorbic acid)

Vitamin C is perhaps most well known for its antiscorbutic properties, meaning that it prevents the onset of scurvy, a bleeding disease affecting those with a deficiency of fresh vegetables and citrus fruits in their diet. Sailors in the Age of Exploration were particularly susceptible to scurvy, and the death toll was high. The Portuguese explorer Vasco da Gama lost more than half his crew to scurvy during his 2-year voyage around the Cape of Good Hope in 1497–1499.

In more recent times, large doses of vitamin C have been claimed to prevent the common cold, cure infertility, delay the onset of symptoms in acquired immunodeficiency syndrome (AIDS), and inhibit the development of gastric and cervical cancers. None of these claims have been backed by medical evidence, however. In the largest study yet done of the effect of vitamin C on the common cold, a meta-analysis of more than 100 separate trials covering 40,000 people found no difference in the incidence of colds between those who took supplemental vitamin C regularly and those who did not. When taken *during* a cold, however, vitamin C does appear to decrease the cold's duration by perhaps a day.

The industrial preparation of vitamin C involves an unusual blend of biological and laboratory organic chemistry, beginning with glucose and following the five-step route shown in **Figure 20.7**. Glucose, a pentahydroxy aldehyde, is first reduced to sorbitol, which

(continued)

Summary

Key words

carboxyl group, 779
carboxylation, 790
carboxylic acid (RCO_2H), 778
Henderson–Hasselbalch equation, 785
nitrile ($RC{\equiv}N$), 781

Carboxylic acids are among the most useful building blocks for synthesizing other molecules, both in nature and in the chemical laboratory. Thus, an understanding of their properties and reactions is fundamental to understanding biological chemistry. In this chapter, we've looked both at acids and at their close relatives, **nitriles ($RC{\equiv}N$)**.

Carboxylic acids are named systematically by replacing the terminal *-e* of the corresponding alkane name with *-oic acid*. Like aldehydes and ketones, the carbonyl carbon atom is sp^2-hybridized; like alcohols, carboxylic acids are associated through hydrogen-bonding and therefore have high boiling points.

The distinguishing characteristic of carboxylic acids is their acidity. Although weaker than mineral acids such as HCl, carboxylic acids dissociate much more readily than alcohols because the resultant carboxylate ions are stabilized by resonance between two equivalent forms.

(continued)

is then oxidized by the microorganism *Acetobacter suboxydans*. No chemical reagent is known that is selective enough to oxidize only one of the six alcohol groups in sorbitol, so an enzymatic reaction is used. Treatment with acetone and an acid catalyst then converts four of the other hydroxyl groups into acetal linkages, and the remaining hydroxyl group is chemically oxidized to a carboxylic acid by reaction with aqueous NaOCl (household bleach). Hydrolysis with acid then removes the two acetal groups and causes an internal ester-forming reaction to take place to give ascorbic acid. Each of the five steps takes place in better than 90% yield.

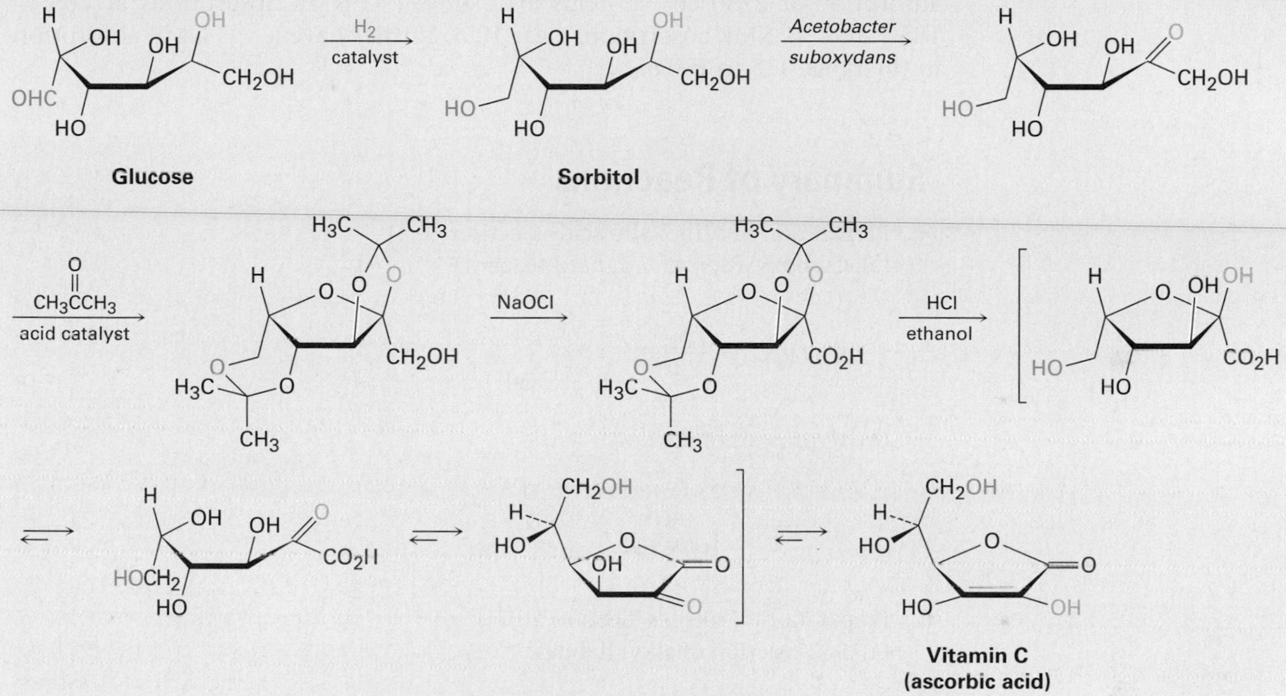

Figure 20.7 The industrial synthesis of ascorbic acid from glucose.

Most carboxylic acids have pK_a values near 5, but the exact pK_a of a given acid depends on structure. Carboxylic acids substituted by electron-withdrawing groups are more acidic (have a lower pK_a) because their carboxylate ions are stabilized. Carboxylic acids substituted by electron-donating groups are less acidic (have a higher pK_a) because their carboxylate ions are destabilized. The extent of dissociation of a carboxylic acid in a buffered solution of a given pH can be calculated with the **Henderson–Hasselbalch equation**. Inside living cells, where the physiological pH = 7.3, carboxylic acids are entirely dissociated and exist as their carboxylate anions.

Methods of synthesis for carboxylic acids include (1) oxidation of alkylbenzenes, (2) oxidation of primary alcohols or aldehydes, (3) reaction of Grignard reagents with CO_2 **(carboxylation)**, and (4) hydrolysis of nitriles. General reactions of carboxylic acids include (1) loss of the acidic proton, (2) nucleophilic acyl substitution at the carbonyl group, (3) substitution on the α carbon, and (4) reduction.

Nitriles are similar in some respects to carboxylic acids and are prepared either by S_N2 reaction of an alkyl halide with cyanide ion or by dehydration of an amide. Nitriles undergo nucleophilic addition to the polar $C{\equiv}N$ bond in the same way that carbonyl compounds do. The most important reactions of nitriles are their hydrolysis to carboxylic acids, reduction to primary amines, and reaction with Grignard reagents to yield ketones.

Carboxylic acids and nitriles are easily distinguished spectroscopically. Acids show a characteristic IR absorption at 2500 to 3300 cm^{-1} due to the O–H bond and another at 1710 to 1760 cm^{-1} due to the C=O bond; nitriles have an absorption at 2250 cm^{-1}. Acids also show ^{13}C NMR absorptions at 165 to 185 δ and ^1H NMR absorptions near 12 δ. Nitriles have a ^{13}C NMR absorption in the range 115 to 130 δ.

Summary of Reactions

1. Preparation of carboxylic acids (Section 20.5)
   (a) Carboxylation of Grignard reagents

$$R{-}MgX \xrightarrow[2.\ H_3O^+]{1.\ CO_2} RCOOH$$

   (b) Hydrolysis of nitriles

$$R{-}C{\equiv}N \xrightarrow[NaOH,\ H_2O]{H_3O^+} RCOOH$$

2. Preparation of nitriles (Section 20.7)
   (a) S_N2 reaction of alkyl halides

$$RCH_2Br \xrightarrow{NaCN} RCH_2C{\equiv}N$$

   (b) Dehydration of amides

$$RCONH_2 \xrightarrow{SOCl_2} R{-}C{\equiv}N + SO_2 + 2\ HCl$$

3. Reactions of nitriles (Section 20.7)
   (a) Hydrolysis to yield carboxylic acids

$$R{-}C{\equiv}N \xrightarrow[2.\ H_3O^+]{1.\ NaOH,\ H_2O} RCOOH + NH_3$$

   (b) Reduction to yield primary amines

$$R{-}C{\equiv}N \xrightarrow[2.\ H_2O]{1.\ LiAlH_4} RCH_2NH_2$$

(c) Reaction with Grignard reagents to yield ketones

$$R{-}C{\equiv}N \xrightarrow[\text{2. } H_3O^+]{\text{1. } R'MgX\text{, ether}} R{-}\overset{O}{\overset{\|}{C}}{-}R' + NH_3$$

Exercises

Visualizing Chemistry

(Problems 20.1–20.16 appear within the chapter.)

OWL Interactive versions of these problems are assignable in OWL for Organic Chemistry.

20.17 Give IUPAC names for the following carboxylic acids (reddish brown = Br):

(a)

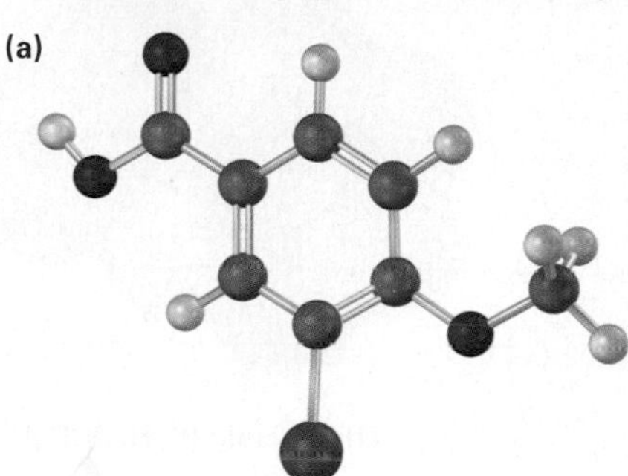

(b)

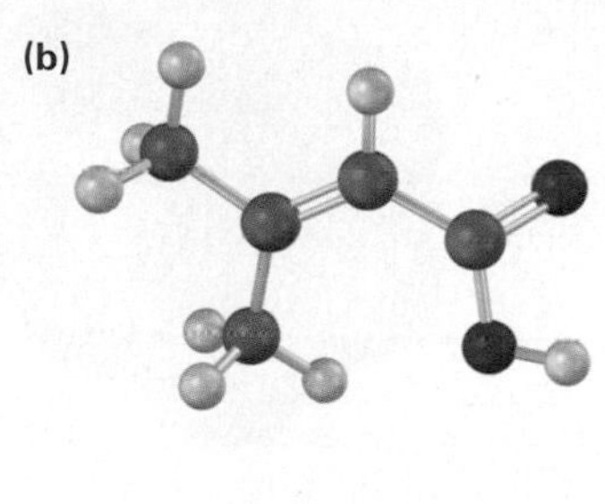

(c)

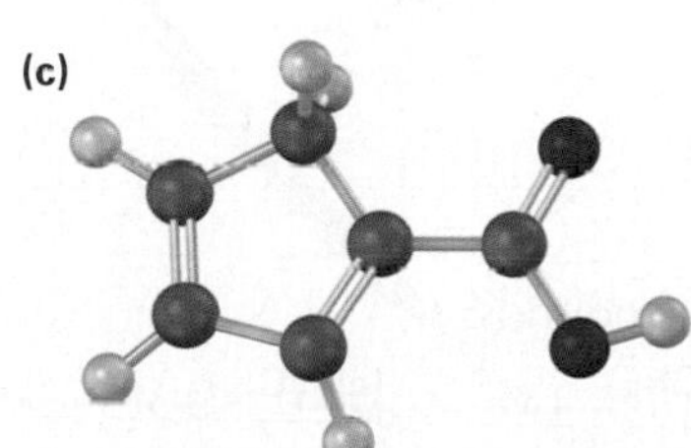

(d)

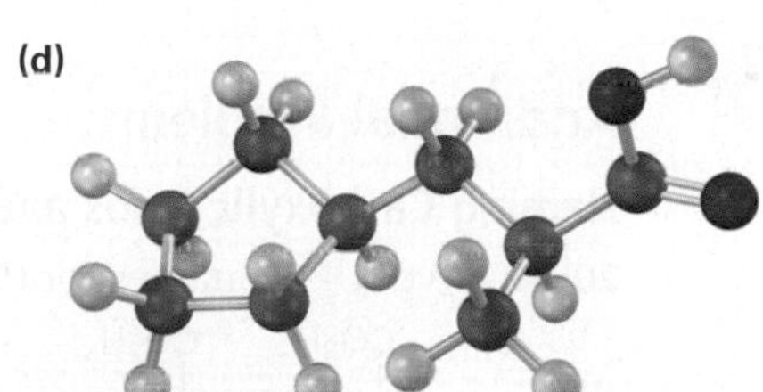

20.18 Would you expect the following carboxylic acids to be more acidic or less acidic than benzoic acid? Explain. (Reddish brown = Br.)

(a)

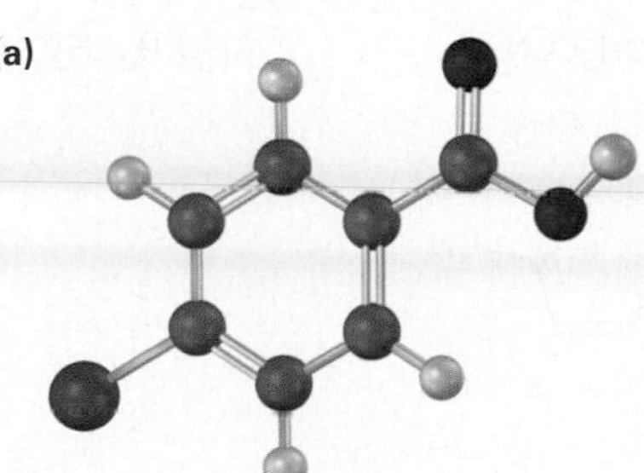

(b)

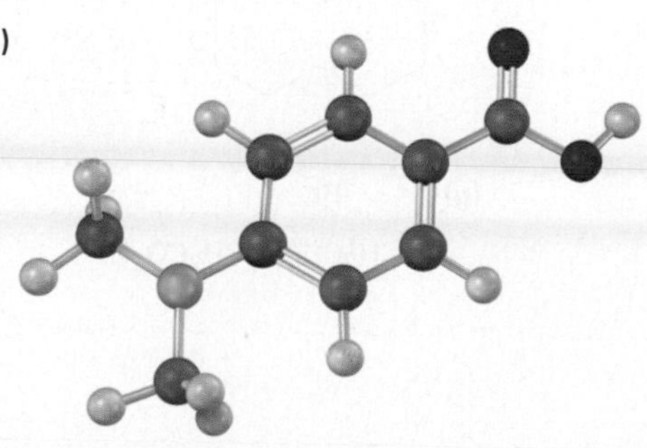

20.19 The following carboxylic acid can't be prepared from an alkyl halide by either the nitrile hydrolysis route or the Grignard carboxylation route. Explain.

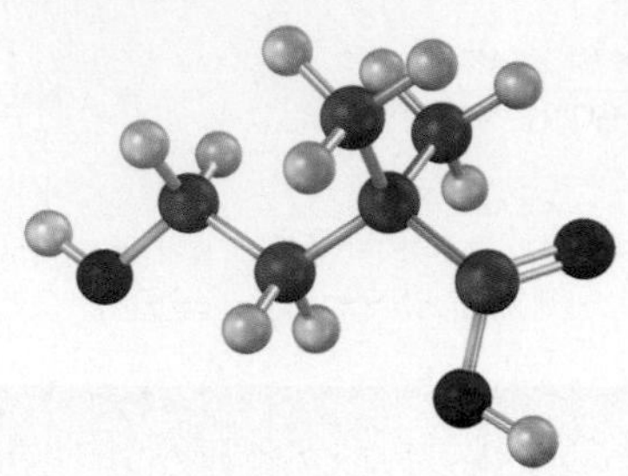

20.20 Electrostatic potential maps of anisole and thioanisole are shown. Which do you think is the stronger acid, *p*-methoxybenzoic acid or *p*-(methylthio)benzoic acid? Explain.

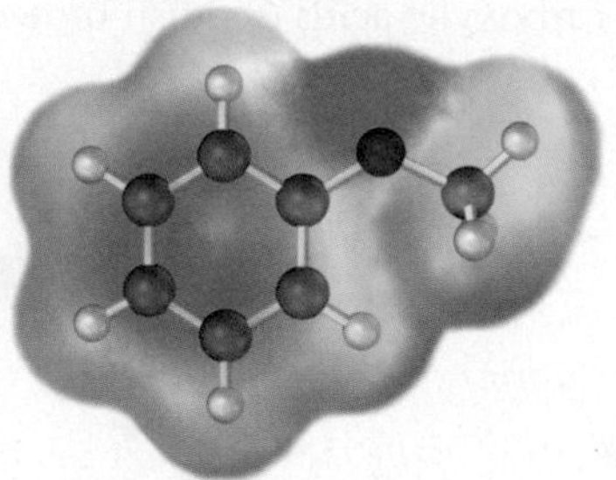

Anisole ($C_6H_5OCH_3$)

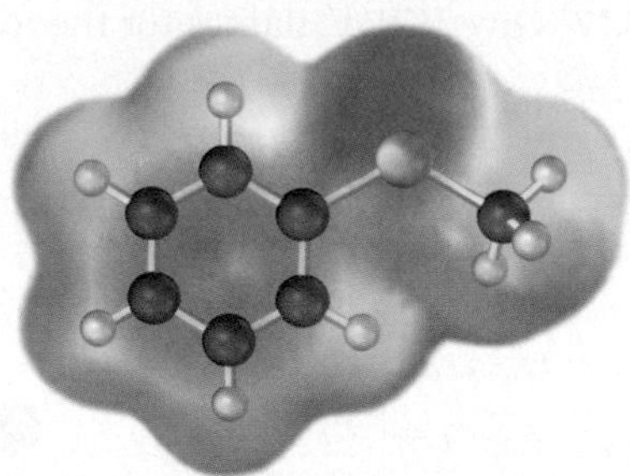

Thioanisole ($C_6H_5SCH_3$)

Additional Problems

Naming Carboxylic Acids and Nitriles

20.21 Give IUPAC names for the following compounds:

(a)
```
   CO2H       CO2H
   |          |
CH3CHCH2CH2CHCH3
```

(b)
```
   CH3
   |
CH3CCO2H
   |
   CH3
```

(c) NC–(benzene ring, meta)–CO_2H

(d) (cyclodecene ring)–CO_2H

(e)
```
   CH3
   |
CH3CCN
   |
   CH3
```

(f)
```
            CH2CO2H
            |
CH3CH2CH2CHCH2CH3
```

(g)
```
       Br
       |
BrCH2CHCH2CH2CO2H
```

(h) (cyclopentene ring)–CN

20.22 Draw structures corresponding to the following IUPAC names:
(a) *cis*-1,2-Cyclohexanedicarboxylic acid **(b)** Heptanedioic acid
(c) 2-Hexen-4-ynoic acid **(d)** 4-Ethyl-2-propyloctanoic acid
(e) 3-Chlorophthalic acid **(f)** Triphenylacetic acid
(g) 2-Cyclobutenecarbonitrile **(h)** *m*-Benzoylbenzonitrile

20.23 Draw and name the following:
(a) The eight carboxylic acids with the formula $C_6H_{12}O_2$
(b) Three nitriles with the formula C_5H_7N

20.24 Pregabalin, marketed as Lyrica, is an anticonvulsant drug that is also effective in treating chronic pain. The IUPAC name of pregabalin is (*S*)-3-(aminomethyl)-5-methylhexanoic acid. (An aminomethyl group is $-CH_2NH_2$.) Draw the structure of pregabalin.

20.25 Isocitric acid, an intermediate in the citric acid cycle of food metabolism, has the systematic name (2*R*,3*S*)-3-carboxy-2-hydroxypentanedioic acid. Draw the structure.

Acidity of Carboxylic Acids

20.26 Order the compounds in each of the following sets with respect to increasing acidity:
(a) Acetic acid, oxalic acid, formic acid
(b) *p*-Bromobenzoic acid, *p*-nitrobenzoic acid, 2,4-dinitrobenzoic acid
(c) Fluoroacetic acid, 3-fluoropropanoic acid, iodoacetic acid

20.27 Arrange the compounds in each of the following sets in order of increasing basicity:
(a) Magnesium acetate, magnesium hydroxide, methylmagnesium bromide
(b) Sodium benzoate, sodium *p*-nitrobenzoate, sodium acetylide
(c) Lithium hydroxide, lithium ethoxide, lithium formate

20.28 Calculate pK_a's for the following acids:
(a) Lactic acid, $K_a = 8.4 \times 10^{-4}$ **(b)** Acrylic acid, $K_a = 5.6 \times 10^{-6}$

20.29 Calculate K_a's for the following acids:
(a) Citric acid, p$K_a = 3.14$ **(b)** Tartaric acid, p$K_a = 2.98$

20.30 Thioglycolic acid, $HSCH_2CO_2H$, a substance used in depilatory agents (hair removers) has p$K_a = 3.42$. What is the percent dissociation of thioglycolic acid in a buffer solution at pH = 3.0?

20.31 In humans, the final product of purine degradation from DNA is uric acid, p$K_a = 5.61$, which is excreted in the urine. What is the percent dissociation of uric acid in urine at a typical pH = 6.0? Why do you think uric acid is acidic even though it does not have a CO_2H group?

Uric acid

20.32 Some pK_a data for simple dibasic acids are shown. How can you account for the fact that the difference between the first and second ionization constants decreases with increasing distance between the carboxyl groups?

Name	Structure	pK_1	pK_2
Oxalic	HO_2CCO_2H	1.2	4.2
Succinic	$HO_2CCH_2CH_2CO_2H$	4.2	5.6
Adipic	$HO_2C(CH_2)_4CO_2H$	4.4	5.4

Reactions of Carboxylic Acids and Nitriles

20.33 How could you convert butanoic acid into the following compounds? Write each step showing the reagents needed.
(a) 1-Butanol (b) 1-Bromobutane (c) Pentanoic acid
(d) 1-Butene (e) Octane

20.34 How could you convert each of the following compounds into butanoic acid? Write each step showing all reagents.
(a) 1-Butanol (b) 1-Bromobutane (c) 1-Butene
(d) 1-Bromopropane (e) 4-Octene

20.35 How could you convert butanenitrile into the following compounds? Write each step showing the reagents needed.
(a) 1-Butanol (b) Butylamine (c) 2-Methyl-3-hexanone

20.36 How would you prepare the following compounds from benzene? More than one step is required in each case.
(a) *m*-Chlorobenzoic acid (b) *p*-Bromobenzoic acid
(c) Phenylacetic acid, $C_6H_5CH_2CO_2H$

20.37 Predict the product of the reaction of *p*-methylbenzoic acid with each of the following:
(a) $LiAlH_4$, then H_3O^+ (b) *N*-Bromosuccinimide in CCl_4
(c) CH_3MgBr in ether, then H_3O^+ (d) $KMnO_4$, H_3O^+

20.38 Using $^{13}CO_2$ as your only source of labeled carbon, along with any other compounds needed, how would you synthesize the following compounds?
(a) $CH_3CH_2{}^{13}CO_2H$ (b) $CH_3{}^{13}CH_2CO_2H$

20.39 How would you carry out the following transformations?

CH2 → ? → CO2H
CH2 → ? → CH3, CO2H

20.40 Which method—Grignard carboxylation or nitrile hydrolysis—would you use for each of the following reactions? Explain.

(a) o-$HOC_6H_4CH_2Br$ $\longrightarrow$ o-$HOC_6H_4CH_2CO_2H$

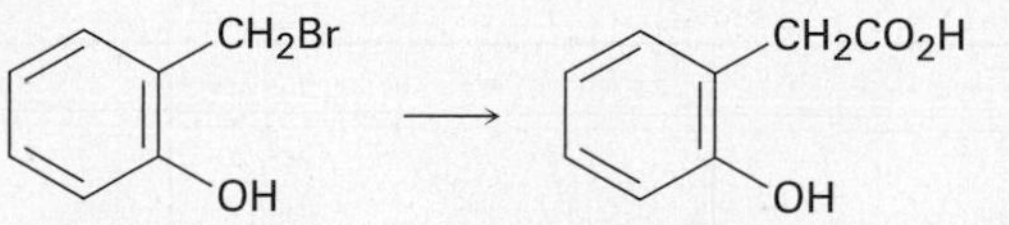

(b) $CH_3CH_2CHBrCH_3$ $\longrightarrow$ $CH_3CH_2CH(CH_3)CO_2H$

(c) $CH_3COCH_2CH_2CH_2I$ $\longrightarrow$ $CH_3COCH_2CH_2CH_2CO_2H$

(d) $HOCH_2CH_2CH_2Br$ $\longrightarrow$ $HOCH_2CH_2CH_2CO_2H$

20.41 1,6-Hexanediamine, a starting material needed for making nylon, can be made from 1,3-butadiene. How would you accomplish the synthesis?

$$H_2C{=}CHCH{=}CH_2 \xrightarrow{?} H_2NCH_2CH_2CH_2CH_2CH_2CH_2NH_2$$

20.42 3-Methyl-2-hexenoic acid (mixture of *E* and *Z* isomers) has been identified as the substance responsible for the odor of human sweat. Synthesize the compound from starting materials having five or fewer carbons.

Spectroscopy

20.43 Propose a structure for a compound $C_6H_{12}O_2$ that dissolves in dilute NaOH and shows the following ^1H NMR spectrum: 1.08 δ (9 H, singlet), 2.2 δ (2 H, singlet), and 11.2 δ (1 H, singlet).

20.44 What spectroscopic method could you use to distinguish among the following three isomeric acids? Tell what characteristic features you would expect for each acid.

$CH_3(CH_2)_3CO_2H$	$(CH_3)_2CHCH_2CO_2H$	$(CH_3)_3CCO_2H$
Pentanoic acid	**3-Methylbutanoic acid**	**2,2-Dimethylpropanoic acid**

20.45 How would you use NMR (either ^{13}C or ^1H) to distinguish between the following pairs of isomers?

(a) m-$C_6H_4(CO_2H)_2$ and p-$C_6H_4(CO_2H)_2$

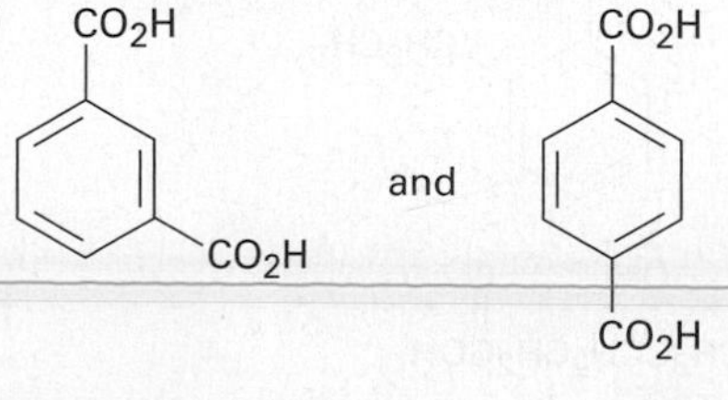

(b) $HO_2CCH_2CH_2CO_2H$ and $CH_3CH(CO_2H)_2$

(c) $CH_3CH_2CH_2CO_2H$ and $HOCH_2CH_2CH_2CHO$

(d) $(CH_3)_2C{=}CHCH_2CO_2H$ and cyclopentane–CO_2H

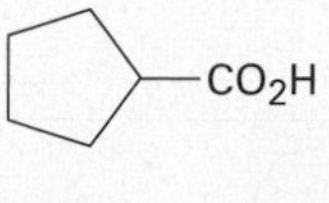

20.46 Compound A, $C_4H_8O_3$, has infrared absorptions at 1710 and 2500 to 3100 cm^{-1} and has the ^1H NMR spectrum shown. Propose a structure for A.

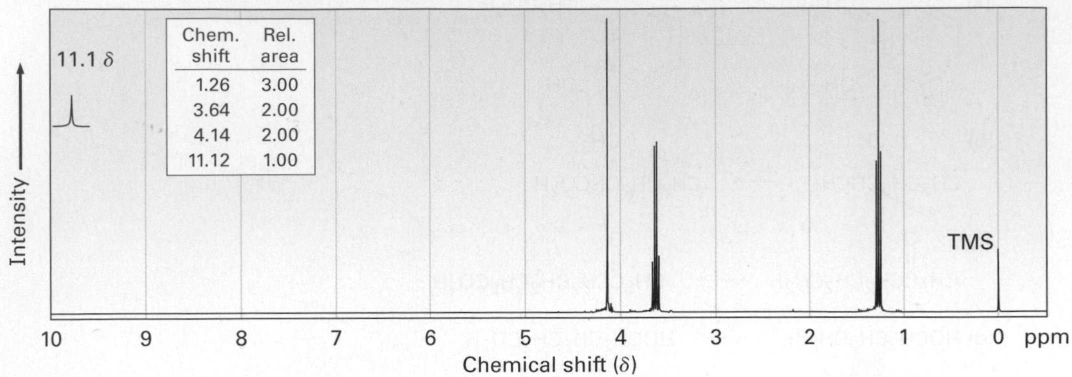

General Problems

20.47 A chemist in need of 2,2-dimethylpentanoic acid decided to synthesize some by reaction of 2-chloro-2-methylpentane with NaCN, followed by hydrolysis of the product. After the reaction sequence was carried out, however, none of the desired product could be found. What do you suppose went wrong?

20.48 Show how you might prepare the anti-inflammatory agent ibuprofen starting from isobutylbenzene. More than one step is needed.

Isobutylbenzene $\Longrightarrow$ Ibuprofen (CO_2H)

20.49 The following synthetic schemes all have at least one flaw in them. What is wrong with each?

(a) $CH_3CH_2CH(Br)CH_2CH_3 \xrightarrow[\text{2. NaCN; 3. } H_3O^+]{\text{1. Mg}} CH_3CH_2CH(CO_2H)CH_2CH_3$

(b) $C_6H_5CH_2CO_2H \xrightarrow[\text{2. } H_3O^+]{\text{1. } LiAlH_4} C_6H_5CH_2CH_3$

(c) $CH_3C(OH)(CH_3)CH_2CH_2Cl \xrightarrow[\text{2. } H_3O^+]{\text{1. NaCN}} CH_3C(OH)(CH_3)CH_2CH_2COH$ (with C=O)

20.50 Naturally occurring compounds called *cyanogenic glycosides,* such as lotaustralin, release hydrogen cyanide, HCN, when treated with aqueous acid. The reaction occurs by hydrolysis of the acetal linkage to form a cyanohydrin, which then expels HCN and gives a carbonyl compound.

(a) Show the mechanism of the acetal hydrolysis and the structure of the cyanohydrin that results.

(b) Propose a mechanism for the loss of HCN, and show the structure of the carbonyl compound that forms.

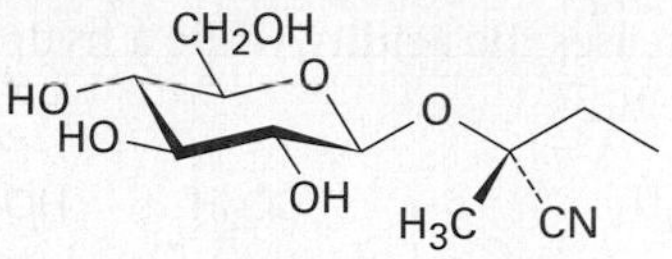

CH_2OH HO HO O OH O H_3C CN **Lotaustralin**

20.51 Acid-catalyzed hydrolysis of a nitrile to give a carboxylic acid occurs by initial protonation of the nitrogen atom, followed by nucleophilic addition of water. Review the mechanism of base-catalyzed nitrile hydrolysis in Section 20.7, and then write all the steps involved in the acid-catalyzed reaction, using curved arrows to represent electron flow in each step.

20.52 *p*-Aminobenzoic acid (PABA) is widely used as a sunscreen agent. Propose a synthesis of PABA starting from toluene.

20.53 Propose a synthesis of the anti-inflammatory drug Fenclorac from phenylcyclohexane.

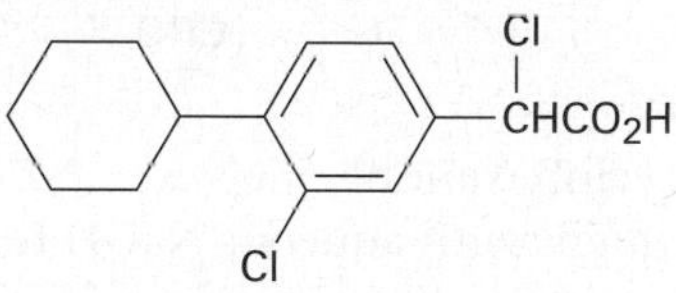

Cl $CHCO_2H$ Cl **Fenclorac**

20.54 The pK_a's of five *p*-substituted benzoic acids ($YC_6H_4CO_2H$) follow. Rank the corresponding substituted benzenes (YC_6H_5) in order of their increasing reactivity toward electrophilic aromatic substitution. If benzoic acid has pK_a = 4.19, which of the substituents are activators and which are deactivators?

Substituent Y	pK_a of Y–C_6H_4–CO_2H
$-Si(CH_3)_3$	4.27
$-CH{=}CHC{\equiv}N$	4.03
$-HgCH_3$	4.10
$-OSO_2CH_3$	3.84
$-PCl_2$	3.59

20.55 How would you carry out the following transformations? More than one step is needed in each case.

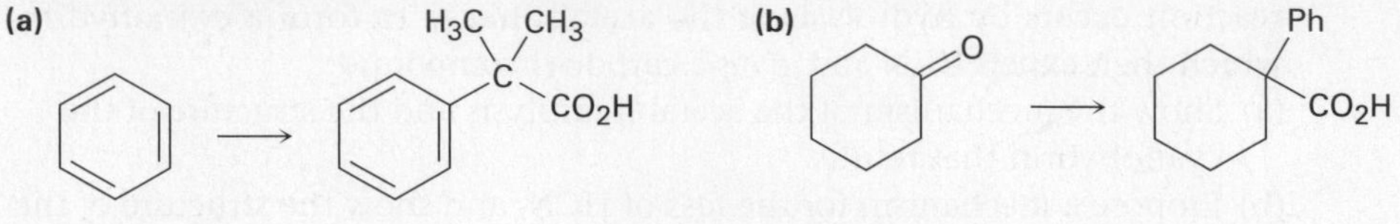

(a) → H_3C CH_3 C CO_2H

(b) O → Ph CO_2H

20.56 The following pK_a values have been measured. Explain why a hydroxyl group in the para position decreases the acidity while a hydroxyl group in the meta position increases the acidity.

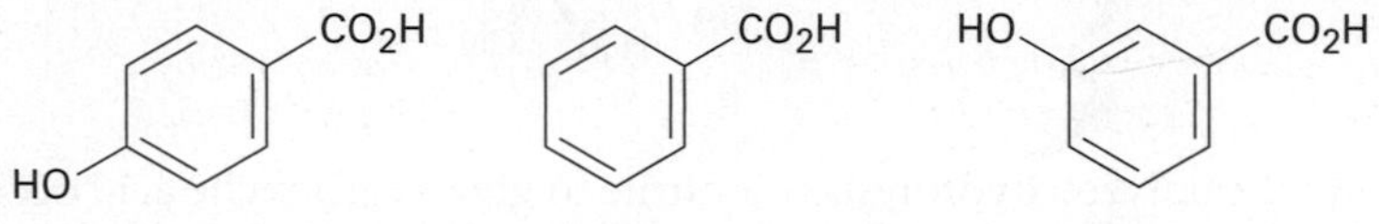

CO_2H HO — **pK_a = 4.48**

CO_2H — **pK_a = 4.19**

HO CO_2H — **pK_a = 4.07**

20.57 Identify the missing reagents a–f in the following scheme:

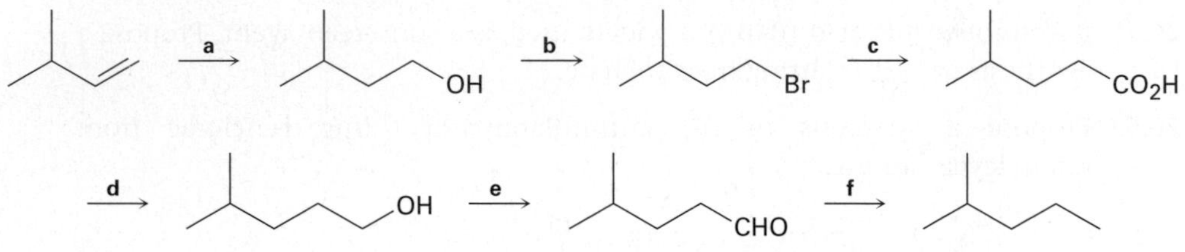

a → OH **b** → Br **c** → CO_2H

d → OH **e** → CHO **f** →

20.58 2-Bromo-6,6-dimethylcyclohexanone gives 2,2-dimethylcyclopentanecarboxylic acid on treatment with aqueous NaOH followed by acidification, a process called the *Favorskii reaction.* Propose a mechanism.

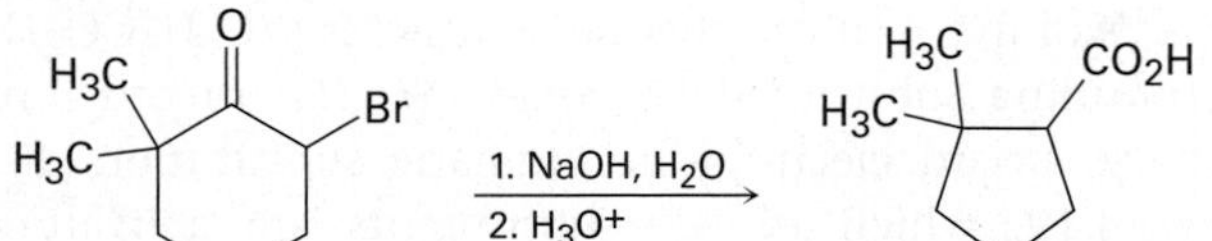

O H_3C H_3C Br — 1. NaOH, H_2O 2. H_3O^+ → H_3C H_3C CO_2H

20.59 Naturally occurring compounds called *terpenoids,* which we'll discuss in Section 27.5, are biosynthesized by a pathway that involves loss of CO_2 from 3-phosphomevalonate 5-diphosphate to yield isopentenyl diphosphate. Use curved arrows to show the mechanism of this reaction.

$$\text{3-Phosphomevalonate 5-diphosphate} \longrightarrow \text{Isopentenyl diphosphate} + CO_2 + PO_4^{3-}$$

3-Phosphomevalonate 5-diphosphate ($^-O_2C{-}CH_2{-}C(CH_3)(OPO_3^{2-}){-}CH_2{-}CH_2OP_2O_5^{3-}$)

Isopentenyl diphosphate ($H_2C{=}C(CH_3){-}CH_2{-}CH_2OP_2O_5^{3-}$)

20.60 Propose a structure for a compound, C_4H_7N, that has the following IR and ^1H NMR spectra:

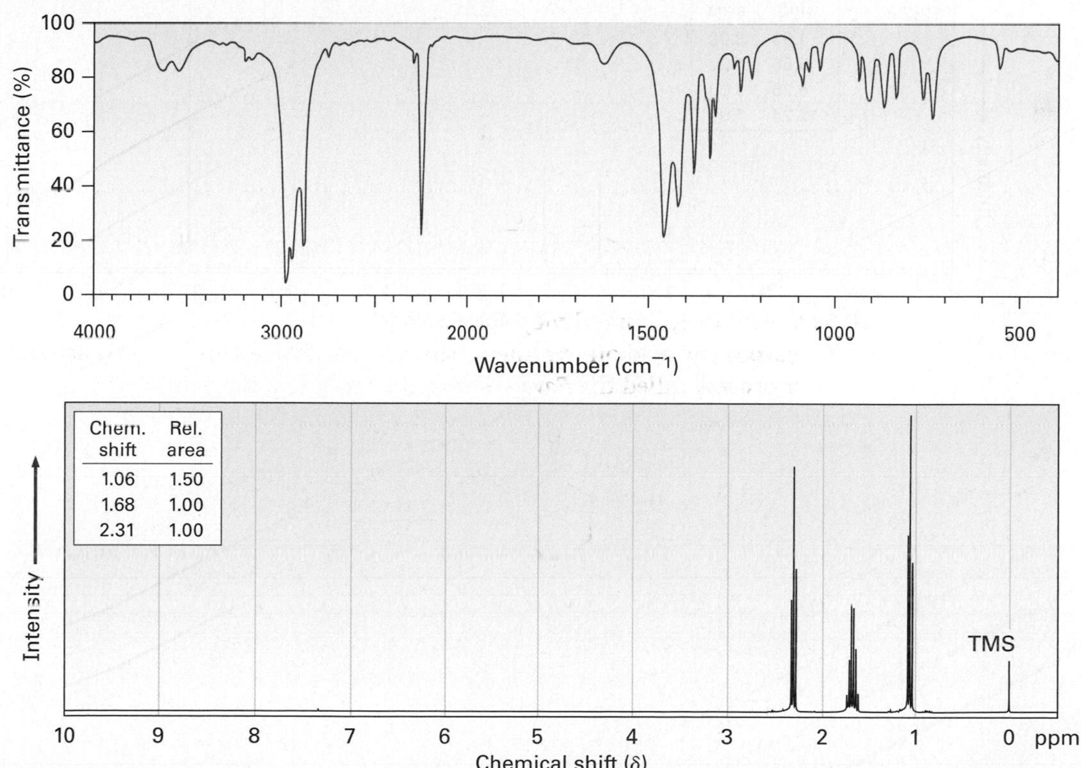

20.61 The two ^{1}H NMR spectra shown here belong to crotonic acid (*trans*-$CH_3CH{=}CHCO_2H$) and methacrylic acid [$H_2C{=}C(CH_3)CO_2H$]. Which spectrum corresponds to which acid? Explain.

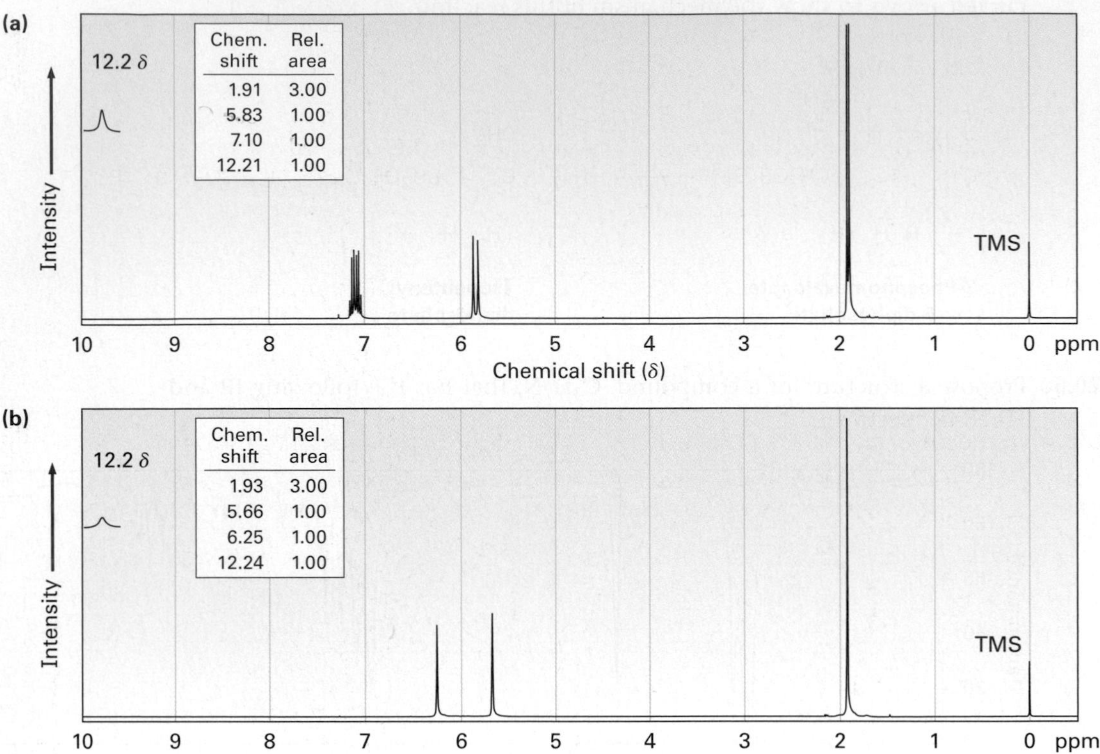

20.62 Propose structures for carboxylic acids that show the following peaks in their ^{13}C NMR spectra. Assume that the kinds of carbons (1°, 2°, 3°, or 4°) have been assigned by DEPT-NMR.

(a) $C_7H_{12}O_2$: 25.5 δ (2°), 25.9 δ (2°), 29.0 δ (2°), 43.1 δ (3°), 183.0 δ (4°)

(b) $C_8H_8O_2$: 21.4 δ (1°), 128.3 δ (4°), 129.0 δ (3°), 129.7 δ (3°), 143.1 δ (4°), 168.2 δ (4°)

20.63 Carboxylic acids having a second carbonyl group two atoms away lose CO_2 *(decarboxylate)* through an intermediate enolate ion when treated with base. Write the mechanism of this decarboxylation reaction using curved arrows to show the electron flow in each step.

$$CH_3\overset{O}{\overset{\|}{C}}CH_2\overset{O}{\overset{\|}{C}}OH \xrightarrow[H_2O]{NaOH} \left[CH_3\overset{O^-}{\overset{|}{C}}{=}CH_2\right] + CO_2 \xrightarrow{H_2O} CH_3\overset{O}{\overset{\|}{C}}CH_3$$

An enolate ion

20.64 In the Ritter reaction, an alkene reacts with a nitrile in the presence of strong aqueous sulfuric acid to yield an amide. Propose a mechanism.

1-Methylcyclohexene (CH_3) $\xrightarrow[H_2O,\ H_2SO_4]{CH_3C{\equiv}N}$ N-(1-methylcyclohexyl)acetamide (H_3C, N–H, $C(=O)CH_3$)

21

The lives of rock climbers depend on their ropes, typically made of a nylon polymer prepared by a nucleophilic acyl substitution reaction. Image copyright ArtmannWitte, 2010. Used under license from Shutterstock.com

Carboxylic Acid Derivatives: Nucleophilic Acyl Substitution Reactions

Closely related to the carboxylic acids and nitriles discussed in the previous chapter are the **carboxylic acid derivatives**, compounds in which an acyl group is bonded to an electronegative atom or substituent that can act as a leaving group in the nucleophilic acyl substitution reaction that we saw briefly in the *Preview of Carbonyl Chemistry:*

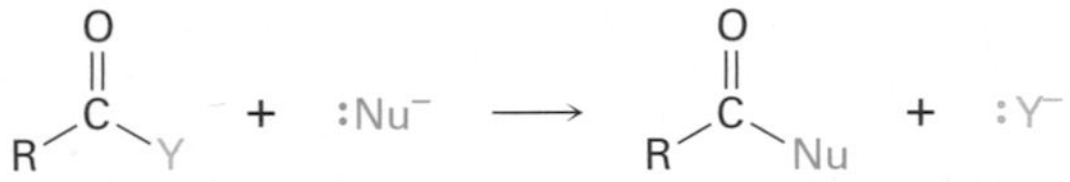

$$RC(=O)Y + :Nu^- \longrightarrow RC(=O)Nu + :Y^-$$

Many kinds of acid derivatives are known, but we'll be concerned primarily with four of the more common ones: **acid halides**, **acid anhydrides**, **esters**, and **amides**. Acid halides and acid anhydrides are used only in the laboratory, while esters and amides are common in both laboratory and biological chemistry. In addition, carboxylic acid derivatives called **thioesters** and **acyl phosphates** are encountered primarily in biological chemistry. Note the structural similarity between acid anhydrides and acyl phosphates.

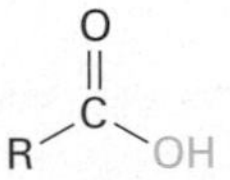

$RC(=O)OH$ — **Carboxylic acid**

$RC(=O)X$ — **Acid halide** (X = Cl, Br)

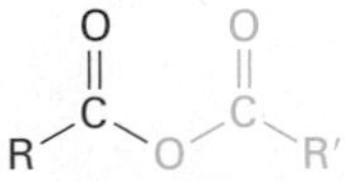

$RC(=O)OC(=O)R'$ — **Acid anhydride**

$RC(=O)OR'$ — **Ester**

$RC(=O)NH_2$ — **Amide**

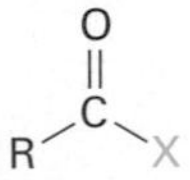

$RC(=O)SR'$ — **Thioester**

$RC(=O)OP(=O)(O^-)O^-$ (or OR′) — **Acyl phosphate**

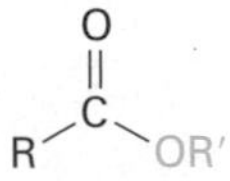

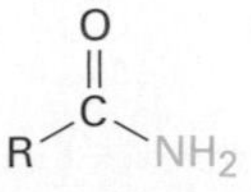

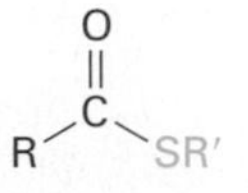

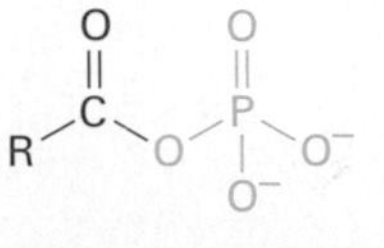

Why This Chapter? Carboxylic acid derivatives are among the most widely occurring of all molecules, both in laboratory chemistry and in biological pathways. Thus, a study of them and their primary reaction—nucleophilic acyl substitution—is fundamental to understanding organic chemistry. We'll begin this chapter by first learning about carboxylic acid derivatives, and we'll then explore the chemistry of acyl substitution reactions.

21.1 Naming Carboxylic Acid Derivatives

Acid Halides, RCOX

Acid halides are named by identifying first the acyl group and then the halide. As described in **Section 20.1** and shown in Table 20.1 on page 780, the acyl group name is derived from the carboxylic acid name by replacing the *-ic acid* or *-oic acid* ending with *-oyl*, or the *-carboxylic acid* ending with *-carbonyl*. To keep things interesting, however, IUPAC recognizes eight exceptions for which a *-yl* rather than an *-oyl* ending is used: formic (formyl), acetic (acetyl), propionic (propionyl), butyric (butyryl), oxalic (oxalyl), malonic (malonyl), succinic (succinyl), and glutaric (glutaryl).

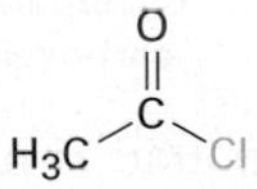

Acetyl chloride 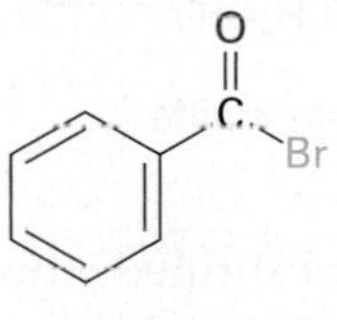 **Benzoyl bromide** 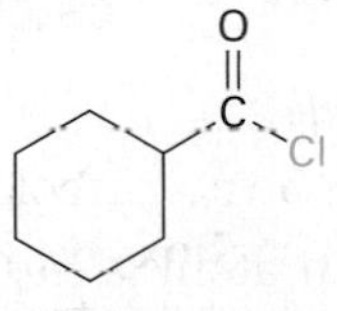 **Cyclohexanecarbonyl chloride**

Acid Anhydrides, RCO_2COR'

Symmetrical anhydrides of unsubstituted monocarboxylic acids and cyclic anhydrides of dicarboxylic acids are named by replacing the word *acid* with *anhydride*.

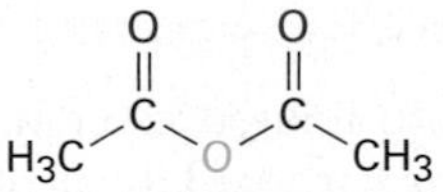

Acetic anhydride 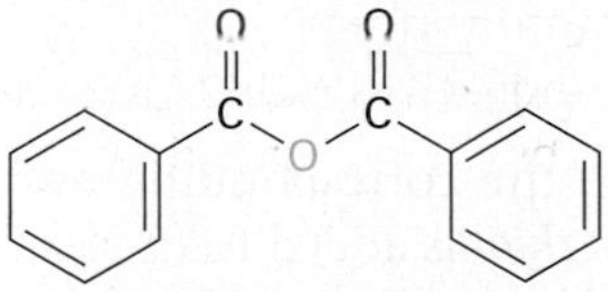 **Benzoic anhydride** 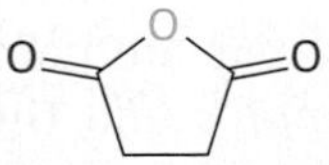 **Succinic anhydride**

Unsymmetrical anhydrides—those prepared from two different carboxylic acids—are named by citing the two acids alphabetically and then adding *anhydride*.

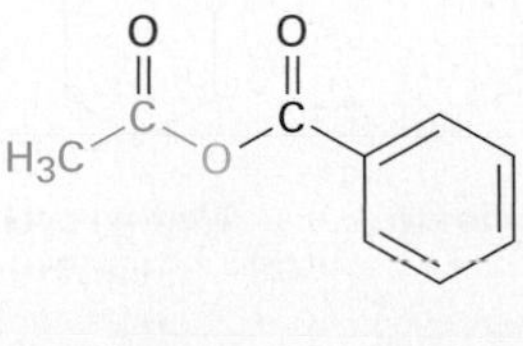

Acetic **benzoic anhydride**

Esters, RCO_2R'

Esters are named by first identifying the alkyl group attached to oxygen and then the carboxylic acid, with the *-ic acid* ending replaced by *-ate*.

$CH_3COOCH_2CH_3$ — **Ethyl acetate**

$CH_3OOC{-}CH_2{-}COOCH_3$ — **Dimethyl malonate**

$C_6H_{11}COOC(CH_3)_3$ — ***tert*-Butyl cyclohexane-carboxylate**

Amides, $RCONH_2$

Amides with an unsubstituted $-NH_2$ group are named by replacing the *-oic acid* or *-ic acid* ending with *-amide*, or by replacing the *-carboxylic acid* ending with *-carboxamide*.

$H_3C{-}CO{-}NH_2$ — **Acetamide**

$CH_3CH_2CH_2CH_2CH_2{-}CO{-}NH_2$ — **Hexanamide**

$C_5H_9{-}CO{-}NH_2$ — **Cyclopentane-carboxamide**

If the nitrogen atom is further substituted, the compound is named by first identifying the substituent groups and then the parent amide. The substituents are preceded by the letter *N* to identify them as being directly attached to nitrogen.

$CH_3CH_2{-}CO{-}NH{-}CH_3$ — ***N*-Methylpropanamide**

$C_6H_{11}{-}CO{-}N(CH_2CH_3)_2$ — ***N,N*-Diethylcyclohexanecarboxamide**

Thioesters, RCOSR′

Thioesters are named like the corresponding esters. If the related ester has a common name, the prefix *thio-* is added to the name of the carboxylate: acetate becomes thioacetate, for instance. If the related ester has a systematic name, the *-oate* or *-carboxylate* ending is replaced by *-thioate* or *-carbothioate:* butanoate becomes butanethioate and cyclohexanecarboxylate becomes cyclohexane-carbothioate, for instance.

$H_3C{-}CO{-}SCH_3$ — **Methyl thioacetate**

$CH_3CH_2CH_2{-}CO{-}SCH_2CH_3$ — **Ethyl butanethioate**

$C_6H_{11}{-}CO{-}SCH_3$ — **Methyl cyclohexane-carbothioate**

Acyl Phosphates, $RCO_2PO_3^{2-}$ and $RCO_2PO_3R'^{-}$

Acyl phosphates are named by citing the acyl group and adding the word *phosphate.* If an alkyl group is attached to one of the phosphate oxygens, it is identified after the name of the acyl group. In biological chemistry, acyl adenosyl phosphates are particularly common.

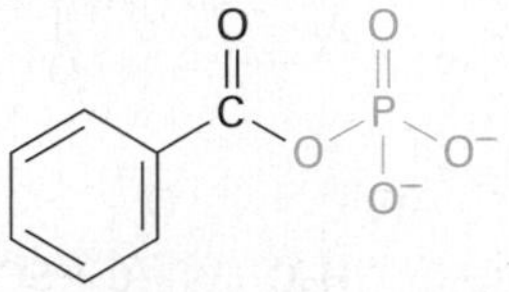

Benzoyl phosphate

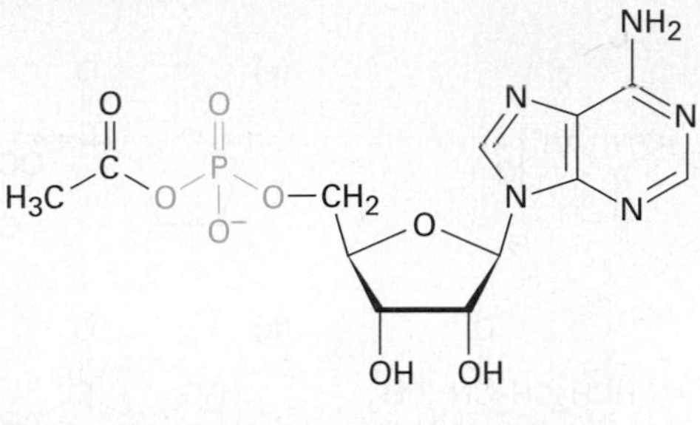

Acetyl adenosyl phosphate

A summary of nomenclature rules for carboxylic acid derivatives is given in Table 21.1.

Table 21.1 Nomenclature of Carboxylic Acid Derivatives

Functional group	Structure	Name ending
Carboxylic acid	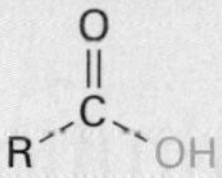 $R{-}C({=}O){-}OH$	*-ic acid* *(-carboxylic acid)*
Acid halide	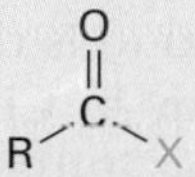 $R{-}C({=}O){-}X$	*-oyl halide* *(-carbonyl halide)*
Acid anhydride	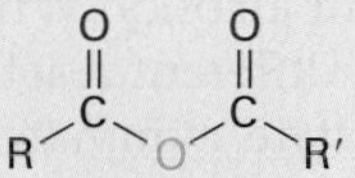 $R{-}C({=}O){-}O{-}C({=}O){-}R'$	*anhydride*
Amide	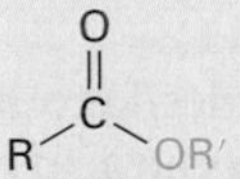 $R{-}C({=}O){-}OR'$	*-amide* *(-carboxamide)*
Ester	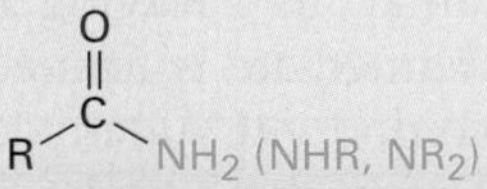 $R{-}C({=}O){-}NH_2$ (NHR, NR_2)	*-oate* *(-carboxylate)*
Thioester	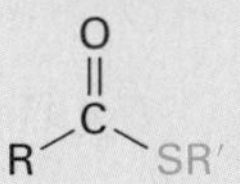 $R{-}C({=}O){-}SR'$	*-thioate* *(-carbothioate)*
Acyl phosphate	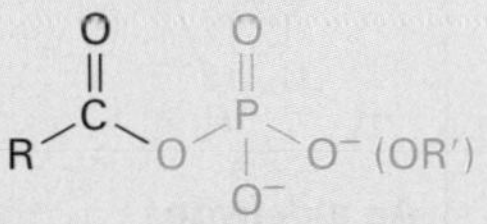 $R{-}C({=}O){-}O{-}P({=}O)(O^-){-}O^-\ (OR')$	*-oyl phosphate*

Problem 21.1

Give IUPAC names for the following substances:

(a) 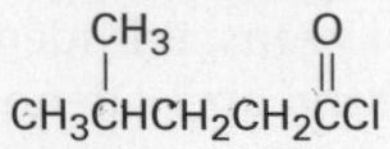 $CH_3CH(CH_3)CH_2CH_2COCl$ (b) 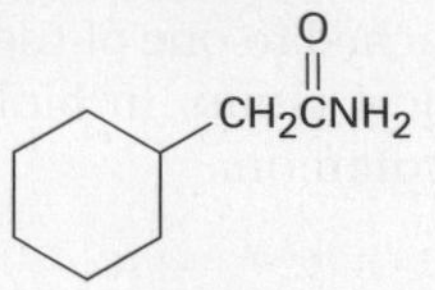 cyclohexyl–CH_2CONH_2 (c) 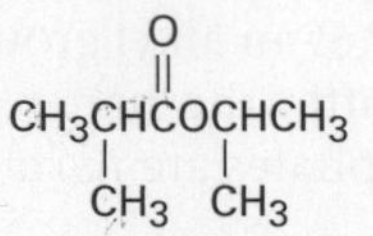 $CH_3CH(CH_3)COOCH(CH_3)CH_3$

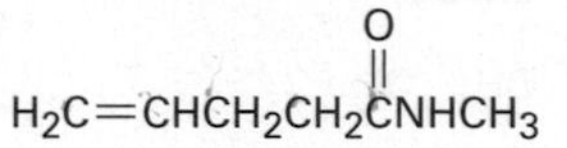

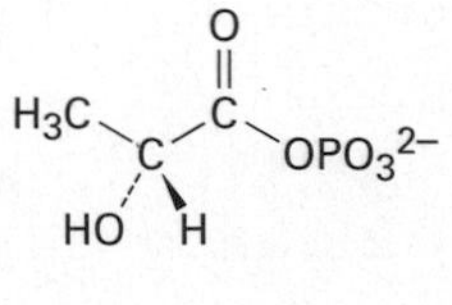

(d) 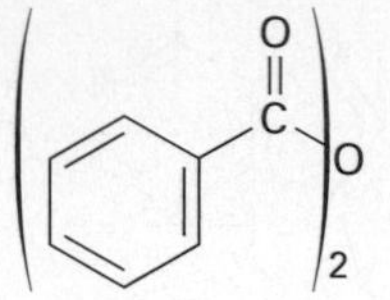 $(C_6H_5CO)_2O$ (e) 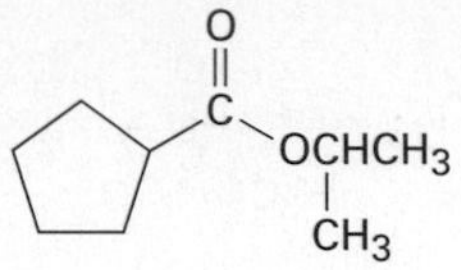 cyclopentyl–$COOCH(CH_3)CH_3$ (f) 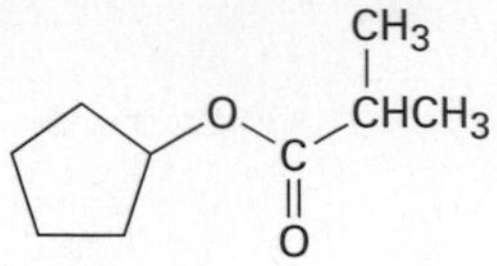 cyclopentyl–$OCOCH(CH_3)CH_3$

(g) $H_2C{=}CHCH_2CH_2CONHCH_3$ (h) H_3C–C(HO)(H)–$COOPO_3^{2-}$ (i) 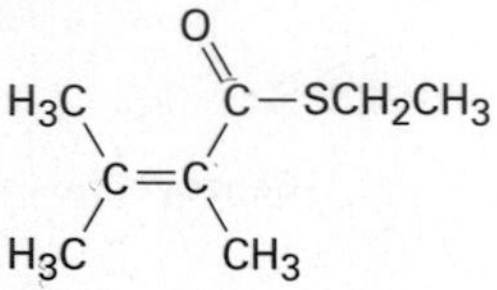 $(H_3C)_2C{=}C(CH_3)COSCH_2CH_3$

Problem 21.2

Draw structures corresponding to the following names:

(a) Phenyl benzoate
(b) *N*-Ethyl-*N*-methylbutanamide
(c) 2,4-Dimethylpentanoyl chloride
(d) Methyl 1-methylcyclohexanecarboxylate
(e) Ethyl 3-oxopentanoate
(f) Methyl *p*-bromobenzenethioate
(g) Formic propanoic anhydride
(h) *cis*-2-Methylcyclopentanecarbonyl bromide

21.2 Nucleophilic Acyl Substitution Reactions

The addition of a nucleophile to a polar C=O bond is the key step in three of the four major carbonyl-group reactions. We saw in Chapter 19 that when a nucleophile adds to an aldehyde or ketone, the initially formed tetrahedral intermediate can be protonated to yield an alcohol. When a nucleophile adds to a carboxylic acid derivative, however, a different reaction course is followed. The initially formed tetrahedral intermediate eliminates one of the two substituents originally bonded to the carbonyl carbon, leading to a net **nucleophilic acyl substitution reaction** **(Figure 21.1)**.

The difference in behavior between aldehydes/ketones and carboxylic acid derivatives is a consequence of structure. Carboxylic acid derivatives have an acyl carbon bonded to a group –Y that can act as a leaving group, often as a stable anion. As soon as the tetrahedral intermediate is formed, the leaving group is expelled to generate a new carbonyl compound. Aldehydes and ketones have no such leaving group, however, and therefore don't undergo substitution.

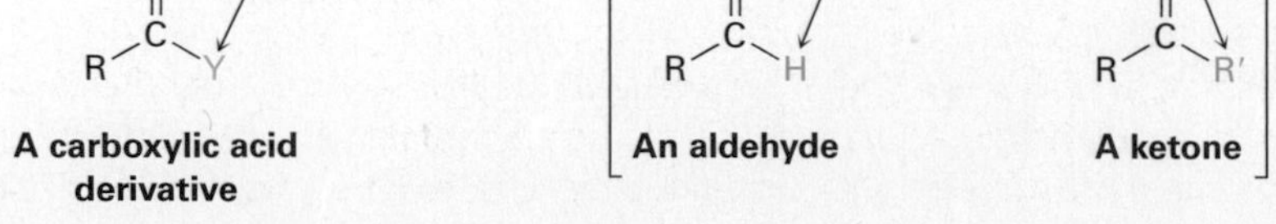

A leaving group → Y in RC(=O)Y — **A carboxylic acid derivative**

Not a leaving group → H in RC(=O)H — **An aldehyde**; R′ in RC(=O)R′ — **A ketone**

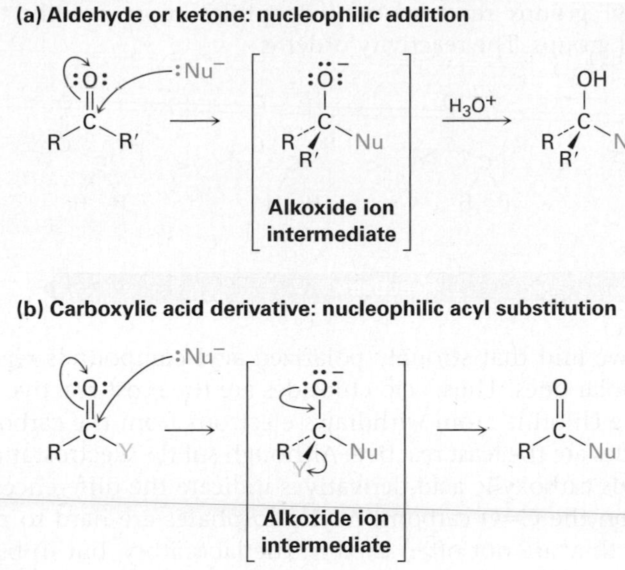

(a) Aldehyde or ketone: nucleophilic addition

(b) Carboxylic acid derivative: nucleophilic acyl substitution

Figure 21.1 The general mechanisms of nucleophilic addition and nucleophilic acyl substitution reactions. Both reactions begin with addition of a nucleophile to a polar C=O bond to give a tetrahedral, alkoxide ion intermediate. **(a)** The intermediate formed from an aldehyde or ketone is protonated to give an alcohol, but **(b)** the intermediate formed from a carboxylic acid derivative expels a leaving group to give a new carbonyl compound.

The net effect of the addition/elimination sequence is a substitution of the nucleophile for the –Y group originally bonded to the acyl carbon. Thus, the overall reaction is superficially similar to the kind of nucleophilic substitution that occurs during an S_N2 reaction **(Section 11.3)**, but the mechanisms of the two reactions are completely different. An S_N2 reaction occurs in a single step by backside displacement of the leaving group, while a nucleophilic acyl substitution takes place in two steps and involves a tetrahedral intermediate.

Problem 21.3

Show the mechanism of the following nucleophilic acyl substitution reaction, using curved arrows to indicate the electron flow in each step:

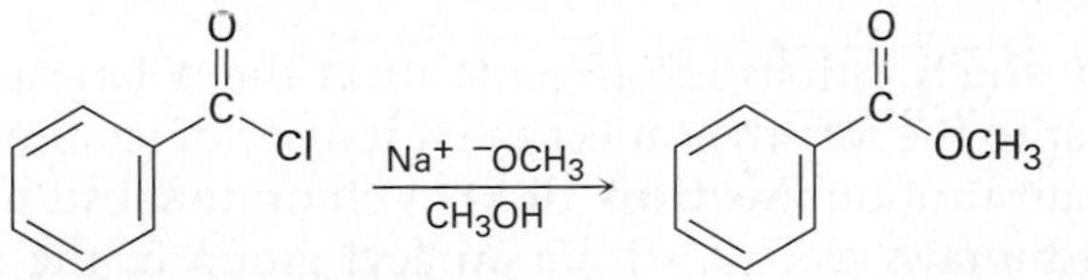

$$C_6H_5COCl \xrightarrow[CH_3OH]{Na^+\ ^-OCH_3} C_6H_5COOCH_3$$

Relative Reactivity of Carboxylic Acid Derivatives

Both the initial addition step and the subsequent elimination step can affect the overall rate of a nucleophilic acyl substitution reaction, but the addition step is generally the rate-limiting one. Thus, any factor that makes the carbonyl group more reactive toward nucleophiles favors the substitution process.

Steric and electronic factors are both important in determining reactivity. Sterically, we find within a series of similar acid derivatives that unhindered, accessible carbonyl groups react with nucleophiles more readily than do sterically hindered groups. The reactivity order is

$R_3C{-}C(=O){-} < R_2HC{-}C(=O){-} < RH_2C{-}C(=O){-} < H_3C{-}C(=O){-}$

Reactivity →

Electronically, we find that strongly polarized acyl compounds react more readily than less polar ones. Thus, acid chlorides are the most reactive because the electronegative chlorine atom withdraws electrons from the carbonyl carbon, whereas amides are the least reactive. Although subtle, electrostatic potential maps of various carboxylic acid derivatives indicate the differences by the relative blueness on the C=O carbons. Acyl phosphates are hard to place on this scale because they are not often used in the laboratory, but in biological systems they appear to be somewhat more reactive than thioesters.

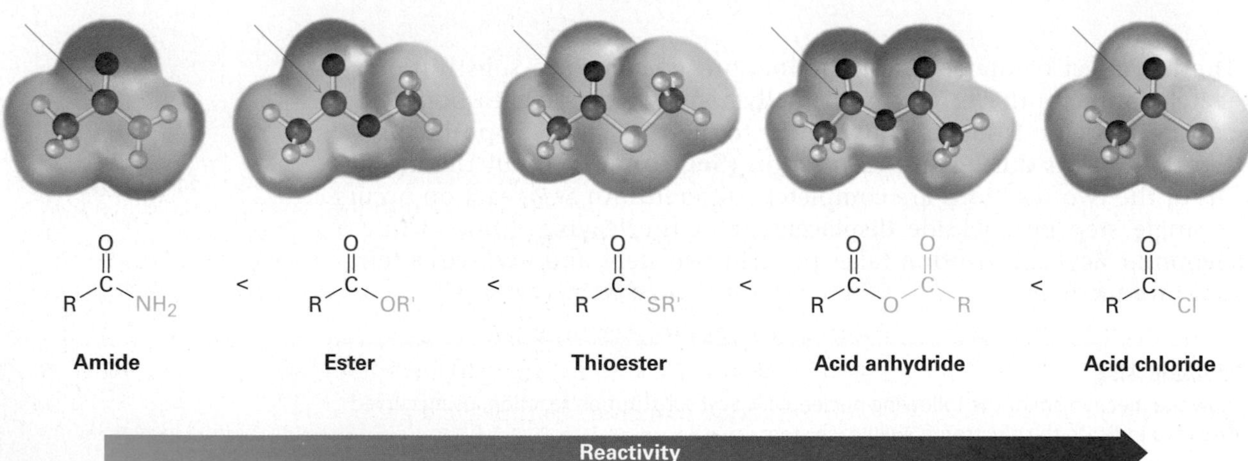

The way in which various substituents affect the polarization of a carbonyl group is similar to the way they affect the reactivity of an aromatic ring toward electrophilic substitution **(Section 16.5)**. A chlorine substituent, for example, inductively withdraws electrons from an acyl group in the same way that it withdraws electrons from and thus deactivates an aromatic ring. Similarly, amino, methoxyl, and methylthio substituents donate electrons to acyl groups by resonance in the same way that they donate electrons to, and thus activate, aromatic rings.

As a consequence of these reactivity differences, it's usually possible to convert a more reactive acid derivative into a less reactive one. Acid chlorides, for instance, can be directly converted into anhydrides, thioesters, esters, and amides, but amides can't be directly converted into esters, thioesters, anhydrides, or acid chlorides. Remembering the reactivity order is therefore a way to keep track of a large number of reactions **(Figure 21.2)**. Another consequence, as noted previously, is that only acyl phosphates, thioesters, esters, and amides are commonly found in nature. Acid halides and acid anhydrides react with water so rapidly that they can't exist for long in living organisms.

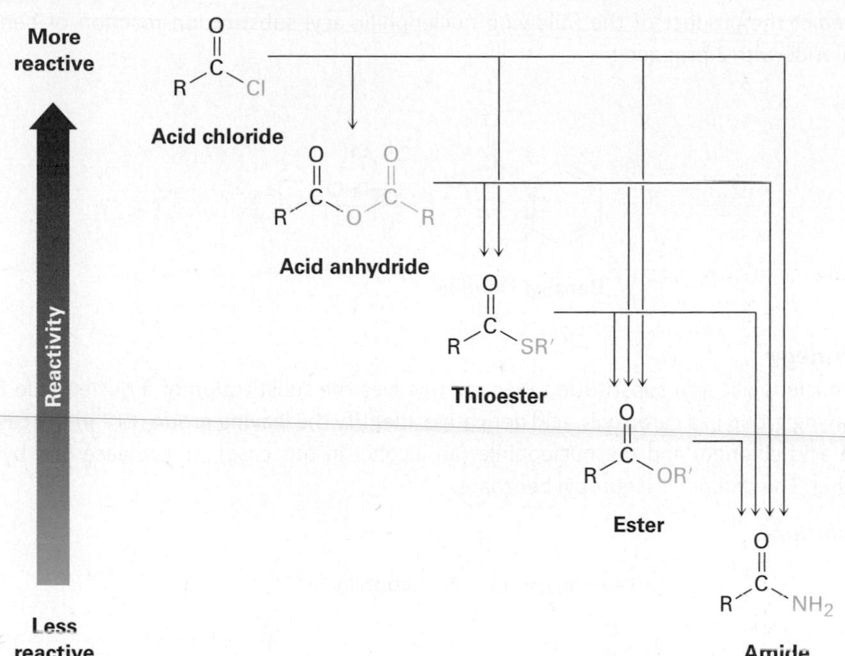

Figure 21.2 Interconversions of carboxylic acid derivatives. A more reactive acid derivative can be converted into a less reactive one, but not vice versa.

In studying the chemistry of carboxylic acid derivatives in the next few sections, we'll be concerned largely with the reactions of just a few nucleophiles and will see that the same kinds of reactions keep occurring **(Figure 21.3)**.

- **Hydrolysis** Reaction with water to yield a carboxylic acid
- **Alcoholysis** Reaction with an alcohol to yield an ester
- **Aminolysis** Reaction with ammonia or an amine to yield an amide
- **Reduction** Reaction with a hydride reducing agent to yield an aldehyde or an alcohol
- **Grignard reaction** Reaction with an organometallic reagent to yield a ketone or an alcohol

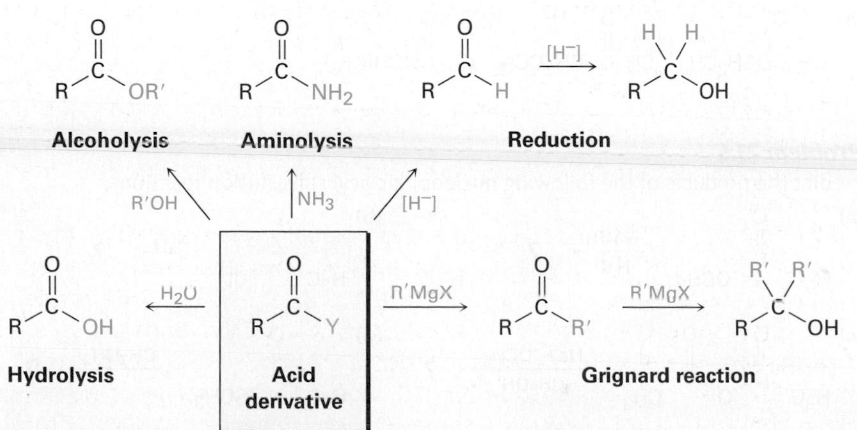

Figure 21.3 Some general reactions of carboxylic acid derivatives.

Worked Example 21.1 Predicting the Product of a Nucleophilic Acyl Substitution Reaction

Predict the product of the following nucleophilic acyl substitution reaction of benzoyl chloride with 2-propanol:

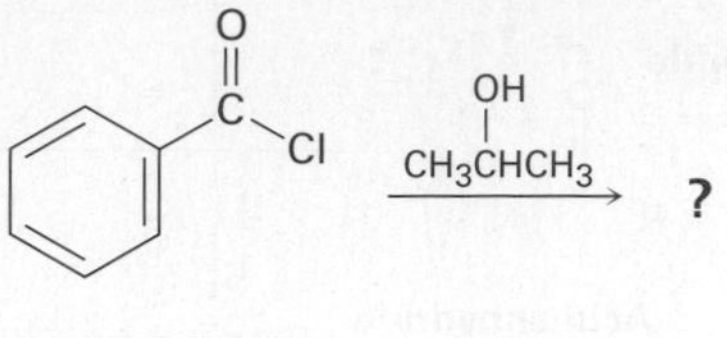

$$\text{Benzoyl chloride } (C_6H_5COCl) \xrightarrow{CH_3CH(OH)CH_3} ?$$

Benzoyl chloride

Strategy

A nucleophilic acyl substitution reaction involves the substitution of a nucleophile for a leaving group in a carboxylic acid derivative. Identify the leaving group (Cl^- in the case of an acid chloride) and the nucleophile (an alcohol in this case), and replace one by the other. The product is isopropyl benzoate.

Solution

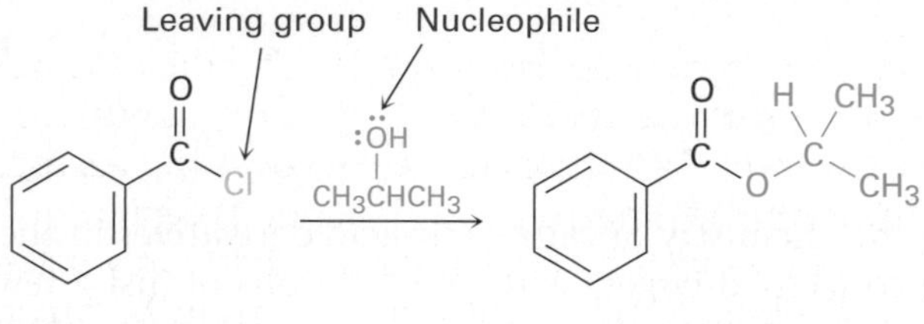

Leaving group Nucleophile

$$C_6H_5COCl \xrightarrow{CH_3CH(\ddot{O}H)CH_3} C_6H_5COOCH(CH_3)_2$$

Benzoyl chloride Isopropyl benzoate

Problem 21.4

Rank the compounds in each of the following sets in order of their expected reactivity toward nucleophilic acyl substitution:

(a) $CH_3\overset{O}{\overset{\|}{C}}Cl$, $CH_3\overset{O}{\overset{\|}{C}}OCH_3$, $CH_3\overset{O}{\overset{\|}{C}}NH_2$

(b) $CH_3\overset{O}{\overset{\|}{C}}OCH_2CH_3$, $CH_3\overset{O}{\overset{\|}{C}}OCH_2CCl_3$, $CH_3\overset{O}{\overset{\|}{C}}OCH(CF_3)_2$

Problem 21.5

Predict the products of the following nucleophilic acyl substitution reactions:

(a) $H_3C\overset{O}{\overset{\|}{C}}OCH_3 \xrightarrow[H_2O]{NaOH} ?$

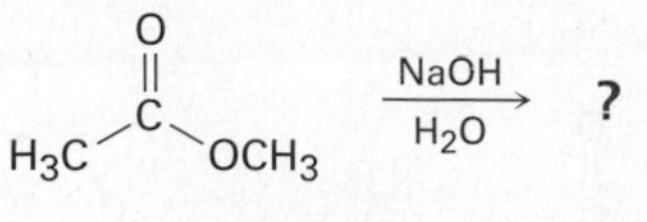

(b) $H_3C\overset{O}{\overset{\|}{C}}Cl \xrightarrow{NH_3} ?$

(c) $H_3C\overset{O}{\overset{\|}{C}}O\overset{O}{\overset{\|}{C}}CH_3 \xrightarrow[CH_3OH]{Na^+\ ^-OCH_3} ?$

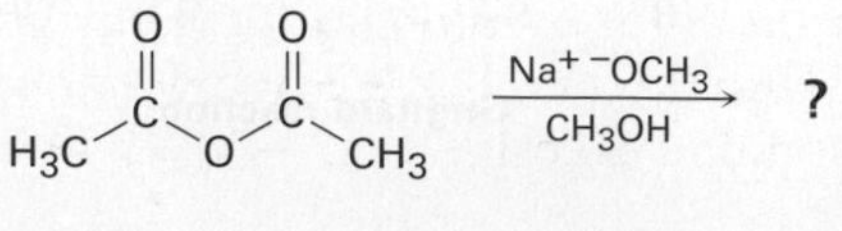

(d) $H_3C\overset{O}{\overset{\|}{C}}SCH_3 \xrightarrow{CH_3NH_2} ?$

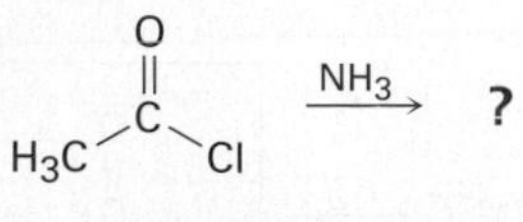

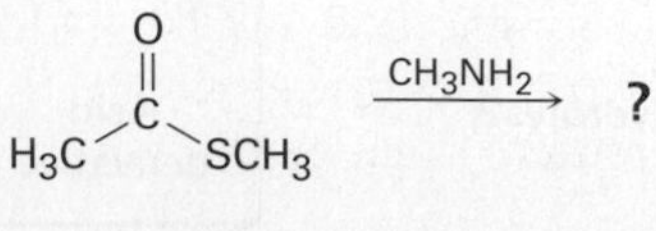

Problem 21.6

The following structure represents a tetrahedral alkoxide ion intermediate formed by addition of a nucleophile to a carboxylic acid derivative. Identify the nucleophile, the leaving group, the starting acid derivative, and the ultimate product.

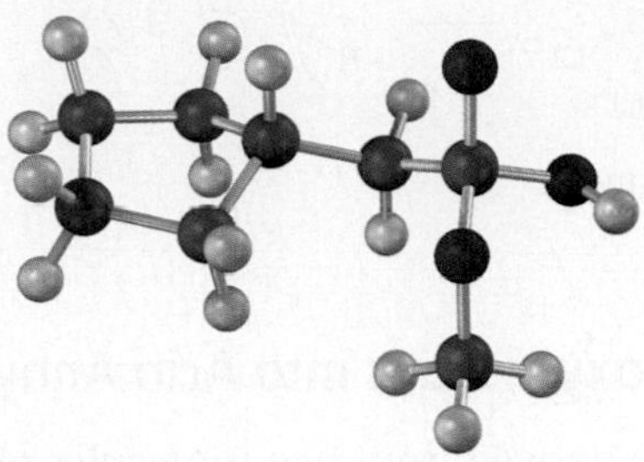

21.3 Nucleophilic Acyl Substitution Reactions of Carboxylic Acids

The direct nucleophilic acyl substitution of a carboxylic acid is difficult because –OH is a poor leaving group **(Section 11.3)**. Thus, it's usually necessary to enhance the reactivity of the acid, either by using a strong acid catalyst to protonate the carboxyl and make it a better acceptor or by converting the –OH into a better leaving group. Under the right circumstances, however, acid chlorides, anhydrides, esters, and amides can all be prepared from carboxylic acids by nucleophilic acyl substitution reactions.

Conversion of Carboxylic Acids into Acid Chlorides

In the laboratory, carboxylic acids are converted into acid chlorides by treatment with thionyl chloride, $SOCl_2$.

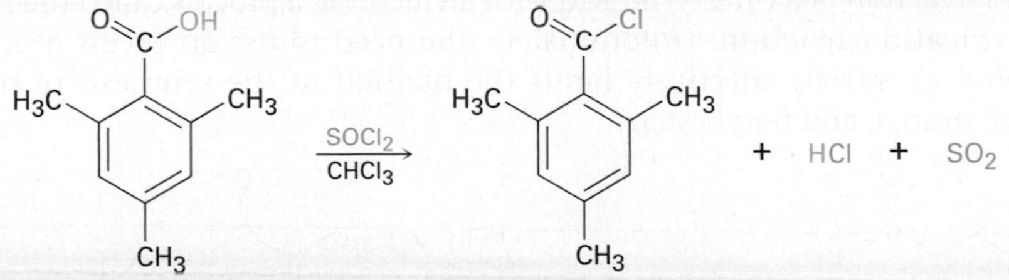

$$\text{2,4,6-Trimethylbenzoic acid} \xrightarrow[\text{CHCl}_3]{\text{SOCl}_2} \text{2,4,6-Trimethylbenzoyl chloride (90\%)} + \text{HCl} + \text{SO}_2$$

2,4,6-Trimethylbenzoic acid **2,4,6-Trimethylbenzoyl chloride (90%)**

The reaction occurs by a nucleophilic acyl substitution pathway in which the carboxylic acid is first converted into an acyl chlorosulfite intermediate, thereby replacing the –OH of the acid with a much better leaving group. The chlorosulfite then reacts with a nucleophilic chloride ion. You might recall from **Section 17.6** that an analogous chlorosulfite is involved in reaction of an alcohol with $SOCl_2$ to yield an alkyl chloride.

$SOCl_2$

Carboxylic acid **A chlorosulfite** **Acid chloride**

Conversion of Carboxylic Acids into Acid Anhydrides

Acid anhydrides can be derived from two molecules of carboxylic acid by heating to remove 1 equivalent of water. Because of the high temperatures needed, however, only acetic anhydride is commonly prepared this way.

800 °C

Acetic acid **Acetic anhydride**

Conversion of Carboxylic Acids into Esters

Perhaps the most useful reaction of carboxylic acids is their conversion into esters. There are many methods for accomplishing the transformation, including the S_N2 reaction of a carboxylate anion with a primary alkyl halide that we saw in **Section 11.3**.

S_N2 reaction

Sodium butanoate **Methyl butanoate (97%)**

Esters can also be synthesized by an acid-catalyzed nucleophilic acyl substitution reaction of a carboxylic acid with an alcohol, a process called the **Fischer esterification reaction**. Unfortunately, the need to use an excess of a liquid alcohol as solvent effectively limits the method to the synthesis of methyl, ethyl, propyl, and butyl esters.

HCl catalyst

Benzoic acid **Ethyl benzoate (91%)**

The mechanism of the Fischer esterification reaction is shown in **Figure 21.4**. Carboxylic acids are not reactive enough to undergo nucleophilic addition directly, but their reactivity is greatly enhanced in the presence of a strong acid such as HCl or H_2SO_4. The mineral acid protonates the carbonyl-group oxygen atom, thereby giving the carboxylic acid a positive charge and rendering it much more reactive. Subsequent loss of water from the tetrahedral intermediate yields the ester product.

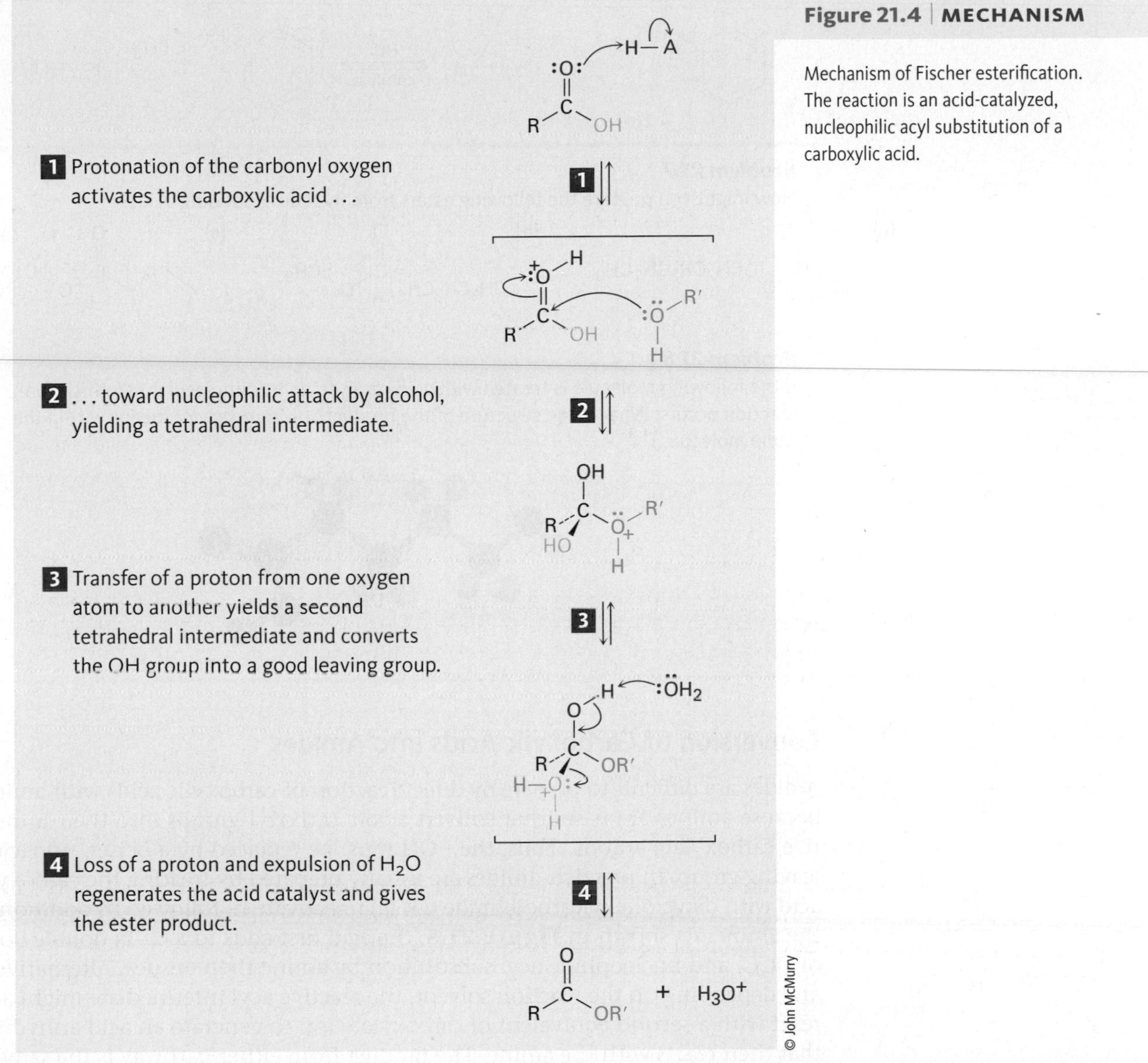

Figure 21.4 | MECHANISM

Mechanism of Fischer esterification. The reaction is an acid-catalyzed, nucleophilic acyl substitution of a carboxylic acid.

The net effect of Fischer esterification is substitution of an –OH group by –OR′. All steps are reversible, and the reaction typically has an equilibrium constant close to 1. Thus, the reaction can be driven in either direction by choice of reaction conditions. Ester formation is favored when a large excess of alcohol is used as solvent, but carboxylic acid formation is favored when a large excess of water is present.

Evidence in support of the mechanism shown in Figure 21.4 comes from isotope-labeling experiments. When ^{18}O-labeled methanol reacts with benzoic acid, the methyl benzoate produced is found to be ^{18}O-labeled but the water produced is unlabeled. Thus, it is the C–OH bond of the carboxylic acid that is broken during the reaction rather than the CO–H bond and the RO–H bond of the alcohol that is broken rather than the R–OH bond.

These bonds are broken

$$C_6H_5COOH + CH_3\overset{*}{O}{-}H \xrightarrow[\text{catalyst}]{HCl} C_6H_5CO\overset{*}{O}CH_3 + HOH$$

Problem 21.7

How might you prepare the following esters from the corresponding acids?

(a) $H_3C{-}C(=O){-}O{-}CH_2CH_2CH_2CH_3$ **(b)** $CH_3CH_2CH_2{-}C(=O){-}O{-}CH_3$ **(c)** cyclopentyl$-C(=O){-}O{-}CH(CH_3)_2$

Problem 21.8

If the following molecule is treated with acid catalyst, an intramolecular esterification reaction occurs. What is the structure of the product? (*Intramolecular* means within the same molecule.)

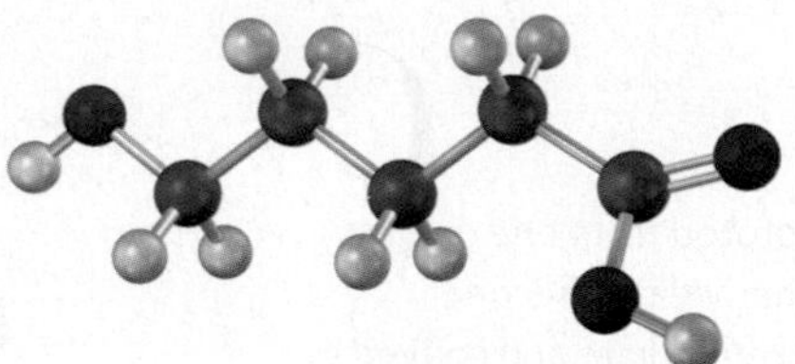

Conversion of Carboxylic Acids into Amides

Amides are difficult to prepare by direct reaction of carboxylic acids with amines because amines are bases that convert acidic carboxyl groups into their unreactive carboxylate anions. Thus, the –OH must be replaced by a better, nonacidic leaving group. In practice, amides are usually prepared by treating the carboxylic acid with dicyclohexylcarbodiimide (DCC) to activate it, followed by addition of the amine. As shown in **Figure 21.5**, the acid first adds to a C=N double bond of DCC, and nucleophilic acyl substitution by amine then ensues. Alternatively, and depending on the reaction solvent, the reactive acyl intermediate might also react with a second equivalent of carboxylate ion to generate an acid anhydride that then reacts with the amine. The product from either pathway is the same.

We'll see in **Section 26.7** that this DCC-induced method of amide formation is the key step in the laboratory synthesis of small proteins, or *peptides*. For instance, when one amino acid with its NH_2 rendered unreactive and a second amino acid with its $-CO_2H$ rendered unreactive are treated with DCC, a dipeptide is formed.

$$\underset{\textbf{Amino acid 1}}{RHN{-}CH(R_1){-}C(=O){-}OH} + \underset{\textbf{Amino acid 2}}{H_2N{-}CH(R_2){-}C(=O){-}OR'} \xrightarrow{DCC} \underset{\textbf{A dipeptide}}{RHN{-}CH(R_1){-}C(=O){-}NH{-}CH(R_2){-}C(=O){-}OR'}$$

Figure 21.5 | MECHANISM

Mechanism of amide formation by reaction of a carboxylic acid and an amine with dicyclohexylcarbodiimide (DCC).

1 Dicyclohexylcarbodiimide is first protonated by the carboxylic acid to make it a better acceptor.

2 The carboxylate then adds to the protonated carbodiimide to yield a reactive acylating agent.

3 Nucleophilic attack of the amine on the acylating agent gives a tetrahedral intermediate.

4 The intermediate loses dicyclohexylurea and gives the amide.

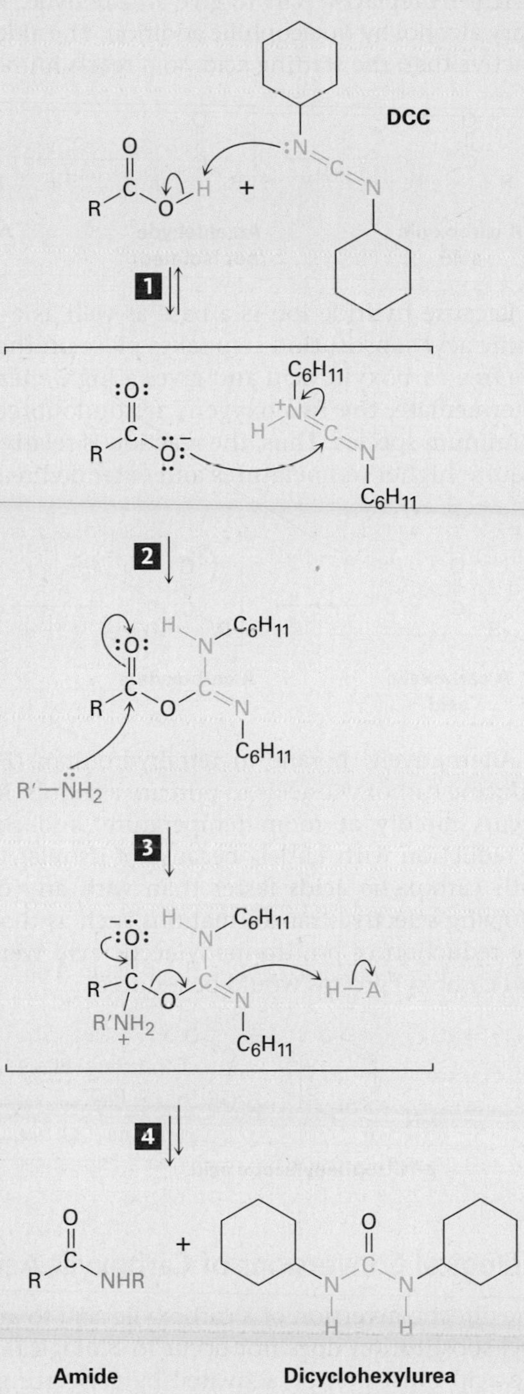

Conversion of Carboxylic Acids into Alcohols

We said in **Section 17.4** that carboxylic acids are reduced by $LiAlH_4$ to give primary alcohols, but we deferred a discussion of the reaction mechanism at that time. In fact, the reduction is a nucleophilic acyl substitution reaction in which –H replaces –OH to give an aldehyde, which is further reduced to a primary alcohol by nucleophilic addition. The aldehyde intermediate is much more reactive than the starting acid, so it reacts immediately and is not isolated.

A carboxylic acid —"H⁻" (LiAlH4)→ [An aldehyde (not isolated) —"H⁻" (LiAlH4)→ An alkoxide ion] —H_3O^+→ A 1° alcohol

Because hydride ion is a base as well as a nucleophile, the actual nucleophilic acyl substitution step takes place on the carboxylate ion rather than on the free carboxylic acid and gives a high-energy *dianion* intermediate. In this intermediate, the two oxygens are undoubtedly complexed to a Lewis acidic aluminum species. Thus, the reaction is relatively difficult, and acid reductions require higher temperatures and extended reaction times.

A carboxylic acid —"H⁻" (LiAlH4)→ [A carboxylate —"H⁻" (LiAlH4)→ A dianion] → An aldehyde

Alternatively, borane in tetrahydrofuran (BH_3/THF) is a useful reagent for reducing carboxylic acids to primary alcohols. Reaction of an acid with BH_3/THF occurs rapidly at room temperature, and the procedure is often preferred to reduction with $LiAlH_4$ because of its relative ease and safety. Borane reacts with carboxylic acids faster than with any other functional group, thereby allowing selective transformations such as that on *p*-nitrophenylacetic acid. If the reduction of *p*-nitrophenylacetic acid were done with $LiAlH_4$, both nitro and carboxyl groups would be reduced.

p-Nitrophenylacetic acid —1. BH_3, THF; 2. H_3O^+→ 2-(*p*-Nitrophenyl)ethanol (94%)

Biological Conversions of Carboxylic Acids

The direct conversion of a carboxylic acid to an acyl derivative by nucleophilic acyl substitution does not occur in biological chemistry. As in the laboratory, the acid must first be activated by converting the –OH into a better leaving group. This activation is often accomplished in living organisms by reaction of the acid with adenosine triphosphate (ATP) to give an acyl adenosyl phosphate, or *acyl adenylate*, a mixed anhydride between a carboxylic acid and adenosine monophosphate (AMP, also known as adenylic acid). In the biosynthesis of fats, for example, a long-chain carboxylic acid reacts with ATP to give an acyl adenylate, followed by subsequent nucleophilic acyl substitution of a thiol group in coenzyme A to give the corresponding acyl CoA **(Figure 21.6)**.

Figure 21.6 | MECHANISM

In fatty-acid biosynthesis, a carboxylic acid is activated by reaction with ATP to give an acyl adenylate, which undergoes nucleophilic acyl substitution with the —SH group on coenzyme A. (ATP = adenosine triphosphate; AMP = adenosine monophosphate.)

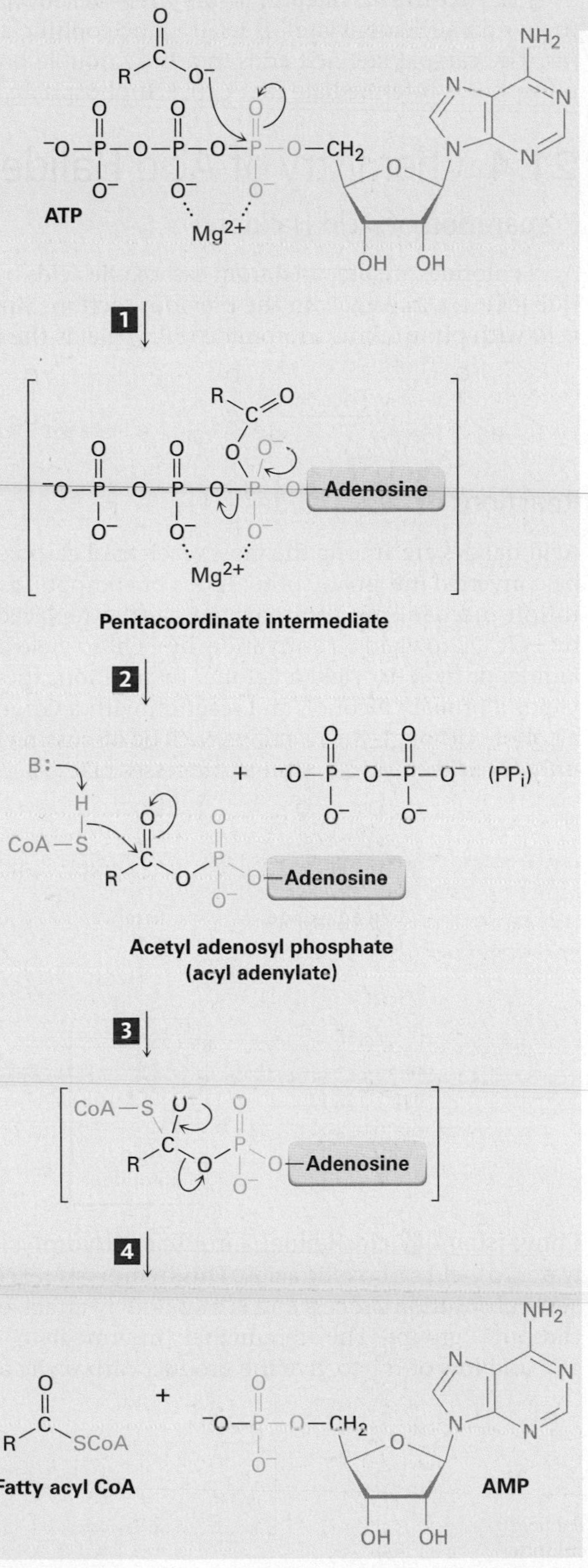

1 ATP is activated by coordination to magnesium ion, and nucleophilic addition of a fatty acid carboxylate to phosphorus then yields a pentacoordinate intermediate . . .

ATP

Mg^{2+}

Pentacoordinate intermediate

2 . . . which expels diphosphate ion (PP_i) as leaving group and gives an acyl adenosyl phosphate in a process analogous to a nucleophilic acyl substitution reaction.

B: CoA—S—H + (PP_i)

Acetyl adenosyl phosphate (acyl adenylate)

3 The -SH group of coenzyme A adds to the acyl adenosyl phosphate, giving a tetrahedral alkoxide intermediate . . .

4 . . . which expels adenosine monophosphate (AMP) as leaving group and yields the fatty acyl CoA.

Fatty acyl CoA + **AMP**

Note that the first step in Figure 21.6—reaction of the carboxylate with ATP to give an acyl adenylate—is itself a nucleophilic acyl substitution on *phosphorus*. The carboxylate first adds to a P=O double bond, giving a five-coordinate phosphorus intermediate that expels diphosphate ion as leaving group.

21.4 Chemistry of Acid Halides

Preparation of Acid Halides

Acid chlorides are prepared from carboxylic acids by reaction with thionyl chloride ($SOCl_2$), as we saw in the previous section. Similar reaction of a carboxylic acid with phosphorus tribromide (PBr_3) yields the acid bromide.

$$RCOOH \xrightarrow{SOCl_2} RCOCl \qquad RCOOH \xrightarrow[\text{Ether}]{PBr_3} RCOBr$$

Reactions of Acid Halides

Acid halides are among the most reactive of carboxylic acid derivatives and can be converted into many other kinds of compounds by nucleophilic acyl substitution mechanisms. The halogen can be replaced by –OH to yield an acid, by –OCOR to yield an anhydride, by –OR to yield an ester, by $-NH_2$ to yield an amide, or by R′ to yield a ketone. In addition, the reduction of an acid halide yields a primary alcohol, and reaction with a Grignard reagent yields a tertiary alcohol. Although the reactions we'll be discussing in this section are illustrated only for acid chlorides, similar processes take place with other acid halides.

Conversion of Acid Halides into Acids: Hydrolysis Acid chlorides react with water to yield carboxylic acids. This hydrolysis reaction is a typical nucleophilic acyl substitution process and is initiated by attack of water on the acid chloride carbonyl group. The tetrahedral intermediate undergoes elimination of Cl^- and loss of H^+ to give the product carboxylic acid plus HCl.

Because HCl is formed during the hydrolysis, the reaction is often carried out in the presence of a base such as pyridine or NaOH to remove the HCl and prevent it from causing side reactions.

Conversion of Acid Halides into Anhydrides Nucleophilic acyl substitution reaction of an acid chloride with a carboxylate anion gives an acid anhydride. Both symmetrical and unsymmetrical acid anhydrides can be prepared.

$$\text{HCOO}^-\,\text{Na}^+ + \text{CH}_3\text{COCl} \xrightarrow[\text{25 °C}]{\text{Ether}} \text{HCOOCOCH}_3$$

Sodium formate **Acetyl chloride** **Acetic formic anhydride (64%)**

Conversion of Acid Halides into Esters: Alcoholysis Acid chlorides react with alcohols to yield esters in a process analogous to their reaction with water to yield acids. In fact, this reaction is probably the most common method for preparing esters in the laboratory. As with hydrolysis, alcoholysis reactions are usually carried out in the presence of pyridine or NaOH to react with the HCl formed.

$$\text{C}_6\text{H}_5\text{COCl} + \text{C}_6\text{H}_{11}\text{OH} \xrightarrow{\text{Pyridine}} \text{C}_6\text{H}_5\text{COOC}_6\text{H}_{11}$$

Benzoyl chloride **Cyclohexanol** **Cyclohexyl benzoate (97%)**

The reaction of an alcohol with an acid chloride is strongly affected by steric hindrance. Bulky groups on either partner slow down the reaction considerably, resulting in a reactivity order among alcohols of primary > secondary > tertiary. As a result, it's often possible to esterify an unhindered alcohol selectively in the presence of a more hindered one. This can be important in complex syntheses in which it's sometimes necessary to distinguish between similar functional groups. For example,

Primary alcohol (less hindered and more reactive)

$$\text{HO–C}_6\text{H}_{10}\text{–CH}_2\text{OH} + \text{H}_3\text{C–COCl} \xrightarrow{\text{Pyridine}} \text{HO–C}_6\text{H}_{10}\text{–CH}_2\text{O–CO–CH}_3$$

Secondary alcohol (more hindered and less reactive)

Problem 21.9

How might you prepare the following esters using a nucleophilic acyl substitution reaction of an acid chloride?

(a) $CH_3CH_2CO_2CH_3$ **(b)** $CH_3CO_2CH_2CH_3$ **(c)** Ethyl benzoate

Problem 21.10

Which method would you choose if you wanted to prepare cyclohexyl benzoate—Fischer esterification or reaction of an acid chloride with an alcohol? Explain.

Conversion of Acid Halides into Amides: Aminolysis Acid chlorides react rapidly with ammonia and amines to give amides. As with the acid chloride-plus-alcohol method for preparing esters, this reaction of acid chlorides with amines is the most commonly used laboratory method for preparing amides. Both monosubstituted and disubstituted amines can be used, but not trisubstituted amines (R_3N).

$$CH_3CH(CH_3)C(=O)Cl + 2\ NH_3 \longrightarrow CH_3CH(CH_3)C(=O)NH_2 + \overset{+}{N}H_4\ Cl^-$$

2-Methylpropanoyl chloride → **2-Methylpropanamide (83%)**

$$C_6H_5C(=O)Cl + 2\ NH(CH_3)_2 \longrightarrow C_6H_5C(=O)N(CH_3)_2 + (CH_3)_2\overset{+}{N}H_2\ Cl^-$$

Benzoyl chloride → ***N,N*-Dimethylbenzamide (92%)**

Because HCl is formed during the reaction, 2 equivalents of the amine must be used. One equivalent reacts with the acid chloride, and one equivalent reacts with the HCl by-product to form an ammonium chloride salt. If, however, the amine component is valuable, amide synthesis is often carried out using 1 equivalent of the amine plus 1 equivalent of an inexpensive base such as NaOH. For example, the sedative trimetozine is prepared commercially by reaction of 3,4,5-trimethoxybenzoyl chloride with the amine morpholine in the presence of 1 equivalent of NaOH.

$$\text{3,4,5-}(CH_3O)_3C_6H_2C(=O)Cl + \text{morpholine (H–N)} \xrightarrow[H_2O]{NaOH} \text{3,4,5-}(CH_3O)_3C_6H_2C(=O)N(\text{morpholino}) + NaCl$$

3,4,5-Trimethoxybenzoyl chloride + **Morpholine** → **Trimetozine (an amide)**

Problem 21.11
Write the mechanism of the reaction just shown between 3,4,5-trimethoxybenzoyl chloride and morpholine to form trimetozine. Use curved arrows to show the electron flow in each step.

Problem 21.12
How could you prepare the following amides using an acid chloride and an amine or ammonia?
(a) $CH_3CH_2CONHCH_3$ **(b)** *N,N*-Diethylbenzamide **(c)** Propanamide

Conversion of Acid Chlorides into Alcohols: Reduction and Grignard Reaction

Acid chlorides are reduced by $LiAlH_4$ to yield primary alcohols. The reaction is of little practical value, however, because the parent carboxylic acids are generally more readily available and can themselves be reduced by $LiAlH_4$ to yield alcohols.

Reduction occurs via a typical nucleophilic acyl substitution mechanism in which a hydride ion (H:$^-$) adds to the carbonyl group, yielding a tetrahedral intermediate that expels Cl^-. The net effect is a substitution of –Cl by –H to yield an aldehyde, which is then further reduced by $LiAlH_4$ in a second step to yield the primary alcohol.

1. $LiAlH_4$, ether
2. H_3O^+

Benzoyl chloride **Benzyl alcohol (96%)**

Grignard reagents react with acid chlorides to yield tertiary alcohols in which two of the substituents are the same. The mechanism of the reaction is similar to that of $LiAlH_4$ reduction. The first equivalent of Grignard reagent adds to the acid chloride, loss of Cl^- from the tetrahedral intermediate yields a ketone, and a second equivalent of Grignard reagent immediately adds to the ketone to produce an alcohol.

CH_3MgBr, Ether
1. CH_3MgBr
2. H_3O^+

Benzoyl chloride **Acetophenone *(Not isolated)*** **2-Phenyl-2-propanol (92%)**

Conversion of Acid Chlorides into Ketones: Diorganocopper Reaction

The ketone intermediate formed in during the reaction of an acid chloride with a Grignard reagent can't usually be isolated because addition of the second equivalent of organomagnesium reagent occurs too rapidly. A ketone *can,* however, be isolated from the reaction of an acid chloride with a lithium diorganocopper (Gilman) reagent, $Li^+ \ R_2Cu^-$. The reaction occurs by initial nucleophilic acyl substitution on the acid chloride by the diorganocopper anion to yield an acyl diorganocopper intermediate, followed by loss of R′Cu and formation of the ketone.

$$\underset{\textbf{An acid chloride}}{R-\overset{O}{\overset{\|}{C}}-Cl} \xrightarrow[\text{Ether}]{R_2'CuLi} \left[\underset{\textbf{An acyl diorganocopper}}{R-\overset{O}{\overset{\|}{C}}-\underset{R'}{\underset{|}{Cu}}-R'}\right] \longrightarrow \underset{\textbf{A ketone}}{R-\overset{O}{\overset{\|}{C}}-R'} + R'Cu$$

The reaction is generally carried out at −78 °C in ether solution, and yields are often excellent. For example, manicone, a substance secreted by male ants to coordinate ant pairing and mating, has been synthesized by reaction of lithium diethylcopper with (*E*)-2,4-dimethyl-2-hexenoyl chloride.

$$\underset{\textbf{2,4-Dimethyl-2-hexenoyl chloride}}{CH_3CH_2\underset{CH_3}{\underset{|}{C}}H-\overset{H}{\overset{|}{C}}=\underset{CH_3}{\underset{|}{C}}-\overset{O}{\overset{\|}{C}}-Cl} \xrightarrow[\text{Ether, −78 °C}]{(CH_3CH_2)_2CuLi} \underset{\textbf{4,6-Dimethyl-4-octen-3-one (manicone, 92\%)}}{CH_3CH_2\underset{CH_3}{\underset{|}{C}}H-\overset{H}{\overset{|}{C}}=\underset{CH_3}{\underset{|}{C}}-\overset{O}{\overset{\|}{C}}-CH_2CH_3}$$

Note that the diorganocopper reaction occurs only with acid chlorides. Carboxylic acids, esters, acid anhydrides, and amides do not react with lithium diorganocopper reagents.

Problem 21.13

How could you prepare the following ketones by reaction of an acid chloride with a lithium diorganocopper reagent?

(a) **(b)**

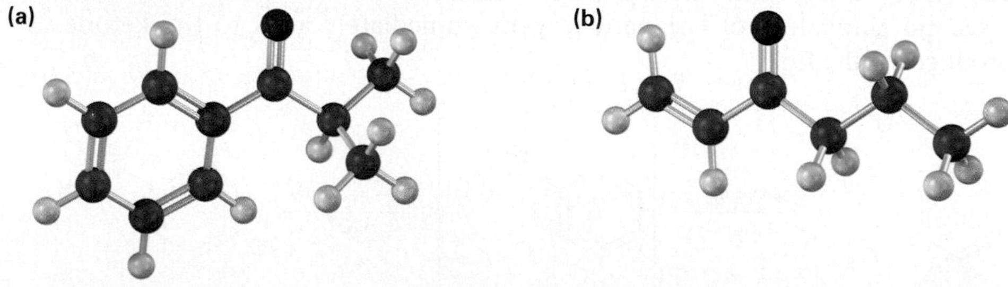

21.5 Chemistry of Acid Anhydrides

Preparation of Acid Anhydrides

Acid anhydrides are typically prepared by nucleophilic acyl substitution reaction of an acid chloride with a carboxylate anion, as we saw in **Section 21.4**.

Both symmetrical and unsymmetrical acid anhydrides can be prepared in this way.

Benzoyl chloride + Sodium acetate $\xrightarrow{\text{Ether}}$ Acetic benzoic anhydride

Reactions of Acid Anhydrides

The chemistry of acid anhydrides is similar to that of acid chlorides, although anhydrides react more slowly. Thus, acid anhydrides react with water to form acids, with alcohols to form esters, with amines to form amides, and with $LiAlH_4$ to form primary alcohols. Only the ester and amide forming reactions are commonly used, however.

Alcoholysis ($RCOOR'$, with $R'OH$); Aminolysis ($RCONH_2$, with NH_3); Hydrolysis ($RCOOH$, with H_2O); Acid anhydride; Reduction ($[H^-]$ → $RCHO$ → $[H^-]$ → RCH_2OH)

Conversion of Acid Anhydrides into Esters Acetic anhydride is often used to prepare acetate esters from alcohols. For example, aspirin (acetylsalicylic acid) is prepared commercially by the acetylation of *o*-hydroxybenzoic acid (salicylic acid) with acetic anhydride.

Salicylic acid (*o*-hydroxybenzoic acid) + Acetic anhydride ($CH_3COOCCH_3$) $\xrightarrow[H_2O]{NaOH}$ Aspirin (an ester) + CH_3CO^- (acetate)

Conversion of Acid Anhydrides into Amides Acetic anhydride is also commonly used to prepare *N*-substituted acetamides from amines. For example, acetaminophen, a drug used in over-the-counter analgesics such as Tylenol, is prepared by reaction of *p*-hydroxyaniline with acetic anhydride. Only the more nucleophilic $-NH_2$ group reacts rather than the less nucleophilic $-OH$ group.

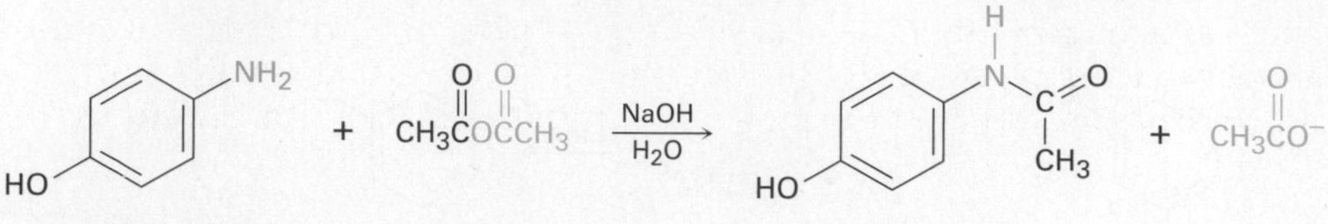

$$\text{HO-}C_6H_4\text{-}NH_2 + CH_3\overset{O}{\overset{\|}{C}}O\overset{O}{\overset{\|}{C}}CH_3 \xrightarrow[H_2O]{NaOH} \text{HO-}C_6H_4\text{-}N(H)\overset{O}{\overset{\|}{C}}CH_3 + CH_3\overset{O}{\overset{\|}{C}}O^-$$

***p*-Hydroxyaniline** **Acetic anhydride** **Acetaminophen**

Notice in both of the previous reactions that only "half" of the anhydride molecule is used, while the other half acts as the leaving group during the nucleophilic acyl substitution step and produces acetate ion as a by-product. Thus, anhydrides are inefficient to use, and acid chlorides are normally preferred for introducing acyl substituents other than acetyl groups.

Problem 21.14

Write the mechanism of the reaction between *p*-hydroxyaniline and acetic anhydride to prepare acetaminophen.

Problem 21.15

What product would you expect from reaction of 1 equivalent of methanol with a cyclic anhydride, such as phthalic anhydride (1,2-benzenedicarboxylic anhydride)? What is the fate of the second "half" of the anhydride?

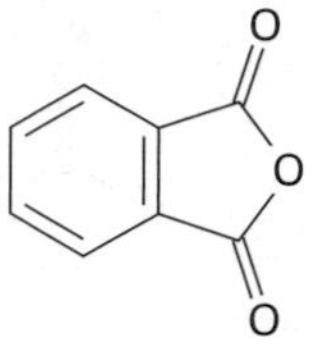

Phthalic anhydride

21.6 Chemistry of Esters

Esters are among the most widespread of all naturally occurring compounds. Many simple esters are pleasant-smelling liquids that are responsible for the fragrant odors of fruits and flowers. For example, methyl butanoate is found in pineapple oil, and isopentyl acetate is a constituent of banana oil. The ester linkage is also present in animal fats and in many biologically important molecules.

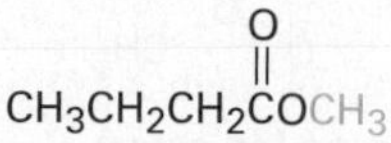

$$CH_3CH_2CH_2\overset{O}{\overset{\|}{C}}OCH_3 \qquad CH_3\overset{O}{\overset{\|}{C}}OCH_2CH_2\overset{CH_3}{\overset{|}{C}}HCH_3$$

Methyl butanoate (from pineapples)  **Isopentyl acetate (from bananas)**

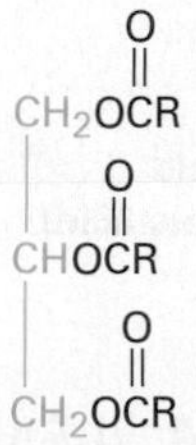

$$\begin{array}{l} CH_2O\overset{O}{\overset{\|}{C}}R \\ | \\ CHO\overset{O}{\overset{\|}{C}}R \\ | \\ CH_2O\overset{O}{\overset{\|}{C}}R \end{array}$$

A fat ($R = C_{11-17}$ chains)

The chemical industry uses esters for a variety of purposes. Ethyl acetate, for instance, is a commonly used solvent, and dialkyl phthalates are used as plasticizers to keep polymers from becoming brittle. You may be aware that there is current concern about possible toxicity of phthalates at high concentrations, although a recent assessment by the U.S. Food and Drug Administration found the risk to be minimal for most people, with the possible exception of male infants.

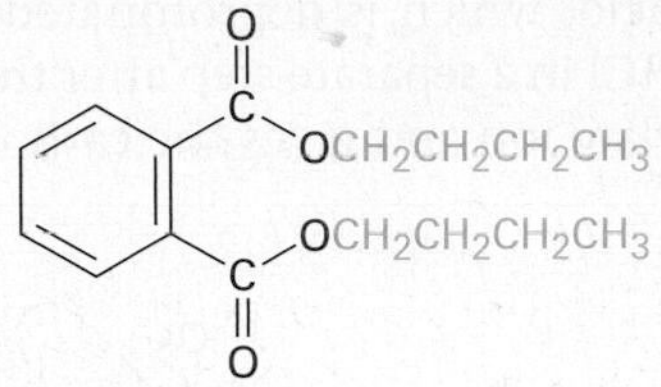

$OCH_2CH_2CH_2CH_3$

$OCH_2CH_2CH_2CH_3$

Dibutyl phthalate (a plasticizer)

Preparation of Esters

Esters are usually prepared from carboxylic acids by the methods already discussed. Thus, carboxylic acids are converted directly into esters by S_N2 reaction of a carboxylate ion with a primary alkyl halide or by Fischer esterification of a carboxylic acid with an alcohol in the presence of a mineral acid catalyst. In addition, acid chlorides are converted into esters by treatment with an alcohol in the presence of base **(Section 21.4)**.

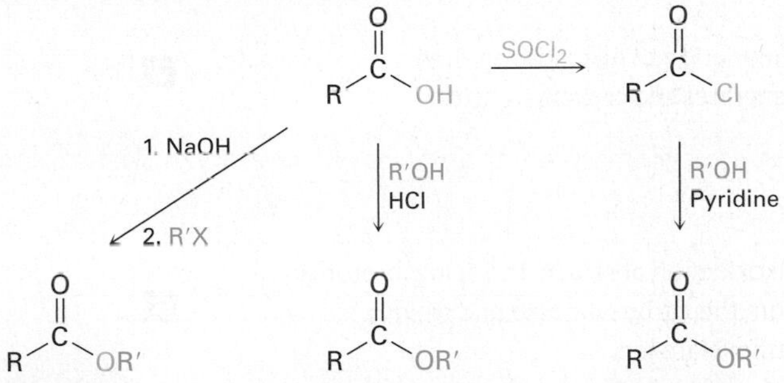

$SOCl_2$

1. NaOH 2. R′X

R′OH HCl

R′OH Pyridine

Method limited to primary alkyl halides	**Method limited to simple alcohols**	**Method is very general**

Reactions of Esters

Esters undergo the same kinds of reactions that we've seen for other carboxylic acid derivatives, but they are less reactive toward nucleophiles than either acid chlorides or anhydrides. All their reactions are applicable to both acyclic and cyclic esters, called **lactones**.

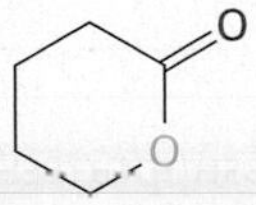

A lactone (cyclic ester)

Conversion of Esters into Carboxylic Acids: Hydrolysis An ester is hydrolyzed, either by aqueous base or by aqueous acid, to yield a carboxylic acid plus an alcohol.

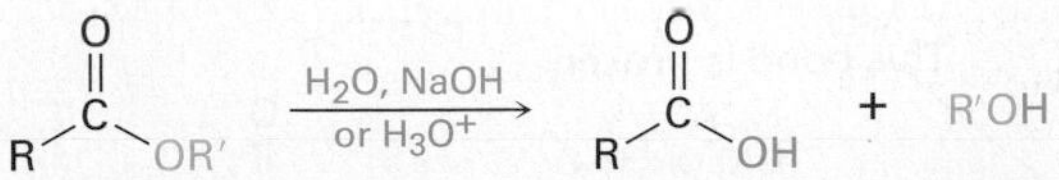

$$RCOOR' \xrightarrow[\text{or } H_3O^+]{H_2O,\ NaOH} RCOOH + R'OH$$

Ester **Acid** **Alcohol**

Ester hydrolysis in basic solution is called **saponification**, after the Latin word *sapo,* meaning "soap." We'll see in **Section 27.2** that soap is in fact made by boiling animal fat with aqueous base to hydrolyze the ester linkages.

As shown in **Figure 21.7**, ester hydrolysis occurs through a typical nucleophilic acyl substitution pathway in which hydroxide ion is the nucleophile that adds to the ester carbonyl group to give a tetrahedral intermediate. Loss of alkoxide ion then gives a carboxylic acid, which is deprotonated to give the carboxylate ion. Addition of aqueous HCl in a separate step after the saponification is complete protonates the carboxylate ion and gives the carboxylic acid.

Figure 21.7 | MECHANISM

Mechanism of base-induced ester hydrolysis (saponification)

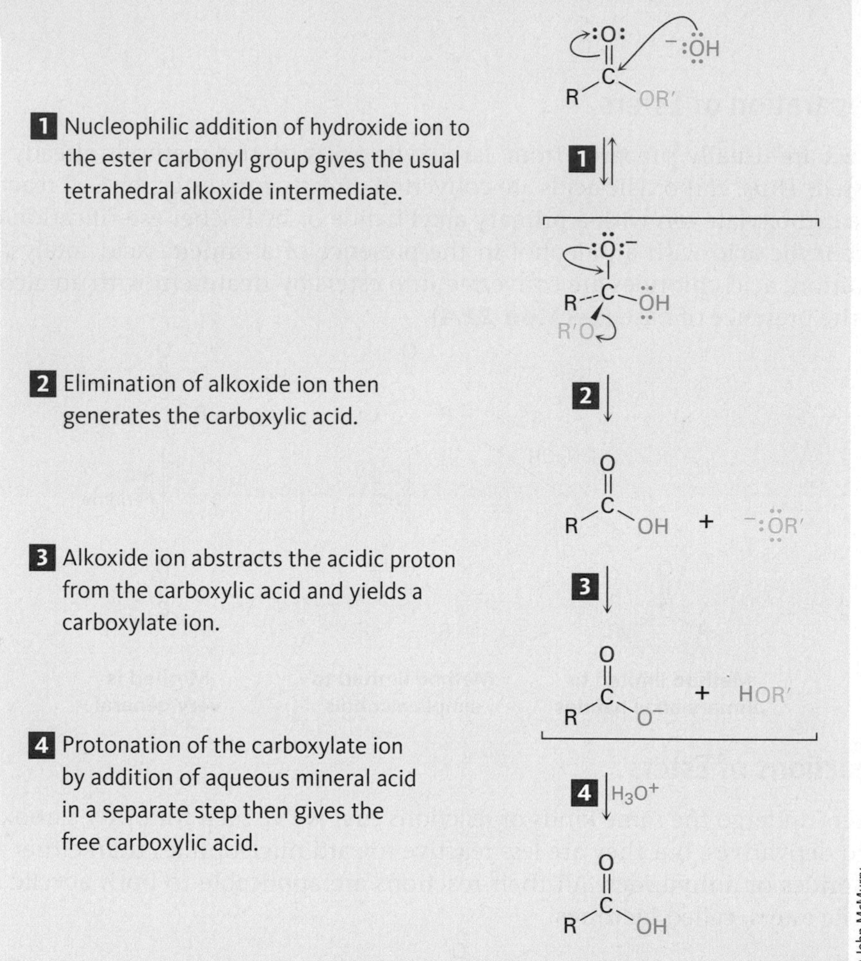

The mechanism shown in Figure 21.7 is supported by isotope-labeling studies. When ethyl propanoate labeled with ^{18}O in the ether-like oxygen is hydrolyzed in aqueous NaOH, the ^{18}O label shows up exclusively in the ethanol product. None of the label remains with the propanoic acid, indicating that saponification occurs by cleavage of the C–OR′ bond rather than the CO–R′ bond.

This bond is broken.

$$CH_3CH_2C(=O)\overset{*}{O}CH_2CH_3 \xrightarrow[\text{2. } H_3O^+]{\text{1. NaOH, } H_2O} CH_3CH_2C(=O)OH + H\overset{*}{O}CH_2CH_3$$

Acid-catalyzed ester hydrolysis can occur by more than one mechanism, depending on the structure of the ester. The usual pathway, however, is just the reverse of a Fischer esterification reaction **(Section 21.3)**. As shown in **Figure 21.8**, the ester is first activated toward nucleophilic attack by protonation of the carboxyl oxygen atom, and nucleophilic addition of water then occurs. Transfer of a proton and elimination of alcohol yields the carboxylic acid. Because this hydrolysis reaction is the reverse of a Fischer esterification reaction, Figure 21.8 is the reverse of Figure 21.4.

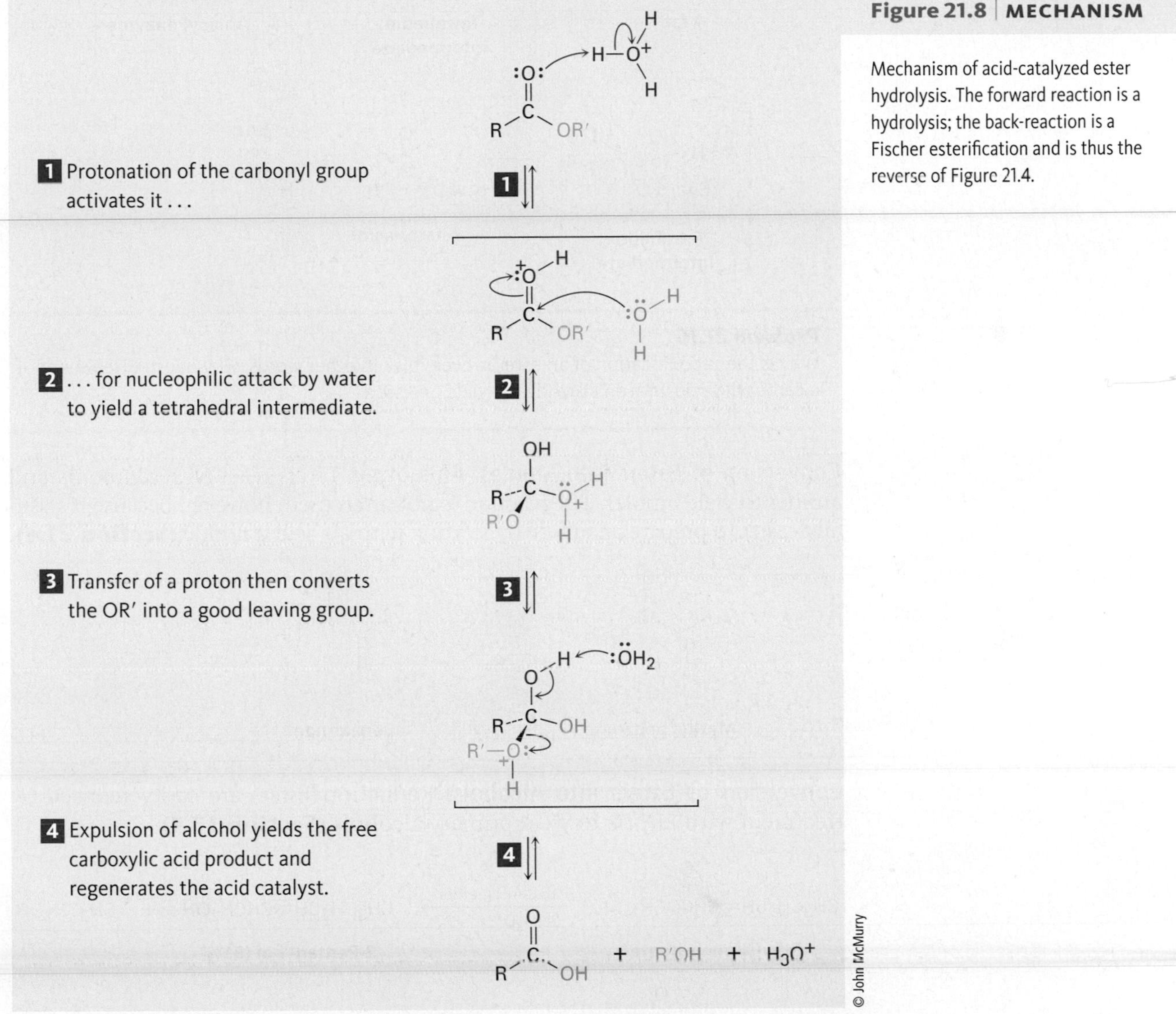

Figure 21.8 | MECHANISM

Mechanism of acid-catalyzed ester hydrolysis. The forward reaction is a hydrolysis; the back-reaction is a Fischer esterification and is thus the reverse of Figure 21.4.

Ester hydrolysis is common in biological chemistry, particularly in the digestion of dietary fats and oils. We'll save a complete discussion of the mechanistic details of fat hydrolysis until **Section 29.2** but will note for now that the reaction is catalyzed by various lipase enzymes and involves two sequential nucleophilic acyl substitution reactions. The first is a *transesterification* reaction in which an alcohol group on the lipase adds to an ester linkage in the fat molecule to give a tetrahedral intermediate that expels alcohol and forms an acyl enzyme intermediate. The second is an addition of water to the acyl enzyme, followed by expulsion of the enzyme to give a hydrolyzed acid plus regenerated enzyme.

A fat → Tetrahedral intermediate → (ROH) → An acyl enzyme → Tetrahedral intermediate → A fatty acid + Enz–OH

Problem 21.16

Why is the saponification of an ester irreversible? In other words, why doesn't treatment of a carboxylic acid with an alkoxide ion yield an ester?

Conversion of Esters into Amides: Aminolysis Esters react with ammonia and amines to yield amides. The reaction is not often used, however, because it's usually easier to prepare an amide by starting with an acid chloride **(Section 21.4)**.

$$\text{Methyl benzoate} \xrightarrow[\text{Ether}]{NH_3} \text{Benzamide} + CH_3OH$$

Conversion of Esters into Alcohols: Reduction Esters are easily reduced by treatment with $LiAlH_4$ to yield primary alcohols **(Section 17.4)**.

$$CH_3CH_2CH{=}CHCOCH_2CH_3 \xrightarrow[\text{2. } H_3O^+]{\text{1. } LiAlH_4\text{, ether}} CH_3CH_2CH{=}CHCH_2OH + CH_3CH_2OH$$

Ethyl 2-pentenoate → **2-Penten-1-ol (91%)**

$$\text{A lactone} \xrightarrow[\text{2. } H_3O^+]{\text{1. } LiAlH_4\text{, ether}} HOCH_2CH_2CH_2CH(OH)CH_3$$

A lactone → **1,4-Pentanediol (86%)**

The mechanism of ester reduction is similar to that of acid chloride reduction in that a hydride ion first adds to the carbonyl group, followed by elimination of alkoxide ion to yield an aldehyde. Further reduction of the aldehyde gives the primary alcohol.

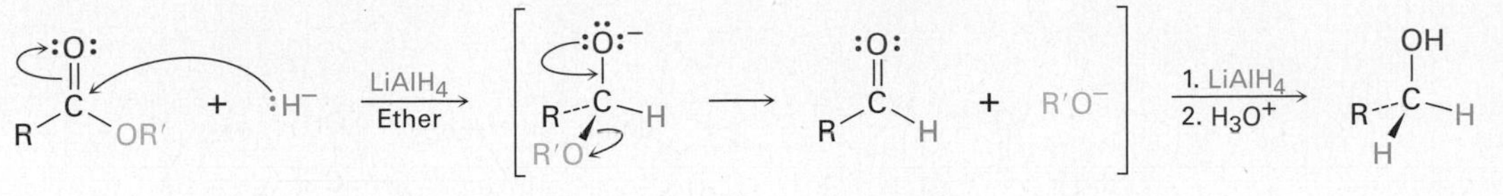

:O: || R–C–OR′ + :H⁻ →($LiAlH_4$, Ether) [:Ö:⁻ | R–C–H | R′O → :O: || R–C–H + R′O⁻] →(1. $LiAlH_4$ 2. H_3O^+) OH | R–C–H | H

Aldehyde

A primary alcohol

The aldehyde intermediate can be isolated if 1 equivalent of diisobutylaluminum hydride (DIBAH, or DIBAL-H) is used as the reducing agent instead of $LiAlH_4$. The reaction has to be carried out at −78 °C to avoid further reduction to the alcohol. Such partial reductions of carboxylic acid derivatives to aldehydes also occur in numerous biological pathways, although the substrate is either a thioester or acyl phosphate rather than an ester.

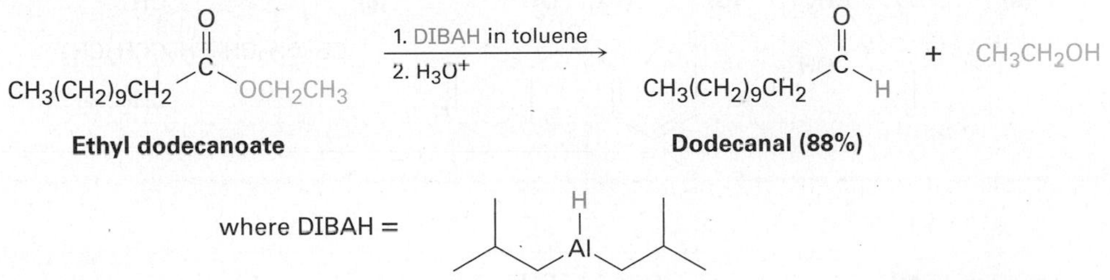

$CH_3(CH_2)_9CH_2COOCH_2CH_3$ →(1. DIBAH in toluene 2. H_3O^+) $CH_3(CH_2)_9CH_2CHO$ + CH_3CH_2OH

Ethyl dodecanoate

Dodecanal (88%)

where DIBAH =

Problem 21.17

What product would you expect from the reaction of butyrolactone with $LiAlH_4$? With DIBAH?

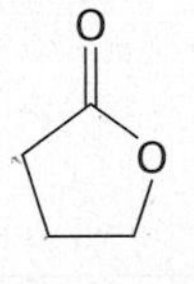

Butyrolactone

Problem 21.18

Show the products you would obtain by reduction of the following esters with $LiAlH_4$:

(a) 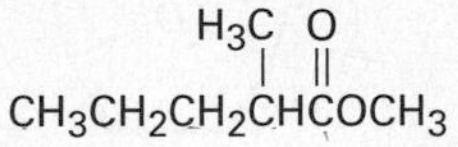 $CH_3CH_2CH_2CH(CH_3)COOCH_3$

(b) 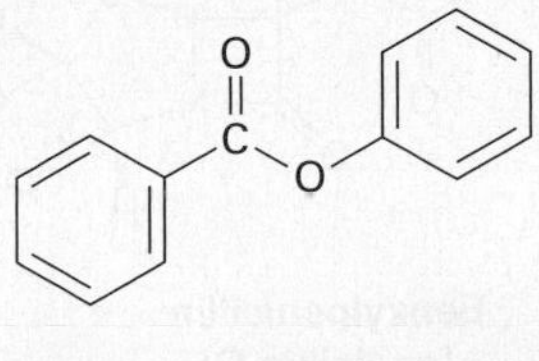 $C_6H_5COOC_6H_5$

Conversion of Esters into Alcohols: Grignard Reaction Esters react with 2 equivalents of a Grignard reagent to yield a tertiary alcohol in which two of the substituents are identical (**Section 17.5**). The reaction occurs by the usual nucleophilic substitution mechanism to give an intermediate ketone, which reacts further with the Grignard reagent to yield a tertiary alcohol.

1. 2 (phenyl)MgBr
2. H_3O^+

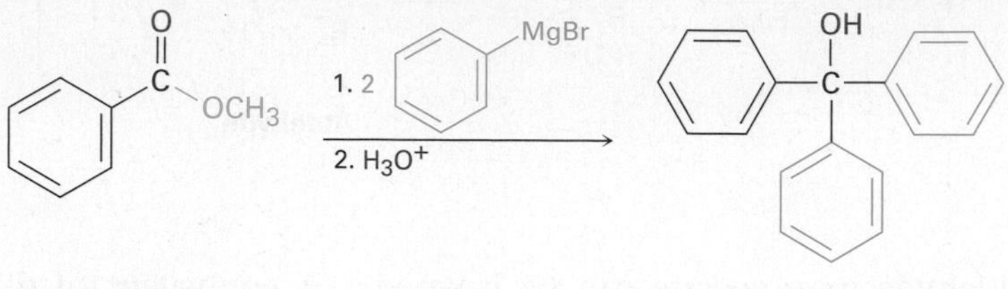

Methyl benzoate → **Triphenylmethanol (96%)**

Problem 21.19

What ester and what Grignard reagent might you start with to prepare the following alcohols?

(a) (phenyl)$C(CH_3)_2OH$ **(b)** $(C_6H_5)_2C(CH_3)OH$ **(c)** $CH_3CH_2CH_2CH_2C(OH)(CH_2CH_3)CH_2CH_3$

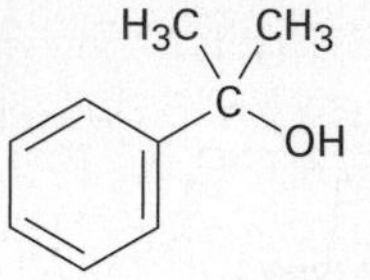

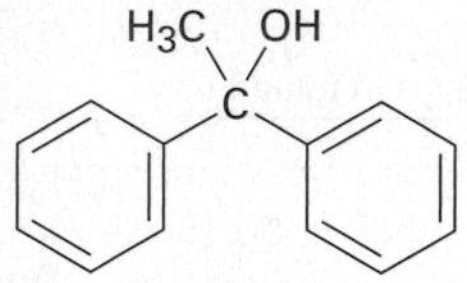

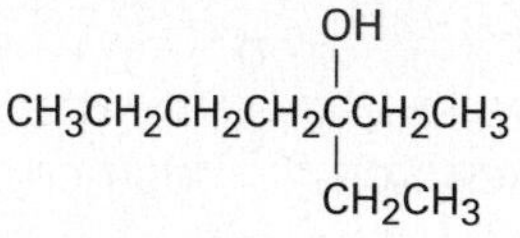

21.7 Chemistry of Amides

Amides, like esters, are abundant in all living organisms. Proteins, nucleic acids, and many pharmaceutical agents have amide functional groups. The reason for this abundance of amides is that they are stable to the aqueous conditions found in living organisms. Amides are the least reactive of the common acid derivatives and undergo relatively few nucleophilic acyl substitution reactions.

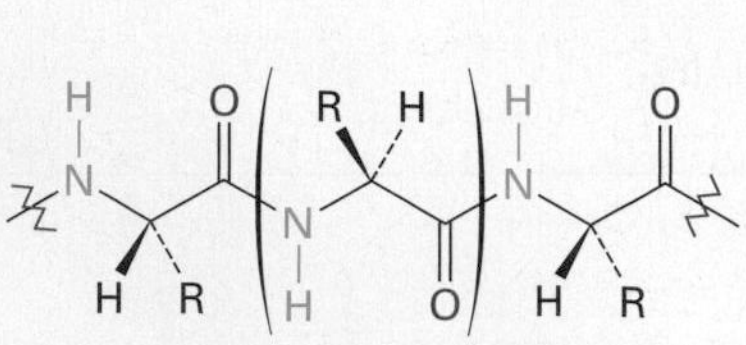

A protein segment 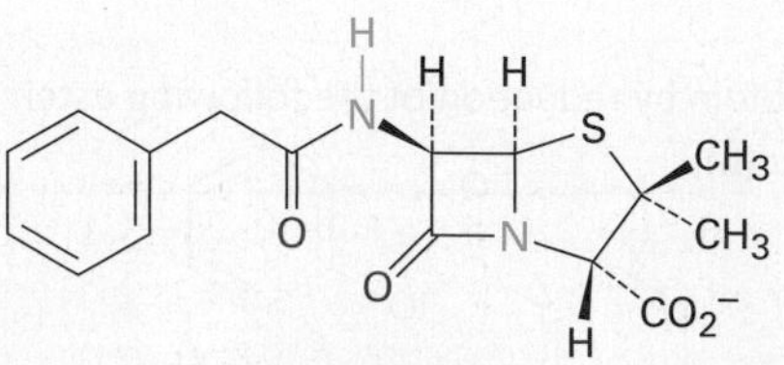 **Benzylpenicillin (penicillin G)** 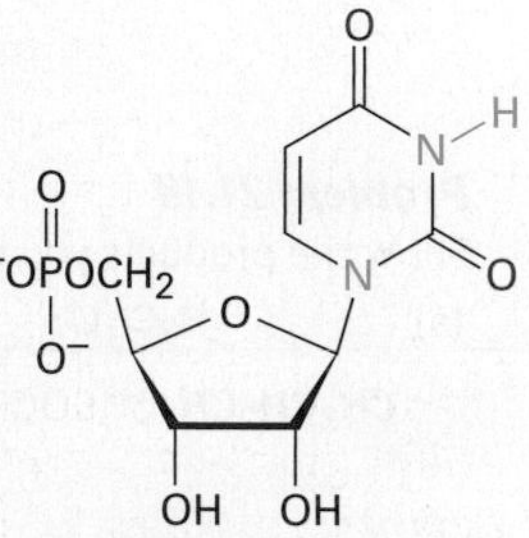 **Uridine 5′-phosphate (a ribonucleotide)**

Preparation of Amides

Amides are usually prepared by reaction of an acid chloride with an amine **(Section 21.4)**. Ammonia, monosubstituted amines, and disubstituted amines all undergo the reaction.

$$RCOCl \xrightarrow{NH_3} RCONH_2 \qquad RCOCl \xrightarrow{R'NH_2} RCONHR' \qquad RCOCl \xrightarrow{R'_2NH} RCONR'_2$$

Reactions of Amides

Conversion of Amides into Carboxylic Acids: Hydrolysis Amides undergo hydrolysis to yield carboxylic acids plus ammonia or an amine on heating in either aqueous acid or aqueous base. The conditions required for amide hydrolysis are more severe than those required for the hydrolysis of acid chlorides or esters, but the mechanisms are similar. Acidic hydrolysis reaction occurs by nucleophilic addition of water to the protonated amide, followed by transfer of a proton from oxygen to nitrogen to make the nitrogen a better leaving group and subsequent elimination. The steps are reversible, with the equilibrium shifted toward product by protonation of NH_3 in the final step.

$$RCONH_2 \xrightarrow{H_3O^+} [RC(=\overset{+}{O}H)NH_2 + H_2O \rightleftharpoons RC(OH)(NH_2)\overset{+}{O}H_2 \rightleftharpoons RC(OH)(\overset{+}{N}H_3)OH] \rightleftharpoons RCOOH + \{H_3O^+, NH_3\} \longrightarrow NH_4^+ + H_2O$$

An amide **A carboxylic acid**

Basic hydrolysis occurs by nucleophilic addition of OH^- to the amide carbonyl group, followed by elimination of amide ion ($^-NH_2$) and subsequent deprotonation of the initially formed carboxylic acid by ammonia. The steps are reversible, with the equilibrium shifted toward product by the final deprotonation of the carboxylic acid. Basic hydrolysis is substantially more difficult than the analogous acid-catalyzed reaction because amide ion is a very poor leaving group, making the elimination step difficult.

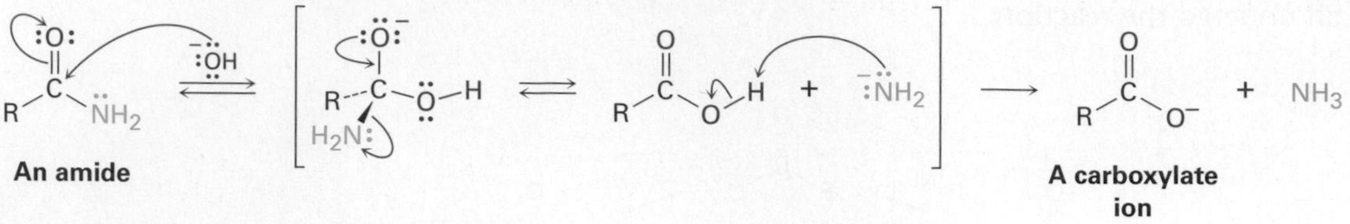

An amide

A carboxylate ion

Amide hydrolysis is common in biological chemistry. Just as the hydrolysis of esters is the initial step in the digestion of dietary fats, the hydrolysis of amides is the initial step in the digestion of dietary proteins. The reaction is catalyzed by protease enzymes and occurs by a mechanism almost identical to that we just saw for fat hydrolysis. That is, an initial nucleophilic acyl substitution of an alcohol group in the enzyme on an amide linkage in the protein gives an acyl enzyme intermediate that then undergoes hydrolysis.

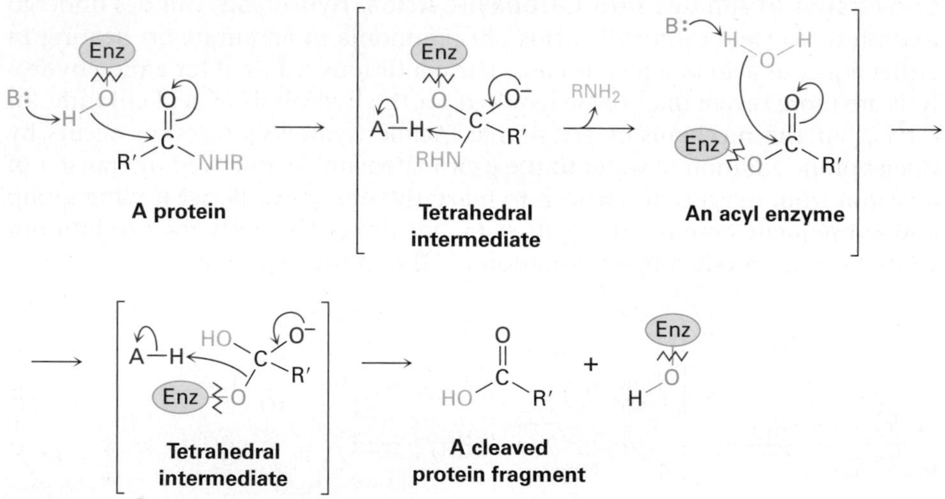

A protein

Tetrahedral intermediate

An acyl enzyme

Tetrahedral intermediate

A cleaved protein fragment

Conversion of Amides into Amines: Reduction Like other carboxylic acid derivatives, amides can be reduced by $LiAlH_4$. The product of the reduction, however, is an amine rather than an alcohol. The net effect of an amide reduction reaction is thus the conversion of the amide carbonyl group into a methylene group ($C{=}O \rightarrow CH_2$). This kind of reaction is specific for amides and does not occur with other carboxylic acid derivatives.

$$CH_3(CH_2)_9CH_2C(=O)NHCH_3 \xrightarrow[2.\ H_2O]{1.\ LiAlH_4\ \text{in ether}} CH_3(CH_2)_9CH_2CH_2NHCH_3$$

N-Methyldodecanamide

Dodecylmethylamine (95%)

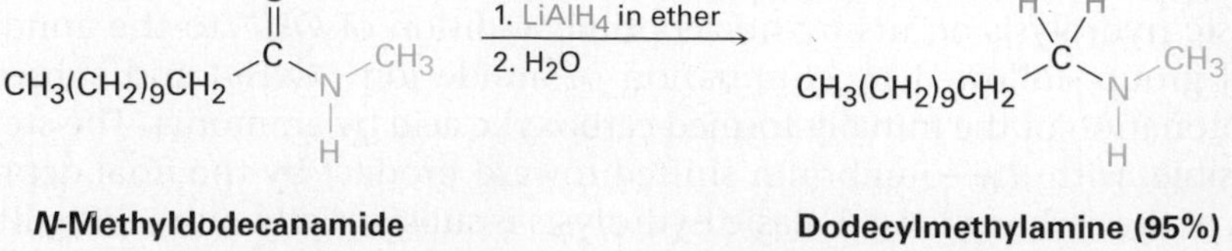

Amide reduction occurs by nucleophilic addition of hydride ion to the amide carbonyl group, followed by expulsion of the *oxygen* atom as an aluminate anion leaving group to give an iminium ion intermediate. The intermediate iminium ion is then further reduced by $LiAlH_4$ to yield the amine.

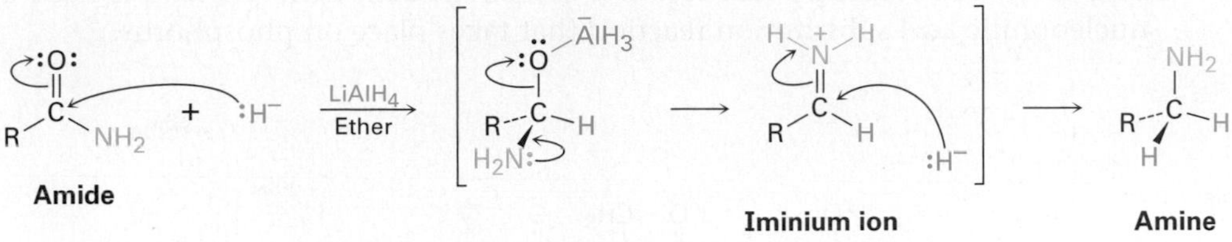

Amide

Iminium ion

Amine

The reaction is effective with both acyclic and cyclic amides, or **lactams**, and is a good method for preparing cyclic amines.

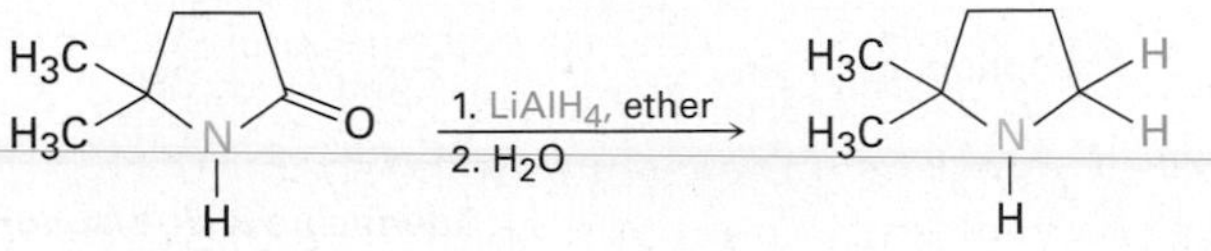

A lactam

A cyclic amine (80%)

Problem 21.20

How would you convert *N*-ethylbenzamide to each of the following products?

(a) Benzoic acid **(b)** Benzyl alcohol **(c)** $C_6H_5CH_2NHCH_2CH_3$

Problem 21.21

How would you use the reaction of an amide with $LiAlH_4$ as the key step in going from bromocyclohexane to (*N,N*-dimethylaminomethyl)cyclohexane? Write all the steps in the reaction sequence.

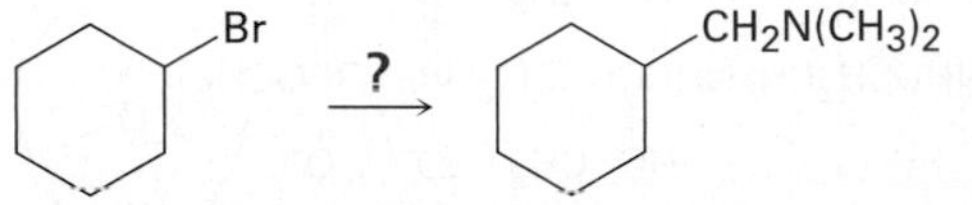

(*N,N*-Dimethylaminomethyl)cyclohexane

21.8 Chemistry of Thioesters and Acyl Phosphates: Biological Carboxylic Acid Derivatives

As mentioned in the chapter introduction, the substrate for a nucleophilic acyl substitution reaction in living organisms is generally either a thioester (RCOSR′) or an acyl phosphate ($RCO_2PO_3^{2-}$ or $RCO_2PO_3R'^-$). Neither is as reactive as an acid chloride or acid anhydride, yet both are stable enough to exist in living organisms while still reactive enough to undergo acyl substitution.

Acyl CoA's, such as acetyl CoA, are the most common thioesters in nature. Coenzyme A, abbreviated CoA, is a thiol formed by a phosphoric anhydride linkage (O=P–O–P=O) between phosphopantetheine and adenosine

3′,5′-bisphosphate. (The prefix *bis-* means "two" and indicates that adenosine 3′,5′-bisphosphate has two phosphate groups, one on C3′ and one on C5′.) Reaction of coenzyme A with an acyl phosphate or acyl adenylate gives the acyl CoA **(Figure 21.9)**. As we saw in **Section 21.3** (Figure 21.6), formation of the acyl adenylate occurs by reaction of a carboxylic acid with ATP and is itself a nucleophilic acyl substitution reaction that takes place on phosphorus.

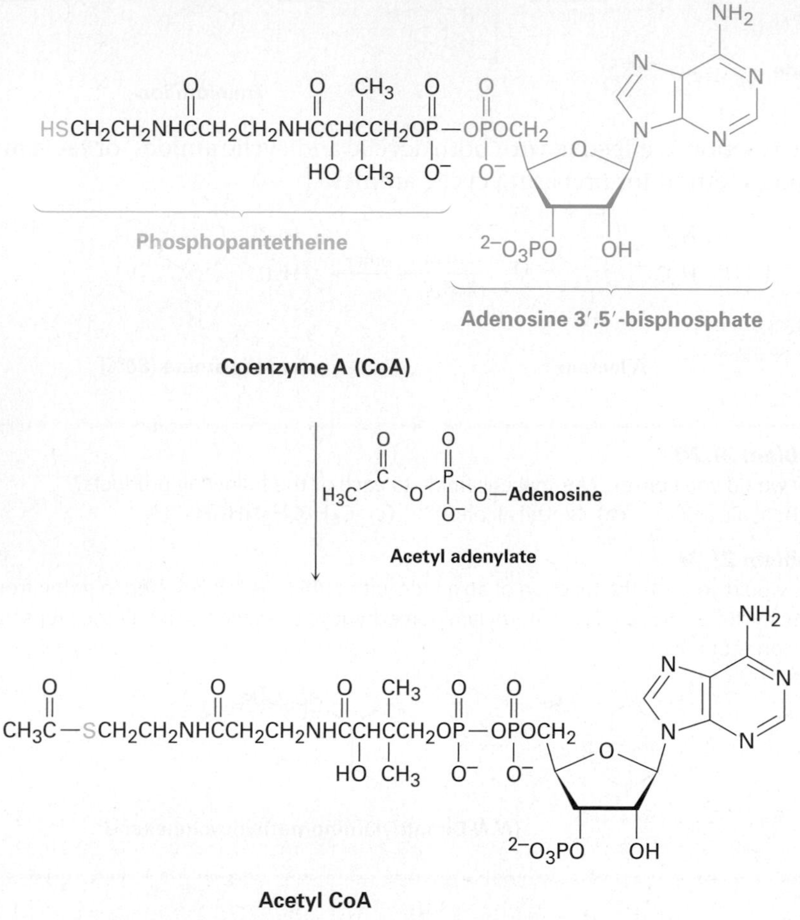

Figure 21.9 Formation of the thioester acetyl CoA by nucleophilic acyl substitution reaction of coenzyme A (CoA) with acetyl adenylate.

Phosphopantetheine

Adenosine 3′,5′-bisphosphate

Coenzyme A (CoA)

Acetyl adenylate

Acetyl CoA

Once formed, an acyl CoA is a substrate for further nucleophilic acyl substitution reactions. For example, *N*-acetylglucosamine, a component of cartilage and other connective tissues, is synthesized by an aminolysis reaction between glucosamine and acetyl CoA.

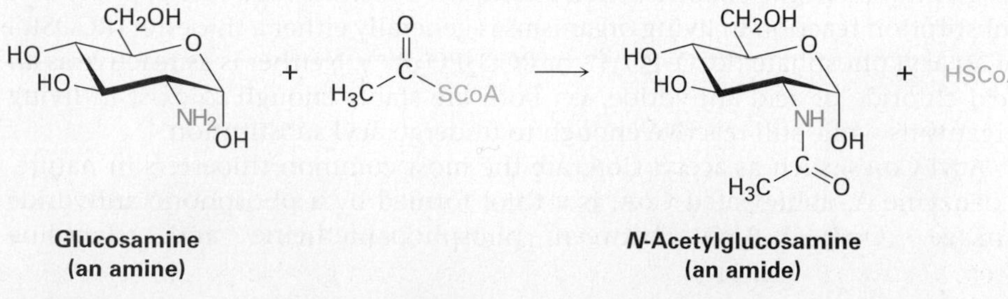

Glucosamine (an amine)

***N*-Acetylglucosamine (an amide)**

Another example of a nucleophilic acyl substitution reaction on a thioester, this one a substitution by hydride ion to effect partial reduction of a thioester to an aldehyde, occurs in the biosynthesis of mevaldehyde, an intermediate in terpenoid synthesis, which we'll discuss in some detail in **Section 27.5**. In this reaction, (3*S*)-3-hydroxy-3-methylglutaryl CoA is reduced by hydride donation from NADPH.

NADPH

$CONH_2$

NADP+

HSCoA

(3*S*)-3-Hydroxy-3-methylglutaryl CoA

(*R*)-Mevaldehyde

Problem 21.22

Write the mechanism of the reaction shown in Figure 21.9 between coenzyme A and acetyl adenylate to give acetyl CoA

21.9 Polyamides and Polyesters: Step-Growth Polymers

When an amine reacts with an acid chloride, an amide is formed. What would happen, though, if a *diamine* and a *diacid chloride* were allowed to react? Each partner could form two amide bonds, linking more and more molecules together until a giant polyamide resulted. In the same way, reaction of a diol with a diacid would lead to a polyester.

$$H_2N(CH_2)_nNH_2 + ClC(=O)(CH_2)_mC(=O)Cl \longrightarrow -[HN(CH_2)_nNH-C(=O)(CH_2)_mC(=O)]-$$

A diamine **A diacid chloride** **A polyamide (nylon)**

$$HO(CH_2)_nOH + HOC(=O)(CH_2)_mC(=O)OH \longrightarrow -[O(CH_2)_nO-C(=O)(CH_2)_mC(=O)]- + H_2O$$

A diol **A diacid** **A polyester**

There are two main classes of synthetic polymers: *chain-growth polymers* and *step-growth polymers*. Polyethylene and other alkene and diene polymers like those we saw in **Sections 8.10 and 14.6** are chain-growth polymers because they are produced in chain-reaction processes. An initiator adds to a C=C bond to give a reactive intermediate, which adds to a second alkene molecule to produce a new intermediate, which adds to a third molecule, and so on. By contrast, polyamides and polyesters are **step-growth polymers** because each bond in the polymer is formed in a discrete step, independent of the others. The key bond-forming step is often a nucleophilic acyl substitution of a carboxylic acid derivative. Some commercially important step-growth polymers are shown in Table 21.2.

Table 21.2 Some Common Step-Growth Polymers and Their Uses

Monomers	Structure	Polymer	Uses
Adipic acid + Hexamethylenediamine	$HOOC(CH_2)_4COOH$; $H_2NCH_2CH_2CH_2CH_2CH_2CH_2NH_2$	Nylon 66	Fibers, clothing, tire cord
Dimethyl terephthalate + Ethylene glycol	$CH_3O_2C{-}C_6H_4{-}CO_2CH_3$; $HOCH_2CH_2OH$	Dacron, Mylar, Terylene	Fibers, clothing, films, tire cord
Caprolactam	(seven-membered cyclic amide, N–H, C=O)	Nylon 6, Perlon	Fibers, castings
Diphenyl carbonate + Bisphenol A	$C_6H_5O{-}CO{-}OC_6H_5$; $HO{-}C_6H_4{-}C(CH_3)_2{-}C_6H_4{-}OH$	Lexan, polycarbonate	Equipment housing, molded articles
Toluene-2,6-diisocyanate + Poly(2-butene-1,4-diol)	$O{=}C{=}N{-}C_6H_3(CH_3){-}N{=}C{=}O$; $HO{+}CH_2CH{=}CHCH_2{+}_nOH$	Polyurethane, Spandex	Fibers, coatings, foams

Polyamides (Nylons)

The best known step-growth polymers are the polyamides, or *nylons,* first prepared by Wallace Carothers at the DuPont Company by heating a diamine with a diacid. For example, nylon 66 is prepared by reaction of adipic acid (hexanedioic acid) with hexamethylenediamine (1,6-hexanediamine) at 280 °C. The designation "66" tells the number of carbon atoms in the diamine (the first 6) and the diacid (the second 6).

$$HO\overset{O}{\overset{\|}{C}}CH_2CH_2CH_2CH_2\overset{O}{\overset{\|}{C}}OH \;+\; H_2NCH_2CH_2CH_2CH_2CH_2CH_2NH_2$$

Adipic acid **Hexamethylenediamine**

$$\downarrow \text{Heat}$$

$$\left(\overset{O}{\overset{\|}{C}}CH_2CH_2CH_2CH_2\overset{O}{\overset{\|}{C}}-NHCH_2CH_2CH_2CH_2CH_2CH_2NH\right)_n \;+\; 2n\,H_2O$$

Nylon 66

Nylons are used both in engineering applications and in making fibers. A combination of high impact strength and abrasion resistance makes nylon an excellent metal substitute for bearings and gears. As fiber, nylon is used in a variety of applications, from clothing to tire cord to ropes.

Polyesters

The most generally useful polyester is that made by reaction between dimethyl terephthalate (dimethyl 1,4-benzenedicarboxylate) and ethylene glycol (1,2-ethanediol). The product is used under the trade name Dacron to make clothing fiber and tire cord and under the name Mylar to make recording tape. The tensile strength of poly(ethylene terephthalate) film is nearly equal to that of steel.

$$CH_3O\overset{O}{\overset{\|}{C}}-C_6H_4-\overset{O}{\overset{\|}{C}}OCH_3 \;+\; HOCH_2CH_2OH \xrightarrow{200\ °C} \left(\overset{O}{\overset{\|}{C}}-C_6H_4-\overset{O}{\overset{\|}{C}}-OCH_2CH_2O\right)_n \;+\; 2n\,CH_3OH$$

Dimethyl terephthalate **Ethylene glycol** **A polyester (Dacron, Mylar)**

Lexan, a polycarbonate prepared from diphenyl carbonate and bisphenol A, is another commercially valuable polyester. Lexan has an unusually high impact strength, making it valuable for use in telephones, bicycle safety helmets, and laptop computer cases.

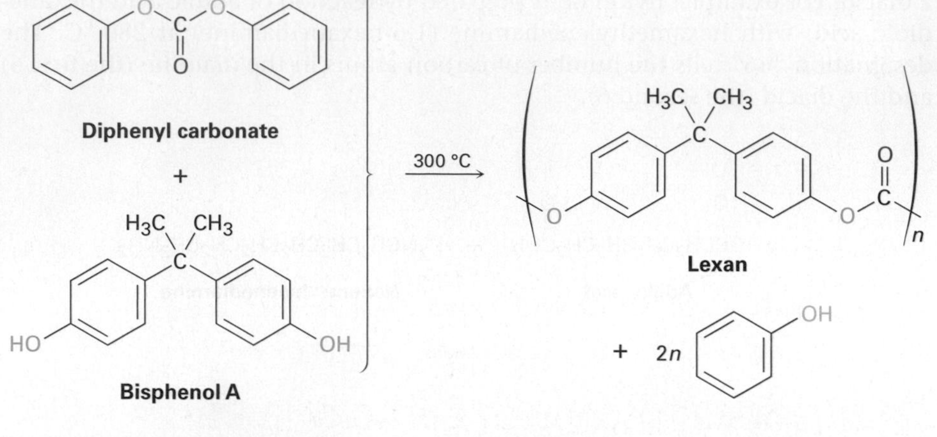

Diphenyl carbonate

+

Bisphenol A

300 °C

Lexan

+ 2*n*

Sutures and Biodegradable Polymers

Because plastics are too often thrown away rather than recycled, much work has been carried out on developing biodegradable polymers, which can be broken down rapidly in landfills by soil microorganisms. Among the most common biodegradable polymers are poly(glycolic acid) (PGA), poly(lactic acid) (PLA), and poly(hydroxybutyrate) (PHB). All are polyesters and are therefore susceptible to hydrolysis of their ester links. Copolymers of PGA with PLA have found a particularly wide range of uses. A 90/10 copolymer of poly(glycolic acid) with poly(lactic acid) is used to make absorbable sutures, for instance. The sutures are entirely hydrolyzed and absorbed by the body within 90 days after surgery.

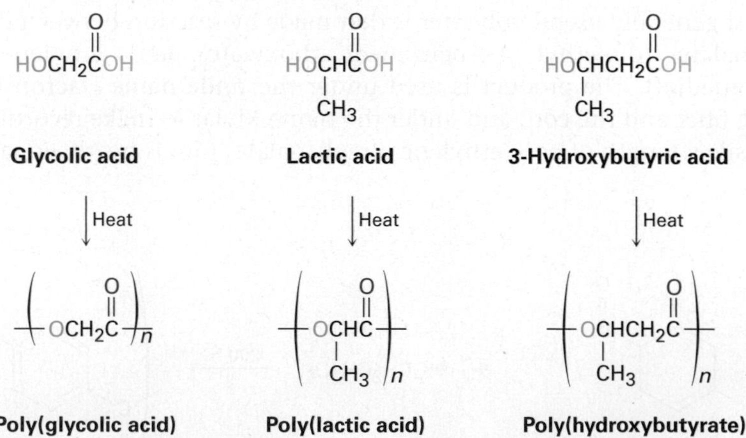

In Europe, interest has centered particularly on poly(hydroxybutyrate), which can be made into films for packaging as well as into molded items. The polymer degrades within 4 weeks in landfills, both by ester hydrolysis and by an E1cB elimination reaction of the oxygen atom β to the carbonyl group. The use of poly(hydroxybutyrate) is limited at present by its cost—about four times that of polypropylene.

Problem 21.23

Draw structures of the step-growth polymers you would expect to obtain from the following reactions:

(a) $BrCH_2CH_2CH_2Br + HOCH_2CH_2CH_2OH \xrightarrow{\text{Base}}$?

(b) $HOCH_2CH_2OH + HO_2C(CH_2)_6CO_2H \xrightarrow{H_2SO_4 \text{ catalyst}}$?

(c) $H_2N(CH_2)_6NH_2 + ClC(=O)(CH_2)_4C(=O)Cl \longrightarrow$?

Problem 21.24

Kevlar, a nylon polymer prepared by reaction of 1,4-benzenedicarboxylic acid (terephthalic acid) with 1,4-benzenediamine (*p*-phenylenediamine), is so strong that it's used to make bulletproof vests. Draw the structure of a segment of Kevlar.

21.10 Spectroscopy of Carboxylic Acid Derivatives

Infrared Spectroscopy

All carbonyl-containing compounds have intense IR absorptions in the range 1650 to 1850 cm^{-1}. As shown in Table 21.3, the exact position of the absorption provides information about the specific kind of carbonyl group. For comparison, the IR absorptions of aldehydes, ketones, and carboxylic acids are included in the table, along with values for carboxylic acid derivatives.

Acid chlorides are easily detected by their characteristic absorption near 1800 cm^{-1}. Acid anhydrides can be identified because they show two absorptions in the carbonyl region, one at 1820 cm^{-1} and another at 1760 cm^{-1}. Esters are detected by their absorption at 1735 cm^{-1}, a position somewhat higher than

Table 21.3 Infrared Absorptions of Some Carbonyl Compounds

Carbonyl type	Example	Absorption (cm^{-1})
Saturated acid chloride	Acetyl chloride	1810
Aromatic acid chloride	Benzoyl chloride	1770
Saturated acid anhydride	Acetic anhydride	1820, 1760
Saturated ester	Ethyl acetate	1735
Aromatic ester	Ethyl benzoate	1720
Saturated amide	Acetamide	1690
Aromatic amide	Benzamide	1675
N-Substituted amide	*N*-Methylacetamide	1680
N,N-Disubstituted amide	*N,N*-Dimethylacetamide	1650
(Saturated aldehyde	Acetaldehyde	1730)
(Saturated ketone	Acetone	1715)
(Saturated carboxylic acid	Acetic acid	1710)

that for either aldehydes or ketones. Amides, by contrast, absorb near the low wavenumber end of the carbonyl region, with the degree of substitution on nitrogen affecting the exact position of the IR band.

Problem 21.25

What kinds of functional groups might compounds have if they show the following IR absorptions?

(a) Absorption at 1735 cm^{-1} **(b)** Absorption at 1810 cm^{-1}

(c) Absorptions at 2500–3300 cm^{-1} and 1710 cm^{-1} **(d)** Absorption at 1715 cm^{-1}

Problem 21.26

Propose structures for compounds that have the following formulas and IR absorptions:

(a) $C_6H_{12}O_2$, 1735 cm^{-1} **(b)** C_4H_9NO, 1650 cm^{-1} **(c)** C_4H_5ClO, 1780 cm^{-1}

Nuclear Magnetic Resonance Spectroscopy

Hydrogens on the carbon next to a carbonyl group are slightly deshielded and absorb near 2 δ in the ^1H NMR spectrum. The identity of the carbonyl group can't be determined by ^1H NMR, however, because the α hydrogens of all acid derivatives absorb in the same range. **Figure 21.10** shows the ^1H NMR spectrum of ethyl acetate.

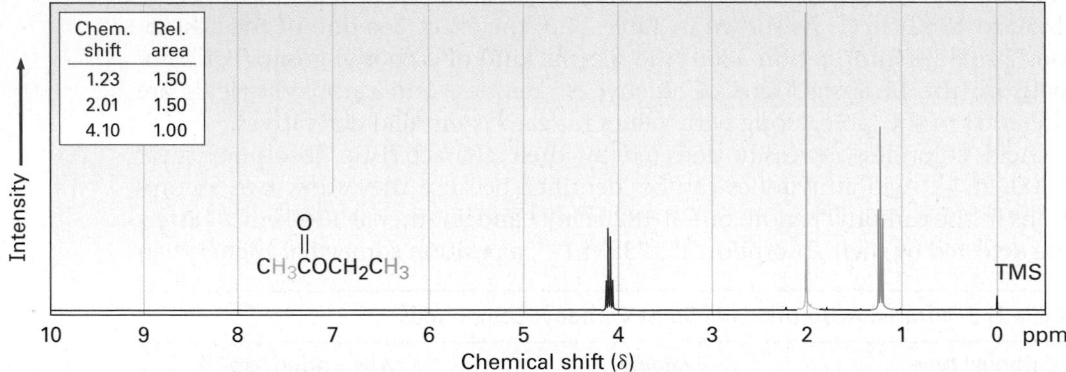

Figure 21.10 Proton NMR spectrum of ethyl acetate.

Although ^{13}C NMR is useful for determining the presence or absence of a carbonyl group in a molecule, the identity of the carbonyl group is difficult to determine. Aldehydes and ketones absorb near 200 δ, while the carbonyl carbon atoms of various acid derivatives absorb in the range 160 to 180 δ (Table 21.4).

Table 21.4 ^{13}C NMR Absorptions in Some Carbonyl Compounds

Compound	Absorption (δ)	Compound	Absorption (δ)
Acetic acid	177.3	Acetic anhydride	166.9
Ethyl acetate	170.7	Acetone	205.6
Acetyl chloride	170.3	Acetaldehyde	201.0
Acetamide	172.6		

A DEEPER LOOK β-Lactam Antibiotics

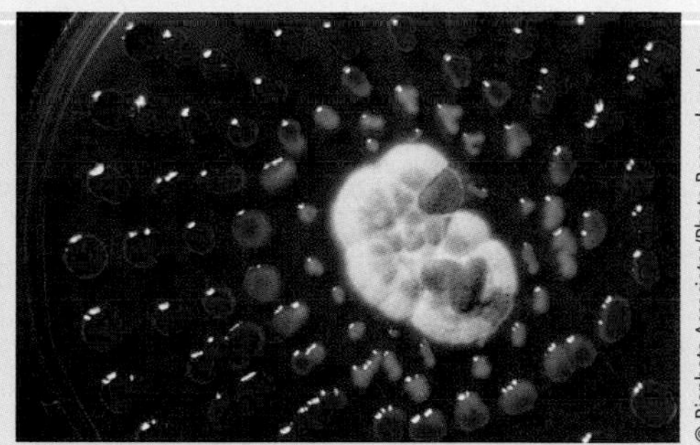

Penicillium mold growing in a petri dish.

You should never underestimate the value of hard work and logical thinking, but it's also true that blind luck plays a role in most real scientific breakthroughs. What has been called "the supreme example of luck in all scientific history" occurred in the late summer of 1928, when the Scottish bacteriologist Alexander Fleming went on vacation, leaving in his lab a culture plate recently inoculated with the bacterium *Staphylococcus aureus.*

While Fleming was away, an extraordinary chain of events occurred. First, a 9-day cold spell lowered the laboratory temperature to a point where the *Staphylococcus* on the plate could not grow. During this time, spores from a colony of the mold *Penicillium notatum* being grown on the floor below wafted up into Fleming's lab and landed in the culture plate. The temperature then rose, and both *Staphylococcus* and *Penicillium* began to grow. On returning from vacation, Fleming discarded the plate into a tray of antiseptic, intending to sterilize it. Evidently, though, the plate did not sink deeply enough into the antiseptic, because when Fleming happened to glance at it a few days later, what he saw changed the course of human history. He noticed that the growing *Penicillium* mold appeared to dissolve the colonies of staphylococci.

Fleming realized that the *Penicillium* mold must be producing a chemical that killed the *Staphylococcus* bacteria, and he spent several years trying to isolate the substance. Finally, in 1939, the Australian pathologist Howard Florey and the German refugee Ernst Chain managed to isolate the active substance, called *penicillin.* The dramatic ability of penicillin to cure infections in mice was soon demonstrated, and successful tests in humans followed shortly thereafter. By 1943, penicillin was being produced on a large scale for military use in World War II, and by 1944 it was being used on civilians. Fleming, Florey, and Chain shared the 1945 Nobel Prize in Medicine.

Now called benzylpenicillin, or penicillin G, the substance first discovered by Fleming is but one member of a large class of so-called β-lactam antibiotics,compounds with a four-membered lactam (cyclic amide) ring. The four-membered lactam ring is fused to a five-membered, sulfur-containing ring, and the carbon atom next to the lactam carbonyl group is bonded to an acylamino substituent, RCONH–. This acylamino side chain can be varied in the laboratory to provide many hundreds of penicillin analogs with different biological activity profiles. Ampicillin, for instance, has an α-aminophenylacetamido substituent [PhCH(NH_2)CONH–].

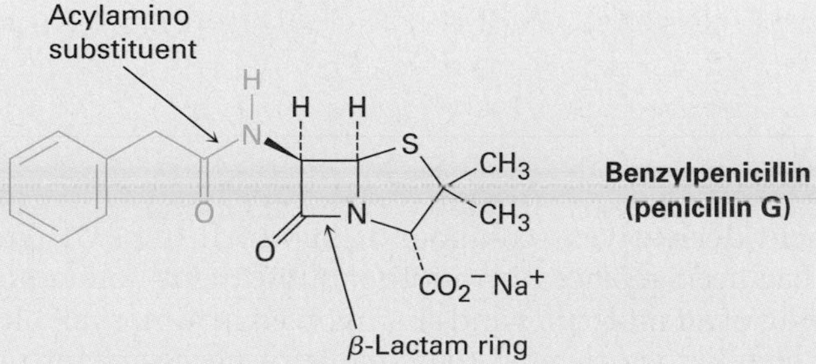

Benzylpenicillin (penicillin G)

Closely related to the penicillins are the *cephalosporins,* a group of β-lactam antibiotics that contain an unsaturated six-membered, sulfur-containing ring. Cephalexin, marketed

(continued)

(continued)

under the trade name Keflex, is an example. Cephalosporins generally have much greater antibacterial activity than penicillins, particularly against resistant strains of bacteria.

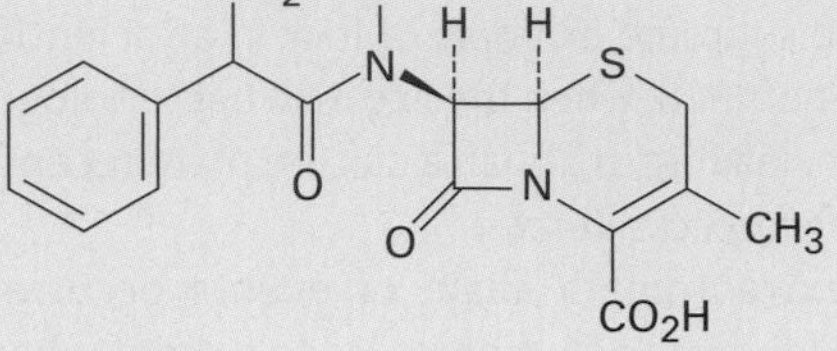

Cephalexin (a cephalosporin)

The biological activity of penicillins and cephalosporins is due to the presence of the strained β-lactam ring, which reacts with and deactivates the transpeptidase enzyme needed to synthesize and repair bacterial cell walls. With the wall either incomplete or weakened, the bacterial cell ruptures and dies.

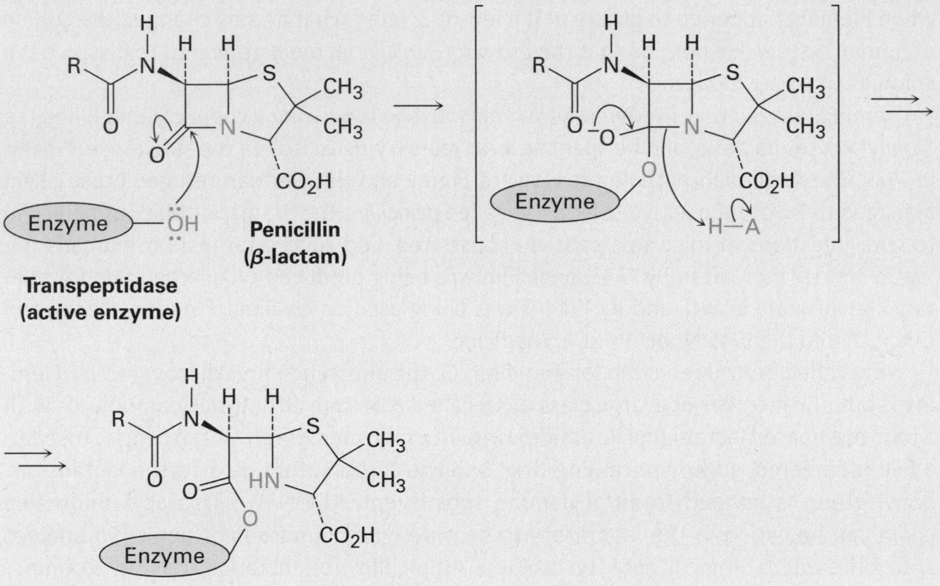

Transpeptidase (active enzyme)

Penicillin (β-lactam)

(inactive enzyme)

Summary

Key words

acid anhydride (RCO_2COR'), 814

acid halide (RCOX), 814

acyl phosphate ($RCOPO_3^{2-}$), 814

amide ($RCONH_2$), 814

Carboxylic acid derivatives—compounds in which the –OH group of a carboxylic acid has been replaced by another substituent—are among the most widely occurring of all molecules and are involved in almost all biological pathways. In this chapter, we covered the chemistry necessary for understanding them and thus also necessary for understanding the chemistry of living organisms. **Acid halides, acid anhydrides, esters,** and **amides** are the most common such derivatives in the laboratory; **thioesters** and **acyl phosphates** are common in biological molecules.

The chemistry of carboxylic acid derivatives is dominated by the **nucleophilic acyl substitution reaction**. Mechanistically, these substitutions take place by addition of a nucleophile to the polar carbonyl group of the acid derivative to give a tetrahedral intermediate, followed by expulsion of a leaving group.

$$RC(=O)Y + :Nu^- \longrightarrow [R{-}C(O^-)(Y)Nu] \longrightarrow RC(=O)Nu + :Y^-$$

The reactivity of an acid derivative toward substitution depends both on the steric environment near the carbonyl group and on the electronic nature of the substituent, Y. The reactivity order is acid halide > acid anhydride > thioester > ester > amide.

The most common reactions of carboxylic acid derivatives are substitution by water to yield an acid (hydrolysis), by an alcohol to yield an ester (alcoholysis), by an amine to yield an amide (aminolysis), by hydride ion to yield an alcohol (reduction), and by an organomagnesium halide to yield an alcohol (Grignard reaction).

Step-growth polymers, such as polyamides and polyesters, are prepared by reactions between difunctional molecules. Polyamides (nylons) are formed by reaction between a diacid and a diamine; polyesters are formed from a diacid and a diol.

IR spectroscopy is a valuable tool for the structural analysis of acid derivatives. Acid chlorides, anhydrides, esters, and amides all show characteristic IR absorptions that can be used to identify these functional groups.

Key words—cont'd

carboxylic acid derivative, 814
ester (RCO_2R'), 814
Fischer esterification reaction, 824
lactam, 845
lactone, 837
nucleophilic acyl substitution reaction, 818
saponification, 838
step-growth polymer, 848
thioester (RCOSR′), 814

Summary of Reactions

1. Reactions of carboxylic acids (Section 21.3)
   (a) Conversion into acid chlorides

$$RC(=O)OH \xrightarrow[CHCl_3]{SOCl_2} RC(=O)Cl + SO_2 + HCl$$

   (b) Conversion into esters

$$RC(=O)O^- + R'X \xrightarrow[S_N2\ reaction]{Via} RC(=O)OR'$$

$$RC(=O)OH + R'OH \xrightarrow[catalyst]{Acid} RC(=O)OR' + H_2O$$

   (c) Conversion into amides

$$RC(=O)OH + RNH_2 \xrightarrow{DCC} RC(=O)NHR$$

(continued)

(d) Reduction to yield primary alcohols

$$RCOOH + LiAlH_4 \longrightarrow RCH_2OH$$

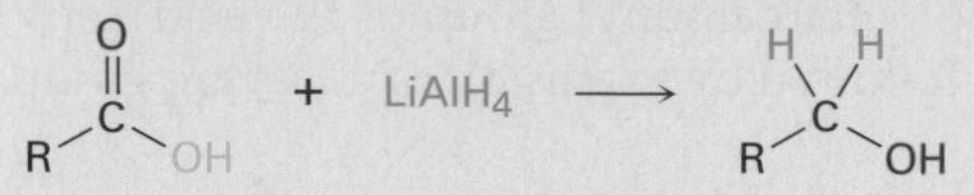

2. Reactions of acid chlorides (Section 21.4)

(a) Hydrolysis to yield acids

$$RCOCl + H_2O \longrightarrow RCOOH + HCl$$

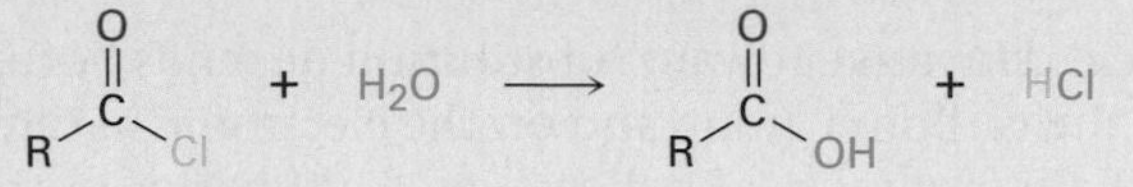

(b) Reaction with carboxylates to yield anhydrides

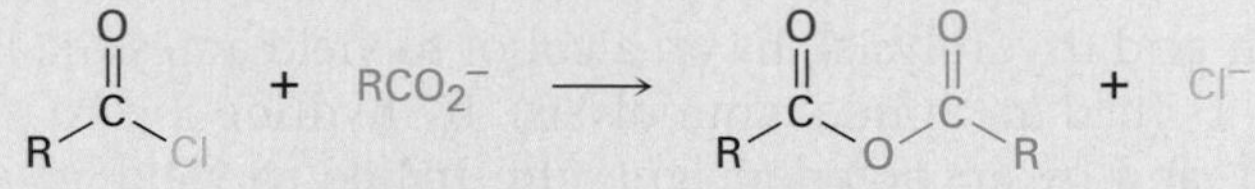

$$RCOCl + RCO_2^- \longrightarrow RCOOCOR + Cl^-$$

(c) Alcoholysis to yield esters

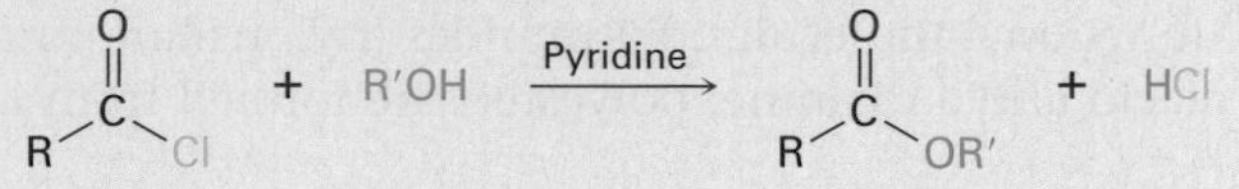

$$RCOCl + R'OH \xrightarrow{\text{Pyridine}} RCOOR' + HCl$$

(d) Aminolysis to yield amides

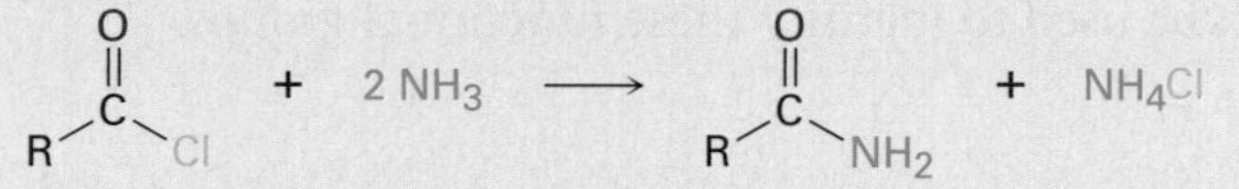

$$RCOCl + 2\ NH_3 \longrightarrow RCONH_2 + NH_4Cl$$

(e) Reduction to yield primary alcohols

$$RCOCl \xrightarrow[\text{2. } H_3O^+]{\text{1. } LiAlH_4\text{, ether}} RCH_2OH$$

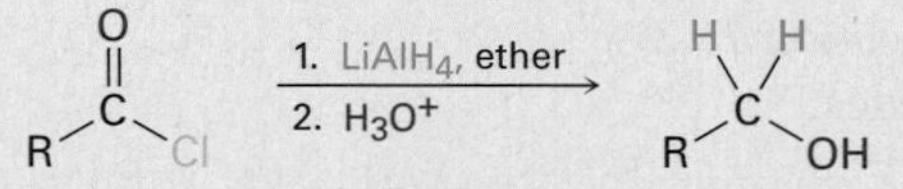

(f) Grignard reaction to yield tertiary alcohols

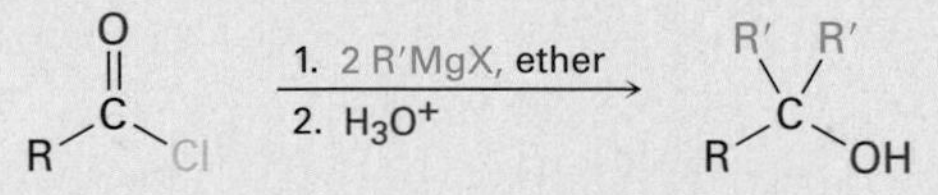

$$RCOCl \xrightarrow[\text{2. } H_3O^+]{\text{1. 2 } R'MgX\text{, ether}} RR'_2COH$$

(g) Diorganocopper reaction to yield ketones

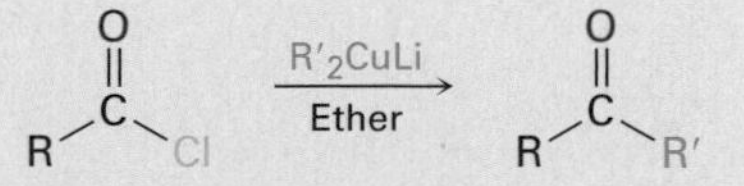

$$RCOCl \xrightarrow[\text{Ether}]{R'_2CuLi} RCOR'$$

3. Reactions of acid anhydrides (Section 21.5)

(a) Hydrolysis to yield acids

$$RCOOCOR + H_2O \longrightarrow 2\ RCOOH$$

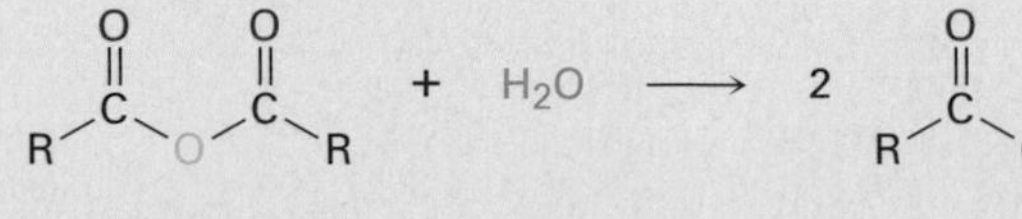

(continued)

(b) Alcoholysis to yield esters

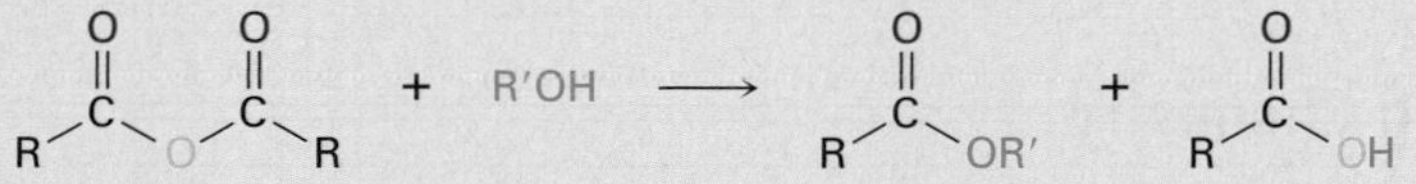

$$RCOOCOR + R'OH \longrightarrow RCOOR' + RCOOH$$

(c) Aminolysis to yield amides

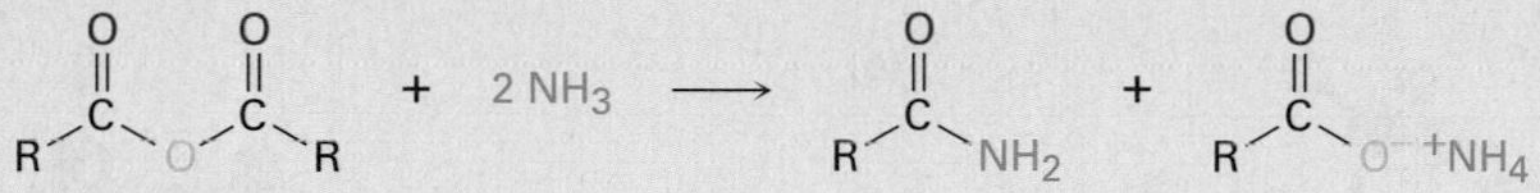

$$RCOOCOR + 2\ NH_3 \longrightarrow RCONH_2 + RCOO^-\ {}^+NH_4$$

4. Reactions of esters (Section 21.6)

(a) Hydrolysis to yield acids

$$RCOOR' \xrightarrow[\text{or NaOH, } H_2O]{H_3O^+} RCOOH + R'OH$$

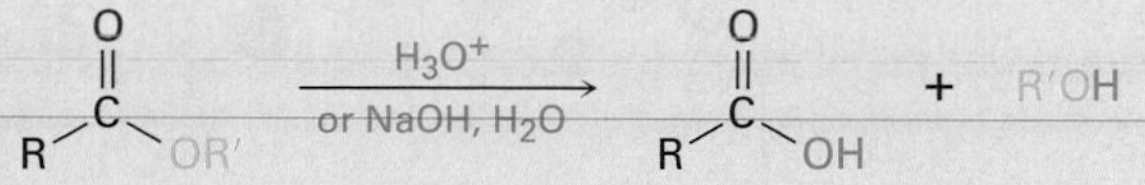

(b) Reduction to yield primary alcohols

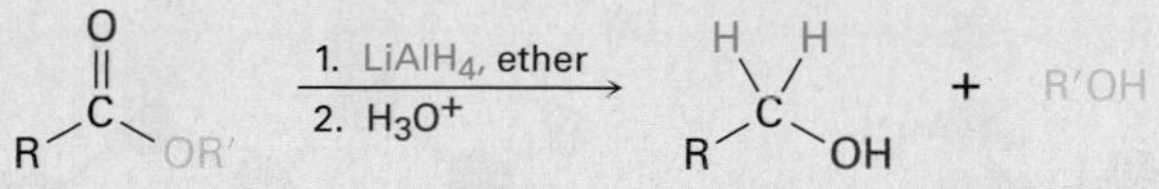

$$RCOOR' \xrightarrow[\text{2. } H_3O^+]{\text{1. } LiAlH_4\text{, ether}} RCH_2OH + R'OH$$

(c) Partial reduction to yield aldehydes

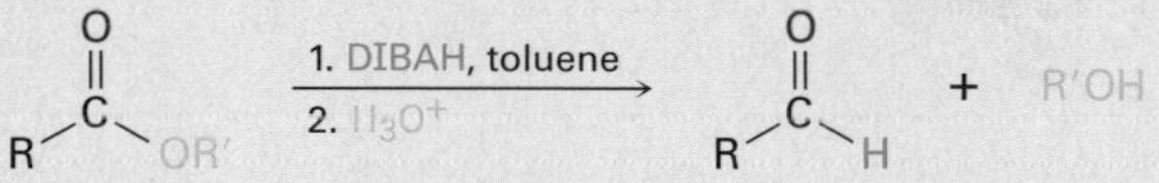

$$RCOOR' \xrightarrow[\text{2. } H_3O^+]{\text{1. DIBAH, toluene}} RCHO + R'OH$$

(d) Grignard reaction to yield tertiary alcohols

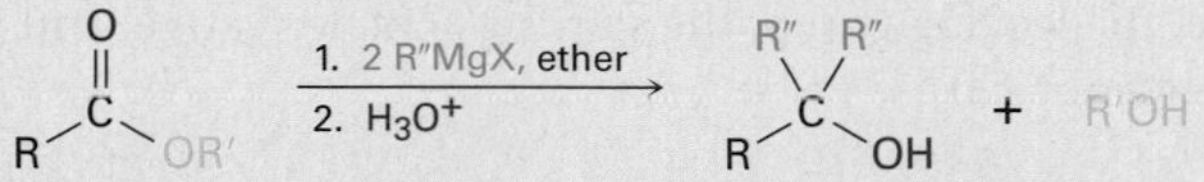

$$RCOOR' \xrightarrow[\text{2. } H_3O^+]{\text{1. 2 R''MgX, ether}} RC(R'')_2OH + R'OH$$

5. Reactions of amides (Section 21.7)

(a) Hydrolysis to yield acids

$$RCONH_2 \xrightarrow[\text{or NaOH, } H_2O]{H_3O^+} RCOOH + NH_3$$

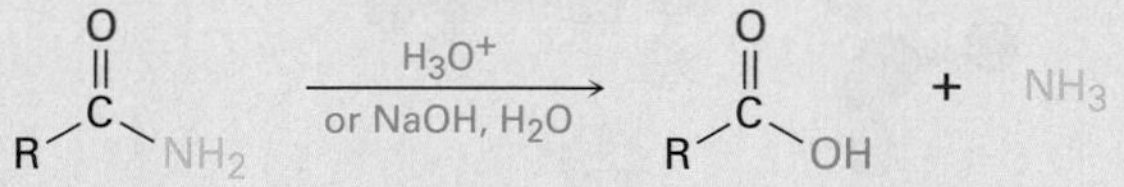

(b) Reduction to yield amines

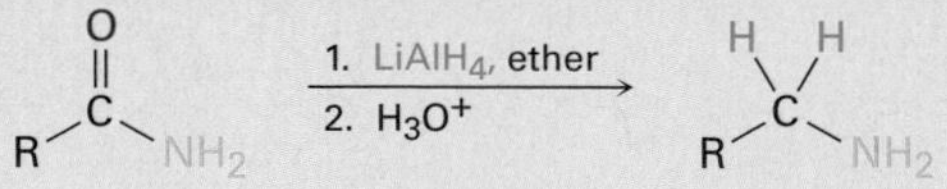

$$RCONH_2 \xrightarrow[\text{2. } H_3O^+]{\text{1. } LiAlH_4\text{, ether}} RCH_2NH_2$$

Exercises

OWL Interactive versions of these problems are assignable in OWL for Organic Chemistry.

Visualizing Chemistry

(Problems 21.1–21.26 appear within the chapter.)

21.27 Name the following compounds:

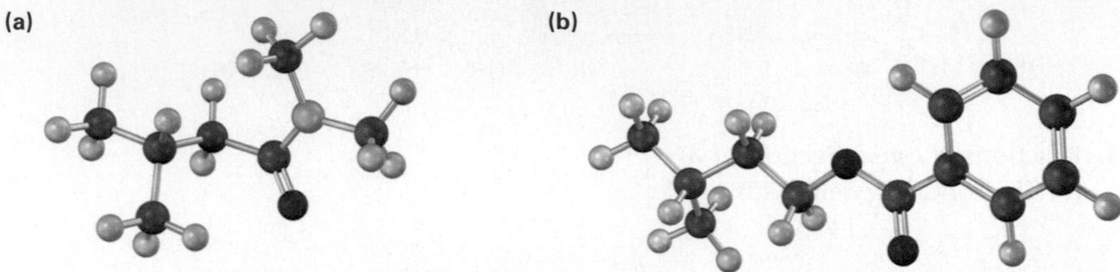

21.28 How would you prepare the following compounds starting with an appropriate carboxylic acid and any other reagents needed? (Reddish brown = Br.)

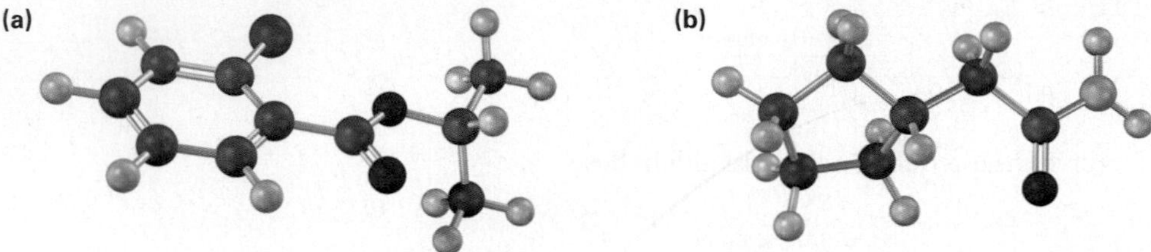

21.29 The following structure represents a tetrahedral alkoxide-ion intermediate formed by addition of a nucleophile to a carboxylic acid derivative. Identify the nucleophile, the leaving group, the starting acid derivative, and the ultimate product (green = Cl).

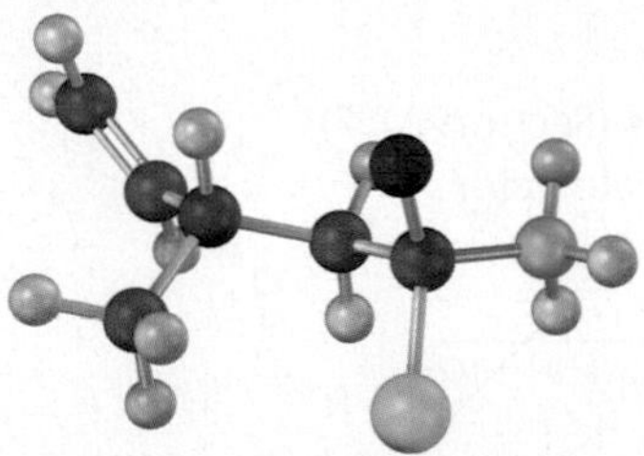

21.30 Electrostatic potential maps of a typical amide (acetamide) and an acyl azide (acetyl azide) are shown. Which of the two do you think is more reactive in nucleophilic acyl substitution reactions? Explain.

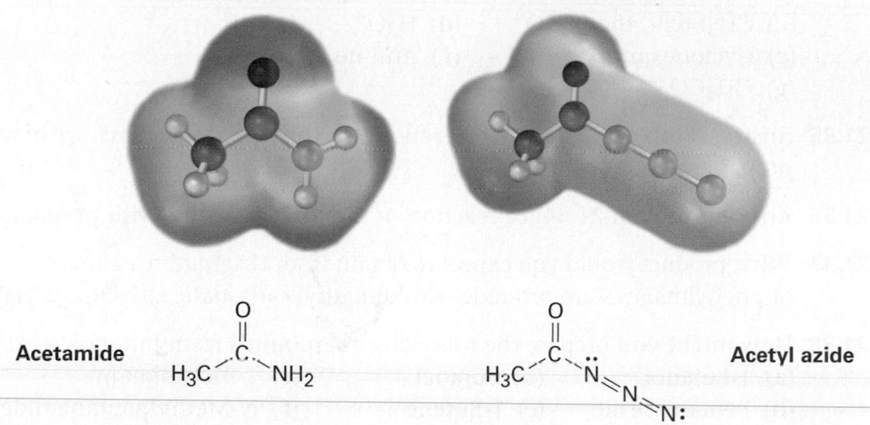

Additional Problems

Naming Carboxylic Acid Derivatives

21.31 Give IUPAC names for the following compounds:

(a) $p\text{-}H_3C\text{-}C_6H_4\text{-}CONH_2$

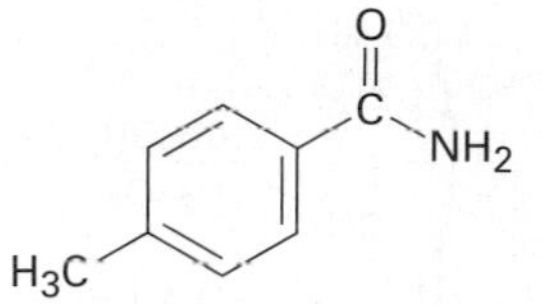

(b) $CH_3CH_2CH(CH_2CH_3)CH{=}CHCOCl$

(c) $CH_3OCOCH_2CH_2COOCH_3$

(d) $C_6H_5CH_2CH_2COOCH(CH_3)CH_3$

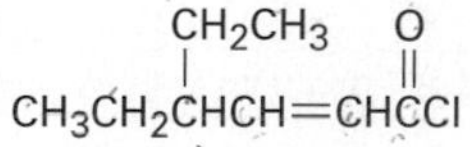

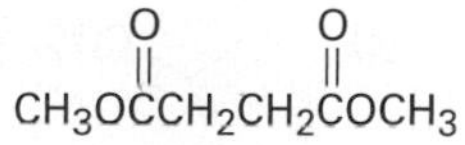

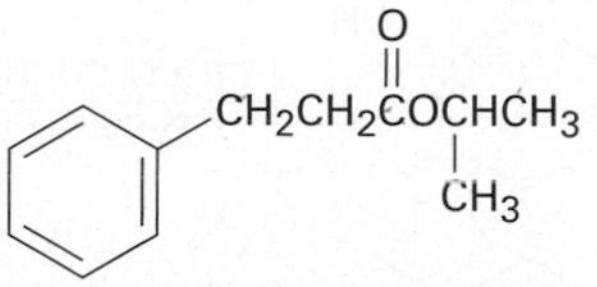

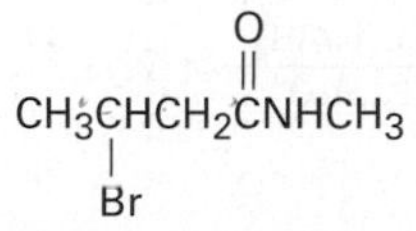

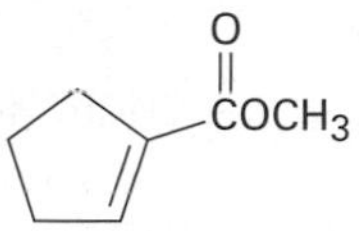

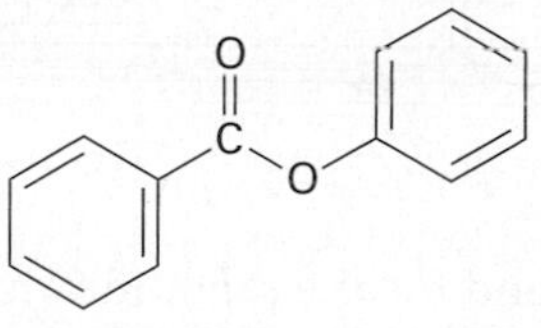

(e) $CH_3CH(Br)CH_2CONHCH_3$

(f) 1-Cyclopentene bearing $COCH_3$ (C=O)

(g) $C_6H_5COOC_6H_5$

(h) $C_6H_5COSCH(CH_3)_2$

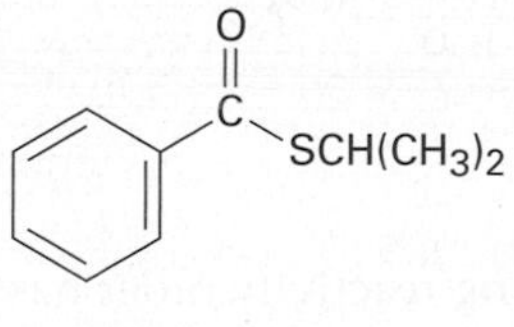

21.32 Draw structures corresponding to the following names:

(a) *p*-Bromophenylacetamide
(b) *m*-Benzoylbenzamide
(c) 2,2-Dimethylhexanamide
(d) Cyclohexyl cyclohexanecarboxylate
(e) Ethyl 2-cyclobutenecarboxylate
(f) Succinic anhydride

21.33 Draw and name compounds that meet the following descriptions:

(a) Three acid chlorides having the formula C_6H_9ClO
(b) Three amides having the formula $C_7H_{11}NO$

Nucleophilic Acyl Substitution Reactions

21.34 Predict the product, if any, of reaction between propanoyl chloride and the following reagents:

(a) $Li(Ph)_2Cu$ in ether
(b) $LiAlH_4$, then H_3O^+
(c) CH_3MgBr, then H_3O^+
(d) H_3O^+
(e) Cyclohexanol
(f) Aniline
(g) $CH_3CO_2^-\ {}^+Na$

21.35 Answer Problem 21.34 for reaction of the listed reagents with methyl propanoate.

21.36 Answer Problem 21.34 for reaction of the listed reagents with propanamide.

21.37 What product would you expect to obtain from Grignard reaction of an excess of phenylmagnesium bromide with dimethyl carbonate, $CH_3OCO_2CH_3$?

21.38 How might you prepare the following compounds from butanoic acid?

(a) 1-Butanol
(b) Butanal
(c) 1-Bromobutane
(d) Pentanenitrile
(e) 1-Butene
(f) *N*-Methylpentanamide
(g) 2-Hexanone
(h) Butylbenzene
(i) Butanenitrile

21.39 Predict the product(s) of the following reactions:

(a) Cyclohexyl–$CO_2CH_2CH_3$ $\xrightarrow[\text{2. } H_3O^+]{\text{1. } CH_3CH_2MgBr}$?

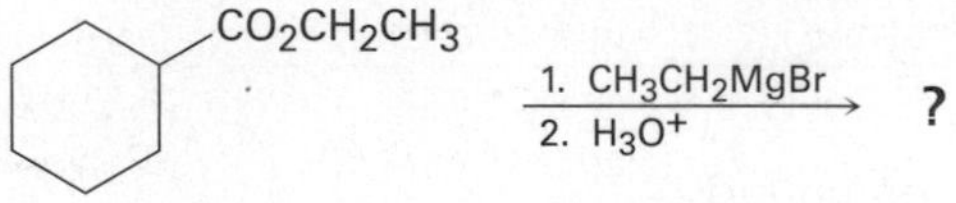

(b) $CH_3CH(CH_3)CH_2CH_2CO_2CH_3$ $\xrightarrow[\text{2. } H_3O^+]{\text{1. DIBAH}}$?

(c) Cyclopentyl–COCl $\xrightarrow{CH_3NH_2}$?

(d) Cyclohexane with CO_2H and H on one carbon, CH_3 and H on the adjacent carbon $\xrightarrow[H_2SO_4]{CH_3OH}$?

(e) $H_2C{=}CHCH(CH_3)CH_2CO_2CH_3$ $\xrightarrow[\text{2. } H_3O^+]{\text{1. } LiAlH_4}$?

(f) Cyclohexyl–OH $\xrightarrow[\text{Pyridine}]{CH_3CO_2COCH_3}$?

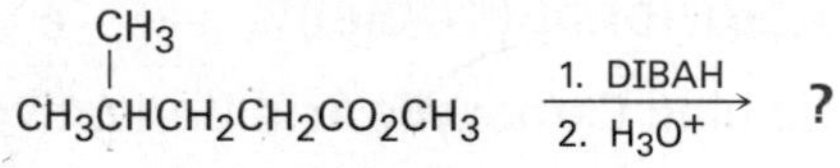

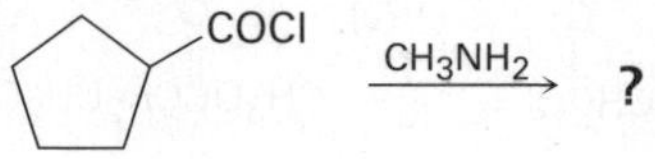

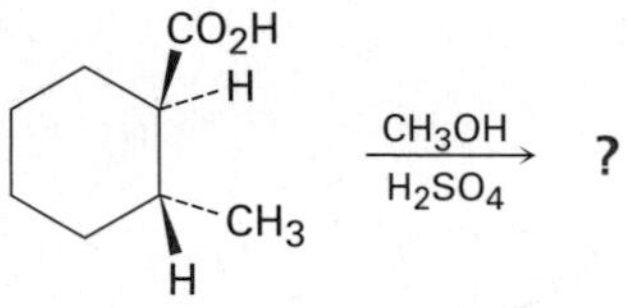

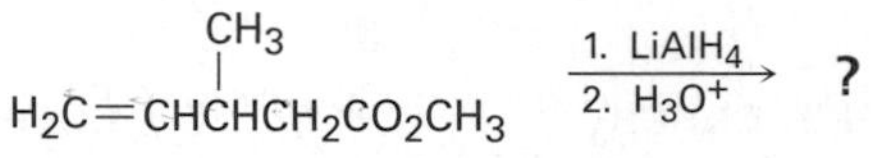

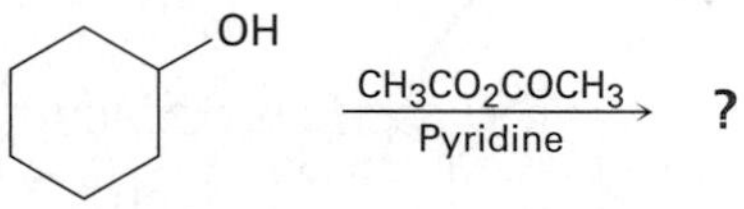

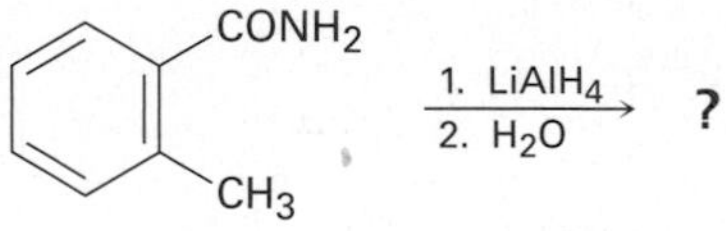

(g) Benzene ring with $CONH_2$ and ortho CH_3 $\xrightarrow[\text{2. } H_2O]{\text{1. } LiAlH_4}$?

(h) Benzene ring with CH_2CO_2H and ortho Br $\xrightarrow{SOCl_2}$?

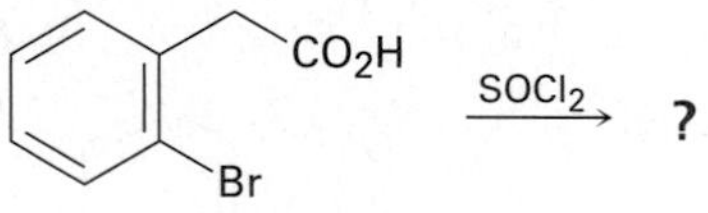

21.40 The following reactivity order has been found for the saponification of alkyl acetates by aqueous NaOH. Explain.

$$CH_3CO_2CH_3 > CH_3CO_2CH_2CH_3 > CH_3CO_2CH(CH_3)_2 > CH_3CO_2C(CH_3)_3$$

21.41 Explain the observation that attempted Fischer esterification of 2,4,6-trimethylbenzoic acid with methanol and HCl is unsuccessful. No ester is obtained, and the acid is recovered unchanged. What alternative method of esterification might be successful?

21.42 Outline methods for the preparation of acetophenone (phenyl methyl ketone) starting from the following:
(a) Benzene (b) Bromobenzene (c) Methyl benzoate
(d) Benzonitrile (e) Styrene

21.43 Treatment of 5-aminopentanoic acid with DCC (dicyclohexylcarbodiimide) yields a lactam. Show the structure of the product and the mechanism of the reaction.

21.44 When *ethyl* benzoate is heated in methanol containing a small amount of HCl, *methyl* benzoate is formed. Propose a mechanism for the reaction.

21.45 *tert*-Butoxycarbonyl azide, a reagent used in protein synthesis, is prepared by treating *tert*-butoxycarbonyl chloride with sodium azide. Propose a mechanism for this reaction.

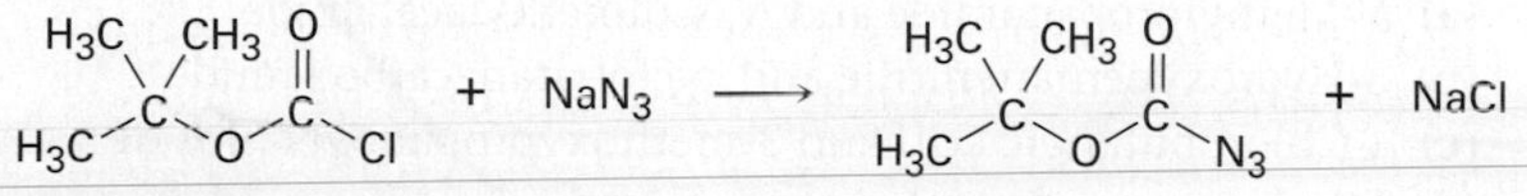

$$(H_3C)_3C{-}O{-}C(=O){-}Cl + NaN_3 \longrightarrow (H_3C)_3C{-}O{-}C(=O){-}N_3 + NaCl$$

Step-Growth Polymers

21.46 The step-growth polymer nylon 6 is prepared from caprolactam. The reaction involves initial reaction of caprolactam with water to give an intermediate open-chain amino acid, followed by heating to form the polymer. Propose mechanisms for both steps, and show the structure of nylon 6.

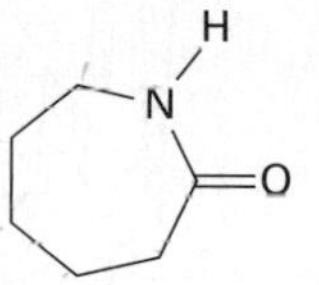

Caprolactam

21.47 *Qiana,* a polyamide fiber with a silky texture, has the following structure. What are the monomer units used in the synthesis of Qiana?

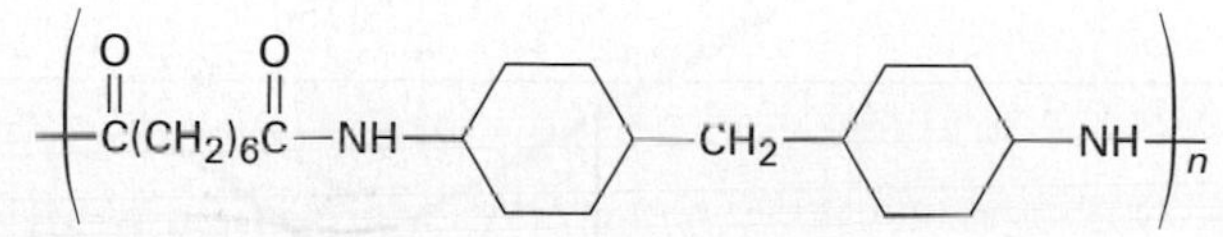

$$\left[\overset{O}{\overset{\|}{C}}(CH_2)_6\overset{O}{\overset{\|}{C}}{-}NH{-}C_6H_{10}{-}CH_2{-}C_6H_{10}{-}NH \right]_n$$

Qiana

21.48 What is the structure of the polymer produced by treatment of β-propiolactone with a small amount of hydroxide ion?

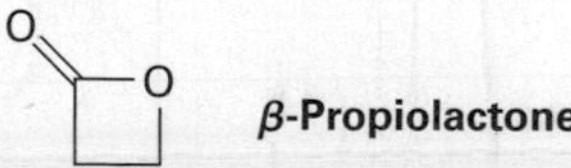

β-Propiolactone

21.49 Polyimides having the structure shown are used as coatings on glass and plastics to improve scratch resistance. How would you synthesize a polyimide? (See Problem 21.59.)

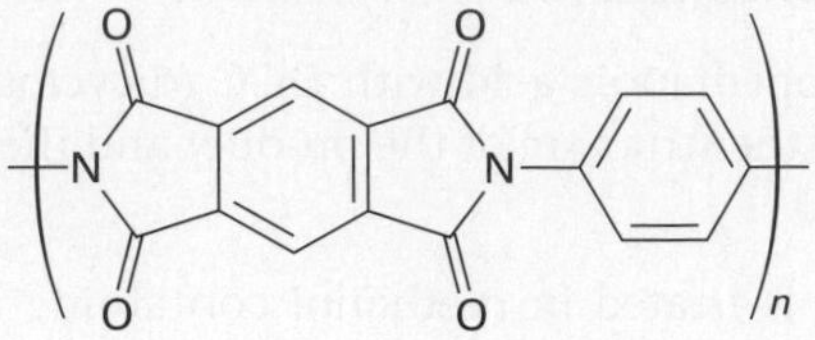

A polyimide

Spectroscopy

21.50 How would you distinguish spectroscopically between the following isomer pairs? Tell what differences you would expect to see.

(a) *N*-Methylpropanamide and *N,N*-dimethylacetamide

(b) 5-Hydroxypentanenitrile and cyclobutanecarboxamide

(c) 4-Chlorobutanoic acid and 3-methoxypropanoyl chloride

(d) Ethyl propanoate and propyl acetate

21.51 Propose a structure for a compound, $C_4H_7ClO_2$, that has the following IR and ^1H NMR spectra:

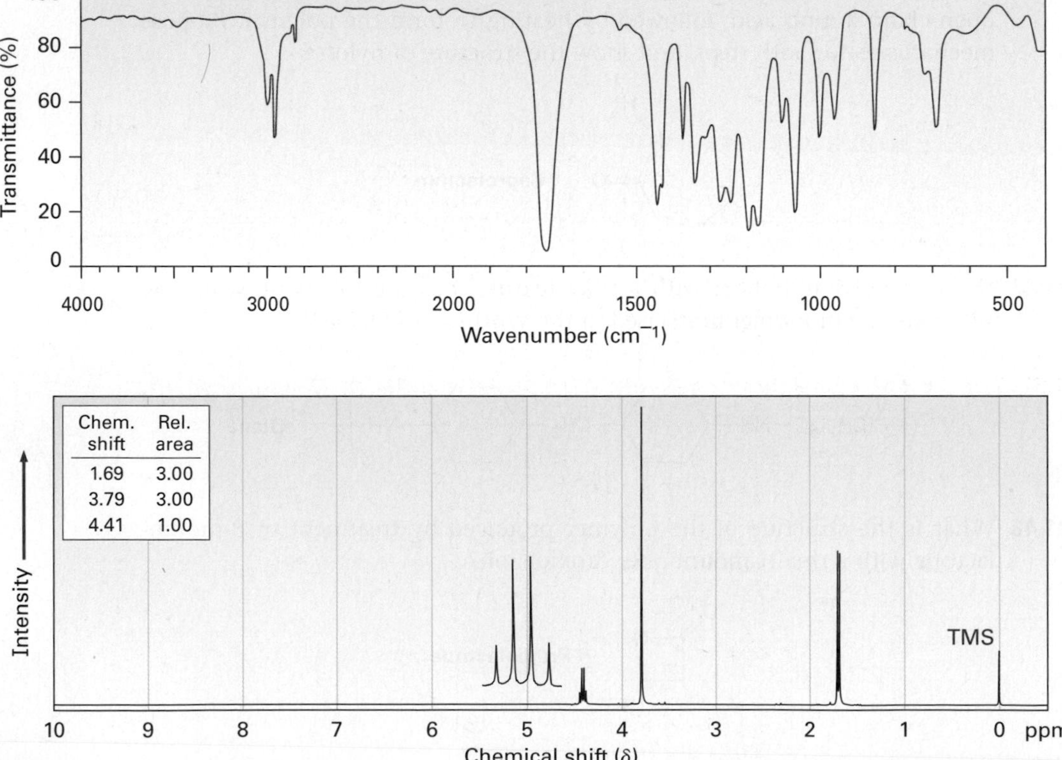

21.52 Assign structures to compounds with the following ^{1}H NMR spectra:

(a) C_4H_7ClO
IR: 1810 cm^{-1}

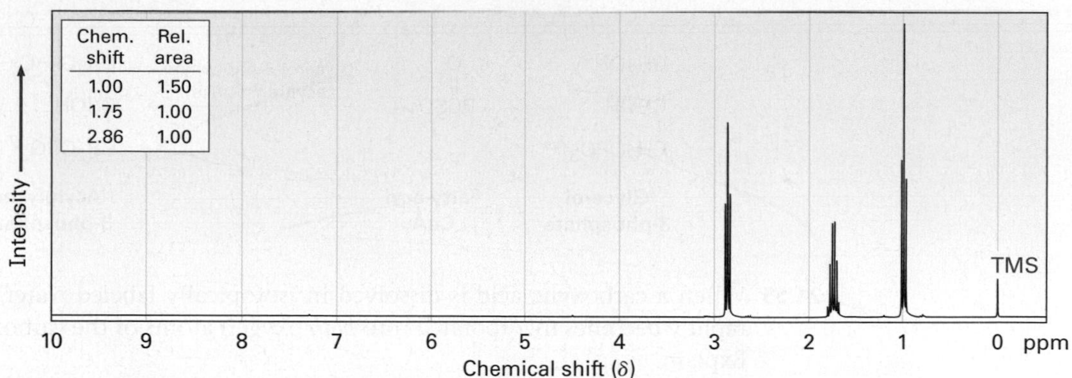

(b) $C_5H_7NO_2$
IR: 2250, 1735 cm^{-1}

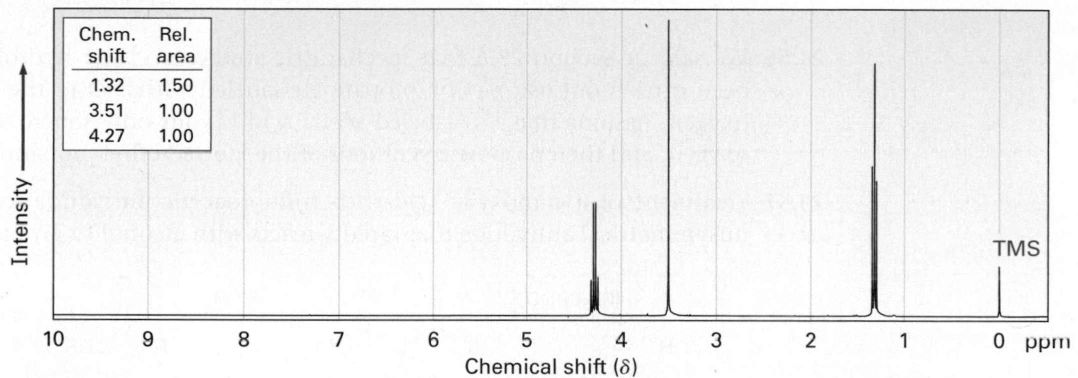

General Problems

21.53 The following reactivity order has been found for the basic hydrolysis of *p*-substituted methyl benzoates:

$$Y = NO_2 > Br > H > CH_3 > OCH_3$$

How can you explain this reactivity order? Where would you expect Y = C≡N, Y = CHO, and Y = NH_2 to be in the reactivity list?

$$p\text{-}Y\text{-}C_6H_4CO_2CH_3 \xrightarrow[H_2O]{^-OH} p\text{-}Y\text{-}C_6H_4CO_2^- + CH_3OH$$

21.54 Fats are biosynthesized from glycerol 3-phosphate and fatty-acyl CoA's by a reaction sequence that begins with the following step. Show the mechanism of the reaction.

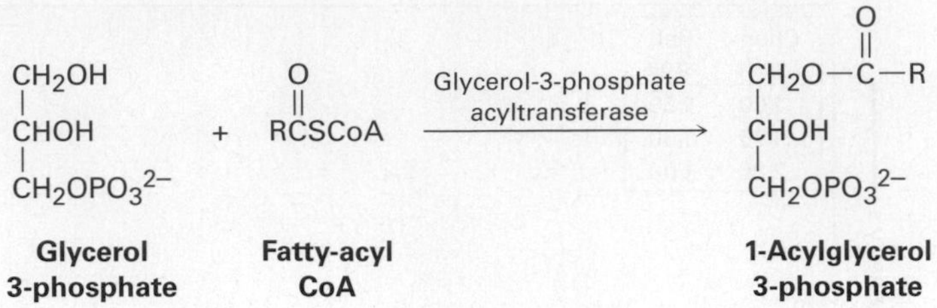

$$\begin{array}{c}CH_2OH\\|\\CHOH\\|\\CH_2OPO_3^{2-}\end{array} + RCSCoA\ (C{=}O) \xrightarrow[\text{acyltransferase}]{\text{Glycerol-3-phosphate}} \begin{array}{c}CH_2O{-}C(=O){-}R\\|\\CHOH\\|\\CH_2OPO_3^{2-}\end{array}$$

Glycerol 3-phosphate **Fatty-acyl CoA** **1-Acylglycerol 3-phosphate**

21.55 When a carboxylic acid is dissolved in isotopically labeled water, the label rapidly becomes incorporated into *both* oxygen atoms of the carboxylic acid. Explain.

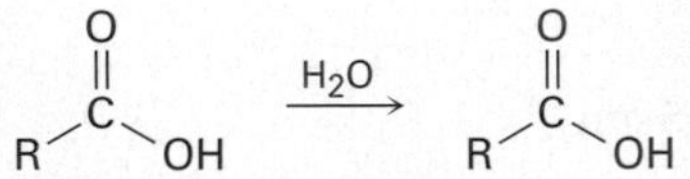

$$RC(=O)OH \xrightarrow{H_2O} RC(=O)OH$$

21.56 We said in Section 21.6 that mechanistic studies on ester hydrolysis have been carried out using ethyl propanoate labeled with ^{18}O in the ether-like oxygen. Assume that ^{18}O-labeled acetic acid is your only source of isotopic oxygen, and then propose a synthesis of the labeled ethyl propanoate.

21.57 Treatment of a carboxylic acid with trifluoroacetic anhydride leads to an unsymmetrical anhydride that rapidly reacts with alcohol to give an ester.

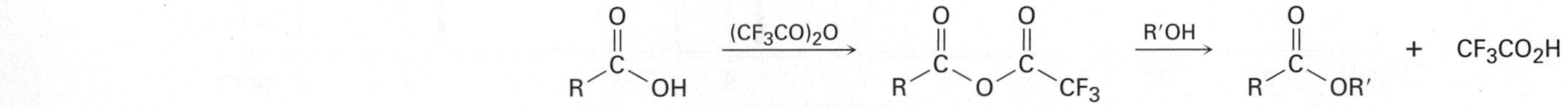

$$RC(=O)OH \xrightarrow{(CF_3CO)_2O} RC(=O)OC(=O)CF_3 \xrightarrow{R'OH} RC(=O)OR' + CF_3CO_2H$$

(a) Propose a mechanism for formation of the unsymmetrical anhydride.
(b) Why is the unsymmetrical anhydride unusually reactive?
(c) Why does the unsymmetrical anhydride react as indicated rather than giving a trifluoroacetate ester plus carboxylic acid?

21.58 Treatment of an α-amino acid with DCC yields a 2,5-diketopiperazine. Propose a mechanism.

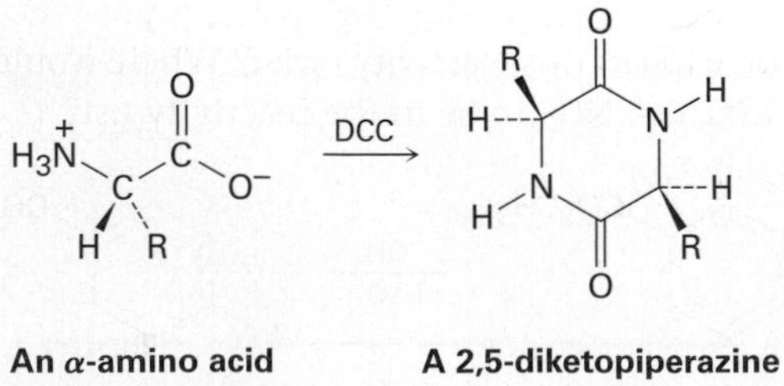

$$H_3\overset{+}{N}CH(R)CO_2^- \xrightarrow{DCC} \text{2,5-diketopiperazine}$$

An α-amino acid **A 2,5-diketopiperazine**

21.59 Succinic anhydride yields the cyclic imide succinimide when heated with ammonium chloride at 200 °C. Propose a mechanism for this reaction. Why do you suppose such a high reaction temperature is required?

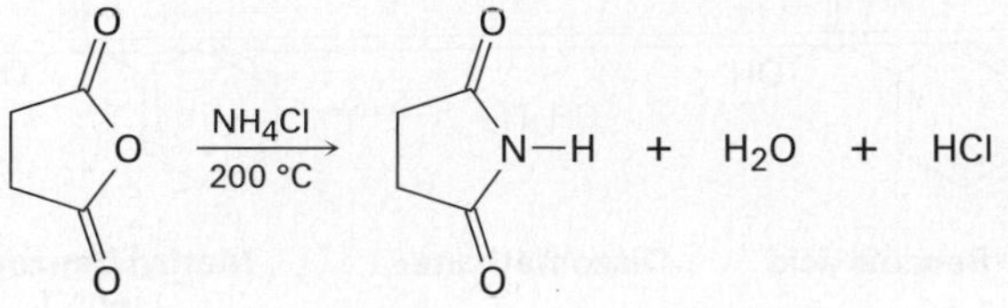

NH_4Cl, 200 °C → N–H + H_2O + HCl

21.60 Butacetin is an analgesic (pain-killing) agent that is synthesized commercially from *p*-fluoronitrobenzene. Propose a synthesis.

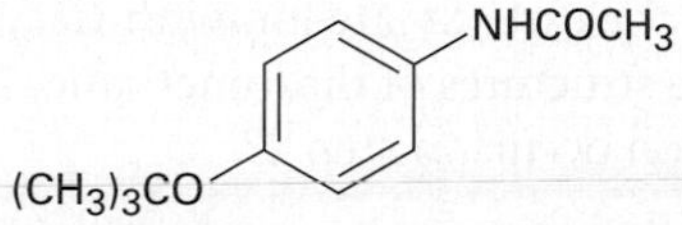

$NHCOCH_3$

Butacetin

$(CH_3)_3CO$

21.61 Phenyl 4-aminosalicylate is a drug used in the treatment of tuberculosis. Propose a synthesis of this compound starting from 4-nitrosalicylic acid.

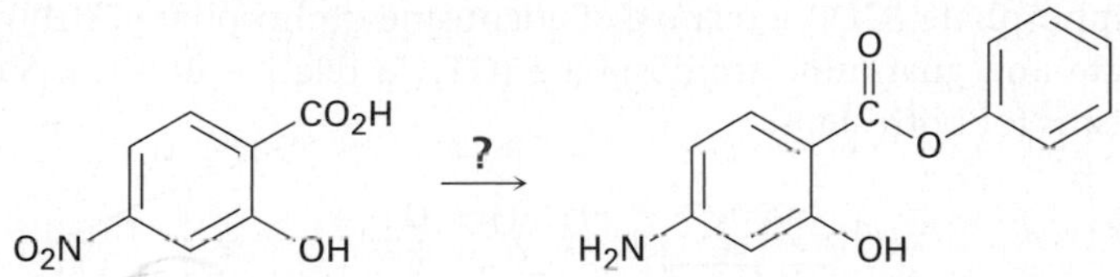

CO_2H ? O_2N OH H_2N OH

4-Nitrosalicylic acid **Phenyl 4-aminosalicylate**

21.62 *N*,*N*-Diethyl-*m*-toluamide (DEET) is the active ingredient in many insect-repellent preparations. How might you synthesize this substance from *m*-bromotoluene?

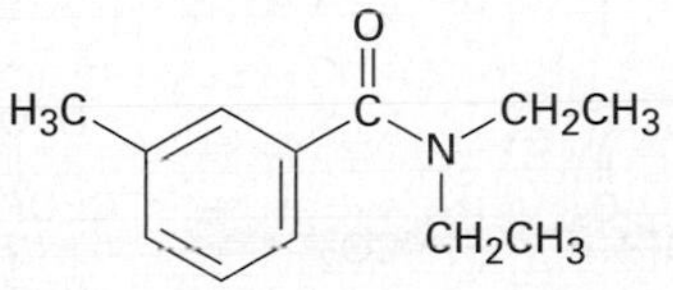

H_3C CH_2CH_3 CH_2CH_3

***N*,*N*-Diethyl-*m*-toluamide**

21.63 Tranexamic acid, a drug useful against blood clotting, is prepared commercially from *p*-methylbenzonitrile. Formulate the steps likely to be used in the synthesis. (Don't worry about cis–trans isomers; heating to 300 °C interconverts the isomers.)

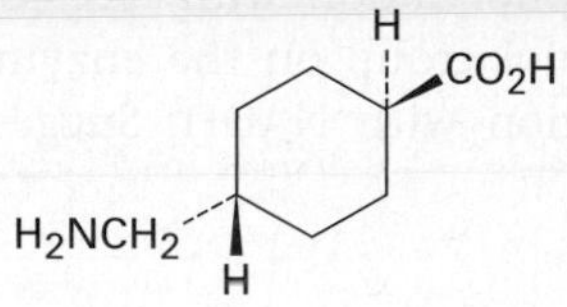

H CO_2H H_2NCH_2 H

Tranexamic acid

21.64 One frequently used method for preparing methyl esters is by reaction of carboxylic acids with diazomethane, CH_2N_2.

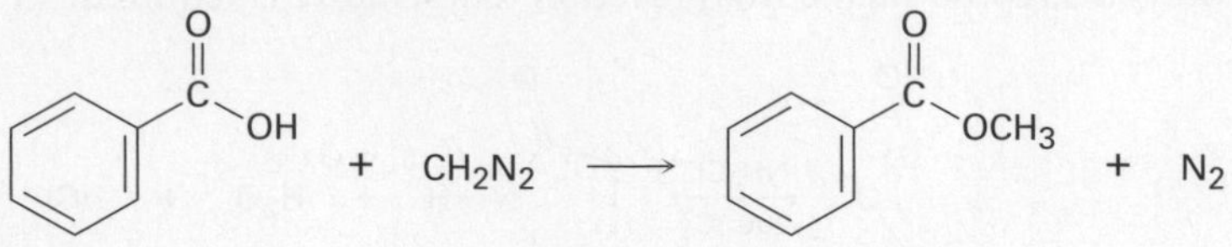

Benzoic acid + CH_2N_2 ⟶ Methyl benzoate (100%) + N_2

Benzoic acid **Diazomethane** **Methyl benzoate (100%)**

The reaction occurs in two steps: (1) protonation of diazomethane by the carboxylic acid to yield methyldiazonium ion, $CH_3N_2^+$, plus a carboxylate ion; and (2) reaction of the carboxylate ion with $CH_3N_2^+$.

(a) Draw two resonance structures of diazomethane, and account for step 1.

(b) What kind of reaction occurs in step 2?

21.65 The hydrolysis of a biological thioester to the corresponding carboxylate is often more complex than the overall result might suggest. The conversion of succinyl CoA to succinate in the citric acid cycle, for instance, occurs by initial formation of an acyl phosphate, followed by reaction with guanosine diphosphate (GDP, a relative of adenosine diphosphate [ADP]) to give succinate and guanosine triphosphate (GTP, a relative of ATP). Suggest mechanisms for both steps.

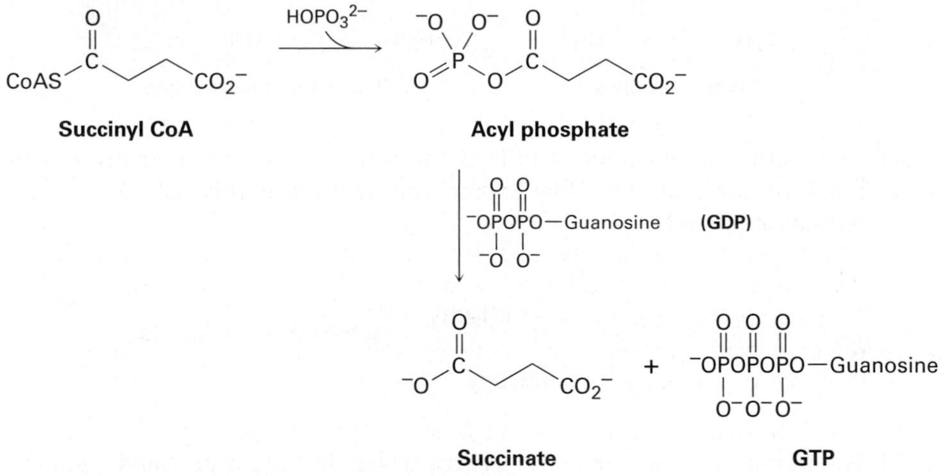

Succinyl CoA —($HOPO_3^{2-}$)→ **Acyl phosphate** —(GDP: $^-$OPOPO—Guanosine)→ **Succinate** + $^-$OPOPOPO—Guanosine **GTP**

21.66 One step in the *gluconeogenesis* pathway for the biosynthesis of glucose is the partial reduction of 3-phosphoglycerate to give glyceraldehyde 3-phosphate. The process occurs by phosphorylation with ATP to give 1,3-bisphosphoglycerate, reaction with a thiol group on the enzyme to give an enzyme-bound thioester, and reduction with NADH. Suggest mechanisms for all three reactions.

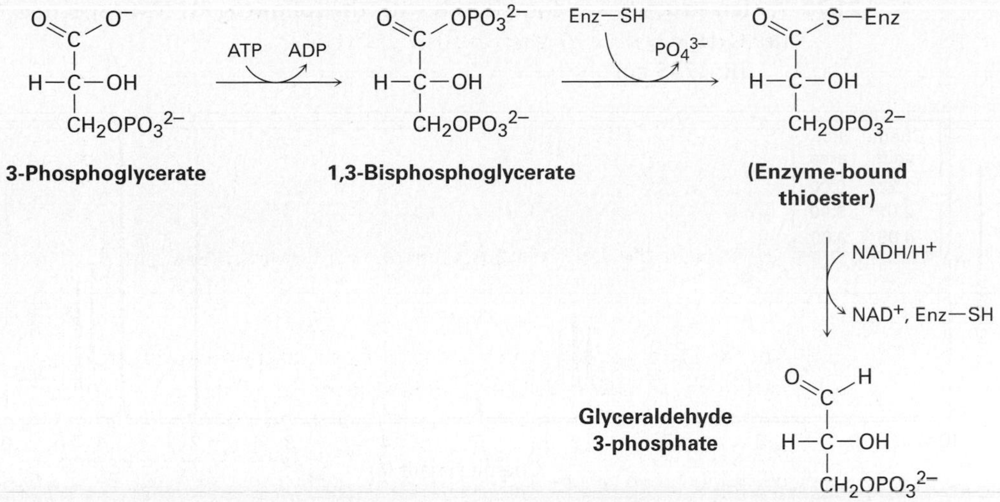

3-Phosphoglycerate → (ATP, ADP) → 1,3-Bisphosphoglycerate → (Enz—SH, PO_4^{3-}) → (Enzyme-bound thioester) → (NADH/H^+; NAD^+, Enz—SH) → Glyceraldehyde 3-phosphate

21.67 Penicillins and other β-lactam antibiotics (see *A Deeper Look* in this chapter) typically develop a resistance to bacteria due to bacterial synthesis of β-lactamase enzymes. Tazobactam, however, is able to inhibit the activity of the β-lactamase by trapping it, thereby preventing resistance from developing.

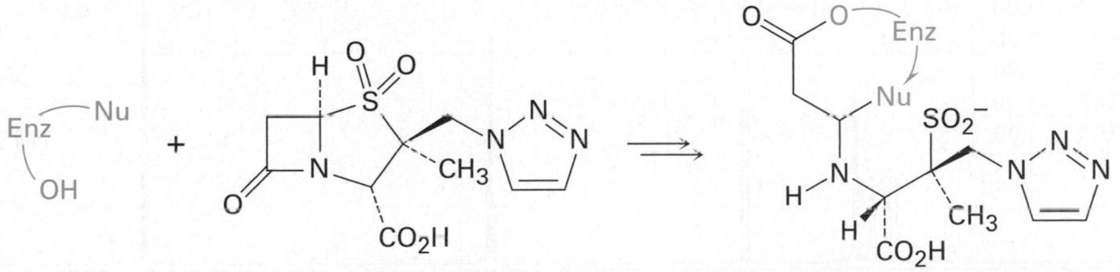

β-Lactamase + **Tazobactam** ⟹ **Trapped β-lactamase**

(a) The first step in trapping is reaction of a hydroxyl group on the β-lactamase to open the β-lactam ring of tazobactam. Show the mechanism.
(b) The second step is opening of the sulfur-containing ring in tazobactam to give an acyclic imine intermediate. Show the mechanism.
(c) Cyclization of the imine intermediate gives the trapped β-lactamase product. Show the mechanism.

21.68 The following reaction, called the *benzylic acid rearrangement,* takes place by typical carbonyl-group reactions. Propose a mechanism (Ph = phenyl).

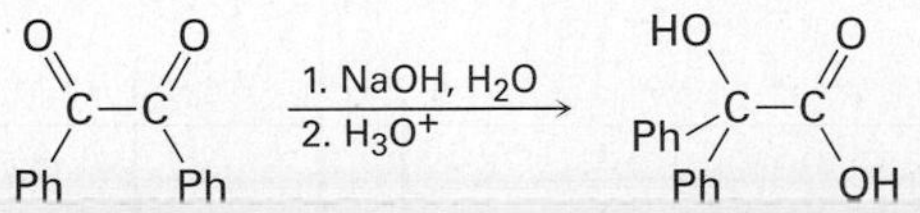

Benzil → 1. NaOH, H_2O; 2. H_3O^+ → **Benzylic acid**

21.69 Draw the structure of the polymer you would expect to obtain from reaction of dimethyl terephthalate with a triol such as glycerol. What structural feature would this new polymer have that was not present in Dacron (Table 21.2)? How do you think this new feature might affect the properties of the polymer?

21.70 Assign structures to compounds with the following ^1H NMR spectra:

(a) $C_5H_{10}O_2$
IR: 1735 cm^{-1}

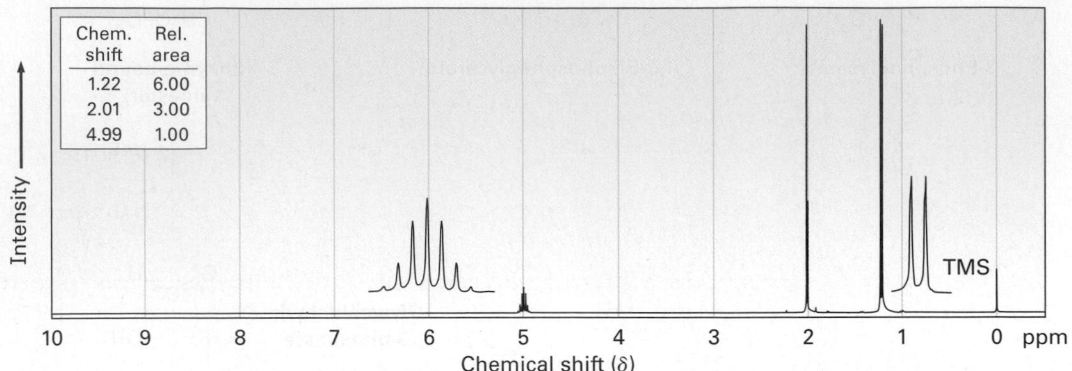

(b) $C_{11}H_{12}O_2$
IR: 1710 cm^{-1}

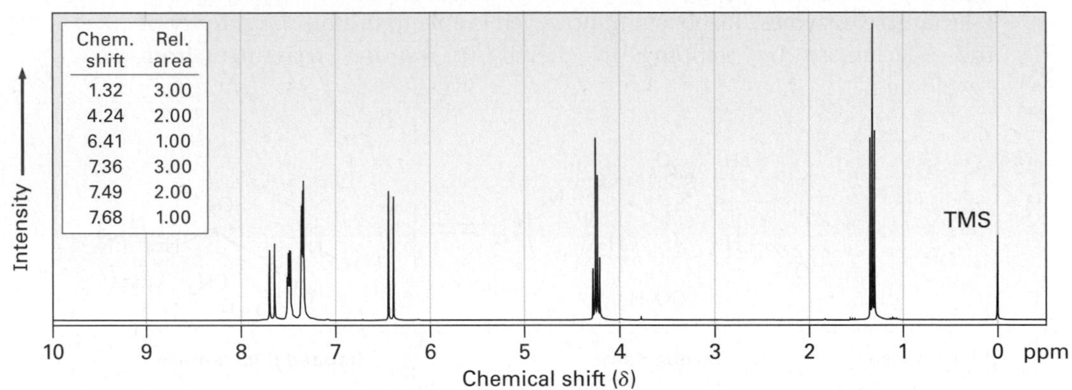

21.71 Propose structures for compounds with the following ^1H NMR spectra:

(a) $C_5H_9ClO_2$
IR: 1735 cm^{-1}

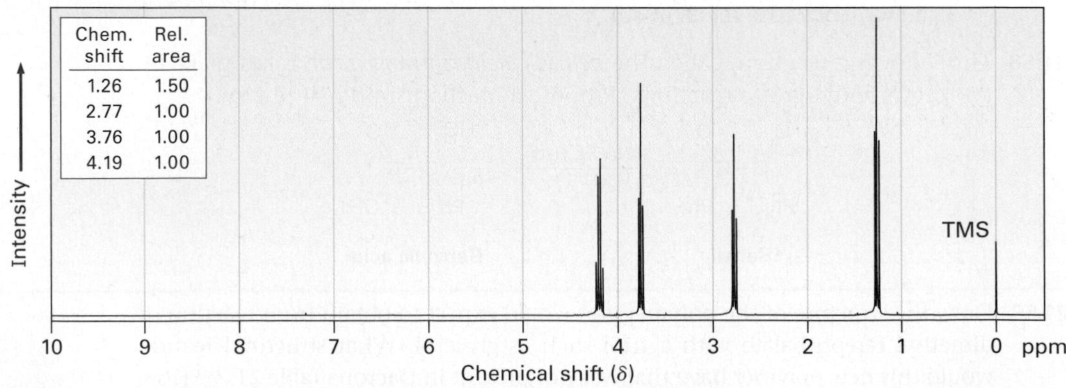

(b) $C_7H_{12}O_4$
IR: 1735 cm^{-1}

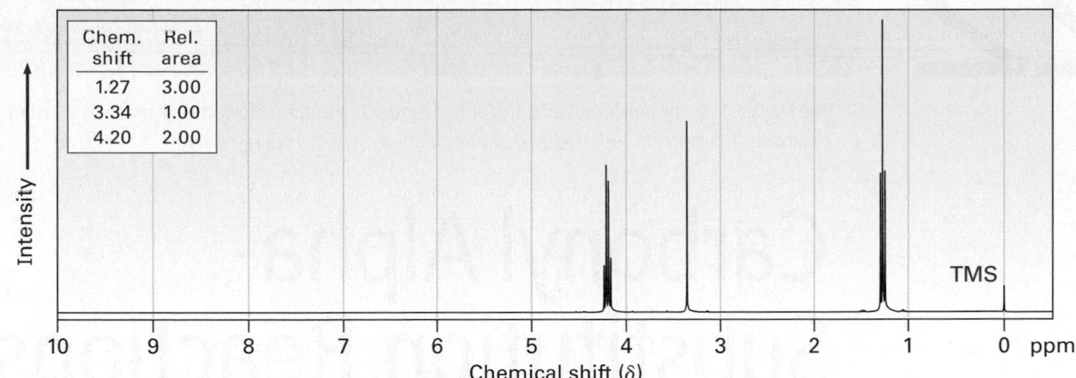

21.72 Epoxy adhesives are prepared in two steps. S_N2 reaction of the disodium salt of bisphenol A with epichlorohydrin forms a "prepolymer," which is then "cured" by treatment with a triamine such as $H_2NCH_2CH_2NHCH_2CH_2NH_2$.

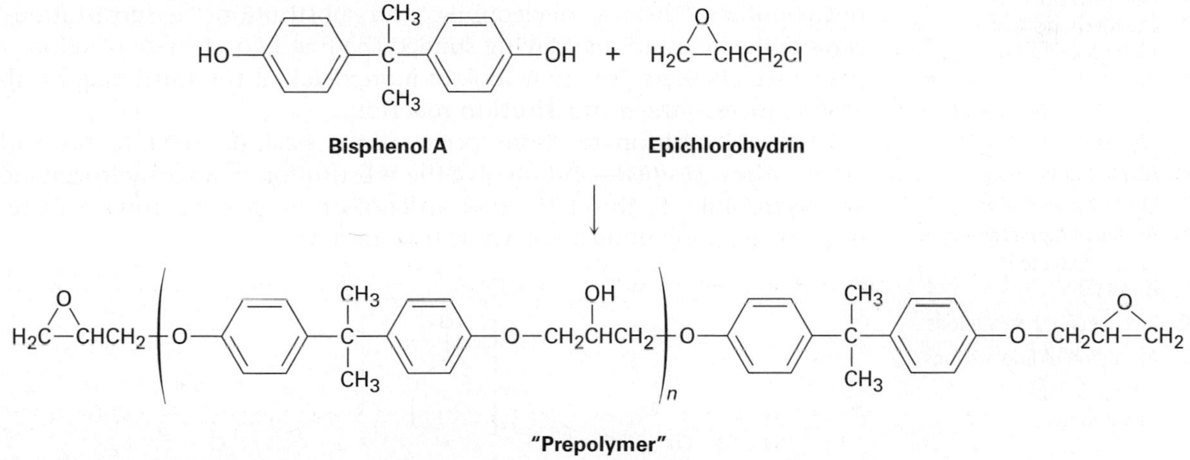

$$HO-C_6H_4-C(CH_3)_2-C_6H_4-OH \quad + \quad H_2C\overset{O}{-}CHCH_2Cl$$

Bisphenol A **Epichlorohydrin**

↓

$$H_2C\overset{O}{-}CHCH_2\left(O-C_6H_4-C(CH_3)_2-C_6H_4-O-CH_2\underset{}{\overset{OH}{C}}HCH_2\right)_n O-C_6H_4-C(CH_3)_2-C_6H_4-O-CH_2CH\overset{O}{-}CH_2$$

"Prepolymer"

Draw structures to show how addition of the triamine results in strengthening the polymer.

21.73 In the *iodoform reaction,* a triiodomethyl ketone reacts with aqueous NaOH to yield a carboxylate ion and iodoform (triiodomethane). Propose a mechanism for this reaction.

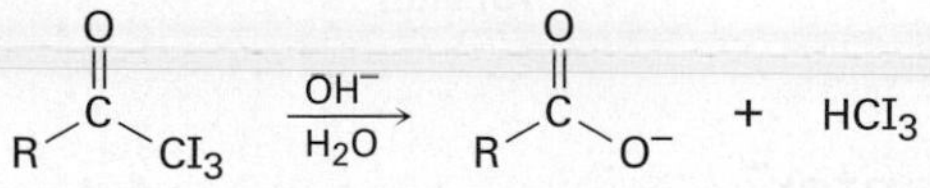

$$R-\overset{O}{\overset{\|}{C}}-CI_3 \xrightarrow[H_2O]{OH^-} R-\overset{O}{\overset{\|}{C}}-O^- \quad + \quad HCI_3$$

22

The tear gas used by police and military for riot control is a simple chloro ketone made by a carbonyl α-substitution reaction. Image copyright JustASC 2010. Used under license from Shutterstock.com

Carbonyl Alpha-Substitution Reactions

We said in the *Preview of Carbonyl Chemistry* that much of the chemistry of carbonyl compounds can be explained by just four fundamental reaction types: nucleophilic additions, nucleophilic acyl substitutions, α substitutions, and carbonyl condensations. Having studied the first two of these reactions in the past three chapters, let's now look in more detail at the third major carbonyl-group process—the **α-substitution reaction**.

Alpha-substitution reactions occur at the position next to the carbonyl group—the *α position*—and involve the substitution of an α hydrogen atom by an electrophile, E, through either an enol or enolate ion intermediate. Let's begin by learning more about these two species.

A carbonyl compound ⇌ [**An enolate ion**] or [**An enol**] —E^+→ **An alpha-substituted carbonyl compound**

Why This Chapter? As with nucleophilic additions and nucleophilic acyl substitutions, many laboratory schemes, pharmaceutical syntheses, and biochemical pathways make frequent use of carbonyl α-substitution reactions. Their great value is that they constitute one of the few general methods for forming carbon–carbon bonds, thereby making it possible to build larger molecules from smaller precursors. We'll see how and why these reactions occur in this chapter.

OWL Sign in to OWL for Organic Chemistry at **www.cengage.com/owl** to view tutorials and simulations, develop problem-solving skills, and complete online homework assigned by your professor.

22.1 Keto–Enol Tautomerism

A carbonyl compound with a hydrogen atom on its α carbon is in an equilibrium with its corresponding **enol** isomer **(Section 9.4)**. This spontaneous interconversion between two isomers, usually with the change in position of a hydrogen, is called *tautomerism,* from the Greek *tauto,* meaning "the same," and *meros,* meaning "part." The individual keto and enol isomers are called **tautomers**.

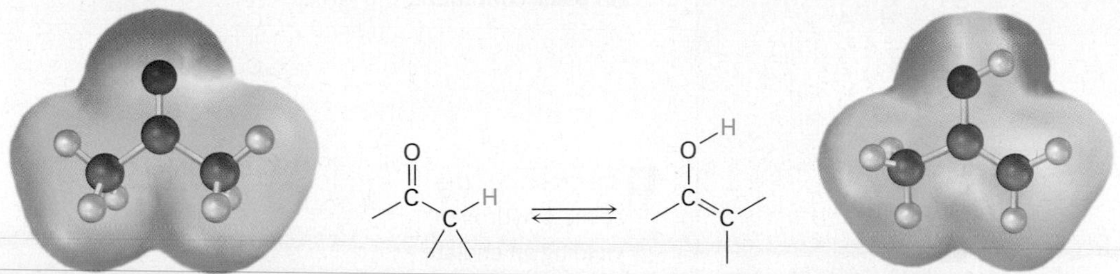

Keto tautomer **Enol tautomer**

Note the difference between tautomers and resonance forms. Tautomers are constitutional isomers—different compounds with different structures—while resonance forms are different representations of a single compound. Tautomers have their *atoms* arranged differently, while resonance forms differ only in the position of their π and nonbonding *electrons*.

Most monocarbonyl compounds exist almost entirely in their keto form at equilibrium, and it's usually difficult to isolate the pure enol. Cyclohexanone, for example, contains only about 0.0001% of its enol tautomer at room temperature. The percentage of enol tautomer is even less for carboxylic acids, esters, and amides. Only when the enol can be stabilized by conjugation or by intramolecular hydrogen bond formation does the enol sometimes predominate. Thus, 2,4-pentanedione is about 76% enol tautomer. Although enols are present only to a small extent at equilibrium, they are nevertheless responsible for much of the chemistry of carbonyl compounds because they are so reactive.

99.999 9% 0.000 1%

Cyclohexanone

99.999 999 9% 0.000 000 1%

Acetone

Keto–enol tautomerism of carbonyl compounds is catalyzed by both acids and bases. Acid catalysis occurs by protonation of the carbonyl oxygen atom to give an intermediate cation that loses H^+ from its α carbon to yield a neutral enol **(Figure 22.1a)**. This proton loss from the cation intermediate is similar to what occurs during an E1 reaction when a carbocation loses H^+ to form an alkene **(Section 11.10)**.

Base-catalyzed enol formation occurs because the presence of a carbonyl group makes the hydrogens on the α carbon weakly acidic. Thus, a carbonyl compound can act as an acid and donate one of its α hydrogens to a sufficiently strong base. The resultant resonance-stabilized anion, an **enolate ion**, is then protonated to yield a neutral compound. If protonation of the enolate ion takes place on the α carbon, the keto tautomer is regenerated and no net change occurs. If, however, protonation takes place on the oxygen atom, then an enol tautomer is formed **(Figure 22.1b)**.

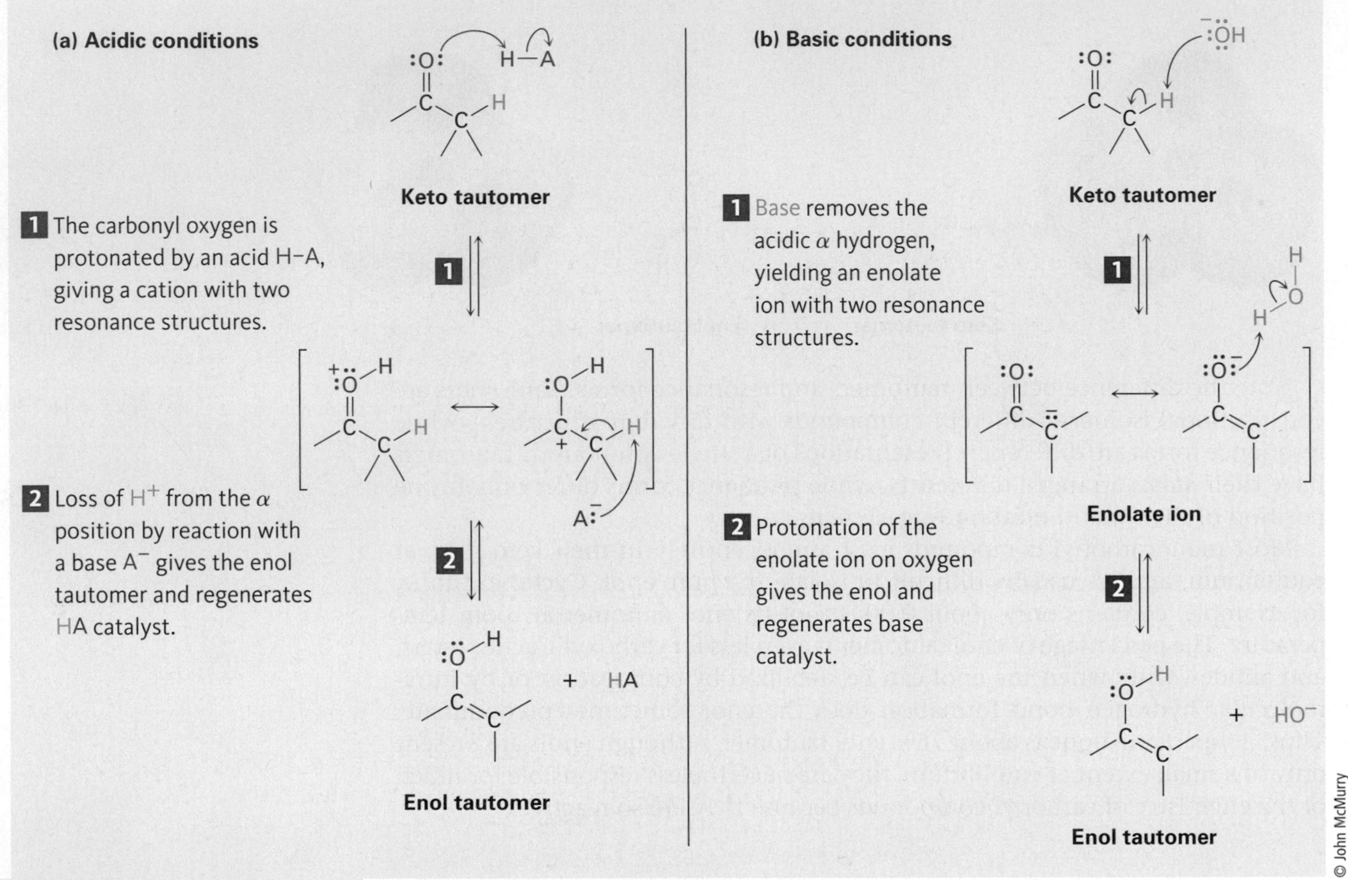

Figure 22.1 | MECHANISM Mechanism of enol formation under both acid-catalyzed and base-catalyzed conditions. **(a)** Acid catalysis involves (**1**) initial protonation of the carbonyl oxygen followed by (**2**) removal of H^+ from the α position. **(b)** Base catalysis involves (**1**) initial deprotonation of the α position to give an enolate ion, followed by (**2**) reprotonation on oxygen.

Note that only the hydrogens on the α position of carbonyl compounds are acidic. Hydrogens at β, γ, δ, and so on, aren't acidic and can't be removed by base because the resulting anions can't be resonance-stabilized by the carbonyl group.

Nonacidic

Base

Acidic

Enolate ion

Problem 22.1

Draw structures for the enol tautomers of the following compounds:

(a) Cyclopentanone **(b)** Methyl thioacetate **(c)** Ethyl acetate
(d) Propanal **(e)** Acetic acid **(f)** Phenylacetone

Problem 22.2

How many acidic hydrogens does each of the molecules listed in Problem 22.1 have? Identify them.

Problem 22.3

Draw structures for all monoenol forms of the following molecule. Which would you expect to be most stable? Explain.

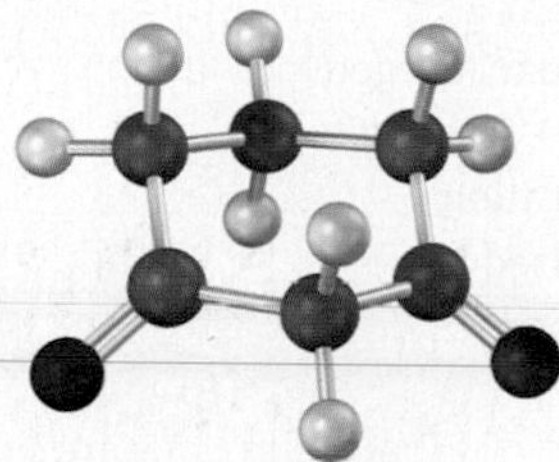

22.2 Reactivity of Enols: The Mechanism of Alpha-Substitution Reactions

What kind of chemistry do enols have? Because their double bonds are electron-rich, enols behave as nucleophiles and react with electrophiles in much the same way that alkenes do. But because of resonance electron donation of a lone-pair of electrons on the neighboring oxygen, enols are more electron-rich and correspondingly more reactive than alkenes. Notice in the following electrostatic potential map of ethenol ($H_2C{=}CHOH$) how there is a substantial amount of electron density (yellow-red) on the α carbon.

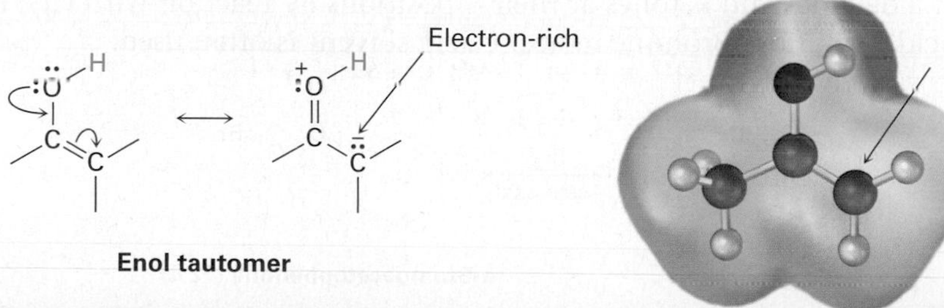

When an *alkene* reacts with an electrophile, E^+, initial addition gives an intermediate cation and subsequent reaction with a nucleophile such as a halide ion yields an addition product **(Section 7.7)**. When an *enol* reacts with an electrophile, however, only the initial addition step is the same **(Figure 22.2)**. Instead of reacting with a nucleophile to give an addition product, the intermediate cation loses the –OH proton to give an α-substituted carbonyl compound.

Figure 22.2 | MECHANISM

General mechanism of a carbonyl α-substitution reaction. In step 3, the initially formed cation loses H^+ to regenerate a carbonyl compound.

1 Acid-catalyzed enol formation occurs by the usual mechanism.

2 An electron pair from the enol oxygen attacks an electrophile (E^+), forming a new bond and leaving a cation intermediate that is stabilized by resonance between two forms.

3 Loss of a proton from oxygen yields the neutral alpha-substitution product as a new C=O bond is formed.

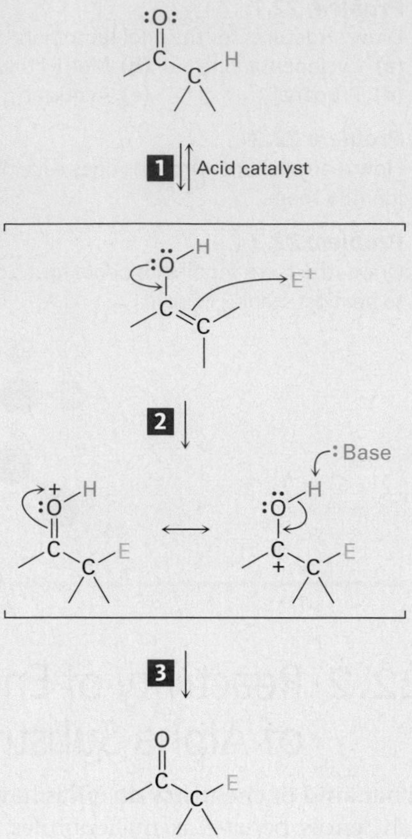

22.3 Alpha Halogenation of Aldehydes and Ketones

A particularly common α-substitution reaction in the laboratory is the halogenation of aldehydes and ketones at their α positions by reaction with Cl_2, Br_2, or I_2 in acidic solution. Bromine in acetic acid solvent is often used.

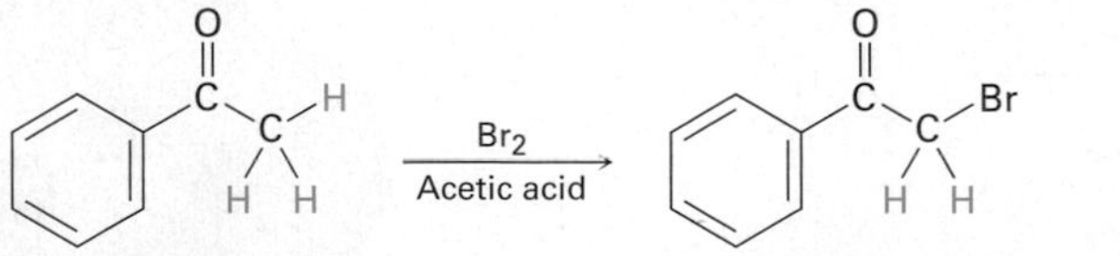

Remarkably, ketone halogenation also occurs in biological systems, particularly in marine alga, where dibromoacetaldehyde, bromoacetone, 1,1,1-tribromoacetone, and other related compounds have been found.

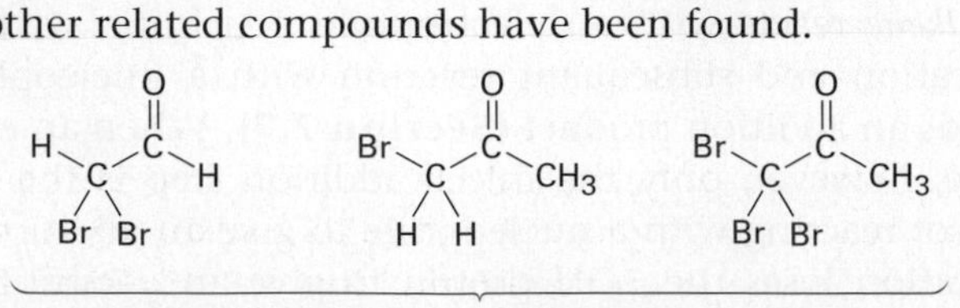

From the Hawaiian alga *Asparagopsis taxiformis*

The halogenation is a typical α-substitution reaction that proceeds by acid-catalyzed formation of an enol intermediate, as shown in **Figure 22.3**.

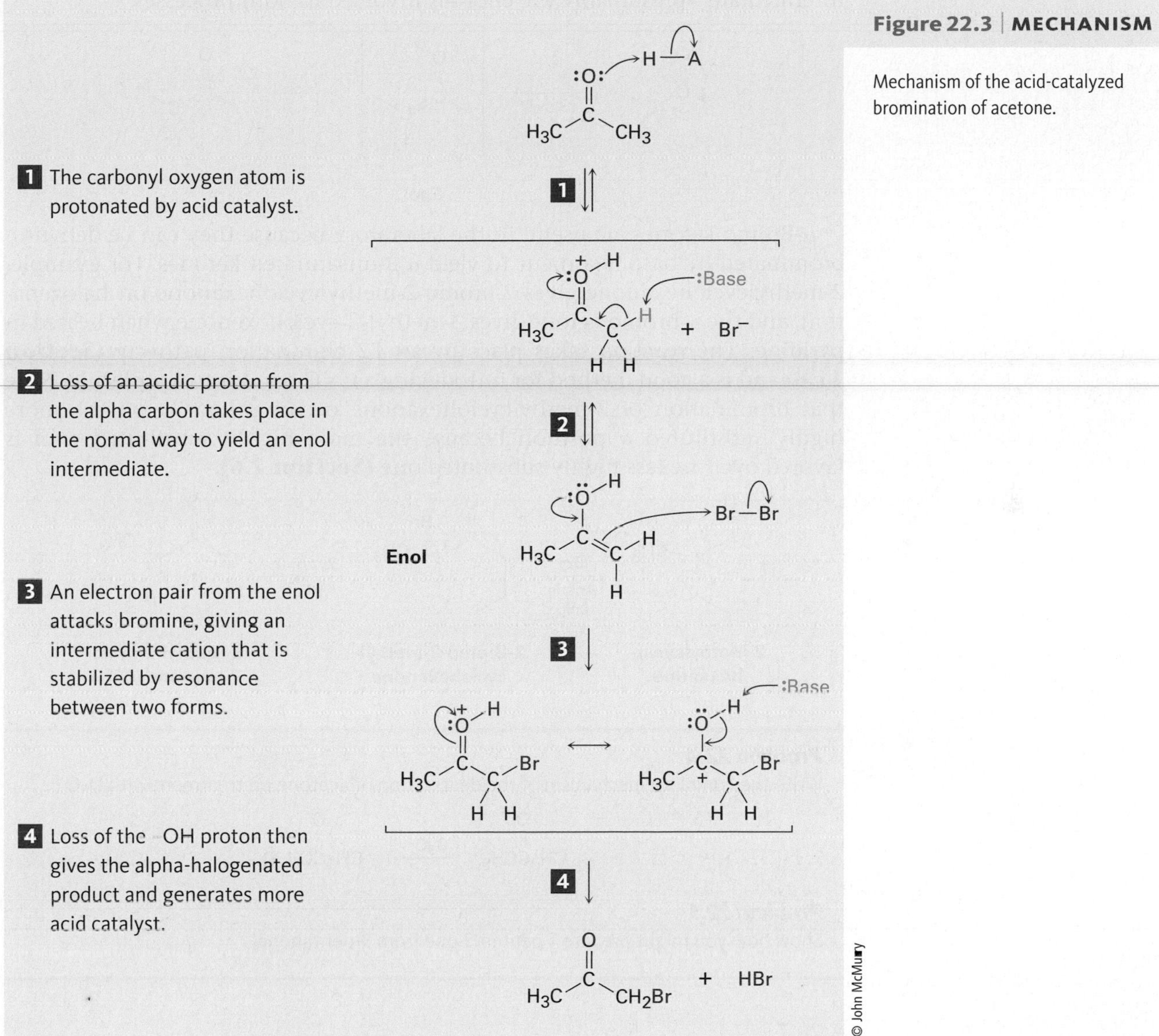

Figure 22.3 | MECHANISM

Mechanism of the acid-catalyzed bromination of acetone.

Evidence for the mechanism shown in Figure 22.3 includes the observation that acid-catalyzed halogenations show second-order kinetics and follow the rate law

$$\text{Reaction rate} = k\,[\text{Ketone}]\,[\text{H}^+]$$

In other words, the rate of halogenation depends only on the concentrations of ketone and acid and is independent of halogen concentration. Halogen is not involved in the rate-limiting step, so chlorination, bromination, and iodination of a given substrate all occur at the same rate.

Furthermore, if an aldehyde or ketone is treated with D_3O^+, the acidic α hydrogens are replaced by deuterium. For a given ketone, the rate of deuterium exchange is identical to the rate of halogenation, implying that a common intermediate—presumably the enol—is involved in both processes.

D_3O^+ (Slow) → Enol → D_3O^+ (Fast)

Enol

α-Bromo ketones are useful in the laboratory because they can be dehydrobrominated by base treatment to yield α,β-unsaturated ketones. For example, 2-methylcyclohexanone gives 2-bromo-2-methylcyclohexanone on halogenation, and the α-bromo ketone gives 2-methyl-2-cyclohexenone when heated in pyridine. The reaction takes place by an E2 elimination pathway **(Section 11.8)** and is a good method for introducing a C=C bond into a molecule. Note that bromination of 2-methylcyclohexanone occurs primarily on the more highly substituted α position because the more highly substituted enol is favored over the less highly substituted one **(Section 7.6)**.

2-Methylcyclohexanone —(Br_2, Acetic acid)→ **2-Bromo-2-methylcyclohexanone** —(Pyridine, Heat)→ **2-Methyl-2-cyclohexenone (63%)**

Problem 22.4

Write the complete mechanism of the deuteration of acetone on treatment with D_3O^+.

$$CH_3\overset{O}{\overset{\|}{C}}CH_3 \xrightarrow{D_3O^+} CH_3\overset{O}{\overset{\|}{C}}CH_2D$$

Problem 22.5

Show how you might prepare 1-penten-3-one from 3-pentanone.

22.4 Alpha Bromination of Carboxylic Acids

The α bromination of carbonyl compounds by Br_2 in acetic acid is limited to aldehydes and ketones because acids, esters, and amides don't enolize to a sufficient extent. Carboxylic acids, however, can be α brominated by a mixture of Br_2 and PBr_3 in the *Hell–Volhard–Zelinskii (HVZ) reaction.*

$$CH_3CH_2CH_2CH_2CH_2CH_2\overset{O}{\overset{\|}{C}}OH \xrightarrow[2.\ H_2O]{1.\ Br_2,\ PBr_3} CH_3CH_2CH_2CH_2CH_2\underset{Br}{\underset{|}{C}}H\overset{O}{\overset{\|}{C}}OH$$

Heptanoic acid → **2-Bromoheptanoic acid (90%)**

The Hell–Volhard–Zelinskii reaction is a bit more complex than it looks and actually involves α substitution of an acid bromide enol rather than a carboxylic acid enol. The process begins with reaction of the carboxylic acid with PBr_3 to form an acid bromide plus HBr (**Section 21.4**). The HBr then catalyzes enolization of the acid bromide, and the resultant enol reacts with Br_2 in an α-substitution reaction to give an α-bromo acid bromide. Addition of water hydrolyzes the acid bromide in a nucleophilic acyl substitution reaction and yields the α-bromo carboxylic acid product.

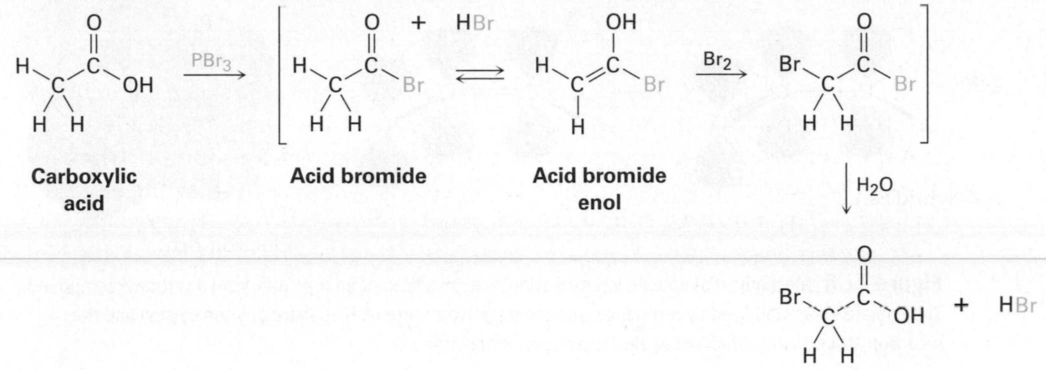

$$\mathrm{H_2(H)C{-}COOH} \xrightarrow{PBr_3} \left[\mathrm{H_2(H)C{-}COBr} + \mathrm{HBr} \rightleftharpoons \mathrm{H(H)C{=}C(OH)Br} \xrightarrow{Br_2} \mathrm{Br(H_2)C{-}COBr} \right] \xrightarrow{H_2O} \mathrm{Br(H_2)C{-}COOH} + \mathrm{HBr}$$

Carboxylic acid — **Acid bromide** — **Acid bromide enol** — **α-Bromo carboxylic acid**

Problem 22.6

If methanol rather than water is added at the end of a Hell–Volhard–Zelinskii reaction, an ester rather than an acid is produced. Show how you could carry out the following transformation, and propose a mechanism for the ester-forming step.

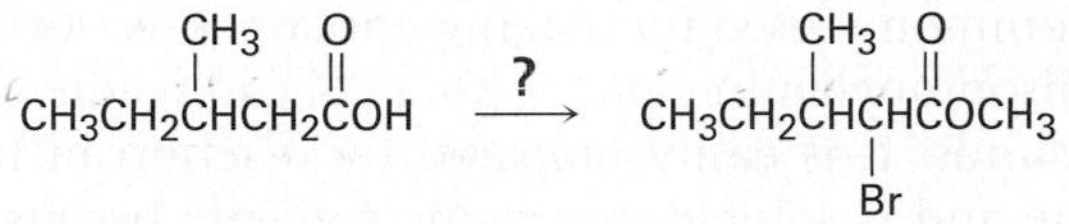

$$\mathrm{CH_3CH_2CH(CH_3)CH_2COOH} \xrightarrow{?} \mathrm{CH_3CH_2CH(CH_3)CH(Br)COOCH_3}$$

22.5 Acidity of Alpha Hydrogen Atoms: Enolate Ion Formation

As noted in **Section 22.1**, a hydrogen on the α position of a carbonyl compound is weakly acidic and can be removed by a strong base to yield an enolate ion. In comparing acetone ($pK_a = 19.3$) with ethane ($pK_a \approx 60$), for instance, the presence of a neighboring carbonyl group increases the acidity of the ketone over the alkane by a factor of 10^{40}.

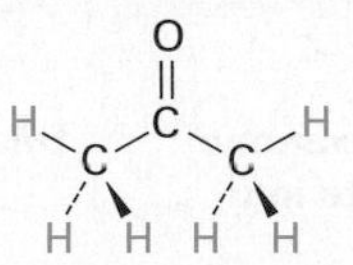

$$\mathrm{CH_3COCH_3} \qquad\qquad \mathrm{CH_3CH_3}$$

Acetone ($pK_a = 19.3$) — **Ethane ($pK_a \approx 60$)**

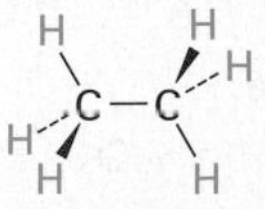

Proton abstraction from a carbonyl compound occurs when the α C–H bond is oriented roughly parallel to the *p* orbitals of the carbonyl group. The α carbon atom of the enolate ion is sp^2-hybridized and has a *p* orbital that overlaps the neighboring carbonyl *p* orbitals. Thus, the negative charge is shared by the electronegative oxygen atom, and the enolate ion is stabilized by resonance (**Figure 22.4**).

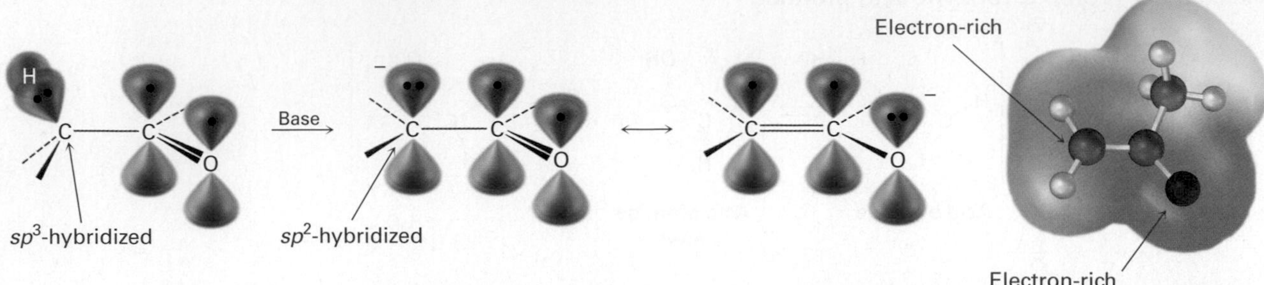

Figure 22.4 Mechanism of enolate ion formation by abstraction of an α proton from a carbonyl compound. The enolate ion is stabilized by resonance, and the negative charge (red) is shared by the oxygen and the α carbon atom, as indicated by the electrostatic potential map.

Because carbonyl compounds are only weakly acidic, a strong base is needed for enolate ion formation. If an alkoxide ion, such as sodium ethoxide, is used as base, deprotonation takes place only to the extent of about 0.1% because acetone is a weaker acid than ethanol ($pK_a = 16$). If, however, a more powerful base is used, then a carbonyl compound is completely converted into its enolate ion.

In practice, the strong base lithium diisopropylamide [$LiN(i\text{-}C_3H_7)_2$; abbreviated LDA] is commonly used for making enolate ions. As the lithium salt of the weak acid diisopropylamine, $pK_a = 36$, LDA can readily deprotonate most carbonyl compounds. It is easily prepared by reaction of butyllithium with diisopropylamine and is soluble in organic solvents because of its two alkyl groups.

$$H{-}N[CH(CH_3)_2]_2 \xrightarrow[\text{THF}]{C_4H_9Li} Li^+\ {}^-N[CH(CH_3)_2]_2 + C_4H_{10}$$

Diisopropylamine **p*K*a = 36** — **Lithium diisopropylamide (LDA)**

$$\text{Cyclohexanone} \xrightarrow[\text{Tetrahydrofuran solvent (THF)}]{\text{LDA}} \text{Cyclohexanone enolate ion } (O^-\ Li^+) + H{-}N[CH(CH_3)_2]_2$$

Cyclohexanone — **Cyclohexanone enolate ion** — **Diisopropylamine**

Many types of carbonyl compounds, including aldehydes, ketones, esters, thioesters, acids, and amides, can be converted into enolate ions by reaction with LDA. Table 22.1 lists the approximate pK_a values of different types of carbonyl compounds and shows how these values compare to other acidic substances we've seen. Note that nitriles, too, are acidic and can be converted into enolate-like anions.

Table 22.1 Acidity Constants for Some Organic Compounds

Functional group	Example	pK_a
Carboxylic acid	$CH_3C(=O)OH$	5
1,3-Diketone	$CH_3C(=O)CH_2C(=O)CH_3$	9
3-Keto ester	$CH_3C(=O)CH_2C(=O)OCH_3$	11
1,3-Diester	$CH_3OC(=O)CH_2C(=O)OCH_3$	13
Alcohol	CH_3OH	16
Acid chloride	$CH_3C(=O)Cl$	16
Aldehyde	$CH_3C(=O)H$	17
Ketone	$CH_3C(=O)CH_3$	19
Thioester	$CH_3C(=O)SCH_3$	21
Ester	$CH_3C(=O)OCH_3$	25
Nitrile	$CH_3C{\equiv}N$	25
N,N-Dialkylamide	$CH_3C(=O)N(CH_3)_2$	30
Dialkylamine	$HN(i\text{-}C_3H_7)_2$	36

When a hydrogen atom is flanked by two carbonyl groups, its acidity is enhanced even more. Table 22.1 thus shows that 1,3-diketones (β-diketones), 3-oxo esters (β-keto esters), and 1,3-diesters (malonic esters) are even more acidic than water. The enolate ions derived from these β-dicarbonyl compounds are stabilized by sharing of the negative charge by both neighboring carbonyl oxygens. The enolate ion of 2,4-pentanedione, for instance, has three resonance forms. Similar resonance forms can be drawn for other doubly stabilized enolate ions.

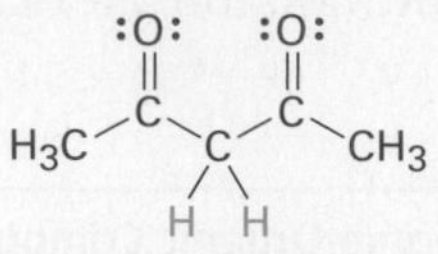

2,4-Pentanedione ($pK_a = 9$)

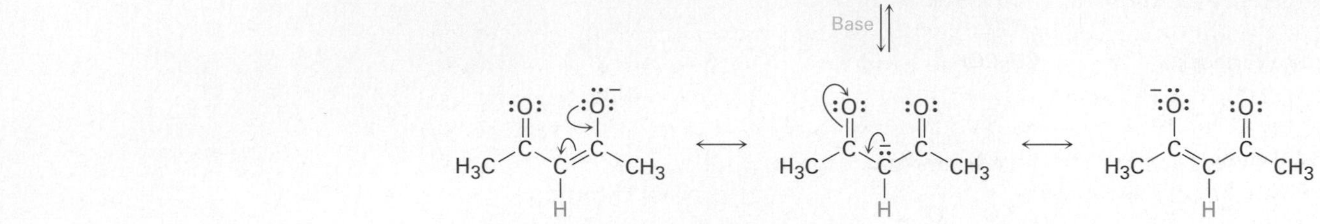

Base

Worked Example 22.1 Identifying the Acidic Hydrogens in a Compound

Identify the most acidic hydrogens in each of the following compounds, and rank the compounds in order of increasing acidity:

(a) **(b)** $CH_3CHCOCH_3$ with CH_3 substituent **(c)**

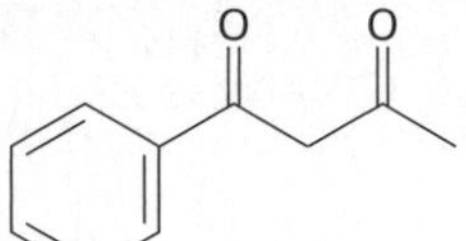

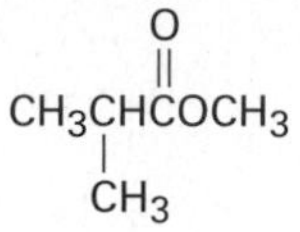

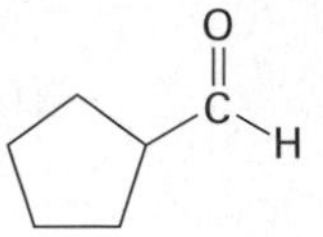

Strategy

Hydrogens on carbon next to a carbonyl group are acidic. In general, a β-dicarbonyl compound is most acidic, a ketone or aldehyde is next most acidic, and a carboxylic acid derivative is least acidic. Remember that alcohols, phenols, and carboxylic acids are also acidic because of their –OH hydrogens.

Solution

The acidity order is (a) > (c) > (b). Acidic hydrogens are shown in red.

(a) More acidic Less acidic **(b)** 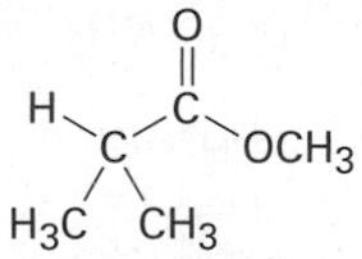 **(c)**

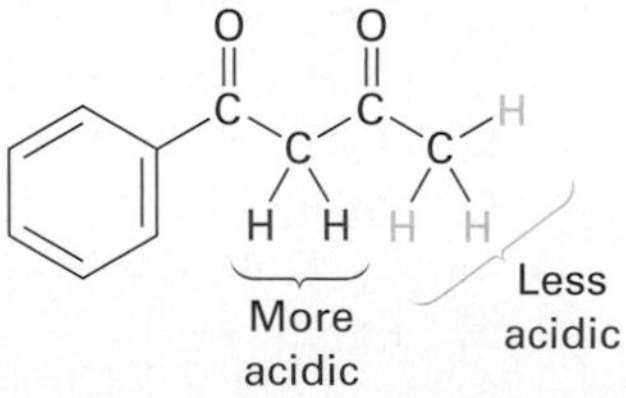

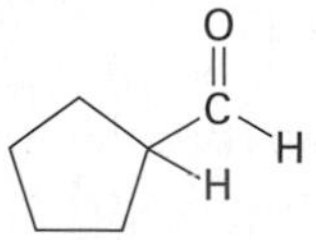

Problem 22.7

Identify the most acidic hydrogens in each of the following molecules:

(a) CH_3CH_2CHO **(b)** $(CH_3)_3CCOCH_3$ **(c)** CH_3CO_2H

(d) Benzamide **(e)** $CH_3CH_2CH_2CN$ **(f)** $CH_3CON(CH_3)_2$

Problem 22.8

Draw a resonance structure of the acetonitrile anion, $^-{:}CH_2C{\equiv}N$, and account for the acidity of nitriles.

22.6 Reactivity of Enolate Ions

Enolate ions are more useful than enols for two reasons. First, pure enols can't normally be isolated but are instead generated only as short-lived intermediates in low concentration. By contrast, stable solutions of pure enolate ions are easily prepared from most carbonyl compounds by reaction with a strong base. Second, enolate ions are more reactive than enols and undergo many reactions that enols don't. Whereas enols are neutral, enolate ions are negatively charged, making them much better nucleophiles.

Because they are resonance hybrids of two, nonequivalent forms, enolate ions can be looked at either as vinylic alkoxides (C=C–O$^-$) or as α-keto carbanions ($^-$C–C=O). Thus, enolate ions can react with electrophiles either on oxygen or on carbon. Reaction on oxygen yields an enol derivative, while reaction on carbon yields an α-substituted carbonyl compound **(Figure 22.5)**. Both kinds of reactivity are known, but reaction on carbon is more common.

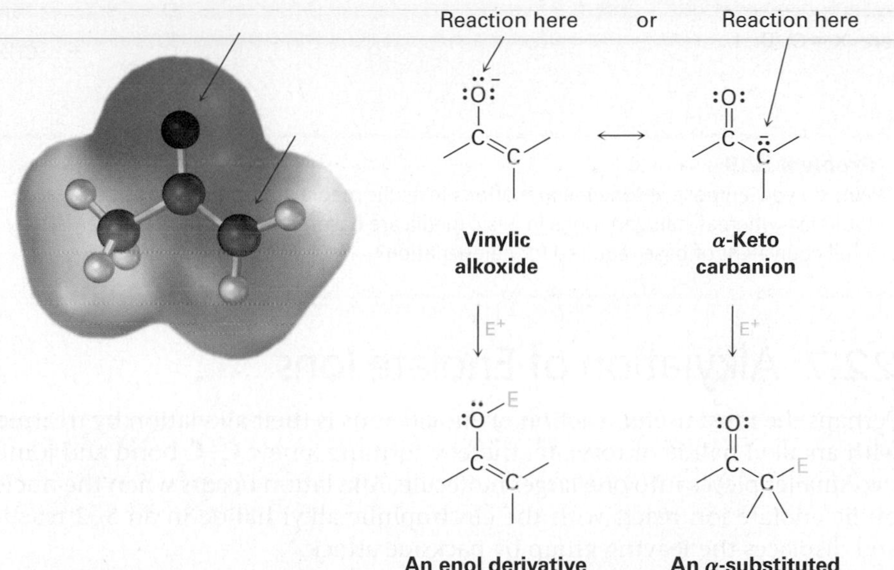

Figure 22.5 The electrostatic potential map of acetone enolate ion shows how the negative charge is delocalized over both the oxygen and the α carbon. As a result, two modes of reaction of an enolate ion with an electrophile E^+ are possible. Reaction on carbon to yield an α-substituted carbonyl product is more common.

As an example of enolate-ion reactivity, aldehydes and ketones undergo base-promoted α halogenation. Even relatively weak bases such as hydroxide ion are effective for halogenation because it's not necessary to convert the ketone completely into its enolate ion. As soon as a small amount of enolate is generated, halogen reacts with it immediately, removing it from the reaction and driving the equilibrium toward further enolate ion formation.

$$\text{O=C–C–H} \underset{H_2O}{\overset{NaOH}{\rightleftharpoons}} \left[\,{:}\ddot{O}{:}^-\text{–C=C}\,\right] \xrightarrow{Br_2} \text{O=C–C–Br} + Br^-$$

Base-promoted halogenation of aldehydes and ketones is little used in practice because it's difficult to stop the reaction at the monosubstituted product. An α-halogenated ketone is generally more acidic than the starting, unsubstituted ketone because of the electron-withdrawing inductive effect of the halogen atom. Thus, the monohalogenated products are themselves rapidly turned into enolate ions and further halogenated.

If excess base and halogen are used, a methyl ketone is triply halogenated and then cleaved by base in the *haloform reaction.* The products are a carboxylic acid plus a so-called haloform (chloroform, $CHCl_3$; bromoform, $CHBr_3$; or iodoform, CHI_3). Note that the second step of the reaction is a nucleophilic acyl substitution of $^-CX_3$ by ^-OH. That is, a halogen-stabilized carbanion acts as a leaving group.

A methyl ketone

where X = Cl, Br, I

Problem 22.9

Why do you suppose ketone halogenations in acidic media are referred to as being acid-*catalyzed*, whereas halogenations in basic media are base-*promoted*? In other words, why is a full equivalent of base required for halogenation?

22.7 Alkylation of Enolate Ions

Perhaps the most useful reaction of enolate ions is their alkylation by treatment with an alkyl halide or tosylate, thereby forming a new C–C bond and joining two smaller pieces into one larger molecule. Alkylation occurs when the nucleophilic enolate ion reacts with the electrophilic alkyl halide in an S_N2 reaction and displaces the leaving group by backside attack.

Base

S_N2 reaction

Enolate ion

Alkylation reactions are subject to the same constraints that affect all S_N2 reactions **(Section 11.3)**. Thus, the leaving group X in the alkylating agent R–X can be chloride, bromide, iodide, or tosylate. The alkyl group R should be primary or methyl, and preferably should be allylic or benzylic. Secondary halides react poorly, and tertiary halides don't react at all because a competing E2 elimination of HX occurs instead. Vinylic and aryl halides are also unreactive because backside approach is sterically prevented.

R–X { –X: Tosylate > –I > –Br > –Cl
R–: Allylic ≈ Benzylic > H_3C– > RCH_2– }

The Malonic Ester Synthesis

One of the oldest and best known carbonyl alkylation reactions is the **malonic ester synthesis**, a method for preparing a carboxylic acid from an alkyl halide while lengthening the carbon chain by two atoms.

$$\text{R–X} \xrightarrow[\text{synthesis}]{\text{Malonic ester}} \text{RCH}_2\text{CO}_2\text{H}$$

Diethyl propanedioate, commonly called diethyl malonate, or *malonic ester*, is relatively acidic ($pK_a = 13$) because its α hydrogens are flanked by two carbonyl groups. Thus, malonic ester is easily converted into its enolate ion by reaction with sodium ethoxide in ethanol. The enolate ion, in turn, is a good nucleophile that reacts rapidly with an alkyl halide to give an α-substituted malonic ester. Note in the following examples that the abbreviation "Et" is used for an ethyl group, $-CH_2CH_3$.

$$\text{EtO}_2\text{C–CH}_2\text{–CO}_2\text{Et} \xrightarrow[\text{EtOH}]{\text{Na}^+\ ^-\text{OEt}} \left[\text{EtO}_2\text{C–}\ddot{\text{C}}^-\text{H–CO}_2\text{Et}\ \ \text{Na}^+\right] \xrightarrow{\text{RX}} \text{EtO}_2\text{C–CHR–CO}_2\text{Et}$$

Diethyl propanedioate (malonic ester) — **Sodio malonic ester** — **An alkylated malonic ester**

The product of a malonic ester alkylation has one acidic α hydrogen remaining, so the alkylation process can be repeated to yield a dialkylated malonic ester.

$$\text{EtO}_2\text{C–CHR–CO}_2\text{Et} \xrightarrow[\text{EtOH}]{\text{Na}^+\ ^-\text{OEt}} \left[\text{EtO}_2\text{C–}\ddot{\text{C}}^-\text{R–CO}_2\text{Et}\ \ \text{Na}^+\right] \xrightarrow{\text{R}'\text{X}} \text{EtO}_2\text{C–CRR}'\text{–CO}_2\text{Et}$$

An alkylated malonic ester — **A dialkylated malonic ester**

On heating with aqueous hydrochloric acid, the alkylated (or dialkylated) malonic ester undergoes hydrolysis of its two ester groups followed by *decarboxylation* (loss of CO_2) to yield a substituted monocarboxylic acid.

$$\text{R–CH(CO}_2\text{Et)}_2 \xrightarrow[\text{Heat}]{\text{H}_3\text{O}^+} \text{RCH}_2\text{CO}_2\text{H} + \text{CO}_2 + 2\ \text{EtOH}$$

An alkylated malonic ester — **A carboxylic acid**

Decarboxylation is not a general reaction of carboxylic acids. Rather, it is unique to compounds that have a second carbonyl group two atoms away from the $-CO_2H$. That is, only substituted malonic acids and β-keto acids undergo loss of CO_2 on heating. The decarboxylation reaction occurs by a cyclic mechanism and involves initial formation of an enol, thereby accounting for the need to have a second carbonyl group appropriately positioned.

A diacid → ($-CO_2$) → An acid enol → A carboxylic acid

A β-keto acid → ($-CO_2$) → An enol → A ketone

As noted previously, the overall effect of the malonic ester synthesis is to convert an alkyl halide into a carboxylic acid while lengthening the carbon chain by two atoms ($RX \rightarrow RCH_2CO_2H$).

1-Bromobutane + diethyl malonate → ($Na^+\ ^-OEt$, EtOH) → (H_3O^+, Heat) → Hexanoic acid (75%)

(1. $Na^+\ ^-OEt$; 2. CH_3I) → (H_3O^+, Heat) → 2-Methylhexanoic acid (74%)

The malonic ester synthesis can also be used to prepare *cyclo*alkanecarboxylic acids. For example, when 1,4-dibromobutane is treated with diethyl malonate in the presence of 2 equivalents of sodium ethoxide base, the second alkylation step occurs intramolecularly to yield a cyclic product. Hydrolysis and decarboxylation then give cyclopentanecarboxylic acid. Three-, four-, five-, and six-membered rings can be prepared in this way, but yields decrease for larger ring sizes.

1,4-Dibromobutane → ($Na^+\ ^-OEt$, EtOH) → ($Na^+\ ^-OEt$, EtOH) → (H_3O^+, Heat) → Cyclopentanecarboxylic acid + CO_2 + 2 EtOH

Using the Malonic Ester Synthesis to Prepare a Carboxylic Acid

Worked Example 22.2

How would you prepare heptanoic acid using a malonic ester synthesis?

Strategy

The malonic ester synthesis converts an alkyl halide into a carboxylic acid having two more carbons. Thus, a seven-carbon acid chain must be derived from the five-carbon alkyl halide 1-bromopentane.

Solution

$$CH_3CH_2CH_2CH_2CH_2Br + CH_2(CO_2Et)_2 \xrightarrow[\text{2. } H_3O^+\text{, heat}]{\text{1. } Na^+ \ ^-OEt} CH_3CH_2CH_2CH_2CH_2CH_2\overset{O}{\overset{\|}{C}}OH$$

Problem 22.10

How could you use a malonic ester synthesis to prepare the following compounds? Show all steps.

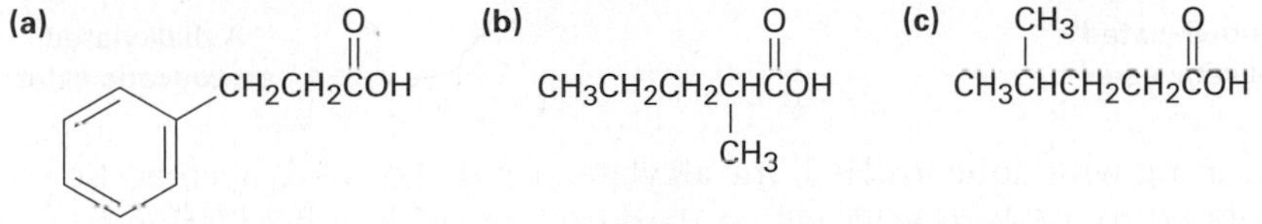

(a) $C_6H_5CH_2CH_2\overset{O}{\overset{\|}{C}}OH$ (b) $CH_3CH_2CH_2\underset{CH_3}{\underset{|}{C}}H\overset{O}{\overset{\|}{C}}OH$ (c) $CH_3\overset{CH_3}{\overset{|}{C}}HCH_2CH_2\overset{O}{\overset{\|}{C}}OH$

Problem 22.11

Monoalkylated and dialkylated acetic acids can be prepared by the malonic ester synthesis, but trialkylated acetic acids (R_3CCO_2H) can't be prepared. Explain.

Problem 22.12

How could you use a malonic ester synthesis to prepare the following compound?

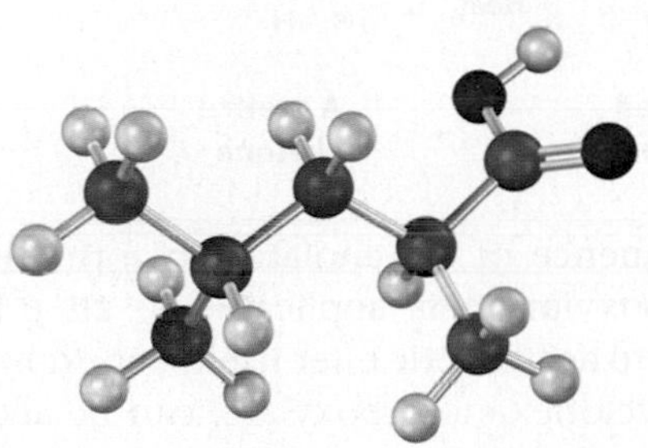

The Acetoacetic Ester Synthesis

Just as the malonic ester synthesis converts an alkyl halide into a carboxylic acid, the **acetoacetic ester synthesis** converts an alkyl halide into a methyl ketone having three more carbons.

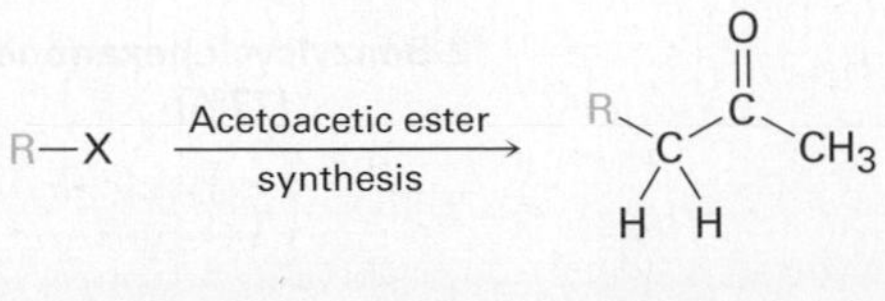

$$R{-}X \xrightarrow[\text{synthesis}]{\text{Acetoacetic ester}} R{-}CH_2{-}\overset{O}{\overset{\|}{C}}{-}CH_3$$

Ethyl 3-oxobutanoate, commonly called ethyl acetoacetate, or *acetoacetic ester,* is much like malonic ester in that its α hydrogens are flanked by two carbonyl groups. It is therefore readily converted into its enolate ion, which can be alkylated by reaction with an alkyl halide. A second alkylation can also be carried out if desired, since acetoacetic ester has two acidic α hydrogens.

$$EtO_2C{-}CH_2{-}C(=O){-}CH_3 \xrightarrow[EtOH]{Na^+\ ^-OEt} \left[EtO_2C{-}\ddot{C}^-H{-}C(=O){-}CH_3\ \ Na^+\right] \xrightarrow{RX} EtO_2C{-}CHR{-}C(=O){-}CH_3$$

Ethyl acetoacetate (acetoacetic ester) — **Sodio acetoacetic ester** — **A monoalkylated acetoacetic ester**

$$EtO_2C{-}CHR{-}C(=O){-}CH_3 \xrightarrow[EtOH]{Na^+\ ^-OEt} \left[EtO_2C{-}\ddot{C}^-R{-}C(=O){-}CH_3\ \ Na^+\right] \xrightarrow{R'X} EtO_2C{-}CRR'{-}C(=O){-}CH_3$$

A monoalkylated acetoacetic ester — **A dialkylated acetoacetic ester**

On heating with aqueous HCl, the alkylated (or dialkylated) acetoacetic ester is hydrolyzed to a β-keto acid, which then undergoes decarboxylation to yield a ketone product. The decarboxylation occurs in the same way as in the malonic ester synthesis and involves a ketone enol as the initial product.

$$R{-}CH(CO_2Et){-}C(=O){-}CH_3 \xrightarrow[Heat]{H_3O^+} R{-}CH_2{-}C(=O){-}CH_3 + CO_2 + EtOH$$

An alkylated acetoacetic ester — **A methyl ketone**

The three-step sequence of (1) enolate ion formation, (2) alkylation, and (3) hydrolysis/decarboxylation is applicable to all β-keto esters with acidic α hydrogens, not just to acetoacetic ester itself. For example, cyclic β-keto esters, such as ethyl 2-oxocyclohexanecarboxylate, can be alkylated and decarboxylated to give 2-substituted cyclohexanones.

Ethyl 2-oxocyclohexanecarboxylate $\xrightarrow[2.\ PhCH_2Br]{1.\ Na^+\ ^-OEt}$ 2-benzyl-2-(ethoxycarbonyl)cyclohexanone $\xrightarrow[Heat]{H_3O^+}$ 2-benzylcyclohexanone + CO_2 + EtOH

Ethyl 2-oxocyclohexanecarboxylate (a cyclic β-keto ester) — **2-Benzylcyclohexanone (77%)**

Using the Acetoacetic Ester Synthesis to Prepare a Ketone

Worked Example 22.3

How would you prepare 2-pentanone by an acetoacetic ester synthesis?

Strategy

The acetoacetic ester synthesis yields a methyl ketone by adding three carbons to an alkyl halide.

$$\underbrace{\text{R}}_{\text{This R group from alkyl halide}}\overset{\text{This bond formed}}{-}\underbrace{\text{CH}_2\overset{\text{O}}{\overset{\|}{\text{C}}}\text{CH}_3}_{\text{These three carbons from acetoacetic ester}}$$

Thus, the acetoacetic ester synthesis of 2-pentanone must involve reaction of bromoethane.

Solution

$$\text{CH}_3\text{CH}_2\text{Br} + \text{EtO}\overset{\text{O}}{\overset{\|}{\text{C}}}\text{CH}_2\overset{\text{O}}{\overset{\|}{\text{C}}}\text{CH}_3 \xrightarrow[\text{2. H}_3\text{O}^+\text{, heat}]{\text{1. Na}^+\ ^-\text{OEt}} \text{CH}_3\text{CH}_2\text{CH}_2\overset{\text{O}}{\overset{\|}{\text{C}}}\text{CH}_3$$

2-Pentanone

Problem 22.13

What alkyl halides would you use to prepare the following ketones by an acetoacetic ester synthesis?

(a) $\text{CH}_3\overset{\text{CH}_3}{\overset{|}{\text{C}}}\text{HCH}_2\text{CH}_2\overset{\text{O}}{\overset{\|}{\text{C}}}\text{CH}_3$ **(b)** $\text{C}_6\text{H}_5\text{–CH}_2\text{CH}_2\text{CH}_2\overset{\text{O}}{\overset{\|}{\text{C}}}\text{CH}_3$

Problem 22.14

Which of the following compounds *cannot* be prepared by an acetoacetic ester synthesis? Explain.

(a) Phenylacetone **(b)** Acetophenone **(c)** 3,3-Dimethyl-2-butanone

Problem 22.15

How would you prepare the following compound using an acetoacetic ester synthesis?

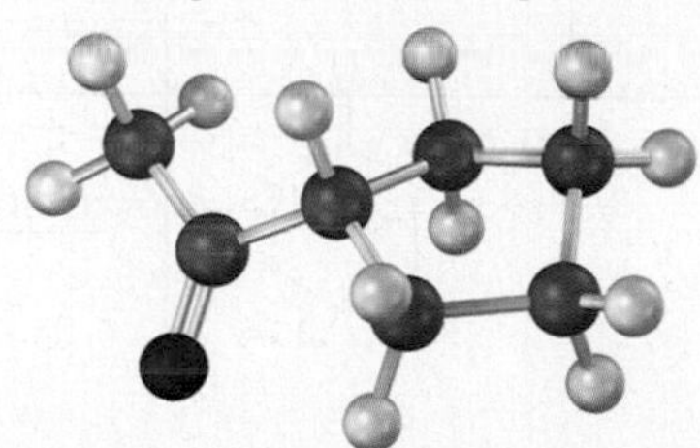

Direct Alkylation of Ketones, Esters, and Nitriles

Both the malonic ester synthesis and the acetoacetic ester synthesis are easy to carry out because they involve relatively acidic dicarbonyl compounds. As a result, sodium ethoxide in ethanol as solvent can be used to prepare the necessary enolate ions. Alternatively, however, it's also possible in many cases to directly alkylate the α position of *mono*carbonyl compounds. A strong, sterically hindered base such as LDA is needed so that complete conversion to the enolate ion takes place rather than a nucleophilic addition, and a nonprotic solvent must be used.

Ketones, esters, and nitriles can all be alkylated using LDA or related dialkylamide bases in THF. Aldehydes, however, rarely give high yields of pure products because their enolate ions undergo carbonyl condensation reactions instead of alkylation. (We'll study this condensation reaction in the next chapter.) Some specific examples of alkylation reactions are shown.

Lactone

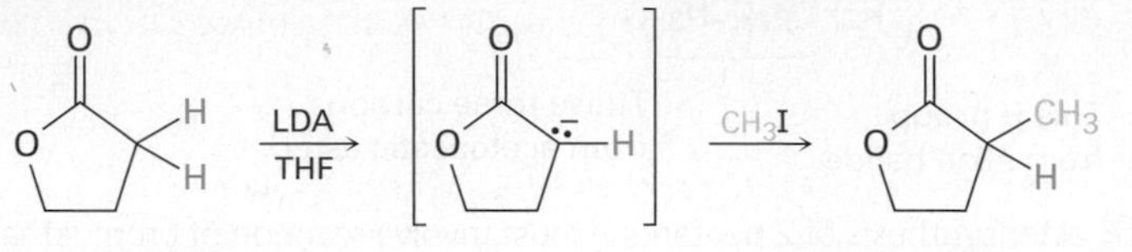

LDA, THF; CH_3I

Butyrolactone → **2-Methylbutyrolactone (88%)**

Ester

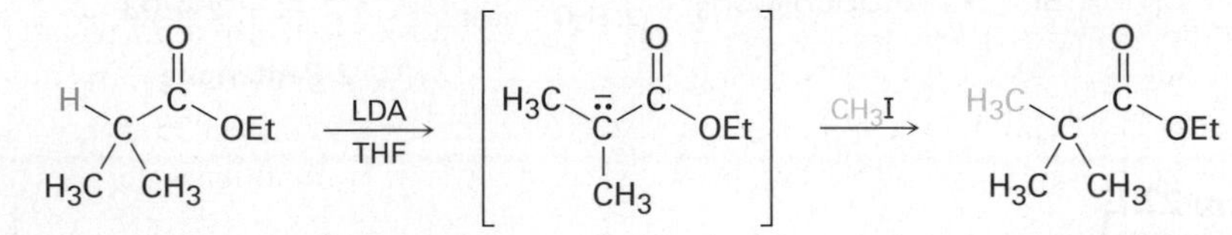

LDA, THF; CH_3I

Ethyl 2-methylpropanoate → **Ethyl 2,2-dimethylpropanoate (87%)**

Ketone

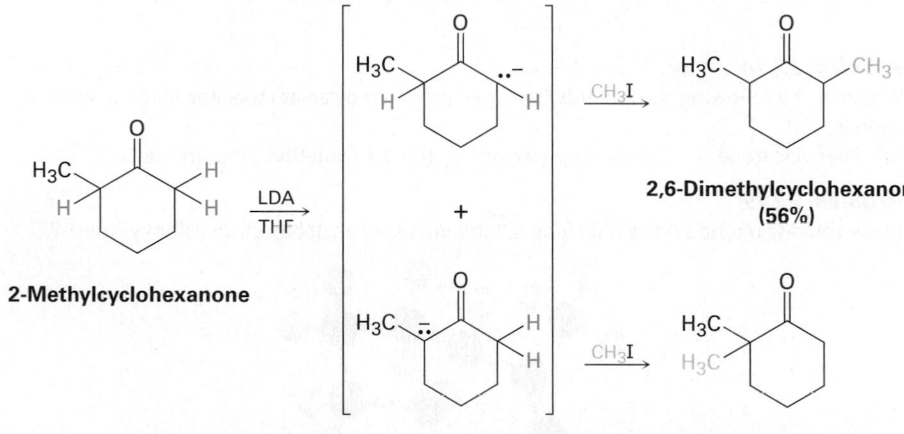

LDA, THF; CH_3I

2-Methylcyclohexanone → **2,6-Dimethylcyclohexanone (56%)** + **2,2-Dimethylcyclohexanone (6%)**

Nitrile

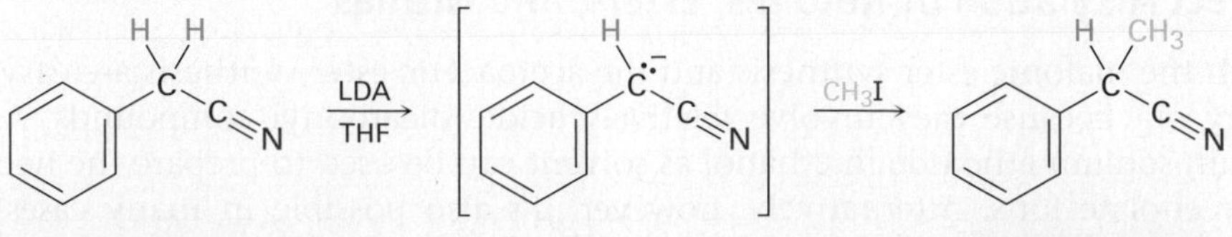

LDA, THF; CH_3I

Phenylacetonitrile → **2-Phenylpropane-nitrile (71%)**

Note in the ketone example that alkylation of 2-methylcyclohexanone leads to a mixture of products because both possible enolate ions are formed. In general, the major product in such cases occurs by alkylation at the less hindered, more accessible position. Thus, alkylation of 2-methylcyclohexanone occurs primarily at C6 (secondary) rather than C2 (tertiary).

Worked Example 22.4 Using an Alkylation Reaction to Prepare a Substituted Ester

How might you use an alkylation reaction to prepare ethyl 1-methylcyclohexanecarboxylate?

CO_2Et / CH_3

Ethyl 1-methylcyclohexanecarboxylate

Strategy

An alkylation reaction is used to introduce a methyl or primary alkyl group onto the α position of a ketone, ester, or nitrile by S_N2 reaction of an enolate ion with an alkyl halide. Thus, we need to look at the target molecule and identify any methyl or primary alkyl groups attached to an α carbon. In the present instance, the target has an α methyl group, which might be introduced by alkylation of an ester enolate ion with iodomethane.

Solution

CO_2Et / H → (1. LDA, THF; 2. CH_3I) → CO_2Et / CH_3

Ethyl cyclohexanecarboxylate **Ethyl 1-methylcyclohexanecarboxylate**

Problem 22.16

Show how you might prepare the following compounds using an alkylation reaction as the key step:

(a) $C_6H_5CH(CH_3)COCH_3$ (CHCCH3 with O and CH_3)

(b) $CH_3CH_2CH_2CH(CH_2CH_3)C{\equiv}N$

(c) 2-($CH_2CH{=}CH_2$)cyclohexanone

(d) 2,2,6,6-tetramethylcyclohexanone (H_3C, H_3C, CH_3, CH_3, O)

(e) $C_6H_5C(O)CH(CH_3)_2$

(f) $CH_3CH(CH_3)CH(CH_2CH_3)COCH_3$ (O‖)

Biological Alkylations

Alkylations are rare but not unknown in biological chemistry. One example occurs during biosynthesis of the antibiotic indolmycin from indolylpyruvate when a base abstracts an acidic hydrogen from an α position and the resultant enolate ion carries out an S_N2 alkylation reaction on the methyl group of *S*-adenosylmethionine (SAM; **Section 11.6**). Although it's convenient to speak of "enolate ion" intermediates in biological pathways, it's unlikely that they exist for long in an aqueous cellular environment. Rather, proton removal and alkylation probably occur at essentially the same time **(Figure 22.6)**.

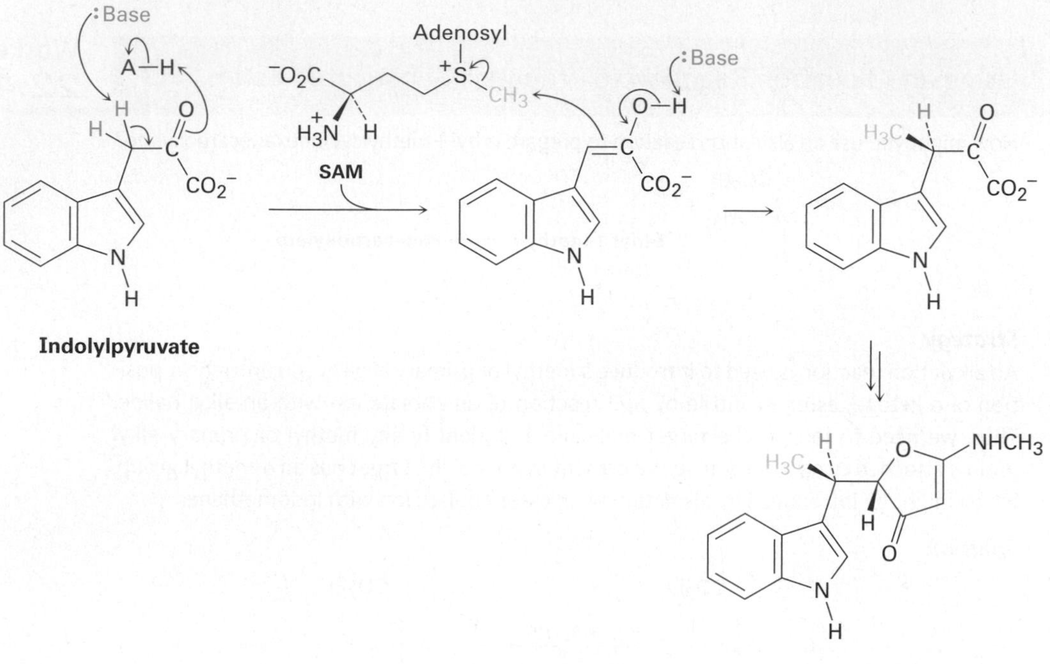

Indolylpyruvate

SAM

Indolmycin (an antibiotic)

Figure 22.6 The biosynthesis of indolmycin from indolylpyruvate occurs through a pathway that includes an alkylation reaction of a short-lived enolate ion intermediate.

Barbiturates | A DEEPER LOOK

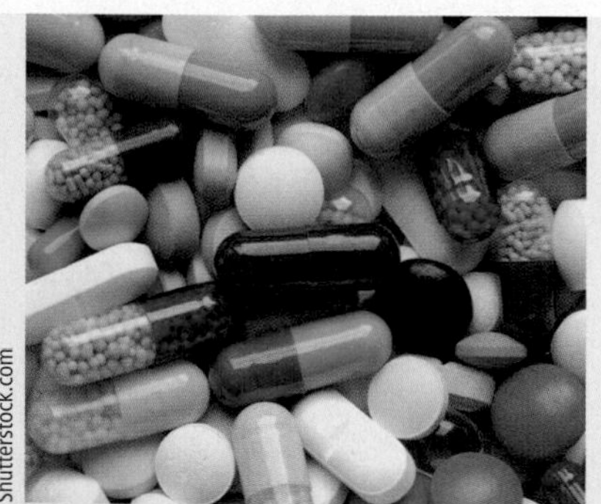

Different barbiturates come in a multitude of colors, giving rise to similarly colorful street names when the drugs are abused.

Using herbal remedies to treat illness and disease goes back thousands of years, but the medical use of chemicals prepared in the laboratory has a much shorter history. The barbiturates, a large class of drugs with a wide variety of uses, constitute one of the earliest successes of medicinal chemistry. The synthesis and medical use of barbiturates goes back to 1904 when Bayer, a German chemical company, first marketed a compound called barbital, trade named Veronal, as a treatment for insomnia. Since that time, more than 2500 different barbiturate analogs have been synthesized by drug companies, more than 50 have been used medicinally, and about a dozen are still in use as anesthetics, anticonvulsants, sedatives, and anxiolytics.

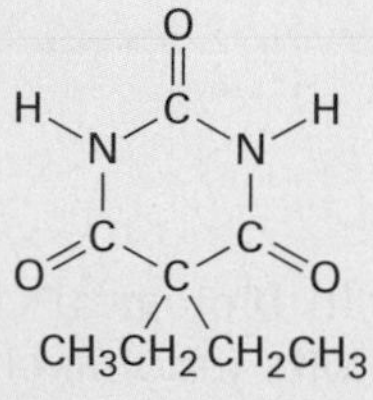

Barbital (Veronal), the first barbiturate

(continued)

(continued)

The synthesis of barbiturates is relatively simple and relies on reactions that are now familiar: enolate alkylations and nucleophilic acyl substitutions. Starting with diethyl malonate, or malonic ester, alkylation of the corresponding enolate ion with simple alkyl halides provides a wealth of different disubstituted malonic esters. Reaction with urea, $(H_2N)_2C{=}O$, then gives the product barbiturates by a twofold nucleophilic acyl substitution reaction of the ester groups with the $-NH_2$ groups of urea **(Figure 22.7)**. Amobarbital (Amytal), pentobarbital (Nembutal), and secobarbital (Seconal) are typical examples.

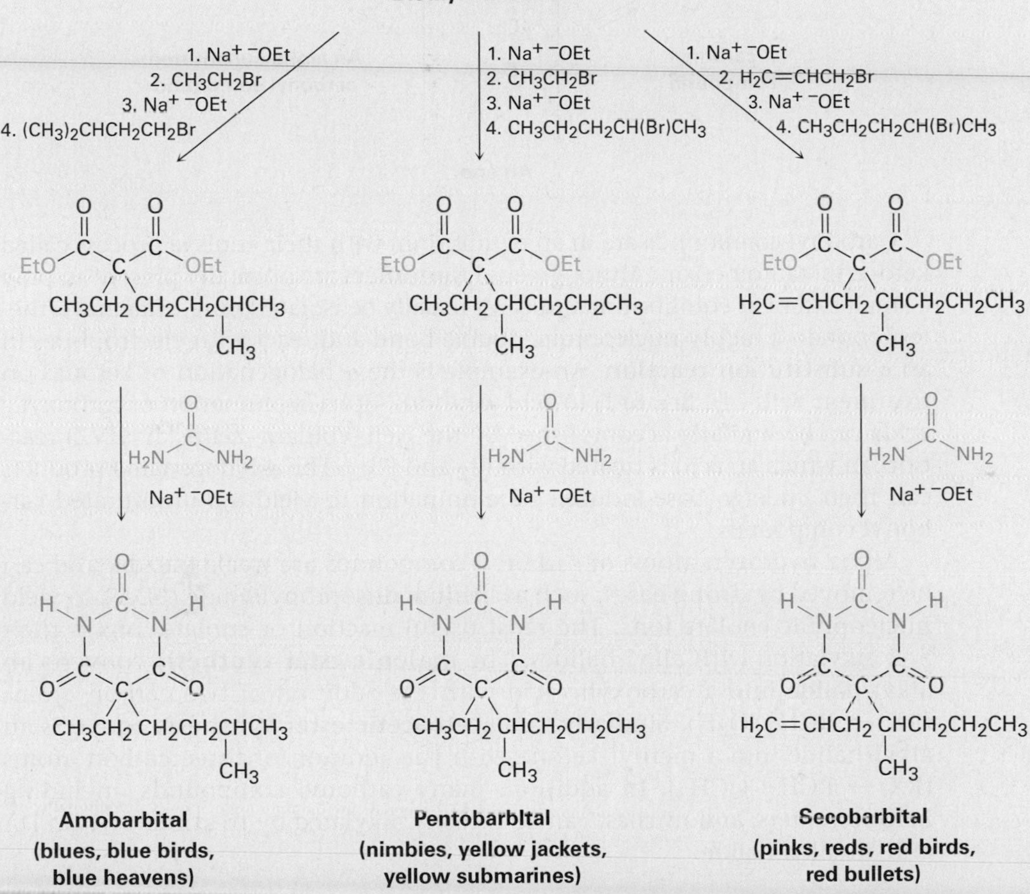

Figure 22.7 The synthesis of barbiturates relies on malonic ester alkylations and nucleophilic acyl substitution reactions. More than 2500 different barbiturates have been synthesized over the past 100 years. In addition to their legal medical uses, some barbiturates are also used illegally as street drugs under many colorful names.

In addition to their prescribed medical uses, many barbiturates have also found widespread illegal use as street drugs. Each barbiturate comes as a tablet of regulated size, shape, and color, and their street names often mimic those colors. Although still used today, most barbiturates have been replaced by safer, more potent alternatives with markedly different structures.

Summary

Key words

acetoacetic ester synthesis, 885
α-substitution reaction, 870
enol, 871
enolate ion, 872
malonic ester synthesis, 883
tautomer, 871

The α-substitution reaction of a carbonyl compound through either an **enol** or **enolate ion** intermediate is one of the four fundamental reaction types in carbonyl-group chemistry.

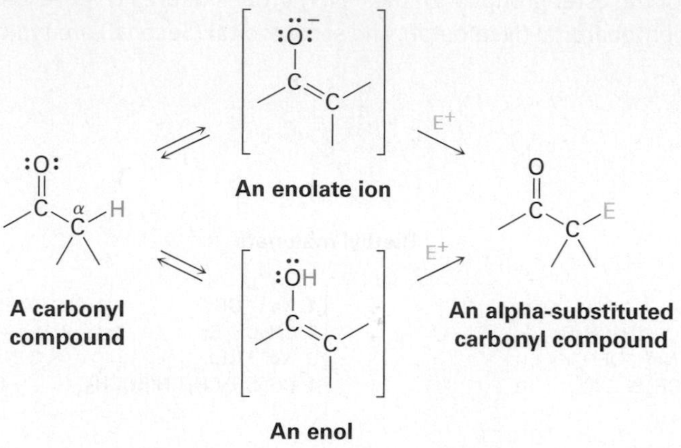

A carbonyl compound

An enolate ion

An enol

An alpha-substituted carbonyl compound

Carbonyl compounds are in an equilibrium with their enols, a process called keto–enol tautomerism. Although enol **tautomers** are normally present to only a small extent at equilibrium and can't usually be isolated pure, they nevertheless contain a highly nucleophilic double bond and react with electrophiles in an ***α*-substitution reaction**. An example is the α halogenation of ketones on treatment with Cl_2, Br_2, or I_2 in acid solution. Alpha bromination of carboxylic acids can be similarly accomplished by the Hell–Volhard–Zelinskii (HVZ) reaction, in which an acid is treated with Br_2 and PBr_3. The α-halogenated products can then undergo base-induced E2 elimination to yield α,β-unsaturated carbonyl compounds.

Alpha hydrogen atoms of carbonyl compounds are weakly acidic and can be removed by strong bases, such as lithium diisopropylamide (LDA), to yield nucleophilic enolate ions. The most useful reaction of enolate ions is their S_N2 alkylation with alkyl halides. The **malonic ester synthesis** converts an alkyl halide into a carboxylic acid with the addition of two carbon atoms ($RX \rightarrow RCH_2CO_2H$). Similarly, the **acetoacetic ester synthesis** converts an alkyl halide into a methyl ketone with the addition of three carbon atoms ($RX \rightarrow RCH_2COCH_3$). In addition, many carbonyl compounds, including ketones, esters, and nitriles, can be directly alkylated by treatment with LDA and an alkyl halide.

Summary of Reactions

1. Aldehyde/ketone halogenation (Section 22.3)

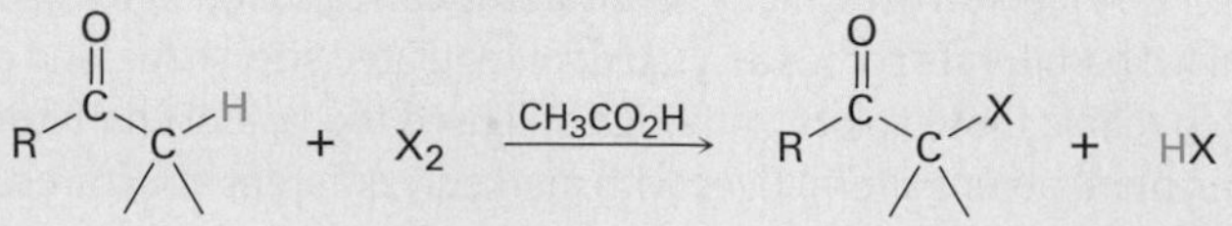

$$RCOCH< + X_2 \xrightarrow{CH_3CO_2H} RCOCX< + HX$$

2. Hell–Volhard–Zelinskii bromination of acids (Section 22.4)

$$HO-C(=O)-C(-)(-)-H \xrightarrow[2.\ H_2O]{1.\ Br_2,\ PBr_3} HO-C(=O)-C(-)(-)-Br$$

3. Dehydrobromination of α-bromo ketones (Section 22.3)

$$R-C(=O)-C(Br)(-)-C(H)< \xrightarrow[\text{Heat}]{\text{Pyridine}} R-C(=O)-C=C<$$

4. Haloform reaction (Section 22.6)

$$R-C(=O)-CH_3 \xrightarrow[\text{NaOH}]{X_2} R-C(=O)-O^- + CHX_3$$

5. Alkylation of enolate ions (Section 22.7)

(a) Malonic ester synthesis

$$EtO_2C-CH_2-CO_2Et \xrightarrow[2.\ RX]{1.\ Na^+\ {}^-OEt,\ \text{ethanol}} EtO_2C-CH(R)-CO_2Et \xrightarrow[\text{Heat}]{H_3O^+} R-CH_2-CO_2H + CO_2 + 2\ EtOH$$

(b) Acetoacetic ester synthesis

$$EtO_2C-CH_2-C(=O)-CH_3 \xrightarrow[2.\ RX]{1.\ Na^+\ {}^-OEt,\ \text{ethanol}} EtO_2C-CH(R)-C(=O)-CH_3 \xrightarrow[\text{Heat}]{H_3O^+} R-CH_2-C(=O)-CH_3 + CO_2 + EtOH$$

(c) Direct alkylation of ketones

$$R-C(=O)-C(-)(-)-H \xrightarrow[2.\ R'X]{1.\ \text{LDA in THF}} R-C(=O)-C(-)(-)-R'$$

(d) Direct alkylation of esters

$$RO-C(=O)-C(-)(-)-H \xrightarrow[2.\ R'X]{1.\ \text{LDA in THF}} RO-C(=O)-C(-)(-)-R'$$

(e) Direct alkylation of nitriles

$$H-C(-)(-)-C\equiv N \xrightarrow[2.\ RX]{1.\ \text{LDA in THF}} R-C(-)(-)-C\equiv N$$

Exercises

OWL Interactive versions of these problems are assignable in OWL for Organic Chemistry.

Visualizing Chemistry

(Problems 22.1–22.16 appear within the chapter.)

22.17 Show the steps in preparing each of the following substances using either a malonic ester synthesis or an acetoacetic ester synthesis:

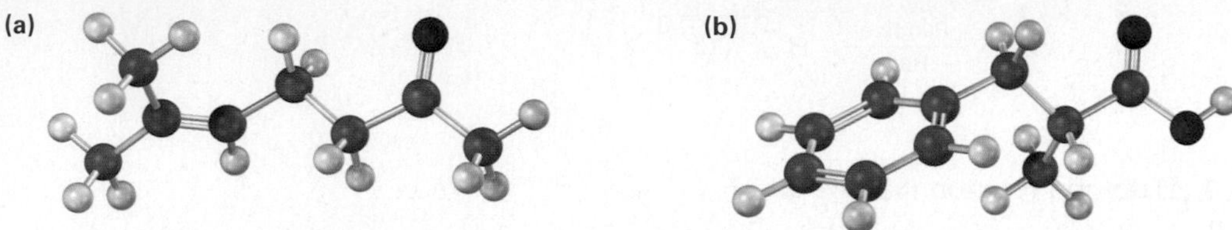

22.18 Unlike most β-diketones, the following β-diketone has no detectable enol content and is about as acidic as acetone. Explain.

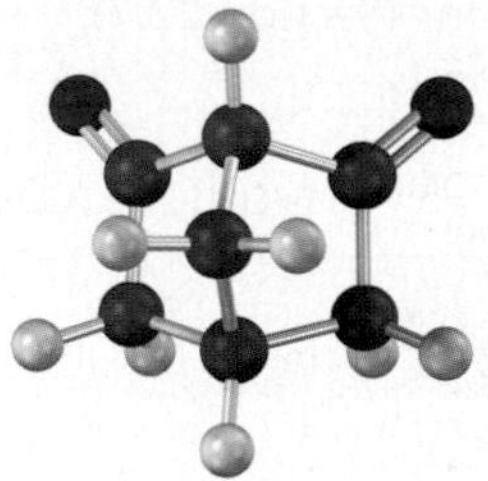

22.19 For a given α hydrogen atom to be acidic, the C–H bond must be parallel to the *p* orbitals of the C=O double bond (that is, perpendicular to the plane of the adjacent carbonyl group). Identify the most acidic hydrogen atom in the conformation shown for the following structure. Is it axial or equatorial?

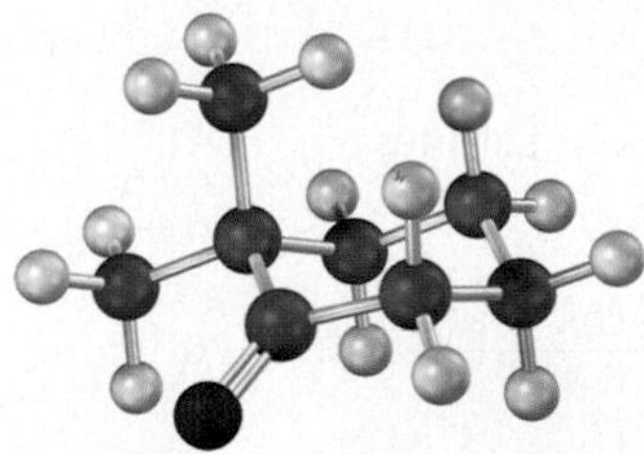

Additional Problems

Acidity of Carbonyl Compounds

22.20 Identify all the acidic hydrogens ($pK_a < 25$) in the following molecules:

(a) $CH_3CH_2CH(CH_3)C(=O)CH_3$ (b) cyclopentane-1,3-dione (c) $HOCH_2CH_2C(=O)C{\equiv}CCH_3$

(d) methyl 2-(cyanomethyl)benzoate (CO_2CH_3, CH_2CN on benzene ring) (e) cyclopentane-COCl (f) $CH_3CH_2C(=O)C(CH_3){=}CH_2$

22.21 Rank the following compounds in order of increasing acidity:

(a) $CH_3CH_2CO_2H$ (b) CH_3CH_2OH (c) $(CH_3CH_2)_2NH$

(d) CH_3COCH_3 (e) $CH_3C(=O)CH_2C(=O)CH_3$ (f) CCl_3CO_2H

22.22 Write resonance structures for the following anions:

(a) $CH_3C(=O)\ddot{C}HC(=O)CH_3$ (b) $CH_3CH{=}CH\ddot{C}HC(=O)CH_3$ (c) $N{\equiv}C\ddot{C}HC(=O)OCH_3$

(d) $C_6H_5\ddot{C}HC(=O)CH_3$ (e) indane-1,3-dione anion bearing $C(=O)OCH_3$ at C-2

22.23 Base treatment of the following α,β-unsaturated carbonyl compound yields an anion by removal of H^+ from the γ carbon. Why are hydrogens on the γ carbon atom acidic?

$$C_6H_5C(=O)CH{=}CHCH_3 \xrightarrow{LDA} C_6H_5C(=O)CH{=}CH\ddot{C}H_2^-$$

22.24 Treatment of 1-phenyl-2-propenone with a strong base such as LDA does not yield an anion, even though it contains a hydrogen on the carbon atom next to the carbonyl group. Explain.

$C_6H_5C(=O)CH{=}CH_2$ 1-Phenyl-2-propenone

α-Substitution Reactions

22.25 Predict the product(s) of the following reactions:

(a) 1,1-Cyclohexanedicarboxylic acid (CO_2H, CO_2H) $\xrightarrow{\text{Heat}}$?

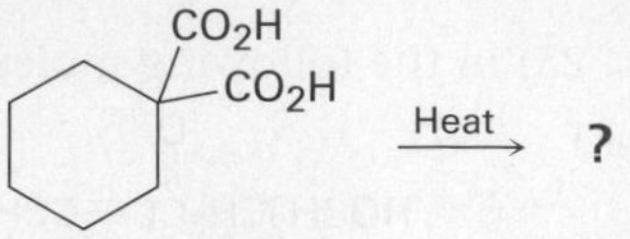

(b) 1,3-Cyclopentanedione (O=, =O) $\xrightarrow[\text{2. } CH_3I]{\text{1. } Na^+ \ ^-OEt}$?

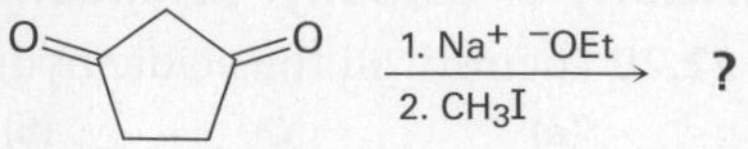

(c) $CH_3CH_2CH_2COH$ (C=O) $\xrightarrow{Br_2,\ PBr_3}$? $\xrightarrow{H_2O}$?

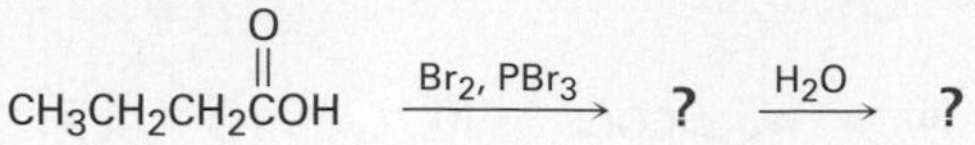

(d) Acetophenone (C(=O)CH_3 on benzene) $\xrightarrow[I_2]{NaOH,\ H_2O}$?

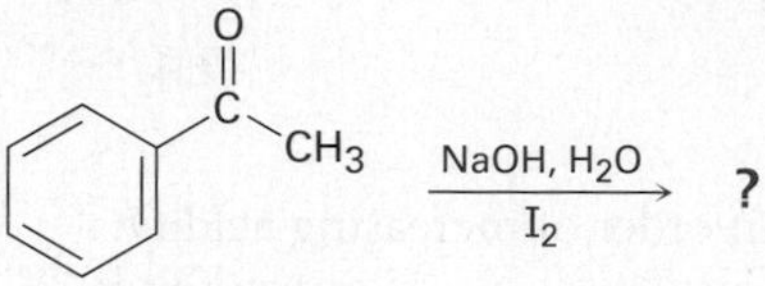

22.26 Which, if any, of the following compounds can be prepared by a malonic ester synthesis? Show the alkyl halide you would use in each case.

(a) Ethyl pentanoate (b) Ethyl 3-methylbutanoate
(c) Ethyl 2-methylbutanoate (d) Ethyl 2,2-dimethylpropanoate

22.27 Which, if any, of the following compounds can be prepared by an acetoacetic ester synthesis? Explain.

(a) Br–C_6H_4–$CH_2C(=O)CH_3$ (meta-bromophenyl) (b) cyclohexyl–$C(=O)CH_3$ (c) CH_3–C(CH_3)(CH_3)–CH_2CCH_3 (C=O)

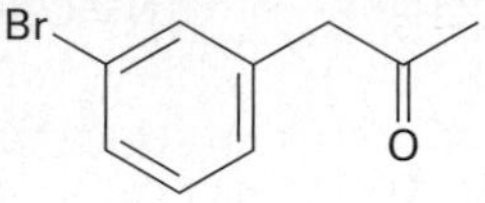

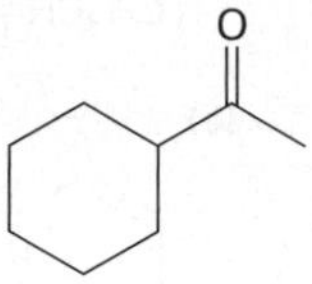

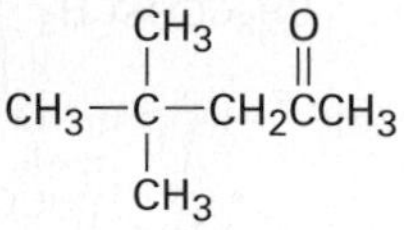

22.28 How would you prepare the following ketones using an acetoacetic ester synthesis?

(a) $CH_3CH_2CHCCH_3$ (C=O; CH_2CH_3 on CH) (b) $CH_3CH_2CH_2CHCCH_3$ (C=O; CH_3 on CH)

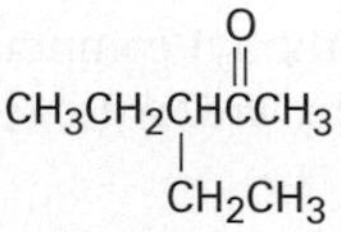

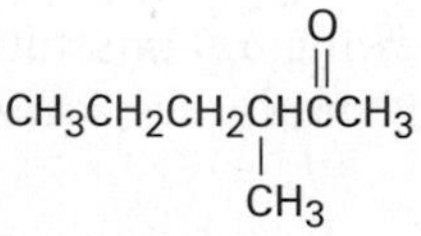

22.29 How would you prepare the following compounds using either an acetoacetic ester synthesis or a malonic ester synthesis?

(a) CH_3CCO_2Et (CH_3 above, CO_2Et below) (b) cycloheptyl–C(=O)CH_3 (c) cyclobutyl–C(=O)OH (d) $H_2C{=}CHCH_2CH_2CCH_3$ (C=O)

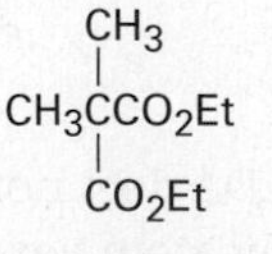

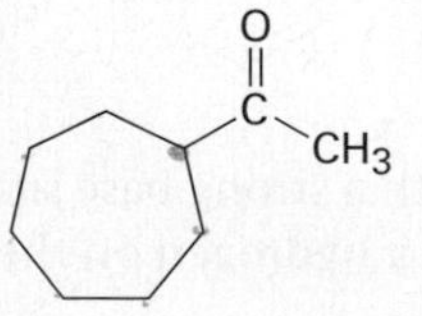

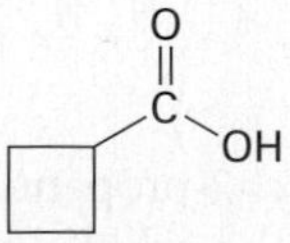

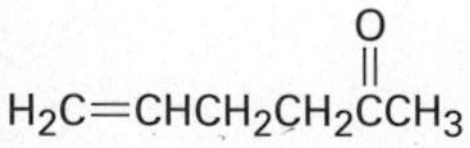

22.30 Which of the following substances would undergo the haloform reaction?

(a) CH_3COCH_3 (b) Acetophenone (c) CH_3CH_2CHO
(d) CH_3CO_2H (e) $CH_3C{\equiv}N$

22.31 How might you convert geraniol into either ethyl geranylacetate or geranylacetone?

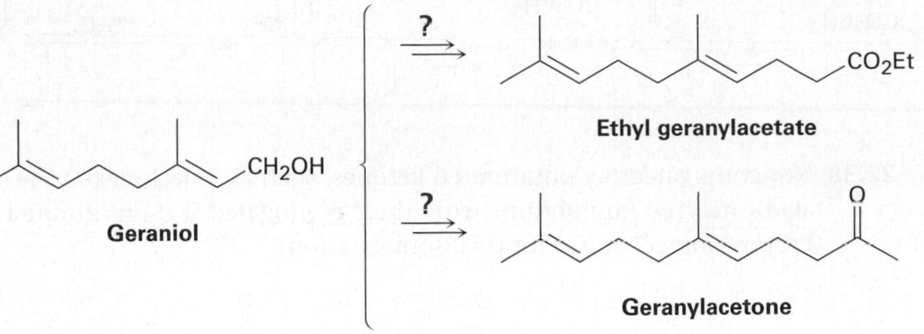

Geraniol

Ethyl geranylacetate

Geranylacetone

22.32 Aprobarbital, a barbiturate once used in treating insomnia, is synthesized in three steps from diethyl malonate. Show how you would synthesize the necessary dialkylated intermediate, and then propose a mechanism for the reaction of that intermediate with urea to give aprobarbital.

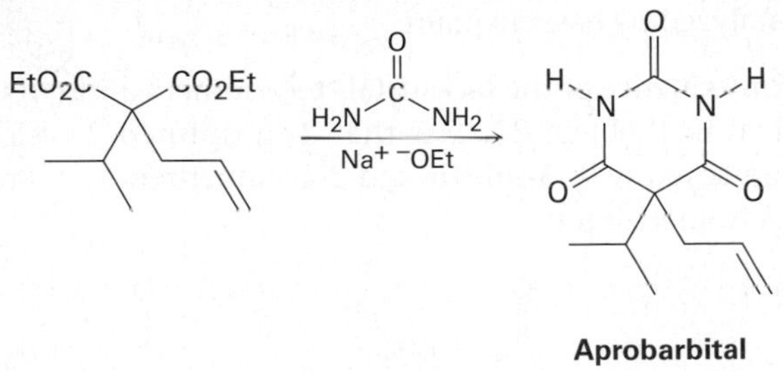

Aprobarbital

General Problems

22.33 One way to determine the number of acidic hydrogens in a molecule is to treat the compound with NaOD in D_2O, isolate the product, and determine its molecular weight by mass spectrometry. For example, if cyclohexanone is treated with NaOD in D_2O, the product has MW = 102. Explain how this method works.

22.34 When optically active (*R*)-2-methylcyclohexanone is treated with either aqueous base or acid, racemization occurs. Explain.

22.35 Would you expect optically active (*S*)-3-methylcyclohexanone to be racemized on acid or base treatment in the same way as 2-methylcyclohexanone (Problem 22.34)? Explain.

22.36 When an optically active carboxylic acid such as (*R*)-2-phenylpropanoic acid is brominated under Hell–Volhard–Zelinskii conditions, is the product optically active or racemic? Explain.

22.37 Fill in the reagents a–c that are missing from the following scheme:

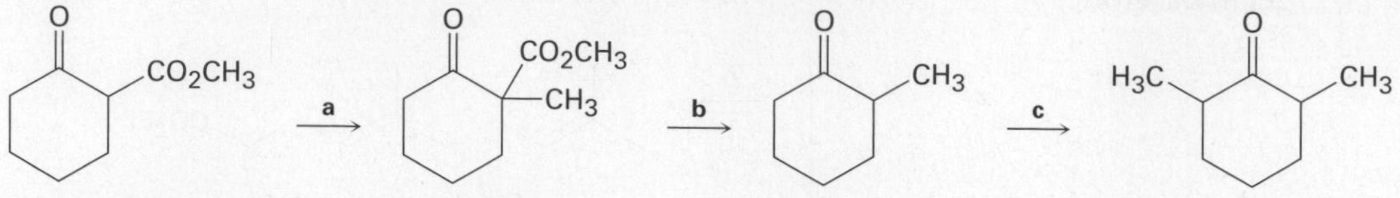

22.38 Nonconjugated β,γ-unsaturated ketones, such as 3-cyclohexenone, are in an acid-catalyzed equilibrium with their conjugated α,β-unsaturated isomers. Propose a mechanism for this isomerization.

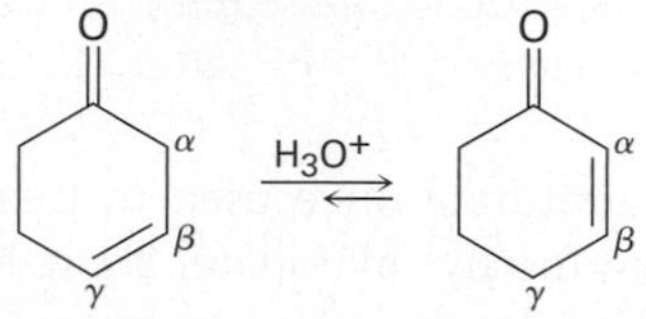

22.39 The interconversion of unsaturated ketones described in Problem 22.38 is also catalyzed by base. Explain.

22.40 One consequence of the base-catalyzed isomerization of unsaturated ketones described in Problem 22.39 is that 2-substituted 2-cyclopentenones can be interconverted with 5-substituted 2-cyclopentenones. Propose a mechanism for this isomerization.

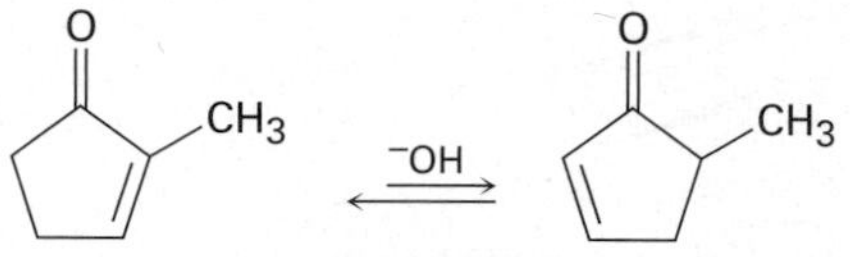

22.41 Although 2-substituted 2-cyclopentenones are in a base-catalyzed equilibrium with their 5-substituted 2-cyclopentenone isomers (Problem 22.40), the analogous isomerization is not observed for 2-substituted 2-cyclohexenones. Explain.

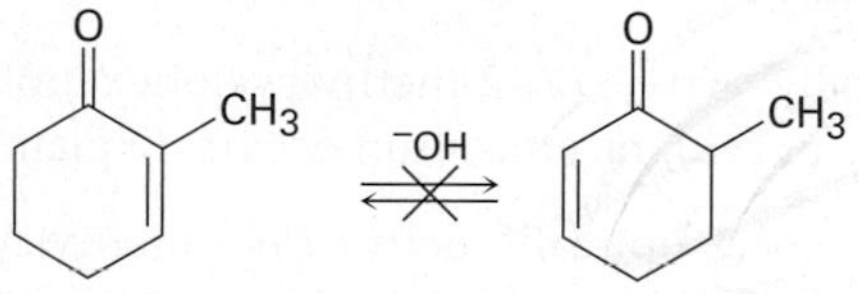

22.42 Using curved arrows, propose a mechanism for the following reaction, one of the steps in the metabolism of the amino acid alanine.

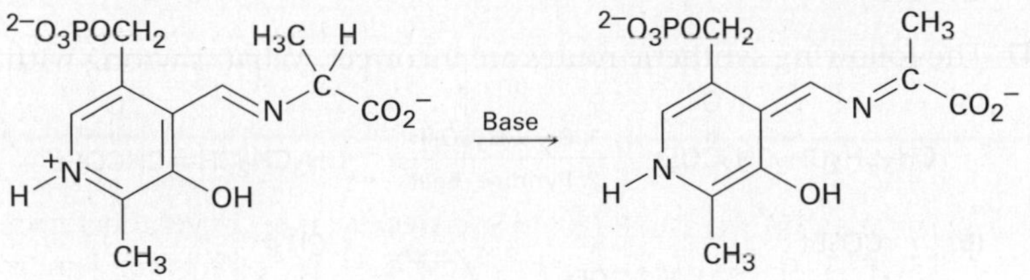

$^{2-}O_3POCH_2$ H$_3$C H C N CO_2^- +N H OH CH_3 —Base→ $^{2-}O_3POCH_2$ CH_3 N=C CO_2^- N H OH CH_3

22.43 Using curved arrows, propose a mechanism for the following reaction, one of the steps in the biosynthesis of the amino acid tyrosine.

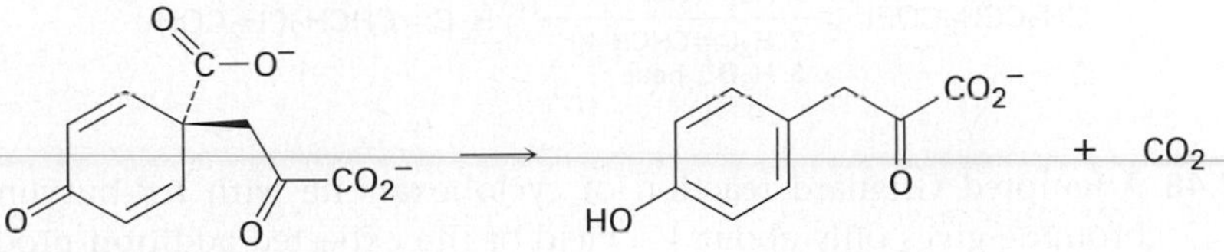

O C—O$^-$ CO_2^- O O → CO_2^- O HO + CO_2

22.44 All attempts to isolate primary and secondary nitroso compounds result only in the formation of oximes. Tertiary nitroso compounds, however, are stable. Explain.

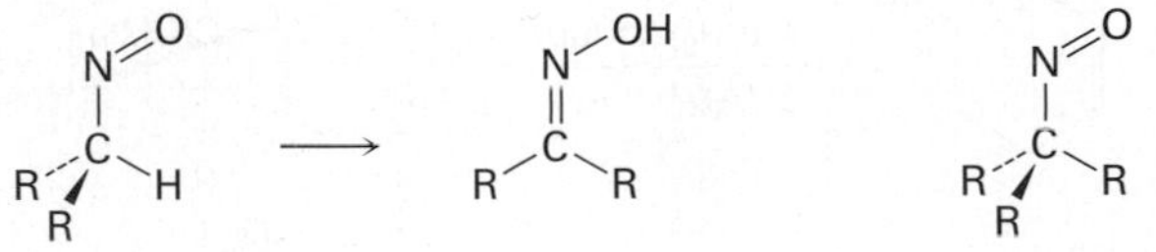

A 1° or 2° nitroso compound (unstable) → **An oxime** **A 3° nitroso compound (stable)**

22.45 How would you synthesize the following compounds from cyclohexanone? More than one step may be required.

(a) CH_2

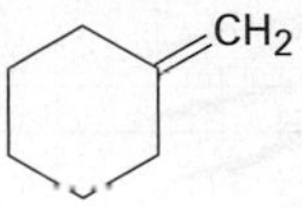

(b) CH_2Br

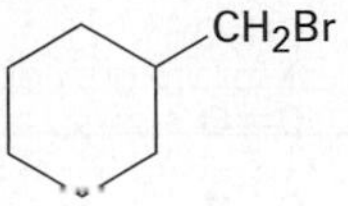

(c)

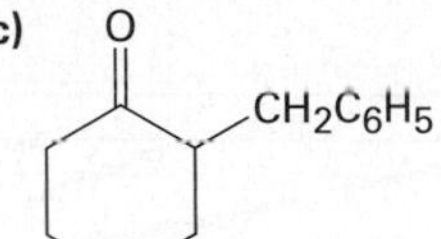

(d) $CH_2CH_2CO_2H$

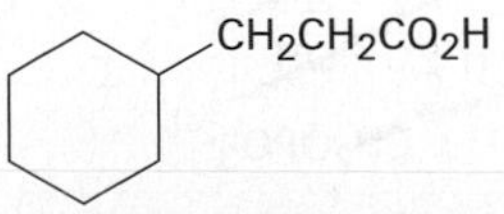

(e) CO_2H

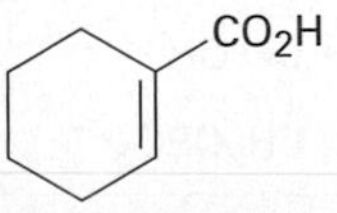

(f) O

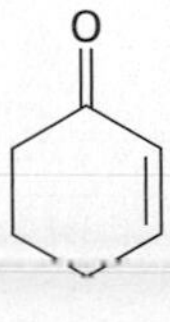

22.46 The two isomers *cis*- and *trans*-4-*tert*-butyl-2-methylcyclohexanone are interconverted by base treatment. Which isomer do you think is more stable, and why?

22.47 The following synthetic routes are incorrect. What is wrong with each?

(a) $CH_3CH_2CH_2CH_2COEt$ (C=O) $\xrightarrow[\text{2. Pyridine, heat}]{\text{1. } Br_2,\ CH_3CO_2H}$ $CH_3CH_2CH{=}CHCOEt$ (C=O)

(b) $CH_3CH(CO_2Et)CO_2Et$ $\xrightarrow[\text{2. PhBr; 3. } H_3O^+\text{, heat}]{\text{1. } Na^+\ ^-OEt}$ $C_6H_5CH(CH_3)CO_2H$

(c) CH_3COCH_2COEt (two C=O) $\xrightarrow[\text{2. } H_2C{=}CHCH_2Br\text{; 3. } H_3O^+\text{, heat}]{\text{1. } Na^+\ ^-OEt}$ $H_2C{=}CHCH_2CH_2COH$ (C=O)

22.48 Attempted Grignard reaction of cyclohexanone with *tert*-butylmagnesium bromide gives only about 1% yield of the expected addition product along with 99% unreacted cyclohexanone. If D_3O^+ is added to the reaction mixture after a suitable period, however, the "unreacted" cyclohexanone is found to have one deuterium atom incorporated into it. Explain.

Cyclohexanone $\xrightarrow[\text{2. } D_3O^+]{\text{1. } (CH_3)_3CMgBr}$ 1-*tert*-butylcyclohexan-1-ol-*O*-d (OD, $C(CH_3)_3$) **1%** + 2-deuteriocyclohexanone **99%**

22.49 One of the later steps in glucose biosynthesis is the isomerization of fructose 6-phosphate to glucose 6-phosphate. Propose a mechanism, using acid or base catalysis as needed.

Fructose 6-phosphate		Glucose 6-phosphate
CH_2OH		$O{=}C{-}H$
$C{=}O$		$H{-}C{-}OH$
$HO{-}C{-}H$	$\longrightarrow$	$HO{-}C{-}H$
$H{-}C{-}OH$		$H{-}C{-}OH$
$H{-}C{-}OH$		$H{-}C{-}OH$
$CH_2OPO_3^{2-}$		$CH_2OPO_3^{2-}$

Fructose 6-phosphate → **Glucose 6-phosphate**

22.50 The *Favorskii reaction* involves treatment of an α-bromo ketone with base to yield a ring-contracted product. For example, reaction of 2-bromocyclohexanone with aqueous NaOH yields cyclopentanecarboxylic acid. Propose a mechanism.

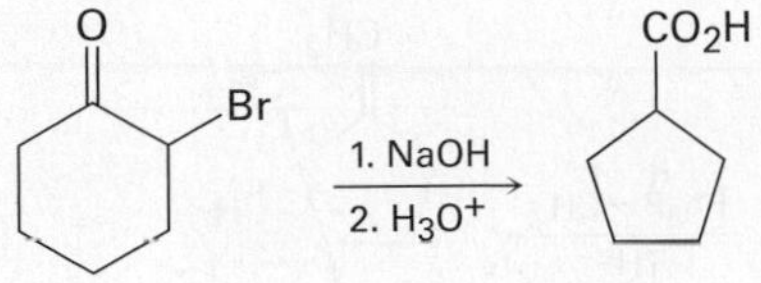

1. NaOH
2. H_3O^+

22.51 Treatment of a cyclic ketone with diazomethane is a method for accomplishing a *ring-expansion reaction*. For example, treatment of cyclohexanone with diazomethane yields cycloheptanone. Propose a mechanism.

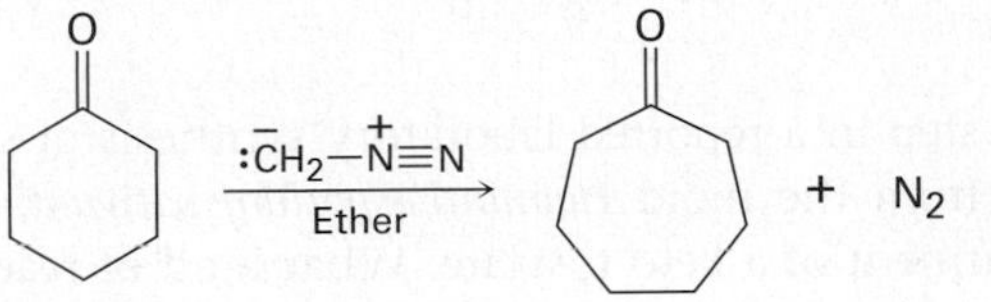

$:\overset{-}{C}H_2-\overset{+}{N}\equiv N$, Ether; + N_2

22.52 Ketones react slowly with benzeneselenenyl chloride in the presence of HCl to yield α-phenylseleno ketones. Propose a mechanism for this acid-catalyzed α-substitution reaction.

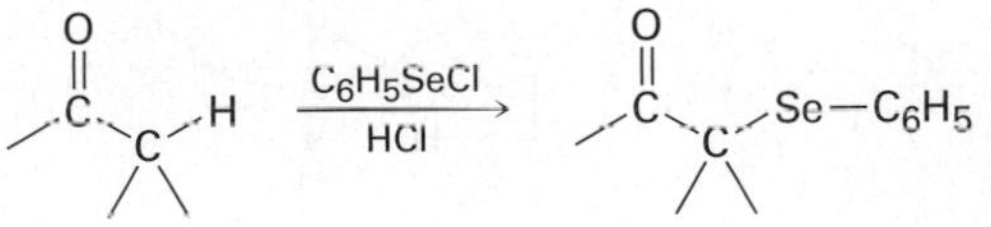

C_6H_5SeCl, HCl; Se$-C_6H_5$

22.53 As far back as the 16th century, South American Incas chewed the leaves of the coca bush, *Erythroxylon coca*, to combat fatigue. Chemical studies of *Erythroxylon coca* by Friedrich Wöhler in 1862 resulted in the discovery of *cocaine*, $C_{17}H_{21}NO_4$, as the active component. Basic hydrolysis of cocaine leads to methanol, benzoic acid, and another compound called *ecgonine*, $C_9H_{15}NO_3$. Oxidation of ecgonine with CrO_3 yields a keto acid that readily loses CO_2 on heating, giving tropinone.

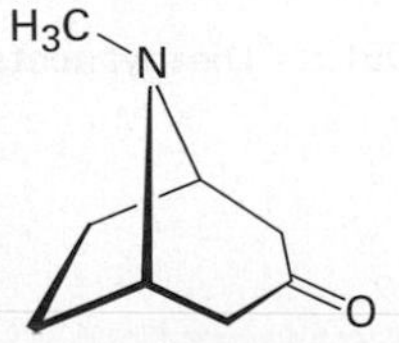

Tropinone

(a) What is a likely structure for the keto acid?
(b) What is a likely structure for ecgonine, neglecting stereochemistry?
(c) What is a likely structure for cocaine, neglecting stereochemistry?

22.54 The final step in an attempted synthesis of laurene, a hydrocarbon isolated from the marine alga *Laurencia glandulifera,* involved the Wittig reaction shown. The product obtained, however, was not laurene but an isomer. Propose a mechanism to account for these unexpected results.

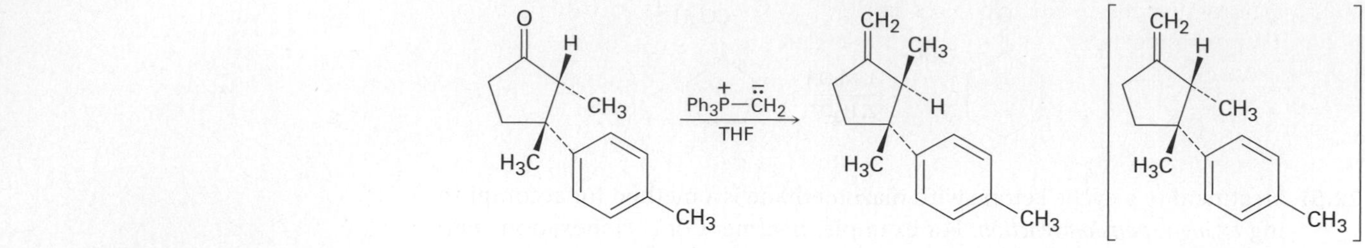

$Ph_3\overset{+}{P}-\ddot{C}H_2$, THF

Laurene
(Not formed)

22.55 The key step in a reported laboratory synthesis of sativene, a hydrocarbon isolated from the mold *Helminthosporium sativum,* involves the following base treatment of a keto tosylate. What kind of reaction is occurring? How would you complete the synthesis?

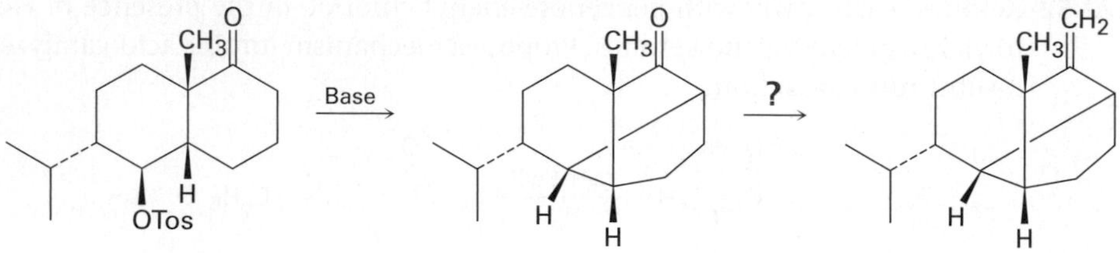

Base → ?

A keto tosylate **Sativene**

22.56 Amino acids can be prepared by reaction of alkyl halides with diethyl acetamidomalonate, followed by heating the initial alkylation product with aqueous HCl. Show how you would prepare alanine, $CH_3CH(NH_2)CO_2H$, one of the twenty amino acids found in proteins, and propose a mechanism for acid-catalyzed conversion of the initial alkylation product to the amino acid.

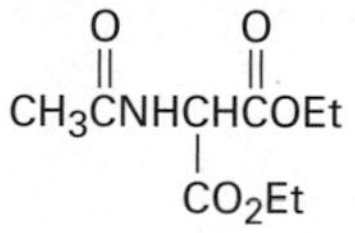

$CH_3\overset{O}{\overset{\|}{C}}NH\underset{CO_2Et}{\underset{|}{C}H}\overset{O}{\overset{\|}{C}}OEt$ **Diethyl acetamidomalonate**

22.57 Amino acids can also be prepared by a two-step sequence that involves Hell–Volhard–Zelinskii reaction of a carboxylic acid followed by treatment with ammonia. Show how you would prepare leucine, $(CH_3)_2CHCH_2CH(NH_2)CO_2H$, and identify the mechanism of the second step.

22.58 Heating carvone with aqueous sulfuric acid converts it into carvacrol. Propose a mechanism for the isomerization.

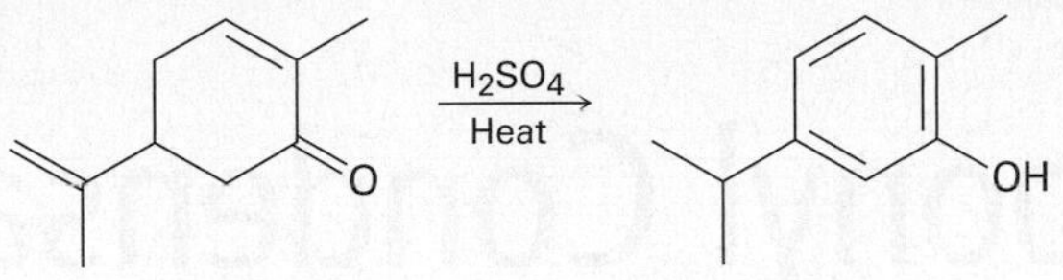

H_2SO_4
Heat

OH

Carvone **Carvacrol**

22.59 Sodium Pentothal is a short-acting barbiturate derivative used as a general anesthetic and known in popular culture as a truth serum. It is synthesized like other barbiturates (see the *A Deeper Look* at the end of this chapter), using thiourea, $(H_2N)_2C{=}S$, in place of urea. How would you synthesize Sodium Pentothal?

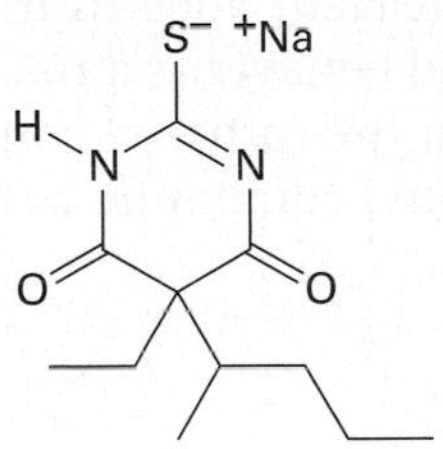

S^- ^+Na
H N N
O O

Sodium Pentothal

23

Many of life's molecules needed by all growing organisms are biosynthesized using carbonyl condensation reactions. © Picturebank/Alamy

Carbonyl Condensation Reactions

We've now studied three of the four general kinds of carbonyl-group reactions and have seen two general kinds of behavior. In nucleophilic addition and nucleophilic acyl substitution reactions, a carbonyl compound behaves as an electrophile when an electron-rich reagent adds to it. In α-substitution reactions, however, a carbonyl compound behaves as a nucleophile when it is converted into its enol or enolate ion. In the carbonyl condensation reaction that we'll study in this chapter, the carbonyl compound behaves *both* as an electrophile and as a nucleophile.

Electrophilic carbonyl group reacts with nucleophiles.

Nucleophilic enolate ion reacts with electrophiles.

Why This Chapter? We'll see later in this chapter and again in Chapter 29 that carbonyl condensation reactions occur in a large number of metabolic pathways. In fact, almost all classes of biomolecules—carbohydrates, lipids, proteins, nucleic acids, and many others—are biosynthesized through pathways that involve carbonyl condensation reactions. As with the α-substitution reaction discussed in the previous chapter, the great value of carbonyl condensations is that they are one of the few general methods for forming carbon–carbon bonds, thereby making it possible to build larger molecules from smaller precursors. We'll see how and why these reactions occur in this chapter.

23.1 Carbonyl Condensations: The Aldol Reaction

Carbonyl condensation reactions take place between two carbonyl partners and involve a combination of nucleophilic addition and α-substitution steps. One partner is converted into an enolate-ion nucleophile and adds to the electrophilic carbonyl group of the second partner. In so doing, the nucleophilic partner undergoes an α-substitution reaction and the electrophilic partner undergoes a nucleophilic addition. The general mechanism of the process is shown in **Figure 23.1**.

Figure 23.1 | MECHANISM

The general mechanism of a carbonyl condensation reaction. One partner becomes a nucleophilic donor and adds to the second partner as an electrophilic acceptor. After protonation, the final product is a β-hydroxy carbonyl compound.

1 A carbonyl compound with an α hydrogen atom is converted by base into its enolate ion.

2 The enolate ion acts as a nucleophilic donor and adds to the electrophilic carbonyl group of a second carbonyl compound.

3 Protonation of the tetrahedral alkoxide ion intermediate gives the neutral condensation product and regenerates the base catalyst.

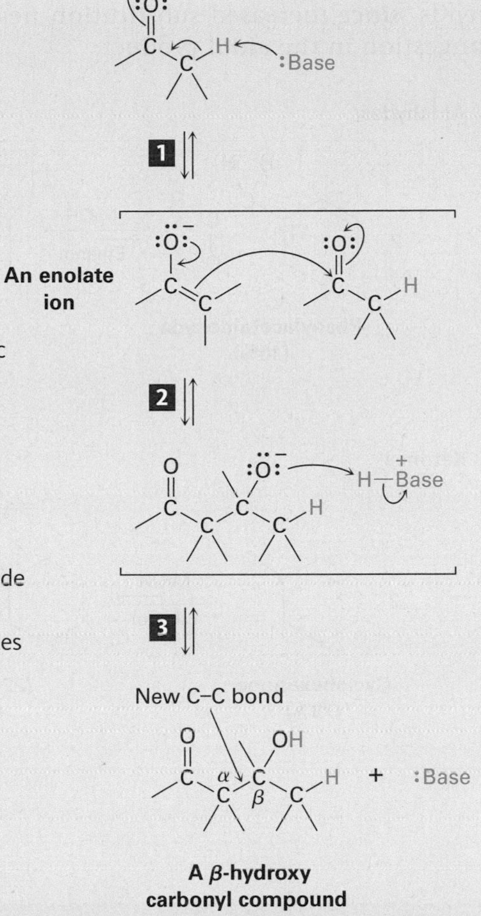

Aldehydes and ketones with an α hydrogen atom undergo a base-catalyzed carbonyl condensation reaction called the **aldol reaction**. For example, treatment of acetaldehyde with a base such as sodium ethoxide or sodium hydroxide in a protic solvent leads to rapid and reversible formation of 3-hydroxybutanal, known commonly as *aldol* (*ald*ehyde + alcoh*ol*), hence the general name of the reaction.

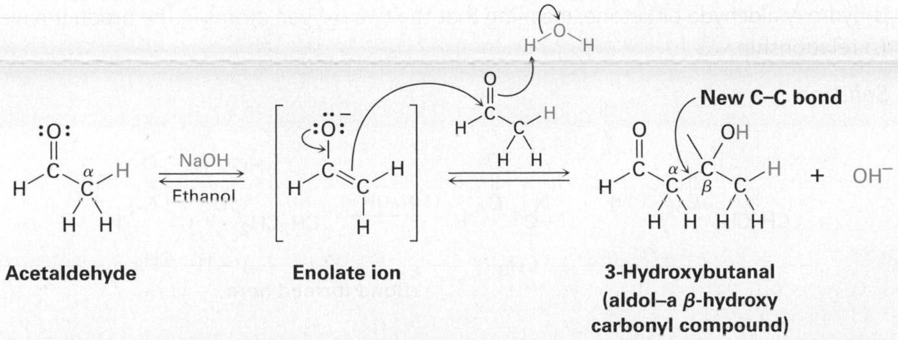

The exact position of the aldol equilibrium depends both on reaction conditions and on substrate structure. The equilibrium generally favors condensation product in the case of aldehydes with no α substituent (RCH_2CHO) but favors reactant for disubstituted aldehydes (R_2CHCHO) and for most ketones. Steric factors are probably responsible for these trends, since increased substitution near the reaction site increases steric congestion in the aldol product.

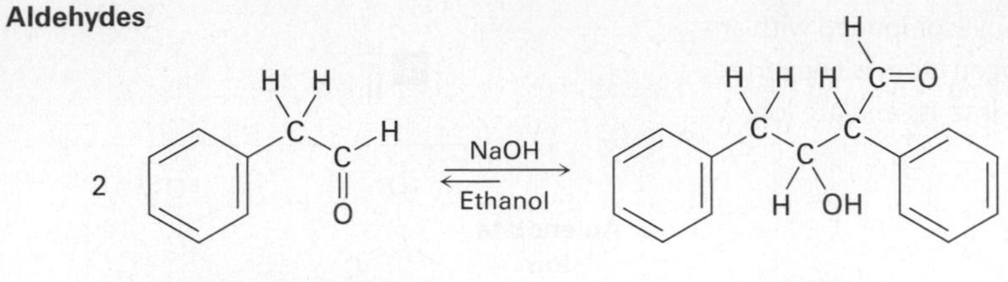

Aldehydes

2 Phenylacetaldehyde $\underset{\text{Ethanol}}{\overset{\text{NaOH}}{\rightleftharpoons}}$ aldol product

Phenylacetaldehyde (10%) — **(90%)**

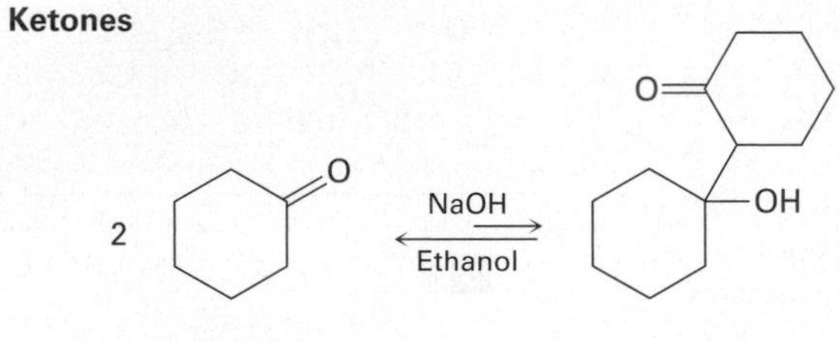

Ketones

2 Cyclohexanone $\underset{\text{Ethanol}}{\overset{\text{NaOH}}{\rightleftharpoons}}$ aldol product (OH)

Cyclohexanone (78%) — **(22%)**

Worked Example 23.1 Predicting the Product of an Aldol Reaction

What is the structure of the aldol product from propanal?

Strategy

An aldol reaction combines two molecules of reactant by forming a bond between the α carbon of one partner and the carbonyl carbon of the second partner. The product is a β-hydroxy aldehyde or ketone, meaning that the two oxygen atoms in the product have a 1,3 relationship.

Solution

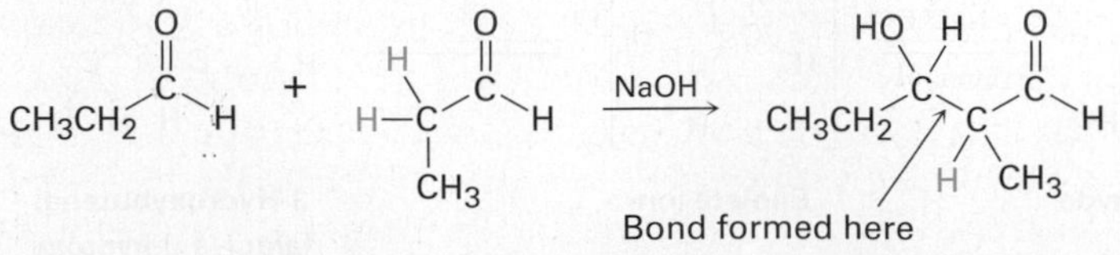

$$CH_3CH_2CHO + CH_3CH_2CHO \xrightarrow{\text{NaOH}} CH_3CH_2CH(OH)CH(CH_3)CHO$$

Bond formed here

Problem 23.1

Predict the aldol reaction product of the following compounds:

(a) $CH_3CH_2CH_2CHO$ (b) $C_6H_5COCH_3$ (c) cyclopentanone

Problem 23.2

Using curved arrows to indicate the electron flow in each step, show how the base-catalyzed retro-aldol reaction of 4-hydroxy-4-methyl-2-pentanone takes place to yield 2 equivalents of acetone.

23.2 Carbonyl Condensations versus Alpha Substitutions

Two of the four general carbonyl-group reactions—carbonyl condensations and α substitutions—take place under basic conditions and involve enolate-ion intermediates. Because the experimental conditions for the two reactions are similar, how can we predict which will occur in a given case? When we generate an enolate ion with the intention of carrying out an α alkylation, how can we be sure that a carbonyl condensation reaction won't occur instead?

There is no simple answer to this question, but the exact experimental conditions usually have much to do with the result. Alpha-substitution reactions require a full equivalent of strong base and are normally carried out so that the carbonyl compound is rapidly and completely converted into its enolate ion at a low temperature. An electrophile is then added rapidly to ensure that the reactive enolate ion is quenched quickly. In a ketone alkylation reaction, for instance, we might use 1 equivalent of lithium diisopropylamide (LDA) in tetrahydrofuran solution at −78 °C. Rapid and complete generation of the ketone enolate ion would occur, and no unreacted ketone would be left so that no condensation reaction could take place. We would then immediately add an alkyl halide to complete the alkylation reaction.

Cyclohexanone $\xrightarrow[\text{THF, −78 °C}]{\text{1 equiv LDA}}$ [lithium enolate, $O^-\ Li^+$] $\xrightarrow{\text{Add } CH_3I}$ 2-methylcyclohexanone

On the other hand, carbonyl condensation reactions require only a catalytic amount of a relatively weak base rather than a full equivalent so that a small amount of enolate ion is generated in the presence of unreacted carbonyl compound. Once a condensation has occurred, the basic catalyst is regenerated. To carry out an aldol reaction on propanal, for instance, we might dissolve the aldehyde in methanol, add 0.05 equivalent of sodium methoxide, and then warm the mixture to give the aldol product.

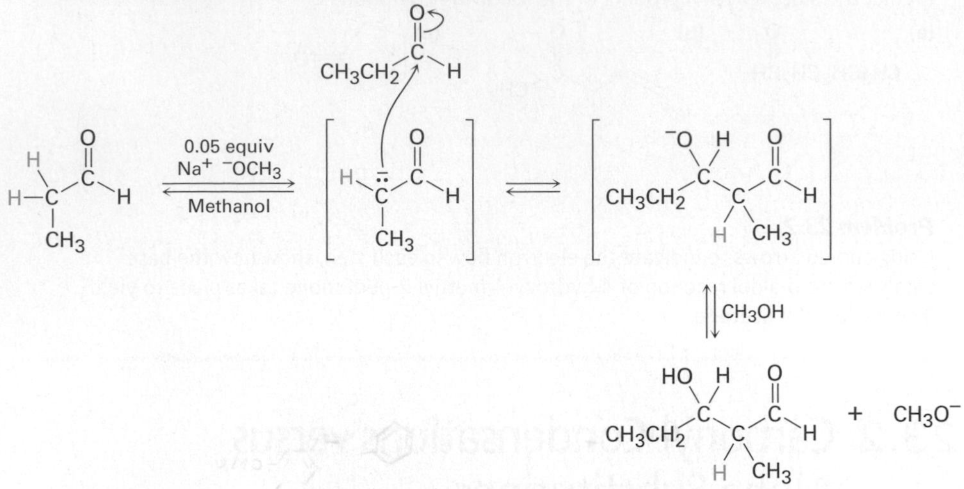

0.05 equiv $Na^+\ ^-OCH_3$

Methanol

CH_3CH_2

CH_3OH

$+\ CH_3O^-$

23.3 Dehydration of Aldol Products: Synthesis of Enones

The β-hydroxy aldehydes or ketones formed in aldol reactions can be easily dehydrated to yield α,β-unsaturated products, or conjugated enones. In fact, it's this loss of water that gives the *condensation* reaction its name, because water condenses out of the reaction when the enone product forms.

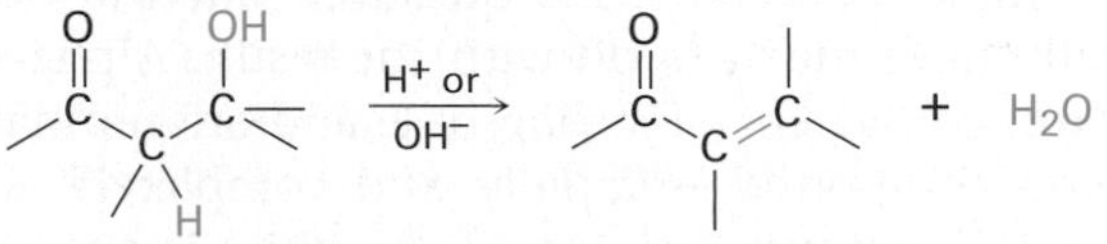

H^+ or OH^- $+\ H_2O$

A β-hydroxy ketone or aldehyde **A conjugated enone**

Most alcohols are resistant to dehydration by base **(Section 17.6)** because hydroxide ion is a poor leaving group, but aldol products dehydrate easily because of the carbonyl group. Under basic conditions, an acidic α hydrogen is removed, yielding an enolate ion that expels the ^-OH leaving group in an E1cB reaction **(Section 11.10)**. Under acidic conditions, an enol is formed, the –OH group is protonated, and water is expelled in an E1 or E2 reaction.

Base-catalyzed

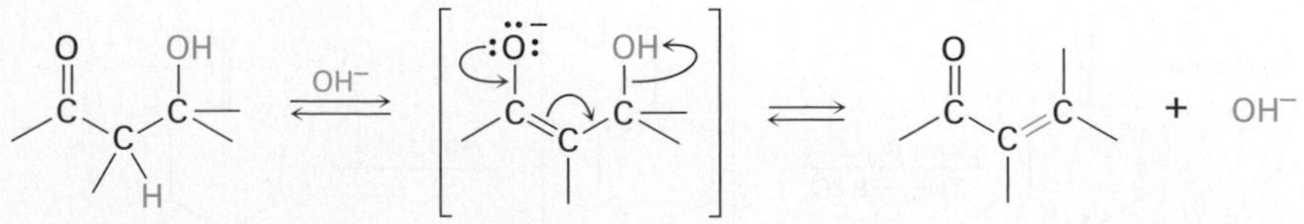

OH^- $+\ OH^-$

Enolate ion

Acid-catalyzed

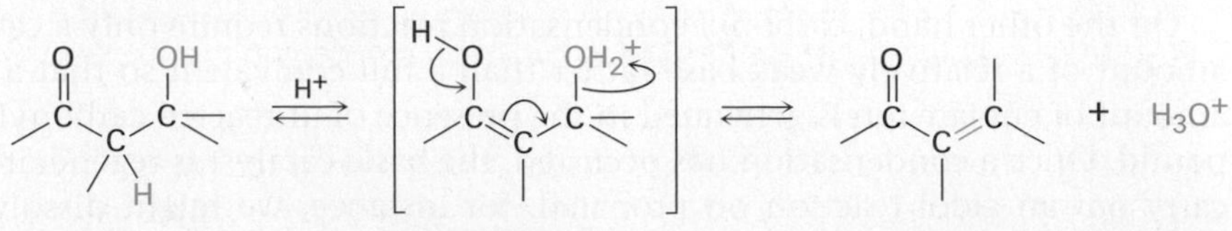

H^+ $+\ H_3O^+$

Enol

The reaction conditions needed for aldol dehydration are often only a bit more vigorous (slightly higher temperature, for instance) than the conditions needed for the aldol formation itself. As a result, conjugated enones are usually obtained directly from aldol reactions without isolating the intermediate β-hydroxy carbonyl compounds.

Conjugated enones are more stable than nonconjugated enones for the same reason that conjugated dienes are more stable than nonconjugated dienes **(Section 14.1)**. Interaction between the π electrons of the C=C bond and the π electrons of the C=O group leads to a molecular orbital description for a conjugated enone that shows an interaction of the π electrons over all four atomic centers **(Figure 23.2)**.

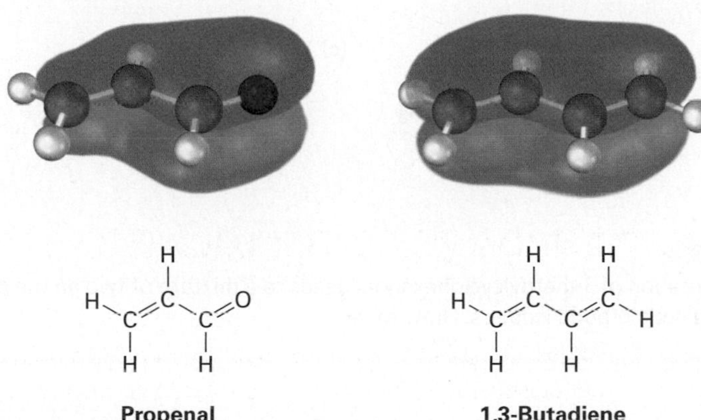

Figure 23.2 The π bonding molecular orbitals of a conjugated enone (propenal) and a conjugated diene (1,3-butadiene) are similar in shape and are spread over the entire π system.

The real value of aldol dehydration is that removal of water from the reaction mixture can be used to drive the aldol equilibrium toward product. Even though the initial aldol step itself may be unfavorable, as it usually is for ketones, the subsequent dehydration step nevertheless allows many aldol condensations to be carried out in good yield. Cyclohexanone, for example, gives cyclohexylidenecyclohexanone in 92% yield even though the initial equilibrium is unfavorable.

NaOH, Ethanol; + H_2O

Cyclohexanone → **Cyclohexylidenecyclohexanone (92%)**

Predicting the Product of an Aldol Reaction

Worked Example 23.2

What is the structure of the enone obtained from aldol condensation of acetaldehyde?

Strategy

In the aldol reaction, H_2O is eliminated and a double bond is formed by removing two hydrogens from the acidic α position of one partner and the carbonyl oxygen from the second partner. The product is thus an α,β-unsaturated aldehyde or ketone.

Solution

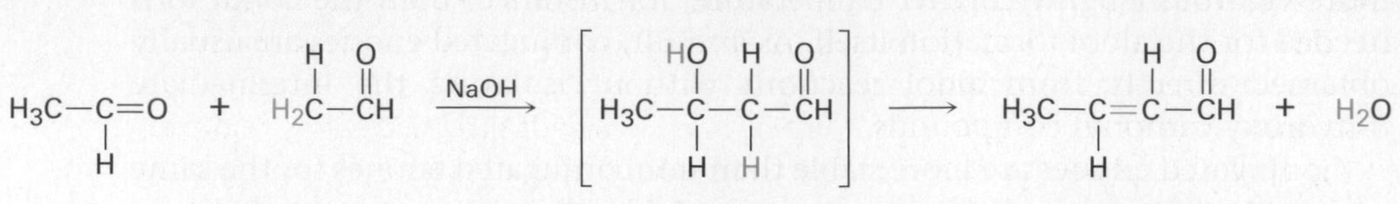

$$H_3C{-}\underset{H}{C}{=}O \; + \; H_2C{-}\overset{O}{\overset{\|}{C}}H \xrightarrow{\text{NaOH}} \left[H_3C{-}\overset{HO}{\underset{H}{C}}{-}\overset{H}{\underset{H}{C}}{-}\overset{O}{\overset{\|}{C}}H \right] \longrightarrow H_3C{-}\underset{H}{C}{=}\overset{H}{C}{-}\overset{O}{\overset{\|}{C}}H \; + \; H_2O$$

2-Butenal

Problem 23.3

What enone product would you expect from aldol condensation of each of the following compounds?

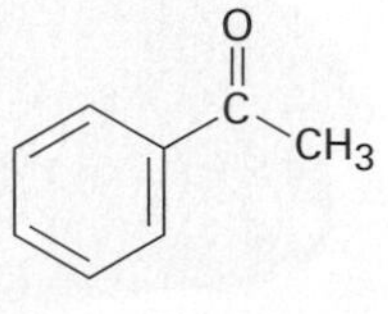

(a) Cyclopentanone **(b)** $C_6H_5COCH_3$ **(c)** $CH_3CH(CH_3)CH_2CHO$

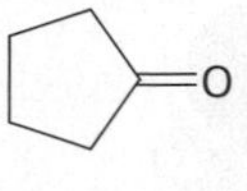

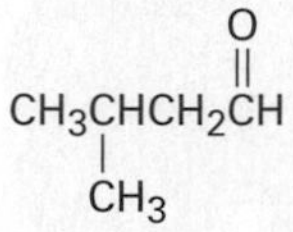

Problem 23.4

Aldol condensation of 3-methylcyclohexanone leads to a mixture of two enone products, not counting double-bond isomers. Draw them.

23.4 Using Aldol Reactions in Synthesis

The aldol reaction yields either a β-hydroxy aldehyde/ketone or an α,β-unsaturated aldehyde/ketone, depending on the experimental conditions. By learning how to think backward, it's possible to predict when the aldol reaction might be useful in synthesis. Whenever the target molecule contains either a β-hydroxy aldehyde/ketone or a conjugated enone functional group, it might come from an aldol reaction.

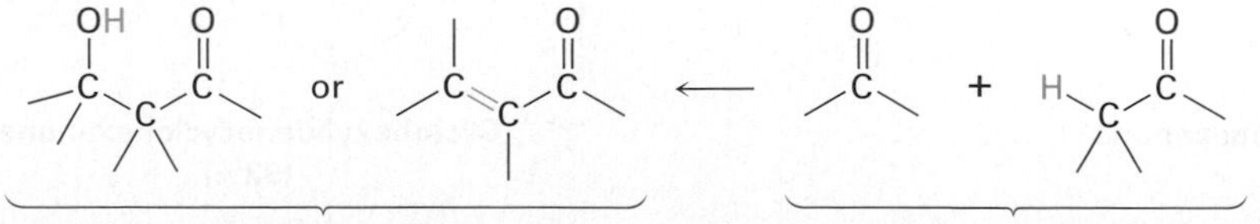

$$\underbrace{\text{β-hydroxy carbonyl} \quad \text{or} \quad \text{α,β-unsaturated carbonyl}}_{\textbf{Aldol products}} \longleftarrow \underbrace{C{=}O \; + \; H{-}C{-}C{=}O}_{\textbf{Aldol reactants}}$$

We can extend this kind of reasoning even further by imagining that subsequent transformations might be carried out on the aldol products. For example, a saturated ketone might be prepared by catalytic hydrogenation of the enone product. A good example can be found in the industrial preparation of 2-ethyl-1-hexanol, an alcohol used in the synthesis of plasticizers for polymers. Although 2-ethyl-1-hexanol bears little resemblance to an aldol product at first glance, it is in fact prepared commercially from butanal by an aldol reaction. Working backward, we can reason that 2-ethyl-1-hexanol might come from 2-ethylhexanal by a reduction. 2-Ethylhexanal, in turn, might be prepared by catalytic reduction of 2-ethyl-2-hexenal, which is the aldol condensation product of butanal. The reactions that follow show the sequence in reverse order.

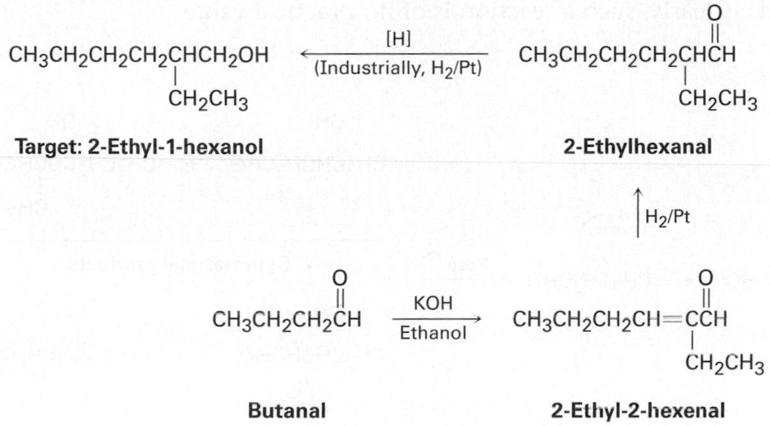

$$\underset{\textbf{Target: 2-Ethyl-1-hexanol}}{CH_3CH_2CH_2CH_2CH(CH_2CH_3)CH_2OH} \xleftarrow[\text{(Industrially, } H_2/Pt)]{[H]} \underset{\textbf{2-Ethylhexanal}}{CH_3CH_2CH_2CH_2CH(CH_2CH_3)CHO}$$

$$\uparrow H_2/Pt$$

$$\underset{\textbf{Butanal}}{CH_3CH_2CH_2CHO} \xrightarrow[\text{Ethanol}]{KOH} \underset{\textbf{2-Ethyl-2-hexenal}}{CH_3CH_2CH_2CH{=}C(CH_2CH_3)CHO}$$

Problem 23.5

Which of the following compounds are aldol condensation products? What is the aldehyde or ketone precursor of each?

(a) 2-Hydroxy-2-methylpentanal **(b)** 5-Ethyl-4-methyl-4-hepten-3-one

Problem 23.6

1-Butanol is prepared commercially by a route that begins with an aldol reaction. Show the steps that are likely to be involved.

Problem 23.7

Show how you would synthesize the following compound using an aldol reaction:

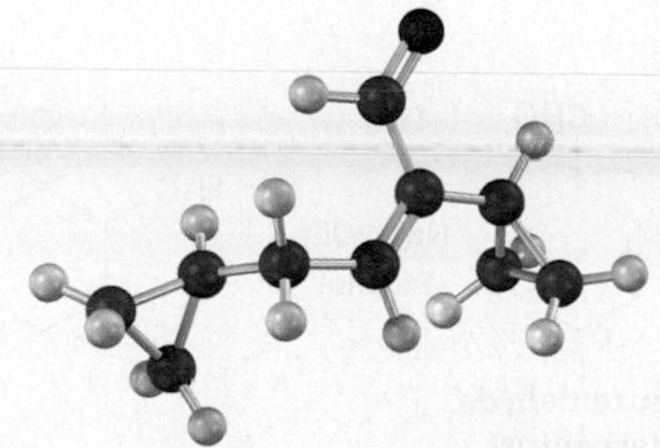

23.5 Mixed Aldol Reactions

Until now, we've considered only symmetrical aldol reactions, in which the two carbonyl components have been the same. What would happen, though, if an aldol reaction were carried out between two different carbonyl partners?

In general, a mixed aldol reaction between two similar aldehyde or ketone partners leads to a mixture of four possible products. For example, base treatment of a mixture of acetaldehyde and propanal gives a complex product mixture containing two "symmetrical" aldol products and two "mixed" aldol products. Clearly, such a reaction is of no practical value.

$$CH_3CHO + CH_3CH_2CHO \xrightarrow{\text{Base}} \underbrace{CH_3CH(OH)CH_2CHO + CH_3CH_2CH(OH)CH(CH_3)CHO}_{\text{Symmetrical products}} \quad \underbrace{CH_3CH(OH)CH(CH_3)CHO + CH_3CH_2CH(OH)CH_2CHO}_{\text{Mixed products}}$$

On the other hand, mixed aldol reactions can lead cleanly to a single product if either of two conditions is met:

- If one of the carbonyl partners contains no α hydrogens, and thus can't form an enolate ion to become a donor, but does contain an unhindered carbonyl group and so is a good acceptor of nucleophiles, then a mixed aldol reaction is likely to be successful. This is the case, for instance, when either benzaldehyde or formaldehyde is used as one of the carbonyl partners.

  Neither benzaldehyde nor formaldehyde can form an enolate ion to add to another partner, yet both compounds have an unhindered carbonyl group. The ketone 2-methylcyclohexanone, for instance, gives the mixed aldol product on reaction with benzaldehyde.

$$\text{2-Methylcyclohexanone (donor)} + \text{Benzaldehyde (acceptor)} \xrightarrow[\text{Ethanol}]{Na^+ \ ^-OEt} \text{2-benzylidene-6-methylcyclohexanone (78\%)} + H_2O$$

- If one of the carbonyl partners is much more acidic than the other and so is transformed into its enolate ion in preference to the other, then a mixed aldol reaction is likely to be successful. Ethyl acetoacetate, for instance, is completely converted into its enolate ion in preference to enolate ion formation from monocarbonyl partners. Thus, aldol condensations of monoketones with ethyl acetoacetate occur preferentially to give the mixed product.

$$\text{Cyclohexanone (acceptor)} + \underset{\text{Ethyl acetoacetate (donor)}}{CH_3\overset{O}{\overset{\|}{C}}CH_2\overset{O}{\overset{\|}{C}}OCH_2CH_3} \xrightarrow[\text{Ethanol}]{Na^+\ ^-OEt} \underset{80\%}{\text{(cyclohexylidene)}{=}C(COCH_3)CO_2CH_2CH_3} + H_2O$$

The situation can be summarized by saying that a mixed aldol reaction leads to a mixture of products unless one of the partners either has no α hydrogens but is a good electrophilic acceptor (such as benzaldehyde) or is an unusually acidic nucleophilic donor (such as ethyl acetoacetate).

Problem 23.8

Which of the following compounds can probably be prepared by a mixed aldol reaction? Show the reactants you would use in each case.

(a) $C_6H_5CH{=}CH\overset{O}{\overset{\|}{C}}CH_3$

(b) $C_6H_5\underset{CH_3}{\underset{|}{C}}{=}CH\overset{O}{\overset{\|}{C}}CH_3$

(c) 2-(cyclohexanone)$=CHCH_2CH_3$

23.6 Intramolecular Aldol Reactions

The aldol reactions we've seen thus far have all been intermolecular, meaning that they have taken place between two different molecules. When certain *di*carbonyl compounds are treated with base, however, an intramolecular aldol reaction can occur, leading to the formation of a cyclic product. For example, base treatment of a 1,4-diketone such as 2,5-hexanedione yields a cyclopentenone product, and base treatment of a 1,5-diketone such as 2,6-heptanedione yields a cyclohexenone.

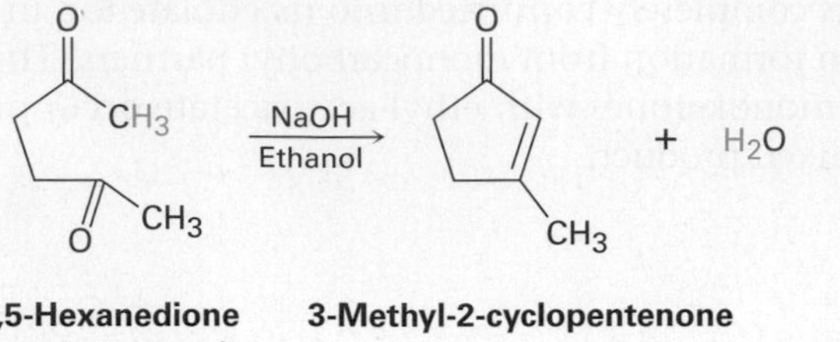

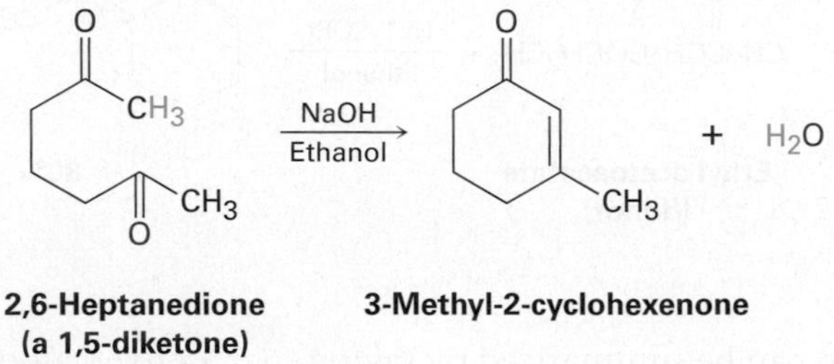

NaOH, Ethanol: 2,5-Hexanedione (a 1,4-diketone) → 3-Methyl-2-cyclopentenone + H_2O

NaOH, Ethanol: 2,6-Heptanedione (a 1,5-diketone) → 3-Methyl-2-cyclohexenone + H_2O

The mechanism of intramolecular aldol reactions is similar to that of intermolecular reactions. The only difference is that both the nucleophilic carbonyl anion donor and the electrophilic carbonyl acceptor are now in the same molecule. One complication, however, is that intramolecular aldol reactions might lead to a mixture of products, depending on which enolate ion is formed. For example, 2,5-hexanedione might yield either the five-membered-ring product 3-methyl-2-cyclopentenone or the three-membered-ring product (2-methylcyclopropenyl)ethanone **(Figure 23.3)**. In practice, though, only the cyclopentenone is formed.

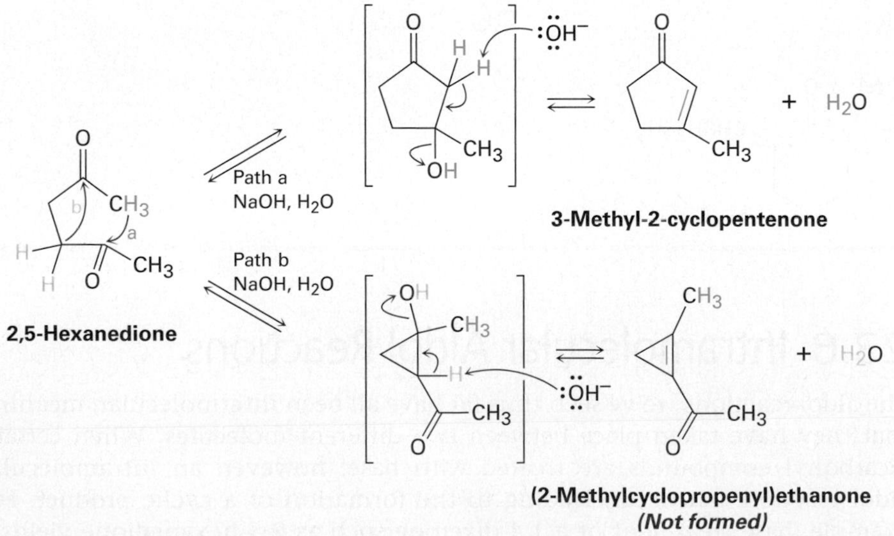

Figure 23.3 Intramolecular aldol reaction of 2,5-hexanedione yields 3-methyl-2-cyclopentenone rather than the alternative cyclopropene.

2,5-Hexanedione — Path a: NaOH, H_2O → 3-Methyl-2-cyclopentenone + H_2O

2,5-Hexanedione — Path b: NaOH, H_2O → (2-Methylcyclopropenyl)ethanone *(Not formed)* + H_2O

The selectivity observed in the intramolecular aldol reaction of 2,5-hexanedione is due to the fact that all steps in the mechanism are reversible, so an equilibrium is reached. Thus, the relatively strain-free cyclopentenone product is considerably more stable than the highly strained cyclopropene alternative. For similar reasons, intramolecular aldol reactions of 1,5-diketones lead only to cyclohexenone products rather than to acylcyclobutenes.

Problem 23.9
Treatment of a 1,3-diketone such as 2,4-pentanedione with base does not give an aldol condensation product. Explain.

Problem 23.10
What product would you expect to obtain from base treatment of 1,6-cyclodecanedione?

$$\text{1,6-Cyclodecanedione} \xrightarrow{\text{Base}} ?$$

1,6-Cyclodecanedione

23.7 The Claisen Condensation Reaction

Esters, like aldehydes and ketones, are weakly acidic. When an ester with an α hydrogen is treated with 1 equivalent of a base such as sodium ethoxide, a reversible carbonyl condensation reaction occurs to yield a β-keto ester. For instance, ethyl acetate yields ethyl acetoacetate on base treatment. This reaction between two ester molecules is known as the **Claisen condensation reaction.** (We'll use ethyl esters, abbreviated "Et," for consistency, but other esters will also work.)

$$H_3C{-}C({=}O){-}OEt + H_3C{-}C({=}O){-}OEt \xrightarrow[\text{2. } H_3O^+]{\text{1. } Na^+\ {}^-OEt\text{, ethanol}} H_3C{-}C({=}O){-}CH_2{-}C({=}O){-}OEt + CH_3CH_2OH$$

2 Ethyl acetate

Ethyl acetoacetate, a β-keto ester (75%)

The mechanism of the Claisen condensation is similar to that of the aldol condensation and involves the nucleophilic addition of an ester enolate ion to the carbonyl group of a second ester molecule **(Figure 23.4)**. The only difference between the aldol condensation of an aldehyde or ketone and the Claisen condensation of an ester involves the fate of the initially formed tetrahedral intermediate. The tetrahedral intermediate in the aldol reaction is protonated to give an alcohol product—exactly the behavior previously seen for aldehydes and ketones **(Section 19.4)**. The tetrahedral intermediate in the Claisen reaction, however, expels an alkoxide leaving group to yield an acyl substitution product—exactly the behavior previously seen for esters **(Section 21.6)**.

Figure 23.4 | MECHANISM

Mechanism of the Claisen condensation reaction.

1 Base abstracts an acidic alpha hydrogen atom from an ester molecule, yielding an ester enolate ion.

2 The enolate ion adds in a nucleophilic addition reaction to a second ester molecule, giving a tetrahedral alkoxide intermediate.

3 The tetrahedral intermediate expels ethoxide ion to yield a new carbonyl compound, ethyl acetoacetate.

4 But ethoxide ion is a strong enough base to deprotonate ethyl acetoacetate, shifting the equilibrium and driving the overall reaction to completion.

5 Protonation of the enolate ion by addition of aqueous acid in a separate step yields the final β-keto ester product.

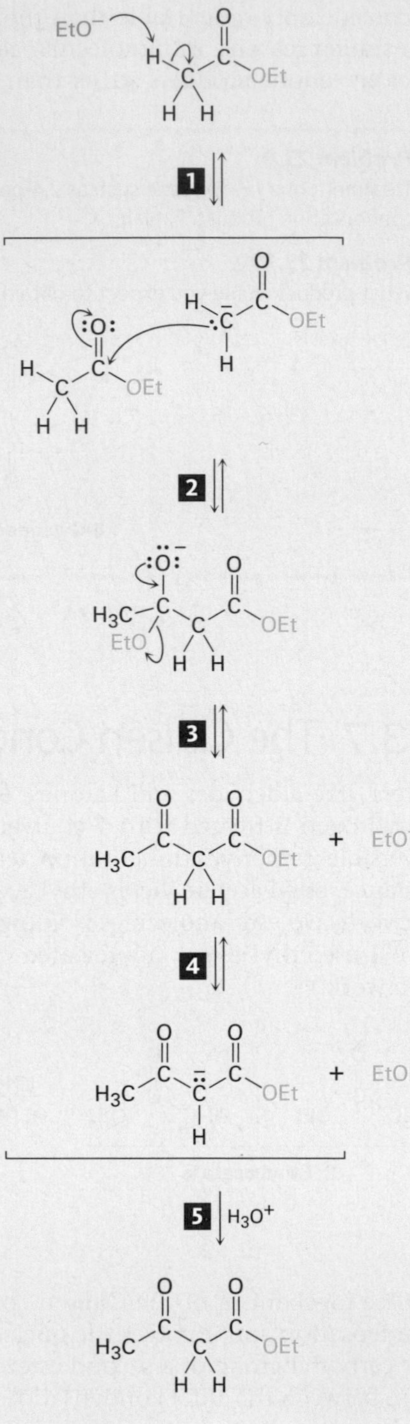

EtO⁻ + H–CH₂–C(=O)OEt

1 ⇅

H_3C–C(=O)OEt + enolate $^{-}$:CH₂–C(=O)OEt

2 ⇅

H_3C–C(O⁻)(OEt)–CH₂–C(=O)OEt

3 ⇅

H_3C–C(=O)–CH₂–C(=O)OEt + EtO⁻

4 ⇅

H_3C–C(=O)–$\ddot{C}H^{-}$–C(=O)OEt + EtOH

5 ↓ H_3O^+

H_3C–C(=O)–CH₂–C(=O)OEt

If the starting ester has more than one acidic α hydrogen, the product β-keto ester has a highly acidic, doubly activated hydrogen atom that can be abstracted by base. This deprotonation of the product requires that a full equivalent of base rather than a catalytic amount be used in the reaction. Furthermore, the deprotonation serves to drive the equilibrium completely to the product side so that high yields are usually obtained in Claisen condensations.

Worked Example 23.3

Predicting the Product of a Claisen Condensation Reaction

What product would you obtain from Claisen condensation of ethyl propanoate?

Strategy

The Claisen condensation of an ester results in loss of one molecule of alcohol and formation of a product in which an acyl group of one reactant bonds to the α carbon of the second reactant. The product is a β-keto ester.

Solution

$$CH_3CH_2\overset{O}{\overset{\|}{C}}{-}OEt + H{-}\underset{CH_3}{\underset{|}{C}}H\overset{O}{\overset{\|}{C}}OEt \xrightarrow[2.\ H_3O^+]{1.\ Na^+\ {}^-OEt} CH_3CH_2\overset{O}{\overset{\|}{C}}{-}\underset{CH_3}{\underset{|}{C}}H\overset{O}{\overset{\|}{C}}OEt + EtOH$$

2 Ethyl propanoate → **Ethyl 2-methyl-3-oxopentanoate**

Problem 23.11

Show the products you would expect to obtain by Claisen condensation of the following esters:
(a) $(CH_3)_2CHCH_2CO_2Et$ **(b)** Ethyl phenylacetate **(c)** Ethyl cyclohexylacetate

Problem 23.12

As shown in Figure 23.4, the Claisen reaction is reversible. That is, a β-keto ester can be cleaved by base into two fragments. Using curved arrows to indicate electron flow, show the mechanism by which this cleavage occurs.

$$C_6H_5\overset{O}{\overset{\|}{C}}CH_2\overset{O}{\overset{\|}{C}}OEt \xrightarrow[\text{Ethanol}]{\text{1 equiv. NaOH}} C_6H_5\overset{O}{\overset{\|}{C}}O^- + H_3C\overset{O}{\overset{\|}{C}}OEt$$

23.8 Mixed Claisen Condensations

The mixed Claisen condensation of two different esters is similar to the mixed aldol condensation of two different aldehydes or ketones **(Section 23.5)**. Mixed Claisen reactions are successful only when one of the two ester components has no α hydrogens and thus can't form an enolate ion. For example, ethyl benzoate and ethyl formate can't form enolate ions and thus can't serve as donors. They can, however, act as the electrophilic acceptor components in reactions with other ester anions to give mixed β-keto ester products.

Ethyl benzoate (acceptor) + Ethyl acetate (donor) → (1. NaH, THF; 2. H_3O^+) Ethyl benzoylacetate + EtOH

Mixed Claisen-like reactions can also be carried out between an ester and a ketone, resulting in the synthesis of a β-diketone. The reaction works best when the ester component has no α hydrogens and thus can't act as the nucleophilic donor. For example, ethyl formate gives high yields in mixed Claisen condensations with ketones.

2,2-Dimethylcyclohexanone (donor) + Ethyl formate (acceptor) → (1. Na^+ ^-OEt, ethanol; 2. H_3O^+) A β-keto aldehyde (91%)

Worked Example 23.4 Predicting the Product of a Mixed Claisen Condensation Reaction

Diethyl oxalate, $(CO_2Et)_2$, often gives high yields in mixed Claisen reactions. What product would you expect to obtain from a mixed Claisen reaction of ethyl acetate with diethyl oxalate?

Strategy

A mixed Claisen reaction is effective when only one of the two partners has an acidic α hydrogen atom. In the present case, ethyl acetate can be converted into its enolate ion, but diethyl oxalate cannot. Thus, ethyl acetate acts as the donor and diethyl oxalate as the acceptor.

Solution

Diethyl oxalate + Ethyl acetate (Acidic H) → (Na^+ ^-OEt, Ethanol) product + EtOH

Problem 23.13

What product would you expect from the following mixed Claisen-like reaction?

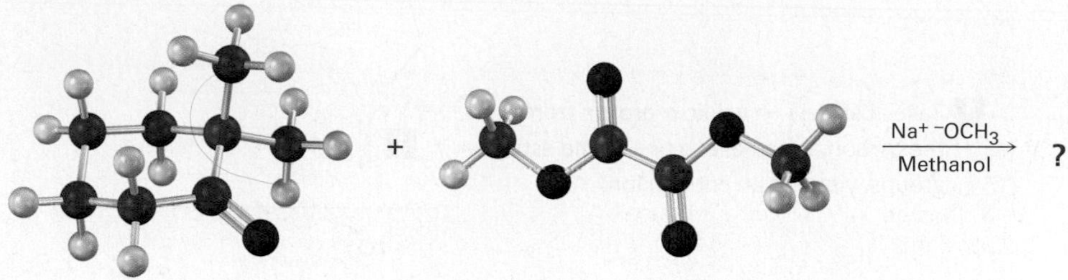

23.9 Intramolecular Claisen Condensations: The Dieckmann Cyclization

Intramolecular Claisen condensations can be carried out with diesters, just as intramolecular aldol condensations can be carried out with diketones (**Section 23.6**). Called the **Dieckmann cyclization**, the reaction works best on 1,6-diesters and 1,7-diesters. Intramolecular Claisen cyclization of a 1,6-diester gives a five-membered cyclic β-keto ester, and cyclization of a 1,7-diester gives a six-membered cyclic β-keto ester.

O OEt O OEt — 1. $Na^+\ ^-OEt$, ethanol; 2. H_3O^+ → O O C OEt + EtOH

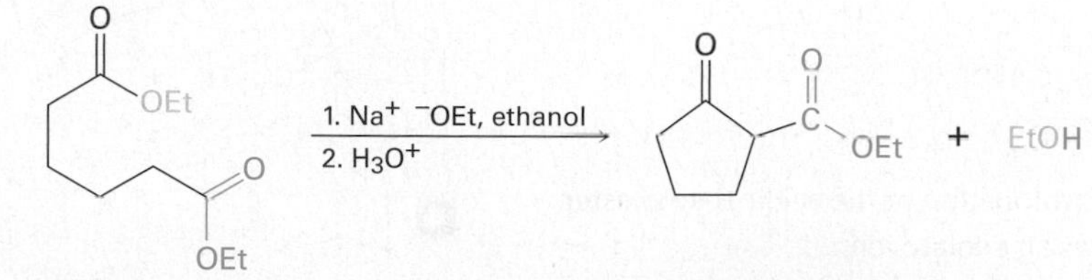

Diethyl hexanedioate (a 1,6-diester) → **Ethyl 2-oxocyclopentanecarboxylate (82%)**

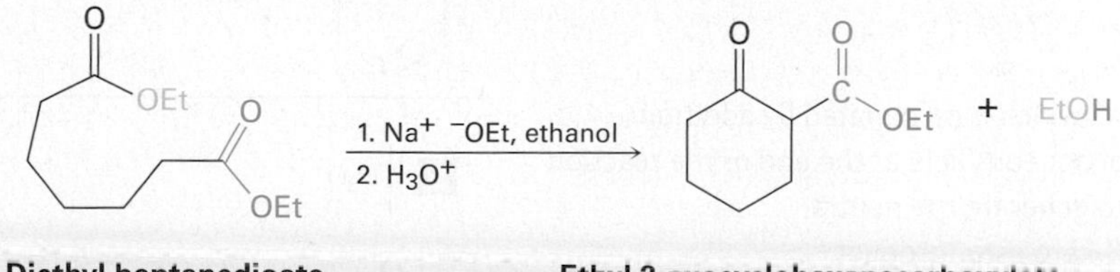

O OEt O OEt — 1. $Na^+\ ^-OEt$, ethanol; 2. H_3O^+ → O O C OEt + EtOH

Diethyl heptanedioate (a 1,7-diester) → **Ethyl 2-oxocyclohexanecarboxylate**

The mechanism of the Dieckmann cyclization, shown in **Figure 23.5**, is the same as that of the Claisen condensation. One of the two ester groups is converted into an enolate ion, which then carries out a nucleophilic acyl substitution on the second ester group at the other end of the molecule. A cyclic β-keto ester product results.

Figure 23.5 | MECHANISM

Mechanism of the Dieckmann cyclization of a 1,7-diester to yield a cyclic β-keto ester product.

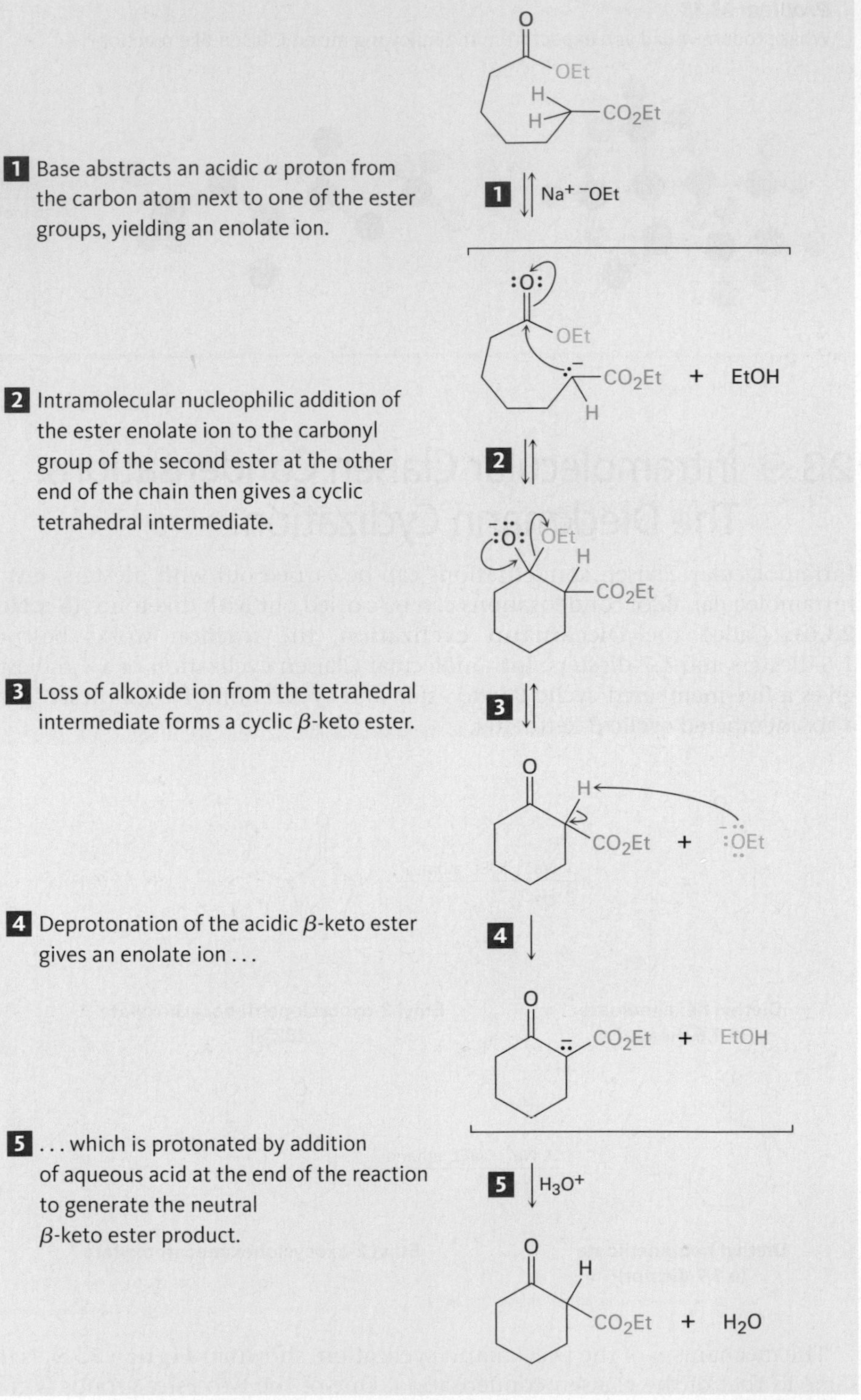

1 Base abstracts an acidic α proton from the carbon atom next to one of the ester groups, yielding an enolate ion.

2 Intramolecular nucleophilic addition of the ester enolate ion to the carbonyl group of the second ester at the other end of the chain then gives a cyclic tetrahedral intermediate.

3 Loss of alkoxide ion from the tetrahedral intermediate forms a cyclic β-keto ester.

4 Deprotonation of the acidic β-keto ester gives an enolate ion . . .

5 . . . which is protonated by addition of aqueous acid at the end of the reaction to generate the neutral β-keto ester product.

The cyclic β-keto ester produced in a Dieckmann cyclization can be further alkylated and decarboxylated by a series of reactions analogous to those used in the acetoacetic ester synthesis **(Section 22.7)**. Alkylation and subsequent decarboxylation of ethyl 2-oxocyclohexanecarboxylate, for instance, yields a 2-alkylcyclohexanone. The overall sequence of (1) Dieckmann cyclization, (2) β-keto ester alkylation, and (3) decarboxylation is a powerful method for preparing 2-substituted cyclopentanones and cyclohexanones.

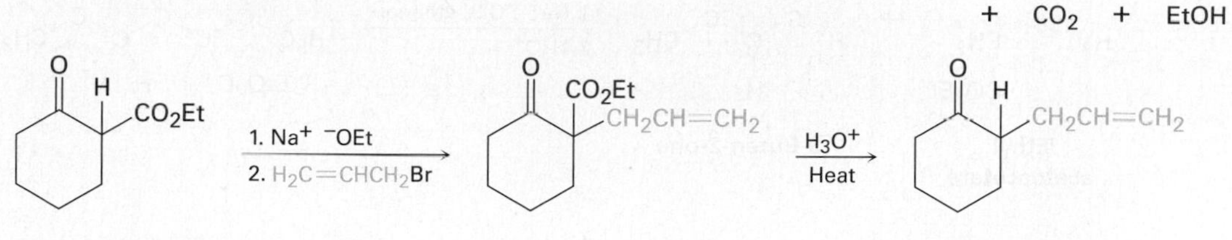

1. Na^+ ^-OEt; 2. $H_2C{=}CHCH_2Br$ → (CO_2Et, $CH_2CH{=}CH_2$) → H_3O^+, Heat → ($CH_2CH{=}CH_2$) + CO_2 + EtOH

Ethyl 2-oxocyclohexanecarboxylate

2-Allylcyclohexanone (83%)

Problem 23.14

What product would you expect from the following reaction?

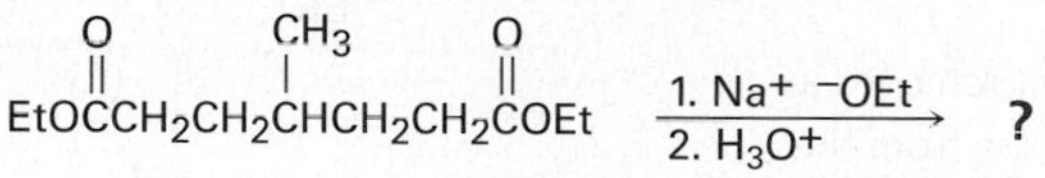

$$\mathrm{EtO\overset{O}{\overset{\|}{C}}CH_2CH_2\overset{CH_3}{\overset{|}{C}}HCH_2CH_2\overset{O}{\overset{\|}{C}}OEt} \xrightarrow[2.\ H_3O^+]{1.\ Na^+\ ^-OEt} \ ?$$

Problem 23.15

Dieckmann cyclization of diethyl 3-methylheptanedioate gives a mixture of two β-keto ester products. What are their structures, and why is a mixture formed?

23.10 Conjugate Carbonyl Additions: The Michael Reaction

We saw in **Section 19.13** that certain nucleophiles, such as amines, react with α,β-unsaturated aldehydes and ketones to give the conjugate addition product, rather than the direct addition product.

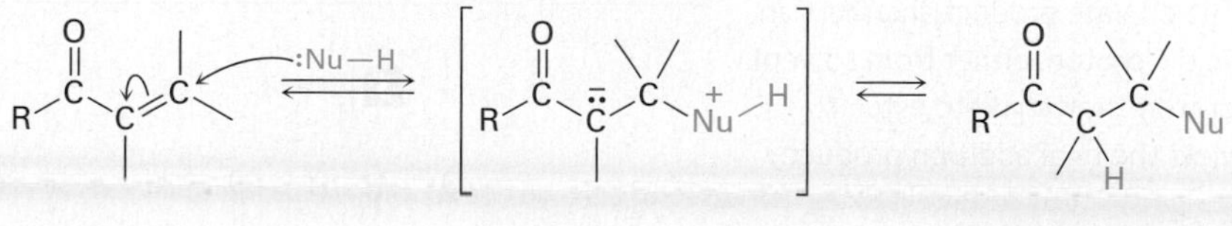

:Nu—H

Conjugate addition product

Exactly the same kind of conjugate addition can occur when a nucleophilic enolate ion reacts with an α,β-unsaturated carbonyl compound—a process known as the **Michael reaction.**

The best Michael reactions are those that take place when a particularly stable enolate ion such as that derived from a β-keto ester or other 1,3-dicarbonyl compound adds to an unhindered α,β-unsaturated ketone. For example, ethyl acetoacetate reacts with 3-buten-2-one in the presence of sodium ethoxide to yield the conjugate addition product.

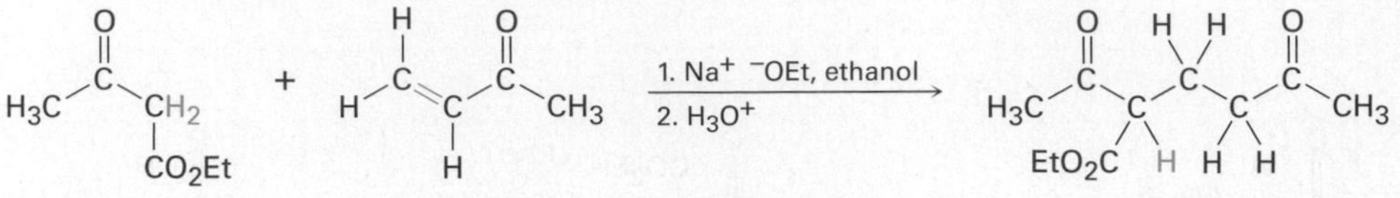

1. Na^+ ^-OEt, ethanol
2. H_3O^+

Ethyl acetoacetate **3-Buten-2-one**

Michael reactions take place by addition of a nucleophilic enolate ion donor to the β carbon of an α,β-unsaturated carbonyl acceptor, according to the mechanism shown in **Figure 23.6**.

Figure 23.6 | MECHANISM

Mechanism of the Michael reaction between a β-keto ester and an α,β-unsaturated ketone. The reaction is a conjugate addition of an enolate ion to the unsaturated carbonyl compound.

1 The base catalyst removes an acidic alpha proton from the starting β-keto ester to generate a stabilized enolate ion nucleophile.

2 The nucleophile adds to the α,β-unsaturated ketone electrophile in a Michael reaction to generate a new enolate as product.

3 The enolate product abstracts an acidic proton, either from solvent or from starting keto ester, to yield the final addition product.

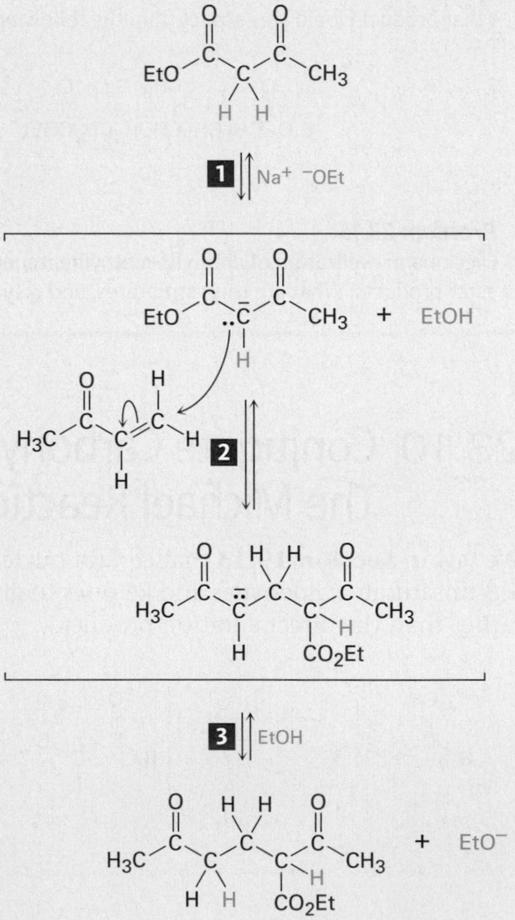

The Michael reaction occurs with a variety of α,β-unsaturated carbonyl compounds, not just conjugated ketones. Unsaturated aldehydes, esters, thioesters, nitriles, amides, and nitro compounds can all act as the electrophilic acceptor component in Michael reactions (Table 23.1). Similarly, a variety of different donors can be used, including β-diketones, β-keto esters, malonic esters, β-keto nitriles, and nitro compounds.

Table 23.1 Some Michael Acceptors and Michael Donors

Michael acceptors		Michael donors	
$H_2C{=}CH\overset{O}{\overset{\parallel}{C}}H$	Propenal	$R\overset{O}{\overset{\parallel}{C}}CH_2\overset{O}{\overset{\parallel}{C}}R'$	β-Diketone
$H_2C{=}CH\overset{O}{\overset{\parallel}{C}}CH_3$	3-Buten-2-one	$R\overset{O}{\overset{\parallel}{C}}CH_2\overset{O}{\overset{\parallel}{C}}OEt$	β-Keto ester
$H_2C{=}CH\overset{O}{\overset{\parallel}{C}}OEt$	Ethyl propenoate	$EtO\overset{O}{\overset{\parallel}{C}}CH_2\overset{O}{\overset{\parallel}{C}}OEt$	Diethyl malonate
$H_2C{=}CH\overset{O}{\overset{\parallel}{C}}NH_2$	Propenamide	$R\overset{O}{\overset{\parallel}{C}}CH_2C{\equiv}N$	β-Keto nitrile
$H_2C{=}CHC{\equiv}N$	Propenenitrile	RCH_2NO_2	Nitro compound
$H_2C{=}\overset{NO_2}{\overset{\vert}{C}H}$	Nitroethylene		

Using the Michael Reaction

Worked Example 23.5

How might you obtain the following compound using a Michael reaction?

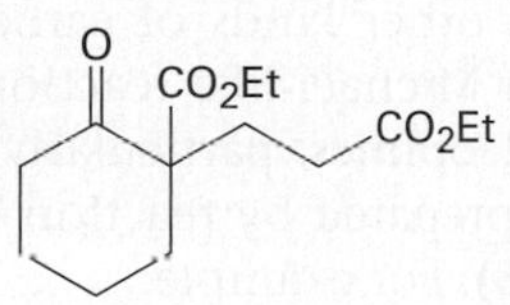

Strategy

A Michael reaction involves the conjugate addition of a stable enolate ion donor to an α,β-unsaturated carbonyl acceptor, yielding a 1,5-dicarbonyl product. Usually, the stable enolate ion is derived from a β-diketone, β-keto ester, malonic ester, or similar compound. The C C bond made in the conjugate addition step is the one between the α carbon of the acidic donor and the β carbon of the unsaturated acceptor.

Solution

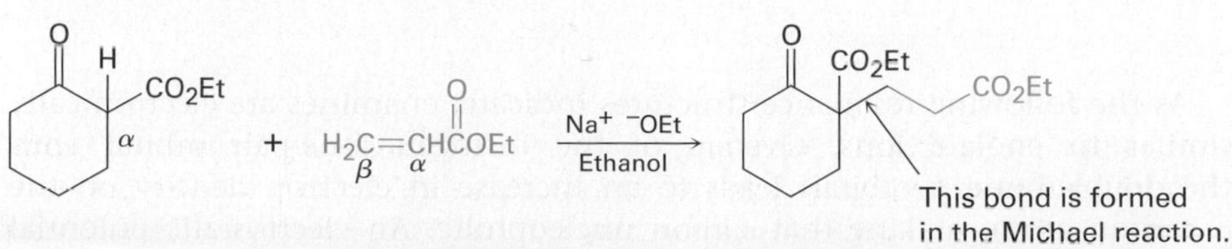

Problem 23.16

What product would you obtain from a base-catalyzed Michael reaction of 2,4-pentanedione with each of the following α,β-unsaturated acceptors?

(a) 2-Cyclohexenone **(b)** Propenenitrile **(c)** Ethyl 2-butenoate

Problem 23.17

What product would you obtain from a base-catalyzed Michael reaction of 3-buten-2-one with each of the following nucleophilic donors?

(a) $EtO\overset{O}{\overset{\|}{C}}CH_2\overset{O}{\overset{\|}{C}}OEt$

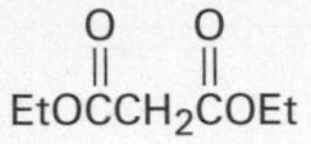

(b) 2-oxocyclopentane ring bearing CO_2Et

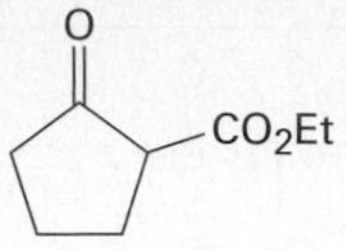

Problem 23.18

How would you prepare the following compound using a Michael reaction?

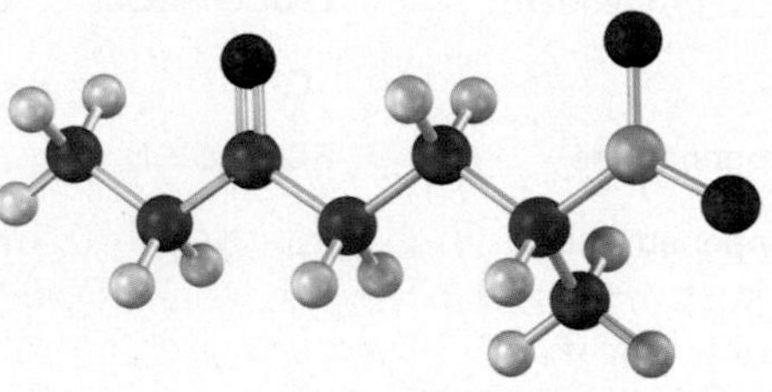

23.11 Carbonyl Condensations with Enamines: The Stork Reaction

In addition to enolate ions, other kinds of carbon nucleophiles also add to α,β-unsaturated acceptors in Michael-like reactions. Among the most important and useful of such nucleophiles, particularly in biological chemistry, are *enamines*, which are readily prepared by reaction between a ketone and a secondary amine **(Section 19.8)**. For example:

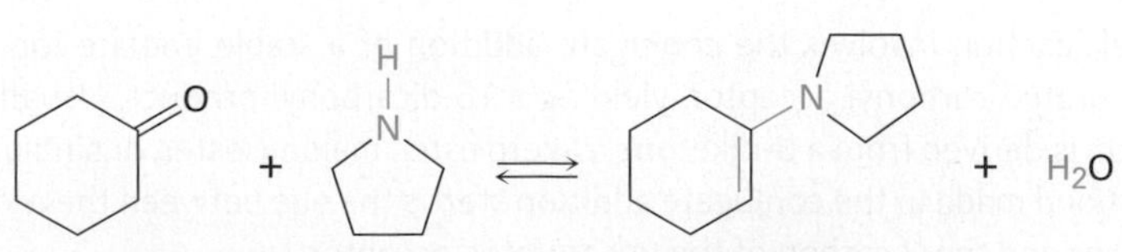

Cyclohexanone + Pyrrolidine $\rightleftharpoons$ 1-Pyrrolidinocyclohexene (87%) + H_2O

Cyclohexanone **Pyrrolidine** **1-Pyrrolidino-cyclohexene (87%)**

As the following resonance structures indicate, enamines are electronically similar to enolate ions. Overlap of the nitrogen lone-pair orbital with the double-bond p orbitals leads to an increase in electron density on the α carbon atom, making that carbon nucleophilic. An electrostatic potential map of *N,N*-dimethylaminoethylene shows this shift of electron density (red) toward the α position.

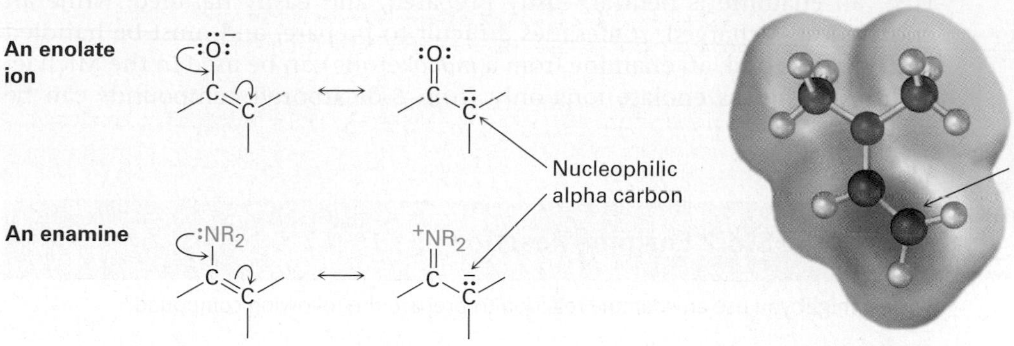

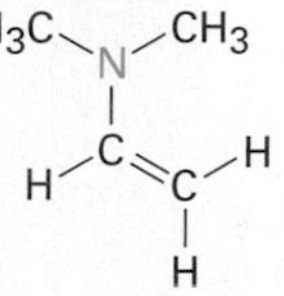

An enolate ion

An enamine

Nucleophilic alpha carbon

Enamines behave in much the same way as enolate ions and enter into many of the same kinds of reactions. In the **Stork reaction**, for example, an enamine adds to an α,β-unsaturated carbonyl acceptor in a Michael-like process. The initial product is then hydrolyzed by aqueous acid to yield a 1,5-dicarbonyl compound. The overall reaction is thus a three-step sequence of (1) enamine formation from a ketone, (2) Michael addition to an α,β-unsaturated carbonyl compound, and (3) enamine hydrolysis back to a ketone.

The net effect of the Stork reaction is the Michael addition of a ketone to an α,β-unsaturated carbonyl compound. For example, cyclohexanone reacts with the cyclic amine pyrrolidine to yield an enamine; further reaction with an enone such as 3-buten-2-one yields a Michael adduct; and aqueous hydrolysis completes the sequence to give a 1,5-diketone **(Figure 23.7)**.

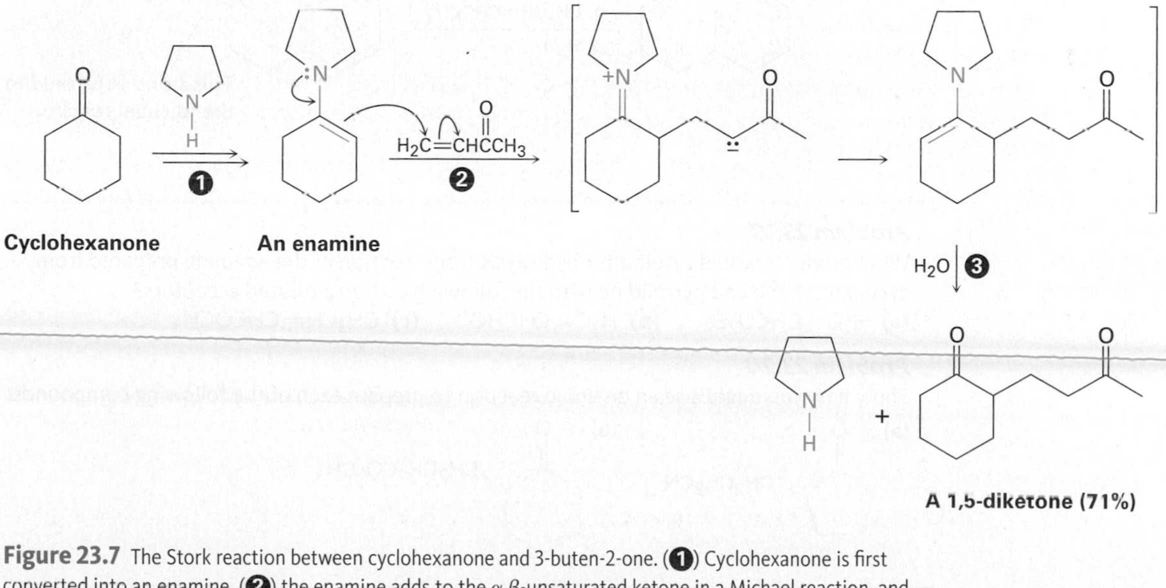

Cyclohexanone

An enamine

$H_2C{=}CHCCH_3$

H_2O 3

A 1,5-diketone (71%)

Figure 23.7 The Stork reaction between cyclohexanone and 3-buten-2-one. (1) Cyclohexanone is first converted into an enamine, (2) the enamine adds to the α,β-unsaturated ketone in a Michael reaction, and (3) the conjugate addition product is hydrolyzed to yield a 1,5-diketone.

The enamine–Michael reaction has two advantages over the enolate-ion–Michael reaction that makes it particularly useful in biological pathways. First, an enamine is neutral, easily prepared, and easily handled, while an enolate ion is charged, sometimes difficult to prepare, and must be handled with care. Second, an enamine from a *mono*ketone can be used in the Michael addition, whereas enolate ions only from β-dicarbonyl compounds can be used.

Worked Example 23.6 Using the Stork Enamine Reaction

How might you use an enamine reaction to prepare the following compound?

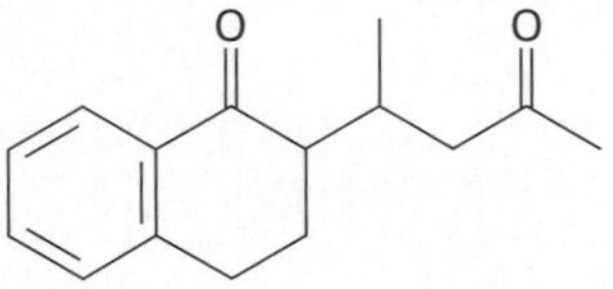

Strategy

The overall result of an enamine reaction is the Michael addition of a ketone as donor to an α,β-unsaturated carbonyl compound as acceptor, yielding a 1,5-dicarbonyl product. The C–C bond made in the Michael addition step is the one between the α carbon of the ketone donor and the β carbon of the unsaturated acceptor.

Solution

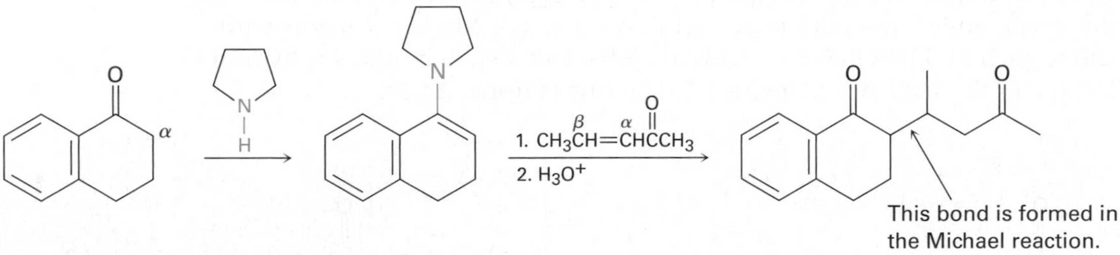

1. $CH_3\overset{\beta}{C}H{=}\overset{\alpha}{C}H\overset{O}{\overset{\|}{C}}CH_3$
2. H_3O^+

This bond is formed in the Michael reaction.

Problem 23.19

What products would result after hydrolysis from reaction of the enamine prepared from cyclopentanone and pyrrolidine with the following α,β-unsaturated acceptors?

(a) $H_2C{=}CHCO_2Et$ **(b)** $H_2C{=}CHCHO$ **(c)** $CH_3CH{=}CHCOCH_3$

Problem 23.20

Show how you might use an enamine reaction to prepare each of the following compounds:

(a) CH_2CH_2CN **(b)** $CH_2CH_2CO_2CH_3$

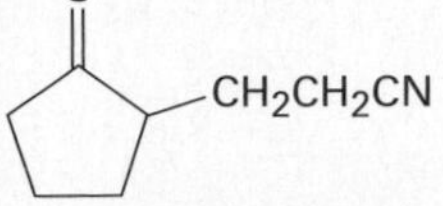

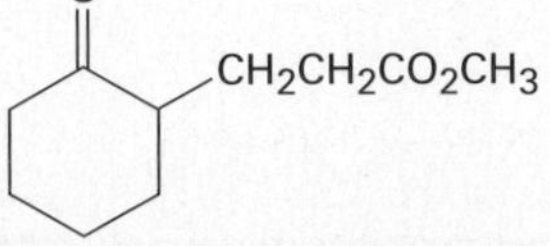

23.12 The Robinson Annulation Reaction

Carbonyl condensation reactions are perhaps the most versatile methods available for synthesizing complex molecules. By putting a few fundamental reactions together in the proper sequence, some remarkably useful transformations can be carried out. One such example is the **Robinson annulation reaction** for the synthesis of polycyclic molecules. The word *annulation* comes from the Latin *annulus,* meaning "ring," so an annulation reaction builds a new ring onto a molecule.

The Robinson annulation is a two-step process that combines a Michael reaction with an intramolecular aldol reaction. It takes place between a nucleophilic donor, such as a β-keto ester, an enamine, or a β-diketone, and an α,β-unsaturated ketone acceptor, such as 3-buten-2-one. The product is a substituted 2-cyclohexenone.

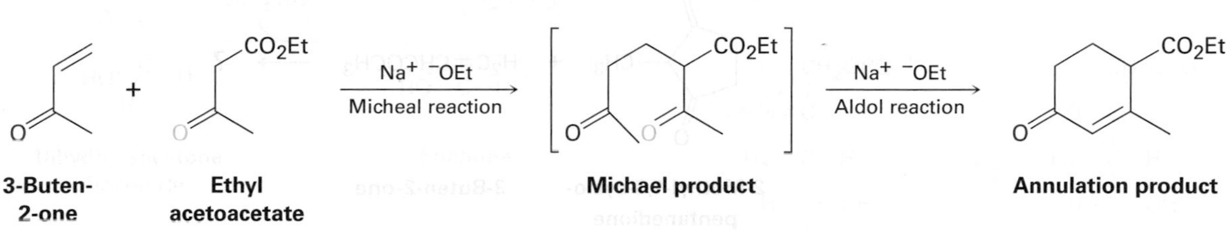

CO_2Et | + | Na^+ ^-OEt | Micheal reaction | CO_2Et | Na^+ ^-OEt | Aldol reaction | CO_2Et

3-Buten-2-one | **Ethyl acetoacetate** | **Michael product** | **Annulation product**

The first step of the Robinson annulation is simply a Michael reaction. An enamine or an enolate ion from a β-keto ester or β-diketone effects a conjugate addition to an α,β-unsaturated ketone, yielding a 1,5-diketone. But as we saw in **Section 23.6**, 1,5-diketones undergo intramolecular aldol condensation to yield cyclohexenones when treated with base. Thus, the final product contains a six-membered ring, and an annulation has been accomplished. An example occurs during a synthesis of the steroid hormone estrone **(Figure 23.8)**.

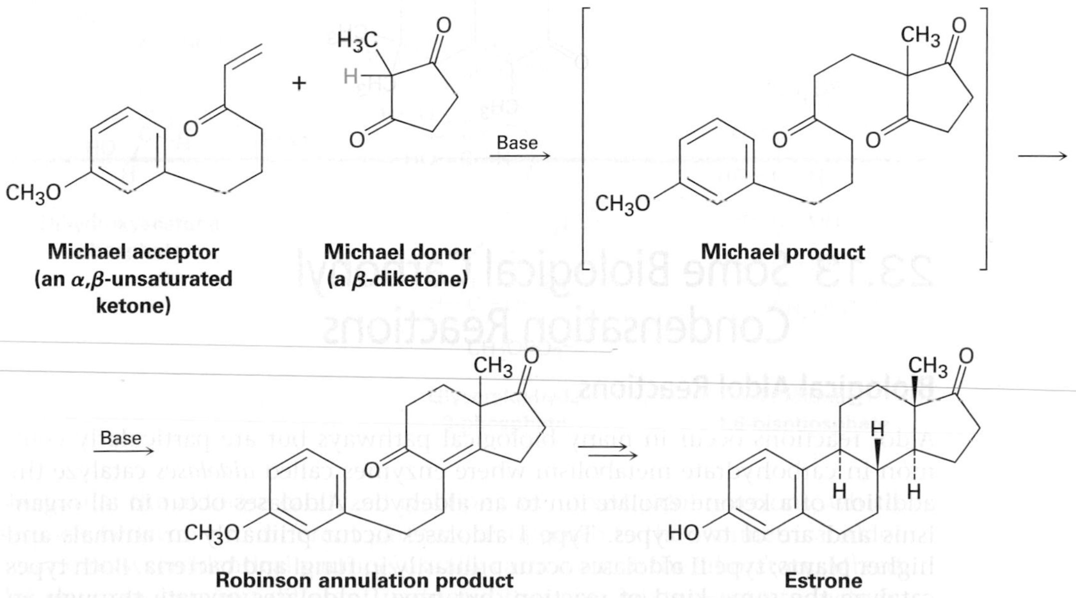

CH_3O | + | H_3C | H | Base | CH_3 | CH_3O

Michael acceptor (an α,β-unsaturated ketone) | **Michael donor (a β-diketone)** | **Michael product**

Base | CH_3 | CH_3O | CH_3 | H | H | H | HO

Robinson annulation product | **Estrone**

Figure 23.8 Synthesis of the steroid hormone estrone using a Robinson annulation reaction. The nucleophilic donor is a β-diketone.

Biological Claisen Condensations

Claisen condensations, like aldol reactions, also occur in a large number of biological pathways. In fatty-acid biosynthesis, for instance, an enolate ion generated by decarboxylation **(Section 22.7)** of malonyl ACP adds to the carbonyl group of another acyl group bonded through a thioester linkage to a synthase enzyme. The tetrahedral intermediate that results then expels the synthase, giving acetoacetyl ACP. (The abbreviation ACP stands for acyl carrier protein, which forms thioester bonds to acyl groups.)

Acetyl synthase

Malonyl ACP → (CO_2) **Enolate ion** → (− Synthase–SH) **Acetoacetyl ACP**

Mixed Claisen condensations also occur frequently in living organisms, particularly in the pathway for fatty-acid biosynthesis that we'll discuss in **Section 29.4**. Butyryl synthase, for instance, reacts with malonyl ACP in a mixed Claisen condensation to give 3-ketohexanoyl ACP.

Butyryl synthase + **Malonyl ACP** → (CO_2, Synthase–SH) **3-Ketohexanoyl ACP**

A Prologue to Metabolism | A DEEPER LOOK

Biochemistry *is* carbonyl chemistry. Almost all metabolic pathways used by living organisms involve one or more of the four fundamental carbonyl-group reactions we've seen in Chapters 19 through 23. The digestion and metabolic breakdown of all the major classes of food molecules—fats, carbohydrates, and proteins—take place by nucleophilic addition reactions, nucleophilic acyl substitutions, α substitutions, and carbonyl condensations. Similarly, hormones and other crucial biological molecules are built up from smaller precursors by these same carbonyl-group reactions.

(continued)

(continued)

Take *glycolysis*, for example, the metabolic pathway by which organisms convert glucose to pyruvate as the first step in extracting energy from carbohydrates.

Glucose $\xrightarrow{\text{Glycolysis}}$ 2 $H_3C-C(=O)-CO_2^-$ **Pyruvate**

You are what you eat. Food molecules are metabolized by pathways that involve the four major carbonyl-group reactions.

Erich Lessing/Art Resource, NY

Glycolysis is a ten-step process that begins with isomerization of glucose from its cyclic hemiacetal form to its open-chain aldehyde form—the reverse of a nucleophilic addition reaction. The aldehyde then undergoes tautomerization to yield an enol, which undergoes yet another tautomerization to give the ketone fructose.

Glucose (hemiacetal) ⇌ (:Base, H—A) **Glucose (aldehyde)** ⇌ **Glucose (enol)** ⇌ **Fructose**

Fructose, a β-hydroxy ketone, is then cleaved by a retro-aldol reaction into two three-carbon molecules—one ketone and one aldehyde. Still further carbonyl-group reactions then occur until pyruvate results.

Fructose (H—A, :Base) ⟶ [CH_2OH, C—OH, HO—C—H] ⟶ CH_2OH—C=O—CH_2OH + H—C(=O)—CH(OH)—CH_2OH

These few examples are only an introduction; we'll look at several of the major metabolic pathways in more detail in Chapter 29. The bottom line is that you haven't seen the end of carbonyl-group chemistry. A solid grasp of carbonyl-group reactions is crucial to an understanding of biochemistry.

Summary

Key words

aldol reaction, 905
carbonyl condensation reaction, 904
Claisen condensation reaction, 915
Dieckmann cyclization, 919
Michael reaction, 921
Robinson annulation reaction, 927
Stork reaction, 925

In this chapter, we've discussed the fourth and last of the common carbonyl-group reactions—the carbonyl condensation. A **carbonyl condensation reaction** takes place between two carbonyl partners and involves both nucleophilic addition and α-substitution processes. One carbonyl partner is converted by base into a nucleophilic enolate ion, which then adds to the electrophilic carbonyl group of the second partner. The first partner thus undergoes an α substitution, while the second undergoes a nucleophilic addition.

Nucleophilic donor **Electrophilic acceptor**

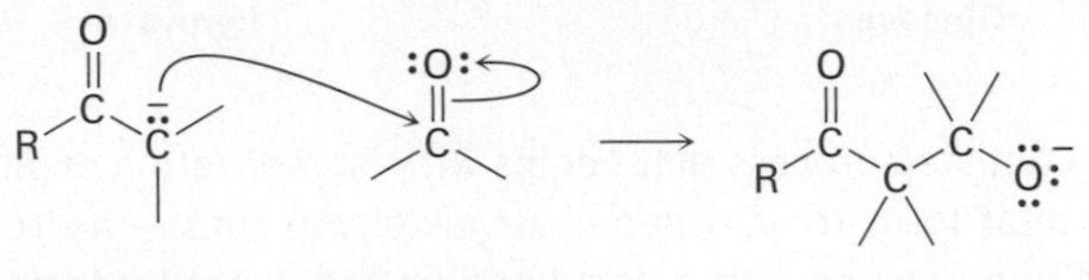

The **aldol reaction** is a carbonyl condensation that occurs between two aldehyde or ketone molecules. Aldol reactions are reversible, leading first to β-hydroxy aldehydes/ketones and then to α,β-unsaturated products after dehydration. Mixed aldol condensations between two different aldehydes or ketones generally give a mixture of all four possible products. A mixed reaction can be successful, however, if one of the two partners is an unusually good donor (ethyl acetoacetate, for instance) or if it can act only as an acceptor (formaldehyde and benzaldehyde, for instance). Intramolecular aldol condensations of 1,4- and 1,5-diketones are also successful and provide a good way to make five- and six-membered rings.

The **Claisen condensation reaction** is a carbonyl condensation that occurs between two ester components and gives a β-keto ester product. Mixed Claisen condensations between two different esters are successful only when one of the two partners has no acidic α hydrogens (ethyl benzoate and ethyl formate, for instance) and thus can function only as the acceptor partner. Intramolecular Claisen condensations, called **Dieckmann cyclization reactions**, yield five- and six-membered cyclic β-keto esters starting from 1,6- and 1,7-diesters.

The conjugate addition of a carbon nucleophile to an α,β-unsaturated acceptor is known as the **Michael reaction**. The best Michael reactions take place between relatively acidic donors (β-keto esters or β-diketones) and unhindered α,β-unsaturated acceptors. Enamines, prepared by reaction of a ketone with a disubstituted amine, are also good Michael donors.

Carbonyl condensation reactions are widely used in synthesis. One example of their versatility is the **Robinson annulation reaction**, which leads to the formation of an substituted cyclohexenone. Treatment of a β-diketone or β-keto ester with an α,β-unsaturated ketone leads first to a Michael addition, which is followed by intramolecular aldol cyclization. Condensation reactions are also used widely in nature for the biosynthesis of such molecules as fats and steroids.

Summary of Reactions

1. Aldol reaction (Section 23.1)

$$2\,RCH_2\overset{O}{\overset{\|}{C}}H \xrightleftharpoons{\text{NaOH, ethanol}} RCH_2\overset{OH}{\overset{|}{C}}H\underset{R}{\underset{|}{C}}H\overset{O}{\overset{\|}{C}}H$$

2. Mixed aldol reaction (Section 23.5)

$$RCH_2\overset{O}{\overset{\|}{C}}R' + PhCHO \xrightleftharpoons{\text{NaOH, ethanol}} Ph\overset{OH}{\overset{|}{C}}H\underset{R}{\underset{|}{C}}H\overset{O}{\overset{\|}{C}}R'$$

$$RCH_2\overset{O}{\overset{\|}{C}}R' + CH_2O \xrightleftharpoons{\text{NaOH, ethanol}} HOCH_2\underset{R}{\underset{|}{C}}H\overset{O}{\overset{\|}{C}}R'$$

3. Intramolecular aldol reaction (Section 23.6)

$$\text{O=C(CH}_2\text{R)–CH}_2\text{CH}_2\text{CH}_2\text{–C(=O)R'} \xrightleftharpoons{\text{NaOH, ethanol}} \text{2-R-3-R'-cyclohex-2-enone} + H_2O$$

4. Dehydration of aldol products (Section 23.3)

$$\overset{OH}{\overset{|}{C}}-\underset{H}{\underset{|}{C}}-\overset{O}{\overset{\|}{C}} \xrightleftharpoons[\text{or } H_3O^+]{\text{NaOH}} C{=}C-\overset{O}{\overset{\|}{C}} + H_2O$$

5. Claisen condensation reaction (Section 23.7)

$$2\,RCH_2\overset{O}{\overset{\|}{C}}OR' \xrightleftharpoons{Na^+\ {}^-OEt,\ \text{ethanol}} RCH_2\overset{O}{\overset{\|}{C}}-\underset{R}{\underset{|}{C}}H\overset{O}{\overset{\|}{C}}OR' + HOR'$$

6. Mixed Claisen condensation reaction (Section 23.8)

$$RCH_2\overset{O}{\overset{\|}{C}}OEt + H\overset{O}{\overset{\|}{C}}OEt \xrightleftharpoons{Na^+\ {}^-OEt,\ \text{ethanol}} H\overset{O}{\overset{\|}{C}}-\underset{R}{\underset{|}{C}}H\overset{O}{\overset{\|}{C}}OEt + HOEt$$

(continued)

7. Intramolecular Claisen condensation (Dieckmann cyclization; Section 23.9)

$EtOC(O)(CH_2)_4COEt$ $\xrightleftharpoons{Na^+\ ^-OEt,\ ethanol}$ 2-(ethoxycarbonyl)cyclopentanone (O, $\overset{O}{\|}$COEt) + HOEt

$EtOC(O)(CH_2)_5COEt$ $\xrightleftharpoons{Na^+\ ^-OEt,\ ethanol}$ 2-(ethoxycarbonyl)cyclohexanone (O, $\overset{O}{\|}$COEt) + HOEt

8. Michael reaction (Section 23.10)

1,3-dicarbonyl (O=C–CH₂–C=O, H H) + α,β-unsaturated carbonyl (C=C–C=O) $\xrightarrow[Ethanol]{Na^+\ ^-OEt}$ Michael adduct (O=C, H, C, C, C, O=C, H)

9. Carbonyl condensations with enamines (Stork reaction; Section 23.11)

Enamine (NR_2, R–C=C) + α,β-unsaturated carbonyl (C=C–C(=O)–R′) $\xrightarrow[2.\ H_3O^+]{1.\ Mix\ in\ THF\ solvent}$ R–C(=O)–C–C–C(H)–C(=O)–R′

Exercises

OWL Interactive versions of these problems are assignable in OWL for Organic Chemistry.

Visualizing Chemistry

(Problems 23.1–23.22 appear within the chapter.)

23.23 What ketones or aldehydes might the following enones have been prepared from by aldol reaction?

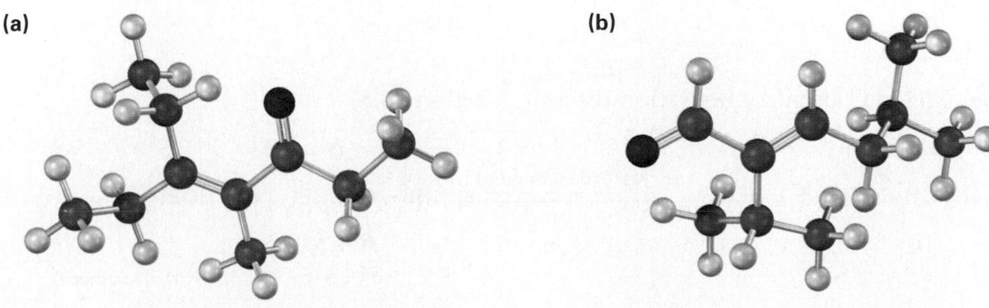
(a) **(b)**

23.24 The following structure represents an intermediate formed by addition of an ester enolate ion to a second ester molecule. Identify the reactant, the leaving group, and the product.

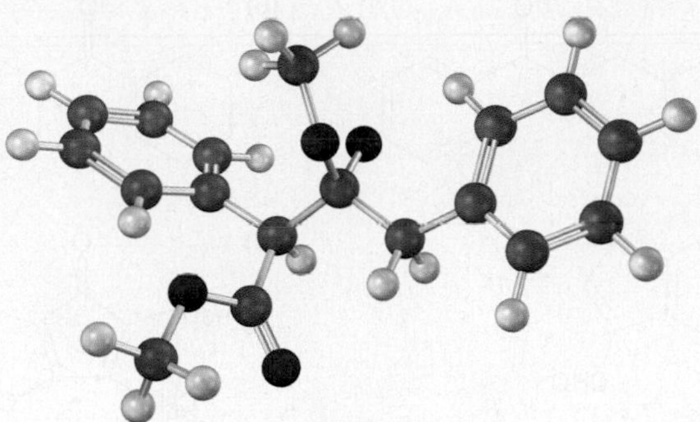

23.25 The following molecule was formed by an intramolecular aldol reaction. What dicarbonyl precursor was used for its preparation?

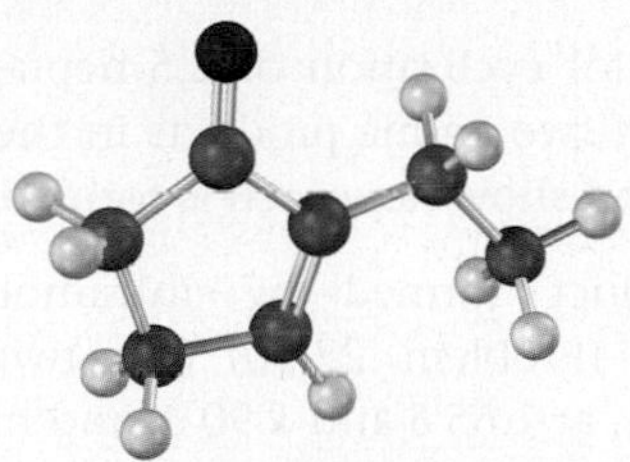

23.26 The following molecule was formed by a Robinson annulation reaction. What reactants were used?

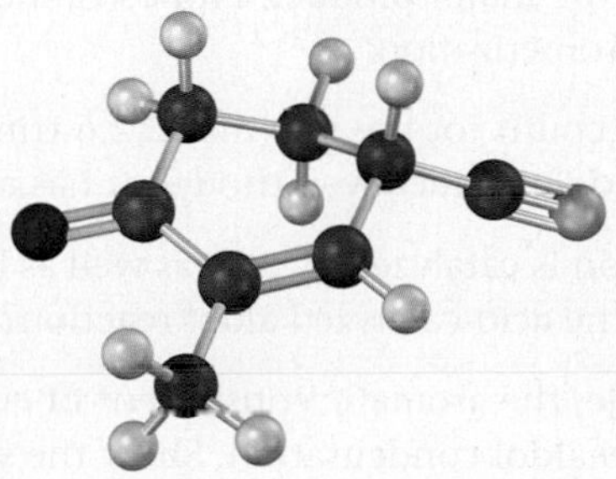

Additional Problems

Aldol Reactions

23.27 Which of the following compounds would you expect to undergo aldol self-condensation? Show the product of each successful reaction.

(a) Trimethylacetaldehyde
(b) Cyclobutanone
(c) Benzophenone (diphenyl ketone)
(d) 3-Pentanone
(e) Decanal
(f) 3-Phenyl-2-propenal

23.28 How might you synthesize each of the following compounds using an aldol reaction? Show the structure of the starting aldehyde(s) or ketone(s) you would use in each case.

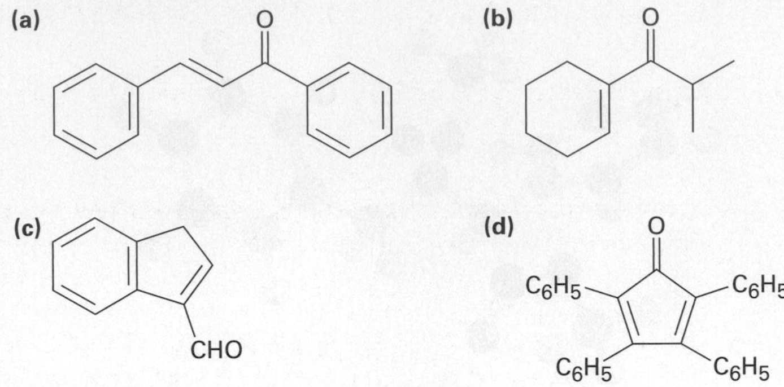

23.29 What product would you expect to obtain from aldol cyclization of hexanedial, $OHCCH_2CH_2CH_2CH_2CHO$?

23.30 Intramolecular aldol cyclization of 2,5-heptanedione with aqueous NaOH yields a mixture of two enone products in the approximate ratio 9:1. Write their structures, and show how each is formed.

23.31 The major product formed by intramolecular aldol cyclization of 2,5-heptanedione (Problem 23.30) has two singlet absorptions in the ^1H NMR spectrum, at 1.65 δ and 1.90 δ, and has no absorptions in the range 3 to 10 δ. What is its structure?

23.32 Treatment of the minor product formed in the intramolecular aldol cyclization of 2,5-heptanedione (Problems 23.30 and 23.31) with aqueous NaOH converts it into the major product. Propose a mechanism to account for this base-catalyzed isomerization.

23.33 How can you account for the fact that 2,2,6-trimethylcyclohexanone yields no detectable aldol product even though it has an acidic α hydrogen?

23.34 The aldol reaction is catalyzed by acid as well as by base. What is the reactive nucleophile in the acid-catalyzed aldol reaction? Propose a mechanism.

23.35 Cinnamaldehyde, the aromatic constituent of cinnamon oil, can be synthesized by a mixed aldol condensation. Show the starting materials you would use, and write the reaction.

CHO

Cinnamaldehyde

23.36 The bicyclic ketone shown below does not undergo aldol self-condensation even though it has two α hydrogen atoms. Explain.

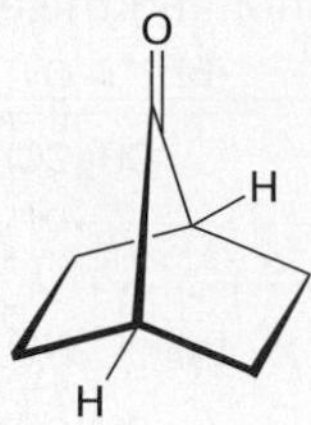

Claisen Condensations

23.37 Give the structures of the possible Claisen condensation products from the following reactions. Tell which, if any, you would expect to predominate in each case.

(a) $CH_3CO_2Et + CH_3CH_2CO_2Et$ **(b)** $C_6H_5CO_2Et + C_6H_5CH_2CO_2Et$
(c) $EtOCO_2Et$ + cyclohexanone **(d)** $C_6H_5CHO + CH_3CO_2Et$

23.38 In the mixed Claisen reaction of cyclopentanone with ethyl formate, a much higher yield of the desired product is obtained by first mixing the two carbonyl components and then adding base rather than by first mixing base with cyclopentanone and then adding ethyl formate. Explain.

23.39 Ethyl dimethylacetoacetate reacts instantly at room temperature when treated with ethoxide ion to yield two products, ethyl acetate and ethyl 2-methylpropanoate. Propose a mechanism for this cleavage reaction.

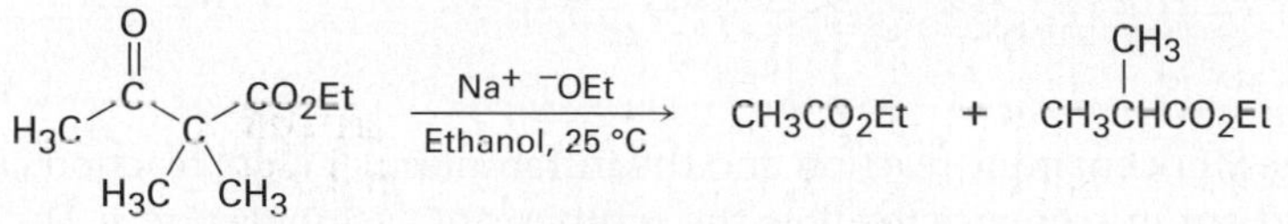

$$H_3C{-}C(=O){-}C(CH_3)_2{-}CO_2Et \xrightarrow[\text{Ethanol, 25 °C}]{Na^+\ {}^-OEt} CH_3CO_2Et + CH_3CH(CH_3)CO_2Et$$

23.40 In contrast to the rapid reaction shown in Problem 23.39, ethyl acetoacetate requires a temperature over 150 °C to undergo the same kind of cleavage reaction. How can you explain the difference in reactivity?

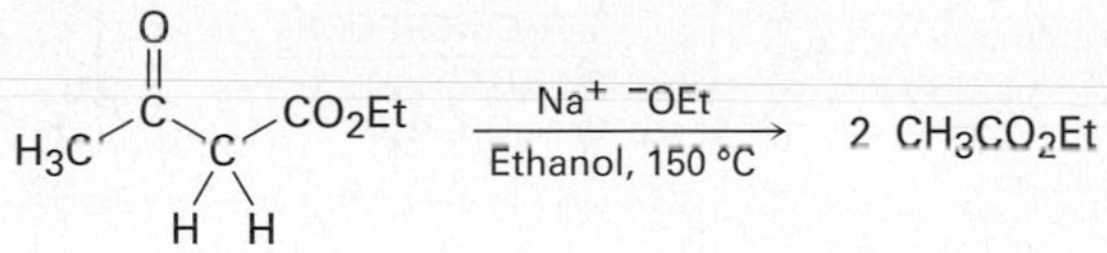

$$H_3C{-}C(=O){-}CH_2{-}CO_2Et \xrightarrow[\text{Ethanol, 150 °C}]{Na^+\ {}^-OEt} 2\ CH_3CO_2Et$$

Michael and Enamine Reactions

23.41 How might the following compounds be prepared using Michael reactions? Show the nucleophilic donor and the electrophilic acceptor in each case.

(a) $CH_3COCH(CO_2Et)CH_2CH_2COC_6H_5$

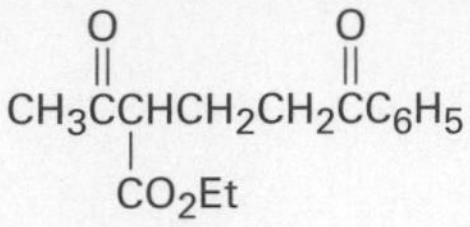

(b) $CH_3COCH_2CH_2CH_2COCH_3$

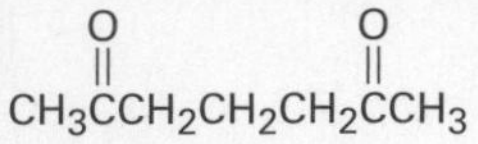

(c) $EtOCOCH(CO_2Et)CH_2CH_2C{\equiv}N$

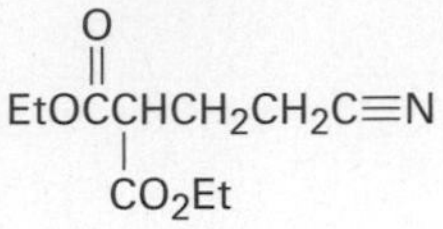

(d) $CH_3CH(NO_2)CH_2CH_2COOEt$

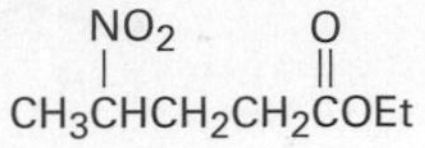

(e) $EtOCOCH(CO_2Et)CH_2CH_2NO_2$

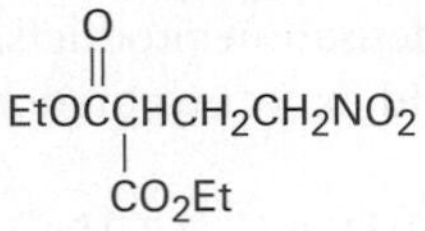

(f) 3-(CH_2NO_2)cyclohexanone

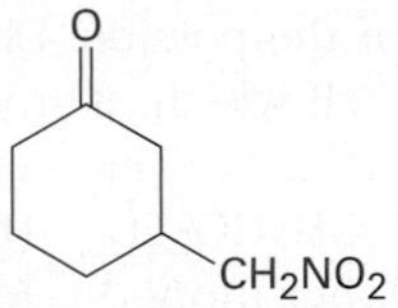

23.42 The so-called Wieland–Miescher ketone is a valuable starting material used in the synthesis of steroid hormones. How might you prepare it from 1,3-cyclohexanedione?

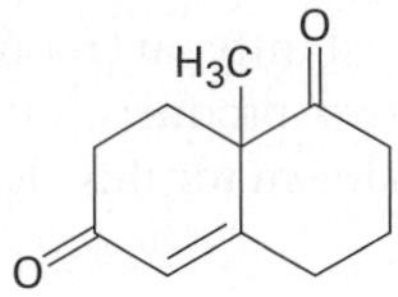

H_3C **Wieland–Miescher ketone**

23.43 The Stork enamine reaction and the intramolecular aldol reaction can be carried out in sequence to allow the synthesis of cyclohexenones. For example, reaction of the pyrrolidine enamine of cyclohexanone with 3-buten-2-one, followed by enamine hydrolysis and base treatment, yields the product indicated. Write each step, and show the mechanism of each.

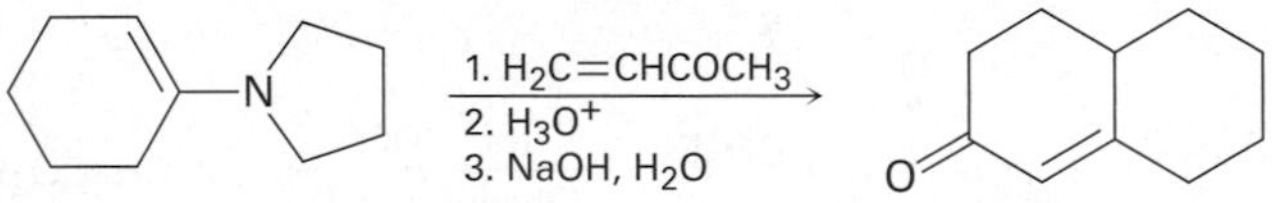

1. $H_2C{=}CHCOCH_3$
2. H_3O^+
3. NaOH, H_2O

23.44 How could you prepare the following cyclohexenones by combining a Stork enamine reaction with an intramolecular aldol condensation? (See Problem 23.43.)

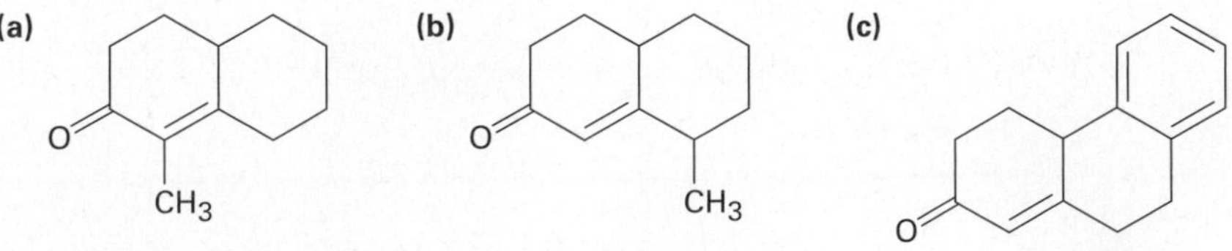

23.45 The following reaction involves two successive intramolecular Michael reactions. Write both steps, and show their mechanisms.

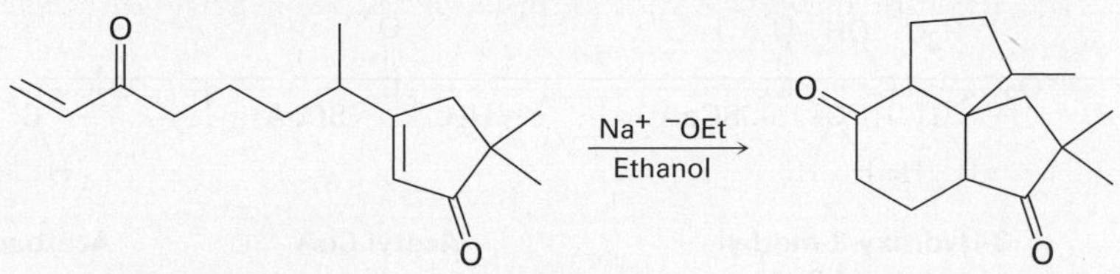

$\xrightarrow[\text{Ethanol}]{Na^+\ ^-OEt}$

General Problems

23.46 What condensation products would you expect to obtain by treatment of the following substances with sodium ethoxide in ethanol?

(a) Ethyl butanoate (b) Cycloheptanone
(c) 3,7-Nonanedione (d) 3-Phenylpropanal

23.47 The following reactions are unlikely to provide the indicated product in high yield. What is wrong with each?

(a) $CH_3CHO + CH_3COCH_3 \xrightarrow[\text{Ethanol}]{Na^+\ ^-OEt} CH_3CH(OH)CH_2COCH_3$

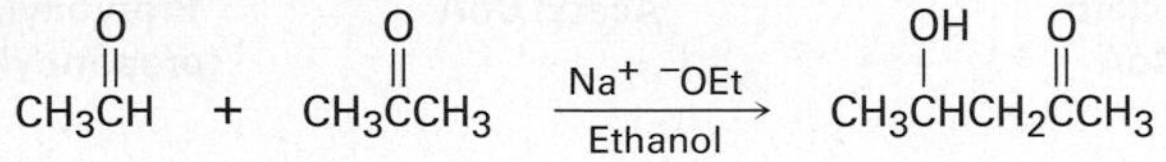

(b) 3-Methylcyclohexanone + $H_2C{=}CHCOCH_3$ $\xrightarrow[\text{Ethanol}]{Na^+\ ^-OEt}$ 2-($CH_2CH_2COCH_3$)-3-CH_3-cyclohexanone

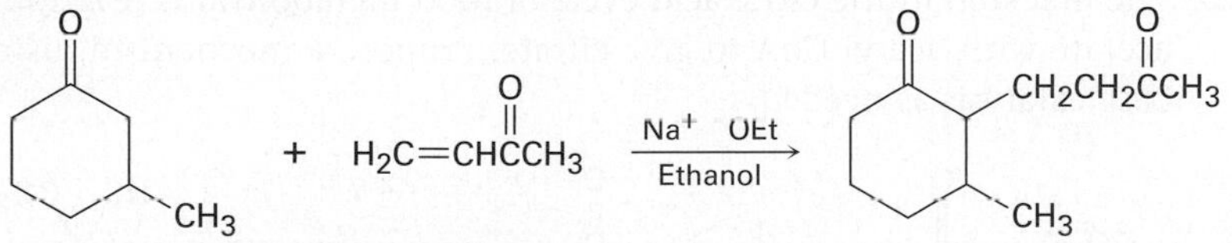

(c) $CH_3COCH_2CH_2CH_2COCH_3 \xrightarrow[\text{Ethanol}]{Na^+\ ^-OEt}$

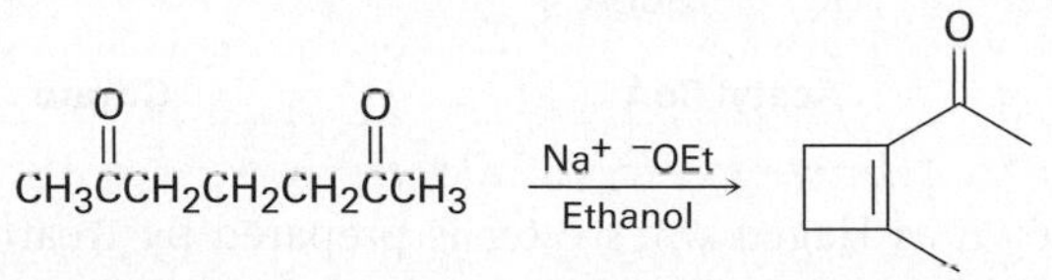

23.48 Fill in the missing reagents **a–h** in the following scheme:

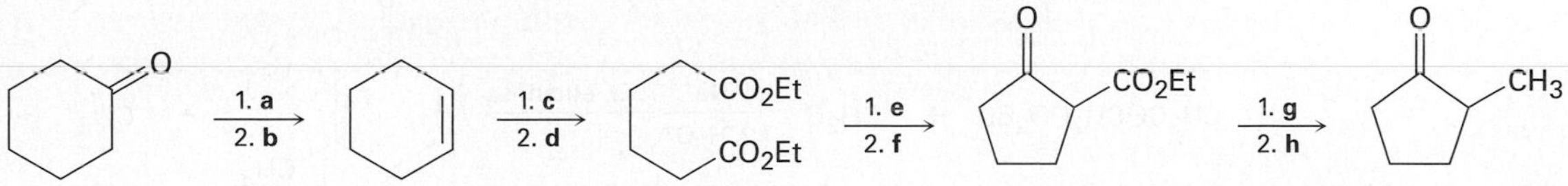

Cyclohexanone $\xrightarrow[\text{2. b}]{\text{1. a}}$ Cyclohexene $\xrightarrow[\text{2. d}]{\text{1. c}}$ (CO_2Et, CO_2Et) $\xrightarrow[\text{2. f}]{\text{1. e}}$ (CO_2Et) $\xrightarrow[\text{2. h}]{\text{1. g}}$ (CH_3)

23.49 How would you prepare the following compounds from cyclohexanone?

(a) C_6H_5CH, CHC_6H_5 (b) CH_2CH_2CN

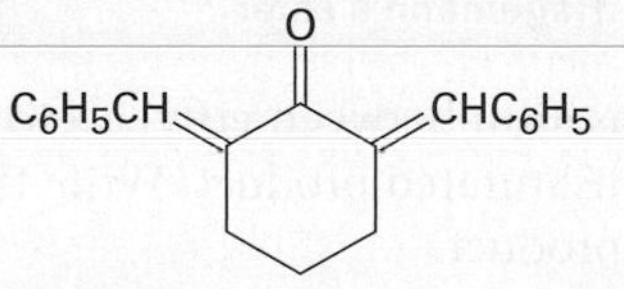

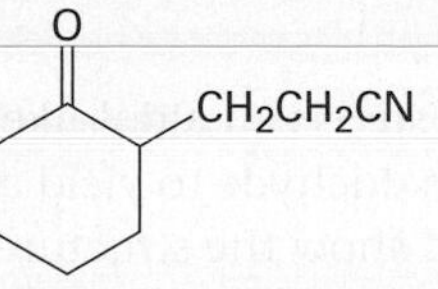

(c) $CH_2CH{=}CH_2$ (d) CO_2Et

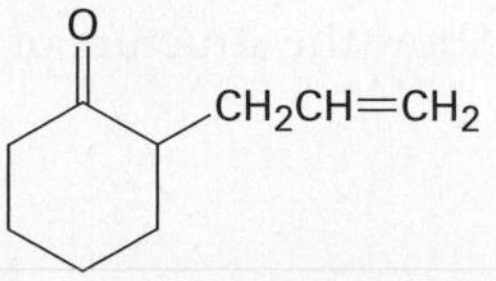

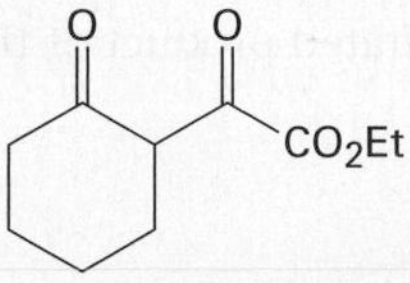

23.50 Leucine, one of the twenty amino acids found in proteins, is metabolized by a pathway that includes the following step. Propose a mechanism.

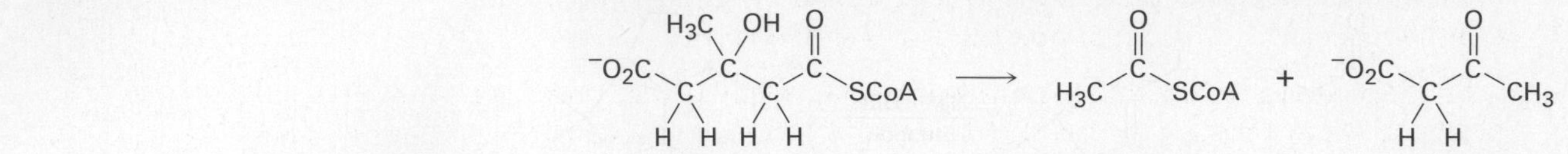

3-Hydroxy-3-methylglutaryl CoA → **Acetyl CoA** + **Acetoacetate**

23.51 Isoleucine, another of the twenty amino acids found in proteins, is metabolized by a pathway that includes the following step. Propose a mechanism.

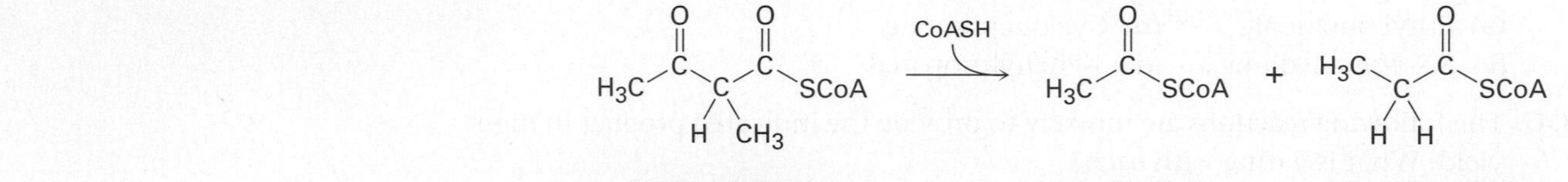

2-Methyl-3-ketobutyryl CoA → (CoASH) **Acetyl CoA** + **Propionyl CoA (propanoyl CoA)**

23.52 The first step in the citric acid cycle of food metabolism is reaction of oxaloacetate with acetyl CoA to give citrate. Propose a mechanism, using acid or base catalysis as needed.

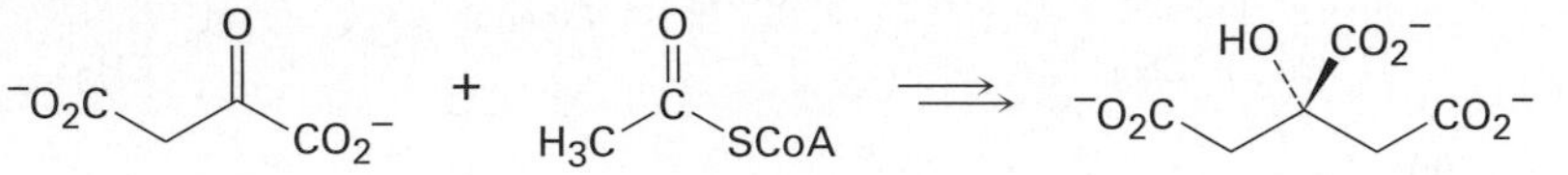

Oxaloacetate + **Acetyl CoA** ⇒ **Citrate**

23.53 The compound known as Hagemann's ester is prepared by treatment of a mixture of formaldehyde and ethyl acetoacetate with base, followed by acid-catalyzed decarboxylation.

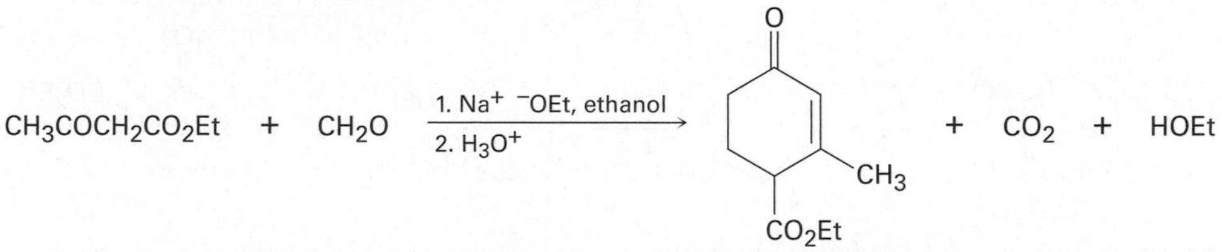

$$CH_3COCH_2CO_2Et + CH_2O \xrightarrow[2.\ H_3O^+]{1.\ Na^+\ ^-OEt,\ ethanol} \text{Hagemann's ester} + CO_2 + HOEt$$

Hagemann's ester

(a) The first step is an aldol-like condensation between ethyl acetoacetate and formaldehyde to yield an α,β-unsaturated product. Write the reaction, and show the structure of the product.

(b) The second step is a Michael reaction between ethyl acetoacetate and the unsaturated product of the first step. Show the structure of the product.

23.54 The third and fourth steps in the synthesis of Hagemann's ester from ethyl acetoacetate and formaldehyde (Problem 23.53) are an intramolecular aldol cyclization to yield a substituted cyclohexenone, and a decarboxylation reaction. Write both reactions, and show the products of each step.

23.55 When 2-methylcyclohexanone is converted into an enamine, only one product is formed despite the fact that the starting ketone is unsymmetrical. Build molecular models of the two possible products and explain the fact that the sole product is the one with the double bond away from the methyl-substituted carbon.

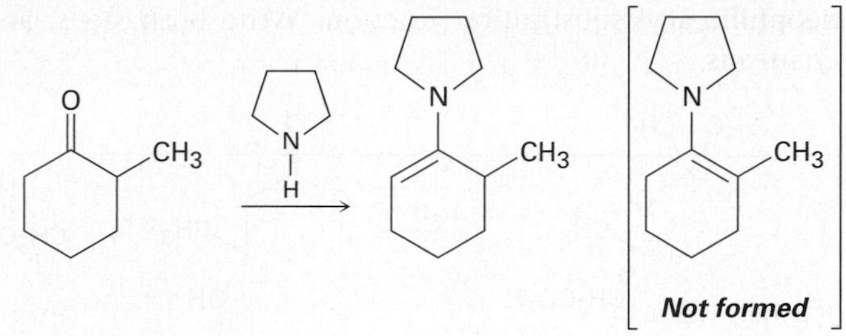

CH_3 N H CH_3 N CH_3 N

Not formed

23.56 The amino acid leucine is biosynthesized from α-ketoisovalerate by the following sequence of steps. Show the mechanism of each.

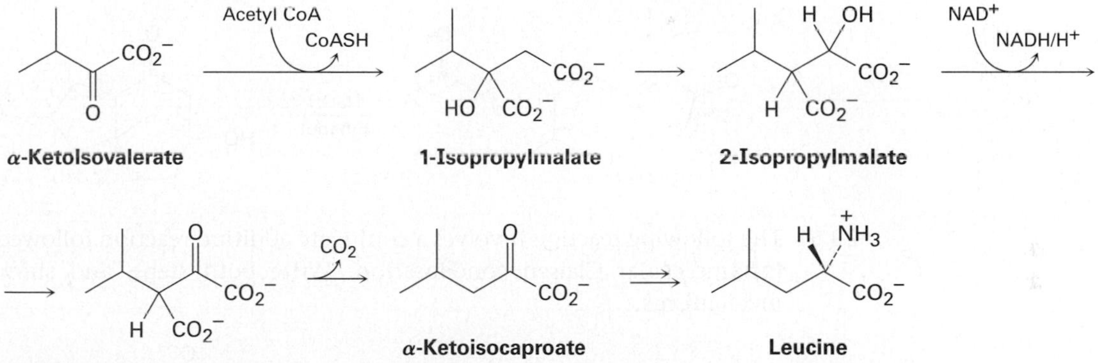

CO_2^- O Acetyl CoA CoASH HO CO_2^- CO_2^- H OH CO_2^- H CO_2^- NAD^+ $NADH/H^+$

α-Ketoisovalerate **1-Isopropylmalate** **2-Isopropylmalate**

O CO_2^- H CO_2^- CO_2 O CO_2^- H $\overset{+}{N}H_3$ CO_2^-

α-Ketoisocaproate **Leucine**

23.57 The Knoevenagel reaction is a carbonyl condensation reaction of an ester with an aldehyde or ketone to yield an α,β-unsaturated product. Show the mechanism of the Knoevenagel reaction of diethyl malonate with benzaldehyde.

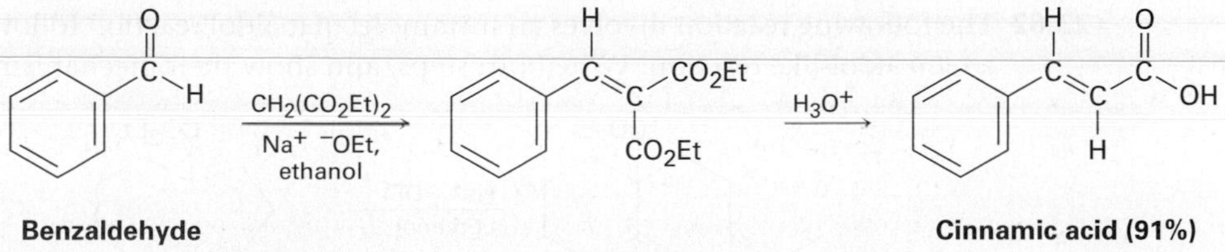

O C H $CH_2(CO_2Et)_2$ Na^+ ^-OEt, ethanol H C C CO_2Et CO_2Et H_3O^+ H O C C C OH H

Benzaldehyde **Cinnamic acid (91%)**

23.58 The Darzens reaction involves a two-step, base-catalyzed condensation of ethyl chloroacetate with a ketone to yield an epoxy ester. The first step is a carbonyl condensation reaction, and the second step is an S_N2 reaction. Write both steps, and show their mechanisms.

+ $ClCH_2CO_2Et$ $\xrightarrow[\text{Ethanol}]{Na^+ \ ^-OEt}$ $CHCO_2Et$

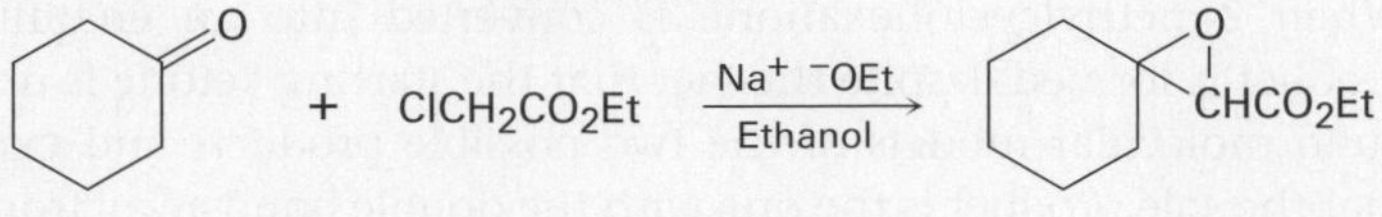

23.59 The following reaction involves a hydrolysis followed by an intramolecular nucleophilic acyl substitution reaction. Write both steps, and show their mechanisms.

H_3C CH_3 O O CH_3 CH_2CO_2H $\xrightarrow{H_3O^+}$ O O CH_3 OH + CH_3CCH_3 (O)

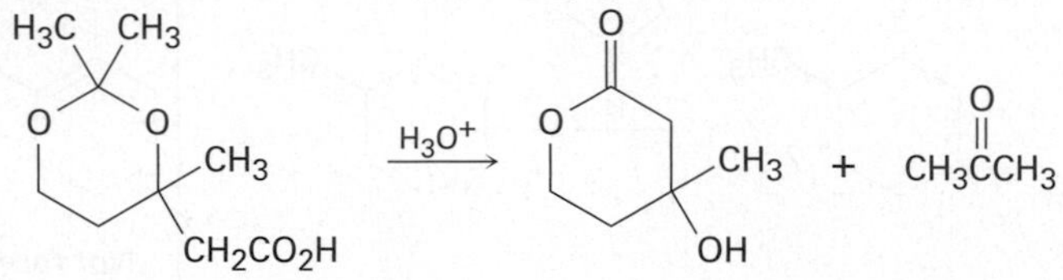

23.60 The following reaction involves an intramolecular Michael reaction followed by an intramolecular aldol reaction. Write both steps, and show their mechanisms.

$\xrightarrow[\text{Ethanol}]{NaOH}$ HO

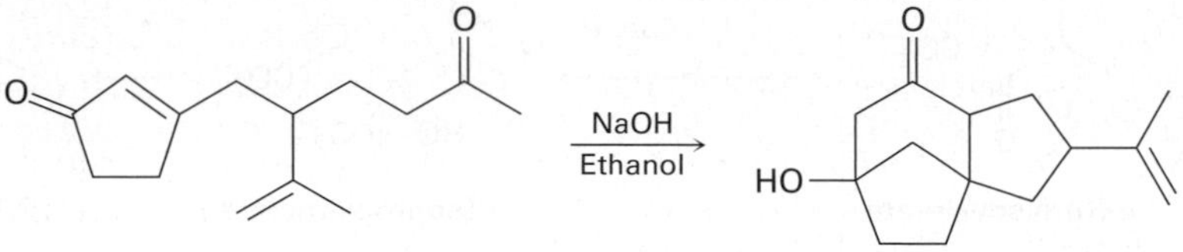

23.61 The following reaction involves a conjugate addition reaction followed by an intramolecular Claisen condensation. Write both steps, and show their mechanisms.

CO_2CH_3 $C{\equiv}C{-}CO_2CH_3$ $\xrightarrow{(CH_3)_2CuLi}$ O CO_2CH_3 CH_3

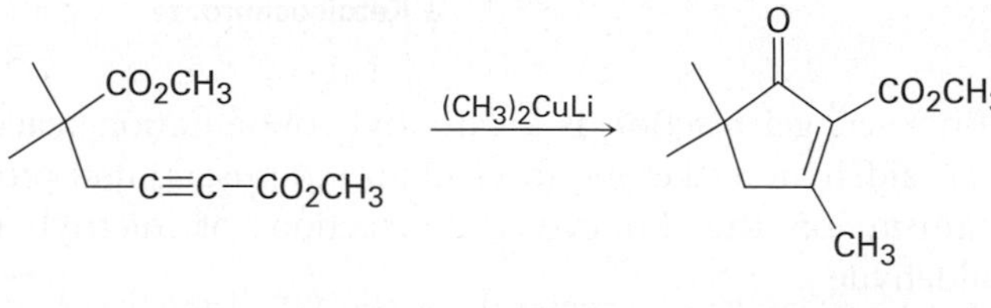

23.62 The following reaction involves an intramolecular aldol reaction followed by a *retro* aldol-like reaction. Write both steps, and show their mechanisms.

CO_2Et O O $\xrightarrow[\text{Ethanol}]{Na^+ \ ^-OEt}$ CO_2Et O O

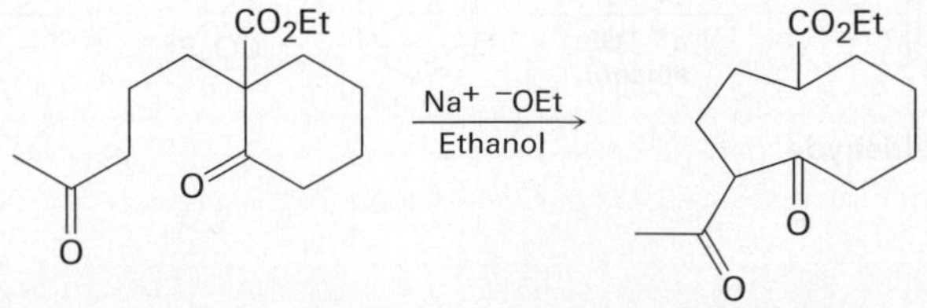

23.63 Propose a mechanism for the following base-catalyzed isomerization:

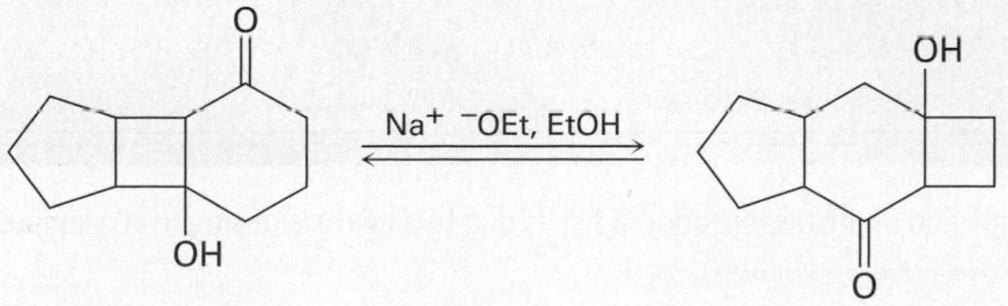

Na^+ ^-OEt, EtOH

23.64 The Mannich reaction of a ketone, an amine, and an aldehyde is one of the few three-component reactions in organic chemistry. Cyclohexanone, for example, reacts with dimethylamine and acetaldehyde to yield an amino ketone. The reaction takes place in two steps, both of which are typical carbonyl-group reactions.

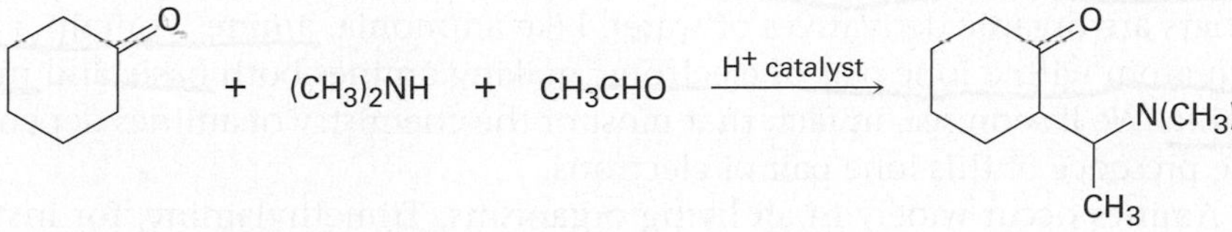

+ $(CH_3)_2NH$ + CH_3CHO $\xrightarrow{H^+ \text{ catalyst}}$ $N(CH_3)_2$, CH_3

(a) The first step is reaction between the aldehyde and the amine to yield an intermediate iminium ion ($R_2C{=}NR_2{}^+$) plus water. Propose a mechanism, and show the structure of the intermediate iminium ion.

(b) The second step is reaction between the iminium ion intermediate and the ketone to yield the final product. Propose a mechanism.

23.65 Cocaine has been prepared by a sequence beginning with a Mannich reaction (Problem 23.64) between dimethyl acetonedicarboxylate, an amine, and a dialdehyde. Show the structures of the amine and dialdehyde.

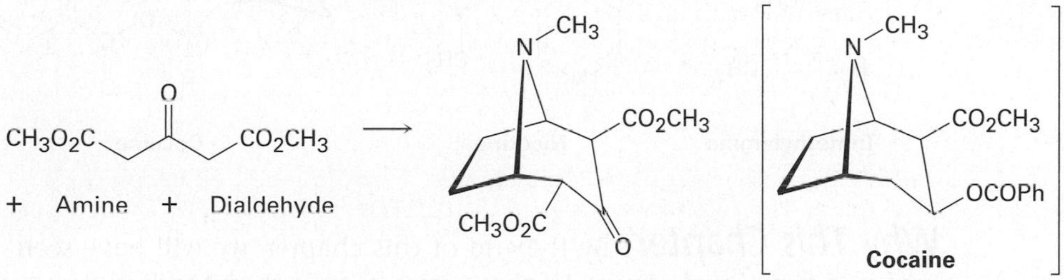

CH_3O_2C, CO_2CH_3 + Amine + Dialdehyde → CH_3, CO_2CH_3, CH_3O_2C

[CH_3, CO_2CH_3, OCOPh **Cocaine**]

23.66 Propose a mechanism to account for the following reaction of an enamine with an alkyl halide:

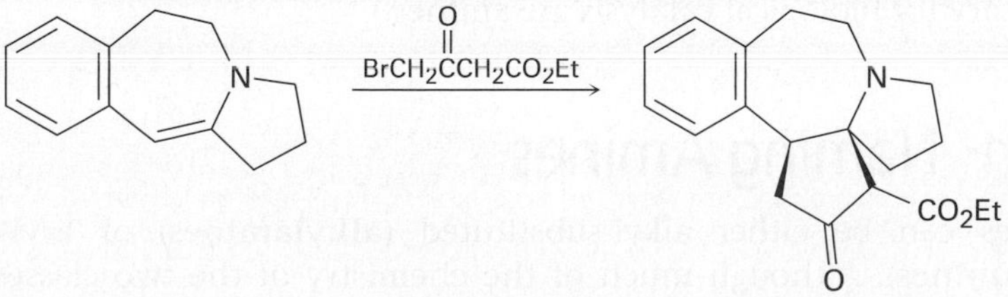

$BrCH_2CCH_2CO_2Et$ (C=O) → CO_2Et

24

The characteristic and unmistakable odor of fish is due to a mixture of simple alkylamines. Image copyright tororo reaction, 2010. Used under license from Shutterstock.com

Amines and Heterocycles

Amines are organic derivatives of ammonia in the same way that alcohols and ethers are organic derivatives of water. Like ammonia, amines contain a nitrogen atom with a lone pair of electrons, making amines both basic and nucleophilic. We'll soon see, in fact, that most of the chemistry of amines depends on the presence of this lone pair of electrons.

Amines occur widely in all living organisms. Trimethylamine, for instance, occurs in animal tissues and is partially responsible for the distinctive odor of fish; nicotine is found in tobacco; and cocaine is a stimulant found in the leaves of the South American coca bush. In addition, amino acids are the building blocks from which all proteins are made, and cyclic amine bases are constituents of nucleic acids.

Trimethylamine **Nicotine** **Cocaine**

Why This Chapter? By the end of this chapter, we will have seen all the common functional groups. Of those groups, amines and carbonyl compounds are the most abundant and have the richest chemistry. In addition to the proteins and nucleic acids already mentioned, the majority of pharmaceutical agents contain amine functional groups, and many of the common coenzymes necessary for biological catalysis are amines.

24.1 Naming Amines

Amines can be either alkyl-substituted (**alkylamines**) or aryl-substituted (**arylamines**). Although much of the chemistry of the two classes is similar, there are also substantial differences. Amines are classified as **primary** **(RNH_2)**, **secondary** **(R_2NH)**, or **tertiary** **(R_3N)**, depending on the number of organic substituents attached to nitrogen. Thus, methylamine (CH_3NH_2) is a primary amine, dimethylamine [$(CH_3)_2NH$] is a secondary amine, and trimethylamine [$(CH_3)_3N$] is a tertiary amine. Note that this usage of the terms *primary, secondary,* and *tertiary* is different from our previous usage. When we speak of a tertiary alcohol or alkyl halide, we refer to the degree of substitution at the alkyl carbon atom, but when we speak of a tertiary amine, we refer to the degree of substitution at the nitrogen atom.

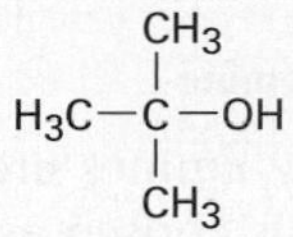

***tert*-Butyl alcohol** **(a tertiary alcohol)** — **Trimethylamine** **(a tertiary amine)** — ***tert*-Butylamine** **(a primary amine)**

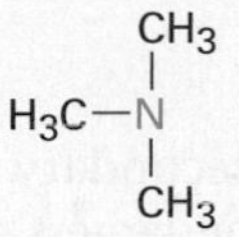

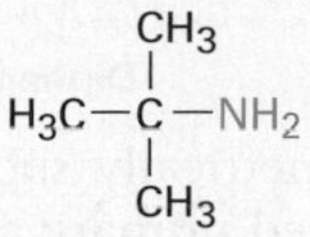

Compounds containing a nitrogen atom with four attached groups also exist, but the nitrogen atom must carry a formal positive charge. Such compounds are called **quaternary ammonium salts.**

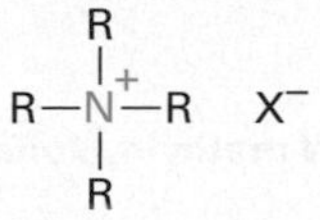

$R_4N^+\ X^-$ **A quaternary ammonium salt**

Primary amines are named in the IUPAC system in several ways. For simple amines, the suffix *-amine* is added to the name of the alkyl substituent. You might also recall from Chapter 15 that phenylamine, $C_6H_5NH_2$, has the common name *aniline*.

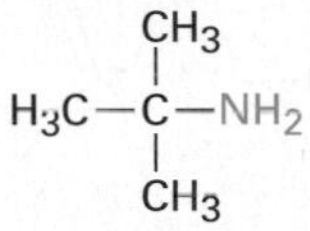

***tert*-Butylamine** 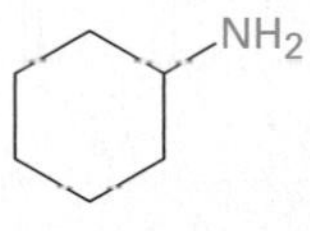 — **Cyclohexylamine** 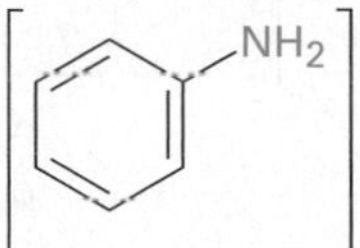 — **Aniline**

Alternatively, the suffix *-amine* can be used in place of the final *-e* in the name of the parent compound.

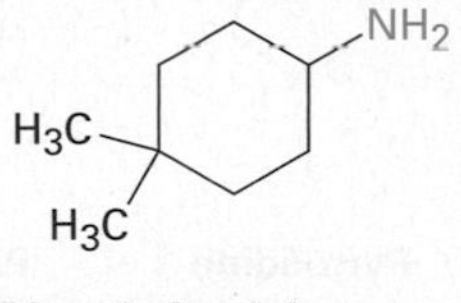

4,4-Dimethylcyclohexanamine — $H_2NCH_2CH_2CH_2CH_2NH_2$ **1,4-Butanediamine**

Amines with more than one functional group are named by considering the $-NH_2$ as an *amino* substituent on the parent molecule.

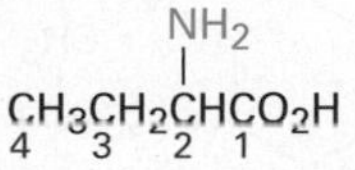

2-Aminobutanoic acid 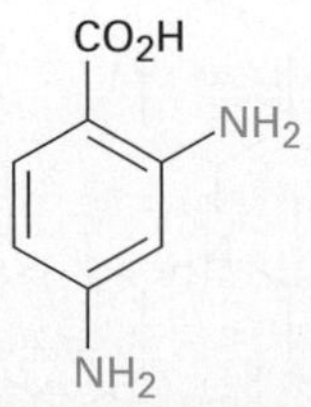 — **2,4-Diaminobenzoic acid** 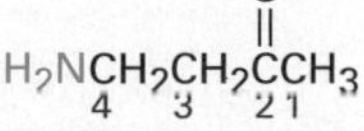 — **4-Amino-2-butanone**

Symmetrical secondary and tertiary amines are named by adding the prefix *di-* or *tri-* to the alkyl group.

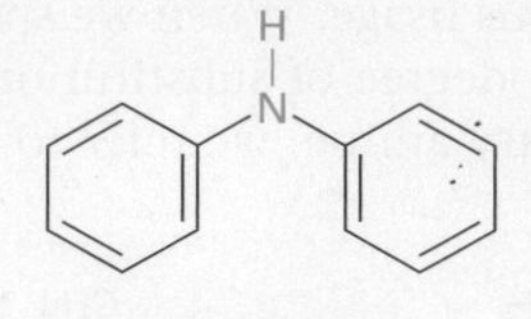

$CH_3CH_2-N(CH_2CH_3)-CH_2CH_3$

Diphenylamine 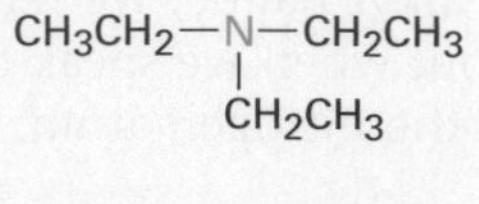 **Triethylamine**

Unsymmetrically substituted secondary and tertiary amines are named as *N*-substituted primary amines. The largest alkyl group is chosen as the parent name, and the other alkyl groups are considered *N*-substituents on the parent (*N* because they're attached to nitrogen).

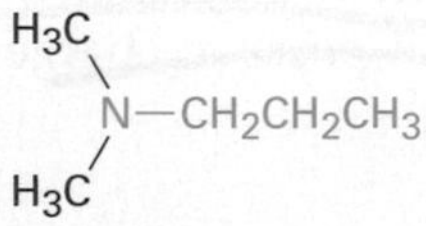

$(H_3C)_2N-CH_2CH_2CH_3$

N,N-Dimethylpropylamine 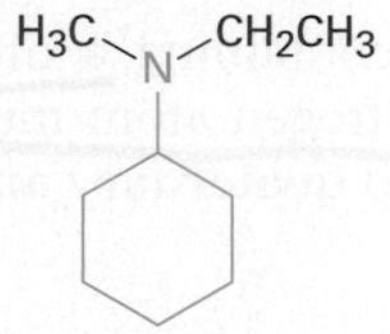 **N-Ethyl-N-methylcyclohexylamine**

Heterocyclic amines—compounds in which the nitrogen atom occurs as part of a ring—are also common, and each different heterocyclic ring system has its own parent name. The heterocyclic nitrogen atom is always numbered as position 1.

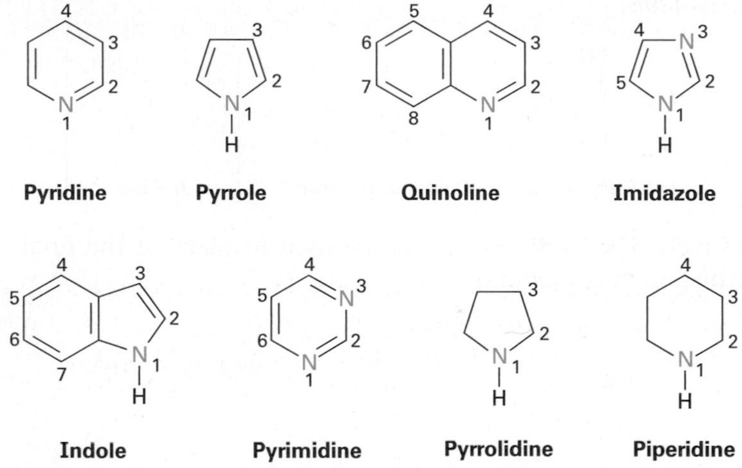

Pyridine **Pyrrole** **Quinoline** **Imidazole**

Indole **Pyrimidine** **Pyrrolidine** **Piperidine**

Problem 24.1

Name the following compounds:

(a) $CH_3NHCH_2CH_3$ **(b)** **(c)**

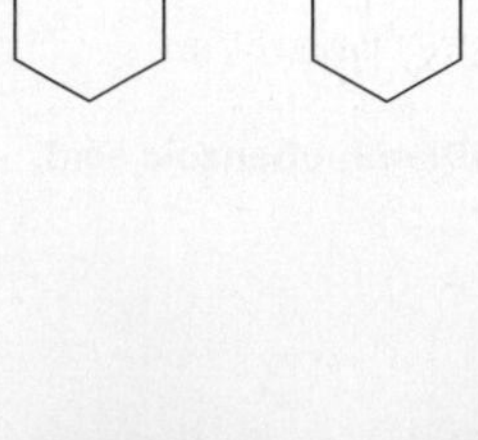

(d) N–CH_3 (e) H–N (f) CH_3 / $H_2NCH_2CH_2CHNH_2$

Problem 24.2

Draw structures corresponding to the following IUPAC names:

(a) Triisopropylamine **(b)** Triallylamine
(c) *N*-Methylaniline **(d)** *N* Ethyl-*N*-methylcyclopentylamine
(e) *N*-Isopropylcyclohexylamine **(f)** *N*-Ethylpyrrole

Problem 24.3

Draw structures for the following heterocyclic amines:

(a) 5-Methoxyindole **(b)** 1,3-Dimethylpyrrole
(c) 4-(*N*,*N*-Dimethylamino)pyridine **(d)** 5-Aminopyrimidine

24.2 Structure and Properties of Amines

The bonding in alkylamines is similar to the bonding in ammonia. The nitrogen atom is sp^3-hybridized, with the three substituents occupying three corners of a regular tetrahedron and the lone pair of electrons occupying the fourth corner. As you might expect, the C–N–C bond angles are close to the 109° tetrahedral value. For trimethylamine, the C–N–C bond angle is 108° and the C–N bond length is 147 pm.

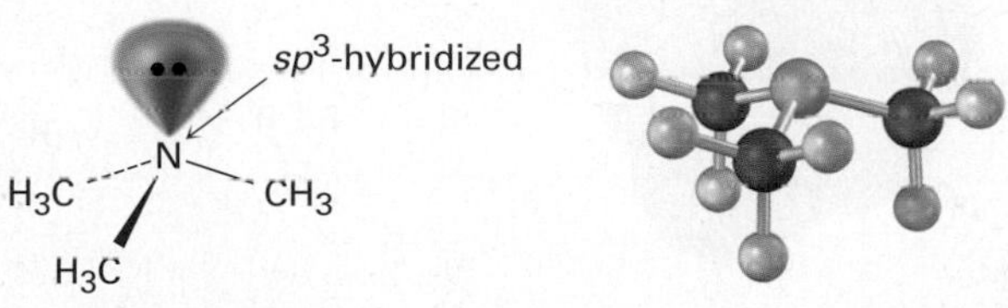

Trimethylamine

One consequence of tetrahedral geometry is that an amine with three different substituents on nitrogen is chiral, as we saw in **Section 5.10**. Unlike chiral carbon compounds, however, chiral amines can't usually be resolved because the two enantiomeric forms rapidly interconvert by a pyramidal inversion, much as an alkyl halide inverts in an S_N2 reaction. Pyramidal inversion occurs by a momentary rehybridization of the nitrogen atom to planar, sp^2 geometry, followed by rehybridization of the planar intermediate to tetrahedral, sp^3 geometry **(Figure 24.1)**. The barrier to inversion is about 25 kJ/mol (6 kcal/mol), an amount only twice as large as the barrier to rotation about a C–C single bond.

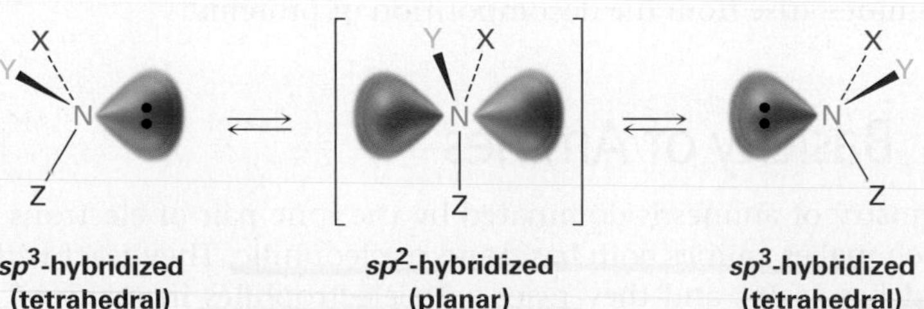

Figure 24.1 Pyramidal inversion rapidly interconverts the two mirror-image (enantiomeric) forms of an amine.

Alkylamines have a variety of applications in the chemical industry as starting materials for the preparation of insecticides and pharmaceuticals. Labetalol, for instance, a so-called β-blocker used for the treatment of high blood pressure, is prepared by S_N2 reaction of an epoxide with a primary amine. The substance marketed for drug use is a mixture of all four possible stereoisomers, but the biological activity derives primarily from the (*R*,*R*) isomer.

Labetalol

Like alcohols, amines with fewer than five carbon atoms are generally water-soluble. Also like alcohols, primary and secondary amines form hydrogen bonds and are highly associated. As a result, amines have higher boiling points than alkanes of similar molecular weight. Diethylamine (MW = 73 amu) boils at 56.3 °C, for instance, while pentane (MW = 72 amu) boils at 36.1 °C.

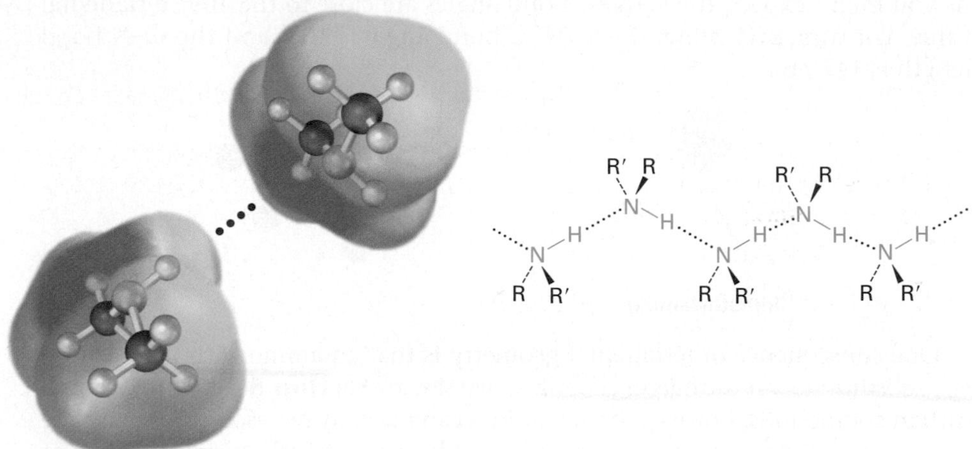

One other characteristic of amines is their odor. Low-molecular-weight amines such as trimethylamine have a distinctive fishlike aroma, while diamines such as cadaverine (1,5-pentanediamine) and putrescine (1,4-butanediamine) have the appalling odors you might expect from their common names. Both these diamines arise from the decomposition of proteins.

24.3 Basicity of Amines

The chemistry of amines is dominated by the lone pair of electrons on nitrogen, which makes amines both basic and nucleophilic. They react with acids to form acid–base salts, and they react with electrophiles in many of the polar reactions seen in past chapters. Note in the following electrostatic potential map of trimethylamine how the negative (red) region corresponds to the lone-pair of electrons on nitrogen.

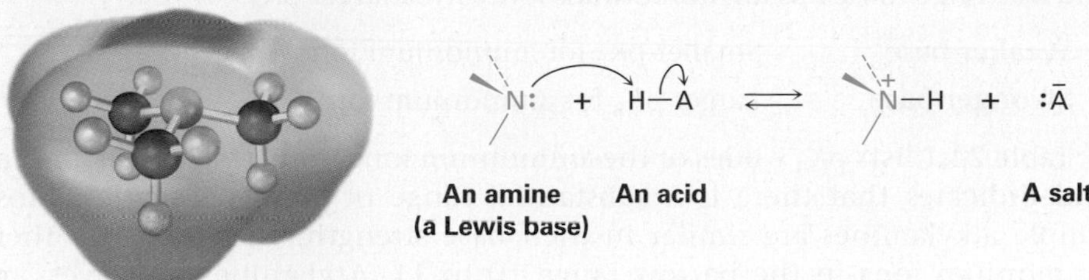

Amines are much stronger bases than alcohols and ethers, their oxygen-containing analogs. When an amine is dissolved in water, an equilibrium is established in which water acts as an acid and transfers a proton to the amine. Just as the acid strength of a carboxylic acid can be measured by defining an acidity constant K_a **(Section 2.8)**, the base strength of an amine can be measured by defining an analogous *basicity constant* K_b. The larger the value of K_b and the smaller the value of pK_b, the more favorable the proton-transfer equilibrium and the stronger the base.

For the reaction

$$RNH_2 + H_2O \rightleftharpoons RNH_3^+ + OH^-$$

$$K_b = \frac{[RNH_3^+][OH^-]}{[RNH_2]}$$

$$pK_b = -\log K_b$$

In practice, K_b values are not often used. Instead, the most convenient way to measure the basicity of an amine (RNH_2) is to look at the acidity of the corresponding ammonium ion (RNH_3^+).

For the reaction

$$RNH_3^+ + H_2O \rightleftharpoons RNH_2 + H_3O^+$$

$$K_a = \frac{[RNH_2][H_3O^+]}{[RNH_3^+]}$$

so $$K_a \cdot K_b = \left[\frac{[RNH_2][H_3O^+]}{[RNH_3^+]}\right]\left[\frac{[RNH_3^+][OH^-]}{[RNH_2]}\right]$$

$$= [H_3O^+][OH^-] = K_w = 1.00 \times 10^{-14}$$

Thus $$K_a = \frac{K_w}{K_b} \quad \text{and} \quad K_b = \frac{K_w}{K_a}$$

and $$pK_a + pK_b = 14$$

These equations say that the K_b of an amine multiplied by the K_a of the corresponding ammonium ion is equal to K_w, the ion-product constant for water (1.00×10^{-14}). Thus, if we know K_a for an ammonium ion, we also know K_b for the corresponding amine base because $K_b = K_w/K_a$. The more acidic the ammonium ion, the less tightly the proton is held and the weaker the corresponding base. That is, a weaker base has an ammonium ion with a smaller pK_a and a stronger base has an ammonium ion with a larger pK_a.

Weaker base	Smaller pK_a for ammonium ion
Stronger base	Larger pK_a for ammonium ion

Table 24.1 lists pK_a values of the ammonium ions from a variety of amines and indicates that there is a substantial range of amine basicities. Most simple alkylamines are similar in their base strength, with pK_a's for their ammonium ions in the narrow range 10 to 11. Arylamines, however, are considerably less basic than alkylamines, as are the heterocyclic amines pyridine and pyrrole.

Table 24.1 Basicity of Some Common Amines

Name	Structure	pK_a of ammonium ion
Ammonia	NH_3	9.26
Primary alkylamine		
Methylamine	CH_3NH_2	10.64
Ethylamine	$CH_3CH_2NH_2$	10.75
Secondary alkylamine		
Diethylamine	$(CH_3CH_2)_2NH$	10.98
Pyrrolidine	NH	11.27
Tertiary alkylamine		
Triethylamine	$(CH_3CH_2)_3N$	10.76
Arylamine		
Aniline	NH_2	4.63
Heterocyclic amine		
Pyridine	N	5.25
Pyrimidine	N, N	1.3
Pyrrole	NH	0.4
Imidazole	N, NH	6.95

In contrast with amines, amides ($RCONH_2$) are nonbasic. Amides aren't protonated by aqueous acids, and they are poor nucleophiles. The main reason for this difference in basicity between amines and amides is that an amide is stabilized by delocalization of the nitrogen lone-pair electrons through orbital overlap with the carbonyl group. In resonance terms, amides are more stable and less reactive than amines because they are hybrids of two resonance forms. This amide resonance stabilization is lost when the nitrogen atom is protonated, so protonation is disfavored. Electrostatic potential maps show clearly the decreased electron density on the amide nitrogen.

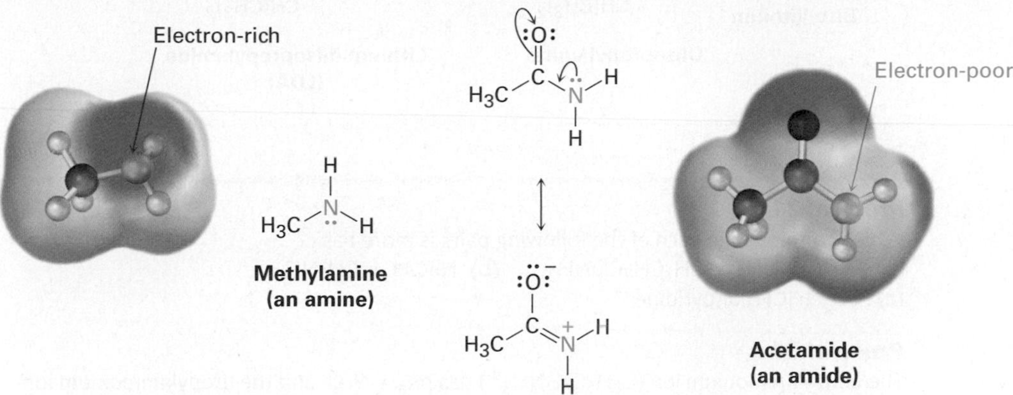

It's often possible to take advantage of their basicity to purify amines. For example, if a mixture of a basic amine and a neutral compound such as a ketone or alcohol is dissolved in an organic solvent and aqueous acid is added, the basic amine dissolves in the water layer as its protonated salt, while the neutral compound remains in the organic solvent layer. Separation of the water layer and neutralization of the ammonium ion by addition of NaOH then provides the pure amine **(Figure 24.2)**.

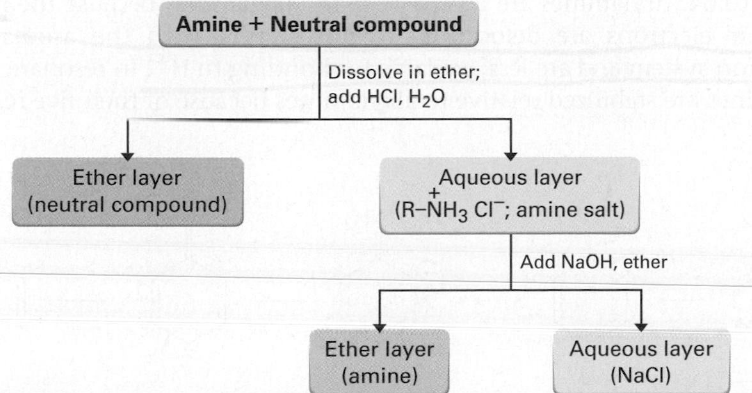

Figure 24.2 Separation and purification of an amine component from a mixture by extraction of its ammonium salt into water.

In addition to their behavior as bases, primary and secondary amines can also act as very weak acids because an N–H proton can be removed by a sufficiently strong base. We've seen, for example, how diisopropylamine ($pK_a \approx 36$) reacts with butyllithium to yield lithium diisopropylamide (LDA; **Section 22.5**). Dialkylamine anions like LDA are very strong bases that are often used in laboratory organic chemistry for the generation of enolate ions from carbonyl compounds **(Section 22.7)**. They are not, however, encountered in biological chemistry.

$$C_4H_9Li + H{-}N(CH(CH_3)_2)_2 \xrightarrow[\text{solvent}]{\text{THF}} Li^{+}\ {}^{-}{:}N(CH(CH_3)_2)_2 + C_4H_{10}$$

Butyllithium **Diisopropylamine** **Lithium diisopropylamide (LDA)**

Problem 24.4

Which compound in each of the following pairs is more basic?

(a) $CH_3CH_2NH_2$ or $CH_3CH_2CONH_2$ **(b)** NaOH or CH_3NH_2

(c) CH_3NHCH_3 or pyridine

Problem 24.5

The benzylammonium ion ($C_6H_5CH_2NH_3^+$) has $pK_a = 9.33$, and the propylammonium ion has $pK_a = 10.71$. Which is the stronger base, benzylamine or propylamine? What are the pK_b's of benzylamine and propylamine?

24.4 Basicity of Arylamines

As noted previously, arylamines are generally less basic than alkylamines. Anilinium ion has $pK_a = 4.63$, for instance, whereas methylammonium ion has $pK_a = 10.64$. Arylamines are less basic than alkylamines because the nitrogen lone-pair electrons are delocalized by interaction with the aromatic ring π electron system and are less available for bonding to H^+. In resonance terms, arylamines are stabilized relative to alkylamines because of their five resonance forms.

:NH₂ ⟷ :NH₂ ⟷ ⁺NH₂ ⟷ ⁺NH₂ ⟷ ⁺NH₂

Much of the resonance stabilization is lost on protonation, however, so the energy difference between protonated and nonprotonated forms is higher for arylamines than it is for alkylamines. As a result, arylamines are less basic. **Figure 24.3** illustrates the difference.

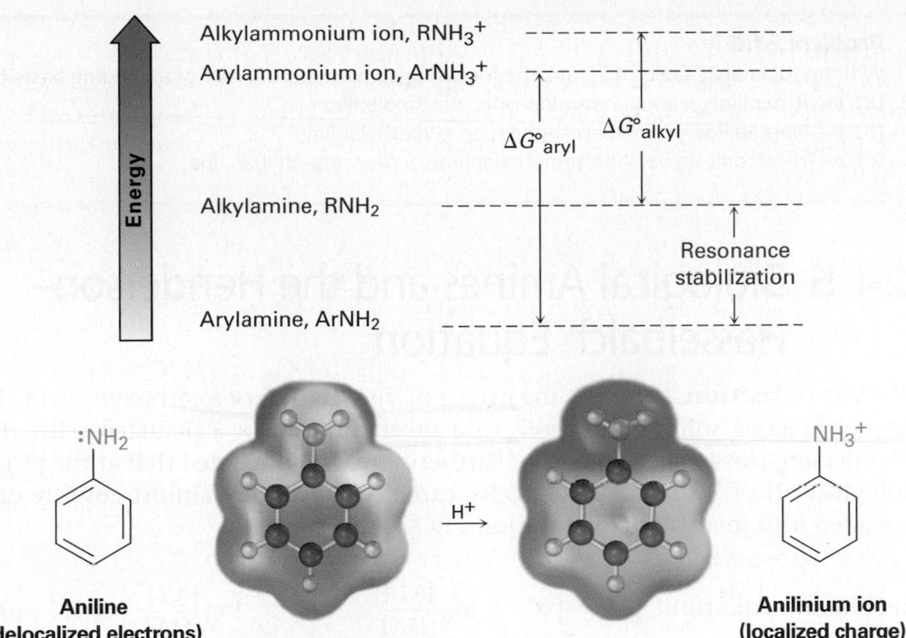

Figure 24.3 Arylamines have a larger positive $\Delta G°$ for protonation and are therefore less basic than alkylamines, primarily because of resonance stabilization of the ground state. Electrostatic potential maps show that lone-pair electron density is delocalized in the amine but the charge is localized in the corresponding ammonium ion.

Substituted arylamines can be either more basic or less basic than aniline, depending on the substituent. Electron-donating substituents, such as $-CH_3$, $-NH_2$, and $-OCH_3$, which increase the reactivity of an aromatic ring toward electrophilic substitution **(Section 16.5)**, also increase the basicity of the corresponding arylamine. Electron-withdrawing substituents, such as $-Cl$, $-NO_2$, and $-CN$, which decrease ring reactivity toward electrophilic substitution, also decrease arylamine basicity. Table 24.2 considers only *p*-substituted anilines, but similar trends are observed for ortho and meta derivatives.

Table 24.2 Base Strength of Some *p*-Substituted Anilines

$$Y-C_6H_4-\ddot{N}H_2 + H_2O \rightleftharpoons Y-C_6H_4-\overset{+}{N}H_3 + {}^-OH$$

	Substituent, Y	pK_a	
Stronger base	$-NH_2$	6.15	Activating groups
	$-OCH_3$	5.34	
	$-CH_3$	5.08	
	$-H$	4.63	
	$-Cl$	3.98	Deactivating groups
	$-Br$	3.86	
	$-CN$	1.74	
Weaker base	$-NO_2$	1.00	

Problem 24.6

Without looking at Table 24.2, rank the following compounds in order of ascending basicity.

(a) *p*-Nitroaniline, *p*-aminobenzaldehyde, *p*-bromoaniline

(b) *p*-Chloroaniline, *p*-aminoacetophenone, *p*-methylaniline

(c) *p*-(Trifluoromethyl)aniline, *p*-methylaniline, *p*-(fluoromethyl)aniline

24.5 Biological Amines and the Henderson–Hasselbalch Equation

We saw in **Section 20.3** that the extent of dissociation of a carboxylic acid HA in an aqueous solution buffered to a given pH can be calculated with the Henderson–Hasselbalch equation. Furthermore, we concluded that at the physiological pH of 7.3 inside living cells, carboxylic acids are almost entirely dissociated into their carboxylate anions, RCO_2^-.

$$\textbf{Henderson–Hasselbalch equation: } \mathrm{pH} = \mathrm{p}K_a + \log\frac{[\mathrm{A^-}]}{[\mathrm{HA}]} \quad \text{so} \quad \log\frac{[\mathrm{A^-}]}{[\mathrm{HA}]} = \mathrm{pH} - \mathrm{p}K_a$$

What about amine bases? In what form do they exist at the physiological pH inside cells? As the amine ($A^- = RNH_2$), or as the ammonium ion ($HA = RNH_3^+$)? Let's take a 0.0010 M solution of methylamine at pH = 7.3, for example. According to Table 24.1, the pK_a of methylammonium ion is 10.64, so from the Henderson–Hasselbalch equation, we have

$$\log\frac{[\mathrm{RNH_2}]}{[\mathrm{RNH_3^+}]} = \mathrm{pH} - \mathrm{p}K_a = 7.3 - 10.64 = -3.34$$

$$\frac{[\mathrm{RNH_2}]}{[\mathrm{RNH_3^+}]} = \mathrm{antilog}(-3.34) = 4.6 \times 10^{-4} \quad \text{so} \quad [\mathrm{RNH_2}] = (4.6 \times 10^{-4})[\mathrm{RNH_3^+}]$$

In addition, we know that

$$[\mathrm{RNH_2}] + [\mathrm{RNH_3^+}] = 0.0010\ \mathrm{M}$$

Solving the two simultaneous equations gives $[RNH_3^+] = 0.0010$ M and $[RNH_2] = 5 \times 10^{-7}$ M. In other words, at a physiological pH of 7.3, essentially 100% of the methylamine in a 0.0010 M solution exists in its protonated form as methylammonium ion. The same is true of other amine bases, so we always write cellular amines in their protonated form and amino acids in their ammonium carboxylate form to reflect their structures at physiological pH.

The amino group is protonated at pH = 7.3.

The carboxylic acid group is dissociated at pH = 7.3.

$H_3C\ \ H$ / $H_3\overset{+}{N}-C-CO_2^-$

Alanine
(an amino acid)

Problem 24.7

Calculate the percentages of neutral and protonated forms present in a solution of 0.0010 M pyrimidine at pH = 7.3. The pK_a of pyrimidinium ion is 1.3.

24.6 Synthesis of Amines

Reduction of Nitriles, Amides, and Nitro Compounds

We've already seen in **Sections 20.7 and 21.7** how amines can be prepared by reduction of nitriles and amides with $LiAlH_4$. The two-step sequence of S_N2 displacement with CN^- followed by reduction thus converts an alkyl halide into a primary alkylamine having one more carbon atom. Amide reduction converts carboxylic acids and their derivatives into amines with the same number of carbon atoms.

$$RCH_2X \xrightarrow{NaCN} RCH_2C{\equiv}N \xrightarrow[2.\ H_2O]{1.\ LiAlH_4,\ ether} RCH_2CH_2NH_2$$

Alkyl halide → **1° amine**

$$RCOOH \xrightarrow[2.\ NH_3]{1.\ SOCl_2} RCONH_2 \xrightarrow[2.\ H_2O]{1.\ LiAlH_4,\ ether} RCH_2NH_2$$

Carboxylic acid → **1° amine**

Arylamines are usually prepared by nitration of an aromatic starting material, followed by reduction of the nitro group (**Section 16.2**). The reduction step can be carried out in many different ways, depending on the circumstances. Catalytic hydrogenation over platinum works well but is often incompatible with the presence elsewhere in the molecule of other reducible groups, such as C=C bonds or carbonyl groups. Iron, zinc, tin, and tin(II) chloride ($SnCl_2$) are also effective when used in acidic aqueous solution. Tin(II) chloride is particularly mild and is often used when other reducible functional groups are present.

$$p\text{-}(CH_3)_3C\text{-}C_6H_4\text{-}NO_2 \xrightarrow[\text{Pt catalyst, ethanol}]{H_2} p\text{-}(CH_3)_3C\text{-}C_6H_4\text{-}NH_2$$

***p-tert*-Butylnitrobenzene** → ***p-tert*-Butylaniline (100%)**

$$m\text{-}OHC\text{-}C_6H_4\text{-}NO_2 \xrightarrow[2.\ NaOH]{1.\ SnCl_2,\ H_3O^+} m\text{-}OHC\text{-}C_6H_4\text{-}NH_2$$

***m*-Nitrobenzaldehyde** → ***m*-Aminobenzaldehyde (90%)**

Problem 24.8

Propose structures for either a nitrile or an amide that might be a precursor of each of the following amines:

(a) $CH_3CH_2CH_2NH_2$ **(b)** $(CH_3CH_2CH_2)_2NH$

(c) Benzylamine, $C_6H_5CH_2NH_2$ **(d)** *N*-Ethylaniline

S_N2 Reactions of Alkyl Halides

Ammonia and other amines are good nucleophiles in S_N2 reactions. As a result, the simplest method of alkylamine synthesis is by S_N2 alkylation of ammonia or an alkylamine with an alkyl halide. If ammonia is used, a primary amine results; if a primary amine is used, a secondary amine results; and so on. Even tertiary amines react rapidly with alkyl halides to yield quaternary ammonium salts, $R_4N^+\ X^-$.

Ammonia	$\ddot{N}H_3 + R{-}X \xrightarrow{S_N2} R\overset{+}{N}H_3\ X^- \xrightarrow{NaOH}$	RNH_2	Primary
Primary	$R\ddot{N}H_2 + R{-}X \xrightarrow{S_N2} R_2\overset{+}{N}H_2\ X^- \xrightarrow{NaOH}$	R_2NH	Secondary
Secondary	$R_2\ddot{N}H + R{-}X \xrightarrow{S_N2} R_3\overset{+}{N}H\ X^- \xrightarrow{NaOH}$	R_3N	Tertiary
Tertiary	$R_3\ddot{N} + R{-}X \xrightarrow{S_N2} R_4\overset{+}{N}\ X^-$		Quaternary ammonium

Unfortunately, these reactions don't stop cleanly after a single alkylation has occurred. Because ammonia and primary amines have similar reactivity, the initially formed monoalkylated substance often undergoes further reaction to yield a mixture of products. Even secondary and tertiary amines undergo further alkylation, although to a lesser extent. For example, treatment of 1-bromooctane with a twofold excess of ammonia leads to a mixture containing only 45% of octylamine. A nearly equal amount of dioctylamine is produced by double alkylation, along with smaller amounts of trioctylamine and tetraoctylammonium bromide.

$$CH_3(CH_2)_6CH_2Br + :NH_3 \longrightarrow CH_3(CH_2)_6CH_2\ddot{N}H_2 + [CH_3(CH_2)_6CH_2]_2\ddot{N}H$$

1-Bromooctane **Octylamine (45%)** **Dioctylamine (43%)**

$$+ [CH_3(CH_2)_6CH_2]_3N: + [CH_3(CH_2)_6CH_2]_4\overset{+}{N}\ \overset{-}{Br}$$

Trace **Trace**

A better method for preparing primary amines is to use azide ion, N_3^-, as the nucleophile rather than ammonia for S_N2 reaction with a primary or secondary alkyl halide. The product is an alkyl azide, which is not nucleophilic, so overalkylation can't occur. Subsequent reduction of the alkyl azide with $LiAlH_4$

then leads to the desired primary amine. Although the method works well, low-molecular-weight alkyl azides are explosive and must be handled carefully.

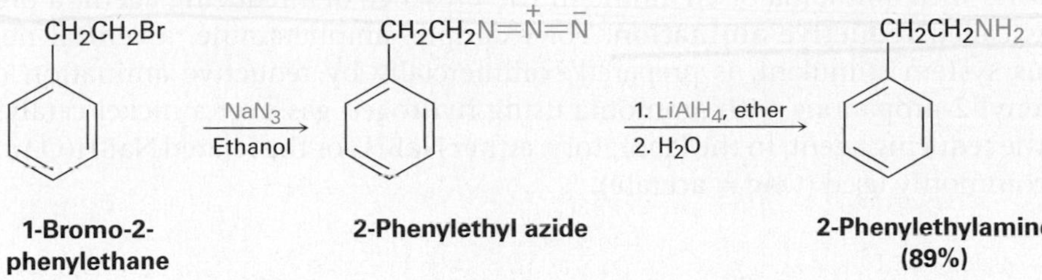

CH_2CH_2Br → (NaN_3, Ethanol) → $CH_2CH_2N{=}\overset{+}{N}{=}\overset{-}{N}$ → (1. $LiAlH_4$, ether; 2. H_2O) → $CH_2CH_2NH_2$

1-Bromo-2-phenylethane | **2-Phenylethyl azide** | **2-Phenylethylamine (89%)**

Another alternative for preparing a primary amine from an alkyl halide is the **Gabriel amine synthesis**, which uses a *phthalimide* alkylation. An **imide** (–CONHCO–) is similar to a β-keto ester in that the acidic N–H hydrogen is flanked by two carbonyl groups. Thus, imides are deprotonated by such bases as KOH, and the resultant anions are readily alkylated in a reaction similar to the acetoacetic ester synthesis **(Section 22.7)**. Basic hydrolysis of the *N*-alkylated imide then yields a primary amine product. The imide hydrolysis step is analogous to the hydrolysis of an amide **(Section 21.7)**.

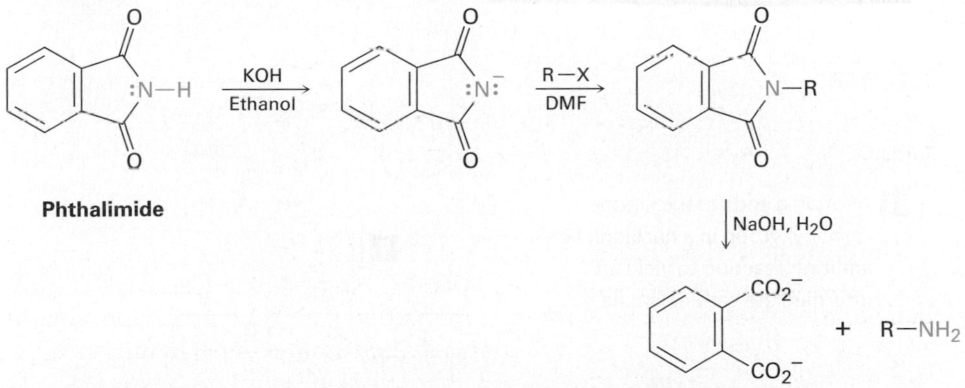

Phthalimide (:N–H) → (KOH, Ethanol) → :N:⁻ → (R–X, DMF) → N–R → (NaOH, H_2O) → phthalate (CO_2^-, CO_2^-) + R–NH_2

Problem 24.9

Write the mechanism of the last step in the Gabriel amine synthesis, the base-promoted hydrolysis of a phthalimide to yield an amine plus phthalate ion.

Problem 24.10

Show two methods for the synthesis of dopamine, a neurotransmitter involved in regulation of the central nervous system. Use any alkyl halide needed.

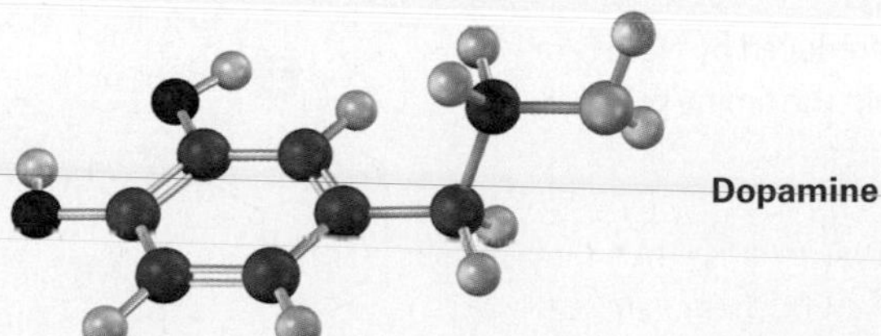

Dopamine

Reductive Amination of Aldehydes and Ketones

Amines can be synthesized in a single step by treatment of an aldehyde or ketone with ammonia or an amine in the presence of a reducing agent, a process called **reductive amination**. For example, amphetamine, a central nervous system stimulant, is prepared commercially by reductive amination of phenyl-2-propanone with ammonia using hydrogen gas over a nickel catalyst as the reducing agent. In the laboratory, either $NaBH_4$ or the related $NaBH(OAc)_3$ is commonly used (OAc = acetate).

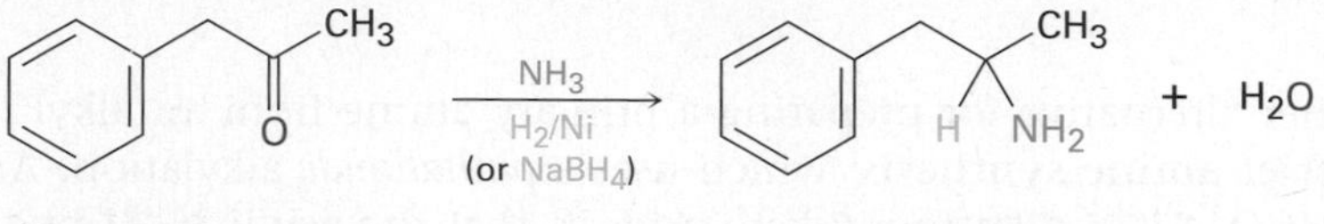

Reductive amination takes place by the pathway shown in **Figure 24.4**. An imine intermediate is first formed by a nucleophilic addition reaction **(Section 19.8)**, and the C=N bond of the imine is then reduced to the amine, much as the C=O bond of a ketone can be reduced to an alcohol.

Figure 24.4 | **MECHANISM**

Mechanism of reductive amination of a ketone to yield an amine. Details of the imine-forming step were shown in Figure 19.6 on page 737.

1 Ammonia adds to the ketone carbonyl group in a nucleophilic addition reaction to yield an intermediate carbinolamine.

2 The carbinolamine loses water to give an imine.

3 The imine is reduced by $NaBH_4$ or H_2/Ni to yield the amine product.

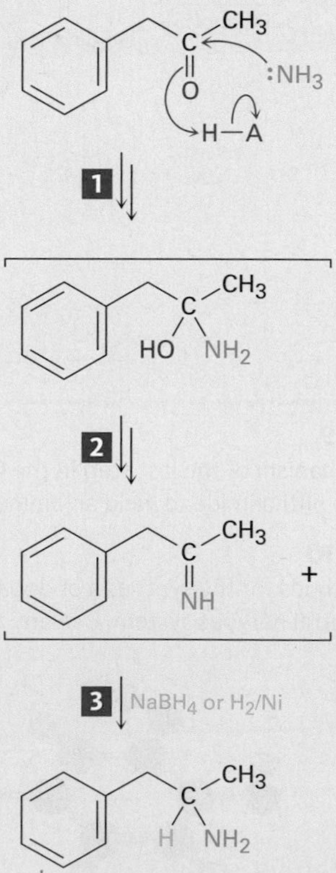

Ammonia, primary amines, and secondary amines can all be used in the reductive amination reaction, yielding primary, secondary, and tertiary amines, respectively.

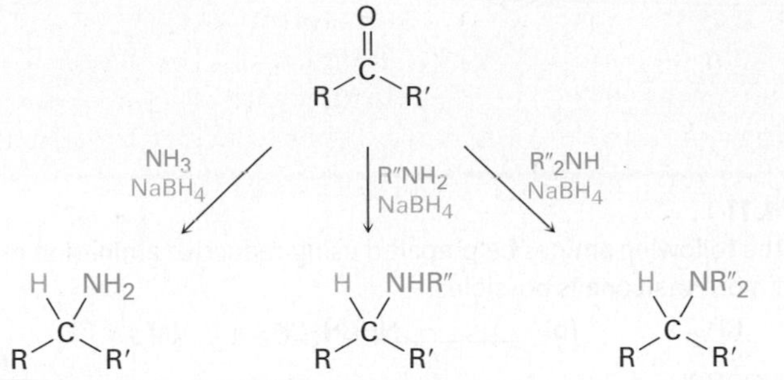

NH_3, $NaBH_4$ → Primary amine; $R''NH_2$, $NaBH_4$ → Secondary amine; R''_2NH, $NaBH_4$ → Tertiary amine

Primary amine **Secondary amine** **Tertiary amine**

Reductive aminations also occur in various biological pathways. In the biosynthesis of the amino acid proline, for instance, glutamate 5-semialdehyde undergoes internal imine formation to give 1-pyrrolinium 5-carboxylate, which is then reduced by nucleophilic addition of hydride ion to the C=N bond. Reduced nicotinamide adenine dinucleotide, NADH, acts as the biological reducing agent.

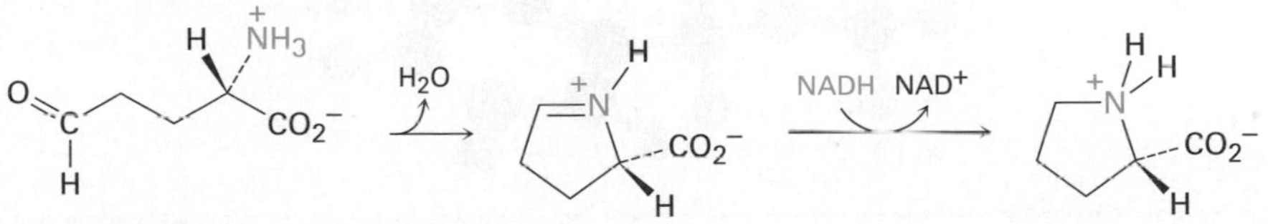

Glutamate 5-semialdehyde → (H_2O) **1-Pyrrolinium 5-carboxylate** → (NADH, NAD^+) **Proline**

Using a Reductive Amination Reaction

Worked Example 24.1

How might you prepare N-methyl-2-phenylethylamine using a reductive amination reaction?

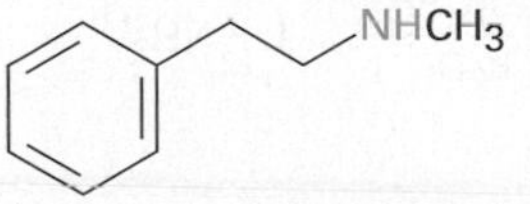

$NHCH_3$

***N*-Methyl-2-phenylethylamine**

Strategy

Look at the target molecule, and identify the groups attached to nitrogen. One of the groups must be derived from the aldehyde or ketone component, and the other must be derived from the amine component. In the case of N-methyl-2-phenylethylamine, two combinations can lead to the product: phenylacetaldehyde plus methylamine or formaldehyde plus 2-phenylethylamine. It's usually better to choose the combination with the simpler amine component—methylamine in this case—and to use an excess of that amine as reactant.

Solution

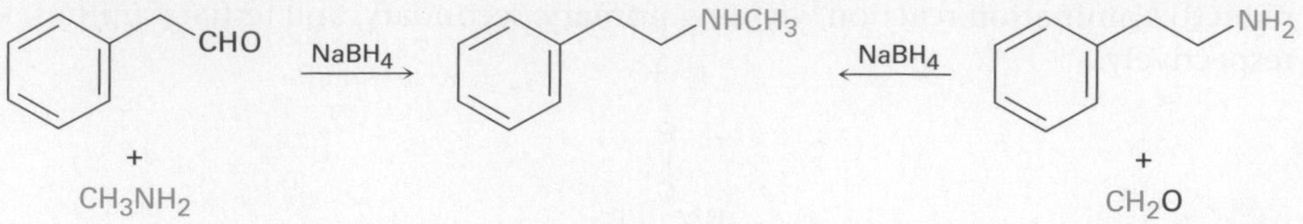

CHO + CH_3NH_2 $\xrightarrow{NaBH_4}$ NHCH3 $\xleftarrow{NaBH_4}$ NH2 + CH_2O

Problem 24.11

How might the following amines be prepared using reductive amination reactions? Show all precursors if more than one is possible.

(a) $CH_3CH_2NHCH(CH_3)CH_3$ (b) $C_6H_5NHCH_2CH_3$ (c) cyclopentyl–$NHCH_3$

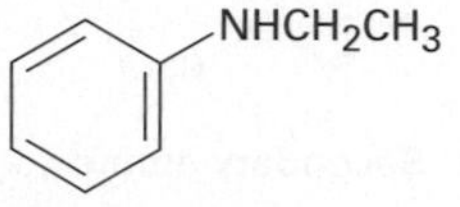

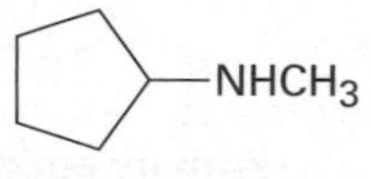

Problem 24.12

How could you prepare the following amine using a reductive amination reaction?

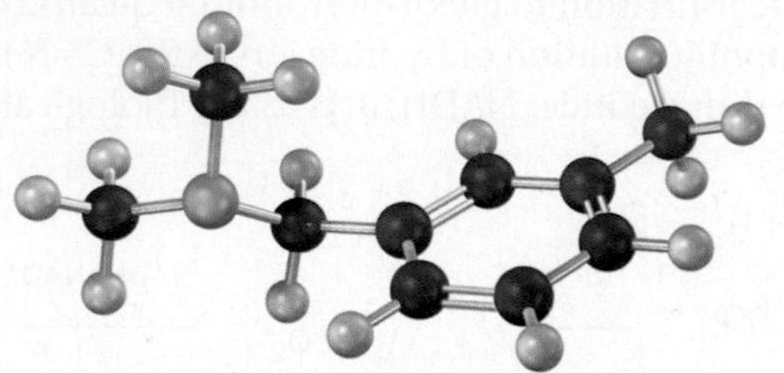

Hofmann and Curtius Rearrangements

Carboxylic acid derivatives can be converted into primary amines with loss of one carbon atom by both the **Hofmann rearrangement** and the **Curtius rearrangement**. Although the Hofmann rearrangement involves a primary amide and the Curtius rearrangement involves an acyl azide, both proceed through similar mechanisms.

Hofmann rearrangement

$$RCONH_2 \xrightarrow[H_2O]{NaOH,\ Br_2} R{-}NH_2 + CO_2$$

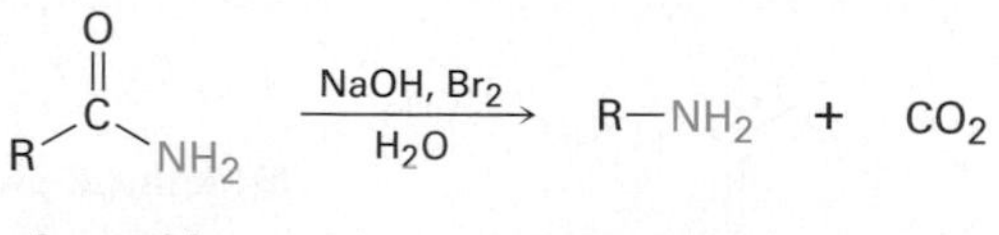

An amide

Curtius rearrangement

$$RCON{=}\overset{+}{N}{=}\overset{-}{N} \xrightarrow[Heat]{H_2O} R{-}NH_2 + CO_2 + N_2$$

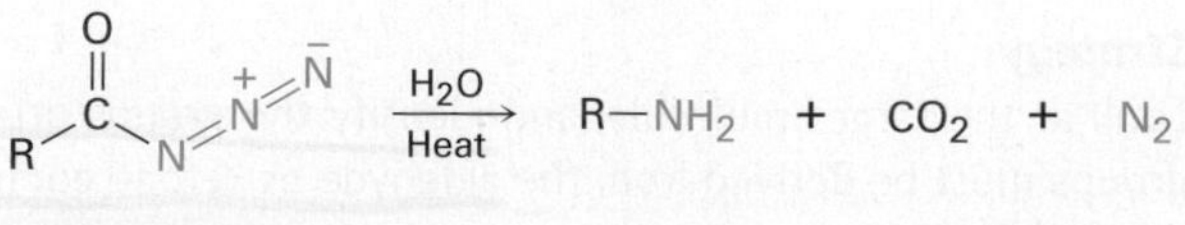

An acyl azide

Hofmann rearrangement occurs when a primary amide, $RCONH_2$, is treated with Br_2 and base **(Figure 24.5)**. The overall mechanism is lengthy, but most of the individual steps have been encountered before. Thus, the bromination of

Figure 24.5 | MECHANISM

Mechanism of the Hofmann rearrangement of an amide to an amine. Each step is analogous to a reaction studied previously.

1 Base abstracts an acidic N–H proton, yielding an amide anion.

2 The anion reacts with bromine in an α-substitution reaction to give an *N*-bromoamide.

3 Abstraction of the remaining N–H proton by base gives a resonance-stabilized bromoamide anion . . .

4 . . . which rearranges when the R group attached to the carbonyl carbon migrates to nitrogen at the same time the bromide ion leaves.

5 The isocyanate formed on rearrangement adds water in a nucleophilic addition step to yield a carbamic acid.

6 The carbamic acid spontaneously loses CO_2 to give an amine.

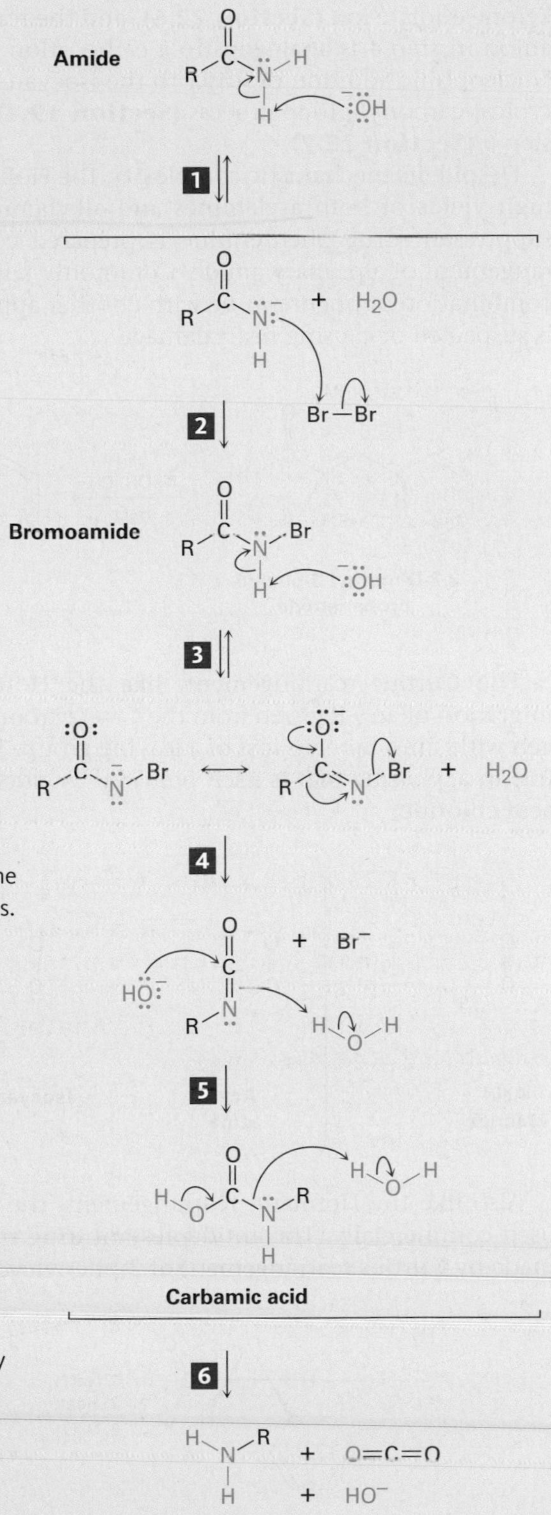

Amide

H_2O

Br—Br

Bromoamide

H_2O

Br^-

Carbamic acid

O=C=O

HO^-

an amide in steps 1 and 2 is analogous to the base-promoted bromination of a ketone enolate ion **(Section 22.6)**, and the rearrangement of the bromoamide anion in step 4 is analogous to a carbocation rearrangement **(Section 7.11).** Nucleophilic addition of water to the isocyanate carbonyl group in step 5 is a typical carbonyl-group process **(Section 19.4)**, as is the final decarboxylation step 6 **(Section 22.7)**.

Despite its mechanistic complexity, the Hofmann rearrangement often gives high yields of both arylamines and alkylamines. For example, the appetite-suppressant drug phentermine is prepared commercially by Hofmann rearrangement of a primary amide. Commonly known by the name Fen-Phen, the combination of phentermine with another appetite-suppressant, fenfluramine, is suspected of causing heart damage.

NaOH, Cl_2, H_2O ⟶ + CO_2

2,2-Dimethyl-3-phenyl-propanamide **Phetermine**

The Curtius rearrangement, like the Hofmann rearrangement, involves migration of an –R group from the C=O carbon atom to the neighboring nitrogen with simultaneous loss of a leaving group. The reaction takes place on heating an acyl azide that is itself prepared by nucleophilic acyl substitution of an acid chloride.

NaN_3 ⟶ [⟶ + N_2] $\xrightarrow{H_2O}$ R—NH_2 + CO_2

Acid chloride **Acyl azide** **Isocyanate** **Amine**

Also like the Hofmann rearrangement, the Curtius rearrangement is often used commercially. The antidepressant drug tranylcypromine, for instance, is made by Curtius rearrangement of 2-phenylcyclopropanecarbonyl chloride.

1. NaN_3 2. Heat 3. H_2O

***trans*-2-Phenylcyclo-propanecarbonyl chloride** **Tranylcypromine**

Worked Example 24.2

Using the Hofmann and Curtius Reactions

How would you prepare o-methylbenzylamine from a carboxylic acid, using both Hofmann and Curtius rearrangements?

Strategy

Both Hofmann and Curtius rearrangements convert a carboxylic acid derivative—either an amide (Hofmann) or an acid chloride (Curtius)—into a primary amine with loss of one carbon, RCOY → RNH_2. Both reactions begin with the same carboxylic acid, which can be identified by replacing the $-NH_2$ group of the amine product by a $-CO_2H$ group. In the present instance, o-methylphenylacetic acid is needed.

Solution

o-Methylphenylacetic acid $\xrightarrow{SOCl_2}$ acid chloride; acid chloride $\xrightarrow{\text{1. } NH_3 \text{; 2. } Br_2\text{, NaOH, } H_2O}$ o-Methylbenzylamine; acid chloride $\xrightarrow{\text{1. } NaN_3 \text{; 2. } H_2O\text{, heat}}$ o-Methylbenzylamine

o-Methylphenyl-acetic acid **o-Methylbenzylamine**

Problem 24.13

How would you prepare the following amines, using both Hofmann and Curtius rearrangements on a carboxylic acid derivative?

(a) $(CH_3)_3CCH_2CH_2NH_2$ (CH_3 above and below the C of $CH_3CCH_2CH_2NH_2$)

(b) p-methylbenzylamine (H_3C–C_6H_4–CH_2NH_2)

24.7 Reactions of Amines

Alkylation and Acylation

We've already studied the two most general reactions of amines—alkylation and acylation. As we saw earlier in this chapter, primary, secondary, and tertiary amines can be alkylated by reaction with a primary alkyl halide. Alkylations of primary and secondary amines are difficult to control and often give mixtures of products, but tertiary amines are cleanly alkylated to give quaternary ammonium salts. Primary and secondary (but not tertiary) amines can also be acylated by nucleophilic acyl substitution reaction with an acid chloride or an acid anhydride to yield an amide **(Sections 21.4 and 21.5)**. Note that overacylation of the nitrogen does not occur because the amide product is much less nucleophilic and less reactive than the starting amine.

$$RCOCl + NH_3 \xrightarrow[\text{solvent}]{\text{Pyridine}} RCONH_2 + HCl$$

$$RCOCl + R'NH_2 \xrightarrow[\text{solvent}]{\text{Pyridine}} RCONHR' + HCl$$

$$RCOCl + R'_2NH \xrightarrow[\text{solvent}]{\text{Pyridine}} RCONR'_2 + HCl$$

Hofmann Elimination

Like alcohols, amines can be converted into alkenes by an elimination reaction. But because an amide ion, NH_2^-, is such a poor leaving group, it must first be converted into a better leaving group. In the **Hofmann elimination reaction**, an amine is completely methylated by reaction with an excess amount of iodomethane to produce the corresponding quaternary ammonium salt. This salt then undergoes elimination to give an alkene on heating with a base, typically silver oxide, Ag_2O. For example, 1-methylpentylamine is converted into 1-hexene.

$$\underset{\textbf{1-Methylpentylamine}}{CH_3CH_2CH_2CH_2CH(NH_2)CH_3} \xrightarrow{\text{Excess } CH_3I} \underset{\textbf{(1-Methylpentyl)trimethylammonium iodide}}{CH_3CH_2CH_2CH_2CH(\overset{+}{N}(CH_3)_3\ I^-)CH_3} \xrightarrow[H_2O,\ \text{heat}]{Ag_2O} \underset{\textbf{1-Hexene (60\%)}}{CH_3CH_2CH_2CH_2CH{=}CH_2} + N(CH_3)_3$$

Silver oxide acts by exchanging iodide ion for hydroxide ion in the quaternary salt, thus providing the base necessary for elimination. The actual elimination step is an E2 reaction **(Section 11.8)** in which hydroxide ion removes a proton at the same time that the positively charged nitrogen atom leaves.

$$\underset{\textbf{Quaternary ammonium salt}}{HO^- + H{-}C{-}C{-}\overset{+}{N}(CH_3)_3} \xrightarrow[\text{reaction}]{\text{E2}} \underset{\textbf{Alkene}}{C{=}C} + H_2O + N(CH_3)_3$$

Unlike what happens in other E2 reactions, the major product of the Hofmann elimination is the less highly substituted alkene rather than the more highly substituted one, as shown by the reaction of (1-methylbutyl)trimethylammonium hydroxide to give 1-pentene rather than the alternative 2-pentene. The reason for this non-Zaitsev result is probably steric. Because of the large size of the trialkylamine leaving group, the base must abstract a hydrogen from the more accessible, least hindered position.

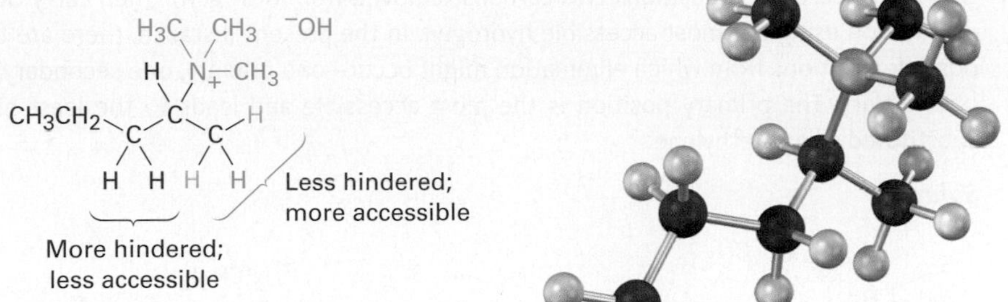

More hindered; less accessible

Less hindered; more accessible

(1-Methylbutyl)trimethylammonium hydroxide

$CH_3CH_2CH_2CH{=}CH_2$ + $CH_3CH_2CH{=}CHCH_3$

1-Pentene (94%)

2-Pentene (6%)

The Hofmann elimination reaction is not often used today in the laboratory, but analogous biological eliminations occur frequently, although usually with protonated ammonium ions rather than quaternary ammonium salts. In the biosynthesis of nucleic acids, for instance, a substance called adenylosuccinate undergoes an elimination of a positively charged nitrogen to give fumarate plus adenosine monophosphate.

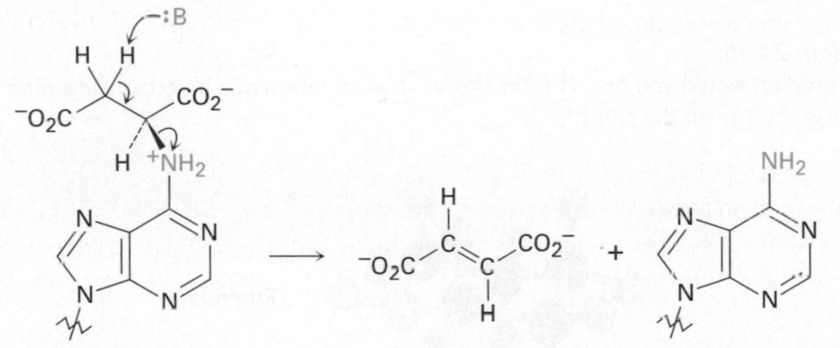

Adenylosuccinate → **Fumarate** + **Adenosine monophosphate**

Predicting the Product of a Hofmann Elimination

Worked Example 24.3

What product would you expect from Hofmann elimination of the following amine?

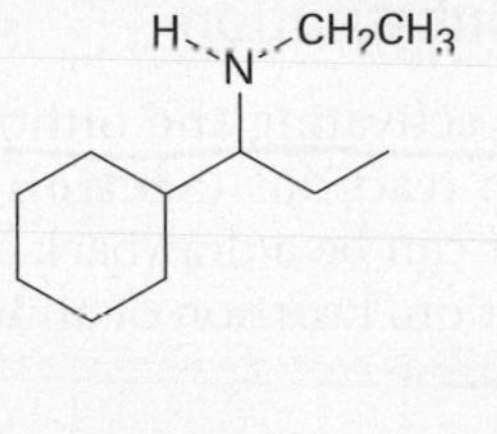

Strategy

The Hofmann elimination is an E2 reaction that converts an amine into an alkene and occurs with non-Zaitsev regiochemistry to form the less highly substituted double bond. To predict the product, look at the reactant and identify the positions from which elimination might occur (the positions two carbons removed from nitrogen). Then carry out an elimination using the most accessible hydrogen. In the present instance, there are three possible positions from which elimination might occur—one primary, one secondary, and one tertiary. The primary position is the most accessible and leads to the least highly substituted alkene, ethylene.

Solution

3° H, N, CH_2CH_3, H, 1°, H H ← 2° $\xrightarrow[\text{2. } Ag_2O,\ H_2O,\ \text{heat}]{\text{1. Excess } CH_3I}$ $N(CH_3)_2$ + $H_2C{=}CH_2$

Problem 24.14

What products would you expect from Hofmann elimination of the following amines? If more than one product is formed, indicate which is major.

(a) $CH_3CH_2CH_2CH(NH_2)CH_2CH_2CH_2CH_3$

(b) Cyclohexyl–NH_2

(c) $CH_3CH_2CH_2CH(NH_2)CH_2CH_2CH_3$

(d) Cyclohexyl–$NHCH_2CH_3$

Problem 24.15

What product would you expect from Hofmann elimination of a heterocyclic amine such as piperidine? Write all the steps.

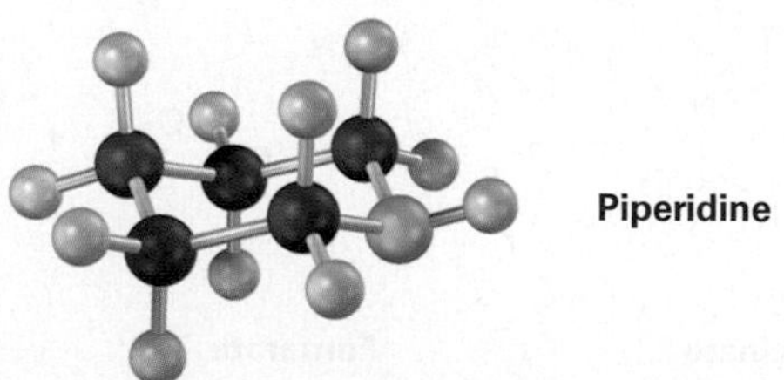

Piperidine

24.8 Reactions of Arylamines

Electrophilic Aromatic Substitution

An amino group is strongly activating and ortho- and para-directing in electrophilic aromatic substitution reactions (**Section 16.5**). This high reactivity of amino-substituted benzenes can be a drawback at times because it's often difficult to prevent polysubstitution. Reaction of aniline with Br_2, for instance, takes place rapidly and yields the 2,4,6-tribrominated product. The amino group is so strongly activating that it's not possible to stop at the monobromo stage.

NH_2 NH_2 Br Br $\xrightarrow[H_2O]{3\ Br_2}$ Br

Aniline **2,4,6-Tribromoaniline (100%)**

Another drawback to the use of amino-substituted benzenes in electrophilic aromatic substitution reactions is that Friedel–Crafts reactions are not successful **(Section 16.3)**. The amino group forms an acid–base complex with the $AlCl_3$ catalyst, which prevents further reaction from occurring. Both drawbacks can be overcome, however, by carrying out electrophilic aromatic substitution reactions on the corresponding amide rather than on the free amine.

As we saw in **Section 21.5**, treatment of an amine with acetic anhydride yields the corresponding acetyl amide, or acetamide. Although still activating and ortho-, para-directing, amido substituents (–NHCOR) are less strongly activating and less basic than amino groups because their nitrogen lone-pair electrons are delocalized by the neighboring carbonyl group. As a result, bromination of an *N*-arylamide occurs cleanly to give a monobromo product, and hydrolysis of the amide with aqueous base then gives the free amine. For example, *p*-toluidine (4-methylaniline) can be acetylated, brominated, and hydrolyzed to yield 2-bromo-4-methylaniline. None of the 2,6-dibrominated product is obtained.

NH_2 CH_3 $\xrightarrow[Pyridine]{(CH_3CO)_2O}$ H–N–C(=O)–CH_3 CH_3 $\xrightarrow{Br_2}$ H–N–C(=O)–CH_3 Br CH_3 $\xrightarrow[H_2O]{NaOH}$ NH_2 Br CH_3 + $CH_3CO_2^-$

***p*-Toluidine** **2-Bromo-4-methylaniline (79%)**

Friedel–Crafts alkylations and acylations of *N*-arylamides also proceed normally. For example, benzoylation of acetanilide (*N*-acetylaniline) under Friedel–Crafts conditions gives 4-aminobenzophenone in 80% yield after hydrolysis.

NH_2 $\xrightarrow[Pyridine]{(CH_3CO)_2O}$ H–N–C(=O)–CH_3 $\xrightarrow[AlCl_3]{C_6H_5COCl}$ H–N–C(=O)–CH_3 O=C $\xrightarrow[H_2O]{NaOH}$ NH_2 O=C

Aniline **4-Aminobenzophenone (80%)**

Modulating the reactivity of an amino-substituted benzene by forming an amide is a useful trick that allows many kinds of electrophilic aromatic substitutions to be carried out that would otherwise be impossible. An example is the preparation of the sulfa drugs, such as sulfanilamide.

Sulfa drugs were among the first pharmaceutical agents to be used clinically against bacterial infection. Although they have largely been replaced today by safer and more powerful antibiotics, sulfa drugs are credited with saving the lives of thousands of wounded during World War II and are still prescribed for urinary tract infections. They are prepared by chlorosulfonation of acetanilide, followed by reaction of *p*-(*N*-acetylamino)benzenesulfonyl chloride with ammonia or some other amine to give a sulfonamide. Hydrolysis of the amide then yields the sulfa drug. Note that hydrolysis of the amide can be carried out in the presence of the sulfonamide group because sulfonamides hydrolyze very slowly.

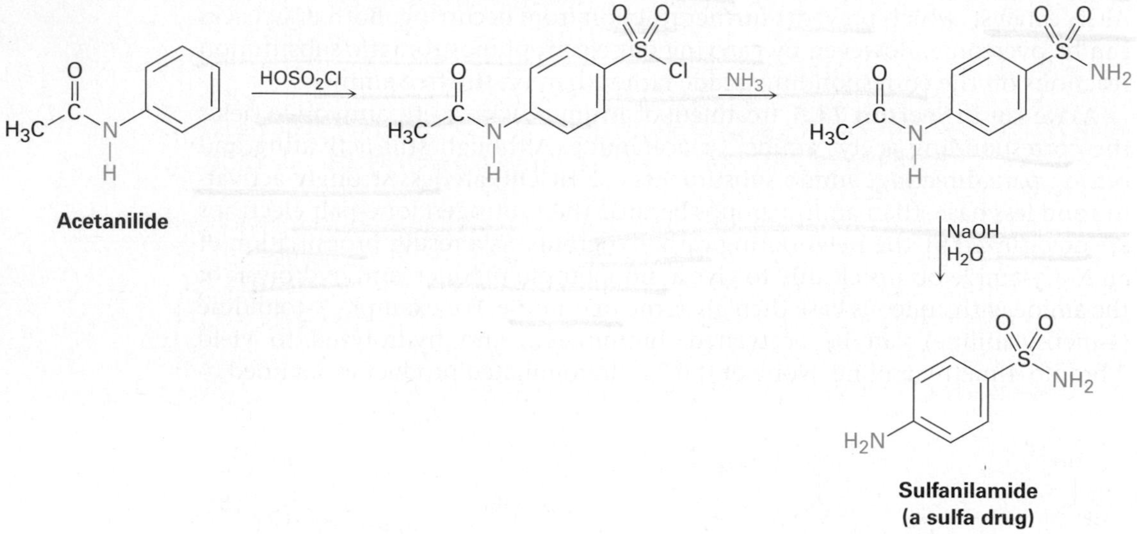

Acetanilide $\xrightarrow{HOSO_2Cl}$ $\xrightarrow{NH_3}$ $\xrightarrow{NaOH,\ H_2O}$ **Sulfanilamide (a sulfa drug)**

Problem 24.16

Propose a synthesis of the drug sulfathiazole from benzene and any necessary amine.

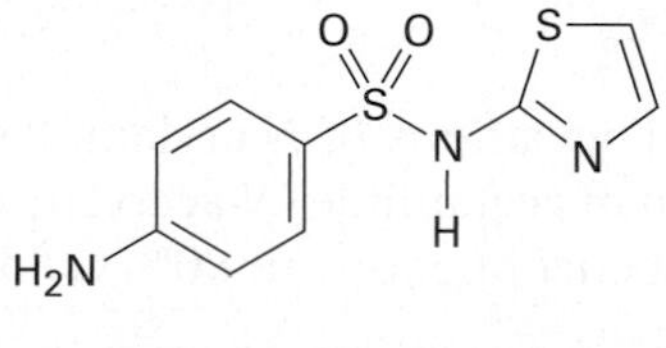

Sulfathiazole

Problem 24.17

Propose syntheses of the following compounds from benzene:

(a) *N,N*-Dimethylaniline **(b)** *p*-Chloroaniline

(c) *m*-Chloroaniline **(d)** 2,4-Dimethylaniline

Diazonium Salts: The Sandmeyer Reaction

Primary arylamines react with nitrous acid, HNO_2, to yield stable *arenediazonium* salts, $Ar—\overset{+}{N}\equiv N\ X^-$, a process called a *diazotization* reaction. Alkylamines also react with nitrous acid, but the corresponding alkanediazonium products are so reactive they can't be isolated. Instead, they lose nitrogen instantly to yield carbocations. The analogous loss of N_2 from an arenediazonium ion to yield an aryl cation is disfavored by the instability of the cation.

$$C_6H_5NH_2 + HNO_2 + H_2SO_4 \longrightarrow C_6H_5\overset{+}{N}{\equiv}N \; HSO_4^- + 2\,H_2O$$

Arenediazonium salts are useful because the diazonio group (N_2) can be replaced by a nucleophile in a substitution reaction.

$$C_6H_5\overset{+}{N}{\equiv}N \; HSO_4^- + :Nu^- \longrightarrow C_6H_5Nu + N_2$$

Many different nucleophiles—halide, hydride, cyanide, and hydroxide among others—react with arenediazonium salts, yielding many different kinds of substituted benzenes. The overall sequence of (1) nitration, (2) reduction, (3) diazotization, and (4) nucleophilic substitution is perhaps the single most versatile method of aromatic substitution.

Aryl chlorides and bromides are prepared by reaction of an arenediazonium salt with the corresponding copper(I) halide, CuX, a process called the **Sandmeyer reaction**. Aryl iodides can be prepared by direct reaction with NaI without using a copper(I) salt. Yields generally fall between 60% and 80%.

$$\text{H}_3\text{C-C}_6\text{H}_4\text{-NH}_2 \xrightarrow[H_2SO_4]{HNO_2} \left[\text{H}_3\text{C-C}_6\text{H}_4\text{-}\overset{+}{N}{\equiv}N \; HSO_4^-\right] \xrightarrow[CuBr]{HBr} \text{H}_3\text{C-C}_6\text{H}_4\text{-Br}$$

***p*-Methylaniline** → ***p*-Bromotoluene (73%)**

$$C_6H_5NH_2 \xrightarrow[H_2SO_4]{HNO_2} \left[C_6H_5\overset{+}{N}{\equiv}N \; HSO_4^-\right] \xrightarrow{NaI} C_6H_5I$$

Aniline → **Iodobenzene (67%)**

Similar treatment of an arenediazonium salt with CuCN yields the nitrile, ArCN, which can then be further converted into other functional groups such as carboxyl. For example, Sandmeyer reaction of *o*-methylbenzenediazonium bisulfate with CuCN yields *o*-methylbenzonitrile, which can be hydrolyzed to give *o*-methylbenzoic acid. This product can't be prepared from *o*-xylene by the usual side-chain oxidation route because both methyl groups would be oxidized.

$$o\text{-CH}_3\text{C}_6\text{H}_4\text{NH}_2 \xrightarrow[H_2SO_4]{HNO_2} o\text{-CH}_3\text{C}_6\text{H}_4\overset{+}{N}{\equiv}N \; HSO_4^- \xrightarrow[CuCN]{KCN} o\text{-CH}_3\text{C}_6\text{H}_4\text{C}{\equiv}\text{N} \xrightarrow{H_3O^+} o\text{-CH}_3\text{C}_6\text{H}_4\text{CO}_2\text{H}$$

***o*-Methylaniline** → ***o*-Methylbenzene-diazonium bisulfate** → ***o*-Methylbenzonitrile** → ***o*-Methylbenzoic acid**

The diazonio group can also be replaced by −OH to yield a phenol and by −H to yield an arene. A phenol is prepared by reaction of the arenediazonium salt with copper(I) oxide in an aqueous solution of copper(II) nitrate, a reaction that is especially useful because few other general methods exist for introducing an −OH group onto an aromatic ring.

p-Methylaniline (p-Toluidine) **p-Cresol (93%)**

Reduction of a diazonium salt to give an arene occurs on treatment with hypophosphorous acid, H_3PO_2. This reaction is used primarily when there is a need for temporarily introducing an amino substituent onto a ring to take advantage of its directing effect. Suppose, for instance, that you needed to make 3,5-dibromotoluene. The product can't be made by direct bromination of toluene because reaction would occur at positions 2 and 4. Starting with *p*-methylaniline (*p*-toluidine), however, dibromination occurs ortho to the strongly directing amino substituent, and diazotization followed by treatment with H_3PO_2 to remove the amino group yields the desired product.

p-Methylaniline **3,5-Dibromotoluene**

Toluene **2,4-Dibromotoluene**

Mechanistically, these diazonio replacement reactions occur through radical rather than polar pathways. In the presence of a copper(I) compound, for instance, it's thought that the arenediazonium ion is first converted to an aryl radical plus copper(II), followed by subsequent reaction to give product plus regenerated copper(I) catalyst.

Diazonium compound **Aryl radical**

Using Diazonium Replacement Reactions

Worked Example 24.4

How would you prepare *m*-hydroxyacetophenone from benzene, using a diazonium replacement reaction in your scheme?

m-Hydroxyacetophenone

Strategy

As always, organic syntheses are planned by working retrosynthetically from the final product, one step at a time. First, identify the functional groups in the product and recall how those groups can be synthesized. *m*-Hydroxyacetophenone has an –OH group and a $-COCH_3$ group in a meta relationship on a benzene ring. A hydroxyl group is generally introduced onto an aromatic ring by a four-step sequence of nitration, reduction, diazotization, and diazonio replacement. An acetyl group is introduced by a Friedel–Crafts acylation reaction.

Next, ask yourself what an immediate precursor of the target might be. Since an acetyl group is a meta director while a hydroxyl group is an ortho and para director, acetophenone might be a precursor of *m*-hydroxyacetophenone. Benzene, in turn, is a precursor of acetophenone.

Solution

Benzene $\xrightarrow[AlCl_3]{CH_3COCl}$ Acetophenone → (1. HNO_3, H_2SO_4; 2. $SnCl_2$, H_3O^+; 3. HNO_2, H_2SO_4; 4. Cu_2O, $Cu(NO_3)_2$, H_2O) *m*-Hydroxyacetophenone

Benzene **Acetophenone** ***m*-Hydroxyacetophenone**

Problem 24.18

How would you prepare the following compounds from benzene, using a diazonium replacement reaction in your scheme?

(a) *p*-Bromobenzoic acid **(b)** *m*-Bromobenzoic acid **(c)** *m*-Bromochlorobenzene
(d) *p*-Methylbenzoic acid **(e)** 1,2,4-Tribromobenzene

Diazonium Coupling Reactions

Arenediazonium salts undergo a coupling reaction with activated aromatic rings such as phenols and arylamines to yield brightly colored **azo compounds, Ar—N=N—Ar′.**

$C_6H_5N^+{\equiv}N\ HSO_4^-$ + $C_6H_5\ddot{Y}$ → $C_6H_5N{=}NC_6H_4\ddot{Y}$

An azo compound

where Y = –OH or $-NR_2$

Diazonium coupling reactions are typical electrophilic aromatic substitutions in which the positively charged diazonium ion is the electrophile that reacts with the electron-rich ring of a phenol or arylamine. Reaction usually occurs at the para position.

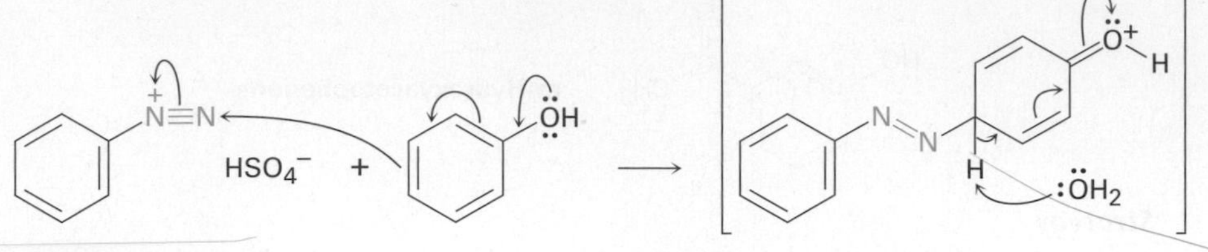

HSO$_4^-$ + ⟶ :ÖH$_2$

Benzenediazonium bisulfate **Phenol**

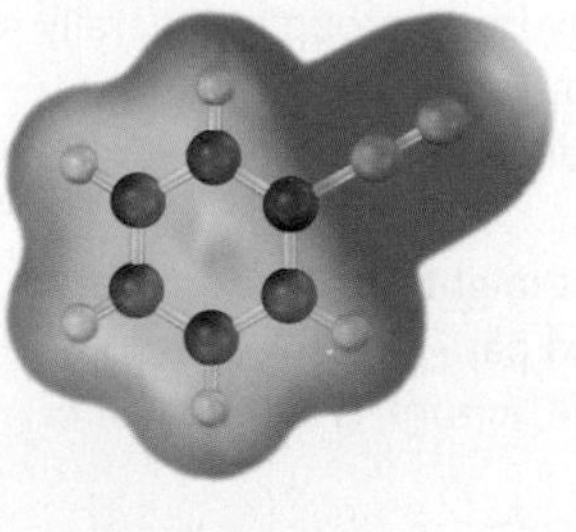

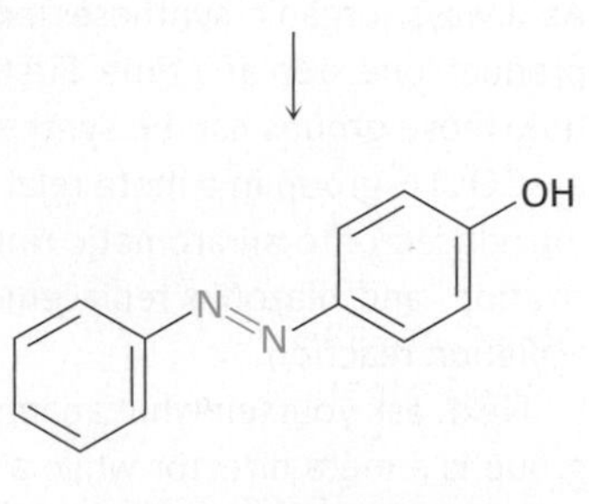

p-Hydroxyazobenzene (orange crystals, mp 152 °C)

Azo-coupled products are widely used as dyes for textiles because their extended conjugated π electron system causes them to absorb in the visible region of the electromagnetic spectrum **(Section 14.9)**. *p*-(Dimethylamino)-azobenzene, for instance, is a bright yellow compound that was at one time used as a coloring agent in margarine.

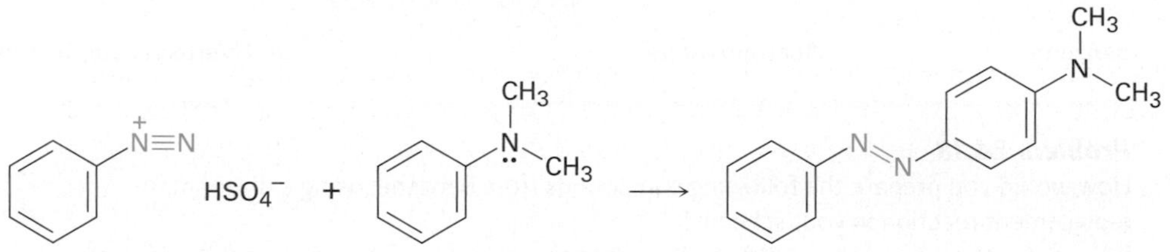

HSO$_4^-$ + ⟶

Benzenediazonium bisulfate ***N,N*-Dimethylaniline** ***p*-(Dimethylamino)azobenzene (yellow crystals, mp 127 °C)**

Problem 24.19

Propose a synthesis of *p*-(dimethylamino)azobenzene from benzene as your only organic starting material.

24.9 Heterocyclic Amines

As noted in **Section 15.5** in connection with a discussion of aromaticity, a cyclic organic compound that contains atoms of two or more elements in its ring is a called a *heterocycle*. Heterocyclic amines are particularly common, and many have important biological properties. Pyridoxal phosphate, a coenzyme; sildenafil (Viagra), a well-known pharmaceutical; and heme, the oxygen carrier in blood, are examples.

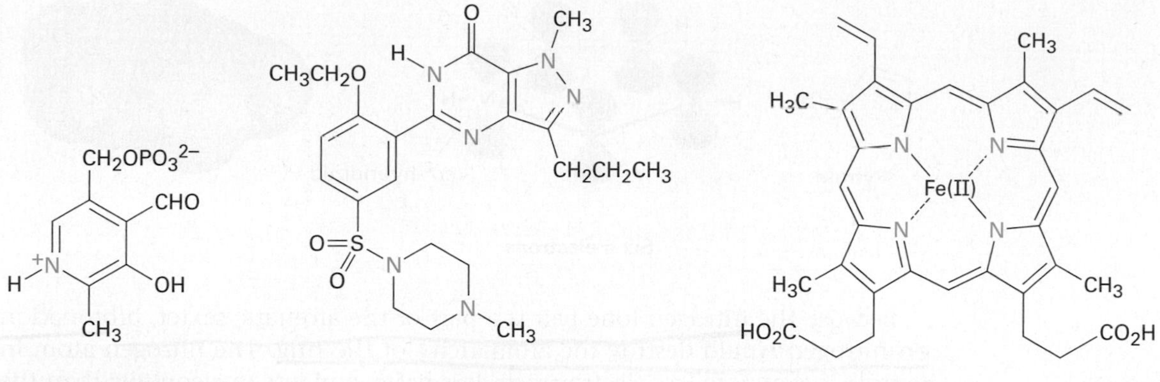

Pyridoxal phosphate (a coenzyme)

Sildenafil (Viagra)

Heme

Most heterocycles have the same chemistry as their open-chain counterparts. Lactones and acyclic esters behave similarly, lactams and acyclic amides behave similarly, and cyclic and acyclic ethers behave similarly. In certain cases, however, particularly when the ring is unsaturated, heterocycles have unique and interesting properties.

Pyrrole and Imidazole

Pyrrole, the simplest five-membered unsaturated heterocyclic amine, is obtained commercially by treatment of furan with ammonia over an alumina catalyst at 400 °C. Furan, the oxygen-containing analog of pyrrole, is obtained by acid-catalyzed dehydration of the five-carbon sugars found in oat hulls and corncobs.

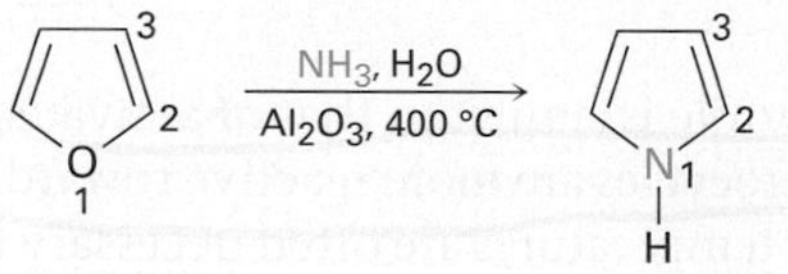

$$\text{Furan} \xrightarrow[\text{Al}_2\text{O}_3,\ 400\ ^\circ\text{C}]{\text{NH}_3,\ \text{H}_2\text{O}} \text{Pyrrole}$$

Furan

Pyrrole

Although pyrrole appears to be both an amine and a conjugated diene, its chemical properties are not consistent with either of these structural features. Unlike most other amines, pyrrole is not basic—the pK_a of the pyrrolinium ion is 0.4; unlike most other conjugated dienes, pyrrole undergoes electrophilic substitution reactions rather than additions. The reason for both these properties, as noted in **Section 15.5** is that pyrrole has six π electrons and is aromatic.

Each of the four carbons contributes one π electron, and the sp^2-hybridized nitrogen contributes two more from its lone pair.

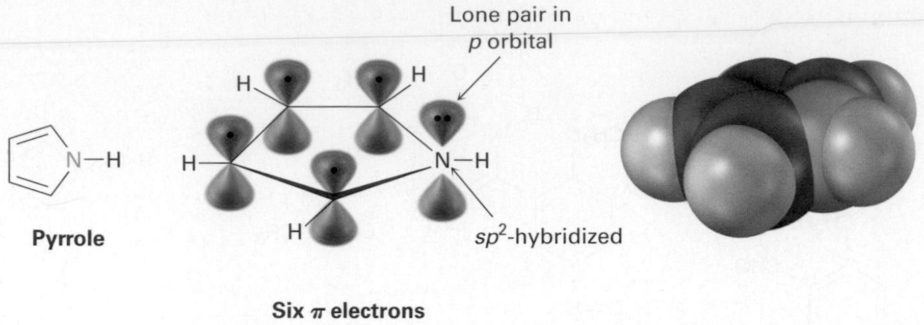

Lone pair in *p* orbital

N–H

Pyrrole

*sp*²-hybridized

Six π electrons

Because the nitrogen lone pair is a part of the aromatic sextet, protonation on nitrogen would destroy the aromaticity of the ring. The nitrogen atom in pyrrole is therefore less electron-rich, less basic, and less nucleophilic than the nitrogen in an aliphatic amine. By the same token, the carbon atoms of pyrrole are more electron-rich and more nucleophilic than typical double-bond carbons. The pyrrole ring is therefore reactive toward electrophiles in the same way that enamines are **(Section 23.11)**. Electrostatic potential maps show how the pyrrole nitrogen is electron-poor (less red) compared with the nitrogen in its saturated counterpart pyrrolidine, while the pyrrole carbon atoms are electron-rich (more red) compared with the carbons in 1,3-cyclopentadiene.

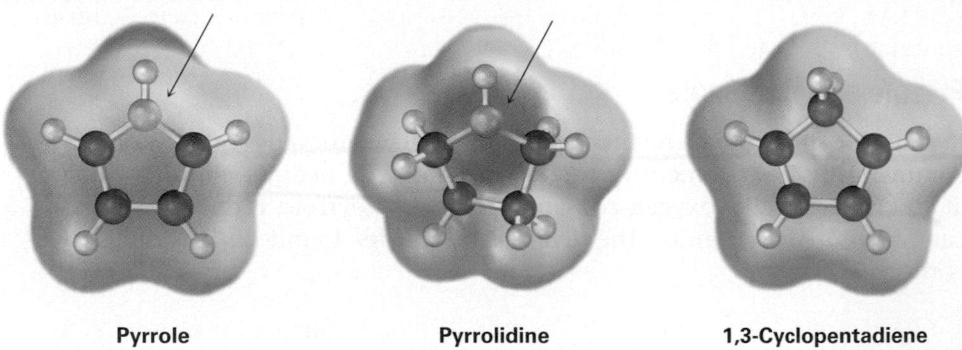

Pyrrole **Pyrrolidine** **1,3-Cyclopentadiene**

The chemistry of pyrrole is similar to that of activated benzene rings. In general, however, the heterocycles are more reactive toward electrophiles than benzene rings are, and low temperatures are often necessary to control the reactions. Halogenation, nitration, sulfonation, and Friedel–Crafts acylation can all be accomplished. For example:

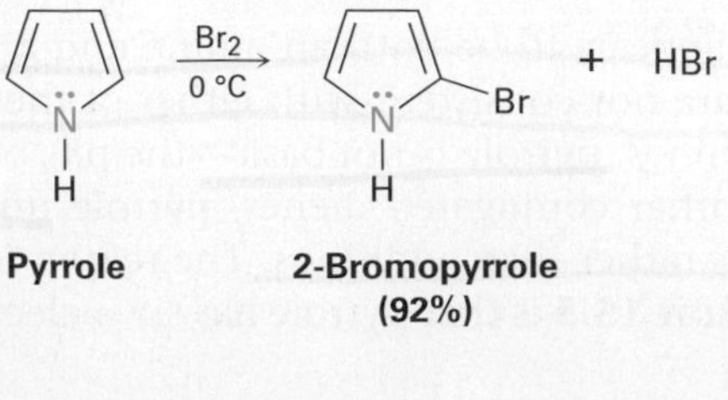

$$\text{Pyrrole} \xrightarrow[0\,°\text{C}]{Br_2} \text{2-Bromopyrrole (92\%)} + HBr$$

Pyrrole **2-Bromopyrrole (92%)**

Electrophilic substitutions normally occur at C2, the position next to the nitrogen, because reaction at this position leads to a more stable intermediate cation having three resonance forms, whereas reaction at C3 gives a less stable cation with only two resonance forms **(Figure 24.6)**.

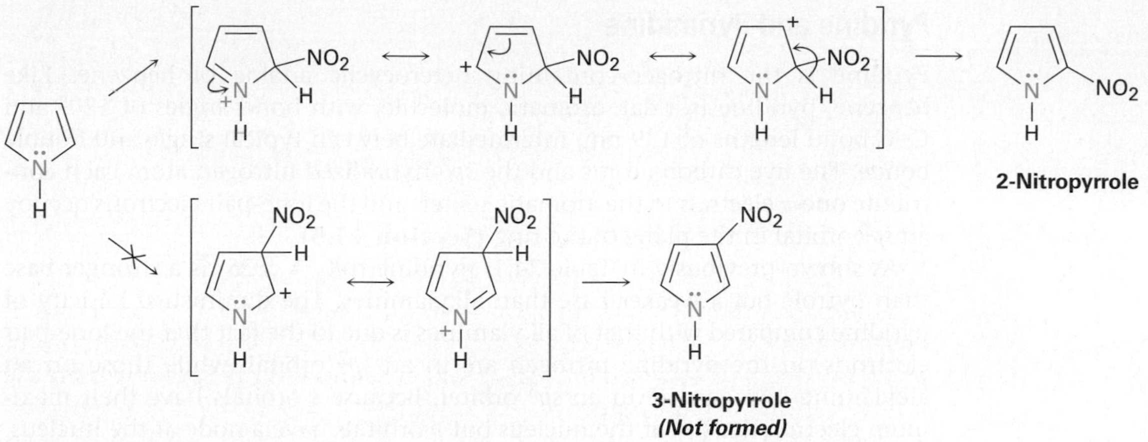

Figure 24.6 Electrophilic nitration of pyrrole. The intermediate produced by reaction at C2 is more stable than that produced by reaction at C3.

Other common five-membered heterocyclic amines include imidazole and thiazole. Imidazole, a constituent of the amino acid histidine, has two nitrogens, only one of which is basic. Thiazole, the five-membered ring system on which the structure of thiamin (vitamin B_1) is based, also contains a basic nitrogen that is alkylated in thiamin to form a quaternary ammonium ion.

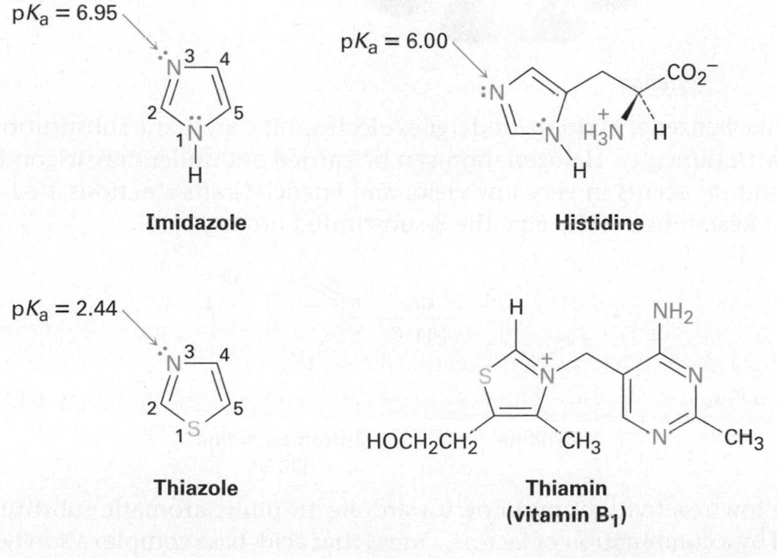

Problem 24.20

Draw an orbital picture of thiazole. Assume that both the nitrogen and sulfur atoms are sp^2-hybridized, and show the orbitals that the lone pairs occupy.

Problem 24.21
What is the percent protonation of the imidazole nitrogen atom in histidine at a physiological pH of 7.3 **(Section 24.5)**?

Pyridine and Pyrimidine

Pyridine is the nitrogen-containing heterocyclic analog of benzene. Like benzene, pyridine is a flat, aromatic molecule, with bond angles of 120° and C–C bond lengths of 139 pm, intermediate between typical single and double bonds. The five carbon atoms and the sp^2-hybridized nitrogen atom each contribute one π electron to the aromatic sextet, and the lone-pair electrons occupy an sp^2 orbital in the plane of the ring **(Section 15.5)**.

As shown previously in Table 24.1, pyridine (pK_a = 5.25) is a stronger base than pyrrole but a weaker base than alkylamines. The diminished basicity of pyridine compared with that of alkylamines is due to the fact that the lone-pair electrons on the pyridine nitrogen are in an sp^2 orbital, while those on an alkylamine nitrogen are in an sp^3 orbital. Because *s* orbitals have their maximum electron density at the nucleus but *p* orbitals have a node at the nucleus, electrons in an orbital with more *s* character are held more closely to the positively charged nucleus and are less available for bonding. As a result, the sp^2-hybridized nitrogen atom (33% *s* character) in pyridine is less basic than the sp^3-hybridized nitrogen in an alkylamine (25% *s* character).

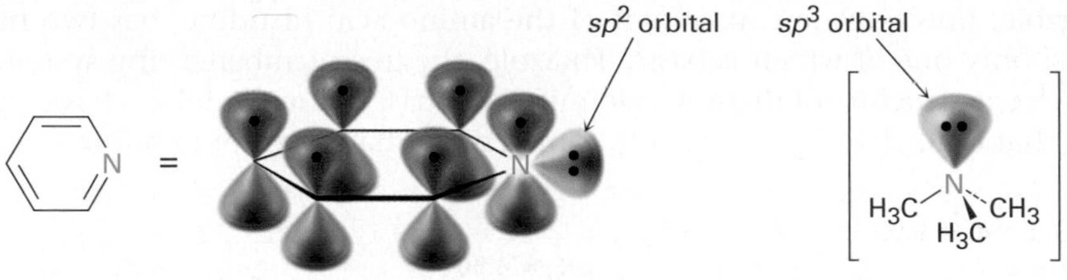

Pyridine

Unlike benzene, pyridine undergoes electrophilic aromatic substitution reactions with difficulty. Halogenation can be carried out under drastic conditions, but nitration occurs in very low yield, and Friedel–Crafts reactions are not successful. Reactions usually give the 3-substituted product.

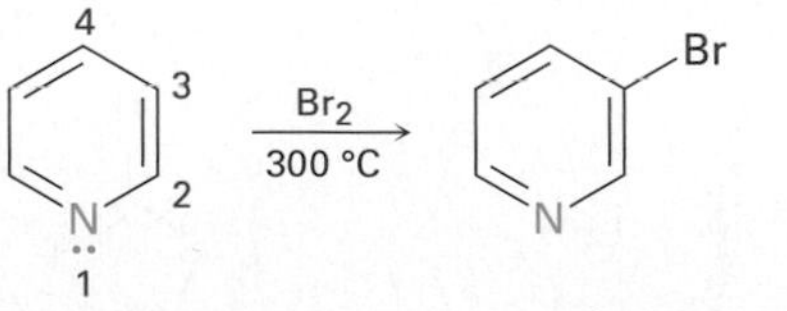

Pyridine **3-Bromopyridine (30%)**

The low reactivity of pyridine toward electrophilic aromatic substitution is caused by a combination of factors. One is that acid–base complexation between the basic ring nitrogen atom and the incoming electrophile places a positive charge on the ring, thereby deactivating it. Equally important is that the electron density of the ring is decreased by the electron-withdrawing inductive effect of the electronegative nitrogen atom. Thus, pyridine has a substantial dipole moment (μ = 2.26 D), with the ring carbons acting as the positive end of the dipole. Reaction of an electrophile with the positively polarized carbon atoms is therefore difficult.

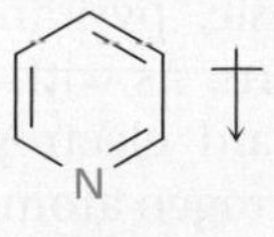

$\mu = 2.26\ \mathrm{D}$

In addition to pyridine, the six-membered diamine pyrimidine is also found commonly in biological molecules, particularly as a constituent of nucleic acids. With a pK_a of 1.3, pyrimidine is substantially less basic than pyridine because of the inductive effect of the second nitrogen.

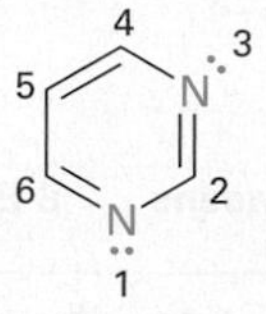

Pyrimidine
pK_a = 1.3

Problem 24.22

Electrophilic aromatic substitution reactions of pyridine normally occur at C3. Draw the carbocation intermediates resulting from reaction of an electrophile at C2, C3, and C4, and explain the observed result.

Polycyclic Heterocycles

As we saw in **Section 15.6**, quinoline, isoquinoline, indole, and purine are common polycyclic heterocycles. The first three contain both a benzene ring and a heterocyclic aromatic ring, while purine contains two heterocyclic rings joined together. All four ring systems occur commonly in nature, and many compounds with these rings have pronounced physiological activity. The quinoline alkaloid quinine, for instance, is widely used as an antimalarial drug; tryptophan is a common amino acid; and the purine adenine is a constituent of nucleic acids.

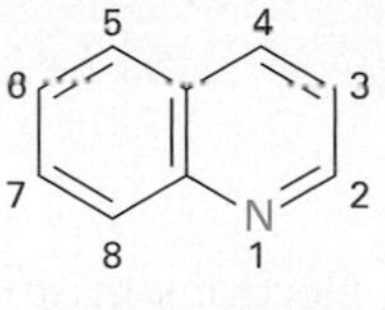

Quinoline 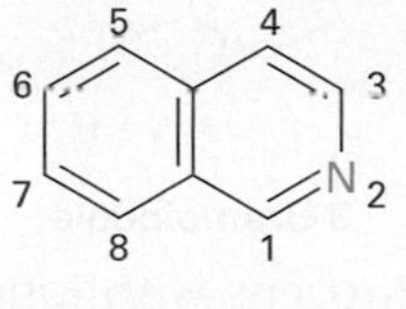 **Isoquinoline**

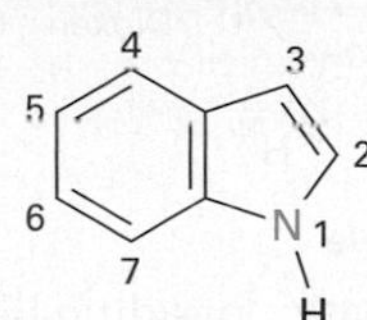

Indole

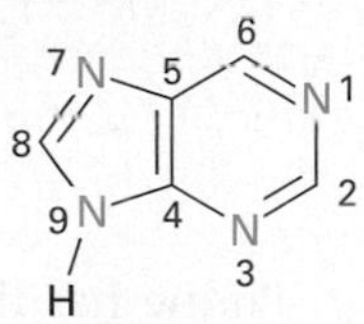

Purine

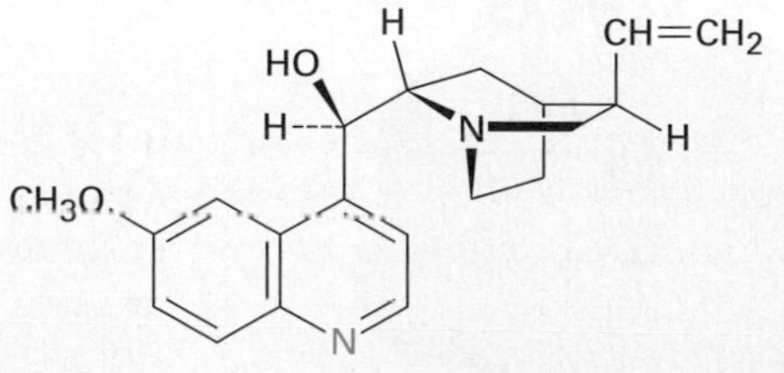

Quinine
(antimalarial)

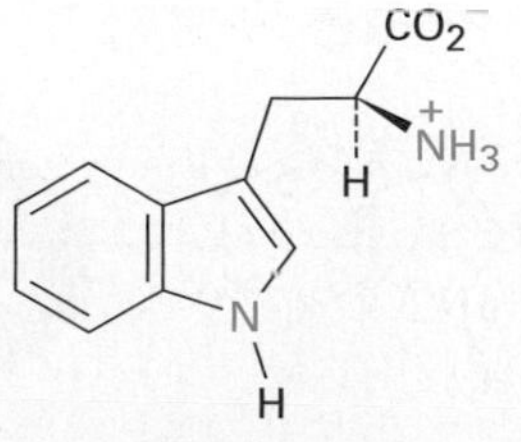

Tryptophan
(amino acid)

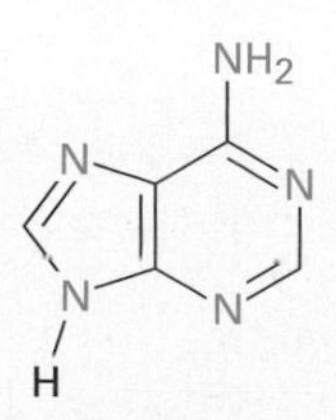

Adenine
(DNA constituent)

The chemistry of these polycyclic heterocycles is just what you might expect from a knowledge of the simpler heterocycles pyridine and pyrrole. Quinoline and isoquinoline both have basic, pyridine-like nitrogen atoms, and both undergo electrophilic substitutions. As with pyridine, both quinoline and isoquinoline are less reactive toward electrophilic substitution than benzene because of the electronegative nitrogen atom that withdraws electrons from the ring. Reaction occurs on the benzene ring rather than on the nitrogen-containing pyridine ring, and a mixture of substitution products is obtained.

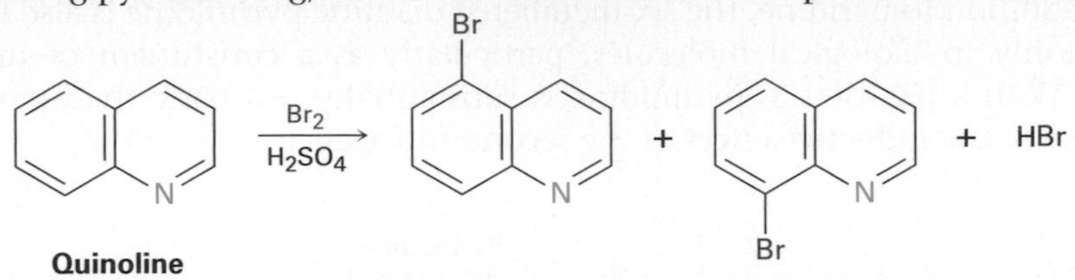

Br
Br_2
H_2SO_4
+ + HBr
N N N
Br
Quinoline
5-Bromoquinoline **8-Bromoquinoline**
A 51 : 49 ratio

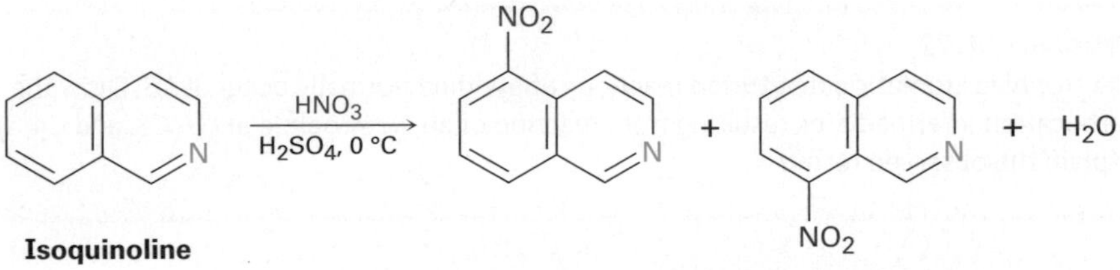

NO_2
HNO_3
H_2SO_4, 0 °C
+ + H_2O
N N N
NO_2
Isoquinoline
5-Nitroisoquinoline **8-Nitroisoquinoline**
A 90 : 10 ratio

Indole has a nonbasic, pyrrole-like nitrogen and undergoes electrophilic substitution more easily than benzene. Substitution occurs at C3 of the electron-rich pyrrole ring rather than on the benzene ring.

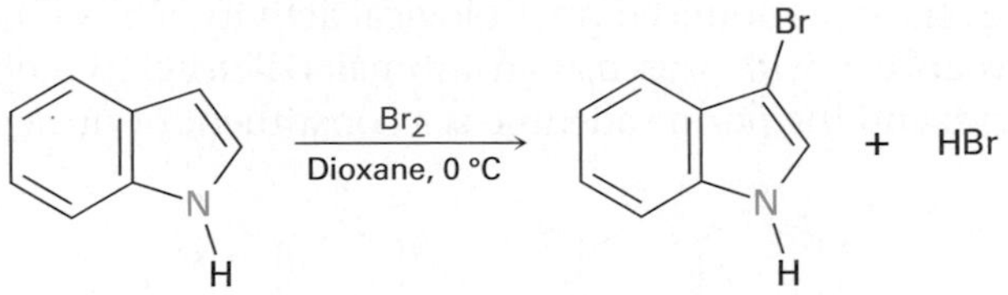

Br
Br_2
Dioxane, 0 °C
+ HBr
N N
H H
Indole **3-Bromoindole**

Purine has three basic, pyridine-like nitrogens with lone-pair electrons in sp^2 orbitals in the plane of the ring. The remaining purine nitrogen is nonbasic and pyrrole-like, with its lone-pair electrons as part of the aromatic π electron system.

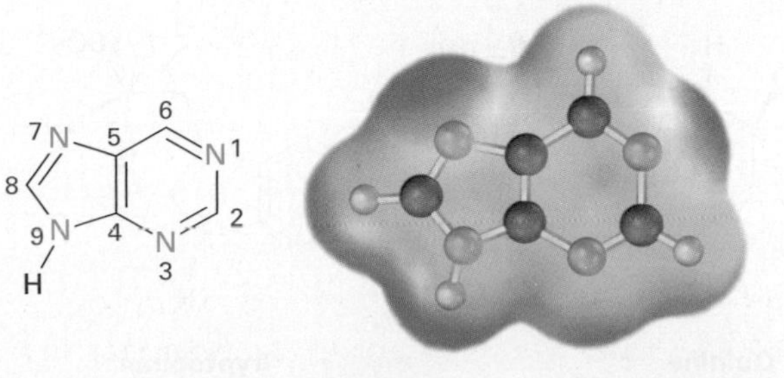

6
7 N 5 N 1
8
9 N 4 N 2
H 3

Purine

Problem 24.23
Which nitrogen atom in the hallucinogenic indole alkaloid *N*,*N*-dimethyltryptamine is more basic? Explain.

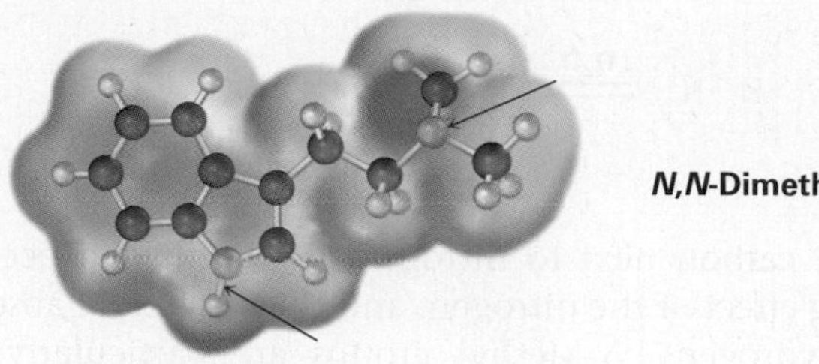

Problem 24.24
Indole reacts with electrophiles at C3 rather than at C2. Draw resonance forms of the intermediate cations resulting from reaction at C2 and C3, and explain the observed results.

24.10 Spectroscopy of Amines

Infrared Spectroscopy

Primary and secondary amines can be identified by a characteristic N–H stretching absorption in the 3300 to 3500 cm^{-1} range of the IR spectrum. Alcohols also absorb in this range **(Section 17.11)**, but amine absorption bands are generally sharper and less intense than hydroxyl bands. Primary amines show a pair of bands at about 3350 and 3450 cm^{-1}, and secondary amines show a single band at 3350 cm^{-1}. Tertiary amines have no absorption in this region because they have no N–H bonds. An IR spectrum of cyclohexylamine is shown in **Figure 24.7**.

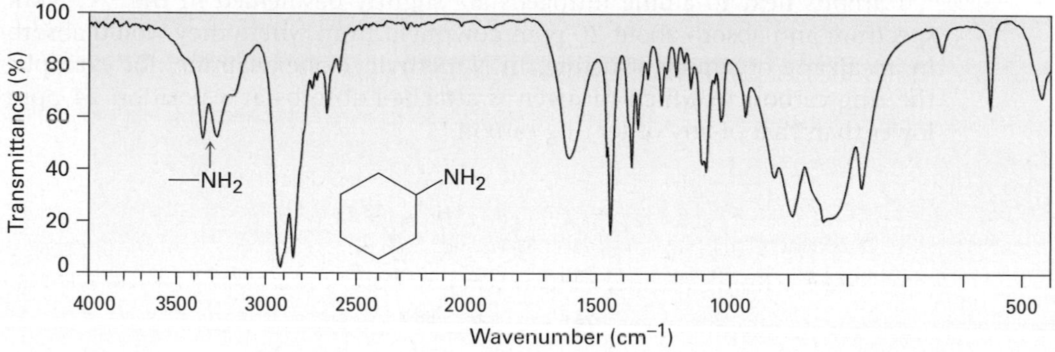

Figure 24.7 IR spectrum of cyclohexylamine.

Nuclear Magnetic Resonance Spectroscopy

Amines are difficult to identify solely by ^{1}H NMR spectroscopy because N–H hydrogens tend to appear as broad signals without clear-cut coupling to neighboring C–H hydrogens. As with O–H absorptions **(Section 17.11)**, amine N–H absorptions can appear over a wide range and are best identified by adding a small amount of D_2O to the sample tube. Exchange of N–D for N–H occurs, and the N–H signal disappears from the NMR spectrum.

$$\gt N{-}H \xrightleftharpoons{D_2O} \gt N{-}D \; + \; HDO$$

Hydrogens on the carbon next to nitrogen are deshielded because of the electron-withdrawing effect of the nitrogen, and they therefore absorb at lower field than alkane hydrogens. *N*-Methyl groups are particularly distinctive because they absorb as a sharp three-proton singlet at 2.2 to 2.6 δ. This *N*-methyl resonance at 2.42 δ is easily seen in the ^1H NMR spectrum of *N*-methylcyclohexylamine **(Figure 24.8)**.

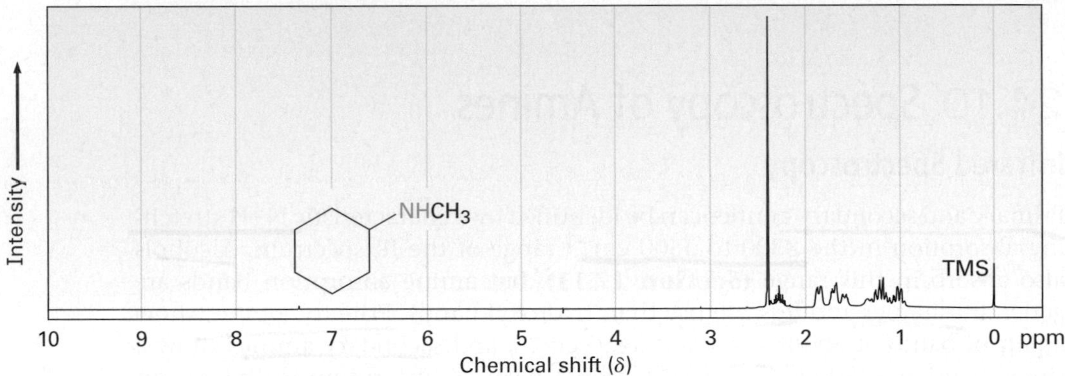

Figure 24.8 Proton NMR spectrum of *N*-methylcyclohexylamine.

Carbons next to amine nitrogens are slightly deshielded in the ^{13}C NMR spectrum and absorb about 20 ppm downfield from where they would absorb in an alkane of similar structure. In *N*-methylcyclohexylamine, for example, the ring carbon to which nitrogen is attached absorbs at a position 24 ppm lower than that of any other ring carbon.

H
33.4
33.3
N
25.2
CH_3
26.5
58.7

Problem 24.25

Compound **A**, $C_6H_{12}O$, has an IR absorption at 1715 cm^{-1} and gives compound **B**, $C_6H_{15}N$, when treated with ammonia and $NaBH_4$. The IR and ^1H NMR spectra of **B** are shown. What are the structures of **A** and **B**?

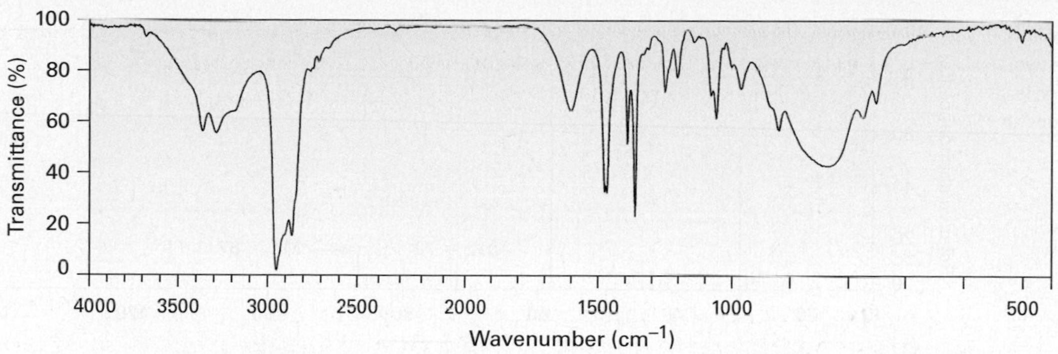

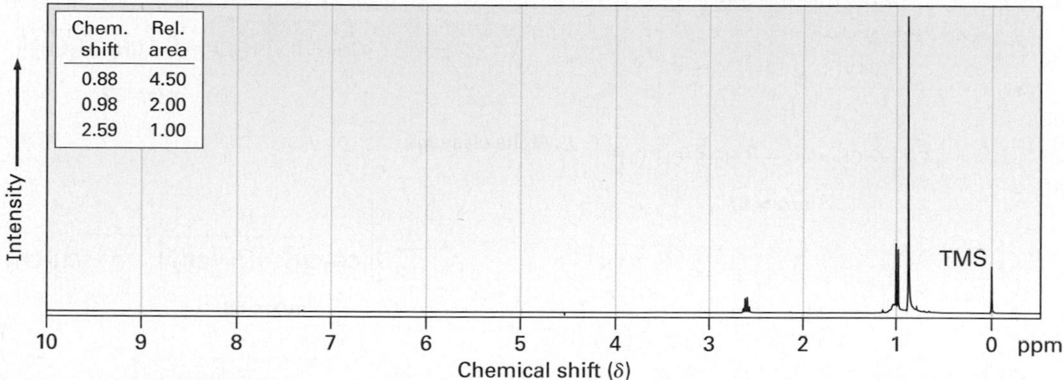

Mass Spectrometry

The *nitrogen rule* of mass spectrometry says that a compound with an odd number of nitrogen atoms has an odd-numbered molecular weight. Thus, the presence of nitrogen in a molecule is detected simply by observing its mass spectrum. An odd-numbered molecular ion usually means that the unknown compound has one or three nitrogen atoms, and an even-numbered molecular ion usually means that a compound has either zero or two nitrogen atoms. The logic behind the rule derives from the fact that nitrogen is trivalent, thus requiring an odd number of hydrogen atoms. For example, morphine has the formula $C_{17}H_{19}NO_3$ and a molecular weight of 285 amu.

Alkylamines undergo a characteristic α cleavage in the mass spectrometer, similar to the cleavage observed for alcohols **(Section 17.11)**. A C–C bond nearest the nitrogen atom is broken, yielding an alkyl radical and a resonance-stabilized, nitrogen-containing cation.

$$\left[RCH_2 \text{-} C(NR_2) \right]^{+\cdot} \xrightarrow[\text{cleavage}]{\text{Alpha}} RCH_2\cdot \; + \; \left[\overset{:NR_2}{\overset{|}{C^+}} \longleftrightarrow \overset{^+NR_2}{\overset{||}{C}} \right]$$

As an example, the mass spectrum of *N*-ethylpropylamine shown in **Figure 24.9** has peaks at m/z = 58 and m/z = 72, corresponding to the two possible modes of α cleavage.

Figure 24.9 Mass spectrum of *N*-ethylpropylamine. The two possible modes of α cleavage lead to the observed fragment ions at $m/z = 58$ and $m/z = 72$.

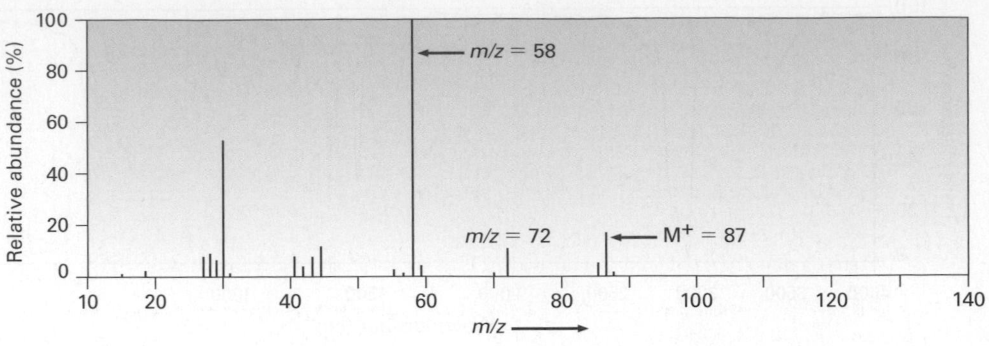

$$\left[\mathrm{CH_3 \wr CH_2{-}\overset{\displaystyle H}{\overset{|}{N}}{-}CH_2 \wr CH_2CH_3}\right]^{+\cdot} \quad m/z = 87$$

Alpha cleavage

$$\rightarrow \cdot\mathrm{CH_3} + \left[\mathrm{H_2C{=}\overset{\displaystyle H}{\overset{|}{N}}{-}CH_2CH_2CH_3}\right]^{+} \quad m/z = 72$$

$$\rightarrow \left[\mathrm{CH_3CH_2{-}\overset{\displaystyle H}{\overset{|}{N}}{=}CH_2}\right]^{+} + \cdot\mathrm{CH_2CH_3} \quad m/z = 58$$

Yes, this liquid really does consist of an ionic rather than a molecular substance.

Green Chemistry II: Ionic Liquids — A DEEPER LOOK

Liquids made of ions? Usually when we think of ionic compounds, we think of high-melting solids: sodium chloride, magnesium sulfate, lithium carbonate, and so forth. But yes, there also ionic compounds that are liquid at room temperature, and they are gaining importance as reaction solvents, particularly for use in green chemistry processes (see the Chapter 11 *A Deeper Look*). More than 1500 ionic liquids are known, and about 500 are available commercially.

Ionic liquids have been known for nearly a century; the first to be discovered was ethylammonium nitrate, $CH_3CH_2NH_3^+ \ NO_3^-$, with a melting point of 12 °C. More generally, however, the ionic liquids in use today are salts in which the cation is unsymmetrical and in which one or both of the ions are bulky so that the charges are dispersed over a large volume. Both factors minimize the crystal lattice energy and disfavor formation of the solid. Typical cations are quaternary ammonium ions from heterocyclic amines, either 1,3-dialkylimidazolium ions, *N*-alkylpyridinium ions, or ring-substituted *N*-alkylpyridinium ions.

$$\left[\mathrm{R = {-}CH_3, {-}CH_2CH_3, {-}CH_2CH_2CH_2CH_3, {-}CH_2CH_2CH_2CH_2CH_2CH_2CH_2CH_3}\right]$$

1,3-Dialkylimidazolium ions

$$\left[\mathrm{R = {-}CH_2CH_3, {-}CH_2CH_2CH_2CH_3, {-}CH_2CH_2CH_2CH_2CH_2CH_3}\right]$$

***N*-Alkylpyridinium ions**

(continued)

(continued)

Anions are just as varied as the cations. Hexafluorophosphate, tetrafluoroborate, alkyl sulfates, trifluoromethanesulfonates (triflates), and halides are some anion possibilities.

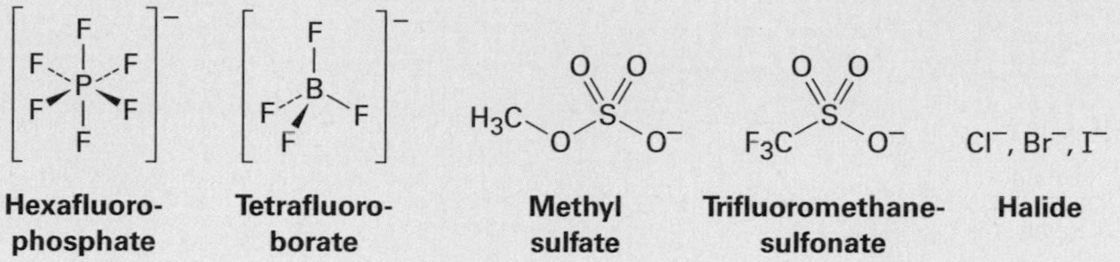

Hexafluoro-phosphate **Tetrafluoro-borate** **Methyl sulfate** **Trifluoromethane-sulfonate** **Halide**

Ionic liquids have several important features that make them attractive for use, both as solvents in green chemistry and as specialty chemicals in such applications as paint additives and refrigerants:

- They dissolve both polar and nonpolar organic compounds, giving high solute concentrations and thereby minimizing the amount of solvent needed.
- They can be optimized for specific reactions by varying cation and anion structures.
- They are nonflammable.
- They are thermally stable.
- They have negligible vapor pressures and do not evaporate.
- They are generally recoverable and can be reused many times.

As an example of their use in organic chemistry, the analgesic drug pravadoline has been synthesized in two steps using 1-butyl-3-methylimidazolium hexafluorophosphate, abbreviated [bmim][PF_6], as the solvent for both steps. The first step is a base-induced S_N2 reaction of 2-methylindole with a primary alkyl halide, and the second is a Friedel–Crafts acylation. Both steps take place in 95% yield, and the ionic solvent is recovered simply by washing the reaction mixture, first with toluene and then with water.

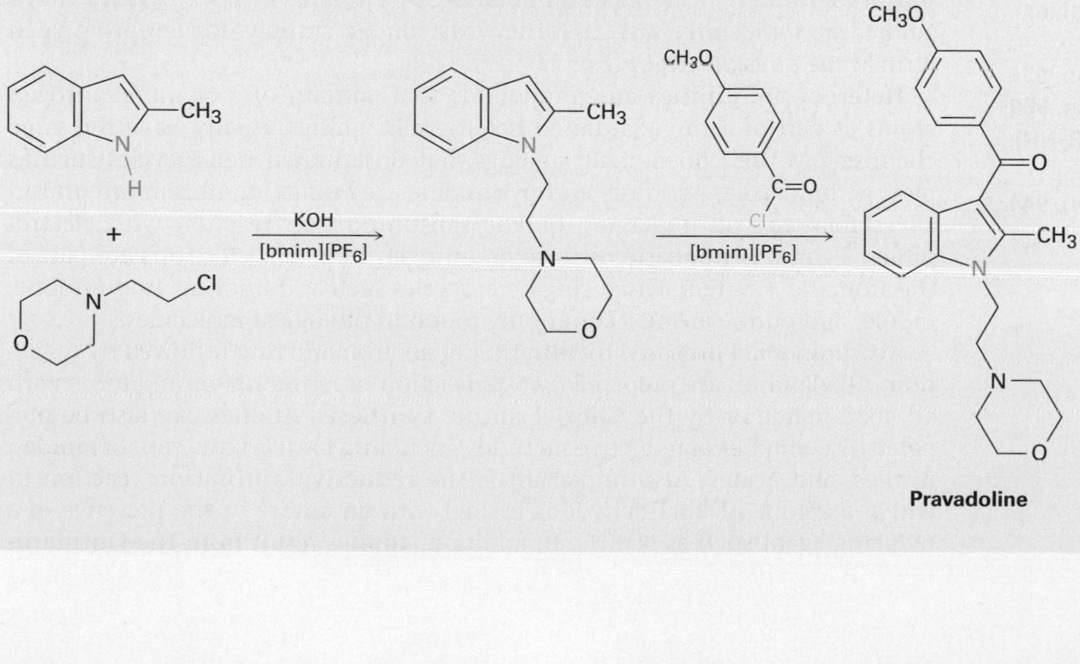

Pravadoline

(continued)

(continued)

The first commercial process using an ionic liquid catalyst was introduced by PetroChina in 2008, when they opened a plant producing 65,000 tons per year of alkylate gasoline from isobutane. The aluminum-based ionic liquid catalyst replaced the sulfuric acid and hydrofluoric acid catalysts that had previously been used.

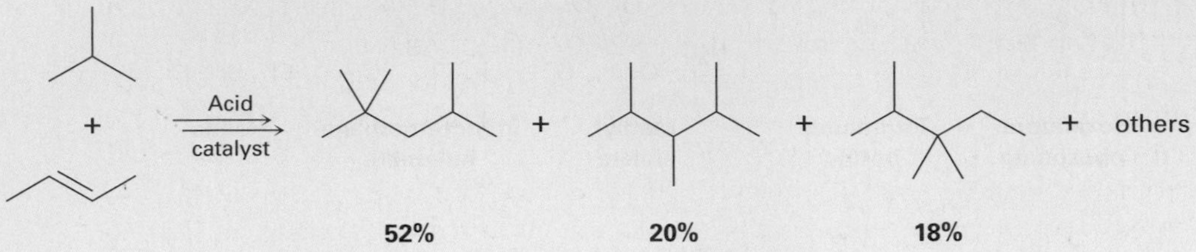

\+ Acid catalyst → + + + others

52% 20% 18%

Summary

Key words

alkylamine, 944
amine, 944
arylamine, 944
azo compound (Ar—N=N—Ar′), 971
Curtius rearrangement, 960
Gabriel amine synthesis, 957
heterocyclic amine, 946
Hofmann elimination reaction, 964
Hofmann rearrangement, 960
imide (–CONHCO–), 957
primary amine (RNH_2), 944
quaternary ammonium salt, 945
reductive amination, 958
Sandmeyer reaction, 969
secondary amine (R_2NH), 944
tertiary amine (R_3N), 944

We've now seen all the common functional groups that occur in organic and biological chemistry. Of those groups, amines are among the most abundant and have among the richest chemistry. In addition to proteins and nucleic acids, the majority of pharmaceutical agents contain amine functional groups and many of the common coenzymes necessary for biological reactions are amines.

Amines are organic derivatives of ammonia. They are named in the IUPAC system either by adding the suffix *-amine* to the name of the alkyl substituent or by considering the amino group as a substituent on a more complex parent molecule.

The chemistry of amines is dominated by the lone-pair electrons on nitrogen, which makes amines both basic and nucleophilic. The basicity of **arylamines** is generally lower than that of **alkylamines** because the nitrogen lone-pair electrons are delocalized by interaction with the aromatic π system. Electron-withdrawing substituents on the aromatic ring further weaken the basicity of a substituted aniline, while electron-donating substituents increase basicity. Alkylamines are sufficiently basic that they exist almost entirely in their protonated form at the physiological pH of 7.3 inside cells.

Heterocyclic amines are compounds that contain one or more nitrogen atoms as part of a ring. Saturated heterocyclic amines usually have the same chemistry as their open-chain analogs, but unsaturated heterocycles such as pyrrole, imidazole, pyridine, and pyrimidine are aromatic. All four are unusually stable, and all undergo aromatic substitution on reaction with electrophiles. Pyrrole is nonbasic because its nitrogen lone-pair electrons are part of the aromatic π system. Fused-ring heterocycles such as quinoline, isoquinoline, indole, and purine are also commonly found in biological molecules.

Arylamines are prepared by nitration of an aromatic ring followed by reduction. Alkylamines are prepared by S_N2 reaction of ammonia or an amine with an alkyl halide or by the **Gabriel amine synthesis**. Amines can also be prepared by a number of reductive methods, including $LiAlH_4$ reduction of amides, nitriles, and azides. Also important is the **reductive amination** reaction in which a ketone or an aldehyde is treated with an amine in the presence of a reducing agent such as $NaBH_4$. In addition, amines result from the **Hofmann** and **Curtius rearrangements** of carboxylic acid derivatives. Both methods involve migration of the –R group bonded to the carbonyl carbon and yield a product that has one less carbon atom than the starting material.

Many of the reactions of amines are familiar from past chapters. Thus, amines react with alkyl halides in S_N2 reactions and with acid chlorides in nucleophilic acyl substitution reactions. Amines also undergo E2 elimination to yield alkenes if they are first quaternized by treatment with iodomethane and then heated with silver oxide, a process called the **Hofmann elimination**.

Arylamines are converted by diazotization with nitrous acid into arenediazonium salts, $ArN_2^+ X^-$. The diazonio group can then be replaced by many other substituents in the **Sandmeyer reaction** to give a wide variety of substituted aromatic compounds. Aryl chlorides, bromides, iodides, and nitriles can be prepared from arenediazonium salts, as can arenes and phenols. In addition to their reactivity toward substitution reactions, diazonium salts undergo coupling with phenols and arylamines to give brightly colored **azo compounds**.

Summary of Reactions

1. Synthesis of amines (Section 24.6)
   (a) Reduction of nitriles

$$RCH_2X \xrightarrow{NaCN} RCH_2C{\equiv}N \xrightarrow[2.\ H_2O]{1.\ LiAlH_4,\ ether} RCH_2CH_2NH_2$$

   (b) Reduction of amides

$$RCONH_2 \xrightarrow[2.\ H_2O]{1.\ LiAlH_4,\ ether} RCH_2NH_2$$

   (c) Reduction of nitrobenzenes

$$C_6H_5NO_2 \xrightarrow[\text{or } Fe,\ H_3O^+ \text{ or } SnCl_2,\ H_3O^+]{H_2,\ Pt} C_6H_5NH_2$$

   (d) S_N2 Alkylation of alkyl halides

Ammonia	$\ddot{N}H_3$ + R–X	⟶	$R\overset{+}{N}H_3\ X^-$	$\xrightarrow{NaOH}$	RNH_2	Primary
Primary	$R\ddot{N}H_2$ + R–X	⟶	$R_2\overset{+}{N}H_2\ X^-$	$\xrightarrow{NaOH}$	R_2NH	Secondary
Secondary	$R_2\ddot{N}H$ + R–X	⟶	$R_3\overset{+}{N}H\ X^-$	$\xrightarrow{NaOH}$	R_3N	Tertiary
Tertiary	$R_3\ddot{N}$ + R–X	⟶	$R_4\overset{+}{N}\ X^-$			Quaternary ammonium

(continued)

(e) Gabriel amine synthesis

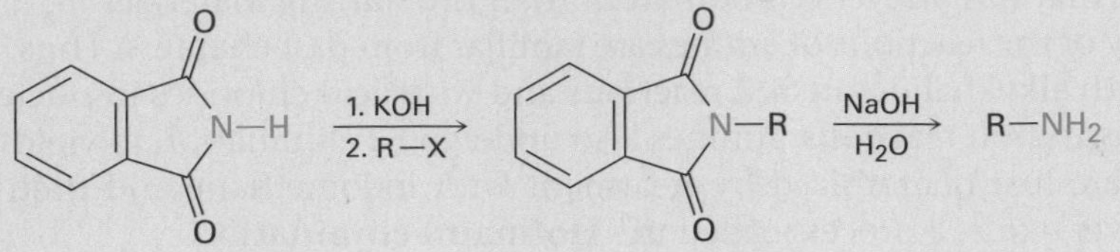

$$\text{Phthalimide (N–H)} \xrightarrow[\text{2. R–X}]{\text{1. KOH}} \text{N-alkylphthalimide (N–R)} \xrightarrow[H_2O]{NaOH} R{-}NH_2$$

(f) Reduction of azides

$$RCH_2{-}X \xrightarrow[\text{ethanol}]{Na^+\ {}^-N_3} RCH_2{-}N{=}\overset{+}{N}{=}\overset{-}{N} \xrightarrow[\text{2. } H_2O]{\text{1. } LiAlH_4\text{, ether}} R{-}NH_2$$

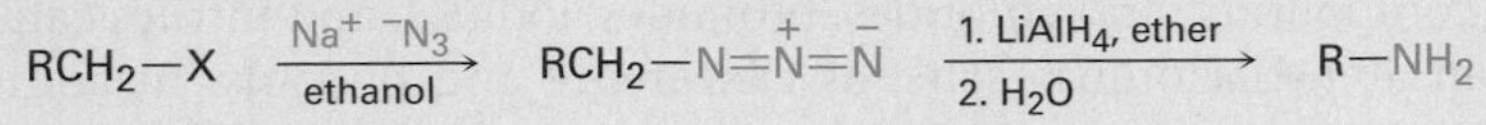

(g) Reductive amination of aldehydes/ketones

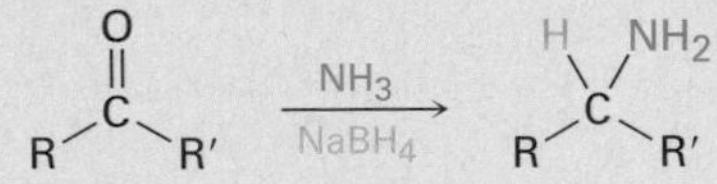

$$RC(=O)R' \xrightarrow[NaBH_4]{NH_3} RCH(NH_2)R'$$

(h) Hofmann rearrangement of amides

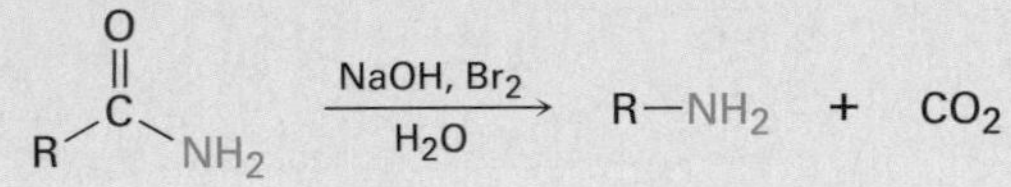

$$RC(=O)NH_2 \xrightarrow[H_2O]{NaOH,\ Br_2} R{-}NH_2 + CO_2$$

(i) Curtius rearrangement of acyl azides

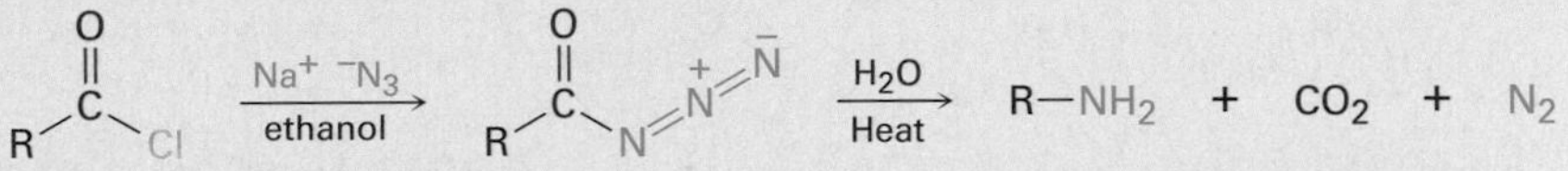

$$RC(=O)Cl \xrightarrow[\text{ethanol}]{Na^+\ {}^-N_3} RC(=O){-}N{=}\overset{+}{N}{=}\overset{-}{N} \xrightarrow[\text{Heat}]{H_2O} R{-}NH_2 + CO_2 + N_2$$

2. Reactions of amines
   (a) Alkylation with alkyl halides; see reaction 1(d)
   (b) Hofmann elimination (Section 24.7)

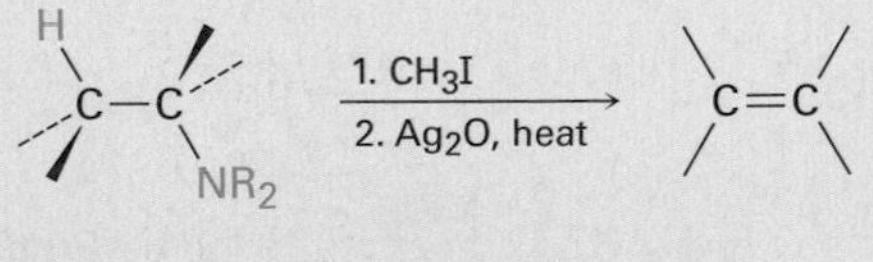

$$H{-}C{-}C{-}NR_2 \xrightarrow[\text{2. } Ag_2O\text{, heat}]{\text{1. } CH_3I} C{=}C$$

   (c) Diazotization (Section 24.8)

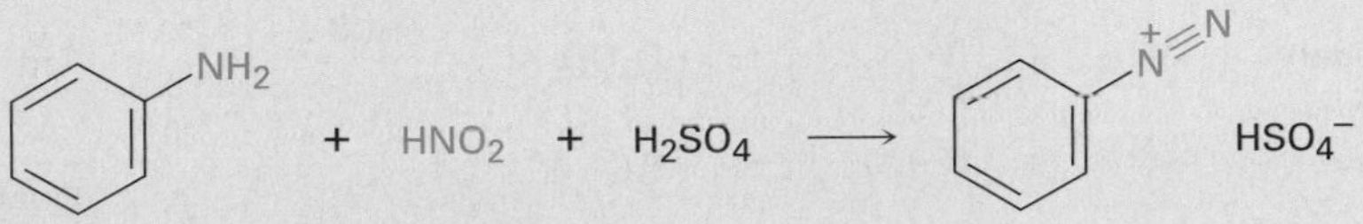

$$C_6H_5NH_2 + HNO_2 + H_2SO_4 \longrightarrow C_6H_5\overset{+}{N}{\equiv}N\ \ HSO_4^-$$

3. Reactions of arenediazonium salts (Section 24.8)
(a) Nucleophilic substitutions

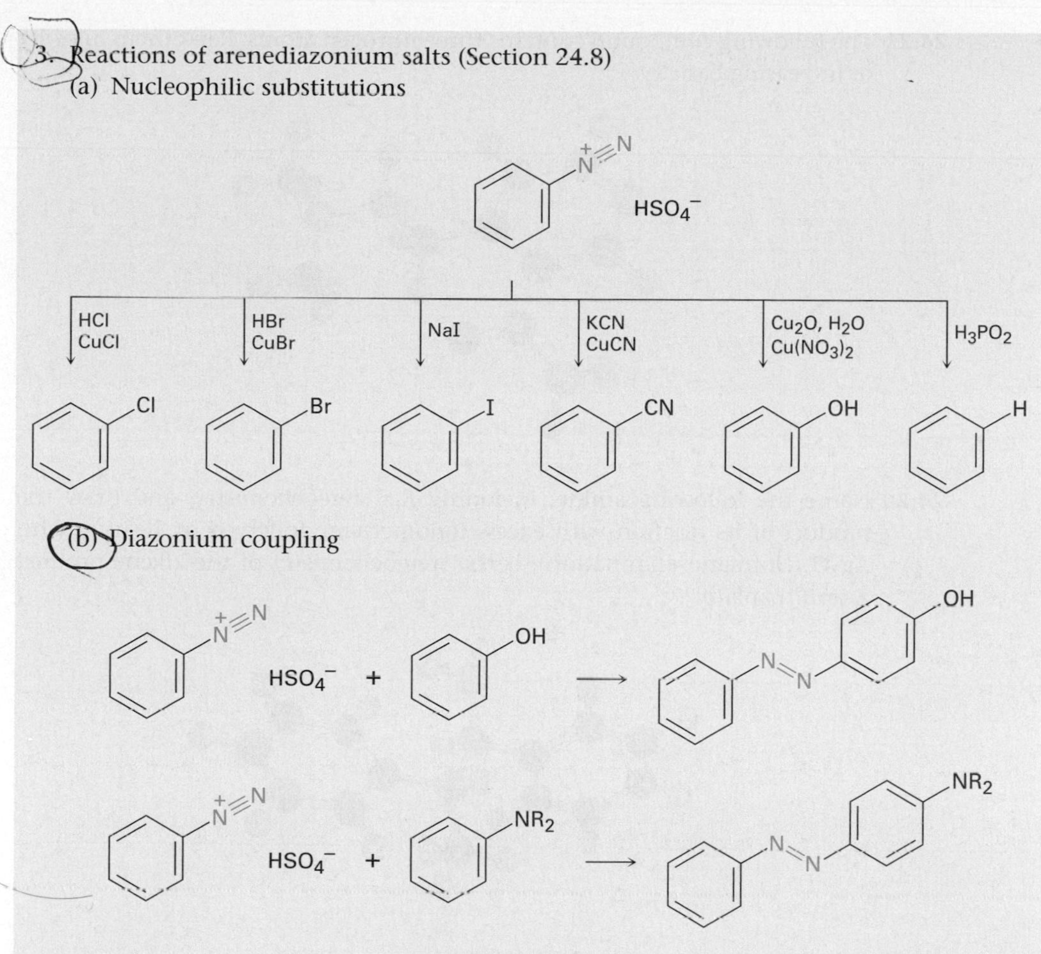

Exercises

Visualizing Chemistry

(Problems 24.1–24.25 appear within the chapter.)

OWL Interactive versions of these problems are assignable in OWL for Organic Chemistry.

24.26 Name the following amines, and identify each as primary, secondary, or tertiary:

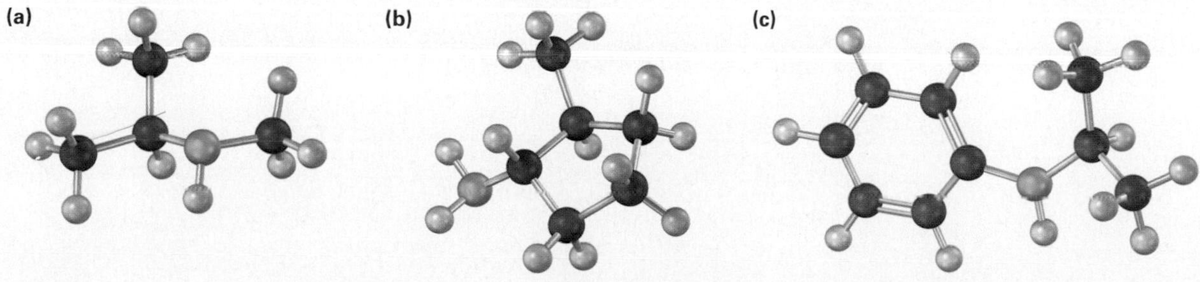

24.27 The following compound contains three nitrogen atoms. Rank them in order of increasing basicity.

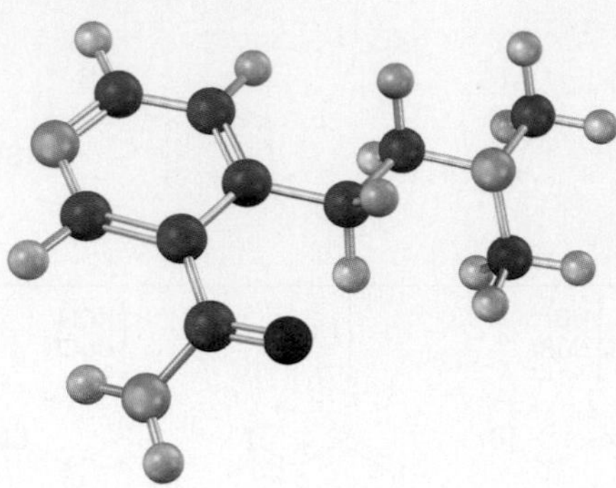

24.28 Name the following amine, including *R,S* stereochemistry, and draw the product of its reaction with excess iodomethane followed by heating with Ag_2O (Hofmann elimination). Is the stereochemistry of the alkene product *Z* or *E*? Explain.

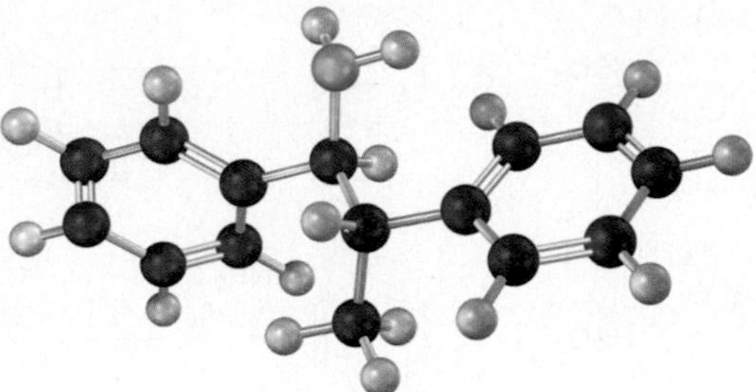

24.29 Which nitrogen atom in the following compound is most basic? Explain.

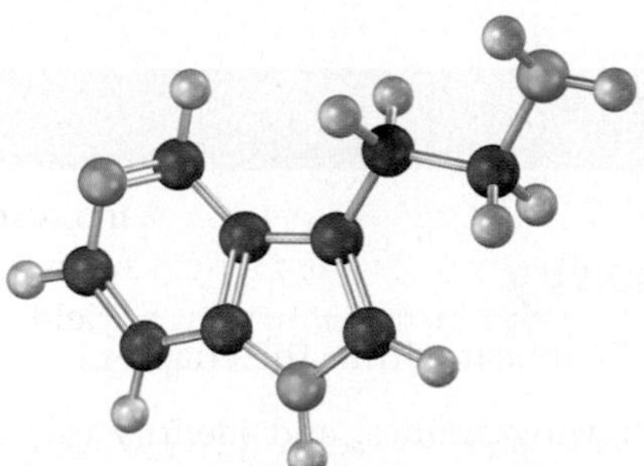

Additional Problems

Naming Amines

24.30 Name the following compounds:

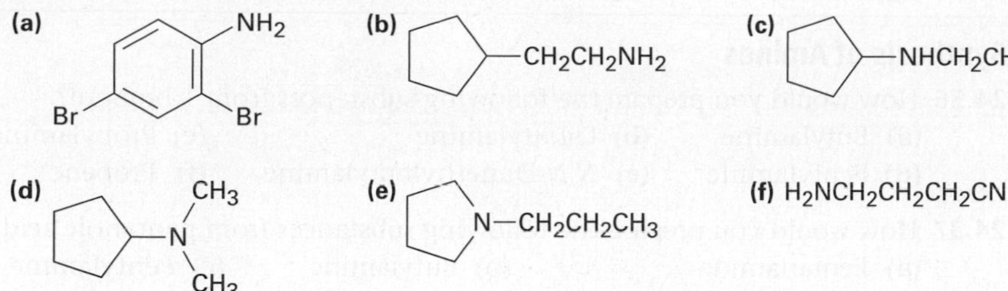

24.31 Draw structures corresponding to the following IUPAC names:

(a) *N,N*-Dimethylaniline
(b) (Cyclohexylmethyl)amine
(c) *N*-Methylcyclohexylamine
(d) (2-Methylcyclohexyl)amine
(e) 3-(*N,N*-Dimethylamino)propanoic acid

24.32 Classify each of the amine nitrogen atoms in the following substances as primary, secondary, or tertiary:

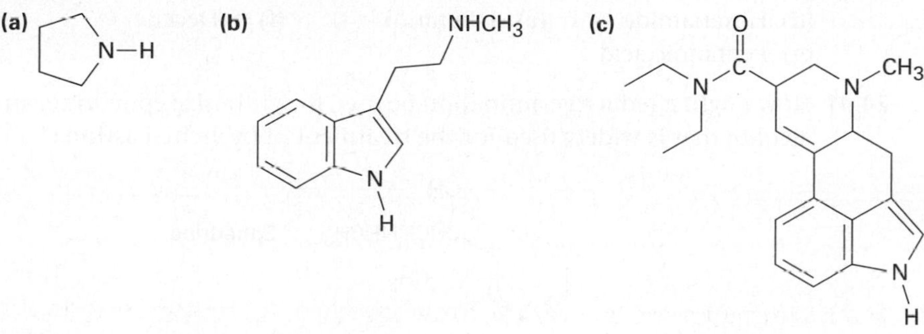

Lysergic acid diethylamide

Amine Basicity

24.33 Although pyrrole is a much weaker base than most other amines, it is a much stronger acid (p$K_a \approx 15$ for the pyrrole versus 35 for diethylamine). The N–H proton is readily abstracted by base to yield the pyrrole anion, $C_4H_4N^-$. Explain.

24.34 Histamine, whose release in the body triggers nasal secretions and constricted airways, has three nitrogen atoms. List them in order of increasing basicity, and explain your ordering.

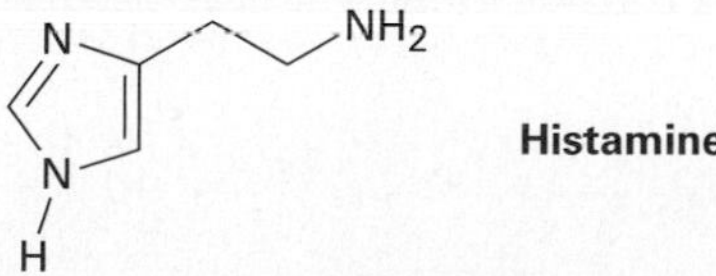

Histamine

24.35 Account for the fact that *p*-nitroaniline ($pK_a = 1.0$) is less basic than *m*-nitroaniline ($pK_a = 2.5$) by a factor of 30. Draw resonance structures to support your argument. (The pK_a values refer to the corresponding ammonium ions.)

Synthesis of Amines

24.36 How would you prepare the following substances from 1-butanol?
(a) Butylamine (b) Dibutylamine (c) Propylamine
(d) Pentylamine (e) *N*,*N*-Dimethylbutylamine (f) Propene

24.37 How would you prepare the following substances from pentanoic acid?
(a) Pentanamide (b) Butylamine (c) Pentylamine
(d) 2-Bromopentanoic acid (e) Hexanenitrile (f) Hexylamine

24.38 How would you prepare aniline from the following starting materials?
(a) Benzene (b) Benzamide (c) Toluene

24.39 How would you prepare benzylamine, $C_6H_5CH_2NH_2$, from benzene? More than one step is needed.

24.40 How might you prepare pentylamine from the following starting materials?
(a) Pentanamide (b) Pentanenitrile (c) 1-Butene
(d) Hexanamide (e) 1-Butanol (f) 5-Decene
(g) Pentanoic acid

24.41 How might a reductive amination be used to synthesize ephedrine, an amino alcohol that is widely used for the treatment of bronchial asthma?

C_6H_5–CH(OH)–CH(CH_3)–NHCH_3 **Ephedrine**

Reactions of Amines

24.42 How would you convert aniline into each of the following products?
(a) Benzene (b) Benzamide (c) Toluene

24.43 Give the structures of the major organic products you would expect from reaction of *m*-toluidine (*m*-methylaniline) with the following reagents:
(a) Br_2 (1 equivalent) (b) CH_3I (excess)
(c) CH_3COCl in pyridine (d) The product of (c), then HSO_3Cl

24.44 Show the products from reaction of *p*-bromoaniline with the following reagents:

(a) CH_3I (excess) (b) HCl (c) HNO_2, H_2SO_4
(d) CH_3COCl (e) CH_3MgBr (f) CH_3CH_2Cl, $AlCl_3$
(g) Product of (c) with CuCl, HCl
(h) Product of (d) with CH_3CH_2Cl, $AlCl_3$

24.45 What are the major products you would expect from Hofmann elimination of the following amines?

(a) 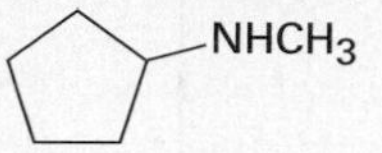 cyclopentyl–$NHCH_3$ (b) C_6H_5–$NHCH(CH_3)CH_2CH_2CH_2CH_3$ (c) $CH_3CH(CH_3)CH(NH_2)CH_2CH_2CH_3$

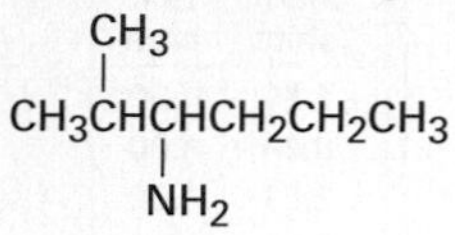

24.46 How would you prepare the following compounds from toluene? A diazonio replacement reaction is needed in some instances.

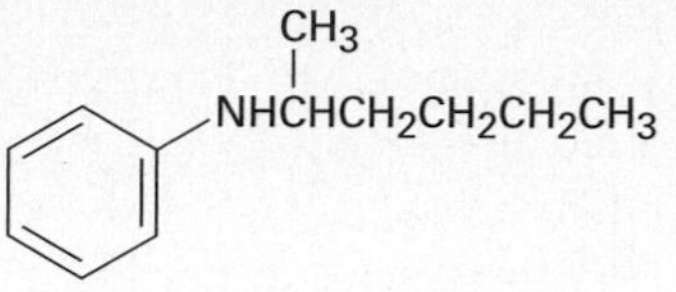

(a) 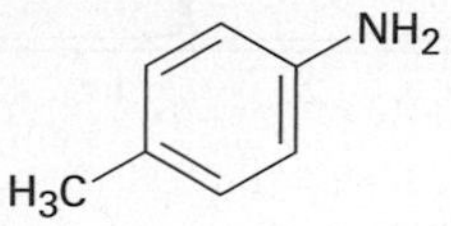 H_3C–C_6H_4–NH_2 (b) H_3C–C_6H_4–CH_2NH_2 (c) I–C_6H_4–$C(=O)OCH_3$

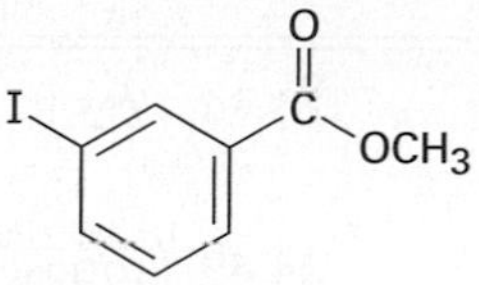

24.47 Predict the product(s) of the following reactions. If more than one product is formed, tell which is major.

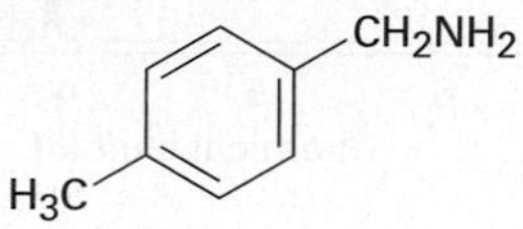

(a) 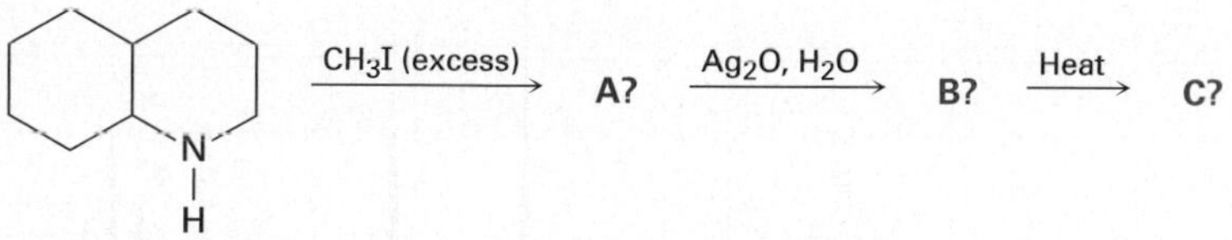 decahydroquinoline (N–H) $\xrightarrow{CH_3I\ (excess)}$ A? $\xrightarrow{Ag_2O,\ H_2O}$ B? $\xrightarrow{Heat}$ C?

(b) C_6H_5COCl $\xrightarrow{NaN_3}$ A? $\xrightarrow{Heat}$ B? $\xrightarrow{H_2O}$ C?

(c) phthalimide (N–H) $\xrightarrow{KOH}$ A? $\xrightarrow{C_6H_5CH_2Br}$ B? $\xrightarrow[H_2O]{KOH}$ C?

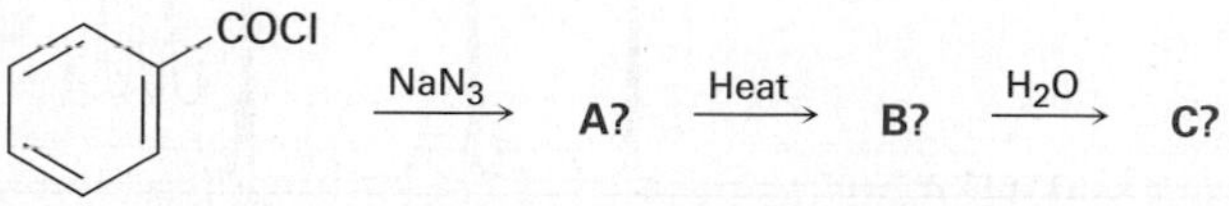

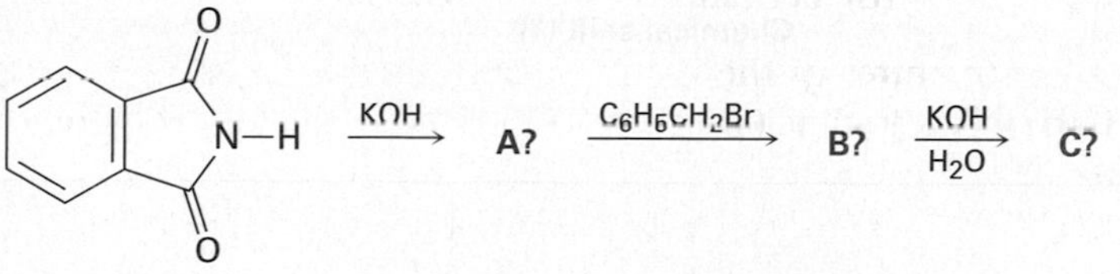

(d) $BrCH_2CH_2CH_2CH_2Br$ + 1 equiv CH_3NH_2 $\xrightarrow[H_2O]{NaOH}$?

Spectroscopy

24.48 Phenacetin, a substance formerly used in over-the-counter headache remedies, has the formula $C_{10}H_{13}NO_2$. Phenacetin is neutral and does not dissolve in either acid or base. When warmed with aqueous NaOH, phenacetin yields an amine, $C_8H_{11}NO$, whose ^1H NMR spectrum is shown. When heated with HI, the amine is cleaved to an aminophenol, C_6H_7NO. What is the structure of phenacetin, and what are the structures of the amine and the aminophenol?

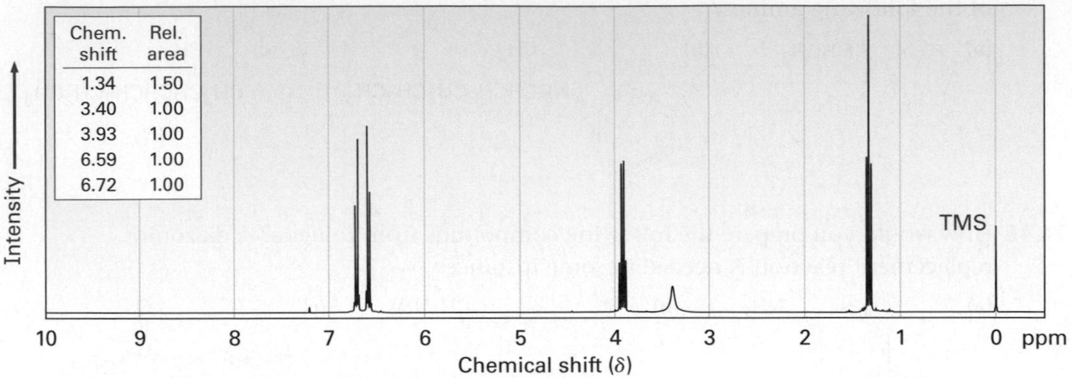

24.49 Propose structures for amines with the following ^1H NMR spectra:

(a) C_3H_9NO

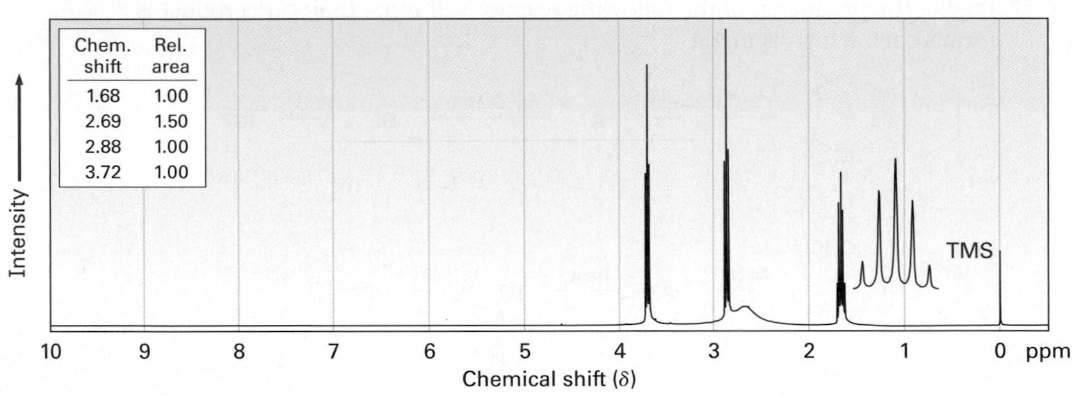

(b) $C_4H_{11}NO_2$

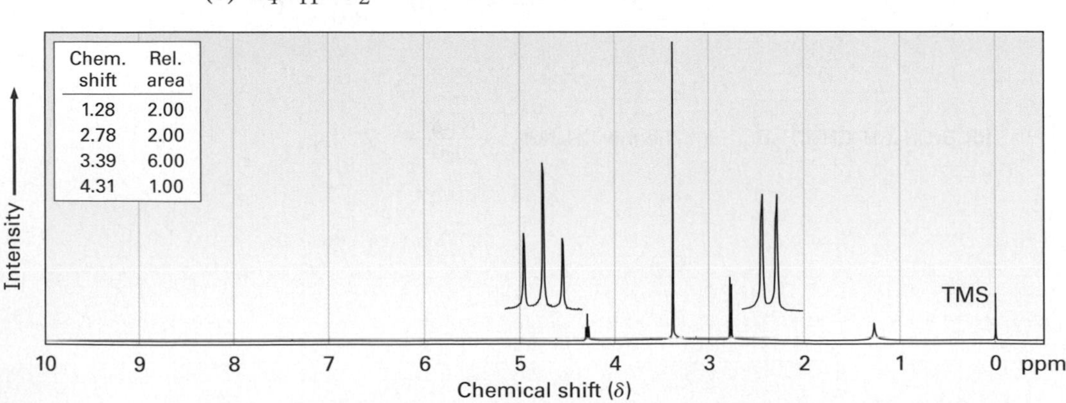

General Problems

24.50 Fill in the missing reagents a–e in the following scheme:

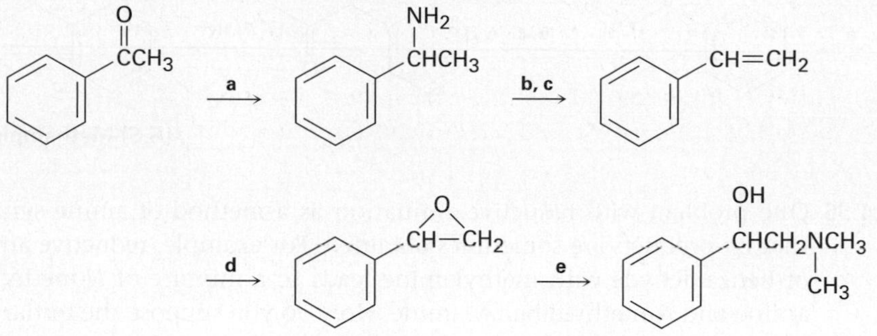

$C_6H_5COCH_3 \xrightarrow{a} C_6H_5CH(NH_2)CH_3 \xrightarrow{b,\ c} C_6H_5CH{=}CH_2 \xrightarrow{d} C_6H_5CH{-}CH_2$ (epoxide, O bridging) $\xrightarrow{e} C_6H_5CH(OH)CH_2N(CH_3)CH_3$

24.51 Oxazole is a five-membered aromatic heterocycle. Would you expect oxazole to be more basic or less basic than pyrrole? Explain.

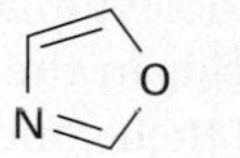

Oxazole

24.52 Protonation of an amide using strong acid occurs on oxygen rather than on nitrogen. Suggest a reason for this behavior, taking resonance into account.

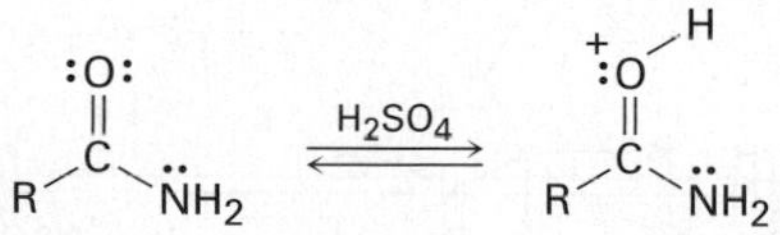

$RC(=O)NH_2 \xrightleftharpoons{H_2SO_4} RC(=\overset{+}{O}H)NH_2$

24.53 Substituted pyrroles are often prepared by treatment of a 1,4-diketone with ammonia. Propose a mechanism.

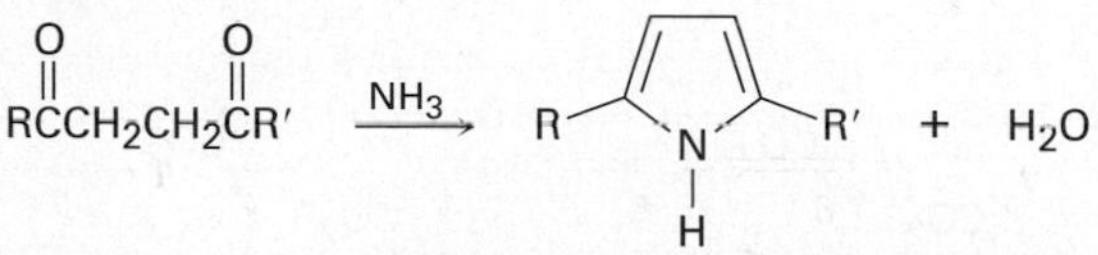

$RCOCH_2CH_2COR' \xrightarrow{NH_3}$ 2-R-5-R'-pyrrole (N–H) $+ H_2O$

24.54 3,5 Dimethylisoxazole is prepared by reaction of 2,4-pentanedione with hydroxylamine. Propose a mechanism.

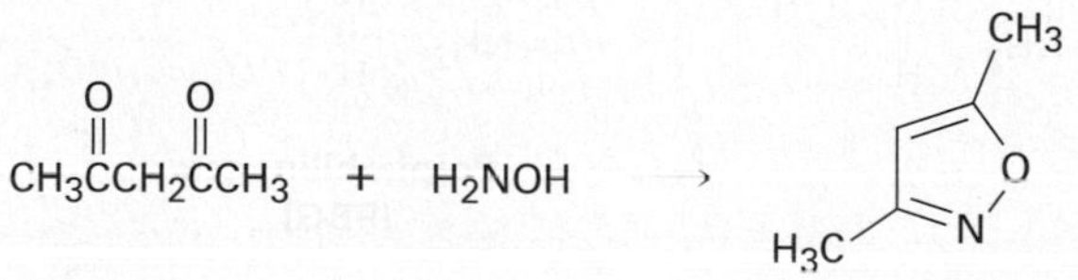

$CH_3COCH_2COCH_3 + H_2NOH \longrightarrow$ 3,5-Dimethylisoxazole

3,5-Dimethylisoxazole

24.55 Fill in the missing reagents **a–d** in the following synthesis of racemic methamphetamine from benzene.

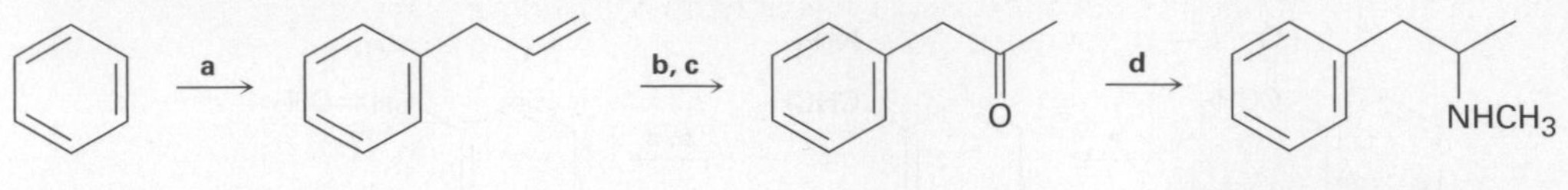

a, b, c, d

$NHCH_3$

***(R,S)*-Methamphetamine**

24.56 One problem with reductive amination as a method of amine synthesis is that by-products are sometimes obtained. For example, reductive amination of benzaldehyde with methylamine leads to a mixture of *N*-methylbenzylamine and *N*-methyldibenzylamine. How do you suppose the tertiary amine by-product is formed? Propose a mechanism.

24.57 Chlorophyll, heme, vitamin B_{12}, and a host of other substances are biosynthesized from porphobilinogen (PBG), which is itself formed from condensation of two molecules of 5-aminolevulinate. The two 5-aminolevulinates are bound to lysine (Lys) amino acids in the enzyme, one in the enamine form and one in the imine form, and their condensation is thought to occur by the following steps. Using curved arrows, show the mechanism of each step.

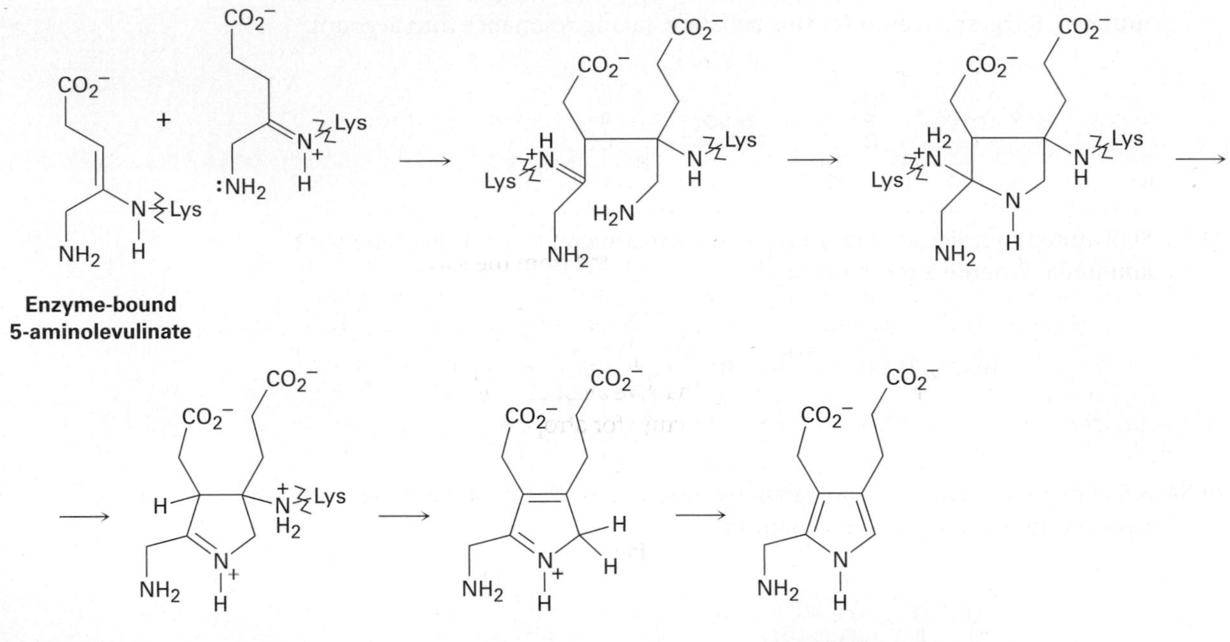

Enzyme-bound 5-aminolevulinate

Porphobilinogen (PBG)

24.58 Choline, a component of the phospholipids in cell membranes, can be prepared by S_N2 reaction of trimethylamine with ethylene oxide. Show the structure of choline, and propose a mechanism for the reaction.

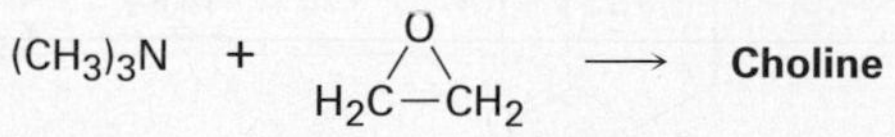

$(CH_3)_3N$ + $H_2C{-}CH_2$ (O bridging; ethylene oxide) ⟶ **Choline**

24.59 Cyclopentamine is an amphetamine-like central nervous system stimulant. Propose a synthesis of cyclopentamine from materials of five carbons or less.

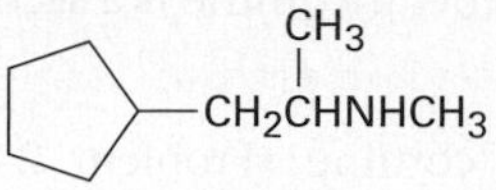

(cyclopentyl)$-CH_2\overset{CH_3}{\overset{|}{C}}HNHCH_3$ **Cyclopentamine**

24.60 Tetracaine is a substance used as a spinal anesthetic.

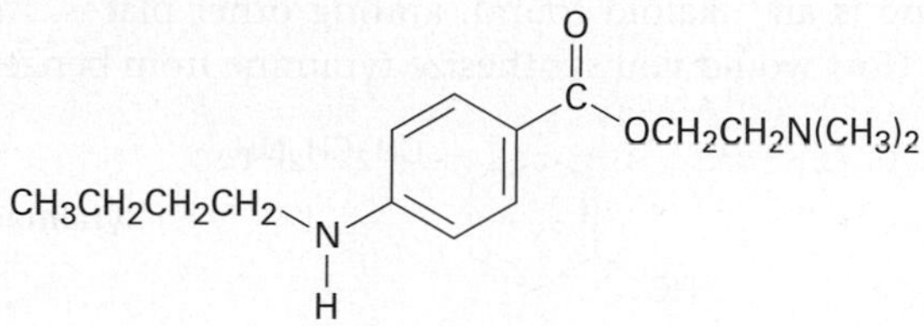

$CH_3CH_2CH_2CH_2-\underset{H}{N}-C_6H_4-\overset{O}{\overset{\|}{C}}-OCH_2CH_2N(CH_3)_2$

Tetracaine

(a) How would you prepare tetracaine from the corresponding aniline derivative, $ArNH_2$?

(b) How would you prepare tetracaine from *p*-nitrobenzoic acid?

(c) How would you prepare tetracaine from benzene?

24.61 Atropine, $C_{17}H_{23}NO_3$, is a poisonous alkaloid isolated from the leaves and roots of *Atropa belladonna,* the deadly nightshade. In small doses, atropine acts as a muscle relaxant; 0.5 ng (nanogram, 10^{-9} g) is sufficient to cause pupil dilation. On basic hydrolysis, atropine yields tropic acid, $C_6H_5CH(CH_2OH)CO_2H$, and tropine, $C_8H_{15}NO$. Tropine is an optically inactive alcohol that yields tropidene on dehydration with H_2SO_4. Propose a structure for atropine.

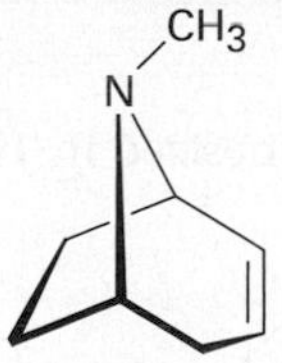

CH_3
N

Tropidene

24.62 Tropidene (Problem 24.61) can be converted by a series of steps into tropilidene (1,3,5-cycloheptatriene). How would you accomplish this conversion?

24.63 Propose a structure for the product with formula $C_9H_{17}N$ that results when 2-(2-cyanoethyl)cyclohexanone is reduced catalytically.

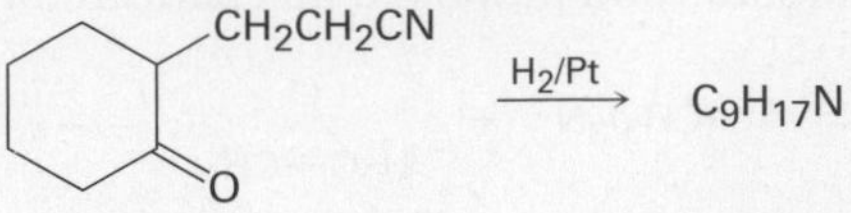

CH_2CH_2CN ; H_2/Pt ; $C_9H_{17}N$; O

24.64 Coniine, $C_8H_{17}N$, is the toxic principle of the poison hemlock drunk by Socrates. When subjected to Hofmann elimination, coniine yields 5-(*N,N*-dimethylamino)-1-octene. If coniine is a secondary amine, what is its structure?

24.65 How would you synthesize coniine (Problem 24.64) from acrylonitrile ($H_2C{=}CHCN$) and ethyl 3-oxohexanoate ($CH_3CH_2CH_2COCH_2CO_2Et$)? (*Hint:* See Problem 24.63.)

24.66 Tyramine is an alkaloid found, among other places, in mistletoe and ripe cheese. How would you synthesize tyramine from benzene? From toluene?

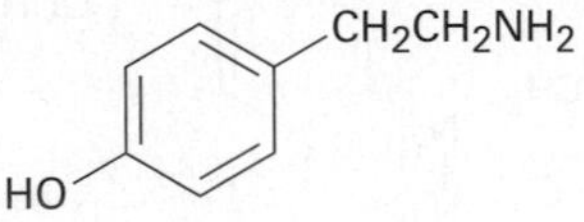

$CH_2CH_2NH_2$; HO ; **Tyramine**

24.67 Reaction of anthranilic acid (*o*-aminobenzoic acid) with HNO_2 and H_2SO_4 yields a diazonium salt that can be treated with base to yield a neutral diazonium carboxylate.

(a) What is the structure of the neutral diazonium carboxylate?

(b) Heating the diazonium carboxylate results in the formation of CO_2, N_2, and an intermediate that reacts with 1,3-cyclopentadiene to yield the following product:

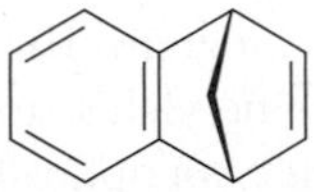

What is the structure of the intermediate, and what kind of reaction does it undergo with cyclopentadiene?

24.68 Cyclooctatetraene was first synthesized in 1911 by a route that involved the following transformation:

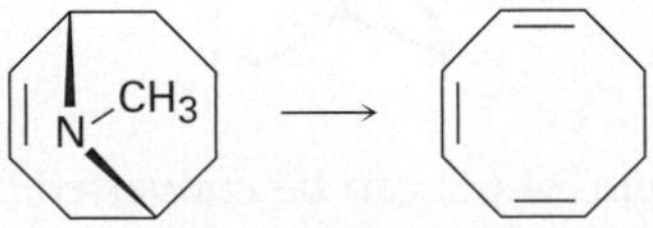

N ; CH_3 ; ⟶

How might you use the Hofmann elimination to accomplish this reaction? How would you finish the synthesis by converting cyclooctatriene into cyclooctatetraene?

24.69 When an α-hydroxy amide is treated with Br_2 in aqueous NaOH under Hofmann rearrangement conditions, loss of CO_2 occurs and a chain-shortened aldehyde is formed. Propose a mechanism.

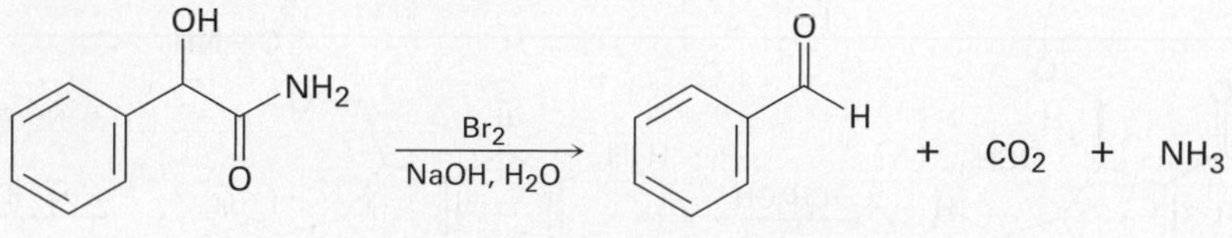

$$\text{PhCH(OH)CONH}_2 \xrightarrow[\text{NaOH, H}_2\text{O}]{\text{Br}_2} \text{PhCHO} + CO_2 + NH_3$$

24.70 The following transformation involves a conjugate nucleophilic addition reaction (Section 19.13) followed by an intramolecular nucleophilic acyl substitution reaction (Section 21.2). Show the mechanism.

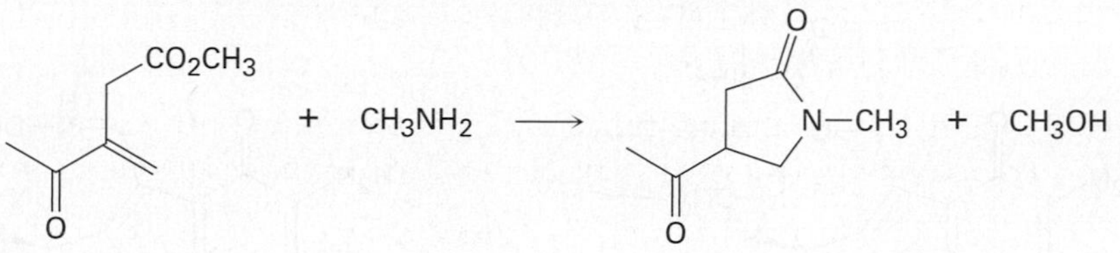

$$\text{(enone ester, } CO_2CH_3\text{)} + CH_3NH_2 \longrightarrow \text{(N–}CH_3 \text{ pyrrolidinone)} + CH_3OH$$

24.71 Propose a mechanism for the following reaction:

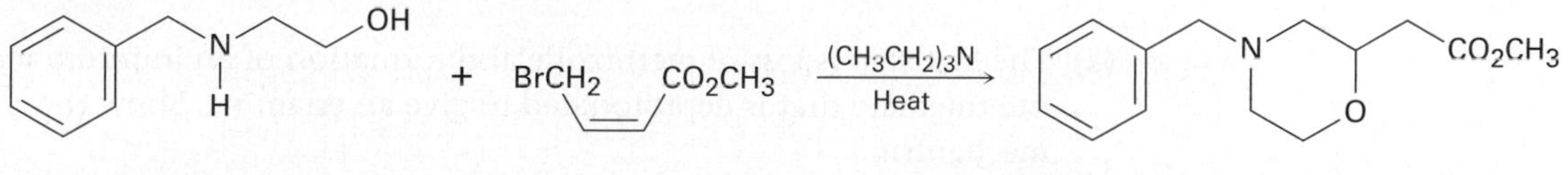

$$\text{(N–H, OH amino alcohol)} + BrCH_2\text{CH=CH}CO_2CH_3 \xrightarrow[\text{Heat}]{(CH_3CH_2)_3N} \text{(morpholine, } CO_2CH_3\text{)}$$

24.72 One step in the biosynthesis of morphine is the reaction of dopamine with *p*-hydroxyphenylacetaldehyde to give (*S*)-norcoclaurine. Assuming that the reaction is acid-catalyzed, propose a mechanism.

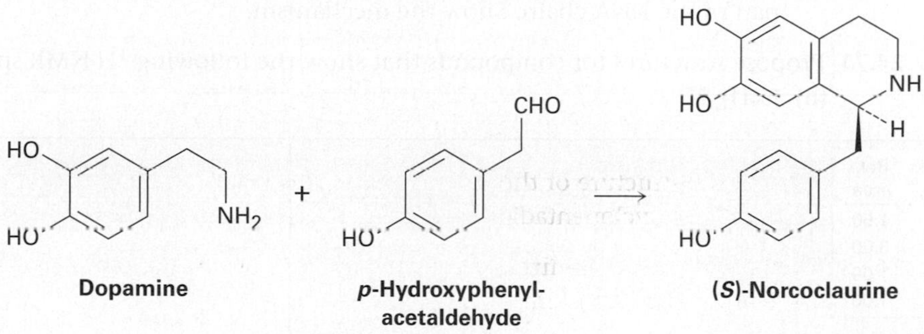

Dopamine + ***p*-Hydroxyphenylacetaldehyde** → **(*S*)-Norcoclaurine**

24.73 The antitumor antibiotic mitomycin C functions by forming cross-links in DNA chains.

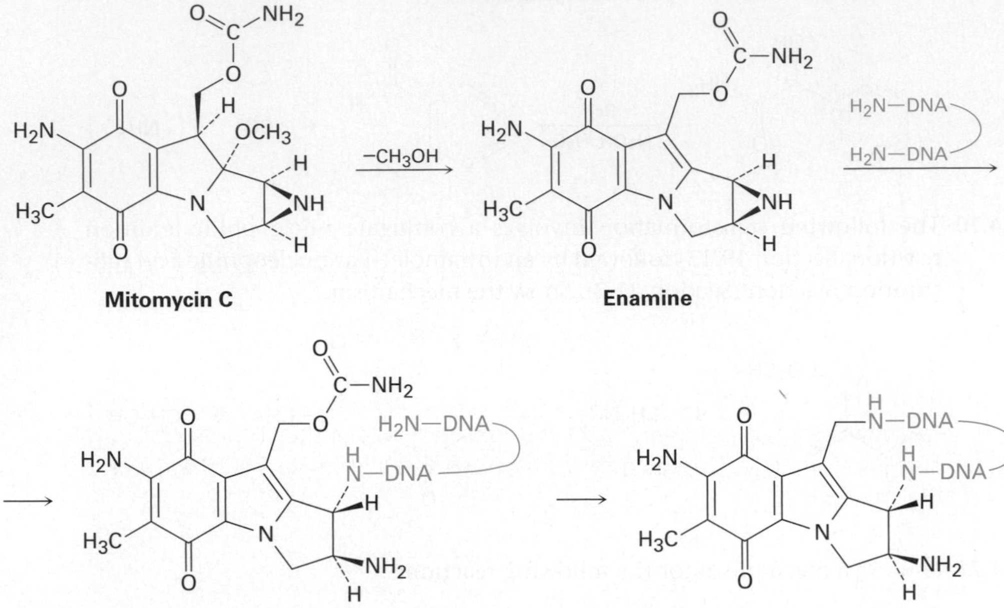

$-CH_3OH$

Mitomycin C **Enamine**

(a) The first step is loss of methoxide and formation of an iminium ion intermediate that is deprotonated to give an enamine. Show the mechanism.

(b) The second step is reaction of the enamine with DNA to open the three-membered, nitrogen-containing (aziridine) ring. Show the mechanism.

(c) The third step is loss of carbamate ($NH_2CO_2^-$) and formation of an unsaturated iminium ion, followed by a conjugate addition of another part of the DNA chain. Show the mechanism.

24.74 Propose structures for compounds that show the following ^1H NMR spectra.

(a) $C_9H_{13}N$

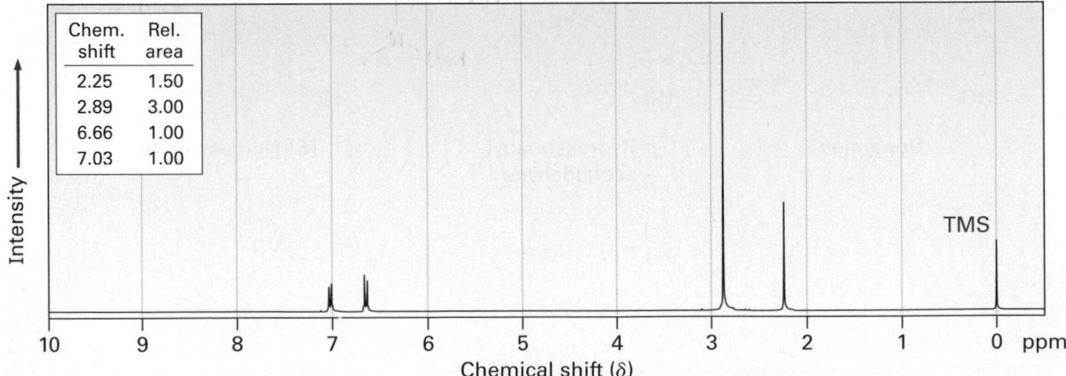

(b) $C_{15}H_{17}N$

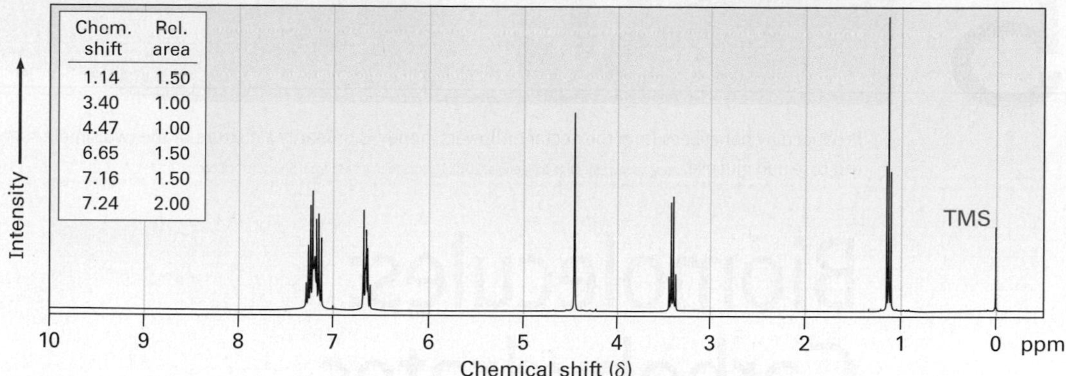

24.75 α-Amino acids can be prepared by the Strecker synthesis, a two-step process in which an aldehyde is treated with ammonium cyanide followed by hydrolysis of the amino nitrile intermediate with aqueous acid. Propose a mechanism for the reaction.

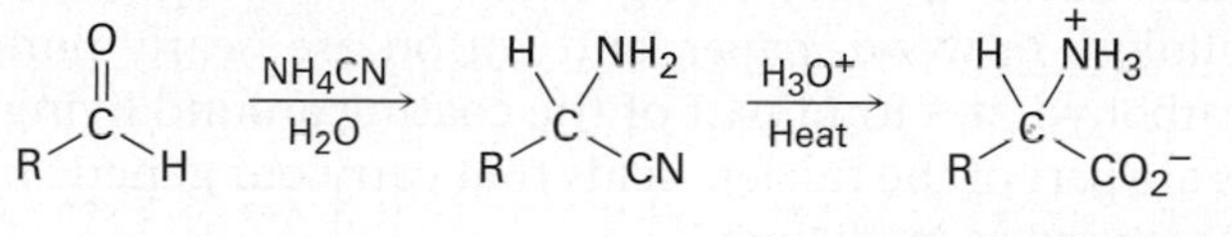

$$RCHO \xrightarrow[H_2O]{NH_4CN} RCH(NH_2)CN \xrightarrow[\text{Heat}]{H_3O^+} RCH(\overset{+}{N}H_3)CO_2^-$$

An α-amino acid

24.76 One of the reactions used in determining the sequence of nucleotides in a strand of DNA is reaction with hydrazine. Propose a mechanism for the following reaction, which occurs by an initial conjugate addition followed by internal amide formation.

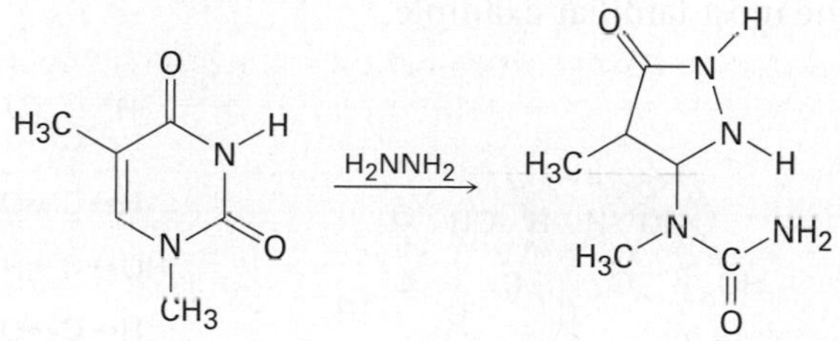

$$\xrightarrow{H_2NNH_2}$$

(Reactant: 1,5-dimethyluracil ring — H_3C, O, N–H, N, O, CH_3. Product: pyrazolidinone ring bearing H_3C, O, N–H, N–H, and H_3C–N–C(=O)–NH_2.)

Produced by honeybees from the nectar of flowers, honey is primarily a mixture of the two simple sugars fructose and glucose. Image copyright Olga Langerova, 2010. Used under license from Shutterstock.com

25 Biomolecules: Carbohydrates

Carbohydrates occur in every living organism. The sugar and starch in food, and the cellulose in wood, paper, and cotton are nearly pure carbohydrates. Modified carbohydrates form part of the coating around living cells, other carbohydrates are part of the nucleic acids that carry our genetic information, and still others are used as medicines.

The word **carbohydrate** derives historically from the fact that glucose, the first simple carbohydrate to be obtained pure, has the molecular formula $C_6H_{12}O_6$ and was originally thought to be a "hydrate of carbon, $C_6(H_2O)_6$." This view was soon abandoned, but the name persisted. Today, the term *carbohydrate* is used to refer loosely to the broad class of polyhydroxylated aldehydes and ketones commonly called sugars. Glucose, also known as dextrose in medical work, is the most familiar example.

Glucose (dextrose), a pentahydroxyhexanal

Carbohydrates are synthesized by green plants during photosynthesis, a complex process in which sunlight provides the energy to convert CO_2 and H_2O into glucose plus oxygen. Many molecules of glucose are then chemically linked for storage by the plant in the form of either cellulose or starch. It has been estimated that more than 50% of the dry weight of the earth's biomass—all plants and animals—consists of glucose polymers. When eaten and metabolized, carbohydrates then provide animals with a source of readily available

OWL Sign in to OWL for Organic Chemistry at **www.cengage.com/owl** to view tutorials and simulations, develop problem-solving skills, and complete online homework assigned by your professor.

energy. Thus, carbohydrates act as the chemical intermediaries by which solar energy is stored and used to support life.

$$6\,CO_2 + 6\,H_2O \xrightarrow{\text{Sunlight}} 6\,O_2 + \underset{\textbf{Glucose}}{C_6H_{12}O_6} \longrightarrow \text{Cellulose, starch}$$

Because humans and most other mammals lack the enzymes needed for digestion of cellulose, they require starch as their dietary source of carbohydrates. Grazing animals such as cows, however, have microorganisms in their first stomach that are able to digest cellulose. The energy stored in cellulose is thus moved up the biological food chain when these ruminant animals eat grass and are themselves used for food.

Why This Chapter? We've now seen all the common functional groups and reaction types that occur in organic and biological chemistry. In this and the next four chapters, we'll focus on the major classes of biological molecules, beginning with a look at the structures and primary biological functions of carbohydrates. Then in Chapter 29, we'll return to the subject to see how carbohydrates are both synthesized and degraded in organisms.

25.1 Classification of Carbohydrates

Carbohydrates are generally classed as either simple or complex. **Simple sugars**, or **monosaccharides**, are carbohydrates like glucose and fructose that can't be converted into smaller sugars by hydrolysis. **Complex carbohydrates** are made of two or more simple sugars linked together by acetal bonds **(Section 19.10)**. Sucrose (table sugar), for example, is made up of one glucose linked to one fructose. Similarly, cellulose is made up of several thousand glucose units linked together. Enzyme-catalyzed hydrolysis of a complex carbohydrate breaks it down into its constituent monosaccharides.

CH_2OH, O, O, HO, $HOCH_2$, OH, O, HO, CH_2OH, HO HO

$\xrightarrow{H_3O^+}$ **1 Glucose + 1 Fructose**

Sucrose
(a disaccharide)

CH_2OH, O, O, HO, OH, O, CH_2OH, O, HO, O, OH

$\xrightarrow{H_3O^+}$ **~3000 Glucose**

Cellulose
(a polysaccharide)

Monosaccharides are further classified as either **aldoses** or **ketoses**. The *-ose* suffix designates a carbohydrate, and the *aldo-* and *keto-* prefixes identify the kind of carbonyl group in the molecule, whether aldehyde or ketone. The number of carbon atoms in the monosaccharide is indicated by the appropriate numerical prefix *tri-*, *tetr-*, *pent-*, *hex-*, and so forth, in the name. Putting it all together, glucose is an aldohexose, a six-carbon aldehydo sugar; fructose is a ketohexose, a six-carbon keto sugar; ribose is an aldopentose, a five-carbon aldehydo sugar; and sedoheptulose is a ketoheptose, a seven-carbon keto sugar. Most of the common simple sugars are either pentoses or hexoses.

Glucose (an aldohexose) **Fructose** (a ketohexose) **Ribose** (an aldopentose) **Sedoheptulose** (a ketoheptose)

Problem 25.1

Classify each of the following monosaccharides:

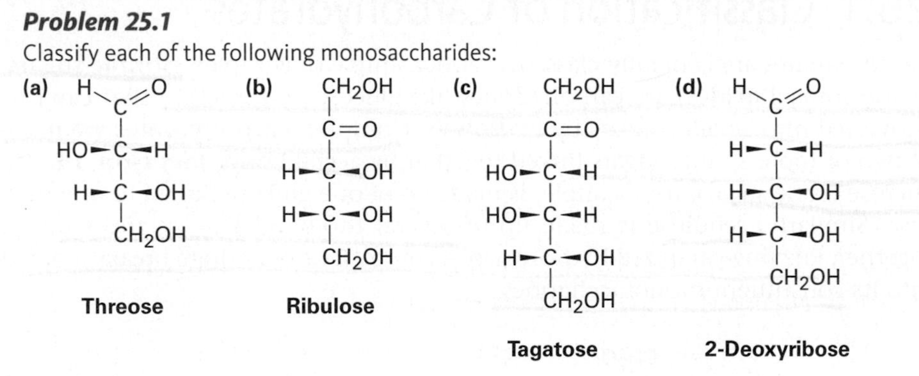

25.2 Depicting Carbohydrate Stereochemistry: Fischer Projections

Because carbohydrates usually have numerous chirality centers, it was recognized long ago that a quick method for representing their stereochemistry is needed. In 1891, the German chemist Emil Fischer suggested a method based on the projection of a tetrahedral carbon atom onto a flat surface. These **Fischer projections** were soon adopted and are now a common means of representing stereochemistry at chirality centers, particularly in carbohydrate chemistry.

A tetrahedral carbon atom is represented in a Fischer projection by two crossed lines. The horizontal lines represent bonds coming out of the page, and the vertical lines represent bonds going into the page.

Press flat

Fischer projection

For example, (*R*)-glyceraldehyde, the simplest monosaccharide, can be drawn as in **Figure 25.1**.

Bonds out of page

Bonds into page

(*R*)-Glyceraldehyde (Fischer projection)

Figure 25.1 A Fischer projection of (*R*) glyceraldehyde.

Because a given chiral molecule can be drawn in many ways, it's sometimes necessary to compare two projections to see if they represent the same or different enantiomers. To test for identity, Fischer projections can be moved around on the paper, but only two kinds of motions are allowed; moving a Fischer projection in any other way inverts its meaning.

- A Fischer projection can be rotated on the page by 180°, but not by 90° or 270°. Only a 180° rotation maintains the Fischer convention by keeping the same substituent groups going into and coming out of the plane. In the following Fischer projection of (*R*)-glyceraldehyde, for example, the –H and –OH groups come out of the plane both before and after a 180° rotation.

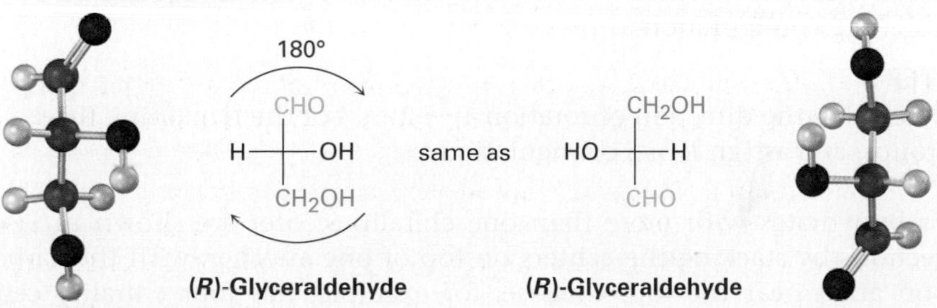

A 90° rotation breaks the Fischer convention by exchanging the groups that go into the plane and those that come out. In the following Fischer projections of (*R*)-glyceraldehyde, the –H and –OH groups come out of the plane before rotation but go into the plane after a 90° rotation. As a result, the rotated projection represents (S)-glyceraldehyde.

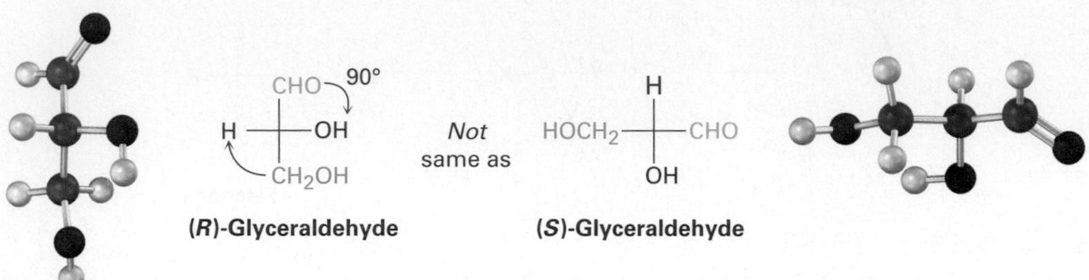

- A Fischer projection can have one group held steady while the other three rotate in either a clockwise or a counterclockwise direction. The effect is simply to rotate around a single bond, which does not change the stereochemistry.

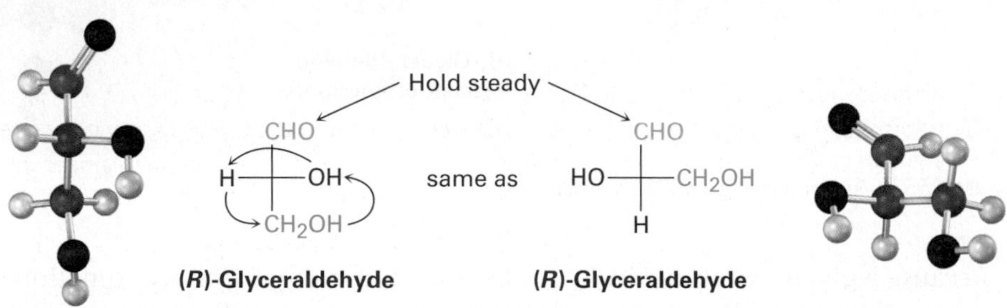

R,S stereochemical designations **(Section 5.5)** can be assigned to the chirality center in a Fischer projection by following three steps, as shown in Worked Example 25.1.

STEP 1
Rank the four substituents in the usual way **(Section 5.5)**.

STEP 2
Place the group of lowest ranking, usually H, at the top of the Fischer projection by using one of the allowed motions. This means that the lowest-ranked group is oriented back, away from the viewer, as required for assigning configuration.

STEP 3
Determine the direction of rotation $1 \rightarrow 2 \rightarrow 3$ of the remaining three groups, and assign *R* or *S* configuration.

Carbohydrates with more than one chirality center are shown in Fischer projections by stacking the centers on top of one another, with the carbonyl carbon at or near the top. Glucose, for example, has four chirality centers stacked on top of one another in a Fischer projection. Such representations don't, however, give an accurate picture of a molecule's true three-dimensional conformation, which is curled around on itself like a bracelet.

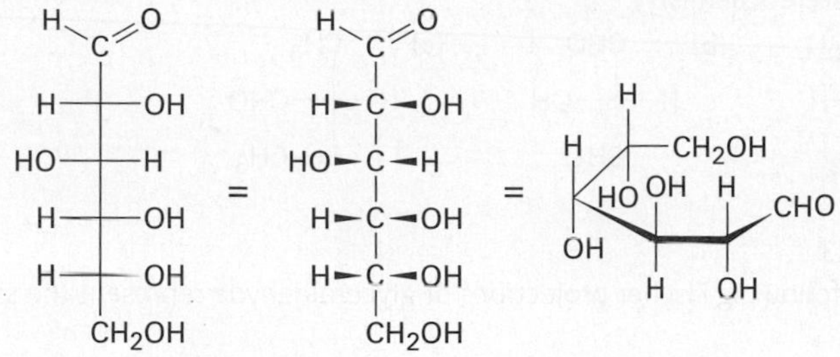

Glucose (carbonyl group at top)

Assigning *R* or *S* Configuration to a Fischer Projection

Worked Example 25.1

Assign *R* or *S* configuration to the following Fischer projection of alanine:

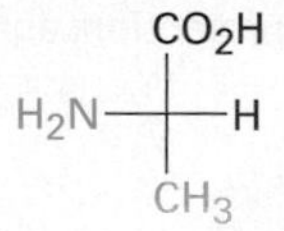

Alanine

Strategy

Follow the steps in the text. (1) Rank the four substituents on the chiral carbon. (2) Manipulate the Fischer projection to place the group of lowest ranking at the top by carrying out one of the allowed motions. (3) Determine the direction 1→2→3 of the remaining three groups.

Solution

The rankings of the groups are (1) $-NH_2$, (2) $-CO_2H$, (3) $-CH_3$, and (4) $-H$. To bring the group of lowest ranking ($-H$) to the top, we might want to hold the $-CH_3$ group steady while rotating the other three groups counterclockwise.

Rotate 3 groups counterclockwise

same as

Hold CH_3 steady

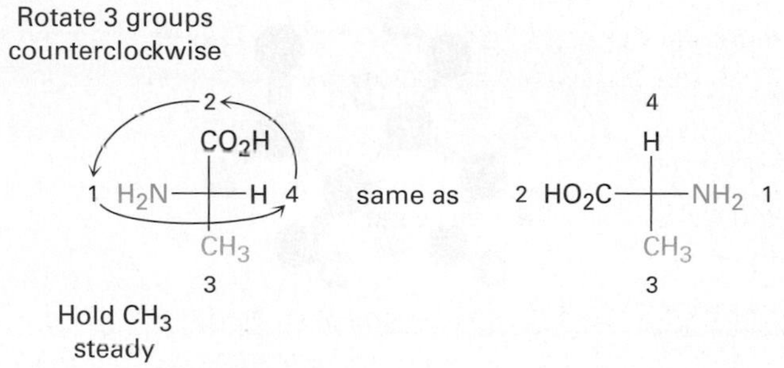

Going from first- to second- to third-highest ranking requires a counterclockwise turn, corresponding to *S* stereochemistry.

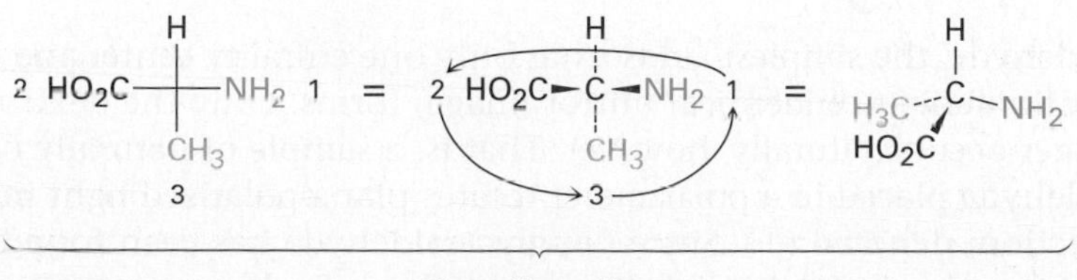

***S* configuration**

Problem 25.2

Convert each of the following Fischer projections into a tetrahedral representation, and assign *R* or *S* stereochemistry:

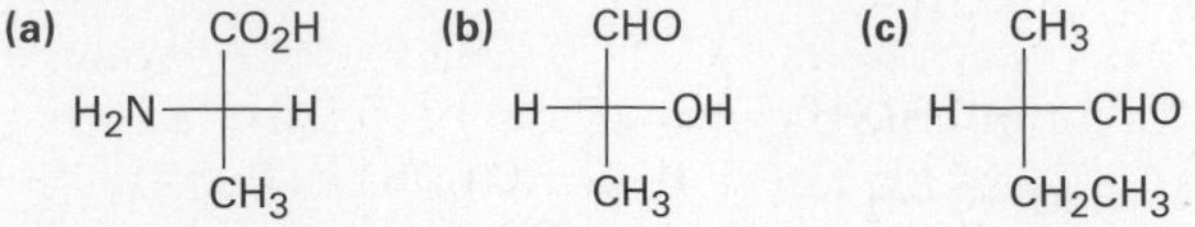

(a) CO_2H (top); H_2N (left); H (right); CH_3 (bottom)

(b) CHO (top); H (left); OH (right); CH_3 (bottom)

(c) CH_3 (top); H (left); CHO (right); CH_2CH_3 (bottom)

Problem 25.3

Which of the following Fischer projections of glyceraldehyde represent the same enantiomer?

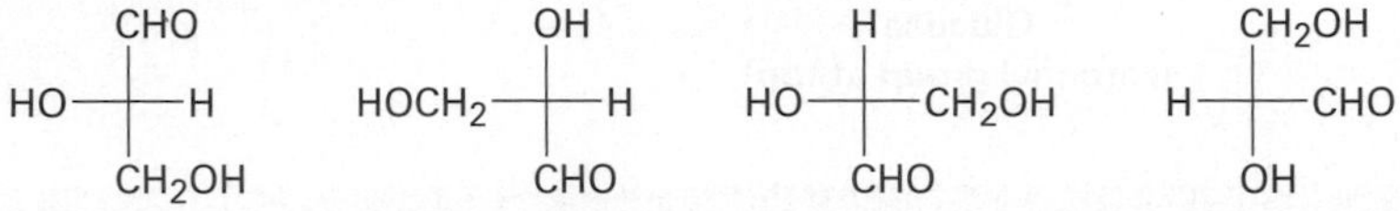

A: CHO (top); HO (left); H (right); CH_2OH (bottom)

B: OH (top); $HOCH_2$ (left); H (right); CHO (bottom)

C: H (top); HO (left); CH_2OH (right); CHO (bottom)

D: CH_2OH (top); H (left); CHO (right); OH (bottom)

Problem 25.4

Redraw the following molecule as a Fischer projection, and assign *R* or *S* configuration to the chirality center (green = Cl):

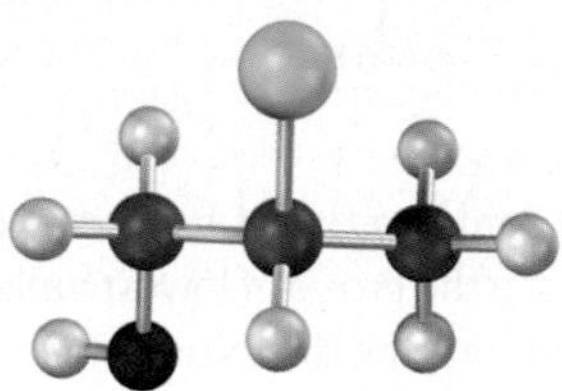

Problem 25.5

Redraw the following aldotetrose as a Fischer projection, and assign *R* or *S* configuration to each chirality center:

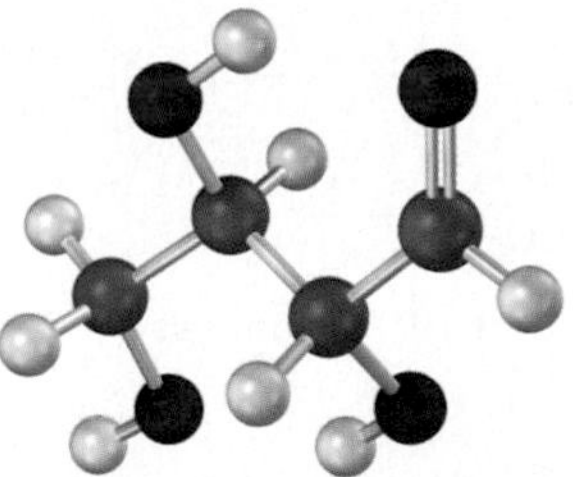

25.3 D,L Sugars

Glyceraldehyde, the simplest aldose, has only one chirality center and thus has two enantiomeric (nonidentical mirror-image) forms. Only the dextrorotatory enantiomer occurs naturally, however. That is, a sample of naturally occurring glyceraldehyde placed in a polarimeter rotates plane-polarized light in a clockwise direction, denoted (+). Since (+)-glyceraldehyde has been found to have an *R* configuration at C2, it can be represented in a Fischer projection as shown in Figure 25.1. For historical reasons dating back long before the adoption of the *R,S* system, (*R*)-(+)-glyceraldehyde is also referred to as D-glyceraldehyde (D for dextrorotatory). The other enantiomer, (*S*)-(−)-glyceraldehyde, is known as L-glyceraldehyde (L for levorotatory).

Because of the way that monosaccharides are biosynthesized in nature, glucose, fructose, and most other naturally occurring monosaccharides have the same *R* stereochemical configuration as D-glyceraldehyde at the chirality center farthest from the carbonyl group. In Fischer projections, therefore, most naturally occurring sugars have the hydroxyl group at the bottom chirality center pointing to the right **(Figure 25.2)**. All such compounds are referred to as **D sugars.**

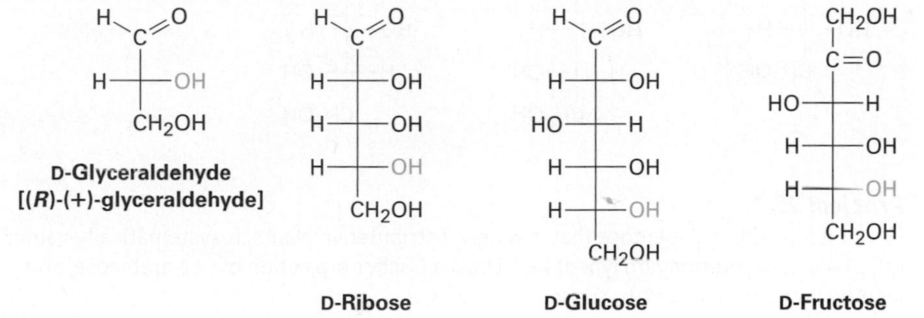

D-Glyceraldehyde
[(*R*)-(+)-glyceraldehyde]

D-Ribose **D-Glucose** **D-Fructose**

Figure 25.2 Some naturally occurring D sugars. The —OH group at the chirality center farthest from the carbonyl group has the same configuration as (*R*)-(+)-glyceraldehyde and points toward the right in Fischer projections.

In contrast with D sugars, **L sugars** have an *S* configuration at the lowest chirality center, with the bottom —OH group pointing to the left in Fischer projections. Thus, an L sugar is the mirror image (enantiomer) of the corresponding D sugar and has the opposite configuration from the D sugar at all chirality centers.

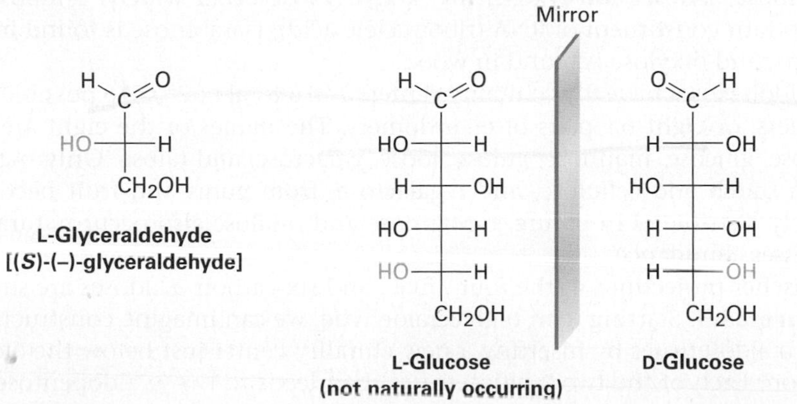

Mirror

L-Glyceraldehyde
[(*S*)-(−)-glyceraldehyde]

L-Glucose
(not naturally occurring)

D-Glucose

Note that the D and L notations have no relation to the direction in which a given sugar rotates plane-polarized light. A D sugar can be either dextrorotatory or levorotatory. The prefix D indicates only that the –OH group at the lowest chirality center has *R* stereochemistry and points to the right when the molecule is drawn in the standard way in a Fischer projection. Note also that the D,L system of carbohydrate nomenclature describes the configuration at only one chirality center and says nothing about the configuration of other chirality centers that may be present.

Problem 25.6

Assign *R* or *S* configuration to each chirality center in the following monosaccharides, and tell whether each is a D sugar or an L sugar:

(a) CHO; HO–C–H; HO–C–H; CH_2OH

(b) CHO; H–C–OH; HO–C–H; H–C–OH; CH_2OH

(c) CH_2OH; C=O; HO–C–H; H–C–OH; CH_2OH

Problem 25.7

(+)-Arabinose, an aldopentose that is widely distributed in plants, is systematically named (2*R*,3*S*,4*S*)-2,3,4,5-tetrahydroxypentanal. Draw a Fischer projection of (+)-arabinose, and identify it as a D sugar or an L sugar.

25.4 Configurations of Aldoses

Aldotetroses are four-carbon sugars with two chirality centers. Thus, there are $2^2 = 4$ possible stereoisomeric aldotetroses, or two D,L pairs of enantiomers named erythrose and threose.

Aldopentoses have three chirality centers and a total of $2^3 = 8$ possible stereoisomers, or four D,L pairs of enantiomers. These four pairs are called ribose, arabinose, xylose, and lyxose. All except lyxose occur widely. D-Ribose is an important constituent of RNA (ribonucleic acid), L-arabinose is found in many plants, and D-xylose is found in wood.

Aldohexoses have four chirality centers and a total of $2^4 = 16$ possible stereoisomers, or eight D,L pairs of enantiomers. The names of the eight are allose, altrose, glucose, mannose, gulose, idose, galactose, and talose. Only D-glucose, from starch and cellulose, and D-galactose, from gums and fruit pectins, are widely distributed in nature. D-Mannose and D-talose also occur naturally but in lesser abundance.

Fischer projections of the four-, five-, and six-carbon D aldoses are shown in **Figure 25.3**. Starting with D-glyceraldehyde, we can imagine constructing the two D aldotetroses by inserting a new chirality center just below the aldehyde carbon. Each of the two D aldotetroses then leads to two D aldopentoses (four total), and each of the four D aldopentoses leads to two D aldohexoses (eight total). In addition, each of the D aldoses in Figure 25.3 has an L enantiomer, which is not shown.

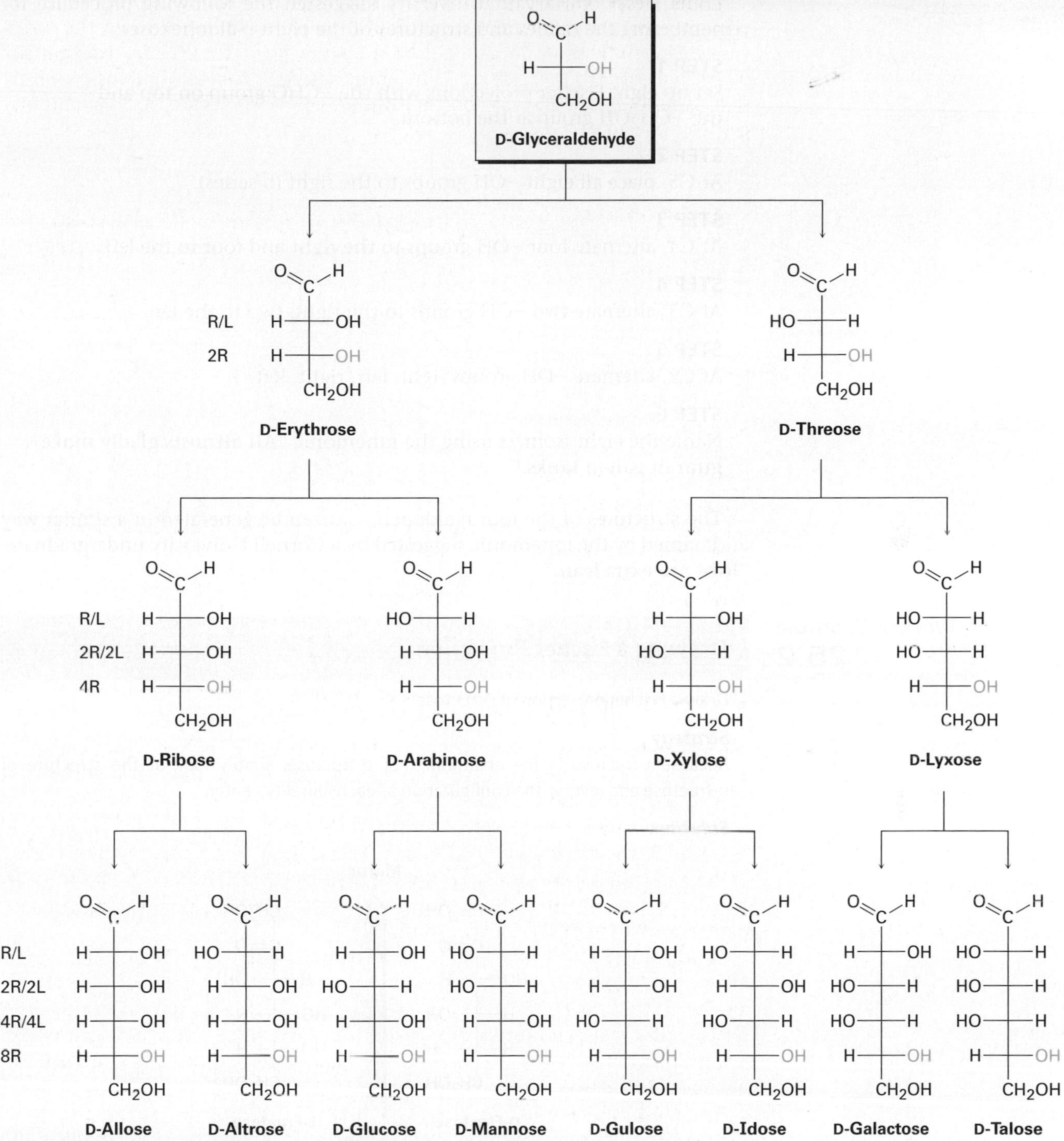

Figure 25.3 Configurations of D aldoses. The structures are arranged from left to right so that the —OH groups on C2 alternate right/left (R/L) in going across a series. Similarly, the —OH groups at C3 alternate two right/two left (2R/2L), the —OH groups at C4 alternate 4R/4L, and the —OH groups at C5 are to the right in all eight (8R). Each D aldose has a corresponding L enantiomer, which is not shown.

Louis Fieser of Harvard University suggested the following procedure for remembering the names and structures of the eight D aldohexoses:

STEP 1
Set up eight Fischer projections with the –CHO group on top and the $-CH_2OH$ group at the bottom.

STEP 2
At C5, place all eight –OH groups to the right (D series).

STEP 3
At C4, alternate four –OH groups to the right and four to the left.

STEP 4
At C3, alternate two –OH groups to the right, two to the left.

STEP 5
At C2, alternate –OH groups right, left, right, left.

STEP 6
Name the eight isomers using the mnemonic "**All altr**uists **gl**adly **ma**ke **gu**m **i**n **gal**lon **ta**nks."

The structures of the four D aldopentoses can be generated in a similar way and named by the mnemonic suggested by a Cornell University undergraduate: "**Rib**s **ar**e e**x**tra **lyx**ean."

Worked Example 25.2 Drawing a Fischer Projection

Draw a Fischer projection of L-fructose.

Strategy

Because L-fructose is the enantiomer of D-fructose, simply look at the structure of D-fructose and reverse the configuration at each chirality center.

Solution

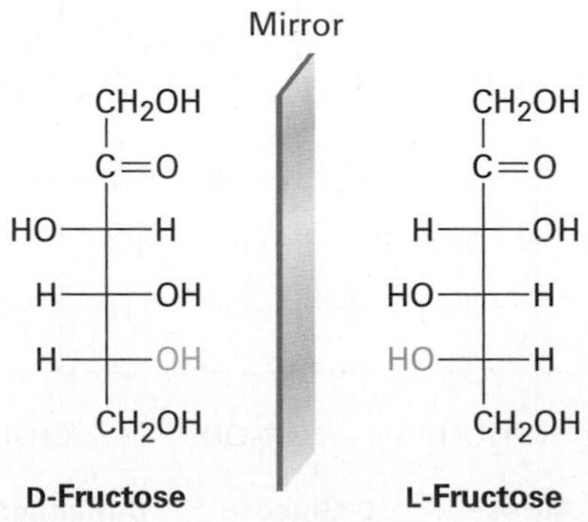

Problem 25.8

Only the D sugars are shown in Figure 25.3. Draw Fischer projections for the following L sugars:

(a) L-Xylose **(b)** L-Galactose **(c)** L-Allose

Problem 25.9

How many aldoheptoses are there? How many are D sugars, and how many are L sugars?

Problem 25.10

The following model is that of an aldopentose. Draw a Fischer projection of the sugar, name it, and identify it as a D sugar or an L sugar.

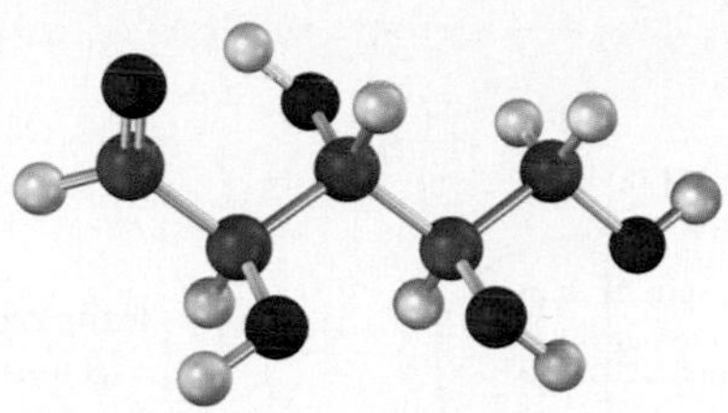

25.5 Cyclic Structures of Monosaccharides: Anomers

We said in **Section 19.10** that aldehydes and ketones undergo a rapid and reversible nucleophilic addition reaction with alcohols to form hemiacetals.

$$\mathrm{RCHO} + \mathrm{R'OH} \underset{}{\overset{\mathrm{H^+}\ \text{catalyst}}{\rightleftharpoons}} \mathrm{RCH(OH)OR'}$$

An aldehyde **A hemiacetal**

If the carbonyl and the hydroxyl group are in the same molecule, an intramolecular nucleophilic addition can take place, leading to the formation of a cyclic hemiacetal. Five- and six-membered cyclic hemiacetals are relatively strain-free and particularly stable, and many carbohydrates therefore exist in an equilibrium between open-chain and cyclic forms. Glucose, for instance, exists in aqueous solution primarily in the six-membered, **pyranose** form resulting from intramolecular nucleophilic addition of the –OH group at C5 to the C1 carbonyl group **(Figure 25.4)**. The word *pyranose* is derived from *pyran*, the name of the unsaturated six-membered cyclic ether.

Like cyclohexane rings **(Section 4.6)**, pyranose rings have a chairlike geometry with axial and equatorial substituents. By convention, the rings are usually drawn by placing the hemiacetal oxygen atom at the right rear, as shown in Figure 25.4. Note that an –OH group on the right in a Fischer projection is on the bottom face of the pyranose ring, and an –OH group on the left in a Fischer projection is on the top face of the ring. For D sugars, the terminal $-CH_2OH$ group is on the top of the ring, whereas for L sugars, the $-CH_2OH$ group is on the bottom.

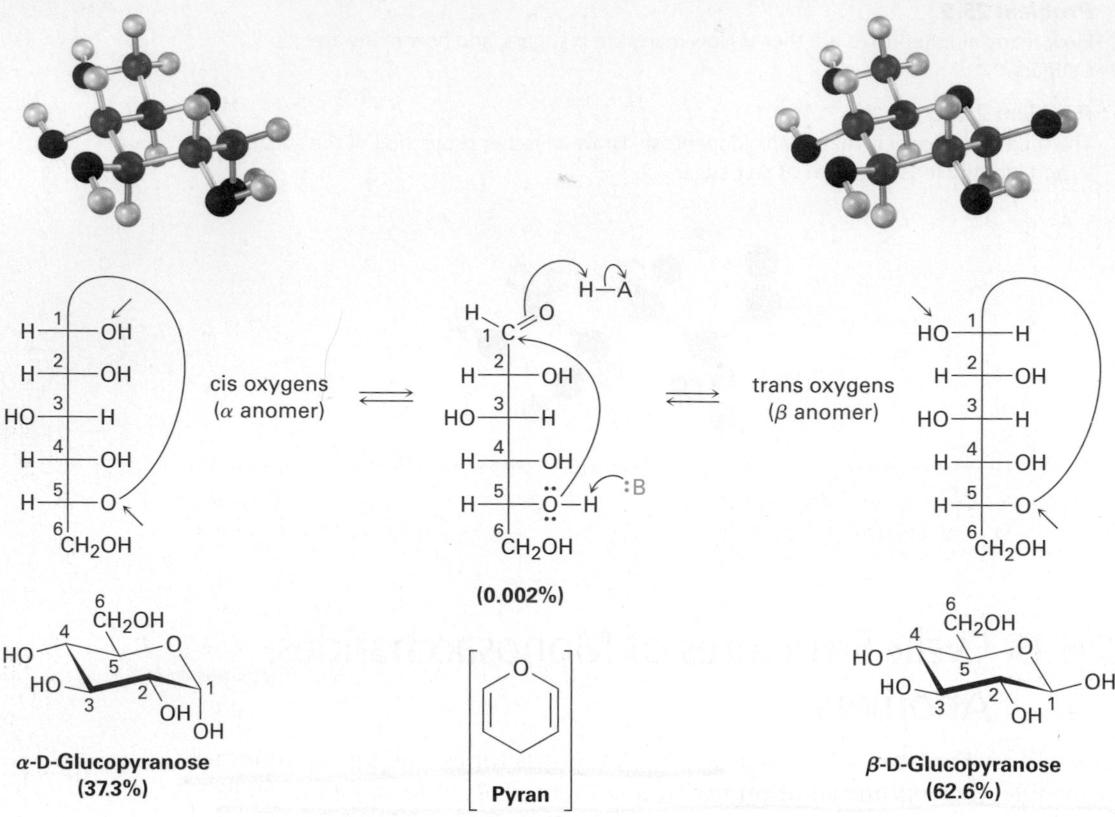

Figure 25.4 Glucose in its cyclic pyranose forms. As explained in the text, two anomers are formed by cyclization of glucose. The molecule whose newly formed –OH group at C1 is cis to the oxygen atom on the lowest chirality center (C5) in a Fischer projection is the α anomer. The molecule whose newly formed –OH group is trans to the oxygen atom on the lowest chirality center in a Fischer projection is the β anomer.

When an open-chain monosaccharide cyclizes to a pyranose form, a new chirality center is generated at the former carbonyl carbon and two diastereomers, called **anomers**, are produced. The hemiacetal carbon atom is referred to as the **anomeric center**. For example, glucose cyclizes reversibly in aqueous solution to a 37:63 mixture of two anomers (Figure 25.4). The compound with its newly generated –OH group at C1 cis to the –OH at the lowest chirality center in a Fischer projection is called the **α anomer**; its full name is α-D-glucopyranose. The compound with its newly generated –OH group trans to the –OH at the lowest chirality center in a Fischer projection is called the **β anomer**; its full name is β-D-glucopyranose. Note that in β-D-glucopyranose, all the substituents on the ring are equatorial. Thus, β-D-glucopyranose is the least sterically crowded and most stable of the eight D aldohexoses.

Some monosaccharides also exist in a five-membered cyclic hemiacetal form called a **furanose**. D-Fructose, for instance, exists in water solution as 70% β-pyranose, 2% α-pyranose, 0.7% open-chain, 23% β-furanose, and 5% α-furanose. The pyranose form results from addition of the –OH at C6 to the carbonyl group, while the furanose form results from addition of the –OH at C5 to the carbonyl group **(Figure 25.5)**.

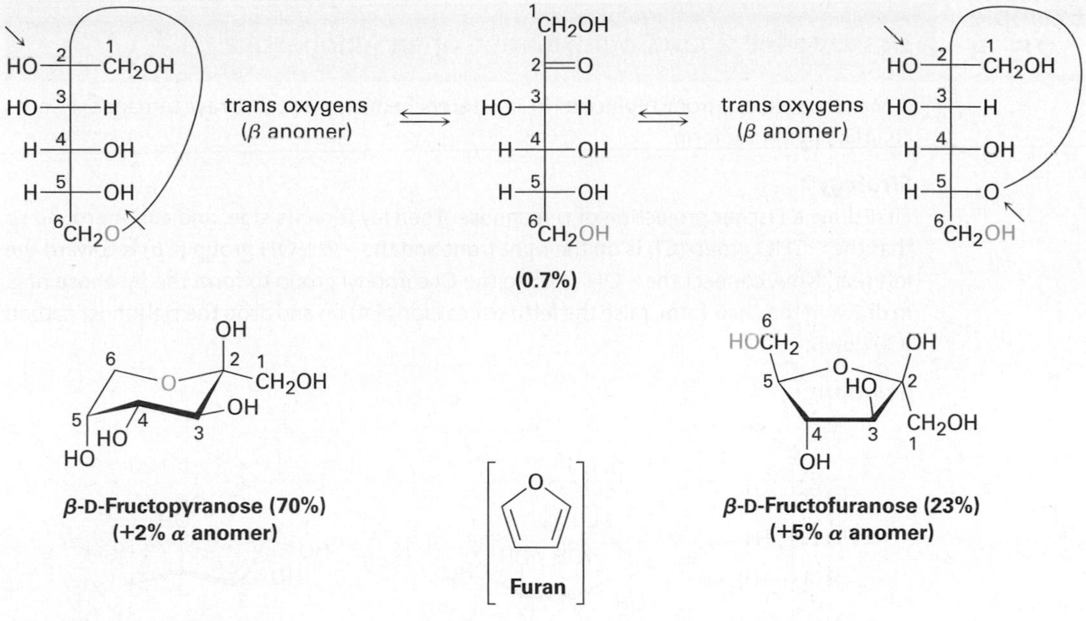

Figure 25.5 Pyranose and furanose forms of fructose in aqueous solution. The two pyranose anomers result from addition of the C6 —OH group to the C2 carbonyl; the two furanose anomers result from addition of the C5 —OH group to the C2 carbonyl.

Both anomers of D-glucopyranose can be crystallized and purified. Pure α-D-glucopyranose has a melting point of 146 °C and a specific rotation $[\alpha]_D = +112.2$; pure β-D-glucopyranose has a melting point of 148 to 155 °C and a specific rotation $[\alpha]_D = +18.7$. When a sample of either pure anomer is dissolved in water, however, its optical rotation slowly changes until it reaches a constant value of +52.6. That is, the specific rotation of the α-anomer solution decreases from +112.2 to +52.6, and the specific rotation of the β-anomer solution increases from +18.7 to +52.6. Called **mutarotation**, this change in optical rotation is due to the slow interconversion of the pure anomers to give a 37:63 equilibrium mixture.

Mutarotation occurs by a reversible ring-opening of each anomer to the open-chain aldehyde, followed by reclosure. Although the equilibration is slow at neutral pH, it is catalyzed by both acid and base.

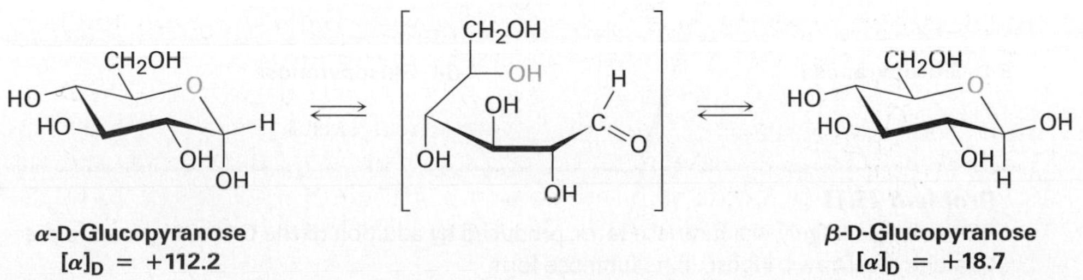

Worked Example 25.3 Drawing the Chair Conformation of an Aldohexose

D-Mannose differs from D-glucose in its stereochemistry at C2. Draw D-mannose in its chairlike pyranose form.

Strategy

First draw a Fischer projection of D-mannose. Then lay it on its side, and curl it around so that the $-CHO$ group (C1) is on the right front and the $-CH_2OH$ group (C6) is toward the left rear. Now, connect the $-OH$ at C5 to the C1 carbonyl group to form the pyranose ring. In drawing the chair form, raise the leftmost carbon (C4) up and drop the rightmost carbon (C1) down.

Solution

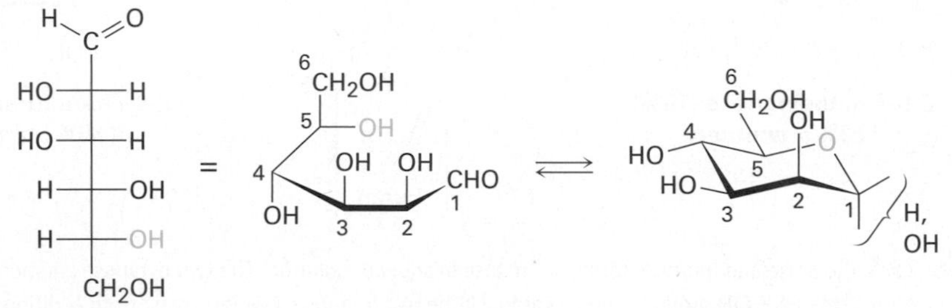

D-Mannose **(Pyranose form)**

Worked Example 25.4 Drawing the Chair Conformation of a Pyranose

Draw β-L-glucopyranose in its more stable chair conformation.

Strategy

It's probably easiest to begin by drawing the chair conformation of β-D-glucopyranose. Then draw its mirror-image L enantiomer by changing the stereochemistry at every position on the ring, and carry out a ring-flip to give the more stable chair conformation. Note that the $-CH_2OH$ group is on the bottom face of the ring in the L enantiomer as is the anomeric $-OH$.

Solution

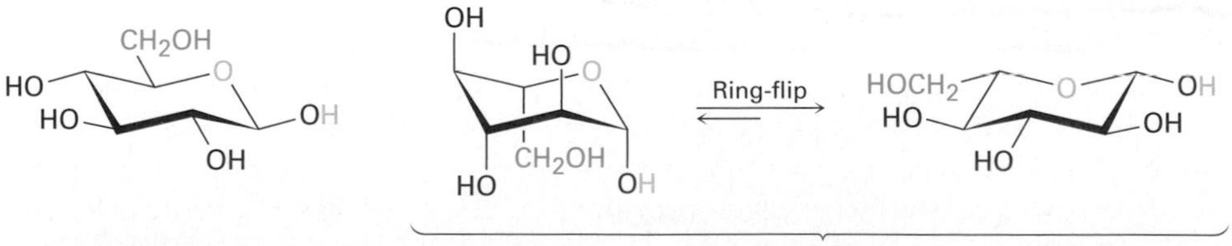

Ring-flip

β-D-Glucopyranose **β-L-Glucopyranose**

Problem 25.11

Ribose exists largely in a furanose form, produced by addition of the C4 $-OH$ group to the C1 aldehyde. Draw D-ribose in its furanose form.

Problem 25.12
Figure 25.5 shows only the β-pyranose and β-furanose anomers of D-fructose. Draw the α-pyranose and α-furanose anomers.

Problem 25.13
Draw β-D-galactopyranose and β-D-mannopyranose in their more stable chair conformations. Label each ring substituent as either axial or equatorial. Which would you expect to be more stable, galactose or mannose?

Problem 25.14
Draw β-L-galactopyranose in its more stable chair conformation, and label the substituents as either axial or equatorial.

Problem 25.15
Identify the following monosaccharide, write its full name, and draw its open-chain form in Fischer projection.

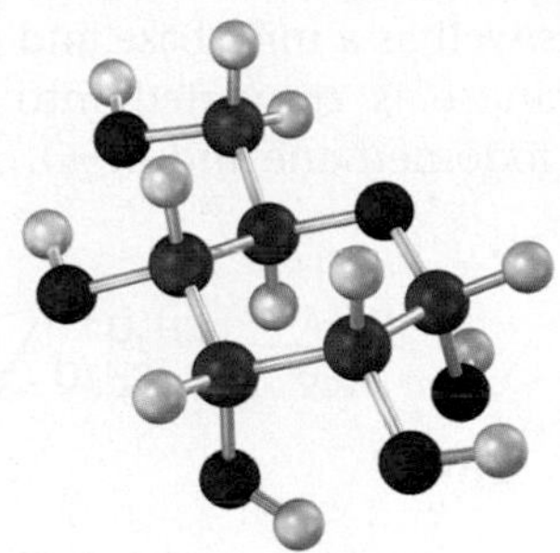

25.6 Reactions of Monosaccharides

Because monosaccharides contain only two kinds of functional groups, hydroxyls and carbonyls, most of the chemistry of monosaccharides is the familiar chemistry of these two groups. As we've seen, alcohols can be converted to esters and ethers and can be oxidized; carbonyl compounds can react with nucleophiles and can be reduced.

Ester and Ether Formation

Monosaccharides behave as simple alcohols in much of their chemistry. For example, carbohydrate –OH groups can be converted into esters and ethers, which are often easier to work with than the free sugars. Because of their many hydroxyl groups, monosaccharides are usually soluble in water but insoluble in organic solvents such as ether. They are also difficult to purify and have a tendency to form syrups rather than crystals when water is removed. Ester and ether derivatives, however, are soluble in organic solvents and are easily purified and crystallized.

Esterification is normally carried out by treating the carbohydrate with an acid chloride or acid anhydride in the presence of a base **(Sections 21.4 and 21.5)**. All the –OH groups react, including the anomeric one. For example, β-D-glucopyranose is converted into its pentaacetate by treatment with acetic anhydride in pyridine solution.

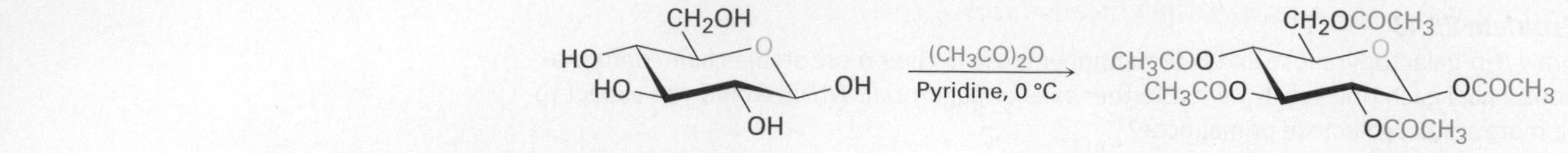

β-D-Glucopyranose → $(CH_3CO)_2O$, Pyridine, 0 °C → **Penta-*O*-acetyl-β-D-glucopyranose (91%)**

Carbohydrates are converted into ethers by treatment with an alkyl halide in the presence of base—the Williamson ether synthesis (**Section 18.2**). Standard Williamson conditions using a strong base tend to degrade sensitive sugar molecules, but silver oxide works well as a mild base and gives high yields of ethers. For example, α-D-glucopyranose is converted into its pentamethyl ether in 85% yield on reaction with iodomethane and Ag_2O.

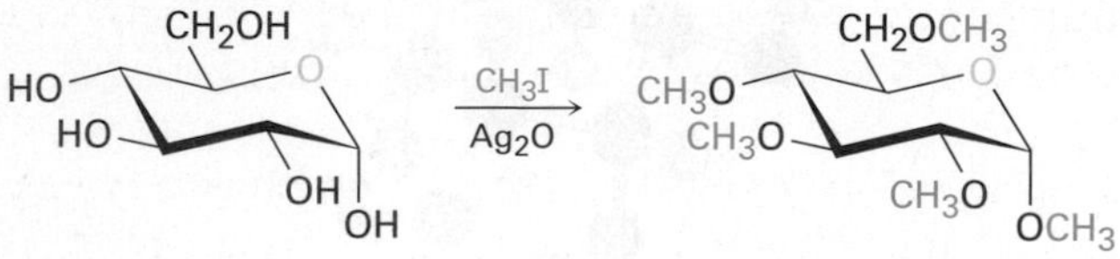

α-D-Glucopyranose → CH_3I, Ag_2O → **α-D-Glucopyranose pentamethyl ether (85%)**

Problem 25.16

Draw the products you would obtain by reaction of β-D-ribofuranose with:

(a) CH_3I, Ag_2O **(b)** $(CH_3CO)_2O$, pyridine

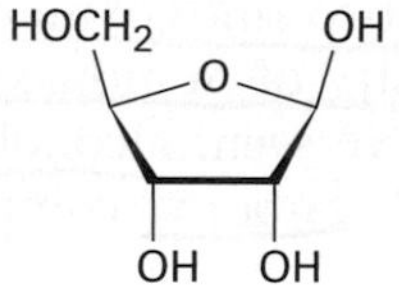

β-D-Ribofuranose

Glycoside Formation

We saw in **Section 19.10** that treatment of a hemiacetal with an alcohol and an acid catalyst yields an acetal.

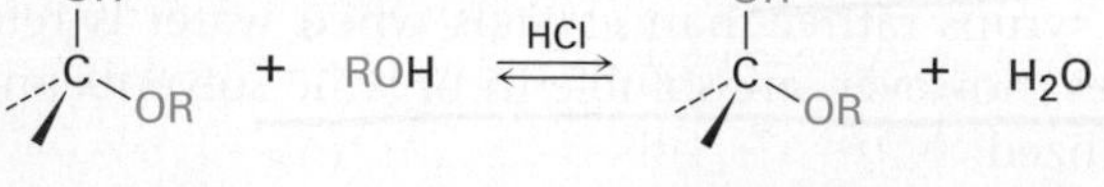

A hemiacetal + ROH ⇌ (HCl) **An acetal** + H_2O

In the same way, treatment of a monosaccharide hemiacetal with an alcohol and an acid catalyst yields an acetal called a **glycoside**, in which the anomeric –OH has been replaced by an –OR group. For example, reaction of β-D-glucopyranose with methanol gives a mixture of α and β methyl D-glucopyranosides. (Note that a *gly*coside is the functional group name for any sugar, whereas a *glu*coside is formed specifically from glucose.)

β-D-Glucopyranose (a cyclic hemiacetal) ⇌ (CH_3OH, HCl) **Methyl α-D-glucopyranoside (66%)** + **Methyl β-D-glucopyranoside (33%)**

Glycosides are named by first citing the alkyl group and then replacing the *-ose* ending of the sugar with *-oside*. Like all acetals, glycosides are stable to neutral water. They aren't in equilibrium with an open-chain form, and they don't show mutarotation. They can, however, be hydrolyzed to give back the free monosaccharide plus alcohol on treatment with aqueous acid **(Section 19.10)**.

Glycosides are abundant in nature, and many biologically important molecules contain glycosidic linkages. For example, digitoxin, the active component of the digitalis preparations used for treatment of heart disease, is a glycoside consisting of a steroid alcohol linked to a trisaccharide. Note also that the three sugars are linked to one another by glycoside bonds.

Trisaccharide — **Steroid**

Digitoxigenin, a glycoside

The laboratory synthesis of glycosides can be difficult because of the numerous –OH groups on the sugar molecule. One method that is particularly suitable for preparing glucose β-glycosides involves treatment of glucose pentaacetate with HBr, followed by addition of the appropriate alcohol in the presence of silver oxide. Called the *Koenigs–Knorr reaction*, the sequence involves formation of a pyranosyl bromide, followed by nucleophilic substitution. For example, methylarbutin, a glycoside found in pears, has been prepared by reaction of tetraacetyl-α-D-glucopyranosyl bromide with *p*-methoxyphenol.

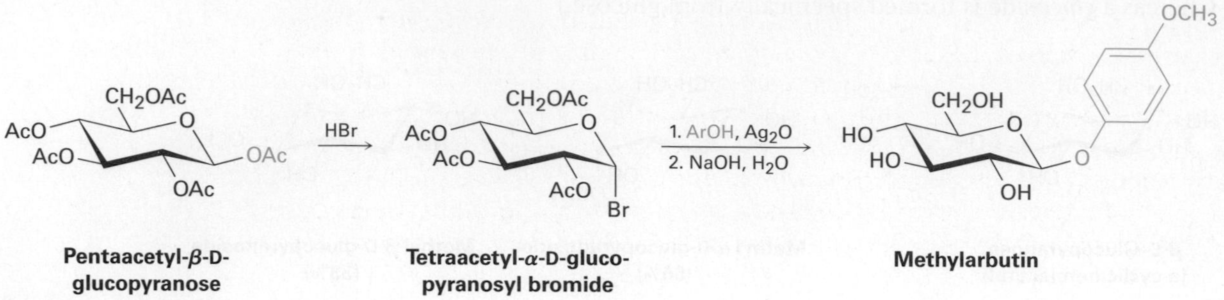

Pentaacetyl-β-D-glucopyranose — HBr → **Tetraacetyl-α-D-glucopyranosyl bromide** — 1. ArOH, Ag_2O; 2. NaOH, H_2O → **Methylarbutin**

Although the Koenigs–Knorr reaction appears to involve a simple backside S_N2 displacement of bromide ion by alkoxide ion, the situation is actually more complex. Both α and β anomers of tetraacetyl-D-glucopyranosyl bromide give the same β-glycoside product, implying that they react by a common pathway.

The results can be understood by assuming that tetraacetyl-D-glucopyranosyl bromide (either α or β anomer) undergoes a spontaneous S_N1-like loss of Br^-, followed by internal reaction with the ester group at C2 to form an oxonium ion. Since the acetate at C2 is on the bottom of the glucose ring, the C–O bond also forms from the bottom. Backside S_N2 displacement of the oxonium ion then occurs with the usual inversion of configuration, yielding a β-glycoside and regenerating the acetate at C2 **(Figure 25.6)**.

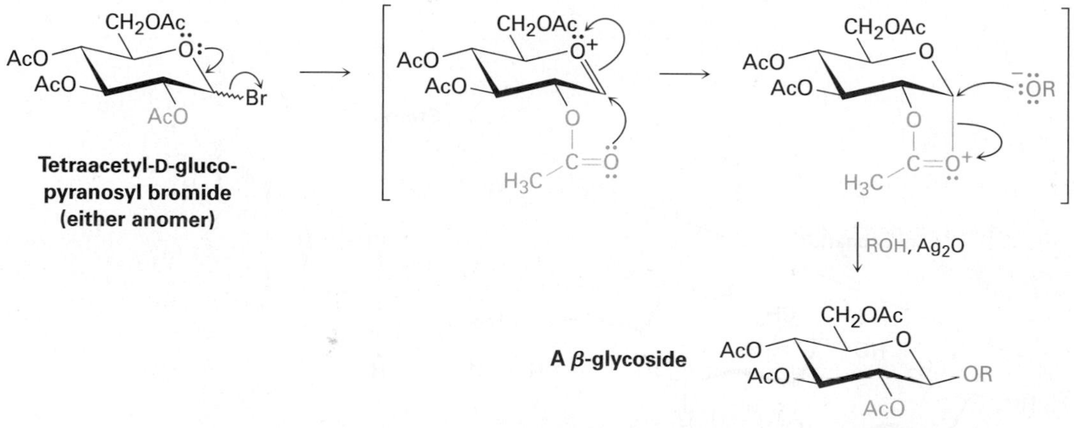

Tetraacetyl-D-glucopyranosyl bromide (either anomer) → [oxonium ion intermediates] — ROH, Ag_2O → **A β-glycoside**

Figure 25.6 Mechanism of the Koenigs–Knorr reaction, showing the neighboring-group effect of a nearby acetate.

The participation shown by the nearby acetate group in the Koenigs–Knorr reaction is referred to as a *neighboring-group effect* and is a common occurrence in organic chemistry. Neighboring-group effects are usually noticeable only because they affect the rate or stereochemistry of a reaction; the nearby group itself does not undergo any evident change during the reaction.

Biological Ester Formation: Phosphorylation

In living organisms, carbohydrates occur not only in the free form but also linked through their anomeric center to other molecules such as lipids (glycolipids) or proteins (glycoproteins). Collectively called *glycoconjugates,* these sugar-linked molecules are components of cell walls and are crucial to the mechanism by which different cell types recognize one another.

Glycoconjugate formation occurs by reaction of the lipid or protein with a glycosyl nucleoside diphosphate. This diphosphate is itself formed by initial reaction of a monosaccharide with adenosine triphosphate (ATP) to give a glycosyl monophosphate, followed by reaction with uridine triphosphate (UTP), to give a glycosyl uridine diphosphate. (We'll see the structures of nucleoside phosphates in **Section 28.1**.) The purpose of the phosphorylation is to activate the anomeric –OH group of the sugar and make it a better leaving group in a nucleophilic substitution reaction by a protein or lipid **(Figure 25.7)**.

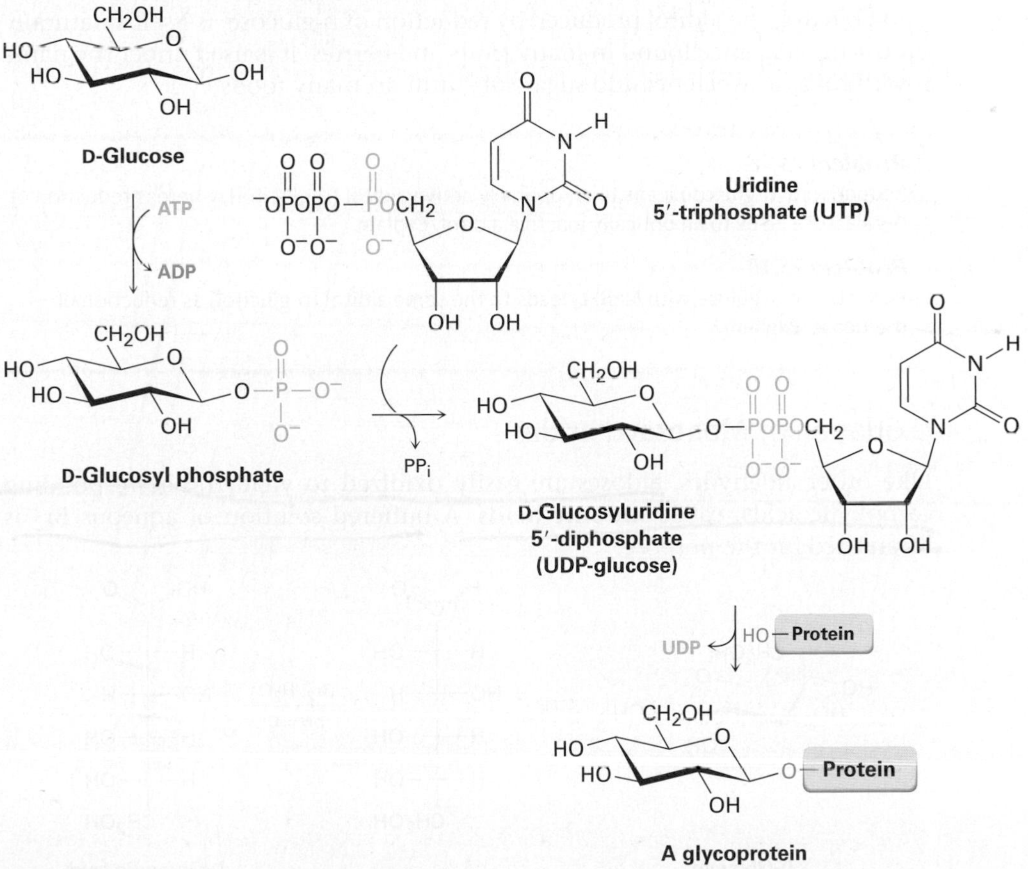

D-Glucose

ATP

ADP

Uridine 5′-triphosphate (UTP)

D-Glucosyl phosphate

PP_i

D-Glucosyluridine 5′-diphosphate (UDP-glucose)

UDP

HO—Protein

A glycoprotein

Figure 25.7 Glycoprotein formation occurs by initial phosphorylation of the starting carbohydrate with ATP to a glycosyl monophosphate, followed by reaction with UTP to form a glycosyl uridine 5′-diphosphate. Nucleophilic substitution by an –OH (or $-NH_2$) group on a protein then gives the glycoprotein.

Reduction of Monosaccharides

Treatment of an aldose or ketose with $NaBH_4$ reduces it to a polyalcohol called an **alditol**. The reduction occurs by reaction of the open-chain form present in the aldehyde/ketone ⇄ hemiacetal equilibrium. Although only a small amount of the open-chain form is present at any given time, that small amount is reduced, more is produced by opening of the pyranose form, that additional amount is reduced, and so on, until the entire sample has undergone reaction.

β-D-Glucopyranose ⇄ [D-Glucose] $\xrightarrow[H_2O]{NaBH_4}$ D-Glucitol (D-sorbitol), an alditol

D-Glucitol, the alditol produced by reduction of D-glucose, is itself a naturally occurring substance found in many fruits and berries. It is used under the name D-sorbitol as a sweetener and sugar substitute in many foods.

Problem 25.17

Reduction of D-glucose leads to an optically active alditol (D-glucitol), whereas reduction of D-galactose leads to an optically inactive alditol. Explain.

Problem 25.18

Reduction of L-gulose with $NaBH_4$ leads to the same alditol (D-glucitol) as reduction of D-glucose. Explain.

Oxidation of Monosaccharides

Like other aldehydes, aldoses are easily oxidized to yield the corresponding carboxylic acids, called **aldonic acids**. A buffered solution of aqueous Br_2 is often used for the purpose.

D-Glucose ⇄ $\xrightarrow[pH = 6]{Br_2,\ H_2O}$ D-Gluconic acid (an aldonic acid)

Historically, the oxidation of an aldose with either Ag^+ in aqueous ammonia (called Tollens' reagent) or Cu^{2+} with aqueous sodium citrate (Benedict's reagent) formed the basis of simple tests for what are called **reducing sugars**. *(Reducing* because the aldose reduces the metal oxidizing agent.) Some simple diabetes self-test kits sold in drugstores still use Benedict's reagent to detect glucose in urine, but more modern methods have largely replaced the chemical test.

All aldoses are reducing sugars because they contain an aldehyde group, but some ketoses are reducing sugars as well. Fructose reduces Tollens' reagent, for example, even though it contains no aldehyde group. Reduction occurs because fructose is readily isomerized to a mixture of aldoses (glucose and mannose) in basic solution by a series of keto–enol tautomeric shifts **(Figure 25.8)**. Glycosides, however, are nonreducing because the acetal group is not hydrolyzed to an aldehyde under basic conditions.

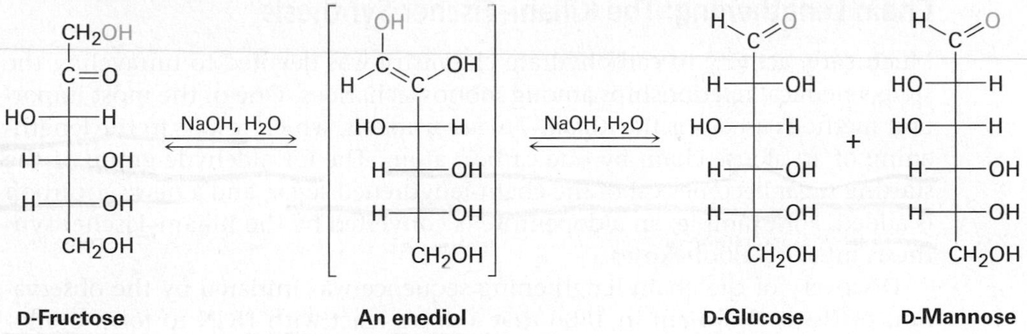

NaOH, H_2O NaOH, H_2O

D-Fructose   An enediol   D-Glucose + D-Mannose

Figure 25.8 Fructose, a ketose, is a reducing sugar because it undergoes two base-catalyzed keto–enol tautomerizations that result in conversion to a mixture of aldoses.

If warm dilute HNO_3 (nitric acid) is used as the oxidizing agent, an aldose is oxidized to a dicarboxylic acid called an **aldaric acid**. Both the aldehyde carbonyl and the terminal $-CH_2OH$ group are oxidized in this reaction.

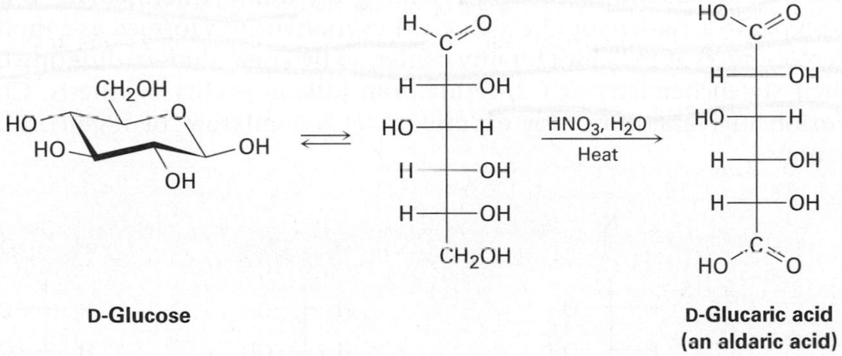

HNO_3, H_2O, Heat

D-Glucose   D-Glucaric acid (an aldaric acid)

Finally, if only the $-CH_2OH$ end of the aldose is oxidized without affecting the $-CHO$ group, the product is a monocarboxylic acid called a **uronic acid**. The reaction can only be done enzymatically; no chemical reagent is known that can accomplish this selective oxidation in the laboratory.

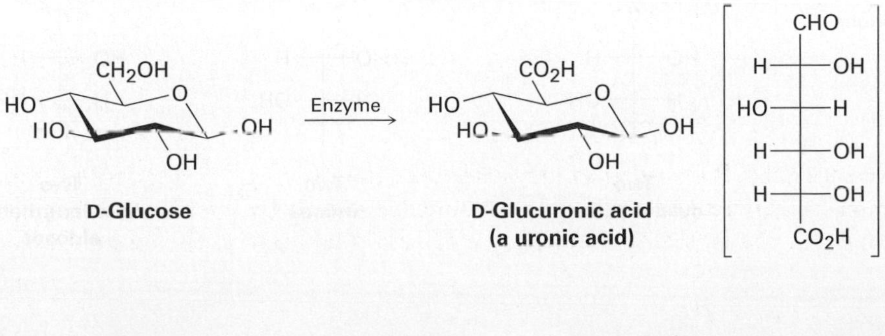

Enzyme

D-Glucose   D-Glucuronic acid (a uronic acid)

Problem 25.19

D-Glucose yields an optically active aldaric acid on treatment with HNO_3, but D-allose yields an optically inactive aldaric acid. Explain.

Problem 25.20

Which of the other six D aldohexoses yield optically active aldaric acids on oxidation, and which yield optically inactive (meso) aldaric acids? (See Problem 25.19.)

Chain Lengthening: The Kiliani–Fischer Synthesis

Much early activity in carbohydrate chemistry was devoted to unraveling the stereochemical relationships among monosaccharides. One of the most important methods used was the *Kiliani–Fischer synthesis,* which results in the lengthening of an aldose chain by one carbon atom. The C1 aldehyde group of the starting sugar becomes C2 of the chain-lengthened sugar, and a new C1 carbon is added. For example, an aldopentose is converted by the Kiliani–Fischer synthesis into two aldohexoses.

Discovery of the chain-lengthening sequence was initiated by the observation of Heinrich Kiliani in 1886 that aldoses react with HCN to form cyanohydrins **(Section 19.6)**. Emil Fischer immediately realized the importance of Kiliani's discovery and devised a method for converting the cyanohydrin nitrile group into an aldehyde.

Fischer's original method for conversion of the nitrile into an aldehyde involved hydrolysis to a carboxylic acid, ring closure to a cyclic ester (lactone), and subsequent reduction. A modern improvement is to reduce the nitrile over a palladium catalyst, yielding an imine intermediate that is hydrolyzed to an aldehyde. Note that the cyanohydrin is formed as a mixture of stereoisomers at the new chirality center, so two new aldoses, differing only in their stereochemistry at C2, result from Kiliani–Fischer synthesis. Chain extension of D-arabinose, for example, yields a mixture of D-glucose and D-mannose.

An aldose $\xrightarrow{HCN}$ Two cyanohydrins $\xrightarrow[\text{Pd catalyst}]{H_2}$ Two imines $\xrightarrow{H_3O^+}$ Two chain-lengthened aldoses

An aldose **Two cyanohydrins** **Two imines** **Two chain-lengthened aldoses**

Problem 25.21
What product(s) would you expect from Kiliani–Fischer reaction of D-ribose?

Problem 25.22
What aldopentose would give a mixture of L-gulose and L-idose on Kiliani–Fischer chain extension?

Chain Shortening: The Wohl Degradation

Just as the Kiliani–Fischer synthesis lengthens an aldose chain by one carbon, the *Wohl degradation* shortens an aldose chain by one carbon. The Wohl degradation is almost the exact opposite of the Kiliani–Fischer sequence. That is, the aldose aldehyde carbonyl group is first converted into a nitrile, and the resulting cyanohydrin loses HCN under basic conditions—the reverse of a nucleophilic addition reaction.

Conversion of the aldehyde into a nitrile is accomplished by treatment of an aldose with hydroxylamine to give an imine called an *oxime* **(Section 19.8)**, followed by dehydration of the oxime with acetic anhydride. The Wohl degradation does not give particularly high yields of chain-shortened aldoses, but the reaction is general for all aldopentoses and aldohexoses. For example, D-galactose is converted by Wohl degradation into D-lyxose.

Problem 25.23
Two of the four D aldopentoses yield D-threose on Wohl degradation. What are their structures?

25.7 The Eight Essential Monosaccharides

Humans need to obtain eight monosaccharides for proper functioning. Although all eight can be biosynthesized from simpler precursors if necessary, it's more energetically efficient to obtain them from the diet. The eight are L-fucose (6-deoxy-L-galactose), D-galactose, D-glucose, D-mannose, *N*-acetyl-D-glucosamine, *N*-acetyl-D-galactosamine, D-xylose, and *N*-acetyl-D-neuraminic acid **(Figure 25.9)**.

All are used for the synthesis of the glycoconjugate components of cell walls, and glucose is also the body's primary source of energy.

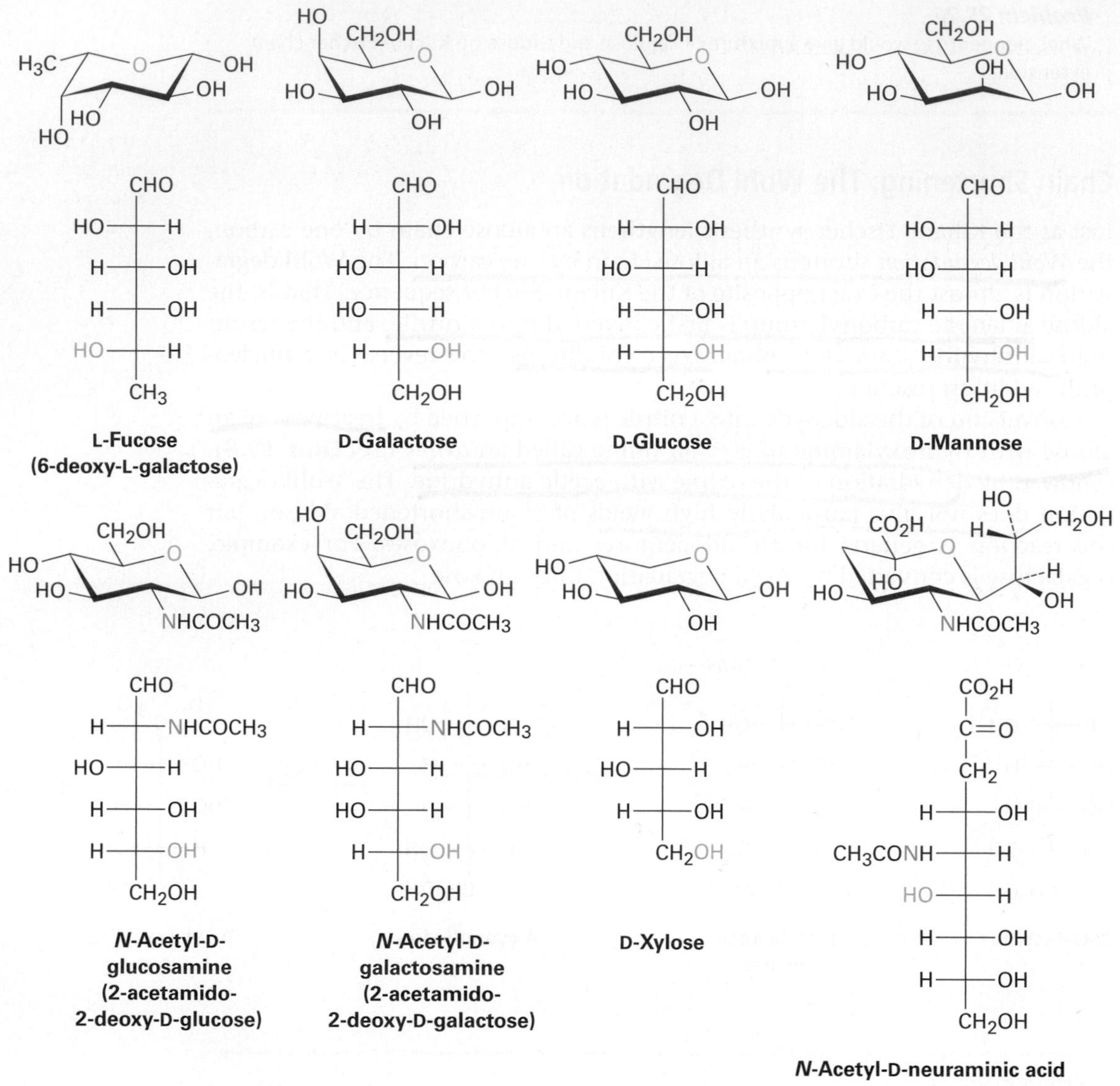

L-Fucose (6-deoxy-L-galactose)

D-Galactose

D-Glucose

D-Mannose

N-Acetyl-D-glucosamine (2-acetamido-2-deoxy-D-glucose)

N-Acetyl-D-galactosamine (2-acetamido-2-deoxy-D-galactose)

D-Xylose

N-Acetyl-D-neuraminic acid

Figure 25.9 Structures of the eight monosaccharides essential to humans.

Of the eight essential monosaccharides, galactose, glucose, and mannose are simple aldohexoses, while xylose is an aldopentose. Fucose is a **deoxy sugar**, meaning that it has an oxygen atom "missing." That is, an –OH group (the one at C6) is replaced by an –H. *N*-Acetylglucosamine and *N*-acetylgalactosamine are amide derivatives of **amino sugars** in which an –OH (the one at C2) is replaced by an $-NH_2$ group. *N*-Acetylneuraminic acid is the parent compound of the sialic acids, a group of more than 30 compounds with different modifications, including various oxidations, acetylations, sulfations, and methylations. Note that neuraminic acid has nine carbons and is an aldol reaction product of *N*-acetylmannosamine with pyruvate ($CH_3COCO_2^-$). We'll see in **Section 25.11** that neuraminic acid is crucially important to the mechanism by which an influenza virus spreads.

All the essential monosaccharides arise from glucose, by the conversions summarized in **Figure 25.10**. We'll not look specifically at these conversions, but might note that Problems 25.54 through 25.56 and 25.71 at the end of the chapter lead you through several of the biosynthetic pathways.

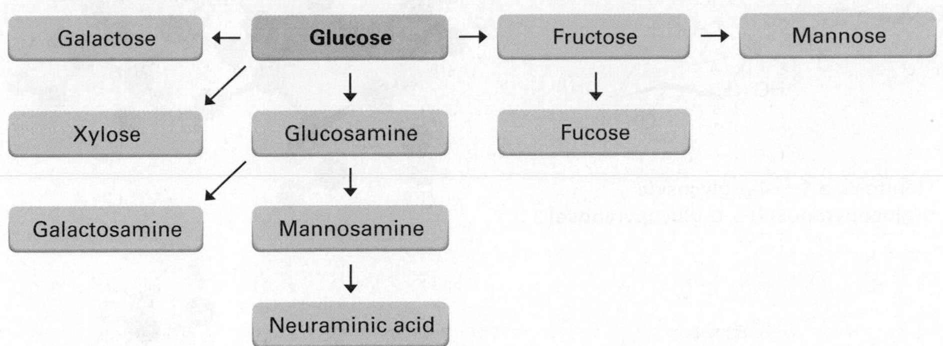

Figure 25.10 An overview of biosynthetic pathways for the eight essential monosaccharides.

Problem 25.24

Show how neuraminic acid can arise by an aldol reaction of *N*-acetylmannosamine with pyruvate ($CH_3COCO_2^-$).

CHO
CH_3CONH—C—H
HO—C—H
H—C—OH
H—C—OH
CH_2OH

***N*-Acetylmannosamine**

25.8 Disaccharides

We saw in **Section 25.6** that reaction of a monosaccharide with an alcohol yields a glycoside in which the anomeric –OH group is replaced by an –OR substituent. If the alcohol is itself a sugar, the glycosidic product is a **disaccharide**.

Maltose and Cellobiose

Disaccharides contain a glycosidic acetal bond between the anomeric carbon of one sugar and an –OH group at any position on the other sugar. A glycosidic bond between C1 of the first sugar and the –OH at C4 of the second sugar is particularly common. Such a bond is called a *1→4 link*.

The glycosidic bond to an anomeric carbon can be either α or β. Maltose, the disaccharide obtained by enzyme-catalyzed hydrolysis of starch, consists of two α-D-glucopyranose units joined by a 1→4-α-glycoside bond. Cellobiose, the disaccharide obtained by partial hydrolysis of cellulose, consists of two β-D-glucopyranose units joined by a 1→4-β-glycoside bond.

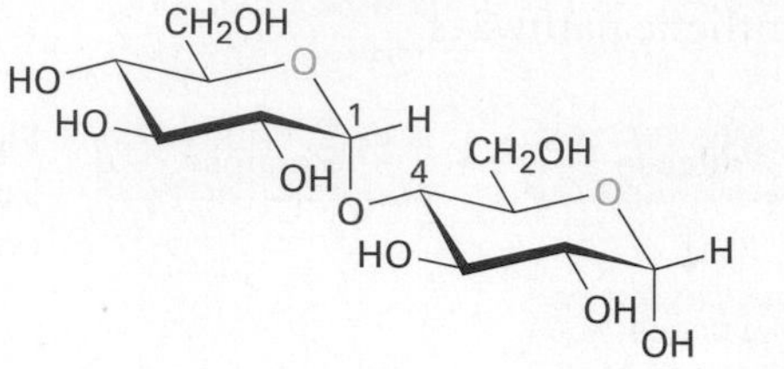

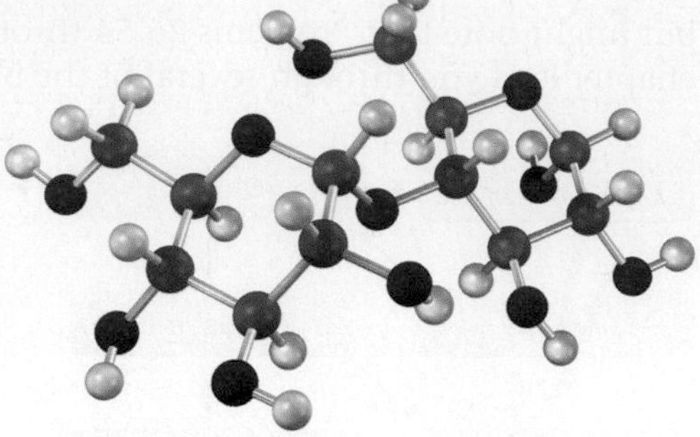

Maltose, a 1→4-α-glycoside
[4-*O*-(α-D-glucopyranosyl)-α-D-glucopyranose]

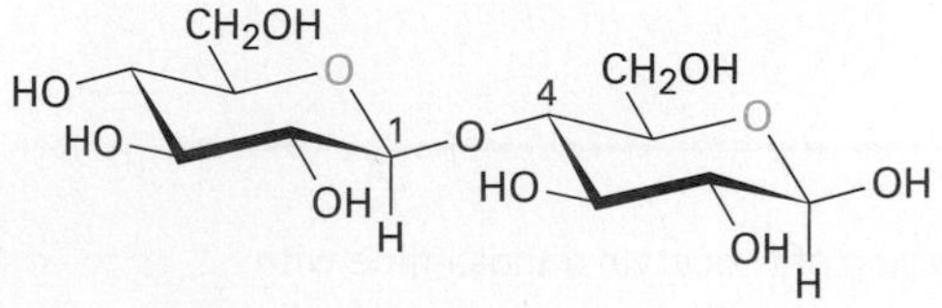

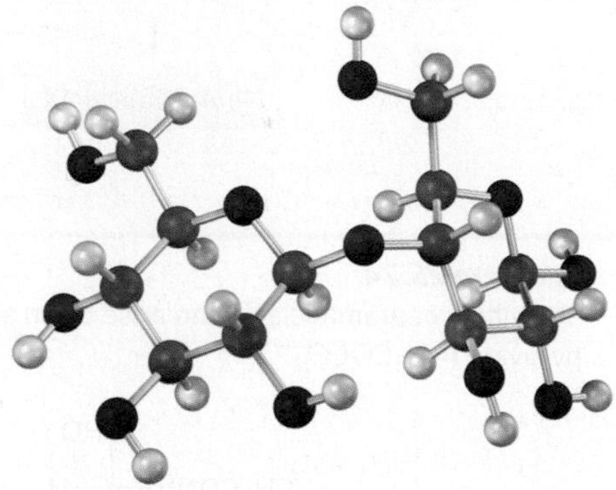

Cellobiose, a 1→4-β-glycoside
[4-*O*-(β-D-glucopyranosyl)-β-D-glucopyranose]

Maltose and cellobiose are both reducing sugars because the anomeric carbons on the right-hand glucopyranose units have hemiacetal groups and are in equilibrium with aldehyde forms. For a similar reason, both maltose and cellobiose exhibit mutarotation of α and β anomers of the glucopyranose unit on the right.

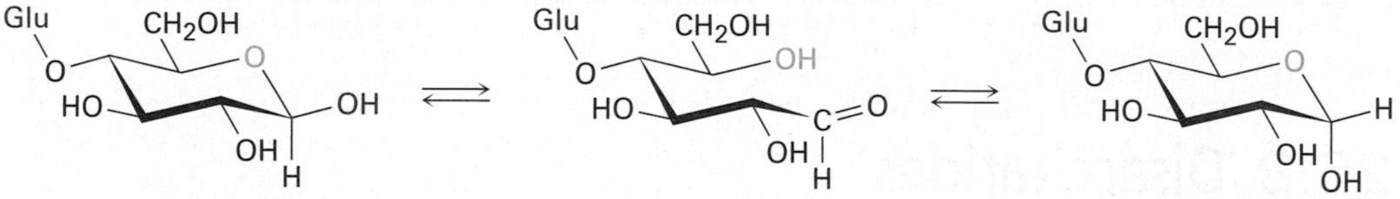

Maltose or cellobiose (β anomers) ⇌ **Maltose or cellobiose (aldehydes)** ⇌ **Maltose or cellobiose (α anomers)**

Despite the similarities of their structures, cellobiose and maltose have dramatically different biological properties. Cellobiose can't be digested by humans and can't be fermented by yeast. Maltose, however, is digested without difficulty and is fermented readily.

Problem 25.25

Show the product you would obtain from the reaction of cellobiose with the following reagents:

(a) $NaBH_4$ **(b)** Br_2, H_2O **(c)** CH_3COCl, pyridine

Lactose

Lactose is a disaccharide that occurs naturally in both human and cow's milk. It is widely used in baking and in commercial milk formulas for infants. Like maltose and cellobiose, lactose is a reducing sugar. It exhibits mutarotation and is a 1→4-β-linked glycoside. Unlike maltose and cellobiose, however, lactose contains two different monosaccharides—D-glucose and D-galactose—joined by a β-glycosidic bond between C1 of galactose and C4 of glucose.

β-Glucopyranose

β-Galactopyranoside

Lactose, a 1→4-β-glycoside
[4-*O*-(β-D-galactopyranosyl)-β-D-glucopyranose]

Sucrose

Sucrose, or ordinary table sugar, is probably the most abundant pure organic chemical in the world. Whether from sugar cane (20% sucrose by weight) or sugar beets (15% by weight), and whether raw or refined, all table sugar is sucrose.

Sucrose is a disaccharide that yields 1 equivalent of glucose and 1 equivalent of fructose on hydrolysis. This 1:1 mixture of glucose and fructose is often referred to as *invert sugar* because the sign of optical rotation changes, or inverts, during the hydrolysis of sucrose ($[\alpha]_D = +66.5$) to a glucose/fructose mixture ($[\alpha]_D = -22.0$). Some insects, such as honeybees, have enzymes called invertases that catalyze the sucrose hydrolysis. Honey, in fact, is primarily a mixture of glucose, fructose, and sucrose.

Unlike most other disaccharides, sucrose is not a reducing sugar and does not undergo mutarotation. These observations imply that sucrose is not a hemiacetal and suggest that glucose and fructose must both be glycosides. This can happen only if the two sugars are joined by a glycoside link between the anomeric carbons of both sugars—C1 of glucose and C2 of fructose.

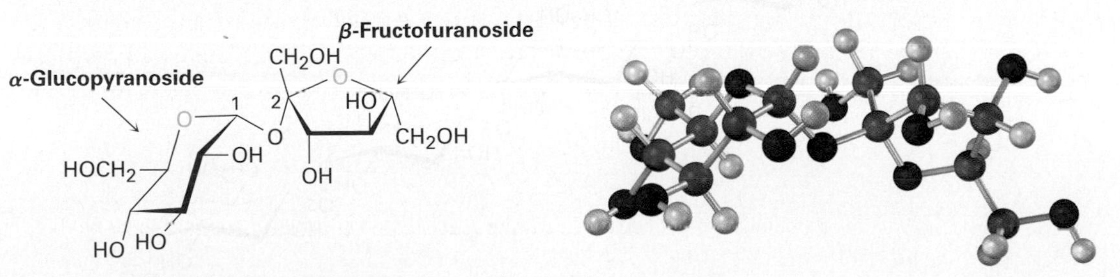

Sucrose, a 1→2-glycoside
[2-*O*-(α-D-glucopyranosyl)-β-D-fructofuranoside]

25.9 Polysaccharides and Their Synthesis

Polysaccharides are complex carbohydrates in which tens, hundreds, or even thousands of simple sugars are linked together through glycoside bonds. Because they have only the one free anomeric –OH group at the end of a very long chain, polysaccharides aren't reducing sugars and don't show noticeable mutarotation. Cellulose and starch are the two most widely occurring polysaccharides.

Cellulose

Cellulose consists of several thousand D-glucose units linked by 1→4-β-glycoside bonds like those in cellobiose. Different cellulose molecules then interact to form a large aggregate structure held together by hydrogen bonds.

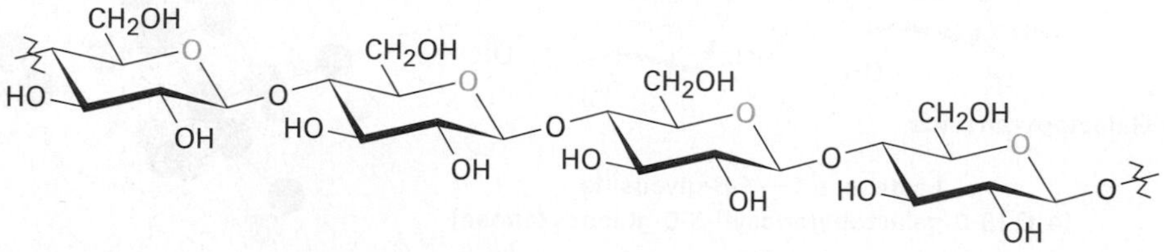

Cellulose, a 1→4-*O*-(β-D-glucopyranoside) polymer

Nature uses cellulose primarily as a structural material to impart strength and rigidity to plants. Leaves, grasses, and cotton, for instance, are primarily cellulose. Cellulose also serves as raw material for the manufacture of cellulose acetate, known commercially as acetate rayon, and cellulose nitrate, known as guncotton. Guncotton is the major ingredient in smokeless powder, the explosive propellant used in artillery shells and in ammunition for firearms.

Starch and Glycogen

Potatoes, corn, and cereal grains contain large amounts of *starch,* a polymer of glucose in which the monosaccharide units are linked by 1→4-α-glycoside bonds like those in maltose. Starch can be separated into two fractions: amylose and amylopectin. Amylose accounts for about 20% by weight of starch and consists of several hundred glucose molecules linked together by 1→4-α-glycoside bonds.

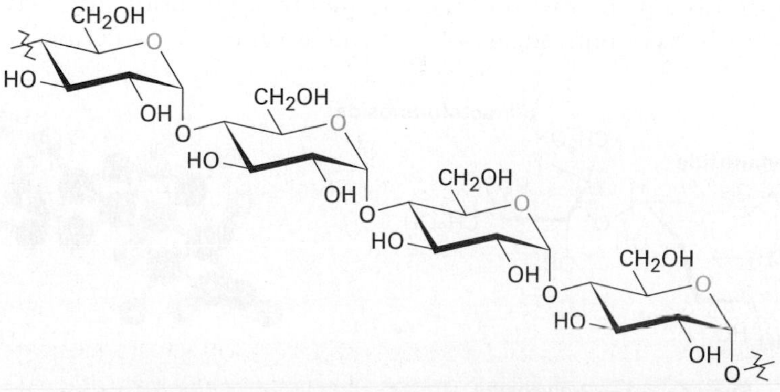

Amylose, a 1→4-*O*-(α-D-glucopyranoside) polymer

Amylopectin accounts for the remaining 80% of starch and is more complex in structure than amylose. Unlike cellulose and amylose, which are linear polymers, amylopectin contains 1→6-α-glycoside branches approximately every 25 glucose units.

α-(1→6) glycoside branch

α-(1→4) glycoside link

Amylopectin: α-(1→4) links with α-(1→6) branches

Starch is digested in the mouth and stomach by α-glycosidases, which catalyze the hydrolysis of glycoside bonds and release individual molecules of glucose. Like most enzymes, α-glycosidases are highly selective in their action. They hydrolyze only the α-glycoside links in starch and leave the β-glycoside links in cellulose untouched. Thus, humans can digest potatoes and grains but not grass and leaves.

Glycogen is a polysaccharide that serves the same energy storage function in animals that starch serves in plants. Dietary carbohydrates not needed for immediate energy are converted by the body to glycogen for long-term storage. Like the amylopectin found in starch, glycogen contains a complex branching structure with both 1→4 and 1→6 links **(Figure 25.11)**. Glycogen molecules are larger than those of amylopectin—up to 100,000 glucose units—and contain even more branches.

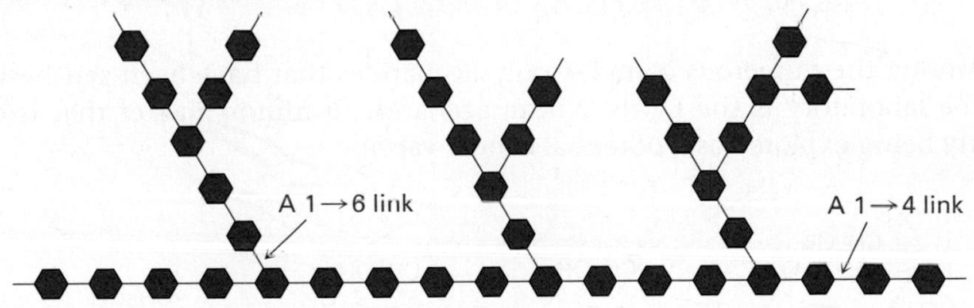

Figure 25.11 A representation of the structure of glycogen. The hexagons represent glucose units linked by 1→4 and 1→6 glycoside bonds.

Polysaccharide Synthesis

With numerous –OH groups of similar reactivity, polysaccharides are so structurally complex that their laboratory synthesis has been a particularly difficult problem. Several methods have recently been devised, however, that have greatly simplified the problem. Among these approaches is the *glycal assembly method.*

Easily prepared from the appropriate monosaccharide, a glycal is an unsaturated sugar with a C1–C2 double bond. To ready it for use in polysaccharide synthesis, the glycal is first protected at its primary –OH group by formation of a silyl ether **(Section 17.8)** and at its two adjacent secondary –OH groups by formation of a cyclic carbonate ester. Then, the protected glycal is epoxidized.

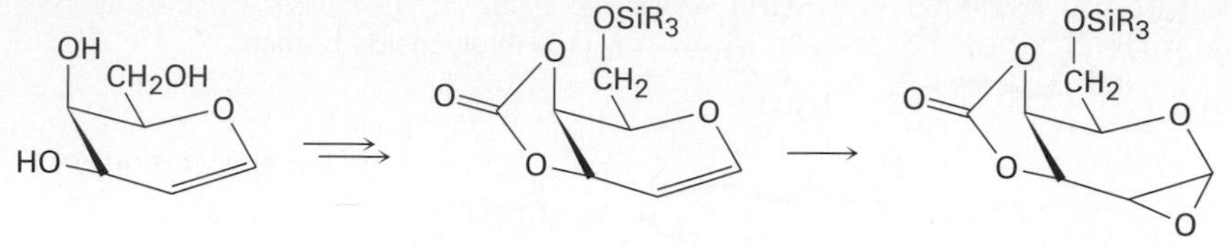

OH, CH_2OH, O, HO — A glycal ⇉ $OSiR_3$, O, CH_2, O, O — A protected glycal ⟶ $OSiR_3$, O, CH_2, O, O, O — An epoxide

A glycal **A protected glycal** **An epoxide**

Treatment of the protected glycal epoxide in the presence of $ZnCl_2$ as a Lewis acid with a second glycal having a free –OH group causes acid-catalyzed opening of the epoxide ring by S_N2 backside attack **(Section 18.6)** and yields a disaccharide. The disaccharide is itself a glycal, so it can be epoxidized and coupled again to yield a trisaccharide, and so on. Using the appropriate sugars at each step, a great variety of polysaccharides can be prepared. After the appropriate sugars are linked, the silyl ethers and cyclic carbonate protecting groups are removed by hydrolysis.

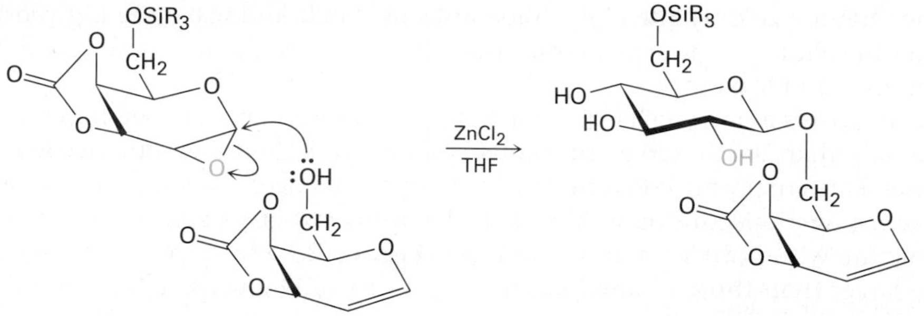

$OSiR_3$, O, CH_2, O, O, O + :ÖH, CH_2, O, O, O $\xrightarrow[\text{THF}]{ZnCl_2}$ $OSiR_3$, CH_2, HO, HO, O, OH, O, O, CH_2, O, O

A disaccharide glycal

Among the numerous complex polysaccharides that have been synthesized in the laboratory is the Lewis Y hexasaccharide, a tumor marker that is currently being explored as a potential cancer vaccine.

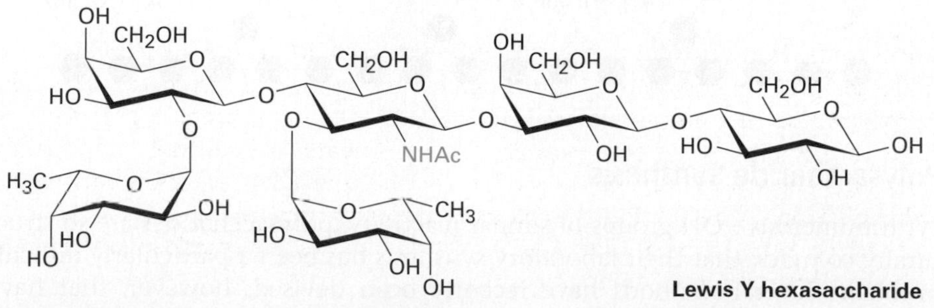

OH, CH_2OH, O, HO, O, CH_2OH, O, O, NHAc, OH, CH_2OH, O, O, OH, O, CH_2OH, O, HO, OH, OH, H_3C, O, OH, HO, HO, O, CH_3, HO, OH, OH

Lewis Y hexasaccharide

25.10 Other Important Carbohydrates

In addition to the common carbohydrates mentioned in previous sections, there are a variety of important carbohydrate-derived materials. Their structural resemblance to sugars is clear, but they aren't simple aldoses or ketoses.

Deoxy sugars, as we saw in **Section 25.7**, have an oxygen atom "missing." That is, an –OH group is replaced by an –H. The most common deoxy sugar is 2-deoxyribose, a monosaccharide found in DNA (deoxyribonucleic acid). Note that 2-deoxyribose exists in water solution as a complex equilibrium mixture of both furanose and pyranose forms.

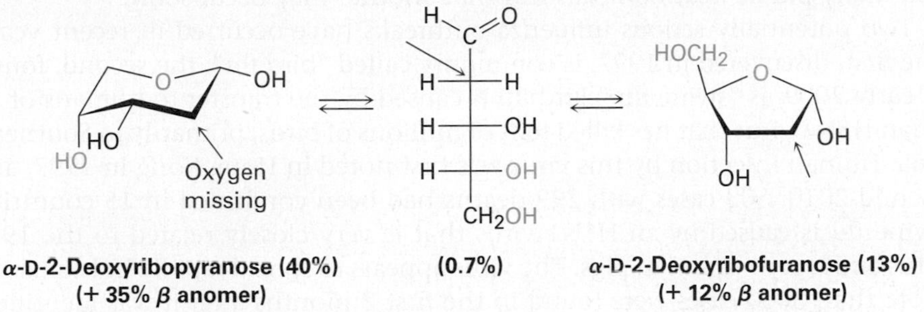

Oxygen missing

α-D-2-Deoxyribopyranose (40%) (+ 35% β anomer) **(0.7%)** **α-D-2-Deoxyribofuranose (13%) (+ 12% β anomer)**

Amino sugars, such as D-glucosamine, have an –OH group replaced by an $-NH_2$. The *N*-acetyl amide derived from D-glucosamine is the monosaccharide unit from which chitin, the hard crust that protects insects and shellfish, is made. Still other amino sugars are found in antibiotics such as streptomycin and gentamicin.

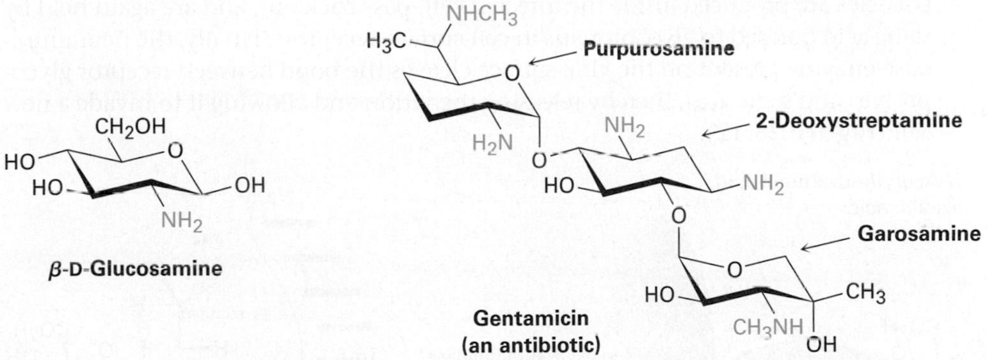

Purpurosamine **2-Deoxystreptamine** **Garosamine**

β-D-Glucosamine

Gentamicin (an antibiotic)

25.11 Cell-Surface Carbohydrates and Influenza Viruses

It was once thought that carbohydrates were useful in nature only as structural materials and energy sources. Although carbohydrates do indeed serve these purposes, they have many other important biochemical functions as well. As noted in **Section 25.6**, for instance, glycoconjugates are centrally involved in cell–cell recognition, the critical process by which one type of cell distinguishes another. Small polysaccharide chains, covalently bound by glycosidic links to –OH or $-NH_2$ groups on proteins, act as biochemical markers on cell surfaces, as illustrated by influenza viruses.

Each year, seasonal outbreaks of influenza occur throughout the world, usually without particular notice. These outbreaks are caused by subtypes of known flu viruses that are already present in the population, and they can usually be controlled or prevented by vaccination. Every 10 to 40 years, however, a new and virulent subtype never before seen in humans appears. The result can be a worldwide pandemic, capable of causing great disruption and killing millions.

Three such pandemics struck in the 20th century, the most serious of which was the 1918–1919 "Spanish flu" that killed an estimated 50 million people worldwide, including many healthy young adults. It has now been more than 40 years since the last pandemic, an outbreak of "Hong Kong flu" in 1968–1969, and many public heath officials fear that another may occur soon.

Two potentially serious influenza outbreaks have occurred in recent years. The first, discovered in 1997, is commonly called "bird flu"; the second, found in early 2009, is "swine flu." Bird flu is caused by the transfer to humans of an avian H5N1 virus that has killed tens of millions of birds, primarily in Southeast Asia. Human infection by this virus was first noted in Hong Kong in 1997, and by mid-2010, 503 cases with 299 deaths had been confirmed in 15 countries. Swine flu is caused by an H1N1 virus that is very closely related to the 1918 virus and is now found in pigs. The virus appears to spread rapidly in humans—more than 3000 cases were found in the first 2 months after it was identified. By mid-2010, 18,449 deaths in 214 countries had been reported.

The classifications H5N1 and H1N1 for the two viral strains are based on the behavior of two kinds of glycoproteins that coat the viral surface—hemagglutinin (H, type 5 or type 1) and neuraminidase (N, type 1), an enzyme. Infection occurs when a viral particle, or *virion,* binds to the sialic acid part **(Section 25.7)** of a receptor glycoprotein on the target cell and is then engulfed by the cell. New viral particles are produced inside the infected cell, pass back out, and are again held by sialic acid bonded to glycoproteins in cell-surface receptors. Finally, the neuraminidase enzyme present on the viral surface cleaves the bond between receptor glycoprotein and sialic acid, thereby releasing the virion and allowing it to invade a new cell **(Figure 25.12)**.

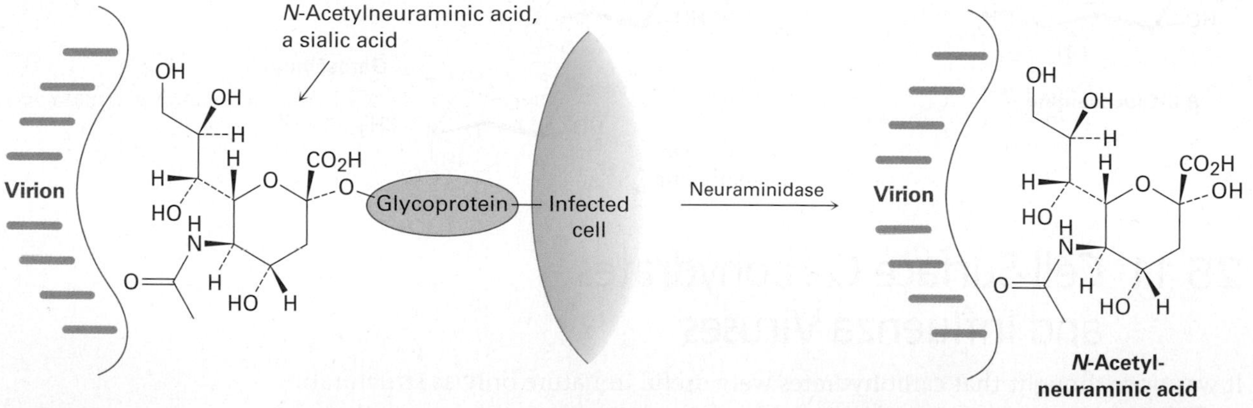

Figure 25.12 Release of a newly formed virion from an infected cell occurs when neuraminidase, present on the surface of the virion, cleaves the bond holding the virion to a sialic acid molecule in a glycoprotein receptor on the infected cell.

So what can be done to limit the severity of an influenza pandemic? Development of a vaccine is the only means to limit the spread of the virus, but work can't begin until the contagious strain of virus has appeared. Until that time, the only hope is that an antiviral drug might limit the severity of infection. Oseltamivir, sold as Tamiflu, and zanamivir, sold as Relenza, are two of only a handful of known substances able to inhibit the neuraminidase enzyme. With the enzyme blocked, newly formed virions are not released, and spread of the infection within the body is thus limited. You might notice in Figure 25.12 the similarity in shape between *N*-acetylneuraminic acid and both oseltamivir and zanamivir, which allows the drugs to bind to and block the action of neuraminidase. Unfortunately, the H1N1 swine flu virus developed almost complete resistance to oseltamivir within a year of appearing, so chemists will have to work hard to keep ahead.

Oseltamivir (Tamiflu)

Zanamivir (Relenza)

A DEEPER LOOK Sweetness

The real thing comes from sugarcane fields like this one.

Say the word *sugar* and most people immediately think of sweet-tasting candies, desserts, and such. In fact, most simple carbohydrates do taste sweet but the degree of sweetness varies greatly from one sugar to another. With sucrose (table sugar) as a reference point, fructose is nearly twice as sweet, but lactose is only about one-sixth as sweet. Comparisons are difficult, though, because perceived sweetness varies depending on the concentration of the solution being tasted and on personal opinion. Nevertheless, the ordering in Table 25.1 is generally accepted.

Table 25.1 Sweetness of Some Sugars and Sugar Substitutes

Name	Type	Sweetness
Lactose	Disaccharide	0.16
Glucose	Monosaccharide	0.75
Sucrose	**Disaccharide**	**1.00**
Fructose	Monosaccharide	1.75
Aspartame	Synthetic	180
Acesulfame-K	Synthetic	200
Saccharin	Synthetic	350
Sucralose	Semisynthetic	600
Alitame	Semisynthetic	2000

(continued)

(continued)

The desire of many people to cut their caloric intake has led to the development of synthetic sweeteners such as saccharin, aspartame, acesulfame, and sucralose. All are far sweeter than natural sugars, so the choice of one or another depends on personal taste, government regulations, and (for baked goods) heat stability. Saccharin, the oldest synthetic sweetener, has been used for more than a century, although it has a somewhat metallic aftertaste. Doubts about its safety and potential carcinogenicity were raised in the early 1970s, but it has now been cleared of suspicion.

Acesulfame potassium, one of the most recently approved sweeteners, is proving to be extremely popular in soft drinks because it has little aftertaste. Sucralose, another recently approved sweetener, is particularly useful in baked goods because of its stability at high temperatures. Alitame, marketed in some countries under the name Aclame, is not approved for sale in the United States. It is some 2000 times as sweet as sucrose and, like acesulfame-K, has no aftertaste. Of the five synthetic sweeteners listed in Table 25.1, only sucralose has clear structural resemblance to a carbohydrate, although it differs dramatically in containing three chlorine atoms. Aspartame and alitame are both dipeptides.

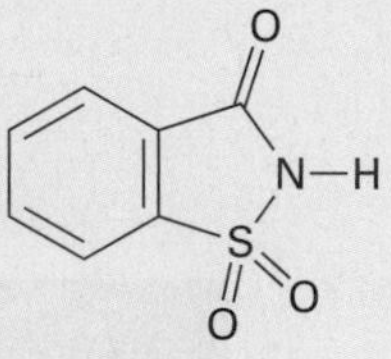

Saccharin 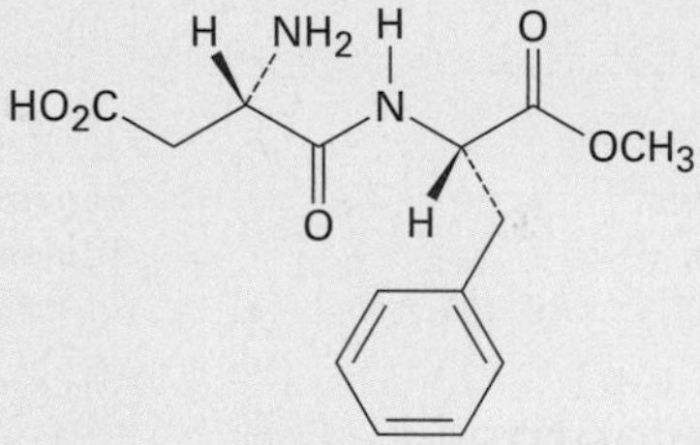 **Aspartame** 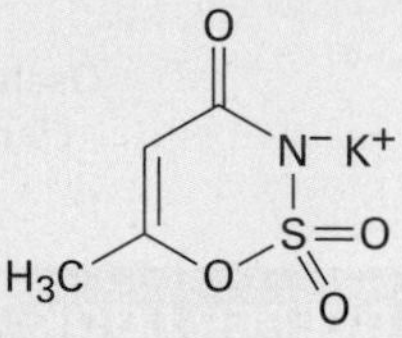 **Acesulfame potassium**

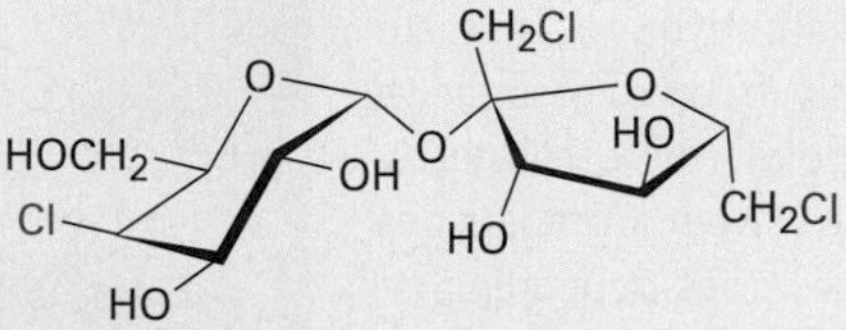

Sucralose 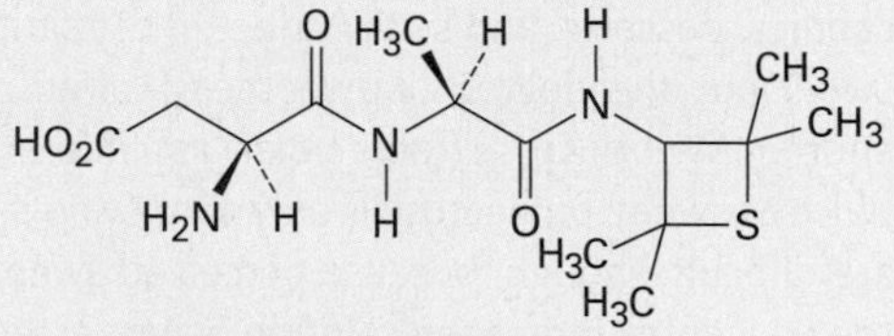 **Alitame**

Summary

Key words

aldaric acid, 1021
alditol, 1020
aldonic acid, 1020
aldose, 1002
amino sugar, 1024
α anomer, β anomer, 1012
anomeric center, 1012

Now that we've now seen all the common functional groups and reaction types, our focus has changed to looking at the major classes of biological molecules. **Carbohydrates** are polyhydroxy aldehydes and ketones. They are classified according to the number of carbon atoms and the kind of carbonyl group they contain. Glucose, for example, is an aldohexose, a six-carbon aldehydo sugar. **Monosaccharides** are further classified as either D **sugars** or L **sugars**, depending on the stereochemistry of the chirality center farthest from the carbonyl group. Carbohydrate stereochemistry is frequently depicted using **Fischer projections**, which represent a chirality center as the intersection of two crossed lines.

Monosaccharides normally exist as cyclic hemiacetals rather than as open-chain aldehydes or ketones. The hemiacetal linkage results from reaction of the carbonyl group with an –OH group three or four carbon atoms away. A five-membered cyclic hemiacetal is called a **furanose**, and a six-membered cyclic hemiacetal is called a **pyranose**. Cyclization leads to the formation of a new chirality center called the **anomeric center** and the production of two diastereomeric hemiacetals called **alpha (α) and beta (β) anomers**.

Much of the chemistry of monosaccharides is the familiar chemistry of alcohols and aldehydes/ketones. Thus, the hydroxyl groups of carbohydrates form esters and ethers. The carbonyl group of a monosaccharide can be reduced with $NaBH_4$ to form an **alditol**, oxidized with aqueous Br_2 to form an **aldonic acid**, oxidized with HNO_3 to form an **aldaric acid**, oxidized enzymatically to form a **uronic acid**, or treated with an alcohol in the presence of acid to form a **glycoside**. Monosaccharides can also be chain-lengthened by the multistep **Kiliani–Fischer synthesis** and can be chain-shortened by the **Wohl degradation**.

Disaccharides are complex carbohydrates in which simple sugars are linked by a glycoside bond between the **anomeric center** of one unit and a hydroxyl of the second unit. The sugars can be the same, as in maltose and cellobiose, or different, as in lactose and sucrose. The glycosidic bond can be either α (maltose) or β (cellobiose, lactose) and can involve any hydroxyl of the second sugar. A 1→4 link is most common (cellobiose, maltose), but others such as 1→2 (sucrose) are also known. **Polysaccharides**, such as cellulose, starch, and glycogen, are used in nature as structural materials, as a means of long-term energy storage, and as cell-surface markers.

Key words—cont'd

carbohydrate, 1000
complex carbohydrate, 1001
D sugar, 1007
deoxy sugar, 1024
disaccharide, 1025
Fischer projection, 1002
furanose, 1012
glycoside, 1016
ketose, 1002
L sugar, 1007
monosaccharide, 1001
mutarotation, 1013
polysaccharide, 1028
pyranose, 1011
reducing sugar, 1020
simple sugar, 1001
uronic acid, 1021

Summary of Reactions

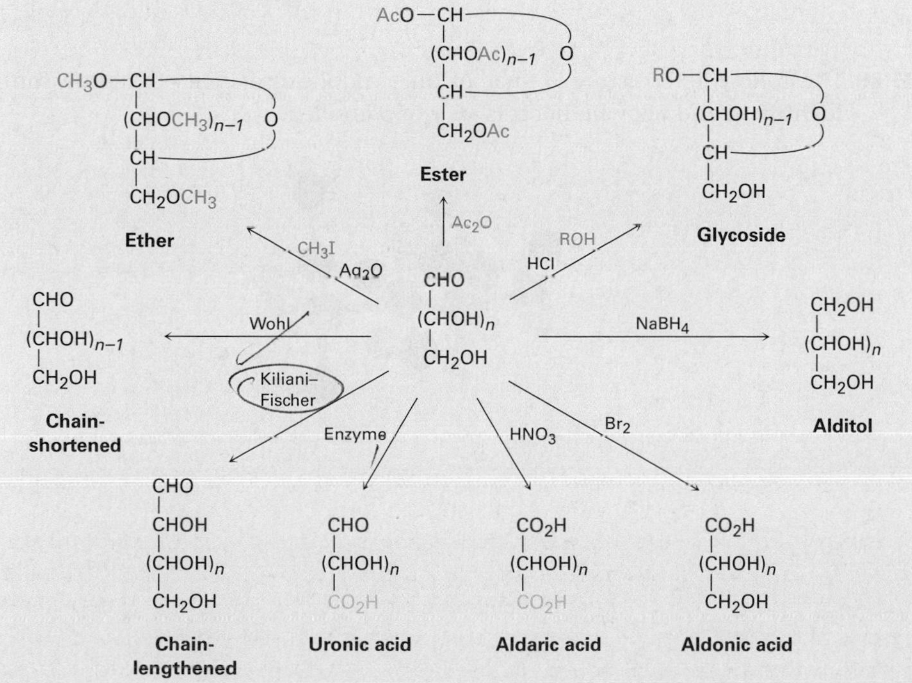

Starting sugar: CHO–(CHOH)$_n$–CH$_2$OH

- Ac_2O → **Ester**: AcO–CH–(CHOAc)$_{n-1}$–CH(–O– ring)–CH$_2$OAc
- CH_3I, Ag_2O → **Ether**: CH$_3$O–CH–(CHOCH$_3$)$_{n-1}$–CH(–O– ring)–CH$_2$OCH$_3$
- ROH, HCl → **Glycoside**: RO–CH–(CHOH)$_{n-1}$–CH(–O– ring)–CH$_2$OH
- Wohl → **Chain-shortened**: CHO–(CHOH)$_{n-1}$–CH$_2$OH
- Kiliani–Fischer → **Chain-lengthened**: CHO–CHOH–(CHOH)$_n$–CH$_2$OH
- $NaBH_4$ → **Alditol**: CH$_2$OH–(CHOH)$_n$–CH$_2$OH
- Enzyme → **Uronic acid**: CHO–(CHOH)$_n$–CO$_2$H
- HNO_3 → **Aldaric acid**: CO$_2$H–(CHOH)$_n$–CO$_2$H
- Br_2 → **Aldonic acid**: CO$_2$H–(CHOH)$_n$–CH$_2$OH

Exercises

OWL Interactive versions of these problems are assignable in OWL for Organic Chemistry.

Visualizing Chemistry

(Problems 25.1–25.25 appear within the chapter.)

25.26 Identify the following aldoses, and tell whether each is a D or L sugar:

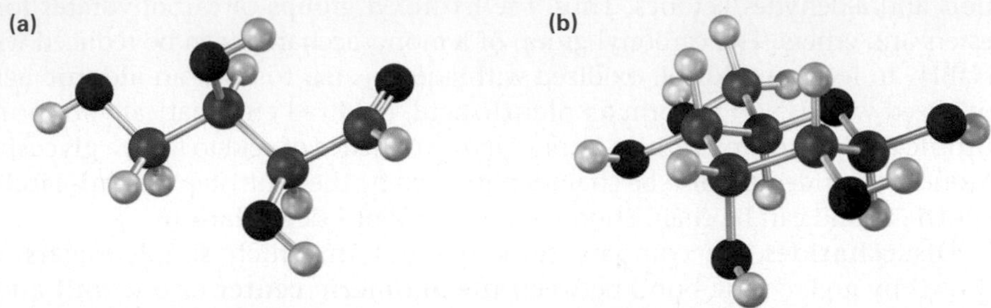

25.27 Draw Fischer projections of the following molecules, placing the carbonyl group at the top in the usual way. Identify each as a D or L sugar.

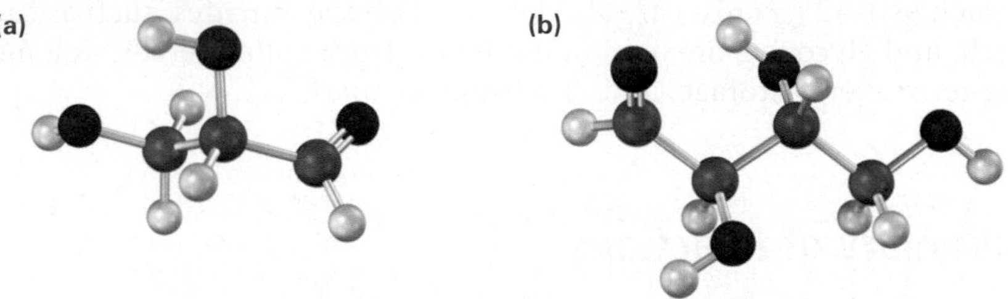

25.28 The following structure is that of an L aldohexose in its pyranose form. Identify it, and tell whether it is an α or β anomer.

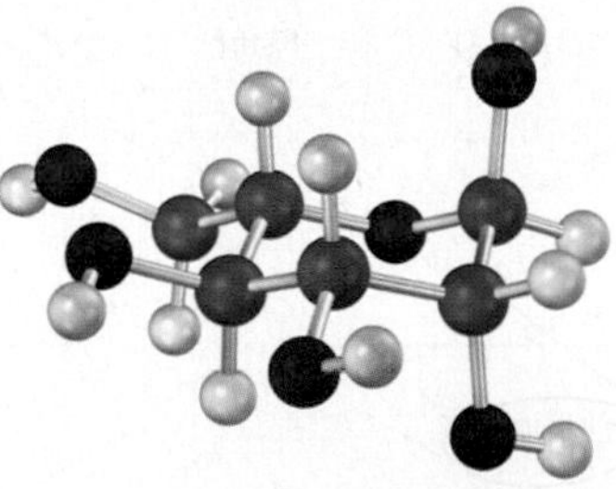

25.29 The following model is that of an aldohexose:

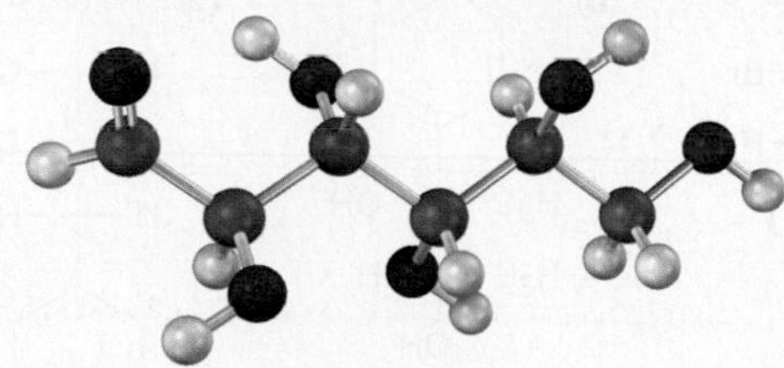

(a) Draw Fischer projections of the sugar, its enantiomer, and a diastereomer.
(b) Is this a D sugar or an L sugar? Explain.
(c) Draw the β anomer of the sugar in its furanose form.

Additional Problems

Carbohydrate Structures

25.30 Classify each of the following sugars. (For example, glucose is an aldohexose.)

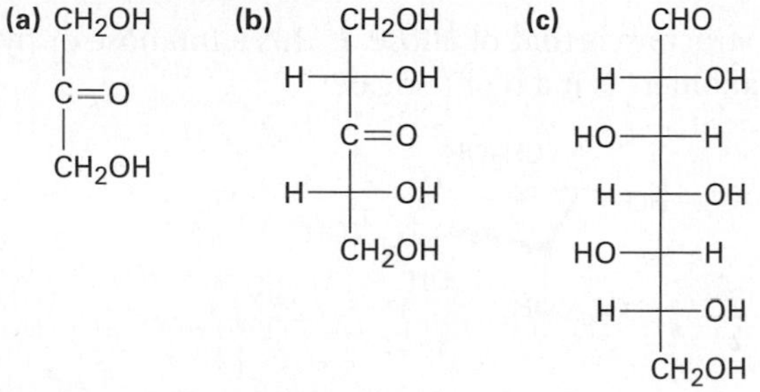

(a)
CH_2OH
C=O
CH_2OH

(b)
CH_2OH
H—OH
C=O
H—OH
CH_2OH

(c)
CHO
H—OH
HO—H
H—OH
HO—H
H—OH
CH_2OH

25.31 Write open-chain structures for the following:
(a) A ketotetrose
(b) A ketopentose
(c) A deoxyaldohexose
(d) A five-carbon amino sugar

25.32 What is the stereochemical relationship of D-ribose to L-xylose? What generalizations can you make about the following properties of the two sugars?
(a) Melting point
(b) Solubility in water
(c) Specific rotation
(d) Density

25.33 Does ascorbic acid (vitamin C) have a D or L configuration?

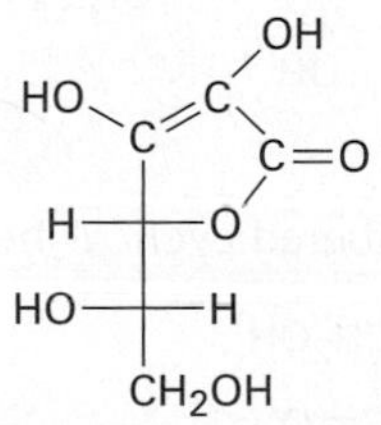

HO—C=C(OH)—C=O (ring closed through O)
H—O
HO—H
CH_2OH

Ascorbic acid

25.34 Draw the three-dimensional furanose form of ascorbic acid (Problem 25.33), and assign *R* or *S* stereochemistry to each chirality center.

25.35 Assign *R* or *S* configuration to each chirality center in the following molecules:

(a) H / H_3C—Br / Br—H / CH_3 (b) phenyl / H_3C—OH / H_3C—H / OH (c) NH_2 / H—CO_2H / H—OH / H—H / phenyl

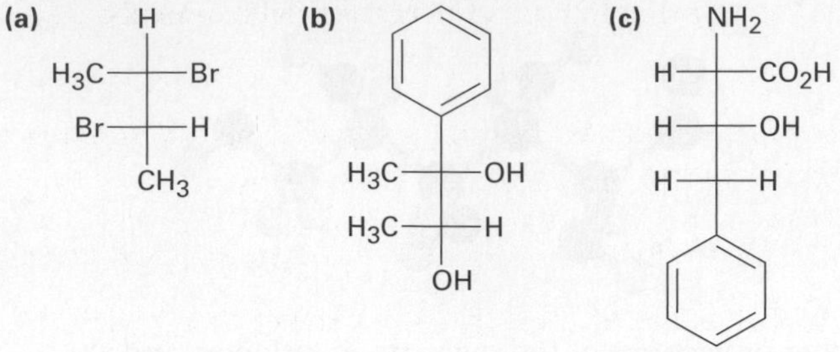

25.36 Draw Fischer projections of the following molecules:

(a) The *S* enantiomer of 2-bromobutane

(b) The *R* enantiomer of alanine, $CH_3CH(NH_2)CO_2H$

(c) The *R* enantiomer of 2-hydroxypropanoic acid

(d) The *S* enantiomer of 3-methylhexane

25.37 Draw Fischer projections for the two D aldoheptoses whose stereochemistry at C3, C4, C5, and C6 is the same as that of D-glucose at C2, C3, C4, and C5.

25.38 The following cyclic structure is that of allose. Is this a furanose or pyranose form? Is it an α or β anomer? Is it a D or L sugar?

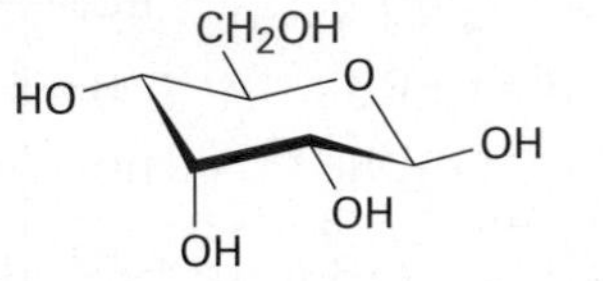

CH_2OH, O, HO, OH, OH, OH

25.39 What is the complete name of the following sugar?

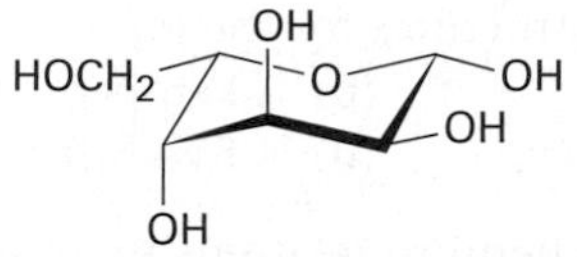

OH, $HOCH_2$, O, OH, OH, OH

25.40 Write the following sugars in their open-chain forms:

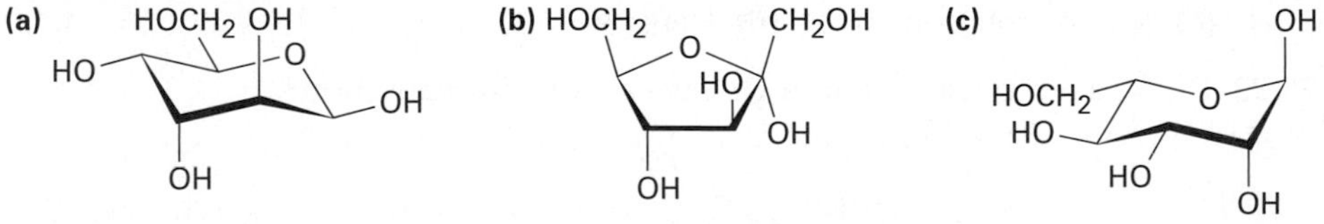

(a) $HOCH_2$ OH, O, HO, OH, OH (b) $HOCH_2$, O, CH_2OH, HO, OH, OH (c) OH, $HOCH_2$, O, HO, HO, OH

25.41 Draw D-ribulose in its five-membered cyclic β-hemiacetal form.

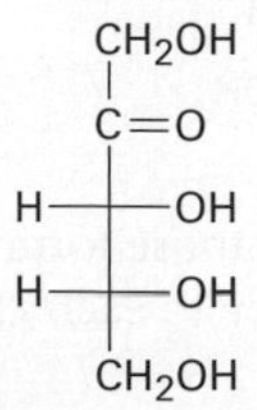

CH_2OH / C=O / H—OH / H—OH / CH_2OH **Ribulose**

25.42 Look up the structure of D-talose in Figure 25.3, and draw the β anomer in its pyranose form. Identify the ring substituents as axial or equatorial.

Carbohydrate Reactions

25.43 Draw structures for the products you would expect to obtain from reaction of β-D-talopyranose with each of the following reagents:

(a) $NaBH_4$ in H_2O (b) Warm dilute HNO_3 (c) Br_2, H_2O
(d) CH_3CH_2OH, HCl (e) CH_3I, Ag_2O (f) $(CH_3CO)_2O$, pyridine

25.44 How many D-2-ketohexoses are possible? Draw them.

25.45 One of the D-2-ketohexoses is called *sorbose.* On treatment with $NaBH_4$, sorbose yields a mixture of gulitol and iditol. What is the structure of sorbose?

25.46 Another D-2-ketohexose, *psicose,* yields a mixture of allitol and altritol when reduced with $NaBH_4$. What is the structure of psicose?

25.47 L-Gulose can be prepared from D-glucose by a route that begins with oxidation to D-glucaric acid, which cyclizes to form two six-membered-ring lactones. Separating the lactones and reducing them with sodium amalgam gives D-glucose and L-gulose. What are the structures of the two lactones, and which one is reduced to L-gulose?

25.48 Gentiobiose, a rare disaccharide found in saffron and gentian, is a reducing sugar and forms only D-glucose on hydrolysis with aqueous acid. Reaction of gentiobiose with iodomethane and Ag_2O yields an octamethyl derivative, which can be hydrolyzed with aqueous acid to give 1 equivalent of 2,3,4,6-tetra-*O*-methyl-D-glucopyranose and 1 equivalent of 2,3,4-tri-*O*-methyl-D-glucopyranose. If gentiobiose contains a β-glycoside link, what is its structure?

General Problems

25.49 All aldoses exhibit mutarotation. For example, α-D-galactopyranose has $[\alpha]_D = +150.7$, and β-D-galactopyranose has $[\alpha]_D = +52.8$. If either anomer is dissolved in water and allowed to reach equilibrium, the specific rotation of the solution is +80.2. What are the percentages of each anomer at equilibrium? Draw the pyranose forms of both anomers.

25.50 What other D aldohexose gives the same alditol as D-talose?

25.51 Which of the eight D aldohexoses give the same aldaric acids as their L enantiomers?

25.52 Which of the other three D aldopentoses gives the same aldaric acid as D-lyxose?

25.53 Draw the structure of L-galactose, and then answer the following questions:

(a) Which other aldohexose gives the same aldaric acid as L-galactose on oxidation with warm HNO_3?

(b) Is this other aldohexose a D sugar or an L sugar?

(c) Draw this other aldohexose in its most stable pyranose conformation.

25.54 Galactose, one of the eight essential monosaccharides (Section 25.7), is biosynthesized from UDP-glucose by galactose 4-epimerase, where UDP = uridylyl diphosphate (a ribonucleotide diphosphate; Section 28.1). The enzyme requires NAD^+ for activity (Section 17.7), but it is not a stoichiometric reactant, and NADH is not a final reaction product. Propose a mechanism.

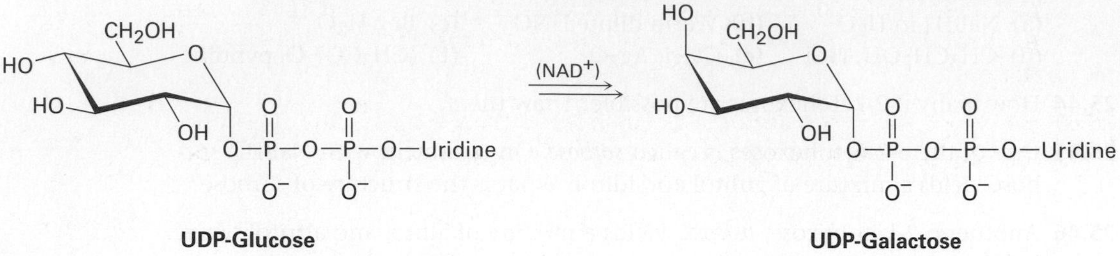

UDP-Glucose $\xrightarrow{(NAD^+)}$ UDP-Galactose

25.55 Mannose, one of the eight essential monosaccharides (Section 25.7), is biosynthesized as its 6-phosphate derivative from fructose 6-phosphate. No enzyme cofactor is required. Propose a mechanism.

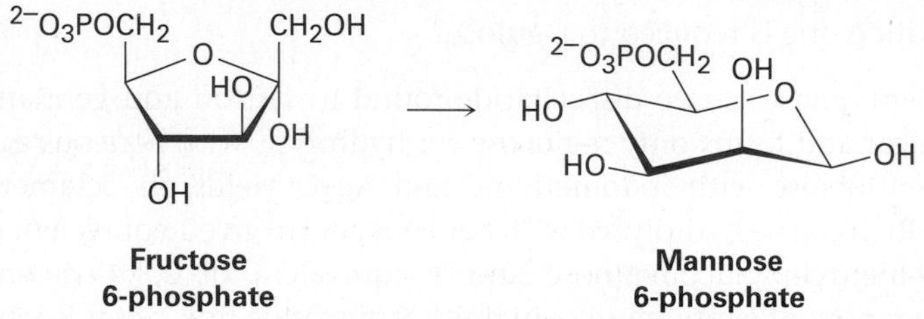

Fructose 6-phosphate $\longrightarrow$ Mannose 6-phosphate

25.56 Glucosamine, one of the eight essential monosaccharides (Section 25.7), is biosynthesized as its 6-phosphate derivative from fructose 6-phosphate by reaction with ammonia. Propose a mechanism.

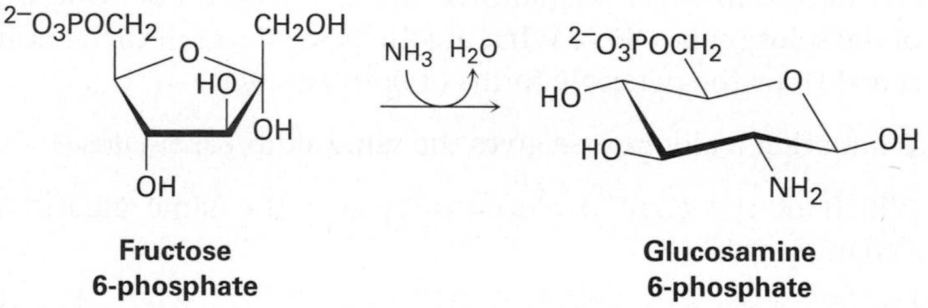

Fructose 6-phosphate $\xrightarrow{NH_3 \quad H_2O}$ Glucosamine 6-phosphate

25.57 Amygdalin, or laetrile, is a cyanogenic glycoside isolated in 1830 from almond and apricot seeds. Acidic hydrolysis of amygdalin liberates HCN, along with benzaldehyde and 2 equivalents of D-glucose. If amygdalin is a β-glycoside of benzaldehyde cyanohydrin with gentiobiose (Problem 21.56), what is its structure?

25.58 Trehalose is a nonreducing disaccharide that is hydrolyzed by aqueous acid to yield 2 equivalents of D-glucose. Methylation followed by hydrolysis yields 2 equivalents of 2,3,4,6-tetra-*O*-methylglucose. How many structures are possible for trehalose?

25.59 Trehalose (Problem 25.58) is cleaved by enzymes that hydrolyze α-glycosides but not by enzymes that hydrolyze β-glycosides. What is the structure and systematic name of trehalose?

25.60 Isotrehalose and neotrehalose are chemically similar to trehalose (Problems 25.58 and 25.59) except that neotrehalose is hydrolyzed only by β-glycosidase enzymes, whereas isotrehalose is hydrolyzed by both α- and β-glycosidase enzymes. What are the structures of isotrehalose and neotrehalose?

25.61 D-Glucose reacts with acetone in the presence of acid to yield the nonreducing 1,2:5,6-diisopropylidene-D-glucofuranose. Propose a mechanism.

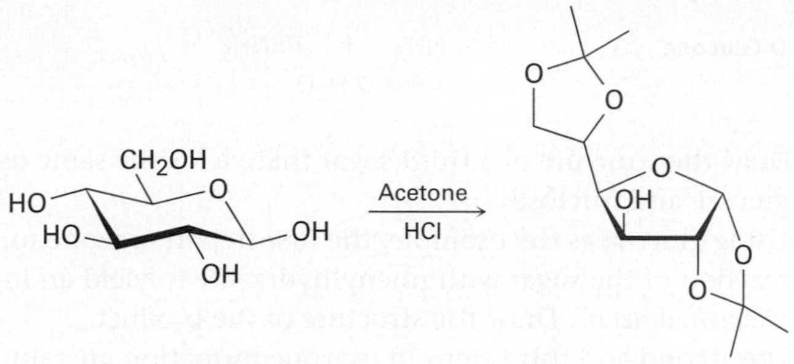

1,2:5,6-Diisopropylidene-D-glucofuranose

25.62 D-Mannose reacts with acetone to give a diisopropylidene derivative (Problem 25.61) that is still reducing toward Tollens' reagent. Propose a likely structure for this derivative.

25.63 Glucose and mannose can be interconverted (in low yield) by treatment with dilute aqueous NaOH. Propose a mechanism.

25.64 Propose a mechanism to account for the fact that D-gluconic acid and D-mannonic acid are interconverted when either is heated in pyridine solvent.

25.65 The *cyclitols* are a group of carbocyclic sugar derivatives having the general formulation 1,2,3,4,5,6-cyclohexanehexol. How many stereoisomeric cyclitols are possible? Draw them in their chair forms.

25.66 Compound A is a D aldopentose that can be oxidized to an optically inactive aldaric acid B. On Kiliani–Fischer chain extension, A is converted into C and D; C can be oxidized to an optically active aldaric acid E, but D is oxidized to an optically inactive aldaric acid F. What are the structures of A–F?

25.67 Simple sugars undergo reaction with phenylhydrazine, $PhNHNH_2$, to yield crystalline derivatives called *osazones*. The reaction is a bit complex, however, as shown by the fact that glucose and fructose yield the same osazone.

CHO; H—OH; HO—H; H—OH; H—OH; CH_2OH — **D-Glucose**

3 equiv $PhNHNH_2$ →

H–C=N—NHPh; C=N—NHPh; HO—H; H—OH; H—OH; CH_2OH

$+ \ NH_3 \ + \ PhNH_2 \ + \ 2\ H_2O$

← 3 equiv $PhNHNH_2$

CH_2OH; C=O; HO—H; H—OH; H—OH; CH_2OH — **D-Fructose**

(a) Draw the structure of a third sugar that yields the same osazone as glucose and fructose.

(b) Using glucose as the example, the first step in osazone formation is reaction of the sugar with phenylhydrazine to yield an imine called a *phenylhydrazone*. Draw the structure of the product.

(c) The second and third steps in osazone formation are tautomerization of the phenylhydrazone to give an enol, followed by elimination of aniline to give a keto imine. Draw the structures of both the enol tautomer and the keto imine.

(d) The final step is reaction of the keto imine with 2 equivalents of phenylhydrazine to yield the osazone plus ammonia. Propose a mechanism for this step.

25.68 When heated to 100 °C, D-idose undergoes a reversible loss of water and exists primarily as 1,6-anhydro-D-idopyranose.

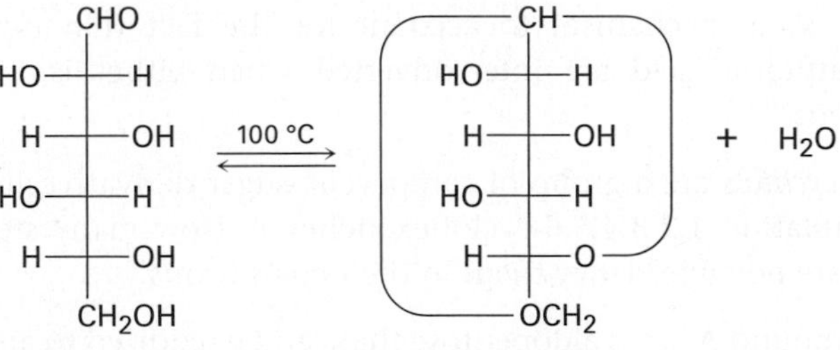

D-Idose **1,6-Anhydro-D-idopyranose**

(a) Draw D-idose in its pyranose form, showing the more stable chair conformation of the ring.

(b) Which is more stable, α-D-idopyranose or β-D-idopyranose? Explain.

(c) Draw 1,6-anhydro-D-idopyranose in its most stable conformation.

(d) When heated to 100 °C under the same conditions as those used for D-idose, D-glucose does not lose water and does not exist in a 1,6-anhydro form. Explain.

25.69 Acetyl coenzyme A (acetyl CoA) is the key intermediate in food metabolism. What sugar is present in acetyl CoA?

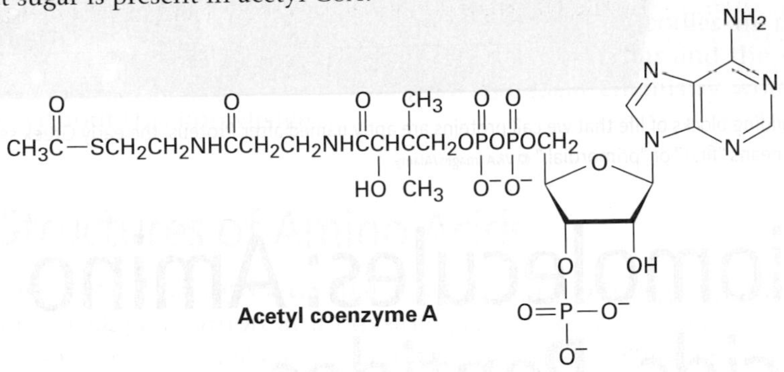

$CH_3C(=O)—SCH_2CH_2NHC(=O)CH_2CH_2NHC(=O)CH(OH)C(CH_3)_2CH_2OP(=O)(O^-)OP(=O)(O^-)OCH_2$ (ribose–adenine, 3′-OPO_3^{2-}, 2′-OH)

Acetyl coenzyme A

25.70 One of the steps in the biological pathway for carbohydrate metabolism is the conversion of fructose 1,6-bisphosphate into dihydroxyacetone phosphate and glyceraldehyde 3-phosphate. Propose a mechanism for the transformation.

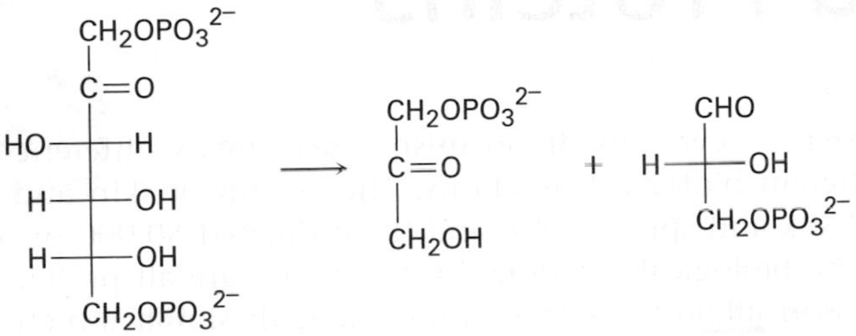

Fructose 1,6-bisphosphate → **Dihydroxyacetone phosphate** + **Glyceraldehyde 3-phosphate**

25.71 L-Fucose, one of the eight essential monosaccharides (Section 25.7), is biosynthesized from GDP-D-mannose by the following three-step reaction sequence, where GDP = guanosine diphosphate (a ribonucleoside diphosphate; Section 28.1):

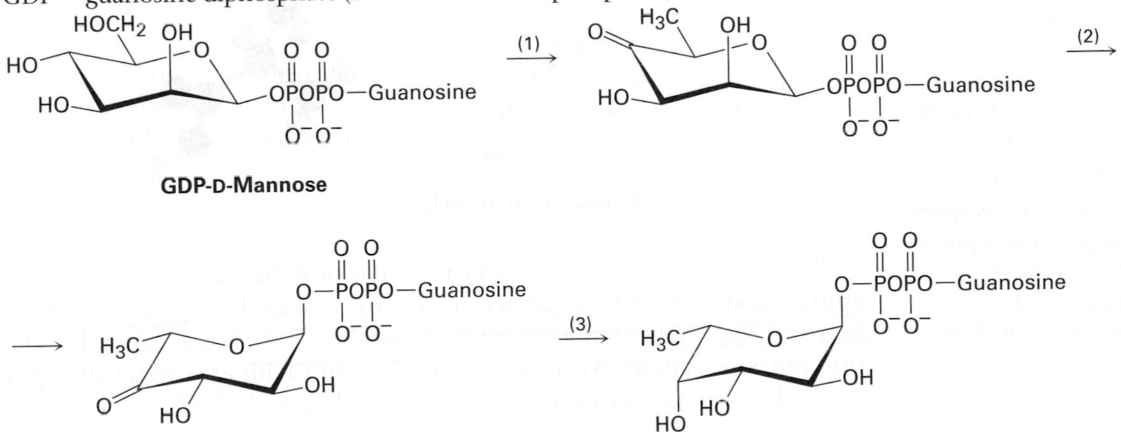

GDP-D-Mannose →(1) →(2) →(3) **GDP-L-Fucose**

(a) Step 1 involves an oxidation to a ketone, a dehydration to an enone, and a conjugate reduction. The step requires $NADP^+$, but no NADPH is formed as a final reaction product. Propose a mechanism.

(b) Step 2 accomplishes two epimerizations and utilizes acidic and basic sites in the enzyme but does not require a coenzyme. Propose a mechanism.

(c) Step 3 requires NADPH as coenzyme. Show the mechanism.

Table 26.1 The 20 Common Amino Acids in Proteins

Name	Abbreviations		MW	Stucture	pK_a α-CO_2H	pK_a α-NH_3^+	pK_a side chain	pI
Neutral Amino Acids								
Alanine	Ala	A	89		2.34	9.69	—	6.01
Asparagine	Asn	N	132		2.02	8.80	—	5.41
Cysteine	Cys	C	121		1.96	10.28	8.18	5.07
Glutamine	Gln	Q	146		2.17	9.13	—	5.65
Glycine	Gly	G	75		2.34	9.60	—	5.97
Isoleucine	Ile	I	131		2.36	9.60	—	6.02
Leucine	Leu	L	131		2.36	9.60	—	5.98
Methionine	Met	M	149		2.28	9.21	—	5.74
Phenylalanine	Phe	F	165		1.83	9.13	—	5.48
Proline	Pro	P	115		1.99	10.60	—	6.30

Table 26.1 The 20 Common Amino Acids in Proteins *(continued)*

Name	Abbreviations		MW	Stucture	pK_a α-CO_2H	pK_a α-NH_3^+	pK_a side chain	p*I*
Neutral Amino Acids *(continued)*								
Serine	Ser	S	105		2.21	9.15	—	5.68
Threonine	Thr	T	119		2.09	9.10	—	5.60
Tryptophan	Trp	W	204		2.83	9.39	—	5.89
Tyrosine	Tyr	Y	181		2.20	9.11	10.07	5.66
Valine	Val	V	117		2.32	9.62	—	5.96
Acidic Amino Acids								
Aspartic acid	Asp	D	133		1.88	9.60	3.65	2.77
Glutamic acid	Glu	E	147		2.19	9.67	4.25	3.22
Basic Amino Acids								
Arginine	Arg	R	174		2.17	9.04	12.48	10.76
Histidine	His	H	155		1.82	9.17	6.00	7.59
Lysine	Lys	K	146		2.18	8.95	10.53	9.74

amino acids are primary amines, RNH_2, and differ only in the nature of the **side chain**—the substituent attached to the α carbon. Proline is a secondary amine whose nitrogen and α carbon atoms are part of a five-membered pyrrolidine ring.

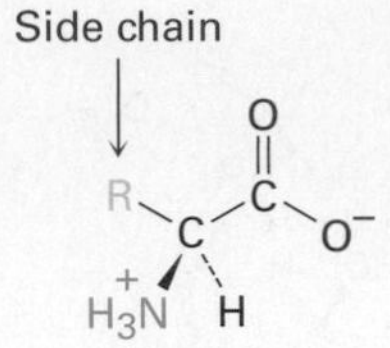

Side chain

A primary α-amino acid

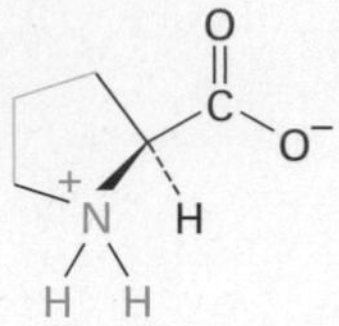

Proline, a secondary α-amino acid

In addition to the 20 amino acids commonly found in proteins, 2 others—selenocysteine and pyrrolysine—are found in some organisms, and more than 700 nonprotein amino acids are also found in nature. γ-Aminobutyric acid (GABA), for instance, is found in the brain and acts as a neurotransmitter; homocysteine is found in blood and is linked to coronary heart disease; and thyroxine is found in the thyroid gland, where it acts as a hormone.

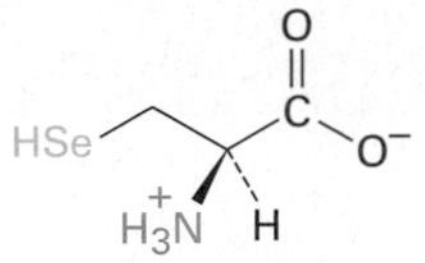

Selenocysteine

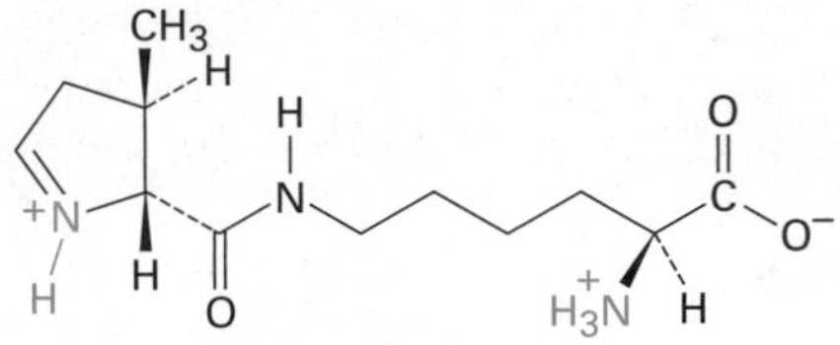

Pyrrolysine

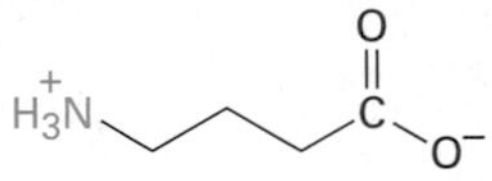

γ-Aminobutyric acid

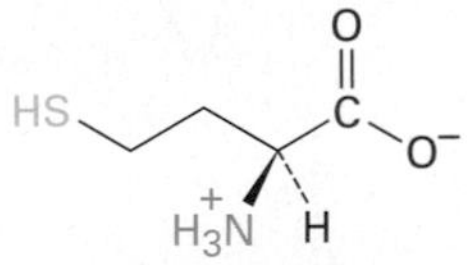

Homocysteine

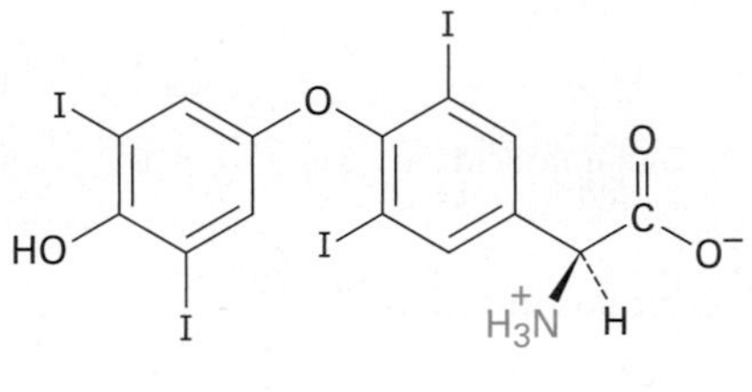

Thyroxine

Except for glycine, $H_2NCH_2CO_2H$, the α carbons of amino acids are chirality centers. Two enantiomers of each are therefore possible, but nature uses only one to build proteins. In Fischer projections, naturally occurring amino acids are represented by placing the $-CO_2^-$ group at the top and the side chain down, as if drawing a carbohydrate **(Section 25.2)** and then placing the $-NH_3^+$ group on the left. Because of their stereochemical similarity to L sugars

(Section 25.3), the naturally occurring α-amino acids are often referred to as L amino acids.

CO_2^- $H_3\overset{+}{N}$—C—H CH_3	CO_2^- $H_3\overset{+}{N}$—C—H CH_2OH	CO_2^- $H_3\overset{+}{N}$—C—H CH_2SH	CHO HO—C—H CH_2OH
L-Alanine **(*S*)-Alanine**	**L-Serine** **(*S*)-Serine**	**L-Cysteine** **(*R*)-Cysteine**	**L-Glyceraldehyde**

The 20 common amino acids can be further classified as neutral, acidic, or basic, depending on the structure of their side chains. Fifteen of the twenty have neutral side chains, two (aspartic acid and glutamic acid) have an extra carboxylic acid function in their side chains, and three (lysine, arginine, and histidine) have basic amino groups in their side chains. Note that both cysteine (a thiol) and tyrosine (a phenol), although usually classified as neutral amino acids, nevertheless have weakly acidic side chains that can be deprotonated in a sufficiently basic solution.

At the physiological pH of 7.3 within cells, the side-chain carboxyl groups of aspartic acid and glutamic acid are deprotonated and the basic side-chain nitrogens of lysine and arginine are protonated. Histidine, however, which contains a heterocyclic imidazole ring in its side chain, is not quite basic enough to be protonated at pH 7.3. Note that only the pyridine-like, doubly bonded nitrogen in histidine is basic. The pyrrole-like singly bonded nitrogen is nonbasic because its lone pair of electrons is part of the six-π-electron aromatic imidazole ring **(Section 24.9)**.

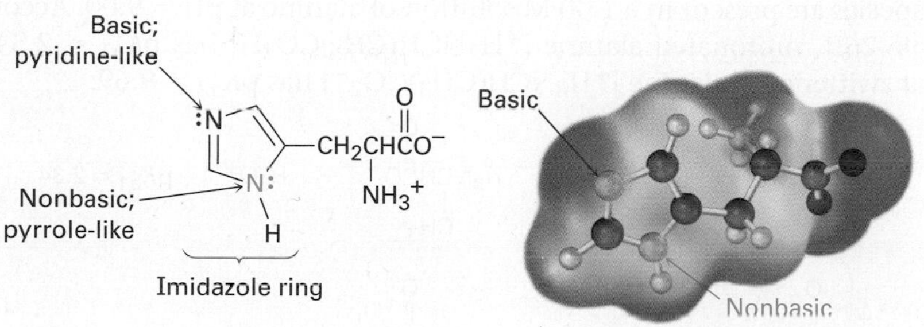

Histidine

Humans are able to synthesize only 11 of the 20 protein amino acids, called nonessential amino acids. The other 9, called essential amino acids, are biosynthesized only in plants and microorganisms and must be obtained in our diet. The division between essential and nonessential amino acids is not clear-cut, however. Tyrosine, for instance, is sometimes considered nonessential because humans can produce it from phenylalanine, but phenylalanine itself is essential and must be obtained in the diet. Arginine can be synthesized by humans, but much of the arginine we need also comes from our diet.

Problem 26.1

How many of the α-amino acids shown in Table 26.1 contain aromatic rings? How many contain sulfur? How many contain alcohols? How many contain hydrocarbon side chains?

Problem 26.2

Eighteen of the nineteen L amino acids have the *S* configuration at the α carbon. Cysteine is the only L amino acid that has an *R* configuration. Explain.

Problem 26.3

The amino acid threonine, (2*S*,3*R*)-2-amino-3-hydroxybutanoic acid, has two chirality centers.

(a) Draw a Fischer projection of threonine.

(b) Draw a Fischer projection of a threonine diastereomer, and label its chirality centers as *R* or *S*.

26.2 Amino Acids and the Henderson–Hasselbalch Equation: Isoelectric Points

According to the Henderson–Hasselbalch equation **(Sections 20.3 and 24.5)**, if we know both the pH of a solution and the pK_a of an acid HA, we can calculate the ratio of $[A^-]$ to [HA] in the solution. Furthermore, when pH = pK_a, the two forms A^- and HA are present in equal amounts because log 1 = 0.

$$\text{pH} = pK_a + \log\frac{[A^-]}{[HA]} \quad \text{or} \quad \log\frac{[A^-]}{[HA]} = \text{pH} - pK_a$$

To apply the Henderson–Hasselbalch equation to an amino acid, let's find out what species are present in a 1.00 M solution of alanine at pH = 9.00. According to Table 26.1, protonated alanine $[^+H_3NCH(CH_3)CO_2H]$ has pK_{a1} = 2.34 and neutral zwitterionic alanine $[^+H_3NCH(CH_3)CO_2^-]$ has pK_{a2} = 9.69:

$$^+H_3NCH(CH_3)COOH + H_2O \rightleftharpoons {}^+H_3NCH(CH_3)CO_2^- + H_3O^+ \qquad pK_{a1} = 2.34$$

$$^+H_3NCH(CH_3)CO_2^- + H_2O \rightleftharpoons H_2NCH(CH_3)CO_2^- + H_3O^+ \qquad pK_{a2} = 9.69$$

Since the pH of the solution is much closer to pK_{a2} than to pK_{a1}, we need to use pK_{a2} for the calculation. From the Henderson–Hasselbalch equation, we have:

$$\log\frac{[A^-]}{[HA]} = \text{pH} - pK_a = 9.00 - 9.69 = -0.69$$

so

$$\frac{[A^-]}{[HA]} = \text{antilog}(-0.69) = 0.20 \quad \text{and} \quad [A^-] = 0.20[HA]$$

In addition, we know that

$$[A^-] + [HA] = 1.00\ M$$

Solving the two simultaneous equations gives [HA] = 0.83 and $[A^-]$ = 0.17. In other words, at pH = 9.00, 83% of alanine molecules in a 1.00 M solution are neutral (zwitterionic) and 17% are deprotonated. Similar calculations can be done at any other pH and the results plotted to give the titration curve shown in **Figure 26.1**.

Each leg of the titration curve is calculated separately. The first leg, from pH 1 to 6, corresponds to the dissociation of protonated alanine, H_2A^+. The second leg, from pH 6 to 11, corresponds to the dissociation of zwitterionic alanine, HA. It's as if we started with H_2A^+ at low pH and then titrated with NaOH. When 0.5 equivalent of NaOH is added, the deprotonation of H_2A^+ is 50% done; when 1.0 equivalent of NaOH is added, the deprotonation of H_2A^+ is complete and HA predominates; when 1.5 equivalents of NaOH is added, the deprotonation of HA is 50% done; and when 2.0 equivalents of NaOH is added, the deprotonation of HA is complete.

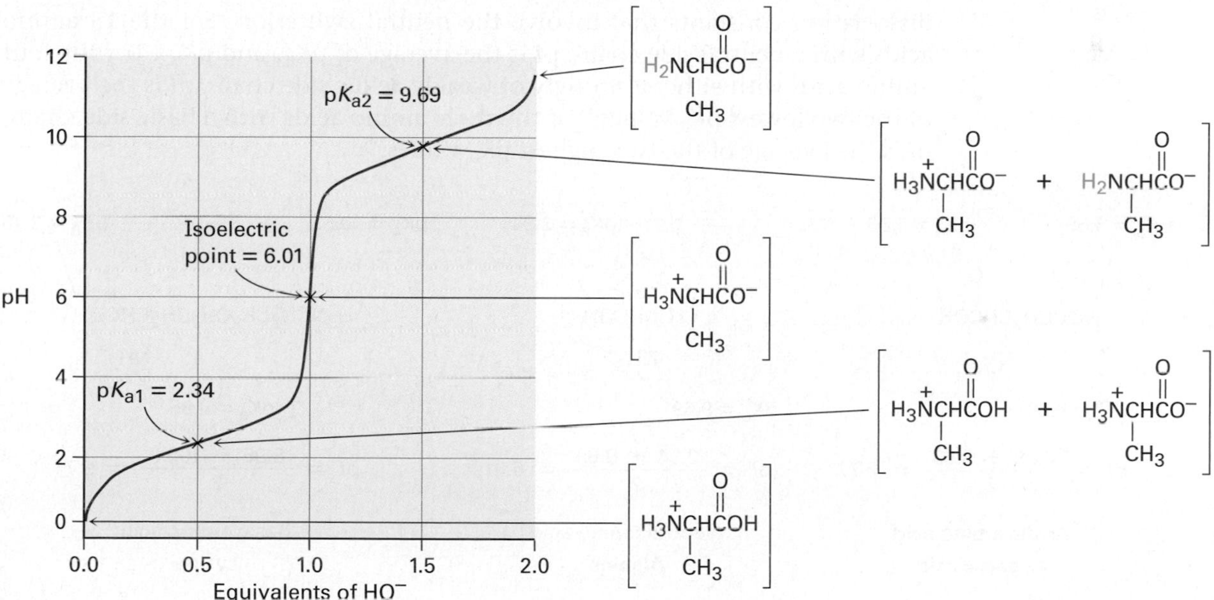

Figure 26.1 A titration curve for alanine, plotted using the Henderson–Hasselbalch equation. Each of the two legs is plotted separately. At pH < 1, alanine is entirely protonated; at pH = 2.34, alanine is a 50:50 mix of protonated and neutral forms; at pH = 6.01, alanine is entirely neutral; at pH = 9.69, alanine is a 50:50 mix of neutral and deprotonated forms; at pH > 11.5, alanine is entirely deprotonated.

Look carefully at the titration curve in Figure 26.1. In acid solution, the amino acid is protonated and exists primarily as a cation. In basic solution, the amino acid is deprotonated and exists primarily as an anion. In between the two is an intermediate pH at which the amino acid is exactly balanced between anionic and cationic forms and exists primarily as the neutral, dipolar zwitterion. This pH is called the amino acid's **isoelectric point (p*I*)** and has a value of 6.01 for alanine.

$$\overset{+}{H_3N}CH(R)COOH \underset{H_3O^+}{\overset{}{\rightleftharpoons}} \overset{+}{H_3N}CH(R)CO_2^- \underset{^-OH}{\rightleftharpoons} H_2NCH(R)CO_2^-$$

Low pH (protonated) — pH → — High pH (deprotonated)

Isoelectric point (neutral zwitterion)

The isoelectric point of an amino acid depends on its structure, with values for the 20 common amino acids given in Table 26.1. The 15 neutral amino acids have isoelectric points near neutrality, in the pH range 5.0 to 6.5. The two acidic amino acids have isoelectric points at lower pH so that deprotonation of the side-chain $-CO_2H$ does not occur at their p*I*, and the three basic amino acids have isoelectric points at higher pH so that protonation of the side-chain amino group does not occur at their p*I*.

More specifically, the p*I* of any amino acid is the average of the two acid-dissociation constants that involve the neutral zwitterion. For the 13 amino acids with a neutral side chain, p*I* is the average of pK_{a1} and pK_{a2}. For the four amino acids with either a strongly or weakly acidic side chain, p*I* is the average of the two lowest pK_a values. For the three amino acids with a basic side chain, p*I* is the average of the two highest pK_a values.

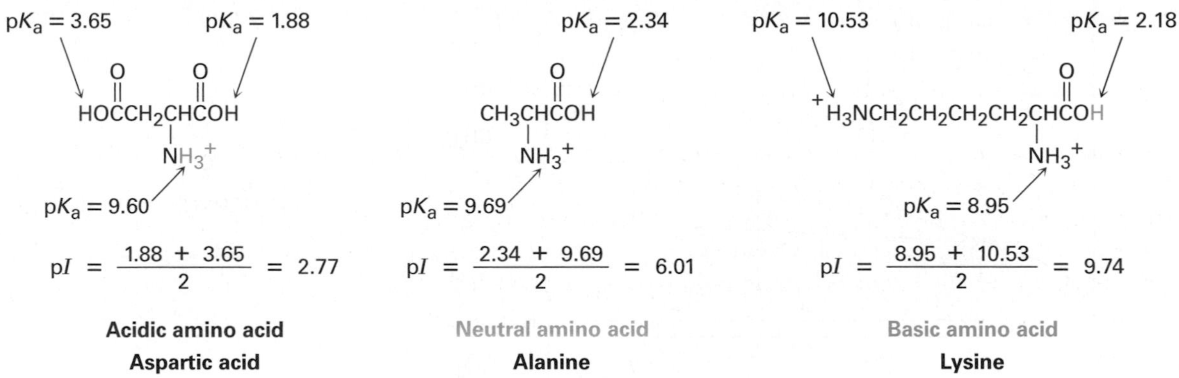

Just as individual amino acids have isoelectric points, proteins have an overall p*I* because of the cumulative effect of all the acidic or basic amino acids they may contain. The enzyme lysozyme, for instance, has a preponderance of basic amino acids and thus has a high isoelectric point (p*I* = 11.0). Pepsin, however, has a preponderance of acidic amino acids and a low isoelectric point (p*I* ~ 1.0). Not surprisingly, the solubilities and properties of proteins with different p*I*'s are strongly affected by the pH of the medium. Solubility in water is usually lowest at the isoelectric point, where the protein has no net charge, and is higher both above and below the p*I*, where the protein is charged.

We can take advantage of the differences in isoelectric points to separate a mixture of proteins into its pure constituents. Using a technique known as *electrophoresis,* a mixture of proteins is placed near the center of a strip of paper or gel. The paper or gel is moistened with an aqueous buffer of a given pH, and electrodes are connected to the ends of the strip. When an electric potential is applied, those proteins with negative charges (those that are deprotonated because the pH of the buffer is above their isoelectric point) migrate slowly toward the positive electrode. At the same time, those amino acids with positive charges (those that are protonated because the pH of the buffer is below their isoelectric point) migrate toward the negative electrode.

Different proteins migrate at different rates, depending on their isoelectric points and on the pH of the aqueous buffer, thereby effecting a separation of the mixture into its components. **Figure 26.2** illustrates this separation for a mixture containing basic, neutral, and acidic components.

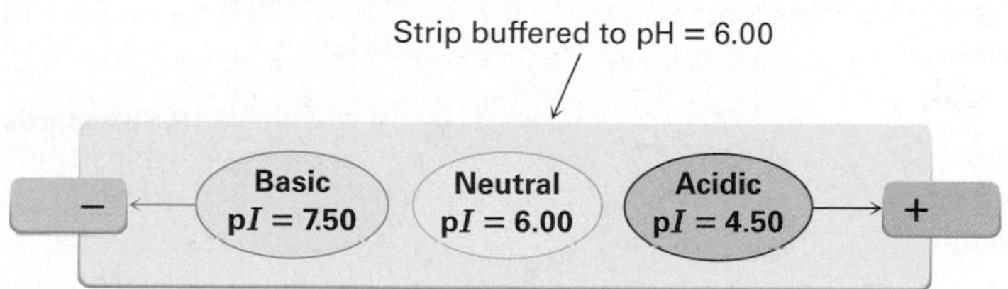

Figure 26.2 Separation of a protein mixture by electrophoresis. At pH = 6.00, a neutral protein does not migrate, a basic protein is protonated and migrates toward the negative electrode, and an acidic protein is deprotonated and migrates toward the positive electrode.

Problem 26.4

Hemoglobin has $pI = 6.8$. Does hemoglobin have a net negative charge or net positive charge at pH = 5.3? At pH = 7.3?

26.3 Synthesis of Amino Acids

α-Amino acids can be synthesized in the laboratory using some of the reactions discussed in previous chapters. One of the oldest methods of α-amino acid synthesis begins with α bromination of a carboxylic acid by treatment with Br_2 and PBr_3 (the Hell–Volhard–Zelinskii reaction; **Section 22.4**). S_N2 substitution of the α-bromo acid with ammonia then yields an α-amino acid.

$$\underset{\textbf{4-Methylpentanoic acid}}{(CH_3)_2CHCH_2CH_2COOH} \xrightarrow[2.\ H_2O]{1.\ Br_2,\ PBr_3} \underset{\textbf{2-Bromo-4-methylpentanoic acid}}{(CH_3)_2CHCH_2CH(Br)COOH} \xrightarrow[\text{(excess)}]{NH_3} \underset{\textbf{(R,S)-Leucine (45\%)}}{(CH_3)_2CHCH_2CH(\overset{+}{N}H_3)CO_2^-}$$

Problem 26.5

Show how you could prepare the following α-amino acids from the appropriate carboxylic acids:

(a) Phenylalanine **(b)** Valine

The Amidomalonate Synthesis

A more general method for preparation of α-amino acids is the *amidomalonate synthesis,* a straightforward extension of the malonic ester synthesis **(Section 22.7)**. The reaction begins with conversion of diethyl acetamidomalonate into an enolate ion by treatment with base, followed by S_N2 alkylation with a primary alkyl halide. Hydrolysis of both the amide protecting group and the esters occurs when the alkylated product is warmed with aqueous acid, and decarboxylation then takes place to yield an α-amino acid. For example, aspartic acid can be prepared from ethyl bromoacetate, $BrCH_2CO_2Et$:

$$\text{H–C(CO}_2\text{Et)}_2\text{–NHCOCH}_3 \xrightarrow[\text{2. BrCH}_2\text{CO}_2\text{Et}]{\text{1. Na}^+\ ^-\text{OEt}} \text{EtOCOCH}_2\text{–C(CO}_2\text{Et)}_2\text{–NHCOCH}_3 \xrightarrow[\text{Heat}]{\text{H}_3\text{O}^+} {}^-\text{O}_2\text{CCH}_2\text{CH(NH}_3^+\text{)CO}_2^-$$

Diethyl acetamidomalonate

(*R,S*)-Aspartic acid (55%)

Problem 26.6

What alkyl halides would you use to prepare the following α-amino acids by the amidomalonate method?

(a) Leucine **(b)** Histidine **(c)** Tryptophan **(d)** Methionine

---

Reductive Amination of α-Keto Acids

Yet another method for the synthesis of α-amino acids is by reductive amination of an α-keto acid with ammonia and a reducing agent. Alanine, for instance, is prepared by treatment of pyruvic acid with ammonia in the presence of $NaBH_4$. As described in **Section 24.6**, the reaction proceeds through formation of an intermediate imine that is then reduced.

$$\text{H}_3\text{C–CO–CO}_2\text{H} \xrightarrow[\text{NaBH}_4]{\text{NH}_3} \left[\text{H}_3\text{C–C(=NH)–CO}_2\text{H}\right] \longrightarrow \text{H}_3\text{C–CH(NH}_3^+\text{)–CO}_2^-$$

Pyruvic acid **Imine intermediate** **(*R,S*)-Alanine**

Enantioselective Synthesis

The synthesis of an α-amino acid from an achiral precursor by any of the methods just described yields a racemic mixture, with equal amounts of *S* and *R* enantiomers. To use an amino acid in the laboratory synthesis of a naturally occurring protein, however, the pure *S* enantiomer must be obtained.

Two methods are used in practice to obtain enantiomerically pure amino acids. One way is to resolve the racemic mixture into its pure enantiomers

(Section 5.8). A more direct approach, however, is to use an enantioselective synthesis to prepare only the desired *S* enantiomer directly. As discussed in the Chapter 19 *A Deeper Look,* the idea behind enantioselective synthesis is to find a chiral reaction catalyst that will temporarily hold a substrate molecule in an unsymmetrical, chiral environment. While in that chiral environment, the substrate may be more open to reaction on one side than on another, leading to an excess of one enantiomeric product over another.

William Knowles at the Monsanto Company discovered some years ago that α-amino acids can be prepared enantioselectively by hydrogenation of a *Z* enamido acid with a chiral hydrogenation catalyst. (*S*)-Phenylalanine, for instance, is prepared in 98.7% purity contaminated by only 1.3% of the (*R*) enantiomer when a chiral rhodium catalyst is used. For this discovery, Knowles shared the 2001 Nobel Prize in Chemistry.

H, CO_2H, C=C, $NHCOCH_3$ — 1. H_2, $[Rh(DiPAMP)(COD)]^+$ BF_4^- ; 2. NaOH, H_2O → CO_2^-, $H_3\overset{+}{N}$ H

A (*Z*) enamido acid **(*S*)-Phenylalanine**

The most effective catalysts for enantioselective amino acid synthesis are coordination complexes of rhodium(I) with 1,5-cyclooctadiene (COD) and a chiral diphosphine such as (*R,R*)-1,2-bis(*o*-anisylphenylphosphino)ethane, the so-called DiPAMP ligand. The complex owes its chirality to the presence of the trisubstituted phosphorus atoms **(Section 5.10)**.

Ph An, P, $\overset{+}{Rh}$, P, An Ph, BF_4^- [An = (2-methoxyphenyl) OCH_3]

$[Rh(R,R\text{-DiPAMP})(COD)]^+$ BF_4^-

Problem 26.7

Show how you could prepare the following amino acid enantioselectively:

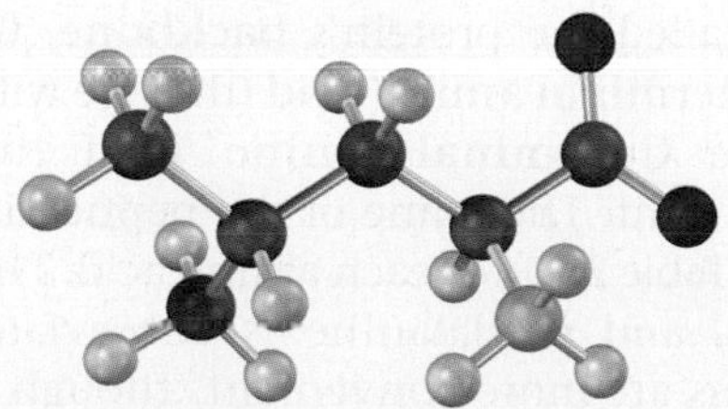

26.4 Peptides and Proteins

Proteins and peptides are amino acid polymers in which the individual amino acids, called **residues**, are linked together by amide bonds, or *peptide bonds*. An amino group from one residue forms an amide bond with the carboxyl of a second residue, the amino group of the second forms an amide bond with the carboxyl of a third, and so on. For example, alanylserine is the dipeptide that results when an amide bond forms between the alanine carboxyl and the serine amino group.

Alanine (Ala) + Serine (Ser) ⟶ Alanylserine (Ala-Ser)

Note that two dipeptides can result from reaction between alanine and serine, depending on which carboxyl group reacts with which amino group. If the alanine amino group reacts with the serine carboxyl, serylalanine results.

Serine (Ser) + Alanine (Ala) ⟶ Serylalanine (Ser-Ala)

The long, repetitive sequence of –N–CH–CO– atoms that makes up a continuous chain is called the protein's **backbone**. By convention, peptides are written with the **N-terminal amino acid** (the one with the free $-NH_3^+$ group) on the left and the **C-terminal amino acid** (the one with the free $-CO_2^-$ group) on the right. The name of the peptide is indicated by using the abbreviations listed in Table 26.1 for each amino acid. Thus, alanylserine is abbreviated Ala-Ser or A-S, and serylalanine is abbreviated Ser-Ala or S-A. The one-letter abbreviations are more convenient, though less immediately recognizable, than the three-letter abbreviations.

The amide bond that links different amino acids together in peptides is no different from any other amide bond **(Section 24.3)**. An amide nitrogen is nonbasic because its unshared electron pair is delocalized by interaction with the carbonyl group. This overlap of the nitrogen *p* orbital with the *p* orbitals of the carbonyl group imparts a certain amount of double-bond character to the C–N bond and restricts rotation around it. The amide bond is therefore planar, and the N–H is oriented 180° to the C=O.

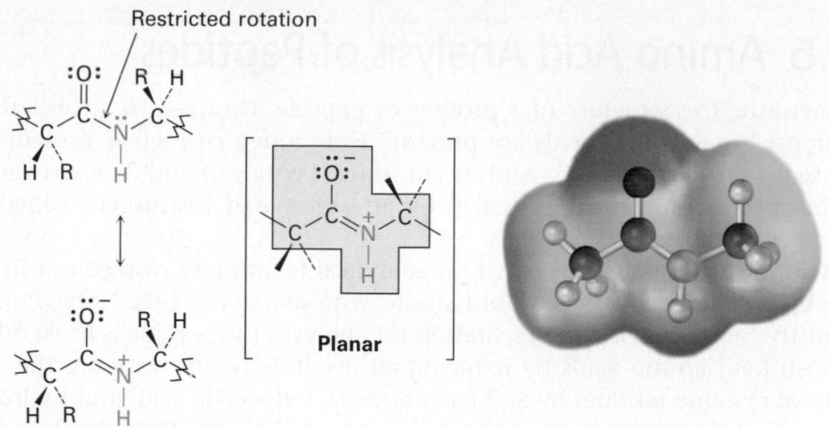

A second kind of covalent bonding in peptides occurs when a disulfide linkage, RS–SR, is formed between two cysteine residues. As we saw in **Section 18.8**, a disulfide is formed by mild oxidation of a thiol, RSH, and is cleaved by mild reduction.

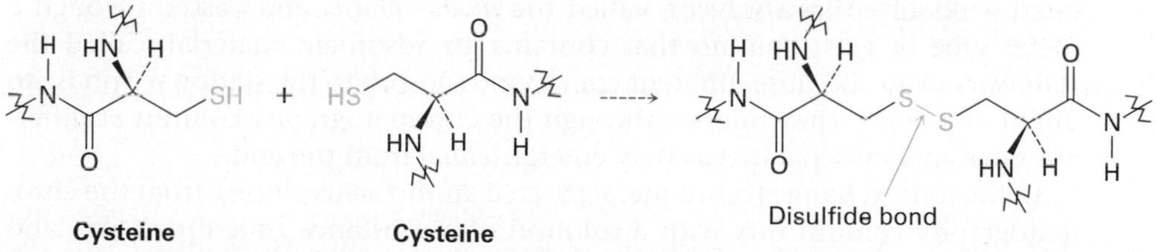

Cysteine + Cysteine ----→ (Disulfide bond)

A disulfide bond between cysteine residues in different peptide chains links the otherwise separate chains together, while a disulfide bond between cysteine residues in the same chain forms a loop. Insulin, for instance, is composed of two chains that total 51 amino acids and are linked by two cysteine disulfide bridges.

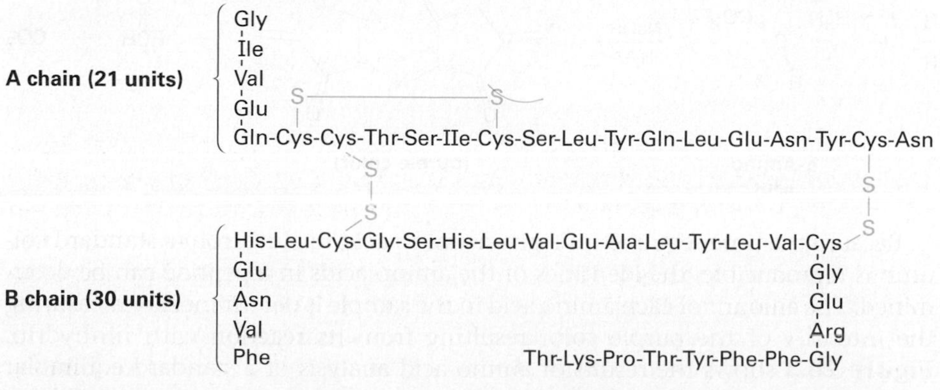

A chain (21 units)

Gly-Ile-Val-Glu-Gln-Cys-Cys-Thr-Ser-Ile-Cys-Ser-Leu-Tyr-Gln-Leu-Glu-Asn-Tyr-Cys-Asn

B chain (30 units)

Phe-Val-Asn-Glu-His-Leu-Cys-Gly-Ser-His-Leu-Val-Glu-Ala-Leu-Tyr-Leu-Val-Cys-Gly-Glu-Arg-Gly-Phe-Phe-Tyr-Thr-Pro-Lys-Thr

Insulin

Problem 26.8

There are six isomeric tripeptides that contain valine, tyrosine, and glycine. Name them using both three- and one-letter abbreviations.

Problem 26.9

Draw the structure of M-P-V-G, and indicate the amide bonds.

26.5 Amino Acid Analysis of Peptides

To determine the structure of a protein or peptide, we need to answer three questions: What amino acids are present? How much of each is present? In what sequence do the amino acids occur in the peptide chain? The answers to the first two questions are provided by an automated instrument called an amino acid analyzer.

An amino acid analyzer is based on analytical techniques worked out in the 1950s by William Stein and Stanford Moore, who shared the 1972 Nobel Prize in Chemistry for their work. In preparation for analysis, the peptide is broken into its constituent amino acids by reducing all disulfide bonds, capping the –SH groups of cysteine residues by S_N2 reaction with iodoacetic acid, and hydrolyzing the amide bonds by heating with aqueous 6 M HCl at 110 °C for 24 hours. The resultant amino acid mixture is then separated into its components by a technique called *chromatography,* either high-pressure liquid chromatography (HPLC) or ion-exchange chromatography.

In both HPLC and ion-exchange chromatography, the mixture to be separated is dissolved in a solvent, called the *mobile phase,* and passed through a metal tube or glass column that contains an adsorbent material, called the *stationary phase.* Because different compounds adsorb to the stationary phase to different extents, they migrate through the chromatography column at different rates and are separated as they emerge *(elute)* from the end.

In the ion-exchange technique, separated amino acids eluting from the chromatography column mix with a solution of a substance called *ninhydrin* and undergo a rapid reaction that produces an intense purple color. The color is detected by a spectrometer, and a plot of elution time versus spectrometer absorbance is obtained.

Ninhydrin + α-Amino acid ($H_3\overset{+}{N}$–CH(R)–CO_2^-) $\xrightarrow[H_2O]{NaOH}$ (purple color) + RCHO + CO_2

Ninhydrin **α-Amino acid** **(purple color)**

Because the time required for a given amino acid to elute from a standard column is reproducible, the identities of the amino acids in a peptide can be determined. The amount of each amino acid in the sample is determined by measuring the intensity of the purple color resulting from its reaction with ninhydrin. **Figure 26.3** shows the results of amino acid analysis of a standard equimolar mixture of 17 α-amino acids. Typically, amino acid analysis requires about 100 picomoles (2–3 μg) of sample for a protein containing about 200 residues.

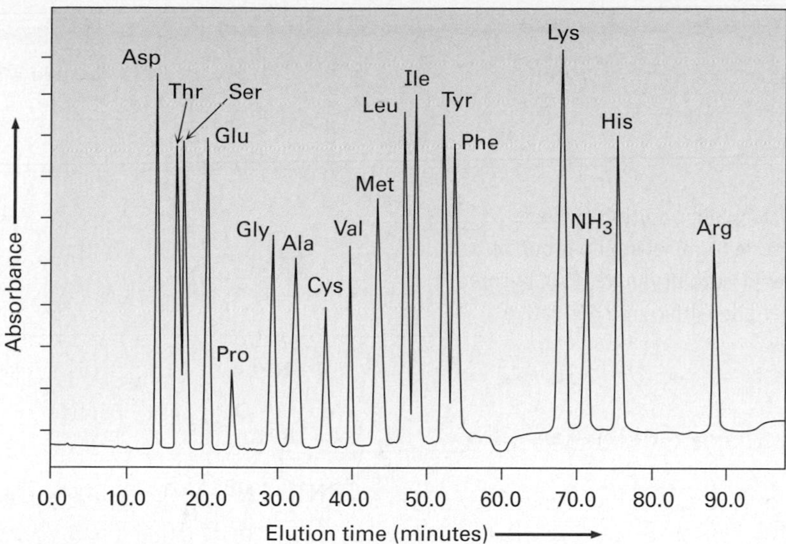

Figure 26.3 Amino acid analysis of an equimolar mixture of 17 amino acids.

Problem 26.10
Show the structure of the product you would expect to obtain by S_N2 reaction of a cysteine residue with iodoacetic acid.

Problem 26.11
Show the structures of the products obtained on reaction of valine with ninhydrin.

26.6 Peptide Sequencing: The Edman Degradation

With the identities and relative amounts of amino acids known, the peptide is then sequenced to find out in what order the amino acids are linked together. Much peptide sequencing is now done by mass spectrometry, using either electrospray ionization (ESI) or matrix-assisted laser desorption ionization (MALDI) linked to a time-of-flight (TOF) mass analyzer, as described in **Section 12.4**. Also in common use is a chemical method of peptide sequencing called the Edman degradation.

The general idea of peptide sequencing by Edman degradation is to cleave one amino acid at a time from an end of the peptide chain. That terminal amino acid is then separated and identified, and the cleavage reactions are repeated on the chain-shortened peptide until the entire peptide sequence is known. Automated protein sequencers are available that allow as many as 50 repetitive sequencing cycles to be carried out before a buildup of unwanted by-products interferes with the results. So efficient are these instruments that sequence information can be obtained from as little as 1 to 5 picomoles of sample—less than 0.1 μg.

As shown in **Figure 26.4**, **Edman degradation** involves treatment of a peptide with phenyl isothiocyanate (PITC), $C_6H_5{-}N{=}C{=}S$, followed by reaction with trifluoroacetic acid. The first step attaches the PITC to the $-NH_2$ group of the N-terminal amino acid, and the second step splits the N-terminal residue from the peptide chain, yielding an anilinothiazolinone (ATZ) derivative plus the

Figure 26.4 | MECHANISM

Mechanism of the Edman degradation for N-terminal analysis of peptides.

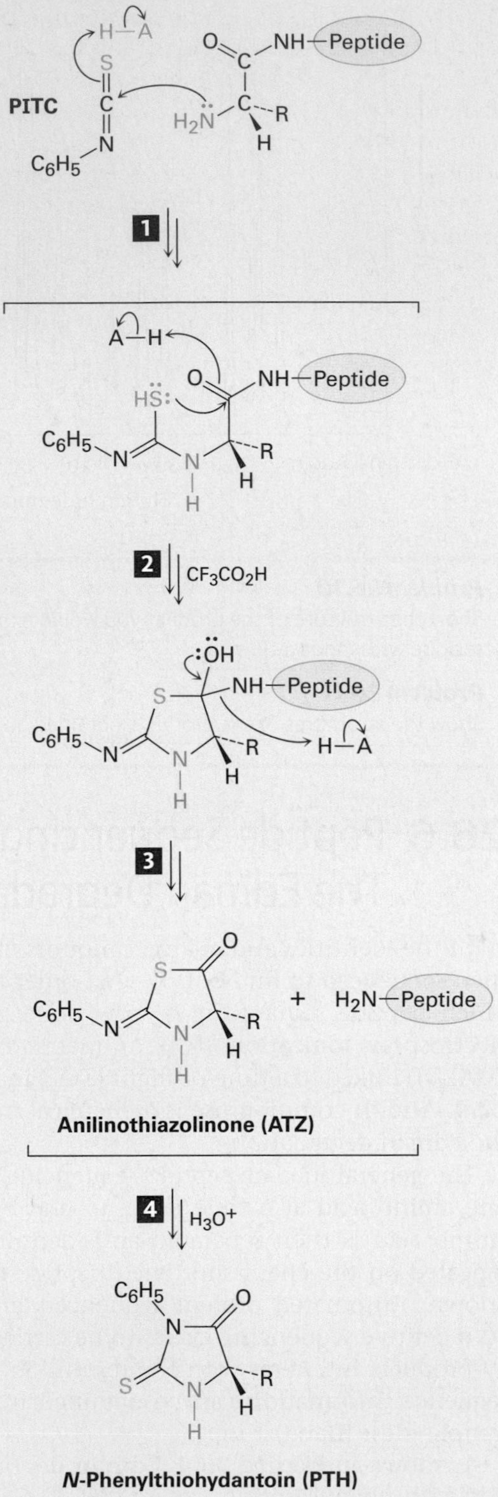

1 Nucleophilic addition of the peptide terminal amino group to phenyl isothiocyanate (PITC) gives an *N*-phenylthiourea derivative.

2 Acid-catalyzed cyclization of the phenylthiourea yields a tetrahedral intermediate . . .

3 . . . which expels the chain-shortened peptide and forms an anilinothiazolinone (ATZ) derivative.

4 The ATZ rearranges in the presence of aqueous acid to an isomeric *N*-phenylthiohydantoin (PTH) as the final product.

chain-shortened peptide. Further acid-catalyzed rearrangement of the ATZ derivative with aqueous acid converts it into a phenylthiohydantoin (PTH), which is identified by comparison of its elution time with the known elution times of PTH derivatives of the 20 common amino acids. The chain-shortened peptide is then automatically resubmitted to another round of Edman degradation.

Complete sequencing of large proteins by Edman degradation is impractical because of the buildup of unwanted by-products. To get around the problem, a large peptide chain is first cleaved by partial hydrolysis into a number of smaller fragments, the sequence of each fragment is determined, and the individual fragments are fitted together by matching the overlapping ends. In this way, protein chains with more than 400 amino acids have been sequenced.

Partial hydrolysis of a peptide can be carried out either chemically with aqueous acid or enzymatically. Acid hydrolysis is unselective and gives a more-or-less random mixture of small fragments, but enzymatic hydrolysis is quite specific. The enzyme trypsin, for instance, catalyzes hydrolysis of peptides only at the carboxyl side of the basic amino acids arginine and lysine; chymotrypsin cleaves only at the carboxyl side of the aryl-substituted amino acids phenylalanine, tyrosine, and tryptophan.

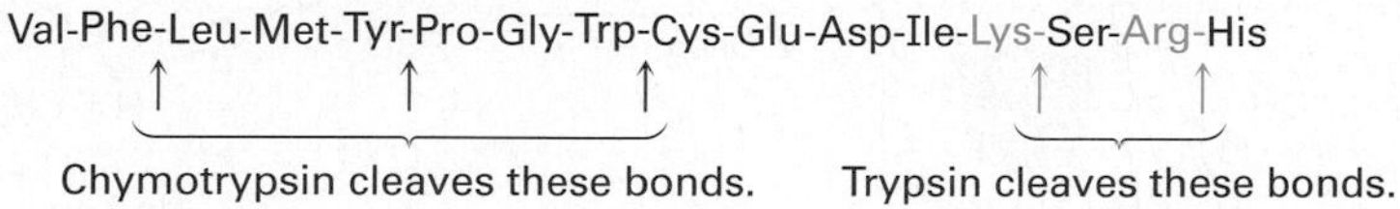

Problem 26.12

The octapeptide angiotensin II has the sequence Asp-Arg-Val-Tyr-Ile-His-Pro-Phe. What fragments would result if angiotensin II were cleaved with trypsin? With chymotrypsin?

Problem 26.13

What is the N-terminal residue on a peptide that gives the following PTH derivative on Edman degradation?

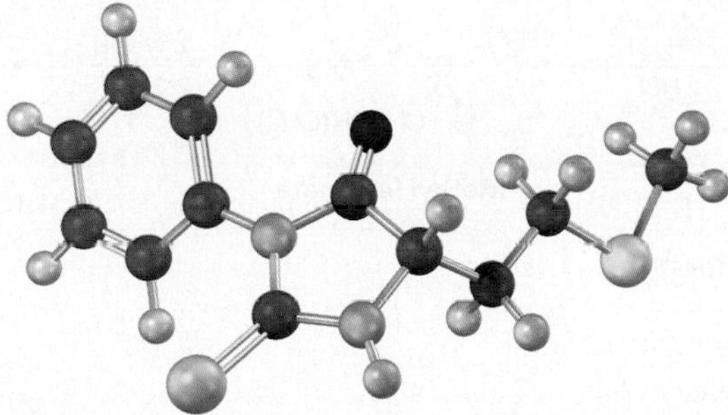

Problem 26.14

Draw the structure of the PTH derivative that would be formed on Edman degradation of angiotensin II (Problem 26.12).

Problem 26.15

Give the amino acid sequence of hexapeptides that produce the following sets of fragments on partial acid hydrolysis:

(a) Arg, Gly, Ile, Leu, Pro, Val gives Pro-Leu-Gly, Arg-Pro, Gly-Ile-Val

(b) N, L, M, W, V_2 gives V-L, V-M-W, W-N-V

26.7 Peptide Synthesis

Once the structure of a peptide is known, its synthesis can then be undertaken—perhaps to obtain a larger amount for biological evaluation. A simple amide might be formed by treating an amine and a carboxylic acid with dicyclohexylcarbodiimide (DCC; **Section 21.7**), but peptide synthesis is a more difficult problem because many different amide bonds must be formed in a specific order rather than at random.

The solution to the specificity problem is protection **(Section 17.8)**. If we want to couple alanine with leucine to synthesize Ala-Leu, for instance, we could protect the $-NH_2$ group of alanine and the $-CO_2H$ group of leucine to shield them from reacting, then form the desired Ala-Leu amide bond by reaction with DCC, and then remove the protecting groups.

Alanine → Protect $-NH_2$ → RHN–CH(CH₃)–CO₂H
Leucine → Protect $-CO_2H$ → H₂N–CH(CH₂CH(CH₃)₂)–CO–O–R
1. DCC (form amide) 2. Deprotect → Ala-Leu

Many different amino- and carboxyl-protecting groups have been devised, but only a few are widely used. Carboxyl groups are often protected simply by converting them into methyl or benzyl esters. Both groups are easily introduced by standard methods of ester formation **(Section 21.6)** and are easily removed by mild hydrolysis with aqueous NaOH. Benzyl esters can also be cleaved by catalytic *hydrogenolysis* of the weak benzylic C–O bond ($RCO_2–CH_2Ph + H_2 \rightarrow RCO_2H + PhCH_3$).

Leucine → CH_3OH, HCl → Methyl leucinate → 1. NaOH 2. H_3O^+ → Leucine
Leucine → $PhCH_2OH$, HCl → Benzyl leucinate → H_2/Pd → Leucine

Amino groups are often protected as their *tert*-butyloxycarbonyl amide (Boc) or fluorenylmethyloxycarbonyl amide (Fmoc) derivatives. The Boc protecting group is introduced by reaction of the amino acid with di-*tert*-butyl dicarbonate in a nucleophilic acyl substitution reaction and is removed by brief treatment with a strong acid such as trifluoroacetic acid, CF_3CO_2H. The Fmoc protecting group is introduced by reaction with an acid chloride and is removed by treatment with base.

Alanine —Di-*tert*-butyl dicarbonate, $(CH_3CH_2)_3N$→ Boc-Ala

Alanine —Fluorenylmethyloxycarbonyl chloride, $(CH_3CH_2)_3N$→ Fmoc-Ala

Thus, five steps are needed to synthesize a dipeptide such as Ala-Leu:

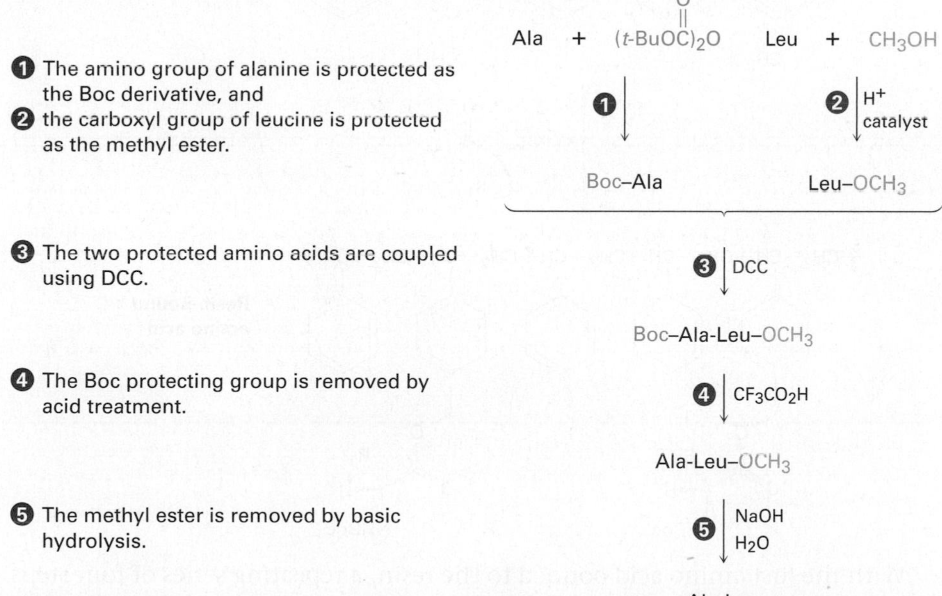

These steps can be repeated to add one amino acid at a time to the growing chain or to link two peptide chains together. Many remarkable achievements in peptide synthesis have been reported, including a complete synthesis of human insulin. Insulin is composed of two chains totaling 51 amino acids linked by two disulfide bridges. Its structure, shown previously on page 1057, was determined by Frederick Sanger, who received the 1958 Nobel Prize in Chemistry for his work.

Problem 26.16

Show the mechanism for formation of a Boc derivative by reaction of an amino acid with di-*tert*-butyl dicarbonate.

Problem 26.17

Write all five steps required for the synthesis of Leu-Ala from alanine and leucine.

26.8 Automated Peptide Synthesis: The Merrifield Solid-Phase Method

As you might imagine, the synthesis of a large peptide chain by sequential addition of one amino acid at a time is a long and arduous process. An immense simplification is possible, however, using methods introduced by R. Bruce Merrifield, who received the 1984 Nobel Prize in Chemistry for his work. In the Merrifield solid-phase method, peptide synthesis is carried out with the growing amino acid chain covalently bonded to small beads of a polymer resin rather than in solution.

In the original procedure, polystyrene resin was used, prepared so that 1 of every 100 or so benzene rings contained a chloromethyl ($-CH_2Cl$) group. A Boc-protected C-terminal amino acid was then attached to the resin through an ester bond formed by S_N2 reaction.

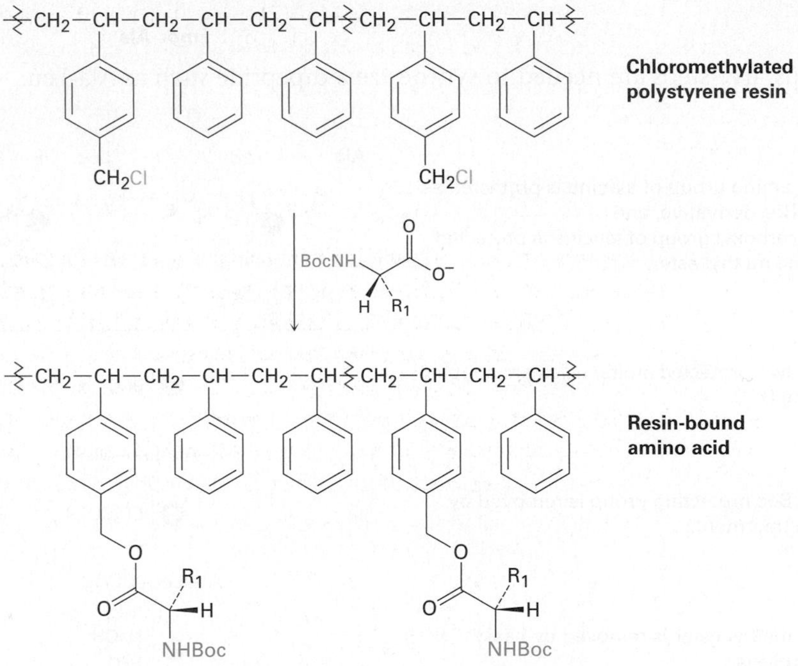

$-CH_2-CH-CH_2-CH-CH_2-CH-CH_2-CH-CH_2-CH-$ (rings; two bearing CH_2Cl)

Chloromethylated polystyrene resin

↓ BocNH–CH(R_1)–CO_2^-

Resin-bound amino acid

(rings bearing CH_2O–C(=O)–CH(R_1)–NHBoc)

With the first amino acid bonded to the resin, a repeating series of four steps is then carried out to build a peptide.

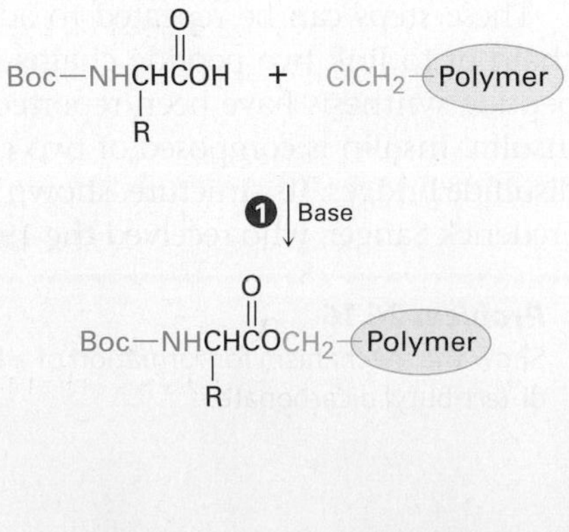

❶ A Boc-protected amino acid is covalently linked to the polystyrene polymer by formation of an ester bond (S_N2 reaction).

$$\text{Boc—NHCH(R)COOH} + \text{ClCH}_2\text{—Polymer} \xrightarrow{\text{Base}} \text{Boc—NHCH(R)COOCH}_2\text{—Polymer}$$

2. The polymer-bonded amino acid is washed free of excess reagent and then treated with trifluoroacetic acid to remove the Boc group.

3. A second Boc-protected amino acid is coupled to the first by reaction with DCC. Excess reagents are removed by washing them from the insoluble polymer.

4. The cycle of deprotection, coupling, and washing is repeated as many times as desired to add amino acid units to the growing chain.

5. After the desired peptide has been made, treatment with anhydrous HF removes the final Boc group and cleaves the ester bond to the polymer, yielding the free peptide.

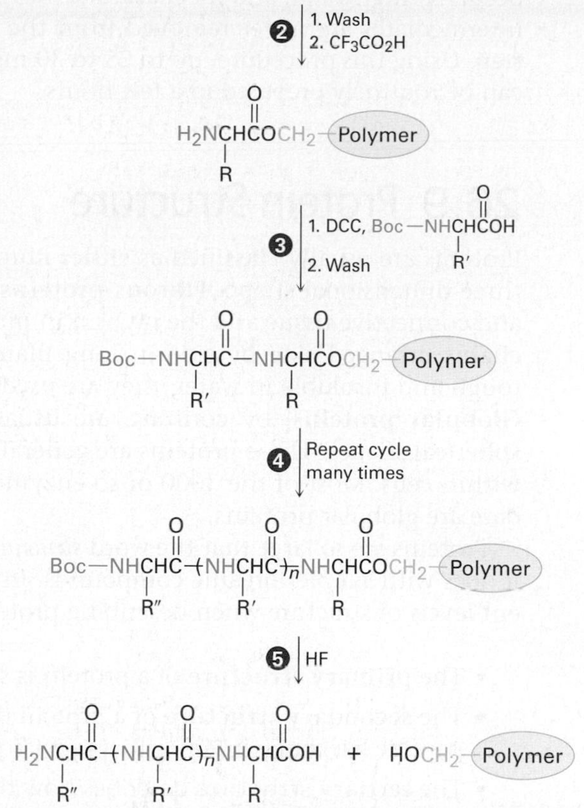

2: 1. Wash; 2. CF_3CO_2H → $H_2NCH(R)COCH_2$–Polymer

3: 1. DCC, Boc–NHCH(R′)COOH; 2. Wash → Boc–NHCH(R′)C(=O)–NHCH(R)COCH_2–Polymer

4: Repeat cycle many times → Boc–NHCH(R″)C(=O)–(NHCH(R′)C(=O))$_n$NHCH(R)COCH_2–Polymer

5: HF → $H_2NCH(R″)C(=O)$–(NHCH(R′)C(=O))$_n$NHCH(R)COOH + $HOCH_2$–Polymer

The steps in the solid-phase procedure have been improved substantially over the years, but the fundamental idea remains the same. The most commonly used resins at present are either the Wang resin or the PAM (phenylacetamidomethyl) resin, and the most commonly used N-protecting group is the Fmoc group rather than Boc.

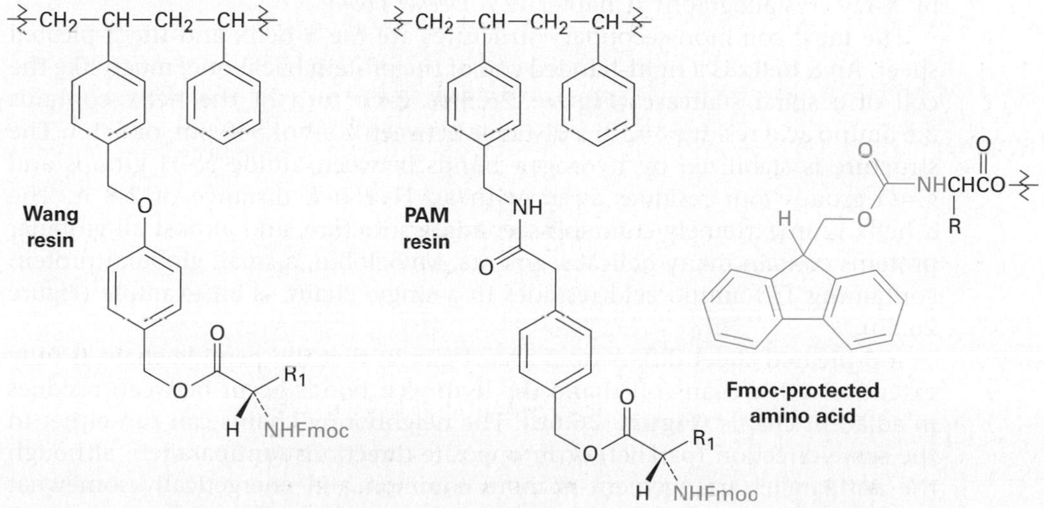

Wang resin

PAM resin

Fmoc-protected amino acid

Robotic peptide synthesizers are now used to automatically repeat the coupling, washing, and deprotection steps with different amino acids. Each step occurs in high yield, and mechanical losses are minimized because the peptide intermediates are never removed from the insoluble polymer until the final step. Using this procedure, up to 25 to 30 mg of a peptide with 20 amino acids can be routinely prepared in a few hours.

26.9 Protein Structure

Proteins are usually classified as either fibrous or globular, according to their three-dimensional shape. **Fibrous proteins**, such as the collagen in tendons and connective tissue and the myosin in muscle tissue, consist of polypeptide chains arranged side by side in long filaments. Because these proteins are tough and insoluble in water, they are used in nature for structural materials. **Globular proteins**, by contrast, are usually coiled into compact, roughly spherical shapes. These proteins are generally soluble in water and are mobile within cells. Most of the 3000 or so enzymes that have been characterized to date are globular proteins.

Proteins are so large that the word *structure* takes on a broader meaning than it does with simpler organic compounds. In fact, chemists speak of four different levels of structure when describing proteins.

- The **primary structure** of a protein is simply the amino acid sequence.
- The **secondary structure** of a protein describes how *segments* of the peptide backbone orient into a regular pattern.
- The **tertiary structure** describes how the *entire* protein molecule coils into an overall three-dimensional shape.
- The **quaternary structure** describes how different protein molecules come together to yield large aggregate structures.

Primary structure is determined, as we've seen, by sequencing the protein. Secondary, tertiary, and quaternary structures are determined either by NMR or by X-ray crystallography (Chapter 12 *A Deeper Look*).

The most common secondary structures are the α helix and the β-pleated sheet. An **α helix** is a right-handed coil of the protein backbone, much like the coil of a spiral staircase **(Figure 26.5a)**. Each turn of the helix contains 3.6 amino acid residues, with a distance between coils of 540 pm, or 5.4 Å. The structure is stabilized by hydrogen bonds between amide N–H groups and C=O groups four residues away, with an N–H····O distance of 2.8 Å. The α helix is an extremely common secondary structure, and almost all globular proteins contain many helical segments. Myoglobin, a small globular protein containing 153 amino acid residues in a single chain, is an example (Figure 26.5b).

A **β-pleated sheet** differs from an α helix in that the peptide chain is fully extended rather than coiled and the hydrogen bonds occur between residues in adjacent chains **(Figure 26.6a)**. The neighboring chains can run either in the same direction (parallel) or in opposite directions (antiparallel), although the antiparallel arrangement is more common and energetically somewhat more favorable. Concanavalin A, for instance, consists of two identical chains of 237 residues, with extensive regions of antiparallel β sheets (Figure 26.6b).

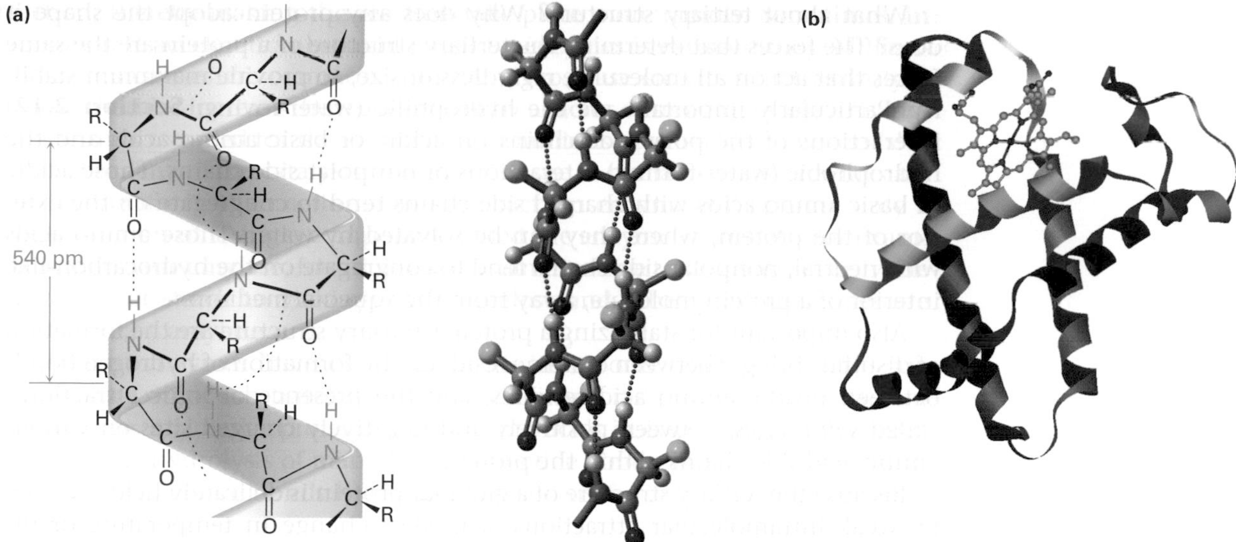

Figure 26.5 **(a)** The α-helical secondary structure of proteins is stabilized by hydrogen bonds between the N—H group of one residue and the C=O group four residues away. **(b)** The structure of myoglobin, a globular protein with extensive helical regions that are shown as coiled ribbons in this representation.

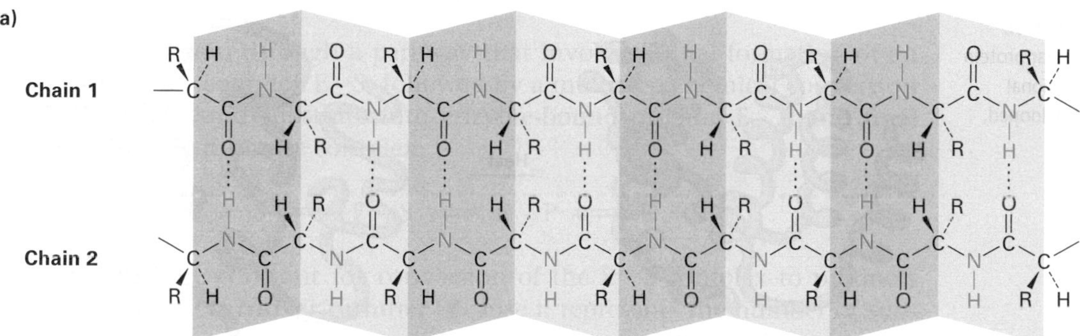

(b)

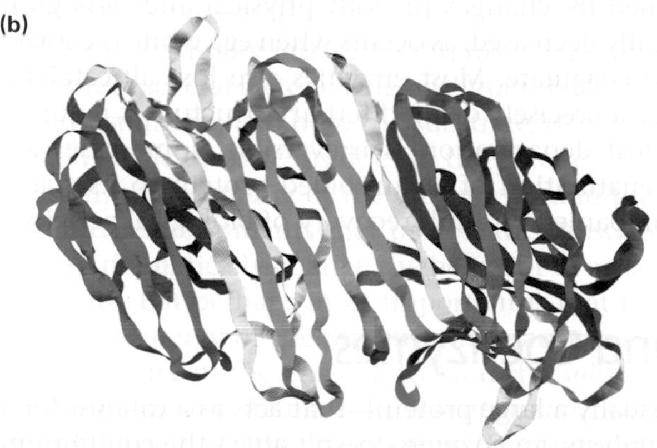

Figure 26.6 **(a)** The β-pleated sheet secondary structure of proteins is stabilized by hydrogen bonds between parallel or antiparallel chains. **(b)** The structure of concanavalin A, a protein with extensive regions of antiparallel β sheets, shown as flat ribbons.

The Protein Data Bank | A DEEPER LOOK

Enzymes are so large, so structurally complex, and so numerous that the use of computer databases and molecular visualization programs has become an essential tool for studying biological chemistry. Of the various databases available online, the Kyoto Encyclopedia of Genes and Genomes (KEGG) database (http://www.genome.ad.jp/kegg), maintained by the Kanehisa Laboratory of Kyoto University Bioinformatics Center, is useful for obtaining information on biosynthetic pathways of the sort we'll be describing in Chapter 29. For obtaining information on a specific enzyme, the BRENDA database (http://www.brenda-enzymes.org/), maintained by the Institute of Biochemistry at the University of Cologne, Germany, is particularly valuable.

Perhaps the most useful of all biological databases is the Protein Data Bank (PDB), operated by the Research Collaboratory for Structural Bioinformatics (RCSB). The PDB is a worldwide repository of X-ray and NMR structural data for biological macromolecules. In mid-2010, data for more than 66,000 structures were available, and more than 6000 new ones were being added yearly. To access the Protein Data Bank, go to http://www.rcsb.org/pdb/ and a home page like that shown in **Figure 26.11** will appear. As with much that is available online, however, the PDB site is changing rapidly, so you may not see quite the same thing.

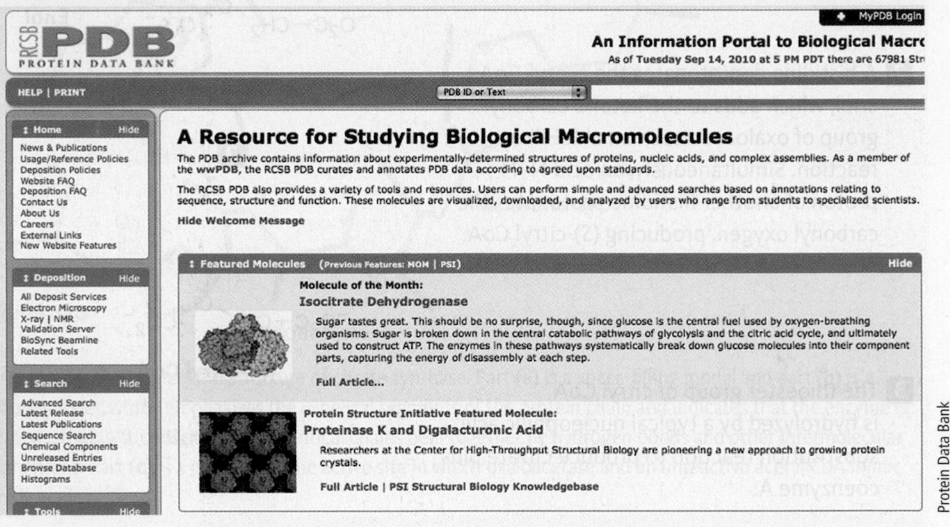

Figure 26.11 The Protein Data Bank home page.

Protein Data Bank

To learn how to use the PDB, begin by running the short tutorial listed under Getting Started at the bottom of the page. After that introduction, start exploring. Let's say you want to view citrate synthase, the enzyme that catalyzes the addition of acetyl CoA to oxaloacetate to give citrate. Type "citrate synthase" (with quotation marks) into the small search box on the top line, click on "Search," and a list of 42 or so structures will appear. Scroll down near the end of the list until you find the entry with a PDB code of 5CTS and the title "Proposed Mechanism for the Condensation Reaction of Citrate Synthase: 1.9 Angstroms Structure of the Ternary Complex with Oxaloacetate and Carboxymethyl Coenzyme A." Alternatively, if you know the code of the enzyme you want, you can enter it directly into the search box. Click on the PDB code of entry 5CTS, and a new page containing information about the enzyme will open.

(continued)

(continued)

If you choose, you can download the structure file to your computer and open it with any of numerous molecular graphics programs to see an image like that in **Figure 26.12**. The biologically active molecule is a dimer of two identical subunits consisting primarily of α-helical regions displayed as coiled ribbons. For now, just click on "View in Jmol" under the enzyme image on the right side of the screen to see some of the options for visualizing and further exploring the enzyme.

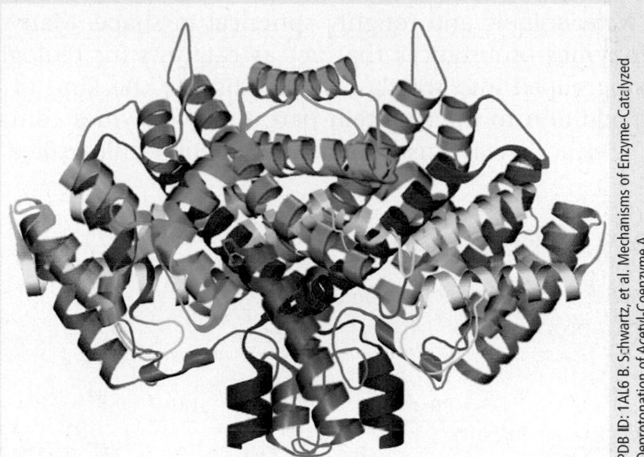

PDB ID: 1AL6 B. Schwartz, et al. Mechanisms of Enzyme-Catalyzed Deprotonation of Acetyl-Coenzyme A

Figure 26.12 An image of citrate synthase, downloaded from the Protein Data Bank.

Summary

Proteins and **peptides** are large biomolecules made of **α-amino acid residues** linked together by amide, or peptide, bonds. Twenty amino acids are commonly found in proteins, and all except glycine have stereochemistry similar to that of L sugars. In neutral solution, amino acids exist as dipolar **zwitterions**.

Amino acids can be synthesized in racemic form by several methods, including ammonolysis of an α-bromo acid, alkylation of diethyl acetamidomalonate, and reductive amination of an α-keto acid. Alternatively, an enantioselective synthesis of amino acids can be carried out using a chiral hydrogenation catalyst.

Determining the structure of a peptide or protein begins with amino acid analysis. The peptide is hydrolyzed to its constituent α-amino acids, which are separated and identified. Next, the peptide is sequenced. **Edman degradation** by treatment with phenyl isothiocyanate (PITC) cleaves one residue from the N terminus of the peptide and forms an easily identifiable phenylthiohydantoin (PTH) derivative of the **N-terminal amino acid**. An automated series of Edman degradations can sequence peptide chains up to 50 residues in length.

Peptide synthesis involves the use of protecting groups. An N-protected amino acid with a free $-CO_2H$ group is coupled using DCC to an O-protected amino acid with a free $-NH_2$ group. Amide formation occurs, the protecting groups are removed, and the sequence is repeated. Amines are usually protected as their *tert*-butyloxycarbonyl (Boc) or fluorenylmethyloxycarbonyl (Fmoc) derivatives; acids are usually protected as esters. The synthesis is often carried out by the Merrifield solid-phase method, in which the peptide is bonded to insoluble polymer beads.

Key words

α-amino acid, 1045
α helix, 1066
backbone, 1056
β-pleated sheet, 1066
C-terminal amino acid, 1056
coenzyme, 1071
cofactor, 1071
denatured, 1068
Edman degradation, 1059
enzyme, 1068
fibrous protein, 1066
globular protein, 1066
isoelectric point, (p*I*), 1052
N-terminal amino acid, 1056
peptide, 1044
primary structure, 1066
protein, 1044
quaternary structure, 1066
residue, 1056
secondary structure, 1066
side chain, 1048
tertiary structure, 1066
turnover number, 1069
zwitterion, 1045

Proteins have four levels of structure. **Primary structure** describes a protein's amino acid sequence; **secondary structure** describes how segments of the protein chain orient into regular patterns—either **α-helix** or **β-pleated sheet**; **tertiary structure** describes how the entire protein molecule coils into an overall three-dimensional shape; and **quaternary structure** describes how individual protein molecules aggregate into larger structures.

Proteins are classified as either globular or fibrous. **Fibrous proteins** such as α-keratin are tough, rigid, and water-insoluble; **globular proteins** such as myoglobin are water-soluble and roughly spherical in shape. Many globular proteins are **enzymes**—substances that act as catalysts for biological reactions. Enzymes are grouped into six classes according to the kind of reaction they catalyze. In addition to their protein part, many enzymes contain **cofactors**, which can be either metal ions or small organic molecules called **coenzymes**.

Summary of Reactions

1. Amino acid synthesis (Section 26.3)
   (a) From α-bromo acids

$$RCH_2CO_2H \xrightarrow[2.\ H_2O]{1.\ Br_2,\ PBr_3} RCHBrCO_2H \xrightarrow{NH_3} RCH(\overset{+}{N}H_3)CO_2^-$$

   (b) Diethyl acetamidomalonate synthesis

$$H_3CCONHCH(CO_2Et)_2 \xrightarrow[3.\ H_3O^+]{1.\ Na^+\ {}^-OEt;\ 2.\ RX} RCH(\overset{+}{N}H_3)CO_2^-$$

   (c) Reductive amination of an α-keto acid

$$RCOCO_2H \xrightarrow[NaBH_4]{NH_3} RCH(\overset{+}{N}H_3)CO_2^-$$

   (d) Enantioselective synthesis

$$RCH{=}C(CO_2H)NHCOCH_3 \xrightarrow[2.\ NaOH,\ H_2O]{1.\ H_2,\ [Rh(DiPAMP)(COD)]^+\ BF_4^-} RCH_2CH(\overset{+}{N}H_3)CO_2^-$$

A (*Z*) enamido acid → **An (*S*)-amino acid**

2. Peptide sequencing by Edman degradation (Section 26.6)

$$C_6H_5N{=}C{=}S + H_2N{-}CHR{-}CO{-}NH{-}\text{Peptide} \longrightarrow \text{phenylthiohydantoin } (C_6H_5,\ R) + H_2N{-}\text{Peptide}$$

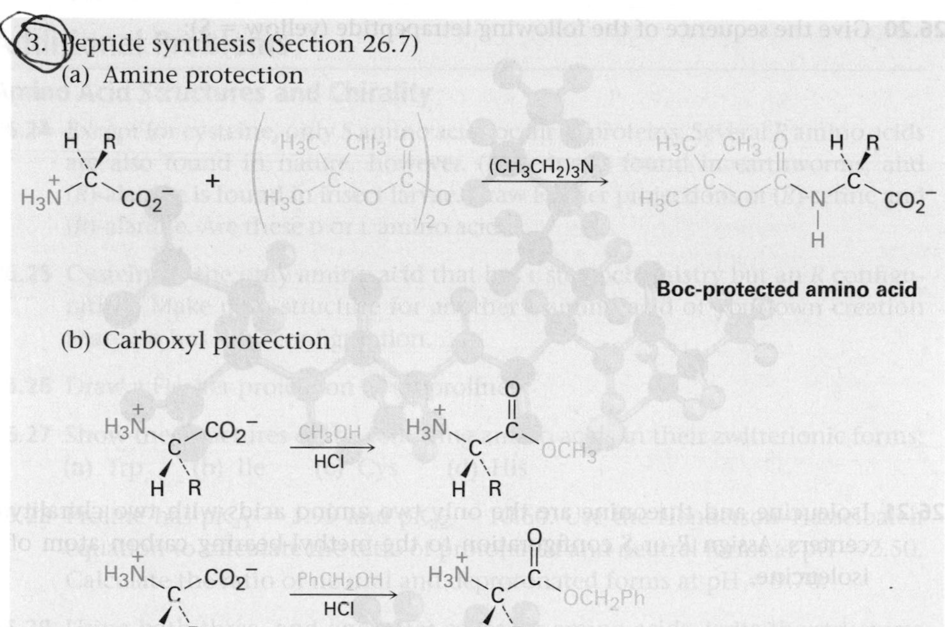

Exercises

Visualizing Chemistry

(Problems 26.1–26.18 appear within the chapter.)

26.19 Identify the following amino acids:

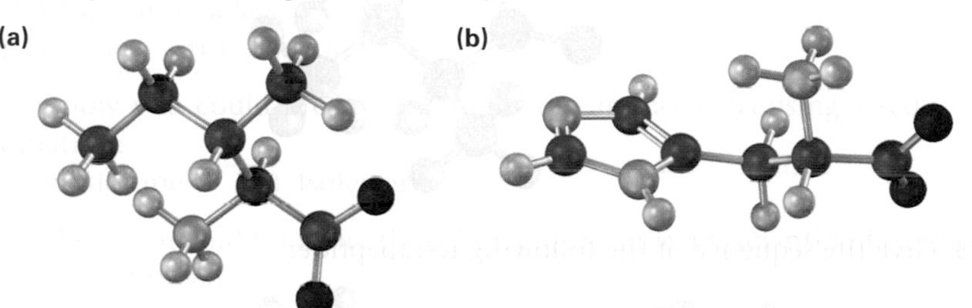

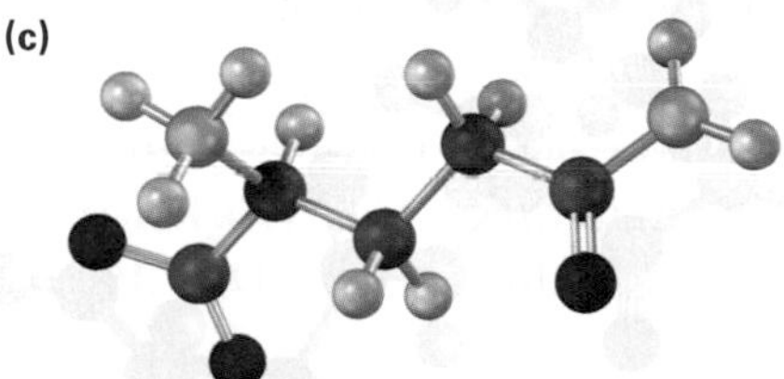

OWL Interactive versions of these problems are assignable in OWL for Organic Chemistry.

26.57 Oxytocin, a nonapeptide hormone secreted by the pituitary gland, functions by stimulating uterine contraction and lactation during childbirth. Its sequence was determined from the following evidence:

1. Oxytocin is a cyclic compound containing a disulfide bridge between two cysteine residues.
2. When the disulfide bridge is reduced, oxytocin has the constitution Asn, Cys_2, Gln, Gly, Ile, Leu, Pro, Tyr.
3. Partial hydrolysis of reduced oxytocin yields seven fragments: Asp-Cys, Ile-Glu, Cys-Tyr, Leu-Gly, Tyr-Ile-Glu, Glu-Asp-Cys, Cys-Pro-Leu.
4. Gly is the C-terminal group.
5. Both Glu and Asp are present as their side-chain amides (Gln and Asn) rather than as free side-chain acids.

What is the amino acid sequence of reduced oxytocin? What is the structure of oxytocin itself?

26.58 *Aspartame,* a nonnutritive sweetener marketed under such trade names as Equal, NutraSweet, and Canderel, is the methyl ester of a simple dipeptide, Asp-Phe-OCH_3.

(a) Draw the structure of aspartame.

(b) The isoelectric point of aspartame is 5.9. Draw the principal structure present in aqueous solution at this pH.

(c) Draw the principal form of aspartame present at physiological pH = 7.3.

26.59 Refer to Figure 26.4 on page 1060 and propose a mechanism for the final step in the Edman degradation—the acid-catalyzed rearrangement of the ATZ derivative to the PTH derivative.

26.60 Amino acids are metabolized by a transamination reaction in which the $-NH_2$ group of the amino acid changes places with the keto group of an α-keto acid. The products are a new amino acid and a new α-keto acid. Show the product from transamination of isoleucine.

26.61 The first step in the biological degradation of histidine is formation of a 4-methylideneimidazol-5-one (MIO) by cyclization of a segment of the peptide chain in the histidine ammonia lyase enzyme. Propose a mechanism.

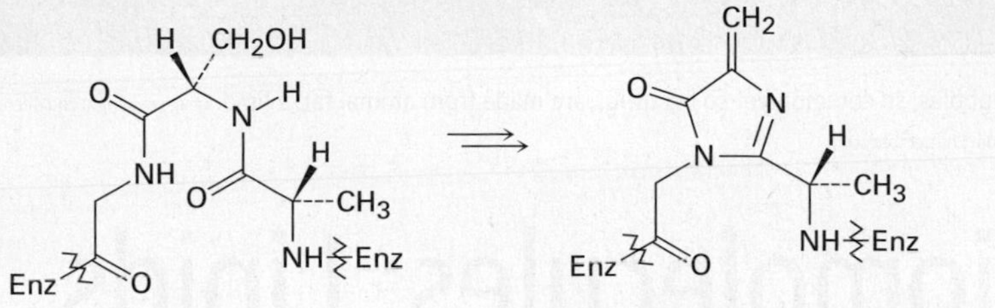

4-Methylidene-5-imidazolone (MIO)

26.62 The first step in the biological degradation of lysine is reductive amination with α-ketoglutarate to give saccharopine. Nicotinamide adenine dinucleotide phosphate (NADPH), a relative of NADH, is the reducing agent. Show the mechanism.

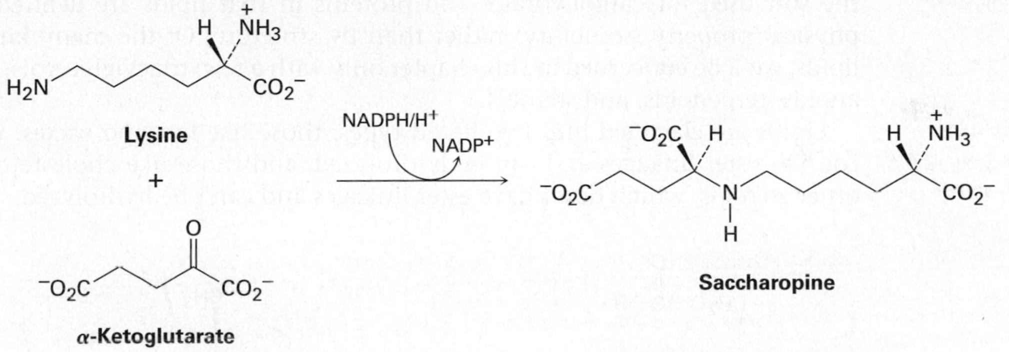

Lysine + **α-Ketoglutarate** $\xrightarrow{NADPH/H^+ \;\; NADP^+}$ **Saccharopine**

27

Soap bubbles, so common yet so beautiful, are made from animal fat, a lipid. Image copyright zhu difeng, 2010. Used under license from Shutterstock.com

Biomolecules: Lipids

Lipids are naturally occurring organic molecules that have limited solubility in water and can be isolated from organisms by extraction with nonpolar organic solvents. Fats, oils, waxes, many vitamins and hormones, and most nonprotein cell-membrane components are examples. Note that this definition differs from the sort used for carbohydrates and proteins in that lipids are defined by a physical property (solubility) rather than by structure. Of the many kinds of lipids, we'll be concerned in this chapter only with a few: triacylglycerols, eicosanoids, terpenoids, and steroids.

Lipids are classified into two broad types: those like fats and waxes, which contain ester linkages and can be hydrolyzed, and those like cholesterol and other steroids, which don't have ester linkages and can't be hydrolyzed.

Animal fat—a triester
(R, R′, R″ = C_{11}–C_{19} chains)

Cholesterol

Why This Chapter? We've now covered two of the four major classes of biomolecules—proteins and carbohydrates—and have two remaining. We'll cover lipids, the largest and most diverse class of biomolecules, in this chapter, looking both at their structure and function and at their metabolism.

27.1 Waxes, Fats, and Oils

Waxes are mixtures of esters of long-chain carboxylic acids with long-chain alcohols. The carboxylic acid usually has an even number of carbons from 16 through 36, while the alcohol has an even number of carbons from 24 through 36. One of the major components of beeswax, for instance, is triacontyl hexadecanoate, the ester of the C_{30} alcohol 1-triacontanol and the

OWL Sign in to OWL for Organic Chemistry at **www.cengage.com/owl** to view tutorials and simulations, develop problem-solving skills, and complete online homework assigned by your professor.

C_{16} acid hexadecanoic acid. The waxy protective coatings on most fruits, berries, leaves, and animal furs have similar structures.

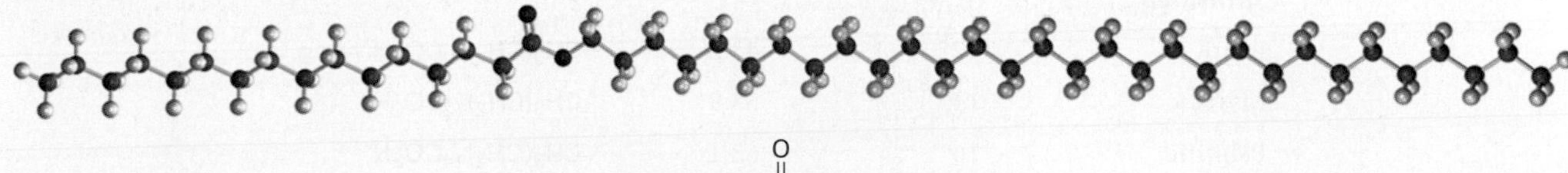

$$CH_3(CH_2)_{14}\overset{O}{\overset{\|}{C}}O(CH_2)_{29}CH_3$$

Triacontyl hexadecanoate (from beeswax)

Animal fats and vegetable oils are the most widely occurring lipids. Although they appear different—animal fats like butter and lard are solids, whereas vegetable oils like corn and peanut oil are liquid—their structures are closely related. Chemically, fats and oils are *triglycerides,* or **triacylglycerols**—triesters of glycerol with three long-chain carboxylic acids called **fatty acids.** Animals use fats for long-term energy storage because they are much less highly oxidized than carbohydrates and provide about six times as much energy as an equal weight of stored, hydrated glycogen.

Glycerol | Fatty acyl

$$CH_2O\overset{O}{\overset{\|}{C}}CH_2CH_2CH_2CH_2CH_2CH_2CH_2CH_2CH_2CH_2CH_2CH_2CH_2CH_2CH_2CH_2CH_3$$ ← Stearoyl (stearic acid)

$$CHO\overset{O}{\overset{\|}{C}}CH_2CH_2CH_2CH_2CH_2CH_2CH_2CH{=}CHCH_2CH_2CH_2CH_2CH_2CH_2CH_2CH_3$$ ← Oleoyl (oleic acid)

$$CH_2O\overset{O}{\overset{\|}{C}}CH_2CH_2CH_2CH_2CH_2CH_2CH_2CH{=}CHCH_2CH{=}CHCH_2CH_2CH_2CH_2CH_3$$ ← Linoleoyl (linoleic acid)

A triacylglycerol

Hydrolysis of a fat or oil with aqueous NaOH yields glycerol and three fatty acids. The fatty acids are generally unbranched and contain an even number of carbon atoms between 12 and 20. If double bonds are present, they have largely, although not entirely, *Z,* or cis, geometry. The three fatty acids of a specific triacylglycerol molecule need not be the same, and the fat or oil from a given source is likely to be a complex mixture of many different triacylglycerols. Table 27.1 lists some of the commonly occurring fatty acids, and Table 27.2 lists the approximate composition of fats and oils from different sources.

More than 100 different fatty acids are known, and about 40 occur widely. Palmitic acid (C_{16}) and stearic acid (C_{18}) are the most abundant saturated fatty acids; oleic and linoleic acids (both C_{18}) are the most abundant unsaturated ones. Oleic acid is monounsaturated because it has only one double bond, whereas linoleic, linolenic, and arachidonic acids are **polyunsaturated fatty acids** because they have more than one double bond. Linoleic and linolenic acids occur in cream and are essential in the human diet; infants grow poorly and develop skin lesions if fed a diet of nonfat milk for prolonged periods. Linolenic acid, in particular, is an example of an *omega-3* fatty acid, which has been found to lower blood triglyceride levels and reduce the risk of heart attack. The

Table 27.1 Structures of Some Common Fatty Acids

Name	No. of carbons	Melting point (°C)	Structure
Saturated			
Lauric	12	43.2	$CH_3(CH_2)_{10}CO_2H$
Myristic	14	53.9	$CH_3(CH_2)_{12}CO_2H$
Palmitic	16	63.1	$CH_3(CH_2)_{14}CO_2H$
Stearic	18	68.8	$CH_3(CH_2)_{16}CO_2H$
Arachidic	20	76.5	$CH_3(CH_2)_{18}CO_2H$
Unsaturated			
Palmitoleic	16	−0.1	(*Z*)-$CH_3(CH_2)_5CH{=}CH(CH_2)_7CO_2H$
Oleic	18	13.4	(*Z*)-$CH_3(CH_2)_7CH{=}CH(CH_2)_7CO_2H$
Linoleic	18	−12	(*Z,Z*)-$CH_3(CH_2)_4(CH{=}CHCH_2)_2(CH_2)_6CO_2H$
Linolenic	18	−11	(all *Z*)-$CH_3CH_2(CH{=}CHCH_2)_3(CH_2)_6CO_2H$
Arachidonic	20	−49.5	(all *Z*)-$CH_3(CH_2)_4(CH{=}CHCH_2)_4CH_2CH_2CO_2H$

Table 27.2 Composition of Some Fats and Oils

	Saturated fatty acids (%)				Unsaturated fatty acids (%)	
Source	C_{12} lauric	C_{14} myristic	C_{16} palmitic	C_{18} stearic	C_{18} oleic	C_{18} linoleic
Animal fat						
Lard	—	1	25	15	50	6
Butter	2	10	25	10	25	5
Human fat	1	3	25	8	46	10
Whale blubber	—	8	12	3	35	10
Vegetable oil						
Coconut	50	18	8	2	6	1
Corn	—	1	10	4	35	45
Olive	—	1	5	5	80	7
Peanut	—	—	7	5	60	20

name omega-3 means that there is a double bond three carbons in from the noncarboxyl end of the chain.

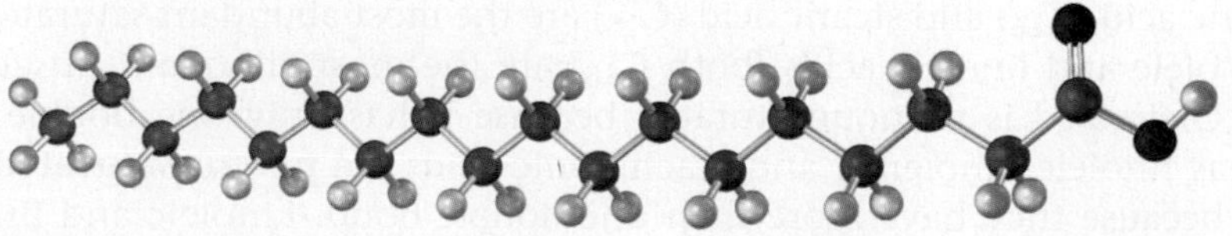

$CH_3CH_2CH_2CH_2CH_2CH_2CH_2CH_2CH_2CH_2CH_2CH_2CH_2CH_2CH_2CH_2CH_2\overset{O}{\overset{\|}{C}}OH$

Stearic acid

$CH_3CH_2CH{=}CHCH_2CH{=}CHCH_2CH{=}CHCH_2CH_2CH_2CH_2CH_2CH_2CH_2\overset{O}{\overset{\|}{C}}OH$

Linolenic acid, an omega-3 polyunsaturated fatty acid

The data in Table 27.1 show that unsaturated fatty acids generally have lower melting points than their saturated counterparts, a trend that is also true for triacylglycerols. Since vegetable oils generally have a higher proportion of unsaturated to saturated fatty acids than animal fats (Table 27.2), they have lower melting points. The difference is a consequence of structure. Saturated fats have a uniform shape that allows them to pack together efficiently in a crystal lattice. In unsaturated vegetable oils, however, the C=C bonds introduce bends and kinks into the hydrocarbon chains, making crystal formation more difficult. The more double bonds there are, the harder it is for the molecules to crystallize and the lower the melting point of the oil.

The C=C bonds in vegetable oils can be reduced by catalytic hydrogenation, typically carried out at high temperature using a nickel catalyst, to produce saturated solid or semisolid fats. Margarine and shortening are produced by hydrogenating soybean, peanut, or cottonseed oil until the proper consistency is obtained. Unfortunately, the hydrogenation reaction is accompanied by some cis–trans isomerization of the double bonds that remain, producing fats with about 10% to 15% trans unsaturated fatty acids. Dietary intake of trans fatty acids increases cholesterol levels in the blood, thereby increasing the risk of heart problems. The conversion of linoleic acid into elaidic acid is an example.

Linoleic acid

Elaidic acid

Problem 27.1
Carnauba wax, used in floor and furniture polishes, contains an ester of a C_{32} straight-chain alcohol with a C_{20} straight-chain carboxylic acid. Draw its structure.

Problem 27.2
Draw structures of glyceryl tripalmitate and glyceryl trioleate. Which would you expect to have a higher melting point?

27.2 Soap

Soap has been known since at least 600 BC, when the Phoenicians prepared a curdy material by boiling goat fat with extracts of wood ash. The cleansing properties of soap weren't generally recognized, however, and the use of soap did not become widespread until the 18th century. Chemically, soap is a mixture of the sodium or potassium salts of the long-chain fatty acids produced by hydrolysis *(saponification)* of animal fat with alkali. Wood ash was used as a source of alkali until the early 1800s, when the development of the LeBlanc process for making Na_2CO_3 by heating sodium sulfate with limestone became available.

$$\begin{array}{l} \text{CH}_2\text{OCOR} \\ | \\ \text{CHOCOR} \\ | \\ \text{CH}_2\text{OCOR} \end{array} \xrightarrow[\text{H}_2\text{O}]{\text{NaOH}} 3\ \text{RCO}_2^-\ \text{Na}^+ + \begin{array}{l} \text{CH}_2\text{OH} \\ | \\ \text{CHOH} \\ | \\ \text{CH}_2\text{OH} \end{array}$$

A fat **(R = C_{11}–C_{19} aliphatic chains)** → **Soap** + **Glycerol**

Crude soap curds contain glycerol and excess alkali as well as soap but can be purified by boiling with water and adding NaCl or KCl to precipitate the pure carboxylate salts. The smooth soap that precipitates is dried, perfumed, and pressed into bars for household use. Dyes are added to make colored soaps, antiseptics are added for medicated soaps, pumice is added for scouring soaps, and air is blown in for soaps that float. Regardless of these extra treatments and regardless of price, though, all soaps are basically the same.

Soaps act as cleansers because the two ends of a soap molecule are so different. The carboxylate end of the long-chain molecule is ionic and therefore hydrophilic **(Section 2.12)**, or attracted to water. The long hydrocarbon portion of the molecule, however, is nonpolar and hydrophobic, avoiding water and therefore more soluble in oils. The net effect of these two opposing tendencies is that soaps are attracted to both oils and water and are therefore useful as cleansers.

When soaps are dispersed in water, the long hydrocarbon tails cluster together on the inside of a tangled, hydrophobic ball, while the ionic heads on the surface of the cluster stick out into the water layer. These spherical clusters, called **micelles**, are shown schematically in **Figure 27.1**. Grease and oil droplets are solubilized in water when they are coated by the nonpolar, hydrophobic tails of soap molecules in the center of micelles. Once solubilized, the grease and dirt can be rinsed away.

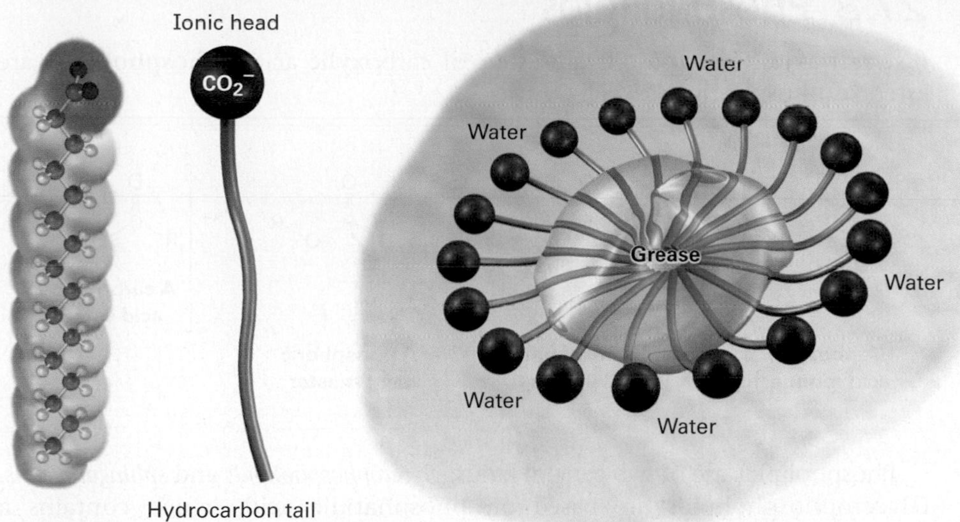

Figure 27.1 A soap micelle solubilizing a grease particle in water. An electrostatic potential map of a fatty acid carboxylate shows how the negative charge is located in the head group.

As useful as they are, soaps also have some drawbacks. In hard water, which contains metal ions, soluble sodium carboxylates are converted into insoluble magnesium and calcium salts, leaving the familiar ring of scum around bathtubs and the gray tinge on white clothes. Chemists have circumvented these problems by synthesizing a class of synthetic detergents based on salts of long-chain alkylbenzenesulfonic acids. The principle of synthetic detergents is the same as that of soaps: the alkylbenzene end of the molecule is attracted to grease, while the anionic sulfonate end is attracted to water. Unlike soaps, though, sulfonate detergents don't form insoluble metal salts in hard water and don't leave an unpleasant scum.

A synthetic detergent
(R = a mixture of C_{12} chains)

Problem 27.3
Draw the structure of magnesium oleate, a component of bathtub scum.

Problem 27.4
Write the saponification reaction of glyceryl dioleate monopalmitate with aqueous NaOH.

27.3 Phospholipids

Just as waxes, fats, and oils are esters of carboxylic acids, **phospholipids** are esters of phosphoric acid, H_3PO_4.

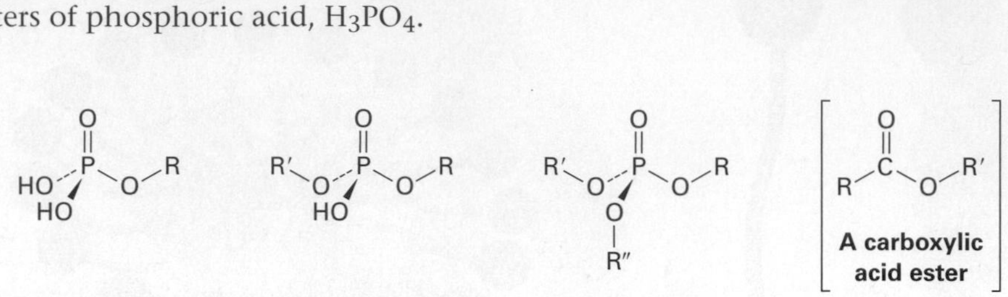

A phosphoric acid monoester **A phosphoric acid diester** **A phosphoric acid triester** **A carboxylic acid ester**

Phospholipids are of two general kinds: *glycerophospholipids* and *sphingomyelins*. Glycerophospholipids are based on phosphatidic acid, which contains a glycerol backbone linked by ester bonds to two fatty acids and one phosphoric acid. Although the fatty-acid residues can be any of the C_{12}–C_{20} units typically present in fats, the acyl group at C1 is usually saturated and the one at C2 is usually unsaturated. The phosphate group at C3 is also bonded to an amino alcohol such as choline $[HOCH_2CH_2N(CH_3)_3]^+$, ethanolamine ($HOCH_2CH_2NH_2$), or serine $[HOCH_2CH(NH_2)CO_2H]$. The compounds are chiral and have an L, or *R*, configuration at C2.

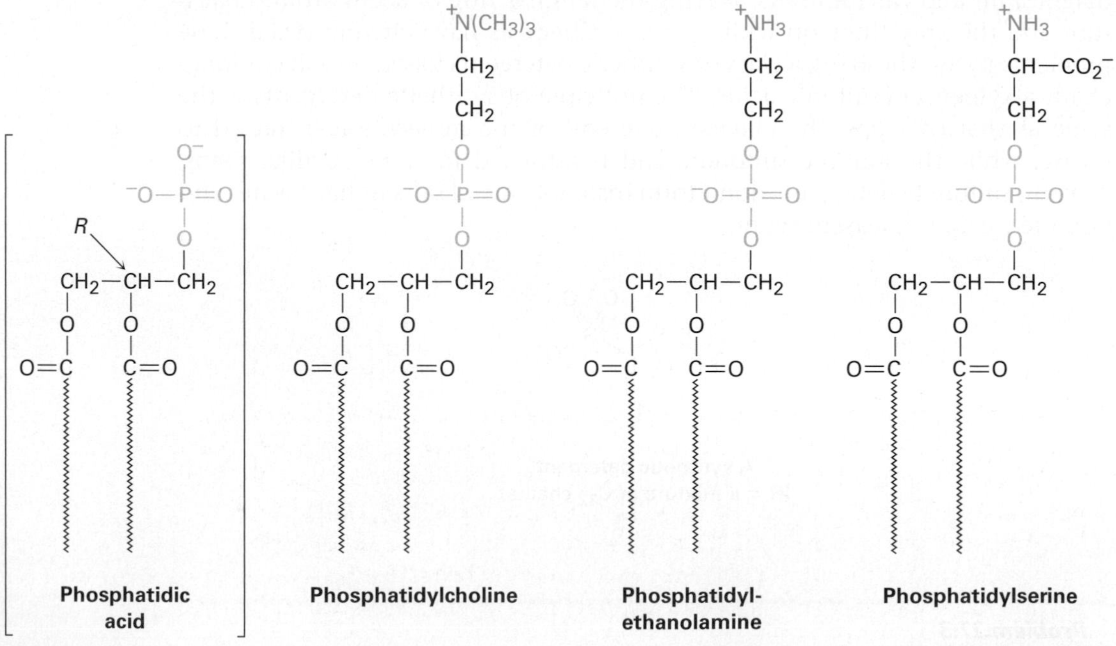

Phosphatidic acid **Phosphatidylcholine** **Phosphatidylethanolamine** **Phosphatidylserine**

Sphingomyelins are the second major group of phospholipids. These compounds have sphingosine or a related dihydroxyamine as their backbone and are particularly abundant in brain and nerve tissue, where they are a major constituent of the coating around nerve fibers.

Sphingosine

A sphingomyelin

Phospholipids are found widely in both plant and animal tissues and make up approximately 50% to 60% of cell membranes. Because they are like soaps in having a long, nonpolar hydrocarbon tail bound to a polar ionic head, phospholipids in the cell membrane organize into a **lipid bilayer** about 5.0 nm (50 Å) thick. As shown in **Figure 27.2**, the nonpolar tails aggregate in the center of the bilayer in much the same way that soap tails aggregate in the center of a micelle. This bilayer serves as an effective barrier to the passage of water, ions, and other components into and out of cells.

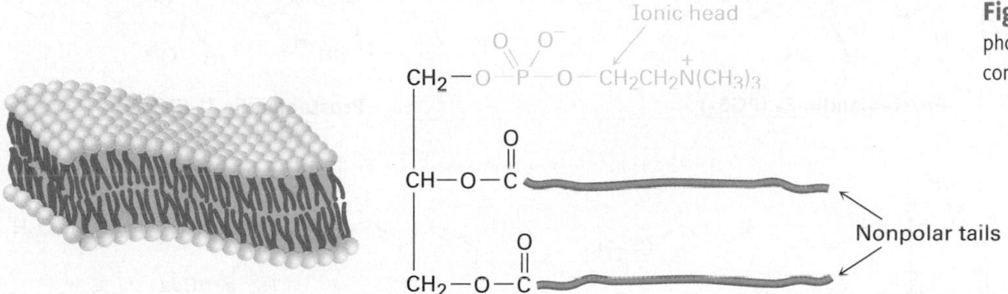

Figure 27.2 Aggregation of glycerophospholipids into the lipid bilayer that composes cell membranes.

27.4 Prostaglandins and Other Eicosanoids

The **prostaglandins** are a group of C_{20} lipids that contain a five-membered ring with two long side chains. First isolated in the 1930s by Ulf von Euler at the Karolinska Institute in Sweden, much of the structural and chemical work on the prostaglandins was carried out by Sune Bergström and Bengt Samuelsson. All three received Nobel Prizes for their work. The name *prostaglandin* derives from the fact that the compounds were first isolated from sheep prostate glands, but they have subsequently been shown to be present in small amounts in all body tissues and fluids.

The several dozen known prostaglandins have an extraordinarily wide range of biological effects. Among their many properties, they can lower blood pressure, affect blood platelet aggregation during clotting, lower gastric secretions, control inflammation, affect kidney function, affect reproductive systems, and stimulate uterine contractions during childbirth.

Prostaglandins, together with related compounds called thromboxanes and leukotrienes, make up a class of compounds called **eicosanoids** because they are derived biologically from 5,8,11,14-eicosatetraenoic acid, or arachidonic acid **(Figure 27.3)**. Prostaglandins (PG) have a cyclopentane ring with two long side chains; thromboxanes (TX) have a six-membered, oxygen-containing ring; and leukotrienes (LT) are acyclic.

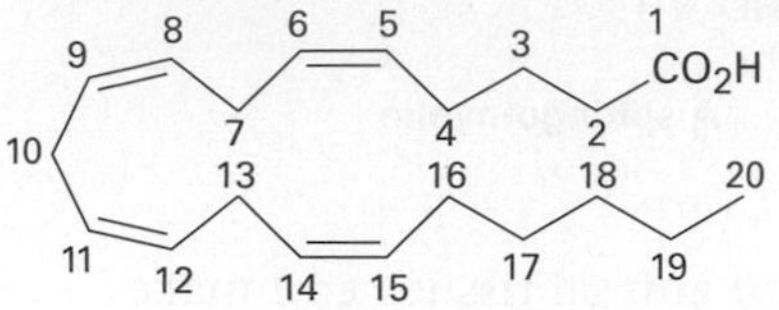

Arachidonic acid

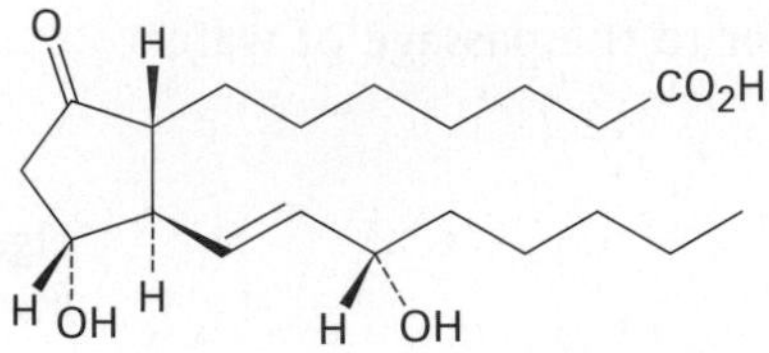

Prostaglandin E_1 (PGE_1)

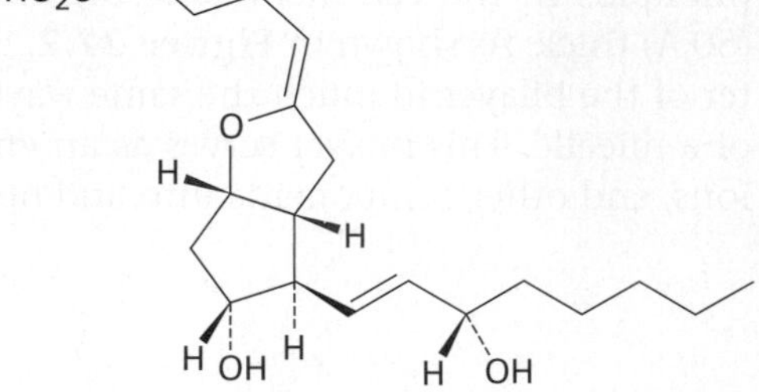

Prostaglandin I_2 (PGI_2) (prostacyclin)

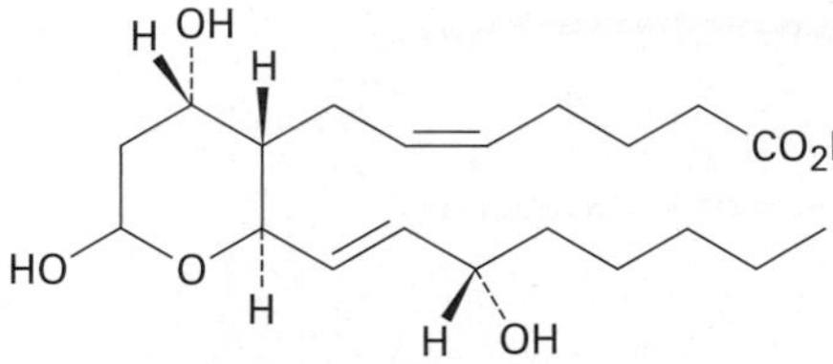

Thromboxane B_2 (TXB_2)

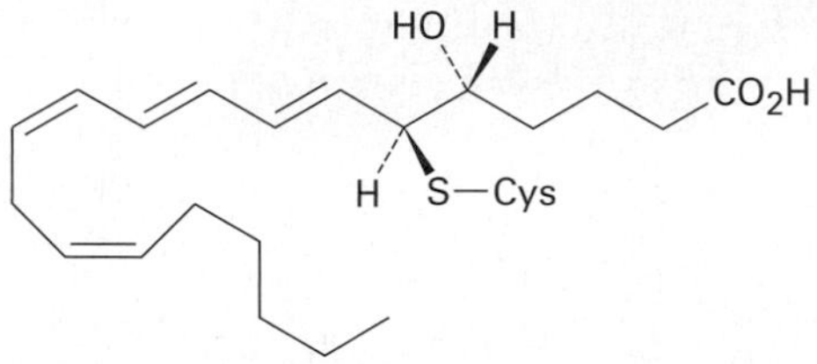

Leukotriene E_4 (LTE_4)

Figure 27.3 Structures of some representative eicosanoids. All are derived biologically from arachidonic acid.

Eicosanoids are named based on their ring system (PG, TX, or LT), substitution pattern, and number of double bonds. The various substitution patterns on the ring are indicated by letter as in **Figure 27.4**, and the number of double bonds is indicated by a subscript. Thus, PGE_1 is a prostaglandin with the "E" substitution pattern and one double bond. The numbering of the atoms in the various eicosanoids is the same as in arachidonic acid, starting with the $-CO_2H$ carbon as C1, continuing around the ring, and ending with the $-CH_3$ carbon at the other end of the chain as C20.

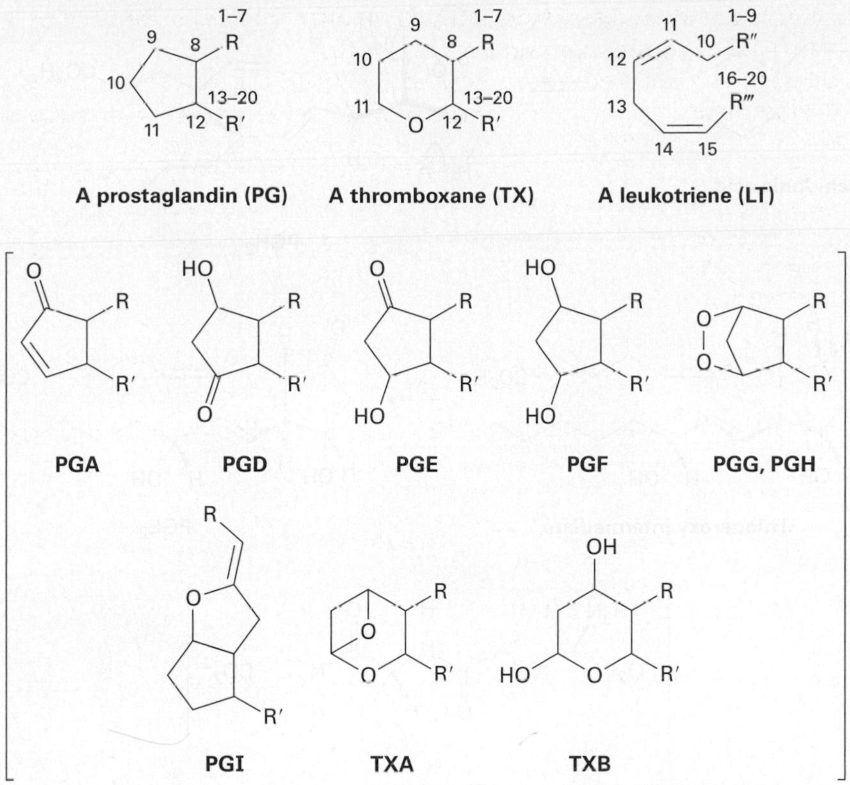

Figure 27.4 The nomenclature system for eicosanoids.

Eicosanoid biosynthesis begins with the conversion of arachidonic acid to PGH_2, catalyzed by the multifunctional PGH synthase (PGHS), also called cyclooxygenase (COX). There are two distinct enzymes, PGHS-1 and PGHS-2 (or COX-1 and COX-2), both of which accomplish the same reaction but appear to function independently. COX-1 carries out the normal physiological production of prostaglandins, and COX-2 produces additional prostaglandin in response to arthritis or other inflammatory conditions. Vioxx, Celebrex, Bextra, and several other drugs selectively inhibit the COX-2 enzyme but also appear to cause potentially serious heart problems in weakened patients. (See the Chapter 15 *A Deeper Look.*)

PGHS accomplishes two transformations, an initial reaction of arachidonic acid with O_2 to yield PGG_2 and a subsequent reduction of the hydroperoxide group (–OOH) to the alcohol PGH_2. The sequence of steps involved in these transformation was shown in Figure 8.10 on page 295.

Further processing of PGH_2 then leads to other eicosanoids. PGE_2, for instance, arises by an isomerization of PGH_2 catalyzed by PGE synthase (PGES). The coenzyme glutathione is needed for enzyme activity, although it is not chemically changed during the isomerization and its role is not fully understood. One possibility is that the glutathione thiolate anion breaks the O–O bond in PGH_2 by an S_N2-like attack on one of the oxygen atoms, giving a thioperoxy intermediate (R—S—O—R′) that eliminates glutathione to give the ketone **(Figure 27.5)**.

Figure 27.5 Mechanism of the conversion of PGH_2 into PGE_2.

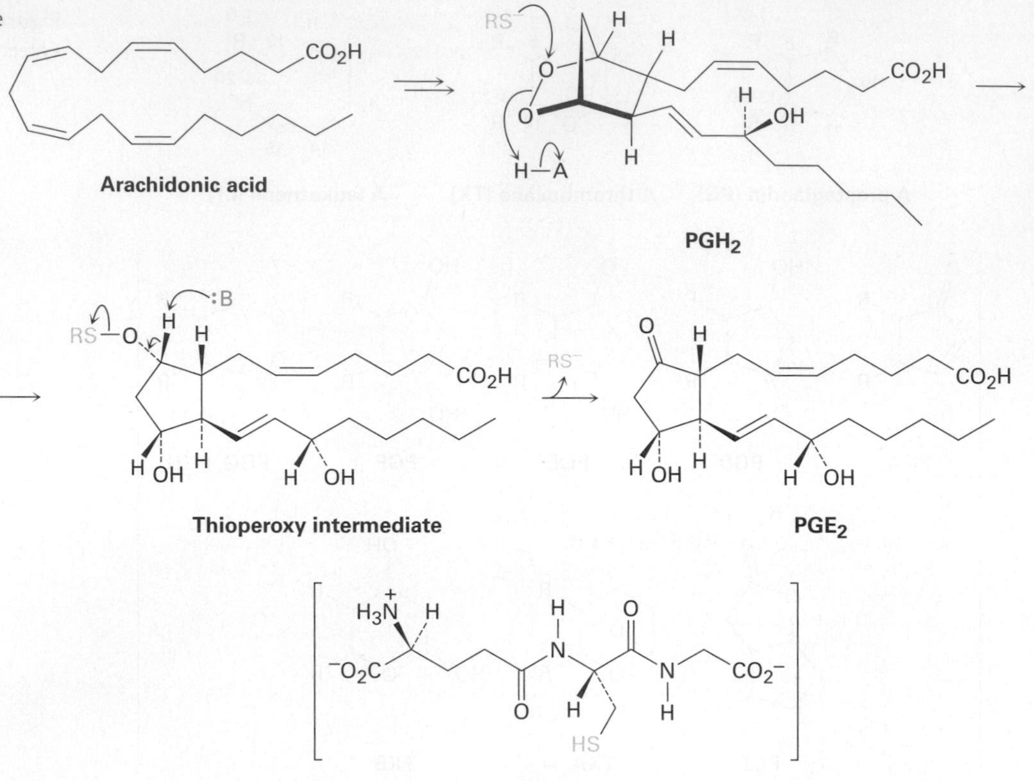

Problem 27.5

Assign *R* or *S* configuration to each chirality center in prostaglandin E_2 (Figure 27.5), the most abundant and biologically potent of mammalian prostaglandins.

27.5 Terpenoids

We saw in the Chapter 8 *A Deeper Look* that **terpenoids** are a vast and diverse group of lipids found in all living organisms. Despite their apparent structural differences, all terpenoids contain a multiple of five carbons and are derived biosynthetically from the five-carbon precursor isopentenyl diphosphate **(Figure 27.6)**. Although formally a *terpenoid* contains oxygen, while a hydrocarbon is called a *terpene,* we'll use the term *terpenoid* to refer to both for simplicity.

You might recall from Chapter 8 that terpenoids are classified according to the number of five-carbon multiples they contain. Monoterpenoids contain 10 carbons and are derived from two isopentenyl diphosphates, sesquiterpenoids contain 15 carbons and are derived from three isopentenyl diphosphates, diterpenoids contain 20 carbons and are derived from four isopentenyl diphosphates, and so on, up to triterpenoids (C_{30}) and tetraterpenoids (C_{40}). Lanosterol, for example, is a triterpenoid from which steroid hormones are made, and β-carotene is a tetraterpenoid that serves as a dietary source of vitamin A (Figure 27.6).

The terpenoid precursor isopentenyl diphosphate, formerly called isopentenyl pyrophosphate and thus abbreviated IPP, is biosynthesized by two different pathways, depending on the organism and the structure of the final product. In animals and higher plants, sesquiterpenoids and triterpenoids arise primarily from the *mevalonate* pathway, whereas monoterpenoids, diterpenoids, and

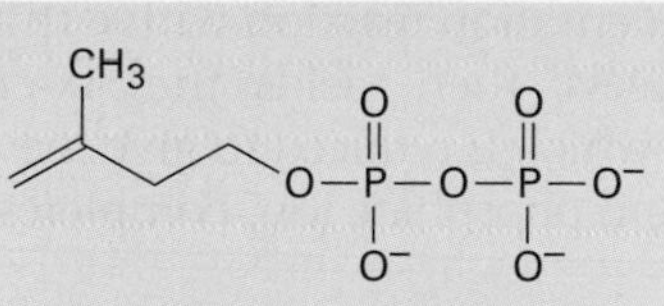

Isopentenyl diphosphate

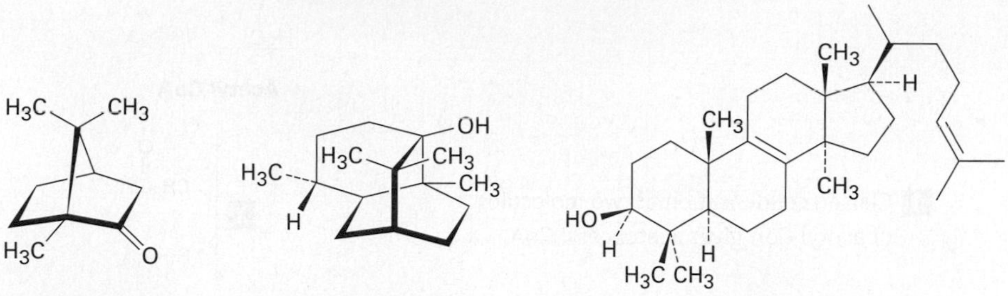

Camphor
(a monoterpenoid—C_{10})

Patchouli alcohol
(a sesquiterpenoid—C_{15})

Lanosterol
(a triterpenoid—C_{30})

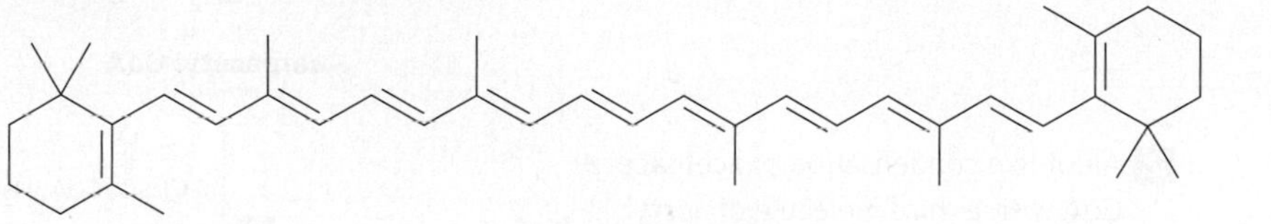

β-Carotene
(a tetraterpenoid—C_{40})

Figure 27.6 Structures of some representative terpenoids.

tetraterpenoids are biosynthesized by the *1-deoxyxylulose 5-phosphate (DXP)* pathway, also called the methylerithritol phosphate, or MEP, pathway. In bacteria, both pathways are used. We'll look only at the mevalonate pathway, which is more common and better understood at present.

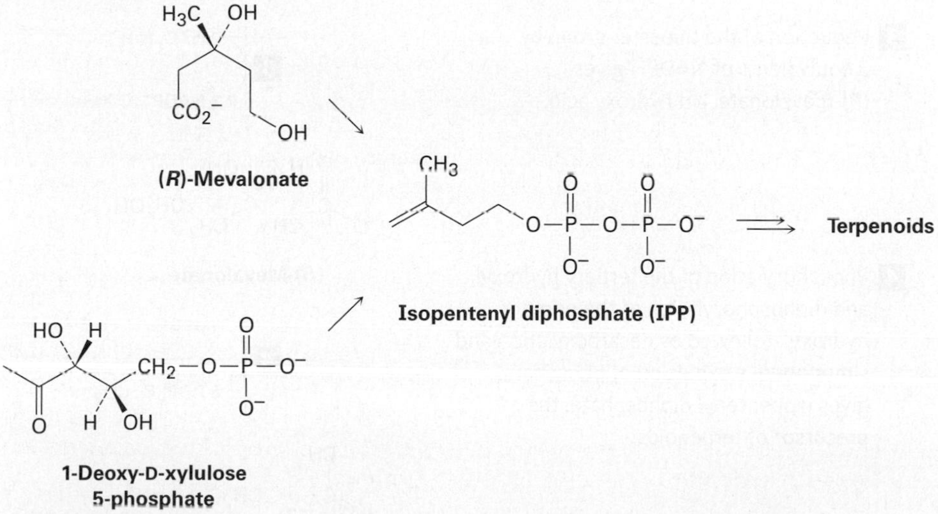

(*R*)-Mevalonate

1-Deoxy-D-xylulose 5-phosphate

Isopentenyl diphosphate (IPP) ⟹ Terpenoids

The Mevalonate Pathway to Isopentenyl Diphosphate

As shown in **Figure 27.7**, the mevalonate pathway begins with the conversion of acetate to acetyl CoA, followed by Claisen condensation to yield acetoacetyl CoA.

A second carbonyl condensation reaction with a third molecule of acetyl CoA, this one an aldol-like process, then yields the six-carbon compound 3-hydroxy-3-methylglutaryl CoA, which is reduced to give mevalonate. Phosphorylation, followed by loss of CO_2 and phosphate ion, completes the process.

Figure 27.7 | MECHANISM

The mevalonate pathway for the biosynthesis of isopentenyl diphosphate from three molecules of acetyl CoA. Individual steps are explained in the text.

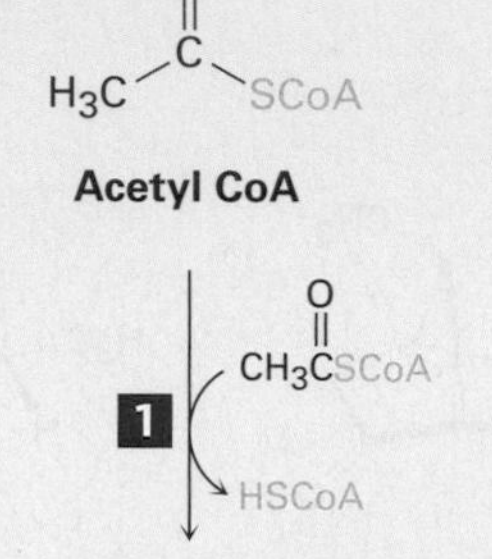

1 Claisen condensation of two molecules of acetyl CoA gives acetoacetyl CoA.

CH_3CSCoA → HSCoA

Acetoacetyl CoA

2 Aldol-like condensation of acetoacetyl CoA with a third molecule of acetyl CoA, followed by hydrolysis, gives (3*S*)-3-hydroxy-3-methylglutaryl CoA.

CH_3CSCoA, H_2O → HSCoA

(3*S*)-3-Hydroxy-3-methylglutaryl CoA

3 Reduction of the thioester group by 2 equivalents of NADPH gives (*R*)-mevalonate, a dihydroxy acid.

2 NADPH/H^+ → 2 $NADP^+$, CoASH

(*R*)-Mevalonate

4 Phosphorylation of the tertiary hydroxyl and diphosphorylation of the primary hydroxyl, followed by decarboxylation and simultaneous expulsion of phosphate, gives isopentenyl diphosphate, the precursor of terpenoids.

3 ATP → 3 ADP, P_i, CO_2

Isopentenyl diphosphate

Step 1 of Figure 27.7: Claisen Condensation The first step in mevalonate biosynthesis is a Claisen condensation to yield acetoacetyl CoA, a reaction catalyzed by acetoacetyl-CoA acetyltransferase. An acetyl group is first bound to the enzyme by a nucleophilic acyl substitution reaction with a cysteine –SH group. Formation of an enolate ion from a second molecule of acetyl CoA, followed by Claisen condensation, then yields the product.

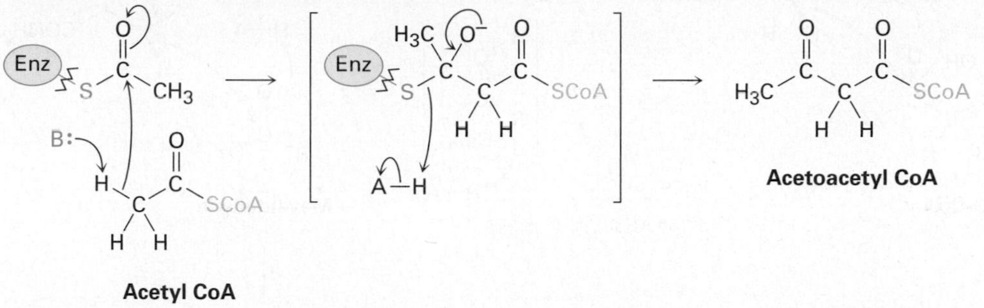

Acetyl CoA

Acetoacetyl CoA

Step 2 of Figure 27.7: Aldol Condensation Acetoacetyl CoA next undergoes an aldol-like addition of an acetyl CoA enolate ion in a reaction catalyzed by 3-hydroxy-3-methylglutaryl-CoA synthase. The reaction occurs by initial binding of the substrate to a cysteine –SH group in the enzyme, followed by enolate-ion addition and subsequent hydrolysis to give (3*S*)-3-hydroxy-3-methylglutaryl CoA (HMG-CoA).

(3*S*)-3-Hydroxy-3-methylglutaryl CoA (HMG-CoA)

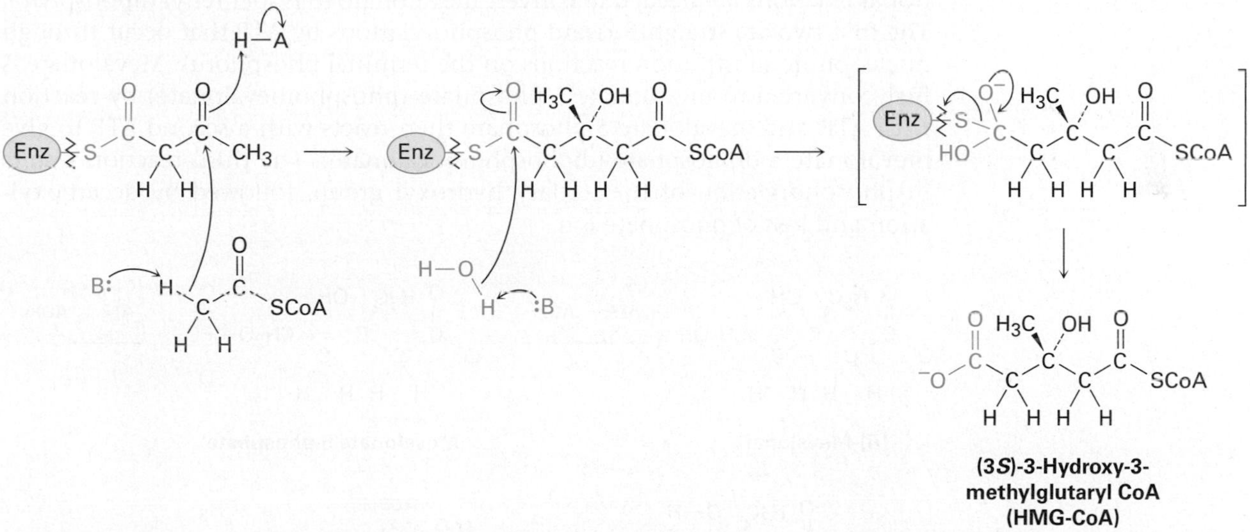

Step 3 of Figure 27.7: Reduction Reduction of HMG-CoA to give (*R*)-mevalonate is catalyzed by 3-hydroxy-3-methylglutaryl-CoA reductase and requires 2 equivalents of reduced nicotinamide adenine dinucleotide phosphate (NADPH), a close relative of NADH **(Section 19.12)**. The reaction occurs in two steps and proceeds through an aldehyde intermediate. The first step is a nucleophilic acyl substitution reaction involving hydride transfer from NADPH to the thioester carbonyl group of HMG-CoA. Following expulsion of HSCoA as leaving group, the aldehyde intermediate undergoes a second hydride addition to give mevalonate.

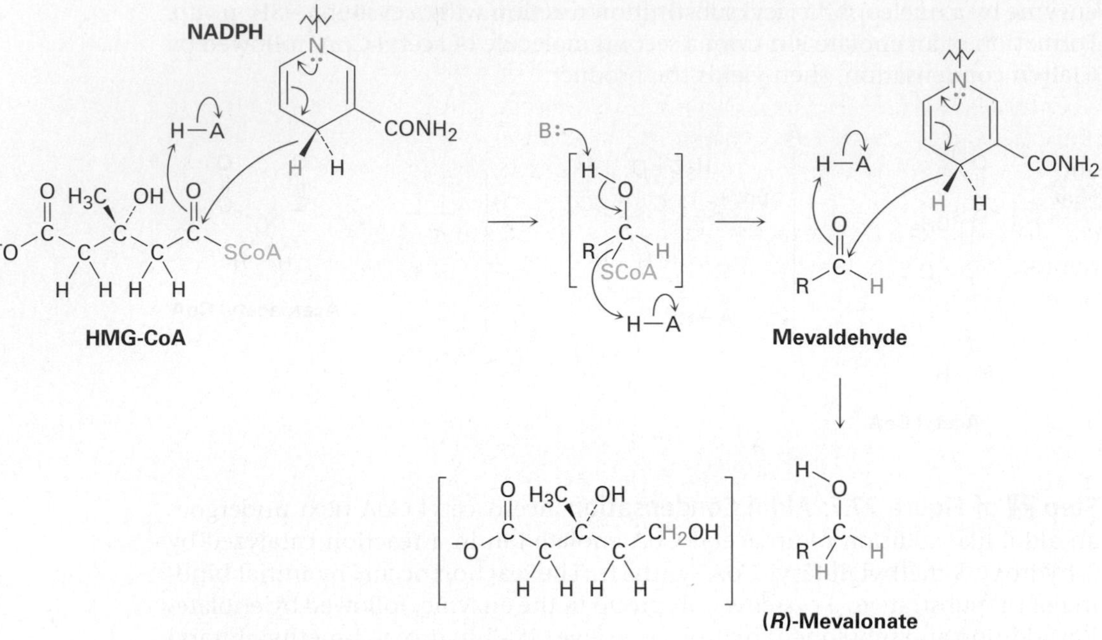

NADPH

HMG-CoA

Mevaldehyde

(*R*)-Mevalonate

Step 4 of Figure 27.7: Phosphorylation and Decarboxylation Three additional reactions are needed to convert mevalonate to isopentenyl diphosphate. The first two are straightforward phosphorylations by ATP that occur through nucleophilic substitution reactions on the terminal phosphorus. Mevalonate is first converted to mevalonate 5-phosphate (phosphomevalonate) by reaction with ATP, and mevalonate 5-phosphate then reacts with a second ATP to give mevalonate 5-diphosphate (diphosphomevalonate). The third reaction results in phosphorylation of the tertiary hydroxyl group, followed by decarboxylation and loss of phosphate ion.

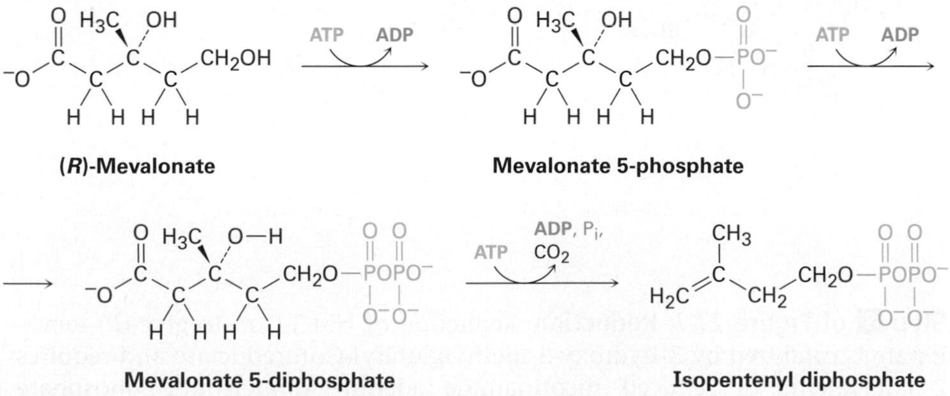

(*R*)-Mevalonate — ATP, ADP → Mevalonate 5-phosphate — ATP, ADP → Mevalonate 5-diphosphate — ATP; ADP, P_i, CO_2 → Isopentenyl diphosphate

The final decarboxylation of mevalonate 5-diphosphate appears unusual because decarboxylations of acids do not typically occur except in β-keto acids and malonic acids, in which the carboxylate group is two atoms away from an additional carbonyl group. As discussed in **Section 22.7**, the function of this second carbonyl group is to act as an electron acceptor and stabilize the charge resulting from loss of CO_2. In fact, though, the decarboxylation of a β-keto acid and the decarboxylation of mevalonate 5-diphosphate are closely related.

Catalyzed by mevalonate-5-diphosphate decarboxylase, the substrate is first phosphorylated on the free –OH group by reaction with ATP to give a tertiary phosphate, which undergoes spontaneous S_N1-like dissociation to give a tertiary carbocation. The positive charge then acts as an electron acceptor to facilitate decarboxylation in exactly the same way a β carbonyl group does, giving isopentenyl diphosphate. (In the following structures, the diphosphate group is abbreviated OPP.)

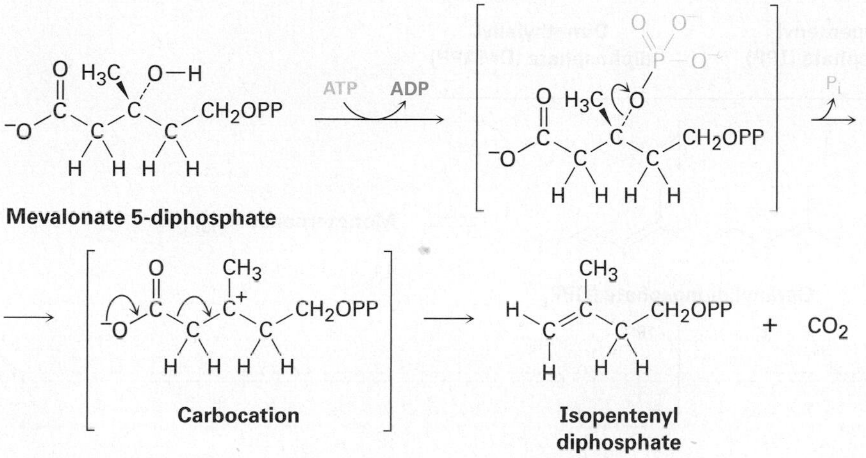

Mevalonate 5-diphosphate —(ATP → ADP)→ [tertiary phosphate] —(P_i)→ [**Carbocation**] → **Isopentenyl diphosphate** + CO_2

Problem 27.6

The conversion of mevalonate 5-phosphate to isopentenyl diphosphate occurs with the following result. Which hydrogen, *pro-R* or *pro-S*, ends up cis to the methyl group, and which ends up trans?

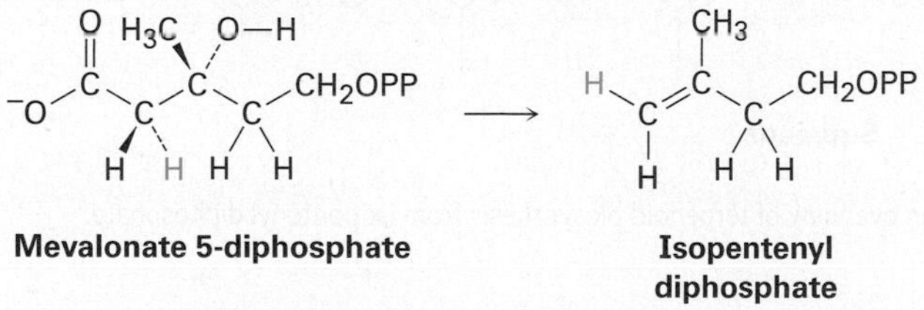

Mevalonate 5-diphosphate → **Isopentenyl diphosphate**

Conversion of Isopentenyl Diphosphate to Terpenoids

The conversion of isopentenyl diphosphate (IPP) to terpenoids begins with its isomerization to dimethylallyl diphosphate, abbreviated DMAPP and formerly called dimethylallyl pyrophosphate. These two C_5 building blocks then combine to give the C_{10} unit geranyl diphosphate (GPP). The corresponding alcohol, geraniol, is itself a fragrant terpenoid that occurs in rose oil.

Further combination of GPP with another IPP gives the C_{15} unit farnesyl diphosphate (FPP), and so on, up to C_{25}. Terpenoids with more than 25 carbons—that is, triterpenoids (C_{30}) and tetraterpenoids (C_{40})—are synthesized by dimerization of C_{15} and C_{20} units, respectively **(Figure 27.8)**. Triterpenoids and steroids, in particular, arise from dimerization of farnesyl diphosphate to give squalene.

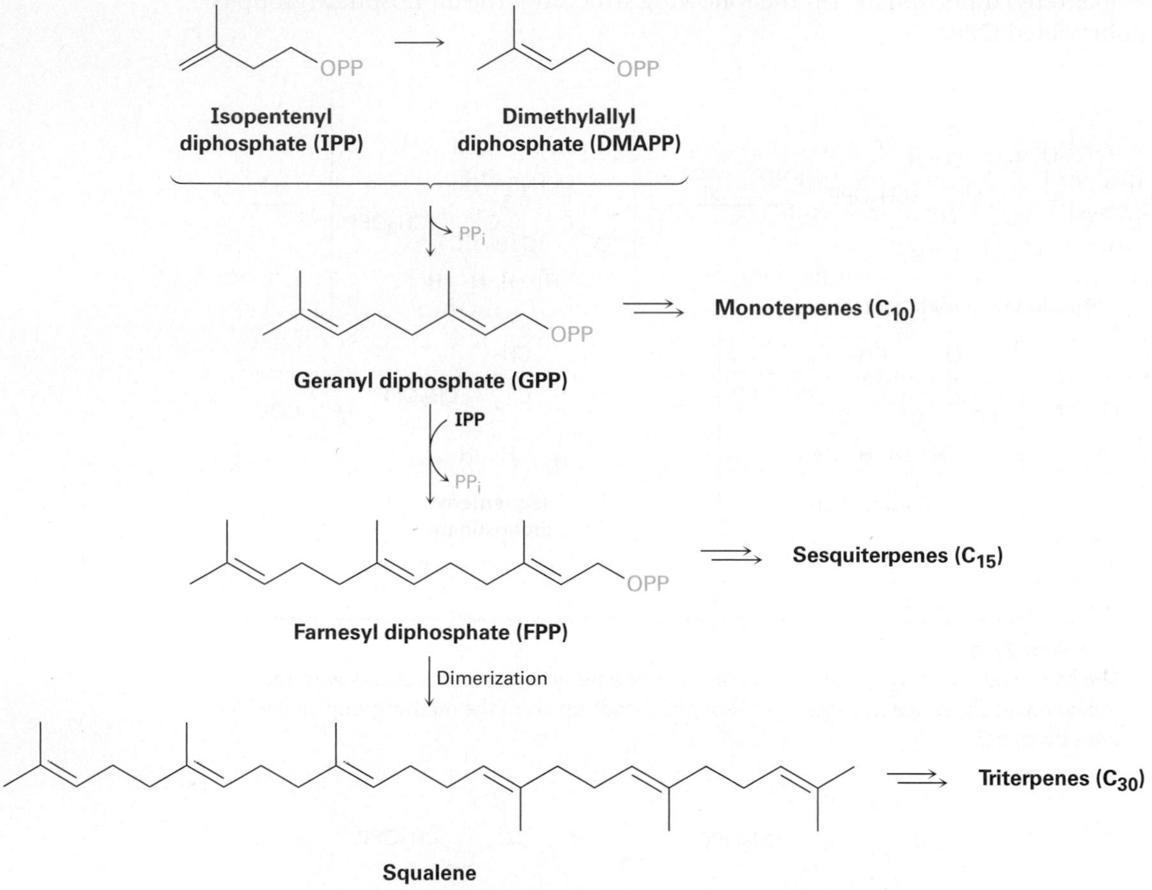

Figure 27.8 An overview of terpenoid biosynthesis from isopentenyl diphosphate.

The isomerization of isopentenyl diphosphate to dimethylallyl diphosphate is catalyzed by IPP isomerase and occurs through a carbocation pathway. Protonation of the IPP double bond by a hydrogen-bonded cysteine residue in the enzyme gives a tertiary carbocation intermediate, which is deprotonated by a glutamate residue as base to yield DMAPP. X-ray structural studies on the enzyme show that it holds the substrate in an unusually deep, well-protected pocket to shield the highly reactive carbocation from reaction with solvent or other external substances.

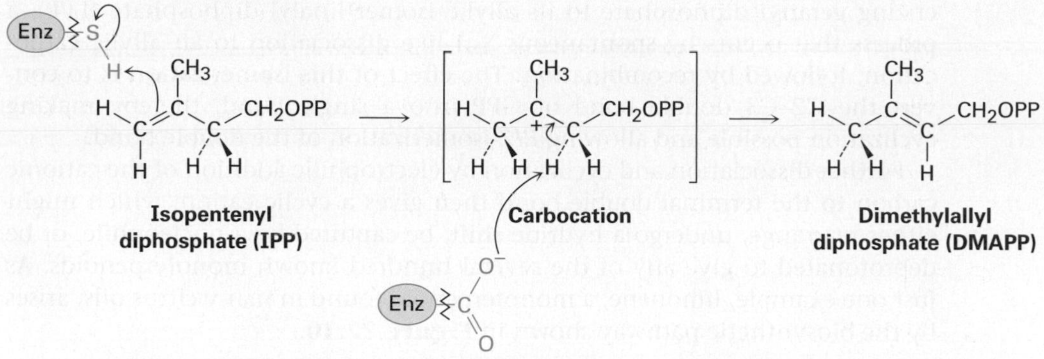

Both the initial coupling of DMAPP with IPP to give geranyl diphosphate and the subsequent coupling of GPP with a second molecule of IPP to give farnesyl diphosphate are catalyzed by farnesyl diphosphate synthase. The process requires Mg^{2+} ion, and the key step is a nucleophilic substitution reaction in which the double bond of IPP behaves as a nucleophile in displacing diphosphate ion leaving group (PP_i) on DMAPP. Evidence suggests that the DMAPP develops considerable cationic character and that spontaneous dissociation of the allylic diphosphate ion in an S_N1-like pathway probably occurs **(Figure 27.9)**.

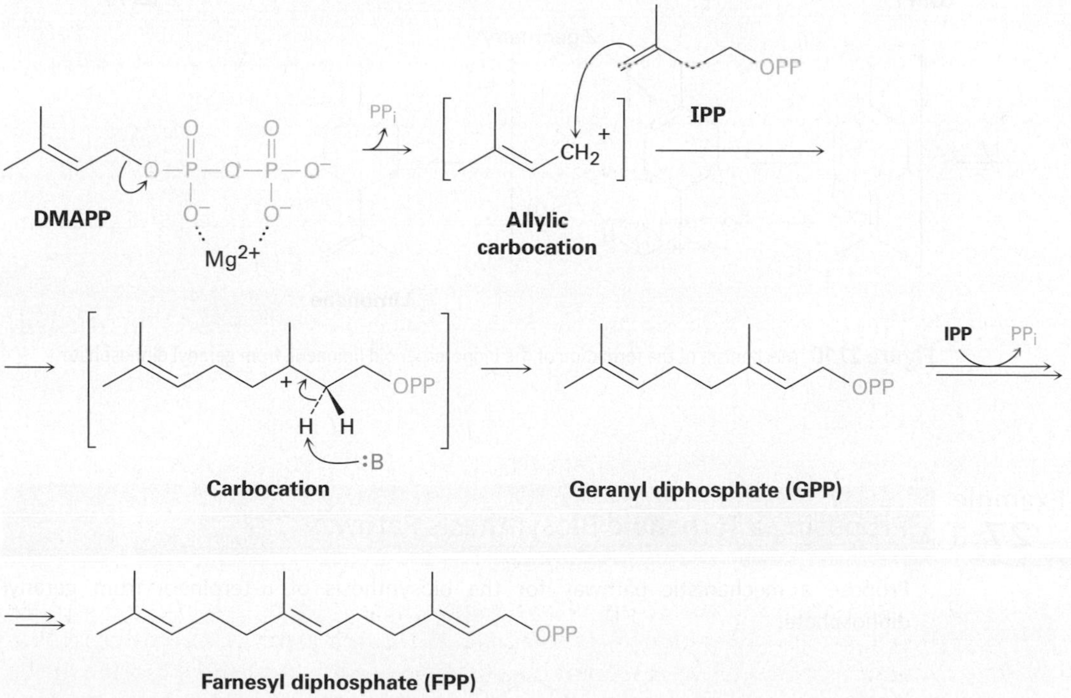

Figure 27.9 Mechanism of the coupling reaction of dimethylallyl diphosphate (DMAPP) and isopentenyl diphosphate (IPP), to give geranyl diphosphate (GPP).

The further conversion of geranyl diphosphate into monoterpenoids typically involves carbocation intermediates and multistep reaction pathways that are catalyzed by terpene cyclases. Monoterpene cyclases function by first isomerizing geranyl diphosphate to its allylic isomer linalyl diphosphate (LPP), a process that occurs by spontaneous S_N1-like dissociation to an allylic carbocation, followed by recombination. The effect of this isomerization is to convert the C2–C3 double bond of GPP into a single bond, thereby making cyclization possible and allowing *E*/*Z* isomerization of the double bond.

Further dissociation and cyclization by electrophilic addition of the cationic carbon to the terminal double bond then gives a cyclic cation, which might either rearrange, undergo a hydride shift, be captured by a nucleophile, or be deprotonated to give any of the several hundred known monoterpenoids. As just one example, limonene, a monoterpenoid found in many citrus oils, arises by the biosynthetic pathway shown in **Figure 27.10**.

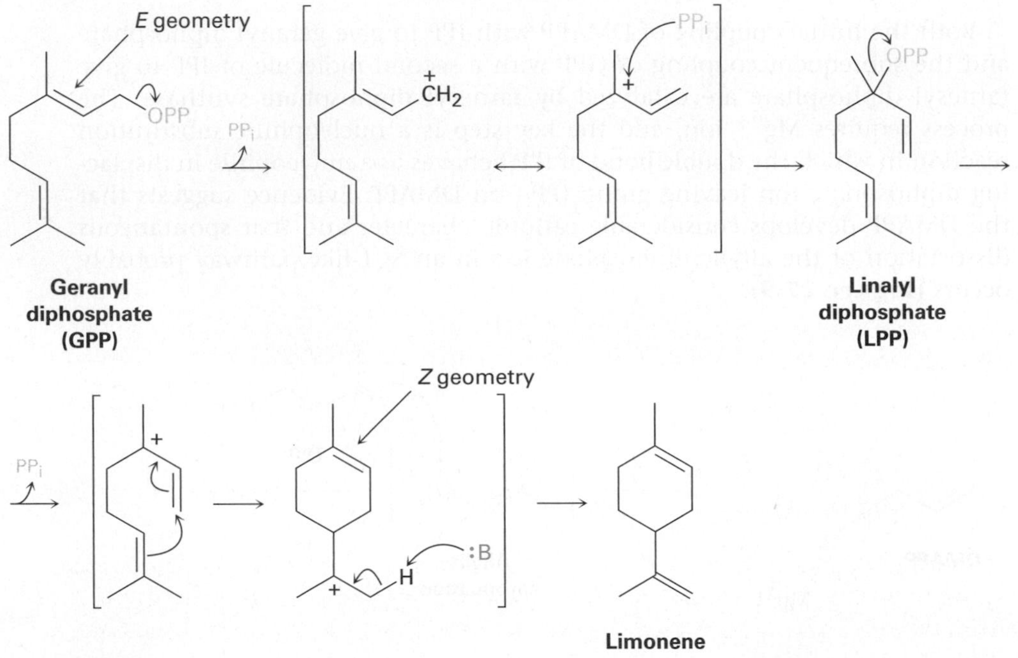

E geometry

OPP

PP_i

$\overset{+}{C}H_2$

PP_i

OPP

Geranyl diphosphate (GPP)

Linalyl diphosphate (LPP)

Z geometry

PP_i

:B

H

Limonene

Figure 27.10 Mechanism of the formation of the monoterpenoid limonene from geranyl diphosphate.

Worked Example 27.1 Proposing a Terpenoid Biosynthesis Pathway

Propose a mechanistic pathway for the biosynthesis of α-terpineol from geranyl diphosphate.

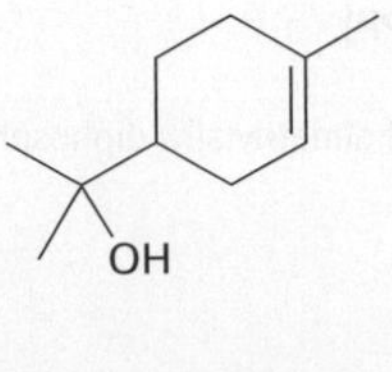

α-Terpineol

OH

Strategy

α-Terpineol, a monoterpenoid, must be derived biologically from geranyl diphosphate through its isomer linalyl diphosphate. Draw the precursor in a conformation that approximates the structure of the target molecule, and then carry out a cationic cyclization, using the appropriate double bond to displace the diphosphate leaving group. Since the target is an alcohol, the carbocation resulting from cyclization evidently reacts with water.

Solution

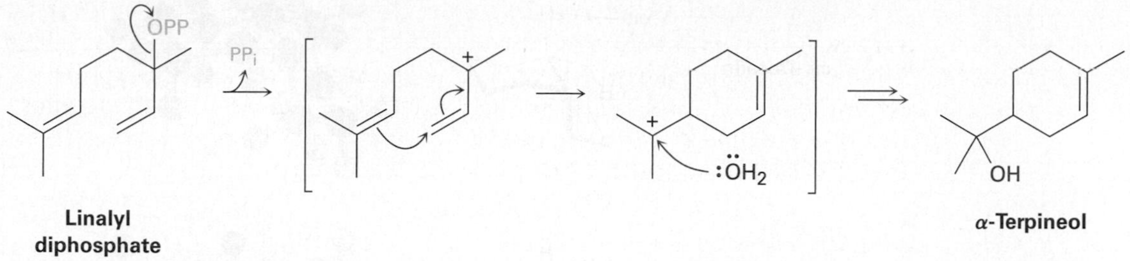

OPP

PP_i

$:\ddot{O}H_2$

OH

Linalyl diphosphate

α-Terpineol

Problem 27.7

Propose mechanistic pathways for the biosynthetic formation of the following terpenoids:

(a)

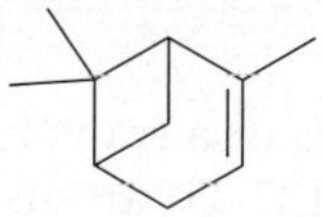

α-Pinene

(b)

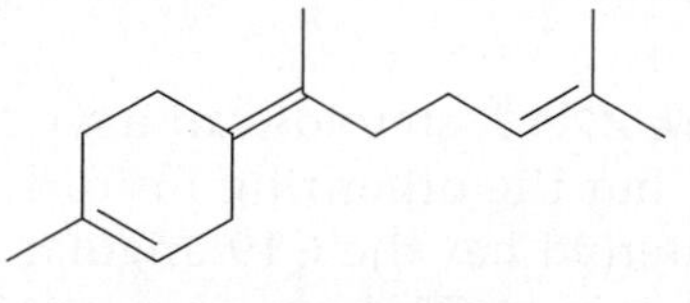

γ-Bisabolene

27.6 Steroids

In addition to fats, phospholipids, eicosanoids, and terpenoids, the lipid extracts of plants and animals also contain **steroids**, molecules that are derived from the triterpenoid lanosterol (Figure 27.6) and whose structures are based on a tetracyclic ring system. The four rings are designated A, B, C, and D, beginning at the lower left, and the carbon atoms are numbered beginning in the A ring. The three 6-membered rings (A, B, and C) adopt chair conformations but are prevented by their rigid geometry from undergoing the usual cyclohexane ring-flips **(Section 4.6)**.

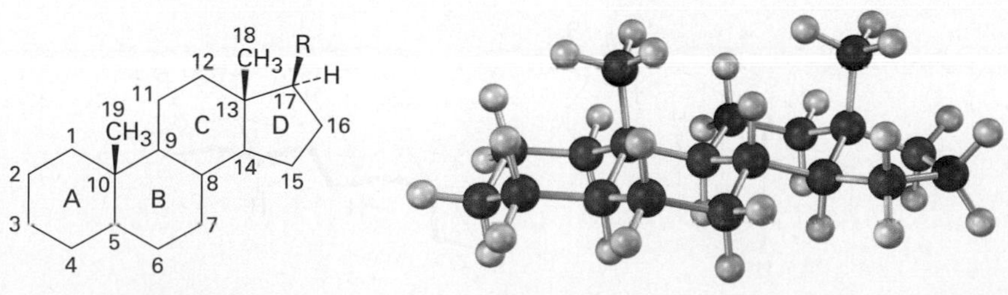

A steroid
(R = various side chains)

Two cyclohexane rings can be joined in either a cis or a trans manner. With cis fusion to give *cis*-decalin, both groups at the ring-junction positions (the *angular* groups) are on the same side of the two rings. With trans fusion to give *trans*-decalin, the groups at the ring junctions are on opposite sides.

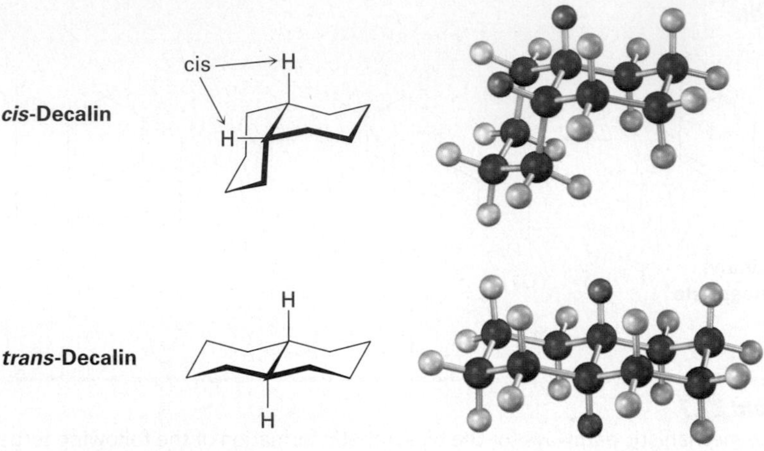

As shown in **Figure 27.11**, steroids can have either a cis or a trans fusion of the A and B rings, but the other ring fusions (B–C and C–D) are usually trans. An A–B trans steroid has the C19 angular methyl group up, denoted β, and the hydrogen atom at C5 down, denoted α, on opposite sides of the molecule. An A–B cis steroid, by contrast, has both the C19 angular methyl

Figure 27.11 Steroid conformations. The three 6-membered rings have chair conformations but are unable to undergo ring-flips. The A and B rings can be either cis-fused or trans-fused.

An A–B trans steroid

An A–B cis steroid

group and the C5 hydrogen atom on the same side (β) of the molecule. Both kinds of steroids are relatively long, flat molecules that have their two methyl groups (C18 and C19) protruding axially above the ring system. The A–B trans steroids are the more common, although A–B cis steroids are found in liver bile.

Substituent groups on the steroid ring system can be either axial or equatorial. As with simple cyclohexanes **(Section 4.7)**, equatorial substitution is generally more favorable than axial substitution for steric reasons. The hydroxyl group at C3 of cholesterol, for example, has the more stable equatorial orientation. Unlike what happens with simple cyclohexanes, however, steroids are rigid molecules whose geometry prevents cyclohexane ring-flips.

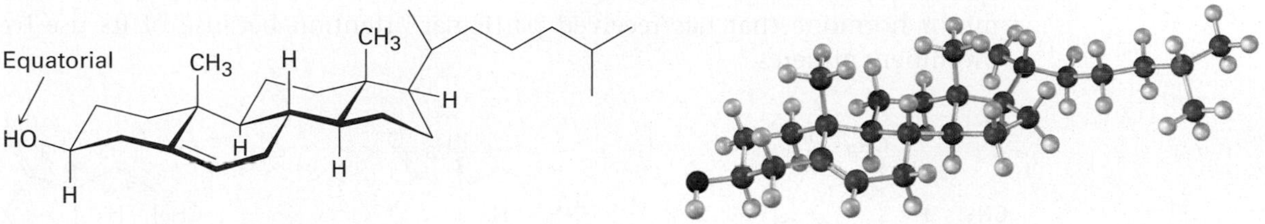

Equatorial HO H CH3 H CH3 H H H

Cholesterol

Problem 27.8

Draw the following molecules in chair conformations, and tell whether the ring substituents are axial or equatorial:

(a) CH3 H H H

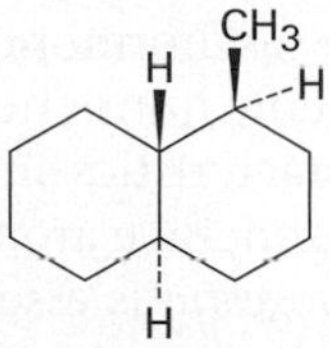

(b) H H CH3 H

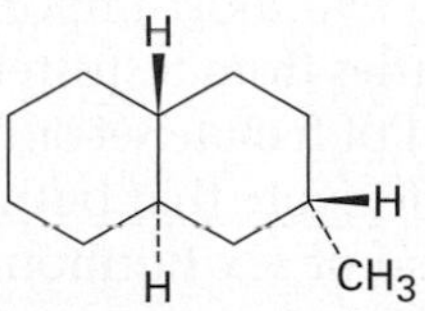

Problem 27.9

Lithocholic acid is an A–B cis steroid found in human bile. Draw lithocholic acid showing chair conformations, as in Figure 27.11, and tell whether the hydroxyl group at C3 is axial or equatorial.

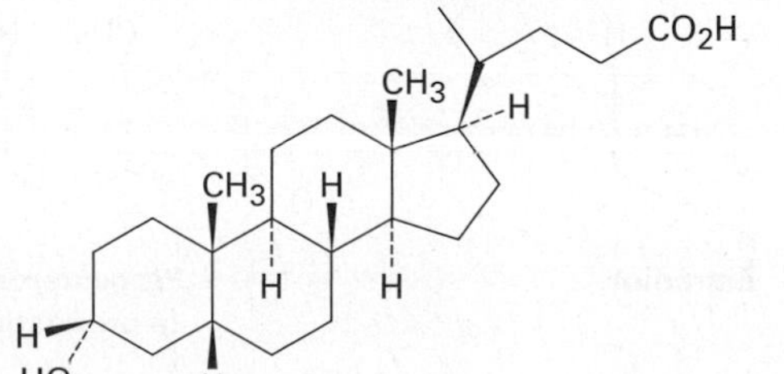

CO_2H CH3 H CH3 H H H H HO H

Lithocholic acid

Steroid Hormones

In humans, most steroids function as **hormones**, chemical messengers that are secreted by endocrine glands and carried through the bloodstream to target tissues. There are two main classes of steroid hormones: the *sex hormones,* which control maturation, tissue growth, and reproduction, and the *adrenocortical hormones,* which regulate a variety of metabolic processes.

Sex Hormones

Testosterone and androsterone are the two most important male sex hormones, or *androgens*. Androgens are responsible for the development of male secondary sex characteristics during puberty and for promoting tissue and muscle growth. Both are synthesized in the testes from cholesterol. Androstenedione is another minor hormone that has received particular attention because of its use by prominent athletes.

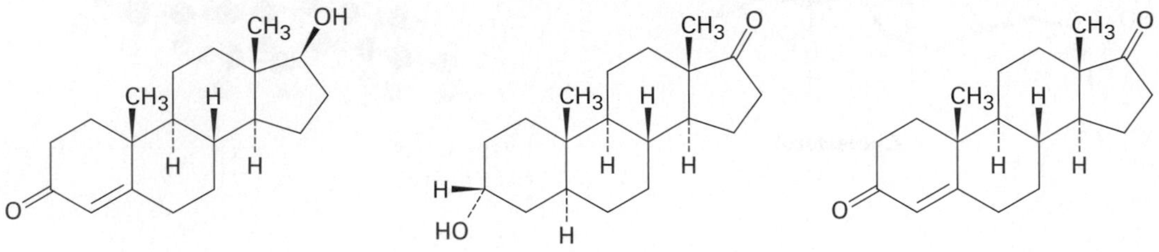

Testosterone **Androsterone** **Androstenedione**

(Androgens)

Estrone and estradiol are the two most important female sex hormones, or *estrogens*. Synthesized in the ovaries from testosterone, estrogenic hormones are responsible for the development of female secondary sex characteristics and for regulation of the menstrual cycle. Note that both have a benzene-like aromatic A ring. In addition, another kind of sex hormone called a *progestin* is essential for preparing the uterus for implantation of a fertilized ovum during pregnancy. Progesterone is the most important progestin.

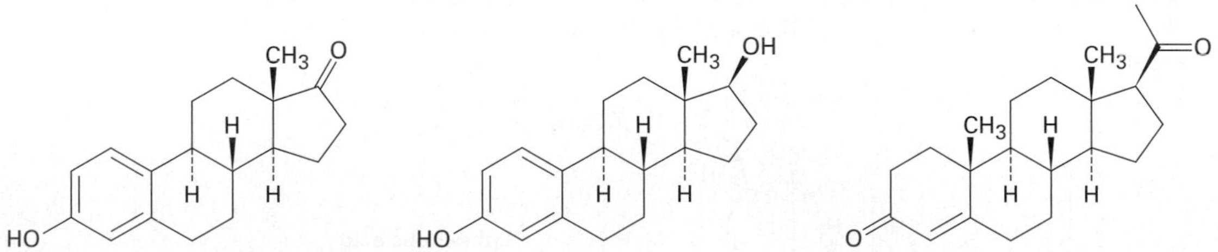

Estrone **Estradiol** **Progesterone (a progestin)**

(Estrogens)

Adrenocortical Hormones Adrenocortical steroids are secreted by the adrenal glands, small organs located near the upper end of each kidney. There are two types of adrenocortical steroids, called *mineralocorticoids* and *glucocorticoids*. Mineralocorticoids, such as aldosterone, control tissue swelling by regulating cellular salt balance between Na^+ and K^+. Glucocorticoids, such as hydrocortisone, are involved in the regulation of glucose metabolism and in the control of inflammation. Glucocorticoid ointments are widely used to bring down the swelling from exposure to poison oak or poison ivy.

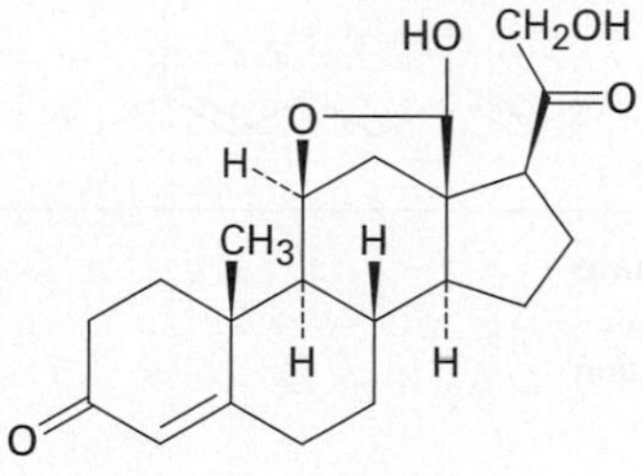

Aldosterone
(a mineralocorticoid)

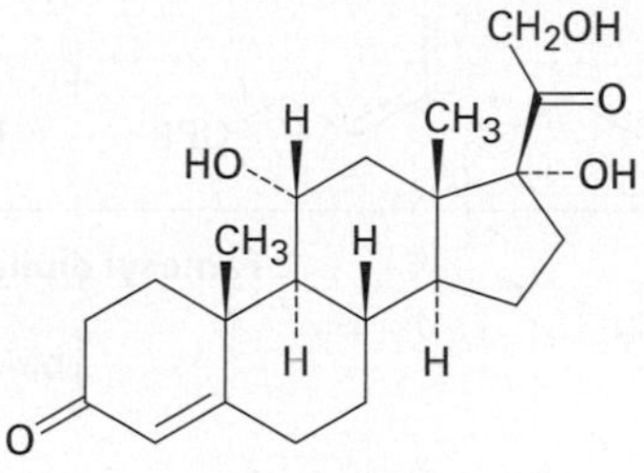

Hydrocortisone
(a glucocorticoid)

Synthetic Steroids In addition to the many hundreds of steroids isolated from plants and animals, thousands more have been synthesized in pharmaceutical laboratories in a search for new drugs. Among the best-known synthetic steroids are the oral contraceptives and anabolic agents. Most birth control pills are a mixture of two compounds, a synthetic estrogen, such as ethynylestradiol, and a synthetic progestin, such as norethindrone. Anabolic steroids, such as methandrostenolone (Dianabol), are synthetic androgens that mimic the tissue-building effects of natural testosterone.

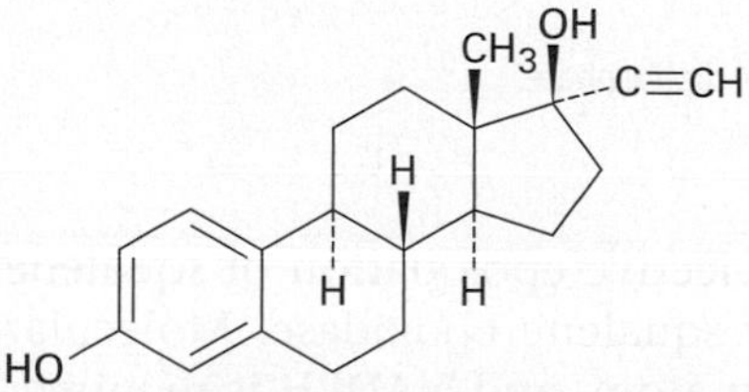

Ethynylestradiol
(a synthetic estrogen)

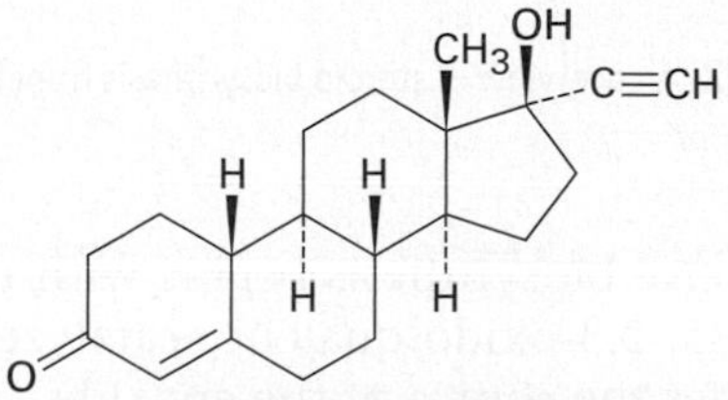

Norethindrone
(a synthetic progestin)

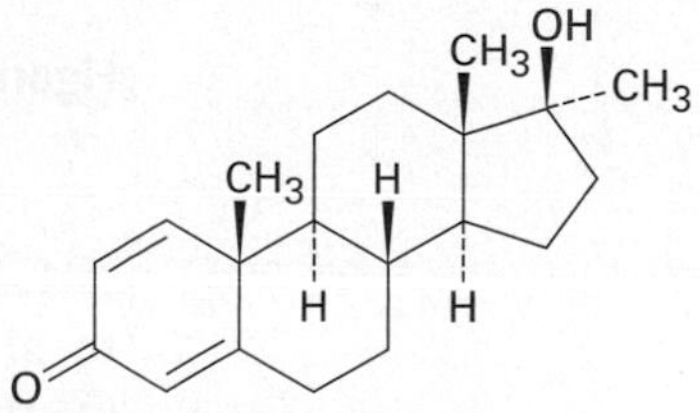

Methandrostenolone
(Dianabol)

27.7 Biosynthesis of Steroids

Steroids are heavily modified triterpenoids that are biosynthesized in living organisms from farnesyl diphosphate (C_{15}). A reductive dimerization first converts farnesyl diphosphate to the acyclic hydrocarbon squalene (C_{30}), which is converted into lanosterol **(Figure 27.12)**. Further rearrangements and degradations then take place to yield various steroids. The conversion of squalene to lanosterol is among the most intensively studied of all biosynthetic transformations. Starting from an achiral, open-chain polyene, the entire process requires only two enzymes and results in the formation of six carbon–carbon bonds, four rings, and seven chirality centers.

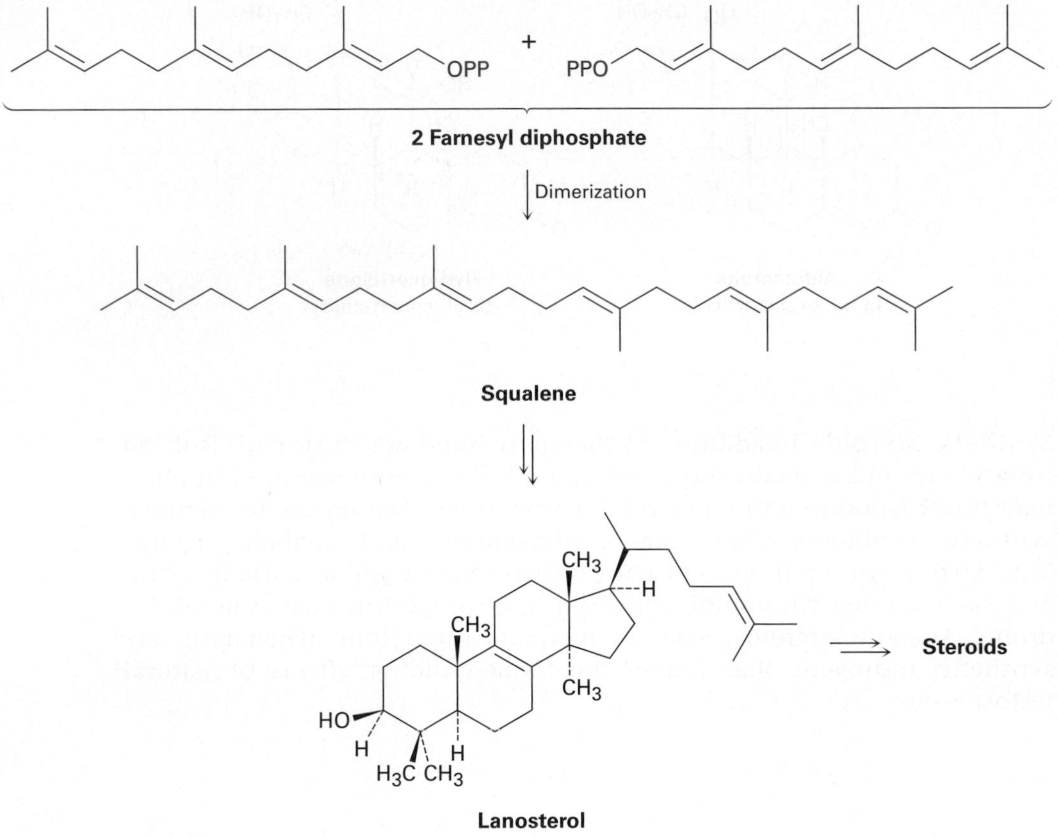

Figure 27.12 An overview of steroid biosynthesis from farnesyl diphosphate.

Lanosterol biosynthesis begins with the selective epoxidation of squalene to give (3*S*)-2,3-oxidosqualene, catalyzed by squalene epoxidase. Molecular O_2 provides the source of the epoxide oxygen atom, and NADPH is required, along with a flavin coenzyme. The proposed mechanism involves reaction of $FADH_2$ with O_2 to produce a flavin hydroperoxide intermediate (ROOH), which transfers an oxygen to squalene in a pathway initiated by nucleophilic attack of the squalene double bond on the terminal hydroperoxide oxygen **(Figure 27.13)**. The flavin alcohol formed as a by-product loses H_2O to give FAD, which is reduced back to $FADH_2$ by NADPH. As noted in **Section 8.7**, this biological epoxidation mechanism is closely analogous to the mechanism by which peroxyacids (RCO_3H) react with alkenes to give epoxides in the laboratory.

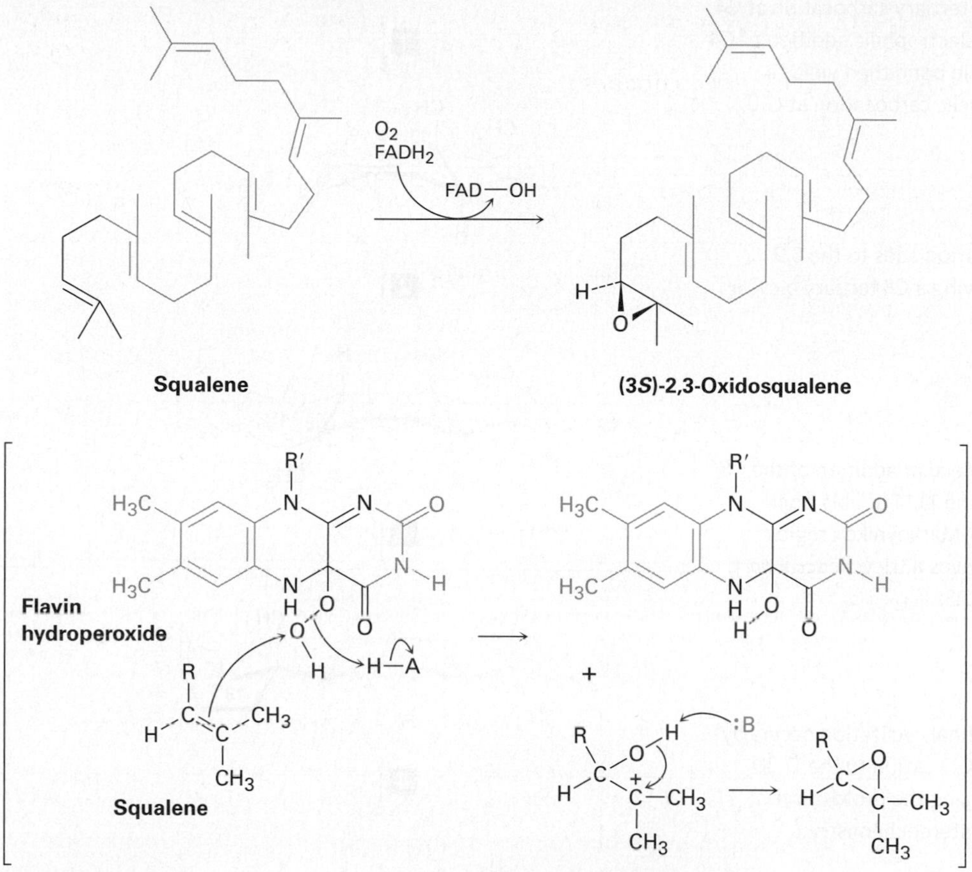

Figure 27.13 Proposed mechanism of the oxidation of squalene by flavin hydroperoxide.

The second part of lanosterol biosynthesis is catalyzed by oxidosqualene: lanosterol cyclase and occurs as shown in **Figure 27.14**. Squalene is folded by the enzyme into a conformation that aligns the various double bonds for undergoing a cascade of successive intramolecular electrophilic additions, followed by a series of hydride and methyl migrations. Except for the initial epoxide protonation/cyclization, the process is probably stepwise and appears to involve discrete carbocation intermediates that are stabilized by electrostatic interactions with electron-rich aromatic amino acids in the enzyme.

1 Protonation on oxygen opens the epoxide ring and gives a tertiary carbocation at C4. Intramolecular electrophilic addition of C4 to the 5,10 double bond then yields a tertiary monocyclic carbocation at C10.

(3*S*)-2,3-Oxidosqualene

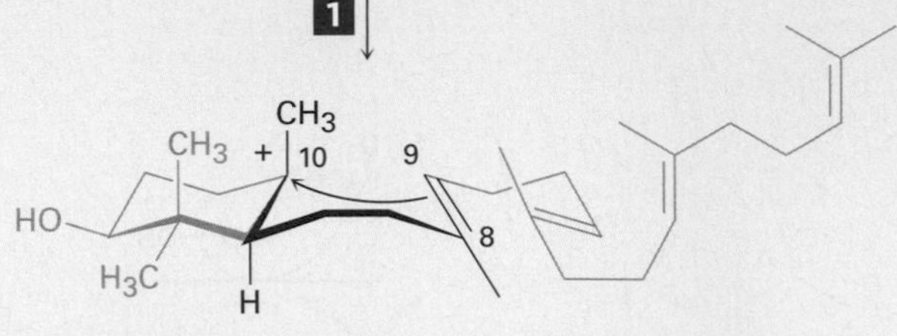

2 The C10 carbocation adds to the 8,9 double bond, giving a C8 tertiary bicyclic carbocation.

3 Further intramolecular addition of the C8 carbocation to the 13,14 double bond occurs with non-Markovnikov regiochemistry and gives a tricyclic *secondary* carbocation at C13.

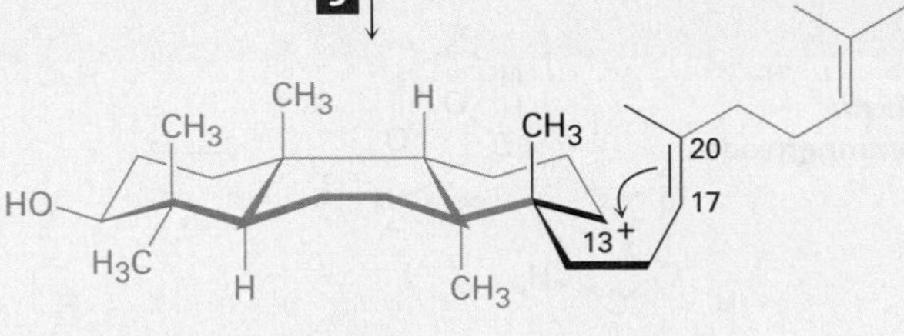

4 The fourth and final cyclization occurs by addition of the C13 cation to the 17,20 double bond, giving the protosteryl cation with 17β stereochemistry.

Protosteryl cation

Figure 27.14 | MECHANISM

Mechanism of the conversion of 2,3-oxidosqualene to lanosterol. Four cationic cyclizations are followed by four rearrangements and a final loss of H^+ from C9. The steroid numbering system is used for referring to specific positions in the intermediates (Section 27.6). Individual steps are explained in the text

Protosteryl cation

5 Hydride migration from C17 to C20 occurs, establishing *R* stereochemistry at C20.

6 A second hydride migration takes place, from C13 to C17, establishing the final 17*β* stereochemistry of the side chain.

7 Methyl migration from C14 to C13 occurs.

8 A second methyl migration occurs, from C8 to C14.

9 Loss of a proton from C9 forms an 8,9 double bond and gives lanosterol.

Lanosterol

Figure 27.14 (Continued)

Steps 1, 2 of Figure 27.14: Epoxide Opening and Initial Cyclizations Cyclization begins in step 1 with protonation of the epoxide ring by an aspartic acid residue in the enzyme. Nucleophilic opening of the protonated epoxide by the nearby 5,10 double bond (steroid numbering; **Section 27.6**) then yields a tertiary carbocation at C10. Further addition of C10 to the 8,9 double bond in step 2 next gives a bicyclic tertiary cation at C8.

(3*S*)-2,3-Oxidosqualene

Step 3 of Figure 27.14: Third Cyclization The third cationic cyclization is somewhat unusual because it occurs with non-Markovnikov regiochemistry and gives a secondary cation at C13 rather than the alternative tertiary cation at C14. There is growing evidence, however, that the tertiary carbocation may in fact be formed initially and that the secondary cation arises by subsequent rearrangement. The secondary cation is probably stabilized in the enzyme pocket by the proximity of an electron-rich aromatic ring.

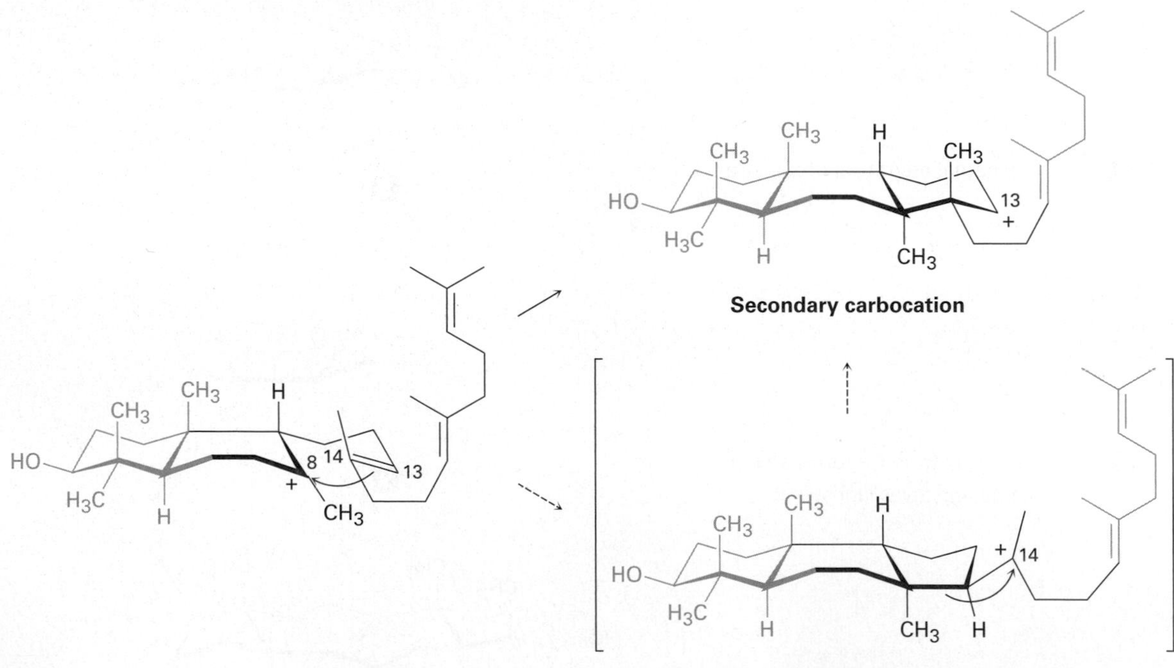

Secondary carbocation

Tertiary carbocation

Step 4 of Figure 27.14: Final Cyclization The fourth and last cyclization occurs in step 4 by addition of the cationic center at C13 to the 17,20 double bond, giving what is known as the *protosteryl* cation. The side-chain alkyl group at C17 has β (up) stereochemistry, although this stereochemistry is lost in step 5 and then reset in step 6.

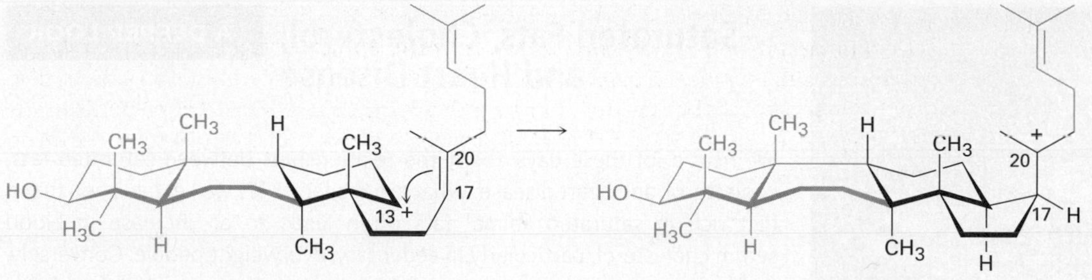

Protosteryl cation

Steps 5–9 of Figure 27.14: Carbocation Rearrangements Once the tetracyclic carbon skeleton of lanosterol has been formed, a series of carbocation rearrangements occur **(Section 7.11)**. The first rearrangement, hydride migration from C17 to C20, occurs in step 5 and results in establishment of *R* stereochemistry at C20 in the side chain. A second hydride migration then occurs from C13 to C17 on the α (bottom) face of the ring in step 6 and reestablishes the 17β orientation of the side chain. Finally, two methyl migrations, the first from C14 to C13 on the top (β) face and the second from C8 to C14 on the bottom (α) face, place the positive charge at C8. A basic histidine residue in the enzyme then removes the neighboring β proton from C9 to give lanosterol.

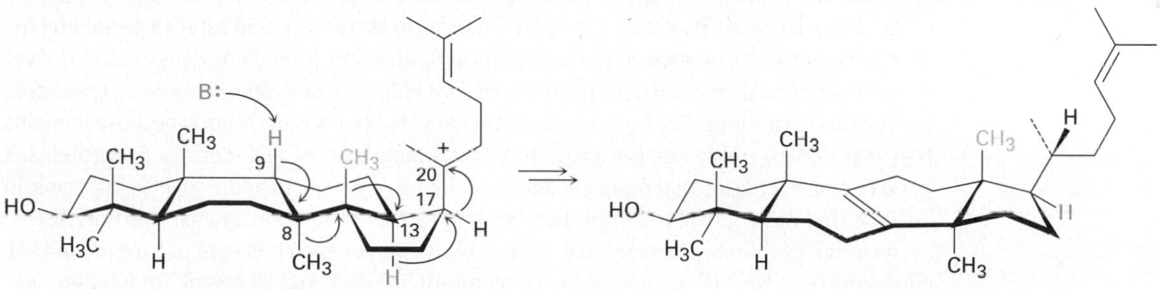

Protosteryl cation

Lanosterol

From lanosterol, the pathway for steroid biosynthesis continues on to yield cholesterol. Cholesterol then becomes a branch point, serving as the common precursor from which all other steroids are derived.

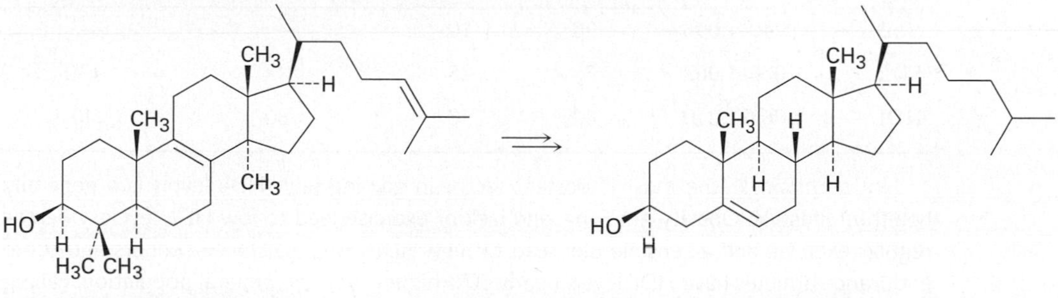

Lanosterol

Cholesterol

Problem 27.10

Compare the structures of lanosterol and cholesterol, and catalog the changes needed for the transformation.

iStockphoto.com/Rob Friedman

It's hard to resist, but a high intake of saturated animal fat doesn't do much for your cholesterol level.

Saturated Fats, Cholesterol, and Heart Disease

A DEEPER LOOK

We hear a lot these days about the relationships between saturated fats, cholesterol, and heart disease. What are the facts? It's well established that a diet rich in saturated animal fats often leads to an increase in blood serum cholesterol, particularly in sedentary, overweight people. Conversely, a diet lower in saturated fats and higher in polyunsaturated fats leads to a lower serum cholesterol level. Studies have shown that a serum cholesterol level greater than 240 mg/dL (a desirable value is <200 mg/dL) is correlated with an increased incidence of coronary artery disease, in which cholesterol deposits build up on the inner walls of coronary arteries, blocking the flow of blood to the heart muscles.

A better indication of a person's risk of heart disease comes from a measurement of blood lipoprotein levels. *Lipoproteins* are complex molecules with both lipid and protein parts that transport lipids through the body. They can be divided into three types according to density, as shown in Table 27.3. Very-low-density lipoproteins (VLDLs) act primarily as carriers of triglycerides from the intestines to peripheral tissues, whereas low-density lipoproteins (LDLs) and high-density lipoproteins (HDLs) act as carriers of cholesterol to and from the liver.

Evidence suggests that LDLs transport cholesterol as its fatty-acid ester to peripheral tissues, whereas HDLs remove cholesterol as its stearate ester from dying cells. If LDLs deliver more cholesterol than is needed, and if insufficient HDLs are present to remove it, the excess is deposited in arteries. Thus, a *low* level of *low*-density lipoproteins is good because it means that less cholesterol is being transported, and a *high* level of *high*-density lipoproteins is good because it means that more cholesterol is being removed. In addition, HDL contains an enzyme that has antioxidant properties, offering further protection against heart disease.

As a rule of thumb, a person's risk drops about 25% for each increase of 5 mg/dL in HDL concentration. Normal values are about 45 mg/dL for men and 55 mg/dL for women, perhaps explaining why premenopausal women appear to be somewhat less susceptible than men to heart disease.

Table 27.3 Serum Lipoproteins

Name	Density (g/mL)	% Lipid	% Protein	Optimal (mg/dL)	Poor (mg/dL)
VLDL	0.940–1.006	90	10	—	—
LDL	1.006–1.063	75	25	<100	>130
HDL	1.063–1.210	60	40	>60	<40

Not surprisingly, the most important factor in gaining high HDL levels is a generally healthful lifestyle. Obesity, smoking, and lack of exercise lead to low HDL levels, whereas regular exercise and a sensible diet lead to high HDL levels. Distance runners and other endurance athletes have HDL levels nearly 50% higher than the general population. Failing that—not everyone wants to run 30 miles or bike 100 miles per week—diet is also important. Diets high in cold-water fish like salmon and whitefish, raise HDL and lower blood cholesterol because these fish contain almost entirely polyunsaturated fat, including a large percentage of omega-3 fatty acids. Animal fat from red meat and cooking fats should be minimized because saturated fats and monounsaturated trans fats raise blood cholesterol.

Summary

Lipids are the naturally occurring materials isolated from plants and animals by extraction with nonpolar organic solvents. Animal fats and vegetable oils are the most widely occurring lipids. Both are **triacylglycerols**—triesters of glycerol with long-chain **fatty acids**. Animal fats are usually saturated, whereas vegetable oils usually have unsaturated fatty acid residues.

Phospholipids are important constituents of cell membranes and are of two kinds. *Glycerophospholipids,* such as phosphatidylcholine and phosphatidylethanolamine, are closely related to fats in that they have a glycerol backbone esterified to two fatty acids (one saturated and one unsaturated) and to one phosphate ester. *Sphingomyelins* have the amino alcohol sphingosine for their backbone.

Eicosanoids and **terpenoids** are still other classes of lipids. Eicosanoids, of which prostaglandins are the most abundant kind, are derived biosynthetically from arachidonic acid, are found in all body tissues, and have a wide range of physiological activity. Terpenoids are often isolated from the essential oils of plants, have an immense diversity of structure, and are produced biosynthetically from the five-carbon precursor isopentenyl diphosphate (IPP). Isopentenyl diphosphate is itself biosynthesized from 3 equivalents of acetate in the mevalonate pathway.

Steroids are plant and animal lipids with a characteristic tetracyclic carbon skeleton. Like the eicosanoids, steroids occur widely in body tissues and have a large variety of physiological activities. Steroids are closely related to terpenoids and arise biosynthetically from the triterpenoid lanosterol. Lanosterol, in turn, arises from cationic cyclization of the acyclic hydrocarbon squalene.

Key words

eicosanoid, 1096
fatty acid, 1089
hormone, 1110
lipid, 1088
lipid bilayer, 1095
micelle, 1092
phospholipid, 1094
polyunsaturated fatty acid, 1089
prostaglandin, 1095
steroid, 1107
terpenoid, 1098
triacylglycerol, 1089
wax, 1088

Exercises

Visualizing Chemistry

(Problems 27.1–27.10 appear within the chapter.)

OWL Interactive versions of these problems are assignable in OWL for Organic Chemistry.

27.11 The following model is that of cholic acid, a constituent of human bile. Locate the three hydroxyl groups, and identify each as axial or equatorial. Is cholic acid an A–B trans steroid or an A–B cis steroid?

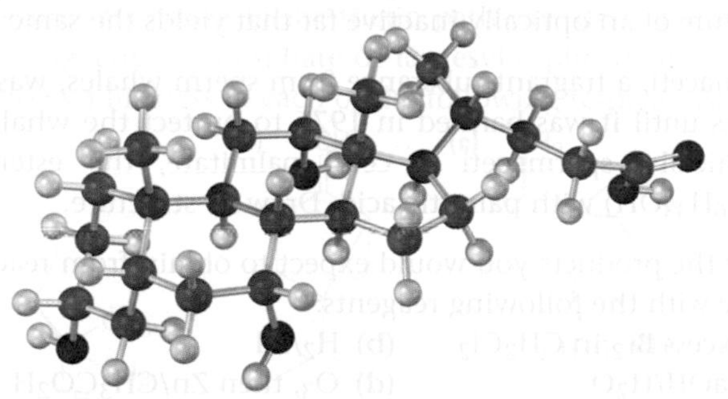

27.35 Draw the most stable chair conformation of menthol, and label each substituent as axial or equatorial.

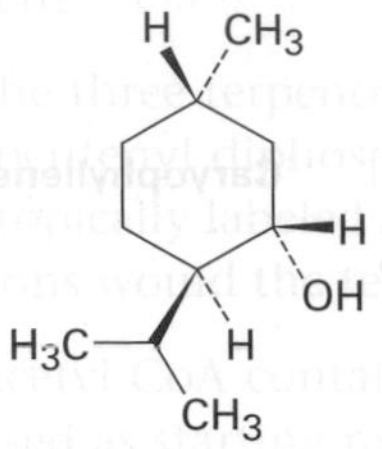

H CH3
H
OH
H3C H
CH3
Menthol (from peppermint oil)

27.36 As a general rule, equatorial alcohols are esterified more readily than axial alcohols. What product would you expect to obtain from reaction of the following two compounds with 1 equivalent of acetic anhydride?

(a) H CH_3 OH HO H H

(b) CH_3 H OH HO H H

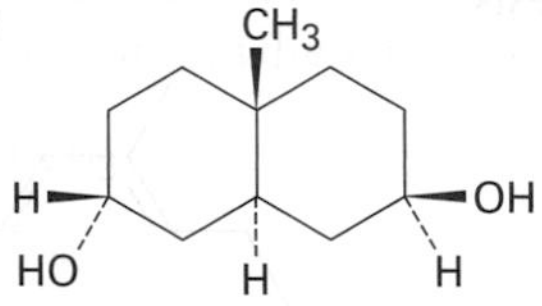

27.37 Propose a mechanistic pathway for the biosynthesis of isoborneol. A carbocation rearrangement is needed at one point in the scheme.

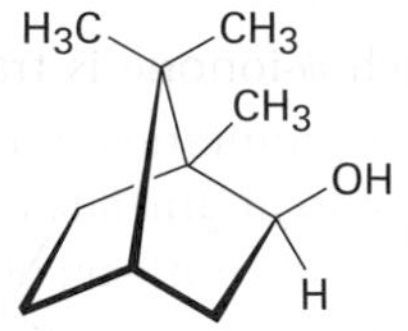

H_3C CH_3 CH_3 OH H

Isoborneol

27.38 Isoborneol (Problem 27.37) is converted into camphene on treatment with dilute sulfuric acid. Propose a mechanism for the reaction, which involves a carbocation rearrangement.

H_3C CH_3 CH_3 OH H $\xrightarrow{H_2SO_4}$ H_2C H_3C CH_3

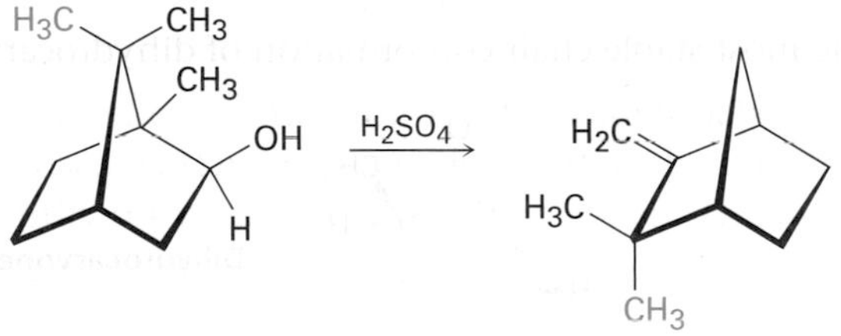

Isoborneol **Camphene**

27.39 Digitoxigenin is a heart stimulant obtained from the purple foxglove *Digitalis purpurea* and used in the treatment of heart disease. Draw the three-dimensional conformation of digitoxigenin, and identify the two –OH groups as axial or equatorial.

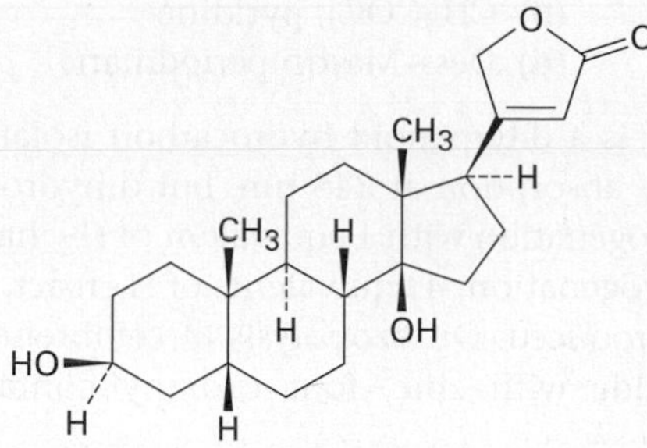

Digitoxigenin

27.40 What product would you obtain by reduction of digitoxigenin (Problem 27.39) with $LiAlH_4$? By oxidation with the Dess–Martin periodinane?

27.41 Vaccenic acid, $C_{18}H_{34}O_2$, is a rare fatty acid that gives heptanal and 11-oxoundecanoic acid [$OHC(CH_2)_9CO_2H$] on ozonolysis followed by zinc treatment. When allowed to react with $CH_2I_2/Zn(Cu)$, vaccenic acid is converted into lactobacillic acid. What are the structures of vaccenic and lactobacillic acids?

27.42 Eleostearic acid, $C_{18}H_{30}O_2$, is a rare fatty acid found in the tung oil used for finishing furniture. On ozonolysis followed by treatment with zinc, eleostearic acid furnishes one part pentanal, two parts glyoxal (OHC—CHO), and one part 9-oxononanoic acid [$OHC(CH_2)_7CO_2H$]. What is the structure of eleostearic acid?

27.43 Diterpenoids are derived biosynthetically from geranylgeranyl diphosphate (GGPP), which is itself biosynthesized by reaction of farnesyl diphosphate with isopentenyl diphosphate. Show the structure of GGPP, and propose a mechanism for its biosynthesis from FPP and IPP.

27.44 Diethylstilbestrol (DES) has estrogenic activity even though it is structurally unrelated to steroids. Once used as an additive in animal feed, DES has been implicated as a causative agent in several types of cancer. Show how DES can be drawn so that it is sterically similar to estradiol.

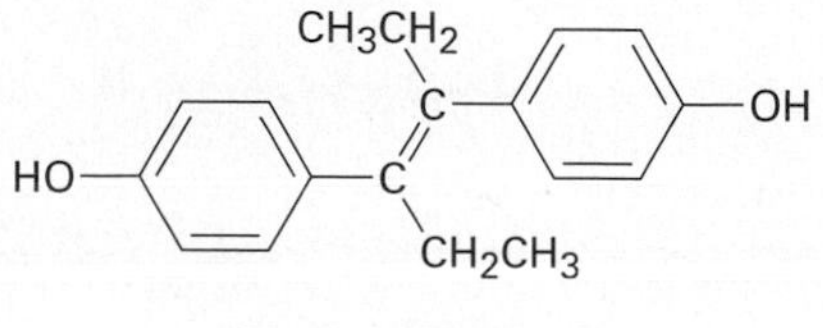

Diethylstilbestrol

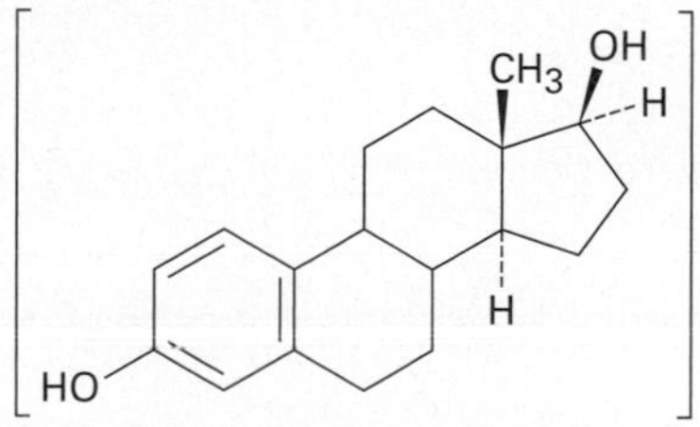

Estradiol

27.45 Propose a synthesis of diethylstilbestrol (Problem 27.44) from phenol and any other organic compound required.

27.46 What products would you expect from reaction of estradiol (Problem 27.44) with the following reagents?

(a) NaH, then CH_3I (b) CH_3COCl, pyridine
(c) Br_2, $FeBr_3$ (d) Dess–Martin periodinane

27.47 Cembrene, $C_{20}H_{32}$, is a diterpenoid hydrocarbon isolated from pine resin. Cembrene has a UV absorption at 245 nm, but dihydrocembrene ($C_{20}H_{34}$), the product of hydrogenation with 1 equivalent of H_2, has no UV absorption. On exhaustive hydrogenation, 4 equivalents of H_2 react, and octahydrocembrene, $C_{20}H_{40}$, is produced. On ozonolysis of cembrene, followed by treatment of the ozonide with zinc, four carbonyl-containing products are obtained:

$$CH_3\overset{O}{\overset{\|}{C}}CH_2CH_2\overset{O}{\overset{\|}{C}}H \;+\; CH_3\overset{O}{\overset{\|}{C}}CHO \;+\; H\overset{O}{\overset{\|}{C}}CH_2\overset{O}{\overset{\|}{C}}H \;+\; CH_3\overset{O}{\overset{\|}{C}}CH_2CH_2\underset{CH_3}{\underset{|}{C}}H\overset{CHO}{\overset{|}{C}}HCH_3$$

Propose a structure for cembrene that is consistent with its formation from geranylgeranyl diphosphate.

27.48 α-Fenchone is a pleasant-smelling terpenoid isolated from oil of lavender. Propose a pathway for the formation of α-fenchone from geranyl diphosphate. A carbocation rearrangement is required.

α-Fenchone

27.49 Fatty acids are synthesized by a multistep route that starts with acetate. The first step is a reaction between protein-bound acetyl and malonyl units to give a protein-bound 3-ketobutyryl unit. Show the mechanism, and tell what kind of reaction is occurring.

$$^{-}O_2C{-}CH_2{-}C(=O){-}S{-}Protein + H_3C{-}C(=O){-}S{-}Protein \longrightarrow \left[H_3C{-}C(=O){-}CH(CO_2^-){-}C(=O){-}S{-}Protein \right] \longrightarrow H_3C{-}C(=O){-}CH_2{-}C(=O){-}S{-}Protein + CO_2$$

Malonyl–protein **Acetyl–protein** **3-Ketobutyryl–protein**

27.50 Propose a mechanism for the biosynthesis of the sesquiterpenoid trichodiene from farnesyl diphosphate. The process involves cyclization to give an intermediate secondary carbocation, followed by several carbocation rearrangements.

OPP

H_3C H H_3C H_3C + H_3C CH_3 H H

Farnesyl diphosphate (FPP) **Trichodiene**

28

If these golden retrievers look similar, that's because they're identical—all cloned from somatic cells of the same donor. Chung Sung-Jun/Getty Images

Biomolecules: Nucleic Acids

The nucleic acids, **deoxyribonucleic acid (DNA)** and **ribonucleic acid (RNA)**, are the chemical carriers of a cell's genetic information. Coded in a cell's DNA is the information that determines the nature of the cell, controls the cell's growth and division, and directs biosynthesis of the enzymes and other proteins required for cellular functions.

In addition to nucleic acids themselves, nucleic acid derivatives such as ATP are involved as phosphorylating agents in many biochemical pathways, and several important coenzymes, including NAD^+, FAD, and coenzyme A, have nucleic acid components. See Table 26.3 on pages 1072 and 1073 for the structures.

Why This Chapter? Nucleic acids are the last of the four major classes of biomolecules we'll consider. So much has been written and spoken about DNA in the media that the basics of DNA replication and transcription are probably known to you. Thus, we'll move fairly quickly through the fundamentals and then look more closely at the chemical details of DNA sequencing, synthesis, and metabolism. The field is moving very rapidly, and there's a lot you may not be familiar with.

28.1 Nucleotides and Nucleic Acids

Just as proteins are biopolymers made of amino acids, nucleic acids are biopolymers made of **nucleotides**, joined together to form a long chain. Each nucleotide is composed of a **nucleoside** bonded to a phosphate group, and each nucleoside is composed of an aldopentose sugar linked through its anomeric carbon to the nitrogen atom of a heterocyclic purine or pyrimidine base.

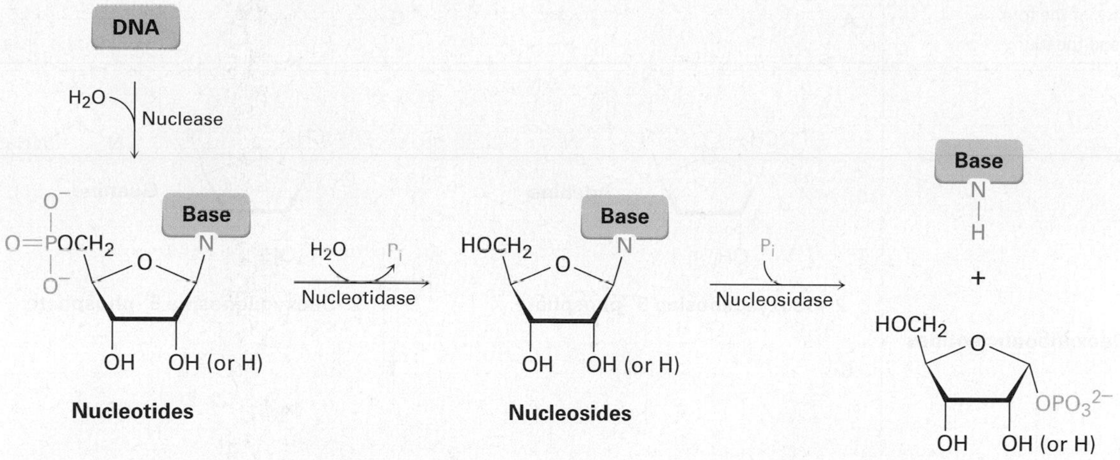

DNA

H_2O Nuclease

Base N, $O=POCH_2$, O^-, OH, OH (or H) — Nucleotides

H_2O, P_i, Nucleotidase

Base N, $HOCH_2$, OH, OH (or H) — Nucleosides

P_i, Nucleosidase

Base N–H + $HOCH_2$, OPO_3^{2-}, OH, OH (or H)

The sugar component in RNA is ribose, and the sugar in DNA is 2′-deoxyribose. (In naming and numbering nucleotides, numbers with a prime superscript refer to positions on the sugar and numbers without a prime superscript refer to positions on the heterocyclic base. Thus, the prefix 2′-deoxy indicates that oxygen is missing from C2′ of ribose.) DNA contains four different amine bases, two substituted purines (adenine and guanine) and two substituted pyrimidines (cytosine and thymine). Adenine, guanine, and cytosine also occur in RNA, but thymine is replaced in RNA by a closely related pyrimidine base called uracil.

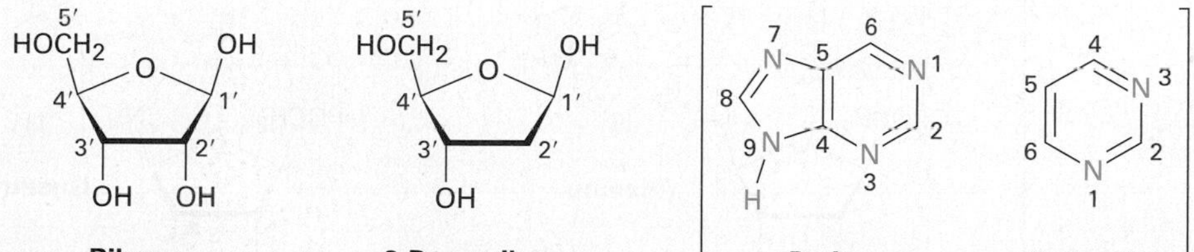

Ribose | 2-Deoxyribose | Purine | Pyrimidine

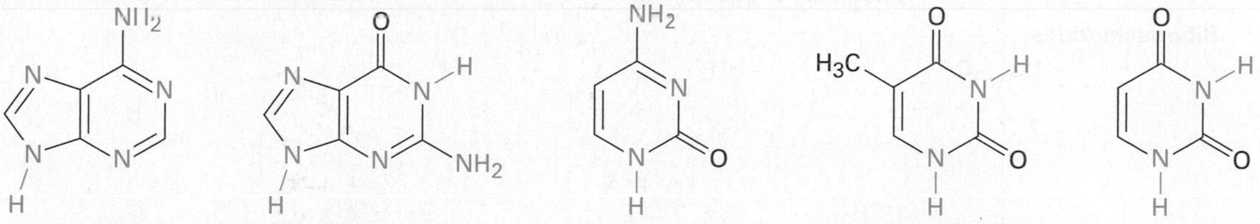

Adenine (A) DNA, RNA | Guanine (G) DNA, RNA | Cytosine (C) DNA, RNA | Thymine (T) DNA | Uracil (U) RNA

The structures of the four deoxyribonucleotides and the four ribonucleotides are shown in **Figure 28.1**. Although similar chemically, DNA and RNA differ dramatically in size. Molecules of DNA are enormous, containing as many as 245 million nucleotides and having molecular weights as high as 75 billion.

Molecules of RNA, by contrast, are much smaller, containing as few as 21 nucleotides and having molecular weights as low as 7000.

Figure 28.1 Structures of the four deoxyribonucleotides and the four ribonucleotides.

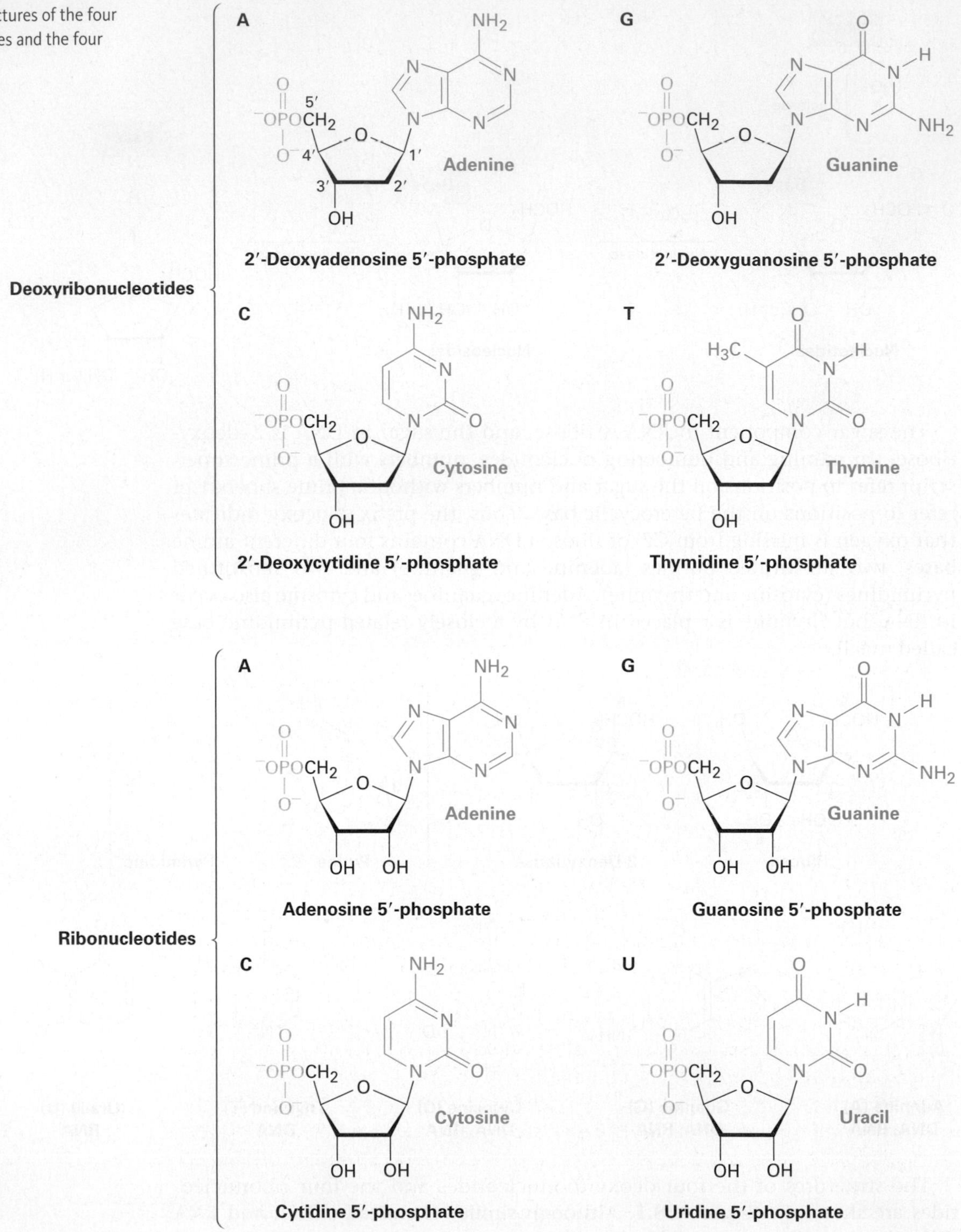

Nucleotides are linked together in DNA and RNA by *phosphodiester* bonds [RO—(PO_2^-)—OR′] between phosphate, the 5′ hydroxyl group on one nucleoside, and the 3′-hydroxyl group on another nucleoside. One end of the nucleic acid polymer has a free hydroxyl at C3′ (the **3′ end**), and the other end has a phosphate at C5′ (the **5′ end**). The sequence of nucleotides in a chain is described by starting at the 5′ end and identifying the bases in order of occurrence, using the abbreviations G, C, A, T (or U for RNA). Thus, a typical DNA sequence might be written as TAGGCT.

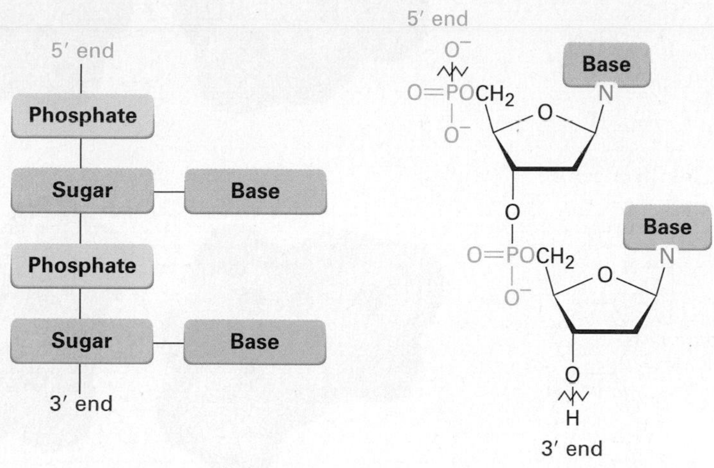

Problem 28.1
Draw the full structure of the DNA dinucleotide AG.

Problem 28.2
Draw the full structure of the RNA dinucleotide UA.

28.2 Base Pairing in DNA: The Watson–Crick Model

Samples of DNA isolated from different tissues of the same species have the same proportions of heterocyclic bases, but samples from different species often have greatly different proportions of bases. Human DNA, for example, contains about 30% each of adenine and thymine and about 20% each of guanine and cytosine. The bacterium *Clostridium perfringens,* however, contains about 37% each of adenine and thymine and only 13% each of guanine and cytosine. Note that in both examples the bases occur in pairs. Adenine and thymine are present in equal amounts, as are cytosine and guanine. Why?

In 1953, James Watson and Francis Crick made their classic proposal for the secondary structure of DNA. According to the Watson–Crick model, DNA under physiological conditions consists of two polynucleotide strands, running in opposite directions and coiled around each other in a **double helix** like the handrails on a spiral staircase. The two strands are complementary rather than identical and are held together by hydrogen bonds between specific pairs of bases, A with T and C with G. That is, whenever an A base occurs in one strand, a T base occurs opposite it in the other strand; when a C base occurs in one, a

G occurs in the other **(Figure 28.2)**. This complementary base-pairing thus explains why A and T are always found in equal amounts, as are G and C.

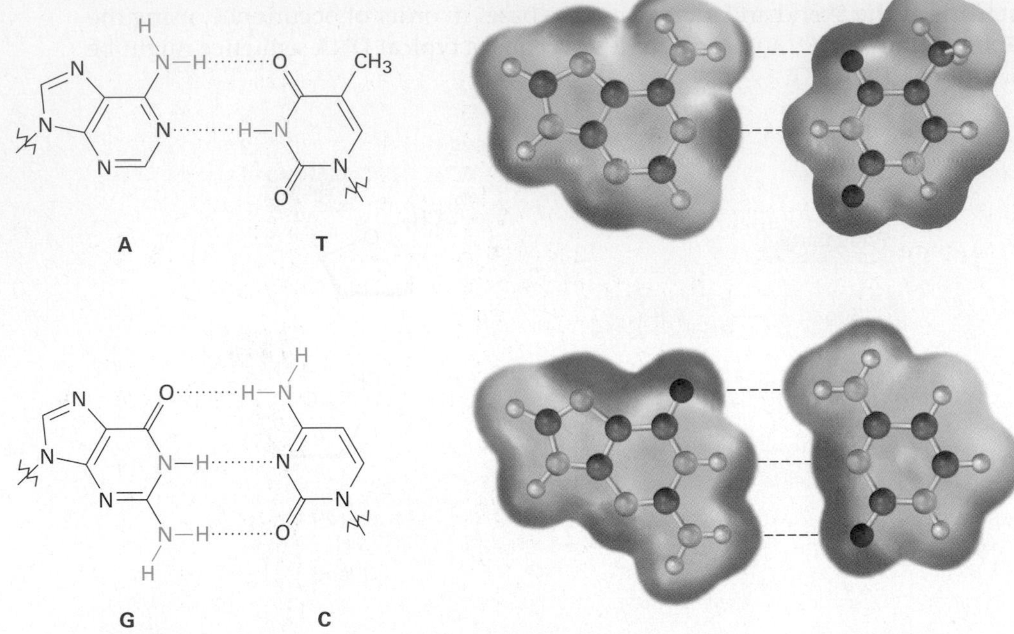

Figure 28.2 Hydrogen-bonding between base pairs in the DNA double helix. Electrostatic potential maps show that the faces of the bases are relatively neutral (green), while the edges have **positive** and **negative** regions. Pairing G with C and A with T brings together oppositely charged regions.

A full turn of the DNA double helix is shown in **Figure 28.3**. The helix is 20 Å wide, there are 10 base pairs per turn, and each turn is 34 Å in length. Notice in Figure 28.3 that the two strands of the double helix coil in such a way that two kinds of "grooves" result, a *major groove* 12 Å wide and a *minor groove* 6 Å wide. The major groove is slightly deeper than the minor groove, and both are lined by flat heterocyclic bases. As a result, a variety of other polycyclic aromatic molecules are able to slip sideways, or *intercalate,* between the stacked bases. Many cancer-causing and cancer-preventing agents function by interacting with DNA in this way.

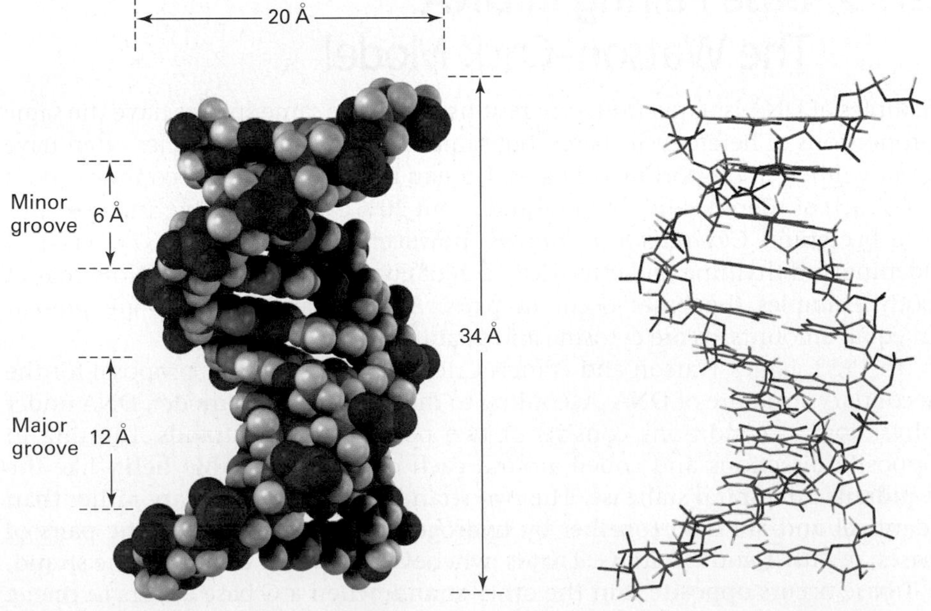

Figure 28.3 A turn of the DNA double helix in both space-filling and wire-frame formats. The sugar–phosphate backbone runs along the outside of the helix, and the amine bases hydrogen bond to one another on the inside. Both major and minor grooves are visible.

An organism's genetic information is stored as a sequence of deoxyribonucleotides strung together in the DNA chain. For the information to be preserved and passed on to future generations, a mechanism must exist for copying DNA. For the information to be used, a mechanism must exist for decoding the DNA message and implementing the instructions it contains.

What Crick called the "central dogma of molecular genetics" says that the function of DNA is to store information and pass it on to RNA. The function of RNA, in turn, is to read, decode, and use the information received from DNA to make proteins. This view is greatly oversimplified but is nevertheless a good place to start. Three fundamental processes take place.

- **Replication**—the process by which identical copies of DNA are made so that information can be preserved and handed down to offspring
- **Transcription**—the process by which the genetic messages are read and carried out of the cell nucleus to ribosomes, where protein synthesis occurs
- **Translation**—the process by which the genetic messages are decoded and used to synthesize proteins

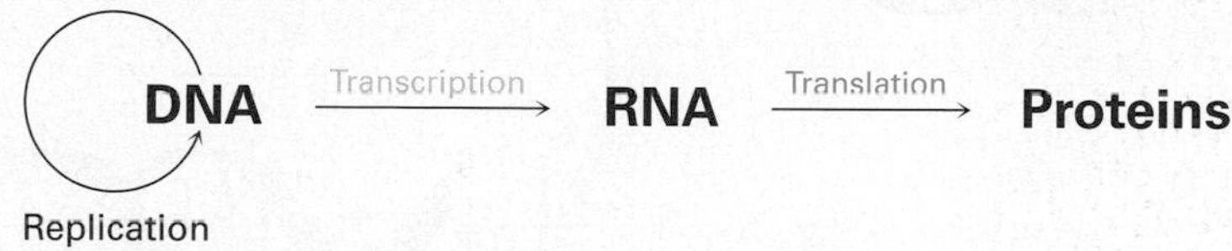

Predicting the Complementary Base Sequence in Double-Stranded DNA

Worked Example 28.1

What sequence of bases on one strand of DNA is complementary to the sequence TATGCAT on another strand?

Strategy

Remember that A and G form complementary pairs with T and C, respectively, and then go through the sequence replacing A by T, G by C, T by A, and C by G. Remember also that the 5′ end is on the left and the 3′ end is on the right in the original strand.

Solution

Original: (5′) TATGCAT (3′)
Complement: (3′) ATACGTA (5′) or (5′) ATGCATA (3′)

Problem 28.3

What sequence of bases on one strand of DNA is complementary to the following sequence on another strand?

(5′) GGCTAATCCGT (3′)

28.3 Replication of DNA

DNA **replication** is an enzyme-catalyzed process that begins with a partial unwinding of the double helix at various points along the chain, brought about by enzymes called *helicases*. Hydrogen bonds are broken, the two strands separate to form a "bubble," and bases are exposed. New nucleotides then line up on each strand in a complementary manner, A to T and G to C, and two new strands begin to grow from the ends of the bubble, called the *replication forks.* Each new strand is complementary to its old template strand, so two identical DNA double helices are produced **(Figure 28.4)**. Because each of the new DNA molecules contains one old strand and one new strand, the process is described as *semiconservative replication.*

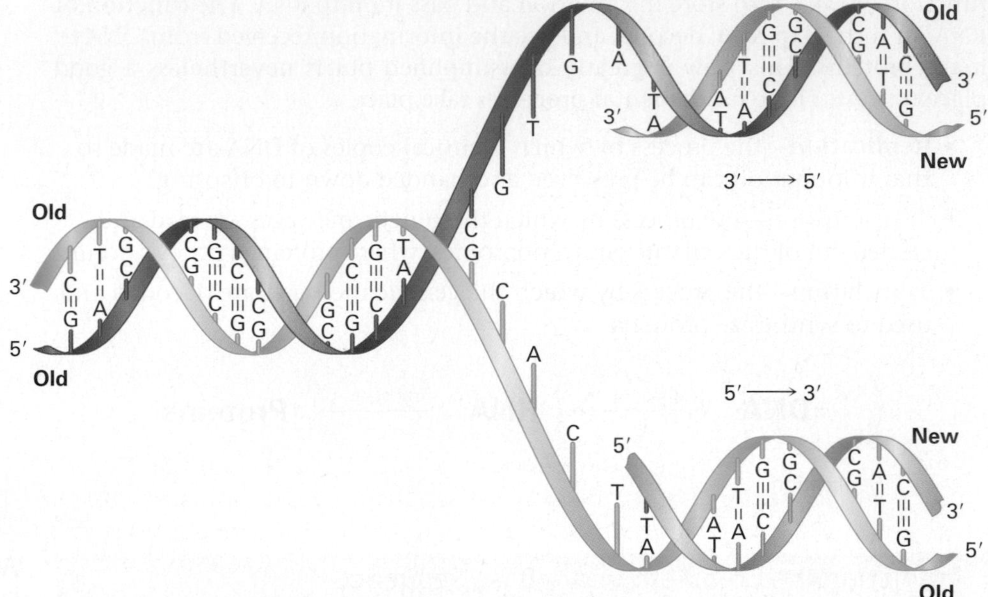

Figure 28.4 A representation of semiconservative DNA replication. The original double-stranded DNA partially unwinds, bases are exposed, nucleotides line up on each strand in a complementary manner, and two new strands begin to grow. Both strands are synthesized in the same 5′ → 3′ direction, one continuously and one in fragments.

Addition of nucleotides to the growing chain takes place in the 5′→3′ direction and is catalyzed by DNA polymerase. The key step is the addition of a nucleoside 5′-triphosphate to the free 3′-hydroxyl group of the growing chain with loss of a diphosphate leaving group.

5′ end
3′ end
Template strand
New strand
5′ end

Because both new DNA strands are synthesized in the 5′ → 3′ direction, they can't be made in exactly the same way. One new strand must have its 3′ end nearer a replication fork, while the other new strand has its 5′ end nearer the replication fork. What happens is that the complement of the original 5′ → 3′ strand is synthesized continuously in a single piece to give a newly synthesized copy called the *leading strand,* while the complement of the original 3′ → 5′ strand is synthesized discontinuously in small pieces called *Okazaki fragments* that are subsequently linked by DNA ligases to form the *lagging strand.*

The magnitude of the replication process is staggering. The nucleus of every human cell contains 2 copies of 22 chromosomes plus an additional 2 sex chromosomes, for a total of 46. Each chromosome consists of one very large DNA molecule, and the sum of the DNA in each of the two sets of chromosomes is estimated to be 3.0 billion base pairs, or 6.0 billion nucleotides. Despite the size of these enormous molecules, their base sequence is faithfully copied during replication. The entire copying process takes only a few hours and, after proofreading and repair, an error gets through only about once each 10 to 100 billion bases. In fact, only about 60 of these random mutations are passed on from parent to child per human generation.

28.4 Transcription of DNA

As noted previously, RNA is structurally similar to DNA but contains ribose rather than deoxyribose and uracil rather than thymine. RNA is of three major kinds, each of which serves a specific purpose. In addition, there are a number of small RNAs that appear to control a wide variety of important cellular functions. All RNA molecules are much smaller than DNA, and all remain single-stranded rather than double-stranded.

- **Messenger RNA (mRNA)** carries genetic messages from DNA to ribosomes, small granular particles in the cytoplasm of a cell where protein synthesis takes place.
- **Ribosomal RNA (rRNA)** complexed with protein provides the physical makeup of the ribosomes.
- **Transfer RNA (tRNA)** transports amino acids to the ribosomes, where they are joined together to make proteins.
- **Small RNAs,** also called *functional RNAs,* have a variety of functions within the cell, including silencing transcription and catalyzing chemical modifications of other RNA molecules.

The genetic information in DNA is contained in segments called *genes,* each of which consists of a specific nucleotide sequence that encodes a specific protein. The conversion of that information from DNA into proteins begins in the nucleus of cells with the synthesis of mRNA by **transcription** of DNA. In bacteria, the process begins when RNA polymerase recognizes and binds to a *promoter sequence* on DNA, typically consisting of around 40 base pairs located upstream (5′) of the transcription start site. Within the promoter are two hexameric *consensus sequences,* one located 10 base pairs upstream of the start and the second located 35 base pairs upstream.

Following formation of the polymerase–promoter complex, several turns of the DNA double helix untwist, forming a bubble and exposing 14 or so base pairs of the two strands. Appropriate ribonucleotides then line up by hydrogen-bonding to their complementary bases on DNA, bond formation occurs in the

5′ → 3′ direction, the RNA polymerase moves along the DNA chain, and the growing RNA molecule unwinds from DNA **(Figure 28.5)**. At any one time, about 12 base pairs of the growing RNA remain hydrogen-bonded to the DNA template.

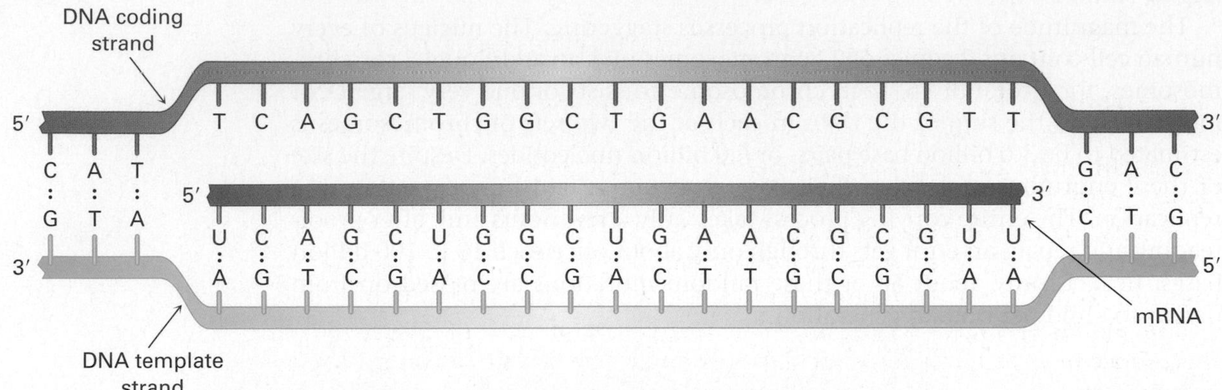

Figure 28.5 Biosynthesis of RNA using a DNA segment as a template.

Unlike what happens in DNA replication, where both strands are copied, only one of the two DNA strands is transcribed into mRNA. The DNA strand that contains the gene is often called the **sense strand**, or *coding strand,* and the DNA strand that gets transcribed to give RNA is called the **antisense strand**, or *noncoding strand.* Because the sense strand and the antisense strand in DNA are complementary, and because the DNA antisense strand and the newly formed RNA strand are also complementary, *the RNA molecule produced during transcription is a copy of the DNA sense strand.* That is, the complement of the complement is the same as the original. The only difference is that the RNA molecule has a U everywhere the DNA sense strand has a T.

Another part of the picture in vertebrates and flowering plants is that genes are often not continuous segments of the DNA chain. Instead, a gene will begin in one small section of DNA called an *exon,* then be interrupted by a noncoding section called an *intron,* and then take up again farther down the chain in another exon. The final mRNA molecule results only after the noncoded sections are cut out of the transcribed mRNA and the remaining pieces are joined together by spliceosomes. The gene for triose phosphate isomerase in maize, for instance, contains eight noncoding introns accounting for approximately 70% of the DNA base pairs and nine coding exons accounting for only 30% of the base pairs.

Problem 28.4

Show how uracil can form strong hydrogen bonds to adenine.

Problem 28.5

What RNA base sequence is complementary to the following DNA base sequence?

(5′) GATTACCGTA (3′)

Problem 28.6
From what DNA base sequence was the following RNA sequence transcribed?

(5′) UUCGCAGAGU (3′)

28.5 Translation of RNA: Protein Biosynthesis

The primary cellular function of mRNA is to direct biosynthesis of the thousands of diverse peptides and proteins required by an organism—as many as 150,000 in a human. The mechanics of protein biosynthesis take place on ribosomes, small granular particles in the cytoplasm of a cell that consist of about 60% ribosomal RNA and 40% protein.

The specific ribonucleotide sequence in mRNA forms a message that determines the order in which amino acid residues are to be joined. Each "word," or **codon**, along the mRNA chain consists of a sequence of three ribonucleotides that is specific for a given amino acid. For example, the series UUC on mRNA is a codon directing incorporation of the amino acid phenylalanine into the growing protein. Of the $4^3 = 64$ possible triplets of the four bases in RNA, 61 code for specific amino acids and 3 code for chain termination. Table 28.1 shows the meaning of each codon.

Table 28.1 Codon Assignments of Base Triplets

		Third base (3′ end)			
First base (5′ end)	**Second base**	**U**	**C**	**A**	**G**
U	U	Phe	Phe	Leu	Leu
	C	Ser	Ser	Ser	Ser
	A	Tyr	Tyr	Stop	Stop
	G	Cys	Cys	Stop	Trp
C	U	Leu	Leu	Leu	Leu
	C	Pro	Pro	Pro	Pro
	A	His	His	Gln	Gln
	G	Arg	Arg	Arg	Arg
A	U	Ile	Ile	Ile	Met
	C	Thr	Thr	Thr	Thr
	A	Asn	Asn	Lys	Lys
	G	Ser	Ser	Arg	Arg
G	U	Val	Val	Val	Val
	C	Ala	Ala	Ala	Ala
	A	Asp	Asp	Glu	Glu
	G	Gly	Gly	Gly	Gly

The message embedded in mRNA is read by transfer RNA (tRNA) in a process called **translation**. There are 61 different tRNAs, one for each of the 61 codons that specifies an amino acid. A typical tRNA is single-stranded and has roughly the shape of a cloverleaf, as shown in **Figure 28.6**. It consists of about 70 to 100 ribonucleotides and is bonded to a specific amino acid by an ester linkage through the 3′ hydroxyl on ribose at the 3′ end of the tRNA. Each tRNA also contains on its middle leaf a segment called an **anticodon**, a sequence of three ribonucleotides complementary to the codon sequence. For example, the codon sequence UUC present on mRNA is read by a phenylalanine-bearing tRNA having the complementary anticodon base sequence GAA. [Remember that nucleotide sequences are written in the 5′ → 3′ direction, so the sequence in an anticodon must be reversed. That is, the complement to (5′)-UUC-(3′) is (3′)-AAG-(5′), which is written as (5′)-GAA-(3′).]

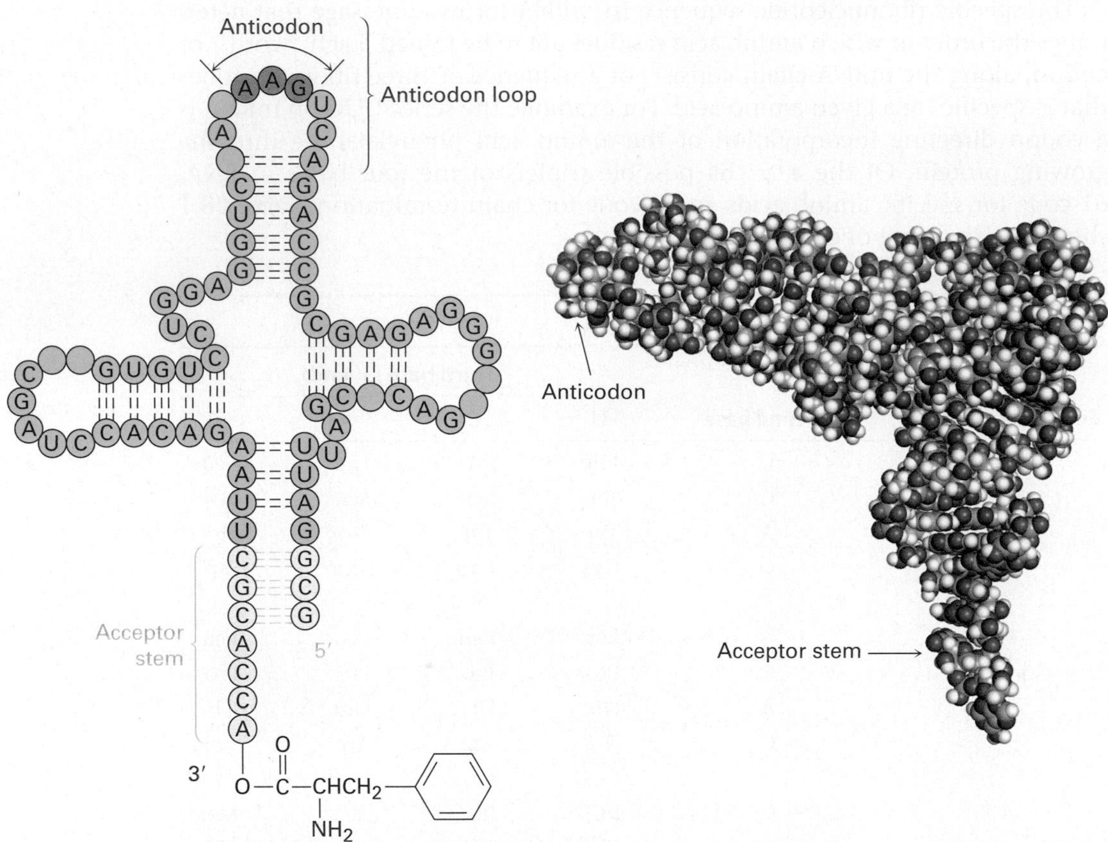

Figure 28.6 Structure of a tRNA molecule. The tRNA is a roughly cloverleaf-shaped molecule containing an anticodon triplet on one "leaf" and an amino acid unit attached covalently at its 3′ end. The example shown is a yeast tRNA that codes for phenylalanine. The nucleotides not specifically identified are chemically modified analogs of the four common nucleotides.

As each successive codon on mRNA is read, different tRNAs bring the correct amino acids into position for enzyme-mediated transfer to the growing peptide. When synthesis of the proper protein is completed, a "stop" codon signals the end and the protein is released from the ribosome. The process is illustrated in **Figure 28.7**.

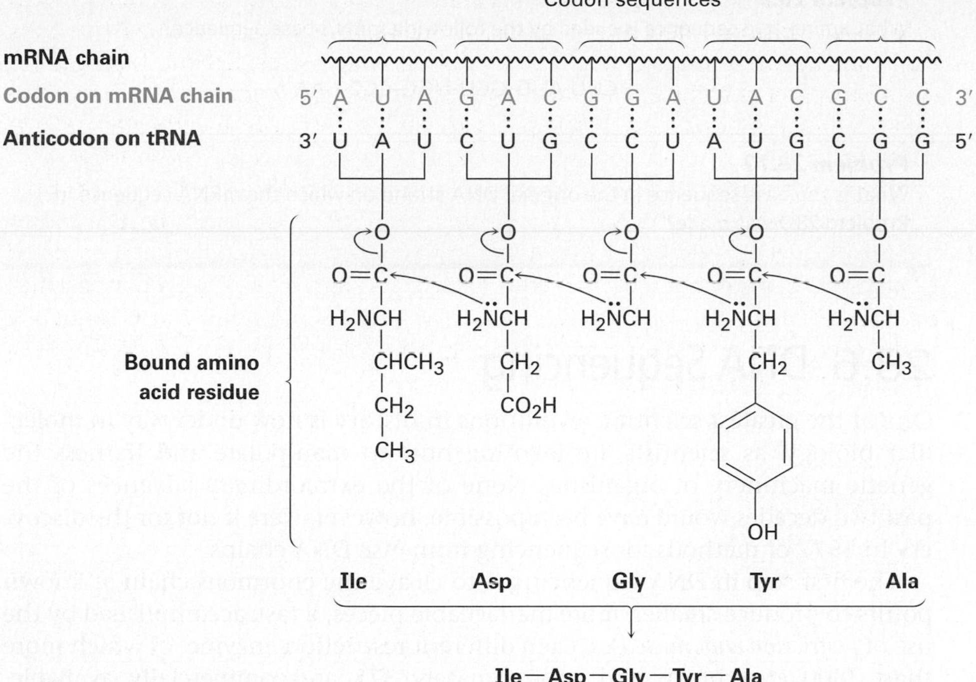

Figure 28.7 A representation of protein biosynthesis. The codon base sequences on mRNA are read by tRNAs containing complementary anticodon base sequences. Transfer RNAs assemble the proper amino acids into position for incorporation into the growing peptide.

Predicting the Amino Acid Sequence Transcribed from DNA

Worked Example 28.2

What amino acid sequence is coded by the following segment of a DNA coding strand (sense strand)?

(5′) CTA-ACT-AGC-GGG-TCG-CCG (3′)

Strategy

The mRNA produced during translation is a copy of the DNA coding strand, with each T replaced by U. Thus, the mRNA has the sequence

(5′) CUA-ACU-AGC-GGG-UCG-CCG (3′)

Each set of three bases forms a codon, whose meaning can be found in Table 28.1.

Solution

Leu-Thr-Ser-Gly-Ser-Pro.

Problem 28.7

List codon sequences for the following amino acids:

(a) Ala **(b)** Phe **(c)** Leu **(d)** Tyr

Problem 28.8

List anticodon sequences on the tRNAs carrying the amino acids shown in Problem 28.7.

Problem 28.9

What amino acid sequence is coded by the following mRNA base sequence?

CUU-AUG-GCU-UGG-CCC-UAA

Problem 28.10

What is the base sequence in the original DNA strand on which the mRNA sequence in Problem 28.9 was made?

28.6 DNA Sequencing

One of the greatest scientific revolutions in history is now under way in molecular biology, as scientists are learning how to manipulate and harness the genetic machinery of organisms. None of the extraordinary advances of the past two decades would have been possible, however, were it not for the discovery in 1977 of methods for sequencing immense DNA chains.

The first step in DNA sequencing is to cleave the enormous chain at known points to produce smaller, more manageable pieces, a task accomplished by the use of *restriction endonucleases*. Each different restriction enzyme, of which more than 3800 are known and approximately 375 are commercially available, cleaves a DNA molecule at a point in the chain where a specific base sequence occurs. For example, the restriction enzyme *AluI* cleaves between G and C in the four-base sequence AG-CT. Note that the sequence is a *palindrome,* meaning that the sequence (5′)-AGCT-(3′) is the same as its complement (3′)-TCGA-(5′) when both are read in the same 5′ → 3′ direction. The same is true for other restriction endonucleases.

If the original DNA molecule is cut with another restriction enzyme having a different specificity for cleavage, still other segments are produced whose sequences partially overlap those produced by the first enzyme. Sequencing of all the segments, followed by identification of the overlapping regions, allows complete DNA sequencing.

Two methods of DNA sequencing are commonly available, and several others are being developed. The *Maxam–Gilbert method* uses chemical techniques, while the **Sanger dideoxy method** uses enzymatic reactions. The Sanger method is the more frequently used of the two and was the method responsible for sequencing the entire human genome of 3.0 billion base pairs. In commercial sequencing instruments, the dideoxy method begins with a mixture of the following:

- The restriction fragment to be sequenced
- A small piece of DNA called a *primer,* whose sequence is complementary to that on the 3′ end of the restriction fragment
- The four 2′-deoxyribonucleoside triphosphates (dNTPs)
- Very small amounts of the four 2′,3′-dideoxyribonucleoside triphosphates (ddNTPs), each of which is labeled with a fluorescent dye of a different color (A 2′,3′-*dideoxy*ribonucleoside triphosphate is one in which both 2′ and 3′ –OH groups are missing from ribose.)

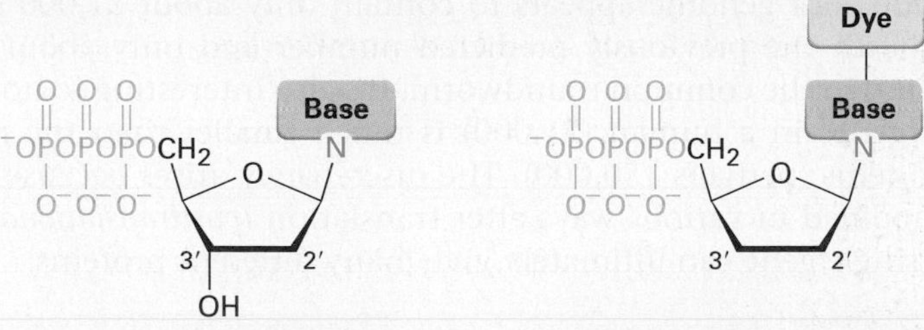

A 2′-deoxyribonucleoside triphosphate (dNTP)

A 2′,3′-dideoxyribonucleoside triphosphate (ddNTP)

DNA polymerase is added to the mixture, and a strand of DNA complementary to the restriction fragment begins to grow from the end of the primer. Most of the time, only normal deoxyribonucleotides are incorporated into the growing chain because of their much higher concentration in the mixture, but every so often, a dideoxyribonucleotide is incorporated. When that happens, DNA synthesis stops because the chain end no longer has a 3′-hydroxyl group for adding further nucleotides.

When reaction is complete, the product consists of a mixture of DNA fragments of all possible lengths, each terminated by one of the four dye-labeled dideoxyribonucleotides. This product mixture is then separated according to the size of the pieces by gel electrophoresis **(Section 26.2)**, and the identity of the terminal dideoxyribonucleotide in each piece—and thus the sequence of the restriction fragment—is determined by noting the color with which the attached dye fluoresces. **Figure 28.8** shows a typical result.

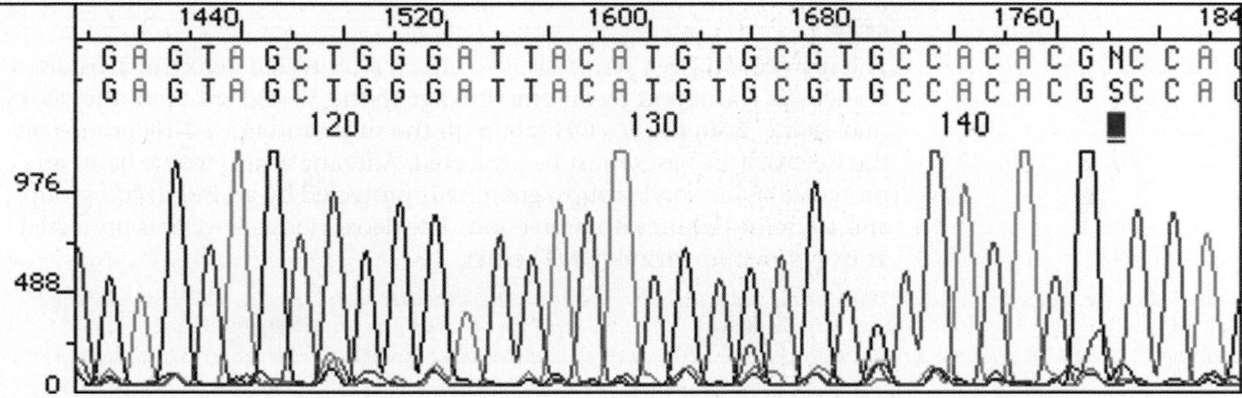

Figure 28.8 The sequence of a restriction fragment determined by the Sanger dideoxy method can be read simply by noting the colors of the dye attached to each of the various terminal nucleotides.

So efficient is the automated dideoxy method that sequences up to 1100 nucleotides in length, with a throughput of up to 19,000 bases per hour, can be sequenced with 98% accuracy. After a decade of work and a cost of about \$500 million, preliminary sequence information for the entire human genome of 3.0 billion base pairs was announced early in 2001 and complete information was released in 2003. More recently, the genome sequencing of specific individuals, including that of James Watson, discoverer of the double helix, has been accomplished. The sequencing price per genome is dropping rapidly and is currently approaching \$10,000, meaning that the routine sequencing of individuals is within reach.

Remarkably, our genome appears to contain only about 21,000 genes, less than one-fourth the previously predicted number and only about twice the number found in the common roundworm. It's also interesting to note that the number of genes in a human (21,000) is much smaller than the number of kinds of proteins (perhaps 150,000). The discrepancy arises because most proteins are modified in various ways after translation *(posttranslational modifications),* so a single gene can ultimately give many different proteins.

28.7 DNA Synthesis

The ongoing revolution in molecular biology has brought with it an increased demand for the efficient chemical synthesis of short DNA segments, called *oligonucleotides,* or simply *oligos.* The problems of DNA synthesis are similar to those of peptide synthesis **(Section 26.7)** but are more difficult because of the complexity of the nucleotide monomers. Each nucleotide has multiple reactive sites that must be selectively protected and deprotected at the proper times, and coupling of the four nucleotides must be carried out in the proper sequence. Automated DNA synthesizers are available, however, that allow the fast and reliable synthesis of DNA segments up to 200 nucleotides in length.

DNA synthesizers operate on a principle similar to that of the Merrifield solid-phase peptide synthesizer **(Section 26.8)**. In essence, a protected nucleotide is covalently bonded to a solid support, and one nucleotide at a time is added to the growing chain by the use of a coupling reagent. After the final nucleotide has been added, all the protecting groups are removed and the synthetic DNA is cleaved from the solid support. Five steps are needed:

STEP 1

The first step in DNA synthesis is to attach a protected deoxynucleoside to a silica (SiO_2) support by an ester linkage to the 3′ –OH group of the deoxynucleoside. Both the 5′ –OH group on the sugar and free $-NH_2$ groups on the heterocyclic bases must be protected. Adenine and cytosine bases are protected by benzoyl groups, guanine is protected by an isobutyryl group, and thymine requires no protection. The deoxyribose 5′ –OH is protected as its *p*-dimethoxytrityl (DMT) ether.

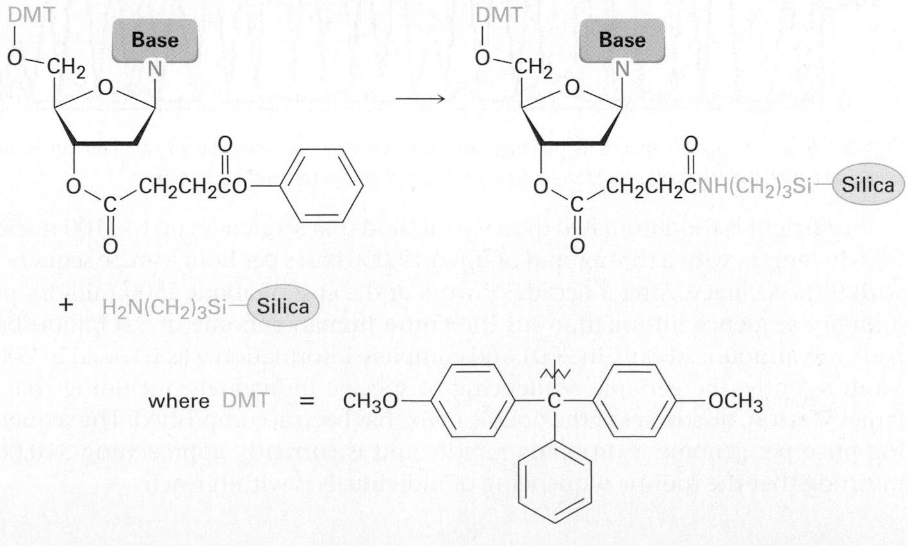

DMT
Base
$+ H_2N(CH_2)_3Si$–Silica
→
$CH_2CH_2CNH(CH_2)_3Si$–Silica

where DMT = CH_3O–C–OCH_3

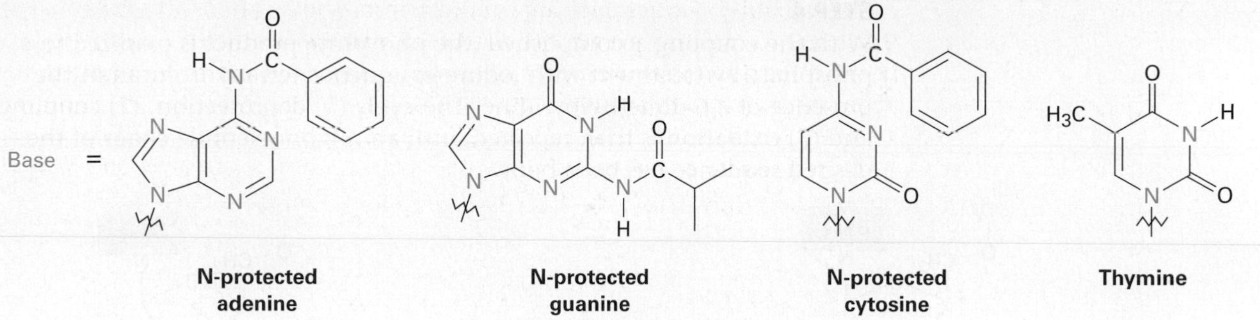

Base =

N-protected adenine

N-protected guanine

N-protected cytosine

Thymine

STEP 2

The second step is removal of the DMT protecting group by treatment with dichloroacetic acid in CH_2Cl_2. The reaction occurs by an S_N1 mechanism and proceeds rapidly because of the stability of the tertiary, benzylic dimethoxytrityl cation.

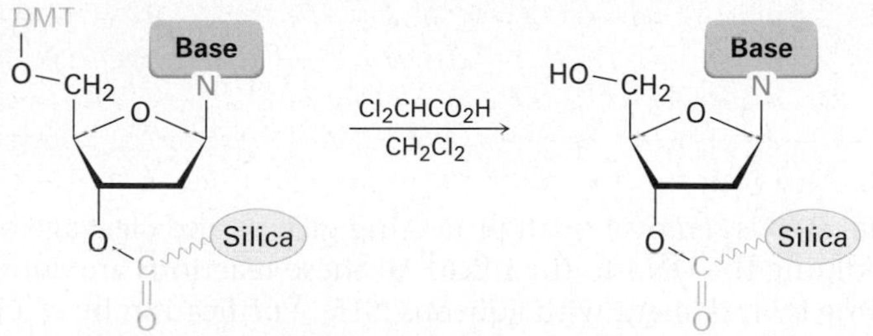

Cl_2CHCO_2H, CH_2Cl_2

STEP 3

The third step is the coupling of the polymer-bonded deoxynucleoside with a protected deoxynucleoside containing a *phosphoramidite* group [$R_2NP(OR)_2$] at its 3′ position. The coupling reaction takes place in the polar aprotic solvent acetonitrile, requires catalysis by the heterocyclic amine tetrazole, and yields a *phosphite,* $P(OR)_3$, as product. Note that one of the phosphorus oxygen atoms is protected by a β-cyanoethyl group, $-OCH_2CH_2C{\equiv}N$. The coupling step takes place in better than 99% yield.

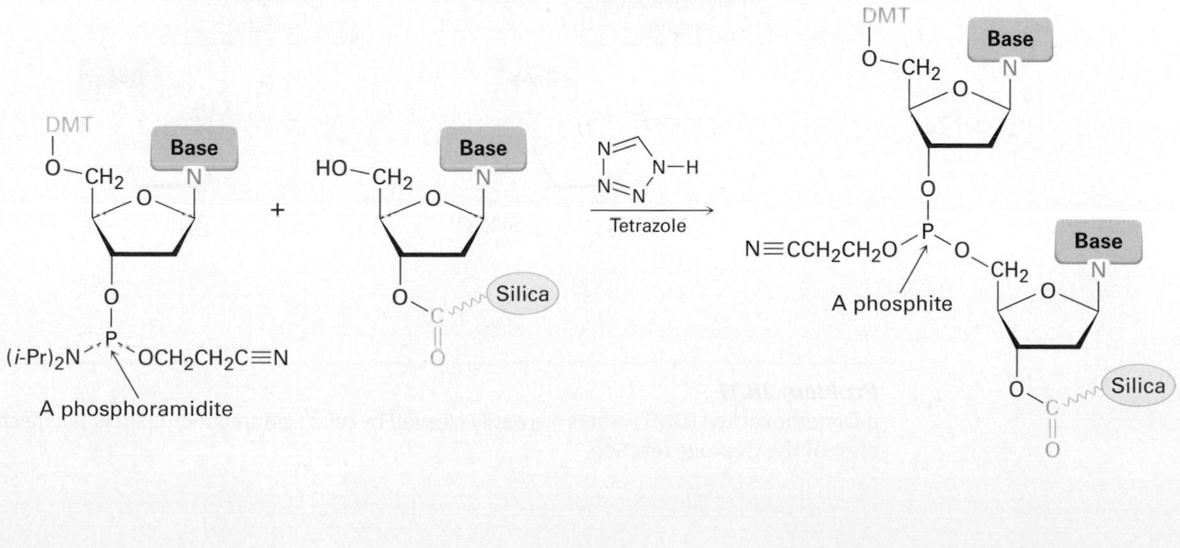

A phosphoramidite

Tetrazole

A phosphite

STEP 4

With the coupling accomplished, the phosphite product is oxidized to a phosphate by treatment with iodine in aqueous tetrahydrofuran in the presence of 2,6-dimethylpyridine. The cycle (1) deprotection, (2) coupling, and (3) oxidation is then repeated until an oligonucleotide chain of the desired sequence has been built.

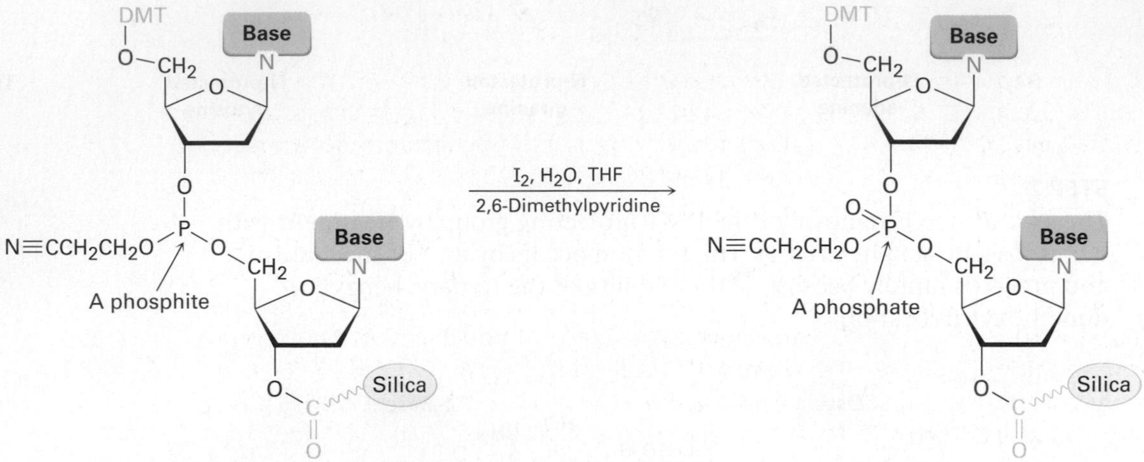

DMT, O—CH_2, Base, N, O, O, P, $N{\equiv}CCH_2CH_2O$, O, CH_2, Base, N, O, A phosphite, O, C, O, Silica

$\xrightarrow[\text{2,6-Dimethylpyridine}]{I_2,\ H_2O,\ THF}$

DMT, O—CH_2, Base, N, O, O, O=P, $N{\equiv}CCH_2CH_2O$, O, CH_2, Base, N, O, A phosphate, O, C, O, Silica

STEP 5

The final step is removal of all protecting groups and cleavage of the ester bond holding the DNA to the silica. All these reactions are done at the same time by treatment with aqueous NH_3. Purification by electrophoresis then yields the synthetic DNA.

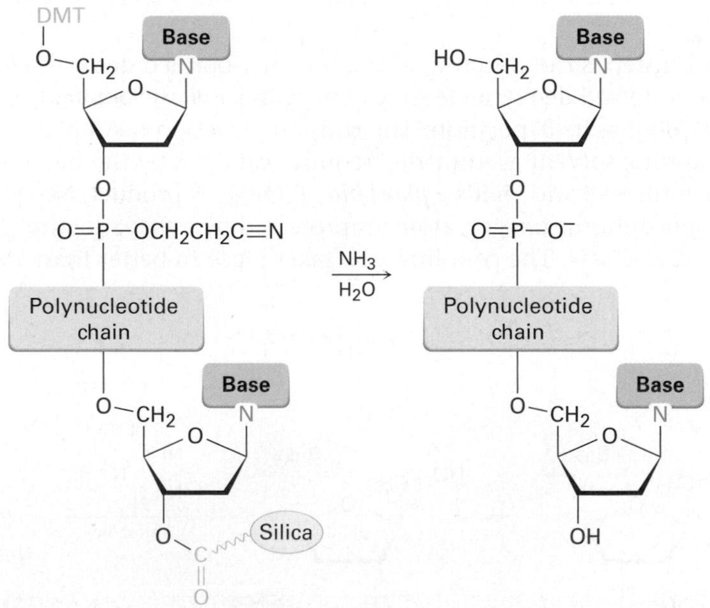

DMT, O—CH_2, Base, N, O, O, $O{=}P{-}OCH_2CH_2C{\equiv}N$, Polynucleotide chain, O—CH_2, Base, N, O, O, C, O, Silica

$\xrightarrow[H_2O]{NH_3}$

HO—CH_2, Base, N, O, O, $O{=}P{-}O^-$, Polynucleotide chain, O—CH_2, Base, N, O, OH

Problem 28.11

p-Dimethoxytrityl (DMT) ethers are easily cleaved by mild acid treatment. Show the mechanism of the cleavage reaction.

Problem 28.12

Propose a mechanism to account for cleavage of the β-cyanoethyl protecting group from the phosphate groups on treatment with aqueous ammonia. (Acrylonitrile, $H_2C{=}CHCN$, is a by-product.) What kind of reaction is occurring?

28.8 The Polymerase Chain Reaction

It often happens that only a tiny amount of DNA can be obtained directly, as might occur at a crime scene, so methods for obtaining larger amounts are sometimes needed to carry out the sequencing and characterization. The invention of the **polymerase chain reaction (PCR)** by Kary Mullis in 1986 has been described as being to genes what Gutenberg's invention of the printing press was to the written word. Just as the printing press produces multiple copies of a book, PCR produces multiple copies of a given DNA sequence. Starting from less than 1 picogram of DNA with a chain length of 10,000 nucleotides (1 pg = 10^{-12} g; about 10^5 molecules), PCR makes it possible to obtain several micrograms (1 μg = 10^{-6} g; about 10^{11} molecules) in just a few hours.

The key to the polymerase chain reaction is *Taq* DNA polymerase, a heat-stable enzyme isolated from the thermophilic bacterium *Thermus aquaticus* found in a hot spring in Yellowstone National Park. *Taq* polymerase is able to take a single strand of DNA that has a short, primer segment of complementary chain at one end and then finish constructing the entire complementary strand. The overall process takes three steps, as shown in **Figure 28.9**. More recently, improved heat-stable DNA polymerases have become available, including Vent polymerase and *Pfu* polymerase, both isolated from bacteria growing near geothermal vents in the ocean floor. The error rate of both enzymes is substantially less than that of *Taq*.

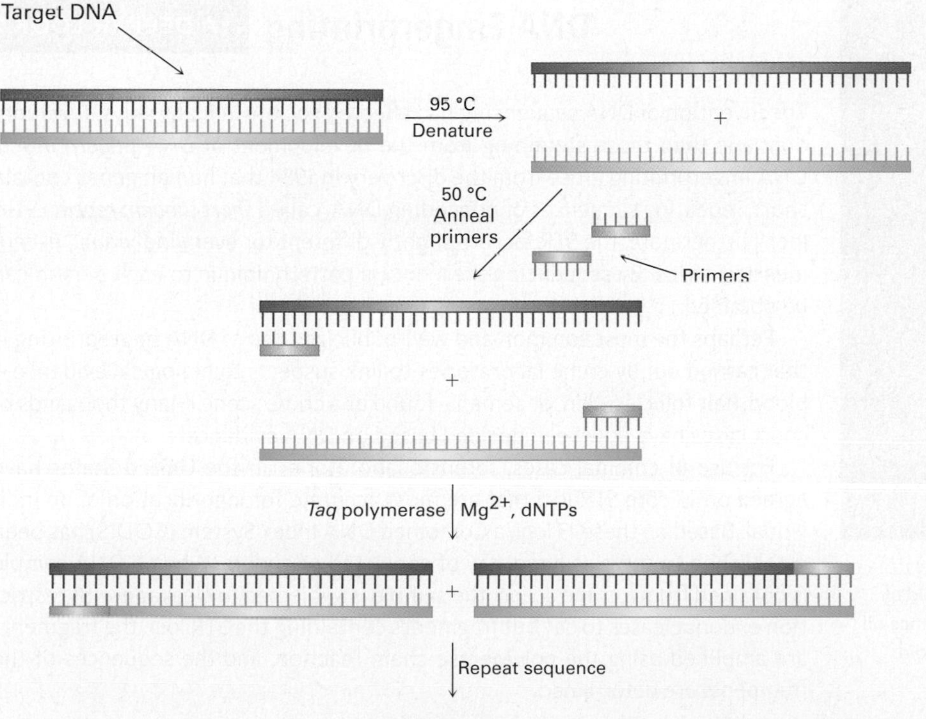

Figure 28.9 The polymerase chain reaction. Details are explained in the text.

STEP 1
The double-stranded DNA to be amplified is heated in the presence of *Taq* polymerase, Mg^{2+} ion, the four deoxynucleotide triphosphate monomers (dNTPs), and a large excess of two short oligonucleotide primers of about 20 bases each. Each primer is complementary to the sequence at the end of one of the target DNA segments. At a temperature of 95 °C, double-stranded DNA denatures, spontaneously breaking apart into two single strands.

STEP 2
The temperature is lowered to between 37 and 50 °C, allowing the primers, because of their relatively high concentration, to anneal by hydrogen-bonding to their complementary sequence at the end of each target strand.

STEP 3
The temperature is then raised to 72 °C, and *Taq* polymerase catalyzes the addition of further nucleotides to the two primed DNA strands. When replication of each strand is finished, *two* copies of the original DNA now exist. Repeating the denature–anneal–synthesize cycle a second time yields four DNA copies, repeating a third time yields eight copies, and so on, in an exponential series.

PCR has been automated, and 30 or so cycles can be carried out in an hour, resulting in a theoretical amplification factor of 2^{30} ($\sim 10^9$). In practice, however, the efficiency of each cycle is less than 100%, and an experimental amplification of about 10^6 to 10^8 is routinely achieved for 30 cycles.

DNA Fingerprinting

A DEEPER LOOK

Rembrandt Peale (1805)/New York Historical Society

Historians have wondered for many years whether Thomas Jefferson fathered a child by Sally Hemings. DNA fingerprinting evidence obtained in 1998 strongly suggests that he did.

The invention of DNA sequencing has affected society in many ways, few more dramatic than those stemming from the development of *DNA fingerprinting*. DNA fingerprinting arose from the discovery in 1984 that human genes contain short, repeating sequences of noncoding DNA, called *short tandem repeat* (STR) loci. Furthermore, the STR loci are slightly different for every individual, except identical twins. By sequencing these loci, a pattern unique to each person can be obtained.

Perhaps the most common and well-publicized use of DNA fingerprinting is that carried out by crime laboratories to link suspects to biological evidence—blood, hair follicles, skin, or semen—found at a crime scene. Many thousands of court cases have now been decided based on DNA evidence.

For use in criminal cases, forensic laboratories in the United States have agreed on 13 core STR loci that are most accurate for identification of an individual. Based on these 13 loci, a Combined DNA Index System (CODIS) has been established to serve as a registry of convicted offenders. When a DNA sample is obtained from a crime scene, the sample is subjected to cleavage with restriction endonucleases to cut out fragments containing the STR loci, the fragments are amplified using the polymerase chain reaction, and the sequences of the fragments are determined.

(continued)

(continued)

If the profile of sequences from a known individual and the profile from DNA obtained at a crime scene match, the probability is approximately 82 billion to 1 that the DNA is from the same individual. In paternity cases, where the DNA of father and offspring are related but not fully identical, the identity of the father can be established with a probability of around 100,000 to 1. Even after several generations have passed, paternity can still be inferred from DNA analysis of the Y chromosome of direct male-line descendants. The most well-known such case is that of Thomas Jefferson, who likely fathered a child by his slave Sally Hemings. Although Jefferson himself has no male-line descendants, DNA analysis of the male-line descendants of Jefferson's paternal uncle contained the same Y chromosome as a male-line descendant of Eston Hemings, the youngest son of Sally Hemings. Thus, a mixing of the two genomes is clear, although the male individual responsible for that mixing can't be conclusively identified.

Among its many other applications, DNA fingerprinting is widely used for the diagnosis of genetic disorders, both prenatally and in newborns. Cystic fibrosis, hemophilia, Huntington's disease, Tay–Sachs disease, sickle cell anemia, and thalassemia are among the many diseases that can be detected, enabling early treatment of an affected child. Furthermore, by studying the DNA fingerprints of relatives with a history of a particular disorder, it's possible to identify DNA patterns associated with the disease and perhaps obtain clues for eventual cure. In addition, the U.S. Department of Defense now requires blood and saliva samples from all military personnel. The samples are stored, and DNA is extracted if the need for identification of a casualty arises.

Summary

DNA (deoxyribonucleic acid) and **RNA (ribonucleic acid)** are biological polymers that act as chemical carriers of an organism's genetic information. Enzyme-catalyzed hydrolysis of nucleic acids yields **nucleotides**, the monomer units from which RNA and DNA are constructed. Further enzyme-catalyzed hydrolysis of the nucleotides yields **nucleosides** plus phosphate. Nucleosides, in turn, consist of a purine or pyrimidine base linked to C1 of an aldopentose sugar—ribose in RNA and 2-deoxyribose in DNA. The nucleotides are joined by phosphate links between the 5′ phosphate of one nucleotide and the 3′ hydroxyl on the sugar of another nucleotide.

Molecules of DNA consist of two complementary polynucleotide strands held together by hydrogen bonds between heterocyclic bases on the different strands and coiled into a **double helix**. Adenine and thymine form hydrogen bonds to each other, as do cytosine and guanine.

Three processes take place in deciphering the genetic information of DNA:

- **Replication** of DNA is the process by which identical DNA copies are made. The DNA double helix unwinds, complementary deoxyribonucleotides line up in order, and two new DNA molecules are produced.
- **Transcription** is the process by which RNA is produced to carry genetic information from the nucleus to the ribosomes. A short segment of the DNA double helix unwinds, and complementary ribonucleotides line up to produce **messenger RNA (mRNA)**.

Key words

anticodon, 1138
antisense strand, 1136
codon, 1137
deoxyribonucleic acid (DNA), 1128
double helix, 1131
3′ end, 1131
5′ end, 1131
messenger RNA (mRNA), 1135
nucleoside, 1128
nucleotide, 1128
polymerase chain reaction (PCR), 1145
replication, 1133
ribonucleic acid (RNA), 1128
ribosomal RNA (rRNA), 1135
Sanger dideoxy method, 1140

Key words—cont'd

- **Translation** is the process by which mRNA directs protein synthesis. Each mRNA is divided into **codons**, ribonucleotide triplets that are recognized by small amino acid–carrying molecules of **transfer RNA (tRNA)**, which deliver the appropriate amino acids needed for protein synthesis.

Sequencing of DNA is carried out by the **Sanger dideoxy method**, and small DNA segments can be synthesized in the laboratory by automated instruments. Small amounts of DNA can be amplified by factors of 10^6 using the **polymerase chain reaction (PCR)**.

Exercises

OWL Interactive versions of these problems are assignable in OWL for Organic Chemistry.

Visualizing Chemistry

(Problems 28.1–28.12 appear within the chapter.)

28.13 Identify the following bases, and tell whether each is found in DNA, RNA, or both:

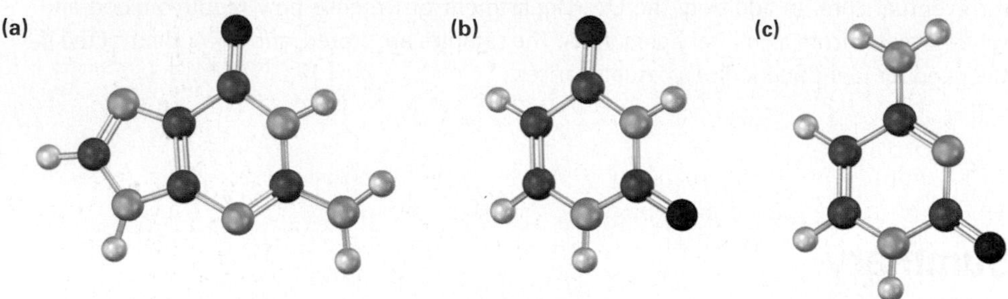

28.14 Identify the following nucleotide, and tell how it is used:

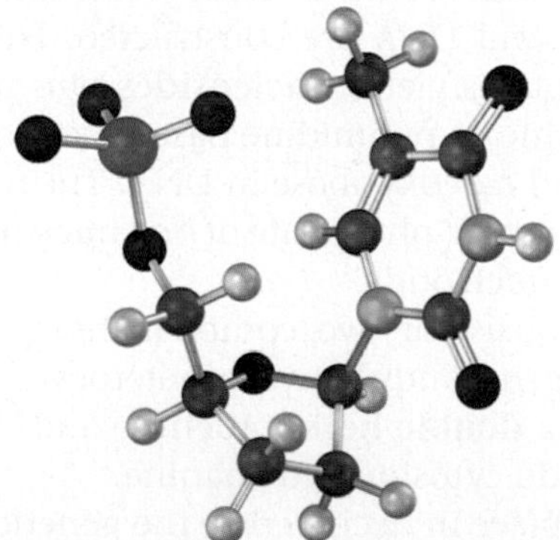

28.15 Amine bases in nucleic acids can react with alkylating agents in typical S_N2 reactions. Look at the following electrostatic potential maps, and tell which is the better nucleophile, guanine or adenine. The reactive positions in each are indicated.

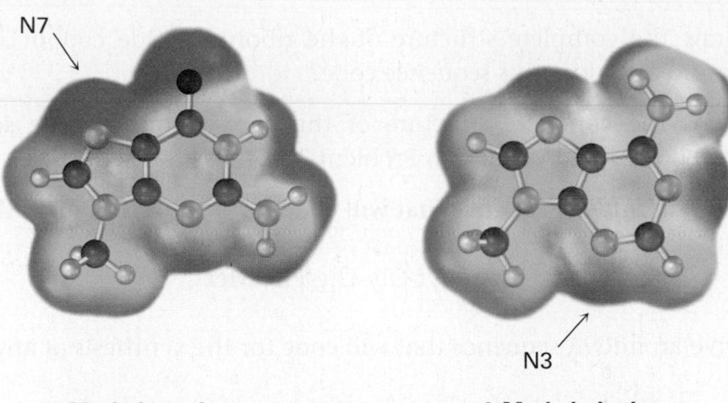

Additional Problems

28.16 Human brain natriuretic peptide (BNP) is a small peptide of 32 amino acids used in the treatment of congestive heart failure. How many nitrogen bases are present in the DNA that codes for BNP?

28.17 Human and horse insulin both have two polypeptide chains, with one chain containing 21 amino acids and the other containing 30 amino acids. They differ in primary structure at two places. At position 9 in one chain, human insulin has Ser and horse insulin has Gly; at position 30 in the other chain, human insulin has Thr and horse insulin has Ala. How must the DNA for the two insulins differ?

28.18 The DNA of sea urchins contains about 32% A. What percentages of the other three bases would you expect in sea urchin DNA? Explain.

28.19 The codon UAA stops protein synthesis. Why does the sequence UAA in the following stretch of mRNA not cause any problems?

-GCA-UUC-GAG-GUA-ACG-CCC-

28.20 Which of the following base sequences would most likely be recognized by a restriction endonuclease? Explain.
(a) GAATTC (b) GATTACA (c) CTCGAG

28.21 For what amino acids do the following ribonucleotide triplets code?
(a) AAU (b) GAG (c) UCC (d) CAU

28.22 From what DNA sequences were each of the mRNA codons in Problem 28.21 transcribed?

28.23 What anticodon sequences of tRNAs are coded for by the codons in Problem 28.21?

28.24 Draw the complete structure of the ribonucleotide codon UAC. For what amino acid does this sequence code?

28.25 Draw the complete structure of the deoxyribonucleotide sequence from which the mRNA codon in Problem 28.24 was transcribed.

28.26 Give an mRNA sequence that will code for synthesis of metenkephalin.

Tyr-Gly-Gly-Phe-Met

28.27 Give an mRNA sequence that will code for the synthesis of angiotensin II.

Asp-Arg-Val-Tyr-Ile-His-Pro-Phe

28.28 What amino acid sequence is coded for by the following DNA coding strand (sense strand)?

(5′) CTT-CGA-CCA-GAC-AGC-TTT (3′)

28.29 What amino acid sequence is coded for by the following mRNA base sequence?

(5′) CUA-GAC-CGU-UCC-AAG-UGA (3′)

28.30 If the DNA coding sequence -CAA-CCG-GAT- were miscopied during replication and became -CGA-CCG-GAT-, what effect would there be on the sequence of the protein produced?

28.31 Show the steps involved in a laboratory synthesis of the DNA fragment with the sequence CTAG.

28.32 The final step in DNA synthesis is deprotection by treatment with aqueous ammonia. Show the mechanisms by which deprotection occurs at the points indicated in the following structure:

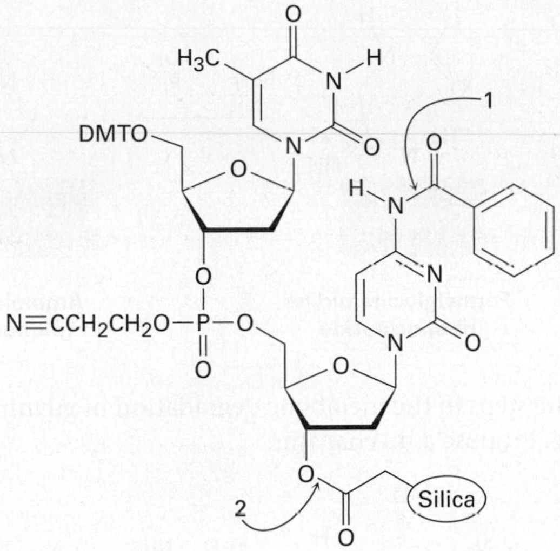

28.33 Draw the structure of cyclic adenosine monophosphate (cAMP), a messenger involved in the regulation of glucose production in the body. Cyclic AMP has a phosphate ring connecting the 3′ and 5′ hydroxyl groups on adenosine.

28.34 The final step in the metabolic degradation of uracil is the oxidation of malonic semialdehyde to give malonyl CoA. Propose a mechanism.

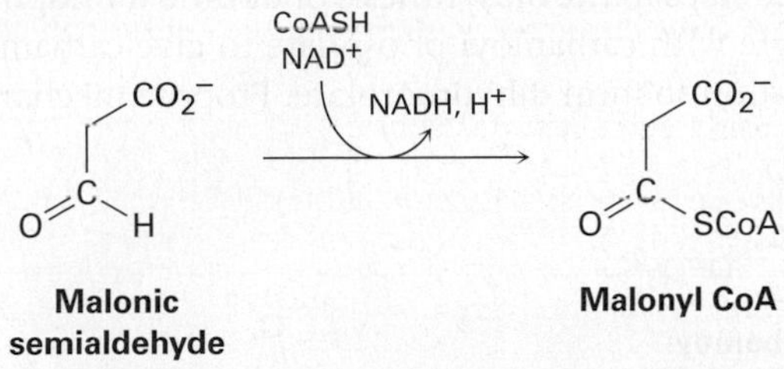

28.35 One of the steps in the biosynthesis of a nucleotide called inosine monophosphate is the formation of aminoimidazole ribonucleotide from formylglycinamidine ribonucleotide. Propose a mechanism.

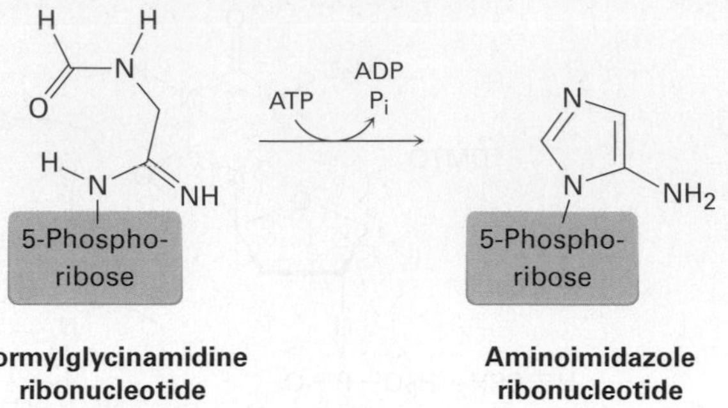

ATP → ADP, P_i

Formylglycinamidine ribonucleotide → **Aminoimidazole ribonucleotide**

28.36 One of the steps in the metabolic degradation of guanine is hydrolysis to give xanthine. Propose a mechanism.

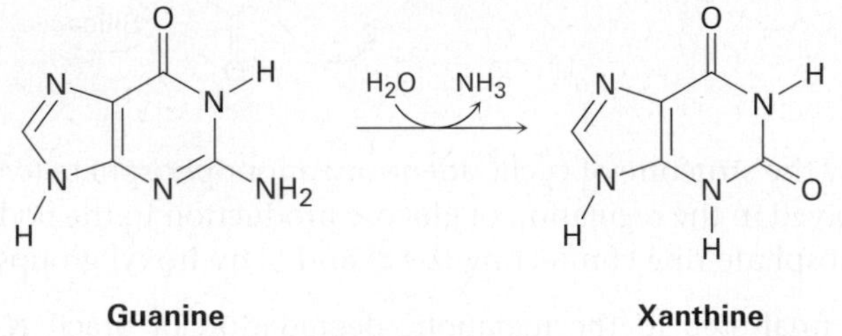

H_2O → NH_3

Guanine → **Xanthine**

28.37 One of the steps in the biosynthesis of uridine monophosphate is the reaction of aspartate with carbamoyl phosphate to give carbamoyl aspartate followed by cyclization to form dihydroorotate. Propose mechanisms for both steps.

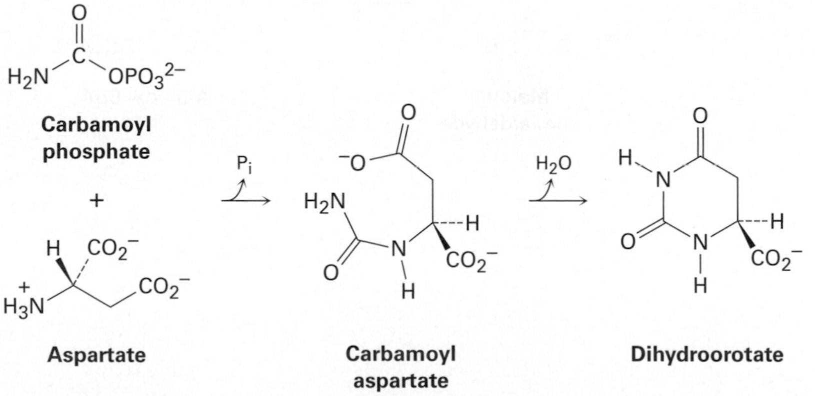

Carbamoyl phosphate + **Aspartate** → (P_i) **Carbamoyl aspartate** → (H_2O) **Dihydroorotate**

28.38 Valganciclovir, marketed as Valcyte, is an antiviral agent used for the treatment of cytomegalovirus. Called a *prodrug*, valganciclovir is inactive by itself but is rapidly converted in the intestine by hydrolysis of its ester bond to produce an active drug, called ganciclovir, along with an amino acid.

Valganciclovir

(a) What amino acid is produced by hydrolysis of the ester bond in valganciclovir?

(b) What is the structure of ganciclovir?

(c) What atoms present in the nucleotide deoxyguanine are missing from ganciclovir?

(d) What role do the atoms missing from deoxyguanine play in DNA replication?

(e) How might valganciclovir interfere with DNA synthesis?

29

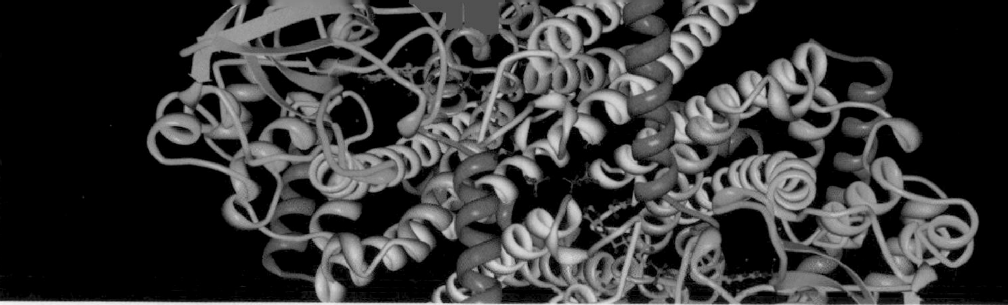

Acyl CoA dehydrogenase is an enzyme that catalyzes the introduction of a C=C double bond into fatty acids during their metabolism. PDB ID: 2WBI. Muniz, J.R.C., Guo, K., Savitsky, P., Roos, A., Yue, W., Pilka, E., Vondelft, F., Edwards, A.M., Bountra, C., Arrowsmith, C.H., Weigelt, J., Oppermann, U. CRYSTAL STRUCTURE OF HUMAN ACYL-COA DEHYDROGENASE 11

The Organic Chemistry of Metabolic Pathways

Anyone who wants to understand or contribute to the revolution now taking place in the biological sciences must first understand life processes at the molecular level. This understanding, in turn, must be based on a detailed knowledge of the chemical reactions and pathways used by living organisms. Just knowing *what* occurs is not enough; it's also necessary to understand *how* and *why* organisms use the chemistry they do.

Biochemical reactions are not mysterious. Even though the biological reactions that take place in living organisms often appear complicated, they follow the same rules of reactivity as laboratory reactions and they take place by the same mechanisms.

A word of caution: some of the molecules we'll be encountering are substantially larger and more complex than those we've been dealing with thus far. But don't be intimidated; keep your focus on the parts of the molecules where changes occur, and ignore the parts where nothing changes. The reactions themselves are exactly the same additions, eliminations, substitutions, carbonyl condensations, and so forth, that we've been dealing with all along. By the end of this chapter, it should be clear that the chemistry of living organisms *is* organic chemistry.

Why This Chapter? In this chapter, we'll look at some of the pathways by which organisms carry out their chemistry, focusing primarily on how they metabolize fats and carbohydrates. The treatment will be far from complete, but it should give you an idea of the kinds of processes that occur.

29.1 An Overview of Metabolism and Biochemical Energy

The many reactions that go on in the cells of living organisms are collectively called **metabolism**. The pathways that break down larger molecules into smaller ones are called **catabolism**, and the pathways that synthesize larger biomolecules from smaller ones are known as **anabolism**. Catabolic reaction pathways are usually exergonic and release energy, while anabolic pathways are often endergonic and absorb energy. Catabolism can be divided into the four stages shown in **Figure 29.1**.

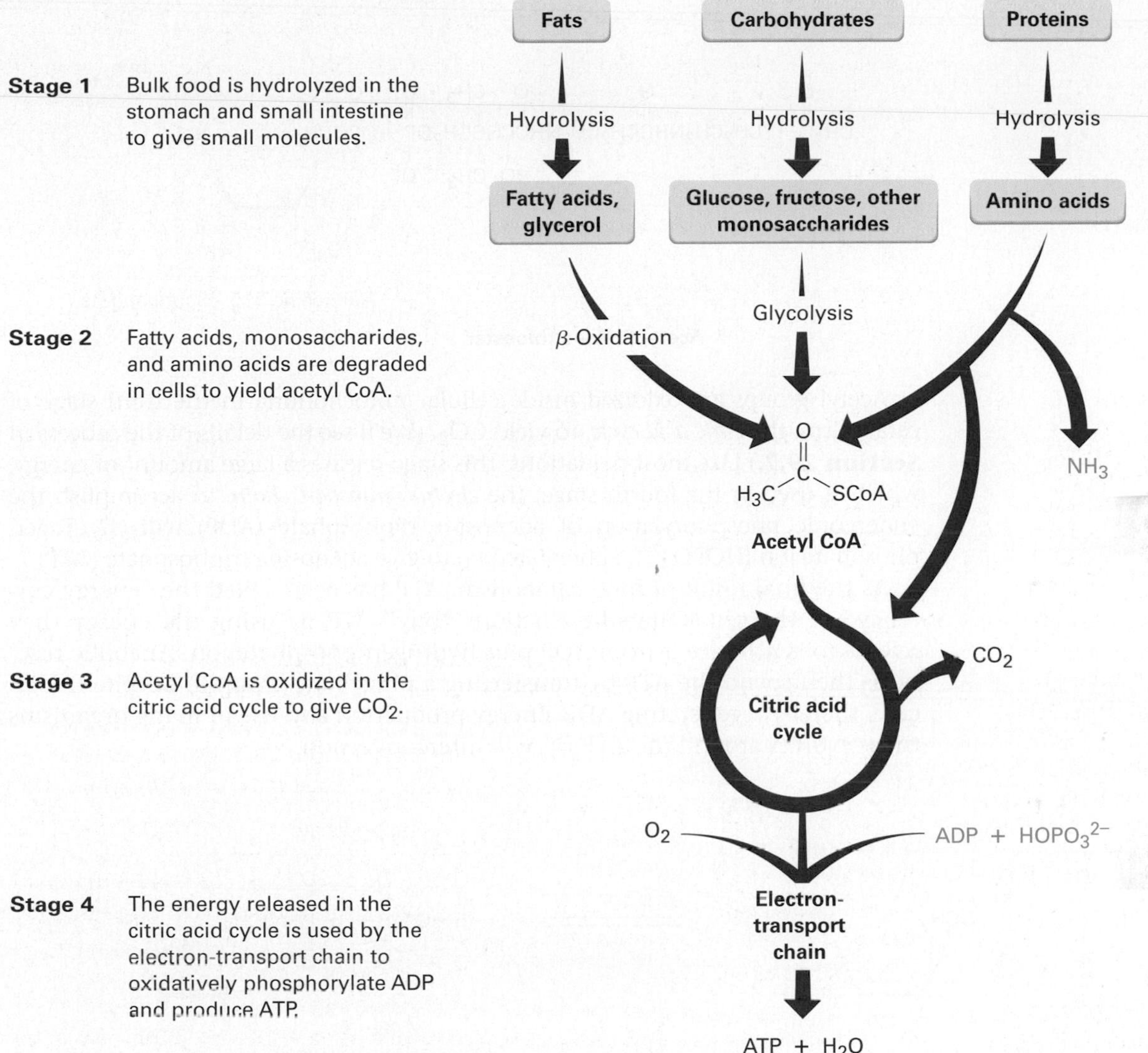

Figure 29.1 An overview of catabolic pathways for the degradation of food and the production of biochemical energy. The ultimate products of food catabolism are CO_2 and H_2O, with the energy released in the citric acid cycle used to drive the endergonic synthesis of adenosine triphosphate (ATP) from adenosine diphosphate (ADP) plus phosphate ion, $HOPO_3^{2-}$.

In the first catabolic stage, commonly called digestion, food is broken down in the mouth, stomach, and small intestine by hydrolysis of ester, acetal (glycoside), and amide (peptide) bonds to yield fatty acids, simple sugars, and amino acids. These smaller molecules are then absorbed and further degraded in the second stage of catabolism to yield acetyl groups attached by a thioester bond to the large carrier molecule coenzyme A. The resultant compound, acetyl coenzyme A (acetyl CoA), is a key substance in the metabolism of food molecules and in many other biological pathways. As noted in **Section 21.8**, the acetyl group in acetyl CoA is linked to the sulfur atom of phosphopantetheine, which is itself linked to adenosine 3′,5′-bisphosphate.

$CH_3C(=O)-SCH_2CH_2NHC(=O)CH_2CH_2NHC(=O)CH(OH)C(CH_3)_2CH_2OP(=O)(O^-)-OP(=O)(O^-)OCH_2$–(ribose 3′-$OPO_3^{2-}$, 2′-OH)–adenine ($NH_2$)

Phosphopantetheine

Adenosine 3′,5′-bisphosphate

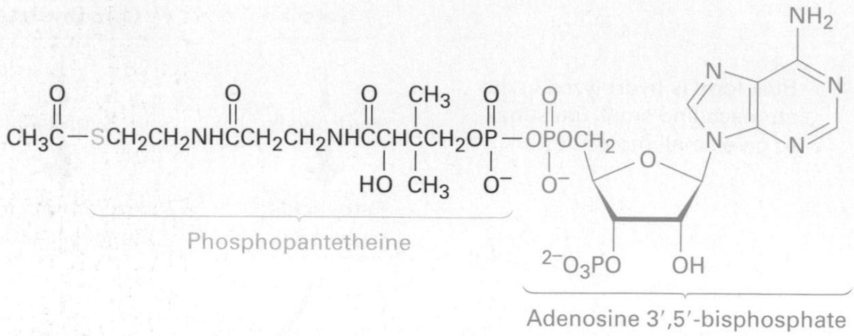

Acetyl CoA—a thioester

Acetyl groups are oxidized inside cellular mitochondria in the third stage of catabolism, the *citric acid cycle*, to yield CO_2. (We'll see the details of the process in **Section 29.7**.) Like most oxidations, this stage releases a large amount of energy, which is used in the fourth stage, the *electron-transport chain*, to accomplish the endergonic phosphorylation of adenosine diphosphate (ADP) with hydrogen phosphate ion ($HOPO_3^{2-}$, abbreviated P_i) to give adenosine triphosphate (ATP).

As the final result of food catabolism, ATP has been called the "energy currency" of the cell. Catabolic reactions "buy" ATP by using the energy they release to synthesize it from ADP plus hydrogen phosphate ion. Anabolic reactions then spend the ATP by transferring a phosphate group to another molecule, thereby regenerating ADP. Energy production and use in living organisms thus revolves around the ATP ⇄ ADP interconversion.

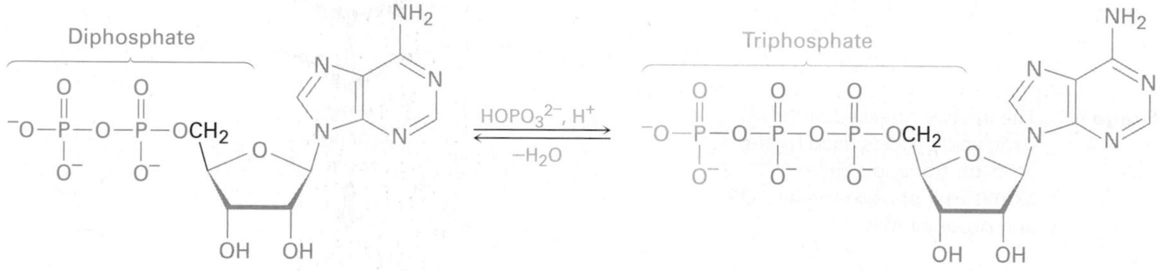

$$^-O-P(=O)(O^-)-O-P(=O)(O^-)-OCH_2\text{-adenosine} \xrightarrow[-H_2O]{HOPO_3^{2-},\ H^+} {}^-O-P(=O)(O^-)-O-P(=O)(O^-)-O-P(=O)(O^-)-OCH_2\text{-adenosine}$$

Diphosphate — Triphosphate

Adenosine diphosphate (ADP) — **Adenosine triphosphate (ATP)**

ADP and ATP are both phosphoric acid anhydrides, which contain $-\overset{O}{\overset{\|}{P}}-O-\overset{O}{\overset{\|}{P}}-$ linkages analogous to the $-\overset{O}{\overset{\|}{C}}-O-\overset{O}{\overset{\|}{C}}-$ linkage in carboxylic acid anhydrides. Just as carboxylic acid anhydrides react with alcohols by breaking a C–O bond and forming a carboxylic ester, ROCOR′, **(Section 21.5)**, phosphoric acid anhydrides react with alcohols by breaking a P–O bond and forming a phosphate ester, $ROPO_3^{2-}$. The reaction is, in effect, a nucleophilic acyl substitution at phosphorus. Note that phosphorylation reactions with ATP generally require the presence of a divalent metal cation in the enzyme, usually Mg^{2+}, to form a Lewis acid–base complex with the phosphate oxygen atoms and neutralize negative charge.

R—O—H + ATP (Mg^{2+}) ⟶ [R—O—P—OPOP—O—Adenosine (Mg^{2+})] ⟶ R—O—P—O^- + $^-$OPOP—O—Adenosine + Mg^{2+}

A phosphate ester ADP

How does the body use ATP? Recall from **Section 6.7** that the free-energy change ΔG must be negative and energy must be released for a reaction to be favorable and occur spontaneously. If ΔG is positive, the reaction is energetically unfavorable and the process can't occur spontaneously.

For an energetically unfavorable reaction to occur, it must be "coupled" to an energetically favorable reaction so that the overall free-energy change for the two reactions together is favorable. To understand what it means for reactions to be coupled, imagine that reaction 1 does not occur to any reasonable extent because it has a small equilibrium constant and is energetically unfavorable; that is, the reaction has $\Delta G > 0$.

$$(1)\ \mathbf{A} + m \rightleftharpoons \mathbf{B} + n \qquad \Delta G > 0$$

where **A** and **B** are the biochemically "important" substances while m and n are enzyme cofactors, H_2O, or other small molecules.

Imagine also that product n can react with substance o to yield p and q in a second, strongly favorable reaction that has a large equilibrium constant and $\Delta G << 0$.

$$(2)\ n + o \rightleftharpoons p + q \qquad \Delta G << 0$$

Taking the two reactions together, they share, or are coupled through, the common intermediate n, which is a product in the first reaction and a reactant in the second. When even a tiny amount of n is formed in reaction 1, it undergoes essentially complete conversion in reaction 2, thereby removing it from the first equilibrium and forcing reaction 1 to continually replenish n until the reactant A is gone. That is, the two reactions added together have a favorable $\Delta G < 0$, and we say that the favorable reaction 2 "drives" the unfavorable reaction 1. Because the two reactions are coupled through n, the transformation of A to B becomes favorable.

$$(1)\ \mathrm{A} + m \rightleftharpoons \mathrm{B} + \cancel{n} \qquad \Delta G > 0$$

$$(2)\ \cancel{n} + o \rightleftharpoons p + q \qquad \Delta G << 0$$

$$\text{Net: } \mathrm{A} + m + o \rightleftharpoons \mathrm{B} + p + q \qquad \Delta G < 0$$

As an example of two reactions that are coupled, look at the phosphorylation reaction of glucose to yield glucose 6-phosphate plus water, an important step in the breakdown of dietary carbohydrates.

$$\underset{\textbf{Glucose}}{\text{HOCH}_2\text{CH(OH)CH(OH)CH(OH)CH(OH)CHO}} \xrightleftharpoons{\text{HOPO}_3^{2-}} \underset{\textbf{Glucose 6-phosphate}}{{}^{-}\text{O}_3\text{POCH}_2\text{CH(OH)CH(OH)CH(OH)CH(OH)CHO}} + \text{H}_2\text{O} \qquad \Delta G^{\circ\prime} = +13.8 \text{ kJ}$$

The reaction of glucose with $HOPO_3^{2-}$ does not occur spontaneously because it is energetically unfavorable, with $\Delta G^{\circ\prime} = +13.8$ kJ/mol. (The standard free-energy change for a biological reaction is denoted $\Delta G^{\circ\prime}$ and refers to a process in which reactants and products have a concentration of 1.0 M in a solution with pH = 7.) At the same time, however, the reaction of water with ATP to yield ADP plus $HOPO_3^{2-}$ is strongly favorable, with $\Delta G^{\circ} = -30.5$ kJ/mol. When the two reactions are coupled, glucose reacts with ATP to yield glucose 6-phosphate plus ADP in a reaction that is favorable by about 16.7 kJ/mol (4.0 kcal/mol). That is, ATP drives the phosphorylation reaction of glucose.

$$\begin{array}{llr}
 & \text{Glucose} + \cancel{\text{HOPO}_3^{2-}} \longrightarrow \text{Glucose 6-phosphate} + \cancel{\text{H}_2\text{O}} & \Delta G^{\circ\prime} = +13.8 \text{ kJ/mol} \\
 & \text{ATP} + \cancel{\text{H}_2\text{O}} \longrightarrow \text{ADP} + \cancel{\text{HOPO}_3^{2-}} + \text{H}^+ & \Delta G^{\circ\prime} = -30.5 \text{ kJ/mol} \\
\hline
\text{Net:} & \text{Glucose} + \text{ATP} \longrightarrow \text{Glucose 6-phosphate} + \text{ADP} + \text{H}^+ & \Delta G^{\circ\prime} = -16.7 \text{ kJ/mol}
\end{array}$$

It's this ability to drive otherwise unfavorable phosphorylation reactions that makes ATP so useful. The resultant phosphates are much more reactive as leaving groups in nucleophilic substitutions and eliminations than the alcohols they're derived from and are therefore more chemically useful.

Problem 29.1

One of the steps in fat metabolism is the reaction of glycerol (1,2,3-propanetriol) with ATP to yield glycerol 1-phosphate. Write the reaction, and draw the structure of glycerol 1-phosphate.

29.2 Catabolism of Triacylglycerols: The Fate of Glycerol

The metabolic breakdown of triacylglycerols begins with their hydrolysis in the stomach and small intestine to yield glycerol plus fatty acids. The reaction is catalyzed by a lipase, whose mechanism is shown in **Figure 29.2**. The active site of the enzyme contains a catalytic triad of aspartic acid, histidine, and serine residues, which act cooperatively to provide the necessary acid and base catalysis for the individual steps. Hydrolysis is accomplished by two sequential nucleophilic acyl substitution reactions, one that covalently binds an acyl group to the side chain –OH of a serine residue on the enzyme and a second that frees the fatty acid from the enzyme.

1 The enzyme active site contains an aspartic acid, a histidine, and a serine. First, histidine acts as a base to deprotonate the –OH group of serine, with the negatively charged carboxylate of aspartic acid stabilizing the nearby histidine cation that results. Serine then adds to the carbonyl group of the triacylglycerol, yielding a tetrahedral intermediate.

2 This intermediate expels a diacylglycerol as leaving group in a nucleophilic acyl substitution reaction, giving an acyl enzyme. The diacylglycerol is protonated by the histidine cation.

3 Histidine deprotonates a water molecule, which adds to the acyl group. A tetrahedral intermediate is again formed, and the histidine cation is again stabilized by the nearby carboxylate.

4 The tetrahedral intermediate expels the serine as leaving group in a second nucleophilic acyl substitution reaction, yielding a free fatty acid. The serine accepts a proton from histidine, and the enzyme has now returned to its starting structure.

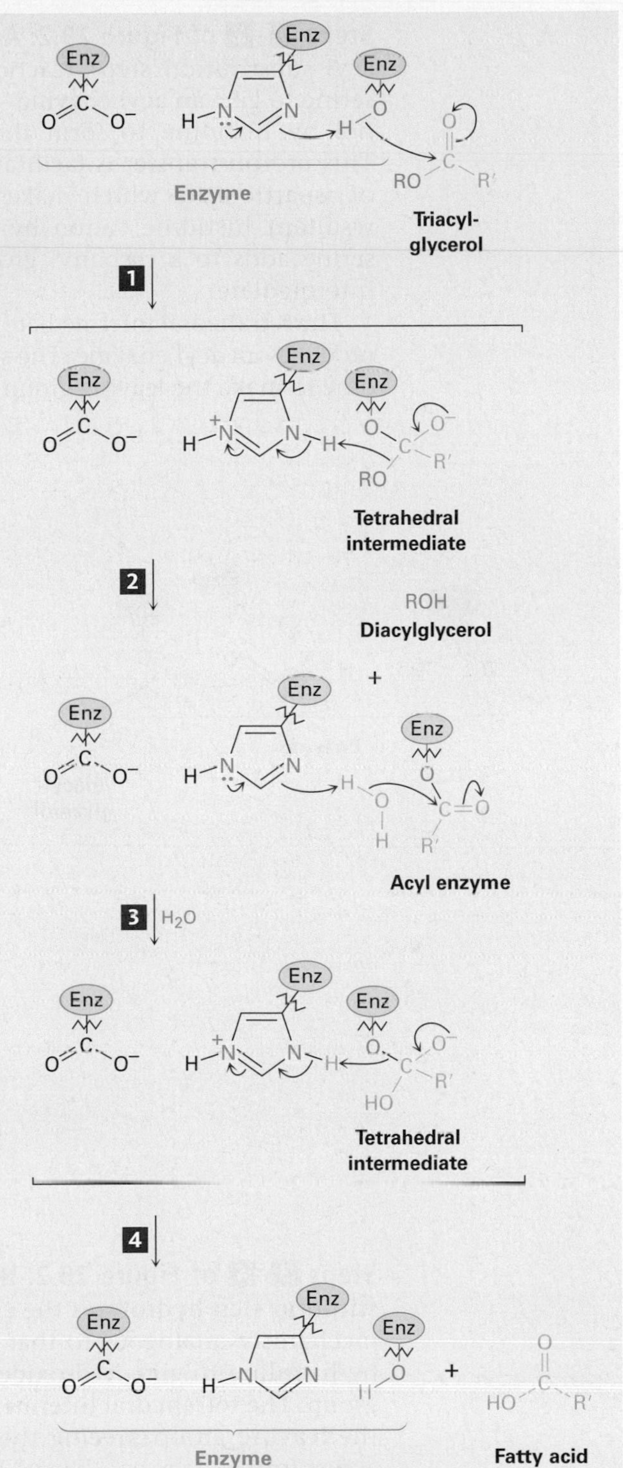

Figure 29.2 | MECHANISM

Mechanism of action of lipase. The active site of the enzyme contains a catalytic triad of aspartic acid, histidine, and serine, which react cooperatively to carry out two nucleophilic acyl substitution reactions. Individual steps are explained in the text.

Steps 1–2 of Figure 29.2: Acyl Enzyme Formation The first nucleophilic acyl substitution step—reaction of the triacylglycerol with the active-site serine to give an acyl enzyme—begins with deprotonation of the serine alcohol by histidine to form the more strongly nucleophilic alkoxide ion. This proton transfer is facilitated by a nearby side-chain carboxylate anion of aspartic acid, which makes the histidine more basic and stabilizes the resultant histidine cation by electrostatic interactions. The deprotonated serine adds to a carbonyl group of a triacylglycerol to give a tetrahedral intermediate.

The tetrahedral intermediate expels a diacylglycerol as the leaving group and produces an acyl enzyme. The step is catalyzed by a proton transfer from histidine to make the leaving group a neutral alcohol.

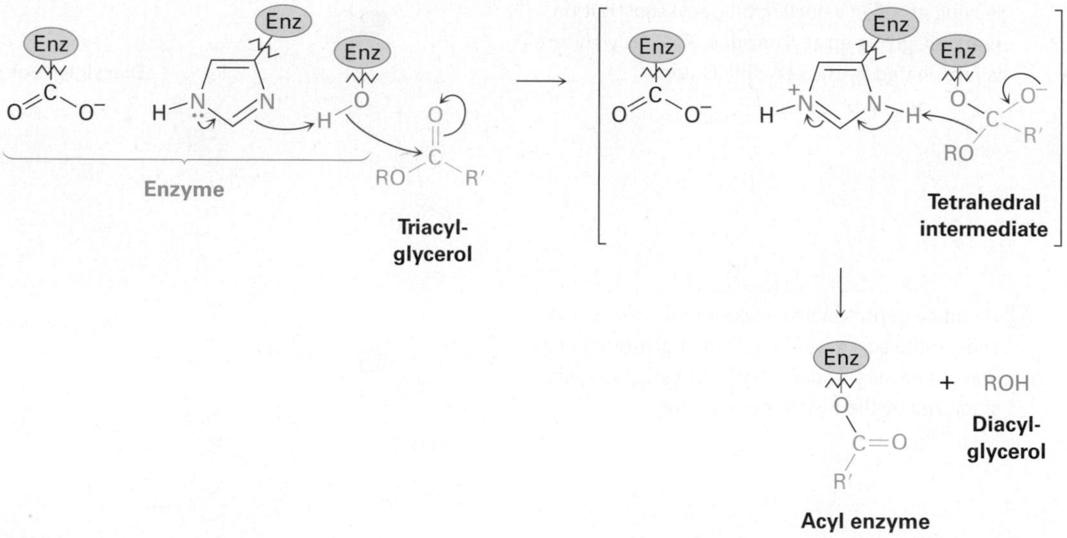

Enzyme

Triacylglycerol

Tetrahedral intermediate

Acyl enzyme + ROH (Diacylglycerol)

Steps 3–4 of Figure 29.2: Hydrolysis The second nucleophilic acyl substitution step hydrolyzes the acyl enzyme and gives the free fatty acid by a mechanism analogous to that of the first two steps. Water is deprotonated by histidine to give hydroxide ion, which adds to the enzyme-bound acyl group. The tetrahedral intermediate then expels the neutral serine residue as the leaving group, freeing the fatty acid and returning the enzyme to its active form.

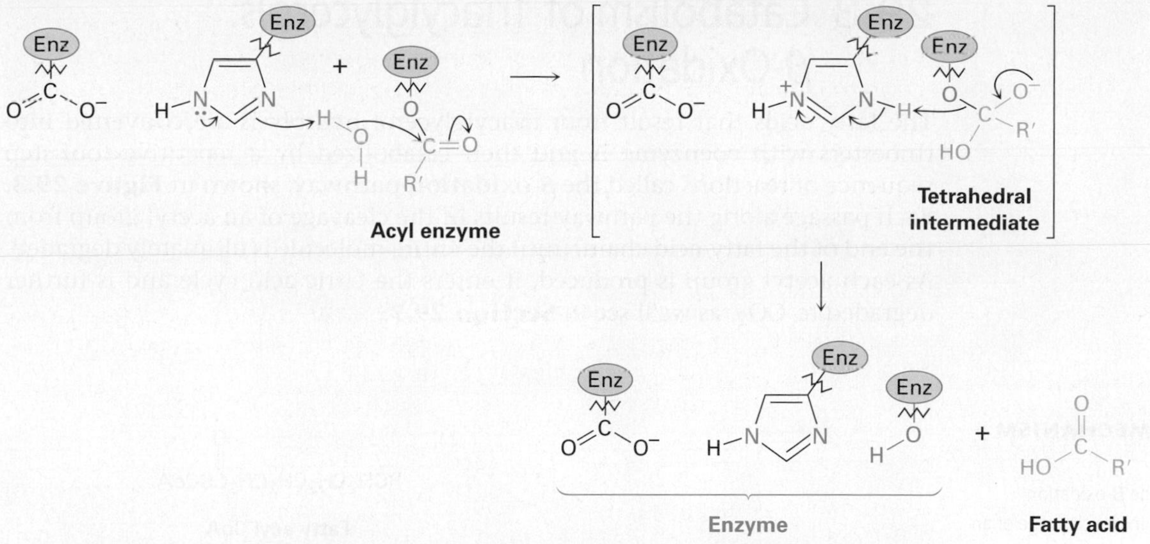

Acyl enzyme

Tetrahedral intermediate

Enzyme

Fatty acid

The fatty acids released on triacylglycerol hydrolysis are transported to mitochondria and degraded to acetyl CoA, while the glycerol is carried to the liver for further metabolism. In the liver, glycerol is first phosphorylated by reaction with ATP and then oxidized by NAD^+. The dihydroxyacetone phosphate (DHAP) that results enters the carbohydrate glycolysis pathway, which we'll discuss in **Section 29.5**.

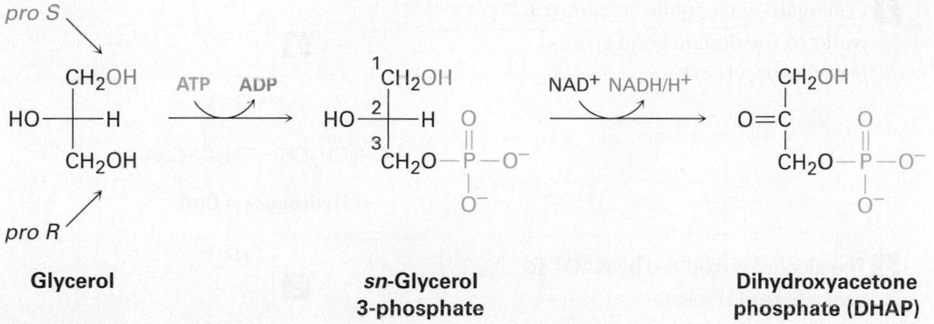

pro S

pro R

ATP ADP

NAD^+ NADH/H^+

Glycerol

***sn*-Glycerol 3-phosphate**

Dihydroxyacetone phosphate (DHAP)

You might note that C2 of glycerol is a prochiral center **(Section 5.11)** with two identical "arms." As is typical for enzyme-catalyzed reactions, the phosphorylation of glycerol is selective. Only the *pro-R* arm undergoes reaction, although this can't be predicted in advance.

Note also that the phosphorylation product is named *sn*-glycerol 3-phosphate, where the *sn-* prefix means "stereospecific numbering." In this convention, the molecule is drawn in Fischer projection with the –OH group at C2 pointing to the left and the glycerol carbon atoms numbered beginning at the top.

29.3 Catabolism of Triacylglycerols: β-Oxidation

The fatty acids that result from triacylglycerol hydrolysis are converted into thioesters with coenzyme A and then catabolized by a repetitive four-step sequence of reactions called the **β-oxidation pathway**, shown in **Figure 29.3**. Each passage along the pathway results in the cleavage of an acetyl group from the end of the fatty-acid chain, until the entire molecule is ultimately degraded. As each acetyl group is produced, it enters the citric acid cycle and is further degraded to CO_2, as we'll see in **Section 29.7**.

Figure 29.3 | MECHANISM

The four steps of the β-oxidation pathway, resulting in the cleavage of an acetyl group from the end of the fatty-acid chain. The key chain-shortening step is a retro-Claisen reaction of a β-keto thioester. Individual steps are explained in the text.

1 A conjugated double bond is introduced by removal of hydrogens from C2 and C3 by the coenzyme flavin adenine dinucleotide (FAD).

2 Conjugate nucleophilic addition of water to the double bond gives a β-hydroxyacyl CoA.

3 The alcohol is oxidized by NAD^+ to give a β-keto thioester.

4 Nucleophilic addition of coenzyme A to the keto group occurs, followed by a retro-Claisen condensation reaction. The products are acetyl CoA and a chain-shortened fatty acyl CoA.

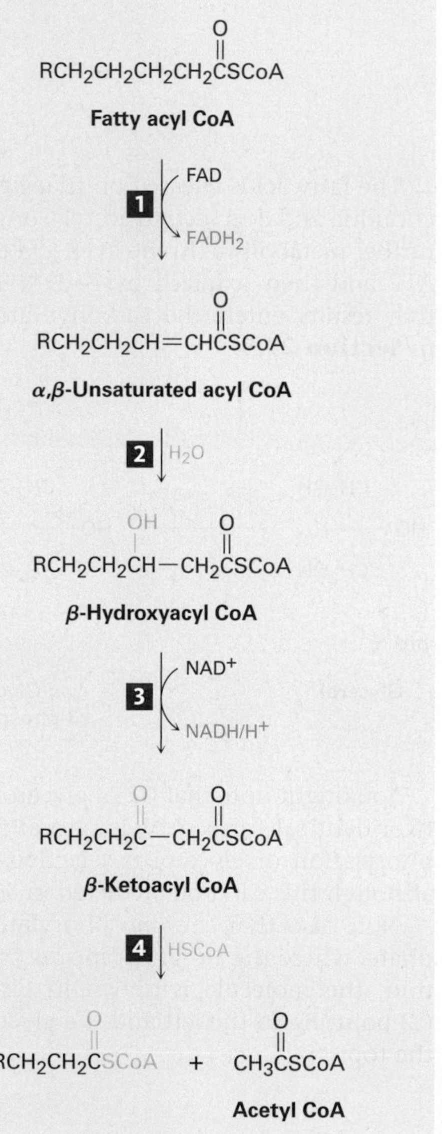

Step 1 of Figure 29.3: Introduction of a Double Bond The β-oxidation pathway begins when two hydrogen atoms are removed from C2 and C3 of the fatty acyl CoA by one of a family of acyl-CoA dehydrogenases to yield an α,β-unsaturated acyl CoA. This kind of oxidation—the introduction of a conjugated double bond into a carbonyl compound—occurs frequently in biochemical pathways and usually involves the coenzyme flavin adenine dinucleotide (FAD). Reduced $FADH_2$ is the by-product.

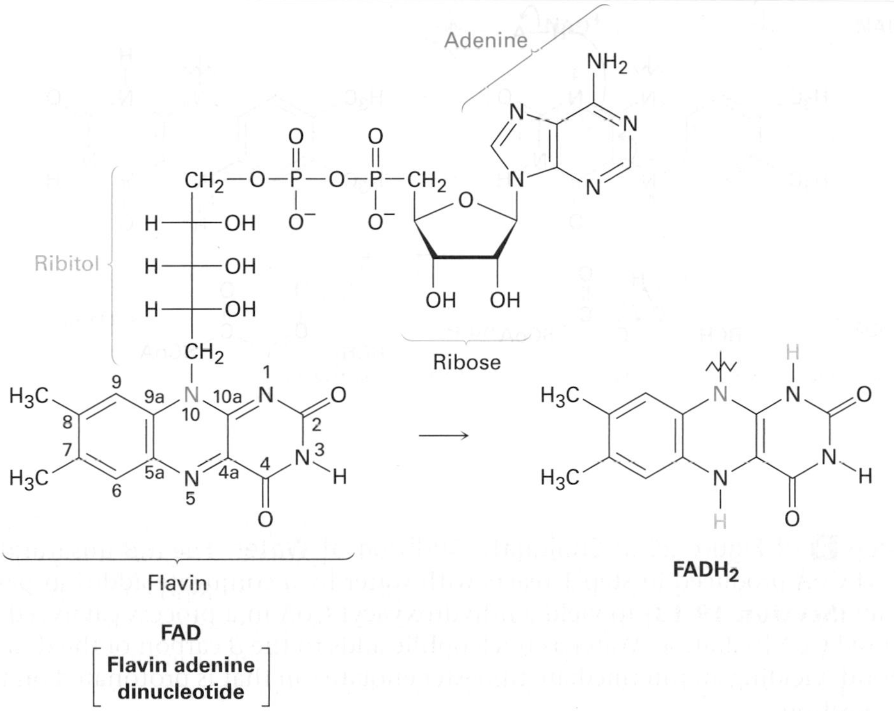

The mechanisms of FAD-catalyzed reactions are often difficult to establish because flavin coenzymes can operate by both two-electron (polar) and one-electron (radical) pathways. As a result, extensive studies of the family of acyl-CoA dehydrogenases have not yet provided a clear picture of how these enzymes function. What is known is that: (1) The first step is abstraction of the *pro-R* hydrogen from the acidic α position of the acyl CoA to give a thioester enolate ion. Hydrogen-bonding between the acyl carbonyl group and the ribitol hydroxyls of FAD increases the acidity of the acyl group. (2) The *pro-R* hydrogen at the β position is transferred to FAD. (3) The α,β-unsaturated acyl CoA that results has a trans double bond.

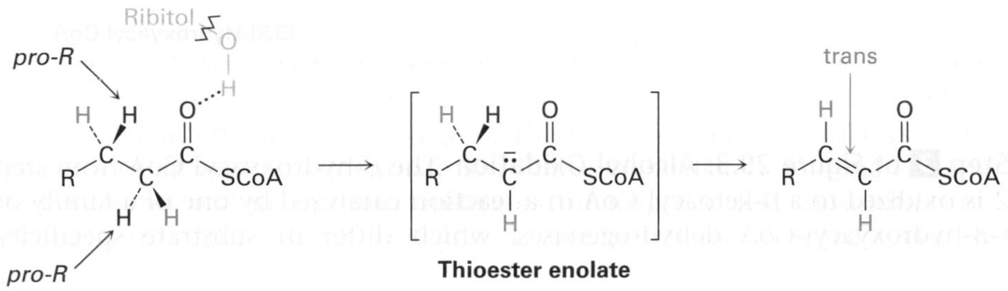

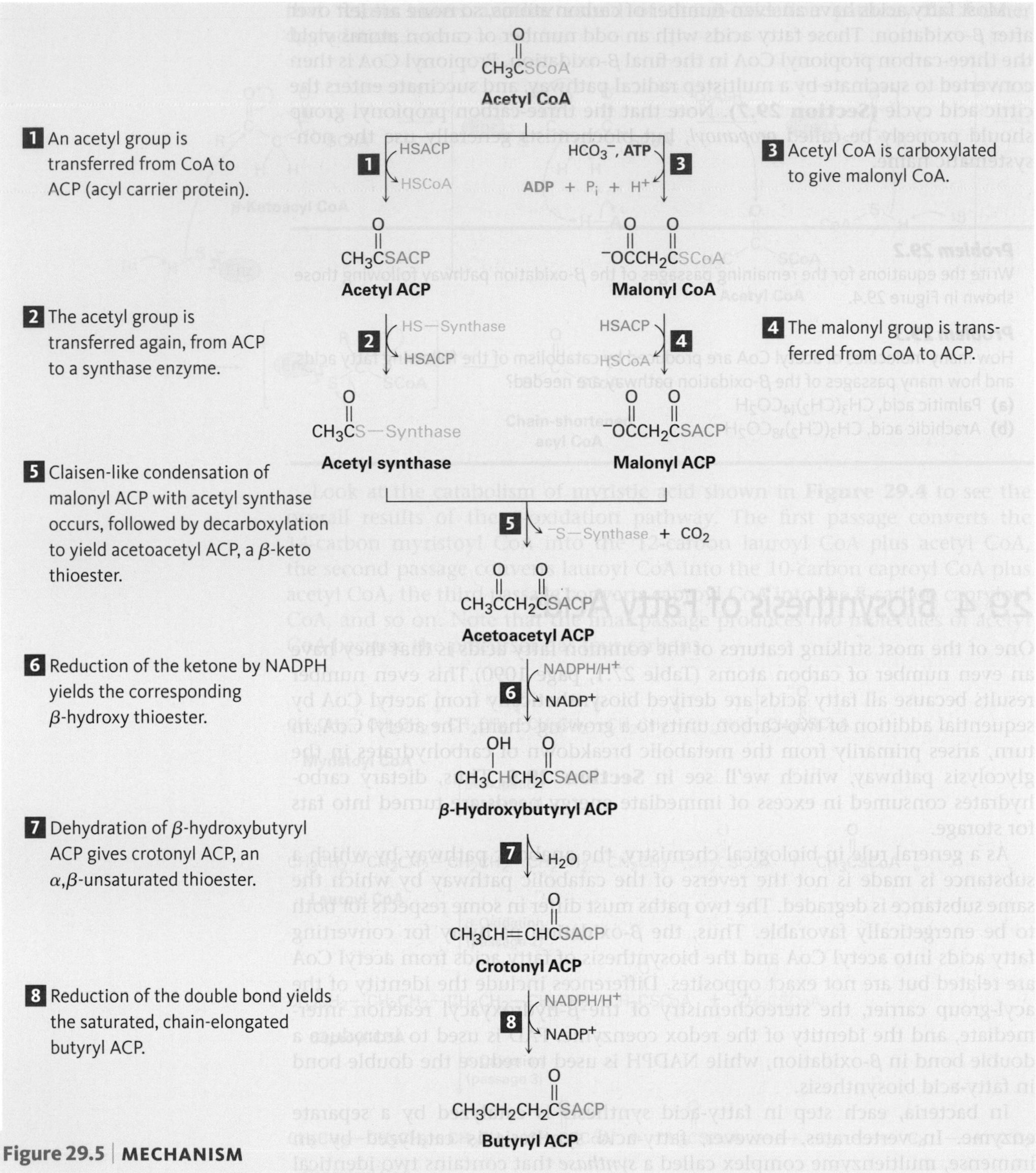

Figure 29.5 | **MECHANISM**

The pathway for fatty-acid biosynthesis from the two-carbon precursor, acetyl CoA. Individual steps are explained in the text.

Steps 1–2 of Figure 29.5: Acyl Transfers The starting material for fatty-acid biosynthesis is the thioester acetyl CoA, the final product of carbohydrate breakdown, as we'll see in **Section 29.6**. The pathway begins with several *priming reactions,* which transport acetyl CoA and convert it into more reactive species. The first priming reaction is a nucleophilic acyl substitution reaction that converts acetyl CoA into acetyl ACP (acyl carrier protein).

Notice that the mechanism of the nucleophilic acyl substitution step can be given in an abbreviated form that saves space by not explicitly showing the tetrahedral reaction intermediate. Instead, electron movement is shown as a heart-shaped path around the carbonyl oxygen to imply the two steps of the full mechanism. Biochemists use this kind of format commonly, and we'll also use it on occasion in the remainder of this chapter.

B: H—SACP H3C SCoA HSCoA SACP H—A

Tetrahedral intermediate

Abbreviated mechanism

In bacteria, ACP is a small protein of 77 residues that transports an acyl group from one enzyme to another. In vertebrates, however, ACP appears to be a long arm on a multienzyme synthase complex, whose apparent function is to shepherd an acyl group from site to site within the complex. As in acetyl CoA, the acyl group in acetyl ACP is linked by a thioester bond to the sulfur atom of phosphopantetheine. The phosphopantetheine is in turn linked to ACP through the side-chain –OH group of a serine residue in the enzyme.

$CH_3C(=O)—SCH_2CH_2NHC(=O)CH_2CH_2NHC(=O)CH(OH)C(CH_3)_2CH_2OP(=O)(O^-)—OCH_2—Ser—ACP$

Phosphopantetheine

Acetyl ACP

Step 2, another priming reaction, involves a further exchange of thioester linkages by another nucleophilic acyl substitution and results in covalent bonding of the acetyl group to a cysteine residue in the synthase complex that will catalyze the upcoming condensation step.

Steps 3–4 of Figure 29.5: Carboxylation and Acyl Transfer Step 3 is a *loading* reaction in which acetyl CoA is carboxylated by reaction with HCO_3^- and ATP to yield malonyl CoA plus ADP. This step requires the coenzyme biotin, which is bonded to the lysine residue of acetyl CoA carboxylase and acts as a carrier of CO_2. Biotin first reacts with bicarbonate ion to give *N*-carboxybiotin, which then reacts with the enolate ion of acetyl CoA and transfers the CO_2 group. Thus, biotin acts as a carrier of CO_2, binding it in one step and releasing it in another.

The mechanism of the CO_2 transfer reaction with acetyl CoA to give malonyl CoA is thought to involve CO_2 as the reactive species. One proposal is that loss of CO_2 is favored by hydrogen-bond formation between the *N*-carboxybiotin carbonyl group and a nearby acidic site in the enzyme. Simultaneous deprotonation of acetyl CoA by a basic site in the enzyme gives a thioester enolate ion that can react with CO_2 as it is formed **(Figure 29.6)**.

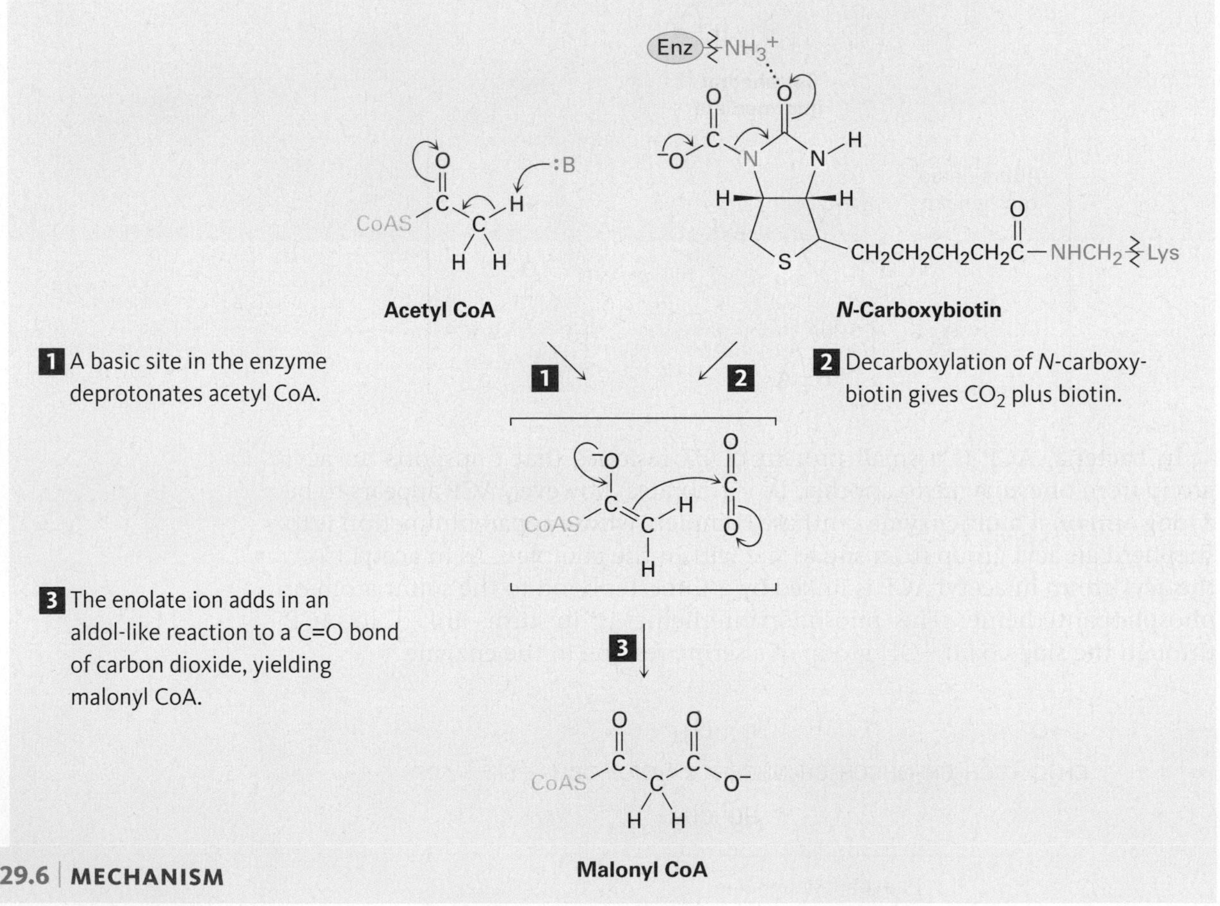

Figure 29.6 | MECHANISM

Mechanism of step 3 in Figure 29.5, the biotin-dependent carboxylation of acetyl CoA to yield malonyl CoA.

Following the formation of malonyl CoA, another nucleophilic acyl substitution reaction occurs in step 4 to form the more reactive malonyl ACP, thereby binding the malonyl group to an ACP arm of the multienzyme synthase. At this point, both acetyl and malonyl groups are bound to the enzyme, and the stage is set for their condensation.

Step 5 of Figure 29.5: Condensation The key carbon–carbon bond-forming reaction that builds the fatty-acid chain occurs in step 5. This step is simply a Claisen condensation between acetyl synthase as the electrophilic acceptor and malonyl ACP as the nucleophilic donor. The mechanism of the condensation is thought to involve decarboxylation of malonyl ACP to give an enolate ion, followed by immediate nucleophilic addition of the enolate ion to the carbonyl group of acetyl synthase. Breakdown of the tetrahedral intermediate then gives the four-carbon condensation product acetoacetyl ACP and frees the synthase binding site for attachment of the chain-elongated acyl group at the end of the sequence.

Synthase—S—C(=O)—CH_3 + ACPS—C(=O)—CH_2—C(=O)—O^- ⟶ [Synthase—S—C(O^-)(H_3C)—CH_2—C(=O)—SACP] + CO_2 ⟶ H_3C—C(=O)—CH_2—C(=O)—SACP + Synthase—S^-

Malonyl ACP **Acetoacetyl ACP**

Steps 6–8 of Figure 29.5: Reduction and Dehydration The ketone carbonyl group in acetoacetyl ACP is next reduced to the alcohol β-hydroxybutyryl ACP by β-keto thioester reductase and NADPH, a reducing coenzyme closely related to NADH. *R* Stereochemistry results at the newly formed chirality center in the β-hydroxy thioester product. (Note that the systematic name of a butyryl group is *butanoyl*.)

A—H + H_3C—C(=O)—CH_2—C(=O)—SACP + NADPH ⟶ H_3C—CH(OH)—CH_2—C(=O)—SACP + $NADP^+$

Acetoacetyl ACP **NADPH** **β-Hydroxybutyryl ACP** **$NADP^+$**

Subsequent dehydration of β-hydroxybutyryl ACP by an E1cB reaction in step 7 yields *trans*-crotonyl ACP, and the carbon–carbon double bond of crotonyl ACP is reduced by NADPH in step 8 to yield butyryl ACP. The double-bond reduction occurs by conjugate nucleophilic addition of a hydride ion from NADPH to the β carbon of *trans*-crotonyl ACP. In vertebrates, the reduction occurs by an overall *syn* addition, but other organisms carry out similar chemistry with different stereochemistry.

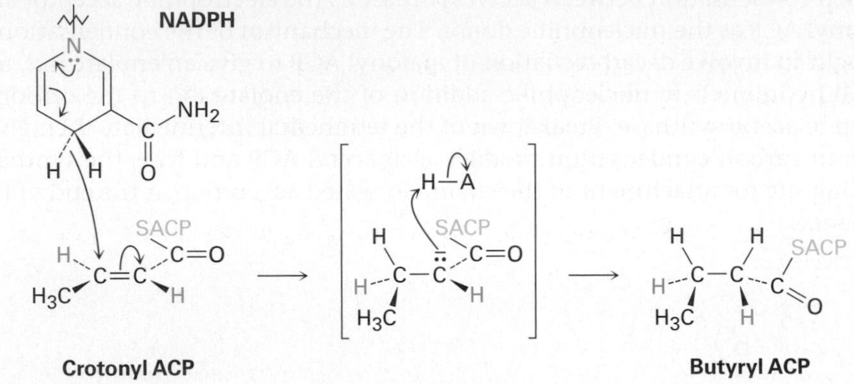

NADPH

Crotonyl ACP → [intermediate] → Butyryl ACP

The net effect of the eight steps in the fatty-acid biosynthesis pathway is to take two 2-carbon acetyl groups and combine them into a 4-carbon butyryl group. Further condensation of the butyryl group with another malonyl ACP yields a 6-carbon unit, and still further repetitions of the pathway add two more carbon atoms to the chain each time until the 16-carbon palmitoyl ACP is reached.

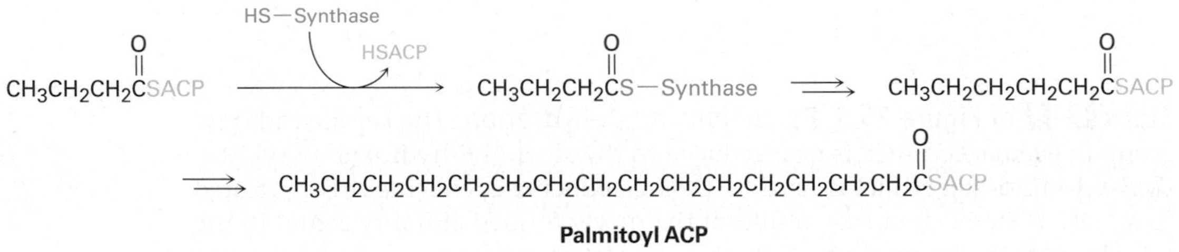

$$CH_3CH_2CH_2C(=O)SACP \xrightarrow[\text{HSACP}]{\text{HS—Synthase}} CH_3CH_2CH_2C(=O)S\text{—Synthase} \rightleftharpoons CH_3CH_2CH_2CH_2CH_2C(=O)SACP$$

$$\rightleftharpoons CH_3CH_2CH_2CH_2CH_2CH_2CH_2CH_2CH_2CH_2CH_2CH_2CH_2CH_2CH_2C(=O)SACP$$

Palmitoyl ACP

Further chain elongation of palmitic acid occurs by reactions similar to those just described, but CoA rather than ACP is the carrier group, and separate enzymes are needed for each step rather than a multienzyme synthase complex.

Problem 29.4

Write a mechanism for the dehydration reaction of β-hydroxybutyryl ACP to yield crotonyl ACP in step 7 of fatty-acid synthesis.

Problem 29.5

Evidence for the role of acetate in fatty-acid biosynthesis comes from isotope-labeling experiments. If acetate labeled with ^{13}C in the methyl group ($^{13}CH_3CO_2H$) were incorporated into fatty acids, at what positions in the fatty-acid chain would you expect the ^{13}C label to appear?

Problem 29.6

Does the reduction of acetoacetyl ACP in step 6 occur on the *Re* face or the *Si* face of the molecule?

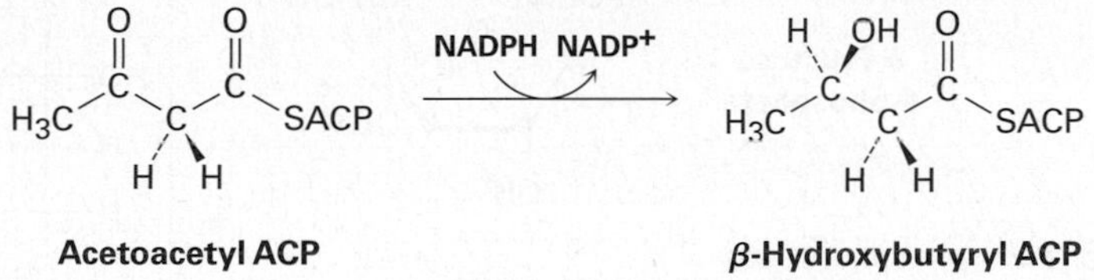

Acetoacetyl ACP —(NADPH → $NADP^+$)→ β-Hydroxybutyryl ACP

29.5 Catabolism of Carbohydrates: Glycolysis

Glucose is the body's primary short-term energy source. Its catabolism begins with **glycolysis**, a series of ten enzyme-catalyzed reactions that break down glucose into 2 equivalents of pyruvate, $CH_3COCO_2^-$. The steps of glycolysis, also called the *Embden–Meyerhoff pathway* after its discoverers, are summarized in **Figure 29.7**.

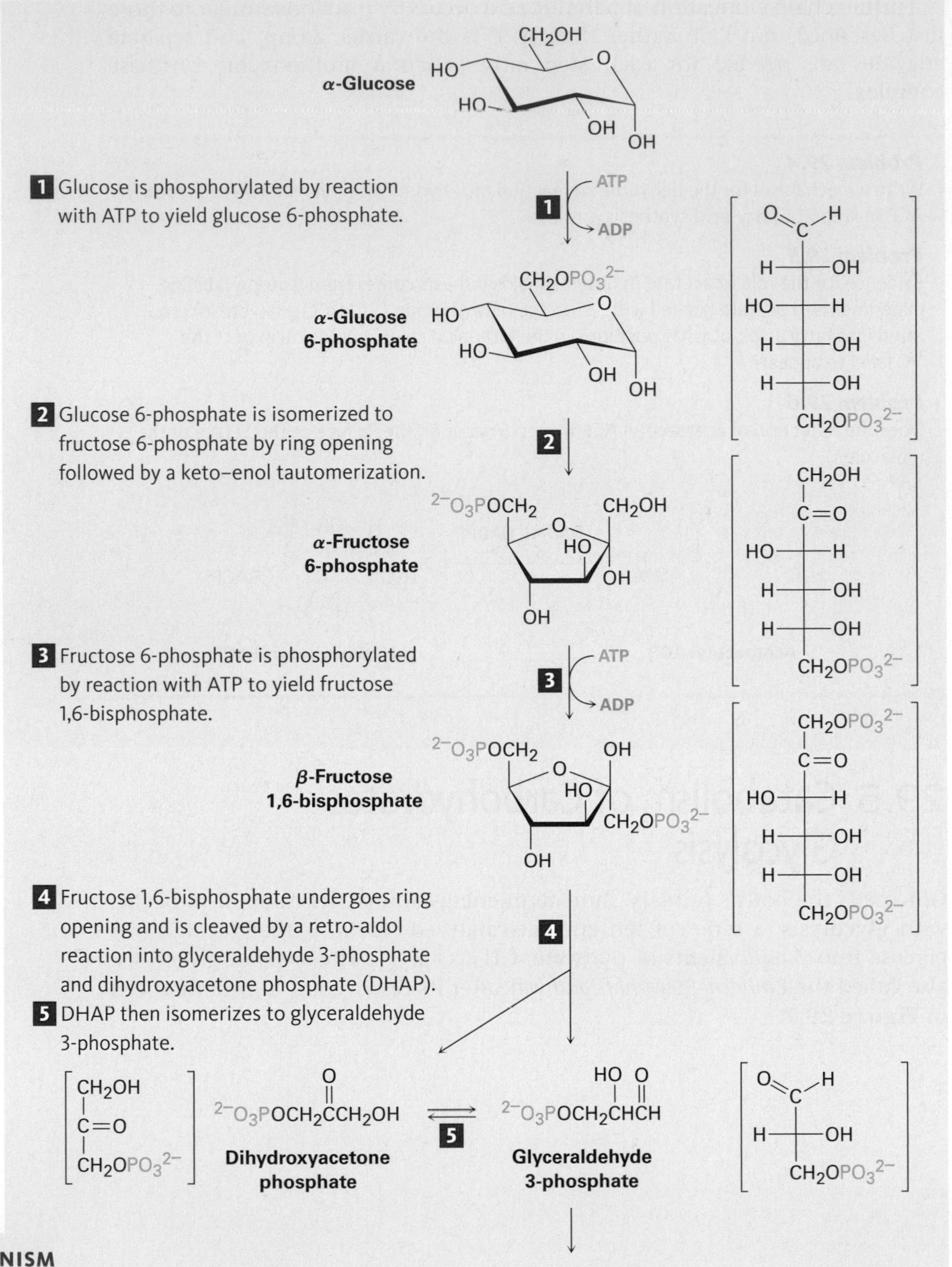

Figure 29.7 | MECHANISM

The ten-step glycolysis pathway for catabolizing glucose to two molecules of pyruvate. Individual steps are described in the text.

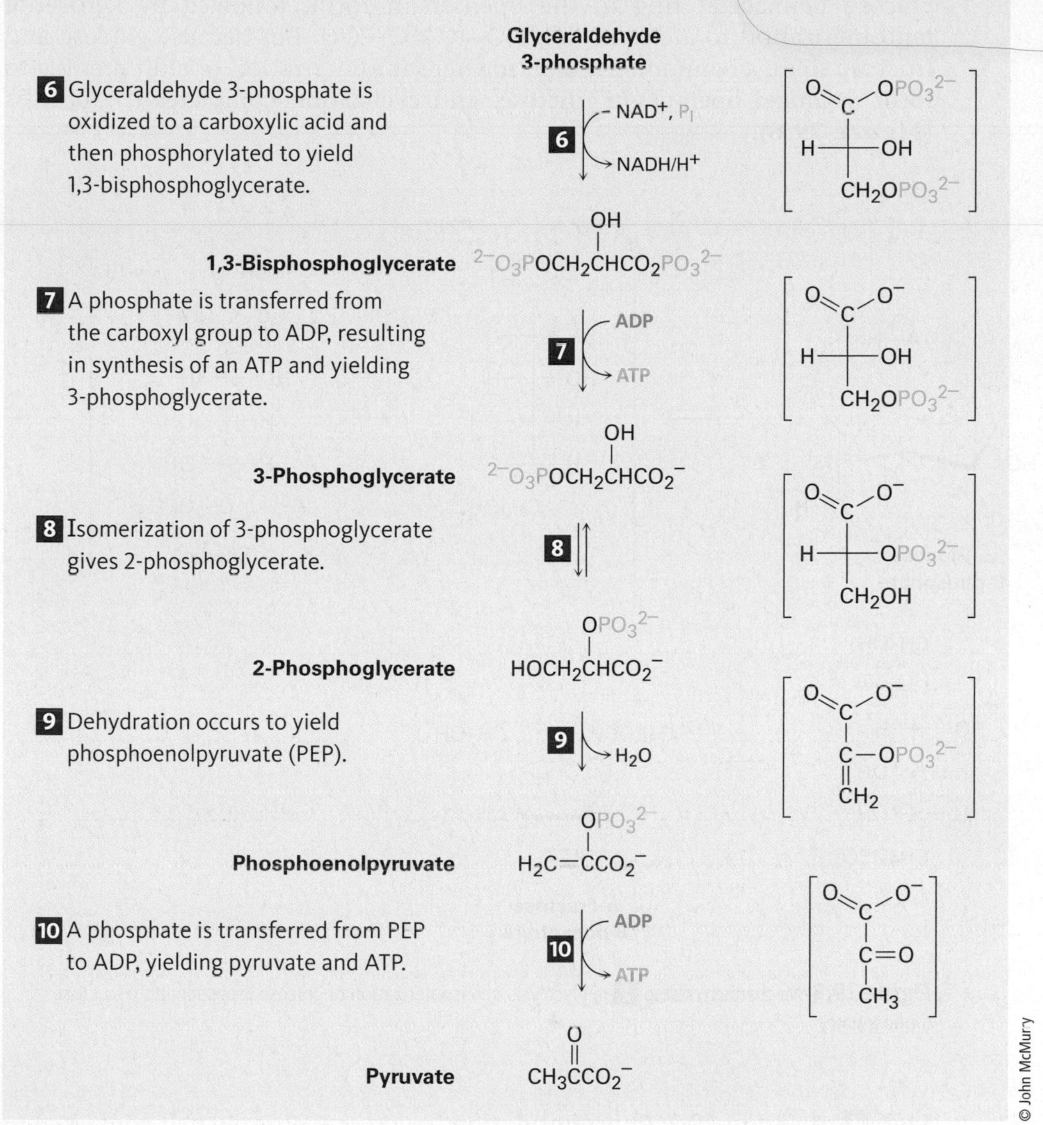

Figure 29.7 (Continued)

Steps 1–2 of Figure 29.7: Phosphorylation and Isomerization Glucose, produced by the digestion of dietary carbohydrates, is phosphorylated at the C6 hydroxyl group by reaction with ATP in a process catalyzed by hexokinase. As noted in **Section 29.1**, the reaction requires Mg^{2+} as a cofactor to complex with the negatively charged phosphate oxygens. The glucose 6-phosphate that results is then isomerized by glucose 6-phosphate isomerase to give fructose 6-phosphate. The isomerization takes place by initial opening of the glucose hemiacetal ring to the open-chain form, followed by keto–enol tautomerization to a cis enediol, HO—C=C—OH. But because glucose and fructose share a common enediol, further tautomerization to a different keto form produces open-chain fructose, and cyclization completes the process **(Figure 29.8)**.

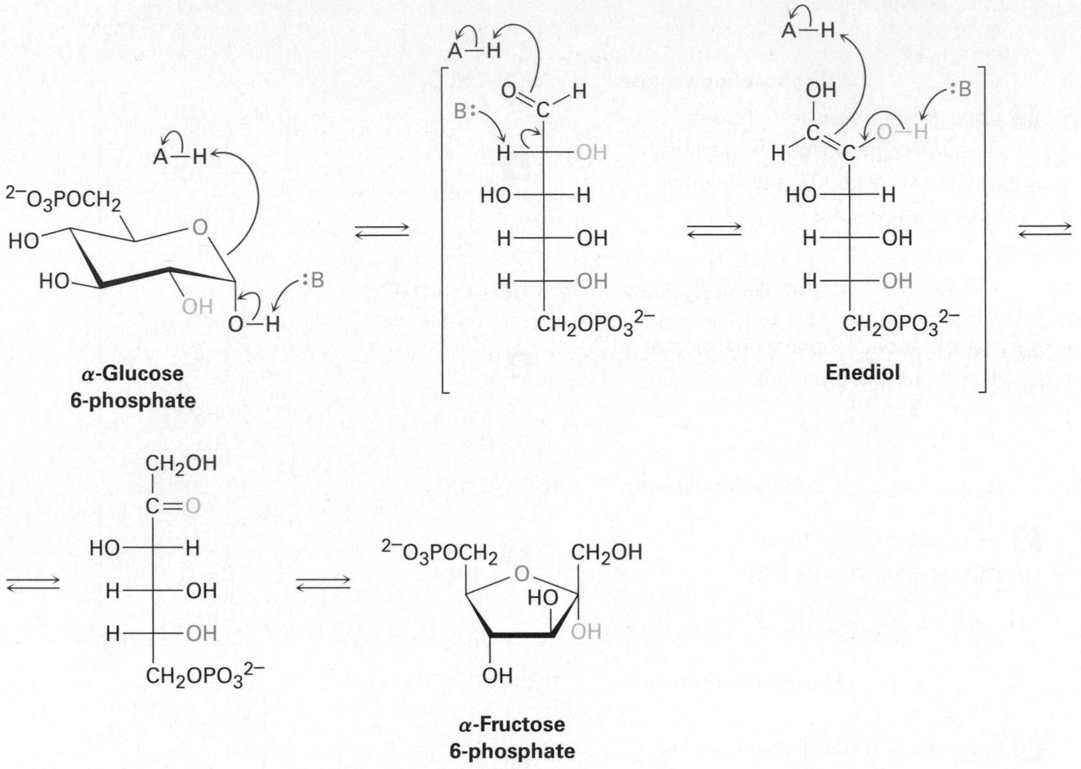

α-Glucose 6-phosphate

Enediol

α-Fructose 6-phosphate

Figure 29.8 Mechanism of step **2** in glycolysis, the isomerization of glucose 6-phosphate to fructose 6-phosphate.

Step 3 of Figure 29.7: Phosphorylation Fructose 6-phosphate is converted in step 3 to fructose 1,6-bisphosphate (FBP) by a phosphofructokinase-catalyzed reaction with ATP (recall that the prefix *bis-* means two). The mechanism is similar to that in step 1, with Mg^{2+} ion again required as cofactor. Interestingly, the product of step 2 is the α anomer of fructose 6-phosphate, but it is the β anomer that is phosphorylated in step 3, implying that the two anomers equilibrate rapidly through the open-chain form. The result is a molecule ready to be split into the two three-carbon intermediates that will ultimately become two molecules of pyruvate.

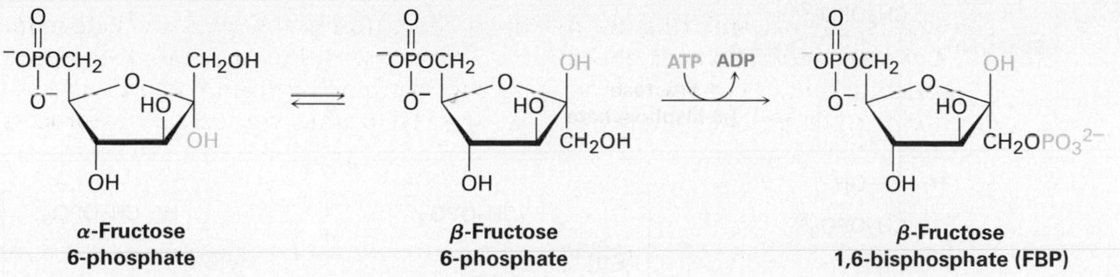

α-Fructose 6-phosphate ⇌ β-Fructose 6-phosphate —(ATP → ADP)→ β-Fructose 1,6-bisphosphate (FBP)

Step 4 of Figure 29.7: Cleavage Fructose 1,6-bisphosphate is cleaved in step 4 into two 3-carbon pieces, dihydroxyacetone phosphate (DHAP) and glyceraldehyde 3-phosphate (GAP). The bond between C3 and C4 of fructose 1,6-bisphosphate breaks, and a C=O group is formed at C4. Mechanistically, the cleavage is the reverse of an aldol reaction **(Section 23.1)** and is catalyzed by an aldolase. A forward aldol reaction joins two aldehydes or ketones to give a β-hydroxy carbonyl compound, while a retro-aldol reaction such as that occurring here cleaves a β-hydroxy carbonyl compound into two aldehydes or ketones.

Fructose 1,6-bisphosphate ⇌ Glyceraldehyde 3-phosphate (GAP) + [enol intermediate] ⇌ Dihydroxyacetone phosphate (DHAP)

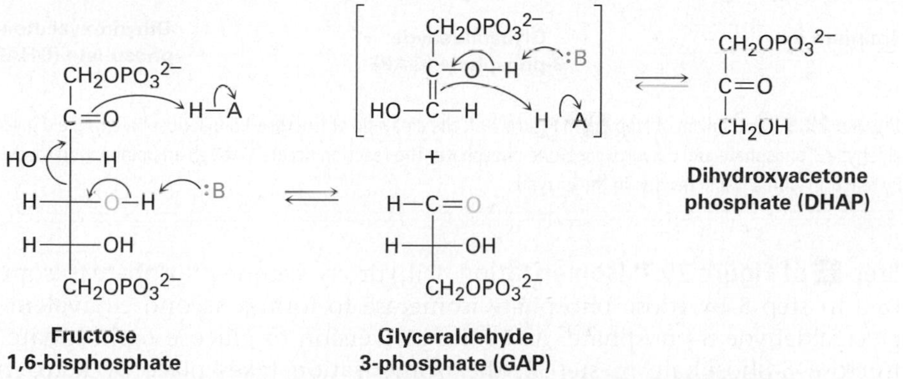

Two classes of aldolases are used by organisms for catalysis of the retro-aldol reaction. In fungi, algae, and some bacteria, the retro-aldol reaction is catalyzed by class II aldolases, which function by coordination of the fructose carbonyl group with Zn^{2+} as Lewis acid. In plants and animals, the reaction is catalyzed by class I aldolases and does not take place on the free ketone. Instead, fructose 1,6-bisphosphate undergoes reaction with the side-chain $-NH_2$ group of a lysine residue on the aldolase to yield a protonated enzyme-bound imine **(Section 19.8)**, often called a **Schiff base** in biochemistry.

Because of its positive charge, the iminium ion is a better electron acceptor than a ketone carbonyl group. Retro-aldol reaction ensues, giving glyceraldehyde 3-phosphate and an enamine, which is protonated to give another iminium ion that is hydrolyzed to yield dihydroxyacetone phosphate **(Figure 29.9)**.

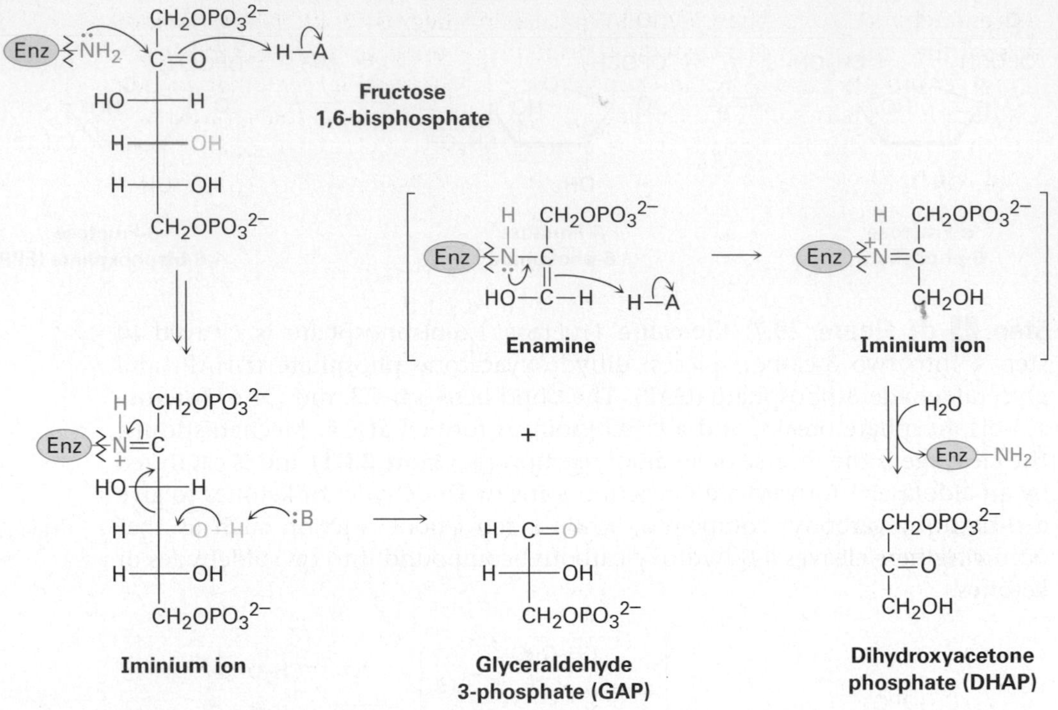

Fructose 1,6-bisphosphate

Enamine → **Iminium ion**

Iminium ion → **Glyceraldehyde 3-phosphate (GAP)** + **Dihydroxyacetone phosphate (DHAP)**

Figure 29.9 Mechanism of step **4** in Figure 29.7, the cleavage of fructose 1,6-bisphosphate to yield glyceraldehyde 3-phosphate and dihydroxyacetone phosphate. The reaction occurs through an iminium ion formed by reaction with a lysine residue in the enzyme.

Step 5 of Figure 29.7: Isomerization Dihydroxyacetone phosphate is isomerized in step 5 by triose phosphate isomerase to form a second equivalent of glyceraldehyde 3-phosphate. As in the conversion of glucose 6-phosphate to fructose 6-phosphate in step 2, the isomerization takes place by keto–enol tautomerization through a common enediol intermediate. A base deprotonates C1 and then reprotonates C2 using the same hydrogen. The net result of steps 4 and 5 together is the production of two glyceraldehyde 3-phosphate molecules, both of which pass down the rest of the pathway. Thus, each of the remaining five steps of glycolysis takes place twice for every glucose molecule that enters at step 1.

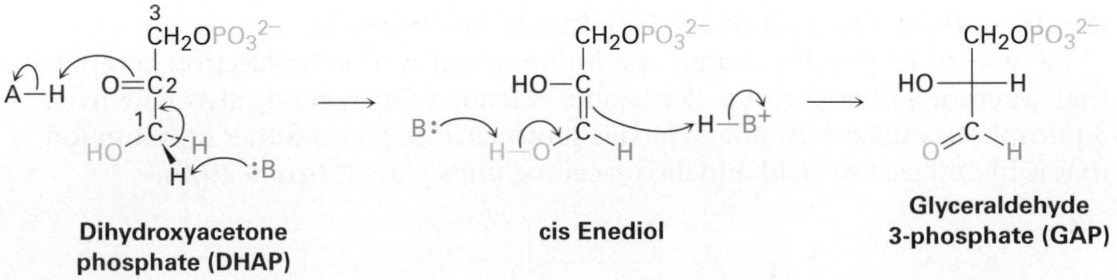

Dihydroxyacetone phosphate (DHAP) → **cis Enediol** → **Glyceraldehyde 3-phosphate (GAP)**

Steps 6–7 of Figure 29.7: Oxidation, Phosphorylation, and Dephosphorylation Glyceraldehyde 3-phosphate is oxidized and phosphorylated in step 6 to give 1,3-bisphosphoglycerate **(Figure 29.10)**. The reaction is catalyzed by glyceraldehyde 3-phosphate dehydrogenase and begins by nucleophilic addition of the –SH group of a cysteine residue in the enzyme to the aldehyde carbonyl group to yield a *hemithioacetal,* the sulfur analog of a hemiacetal. Oxidation of the hemithioacetal –OH group by NAD^+ then yields a thioester, which reacts with phosphate ion in a nucleophilic acyl substitution step to yield 1,3-bisphosphoglycerate, a mixed anhydride between a carboxylic acid and phosphoric acid.

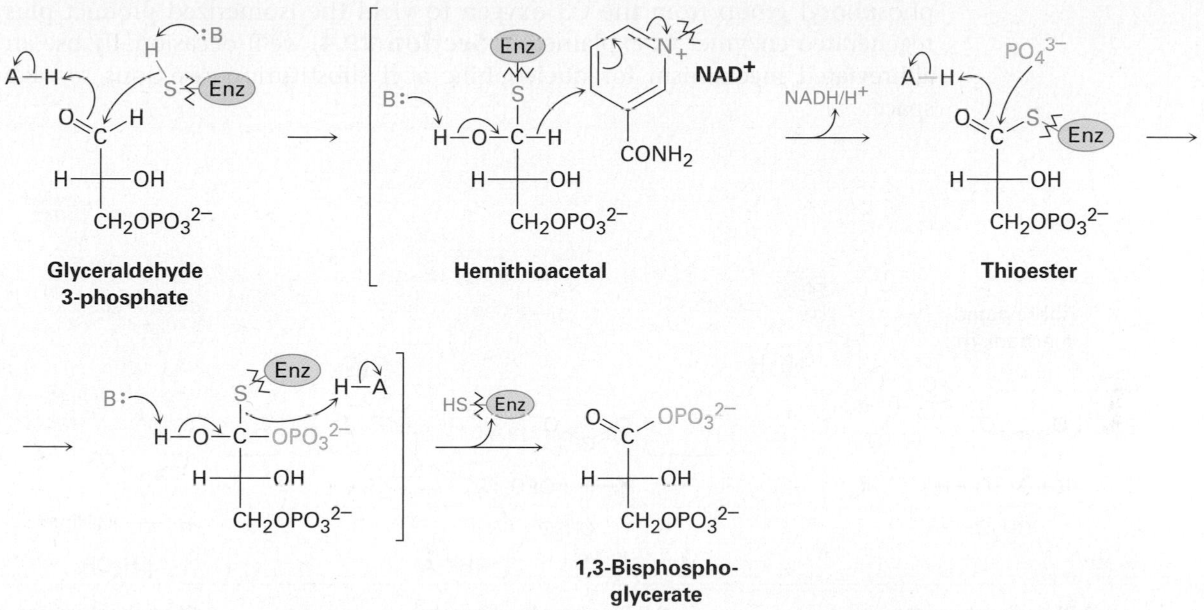

Figure 29.10 Mechanism of step 6 in Figure 29.7, the oxidation and phosphorylation of glyceraldehyde 3-phosphate to give 1,3-bisphosphoglycerate.The process occurs through initial formation of a hemiacetal that is oxidized to a thioester and converted into an acyl phosphate.

Like all anhydrides **(Section 21.5)**, the mixed carboxylic–phosphoric anhydride is a reactive substrate in nucleophilic acyl (or phosphoryl) substitution reactions. Reaction of 1,3-bisphosphoglycerate with ADP occurs in step 7 by substitution on phosphorus, resulting in transfer of a phosphate group to ADP and giving ATP plus 3-phosphoglycerate. The process is catalyzed by phosphoglycerate kinase and requires Mg^{2+} as cofactor. Together, steps 6 and 7 accomplish the oxidation of an aldehyde to a carboxylic acid.

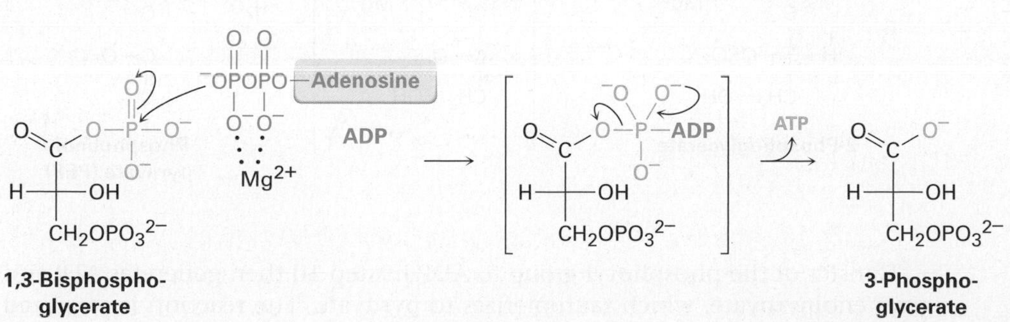

Step 8 of Figure 29.7: Isomerization 3-Phosphoglycerate isomerizes to 2-phosphoglycerate in a step catalyzed by phosphoglycerate mutase. In plants, 3-phosphoglycerate transfers its phosphoryl group from its C3 oxygen to a histidine residue on the enzyme in one step and then accepts the same phosphoryl group back onto the C2 oxygen in a second step. In animals and yeast, however, the enzyme contains a phosphorylated histidine, which transfers its phosphoryl group to the C2 oxygen of 3-phosphoglycerate and forms 2,3-bisphosphoglycerate as intermediate. The same histidine then accepts a phosphoryl group from the C3 oxygen to yield the isomerized product plus regenerated enzyme. As explained in **Section 29.4**, we'll occasionally use an abbreviated mechanism for nucleophilic acyl substitution reactions to save space.

Abbreviated mechanism

3-Phosphoglycerate → 2,3-Bisphosphoglycerate → 2-Phosphoglycerate

Steps 9–10 of Figure 29.7: Dehydration and Dephosphorylation Like most β-hydroxy carbonyl compounds, 2-phosphoglycerate undergoes a ready dehydration in step 9 by an E1cB mechanism **(Section 23.3)**. The process is catalyzed by enolase, and the product is phosphoenolpyruvate, abbreviated PEP. Two Mg^{2+} ions are associated with the 2-phosphoglycerate to neutralize the negative charges.

2-Phosphoglycerate → [intermediate] → Phosphoenolpyruvate (PEP)

Transfer of the phosphoryl group to ADP in step 10 then generates ATP and gives enolpyruvate, which tautomerizes to pyruvate. The reaction is catalyzed by pyruvate kinase and requires that a molecule of fructose 1,6-bisphosphate also be present, as well as 2 equivalents of Mg^{2+}. One Mg^{2+} ion coordinates to ADP, and the other increases the acidity of a water molecule necessary for protonation of the enolate ion.

Phosphoenol-pyruvate (PEP) + ADP (Mg^{2+}) → ATP + [Enolpyruvate] ⇌ Pyruvate

The overall result of glycolysis can be summarized by the following equation:

$$\text{Glucose} + 2\ NAD^+ + 2\ P_i + 2\ ADP \longrightarrow 2\ \text{Pyruvate} + 2\ NADH + 2\ ATP + 2\ H_2O$$

Problem 29.7
Identify the two steps in glycolysis in which ATP is produced.

Problem 29.8
Look at the entire glycolysis pathway, and make a list of the kinds of organic reactions that take place—nucleophilic acyl substitutions, aldol reactions, E1cB reactions, and so forth.

29.6 Conversion of Pyruvate to Acetyl CoA

Pyruvate, produced by catabolism of glucose (and by degradation of several amino acids), can undergo several further transformations depending on the conditions and on the organism. In the absence of oxygen, pyruvate can be either reduced by NADH to yield lactate [$CH_3CH(OH)CO_2^-$] or, in yeast, fermented to give ethanol. Under typical aerobic conditions in mammals, however, pyruvate is converted by a process called *oxidative decarboxylation* to give acetyl CoA plus CO_2. (*Oxidative* because the oxidation state of the carbonyl carbon rises from that of a ketone to that of a thioester.)

The conversion occurs through a multistep sequence of reactions catalyzed by a complex of enzymes and cofactors called the *pyruvate dehydrogenase complex*. The process occurs in three stages, each catalyzed by one of the enzymes in the complex, as outlined in **Figure 29.11**. Acetyl CoA, the ultimate product, then acts as fuel for the final stage of catabolism, the citric acid cycle.

1 Nucleophilic addition of thiamin diphosphate (TPP) ylide to pyruvate gives an alcohol addition product.

2 Decarboxylation occurs in a step analgous to the loss of CO_2 from a β-keto acid, yielding the enamine hydroxyethylthiamin diphosphate (HETPP).

3 The enamine double bond attacks a sulfur atom of lipoamide and carries out an S_N2-like displacement of the second sulfur to yield a hemithioacetal.

4 Elimination of thiamin diphosphate ylide from the hemithioacetal intermediate yields acetyl dihydrolipoamide . . .

5 . . . which reacts with coenzyme A in a nucleophilic acyl substitution reaction to exchange one thioester for another and give acetyl CoA plus dihydrolipoamide.

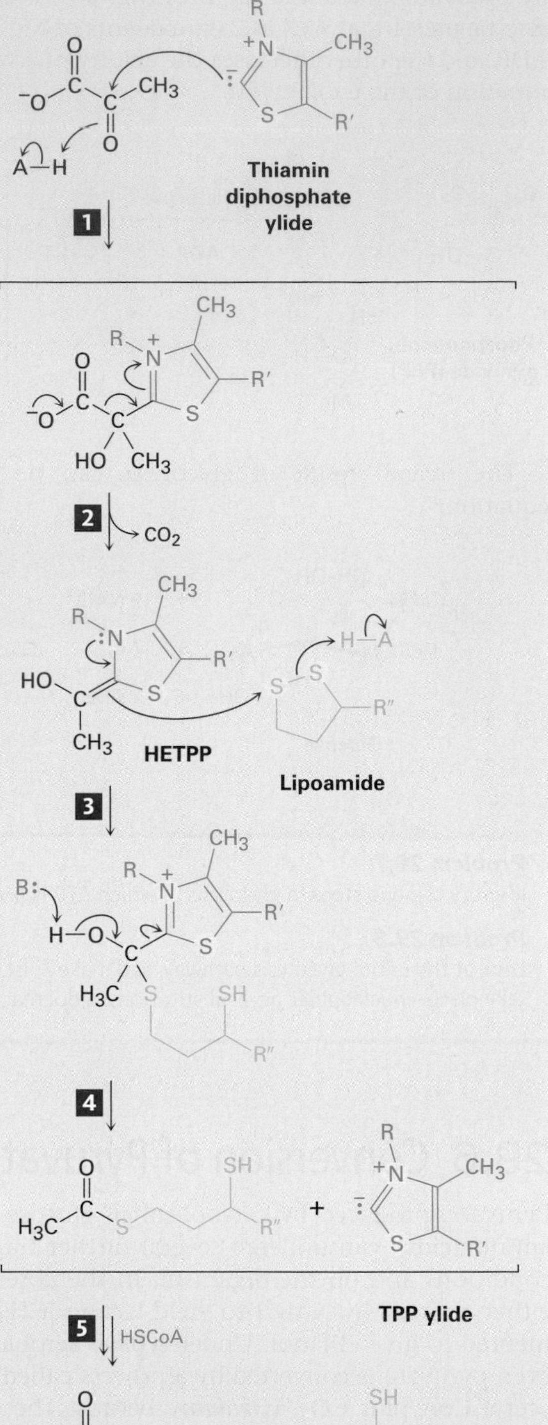

Figure 29.11 | MECHANISM

Mechanism of the conversion of pyruvate to acetyl CoA through a multistep sequence of reactions that requires three different enzymes and four different coenzymes. The individual steps are explained in the text.

Step 1 of Figure 29.11: Addition of Thiamin Diphosphate The conversion of pyruvate to acetyl CoA begins by reaction of pyruvate with thiamin diphosphate, a derivative of vitamin B_1. Formerly called thiamin *pyro*phosphate, thiamin diphosphate is usually abbreviated as TPP. The spelling *thiamine* is also correct and frequently used.

The key structural element in thiamin diphosphate is the presence of a thiazolium ring—a five-membered, unsaturated heterocycle containing a sulfur atom and a positively charged nitrogen atom. The thiazolium ring is weakly acidic, with a pK_a of approximately 18 for the ring hydrogen between N and S. Bases can therefore deprotonate thiamin diphosphate, leading to formation of an ylide much like the phosphonium ylides used in Wittig reactions **(Section 19.11)**. As in the Wittig reaction, the TPP ylide is a nucleophile and adds to the ketone carbonyl group of pyruvate to yield an alcohol addition product.

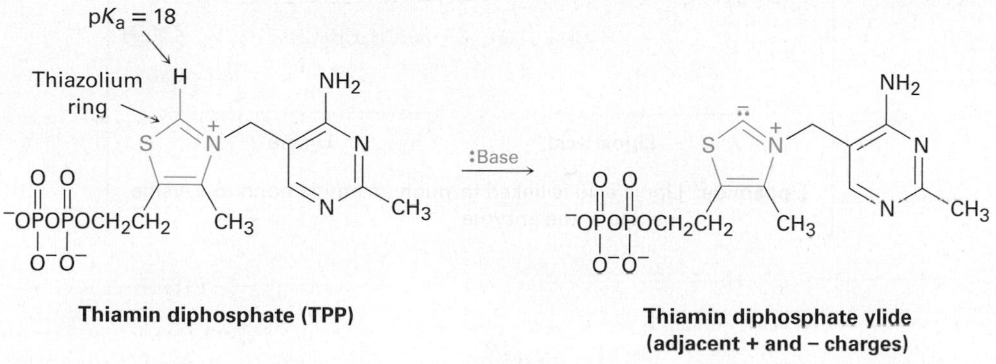

$pK_a = 18$

Thiazolium ring

:Base

Thiamin diphosphate (TPP)

Thiamin diphosphate ylide (adjacent + and − charges)

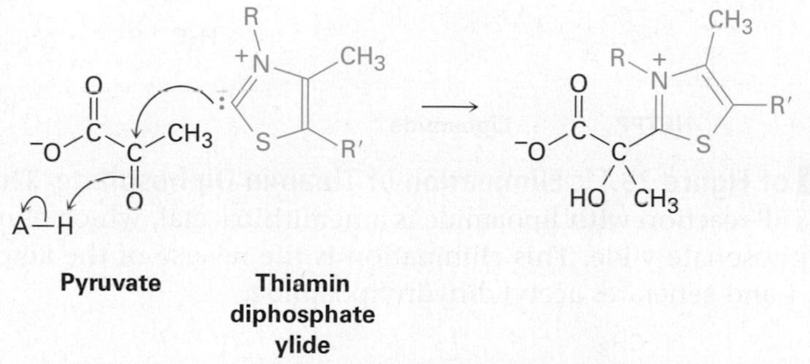

Pyruvate

Thiamin diphosphate ylide

Step 2 of Figure 29.11: Decarboxylation The TPP addition product, which contains an iminium ion β to a carboxylate anion, undergoes decarboxylation in much the same way that a β-keto acid decarboxylates in the acetoacetic ester synthesis **(Section 22.7)**. The $C{-}N^+$ bond of the pyruvate addition product acts like the C=O bond of a β-keto acid to accept electrons as CO_2 leaves, giving hydroxyethylthiamin diphosphate (HETPP).

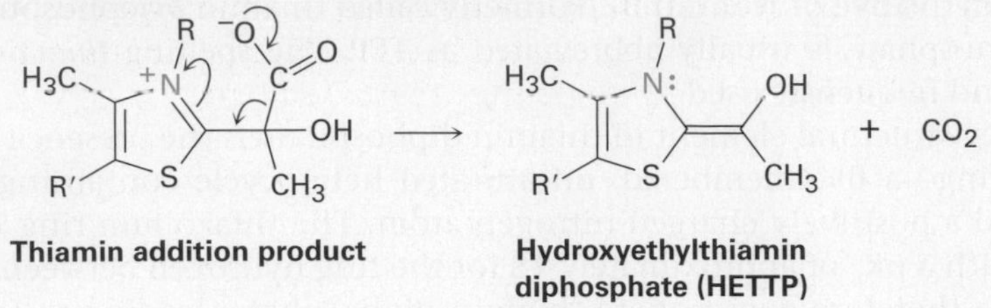

Thiamin addition product → Hydroxyethylthiamin diphosphate (HETTP) + CO_2

Step 3 of Figure 29.11: Reaction with Lipoamide Hydroxyethylthiamin diphosphate is an enamine ($R_2N—C=C$), which, like all enamines, is nucleophilic **(Section 23.11)**. It therefore reacts with the enzyme-bound disulfide lipoamide by nucleophilic attack on a sulfur atom, displacing the second sulfur in an S_N2-like process.

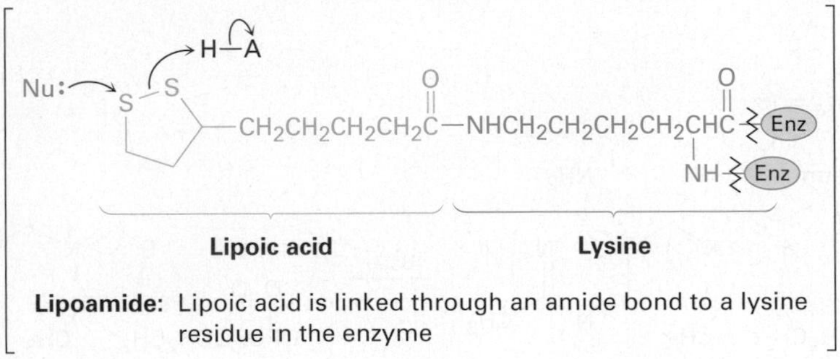

Lipoic acid — Lysine

$Nu:$ ... $-CH_2CH_2CH_2CH_2C(=O)—NHCH_2CH_2CH_2CH_2CHC(=O)$-Enz, NH-Enz

Lipoamide: Lipoic acid is linked through an amide bond to a lysine residue in the enzyme

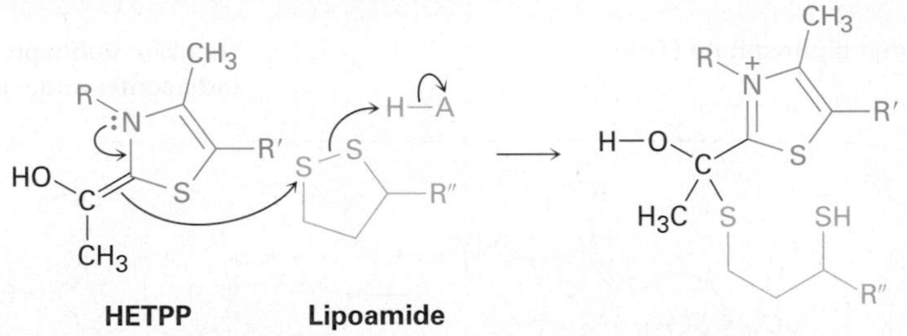

HETPP + Lipoamide →

Step 4 of Figure 29.11: Elimination of Thiamin Diphosphate The product of the HETPP reaction with lipoamide is a hemithioacetal, which eliminates thiamin diphosphate ylide. This elimination is the reverse of the ketone addition in step 1 and generates acetyl dihydrolipoamide.

→ Acetyl dihydrolipoamide + TPP ylide

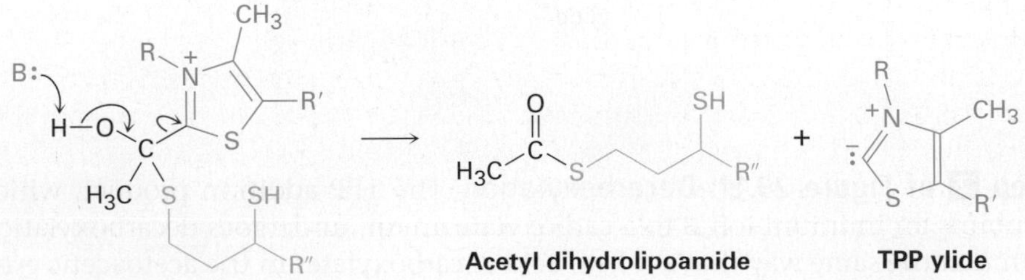

Step 5 of Figure 29.11: Acyl Transfer Acetyl dihydrolipoamide, a thioester, undergoes a nucleophilic acyl substitution reaction with coenzyme A to yield acetyl CoA plus dihydrolipoamide. The dihydrolipoamide is then oxidized back to lipoamide by FAD **(Section 29.3)**, and the $FADH_2$ that results is in turn oxidized back to FAD by NAD^+, completing the catalytic cycle.

Problem 29.9
Which carbon atoms in glucose end up as $-CH_3$ carbons in acetyl CoA? Which carbons end up as CO_2?

29.7 The Citric Acid Cycle

The initial stages of catabolism result in the conversion of both fats and carbohydrates into acetyl groups that are bonded through a thioester link to coenzyme A. Acetyl CoA then enters the next stage of catabolism—the **citric acid cycle**, also called the *tricarboxylic acid (TCA) cycle,* or *Krebs cycle,* after Hans Krebs, who unraveled its complexities in 1937. The overall result of the cycle is the conversion of an acetyl group into two molecules of CO_2 plus reduced coenzymes by the eight-step sequence of reactions shown in **Figure 29.12**.

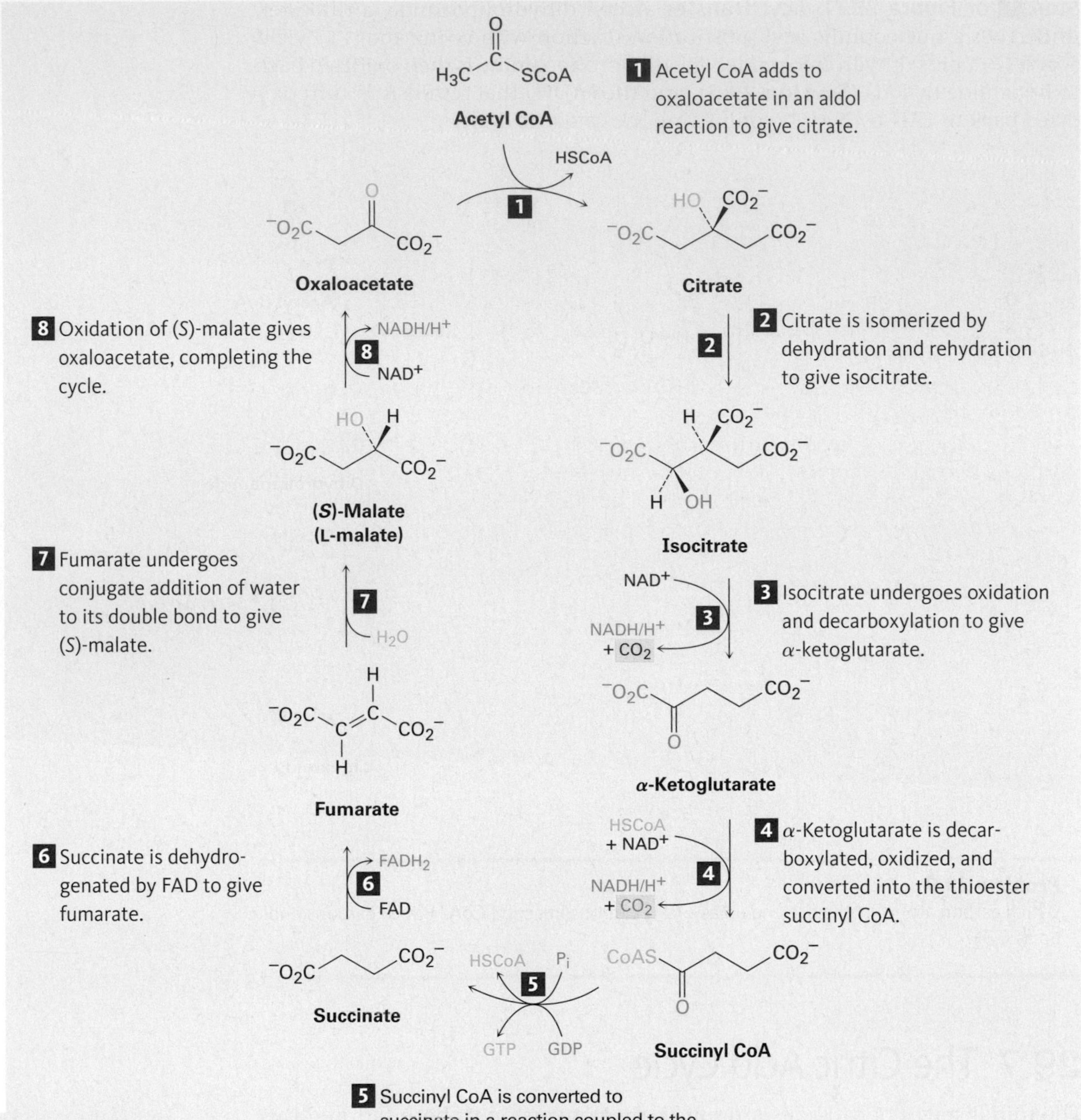

Figure 29.12 | **MECHANISM**

The citric acid cycle is an eight-step series of reactions that results in the conversion of an acetyl group into two molecules of CO_2 plus reduced coenzymes. Individual steps are explained in the text.

As its name implies, the citric acid *cycle* is a closed loop of reactions in which the product of the final step (oxaloacetate) is a reactant in the first step. The intermediates are constantly regenerated and flow continuously through the cycle, which operates as long as the oxidizing coenzymes NAD^+ and FAD are available. To meet this condition, the reduced coenzymes NADH and $FADH_2$ must be reoxidized via the electron-transport chain, which in turn relies on oxygen as the ultimate electron acceptor. Thus, the cycle is dependent on the availability of oxygen and on the operation of the electron-transport chain.

Step 1 of Figure 29.12: Addition to Oxaloacetate Acetyl CoA enters the citric acid cycle in step 1 by nucleophilic addition to the oxaloacetate carbonyl group, to give (*S*)-citryl CoA. The addition is an aldol reaction and is catalyzed by citrate synthase, as discussed in **Section 26.11**. (*S*)-Citryl CoA is then hydrolyzed to citrate by a typical nucleophilic acyl substitution reaction with water, catalyzed by the same citrate synthase enzyme.

Note that the hydroxyl-bearing carbon of citrate is a prochirality center and contains two identical arms. Because the initial aldol reaction of acetyl CoA to oxaloacetate occurs specifically from the *Si* face of the ketone carbonyl group, the *pro-S* arm of citrate is derived from acetyl CoA and the *pro-R* arm is derived from oxaloacetate.

Acetyl CoA → [Oxaloacetate] → (*S*)-Citryl CoA → (H_2O, HSCoA) → Citrate

Step 2 of Figure 29.12: Isomerization Citrate, a prochiral tertiary alcohol, is next converted into its isomer, (2*R*,3*S*)-isocitrate, a chiral secondary alcohol. The isomerization occurs in two steps, both of which are catalyzed by the same aconitase enzyme. The initial step is an E1cB dehydration of a β-hydroxy acid to give *cis*-aconitate, the same sort of reaction that occurs in step 9 of glycolysis (Figure 29.7 on page 1175). The second step is a conjugate nucleophilic addition of water to the C=C bond **(Section 19.13)**. The dehydration of citrate takes place specifically on the *pro-R* arm—the one derived from oxaloacetate—rather than on the *pro-S* arm derived from acetyl CoA.

Citrate → (H_2O) → *cis*-Aconitate → (H_2O) → (2*R*,3*S*)-Isocitrate

Step 3 of Figure 29.12: Oxidation and Decarboxylation (2*R*,3*S*)-Isocitrate, a secondary alcohol, is oxidized by NAD^+ in step 3 to give the ketone oxalosuccinate, which loses CO_2 to give α-ketoglutarate. Catalyzed by isocitrate dehydrogenase, the decarboxylation is a typical reaction of a β-keto acid, just like that in the acetoacetic ester synthesis **(Section 22.7)**. The enzyme requires a divalent cation as cofactor to polarize the ketone carbonyl group and make it a better electron acceptor.

(2*R*,3*S*)-Isocitrate → (NAD^+ → $NADH/H^+$) → [Oxalosuccinate → (CO_2)] → α-Ketoglutarate

Step 4 of Figure 29.12: Oxidative Decarboxylation The transformation of α-ketoglutarate to succinyl CoA in step 4 is a multistep process just like the transformation of pyruvate to acetyl CoA that we saw in Figure 29.11 on page 1182. In both cases, an α-keto acid loses CO_2 and is oxidized to a thioester in a series of steps catalyzed by a multienzyme dehydrogenase complex. As in the conversion of pyruvate to acetyl CoA, the reaction involves an initial nucleophilic addition reaction of thiamin diphosphate ylide to α-ketoglutarate, followed by decarboxylation. Reaction with lipoamide, elimination of TPP ylide, and finally a transesterification of the dihydrolipoamide thioester with coenzyme A yields succinyl CoA.

α-Ketoglutarate → (HSCoA + NAD^+ → $NADH/H^+$ + CO_2) → Succinyl CoA

Step 5 of Figure 29.12: Acyl CoA Cleavage Succinyl CoA is converted to succinate in step 5. The reaction is catalyzed by succinyl CoA synthetase and is coupled with phosphorylation of guanosine diphosphate (GDP) to give guanosine triphosphate (GTP). The overall transformation is similar to that of steps 6 through 8 in glycolysis (Figure 29.7), in which a thioester is converted into an acyl phosphate and a phosphate group is then transferred to ADP. The overall result is a "hydrolysis" of the thioester group without involvement of water.

Abbreviated mechanism

Succinyl CoA → Acyl phosphate → Succinate + GTP (from GDP)

Step 6 of Figure 29.12: Dehydrogenation Succinate is dehydrogenated in step 6 by the FAD dependent succinate dehydrogenase to give fumarate. The process is analogous to what occurs during the β-oxidation pathway of fatty-acid catabolism **(Section 29.3)**. The reaction is stereospecific, removing the *pro-S* hydrogen from one carbon and the *pro-R* hydrogen from the other.

Succinate —(Enz-FAD → Enz-FADH$_2$)→ Fumarate

Steps 7–8 of Figure 29.12: Hydration and Oxidation The final two steps in the citric acid cycle are the conjugate nucleophilic addition of water to fumarate to yield (*S*)-malate and the oxidation of (*S*)-malate by NAD$^+$ to give oxaloacetate. The addition is catalyzed by fumarase and is mechanistically similar to the addition of water to *cis*-aconitate in step 2. The reaction occurs through an enolate-ion intermediate, which is protonated on the side opposite the OH, leading to a net anti addition.

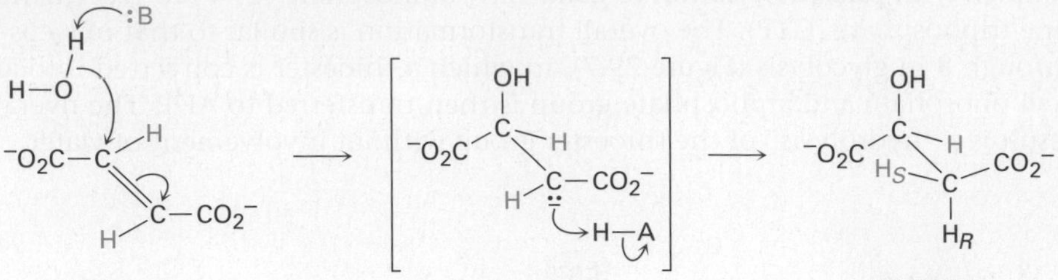

Fumarate

(*S*)-Malate

The final step is the oxidation of (*S*)-malate by NAD^+ to give oxaloacetate, a reaction catalyzed by malate dehydrogenase. The citric acid cycle has now returned to its starting point, ready to revolve again. The overall result of the cycle is

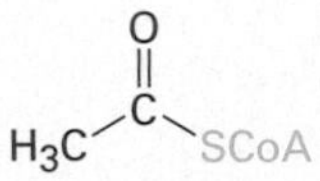

$$\text{Acetyl CoA} + 3\ NAD^+ + FAD + GDP + P_i + 2\ H_2O \longrightarrow 2\ CO_2 + HSCoA + 3\ NADH + 2\ H^+ + FADH_2 + GTP$$

Acetyl CoA

Problem 29.10

Which of the substances in the citric acid cycle are tricarboxylic acids, thus giving the cycle its alternative name?

Problem 29.11

Write mechanisms for step 2 of the citric acid cycle, the dehydration of citrate and the addition of water to aconitate.

Problem 29.12

Is the *pro-R* or *pro-S* hydrogen removed from citrate during the dehydration in step 2 of the citric acid cycle? Does the elimination reaction occur with syn or anti geometry?

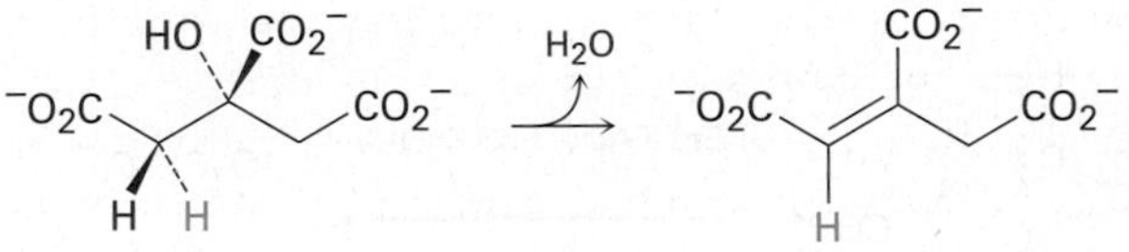

Citrate

cis-Aconitate

29.8 Carbohydrate Biosynthesis: Gluconeogenesis

Glucose is the body's primary fuel when food is plentiful, but in times of fasting or prolonged exercise, glucose stores can become depleted. Most tissues then begin metabolizing fats as their source of acetyl CoA, but the brain is different. The brain relies almost entirely on glucose for fuel and is dependent on receiving a continuous supply in the blood. When the supply of glucose fails, even for a brief time, irreversible damage can occur. Thus, a pathway for synthesizing glucose from simple precursors is crucial.

Higher organisms are not able to synthesize glucose from acetyl CoA but must instead use one of the three-carbon precursors lactate, glycerol, or alanine, all of which are readily converted into pyruvate.

(*S*)-Lactate **Alanine** **Glycerol**

Pyruvate *Gluconeogenesis* → **Glucose**

Pyruvate then becomes the starting point for **gluconeogenesis**, the 11-step biosynthetic pathway by which organisms make glucose **(Figure 29.13)**. The gluconeogenesis pathway by which glucose is made, however, is not the reverse of the glycolysis pathway by which it is degraded. As with the catabolic and anabolic pathways for fatty acids **(Sections 29.3 and 29.4)**, the catabolic and anabolic pathways for carbohydrates differ in some details so that both are energetically favorable.

Figure 29.13 | MECHANISM

The gluconeogenesis pathway for the biosynthesis of glucose from pyruvate. Individual steps are explained in the text.

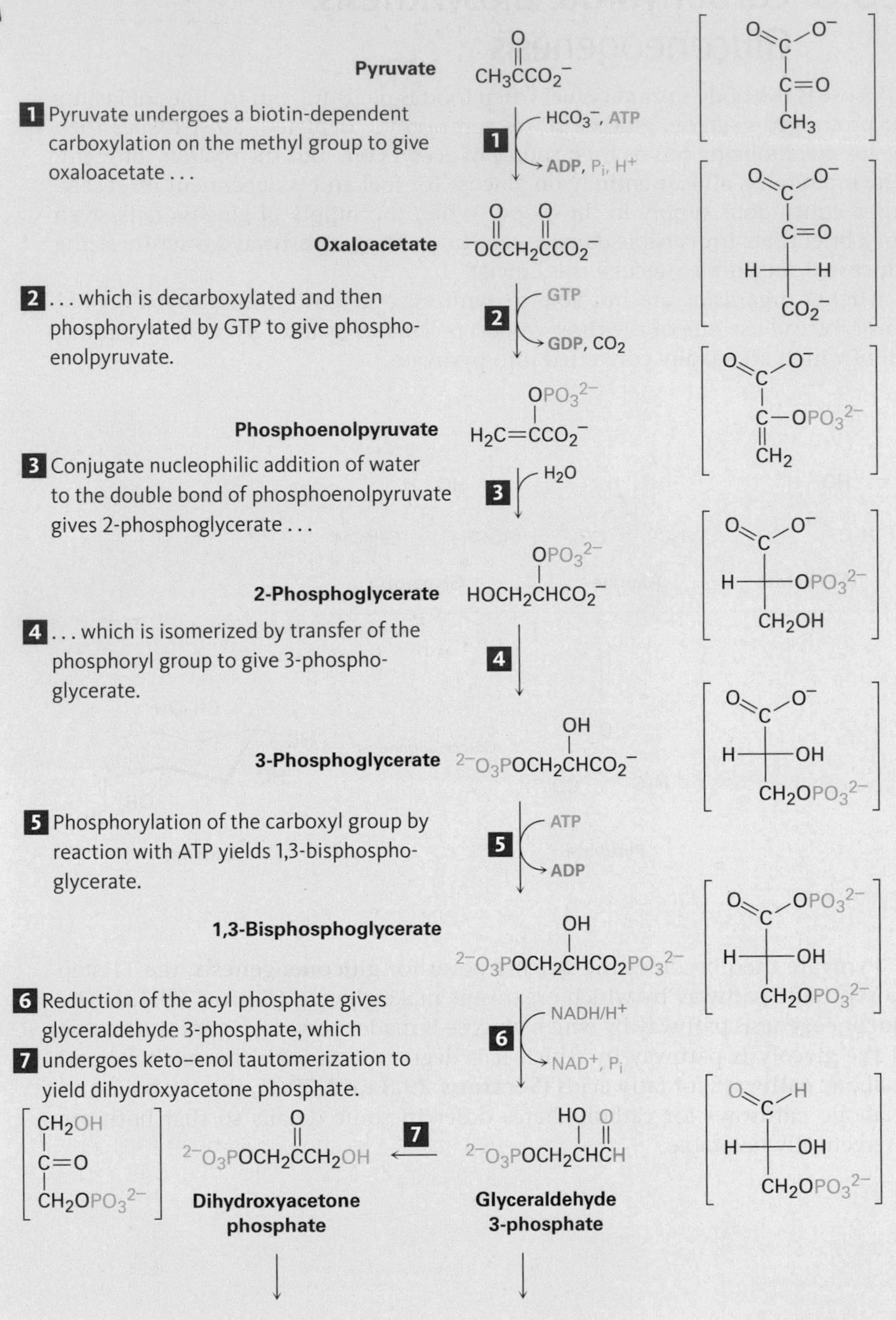

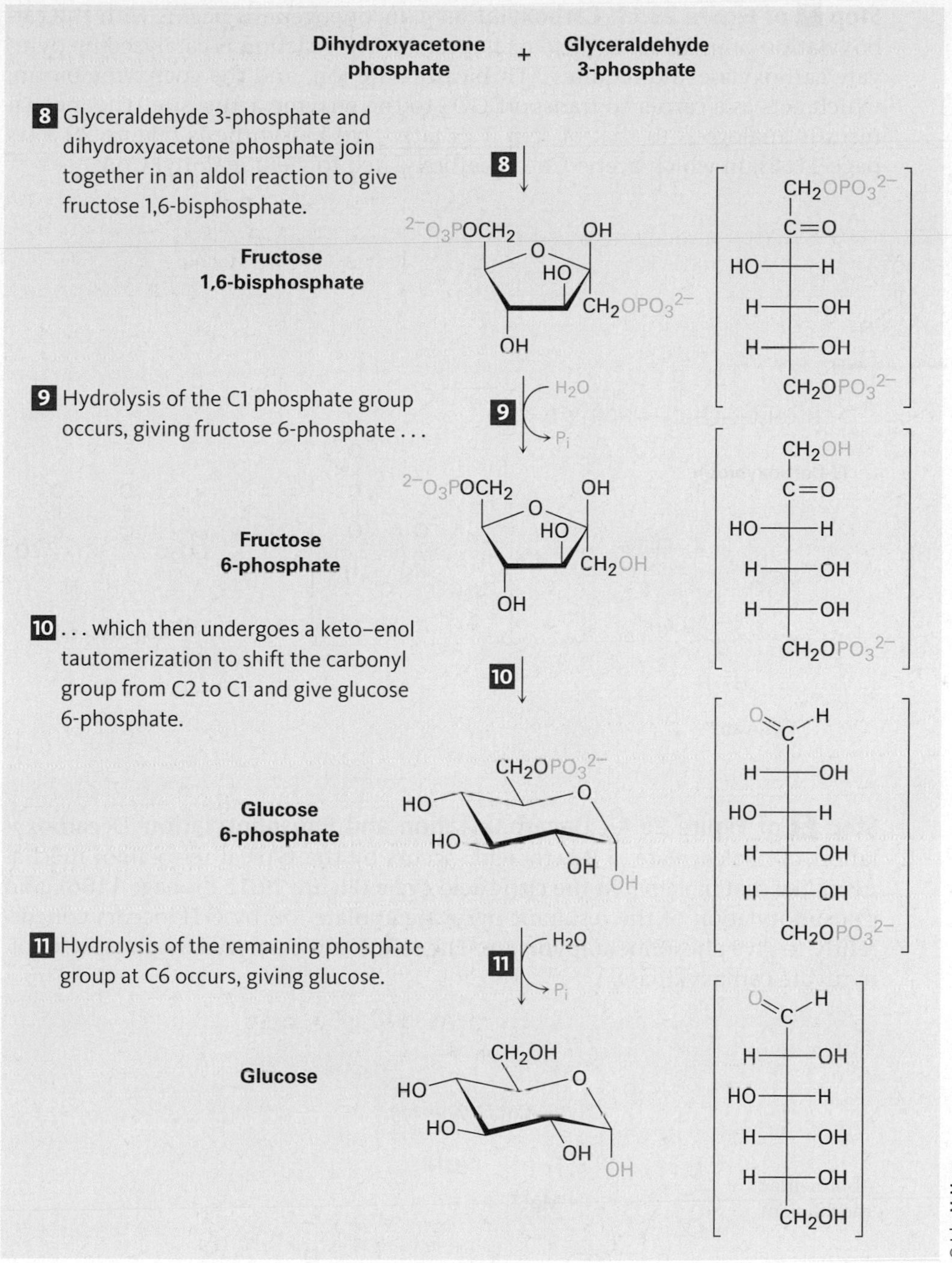

Figure 29.13 (Continued)

Step 1 of Figure 29.13: Carboxylation Gluconeogenesis begins with the carboxylation of pyruvate to yield oxaloacetate. The reaction is catalyzed by pyruvate carboxylase and requires ATP, bicarbonate ion, and the coenzyme biotin, which acts as a carrier to transport CO_2 to the enzyme active site. The mechanism is analogous to that of step 3 in fatty-acid biosynthesis (Figure 29.5 on page 1168), in which acetyl CoA is carboxylated to yield malonyl CoA.

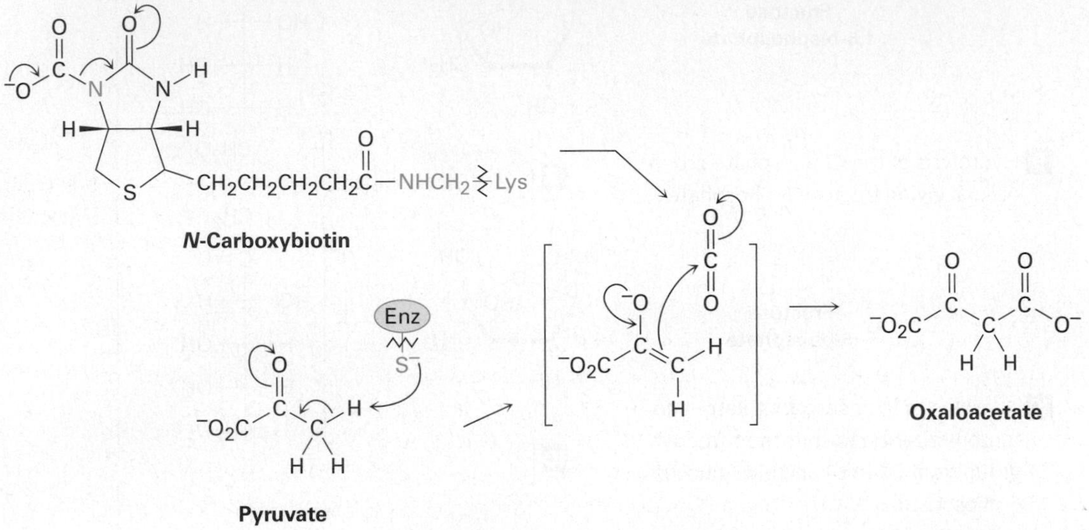

N-Carboxybiotin

Enz

Pyruvate

Oxaloacetate

Step 2 of Figure 29.13: Decarboxylation and Phosphorylation Decarboxylation of oxaloacetate, a β-keto acid, occurs by the typical retro-aldol mechanism like that in step 3 in the citric acid cycle (Figure 29.12 on page 1186), and phosphorylation of the resultant pyruvate enolate ion by GTP occurs concurrently to give phosphoenolpyruvate. The reaction is catalyzed by phosphoenolpyruvate carboxykinase.

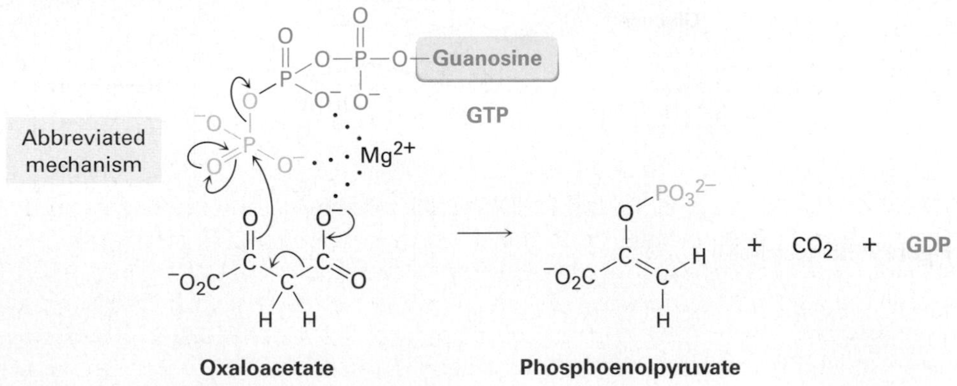

Abbreviated mechanism

Guanosine

GTP

Mg^{2+}

Oxaloacetate

Phosphoenolpyruvate + CO_2 + GDP

Steps 3–4 of Figure 29.13: Hydration and Isomerization Conjugate nucleophilic addition of water to the double bond of phosphoenolpyruvate gives 2-phosphoglycerate by a process similar to that of step 7 in the citric acid cycle. Phosphorylation of C3 and dephosphorylation of C2 then yields 3-phosphoglycerate. Mechanistically, these steps are the reverse of steps 9 and 8 in glycolysis (Figure 29.7), which have equilibrium constants near 1 so that substantial amounts of reactant and product are both present.

$H_2C{=}C(OPO_3^{2-})CO_2^-$ (**Phosphoenolpyruvate**) $\xrightarrow{H_2O}$ $HOCH_2CH(OPO_3^{2-})CO_2^-$ (**2-Phosphoglycerate**) $\xrightarrow{Enz{-}PO_3^{2-} \;\to\; Enz}$ ${}^{2-}O_3POCH_2CH(OPO_3^{2-})CO_2^-$ (**2,3-Bisphosphoglycerate**) $\xrightarrow{H_2O \;\to\; P_i}$ ${}^{2-}O_3POCH_2CH(OH)CO_2^-$ (**3-Phosphoglycerate**)

Steps 5–7 of Figure 29.13: Phosphorylation, Reduction, and Tautomerization Reaction of 3-phosphoglycerate with ATP generates the corresponding acyl phosphate, 1,3-bisphosphoglycerate, which binds to the glyceraldehyde 3-phosphate dehydrogenase by a thioester bond to a cysteine residue. Reduction of the thioester by NADH/H$^+$ yields the corresponding aldehyde, and keto–enol tautomerization of the aldehyde gives dihydroxyacetone phosphate. All three steps are mechanistically the reverse of the corresponding steps 7, 6, and 5 of glycolysis and have equilibrium constants near 1.

${}^{2-}O_3POCH_2CH(OH)CO_2^-$ (**3-Phosphoglycerate**) $\xrightarrow{ATP \;\to\; ADP}$ ${}^{2-}O_3POCH_2CH(OH)C(=O)OPO_3^{2-}$ (**1,3-Bisphosphoglycerate**) $\xrightarrow{Enz{-}SH \;\to\; P_i}$ ${}^{2-}O_3POCH_2CH(OH)C(=O)S{-}Cys$ (**(Enzyme-bound thioester)**) $\xrightarrow{NADH/H^+ \;\to\; NAD^+}$ ${}^{2-}O_3POCH_2CH(OH)CHO$ (**Glyceraldehyde 3-phosphate**) $\rightleftharpoons$ ${}^{2-}O_3POCH_2C(=O)CH_2OH$ (**Dihydroxyacetone phosphate**)

Step 8 of Figure 29.13: Aldol Reaction Dihydroxyacetone phosphate and glyceraldehyde 3-phosphate, the two 3-carbon units produced in step 7, join by an aldol reaction to give fructose 1,6-bisphosphate, the reverse of step 4 in glycolysis (Figure 29.9 on page 1178). As in glycolysis, the reaction is catalyzed in plants and animals by a class I aldolase and takes place on an iminium ion formed by reaction of dihydroxyacetone phosphate with a side-chain lysine $-NH_2$ group on the enzyme. Loss of a proton from the neighboring carbon then generates an enamine, an aldol-like reaction ensues, and the product is hydrolyzed.

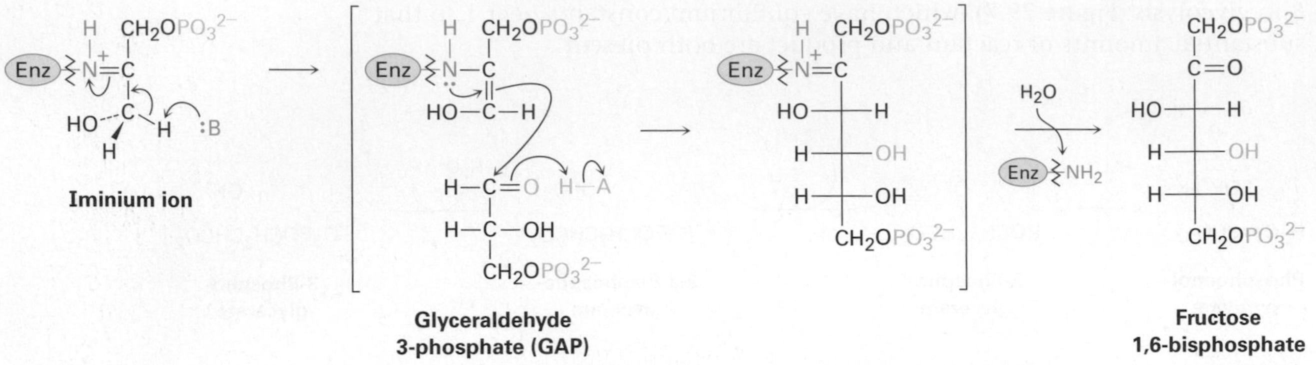

Iminium ion

Glyceraldehyde 3-phosphate (GAP)

Fructose 1,6-bisphosphate

Steps 9–10 of Figure 29.13: Hydrolysis and Isomerization Hydrolysis of the phosphate group at C1 of fructose 1,6-bisphosphate gives fructose 6-phosphate. Although the result of the reaction is the exact opposite of step 3 in glycolysis, the mechanism is not. In glycolysis, the phosphorylation is accomplished by reaction of fructose with ATP, with formation of ADP as by-product. The reverse of that process, however—the reaction of fructose 1,6-bisphosphate with ADP to give fructose 6-phosphate and ATP—is energetically unfavorable because ATP is too high in energy. Thus, an alternative pathway is used in which the C1 phosphate group is removed by a direct hydrolysis reaction, catalyzed by fructose 1,6-bisphosphatase.

Following hydrolysis, keto–enol tautomerization of the carbonyl group from C2 to C1 gives glucose 6-phosphate. The isomerization is the reverse of step 2 in glycolysis.

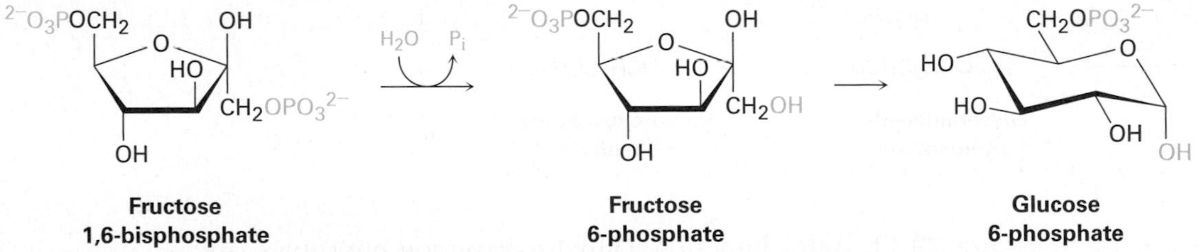

Fructose 1,6-bisphosphate

Fructose 6-phosphate

Glucose 6-phosphate

Step 11 of Figure 29.13: Hydrolysis The final step in gluconeogenesis is the conversion of glucose 6-phosphate to glucose by a second phosphatase-catalyzed hydrolysis reaction. As just discussed for the hydrolysis of fructose 1,6-bisphosphate in step 9, and for the same energetic reasons, the mechanism of the glucose 6-phosphate hydrolysis is not the exact opposite of the corresponding step 1 in glycolysis.

Interestingly, however, the mechanisms of the two phosphate hydrolysis reactions in steps 9 and 11 are not the same. In step 9, water is the nucleophile, but in the glucose 6-phosphate reaction of step 11, a histidine residue on the enzyme attacks phosphorus, giving a phosphoryl enzyme intermediate that subsequently reacts with water.

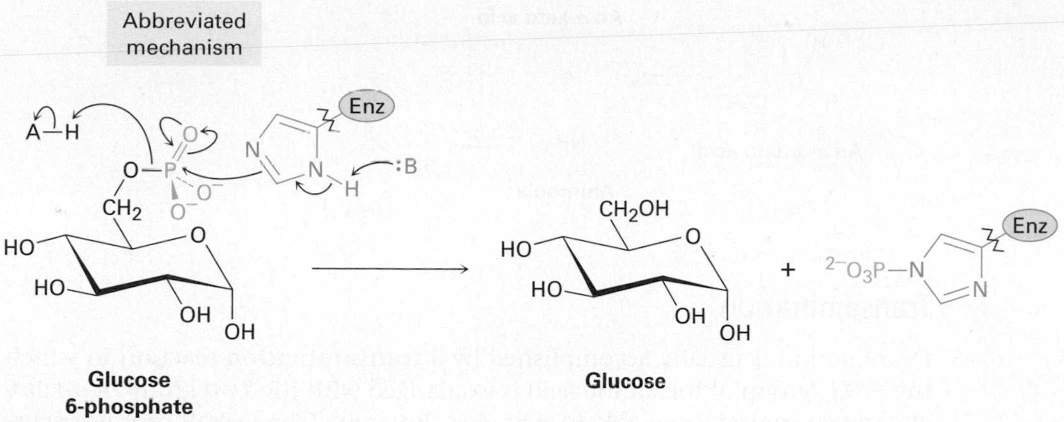

Abbreviated mechanism

Glucose 6-phosphate → Glucose + $^{2-}O_3P$–N(imidazole)–Enz

The overall result of gluconeogenesis is summarized by the following equation:

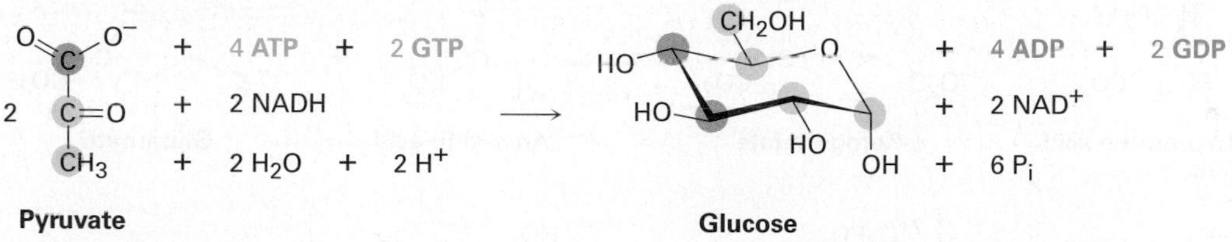

$$2\ \text{Pyruvate} + 4\ ATP + 2\ GTP + 2\ NADH + 2\ H_2O + 2\ H^+ \longrightarrow \text{Glucose} + 4\ ADP + 2\ GDP + 2\ NAD^+ + 6\ P_i$$

Pyruvate Glucose

Problem 29.13

Write a mechanism for step 6 of gluconeogenesis, the reduction of 3-phosphoglyceryl phosphate with NADH/H^+ to yield glyceraldehyde 3-phosphate.

29.9 Catabolism of Proteins: Deamination

The catabolism of proteins is much more complex than that of fats and carbohydrates because each of the 20 α-amino acids is degraded through its own unique pathway. The general idea, however, is that (1) the α amino group is first removed as ammonia by a *deamination* process, (2) the ammonia is converted into urea, and (3) the remaining amino acid carbon skeleton (usually an α-keto acid) is converted into a compound that enters the citric acid cycle.

Transamination

Deamination is usually accomplished by a **transamination** reaction in which the $-NH_2$ group of the amino acid is exchanged with the keto group of α-keto-glutarate, forming a new α-keto acid plus glutamate. The overall process occurs in two parts, is catalyzed by aminotransferases, and involves participation of the coenzyme pyridoxal phosphate, abbreviated PLP, a derivative of pyridoxine (vitamin B_6). Different aminotransferases differ in their specificity for amino acids, but the mechanism remains the same.

The mechanism of the first part of transamination is shown in **Figure 29.14**. The process begins with reaction between the α-amino acid and pyridoxal phosphate, which is covalently bonded to the aminotransferase by an imine linkage between the side-chain $-NH_2$ group of a lysine residue in the enzyme and the PLP aldehyde group. Deprotonation/reprotonation of the PLP–amino acid imine effects tautomerization of the imine C=N bond, and hydrolysis of the tautomerized imine gives an α-keto acid plus pyridoxamine phosphate (PMP).

1 An amino acid reacts with the enzyme-bound PLP imine by nucleophilic addition of its $-NH_2$ group to the C=N bond of the imine, giving a PLP–amino acid imine and releasing the enzyme amino group.

2 Deprotonation of the acidic α carbon of the amino acid gives an intermediate α-keto acid imine . . .

3 . . . that is reprotonated on the PLP carbon. The net result of this deprotonation/reprotonation sequence is tautomerization of the imine C=N bond.

4 Hydrolysis of the α-keto acid imine by nucleophilic addition of water to the C=N bond gives the transamination products pyridoxamine phosphate (PMP) and α-keto acid.

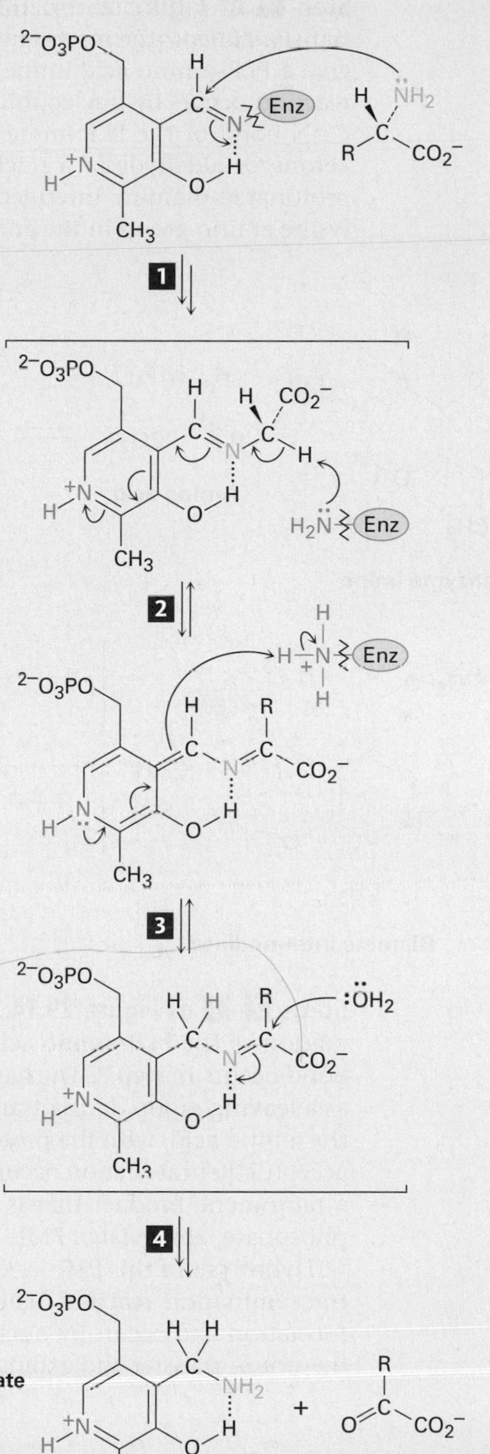

Figure 29.14 | MECHANISM

Mechanism of the enzyme-catalyzed, PLP-dependent transamination of an α-amino acid to give an α-keto acid. Individual steps are explained in the text.

Step 1 of Figure 29.14: Transimination The first step in trans*amination* is trans*imination*—the reaction of the PLP–enzyme imine with an α-amino acid to give a PLP–amino acid imine plus expelled enzyme as the leaving group. The reaction occurs by nucleophilic addition of the amino acid $-NH_2$ group to the C–N bond of the PLP imine, much as an amine adds to the C=O bond of a ketone or aldehyde in a nucleophilic addition reaction **(Section 19.8)**. The protonated diamine intermediate undergoes a proton transfer and expels the lysine amino group in the enzyme to complete the step.

PLP–enzyme imine **Amino acid** **Diamine intermediate**

Diamine intermediate **PLP–amino acid imine**

Steps 2–4 of Figure 29.14: Tautomerization and Hydrolysis Following formation of the PLP–amino acid imine in step 1, a tautomerization of the C=N bond occurs in step 2. The basic lysine residue in the enzyme that was expelled as a leaving group during transimination deprotonates the acidic α position of the amino acid, with the protonated pyridine ring of PLP acting as the electron acceptor. Reprotonation occurs on the carbon atom next to the ring, generating a tautomeric product that is the imine of an α-keto acid with pyridoxamine phosphate, abbreviated PMP.

Hydrolysis of this PMP–α-keto acid imine then completes the first part of the transamination reaction. The hydrolysis is the mechanistic reverse of imine formation and occurs by nucleophilic addition of water to the imine, followed by proton transfer and expulsion of PMP as leaving group.

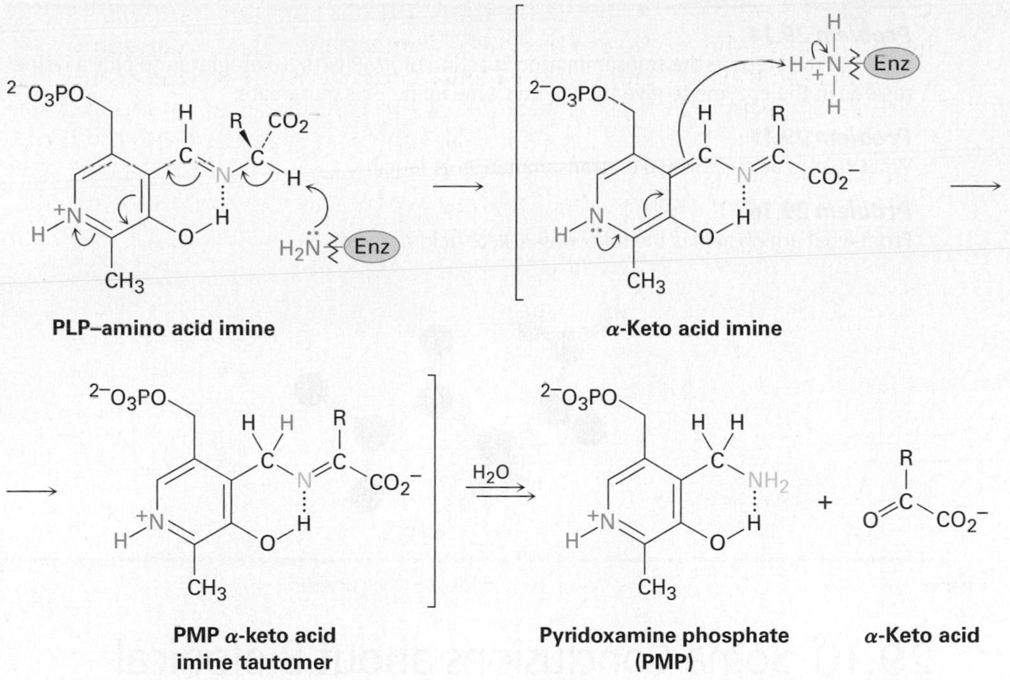

PLP–amino acid imine

α-Keto acid imine

PMP α-keto acid imine tautomer

Pyridoxamine phosphate (PMP)

α-Keto acid

Regeneration of PLP from PMP

With PLP plus the α-amino acid now converted into PMP plus an α-keto acid, PMP must be transformed back into PLP to complete the catalytic cycle. The conversion occurs by another transamination reaction, this one between PMP and an α-keto acid, usually α-ketoglutarate. The products are PLP plus glutamate, and the mechanism of the process is the exact reverse of that shown in Figure 29.14. That is, PMP and α-ketoglutarate give an imine; the PMP–α-ketoglutarate imine undergoes tautomerization of the C=N bond to give a PLP–glutamate imine; and the PLP–glutamate imine reacts with a lysine residue on the enzyme in a transimination process to yield PLP–enzyme imine plus glutamate.

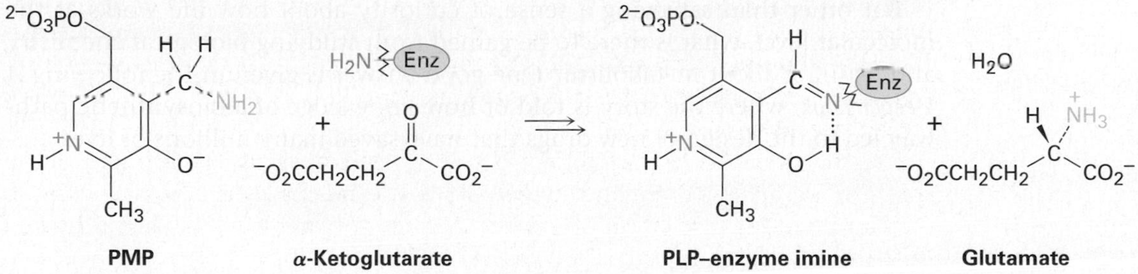

PMP

α-Ketoglutarate

PLP–enzyme imine

Glutamate

Problem 29.14

Write all the steps in the transamination reaction of PMP with α-ketoglutarate plus a lysine residue in the enzyme to give the PLP–enzyme imine plus glutamate.

Problem 29.15

What α-keto acid is formed on transamination of leucine?

Problem 29.16

From what amino acid is the following α-keto acid derived?

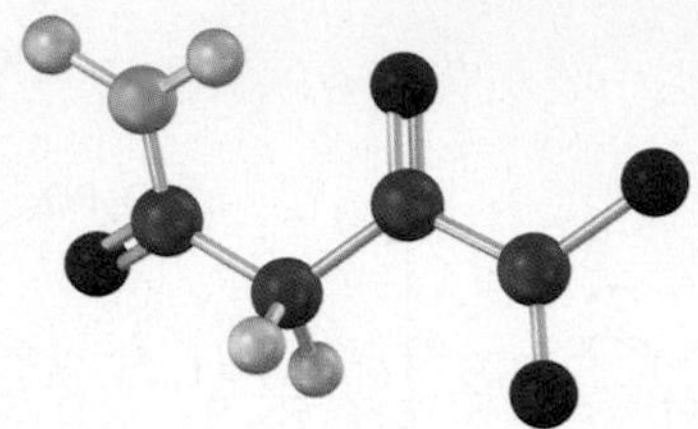

29.10 Some Conclusions about Biological Chemistry

As promised in the chapter introduction, the past few sections have been a fast-paced tour of a large number of reactions. Following it all undoubtedly required a lot of work and a lot of page-turning to look at earlier sections.

After examining the various metabolic pathways, perhaps the main conclusion about biological chemistry is the remarkable similarity between the mechanisms of biological reactions and the mechanisms of laboratory reactions. In all the pathways described in this chapter, terms like *imine formation, aldol reaction, nucleophilic acyl substitution reaction, E1cB reaction,* and *Claisen reaction* appear constantly. Biological reactions aren't mysterious; there are clear, understandable explanations for the reactions carried out by living organisms. Biological chemistry *is* organic chemistry.

But other than satisfying a sense of curiosity about how life works at the molecular level, what is there to be gained from studying biological chemistry and learning about metabolism? One good answer is given in the following *A Deeper Look,* where the story is told of how knowledge of a biosynthetic pathway led to the design of new drugs that have saved many millions of lives.

A DEEPER LOOK Statin Drugs

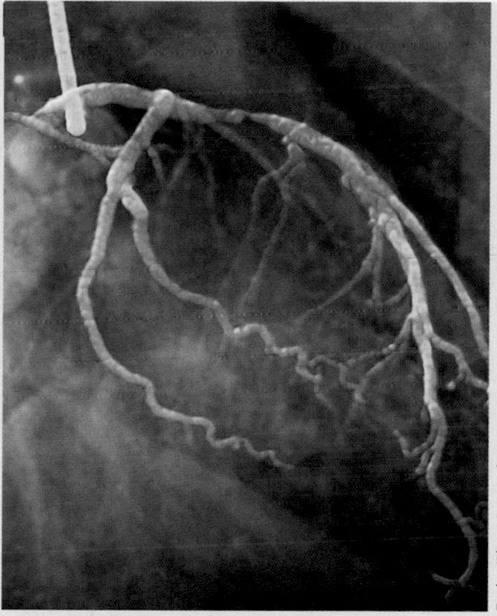

The buildup of cholesterol deposits inside arteries can cause coronary heart disease, a leading cause of death for both men and women.

Coronary heart disease—the buildup of cholesterol-containing plaques on the walls of heart arteries—is the leading cause of death for both men and women older than age 20 in industrialized countries. It's estimated that up to one-third of women and one-half of men will develop the disease at some point in their lives.

The onset of coronary heart disease is directly correlated with blood cholesterol levels (see the Chapter 27 *A Deeper Look*), and the first step in disease prevention is to lower those levels. It turns out that only about 25% of your blood cholesterol comes from what you eat; the remaining 75%—about 1000 mg each day—is biosynthesized in your liver from dietary fats and carbohydrates. Thus, any effective plan for lowering your cholesterol level means limiting the amount that your body makes, which is where a detailed chemical knowledge of cholesterol biosynthesis comes in.

We saw in **Sections 27.5 and 27.7** that all steroids, including cholesterol, are biosynthesized from the triterpenoid lanosterol, which in turn comes from acetyl CoA through isopentenyl diphosphate. If you knew all the mechanisms for all the chemical steps in cholesterol biosynthesis, you might be able to devise a drug that would block one of those steps, thereby short-circuiting the biosynthetic process and controlling the amount of cholesterol produced.

But we do know those mechanisms! Look back at the pathway for the biosynthesis of isopentenyl diphosphate from acetyl CoA, shown in Figure 27.7 on page 1100. It turns out that the rate-limiting step in the pathway is the reduction of 3-hydroxy-3-methylglutaryl CoA (abbreviated HMG-CoA) to mevalonate, brought about by the enzyme HMG-CoA reductase. If that enzyme could be stopped from functioning, cholesterol biosynthesis would also be stopped.

Rate limiting

H_3C OH, ^-O_2C, O, C, SCoA — **3-Hydroxy-3-methylglutaryl coenzyme A (HMG-CoA)** → H_3C OH, CO_2^-, OH — **Mevalonate** → CH_3, H, CH_3, H, H, H, HO, H — **Cholesterol**

To find a drug that blocks HMG-CoA reductase, chemists did two simultaneous experiments on a large number of potential drug candidates isolated from soil microbes. In one experiment, the drug candidate and mevalonate were added to liver extract; in the second experiment, only the drug candidate was added without mevalonate. If cholesterol was produced only in the presence of added mevalonate but not in the absence of mevalonate, the drug candidate must have blocked the enzyme for mevalonate synthesis.

The drugs that block HMG-CoA reductase, and thus control cholesterol synthesis in the body, are called *statins*. They are the most widely prescribed drugs in the world, with an estimated $15 billion in annual sales. So effective are they that in the 10-year period following their introduction in 1994, the death rate from coronary heart disease decreased by

(continued)

(continued)

33% in the United States. Atorvastatin (Lipitor), simvastatin (Zocor), rosuvastatin (Crestor), pravastatin (Pravachol), and lovastatin (Mevacor) are examples. An X-ray crystal structure of the active site in the HMG-CoA reductase enzyme is shown in the accompanying graphic, along with a molecule of atorvastatin (blue) that is tightly bound in the active site and stops the enzyme from functioning. A good understanding of organic chemistry certainly paid off in this instance.

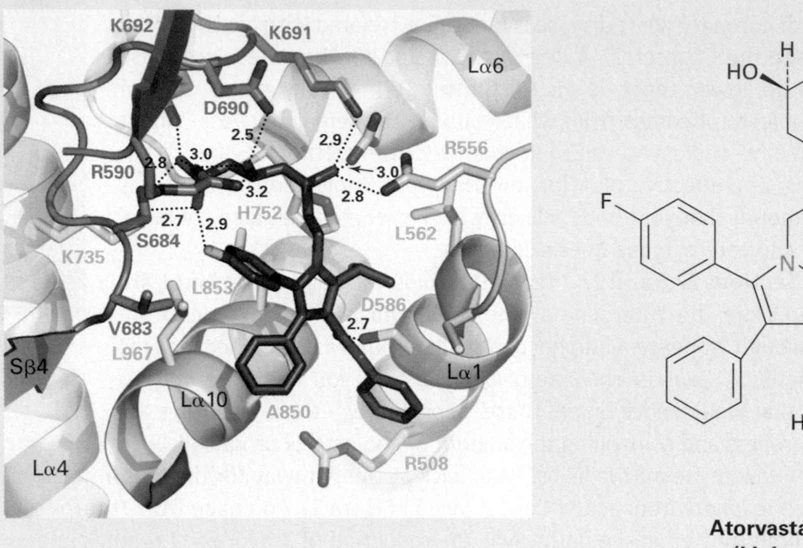

Atorvastatin (Lipitor)

Summary

Key words

anabolism, 1154
β-oxidation pathway, 1162
catabolism, 1154
citric acid cycle, 1185
gluconeogenesis, 1191
glycolysis, 1173
metabolism, 1154
Schiff base, 1177
transamination, 1198

Metabolism is the sum of all chemical reactions in the body. Reactions that break down large molecules into smaller fragments are called **catabolism**, and those that build up large molecules from small pieces are called **anabolism**. Although the details of specific biochemical pathways are sometimes complex, all the reactions that occur follow the normal rules of organic chemical reactivity.

The catabolism of fats begins with digestion, in which ester bonds are hydrolyzed to give glycerol and fatty acids. The fatty acids are degraded in the four-step ***β*-oxidation pathway** by removal of two carbons at a time, yielding acetyl CoA. Catabolism of carbohydrates begins with the hydrolysis of glycoside bonds to give glucose, which is degraded in the ten-step **glycolysis** pathway. Pyruvate, the initial product of glycolysis, is then converted into acetyl CoA. Acetyl CoA next enters the eight-step **citric acid cycle**, where it is further degraded into CO_2. The cycle is a closed loop of reactions in which the product of the final step (oxaloacetate) is a reactant in the first step.

Catabolism of proteins is more complex than that of fats or carbohydrates because each of the 20 different amino acids is degraded by its own unique pathway. In general, though, the amino nitrogen atoms are removed and the substances that remain are converted into compounds that enter the citric acid cycle. Most amino acids lose their nitrogen atom by **transamination**, a reaction in which the $-NH_2$ group of the amino acid changes places with the keto group of an α-keto acid such as α-ketoglutarate. The products are a new α-keto acid and glutamate.

The energy released in catabolic pathways is used in the *electron-transport chain* to make molecules of adenosine triphosphate, ATP. ATP, the final result of food catabolism, couples to and drives many otherwise unfavorable reactions.

Biomolecules are synthesized as well as degraded, but the pathways for anabolism and catabolism are not the exact reverse of one another. Fatty acids are biosynthesized from acetate by an 8-step pathway, and carbohydrates are made from pyruvate by the 11-step **gluconeogenesis** pathway.

Exercises

Visualizing Chemistry

OWL Interactive versions of these problems are assignable in OWL for Organic Chemistry.

(Problems 29.1–29.16 appear within the chapter.)

29.17 Identify the amino acid that is a catabolic precursor of each of the following α-keto acids:

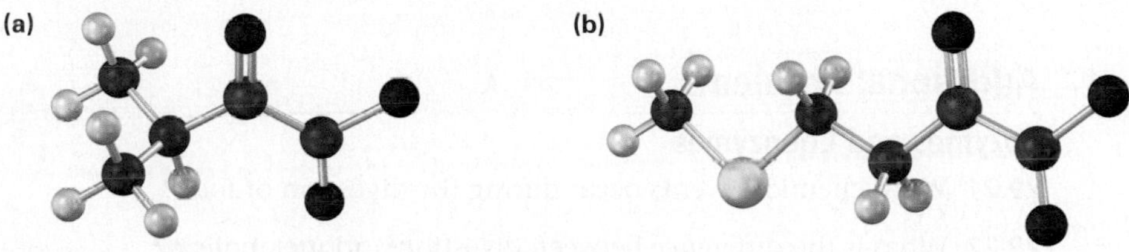

29.18 Identify the following intermediate in the citric acid cycle, and tell whether it has *R* or *S* stereochemistry:

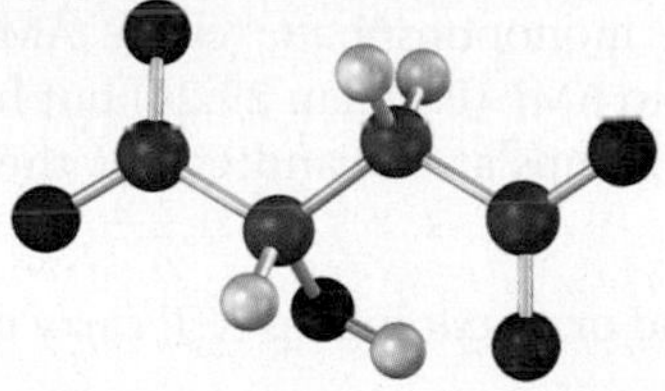

29.19 The following compound is an intermediate in the biosynthesis of one of the 20 common α-amino acids. Which one is it likely to be, and what kind of chemical change must take place to complete the biosynthesis?

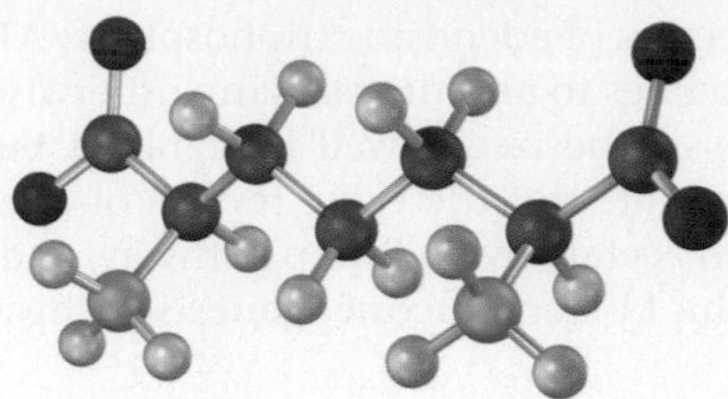

29.20 The following compound is an intermediate in the pentose phosphate pathway, an alternative route for glucose metabolism. Identify the sugar it is derived from.

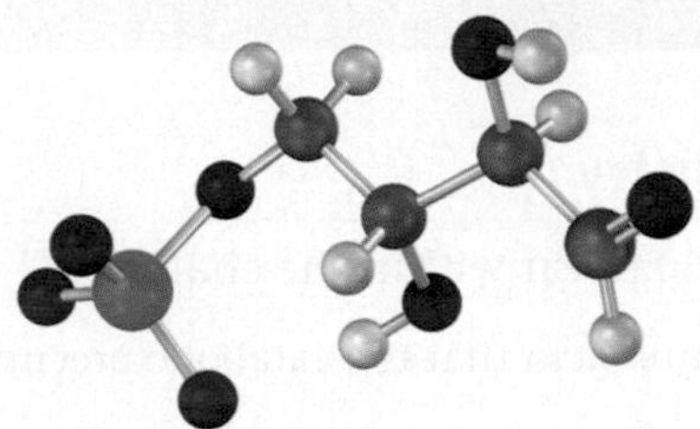

Additional Problems

Enzymes and Coenzymes

29.21 What chemical events occur during the digestion of food?

29.22 What is the difference between digestion and metabolism?

29.23 What is the difference between anabolism and catabolism?

29.24 Draw the structure of adenosine 5′-monophosphate (AMP), an intermediate in some biochemical pathways.

29.25 Cyclic adenosine monophosphate (cyclic AMP), a modulator of hormone action, is related to AMP (Problem 29.24) but has its phosphate group linked to *two* hydroxyl groups at C3′ and C5′ of the sugar. Draw the structure of cyclic AMP.

29.26 What general kind of reaction does ATP carry out?

29.27 What general kind of reaction does NAD^+ carry out?

29.28 What general kind of reaction does FAD carry out?

29.29 What enzyme cofactor is associated with each of the following kinds of reactions?

(a) Transamination

(b) Carboxylation of a ketone

(c) Decarboxylation of an α-keto acid

29.30 Lactate, a product of glucose catabolism in oxygen-starved muscles, can be converted into pyruvate by oxidation. What coenzyme do you think is needed? Write the equation in the normal biochemical format using a curved arrow.

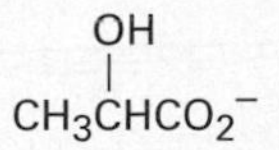

$$\underset{\displaystyle CH_3CHCO_2^-}{\overset{OH}{|}} \quad \textbf{Lactate}$$

Metabolism

29.31 Write the equation for the final step in the β-oxidation pathway of any fatty acid with an even number of carbon atoms.

29.32 Show the products of each of the following reactions:

(a)

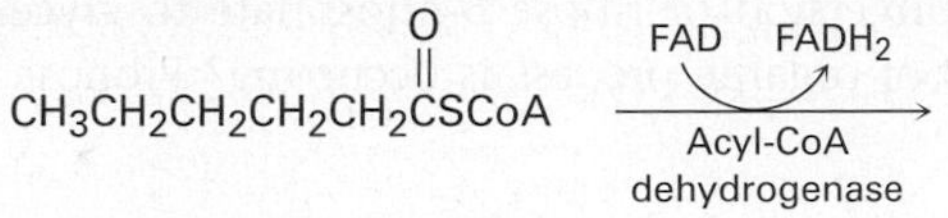

$$CH_3CH_2CH_2CH_2CH_2\overset{O}{\overset{\|}{C}}SCoA \xrightarrow[\text{Acyl-CoA dehydrogenase}]{FAD \quad FADH_2} ?$$

(b)

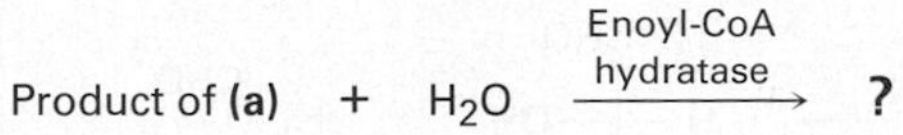

$$\text{Product of (a)} + H_2O \xrightarrow{\text{Enoyl-CoA hydratase}} ?$$

(c)

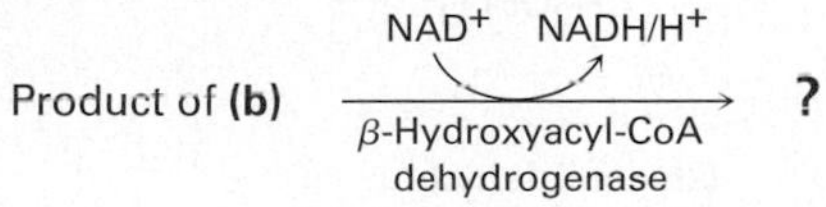

$$\text{Product of (b)} \xrightarrow[\beta\text{-Hydroxyacyl-CoA dehydrogenase}]{NAD^+ \quad NADH/H^+} ?$$

29.33 Why aren't the glycolysis and gluconeogenesis pathways the exact reverse of each other?

29.34 How many moles of acetyl CoA are produced by catabolism of the following substances?
(a) 1.0 mol of glucose (b) 1.0 mol of palmitic acid (c) 1.0 mol of maltose

29.35 How many grams of acetyl CoA (MW = 809.6 amu) are produced by catabolism of the following substances? Which substance is the most efficient precursor of acetyl CoA on a weight basis?
(a) 100.0 g of glucose
(b) 100.0 g of palmitic acid
(c) 100.0 g of maltose

29.36 What is the structure of the α-keto acid formed by transamination of each of the following amino acids?
(a) Threonine (b) Phenylalanine (c) Asparagine

29.37 The glycolysis pathway shown in Figure 29.7 has a number of intermediates that contain phosphate groups. Why can 3-phosphoglyceryl phosphate and phosphoenolpyruvate transfer a phosphate group to ADP while glucose 6-phosphate cannot?

29.38 In the *pentose phosphate* pathway for degrading sugars, ribulose 5-phosphate is converted to ribose 5-phosphate. Propose a mechanism for the isomerization.

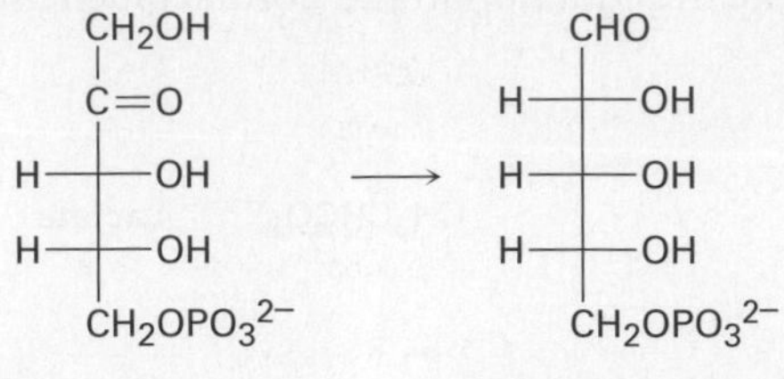

$$\begin{array}{c} CH_2OH \\ | \\ C{=}O \\ H{-}\!\!\!+\!\!\!{-}OH \\ H{-}\!\!\!+\!\!\!{-}OH \\ CH_2OPO_3^{2-} \end{array} \longrightarrow \begin{array}{c} CHO \\ H{-}\!\!\!+\!\!\!{-}OH \\ H{-}\!\!\!+\!\!\!{-}OH \\ H{-}\!\!\!+\!\!\!{-}OH \\ CH_2OPO_3^{2-} \end{array}$$

Ribulose 5-phosphate **Ribose 5-phosphate**

29.39 Another step in the pentose phosphate pathway for degrading sugars (see Problem 29.38) is the conversion of ribose 5-phosphate to glyceraldehyde 3-phosphate. What kind of organic process is occurring? Propose a mechanism for the conversion.

$$\begin{array}{c} CHO \\ H{-}\!\!\!+\!\!\!{-}OH \\ H{-}\!\!\!+\!\!\!{-}OH \\ H{-}\!\!\!+\!\!\!{-}OH \\ CH_2OPO_3^{2-} \end{array} \rightleftharpoons \begin{array}{c} CHO \\ H{-}\!\!\!+\!\!\!{-}OH \\ CH_2OPO_3^{2-} \end{array} + \begin{array}{c} CHO \\ | \\ CH_2OPO_3^{2-} \end{array}$$

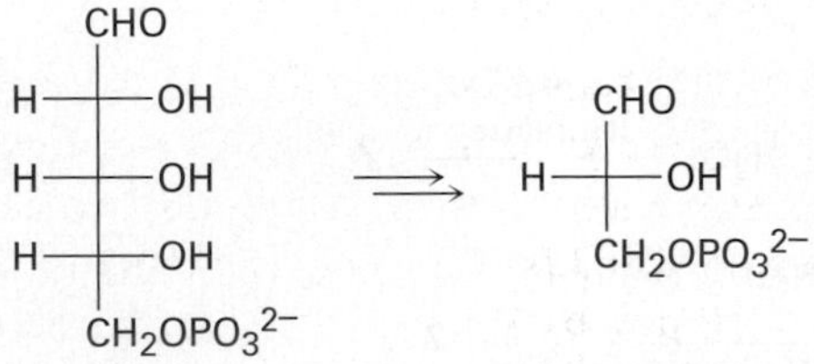

Ribose 5-phosphate **Glyceraldehyde 3-phosphate**

29.40 Write a mechanism for the conversion of α-ketoglutarate to succinyl CoA in step 4 of the citric acid cycle (Figure 29.12).

29.41 In step 2 of the citric acid cycle (Figure 29.12), *cis*-aconitate reacts with water to give (2*R*,3*S*)-isocitrate. Does –OH add from the *Re* face of the double bond or from the *Si* face? What about –H? Does the addition of water occur with syn or anti geometry?

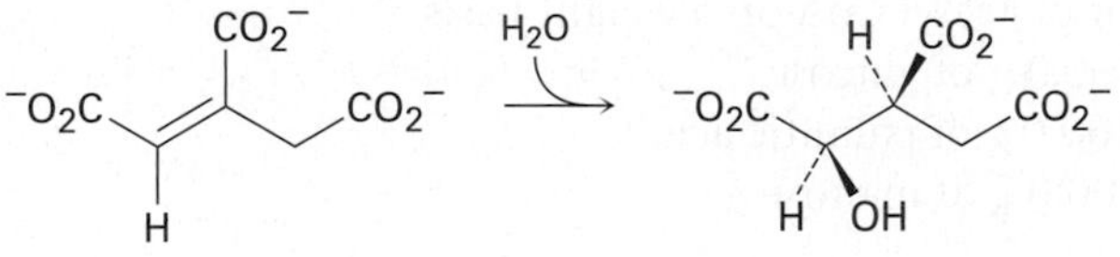

CO_2^-, ^-O_2C, CO_2^-, H → (H_2O) → H CO_2^-, ^-O_2C, CO_2^-, H OH

***cis*-Aconitate** **(2*R*,3*S*)-Isocitrate**

General Problems

29.42 In glycerol metabolism, the oxidation of *sn*-glycerol 3-phosphate to give dihydroxyacetone phosphate is catalyzed by *sn*-glycerol-3-phosphate dehydrogenase, with NAD^+ as cofactor. The reaction is stereospecific, occurring exclusively on the *Re* face of the nicotinamide ring.

NAD^+ $\xrightarrow{\text{sn-Glycerol 3-phosphate} \rightarrow \text{Dihydroxyacetone phosphate}}$ NADH

Which hydrogen in the NADH product comes from *sn*-glycerol 3-phosphate? Does it have *pro-R* or *pro-S* stereochemistry?

29.43 The primary fate of acetyl CoA under normal metabolic conditions is degradation in the citric acid cycle to yield CO_2. When the body is stressed by prolonged starvation, however, acetyl CoA is converted into compounds called *ketone bodies,* which can be used by the brain as a temporary fuel. Fill in the missing information indicated by the four question marks in the following biochemical pathway for the synthesis of ketone bodies from acetyl CoA:

$$2\ CH_3\overset{O}{\overset{\|}{C}}SCoA \xrightarrow{-HSCoA} CH_3\overset{O}{\overset{\|}{C}}CH_2\overset{O}{\overset{\|}{C}}SCoA \xrightarrow{H_2O \;\; ?} CH_3\overset{O}{\overset{\|}{C}}CH_2\overset{O}{\overset{\|}{C}}O^-$$

Acetyl CoA → Acetoacetyl CoA → Acetoacetate

Acetoacetate $\xrightarrow{?}$ $CH_3\overset{O}{\overset{\|}{C}}CH_3$ (Acetone)

Acetoacetate $\xrightarrow{? \;\to\; ?}$ $CH_3\overset{OH}{\overset{|}{C}}HCH_2\overset{O}{\overset{\|}{C}}O^-$ (3-Hydroxybutyrate)

Ketone bodies

29.44 The initial reaction in Problem 29.43, conversion of two molecules of acetyl CoA to one molecule of acetoacetyl CoA, is a Claisen reaction. Assuming that there is a base present, show the mechanism of the reaction.

29.45 In step 6 of fatty-acid biosynthesis (Figure 29.5), acetoacetyl ACP is reduced stereospecifically by NADPH to yield an alcohol. Does hydride ion add to the *Si* face or the *Re* face of acetoacetyl ACP?

$$H_3C-C(=O)-CH_2-C(=O)-SACP \xrightarrow{NADPH \;\; NADP^+} H_3C-CH(OH)-CH_2-C(=O)-SACP$$

Acetoacetyl ACP → **β-Hydroxybutyryl ACP**

29.46 In step 7 of fatty-acid biosynthesis (Figure 29.5), dehydration of a *β*-hydroxy thioester occurs to give *trans*-crotonyl ACP. Is the dehydration a syn elimination or an anti elimination?

$$H_3C-CH(OH)-CH_2-C(=O)-SACP \xrightarrow{H_2O} H_3C-CH=CH-C(=O)-SACP$$

***trans*-Crotonyl ACP**

29.47 In step 8 of fatty-acid biosynthesis (Figure 29.5), reduction of *trans*-crotonyl ACP gives butyryl ACP. A hydride from NADPH adds to C3 of the crotonyl group from the *Re* face, and protonation on C2 occurs on the *Si* face. Is the reduction a syn addition or an anti addition?

$$H_3C-CH=CH-C(=O)SACP \longrightarrow CH_3CH_2CH_2\overset{O}{\overset{\|}{C}}SACP$$

Crotonyl ACP → **Butyryl ACP**

29.48 One of the steps in the pentose phosphate pathway for glucose catabolism is the reaction of sedoheptulose 7-phosphate with glyceraldehyde 3-phosphate in the presence of a transaldolase to yield erythrose 4-phosphate and fructose 6-phosphate.

Sedoheptulose 7-phosphate (CH_2OH; C=O; HO–C–H; H–C–OH; H–C–OH; H–C–OH; $CH_2OPO_3^{2-}$) + Glyceraldehyde 3-phosphate (CHO; H–C–OH; $CH_2OPO_3^{2-}$) ⟶ Erythrose 4-phosphate (CHO; H–C–OH; H–C–OH; $CH_2OPO_3^{2-}$) + Fructose 6-phosphate (CH_2OH; C=O; HO–C–H; H–C–OH; H–C–OH; $CH_2OPO_3^{2-}$)

Sedoheptulose 7-phosphate **Glyceraldehyde 3-phosphate** **Erythrose 4-phosphate** **Fructose 6-phosphate**

(a) The first part of the reaction is formation of a protonated Schiff base of sedoheptulose 7-phosphate with a lysine residue in the enzyme followed by a retro-aldol cleavage to give an enamine plus erythrose 4-phosphate. Show the structure of the enamine and the mechanism by which it is formed.

(b) The second part of the reaction is nucleophilic addition of the enamine to glyceraldehyde 3-phosphate followed by hydrolysis of the Schiff base to give fructose 6-phosphate. Show the mechanism.

29.49 One of the steps in the pentose phosphate pathway for glucose catabolism is the reaction of xylulose 5-phosphate with ribose 5-phosphate in the presence of a transketolase to give glyceraldehyde 3-phosphate and sedoheptulose 7-phosphate.

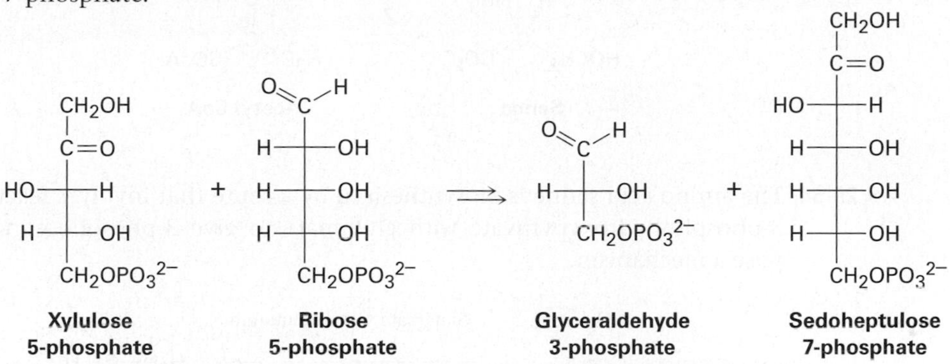

Xylulose 5-phosphate + **Ribose 5-phosphate** ⟶ **Glyceraldehyde 3-phosphate** + **Sedoheptulose 7-phosphate**

(a) The first part of the reaction is nucleophilic addition of thiamin diphosphate (TPP) ylide to xylulose 5-phosphate, followed by a retro-aldol cleavage to give glyceraldehyde 3-phosphate and a TPP-containing enamine. Show the structure of the enamine and the mechanism by which it is formed.

(b) The second part of the reaction is addition of the enamine to ribose 5-phosphate followed by loss of TPP ylide to give sedoheptulose 7-phosphate. Show the mechanism.

29.50 The amino acid tyrosine is biologically degraded by a series of steps that include the following transformations:

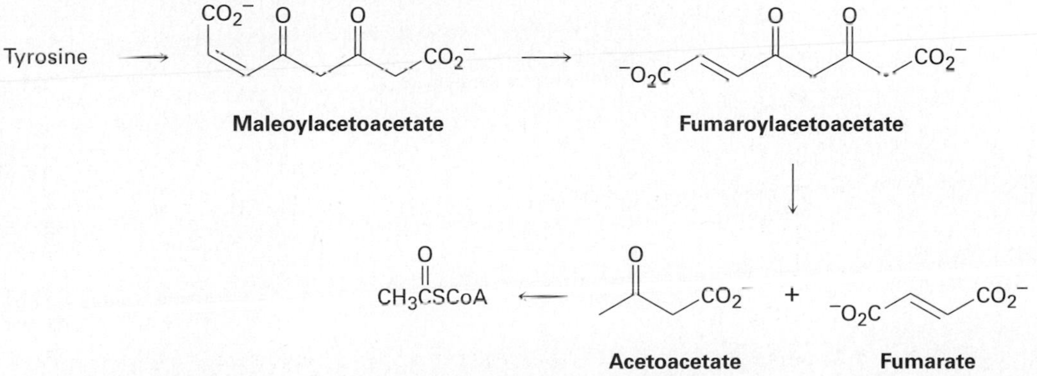

Tyrosine ⟶ **Maleoylacetoacetate** ⟶ **Fumaroylacetoacetate** ⟶ **Acetoacetate** + **Fumarate**; Acetoacetate ⟶ CH_3CSCoA (O)

The double-bond isomerization of maleoylacetoacetate to fumaroylacetoacetate is catalyzed by practically any nucleophile, :Nu^-. Propose a mechanism.

29.51 Propose a mechanism for the conversion of fumaroylacetoacetate to fumarate plus acetoacetate (Problem 29.50).

29.52 Propose a mechanism for the conversion of acetoacetate to acetyl CoA (Problem 29.50).

29.53 Design your own degradative pathway. You know the rules (organic mechanisms), and you've seen the kinds of reactions that occur in the biological degradation of fats and carbohydrates into acetyl CoA. If you were Mother Nature, what series of steps would you use to degrade the amino acid serine into acetyl CoA?

$$HOCH_2CH(\overset{+}{N}H_3)CO_2^- \xrightarrow{?} H_3C{-}C(=O){-}SCoA$$

Serine → **Acetyl CoA**

29.54 The amino acid serine is biosynthesized by a route that involves reaction of 3-phosphohydroxypyruvate with glutamate to give 3-phosphoserine. Propose a mechanism.

$$^-O_2C{-}C(=O){-}CH_2OPO_3^{2-} \xrightarrow{\text{Glutamate} \;\; \alpha\text{-Ketoglutarate}} {}^-O_2C{-}CH(\overset{+}{N}H_3){-}CH_2OPO_3^{2-}$$

3-Phosphohydroxypyruvate → **3-Phosphoserine**

29.55 The amino acid leucine is biosynthesized from α-ketoisocaproate, which is itself prepared from α-ketoisovalerate by a multistep route that involves (1) reaction with acetyl CoA, (2) hydrolysis, (3) dehydration, (4) hydration, (5) oxidation, and (6) decarboxylation. Show the steps in the transformation, and propose a mechanism for each.

Acetyl CoA, H_2O, NAD^+ → HSCoA, CO_2, NADH/H^+

α-Ketoisovalerate → **α-Ketoisocaproate**

29.56 The amino acid cysteine, $C_3H_7NO_2S$, is biosynthesized from a substance called cystathionine by a multistep pathway.

$$^{-}OC(=O)CH(NH_3^{+})CH_2CH_2SCH_2CH(NH_3^{+})C(=O)O^{-} \longrightarrow NH_4^{+} + \ ? \ + \text{Cysteine}$$

Cystathionine

(a) The first step is a transamination. What is the product?

(b) The second step is an E1cB reaction. Show the products and the mechanism of the reaction.

(c) The final step is a double-bond reduction. What organic cofactor is required for this reaction, and what is the product represented by the question mark in the equation?

30

All vertebrates need vitamin D, which is synthesized by a pericyclic reaction when skin oils are exposed to sunlight. If the animal has no exposed skin, however, vitamin D is made from oily skin secretions that are deposited onto fur and then ingested during grooming. Image copyright Krylova Ksenia, 2010. Used under license from Shutterstock.com

Orbitals and Organic Chemistry: Pericyclic Reactions

Most organic reactions take place by polar mechanisms, in which a nucleophile donates two electrons to an electrophile in forming a new bond. Other reactions take place by radical mechanisms, in which each of two reactants donates one electron in forming a new bond. Both kinds of reactions occur frequently in the laboratory and in living organisms. Less common, however, is the third major class of organic reactions—*pericyclic reactions.*

A **pericyclic reaction** is one that occurs by a concerted process through a cyclic transition state. The word *concerted* means that all bonding changes occur simultaneously; no intermediates are involved. Rather than try to expand this definition now, we'll begin by briefly reviewing some of the ideas of molecular orbital theory introduced in Chapters 1 and 14 and then looking individually at the three main classes of pericyclic reactions: *electrocyclic reactions, cycloadditions,* and *sigmatropic rearrangements.*

Why This Chapter? The broad outlines of both polar and radical reactions have been known for more than a century, but our understanding of pericyclic reactions has emerged more recently. Prior to the mid-1960s, in fact, they were even referred to on occasion as "no-mechanism reactions." They occur largely in the laboratory rather than in biological processes, but a knowledge of them is necessary, both for completeness in studying organic chemistry and in understanding those biological pathways where they do occur.

30.1 Molecular Orbitals of Conjugated Pi Systems

A conjugated polyene, as we saw in **Section 14.1**, is one with alternating double and single bonds. According to molecular orbital (MO) theory, the *p* orbitals on the sp^2-hybridized carbons of a conjugated polyene interact to form a set of π molecular orbitals whose energies depend on the number of nodes they have between nuclei. Those molecular orbitals with fewer nodes are lower in energy than the isolated *p* atomic orbitals and are *bonding MOs;* those molecular orbitals with more nodes are higher in energy than the isolated *p* orbitals and are *antibonding MOs*. Pi molecular orbitals of ethylene and 1,3-butadiene are shown in **Figure 30.1**.

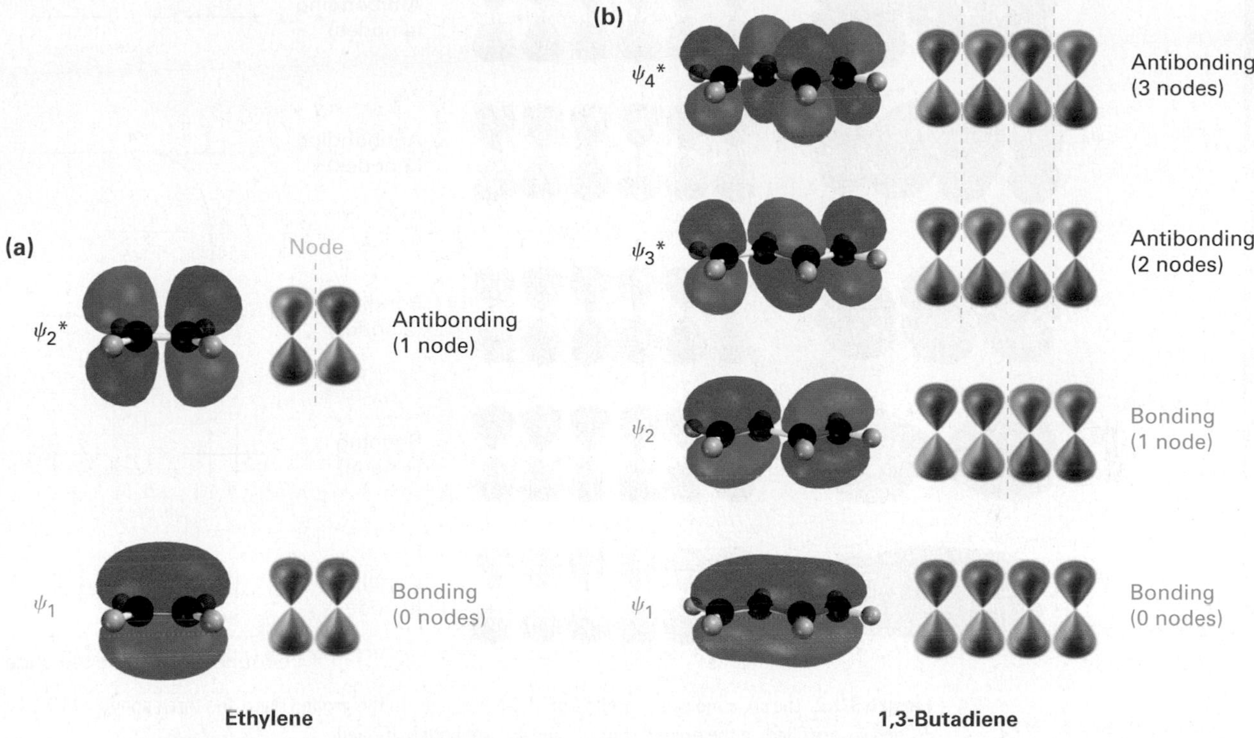

Figure 30.1 Pi molecular orbitals of **(a)** ethylene and **(b)** 1,3-butadiene.

A similar sort of molecular orbital description can be derived for any conjugated π electron system. 1,3,5-Hexatriene, for example, has three double bonds and six π MOs, as shown in **Figure 30.2**. In the ground state, only the three bonding orbitals, ψ_1, ψ_2, and ψ_3, are filled. On irradiation with ultraviolet light, however, an electron is promoted from the highest-energy filled orbital (ψ_3) to the lowest-energy unfilled orbital (ψ_4*) to give an excited state **(Section 14.7)**, in which ψ_3 and ψ_4* are each half-filled. (An asterisk denotes an antibonding orbital.)

What do molecular orbitals and their nodes have to do with pericyclic reactions? The answer is, *everything*. According to a series of rules formulated in the mid-1960s by R. B. Woodward and Roald Hoffmann, a pericyclic reaction can take place only if the symmetries of the reactant MOs are the same as the symmetries of the product MOs. In other words, the lobes of reactant MOs must be of the correct algebraic sign for bonding to occur in the transition state leading to product.

If the symmetries of reactant and product orbitals match up, or correlate, the reaction is said to be **symmetry-allowed**. If the symmetries of reactant and product orbitals don't correlate, the reaction is **symmetry-disallowed**.

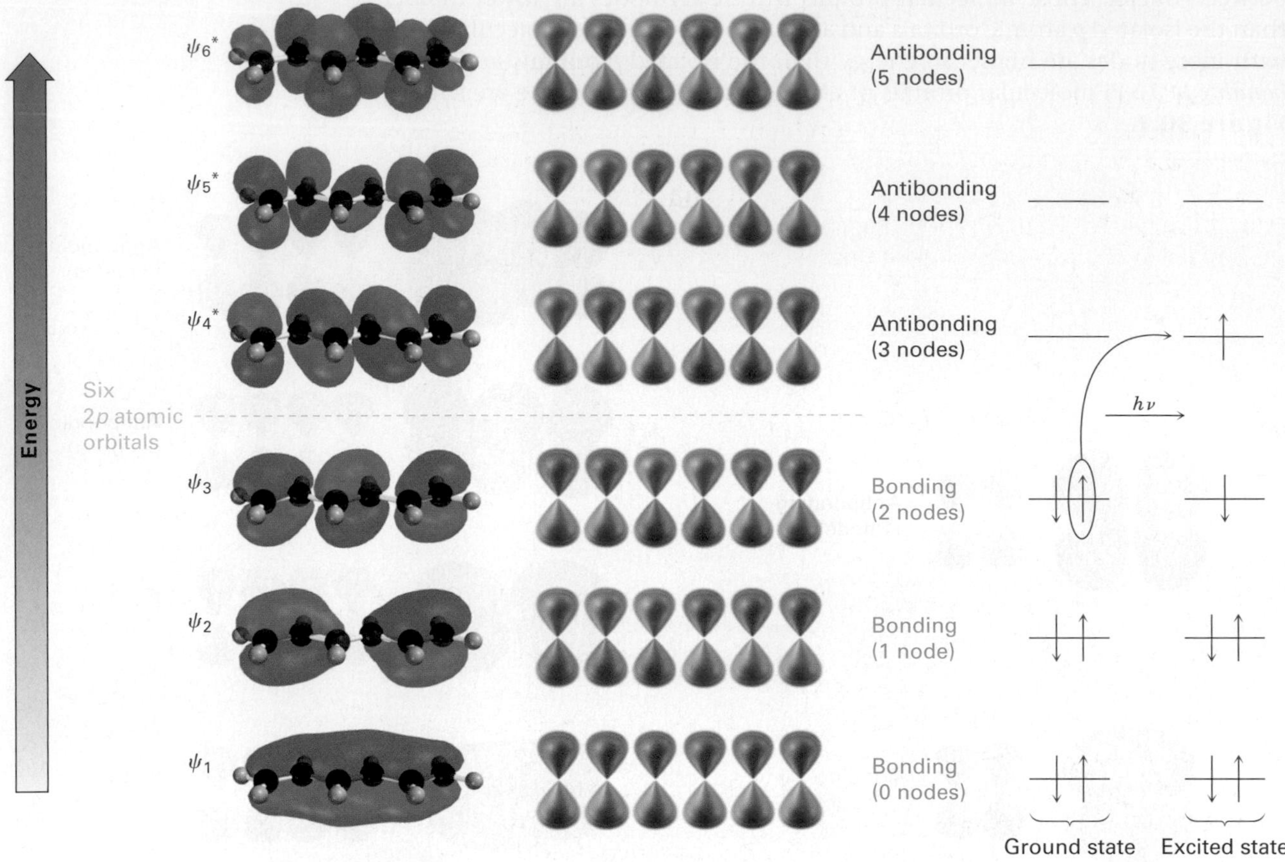

Figure 30.2 The six π molecular orbitals of 1,3,5-hexatriene. In the ground state, the three bonding MOs, ψ_1, ψ_2, and ψ_3, are filled. In the excited state, ψ_3 and $\psi_4{}^*$ are both half-filled.

Symmetry-allowed reactions often occur under relatively mild conditions, but symmetry-disallowed reactions can't occur by concerted paths. Either they take place by nonconcerted, higher-energy pathways, or they don't take place at all.

The Woodward–Hoffmann rules for pericyclic reactions require an analysis of all reactant and product molecular orbitals, but Kenichi Fukui at Kyoto Imperial University in Japan introduced a simplified version. According to Fukui, we need to consider only two molecular orbitals, called the **frontier orbitals**. These frontier orbitals are the **highest occupied molecular orbital (HOMO)** and the **lowest unoccupied molecular orbital (LUMO)**. In ground-state 1,3,5-hexatriene, for example, ψ_3 is the HOMO and $\psi_4{}^*$ is the LUMO (Figure 30.2). In excited-state 1,3,5-hexatriene, however, $\psi_4{}^*$ is the HOMO and $\psi_5{}^*$ is the LUMO.

Problem 30.1

Look at Figure 30.1, and tell which molecular orbital is the HOMO and which is the LUMO for both ground and excited states of ethylene and 1,3-butadiene.

30.2 Electrocyclic Reactions

The best way to understand how orbital symmetry affects pericyclic reactions is to look at some examples. Let's look first at a group of polyene rearrangements called *electrocyclic reactions*. An **electrocyclic reaction** is a pericyclic process that involves the cyclization of a conjugated acyclic polyene. One π bond is broken, the other π bonds change position, a new σ bond is formed, and a cyclic compound results. For example, a conjugated triene can be converted into a cyclohexadiene, and a conjugated diene can be converted into a cyclobutene.

Heat

A conjugated triene **A cyclohexadiene**

Heat

A conjugated diene **A cyclobutene**

Pericyclic reactions are reversible, and the position of the equilibrium depends on the specific case. In general, the triene ⇄ cyclohexadiene equilibrium favors the cyclic product, whereas the diene ⇄ cyclobutene equilibrium favors less strained open-chain product.

The most striking feature of electrocyclic reactions is their stereochemistry. For example, (2*E*,4*Z*,6*E*)-2,4,6-octatriene yields only *cis*-5,6-dimethyl-1,3-cyclohexadiene when heated, and (2*E*,4*Z*,6*Z*)-2,4,6-octatriene yields only *trans*-5,6-dimethyl-1,3 cyclohexadiene. Remarkably, however, the stereochemical results change completely when the reactions are carried out under what are called **photochemical**, rather than thermal, conditions. Irradiation, or *photolysis*, of (2*E*,4*Z*,6*E*)-2,4,6-octatriene with ultraviolet light yields *trans*-5,6-dimethyl-1,3-cyclohexadiene **(Figure 30.3)**.

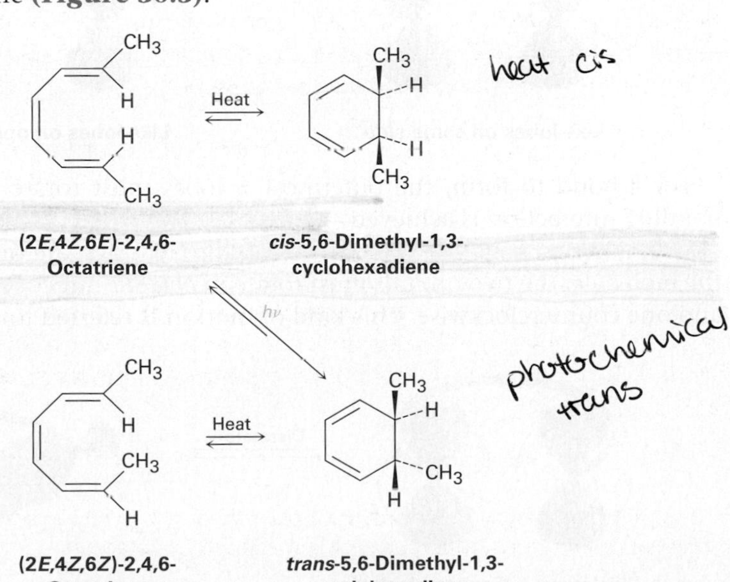

Figure 30.3 Electrocyclic interconversions of 2,4,6-octatriene isomers and 5,6-dimethyl-1,3-cyclohexadiene isomers.

A similar result is obtained for the thermal electrocyclic ring-opening of 3,4-dimethylcyclobutene. The trans isomer yields only (2*E*,4*E*)-2,4-hexadiene when heated, and the cis isomer yields only (2*E*,4*Z*)-2,4-hexadiene. On UV irradiation, however, the results are opposite. Cyclization of the 2*E*,4*E* isomer under photochemical conditions yields cis product **(Figure 30.4)**.

Figure 30.4 Electrocyclic interconversions of 2,4-hexadiene isomers and 3,4-dimethylcyclobutene isomers.

cis-3,4-Dimethylcyclobutene ⇌ (Heat) (2*E*,4*Z*)-2,4-Hexadiene

trans-3,4-Dimethylcyclobutene ⇌ (Heat) (2*E*,4*E*)-2,4-Hexadiene

(2*E*,4*E*)-2,4-Hexadiene → ($h\nu$) *cis*-3,4-Dimethylcyclobutene

To account for these results, we need to look at the two outermost lobes of the polyene MOs—the lobes that interact when cyclization occurs. There are two possibilities: the lobes of like sign can be either on the same side or on opposite sides of the molecule.

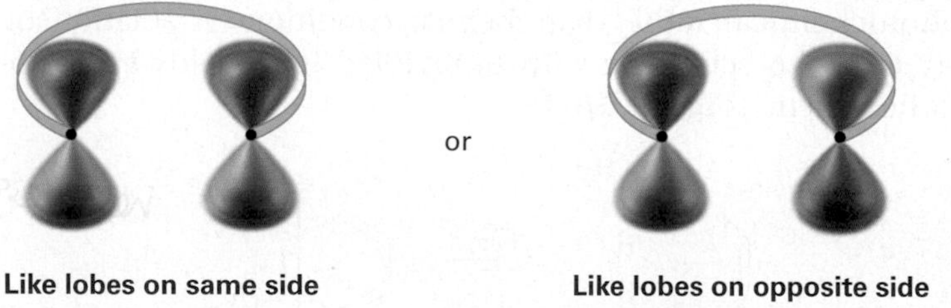

For a bond to form, the outermost π lobes must rotate so that favorable bonding interaction is achieved—a positive lobe with a positive lobe or a negative lobe with a negative lobe. If two lobes of like sign are on the *same* side of the molecule, the two orbitals must rotate in *opposite* directions—one clockwise and one counterclockwise. This kind of motion is referred to as **disrotatory**.

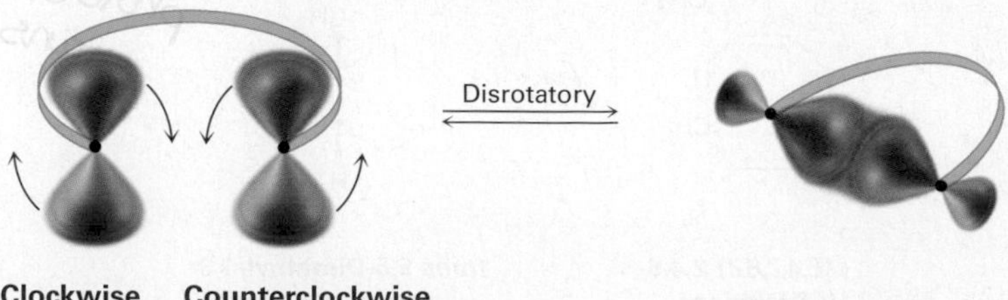

Conversely, if lobes of like sign are on *opposite* sides of the molecule, both orbitals must rotate in the *same* direction, either both clockwise or both counterclockwise. This kind of motion is called **conrotatory.**

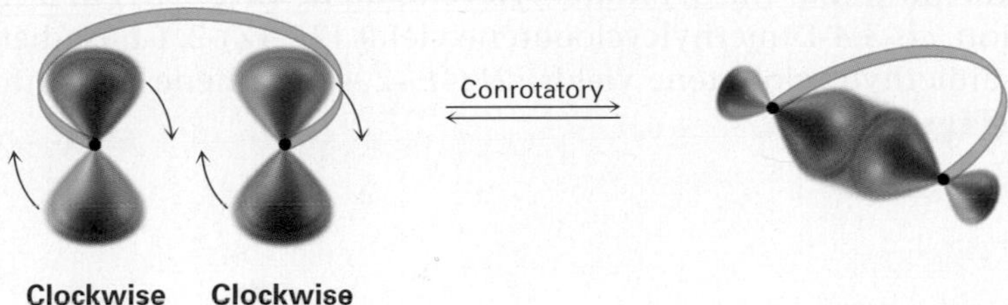

30.3 Stereochemistry of Thermal Electrocyclic Reactions

How can we predict whether conrotatory or disrotatory motion will occur in a given case? According to frontier orbital theory, *the stereochemistry of an electrocyclic reaction is determined by the symmetry of the polyene HOMO.* The electrons in the HOMO are the highest-energy, most loosely held electrons and are therefore most easily moved during reaction. For thermal reactions, the ground-state electron configuration is used to identify the HOMO; for photochemical reactions, the excited-state electron configuration is used.

Let's look again at the thermal ring-closure of conjugated *trienes*. According to Figure 30.2, the HOMO of a conjugated triene in its ground state has lobes of like sign on the same side of the molecule, a symmetry that predicts disrotatory ring-closure. This disrotatory cyclization is exactly what is observed in the thermal cyclization of 2,4,6-octatriene. The 2*E*,4*Z*,6*E* isomer yields cis product; the 2*E*,4*Z*,6*Z* isomer yields trans product **(Figure 30.5)**.

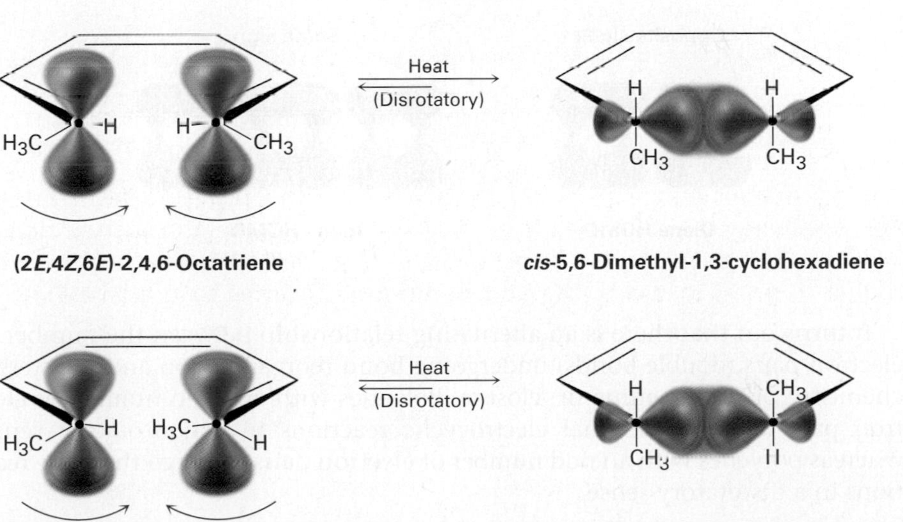

Figure 30.5 Thermal cyclizations of 2,4,6-octatrienes occur by disrotatory ring-closures.

In the same way, the ground-state HOMO of conjugated *dienes* (Figure 30.1) has a symmetry that predicts conrotatory ring-closure. In practice, however, the conjugated diene reaction can be observed only in the reverse direction (cyclobutene → diene) because of the position of the equilibrium. We therefore find that the 3,4-dimethylcyclobutene ring *opens* in a conrotatory fashion. *cis*-3,4-Dimethylcyclobutene yields (2*E*,4*Z*)-2,4-hexadiene, and *trans*-3,4-dimethylcyclobutene yields (2*E*,4*E*)-2,4-hexadiene by conrotatory opening **(Figure 30.6)**.

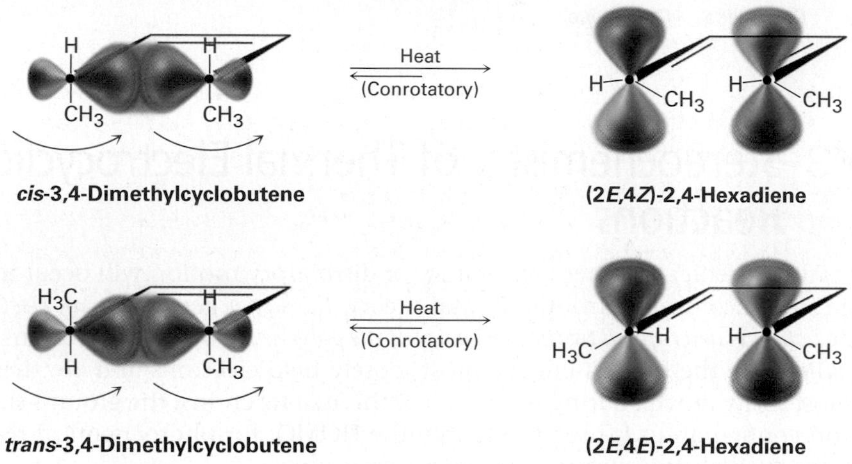

Figure 30.6 Thermal ring-openings of *cis*- and *trans*-dimethylcyclobutene occur by conrotatory paths.

Note that a conjugated diene and a conjugated triene react with opposite stereochemistry. The diene opens and closes by a conrotatory path, whereas the triene opens and closes by a disrotatory path. The difference is due to the different symmetries of the diene and triene HOMOs.

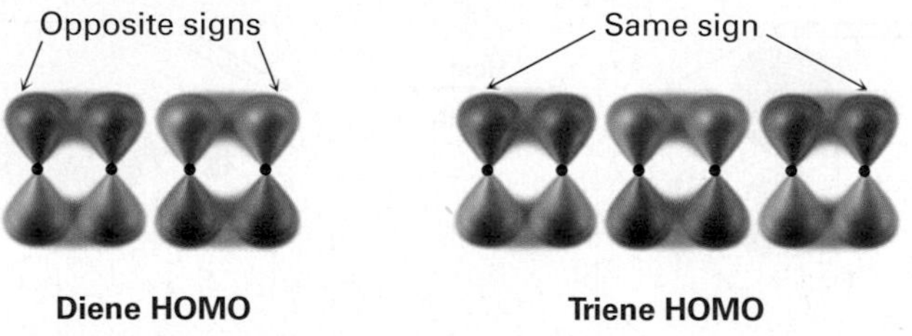

It turns out that there is an alternating relationship between the number of electron pairs (double bonds) undergoing bond reorganization and the stereochemistry of ring-opening or -closure. Polyenes with an even number of electron pairs undergo thermal electrocyclic reactions in a conrotatory sense, whereas polyenes with an odd number of electron pairs undergo the same reactions in a disrotatory sense.

Problem 30.2

Draw the products you would expect from conrotatory and disrotatory cyclizations of (2*Z*,4*Z*,6*Z*)-2,4,6-octatriene. Which of the two paths would you expect the thermal reaction to follow?

Problem 30.3

trans-3,4-Dimethylcyclobutene can open by two conrotatory paths to give either (2*E*,4*E*)-2,4-hexadiene or (2*Z*,4*Z*)-2,4-hexadiene. Explain why both products are symmetry-allowed, and then account for the fact that only the 2*E*,4*E* isomer is obtained in practice.

30.4 Photochemical Electrocyclic Reactions

We noted previously that photochemical electrocyclic reactions take a different stereochemical course than their thermal counterparts, and we can now explain this difference. Ultraviolet irradiation of a polyene causes an excitation of one electron from the ground-state HOMO to the ground-state LUMO, thus changing their symmetries. But because electronic excitation changes the symmetries of HOMO and LUMO, it also changes the reaction stereochemistry. (2*E*,4*E*)-2,4-Hexadiene, for instance, undergoes photochemical cyclization by a disrotatory path, whereas the thermal reaction is conrotatory. Similarly, (2*E*,4*Z*,6*E*)-2,4,6-octatriene undergoes photochemical cyclization by a conrotatory path, whereas the thermal reaction is disrotatory **(Figure 30.7)**.

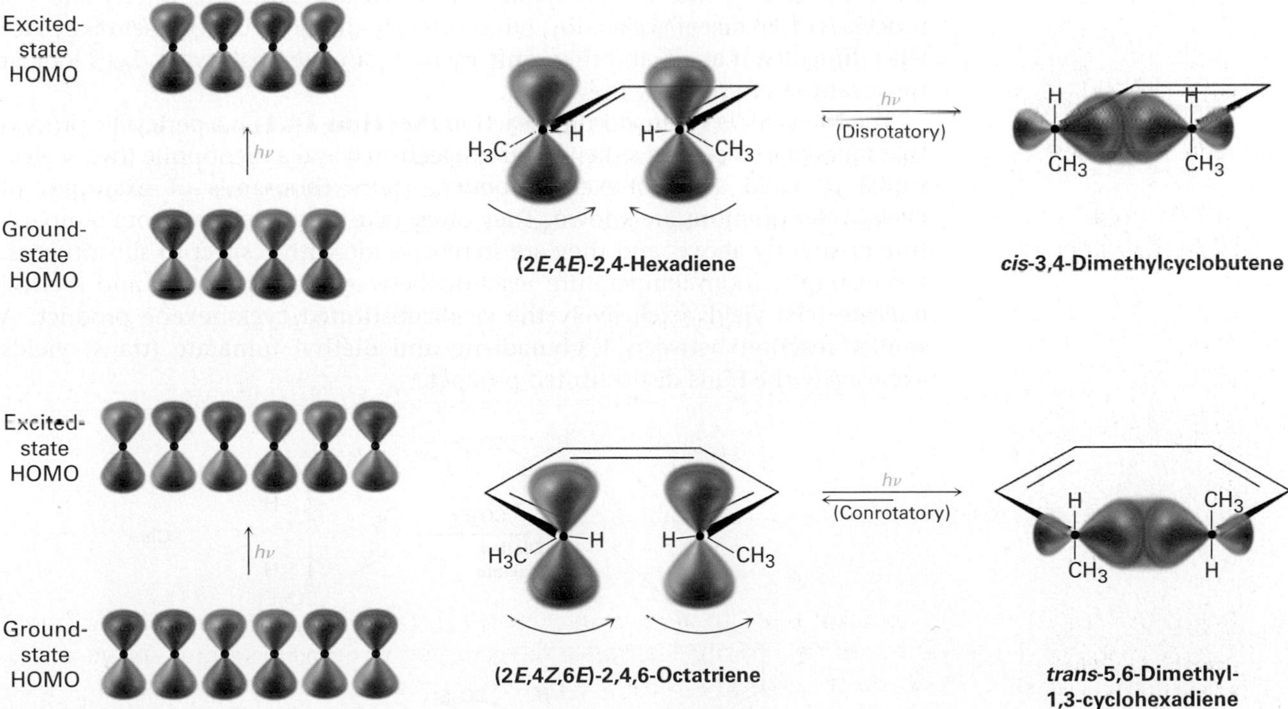

Figure 30.7 Photochemical cyclizations of conjugated dienes and trienes. The two processes occur with different stereochemistry because of their different orbital symmetries.

Thermal and photochemical electrocyclic reactions always take place with opposite stereochemistry because the symmetries of the frontier orbitals are always different. Table 30.1 gives some simple rules that make it possible to predict the stereochemistry of electrocyclic reactions.

Table 30.1 Stereochemical Rules for Electrocyclic Reactions

Electron pairs (double bonds)	Thermal reaction	Photochemical reaction
Even number	Conrotatory	Disrotatory
Odd number	Disrotatory	Conrotatory

Problem 30.4

What product would you expect to obtain from the photochemical cyclization of (2*E*,4*Z*,6*E*)-2,4,6-octatriene? Of (2*E*,4*Z*,6*Z*)-2,4,6-octatriene?

30.5 Cycloaddition Reactions

A **cycloaddition reaction** is one in which two unsaturated molecules add to one another to yield a cyclic product. As with electrocyclic reactions, cycloadditions are controlled by the orbital symmetry of the reactants. Symmetry-allowed processes often take place readily, but symmetry-disallowed processes take place with difficulty, if at all, and then only by nonconcerted pathways. Let's look at two examples to see how they differ.

The Diels–Alder cycloaddition reaction **(Section 14.4)** is a pericyclic process that takes place between a diene (four π electrons) and a dienophile (two π electrons) to yield a cyclohexene product. Many thousands of examples of Diels–Alder reactions are known. They often take place easily at room temperature or slightly above, and they are stereospecific with respect to substituents. For example, room-temperature reaction between 1,3-butadiene and diethyl maleate (cis) yields exclusively the cis-disubstituted cyclohexene product. A similar reaction between 1,3-butadiene and diethyl fumarate (trans) yields exclusively the trans-disubstituted product.

H CO2Et H CO2Et Diethyl maleate → CO2Et H H CO2Et Cis

1,3-Butadiene

H CO2Et EtO2C H Diethyl fumarate → CO2Et H CO2Et H Trans

In contrast to the [4 + 2]-π-electron Diels–Alder reaction, the [2 + 2]-π-electron cycloaddition between two alkenes does not occur thermally. The [2 + 2] cycloaddition takes place only on irradiation, yielding cyclobutane products.

Two alkenes $\xrightarrow{\text{Heat}}$ *No reaction*

Two alkenes $\xrightarrow{h\nu}$ A cyclobutane

For a successful cycloaddition to take place, the terminal π lobes of the two reactants must have the correct symmetry for bonding to occur. This can happen in either of two ways, called *suprafacial* and *antarafacial*. **Suprafacial** cycloadditions take place when a bonding interaction occurs between lobes on the same face of one reactant and lobes on the same face of the other reactant. **Antarafacial** cycloadditions take place when a bonding interaction occurs between lobes on the same face of one reactant and lobes on *opposite* faces of the other reactant **(Figure 30.8)**.

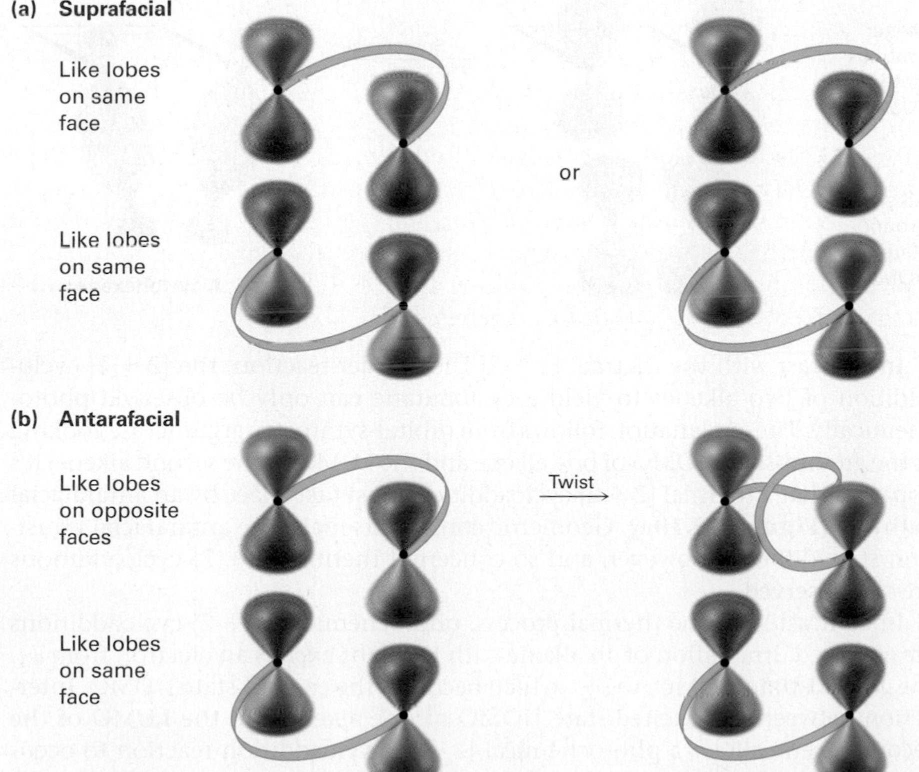

Figure 30.8 **(a)** Suprafacial cycloaddition occurs when there is bonding between lobes on the same face of one reactant and lobes on the same face of the other reactant. **(b)** Antarafacial cycloaddition occurs when there is bonding between lobes on the same face of one reactant and lobes on opposite faces of the other, which requires a twist in one π system.

Note that both suprafacial and antarafacial cycloadditions are symmetry-allowed. Geometric constraints often make antarafacial reactions difficult, however, because there must be a twisting of the π orbital system in one of the reactants. Thus, suprafacial cycloadditions are much more common for small π systems.

30.6 Stereochemistry of Cycloadditions

How can we predict whether a given cycloaddition reaction will occur with suprafacial or with antarafacial geometry? According to frontier orbital theory, a cycloaddition reaction takes place when a bonding interaction occurs between the HOMO of one reactant and the LUMO of the other. An intuitive explanation of this rule is to imagine that one reactant donates electrons to the other. As with electrocyclic reactions, it's the electrons in the HOMO of the first reactant that are least tightly held and most likely to be donated. But of course when the second reactant accepts those electrons, they must go into a vacant, unoccupied orbital—the LUMO.

For a [4 + 2] cycloaddition (Diels–Alder reaction), let's arbitrarily select the diene LUMO and the alkene HOMO. The symmetries of the two ground-state orbitals are such that bonding of the terminal lobes can occur with suprafacial geometry **(Figure 30.9)**, so the Diels–Alder reaction takes place readily under thermal conditions. Note that, as with electrocyclic reactions, we need be concerned only with the terminal lobes. For purposes of prediction, interactions among the interior lobes need not be considered.

Figure 30.9 Interaction of diene LUMO and alkene HOMO in a suprafacial [4 + 2] cycloaddition reaction (Diels–Alder reaction).

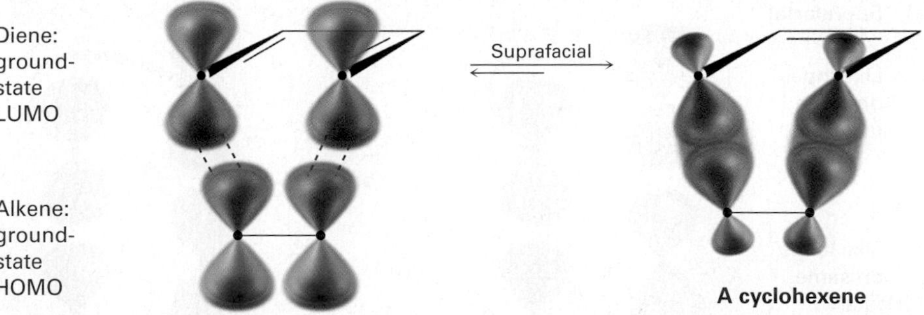

In contrast with the thermal [4 + 2] Diels–Alder reaction, the [2 + 2] cycloaddition of two alkenes to yield a cyclobutane can only be observed photochemically. The explanation follows from orbital-symmetry arguments. Looking at the ground-state HOMO of one alkene and the LUMO of the second alkene, it's apparent that a thermal [2 + 2] cycloaddition must take place by an antarafacial pathway **(Figure 30.10a)**. Geometric constraints make the antarafacial transition state difficult, however, and so concerted thermal [2 + 2] cycloadditions are not observed.

In contrast with the thermal process, photochemical [2 + 2] cycloadditions *are* observed. Irradiation of an alkene with UV light excites an electron from ψ_1, the ground-state HOMO, to $\psi_2{}^*$, which becomes the excited-state HOMO. Interaction between the excited-state HOMO of one alkene and the LUMO of the second alkene allows a photochemical [2 + 2] cycloaddition reaction to occur by a suprafacial pathway (Figure 30.10b).

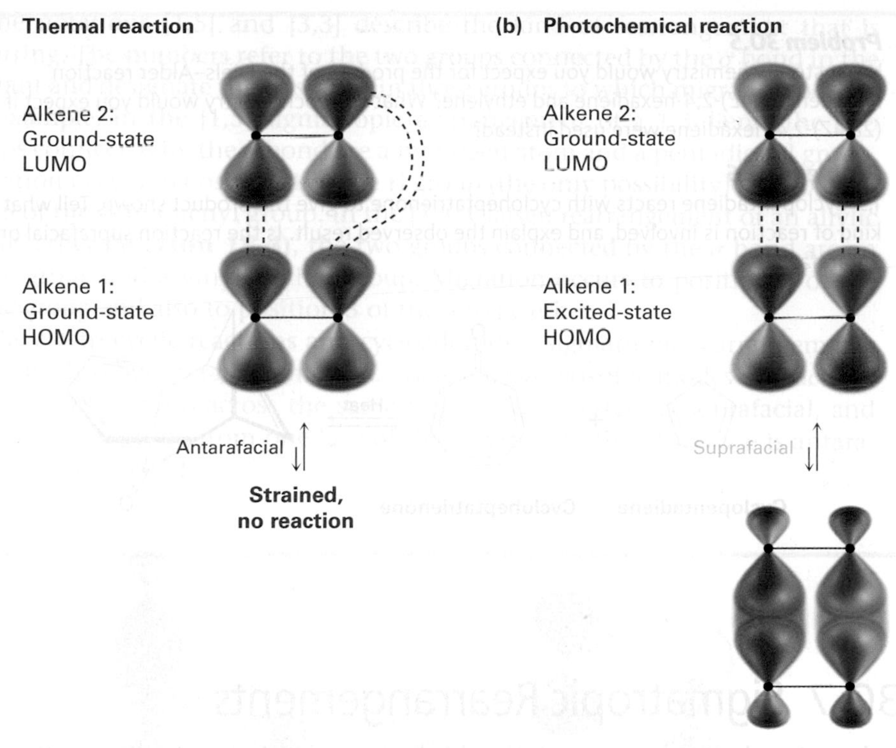

Figure 30.10 **(a)** Interaction of a ground-state HOMO and a ground-state LUMO in a potential [2 + 2] cycloaddition does not occur thermally because the antarafacial geometry is too strained. **(b)** Interaction of an excited-state HOMO and a ground-state LUMO in a photochemical [2 + 2] cycloaddition reaction is less strained, however, and occurs with suprafacial geometry.

The photochemical [2 + 2] cycloaddition reaction occurs smoothly, particularly with α,β-unsaturated carbonyl compounds, and represents one of the best methods known for synthesizing cyclobutane rings. For example:

2-Cyclohexenone + 2-Methylpropene $\xrightarrow{h\nu}$ (40%)

Thermal and photochemical cycloaddition reactions always take place with opposite stereochemistry. As with electrocyclic reactions, we can categorize cycloadditions according to the total number of electron pairs (double bonds) involved in the rearrangement. Thus, a thermal [4 + 2] Diels–Alder reaction between a diene and a dienophile involves an odd number (three) of electron pairs and takes place by a suprafacial pathway. A thermal [2 + 2] reaction between two alkenes involves an even number (two) of electron pairs and must take place by an antarafacial pathway. For photochemical cyclizations, these selectivities are reversed. The general rules are given in Table 30.2.

Table 30.2 Stereochemical Rules for Cycloaddition Reactions

Electron pairs (double bonds)	Thermal reaction	Photochemical reaction
Even number	Antarafacial	Suprafacial
Odd number	Suprafacial	Antarafacial

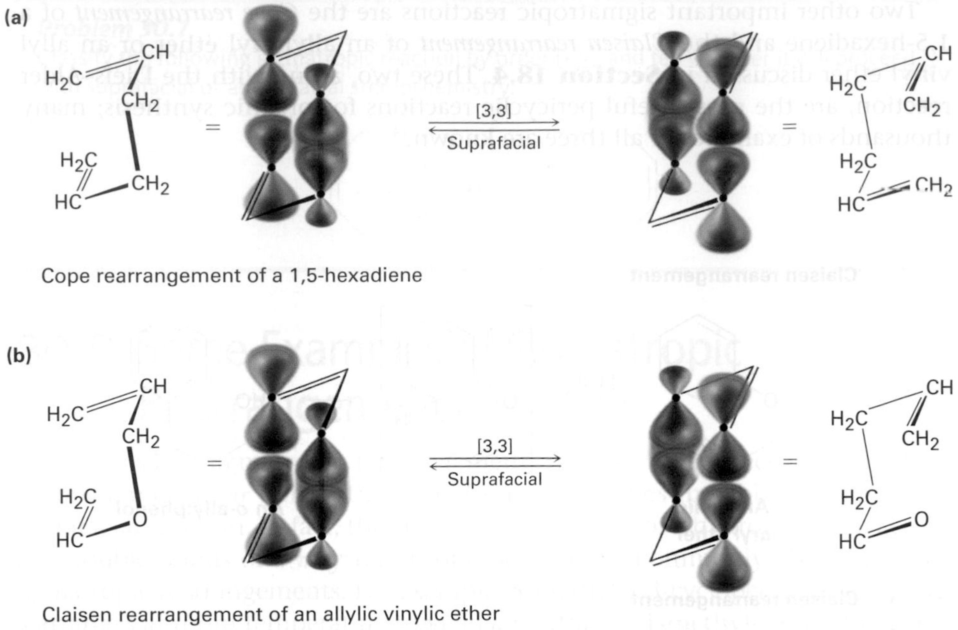

(a)

Cope rearrangement of a 1,5-hexadiene

(b)

Claisen rearrangement of an allylic vinylic ether

Figure 30.13 Suprafacial [3,3] **(a)** Cope and **(b)** Claisen rearrangements.

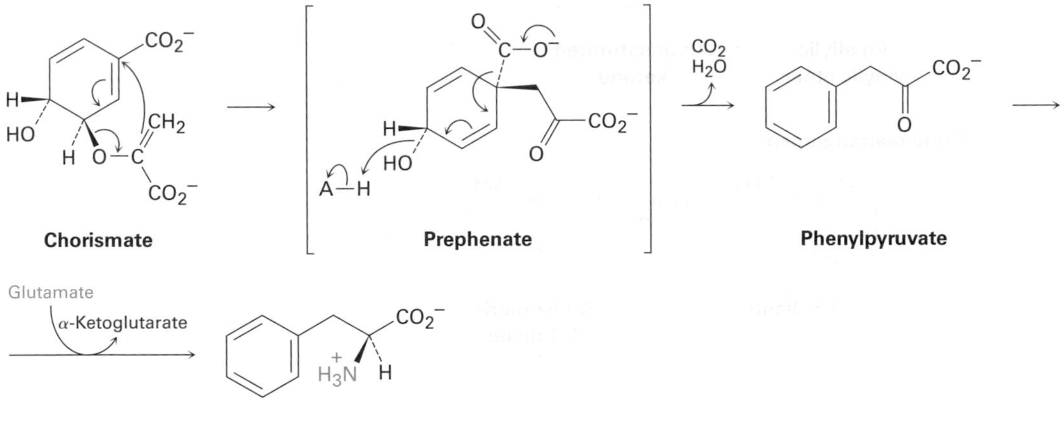

Chorismate **Prephenate** **Phenylpyruvate** **Phenylalanine**

Figure 30.14 Pathway for the bacterial biosynthesis of phenylalanine from chorismate, involving a Claisen rearrangement.

Problem 30.8

Propose a mechanism to account for the fact that heating 1-deuterioindene scrambles the isotope label to all three positions on the five-membered ring.

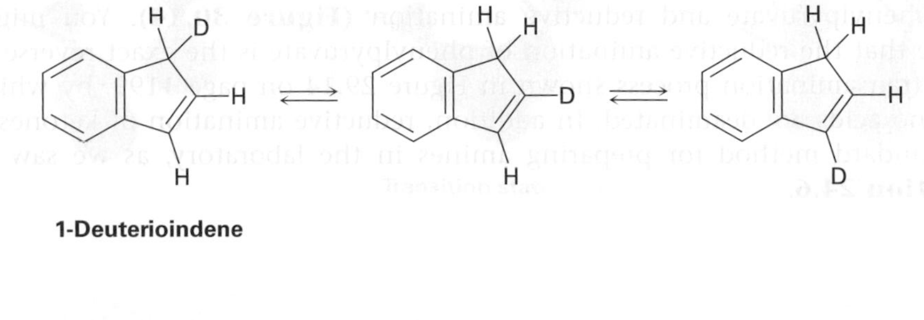

1-Deuterioindene

Problem 30.9

When a 2,6-disubstituted allyl phenyl ether is heated in an attempted Claisen rearrangement, migration occurs to give the *p*-allyl product as the result of two sequential pericyclic reactions. Explain.

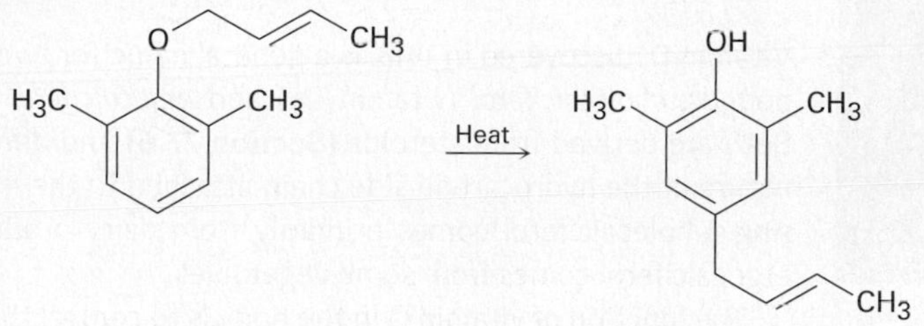

30.9 A Summary of Rules for Pericyclic Reactions

How can you keep straight all the rules about pericyclic reactions? The summary information in Tables 30.1 to 30.3 can be distilled into one mnemonic phrase that provides an easy way to predict the stereochemical outcome of any pericyclic reaction:

The Electrons Circle Around (TECA)

*T*hermal reactions with an *E*ven number of electron pairs are *C*onrotatory or *A*ntarafacial.

A change either from thermal to photochemical or from an even to an odd number of electron pairs changes the outcome from conrotatory/antarafacial to disrotatory/suprafacial. A change from both thermal and even to photochemical and odd causes no change because two negatives make a positive.

These selection rules are summarized in Table 30.4; knowing them gives you the ability to predict the stereochemistry of literally thousands of pericyclic reactions.

Table 30.4 Stereochemical Rules for Pericyclic Reactions

Electronic state	Electron pairs	Stereochemistry
Ground state (thermal)	Even number	Antara–con
	Odd number	Supra–dis
Excited state (photochemical)	Even number	Supra–dis
	Odd number	Antara–con

Problem 30.10

Predict the stereochemistry of the following pericyclic reactions:

(a) The thermal cyclization of a conjugated tetraene
(b) The photochemical cyclization of a conjugated tetraene
(c) A photochemical [4 + 4] cycloaddition
(d) A thermal [2 + 6] cycloaddition
(e) A photochemical [3,5] sigmatropic rearrangement

Synthesizing vitamin D takes dedication and hard work.

Vitamin D, the Sunshine Vitamin

A DEEPER LOOK

Vitamin D, discovered in 1918, is a general name for two related compounds, *cholecalciferol* (vitamin D_3) and *ergocalciferol* (vitamin D_2). Both are derived from steroids **(Section 27.6)** and differ only in the nature of the hydrocarbon side chain attached to the five-membered ring. Cholecalciferol comes primarily from dairy products and fish; ergocalciferol comes from some vegetables.

The function of vitamin D in the body is to control the calcification of bones by increasing intestinal absorption of calcium. When sufficient vitamin D is present, approximately 30% of ingested calcium is absorbed, but in the absence of vitamin D, calcium absorption falls to about 10%. A deficiency of vitamin D thus leads to poor bone growth and to the diseases *rickets* in children and *osteoporosis* in adults.

Actually, neither vitamin D_2 nor D_3 is present in foods. Rather, foods contain the precursor molecules 7-dehydrocholesterol and ergosterol. In the presence of sunlight, both precursors are converted in the outer, epidermal layer of skin to the active vitamins, hence the nickname for vitamin D, the "sunshine vitamin."

CH_3, R, CH_3, HO — *hν* → [CH_3, R, H, H_2C, OH] — [1,7] H shift → CH_3, R, CH_2, HO

7-Dehydrocholesterol — R = $CH(CH_3)CH_2CH_2CH_2CH(CH_3)_2$ — **Cholecalciferol**

Ergosterol — R = $CH(CH_3)CH{=}CHCH(CH_3)CH(CH_3)_2$ — **Ergocalciferol**

Pericyclic reactions are unusual in living organisms, and the photochemical synthesis of vitamin D is one of only a few well-studied examples. The reaction takes place in two steps, an electrocyclic ring-opening of a cyclohexadiene to yield an open-chain hexatriene, followed by a sigmatropic [1,7] H shift to yield an isomeric hexatriene. Only the initial, electrocyclic ring-opening requires irradiation, with so-called UVB light of 295 to 300 nm wavelength required. The subsequent sigmatropic [1,7] H shift occurs spontaneously by a thermal isomerization.

Following synthesis under the skin, further metabolic processing of cholecalciferol and ergocalciferol in the liver and kidney introduces two additional –OH groups to give the active forms of the vitamin, calcitriol and ergocalcitriol.

Summary

A **pericyclic reaction** takes place in a single step through a cyclic transition state without intermediates. There are three major classes of pericyclic processes: electrocyclic reactions, cycloaddition reactions, and sigmatropic rearrangements. The stereochemistry of these reactions is controlled by the symmetry of the orbitals involved in bond reorganization.

Electrocyclic reactions involve the cyclization of conjugated acyclic polyenes. For example, 1,3,5-hexatriene cyclizes to 1,3-cyclohexadiene on heating. Electrocyclic reactions can occur by either **conrotatory** or **disrotatory** pathways, depending on the symmetry of the terminal lobes of the π system. Conrotatory cyclization requires that both lobes rotate in the same direction, whereas disrotatory cyclization requires that the lobes rotate in opposite directions. The reaction course in a specific case can be found by looking at the symmetry of the **highest occupied molecular orbital (HOMO)**.

Cycloaddition reactions are those in which two unsaturated molecules add together to yield a cyclic product. For example, Diels–Alder reaction between a diene (four π electrons) and a dienophile (two π electrons) yields a cyclohexene. Cycloadditions can take place either by **suprafacial** or **antarafacial** pathways. Suprafacial cycloaddition involves interaction between lobes on the same face of one component and on the same face of the second component. Antarafacial cycloaddition involves interaction between lobes on the same face of one component and on opposite faces of the other component. The reaction course in a specific case can be found by looking at the symmetry of the HOMO of one component and the **lowest unoccupied molecular orbital (LUMO)** of the other component.

Sigmatropic rearrangements involve the migration of a σ-bonded group across a π electron system. For example, Claisen rearrangement of an allylic vinylic ether yields an unsaturated carbonyl compound, and Cope rearrangement of a 1,5-hexadiene yields an isomeric 1,5-hexadiene. Sigmatropic rearrangements can occur with either suprafacial or antarafacial stereochemistry; the selection rules for a given case are the same as those for cycloaddition reactions.

The stereochemistry of any pericyclic reaction can be predicted by counting the total number of electron pairs (bonds) involved in bond reorganization and then applying the mnemonic "The Electrons Circle Around." That is, **thermal** (ground-state) reactions involving an even number of electron pairs occur with either conrotatory or antarafacial stereochemistry. Exactly the opposite rules apply to **photochemical** (excited-state) reactions.

Key words

antarafacial, 1223
conrotatory, 1219
cycloaddition reaction, 1222
disrotatory, 1218
electrocyclic reaction, 1217
frontier orbital, 1216
highest occupied molecular orbital (HOMO), 1216
lowest unoccupied molecular orbital (LUMO), 1216
pericyclic reaction, 1214
photochemical reaction, 1217
sigmatropic rearrangement, 1226
suprafacial, 1223
symmetry-allowed, 1215
symmetry-disallowed, 1215

Exercises

OWL Interactive versions of these problems are assignable in OWL for Organic Chemistry.

Visualizing Chemistry

(Problems 30.1–30.10 appear within the chapter.)

30.11 Predict the product obtained when the following substance is heated:

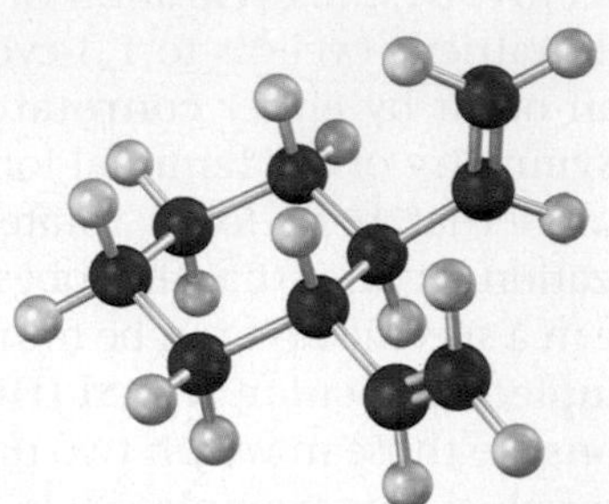

30.12 The ^{13}C NMR spectrum of homotropilidene taken at room temperature shows only three peaks. Explain.

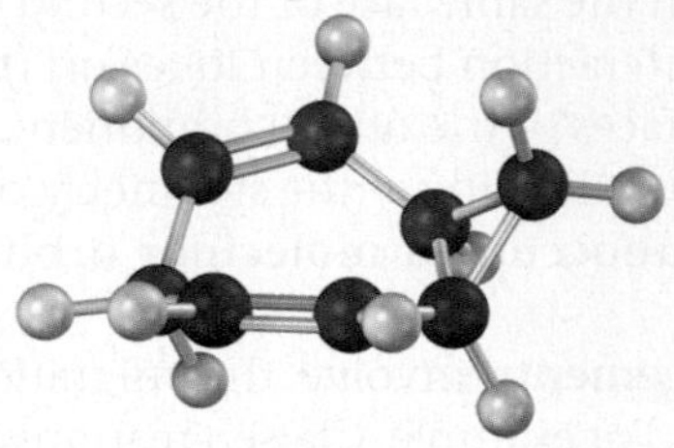

Additional Problems

Electrocyclic Reactions

30.13 Have the following electrocyclic reactions taken place in a conrotatory or disrotatory manner? Under what conditions, thermal or photochemical, would you carry out each reaction?

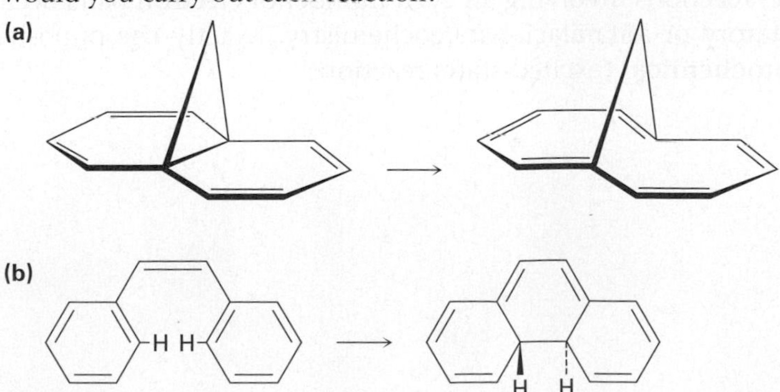

30.14 The following thermal isomerization occurs under relatively mild conditions. Identify the pericyclic reactions involved, and show how the rearrangement occurs.

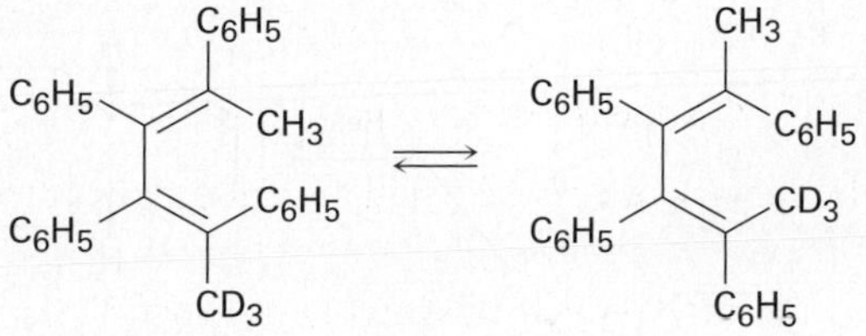

30.15 Would you expect the following reaction to proceed in a conrotatory or disrotatory manner? Show the stereochemistry of the cyclobutene product, and explain your answer.

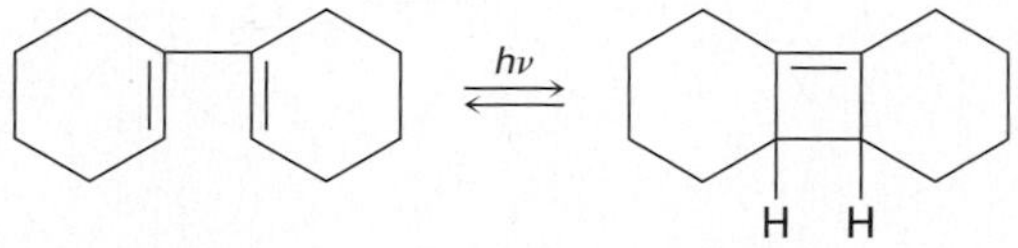

30.16 Heating (1*Z*,3*Z*,5*Z*)-1,3,5-cyclononatriene to 100 °C causes cyclization and formation of a bicyclic product. Is the reaction conrotatory or disrotatory? What is the stereochemical relationship of the two hydrogens at the ring junctions, cis or trans?

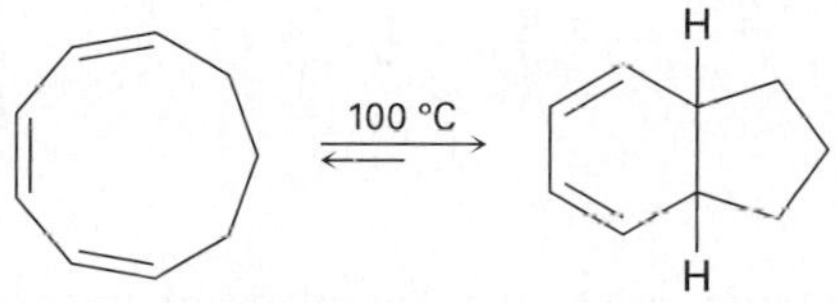

(1*Z*,3*Z*,5*Z*)-1,3,5-Cyclononatriene

30.17 (2*E*,4*Z*,6*Z*,8*E*)-2,4,6,8-Decatetraene has been cyclized to give 7,8-dimethyl-1,3,5-cyclooctatriene. Predict the manner of ring-closure—conrotatory or disrotatory—for both thermal and photochemical reactions, and predict the stereochemistry of the product in each case.

30.18 Answer Problem 30.17 for the thermal and photochemical cyclizations of (2*E*,4*Z*,6*Z*,8*Z*)-2,4,6,8-decatetraene.

30.19 The cyclohexadecaoctaene shown isomerizes to two different isomers, depending on reaction conditions. Explain the observed results, and indicate whether each reaction is conrotatory or disrotatory.

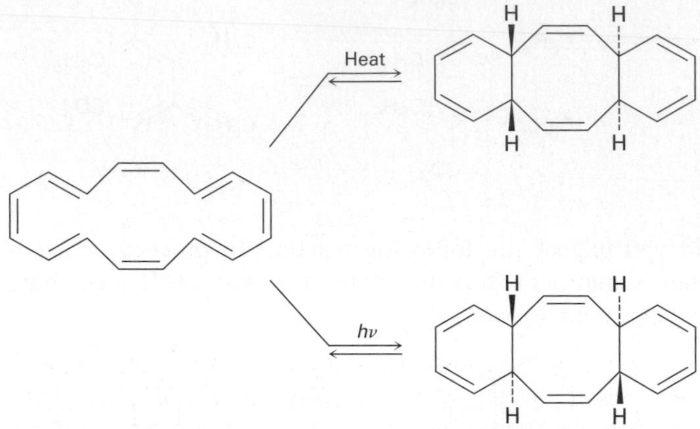

Heat

$h\nu$

Cycloaddition Reactions

30.20 Which of the following reactions is more likely to occur? Explain.

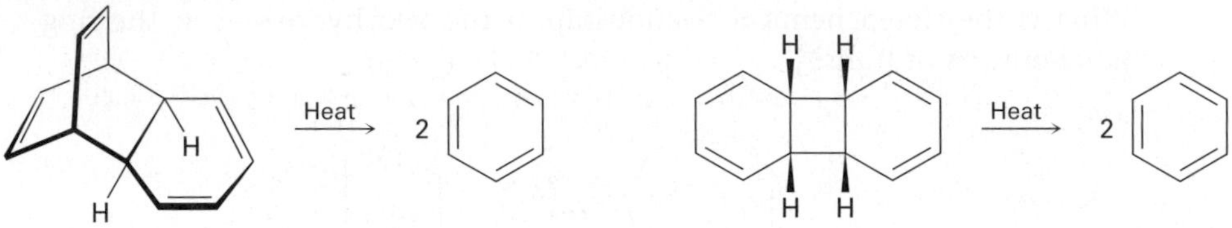

Heat → 2

Heat → 2

30.21 The following reaction takes place in two steps, one of which is a cycloaddition and the other of which is a *reverse* cycloaddition. Identify the two pericyclic reactions, and show how they occur.

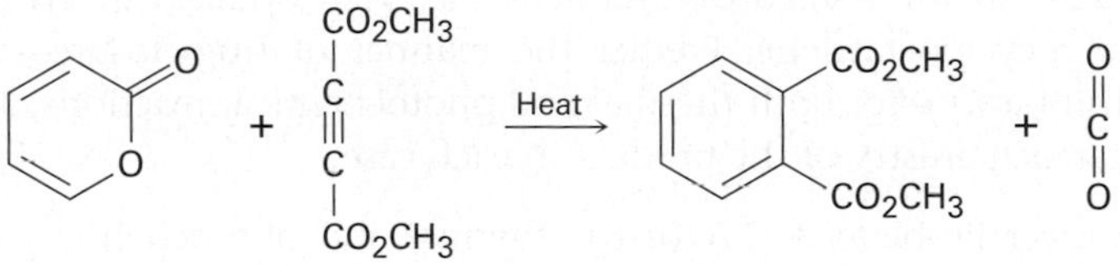

CO_2CH_3 Heat CO_2CH_3 + $O=C=O$

30.22 Two sequential pericyclic reactions are involved in the following furan synthesis. Identify them, and propose a mechanism for the transformation.

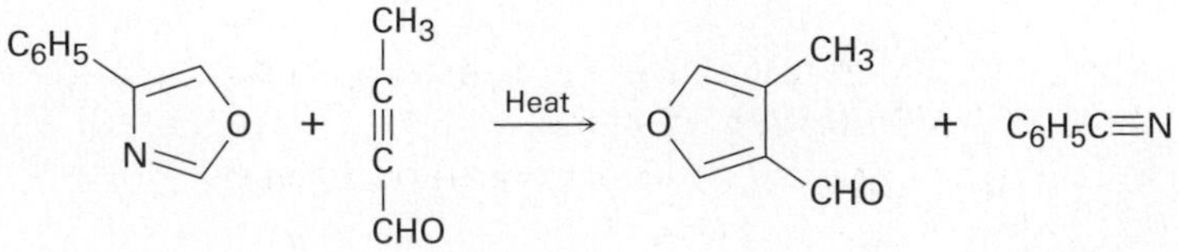

C_6H_5 + CH_3, CHO Heat → CH_3, CHO + $C_6H_5C{\equiv}N$

Sigmatropic Rearrangements

30.23 Predict the product of the following pericyclic reaction. Is this [5,5] shift a suprafacial or an antarafacial process?

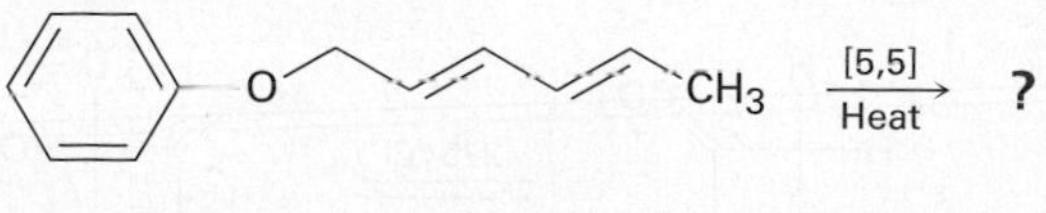

[5,5] Heat → ?

30.24 Propose a pericyclic mechanism to account for the following transformation:

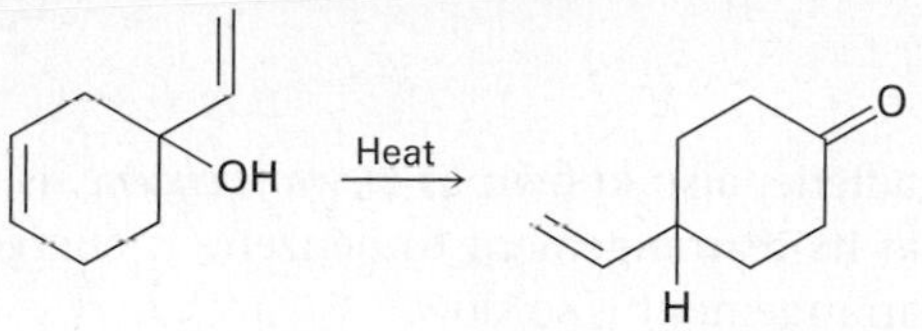

Heat →

30.25 Vinyl-substituted cyclopropanes undergo thermal rearrangement to yield cyclopentenes. Propose a mechanism for the reaction, and identify the pericyclic process involved.

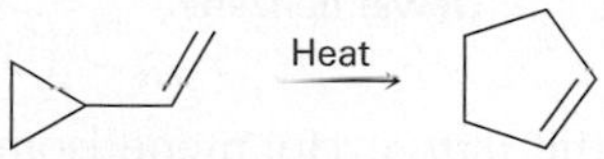

Heat →

Vinylcyclopropane **Cyclopentene**

30.26 The following synthesis of dienones occurs readily. Propose a mechanism to account for the results, and identify the kind of pericyclic reaction involved.

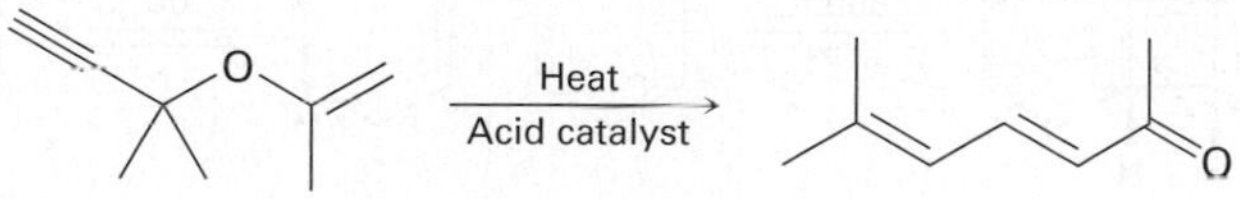

Heat, Acid catalyst →

30.27 Karahanaenone, a terpenoid isolated from oil of hops, has been synthesized by the thermal reaction shown. Identify the kind of pericyclic reaction, and explain how karahanaenone is formed.

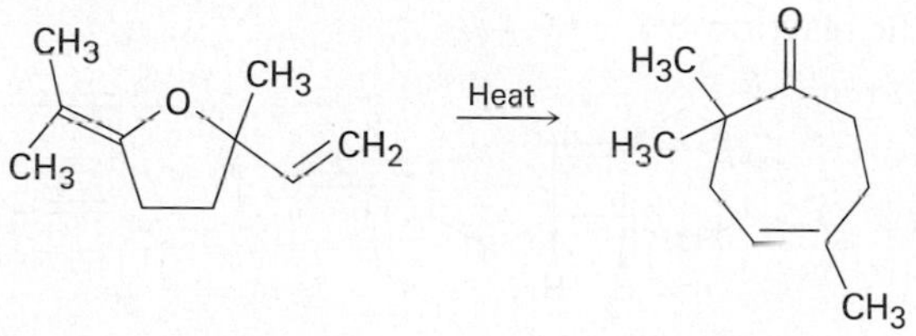

Heat →

Karahanaenone

General Problems

30.28 What stereochemistry—antarafacial or suprafacial—would you expect to observe in the following reactions?

(a) A photochemical [1,5] sigmatropic rearrangement
(b) A thermal [4 + 6] cycloaddition
(c) A thermal [1,7] sigmatropic rearrangement
(d) A photochemical [2 + 6] cycloaddition

30.29 The following thermal rearrangement involves two pericyclic reactions in sequence. Identify them, and propose a mechanism to account for the observed result.

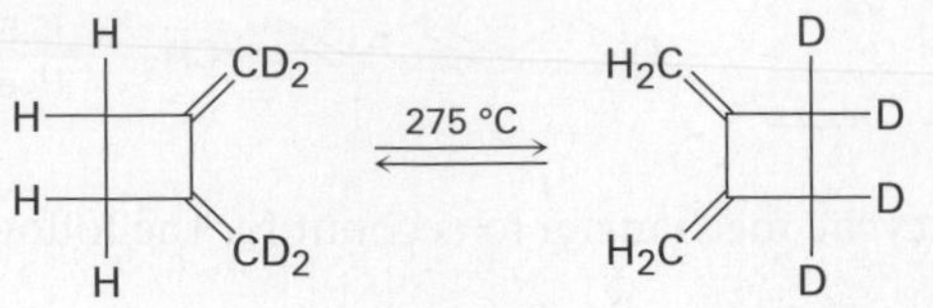

H, H, H, H, CD_2, CD_2 —275 °C→ H_2C, H_2C, D, D, D, D

30.30 Bicyclohexadiene, also known as *Dewar benzene,* is extremely stable despite the fact that its rearrangement to benzene is energetically favored. Explain why the rearrangement is so slow.

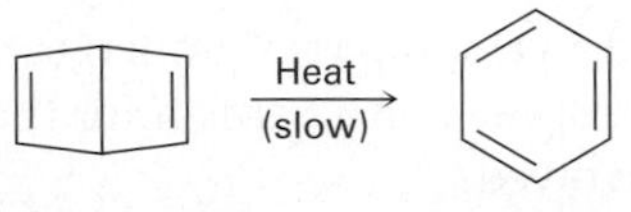

Heat (slow)

Dewar benzene **Benzene**

30.31 Ring-opening of the *trans*-cyclobutene isomer shown takes place at much lower temperature than a similar ring-opening of the *cis*-cyclobutene isomer. Explain the temperature effect, and identify the stereochemistry of each reaction as either conrotatory or disrotatory.

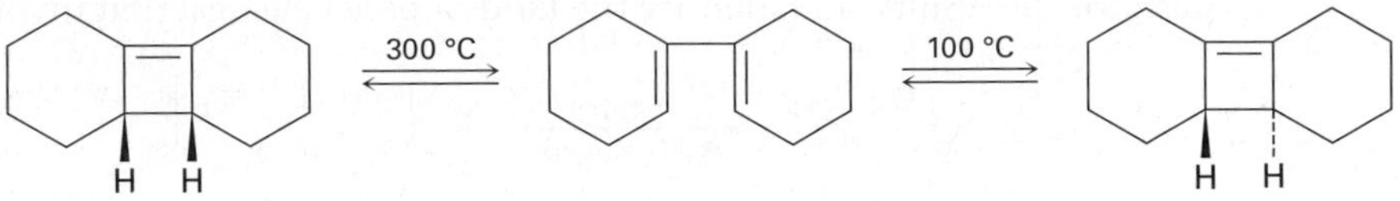

H H —300 °C→ ←100 °C— H H

30.32 Photolysis of the *cis*-cyclobutene isomer in Problem 30.31 yields *cis*-cyclododecaen-7-yne, but photolysis of the trans isomer yields *trans*-cyclododecaen-7-yne. Explain these results, and identify the type and stereochemistry of the pericyclic reaction.

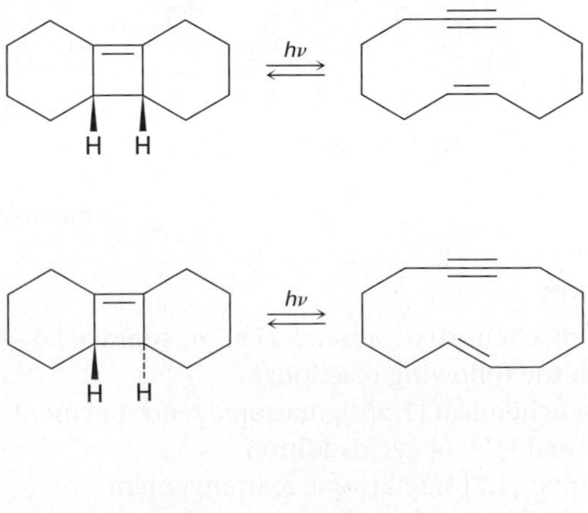

H H —$h\nu$→

H H —$h\nu$→

30.33 The ^1H NMR spectrum of bullvalene at 100 °C consists only of a single peak at 4.22 δ. Explain.

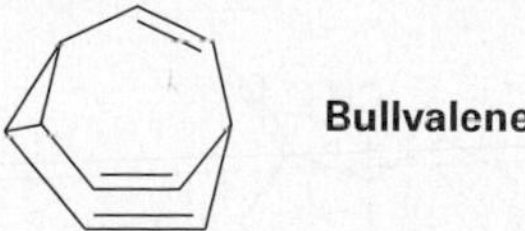

30.34 The following rearrangement was devised and carried out to prove the stereochemistry of [1,5] sigmatropic hydrogen shifts. Explain how the observed result confirms the predictions of orbital symmetry.

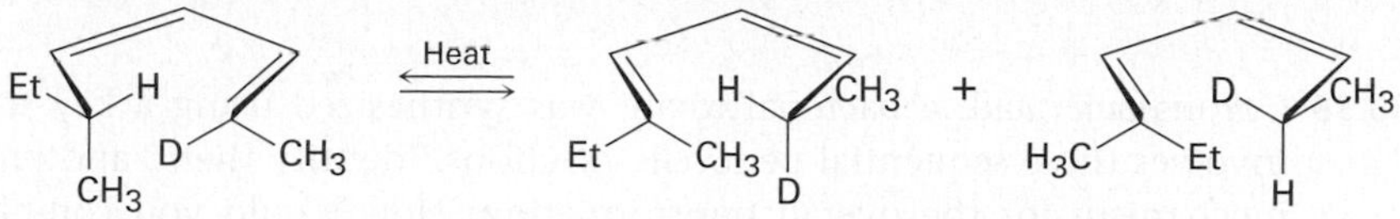

30.35 The following reaction is an example of a [2,3] sigmatropic rearrangement. Would you expect the reaction to be suprafacial or antarafacial? Explain.

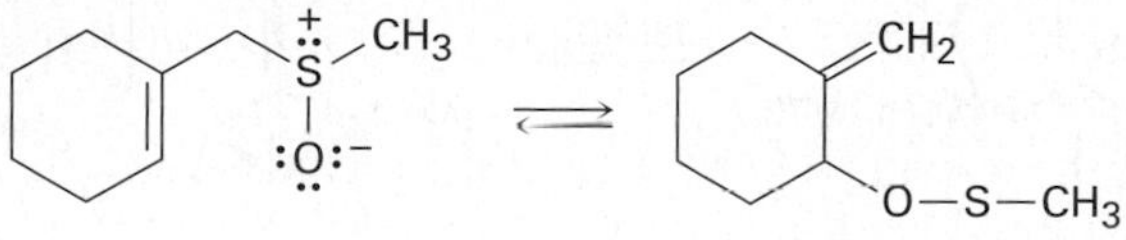

30.36 When the compound having a cyclobutene fused to a five-membered ring is heated, (1*Z*,3*Z*)-1,3-cycloheptadiene is formed. When the related compound having a cyclobutene fused to an eight-membered ring is heated, however, (1*E*,3*Z*)-1,3-cyclodecadiene is formed. Explain these results, and suggest a reason why opening of the eight-membered ring occurs at a lower temperature.

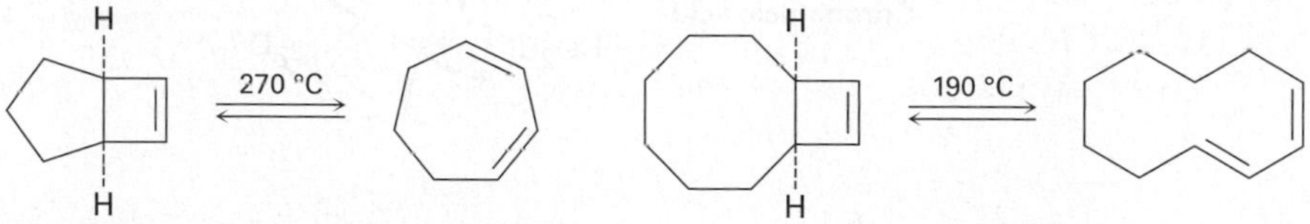

30.37 In light of your answer to Problem 30.36, explain why a mixture of products occurs in the following reaction:

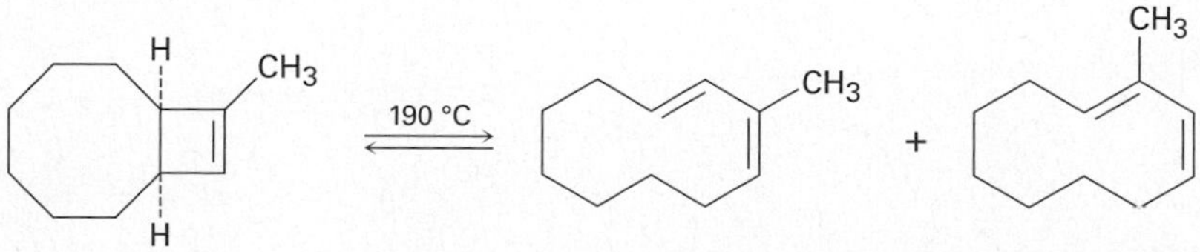

30.38 The sex hormone estrone has been synthesized by a route that involves the following step. Identify the pericyclic reactions involved, and propose a mechanism.

Heat

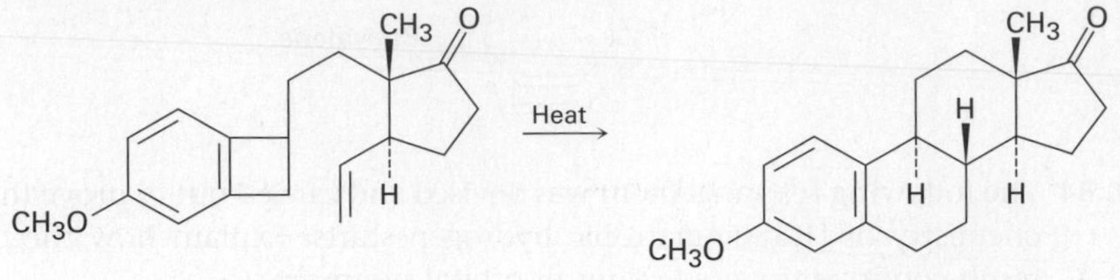

Estrone methyl ether

30.39 Coronafacic acid, a bacterial toxin, was synthesized using a key step that involves three sequential pericyclic reactions. Identify them, and propose a mechanism for the overall transformation. How would you complete the synthesis?

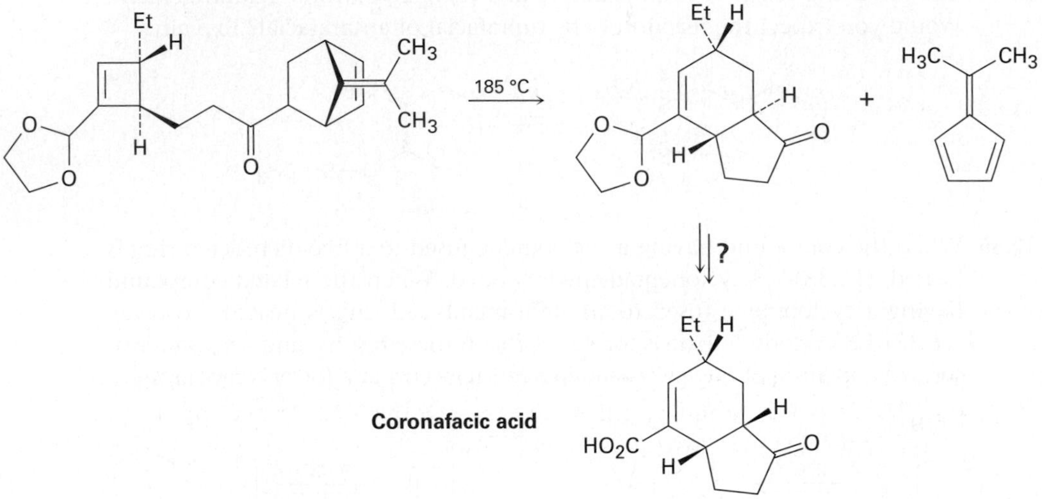

185 °C

Coronafacic acid

30.40 The following rearrangement of *N*-allyl-*N*,*N*-dimethylanilinium ion has been observed. Propose a mechanism.

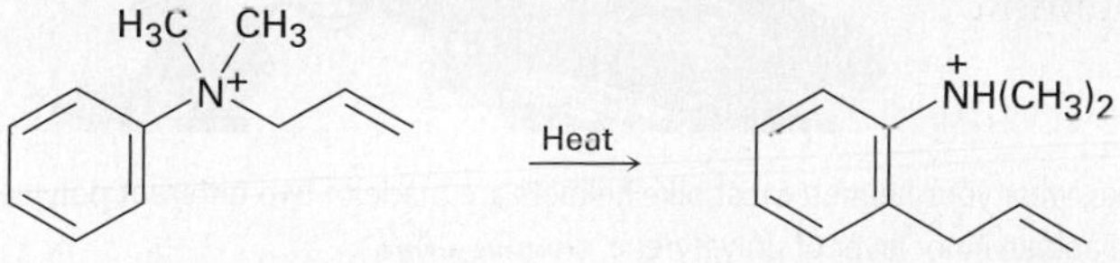

H$_3$C CH$_3$ N$^+$ Heat $\overset{+}{N}H(CH_3)_2$

***N*-Allyl-*N*,*N*-dimethylanilinium ion** ***o*-Allyl-*N*,*N*-dimethylanilinium ion**

30.41 Plastic photochromic sunglasses are based on the following reversible rearrangement of a dye inside the lenses that occurs when the lenses are exposed to sunlight. The original dye absorbs UV light but not visible light and is thus colorless, while the rearrangement product absorbs visible light and is thus darkened.

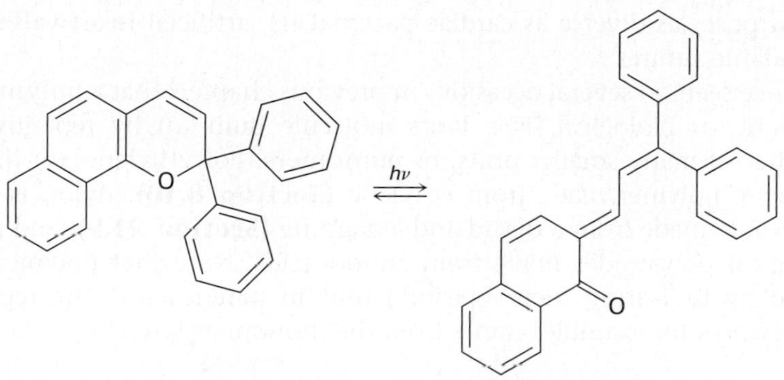

O $h\nu$ O

(Colorless) **(Dark)**

(a) Show the mechanism of the rearrangement.
(b) Why does the rearrangement product absorb at a longer wavelength (visible light) than the original dye (UV)?

31

If you ride a bike, wear your helmet! Most bike helmets are made of two different polymers, a hard polycarbonate shell and an inner layer of polystyrene. © Dattatreya/Alamy

Synthetic Polymers

Polymers are a fundamental part of the modern world, used in everything from coffee cups to cars to clothing. In medicine, too, their importance is growing for purposes as diverse as cardiac pacemakers, artificial heart valves, and biodegradable sutures.

We've seen on several occasions in previous chapters that a **polymer**, whether synthetic or biological, is a large molecule built up by repetitive bonding together of many smaller units, or **monomers**. Polyethylene, for instance, is a synthetic polymer made from ethylene **(Section 8.10)**, nylon is a synthetic polyamide made from a diacid and a diamine **(Section 21.9)**, and proteins are biological polyamides made from amino acids. Note that polymers are often drawn by indicating their repeating unit in parentheses. The repeat unit in polystyrene, for example, comes from the monomer styrene.

$H_2C{=}CH$ (with phenyl) $\Longrightarrow$ $\left(CH_2{-}CH \right)_n$ (with phenyl)

Styrene **Polystyrene**

Why This Chapter? Our treatment of polymers has thus far been dispersed over several chapters, but it's also important to take a more comprehensive view. In the present chapter, we'll look further at how polymers are made, and we'll see how polymer structure correlates with physical properties. No course in organic chemistry would be complete without a look at polymers.

31.1 Chain-Growth Polymers

Synthetic polymers are classified by their method of synthesis as either *chain-growth* or *step-growth*. The categories are somewhat imprecise but nevertheless provide a useful distinction. Chain-growth polymers are produced by chain-reaction polymerization in which an initiator adds to a carbon–carbon double bond of an unsaturated substrate (a *vinyl monomer*) to yield a reactive intermediate. This intermediate reacts with a second molecule of monomer to yield a new intermediate, which reacts with a third monomer unit, and so on.

OWL Sign in to OWL for Organic Chemistry at **www.cengage.com/owl** to view tutorials and simulations, develop problem-solving skills, and complete online homework assigned by your professor.

The initiator can be a radical, an acid, or a base. Historically, as we saw in **Section 8.10**, radical polymerization was the most common method because it can be carried out with practically any vinyl monomer.

Benzoyl peroxide $\xrightarrow{\text{Heat}}$ 2 [Benzoyloxy radical] $\longrightarrow$ 2 Phenyl radical (Ph·) + 2 CO_2

$$\text{Ph}\cdot \;+\; H_2C{=}CH_2 \longrightarrow \text{Ph}{-}CH_2CH_2\cdot \xrightarrow[\text{many times}]{\text{Repeat}} \text{Ph}{-}(CH_2CH_2)_nCH_2CH_2\cdot$$

Acid-catalyzed (cationic) polymerization, by contrast, is effective only with vinyl monomers that contain an electron-donating group (EDG) capable of stabilizing the chain-carrying carbocation intermediate.

$$\text{A}{-}\text{H} + H_2C{=}CH(\text{EDG}) \longrightarrow \left[\text{H}{-}CH_2{-}CH^+(\text{EDG}) \xrightarrow{H_2C=CH(\text{EDG})} CH_3CH(\text{EDG}){-}CH_2CH^+(\text{EDG})\right] \xrightarrow{\text{Repeat}} {+}CH_2{-}CH(\text{EDG}){+}_n$$

where EDG = an electron-donating group

Isobutylene (2-methylpropene) is a good example of a monomer that polymerizes rapidly under cationic conditions. The reaction is carried out commercially at −80 °C, using BF_3 and a small amount of water to generate $BF_3OH^-\ H^+$ catalyst. The product is used in the manufacture of truck and bicycle inner tubes.

$$H_2C{=}CH(CH_3) \xrightarrow{BF_3OH^-\ H^+} {+}CH_2{-}CH(CH_3){+}_n$$

Isobutylene — Polyisobutylene

Vinyl monomers with electron-withdrawing groups (EWG) can be polymerized by basic (anionic) catalysts. The chain-carrying step is conjugate nucleophilic addition of an anion to the unsaturated monomer **(Section 19.13)**.

$$\text{Nu:}^- + H_2C{=}CH(\text{EWG}) \longrightarrow \left[\text{Nu}{-}CH_2{-}CH(\text{EWG}){:}^- \xrightarrow{H_2C=CH(\text{EWG})} \text{Nu}CH_2CH(\text{EWG}){-}CH_2CH(\text{EWG}){:}^-\right] \xrightarrow{\text{Repeat}} {+}CH_2{-}CH(\text{EWG}){+}_n$$

where EWG = an electron-withdrawing group

Acrylonitrile ($H_2C{=}CHCN$), methyl methacrylate [$H_2C{=}C(CH_3)CO_2CH_3$], and styrene ($H_2C{=}CHC_6H_5$) can all be polymerized anionically. The polystyrene used in foam coffee cups, for example, is prepared by anionic polymerization of styrene using butyllithium as catalyst.

Bu:$^-$ Li$^+$ $H_2C{=}CH$ ⟶ [Bu—CH_2—CH:$^-$ $\xrightarrow{H_2C=CH}$] $\xrightarrow{\text{Repeat}}$ $(CH_2$—$CH)_n$

Styrene **Polystyrene**

An interesting example of anionic polymerization accounts for the remarkable properties of "super glue," one drop of which can support up to 2000 lb. Super glue is simply a solution of pure methyl α-cyanoacrylate, which has two electron-withdrawing groups that make anionic addition particularly easy. Trace amounts of water or bases on the surface of an object are sufficient to initiate polymerization of the cyanoacrylate and bind articles together. Skin is a good source of the necessary basic initiators, and many people have found their fingers stuck together after inadvertently touching super glue. So good is super glue at binding tissues together that related cyanoacrylate esters such as Dermabond are often used in place of sutures to close wounds.

Nu: $H_2C{=}C(C{\equiv}N)CO_2CH_3$ ⟶ [Nu—CH_2—C:$^-$(C≡N)(C(=O)—OCH_3)] ⟹ $(CH_2$—C(CN)(CO_2CH_3)$)_n$

Methyl α-cyanoacrylate

$H_2C{=}C(C{\equiv}N)$—C(=O)—$OCH_2CH(CH_2CH_3)CH_2CH_2CH_2CH_3$

Dermabond
(2-ethylhexyl α-cyanoacrylate)

Problem 31.1

Order the following monomers with respect to their expected reactivity toward cationic polymerization, and explain your answer:

$$H_2C{=}CHCH_3,\ H_2C{=}CHCl,\ H_2C{=}CH{-}C_6H_5,\ H_2C{=}CHCO_2CH_3$$

Problem 31.2

Order the following monomers with respect to their expected reactivity toward anionic polymerization, and explain your answer:

$$H_2C{=}CHCH_3,\ H_2C{=}CHC{\equiv}N,\ H_2C{=}CHC_6H_5$$

Problem 31.3

Polystyrene is produced commercially by reaction of styrene with butyllithium as an anionic initiator. Using resonance structures, explain how the chain-carrying intermediate is stabilized.

31.2 Stereochemistry of Polymerization: Ziegler–Natta Catalysts

Although we didn't point it out when discussing chain-growth polymers in **Section 8.10**, the polymerization of a substituted vinyl monomer can lead to a polymer with numerous chirality centers in its chain. Propylene, for example, might polymerize with any of the three stereochemical outcomes shown in **Figure 31.1**. The polymer having all methyl groups on the same side of the zigzag backbone is called **isotactic**, the one in which the methyl groups alternate regularly on opposite sides of the backbone is called **syndiotactic**, and the one having the methyl groups randomly oriented is called **atactic**.

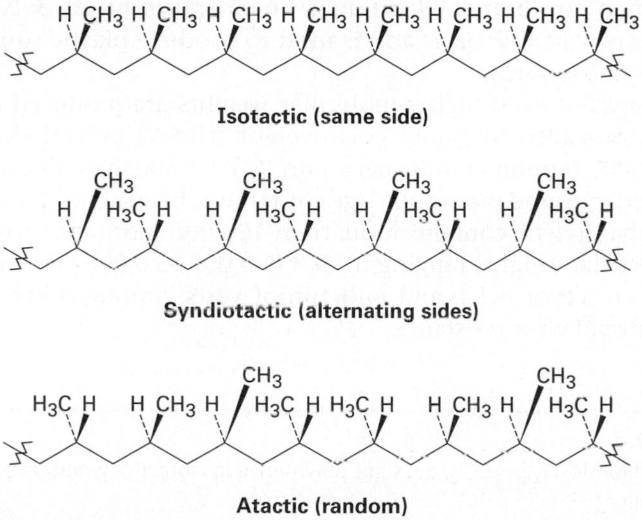

H CH_3 H CH_3 H CH_3 H CH_3 H CH_3 H CH_3 H CH_3 H CH_3

Isotactic (same side)

CH_3 H H_3C H CH_3 H H_3C H CH_3 H H_3C H CH_3 H H_3C H

Syndiotactic (alternating sides)

H_3C H H CH_3 H CH_3 H_3C H H_3C H H CH_3 H CH_3 H_3C H

Atactic (random)

Figure 31.1 Isotactic, syndiotactic, and atactic forms of polypropylene.

The three different stereochemical forms of polypropylene all have somewhat different properties, and all can be made by using the right polymerization catalyst. Propylene polymerization using radical initiators does not work well, but polymerization using *Ziegler–Natta catalysts* allows preparation of isotactic, syndiotactic, and atactic polypropylene.

Ziegler–Natta catalysts—there are many different formulations—are organometallic transition-metal complexes prepared by treatment of an alkylaluminum with a titanium compound. Triethylaluminum and titanium tetrachloride form a typical preparation.

$$(CH_3CH_2)_3Al + TiCl_4 \rightarrow \text{A Ziegler–Natta catalyst}$$

Following their introduction in 1953, Ziegler–Natta catalysts revolutionized the field of polymer chemistry because of two advantages: first, the resultant polymers are linear, with practically no chain branching, and second, they are stereochemically controllable. Isotactic, syndiotactic, and atactic forms can all be produced, depending on the catalyst system used.

The active form of a Ziegler–Natta catalyst is an alkyltitanium intermediate with a vacant coordination site on the metal. Coordination of alkene monomer to the titanium occurs, and the coordinated alkene then inserts into the carbon–titanium bond to extend the alkyl chain. A new coordination site opens up during the insertion step, so the process repeats indefinitely.

$Ti-CH_2CH_2-R$ (Vacant coordination site) $\xrightarrow{H_2C=CH_2}$ $Ti(-CH_2CH_2-R)(\leftarrow CH_2{=}CH_2)$ $\longrightarrow$ $Ti-CH_2CH_2CH_2CH_2-R$ (Vacant coordination site)

The linear polyethylene produced by the Ziegler–Natta process, called *high-density polyethylene,* is a highly crystalline polymer with 4000 to 7000 ethylene units per chain and molecular weights in the range 100,000 to 200,000 amu. High-density polyethylene has greater strength and heat resistance than the branched product of radical-induced polymerization, called *low-density polyethylene,* and is used to produce plastic squeeze bottles and molded housewares.

Polyethylenes of even higher molecular weights are produced for specialty applications. So-called high-molecular-weight (HMW) polyethylene contains 10,000 to 18,000 monomer units per chain (MW = 300,000–500,000 amu) and is used for underground pipes and large containers. Ultrahigh-molecular-weight (UHMW) polyethylene contains more than 100,000 monomer units per chain and has molecular weights ranging from 3,000,000 to 6,000,000 amu. It is used in bearings, conveyor belts, and bulletproof vests, among other applications requiring unusual wear resistance.

Problem 31.4

Vinylidene chloride, $H_2C{=}CCl_2$, does not polymerize in isotactic, syndiotactic, and atactic forms. Explain.

Problem 31.5

Polymers such as polypropylene contain a large number of chirality centers. Would you therefore expect samples of isotactic, syndiotactic, or atactic polypropylene to rotate plane-polarized light? Explain.

31.3 Copolymers

Up to this point we've discussed only **homopolymers**—polymers that are made up of identical repeating units. In practice, however, *copolymers* are more important commercially. **Copolymers** are obtained when two or more different monomers are allowed to polymerize together. For example, copolymerization of vinyl chloride with vinylidene chloride (1,1-dichloroethylene) in a 1:4 ratio leads to the polymer Saran.

$$\underset{\textbf{Vinyl chloride}}{H_2C{=}CHCl} + \underset{\textbf{Vinylidene chloride}}{H_2C{=}CCl_2} \longrightarrow \underset{\textbf{Saran}}{\left(\left(CH_2CHCl\right)_m\left(CH_2CCl_2\right)\right)_n}$$

Copolymerization of monomer mixtures often leads to materials with properties quite different from those of either corresponding homopolymer, giving the polymer chemist a vast amount of flexibility for devising new materials. Table 31.1 lists some common copolymers and their commercial applications.

Table 31.1 Some Common Copolymers and Their Uses

Monomers	Structures	Trade name	Uses
Vinyl chloride Vinylidene chloride	$H_2C{=}CHCl$ + $H_2C{=}CCl_2$	Saran	Fibers, food packaging
Styrene 1,3-Butadiene	$H_2C{=}CHC_6H_5$ + $H_2C{=}CH{-}CH{=}CH_2$	SBR (styrene–butadiene rubber)	Tires, rubber articles
Hexafluoropropene Vinylidene fluoride	$F_2C{=}CFCF_3$ + $H_2C{=}CF_2$	Viton	Gaskets, seals
Acrylonitrile 1,3-Butadiene	$H_2C{=}CHCN$ + $H_2C{=}CH{-}CH{=}CH_2$	Nitrile rubber	Adhesives, hoses
Isobutylene Isoprene	$H_2C{=}C(CH_3)_2$ + $H_2C{=}C(CH_3){-}CH{=}CH_2$	Butyl rubber	Inner tubes
Acrylonitrile 1,3-Butadiene Styrene	$H_2C{=}CHCN$ + $H_2C{=}CH{-}CH{=}CH_2$ + $H_2C{=}CHC_6H_5$	ABS (monomer initials)	Pipes, high-impact applications

Several different types of copolymers can be defined, depending on the distribution of monomer units in the chain. If monomer A is copolymerized with monomer B, for instance, the resultant product might have a random distribution of the two units throughout the chain, or it might have an alternating distribution.

—A—A—A—B—A—B—B—A—B—A—A—A—B—B—B—

Random copolymer

—A—B—A—B—A—B—A—B—A—B—A—B—A—B—A—

Alternating copolymer

The exact distribution of monomer units depends on the initial proportions of the two reactant monomers and their relative reactivities. In practice, neither perfectly random nor perfectly alternating copolymers are usually found. Most copolymers have many random imperfections.

Two other forms of copolymers that can be prepared under certain conditions are called *block copolymers* and *graft copolymers*. **Block copolymers** are those in which different blocks of identical monomer units alternate with each other; **graft copolymers** are those in which homopolymer branches of one monomer unit are "grafted" onto a homopolymer chain of another monomer unit.

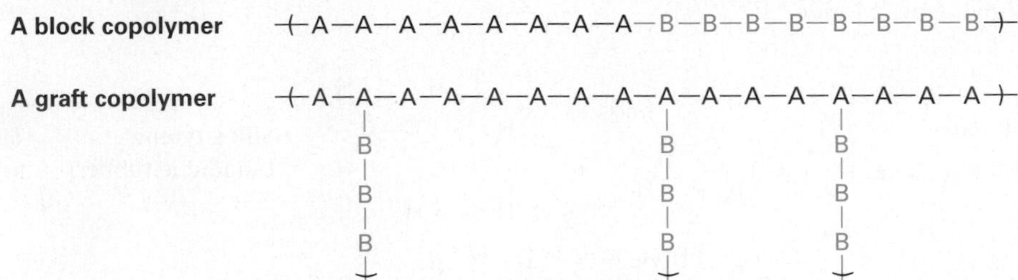

Block copolymers are prepared by initiating the polymerization of one monomer as if growing a homopolymer chain and then adding an excess of the second monomer to the still-active reaction mix. Graft copolymers are made by gamma irradiation of a completed homopolymer chain in the presence of the second monomer. The high-energy irradiation knocks hydrogen atoms off the homopolymer chain at random points, thus generating new radical sites that can initiate polymerization of the added monomer.

Problem 31.6
Draw the structure of an alternating segment of butyl rubber, a copolymer of isoprene (2-methyl-1,3-butadiene) and isobutylene (2-methylpropene) prepared using a cationic initiator.

Problem 31.7
Irradiation of poly(1,3-butadiene), followed by addition of styrene, yields a graft copolymer that is used to make rubber soles for shoes. Draw the structure of a representative segment of this styrene–butadiene graft copolymer.

31.4 Step-Growth Polymers

Step-growth polymers are produced by reactions in which each bond in the polymer is formed stepwise, independently of the others. Like the polyamides (nylons) and polyesters that we saw in **Section 21.9**, most step-growth polymers are produced by reaction between two difunctional reactants. Nylon 66, for instance, is made by reaction between the six-carbon adipic acid and the six-carbon hexamethylenediamine (1,6-hexanediamine). Alternatively, a single reactant with two different functional groups can polymerize. Nylon 6 is made by polymerization of the six-carbon caprolactam. The reaction is initiated by addition of a small amount of water, which hydrolyzes some caprolactam to 6-aminohexanoic acid. Nucleophilic addition of the amino group to caprolactam then propagates the polymerization.

$HOCCH_2CH_2CH_2CH_2COH$ (with C=O at both carbonyls) + $H_2NCH_2CH_2CH_2CH_2CH_2CH_2NH_2$

Adipic acid (hexanedioic acid) **Hexamethylenediamine (1,6-hexanediamine)**

↓ Heat

$\left[CCH_2CH_2CH_2CH_2C-NHCH_2CH_2CH_2CH_2CH_2CH_2NH \right]_n$ (with C=O at both carbonyls)

Nylon 66

Caprolactam $\xrightarrow[\text{Heat}]{H_2O}$ $[HOCCH_2CH_2CH_2CH_2CH_2\ddot{N}H_2]$ **6-Aminohexanoic acid** $\xrightarrow{\text{Heat}}$ $\left[CCH_2CH_2CH_2CH_2CH_2NH \right]_n$ **Nylon 66**

Polycarbonates

Polycarbonates are like polyesters, but their carbonyl group is linked to two –OR groups, [O=C(OR)$_2$]. Lexan, for instance, is a polycarbonate prepared from diphenyl carbonate and a diphenol called bisphenol A. Lexan has unusually high impact strength, making it valuable for use in machinery housings, telephones, bicycle safety helmets, and bulletproof glass.

Diphenyl carbonate + **Bisphenol A** $\xrightarrow{300\ °C}$ **Lexan** + 2*n* phenol (OH)

Polyurethanes

A *urethane* is a carbonyl-containing functional group in which the carbonyl carbon is bonded to both an $-OR$ group and an $-NR_2$ group. As such, a urethane is halfway between a carbonate and a urea.

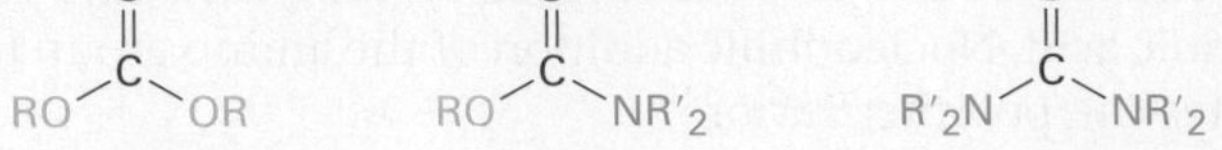

A carbonate **A urethane** **A urea**

A urethane is typically prepared by nucleophilic addition reaction between an alcohol and an isocyanate (R—N=C=O), so a **polyurethane** is prepared by reaction between a diol and a diisocyanate. The diol is usually a low-molecular-weight polymer (MW ≈ 1000 amu) with hydroxyl end-groups; the diisocyanate is often toluene-2,4-diisocyanate.

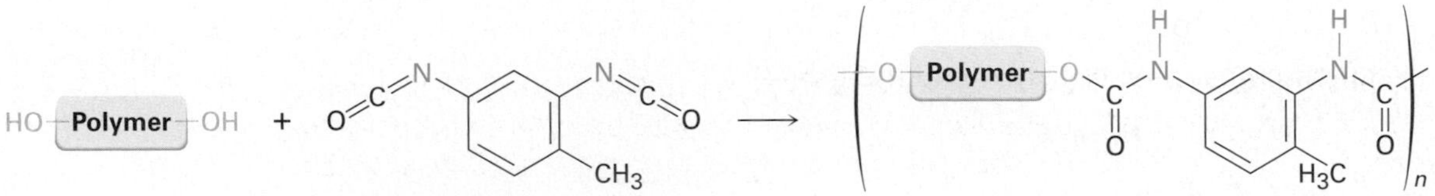

Toluene-2,4-diisocyanate **A polyurethane**

Several different kinds of polyurethanes are produced, depending on the nature of the polymeric alcohol used. One major use of polyurethane is in the stretchable spandex fibers used for bathing suits and athletic gear. These polyurethanes have a fairly low degree of cross-linking so that the resultant polymer is soft and elastic. A second major use of polyurethanes is in the foams used for insulation. Foaming occurs when a small amount of water is added during polymerization, giving a carbamic acid intermediate that spontaneously loses bubbles of CO_2.

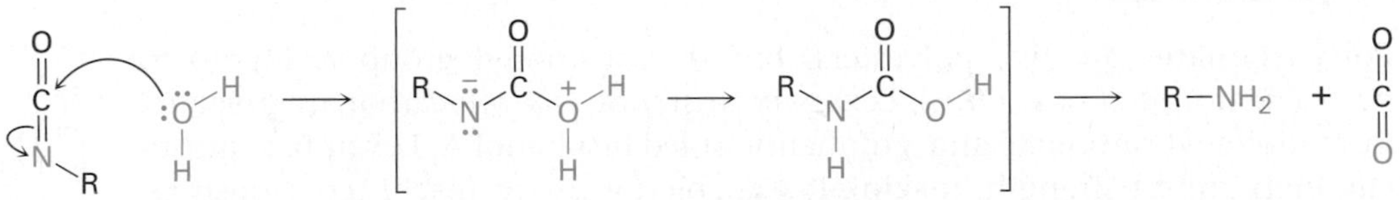

A carbamic acid

Polyurethane foams are generally made using a *poly*alcohol rather than a diol as the monomer so that the polymer has a high amount of three-dimensional cross-linking. The result is a rigid but very light foam suitable for use as thermal insulation in building construction and portable ice chests.

Problem 31.8

Poly(ethylene terephthalate), or PET, is a polyester used to make soft-drink bottles. It is prepared by reaction of ethylene glycol with 1,4-benzenedicarboxylic acid (terephthalic acid). Draw the structure of PET.

Problem 31.9

Show the mechanism of the nucleophilic addition reaction of an alcohol with an isocyanate to yield a urethane.

31.5 Olefin Metathesis Polymerization

Perhaps the most important advance in polymer synthesis in recent years has been the development of *olefin metathesis polymerization.* At its simplest, an olefin metathesis reaction is one in which two olefins (alkenes) exchange substituents on their double bonds.

An olefin metathesis reaction

R–CH=CH–R + R′–CH=CH–R′ ⇌ (Catalyst) R–CH=CH–R′ + R–CH=CH–R′

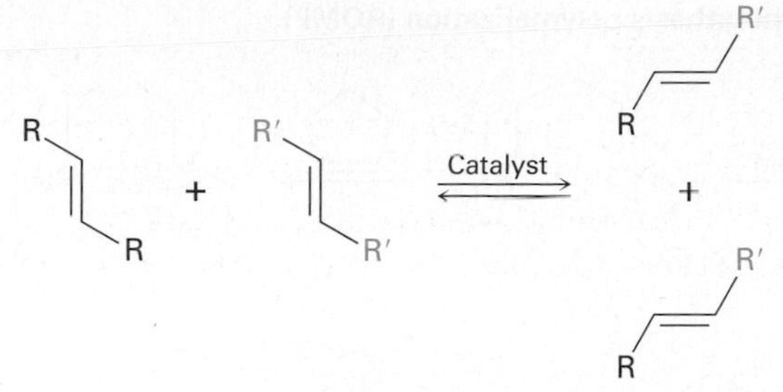

Olefin metathesis catalysts, such as the Grubbs catalyst now in common use, contain a carbon–metal double bond (usually to ruthenium, Ru) and have the general structure M=CHR. They function by reacting reversibly with an alkene to form a four-membered, metal-containing intermediate called a *metallacycle,* which immediately opens to give a different catalyst and a different alkene. The mechanism is shown in **Figure 31.2**.

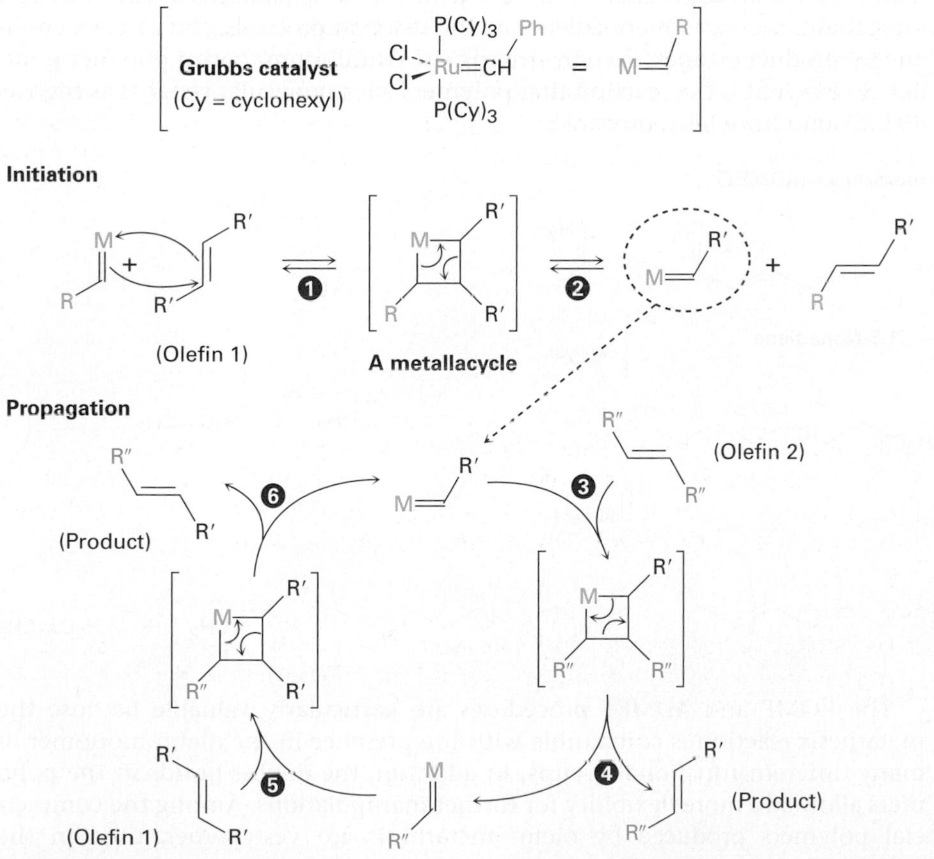

Grubbs catalyst (Cy = cyclohexyl): $Cl_2(P(Cy)_3)_2Ru{=}CHPh$ = M=CHR

Initiation

M=CHR + R′CH=CHR′ (Olefin 1) ⇌ (1) [A metallacycle] ⇌ (2) M=CHR′ + RCH=CHR′

Propagation

M=CHR′ + R″CH=CHR″ (Olefin 2) → (3) [metallacycle] → (4) R′CH=CHR″ (Product) + M=CHR″; M=CHR″ + R′CH=CHR′ (Olefin 1) → (5) [metallacycle] → (6) R″CH=CHR′ (Product) + M=CHR′

Figure 31.2 Mechanism of the olefin metathesis reaction. The process is initiated by a two-step sequence that involves (1) reaction of the catalyst and olefin 1 to give a four-membered metallacycle intermediate, followed by (2) ring-opening to give a different form of catalyst that contains part of olefin 1. (3) Reaction of this new catalyst with olefin 2 gives another metallacycle intermediate, (4) which opens to give metathesis product and another form of catalyst. (5, 6) The repeating ring-forming and ring-opening steps then continue.

There are several methods for implementing the olefin metathesis reaction to prepare polymers. One method, called *ring-opening metathesis polymerization,* or ROMP, involves use of a moderately strained cycloalkene, such as cyclopentene. The strain of the ring favors ring-opening, thereby driving formation of the open-chain product. The polymer that results has double bonds spaced regularly along the chain, allowing for either hydrogenation or further functionalization if desired.

Ring-opening metathesis polymerization (ROMP)

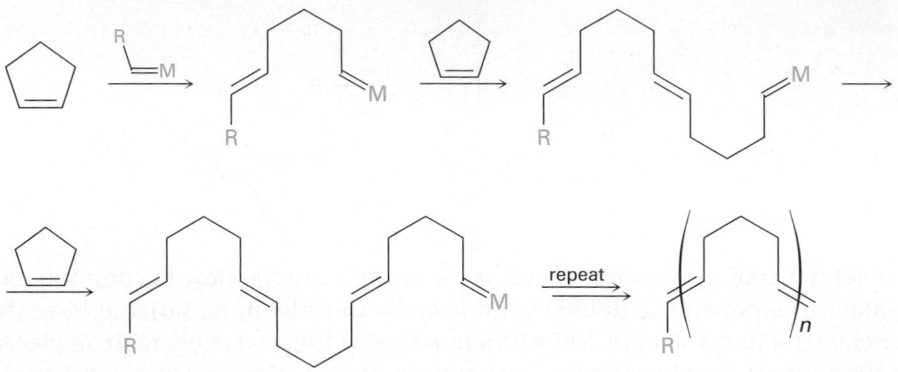

A second method of using olefin metathesis to prepare polymers is by *acyclic diene metathesis,* or ADMET. As the name suggests, ADMET involves olefin metathesis of an open-chain substrate with two double bonds at the ends of a long chain, such as 1,8-nonadiene. As the reaction proceeds, the gaseous ethylene by-product escapes, thereby driving the equilibrium toward polymer product. So efficient is the reaction that polymers with molecular weights as high as 80,000 amu have been prepared.

Acyclic diene metathesis (ADMET)

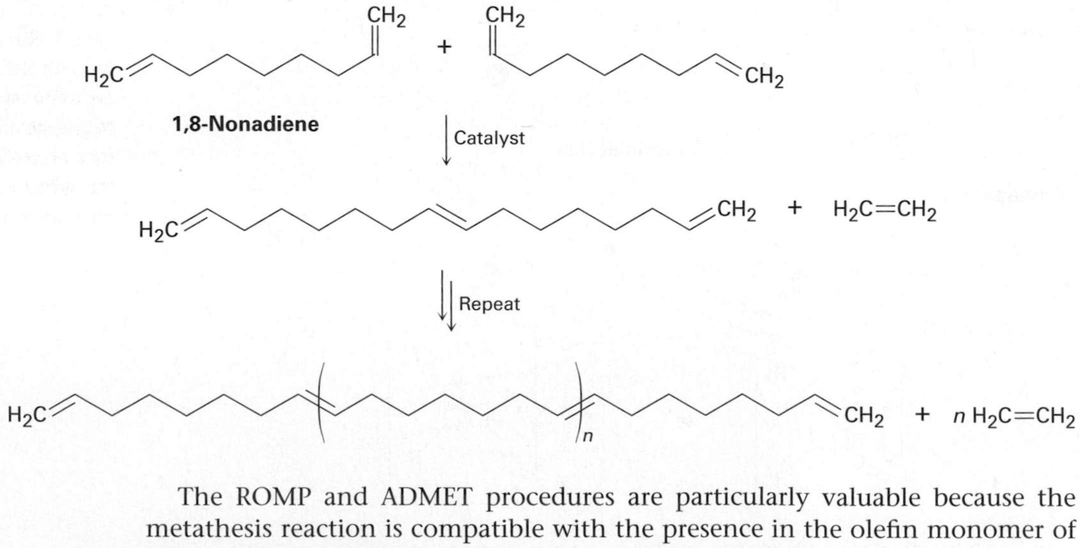

The ROMP and ADMET procedures are particularly valuable because the metathesis reaction is compatible with the presence in the olefin monomer of many different functional groups. In addition, the double bonds in the polymers allow still more flexibility for further manipulations. Among the commercial polymers produced by olefin metathesis are Vestenamer, used in the manufacture of tires and other molded rubber objects, and Norsorex, used in the automobile industry as a sealing material.

Vestenamer   Norsorex

Problem 31.10

Look at the structures of Vestenamer and Norsorex, and show how they might be made by olefin metathesis polymerization.

31.6 Polymer Structure and Physical Properties

Polymers aren't really that different from other organic molecules. They're much larger, of course, but their chemistry is similar to that of analogous small molecules. Thus, the alkane chains of polyethylene undergo radical-initiated halogenation, the aromatic rings of polystyrene undergo typical electrophilic aromatic substitution reactions, and the amide linkages of nylon are hydrolyzed by aqueous base.

The major difference between small and large organic molecules is in their physical properties. For instance, their large size means that polymers experience substantially larger van der Waals forces than do small molecules **(Section 2.12)**. But because van der Waals forces operate only at close distances, they are strongest in polymers like high-density polyethylene, in which chains can pack together closely in a regular way. Many polymers, in fact, have regions that are essentially crystalline. These regions, called **crystallites**, consist of highly ordered portions in which the zigzag polymer chains are held together by van der Waals forces **(Figure 31.3)**.

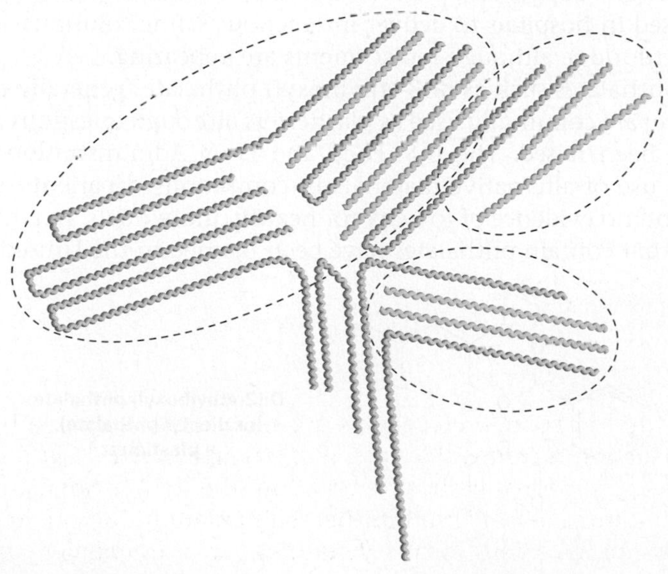

Figure 31.3 Crystallites in linear polyethylene. The long polymer chains are arranged in parallel lines in the crystallite regions.

As you might expect, polymer crystallinity is strongly affected by the steric requirements of substituent groups on the chains. Linear polyethylene is highly crystalline, but poly(methyl methacrylate) is noncrystalline because the chains can't pack closely together in a regular way. Polymers with a high degree of crystallinity are generally hard and durable. When heated, the crystalline regions melt at the **melt transition temperature**, T_m, to give an amorphous material.

Noncrystalline, amorphous polymers like poly(methyl methacrylate), sold under the trade name Plexiglas, have little or no long-range ordering among chains but can nevertheless be very hard at room temperature. When heated, the hard amorphous polymer becomes soft and flexible at a point called the **glass transition temperature**, T_g. Much of the art in polymer synthesis lies in finding methods for controlling the degree of crystallinity and the glass transition temperature, thereby imparting useful properties to the polymer.

In general, polymers can be divided into four major categories, depending on their physical behavior: *thermoplastics, fibers, elastomers,* and *thermosetting resins*. **Thermoplastics** are the polymers most people think of when the word *plastic* is mentioned. These polymers have a high T_g and are therefore hard at room temperature but become soft and viscous when heated. As a result, they can be molded into toys, beads, telephone housings, or any of a thousand other items. Because thermoplastics have little or no cross-linking, the individual chains can slip past one another in the melt. Some thermoplastic polymers, such as poly(methyl methacrylate) and polystyrene, are amorphous and noncrystalline; others, such as polyethylene and nylon, are partially crystalline. Among the better-known thermoplastics is poly(ethylene terephthalate), or PET, used for making plastic soft-drink bottles.

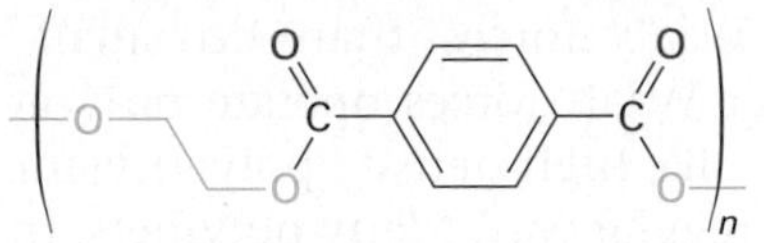

Poly(ethylene terephthalate)

Plasticizers—small organic molecules that act as lubricants between chains—are usually added to thermoplastics to keep them from becoming brittle at room temperature. An example is poly(vinyl chloride), which is brittle when pure but becomes supple and pliable when a plasticizer is added. In fact, most drip bags used in hospitals to deliver intravenous saline solutions are made of poly(vinyl chloride), although replacements are appearing.

Dialkyl phthalates such as di(2-ethylhexyl) phthalate (generally called dioctyl phthalate) are commonly used as plasticizers although questions about their safety have been raised. The U.S. Food and Drug Administration (FDA) has advised the use of alternative materials in compromised patients and infants but has found no evidence of toxicity for healthy individuals. In addition, children's toys that contain phthalates have been banned in the United States.

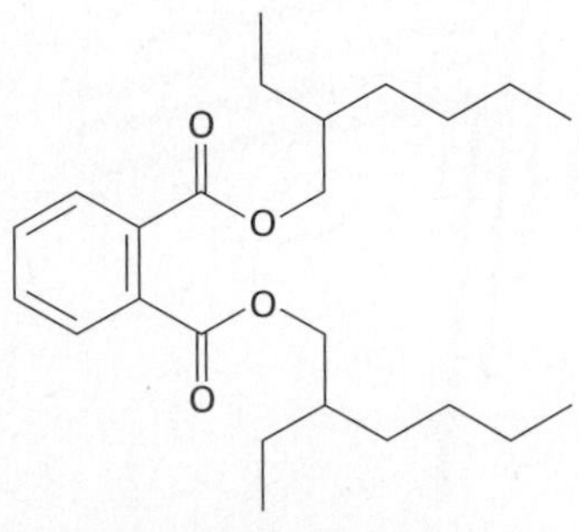

Di(2-ethylhexyl) phthalate (or dioctyl phthalate), a plasticizer

Fibers are thin threads produced by extruding a molten polymer through small holes in a die, or spinneret. The fibers are then cooled and drawn out, which orients the crystallite regions along the axis of the fiber and adds considerable tensile strength **(Figure 31.4)**. Nylon, Dacron, and polyethylene all have the semicrystalline structure necessary for drawing into oriented fibers.

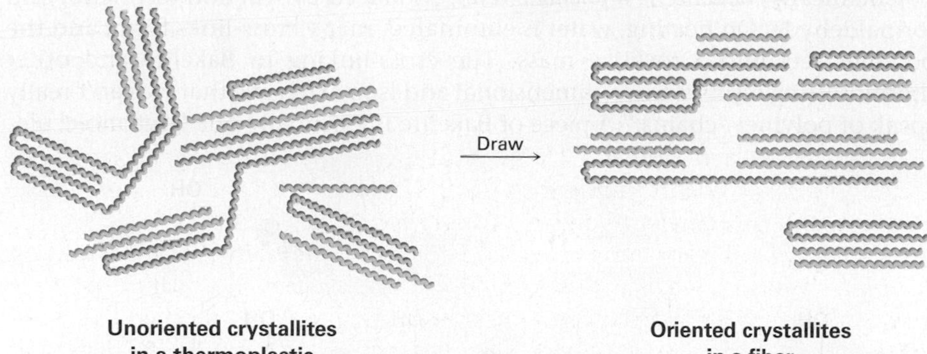

Figure 31.4 Oriented crystallite regions in a polymer fiber.

Elastomers are amorphous polymers that have the ability to stretch out and spring back to their original shapes. These polymers must have low T_g values and a small amount of cross-linking to prevent the chains from slipping over one another. In addition, the chains must have an irregular shape to prevent crystallite formation. When stretched, the randomly coiled chains straighten out and orient along the direction of the pull. Van der Waals forces are too weak and too few to maintain this orientation, however, and the elastomer therefore reverts to its random coiled state when the stretching force is released **(Figure 31.5)**.

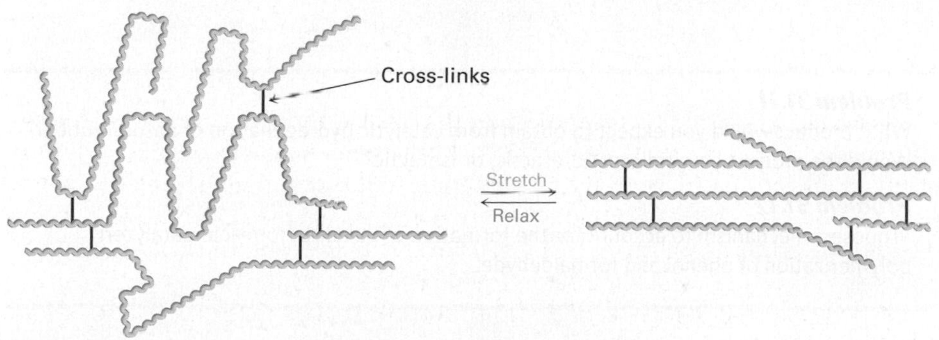

Figure 31.5 Unstretched and stretched forms of an elastomer.

Natural rubber **(Section 14.6)** is the most common example of an elastomer. Rubber has the long chains and occasional cross-links needed for elasticity, but its irregular geometry prevents close packing of the chains into crystallites. Gutta-percha, by contrast, is highly crystalline and is not an elastomer **(Figure 31.6)**.

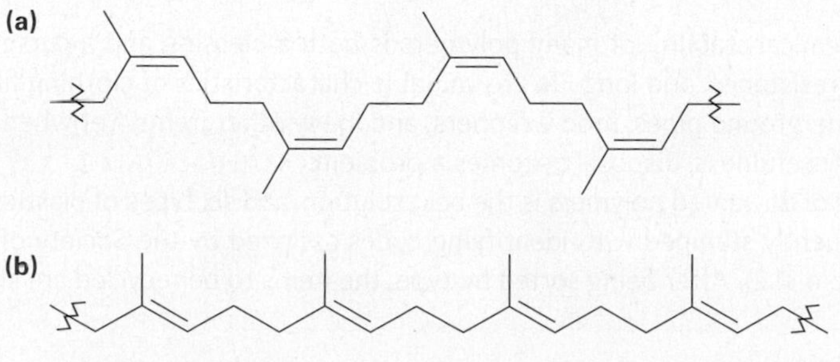

(a)

(b)

Figure 31.6 **(a)** Natural rubber is elastic and noncrystalline because of its cis double-bond geometry, but **(b)** gutta-percha is nonelastic and crystalline because its geometry allows for better packing together of chains.

Thermosetting resins are polymers that become highly cross-linked and solidify into a hard, insoluble mass when heated. *Bakelite,* a thermosetting resin first produced in 1907, has been in commercial use longer than any other synthetic polymer. It is widely used for molded parts, adhesives, coatings, and even high-temperature applications such as missile nose cones.

Chemically, Bakelite is a *phenolic resin,* produced by reaction of phenol and formaldehyde. On heating, water is eliminated, many cross-links form, and the polymer sets into a rocklike mass. The cross-linking in Bakelite and other thermosetting resins is three-dimensional and is so extensive that we can't really speak of polymer "chains." A piece of Bakelite is essentially one large molecule.

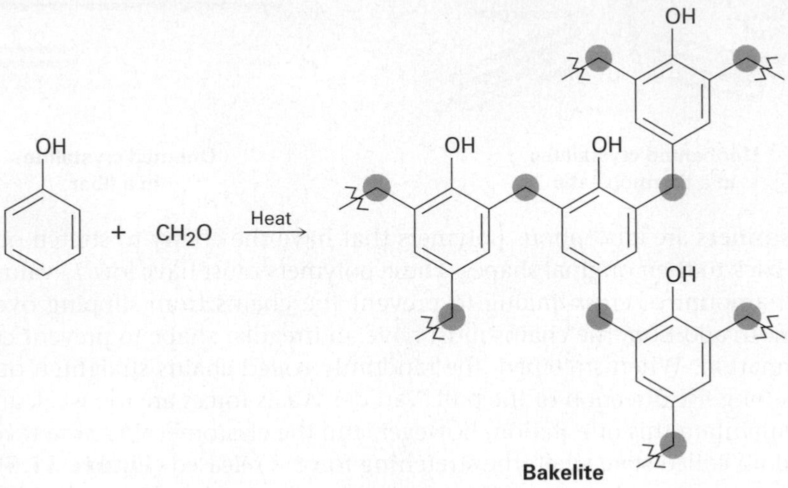

OH + CH_2O Heat → OH OH OH OH

Bakelite

Problem 31.11

What product would you expect to obtain from catalytic hydrogenation of natural rubber? Would the product be syndiotactic, atactic, or isotactic?

Problem 31.12

Propose a mechanism to account for the formation of Bakelite from acid-catalyzed polymerization of phenol and formaldehyde.

Biodegradable Polymers | A DEEPER LOOK

The high chemical stability of many polymers is both a blessing and a curse. Heat resistance, wear resistance, and long life are valuable characteristics of clothing fibers, bicycle helmets, underground pipes, food wrappers, and many other items. Yet when those items outlive their usefulness, disposal becomes a problem.

Recycling of unwanted polymers is the best solution, and six types of plastics in common use are frequently stamped with identifying codes assigned by the Society of the Plastics Industry (Table 31.2). After being sorted by type, the items to be recycled are shredded into

(continued)

(continued)

small chips, washed, dried, and melted for reuse. Soft-drink bottles, for instance, are made from recycled poly(ethylene terephthalate), trash bags are made from recycled low-density polyethylene, and garden furniture is made from recycled polypropylene and mixed plastics.

What happens to the plastics that end up here?

Table 31.2 Recyclable Plastics

Polymer	Recycling code	Use
Poly(ethylene terephthalate)	1—PET	Soft-drink bottles
High-density polyethylene	2—HDPE	Bottles
Poly(vinyl chloride)	3—V	Floor mats
Low-density polyethylene	4—LDPE	Grocery bags
Polypropylene	5—PP	Furniture
Polystyrene	6—PS	Molded articles
Mixed plastics	7	Benches, plastic lumber

Frequently, however, plastics are simply thrown away rather than recycled, and much work has therefore been carried out on developing *biodegradable* polymers, which can be broken down rapidly by soil microorganisms. Among the most common biodegradable polymers are polyglycolic acid (PGA), polylactic acid (PLA), and polyhydroxybutyrate (PHB). All are polyesters and are therefore susceptible to hydrolysis of their ester links. Copolymers of PGA with PLA have found a particularly wide range of uses. A 90/10 copolymer of polyglycolic acid with polylactic acid is used to make absorbable sutures that are degraded and absorbed by the body within 90 days after surgery.

$$\mathrm{HOCH_2\overset{O}{\overset{\|}{C}}OH} \qquad \mathrm{HO\underset{CH_3}{\underset{|}{C}}H\overset{O}{\overset{\|}{C}}OH} \qquad \mathrm{HO\underset{CH_3}{\underset{|}{C}}HCH_2\overset{O}{\overset{\|}{C}}OH}$$

Glycolic acid **Lactic acid** **3-Hydroxybutyric acid**

$$\downarrow \text{Heat} \qquad \downarrow \text{Heat} \qquad \downarrow \text{Heat}$$

$$\left(\mathrm{OCH_2\overset{O}{\overset{\|}{C}}}\right)_n \qquad \left(\mathrm{O\underset{CH_3}{\underset{|}{C}}H\overset{O}{\overset{\|}{C}}}\right)_n \qquad \left(\mathrm{O\underset{CH_3}{\underset{|}{C}}HCH_2\overset{O}{\overset{\|}{C}}}\right)_n$$

Poly(glycolic acid) **Poly(lactic acid)** **Poly(hydroxybutyrate)**

Summary

Key words

atactic, 1245
block copolymer, 1248
copolymer, 1246
crystallite, 1253
elastomer, 1255
fiber, 1255
glass transition temperature (T_g), 1254
graft copolymer, 1248
homopolymer, 1246
isotactic, 1245
melt transition temperature (T_m), 1254
monomer, 1242
plasticizer, 1254
polycarbonate, 1249
polymer, 1242
polyurethane, 1250
syndiotactic, 1245
thermoplastic, 1254
thermosetting resin, 1256
Ziegler–Natta catalyst, 1245

Synthetic polymers can be classified as either chain-growth or step-growth. Chain-growth polymers are prepared by chain-reaction polymerization of vinyl monomers in the presence of a radical, an anion, or a cation initiator. Radical polymerization is sometimes used, but alkenes such as 2-methylpropene that have electron-donating substituents on the double bond polymerize easily by a cationic route through carbocation intermediates. Similarly, monomers such as methyl α-cyanoacrylate that have electron-withdrawing substituents on the double bond polymerize by an anionic, conjugate addition pathway.

Copolymerization of two monomers gives a product with properties different from those of either homopolymer. **Graft copolymers** and **block copolymers** are two examples.

Alkene polymerization can be carried out in a controlled manner using a **Ziegler–Natta catalyst**. Ziegler–Natta polymerization minimizes the amount of chain branching in the polymer and leads to stereoregular chains—either **isotactic** (substituents on the same side of the chain) or **syndiotactic** (substituents on alternate sides of the chain), rather than **atactic** (substituents randomly disposed).

Step-growth polymers, the second major class of polymers, are prepared by reactions between difunctional molecules, with the individual bonds in the polymer formed independently of one another. **Polycarbonates** are formed from a diester and a diol, and **polyurethanes** are formed from a diisocyanate and a diol.

The chemistry of synthetic polymers is similar to the chemistry of small molecules with the same functional groups, but the physical properties of polymers are greatly affected by size. Polymers can be classified by physical property into four groups: **thermoplastics**, **fibers**, **elastomers**, and **thermosetting resins**. The properties of each group can be accounted for by the structure, the degree of crystallinity, and the amount of cross-linking they contain.

Exercises

OWL Interactive versions of these problems are assignable in OWL for Organic Chemistry.

Visualizing Chemistry

(Problems 31.1–31.12 appear within the chapter.)

31.13 Identify the structural class to which the following polymer belongs, and show the structure of the monomer units used to make it:

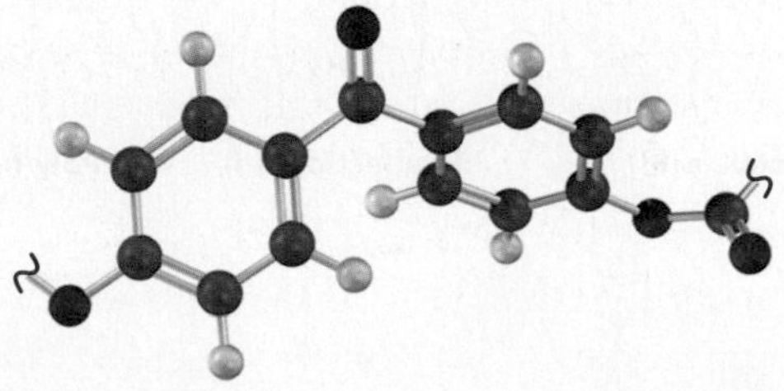

31.14 Show the structures of the polymers that could be made from the following monomers (green = Cl):

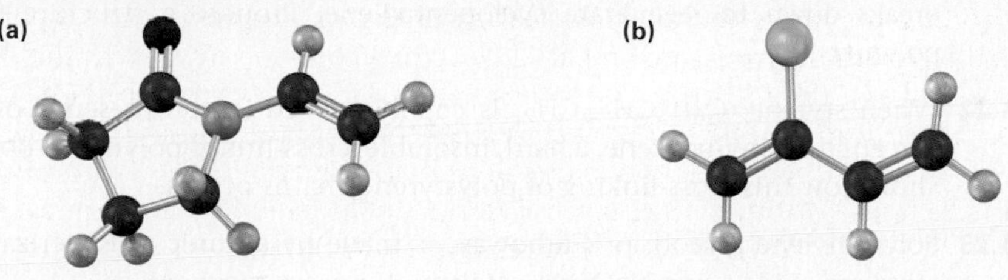

Additional Problems

31.15 Identify the monomer units from which each of the following polymers is made, and tell whether each is a chain-growth or a step-growth polymer:

(a) $-\!\!(\!-CH_2-O-\!)\!_n$ (b) $-\!\!(\!-CF_2-CFCl-\!)\!_n$ (c) $\left(-NHCH_2CH_2CH_2\overset{O}{\overset{\|}{C}}-\right)_n$

(d) $\left(-O-C_6H_4-\overset{O}{\overset{\|}{C}}-\right)_n$ (e) $\left(-O-C_6H_{10}-O-\overset{O}{\overset{\|}{C}}-\right)_n$

31.16 Draw a three-dimensional representation of segments of the following polymers:

(a) Syndiotactic polyacrylonitrile (b) Atactic poly(methyl methacrylate)
(c) Isotactic poly(vinyl chloride)

31.17 Draw the structure of Kodel, a polyester prepared by heating dimethyl 1,4-benzenedicarboxylate with 1,4-bis(hydroxymethyl)cyclohexane.

$HOCH_2-C_6H_{10}-CH_2OH$ **1,4-Bis(hydroxymethyl)cyclohexane**

31.18 Show the structure of the polymer that results from heating the following diepoxide and diamine:

[diepoxide structure] + $H_2N-C_6H_4-NH_2 \xrightarrow{\text{Heat}}$?

31.19 Nomex, a polyamide used in such applications as fire-retardant clothing, is prepared by reaction of 1,3-benzenediamine with 1,3-benzenedicarbonyl chloride. Show the structure of Nomex.

31.20 Nylon 10,10 is an extremely tough, strong polymer used to make reinforcing rods for concrete. Draw a segment of nylon 10,10, and show its monomer units.

31.21 1,3-Cyclopentadiene undergoes thermal polymerization to yield a polymer that has no double bonds in the chain. On strong heating, the polymer breaks down to regenerate cyclopentadiene. Propose a structure for the polymer.

31.22 When styrene, $C_6H_5CH{=}CH_2$, is copolymerized in the presence of a few percent *p*-divinylbenzene, a hard, insoluble, cross-linked polymer is obtained. Show how this cross-linking of polystyrene chains occurs.

31.23 Poly(ethylene glycol), or Carbowax, is made by anionic polymerization of ethylene oxide using NaOH as catalyst. Propose a mechanism.

$+O{-}CH_2CH_2+_n$ **Poly(ethylene glycol)**

31.24 Nitroethylene, $H_2C{=}CHNO_2$, is a sensitive compound that must be prepared with great care. Attempted purification of nitroethylene by distillation often results in low recovery of product and a white coating on the inner walls of the distillation apparatus. Explain.

31.25 Poly(vinyl butyral) is used as the plastic laminate in the preparation of automobile windshield safety glass. How would you synthesize this polymer?

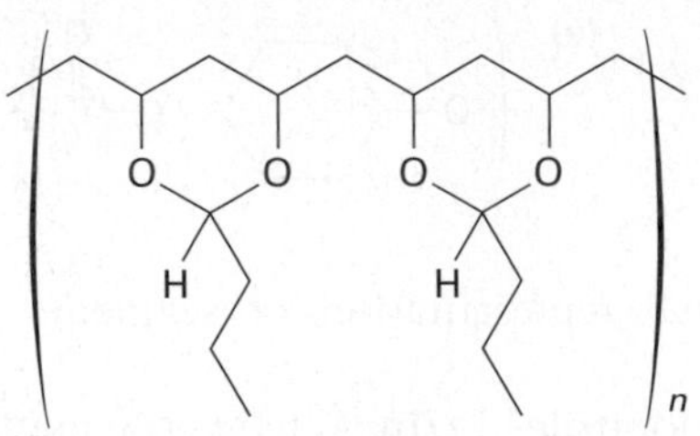

Poly(vinyl butyral)

31.26 What is the structure of the polymer produced by anionic polymerization of *β*-propiolactone using NaOH as catalyst?

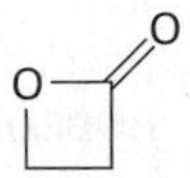

***β*-Propiolactone**

31.27 Glyptal is a highly cross-linked thermosetting resin produced by heating glycerol and phthalic anhydride (1,2-benzenedicarboxylic acid anhydride). Show the structure of a representative segment of glyptal.

31.28 Melmac, a thermosetting resin often used to make plastic dishes, is prepared by heating melamine with formaldehyde. Look at the structure of Bakelite shown in Section 31.6, and then propose a structure for Melmac.

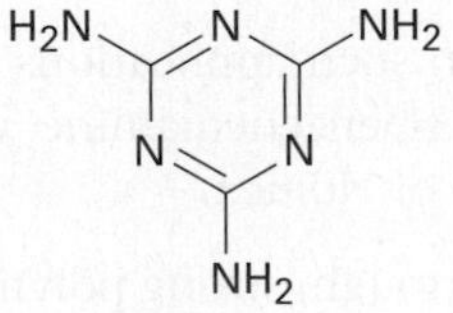

Melamine

31.29 Epoxy adhesives are cross-linked resins prepared in two steps. The first step involves S_N2 reaction of the disodium salt of bisphenol A with epichlorohydrin to form a low-molecular-weight prepolymer. This prepolymer is then "cured" into a cross-linked resin by treatment with a triamine such as $H_2NCH_2CH_2NHCH_2CH_2NH_2$.

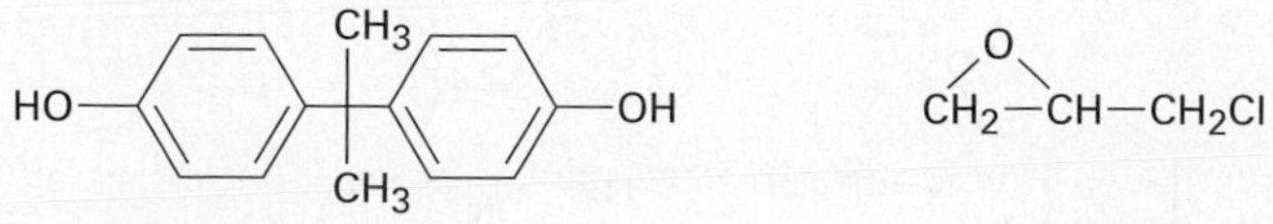

HO–C₆H₄–C(CH₃)₂–C₆H₄–OH   **Bisphenol A**

CH_2–CH–CH_2Cl (epoxide O bridging CH_2 and CH)   **Epichlorohydrin**

(a) What is the structure of the prepolymer?

(b) How does addition of the triamine to the prepolymer result in cross-linking?

31.30 The polyurethane foam used for home insulation uses methanediphenyldiisocyanate (MDI) as monomer. The MDI is prepared by acid-catalyzed reaction of aniline with formaldehyde, followed by treatment with phosgene, $COCl_2$. Propose mechanisms for both steps.

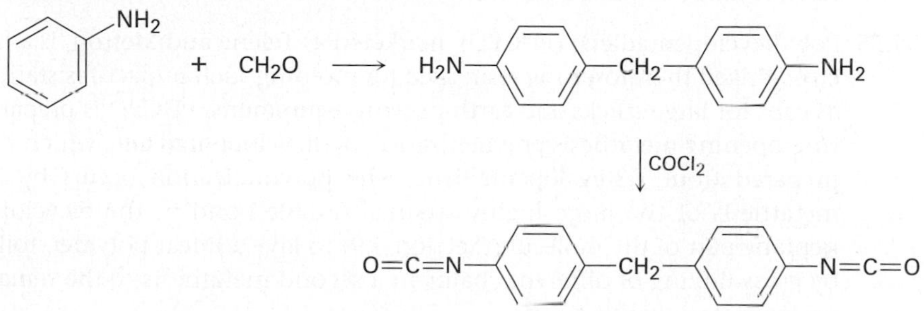

$$C_6H_5NH_2 + CH_2O \longrightarrow H_2N{-}C_6H_4{-}CH_2{-}C_6H_4{-}NH_2 \xrightarrow{COCl_2} O{=}C{=}N{-}C_6H_4{-}CH_2{-}C_6H_4{-}N{=}C{=}O$$

MDI

31.31 Write the structure of a representative segment of polyurethane prepared by reaction of ethylene glycol with MDI (Problem 31.30).

31.32 The smoking salons of the Hindenburg and other hydrogen-filled dirigibles of the 1930s were insulated with urea–formaldehyde polymer foams. The structure of this polymer is highly cross-linked, like that of Bakelite (Section 31.6). Propose a structure.

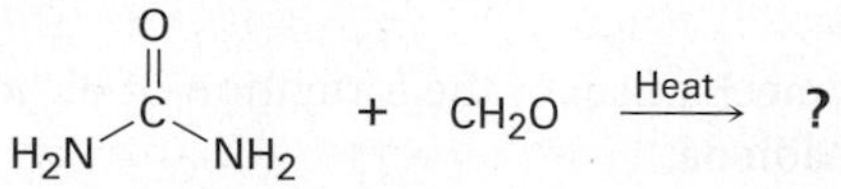

$$H_2N{-}C({=}O){-}NH_2 + CH_2O \xrightarrow{\text{Heat}} ?$$

31.33 The polymeric resin used for Merrifield solid-phase peptide synthesis (Section 26.8) is prepared by treating polystyrene with *N*-(hydroxymethyl) phthalimide and trifluoromethanesulfonic acid, followed by reaction with hydrazine. Propose a mechanism for both steps.

N—CH_2OH

CF_3SO_3H

CH_2

H_2NNH_2

CH_2NH_2

—CH_2—CH—$_n$

Polystyrene

31.34 2-Ethyl-1-hexanol, used in the synthesis of di(2-ethylhexyl) phthalate plasticizer, is made commercially from butanal. Show the likely synthesis route.

31.35 Polydicyclopentadiene (PDCPD), marketed as Telene and Metton, is a highly cross-linked thermosetting resin used for molding such impact-resistant parts as cabs for large trucks and earth-moving equipment. PDCPD is prepared by ring-opening metathesis polymerization of dicyclopentadiene, which is itself prepared from 1,3-cyclopentadiene. The polymerization occurs by initial metathesis of the more highly strained double bond in the bicyclo[2.2.1] heptane part of the molecule (Section 4.9) to give a linear polymer, followed by cross-linking of different chains in a second metathesis of the remaining cyclopentene double bond.

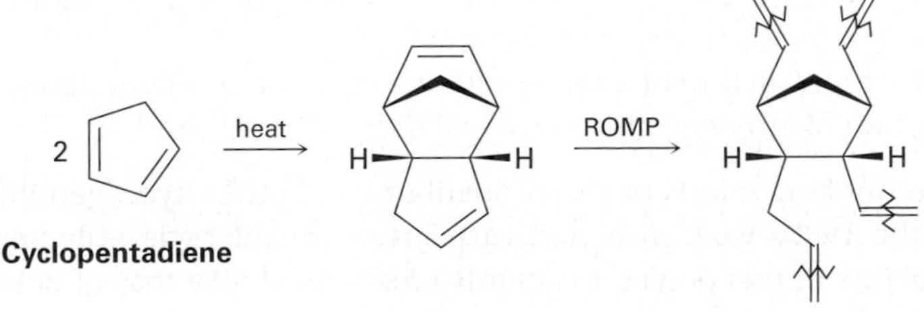

Dicyclopentadiene **Polydicyclopentadiene**

(a) Show the mechanism of the formation of dicyclopentadiene from cyclopentadiene.

(b) Draw the structure of a representative sample of the initially formed linear polymer containing three monomer units.

(c) Draw the structure of a representative sample of PDCPD that shows how cross-linking of the linear chains takes place.

APPENDIX A

Nomenclature of Polyfunctional Organic Compounds

With more than 40 million organic compounds now known and thousands more being created daily, naming them all is a real problem. Part of the problem is due to the sheer complexity of organic structures, but part is also due to the fact that chemical names have more than one purpose. For Chemical Abstracts Service (CAS), which catalogs and indexes the worldwide chemical literature, each compound must have only one correct name. It would be chaos if half the entries for CH_3Br were indexed under "M" for methyl bromide and half under "B" for bromomethane. Furthermore, a CAS name must be strictly systematic so that it can be assigned and interpreted by computers; common names are not allowed.

People, however, have different requirements than computers. For people—which is to say students and professional chemists in their spoken and written communications—it's best that a chemical name be pronounceable and that it be as easy as possible to assign and interpret. Furthermore, it's convenient if names follow historical precedents, even if that means a particularly well-known compound might have more than one name. People can readily understand that bromomethane and methyl bromide both refer to CH_3Br.

As noted in the text, chemists overwhelmingly use the nomenclature system devised and maintained by the International Union of Pure and Applied Chemistry, or IUPAC. Rules for naming monofunctional compounds were given throughout the text as each new functional group was introduced, and a list of where these rules can be found is given in Table A.1.

Table A.1 Nomenclature Rules for Functional Groups

Functional group	Text section	Functional group	Text section
Acid anhydrides	21.1	Aromatic compounds	15.1
Acid halides	21.1	Carboxylic acids	20.1
Acyl phosphates	21.1	Cycloalkanes	4.1
Alcohols	17.1	Esters	21.1
Aldehydes	19.1	Ethers	18.1
Alkanes	3.4	Ketones	19.1
Alkenes	7.3	Nitriles	20.1
Alkyl halides	10.1	Phenols	17.1
Alkynes	9.1	Sulfides	18.8
Amides	21.1	Thiols	18.8
Amines	24.1	Thioesters	21.1

Naming a monofunctional compound is reasonably straightforward, but even experienced chemists often encounter problems when faced with naming a complex polyfunctional compound. Take the following compound, for instance. It has three functional groups, ester, ketone, and C=C, but how should it be named? As an ester with an *-oate* ending, a ketone with an *-one* ending, or an alkene with an *-ene* ending? It's actually named methyl 3-(2-oxo-6-cyclohexenyl)propanoate.

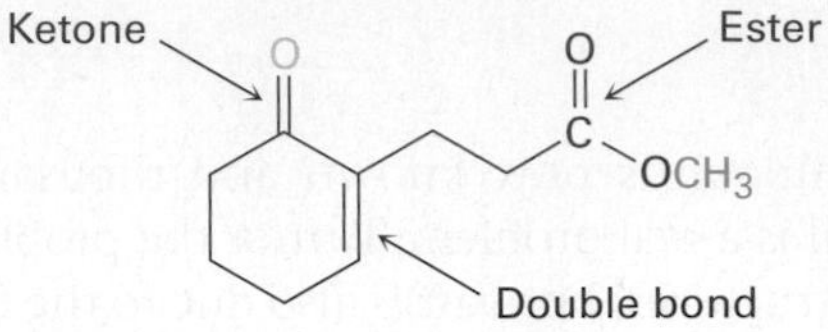

Ketone Ester Double bond

Methyl 3-(2-oxo-6-cylohexenyl)propanoate

The name of a polyfunctional organic molecule has four parts—suffix, parent, prefixes, and locants—which must be identified and expressed in the proper order and format. Let's look at each of the four.

Name Part 1. The Suffix: Functional-Group Precedence

Although a polyfunctional organic molecule might contain several different functional groups, we must choose just one suffix for nomenclature purposes. It's not correct to use two suffixes. Thus, keto ester **1** must be named either as a ketone with an *-one* suffix or as an ester with an *-oate* suffix, but it can't be named as an *-onoate*. Similarly, amino alcohol **2** must be named either as an alcohol (*-ol*) or as an amine (*-amine*), but it can't be named as an *-olamine* or *-aminol*.

1. $CH_3C(=O)CH_2CH_2C(=O)OCH_3$ **2.** $CH_3CH(OH)CH_2CH_2CH_2NH_2$

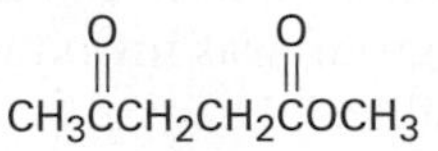

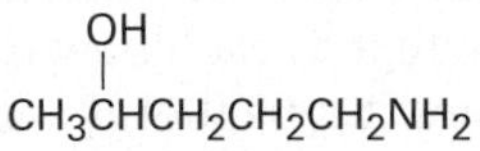

The only exception to the rule requiring a single suffix is when naming compounds that have double or triple bonds. Thus, the unsaturated acid $H_2C{=}CHCH_2CO_2H$ is 3-butenoic acid, and the acetylenic alcohol $HC{\equiv}CCH_2CH_2CH_2OH$ is 5-pentyn-1-ol.

How do we choose which suffix to use? Functional groups are divided into two classes, **principal groups** and **subordinate groups**, as shown in Table A.2. Principal groups can be cited either as prefixes or as suffixes, while subordinate groups are cited only as prefixes. Within the principal groups, an order of priority has been established, with the proper suffix for a given compound determined by choosing the principal group of highest priority. For example, Table A.2 indicates that keto ester **1** should be named as an ester rather than as a ketone because an ester functional group is higher in priority than a ketone. Similarly, amino alcohol **2** should be named as an alcohol rather than as an amine.

Table A.2 Classification of Functional Groups[a]

Functional group	Name as suffix	Name as prefix
Principal groups		
Carboxylic acids	-oic acid -carboxylic acid	carboxy
Acid anhydrides	-oic anhydride -carboxylic anhydride	—
Esters	-oate -carboxylate	alkoxycarbonyl
Thioesters	-thioate -carbothioate	alkylthiocarbonyl
Acid halides	-oyl halide -carbonyl halide	halocarbonyl
Amides	-amide -carboxamide	carbamoyl
Nitriles	-nitrile -carbonitrile	cyano
Aldehydes	-al -carbaldehyde	oxo
Ketones	-one	oxo
Alcohols	-ol	hydroxy
Phenols	-ol	hydroxy
Thiols	-thiol	mercapto
Amines	-amine	amino
Imines	-imine	imino
Ethers	ether	alkoxy
Sulfides	sulfide	alkylthio
Disulfides	disulfide	—
Alkenes	-ene	—
Alkynes	-yne	—
Alkanes	-ane	—
Subordinate groups		
Azides	—	azido
Halides	—	halo
Nitro compounds	—	nitro

[a]Principal groups are listed in order of decreasing priority; subordinate groups have no priority order.

Thus, the name of **1** is methyl 4-oxopentanoate and the name of **2** is 5-amino-2-pentanol. Further examples are shown:

$CH_3C(=O)CH_2CH_2C(=O)OCH_3$

1. Methyl 4-oxopentanoate
(an ester with a ketone group)

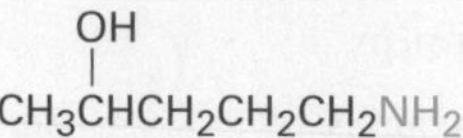

2. 5-Amino-2-pentanol
(an alcohol with an amine group)

$CH_3CH(CHO)CH_2CH_2CH_2C(=O)OCH_3$

3. Methyl 5-methyl-6-oxohexanoate
(an ester with an aldehyde group)

$H_2NC(=O)CH_2CH(OH)CH_2CH_2C(=O)OH$

4. 5-Carbamoyl-4-hydroxypentanoic acid
(a carboxylic acid with amide and alcohol groups)

(cyclohexanone ring with CHO substituent at C3)

5. 3-Oxocyclohexanecarbaldehyde
(an aldehyde with a ketone group)

Name Part 2. The Parent: Selecting the Main Chain or Ring

The parent, or base, name of a polyfunctional organic compound is usually easy to identify. If the principal group of highest priority is part of an open chain, the parent name is that of the longest chain containing the largest number of principal groups. For example, compounds **6** and **7** are isomeric aldehydo amides, which must be named as amides rather than as aldehydes according to Table A.2. The longest chain in compound **6** has six carbons, and the substance is named 5-methyl-6-oxohexanamide. Compound **7** also has a chain of six carbons, but the longest chain that contains both principal functional groups has only four carbons. Thus, compound **7** is named 4-oxo-3-propylbutanamide.

$HC(=O)CH(CH_3)CH_2CH_2CH_2C(=O)NH_2$

6. 5-Methyl-6-oxohexanamide

$CH_3CH_2CH_2CH(CHO)CH_2C(=O)NH_2$

7. 4-Oxo-3-propylbutanamide

If the highest-priority principal group is attached to a ring, the parent name is that of the ring system. Compounds **8** and **9**, for instance, are isomeric keto nitriles and must both be named as nitriles according to Table A.2. Substance **8** is named as a benzonitrile because the –CN functional group is a substituent on the aromatic ring, but substance **9** is named as an acetonitrile because the –CN functional group is on an open chain. The names are 2-acetyl-(4-bromomethyl)benzonitrile (**8**) and (2-acetyl-4-bromophenyl)acetonitrile (**9**). As further examples, compounds **10** and **11** are both keto acids and must be named as acids, but the parent name in **10** is that of a ring system (cyclohexanecarboxylic acid) and the parent name in **11** is that of an open chain (propanoic acid). The names are *trans*-2-(3-oxopropyl)cyclohexanecarboxylic acid (**10**) and 3-(2-oxocyclohexyl)propanoic acid (**11**).

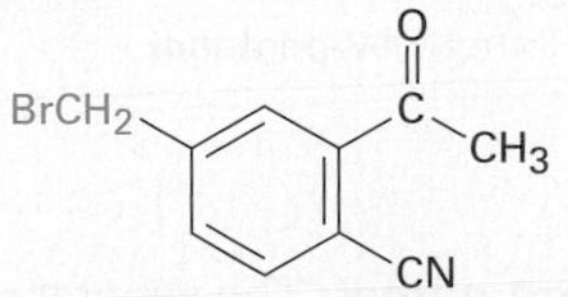

8. 2-Acetyl-(4-bromomethyl)benzonitrile

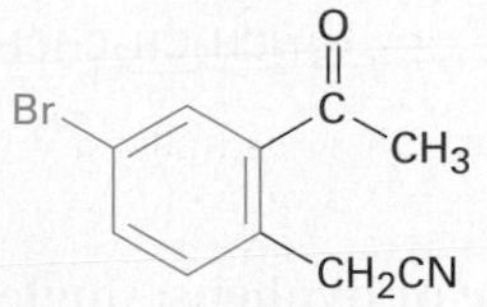

9. (2-Acetyl-4-bromophenyl)acetonitrile

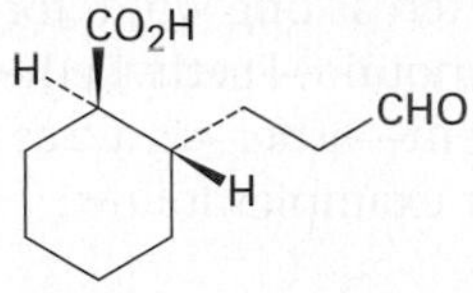

10. *trans*-2-(3-oxopropyl)cyclohexanecarboxylic acid

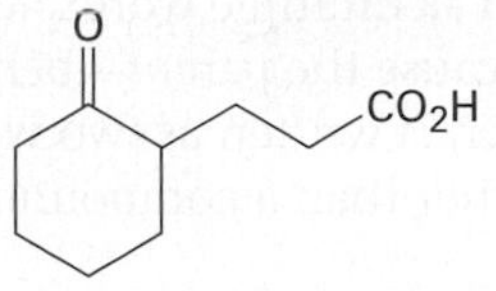

11. 3-(2-Oxocyclohexyl)propanoic acid

Name Parts 3 and 4. The Prefixes and Locants

With the parent name and the suffix established, the next step is to identify and give numbers, or *locants,* to all substituents on the parent chain or ring. The substituents include all alkyl groups and all functional groups other than the one cited in the suffix. For example, compound **12** contains three different functional groups (carboxyl, keto, and double bond). Because the carboxyl group is highest in priority and the longest chain containing the functional groups has seven carbons, compound **12** is a heptenoic acid. In addition, the parent chain has a keto (oxo) substituent and three methyl groups. Numbering from the end nearer the highest-priority functional group gives the name (*E*)-2,5,5-trimethyl-4-oxo-2-heptenoic acid. Look back at some of the other compounds we've named to see other examples of how prefixes and locants are assigned.

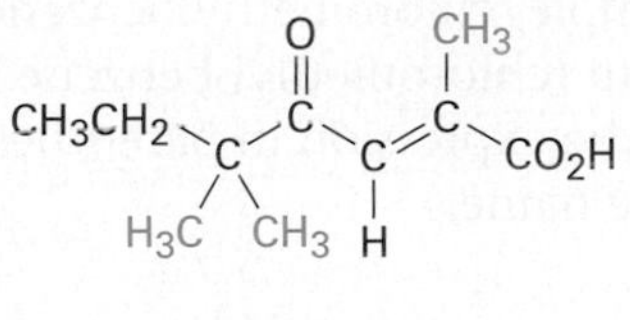

12. (*E*)-2,5,5-Trimethyl-4-oxo-2-heptenoic acid

Writing the Name

With the name parts established, the entire name is then written out. Several additional rules apply:

1. **Order of prefixes.** When the substituents have been identified, the parent chain has been numbered, and the proper multipliers such as *di-* and *tri-* have been assigned, the name is written with the substituents listed in alphabetical, rather than numerical, order. Multipliers such as *di-* and

tri- are not used for alphabetization, but the italicized prefixes *iso-* and *sec- are* used.

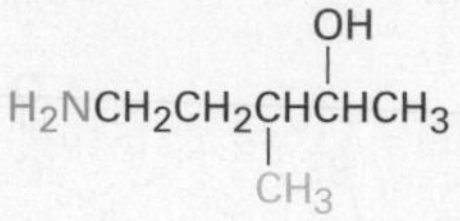

$$H_2NCH_2CH_2CH(CH_3)CH(OH)CH_3$$

13. 5-Amino-3-methyl-**2-pentanol**

2. **Use of hyphens; single- and multiple-word names.** The general rule is to determine whether the parent is itself an element or compound. If it is, then the name is written as a single word; if it isn't, then the name is written as multiple words. Methylbenzene is written as one word, for instance, because the parent—benzene—is itself a compound. Diethyl ether, however, is written as two words because the parent—ether—is a class name rather than a compound name. Some further examples follow:

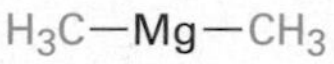

$$H_3C{-}Mg{-}CH_3$$

14. Dimethyl**magnesium** (one word, because magnesium is an element)

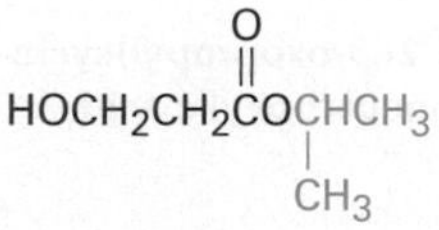

$$HOCH_2CH_2COOCH(CH_3)CH_3$$

15. Isopropyl **3-hydroxypropanoate** (two words, because "propanoate" is not a compound)

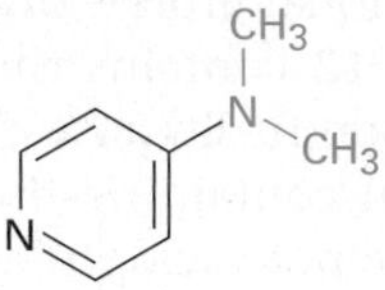

16. 4-(Dimethylamino)**pyridine** (one word, because pyridine is a compound)

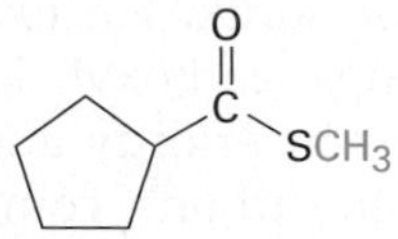

17. Methyl **cyclopentanecarbothioate** (two words, because "cyclopentanecarbothioate" is not a compound)

3. **Parentheses.** Parentheses are used to denote complex substituents when ambiguity would otherwise arise. For example, chloromethylbenzene has two substituents on a benzene ring, but (chloromethyl)benzene has only one complex substituent. Note that the expression in parentheses is not set off by hyphens from the rest of the name.

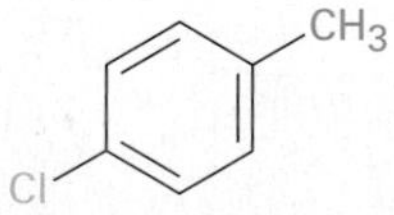

18. *p*-Chloromethyl**benzene**

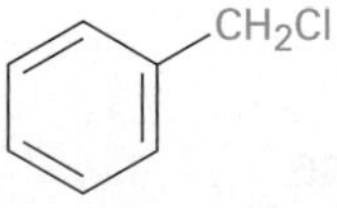

19. (Chloromethyl)**benzene**

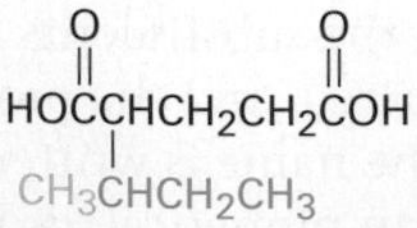

$$HOOCCH(CH(CH_3)CH_2CH_3)CH_2CH_2COOH$$

20. 2-(1-Methylpropyl)**pentanedioic acid**

Additional Reading

Further explanations of the rules of organic nomenclature can be found online at http://www.acdlabs.com/iupac/nomenclature/ (accessed September 2010) and in the following references:

1. "A Guide to IUPAC Nomenclature of Organic Compounds," CRC Press, Boca Raton, FL, 1993.
2. "Nomenclature of Organic Chemistry, Sections A, B, C, D, E, F, and H," International Union of Pure and Applied Chemistry, Pergamon Press, Oxford, 1979.

APPENDIX B

Acidity Constants for Some Organic Compounds

Compound	pK_a
CH_3SO_3H	−1.8
$CH(NO_2)_3$	0.1
NO_2 / O_2N—(ring)—OH / NO_2	0.3
CCl_3CO_2H	0.5
CF_3CO_2H	0.5
CBr_3CO_2H	0.7
$HO_2CC{\equiv}CCO_2H$	1.2; 2.5
HO_2CCO_2H	1.2; 3.7
$CHCl_2CO_2H$	1.3
$CH_2(NO_2)CO_2H$	1.3
$HC{\equiv}CCO_2H$	1.9
(Z) $HO_2CCH{=}CHCO_2H$	1.9; 6.3
(ring) CO_2H / NO_2	2.4
CH_3COCO_2H	2.4
$NCCH_2CO_2H$	2.5
$CH_3C{\equiv}CCO_2H$	2.6
CH_2FCO_2H	2.7
CH_2ClCO_2H	2.8
$HO_2CCH_2CO_2H$	2.8; 5.6
CH_2BrCO_2H	2.9
(ring) CO_2H / Cl	3.0
(ring) CO_2H / OH	3.0
CH_2ICO_2H	3.2
$CHOCO_2H$	3.2
O_2N—(ring)—CO_2H	3.4
O_2N / O_2N—(ring)—CO_2H	3.5
$HSCH_2CO_2H$	3.5; 10.2
$CH_2(NO_2)_2$	3.6
$CH_3OCH_2CO_2H$	3.6
$CH_3COCH_2CO_2H$	3.6
$HOCH_2CO_2H$	3.7
HCO_2H	3.7
Cl / (ring)—CO_2H	3.8
Cl—(ring)—CO_2H	4.0
$CH_2BrCH_2CO_2H$	4.0
O_2N (ring) NO_2 / OH	4.1
(ring) CO_2H	4.2
$H_2C{=}CHCO_2H$	4.2
$HO_2CCH_2CH_2CO_2H$	4.2; 5.7
$HO_2CCH_2CH_2CH_2CO_2H$	4.3; 5.4
Cl Cl / Cl—(ring)—OH / Cl Cl	4.5
$H_2C{=}C(CH_3)CO_2H$	4.7
CH_3CO_2H	4.8
$CH_3CH_2CO_2H$	4.8
$(CH_3)_3CCO_2H$	5.0
$CH_3COCH_2NO_2$	5.1
O / (ring) / O	5.3
$O_2NCH_2CO_2CH_3$	5.8

Compound	pK_a
O, CHO	5.8
Cl, Cl–, –OH, Cl	6.2
–SH	6.6
HCO_3H	7.1
NO_2, OH	7.2
$(CH_3)_2CHNO_2$	7.7
Cl–, –OH, Cl	7.8
CH_3CO_3H	8.2
OH, Cl	8.5
$CH_3CH_2NO_2$	8.5
F_3C–, –OH	8.7
$CH_3COCH_2COCH_3$	9.0
HO, OH	9.3; 11.1
OH, OH	9.3; 12.6

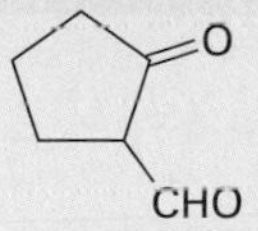

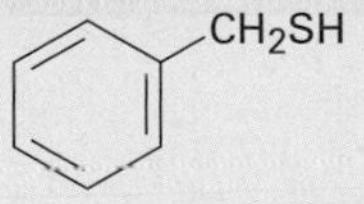

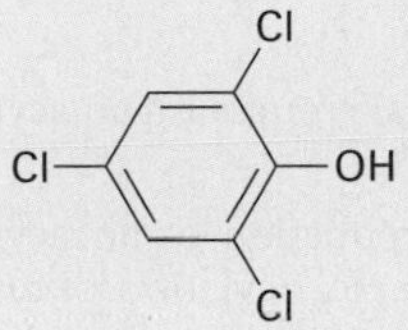

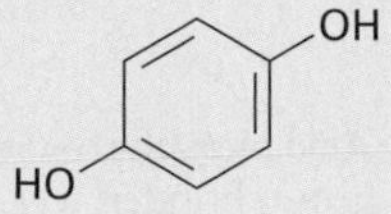

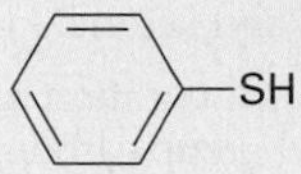

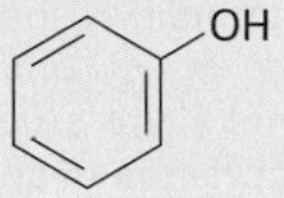

Compound	pK_a
CH_2SH	9.4
OH, HO	9.9; 11.5
OH	9.9
$CH_3COCH_2SOCH_3$	10.0
OH, CH_3	10.3
CH_3NO_2	10.3
CH_3SH	10.3
$CH_3COCH_2CO_2CH_3$	10.6
CH_3COCHO	11.0
$CH_2(CN)_2$	11.2
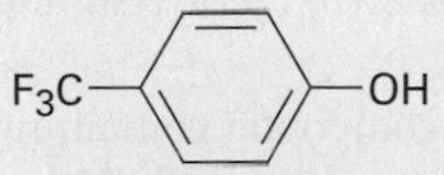 CCl_3CH_2OH	12.2 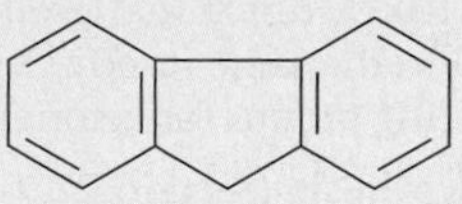
Glucose	12.3
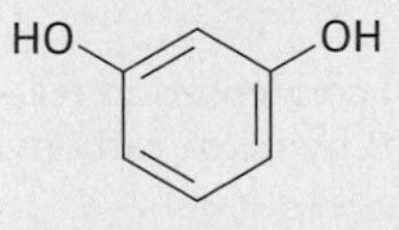 $(CH_3)_2C{=}NOH$	12.4
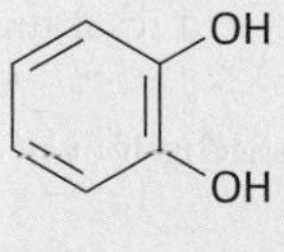 $CH_2(CO_2CH_3)_2$	12.9
$CHCl_2CH_2OH$	12.9
$CH_2(OH)_2$	13.3
$HOCH_2CH(OH)CH_2OH$	14.1
CH_2ClCH_2OH	14.3
	15.0
–CH_2OH	15.4
CH_3OH	15.5

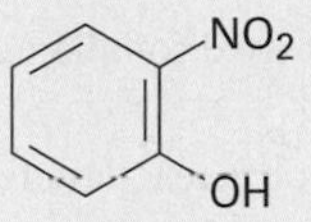

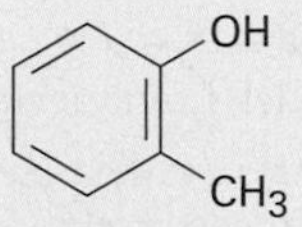

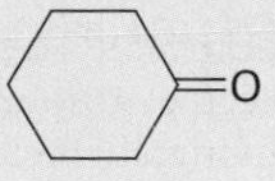

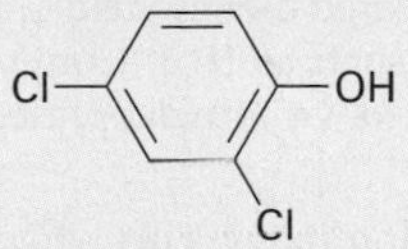

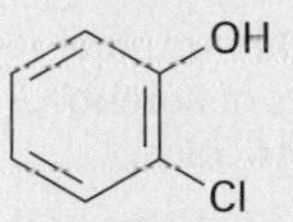

Compound	pK_a
$H_2C{=}CHCH_2OH$	15.5
CH_3CH_2OH	16.0
$CH_3CH_2CH_2OH$	16.1
CH_3COCH_2Br	16.1
=O	16.7
CH_3CHO	17
$(CH_3)_2CHCHO$	17
$(CH_3)_2CHOH$	17.1
$(CH_3)_3COH$	18.0
CH_3COCH_3	19.3
	23
$CH_3CO_2CH_2CH_3$	25
$HC{\equiv}CH$	25
CH_3CN	25
$CH_3SO_2CH_3$	28
$(C_6H_5)_3CH$	32
$(C_6H_5)_2CH_2$	34
CH_3SOCH_3	35
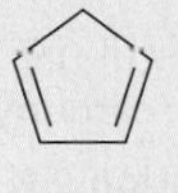 NH_3	36
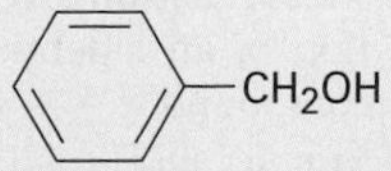 $CH_3CH_2NH_2$	36
$(CH_3CH_2)_2NH$	40
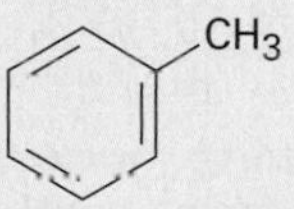 CH_3	41
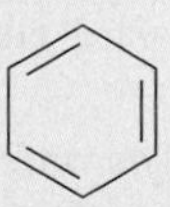	43
$H_2C{=}CH_2$	44
CH_4	~60

An acidity list covering more than 5000 organic compounds has been published: E.P. Serjeant and B. Dempsey (eds.), "Ionization Constants of Organic Acids in Aqueous Solution," IUPAC Chemical Data Series No. 23, Pergamon Press, Oxford, 1979.

APPENDIX C

Glossary

Absolute configuration (Section 5.5): The exact three-dimensional structure of a chiral molecule. Absolute configurations are specified verbally by the Cahn–Ingold–Prelog *R,S* convention.

Absorbance (Section 14.7): In optical spectroscopy, the logarithm of the intensity of the incident light divided by the intensity of the light transmitted through a sample; $A = \log I_0/I$.

Absorption spectrum (Section 12.5): A plot of wavelength of incident light versus amount of light absorbed. Organic molecules show absorption spectra in both the infrared and the ultraviolet regions of the electromagnetic spectrum.

Acetal (Section 19.10): A functional group consisting of two –OR groups bonded to the same carbon, $R_2C(OR')_2$. Acetals are often used as protecting groups for ketones and aldehydes.

Acetoacetic ester synthesis (Section 22.7): The synthesis of a methyl ketone by alkylation of an alkyl halide with ethyl acetoacetate, followed by hydrolysis and decarboxylation.

Acetyl group (Section 19.1): The CH_3CO– group.

Acetylide anion (Section 9.7): The anion formed by removal of a proton from a terminal alkyne, $R{-}C{\equiv}C{:}^-$.

Achiral (Section 5.2): Having a lack of handedness. A molecule is achiral if it has a plane of symmetry and is thus superimposable on its mirror image.

Acid anhydride (Section 21.1): A functional group with two acyl groups bonded to a common oxygen atom, RCO_2COR'.

Acid halide (Section 21.1): A functional group with an acyl group bonded to a halogen atom, RCOX.

Acidity constant, K_a (Section 2.8): A measure of acid strength. For any acid HA, the acidity constant is given by the expression

$$K_a = \frac{[H_3O^+][A^-]}{[HA]}.$$

Activating group (Section 16.4): An electron-donating group such as hydroxyl (–OH) or amino ($-NH_2$) that increases the reactivity of an aromatic ring toward electrophilic aromatic substitution.

Activation energy (Section 6.9): The difference in energy between ground state and transition state in a reaction. The amount of activation energy determines the rate at which the reaction proceeds. Most organic reactions have activation energies of 40–100 kJ/mol.

Active site (Sections 6.11, 26.11): The pocket in an enzyme where a substrate is bound and undergoes reaction.

Acyclic diene metathesis (ADMET) (Section 31.5): A method of polymer synthesis that uses the olefin metathesis reaction of an open-chain diene.

Acyl group (Sections 16.3, 19.1): A –COR group.

Acyl phosphate (Section 21.8): A functional group with an acyl group bonded to a phosphate, $RCO_2PO_3^{2-}$.

Acylation (Sections 16.3, 21.4): The introduction of an acyl group, –COR, onto a molecule. For example, acylation of an alcohol yields an ester, acylation of an amine yields an amide, and acylation of an aromatic ring yields an alkyl aryl ketone.

Acylium ion (Section 16.3): A resonance-stabilized carbocation in which the positive charge is located at a carbonyl-group carbon, $R{-}\overset{+}{C}{=}O \leftrightarrow R{-}C{\equiv}O^+$. Acylium ions are intermediates in Friedel–Crafts acylation reactions.

Adams catalyst (Section 8.6): The PtO_2 catalyst used for alkene hydrogenations.

1,2 Addition (Sections 14.2, 19.13): Addition of a reactant to the two ends of a double bond.

1,4 Addition (Sections 14.2, 19.13): Addition of a reactant to the ends of a conjugated π system. Conjugated dienes yield 1,4 adducts when treated with electrophiles such as HCl. Conjugated enones yield 1,4 adducts when treated with nucleophiles such as amines.

Addition reaction (Section 6.1): The reaction that occurs when two reactants add together to form a single product with no atoms left over.

Adrenocortical hormone (Section 27.6): A steroid hormone secreted by the adrenal glands. There are two types of adrenocortical hormones: mineralocorticoids and glucocorticoids.

Alcohol (Chapter 17 Introduction): A compound with an –OH group bonded to a saturated, sp^3-hybridized carbon, ROH.

Aldaric acid (Section 25.6): The dicarboxylic acid resulting from oxidation of an aldose.

Aldehyde (Chapter 19 Introduction): A compound containing the –CHO functional group.

Alditol (Section 25.6): The polyalcohol resulting from reduction of the carbonyl group of a sugar.

Aldol reaction (Section 23.1): The carbonyl condensation reaction of an aldehyde or ketone to give a β-hydroxy carbonyl compound.

Aldonic acid (Section 25.6): The monocarboxylic acid resulting from oxidation of the –CHO group of an aldose.

Aldose (Section 25.1): A carbohydrate with an aldehyde functional group.

Alicyclic (Section 4.1): A nonaromatic cyclic hydrocarbon such as a cycloalkane or cycloalkene.

Aliphatic (Section 3.2): A nonaromatic hydrocarbon such as a simple alkane, alkene, or alkyne.

Alkaloid (Chapter 2 *A Deeper Look*): A naturally occurring organic base, such as morphine.

Alkane (Section 3.2): A compound of carbon and hydrogen that contains only single bonds.

Alkene (Chapter 7 Introduction): A hydrocarbon that contains a carbon–carbon double bond, $R_2C{=}CR_2$.

Alkoxide ion (Section 17.2): The anion RO^- formed by deprotonation of an alcohol.

Alkoxymercuration reaction (Section 18.2): A method for synthesizing ethers by mercuric-ion catalyzed addition of an alcohol to an alkene followed by demercuration on treatment with $NaBH_4$.

Alkyl group (Section 3.3): The partial structure that remains when a hydrogen atom is removed from an alkane.

Alkyl halide (Chapter 10 Introduction): A compound with a halogen atom bonded to a saturated, sp^3-hybridized carbon atom.

Alkylamine (Section 24.1): An amino-substituted alkane, RNH_2, R_2NH, or R_3N.

Alkylation (Sections 9.8, 16.3, 18.2, 22.7): Introduction of an alkyl group onto a molecule. For example, aromatic rings can be alkylated to yield arenes, and enolate anions can be alkylated to yield α-substituted carbonyl compounds.

Alkyne (Chapter 9 Introduction): A hydrocarbon that contains a carbon–carbon triple bond, $RC{\equiv}CR$.

Allyl group (Section 7.3): A $H_2C{=}CHCH_2{-}$ substituent.

Allylic (Section 10.3): The position next to a double bond. For example, $H_2C{=}CHCH_2Br$ is an allylic bromide.

α-Amino acid (Section 26.1): A difunctional compound with an amino group on the carbon atom next to a carboxyl group, $RCH(NH_2)CO_2H$.

α Anomer (Section 25.5): The cyclic hemiacetal form of a sugar that has the hemiacetal –OH group cis to the –OH at the lowest chirality center in a Fischer projection.

α Helix (Section 26.9): The coiled secondary structure of a protein.

α Position (Chapter 22 Introduction): The position next to a carbonyl group.

α-Substitution reaction (Section 22.2): The substitution of the α hydrogen atom of a carbonyl compound by reaction with an electrophile.

Amide (Chapter 21 Introduction): A compound containing the $-CONR_2$ functional group.

Amidomalonate synthesis (Section 26.3): A method for preparing α-amino acids by alkylation of diethyl amidomalonate with an alkyl halide followed by deprotection and decarboxylation.

Amine (Chapter 24 Introduction): A compound containing one or more organic substituents bonded to a nitrogen atom, RNH_2, R_2NH, or R_3N.

Amino acid (Section 26.1): *See* α-Amino acid.

Amino sugar (Section 25.7): A sugar with one of its –OH groups replaced by $-NH_2$.

Amphiprotic (Section 26.1): Capable of acting either as an acid or as a base. Amino acids are amphiprotic.

Amplitude (Section 12.5): The height of a wave measured from the midpoint to the maximum. The intensity of radiant energy is proportional to the square of the wave's amplitude.

Amyl group (Section 3.3): An alternative name for a pentyl group.

Anabolic steroid (Section 27.6): A synthetic androgen that mimics the tissue-building effects of natural testosterone.

Anabolism (Section 29.1): The group of metabolic pathways that build up larger molecules from smaller ones.

Androgen (Section 27.6): A male steroid sex hormone.

Angle strain (Section 4.3): The strain introduced into a molecule when a bond angle is deformed from its ideal value. Angle strain is particularly important in small-ring cycloalkanes, where it results from compression of bond angles to less than their ideal tetrahedral values.

Annulation (Section 23.12): The building of a new ring onto an existing molecule.

Anomeric center (Section 25.5): The hemiacetal carbon atom in the cyclic pyranose or furanose form of a sugar.

Anomers (Section 25.5): Cyclic stereoisomers of sugars that differ only in their configuration at the hemiacetal (anomeric) carbon.

Antarafacial (Section 30.5): A pericyclic reaction that takes place on opposite faces of the two ends of a π electron system.

Anti conformation (Section 3.7): The geometric arrangement around a carbon–carbon single bond in which the two largest substituents are 180° apart as viewed in a Newman projection.

Anti periplanar (Section 11.8): Describing the stereochemical relationship in which two bonds on adjacent carbons lie in the same plane at an angle of 180°.

Anti stereochemistry (Section 8.2): The opposite of syn. An anti addition reaction is one in which the two ends of the double bond are attacked from different sides. An anti elimination reaction is one in which the two groups leave from opposite sides of the molecule.

Antiaromatic (Section 15.3): Referring to a planar, conjugated molecule with $4n$ π electrons. Delocalization of the π electrons leads to an increase in energy.

Antibonding MO (Section 1.11): A molecular orbital that is higher in energy than the atomic orbitals from which it is formed.

Anticodon (Section 28.5): A sequence of three bases on tRNA that reads the codons on mRNA and brings the correct amino acids into position for protein synthesis.

Antisense strand (Section 28.4): The template, noncoding strand of double-helical DNA that does not contain the gene.

Arene (Section 15.1): An alkyl-substituted benzene.

Arenediazonium salt (Section 24.8): An aromatic compound $Ar{-}\overset{+}{N}{\equiv}N\ X^-$; used in the Sandmeyer reaction.

Aromaticity (Chapter 15 Introduction): The special characteristics of cyclic conjugated molecules, including unusual stability and a tendency to undergo substitution reactions rather than addition reactions on treatment with electrophiles. Aromatic molecules are planar, cyclic, conjugated species with $4n + 2$ π electrons.

Arylamine (Section 24.1): An amino-substituted aromatic compound, $ArNH_2$.

Atactic (Section 31.2): A chain-growth polymer in which the stereochemistry of the substituents is oriented randomly along the backbone.

Atomic mass (Section 1.1): The weighted average mass of an element's naturally occurring isotopes.

Atomic number, *Z* (Section 1.1): The number of protons in the nucleus of an atom.

ATZ Derivative (Section 26.6): An anilinothiazolinone, formed from an amino acid during Edman degradation of a peptide.

Aufbau principle (Section 1.3): The rules for determining the electron configuration of an atom.

Axial bond (Section 4.6): A bond to chair cyclohexane that lies along the ring axis, perpendicular to the rough plane of the ring.

Azide synthesis (Section 24.6): A method for preparing amines by S_N2 reaction of an alkyl halide with azide ion, followed by reduction.

Azo compound (Section 24.8): A compound with the general structure R—N=N—R′.

Backbone (Section 26.4): The continuous chain of atoms running the length of a protein or other polymer.

Base peak (Section 12.1): The most intense peak in a mass spectrum.

Basicity constant, K_b (Section 24.3): A measure of base strength in water. For any base B, the basicity constant is given by the expression

$$B + H_2O \rightleftharpoons BH^+ + OH^-$$

$$K_b = \frac{[BH^+]\,[OH^-]}{[B]}$$

Bent bonds (Section 4.4): The bonds in small rings such as cyclopropane that bend away from the internuclear line and overlap at a slight angle, rather than head-on. Bent bonds are highly strained and highly reactive.

Benzoyl group (Section 19.1): The C_6H_5CO- group.

Benzyl group (Section 15.1): The $C_6H_5CH_2-$ group.

Benzylic (Section 11.5): The position next to an aromatic ring.

Benzyne (Section 16.8): An unstable compound having a triple bond in a benzene ring.

β Anomer (Section 25.5): The cyclic hemiacetal form of a sugar that has the hemiacetal –OH group trans to the –OH at the lowest chirality center in a Fischer projection.

β Diketone (Section 22.5): A 1,3-diketone.

β-Keto ester (Section 22.5): A 3-oxoester.

β Lactam (Chapter 21 *A Deeper Look*): A four-membered lactam, or cyclic amide. Penicillin and cephalosporin antibiotics contain β-lactam rings.

β-Oxidation pathway (Section 29.3): The metabolic pathway for degrading fatty acids.

β-Pleated sheet (Section 26.9): A type of secondary structure of a protein.

Betaine (Section 19.11): A neutral dipolar molecule with nonadjacent positive and negative charges. For example, the adduct of a Wittig reagent with a carbonyl compound is a betaine.

Bicycloalkane (Section 4.9): A cycloalkane that contains two rings.

Bimolecular reaction (Section 11.2): A reaction whose rate-limiting step occurs between two reactants.

Block copolymer (Section 31.3): A polymer in which different blocks of identical monomer units alternate with one another.

Boat cyclohexane (Section 4.5): A conformation of cyclohexane that bears a slight resemblance to a boat. Boat cyclohexane has no angle strain but has a large number of eclipsing interactions that make it less stable than chair cyclohexane.

Boc derivative (Section 26.7): A butyloxycarbonyl N-protected amino acid.

Bond angle (Section 1.6): The angle formed between two adjacent bonds.

Bond dissociation energy, *D* (Section 6.8): The amount of energy needed to break a bond and produce two radical fragments.

Bond length (Section 1.5): The equilibrium distance between the nuclei of two atoms that are bonded to each other.

Bond strength (Section 1.5): An alternative name for bond dissociation energy.

Bonding MO (Section 1.11): A molecular orbital that is lower in energy than the atomic orbitals from which it is formed.

Branched-chain alkane (Section 3.2): An alkane that contains a branching connection of carbons as opposed to a straight-chain alkane.

Bridgehead atom (Section 4.9): An atom that is shared by more than one ring in a polycyclic molecule.

Bromohydrin (Section 8.3): A 1,2-bromoalcohol; obtained by addition of HOBr to an alkene.

Bromonium ion (Section 8.2): A species with a divalent, positively charged bromine, R_2Br^+.

Brønsted–Lowry acid (Section 2.7): A substance that donates a hydrogen ion (proton; H^+) to a base.

Brønsted–Lowry base (Section 2.7): A substance that accepts H^+ from an acid.

C-terminal amino acid (Section 26.4): The amino acid with a free $-CO_2H$ group at the end of a protein chain.

Cahn–Ingold–Prelog sequence rules (Sections 5.5, 7.5): A series of rules for assigning relative rankings to substituent groups on a chirality center or a double-bond carbon atom.

Cannizzaro reaction (Section 19.12): The disproportionation reaction of an aldehyde on treatment with base to yield an alcohol and a carboxylic acid.

Carbanion (Sections 10.6, 19.7): A carbon anion, or substance that contains a trivalent, negatively charged carbon atom ($R_3C{:}^-$). Alkyl carbanions are sp^3-hybridized and have eight electrons in the outer shell of the negatively charged carbon.

Carbene (Section 8.9): A neutral substance that contains a divalent carbon atom having only six electrons in its outer shell ($R_2C{:}$).

Carbinolamine (Section 19.8): A molecule that contains the $R_2C(OH)NH_2$ functional group. Carbinolamines are produced as intermediates during the nucleophilic addition of amines to carbonyl compounds.

Carbocation (Sections 6.5, 7.9): A carbon cation, or substance that contains a trivalent, positively charged carbon atom having six electrons in its outer shell (R_3C^+).

Carbohydrate (Chapter 25 Introduction): A polyhydroxy aldehyde or ketone. Carbohydrates can be either simple sugars, such as glucose, or complex sugars, such as cellulose.

Carbonyl condensation reaction (Section 23.1): A reaction that joins two carbonyl compounds together by a combination of α-substitution and nucleophilic addition reactions.

Carbonyl group (Preview of Carbonyl Chemistry): The C=O functional group.

Carboxyl group (Section 20.1): The $-CO_2H$ functional group.

Carboxylation (Section 20.5): The addition of CO_2 to a molecule.

Carboxylic acid (Chapter 20 Introduction): A compound containing the $-CO_2H$ functional group.

Carboxylic acid derivative (Chapter 21 Introduction): A compound in which an acyl group is bonded to an electronegative atom or substituent that can act as a leaving group in a substitution reaction. Esters, amides, and acid halides are examples.

Catabolism (Section 29.1): The group of metabolic pathways that break down larger molecules into smaller ones.

Catalyst (Section 6.11): A substance that increases the rate of a chemical transformation by providing an alternative mechanism but is not itself changed in the reaction.

Cation radical (Section 12.1): A reactive species, typically formed in a mass spectrometer by loss of an electron from a neutral molecule and having both a positive charge and an odd number of electrons.

Chain-growth polymer (Sections 8.10, 31.1): A polymer whose bonds are produced by chain reaction mechanisms. Polyethylene and other alkene polymers are examples.

Chain reaction (Section 6.3): A reaction that, once initiated, sustains itself in an endlessly repeating cycle of propagation steps. The radical chlorination of alkanes is an example of a chain reaction that is initiated by irradiation with light and then continues in a series of propagation steps.

Chair conformation (Section 4.5): A three-dimensional conformation of cyclohexane that resembles the rough shape of a chair. The chair form of cyclohexane is the lowest-energy conformation of the molecule.

Chemical shift (Section 13.3): The position on the NMR chart where a nucleus absorbs. By convention, the chemical shift of tetramethylsilane (TMS) is set at zero, and all other absorptions usually occur downfield (to the left on the chart). Chemical shifts are expressed in delta units (δ), where 1 δ equals 1 ppm of the spectrometer operating frequency.

Chiral (Section 5.2): Having handedness. Chiral molecules are those that do not have a plane of symmetry and are therefore not superimposable on their mirror image. A chiral molecule thus exists in two forms, one right-handed and one left-handed. The most common cause of chirality in a molecule is the presence of a carbon atom that is bonded to four different substituents.

Chiral environment (Section 5.12): The chiral surroundings or conditions in which a molecule resides.

Chirality center (Section 5.2): An atom (usually carbon) that is bonded to four different groups.

Chlorohydrin (Section 8.3): A 1,2-chloroalcohol; obtained by addition of HOCl to an alkene.

Chromatography (Section 26.5): A technique for separating a mixture of compounds into pure components. Different compounds adsorb to a stationary support phase and are then carried along it at different rates by a mobile phase.

Cis–trans isomers (Sections 4.2, 7.4): Stereoisomers that differ in their stereochemistry about a ring or double bond.

Citric acid cycle (Section 29.7): The metabolic pathway by which acetyl CoA is degraded to CO_2.

Claisen condensation reaction (Section 23.7): The carbonyl condensation reaction of two ester molecules to give a β-keto ester product.

Claisen rearrangement reaction (Sections 18.4, 30.8): The pericyclic conversion of an allyl phenyl ether to an *o*-allylphenol or an allyl vinyl ether to a γ,δ-unsaturated ketone by heating.

Coding strand (Section 28.4): The sense strand of double-helical DNA that contains the gene.

Codon (Section 28.5): A three-base sequence on a messenger RNA chain that encodes the genetic information necessary to cause a specific amino acid to be incorporated into a protein. Codons on mRNA are read by complementary anticodons on tRNA.

Coenzyme (Section 26.10): A small organic molecule that acts as a cofactor in a biological reaction.

Cofactor (Section 26.10): A small nonprotein part of an enzyme that is necessary for biological activity.

Combinatorial chemistry (Chapter 16 *A Deeper Look*): A procedure in which anywhere from a few dozen to several hundred thousand substances are prepared simultaneously.

Complex carbohydrate (Section 25.1): A carbohydrate that is made of two or more simple sugars linked together by glycoside bonds.

Concerted reaction (Section 30.1): A reaction that takes place in a single step without intermediates. For example, the Diels–Alder cycloaddition reaction is a concerted process.

Condensed structure (Section 1.12): A shorthand way of writing structures in which carbon–hydrogen and carbon–carbon bonds are understood rather than shown explicitly. Propane, for example, has the condensed structure $CH_3CH_2CH_3$.

Configuration (Section 5.5): The three-dimensional arrangement of atoms bonded to a chirality center.

Conformation (Section 3.6): The three-dimensional shape of a molecule at any given instant, assuming that rotation around single bonds is frozen.

Conformational analysis (Section 4.8): A means of assessing the energy of a substituted cycloalkane by totaling the steric interactions present in the molecule.

Conformer (Section 3.6): A conformational isomer.

Conjugate acid (Section 2.7): The product that results from protonation of a Brønsted–Lowry base.

Conjugate addition (Section 19.13): Addition of a nucleophile to the β carbon atom of an α,β-unsaturated carbonyl compound.

Conjugate base (Section 2.7): The product that results from deprotonation of a Brønsted–Lowry acid.

Conjugation (Chapter 14 Introduction): A series of overlapping *p* orbitals, usually in alternating single and multiple bonds. For example, 1,3-butadiene is a conjugated diene, 3-buten-2-one is a conjugated enone, and benzene is a cyclic conjugated triene.

Conrotatory (Section 30.2): A term used to indicate that *p* orbitals must rotate in the same direction during electrocyclic ring-opening or ring-closure.

Constitutional isomers (Sections 3.2, 5.9): Isomers that have their atoms connected in a different order. For example, butane and 2-methylpropane are constitutional isomers.

Cope rearrangement (Section 30.8): The sigmatropic rearrangement of a 1,5-hexadiene.

Copolymer (Section 31.3): A polymer obtained when two or more different monomers are allowed to polymerize together.

Coupled reactions (Section 29.1): Two reactions that share a common intermediate so that the energy released in the favorable step allows the unfavorable step to occur.

Coupling constant, J (Section 13.11): The magnitude (expressed in hertz) of the interaction between nuclei whose spins are coupled.

Covalent bond (Section 1.5): A bond formed by sharing electrons between atoms.

Cracking (Chapter 3 *A Deeper Look*): A process used in petroleum refining in which large alkanes are thermally cracked into smaller fragments.

Crown ether (Section 18.7): A large-ring polyether; used as a phase-transfer catalyst.

Crystallite (Section 31.6): A highly ordered crystal-like region within a long polymer chain.

Curtius rearrangement (Section 24.6): The conversion of an acid chloride into an amine by reaction with azide ion, followed by heating with water.

Cyanohydrin (Section 19.6): A compound with an –OH group and a –CN group bonded to the same carbon atom; formed by addition of HCN to an aldehyde or ketone.

Cycloaddition reaction (Sections 14.4, 30.5): A pericyclic reaction in which two reactants add together in a single step to yield a cyclic product. The Diels–Alder reaction between a diene and a dienophile to give a cyclohexene is an example.

Cycloalkane (Section 4.1): An alkane that contains a ring of carbons.

D Sugar (Section 25.3): A sugar whose hydroxyl group at the chirality center farthest from the carbonyl group has the same configuration as D-glyceraldehyde and points to the right when drawn in Fischer projection.

***d,l* form** (Section 5.8): The racemic mixture of a chiral compound.

Deactivating group (Section 16.4): An electron-withdrawing substituent that decreases the reactivity of an aromatic ring toward electrophilic aromatic substitution.

Deamination (Section 29.9): The removal of an amino group from a molecule, as occurs with amino acids during metabolic degradation.

Debye, D (Section 2.2): The unit for measuring dipole moments; $1\text{ D} = 3.336 \times 10^{-30}$ coulomb meter (C · m).

Decarboxylation (Section 22.7): The loss of carbon dioxide from a molecule. β-Keto acids decarboxylate readily on heating.

Degenerate orbitals (Section 15.2): Two or more orbitals that have the same energy level.

Degree of unsaturation (Section 7.2): The number of rings and/or multiple bonds in a molecule.

Dehydration (Sections 8.1, 11.10, 17.6): The loss of water from an alcohol to yield an alkene.

Dehydrohalogenation (Sections 8.1, 11.8): The loss of HX from an alkyl halide. Alkyl halides undergo dehydrohalogenation to yield alkenes on treatment with strong base.

Delocalization (Sections 10.4, 15.2): A spreading out of electron density over a conjugated π electron system. For example, allylic cations and allylic anions are delocalized because their charges are spread out over the entire π electron system. Aromatic compounds have $4n + 2$ π electrons delocalized over their ring.

Delta scale (Section 13.3): An arbitrary scale used to calibrate NMR charts. One delta unit (δ) is equal to 1 part per million (ppm) of the spectrometer operating frequency.

Denaturation (Section 26.9): The physical changes that occur in a protein when secondary and tertiary structures are disrupted.

Deoxy sugar (Section 25.7): A sugar with one of its –OH groups replaced by an –H.

Deoxyribonucleic acid (DNA) (Section 28.1): The biopolymer consisting of deoxyribonucleotide units linked together through phosphate–sugar bonds. Found in the nucleus of cells, DNA contains an organism's genetic information.

DEPT-NMR (Section 13.6): An NMR method for distinguishing among signals due to CH_3, CH_2, CH, and quaternary carbons. That is, the number of hydrogens attached to each carbon can be determined.

Deshielding (Section 13.2): An effect observed in NMR that causes a nucleus to absorb toward the left (downfield) side of the chart. Deshielding is caused by a withdrawal of electron density from the nucleus.

Dess–Martin periodinane (Section 17.7): An iodine-based reagent commonly used for the laboratory oxidation of a primary alcohol to an aldehyde or a secondary alcohol to a ketone.

Deuterium isotope effect (Section 11.8): A tool used in mechanistic investigations to establish whether a C–H bond is broken in the rate-limiting step of a reaction.

Dextrorotatory (Section 5.3): A word used to describe an optically active substance that rotates the plane of polarization of plane-polarized light in a right-handed (clockwise) direction.

Diastereomers (Section 5.6): Non–mirror-image stereoisomers; diastereomers have the same configuration at one or more chirality centers but differ at other chirality centers.

Diastereotopic (Section 13.8): Hydrogens in a molecule whose replacement by some other group leads to different diastereomers.

1,3-Diaxial interaction (Section 4.7): The strain energy caused by a steric interaction between axial groups three carbon atoms apart in chair cyclohexane.

Diazonium salt (Section 24.8): A compound with the general structure $RN_2^+ X^-$.

Diazotization (Section 24.8): The conversion of a primary amine, RNH_2, into a diazonium ion, RN_2^+, by treatment with nitrous acid.

Dideoxy DNA sequencing (Section 28.6): A biochemical method for sequencing DNA strands.

Dieckmann cyclization reaction (Section 23.9): An intramolecular Claisen condensation reaction of a diester to give a cyclic β-keto ester.

Diels–Alder reaction (Sections 14.4, 30.5): The cycloaddition reaction of a diene with a dienophile to yield a cyclohexene.

Dienophile (Section 14.5): A compound containing a double bond that can take part in the Diels–Alder cycloaddition reaction. The most reactive dienophiles are those that have electron-withdrawing groups on the double bond.

Digestion (Section 29.1): The first stage of catabolism, in which food is broken down by hydrolysis of ester, glycoside (acetal), and peptide (amide) bonds to yield fatty acids, simple sugars, and amino acids.

Dihedral angle (Section 3.6): The angle between two bonds on adjacent carbons as viewed along the C–C bond.

Dipole moment, μ (Section 2.2): A measure of the net polarity of a molecule. A dipole moment arises when the centers of mass of positive and negative charges within a molecule do not coincide.

Dipole–dipole force (Section 2.12): A noncovalent electrostatic interaction between dipolar molecules.

Disaccharide (Section 25.8): A carbohydrate formed by linking two simple sugars through an acetal bond.

Dispersion force (Section 2.12): A noncovalent interaction between molecules that arises because of constantly changing electron distributions within the molecules.

Disrotatory (Section 30.2): A term used to indicate that *p* orbitals rotate in opposite directions during electrocyclic ring-opening or ring-closing reactions.

Disulfide (Section 18.8): A compound of the general structure RSSR′.

DNA (Section 28.1): Deoxyribonucleic acid.

Double bond (Section 1.8): A covalent bond formed by sharing two electron pairs between atoms.

Double helix (Section 28.2): The structure of DNA in which two polynucleotide strands coil around each other.

Doublet (Section 13.11): A two-line NMR absorption caused by spin–spin splitting when the spin of the nucleus under observation couples with the spin of a neighboring magnetic nucleus.

Downfield (Section 13.3): Referring to the left-hand portion of the NMR chart.

***E* geometry** (Section 7.5): A term used to describe the stereochemistry of a carbon–carbon double bond. The two groups on each carbon are ranked according to the Cahn–Ingold–Prelog sequence rules, and the two carbons are compared. If the higher-ranked groups on each carbon are on opposite sides of the double bond, the bond has *E* geometry.

E1 reaction (Section 11.10): A unimolecular elimination reaction in which the substrate spontaneously dissociates to give a carbocation intermediate, which loses a proton in a separate step.

E1cB reaction (Section 11.10): A unimolecular elimination reaction in which a proton is first removed to give a carbanion intermediate, which then expels the leaving group in a separate step.

E2 reaction (Section 11.8): A bimolecular elimination reaction in which C–H and C–X bond cleavages are simultaneous.

Eclipsed conformation (Section 3.6): The geometric arrangement around a carbon–carbon single bond in which the bonds to substituents on one carbon are parallel to the bonds to substituents on the neighboring carbon as viewed in a Newman projection.

Eclipsing strain (Section 3.6): The strain energy in a molecule caused by electron repulsions between eclipsed bonds. Eclipsing strain is also called torsional strain.

Edman degradation (Section 26.6): A method for N-terminal sequencing of peptide chains by treatment with *N*-phenylisothiocyanate.

Eicosanoid (Section 27.4): A lipid derived biologically from 5,8,11,14-eicosatetraenoic acid, or arachidonic acid. Prostaglandins, thromboxanes, and leukotrienes are examples.

Elastomer (Section 31.6): An amorphous polymer that has the ability to stretch out and spring back to its original shape.

Electrocyclic reaction (Section 30.2): A unimolecular pericyclic reaction in which a ring is formed or broken by a concerted reorganization of electrons through a cyclic transition state. For example, the cyclization of 1,3,5-hexatriene to yield 1,3-cyclohexadiene is an electrocyclic reaction.

Electromagnetic spectrum (Section 12.5): The range of electromagnetic energy, including infrared, ultraviolet, and visible radiation.

Electron configuration (Section 1.3): A list of the orbitals occupied by electrons in an atom.

Electron-dot structure (Section 1.4): A representation of a molecule showing valence electrons as dots.

Electron-transport chain (Section 29.1): The final stage of catabolism in which ATP is produced.

Electronegativity (Section 2.1): The ability of an atom to attract electrons in a covalent bond. Electronegativity increases across the periodic table from right to left and from bottom to top.

Electrophile (Section 6.4): An "electron-lover," or substance that accepts an electron pair from a nucleophile in a polar bond-forming reaction.

Electrophilic addition reaction (Section 7.7): The addition of an electrophile to a carbon–carbon double bond to yield a saturated product.

Electrophilic aromatic substitution reaction (Chapter 16 Introduction): A reaction in which an electrophile (E^+) reacts with an aromatic ring and substitutes for one of the ring hydrogens.

Electrophoresis (Sections 26.2, 28.6): A technique used for separating charged organic molecules, particularly proteins and DNA fragments. The mixture to be separated is placed on a buffered gel or paper, and an electric potential is applied across the ends of the apparatus. Negatively charged molecules migrate toward the positive electrode, and positively charged molecules migrate toward the negative electrode.

Electrostatic potential map (Section 2.1): A molecular representation that uses color to indicate the charge distribution in the molecule as derived from quantum-mechanical calculations.

Elimination reaction (Section 6.1): What occurs when a single reactant splits into two products.

Elution (Section 26.5): The passage of a substance from a chromatography column.

Embden–Meyerhof pathway (Section 29.5): An alternative name for glycolysis.

Enamine (Section 19.8): A compound with the $R_2N—CR{=}CR_2$ functional group.

Enantiomers (Section 5.1): Stereoisomers of a chiral substance that have a mirror-image relationship. Enantiomers have opposite configurations at all chirality centers.

Enantioselective synthesis (Chapter 19 *A Deeper Look*): A reaction method that yields only a single enantiomer of a chiral product starting from an achiral reactant.

Enantiotopic (Section 13.8): Hydrogens in a molecule whose replacement by some other group leads to different enantiomers.

3′ End (Section 28.1): The end of a nucleic acid chain with a free hydroxyl group at C3′.

5′ End (Section 28.1): The end of a nucleic acid chain with a free hydroxyl group at C5′.

Endergonic (Section 6.7): A reaction that has a positive free-energy change and is therefore nonspontaneous. In an energy diagram, the product of an endergonic reaction has a higher energy level than the reactants.

Endo (Section 14.5): A term indicating the stereochemistry of a substituent in a bridged bicycloalkane. An endo substituent is syn to the larger of the two bridges.

Endothermic (Section 6.7): A reaction that absorbs heat and therefore has a positive enthalpy change.

Energy diagram (Section 6.9): A representation of the course of a reaction, in which free energy is plotted as a function of reaction progress. Reactants, transition states, intermediates, and products are represented, and their appropriate energy levels are indicated.

Enol (Sections 9.4, 22.1): A vinylic alcohol that is in equilibrium with a carbonyl compound, C=C–OH.

Enolate ion (Section 22.1): The anion of an enol, $C{=}C{-}O^-$.

Enthalpy change, ΔH (Section 6.7): The heat of reaction. The enthalpy change that occurs during a reaction is a measure of the difference in total bond energy between reactants and products.

Entropy change, ΔS (Section 6.7): The change in amount of molecular randomness. The entropy change that occurs during a reaction is a measure of the difference in randomness between reactants and products.

Enzyme (Sections 6.11, 26.10): A biological catalyst. Enzymes are large proteins that catalyze specific biochemical reactions.

Epimers (Section 5.6): Diastereomers that differ in configuration at only one chirality center but are the same at all others.

Epoxide (Section 8.7): A three-membered-ring ether functional group.

Equatorial bond (Section 4.6): A bond to cyclohexane that lies along the rough equator of the ring.

ESI (Section 12.4): Electrospray ionization; a "soft" ionization method used for mass spectrometry of biological samples of very high molecular weight.

Essential amino acid (Section 26.1): One of nine amino acids that are biosynthesized only in plants and microorganisms and must be obtained by humans in the diet.

Essential monosaccharide (Section 25.7): One of eight simple sugars that is best obtained in the diet rather than by biosynthesis.

Essential oil (Chapter 8 *A Deeper Look*): The volatile oil obtained by steam distillation of a plant extract.

Ester (Chapter 21 Introduction): A compound containing the $-CO_2R$ functional group.

Estrogen (Section 27.6): A female steroid sex hormone.

Ether (Chapter 18 Introduction): A compound that has two organic substituents bonded to the same oxygen atom, ROR′.

Exergonic (Section 6.7): A reaction that has a negative free-energy change and is therefore spontaneous. On an energy diagram, the product of an exergonic reaction has a lower energy level than that of the reactants.

Exo (Section 14.5): A term indicating the stereochemistry of a substituent in a bridged bicycloalkane. An exo substituent is anti to the larger of the two bridges.

Exon (Section 28.4): A section of DNA that contains genetic information.

Exothermic (Section 6.7): A reaction that releases heat and therefore has a negative enthalpy change.

Fat (Section 27.1): A solid triacylglycerol derived from an animal source.

Fatty acid (Section 27.1): A long, straight-chain carboxylic acid found in fats and oils.

Fiber (Section 31.6): A thin thread produced by extruding a molten polymer through small holes in a die.

Fibrous protein (Section 26.9): A protein that consists of polypeptide chains arranged side by side in long threads. Such proteins are tough, insoluble in water, and used in nature for structural materials such as hair, hooves, and fingernails.

Fingerprint region (Section 12.7): The complex region of the infrared spectrum from 1500–400 cm^{-1}.

First-order reaction (Section 11.4): A reaction whose rate-limiting step is unimolecular and whose kinetics therefore depend on the concentration of only one reactant.

Fischer esterification reaction (Section 21.3): The acid-catalyzed nucleophilic acyl substitution reaction of a carboxylic acid with an alcohol to yield an ester.

Fischer projection (Section 25.2): A means of depicting the absolute configuration of a chiral molecule on a flat page. A Fischer projection uses a cross to represent the chirality center. The horizontal arms of the cross represent bonds coming out of the plane of the page, and the vertical arms of the cross represent bonds going back into the plane of the page.

Fmoc derivative (Section 26.7): A fluorenylmethyloxycarbonyl N-protected amino acid.

Formal charge (Section 2.3): The difference in the number of electrons owned by an atom in a molecule and by the same atom in its elemental state.

Formyl group (Section 19.1): A –CHO group.

Frequency, ν (Section 12.5): The number of electromagnetic wave cycles that travel past a fixed point in a given unit of time. Frequencies are expressed in units of cycles per second, or hertz.

Friedel–Crafts reaction (Section 16.3): An electrophilic aromatic substitution reaction to alkylate or acylate an aromatic ring.

Frontier orbitals (Section 30.1): The highest occupied (HOMO) and lowest unoccupied (LUMO) molecular orbitals.

FT-NMR (Section 13.4): Fourier-transform NMR; a rapid technique for recording NMR spectra in which all magnetic nuclei absorb at the same time.

Functional group (Section 3.1): An atom or group of atoms that is part of a larger molecule and has a characteristic chemical reactivity.

Functional RNA (Section 28.4): An alternative name for small RNAs.

Furanose (Section 25.5): The five-membered-ring form of a simple sugar.

Gabriel amine synthesis (Section 24.6): A method for preparing an amine by S_N2 reaction of an alkyl halide with potassium phthalimide, followed by hydrolysis.

Gauche conformation (Section 3.7): The conformation of butane in which the two methyl groups lie 60° apart as viewed in a Newman projection. This conformation has 3.8 kJ/mol steric strain.

Geminal (Section 19.5): Referring to two groups attached to the same carbon atom. For example, the hydrate formed by nucleophilic addition of water to an aldehyde or ketone is a geminal diol.

Gibbs free-energy change, D*G* (Section 6.7): The free-energy change that occurs during a reaction, given by the equation $\Delta G = \Delta H - T\Delta S$. A reaction with a negative free-energy change is spontaneous, and a reaction with a positive free-energy change is nonspontaneous.

Gilman reagent (Section 10.7): A diorganocopper reagent, R_2CuLi.

Glass transition temperature, T_g (Section 31.6): The temperature at which a hard, amorphous polymer becomes soft and flexible.

Globular protein (Section 26.9): A protein that is coiled into a compact, nearly spherical shape. Globular proteins, which are generally water-soluble and mobile within the cell, are the structural class to which enzymes belong.

Gluconeogenesis (Section 29.8): The anabolic pathway by which organisms make glucose from simple three-carbon precursors.

Glycal (Section 25.9): An unsaturated sugar with a C1–C2 double bond.

Glycal assembly method (Section 25.9): A method for linking monosaccharides together to synthesize polysaccharides.

Glycerophospholipid (Section 27.3): A lipid that contains a glycerol backbone linked to two fatty acids and a phosphoric acid.

Glycoconjugate (Section 25.6): A molecule in which a carbohydrate is linked through its anomeric center to another biological molecule such as a lipid or protein.

Glycol (Section 8.7): A diol, such as ethylene glycol, $HOCH_2CH_2OH$.

Glycolipid (Section 25.6): A biological molecule in which a carbohydrate is linked through a glycoside bond to a lipid.

Glycolysis (Section 29.5): A series of ten enzyme-catalyzed reactions that break down glucose into 2 equivalents of pyruvate, $CH_3COCO_2^-$.

Glycoprotein (Section 25.6): A biological molecule in which a carbohydrate is linked through a glycoside bond to a protein.

Glycoside (Section 25.6): A cyclic acetal formed by reaction of a sugar with another alcohol.

Graft copolymer (Section 31.3): A copolymer in which homopolymer branches of one monomer unit are "grafted" onto a homopolymer chain of another monomer unit.

Green chemistry (Chapters 11, 24 *A Deeper Look*): The design and implementation of chemical products and processes that reduce waste and minimize or eliminate the generation of hazardous substances.

Grignard reagent (Section 10.6): An organomagnesium halide, RMgX.

Ground state (Section 1.3): The most stable, lowest-energy electron configuration of a molecule or atom.

Haloform reaction (Section 22.6): The reaction of a methyl ketone with halogen and base to yield a haloform (CHX_3) and a carboxylic acid.

Halogenation (Sections 8.2, 16.1): The reaction of halogen with an alkene to yield a 1,2-dihalide addition product or with an aromatic compound to yield a substitution product.

Halohydrin (Section 8.3): A 1,2-haloalcohol, such as that obtained on addition of HOBr to an alkene.

Halonium ion (Section 8.2): A species containing a positively charged, divalent halogen. Three-membered-ring bromonium ions are intermediates in the electrophilic addition of Br_2 to alkenes.

Hammond postulate (Section 7.10): A postulate stating that we can get a picture of what a given transition state looks like by looking at the structure of the nearest stable species. Exergonic reactions have transition states that resemble reactant; endergonic reactions have transition states that resemble product.

Heat of combustion (Section 4.3): The amount of heat released when a compound burns completely in oxygen.

Heat of hydrogenation (Section 7.6): The amount of heat released when a carbon–carbon double bond is hydrogenated.

Heat of reaction (Section 6.7): An alternative name for the enthalpy change in a reaction, ΔH.

Hell–Volhard–Zelinskii (HVZ) reaction (Section 22.4): The reaction of a carboxylic acid with Br_2 and phosphorus to give an α-bromo carboxylic acid.

Hemiacetal (Section 19.10): A functional group having one –OR and one –OH group bonded to the same carbon.

Henderson–Hasselbalch equation (Sections 20.3, 24.5, 26.2): An equation for determining the extent of dissociation of a weak acid at various pH values.

Hertz, Hz (Section 12.5): A unit of measure of electromagnetic frequency, the number of waves that pass by a fixed point per second.

Heterocycle (Sections 15.5, 24.9): A cyclic molecule whose ring contains more than one kind of atom. For example, pyridine is a heterocycle that contains five carbon atoms and one nitrogen atom in its ring.

Heterolytic bond breakage (Section 6.2): The kind of bond-breaking that occurs in polar reactions when one fragment leaves with both of the bonding electrons: $A{:}B \rightarrow A^+ + B{:}^-$.

Hofmann elimination reaction (Section 24.7): The elimination reaction of an amine to yield an alkene by reaction with iodomethane followed by heating with Ag_2O.

Hofmann rearrangement (Section 24.6): The conversion of an amide into an amine by reaction with Br_2 and base.

HOMO (Sections 14.7, 30.1): The highest occupied molecular orbital. The symmetries of the HOMO and LUMO are important in pericyclic reactions.

Homolytic bond breakage (Section 6.2): The kind of bond-breaking that occurs in radical reactions when each fragment leaves with one bonding electron: $A{:}B \rightarrow A\cdot + B\cdot$.

Homopolymer (Section 31.3): A polymer made up of identical repeating units.

Homotopic (Section 13.8): Hydrogens in a molecule that give the identical structure on replacement by X and thus show identical NMR absorptions.

Hormone (Section 27.6): A chemical messenger that is secreted by an endocrine gland and carried through the bloodstream to a target tissue.

HPLC (Section 26.5): High-pressure liquid chromatography; a variant of column chromatography using high pressure to force solvent through very small absorbent particles.

Hückel's rule (Section 15.3): A rule stating that monocyclic conjugated molecules having $4n + 2$ π electrons (n = an integer) are aromatic.

Hund's rule (Section 1.3): If two or more empty orbitals of equal energy are available, one electron occupies each, with their spins parallel, until all are half-full.

Hybrid orbital (Section 1.6): An orbital derived from a combination of atomic orbitals. Hybrid orbitals, such as the sp^3, sp^2, and sp hybrids of carbon, are strongly directed and form stronger bonds than atomic orbitals do.

Hydration (Section 8.4): Addition of water to a molecule, such as occurs when alkenes are treated with aqueous sulfuric acid to give alcohols.

Hydride shift (Section 7.11): The shift of a hydrogen atom and its electron pair to a nearby cationic center.

Hydroboration (Section 8.5): Addition of borane (BH_3) or an alkylborane to an alkene. The resultant trialkylborane products can be oxidized to yield alcohols.

Hydrocarbon (Section 3.2): A compound that contains only carbon and hydrogen.

Hydrogen bond (Sections 2.12, 17.2): A weak attraction between a hydrogen atom bonded to an electronegative atom and an electron lone pair on another electronegative atom.

Hydrogenation (Section 8.6): Addition of hydrogen to a double or triple bond to yield a saturated product.

Hydrogenolysis (Section 26.7): Cleavage of a bond by reaction with hydrogen. Benzylic ethers and esters, for instance, are cleaved by hydrogenolysis.

Hydrophilic (Section 2.12): Water-loving; attracted to water.

Hydrophobic (Section 2.12): Water-fearing; repelled by water.

Hydroquinone (Section 17.10): A 1,4-dihydroxybenzene.

Hydroxylation (Section 8.7): Addition of two –OH groups to a double bond.

Hyperconjugation (Sections 7.6, 7.9): An electronic interaction that results from overlap of a vacant p orbital on one atom with a neighboring C–H σ bond. Hyperconjugation is important in stabilizing carbocations and substituted alkenes.

Imide (Section 24.6): A compound with the –CONHCO– functional group.

Imine (Section 19.8): A compound with the $R_2C{=}NR$ functional group.

Inductive effect (Sections 2.1, 7.9, 16.5): The electron-attracting or electron-withdrawing effect transmitted through σ bonds. Electronegative elements have an electron-withdrawing inductive effect.

Infrared (IR) spectroscopy (Section 12.6): A kind of optical spectroscopy that uses infrared energy. IR spectroscopy is particularly useful in organic chemistry for determining the kinds of functional groups present in molecules.

Initiator (Sections 6.3, 31.1): A substance that is used to initiate a radical chain reaction or polymerization. For example, radical chlorination of alkanes is initiated when light energy breaks the weak Cl–Cl bond to form Cl· radicals.

Integration (Section 13.10): A technique for measuring the area under an NMR peak to determine the relative number of each kind of proton in a molecule.

Intermediate (Section 6.10): A species that is formed during the course of a multistep reaction but is not the final product. Intermediates are more stable than transition states but may or may not be stable enough to isolate.

Intramolecular, intermolecular (Section 23.6): A reaction that occurs within the same molecule is intramolecular; a reaction that occurs between two molecules is intermolecular.

Intron (Section 28.4): A section of DNA that does not contain genetic information.

Ion pair (Section 11.5): A loose association between two ions in solution. Ion pairs are implicated as intermediates in S_N1 reactions to account for the partial retention of stereochemistry that is often observed.

Ionic bond (Section 1.4): The electrostatic attraction between ions of unlike charge.

Isoelectric point, p*I* (Section 26.2): The pH at which the number of positive charges and the number of negative charges on a protein or an amino acid are equal.

Isomers (Sections 3.2, 5.9): Compounds that have the same molecular formula but different structures.

Isoprene rule (Chapter 8 *A Deeper Look*): An observation to the effect that terpenoids appear to be made up of isoprene (2-methyl-1,3-butadiene) units connected head-to-tail.

Isotactic (Section 31.2): A chain-growth polymer in which the stereochemistry of the substituents is oriented regularly along the backbone.

Isotopes (Section 1.1): Atoms of the same element that have different mass numbers.

IUPAC system of nomenclature (Section 3.4): Rules for naming compounds, devised by the International Union of Pure and Applied Chemistry.

Kekulé structure (Section 1.4): An alternative name for a line-bond structure, which represents a molecule by showing covalent bonds as lines between atoms.

Ketal (Section 19.10): An alternative name for an acetal derived from a ketone rather than an aldehyde and consisting of two –OR groups bonded to the same carbon, $R_2C(OR')_2$. Ketals are often used as protecting groups for ketones.

Keto–enol tautomerism (Sections 9.4, 22.1): The equilibration between a carbonyl form and vinylic alcohol form of a molecule.

Ketone (Chapter 19 Introduction): A compound with two organic substituents bonded to a carbonyl group, $R_2C{=}O$.

Ketose (Section 25.1): A carbohydrate with a ketone functional group.

Kiliani–Fischer synthesis (Section 25.6): A method for lengthening the chain of an aldose sugar.

Kinetic control (Section 11.3): A reaction that follows the lowest activation energy pathway is said to be kinetically controlled. The product is the most rapidly formed but is not necessarily the most stable.

Kinetics (Section 11.2): Referring to reaction rates. Kinetic measurements are useful for helping to determine reaction mechanisms.

Koenigs–Knorr reaction (Section 25.6): A method for the synthesis of glycosides by reaction of an alcohol with a pyranosyl bromide.

Krebs cycle (Section 29.7): An alternative name for the citric acid cycle, by which acetyl CoA is degraded to CO_2.

L Sugar (Section 25.3): A sugar whose hydroxyl group at the chirality center farthest from the carbonyl group points to the left when drawn in Fischer projection.

Lactam (Section 21.7): A cyclic amide.

Lactone (Section 21.6): A cyclic ester.

Lagging strand (Section 28.3): The complement of the original $3' \rightarrow 5'$ DNA strand that is synthesized discontinuously in small pieces that are subsequently linked by DNA ligases.

LDA (Section 22.5): Lithium diisopropylamide, $LiN(i\text{-}C_3H_7)_2$, a strong base commonly used to convert carbonyl compounds into their enolate ions.

LD_{50} (Chapter 1 *A Deeper Look*): The amount of a substance per kilogram body weight that is lethal to 50% of test animals.

Leading strand (Section 28.3): The complement of the original $5' \rightarrow 3'$ DNA strand that is synthesized continuously in a single piece.

Leaving group (Section 11.2): The group that is replaced in a substitution reaction.

Levorotatory (Section 5.3): An optically active substance that rotates the plane of polarization of plane-polarized light in a left-handed (counterclockwise) direction.

Lewis acid (Section 2.11): A substance with a vacant low-energy orbital that can accept an electron pair from a base. All electrophiles are Lewis acids.

Lewis base (Section 2.11): A substance that donates an electron lone pair to an acid. All nucleophiles are Lewis bases.

Lewis structure (Section 1.4): A representation of a molecule showing valence electrons as dots.

Lindlar catalyst (Section 9.5): A hydrogenation catalyst used to convert alkynes to cis alkenes.

Line-bond structure (Section 1.4): An alternative name for a Kekulé structure, which represents a molecule by showing covalent bonds as lines between atoms.

1→4 Link (Section 25.8): A glycoside link between the C1 –OH group of one sugar and the C4 –OH group of another sugar.

Lipid (Chapter 27 Introduction): A naturally occurring substance isolated from cells and tissues by extraction with a nonpolar solvent. Lipids belong to many different structural classes, including fats, terpenoids, prostaglandins, and steroids.

Lipid bilayer (Section 27.3): The ordered lipid structure that forms a cell membrane.

Lipoprotein (Chapter 27 *A Deeper Look*): A complex molecule with both lipid and protein parts that transports lipids through the body.

Locant (Section 3.4): A number in a chemical name that locates the positions of the functional groups and substituents in the molecule.

Lone-pair electrons (Section 1.4): Nonbonding valence-shell electron pairs. Lone-pair electrons are used by nucleophiles in their reactions with electrophiles.

LUMO (Sections 14.7, 30.1): The lowest unoccupied molecular orbital. The symmetries of the LUMO and the HOMO are important in determining the stereochemistry of pericyclic reactions.

Magnetic resonance imaging, MRI (Chapter 13 *A Deeper Look*): A medical diagnostic technique based on nuclear magnetic resonance.

MALDI (Section 12.4): Matrix-assisted laser desorption ionization; a soft ionization method used for mass spectrometry of biological samples of very high molecular weight.

Malonic ester synthesis (Section 22.7): The synthesis of a carboxylic acid by alkylation of an alkyl halide with diethyl malonate, followed by hydrolysis and decarboxylation.

Markovnikov's rule (Section 7.8): A guide for determining the regiochemistry (orientation) of electrophilic addition reactions. In the addition of HX to an alkene, the hydrogen atom bonds to the alkene carbon that has fewer alkyl substituents.

Mass number, *A* (Section 1.1): The total of protons plus neutrons in an atom.

Mass spectrometry (Section 12.1): A technique for measuring the mass, and therefore the molecular weight (MW), of ions.

McLafferty rearrangement (Section 12.3): A mass-spectral fragmentation pathway for carbonyl compounds.

Mechanism (Section 6.2): A complete description of how a reaction occurs. A mechanism accounts for all starting materials and all products and describes the details of each individual step in the overall reaction process.

Meisenheimer complex (Section 16.7): An intermediate formed by addition of a nucleophile to a halo-substituted aromatic ring.

Melt transition temperature, T_m (Section 31.6): The temperature at which crystalline regions of a polymer melt to give an amorphous material.

Mercapto group (Section 18.8): An alternative name for the thiol group, –SH.

Meso compound (Section 5.7): A compound that contains chirality centers but is nevertheless achiral because it contains a symmetry plane.

Messenger RNA (Section 28.4): A kind of RNA formed by transcription of DNA and used to carry genetic messages from DNA to ribosomes.

Meta, *m*- (Section 15.1): A naming prefix used for 1,3-disubstituted benzenes.

Metabolism (Section 29.1): A collective name for the many reactions that go on in the cells of living organisms.

Metallacycle (Section 31.5): A cyclic compound that contains a metal atom in its ring.

Methylene group (Section 7.3): A $-CH_2-$ or $=CH_2$ group.

Micelle (Section 27.2): A spherical cluster of soaplike molecules that aggregate in aqueous solution. The ionic heads of the molecules lie on the outside, where they are solvated by water, and the organic tails bunch together on the inside of the micelle.

Michael reaction (Section 23.10): The conjugate addition reaction of an enolate ion to an unsaturated carbonyl compound.

Molar absorptivity (Section 14.7): A quantitative measure of the amount of UV light absorbed by a sample.

Molecular ion (Section 12.1): The cation produced in a mass spectrometer by loss of an electron from the parent molecule. The mass of the molecular ion corresponds to the molecular weight of the sample.

Molecular mechanics (Chapter 4 *A Deeper Look*): A computer-based method for calculating the minimum-energy conformation of a molecule.

Molecular orbital (MO) theory (Sections 1.11, 14.1): A description of covalent bond formation as resulting from a mathematical combination of atomic orbitals (wave functions) to form molecular orbitals.

Molecule (Section 1.4): A neutral collection of atoms held together by covalent bonds.

Molozonide (Section 8.8): The initial addition product of ozone with an alkene.

Monomer (Sections 8.10, 21.9; Chapter 31 Introduction): The simple starting unit from which a polymer is made.

Monosaccharide (Section 25.1): A simple sugar.

Monoterpenoid (Chapter 8 *A Deeper Look,* Section 27.5): A ten-carbon lipid.

Multiplet (Section 13.11): A pattern of peaks in an NMR spectrum that arises by spin–spin splitting of a single absorption because of coupling between neighboring magnetic nuclei.

Mutarotation (Section 25.5): The change in optical rotation observed when a pure anomer of a sugar is dissolved in water. Mutarotation is caused by the reversible opening and closing of the acetal linkage, which yields an equilibrium mixture of anomers.

$n + 1$ rule (Section 13.11): A hydrogen with n other hydrogens on neighboring carbons shows $n + 1$ peaks in its ^1H NMR spectrum.

N-terminal amino acid (Section 26.4): The amino acid with a free $-NH_2$ group at the end of a protein chain.

Natural gas (Chapter 3 *A Deeper Look*): A naturally occurring hydrocarbon mixture consisting chiefly of methane, along with smaller amounts of ethane, propane, and butane.

Natural product (Chapter 7 *A Deeper Look*): A catchall term generally taken to mean a secondary metabolite found in bacteria, plants, and other living organisms.

Neopentyl group (Section 3.4): The 2,2-dimethylpropyl group, $(CH_3)_3CCH_2-$.

Neuraminidase (Section 25.11): An enzyme present on the surface of viral particles that cleaves the bond holding the newly formed viral particles to host cells.

New molecular entity, NME (Chapter 6 *A Deeper Look*): A new biologically active chemical substance approved for sale as a drug by the U.S. Food and Drug Administration.

Newman projection (Section 3.6): A means of indicating stereochemical relationships between substituent groups on neighboring carbons. The carbon–carbon bond is viewed end-on, and the carbons are indicated by a circle. Bonds radiating from the center of the circle are attached to the front carbon, and bonds radiating from the edge of the circle are attached to the rear carbon.

Nitration (Section 16.2): The substitution of a nitro group onto an aromatic ring.

Nitrile (Section 20.1): A compound containing the $C\equiv N$ functional group.

Nitrogen rule (Section 24.10): A compound with an odd number of nitrogen atoms has an odd-numbered molecular weight.

Node (Section 1.2): A surface of zero electron density within an orbital. For example, a *p* orbital has a nodal plane passing through the center of the nucleus, perpendicular to the axis of the orbital.

Nonbonding electrons (Section 1.4): Valence electrons that are not used in forming covalent bonds.

Noncoding strand (Section 28.4): An alternative name for the antisense strand of DNA.

Noncovalent interaction (Section 2.12): One of a variety of nonbonding interactions between molecules, such as dipole–dipole forces, dispersion forces, and hydrogen bonds.

Nonessential amino acid (Section 26.1): One of the eleven amino acids that are biosynthesized by humans.

Normal alkane (Section 3.2): A straight-chain alkane, as opposed to a branched alkane. Normal alkanes are denoted by the suffix *n,* as in n-C_4H_{10} (*n*-butane).

NSAID (Chapter 15 *A Deeper Look*): A nonsteroidal anti-inflammatory drug, such as aspirin or ibuprofen.

Nuclear magnetic resonance, NMR (Chapter 13 Introduction): A spectroscopic technique that provides information about the carbon–hydrogen framework of a molecule. NMR works by detecting the energy absorptions accompanying the transitions between nuclear spin states that occur when a molecule is placed in a strong magnetic field and irradiated with radiofrequency waves.

Nucleic acid (Section 28.1): Deoxyribonucleic acid (DNA) and ribonucleic acid (RNA); biological polymers made of nucleotides joined together to form long chains.

Nucleophile (Section 6.4): An electron-rich species that donates an electron pair to an electrophile in a polar bond-forming reaction. Nucleophiles are also Lewis bases.

Nucleophilic acyl substitution reaction (Section 21.2): A reaction in which a nucleophile attacks a carbonyl compound and substitutes for a leaving group bonded to the carbonyl carbon.

Nucleophilic addition reaction (Section 19.4): A reaction in which a nucleophile adds to the electrophilic carbonyl group of a ketone or aldehyde to give an alcohol.

Nucleophilic aromatic substitution reaction (Section 16.7): The substitution reaction of an aryl halide by a nucleophile.

Nucleophilic substitution reaction (Section 11.1): A reaction in which one nucleophile replaces another attached to a saturated carbon atom.

Nucleophilicity (Section 11.3): The ability of a substance to act as a nucleophile in an S_N2 reaction.

Nucleoside (Section 28.1): A nucleic acid constituent, consisting of a sugar residue bonded to a heterocyclic purine or pyrimidine base.

Nucleotide (Section 28.1): A nucleic acid constituent, consisting of a sugar residue bonded both to a heterocyclic purine or pyrimidine base and to a phosphoric acid. Nucleotides are the monomer units from which DNA and RNA are constructed.

Nylon (Section 21.9): A synthetic polyamide step-growth polymer.

Okazaki fragment (Section 28.3): A short segment of a DNA lagging strand that is biosynthesized discontinuously and then linked by DNA ligases.

Olefin (Chapter 7 Introduction): An alternative name for an alkene.

Olefin metathesis polymerization (Section 31.5): A method of polymer synthesis based on using an olefin metathesis reaction.

Olefin metathesis reaction (Section 31.5): A reaction in which two olefins (alkenes) exchange substituents on their double bonds.

Oligonucleotide (Section 28.7): A short segment of DNA.

Optical activity (Section 5.3): The rotation of the plane of polarization of plane-polarized light by a chiral substance in solution.

Optical isomers (Section 5.4): An alternative name for enantiomers. Optical isomers are isomers that have a mirror-image relationship.

Orbital (Section 1.2): A wave function, which describes the volume of space around a nucleus in which an electron is most likely to be found.

Organic chemistry (Chapter 1 Introduction): The study of carbon compounds.

Organohalide (Chapter 10 Introduction): A compound that contains one or more halogen atoms bonded to carbon.

Organometallic compound (Section 10.6): A compound that contains a carbon–metal bond. Grignard reagents, RMgX, are examples.

Organophosphate (Section 1.10): A compound that contains a phosphorus atom bonded to four oxygens, with one of the oxygens also bonded to carbon.

Ortho, *o*- (Section 15.1): A naming prefix used for 1,2-disubstituted benzenes.

Oxidation (Section 10.8): A reaction that causes a decrease in electron ownership by carbon, either by bond formation between carbon and a more electronegative atom (usually oxygen, nitrogen, or a halogen) or by bond-breaking between carbon and a less electronegative atom (usually hydrogen).

Oxime (Section 19.8): A compound with the $R_2C{=}NOH$ functional group.

Oxirane (Section 8.7): An alternative name for an epoxide.

Oxymercuration (Section 8.4): A method for double-bond hydration by reaction of an alkene with aqueous mercuric acetate followed by treatment with $NaBH_4$.

Ozonide (Section 8.9): The product initially formed by addition of ozone to a carbon–carbon double bond. Ozonides are usually treated with a reducing agent, such as zinc in acetic acid, to produce carbonyl compounds.

Para, *p*- (Section 15.1): A naming prefix used for 1,4-disubstituted benzenes.

Paraffin (Section 3.5): A common name for alkanes.

Parent peak (Section 12.1): The peak in a mass spectrum corresponding to the molecular ion. The mass of the parent peak therefore represents the molecular weight of the compound.

Pauli exclusion principle (Section 1.3): No more than two electrons can occupy the same orbital, and those two must have spins of opposite sign.

Peptide (Chapter 26 Introduction): A short amino acid polymer in which the individual amino acid residues are linked by amide bonds.

Peptide bond (Section 26.4): An amide bond in a peptide chain.

Pericyclic reaction (Chapter 30 Introduction): A reaction that occurs in a single step by a reorganization of bonding electrons in a cyclic transition state.

Periplanar (Section 11.8): A conformation in which bonds to neighboring atoms have a parallel arrangement. In an eclipsed conformation, the neighboring bonds are syn periplanar; in a staggered conformation, the bonds are anti periplanar.

Peroxide (Section 18.1): A molecule containing an oxygen–oxygen bond functional group, ROOR′ or ROOH.

Peroxyacid (Section 8.7): A compound with the $-CO_3H$ functional group. Peroxyacids react with alkenes to give epoxides.

Phenol (Chapter 17 Introduction): A compound with an –OH group directly bonded to an aromatic ring, ArOH.

Phenoxide ion (Section 17.2): The anion of a phenol, ArO^-.

Phenyl group (Section 15.1): The name for the $-C_6H_5$ unit when the benzene ring is considered as a substituent. A phenyl group is abbreviated as –Ph.

Phosphine (Section 5.10): A trivalent phosphorus compound, R_3P.

Phosphite (Section 28.7): A compound with the structure $P(OR)_3$.

Phospholipid (Section 27.3): A lipid that contains a phosphate residue. For example, glycerophospholipids contain a glycerol backbone linked to two fatty acids and a phosphoric acid.

Phosphoramidite (Section 28.7): A compound with the structure $R_2NP(OR)_2$.

Phosphoric acid anhydride (Section 29.1): A substance that contains PO_2PO link, analogous to the CO_2CO link in carboxylic acid anhydrides.

Physiological pH (Section 20.3): The pH of 7.3 that exists inside cells.

Photochemical reaction (Section 30.2): A reaction carried out by irradiating the reactants with light.

Pi (π) bond (Section 1.8): The covalent bond formed by sideways overlap of atomic orbitals. For example, carbon–carbon double bonds contain a π bond formed by sideways overlap of two *p* orbitals.

PITC (Section 26.6): Phenylisothiocyanate; used in the Edman degradation.

pK_a (Section 2.8): The negative common logarithm of the K_a; used to express acid strength.

Plane of symmetry (Section 5.2): A plane that bisects a molecule such that one half of the molecule is the mirror image of the other half. Molecules containing a plane of symmetry are achiral.

Plane-polarized light (Section 5.3): Light that has its electromagnetic waves oscillating in a single plane rather than in random planes. The plane of polarization is rotated when the light is passed through a solution of a chiral substance.

Plasticizer (Section 31.6): A small organic molecule added to polymers to act as a lubricant between polymer chains.

Polar aprotic solvent (Section 11.3): A polar solvent that can't function as a hydrogen ion donor. Polar aprotic solvents such as dimethyl sulfoxide (DMSO) and dimethylformamide (DMF) are particularly useful in S_N2 reactions because of their ability to solvate cations.

Polar covalent bond (Section 2.1): A covalent bond in which the electron distribution between atoms is unsymmetrical.

Polar reaction (Section 6.4): A reaction in which bonds are made when a nucleophile donates two electrons to an electrophile and in which bonds are broken when one fragment leaves with both electrons from the bond.

Polarity (Section 2.1): The unsymmetrical distribution of electrons in a molecule that results when one atom attracts electrons more strongly than another.

Polarizability (Section 6.4): The measure of the change in a molecule's electron distribution in response to changing electrostatic interactions with solvents or ionic reagents.

Polycarbonate (Section 31.4): A polyester in which the carbonyl groups are linked to two –OR groups, $[O{=}C(OR)_2]$.

Polycyclic aromatic compound (Section 15.6): A compound with two or more benzene-like aromatic rings fused together.

Polycyclic compound (Section 4.9): A compound that contains more than one ring.

Polymer (Sections 8.10, 21.9; Chapter 31 Introduction): A large molecule made up of repeating smaller units. For example, polyethylene is a synthetic polymer made from repeating ethylene units and DNA is a biopolymer made of repeating deoxyribonucleotide units.

Polymerase chain reaction, PCR (Section 28.8): A method for amplifying small amounts of DNA to produce larger amounts.

Polysaccharide (Section 25.9): A carbohydrate that is made of many simple sugars linked together by glycoside (acetal) bonds.

Polyunsaturated fatty acid (Section 27.1): A fatty acid that contains more than one double bond.

Polyurethane (Section 31.4): A step-growth polymer prepared by reaction between a diol and a diisocyanate.

Posttranslational modification (Section 28.6): A chemical modification of a protein that occurs after translation from DNA.

Primary, secondary, tertiary, quaternary (Section 3.3): Terms used to describe the substitution pattern at a specific site. A primary site has one organic substituent attached to it, a secondary site has two organic substituents, a tertiary site has three, and a quaternary site has four.

	Carbon	Carbocation	Hydrogen	Alcohol	Amine
Primary	RCH_3	RCH_2^+	RCH_3	RCH_2OH	RNH_2
Secondary	R_2CH_2	R_2CH^+	R_2CH_2	R_2CHOH	R_2NH
Tertiary	R_3CH	R_3C^+	R_3CH	R_3COH	R_3N
Quaternary	R_4C				

Primary structure (Section 26.9): The amino acid sequence in a protein.

pro-R (Section 5.11): One of two identical atoms or groups of atoms in a compound whose replacement leads to an *R* chirality center.

pro-S (Section 5.11): One of two identical atoms or groups of atoms in a compound whose replacement leads to an *S* chirality center.

Prochiral (Section 5.11): A molecule that can be converted from achiral to chiral in a single chemical step.

Prochirality center (Section 5.11): An atom in a compound that can be converted into a chirality center by changing one of its attached substituents.

Promotor sequence (Section 28.4): A short sequence on DNA located upstream of the transcription start site and recognized by RNA polymerase.

Propagation step (Section 6.3): A step in a radical chain reaction that carries on the chain. The propagation steps must yield both product and a reactive intermediate.

Prostaglandin (Section 27.4): A lipid derived from arachidonic acid. Prostaglandins are present in nearly all body tissues and fluids, where they serve many important hormonal functions.

Protecting group (Sections 17.8, 19.10, 26.7): A group that is introduced to protect a sensitive functional group toward reaction elsewhere in the molecule. After serving its protective function, the group is removed.

Protein (Chapter 26 Introduction): A large peptide containing 50 or more amino acid residues. Proteins serve both as structural materials and as enzymes that control an organism's chemistry.

Protein Data Bank (Chapter 19 *A Deeper Look*): A worldwide online repository of X-ray and NMR structural data for biological macromolecules. To access the Protein Data Bank, go to http://www.rcsb.org/pdb/.

Protic solvent (Section 11.3): A solvent such as water or alcohol that can act as a proton donor.

Pyramidal inversion (Section 24.2): The rapid stereochemical inversion of a trivalent nitrogen compound.

Pyranose (Section 25.5): The six-membered, cyclic hemiacetal form of a simple sugar.

Quartet (Section 13.11): A set of four peaks in an NMR spectrum, caused by spin–spin splitting of a signal by three adjacent nuclear spins.

Quaternary: *See* **Primary.**

Quaternary ammonium salt (Section 24.1): An ionic compound containing a positively charged nitrogen atom with four attached groups, $R_4N^+ X^-$.

Quaternary structure (Section 26.9): The highest level of protein structure, involving an ordered aggregation of individual proteins into a larger cluster.

Quinone (Section 17.10): A 2,5-cyclohexadiene-1,4-dione.

***R* configuration** (Section 5.5): The configuration at a chirality center as specified using the Cahn–Ingold–Prelog sequence rules.

R group (Section 3.3): A generalized abbreviation for an organic partial structure.

Racemate (Section 5.8): A mixture consisting of equal parts (+) and (−) enantiomers of a chiral substance; also called a racemic mixture.

Radical (Section 6.2): A species that has an odd number of electrons, such as the chlorine radical, Cl·.

Radical reaction (Section 6.3): A reaction in which bonds are made by donation of one electron from each of two reactants and in which bonds are broken when each fragment leaves with one electron.

Rate constant (Section 11.2): The constant *k* in a rate equation.

Rate equation (Section 11.2): An equation that expresses the dependence of a reaction's rate on the concentration of reactants.

Rate-limiting step (Section 11.4): The slowest step in a multistep reaction sequence; also called the rate-determining step. The rate-limiting step acts as a kind of bottleneck in multistep reactions.

***Re* face** (Section 5.11): One of two faces of a planar, sp^2-hybridized atom.

Rearrangement reaction (Section 6.1): What occurs when a single reactant undergoes a reorganization of bonds and atoms to yield an isomeric product.

Reducing sugar (Section 25.6): A sugar that reduces silver ion in the Tollens test or cupric ion in the Fehling or Benedict tests.

Reduction (Section 10.8): A reaction that causes an increase of electron ownership by carbon, either by bond-breaking between carbon and a more electronegative atom or by bond formation between carbon and a less electronegative atom.

Reductive amination (Sections 24.6, 26.3): A method for preparing an amine by reaction of an aldehyde or ketone with ammonia and a reducing agent.

Refining (Chapter 3 *A Deeper Look*): The process by which petroleum is converted into gasoline and other useful products.

Regiochemistry (Section 7.8): A term describing the orientation of a reaction that occurs on an unsymmetrical substrate.

Regiospecific (Section 7.8): A term describing a reaction that occurs with a specific regiochemistry to give a single product rather than a mixture of products.

Replication (Section 28.3): The process by which double-stranded DNA uncoils and is replicated to produce two new copies.

Replication fork (Section 28.3): The point of unraveling in a DNA chain where replication occurs.

Residue (Section 26.4): An amino acid in a protein chain.

Resolution (Section 5.8): The process by which a racemate is separated into its two pure enantiomers.

Resonance effect (Section 16.4): The donation or withdrawal of electrons through orbital overlap with neighboring π bonds. For example, an oxygen or nitrogen substituent donates electrons to an aromatic ring by overlap of the O or N orbital with the aromatic ring *p* orbitals.

Resonance form (Section 2.4): An individual structural form of a resonance hybrid.

Resonance hybrid (Section 2.4): A molecule, such as benzene, that can't be represented adequately by a single Kekulé structure but must instead be considered as an average of two or more resonance forms. The resonance forms themselves differ only in the positions of their electrons, not their nuclei.

Restriction endonuclease (Section 28.6): An enzyme that is able to cleave a DNA molecule at points in the chain where a specific base sequence occurs.

Retrosynthetic (Sections 9.9, 16.11): Planning an organic synthesis by working backward from the final product to the starting material.

Ribonucleic acid (RNA) (Section 28.1): The biopolymer found in cells that serves to transcribe the genetic information found in DNA and uses that information to direct the synthesis of proteins.

Ribosomal RNA (Section 28.4): A kind of RNA used in the physical makeup of ribosomes.

Ring current (Section 15.7): The circulation of π electrons induced in aromatic rings by an external magnetic field. This effect accounts for the downfield shift of aromatic ring protons in the ^{1}H NMR spectrum.

Ring-flip (Section 4.6): A molecular motion that interconverts two chair conformations of cyclohexane. The effect of a ring-flip is to convert an axial substituent into an equatorial substituent.

Ring-opening metathesis polymerization (ROMP): A method of polymer synthesis that uses an olefin metathesis reaction of a cycloalkene.

RNA (Section 28.1): Ribonucleic acid.

Robinson annulation reaction (Section 23.12): A method for synthesis of cyclohexenones by sequential Michael reaction and intramolecular aldol reaction.

***S* configuration** (Section 5.5): The configuration at a chirality center as specified using the Cahn–Ingold–Prelog sequence rules.

***s*-Cis conformation** (Section 14.5): The conformation of a conjugated diene that is cis-like around the single bond.

Saccharide (Section 25.1): A sugar.

Salt bridge (Section 26.9): An ionic attraction between two oppositely charged groups in a protein chain.

Sandmeyer reaction (Section 24.8): The nucleophilic substitution reaction of an arenediazonium salt with a cuprous halide to yield an aryl halide.

Sanger dideoxy method (Section 28.6): A commonly used method of DNA sequencing.

Saponification (Section 21.6): An old term for the base-induced hydrolysis of an ester to yield a carboxylic acid salt.

Saturated (Section 3.2): A molecule that has only single bonds and thus can't undergo addition reactions. Alkanes are saturated, but alkenes are unsaturated.

Sawhorse structure (Section 3.6): A manner of representing stereochemistry that uses a stick drawing and gives a perspective view of the conformation around a single bond.

Schiff base (Sections 19.8, 29.5): An alternative name for an imine, $R_2C{=}NR'$, used primarily in biochemistry.

Second-order reaction (Section 11.2): A reaction whose rate-limiting step is bimolecular and whose kinetics are therefore dependent on the concentration of two reactants.

Secondary: *See* **Primary.**

Secondary metabolite (Chapter 7 *A Deeper Look*): A small naturally occurring molecule that is not essential to the growth and development of the producing organism and is not classified by structure.

Secondary structure (Section 26.9): The level of protein substructure that involves organization of chain sections into ordered arrangements such as β-pleated sheets or α helices.

Semiconservative replication (Section 28.3): The process by which DNA molecules are made containing one strand of old DNA and one strand of new DNA.

Sense strand (Section 28.4): The coding strand of double-helical DNA that contains the gene.

Sequence rules (Sections 5.5, 7.5): A series of rules for assigning relative rankings to substituent groups on a double-bond carbon atom or on a chirality center.

Sesquiterpenoid (Section 27.5): A 15-carbon lipid.

Sharpless epoxidation (Chapter 19 *A Deeper Look*): A method for enantioselective synthesis of a chiral epoxide by treatment of an allylic alcohol with *tert*-butyl hydroperoxide, $(CH_3)_3C{-}OOH$, in the presence of titanium tetraisopropoxide and diethyl tartrate.

Shell (electron) (Section 1.2): A group of an atom's electrons with the same principal quantum number.

Shielding (Section 13.2): An effect observed in NMR that causes a nucleus to absorb toward the right (upfield) side of the chart. Shielding is caused by donation of electron density to the nucleus.

***Si* face** (Section 5.11): One of two faces of a planar, sp^2-hybridized atom.

Sialic acid (Section 25.7): One of a group of more than 300 carbohydrates based on acetylneuramic acid.

Side chain (Section 26.1): The substituent attached to the α carbon of an amino acid.

Sigma (σ) bond (Section 1.5): A covalent bond formed by head-on overlap of atomic orbitals.

Sigmatropic reaction (Section 30.8): A pericyclic reaction that involves the migration of a group from one end of a π electron system to the other.

Silyl ether (Section 17.8): A substance with the structure $R_3Si{-}O{-}R$. The silyl ether acts as a protecting group for alcohols.

Simmons–Smith reaction (Section 8.9): The reaction of an alkene with CH_2I_2 and Zn–Cu to yield a cyclopropane.

Simple sugar (Section 25.1): A carbohydrate that cannot be broken down into smaller sugars by hydrolysis.

Single bond (Section 1.8): A covalent bond formed by sharing one electron pair between atoms.

Skeletal structure (Section 1.12): A shorthand way of writing structures in which carbon atoms are assumed to be at each intersection of two lines (bonds) and at the end of each line.

Small RNAs (Section 28.4): A type of RNA that has a variety of functions within the cell, including silencing transcription and catalyzing chemical modifications of other RNA molecules.

S_N1 reaction (Section 11.4): A unimolecular nucleophilic substitution reaction.

S_N2 reaction (Section 11.2): A bimolecular nucleophilic substitution reaction.

Solid-phase synthesis (Section 26.8): A technique of synthesis whereby the starting material is covalently bound to a solid polymer bead and reactions are carried out on the bound substrate. After the desired transformations have been effected, the product is cleaved from the polymer.

Solvation (Section 11.3): The clustering of solvent molecules around a solute particle to stabilize it.

***sp* Hybrid orbital** (Section 1.9): A hybrid orbital derived from the combination of an *s* and a *p* atomic orbital. The two *sp* orbitals that result from hybridization are oriented at an angle of 180° to each other.

sp^2 Hybrid orbital (Section 1.8): A hybrid orbital derived by combination of an *s* atomic orbital with two *p* atomic orbitals. The three sp^2 hybrid orbitals that result lie in a plane at angles of 120° to each other.

sp^3 Hybrid orbital (Section 1.6): A hybrid orbital derived by combination of an *s* atomic orbital with three *p* atomic orbitals. The four sp^3 hybrid orbitals that result are directed toward the corners of a regular tetrahedron at angles of 109° to each other.

Specific rotation, $[\alpha]_D$ (Section 5.3): The optical rotation of a chiral compound under standard conditions.

Sphingomyelin (Section 27.3): A phospholipid that has sphingosine as its backbone rather than glycerol.

Spin–spin splitting (Section 13.11): The splitting of an NMR signal into a multiplet because of an interaction between nearby magnetic nuclei whose spins are coupled. The magnitude of spin–spin splitting is given by the coupling constant, *J*.

Staggered conformation (Section 3.6): The three-dimensional arrangement of atoms around a carbon–carbon single bond in which the bonds on one carbon bisect the bond angles on the second carbon as viewed end-on.

Statin (Chapter 29 *A Deeper Look*): A drug that controls cholesterol biosynthesis in the body by blocking the HMG-CoA reductase enzyme.

Step-growth polymer (Sections 21.9, 31.4): A polymer in which each bond is formed independently of the others. Polyesters and polyamides (nylons) are examples.

Stereocenter (Section 5.2): An alternative name for a chirality center.

Stereochemistry (Chapters 3, 4, 5): The branch of chemistry concerned with the three-dimensional arrangement of atoms in molecules.

Stereogenic center (Section 5.2): An alternative name for a chirality center.

Stereoisomers (Section 4.2): Isomers that have their atoms connected in the same order but have different three-dimensional arrangements. The term *stereoisomer* includes both enantiomers and diastereomers.

Stereospecific (Section 8.9): A term indicating that only a single stereoisomer is produced in a given reaction rather than a mixture.

Steric strain (Sections 3.7, 4.7): The strain imposed on a molecule when two groups are too close together and try to occupy the same space. Steric strain is responsible both for the greater stability of trans versus cis alkenes and for the greater stability of equatorially substituted versus axially substituted cyclohexanes.

Steroid (Section 27.6): A lipid whose structure is based on a tetracyclic carbon skeleton with three 6-membered and one 5-membered ring. Steroids occur in both plants and animals and have a variety of important hormonal functions.

Stork enamine reaction (Section 23.11): The conjugate addition of an enamine to an α,β-unsaturated carbonyl compound, followed by hydrolysis to yield a 1,5-dicarbonyl product.

STR loci (Chapter 28 *A Deeper Look*): Short tandem repeat sequences of noncoding DNA that are unique to every individual and allow DNA fingerprinting.

Straight-chain alkane (Section 3.2): An alkane whose carbon atoms are connected without branching.

Substitution reaction (Section 6.1): What occurs when two reactants exchange parts to give two new products. S_N1 and S_N2 reactions are examples.

Sulfide (Section 18.8): A compound that has two organic substituents bonded to the same sulfur atom, RSR′.

Sulfonation (Section 16.2): The substitution of a sulfonic acid group ($-SO_3H$) onto an aromatic ring.

Sulfone (Section 18.8): A compound of the general structure RSO_2R'.

Sulfonium ion (Section 18.8): A species containing a positively charged, trivalent sulfur atom, R_3S^+.

Sulfoxide (Section 18.8): A compound of the general structure RSOR′.

Suprafacial (Section 30.5): A word used to describe the geometry of pericyclic reactions. Suprafacial reactions take place on the same side of the two ends of a π electron system.

Suzuki–Miyaura reaction (Section 10.7): The palladium-catalyzed coupling reaction of an aromatic or vinylic halide with an aromatic or vinylic boronic acid.

Symmetry-allowed, symmetry-disallowed (Section 30.2): A symmetry-allowed reaction is a pericyclic process that has a favorable orbital symmetry for reaction through a concerted pathway. A symmetry-disallowed reaction is one that does not have favorable orbital symmetry for reaction through a concerted pathway.

Symmetry plane (Section 5.2): A plane that bisects a molecule such that one half of the molecule is the mirror image of the other half. Molecules containing a plane of symmetry are achiral.

Syn periplanar (Section 11.8): Describing a stereochemical relationship in which two bonds on adjacent carbons lie in the same plane and are eclipsed.

Syn stereochemistry (Section 8.5): The opposite of anti. A syn addition reaction is one in which the two ends of the double bond react from the same side. A syn elimination is one in which the two groups leave from the same side of the molecule.

Syndiotactic (Section 31.2): A chain-growth polymer in which the stereochemistry of the substituents alternates regularly on opposite sides of the backbone.

Tautomers (Sections 9.4, 22.1): Isomers that interconvert spontaneously, usually with the change in position of a hydrogen.

Template strand (Section 28.4): The strand of double-helical DNA that does not contain the gene.

Terpenoid (Chapter 8 *A Deeper Look,* Section 27.5): A lipid that is formally derived by head-to-tail polymerization of isoprene units.

Tertiary: *See* **Primary.**

Tertiary structure (Section 26.9): The level of protein structure that involves the manner in which the entire protein chain is folded into a specific three-dimensional arrangement.

Thermodynamic control (Section 14.3): An equilibrium reaction that yields the lowest-energy, most stable product is said to be thermodynamically controlled.

Thermoplastic (Section 31.6): A polymer that has a high T_g and is hard at room temperature but becomes soft and viscous when heated.

Thermosetting resin (Section 31.6): A polymer that becomes highly cross-linked and solidifies into a hard, insoluble mass when heated.

Thioester (Section 21.8): A compound with the RCOSR′ functional group.

Thiol (Section 18.8): A compound containing the –SH functional group.

Thiolate ion (Section 18.8): The anion of a thiol, RS^-.

TMS (Section 13.3): Tetramethylsilane; used as an NMR calibration standard.

TOF (Section 12.4): Time-of-flight mass spectrometry; a sensitive method of mass detection accurate to about 3 ppm.

Tollens' reagent (Section 25.6): A solution of Ag_2O in aqueous ammonia; used to oxidize aldehydes to carboxylic acids.

Torsional strain (Section 3.6): The strain in a molecule caused by electron repulsion between eclipsed bonds. Torsional strain is also called eclipsing strain.

Tosylate (Section 11.1): A *p*-toluenesulfonate ester; useful as a leaving group in nucleophilic substitution reactions.

Transamination (Section 29.9): The exchange of an amino group and a keto group between reactants.

Transcription (Section 28.4): The process by which the genetic information encoded in DNA is read and used to synthesize RNA in the nucleus of the cell. A small portion of double-stranded DNA uncoils, and complementary ribonucleotides line up in the correct sequence for RNA synthesis.

Transfer RNA (Section 28.4): A kind of RNA that transports amino acids to the ribosomes, where they are joined together to make proteins.

Transimination (Section 29.9): The exchange of an amino group and an imine group between reactants.

Transition state (Section 6.9): An activated complex between reactants, representing the highest energy point on a reaction curve. Transition states are unstable complexes that can't be isolated.

Translation (Section 28.5): The process by which the genetic information transcribed from DNA onto mRNA is read by tRNA and used to direct protein synthesis.

Tree diagram (Section 13.12): A diagram used in NMR to sort out the complicated splitting patterns that can arise from multiple couplings.

Triacylglycerol (Section 27.1): A lipid, such as those found in animal fat and vegetable oil, that is a triester of glycerol with long-chain fatty acids.

Tricarboxylic acid cycle (Section 29.7): An alternative name for the citric acid cycle by which acetyl CoA is degraded to CO_2.

Triple bond (Section 1.9): A covalent bond formed by sharing three electron pairs between atoms.

Triplet (Section 13.11): A symmetrical three-line splitting pattern observed in the ^1H NMR spectrum when a proton has two equivalent neighbor protons.

Turnover number (Section 26.10): The number of substrate molecules acted on by an enzyme molecule per unit time.

Twist-boat conformation (Section 4.5): A conformation of cyclohexane that is somewhat more stable than a pure boat conformation.

Ultraviolet (UV) spectroscopy (Section 14.7): An optical spectroscopy employing ultraviolet irradiation. UV spectroscopy provides structural information about the extent of π electron conjugation in organic molecules.

Unimolecular reaction (Section 11.4): A reaction that occurs by spontaneous transformation of the starting material without the intervention of other reactants. For example, the dissociation of a tertiary alkyl halide in the S_N1 reaction is a unimolecular process.

Unsaturated (Section 7.2): A molecule that has one or more multiple bonds.

Upfield (Section 13.3): The right-hand portion of the NMR chart.

Urethane (Section 31.4): A functional group in which a carbonyl group is bonded to both an –OR and an $-NR_2$.

Uronic acid (Section 25.6): A monocarboxylic acid formed by oxidizing the $-CH_2OH$ end of an aldose without affecting the –CHO end.

Valence bond theory (Section 1.5): A bonding theory that describes a covalent bond as resulting from the overlap of two atomic orbitals.

Valence shell (Section 1.4): The outermost electron shell of an atom.

van der Waals forces (Section 2.12): Intermolecular forces that are responsible for holding molecules together in the liquid and solid states.

Vegetable oil (Section 27.1): A liquid triacylglycerol derived from a plant source.

Vicinal (Section 9.2): A term used to refer to a 1,2-disubstitution pattern. For example, 1,2-dibromoethane is a vicinal dibromide.

Vinyl group (Section 7.3): A $H_2C{=}CH{-}$ substituent.

Vinyl monomer (Sections 8.10, 31.1): A substituted alkene monomer used to make a chain-growth polymer.

Vinylic (Section 9.3): A term that refers to a substituent at a double-bond carbon atom. For example, chloroethylene is a vinylic chloride, and enols are vinylic alcohols.

Virion (Section 25.11): A viral particle.

Vitamin (Section 26.10): A small organic molecule that must be obtained in the diet and is required in trace amounts for proper growth and function.

Vulcanization (Section 14.6): A technique for cross-linking and hardening a diene polymer by heating with a few percent by weight of sulfur.

Walden inversion (Section 11.1): The inversion of configuration at a chirality center that accompanies an S_N2 reaction.

Wave equation (Section 1.2): A mathematical expression that defines the behavior of an electron in an atom.

Wave function (Section 1.2): A solution to the wave equation for defining the behavior of an electron in an atom. The square of the wave function defines the shape of an orbital.

Wavelength, λ (Section 12.5): The length of a wave from peak to peak. The wavelength of electromagnetic radiation is inversely proportional to frequency and inversely proportional to energy.

Wavenumber, $\tilde{\nu}$ (Section 12.6): The reciprocal of the wavelength in centimeters.

Wax (Section 27.1): A mixture of esters of long-chain carboxylic acids with long-chain alcohols.

Williamson ether synthesis (Section 18.2): A method for synthesizing ethers by S_N2 reaction of an alkyl halide with an alkoxide ion.

Wittig reaction (Section 19.11): The reaction of a phosphorus ylide with a ketone or aldehyde to yield an alkene.

Wohl degradation (Section 25.6): A method for shortening the chain of an aldose sugar by one carbon.

Wolff–Kishner reaction (Section 19.9): The conversion of an aldehyde or ketone into an alkane by reaction with hydrazine and base.

X-ray crystallography (Chapter 12 *A Deeper Look*): A technique that uses X rays to determine the structure of molecules.

Ylide (Section 19.11): A neutral species with adjacent + and − charges, such as the phosphoranes used in Wittig reactions.

***Z* geometry** (Section 7.5): A term used to describe the stereochemistry of a carbon–carbon double bond. The two groups on each carbon are ranked according to the Cahn–Ingold–Prelog sequence rules, and the two carbons are compared. If the higher ranked groups on each carbon are on the same side of the double bond, the bond has *Z* geometry.

Zaitsev's rule (Section 11.7): A rule stating that E2 elimination reactions normally yield the more highly substituted alkene as major product.

Ziegler–Natta catalyst (Section 31.2): A catalyst of an alkylaluminum and a titanium compound used for preparing alkene polymers.

Zwitterion (Section 26.1): A neutral dipolar molecule in which the positive and negative charges are not adjacent. For example, amino acids exist as zwitterions, $H_3\overset{+}{N}{-}CHR{-}CO_2^-$.

APPENDIX D

Answers to In-Text Problems

The following answers are meant only as a quick check while you study. Full answers for all problems are provided in the accompanying *Study Guide and Solutions Manual.*

Chapter 1

1.1 (a) $1s^2\ 2s^2\ 2p^4$ (b) $1s^2\ 2s^2\ 2p^3$
(c) $1s^2\ 2s^2\ 2p^6\ 3s^2\ 3p^4$

1.2 (a) 2 (b) 2 (c) 6

1.3

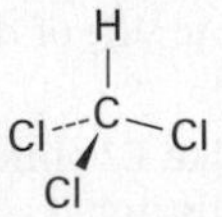

```
     H
     |
Cl---C---Cl
     |
    Cl
```

1.4

```
H       H H
  \     |/
   C — C
  /|     \
 H H      H
```

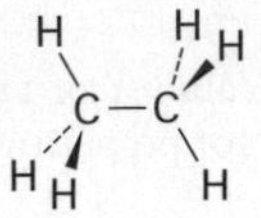

1.5 (a) CCl_4 (b) AlH_3 (c) CH_2Cl_2
(d) SiF_4 (e) CH_3NH_2

1.6 (a)

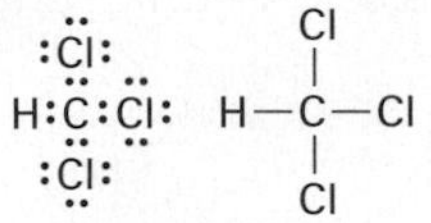

```
   ..                 Cl
  :Cl:                |
 ..  ..  ..       H—C—Cl
H:C:Cl:               |
  ..  ..              Cl
  :Cl:
   ..
```

(b)

```
  H         H
  ..        |
 :S:H       S—H
  ..
```

(c)

```
  H  ..       H
  ..          |
H:C:N:H   H—C—N—H
  ..  ..      |  |
  H  H        H  H
```

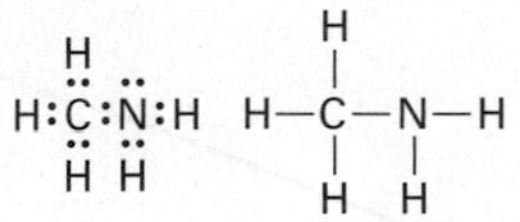

(d)

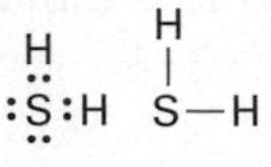

```
  H          H
  ..         |
H:C:Li   H—C—Li
  ..         |
  H          H
```

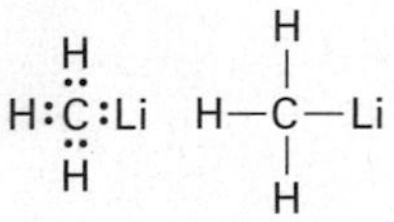

1.7 C_2H_7 has too many hydrogens for a compound with two carbons.

1.8

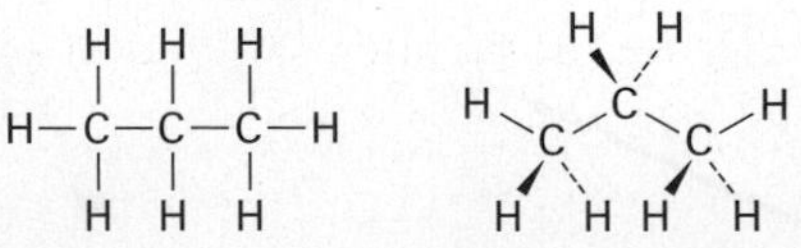

```
  H  H  H
  |  |  |
H—C—C—C—H
  |  |  |
  H  H  H
```

```
      H  H
       \/
  H    C    H
   \  / \  /
    C     C
   / \   / \
  H   H H   H
```

All bond angles are near 109°.

1.9

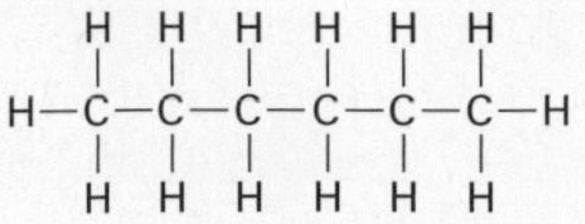

```
  H  H  H  H  H  H
  |  |  |  |  |  |
H—C—C—C—C—C—C—H
  |  |  |  |  |  |
  H  H  H  H  H  H
```

1.10 The CH_3 carbon is sp^3; the double-bond carbons are sp^2; the C=C–C and C=C–H bond angles are approximately 120°; other bond angles are near 109°.

```
  H H
   \|
 H—C    H
     \  /
     C=C
    /   \
   H     H
```

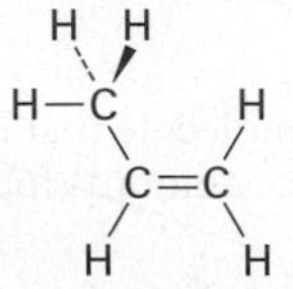

1.11 All carbons are sp^2, and all bond angles are near 120°.

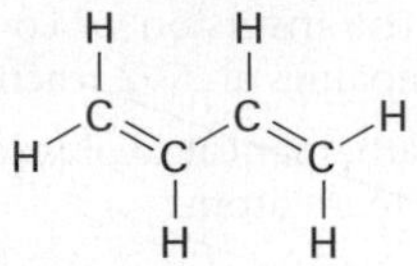

```
   H    H
   |    |
   C    C    H
 /  \\ /  \\ /
H    C    C
     |    |
     H    H
```

1.12 All carbons except CH_3 are sp^2.

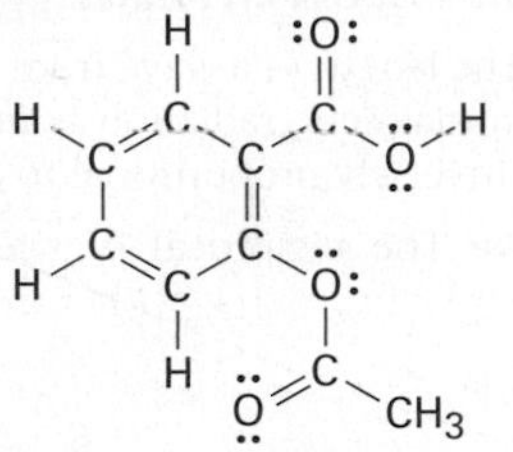

```
   H        :O:
   |        ||
 H  C      C   ..
   / \\   /  \ O—H
  C     C       ..
  |     ||
  C     C   ..
 / \\  /  \ O:
H    C      |
     |      C
     H   ..//  \
         :O      CH3
```

1.13 The CH_3 carbon is sp^3; the triple-bond carbons are sp; the C≡C–C and H–C≡C bond angles are approximately 180°.

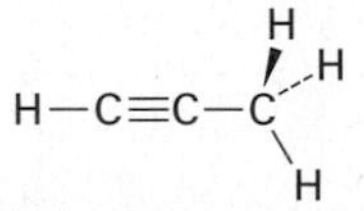

```
             H
             |  H
H—C≡C—C
             |
             H
```

1.14 (a) O has 2 lone pairs and is sp^3-hybridized.
(b) N has 1 lone pair and is sp^3-hybridized.
(c) P has 1 lone pair and is sp^3-hybridized.
(d) S has 2 lone pairs and is sp^3-hybridized.

1.15 **(a)**

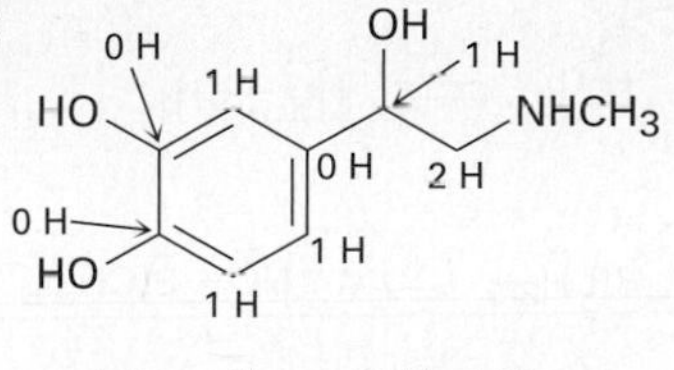

Adrenaline—$C_9H_{13}NO_3$

(b)

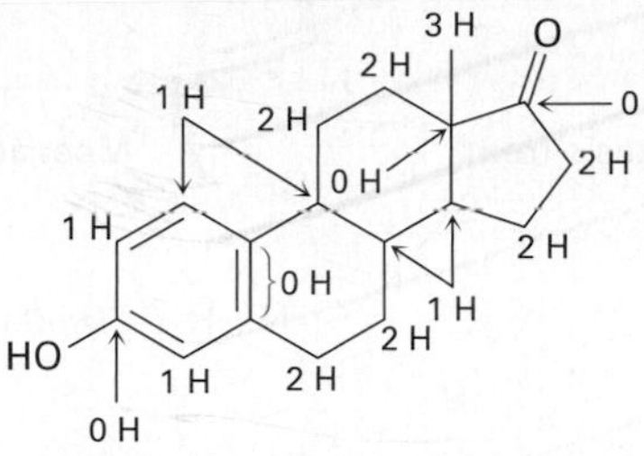

Estrone—$C_{18}H_{22}O_2$

1.16 There are numerous possibilities, such as:

(a) C_5H_{12} $CH_3CH_2CH_2CH_2CH_3$ $CH_3CH_2CH(CH_3)CH_3$ $CH_3C(CH_3)_2CH_3$

(b) C_2H_7N $CH_3CH_2NH_2$ CH_3NHCH_3

(c) C_3H_6O CH_3CH_2CHO $H_2C{=}CHCH_2OH$ $H_2C{=}CHOCH_3$

(d) C_4H_9Cl $CH_3CH_2CH_2CH_2Cl$ $CH_3CH_2CHClCH_3$ $CH_3CH(CH_3)CH_2Cl$

1.17

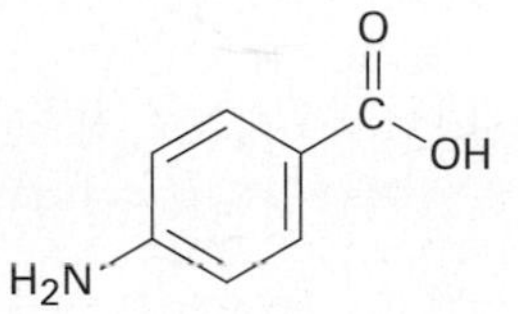

Chapter 2

2.1 (a) H (b) Br (c) Cl (d) C

2.2

(a) $\overset{\delta+}{H_3C}-\overset{\delta-}{Cl}$ **(b)** $\overset{\delta+}{H_3C}-\overset{\delta-}{NH_2}$ **(c)** $\overset{\delta-}{H_2N}-\overset{\delta+}{H}$

(d) H_3C-SH **(e)** $\overset{\delta-}{H_3C}-\overset{\delta+}{MgBr}$ **(f)** $\overset{\delta+}{H_3C}-\overset{\delta-}{F}$

Carbon and sulfur have identical electronegativities.

2.3 $H_3C{-}OH < H_3C{-}MgBr < H_3C{-}Li = H_3C{-}F < H_3C{-}K$

2.4 The chlorine is electron-rich, and the carbon is electron-poor.

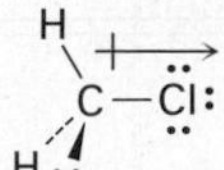

2.5 The two C–O dipoles cancel because of the symmetry of the molecule:

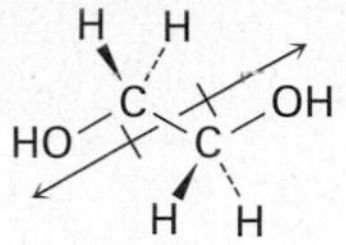

2.6 **(a)** $H_2C{=}CH_2$ No dipole moment **(b)** $CHCl_3$

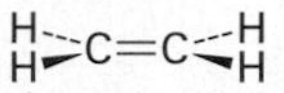

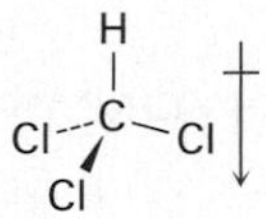

(c) CH_2Cl_2 **(d)** $H_2C{=}CCl_2$

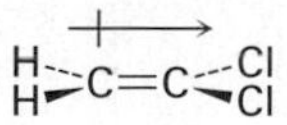

2.7 (a) For carbon: FC = 4 − 8/2 − 0 = 0
For the middle nitrogen: FC = 5 − 8/2 − 0 = +1
For the end nitrogen: FC = 5 − 4/2 − 4 = −1

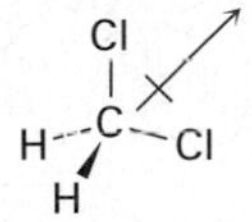

(b) For nitrogen: FC = 5 − 8/2 − 0 = +1
For oxygen: FC = 6 − 2/2 − 6 = −1

(c) For nitrogen: FC = 5 − 8/2 − 0 = +1
For the triply bonded carbon: FC = 4 − 6/2 − 2 = −1

2.8

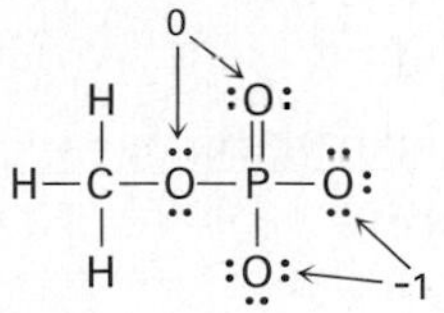

2.9 The structures in (a) are resonance forms.

2.10

(a) Resonance structures of $CH_3OPO_3^{2-}$ (structures not reproducible as text)

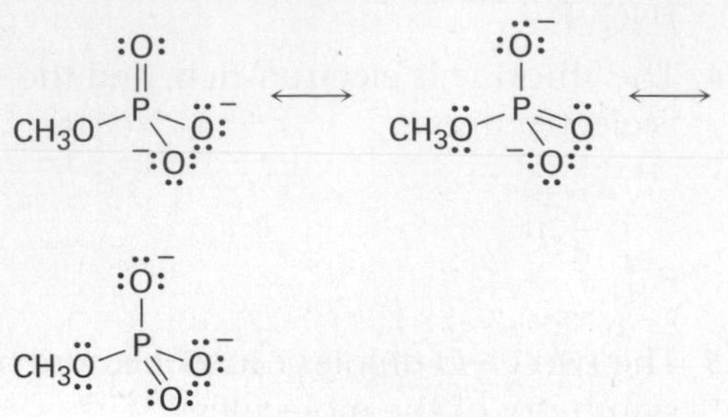

(b) Resonance structures of NO_3^-

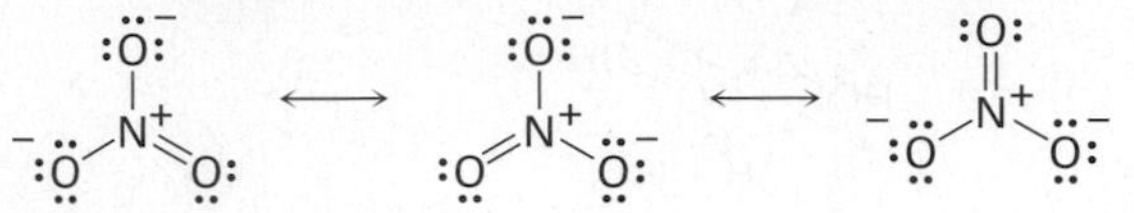

(c) $H_2C{=}CH{-}CH_2^+ \longleftrightarrow H_2\overset{+}{C}{-}CH{=}CH_2$

(d) Resonance structures of benzoate

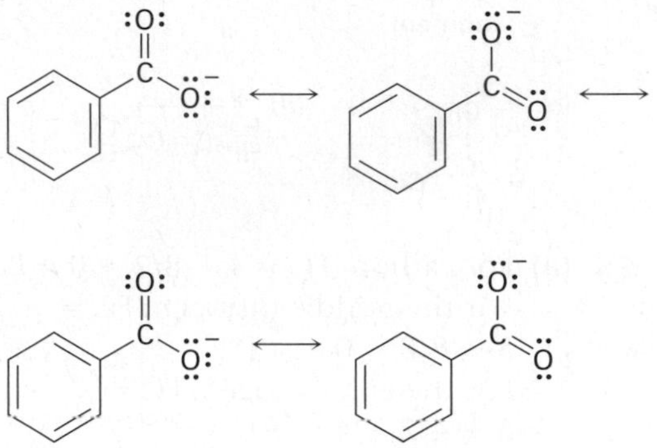

2.11 HNO_3 + NH_3 ⟶ NH_4^+ + NO_3^-

HNO_3	NH_3	NH_4^+	NO_3^-
Acid	Base	Conjugate acid	Conjugate base

2.12 Phenylalanine is stronger.

2.13 Water is a stronger acid.

2.14 Neither reaction will take place.

2.15 Reaction will take place.

2.16 $K_a = 4.9 \times 10^{-10}$

2.17

(a)

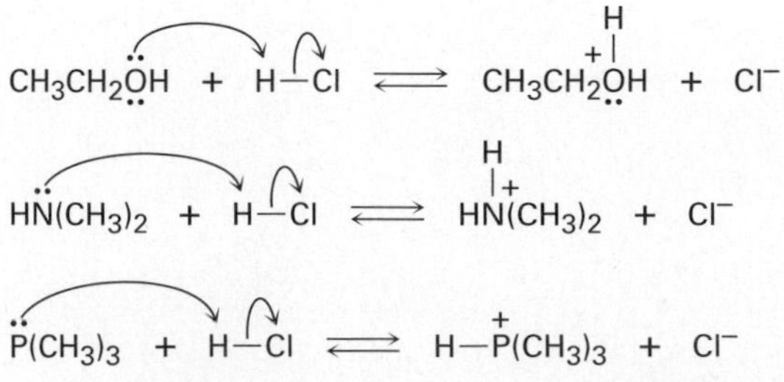

$CH_3CH_2\ddot{O}H + H{-}Cl \rightleftharpoons CH_3CH_2\overset{+}{O}H_2 + Cl^-$

$H\ddot{N}(CH_3)_2 + H{-}Cl \rightleftharpoons H_2\overset{+}{N}(CH_3)_2 + Cl^-$

$\ddot{P}(CH_3)_3 + H{-}Cl \rightleftharpoons H{-}\overset{+}{P}(CH_3)_3 + Cl^-$

(b)

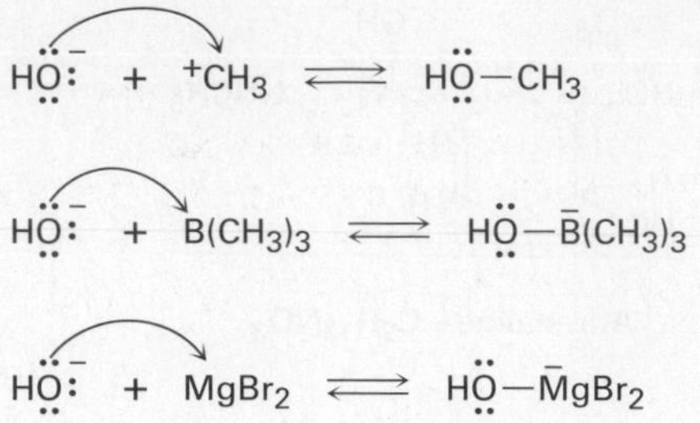

$H\ddot{O}:^- + {}^+CH_3 \rightleftharpoons H\ddot{O}{-}CH_3$

$H\ddot{O}:^- + B(CH_3)_3 \rightleftharpoons H\ddot{O}{-}\bar{B}(CH_3)_3$

$H\ddot{O}:^- + MgBr_2 \rightleftharpoons H\ddot{O}{-}\bar{M}gBr_2$

2.18

(a) More basic (red) — N=; Most acidic (blue) — N–H. Imidazole

(b) Protonation of imidazole by A–H and deprotonation by :B, with resonance structures (structures not reproducible as text)

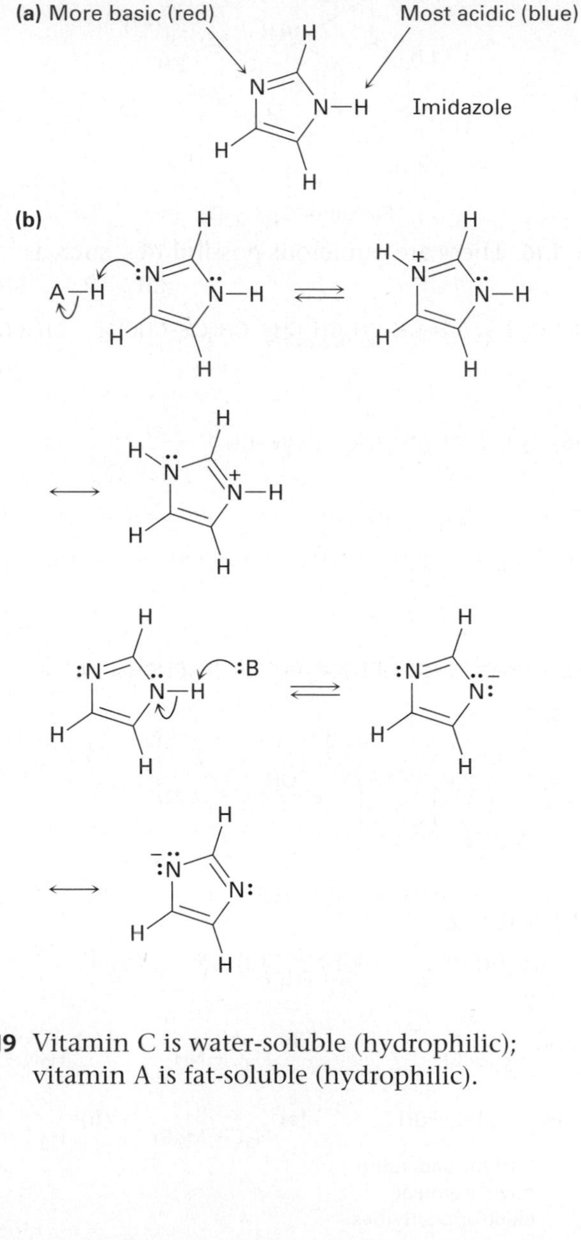

2.19 Vitamin C is water-soluble (hydrophilic); vitamin A is fat-soluble (hydrophilic).

Chapter 3

3.1 (a) Sulfide, carboxylic acid, amine
(b) Aromatic ring, carboxylic acid
(c) Ether, alcohol, aromatic ring, amide, C=C bond

3.2 (a) CH_3OH (b) 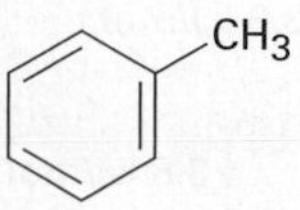 (c) 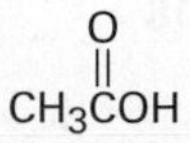
(d) CH_3NH_2 (e) 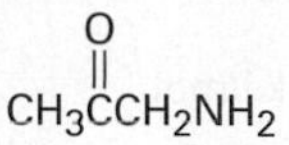 (f) 

3.3

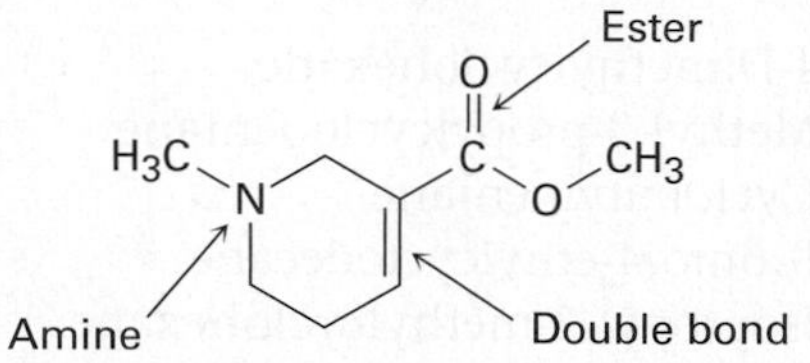

$C_8H_{13}NO_2$

3.4

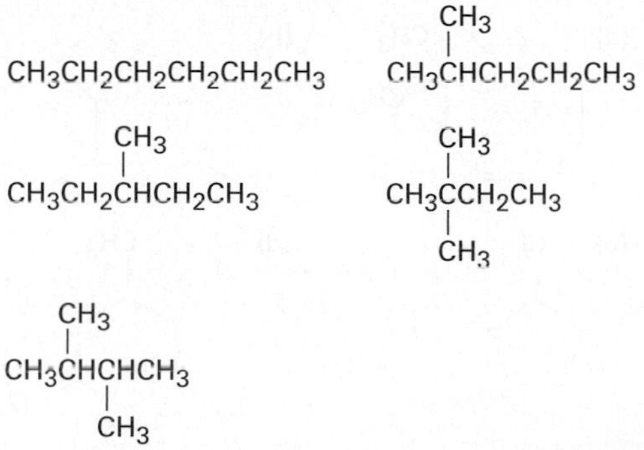

3.5 Part (a) has nine possible answers.

(a) 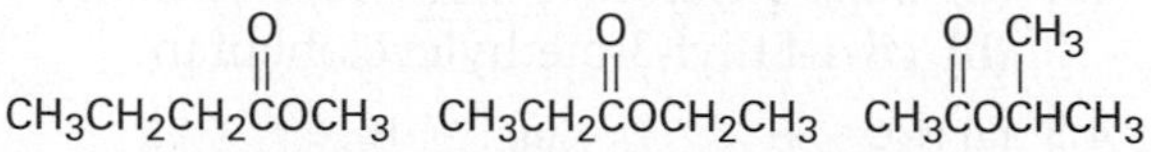

(b) 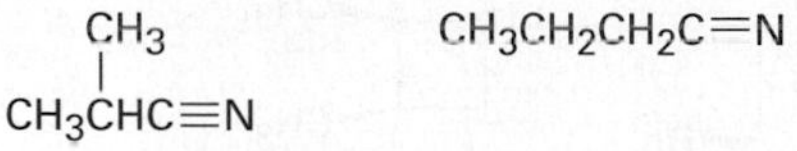

(c) $CH_3CH_2SSCH_2CH_3$ $CH_3SSCH_2CH_2CH_3$ 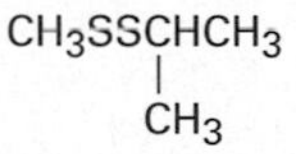

3.6 (a) Two (b) Four (c) Four

3.7

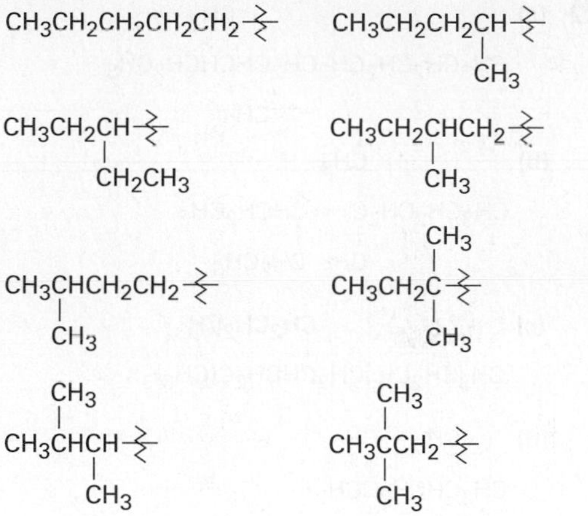

3.8 (a) (b) (c)

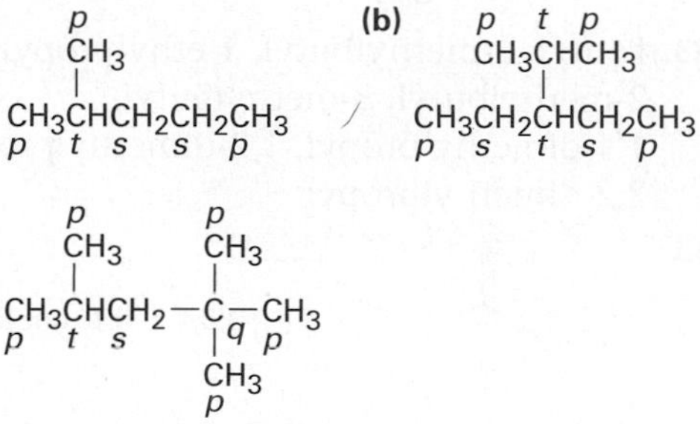

3.9 Primary carbons have primary hydrogens, secondary carbons have secondary hydrogens, and tertiary carbons have tertiary hydrogens.

3.10 (a) (b) (c)

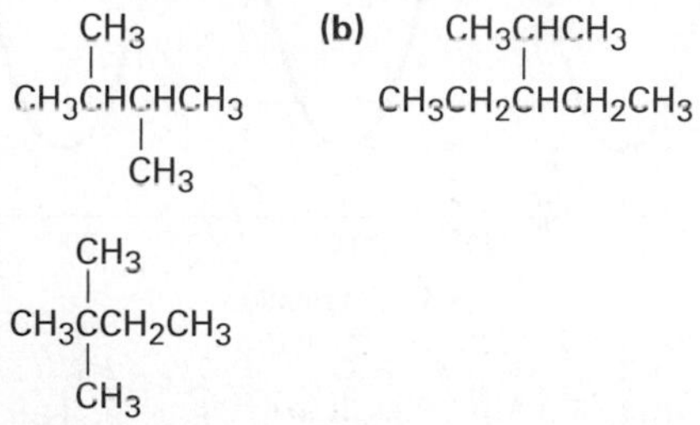

3.11 (a) Pentane, 2 methylbutane, 2,2-dimethylpropane
(b) 2,3-Dimethylpentane
(c) 2,4-Dimethylpentane
(d) 2,2,5-Trimethylhexane

3.12 (a)

$$\begin{array}{c} \qquad\qquad\qquad CH_3 \\ CH_3CH_2CH_2CH_2CH_2CHCHCH_2CH_3 \\ \qquad\qquad\qquad CH_3 \end{array}$$

(b)

$$\begin{array}{c} CH_3 \\ CH_3CH_2CH_2C—CHCH_2CH_3 \\ CH_3\ \ CH_2CH_3 \end{array}$$

(c)

$$\begin{array}{c} CH_2CH_2CH_3 \\ CH_3CH_2CH_2CH_2CHCH_2C(CH_3)_3 \end{array}$$

(d)

$$\begin{array}{c} CH_3\quad CH_3 \\ CH_3CHCH_2CCH_3 \\ \qquad\qquad CH_3 \end{array}$$

3.13 Pentyl, 1-methylbutyl, 1-ethylpropyl, 2-methylbutyl, 3-methylbutyl, 1,1-dimethylpropyl, 1,2-dimethylpropyl, 2,2-dimethylpropyl

3.14 3,3,4,5-Tetramethylheptane

3.15

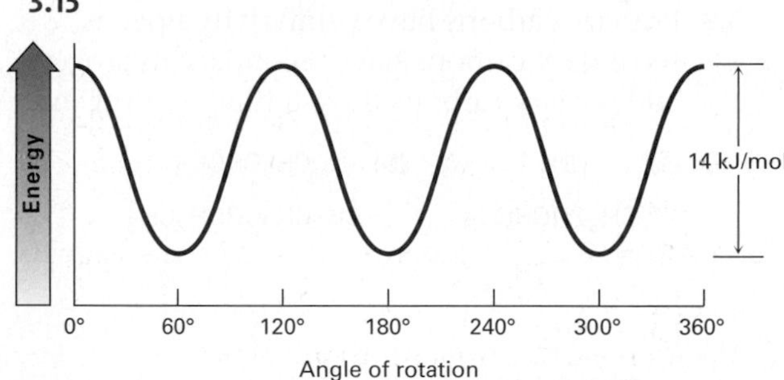

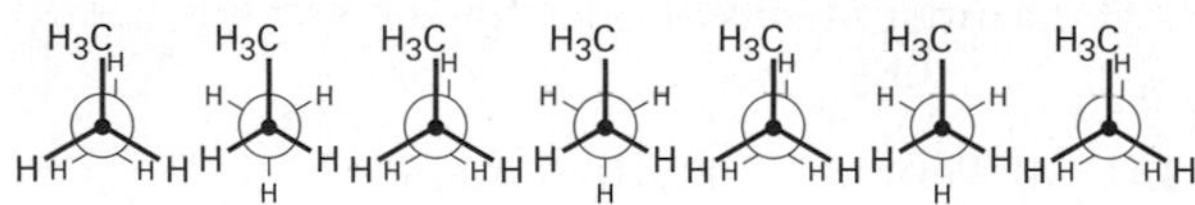

3.16

(a) CH_3, H, H, H, CH_3, H

(b) 4.0 kJ/mol; 6.0 kJ/mol; CH_3, H, H H, H CH_3

(c), (d)

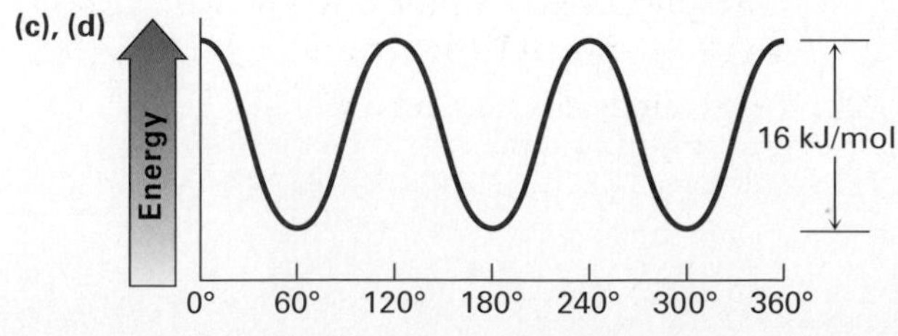

3.17 CH_3, H_3C, H, H, CH_3, CH_3

3.18 CH_3 3.8 kJ/mol; H, CH_3 3.8 kJ/mol; H, CH_3; CH_3 3.8 kJ/mol. Total: 11.4 kJ/mol

Chapter 4

4.1 (a) 1,4-Dimethylcyclohexane
(b) 1-Methyl-3-propylcyclopentane
(c) 3-Cyclobutylpentane
(d) 1-Bromo-4-ethylcyclodecane
(e) 1-Isopropyl-2-methylcyclohexane
(f) 4-Bromo-1-*tert*-butyl-2-methylcycloheptane

4.2

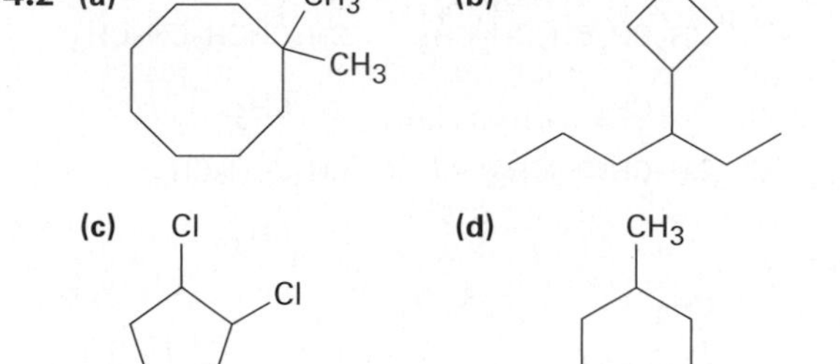

4.3 3-Ethyl-1,1-dimethylcyclopentane

4.4 (a) *trans*-1-Chloro-4-methylcyclohexane
(b) *cis*-1-Ethyl-3-methylcycloheptane

4.5 (a) H_3C, H, Br, H (b) H, CH_3, CH_3, H

(c) CH_2CH_3, H, $C(CH_3)_3$, H

4.6 The two hydroxyl groups are cis. The two side chains are trans.

4.7 (a) *cis*-1,2-Dimethylcyclopentane
(b) *cis*-1-Bromo-3-methylcyclobutane

4.8 Six interactions; 21% of strain

4.9 The cis isomer is less stable because the methyl groups nearly eclipse each other.

4.10 Ten eclipsing interactions; 40 kJ/mol; 35% is relieved.

4.11 Conformation **(a)** is more stable because the methyl groups are farther apart.

4.12

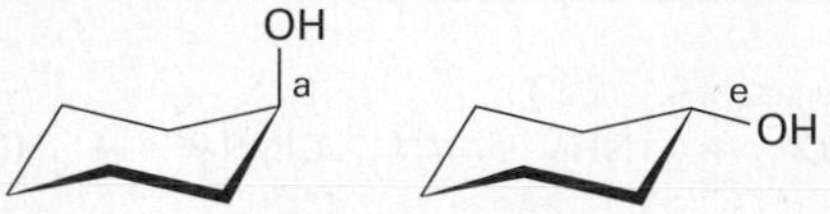

4.13

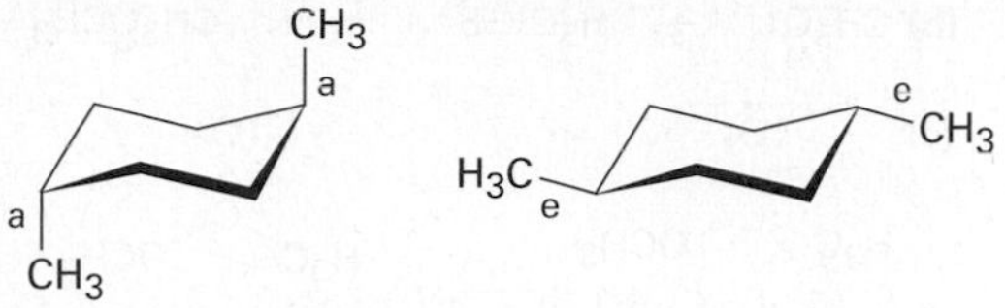

4.14 Before ring-flip, red and blue are equatorial and green is axial. After ring-flip, red and blue are axial and green is equatorial.

4.15 4.2 kJ/mol

4.16 Cyano group points straight up.

4.17 Equatorial = 70%; axial = 30%

4.18 **(a)** 2.0 kJ/mol (axial Cl)
(b) 11.4 kJ/mol (axial CH_3)
(c) 2.0 kJ/mol (axial Br)
(d) 8.0 kJ/mol (axial CH_2CH_3)

4.19

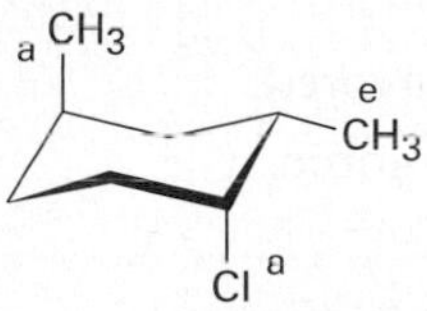

1-Chloro-2,4-dimethylcyclohexane (less stable chair form)

4.20 *trans*-Decalin is more stable because it has no 1,3-diaxial interactions.

4.21 Both ring-fusions are trans.

Chapter 5

5.1 Chiral: screw, shoe

5.2 **(a)** **(b)** 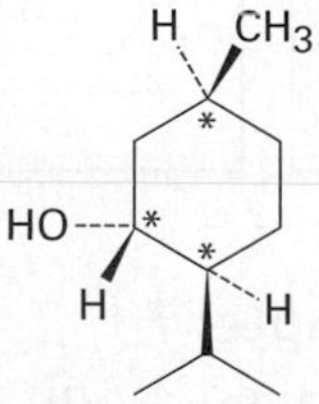 **(c)**

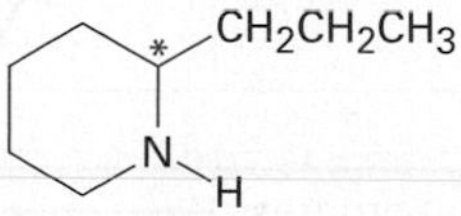

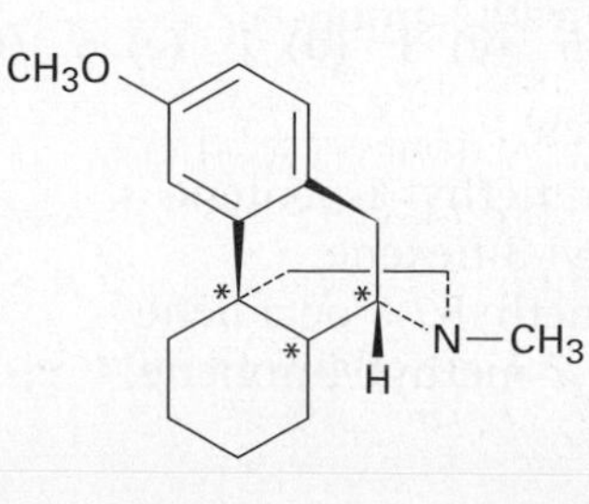

5.3 and

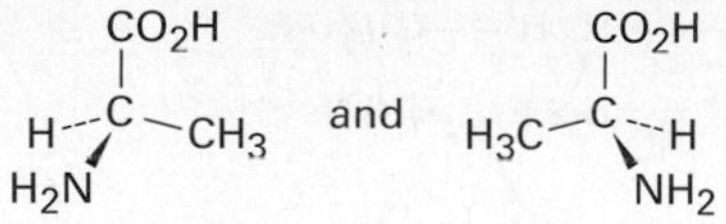

5.4 **(a)** 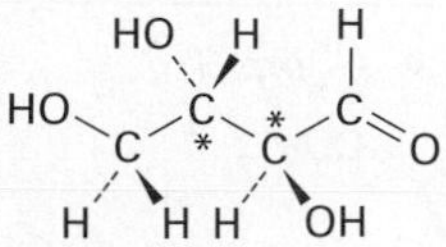 **(b)**

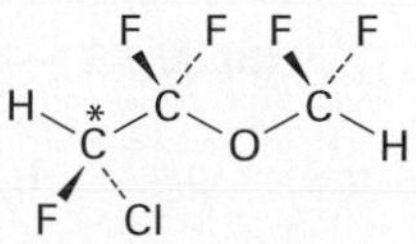

5.5 Levorotatory

5.6 +16.1°

5.7 **(a)** $-Br$ **(b)** $-Br$
(c) $-CH_2CH_3$ **(d)** $-OH$
(e) $-CH_2OH$ **(f)** $-CH{=}O$

5.8 **(a)** $-OH$, $-CH_2CH_2OH$, $-CH_2CH_3$, $-H$
(b) $-OH$, $-CO_2CH_3$, $-CO_2H$, $-CH_2OH$
(c) $-NH_2$, $-CN$, $-CH_2NHCH_3$, $-CH_2NH_2$
(d) $-SSCH_3$, $-SH$, $-CH_2SCH_3$, $-CH_3$

5.9 **(a)** *S* **(b)** *R* **(c)** *S*

5.10 **(a)** *S* **(b)** *S* **(c)** *R*

5.11

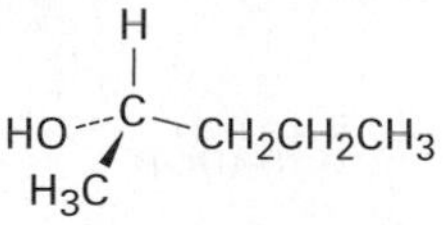

5.12 *S*

5.13 Compound **(a)** is D-erythrose 4-phosphate, **(d)** is its enantiomer, and **(b)** and **(c)** are diastereomers.

5.14 Five chirality centers and $2^5 = 32$ stereoisomers

5.15 *S*,*S*

5.16 Compounds **(a)** and **(d)** are meso.

5.17 Compounds **(a)** and **(c)** have meso forms.

5.18 Meso

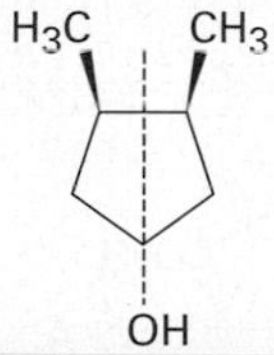

5.19 The product retains its *S* stereochemistry because the chirality center is not affected.

5.20 Two diastereomeric salts: (*R*)-lactic acid plus (*S*)-1-phenylethylamine and (*S*)-lactic acid plus (*S*)-1-phenylethylamine

5.21 **(a)** Constitutional isomers
(b) Diastereomers

5.22 (a) 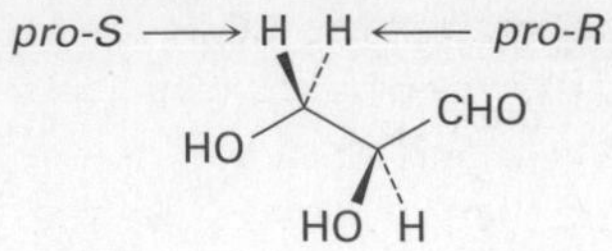 *pro-S* ⟶ H H ⟵ *pro-R*

HO, CHO, HO H

(b) 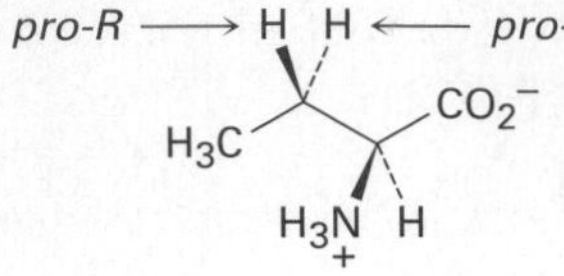 *pro-R* ⟶ H H ⟵ *pro-S*

H_3C, CO_2^-, $H_3\overset{+}{N}$ H

5.23 (a) *Re* face, $H_3C-C(=O)CH_2OH$, *Si* face (b) *Re* face, $H_3C-C(H)=C(H)-CH_2OH$, *Si* face

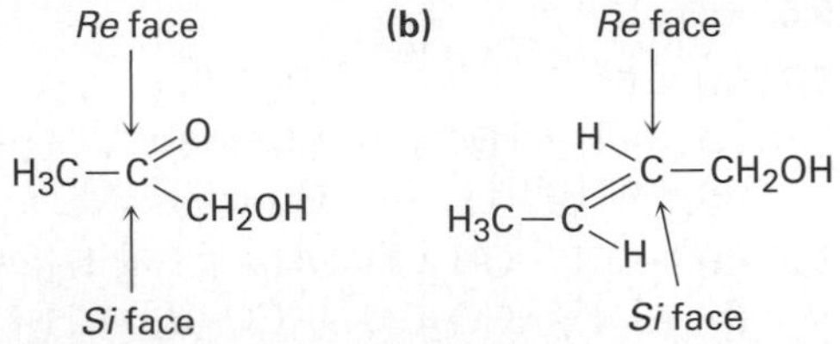

5.24 (*S*)-Lactate

5.25 The –OH adds to the *Re* face of C2, and –H adds to the *Re* face of C3. The overall addition has anti stereochemistry.

Chapter 6

6.1 (a) Substitution (b) Elimination (c) Addition

6.2 1-Chloro-2-methylpentane, 2-chloro-2-methylpentane, 3-chloro-2-methylpentane, 2-chloro-4-methylpentane, 1-chloro-4-methylpentane

6.3 H, O, O, H, CO_2H ↓ H, O, O, H, CO_2H

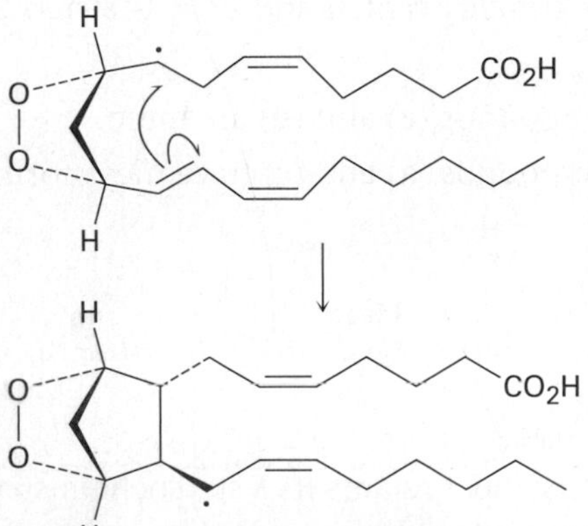

6.4 (a) Carbon is electrophilic.
(b) Sulfur is nucleophilic.
(c) Nitrogens are nucleophilic.
(d) Oxygen is nucleophilic; carbon is electrophilic.

6.5 BF_3 — Electrophilic; vacant *p* orbital

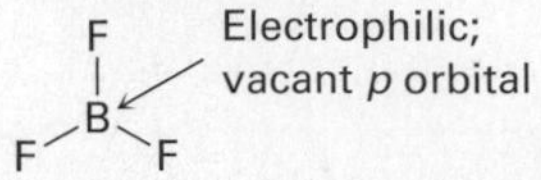

6.6 Bromocyclohexane; chlorocyclohexane

6.7 $(CH_3)_3C^+$

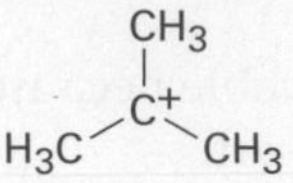

6.8

(a) $Cl-Cl$ + $:NH_3$ ⇌ $ClNH_3^+$ + Cl^-

(b) $CH_3O:^-$ + H_3C-Br ⟶ CH_3OCH_3 + Br^-

(c) ⟶ $H_3C-C(=O)-OCH_3$ + Cl^-

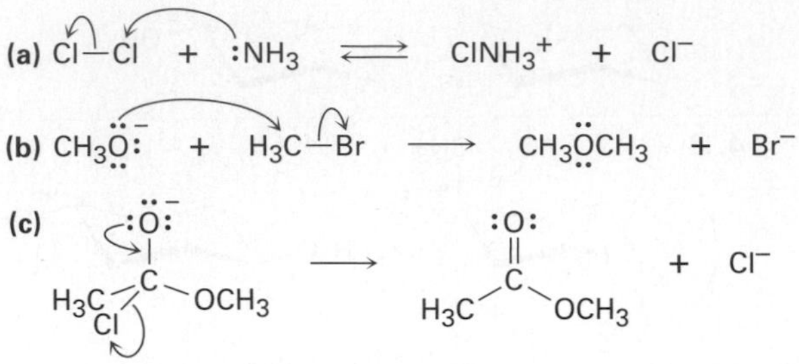

6.9

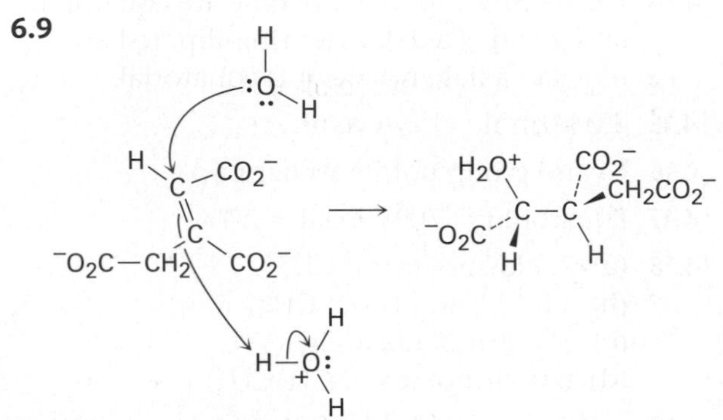

6.10 Negative $\Delta G°$ is more favored.

6.11 Larger K_{eq} is more exergonic.

6.12 Lower $\Delta G^‡$ is faster.

6.13

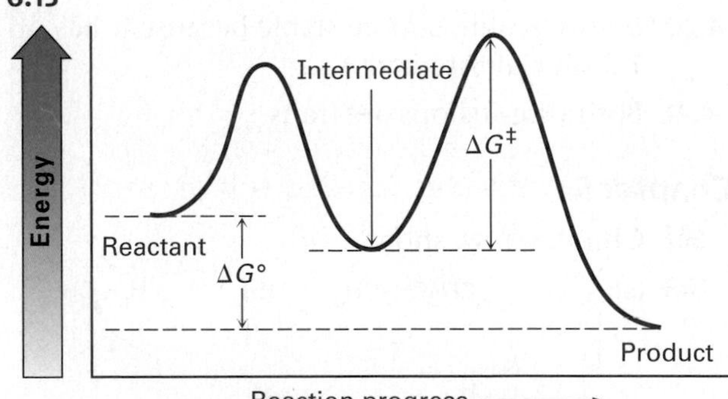

Chapter 7

7.1 (a) 1 (b) 2 (c) 2

7.2 (a) 5 (b) 5 (c) 3 (d) 1 (e) 6 (f) 5

7.3 $C_{16}H_{13}ClN_2O$

7.4 (a) 3,4,4-Trimethyl-1-pentene
(b) 3-Methyl-3-hexene
(c) 4,7-Dimethyl-2,5-octadiene
(d) 6-Ethyl-7-methyl-4-nonene

7.5 (a)
$$H_2C{=}CHCH_2CH_2\overset{CH_3}{\overset{|}{C}}{=}CH_2$$
(b)
$$CH_3CH_2CH_2CH{=}\overset{CH_2CH_3}{\overset{|}{C}}C(CH_3)_3$$
(c)
$$CH_3CH{=}CHCH{=}CH\overset{CH_3}{\overset{|}{\underset{CH_3}{\underset{|}{C}}}}{-}\overset{CH_3}{\overset{|}{C}}{=}CH_2$$
(d)
$$(CH_3)_2CH(CH_3)CH\text{, }(CH_3)_2CH(CH_3)CH \quad C{=}C \quad CH(CH_3)CH(CH_3)_2\text{, }CH(CH_3)CH(CH_3)_2$$

7.6 (a) 1,2-Dimethylcyclohexene
(b) 4,4-Dimethylcycloheptene
(c) 3-Isopropylcyclopentene

7.7 (a) 2,5,5-Trimethylhex-2-ene
(b) 2,3-Dimethylcyclohexa-1,3-diene

7.8 [structure]

7.9 Compounds (c), (e), and (f) have cis–trans isomers.

7.10 (a) *cis*-4,5-Dimethyl-2-hexene
(b) *trans*-6-Methyl-3-heptene

7.11 (a) $-CH_3$ (b) $-Cl$ (c) $-CH{=}CH_2$
(d) $-OCH_3$ (e) $-CH{=}O$ (f) $-CH{=}O$

7.12 (a) $-Cl$, $-OH$, $-CH_3$, $-H$
(b) $-CH_2OH$, $-CH{=}CH_2$, $-CH_2CH_3$, $-CH_3$
(c) $-CO_2H$, $-CH_2OH$, $-C{\equiv}N$, $-CH_2NH_2$
(d) $-CH_2OCH_3$, $-C{\equiv}N$, $-C{\equiv}CH$, $-CH_2CH_3$

7.13 (a) *Z* (b) *E* (c) *Z* (d) *E*

7.14 [structure: CO_2CH_3, CH_2OH] *Z*

7.15 (a) 2-Methylpropene is more stable than 1-butene.
(b) *trans*-2-Hexene is more stable than *cis*-2-hexene.
(c) 1-Methylcyclohexene is more stable than 3-methylcyclohexene.

7.16 (a) Chlorocyclohexane
(b) 2-Bromo-2-methylpentane
(c) 4-Methyl-2-pentanol
(d) 1-Bromo-1-methylcyclohexane

7.17 (a) Cyclopentene
(b) 1-Ethylcyclohexene or ethylidenecyclohexane
(c) 3-Hexene
(d) Vinylcyclohexane (cyclohexylethylene)

7.18 (a)
$$CH_3CH_2\overset{CH_3}{\overset{|}{\underset{+}{C}}}CH_2\overset{CH_3}{\overset{|}{C}}HCH_3$$
(b) [structure: cyclopentyl cation with CH_2CH_3]

7.19 In the conformation shown, only the methyl-group C–H that is parallel to the carbocation *p* orbital can show hyperconjugation.

7.20 The second step is exergonic; the transition state resembles the carbocation.

7.21 [structures: H, H, C, CH_2, H–Br; H, H, +, :Br:⁻, H, H, +; Br]

Chapter 8

8.1 2-Methyl-2-butene and 2-methyl-1-butene

8.2 Five

8.3 *trans*-1,2-Dichloro-1,2-dimethylcyclohexane

8.4 [structures: Cl, CH_3, CH_3, H and CH_3, Cl, CH_3, H]

8.5 *trans*-2-Bromocyclopentanol

8.6 Markovnikov

8.7 (a) 2-Pentanol (b) 2-Methyl-2-pentanol

11.11 $H_2C{=}CHCH(Br)CH_3 > CH_3CH(Br)CH_3 > CH_3CH_2Br > H_2C{=}CHBr$

11.12 The same allylic carbocation intermediate is formed.

11.13 (a) S_N1 (b) S_N2

11.14

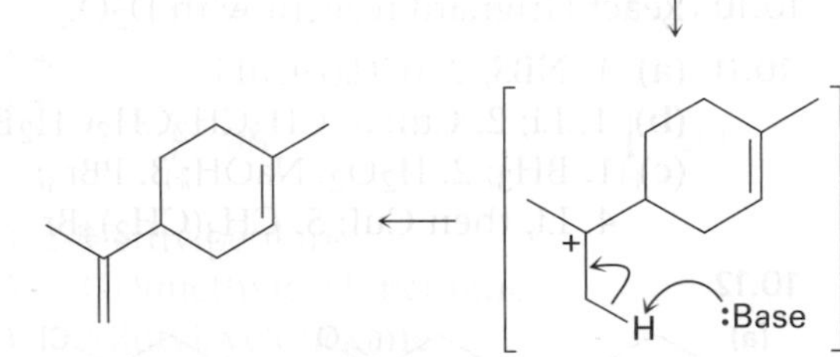

:Base

H

Limonene

11.15 (a) Major: 2-methyl-2-pentene; minor: 4-methyl-2-pentene
(b) Major: 2,3,5-trimethyl-2-hexene; minor: 2,3,5-trimethyl-3-hexene and 2-isopropyl-4-methyl-1-pentene
(c) Major: ethylidenecyclohexane; minor: cyclohexylethylene

11.16 (a) 1-Bromo-3,6-dimethylheptane
(b) 4-Bromo-1,2-dimethylcyclopentane

11.17 (*Z*)-1-Bromo-1,2-diphenylethylene

11.18 (*Z*)-3-Methyl-2-pentene

11.19 Cis isomer reacts faster because the bromine is axial.

11.20 (a) S_N2 (b) E2 (c) S_N1 (d) E1cB

Chapter 12

12.1 $C_{19}H_{28}O_2$

12.2 (a) 2-Methyl-2-pentene (b) 2-Hexene

12.3 (a) 43, 71 (b) 82 (c) 58 (d) 86

12.4 102 (M^+), 84 (dehydration), 87 (alpha cleavage), 59 (alpha cleavage)

12.5 X-ray energy is higher; $\lambda = 9.0 \times 10^{-6}$ m is higher in energy.

12.6 (a) 2.4×10^6 kJ/mol (b) 4.0×10^4 kJ/mol (c) 2.4×10^3 kJ/mol (d) 2.8×10^2 kJ/mol (e) 6.0 kJ/mol (f) 4.0×10^{-2} kJ/mol

12.7 (a) Ketone or aldehyde (b) Nitro compound (c) Carboxylic acid

12.8 (a) CH_3CH_2OH has an –OH absorption.
(b) 1-Hexene has a double-bond absorption.
(c) $CH_3CH_2CO_2H$ has a very broad –OH absorption.

12.9 1450–1600 cm^{-1}: aromatic ring; 2100 cm^{-1}: C≡C; 3300 cm^{-1}: C≡C–H

12.10 (a) 1715 cm^{-1} (b) 1730, 2100, 3300 cm^{-1} (c) 1720, 2500–3100, 3400–3650 cm^{-1}

12.11 1690, 1650, 2230 cm^{-1}

Chapter 13

13.1 7.5×10^{-5} kJ/mol for ^{19}F; 8.0×10^{-5} kJ/mol for ^1H

13.2 1.2×10^{-4} kJ/mol

13.3 The vinylic C–H protons are nonequivalent.

b H, CH_3 a, C=C, c H, Cl

13.4 (a) 7.27 δ (b) 3.05 δ (c) 3.46 δ (d) 5.30 δ

13.5 (a) 420 Hz (b) 2.1 δ (c) 1050 Hz

13.6 (a) 4 (b) 7 (c) 4 (d) 5 (e) 5 (f) 7

13.7 (a) 1,3-Dimethylcyclopentene
(b) 2-Methylpentane
(c) 1-Chloro-2-methylpropane

13.8 $-CH_3$, 9.3 δ; $-CH_2-$, 27.6 δ; C=O, 174.6 δ; $-OCH_3$, 51.4 δ

13.9

23, 26 δ; OH; 124 δ; 24 δ; 18 δ; 132 δ; 39 δ; 68 δ

13.10

DEPT-135 (+) → $H_3C{-}O$; DEPT-135 (–) → $-CH_2-$; DEPT-135 (+) → $-CH_3$; O, $C{-}CH_2{-}CH_3$, C=C; DEPT-135 (+) → H_3C; H ← DEPT-90, DEPT-135 (+)

13.11

$C_6H_5{-}CH_2{-}C(CH_3)_2{-}CH_3$

13.12 A DEPT-90 spectrum would show two absorptions for the non-Markovnikov product (RCH=CHBr) but no absorptions for the Markovnikov product ($RBrC{=}CH_2$).

13.13 (a) Enantiotopic (b) Diastereotopic
(c) Diastereotopic (d) Diastereotopic
(e) Diastereotopic (f) Homotopic

13.14 (a) 2 (b) 4 (c) 3 (d) 4 (e) 5 (f) 3

13.15 4

13.16 (a) 1.43 δ (b) 2.17 δ (c) 7.37 δ
(d) 5.30 δ (e) 9.70 δ (f) 2.12 δ

13.17 Seven kinds of protons

13.18 Two peaks; 3:2 ratio

13.19 (a) $-CHBr_2$, quartet; $-CH_3$, doublet
(b) CH_3O-, singlet; $-OCH_2-$, triplet; $-CH_2Br$, triplet
(c) $ClCH_2-$, triplet; $-CH_2-$, quintet
(d) CH_3-, triplet; $-CH_2-$, quartet; $-CH-$, septet; $(CH_3)_2$, doublet
(e) CH_3-, triplet; $-CH_2-$, quartet; $-CH-$, septet; $(CH_3)_2$, doublet
(f) $=CH$, triplet, $-CH_2-$, doublet, aromatic C–H, two multiplets

13.20 (a) CH_3OCH_3 (b) $CH_3CH(Cl)CH_3$
(c) $ClCH_2CH_2OCH_2CH_2Cl$
(d) $CH_3CH_2CO_2CH_3$ or $CH_3CO_2CH_2CH_3$

13.21 $CH_3CH_2OCH_2CH_3$

13.22 J_{1-2} = 16 Hz; J_{2-3} = 8 Hz

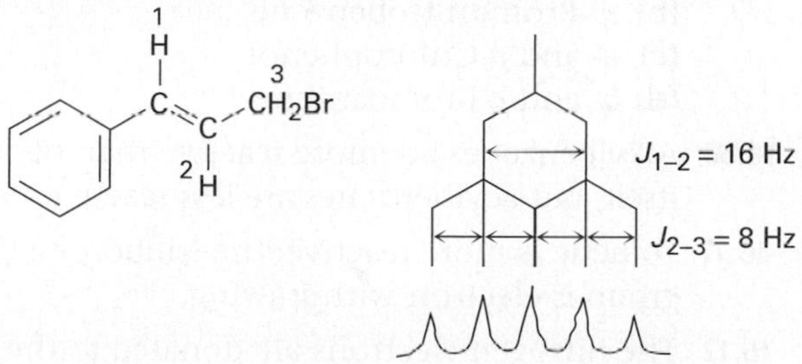

1 H; 3 CH_2Br; 2 H

J_{1-2} = 16 Hz

J_{2-3} = 8 Hz

13.23 1-Chloro-1-methylcyclohexane has a singlet methyl absorption.

Chapter 14

14.1 Expected ΔH°_{hydrog} for allene is -252 kJ/mol. Allene is less stable than a nonconjugated diene, which is less stable than a conjugated diene.

14.2 1-Chloro-2-pentene, 3-chloro-1-pentene, 4-chloro-2-pentene

14.3 4-Chloro-2-pentene predominates in both.

14.4 1,2 Addition: 6-bromo-1,6-dimethylcyclohexene
1,4 Addition: 3-bromo-1,2-dimethylcyclohexene

14.5 Interconversion occurs by S_N1 dissociation to a common intermediate cation.

14.6 The double bond is more highly substituted.

14.7

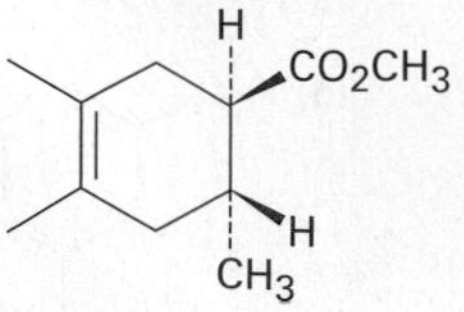

H, CO_2CH_3, H, CH_3

14.8 Good dienophiles: (a), (d)

14.9 Compound (a) is *s*-cis. Compound (c) can rotate to *s*-cis.

14.10

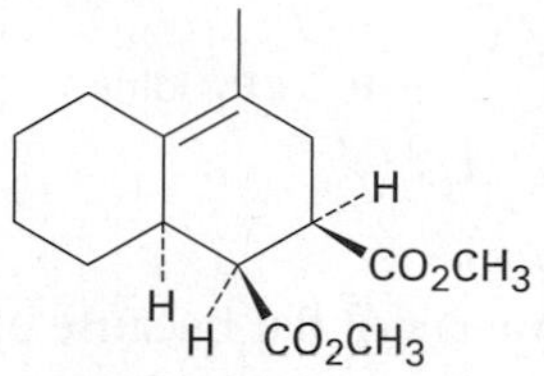

H, CO_2CH_3, H, H, CO_2CH_3

14.11

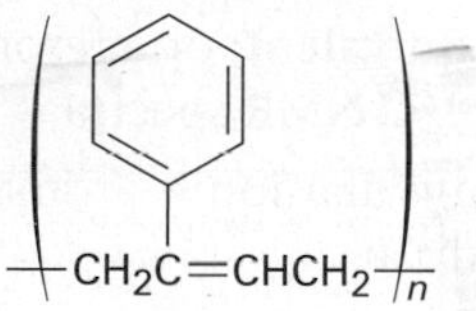

$-\!\!(CH_2C{=}CHCH_2)\!\!-_n$

14.12

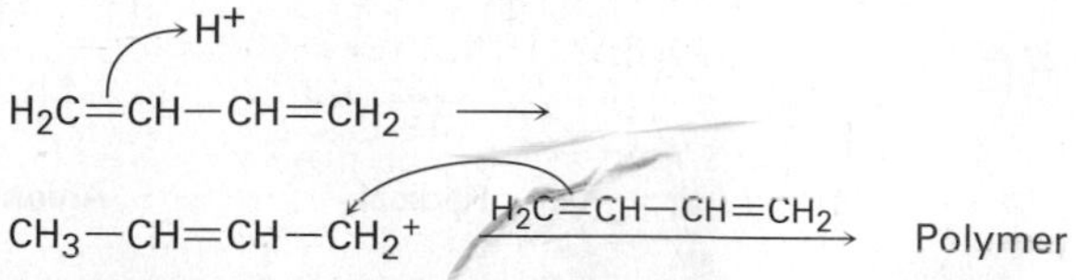

H^+

$H_2C{=}CH{-}CH{=}CH_2 \longrightarrow$

$CH_3{-}CH{=}CH{-}CH_2^+ \xrightarrow{H_2C{=}CH{-}CH{=}CH_2}$ Polymer

14.13 300–600 kJ/mol; UV energy is greater than IR or NMR energy.

14.14 1.46×10^{-5} M

14.15 All except (a) have UV absorptions.

Chapter 15

15.1 (a) Meta (b) Para (c) Ortho

15.2 (a) *m*-Bromochlorobenzene
(b) (3-Methylbutyl)benzene
(c) *p*-Bromoaniline
(d) 2,5-Dichlorotoluene
(e) 1-Ethyl-2,4-dinitrobenzene
(f) 1,2,3,5-Tetramethylbenzene

15.3 (a) Br, Cl (b) Br, CH_3

(c) Cl, NH_2 (d) H_3C, Cl, CH_3

15.4 Pyridine has an aromatic sextet of electrons.

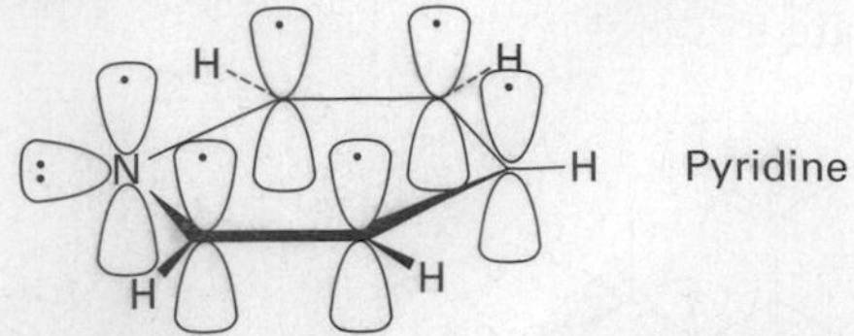

15.5 Cyclodecapentaene is not flat because of steric interactions.

15.6 All C–C bonds are equivalent; one resonance line in both ^{1}H and ^{13}C NMR spectra.

15.7 The cyclooctatetraenyl dianion is aromatic (ten π electrons) and flat.

15.8

Cation Radical Anion

15.9

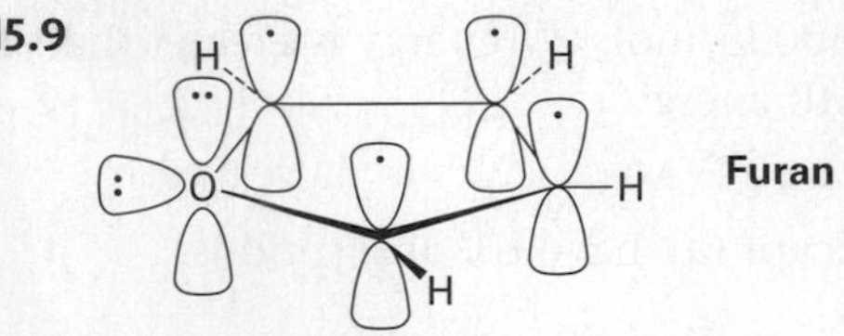

15.10 The thiazolium ring has six π electrons.

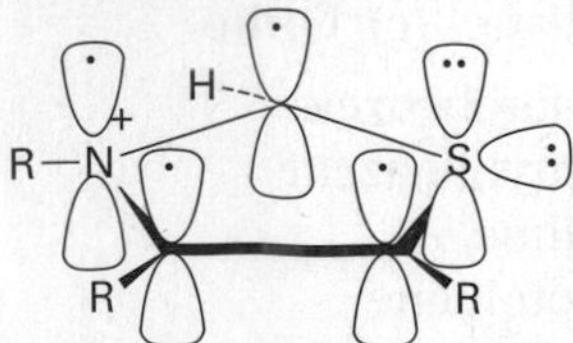

15.11

15.12 The three nitrogens in double bonds each contribute one; the remaining nitrogen contributes two.

Chapter 16

16.1 *o*-, *m*-, and *p*-Bromotoluene

16.2

2 BF_4^- :Base F–N N–CH_2Cl (F-TEDA-BF_4) H F H

F

16.3 *o*-Xylene: 2; *p*-xylene: 1; *m*-xylene: 3

16.4 D^+ does electrophilic substitutions on the ring.

16.5 No rearrangement: (a), (b), (e)

16.6 *tert*-Butylbenzene

16.7 (a) $(CH_3)_2CHCOCl$ (b) PhCOCl

16.8 (a) Phenol > Toluene > Benzene > Nitrobenzene
(b) Phenol > Benzene > Chlorobenzene > Benzoic acid
(c) Aniline > Benzene > Bromobenzene > Benzaldehyde

16.9 (a) *o*- and *p*-Bromonitrobenzene
(b) *m*-Bromonitrobenzene
(c) *o*- and *p*-Chlorophenol
(d) *o*- and *p*-Bromoaniline

16.10 Alkylbenzenes are more reactive than benzene itself, but acylbenzenes are less reactive.

16.11 Toluene is more reactive; the trifluoromethyl group is electron-withdrawing.

16.12 The nitrogen electrons are donated to the nearby carbonyl group by resonance and are less available to the ring.

16.13 The meta intermediate is most favored.

16.14 (a) Ortho and para to $-OCH_3$
(b) Ortho and para to $-NH_2$
(c) Ortho and para to $-Cl$

16.15 (a) Reaction occurs ortho and para to the $-CH_3$ group.
(b) Reaction occurs ortho and para to the $-OCH_3$ group.

16.16 The phenol is deprotonated by KOH to give an anion that carries out a nucleophilic acyl substitution reaction on the fluoronitrobenzene.

16.17 Only one benzyne intermediate can form from *p* bromotoluene; two different benzyne intermediates can form from *m*-bromotoluene.

16.18 (a) *m*-Nitrobenzoic acid
(b) *p*-*tert*-Butylbenzoic acid

16.19 A benzyl radical is more stable than a primary alkyl radical by 52 kJ/mol and is similar in stability to an allyl radical.

16.20 1. CH_3CH_2Cl, $AlCl_3$; 2. NBS; 3. KOH, ethanol

16.21 1. PhCOCl, $AlCl_3$; 2. H_2/Pd

16.22 (a) 1. HNO_3, H_2SO_4; 2. Cl_2, $FeCl_3$
(b) 1. CH_3COCl, $AlCl_3$; 2. Cl_2, $FeCl_3$; 3. H_2/Pd
(c) 1. CH_3CH_2COCl, $AlCl_3$; 2. Cl_2, $FeCl_3$; 3. H_2/Pd; 4. HNO_3, H_2SO_4
(d) 1. CH_3Cl, $AlCl_3$; 2. Br_2, $FeBr_3$; 3. SO_3, H_2SO_4

16.23 (a) Friedel–Crafts acylation does not occur on a deactivated ring.
(b) Rearrangement occurs during Friedel–Crafts alkylation with primary halides; chlorination occurs ortho to the alkyl group.

Chapter 17

17.1 (a) 5-Methyl-2,4-hexanediol
(b) 2-Methyl-4-phenyl-2-butanol
(c) 4,4-Dimethylcyclohexanol
(d) *trans*-2-Bromocyclopentanol
(e) 4-Bromo-3-methylphenol
(f) 2-Cyclopenten-1-ol

17.2 (a) H_3C, CH_2OH, C=C, H, CH_2CH_3
(b) OH
(c) OH, H, Cl, H
(d) $CH_3CH(OH)CH_2CH_2CH_2OH$
(e) OH, H_3C, CH_3
(f) OH, CH_2CH_2OH

17.3 Hydrogen-bonding is more difficult in hindered alcohols.

17.4 (a) $HC{\equiv}CH < (CH_3)_2CHOH < CH_3OH < (CF_3)_2CHOH$
(b) *p*-Methylphenol < Phenol < *p*-(Trifluoromethyl)phenol
(c) Benzyl alcohol < Phenol < *p*-Hydroxybenzoic acid

17.5 The electron-withdrawing nitro group stabilizes an alkoxide ion, but the electron-donating methoxyl group destabilizes the anion.

17.6 (a) 2-Methyl-3-pentanol
(b) 2-Methyl-4-phenyl-2-butanol
(c) *meso*-5,6-Decanediol

17.7 (a) $NaBH_4$ (b) $LiAlH_4$ (c) $LiAlH_4$

17.8 (a) Benzaldehyde or benzoic acid (or ester)
(b) Acetophenone (c) Cyclohexanone
(d) 2-Methylpropanal or 2-methylpropanoic acid (or ester)

17.9 (a) 1-Methylcyclopentanol
(b) 1,1-Diphenylethanol
(c) 3-Methyl-3-hexanol

17.10 (a) Acetone + CH_3MgBr, or ethyl acetate + 2 CH_3MgBr
(b) Cyclohexanone + CH_3MgBr
(c) 3-Pentanone + CH_3MgBr, or 2-butanone + CH_3CH_2MgBr, or ethyl acetate + 2 CH_3CH_2MgBr
(d) 2-Butanone + PhMgBr, or ethyl phenyl ketone + CH_3MgBr, or acetophenone + CH_3CH_2MgBr
(e) Formaldehyde + PhMgBr
(f) Formaldehyde + $(CH_3)_2CHCH_2MgBr$

17.11 Cyclohexanone + CH_3CH_2MgBr

17.12 1. *p*-TosCl, pyridine; 2. NaCN

17.13 (a) 2-Methyl-2-pentene
(b) 3-Methylcyclohexene
(c) 1-Methylcyclohexene
(d) 2,3-Dimethyl-2-pentene
(e) 2-Methyl-2-pentene

17.14 (a) 1-Phenylethanol
(b) 2-Methyl-1-propanol
(c) Cyclopentanol

17.15 (a) Hexanoic acid, hexanal
(b) 2-Hexanone
(c) Hexanoic acid, no reaction

17.16 S_N2 reaction of F^- on silicon with displacement of alkoxide ion.

17.17 Protonation of 2-methylpropene gives the *tert*-butyl cation, which carries out an electrophilic aromatic substitution reaction.

17.18 Disappearance of –OH absorption; appearance of C=O

17.19 (a) Singlet (b) Doublet (c) Triplet (d) Doublet (e) Doublet (f) Singlet

Chapter 18

18.1 (a) Diisopropyl ether
(b) Cyclopentyl propyl ether
(c) *p*-Bromoanisole or 4-bromo-1-methoxybenzene
(d) 1-Methoxycyclohexene
(e) Ethyl isobutyl ether
(f) Allyl vinyl ether

18.2 A mixture of diethyl ether, dipropyl ether, and ethyl propyl ether is formed in a 1:1:2 ratio.

18.3 (a) $CH_3CH_2CH_2O^- + CH_3Br$
(b) $PhO^- + CH_3Br$
(c) $(CH_3)_2CHO^- + PhCH_2Br$
(d) $(CH_3)_3CCH_2O^- + CH_3CH_2Br$

18.4

CH_3 (1-methylcyclopentene) $\xrightarrow{Hg(O_2CCF_3)_2}$ [H_3C, $^+HgOCOCF_3$ bridged cation] $\xrightarrow{H\ddot{O}CH_2CH_3}$ [H_3C OCH_2CH_3, HgO_2CCF_3] $\xrightarrow{NaBH_4}$ H_3C OCH_2CH_3

18.5 (a) Either method (b) Williamson (c) Alkoxymercuration (d) Williamson

18.6 (a) Bromoethane > 2-Bromopropane > Bromobenzene
(b) Bromoethane > Chloroethane > 1-Iodopropene

18.7 (a) (2-bromo-2-propyl)benzene, Br + CH_3OH

(b) $CH_3CH_2CH(CH_3)OH$ + $CH_3CH_2CH_2Br$

18.8 Protonation of the oxygen atom, followed by E1 reaction

18.9 Br^- and I^- are better nucleophiles than Cl^-.

18.10 *o*-(1-Methylallyl)phenol

18.11 Epoxidation of *cis*-2-butene yields *cis*-2,3-epoxybutane, while epoxidation of *trans*-2-butene yields *trans*-2,3-epoxybutane.

18.12 (a) cyclohexyl–CH(OH)–C(CH₃)₂Cl (OH, Cl) (b) cyclohexane with Cl and CH_2OH on the same carbon

18.13 (a) 1-Methylcyclohexene + OsO_4; then $NaHSO_3$
(b) 1-Methylcyclohexene + *m*-chloroperoxybenzoic acid, then H_3O^+

18.14 (a) $CH_3CH_2C(CH_3)(OH)-CH_2(^*OH)$ (HO, *OH) (b) $CH_3CH_2C(CH_3)(HO^*)-CH_2OH$ (HO*, OH)

(c) $C_6H_5-CH(CH_3)-C(OH)(CH_3)CH_2CH_3$

18.16 (a) 2-Butanethiol
(b) 2,2,6-Trimethyl-4-heptanethiol
(c) 2-Cyclopentene-1-thiol
(d) Ethyl isopropyl sulfide
(e) *o*-Di(methylthio)benzene
(f) 3-(Ethylthio)cyclohexanone

18.17 (a) 1. $LiAlH_4$; 2. PBr_3; 3. $(H_2N)_2C{=}S$; 4. H_2O, NaOH
(b) 1. HBr; 2. $(H_2N)_2C{=}S$; 3. H_2O, NaOH

18.18 1,2-Epoxybutane

Preview of Carbonyl Chemistry

1. Acetyl chloride is more electrophilic than acetone.

2. $H_3C-C(=O)-CH_3 \xrightarrow{^-CN} [(H_3C)_2C(O^-)CN] \xrightarrow{H_3O^+} (H_3C)_2C(OH)CN$

3. (a) Nucleophilic acyl substitution
(b) Nucleophilic addition
(c) Carbonyl condensation

Chapter 19

19.1 (a) 2-Methyl-3-pentanone
(b) 3-Phenylpropanal
(c) 2,6-Octanedione
(d) *trans*-2-Methylcyclohexanecarbaldehyde
(e) 4-Hexenal
(f) *cis*-2,5-Dimethylcyclohexanone

19.2
(a) $CH_3CH(CH_3)CH_2CHO$
(b) $CH_3CH(Cl)CH_2C(=O)CH_3$
(c) $C_6H_5CH_2CHO$
(d) $(CH_3)_3C$–cyclohexane–CHO (H, H)
(e) $H_2C{=}C(CH_3)CH_2CHO$
(f) $CH_3CH_2CH(CH_3)CH_2CH_2CH(CH_2CHCl)CHO$

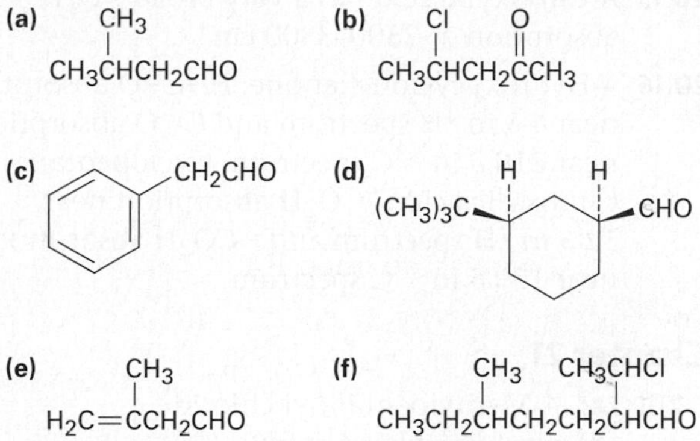

19.3 (a) Dess–Martin periodinane (b) 1. O_3; 2. Zn
(c) DIBAH
(d) 1. BH_3, then H_2O_2, NaOH;
2. Dess–Martin periodinane

19.4 (a) $HgSO_4$, H_3O^+
(b) 1. CH_3COCl, $AlCl_3$; 2. Br_2, $FeBr_3$
(c) 1. Mg; 2. CH_3CHO; 3. H_3O^+; 4. CrO_3
(d) 1. BH_3; 2. H_2O_2, NaOH; 3. CrO_3

19.5 Cyclohexane with CN and OH on the same carbon

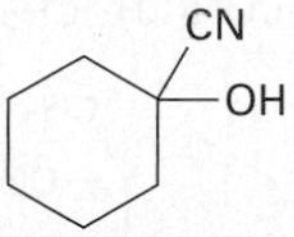

19.6 The electron withdrawing nitro group in *p*-nitrobenzaldehyde polarizes the carbonyl group.

19.7 $CCl_3CH(OH)_2$

19.8 Labeled water adds reversibly to the carbonyl group.

19.9 The equilibrium is unfavorable for sterically hindered ketones.

19.10 Cyclohexylidene=NCH_2CH_3 and cyclohexenyl–$N(CH_2CH_3)_2$

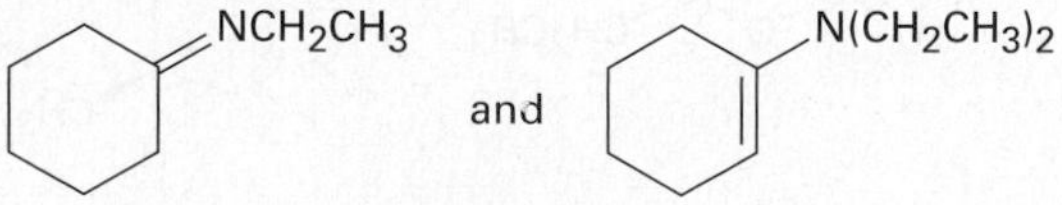

19.11 The steps are the exact reverse of the forward reaction, shown in Figure 19.6.

19.12 Cyclopentanone + $(CH_3CH_2)_2NH$ ⟶ cyclopentenyl–$N(CH_2CH_3)_2$

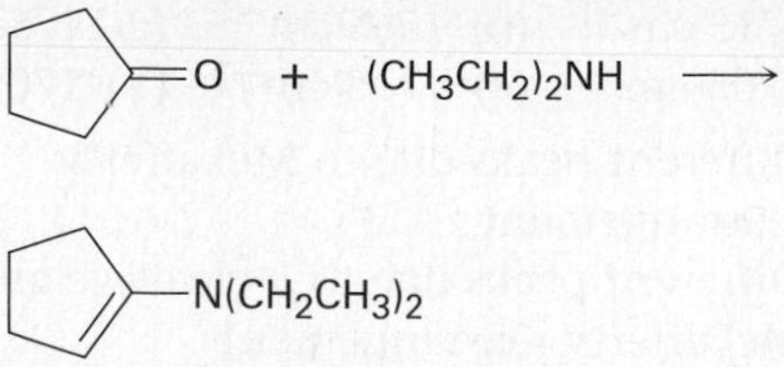

19.13 (a) H_2/Pd (b) N_2H_4, KOH
(c) 1. H_2/Pd; 2. N_2H_4, KOH

19.14 The mechanism is identical to that between a ketone and 2 equivalents of a monoalcohol, shown in Figure 19.10.

19.15 CH_3O_2C–C_6H_4–$CH(CH_3)CHO$ + CH_3OH

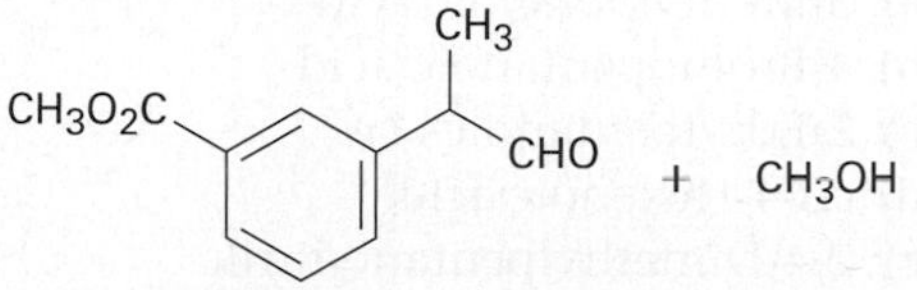

19.16 (a) Cyclohexanone + $(Ph)_3P{=}CHCH_3$
(b) Cyclohexanecarbaldehyde + $(Ph)_3P{=}CH_2$
(c) Acetone + $(Ph)_3P{=}CHCH_2CH_2CH_3$
(d) Acetone + $(Ph)_3P{=}CHPh$
(e) $PhCOCH_3$ + $(Ph)_3P{=}CHPh$
(f) 2-Cyclohexenone + $(Ph)_3P{=}CH_2$

19.17

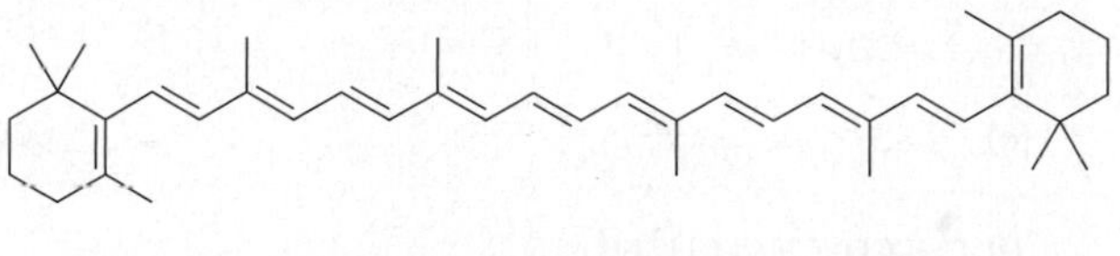

β-Carotene

19.18 Intramolecular Cannizzaro reaction

19.19 Addition of the *pro-R* hydrogen of NADH takes place on the *Re* face of pyruvate.

19.20 The –OH group adds to the *Re* face at C2, and –H adds to the *Re* face at C3, to yield (2*R*,3*S*)-isocitrate.

19.21 3-Oxocyclohexanecarbonitrile (O=, CN)

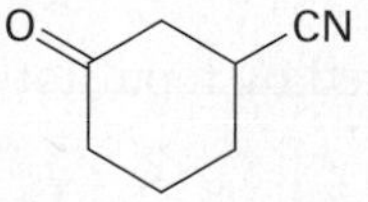

19.22 (a) 3-Buten-2-one + $(CH_3CH_2CH_2)_2CuLi$
(b) 3-Methyl-2-cyclohexenone + $(CH_3)_2CuLi$
(c) 4-*tert*-Butyl-2-cyclohexenone + $(CH_3CH_2)_2CuLi$
(d) Unsaturated ketone + $(H_2C{=}CH)_2CuLi$

19.23 Look for appearance of either an alcohol or a saturated ketone in the product.

19.24 (a) 1715 cm^{-1} (b) 1685 cm^{-1} (c) 1750 cm^{-1} (d) 1705 cm^{-1} (e) 1715 cm^{-1} (f) 1705 cm^{-1}

19.25 (a) Different peaks due to McLafferty rearrangement
(b) Different peaks due to α cleavage and McLafferty rearrangement
(c) Different peaks due to McLafferty rearrangement

19.26 IR: 1750 cm^{-1}; MS: 140, 84

Chapter 20

20.1 (a) 3-Methylbutanoic acid
(b) 4-Bromopentanoic acid
(c) 2-Ethylpentanoic acid
(d) *cis*-4-Hexenoic acid
(e) 2,4-Dimethylpentanenitrile
(f) *cis*-1,3-Cyclopentanedicarboxylic acid

20.2

(a) $CH_3CH_2CH_2CH(CH_3)CH(CH_3)CO_2H$ (b) $CH_3CH(CH_3)CH_2CH_2CO_2H$

(c) cyclobutane with H, CO_2H and H, CO_2H (d) benzene ring with CO_2H and OH

(e) skeletal chain with two C=C bonds ending in CO_2H

(f) $CH_3CH_2CH{=}CHCN$

20.3 Dissolve the mixture in ether, extract with aqueous NaOH, separate and acidify the aqueous layer, and extract with ether.

20.4 43%

20.5 (a) 82% dissociation (b) 73% dissociation

20.6 Lactic acid is stronger because of the inductive effect of the –OH group.

20.7 The dianion is destabilized by repulsion between charges.

20.8 More reactive

20.9 (a) *p*-Methylbenzoic acid < Benzoic acid < *p*-Chlorobenzoic acid
(b) Acetic acid < Benzoic acid < *p*-Nitrobenzoic acid

20.10 (a) 1. Mg; 2. CO_2; 3. H_3O^+
(b) 1. Mg; 2. CO_2; 3. H_3O^+ or 1. NaCN; 2. H_3O^+

20.11 1. NaCN; 2. H_3O^+; 3. $LiAlH_4$

20.12 1. PBr_3; 2. NaCN; 3. H_3O^+; 4. $LiAlH_4$

20.13 (a) Propanenitrile + CH_3CH_2MgBr, then H_3O^+
(b) *p*-Nitrobenzonitrile + CH_3MgBr, then H_3O^+

20.14 1. NaCN; 2. CH_3CH_2MgBr, then H_3O^+

20.15 A carboxylic acid has a very broad –OH absorption at 2500–3300 cm^{-1}.

20.16 4-Hydroxycyclohexanone: **H**–C–O absorption near 4 δ in ^1H spectrum and **C**=O absorption near 210 δ in ^{13}C spectrum. Cyclopentanecarboxylic acid: $-CO_2\mathbf{H}$ absorption near 12 δ in ^1H spectrum and $-\mathbf{C}O_2H$ absorption near 170 δ in ^{13}C spectrum.

Chapter 21

21.1 (a) 4-Methylpentanoyl chloride
(b) Cyclohexylacetamide
(c) Isopropyl 2-methylpropanoate
(d) Benzoic anhydride
(e) Isopropyl cyclopentanecarboxylate
(f) Cyclopentyl 2-methylpropanoate
(g) *N*-Methyl-4-pentenamide
(h) (*R*)-2-Hydroxypropanoyl phosphate
(i) Ethyl 2,3-Dimethyl-2-butenethioate

21.2

(a) $C_6H_5CO_2C_6H_5$ (b) $CH_3CH_2CH_2CON(CH_3)CH_2CH_3$

(c) $(CH_3)_2CHCH_2CH(CH_3)COCl$ (d) cyclohexane with CH_3 and CO_2CH_3 on the same carbon

(e) $CH_3CH_2C(=O)CH_2C(=O)OCH_2CH_3$ (f) *p*-bromophenyl–$C(=O)SCH_3$

(g) $HC(=O)OC(=O)CH_2CH_3$ (h) cyclopentane with H, COBr and H, CH_3

21.3

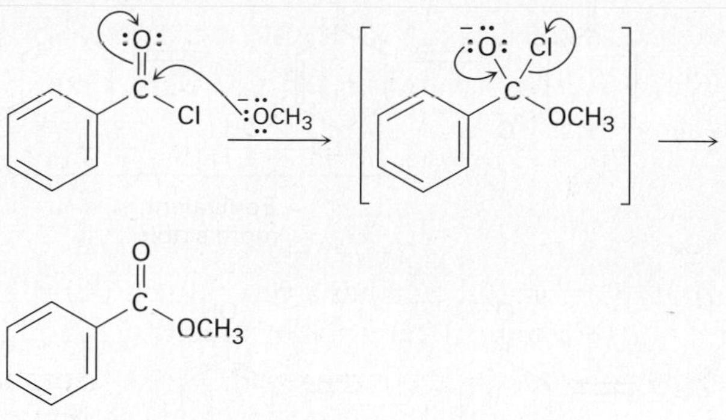

21.4 (a) Acetyl chloride > Methyl acetate > Acetamide
(b) Hexafluoroisopropyl acetate > 2,2,2-Trichloroethyl acetate > Ethyl acetate

21.5 (a) $CH_3CO_2^-\ Na^+$ (b) CH_3CONH_2
(c) $CH_3CO_2CH_3 + CH_3CO_2^-\ Na^+$
(d) $CH_3CONHCH_3$

21.6

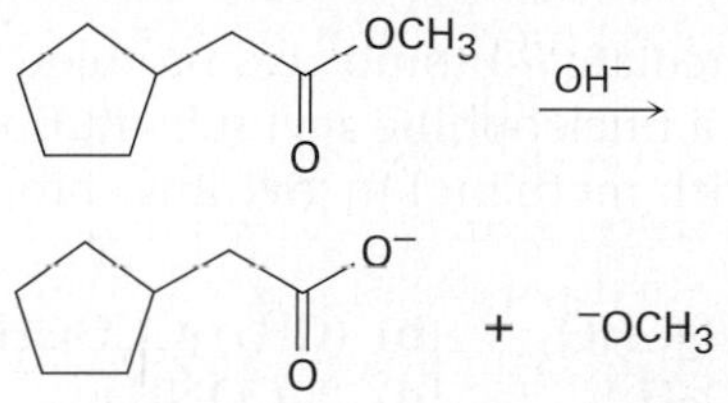

21.7 (a) Acetic acid + 1-butanol
(b) Butanoic acid + methanol
(c) Cyclopentanecarboxylic acid + isopropyl alcohol

21.8

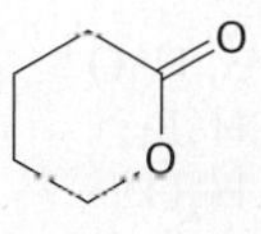

21.9 (a) Propanoyl chloride + methanol
(b) Acetyl chloride + ethanol
(c) Benzoyl chloride + ethanol

21.10 Benzoyl chloride + cyclohexanol

21.11 This is a typical nucleophilic acyl substitution reaction, with morpholine as the nucleophile and chloride as the leaving group.

21.12 (a) Propanoyl chloride + methylamine
(b) Benzoyl chloride + diethylamine
(c) Propanoyl chloride + ammonia

21.13 (a) Benzoyl chloride + $[(CH_3)_2CH]_2CuLi$, or 2-methylpropanoyl chloride + Ph_2CuLi
(b) 2-Propenoyl chloride + $(CH_3CH_2CH_2)_2CuLi$, or butanoyl chloride + $(H_2C{=}CH)_2CuLi$

21.14 This is a typical nucleophilic acyl substitution reaction, with *p*-hydroxyaniline as the nucleophile and acetate ion as the leaving group.

21.15 Monomethyl ester of benzene-1,2-dicarboxylic acid

21.16 Reaction of a carboxylic acid with an alkoxide ion gives the carboxylate ion.

21.17 $LiAlH_4$ gives $HOCH_2CH_2CH_2CH_2OH$; DIBAH gives $HOCH_2CH_2CH_2CHO$

21.18 (a) $CH_3CH_2CH_2CH(CH_3)CH_2OH + CH_3OH$
(b) $PhOH + PhCH_2OH$

21.19 (a) Ethyl benzoate + 2 CH_3MgBr
(b) Ethyl acetate + 2 PhMgBr
(c) Ethyl pentanoate + 2 CH_3CH_2MgBr

21.20 (a) H_2O, NaOH (b) Benzoic acid + $LiAlH_4$
(c) $LiAlH_4$

21.21 1. Mg; 2. CO_2, then H_3O^+; 3. $SOCl_2$; 4. $(CH_3)_2NH$; 5. $LiAlH_4$

21.22

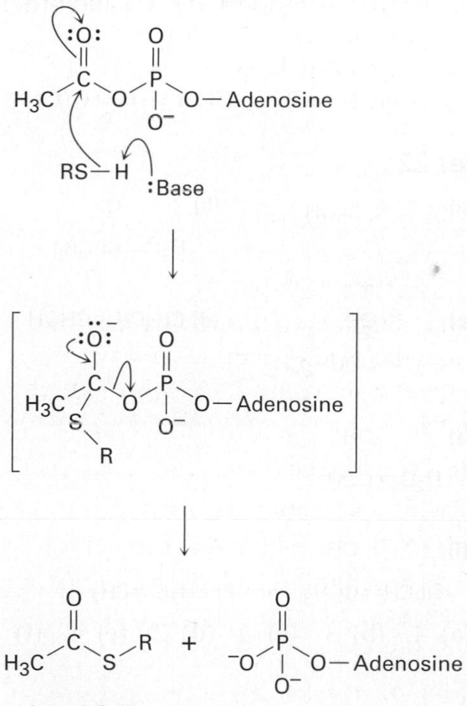

Acetyl CoA

21.23 (a) $-\!\!(OCH_2CH_2CH_2OCH_2CH_2CH_2)\!\!-_n$

(b) $-\!\!(OCH_2CH_2OC(=O)(CH_2)_6C(=O))\!\!-_n$

(c) $-\!\!(NH(CH_2)_6NHC(=O)(CH_2)_4C(=O))\!\!-_n$

21.24 $-\!\!(NH{-}C_6H_4{-}NH{-}C(=O){-}C_6H_4{-}C(=O))\!\!-_n$

21.25 (a) Ester (b) Acid chloride
(c) Carboxylic acid
(d) Aliphatic ketone or cyclohexanone

21.26 (a) $CH_3CH_2CH_2CO_2CH_2CH_3$ and other possibilities
(b) $CH_3CON(CH_3)_2$
(c) $CH_3CH{=}CHCOCl$ or $H_2C{=}C(CH_3)COCl$

Chapter 22

22.1 (a) 1-cyclopentenol (cyclopentene ring with OH on the double-bond carbon) (b) $H_2C{=}C(OH)SCH_3$

(c) $H_2C{=}C(OH)OCH_2CH_3$ (d) $CH_3CH{=}CHOH$

(e) $H_2C{=}C(OH)OH$

(f) $PhCH{=}C(OH)CH_3$ or $PhCH_2C(OH){=}CH_2$

22.2 (a) 4 (b) 3 (c) 3 (d) 2 (e) 4 (f) 5

22.3 1,3-cyclohexanedione ⇌ enol (OH at C3, C2=C3 double bond) ⇌ enol (OH at C1, C1=C2 double bond)

Equivalent; more stable

⇌ enol (C3=C4 double bond, OH at C3) ⇌ enol (C1=C6 double bond, OH at C1)

Equivalent; less stable

22.4 Acid-catalyzed formation of an enol is followed by deuteronation of the enol double bond and dedeuteronation of oxygen.

22.5 1. Br_2; 2. Pyridine, heat

22.6 The intermediate α-bromo acid bromide undergoes a nucleophilic acyl substitution reaction with methanol to give an α-bromo ester.

22.7 (a) CH_3CH_2CHO (b) $(CH_3)_3CCOCH_3$
(c) CH_3CO_2H (d) $PhCONH_2$
(e) $CH_3CH_2CH_2CN$ (f) $CH_3CON(CH_3)_2$

22.8 $^{-}{:}CH_2C{\equiv}N{:} \longleftrightarrow H_2C{=}C{=}\ddot{N}{:}^{-}$

22.9 Acid is regenerated, but base is used stoichiometrically.

22.10 (a) 1. $Na^+\ ^-OEt$; 2. $PhCH_2Br$; 3. H_3O^+
(b) 1. $Na^+\ ^-OEt$; 2. $CH_3CH_2CH_2Br$; 3. $Na^+\ ^-OEt$; 4. CH_3Br; 5. H_3O^+
(c) 1. $Na^+\ ^-OEt$; 2. $(CH_3)_2CHCH_2Br$; 3. H_3O^+

22.11 Malonic ester has only two acidic hydrogens to be replaced.

22.12 1. $Na^+\ ^-OEt$; 2. $(CH_3)_2CHCH_2Br$; 3. $Na^+\ ^-OEt$; 4. CH_3Br; 5. H_3O^+

22.13 (a) $(CH_3)_2CHCH_2Br$ (b) $PhCH_2CH_2Br$

22.14 None can be prepared.

22.15 1. 2 $Na^+\ ^-OEt$; 2. $BrCH_2CH_2CH_2CH_2Br$; 3. H_3O^+

22.16 (a) Alkylate phenylacetone with CH_3I
(b) Alkylate pentanenitrile with CH_3CH_2I
(c) Alkylate cyclohexanone with $H_2C{=}CHCH_2Br$
(d) Alkylate cyclohexanone with excess CH_3I
(e) Alkylate $C_6H_5COCH_2CH_3$ with CH_3I
(f) Alkylate methyl 3-methylbutanoate with CH_3CH_2I

Chapter 23

23.1 (a) $CH_3CH_2CH_2CH(OH)CH(CH_2CH_3)CHO$

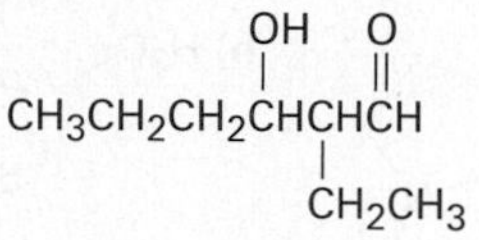

(b) PhCOCH₂C(OH)(CH₃)Ph

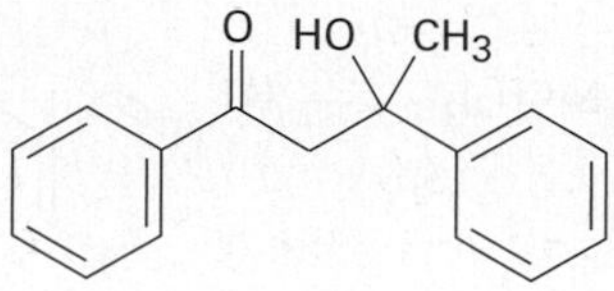

(c) [structure: cyclopentanol–cyclopentanone aldol product; labels OH, O]

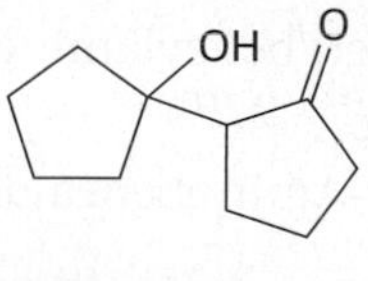

23.2 The reverse reaction is the exact opposite of the forward reaction, shown in Figure 23.1.

23.3

(a) [structure: 2-cyclopentylidenecyclopentanone; label O]

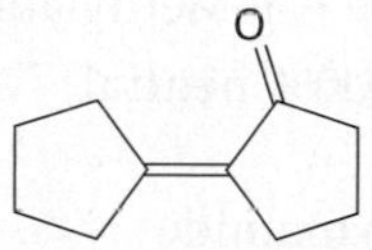

(b) $C_6H_5C(CH_3){=}CHCOC_6H_5$

(c) $(CH_3)_2CHCH_2CH{=}C(CH(CH_3)_2)CHO$

23.4 [structures; labels O, CH₃, H₃C]

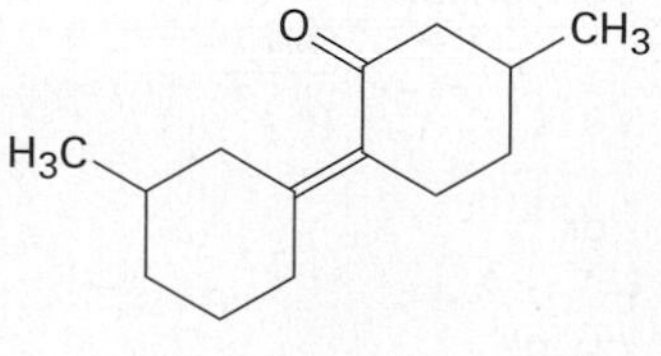

and

[structure; labels O, H₃C, CH₃]

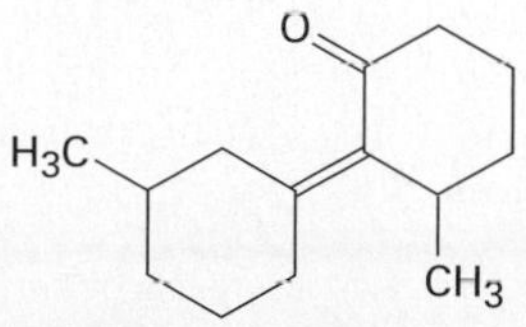

23.5 (a) Not an aldol product (b) 3-Pentanone

23.6 1. NaOH; 2. $LiAlH_4$; 3. H_2/Pd

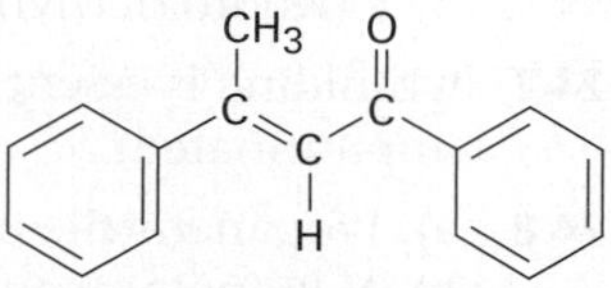

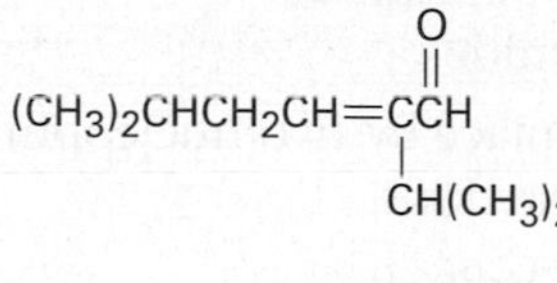

23.7 [structure: cyclopropyl acetaldehyde →(NaOH) aldol condensation product; labels CHO, H, NaOH, CHO, H, H]

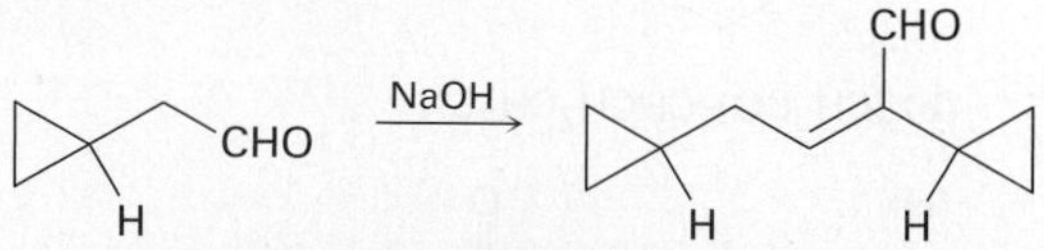

23.8 (a) $C_6H_5CHO + CH_3COCH_3$
(b), (c) Not easily prepared

23.9 The CH_2 position between the two carbonyl groups is so acidic that it is completely deprotonated to give a stable enolate ion.

23.10 [structure: bicyclic enone; label O]

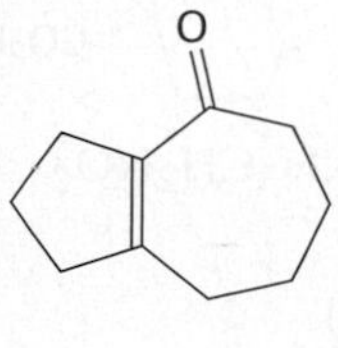

23.11 (a) $(CH_3)_2CHCH_2COCH(CH(CH_3)_2)COOEt$ (b) $PhCH_2COCH(Ph)COOEt$

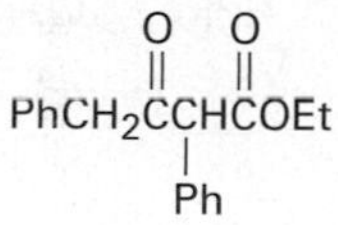

(c) $C_6H_{11}CH_2COCH(C_6H_{11})COOEt$

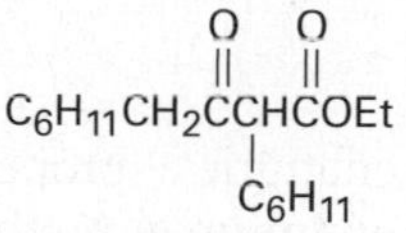

23.12 The cleavage reaction is the exact reverse of the forward reaction.

23.13 [structure: 3,3-dimethyl-substituted cyclohexanone bearing C(=O)–COCH₃; labels O, O O, C–COCH₃]

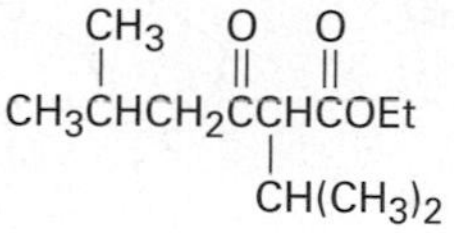

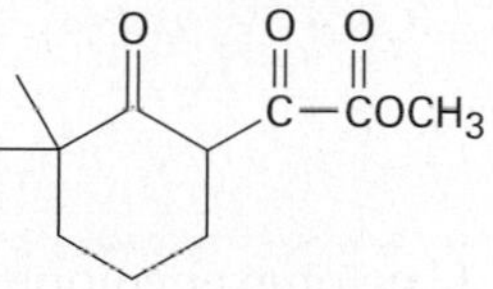

23.14 [structure; labels O, H₃C, CO₂Et]

23.15 [structures; labels H₃C, O, CO₂Et] + [structure; labels O, CO₂Et, CH₃]

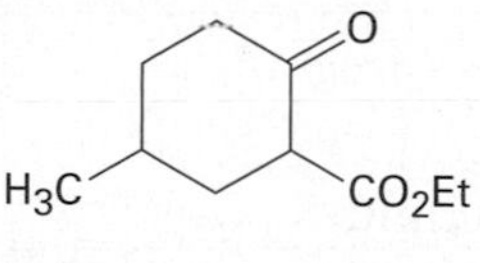

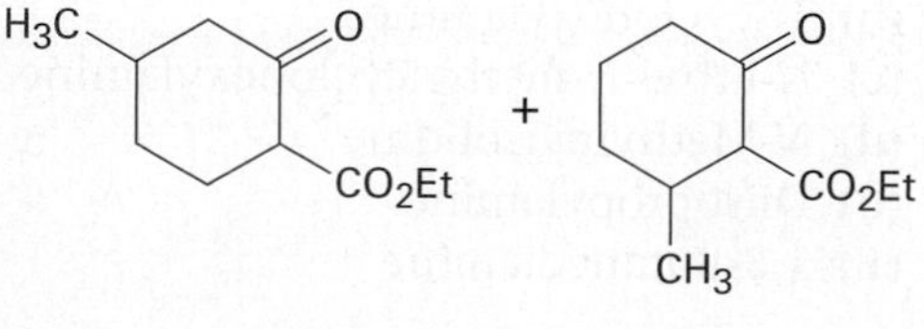

23.16 (a) O= (cyclohexane ring) $CH(COCH_3)_2$

(b) $(CH_3CO)_2CHCH_2CH_2CN$

(c) $(CH_3CO)_2CHCH(CH_3)CH_2COEt$ (with C=O on the COEt carbon)

23.17

(a) $(EtO_2C)_2CHCH_2CH_2CCH_3$ (with C=O on the CCH_3 carbon)

(b) O (cyclopentanone ring) $CH_2CH_2CCH_3$ (with C=O on the CCH_3 carbon), CO_2Et

23.18 $CH_3CH_2COCH{=}CH_2 + CH_3CH_2NO_2$

23.19

(a) O (cyclopentanone ring) $CH_2CH_2CO_2Et$

(b)

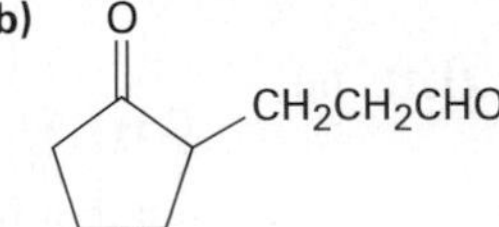

(c) O (cyclopentanone ring), O

23.20 (a) Cyclopentanone enamine + propenenitrile
(b) Cyclohexanone enamine + methyl propenoate

23.21 O, O

23.22 2,5,5-Trimethyl-1,3-cyclohexanedione + 1-penten-3-one

Chapter 24

24.1 (a) *N*-Methylethylamine
(b) Tricyclohexylamine
(c) *N*-Ethyl-*N*-methylcyclohexylamine
(d) *N*-Methylpyrrolidine
(e) Diisopropylamine
(f) 1,3-Butanediamine

24.2 (a) $[(CH_3)_2CH]_3N$ (b) $(H_2C{=}CHCH_2)_3N$

(c) (benzene ring) $NHCH_3$ (d) (cyclopentane ring) $N(CH_3)CH_2CH_3$

(e) (cyclohexane ring) $NHCH(CH_3)_2$ (f) (pyrrole ring) $N{-}CH_2CH_3$

24.3 (a) CH_3O (indole ring), N, H (b) H_3C (pyrrole ring) $N{-}CH_3$

(c) N (pyridine ring) $N(CH_3)_2$ (d) N, N (pyrimidine ring) NH_2

24.4 (a) $CH_3CH_2NH_2$ (b) NaOH (c) CH_3NHCH_3

24.5 Propylamine is stronger; benzylamine pK_b = 4.67; propylamine pK_b = 3.29

24.6 (a) *p*-Nitroaniline < *p*-Aminobenzaldehyde < *p*-Bromoaniline
(b) *p*-Aminoacetophenone < *p*-Chloroaniline < *p*-Methylaniline
(c) *p*-(Trifluoromethyl)aniline < *p*-(Fluoromethyl)aniline < *p*-Methylaniline

24.7 Pyrimidine is essentially 100% neutral (unprotonated).

24.8 (a) Propanenitrile or propanamide
(b) *N*-Propylpropanamide
(c) Benzonitrile or benzamide
(d) *N*-Phenylacetamide

24.9 The reaction takes place by two nucleophilic acyl substitution reactions.

24.10 HO, HO (benzene ring) CH_2CH_2Br $\xrightarrow{NH_3}$

or

HO, HO (benzene ring) CH_2Br $\xrightarrow[\text{2. LiAlH}_4]{\text{1. NaCN}}$

24.11 (a) Ethylamine + acetone, or isopropylamine + acetaldehyde
(b) Aniline + acetaldehyde
(c) Cyclopentylamine + formaldehyde, or methylamine + cyclopentanone

24.12 H_3C–C_6H_4–CHO (meta) + $(CH_3)_2NH$ $\xrightarrow{NaBH_4}$

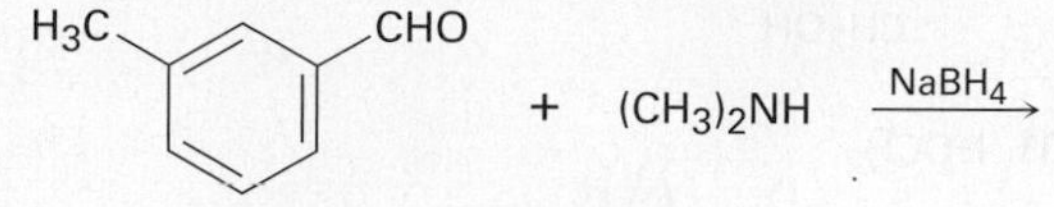

24.13 (a) 4,4-Dimethylpentanamide or 4,4-dimethylpentanoyl azide
(b) *p*-Methylbenzamide or *p*-methylbenzoyl azide

24.14 (a) 3-Octene and 4-octene
(b) Cyclohexene (c) 3-Heptene
(d) Ethylene and cyclohexene

24.15 $H_2C{=}CHCH_2CH_2CH_2N(CH_3)_2$

24.16 1. HNO_3, H_2SO_4; 2. H_2/PtO_2; 3. $(CH_3CO)_2O$; 4. $HOSO_2Cl$; 5. aminothiazole; 6. H_2O, NaOH

24.17 (a) 1. HNO_3, H_2SO_4; 2. H_2/PtO_2; 3. 2 CH_3Br
(b) 1. HNO_3, H_2SO_4; 2. H_2/PtO_2; 3. $(CH_3CO)_2O$; 4. Cl_2; 5. H_2O, NaOH
(c) 1. HNO_3, H_2SO_4; 2. Cl_2, $FeCl_3$; 3. $SnCl_2$
(d) 1. HNO_3, H_2SO_4; 2. H_2/PtO_2; 3. $(CH_3CO)_2O$; 4. 2 CH_3Cl, $AlCl_3$; 5. H_2O, NaOH

24.18 (a) 1. CH_3Cl, $AlCl_3$; 2. HNO_3, H_2SO_4; 3. $SnCl_2$; 4. $NaNO_2$, H_2SO_4; 5. CuBr; 6. $KMnO_4$, H_2O
(b) 1. HNO_3, H_2SO_4; 2. Br_2, $FeBr_3$; 3. $SnCl_2$, H_3O^+; 4. $NaNO_2$, H_2SO_4; 5. CuCN; 6. H_3O^+
(c) 1. HNO_3, H_2SO_4; 2. Cl_2, $FeCl_3$; 3. $SnCl_2$; 4. $NaNO_2$, H_2SO_4; 5. CuBr
(d) 1. CH_3Cl, $AlCl_3$; 2. HNO_3, H_2SO_4; 3. $SnCl_2$; 4. $NaNO_2$, H_2SO_4; 5. CuCN; 6. H_3O^+
(e) 1. HNO_3, H_2SO_4; 2. H_2/PtO_2; 3. $(CH_3CO)_2O$; 4. 2 Br_2; 5. H_2O, NaOH; 6. $NaNO_2$, H_2SO_4; 7. CuBr

24.19 1. HNO_3, H_2SO_4; 2. $SnCl_2$; 3a. 2 equiv. CH_3I; 3b. $NaNO_2$, H_2SO_4; 4. product of 3a + product of 3b

24.20

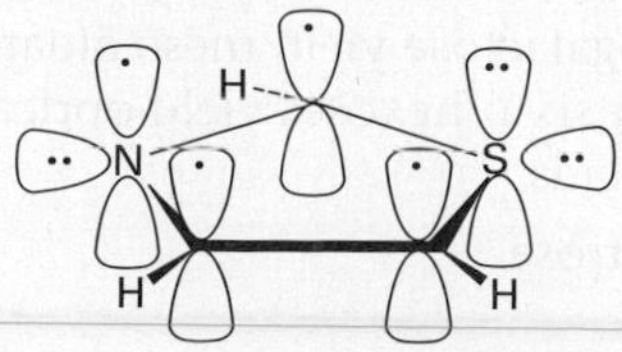

N H S H H

24.21 4.1% protonated

24.22

Attack at C2:

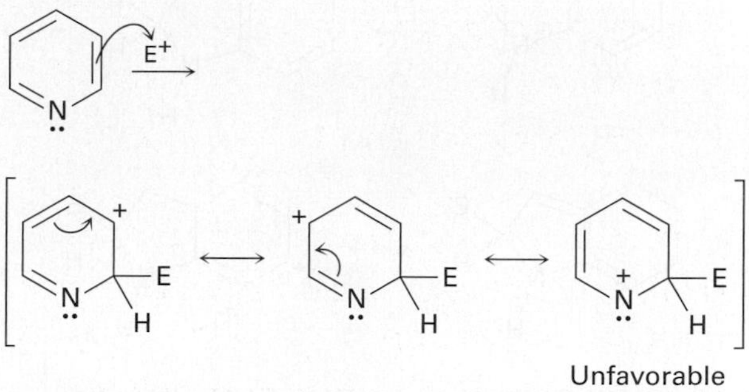

E^+

E H E H E H

Unfavorable

Attack at C3:

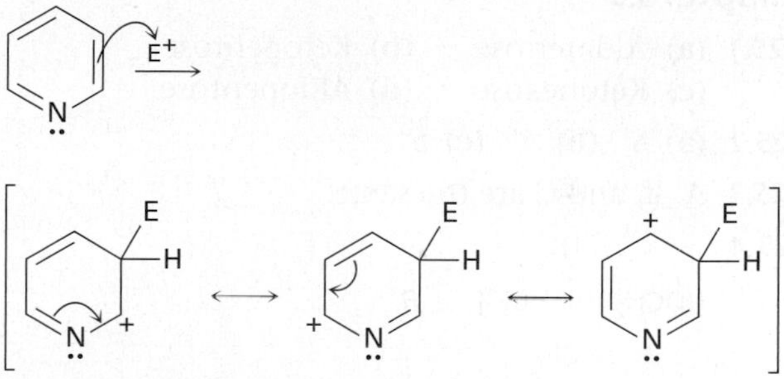

E^+

E H E H E H

Attack at C4:

E^+

E H E H E H

Unfavorable

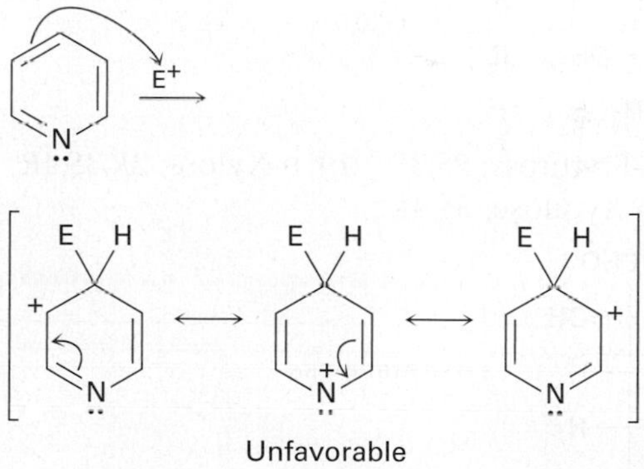

24.23 The side-chain nitrogen is more basic than the ring nitrogen.

24.24 Reaction at C2 is disfavored because the aromaticity of the benzene ring is lost.

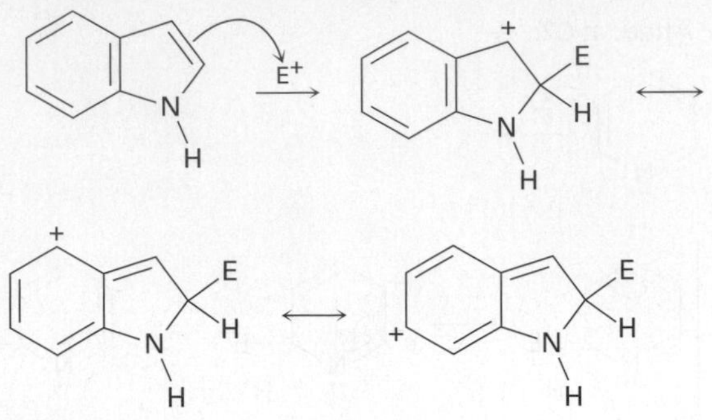

24.25 $(CH_3)_3CCOCH_3 \rightarrow (CH_3)_3CCH(NH_2)CH_3$

Chapter 25

25.1 (a) Aldotetrose (b) Ketopentose
(c) Ketohexose (d) Aldopentose

25.2 (a) *S* (b) *R* (c) *S*

25.3 A, B, and C are the same.

25.4

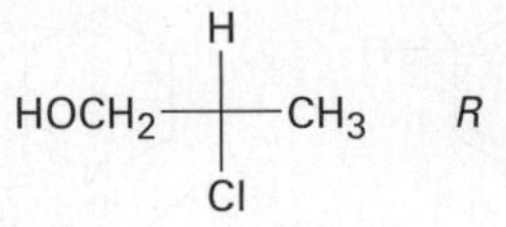

25.5

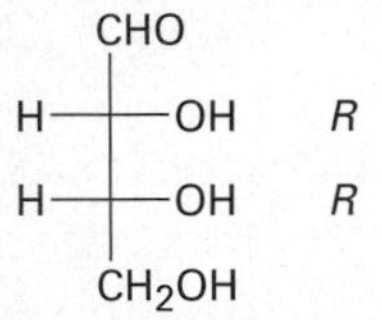

25.6 (a) L-Erythrose; 2*S*,3*S* (b) D-Xylose; 2*R*,3*S*,4*R*
(c) D-Xylulose; 3*S*,4*R*

25.7 L-(+)-Arabinose

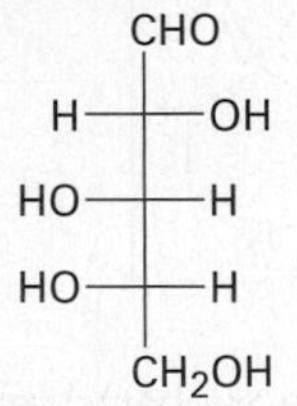

25.8

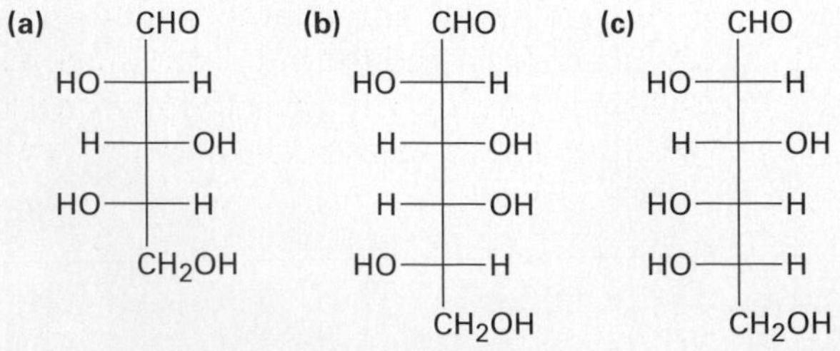

25.9 16 D and 16 L aldoheptoses

25.10 D-Ribose

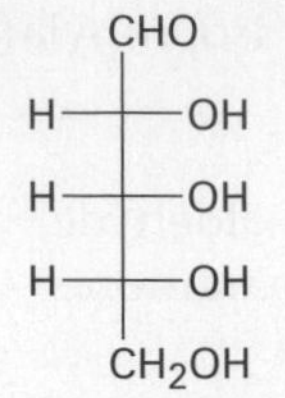

25.11

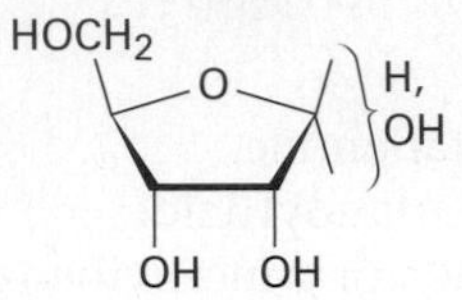

25.12

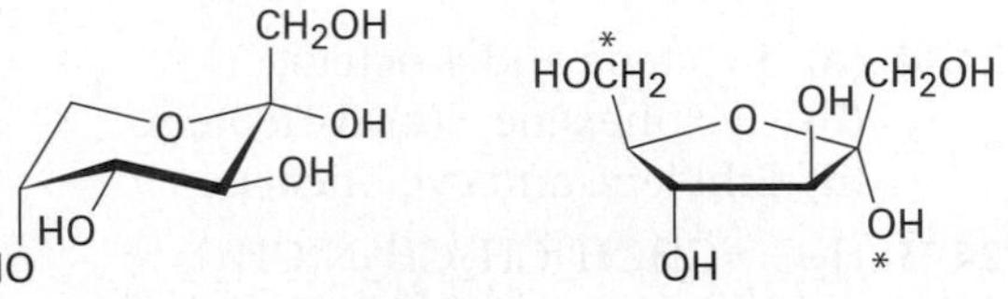

α-D-Fructopyranose α-D-Fructofuranose

25.13

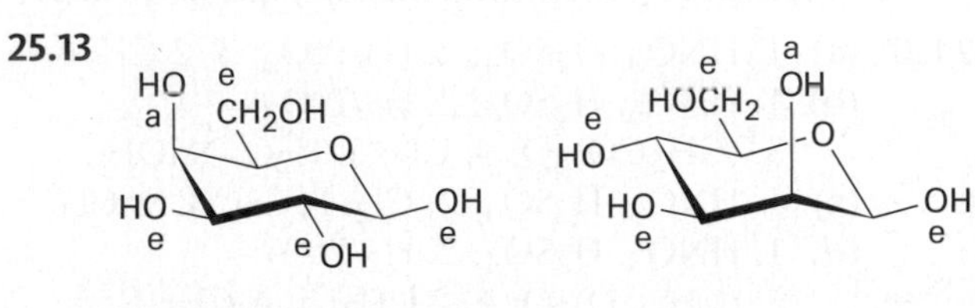

β-D-Galactopyranose β-D-Mannopyranose

25.14

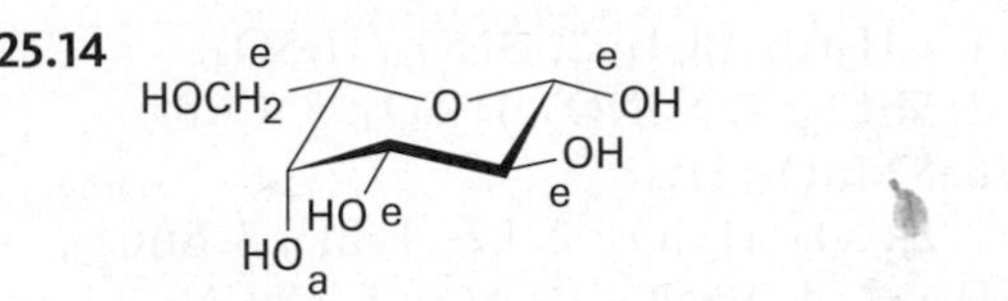

25.15 α-D-Allopyranose

25.16

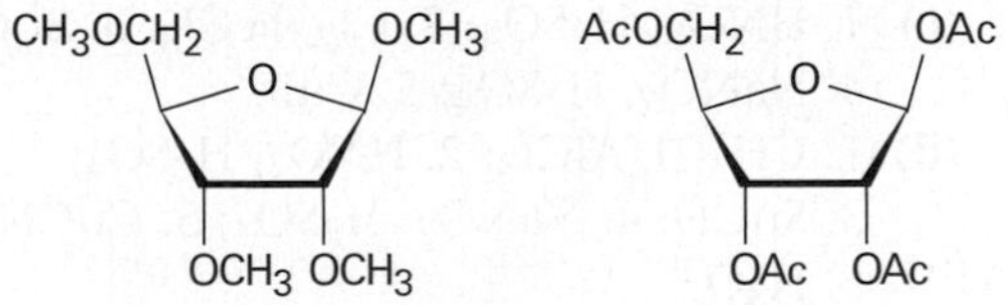

25.17 D-Galactitol has a plane of symmetry and is a meso compound, whereas D-glucitol is chiral.

25.18 The –CHO end of L-gulose corresponds to the $-CH_2OH$ end of D-glucose after reduction.

25.19 D-Allaric acid has a symmetry plane and is a meso compound, but D-glucaric acid is chiral.

25.20 D-Allose and D-galactose yield meso aldaric acids; the other six D-hexoses yield optically active aldaric acids.

25.21 D-Allose + D-altrose

25.22 L-Xylose

25.23 D-Xylose and D-lyxose

25.24

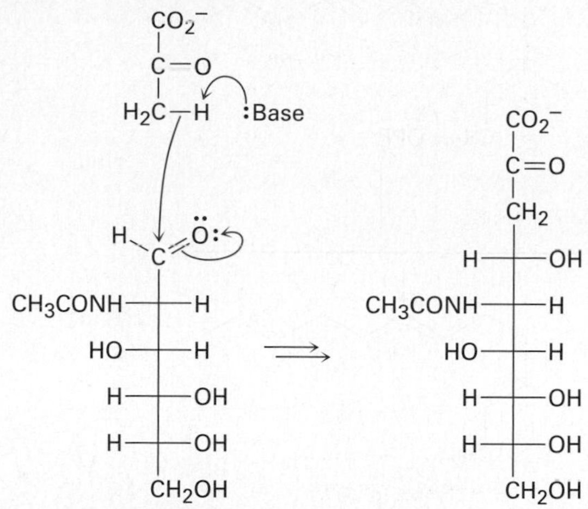

CH3CONH—H; HO—H; H—OH; H—OH; CH_2OH (with CO_2^-, C=O, H_2C—H :Base, H—C=Ö:) ⟶ CO_2^-, C=O, CH_2, H—OH, CH_3CONH—H, HO—H, H—OH, H—OH, CH_2OH

25.25 **(a)** The hemiacetal ring is reduced.
(b) The hemiacetal ring is oxidized.
(c) All hydroxyl groups are acetylated.

Chapter 26

26.1 Aromatic: Phe, Tyr, Trp, His; sulfur-containing: Cys, Met; alcohols: Ser, Thr; hydrocarbon side chains: Ala, Ile, Leu, Val, Phe

26.2 The sulfur atom in the $-CH_2SH$ group of cysteine makes the side chain higher in ranking than the $-CO_2H$ group.

26.3

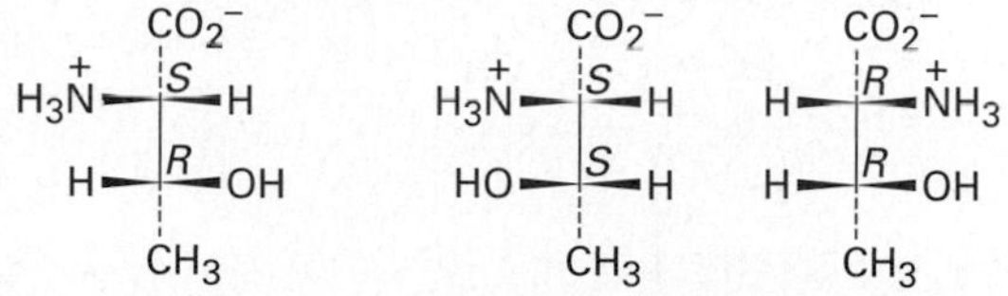

L-Threonine: CO_2^-; $H_3\overset{+}{N}$—S—H; H—R—OH; CH_3

Diastereomers of L-threonine: CO_2^-; $H_3\overset{+}{N}$—S—H; HO—S—H; CH_3 and CO_2^-; H—R—$\overset{+}{N}H_3$; H—R—OH; CH_3

26.4 Net positive at pH = 5.3;
net negative at pH = 7.3

26.5 **(a)** Start with 3-phenylpropanoic acid:
1. Br_2, PBr_3; 2. NH_3
(b) Start with 3-methylbutanoic acid:
1. Br_2, PBr_3; 2. NH_3

26.6 **(a)** $(CH_3)_2CHCH_2Br$ **(b)** 4-(bromomethyl)imidazole (CH_2Br, N, N—H)

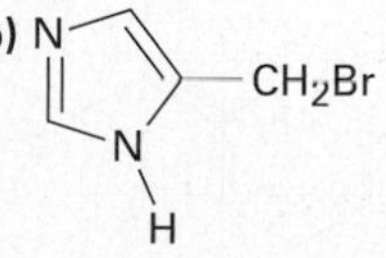

(c) 3-(bromomethyl)indole (CH_2Br, N—H) **(d)** $CH_3SCH_2CH_2Br$

26.7

$(CH_3)_2CH(H)C=C(CO_2H)NHCOCH_3$ $\xrightarrow[\text{2. NaOH, } H_2O]{\text{1. } H_2\text{, [Rh(DiPAMP)(COD)]}^+ BF_4^-}$ $(CH_3)_2CHCH_2CH(\overset{+}{N}H_3)CO_2^-$ (H, $H_3\overset{+}{N}$, CO_2^-)

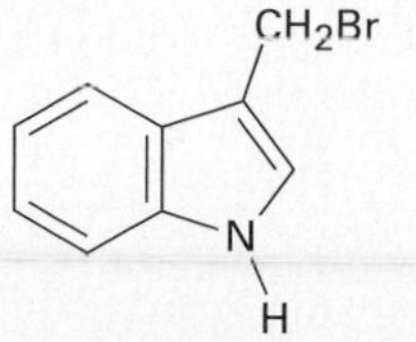

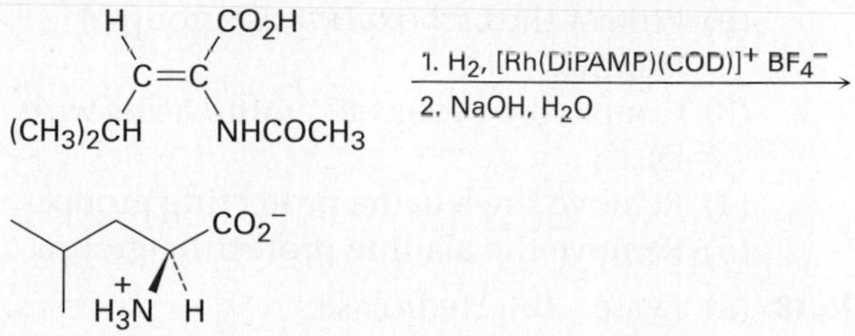

26.8 Val-Tyr-Gly (VYG), Tyr-Gly-Val (YGV), Gly-Val-Tyr (GVY), Val-Gly-Tyr (VGY), Tyr-Val-Gly (YVG), Gly-Tyr-Val (GYV)

26.9 $H_3\overset{+}{N}CH(CH_2CH_2SCH_3)C(=O)—N—CHC(=O)—NHCH(CH(CH_3)_2)C(=O)—NHCH_2CO_2^-$ (proline ring)

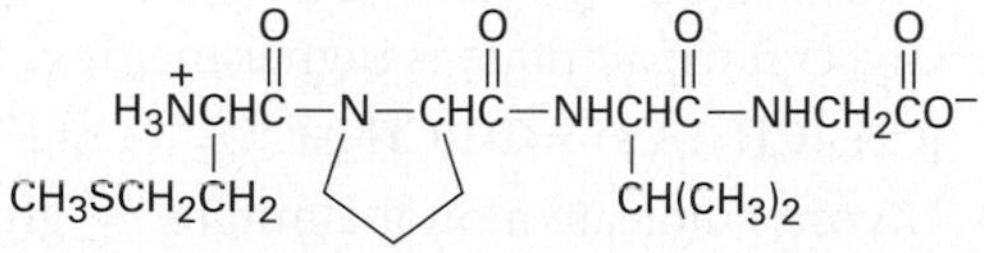

26.10 $HOC(=O)CH_2—SCH_2CH(\overset{+}{N}H_3)C(=O)O^-$

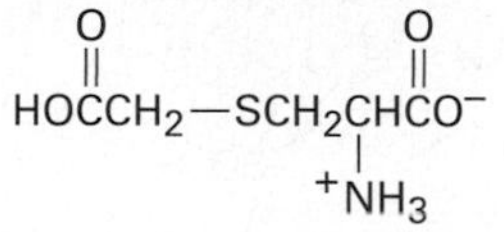

26.11

(ninhydrin-derived purple anion: O, =N, $^-$O, O, O) + $(CH_3)_2CHCHO$ + CO_2

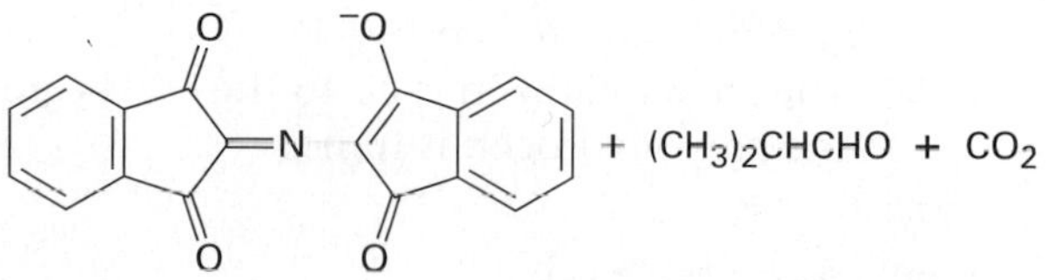

26.12 Trypsin: Asp-Arg + Val-Tyr-Ile-His-Pro-Phe
Chymotrypsin: Asp-Arg-Val-Tyr + Ile-His-Pro-Phe

26.13 Methionine

26.14 C_6H_5—N—C(=O); S=C; C---CH_2CO_2H; N—H; H

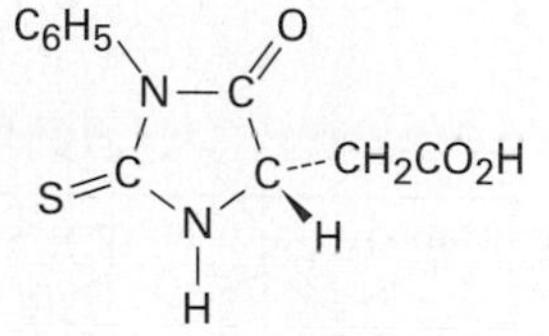

26.15 **(a)** Arg-Pro-Leu-Gly-Ile-Val
(b) Val-Met-Trp-Asp-Val-Leu (VMWNVL)

26.16 This is a typical nucleophilic acyl substitution reaction, with the amine of the amino acid as the nucleophile and *tert*-butyl carbonate as the leaving group. The *tert*-butyl carbonate then loses CO_2 and gives *tert*-butoxide, which is protonated.

26.17 (1) Protect the amino group of leucine.
(2) Protect the carboxylic acid group of alanine.
(3) Couple the protected amino acids with DCC.
(4) Remove the leucine protecting group.
(5) Remove the alanine protecting group.

26.18 (a) Lyase (b) Hydrolase
(c) Oxidoreductase

Chapter 27

27.1 $CH_3(CH_2)_{18}CO_2CH_2(CH_2)_{30}CH_3$

27.2 Glyceryl tripalmitate is higher melting.

27.3 $[CH_3(CH_2)_7CH{=}CH(CH_2)_7CO_2^-]_2\ Mg^{2+}$

27.4 Glyceryl dioleate monopalmitate → glycerol + 2 sodium oleate + sodium palmitate

27.5

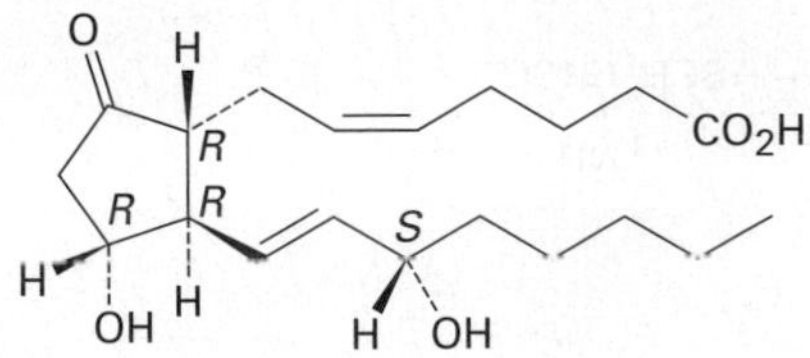

O H R CO_2H R R S H H OH H OH

27.6 The *pro-S* hydrogen is cis to the $-CH_3$ group; the *pro-R* hydrogen is trans.

27.7

(a)

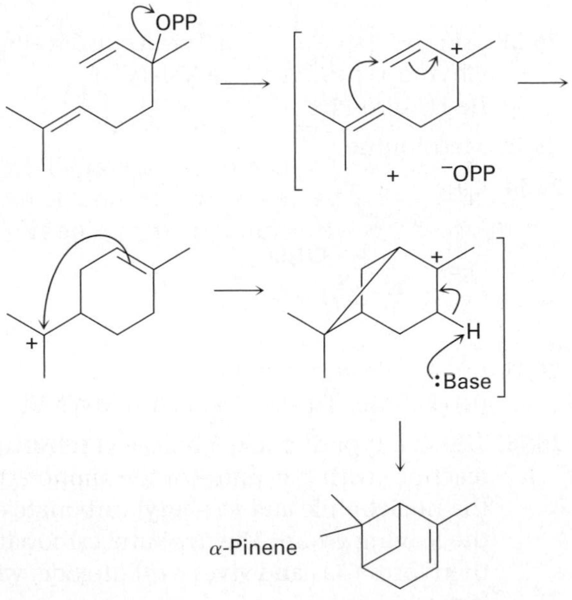

OPP → + $^-$OPP →

→ H :Base

α-Pinene

(b)

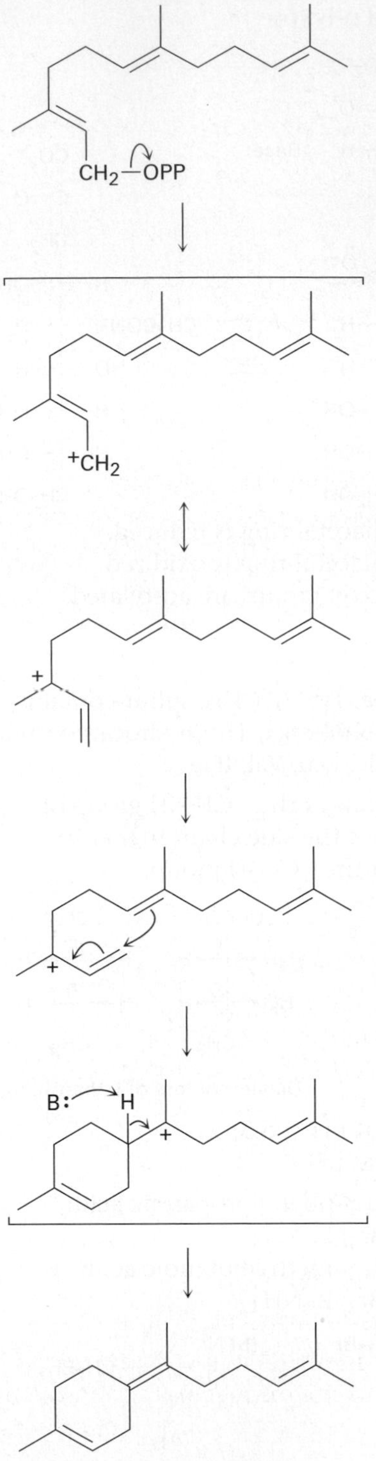

CH_2—OPP

$^+CH_2$

B: H

γ-Bisabolene

27.8

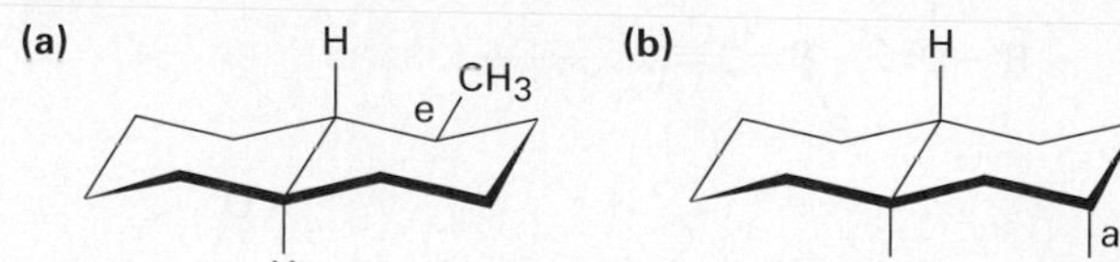

27.9

CH3 CH3 CH3 CO2H H e OH

27.10 Three methyl groups are removed, the side-chain double bond is reduced, and the double bond in the B ring is migrated.

Chapter 28

28.3 (5′) ACGGATTAGCC (3′)

28.4

28.5 (3′) CUAAUGGCAU (5′)

28.6 (5′) ACTCTGCGAA (3′)

28.7 (a) GCU, GCC, GCA, GCG
(b) UUU, UUC
(c) UUA, UUG, CUU, CUC, CUA, CUG
(d) UAU, UAC

28.8 (a) AGC, GGC, UGC, CGC
(b) AAA, GAA
(c) UAA, CAA, GAA, GAG, UAG, CAG
(d) AUA, GUA

28.9 Leu-Met-Ala-Trp-Pro-Stop

28.10 (5′) TTA-GGG-CCA-AGC-CAT-AAG (3′)

28.11 The cleavage is an S_N1 reaction that occurs by protonation of the oxygen atom followed by loss of the stable triarylmethyl carbocation.

28.12

$RO{-}P(=O)(OR'){-}O{-}CH_2{-}CH(H)C{\equiv}N$ + $:NH_3$ E2 reaction

Chapter 29

29.1 $HOCH_2CH(OH)CH_2OH$ + ATP → $HOCH_2CH(OH)CH_2OPO_3^{2-}$ + ADP

29.2 Caprylyl CoA → Hexanoyl CoA → Butyryl CoA → 2 Acetyl CoA

29.3 (a) 8 acetyl CoA; 7 passages
(b) 10 acetyl CoA; 9 passages

29.4 The dehydration is an E1cB reaction.

29.5 At C2, C4, C6, C8, and so forth

29.6 The *Si* face

29.7 Steps 7 and 10

29.8 Steps 1, 3: Phosphate transfers; steps 2, 5, 8: isomerizations; step 4: retro-aldol reaction; step 5: oxidation and nucleophilic acyl substitution; steps 7, 10: phosphate transfers; step 9: E1cB dehydration

29.9 C1 and C6 of glucose become $-CH_3$ groups; C3 and C4 become CO_2.

29.10 Citrate and isocitrate

29.11 E1cB elimination of water, followed by conjugate addition

29.12 *pro-R;* anti geometry

29.13 The reaction occurs by two sequential nucleophilic acyl substitutions, the first by a cysteine residue in the enzyme, with phosphate as leaving group, and the second by hydride donation from NADH, with the cysteine residue as leaving group.

29.14 Initial imine formation between PMP and α-ketoglutarate is followed by double-bond rearrangement to an isomeric imine and hydrolysis.

29.15 $(CH_3)_2CHCH_2COCO_2^-$

29.16 Asparagine

Chapter 30

30.1 Ethylene: ψ_1 is the HOMO and ψ_2^* is the LUMO in the ground state; ψ_2^* is the HOMO and there is no LUMO in the excited state. 1,3-Butadiene: ψ_2 is the HOMO and ψ_3^* is the LUMO in the ground state; ψ_3^* is the HOMO and ψ_4^* is the LUMO in the excited state.

30.2 Disrotatory: *cis*-5,6-dimethyl-1,3-cyclohexadiene; conrotatory: *trans*-5,6-dimethyl-1,3-cyclohexadiene. Disrotatory closure occurs.

30.3 The more stable of two allowed products is formed.

30.4 *trans*-5,6-Dimethyl-1,3-cyclohexadiene; *cis*-5,6-dimethyl-1,3-cyclohexadiene

30.5 *cis*-3,6-Dimethylcyclohexene; *trans*-3,6-dimethylcyclohexene

30.6 A [6 + 4] suprafacial cycloaddition

30.7 An antarafacial [1,7] sigmatropic rearrangement

30.8 A series of [1,5] hydrogen shifts occur.

30.9 Claisen rearrangement is followed by a Cope rearrangement.

30.10 (a) Conrotatory (b) Disrotatory
(c) Suprafacial (d) Antarafacial
(e) Suprafacial

Chapter 31

31.1 $H_2C{=}CHCO_2CH_3 < H_2C{=}CHCl < H_2C{=}CHCH_3 < H_2C{=}CH{-}C_6H_5$

31.2 $H_2C{=}CHCH_3 < H_2C{=}CHC_6H_5 < H_2C{=}CHC{\equiv}N$

31.3 The intermediate is a resonance-stabilized benzylic carbanion, $Ph{-}\ddot{C}HR$.

31.4 The polymer has no chirality centers.

31.5 The polymers are racemic and have no optical rotation.

31.6

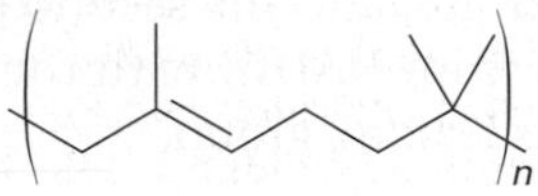

31.7

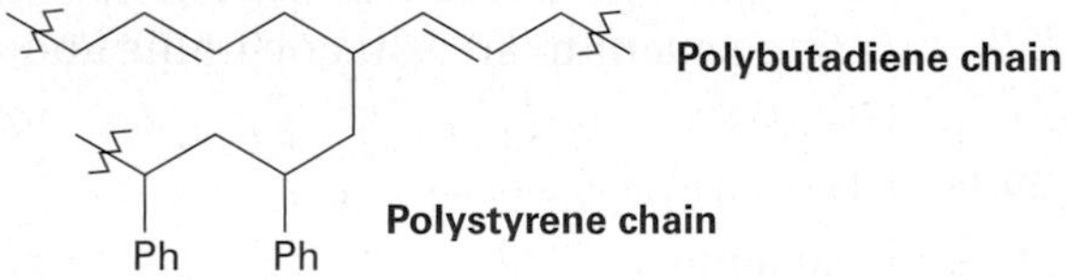

Polybutadiene chain

Polystyrene chain

Ph Ph

31.8

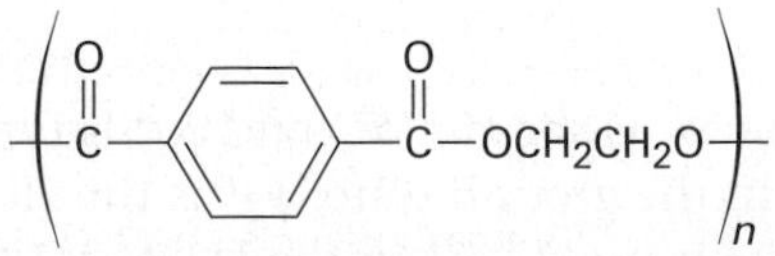

$$\left(\overset{O}{\overset{\|}{C}}-C_6H_4-\overset{O}{\overset{\|}{C}}-OCH_2CH_2O\right)_n$$

31.9

$R'{-}\ddot{O}{:}(H) + R{-}N{=}C{=}\ddot{O} \longrightarrow$

$R'{-}O^{+}(H){-}C({-}\ddot{O}{:}^{-}){=}N{-}R$, $H^{+} \longrightarrow$ $R'O{-}C({=}O){-}RNH$

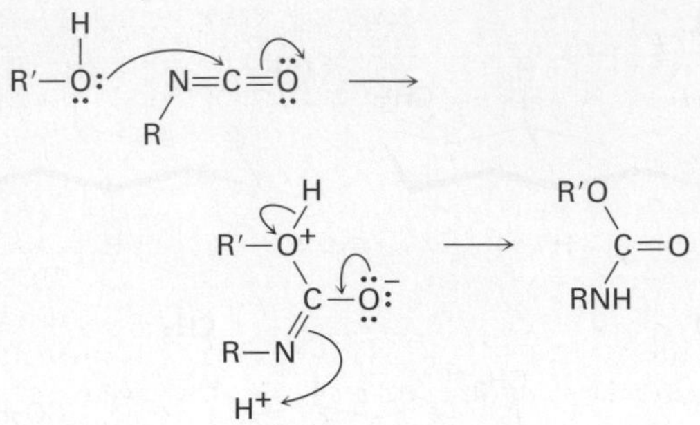

31.10 Vestenamer: ADMET polymerization of 1,9-decadiene or ROMP of cyclooctene; Norsorex: ROMP of norbornene.

Norbornene

31.11

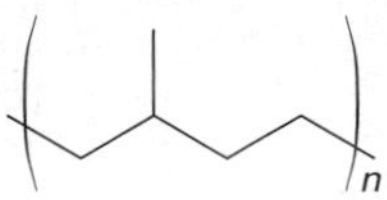

Atactic

31.12

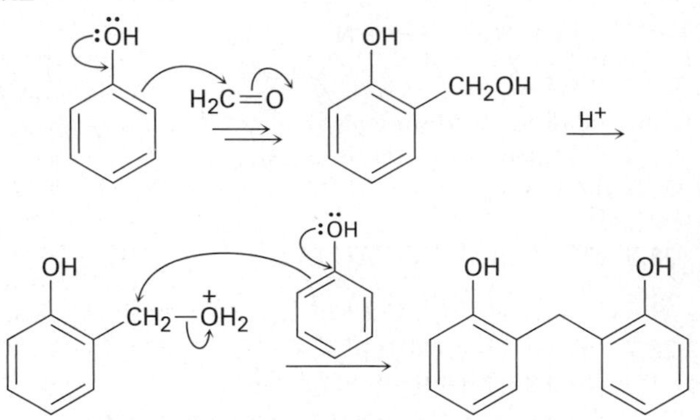

$:\ddot{O}H$ $H_2C{=}O$ OH CH_2OH H^+

OH $CH_2{-}\overset{+}{O}H_2$ $:\ddot{O}H$ OH OH

Index

Boldface references refer to pages where terms are defined.

Structures of Some Common Functional Groups

Name	Structure*	Name ending	Example
Alkene (double bond)	$>C{=}C<$	*-ene*	$H_2C{=}CH_2$ Ethene
Alkyne (triple bond)	$-C{\equiv}C-$	*-yne*	$HC{\equiv}CH$ Ethyne
Arene (aromatic ring)	(substituted benzene ring)	None	Benzene
Halide	C–X (X = F, Cl, Br, I)	None	CH_3Cl Chloromethane
Alcohol	C–OH	*-ol*	CH_3OH Methanol
Ether	C–O–C	*ether*	CH_3OCH_3 Dimethyl ether
Monophosphate	C–O–P(=O)(O^-)–O^-	*phosphate*	$CH_3OPO_3{}^{2-}$ Methyl phosphate
Diphosphate	C–O–P(=O)(O^-)–O–P(=O)(O^-)–O^-	*diphosphate*	$CH_3OP_2O_6{}^{3-}$ Methyl diphosphate
Amine	C–N:	*-amine*	CH_3NH_2 Methylamine
Imine (Schiff base)	C–C(=N:)–C	None	$CH_3C(=NH)CH_3$ Acetone imine
Nitrile	$-C{\equiv}N$	*-nitrile*	$CH_3C{\equiv}N$ Ethanenitrile
Thiol	C–SH	*-thiol*	CH_3SH Methanethiol

*The bonds whose connections aren't specified are assumed to be attached to carbon or hydrogen atoms in the rest of the molecule.